CELL BIOLOGY

(Cytology, Biomolecules and Molecular Biology)

About the Authors

Dr. P. S. Verma, M.Sc., PhD, F.E.S.I. and F.A.Z, has immense teaching experience. He taught Zoology to undergraduate and postgraduate students at Meerut College, Meerut till 2000. Dr. Verma has engaged in active research, and also successfully guided several PhD students. He has to his credit publication of several research papers.

A celebrated author, Dr. Verma has written several books on Zoology. His books include Biology for Schools, and Biology for classes XI and XII (CBSE and ISC). He has also written several books for undergraduate and postgraduate courses. His books have been very popular among the student community.

His famous books include Invertebrate Zoology, Chordate Zoology, Practical Zoology (Invertebrate), Practical Zoology (Chordate), Cell Biology, Genetics, Cytology, Chordate Embryology, Textbook of Biology for class IX, Textbook of Biology for classes XI and XII and ISC Biology for classes XI and XII.

Dr. V. K. Agarwal, M.Sc., PhD, has rich teaching experience of more than four decades at Meerut College, Meerut. He has taught subjects such as Cell Biology, Genetics, Molecular Biology, Chordate Embryology (Developmental Biology), Invertebrate Zoology, Evolution, Ecology, Animal Behaviour (Ethology) to undergraduate and postgraduate students.

Dr. Agarwal has to his credit the authorship of several widely acclaimed books such as Cytology, Genetics, Chordate Embryology (Developmental Biology), Ecology (Environmental Biology), Ethology (Animal Behaviour), Zoology for B.Sc. - I (includes Invertebrate Zoology), Molecular Biology, Genetic Engineering, and a multi-coloured edition of Cell Biology, Genetics, Molecular Biology, Evolution and Ecology. The author has also co-authored CBSE – IX Biology. He has also published few research papers on Insect Physiology and Toxicology.

CELL BIOLOGY

Cytology, Biomolecules and Molecular Biology

For Undergraduate and Postgraduate Students of all Indian Universities

Dr. P.S. VERMA
M.Sc., Ph.D., F.E.S.I., F.A.Z.
Formerly Reader

Dr. V.K. AGARWAL
M.Sc., Ph.D.,
Formerly Reader and Head

Department of Zoology
Meerut College
Meerut

S. CHAND
PUBLISHING

S Chand And Company Limited
(ISO 9001 Certified Company)

S Chand And Company Limited
(ISO 9001 Certified Company)
Head Office: D-92, Sector–2, Noida – 201301, U.P. (India), Ph. 91-120-4682700
Registered Office: A-27, 2nd Floor, Mohan Co-operative Industrial Estate, New Delhi – 110 044, Phone: 011-49731800
www.**schandpublishing.com;** e-mail: **info@schandpublishing.com**

Marketing Offices:

Chennai	:	Ph: 23632120; chennai@schandpublishing.com
Guwahati	:	Ph: 2738811, 2735640; guwahati@schandpublishing.com
Hyderabad	:	Ph: 40186018; hyderabad@schandpublishing.com
Jalandhar	:	Ph: 4645630; jalandhar@schandpublishing.com
Kolkata	:	Ph: 23357458, 23353914; kolkata@schandpublishing.com
Lucknow	:	Ph: 4003633; lucknow@schandpublishing.com
Mumbai	:	Ph: 25000297; mumbai@schandpublishing.com
Patna	:	Ph: 4011400; patna@schandpublishing.com

First Edition 2016; Reprints 2016, 2019, 2020, 2021

Low Priced Students' Paperback Edition (LPSPE), 2022; *Reprint 2022 (Twice), 2024, 2025*

Reprint 2026

ISBN: 978-93-550-1102-2 **Product Code:** H6CBY68ZOOL10ENAB21L

PRINTED IN INDIA

By Shri Shyamlal Printing Press Private Limited, Plot 20/4, Site-IV, Industrial Area Sahibabad, Ghaziabad – 201 010 and Published by S Chand And Company Limited, A-27, 2nd Floor, Mohan Co-operative Industrial Estate, New Delhi – 110 044.

PREFACE

Cell Biology deals with cells, biomolecules, genomes and proteoms. Myriad of biomolecules have interacted to control the morphology, physiology, molecular biology, evolution and growth of the cells. Cell Biology tries to understand various intricacies of functions which govern the life of the cell. In this tune, readers may come across in this textbook the topics of Induction, Differentiation, Stem cells, Cancer, Aging, Cell death, etc. Cell Biology has touched many vital aspects of human life. It has proved to be a boon as well as a curse in the hands of human beings as the following paradox indicates.

In the year 2012, Jennifer A. Doudna, a biochemist at the University of California, Berkeley, U.S.A., helped make one of the most monumental discoveries in Cell Biology regarding cellular defense system; a relatively easy way to alter any organism's DNA just as a computer user can edit a word in a document. Doudna's CRISPR-Cas9 genome editing technique (using the guide RNA) is already widely used in laboratory studies, and scientists hope it may one day help rewrite flawed genes in people, opening tremendous new possibilities for treating, even curing the dreaded diseases. (CRISPR = clustered regularly interspaced short palindromic repeats; Cas9 is the name of bacterium which has synthesized the protein for cutting the DNA.) However, many scientists including Ms Doudna are now worried that this technique could also be used to alter genes in human embryos, sperms and eggs in ways that can passed from generation to generation. The prospect raises fears of a dystopian future in which scientists create an elite population of designer babies with enhanced intelligence, beauty or other traits (Andrew Pollack, in Indian Express, 17th May 2015).

This book of Cell Biology has been developed according to the latest U.G.C. syllabi and is meant to cater to the needs of B.Sc. and M.Sc. students of almost all Indian Universities.

This book of **Cell Biology**, has been written in a simple and lucid language. Illustrations of the book have been redrawn from various standard resources. Main motives of the authors have remained to be quite authentic, modern, yet straightforward in their approach. For revision and practice of our readers, a variety of exercise questions such as Long Answer Questions, Short Answer Questions, Very Short Answer Questions, Fill in the Blanks, Multiple Choice Questions, etc., have been provided at the end of each chapter of this book. Appropriate answers to Very Short Answer Questions, Fill in the Blanks and Multiple Choice Questions have also been provided in the book.

At the end of the book, our readers may go through an exhaustive Glossary.

We hope our readers may appreciate our painstaking efforts to serve them in every cognitive ways.

Errors and shortcomings in the book are regretted and feedbacks from our readers will help the authors for corrections and improvements in the future editions of this book.

Our heartful thanks to our respective families for their cooperation and support to this endeavour.

Our sincere thanks to the Management and the team of young wizards of Editorial Department and D.T.P. Section of S.Chand & Company Pvt. Ltd., New Delhi, for help and support in the publication of this book.

Our thanks to Ms. Surabhi Khare, Ms. Shalini Negi and Mr. Rupesh Gupta of Editorial Department and also Mr. Benny and Mr. Rajesh of D.T.P., S.Chand & Company Pvt. Ltd., Delhi for speedy release of the book.

Dr. P.S. VERMA
"Preeti"
121, New Mohanpuri
Meerut-250 001

Dr. V.K. AGARWAL
Villa–20,
Green Wood City
Bypass Road
Meerut-250 005

Contents

1 Introduction

The **cytology** or **cell biology** is a biological science which deals with the study of cells from morphological, biochemical, physiological, developmental, genetical, pathological and evolutionary point of views. Recently, it has shown rapid progress and has become fundamental in the study of structure and function of living organism. The cell itself can be regarded as the vital unit of organisms and the anatomic and physiologic substrate of biologic phenomena. In its morphological aspect, modern cell biology has gone beyond simple description of structures visible to the light microscope ; by the application of new methods, an analysis has been started of sub-microscopic organization—the architectural arrangement of the molecules and micelles comprising living matter. In this functional aspect, it has transcended the stage of pure description of physiologic changes and seeks an explanation of them in the intimate physico-chemical and metabolic processes of protoplasm. Finally, we can say that the modern cytology or cell biology is attempting to interpret and explain the phenomena of metabolism, biosynthesis, heredity, sex, variation, mutation and evolution of living organisms in terms of molecules or macromolecules such as proteins, ribonucleic acids (RNAs) and deoxyribonucleic acid (DNA).

1.1 ADVENT OF CELL BIOLOGY: THREE HISTORICAL STRANDS

Modern cell biology involves the weaving togather of three distinctly different strands into a single cord (**Becker** *et al.*, 2006). Each of the strands had its own historical origins and most of the intertwining had occurred only within the last 75 years. Each strand should be appreciated in its own right, because each makes its own unique and significant contribution.

1. The first of these historical strands is **cytology** which is concerned primarily with cellular structure. The Greek prefix *cyto*—means "cell" as does the suffix—*cyte* + *logous* = to discourse. (Actually, the literal meaning of the Greek word *cytos* is "hollow vessel" which fits well with Hooke's initial impression of cells). As we will describe later on, cytology has its origin more than three centuries ago and depended heavily on the **light microscope** for its initial motivation. The advent of **electron microscopy** and several related optical techniques has led to considerable additional cytological activity and understanding.

2. The second strand represents the contributions of **biochemistry** to our understanding of cellular function. Most of the developments in this field have occurred within the last 80 years, though again the roots go back much further. Especially important has been the development of techniques such as **ultracentrifugation**, **chromatography** and **electrophoresis** for the separation of cellular components and molecules. The use of **radio-actively labeled compounds** in the study of enzyme-catalyzed reactions and metabolic pathways is another very significant contribution of biochemistry to our understanding of how cells function.

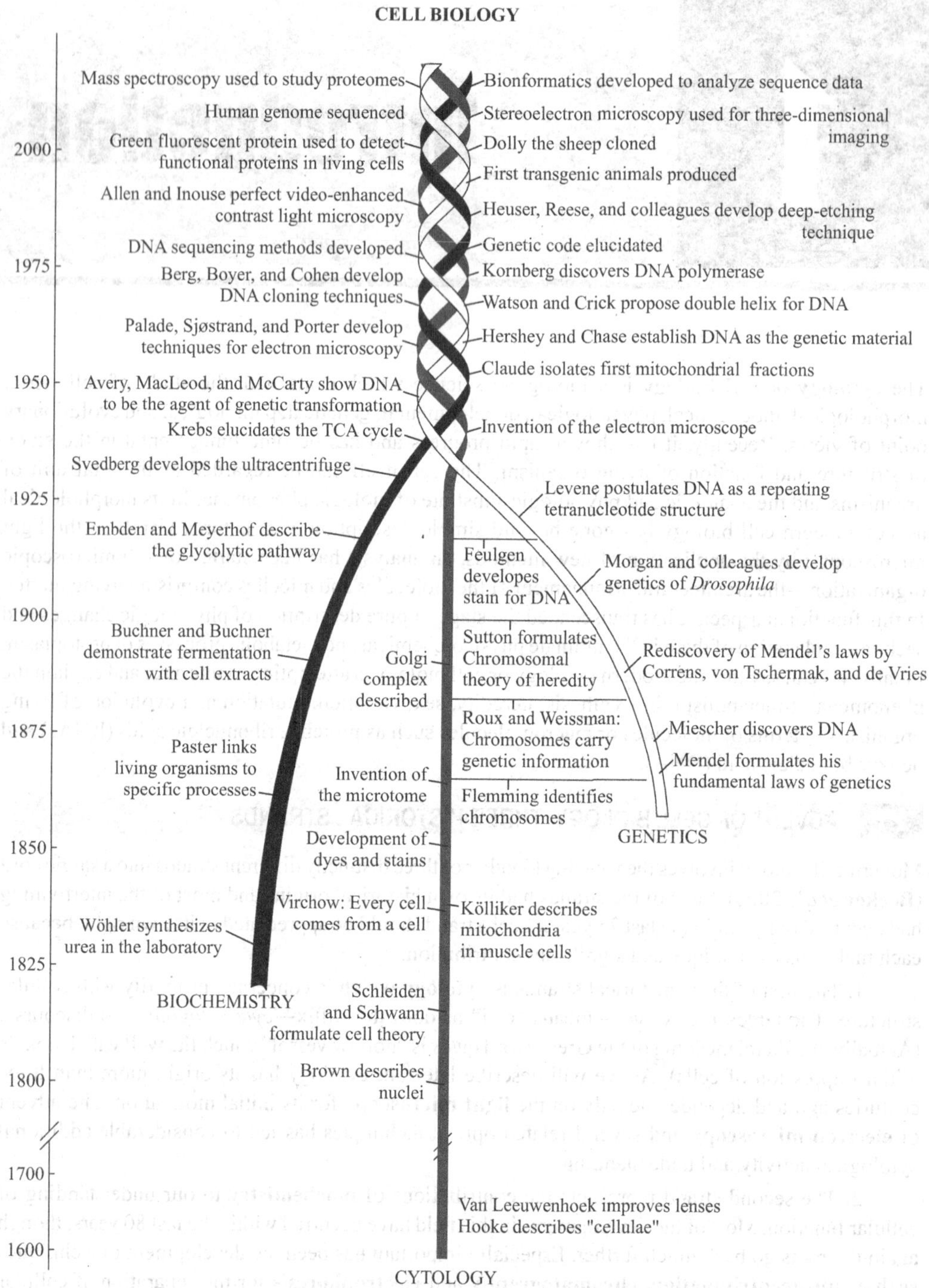

Fig. 1.1. A single cord of cell biology involves weaving together of three strands: Cytology, Biochemistry and Genetics during last 300 or more years.

3. The third strand is **genetics**. Here, the historical continuum stretches back more than 155 years to **Gregor Mendel**. Again, however, much of our present understanding has come within the last 80 years. An especially important landmark on the genetic strand came with the demonstration that DNA (deoxyribonucleic acid) is the bearer of genetic information in most life forms, specifying the order of subunits (= amino acids) and hence the properties, of the proteins that are responsible for most of the functional and structural features of the cells. Recent accomplishments on the genetic strand include the **sequencing** of the entire genomes (all of the DNA) of humans and other species and the **cloning** (production of genetically identical organisms) of mammals, including sheep, goat, cattle, camels and cats.

To understand present-day cell biology therefore means to appreciate its diverse roots and the important contributions that each of its component strands has made to our current understanding of what a cell is and what it can do.

1.2. HISTORICAL REVIEW

The cytology has taken its individual entity by the end of the nineteenth century. Its gradual development in the last centuries can be traced as follows:

Cytology before 19th Century

Ancient philosophers and naturalists, particularly **Aristotle** and **Paracelsus** concluded that "all animals and plants, however, complicated are constituted by few elements which are repeated in each one of them". They were referring to the macroscopic structures of an organism, such as roots, leaves and flowers common to different plants or segments and organs that are repeated in the animal kingdom. Many centuries later, owing to the invention of magnifying lenses, the world of microscopic dimensions was discovered.

The beginning of cytology can be visualized in the 15th century, when **da Vinci** in 1485 has stressed upon the use of lenses in viewing small objects. In the year 1658, **Jan Swammerdam** gave the first description of the cell in his account of the red blood cells of the frog. The cytology came in its actual existence with the discovery of cell in 1665, by an English botanist, **Robert Hooke** (1635–1703). While examining a thin slice of cork under his crude compound microscope, **Hooke** observed its honey-combed or porous structure. He found similar alveoli in other plant tissues and gave them the name "cells" (L., *cellula*—little room). What Hooke had seen of the cell was mainly its walls of thickened cellulose, and he ignored the existence of the substance occurring within the cell wall and thus, failed to grasp its true nature and significance.

In the year 1672, an Italian anatomist, **Marcello Malpighi** (1688–1694) and an English physician **Nehmian Grew**, without using the word cell mentioned that certain parts of plants were made up of minute elementary organisms—"utricles", "sacs", and "vesicles". **Malpighi** also studied a variety of animal tissues microscopically and therefore, he is generally considered as **the father of 'microscopic anatomy'**. During the 17th century, a Dutch microscopist, **A.V. Leeuwenhoek** (1632–1723) discovered the animalcules, infusoria (Protozoa), bacteria, etc., and made microscopical observations on Protozoa, ants, aphids, spermatozoa, red blood cells, muscles, nerves, skin, teeth and certain plants. Leeuwenhoek discovered the free living cells in pond water for the first time (1674) and in 1678 identified the sperm cells of humans, dogs, rabbits, frogs, fish and insects.

During the 18th century, little knowledge was added to cytology. However, some workers produced the accounts of plant cells and referred them by different names such as 'fibres' by **Haller** (1757) and **Bonnet**; 'cylinders' by **Fontana** (1781); 'otricles' by **Brisseau-Mirabel** and 'elementary vesicles' by **Oken**. In 1781, **F. Fontana** also described the nucleolus from the skin of an eel.

Cytology in 19th Century

The 19th century can be considered as the "classical era of cytology" because most important cytological investigations were made extensively in this century. By the beginning of the 19th century it was known that all the organs of animals were made up of tissues such as muscle, bone, cartilage and fat; and similarly that the stems, roots, leaves, and other organs of higher plants were also composed of distinct tissues. The cytological contributions of 19th century can be represented chronologically in the following Table 1.1:

Table 1.1. Chronological representation of certain significant cytological investigations of 19th century.

Year	Name of contributor	Cytological contribution
1808	Mirabel	Reported that plants consist of membranous cellular tissues.
1809	J. B. Lamarck	Stated that in the living organisms the cell has certain important functions.
1824	R. J. H. Dutrochet	Showed that all animals and plants composed of cells and these cells remain united by simple adhesive forces.
1824	P. Prerost and J. B. A. Dumas	Described cell division first of all by studying cleavage of the frog's egg.
1825	F. V. Raspail	Used iodine in the starch detection and developed the frozen-section technique; founder of cytochemistry.
1826	Turpin	Reported the occurrence of cell division.
1827	Karl von Baer	Discovered the mammalian ovum.
1828	Wohler	Synthesized urea and discredited the view that organic compounds could only be made by living things and thus paved the way for a systematic investigation of cellular reactions.
1828	R. Brown	Reported Brownian movement in cellular particles.
1830	G.B. Amici	Discovered fertilization in plants.
1830	Meyen	Suggested that each plant cell is an independent, isolated unit capable of receiving nourishment and building its own internal structures.
1831	R. Brown	Described the nucleus as a central feature in plant cells.
1835	Felix Dujardin	Described protoplasm ("sarcode") in Protozoa and considered it to be living matter.
1835	H. von Mohl	Described the cell division.
1838	M. J. Schleiden	Described the nucleoli and proposed "cell theory" along with T. Schwann.
1839	T. Schwann	Applied "cell theory" to animals.
1840	J. E. Purkinje	Named the cell contents as protoplasm.
1841	Robert Remak	Described amitotic cell division in the R.B.C. of the chick embryo.
1845	A. Donne	Studied spermatozoa and used photomicroscopy first time.
1845	Kolliker	Reported that the ova and spermatozoa are single cell products.
1846	H. von Mohl	Used the term protoplasm for the cytoplasm of modern time.
1846	K. Nageli	Reported that new plant cells arise from the pre-existing cells.
1849	W. Hofmeister	Studied nuclear divisions in stamens and fertilization. He drew figures of chromosomes of the nuclei of pollen mother cells of *Tradescantia*.
1855	R. Virchow	Stated that all cells arise from pre-existing cells.
1855	Pringsheim	Described the complete fertilization process in the alga.

Year	Name of contributor	Cytological contribution
1857	Kolliker	Discovered mitochondria in muscle.
1858	R. Virchow	Correctly asserted that as the functional units of life, cells are the primary site of disease.
1861	Schultze	Said that the cell is a living substance which possesses a nucleus and cell membrance. He used the term protoplasm for the living substances of cell and stated protoplasm as "physical basis of life".
1863	Waldeyer	Reported the utility of haematoxylin in staining the chromosomes of cell.
1865	G. Mendel	Discovered fundamental principles of genetics.
1866	Haeckel	Named plastids.
1867	L. St. George	Discovered what was later called the "Golgi-complex".
1870	W. His	Developed the microtome for cutting serial sections of tissue for cell study.
1871	F. Miescher	Discovered nucleoproteins and nucleic acid (Nuclein).
1873	H. Fol	Described the spindle and astral rays.
1875	Strasburger	Described chromosomes.
1876	L. Pasteur	Discovered anaerobic release of energy from cells (fermentation) in yeasts and moulds.
1876	O. Hertwig	Studied reproduction in sea urchin and concluded that fertilization involves the union of sperms and egg nuclei.
1877	E. Abbe	Discovered the oil-immersion objective.
1879	H. Fol	Observed the penetration of single sperm in the egg during fertilization.
1879	W. Flemming	Introduced the term chromatin and described splitting of chromosomes.
1881	E. G. Balbiani	Discovered the giant chromosomes of salivary glands of the larva of *Chironomus*.
1882	W. Flemming	Introduced the term mitosis, improved the techniques of fixation and staining and suggested relations of nucleic acid with the chromatin.
1882	Strasburger	Described mitosis in plant cells and introduced the terms cytoplasm and nucleoplasm.
1882	W. Pfitzner	Discovered chromomeres, the 'granules' on the chromosomes.
1883	W. Roux	Proposed that chromosomes contain hereditary units.
1883	E. van Beneden	Showed in *Ascaris* that gametes contain half of chromosomes than the body cells.
1883	Schimper	Introduced the term chloroplast.
1883	Meyer	Described details of chloroplast structure.
1883	Metschnikoff	Observed and named phagocytosis in cells.
1886	R. Altmann	Stained mitochondria and other granular components of cell and suggested their role in respiratory metabolism of the cell.
1886	C. A. MacMunn	Discovered cytochromes (originally "histohematins").
1887	E. van Beneden	Discovered centrioles.
1888	T. Boveri	Described the centrioles.
1888	Waldeyer	Introduced the term chromosome.
1888	Kolliker	Isolated mitochondria.

Year	Name of contributor	Cytological contribution
1890	Boveri and Guignard	Proved the re-establishment of a diploid chromosome number at fertilization, by joining of equal sets from the male and female gametes.
1892	A. Weismann	Proposed the germplasm theory and showed the chromosomes as the most important components of the nucleus.
1892	T. Boveri	Described spermatogenesis and oogenesis in *Ascaris.*
1897	Granier	Named and described the ergastoplasm.
1898	C. Benda	Named the mitochondrion.
1898	C. Golgi	Described the Golgi complex in nerve cells.

Cytology After 19th Century

In the 20th century various modern microtechniques have been employed in cytological (cell biological) investigations. For instance, new histochemical and cytochemical methods have been developed to detect various molecular components of the cell; various cellular components have been separated by ultracentrifugation; different biochemical events of the cell could be known in detail by autoradiography; and methods of tissue culturing have made possible the study of living cells. Moreover, micromanipulators, ultra-microtomes, chromatography, electrophoresis, electron microscopy, X-ray microscopy, phase contrast microscopy, spectrophotometry, X-ray structural analysis, etc., have provided new opportunities to cell biologists to investigate minute details of cell and its components.

Various important cytological investigations have been made in 20th century and many cytologists have received Nobel Prizes for their landmark cytological investigations. The cytological investigations of 20th century can be represented chronologically in the Table 1.2.

Table 1.2. Chronological tabulation of certain important cytological investigations of 20th century.

Year	Name of contributor	Cytological contribution
1900	J. Loeb	Discovered artificial parthenogenesis of eggs by chemical and mechanical methods.
1901	T. H. Montgomery	Showed that homologous chromosomes undergo pairing (synapsis) during reduction division.
1901	Strasburger	Introduced the term plasmodesmata.
1902	C. E. McClung	Identified the sex chromosomes in the Hemiptera.
1902	W. S. Sutton	Recognized the significance of reduction division and proposed the "chromosome theory" of heredity. He pointed out the parallels between genes and chromosome behaviour and suggested a chromosomal location for genes.
1903	E. Buchner	Discovered the enzyme and got Nobel Prize.
1904	F. Meves	Demonstrated the presence of mitochondria in plant cells.
1905	J. B. Farmer	Coined tbe term meiosis along with J. E. Moore.
1906	M. Tswett	Discovered chromatography.
1907	R. G. Harrison	Developed the technique for growing tissues in culture.
1909	F. A. Janssens	Stated chiasmata formed by the exchanges between two chromatids of non-homologous chromosomes.
1910	A. Kossel	Investigated the chemistry of the nucleus and got Nobel Prize.

Year	Name of contributor	Cytological contribution
1911	T. H. Morgan	Advanced his gene theory by proposing that genes are linearly arranged along chromosomes in a definite order.
1912	Warburg	Demonstrated that iron is essential to respiration.
1913	Wilstatter and Stoll	Isolated chlorophyll and later determined its structure.
1915	R. M. Wilstatter	Got Nobel Prize for the research on chlorophyll.
1923	F. Pregl	Got Nobel Prize for microanalysis of organic substances.
1923	G. Hevesy	Discovered the technique of isotopic tracing in which the fate of labelled isotope molecules can be traced through a metabolic pathway.
1923	O. Warburg	Discovered a method of measuring gaseous exchange in living tissue by manometry.
1924	R. Feulgen and H. Rossenbeck	Described a test for DNA.
1926	T. Svedberg	Got Nobel Prize for the discovery of ultracentrifuge.
1928	F. Griffith	Discovered genetic transduction in bacteria.
1929	K. Lohmann	Discovered ATP, the source of energy in biochemical reactions.
1931	C. Stern, H. Creighton and B. McClintock	Demonstrated cytologically the process of crossing over.
1931	O. H. Warburg	Got Nobel Prize for the discovery of respiratory enzymes and their actions.
1931	Engelhardt	Demonstrated that phosphorylation is coupled to oxygen consumption.
1931	W. H. Lewis	Discovered pinocytosis.
1932	M. Knoll and E. Ruska	Produced first electron microscope.
1933	T. H. Morgan	Received Nobel Prize for the discovery of function of chromosomes in transmission of heredity.
1933	Keilin	Partially reconstituted an electron transport chain.
1935	W. M. Stanley	Isolated tobacco mosaic virus in crystalline form.
1937	Hans A.Krebs	Formulated the citric acid cycle.
1937	W. N. Haworth and P. Karrer	Received Nobel Prize for the discovery of molecular structures of carbohydrates, vitamin C, and carotenoids, flavins, Vitamins A and B, respectively.
1938	T. Caspersson	Developed ultraviolet photomicrography for the study of nucleic acids.
1939–1941	F. Lipmann	Proposed a central metabolic role for ATP.
1943	A. Claude	Isolated cell components such as ribosomes, mitochondria and nuclei in relatively pure form by differential ultracentrifugation.
1944	O. T. Avery, C. H. Mcleod and H. McCarty	Showed the significance of DNA in hereditary transmission in bacteria.
1945	K. R. Porter	Discovered the endoplasmic reticulum.
1945	F. Lipmann	Discovered coenzyme A and got the Nobel Prize for this discovery in 1953.

Year	Name of contributor	Cytological contribution
1947–1950	Lipmann and Kaplan	Determined the structure of coenzyme A.
1948–1950	Kennedy and Lehninger	Showed that the citric acid cycle, oxidative phosphorylation and fatty acid oxidation take place in the mitochondria.
1948	A. Boivin, R. Vendrely and C. Vendrely	Demonstrated quantitative constancy of DNA material in various cells of same organism.
1952	C. Du Duve	Identified lysosomes.
1952	G. E. Palade	Analysed the fine structure of mitochondria and showed the presence of mitochondrial cristae.
1952	M. Chase and A. D. Hershey	Showed that the gene was DNA.
1953	F. Zemike	Received Nobel Prize for the discovery of phase contrast microscope.
1953	H. A. Krebs	Received Nobel Prize for the discovery of citric acid cycle.
1953	J. D. Watson and F. H. C. Crick	Proposed the double helix model for the DNA molecule and got Nobel Prize in 1962.
1954	F. Sanger	Gave the first complete structure of a protein molecule when he worked out the structure of the insulin molecule.
1956	J. H. Tijo and A. Levan	Gave the first correct human chromosome count (46 chromosomes in the diploid condition).
1957	Seymour Benzer	Gave the concept of the cistron (gene), the unit of function.
1957	A. R. Todd	Got Nobel Prize for the discovery of nucleotides and nucleotidic coenzymes.
1958	F. Sanger	Received Nobel Prize for thc discovery of the structure of insulin.
1958	G. W. Beadle and E. L. Tatum	Received the Nobel Prize for the discovery that one gene regulates one definite chemical process.
1958	F. H. C. Crick	Proposed the central dogma of molecular biology that DNA determines thc sequence of amino acids in polypeptide.
1959	S. Ochoa	Received the Nobel Prize for the synthesis of polyribonucleotide *in vitro*.
1959	A. Kornberg	Received the Nobel Prize for *in vitro* synthesis of polydeoxyribonucleotides.
1960	Park and Pon	Discovered quantosomes in the chloroplast.
1960	J. Hurwitz, A. Stevens and S. Weiss	Showed the role of the enzyme RNA polymerase in the synthesis of RNA from a DNA template.
1961	F. H. C. Crick *et al.*	Produced direct evidence that the genetic code is a triplet one.
1961	F. Jacob and J. Monod	Discovered regulatory genes.
1961	M. Calvin	Received the Nobel Prize for the work on photosynthetic activity.
1964	R. W. Holley	Described the nucleotide sequence of alanine tRNA molecule of yeast.
1968	M. W. Nirenberg and H. G. Khorana	Received the Nobel Prize for the work on genetic code.
1968	R. H. Holley	Received the Nobel Prize for the discovery of base sequence of tRNA.

Year	Name of contributor	Cytological contribution
1969	M. Delbrueck, A. D. Mershey and S. E. Luria	Received thet Nobel Prize for the discovery of reproductive patterns in viruses.
1970	Knippers; Kornberg and Gefter; Moses and Richardson	Isolated DNA-polymerase-II enzyme.
1972	C. B. Anfinsen; S. Moore and W. H. Stein	Got Nobel Prize (chemistry) for the discovery of chemical structure and activity of the enzyme ribonuclease.
1974	A. Claude, G. E. Palade and C. de Duve	Were awarded Nobel Prize for their innovations that started the modern science of cell biology nearly 39 years ago, and for their major contributions to the understanding of the inner working of living cell.
1975	R. Dulbecco, H. M.Temin and D. Baltimore	Were awarded Nobel Prize for their discoveries concerning the interaction between tumour viruses and the genetic material of the cell.
1976	D. C. Gajdusek	Got Nobel Prize for his finding of the cause of a puzzling, fatal degenerative disease called Kuru. The agent responsible for Kuru was found to be a slow virus and this discovery has been considered a great achievement in medical genetics and neurobiology.
1977	Rosalyn Talow, Roger Guillemin and Andrew Schally	These three American endocrinologists got Nobel Prize in Physiology and Medicine for developing the techniques of radioimmunoassay for the estimation of hormones in blood and for isolating, characterising and synthesising three polypeptides involved in the control of the anterior pituitary gland from brain.
1978	Daniel Nathans, Hamilton O. Smith and Werner Arber	These three microbiologists were awarded Nobel Prize in Physiology and Medicine for their "discovery of restriction enzymes and their application to problem of molecular biology". The restriction enzymes provide the "chemical knives" that split chromosomes in their component genes.
1978	Peter Dennis Mitchell	Got Nobel Prize in Chemistry for his work in bioenergetics which concerns the chemical processes responsible for the energy supply in living cells.
1980	Paul Berg, Frederick Sanger and Walter Gilbert	Got Nobel Prize in Chemistry for their investigations in genetic engineering. Half of this year's Prize was shared by Sanger and Gilbert for developing techniques for decoding about 1000 alphabets of heredity from a DNA strand in one day. Another half of Prize was awarded to Berg for his invention of technique of gene-splicing which literally means joining together of the various pieces of gene or DNA.
1980	George Snell, Jean Dausset and B. Benacerraf	Got Nobel Prize in Physiology and Medicine for their research work in establishing links between genetics and immunology and to show ways to cure many human genetic diseases.

Discovery of Cell and its Organelles. Cells were discovered in 1665 by an English scientist, **Robert Hooke**. He saw cells for the first time in a thin slice of cork with his crude microscope. Hooke described the cork as made up of hundreds of little hexagonal boxes, giving a kind of honey-comb appearance. He called these little boxes **cells**. In 1833, a Scottish botanist, **Robert Brown** discovered and named the **nucleus** in a cell. **J.E. Purkinje** in 1839, used the term **protoplasm** for a living substance present inside the cell. For coinage of other terms in cell biology see Box 1.1.

Box 1.1 The Naming of Various Cell Components

1. Cell	*Robert Hooke* (1665)
2. Cell membrane	*C. Nageli* and *C. Cramer* (1855)
3. Plastids	*Haeckel* (1866)
4. Karyokinesis	*Schleicher* (1878)
5. Chromatin	*W. Flemming* (1879)
6. Mitosis	*W. Flemming* (1882)
7. Chloroplast	*Schimper* (1883)
8. Phagocytosis	*Metchinikoff* (1883)
9. Cytoplasm and nucleoplasm	*E. Strasburger* (1884)
10. Centrosome	*T. Boveri* (1888)
11. Chromosome	*Waldeyer* (1888)
12. Nucleic acid	*Richard Altmann* (1889)
13. Mitochondrion	*C. Benda* (1897)
14. Golgi complex	*Camillo Golgi* (1898)
15. Plasmodesmata	*E. Strasburger* (1901)
16. Meiosis	*J.B. Farmer* and *J.B. Moore* (1905)
17. Plasmalemma	*J.Q. Plowe* (1931)
18. Endoplasmic reticulum	*K.R. Porter* (1945)
19. Lysosome	*C.de. Duve* (1948)
20. Microbody	*J. Rhodin* (1954)
21. Prokaryote and Eukaryote	*Hans Ris* (1960)
22. Peroxisomes	*De Duve* (1965)
23. Glyoxisomes	*R.W. Breidenbach* (1967)
24. Viroids	*T.O. Diener* (1971)
25. Prion	*Prusiner* (1982)

1.3. CELL THEORY

In 1838, a German botanist **Mathias Jacob Schleiden** (1804–1881) put forth the idea that cells were the units of structure in the plants. In 1839, his co-worker, a German zoologist, **Theador Schwann** (1810–1882) applied Schleiden's thesis to the animals. Both of them, thus, postulated that the cell is the basic unit of structure and function in all life. This simple, basic and formal biological generalization is known as **cell theory** or **cell doctrine**. In fact, both **Schleiden** and **Schwann** are incorrectly credited for the formulation of the cell theory; they merely made the generalizations which were based on the works of their predecessors such as **Oken** (1805), **Mirbel** (1807), **Lamarck** (1809), **Dutrochet** (1824), **Turpin** (1826), etc. However, **Schleiden** was the first to describe the nucleoli and to appreciate the fact that each cell leads a double life—one independent, pertaining to its own development, and another, as integral part of a multicellular plant. **Schwann** studied both plant and animal tissues and his work with the connective tissues such as bone and cartilage led him to modify the evolving cell theory to include the idea that living things are composed of both cells and the products or secretions of the cells. **Schwann** also introduced the term **metabolism** to describe the activities of the cells.

In the coming years, the cell theory was to be extended and refined further. **K. Negeli** (1817–1891) showed in 1846 that plant cells arise from the division of pre-existing cells. In 1855, a German pathologist **Rudolf Virchow** (1821–1902) confirmed the Nageli's principle of the cellular basis of life's continuity. He stated in Latin that the cells arise only from the pre-existing cells (*viz.*, his actual aphorism was "*omnis cellula e cellula*"—every cell from a cell). **Virchow**, thus, established the significance of cell division in the reproduction of organisms. In 1858, **Virchow** published his classical textbook *Cellular Pathology* and in it he correctly asserted that as functional units of life, the cells were the primary sites of disease and cancer. Later, in 1865, **Louis Pasteur** (1822–1895) in France gave experimental evidence to support Virchow's extension of the cell theory.

The modern version of cell theory states that (1) All living organisms (animals, plants and microorganisms) are made up of one or more cells and cell products. (2) All metabolic reactions in unicellular and multicellular organisms take place in cells. (3) Cells originate only from other cells, *i.e.*, no cell can originate spontaneously or *de novo*, but comes into being only by division and duplication of already existing cells. (4) The smallest clearly defined unit of life is the cell.

The cell theory had its wide biological applications. With the progress of biochemistry, it was shown that there were fundamental similarities in the chemical composition and metabolic activities of all cells. **Kolliker** applied the cell theory to embryology—after it was demonstrated that the organisms developed from the fusion of two cells—the spermatozoon and the ovum. However, in the recent years, large number of sub-cellular structures such as ribosomes, lysosomes, mitochondria, chloroplasts, etc., have been discovered and studied in detail. Consequently, it may appear that cell is no longer a basic unit of life, because life may exist without cells also. Even then, the cell theory remains a useful concept.

Exception to cell theory. Cell theory does not have universal application, *i.e.*, there are certain living organisms which do not have true cells. All kinds of true cells share the following three basic characteristics: 1. **A set of genes** which constitute the blueprints for regulating cellular activities and making new cells. 2. A limiting **plasma membrane** that permits controlled exchange of matter and energy with the external world. 3. **A metabolic machinery** for sustaining life activities such as growth, reproduction and repair of parts. **Viruses** do not easily fit in these parameters of a true cell. Thus, they lack a plasma membrane and a metabolic machinery for energy production and for the synthesis of proteins. However, like any other cellular organism, viruses have (1) a definite genetically determined macromolecular organization; (2) a genetic or hereditary material in the form of either DNA or RNA; (3) a capacity of auto-reproduction; and (4) a capacity of mutation in their genetic substance. In consequence, viruses can only reproduce inside the host cells which may belong to animals, plants or bacteria (*i.e.*, microorganisms). They use their own genetic programme for reproduction but rely on the raw materials (*i.e.*, amino acids, nucleotides) and biosynthetic machinery of the host cells (*i.e.*, ribosomes, tRNA, enzymes) for their multiplication. Thus, a virus may be defined as an infectious, subcellular and ultramicroscopic particle representing an obligate cellular parasite and a potential pathogen whose reproduction (replication) in the host cell and transmission by infection cause characteristic reaction in the host cells. Outside the host cells, viruses are just like non-living inert particles and like the salt or sugar, they can be purified, crystallized and placed into jars on a shelf for years. Due to this fact, viruses have been variously described as "*naked genes that had somehow acquired the ability to move from one cell to another*" or as "*cellular forms that have degenerated through parasitism*" or as "*primitive organisms that have not reached a cellular state*".

There are certain other organisms such as the protozoan *Paramecium*, the fungus *Rhizopus* and the alga *Vaucheria* (Fig. 1.2B) which do not fit into the purview of the cell theory. All of these organisms have bodies containing undivided mass of protoplasm which lacks cell-like organization and has more than one nucleus. They tend to raise the question that whether cell is a basic unit of structure in them.

1.4. PROTOPLASM THEORY

Up to middle of the 19th century, greater emphasis was given to the cell wall and less to the cellular content. But soon cell biologists started to recognize the importance of "juicy" or "slimy" contents of the cells. In 1835, **Felix Dujardin** termed the jelly-like material within protozoans as sarcode. In 1835, **H.von Mohl** (1805–1875) described cell division. In 1839, the Czech biologist **J.E. Purkinje** (1787–1869) coined the term protoplasm to describe the contents of cells (animal embryos). **Von Mohl**, in 1846, applied the name protoplasm to the contents of embryonic cells of the plants. **Max Schultze**, in 1861, established similarity between sarcode and protoplasm of animal and plant cells and, thus, offering a theory which later on was improved and called protoplasm theory by **O. Hertwig** (1849–1922) in 1892.

Protoplasm theory holds that all living matter, out of which animals and plants are formed, is the protoplasm. The cell is an accumulation of living substance or protoplasm which is limited in space by outer membrane and possesses a nucleus. The protoplasm which is filled in the nucleus is called **nucleoplasm** and that exists between the nucleus and the plasma membrane is called **cytoplasm**.

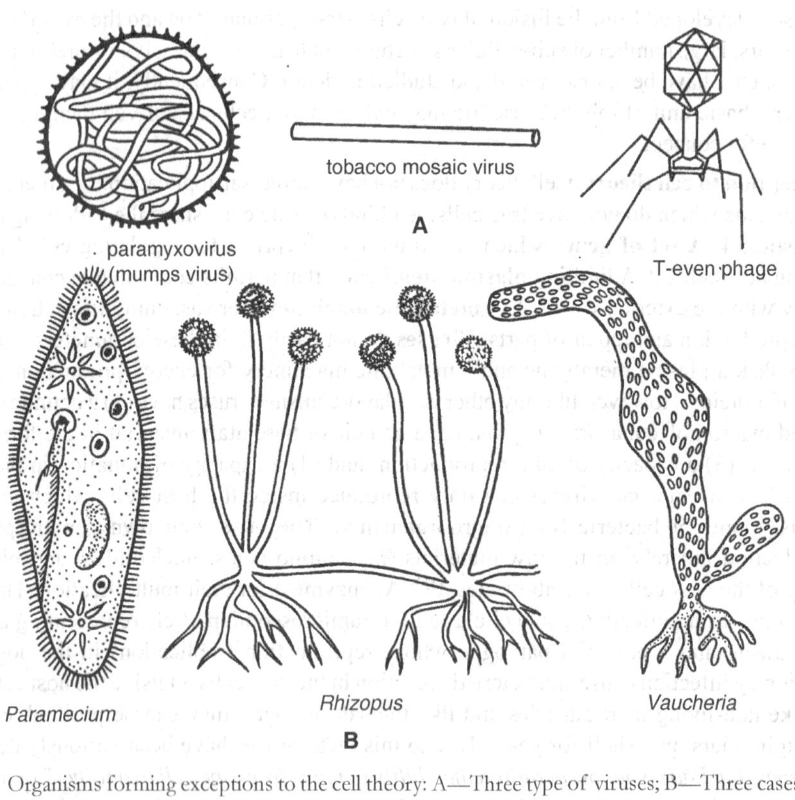

Fig. 1.2. Organisms forming exceptions to the cell theory: A—Three type of viruses; B—Three cases of cellular organization.

1.5 ORGANISMAL THEORY

The organismal theory holds that the body of all multicellular organisms is a continuous mass of protoplasm which remains divided incompletely into small centres, the cells, for the various biological activities. Thus, a multicellular organism is a highly differentiated protoplasmic individual, differing with a unicellular Protozoa only in size and degree of differentiation of the protoplasm. The differentiation involves separation of the protoplasm into subordinate semi-independent compartments, the so-called cells. Even the embryological development of a multicellular individual includes only growth and progressive internal differentiation of a small single protoplasmic individual (egg). Organismal theory too fails to ascertain the position of viruses.

1.6. GROWTH OF CYTOLOGY IN LAST FIVE DECADES

Table 1.3 Nobel Prize awarded for research in Cytology and Molecular biology since 1958 (Source: Karp 2010).

Year	Recipient	Prize	Area of Research
1958	George W. Beadle, Joshua Lederberg and Edward L. Tatum	M&P	Gene expression
	Frederick Sanger	Chemistry	Primary structure of protein
1959	Arthur Kornberg and Severo Ochoa	M&P	Synthesis of DNA and RNA
1960	F. MacFarlane Burnet and Peter B. Medawar	M&P	Clonal selection theory of antibody formation
1961	Melvin Calvin	Chemistry	Biochemistry of CO_2 assimilation during photosynthesis
1962	Francis H.C. Crick, James D. Watson and Maurice H.F. Wilkins	M&P	Three-dimensional structure of DNA molecule
	John C. Kendrew and Max Perutz	Chemistry	Three-dimensional structure of globular proteins
1963	John C. Eccles, Alan L. Hodgkin and Andrew F. Huxley	M&P	Ionic basis of nerve membrane potential
1964	Dorothy C. Hodgkin	Chemistry	X-ray structure of complex biological molecules
1965	Francois Jacob, Andre M. Lwoff and Jacques L. Monod	M&P	Bacterial operons and messenger RNA
1966	Peyton Rous	M&P	Tumor viruses
1968	H. Gobind Khorana, Marshall W. Nirenberg and Robert W. Holley	M&P	Genetic code
1969	Max Delbruck, Alfred D. Hershey and Salvador E. Luria	M&P	Genetic structure of viruses
1970	Bernard Katz and Ulf von Euler	M&P	Nerve impulse propagation
	Leloir	Chemistry	Role of sugar in carbohydrate synthesis
1971	Earl W. Sutherland	M&P	Mechanism of hormone action and cyclic AMP
1972	Gerald Edelman and Rodney R. Porter	M&P	Immunoglobulin structure
	Christian B. Anfinsen	Chemistry	Relationship between primary and tertiary structure of proteins
1974	Albert Claude, Christian de Duve and George F. Palade	M&P	Structure and function of internal components of cells
1975	David Baltimore, Renato Dulbecco and Howard M. Temin	M&P	Reverse transcriptase and tumor virus activity
1976	D. Carleton Gajdusek	M&P	Prion-based diseases

Year	Recipient	Prize	Area of Research
1978	Werner Arber, Daniel Nathans and Hamilton O. Smith	M&P	Restriction endonuclease technology
	Peter Mitchell	Chemistry	Chemiosmotic mechanism of oxidative phosphorylation
1980	Paul Berg, Walter Gilbert and Frederick Sanger	Chemistry	Recombinant DNA technology and DNA sequencing technology
	Baruj Bennacerraf, Jean Dausset and George D. Snell	M&P	Major histocompatibility complex
1982	Aaron Klug	Chemistry	Structure of nucleic acid-protein complexes (TMV, Nucleosomes)
1983	Barbara McClintok	M&P	Mobile elements in the genome
1984	Georges Kohler, Cesar Milstein and Niels K. Jerne	M&P	Monoclonal antibodies
	Bruce Merrifield	Chemistry	Chemical synthesis of peptides
1985	Michael S. Brown and Joseph L. Goldstein	M&P	Regulation of cholesterol metabolism and endocytosis
1986	Rita Levi-Montalcini and Stanley Cohen	M&P	Factors that affect nerve outgrowth
1987	Susuma Tonegawa	M&P	DNA rearrangements responsible for antibody diversity
1988	Johann Deisenhofer, Robert Huber and Hartmut Michel	Chemistry	Bacterial photosynthetic reaction center
1989	J. Michael Bishop and Harold Varmus	M&P	Cellular genes capable of causing malignant transformation
	Thomas R.Cech and Sidney Altman	Chemistry	Ability of RNA to catalyze reactions
1991	Erwin Neher and Bert Sakmann	M&P	Measurement of ion flux by patch clamp recording (cell membrane function)
1992	Edmond Fischer and Edwin Krebs	M&P	Alteration of enzyme activity by phosphorylation/dephosphorylation
1993	Kary Mullis and Michael Smith	Chemistry	Polymerase chain reaction (PCR); Site directed mutagenesis (SDM)
	Richard J. Roberts and Phillip A. Sharp	M&P	Intervening sequences
1994	Alfred Gilman and Martin Rodbell	M&P	Structure and function of GTP-binding (G) proteins
1995	Edward B. Lewis, Christiane Nusslein-Volhard and Eric Wieschaus	M&P	Genetic control of embryonic development
1996	Rolf M. Zinkernagel, and Peter C. Doherty	M&P	Recognition of virus infected cells by the immune system
1997	Jens C. Skou, Paul Boyer, John Walker and Stanley B. Prusiner	Chemistry	Na^+/K^+-ATPase and mechanism of ATP synthesis
1998	Robert Furchgott Louis Ignarro and Ferid Murad	M&P	Nitric oxide (NO) as intercellular messenger
1999	Gunther Blobel	M&P	Protein trafficking
2000	Arvid Carlsson, Paul Greengard and Eric Kandel	M&P	Synaptic transmission and signal transduction

Year	Recipient	Prize	Area of Research
2001	Leland H. Hartwell, Tim Hunt and Paul Nurse	M&P	Control of cell cycle
2002	Sydney Brenner and John Sulston	M&P	Introduction of *Caenorhabditis elegans* as a model organism
	H. Robert Horvitz		Apoptosis in *C. elegans*
	John B. Fenn	Chemistry	Electrospray ionization in mass spectometry (MS)
	Koichi Tanaka		MALDI in MS (MALDI = Matrix-assisted laser desorption ionization)
	Kurt Wuthrich		NMR analysis of proteins (NMR = Nuclear magnetic resonance spectroscopy – a method for determination of three dimensional structure of small proteins)
2003	Peter Agre and Roderick Mackinnon	Chemistry	Structure of membrane channels
2004	Richard Axel and Linda B. Buck	M&P	Olfactory receptors
	Aaron Ciechanover and Avram Hershko	Chemistry	Ubiquitin and proteasome
2006	Andrew Z. Fire, Craig C. Mello and Roger D. Kornberg	M&P	RNA interference (RNAi)
2007	Mario R. Capecchi, Martin J. Evans and Oliver Smithies	M&P	Development techniques for knockout mice
2008	Francoise Barre-Sinoussi, Luc Montagnier	M&P	Discovery of HIV
	Herald Zur Hausen		Role of HIV in cancer
	Martin Chalfie, Osamu Shimomura and Roger Tsien	Chemistry	Discovery and development of GFP (green fluorescent protein)
	Bruce A. Butler and Jules A. Hoffman	M&P	Discoveries concerning activation of innate immunity
	Ralph M. Steiman	M&P	Dendritic cells and its role in adaptive immunity
2010	Ralph M. Steiman	M&P	Dendritic cells and its role in adaptive immunity
2011	Dan Schectman	Chemistry	Discovery of quasi crystals
2012	Robert J. Lefkowitz and Brian K. Kubilha	Chemistry	Studies of G-proteins coupled receptors
	Sir John B. Govdon and Shinya Yamanaka	M&P	Discovery that mature cells can be reprogrammed to become pluripotent

1.7. RELATIONSHIP OF CYTOLOGY WITH OTHER BIOLOGICAL SCIENCES

The cytology has helped the biologists to understand various complicated life activities such as metabolism, growth, differentiation, heredity and evolution at the cellular and molecular levels. Due to its wide application in various branches of biological science, many new hybrid biological sciences have sprang up. Some of them are as follows:

1. Cytotaxonomy (Cytology and Taxonomy). Each plant and animal species has a definite number of chromosomes in its cells and the chromosomes of the individuals of a species resemble closely with one another in shape and size. These characteristics of the chromosomes are helpful to a taxonomist in determining the taxonomical position of a species. Further cytology furnishes strong support to the manner of origin of certain taxonomic units. Therefore, the cytotaxonomy can be defined as a cytological science which provides cytological support to the taxonomic position of any species.

2. Cytogenetics (Cytology and Genetics). Cytogenetics is that branch of cytology which is concerned with the cytological and molecular bases of heredity, variation, mutation, phylogeny, morphogenesis and evolution of organisms. The Weismann's germ plasm theory, Mendel's Laws of inheritance and the concept of gene could be well understood only after the application of cytological concept to the genetics.

3. Cell physiology (Cytology and Physiology). The cell physiology is the study of life activities, *viz.*, nutrition, metabolism, excitability, growth, reproduction or cell division and differentiation of the cells. The cell physiology has helped in understanding various complicated physiological activities at the cellular level.

4. Cytochemistry (Cytology and Biochemistry). The cytochemistry is that branch of cytology which deals with the chemical and physico-chemical analysis of living matter. For example, the cytochemical analysis has revealed the presence of carbohydrates, lipids, proteins, nucleic acids and other organic and inorganic chemical compounds in the cells.

5. Ultrastructure and Molecular Biology. These are the most modem branches of biology in which the merging of cytology with biochemistry, physiochemistry and especially macromolecular and colloidal chemistry becomes increasingly complex. Knowledge of the submicroscopic organisation or ultrastructure of the cell is of fundamental importance because practically all the functional and physiochemical transformations take place with the molecular architecture of the cell and at a molecular level. The recent discoveries in molecular biology such as the discovery of molecular model of **DNA** by **Watson** and **Crick** in 1953, molecular interpretation of protein synthetic mechanism, genetic code, etc., having an extraordinary impact on modern cytology and biology.

6. Cytopathology (Cytology and Pathology). The application of molecular biology to pathological sciences has helped in understanding various human diseases at molecular level. Because most diseases are caused due to disorder of genetic codes in DNA molecule which alter the synthetic process of enzymes and ultimately disturbs metabolic activities of the cell.

7. Cytoecology (Cytology and Ecology). The cytoecology is the science in which one studies the effects of ecological changes on the chromosome number of the cell. The cytological studies on plants and animals have revealed that the ecological habitat and geographical distribution have the correlation with chromosome numbers.

Now we can say that cytology or cell biology has given a great impetus to early and modern biologists to explore new vistas in biological sciences. The modern cytological studies are helpful in undersanding various life activities in terms of molecules, in the curing of various chronic human diseases and in improving the breeds of plants and animals.

1.8. APPLICATIONS OF MOLECULAR CELL BIOLOGY

Understanding the molecular biology of cells is an active area of research that is fundamental to all of the biological sciences. This is true not only from the standpoint of basic science, but also with respect to a growing number of applications in medicine, agriculture, biotechnolgy and biomedical engineering. Especially with the completion of the sequence of the human genome (Box 1.2), progress in cell and molecular biology has opened new horizons in the practice of medicine. Striking examples include the development of new drugs specifically targeted to interfere with the growth of cancer cells and the potential use of stem cells to replace damage tissues and treat patients suffering conditions such as diabetes, Parkinson's disease, Alzheimer's disease, spinal cord injuries and heart disease (**Cooper** and **Hausman**, 2007).

Box 1.2

Entire human genome contains about 3.2 billion bases. Sequencing of entire human genome was accomplished by the **Human Genome Project (HGP)**, a cooperative international effort that began in 1990, involved hundreds of scientists and established the complete sequence of the human genome by 2003 (**Becker *et al.***, 2006).

The first truly complete genome sequence from a single individual is now available. This belongs to **Dr J. Craig Venter** of Celera Genomics himself; the paper describing Dr Venter's genome appearing in the September 3, 2007 issue of an open access journal named ***Plos Biology*** (see **Jain** 2012).

QUESTIONS

Long Answer Questions

1. Give a detailed accout of cell theory.
2. What are the landmark cytological discoveries of 19th and 20th centuries?
3. Write a brief essay on the scope of cytology.

Short Answer Questions

1. What is cell biology? Who had discovered the cell?
2. Who had discovered the cell? What was the contribution of Leeuwenhoek?
3. How would the modern "cell theory" be started?
4. Write short notes on the following:
   (*i*) Cell theory
   (*ii*) Protoplasm theory
   (*iii*) Organismal theory.

Very Short Answer Question

1. Explain the contribution of the following in the discipline of Cell Biology.
   (*i*) Robert Hooke
   (*ii*) C. Benda
   (*iii*) R. Altman
   (*iv*) M. Knoll and Ruska
   (*v*) C. Golgi.
2. Explain /Define the following:
   (*a*) Cell theory

   (*Madras 2011*)

Fill in the Blanks

1. .... are an exception to cell theory.
2. Cell theory was first given by .... and ....

Matching of Columns

	Column A		Column B
(*i*)	Robert Hooke	(*a*)	Protoplasm
(*ii*)	Robert Brown	(*b*)	Cell theory
(*iii*)	Schleiden and Schwann	(*c*)	The term "cell"
(*iv*)	Von Mohl	(*d*)	Nucleus

Multiple Choice Questions

1. Which of the following is an exception to cell theory?
   (*a*) viruses (*b*) mycoplasma
   (*c*) bacteria (*d*) RBC
2. The cell was seen for the first time under the microscope by
   (*a*) Robert Hooke (*b*) Malpighi
   (*c*) Leeuwenhoek (*d*) Grew
3. Cell theory that all plants and animals are made up of small cells was proposed by
   (*a*) Robert Hooke
   (*b*) Knoll and Ruska
   (*c*) Schleiden and Schwann
   (*d*) Watson and Crick
4. The cell theory states that
   (*a*) cells are fundamental structural units of plants and animals
   (*b*) cells reproduce by mitosis and meiosis
   (*c*) all cells are living
   (*d*) all cells have nucleus
5. Which German Physicist invented the electron microscope which won him the 1986 Nobel Prize in Physics?
   (*a*) Ernst Ruska (*b*) Van't Hoff
   (*c*) J.H.D. Jensen (*d*) Eugene P. Winger
6. The term cytology was proposed by
   (*a*) Hertwig (*b*) Robert Hooke
   (*c*) Swanson (*d*) Robert Brown

ANSWERS

Very Short Answer Questions

1. (*i*) Discovery of cell in 1665; (*ii*) Discovery of mitochondria in 1897; (*iii*) Used the term nucleic acid in 1889; (*iv*) Production of first electron microscope in 1932; (*v*) Discovery of Golgi complex in 1898.
2. (*i*) Cell theory is a theory of cellular organization that states that all organisms consist of one or more cells, that the cell is the basic unit of structure for all organisms, and that all cells arise only from preexisting cells.

Fill in the Blanks

1. Viruses 2. Schleiden, Schwann.

Matching of the Columns

(*i*) *c*; (*ii*) *d*; (*iii*) *b*; (*iv*) *a*.

Mutiple Choice Questions

1. (*a*) 2. (*a*) 3. (*c*) 4. (*a*) 5. (*a*) 6. (*a*).

2 Microscopy

A **microscope** is an optical instrument consisting of a single lens or a combination of lenses for magnifying objects too small to be seen or clearly observed by the naked eye. Thus, it is used to obtain magnified images of small objects such as cells of living organisms. Investigations or studies by means of the microscope is called **microscopy**. In cytology or cell biology, various types of microscopes are used to study the structure of cell; some important ones are the following:

1. Optical or light microscopes (simple, monocular, binocular and compound microscopes).
2. Electron microscope.
3. Phase contrast microscope.
4. X-ray diffraction microscope.
5. Interference microscope.
6. Polarized light microscope.
7. Ultra-violet microscope.
8. Ultra-microscope or Dark field microscope.

A **simple microscope** (also called dissecting microscope) has a single lens system. It is composed of a single convex lens. A **compound microscope** consists of two or more **lens systems** or **elements**. A **monocular microscope** contains one ocular or eyepiece while a **binocular microscope** has two oculars.

Here, only two types of microscopes, namely light microscope and electron microscope will be described. (For some historical microscopes see Box 2.1).

Box 2.1 Some Historical Microscopes

1. First simplest compound microscope was designed by **Francis** and **Janssen** in 1590.
2. **Robert Hooke** (1665) constructed the compound microscope (Fig. 2.1).
3. **Anton van Leeuwenhoek** (1675) was the first to use microscope for biological studies. He was first to observe living free cells. His microscope (Fig. 2.2) consisted of a single lens with a high power of magnification and was called **simple microscope**. He made 247 microscopes.

2.1. RESOLVING POWER OF MICROSCOPE

In the search for information about the structure and composition of cells, the cell biologists immediately face two limitations: the exceedingly small dimensions of cells and their component parts and the transparent nature of cells. The diameters of the majority of cells fall within a range of 0.2 and 50 μm. The human eyes have limited distinguishing or resolving power. The ability of an observational instrument such as a human eye or a microscope to reveal details of structure is expressed in terms of **limit of resolution** (l) which is defined as *the smallest distance that may separate two points on an object and still permit their observation as distinct separate points.* The unaided human eye under optimal conditions in green light (to which it is most sensitive) cannot distinguish between points less than about 0.1 µm or 100 µm apart. Structural details smaller than this, *e.g.*, cell, is unresolved unless some instrument capable of higher resolution is used. **Magnification**, *the increase in size of optical image over the size of the object being viewed*, is of no use unless the observational system can resolve the various parts of the structure being examined. Increased magnification without improved resolution results only in a large blurred image. The human eye has no power of magnification, so magnifying glasses may be used to magnify images up to about 10 times. A light compound microscope in which many lenses are combined together has a useful magnification of about 1,500 times.

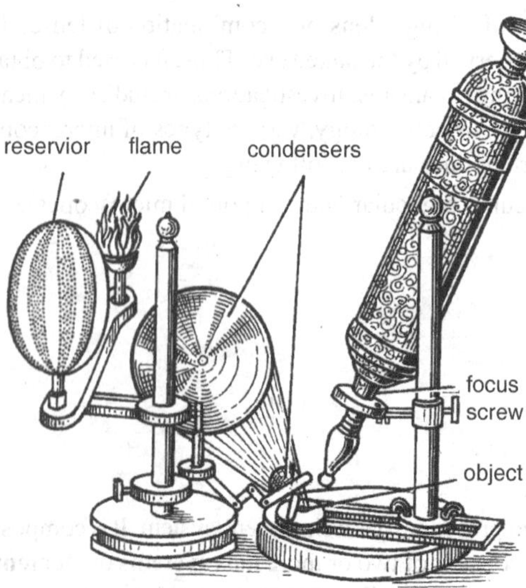

Fig. 2.1. Hooke's compound microscope.

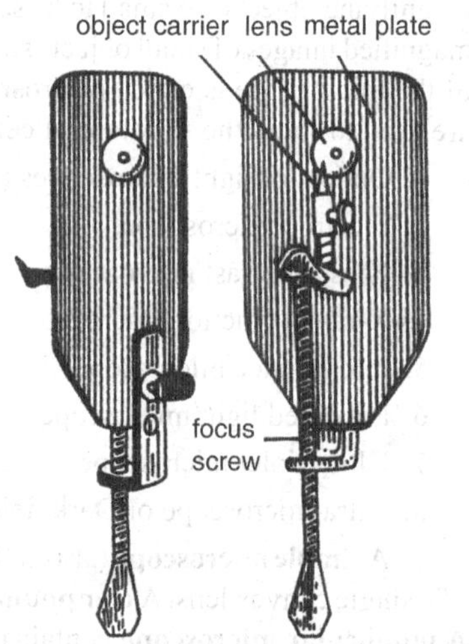

Fig. 2.2. Leeuwenhoek's microscope.

The limit of resolution (l) of any optical instrument (*i.e.*, eye or microscope) is given approximately by the Abbe's relationships:

$$\text{Resolution } (l) = \frac{\text{wevelength } (\lambda)}{\text{numerical aperture } (n \sin \alpha)}$$

where λ (lambda) is the wavelength ("colour") of the illumination or radiation used to form the image, n is refractive index (a function of density) of the material (*i.e.*, mostly air or water) between the specimen and the first lens (or objective lens), and $\sin \alpha$ is sine of the semi-angle of aperture of the first lens as viewed from the specimen. The quantity "$n \sin \alpha$" is often called the **numerical aperture (NA)**.

Abbe's relationships make it clear that high resolution in a microscope can only be achieved by manipulating a small number of variables: the wavelength of the illuminating radiation, the refractive index and the aperture. The **aperture** is limited to something less than 90° since that would have the lens and specimen in contact with one another. In fact, 85° is about the limit in good optical microscopes. Such angles require an excellent lens. In most cases, the aperture is less because the edges of the lens introduce distortions and so cannot be used. **Refractive index** is easy to alter, but only within narrow limits. It can be increased by using oils to fill the space between the specimen and the objective lens. Transparent immersion oils used in today's microscopes (*i.e.*, **oil immersion lens**) have n up to about 1.6. Still 1.6 is big improvement over air or water ($n = 1$). In a microscope, the smallest detectable detail is equal to about one-half the **wavelength** of light with which it is observed. The smaller the object, the shorter the wavelength of light required. Hence, the wavelength of light is the area which has great chances of improvement. One can, for example, use ultraviolet light instead of visible light, thus, improving resolution as much as two-fold. In order to do that however, special lenses (*e.g.*, of quartz) must be used since ordinary glass blocks much ultraviolet light. In such a microscope, called ultraviolet microscope, the eyes cannot be used to view the image directly, for they are insensitive to ultraviolet light. Lastly, a specimen cannot absorb light of wavelength below 0.3 μm.

Thus, a good light microscope, with a numerical aperture of 1.4 and using light of short wavelength (0.4 μm) will resolve two points at about 0.17 μm separations. By such a microscope though, one can see considerable details in most cells, there is also a great deal that cannot be seen. For instance, ribosomes and chromatin threads of nucleus are about 0.02 mm in diameter and quite invisible to the light microscope. For them electron microscope is used. In cell biological studies, the following two types of microscopes are most extensively used:

2.2. LIGHT MICROSCOPES (OPTICAL MICROSCOPES)

The compound light microscope (Fig. 2.3) is the simplest and most commonly used microscope. It consists of the following three lens system.

(*i*) **Condenser lens system** occurs beneath the specimen and its function is the collection and focussing of the light rays on the object or specimen which is placed on the stage of the microscope.

(*ii*) **Objective lens system** remains near and above the specimen. It produces and magnifies the image of the specimen.

(*iii*) **Eye piece lens system** or ocular lens system remains near the eyes of the observer and it magnifies and forms the image (secondary) of the (primary) image previously produced by the objective.

Lenses. A simple lens can magnify an object only three times (3X). For getting a magnification of more than 3X, a combination of several lenses is used. Such a combination of several lenses (called **elements**) function as a single convex lens and a magnification of about 20X can be obtained. For example, an **eye piece lens** contains 2 to 3 lenses or elements. Likewise, the objective lens is made by placing 8 to 10 lenses or elements. Individual lenses are placed so close to each other that they act as a single lens. Only achromatic lenses are used in a microscope. (For certain other facts see Box 2.2).

Box 2.2 Magnification Power of Compound Microscope

1. In compound light microscope natural light or light rays from a source (bulb, etc.) are used.
2. The mirror of the microscope is **plano-concave** and is used to focus light rays on the object through substage condenser. Its concave surface is used without condenser and plain surface is used with condenser.
3. An eye piece usually has either 5X, 10X or 15X lenses. It means, it can magnify an image 5 times, 10 times or 15 times, respectively.
4. Each objective too has either 10X, 40X or 100X magnification power.
5. By using 15X eye piece and 100X objective, a maximum magnification of 1500 time (*i.e.*, 15 × 100 = 1500X) may be obtained under a best type of compound microscope (*e.g.*, research microscope).

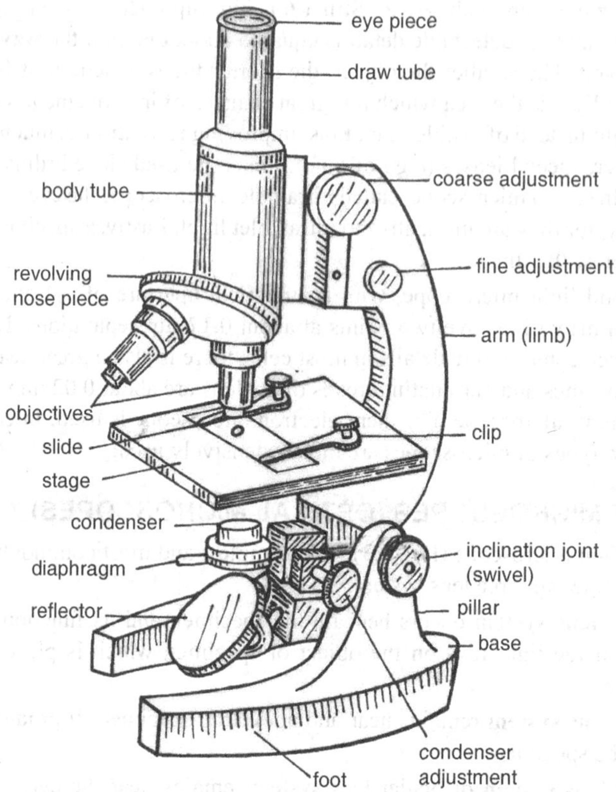

Fig. 2.3. Light or compound microscope.

Working. Light rays from the light source (sunlight or bulb-light) are focussed on the object by the condenser lens. The objective lens produces the enlarged image, called **primary image**, of the object or material. The eye piece lens further magnifies the primary image into secondary image which is seen by our eye. Cellular organelles such as mitochondria and bacteria, both of which are 0.5 μm wide, are the smallest objects that can be clearly seen in the light microscope.

2.3. ELECTRON MICROSCOPE

Two basically different types of electron microscopes (Fig. 2.4) have been developed. **Transmission electron microscopes (TEMs)** form images from electrons that are transmitted through a specimen. Whereas **scanning electron microscopes (SEMs)** utilize electrons that have bounced off the surface of the specimen. A transmission electron microscope has a very high magnification (500,000 times). The TEM was designed by **Knoll** and **Ruska** of Germany in 1932. It permits direct study of biological ultrastructures (*e.g.*, structure of cell organelles). The practical limit of resolution of electron microscope is about 3 to 5 Å.

In TEM, the illuminating agent is not light but the **electrons** of short wavelength (0.500 Å). The wavelength of electrons is determined by the voltage at which these are generated. For example, at 50,000 volts, the electrons have 0.50 Å wavelength and resolution power of electron microscope can be 0.50 Å/2 = 0.25 Å. Similarly, in an electron microscope with a voltage of 100,000 volts, the wavelength of an electron is 0.004 nm (or .04 Å). Theoretically the resolution of such a microscope should be 0.02 Å. But practical resolving power of most modern electron microscope is 0.5 μm or 5 Å. This is 400 times greater than that of a light microscope.

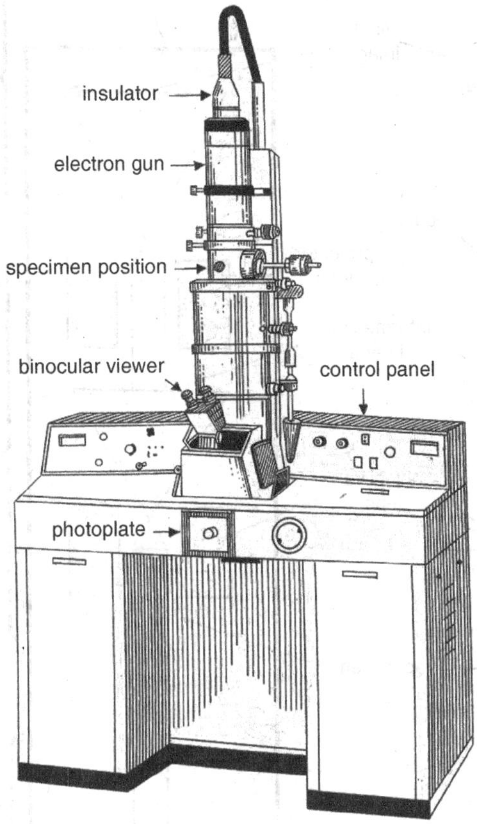

Fig. 2.4. An electron microscope.

Structure

The transmission electron microscope comprises the following parts:

1. Electron gun and anode. In TEM, the source of illumination is electrons which are produced and concentrated into an electron beam by **electron gun**. Electron gun consists of an electrically heated **tungsten filament** or **cathode** that emits electrons. Outside the gun is an anode which attracts electrons. Thus, between the filament and anode the electrons attain a high velocity with wavelength of .005 to .003 nm.

2. Microscope column. Electrons can travel in straight line in vacuum only, because in the air, these collide with oxygen or nitrogen atoms. The electron microscope, therefore, is enclosed in a evacuated metal tube.

3. Condenser lens. It is the **electromagnetic coil** which focusses or condenses the electron beam in the plane of object.

4. Objective lens. It is the **electromagnetic coil** which produces first magnified image of the object. It focusses the electrons which are reflected by the object and forms the first image.

5. Projector lens. It is also the electromagnetic coil which magnifies the first image formed by objective lens. It produces the final image.

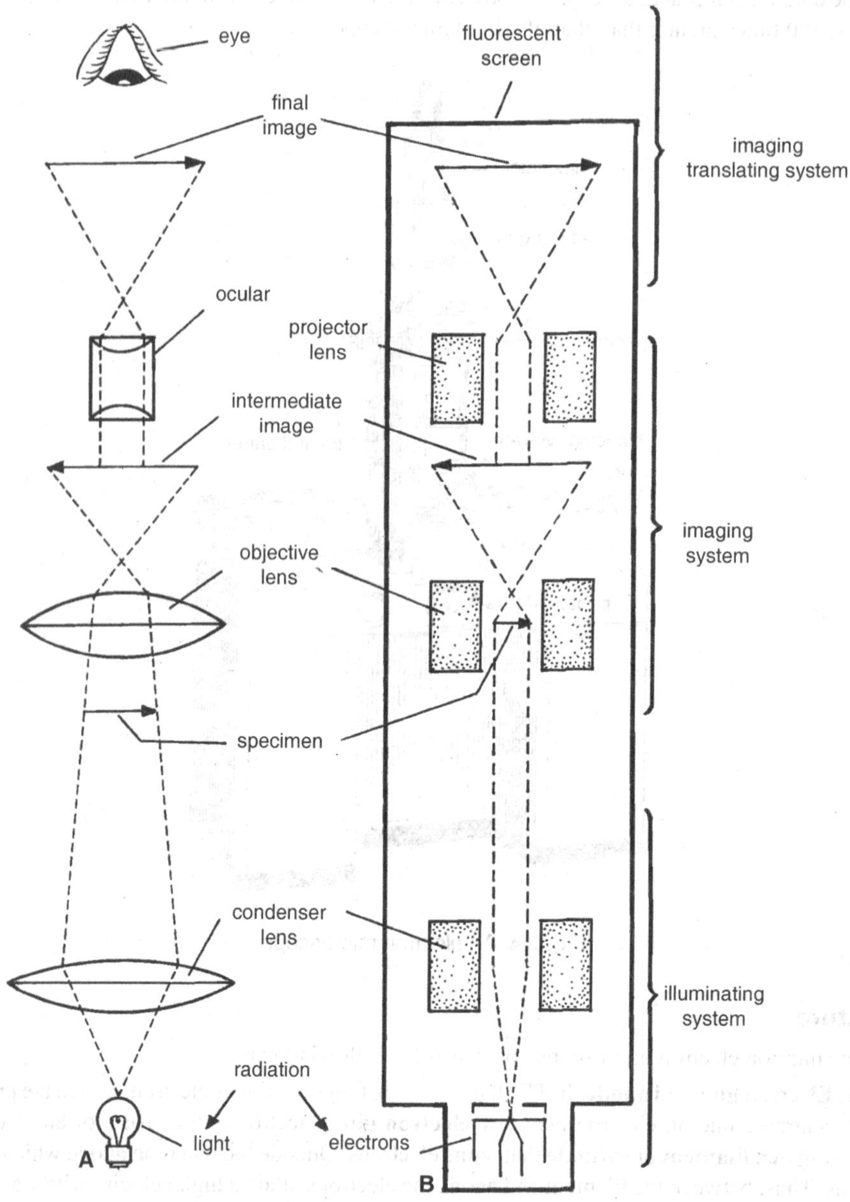

Fig. 2.5. A comparison of the lens systems of light (A) and electron microscope (B).

6. Fluorescent screen or photographic plate. Fluorescent screen is used for observing the magnified image of the object. It remains coated with a chemical (*e.g.*, zinc sulphide) which on being excited form the image as on the screen of television. The final image can also be captured on photographic film; such photographs are known as **electron micrographs**.

In the light microscope, the focussing is done by moving the lens tube up and down. In TEM, focussing is done by changing the flow of current through the coils of the electromagnetic lenses.

Further, in electron microscopy only dead and dried specimens are studied. With this microscope, living cells cannot be studied, since they possess water which causes large scattering of the electrons.

Differences Between Electron and Light Microscope

Table 2.1. Main differences in electron and light microscope.

	Feature	Electron microscope	Light microscope
1.	Radiation source	Electrons; their source is located on the top of the microscope	Light, its source is located at the bottom of microscope
2.	Wavelength	0.005 nm at 50 kV	400–700 nm
3.	Maximum useful magnification	× 250,000 (on screen)	× 1500
4.	Maximum resolution		
	(*i*) in practice	0.5 nm	200–250 nm
	(*ii*) in theory	0.2 nm	200 nm
5.	Lenses	Electromagnets	Glass (quartz for ultraviolet radiation)
6.	Specimen	Non-living, dehydrated, relatively small or thin.	Living or non-living
7.	Support of object	Supported on a small copper grid in a vacuum	Supported on a glass slide
8.	Common stains	They include heavy metals to reflect electrons such as lead acetate, lead citrate, lead hydroxide, uranyl acetate and phosphotungistic acid	Coloured dyes such as borax carmine, haematoxylin, eosin
9.	Image	Black and white	Usually coloured
10.	Mode of observation or recording	Image is either seen on a fluorescent screen or recorded on a photographic film	Image can be seen with eye or recorded on photographic film by camera

2.4. EXAMPLES OF OPTICAL MICROSCOPES

1. Phase Contrast Microscope

Phase contrast microscope was invented by **Frits Zernike** of Netherland in 1932. He got Nobel Prize for this invention in 1953.

Principle and working. Small, unstained specimens, such as a living cell, can be very difficult to see in a bright-field microscope. The phase contrast microscope has solved this problem by making highly transparent objects more visible. The ability to see different parts of an object depends on their capacity to affect light differently from one another. One basis upon which intracellular organelles differ is their refractive index. Cell organelles are made up of different proportions of various molecules: DNA, RNA, protein, lipid, carbohydrate, salts and water. Regions of different composition are likely to have different refractive indexes. Normally, however, such differences cannot be detected by our eyes. The phase contrast microscope converts differences in a refractive index into differences in intensity (relative brightness and darkness) which are then visible to the eye. The basis for this conversion centres on the ability to light waves to interact with one another, a property termed **interference**.

In contrast to bright-field microscope, phase contrast microscope performs two functions:

1. It separates the direct light (the background light of the field) from the light diffracted by the object, and,
2. It causes these two types of waves to be approximately one-half wavelength out of phase with one another so that they can destructively interact and cause changes in intensity.

Suppose that all parts of the object being examined have exactly the same refractive index as the medium in which they are suspended. If this were the case, all the light from the object and the direct light would be one-half wave-length out of phase with one another, and **interference** would result in a uniform decrease in intensity (Fig. 2.6). Now consider an object whose various parts have different refractive indexes, all of which are greater than the surrounding medium. Differences in refractive index result in **phase shifts** of light rays passing through different parts of the specimen (Fig. 2.7). The relative brightness (**negative contrast**) or darkness (**positive contrast**) of each part of the image reflects the way in which the light from that part of the specimen interferes with the direct light. Since parts of the specimen have different refractive indexes, the various parts of the specimen appear in the image with varying intensity, thereby providing the necessary contrast to be visualized.

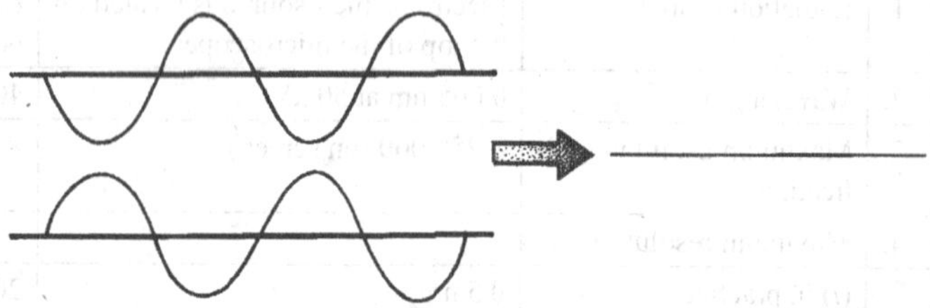

Fig. 2.6. Destructive interference. The two waves of light are one-half wavelength out of phase. When the waves meet, they combine to cancel each other out.

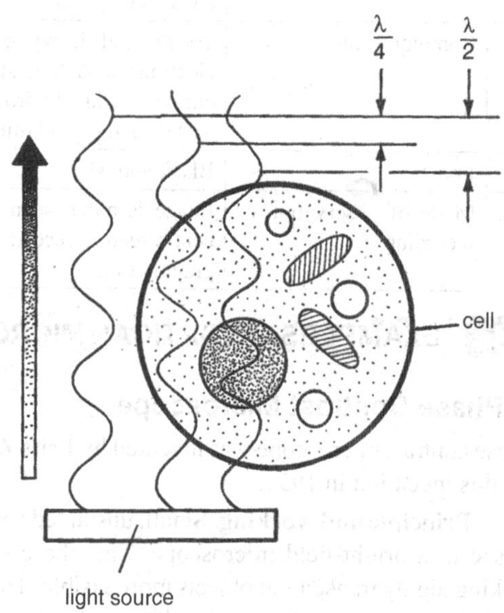

Fig. 2.7. Schematic diagram of retardation of light rays to varying degrees as they pass through different parts of a cell. Because the light rays that emerge from different parts of the cell are in *different phases* of their cycle, they interfere with the direct light to different degrees, producing different degrees of lightness and darkness in the final image.

Structure. The phase contrast microscope contains an annular phase plate in the objective and an annular diaphragm in the condenser (Fig. 2.8). The annular diaphragm of phase contrast microscope permits light to pass through the condenser as a hollow cone, the remaining light being absorbed. This cone is focussed on the object. The phase plate which is placed at the back focal plane of the objective lens, is a transparent disc containing a groove (or elevation) of such a size and shape as to coincide with the direct image of the substage annular diaphragm which is formed at the back focal plane when no object is viewed. If an object is placed between the condenser and the objective, in addition to direct image a number of overlapping diffracted images of the diaphragm then appear at the back focal plane of the objective. The depth of the groove (or the elevation) in the phase plate of the objective is so made that the two sets of rays forming the direct image and the diffracted image differ in optical path by a quarter wavelength of the illuminating beam of light. Under these conditions the phase difference, which is not seen by the eye, is converted to intensify difference which we see. In the bright contrast phase optical system the two sets of light rays are added to make a brighter image and in the dark contrast phase they partially cancel one another making a more contrasting darker image. The phase contrast microscope helps in the observation of behaviour of chromosomes of living cells during mitosis or meiosis and various other cellular components such as mitochondria and vacuoles. Observations of living cells on a television screen using a video camera provide especially dramatic portraits of the dynamic activities of a cell (Table 2.2).

2. Dark Field Microscope

Any ordinary light microscope in which light from the illuminating source is caused to converge on the specimen by the substage condenser, thereby forming a cone of bright light that can enter the objective lens, is called a **bright field microscope**. The cone of illuminating light is seen as a bright background against which the image of the specimen must be contrasted. Bright field microscopy is ideally suited only for specimens of high contrast, such as stained sections of tissues (Fig. 2.9).

The dark field microscope is an ordinary compound microscope which has a special condenser. In **dark field microscope**, an opaque disc is placed in the centre of the condenser. The light rays are thus removed from the centre and no direct

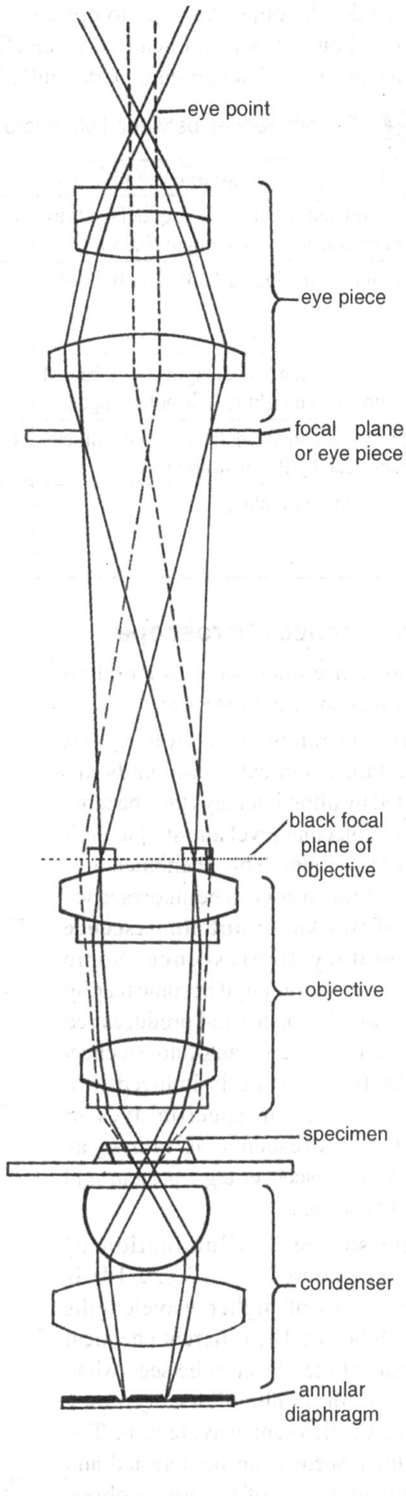

Fig. 2.8. Light path in phase contrast microscope.

light rays enter the objective. Due to opaque disc, light rays are directed from the sides and only scattered (oblique) rays tend to enter the objective lens. As a result, the object appears illuminated (bright) against black background. Dark field microscopy improves visibility and resolution.

Table 2.2 The differences between light microscope and phase contrast microscope.

	Light microscope		Phase contrast microscope
1.	The refracted light rays are delayed by about one-quarter wavelength (0.25 λ).	1.	The refracted light rays are delayed by one-half wavelength (0.50 λ).
2.	All points of the image look equally bright.	2.	The brightness of different points of image differs. It corresponds to differences in the refractive indexes of various cellular organelles.
3.	Difference in the wavelengths of refracted and unrefracted light rays is not changed.	3.	This increases the difference to produce maximum contrast (*i.e.*, interference) in the image.
4.	Both living and dead (stained) specimens may be seen by this instrument.	4.	Living (unstained) structures can be seen by this instrument.
5.	There is no phase plate.	5.	In this microscope, a glass phase plate is placed above the objective lens to increase the phase difference between the refracted and unrefracted light rays.

3. Fluorescence Microscope

The fluorescence microscope is based on the phenomenon of fluorescence.

Fluorescence. It is found that when certain chemical compounds are illuminated by ultraviolet rays they become excited at molecular level and start to emit light of visible range. This phenomenon is known as **fluorescence**. The fluorescence may be of two kinds: **autofluorescence** and **secondary fluorescence**. Some substance, *e.g.*, chlorophyll become readily fluoresces or illuminates and produces red light, is an example of autofluorescence. When the fluorescence is induced in a substance by certain specific dyes or stains, the fluorescence is known as secondary fluorescence, *e.g.*, protein and carbohydrates, etc.

The source of illumination of fluorescence microscope (Fig. 2.10) is ultraviolet rays of higher wavelengths (3500 to 4000 Å). The different chemical compounds of the object to be seen when fluoresce by these ultraviolet rays, they emit rays of different wavelength. The rays of fluorescence can be detected and the chemical nature of the given object can be tested.

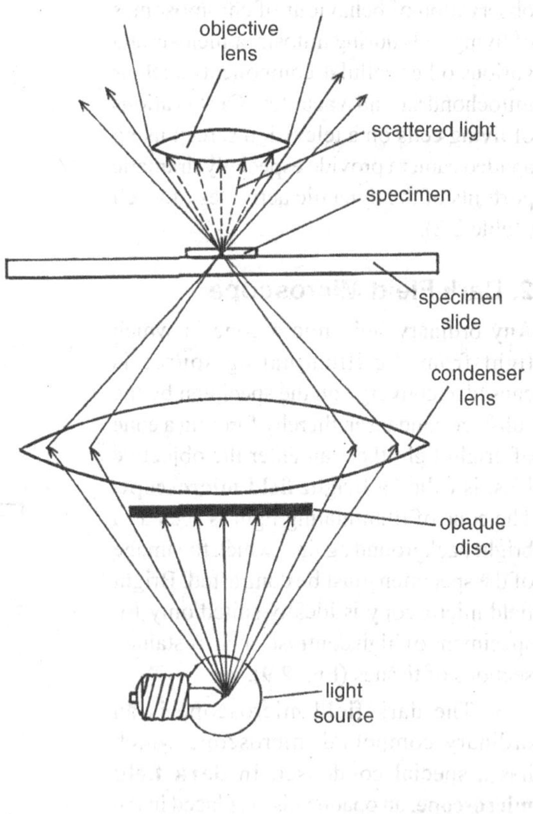

Fig. 2.9. Dark field microscope. Light path from the condenser to the objective lens.

The ordinary compound microscope can be easily converted into the fluorescence microscope by placing a special filter in between the source of illumination and eye piece.

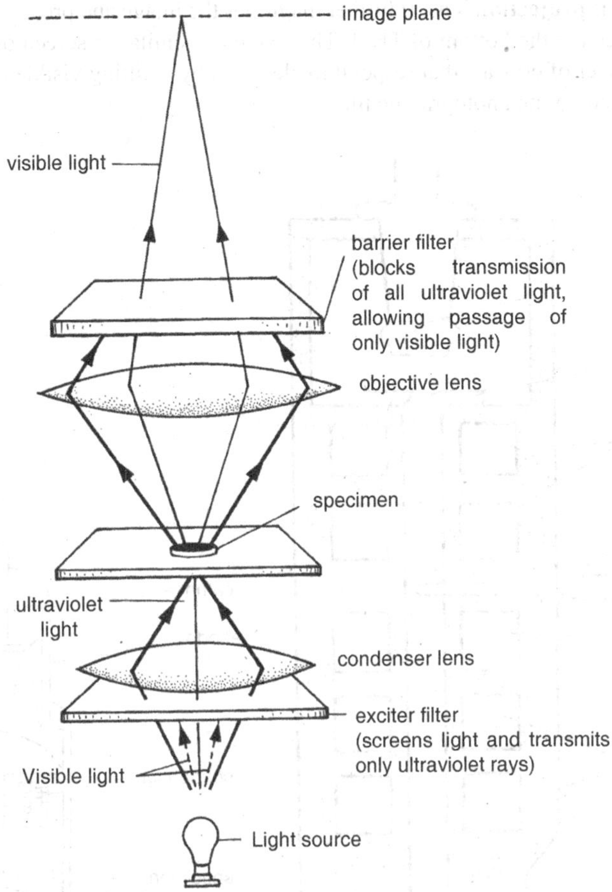

Fig. 2.10. Optics of the fluorescence microscope.

2.5. EXAMPLES OF ELECTRON MICROSCOPES

There are two types of electron microscopes:

1. Transmission electron microscope (TEM)
2. Scanning electron microscope (SEM)

The construction and operation of the TEM are very different from that of the SEM. The image formed by these two types of electron microscopes is also quite different.

1. Transmission Electron Microscope (TEM)

TEM (Fig. 2.11) is the most commonly used electron microscope. It is widely used to study the internal structure of the cell and has great resolving power. In TEM, an electric field propels electrons from a negatively charged electrode (**cathode** or electron gun). The electrons emitted from the gun are attracted towards the positively charged anode and then are focussed into beam by the electromagnetic **condenser**. This beam of electrons illuminates entire stained specimen. Many of these electrons pass right through the specimen and reach the electromagnetic objective. But some are absorbed or scattered by atoms of metallic stains that have combined with certain atoms of specimen. The **objective** magnifies the image of the specimen. This image is received by the third electromagnetic lens which works as **projection lens.** It further magnifies the image and projects it on a zinc sulphide or fluorescent screen at the bottom of TEM. This screen is similar to screen of a television tube. It is coated with a layer of crystals that respond to electrons by emitting visible light. The final image can also be captured on the photographic film.

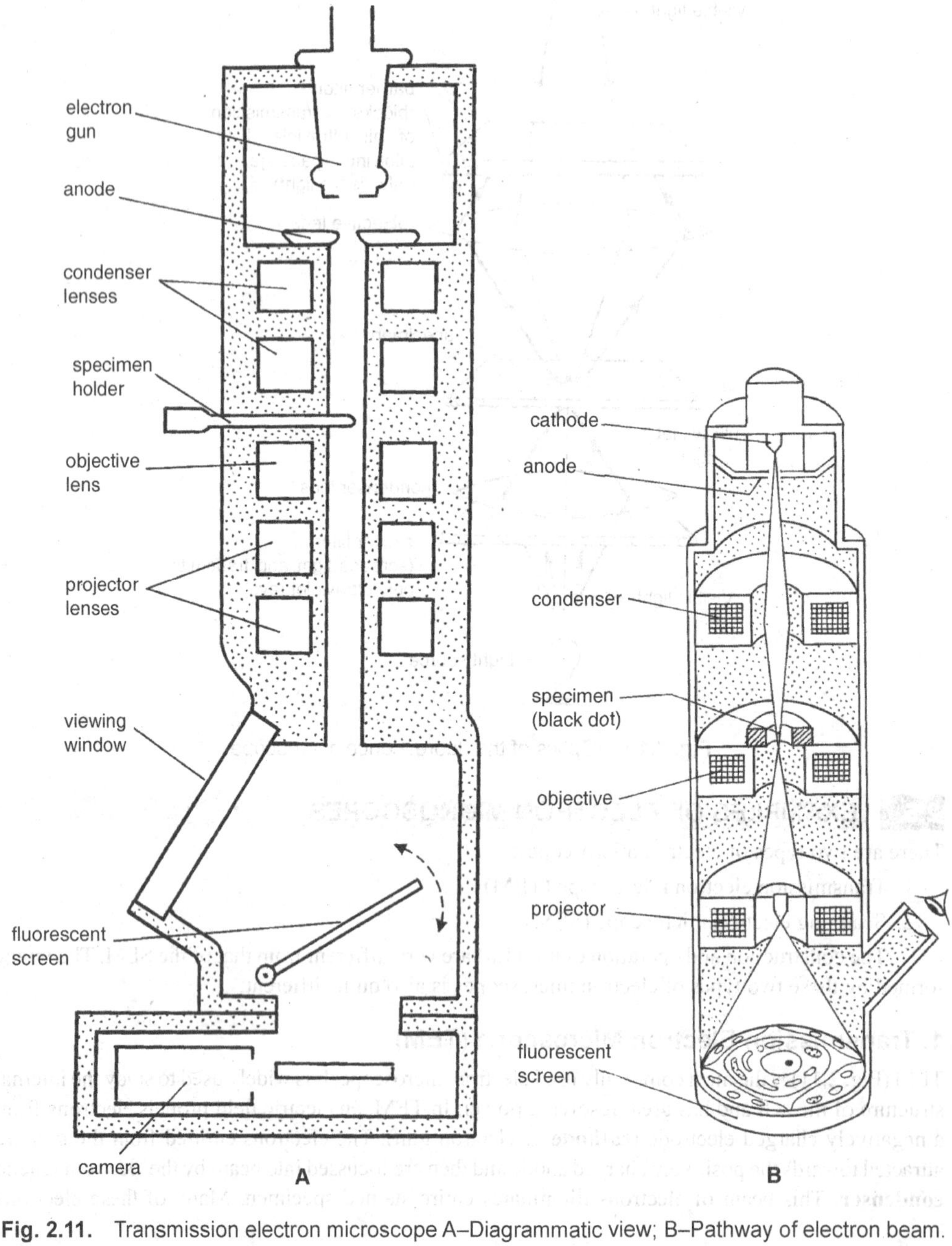

Fig. 2.11. Transmission electron microscope A–Diagrammatic view; B–Pathway of electron beam.

Electrons that are not transmitted by the specimen leave correspondingly dark regions on the viewing plate. Thus, image from a TEM is a pattern of bright and dark areas corresponding to the areas of greater or lesser electron density.

2. Scanning Electron Microscope (SEM)

The electron gun and lens system of SEM is like TEM but its operation is different.

The specimen to be examined is fixed and dried (by the technique of **critical point drying; Karp** 1996) and then coated with a layer of carbon and heavy metal such as gold or gold-palladium—a process called **shadowing**. This step makes specimen suitable as a target for an electron beam. SEM is used to examine the surface of specimen, *i.e.*, outer cell surface and various processes, extensions and extra-cellular materials. SEM provides a three-dimensional image of a specimen.

In SEM, an extremely fine beam of electrons (5 to 20 nm in diameter) is made at 3–30 kV for scanning a selected area of specimen. In SEM, the electron beam does not pass through the specimen. The condenser lens focusses a fine electron beam on the surface of the specimen. The beam is moved rapidly back and forth by **beam deflectors** to scan the specimen surface. As the electron beam hits the surface of specimen it excites the specimen molecules to high energy levels. Due to the this, **secondary electrons** are emitted from the metallic surface. These secondary electrons are collected by the positively charged grid. The **collector** gives rise to a flash of light in a solid scintillator. The light output is amplified in a **photo-multiplier** or **video amplifier** (Fig. 2.12).

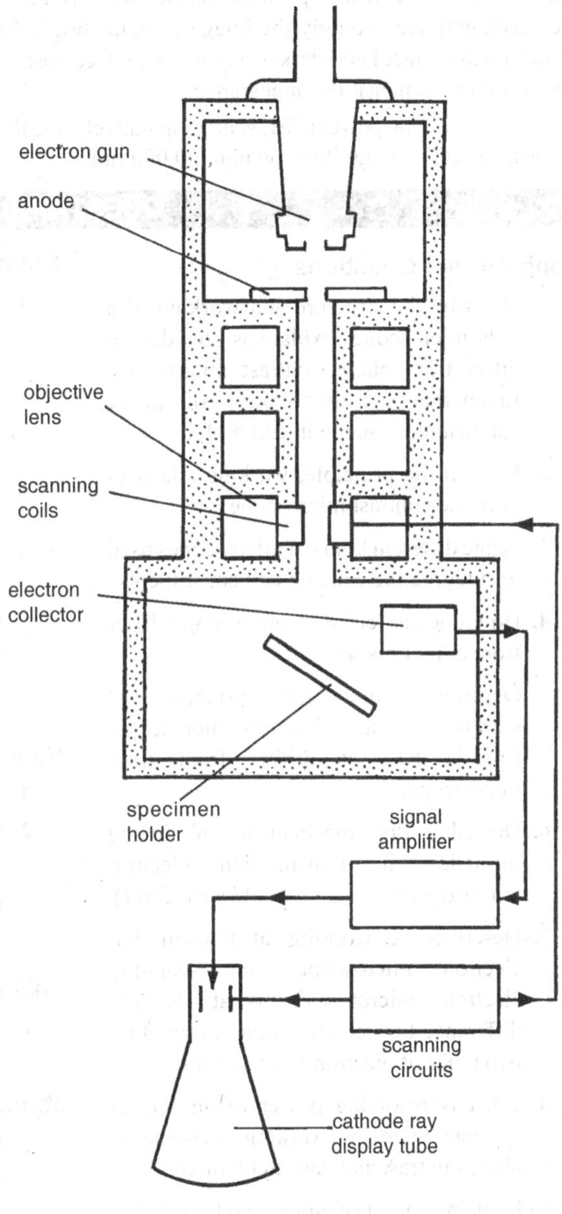

Fig. 2.12. A scanning electron microscope.

The signal from the grid is transferred to a television tube which scans and forms the image on the screen. Thus image formation in the SEM is indirect compared to that in the TEM.

The amount of secondary electrons produced depends on the angle of specimen points with the scanning beam. Surface perpendicular to the beam produces maximum electrons while surfaces at greater angles release less electrons. Thus, the number of electrons produced depends on the three-dimensional shape of the specimen surface (or surface topology, *i.e.*, the crevices, hills and pits of the specimen). Accordingly the image contains bright (which correspond the elevations or ridges in specimen surface) and darker regions (which correspond to the valleys). These shadows give the image a three-dimensional appearance.

The resolving power of SEM is comparatively less than that of transmission electron microscope. It has an effective magnification upto 20,000 times.

QUESTIONS

Long Answer Questions

1. Describe the structure and working of a light microscope. Explain as how does it differ from phase contrast microscope. Enumerate the importance of phase contrast microscope in cell biology.
2. Explain the principle, working and uses of phase contrast microscope.
3. Name different kinds of microscopes used in studying the cell and cell constituents.
4. How does an electron microscope differ from a light microscope?
5. Describe composition, principle of working of an electron microscope. Explain how it differs from light microscope.
6. Describe the mechanism of image formation in transmission electron microsope. (*Madras 2011*)
7. Describe the working of transmission electron microscope and scanning electron microscope. What is the difference between the image formed by two types of electron microscopes?
8. What is resolving power of an optical system? State the working principle of phase contrast and dark field microscopy.
9. Explain the principles and working mechanism of fluorescence microscope.
10. What is a dark field microscope?

Short Answer Questions

1. What is the limitation of a microscope? Describe the difference between light microscopy and electron microscopy with the help of schematic diagrams.
2. Describe simple microscope. (*Madras 2011*)
3. Describe the principle of working of transmission electron microscope.
4. Write short note on the following.
   (*i*) Electron microscopy
   (*ii*) Resolving power of microscope
   (*iii*) Compound microscope
   (*iv*) Fundamentals of TEM
   (*v*) Phase contrast microscope

Very Short Answer Questions

1. What is electron microscope?
2. Explain
   (*i*) SEM (*Madras 2011*)
   (*ii*) Compound microscope (*Madras 2011*)

Fill in the Blanks

1. Ultrastructure of cell organelles can be studied by ....

Multiple Choice Questions

1. Bacterial cells were first seen by
   (*a*) Robert Brown
   (*b*) Robert Hooke
   (*c*) A. Leeuvenhoek
   (*d*) R. Virchow

2. Living cells can be studied by
   (*a*) phase contract microscope
   (*b*) flurorescent microscope
   (*c*) electron microscope
   (*d*) light microscope
3. The number of lenses in compound light microscope is
   (*a*) 2
   (*b*) 3
   (*c*) 4
   (*d*) 1
4. To determine the ultrastructure of a cell organelle the most likely method to be used would be
   (*a*) phase contrast microscope
   (*b*) light microscope
   (*c*) electron microscope
   (*d*) microdissection

ANSWERS

Very Short Answer Questions

1. It is an instrument that uses a beam of electrons to visualize cellular structures and thereby examine cellular architecture; the resolution is much greater than that of the light microscope, allowing detailed ultrastructural examination.
2. (*i*) SEM stands for scanning electron microscope. It is a microscope in which an electron beam scans across the surface of a specimen and forms an image from electrons that deflected from the outer surface of the specimen.
2. (*ii*) Compound microscope is the light microscope that uses several lenses in combination; usually has a condenser lens, an objective lens and an ocular lens.

Fill in the Blanks

1. Electron microscope.

Multiple Choice Questions

1. (*c*) 2. (*a*) 3. (*a*) 4. (*c*).

3 Tools and Techniques in Cell Biology

3.1. CELL FRACTIONATION AND CENTRIFUGATION

Sometimes it becomes necessary to break up tissues and cells and to isolate various parts of the cell for structural or biochemical analysis. For this purpose, the technique of cell fractionation is employed. Cell fractionation method involves, essentially the homogenisation or destruction of cell boundaries by different mechanical or chemical procedures, followed by the separation of the subcellular fractions according to mass, surface and specific gravity by centrifuges.

The centrifuge is the most important tool available to cell biologists for the separation and analysis of subcellular organelles and their constituent macromolecules. Centrifugal force is generated by placing test tubes containing samples to be fractionated in a holder called a **rotor** and then spinning the rotor at high speeds by the help of a **motor** in a centrifuge. The ultra-centrifuge, developed by the Swedish chemist, **Theodor Svedberg** between 1920 to 1940, is a special high-speed instrument capable of spinning rotors at speeds of 100,000 revolutions per minute or more. The large forces generated by these high speeds make it possible to isolate not just organelles but macromolecules such as proteins and nucleic acids as well.

Stokes formula. The rate of movement of any particle subjected to a centrifugal force is given by **Stokes formula**:

$$\frac{dx}{dt} = \frac{2r^2(\rho_p - \rho_M)}{9\eta}.g$$

where dx/dt is the velocity with which the particle moves towards the bottom of the tube, r is the particle radius, ρ_p and ρ_M are the respective densities of the particle and the suspending medium and g is the centrifugal force exerted on the particle. This formula tells us that when the density of a particle exceeds that of the medium, its velocity will be positive and the particle will migrate in the direction of the force field (towards the bottom of tube). If the particle is less dense than the medium, its velocity will be negative, meaning that it will move towards the top of the tube. Finally, if the density of particle is exactly equal to the density of the suspending medium, the particle will not move because the term, $\rho_p - \rho_M = 0$.

Types of Centrifugation

(*a*) **Velocity centrifugation** (It separates organelles and molecules based on differences in size)

(*i*) **Differential centrifugation**

(*ii*) **Moving zone centrifugation**

(*b*) **Isodensity centrifugation** (It separates organelles and molecules on the basis of differences in density).

For these centrifugations, the following three centrifugation processes are employed:

1. **High speed centrifugation.** In high speed centrifugation rotor runs at a speed of about 20,000 rpm and generates a gravitational pull of about 50,000 g. At this speed most microorganisms, cell debris and cell organelles form sediments. It is used in **differential centrifugation.**
2. **Ultracentrifugation.** Ultracentrifuge (Fig. 3.1) of **Svedburg** operates at extremely high speed (about 75,000 rpm) and produces a gravitational pull of about 500,000 g. It is used for **differential centrifugation**, **zonal centrifugation**, **density gradient centrifugation** and **isopycnic centrifugation**.
3. **Analytical centrifugation.** This is the latest and most sophisticated instrument. Its sample container has a quartz window and the rotor has an optical system. It is possible to observe the material sedimenting and measure the sedimentation rate during the run.

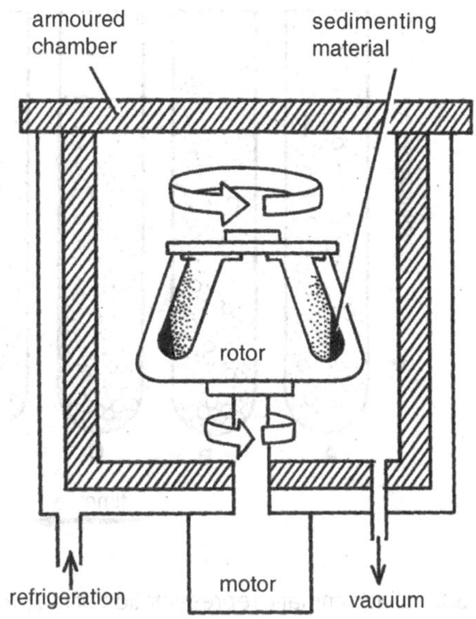

Fig. 3.1. The ultracentrifuge. The sample (*i.e.*, homogenate) is contained in tubes that are inserted into a ring of cylindrical holes in a metal rotor. Rapid rotation of the rotor generates immense centrifugal forces, which cause particles in sample to sediment. The vacuum tends to reduce friction, preventing heating of the rotor and allowing the refrigeration system to maintain the sample at 4°C.

Mechanism of Centrifugation (Cell Fractionation)

1. In the cell fractionation, the cells are gently broken by grinding a small piece of tissue in a homogeniser having a moving close-fitting glass or plastic pestle within a tube that contains a medium such as sucrose solution to preserve the cellular organelles (Fig. 3.3). The solution containing homogenised or disrupted cells, is called **homogenate**. In a homogenate the particles of different sizes are distributed homogeneously. The homogenate is subjected differential centrifugation of increasing velocity.
2. The tissue/liver homogenate is subjected first to low speed centrifugation at 700X g for 10 minutes. This sediments cell nuclei to the bottom of centrifuge tube. This **sediment-I** is called **nuclear fraction**. Nuclear fraction includes nucleus, intact cells, tissue debris and chloroplasts (in case of plant cells).
3. The overlying **supernatant-I** is transferred to another tube for second centrifugation at 10,000Xg for 20 minutes. The **sediment-II** is called **mitochondrial fraction**. It contains mitochondria and also lysosomes and peroxisomes.
4. Again the **supernatant-II** is removed and subjected to third centrifugation at (~ 100,000Xg) for 60 minutes. This causes the sedimentation of **microsomal fraction** which includes free ribosomes, fragments of smooth ER, rough ER, Golgi apparatus and plasma membrane.

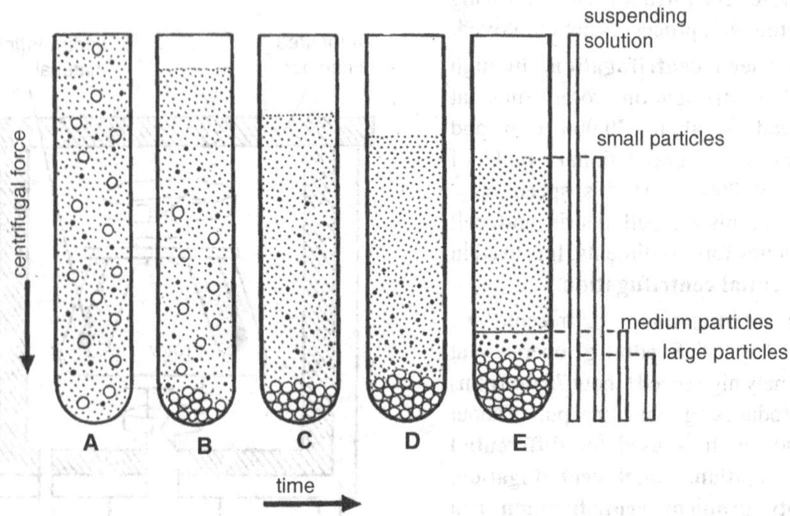

Fig. 3.2. Diagrammatic representation of the process of centrifugation and sedimentation of particles of different sizes.

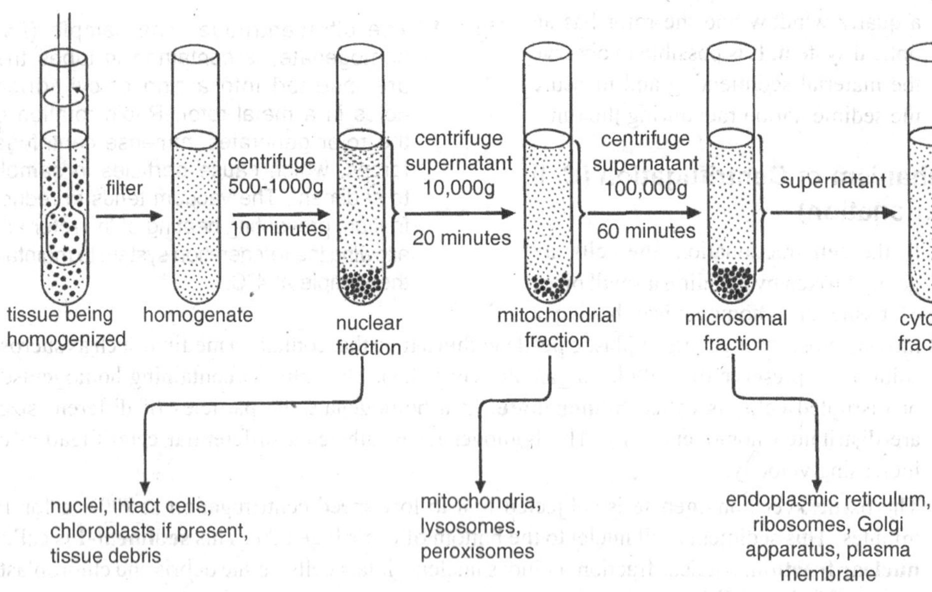

Fig. 3.3. Subcellular fractionation by differential centrifugation.

5. The **final supernatant** contains soluble molecules of cytoplasm and is called **cytosol fractions**. It contains proteins, soluble nucleic acids, soluble polysaccharides, lipid droplets and other particles. These are separated by **chromatography**, **dialysis** and **electrophoresis**.

The technique of cell fractionation has been improved through the use of **density gradient centrifugation**. In this method, centrifuge tube is loaded with layers of solution of varying densities of either sucrose, heavy water, cesium chloride or albumin, in a gradient from top to bottom. Once the gradient is formed, the homogenate is layered on the top and centrifuged until the particles reach equilibrium with the gradient. For this reason, this type of separation is called **equilibrium density** or **isopycnic centrifugation**.

3.2. AUTORADIOGRAPHY

Autoradiography is a technique which is used to locate radioactive isotopes in cells, tissues, organs and whole organisms. A specimen is exposed to a solution containing molecules that have been made radioactive by the incorporation of radioactive isotopes, such as tritium (^3H), carbon 14 (^{14}C), phosphorus (^{32}P) and sulphur (^{35}S). The tagged molecules are often precursor molecules used by the cell in the synthesis of other needed molecules. At intervals, samples are removed from the solution; in case of smaller tissues, the samples are sectioned and mounted on glass slides or grids. The sections are then coated with a photographic emulsion and stored in the dark for periods ranging up to several months.

When a radioactive atom emits a beta particle (*i.e.*, electron) the photographic emulsion is affected in a manner similar to the exposure of a photographic emulsion to light. Over a period of time sufficient radioactive emissions occur to affect the silver grains of the emulsion. Black spots will appear at those sites when the emulsion is developed. Such spots will mark sites in the tissues where the radioactive atoms have accumulated. These sites can be identified by examining the stained tissue sections under the light microscope.

In the technique of autoradiography, for the study of DNA metabolism of cell ^3H-thymidine is used; for RNA metabolism ^3H-uridine is used; for protein synthesis various tritiated (^3H-tagged) amino acids are used; and for polysaccharides and glycoproteins tritiated monosaccharides such as ^3H-mannose and ^3H-fucose are employed.

Pulse-labelling technique. This technique is used for those cases where biological molecules undergo considerable modifications after their synthesis (*e.g.*, ribosomal RNA). Here, actively growing cells are exposed to a radioactive precursor for a short period. The labelled precursor is then removed and replaced by 'cold' (unlabelled) precursor molecules. The unlabelled precursors are incorporated into the newly synthesized molecules and have the effect of 'chasing' the previously synthesized molecules containing the radioactive precursor through any maturation process. If the molecular species under investigation is sampled shortly after the start of the experiment, only the primary synthetic product will contain radioactivity. After longer time intervals, the original radioactive molecules will have been replaced with non-radioactive molecules. In this way the flow of radioactivity through a maturation process can be followed, together with any movement of molecular species within the cell.

3.3. CHROMATOGRAPHY

The separation of molecules from biological materials is an important part of biochemical work. Various methods are used for separating molecules. **Chromatography** is a method of separation which is utilised for separation and purification of both organic and inorganic molecules. The word 'chromatography' has its origin in Greek words '*chromo*' meaning colour and '*graphy*' meaning to measure. The Russian botatinst **Mikhail Tswett** is credited with the original development of this technique. In 1903, he reported the successful separation of a mixture of plant pigments using a column of calcium carbonate. The early methods of isolation and purification of compounds of mixtures were empirical, slow and laborious. But with the advancements in separation procedures over the years, chromatography has become a highly efficient technique for separation and purification of compounds.

Chromatography is based on the phenomenon of adsorption and differential migration of components of a mixture or a compound. This needs the following two basic things:

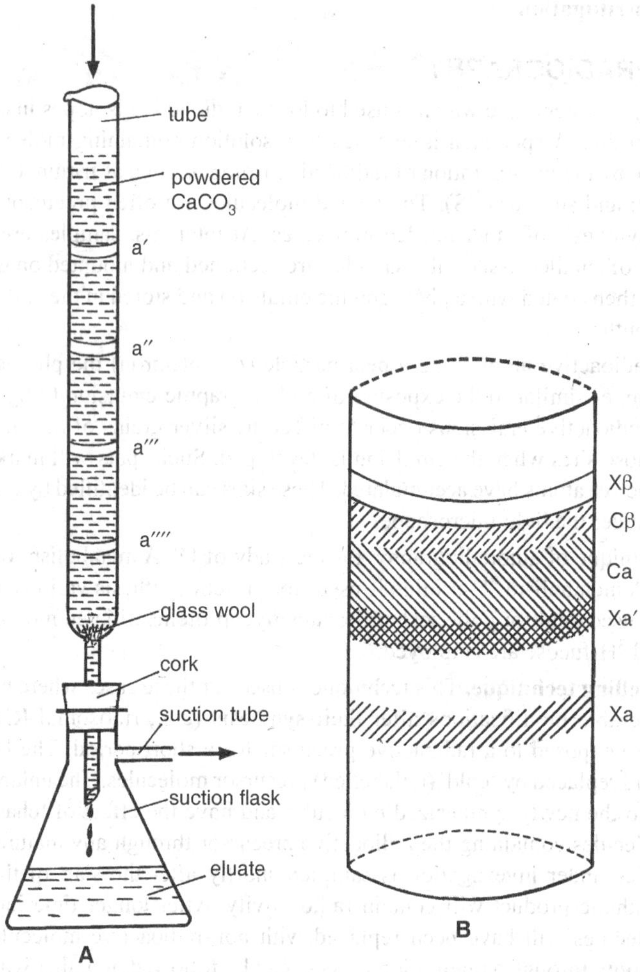

Fig. 3.4. A—An apparatus for column chromatography: B—Chromatogram of chlorophyll.

1. A force for propelling the molecules of any solution, mixture or cytosol. It is called **mobile phase** and it involves moving solvent.
2. A selective impedance causing differential movement of various molecules (*i.e.*, called **impeding medium** or **adsorbent**). It is called **immobile phase** and it involves the matrix (filter paper, $CaCO_3$, etc.) through which the solvent is moving.

During chromatography the mixture or solution is allowed to percolate in an insoluble medium or adsorbent or matrix which has different affinity for molecules of different substances. The molecules migrate through the medium (matrix) at different rates and are separated. For example, if a solution of leaf pigment, chlorophyll in petroleum ether is allowed to percolate through a column of powdered calcium carbonate, the pigments are initially absorbed by pure solvent (*i.e.*, petroleum ether), the individual pigments—carotene, chlorophyll and xanthophyll separate and appear as the coloured bands or zones on the calcium carbonate column like the light rays in spectrum. Such a preparation is called **chromatogram**.

The impeding medium (or immobile phase) is called **adsorbent**. The liquid that passes through adsorbent is **filterate** and the solution obtained after eluting adsorbed substances is **eluate.** The substances that are separated are **samples** or **solutes** and preparation having these solutes is called **chromatogram**.

Principle of Chromatography

Chromatography is based on **partition** or **distribution coefficient (Kd)**. For two such immiscible (*i.e.*, liquids which are not able to be mixed together) phases A and B, the value for this coefficient is constant at a given temperature and is given as:

$$\text{Kd} = \frac{\text{Concentration in phase A}}{\text{Concentration in phase B}}$$

The term **effective distribution coefficient** is defined as the total amount of substance in one phase divided by the total amount present in the other phase.

$$\text{Effective distribution coefficient} = \frac{\text{Total amount in phase A}}{\text{Total amount in phase B}}$$

Basically all chromatographic systems consist of the **stationary phase** which may be a solid, gel, liquid or solid/liquid mixture that is immobilised, and the **mobile phase** which may be liquid or gas. The mobile phase flows over or through the stationary phase. The choice of stationary and mobile phase is made that the compounds to be separated have different distribution coefficient.

Types of Chromatography

Chromatography can be classified into various types depending upon the type of solid support, stationary phase and the mobile phase. Depending upon the physical state of the stationary phase, whether the stationary phase is solid or liquid, chromatography can be divided into two types:

1. **Adsorption chromatography.** The chromatography in which stationary phase is solid, *e.g.*, thin-layer chromatography and ion exchange chromatography are adsorption chromatography.

Table 3.1 Phase of various types of chromatographic techniques.

	Technique	Stationary phase	Mobile phase
1.	Adsorption chromatography	Solid	Liquid
	(*i*) Thin-layer chromatography	Solid	Liquid
	(*ii*) Ion-exchange chromatography	Solid	Liquid
	(*iii*) Affinity chromatography	Solid	Liquid
2.	Partition chromatography	Liquid	Liquid
	(*i*) Paper chromatography	Liquid (supported by cellulose fibers)	Liquid
	(*ii*) Gas-liquid chromatography	Liquid or solid	Gas

2. **Partition chromatography.** When the stationary phase is liquid, it is called partition chromatography, *e.g.*, paper chromatography.

In an alternative method, chromatography is divided into following two main types:

I. Column Chromatography

In column chromatography, the stationary phase is attached to a suitable matrix and it is packed into a glass or metal column (Fig. 3.4). The mobile phase is passed through the column either by gravity or by use of a pumping system. Adsorption, ion exchange, molecular exclusion, affinity and gas-liquid chromatographies are examples of column chromatography.

In column chromatography an insoluble medium is packed into a glass tube; the length and width of this tube influence the separation of the molecules. The molecules to be separated are applied to the top of the column and their migration is started by adding a solvent. The characteristic separation which results depends on the choice of solvent and carrier material. A positively charged carrier binds negatively charged molecules; other carriers contain pores which are penetrated by the smaller molecules, which are, therefore, slowed down. The solvent flows through the column, the *eluate* (the liquid emerging from the bottom of the column) is collected in many fractions (Fig. 3.5). The eluate fractions contain the separated classes of molecules. Column chromatography is important for the separations of mixtures of proteins, that is, for the isolation of enzymes such as cytochrome c or RNA polymerase.

Adsorbents and solvents. A few of the adsorbents used in column chromatography are carbonates of calcium, magnesium and sodium; various forms of charcoal activated in special ways; Fuller's earth, bentonite, Lloyd's reagent, talcum, clays of different types; alumina, silica gel, cellulose, resin, starch, sucrose, inulin and benzoic acid, etc. The adsorbent column neither dissolves in the developing solvent nor reacts chemically with the solvent or the substance to be separated.

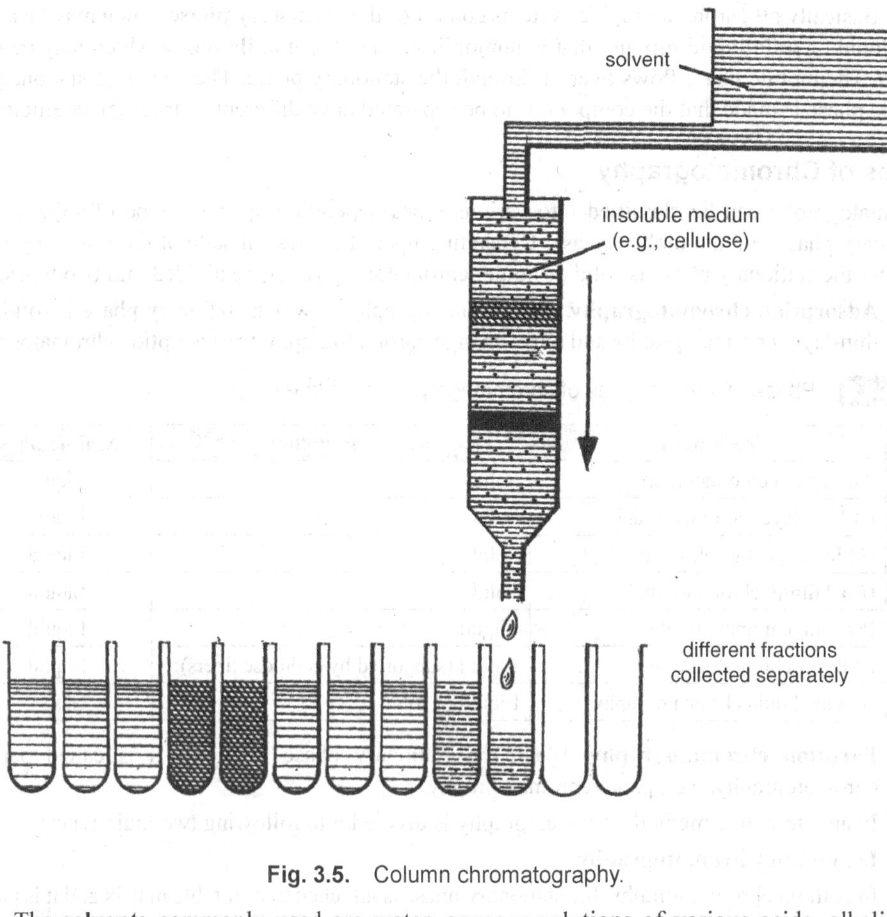

Fig. 3.5. Column chromatography.

The **solvents** commonly used are water, aqueous solutions of various acids, alkalis and salts; alcohols, ketones and cyclohexanone, hydrocarbons such as benzene, chloroform and carbon tetrachloride, etc.

Table 3.2 Differences between column and partition chromatography.

	Column chromatography		Partition chromatography
1.	It was discovered by **Tswett.**	1.	It was discovered by **Martin Synge.**
2.	The adsorbent column may be of carbonates of calcium, magnesium or sodium; activated charcoal; Fuller's earth, resin, starch, etc.	2.	The adsorbent column may be of silica gel, cellulose or starch which are capable of holding water.
3.	Solvent used may be water, aqueous solution of inorganic acids or alkalies or salts, alcohol, ketones, hydrocarbons, benzene, chloroform or carbon tetrachloride.	3.	More than one solvents are used at a time. The substance to be separated are dissolved in alcohol and are then passed through the adsorbent column, already moistened with water.
4.	Column chromatography is based on the principle of differential rate of diffusion of the substances to be separated.	4.	Partition chromatography is based on the principle of differential solubility of substances to be separated in two or more immiscible solvents.
5.	This technique is used to separate any type of substances from any solvent.	5.	This technique is used to separate water soluble compounds from a mixture of two or more immiscible solvents.

II. Thin Layer Chromatography or Planer Chromatography

In this case, the stationary phase, attached to a suitable matrix is coated thinly on to a glass, plastic or metal foil plate. The mobile liquid phase passes across the thin-layer plate by capillary action.

Paper chromatography is a planer chromatography in which the stationary phase is supported by cellulose fibers of a paper sheet. The mobile phase passes over by capillary action over the stationary phase.

A. Adsorption chromatography. In includes the following five types of chromatographies:

1. Adsorption chromatography. This is most commonly used to separate non-ionic and water-insoluble compounds such as triglycerides, vitamins and some drugs.

In this technique, column is packed with an adsorbent (Box 3.1). When the sample is passed through the column, the molecules with specific groups bind with adsorbent. The strength of binding of a particular anylate depends upon the functional groups present in its structure. Hydroxyl groups and aromatic groups tend to increase interaction depends upon the relative strength of the interaction of various molecules present in the eluent to the specific adsorption sites. As eluent is constantly passed down the column, differences in the binding strengths eventually lead to the separation of the analytes. The adsorption chromatography is influenced more by the presence of specific groups than by simple molecular size because only a specific group rather than the whole molecule can interact with the adsorption site.

Box 3.1

Certain solid materials have the ability to hold molecules at their surfaces. Such molecules are known as **adsorbents** and the process is known as **adsorption**. This involves weak, non-ionic attractive forces of van der Walls and hydrogen bonding type. The binding of the molecules occur at specific adsorption sites on the adsorption. Various materials such as silica, alumina and activated charcol are used as adsorbents.

2. Ion-exchange chromatography. In ion exchange chromatography molecules are separated by their difference in charge. The ion-exchange resin is used as chromatographic adsorbent which consists of porous polystyrene beads (polymers) to which ionizable groups are added chemically. Ion-exchange chromatography depends on the attraction between oppositely charged particles. The technique is used for the separation of many biological materials having ionizable groups such as amino acids and proteins.

Column is packed with an ion-exchanger which may be either cation or anion exchanger. **Cation exchangers** (*e.g.*, **DEAE-** or **diethylaminoethyl-cellulose**) possess negatively charged groups and attract positively charged anions. These are also known as **acidic ion-exchanger** as their negatively charges result from the ionization of acidic groups. **Anion-exchangers (*e.g.*, CM or carboxy methyl cellulose)** have positively charged groups that attract negative charged cations. These are also called **basic-ion exchange materials** since positively charges generally result from the association of protons with basic groups.

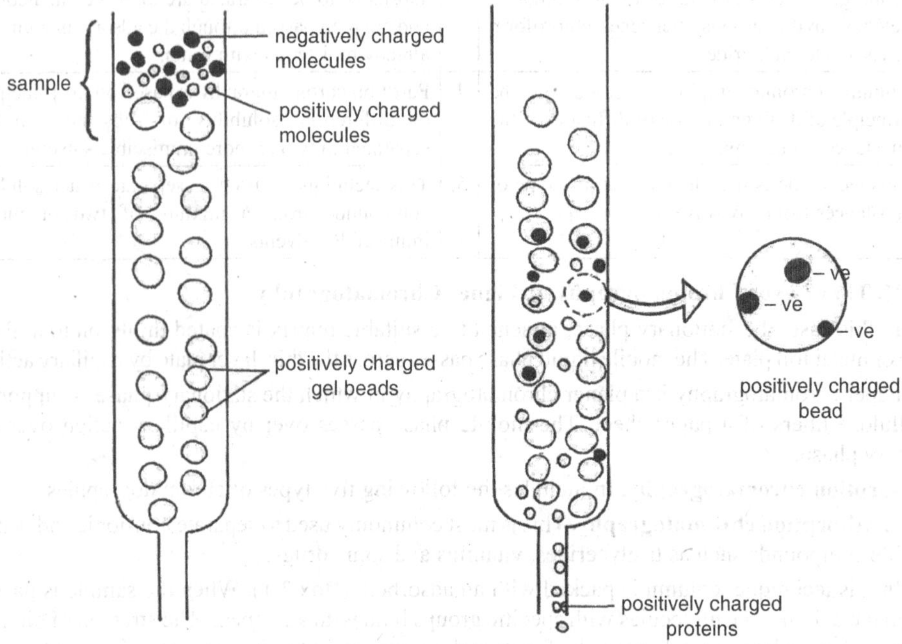

Fig. 3.6. Ion-exchange chromatography.

When a sample (a mixture of charged molecules such as proteins) is passed through the column carrying the ion-exchange material (*e.g.*, cation exchanger), the molecules with greater positive charges interact with the ion-exchange material and bind tighly to the molecules carrying lesser number of positive charges. Molecules with negative charge may just be washed out without any binding. Then, the bound molecules are eluted out one by one with an eluting buffer of a suitable pH. A number of ion-exchanger depends upon the stability of the sample components, their relative molecular mass and the specific requirements of the separation.

3. Molecular exclusion (permeation) chromatography. In molecular exclusion chromatography, the molecules are separated on the basis of their molecular size and shapes. A variety of porous materials act as molecular sieve, *e.g.*, glass granules and a number of the gels are used as molecular sieves. When the separation is carried out by using the gels as the molecular sieve, the technique is also called **gel filteration chromatography**. If porous glass granules are utilised as molecular sieve, the technique is called **controlled pore glass chromatography**. The term exclusion (permeation) chromatography describes all molecular separation processes using molecular sieves. When the mixture is poured over a column filled with gel particles or porous glass granules, large molecules in the mixture are completely excluded from the pores and pass through the interstitial spaces and appear in the eluate first. The smaller molecules are distributed between the mobile phase inside and outside the molecular sieve and then pass through the column at a slower rate, hence appear last in the eluate (Fig. 3.6).

4. Affinity chromatography. Affinity chromatography is the method to separate all the molecules of a particular specificity from a mixture. It is basically a type of adsorption chromatography in which the molecules to be purified are specifically and reversibly adsorbed by a complimentary binding substance or ligand immobilised on insoluble support matrix. The technique is used for the separation of specific antibodies from the blood serum. For the purification of antibodies an immunoadsorbent is prepared which consists of a solid matrix to which antigen 'ag' has been coupled. The matrix used must be stable during binding of macromolecules and its subsequent elution. Agarose, sephadex and some derivatives of cellulose can be used as matrix. When the serum is passed over the immuno-adsorbent, the specific antibodies in the mixture (serum) will bind with the 'ag' and retained. Other antibodies and serum proteins will pass through unimpeded. Then a reagent is passed through the column to release the bound antibodies from the immunoadsorbent. The eluate is then dialyzed to remove the reagent used for elution (Fig. 3.7).

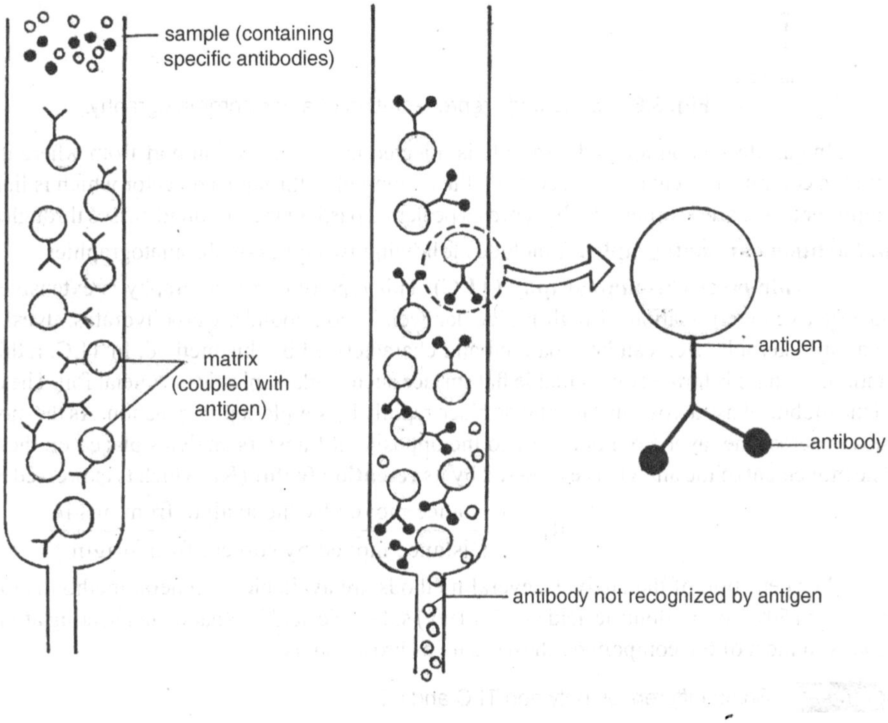

Fig. 3.7. Affinity chromatography.

5. Gas chromatography. This technique is used to separate volatile organic compounds. Gas chromatography is an efficient technique for identification of gases, pollutants, drugs, vitamins, alkaloids, etc. This technique was introduced by **Martin** and **Synge** (1952). This type of chromatography involves injecting the volatile materials into a column contaning a liquid or solid stationary phase. A gas chromatograph consists of a flowing mobile phase, an injection port, a separation column containing the stationary phase and a detector. The stationary phase may be either a liquid or solid. In gas-liquid chromatography (GLC), the mobile phase is a gas and the stationary phase is a thin layer of non-volatile liquid bound to a solid support. In gas-solid chromatography a solid adsorbent acts as the stationary phase and an adsorption process occurs. The organic compounds are separated due to differences in their partitioning behaviour between the mobile gas phase and stationary phase in the column.

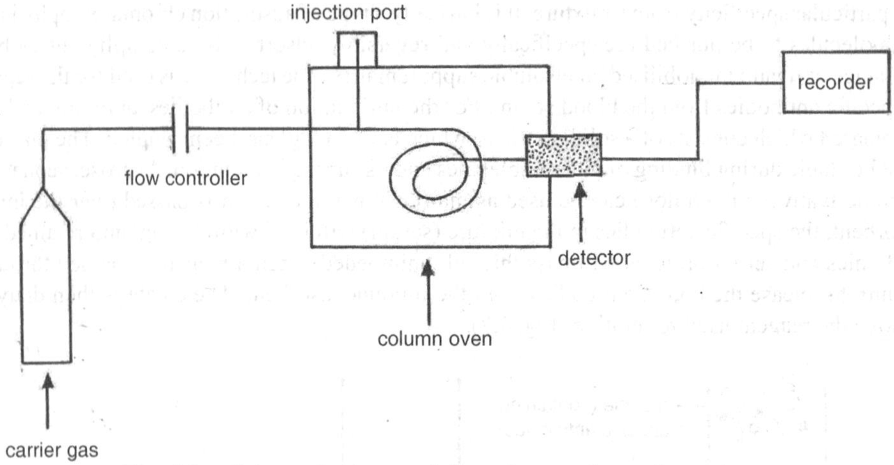

Fig. 3.8. Schematic representation of a gas chromatography.

In gas chromatography, the sample is injected into the injection port from where it is carried into the column. The eluate released from the column pass through a detector which is linked via an amplifier to a chart recorder which records a peak as a particular compound passes through a detector.

B. Partition chromatography. It includes following two types of chromatographies:

1. Thin layer chromatography (TLC). This type of chromatography is extensively used in quantitative determination of high molecular weight compounds. Carbohydrates, dyes, pigments, vitamins, steroids, etc., can be separated and characterized by this method. In TLC, a thin layer of stationary phase is formed on a suitable flat surface such as glass, plastic or a metal foil. The movement of the mobile phase across the layer takes place rapidly by simple capillary action. As the mobile phase moves across the layer from one edge to the opposite, it transfers analytes placed on the thin layer. The movement of the analyte is expressed by its **retention factor (Rf)** which is expressed as follows:

$$\text{Rf} = \frac{\text{Distance moved by the analyte from origin}}{\text{Distance moved by solvent from origin}}$$

For detection of the analyte, several methods are available. A general method is to spray the plate with 50% (v/v) sulphuric acid or 25% (v/v) sulphuric acid in ethanol and heating at 110°C. This results in most of the compounds showing up as brown spots.

Table 3.3 Some differences between TLC and GLC.

	TLC or Thin-layer chromatography		GLC or Gas-liquid chromatography
1.	Adsorbent may be cellulose powder, alumina or cellulose.	1.	Adsorbent or the immobile phase is represented by liquid.
2.	The mobile phase is the solvent (liquid).	2.	The mobile phase is inert gas such as nitrogen, carbon dioxide, helium or argon.

2. Paper chromatography. Paper chromatography is a widely used technique in the separation of organic and biochemical products in laboratories. In this technique, stationary liquid phase is supported by cellulose fibers of a paper sheet. The mobile phase is developing solvent. The mobile phase passes over the stationary phase by capillary action. In paper chromatography the test solution is applied as a small spot on the chromatography paper (Whatman's No. 1 paper) and dried. The solvent is allowed to travel on the paper. As the solvent moves it carries the mixture components along with it. There are two types of paper chromatography: (*a*) **Ascending** (Fig. 3.9) and (*b*) **descending chromatography**. In ascending chromatography, the mobile phase (solvent) is at the bottom of the chromatographic chamber. The sample spot (loading spot) is kept in position just above the surface of the solvent. As the solvent moves vertically up on the paper, separation of the sample is achieved.

In descending chromatography, the mobile phase (solvent) is kept in the upper position of the chromatography chamber. The paper is inserted with the upper end with sample spot close to the mobile phase kept at the upper position of chromatographic chamber. Separation of the sample occurs as the solvent moves downwards due to capillary action and gravity.

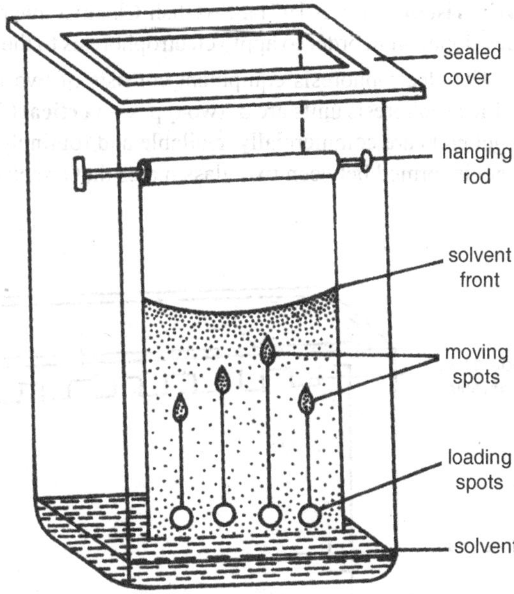

Fig. 3.9. Ascending paper chromatography.

A number of methods are available for the detection of eluate. Spraying with aqueous solution of 0.2% ninhydrin saturated with butyl alcohol is commonly employed for identification. The identification of a given compound is made on the basis of its Rf value.

3.4. ELECTROPHORESIS

Electrophoresis is a biochemical technique used to separate charged particles from a mixture (Fig. 3.10). The term 'electrophoresis' describes the migration of a charged particle under the influence of an electric field. This is a simple, rapid, sensitive and versatile analytical tool used to study and purify the charged molecules such as amino acids, peptides, proteins, nucleotides and nucleic acids. These molecules exist in solution as electrically charged species either as cations (–) and anions (+). Under the influence of an electric current these particles will migrate either to the cathode or anode, depending on the nature of their net charge.

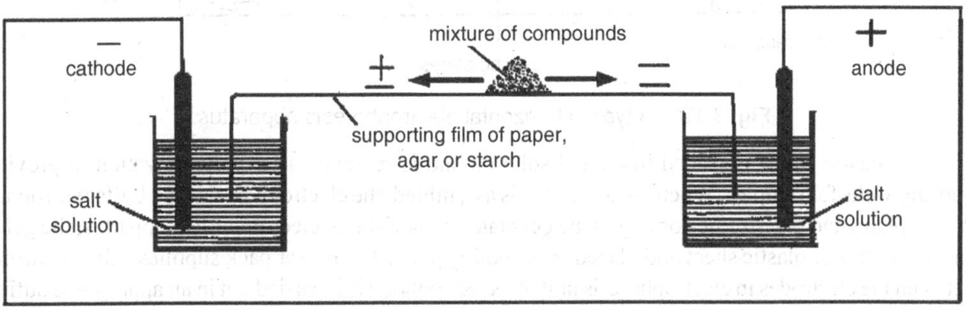

Fig. 3.10. Electrophoresis.

Electrophoresis is a technique, developed by **Tiselius** in 1937, for the analysis and separation of colloids and is based on the movement of charged colloidal particles in an electric field. **Arne Wilhelm Kaurin Tiselius** (1902–1971), Swedish Chemist awarded the 1948 Nobel Prize for Chemistry for his development in order to apply electrophoresis to the study of proteins, notably those of the blood.

The electrophoresis equipment consists of two units—a power pack and an electrophoresis unit. Electrophoresis units are of two types—**vertical** (Fig. 3.11) and **horizontal** (Fig. 3.12). Vertical slab gel units are commercially available and routinely used to separate proteins in acrylamide gel. The gel is formed between two glass plates that are clamped together.

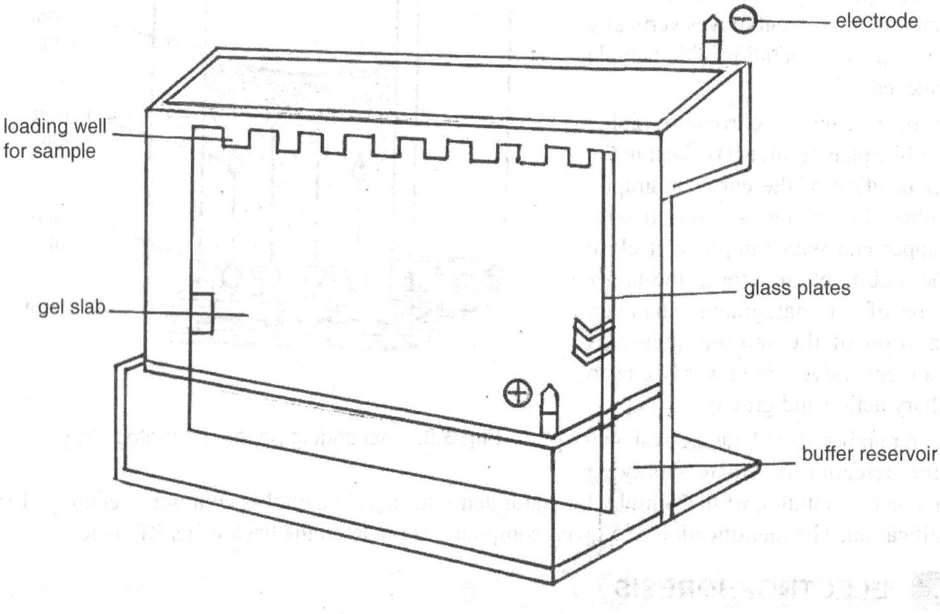

Fig. 3.11. A vertical gel electrophoresis apparatus.

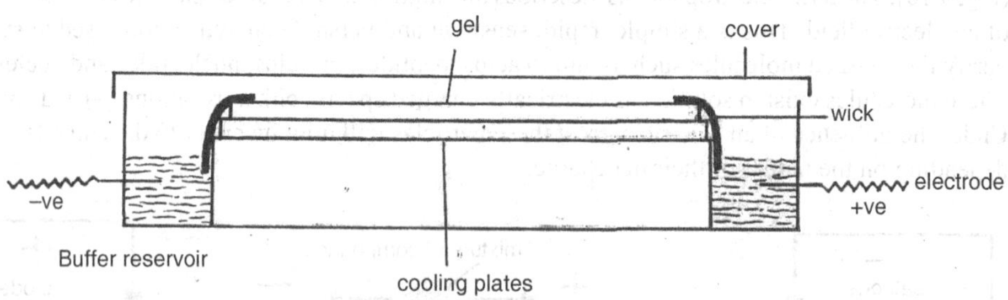

Fig. 3.12. A typical horizontal electrophoresis apparatus.

A **plastic comb** is placed in the gel solution and is removed after polymerization to provide **loading wells** for samples. When the apparatus is assembled, the electrophoresis tank buffer surrounds the gel plates and gives some cooling of the gel plates. In horizontal electrophoresis appartus, the gel is cast on a glass or plastic sheet and placed on a cooling plate. The power pack supplies a direct current between the electrodes in electrophoresis unit. All electrophoresis is carried out in an appropriate buffer. The buffer is essential to maintain a constant state of ionization of the molecules being separated. The sample moves under the influence of current. At the end of run, the position of the molecules on the gel is fixed with a fixative to prevent simple diffusion. Then the separated components are stained to visualize them. The **bands** can be quantified (by elution or by scanning with densitometer) as the uptake of the dye is directly proportional to the concentration of the molecules in each band.

The rate of electrophoretic migration (or 'mobility') depends on the pH of the medium, strength of electric field, magnitude of the net charge on the molecule, and the size and shape of the molecule.

Types of Electrophoresis

1. Paper electrophoresis. It is most common type of electrophoresis which is run in clinical laboratories. This technique is used for separation of small molecules such as amino acids, peptides and carbohydrates.

2. Agarose gel electrophoresis. Starch, agarose or polysaccharide gels are used for electrophoretic techniques. Agarose is a linear polysaccharide made up of the basic repeat unit agarobiose. Agarose gel is formed by boiling dry agarose in aqueous buffer until a clean solution forms. This is allowed to cool at room temperature to form a rigid gel. The pore size in the gel depends upon the initial concentration of agarose. If the concentration is low large pore sizes are formed. Agarose is usually used at concentration between 1% and 3%. Agarose gels are used for electrophoresis of both proteins and nucleic acids. Agarose gel electrophoresis run by placing the gel under the level of buffer is most commonly employed in all the nucleic acid research applications such as DNA fingerprinting, sequencing and recombinant DNA technology. Agarose gel is also used in techniques such as immuno-electrophoresis (used for antigens and antibodies).

3. SDS-Page. SDS-polyacrylamide gel electrophoresis is a widely used technique for the qualitative analysis of protein mixtures. The relative molecular mass of proteins can also be determined by this method. Sodium dodecyl sulphate (SDS) is an anionic detergent. In this method the protein mixture (sample) is first boiled in a buffer containing β-merceptoethanol and SDS. This denatures the proteins and SDS binds with the proteins. The sample buffer also contains an ionizable tracking dye which allows the electrophoretic run to be monitored. It also contains sucrose and glycerol. These allow the sample to settle easily through the electrophoresis buffer to the bottom when injected into the loading wells. Once the samples are all loaded, a current is passed through the gel. Normally 15% polyacrylamide gel is used as the separating gel for separating proteins in the range of molecular mass of 100,000 to 10,000.

3.5. DIALYSIS

It is a sensitive method for separating lower molecular-weight component from macromolecules. A thin membrane in the form of a tube is filled with the solution containing the molecules to be separated. The pore size of the membrane allows the diffusion of small molecules such as salt or amino acids; larger molecules such as proteins or nucleic acids cannot pass through the pores and so remain inside the dialysis tube.

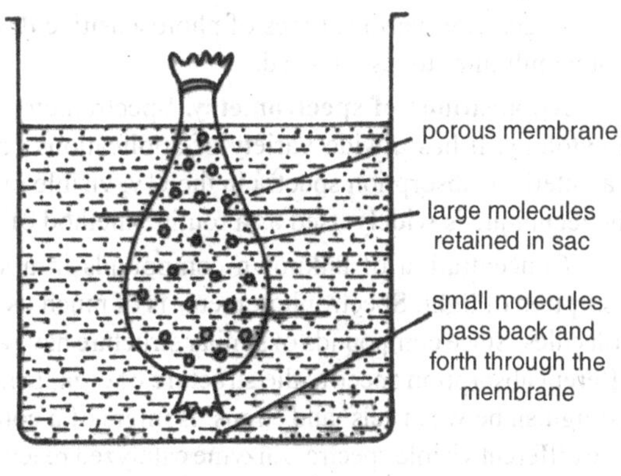

Fig. 3.13. Dialysis.

In dialysis, a protein solution is placed in a cylinder of cellophane tubing sealed at one end. The tubing is then sealed at the other end and suspended in a large volume of buffer. The cellophane membrane is semipermeable: high molecular weight proteins are too large to pass through the pores of the membrane so proteins remain inside the tubing while low molecular weight solutes including, in this case, ammonium and sulphate ions diffuse out and are replaced by solutes in the buffer.

3.6. SPECTROPHOTOMETRY

Spectrophotometry is the measurement of the intensity of different colours (wavelengths) present in a beam of light, normally using a prism or diffraction grating and a photoelectrical cell. **Spectrophotometer** is an instrument for producing a spectrum and measuring the relation between absorption of electromagnetic radiation and frequency (or wavelength) of that radiation. It is a device used by researchers to determine how much radiation at various wavelengths a sample can absorb. The amount of absorption can be used to determine the concentration of particular molecules within a sample, because each type of molecule or macromolecule has its own characteristic wavelength(s) of absorption, called **absorption spectrum.** This instrument is used for colour comparison and in chemical analysis. Spectrophotometer is based on the principle of Beer–Lambert law.

A spectrophotometer typically has two light sources, which can emit ultraviolet or visible light. The light source is passed through a monochrometer, which emits light at a desired wavelength. This incident light then strikes a sample contained within a cuvette. Some of the incident light will be absorbed, and some will not. The unabsorbed light passes through the sample and is detected by the spectrophotometer. The amount of light which strikes the detector is substracted from the amount of incident light, yielding the measure of absorption. In this way, the spectrophotometer provides an absorption reading for the sample. This reading can be used to calculate the concentration of particular molecules or macromolecules in a sample.

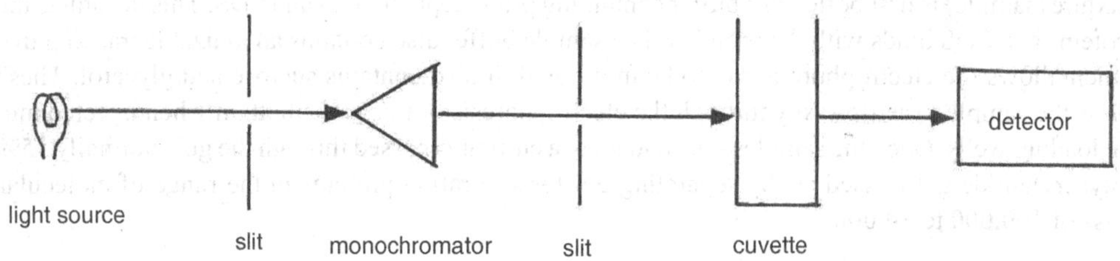

Fig. 3.14. Schematic representation of single beam spectrophotometer.

In detector, various types of photosensitive devices such as photovoltaic cells, phototubes or photoamultiplier tubes are used.

Applications of spectrometry. Spectrometry is one of most valuable analytical techniques of cytology. It has a wide variety of applications or uses. Compounds may be identified by their characteristic absorption spectra in the UV, visible or infrared regions of electromagnetic spectrum. The technique is widely used as a routine method of quantitative analysis of compounds.

Concentration of unknown compounds in a solution can be determined by measuring the absorption of light. Spectrophotometry is of much use in elucidating the structure of various organic molecules. Since compounds of similar structure have analogous absorption spectra, thus by comparing different absorption spectra, the structure of organic molecules can be elucidated. This technique can distinguish between 'cis' and 'trans' isomers of a complex. The geometrical isomers of compounds have different visible spectra. Enzyme catalyzed reactions can be studied by measuring the appearance of produces or disappearance of substrate with the help of spectrophotometry.

3.7. SAMPLE PREPARATION FOR MICROSCOPY

Cells are transparent and optically homogeneous: so either they are viewed as such by instruments such as phase contrast microscope or to produce necessary contrast, the cells are passed through various steps of slide preparation such as killing, fixation, dehydration, embedding, sectioning, staining and mounting.

Thus, superior specimens for microscopic examination can be obtained by **killing** the cells and coagulating or **fixing** the protoplasm by preservatives, called **fixatives** such as alcohols, formaldehyde, mercuric chloride, picric acid, acetic acid and mixture of these. The process of fixation involves the following events:

1. The proteins and other macromolecules are precipitated.
2. The intracellular hydrolytic enzymes are denatured, preventing autolysis.
3. Cross links are formed between macromolecules, making the preparation more stable and minimizing shrinkage upon drying.
4. Substances are introduced which prevent attack by microorganisms.
5. The tissues become stiffer, making their sectioning easier.
6. The affinity of the tissue for dyes (stains) is increased.

Fixation is generally followed by **dehydration** (*i.e.*, gradual removal of water vapours from the tissue) by the organic solvents such as ethanol. The dehydrated specimens are **embedded**, *i.e.*, they are infiltrated with molten paraffin which hardens upon cooling and provides enough support to allow thin sections to be cut with a **microtome**. By the microtome, serial sections, 5 to 10 μm thick can be cut and placed on slides in the order of cutting and permitting a sequence of specimens for observation. These sections are stained with a non-vital-stain to increase the contrast.

Stains are the chemicals that can selectively attach to particular molecules of specific cellular structures and make them stand out from other parts of the cell. The non-vital stains fall into two main classes: **acid stains** such as eosin, orange *G*, aniline blue and fast green, all of which combine with basic molecules such as proteins of the fixed cells; and **basic stains** such as methylene blue, crystal violet, haematoxylin, basic fuchsin, etc., all of which combine with nucleic acids and other acidic molecules of the fixed cells. The cellular structures that stain with acid stains are called **acidophilic** and those that stain with basic dyes are called **basophilic**.

Methods of Sample Preparation for Transmission Electron Microscopy

The standard procedure for the preparation of specimen for TEM entails fixation, dehydration, staining and sectioning similar to light microscopy (Table 3.4). However, the most significant difference being the need for ultra-thin sections. Following techniques of sample preparation are generally used for different types of methods of studying ultrastructure of the cell:

Table 3.4. Comparison between light and electron microscopy.

Steps	Electron microscopy	Light microscopy
Fixation	Osmium tetroxide, postassium permanganate, formalin, glutaraldehyde.	Bouin's solution; formalin; Zenker's fluid.
Dehydration	Increasing concentration of ethanol (or acetone) followed by propylene oxide.	Increasing concentration of ethanol followed by benzene.
Embedding	Araldite; Vestoplaw, Epan 812; Maraglas; Durcopan.	Paraffin.
Sectioning	Usually 10–100 nm thick sections cut with a glass or diamond knife on an ultramicrotome.	Usually 6 μm thick sections cut with a razor blade on a microtome.

Steps	Electron microscopy	Light microscopy
Mounting	On a perforated metal disc (grid) usually covered with formvar or paralodian.	On a glass slide with an egg albumin adhesive. Deparaffinized in xylol for staining.
Staining	With salts of heavy metals such as lead acetate, lead citrate, lead hydroxide, urynyl acetate; phosphotungstic acid.	Selective chromatic stains (as haematoxylin and eosin), dehydrated in ethanol series, cleared in xylol and mounted for viewing in Canada balsam or Permount.
Viewing	Grid is placed between the condenser and objective lenses in vacuum and the image is viewed on phosphorescent screen.	Slide is placed between the condenser and objective lenses and the image viewed in the ocular lens.

1. Monolayer technique. Macromolecules such as DNA and RNA are studied by monolayer technique in which the macromolecules are extended on the air-water interface before being collected on a film.

2. Thin sectioning. This method uses a cutting device known as **ultramicrotome** to remove ultrathin (*i.e.*, 10 nm to 100 nm thick) sections from the specimen. To withstand the passage of ultrafine diamond or glass knife without tearing, the specimen is first embedded in a hard plastic such as, epoxy resin (Table 3.4). The resin is allowed to penetrate the sample before it is polymerized. Sections are floated from the knife of ultramicrotome onto the surface of water and picked up by touching them with a fine wire mesh or small circular copper grid (*i.e.*, small discs perforated with numerous openings). Prior to its use mesh or grid is coated with a thin monolayer film (7.5 to 15 nm thick) of plastic (such as formvar or collodion) or carbon to provide a support to the sections (or sample) (Fig. 3.15). The specimen is visualized through the holes of screen.

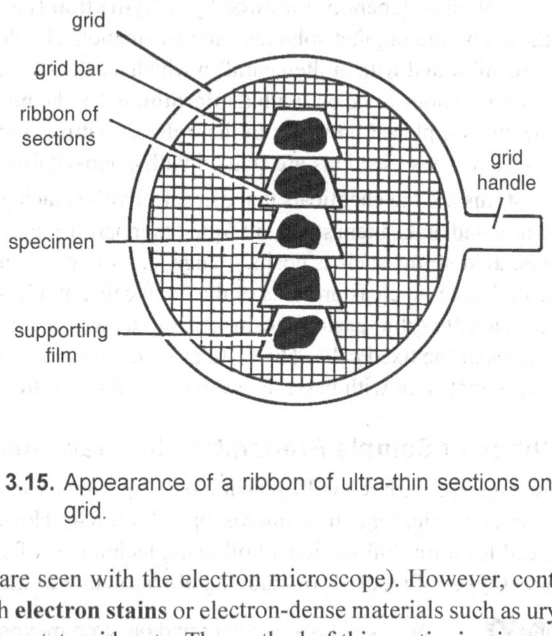

Fig. 3.15. Appearance of a ribbon of ultra-thin sections on the grid.

Sections to be examined with the electron microscope are generally not stained (since no colours are seen with the electron microscope). However, contrast may be improved by "post-staining" with **electron stains** or electron-dense materials such as urynyl acetate, urynyl citrate, lead citrate, osmium tetroxide, etc. The method of thin sectioning is used to study morphology of cell.

3. Negative staining. This technique is used to study small particles such as viruses or macromolecules. Here, the specimen is embedded in a droplet of electron dense material, such as, phosphotungstic acid ($H_3PW_{12}O_{40}$). The electron stain penetrates into all the empty spaces (*i.e.*, openings and crevices) between the macromolecules. The spaces appear well defined in negative contrast. The portions of sepecimen that exclude stain transmit electrons readily, so their images can be seen.

4. Shadow casting or heavy metal shadowing. This technique is used to study three-dimensional appearance of viruses and certain macromolecules such as DNA molecules and collagen fibres. It involves placing of specimen in an evacuated chamber and evaporating at an angle, a heavy metal such as chromium, palladium, platinum or uranium from a filament of incandescent tungsten.

The vapour of heavy metal is deposited on one side of the surface of the elevated particles; on the other side a shadow forms, the length of which permits determination of the height of the particle. In such a specimen, during electron microscopy, the electrons pass readily through the area of light metal content, less readily through the plane on which the particle sits, and are scattered more severely by the side of the particle on which metal has accumulated. Thus, by shadow casting, shape and profile of a particle can be observed.

5. Tracers. Several biological processes such as pinocytosis, phagocytosis and transport of molecules across plasma membrane can be studied by the use of appropriate tracers (*e.g.*, gold, mercuric sulphide, iron oxide, etc.). These tracers are detected by their electron opacity. An ideal tracer should be non-toxic, physiologically inert, composed of small-sized particles of uniform and known size and preserved *in situ* during the processing of the tissue.

6. Freeze-fracture. This technique is used to study the molecular arrangement in the plasma membrane and other cellular membranes. It is carried out by rapidly cooling or freezing the sample (**cryofixation**) and then **fracturing** (cracking) it in the vacuum while it is still at –100°C. The knife does not cut cleanly under those conditions, but tends to **fracture** (crack) the specimen along the lines of natural weakness, such as the middle of a membrane that runs parallel to the cut (Fig. 3.16). After fracture, the sample is left in the vacuum long enough to allow some water to evaporate from the exposed surfaces, a process called **freeze etching.** The exposed face is then shadowed with electron-dense combination of carbon and metal such as platinum to provide the necessary contrast, after which organic material (*i.e.*, specimen itself) is removed by acids to leave a metal **replica** for examination in the electron microscope. Replica reveals a natural-looking representation of the surface of the freeze-etched object and is the only way of seeing membrane interior and certain other features of the cells.

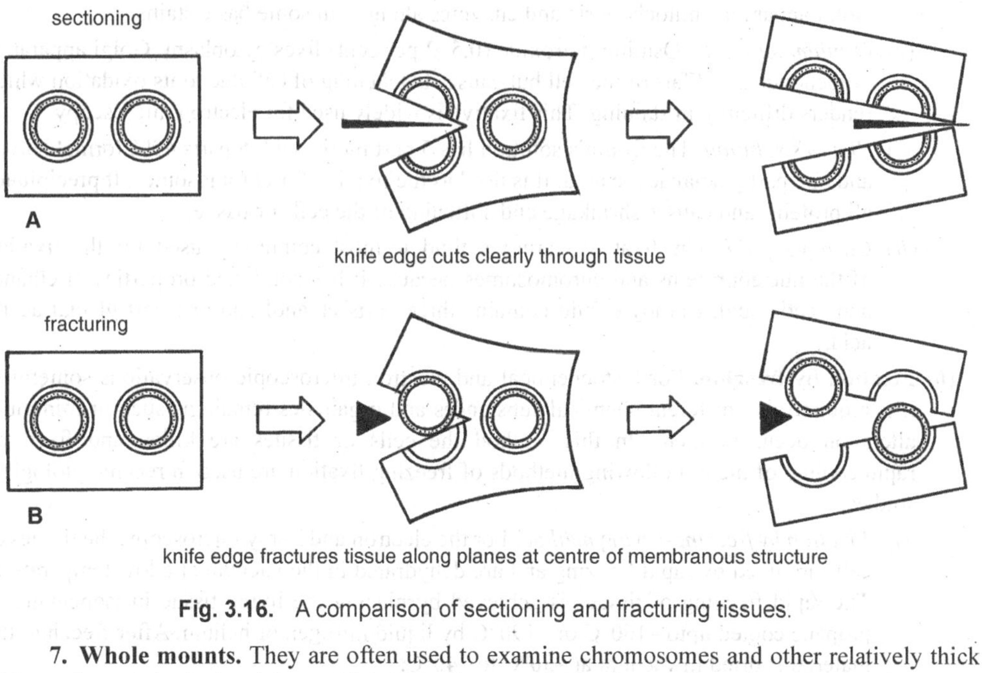

Fig. 3.16. A comparison of sectioning and fracturing tissues.

7. Whole mounts. They are often used to examine chromosomes and other relatively thick objects that can be isolated free of debris. In these methods, the specimen is neither sectioned nor stained. Thick areas will scatter electrons more strongly than thin areas, providing enough contrast to form an image.

Fixation

For microscopical study of dead cells, the process of killing and **fixation** brings about the sudden death of cells or tissues in such a manner that morphological organization and chemical composition of the cell remain unaltered up to greatest extent. The fixative or fixing solution performs following functions: it prevents bacterial decay and autolysis (self-digestion) of the cells; renders the components of the cell insoluble, reduces the cell distortion and shrinkage, increases the visibility of different cellular components and prepares the cells for the staining. The fixation of the cells can be achieved either by chemicals or cooling which is as follows:

(*a*) **Chemical fixation.** The choice of suitable fixative generally depends on the type of analysis and the chemical nature of different cells. For instance, for the study of the nucleus and chromosomes the **acid fixative** such as **Carnoy's solution** or **Bouin's fluid** is used. The commonly used fixatives are as follows:

(*i*) *Acetic acid.* The acetic acid is used from 0.3 to 50 per cent, alone or in mixtures. It does not fix cytoplasmic proteins, but precipitates nucleoproteins, destroys Golgi apparatus and mitochondria of the cells. Morever, the cells become swell up and soft. The acetic acid is used with basic dyes for the fixation of **nucleus** and **chromosomes.**

(*ii*) *Potassium dichromate.* Potassium dichromate is generally used in mixtures and in the percentage of 2–7. It renders proteins insoluble in water and fixes lipids. Potassium dichromate is used for the fixation of chromosomes and some cytoplasmic structures.

(*iii*) *Ethanol.* 70 to 100 per cent ethanol is used either alone or in mixture for the fixation of nucleus along with certain basic dyes. It does not precipitate proteins but dissolves lipids. Ethanol makes the cells to become hard but causes shrinkage.

(*iv*) *Formaldehyde.* The formaldehyde (4–10 per cent) is used in mixtures for the fixation of Golgi apparatus, mitochondria and enzymes along with some basic stains.

(*v*) *Osmium tetroxide.* Osmium tetroxide (0.5–2 per cent) fixes cytoplasm, Golgi apparatus, mitochondria and fats of the cell but causes blackening of cell due to its oxidation which renders difficulty in staining. This fixative is widely used for electron microscopy.

(*vi*) *Bouin's solution.* The Bouin's solution has 5 part picric acid, 5 parts 40% formaldehyde and one part glacial acetic acid. It is used as the fixative for chromosomes. It precipitates all proteins and causes shrinkage and softening of the cells or tissue.

(*vii*) *Carnoy's fluid.* The fixative Carnoy's fluid is most commonly used for the fixation of the nucleoproteins and chromosomes because it has combined properties of ethanol and acetic acid. Carnoy's fluid contains three parts ethanol and one part glacial acetic acid.

(*b*) **Fixation by freezing.** For histochemical and electron microscopic observations sometimes it is required that different chemical substances and organelles remain as such or minimum alteration occur in them. In this method the cells or tissues are killed and fixed by rapid cooling of them. Following methods of freezing fixation are used in recent cytological studies.

(*i*) *Fixation by freezing-drying method.* For the electron and X-ray microscopy, the tissues or cells are fixed by rapid freezing and are dehydrated in the vacuum at a low temperature. The rapid freezing of tissues is achieved by placing the living tissue in isopentane or propane cooled upto –160°C or –190°C by liquid nitrogen or helium. After freezing, the material is dried in vacuum at –30°C to –40°C.

By freezing-drying fixation method, the tissue does not shrink, homogeneous fixation of cell takes place, soluble substances are not extracted, the chemical composition of the cell is maintained without any alteration and the fixation takes place very quickly.

(*ii*) *Fixation by freezing-substitution method.* In the freezing-substitution fixation method, the living tissue is rapidly frozen and then kept frozen at a low temperature (–20 to –60°C) in a reagent that dissolves the ice-crystals. The reagent may be ethanol, methanol or acetone. This method of fixation is also used in electron microscopy and X-ray microscopy.

(*iii*) *Fixation by freezing-etching method.* In the freezing-etching method of fixation the specimen is placed in 20 per cent glycerol and is frozen at –100°C. It is mounted on a chilled holder and splintered with a knife along natural cleavage planes, usually along surfaces of membranes. Occasionally the structure is cross-fractured, giving cross sections of organelles. The splintered preparation is freeze-dried and covered with a platinum or carbon coating in the high vacuum of a freezing ultramicrotome. When the preparation is dried the vacuum is broken and the preparation is placed in water to float the replica off the carrier. The replica is then washed in basic solution to remove the cellular material and the replica is mounted on a grid and dried for elctron microscopy. Such replica gives the outlines of various cell structures and verify the structures seen in thin fixed and stained sections of various cellular organelles in various types of microscopy.

Staining

The process of colouring the cells, tissues of animal or plant bodies by certain inorganic or organic dyes is known as **staining**. The selection of dye or stain for a particular material usually depends on its chemical nature, the pH value of the fixative used and the chemical reactivity of stain to the material.

Stains. Most cytological stains are solutions of dyes of aromatic organic compounds which have two kinds of active chemical groups such as, **chromophoric** and **auxochromic** groups. The **chromophoric groups** gives the colour to the dye, *e.g.*, carboxyl (–COOH), azo (= N–), nitro ($-NO_2$), quinoid (O = ⬡ = O) and indamin (– N=) groups. The **auxochromic group** gives to the dye the ability to attach to the tissue or the material and to dissolve and dissociate, in the water, *e.g.*, *Hydroxy* (–OH) group.

The organic stains are classified into three groups:

1. acidic stains,
2. basic stains, and
3. neutral stains.

1. Acidic stains. The acidic stains are usually used for the cytoplasm and proteins. These have great capacity for combining with the tissue at low pH than basic dyes. The most common cytological acidic stains are picric acid, acid fuchsin, Congo red, Janus green *B*, orange *G*, methyl blue, eosin, aniline blue, Bismark brown and fluorescein.

2. Basic stains. The basic stains are used to stain the nucleus, chromosomes and particularly the nucleic acids. The most common cytological basic stains are basic fuchsin, crystal violet, methyl green, safranine, acridine red, azures, methylene blue, thionine and haematoxylin.

3. Neutral stains. The neutral stains have both the properties of acidic and basic stains.

Acidiphilic and basiphilic tissues. The tissue, cell or cellular component taking acidic stains is known as **acidiphilic** tissue, cell or cellular component, *e.g.*, the cytoplasm. The nucleus, chromosomes and DNA have affinity for the basic stains and known as **basiphilic** organelles.

Metachromasia. Some basic dyes stain certain cell components with a different colour than their original colour. This property of stain is known as **metachromasia** and is very useful for **histochemical** and **physiochemical** tests. The property of metachromasia is displayed by the basic stains such as thionine, azure A and toluidine blue which react with mucopolysaccharides, nucleic acids and acidic lipids.

Mordant and lake. Certain dyes stain the proteins and cytoplasm in the presence of some metal or metallic compound which is known as **mordant**. Usually chemically a mordant is a double salt of potassium or ammonium or ferric sulphate. The mordant and the stain are collectively known as the **lake**. The most important mordant is ferric ammonium sulphate (iron alum), which is used along with the stain haematoxylin and carmine.

Staining for Light Microscopy

For light microscopy both acidic and basic stains are used for the staining of cells or tissue. The stains are specific for different types of cells and also for different organelles of cell. The cytoplasmic proteins and carbohydrates are stained by acidic stains, while the nucleus, chromosomes, etc., are stained by basic stains.

Staining for Electron Microscopy

Electron microscopy usually requires no stain because beside observing visually, the image is photographed and due to the contrast between cellular components a black and white photograph is received. But certain inorganic stains such as lead acetate, lead citrate, lead hydroxide, urynyl acetate, phosphotungstic acid, osmium tetroxide and potassium permanganate are used to increase contrast between cellular components. Sometimes certain organic stains as azure II, leucofuchsin, orcein, etc., are also used to increase the contrast.

Cytochemical Staining

In addition, there are certain specific stains, called **cytochemical** stains that bind selectively to some specific groups of cellular macromolecules such as proteins, nucleic acids, polysaccharides and lipids. For example, Millon reaction, diazonium reaction and Napthol Yellow 5 stain are used for the proteins; alkaline fast green is used for histone (basic protein); Feulgen reaction (using Schiff's reagent) is used for DNA; methyl green-pyronine stain (Unna-Pappenheim stain) is used in distinguishing between DNA and RNA and it stains DNA green and RNA red; acetocarmine and acetoorcein stains are used to stain chromosomes of dividing cells; periodic acid-Schiff (PAS) reaction is used for the demonstration of polysaccharide materials such as starch, cellulose, hemicellulose, and pectin in the plant cells and mucoproteins (glycoproteins), hyaluronic acid and chitin in animal cells; and fat soluble dyes such as Sudan Red and Sudan Black B are used for the lipids. The Sudan Black B is a specific stain for phospholipids and is used to stain Golgi apparatus.

Vital Staining

Vital stains selectively stain the intracellular structures of living cells without serious alteration of cellular metabolism and function. For example, Janus green B selectively stains mitochondria; neutral red stains plant vacuoles and methylene blue stains Golgi apparatus and also nuclear chromatin of dividing cells.

Immunostaining or usage of antibodies in Cell Biology. To use fluorescence microscopy for locating specific molecules or ions within cells, researchers must employ special indicator molecules called **fluorescent probe**. A fluorescent probe is a molecule capable of emitting fluorescent light that can be used to indicate the presence of a specific molecule or ion.

One of the most common application of fluorescent probes is **immunostaining**, a technique based on ability of **antibodies** to recognize and bind to specific molecules (the molecules to which antibodies bind are called **antigens**). Antibodies are proteins produced naturally by the immune

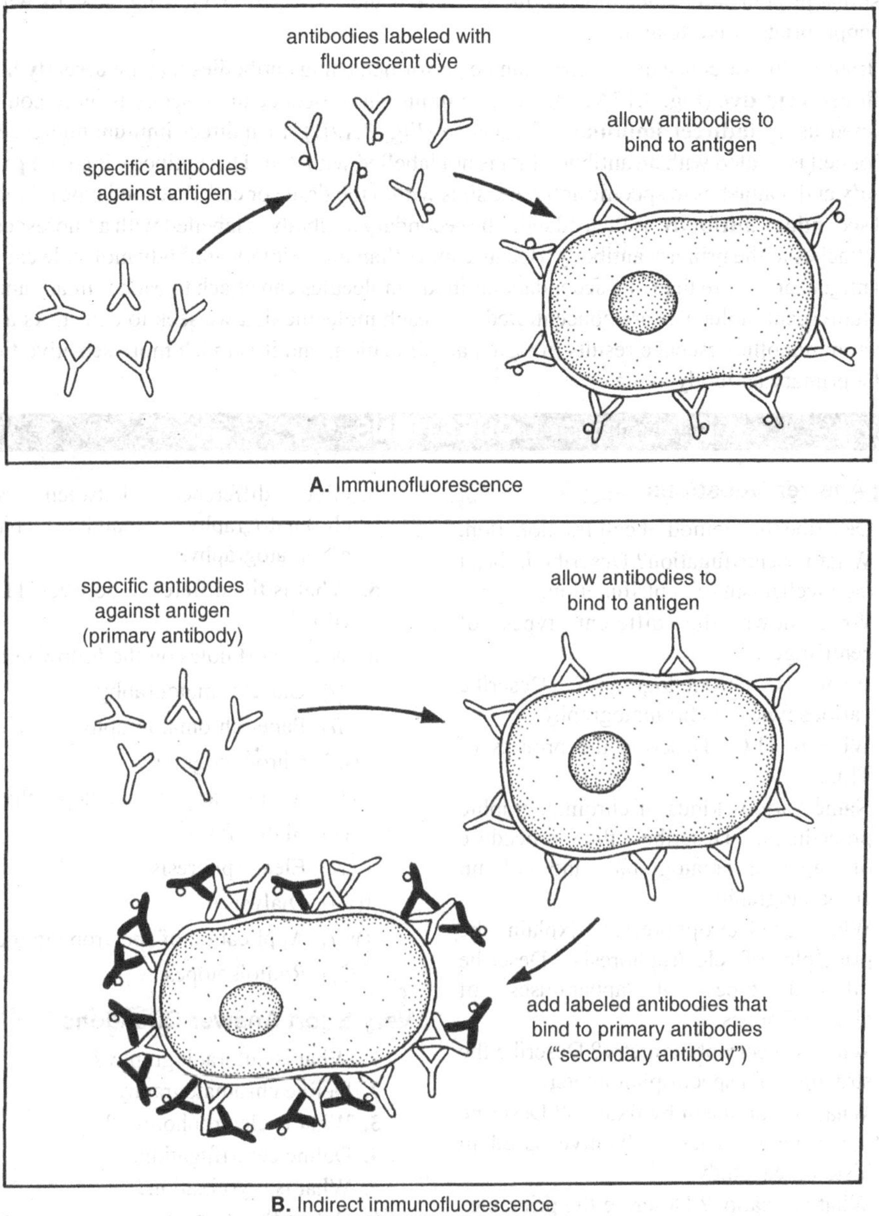

Fig. 3.17. Immunostaining using fluorescent antibodies (after Becker *et al.*, 2006).

system in response to invading microorganisms, but they can also be generated in the laboratory by injecting a foreign protein or macromolecule into an animal such as a rabbit or mouse. In this way, it is possible to produce antibodies that will bind selectively to virtually any protein that a scientist wishes to study. Antibodies are not directly visible using light microscopy, however, so they are linked to a fluorescent dye such as **fluorescein**, they emits a green fluorescence, or **rhodamine**, they emit a red fluorescence. More recently, antibodies have been linked to "**quantum dots**", which are tiny light-emitting crystals that are chemically more stable than traditional dyes. To identify the subcellular location of a specific protein, cell are simply stained with a fluorescent antibody directed against that protein and the location of fluorescence is then detected by viewing the cells with light of the appropriate wave-length.

Immunofluorescence microscopy can be performed using antibodies that are **directly labelled with fluorescent dye** (Fig. 3.17A). However, immunofluorescence microscopy is more commonly performed using **indirect immunofluorescence** (Fig. 3.17B). In indirect immunofluorescence, a tissue or cell is treated with an antibody that is not labelled with dye. This antibody is called **primary antibody** and it attaches to specific antigenic sites within the tissue or cell. A second type of antibody, called **secondary antibody**, is then added. The secondary antibody is labelled with a fluorescent dye, and it attaches to the primary antibody. Because more than one primary antibody molecule can attach to an antigen, and more than one secondary antibody molecules can attach to each primary antibody, more fluorescent molecules are concentrated near each molecule that we seek to detect. As a result, indirect immunofluorescence results in signal amplification, and it is much more sensitive than the use of a primary antibody alone.

QUESTIONS

Long Answer Questions

1. Describe the method of cell fractionation.
2. What is centrifugation? Describe in brief the mechanism of centrifugation.
3. Write down the different types of centrifuges.
4. What is chromatography? Describe various types of chromatography.
5. What is TLC? Describe the process of TLC.
6. Name various kinds of chromatographic procedures. Describe the procedure of paper chromatography and column chromatography.
7. What is electrophoresis? Explain the principle of electrophoresis. Describe different types of apparatuses of electrophoresis.
8. What is spectrophotometry? Describe the structure of a spectrophotometer.
9. What do you mean by fixation? Describe the different types of fixatives used in histological study.
10. What is fixation? Describe the principles of staining with reference to different types of stains used in the study of cell.

Short Answer Questions

1. Write a short note on the principle of autoradiography.
2. Discuss the principle of autoradiography and its application.
3. What is chromatography? Write down the principle.
4. Give differences between column chromatography and partition chromatography.
5. What is the difference between TLC and GLC?
6. Write short notes on the following:
   (*i*) Gas chromatography
   (*ii*) Paper chromatography
   (*iii*) Chromatography
   (*iv*) Ion exchange chromatography
   (*v*) SDS – PAGE
   (*vi*) Electrophoresis
   (*vii*) Dialysis
   (*viii*) Application of spectrophotometry
   (*ix*) Radioisotopes

Very Short Answer Questions

1. What is autoradiography?
2. Define chromatography.
3. What is electrophoresis?
4. Define centrifugation.
5. What is cryofixation?
6. Define the staining.
7. Define the term colorimetry.
8. Give the names of two type of centrifuges.

Multiple Choice Questions

1. The physical measurement of a molecule in chromatography is
   (*a*) Rg value (*b*) Rf value
   (*c*) half life (*d*) Rs value

2. In electrophoresis, the buffer establishes contact between
   (*a*) two electrodes
   (*b*) the sample and electrodes
   (*c*) the sample and supporting medium
   (*d*) the supporting medium and electrode

3. Separation by electrophoresis in based on
   (*a*) charge
   (*b*) density
   (*c*) solubility
   (*d*) refraction

ANSWERS

Very Short Answer Questions

1. Autoradiography is a procedure for detecting the location of radioactive molecules by overlaying a sample with photographic film, which becomes darkened upon exposure to radioactivity.
2. It is a group of related techniques that utilize the flow of a fluid phase over a non-mobile absorbing phase to separate molecules based on their relative affinities for the two phases, which in turn reflect differences in size, charge, hydrophobicity or affinity for a particular chemical group.
3. It is a group of related techniques that utilize an electrical field to separate electrically charged molecules.
4. It is a process of rapidly spinning a tube containing a fluid to subject its contents to a centrifugal force.
5. It is rapid freezing of small samples so that cellular structures can be immobilized in milliseconds. It is often followed by freeze substitution, in which an organic solvent replaces the frozen water in the sample.
6. It is incubation of tissue specimens in a solution of dye, heavy metal, or other substance that binds specifically to selected cellular constituents imparting a distinctive colour or electron density.
7. When light is incident upon a substance, a part of it is emitted and a part is absorbed by the substance. Many instruments are used for measuring the emission or absorption of radiant energy from substances such as *photometer*, *colorimeter* and *spectrophotometer*. *Colorimetry* is the technique for measuring and describing the phenomena of colour.
8. Ultracentrifuge and analytical centrifuge.

Multiple Choice Questions

1. (*b*) 2. (*d*) 3. (*a*)

4 The Origin of Life

Origin of life is one of the most interesting topics that has occupied the attention of humankind since antiquity. Emergence of a highly ordered self-organising system capable of self-adjustment and self-perpetuation may be considered a miraculous fate of nature. Can we describe origin of life as a spontaneous process? To answer these questions we have to trace the early chemical history of the Earth and the conditions necessary for the formation of organic compounds which might have given a start to the evolution of life?

4.1. PRIMITIVE EARTH

The origin of life or **biopoiesis** cannot be explained unless we trace the geological history of the Earth. According to a reliable estimate (**Urey** 1952), the Earth is believed to be formed about 4.5 billion years ago. Enormous heat was generated in the Earth and the atmosphere consisted of many lighter elements in a gaseous state. The earliest atmosphere of the Earth probably consisted of *water*, *ammonia*, *methane*, *hydrogen sulphide*, *hydrogen*, *nitrogen* and *carbon dioxide*. As the Earth gradually cooled, most of these lighter elements (such as hydrogen, ammonia and methane) eventually get lost.

The upper crust of the Earth became hard and rocky, but below the surface the trapped hydrogen and oxygen molecules were subjected to great pressure and heat, which eventually condendensed into water. This water escaped from the Earth and formed an atmosphere of vapour. Gradually the Earth's surface got further cooled, condensing the vapour into water which poured down in the form of rains. The rains continued ceaselessly for thousands of years. This unending downpour resulted in erosion of rocks on the Earth, forming vast oceans carrying dissolved minerals in it. Most of the minerals commonly found in the oceans formed a **prebiotic soup** (this lacked in oxides and sulphates which were not formed then).

This is generally believed that the primitive forms of life must have depended on some kind of energy, as the present forms do. This suggests many possibilities. The *sun-light, ultraviolet radiations, electric discharges* and *heat* must have been the agents responsible for catalytic reactions in the prebiotic Earth. Electric discharges catalyse reactions to synthesize molecules, a phenomenon that has been experimentally proved. Sunlight can produce large amount of oxygen by splitting water and producing simultaneously hydrogen, which is capable of enhancing the reducing power of the atmosphere. **Haldane** long ago suggested that in the prebiological Earth ultraviolet radiations were responsible for generation of complex molecules. The reactions of this type do exist, as evidenced by the presence of formaldehyde and glyoxal in mixtures irradiated with ultraviolet rays.

4.2. SETTING STAGE FOR CHEMICAL EVOLUTION

Chemical evolution of the Earth has been much debated by **Haldane** (1954), **Oparin** (1953), **Pringle** (1953) and **Bernal** (1965). The early Earth probably contained little oxygen but carbon existed either in combination with metals of the Earth's core or as hydrocarbons. Solar radiations might have caused splitting of water and other compounds into free radicals. Free hydrogen was lost into the atmosphere but oxygen and peroxides were freely available on the Earth, associated with humidity. **Haldane** attributed the production of the complex organic compounds to the direct action of radiation. Biological arguments, however, are contrary to Haldane's hypothesis. It is true that ultra-violet radiations favour the synthesis of compounds, but they help in the breakdown also. The ultraviolet light is found to be deterimental to the present day organisms, then how is it possible that the early forms (protobionts) were able to use the same energy for a building up process.

Pringle asserted that the Earth's atmosphere did not allow penetration of short wave radiations through the lower atmosphere, but the action was limited to providing small amounts of oxygen and as a rain from above. This might have produced an appropriate condition for chemical activity to slowly oxidise the hydrocarbons.

Ample evidence exists that suggests that the organic compounds were formed from inorganic components of the atmosphere under the influence of ultraviolet radiations. In this process, slow oxidation reactions may have proceeded at low temperatures, giving rise to chain reactions generated spontaneously or photochemically. Chain reactions between molecules dissolved in the primitive sea appear to be the first step towards organic synthesis, resulting in the emergence of a primitive organisms or **protobiont**. Through this process amino acids and other simple precursors may have originated that are believed to form primitive polypeptides, nucleic acids (RNA, DNA), sugars and lipids.

Origin of life was a gradual but slow process rather than a sudden phenomenon. This naturally involved some kind of selection and an evolutionary process in a non-biological system. This has been adequately discussed by **Oparin**.

4.3. OPARIN'S COACERVATE THEORY

In 1938 **Alexander I. Oparin** suggested that life on Earth emerged at the molecular level in the prebiotic seas through natural selection. He proposed the formation of stable molecular aggregates from the unstable molecular structures. These stable aggregates accumulated in the seas of the early Earth and by chance gave rise to colloidal particles. Evidence from physical chemistry suggests the possibility of formation of such colloidal particles. These particles were submicroscopic, composed of two or more colloids, and termed **coacervates** (Fig. 4.1). These coacervates could give rise to larger aggregates. They could not be called **living** but they did behave like primitive biological systems. They showed primary characteristics, *i.e.*, (*i*) they were subjected natural selection in which the stable molecular complexes survived and the unstable ones suffered spontaneous degradation;

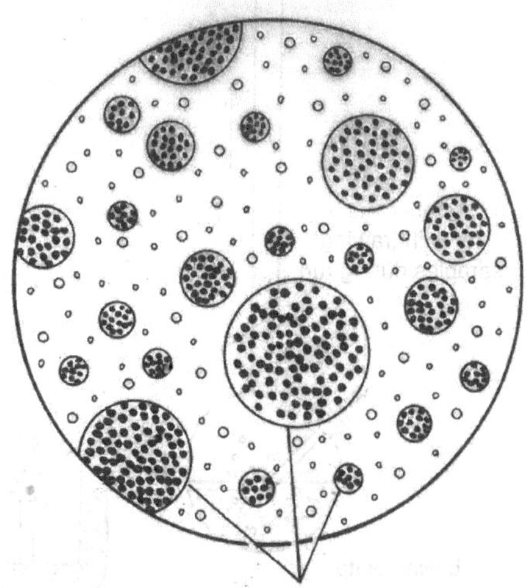

Fig. 4.1. Coacervate particles.

and (*ii*) they could selectively accumulate organic material. The larger complexes could give rise to more and more smaller complexes by fragmentation.

4.4. FIRST SIMULATION EXPERIMENT

Oparin's ideas were well received and generated much excitement in the scienticfic world. In 1952, **Stanley Miller** and **Harold Urey** devised an experimental simulating prebiological conditions of Earth to synthesize molecular forms of complexes. Their apparatus (Fig. 4.2) consisted of exposing water vapour to electric shock of 75,000 volts in an atmosphere of hydrogen, ammonia and methane that represented the atmosphere of the early Earth. The treatment continued for two weeks. There after the final products were subjected to chromatographic analysis and it was found that the organic compounds so synthesized included amino acids and carbohydrates, which serve as important metabolic intermediates in the present-day organisms. Miller's experiment proved that in order to produce biologically active molecules, three factors are necessary: (*i*) a high-energy source; (*ii*) reducing conditions (ammonia is a reducing agent and is homologous to NADH or cytochromes), and (*iii*) an aquatic medium homologous to the primitive seas of the early Earth.

Miller's success demonstrated that it is possible to synthesize organic compounds from inorganic matter, but the vexed question still remained that whether formation of molecular aggregates as suggested by Oparin is possible or not.

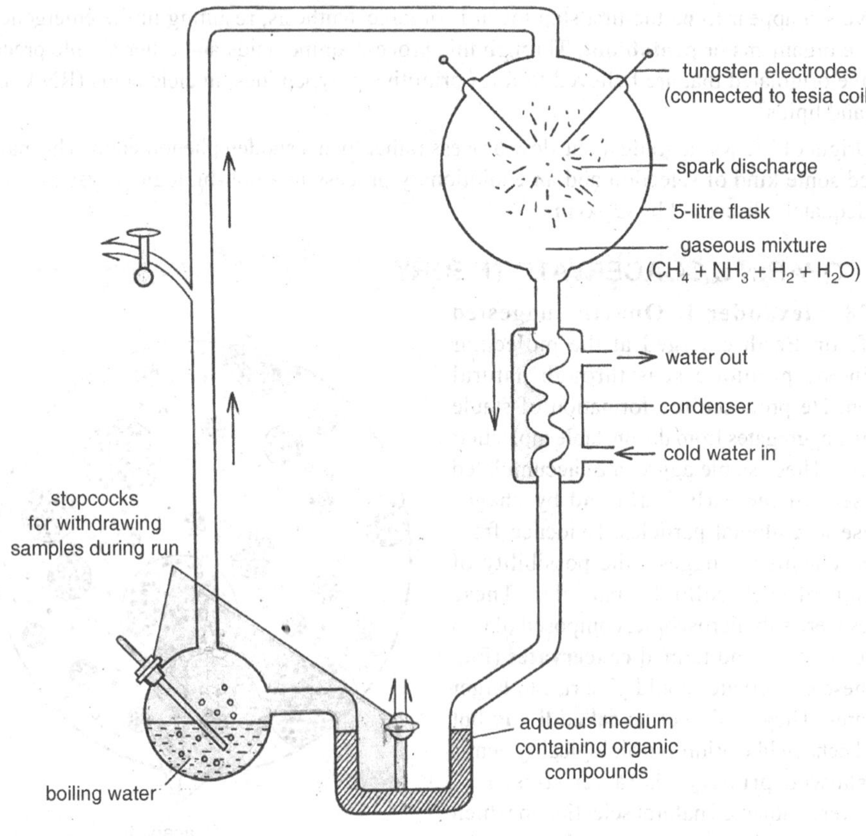

Fig. 4.2. The spark-charged apparatus of Miller for producing organic compounds from a reducing atmosphere.

4.5. FORMATION OF POLYMERS

1. Origin of proteinoids. Several investigators have reported the formation of polymers when amino acids are subjected to heat or electric discharges. **Sydney Fox** and his colloborators in 1957 heated a mixture of 20 amino acids at 170°C for a few hours and then cooled it. Since the mixture of amino acids was heated on volcanic rocks in an oven, cooling was accomplished by sprinkling water on the rock. The residue in the oven was examined under the microscope. It was found that millions of sphere-like structures, varying in size are formed. These spheres resembled yeast cells. **Fox** (1969) named them **protenoids** or **microspheres** (Fig. 4.3).

Fig. 4.3. Proteinoids or microspheres.

2. Origin of polypeptides. On chemical examination, the proteinoids were found to contain polypeptides, which on hydrolysis yielded amino acids. Proteinoids contain some non-amino acid components also, formed as by-product during thermal polymerisation. They could also be hydrolysed by proteolytic enzymes, and possess all the properties of protein molecules. The polypeptide chains in proteinoids consisted of about 18 amino acid residues and were formed spontaneously, each having a different amino acid sequence.

3. Natural slection of protenoids. There are few other properties of protenoids that need critical analysis. Proteinoids thus formed were probably abounding in the aquatic medium and may have been subjected to hydrolytic breakdown, since the peptide bond in aquatic medium is unstable. This means that proteinoids had a very short span of life. Unless they had some self-replicating power, the polypeptide sequences must have undergone destruction. If natural selection was also at work, better protenoids might have resulted.

4. Origin of enzymes. Proteinoids were able to carry out a variety of enzymatic reactions, especially *oxidation, decarboxylation, amination* and *hydrolysis*. Some of the proteinoids contained zinc along with ATPase activity. It has been postulated that catalytic activity appeared first, followed by substrate specificity. The proteinoids could grow in size by accumulating more organic materials, which ultimately gave rise to small protuberances in yeast-like fashion (Budding). These protuberances could detach from the parent proteinoids to lead an independent existence.

5. Origin of membranes. The most important structure of the proteinoids was the double membrane envelope around the protenoids, which behaved exactly like a semi-permeable membrane. The double-layered membrane has a unique biological significance since it contains specific amino acid sequences, predetermined by some genetic mehcanism. In the present case, however, one would think that its origin was spontaneous.

6. Origin of nucleic acids. There is yet another polymer, nucleic acid, that imparts the character of self-regulation to the living organisms. The building blocks of nucleic acids are mononucleotides, consisting of nitrogenous bases, ribose or deoxyribose sugar and a phosphoric acid group. We have now evidence to suggest that nucleic acids and nucleotides could be obtained by simulating conditions prevalent on the primeval Earth. Simulation experiments using appropriate bases and pentose sugars have demonstrated that nucleosides can be synthesized by heating or subjecting the nucleosides to ultraviolet radiation in the presence of phosphoric esters. **Loharmann, Bridson** and **Orgel** (1980) were able to demonstrate polymerisation of nucleotides on nucleic acid templates in enzyme-free systems. Thus nucleotides could have originated abiotically on the primitive Earth.

The next step in the evolutionary process is the formation of internucleotide linkages (bonds), so that monomers become interlinked to produce a strand-like structure. When mononucleotides are heated in the presence of phosphoric acid molecules (50° to 65°C), ester linkages are formed, which are 2′, 5′ linkages. In nucleic acids 3′, 5′ linkages are present. The 2′, 5′ linkages may not be disadvantageous and may not interfere with the properties of nucleic acids.

Based on these findings scientists concluded that the conditions on the prebiotic Earth may have created suitable environment for the abiotic origin of protein and nucleic acid chains. Their formation may have been purely accidental. It has been assumed that primitive nucleic acid molecule existed as short strand with the progress of chemical evolution, this strand functioned as a template for the synthesis of complementary strands. This claim is justified, since mononucleotides can be interlinked in the presence of a phosphorylating agent using nucleic acid template as a primer.

Origin of DNA was perhaps the greatest feat of nature gifted with self-replicating property. DNA is a double-helix, whose both the strands may fall apart on heating a solution containing the DNA. We may thus imagine a short strand of nucleic acid on which free nucleotides can associate in a manner that would be complementary to the original strand that is being copied. The base pairs would be held together by hydrogen bonding. Polypeptide chains do not appear to serve as templates, hence, cannot synthesize complementary peptide sequences.

4.6. THE FIRST LIVING FORM

Thus the conditions on the prebiotic Earth were favourable to a sort of continuing chemical evolution that gave rise to a special class of organic molecules. Oparin's coacervates and Fox's proteinoids were not living. It is perhaps, not easy to imagine, in time scale, of the events when the first symptoms of life appeared.

Some people have asserted that the molecular viruses existing today may be regarded as the most primitive creatures of Earth. To most of us viruses do not appear to be living at all, since they require a living cell to multiply their race. It is, therefore, necessary that we define the parameters of life.

1. Parameters of life. It is not possible to define life; however biologists suggest a set of following properties that characterise life.

***(i)* Self-organisation.** Living systems are self-organising entities whereby specialised structures would have metabolic cooperation in an organised manner.

***(ii)* Capacity of self-regulation.** Living systems must exhibit a property of self-regulation of metabolic processes. Proteins, which also function, as enzymes, are capable of carrying out polymeric synthesis and degradation.

***(iii)* Energy-transfer mechanism.** Living systems follow the laws of thermodynamics. They have the capacity to liberate energy resulting from the degradation of food materials, which is conserved in the phosphate groups of energy rich compounds such as ATP. The energy of these phosphate groups is transferred to other intermediates, which transmit it to energy consuming system.

***(iv)* Capacity of self-perpetuation.** All living systems live within a time scale. Hence, they would be able to perpetuate their own kind during their life span. DNA is the informational molecule, which is not only capable of self-replication but also able to transfer genetic information to daughter cells when placed in a proper environment.

***(v)* Isothermal open system.** Living systems have the capacity to exchange matter and energy with the environment, hence, they are regarded as an *open system* in a steady state. All chemical transformation of the matter occur at a physiological temperature (around 37°C in most cases), which otherwise would require considerable high temperature.

***(vi)* Capacity to adapt.** Living systems are genetically able to adapt to new environments or improve their efficiency in the same environment by random changes in their information content, causing evolutionary changes.

The main problem before us is to figure out the existence of an organism that could survive and evolve in a primeval environment when the oceans were full of organic soup.

2. Protobiont: The Primitive Form of Life. The simplest molecular aggregate or the primitive organism must have been far from the modern cell in structural organisation. A number of models have been suggested, but we have to look for an acceptable model that fits in the framework of a set of properties outlined above.

(*i*) The primitive organism, as **Haldane** suggested might have been an anaerobe because the biochemical mechanisms in anaerobic metabolism are similar in all organisms. This was possible in a primitive reducing atmosphere.

(*ii*) Many evolutionists belived that the first form was a photosynthesising organism. This suggestion was based on the plea that photosynthetic organisms synthesise their own food molecules and require a very simple environment. This idea has been rejected on two grounds. First, the photosynthetic organisms have much complex organisation and for synthesis they would require a very large number of enzymes. Second, it is accepted that life started in seas which were dilute organic soup and could not have sustained a complex photosynthetic organisation. In fact, scientists have much clearer ideas about the period when photosynthesis began.

Sulphur exists in two isotopes, S^{32} and S^{34} and their ratios in the early pre-Cambrian were about 22 : 1. About 800 million years ago the amount of light sulphur increased and diminished sulphates. this could be possible only if sulphur bacteria were present, because they oxidise sulphide to sulphate. **Thode**, **Macnamara** and **Flemming** have suggested that sulphur bacteria are about 8 million years old, a period earlier to the beginning of photosynthetic activity.

(*iii*) **Pringle** has suggested that primitive organism may have evolved at a great depth in the sea. Oxidised compounds probably settled down at great depths, diffusing hydrogen and other reducing substances, favouring the formation of complex molecules.

(*iv*) **Oparin** has suggested that the coacervate proposed by him was perhaps the primitive organism or the **protobiont**. Formation of stable complexes through conservation was possible as a chance event in the aquatic media containing polymers, which were probably proteins. The molecular aggregates could have been easily surrounded by a boundary layer, akin to a membrane. They were also able to trap small molecules of glucose, amino acids and catalysts that enabled them to carry on metabolic activity. *Nucleic acid polymers were not associated with coacervates*, hence oparin suggested that the self-replicating genetic system is a much later development in the biological evolution.

(*v*) It has been suggested that life without nucleic acids would be impossible because they represent the information molecules that direct the self-regulatory behaviour of the living system. Since cocervates were devoid of nucleic acid component, they could not be considered the fore-runners of the living cell. There has been much argument concerning the origin of nucleic acids and proteins. According to one view, proteins were to come first, where as the other view suggests that nucleic acids were arose first.

(*vi*) Studies indicate that RNA was the first biologically active molecule, that in time, associated the molecular aggregates (Fig. 4.4) to form the first living unit. Another problem that confronted the scientists was to provide a clue, whether the DNA that is blue-print for proteins or without which neither DNA nor proteins can be synthesized. It has been shown that certain reactions are catalysed by RNA molecules that are transcribed from introns of certain eukaryotic genes (Box 4.1).

Box 4.1 RNA World

A critical step in understanding molecular evolution was reached in the early 1980s, when it was discovered in the laboratories of **Sid Altman** and **Tom Cech** that RNA is capable of catalyzing a number of chemical reactions, including the polymerisation of nucleotides. Further studies have extended the known catalytic activities of RNA, including the description of RNA molecules that direct the synthesis of a new RNA strand from a RNA template. RNA is thus uniquely able to both serve as template for and to catalyze its own replication. Consequently, RNA is generally believed to have been the **initial genetic system** and an early stage of chemical evolution thought to have been based on self-replicating RNA molecules—a period of evolution known as **RNA World**. Ordered interactions between RNA and amino acids then involved into the present day genetic code, and DNA eventually replaced RNA as the genetic materials (**Cooper** and **Hausman**, 2007). Studies have shown that short stretches of RNA, about 10 bases long, can be synthesized with inorganic chemicals acting as catalysts. The process is extremely slow as compared to enzyme-catalyzed reactions. So it is believed that short stretches of RNA may have been spontaneously generated which later joined to form longer stretches.

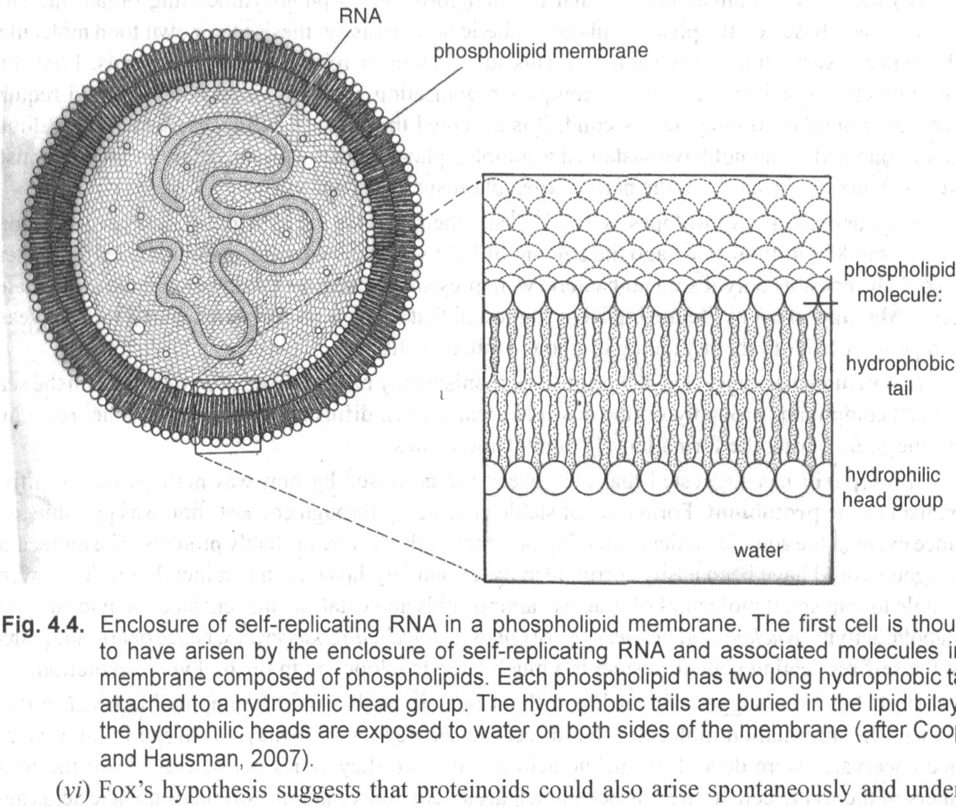

Fig. 4.4. Enclosure of self-replicating RNA in a phospholipid membrane. The first cell is thought to have arisen by the enclosure of self-replicating RNA and associated molecules in a membrane composed of phospholipids. Each phospholipid has two long hydrophobic tails attached to a hydrophilic head group. The hydrophobic tails are buried in the lipid bilayer; the hydrophilic heads are exposed to water on both sides of the membrane (after Cooper and Hausman, 2007).

(*vi*) Fox's hypothesis suggests that proteinoids could also arise spontaneously and undergo division of budding for multiplication. These proteinoids also have a double-membrane covering to limit the boundary of complex materials. But this model does not fit in the framework of life, because the absence of genetic material would not allow a system to evolve. Basic properties to carry out metabolism and self-replication are inherent in the DNA molecules only. It is therefore logical to think that the emergence of the information molecule was the prime necessity of life. Hence scientist pursued the search for an organism that had the capacity to store information, decode it and self-replicate.

3. Search for a Protobiont Organisation. Coacervates and proteinoids could not be the likely candidates, since they did not have the potential capacity to code for proteins to undergo self-replication and to evolve through natural selection. As the nucleic acids (consisting of genes) can undergo **mutations**, only the cells possessing them can have the tendency to evolve. Therefore, thinking of biologist has the natural leaning toward nucleic acid hypothesis. This hypothesis have following three main offshoots:

(a) **Virus hypothesis.** Viruses, which are at the threshold of life, may be considered as the forerunners of a prokaryotic organization. All viruses contain either DNA or RNA, a protein coat and are conspicuously devoid of an ATP-ADP system. However, they are unable to multiply unless allowed to infect a host cell. Inability of viruses to multiply independently is due to the absence of an enzyme system needed for the purpose. A host cell could not have been present in the prebiotic Earth, hence viruses cannot be considered for a protobiont.

(b) **Nucleotide hypothesis.** Another line of thought supports the role of nucleic acids in the origin of life. The monomeric units (nucleotides) function as carriers and coenzymes, which suggests that nucleotides were the prime requirements for evolving a genetic apparatus and suitable metabolic machinery.

(c) **Hypercycle hypothesis.** This concept was developed by **Eigene** and **Schuster** and it takes recourse to the application of irreversible thermodynamics in chemical reactions that are far away from equilibrium. The hypercycle concept imagines the formation of polymeric molecules, such as nucleic acids and proteins, that could provide a structural and functional organisation. These molecules could have been formed in the prebiotic soup and their interactions probably gave rise to more stable complexes that became integral part of the structural organisation. The hypercycles concept has found some support among the advocates supporting the role of nucleic acids in the origin of life.

(d) **RNA hypothesis.** We have earlier discussed spontaneous formation of short RNA segments, which might have joined to form longer molecules. Aquisition of catalytic property by certain RNA configurations probably induced the synthesis of short peptides from the available amino acids. Thus, the key molecules appear to be RNA that developed the means to self-replicate. Such RNA molecules were under great selective pressure, and out of several RNA molecules one may have become selected. Self-replication may not be the only criterion for RNA to be biologically active for the initiation of a cell, it should also serve as a template for synthesis of proteins.

There is, however, one plea against RNA molecules. Compared to double stranded DNA RNA molecules are single stranded and very unstable. Therefore, it appears that the information contained in the RNA molecules, must have been transferred to DNA strands synthesized on RNA templates. To confirm this view, we can cite an evidence from **retroviruses** which synthesize DNA from RNA templates with the help of the enzyme **reverse transcriptase**. In earliest stages, single stranded DNA were synthesized, and in time, they acquired a complementary strand to become double stranded. Once the first organism (cell) containing DNA was evolved, it was subjected to intense biological evolution.

4. Concept of Non-informational RNA and Ribosomes. If we consider for a while that nucleic acid (DNA) arose first, then how did the proteins come about? **Crick** and **Orgel** have suggested that the real answer to the origin of life may be found in the evolution and organisation of transcription and translation machinery of the primeval cell.

Protein synthesis involves two types of RNAs which are **non-informational** and do not serve as templates. These are **transfer RNA (tRNA)** and **ribosomal RNA (rRNA)**. These possess a folded three-dimensional structure that may have participated in making informational proteins. Thus, tRNAs and rRNAs were most important entities in the primitive organisation of the cell.

The next step may have been the evolution of a primitive peptide-forming machinery. Evidently prokaryotes possess more RNA than the eukaryotes. It has been suggested that in the primitive ribosome no protein is associated. Such a ribosome was probably capable of making peptides with the help of tRNA that could bind with rRNA. At this stage the enzyme system had not been evolved. Hence it was possible that the tRNA molecule could link with a given amino acid at the anticodon site through non-covalent linkage. It indicates an enzyme-like role of the primitive tRNA and rRNA.

An enzymeless non-covalent linkage of amino acid to the tRNA molecule could not be error-free and may have led to many copying errors. Therefore, *development of a template system was the prime necessity to specify a correct sequence*. How could this be achieved? Perhaps this requirement could lead to the formation of first **messenger RNA (mRNA)** in the organisation of a transcription and translation machinery. The solution may have been provided by linking another RNA strand with tRNA and also to the ribosome. This gave rise to the mRNA, an intermediate transcription molecule.

The formation of catalytic protein may have followed the synthesis of nucleo-protein complexes. At some stage the proteins may have acquired to synthesize enzymes catalyzing the formation of peptide bonds. There are a number of peptides that could be synthesized under simulated conditions prevailing on the prebiotic Earth. These peptides perhaps had the property of behaving like catalysts. As a cosequence, it is not unlikely that other enzymes such as **ribonuclease** (Box 4.2) may have appeared for efficient translation. Formation of long-chain proteins may have influenced the size of RNA also. As nucleic acids can serve as template in the absence of enzymes and proteins, a complementary nucleic acid may have formed in the presence of suitable condensing agent. This may have led to the formation of DNA, which had the self-replicating property and permanent information storage capacity.

Box 4.2

Ribonuclease or **RNAase** is a nuclease enzyme which have systematic name as E.C. No. 3.1.27.5. It catalyzes the hydrolysis of certain internucleotide linkages (=bonds) of RNA.

5. Evolution of Metabolism. Because the cells originated in a sea of organic molecules, they being heterotrophic were able to obtain food and energy directly from their environment. But such a situation is self-limiting, so cells needed to evolve their own mechanism for generating energy and synthesizing the molecules necessary for their replication. The generation and controlled utilisation of metabolic energy is central to all cell activities, and the principlal pathways of energy metabolism are highly conserved in present-day cells. All cells use **adenosine 5′-triphosphate (ATP)** as their source of metabolic energy to carry to drive the synthesis of cell constituents and carry out other energy requiring activities, such as movement (*e.g.*, muscle contraction). The mechanisms used by cells for the generation of ATP are thought to have evolved in *three* stages, corresponding to the (*i*) evolution of glycolysis; (*ii*) evolution of photosynthesis and (*iii*) evolution of oxidative metabolism (Fig. 4.5). The development of these metabolic pathwys changed Earth's atmosphere, thereby altering the course of further evolution.

1. Glycolysis

$C_6H_{12}O_6$ (Glucose) ⟶ $2C_3H_6O_3$ (Lactic acid) — Generates 2ATP

2. Photosynthesis

$6CO_2 + 6H_2O$ —Sun's radiant energy⟶ $C_6H_{12}O_6$ (Glucose) $+ 6O_2$

3. Oxidative metabolism

$C_6H_{12}O_6$ (Glucose) $+ 6O_2$ ⟶ $6CO_2 + 6H_2O$ — Generates 36–38 ATP

Fig. 4.5. Generation of metabolic energy. 1. Glycolysis is the anaerobic breakdown of glucose to lactic acid. 2. Photosynthesis utilises energy from sunlight to drive the synthesis of glucose from CO_2 and H_2O, with the release of O_2 as a by-product. 3. The O_2 released by photosynthesis is used in oxidative metabolism, in which glucose is broken down to CO_2 and H_2O, releasing much more energy than is obtained from glycolysis (after Cooper and Hausman, 2007).

***(i)* Evolution of glycolysis.** In the initially anaerobic atmosphere of Earth, the first energy-generating reactions presumably involved the breakdown of organic molecules in the absence of oxygen. These reactions are likely to have a form of present-day **glycolysis**—the anaerobic breakdown of glucose to lactic acid, with the net energy gain of two molecules of ATP. In addition to using ATP as their source of intracellular chemical energy, all present-day cells carry out glycolysis, consistent with the notion that these reactions arose very early in evolution.

***(ii)* Evolution of photosynthesis.** Glycolysis provided a mechanism by which the energy in preformed organic molecules (*e.g.*, glucose) could be converted to ATP which could then be used as a source of energy to drive other metabolic reactions. The development of **photosynthesis** is generally thought to have been the next major evolutionary step, which allowed the cell to harness energy from sunlight and provided independence from the utilisation of preformed organic molecules. The first photosynthetic bacteria (the prokaryotic organisations), which evolved more than 3 billion years ago, probably utilising H_2S to convert CO_2 to organic molecules—a pathway of photosynthesis still used by some bacteia. The use of H_2O as a donor of electrons and hydrogen for the conversion of CO_2 to organic compounds evolved later and had the important consequence of changing Earth's atmosphere. The use of H_2O in photosynthetic reactions produces the by-product free O_2, this mechanism is thought to have been responsible for making O_2 abundant in Earth's atmosphere.

***(iii)* Evolution of oxidative metabolism.** The release of O_2 as a consequence of photosynthesis changed the environment in which cells evolved and is commonly thought to have led to the development of **oxidative metabolism**. Alternatively, oxidative metabolism may have evolved before photosynthesis, with the increase in atmospheric O_2 then providing a strong selective advantage for organisms capable of using O_2 in energy producing reactions. In either case, molecular oxygen (O_2) is highly reactive molecule, and oxidative metabolism, utilising this reactivity, has provided a mechanism for generating energy from organic molecules that is much more efficient than anaerobic glycolysis. For example, the complete oxidative breakdown of glucose into CO_2 and H_2O yields energy equivalent to that of 36 to 38 molecules of ATP, in contrast to the 2 ATP molecules formed by anaerobic glycolysis. With few exceptions, present-day cells use oxidative reactions as their principal source of energy.

6. Evolution of protobiont. Chemical evolution on prebiotic Earth thus gave rise to organic molecules which were asymmetrical and included proteins and nucleic acids. Once the template was established there was no difficulty in evolving enzyme systems and a surrounding membrane around the molecular complexes. *This may have been the beginning of a structural organization.* Being heterotrophic these primitive structures utilised the organic molecules from the environment that may have been depleted in the primeval broth. Thus, the first organism was probably an anaerobe. We do not know whether this organism had a form of a biological cell. It is almost certain that energy interactions were established for the biological evolution to proceed.

A very important step in the biological evolution of cell may have been the formation of lipid membranes. Some biologists, however, believe that polar lipids were perhaps the first requirement to form a boundary to limit the molecular aggregates before commencement of the synthesis of nucleic acids and proteins. A continuous supply of free energy is necessary for the maintenance of a biological organisation, which was abundantly available as radiation from the sun. ATP is perhaps the only life-like molecule that may have arisen in the begining a life to establish an energy-transfer mechanism. Since the early Earth did not contain oxygen, the metabolic pathways would have been similar to those of present-day anaerobes.

Origin of progenote. The precellular stage of organization has been termed **progenote** (**Woese**, 1987). The nature of progenote is uncertain. It might have been functional, slow replicating, independently evolving organism, or it might have been an amorphous group of primitive transcription units plus a primitive transcription and translation apparatus from which many different cell lineages evolved; three of which have survived to the present day (*i.e.*, archaebacteria, urkaryote and eubacteria). The progenote contained DNA, most genes with introns, a slow growth and heterotrophic nutrition.

Any search for an anaerobic living system possessing a cellular organisation leads to the microbial world. Bacteria are ubiquitous and occur everywhere and some of them have a very complex organization. Many of the bacteria are extincted because of their structural complexity. There is one group of bacteria that are about one-tenth the size of most bacteira, called **PPLO** (pleuropneumonia-like organisms), form. The PPLO forms have been grown in artificial media, and require no macromolecules or high unstable metabiolic intermediates.

PPLO forms (Forerunners of Prokaryotes). The PPLO forms fall in the category of prokaryotes since they have no nucleus, have 70S type ribosomes, can be grown on artificial media and exhibit a very simple type of metabolism. The PPLO forms are made up of about 4% DNA, 8% RNA and about 40 different kinds of enzymes. It is, however, not clear how many different types of proteins PPLOs were capable of producing. The **protein potential** of the PPLO is about 80, which is calculated by using the value of the DNA content. The 4% DNA represents about 75,000 nucleotide pairs. Each protein contaings about 300 amino acids, and one amino acid is coded by three bases, the protein potential can be calculated as follows:

$$\frac{\text{Nucleotide pairs of DNA} = 75,000}{\text{Triplets} \times \text{Amino acids} = 3 \times 300} = \text{approximately } 80$$

The protein potential indicates that PPLO forms were capable of making at least 80 types of proteins, half of which are the enzymes. A cell in the prebiotic environment would require almost no enzymes for metabolism. The only enzymatic controls that would be necessary are those for electron transport system for generation of ATP, RNA and nucleotides from nitrogen bases. From theoretical consideration, it has been estimated that the primitive organisms could not have a protein potential of more than 30. It is, therefore, reasonable to assume that the protobiont must have had a very low grade of organismal complexity. It probably had some but very few control mechanisms. Its reproduction was hardly beyond fragmentation as in case of protenoids, and its existence was dependent on the metabolic precursors from the environment. It was an organism capable of synthesis of utilising ATP (coupling of adenosine and phosphate by phosphorylating enzyme was possible in the environment). It could control macromolecular synthesis by a DNA-RNA-protein sequence as found in present-day forms, and it underwent evolutionary advance by natural selection. Operating among the protobiont population. It may then be concluded that the molecular aggregate with the minimum biological level must have been produced through a succession of events influenced by selection pressure of the total environment. This may have taken at least a billion years of evolution to reach the level of PPLO-like form. Thus, the protobiont could not have been PPLO-like form.

7. Emergence of Protobiont. The protobiont, as outlined above, was a complicated little creature, which did not come about by reproduction, but was spontaneously generated through the combinations of monomers available in the environment. This was endowed with the properties of life. Through the process of natural selection the protobiont type was formed, not a single chance event, but as the outcome of a stepwise accumulation of traits, directed by the selection pressures of the total environment.

On their evolution toward the probiont (or progenote) molecular aggregates acquired enzyme systems. It is considered that the enzymes evolved before the DNA code for protein structure. This assumption is based on the belief that the early seas had abundant coenzymes, which accelerate chemical reactions even without protein component usually associated with enzymes. When these coenzymes became scarce threatening the survival of the organism, presence of an information system became important.

It is assumed that some molecular aggregates were capable of surrounding the coenzyme with organic material that limited the shapes of acceptable substrates, giving rise to **cloaked coenzymes.** Cloaked coenzymes were responsible for more specific reactions, resulting in increased stability of molecular aggregates or fragmentation of aggregates. Fragmentation would ensure multiplication of a successful type. Subsequently, information molecules (DNA), a decoding system, and enzymes for DNA replication must have been appeared. It is possible that all these three components of the information system arose simultaneously. Once DNA-RNA-protein sequence was established the protobiont or progenote has been formed.

Evolution of Prokaryotes and Origin of Eukaryotes

The progenote evolved into three types of cells: 1. Archaebacteria, 2. Eubacteria and 3. Urkaryote. **Archaebacteria** are the prokaryotes which appear to have diverged from the true bacteria (or eubacteria) very early in evolution (Fig. 4.6). The strange class of archaebacteria consists of bacteria that are tolerant of acid (**thermophiles** or **thermoacidophiles**), bacteria that thrive in salt (**halophiles**) and bacteria that generate methane (**methanogens**). Recent developments in molecular biology reveal that these bacteia share some striking biochemical features (such as 16S rRNA sequence) that are absent from other bacteria. **Eubacteria** includes **cyanobacteria** (aerobic, H_2O photosynthetic); **chloroxybacteira** (aerobic, H_2O photosynthetic); **paracoccus** (aerobic, heterotrophic), **non-sulphur bacteria** (aerobic, H_2S photosynthetic), **sulphur bacteria** (anaerobic, H_2S photosynthetic), **green-filamentous bacteria** (aerobic, H_2S photosynthetic), **green sulphur bacteria** (anaerobic, H_2S photosynthetic), **spirochaetes** (anaerobic, heterotrophic), *Clostridia* (anaerobic, heterotrophic) and *Desulphovibrio* (sulphate respiration). The **urkaryote** is the ancestor of modern eukaryote.

In contrast to the prokaryotes, the cells of the morphologically more complex plants and animals, or **eukaryotic cells**, have a distinct nuclear membrane that encloses strands of DNA, or discrete chromosomes. Eukaryotic cells also have an elaborate system of membrane bound cytoplasmic organelles. The enzymes that are responsible for releasing energy from complex organic molecules are packed inside sausage-shaped organelles called **mitochondria**. In the eukaryotic plant cell, a prominent organelle is the **chloroplast**, the photosynthesizing particle that converts light energy into chemical energy. The following two hypotheses have been proposed regarding the origin of semiautonomous organelles of eukaryotes, *i.e.*, mitochondria and chloroplasts:

A. Endosymbiotic hypothesis. One of the most compelling speculations is that mitochondria and chloroplasts arose not by a gradual evolutionary process but abruptly in an unusually striking manner, *i.e.*, **symbiosis**. In fact, mitochondria have a number of interesting properties that advocate that they were once free-living or independent, bacteria-like (*i.e.*, Prokaryotic) organisms. Mitochondria have small amounts of their own DNA; this DNA exists as a loop-shaped molecule like the DNA of bacteria. Mitochondria also possess the genetic capacity to incorporate amino acids into proteins (on their own ribosomes). These properties have led, in 1970, **Lynn Margulis** of Boston University to propose the **serial symbiotic hypothesis** or **endosymbiotic hypothesis**. She hypothesized that mitochondria may have been derived from primitive aerobic bacteria [*i.e.*, Paracoccus (Box 4.3)] that were engulfed by predatory organisms, probably fermentative bacteria, destined to become eukaryotic (Fig. 4.7). The predatory hosts became dependent on their enslaved mitochondria and the latter, in turn, became dependent on their hosts. Thus, the association was of mutual advantage to the predatory cell and the engulfed prey. Such a close association or partnership is called **symbiosis**. In principle, the mitochondria (formerly oxygen-respiring bacteira) established permanent residence within the hosts. These predatory hosts became the **first eukaryotic cells** (*i.e.*, animal cells).

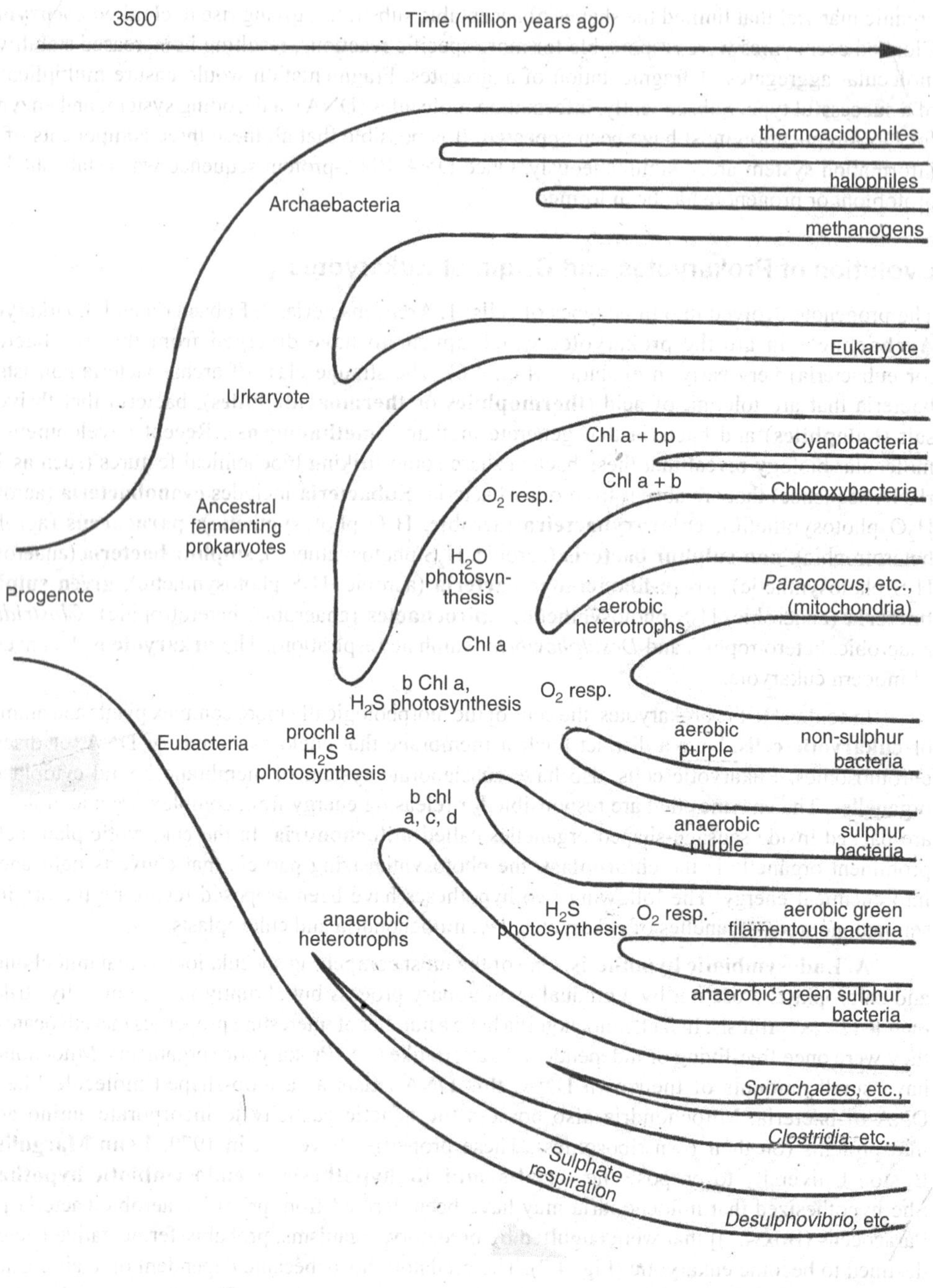

Fig. 4.6. A phylogenetic tree of the main groups of prokaryotes mentioned in the text. chl is chlorophyll; bchl; bacterio-chlorophyll; pb, phycobilin; prochl a, photo chlorophyll a and resp., is respiration.

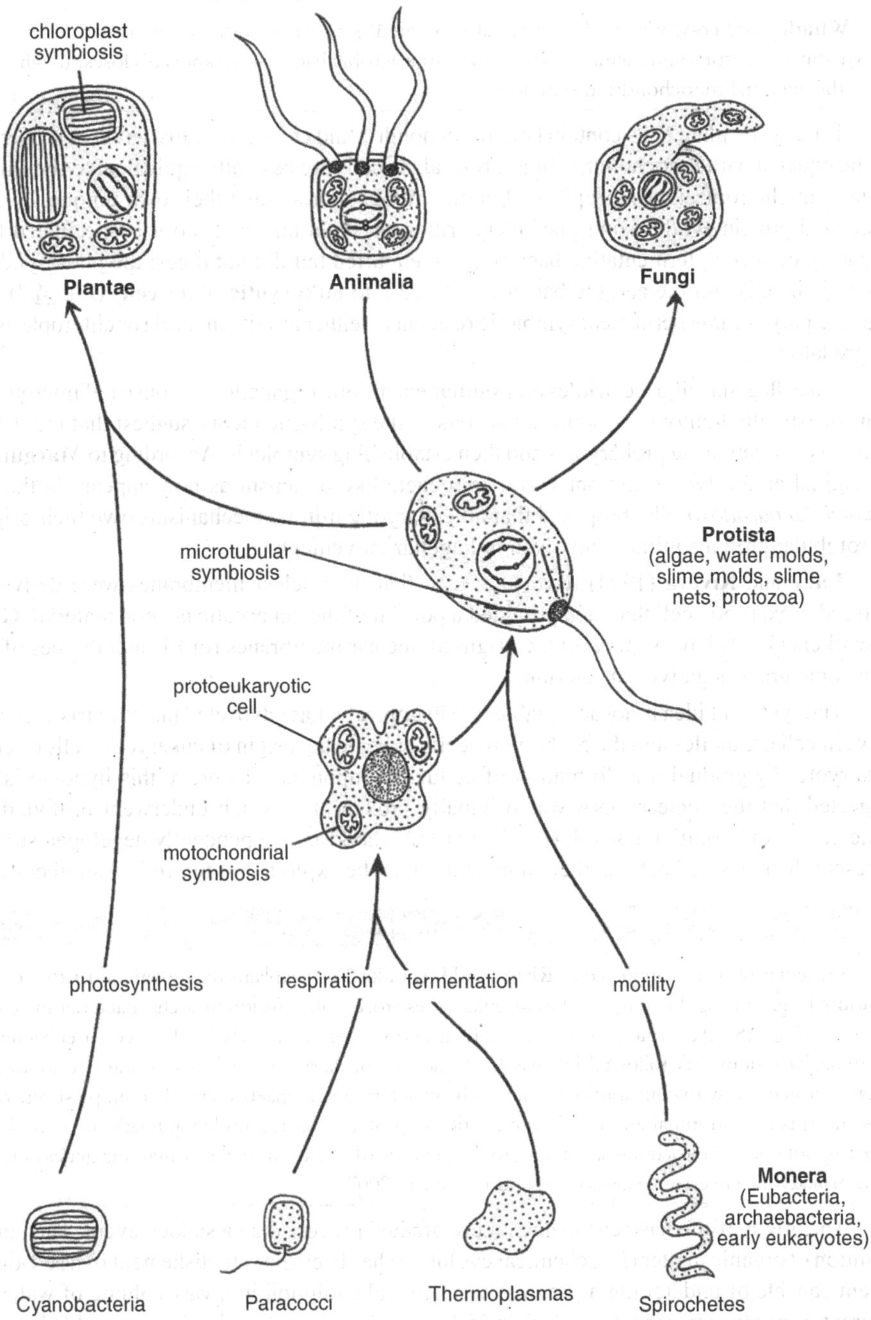

Fig. 4.7. Symbiotic events during the evolution of eukaryotic cells according to Margulis. Mitochondrial symbiosis arose from an early invasion of aerobic respiring prokaryotes such as *Paracoccus*. Eukaryotic motility as well as the microtubular structures involved in mitosis and meiosis was then gained by symbiosis with a prokaryotic spirochete-like form. In a later event, photosynthetic cyanobacteria invaded some eukaryotic cells, then giving rise to plant chloroplasts.

Box 4.3

Whatley and co-workers (1979) have also pointed out that because of its many mitochondrion-like enzymatic, respiratory, and membranous features, the aerobic bacterium *z* appears closest to what may have been the ancestral mitochondrion symbiont.

Eukaryotic plant cells contain both mitochondria and chlorplasts. Just as aerobic bacterial cells can be equated with mitochondria, blue-green algae cells are basically equivalent to the chlorophyll-containing chloroplasts. Chloroplasts, like the mitochondria, have their own unique DNA and the associated protein-synthesizing machinery (ribosomes). In this case too it is speculated that those predatory cells (*i.e.*, fermentative bacteria) that engulfed but did not digest the photosynthetic blue green algal cells and the aerobic bacteria evolved into **eukaryotic plant cells** (Fig. 4.7). Thus the engulfed prey became permanent symbiotic residents—either as mitochondria or chloroplasts—within the predatory cell.

Since flagella, cilia, centrioles and similar eukaryotic organelles all consist of microtubules and seem structurally homologus, some supporters of the symbiotic theory suggest that these also arose by eukaryotic capturing prokaryotes and then establishing symbiosis. According to **Margulis** (1993), an original cilium-type symbiont was a spirochete-like organism as now appears in the protistan *Myxotricha paradoxa*. She proposed that all eukaryotic mitotic mechanisms owe their origin to the microtubular proteins initially involved in flagellar movement.

Lake and **Rivera** (1994) have suggested that the nuclear membranes were derived from a captured prokaryotic cell that then provided a portion of the eukaryotic genetic material. **Gupta** and co-workers (1994) have supported the origin of nuclear membranes (or ER membranes of *Giardia*) from some gram-negative eubacterium.

The symbiont idea is not acceptable to **Allsopp**, who has advocated that the striking similarities between cell organelles and the prokaryotic cells indicate the origin of eukaryotic cells directly from prokaryotes by gradual transformation of cellular components. To prove this hypothesis, Allsopp suggested that the nuclear DNA was originally prokaryotic, which underwent profound changes in the course of evolution (Box 4.4). Whether mitochondria independently developed structures or represent the ingested bacteria, their similarities may be exploited in future in a number of ways.

Box 4.4

A recent hypothesis (proposed by **Rivera** and **J.A. Lake** 2004) explains the **mosaic nature of eukaryotic genome** by proposing that the genomes of eukaryotes arose from a fusion of archaebacterial and eubacterial genomes (Fig. 4.8). According to theis model, an endosymbiotic association between a eubacterium and an archaebacterium was followed by fusion of the two prokaryotic genomes, giving rise to an ancestral eukaryotic genome with contributions from both eubacteria and archaebacteria. The simplest version of this hypothesis is that an initial endosymbiotic relationship of a eubacterium living inside an archaebacterium gave rise not only to mitochondria but also to the genome of eukaryotic cells, containing genes derived from both prokaryotic ancestors (see Cooper and Hausman, 2007).

The origin of life has been a much more gradual process than a sudden event. The outcome of evolution of organic matter (*i.e.*, chemical evolution) has been the establishement of a self-organising system capable of undergoing a progressive chemical evolution in a vast volume of water. Once a coherent organism emerged through chemical reactivity, several properties were added step by step through natural selection. After a long history of chemical evolution, biological evolution must have taken place to give rise to complex biological organisms that we know today. It is left to the future generations to evaluate the present and future evidence before final conclusion can be drawn about the actual sequence of events.

QUESTIONS

Long Answer Questions

1. Summarise Oparin's concept of origin of life on Earth.
2. Describe Stanley Miller's experiment and explain how does it prove the biochemical theory of origin of life.
3. Enumerate the different steps in chemical evolution leadng to the synthesis of self-duplicating organic molecules.
4. Explain following with reasons:
   (*a*) The primitive atmosphere of Earth was without oxygen.
   (*b*) All organic compounds formed in the primitive Earth had hydrogen and carbon.
   (*c*) The sea of primitive Earth was a broth of numerous organic compounds but it was not exposed to any disintegrating activities.
5. Enumerate the energy sources in the atmosphere of primitive Earth which led to the synthesis of various organic molecules.
6. According to latest thinking, nucleic acids are the basis of the origin of life. Discuss this in the light of various theories.

Short Answer Questions

1. What is Miller-Urey experiment?

   Ans. Miller-Urey experiment is an experiment, suggested in 1952 by **H.C. Urey** and performed by **S.L. Miller** to explore the possible origins of life on Earth. An electric spark discharge was passed through a mixture of water vapour, methane, ammonia and hydrogen, gases meant to simulate the Earth's prebiotic atmosphere. After several days hydrogen cyanide, methanol, urea, methanoic, ethanoic and other acids, including amino acids had formed.
2. Write short notes on the following:
   (*a*) Concept of hypercycles;
   (*b*) Evolution of progenote;
   (*c*) Margulis hypothesis of origin of life.
3. Describe concept of non-informational RNA and ribosomes about origin of life.

Very Short Answer Questions

1. What are coacervates?
2. Define biogenesis.
3. Who created microspheres experimentally?
4. Which are forerunners of prokaryotes?
5. Name the ancestor of eukaryote.
6. Who proposed endosymbiotic theory of eukaryotic cells?

Multiple Choice Questions

1. Modern theory of origin of life was propounded by
   (*a*) Miller (*b*) Darwin
   (*c*) Khorana (*d*) Oparin
2. Who said that organisms develop from pre-existing organisms?
   (*a*) Aristotle (*b*) Louis Pasteur
   (*c*) Oparin (*d*) Morgan
3. The most primitive cell-like chemical aggresates capable of growth and division were
   (*a*) chemoautotrophs
   (*b*) eobionts
   (*c*) prokaryotes
   (*d*) coacervates
4. Due to discovery of which of the following in 1980's the evolution was termed as RNA world?
   (*a*) mRNA, tRNA, rRNA synthesize proteins
   (*b*) in some viruses, RNA is the genetic material
   (*c*) some RNAs have enzymatic property
   (*d*) RNA is not found in all cells
5. In the experiment of Stanely Miller, the end product was
   (*a*) amino acids (*b*) nucleotides
   (*c*) acetic acid (*d*) oxaloacetic acid

6. The term 'microsphere' was given by
   (*a*) Haldane
   (*b*) Oparin
   (*c*) Sidney Fox
   (*d*) Miller

7. One of the possible early sources of energy was/were
   (*a*) carbon dioxide
   (*b*) chlorophyll
   (*c*) green plants
   (*d*) UV rays and lightening

ANSWERS

Very Short Answer Questions

1. Coacervates are discrete tiny droplets into which polysaccharides and proteins can spontaneously concentrate. They were proposed by Oparin.
2. Biogenesis is the principle that all living organisms are derived from other living organisms. It is opposite of spontaneous generation. Biogenesis is widely accepted at the present time, although the evolution of life on Earth may have begun with the generation of organisms from abiotic precursors (abiogenesis).
3. Sydney Fox in 1957.
4. PPLO forms.
5. Urkaryote.
6. Lynn Margulis in 1970.

Multiple Choice Questions

1. (*d*) **2.** (*b*) **3.** (*d*) **4.** (*c*) **5.** (*a*) **6.** (*c*) **7.** (*d*)

5 Prokaryotic Cell and Eukaryotic Cell

The body of all living organisms (archaea, bacteria, blue green algae, plants and animals) except viruses has **cellular organization** and may contain one or many cells (Box 5.1). A **cell** is the smallest and complete expression of the fundamental structure and function of all living organisms. It is a unit of structure, function and heredity. The cell is the basic unit of organization or structure, of all living matter. Within a selective and retentive semipermeable membrane, it contains a complete set of different kinds of units necessary to permit its own growth and reproduction from simple nutrients. It has always been quite difficult to define a cell. Different cell biologists have defined the cell differently as follows: **A.G. Loewy** and **P. Siekevitz** (1963) have defined a cell as *a unit of biological activity delimited by a semipermeable membrane and capable of self-reproduction in a medium free of other living systems*. **Wilson** and **Morrison** (1966) have defined the cell as *an integrated and continuously changing system*.**John Paul** (1970) has defined the cell as *the simplest integrated orgainization in living systems, capable of independent survival*. The organisms with only one cell in their body are called **unicellular organisms** (e.g., archaea, bacteria, blue green algae, some algae, Protozoa, etc.). The organisms having many cells in their body are called **multicellular organisms** (e.g., most plants and animals). Any cellular orgainsm may contain only one type of cell from the following types of cells:

A. Prokaryotic cell; B. Eukaryotic cell.

Box 5.1 Why viruses are not true cells?

A virus is neither an organism nor a cell, yet it consists of a core of nucleic acid (DNA or RNA) enclosed in an external mantle of protein. In the free state viruses are quite inert. They become activated only when they infect a living host cell and in the process only the nucleic acid core enter the host's cell. The nucleic acid which is the genetic substance, takes over the metabolic activity of the host cell and utilises the cell machinery for the formation of more viruses, ultimately killing the host cell. In a way, thus, viruses are cellular parasites that cannot reproduce by itself. They are primitive and simpler units of life.

5.1. THREE CELLULAR DOMAINS

The body of all living organisms except viruses has cellular organization. With the advent of electron microscopy, biologists came to recognize two fundamentally different plans of cellular organizations: the simpler one characteristic of bacteria and the more complex one found in all other kinds of cells. Based on structural differences of their cells, organisms were initially divided into two broad groups, the **prokaryotes** (bacteria) and the **eukaryotes** (all other forms of life such as protozoa, plants and animals). The most fundamental distinction between the two groups is that eukaryotic cells have a true, membrane-bounded nucleus (*eu*-is a Greek word for "true" or "genuine", *karyon*-means "nucleus", where as prokaryotic cells do not (*pro*-means "before" suggesting an evolutionarily earlier form of life). The terms prokaryotic and eukaryotic were suggested by **Hans Ris** in the 1960's. On the basis of molecular and biochemical criteria, especially sequence analysis of ribosomal RNAs (rRNAs), cell biologists now recognize that the prokaryotes can be further differentiated as either **eubacteria** and **archaea**. The eubacteria ("true bacteria") include most present-day **bacteria** and **cyanobacteria** (also known as blue-green algae)—most of the commonly encountered bacteria, in other words. The archaea (also called **archaebacteria**) are similar to eubacteria in cellular structure, but are as different from eubacteria as they are from eukaryotes in terms of molecular and biochemical distinctions. Archaea are regarded as modern descendants of an evolutionarily ancient form of prokaryote that differed fundamentally from the ancestors of present eubacteria (*Archae*-is a Greek prefix meaning "ancient" or "original"). Present day archaea can be subdivided into three main groups: the **methanobacteria**, which obtain energy by converting carbon dioxide and hydrogen into methane; the **halobacteria**, which can grow in salty environments (upto 5.5 M NaCl) and the **sulphobacteria**, which obtain energy from sulphur-containing compounds. Some of these archaea are also **thermacidophiles,** which thrive in acidic hot springs with pH as low as 2 and temperatures that can exceed 100°C.

Based on the RNA sequencing work of **Carl Woese, C. Fred Fox** and others (1977, 1990), cell biologists now recognize **eukaryotes, eubacteria** and **archaea** as three **fundamentally different groups** or **domains,** of organisms. In addition to the characteristics of their tRNAs, the cells in each group have other distinctive molecular, biochemical and cellular entities and properties in common, including RNA polymerases (the enzymes that synthesize RNA), sensitivity to specific inhibitors of protein and nucleic acid synthesis, and the presence or absence of memberane-bound intracellular compartments (organelles).

Origin of Eukaryotes

The crucial step in the evolution of eukaryotic cells was the acquirement of memberane-enclosed subcellular organelles allowing the development of the **complexity characteristic** of these cells. The organelles of eukaryotes are thought to have arisen by **endosymbiosis**—one cell living inside another. In particular eukaryotic organelles are thought to have evolved from prokaryotic cells living inside the ancestors of eukaryotes.

The hypothesis that eukaryotic cells evolved by endosymbiosis is particularly well supported by studies of mitochondria and chloroplasts, which are thought to have evolved from eubacteria living in larger cells (**Margulis** 1970, 1981). Both mitochondria and chloroplasts are similar to bacteria in size, and like bacteria, they reproduce by dividing in two. Most important, both mitochondria and chloroplasts contain their own DNA, which encodes some of their components. The mitochondrial and chloroplast DNAs are replicated each time the organelle divides, and the genes they encode are transcribed within the organelle and translated on organelle ribosomes. Mitochondria and chloroplasts thus contain their own genetic systems, which are distinct from the nuclear genome of the cells. Furthermore, the ribosomes and ribosomal RNAs of these organelles are more closely related to those of bacteria than to those encoded by the nuclear genomes of eukaryotes.

An endosymbiotic origin of these organelles is now generally accepted with mitochondria thought to have evolved from aerobic eubacteria and chloroplasts from photosynthetic eubacteria such as the cyanobacteria (blue-green algae) (see Chapter 4). The acquisition of aerobic bacteria would have provided an anaerobic cell with the ability to carry out oxidative metabolism. The acquisition of photosynthetic bacteria would have provided the nutritional independence afforded by the ability to perform photosynthesis. Thus, these endosymbiotic associations were highly advantageous to their partners and were selected for in the course of evolution. Through time, most of the genes originally present in these bacteria evidently became incorporated into the nuclear genome of the cell, so only a few components of mitochondria and chloroplasts are still encoded by the organelle genomes (see **Cooper** and **Hausman**, 2007).

A recent hypothesis (**Rivera** and **Lake**, 2004) explained the mosaic nature of eukaryotic genomes by proposing that the genome of eukaryotes arose from a fusion of archaebacterial and eubacterial genomes (Fig. 5.1). According to this model, an endosymbiotic association between an eubacterium and an archaebacterium was followed by fusion of the two prokaryotic genomes, giving rise to an ancestral eukaryotic genome with contributions from both eubacteria and archaebacteria. The simplest version of this hypothesis is that an initial endosymbiotic relationship of an eubacterium living inside an archaebacterium gave rise not only to mitochondria but also to the genome of eukaryotic cells, containing genes derived from both prokaryotic ancestors.

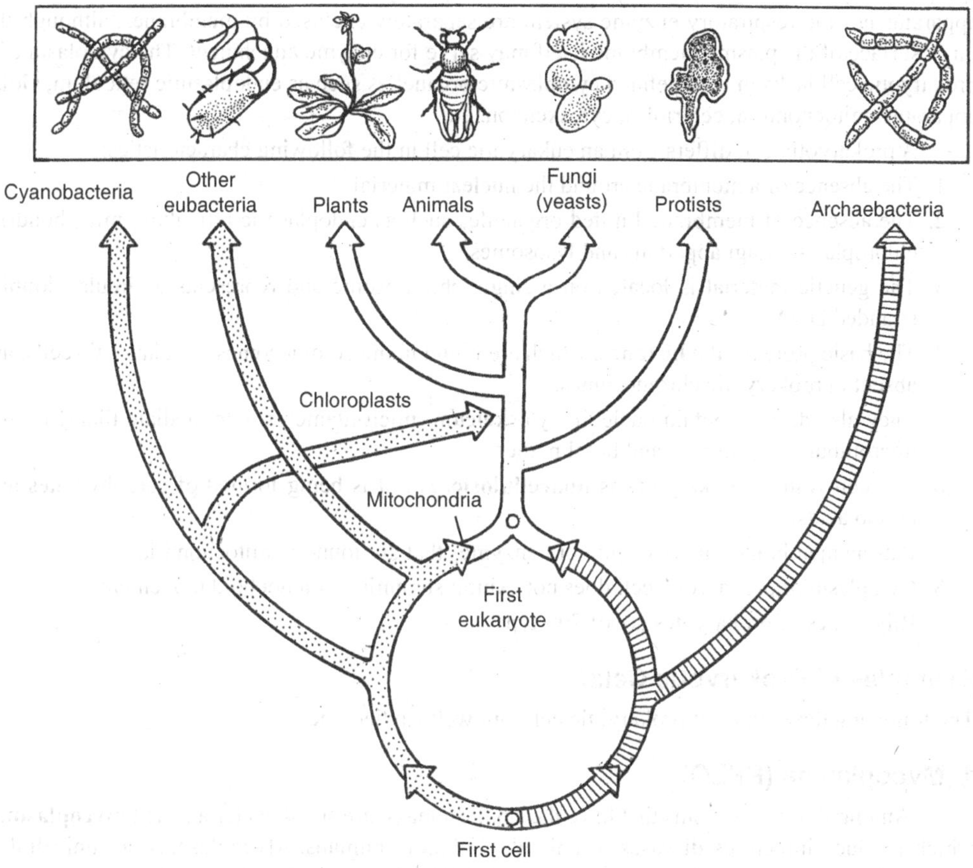

Fig. 5.1. Evolution of cells. Present day cells evolved from a common prokaryotic ancestor which diverged along two lines of descent, giving rise to archaebacteria and eubacteria. Eukaryotic cell may have arisen by endosymbiotic association of an aerobic eubacterium with an archaebacterium, leading to the development of mitochondria as well as formation of a eukaryotic genome with genes derived from both eubacteria and archaebacteria. Chloroplasts subsequently evolved as a result of the endosymbiotic association of a cyanobacterium with the ancestor of plants (after Cooper and Hausman, 2007).

5.2. PROKARYOTIC CELL

The prokaryotic (Gr.,*pro* = primitive or before; *karyon* = nucleus) cells are small, simple and most primitive. They are probably the first to come into existence perhaps 3.5 billion years ago. For example, the **stromatolites** (*i.e.*, giant colonies of extinct cyanobacteria or blue green algae) of Western Australia are known to be at least 3.5 billion years old. The **eukaryotic** (Gr.,*eu* = well; *karyon* = nucleus) cells have evolved from the prokaryotic cells and the first eukaryotic (nucleated) cells may have arisen 1.4 billion years ago (**Vidal,** 1983).

The prokaryotic cells are the most primitive cells from the morphological point of view. They occur in the bacteria (*i.e.*, mycoplasma, bacteria and cyanobacteria or blue-green algae) and archaea. A prokaryotic cell is essentially a **one-envelope system** organized in depth. It consists of central nuclear components (viz., DNA molecules, RNA molecules and nuclear proteins) surrounded by cytoplasmic ground substance, with the whole enveloped by a plasma membrane. Neither the nuclear apparatus nor the respiratory enzyme system are separately enclosed by membranes, although the inner surface of the plasma membrane itself may serve for enzyme attachment. The cytoplasm of a prokaryotic cell lacks in well defined cytoplasmic organelles such as endoplasmic reticulum, Golgi apparatus, mitochondria, centrioles, cytoskeleton, etc.

A prokaryotic cell differs from an eukaryotic cell in the following characteristics:

1. The absence of a membrane around the nuclear material.
2. The absence of membrane limited organelles such as endoplasmic reticulum, mitochondria, chloroplasts, Golgi appartaus and lysosomes.
3. The genetic material is located on a single chromosome and represents a circular double-stranded DNA.
4. The basic proteins, the **histones**, which are found in the chromosomes of eukaryotic cells, are absent in prokaryotic chromosomes.
5. They also do not contain nucleoli, cytoskeleton (microfilaments, intermediate filaments and microtubules), centrioles and basal bodies.
6. The cell wall of prokaryotes is **noncellulosic,** *i.e.*, it is being formed of carbohydrates and amino acids.
7. Plasma membrane carries respiratory enzymes that are found in mitochondria.
8. Cytoplasm of prokaryotic cell does not exhibit streaming or amoeboid movement.
9. Ribosomes of prokaryotes are of 70S type.

Examples of Prokaryotic Cells

The following three types of prokaryotic cells are well studied ones:

1. Mycoplasma (PPLO)

Among living organisms that have the smallest mass, are small bacteria called **mycoplasmas** which produce infectious diseases in animals including humans. Mycoplasmas are unicellular, prokaryotes, containing a plasma membrane, DNA, RNA and a metabolic machinery to grow and multiply in the absence of other cells (*i.e.*, they are capable of autonomous growth). They can be cultured *in vitro* like any bacteria, forming **pleomorphic** (Gr., *pleo* = many; *morphe* = forms) **colonies,** *i.e.*, depending on the type of culture medium, mycoplasmas tend to form different shaped colonies such as spheroid (fried-egg-shaped), thin, branching filaments, stellate, asteroid or irregular. They differ from the bacteria in the following respects:

1. Mycoplasmas are filterable through the bacterial filters (this fact was first demonstrated by **Iwanowsky** in 1892).
2. They do not contain cell wall and mesosomes.
3. Like the viruses and animal cells, they are resistant to antibiotics such as penicillin which kills bacteria by interfering with cell wall synthesis.
4. Their growth is inhibited by tetracyclines and similar antibiotics that act on metabolic pathways.

Mycoplasmas were discovered by French scientists, **E. Nocard** and **E. R. Roux** in 1898 while studying pleural fluids of cattle suffering from the disease **pleuropneumonia** (*i.e.*, an infectious disease of warm blooded animals producing pleural and lung inflammation). Similar organisms were later isolated from other animals such as sheep, goats, dogs, rats, mice and human beings and were named as **pleuropneumonia-like organisms (PPLO).** PPLO were later on included under the genus *Mycoplasma* by **Nowak** (1929) and these organisms are now commonly called **mycoplasmas. W.V. Iterson** (1969) has placed PPLO in the group **Mycoplasmatceae** of bacteria. Currently mycoplasmas are considered as the simplest bacteria. However, some cell biologists still prefer to place PPLO in between the viruses and bacteria.

Mycoplasmas are mostly free-living, saprophytes or parasites. For example, *Mycoplasma laidlawii* (0.1 μm in diameter) is saprophytic and is found in sewage, compost, soil, etc. *Mycoplasma gallisepticum* (0.25 μm in diameter) is parasitic and pathogenic; it is the parasite of cells and cell exudates of respiratory organs of warm-blooded animals causing in them various chronic respiratory diseases.

Mycoplasmas range in size (diameter) from 0.25 to 0.1 μm. They correspond in size to some of the large viruses. The spherical cell of a mycoplasma is bounded at its surface by a 75 Å thick plasma membrane which is composed of molecules of proteins and lipids, but there is no cell wall. Internally the cell's composition is more or less diffuse. The only micro-scopically discernible features within the cell are its genetic component, the DNA and the ribosomes. The DNA molecule is contained in a membraneless and clear-nucleus-like region and it is a double helix which may exit either as the linear strands or a single circular molecule. The nuclear region is surrounded by numerous (50 to 100) 70S type ribosomes existing either freely or in the polysomes (Fig. 5.2). A variety of other cytoplasmic inclusions, such as vacuoles and granules, have also been detected, but their functions are not known. At one side of the cell occurs **bleb** (localised collection of fluid) of ill understood function. As in other prokaryotes, there is no intracellular membranous structure.

The PPLO cells contain many enzymes which may be required for DNA replication, the transcription of different kinds of RNA molecules and translation involved in protein synthesis, and also in the biosynthesis of adenosine triphosphate (ATP) by anaerobic breakdown of sugars. Unlike viruses, they are free living and do not require host cells for their duplication. PPLO reproduce by binary fission, budding, formation of small spore-like bodies and by growth of large branched filaments that ultimately fragment.

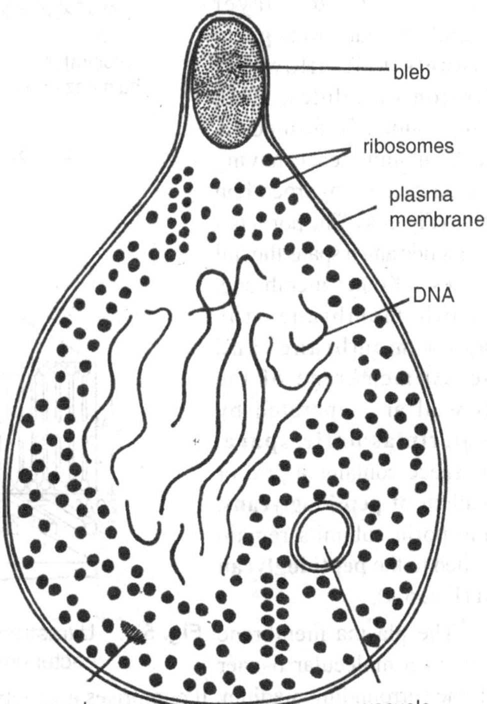

Fig. 5.2. A schematic diagram of typical PPLO cell.

2. *Escherichia coli* (*E. coli*)

E. coli is a Gram negative, monotrichous, symbiotic bacillus of colon of human beings and other vertebrates. It is heterotrophic and non-pathogenic bacteria producing some vitamins (*e.g.*, vitamin K) for human use. Some strains of *E. coli* are known to recognise and bind specifically to sugar-containing target cells on the surface of gut lining of mammals (*e.g.*, D-mannose residues of epithelial cells of human gut or colon). *E. coli* is one of the best studied bacteria. It has served well in the field of molecular biology, since this bacterium is particularly easy to grow in an artificial medium where it divides every 20 minutes at 37°C under optimal conditions. Thus, a single cell becomes 10^9 bacteria in about 20 hours.

The prokaryotic cell of *E. coli* (Fig. 5.3) is about 2 µm long and 1 µm wide. The cytoplasm of the bacterium is bounded by a typical fluid mosaic **plasma membrane**. External to the plasma membrane occurs the rigid and protective **cell wall** which has a complex organization; it comprises following two structures:

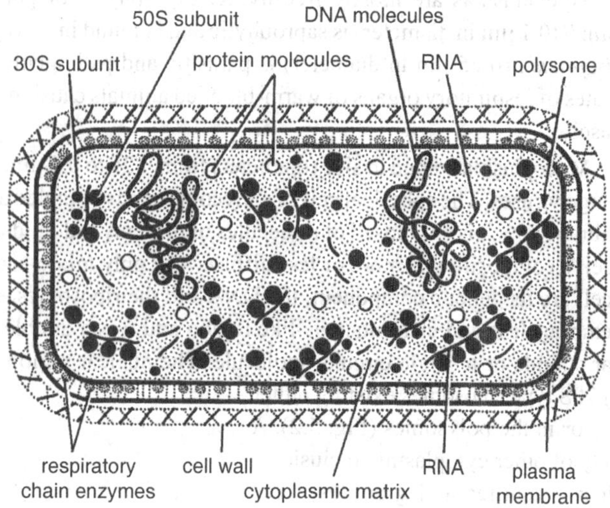

Fig. 5.3. A prokaryotic cell of *Escherichia coli.*

1. External membrane which is a lipid bilayer traversed by numerous **porin channels** that allow the diffusion of solutes. Each porin channel is formed by 6 to 8 subunits, each having three suspended hydrocarbon chains (Fig. 5.4). The porin is a polypeptide and it spans the full thickness of outer membrane.

2. Both membranes-the plasma membrane and external membrane of the cell wall are separated by the **periplasmatic space**. This space contains a grid or reticulum of **peptidoglycans**. Some porin subunits remain attached to the **peptidoglycan** grid (Fig. 5.4).

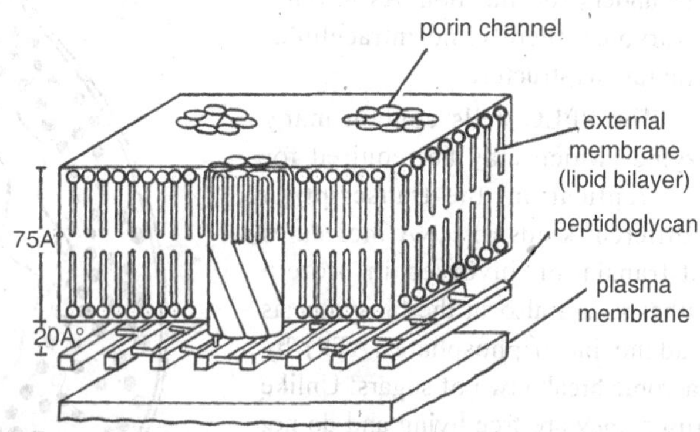

Fig. 5.4. Ultrastructure of the cell wall of a Gram-negative becterium.

The plasma membrane serves as a molecular barrier with the surrounding medium. It comprises a variety of transport proteins, called **permeases** which control the entrance and exit of small molecules and ions. It contributes to the establishment of bacterial protoplasm. *E. coli* has both oxygen-requiring (aerobic) and non-oxygen-requiring (anaerobic) respiratory machinery for the breakdown of sugar and contains a special group of proteins called the

electron transport chain for the generation of stored energy in the form of ATP molecules. *E. coli* lacks mitochondria, and respiratory chain enzymes such as cytochromes, enzymes of Krebs cycle, NADH, acid phosphatase, etc., are attached to inner face of the plasma membrane.

All genes of *E. coli* are contained on a single supercoiled, double-stranded, circular **DNA molecule**, which occurs in a clear zone of cytoplasm, called **nucleoid**, and is attached to the plasma membrane at one point. The total length of the DNA circle is about 1300 μm, comprising about 4.7×10^6 nucleotide pairs; this is enough DNA to code for about 4000 different proteins. The DNA of *E.coli* is naked, lacking histones, but certain polyamines are bound to some of its phosphates. Electron microscopy of isolated chromosome of *E. coli* has shown that DNA is folded into a series of **looped domains**, *i.e.*, about 45 loops radiate out from a dense proteinaceous **scaffold** (Fig. 5.5). The DNA of loops is in the so-called supercoiled conformation in which the double helix is itself twisted. The enzyme **DNA gyrase** is responsible for the DNA supercoiling (it is inhibited by the drug called **coumermycin**).

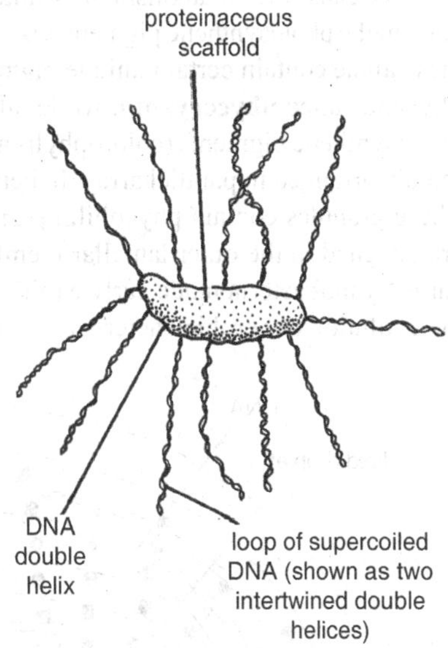

Fig. 5.5. Schematic representation of the chromosome of *E. coli*, showing only 12 of the 45 supercoiled loops.

The colloidal cytoplasmic matrix of *E.coli* contains about 5000 distinguishable components, ranging from water to DNA (*i.e.*, three types of RNA, enzymes, glycogen, amino acids, monosaccharides and various other small molecules). Surrounding the DNA is dark dense region of matrix containing 20,000 to 30,000 70S type **ribosomes**, each existing in the form of their two subunits. During protein synthesis numerous complete ribosomes read the codes of mRNA molecules to form the polysomes (Fig. 5.3).

3. Cyanobacteria (Blue Green Algae)

The Gram-negative cyanobacteria or oxyphotobacteria (*i.e.*, oxygen yielding photosynthetic blue green algae) are one of the most successful and primitive (3.5 billion year old) groups of organisms on earth. They even inhabit the steaming hot springs and the undersides of icebergs. Cyanobacteria form another group of prokaryotes which include about 1500 species (85 genera and 750 species are found in India).

Cyanobacteria occur as individual cells, as small clusters or colonies of cells, or as long, filamentous chains. They lack flagella but are able to perform movement by rotatory motion or gliding over a gelatinous layer secreted through the cell surface.

A typical cell of a blue green algae is composed of outer cellular coverings and cytoplasm. The **outer cellular coverings** include an outer most **gelatinous** or **slimy layer, the capsule**, a middle **cell wall** and an innermost lipoproteinous **plasma membrane**. The cell wall of blue green algae resembles the cell wall of bacteria and contains an **outer bimolecular membrane** of phospholipids, lipoproteins and lipopolysaccharides, and a grid of peptidoglycans (muramic acid) in the **periplasmatic space** existing in between cell wall and plasma membrane. The **cytoplasm** of cyanobacteria appears more organized than that of other bacteria. The matrix extends throughout the cell. The cytoplasm (or protoplast) is differentiated in two regions:

1. Outer or peripheral pigmented region, the **chromoplasm** having photosynthetic lamellae or **thylakoids**.
2. Inner or central colourless region called **centroplasm** or **DNA plasm** having DNA and crystalline granules (Fig. 5.6).

Because the metabolism of the blue green algae is based on photosynthesis, therefore, their cells contain the photosynthetic pigment, viz., the **chlorophyll** and **carotenoid**. In addition to these pigments, these algae contain certain unique pigments collectively called **phycobilin**; one of the phycobilin is blue and called **phycocyanin**, while other type of phycobilin is red and called **phycoerythrin**. The photosynthetic pigments (chlorophylls and carotenoids) occur in flattened sacs called **lamellae** which remain arranged in parallel array. In between the lamellae occur certain granules of 400A° diameter. These granules contain phycobilin pigments and are called **cyanosomes** or **phycobilisomes**. They are attached to the outer lamellar membrane surface. Being earliest oxygenic photosynthesizers of earth, cyanobacteria made early earth's atmosphere aerobic providing the conditions favourable for the evolution of aerobic bacteria and eukaryotes.

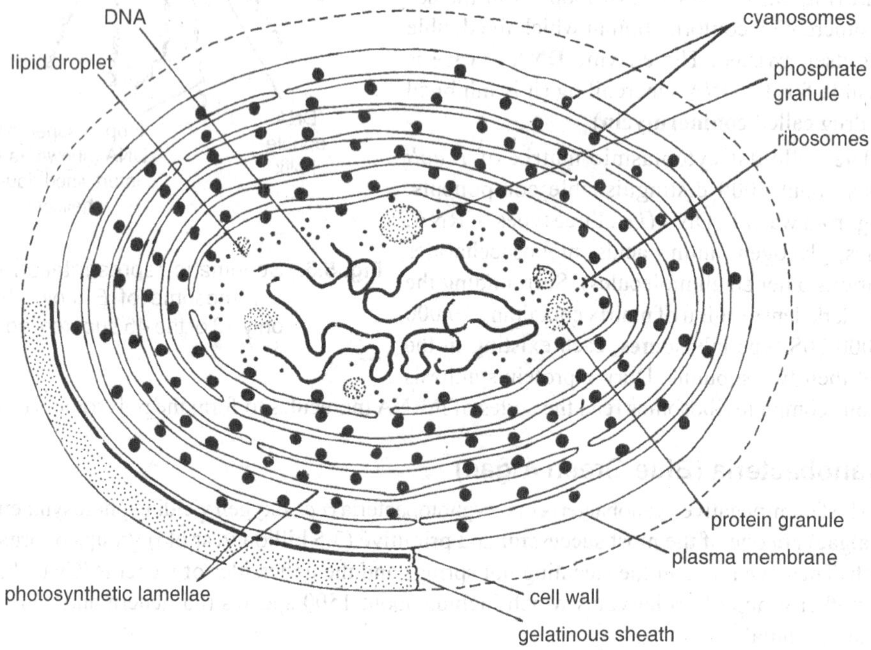

Fig. 5.6. A prokaryotic cell of cyanobacteria (electron microscopic view).

The two subunits of 70S ribosomes of cyanobacteria are freely distributed in the cytoplasm and form polyribosomes during protein synthesis. As in all prokaryotes, the DNA molecule of blue green algae is circular, double-stranded helix and occurs in the centroplasm. This area (nucleoid) is not bound by the nuclear membrane and it does not contain a nucleolus.

Cyanobacteria also contain a variety of **inclusions** in its cytoplasm. Membrane-bound inclusions are the gas vacuoles and the carboxysomes. **Gas vacuoles** are gas-filled cavities which are located in the inner part of chromatoplasm. They occur commonly in planktonic species such as *Nostoc*, *Anabaena*, *Phormidium*, *Calothrix*, *Galaeotrichia*, etc. Gas vacuoles serve the function of flotation or buoyancy. **Carboxysomes** contain enzymes involved in carbon dioxide fixation.

The cytoplasm of blue green algae also contains a variety of membrane-free inclusions such as (1) **cyanophycin granules** which are located in chromatoplasm and are protein storage products, containing large amount of arginine amino acid or copolymers of alanine and aspartic acid; (2) **myxophycean starch** which is the main food storage compound; (3) **polyglucon granules**, polyhedral bodies, lipid droplets, polyphosphate bodies, etc., are some other cytoplasmic inclusions of cyanobacteria.

Lastly, many cyanobacteria (about 20 species) tend to fix atmospheric nitrogen as ammonia, *e.g.*, *Anabaena*, *Nostoc*, *Mastigocladus*, etc. Under aerobic condition nitrogen fixation is done principally in special type of cells called **heterocysts**, as in *Nostoc*.

5.3. EUKARYOTIC CELL

The eukaryotic cells (Gr., *eu* = good, *karyotic* = nucleated) are essentially **two envelope systems** and they are very much larger than prokaryotic cells. Secondary membranes envelop the nucleus and other internal organelles and to a great extent they pervade the cytoplasm as the endoplasmic reticulum. The eukaryotic cells are the true cells which occur in the Protista (Protozoa and Protophyta), fungi, plants (from algae to angiosperms) and the animals (from poriferans to mammals). Though the eukaryotic cells have different shape, size, and physiology; all the cells are typically composed of plasma membrane, cytoplasm and its organelles, viz., mitochondria, endoplasmic reticulum, ribosomes, Golgi apparatus, etc., and a true nucleus. Here the nuclear contents, such as DNA, RNA, nucleoproteins and nucleolus remain separated from the cytoplasm by the thin, perfotated nuclear membranes.

Cell Shape

The basic shape of the eukaryotic cell is **spherical**, however, the shape is ultimately determined by the specific function of the cell. Thus, the shape of the cell may be **variable** (*i.e.*, frequently changing the shape) or **fixed**. Variable or irregular shape occurs in *Amoeba* and white blood cells or leucocytes (In fact, leucocytes are spherical in the circulating blood, but in other conditions they may produce pseudopodia and become irregular in shape). Fixed shape of the cell occurs in almost all protists (*e.g.*, *Euglena*, *Paramecium*), plants and animals. In unicellular organisms the cell shape is maintained by tough plasma membrane and exoskeleton. In a multicellular organism, the shape of the cell depends mainly on its functional adaptations and partly on the surface tension, viscosity of the protoplasm, cytoskeleton of microtubules, microfilaments and intermediate filaments, the mechanical action exerted by adjoining cells and rigidity of the plasma membrane (*i.e.*, presence of rigid cell wall in plant cells). The shape of the cell may vary from animal to animal and from organ to organ. Even the cells of the same organ may display variations in the shape. Thus, cells may have diverse shapes such as **polyhedral** (with 8, 12 or 14 sides; *e.g.*, squamous epithelium); **flattened** (*e.g.*, squamous epithelium, endothelium and the upper layers of the epidermis); **cuboidal** (*e.g.*, in thyroid gland follicles); **columnar** (*e.g.*, the cells lining the intestine); **discoidal** (*e.g.*, red blood cells or erythrocytes); **spherical** (*e.g.*, eggs of many animals); **spindle shaped** (*e.g.* smooth-muscle fibres); **elongated** (*e.g.*, nerve cells or neurons); or **branched** (*e.g.*, chromatophores or pigment cells of skin). Among plants, the cell shape also depends upon the function of the cell. For example, cells such as glandular hairs on a leaf, the guard cells of stomata and root hair cells have their special shape.

Cell Size

The eukaryotic cells are typically larger (mostly ranging between 10 to 100 μm) than the prokaryotic cells (mostly ranging between 1 to 10 μm). Size of the cells of the unicellular organisms is larger than a typical multicellular organism's cells. For example, *Amoeba proteus* is biggest among the unicellular organisms; its length being 1000 μm (1 mm). One species of *Euglena* is found up to 500 μm (0.5mm) in length. *Euplotes* (a freshwater ciliate) is 120 μm in length. Another ciliate, *Paramecium caudatum* is from 150 to 300 μm (0.15 to 0.3 mm) in length. Diatoms have a length of 200 μm or more. The single-celled alga, *Acetabularia* which consists of a stalk and a cap is exceptionally large-sized and measures up to 10 cm in height.

The size of the cells of multicellular organisms ranges between 20 to 30 μm. Among animals, the smallest cells have a diameter of 4 μm (*e.g.*, polocytes); human erythrocytes being 7 to 8 μm in diameter. Largest animal cell is the egg of ostrich, having a diameter of 18 cm (its yolk or deutoplasm is about 5 cm in diameter); though, some nerve cells of human beings have a meter long "tails" or axons. Among the multicellular plants, the largest cell is the ovule of *Cycas*. The fibre cells (*i.e.*, sclerenchyma cells) of Manila hemp are over 100 cm in length.

Cell Volume

The volume of a cell is fairly constant for a particular cell type and is independent of the size of the organism. (This is called the **law of constant volume**). For example, kidney or liver cells are about the same size in the bull, horse and mouse. The difference in the total mass of the organ or organism depends on the number, not on the volume of the cells. Thus, the cells of an elephant are not necessarily larger than those of other tiny animals or plants. The large size of the elephant is due to the larger number of cells present in its body.

If a cell is to be efficient, the ratio of volume to surface should be within a limited range. An increase in cell volume is accompanied by a much smaller expansion in the surface area of the cell (In fact, volume increases as cube of radius, while surface area increases as square of radius). In other words, a large cell has a proportionately smaller surface and a higher volume: surface ratio than a smaller cell. Further, a large cell volume has to accommodate many organelles simultaneously limiting the exchange of information and materials through the surface. This problem is partially overcome by developing a cylindrical shape or by forming numerous extensions (*e.g.*, microvilli) of the plasma membrane. It is also for this reason that metabolically active cells, tend to be smaller in size.

Cell Number

The number of cells present in an organism varies from a single cell in a **unicellular organism** (Protists such as Protozoa and Protophyta) to many cells in multicellular organisms (most plants, fungi and animals). The number of cells in the multicellular organisms usually remains correlated with size of the organisms and, therefore, small-sized organisms have less number of cells in comparison to large-sized organisms. For example, a human being weighing about 80 kg may contain about 60 thousand billion cells in his body. This number would be more in certain other multicellular organisms.

Further, the number of cells in most multicellular organisms is indefinite, but the number of cells may be fixed in some multicellular organisms (*e.g.*, nematodes, rotifers). For example, in rotifers, number of nuclei in the various organs are found to be constant in any given species. This phenomenon of cells or nuclear constancy is called **eutely**. In one species of rotifer, **Martini** (1912) always found 183 nuclei in the brain, 39 in the stomach, 172 in the cornea epithelium, and so on. Among plants, colonial green algae exhibit cell constancy. For example, the green alga, *Pandorina* has a colony consisting of 8, 16 or 32 cells. Likewise, another green alga, *Eudornia*, has 16, 32 or 64 cells in its colony.

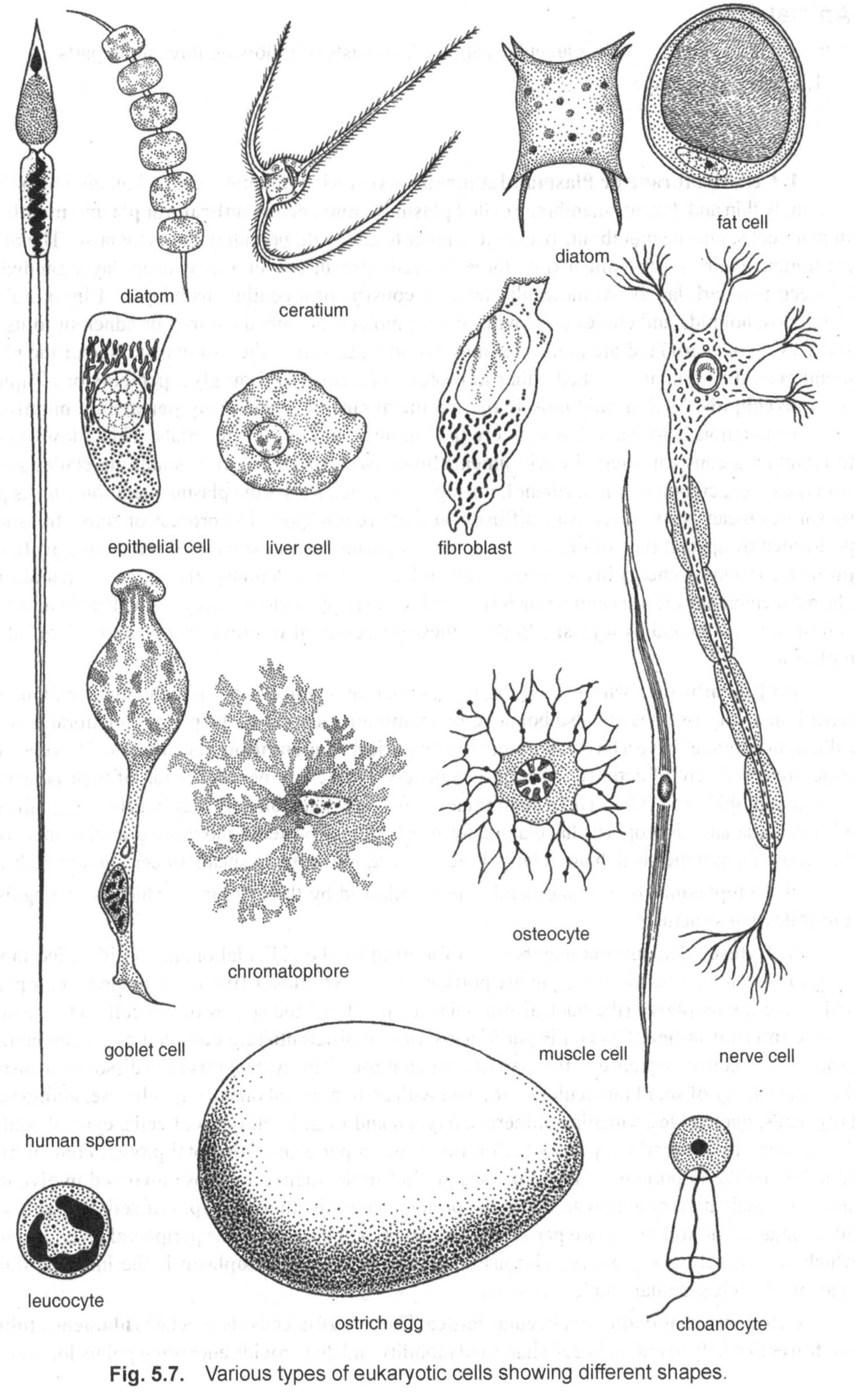

Fig. 5.7. Various types of eukaryotic cells showing different shapes.

Animal Cell

A typical animal cell, which is an eukaryotic cell, consists of following three main parts:

1. Plasma membrane
2. Cytoplasm
3. Nucleus

I. Cell Membrane or Plasma Membrane. Every kind of animal cell is bounded by a living, extremely thin and delicate membrane called **plasmalemma, cell membrane** or **plasma membrane**. In plant cells, plasma membrane occurs just inner to cell wall, bounding the cytoplasm. The plasma membrane exhibits a trilaminar (*i.e.*, three-layered) structure with a translucent layer sandwiched between two dark layers. At molecular level, it consists of a continuous bilayer of lipid molecule (*i.e.*, phospholipids and cholesterol) with protein molecules embedded in it or adherent to its both surfaces. Some carbohydrate molecules may also be attached to the external surface of the plasma membrane, they remain attached either to protein molecules to form **glycoproteins** or to lipids to form **glycolipids**. Such a fluid mosaic plasma membrane is a **selectively permeable membrane;** its main function is to control selectively the entrance and exit of materials. This allows the cell to maintain a constant internal environment **(homeostasis)**. Transport of small molecules such as water, oxygen, carbon dioxide, ethanol, ions; glucose, etc., across the plasma membrane takes place by various means such as osmosis, diffusion and active transport. The process of active transport is performed by special type of protein molecules of plasma membrane called **transport proteins** or **pumps,** consuming energy in the form of ATP molecules. For bulk transport of large-sized molecules, plasma membrane performs **endocytosis** (*i.e.*, endocytosis, pinocytosis, receptor-mediated endocytosis and phagocytosis) and **exocytosis**. Both of these processes also utilise energy in the form of ATP molecules.

Unit membrane. Various cell organelles such as chloroplasts, mitochondria, endoplasmic reticulum and lysosomes are also bounded by membranes similar to the plasma membrane. All the cellular membranes have a basic trilaminar fluid mosaic **unit membrane** construction. However, their structure and extent of activity are mainly depended on the relative proportion of their constituent protein and lipid molecules. Thus, membranes which are metabolically highly active, *e.g.*, those of mitochondria and chloroplasts have a greater proportion of proteins and more granular appearance than those membranes which are relatively less active, *e.g.*, myelin sheath of certain nerve fibres.

II. Cytoplasm. The plasma membrane is followed by the cytoplasm which is distinguished into following structures:

A. Cytosol. The plasma membrane is followed by the colloidal organic fluid called **matrix** or **cytosol**. The cytosol is the aqueous portion of the **cytoplasm** (the extra nuclear protoplasm) and of the **nucleoplasm** (the nuclear protoplasm). It fills all the spaces of the cell and constitutes its true **internal milieu**. Cytosol is particularly rich in differentiating cells and many fundamental properties of cell are because of this part of the cytoplasm. The cytosol serves to dissolve or suspend the great variety of small molecules concerned with cellular metabolism, *e.g.*, glucose, amino acids, fatty acids, nucleotides, vitamins, minerals, oxygen and ions. In all types of cells, cytosol contains the soluble proteins and enzymes which form 20 to 25 per cent of the total protein content of the cell. Among the important soluble enzymes present in the matrix are those involved in glycolysis and in the activation of amino acids for the protein synthesis. In many types of cells, the cytosol is differentiated into following two parts: (*i*) **Ectoplasm** or **cell cortex** is the peripheral layer of cytosol which is relatively non-granular, viscous, clear and rigid. (*ii*) **Endoplasm** is the inner portion of cytosol which is granular and less viscous.

Cytoskeleton and microtrabecular lattice. The cytosol of cells also contains **filaments, tubules** and **fibres** that help to maintain cell shape and mobility and that provide anchoring points for the other cellular structures. Collectively, these protein structures or fibres are termed as the **cytoskeleton.** At least three general classes of such fibres have been identified. 1. The thickest are the **microtubules** (20 nm in diameter) which consist primarily of the **tubulin** protein. The function of microtubules is the transportation of water, ions or small molecules, cytoplasmic streaming (cyclosis), and thc formation of fibres or asters of the mitotic or meitotic spindle during cell division. Moreover, they form the structural units of the centrioles, basal granules, cilia and flagella. 2. The thinnest are the **microfilaments** (7 nm in diameter) which are solid and are principally formed of **actin** protein. They maintain the shape of cell and form contractile component of cells, mainly of the muscle cells. 3. The fibres of middle order are called the **intermediate filaments (IFs)** having a diameter of 10 nm. They have been classified according to their constituent protein such as **desmin filaments, keratin filaments, neurofilaments, vimentin** and **glial filaments**.

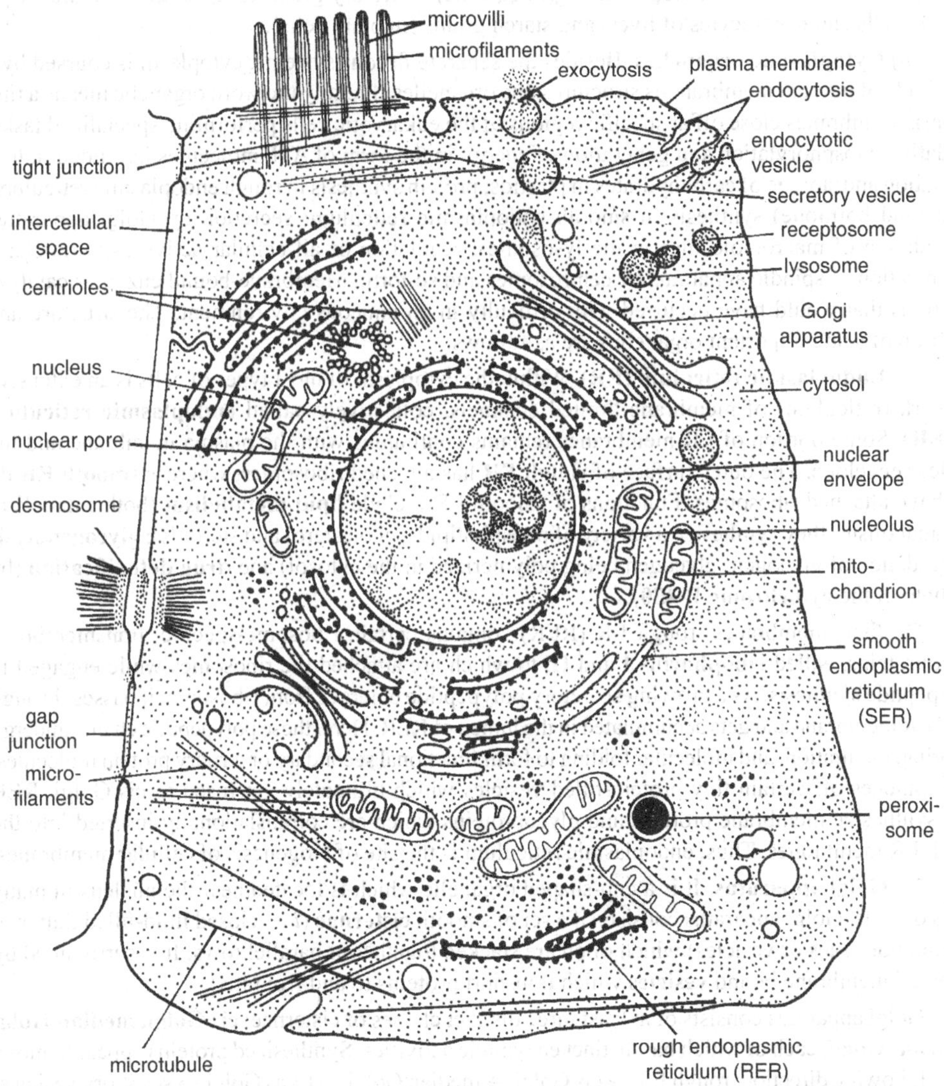

Fig. 5.8. Ultrastructure of a typical animal cell as seen in the electron microscope.

Recently, cytoplasm has been found to be filled with a three-dimensional network of interlinked filaments of cytoskeletal fibres, called **microtrabecular lattice.** Various cellular organelles such as ribosomes, lysosomes, etc., and enzymes are found anchored to this lattice. The microtrabecular lattice being flexible, changes its shape and results in the change of cell shape during cell movement.

B. Cytoplasmic structures. In the cytoplasmic matrix (cytosol) certain non-living and living structures remain suspended. The non-living structures are called **paraplasm** or **inclusions,** while the living structures are membrane bounded and are called **organoids** or **organelles.** Both kinds of cytoplasmic structures can be studied under the following headings:

(*a*) Cytoplasmic inclusions. The stored food and secretory substances of the cell remain suspended in the cytoplasmic matrix in the form of refractile granules forming the cytoplasmic inclusions. The cytoplasmic inclusions include oil drops, triacylglycerols (*e.g.*, fat cells of adipose tissue), yolk granules (or **deutoplasm,** *e.g.*, egg cells), secretory granules, glycogen granules (*e.g.*, muscle cells and hepatocytes of liver) and starch grains (in plant cells).

(*b*) Cytoplasmic organelles. Besides the separate fibrous systems, cytoplasm is coursed by a multitude of internal membranous structures, the organelles (literally the word organelle means a tiny organ). Membranes close off at specific regions of the eukaryotic cells performing specialized tasks: oxidative phosphorylation and generation of energy in the form of ATP molecules in mitochondria; formation and storage of carbohydrates in plastids; protein synthesis in rough endoplasmic reticulum; lipid (and hormone) synthesis in smooth endoplasmic reticulum; secretion by Golgi apparatus; degradation of macromolecules in the lysosomes; regulation of all cellular activities by nucleus; organization of spindle apparatus by centrosomes and so forth. Membrane-bound enzymes catalyze reactions that would have occurred with difficulty in an aqueous environment. The structure and function of some important organelles are as follows:

1. Endoplasmic reticulum (ER). Within the cytoplasm of most types of cells is an extensive network (reticulum) of membrane-limited channels, collectively called **endoplasmic reticulum** (or **ER**). Some portion of ER membranes remains continuous with the plasma membrane and the nuclear envelope. The outer surface of **rough ER** has attached ribosomes, whereas **smooth ER** do not have attached ribosomes. Functions of smooth ER include **lipid metabolism** (both catabolism and anabolism; they synthesize a variety of phospholipids, cholesterol and steroids); **glycogenolysis** (degradation of glycogen; glycogen being polymerized in the cytosol) and **drug detoxification** (by the help of the **cytochrome P-450**).

On their membranes, rough ER (RER) contains certain ribosome-specific, transmembrane glycoproteins, called **ribophorins I** and **II**, to which are attached the ribosomes while engaged in polypeptide synthesis. As a growing secretory polypeptide emerges from ribosome, it passes through the RER membrane and gets accumulated in the lumen of RER. Here, these polypeptide chains undergo tailoring, maturation, and molecular folding to form functional secondary or tertiary protein molecules. RER pinches off certain tiny protein-filled vesicles which ultimately get fused to cis Golgi. RER also synthesize membrane proteins and glycoproteins which are cotranslationally inserted into the rough ER membranes. Thus, endoplasmic reticulum is the site of biogenesis of cellular membranes.

2. Golgi apparatus. It is a cup-shaped organelle which is located near the nucleus in many types of cells. Golgi apparatus consists of a set of smooth **cisternae** (*i.e.*, closed fluid-filled flattened membranous sacs or vesicles) which often are stacked together in parallel rows. It is surrounded by spherical membrane bound **vesicles** which transport proteins to and from it.

Golgi apparatus consists of at least three distinct classes of cisternae: **cis Golgi, median Golgi** and **trans Golgi**, each of which has distinct enzymatic activities. Synthesized proteins appear to move in the following direction: rough ER → cis Golgi → median Golgi → trans Golgi → secretory vesicles/cortical granules of egg/lysosomes or peroxisomes. Thus, the size and number of Golgi apparatus in a cell indicate the active metabolic, mainly synthetic, state of that cell. Plant cells contain many freely distributed sub-units of Golgi apparatus, called **dictyosomes**, secreting cellulose and pectin for cell wall formation during the cell division.

Generally, Golgi apparatus performs the following important functions:

1. The packaging of secretory materials (*e.g.*, enzymes, mucin, lactoprotein of milk, melanin pigment, etc.) that are to be discharged from the cell.
2. The **processing** of proteins, *i.e.*, glycosylation, phosphorylation, sulphation and selective proteolysis.
3. The synthesis of certain polysaccharides and glycolipids.
4. The sorting of proteins destined for various locations (*e.g.*, lysosomes, peroxisomes, etc.) in the cell.
5. The proliferation of membranous element for the plasma membrane.
6. Formation of the acrosome of the spermatozoa and cortical granules of an ovum.

3. Lysosomes. The cytoplasm of animal cells contains many tiny, spheroid or irregular-shaped, membrane-bounded vesicles known as **lysosomes**. The lysosomes are originated from Golgi apparatus and contain numerous (about 50) hydrolytic enzymes (*e.g.*, **acid phosphatase** that is cytochemically identified) for intracellular and extracellular digestion. They digest the material taken in by endocytosis (such as phagocytosis, endocytosis and pinocytosis), parts of the cell (by autophagy) and extracellular substances. Lysosomes have a high acidic medium (pH 5) and this acidification depends on ATP-dependent **proton pumps** which are present in the membrane of lysosomes and which accumulate protons (H^+) inside the lysosomes. Lysosomes exhibit great **polymorphism,** *i.e.*, there are following four types of lysosomes: primary lysosomes (storage granules), secondary lysosomes (digestive vacuoles), residual bodies and autophagic vacuoles. The lysosomes of plant cells are membrane-bounded storage granules containing hydrolytic digestive enzymes, *e.g.*, large **vacuoles** of parenchymatous cells of corn seedlings, **protein** or **aleurone bodies** and **starch granules** of cereal and other seeds.

4. Cytoplasmic vacuoles. The cytoplasm of some animal cells (*e.g.*, ciliate protozoans) contains numerous small or large-sized, hollow, liquid-filled structures, the **vacuoles.** These vacuoles are supposed to be greatly expanded endoplasmic reticulum or Golgi apparatus. The vacuoles of animal cells are bounded by a lipoproteinous membrane and their function is the storage, transmission of the materials and the maintenance of internal pressure of the cell.

5. Peroxisomes. These are tiny circular membrane-bound organelles containing a crystal-core of enzymes (such as urate oxidase, peroxidase, D-amino oxidase and catalase) of certain cells (*e.g.*, liver cells and kidney cells). These enzymes are required by peroxisomes in **detoxification** activity, *i.e.*, in the metabolism or production and decomposition, of hydrogen peroxide or H_2O_2 molecules which are produced during neutralization of certain superoxides—the end products of mitochondrial or cytosolic reactions. Peroxisomes are also related with β-oxidation of fatty acids and thermogenesis like the mitochondria and also in degradation of the amino acids. In green leaves of plants, peroxisomes carry out the process of **photorespiration**.

6. Mitochondria. Mitochondria are oxygen-consuming ribbon-shaped cellular organelles of immense importance. Each mitochondrion is bounded by two unit membranes. The outer mitochondrial membrane resembles more with the plasma membrane in structure and chemical composition. It contains **porins,** proteins that render the membrane permeable to molecules having molecular weight as high as 10,000. Inner mitochondrial membrane is rich in many enzymes, coenzymes and other components of electron transport chain. It also contains **proton pumps** and many **permease** proteins for the transport of various molecules such as citrates, ADP, phosphate and ATP. Inner mitochondrial membrane gives out finger-like outgrowths **(cristae)** towards the lumen of mitochondrion and contains tennis-racket shaped F_1 **particles** which contain ATPase enzyme for ATP synthesis.

Mitochondrial matrix which is the liquid (colloidal) area encircled by the inner membrane, contains the soluble enzymes of Krebs cycle which completely oxidize the **acetyl-CoA** (an end product of cytosolic glycolysis and mitochondrial oxidative decarboxylation) to produce CO_2, H_2O and hydrogen ions. Hydrogen ions reduce the molecules of NAD and FAD, both of which pass on hydrogen ions to respiratory or electron transport chain where oxidative phosphorylation takes place to generate energy-rich ATP molecules. Since mitochondria act as the 'power-houses' of cells, they are abundantly found on those sites where energy is earnestly required such as sperm tail, muscle cell, liver cell (up to 1600 mitochondria), microvilli, oocyte (more than 300,000 mitochondria), etc. Mitochondria also contain in their matrix single or double circular and double stranded DNA molecules, called **mtDNA (or mitochondrial DNA)** and also the 55S ribosomes, called **mitoribosomes**. Since mitochondria can synthesize 10 per cent of their proteins in their own protein-synthetic machinery, they are considered as **semi-autonomous, organelles.** Mitochondria may also produce heat (brown fat), and accumulate iron containing pigments (Heme, ferritin), ions of Ca^{2+} and HPO_4^{2-} (or phosphate; *e.g.*, osteoblasts of bones or yolk proteins).

7. Ribosomes. Ribosomes are tiny spheroidal dense particles (of 150 to 200 A° diameter) that contain approximately equal amounts of RNA and proteins. They are primarily found in all cells and serve as a scaffold for the ordered interaction of the numerous molecules involved in protein synthesis. Ribosome granules may exist either in the **free state** in the cytosol (*e.g.*, basal epidermal cells) or **attached** to RER (*e.g.*, pancreatic acinar cells, plasma cells or antibodies-secreting lymphocytes, osteoblasts, etc.). Ribosomes have a sedimentation coefficient of about **80S** and are composed of two subunits namely **40S** and **60S**. The smaller 40S ribosomal subunit is prolate ellipsoid in shape and consists of one molecule of 18S ribosomal RNA (or rRNA) and 30 types of proteins (named as S_1, S_2, S_3, and so on). The larger 60S ribosomal subunit is round in shape and contains a **channel** through which growing polypeptide chain makes its exit. It consists of three types of rRNA molecules, *i.e.*, 28S rRNA, 5.8 rRNA and 5S rRNA, and 40 types of proteins (named as L_1, L_2, L_3 and so on).

8. Microtubules and microtubular organelles. With rare exceptions, such as human erythrocyte, microtubules are found in the cytoplasm of all types of eukaryotic cells. They are long fibres (of indefinite length) about 24 nm in diameter. In cross section each microtubule appears to have a dense wall of 6 nm thickness and a light or hollow centre. In cross section, the wall of a microtubule is made up of 13 globular subunits, called **protofilaments,** about 4 to 5 nm in diameter. Chemically, microtubules are composed of two kinds of protein subunits: **α-tubulin (tubulin A)** and **β-tubulin (tubulin B)**, each of M.W. 55,000 daltons. The wall of a microtubule is made up of a helical array of repeating α and β tubulin subunits. Assembly studies have indicated that the structural unit is an αβ **dimer** of 8 nm length. Thus, in each microtubule, there are 13 protofilaments, each composed of αβ dimers that run parallel to the long axis of the tubule: The repeating unit is an αβ heterodimer which is arranged 'head to tail' within the microtubule, that is $\alpha\beta \rightarrow \alpha\beta \rightarrow \alpha\beta$. Thus, all microtubules have a defined **polarity:** their two ends are not structurally equivalent.

Microtubules undergo reversible assembly-disassembly (*i.e.*, polymerization and depolymerization), depending on the need of the cell or organelles. Their polymerization is regulated by certain **MAPs** or **microtubule-associated proteins** (*e.g.*, Tau protein). The assembly of microtubules involves preferential addition of subunits (αβ dimers) to one end of tubule, called **A end** (or **net assembly end**); the other end of the tubule is called **D end** (or **net disassembly end**). Such an assembly involves the hydrolysis of GTP to GDP. Thus, assembly of tubulin in the formation of microtubules is a specifically oriented and programmed process. Centrioles, basal bodies and centromeres of chromosomes are the sites of orientation for this assembly. Calcium and **calmodulin** (an acidic protein having four Ca^{2+} binding sites) are some other regulating factors in the *in vivo* polymerization of tubulin. Certain drugs such as **colchicine** and **vinblastin,** are found to block the polymerization of tubulin.

The following cell organelles are derived from special assemblies of microtubules:

1. Cilia and Flagella. Ciliary and flagellar cell motility is adapted to liquid media and is executed by minute, specially differentiated appendices, called **cilia** and **flagella.** Both of these organelles have very similar structure; they differ mainly in size and number (*i.e.*, flagella are longer and fewer in number, while cilia are short and numerous). Cilia are used for locomotion in isolated cells, such as certain protozoans (*e.g.*, *Paramecium*) or to move particles in the medium, as in air passages and oviduct. Flagella are generally used for locomotion of cells, such as the spermatozoon and *Euglena* (protozoan). All cilia and flagella are built on a common fundamental plan: a bundle of microtubules called the **axoneme** (1 to 2 nm in length and 0.2 μm in diameter) is surrounded by a membrane that is part of the plasma membrane. The axoneme is connected with the basal body which is an intracellular granule lying in the cell cortex and which originates from the centrioles. Each axoneme is filled with **ciliary matrix,** in which are embedded two central **singlet** microtubules, each with the 13 protofilaments and nine outer pairs of microtubules, called **doublets**. This recurring motif is known as the 9 + 2 array. Each doublet contains one complete microtubule, called the **A subfibre**, containing all the 13 protofilaments. Attached to each A subfibre is a **B subfibre** with 10 protofilaments. Subfibre A has two **dynein arms** which are oriented in a clockwise direction. Doublets are linked together by **nexin links.** Each subfibre A is also connected to the central microtubules by **radial spokes** terminating in fork-like structures, called **spoke knobs** or **heads**.

Propulsion by both cilia and flagella is caused by bending at their base. Cilia move by a whip-like power stroke fueled by hydrolysis of ATP, followed by **a recovery stroke.** Flagellar movement is also powered by ATP hydrolysis. In contrast to cilia, flagella generally move by waves that emanate from the base and spread outward toward the tip.

2. Basal bodies and Centrioles. Basal bodies and centrioles are similar in structure and function; both act as nucleating centres from which microtubules grow. **Centrioles** are cylinders that measure 0.2 μm × 0.5 μm. This cylinder is open on both ends, unless it carries a cilium or flagellum (then it is called **basal body** or **kinetosome**). The wall of a centriole has nine groups of microtubules arranged in a circle. Each group, called **blade** is a **triplet** formed of three tubules—*A*, *B*, and *C* that are skewed toward the centre. Tubule *A* has 13 protofilaments, while tubules *B* and *C* have only 10 protofilaments each. There are no central microtubules in the centrioles and no dynein arms like the cilia; however, triplets are linked by connectives. The **procentriole** (or daughter centriole) is formed at right angles to the centriole and is located near the proximal end of the centriole. Both centrioles are found in a specially differentiated region the **centrosome, cell centre** or **centrosphere.** The centrosome is juxtanuclear (L., juxta = near) and firmly attached to the nuclear envelope. At the time of cell division two pairs of centrioles are formed and form the spindle of micro-tubules which help in the separation and movement of chromosomes during concluding stages of cell divisions.

3. Microfilaments and Motor Proteins. Cells are capable of remarkable **motility**. The *neural crest cells* in a vertebrate embryo leave the developing nervous system and migrate across the entire width of the embryo, forming such diverse products as the pigment cells of the skin, teeth, and the cartilage of the jaws. Horde of white blood cells patrol the tissues of the body searching for debris and microorganisms. Certain parts of the cells can also be motile; broad projections of epithelial cells at the edge of a wound act as motile devices that pull the sheet of cells over the damaged area, sealing the wound. Similarly, the leading edge of a growing axon sends out microscopic processes that survey the substratum and guide the cell towards a synaptic target. All of these various examples of motility share at least one component: they all depend on **microfilaments**, the second major type of cytoskeletal element. Microfilaments are also involved in intra-cellular motile processes, such as movement of vesicles, phagocytosis and cytokinesis. In fact, plant cells rely primarily on microfilaments, rather than microtubules, for the long-distance transport of cytoplasmic vesicles and organelles.

Actin filaments or microfilaments are 8 nm in diameter, composed of a double-helix polymer of the protein actin, and play a key role in virtually all types of contractility and motility within cell. The role of actin filaments in a particular process is most readily tested by treating the cells with **cytochalasin**, which promotes the depolymerization of the filament, or with **phalloidin**, which prevents it from disassembling and participating in dynamic activities. The forces responsible for microfilament-dependent processes may be generated by the assembly of the actin filament or as the result of interaction with motor proteins such as **myosin** and **kinesin** (**Karp** 2010).

III. Nucleus. The nucleus is centrally located and spherical cellular component which controls all the vital activities of the cytoplasm and carries the hereditary material the DNA in it. The nucleus consists of the following three structures:

1. Chromatin. Nucleus being the heart of every type of eukaryotic cell, contains the **genes,** the hereditary units. Genes are located on the **chromosomes** which exist as **chromatin network** in the non-dividing cell, *i.e.*, during interphase. The chromatin has two forms:

(*i*) **Euchromatin** is the well-dispersed form of chromatin which takes lighter DNA-stain and is genetically active, *i.e.*, it is involved in gene duplication, gene transcription (DNA-dependent RNA synthesis) and **phenogenesis** or phenotypic expression of a gene through some type of protein synthesis.

(*ii*) **Heterochromatin** is the highly condensed form of chromatin which takes dark DNA-stain and is genetically inert. Such type of chromatin exists both in the region of centromere (called **constitutive heterochromatin**) and in the sex chromatin (called **facultative heterochromatin**) and is late-replicating one.

Chemically, the chromatin contains a single DNA molecule, equal amount of histone proteins, (which are of five basic types), some RNA molecules and variable amount of different types of acidic proteins. In fact, the chromatin has its unit structures in the form of **nucleosomes**. The chromatin binds strongly to the inner part of **nuclear lamina,** a 50 to 80 nm thick fibrous lamina lining the inner side of the nuclear envelope. Nuclear lamina is made up of three types of proteins, namely **lamin A, B** and **C**. Lamin proteins are homologous in structure to IF proteins and serve the following functions:

(*i*) They anchor parts of interphase chromatin to the nuclear membrane. They tend to interfere with chromatin condensation during interphase of cell cycle.

(*ii*) Lamins may play a crucial role in the assembly of interphase nuclei after each mitosis.

2. Nuclear envelope and nucleoplasm. Nuclear envelope comprises two nuclear membranes—an **inner nuclear membrane** which is lined by nuclear lamina and an **outer nuclear membrane** which is continuous with rough ER. At certain points the nuclear envelope is interrupted by structures called **pores** or **nucleopores**. Nuclear pores contain octagonal **pore complexes** which regulate exchange between the nucleus and cytoplasm. The number of nucleopores is found to be correlated with the transcriptional activity of the cell. For example, in the frog *Xenopus laevis* oocytes (which are very active in transcription) have 60 pores/μm^2 (and upto 30 million pore complexes per nucleus), whereas frog's mature erythrocytes (inactive in transcription) have only about 3 pores/μm^2 (and a total of only 150 to 300 pores per nucleus).

The nuclear envelope binds the **nucleoplasm** which is rich in those molecules which are needed for DNA replication, transcription, regulation of gene actions and processing of various types of newly transcribed RNA molecules (*i.e.*, tRNA, mRNA and other types of RNA).

3. Nucleolus. Nucleus contains in its nucleoplasm a conspicuous, darkly stained, circular sub-organelle, called **nucleolus.** Nucleolus lacks any limiting membrane and is formed during interphase by the ribosomal DNA (rDNA) of **nucleolar organizer (NO)**. Nucleolus is the site where ribosomes are manufactured. It is here where ribosomal DNA transcribes most of rRNA molecules and these molecules undergo processing before their step-wise addition to 70 types of ribosomal proteins to form the ribosomal subunits.

Plant Cell

All the cytoplasmic and nuclear components found in the animal cells are found in the plant cells (Fig. 5.9). In addition to these cell components, plant cells contain certain structures not found in animal cells. These are **cell wall, plasmodesmata, plastids, large vacuoles** and **glyoxysomes**.

1. Cell wall. The outermost structure of most plant cells is a dynamic and rigid layer called **cell wall** (see **Buchanan** *et al.*, 2000). It is mainly composed of carbohydrates such as cellulose, pectin, hemicellulose and lignin and certain fatty substances such as waxes. Ultrastructurally cell wall is found to consist of a microfibrilar network lying in a gel-like matrix. The microfibrils are mostly made up of cellulose. There is a pectin rich cementing substance between the walls of adjacent cells which is called **middle lamella**. The cell wall which is formed immediately after the division of cell, constitutes the **primary cell wall.** Many kinds of plant cells have only primary cell wall around them. Primary cell wall is composed of pectin, hemicellulose and loose network of cellulose microfibrils.

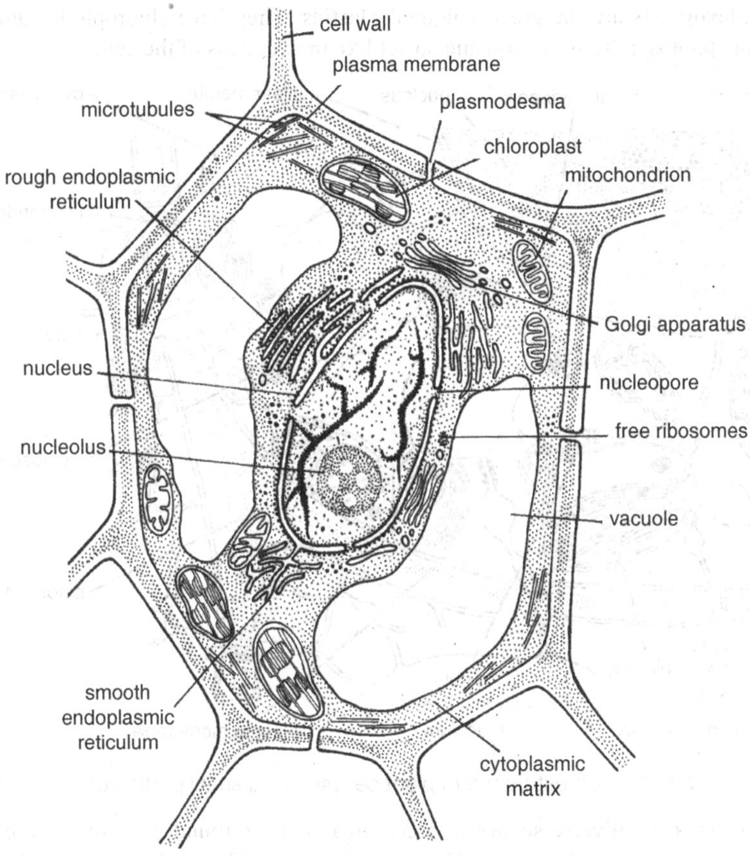

Fig. 5.9. Ultrastructure of a typical eukaryotic cell of a higher plant.

In certain types of cells such as phloem and xylem, an additional layer is added to the inner surface of the primary cell wall at a later stage. This layer is called **secondary cell wall** and it consists mainly of cellulose, hemicellulose and lignin. In many plant cells, there are tunnels running through the cell wall called **plasmodesmata** which allow communication with the other cells in a tissue.

The cell wall constitutes a kind of exoskeleton that provides protection and mechanical support to the plant cell. It determines the shape of plant cell and prevents it from desiccation.

2. Plastids. Plastids occur only in the plant cells. They contain pigments and may synthesize and accumulate various substances. Plastids are of the following types:

(*i*) **Leucoplasts** are colourless plastids of embryonic and germ cells lacking thylakoids and ribosomes.

(*ii*) **Amyloplasts** produce starch.

(*iii*) **Proteinoplasts** accumulate protein.

(*iv*) **Oleosomes** or **elaioplasts** store fats and essential oils.

(*v*) **Chromoplasts** contain pigment molecules and are coloured organelles. Chromoplasts impart a variety of colours to plant cells, such as red colour in tomatoes, red chillies and carrots, various colours to petals of flowers and green colour to many plant cells.

(*vi*) **Chloroplasts** are the green coloured plastids. They have chlorophyll pigment and are involved in the photosynthesis of food and so act like the kitchens of the cell.

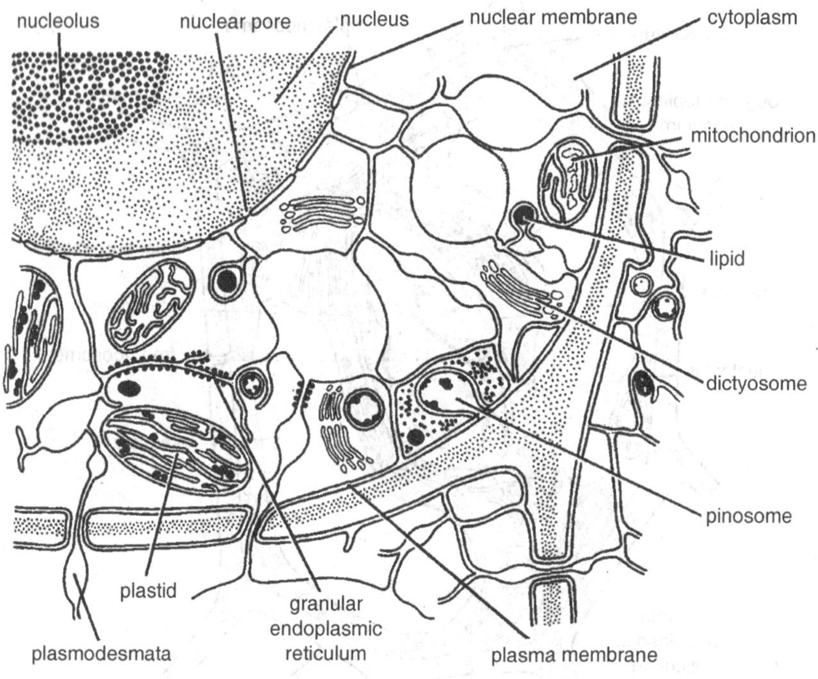

Fig. 5.10. Ultrastructure of plant cell (showing small portion of cell).

Chloroplasts have diverse shapes in green algae but are round, oval or discoid in shape in higher plants. Like mitochondria, each chloroplast is bounded by two membranes, both of which have no chlorophyll pigment. However, unlike mitochondria there occurs third system of membranes within the boundary of inner membrane, called **grana**. The grana form the main functional units of chloroplast and are bathed in the homogeneous matrix, called the **stroma**. Stroma contains a variety of photosynthetic enzymes and starch grains. Grana are stacks of membrane-bounded, flattened discoid sacs, arranged like neat piles of coins. A chloroplast contains many such interconnected grana on which are located various photosynthetic enzymes and the molecules of green pigment chlorophyll and other photosynthetic pigments to trap the light energy. The stroma of chloroplast contains DNA, ribosomes and complete protein synthetic machinery.

3. Vacuoles. Often most of the volume (up to 90 per cent) of a mature cell is occupied by one or more large vacuoles. The vacuoles of the plant cells are bounded by a single, semipermeable membrane known as **tonoplast**. These vacuoles contain water, phenol, flavonols, anthocyanins (blue and red pigment), alkaloids and storage products, such as sugars and proteins.

4. Glyoxysomes. These organelles develop in a germinating plant seed (*e.g.*, castor bean or *Ricinus*) to utilize stored fat of the seed (*i.e.*, to metabolise the triglycerides). Glyoxysomes consist of an amorphous protein matrix surrounded by a limiting membrane. The membrane of glyoxysomes originates from the ER and their enzymes are synthesized in the free ribosomes in the cytosol. Enzymes of glyoxysomes are used to transform the fat stores of the seed into carbohydrates by way of **glyoxylate cycle.**

Table 5.1 Functions of cellular organelles (of an eukaryotic cell).

Cell organelle	Functions
Plasma membrane	It is differentially permeable membrane through which extracellular substances may be selectively sampled and cell products may be liberated.
Cell wall (plant cells)	It forms a thick cellulose wall around the plasma membrane giving strength and rigidity to the cell.
Cytoplasm	It contains machinery for carrying out the instructions sent from the nucleus.
Hyaloplasm or matrix	It contains enzymes for glycolysis and structural materials such as sugars, fatty acids, amino acids, water, vitamins, nucleotides, etc.
Microtubules	They transport water, ions and small molecules of various substances, form structural unit of centrioles, basal granules, cilia, flagella and mitotic spindle and play active role in the cyclosis (amoeboid movement) of cytoplasm and in movement of chromosomes during the cell divisions.
Microfilaments	They maintain the structure of cell and form contractile components of muscle cell and cells performing amoeboid movement.
Centrioles	They form the spindle and have significant role in the movement of chromosomes during cell divisions.
Basal granules	They originate the cilia or flagella and have their importance in ciliary or flagellar movement.
Cilia and flagella	These are microtubule-containing extracellular projections of the cell which provide movement to the cells and serve various other roles of cells.
Endoplasmic reticulum	It provides greatly expanded surface area for biochemical reactions which normally occur at or across membrane surfaces and provide skeletal support to the colloidal complex of hyaloplasm. It also has transporting business of extracellular and intracellular chemical molecules.
Ribosomes	These are the sites for synthesis of protein molecules.
Golgi apparatus	It stores and modifies the synthetic proteins and enzymes of cells and forms secretory vesicles and lysosomes.
Lysosomes (animals only)	They contain intracellular digestive enzymes which aid in disposal of bacteria and other foreign bodies.
Peroxisomes (Microbodies)	They contain enzymes for hydrogen peroxide metabolism, purine metabolism, gluconeogenesis and photorespiration.
Sphaerosomes	They store fats.
Vacuoles	They are storage depots for excess water, waste products, soluble pigments, etc.

Cell organelle	Functions
Mitochondria	They are the organelles for the oxidation of food by Krebs cycle, production of energy by electron transport chain and beta-oxidation of fatty acids.
Plastids (Plants only)	They are the structures for storage of starch, pigments, and other cellular products. Photosynthesis occurs in chloroplasts.
Nucleus	It regulates growth and reproduction of cell.
Nuclear membrane	It provides selective continuity between nuclear and cytoplasmic materials.
Nucleoplasm	It contains materials for building DNA and messenger molecules (different kinds of RNA molecules, *viz.*, rRNA, mRNA and tRNA) which act as intermediates between nucleus and cytoplasm.
Chromosomes	They are the bearers of hereditary instructions and regulations of cellular processes.
Nucleolus	It disappears during cellular replication and synthesizes the ribosomes.

Differences between Prokaryotic and Eukaryotic cells

Main differences between prokaryotic cells (of archaea, bacteria, blue green algae and mycoplasma) and eukaryotic cells (of protozoa, fungi, plants and animals) have been tabulated in Table 5.2.

Table 5.2. Differences between prokaryotic and eukaryotic cells.

	Feature	Prokaryotic cell	Eukaryotic cell
1.	Size	Mostly 1–10 μm	Mostly 10–100 μm
2.	Multicellular forms	Rare	Common, with extensive tissue formation
3.	Cell wall	Present in most but not in all cells	Present in plant and fungal cell only
4.	Plasma membrane	Present	Present
5.	Nucleus	Absent	Present
6.	Nuclear membranes	Absent	Present
7.	Chromatin with histone	Absent	Present
8.	Genetic material	Circular or linear, double-stranded DNA: genes are not interrupted by intron* (notable exception is archaea)	Linear double-stranded DNA: genes frequently interrupted by intron sequences, especially in higher eukaryotes (such genes are called **split genes**)
9.	Nucleoli and mitotic apparatus	Absent	Present

Feature			Prokaryotic cell	Eukaryotic cell
10.	Plasmids		Commonly present	Rare
11.	Cellular organelles:			
	(i)	Mitochondria	Absent	Present
	(ii)	Endoplasmic reticulum	Absent	Present
	(iii)	Vacuoles	Absent	Present
	(iv)	Lysosomes	Absent	Present
	(v)	Chloroplasts	Absent	Present (only in plants)
	(vi)	Centrioles	Absent	Present (absent in higher plants)
	(vii)	Ribosomes	Present (70S)	Present (80S)
	(viii)	Microtubules	Absent	Present
	(ix)	Flagellae	Simple structure composed of the protein flagellin	Complex 9 + 2 structure of tubulin and other proteins
12.	Respiration		Many strict anaerobes (oxygen fatal)	All aerobic, but some facultative anaerobes by secondary modifications
13.	Metabolic patterns		Great variations	All share cytochrome electron transport chains, Krebs cycle oxidation, Embden-Meyerhof glucose metabolism or glycolysis
14.	Photosynthetic enzymes		Bound to plasma membrane as composite chromatophores	Enzymes packaged in plastids bound by membrane
15.	Sexual system		Rare: if present one way (and usually partial); transfer of DNA from donor to recipient cell occurs	Both sexes involved in sexual participation and entire genomes transferred; alternation of haploid and diploid generations is also evident

*Intron is an intervening sequence of nucleotides in DNA, located within a gene that is not included in the mature mRNA.

Differences between Plant and Animal Cells

Differences between typical animal and plant cells have been tabulated in Table 5.3.

Table 5.3 Differences between animal and plant cells.

	Animal cell		Plant cell
1.	Animal cells are generally small in size.	1.	Plant cells are larger than animal cells.
2.	Cell wall is absent.	2.	The plasma membrane of plant cells is surrounded by a rigid cell wall of cellulose.
3.	Except the protozoan *Euglena* no animal cell possesses plastids.	3.	Plastids are present.
4.	Vacuoles in animal cells are many and small.	4.	Most mature plant cells have a large central sap vacuole.
5.	Animal cells have a single highly complex and prominent Golgi apparatus.	5.	Plant cells have many simpler units of Golgi apparatus, called dictyosomes.
6.	Animal cells have centrosome and centrioles.	6.	Plant cells lack centrosome and centrioles.

QUESTIONS

Long Answer Questions

1. Describe the structure or a prokaryotic cell.
2. Describe generalised structure of bacterial cell.
3. Describe the ultrastructure of *E.coli.*
4. What are the chief characteristics of a cyanobacteria?
5. Prokaryotes are considered to be ancestors of eukaryotes. Discuss.
6. Define cell. Give electron microscopic structure of PPLO cell. Give a well labeled diagram of a eukaryotic animal cell.
7. Define the cell. What is an eukaryotic cell? Describe the general shape and size, number of eukaryotic cells in the living organism.
8. Give an account of the structure of an animal cell and state the functions which are attributed to its component parts.
9. Give an illustrated account of the functional anatomy of animal cell.
10. Enumerate the various organelles found in a typical animal cell and briefly describe the functions of Golgi body and ribosomes.
11. Give an illustrated account of ultrastructure of animal cell.
12. Give a detailed labeled diagram of electron microscopic structure of animal cell along with important cell organelles.
13. Give distinguishing characteristics of an eukaryotic plant cell.
14. Describe structure of a typical animal cell and mention the differences between a plant and animal cell.

Short Answer Questions

1. Summarize important diagnostic features of prokaryotes.
2. Write a short note on PPLOs.
3. Write a short note on 'Prokaryotic cell'.
4. Mention the functions of the following cellular companents:
   (*a*) Cell wall (*b*) Cellmembrane
   (*c*) Golgi apparatus (*d*) Lysosomes
   (*e*) Ribosomes (*f*) Mitochondria
   (*g*) Centrioles (*h*) Plastids
   (*i*) Chromatin.
5. Discuss the factors which control the shape of the cells.
6. What are the conditions which set upper limit in cell size?
7. What are the differences between prokaryotic and eukaryotic cells?
8. Draw a neat and labeled diagram of animal cell. Explain differences between prokaryotic and eukaryotic cells.
9. Differentiate between 'Animal cell and Plant cell'.
10. Draw neat and labeled diagram of the following:
    (*i*) Prokaryotic cell.
11. Give the structure of Eukaryotic cell.

Very Short Answer Questions

1. Name the smallest known cell. Give its size.
2. Enumerate cell organelles that fall under following categories:
   (*i*) Cytomembranes
   (*ii*) Membrane bound
   (*iii*) Membraneless
3. What will happen if all the mitochondria of a cell are destroyed?
4. How can you change an animal cell into a plant cell?
5. Name the organelle where photosynthesis occurs in the cell.
6. A cell wall is a part of all plant cells. Is it living or nonliving?
7. Apart from nucleus, which two other cell organelles have their independent DNA?
8. Name the nitrogen fixing cells of cyanobacteria.

Fill in the Blanks

1. The longest animal cell..............
2. Vacuole is generally present in .... cells but absent in .... cells.
3. .... are the fundamental units of life.
4. .... are not surrounded by cell membrane.
5. The nuclear region of prokaryotic cells is called ....

Matching of Columns

Match appropriately the items in column A with those of column B

	Column A		Column B
1.	Mitochondria	(*a*)	Animal cell
2.	Centriole	(*b*)	Bacteria
3.	Nucleus	(*c*)	Eukaryotic cell
4.	Chromosome without histone	(*d*)	Protoplasm
5.	The nerve cell of human beings	(*e*)	Hyaloplasm
6.	The ground substance of cytoplasm	(*f*)	Control room of cell
7.	DNA	(*g*)	90 cm
8.	Colloidal substance	(*h*)	Hereditary material

Multiple Choice Questions

1. Which of the following structures is present only in animal cell?
   (*a*) cell membrane
   (*b*) lysosomes
   (*c*) centrioles
   (*d*) ribosomes
2. Which of the following structures is associated with the formation of aster during nuclear division?
   (*a*) endoplasmic reticulum
   (*b*) centrioles
   (*c*) sphaerosome
   (*d*) ribosome
3. Which of the following statement is not correct?
   (*a*) ribosomes are power house of cell
   (*b*) lysosomes are called suicide bags
   (*c*) viruses lack their own enzyme system
   (*d*) bacteria lack a definite cell wall
4. Plant cells differ from animal cells in the absence of
   (*a*) endoplasmic reticulum
   (*b*) ribosomes
   (*c*) mitochondria
   (*d*) centrioles
5. A well organized nucleus with a distinct nuclear membrane is absent in
   (*a*) bacterial cells
   (*b*) eukaryotic cells
   (*c*) protozoan cells
   (*d*) fungal cells
6. Eukaryotic cells differ from prokaryotic cells in having
   (*a*) endoplasmic reticulum
   (*b*) ribosomes
   (*c*) true nucleus
   (*d*) cytoplasm
7. Which of the cell organelle is considered to be rich in catabolic enzymes?
   (*a*) lysosomes
   (*b*) Golgi apparatus
   (*c*) mitochondria
   (*d*) endoplasmic reticulum
8. Lysosomes are called suicide bags because of
   (*a*) hydrolytic enzymes
   (*b*) phagocytic activity
   (*c*) proteolytic activity
   (*d*) respiratory enzymes
9. Cellular organelles containing hydrolytic enzymes are called
   (*a*) microsomes
   (*b*) lysosomes
   (*c*) oxysomes
   (*d*) ribosomes

10. Single envelope system is characteristic of
    (*a*) prokaryotic cell
    (*b*) eukaryotic cell
    (*c*) none
    (*d*) both
11. Cell boundary of animal cell is
    (*a*) cell wall
    (*b*) plasma membrane
    (*c*) both
    (*d*) none of these
12. The granular structures found attached with endoplasmic reticulum are
    (*a*) ribosomes (*b*) lysosome
    (*c*) desmosome (*d*) kinetosome
13. The power house of cell is
    (*a*) Golgi complex (*b*) histone
    (*c*) mitochondria (*d*) chloroplasts
14. Prokaryotes and eukaryotes have the common
    (*a*) mitotic apparatus (*b*) histone
    (*c*) genetic code (*d*) mitochondria

ANSWERS

Very Short Answer Questions

1. Smallest cells are bacterium *Dialister pneumogintes and Mycoplasma gallisepticum.*
2. (*i*) Endoplasmic reticulum, Golgi apparatus and nuclear envelope.
   (*ii*) Lysosomes, mitochondria and chloroplasts.
   (*iii*) Ribosomes and centrioles.
3. There will be no metabolic energy for metabolic activities and cell will die.
4. By introducing plastids, vacuoles and cell wall.
5. Chloroplast.
6. Nonliving (Currently cell wall is regarded living or dynamic; see **Buchanan** *et al.*, 2000).
7. Mitochondria and chloroplasts.
8. Heterocysts.

Fill in the Blank

1. Nerve cell; 2. Plant, animal; 3. Cells; 4. Ribosomes;
5. Nucleoid.

Matching of Columns

1. (*c*) 2. (*a*) 3. (*f*) 4. (*b*) 5. (*g*) 6. (*e*) 7. (*h*)
8. (*d*)

Multiple Choice Questions

1. (*c*) 2. (*b*) 3. (*a*) 4. (*d*) 5. (*a*) 6. (*c*) 7. (*a*)
8. (*b*) 9. (*b*) 10. (*a*) 11. (*b*) 12. (*a*) 13. (*c*) 14. (*c*)

6 Cytosol: Elements, Biomolecules and Properties

Within the cells of any organism, the living substance, or **protoplasm,** is itself comprised of a multitude of non-living constituents: proteins, nucleic acids, fats (lipids), carbohydrates, vitamins, minerals, waste metabolites, crystalline aggregates, pigments, and many others, all of which are composed of molecules and their constituent atoms. The *protoplasm is alive because of the highly complex organization of these non-living substances and the way they interact with one another.*

6.1. PHYSICAL NATURE OF CYTOSOL

The fluid and soluble portion of the cytoplasm that exists outside the organelles is called **cytoplasmic matrix** or **cytosol.** The cytosol (cytoplasmic matrix) is a colourless or grayish, translucent, viscid, gelatinous or jelly-like **colloidal** substance. It is heavier than water and capable of flowing. In past, there has been a lot of controversy about the physical nature of the matrix. Different workers advanced different theories about the physical characteristics of the matrix. Their theories can be represented as follows:

1. **Reticular theory** suggests that the matrix is composed of reticulum of fibres or particles in the ground substances (Fig. 6.1A).

2. **Alveolar theory** was proposed by **Butschili** in 1892 and according to it, the matrix consists of many suspended droplets or alveoli or minute bubbles resembling the foams of emulsion (Fig. 6.1B).

3. **Granular theory** was propounded by **Altmann** in 1893. This theory supports the view that the matrix contains many granules of smaller and larger size arranged differently. These granules were known as **bioplasts** (Fig. 6.1C).

4. **Fibrillar theory** was proposed by **Fleming** and it holds that the matrix is fibrillar in nature (Fig. 6.1D).

5. **Colloidal theory** has been forwarded very recently after the electron microscopical observations of the matrix. According to this recent concept, the matrix is partly a true **solution**, partly a **colloidal system.**

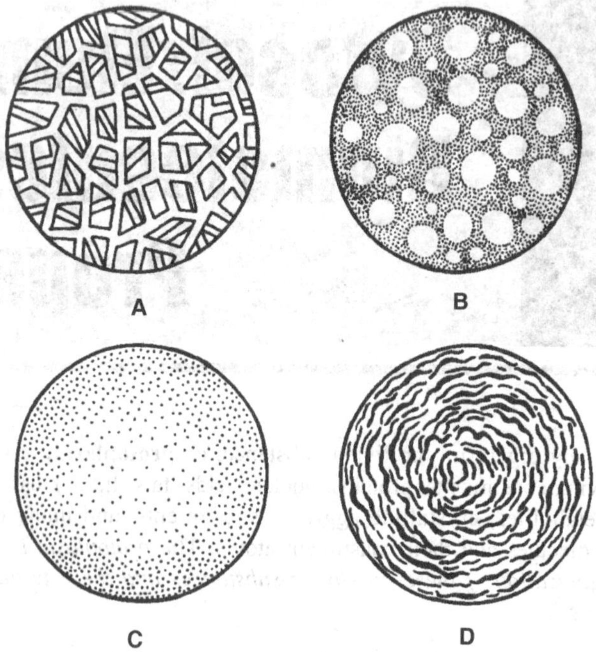

Fig. 6.1. Physical appearance of protoplasm. A—Reticular, B—Alveolar, C—Granular, and D—Fibrillar.

Phase Reversal

Cytosol (cytoplasmic matrix) like many colloidal systems, shows the property of **phase reversal**. For example, gelatin particles (discontinuous phase) are dispersed through water (continuous phase) in a thin consistency that is freely shakable (Fig. 6.2 A). Such a condition is called a **sol**. When the solution cools, gelatin now becomes the continuous phase and the water is in the discontinuous phase. Moreover, now the solution has stiffened and becomes semisolid and is called a **gel**.

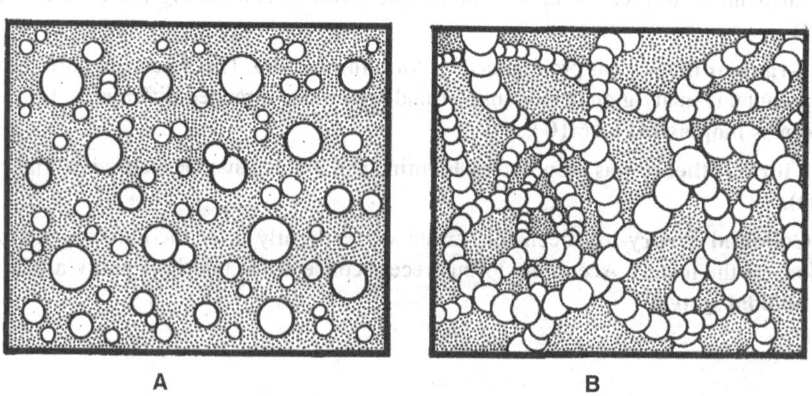

Fig. 6.2. Sol and gel state of the cytosol. A—Sol condition in which gelatin particles are the discontinuous phase, water the continuous phase; B—Gel condition in which gelatin particles form continuous phase (net work), enclosing water as discontinuous phase.

6.2. CHEMICAL ORGANIZATION OF CYTOSOL

Chemically, the cytoplasmic matrix (cytosol) is composed of many chemical elements in the form of atoms, ions and molecules.

Chemical Elements

Of the 92 naturally occurring elements, perhaps 46 are found in the cytosol (cytoplasmic matrix). Twenty four of these are considered essential for life (called **essential elements**), while others are present in cytosol only because they exist in the environment with which the organism interacts. Of the 24 essential elements, six play especially important roles in living systems. These **major elements** are carbon (C, 20 per cent), hydrogen (H, 10 per cent), nitrogen (N, 3 per cent), oxygen (O, 62 per cent), phosphorus (P, 1.14 percent) and sulphur (S, 0.14 per cent). Most organic moleceules are built with these six elements. Another five essential elements found in less abundance in living systems are calcium (Ca, 2.5 per cent), potassium (K, 0.11 per cent), sodium (Na, 0.10 per cent), chlorine (Cl, 0.16 per cent) and magnesium (Mg, 0.07 per cent). Several other elements, called **trace elements,** are also found in minute amounts in animals and plants, but are nevertheless essential for life. These are iron (Fe, 0.10 per cent), iodine (I, 0.014 per cent), molybdenum (Mo), manganese (Mn), Cobalt (Co), zinc (Zn), selenium (Se), copper (Cu), chromium (Cr), tin (Sn), vanadium (V), silicon (Si), nickel (Ni), fluorine (F) and boron (B).

Ions

The cytoplasmic matrix consists of various kinds of ions. The ions are important in maintaining osmotic pressure and acid-base balance in the cells. Retention of ions in the matrix produces an increase in osmotic pressure and, thus, the entrance of water in the cell. The concentration of various ions in the intracellular fluid (matrix) differs from that in the interstitial fluid. For example, in the cell K^+ and Mg^{++} can be high, and Na^+ and Cl^- high outside the cell. In muscle and nerve cells a high order of difference exists between intracellular K^+ and extracellular Na^+. Free calcium ions (Ca^{++}) may occur in cells or circulating blood. Silicon ions occur in the epithelium cells of grasses. The free ions of phosphate (primary, $H_2PO_4^-$ and secondary, $HPO_4^=$) occur in the matrix and blood. These ions act as a buffering system and tend to stabilize pH of blood and cellular fluids. The ions of different cells also include sulphate ($SO_4^=$), carbonate ($CO_3^=$), bicarbonate (HCO_3^-), magnesium (Mg^{++}) and amino acids. Cellular functions of certain ions have been tabulated in Table 6.1.

Table 6.1. Cellular functions of certain ions.

	Element	Ionic form present	Functions
1.	Molybdenum	MoO_4^{2-}	Cofactor or activator of certain enzymes (*e.g.*, nitrogen fixation, nucleic acid metabolism, aldehyde oxidation).
2.	Cobalt	Co^{2+}	Constituent of vitamin B_{12}.
3.	Copper	Cu^+, Cu^{2+}	Constituent of plastocyanin and cofactor of respiratory enzymes.
4.	Iodine (Heaviest trace element)	I^-	Constituent of thyroxin, triiodothyronine and other thyroid hormones.
5.	Boron	BO_3^{3-}, $B_4O_7^{2-}$	Activates arabinose isomerase.
6.	Zinc	Zn^{2+}	Cofactor of certain enzymes (*e.g.*, carbonic anhydrase, carboxypeptidase).
7.	Manganese	Mn^{2+}	Cofactor of certain enzymes (*e.g.*, several kinases, isocitric decarboxylase).

	Element	Ionic form present	Functions
8.	Iron	Fe^{2+}, Fe^{3+}	Constituent of haemoglobin, myoglobin and cytochromes.
9.	Magnesium	Mg^{2+}	Constituent of chlorophyll; activates ATPase enzyme.
10.	Sulphur	SO_4^{2-}	Constituent of coenzyme A, biotin, thiamine, proteins.
11.	Phosphorus	PO_4^{3-}, $H_2PO_4^-$	Constituent of lipids, proteins, nucleic acids, sugar phosphates, nucleoside phosphates.
12.	Calcium	Ca^{2+}	Constituent of plant cell walls; matrix component of bone tissue; cofactor of coagulation enzymes.
13.	Potassium	K^+	Cofactor for pyruvate kinase and K^+– stimulated ATPase.

Types of Compounds of Cytosol

Chemical compounds are conventionally divided into two groups: **organic** and **inorganic**. Organic compounds form 30 per cent of a typical cell, rest are the inorganic substances such as water and other substances.

Table 6.2. The approximate percentage composition of the human body.

	Substance	Percentage
1.	Water	65
2.	Protein	18
3.	Fat	10
4.	Carbohydrate	5
5.	Other organic	1
6.	Inorganic	1

I. Inorganic Compounds

The inorganic compounds are those compounds which normally found in the bulk in the physical, non-living universe, such as elements, metals, non-metals, and their compounds such as water, salts and variety of electrolytes and non-electrolytes. In the previous section, we have discussed a lot about the inorganic substances except the water which will be discussed in the following paragraph.

Water. The most abundant inorganic component of the cytosol is the water (the notable exceptions are seeds, bone and enamel). Water constitutes about 65 to 80 per cent of the matrix. In the matrix the water occurs in two forms, *viz.*, **free water** and **bound water**. The 95 per cent of the total cellular water is used by the matrix as the solvent for various inorganic substances and organic compounds and is known as **free water**. The remaining 5 per cent of the total cellular water remains loosely linked with protein molecules by hydrogen bonds or other forces and is known as **bound water**.

The water contents of the cellular matrix of an organism depend directly on the age, habitat and metabolic activities. For instance, the cells of the embryo have 90 to 95 per cent water which decreases progressively in the cells of the adult organism. The cells of lower aquatic animals contain comparative high percentage of the water than the cells of higher terrestrial animals. Further the percentage of water in the matrix also varies from cell to cell according to the rate of the metabolism.

Unique physical properties of water and their biological utility. There are several extraordinary properties of water that make it especially fit for its essential role in the protoplasmic systems (*i.e.*, cytosol or matrix). Some of the unique properties of water are the following:

1. Water as a solvent. Water is most stable yet versatile of all solvents. Water's properties as a solvent for inorganic substances as mineral ions, solids, etc., and organic compounds such as carbohydrates and proteins, depend on water's dipole nature. Because of this polarity, water can bind electrostatically to both positively and negatively charged groups in the protein. Thus, each amino group in a protein molecule is capable of binding 2.6 molecules of water. The solvency is of great biological importance because all the chemical reactions that take place in the cells do so in aqueous solution. The water also forms the good dispersion medium for the colloidal system of the matrix.

2. Water's thermal properties. Water is the only substance that occurs in nature in the three phases of solid, liquid and vapour within the ordinary range of Earth's temperatures. Water has a **high specific heat:** it requires 1 calorie (4.185 joules) to elevate the temperature of 1 gram of water by 1°C (such as from 15 to 16°C). Such a high thermal capacity of water has a great moderating effect on environmental temperature changes and is a great protective agent for all life.

Water also has a **high heat of vaporization.** It requires more than 540 calories (2259 joules) to change 1 gram of liquid water into water vapour. Thus, water tends to have a remarkably high boiling point (100°C) for a substance of such low relative molecular mass. Were it not for this lucky accident, it is likely that liquid water would never have existed on Earth and would have been lost to outer space. Further, for terrestrial plants and animals, cooling produced by the evaporation of water is an important means of getting rid of excess heat. Moreover, at the other temperature extreme, large amounts of energy (335 joules or 80 cal per gram) must be lost for water to be converted from the liquid to the solid state. This is called **heat of fusion.** Water's melting point being 0°C.

Another important property of water from a biological standpoint is its unique **density behaviour** during change of temperature. Most liquids become continually more dense with cooling. Water, however, reaches its maximum density at 4°C and then becomes lighter with further cooling. Therefore, ice floats rather than settling on the bottom of lakes and ponds. This protects the aquatic life from freezing.

3. Surface tension. Water has a **high surface tension**. This property, caused by the great cohesiveness of water molecules, is important in the maintenance of protoplasmic form and movement. Despite its high surface tension, water has **low viscosity,** a property that favours the movement of blood through minute capillaries and of cytoplasm inside cellular boundaries.

Molecules dissolved in water, lower its surface tension and tend to collect at the interface between its liquid phase and other phases. This may have been important in the development of the plasma membrane, and certainly plays an important role in the movement of molecules across it.

4. Transparency. The water is transparent to light, enabling the specialized photosynthetic organelles, the chloroplast, inside the plant cell to absorb the sunlight for the process of photosynthesis.

II. Organic Compounds (Micromolecules and Macromolecules)

The chemical substances which contain carbon (C) in combination with one or more other elements as hydrogen (H), nitrogen (N), sulphur (S), etc., are called **organic compounds.** The organic compounds usually contain large molecules which are formed by the similar or dissimilar unit structure known as the **monomers.** A monomer (Gr., *mono* = one, *meros* = part) is the simplest unit of the organic molecule which can exist freely. Some organic compounds such as carbohydrates occur in the matrix as the monomers. The monomers usually link with other monomers to form **oligomers** (Gr., *oligo* = few or little, *meros* = part) and **polymers** (Gr., *poly* = many, *meros* = part). The oligomers contain small number of monomers, while the polymers contain large number of monomers. The oligomers and polymers contain large-sized molecules or macromolecules. When a polymer contains similar kinds of monomers in its macromolecule it is known as **homopolymer** and when the polymer is composed of different kinds of monomers it is known as the **heteropolymer**.

The main organic compounds of the matrix are the carbohydrates, lipids, proteins, vitamins, hormones and nucleotides.

6.3. PHYSICAL AND BIOLOGICAL PROPERTIES OF CYTOPLASMIC MATRIX

The matrix is a living substance and it has following physical and biological properties:

Physical Properties. The most of the physical properties of the matrix are due to its colloidal nature and these are as follows:

1. Tyndall's effect. When a beam of strong light is passed through the colloidal system, of the matrix at right angles in the dark room, the small colloidal particles which remain suspended in the colloidal system reflect the light. The path of the light appears like a cone. This light cone is known as **Tyndall's cone** because, this phenomenon has been first of all reported by **Tyndall** (1820–1893) in colloids.

2. Brownian movement. The suspended colloidal particles of the matrix always move in zig-zag fashion. This movement of molecules is caused by moving water molecules which strike with the colloidal molecules to provide motion to them. This type of movement was first of all observed by Scottish botanist **Robert Brown** in 1827 in the colloidal solution. Therefore, such movements are known as **Brownian movement**. The Brownian movement is the peculiarity of all colloidal solutions and depends on the size of the particles and temperature.

3. Cyclosis and amoeboid movement. Due to the phase reversal property of the cytoplasmic matrix, the intracellular streaming or movement of the matrix takes place. This property of intracullular movement of matrix is known as the **cyclosis**. The cyclosis usually occurs in the sol-phase of the matrix and is effected by the hydrostatic pressure, temperature, pH, viscosity, etc. The intracellular movements of the pinosomes, phagosomes and various cytoplasmic organelles such as the lysosomes, mitochondria, chromosomes, centrioles, etc., occur only due to cyclosis of the matrix. The cyclosis has been observed, in most animal and plant cells.

The amoeboid movement depends directly on the cyclosis. The amoeboid movement occurs in the protozoans, leucocytes, epithelia, mesenchymal and other cells. In the amoeboid movement the cell changes its shape actively and gives out cytoplasmic projections known as **pseudopodia**. Due to cyclosis matrix moves these pseudopodia and this causes forward motion of the cell.

4. Surface tension. The molecules in the interior of a homogeneous liquid are free to move and are attracted by surrounding molecules equally in all directions. At the surface of the liquid where it touches air or some other liquid, however, they are attracted downward and sideways or inward, more than upward; consequently they are subjected to unequal stress and are held together to form a membrane. The force by which the molecules are bound is called the **surface tension** of the liquid. The cytoplasmic matrix being a liquid possesses the property of surface tension. The proteins and lipids of matrix have less surface tension, therefore, occur at the surface and form the membrane, while the chemical substances such as NaCl have high surface tension, therefore, occur in deeper part of the matrix.

5. Adsorption. The increase in the concentration of a substance at the surface of a solution is known as adsorption (*L.*, *ad* = to, *sorbex* = to draw in). The phenomenon of adsorption helps the matrix to form protein boundaries.

6. Other mechanical or physical properties of matrix. Besides surface tension and adsorption, the matrix possesses other mechanical properties, *e.g.*, elasticity, contractility, rigidity and viscosity which provide to the matrix many physiological utilities.

7. Polarity of the egg. The colloidal system due to its stable phase determines the polarity of the cell matrix which cannot be altered by centrifugation of other mechanical means.

8. pH. Hydrogen ion concentration of a solution is the number of gram ions of hydrogen in one liter of a solution. The *acidity* or *alkalinity* of a solution is expressed in terms of hydrogen ion concentration.

Box 6.1 The little 'p' in pH

The term pH was first used in 1909 by **Soren Peter Lauritz Sorensen** (1868–1939), director of the Carlsberg Laboratories in Denmark. Sorensen never mentioned what the little "p" stood for (the "H" is obviously hydrogen). Many years later some of the scientists who write chemistry textbooks began to associate the little "p" with the words "power" or "potential". This association, as it turns out, is based on a rather tenuous connection in some of Sorensen's early papers. A recent investigation of the historical records by **Jens G. Noby** suggests that the little "pH" was an arbitrary choice based on Sorensen's use of **p** and **q** to to stand for unknown variables in much the same way that we might use **x** and **y** today.

No matter what the historical origin, it is important to remember that the symbol pH now stands for the negative logarithm of the hydrogen ion concentration (**Horton et al.**, 2006).

pH of pure water is 7 (neutral). When pH is less than 7, *i.e.*, 0–7, the solution is acidic and when pH is more than 7, *i.e.*, 7–14, the solution is alkaline in nature.

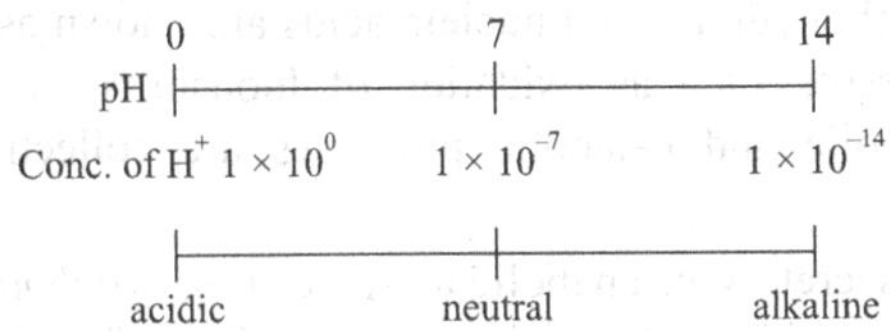

A lower pH value indicates increasing strength of acidity and a higher pH indicates increasing strength of alkalinity.

Box 6.2 Determination of pH

pH is determined by two methods: 1. pH–indicator method (in which chemicals called **pH indicators** such as methyl blue, etc., are used); 2. **Potentiometric** method (in which an electrode or pH meter is used). The latter method uses different types of electrodes such as glass electrode, hydrogen electrode, quinhydrone electrode, antimony-antimony electrode, oxide electrode and ion-selective electrodes (**Sharma** 2010).

pH has great significance in biological systems. In the body, metabolic reactions take place at an optimal pH. The enzymes act on optimal pH. Any deviation from this pH affects the rate of metabolic reactions. In body different fluids have specific pH and the body follows various mechanisms to maintain pH.

9. Buffer solutions. Generally, when an acid or base is added to a solution it changes its pH. But there are certain solutions which resist any change in pH when a little amount of acid or base is added to them. Such solutions are called buffer solutions.

A buffer may be an acid buffer or basic buffer. **Acid buffers** contain a weak acid and its salts, *e.g.*, acetic acid and sodium acetate. **Basic buffers** contain a weak base and its salts, *e.g.*, ammonium hydroxide and ammonium chloride.

Buffers are highly significant. Biological fluids such as blood, urine, etc., have a definite pH. In human blood, a constant pH of 7.4 is maintained by the carbon dioxide-carbonic acid-bicarbonate buffer system.

Enzymes act at a specific pH. Any deviation from this normal pH adversely affects the metabolic reactions in the body. Buffer systems help in carrying out the biochemical reactions at normal level. pH is maintained by different buffer systems in the body. Buffers provide protection to the cells and tissue against sudden changes in pH. Bicarbonate, phosphate, protein, haemoglobin are some major buffer systems of blood.

Biological Properties

The matrix is a living substance and it has following biological properties:

1. **Irritability.** The irritability is the fundamental and inherent property of the matrix. It possesses a sensitivity to stimulation, an ability to transmission of excitation and ability to react according to stimuli. The heat, light, chemical substances and other factors stimulate the cytoplasmic matrix to contract.
2. **Conductivity.** The conductivity is the process of conduction or transmission of excitation from the place of its origin to the region of its reaction. The matrix of nerve cells possesses the property of the conductivity.
3. **Movement.** The cytoplasmic matrix can perform movement due to cyclosis. The cyclosis depends on the age, water contents, heredity factors and composition of the cells.
4. **Metabolism.** The matrix is the seat of various chemical activities. These activities may be either constructive or destructive in nature. The constructive processes such as biosynthesis of proteins, lipids, carbohydrates and nucleic acids are known as **anabolic processes,** while the destructive processes such as oxidation of foodstuffs, *etc.*, are known as **catabolic processes.** The anabolic and catabolic processes are collectively known as **metabolic process.**
5. **Growth.** Due to the secretory or anabolic activities (*Gr.*, *anabolism* = a throwing up) of the cell, new protoplasm continuously increases in its volume. The increase in the volume of the matrix causes into the growth of the cell which ultimately divides into daughter cells by the cell division.
6. **Reproduction.** The cytoplasm has the property of asexual and sexual reproduction.

QUESTIONS

Long Answer Questions

1. Write in brief the physical properties of the cytosol.
2. Write in brief the chemical properties of the cytosol.
3. Describe biological properties of cytoplasmic matrix.
4. Write functions of ions of cytosol.

Short Answer Questions

1. Define phase reversal.
2. Write about unique properties of water.

Very Short Answer Questions

1. Give example of biologically important trace element.
2. How many elements are essential for life?
3. Define buffer.

Mutiple Choice Quesions

1. The living part of a cell is called
   (*a*) cell wall (*b*) protoplasm
   (*c*) hyaloplasm (*d*) cell sap
2. The difference between protoplasm and cell sap was given by
   (*a*) Huxley (*b*) Purkinje
   (*c*) Hugo von Mohl (*d*) none of these
3. The term protoplasm was coined by
   (*a*) Dujardin (*b*) Purkinje
   (*c*) Robert Hooke (*d*) Robert Brown
4. Which term is more precise to describe photoplasm?
   (*a*) colloid (*b*) fibrous
   (*c*) cellular (*d*) solution
5. Zinc is an essential micronutrient of
   (*a*) vitamin
   (*b*) hormone
   (*c*) cofactor of vitamins
   (*d*) cofactor for enzyme
6. A plant cell requires calcium for
   (*a*) holding its cells together
   (*b*) synthesizing chlorophyll
   (*c*) DNA replication
   (*d*) opening and closing of stomata

7. Which one of the following mineral elements plays an important role in biological nitrogen fixation?
   (*a*) copper (*b*) manganese
   (*c*) zinc (*d*) molybdenum
8. The most abundant component of a cell is
   (*a*) water (*b*) cellulose
   (*c*) fats (*d*) carbohydrates
9. Ca^{2+} ions are essential for
   (*a*) muscle contraction
   (*b*) clotting of blood
   (*c*) formation of bone
   (*d*) all above
10. Cobalt as a rare element is essential in the synthesis of this vitamin
    (*a*) vitamin B_1 (*b*) vitamin B_6
    (*c*) vitamin B_{12} (*d*) vitamin C
11. Endoenzymes act at
    (*a*) alkaline pH (*b*) neutral pH
    (*c*) acidic pH (*d*) at any pH
12. Cholera patients are given saline drips because
    (*a*) NaCl maintains RBC
    (*b*) Cl ions are major components of plasma
    (*c*) Cl ions are used to make HCl in stomach
    (*d*) Na helps to retain water in the body and is a component of selective membrane transport
13. The four elements that form the 99% of the protoplasm are
    (*a*) N, O, P, S (*b*) C, H, N, O
    (*c*) H, N, Ca, Cl (*d*) C, O, Ca, S
14. Which of the following contain iodine?
    (*a*) insulin
    (*b*) thyroxine
    (*c*) oxytocin
    (*d*) adrenalin

ANSWERS

Very Short Answer Questions

1. Iron (Fe).
2. Twenty four.
3. A substance or solution that resists change in pH on addition of the acid (*i.e.*, H^+ ions) or alkali (*i.e.*, OH^- ions), absorbing protons from acids and releasing them on addition of alkali.

Multiple Choice Questions

1. (*b*) 2. (*c*) 3. (*b*) 4. (*a*) 5. (*c*) 6. (*a*) 7. (*d*)
8. (*a*) 9. (*d*) 10. (*c*) 11. (*b*) 12. (*d*) 13. (*b*) 14. (*b*)

7 Amino Acids, Peptides and Proteins

Of all the macromolecules found in the cell, the **proteins** are chemically and physically more diverse. They are important constituents of the cell forming more than 50 percent of the cell's dry weight. The term protein was coined by Swedish chemist **Berzelius** in 1838 and was used first (in 1838) by Dutch chemist **G.J. Mulder** (1802–1880) and is derived from Greek word *proteios*, which means "of the first rank".

Proteins serve as the chief structural material of protoplasm and play numerous other essential roles in living systems. They form enzymes—globular proteins specialized to serve as catalysts in virtually all biochemical activities of the cells. Other proteins are antibodies (immunoglobulins), transport proteins, storage proteins, contractile proteins, and some hormones. In every living organism, there are thousands of different proteins, each fitted to perform a specific functional or structural role. Indeed, a single human cell may contain more than 10,000 different types of protein molecules. Chemically, proteins are polymers of amino acids.

Nobel Laureate **Emil Fischer** (1902) discovered that all proteins consist of chains (linear sequence) of smaller units that he named **amino acids**. There are about 20 different amino acids which occur regularly as constituents of naturally occurring proteins (Table 7.2). An organic compound containing one or more amino groups ($-NH_2$) and one or more carboxyl groups (–COOH) is known as amino acid. The amino acids occur freely in the cytoplasmic matrix and constitute the so called **amino acid pool**. Of the 20 commonly occurring amino acids, 19 may be represented by the following general formula (Fig. 7.1).

The sole exception is proline, where the amino group forms part of a ring structure. The central or **alpha** carbon atom of each amino acid is covalently bonded to four groups:

(*i*) A hydrogen atom,

(*ii*) an amino group ($-NH_2$),

(*iii*) an acid (or carboxyl) group, and

(*iv*) a side chain called an **R-group**. It is the particular chemical structure of the R-group that distinguishes one amino acid from another.

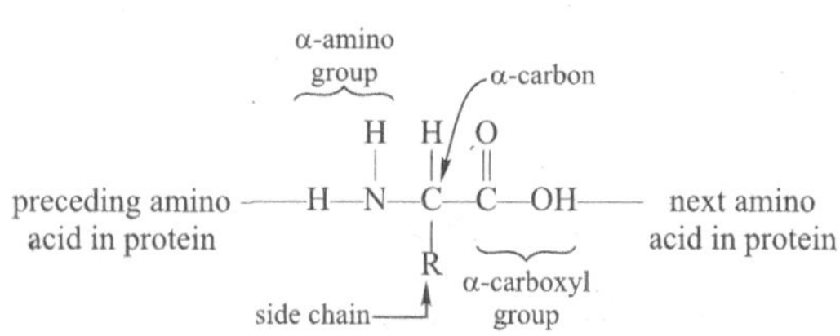

Fig. 7.1. Basic structure of an amino acid.

7.1. ACID DISSOCIATION CONSTANTS (*Ka*) OF WEAK ACIDS

Acids and bases which dissociate completely in water, such as hydrochloric acid and sodium hydroxide, are called **strong acids** and **strong bases**. Many other acids and bases, such as the amino acids from which proteins are made and the purines and pyrimidines from DNA and RNA, do not dissociate completely in water. These substances are known as **weak acids** and **weak bases.**

Acetic acid is the weak acid present in vinegar. The equilibrium reaction for the ionization of acetic acid is

$$\underset{\substack{\text{Acetic acid}\\\text{(Weak acid)}}}{CH_3COOH} \xrightleftharpoons{K_2} \underset{\text{Proton}}{H^+} + \underset{\substack{\text{Acetate ion}\\\text{(Conjugate base)}}}{CH_3COO^-} \quad ...(7.1)$$

The equilibrium constant for the dissociation of a proton from an acid in water is called the **acid dissociation constant, *Ka*.** When it reaches equilibrium, which happens very rapidly, the acid dissociation constant is equal to the concentration of the products divided by the concentration of the reactants. For reaction 7.1 the acid dissociation constant is

$$Ka = \frac{[H^+][CH_3COO^-]}{[CH_3COOH]} \quad ...(7.2)$$

The *Ka* for acetic acid at 25°C is 1.76×10^{-5} M. Because *Ka* values are numerically small and inconvienient in calculations it is useful to place them on a logarithmic scale. The parameter **p*Ka*** is defined by analogy with pH

$$pKa = -\log Ka = \log \frac{1}{Ka} \quad ...(7.3)$$

A pH value is a measure of the acidity of a solution and p*Ka* value is a measure of the strength of the base (ammonia, NH_3) in aqueous solution. The *Ka* values for some common substances are listed in Table 7.1.

Table 7.1. Dissociation constant [*Ka*(M)] and p*Ka* values of weak acids in aqueous solutions at 25°C *Source:* Hurton *et al.*, 2006).

	Acid	*Ka* (M)	p*Ka*
1.	HCOOH (Formic acid)	1.77×10^{-4}	3.8
2.	CH_3COOH (Acetic acid)	1.76×10^{-5}	4.8
3.	$CH_3CHOHCOOH$ (Lactic acid)	1.36×10^{-4}	3.9
4.	H_3PO_4 (Phosphoric acid)	7.52×10^{-3}	2.2
5.	$H_2PO_4^-$ (Dihydrogen phosphate ion)	6.23×10^{-8}	7.2
6.	HPO_4^- (Monohydrogen phosphate ion)	2.20×10^{-13}	12.7
7.	H_2CO_3 (Carbonic acid)	4.30×10^{-7}	6.4
8.	HCO_3 (Bicarbonate ion)	5.61×10^{-11}	10.2
9.	NH_4^+ (Ammonium ion)	5.62×10^{-10}	9.2
10.	$CH_3NH_3^+$ (Methylammonium ion)	2.70×10^{-11}	10.7

7.2. AMINO ACIDS

All organisms use the same 20 amino acids (Table 7.2) as building blocks for the assembly of protein molecules. These 20 amino acids are called the **common**, or **standard**, amino acids. Despite the limited number of amino acids an enormous variety of different polypeptides can be produced by connecting the 20 common amino acids in various combinations.

Table 7.2. The 20 amino acids which occur naturally.

	Group of amino acid		No.	Name of amino acid	Symbol of amino acid	Single letter code of amino acid
A.	**Aliphatic amino acids**					
	1.	Simple amino acids or Monoamino-monocarboxylic amino acids	1.	Glycine	Gly	G
			2.	Alanine	Ala	A
			3.	Valine	Val	V
			4.	Leucine	Leu	L
			5.	Isoleucine	Ile	I
	2.	Acidic amino acids or Monoamino-monocarboxylic acids	6.	Aspartic acid	Asp	D
			7.	Glutamic acid	Glm	E
	3.	Basic amino acids or Diamino-monocarboxylic amino acids	8.	Lysine	Lys	K
			9.	Arginine	Arg	R
			10.	Histidine	His	H
	4.	Hydroxyl containing amino acids	11.	Serine	Ser	S
			12.	Threonine	Thr	T
	5.	Sulphur-containing amino acids	13.	Cysteine	Cys	C
			14.	Methionine	Met	M
B.	**Aromatic amino acids**					
			15.	Phenylalanine	Phe	F
			16.	Tyrosine	Tyr	Y
			17.	Tryptophan	Try	W
C.	**Secondary amino acids**					
			18.	Proline	Pro	P
D.	**Amino acid amides**					
			19.	Aspargine	Asn	N
			20.	Glutamine	Glu	Q

1. Nomenclature of Amino Acids

Amino acids are called amino acids because they are amino derivatives of carboxylic acid. In the 20 common amino acids the amino group and carboxyl group are bonded to the same carbon atom: the α-carbon atom (Fig. 7.1). Thus, all of the standard amino acids found in proteins are α-amino acids. Two other substituents are bound to the α-carbon—a hydrogen atom and a side chain (R group) that is distinctive for each amino acid. In the **chemical names** or **systematic names**, carbon atoms are identified by numbers, beginning with the carbon atom of the carboxyl group. [the correct chemical name follows rules established by **International Union of Pure and Applied Chemistry (IUPAC)** and the **International Union of Biochemistry and Molecular Biology (IUBMB)**]. If the R group is $–CH_3$, then the systematic name for that amino acid would be **2-aminopropanoic acid**. (Propanoic acid is $CH_3–CH_2–COOH$). The trivial name for $CH_3–CH(NH_2)–COOH$ is alanine. An alternate nomenclature uses Greek letters to identify the α-carbon atom and carbon atom of the side chain. The alternate nomenclature identifies the carbon atom relative to carboxyl group so the carbon atom of the carboxyl group is not specified, unlike the systematic nomenclature where this carbon atom is number 1 in the numbering system. Biochemists have traditionally used the old alternate nomenclature.

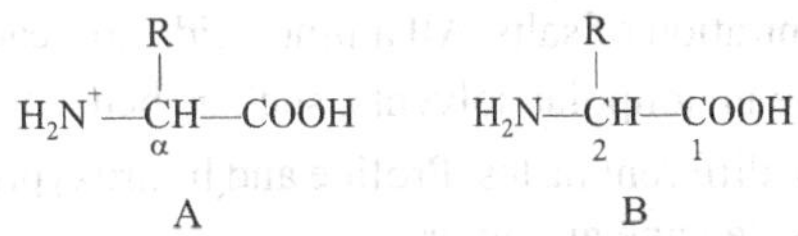

Fig. 7.2. Numbering conventions for amino acids. A—In traditional names, the carbon atoms adjacent to the carboxyl group are identified by the Greek letters α, β, γ, etc. B—In the official IUPAC/IUBMB chemical names or systematic names, the carbon atom in the carboxyl group is number 1 and the adjacent carbons are numbered sequentially. Thus, the α-carbon atom in traditional names is the carbon 2 atom in systematic names (after Horton *et al.*, 2006).

Figure 7.3A, shows the general structure of an amino acid in perspective. Figure 7.3B, shows a ball-and-stick model of representative amino acid, serine, whose side chain is $-CH_2OH$. The carbon atoms of a side chain are sequentially labelled β, γ, δ and ε, which refers to carbons 3, 4, 5 and 6, respectively. The systematic name for serine is 2-amino-3-hydroxypropanoic acid.

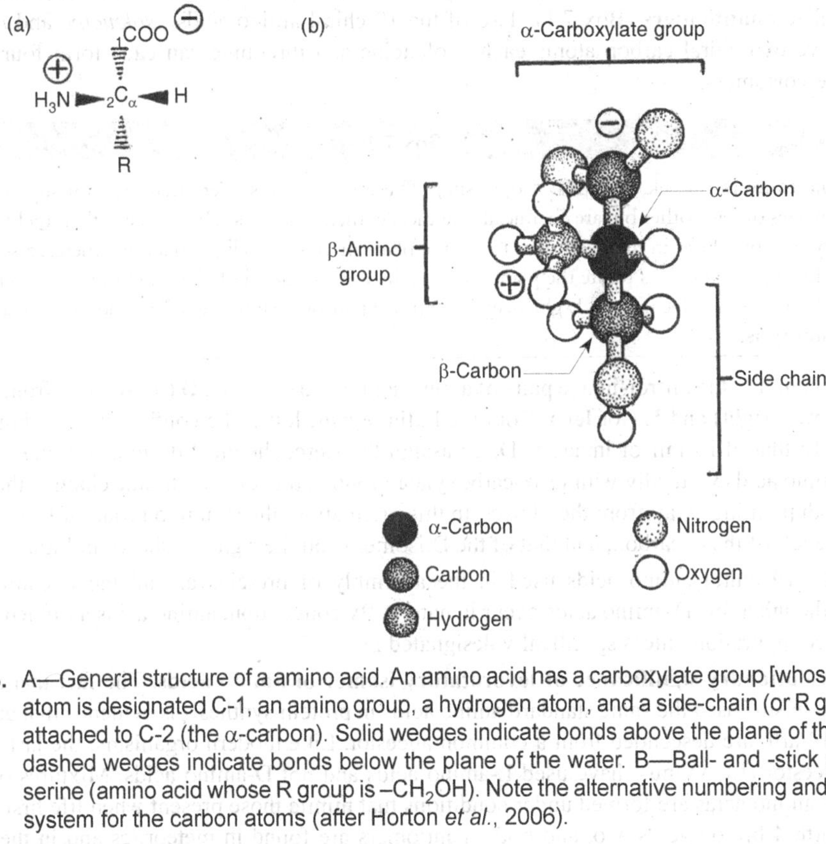

Fig. 7.3. A—General structure of a amino acid. An amino acid has a carboxylate group [whose carbon atom is designated C-1, an amino group, a hydrogen atom, and a side-chain (or R group), all attached to C-2 (the α-carbon). Solid wedges indicate bonds above the plane of the paper; dashed wedges indicate bonds below the plane of the water. B—Ball- and -stick model of serine (amino acid whose R group is $-CH_2OH$). Note the alternative numbering and lettering system for the carbon atoms (after Horton *et al.*, 2006).

2. Physical Properties of Amino Acids

(*i*) **Appearance.** Amino acids are colourless, crystalline compounds which appear as white powder when in bulk. The crystals of amino acids have different appearance.

(*ii*) **Solubility.** Amino acids are soluble in polar solvents such as water but their solubility varies greatly. The solubility of amino acids is influenced by the presence of neutral salts in the solution. Solubility of amino acids increases in the presence of low salt concentration, but decreases in high concentration of salts. All amino acids are readily soluble in dilute acids and alkalies but are insoluble in nonpolar solvents, such as benzene, ether or alcohol.

(*iii*) **Taste.** Amino acids have different tastes. Proline and hydroxyproline are sweet, valine is bitter sweet, tryptophan and isoleucine are bitter.

(*iv*) **Melting point.** The melting point of amino acids is above 200°C.

(*v*) **Stereoisomerism.** In 19 of the 20 amino acids used for the biosynthesis of proteins, the α-carbon atoms is **chiral**, or asymmetric, since it has four different groups bonded in it. The exception is glycine, whose R group is simply a hydrogen atom (the molecule is not chiral because the α-carbon atom is bonded to two identical hydrogen atoms). The 19 chiral amino acids can therefore exist as stereoisomers. **Stereoisomers** are compounds that have the same molecular formula but differ in the arrangement or configuration, of their atoms in space. The two stereoisomers are distinct molecules that cannot be easily converted from one form to other since a change in configuration requires the breaking of one or more bonds. Amino acid stereoisomers are non-superimposable mirror images. Such stereoisomers are called **enantiomers** (Box 7.1). Two of the 19 chiral amino acids, *isoleucine* and *threonine*, have two chiral carbon atoms each. Isoleucine and threonine can each form four different stereoisomers.

Box 7.1

Enantiomers (Greek, enantios = opposite). These are stereoisomers that are non-superimposable mirror images of each other but are chemically identical in their reactions. They agree in their melting points, solubility, etc., but differ in their ability to rotate the plane of polarized light in a polarimeter; a solution of one of the two enantiomers rotate the plane to the right and a solution of the other to the left. Thus, the 2 forms of glucose (D-glucose and L-glucose) constitute a pair of enantiomers. Enantiomers are also called **enantiomorphs.**

Commonly the mirror-image pairs of amino acids are designated **D** (for **dextro**, from the Latin *dexter*, right) and **L** (for **levo**, from the Latin *laevus*, left). The configuration in Figure 7.3A is L; that of its mirror image is D. To assign the stereochemical designation, one draws the amino acids vertically with its α-carboxylate group at the top and its side chain at the bottom, both pointing away from the viewer. In this orientation, the α-amino group of L isomer is on the left of the α-carbon, and that of the D isomer is on the right, as shown in Figure 7.4.

The 19 chiral amino acids used in the assembly of proteins are all the L configuration, although a few D-amino acids occur in nature. By convention, amino acids assumed to be the L configuration unless specifically designated D.

Evolutionary significance of levorotatory isomer of amino acids. The fact that all living organisms used the same standard amino acids in protein synthesis is evidence that all species on Earth are descended from a common ancestor. Like modern organisms, the last common ancestor (LCA) must have used L-amino acids and not D-amino acids. Mixtures of L- and D- amino acids are formed under conditions that mimic those present when life first arose on Earth 4 billion years ago, and both enantiomers are found in meteorites and in the vicinity of stars. It is not known how or why primitive life forms selected L-amino acids from the presumed mixture of the enantiomers present when life first arose. It is likely that the first proteins were composed of a small number of simple amino acids and selection of L-amino acids over D-amino acids was a chance event. Modern living organisms do not select L-amino acids from a mixture since only L-amino acids are synthesized in sufficient quantities. Thus, the predominance of L-amino acids in modern species is due to the evolution of metabolic pathways that produce L-amino acids and not D-amino acids.

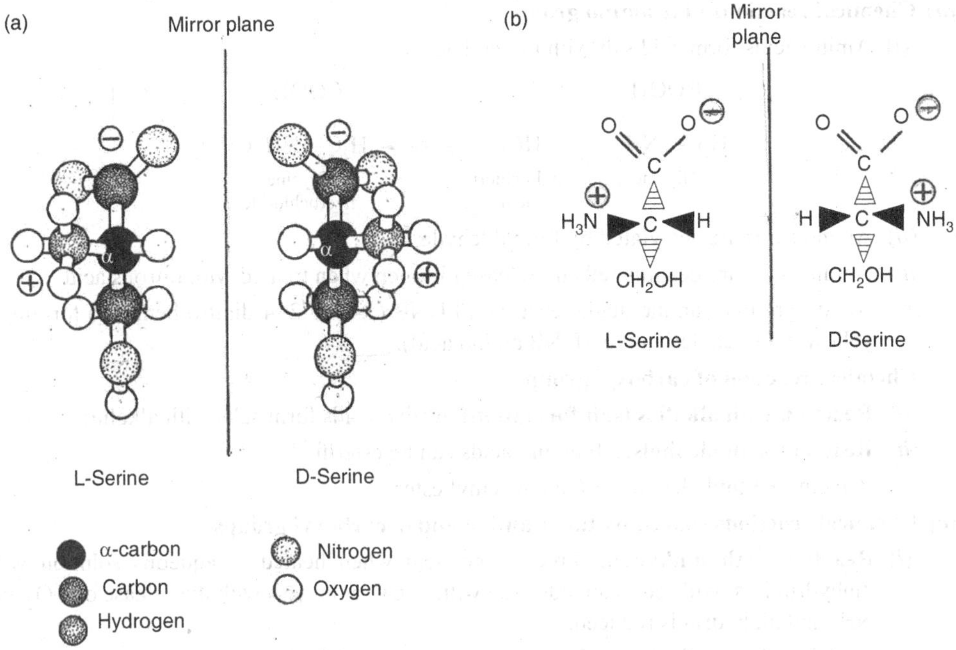

Fig. 7.4. Mirror-image pairs of amino acids. A—Ball-and-stick models of L-serine and D-serine. Note that the two molecules are not identical; they cannot be superimposed. B—L-serine and D-serine.

(*vi*) **Zwitter ions.** Inside a cell, under normal physiological conditions, the amino group is protonated ($-NH_3^+$) because p*Ka* of this group is close to 9. The carboxyl group is ionized ($-COO^-$) because the p*Ka* of the group is below 3 (Table 7.1). Thus, in the physiological pH range of 6.8 to 7.4 amino acids are **zwitter ions**, or **dipolar ions**, even though their net charges may be zero. Like a zwitter ion an amino acid has following characteristics:

(*a*) At its isoelectric pH, each amino acid has an equal number of plus (+) and minus (–) charges.

(*b*) In **acidic solution** an amino acid acts as a base and combines with H^+ ions of the solution.

(*c*) In **alkaline solution** amino acid acts as a acid and combines with OH^- ions of the solution.

(*d*) Zwitter ion is *ampholyte* because it acts as proton donor and proton acceptor.

(*e*) In a pure aqueous solution zwitter ions are not neutral because their acidic and basic strength are different.

3. Chemical Properties of Amino Acids

Amino acids are **amphoteric** having both amino and carboxyl groups (Box 7.2). Their chemical properties can be discussed under following three main headings:

Box 7.2 Amphoteric and Ampholytic Nature of Amino Acids

A compound is called amphoteric if it can react with both acids and bases. Both amino ($-NH_2$) and carboxyl ($-COOH$) groups of amino acids are ionizable and readily form salts either with acids or alkalies. Depending upon H^+ ions concentration of the medium, amino acids behave either negatively or positively charged.

1. **In acidic** medium, amino acids carry positive charge and act as base, yielding cations. In electric field, they migrate to cathode or positive pole.
2. **In alkaline medium**, amino acids carry negative charge and act as acid, yielding anions. In electric field, they migrate to anode or negative pole.

(a) Chemical reactions of α-amino group

(i) Amino acids form acid salt with mineral acids

$$\underset{\text{Glycine}}{\overset{\text{COOH}}{\overset{|}{H_2C{-}NH_2}}} + \underset{\text{Hydrochloric acid}}{HCl} \longrightarrow \underset{\text{Glycine hydrochloride}}{\overset{\text{COOH}}{\overset{|}{H_2C{-}NH_2\,Cl}}}$$

(ii) Amino acids are acetylated by formaldehyde.

(iii) Amino acids are deaminated and release nitrogen when treated with nitrous acid.

(iv) Amino group of amino acids react with FDNB (1-fluro 2, 4-dinitro benzene) forming a yellow coloured derivative (DNB amino acid).

(b) Chemical reaction of carboxyl group

(i) **Reaction with alkalies (salt formation).** Amino acids form salts with alkalies.

(ii) **Reaction with alcohols.** All amino acids can be esterified.

Glycine + Ethyl alcohol → Glycine ethyl ester

(c) Chemical reactions shared by the α-amino and α-carboxyl groups

(i) **Reaction with ninhydrin.** Free amino acid when heated in aqueous solution with ninhydrin it is oxidised to an aldehyde with one C-atom less with the release of CO_2 and NH_3 and ninhydrin is reduced.

Ninhydrin + Glycine → Reduced ninhydrin + aldehyde + CO_2 + NH_3

(ii) **Peptide linkage.** Two amino acids join by the union of α-carboxyl group of one amino acid with the α-amino group of other amino acid forming a **dipeptide** and a molecule of water is eliminated.

Alanine + Glycine → Anylglycine + H_2O

(iii) **Chelation.** Amino acids also acts with ions of certain heavy metals by the process of *chelation* (*i.e.*, the binding of a metal ion to an organic molecule from which it can later be released). Such metalic ions are Cu^{2+}, Co^{2+} and Ca^{2+}.

4. Classification of Amino Acids

Examination of the structures reveals considerable variation in the side chains (*i.e.*, R group) of the 20 amino acids. Some side chains are nonpolar and thus hydrophobic, whereas others are polar or ionized at neutral pH and therefore are hydrophilic. The properties of the side chains greatly influence the overall three-dimensional shape, or conformation of a protein. For example, most of the hydrophobic side chains of a water-soluble protein fold into the interior, giving the protein a compact, globular shape.

Both the three-letter and one-letter abbreviations for each amino acid are shown in figures. The three-letter abbreviation is self-evident, but the simple one-letter abbreviation is less obvious. Several amino acids begin with the same letter so other letters of the alphabet have to be used in order to provide a unique label: for example, threonine = T, tyrosine = Y, and tryptophan = W.

Standard amino acids are grouped by their general properties and the chemical structures of their side chains. The side chains or R groups fall into following chemical classes: 1. Aliphatic; 2. Aromatic; 3. Sulphur-containing; 4. Alcohols; 5. Basic; 6. Acidic and 7. Amides.

A. Aliphatic R Groups

Glycine (Gly, G) (Fig. 7.5) is the smallest amino acid since its R group is simply a hydrogen atom: consequently, the α-carbon of glycine is not chiral. The two hydrogen atoms of the α-carbon of glycine impart little hydrophobic character to the molecule. Glycine plays a unique role in the structure of many proteins becuase its side chain (R group) is small enough to fit into niches that can accommodate no other amino acid.

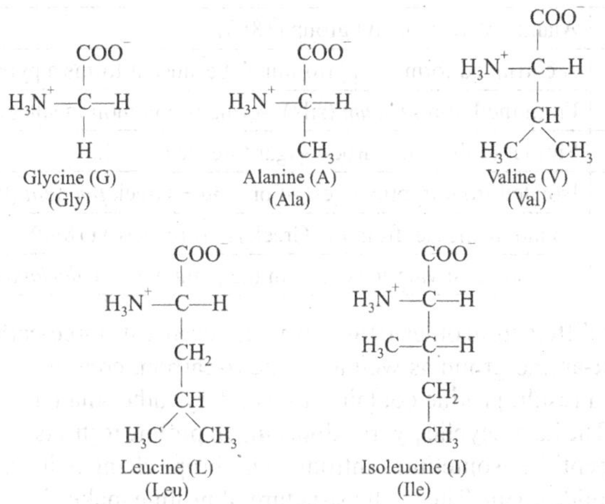

Fig. 7.5. Amino acids with aliphatic (R) groups.

Four amino acids, **alanine (Ala, A), valine (Val, V), leucine (Leu, L)**, and structural isomer of leucine, **isoleucine (Ile, I)**, have saturated aliphatic side chains. The side chain of alanine is a methyl group, whereas valine has a three-carbon branched side chain and leucine and isoleucine each contain a four-carbon branch side chain. Both α- and β-carbon atoms of isoleucine are asymmetric.

Box 7.3

Because isoleucine has two chiral centers, it has four possible stereoisomers. The stereoisomer used in proteins is called L-isoleucine, and the amino acid that differs at the β-carbon is called **L-alloisoleucine.** The other two stereoisomers are **D-isoleucine** and **D-alloleucine**.

Alanine, valine and leucine play an important role in establishing and maintaining the three-dimensional structures of proteins because of their tendency to cluster away from water. Valine, leucine and isoleucine are called **branched-chain amino acids** because their side chains (R groups) of carbon atoms contain branches. All three amino acids are highly hydrophobic.

Table 7.3. Common names of amino acids (*Source:* Horton *et al.*, 2006).

	Amino acid	Derivation of common name of amino acid and year
1.	Alanine	Probably from *a*ldehyde + "an" (for convience) + am*ine* (1849).
2.	Arginine	Crystallizes as a silver salt, from Latin *argentum* (silver) (1886).
3.	Asparagine	First isolated from asparagus (1813).
4.	Aspartate	Similar to asparagine (1836).
5.	Glutamate	In the plant protein gluten (1866).
6.	Glutamine	Similar to glutamate (1866).
7.	Glycine	From the Greek *glykys* (sweet), tastes sweet (1848).
8.	Cysteine	From the Greek *kystis* (bladder), discovered in bladder stones (1882).
9.	Histidine	First isolated from sturgeon sperm, named for the Greek *histidin* (tissue) (1896).
10.	Isoleucine	Isomer of leucine.
11.	Lysine	Product of protein hydrolysis, from the Greek *lysis* (loosening) (1891).

	Amino acid	Derivation of common name of amino acid and year
12.	Methionine	Side chain is a sulphur (Greek *theion*) atom with a *methyl* group (1928).
13.	Phenylalanine	Alanine with a phenyl group (1883).
14.	Proline	A corrupted form of "pyrrolidine" because it forms a pyrrolidine ring (1904).
15.	Serine	From the Latin *sericum* (silk); serine is common in silk (1865).
16.	Threonine	Similar to the four-carbon sugar threose (1936).
17.	Tryptophan	Isolated from tryptic digest of protein + Greek *phanein* (to appear) (1890).
18.	Tyrosine	Found in cheese, from the Greek *tyros* (cheese) (1890).
19.	Valine	Derivative of valeric acid from the plant genus *Valeriana* (1906).

Proline (Pro, P) differs from other 19 amino acids because its three-carbon side chain is bonded to the nitrogen of its α-amino group as well as to the α-carbon, creating a cyclic molecule. As a result, proline contains a secondary rather than a primary amino group. The heterocyclic **pyrrolidine ring** of proline restricts the geometry of polypeptides, sometimes introducing abrupt changes in the direction of the peptide chain. The cyclic structure of proline makes it much less hydrophobic than valine, leucine and isoleucine.

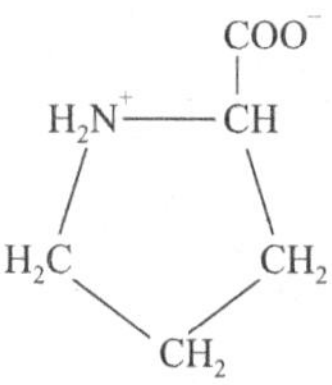

Fig. 7.6. Proline (P) (Pro).

B. Aromatic Groups

Phenylalanine (Phe, F), tyrosine (Tyr, Y), and tryptophan (Trp, W), have side chains with aromatic groups. Phenylalanine has a hydrophobic benzyl side chain. Tyrosine is structurally similar to phenylalanine; the *para* hydrogen of phenylalanine is replaced in tyrosine by a hydroxyl group (–OH) making tyrosine a phenol. The side chain of tryptophan contains a bicyclic indole group. Tyrosine and tryptophan are not as hydrophobic as phenylalanine because their side chains include polar groups.

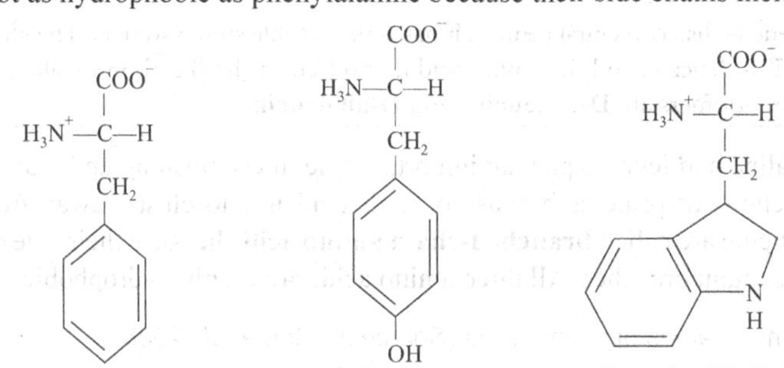

Phenylalanine (F) (Phe) Tyrosine (Y) (Tyr) Tryptophan (W) (Trp)

Fig. 7.7. Amino acids with aromatic R groups.

Box 7.4

All three aromatic amino acids absorb ultraviolet (UV) light. At neutral pH, both tryptophan and tyrosine absorb light at 280 nm. Such absorbance at 280 nm is routinely used to estimate the concentration of proteins in solution.

C. Sulphur-containing R Groups

Methionine (Met, M) and **cysteine (Cys, C)** are the two sulphur-containing amino acids. Methionine contains a nonpolar methyl thioether group in its side chain, and this makes it one of the most hydrophobic amino acids. Methionine plays a special role in protein synthesis since it is almost always the first amino acid in a polypeptide chain. The structure of cysteine resembles that of alanine with a hydrogen atom replaced by sulphydryl group (–SH).

Although the side chain of cysteine is somewhat hydrophobic, it is also highly reactive. Because the sulphur atom is polarizable, the sulphydryl group of cysteine can form weak hydrogen bonds with oxygen and nitrogen. Moreover, the sulphydryl group of cysteine is a weak acid, which allows it to lose its proton to become negatively charged thiolate ion.

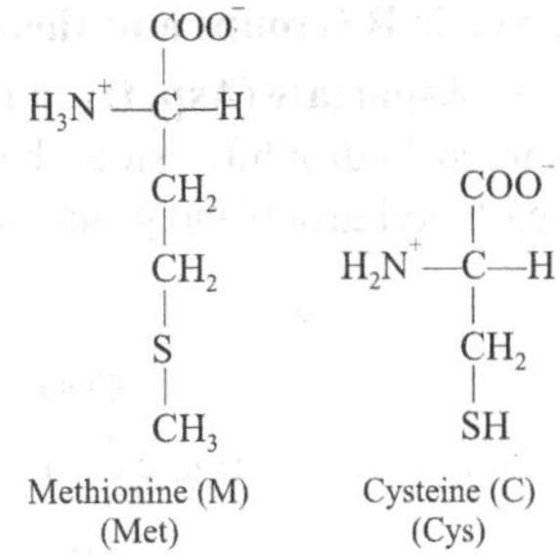

Fig. 7.8. Amino acids with sulphur containing R groups.

A compound called **cystine** can be isolated when some proteins are hydrolyzed. Cystine is formed from two cysteine molecules that are linked by a **disulphide bonds** or **disulphide bridges**. Most proteins do not contain disulphide bonds because conditions inside the cell do not favour oxidation. However, many secreted or extracellular proteins contain disulphide bridges.

D. Side Chains with Alcohol Groups

Serine (Ser, S) and **threonine (Thr, T)** have uncharged polar side chains containing β-hydroxyl groups. These alcohol groups give hydrophilic character to the aliphatic side chains. Unlike the more acidic phenolic side chain of tyrosine, the hydroxyl groups of serine and threonine have the weak ionization properties of primary and secondary alcohols. Threonine, like isoleucine, has two chiral centers, the α- and β-carbon atoms. L-Threonine is the only one of the four stereoisomers that commonly occurs in proteins. (The other stereoisomers are called D-threonine, L-allothreonine and D-allothreonine).

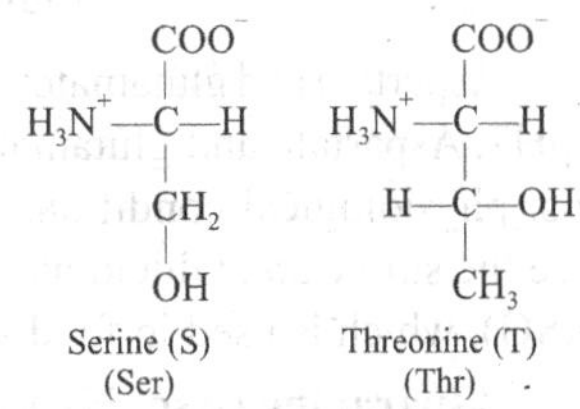

Fig. 7.9. Amino acids having side chains with alcohol groups.

E. Basic R Groups

Histidine (His, H), **lysine (Lys, K)** and **arginine (Arg, R)** have hydrophilic side chains that are nitrogenous bases and are positively charged at pH 7.

The side chains of histidine contains an **imidazole ring** substitute. The protonated form of this ring is called an **imidazolium ion**. Lysine is a diamino acid, having both α- and ε-amino groups. The ε-amino group exists as an alkylammonium ion ($-CH_2-NH_3^+$) at neutral pH and confers positive charge on proteins. Arginine is the most basic of the 20 amino acids because its side-chain guanidinium ion is protonated under all conditions normally found within a cell. Arginine side chains also contribute positive charges to proteins. (Note. Guanidinium ions means protonated form of arginine side chain).

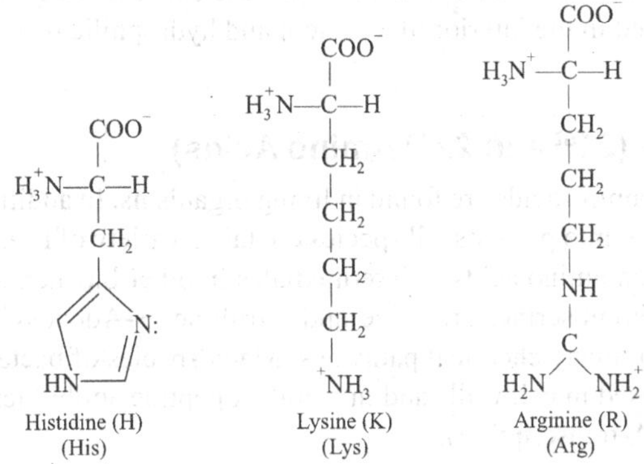

Fig. 7.10. Amino acids with basic R groups.

F. Acidic R Groups and their Amide Derivatives

Aspartate (Asp, D) and **glutamate (Glu, E)** are dicarboxylic amino acids and have negatively charged hydrophilic side chains at pH7. In addition to α-carboxyl groups glutamate possess a β-carboxyl group and glutamate possesses a γ-carboxyl group and their amides.

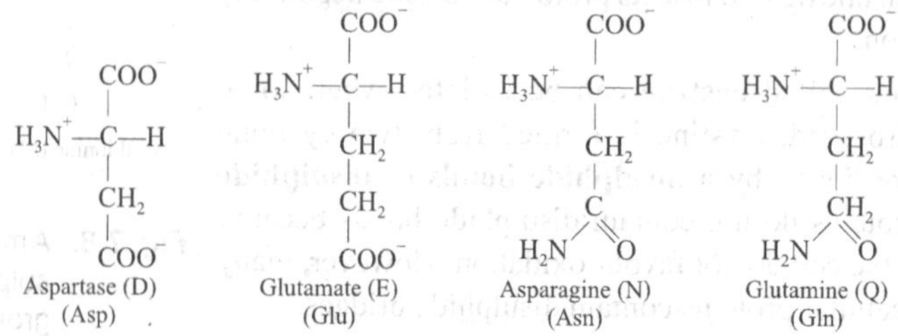

Fig. 7.11. Amino acids with acidic R groups.

Aspartate and glutamate confer negative charges on proteins because their side chains are ionized at pH7. Aspartate and glutamate are sometimes called **aspartic acid** and **glutamic acid**, but under most physiological conditions they are found as the conjugate bases and, like other carboxylates, have the suffix-*ate*. Glutamate is probably familiar as its monosodium salt, monosodium glutamate (MSG), which is used in food as a flavor enhancer.

Asparagine (Asn, N) and **glutamine (Gln, Q)** are the *amides* of aspartic acid and glutamic acid, respectively. Although the side chains of asparagine and glutamine are uncharged, these amino acids are highly polar and often found on the surfaces of proteins, where they can interact with water molecules.

Hydropathy

Various side chains of amino acids range from highly *hydrophobic*, through weakly polar, to highly *hydrophilic*. The relative hydrophobicity or hydrophilicity of each amino acid is called its **hydropathy**. There are several ways of measuring hydropathy, but most of them rely on calculating the tendency of an amino acid to prefer a hydrophobic environment. Example of **highly hydrophobic** amino acids (hydropath range: 3.1 to 1) are *five* amino acids such as isoleucine, phenylalanine, valine, leucine and methionine; **less hydrophibic** amino acids (Hydropathy range: 1.5 to 1.1) include *eight* amino acids such as tryptophan, alanine, glycine, cysteine, tyrosine, proline, threonine and serine. **Highly hydrophilic** amino acids (Hydropathy range: –1.7 to –7.5) include *seven* amino acids such as histidine, glutamate, asparagine, glutamine, aspartate, lysine and arginine.

Hydropathy is an important determinant of protein-chain folding because hydrophobic side chains tend to be clustered in the interior of a protein and hydrophilic residues are usually found on the surface.

Other Amino Acids (21st and 22nd Amino Acids)

More than 200 different amino acids are found in living organisms. In addition to the standard amino acids that are incorporated into proteins, all species contain a variety of L-amino acids that are either precursors of the common amino acids or intermediates in other biochemical pathways. Examples include homocysteine, homoserine, ornithine and citrulline. S-Adenosylmethionine (SAM) is a common methyl donor in many chemical pathways. Many species of bacteria and fungi synthesize D-amino acids that are used in cell walls and in complex peptide antibiotics such as actinomycin D (*i.e.*, an inhibitor of DNA-transcription).

Several common amino acids are chemically modified to produce biologically important amines. These are synthesized by enzyme-catalyzed reactions that include decarboxylation and deamination. In the mammalian brain, for example, glutamate is converted to the neurotransmitter γ-amino butyrate (GABA). Mammals can also synthesize **histamine** from histidine. Histamine controls the constriction of certain blood vessels and also the secretion of hydrochloric acid by the stomach. In the adrenal medulla, tyrosine is metabolized to **epinephrine** or **adrenaline**. Epinephrine and its precursor, norepinephrine are hormones that help regulate metabolism in mammals. Tyrosine is also the precursor of the thyroid hormones such as **thyroxine** and **triiodothyronine**.

Some amino acids are chemically modified after they have been incorporated into polypeptides. For example, some proline residues in the protein collagen are oxidized to form **hydroxyproline** residues. Another common modification is the addition of complex carbohydrate chains, a process called **glycosylation**. Many proteins are **phosphorylated** by the addition of phosphoryl groups to the side chains of serine, threonine or tyrosine. The oxidation of pairs of cysteine residues to form cystine also occurs after a polypeptide has been synthesized. In bacteria, the first amino acid in a protein is usually methionine, which is modified by the addition of a formyl group to form *N*-formylmethionine.

An unexpected discovery was that a *21st amino acid*, **selenocysteine** (which contains selenium in place of the sulphur of cysteine), is incorporated into a few proteins in a wide variety of species. Selenocysteine is formed from *serine* during protein synthesis. The *22nd amino acid* is **pyrrolysine**, found in some species of archaebacteria. Pyrrolysine is a modified form of lysine that is synthesized before being added to a growing polypeptide chain by the translation machinery. Selenocysteine and pyrrolysine have their own codons, which is why they are considered additions to the standard reportoir of protein precursors.

5. Functions of Amino Acids

1. **Formation of proteins.** Amino acids are linked into long chains through their peptide linkage forming into polypeptides and proteins. Proteins form structural and functional components of the cell.
2. **Formation of other compounds.** Some amino acids give rise to other compounds, such as
   (*i*) **Tyrosine** is part of hormone thyroxine and adrenaline.
   (*ii*) **Glycine** participates in the synthesis of heme.
   (*iii*) **Tryptophan** produces vitamin nicotinamide and plant hormone indole-3-acetic acid (IAA).
   (*iv*) Amino acid **β-alanine** is needed in the synthesis of coenzyme-A and vitamin pantothenic acid.
3. **Formation of antibiotics.** Nonprotein amino acids are components of antibiotics.
4. **Formation of enkephalins.** These are small peptides (pentapeptides) formed in certain part of brain. These are associated with perception of pain and pleasure.
5. **Biological buffers.** Because of the amphoteric nature amino acids act as buffers in the body fluids. They maintain pH either by donating H^+ ions when pH increases or by accepting H^+ ions when pH falls.
6. **Synthesis of glucose.** When glucose level in blood comes down, glucose is synthesized from amino acids in the hepatic cells.
7. **Formation of histamine.** Amines in the body are derived from some amino acids by the loss of carboxyl group. For example histamine is derived from amino acid histidine. Histamine is an important vasodilator. It also causes constriction of bronchiole smooth muscles and stimulate a gastric secretion.
8. **Prokaryotic cell wall.** Some peptides are components of **peptidoglycans** which form bacterial cell wall.

7.3. PEPTIDES

A **peptide** is a chain of amino acids linked by **peptide bond** or **amide bond**. The role of peptide bond in protein structure was understood by **Hofmeister** (1902). Peptide bond is a special type of amide bond formed between carboxyl (–COOH) group of one amino acid and amino (–NH_2) group of next amino acid with the removal of one molecule of water.

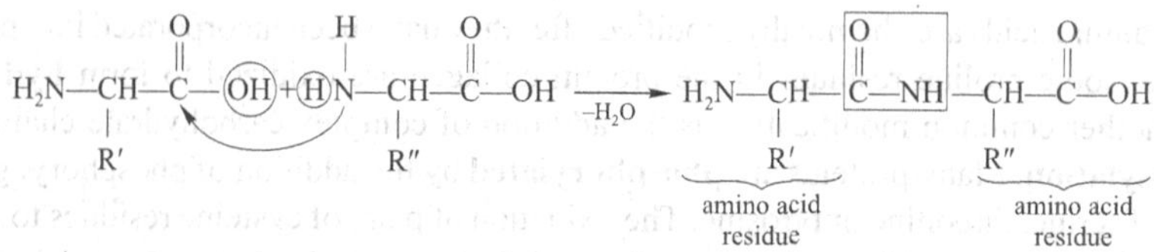

Fig. 7.12. Formation of a peptide bond or amide linkage to join two molecules of amino acids.

A peptide chain may have upto millions of amino acids. Each amino acid in the chain is called a **residue**. Depending on the number of amino acid residues, the peptides are called:

1. **Dipeptide.** A chain of two amino acids.
2. **Tripeptide.** A chain of three amino acids.
3. **Oligopeptide.** A chain of amino acid upto 10 amino acids.
4. **Polypeptide.** A peptide containing more than 10 amino acids.
5. **Macropeptide.** A polypeptide chain of more than 100 amino acids (Box 7.5).

Box 7.5

Strictly speaking, the **proteins** are polypeptides with more than 100 amino acids. The simplest peptide is glycylglicine. All naturally-occurring important peptides, however, possess a shorter individual name, such as tripeptide glutathione which consists of 3 amino acids—glutamic acid, cysteine and glycine. (Glutathione acts as sulphydryl buffer that maintains the cysteine residues of haemoglobin and other proteins of RBCs in the redued state. It also has a role in detoxifying H_2O_2 and organic peroxides, the harmful products of aerobic life.) The two pituitary hormones—oxytocin and vasopressin—are nonapeptides (having 9 amino acid residues). Peptides arise as partial hydrolysis products of proteins, e.g., during digestion (**Jain** 2012).

1. Primary Structure of Peptides

The sequence of amino acids in a polypeptide chain is referred to as its **primary structure**. The two ends of peptide chains are marked, the **N-terminal** on the left and **C-terminal** at the right end of the polypeptide. The N-terminal has amino (–NH_2) group of left terminal amino acid and is basic, C-terminal has carboxyl (–COOH) group of the last amino acid and is acidic. These two terminal groups (one basic and one acidic are ionisable).

The primary structure of a polypeptide is very important. The smallest change of even single amino acid in the primary structure often produces remarkable physiological effect. For example, substitution of a single amino acid in the sequence of 100 or more amino acids may change the biological activity having serious consequences.

2. Biological Importance of Peptides

1. Peptides are intermediates in the formation of proteins.
2. Certain peptides serve as growth factors, *e.g.*, folic acid. **Streptogenins** is another peptide which serves as a growth factor for microorganisms.

3. Certain peptides act as hormones, *e.g.*, oxytocin, vasopressin, prolactin, etc.
4. Certain peptides such as **glutathione** participate in controlling oxidation reduction potential.
5. Many peptides such as **penicilin G** have antibiotic activities.
6. Some peptides are constituents of alkaloids. For example, **ergotamine** is a peptide alkaloid from rye ergot (fungi).
7. Some peptides introduce psychiatric disturbances in human beings.

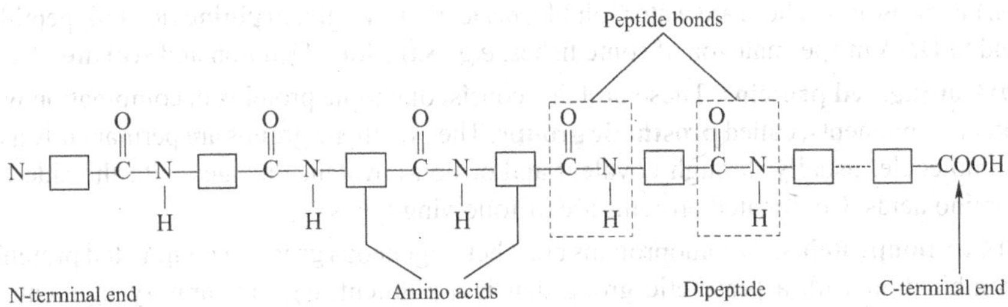

Fig. 7.13. A peptide chain showing peptide bonds between amino acids.

7.4. PROTEINS

Proteins are the most abundant organic molecules of the living system. They occupy a central position in the architecture and functioning of living matter. On this basis, proteins can be broadly grouped as **static** or **structural proteins** (*e.g.*, collagen, elastin, α-Keratin) and **dynamic** or **functional proteins** (*e.g.*, enzymes, hormones, blood clotting factors, immunoglobulins, membrane receptors, storage proteins, motor proteins, etc.).

I. Types of Proteins

A. Classification based on shape of proteins. According to the shape or conformation, two major types of proteins have been recognized:

(*a*) **Fibrous proteins.** Fibrous proteins are water-insoluble, thread-like proteins having greater length than their diameter. They contain secondary protein structure and occur in those cellular or extracellular structures, where strength, elasticity and rigidity are required, *e.g.*, collagen, elastin, keratin, fibrin (blood-clot proteins) and myosin (muscle contractile proteins).

(*b*) **Globular proteins.** Globular proteins are water-soluble roughly spheroidal or ovoidal in shape. They readily go into colloidal suspension. They have tertiary protein structure and are usually functional proteins, *e.g.*, enzymes, hormones and immunoglobulins (antibodies). **Actin** of microfilaments and **tubulins** of microtubules are also globular proteins.

B. Classification based on solubility characteristics. According to this criterion proteins can be classified into two main types:

(*a*) **Simple proteins.** These proteins contain only amino acids in their molecules and they are of following types:

(*i*) **Albumins.** These are water soluble proteins found in all body cells and also in blood stream, *e.g.*, **lactalbumin**, found in milk and **serum albumin** found in blood.

(*ii*) **Globulins.** These are insoluble in water but are soluble in dilute salt solutions of strong acids and bases, *e.g.*, **lactoglobulin** found in milk and **ovoglobulin**.

(*iii*) **Glutelins.** These plant proteins are soluble in dilute acids and alkalis, *e.g.*, **glutenin** of wheat.

(*iv*) **Prolamines.** These plant proteins are soluble in 70 to 80 per cent alcohol, *e.g.*, **gliadin** of wheat and **zein** of corn.

(*v*) **Scleroproteins.** They are insoluble in all neutral solvents and in dilute alkalis and acids, *e.g.*, keratin and collagen.

(*vi*) **Histones.** These are water soluble proteins which are rich in basic amino acids such as arginine and lysine. In eukaryotes histones are associated with DNA of chromosomes to form nucleoproteins.

(*vii*) **Protamines.** These are water soluble, basic, light weight, **arginine** rich polypeptides. They are bound to DNA in spermatozoa of some fishes, *e.g.*, **salmine** of salmon and **sturine** of sturgeons.

(*b*) **Conjugated proteins.** These proteins consist of simple proteins in combination with some non-protein components, called **prosthetic groups**. The prosthetic groups are permanently associated with the molecule, usually through covalent and/or non-covalent linkages with the side chains of certain amino acids. Conjugated proteins are of following types:

(*i*) **Chromoproteins.** Chromoproteins are a heterogeneous group of conjugated proteins which are in combination with a prosthetic group that is a pigment, *e.g.*, respiratory pigments such as **haemoglobin**, **myoglobin** and **haemocyanin**; **catalase**, **cytochromes**, **haemerythrins**; visual purple or **rhodopsin** of rods of retina of eye and yellow enzymes or **flavoproteins.**

(*ii*) **Glycoproteins.** Glycoproteins are proteins that contain various amounts (1 to 85 per cent) of carbohydrates. Of the known 100 monosaccharides, only nine are found to occur as regular constituents of glycoproteins (*e.g.*, glucose, galactose, mannose, fucose, acetylglucosamine, acetylgalactosamine, acetylneuraminic acid, arabinose and xylose). Glycoproteins are of two main types: **1. Intracellular glycoproteins** which are present in cell membranes and have an important role in membrane interaction and recognition. They also serve as antigenic determinants and receptor sites. **2. Secretory glycoproteins** are **plasma glycoproteins** secreted by the liver; **thyroglobulin**, secreted by the thyroid gland; **immunoglobulins** secreted by the plasma cells; **ovoalbumins** secreted by the cells of oviduct of hen; **ribonucleases** and **deoxyribonucleases**. **Mucus** and **synovial fluid** are also glycoproteins with lubricative properties.

(*iii*) **Lipoproteins.** Lipid containing proteins are called **lipoproteins**. Their lipid contents are 40 to 90 per cent of their molecular weight and this tends to affect the density of the molecule. There are four types of lipoproteins: **1. High density lipoproteins (HDL)** or α-lipoproteins; **2. Low density lipoproteins (LDL)** or β-lipoprotiens; **3. Very low density lipoproteins (VLDL)** or pre-β-lipoproteins; and **4. Chylomicrons.** Lipoproteins include some of the blood plasma proteins, various types of membrane proteins, lipovitellin of egg yolk and proteins of brain and nerve tissue.

(*iv*) **Nucleoproteins.** Nucleoproteins are proteins in combination with nucleic acids (DNA and RNA). However, these proteins are not true conjugated proteins since the nucleic acid involved cannot be regarded as prosthetic groups. Nucleoproteins are of two types: **1. Histones** which are quite similar in all plants and animals. Their highly basic nature accounts for the close associations histones form with the nucleic acids. Histones are involved in the tight packing of DNA molecules during the condensation of chromatin into chromosomes for the mitosis. **2. Nonhistones** have great heterogeneous amino acid composition and are acidic in nature. They have selective combination with certain stretches of nuclear DNA and, thus, are involved in the regulation of gene expression.

(*v*) **Metalloproteins.** Metalloproteins are proteins conjugated to metal ions which are not part of the prosthetic group, *e.g.*, **carbonic anhydrase** enzyme contains zinc ions and amino acids in its molecule; **caeruloplasmin**, an oxidase enzyme containing copper; and **siderophilin** contains iron.

(*vi*) **Phosphoproteins.** Phosphoproteins are proteins in combination with a phosphate group, e.g., casein of the milk and ovovitellin of eggs.

II. Formation of Proteins

Because a molecule of the amino acid contains both basic or amino ($-NH_2$) and acidic or carboxyl ($-COOH$) group, it can behave as an acid and base at a time. The molecules of such organic compounds which contain both acidic and basic properties are known as **amphoteric molecules**. Due to amphoteric molecules, the amino acids unite with one another to form complex and large protein molecules. When two molecules of amino acids are combined then the basic group ($-NH_2$) of one amino acid molecule combines with the carboxylic ($-COOH$) group of other amino acid and the loss of a water molecule takes place. This sort of condensation of two amino acid molecules by –NH–CO linkage or bond is known as **peptide linkage** or **peptide bond**. A combination of two amino acids by the peptide bond is known as **dipeptide**. When three amino acids are united by two peptide bonds, they form **tripeptide**. Likewise, by condensation of few or many amino acids by the peptide bonds the **oligopeptides** and **polypeptides** are formed respectively. The various molecules of polypeptides unite to form the **peptones, proteases** and **proteins**. Thus, protein macromolecules are the polymers of many amino acid monomers. The size (molecular weight), shape, and function of proteins are determined by the number, type and distribution of the amino acids present in the molecule. Proteins occur in a wide spectrum of molecular sizes from small molecules such as the hormone ACTH (or adrenocorticotrophic hormone) which consists of only 39 amino acids and has a molecular weight of 4500, to extremely large proteins such as haemocyanin (an invertebrate blood pigment) which consists of 8200 amino acids and has a molecular weight greater than 900,000.

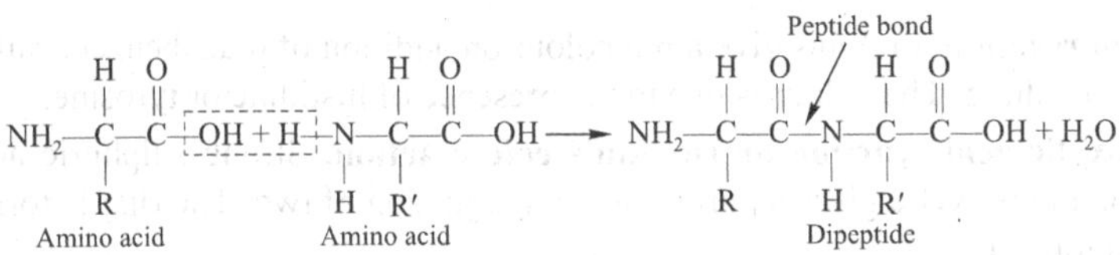

Fig. 7.14. A chemical reaction showing formation of a dipeptide.

III. General Properties of Proteins

Proteins exhibit following general properties:

1. Physical properties

***(i)* Colloid.** Proteins exist in colloidal state. Their molecules are of large size. Therefore, these are unable to diffuse through plasma membrane.

***(ii)* Solubility.** Because proteins are colloids of large-sized molecules, these form turbid solution in water. Proteins are insoluble in alcohol. They are precipitated by acids in certain concentration.

2. Chemical properties

***(i)* Amphoteric nature.** Proteins are **amphoteric**. These behave as acid to alkaline solutions and alkaline to acidic solutions and form salts with them.

***(ii)* Coagulation.** On heating proteins are coagulated but temperature for coagulation differs for different proteins.

***(iii)* Optical properties.** It is already mentioned that amino acids are optically active except glycine. These are mostly **laevorotatory**. Some of the proteins, haemoglobin and nucleoproteins are dextrorotory.

***(iv)* Hydrolysis.** When proteins are boiled with dilute mineral acids in a reflux condenser, the protein molecules gradually break up into simple ones until these are reduced to amino acids. The process of hydrolysis of proteins involves the following steps:

$$\text{Proteins} \rightarrow \text{Proteoses} \rightarrow \text{Peptones} \rightarrow \text{Amino acids}$$

(*v*) **Denaturation.** Denaturation of protein is the disruption of secondary, tertiary and quaternary structure of the functional protein resulting in the changes of its physical, chemical and biological characteristics. Various physical and chemical agents such as heat, UV light, ultrasonic waves, high pressure, acids, alkalies, urea, guanidine, ethanol and detergents can cause denaturation of proteins.

The denatured protein becomes insoluble and precipitate out. Their digestability by proteolytic enzymes increases, but their enzymatic or antibody functions are altered.

(*vi*) Colour reactions

(*a*) **Biuret reaction.** All proteins give **biuret test** with caustic soda and a drop of dilute copper sulphate giving pinkish violet colour.

(*b*) **Millon's reagent.** Mercury dissolved in strong nitric acid gives a white precipitate, which turns red on heating. The reaction occurs on account of amino acid tyrosine.

(*c*) **Rosenheim's reaction.** Add a drop of formaldehyde solution to one volume of protein and shake thoroughly. Dilut with one volume of water. When concentrated sulphuric acid, is added slowly a beautiful purple colour appears at the junction. If a drop of ferric chloride solution is added to it the colour is accentuated.

(*d*) **Xanthoproteic reaction.** Strong nitric acid (HNO_3) gives a white precipitate which turns yellow. The yellow colour changes to orange on addition of alkali. This is due to the presence of phenyl radical.

(*e*) **Diazo reaction.** Proteins give a red colour on addition of diazobenzene sulphonic acid in a mild alkaline medium. The colour is due to the presence of histidine or tyrosine.

(*f*) **Glyoxylic acid reaction or Hopkin's cole reaction.** Strong sulphuric acid added to a mixture of protein glyoxylic gives a purple ring at the juction of two. The ring is formed due to the reaction of tryptophan.

IV. Structural Levels of Proteins

The protein as synthesized on the ribosome is a linear sequence of amino acids, polymerized by the elimination of water between successive amino acids to form the peptide bond, and existing as a randomly coiled chain without specific shape and possessing no biological (*i.e.*, catalytic) acitivity. Within seconds of synthesis being completed, the protein folds into a specific three-dimensional form, which is the same for all molecules of the same type of protein and which now is capable of doing catalysis. According to their mode of folding the following four levels of protein organization have been recognized (Fig. 7.15):

(*a*) **Primary structure of protein**, *e.g.*, insulin.

(*b*) **Secondary structure of protein**, *e.g.*, α-Keratin, collagen, globular proteins.

(*c*) **Tertiary structure of protein**, *e.g.*, myoglobin.

(*d*) **Quaternary structure of protein**, *e.g.*, haemoglobin.

(*a*) **Primary protein structure.** The primary protein structure is defined as the particular sequence of amino acids found in the protein. It is determined by the covalent peptide bondings between amino acids. Primary structure also includes other covalent linkages in proteins, for example the linkages that may exist between sulphur atoms of cysteine amino acids located in the chain of the protein insulin. The first protein to have its primary structure determined was of insulin, the pancreatic hormone that regulates glucose metabolism in mammals. Insulin has a molecular weight of 5,800 daltons and contains 51 amino acids. Insulin consists of two polypeptide chains of 21 and 30 amino acid residues, called the A and B chains, respectively (Fig. 7.16). (An amino acid residue is that which is left when the elements of water are split out during polymerization).

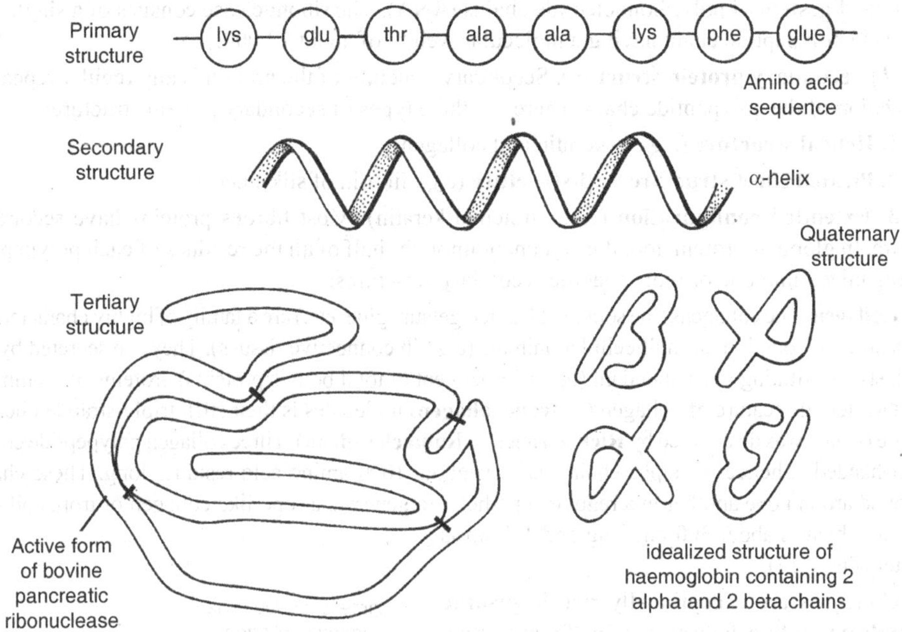

Fig. 7.15. The four levels of structural organization in protein molecules.

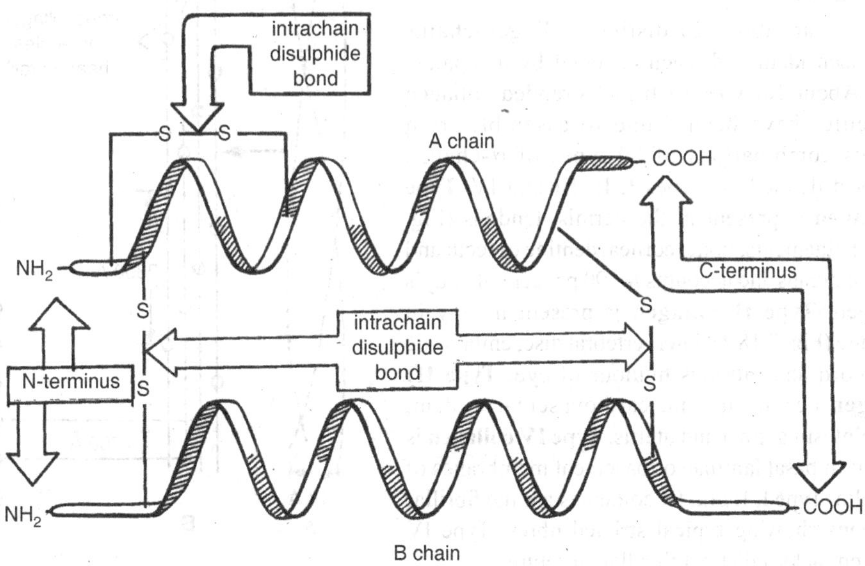

Fig. 7.16. Molecular structure of insulin (After Sheeler and Bianchi, 1987).

Since the elucidation of the primary structure of insulin in 1953 by **F. Sanger** (for which Sanger received a Nobel Prize), several hundred proteins have been fully sequenced. Among the fully sequenced proteins are ribonuclease and nearly 100 types of haemoglobin. For example; **Stein** and his coworkers established the amino acid sequence (*i.e.*, primary structure) of the enzyme **ribonuclease**. This enzyme is produced by the pancreas and secreted into the small intestine where it catalyzes the hydrolytic digestion of polyribonucleotide chains (RNA). The ribonuclease consists of a single 124 amino acid polypeptide chain having a molecular weight of about 12,000.

(*b*) **Secondary protein structure.** Secondary structure of the protein is any regular repeating organization of the polypeptide chain. There are three types of secondary protein structure:

1. **Helical structure** (*e.g.*, α-keratin and collagen);

2. **Pleated sheet structure or β-structure** (*e.g.*, fibroin of silk); and

3. **Extended configuration** (*e.g.*, stretched keratin). Most fibrous proteins have secondary structure. In globular protein, too, it is not uncommon for half of all the residues of each polypeptide to be organized into one or more specific secondary structures.

Collagen. The collagens (the source of leather, gelatin, glue, etc.) are a family of highly characteristic fibrous proteins found in all multicellular animals (*e.g.*, in connective tissues). They are secreted by the fibroblasts constituting most abundant (up to 25 per cent of total body's proteins) proteins of mammals. The characteristic feature of collagen (or **tropocollagen**) molecules is their stiff, triple-stranded helical structure (which was discovered by **Rich**, **Crick** and **Ramachandran**). Three collagen polypeptide chains are left-handed α-helices or alpha chains, each is about 1000 amino acid residues long. These chains are wound around one another in a regular superhelix to generate a rope-like collagen or tropocollagen molecule which is about 300 nm long and 1.5 nm is diameter (Fig. 7.17).

Collagens are exceptionally rich in **proline** (and **hydroxyproline**; both accounting for more than 20 percent of collagen's amino acids) and **glycine**. Other dominant amino acids of collagens are **lysine** and **alanine**.

So far, about 20 distinct collagen-chains have been identified, each encoded by a separate gene. About 10 types of triplet-stranded collagen molecules have been found to assemble from various combinations of 20 types of α-chains. The best defined are types I, II, III and IV. **Type I collagen** is present in the dermis, tendons (Fig. 7.18A), ligaments, bone, cornea, dentine of teeth and internal organs and accounts for 90 per cent of body's collagen. **Type II collagen** is present mainly in cartilage (Fig. 7.18B), intervertebral disc, embryonic notochord and vitreous humour of eye. **Type III collagen** occurs in skin, cardiovascular system, gastrointestinal tract and uterus. **Type IV collagen** is present in basal laminae or basement membranes of epithelia. Type I, II and III collagens are the fibrillar collagens showing typical striated fibres. Type IV collagen lacks a distinct fibrillar structure.

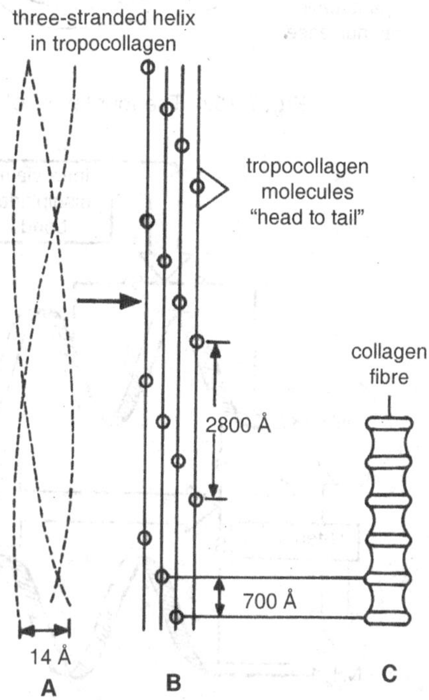

Fig. 7.17. A—A tropocollagen molecule (collagen or superhelix) with three intertwined left-handed α-helices; B—Staggered arrangement of tropocollagen molecules (super helices) in a collagen fibril; C—Striated appearance of a collagen fibril under the electron microscope (after Dyson, 1978).

The individual collagen polypeptide chains (α-chains) are synthesized on membrane bound ribosomes and injected in the lumen of ER as larger precursors, called pro-α-chains. These precursors have distinct polarity, containing terminal propeptides at their *N*-and *C*-terminus. Each pro-α-chain then combines with two others to form a hydrogen-bonded, triple stranded helical molecule, called **procollagen**. Such a procollagen of fibrillar collagens (I, II, and III) is serceted from the fibroblast in the extracellular space. Due to enzyme action its terminal propeptides are removed and the procollagen is converted into a **tropocollagen molecule**. Many tropocollagen molecules spontaneously assemble into the ordered arrays, called **collagen fibrils**. The collagen fibrils are thin (10 to 300 nm in diameter), cable-like structures, many micrometres long, exhibiting cross-striations every 67 nm and are clearly visible in the electron micrographs. The collagen fibrils often aggregate into larger bundles which can be seen in light microscope as **collagen fibres**. Type IV collagen molecules assemble to form a sheet-like **meshwork** that constitutes a major part of all basal laminae (**Martin** *et al.*, 1985, **Burgeson**, 1988).

***(c)* Tertiary protein structure.** Tertiary protein structure refers to a more compact structure in which the helical and non-helical regions of a polypeptide chain are folded back on themselves. This structure is typical of globular protein structure, in which it is the non-helical region that permits the folding. The folding of a polypeptide chain is not random but occurs in a specific fashion, thereby imparting certain steric (*i.e.*, three-dimensional) properties to the protein. For example, in enzymes folding brings together **active amino acids**, which are otherwise scattered along the chain, and may form a distinctive cavity or cleft in which the substrate is bound.

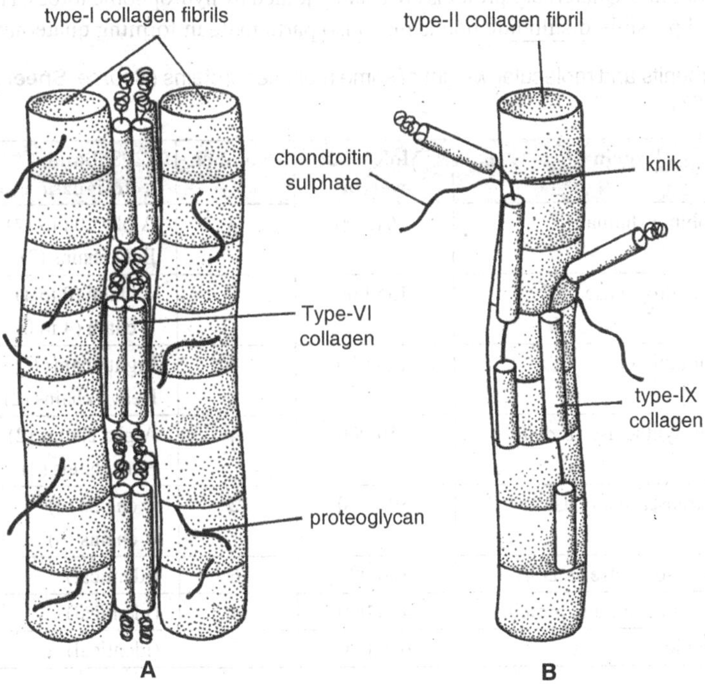

Fig. 7.18. Interactions of fibrous collagens with nonfibrous collagens. A—In tendons, type I collagen fibrils are associated with type-VI collagen. B—In cartilage, type II collagen fibrils are associated with type IX collagen (after Lodish *et al.*, 2004).

The complete tertiary structure of a protein can only be deducted by a laborious analysis of X-ray scattering patterns from crystals. The first protein to have its secondary and tertiary structure determined was **myoglobin** (Fig. 7.21B), a 153-amino acid, oxygen-binding protein found primarily in red muscle and largely responsible for the colour of that tissue. The work was done at Cambridge under the direction of **J.C. Kendrew** (1961). Although at some points the polypeptide chain does have secondary structure (alpha-helical structure), the chain is mainly characterized by seemingly random loops and folds.

In a tertiary protein the polypeptide chain is held in position by weak secondary bonds which are of different type such as **ionic bonds** (or electrostatic bonds or salt or salt bridges); **hydrogen bonds; hydrophobic bonds** and **disulphide bonds.**

(*d*) Quaternary protein structure. In proteins that are composed of two or more polypeptide chains, the quaternary structure refers to the specific orientation of these chains with respect to one another and the nature of the interactions that stabilize this orientation. The individual polypeptide chains of the protein are called **sub-units** and the active protein itself is called **multimer**. While multimeric proteins containing up to 32 subunits have been described, the most common multimers are **dimers, trimers, tetramers, pentamers** (*e.g.*, RNA polymerase) and **decamers** (*e.g.*, DNA polymerase III) (Table 7.4). If the protein consists of identical sub-units, it is called **homopolymer** and is said to have **homogeneous quaternary structure,** *e.g.*, the isozymes H_4 and M_4 of lactic dehydrogenase (LDH), enzyme phosphorylase and L-arabinose isomerase. The enzyme β-galactosidase, consists of four identical polypeptide chains. Lastly, when the sub-units of the protein are different, the protein is called **heteropolymer** and is said to have a **heterogeneous quaternary structure,** *e.g.*, haemoglobin and immunoglobulins. Quaternary proteins are usually joined by hydrophobic forces. Hydrogen bonds, ionic bonds and possibly disulphide bonds may also participate in forming quaternary structures.

Table 7.4. Subunits and molecular weight of some multimer proteins (*Source:* **Sheeler** and **Bianchi**, 1987).

	Protein	Molecular weight	Number	Subunits designation	Molecular weight
1.	Haemoglobin A (human)	64,500	4	Alpha chains (2) Beta chains (2)	15,700 16,500
2.	Lactate dehydrogenase	135,000	4	A chain (0 to 4) B chain (4 to 0)	33,600 33,600
3.	Immunoglobulin G	150,000	4	Light chains (2) Heavy chains (2)	25,000 50,000
4.	Tryptophan synthetase (*E.coli*)	150,000	4	Alpha chains (2) Beta chains (2)	29,500 45,000
5.	Aspartate transcarbamylase	306,000	12	C chains (6) R chains (6)	34,000 17,000
6.	L-arabinose isomerase (*E.coli*)	360,000	6	(identical)	60,000
7.	Apoferritin (iron storage protein)	456,000	24	(identical)	19,000
8.	Thyroglobulin	670,000	2	(identical)	3,35,000

Some Examples of Tertiary and Quaternary Proteins

(*i*) Ribonuclease. C.B. Anfinsen initiated and confirmed the notion that, acting in concert, the specific primary structure of a polypeptide and the innate properties of the side chains of its amino acids cause the polypeptide to spontaneously assume its biologically active tertiary structure. In 1972, he got the Nobel Prize for this definitive work. **Anfinsen** identified four disulphide bridges in the ribonuclease protein, suggesting that the enzyme is highly folded. As is the case with almost all enzymes, the catalytic activity of ribonuclease depends on the maintenance of a particular three-dimensional shape. In concentrated solutions of β-mercaptoethanol and urea, the disulphide bridges of the enzyme are broken and the resultant unfolding of the polypeptide chain is accompanied by a loss of enzyme activity. The enzyme is said to be **denatured**. If the β-mercaptoethanol and urea are removed by dialysis and the denatured ribonuclease reacted with oxygen, the four disulphide bridges re-form spontaneously, and essentially all the catalytic activity of the protein is restored. Similar observations have been made with other proteins, that is, they are capable of spontaneously reestablishing their biologically active tertiary (or even quaternary, *e.g.*, haemoglobin) structure after having undergone extensive molecular disorganization.

***(ii)* Haemoglobin.** Haemoglobin is one of the fully sequenced protein. Our present understanding of the structure and function of haemoglobin is the outcome of 50 years of research of **M.F. Perutz**. He got the Nobel Prize in 1962, along with **J.C. Kendrew**, for their studies of haemoglobin and myoglobin.

The haemoglobin (Fig. 7.21A) is a conjugated globular protein, that is, it contains some non-protein part. In all but the lowest vertebrates, haemoglobin is a tetramer (a heteropolymer). In lampreys, however, haemoglobin is monomeric, that is, it contains a single globin chain like the myoglobin. In humans, most common type of haemoglobin is haemoglobin A (HbA), which consists of 574 amino acid residues and has a molecular weight of 64,500. Its secondary, tertiary and quaternary structure is typical of all higher vertebrate haemoglobins. The protein portion of the haemoglobin molecule, called **globin**, is composed, of four polypeptide chains, each of which is also globular in shape. The four globin chains consist of, two identical pairs: two **alpha chains** (141 amino acids each) and two **beta chains** (146 amino acids each). The non-protein portion of haemoglobin consists of four iron-containing haem groups, one associated with each of the four globin chains. Nineteen of the twenty biologically important amino acids are included in the globin of haemoglobin.

Haemoglobin molecule is highly symmetric; it can be divided into two identical halves, each consisting of an αβ-dimer. The complete tetramer is similar to a mildly flattened sphere having a maximum diameter of about 5.5 nm. The four polypeptide chains are arranged in such a way that unlike chains have numerous stabilizing interactions, whereas, like chains have few. A cavity about 2.5 nm long and varying in width from about 5 to 10 Å passes through the molecule along the axis. Each globin chain envelops its haem group in a deep cleft.

***(iii)* Immunoglobulins.** The ability of animals to resist infection by pathogens (viruses, bacteria and other unicellular parasites) and by multicellular endoparasites, is called **immunity**. Specific immune response function by recognizing particular chemical structures, known as **antigens**—on the surface of invading cells. An antigen can be a protein, lipid, carbohydrate or any other molecule. These antigens interact with protein molecules produced by the host, the **immunoglobulins**, which bind the antigen in much the same way as an enzyme binds its substrate. Specific immune responses involve many different types of cells. One type, the **B-lymphocyte** or **B-cell**, is capable of producing **free immunoglobulins**, called **antibodies**.

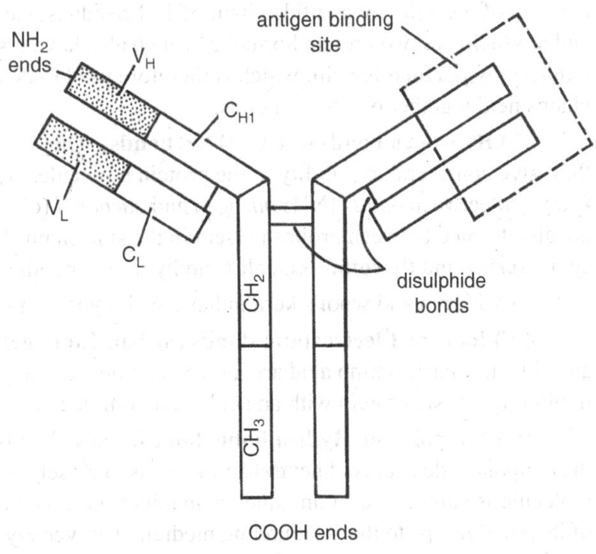

Fig. 7.19. Molecular structure of an immunoglobulin molecule (after Stansfield, 1986).

An immunoglobulin (Ig) molecule is a Y-shaped hetero-polymer and is composed of two identical **H (heavy)** polypeptide chains and two smaller identical **L (Light)** polypeptide chains (Fig. 7.19). Heavy chains contain antigenic determinants in the "tail" (carboxyl) segments by which they can be classified as IgG, IgM, IgA, IgD or IgE. Light chains can likewise be typed as kappa or lambda. Within a H chain class or L chain type, these segments exhibit very little variation in primary structure from one individual to another and are called **constant regions (C)**. The amino ($-NH_2$) ends, however, are extremely diverse in primary structure, even within a class and are called **variable (V) regions**. The V_H and V_L regions together form two **antibody-combining sites** (called **antigen-binding sites**) for specific interactions with homologous antigen molecules. The C_H region consists of three or four similar segments, presumably derived evolutionarily by duplication of an ancestral gene and subsequent modification by mutations; the similar segments are called **domains** and are labelled CH_1, CH_2, CH_3, etc. A mature lymphocyte (plasma cell) produces antibodies with a single class H chain and a single type of L chain, hence, also a single antigen-binding specificity. The first antibodies produced by a developing plasma cells are usually of class IgM.

V. Chemical Bonds in Protein Structure

In a protein molecule amino acid residues are joined to each other by bonds forming a chain. Amino acids of different chains of a proteins are also joined together by different types of bonds or forces. These chemical bonds are of the following types:

1. Primary bonds. These are covalent **peptide bonds** between different amino acids of polypeptide chain. Peptide bonds are specialized amide bonds or amide linkages between –COOH (carboxyl) and $-NH_2$ (amino) groups. These bonds form the backbone of peptide chain.

2. Secondary bonds. These bonds grant natural configuration to a polypeptide chain and help in maintaining the secondary structure of the proteins. These are of following types:

***(i)* Disulphide bonds (–S–S– bonds).** These are covalent bonds which give some rigidity to the protein molecule. They are relatively more stable than the other types of bonds. A disulphide bond is formed between two *cysteine* residues having sulphydryl (–SH) groups. The enzyme **ribonuclease** consists of a single polypeptide chain of 124 residues and contains four disulphide bonds. **Insulin** and **oxytocin** are two protein hormones which also have disulphide bonds. The disulphide bonds are oxidised by **performic acid**, which is therefore used to denature proteins or to separate polypeptide chains held together by –S–S– bonds.

***(ii)* Hydrogen bonds (> CO–HN < bonds).** They are weak bonds but since they are numerous they give considerable stability to the protein molecule. Hydrogen bonds can be formed by sharing *hydrogen* between *amide* (NH) *nitrogen* and *carbonyl* (*C=O*) *oxygen* of the peptide backbone. They are also formed between groups present in the side chain. The side chains of **aspartic** and **glutamic** acids, **serine** and **threonine** can all form hydrogen bonds.

Silk fibroin and wool's keratin have such hydrogen bonds.

***(iii)* Ionic or Electrostatic Bonds (or Salt Linkage/Bridges).** These bonds are formed when an acid and a basic amino acid are ionized and lie close together. Ionic bonds play an important role in binding a basic protein with an acidic macromolecule, *e.g.*, in the formation of nucleoproteins.

***(iv)* Nonpolar or Hydrophobic bonds.** These bonds are formed because of the tendency of the nonpolar side chains of neutral amino acids to closely associate with one another. When a protein molecule is submerged in an aqueous medium there is a tendency to expose a maximum number of its polar groups to the surrounding medium. Conversely a maximum number of nonpolar groups are directed internally. Although hydrophobic interactions do not form true bonds they are of great importance in the formation of tertiary structure.

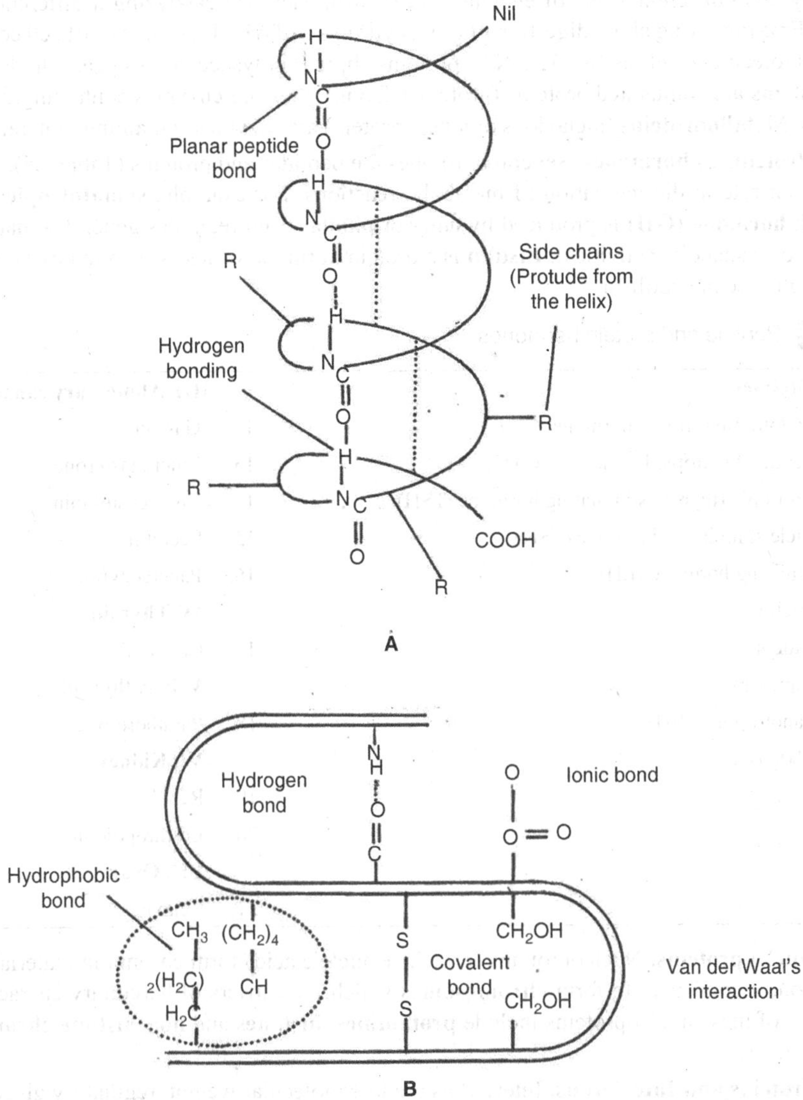

Fig. 7.20. Types of chemical bonds in protein structure: A—showing hydrogen bond and peptide bond; B—showing hydrophobic bond, ionic bond, covalent bond and Van der Waal's interactions. (**Note.** *Van der Waal's interactions.* Attractive forces that exist between atoms or molecules of all substances. Van der Waal's interactions are weak, non-specific, interatomic interactions that come into play when any two uncharged atoms are 3 to 4 A° apart. Since Van der Waals interactions are very weak, it is customary to call them as a force rather than a bond, Jain 2010).

VI. Biological Importance of Proteins

Proteins form a large part of the structure of cells and are present in all tissues. Many proteins also have special physiological functions:

1. Proteins as enzymes. Enzymes are special proteins produced within an organism which are capable of catalyzing specific catalytic reactions. They are **biocatalysts** which influence the rate of chemical reactions, usually without undergoing any change themselves. All enzymes are proteins and nearly 2000 different kinds of enzymes are known, each one catalyzing a different chemical reaction. Enzymes bring about digestion of food, hydrolysis of ATP, biosynthesis of cell components or macromolecules such as DNA, RNA, proteins, lipids, polysaccharides (carbohydrates), etc. **Flavoproteins** are conjugated proteins (protein + flavin); they are enzymes containing the vitamin riboflavin. **Metalloproteins** (metal ions or ions + protein) are enzymes containing mineral elements.

2. Proteins as hormones. Several hormones are peptides and proteins (Table 7.5). They play an important role in the regulation of metabolic reactions. For example, **somatotropic hormone** or **growth hormone (GH)** is produced by anterior pituitary and regulates general somatic growth (*i.e.*, promotes anabolic activities). **Insulin** is a protein hormone which is secreted by pancreas and regulates glucose metabolism.

Table 7.5. Peptide and protein hormones.

	I. Pituitary		**III. Alimentary canal**
1.	Somatotropin (growth hormone, GH)	12.	Gastrin
2.	Adrenocorticotropic hormone (ACTH)	13.	Enterogasterone
3.	Thyrotropic (thyroid stimulating hormone, TSH)	14.	Cholecystokinin
4.	Follicle stimulating hormone (FSH)	15.	Secretin
5.	Luteinizing hormone (LH)	16.	Pancreozymin
6.	Prolactin		**IV. Thyroid**
7.	Oxytocin	17.	Calcitonin
8.	Vasopressin		**V. Parathyroid**
9.	Melanotropin (MSH)	18.	Parathormone
	II. Pancreas		**VI. Kidney**
10.	Insulin	19.	Renin
11.	Glucagon	20.	Erythropoietin
			VII. Ovary
		21.	Relaxin

3. Nucleoproteins. Nucleoproteins (protein + nucleic acid) form chromatin material. During cell division, it condenses to form **chromosomes** which are carriers of hereditary characters. The protein part of these nucleoproteins include **protamines**, **histones** and **non-histone chromosomal (NHC)** proteins.

4. Proteins and Interferons. Interferons are low molecular weight, regulatory glycoproteins produced in response to viral infection, endotoxins, antigenic stimuli, rickettesias and protozoan parasites. Interferons are important therapeutic agents for treating viral diseases, such as hepatitis, encephalitis, cancer and common cold.

5. Proteins and Antibiotics. Some antibiotics such as gramicidins, tyrocidin and penicillin G are peptides.

6. Proteins and protection. Protective proteins are **antibodies** or **immunoglobulins**. These combine and neutralize foreign proteins (**antigens**). Immunoglobulins are **gamma (γ) globulins** which are produced in the spleen and lymphatic cells in response to foreign substances (antigens). There are five classes of immunoglobulins, IgG, IgA, IgM, IgD and IgE.

The blood proteins, **thrombin** and **fibrinogen** are responsible for blood clotting and thus prevent loss of blood at the time of injury.

7. Proteins and Biochemical individuality. Proteins confer specific individuality to the living organisms and species. The protein individuality is well illustrated by **antigen-antibody reactions** such as ABO blood group systems and MHC proteins.

(*i*) ABO blood group system. The blood of one organism or of individuals of one species cannot be transfused into the circulation of other animals. ABO blood group is a classification of blood which is based on natural variation in human blood types. There are 4 groups: A, B, AB and O, each classified by a particular combination of antigen(s) on the red blood cells (RBCs) and naturally occurring antibodies in the blood plasma.

(*ii*) MHC proteins. Histocompatibility is a state of mutual tolerance between tissues that allow them to be grafted effectively. **Histocompatibility antigen** is any of the antigenic glycoproteins on the surface membranes of cells that enable the body's immune system to recognise a cell as a native or foreign and that are determined by the *major histocompatibility complexes* (*MHC*) *genes* (Box 7.6).

Box 7.6 MHC

During the first part of 20th century, clinical researches discovered that blood could be transfused from one person to another, as long as two individuals were compatiable for the ABO blood group system. The success of blood transfusion led to the proposal that skin might also be grafted between individuals. This idea was tested during World War II when skin grafts were attempted on piolets and other military personnel who had received serious burns. The grafts were rapidly and completely rejected. After the war, researchers set out to determine the basis for tissue rejection. It was discovered that skin could be grafted successfully between mice of same inbred strain, but that grafts between mice of different strains were rapidly rejected. Mice of the same inbred strain are like identical twins; they are genetically indentical. Subsequent studies revealed that the genes that governed this graft rejection were clustered in a region of the genome that was named **major histocompatibility complex (MHC)**. Approximately 20 different MHC genes have been characterized, most of which are highly polymorphic: over 2000 different alleles of MHC genes have been identified, for more than any other loci in the human genome. So, it is very unlikely that two individuals in a population have the same combination of MHC alleles. This is the reason that transplanted organs are so likely to be rejected and why transplant patients are given drugs, such as **cyclosporin A**, to suppress the immune system following surgery. Cyclosporin A is a cyclic peptide produced by a soil fungus. Cyclosporin A inhibits a particular phosphatase enzyme in the signaling pathway leading to the production of cytokines required for T-cell activation (**Karp** 2010).

MHC proteins can be subdivided into two major groups: **MHC class I** and **MHC class II** molecules. MHC class I molecules consist of one polypeptide chain encoded by an MHC allele (known as heavy chain) associated noncovalently with a non-MHC polypeptide known as **β_2-microglobulin**. MHC class II molecules also consist of heterodimers but both subunits are encoded by MHC alleles. Both classes of MHC molecules as well as β_2-microglobulin contain Ig-like domains and are thus members of immunoglobulin super family (see **Karp** 2010).

8. Blood Plasma Proteins. Electrophoresis at pH 8.6 in barbital buffer separates six major blood's plasma components: **albumin**, **α_1-globulin**, **α_2-globulin**, **β-globulin**, **γ-globulin** and **fibrinogen**. Starch gel electrophoresis breaks up the bands into 20 zones containing 50 plasma proteins.

9. Proteins and transport. Some proteins bind and transport specific types of molecules via blood. For example

(*i*) **Albumin** (MW 76,000) is synthesized in the liver and is the most abundant plasma protein. It maintains the colloid osmotic pressure of plasma and takes part in the transport of a variety of materials. It carries substances such as bilrubin and some drugs which are otherwise insoluble in plasma.

Serum albumin binds free fatty acids molecules and transport them between adipose tissue and other cells of the body.

(ii) **Lipoproteins** (*e.g.*, HDL, LDL and VLDL) of blood plasma transport lipids (fats) between the intestine, liver and adipose tissue.

(iii) **Haemoglobin** is a conjugated protein consisting of globin plus haeme (an Fe porphyrin prosthetic group). It consists of 4 polypeptide chains, 2 alpha chains and 2 beta chains. Each chain has a molecular weight of 16,750 and is attached to one atom of iron.

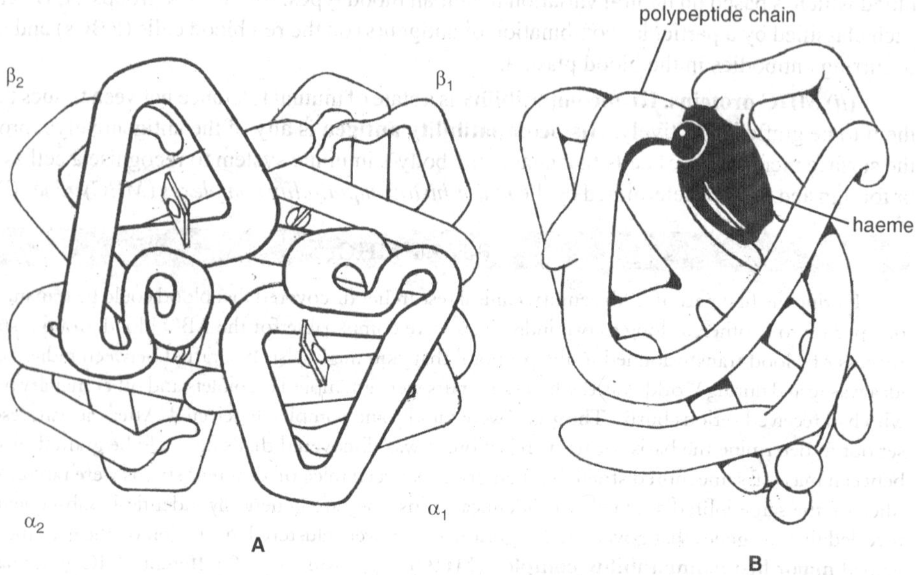

Fig. 7.21. A—Haemoglobin; B—Myoglobin.

Haemoglobin in vertebrates and **haemocyanin** in some invertebrates transport oxygen from a respiratory organ, *e.g.*, lungs to tissue cells (oxygen carrier of blood).

(iv) **Myoglobin** stores oxygen in muscles.

(v) **Transferrin** (MW 90,000) is β-globulin involved in the transfer of Fe to plasma.

(vi) **Hepatoglobin** is an α-2-globulin. When RBCs are damaged, haemoglobin released into the plasma is bound by hepatoglobin and thus prevented from being excreted through the urine.

α_1**-chymotrypsin** (MW 45,000) inhibits the proteolytic action of protein-digesting enzymes such as trypsin, chymotrypsin and plasmin.

α_2**-macroglobulin** (MW 820,000) is an inhibitor protein of proteolytic enzymes.

(vii) **Complement** is thermolabile group of proteins in normal blood serum and plasma that in combination with antibodies causes the destruction of particular antigens (as bacteria and foreign blood corpuscles). There are 9 complements, C′1 to C′9.

(viii) **Permeases.** These are transport proteins. They transport molecules (Na, K, Ca, glucose, etc.) through cell membrane.

(ix) Proteins help in the transport of organic food through phloem.

(x) **Cytoplasmic motor proteins.** Proteins that utilize energy of ATP hydrolysis to generate mechanical forces that propel the proteins, as well as attached **cargo**, along one of the components of the cytoskeleton. Three families of motor proteins are known: **kinesins** and **dyneins** move along the microtubules and **myosins** move along the microfilaments (**Karp** 2010).

(a) **Kinesin** is a plus end-directed motor protein that moves membranous vesicles and other organelles along microtubules through the cytoplasm.

(*b*) **Dynein** is an exceptionally large, cargo-carrying multisubunit motor protein that moves along microtubules toward their minus end. This family of proteins occurs as **cytoplasmic dyneins** and **ciliary** or **axonemal dyneins**.

(*c*) **Myosin** is a large family of motor proteins that moves along actin-containing microfilaments. Most myosins are plus- end directed motors. **Conventional myosin** (*i.e.*, myosin II) is the protein that mediates muscle *contractility* as well as certain types of nonmuscle motility, such as **cytokinesis.** **Unconventional myosins** (*i.e.*, myosin I and III to XVIII) have many diverse roles including **organelle transport**.

10. Toxins. Some proteins are highly poisonous to higher animals, such as **snake venoms, gossypin** (the toxin protein of cotton seed), **ricin** (the protein of castor bean), proteins of pathogenic bacteria (diphtheria toxin), etc.

11. Contraction. Muscle fibers consist of thin filaments of **actin, tropomyosin** and **troponin** which make up the I bands or light bands and thick filaments of **myosin** which make up the A or dark bands.

12. Movement. Locomotory organs of cells such as the **cilia** and **flagella** are formed of fine tubules of **tubulin** proteins. Tubulins are dimeric proteins having α and β subunits.

13. Receptor proteins. Some proteins are present in the plasma membrane and other cell membranes to receive sensory stimuli from the surroundings.

14. Structural proteins.

(*i*) Proteins form an important part of all membranes and membrane bound organelles of the cell [*i.e.*, plasma membrane, ER, nuclear membranes, lysosome, mitochondria, chloroplast, vacuole (tonoplast), etc.].

(*ii*) Collagen fibers and elastin fibers are found in connective tissue (*e.g.*, cartilage) and bone; both form the fibrous proteins.

Elastin is a principal fibrous, highly extensible and elastic protein that is the major constituent of yellow elastic fibers of animal connective tissue. It originates from mesenchyma. Elastin is especially rich in glycine (Gly), alanine (Ala), proline (Pro) and other nonpolar amino acids and are cross-linked, making the protein insoluble. It is particularly abundant in elastic cartilage, lungs, walls of large arteries, ligaments and the heart.

Elastin fibers can stretch to many times their length and then return to their original size. It differs from collagen in not being converted to soluble gelatine.

Collagen is the most abundant fibrous (structural) protein of connective tissue, occurring in **white fibers**. It accounts for about 25 – 33% of total body protein of higher vertebrates, including human beings. Collagen found extensively in all connective tissue, *e.g.*, skin, cartilage, tendons, ligaments and bones. Collagen is also present in ECM (of epithelium). Collagen fibers are composed of masses of tropocolagen molecules, each a triple helix of collagen monomers; the three polypeptide chains are wound tightly around each other to give a **helix**. It is especially rich in glycine (35%), alanine (11%) and proline + hydroxyproline (25%). Collagen is a rod-shaped molecule about 3,000 A° (= 300 nm) long and 15 A° (1.5 nm) in diameter. The helical motif of its three chains is entirely different from the α helix. Tropocollagen subunits are synthesized and released by fibroblasts.

Collagen fibers have **high tensile strength** (*i.e.*, resist stretching forces, as in tendon) and limited **elasticity** (unlike elastin). Collagen is essentially insoluble in water, but when boiled with water, the strands separate and undergo some hydrolysis, producing **gelatin**. Although a protein, collagen has little nutritional value because it lacks significant amount of many amino acids that are essential in the human diet (**Jain** 2012).

(*iii*) Protein **chondrin** and **ossein** form matrix of cartilage and bone respectively.

(*iv*) Scales, feathers, hair, fur wool, hoof, nail, horns, etc., formed of keratin sulphate, a protein derivative.

15. Special secretions. Spiders and silkworms secrete a thick solution of protein **fibroin** which solidifies into the thread of exceptional tensile strength and is used for weaving a web or cocoon.

16. Storage proteins. Proteins are stored as reserve food such as egg albumen or protein stored in various seeds (**glutellin** in rice), **zeins** in maize and **phaseolin** in pea.

17. Exotic proteins. These are proteins which are secreted by different animals or plants for specific purpose. For example:

(*i*) Some marine and aquatic animals secrete sticky proteinaceous substance for adhesion to the rock or substratum.

(*ii*) Fishes live in freezing cold water of Arctic and Antarctic oceans synthesize a **anti-freeze protein** to protect their blood from freezing.

(*iii*) Some african plants synthesize protein **monellin** which is very sweet and is used as artificial sweetner for diabetic patients.

QUESTIONS

Long Answer Questions

1. What are amino acids? Describe the physical and chemical properties of amino acids.
2. List the acidic, basic and nonpolar amino acids.
3. What are proteins? Explain their biological significance.
4. Classify proteins on the basis of their structure and complexity.
5. Describe the helix and pleated structures of proteins.
6. Describe various structural levels of proteins.
7. Discuss the functions of proteins.

Short Answer Questions

1. Describe structure of amino acids.
2. Write short notes on the following:
   (*i*) Denaturation of proteins
   (*ii*) Structure of haemoglobin
   (*iii*) Glycoproteins

Very Short Answer Questions

1. Which protein molecule has antiparallel β-sheet structure?
2. Which type of structure is found in haemoglobin?
3. Which protein is found in rice?
4. Which protein is found in hair?
5. Which animal protein changes into gelatin on boiling?
6. Which protein occurs in tendons?
7. Give an example of zinc containing protein.
8. Give an example of essential amino acid.
9. Give two examples of complete proteins.
10. Define complete proteins.

Multiple Choice Questions

1. Which of the amino acids is essential for metabolism?
   (*a*) serine (*b*) creatine
   (*c*) phenylalanine (*d*) aspartic acid
2. The amino acid found only in bacteria and blue green algae is
   (*a*) methionine
   (*b*) muramic acid
   (*c*) diamino-palmic acid
   (*d*) aspargine
3. Which amino acid is denoted by symbol F?
   (*a*) phenylalanine (*b*) proline
   (*c*) tryptophan (*d*) methionine
4. Histones are
   (*a*) acidic proteins (*b*) basic proteins
   (*c*) mucoproteins (*d*) glycoproteins

5. Peptide bond binds molecules of
   (*a*) sucrose (*b*) amino acids
   (*c*) glucose (*d*) none of these
6. The primary structure of protein is made up of
   (*a*) peptide bonds
   (*b*) hydrogen bonds
   (*c*) ionic bonds
   (*d*) Van der waals interactions
7. Collagen contains
   (*a*) hydroxyproline
   (*b*) tryptophan
   (*c*) arginine
   (*d*) peptone
8. Proteins in silk fiber are
   (*a*) fibrin and serine
   (*b*) chondrin and mucin
   (*c*) sericin and fibroin
   (*d*) collagen and elastin
9. The example of dynamic fibrous protein is
   (*a*) gliadin (*b*) collagen
   (*c*) haemoglobin (*d*) hordein
10. Avidin is a protein found in
    (*a*) egg yolk (*b*) albumin
    (*c*) raw egg white (*d*) liver
11. Scleroproteins are
    (*a*) keratin (*b*) collagen
    (*c*) both (*a*) and (*b*) (*d*) alcoprotein
12. Haemoglobin is a
    (*a*) protein
    (*b*) fibrous protein
    (*c*) globular protein
    (*d*) enzyme
13. Which of the following is a conjugated protein
    (*a*) collagen (*b*) albumin
    (*c*) casein (*d*) keratin

ANSWERS

Very Short Answer Questions

1. Fibroin of silk.
2. Quaternary.
3. Oryzenin.
4. Keratin.
5. Collagen.
6. Elastin.
7. Carbonic anhydrase.
8. Tryptophan.
9. Egg albumin and milk casein.
10. The complete proteins are the proteins which have all the ten essential amino acids in the required portion by the human body to promote good growth.

Multiple Choice Questions

1. (*c*) **2.** (*a*) **3.** (*a*) **4.** (*b*) **5.** (*b*) **6.** (*b*) **7.** (*a*)
8. (*c*) **9.** (*b*) **10.** (*c*) **11.** (*c*) **12.** (*c*) **13.** (*c*)

8 Carbohydrates

Carbohydrates (L. *carbo* = coal; G. *hydro* = water) are compounds of carbon, hydrogen and oxygen *e.g.*, glucose, fructose, sugar, starch, glycogen, cellulose, chitin, etc. They are also called **saccharides** (G. *sakcharon* = sugar) because they are sugars or polymers of sugars. On the basis of mass, carbohydrates are the most abundant class of biological molecules on Earth. Although all organisms can synthesize carbohydrate, much of it is produced by photosynthetic organisms, including bacteria, algae and plants. These organisms convert solar energy to chemical energy, which is then used to make carbohydrate from carbon dioxide. Carbohydrates play several crucial roles in living organisms. In animals and plants, carbohydrate polymers act as energy storage molecules (*e.g.*, starch in plants and glycogen in animals). Animals can ingest carbohydrates (*e.g.*, glucose) that can then be oxidized to yield energy for metabolic processes. Polymeric carbohydrates (*e.g.*, cellulose) are also found in cell walls and in the protective coatings (*e.g.*, chitin) of many organisms. Some other carbohydrate polymers are marker molecules (*e.g.*, antigens A, B, AB and O of RBCs) that allow one type of cell to recognize and interact with another type. Carbohydrate derivatives are found in a number of biological molecules, including some coenzymes and the nucleic acids DNA and RNA (*i.e.*, Deoxyribose and ribose sugars) of DNA and RNA.

8.1. CLASSIFICATION

Carbohydrates can be described by the number of monomeric units they contain: Monosaccharides, Oligosaccharides and Polysaccharides.

1. **Monosaccharides.** They are the smallest units of carbohydrate structure. The name carbohydrate, "hydrate of carbon", refers to their empirical formula $(CH_2O)n$, where n is 3 or greater (n is usually 5 or 6 but can be upto 9).
2. **Oligosaccharides.** They are polymers of 2 to about 20 monosaccharide residues. In them, monomers are joined together by **glucosidic bonds**. The most common oligosaccharides are disaccharides such as sucrose and maltose in plants and lactose in animals; they tend to have two linked monosaccharide residues.
3. **Polysaccharides.** They are polymers that contain many (usually more than 20) monosaccharide residues.

Monosaccharides and disaccharides are commonly known as **sugars** and are readily soluble in water. They are considered to be **micromolecules** whereas polysaccharides are considered to be **macromolecules**.

Oligosaccharides and polysaccharides do not have the empirical formula $(CH_2O)n$ because water is eliminated during polymer formation. The term **glycan** is more general term for carbohydrate polymers. It can refer to a polymer of identical sugars (called **homoglycan**) or of different sugars (called **heteroglycan**).

Glycoconjugates are carbohydrate derivatives in which one or more carbodydrate chains are linked covalently to a peptide, protein or lipid. These derivatives include **proteoglycans**, **glycoproteins** and **glycolipids**.

8.2. MONOSACCHARIDES

Monosaccharides (G. *mono* = one) are water-soluble, white, crystalline solids that have a sweet taste. Examples include glucose and fructose. Chemically, most monosaccharides are **chiral** compounds (see Box 8.1). Monosaccharides are polyhydroxy aldehydes or **aldoses**, or polyhydroxy ketones or **ketoses** (Fig. 8.1). Monosaccharides are the monomers and form structural units of oligo- and polysaccharides. They are classified by their type of carbony group (Box 8.1) and their number of carbon atoms. As a rule, the suffix—*ose* is used in naming carbohydrates, although there are a number of exceptions. All monosaccharides contain at least three cabon atoms (Table 8.1).

Box 8.1

Chiral atom. An atom with asymmetric substitutions that can exist in two different configurations. For example, carbon is a chiral atom since it has 4 different substituents, ~~, in a complex organic compound, the tetrahedrally bonded carbon atom is attached to 4 different atoms or groups of atoms (see **Jain** 2012).

Carbonyl group. $\left(\begin{array}{c} O \\ \| \\ -C- \end{array}\right)$ A carbonyl group is a pair of atoms consisting of a carbon atom linked to an oxygen atom by a double bond (–C = O).

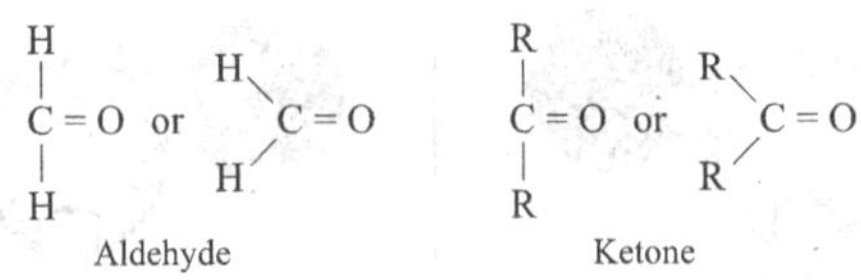

Fig. 8.1. General formula of aldehyde and ketone.

One of these is carbonyl carbon, and each of the remaining carbon atoms bear a hydroxyl group. In aldoses, the most oxidised carbon atom is designated C-1 and is drawn at the top of Fisher projection. In ketoses, the most oxidised carbon atom is usually C-2.

Table 8.1. Some monosaccharides.

	Carbohydrates	Number of carbon atoms	Aldoses	Ketoses
1.	Trioses ($C_3H_6O_3$)	3	Glycerose	Dihydroxyacetone
2.	Tetroses ($C_4H_8O_4$)	4	Erythrose and Threoses	Erythrulose
3.	Pentoses ($C_5H_{10}O_5$)	5	Ribose, Xylose, Lyxose, Arabinose	Ribulose, Xylulose
4.	Hexoses ($C_6H_{12}O_6$)	6	Glucose, Galactose, Mannose, Allose, Altrose Talose	Fructose (levulose), Sorbose, Psicose, Tagatose
5.	Heptoses ($C_7H_{14}O_7$)	7	Mannoheptulose	Sedoheptulose

The smallest monosaccharides are **trioses**, or three-carbon sugars. One- or two-carbon compounds having the general formula $(CH_2O)n$ do not have properties typical of carbohydrates (such as sweet taste and the ability to crystalize). The aldehyde triose or aldotriose, is glyceraldehyde (Fig. 8.2A). Because its central carbon, C-2 has four different group attached to it, glyceraldehyde is chiral (Box 8.1). The ketonic triose, or ketotriose is dihydroxyacetone (Fig. 8.2B), which is achiral—it has no asymmetric carbon atom. All other monosaccharides, longer chain versions of these two sugars are chiral.

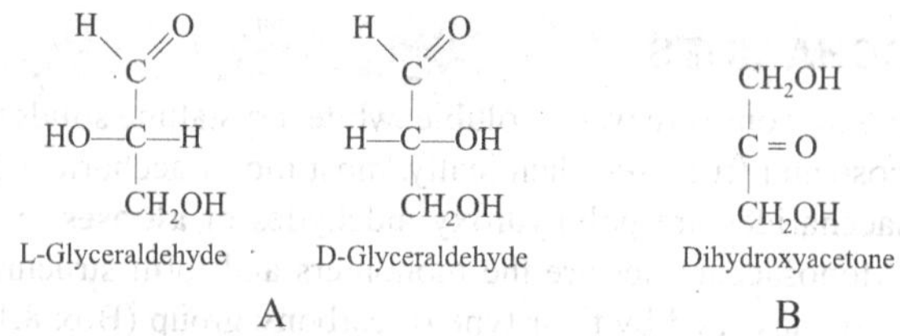

Fig. 8.2. Fischer projections of glyceraldehyde (A) and dihydroxyacetone. The designation L (for left) and D (for right) for glyceraldehyde refers to the configuration of the hydroxyl group of the chiral carbon (C-2). Dihydroxyacetone is achiral.

In Figure 8.2, the stereoisomers D- and L- glyceraldehyde are shown as ball- and -stick models. Chiral molecules are optically active; that is, they rotate the plane of polarised light. The convention for designating D and L isomers was originally base on the optical properties of glyceraldehyde. The form of glyceraldehyde that caused rotation to the right (dextrorotory) was designated D; the form that caused rotation to the left (levoratory) was designated L. X-ray crystallographic studies have confirmed these structural assignments.

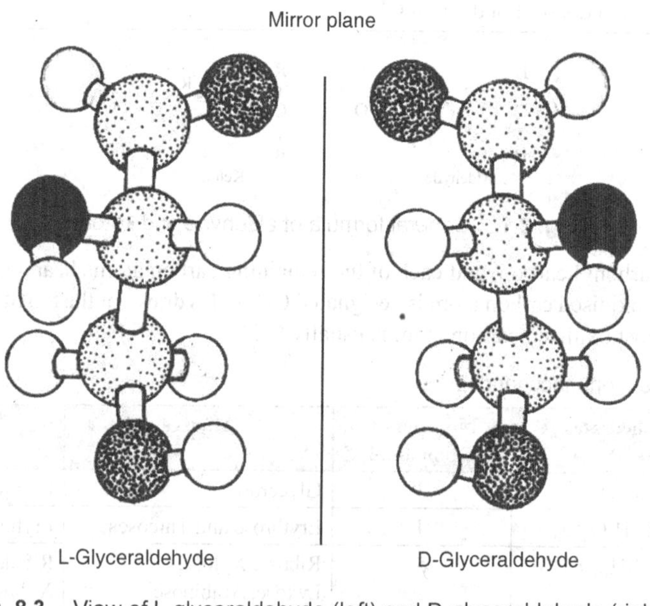

Fig. 8.3. View of L-glyceraldehyde (left) and D-glyceraldehyde (right).

Types of Monosaccharides

1. Trioses ($C_3H_6O_3$). Trioses are formed only during metabolic breakdown of hexoses. They do not occur in nature. They are formed of 3 carbon atom chains and are smallest carbohydrate molecules. *Glyceraldehyde* and *dihydroxyacetone* are two trioses formed during glycolysis of glucose.

2. Tetroses ($C_4H_8O_4$). They are four carbon compounds. Examples are **erythrose** and **threose.**

3. Pentoses ($C_5H_{10}O_5$). These are 5 carbon sugars. Examples of pentoses are **ribose, deoxyribose, ribulose, xylulose, arabinose.**

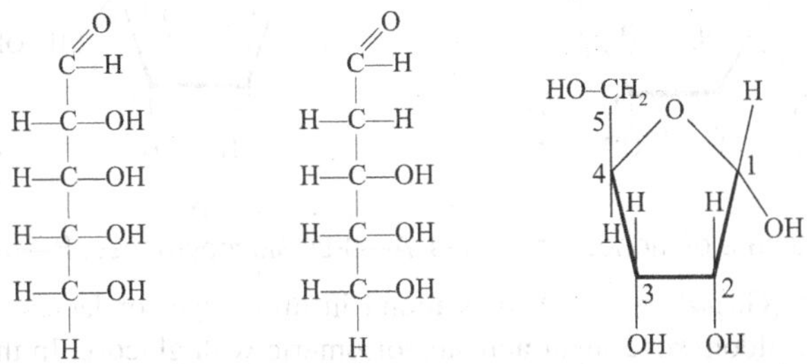

Straight chain form of ribose — Straight chain form of deoxyribose — Ring form of D-ribose (Furanose form)

Fig. 8.4. Pentose sugars: ribose and deoxyribose.

The pentose sugar, **ribose** is the important constituent molecule of the ribonucleic acid (RNA) and certain coenzymes as nicotinamide adenine dinucleotide (NAD), NAD phosphate (NADP), adenosine triphosphate (ATP) and coenzyme A (CoA). Another pentose sugar the **deoxyribose** is the important constituent of the deoxyribonucleic acid (DNA). The **ribulose** is a pentose sugar which is necessary for photosynthetic mechanism.

4. Hexoses ($C_6H_{12}O_6$). These are compounds of 6-carbon atoms and are found as **glucose, fructose, galactose** and **mannose.** All of them are represented by the same formula $C_6H_{12}O_6$. These sugars are sweet, neutral in reaction, crystallizable, soluble in water and readily dialysable. They can undergo alcoholic fermentation. Hexoses form a major source of energy and also act as a raw material for other carbohydrates. All the carbohydrates taken into body (by feeding) are ultimately hydrolyzed into glucose which is absorbed in blood and transported to the cells. Only glucose and fructose occur in solution.

Glucose, fructose and mannose differ only in their first and second carbon atoms. The carbonyl (C = O) group in glucose is attached to the first carbon atom and forms **aldehyde group**. But in fructose it is found with second carbon and forms **keto group**.

***(i)* Glucose** (G. *gleukos* = sweet new wine, sweetness). **Grape sugar** or **dextrose** occurs in fruits and honey along with fructose. In the body, glucose is present in the blood, in the intestine during digestion and in the urine of dibetics. It occurs in the sap of plants. Glucose is natural component of many disaccharides (*e.g.*, sucrose, lactose, maltose, cellobiose, trehalose) and polysaccharides (*e.g.*, starch, glycogen, cellulose). Glucose occurs in two isomeric forms: D- glucose and L- glucose. Glucose is main metabolic fuel.

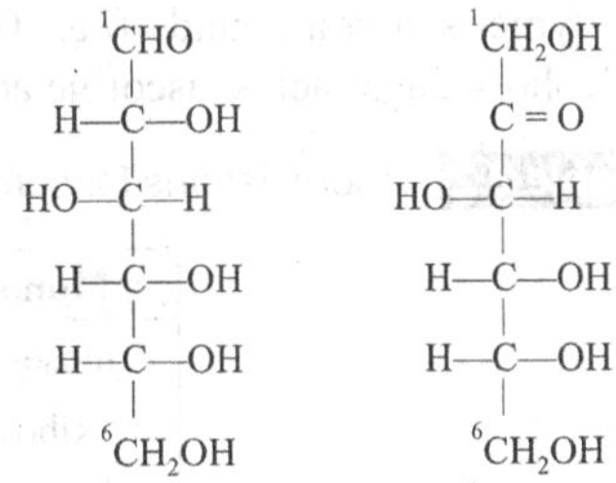

Glucose — Fructose

Fig. 8.5. Glucose and Fructose.

***(ii)* Fructose.** Fructose or fruit sugar is also known as **levulose.** It occurs in green plants, fruits and honey. Fructose has a composition similar to glucose but it is **keto-hexose** and is less readily absorbed by tissue cells. This is due to the difference in position of OH group. In liver and intestine it is changed to glucose.

Fructose is sweetest of all sugars and taste 1.73 times sweeter than sucrose (cane sugar), of which it is constituent. It is very soluble in water and crystallizes in large needles. Its some polysaccharide derivatives (such as **inulin**, which is made up of 30–35 β-D-fructose units per molecule) are energy stores of some plants.

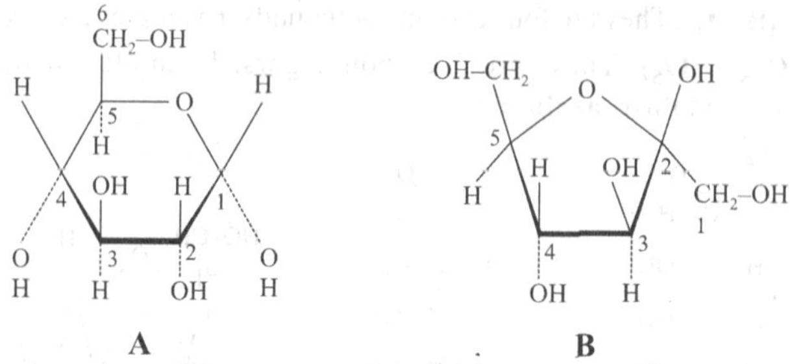

Fig. 8.6. Ring structure of monosaccharides. A—β-D-Glucopyranose; B—α-D-Fructofuranose.

(*iii*) **Galactose** (G. *gala* = milk). It is found in milk-sugar or lactose along with glucose. Galactose is a simple aldohexose sugar and stereoisomeric with glucose. In the body it is changed into glucose to be utilised. Galactose does not occur freely in nature and is constituent of many plant polysaccharides (many gums such as gum arabic, mucilages, pectins), animal glycolipids (*i.e.*, occur in human nervous system) and glycoproteins.

Some human infants do not have the enzyme **uridyltransferase** to metabolise galactose of the milk sugar (of mother's milk), so they store galactose in their different tissues such as brain causing malnutrition and mental retardation. This inborn error of metabolism is called **galactosemia** (**Jain** 2012).

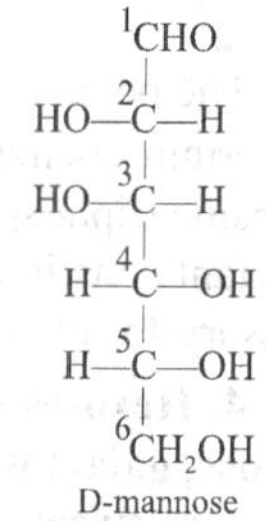

Fig. 8.7. D-mannose.

(*iv*) **Mannose.** It is a 6-carbon aldo sugar that occurs naturally only in polymerised forms called **mannans**. It occurs in plants, yeasts and other fungi and bacteria. Mannose is a constituent of prosthetic polysaccharides of albumins, globulins, mucoproteins. It also occurs in glycoproteins. In animals, mannose changes to glucose before being used in metabolism. It is an epimer of glucose and serves as food energy store.

5. **Heptoses ($C_7H_{14}O_7$).** They include sedoheptulose, glucoheptose, etc. **Sedoheptulose** is an amorphous ketose sugar that plays a role in carbohydrate metabolism but is not fermented by yeast and it is a laboratory source of D-altrose (a synthetic syrupy hexose sugar) and D-ribose.

Derived Monosaccharides

There are many derivatives of the basic monosaccharide. These derivatives include polymerised monosaccharides, such as oligosaccharides and polysaccharides, as well as several classes of non-polymerised compounds (*i.e.*, Glycosides, sugar phosphates, deoxy-sugars, amino-sugars, sugar alcohols, sugar acids, ascorbic acids) (Table 8.2).

Table 8.2. Abbreviations for some monosaccharides and their derivatives (after Horton *et al.*, 2006).

Monosaccharide derivative	Abbreviation
Pentoses	
Ribose	Rib
Xylose	Xyl
Hexoses	
Fructose	Fru
Galactose	Gal
Glucose	Glc
Mannose	Man

Monosaccharide derivative	Abbreviation
Deoxy sugars	
Abequose	Abe
Fucose	Fuc
Amino sugars	
Glucosamine	GlcN
Galactosamine	GalN
N-Acetylglucosamine	GlcNAc
N-Acetylgalactosamine	GalNAc
N-Acetylmuramic acid	MurNAc
Sugar acids	
Glucuronic acid	GlcUA
Iduronic acid	IdoA

1. Glycosides. The hydroxyl group on carbon 1 in the sugar molecule can be replaced by other radicals, forming compounds konwn as **glycosides**. Thus, if a *methyl group* replaces the hydroxyl group of glucose it results in the formation of **methyl glucosides**. There are two methyl glucosides possible, α and β, corresponding to α and β forms of glucose. Glucosides formed from *galactose* are known as **galactosides**. Among the complex glucosides occurring in nature are **digoxin** ($C_{41}H_{64}O_{14}$) which is obtained from the leaves of a foxglove plant (*Digitalis lanata*) and acts on the heart (a cardiotonic) and the antibiotic **streptomycin**.

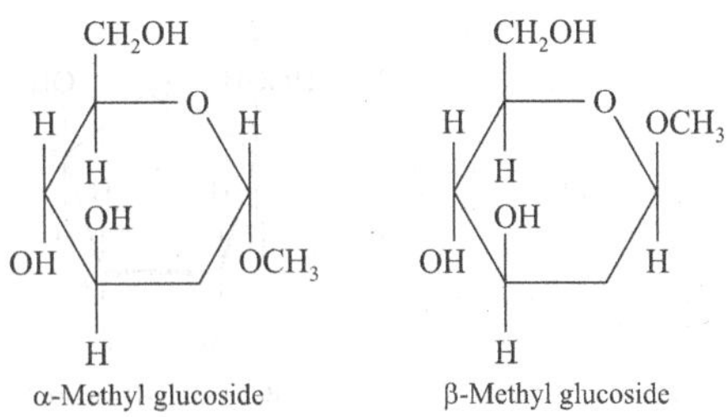

Fig. 8.8. Glycosides.

2. Sugar phosphates. Monosaccharides in metabolic pathways are often converted to phosphate esters. The triose phosphates, ribose 5-phosphate, and glucose 6-phosphate is a hemiacetal phosphate. The ability of UDP-glucose to act as a glycosyl donor is evidence of this reactivity.

Box 8.2

The aldehyde and ketone groups of carbohydrates with five and six carbon atoms react with alcohol group present in the neighbouring carbons to form **hemiacetal** and **hemiketals** respectively. This results in the formation of five to six membered ring (*i.e.*, pyranoses or furanoses).

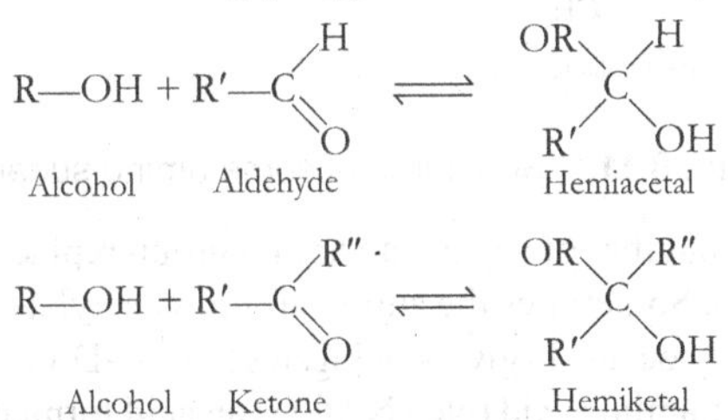

3. Deoxy sugars. In these derivatives, a hydrogen atom replaces one of the hydroxyl groups in the parent monosaccharide. 2 Deoxy-D-ribose is an important building block for DNA (Fig. 8.10). L-fucose (6-deoxy-L-galactose) is widely distributed in plants, animals and micro-organisms. Fucose is derived metabolically from D-mannose.

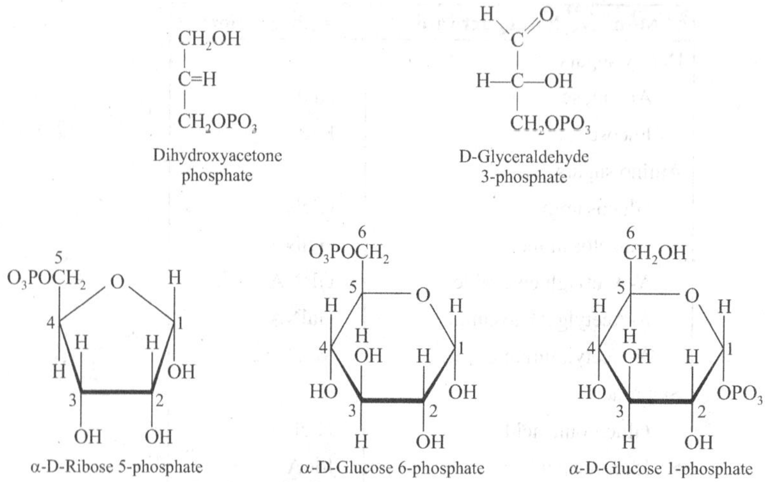

α-D-Ribose 5-phosphate, α-D-Glucose 6-phosphate, α-D-Glucose 1-phosphate, Dihydroxyacetone phosphate, D-Glyceraldehyde 3-phosphate

Fig. 8.9. Some sugar phosphates.

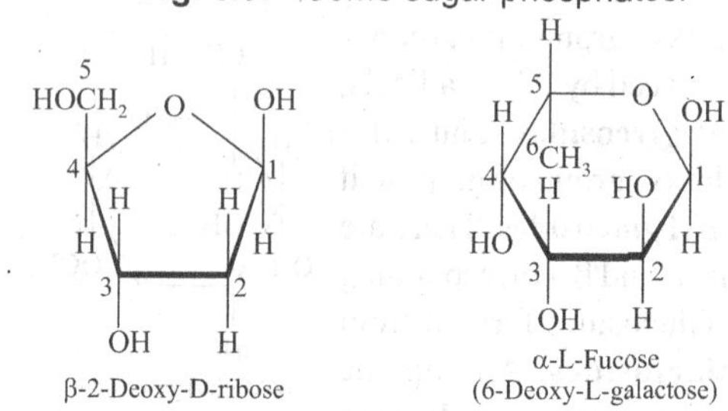

β-2-Deoxy-D-ribose, α-L-Fucose (6-Deoxy-L-galactose)

Fig. 8.10. Structure of deoxysugars.

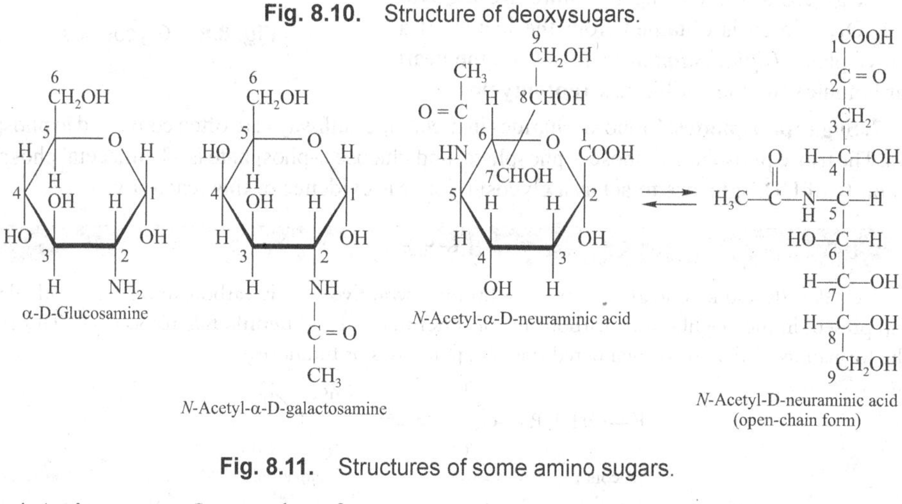

α-D-Glucosamine, *N*-Acetyl-α-D-galactosamine, *N*-Acetyl-α-D-neuraminic acid, *N*-Acetyl-D-neuraminic acid (open-chain form)

Fig. 8.11. Structures of some amino sugars.

4. Amino sugars. In a number of sugars, an amino group replaces one of the hydroxyl groups in the parent monosaccharide. Sometimes the amino group is acetylated. Amino sugars formed from glucose and galactose commnly occurs in glycoconjugates (*e.g.*, α-D-Glucosamine and N-Acetyl-α-D-galactosamine). N-Acetylneuraminic acid (Neu NAC) is an acid formed from N-acetylmannosamine and pyruvate. When this compound cyclizes to form a pyranose, the carbonyl group at C-2 (from the pyruvate moiety) reacts with the hydroxyl group of C-6. NeuNAC is an important constituent of many glycoproteins and of a family of lipid called **gangliosides**. Neuraminic acid and its derivatives, including NeuNAC, are collectively known as **sialic acids**.

5. Sugar alcohols. In a sugar alcohol, the carbonyl oxygen of the parent monosaccharide has been reduced, producing a polyhydroxy alcohol, *e.g.*, **glycerol, myo-Inositol** and **D-ribitol** (Fig. 8.12). Glycerol and *myo*-inositol are important component of lipids. Ribitol is a component of flavin mononucleotide (FMN) and flavin adenine dinucleotide (FAD). In general, sugar alcohols are named by replacing the suffix *-ose* of the parent monosaccharides with *-itol*.

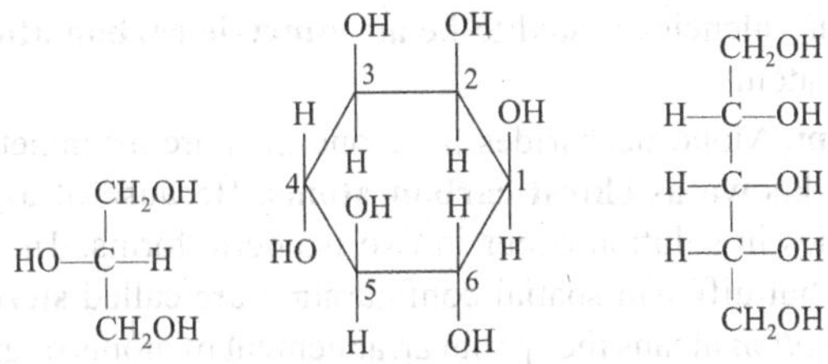

Fig. 8.12. Structure of some sugar alcohols. Glycerol (a reduced form of glyceraldehyde) and *myo*-inositol (metabolically derived from glucose) are important constituents of many lipids. Ribitol (a reduced form of ribose) is a constituent of the vitamin riboflavin and its coenzyme.

6. Sugar acids. Sugar acids are carboxylic acids derived from aldoses, either by oxidation of C-1 (the aldehydic carbon) to yield an aldonic acid or by oxidation of the highest-numbered carbon (the carbon bearing the primary alcohol) to yield an **alduronic acid**. The structures of the aldonic and alduronic derivatives of glucose-gluconate and glucuronate—are shown in Fig. 8.13. Aldonic acid exists in the open chain form in alkaline solution form lactones (intramolecular esters) on acidification. Alduronic acids can exist as pyranose, so they possess an anomeric carbon. N-Acetylneuraminic acid is an aldonic acid as well as an amino sugar. Sugar acids are important component of many polysaccharides.

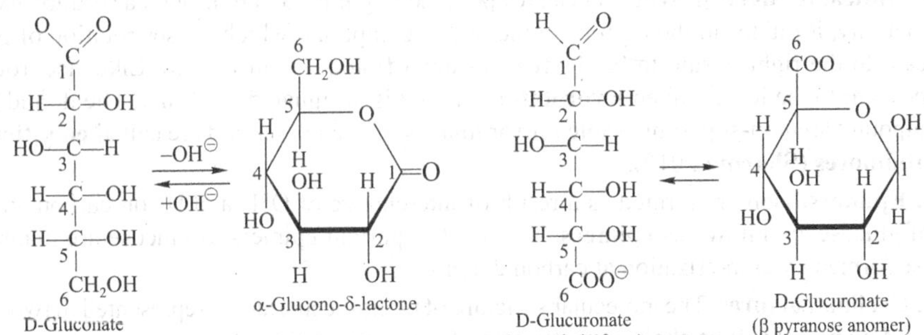

Fig. 8.13. Structures of sugar acids derived from D-glucose. A—Gluconate and its D-lactone. B—The open-chain and pyranose forms of glucuronate.

Many drugs, pesticides and environmental pollutants and hormones are coupled with glucuronic acid (*i.e.*, glucuronate) and excreted in urine or bile as glycuronides.

7. Ascorbic acid (Vitamin C). L-Ascorbic acid (Fig. 8.14) or vitamin C is an enediol of a lactone derived from D-glucuronate. Primates cannot convert-glucuronate to ascorbic acid and must therefore obtain ascorbic acid from the diet. Ascorbic acid is an essential cofactor for the enzymes that catalyze the hydroxylation of proline and lysine residues during collagen synthesis.

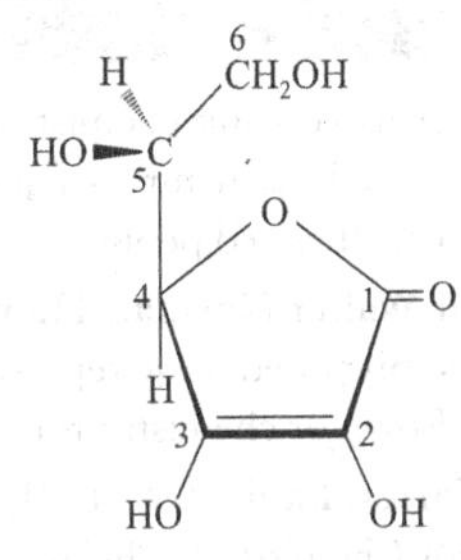

Fig. 8.14. L-Ascorbic acid (vitamin C).

Structural Specialities of Monosaccharides

1. Aldose and ketose form. Monosaccharides or carbohydrate monomers are structural units of carbohydrates. They are mainly hexose sugars. They either occur as aldoses or ketoses and are called **aldohexoses** or **ketohexoses**.

2. Asymmetry of carbon atom. A carbon atom which has four different atoms or groups of atoms attached to its four valencies is said to be **asymmetric carbon atom**. All monosaccharides have asymmetric carbon atoms.

3. Stereoisomerism. Monosaccharides have one or more asymmetric carbon atoms. These asymmetric carbons are known as **chiral carbon atoms**. Because of asymmetric carbon atoms, monosaccharide molecules in solution occur in two isomeric forms. The compounds which have same structural formula but differ in spatial configuration are called **stereoisomers** or **geometric isomers**. (**Note.** *Configuration* means the spatial arrangement of atoms or groups in a molecule.) For example, glucose with four asymmetric carbon atoms have 16 isomeric forms.

4. D- and L-isomers. In monosaccharide molecules, when OH group on the second carbon, *i.e.*, adjacent to the carbon of terminal primary alcohol is on the right, the molecule is represented as **D-isomer**. When OH is on the left the monosaccharide molecule is **L-isomer**. The naturally occurring monosaccharides are mainly D-isomers.

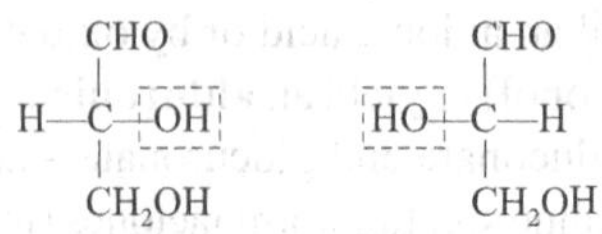

D-Glyceraldehyde L-Glyceraldehyde

Fig. 8.15. Two isomers of glyceraldehyde: D-isomer and L-isomer.

5. Optical isomerism. When a beam of polarised light is passed through a solution exhibiting optical activity, it rotates to the right or to the left. A compound which causes rotation of polarised light beam to the right is said to be **dextrorotatory** (d) and + sign is used. Likewise, rotation of polarised light to the left is called **levorotatory** (l) and is designated by – sign. The d- and l- forms of a compound are non-superimposable mirror images of each other and are called **enantiomorphs** or **enantiomores (Sharma** 2010).

6. Epimers. Isomers formed as a result of interchange of OH^- and H^+ on carbon atoms 2, 3 and 4 of glucose are known as **epimers**. The most important epimers of glucose are mannose and galactose formed by epimerization at carbon 2 and 4.

7. Cyclic structures. The molecular structure of monosaccharides is represented in two different forms: (*i*) **Open** or **straight chain structure** and (*ii*) **Cyclic** or **ring structure.**

The straight chain structural formula accounts for its aldohexose character. This was proposed by a **E.H. Fischer** (see Box 8.3) and is called **Fischer formula** or **Fischer projection.**

Box 8.3. Fischer, Emil Hermann (1852–1919)

German organic chemist and pioneer of biochemistry, awarded the 1902 Nobel Prize for Chemistry for his work on the structures of sugars and purines, simple members of both of which families he synthesized. He also synthesized peptides and studied enzyme action in the breaking down of proteins.

Sir Walter Norman Haworth (1883–1950) worked on the **cyclization reactions** of carbohydrates and first proposed these representations (*i.e.*, Haworth projections). He was awarded with **Karrer** the 1937 Noble Prize for chemistry for his work on the structures of carbohydrates and vitamin C.

Karrer, Paul (1889–1971). Russian-born Swiss chemist awarded with W.N. Haworth the 1937 Nobel Prize for Chemistry for his work on the carotenoids and flavins, and on vitamins A and B.

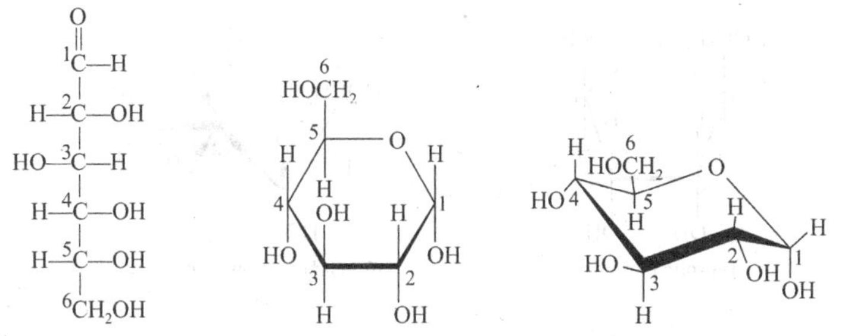

Fig. 8.16. Three types of molecule structures of D-glucose. A—Straight chain form; B—α-D-glucose-Haworth projection; C—α-D-glucose; chair form.

Monosaccharide sugars of 5 or more carbon atoms have **cyclic structures** or **Hawroth projections**. In case of cyclic structure, the carbonyl group (–C = O) forms a covalent bond with the oxygen of –OH group. In cyclic representation the molecule is viewed from the side and above the plane of the ring. By convention, the bonds nearest to the viewer are bold and thickened. X-ray diffraction analysis shows that the six-membered ring containing one oxygen atom is actually in the form of a **chair** or **boat** (**Murray** *et al.*, 2000; **Horton** *et al.*, 2006).

8. Mutarotation. The cyclic rings of monosaccharides occur in two stereoisomeric forms. These are named as α and β **anomers**. Interconversion of α and β forms in solution to maintain a balance in their concentration is called **mutarotation**. In mutarotation the hemiacetal ring opens and is reformed with change in the position of –H and –OH groups. Two anomers of D-glucose α and β forms occur naturally. Specific rotatin of α-D glucose is +112.2° while that of β-D glucose is 18.7°. In a solution D-glucose α anomer is 36% and β anomer is 64%. The specific rotation of glucose solution is always maintained at 52.7%. When α and β D-glucose is dissolved in water, it gradually changes into another form till ratio is achieved.

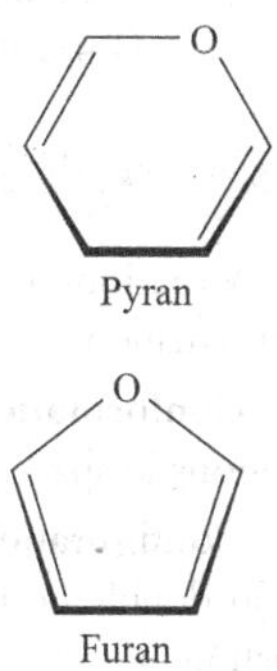

Fig. 8.17. A—Pyran and B—Furan.

9. Pyranose and furanose ring structure. The cyclic or ring like structure of 6 carbon sugars is described as a **pyranose ring**. A pyranose ring has pyran-like structure, *i.e.*, a 6-membered ring with a linkage between C-1 and C-5 carbons and 6th carbon remain outside the ring (Fig. 8.17). The **furanose ring** of monosaccharide has a furan-like structure, *i.e.*, a 5-membered ring with 4 carbon atoms and one oxygen, the linkage is between C-1 and C-3 carbons and the 5th and 6th carbon remain outside. Thus pyranose ring is hexagonal while furanose ring is pentagonal.

Both pyranose and furanose rings have two isomeric forms: α and β. In a solution, the furanose forms are less stable than pyranose forms.

Envelope and Twist Conformations of Furanose

Since the geometry of the carbon atoms of a monosaccharide ring is **tetrahedral** (bond angles near 110°) so monosaccharide rings are not actually planar. Cyclic monosaccharides can exist in a variety of conformations (see Box 8.4). Furanose rings adopt **envelope conformation** in which one of the five ring atoms (either C-2 or C-3) is out-of-plane and the remaining four are approximately coplanar (Fig. 8.18). Furanoses can also form **twist conformations**, in which two of the five ring atoms are out-of-plane, one on either side of the plane formed by the other three atoms. Various conformers of unsubstituted monosaccharides can rapidly interconvert.

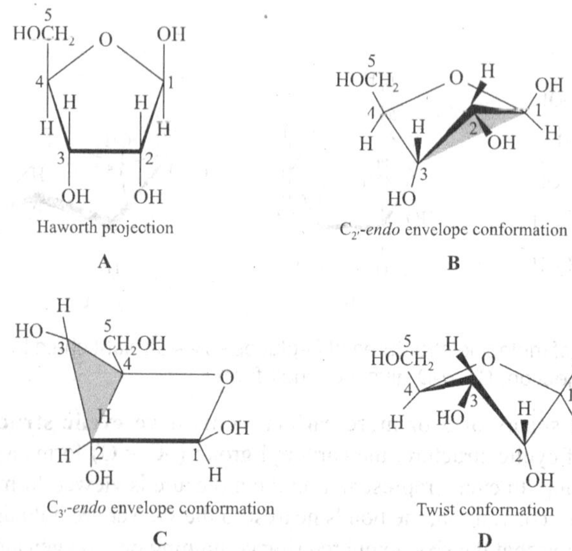

Fig. 8.18. Conformation of β-D-ribofuranose. A—Haworth projection; B C2′β-endo envelope conformation; C—C3′-endo envelope conformation; D—Twist conformation. In the C2′-endo conformation, C-2 lies above the plane defined by C-1, C-3, C-4, and the ring oxygen. In the C3′-endo conformation, C-3 lies above the plane defined by C-1, C-2, C-4, and the ring oxygen. In the twist conformation shown, C-3 lies above and C-2 lies below the plane defined C-1, C-4 and the ring oxygen (after Horton *et al.*, 2006).

Box 8.4 Conformation and Configuration

Variations in the 3-'D' structure of biomolecules are described in terms of configuration and conformation.

Conformation refers to the spatial arrangement of substituent groups that are free to assume different positions in space, without breaking any bonds, because of the freedom of bond rotation.

Configuration denotes the spatial arrangement of an organic molecule that is conferred by the presence of (a) double bonds, around which there is no freedom of rotation, or (b) chiral centers, around which substituent groups are arranged in a specific sequence.

Chair and Boat Conformations of Pyranose

Pyranose rings tend to assume one of the two conformations, the **chair conformations** or the **boat conformations** (Fig. 8.19). For each pyranose, there are two distinct conformers and six boat conformers. The chair conformations minimize steric repulsions among the ring substituents and are generally more stable than boat conformations. The –H, –OH and $–CH_2OH$ substituent of a pyranose ring in the chair conformation may occupy two different positions. In the axial position the substituent is above or below the plane of the ring, while in the equatorial position the substituent lies in the plane of the ring. In pyranoses, five substituents are axial and five are equatorial. Whether a group is axial or equatorial depends on which carbon atom (C-1 or C-4) extends above the plane of the ring when the ring is in the chair conformation (Fig. 8.20) shows the two different chair conformers of β-D-glucopyranose. The more stable conformation is the one in which the bulkiest ring substituents are equatorial (Top structure). In fact, this conformation of β-D-glucose has the least strict strain of any aldohexose. Pyranose rings are occasionally forced to adopt slightly different conformations, such as the unstable half-chair adopted by a polysaccharide residue in the active site of lysozyme.

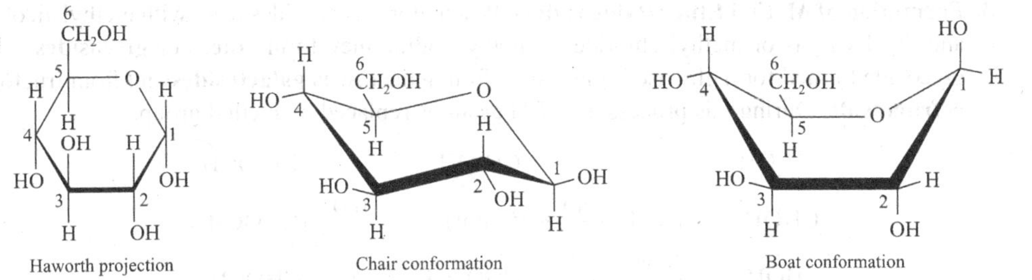

Fig. 8.19. Conformations of β-D-glucopyranose. Haworth projection, a chair conformation and a boat conformation.

Properties of Monosaccharides

1. Physical Properties

(*i*) Monosaccharides are crystalline.

(*ii*) They are sweet and readily soluble in water.

(*iii*) In solution monosaccharides occur in two forms: in D-form and in L-form.

(*iv*) They exhibit stereo-isomerism.

(*v*) Monosaccharides occur in two forms: in the form of straight open chain and in ring form.

(*vi*) They occur in two types of rings: pyranose ring form and furanose ring form.

(*vii*) Monosaccharides can not be hydrolysed into smaller molecules.

(*viii*) They can diffuse through semipermeable cell or plasma membrane.

(*ix*) Molecules of monosa-ccharides also exhibit the optic isomerism.

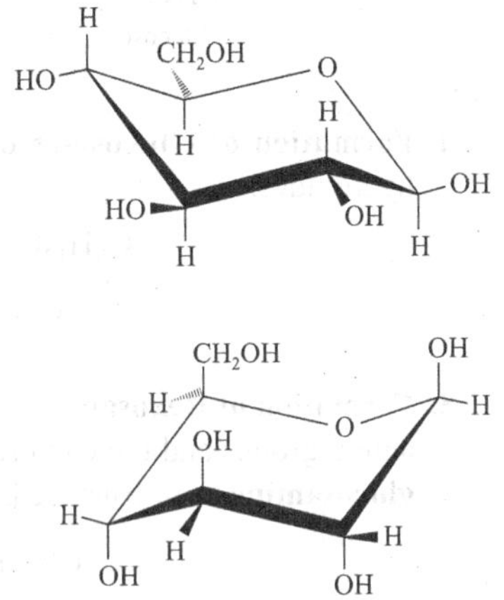

Fig. 8.20. The two chair conformers of β-D-glucopyranose. The top conformer is more stable.

2. Chemical Properties of Monosaccharides. Chemically, monosaccharides are derivatives of **polyhydric alcohols**. Therefore, these exhibit properties of both hydroxyl and carbonyl groups. Since monosaccharides also contain either a free aldehyde or a ketone group, so exhibit reducing property. They are also called **reducing sugars**.

A. Properties of Hydroxyl Group of the Monosaccharide

1. Esterification (Formation of Esters). Due to the presence of hydroxyl group (–OH) monosaccharides form esters with organic and inorganic acids. Acetates, benzoates and phosphates are the common esters of monosaccharides. Hexose phosphates (glucose and fructose-phosphates play important role in carbohydrate metabolism).

2. Acetylation. When monosaccharides are heated with acetic anhydride or acecyl chloride in presence of zinc chloride, their –OH groups are acetylated. The number of acetyl groups incorporated in a molecule of monosaccharide is equal to the number of –OH groups present in it.

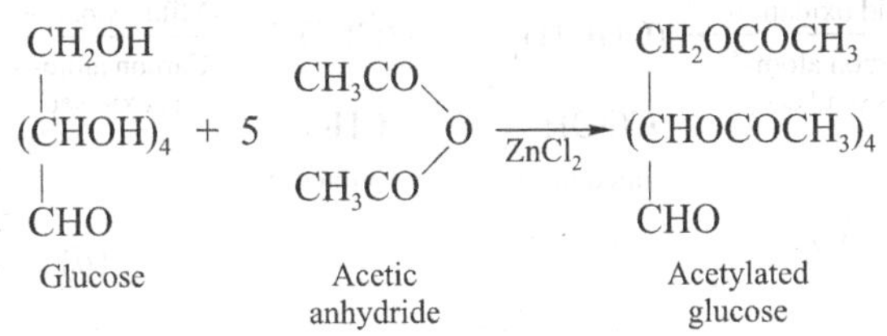

3. **Formation of Methyl Ether (Glucosides).** When monosaccharides react with methyl alcohol and dry HCl gas or methyl chloride or methyl iodide they form **ethers** or **glycosides**. The glycoside formed form glucose is glucoside, from galactose is **galactosides** and from fructose is **fructoside**. During this process the –OH group is replaced by methyl group.

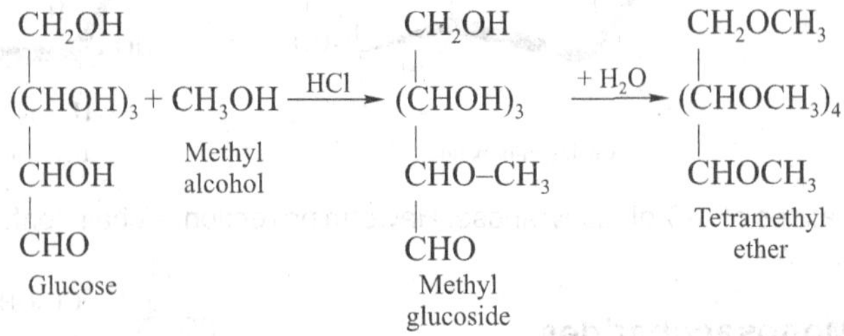

4. **Formation of Glucosates or Fructosates.** Hexoses form hexosates with certain metallic hydroxides:

$$\underset{\text{Glucose or Fructose}}{C_6H_{12}O_6} + \underset{\text{Calcium hydroxide}}{Ca(OH)_2} \rightarrow \underset{\text{Calcium glucosate or calcium fructosate}}{C_6H_{12}O_6CaO} + \underset{\text{Water}}{H_2O}$$

5. **Formation of Hexosamine.** The hydroxyl group in hexose sugars can easily be replaced with amino groups and the end product is **hexosamine** or amino sugar. Hexosamine of glucose is **glucosamine** or of fructose is **fructosamine**.

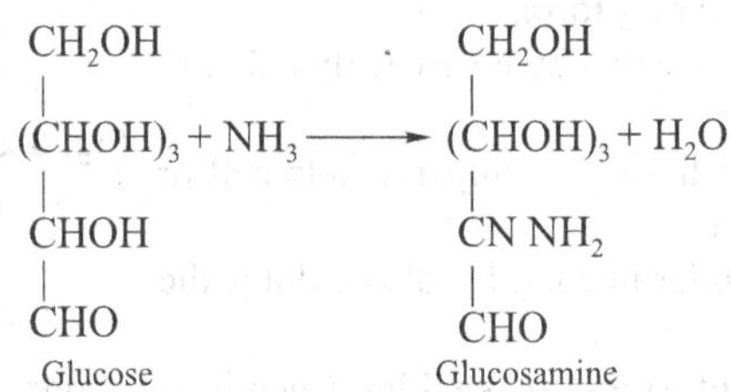

B. Reactions of Carboxyl Group

Monosaccharides contain a carboxyl group which may be an aldehyde or ketone and exhibit following reactions:

1. **Oxidation.** Aldoses can be oxidized with mild oxidizing agents such as Fehling's solution, Benedict's solution, bromine water, or Tollen's reagent. These form monocarboxylic hydroxy acids. If carbon atom one (C-1) is oxidised in a hexose, the oxidation product is **aldonic acid** (*e.g.*, glucose → gluconic acid), but if carbon atom six (C-6) is oxidised it is called **uronic acid** (glucose → **glucuronic acid**).

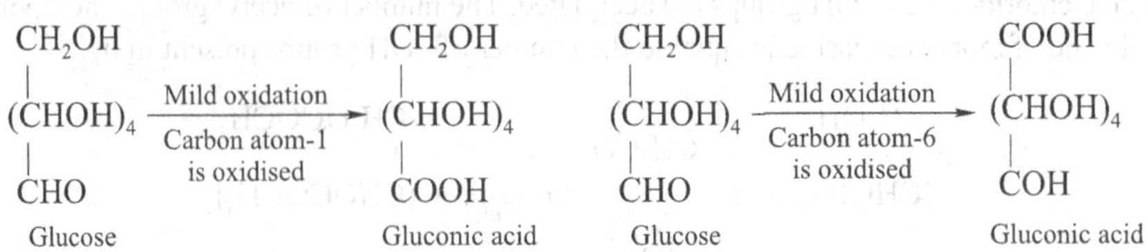

(*i*) (*ii*)

Aldoses, on being oxidised with dilute nitric acid produce dicarboxylic hydroxy acid or **aldaric acid** and with strong oxidising agent (glucose → glucaric acid or saccharic acid).

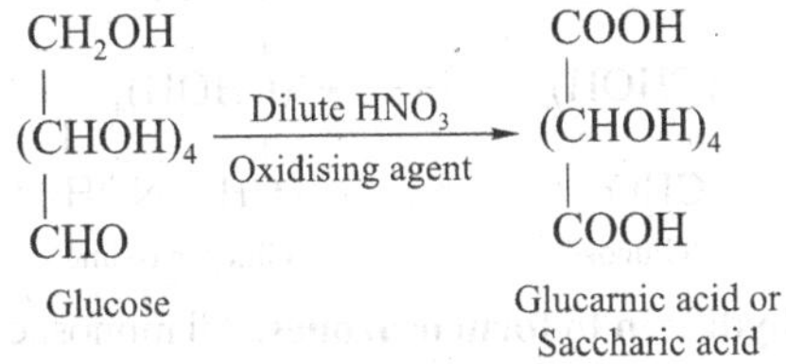

Ketoses when oxidised with nitric acid, undergoes cleavage at C = O group and form acids with a smaller number of carbon atoms.

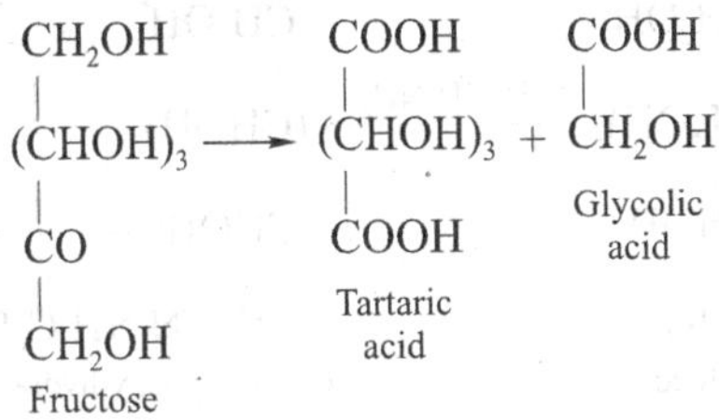

2. **Reduction.** Free aldehyde and ketone groups of monosaccharides are reduced to their corresponding polyhydroxy alcohols by reducing agents such as sodium amalgam (sodium-mercury amalgam), electrolytic and catalytic reduction.

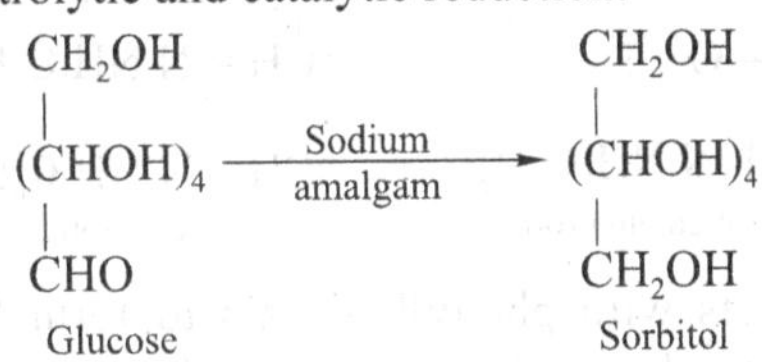

Thus, by reaction of reduction glucose yields **sorbitol**, galactose yields **gulcitol**, mannose yields **mannitol** and fructose produces mannitol and sorbitol.

3. **Iodo compounds.** When an aldose monosaccharide is heated with concentrated hydroiodic acid (H1) and red phosphorus at 100°C, it loses all its oxygen and is converted into iodo compound.

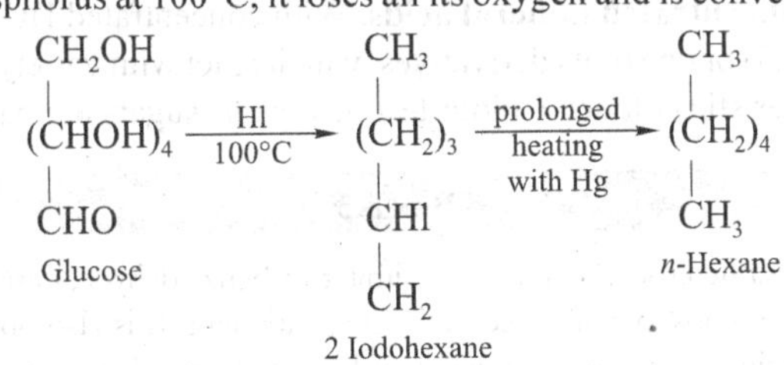

4. **Reaction with hydrogen cyanide (HCN).** During this reaction, the HCN is added to free aldehyde or ketone groups of monosaccharides to form cyanohydrins.

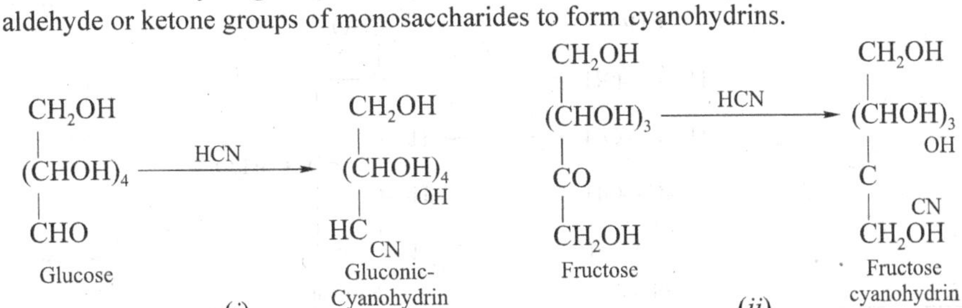

5. **Reaction with Hydroxylamine.** Due to this reaction, saccharides condense to form oximes.

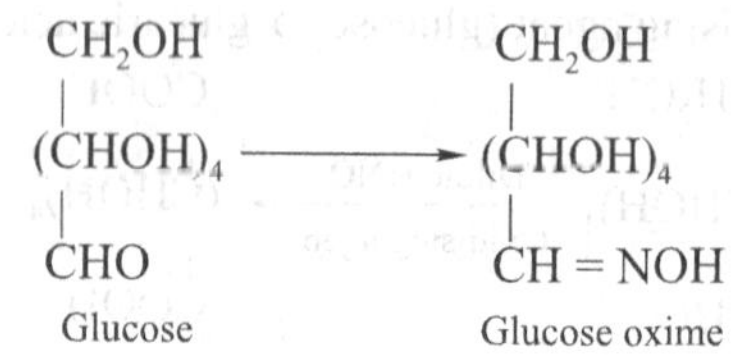

6. **Reaction with phenylhydrazin to form osazones.** All monosaccharides or reducing sugars in the slightly acidic medium and excess of hydrazine form **phenyl osazone** compounds. These have crystalline appearance. Phenylhydrazine ($C_6H_5NH.NH_2$) is a colourless toxic liquid.

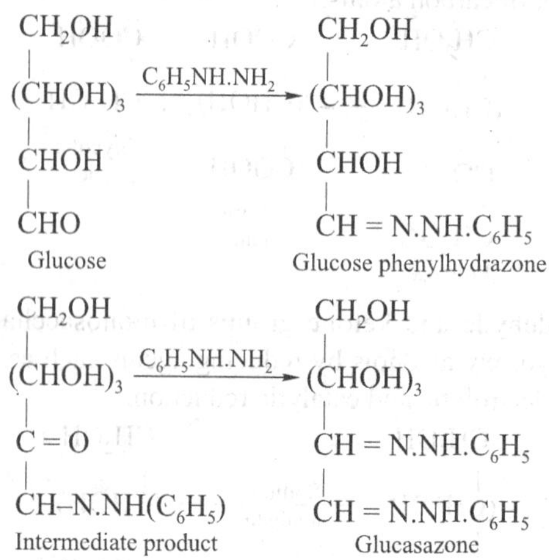

Similarly, fructose reacts with phenylhydrazin to form fructosazone. Reaction with phenylhydrazine involves only 2 carbon atoms *viz.*, the carbonyl (*i.e.*, the aldehyde or ketone group) and the adjacent one.

C. Other Reactions

1. **Reactions with concentrated mineral acids.** With concentrated HCl monosaccharides yield furfural (see Box 8.5) or one of its derivatives, which react with variety of phenolic compounds to produce characteristic colours. Colour test of various sugars are based on such reaction.

Box 8.5

Furfural ($C_5H_4O_2$) is a heterocyclic aldehyde, similar to benzaldehyde, derived from furan (C_4H_4O). It is made by digesting corn cobs, oat and rice hulls, etc., with acid. It is also obtained on heating ribose. Furfural is a colourless liquid. It is used as a solvent and as an intermediate in the manufacture of plastics and other compounds.

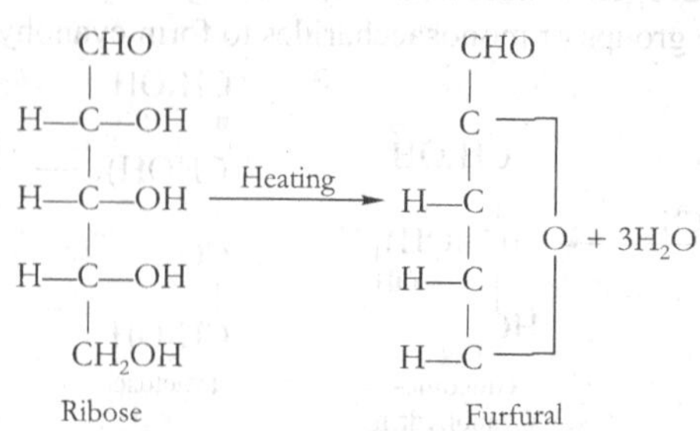

2. Reaction with alkali.

(*a*) In dilute alkalies the monosaccharides change to cyclic α and β structures and the two isomeric forms are in equilibrium.

On standing, these produce a mixture of glucose, fructose and mannose (inter-conversion).

(*b*) If this mixture is heated to 37°C, a series of enols are formed. (Here double bonds shifts from oxygen to carbon atom).

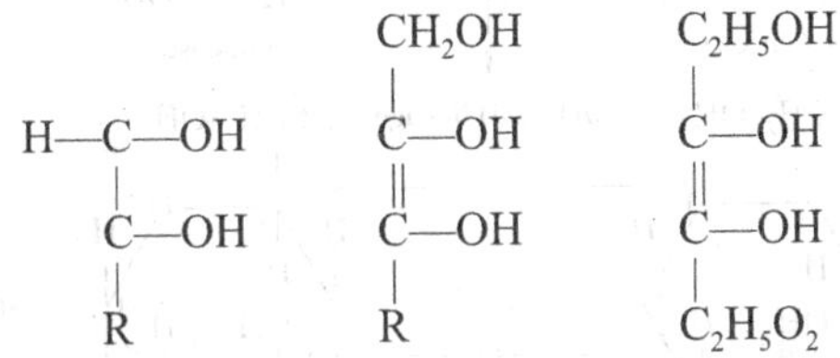

1, 2 Enediol   2, 3 Enediol   3, 4 Enediol

(*c*) In the presence of concentrated alkali sugar caramelizes and produces a series of decomposition products. Yellow and brown pigments are developed.

3. Fermentation. Monosaccharides are readily fermented by yeast to alcohols:

$$\underset{\substack{\text{Glucose or}\\ \text{Fructose}}}{C_6H_{12}O_6} \xrightarrow{\text{Zymase}} \underset{\text{Ethanol}}{2C_2H_5OH} + 2CO_2$$

Reducing Property

1. Due to a free aldehyde or ketone group, they reduce Cu^{++} to Cu^{+}. This property is the basis for Benedict's test and Fehling's test used to detect the presence of sugar (glucose in urine).
2. Aldehyde or ketone group of monosaccharide binds with the alcoholic group of another organic compound thus binding the two compounds togather. This bond is called **glycosidic bond**. On hydrolysis, the original compounds are formed.

Uses of Monosaccharides

1. Glyceraldehyde is fundamental in the process of glycolysis.
2. Deoxyribose and ribose (pentose) sugars are constituents of nucleic acids.
3. Ribose is a constituent of several coezymes.
4. Glucose is the source of energy and is the major product of photosynthesis.
5. Glucose is the easily accessible food for children and invalids.
6. Glucose is used in making sweets, candies, jellies, syrups and cakes.
7. Glucose is used for the manufacture of wines and glycerol.
8. Glucose forms raw material in the synthesis of vitamin C (ascorbic acid).
9. Dextroses (glucose in solution is in dextrorotary form) is frequently used in medical practice, *i.e.*, D-glucose solution is given intravenously to the patients.
10. Glucose is used as cheap reducing agent in the industry in silvering of mirrors and in vat dyeing with indigo.
11. Fructose is chiefly used as food, specially the diabetic patients use it in place of cane sugar.
12. Fructose is abundantly found in semen which is utilised by the sperms for energy.

8.3. OLIGOSACCHARIDES

The oligosaccharides (G. *oligos* = few) are formed by joining together of 2–10 monosaccharide molecules. Linkage of two glucose molecules of form maltose illustrates the general pattern of the reactions forming oligosaccharides, also polysaccharides (Fig. 8.21).

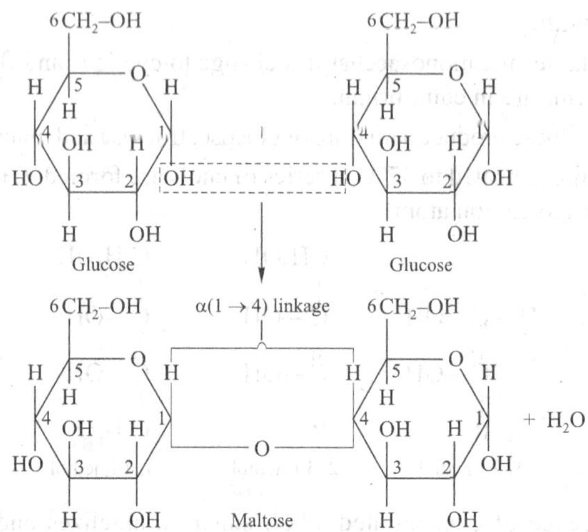

Fig. 8.21. Synthesis of the disaccharide maltose from two glucose molecules. The linkage formed by oxygen bridge is an alpha 1, 4 glycosidic.

Linkage (or bond) is established between the number 1 carbon of one glucose molecule and the number 4 carbon of the other glucose molecule by an oxygen bridge (C–O–C). The process involves loss of water and is called **condensation** or **dehydration**. The compounds of water, H^+ and OH^-, are split from the reacting chemical groups of the combining molecules. The chemical linkages joining the monosaccharide units are called **glycosidic bonds**. These bonds are designated as $1 \rightarrow 4$ or 1, 4 linkages, which may be alpha (α) or beta (β) depending on the orientation of the hydrogen at the number 1 carbon forming the bond. In a maltose molecule, the linkage is α $(1 \rightarrow 4)$ glycosidic because the hydrogen at the number 1 carbon of the bond is in the α position (see Box 8.6). The oligosaccharides are hydrolysed into component monosaccharides by an acid or enzyme.

Box 8.6

It may be noted that the linkages (*i.e.*, α-1-4 glycosidic bonds) are characteristic of energy providing carbohydrates, namely, sugars and starches, which can be digested by organisms. The structural carbohydrate cellulose has β-1, 4 glycosidic linkages between glucose units, and cannot be digested by most eukaryotic organisms.

Classification of Oligosaccharides

The oligosaccharides are classified according to the number of their monosaccharides units or monomers as follows (see Box 8.7):

Box 8.7

Disaccharides

- **Nonreducing** (or **Glycosylglycosides**)
  - C_1–C_1 *glycosidic linkage*
  - C_1–C_2 *glycosidic linkage*
- **Reducing** (or **Glycosylaldoses** or **Glycosylketoses**)
  - C_1–C_4 *glycosidic linkage* *e.g.*, Lactose Maltose
  - C_1–C_6 *glucosidic linkage* *e.g.*, Gentiobiose Malibiose

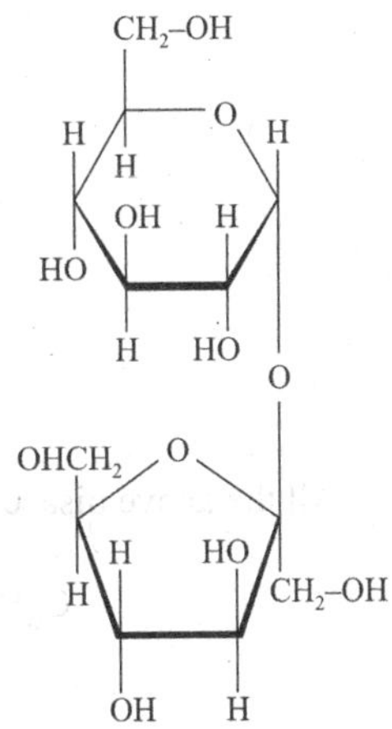

Fig. 8.22. Chemical formula of Sucrose.

1. **Disaccharides.** These have 2 monosaccharide monomers per molecule, *e.g.*, maltose, sucrose, lactose, etc.
2. **Trisaccharides.** These contain 3 monosaccharide units per molecule. These include
   (*a*) **Mannotriose.** It is formed of two molecules of galactose and one molecule of glucose.
   (*b*) **Robinose.** It is formed by the condensation of one molecule of galactose and two molecules of rhamnose.
   (*c*) **Raffinose.** It is formed of a molcule of glucose, a molecule of galactose and one molecule of fructose. This crystalline sugar is obtained commercially from cottonsced meal.
   (*d*) **Gentinose.** It is formed of one molecule of fructose and two molecules of glucose. It is obtained from the gentian roots.
   (*e*) **Melezitose.** It is formed of one molecule of fructose and two molecules of glucose. It is obtained from exudation of various trees (such as larch).
3. **Tetrasaccharides.** These have 4 monosaccharide monomers per molecule. Only following to tetrasaccharides are known:
   (*a*) **Stachyose.** It is composed of D-glucose, fructose and two molecules of galactose. It is found in the haricot beans.
   (*b*) **Scorodose.** It is found in the bulbs of garlic and onion.
4. **Pentasaccharides.** These have 5 monosaccharide units per molecule.

Examples of Biologically Important Disaccharides

The disaccharides are the most abundant oligosaccharides in the cells. They, as already mentioned, have two monosaccharide monomers (or units) joined by glycosidic bond. They are also called **double sugars**. Biologically important disaccharides are sucrose, lactose, maltose, trehalose and cellobiose.

1. **Sucrose.** It is the simple **cane sugar** or **table sugar**. Sucrose is common in higher plants and its commerical sources are sugarcane and sugarbeet (*Beta vulgaris*). It is the storage product of photosynthesis in these plants. Sucrose is formed due to union of a glucose molecule with a fructose molecule (Fig. 8.22).
2. **Lactose.** It is called **milk sugar**. It is found in the milk of mammals. It is synthesized in mammary glands. Lactose is composed of one glucose molecule and one galactose molecule (Fig. 8.23). Souring of milk occurs when bacteria convert lactose in lactic acid.

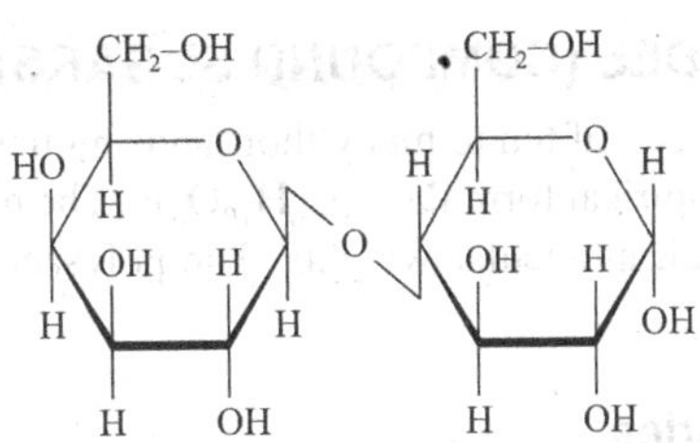

Fig. 8.23. Chemical formula of Lactose.

3. **Maltose.** It is formed in the germinating starchy seeds and malt. It is called the **malt sugar**. Maltose is degradation product of larger polysaccharide molecule (starch). Maltose is composed of two glucose molecules (Fig. 8.24).

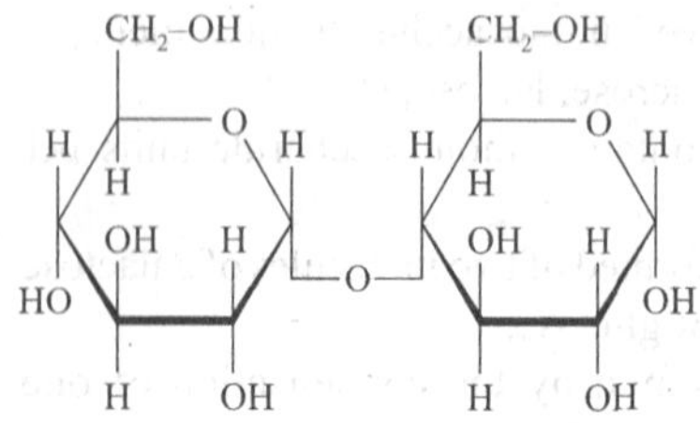

Fig. 8.24. Chemical formula of Maltose.

All the above disaccharides have an identical chemical formula $C_{12}H_{22}O_{11}$ derived as under:

$$\underset{\substack{\text{Glucose}\\ \text{Fructose}\\ \text{Galactose}}}{C_6H_{12}O_6 + C_6H_{12}O_6} \xrightarrow{\text{Condensation}} \underset{\substack{\text{Maltose}\\ \text{Sucrose}\\ \text{Lactose}}}{C_{12}H_{22}O_{11}} + \underset{\text{Water}}{H_2O}$$

4. Trehalose. It is the major sugar of insect haemolymph, in which it serves as an energy storage compound. Trehalose has no free aldehyde group and does not form an osazone. On hydrolysis, it yields glucose.

5. Cellobiose. It is formed during hydrolysis of cellulose. Cellobiose is composed of two molecules of D-glucose, and is identical to maltose except that it has β-1, 4 glycosidic linkage instead of α-1, 4 linkage in maltose.

Properties of Disaccharides

Like the monosaccharides, the disaccharides are also sweetish to taste and form a true solution in water. They, however, diffuse very slowly through the plasma membrane. They hydrolyze into free monosaccharide units when treated with a dilute acid or with an enzyme. Maltose and lactose are reducing disaccharides but sucrose is not.

Oligosaccharides may form unbranched or branched chains. Larger branched or unbranched oligosaccharide chains are found attached to plasma membrane.

Functions of Oligosaccharides

Some common functions of the oligosaccharides are the following:

1. Cellular interactions. The oligosaccharides attached to the plasma membrane help in recognizing cells of their own kind or species.

2. Storage products. Sucrose is a reserve food in sugarcane and sugarbeet.

3. Fuel. Disaccharides also serve as fuel to provide energy.

8.4. POLYSACCHARIDES (COMPOUND SUGARS)

The polysaccharides are composed of ten to many thousands monosaccharides as the monomers in their macromolecules. Their empirical formula is $(C_6H_{10}O_6)n$. The molecules of the polysaccharides are of colloidal size having high molecular weights. The polysaccharides can be hydrolysed into simple sugars.

Naming of Polysaccharides

Polysaccharides are named after their monosaccharide monomers by changing the *–ose* ending of the monosaccharide to *–an*. Thus **cellulose**, a polysaccharide formed of *glucose* units, is called **glucan**. Fructose polymers are called **fructans** (previously fructosans). Likewise, galactose polysaccharides are called **galactans**; xylose polysaccharides are called **xylans**, pentose polymers are called **pentans** (formerly pentosans). Since the term glycose is used for any monosaccharide, the general name for any polysaccharide is **glycan.** If a polysaccharide contains more than one type of sugar unit the two sugars are named in alphabetical order. Thus a polysaccharide consisting of D-glucose and D-mannose units is termed D-gluco-D-mannoglycan. Some old names which have not been changed to conform with *–an* ending include starch, cellulose, pectin, amylopectin, inulin, chitin and heparin.

Classification of Polysaccharides

Polysaccharides have been classified as **homopolysaccharides (homoglycans)** and **heteropolysaccharides (heteroglycans)**. Homoglycans contain only one type of monosaccharide while heteroglycans contain at least two types of glycose monomers. Most polysaccharides contain one or two types of glycose units. Polysaccharides containing three types of glycose units are less common.

Some homopolysaccharides are formed of amino sugars such as *glucosamine* and *galactosamine* monosaccharides or of simple glucoronic acid and iduronic acid molecules.

Polysaccharides are formed by condensation of many molecules of monosaccharides with corresponding elimination of water molecules. The monosaccharide units are joined by **glycosidic bonds.**

1. Homopolysaccharides

(*i*) **Glucans** (of glucose monomers), *e.g.*, starch, glycogen, cellulose, etc.
(*ii*) **Galactans** (of galactose monomers), *e.g.*, pectin, galactan from snails.
(*iii*) **Mannans** (of mannose monomers), *e.g.*, yeast mannan.
(*iv*) **Xylans** (of xylose monomers), *e.g.*, hemicellulose, xylan.
(*v*) **Fructans or fructosans** (of fructose monomers), *e.g.*, inulin.

(*a*) **Inulin** is the reserve carbohydrate of many plants of family compositae such as artichock, *Dahalia, Dendaleon*, garlic, onion, etc. It consists of 30 fructose units linked by $\beta(1 \rightarrow 2)$ bonds and has a molecular weight of 5000. Inulin is not metabolised in human body and is readily filtered through the kidney. It is therefore used in testing kidney function, especially glomerular filteration rate (GFR).
Paramylum. A reserve starch like unbranched nutrient homopolysaccharide of glucose found in various protozoans (*e.g.*, *Euglena*) and algae.

(*b*) **Starch.** Starch is a white soft amorphous powder and lacks sweetness. It is insoluble in water, alcohol and ether at ordinary temperature. Starch is a storage polysaccharide in green plants and the most important source of carbohydrate in our food. It is stored (see Box 8.8) in cereals, potatoes, legumes, sweet potatoes, tapioca and fruits (*e.g.*, banana). It is a polymer of D-glycopyranose units linked by α-1-4 linkage or α-glycosidic bonds. Starch consists of a mixture of **amylose** and **amylopectin** in the proportion of 1 : 4 (see Box 8.8). Both are high molecular weight compounds. Amylose is linear while amylopectin is branched. Treating starch with hot water dissolves amylose, while amylopectin remain.

Box 8.8 Starch grains

Starch is stored as starch grains in chloroplasts or in special leucoplasts called **amyloplasts**. The diameter of starch grains (= granules) range from 3 to 100 μm. The starch grains are of two types: **simple** that occur singly and **compound** which occur in groups. A starch grain consists of several layers, the **shells**, arranged in concentric or eccentric around a point, the **hilum**. The microscopic form of the starch grains is characteristic of the source of starch. The starch grains from a particular source have a characteristic shape and can be identified by examination under the light microscope (Fig. 8.25). Starch gives blue colour with iodine solution.

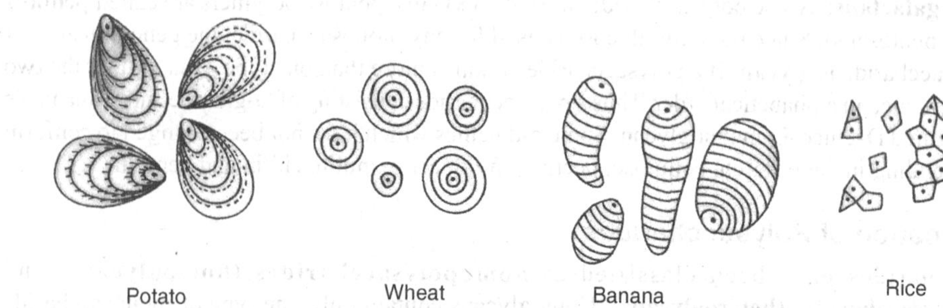

Fig. 8.25. Starch grains from different sources.

Box 8.9

Natural starches consists of two components: **amylose** (15–20%), a long unbranched straight chain component and **amylopectin** (80–85%), a branched chain polysaccharide. Potato and cereal starches are 20–30% amylose and 70–80% amylopectin. Starch from waxy corn is notable as it consists practically of amylopectin component, there being no amylose. However, starch of some varieties of pea having wrinkled surface may have as much as 98% of amylose.

***(i)* Amylose** (Fig. 8.26). Amylose is an unbranched polymer of about 100 to 1000 D-glucose residues connected by α-(1–4) glucosidic linkages, specifically termed **α(1 → 4) glycosidic bonds** because the anomeric carbons belong to glucose residues. Although it is not truely soluble in water, amylose forms hydrated micelles in water and can assume a helical structure under some conditions (Fig. 8.26). Amylose molecule is coiled like a watch spring. Each turn of amylose helix contains about six glucose units. Amylose gives blue colour with iodine.

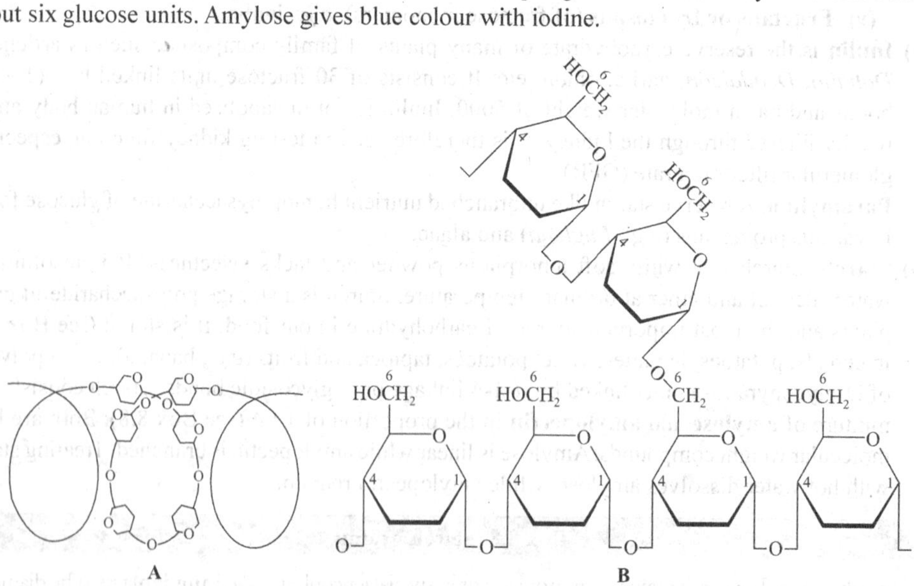

Fig. 8.26. Structure of starch. A—Amylose, showing helical coil structure (on suspension in water); B—Amylopectin, showing 1 → 6 branch (after Murray *et al.*, 2000).

***(ii)* Amylopectin.** Amylopectin is a branched version of amylose (Fig. 8.26). Branches, or polymeric side chains, are attached via α-(1 → 6) glucosidic bonds to linear chains of residues linked by α-(1 → 6) glucosidic bonds. Branching occurs, on average, once every 25 residues and the side chains contain about 15 to 25 glucose residues. Some side chains themselves are branched. Amylopectin molecules isolated from living cells may contain 300 to 6000 glucose residues (**Horton** *et al.*, 2006).

Molecule of amylopectin is coiled concentrically. Because of its more complex structure, amylopectin is less soluble in water than amylose. It gives red-purple colour with iodine solution.

Dextrins and Limit Dextrins

Starch on hydrolysis yields lower molecular weight polysaccharides and finally maltose or glucose. Partial hydrolysis results in substances called **dextrins**. At first **amylodextrins** are formed which give rise to violet colour with iodine. Then **erythrodextrins** are formed which gives red colour with iodine. Further hydrolysis results in the formation of **achrodextrins** which do not give any colour with iodine. Finally reducing sugars (maltose and glucose) appear. Dextrins are easily digested and are therefore used for feeding infants. The enzymes which bring about hydrolysis of starch are called **amylases**.

Dietary starch is degraded in the gastrointestinal tract by the actions. of **α-amylase** and debranching enzyme. α-Amylase which is present in both animals and plants, is an **endoglycosidase** (it acts on internal glycosidic bonds). The enzyme catalyzes random hydrolysis of the α-(1 → 4) glucosidic bonds of amylose and amylopectin.

Another hydrolase, **β-amylase**, is found in seeds and tubers of some plants. **β-amylase** is an **exoglycosidase** (it acts on terminal glucosidic bonds). It catalyzes sequential hydrolytic release of maltose from the free, nonreducing ends of amylopectin.

Despite their α and β designations, both types of amylases act only on α-(1 → 4)-D-glycoside bonds. Figure 8.27, shows the action of α-amylase and β-amylase on amylopectin. The α-(1 → 6) linkages at branch points are not substrates for either α- or β-amylase. After amylase catalyzed hydrolysis of amylopectin, highly branched cores resistant to further hydrolysis, called **limit dextrins**, remain. Limit dextrins can be further degraded only after debranching enzymes have catalyzed hydrolysis of the α-(1 → 6) linkages at branch points.

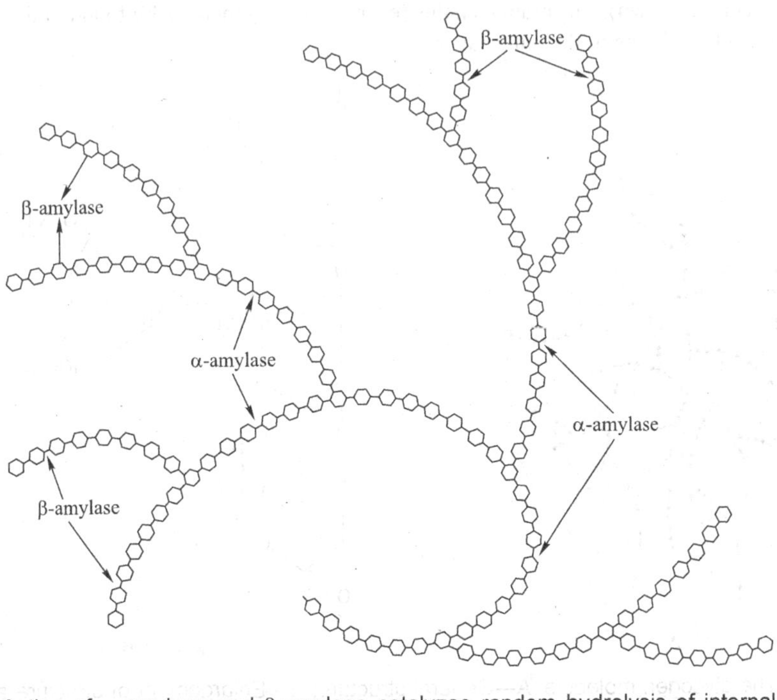

Fig. 8.27. Action of α-amylase and β-amylase catalyzes random hydrolysis of internal α-(1 → 4) glycosidic bonds; β-amylase acts on the nonreducing ends. Each hexagon, represents a glucose residue; the single reducing end of the branched polymer (after Horton *et al.*, 2006).

The branched structures of amylopectin (and glycogen) possess only one reducing end but many nonreducing ends. It is at these nonreducing ends that enzymatic lengthening and degradation occur.

Table 8.3. Differences between amylose and amylopectin.

Amylose	Amylopectin
1. It forms 20% of the starch.	1. It forms 80% part of starch.
2. It is a straight chained polysaccharide. The chain is unbranched.	2. It is a branched polysaccharide.
3. Each amylose molecule is formed of 100 to 1000 glucose residues.	3. An amylopectin molecule is formed of 3000 to 6000 glucose units.
4. Amylose is soluble in hot water.	4. Amylopectin is insoluble in water.
5. Its molecular weight ranges 50,000 to 60,000.	5. Its molecular weight is more than 500,000.
6. It gives blue colour with iodine reagent.	6. It gives pink purple (or red) colour with iodine.
7. Amylose is hydrolysed with β-amylase enzyme.	7. Amylopectin is hydrolysed into dextrin by enzyme β-amylase.

Glycogen. Glycogen is the main reserve food material of animal cells. It is known as **animal starch**. Glycogen is found mainly in liver, muscles and brain. It also occurs in cells of fungi and bacteria. The glycogen in the liver supplies glucose to all tissues through the blood. Muscle glycogen on the other hand is available during contraction of muscle.

Glycogen is a soluble polysaccharide. The molecule of glycogen is a **sphere** (Fig. 8.28) approximately 21 nm in diameter that can be visualised in electron microscope. It has a molecular mass of 10^7 Da and consists of polysaccharide chains each containing about 13 glucose residues. The chains are either branched or unbranched and are arranged in 12 concentric layers (in Fig. 8.28, only four layers are shown). The branch chains (each has two branches) are found in the inner layers and the unbranched chains in the outer layers.

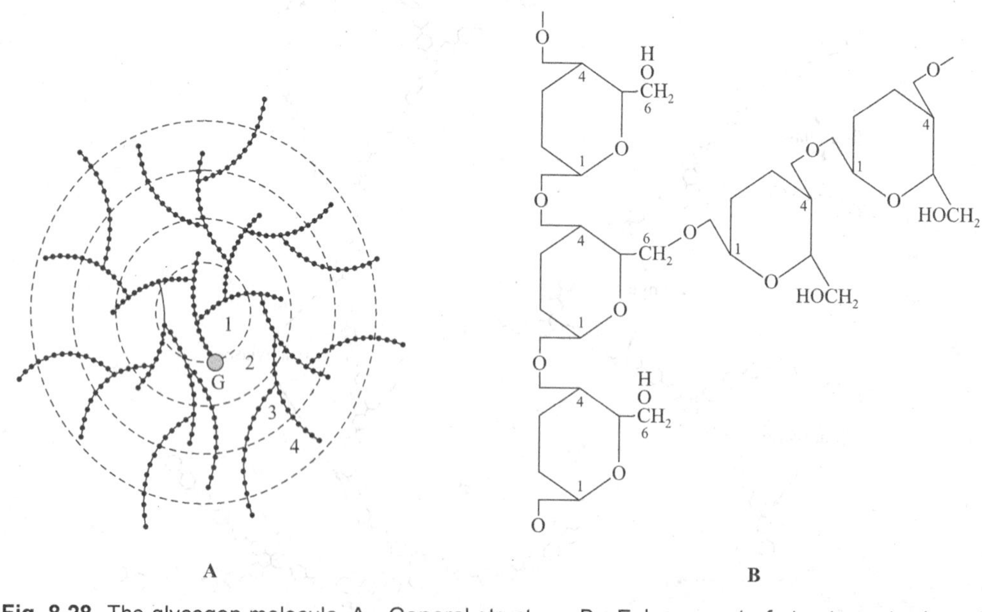

Fig. 8.28. The glycogen molecule. A—General structure; B—Enlargement of structure at a branch point. G=glycogenin, the primer molecule for glycogen sythesis (after Murray *et al.*, 2000).

Chemically glycogen is analogous to starch. Both are glucans consisting of glucose units linked by α-(1 → 4) glycosidic linkages, with branch chains formed at α-(1 → 6) glycosidic bonds. They differ mainly in their molecular weight and degree of branching. Glycogen's molecular weights is about 4,000,000. Each glycogen molecule contains about 50,000 glucose residues. The branching chains of glycogen average about 12 glucose units as compared to 25 or so for starch. Glycogen is non-reducing and gives a red colour with iodine.

Glycogen is produced from glucose in the mammalian liver and muscles when blood sugar levels are high. a process called **glycogenesis**, which is under the influence of insulin hormone. Glycogen in the liver can be broken down to glucose when blood sugar levels are low, a process called **glycogenolysis**, which is under the influence of **glucagon** hormone. Glycogen in the muscle, however, is broken down to lactic acid in glycolysis (**Jain** 2012).

Table 8.4. Differences in starch and glycogen.

Starch	Glycogen
1. It is a storage polysaccharide in the plants.	1. It is called animal starch. It is stored in animal cells such as liver and muscles.
2. It is formed of amylose and amylopectin. Both contain glucose units.	2. It is formed of glucose residues.
3. Amylose chains are unbranched while amylopectin chains are branched.	3. Glycogen molecules are branched similar to amylopectin.
4. The chains of amylopectin are branched after every 20–24 glucose units.	4. The glycogen chains are branched after every 8–10 glucose units.

Cellulose. It is most abundant carbohydrate, in fact, the most abundant organic compound on Earth. It is the main structural polysaccharide of the plants. Cellulose is totally absent in animals. Cellulose is common in the cell walls of most algae, all higher plants, certain fungi and some protists. It is especially abundant in wood (25–50%) and cotton fibers (90%). Cellulose molecule is a long unbranched chain of about 300 to more than 15000 glucose units (see **Horton** *et al.*, 2006) with molecular weight between 0.5 to 2.5 million (Fig. 8.29). The molecules of cellulose are folded in such a way so as to form long, highly tensile fibers that are aggregated into bundles, called **microfibrils**. The molecules in a microfibril are held together by hydrogen bonds. Cellulose gives no colour with iodine solution.

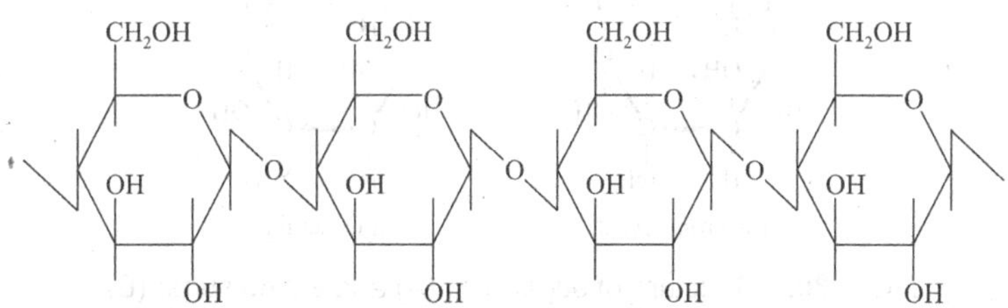

Fig. 8.29. Chemical formula of cellulose.

Cellulose is a linear polymer of β-D-glucose units which are connected through β-1 → 4 glycosidic bonds. Partial hydrolysis of cellulose yields **cellotetrose**, **cellotriose** and **cellobiose** consisting of four, three or two monosaccharide units, respectively. Strong mineral acid is required for complete hydrolysis to D-glucose. The β-(1 → 4) glycosidic linkages can withstand hydrolysis by glycosidases (enzymes) found in digestive tract of human beings and higher animals. Since, it is not broken down to glucose by animals, it is not useful as a food source. It simply forms "bulk" or "roughage" of the food necessary for proper functioning of the alimentary canal. Some animals, such as the farm animals ("ruminants" like cow and sheep) with the help of digestive system bacteria and the termites with the aid of intestinal flagellates can hydrolyse cellulose. These symbiotic organisms produce an enzyme **cellulase** for digesting cellulose. Some snails are said to produce the cellulase enzyme.

Uses of cellulose. Cellulose provides many useful materials. For example, cotton, linen, jute, rayon, cellulose acetate (used in fabrics, plastics and unbreakable glass), cellulose nitrate (used in propellant explosives), cellophane and carboxymethyl cellulose (added to icecreams, cosmetics and medicine). Cattle and buffaloes use cellulose as food because they can digest it by the help their symbiotic microorganisms. Cellulose serve as roughage in the human digestive tract. Cellulose rich wood is used in the manufacture of furniture, sports goods, tool handles and many more articles. Cellulose is also used as fuel (*e.g.*, fire wood) and in the production of paper.

Agar (Agar-Agar). It is the gelatinous extract from the cell wall of certain red algae (*e.g.*, *Gelidium*, etc.) and consists of the polysaccharide **agarose** (β-1, 3 linked D-galactose and 1, 4 linked anhydro-L-galactose) and **agropectin** (a sulphated galactan mixture) in a ratio of 7 : 3. Thus, this homopolysaccharide is composed of galactose monomer. Japan is the major producer of agar. Agar forms highly viscous gels. Once solidifies, agar gel does not melt below 100°C. Agar is an inert substance and hence provides no nutrients. Different nutrients may be added to make media suitable for different microorganisms.

Agar is used to solidify culture media on which bacteia are grown in the laboratory. Agar is also used as laxative and as food (deserts) in the East.

Heteropolysaccarides

These are composed of modified monosaccharide molecules of different types. Examples of heteropolysaccharides include mucopolysaccharides, glycoproteins, chitin, peptidoglycans, pectin, hemicellulose, etc.

I. Mucopolysaccharides. These are polymers of modified sugars (galactose and mannose) called sugar acids and amino sugars. Sugar acid has at least one hydroxyl group replaced by an acid group. Amino sugar has an amino group in place of one of the hydroxyl groups (Fig. 8.30).

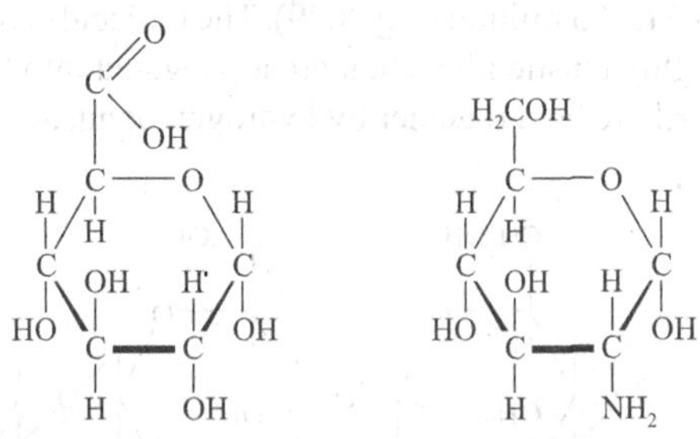

Fig. 8.30. Structure of sugar acid (A) and amino sugar (B).

The mucopolysaccharides are slimy compounds like that noticed in okra (bhindi) fruit. Important examples of mucopolysaccharides are hyaluronic acid, heparin, chondroitin sulphate and keratin sulphate.

Proteoglycans. The molecules of proteoglycans consist of much longer portion of polysaccharide and a small portion of protein. They are also called **mucoproteins**. The proteoglycans are amorphous and form gels which are able to hold large amounts of water.

The **cartilage proteoglycan** is found extracellularly in cartilage and bone. In its molecule, strands of protein, called **core protien**, extend radially from a long, central hyaluronic acid molecule. In each core protein strand, three carbohydrate bearing regions may be identified. The first region contains numerous oligosaccharides, the second region contains keratin sulphate chains and the third region contains chondroitin sulphate chains. This arrangement gives cartilage its resilience and tensile strength.

(*i*) **Hyaluronic acid.** It is polymer of D-glucuronic acid and N-acetyl D-glucosamine units arranged alternately (Fig. 8.31). Hyaluronic acid is present in the various lubricating fluids of the body such as synovial fluid of bone joints, cerebrospinal fluid of brain and spinal cord, Viterous humour of the eye. It also occurs as a coat around the ovum and in the skin and between the cells.

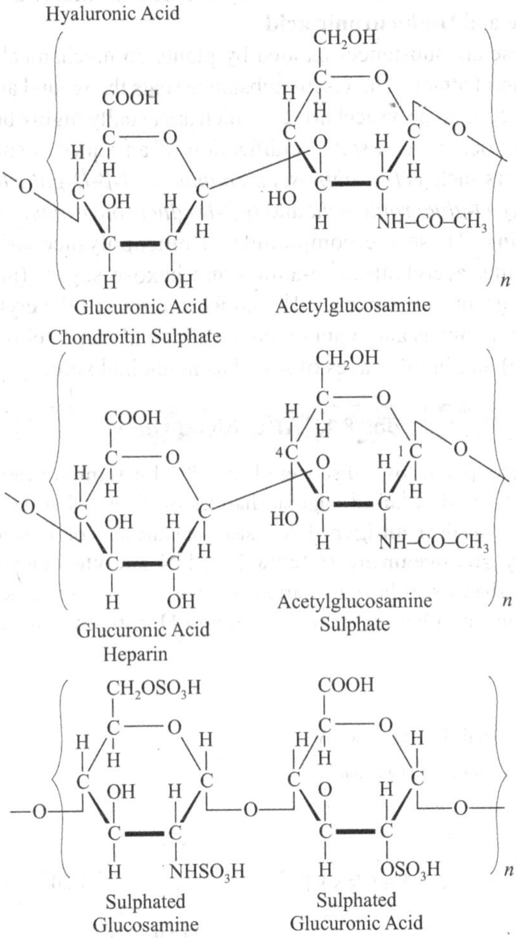

Fig. 8.31. Structure of some mucopolysaccharides.

(*ii*) **Heparin.** It is an anticoagulant and is found in the blood. It is used to prevent clotting of blood. Heparin is produced by the mast cells of the liver, intestinal muscosa, lung, spleen and kidney. Heparin is polymer of sulphated glucuronic acid and sulphated glucosamine units (Fig. 8.31).

(*iii*) **Chondroitin sulphate.** It is present in the ground substance of connective tissue and is predominant in cornea, cartilage, tendons, skin, heart valves and saliva. The repeating unit is a disaccharide consisting of **glucuronic acid** linked to sulphate ester of **N-acetyl**

galactosamine through β-1, 3 glucosidic bond. The disaccharides are linked through. β-1, 4 linkage. The sulphate esters are of two types: **chondroitin-4-sulphate** (A) and **chondroitin-6-sulphate** (C). A third type of sulphate contains **iduronic acid** instead of glucuronic acid and is labelled B.

(*iv*) **Keratin sulphate.** It is a component of cartilage and cornea. It consists of **N-acetylglucosamine, galactose** and **sulphuric acid.**

II. Neutral Sugars:

(*a*) **Hemicellulose.** It is found in association with cellulose in cell walls. The commonly found sugars in hemicellulose are **D-xylose, L-arabinose, D-galactose, L-rhamnose, D-mannose** and **D-glucuronic acid**.

(*b*) **Gums.** These are substances exuded by plants on mechanical injury or after bacterial, fungal or insect attack. The viscus substance seals the wound and thus protects the plant. Gums often contain polysaccharides, which are usually highly branched. The commercial Gum-Arabic occurs as a salt. Acidification of an aqueous solution of the gum yields various sugars such as *L-arabinose, L-rhamnose, 3-β-D-galactopyranosyl-L-arabinose, D-galactose, D-glucuronic acid* and *6-β-D-glucuropyranosyl-D-galactose*.

III. Glycoproteins. These are compounds of mucopolysaccharides and proteins. They contain acetylgucosamine, acetylgalactose-amine and hexose sugars (mannose or galactose) or their derivatives. Blood group substances A, B, and Rh antigens of the erythrocytes (see Box 8.10), luteinizing hormone (LH), mucus and plant mucilage all are glycoproteins. Mucilage in the husk of isabgol (*Plantago ovata*) and in the leaves of *Aloe* has medicinal value.

Box 8.10 ABO Blood Group

The ABO blood group was first discovered in 1901 by **Landsteiner**. Most primates display three different kinds of O- and N-linked oligosaccharides on their cell surfaces. The core structure of these oligosaccharides is called **H antigen**. It consists of various combinations of **galactose (Gal), fucose (Fuc), N-acetylglucoseamine (GlcNac)** and **N-acetyle-neuraminic acid** (sialic acid, NeuNac). These monosaccharides are linked in various ways to form a short branched structure that exhibits considerable microhetrogeneity. One of the most common **H antigen** structures is shown in following Fig. 8.32.

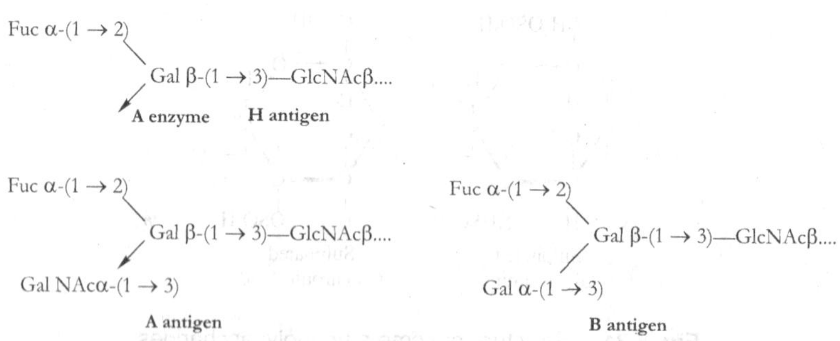

Fig. 8.32. Mode of conversion of H antigen into A antigen and B antigen (after Horton *et al.*, 2006).

The core structure (H antigen) can be modified in various ways. The addition of a GalNAc residue in α-(1 → 3) linkage forms A antigen. This reaction is catalyzed by A enzyme. The addition of Gal in α-(1 → 3) linkage is catalyzed by B enzyme.

If only A antigen is present a person will have **A blood group**. If only B antigen is present their blood type will be B (*i.e.*, **B blood group**). The **AB blood type** indicates that both A antigen and B antigen are present on cell surfaces. If neither GalNAc nor Gal have been added to the H antigen structure, then neither A antigen nor B antigen will be present and the blood type is O (*i.e.*, O blood group).

The ABO blood group is determined by a single gene on autosomal chromosome 9. Human (and other primate) populations contain many alleles of this gene. The original encoded A enzyme, which transfers GalNAc. Variants of this gene have altered specificity of the enzyme so that it no longer recognizes GalNAc but instead transfers Gal. These B enzymes differ by several amino acid residues from the allele that encodes the A enzyme. The structure of both types of **glycosyltransferase** enzymes have been solved and they reveal that only a single amino acid **substitution** is required to change the specificity from *N*-acetyl-aminogalactosyltransferase enzyme to galactosyltransferase enzyme.

The chromosome 9 locus can also contain several alleles that encode non-functional proteins one of the most common mutations is a single base pair deletion near the N-terminus of the coding region. This **deletion** shifts the reading frame for translation, making it impossible to synthesize a functional enzyme of type. People who are homozygous for these non-functional O allele will not synthesize either A antigen or B antigen and their blood type will be O.

All cells contain some of the core oligosaccharide (H antigen) even if your blood type is A, B, or AB. This is because not all of the H antigen structures are modified. Under normal circumstances human plasma will not contain antibodies against H antigen. However, O-type individuals will have antibodies against A antigen and B antigen because these structures are recognised as non-self. If O-type individuals receive a blood transfusion from someone with A, B or AB blood, they will mount an immune response and reject it. Similarly, if one has A-type blood he will have anti-B antibodies and cannot receive a transfusion from someone with B or AB blood type.

The O allele (non-functional enzyme) is the most common allele in most human populations and the B allele is the rarest. Some Native American populations has type O blood type. Type O individuals are perfectly normal, indicating that the absence of the A and B oligosaccharide structures has no effect on normal growth and development (Horton *et al.*, 2006).

Glycoproteins are associated with following functions:

1. They form viscous solutions which function as lubricants and form a protective covering on the cell surface.
2. They provide support to the cells and bind them together.
3. They ensure exchange of substances between blood and tissue cells.
4. They form a lubricating medium to the juction of cartilage, tendon, ligaments and bones.

IV. Chitin. It is present in the exoskeleton of arthropods (crustaceans and insects) and in the cell wall of fungi. It is a nitrogen-containing polysaccharide. Chitin is a polyglycan consisting of both **N-acetyl-D-glucosamine** units connected through β-1, 4-glycosidic linkages. Like cellulose it consists of parallel chains of molecules held together in bundles by hydrogen bonds.

V. Glycolipids. These molecules are covalent combinations of carbohydrate and lipid. The carbohydrate portion may be a single monosaccharide or a linear or branched chain. Glycolipids form the component of most cell membranes, *e.g.*, cerebrosides and gangliosides.

Gangliosides. They were isolated from brain tissue and named by **Klenk** in 1955. Gangliosides are complex lipids found in significant concentrations in ganglion cells of nervous tissue and also in most parenchymtous tissues such as spleen and erythrocytes. They make up about 6% of the membrane lipids in the gray matter of the brain. The structure of gangliosides is complex and related to that of cerebrosides in that they contain a ceramide (N-ancylspinosine) linked to a carbohydrate (galactose or glucose). In **Tay-Sachs disease**, the gangliosides are stored in relatively large amounts in the brain and spleen. Gangliosides are thought to act as receptors for toxic agents such as pathogens, *Vibrio cholerae*, influenza virus and tetanus toxin. They are also implicated to play a role in cell to cell interaction (**Jain** 2012).

VI. Peptidoglycans (Murein). Murein is a polysaccharide linked to short peptide chains. It is found in the cell walls of many bacteria. It consists of alternating units of **N-acetylglucosamine (NAG)** and **N-acetylmuramic acid (NAM)** linked by its carboxyl group to the peptide chain consisting of four amino acids (L-alanine, D-glutamic acid, L-lysine and D-alanine). Five glycine residues link peptide chain to its neighbour. The enzyme **lysozyme** hydrolyzes the chain into NAG-NAM disaccharides, resulting in rupture of cell wall.

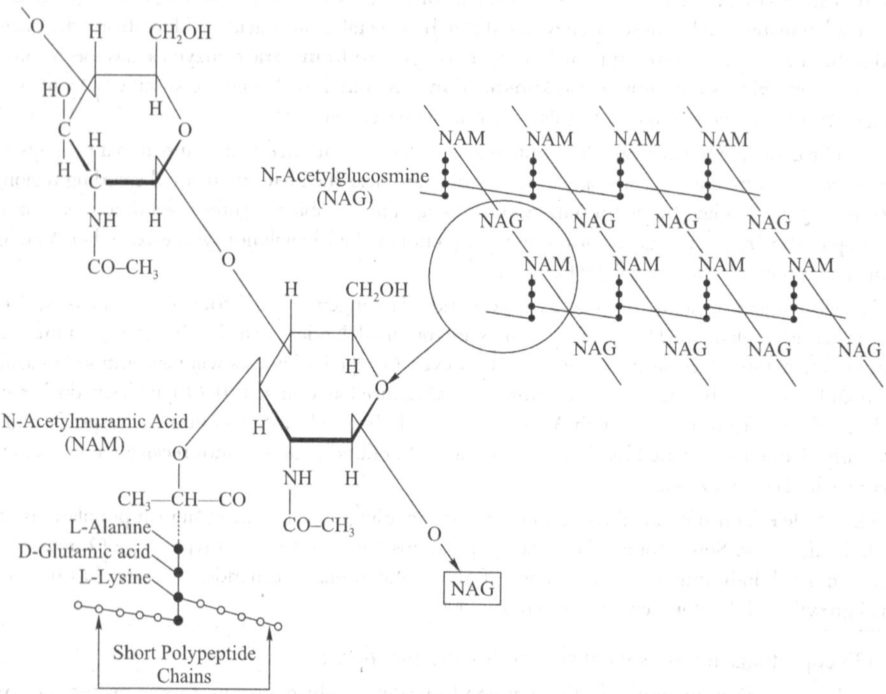

Fig. 8.33. Peptidoglycan sheet structure of bacterial cell wall.

Table 8.5. Differences between oliogosaccharides and polysaccharides.

Oligosaccharides	Polysaccharides
1. They contain 2 – 10 monosaccharide residues.	1. Polysaccharides possess a large number of monosaccharide residues.
2. Oligosaccharides are soluble in water.	2. Polysaccharides are usually insoluble.
3. They are commonly sweet to taste.	3. They are tasteless.
4. Transport of carbohydrates (in plants) occurs in the oligosaccharide state.	4. Polysaccharides are not involved in transport.
5. Food storage in oligosaccharide state is less common.	5. Food storage occurs in the state of polysaccharide.
6. They form component of external surface of plasma membrane and cell coat.	6. They form component of cell wall.
Examples: Sucrose, Maltose, Raffinose	Examples: Starch, Glycogen, Cellulose

8.5. BIOLOGICAL FUNCTIONS OF CARBOHYDRATES

1. Storage substances of potential energy. The most important function of carbohydrates is to provide energy to the body. They are storage substances of potential energy in animals. About 60% of total energy requirement of human beings is provided by the breakdown of carbohydrates. One gram of carbohydrate on oxidation yields on average four calories. Glucose supplies the immediate energy needed by tissues. Glucose is the only form of energy for the brain and other nervous tissues. Lack of glucose or of oxygen for the metabolism of brain leads to rapid damage to the brain. Carbohydrate is stored in the body in the form of glycogen. About 100 gm of glycogen is stored in the human liver and by its breakdown maintains the glucose level in the blood. Human muscles contain about 200 to 240 gm glycogen. This glycogen is, however, utilised only by the muscles themselves and it is not available for regulating the blood sugar level. In plants, storage carbohydrate is starch. In yeast and other fungi storage carbohydrate is glycogen.

2. Structural components. Carbohydrates are important components of some structural materials of living materials. In plant cells carbohydrate (cellulose) constitute the structural framework (cell wall). In bacteria, polysaccharides form the capsule.

Monosaccharides are important constituents of nucleic acids, coenzymes, flavoproteins and blood group substances (A, B antigens). *Ascorbic acid* (vitamin C) is related to sugars. *Immunopolysaccharides* play a vital part in resistance of infections. *Hyaluronic acid* is the viscous substance present in the matrix of connective tissue. *Heparin* prevents the clotting of blood. *Glucuronic acid*, which occurs in liver, acts as detoxifying agent by combining with toxic substances and bacterial byproducts. *Chondroitin sulphates* are present in cornea, cartilage, tendons, skin, heart valves and saliva. *Glycosides* are components of steroid hormones. *Galactolipins* are constituents of nervous tissue.

3. Regulation of fat metabolism. Some carbohydrates are essential for normal oxidation of fats. When carbohydrates are restricted in the diet there is more rapid metabolization of fats. This results in the accumulation of incompletely oxidized intermediate products leading to **ketosis**. This is common in uncontrolled diabetes mellitus.

4. Protein sparing function. Carbohydrate is preferentially metabolized in the body as a source of energy as long as its is present in the required quantity. This spares protein for building of tissue. When there is deficiency of calories in the diet, however, fat and then protein are utilized for supplying energy.

5. Role of carbohydrates in the normal functioning of gastrointestinal tract. Indigestible substances such as cellulose, hemicellulose and proteins provide the bulk of roughage.

6. Survival of Antarctic fish. The survival of Antarctic fish below –2° is attributed to the antifreeze glycoproteins. This glycoprotein consists of 50 repeating units of the tripeptide, alanine-alanine-threonine.

8.6. CLINICAL IMPORTANCE OF CARBOHYDRATES

1. **Glucose.** Glucose is the ultimate source of energy for all body cells. Its continuous supply is essential for following purposes:
   (*i*) As source of energy for the nervous system and for erythrocytes.
   (*ii*) As source of glyceride-glycerol in adipose tissue.
   (*iii*) For maintaning the level of intermediates of the citric acid cycle.
   (*iv*) As precursor of milk sugar, lactose, in the mammary gland.
   (*v*) As fuel for the supply of energy to the skeletal muscles especially under aerobic condition.

   Glycolysis and glyconeogenesis are two main metabolic activities that maintain an non-stop glucose supply to all body cells. Lowering the glucose supply below a critical level leads to brain dysfunction and may result in coma and death.
2. **Fructosurea.** Fructosurea is the appearance of fructose in urine. This clinical condition may occur under following circumstances:
   (*i*) Due to deficiency of **fructokinase** enzyme, fructose is not utilized completely and is excreted in the urine. This is a rare congenital disorder.

(*ii*) When large quantities of fructose are ingested, some of it is excreted in urine. This is due to hepatic insufficiency.

(*iii*) Due to deficiency of enzyme **aldolase**, fructose-1-phosphate is accumulated in the liver. This blocks pathways of fructose utilisation.

3. **Clinical significance of sorbitol.** Accumulation of sorbitol and dulicol in tissues may cause certain pathological conditions, *e.g.*, cataract, nephropathy.
4. **Clinical significance of glycogen.** Glycogen is animal starch. It is stored in the muscles and liver cells. **Muscle glycogen** provides hexose sugars to muscle cells for energy by glycolysis. **Liver glycogen** provides hexose sugars to all body cells for energy and maintains blood glucose level between meals. Liver glycogen is exhausted after 12 – 18 hours of fasting, whereas muscle glycogen is exhausted after prolonged vigorous exercise.

Inherited disorders result in **glycogen storage diseases**. They are caused due to in adequate mobilisation of glycogen and resultant deposition of glycogen in liver, heart, kidney and muscles. This condition is called **glycogenosis** and causes muscular weakness and may ultimately lead to death.

Types of Glycogen Storage Diseases

I. Von Gierke's Disease (Type I Glycogenosis). This disease is caused by the deficiency of enzyme **glucose-6-phosphatase**. In the absence of this enzyme glycogen is not broken down to glucose. Hence stored glycogen is not utilised, but the synthesis of glycogen from glucose-6-phosphate continues. This disease has clinical onset at birth or during infancy and has the following effects:

1. Enlargement of liver and kidney.
2. Children suffering from this disease develop **hypoglycemia** (low blood glucose level).
3. Hypoglycemia inhibits insulin secretion which in turn inhibit protein synthesis. This leads to retarded growth.
4. Hypoglycemia stimulates production of epinephrine. This causes breakdown of muscle glycogen forming lactate. The lactate interferes with the excretion of urate by the kidney raising blood urate level.
5. The persons suffering from this disease tend to use fat as energy source and develop **lipemia** (*i.e.*, the presence of an excess of fats or lipids in the blood); **acidemia** (*i.e.*, increased hydrogen ion concentration is the blood); **ketosis** (*i.e.*, abnormal increase of ketone bodies in the body) and **nosebleeds**.
6. The liver shows fatty infiltration.

This disease is inherited as an autosomal recessive trait.

II. Pompe's disease (Type II Glycogenosis). This fatal disease occurs due to deficiency of a lysosomal enzyme called **acid maltase**. Hence, this disease is also called **acid maltase deficiency**. Absence of acid maltase enzyme causes excessive accumulation of glycogen in the cells of liver, heart and muscles. The heart is enlarged. The patient shows extreamly weak muscles. Death occurs usually before the age of nine months.

III. Limit dextrinosis (Type III Glycogenolysis). This disease is caused by the deficiency of limit dextrin in liver and muscles. This condition is produced by the action of enzyme **phosphorylase** on glycogen.

IV. Amylopectinosis (Type IV Glycogenosis). This disease is caused by the deficiency of branching enzyme in the liver. The disease is fatal. In the absence of this enzyme amylopectins are formed in the liver, heart, kidneys and muscles.

V. Mc Ardie disease (Type V Glycogenosis). This glycogen storage disease is caused by the deficiency of **muscle phosphorylase** and affects the skeletal muscles. This leads to cramps in the muscles.

VI. Type VI Glycogenosis. This disease is due to deficinecy of **liver phosphorylase** resulting in increased liver glycogen.

VII. Type VII Glycogenosis. Deficiency of enzyme **phosphofructokinase** causes accumulation of glucose-6-phosphate and fructose-6-phosphate.

QUESTIONS

Long Answer Questions

1. Classify carbohydrates and give at least one example of each kind.
2. What are carbohydrates? What functions do they serve in the living system?
3. Describe the structure of monosaccharides. Give their classification also.
4. Write down the structure, types and functions of oligosaccharides.
5. Write down the functions of polysaccharides.
6. Describe the isomerism in carbohydrates.

Short Answer Questions

1. What are gangliosides? Give their importance.
2. What are heteropolysaccharides? Explain one heteropolysaccharide of each category.
3. Give structure of following:
   (*i*) Glycogen and
   (*ii*) Amylopectin.

Very Short Answer Questions

1. What are dextrins?
2. Name the compound that forms matrix of connective tissue and cartilage.
3. What is the function of hyaluronic acid?
4. Why are monosaccharides called reducing sugars?
5. Name any two amino sugars.
6. What is heparin?
7. Lists the types and one example of each type of heteropolysaccharides.

Multiple Choice Questions

1. Reducing sugar among the following is
   (*a*) galactose (*b*) gluconic acid
   (*c*) glycogen (*d*) sucrose
2. What is structure of glucose?
   (*a*) $C_3H_8O_3$ (*b*) $C_6H_{12}O_6$
   (*c*) $C_{55}H_{70}O_6$ (*d*) $C_6H_{10}O_6$
3. In which of the following groups all are polysaccharides?
   (*a*) sucrose, glucose and fructose
   (*b*) maltose, lactose and fructose
   (*c*) glycogen, sucrose and maltose
   (*d*) glycogen, cellulose and starch
4. Which one of the following is an example of glycoprotein?
   (*a*) haemoglobin
   (*b*) lecithin
   (*c*) mucin
   (*d*) casein
5. Animal cells store food in the form of
   (*a*) starch
   (*b*) glycogen
   (*c*) paramylum
   (*d*) glucose
6. Cellulose is
   (*a*) hexose polysaccharide
   (*b*) heptose polysaccharide
   (*c*) heteropolysaccharide
   (*d*) pentose polysaccharide
7. Which one is a polymer of α-d glucose?
   (*a*) cellulose
   (*b*) glycogen
   (*c*) inulin
   (*d*) chitin
8. Carbohydrate is stored in both liver and muscle as
   (*a*) glucose
   (*b*) starch
   (*c*) glycogen
   (*d*) sucrose
9. The dye ruthenium red stains
   (*a*) pectin
   (*b*) cellulose
   (*c*) mitochondria
   (*d*) Golgi complex

10. Identify the trisaccharide in the following
(*a*) mannose
(*b*) maltose
(*c*) galactose
(*d*) raffinose

11. Mucus secretions are rich in
(*a*) glycoproteins
(*b*) mucopolysaccharides
(*c*) uric acid
(*d*) all above

ANSWERS

Very Short Answer Questions

1. Dextrins are formed by the partial digestion of starch. It is also formed when starch is heated dry.
2. The matrix of connective tissue and cartilage is formed of heteropolysaccharide, chondroitin-sulphate.
3. Hyaluronic acid is glycosaminoglycan compound found at the joints of bones. It lubricates the joints and absorbs shocks.
4. Monosaccharides have either aldehyde –CHO or keto (C = O) group, both of which have the reducing property.
5. (*i*) N-acetyl glucosamine and (*ii*) Glucosaminoglycans or mucopolysaccharides.
6. Heparin is a complex mucopolysaccharide, found in blood and connective tissue. It helps in blood clotting.
7. (*i*) Mucopolysaccharides, *e.g.*, hyaluronic acid.
(*ii*) Neutral sugars, *e.g.*, D-xylose.
(*iii*) Glycoproteins, *e.g.*, A, B, Rh antigens.
(*iv*) Chitin.
(*v*) Glycolipids, *e.g.*, ganglioside.
(*vi*) Peptidoglycans (Murein).

Multiple Choice Questions

1. (*a*)	**2.** (*b*)	**3.** (*d*)	**4.** (*c*)	**5.** (*b*)	**6.** (*a*)	**7.** (*b*)
8. (*c*)	**9.** (*a*)	**10.** (*d*)	**11.** (*b*)			

9 Lipids

Lipids (Gr. *lipos* = fat) are oily, greasy and wax-like substances which are insoluble in water (a polar solvent) but soluble in non-polar organic solvents such as ether, chloroform, hot alcohol, benzene, petroleum or carbon disulphide. The lipids are a heterogenous group of organic compounds which include cooking oil, butter, ghee, waxes, cholesterol, essential oils, natural rubber, some plant pigments (carotene of carrots, lycopene of tomatoes), menthol (mint oil), vitamins A, E and K and eucalyptus oil. Lipids form major part of stored food material in animal body and a major structural component. The term **lipid** was used by German biochemist **Wilhelm Bloor** in the year 1943 to a group of fats and fat-like substances.

Percentage. The lipids form about 3.5 per cent of the cell content. In plants they are present in seeds, nuts and fruits (*e.g.*, olive), while in animals they are stored in adipose tissues, yellow bone marrow and nervous tissue.

9.1. SPECIAL FEATURES OF LIPIDS

1. Lipids are compounds of carbon, hydrogen and oxygen. In them the number of oxygen atoms is much less in comparison to carbon and hydrogen.
2. Chemically they are esters of fatty acids and polyhydric alcohols. Esters are formed when organic acids and alcohols are reacted just as inorganic acids and bases react to form salts. In one lipid molecule three molecules of fatty acids are joined to one glycerol molecule. The carboxyl groups (–COOH) of fatty acids form ester bonds with alcoholic groups (–OH) of glycerol molecule (Fig. 9.1). The characteristic of lipids depends on the nature of fatty acids.

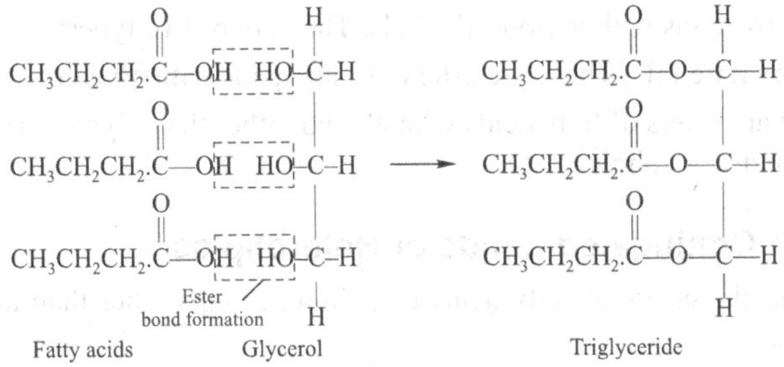

Fig. 9.1. Formation of a triglyceride or lipid.

3. Lipid molecules are **polar** and **hydrophobic**.
4. Generally the lipids do not polymerise to form macromolecules. However they may combine with carbohydrates and protein macromolecules.
5. Lipids can be hydrolysed by heating with alkalis. This process is called **saponification** (Fig. 9.2).

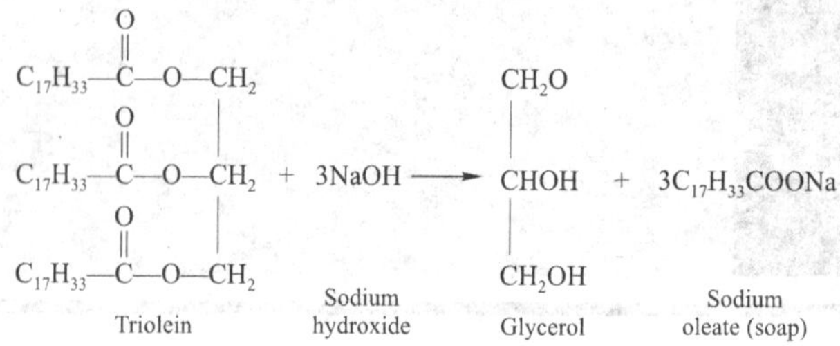

Fig. 9.2. Saponification of a neutral lipid.

9.2. CLASSIFICATION OF LIPIDS

Lipids are generally classified into the following groups:

I. Simple lipids
II. Compound lipids
III. Derived lipids

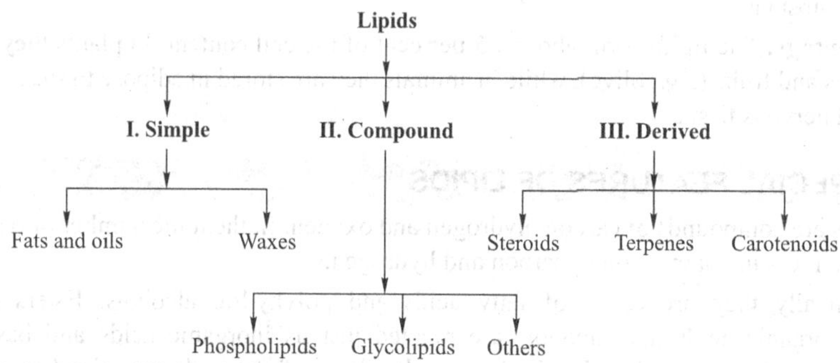

Fig. 9.3. A simple (classical) classification of lipids.

I. Simple Lipids or Homolipids

They are esters of fatty acids with various alcohols. They are of two types:

1. **Neutral fats or true fats.** These are esters of fatty acids with glycerol.
2. **Waxes.** These are esters of fatty acids with alcohol other than glycerol (*i.e.*, higher molecular weight monohydric alcohols).

II. Compound or Conjugated Lipids or Heterolipids

Compound lipids are the esters of fatty acids containing groups other than and in addition to an alcohol and fatty acids.

1. **Phospholipids.** They have a nitrogen-containing base and a phosphate group in addition to fatty acids and glycerol, *e.g.*, lecithins, cephalins, sphinomyelins and plasmalogens.
2. **Glycolipids.** They have fatty acids, an amino alcohol and a carbohydrate, *e.g.*, phrenosin,

kerasin, nervone and oxynervone.

3. **Lipoproteins.** They have lipids (mainly phospholipids) and proteins.

4. **Chromolipids,** *e.g.*, carotenoids and related pigments.

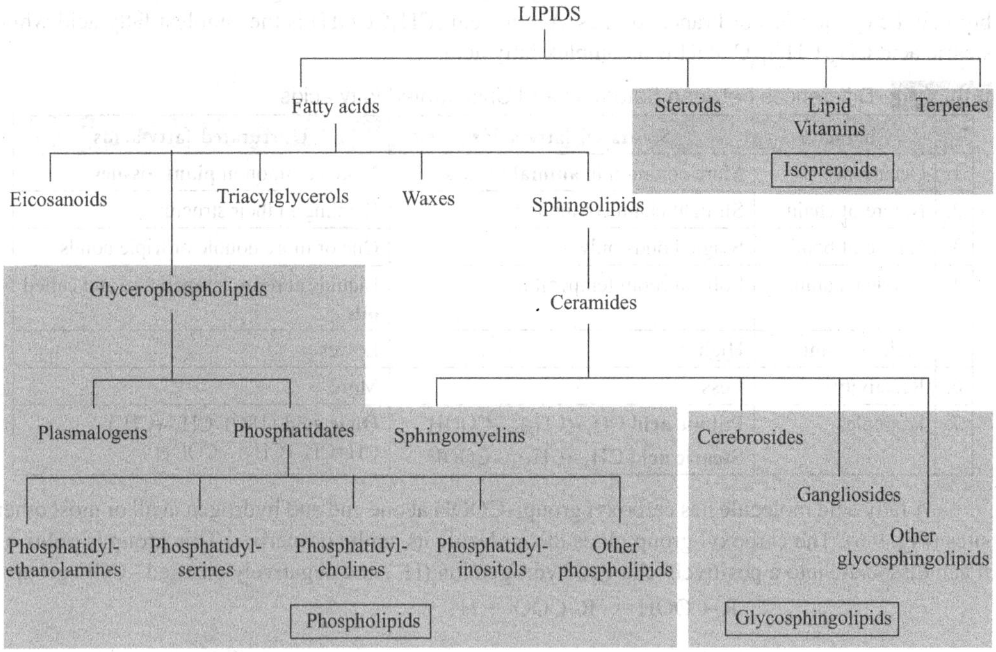

Fig. 9.4. Structural relationships of the major classes of lipids (after **Horton** *et al.*, 2006)

III. Derived Lipids

These are formed from the above lipids by hydrolysis. They include fatty acids, glycerol, vitamin D, cholesterol, bile acids, steroids, sterones, terpenes and prostaglandins.

IV. Unclassified Lipids

These include tocopherol (vitamin E) and vitamin K.

9.3. SIMPLE LIPIDS

A. Neutral Fats or True Fats (Glycerides)

Neutral fats are also called **triglycerides** and are the main energy storing compounds in the body.

Composition. The neutral or true fats are composed of carbon, hydrogen and oxygen like the carbohydrates, but have far fewer oxygen atoms than carbon atoms unlike carbohydrates. Some lipids have phosphorus, nitrogen and sulphur also. A neutral fat molecule consists of two components: one molecule of alcohol called **glycerol** (or glycerine) and one or three molecules of the same or different **fatty acids**.

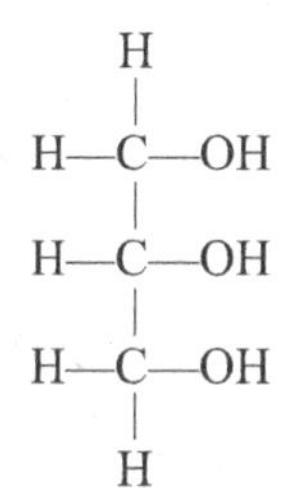

Fig. 9.5. Glycerol.

1. Glycerol. It is trihydroxy alcohol (with 3-OH groups). Glycerol has a molecular formula of $C_3H_5(OH)$, and it can form 3 ester bonds with 3 similar or dissimilar fatty acid molecules (Fig. 9.5). Glycerol is optically inactive. However, when it is esterified by two different fatty acids at 3 positions, the carbon 2 becomes asymmetric so the glycerol acts as a binder or carrier for fatty acids.

2. Fatty acids. These are organic molecules made up of long hydrocarbon chain which contain a terminal carboxyl, –COOH group, so fatty acids are always **monocarboxylic**. On the basis of nature of bond present, these fatty acids may be *saturated* or *unsaturated* (Table 9.1), but alway have even number of carbon (C) atoms (varies from 2–88). These are generally straight chain organic compounds but may be cyclic ones or branched ones. Acetic acid (CH_3COOH) is the simplest fatty acid while stearic acid $CH_3(CH_2)_{16}COOH$ is a complex fatty acid.

Table 9.1. Differences between Saturated and Unsaturated fatty acids.

	Character	Saturated fatty acids	Unsturated fatty acids
1.	Occurrence	More common in **animal tissues**	More common in **plant tissues**
2.	Nature of chain	Straight chain compounds	Bending in their structure
3.	Nature of bonds	Single bonds only	One or more double or triple bonds
4.	Physical nature	Solid at room temperature	Liquids at room temperature and called **oils**
5.	Melting point	Higher	Lower
6.	Reactivity	Less	More
7.	Examples	**Palmic acid** CH_3–$(CH_2)_{14}$–COOH **Stearic acid** CH_3–$(CH_2)_{16}$–COOH	**Oleic acid** (18C)–CH_3–$(CH_2)_7$–CH=CH–$(CH_2)_7$–COOH

A fatty acid molecule has carboxyl group –COOH at one end and hydrogen at all or most other sites (Fig. 9.6). The carboxyl group gives the molecule its acidic properties. This group is **polar** as it can dissociate into a positively charged hydrogen ion (H^+) and negatively charged –COO group:

$$R\text{–}COOH = \text{–}R\text{–}COO^- + H^+$$

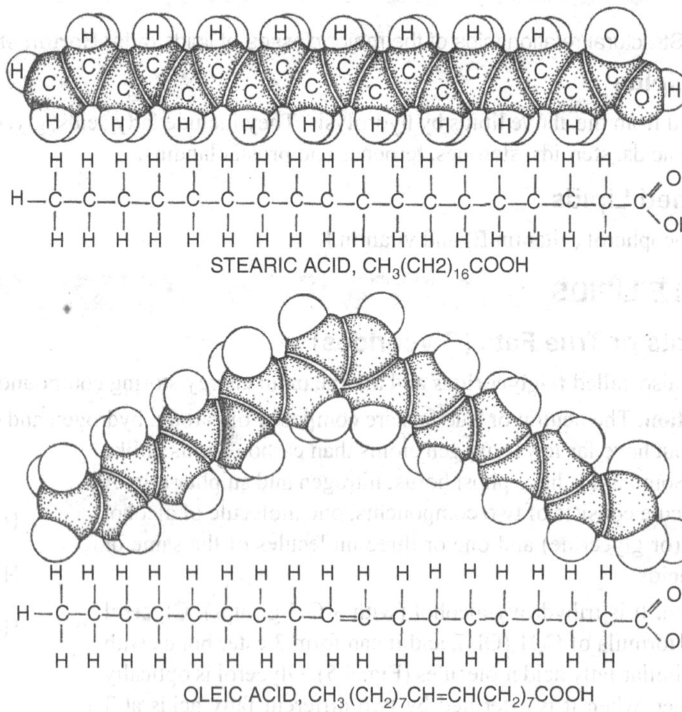

Fig. 9.6. Structure of stearic acid (a saturated fatty acid) and oleic acid (an unsalturated fatty acid). Note the kink produced by the double bond.

Dissociation cannot occur when the fatty acid is a part of a fat molecule, hence, the name "neutral fat". *Water insolubility* of lipids is due to the repulsion present between the nonpolar lipid molecule and the polar water molecule.

(*a*) Saturated fatty acids. In saturated fatty acids all carbon atoms are interlinked by a single bond and each carbon atom carries a hydrogen atom, *e.g.*, palmitic acid and stearic acid. The general chemical (molecular) formula for saturated fatty acids is $C_nH_{2n}O_2$. The chemical formula for stearic acid is $C_{18}H_{36}O_6$ and for palmitic acid is $C_{16}H_{32}O_2$ (Table 9.2).

General formula for saturated fatty acids is R–COOH where R represent CH_3–$(CH_2)_n$–(called **acyl radical**) and *n* varies from zero in acetic acid and 86 in mycolic acid.

Table 9.2. Important saturated fatty acids ($C_nH_{2n+1}COOH$).

	Name of fatty acid	Molecular formula	Number of C-atoms
1.	Acetic acid	CH_3–COOH	2
2.	Butyric acid	$CH_3(CH_2)_2COOH$	4
3.	Caproic acid	$CH_3(CH_2)_4COOH$	6
4.	Caprylic acid	$CH_3(CH_2)_6COOH$	8
5.	Capric acid	$CH_3(CH_2)_8COOH$	10
6.	Lauric acid	$CH_3(CH_2)_{10}COOH$	12
7.	Myristic acid	$CH_3(CH_2)_{12}COOH$	14
8.	Palmitic acid	$CH_3(CH_2)_{14}COOH$	16
9.	Stearic acid	$CH_3(CH_2)_{16}COOH$	18
10.	Arachidic acid	$CH_3(CH_2)_{18}COOH$	20
11.	Behinic acid	$CH_3(CH_2)_{20}COOH$	22
12.	Lignoceric acid	$CH_3(CH_2)_{22}COOH$	24

(*b*) Unsaturated fatty acids. In unsaturated fatty acids, two adjacent carbon atoms at one or more places lack hydrogen atoms and are interlinked by double bonds. These include 18 carbon oleic, linoleic and linolenic acid (Table 9.3). They have 1, 2 and 3 double bonds respectively. A double bond

Table 9.3. Important unsaturated and special fatty acids.

I. Unsaturated acids		
A. Monoethenic, $C_nH_{2n-1}COOH$, one double bond		
(a) Palmitoleic Δ^9	$C_{15}H_{29}COOH$	Animal (Milk) and Sardine fats
(b) Oleic acid Δ^9	$C_{17}H_{33}COOH$	
(c) Eracic acid Δ^{13}	$C_{21}H_{41}COOH$	
(d) Nervonic acid Δ^{15}	$C_{23}H_{45}COOH$	
B. Dienoic C_n–$H_{2n-3}COOH$, two double bonds		
Linoleic acid $D^{9,12}$	$C_{17}H_{31}COOH$	Linseed oil, animal and plant fats, corn oil, soyabean oil, peanut oil etc.
C. Trienoic $C_nH_{2n-5}COOH$, three double bonds		
Linolenic acid $\Delta^{9,12,15}$	$C_{17}H_{29}COOH$	Linseed oil, rapeseed oil, soyabean oil, liver oils etc.
D. Polyenoic, $C_nH_{2n-7}COOH$, more than three double bonds		
Arachidonic acid $\Delta^{5,8,11,14}$	$C_{19}H_{31}COOH$	Fats, phospholipids etc.

II. Branched-chain acids		
(a) Isobutyric acid	C_3H_7COOH	Waxes
(b) Tuberculostearic acid	$C_{18}H_{37}COOH$	Wax of tubercle bacillus
III. Hydroxy acids		
(a) Cerebronic acid	$C_{23}H_{46}(OH)COOH$	Animal and vegetable fats
(b) Ricinolic acid	$C_{17}H_{32}(OH)COOH$	Animal and vegetable fats
(c) Hydroxynervonic acid	$C_{23}H_{44}(OH)COOH$	Animal and vegetable fats
IV. Cyclic acids		
Chaulmoogric acid	$*R—(CH_2)_{12}—COOH$	Seed oils
Hydnocarpic acid	$*R—(CH_2)_{10}—COOH$	Seed oils
*where R is	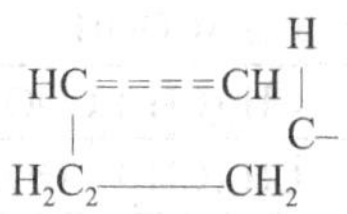	

in an unsaturated fatty acid may have a **cis** or **trans configuration** (Fig. 9.7). The double bond in most unsaturated fatty acids have the cis orientation, which introduces a bend (109° bond angle) in the hydrocarbon chain.

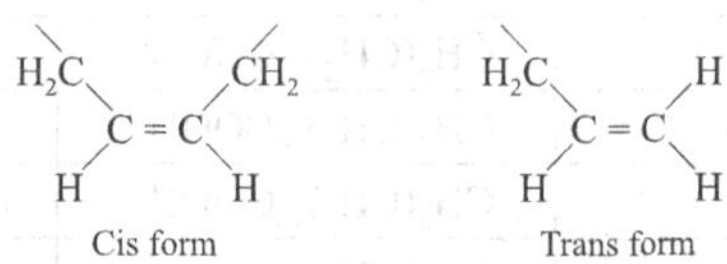

Fig. 9.7. Two configurations of double bond in an unsaturated fatty acid.

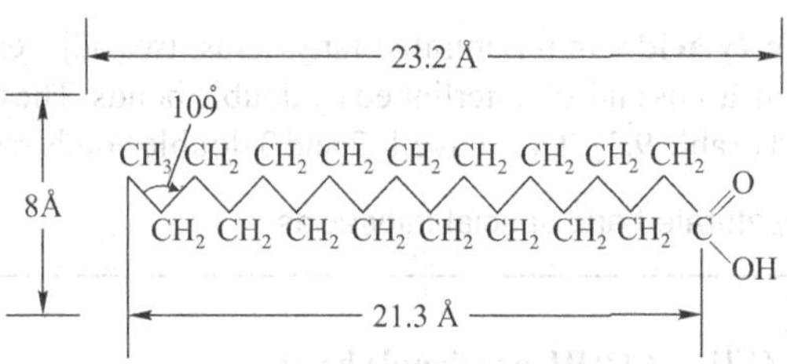

Fig. 9.8. Configuration of a saturated fatty acid.

Box 9.1. Trans Fatty Acids and Margarine

(**Trans.** (L. *trans* = through, across). *In organic chemistry.* In italics, a form of geometric isomerism in which the atoms attached to carbon atoms, joined by double bonds, are located on opposite sides of the molecule. *In Biochemistry.* A prefix to a group name in an enzyme name or reaction denoting transfer of that group from one compound to another, *e.g.*, transformylase.)

While the configuration of most double bonds in unsaturated fatty acids is *cis*, some fatty acids in the human diet have the *trans* configuration. *Trans* fatty acids can come from animal sources such as dairy products and ruminant meats. Most edible trans fatty acids, though are present as hydrogenated vegetable oils in some margarines or shortenings (*i.e.*, fat used for making pastry). Dietery *trans* monosaturated fatty acids can increase plasma levels of cholesterol and triglycerides, and thus their ingestion may increase the risk of cardiovascular disease.

Plant oils such as corn oil, sunflower oil, peanut oil, etc., can be converted to "spreadable" semisolid substances such as *margarines*. Margarines can be produced by the partial or complete hydrogenation of the double bonds in plant oils. The hydrogenation process itself not only saturates the carbon-carbon double bonds of fatty acid esters, but can also change the configuration of the remaining double bonds from *cis* to *trans*. The physical properties of these *trans* fatty acids are similar to those of saturated fatty acids. In order to reduce consumption of *trans* fatty acids, many margarines are now produced from plant oils without hydrogenation by adding other edible components such as skim milk powder (**Horton** *et al.*, 2006; **Jain** *et al.*, 2012).

The unsaturated fatty acids have lower melting points than the saturated fatty acids and are more abundant in the living organisms. Saturated fatty acids (SFA) have straight chains whereas unsaturated fatty acids (UFA) have a bend or kink at each double bond. The kinks affects the packing of UFA chains in cell organelles, making the chains more disordered and hence more fluid at ordinary temperatures. This property maintains the flexibility and fluidity of cell membranes. The general chemical formula for unsaturated fatty acids is $C_nH_{2n-2x}O_2$. The chemical formula of oleic acid with a single double bond is $C_{18}H_{34}O_2$ and of linoleic acid with two double bonds is $C_{18}H_{32}O_2$. Most plant fats contain unsaturated fatty acids and most animal fats have saturated fatty acids. Unsaturated fatty acids are of following types:

(*a*) **Monounsaturated (or Monoenoic) Fatty Acids.** These fatty acids have only one unsaturation (*i.e.*, one double bond).

Examples:

(*i*) **Palmitoleic acid (16C).** It is derived from palmitic acid and has a double bond between 9th and 10th carbon (hence also called **9-hexadeconic**). It is found in milk fat and sardine oil. Its molecular formula is:

$$\underset{\text{Palmitoleic acid}}{CH_3-(CH_2)_5-CH=CH-(CH_3)_7-COOH}$$

(*ii*) **Oleic acid (18C).** It is a derivative of stearic acid and has a double bond between 9th and 10th carbon atom. It is also called **9-octadecenoic**. It is found in olive oil, pork fat. Oleic acid is most abundant fatty acid of human beings and is present in adipose tissue and milk. Its molecular formula is:

$$\underset{\text{Oleic acid}}{CH_3-(CH_2)_9-CH=CH-(CH_2)_7-COOH}$$

(*iii*) **Euracic acid (22C).** It is found in mustard oil and has one double bond between 13th and 14th carbon. Its molecular formula is:

$$\underset{\text{Euracic acid}}{CH_3(CH_2)_7CH=CH(CH_2)_{11}COOH}$$

(*b*) **Polyunsaturated (Polyenoic) fatty acids.** These fatty acids contain more than one double bond. Polyunsaturate oils are recommended by physicians for persons who have high blood **cholesterol** or **cardiovascular diseases**. There use lowers the blood cholesterol level. Polyunsaturated fatty acids include the following:

(*i*) **Linoleic acid ($C_{18}H_{32}O_2$).** It is derived from stearic acid and is found in many seed oils, *e.g.*, linseed oil, soyabean oil, ground nut oil, cotton seed oil, etc. It has two double bonds which are present between 9th and 10th carbon atoms and between 12th and 13th carbon atoms, so it is also called **dienoic fatty acid**. Its systematic name is **9, 12-octadecadienoic acid**. Linoleic acid has the following molecular formula:

$$\underset{\text{Linoleic acid}}{CH_3-(CH_2)_4-CH=CH-CH_2-CH=CH-(CH_2)_7-COOH}$$

Linoleic acid is an *essential fatty acid* and its deficiency in diet results in increased metabolic activity, failure in growth and even death.

(*ii*) **Linolenic acid ($C_{18}H_{30}O_2$).** This *essential fatty acid* is also derived from stearic acid and is present in linseed oil, soya bean oil and hempseed oil. It has three double bonds in its molecule; these are present between 9th/10th, 12th/13th and 15th/16th carbon atoms. Thus it is a **trienoic** and its systematic name is **9, 12, 15-octadecatrienoic**. Linolenic acid has the following molecular formula:

$$\underset{\text{Linolenic acid}}{CH_3-CH_2-CH=CH-CH_2-CH=CH-CH_2-CH=CH-(CH_2)_7-COOH}$$

(*iii*) **Arachidonic acid (20C).** It is an *essential fatty acid*, derived from arachidic acid and is found in mammalian tissues. It is a **tetraenoic** having four double bonds between 5th/6th, 8th/9th, 11th/12th and 14th/15th carbon atoms. Its systematic name is **5, 8, 11, 14-eicosatetraenoic acid** and biologic precursor of eicosanoids (*i.e.*, the prostaglandins, the thromboxanes, and the leukotrienes. **Thromboxanes** or **Txs** is a class of molecules involved in platelet aggregation during blood clotting. **Leukotrienes** or **LTs** are of 6 types and secreted by white blood cells and lung tissues and involved in asthmatic attacks and stimulation of mucus secretion (**Jain** 2012). It has the following molecular formula:

$$CH_3-(CH_2)_4-CH=CH-CH_2-CH=CH-CH_2-CH=CH-CH_2-CH$$
$$=CH-(CH_2)_3-COOH$$

Essential and Non-essential Fatty Acids

On the basis of their synthesis, the fatty acids are of following two types:

1. **Essential fatty acid (EFAs).** These fatty acids are not synthesize in the body tissues of humans. So they must be present in human diet. These EFAs include **linoleic acid**, **linolenic acid** and **arachidonic acid**. These fatty acids are present in a number of vegetable cooking oils, *e.g.*, groundnut oil, sunflower oil and safflower oil.
2. **Non-essential fatty acids (NEFAs).** These fatty acids can be synthesized inside the human's body tissues. They may or may not be present in human food acid, *e.g.*, palmitic acid, stearic acid and oleic acid.

Formation of Fats

The three hydroxyl (–OH) groups of a glycerol molecule join with the carboxyl (–COOH) groups of three fatty acid molecules to form a fat molecule (Fig. 9.9). The chemical linkage between the glycerol and fatty acids is called the **ester bond**. It results from the elimination of 3 molecules of water, one for each linkage (condensation or dehydration). In a neutral fat the three fatty acids may be same or different. The fatty acid components of most of the animal fats are 16-carbon palmitic acid and 18-carbon stearic acid. The neutral fats formed by these fatty acids are named **tripalmitin** and **tristearin** respectively. Both are **pure neutral fats**. A fat having a molecule each of palmitic, oleic and stearic acid is called **palmito-oleio-stearin**. It is a **mixed neutral fat**. Butter is an example of the mixed fat.

The process that joins the glycerol and fatty acid molecules is called **esterification**.

Type of Fats

The neutral fats are of three types: monoglycerides, diglycerides and triglycerides. A true fat having 3 molecules of fatty acids is called **triglyceride**. Triglycerides are common in cells. Different species of animals usually have distinctive mixtures of fatty acids in their triglycerides, but these may change to some extent with the diet of the animal.

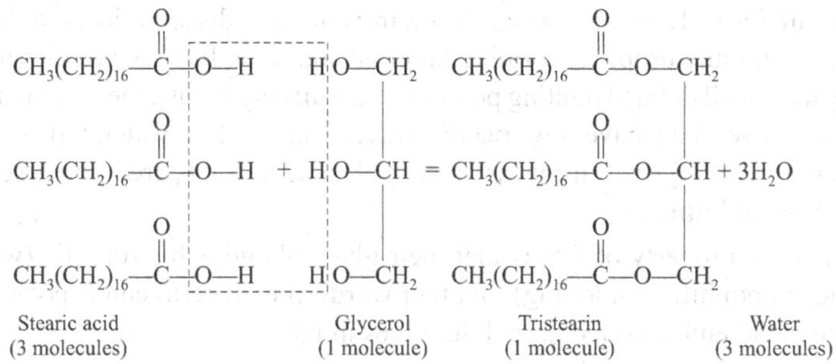

Fig. 9.9. Structural formula of tristearin.

A neutral fat having only one or two fatty acid molecules is termed a **monoglyceride** or a **diglyceride**. These two types of fats occur as intermediates in certain biosynthetic reactions. Out of these three fats, triglycerides form about 98% of our dietary lipids while remaining 2% of lipids are formed of phospholipids and cholesterol. All the three types of neutral fats are given a common name **glyceride**.

Physical Nature of Fats

The physical nature of a neutral fat depends upon the kind of fatty acid present. Some fats have saturated fatty acids, and are solid at ordinary temperature, *e.g.*, butter, lard, beef tallow and bacon fat. These are called **hard fats**. Other neutral fats have unsaturated fatty acids and are liquid at ordinary temperature, *e.g.*, cod liver oil, shark liver oil, whale oil, groundnut oil, rape seed oil, mustard oil, sesame oil (or til oil), linseed oil, sunflower oil, safflower oil, corn oil, olive oil, castor oil, soyabean oil and cotton seed oil. These are termed **oils.** Most plant lipids (oils) have unsaturated fatty acids and most animal fats have saturated fatty acids. There is clinical evidence that saturated fats (animal fats) cause increased production of cholesterol and arteriosclerosis (hardening of arteries) in humans. This condition is dangerous, often fatal, for heart patients as it reduces the flow of blood through arteries.

Box. 9.2.

Triglycerides having the saturated fatty acids are *solid* at room temperature because the aliphatic chains of fatty acids are arranged parallel to one another so that molecules become compact and need high amount of energy in order to melt them (Fig. 9.10A).

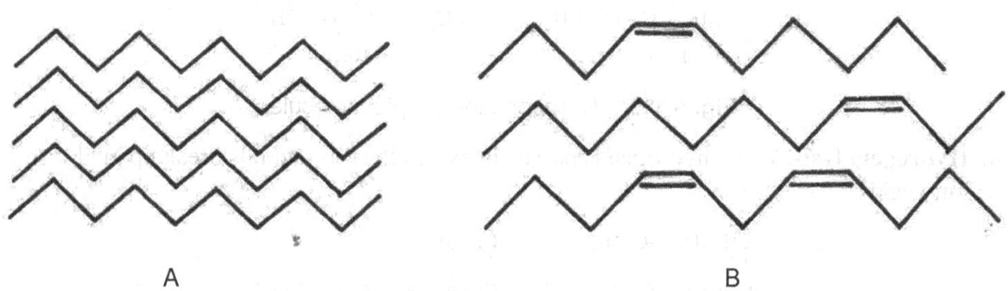

Fig. 9.10. Interactions between fatty acid chains. A—Saturated fatty acid chains. B—Unsaturated fatty acid chains.

Triglycerides having the unsturated fatty acids are *liquid* at room temperature and are called *oils.* It is so as the unsaturated fatty acids are cis-isomers and cis double bonds interrupt the regular packing of their chains so these need less energy to melt (Fig. 9.10B). This disorganisation increases with the increase in the number of double bonds which lowers their melting point.

Properties of Fats. True fats are insoluble in water but dissolve in organic solvents. They disperse in water as minute droplets, forming an emulsion. In cells they are dispersed through the cytoplasm. They have well defined melting points and solidifying points, the latter temperatures being only a few degrees lower. Fats have low specific gravity and float on water. Oils spread on water to form thin monomolecular layers. All fats are greasy. The neutral fats are hydrolysed in cells by the action of enzymes called **lipases.**

A rather unusual property of fats is that their glycerol end is hydrophilic (water loving) and the fatty acid end is lipophilic (fat loving). In other words, the glycerol end is polar and dissolves in water and the fatty acid end is nonpolar and dissolves in fat.

Fats are bad conductors of heat, so act as heat insulators. A layer of fats below skin provides an insulting blanket in all the warm blooded animals.

Chemical Properties of Fats

1. **Hydrolysis.** The fats are hydrolysed into glycerol and constituent fatty acids by superheated steam, alkalies and fat splitting enzymes. Water is utilised during this process. Fat splitting enzyme **lipase** acts at a slight alkaline pH (7.5–8.5) in a step wise fashion.

$$\text{Triglyceride} + H_2O \xrightarrow{\text{Lipase}} \text{Diglyceride} + \text{Fatty acid}$$
$$\text{Diglyceride} + H_2O \xrightarrow{\text{Lipase}} \text{Monoglyceride} + \text{Fatty acid}$$
$$\text{Monoglyceride} + H_2O \xrightarrow{\text{Lipase}} \text{Glycerol} + \text{Fatty acid}$$

2. **Saponification.** The splitting of fat (triglyceride) by alkali is called **saponification**. The fatty acids formed as a result of hydrolysis of fats are neutralized by the alkali. These form the soap.

Triolein + Sodium hydroxide $\rightarrow$ Glycerol + Sodium oleate (soap)

3. **Hydrogenation.** When hydrogen is passed through unsaturated fats (or oils), these are converted into saturated fats. The process is completed rapidly in the presence of nickle (Ni) which acts as a catalyst. This process is called **hydrogenation** and is utilised for the manufacture of ghee (vanaspati) from vegetable oils.

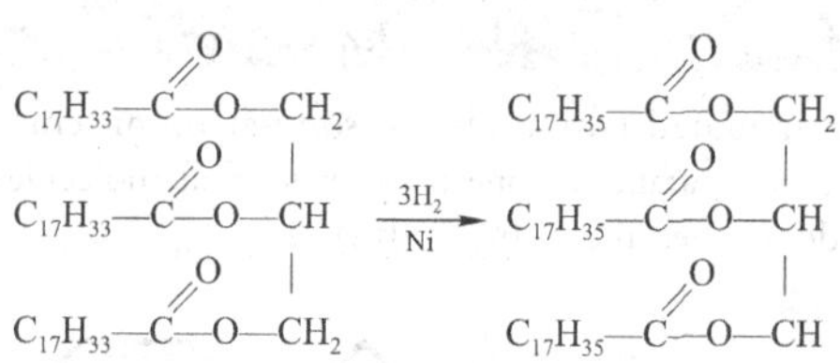

Triolein (oil) Tristearin (fat)

Fig. 9.11. Hydrogenation of lipid molecule.

4. **Hydrogenolysis.** When hydrogen is passed in excess the oils and fats break up in glycerol and fatty acids.

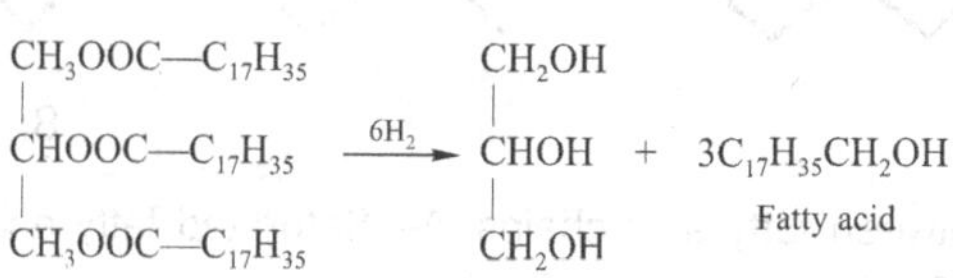

Fat Glycerol

Fig. 9.12. Hydrogenolysis of a fat molecule.

5. **Rancidity.** An exposure of air, fats develop unpleasant odour and flavour. This occurs due to oxidation of fat which releases volatile fatty acids having unpleasant odour. This is known as

rancidity. It is mostly found in vegetable oils and unsaturated fats.

Rancidity of oils also takes place due to bacterial growth, moisture, heat and light. During rancidity, oxidation of fatty acids takes place. This is also known as **auto-oxidation** of fat. The incomplete hydrolysis of ester bonds in the fat produces fatty acid which also leads to rancidity.

The process of rancidity and auto-oxidation of some fats is prevented by certain substances such as vitamin E (which is an **antirancid agent**) which are naturally present in them. Examples of other antirancid agents include phenols, naphthols and quinones.

6. **Fat analysis.** For the analysis of fat the following chemical contacts are used:
   (*i*) **Acetyl number.** Milligram (mg) of KOH required to neutralize the acetic acid obtained by saponification of one gram (g) of acetylated fat. This number indicates the number of hydroxyl groups in fatty acids.
   (*ii*) **Acid number.** The number of mg of KOH required to neutralize the free fatty acids present in 1 g fat or oil. High acid value indicates impurity.
   (*iii*) **Saponification number.** It is the number of milligrams of KOH required to saponify completely one gram of fat. If a fat is composed of fatty acid of higher molecular weight (such as palmitic, stearic and oleic acid), its saponification number will be low and will be high if the fat has shorter chain fatty acids.
   (*iv*) **Iodine number.** It is number of grams of iodine absorbed by 100 g of fat. It is used to determine the total amount of unsaturation in fat, but it does not indicate the number of double bonds.
   (*v*) **Reichet–Missel number (R–M value).** The number of ml of 0.1 N alkali required to neutralise the volatile fatty acids contained in 5 g of fat. R–M value is of special significance in characterisation of fat and testing its purity.

Functions of Neutral Fats or Oils

True fats have the following functions in plants and animals.

1. Reserve food materials. The neutral fats serve as food reserves in both plants and animals. They are stored in special regions—connective tissues (*e.g.*, adipose tissue) around the kidneys, under the dermis in mammals, in the fat bodies in frogs and lizards, and in the liver of fish. These fat deposits are not long term reserves of stored food used only in starvation, but are constantly being used up and reformed.

In plants fat is stored in the seeds to provide nourishment to the germinating embryo. The oil extracted from seeds of groundnut, mustard, soyabean, sunflower and coconut is used for cooking and other purpose (*e.g.*, castor-oil).

2. Fuel. The neutral fats form a concentrated fuel for the working of the animals and plants. They produce more than twice as much energy for gram as do the carbohydrates. They, thus, represent an economical food reserve in the body.

3. Insulation coats. The neutral fats form an insulating layer under the skin (the subcutaneous fat). Women tend to have thicker layer of adipose tissue than men have, and whales have a very thick layer of adipose tissue, called **blubber**, under the skin for insulation.

4. Shock-absorbing cushions. The neutral fats serve as shock-absorbing cushion around the eyeballs, kidneys and gonads.

5. Improvement of physique. The subcutaneous fat in human beings keeps the skin firm and helps in rounding off the contour of the body. These features help in ethology of the animals.

B. Waxes

Waxes are those simple lipids in which one molecule of higher fatty acid (with longer side chains) is esterified with a higher monohydric aliphatic alcohol other than glycerol. The constituent fatty acid and alcohol usually have 24 to 36 carbon atoms. Waxes are chemically inert so are not digested by lipase enzymes. Molecules of waxes do not contain any double bond in their hydrocarbon chain and are insoluble in water, so are resistant to atmospheric oxidation. They have high melting points (60°C to 100°C).

Waxes are saponified with great difficulty (as they contain 31 to 55% unsaponified matter) than fats (which contain only 1 to 2% unsaponified matter).

Types of Waxes

1. Bee wax. It is a simple lipid formed by esterification of one molecule of **palmitic acid** ($C_{15}H_{31}COOH$) and one molecule of mericyl alcohol ($C_{30}H_{61}OH$) or *n-hexacosanol* ($C_{26}H_{53}OH$; Fig. 9.13). The hydrophobicity of **mericyl palmitate** (Fig. 9.14) makes bee wax very water-insoluble and its higher melting point (due to the long, saturate hydrocarbon chains) bees wax is hard and solid at typical outdoor temperatures.

$$\underset{\text{Palmitic acid}}{CH_3(CH_2)_{14}CO-OH} + \underset{\text{Hexacosanol}}{H-OCH_2(CH_2)_{24}-CH_3} \rightarrow \underset{\text{Hexacosyl palmitate (Bee wax)}}{CH_3(CH_2)_{14}-\overset{O}{\overset{\|}{C}}-O-CH_2-(CH_2)_{24}CH_3}$$

Fig. 9.13. Formation of bee wax.

$$H_3C-(CH_2)_{14}-\overset{O}{\overset{\|}{C}}-O-(CH_2)_{29}-CH_3$$

Fig. 9.14. Mericyl palmitate, a wax.

2. Ear wax. It is also known as **cerumen** (L. *cera* = wax). It is secreted by cells lining the auditory canal. Cerumen lubricates the canal and traps particles that could damage the eardrum. Ear wax is a complex mixture made up mostly of long chain fatty acids, cholesterol and ceramides. It also contains sequalene, triacylglycerols and true waxes (about 10% of weight) (**Horton** *et al.*, 2006).

$$CH_3(CH_2)_{14}-\overset{O}{\overset{\|}{C}}-O-(CH_2)_{15}-CH_3$$

Fig. 9.15. Cetyl palmitate (spermaceti).

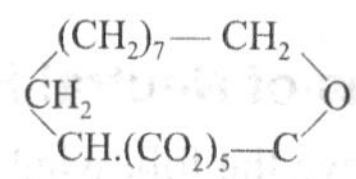

Fig. 9.16. Ambretolide.

3. Lanoline or wool fat. It is formed by esterification of one molecule of either palmitic acid ($C_{15}H_{31}COOH$) or stearic acid ($C_{17}H_{35}COOH$) or oleic acid ($C_{17}H_{33}COOH$) and one molecule of cholesterol. It is secreted by cutaneous glands and closely resembles the sebum.

4. Carnauba wax. It is formed by the esterification of cerotate acid (having 26C) with one molecule of mericyl alcohol (having 30C). It is extracted from a Brazilian palm tree.

5. Spermaceti (Fig. 9.15). It is palmitic ester of cetyl alcohol ($C_{16}H_{33}OH$). It is an oil extracted from the head of sperm whale.

6. Ambretolide (Fig. 9.16). It is extracted from the seeds of *Abelmoschus esculentus*. It is hydroxyl acid and gives a characteristic smell to the seeds.

Functions of Waxes

1. Waxes form a water-proof coating on plant parts such as leaves, etc., to reduce the rate of transpiration and also prevent their wetting.
2. Wax esters act as superior machine lubricant.
3. These form water resistant layers in insects, birds such as ducks and furred mammals.
4. Waxes form protective sheaths on fruits.
5. Waxes are used in polishing the furniture.
6. Lanolin fat is used in manufacture of cosmetic creams, ointments, etc.
7. Spermaceti is used in manufacture of polishes, ointments, candles, etc.

8. Ear wax or cerumen lubricates and protects the ear drum.
9. Sperm oil which is found along with spermaceti in head of sperm whale, acts as a valuable lubricant for instruments such as watches.
10. Waxy coating of fruits such as apple and citrous fruits prevent them from dyring so these can be stored for long period of time.
11. Wax papers are commonly used for packing perishable materials.

9.4. COMPOUND LIPIDS

I. Phospholipids

Phospholipids are compounds of glycerol, fatty acids and phosphoric acid.

Structure. The phospholipids are composed of a molecule of glycerol or other alcohol having (*i*) a phosphate group joined to either of its outer –OH groups (never to the middle one), (*ii*) two fatty acid molecules linked to the other two –OH groups, and (*iii*) a base (alcohol) molecule usually containing one or more nitrogen atoms bound to the phosphate group (Fig. 9.17). Thus, in a phospholipid the phosphate group forms a linking bridge between the glycerol and the base. Both the fatty acid molecules of a phospholipid may be saturated or unsaturated, or one may be saturated and the other unsaturated (Fig. 9.18).

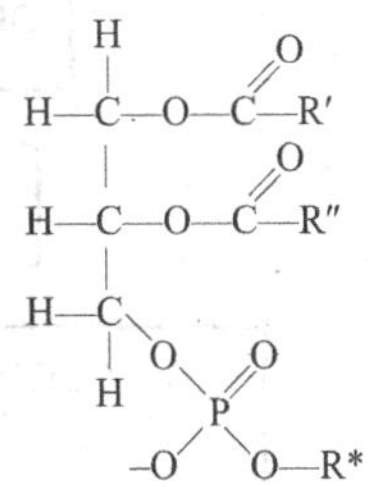

Fig. 9.17. Structure of a phospholipid molecule. R′ and R″ are fatty acid chains and R* a nitrogen containing base.

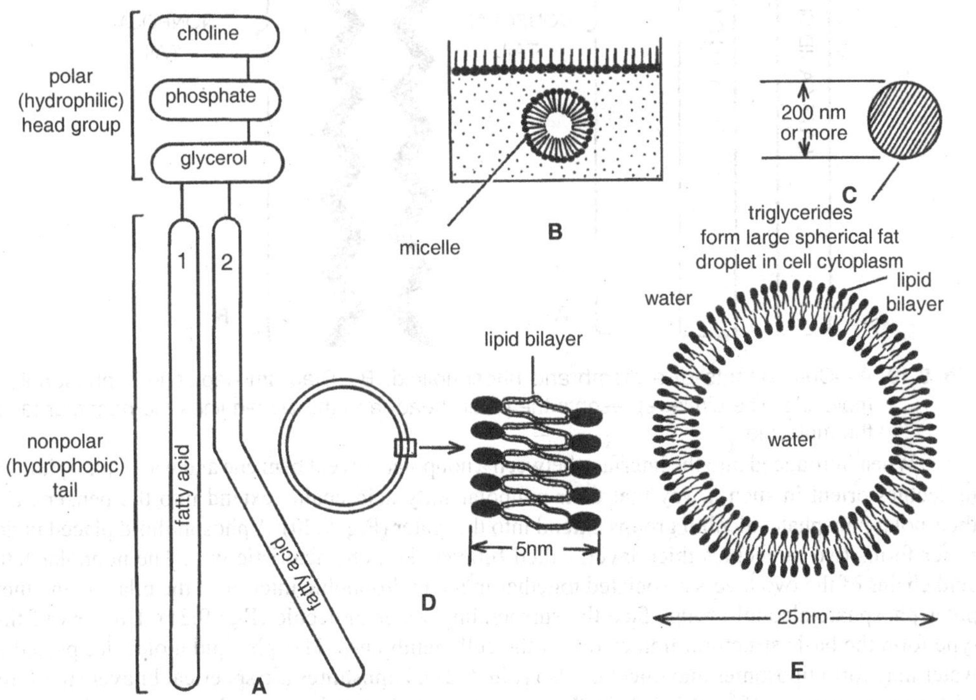

Fig. 9.18. Formation of various types of lipid aggregates. A—Schematic representation of a phospholipid molecule; B—Formation of micelle and monolayer film; C—Formation of a fat droplet by triglycerides; D—Formation of self-sealing lipid bilayer (*e.g.*, liposome); E—Cross-section of a liposome.

Special property. The phospholipids have a special property of dual solubility. A phospholipid molecule has a strongly nonpolar and hydrophobic (water hating or water insoluble) "tail" consisting of fatty acid chains, and a polar and hydrophilic (water loving or water soluble) "head" comprising a negatively charged phosphate group and a positively charged base (Fig. 9.19). The phospholipids are, thus, **amphipathic** (Gr. *amphi* – both + *pathos* = disease, meaning "tolerant to both") lipids, one end of a molecule being hydrophobic and the other hydrophilic. It is this dual nature of the phospholipids that makes them very important for the structure and functioning of the membranes of the cell.

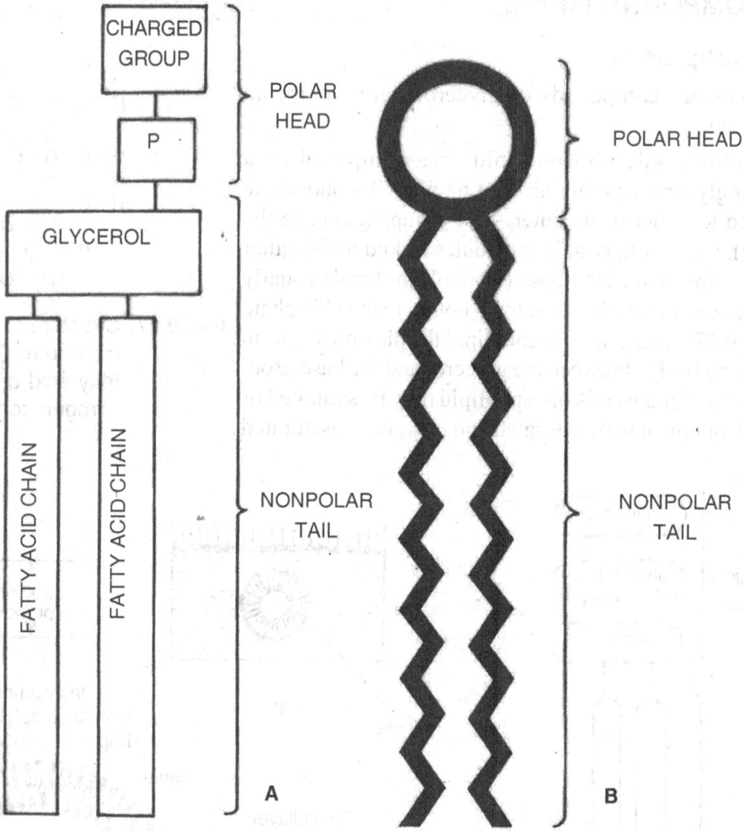

Fig. 9.19. A—Components of a membrane phospholipid. B—Diagram depicting a phospholipid molecule. The circle represents the polar "head" and the zigzag lines the nonpolar tail of the molecule.

When introduced into the interface between a nonpolar solvent benzene and water, phospholipid molecules orient in such a way that their nonpolar fatty acid chains extend into the benzene and their polar phosphate–alcohol groups extend into the water (Fig. 9.20). A phospholipid placed under water forms two molecules thick layer, called **bilayer**, in a characteristic way. The nonpolar fatty acid chains of the two layers associated together in the hydrophobic interior of the bilayer, and their polar phosphate-alcohol groups face the surrounding water molecule (Fig. 9.21). Bilayers of this type form the basic structural framework of the cell membranes. Phospholipid molecules placed in water may form liposomes and micelles also (Fig. 9.21). **Liposomes** are spherical bilayer structures with aqueous interior (Fig. 9.21). **Micelles** are spherical clusters having polar heads on the surface in contant with water and nonpolar fatty acid chains directed towards the center (Fig. 9.21).

Liposomes. When aqueous suspensions of phospholipids are subjected to rapid agitation by using ultrasound (*i.e.*, insonation), the lipid disperses in the water and forms **liposomes** or lipid **vesicles**.

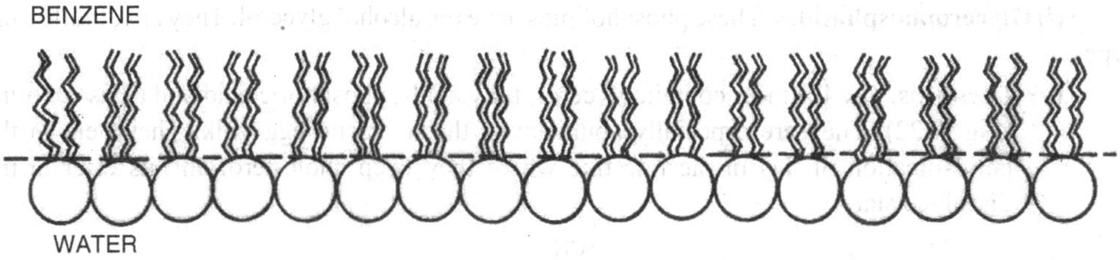

Fig. 9.20. Orientation of phospholipid molecules in the inter face between benzene and water.

Liposomes are small spherical bodies (25 nm to 1 μm in diameter) whose surface is formed by a bilayer of phospholipid molecules enclosing a small volume of the aqueous medium. They exhibit many of the permeability properties of natural membranes, *i.e.*, water soluble small molecules or ions can be enclosed by the liposomes and they can also traverse the lipid bilayer of latter. Recently, liposomes have been found to have great therapeutic promise, since, they can be used as vectors for the transfer of specific drugs, proteins, hormones, nucleic acids, ions or any other molecule into the specific types of animal cells. The contents of the liposomes can enter the target cells by two routes:

1. The liposomes can attach to the surface of target cells and may fuse with the plasma membrane, following which their contents are released into the cytosol or cytoplasmic matrix.
2. The entire liposomes may be endocytosed and degraded intracellularly.

Types of Phospholipids. The phospholipids are classified on the basis of alcohol present in them. The more common types are glycerophosphatides phosphoinositides or phosphosphingosides.

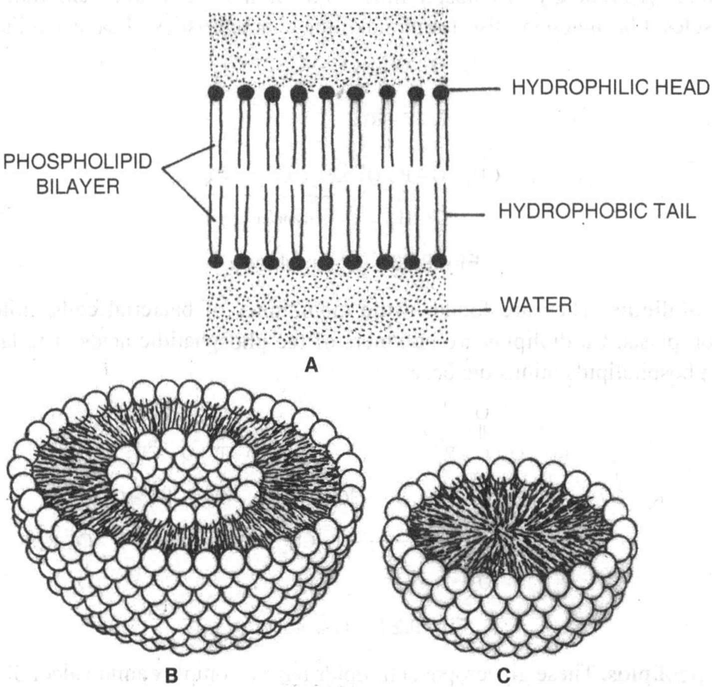

Fig. 9.21. A—Bilayer formed by phospholipid molecules placed under water. Nonpolar tails are associated together in the bilayer interior. Polar heads face the surrounding water molecules. B—A liposome. C—A micelle.

(*i*) **Glycerophosphatides.** These phospholipids have the alcohol glycerol. They are of following types:

(*a*) **Lecithins.** The lecithins contain glycerol, fatty acids, phosphoric acid and a base choline (Fig. 9.22). They are especially abundant in the liver and egg yolk. They help in the emulsification of fats in the intestine where they keep cholesterol and its ester in the dissolved state.

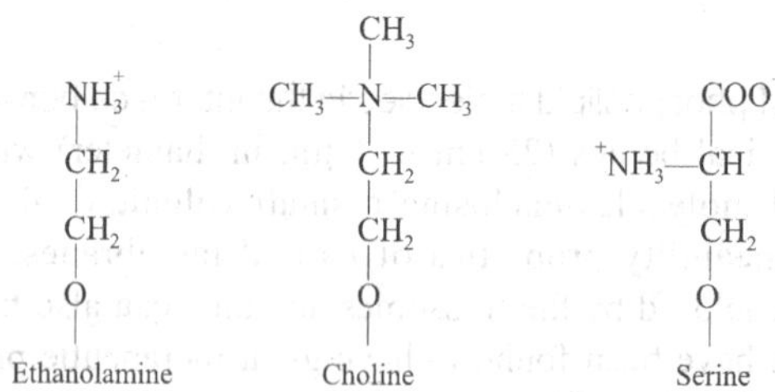

Fig. 9.22. Chemical groups linked to phosphate group in phospholipids.

Part of toxicity of snake venom is found to be related to the presence of enzyme named lecithinases, which hydrolyze the lecithins.

(*b*) **Cephalins.** The cephalins differ from the lecithins only in that the base is ethanolamine ($NH_2CH_2CH_2OH$) instead of choline (Fig. 9.22).

(*c*) **Plasmalogens.** The plasmalogens are glycerophosphospholipids and have one fatty acid replaced by an aldehyde. The base is choline or ethanolamine. Plasmalogens form about 23% of the glycerophospholipids of the human brain and membranes of nerves and muscles. Plasmalogens also form one of the characteristics of cancer cells.

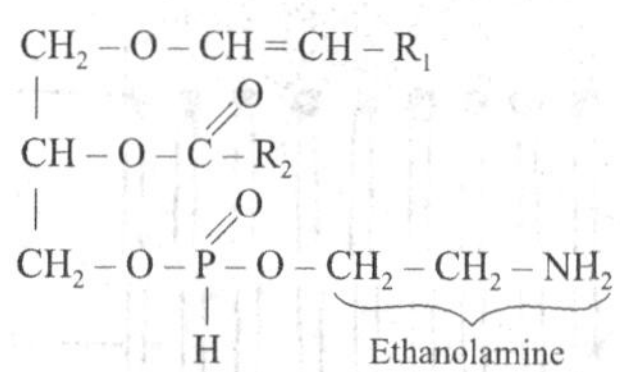

Fig. 9.23. Plasmalogen.

(*d*) **Cardiolipins.** They are found in the membranes of bacterial cells, mitochondria and chloroplasts. Cardiolipins are polymers of the phosphatidic acids. The latter are any of the phospholipids minus the base.

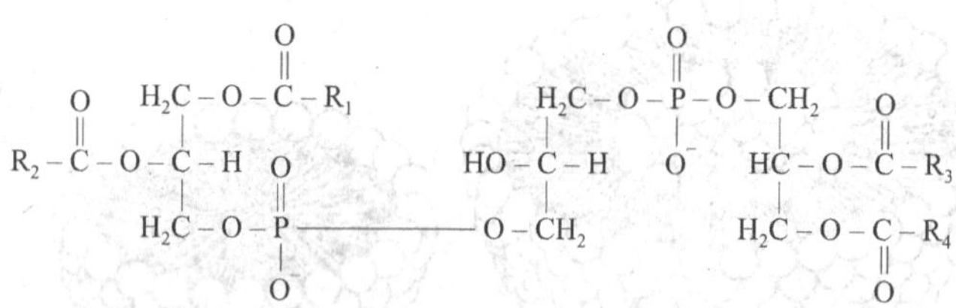

Fig. 9.24. Cardiolipin.

(*ii*) **Sphingolipids.** These glycerophospholipids have a complex amino alcohol sphingosine in place of glycerol. Their other constituents are fatty acids, phosphoric acid and choline. **Sphinomyelins** belong to their group. They are specially prominent components of cell membranes in the brain and nervous tissue (*i.e.*, present in the myelin sheath).

Box 9.3.

Three major families of sphingolipids are the **sphingomyelins**, the **cerebrosides** and the **gangliosides**. Of these, only sphingomyelins contain phosphate and are classified as phospholipids. Cerebrosides and gangliosides contain carbohydrate residues and are classified as glycosphingolipids (**Horton** *et al.*, 2006).

***(iii)* Phosphoinositides.** These phospholipids have the cyclic hexahydric alcohol named **inositol** in place of the base. Inositol $(CH_2O)_6$ exists in the inner membrane of mitochondria. It is a carbocyclic, hexahydric alcohol which is sweet in taste but is not sugar (carbohydrate). It is found in muscles (called **musucle sugar**), liver, kidneys, brain, cerebrospinal fluid (CSF), erythrocytes and tissues of eye. It is a fundamental ingradient of cell membranes. Among plants, inositol occurs in fruits, vegetables, whole grains and nuts as phytic acid esters. Milk and yeast too contain appreciable quantities of inositol.

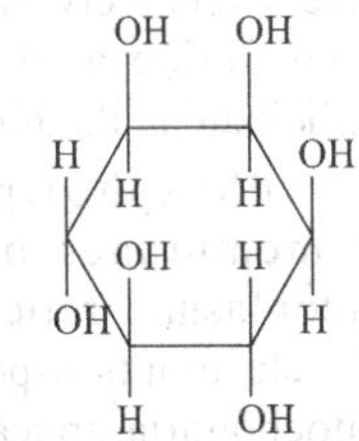

Fig. 9.25. Inositol.

Functions of Phospholipids. Phospholipids serve the following functions:

1. Phospholipids are main components of plasma membrane and other cell membranes (*i.e.*, nuclear, mitochondrial, ER). They are thought to regulate the passage of materials into and out of the cell and cell organelles. Galactolipids and sulpholipids occur in plastid membranes.
2. Some phospholipids such as lecithins form an important component of yolk and protoplasm.
3. Small amount of phospholipids may occur in some storage fats.
4. Spingomyelins and cephalins insulate the nerve fibers (axons, *i.e.*, impulse) from its surrounding structures.
5. Phospholipids (*i.e.*, lipoproteins of serum) form important intermediates in the transport of lipids to and from the liver.
6. Lecithin also act as a component of acetylcholine—neurotransmitters in synaptic conduction of nerve impulses.
7. Glycosphinolipids are the constituents of cell membranes so these are responsible for tissue immunity and cellular recognition.

2. Glycolipids

The glycolipids contain fatty acid, amino alcohol sphingosine (a high molecular weight fatty acid) and sugar. The sugar replaces one fatty acid molecule with galactose (Fig. 9.26). Their name refers to the presence of a carbohydrate in them. Glycolipids are of two types:

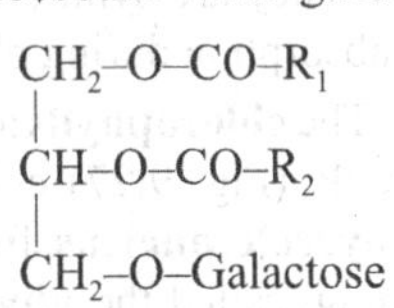

Fig. 9.26. Structure of a glycolipid molecule, R_1 and R_2 are fatty acid.

(i) **Cerebrosides:** These are found particularly in the myelin sheath of medullated nerve fibers and the white matter of the brain.

(ii) **Gangliosides:** These are rich in sialic acid (sugar) and ceramide (fatty acid) and found in ganglion cells of nervous tissue (forming 6% of gray matter). They are also found in parenchymatous tissues such as spleen and in RBCs. Gangliosides exist even in outer surface of plasma membrane. They have considerable physiological significance. Gangliosides are the receptors for certain bacterial protein toxins such as cholera toxins and certain pituitary glycoprotein hormones (**Sharma** 2010).

Other Compound Lipids

1. Sulpholipids: They are widely distributed in plants and are localised in chloroplasts. Sulpholipids are also found in the chromatophores of photosynthetic bacteria.

2. Lipoproteins: The lipoproteins contain lipids (mainly phospholipids) and proteins in their molecules. The lipid components consist of triacyl glycerol, phospholipids and cholesterol. The protein components of lipoproteins have a relatively high portion of non-polar amino acid residues that can participate in the binding of the lipids.

Phospholipids are found in membranes of mitochondria, endoplasmic reticulum and nuclei. The electron transport chain system of mitochondria appear to contain large amounts of lipoproteins. Lipoproteins are also found in mammalian blood plasma, milk and egg-yolk, etc. Some important examples of lipoproteins are:

(*i*) **Rhodopsin** from the retina; (*ii*) **chloroplastin** from leaves; (*iii*) **lipovetellin** and (*iv*) **lipovitellenin** from egg yolk.

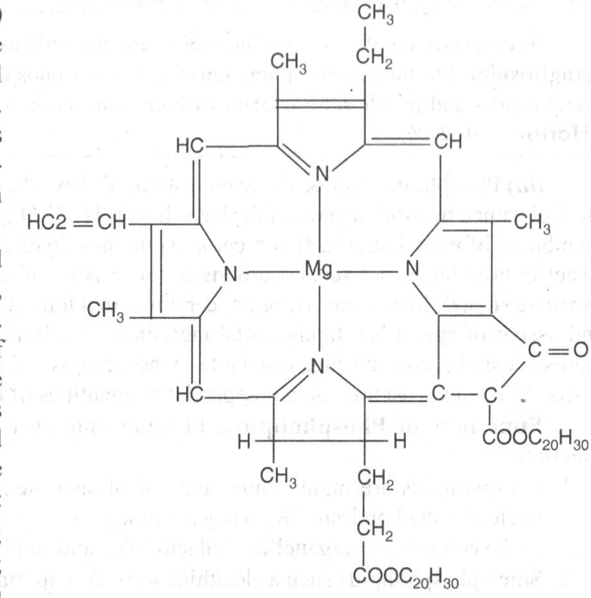

Fig. 9.27. Chemical formula of chlorophyll.

Further, on the basis of centrifugational properties lipoproteins include following four types: (1) Chilomicrons; (2) Very low density lipoprotein (VLDL); (3) High density lipoproteins (HDL) and (4) Low density lipoproteins (LDL).

3. Chromolipids: The **carotenoids** are the compound lipids and they form the pigments of the animal and plant cells. There are about 70 carotenoids occurring in both types of cells. The important carotenoids of cells are the α, β and γ carotenes, retinene, xanthophylls, lactoflavin in milk, riboflavin (vitamin B_2), xanthocyanins, coenzyme Q, anthocyanins, flavones, flavonols and flavonones, etc. Chemically all carotenoids are long-chain isoprenoids having an alternating series of double bonds. They are synthesized by plant tissues and are located in the chloroplast lamellae to help in light absorption during photosynthesis. In animal cells, carotenoids serve as precursors of vitamin A.

The **chlorophylls** are essential photosynthetic green pigments of the chloroplasts. A chlorophyll molecule (Fig. 9.27) consists of a **head** and a **tail**. The head consists of a **porphyrin ring** or **tetrapyrrole nucleus** from which extends a hydrophobic tail which is made up of a 20-carbon grouping, called the **phytol**. Phytol ($C_{20}H_{39}$) is a long straight-chain alcohol containining a single double bond. It may be regarded as a hydrogenated carotene (vitamin A). The porphyrins (Gr., *porphyra* = purple) are complex cabron-nitrogen molecules that usually surround a metal, *i.e.*, it is formed from four pyrrol rings linked together by methane bridges and metal atom (Mg or Fe) is linked to pyrrol rings. In chlorophyll molecule, the porphyrin surrounds a magnesium ion, while in haeme of **haemoglobin**, it surrounds an iron ion (Fig. 9.28). Many other pigments of animal cells such as **myoglobin** and **cytochromes** have porphyrin rings in their molecules.

9.5. DERIVED LIPIDS

1. Steroids

The steroids (Gr., *stereos* = solid + *eidos* = form) do not contain fatty acids but are included in lipids because they have lipid-like properties. Unlike other lipids, they are not straight-chain compounds. They consist of a complex carbon framework of fused four rings, three of which have six carbon atoms each and the fourth have five (Fig. 9.29). The conventional method of naming the rings and numbering of seventeen positions in the rings have been shown in Fig. 9.29. The various steroids differ in the number and position of double bonds between the carbon atoms in the rings, and also in the side groups linked to the rings.

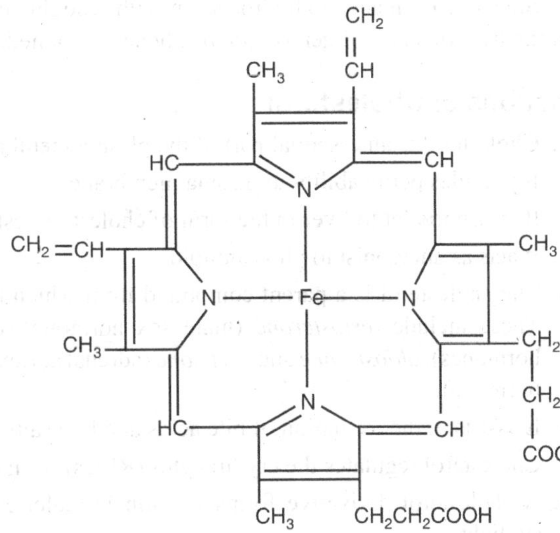

Fig. 9.28. Chemical formula of haeme portion of haemoglobin.

Examples of steroids. The most abundant group of steroids includes the **sterols**. The sterols are complex alcohols (Gr., *stereos* = solid + *-ol* = alcohol), having a polar or hydrophilic hydroxyl (–OH) group linked at one end and a complex non-polar or hydrophobic carbon chain joined at the other end of the ring structure (Fig. 9.30). Thus, sterols are oriented in a way that they show their dual **solubility** properties. The sterols occur in both animal and plant cells. The common sterols include *cholesterol* and *ergosterol*. Other steroids are *bile acids*, vitamin D, male and female sex hormones and adrenal cortical hormones (such as cortisone, aldosterone) in vertebrates and *ecdysone* or moulting hormone of insects.

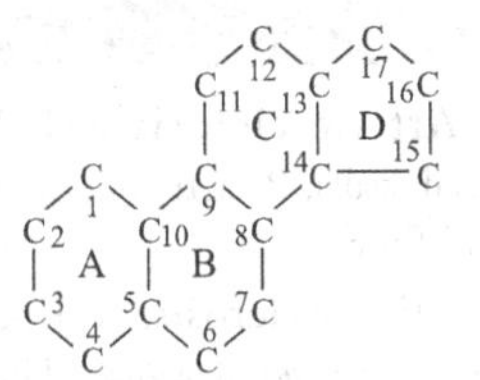

Fig. 9.29. Arrangement of four carbon rings in a steroid molecule.

(*i*) **Cholesterol.** Cholesterol is an animal sterol. It is a white, crystalline and levo-rotatory substance having a melting point of 149–150°C. It is tasteless and odourless substance. Cholestrol is insoluble in water, acids and alkalies but highly soluble in solutions of bile salts and organic solvents such as chloroform, petroleum ether, etc. Cholesterol is found in human body in large quantities specially in brain and nervous tissue. It is an important component of the plasma membrane of animal cells (*e.g.*, erythrocytes). Various human organs such as adrenal glands, liver, kidney, spleen and skin also contain cholesterol. Egg yolk is rich in cholesterol. The normal concentration of cholesterol is 140–250 mg per 100 ml of blood. Its increase in the blood is harmful and may cause **atherosclerosis** (*i.e.*, narrowing of arteries) by deposition on their lining (such a deposition is called **atheroma**). Deposition occurs since both cholesterol and its esters are insoluble in water. Narrowing of cerebral arteries may lead to **stroke** while that of coronary arteries causes **heart attack**.

Cholesterol was first isolated from gall stones, hence its name which means 'a solid alcohol from bile' (Gr., *chole* = bile + *steros* = solid + ol = alcohol). Its formula is $C_{27}H_{45}OH$ (Fig. 9.30). It has a double bond between carbon 5 and 6. Cholesterol is synthesized in

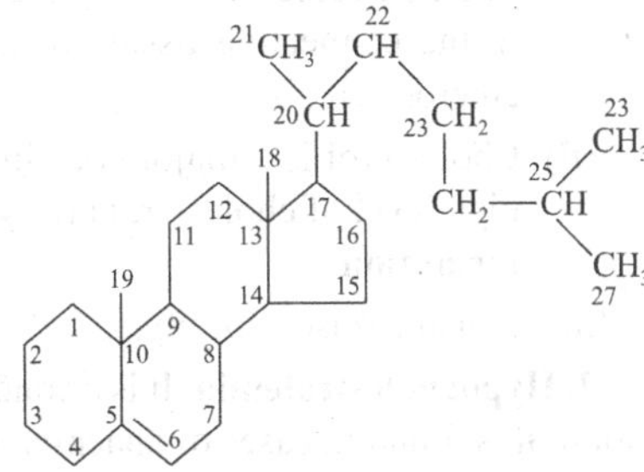

Fig. 9.30. Structure of a cholesterol molecule.

10 steps. **Squalene** ($C_{30}H_{50}$) is one of the intermediates in its synthesis.

Squalene is an acyclic hydrocarbon with 6 double bonds in its molecule. It was first isolated from the liver of shark of genus *Squalus*, hence so named.

Functions of Cholesterol

1. Cholesterol is an essential part of the plasma membrane.
2. It provides permeability to plasma membrane.
3. It transports fat to liver in the form of cholesterol ester.
4. It acts as antagonist to phospholipid.
5. The cholesterol is a parent compound from which many hormones are synthesized in cells. These include *testosterone* (male sex hormone), *estradiol* and *progesterone* (female sex hormones), *aldosterone* and *cortisone* (adrenal cortex hormone) and *ecdysone* (insect moulting hormone).
6. It assists in the formation of bile acids and bile salts.
7. Cholesterol regulates the erythrocytes (RBCs) from being haemolized.
8. A cholesterol derivative forms vitamin D (calciferol) by the action of ultraviolet rays of sunlight.
9. Cholesterol is necessary for human life but the excess of it is known to cause cardiovascular problems such as arteriosclerosis, atherosclerosis, hypertension, heart attack, stroke, etc.

Box 9.4.

Arteriosclerosis includes three pathological conditions: 1. an increase in the thickness of the arterial walls; 2. a reduction in elasticity of the vessel and 3. a constriction of diameter, which affects blood flow.

Atherosclerosis. A degenerative disease of the arteries caused by a atheroma (*i.e.*, fatty deposits or plaques in arterial walls). These either can block an artery directly or increase the chance of its being blocked by a blood clot. Eventually, the arteries become hardened and lose elasticity.

Clinical Effects of Cholesterol

1. Hypercholesterolemia. It is caused when cholesterol level in blood is raised beyond the threshold limit. It is caused due to uncontrolled diabetes mellitus, impairment of liver, obstructive jaundice, glomerulonephritis, nephrosis and hypothyroidism. It leads to following types of diseases:

(*i*) Hypercholesterolemia leads to **atherosclerosis**, which is characterized by the hardening of blood vessels caused by the deposition of cholesterol, cholesteryl ester and other lipids in the connective tissue of areterial wall. Elevated level of LDL and VLDL leads to atherosclerosis.

(*ii*) Cholesterol is a major constituent of gall stone. Therefore, high level of cholesterol always poses a threat of **gall stone formation**.

(*iii*) It also casuses obesity.

2. Hypocholestrolemia. It is a condition of low cholesterol level in blood. It is found in cases of anaemia, hyperthyroidism, carcinoma, acute pancreatitis and in hepatic diseases. A decreased percentage of cholesterol ester is also known to cause hepatitis, cirrhosis and most types of liver damage.

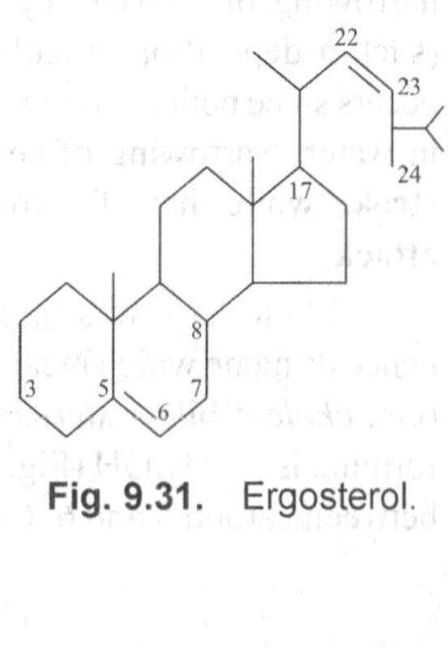

Fig. 9.31. Ergosterol.

(*b*) **Ergosterol.** It is a plant sterol. It is also found in the parasitic fungus, *Claviceps purpurea* (ergot) growing on the rye plant. It is present in yeast and vegetables. Its formula is $C_{28}H_{43}OH$ (Fig. 9.31). Ergosterol becomes vitamin D (calciferol) when activated by ultraviolet rays.

(*c*) **Bile acids and bile salts.** The bile acids are steroids having 24 carbon atoms. These include cholic acid, deoxycholic acid, lithocholic acid, etc. The bile acids combine in liver with 'glycine' or 'taurine' to form 'glycocholic' or 'taurocholic' acid, which is then joined by sodium to form bile salt, namely, **sodium glycholate** or **sodium taurocholate**.

Functions:

1. Bile acids represent the only significant way of excretion of cholesterol. The bile acids and phospholipids solubilise cholesterol in the bile.
2. The bile salts lower the surface tension and thus helps in the emulsification of fats for their digestion in the intestine.
3. Bile acids increase the absorption of fat-soluble vitamins, particularly vitamin D, from the intenstine.

(ii) Steroid Hormones

Steroid hormones are fat soluble hormones having a sterol ring and are derived from the cholesterol. They can be classified according to the number of carbon atoms contained in their molecules:

1. **C_{18} hormones** which include **estrogens** such as esteradiole and esterone of ovary.
2. **C_{19} hormones** which include **androgens** such as testosterone of testes.
3. **C_{21} hormones** which include **progesterone** of corpus luteum and **mineralocorticoids** such as aldosterone and **glucocorticoids** such as cortisol and cortisone of adrenal cortex.

A. Estrogens (C_{18} Hormone). Estrogens are produced by the follicular cells of membrana granulosa of Graafian follicle of ovary, corpus luteum of the ovary and by the placenta during the second and third trimesters of pregnancy. These include esteradiol, esterone and estriol. Out of these most important is **β-esteradiol**. It is found to act at a cellular level by binding to the nucleus of cells in its target organ, where it seems to stimulate the synthesis of RNA (transcription). Estradiol ($C_{18}H_{24}O_2$) promotes the development of endometrium (of uterus) and stimulate LH production in anterior lobe of pituitary gland. LH or **luteinizing hormone** in male humans stimulate interestial cells in testes to produce *testosterone*; in females, LH stimulates ovary to perform ovulation and causes ruptured ovarian follicle to develop into *progesterone*-secreting corpus luteum.

Functions of Estrogens. Estrogens stimulate the growth, development and normal functioning of **female secondary sex organs** such as Fallopian tubes, uterus, vagina, etc. These hormones regulate the cyclic changes in uterus and ovaries during the menstrual cycle.

Estrogens also control the development of **female secondary sexual characters** such as breast-development, broadening of pelvis, growth of pubic and axillary hair and beginning of menstrual cycle. Esstrogens are also related with development of sexual desire in the female.

B. Androgens (C_{19} Hormone). Androgens are produced by the Leydig cells (interstitial cells), of the testes and the adrenal glands in both the sexes. The ovary also produces androgen in small amounts. **Testosterone** is the major androgen produced in the testes at the rate of 6–7 mg/day in adult male. Active form of testosterone is its metabolite called **5a-dihydrotestosterone (DHT)**. Other androgens are **androsterone, epiandrosterone** and **dehydro-epiandrosterone**.

Functions of Testosterone. In humans and other mammals, testosterone controls the growth and development of **male secondary sex organs** such as epididymis, prostate gland, seminal vesicles and penis. It regulates the development of beared, moustaches, pubic hair, axial hair, deepening of voice, broadening of shoulders, increased height (due to elongation of bones and increased development of limbs). It also initiates sexual desire.

Testosterone also regulates spermatogenesis and maturation of sperms. It also increases the fertilizing power of the sperm.

C. Progesterone (C_{21} hormone). This steroid hormone is secreted by yellowish endocrine gland of ovary called **corpus luteum** which originates from ruptured ovarian follicle after the ovulation. Small amounts of progesterone is also secreted by adrenal cortex and placenta. Secretion of progesterone is stimulated by leuteinizing hormone (LH) of anterior lobe of pituitary gland.

Functions:

1. Progesterone stimulates the proliferation of the endometrium of uterus and prepare it for implantation.
2. It desensitizes the uterine muscles to the action of oxytocin.
3. It helps in implantation, placenta formation and normal development of foetus in the womb (uterus).
4. It suspends ovulation during pregnancy.
5. It stimulates the proliferation of mammary glands and enlargement of breasts.

D. Aldosterone (C_{21} hormone). It is the major **mineralo corticoid hormone** secreted by *zona glomerulosa* of adrenal cortex. Its secretion is stimulated by *decrease in sodium level* and *increase in potassium level* in the blood. Its secretion is controlled by ACTH of anterior lobe of pituitary.

Functions. Aldosterone maintains **hypernatraemia** (high Na^+ level in blood) and **hypokalaemia** (low K^+ level in blood) by increased sodium reabsorption from nephric filterate and decreased loss of sodium in sweat. This Na^+/K^+ balance helps in nerve impulse conduction, muscle contraction, maintenance of blood volume and blood pressure.

E. Cortisol (C_{21} Hormone). It is a **glucocorticoid hormone** secreted by *zona fasciculata* of adrenal cortex. Its secretion is stimulated by low sugar level in the blood and is regulated through ACTH of anterior lobe of pituitary.

Functions. 1. It regulates **carbohydrate metabolism** and causes increase in sugar level by following ways: (*i*) increase rate of **gluconeogenesis** (conversion of proteins in liver into sugar), (*ii*) decrease in peripheral utilisation of glucose. 2. It has **anti-inflammatory** and **anti-allergic** processes.

2. Terpenes

The terpenes include certain **fat-soluble vitamins** (*e.g.*, vitamins A, E and K) **carotenoids** (*e.g.*, photosynthetic pigments of plants), and certain **coenzymes** (such as coenzyme Q or ubiquinone). All the terpenes are synthesized from various numbers of a five-carbon building block, called **isoprene unit**. The isoprene units are bonded together in a head-to-tail organization. Two isoprene units form a **monoterpene**, four form a **diterpene**, six form a **triterpene**, and so on. The monoterpenes are responsible for the characteristic odours and flavours of plants (*e.g.*, **geraniol** from geraniums, **menthol** from mint and **limoneme** from lemons). **Dolicol phosphate** is a **polyisoprenoid** (*i.e.*, long chain polymer of isoprene) and is used to carry activated sugars in the membrane-associated synthesis of glycoproteins and some polysaccharides.

3. Icosanoids

Icosanoids are **paracrine hormones**, which influence the nearby cells and are not circulated by blood to distant body parts. They are of following three types:

1. Prostaglandins (PGs). The prostaglandins are recently discovered, hormone-like compounds. The name 'prostaglandin' was given by **U.S.V. Euler** in 1935. PGs are derivatives of 20-carbon polyunsaturated hydroxylated fatty acids such as **arachidonic acid**. They have a characteristic five-carbon ring in the middle of the chain and hydroxyl or ketone groups or certain carbons. Many different prostaglandins have been discovered.

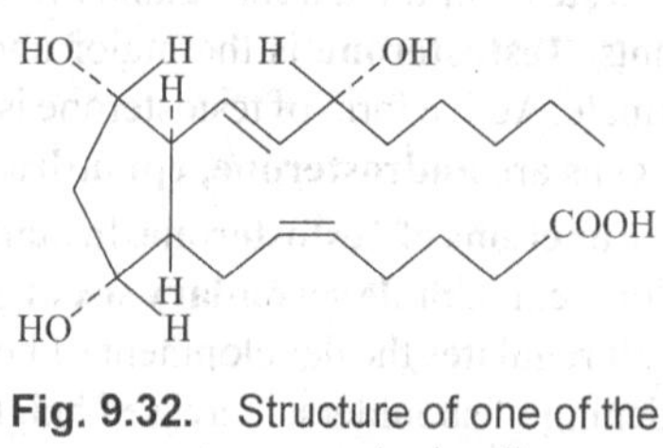

Fig. 9.32. Structure of one of the prostaglandins.

The prostaglandins were first found in the human semen, hence their name. It is now known that they are produced in many types of human tissues. They occur in lungs, thymus, kidneys, thyroid, brain, spinal cord, gastrointestinal mucosa, pancreas, endometrium, amniotic fluid, placenta and menstrual fluid. Many cells have minute amounts of prostaglandins in their membranes, but the richest sources are the secretion of seminal vesicles in males and menstrual fluid in females. Low-back pain in women during menstrual period may be due to prostaglandins released by disintegration of endometrial cells.

Prostaglandins are classified into 9 groups designated as A through I (*i.e.*, PGA through PGI). PGs have a half-life of 5 minute or less, hence they are destroyed very rapidly in the body. The short half-life of these tissue-hormones is thought to ensure their transient and limited response at the intermediate site of production (see **Jain** *et al.*, 2012).

Functions. The prostaglandins have a variety of effect on smooth visceral muscles, nervous system and blood pressure. They are thought to control many biological processes, perhaps by regulating the production cyclic AMP by adenyl cyclase. PGs increase uterine contractions. Hence, they are used to induce *parturition* and **abortion**. By increasing the activity of uterine muscles, the PGs may improve the chances of the union of sperm and egg. It has been found that the deficiency of prostaglandins in semen may cause male sterility. Prostaglandins also increase the motility of intestinal muscles, and may cause severe cramps, nausea, vomiting and diarrhoea. Prostaglandins inhibit gastric secretion, decrease arterial blood pressure, clear the nasal passages and relax the bronchioles. They have a sedative effect on many animals.

Clinical significance. From medical viewpoint, PGs are potentially most revolutionary therapeutic substances yet discovered. They have the following clinical applications:

1. PGs exert stimulatory effect on the contractions of human uterus. Therefore, PGE, PGE_2 and $PGE_2\alpha$ are given intravenously to induce labour.
2. PGE_1 is also given to asthmatic patients through inhalation for vasodilation.
3. PGE_1 is also an effective nasal decongestant.
4. PGA_1, PGE_1, PGE_2 and $PGE_2\alpha$ act as vasoconstrictor on blood vessels of nasal mucosa.
5. Prostaglandins are also used to control inflammation.
6. PGs lower down the blood pressure.

2. Leucotrienes (LTs). They are secreted by WBCs and are derived via lipoxygenase pathway. They have three conjugated double bonds. Leucotrienes stimulate mucus secretion and carry out vasoconstriction of bronchial muscles. Their excessive release causes **asthama**.

3. Thromboxanes (TXs). They are a class of molecules derived from arachidonate and are released by blood platelets or thrombocytes. They produce vasoconstriction and increase blood pressure. They cause release of serotonin and calcium ions (Ca^{2+}) for blood clotting.

QUESTIONS

Long Answer Questions

1. What are lipids? Classify lipids with suitable examples. Describe their importance for living beings.
2. What are lipids? Discuss their structure. Classify lipids with suitable examples.
3. What are lipids? Discuss their structure. Classify lipids with suitable examples and explain their hydrolysis.
4. What are lipids? Give structure and classification of lipids.
5. What are lipids? Describe their importance for living beings.

6. Discuss the importance of lipids in human beings.
7. What are lipids? Give examples of important lipids and lipid derivatives. Describe their chemical composition and role played by them in a living system.
8. What is significance of lipids in human body? Write about any two lipids present in plasma membrane.
9. Describe derived lipids with examples.
10. Describe structure and properties of fats and oils.
11. What are fatty acids? Describe different types of fatty acids.
12. Discuss importance of prostaglandins.
13. What is ketosis? How are ketone bodies produced in the body?
14. Describe physiological or biological importance of cholesterol in the body. Summarise its clinical effect.

Short Answer Questions

1. What are lipids? Classify lipids with suitable examples.
2. Write a short note on hydrolysis of fats.
3. Describe biological significance of lipids.
4. Describe special properties of phospholipids. What is their significance or biological importance in living beings?
5. What is the reason that phospholipid molecules form a bilayer in living cells?
6. What are glycolipids? Give their functions.
7. Give functions of lipoproteins or proteiolipids.
8. Name the plasma lipoproteins found in blood plasma. Give their chemical composition.
9. What do you mean by iodine number of fat?
10. What is rancidity?
11. Write short notes on the following:
    (*i*) Lipids
    (*ii*) Lipolysis
    (*iii*) Phospholipids
    (*iv*) Saturated fatty acids
    (*v*) Fatty acids
    (*vi*) Esterification
    (*vii*) Cholesterol
    (*viii*) Sterols
    (*ix*) Hydrolysis of fat
    (*x*) Glycoproteins
    (*xi*) Prostaglandins
12. Give differences between the following:
    (*a*) Choline and cholesterol
    (*b*) Phospholipids and glycolipids
    (*c*) Saturated and unsaturated fatty acids

Very Short Answer Questions

1. What are fatty acids?
2. Which fatty acids are called saturated?
3. What are monosaturated fatty acids?
4. Give two examples of polysaturated fatty acids.
5. What are carotenoids?
6. What is difference between wax and paraffin wax?
7. What are chilomicrons?
8. Name two groups of derived lipids.
9. Give full names of following: PUFA, HDL, VLDL.
10. What is scientific name of lecithin?
11. Define saponification number.
12. If the saponification number of a fat is high, what does it indicate?
13. What types of lipids are present in the plasma membrane?
14. Which type of lipids are found in animal fats?
15. Who discovered essential and non-essential fatty acids?
16. Give empirical formula of fatty acids.
17. Name sources of essential fatty acids.
18. What is the number of fatty acid in a phospholipid volume?
19. What material is used for making the candles?
20. Which lipid forms essential part of our food?

21. Which fatty acid is most abundant in sunflower oil?
22. Name any one polyunsaturated fatty acid.
23. Who coin the term lipids?
24. Which fatty acid causes sterility?
25. Which compound is being used as an antifertility agent?
26. How many molecules of glycerol and fatty acids form one fat molecule?
27. What is the effect of deficiency of essential fatty acids (EFA)?
28. Give clinical importance of LDL and VLDL.
29. What are prostaglandins?
30. What is meant by rancidity of fat?
31. How the fat can be protected from rancidity?
32. Give an example of following (with structure) : A lipid containing glycerol, fatty acids, phosphate and ethanolamine.
33. What is cause of Norum's disease?
34. What is cause of Tay-Sach's disease?
35. What disease is caused by accumulation of sphingomyelins?
36. What is Gaucher's disease?

Multiple Choice Questions

1. Which of the following is incorrect about fatty acids?
   (*a*) they are straight chain compounds
   (*b*) all the carbon atoms are interlinked by single bonds
   (*c*) they have low melting points
   (*d*) they have general formula $C_nH_{2n}O_2$
2. The chemical linkage between glycerol and fatty acids is called
   (*a*) ester bond
   (*b*) glycosidic bond
   (*c*) phosphodiester bond
   (*d*) none of these
3. Atherosclerosis occurs due to high consumption of food rich in
   (*a*) proteins
   (*b*) carbohydrates
   (*c*) saturated fats
   (*d*) all of them
4. The most common lipids in the cell are
   (*a*) monoglycerides
   (*b*) diglycerides
   (*c*) triglycerides
   (*d*) polyglycerides
5. Which of the following represents the property of true fats?
   (*a*) they have high specific gravity
   (*b*) they have ill-defined melting points
   (*c*) they are insolube in organic solvent
   (*d*) they disperse in water as minute droplets
6. Unusual property of fat is that
   (*a*) their glycerol end is lipophilic and fatty acid end is hydrophobic
   (*b*) their glycerol end is hydrophobic and fatty acid is hydrophilic
   (*c*) their glycerol end is hydrophilic and fatty acid end is lipophilic
   (*d*) their glycerol end is lipophilic and fatty acid end is hydrophilic
7. ____________ molecule has strongly nonpolar and hydrophobic tail consisting fatty acid chains and a polar hydrophilic head comprising a phosphate group
   (*a*) phospholipid
   (*b*) glycolipid
   (*c*) glyceride
   (*d*) tristearin
8. The nitrogenous base present in lecithin is
   (*a*) choline
   (*b*) ethanolamine
   (*c*) inositol
   (*d*) serine
9. One of the following is an amphipathic lipid
   (*a*) phospholipids
   (*b*) fatty acid
   (*c*) bile salts
   (*d*) all of these

ANSWERS

Very Short Answer Questions

1. Fatty acids are long-chain (4–30 C atoms), aliphatic organic acids (Box 9.5) containing only C, H and O. They have a single carboxyl group attached to a long nonpolar hydrocarbon tail which gives most lipids their hydrophobic and oily or greasy nature. They do not occur in free (*i.e.*, uncombined) state in cells or tissues but are present in covalently bound form in different classes of lipids.

Box 9.5

Aliphatic compounds. A major class of organic compounds with carbon atoms linked in straight or branched open chains. The other classes are alicyclic, aromatic and heterocyclic compounds.

2. In saturated fatty acids all the carbon atoms in the hydrocarbon chain of a fatty acid are saturated. It means all the carbon atoms in the chain are connected through a single bond, *e.g.*, Palmitic acid.
3. Fatty acids with single double bond in their hydrocarbon chain are called monosaturated fatty acids, *e.g.*, Oleic acid.
4. Linoleic acid and archidonic acid.
5. Carotenoids are fat soluble photosynthetic pigments. These are formed of **isoprene units**.
6. Waxes are esters of monohydro alcohols and fatty acids with long hydrocarbon chains. They have an odd number of carbon ranging from C_{25} to C_{35}.
   Paraffin wax consists of a mixture of hydrocarbons of alkane series (*i.e.*, ethane series).
7. Chilomicrons are droplets of lipoproteins present in blood plasma. They are synthesized by smooth endoplasmic reticulum of epithelial cells linining the small intestine and serve to transport dietary lipids in the circulation.
   Thus, chilomicrons are formed of triglycerols and cholesteryl esters and are sorrounded by proteins, phospholipids and cholesterols.
8. (*i*) Terpenes and (*ii*) Elcosanoids.
9. (*i*) PUFA : Polyunsaturated fatty acids;
   (*ii*) HDL : High density lipids;
   (*iii*) VLDL : Very low density lipids.
10. Phosphatidylcholine.
11. Saponification number is the amount of KOH (in milligrams) which is used for the complete saponification of one gram of fat or oil.
12. High saponification number of fat indicates that it is formed of short hydrocarbon chain.
13. The lipids present in the plasma membrane are mostly unsaturated.
14. Animal fats are rich in saturated fatty acids.
15. Evans and Burr.
16. $C_8H_{2n}O_2$.
17. Vegetable oils and fish or fish liver oils.
18. Two fatty acids that form tail part of phospholipid.
19. Paraffin wax and stearic acid.
20. Linolenic acid.
21. Linolenic acid.

22. Arachidonic acid.
23. Wilhelm Bloor.
24. Deficiency of essential fatty acids causes sterility.
25. **Diosgenin** which is a crystalline steroid sapogenin $C_{27}H_{42}O_3$ obtained from yams (*Dioscorea* sp.) and used as a starting material for the synthesis of steroid hormones (as cortisone and progesterone).
26. In one fat molecule three molecules of fatty acids are joined to the glycerol molecule.
27. Deficiency of EFA causes swelling of mitochondrial membrane and reduction in the efficiency of oxidative phosphorylation.
28. Both LDL and VLDL are responsible for **atherosclerosis** and **coronary thrombosis**.
29. **Prostaglandins** are a group of cyclic fatty acids derived from essential fatty acids. They are present in many human's body cells and in semen and exhibit hormone-like activity, *e.g.*, promotion of uterine contraction in human females.
30. The change in colour and odour of fats on standing, exposure to light, moisture and air is known as **rancidity**. It occurs due to action of enzyme lipase and changes in atmospheric moisture and temperature to bring partial hydrolysis of the fat. Rancidity is also accompanied with some degree of oxidation of the unsaturated fatty acids in the fat.
31. Rancidity of fats can be prevented in the presence of α-tocopherol (vitamin E) which acts as a radical chain breaker and thereby inhibits the destructive peroxidation of polysaturated fatty acids.
32. **Plasmnalogen** is an example of a lipid containing glycerol, fatty acids and ethanolamine.

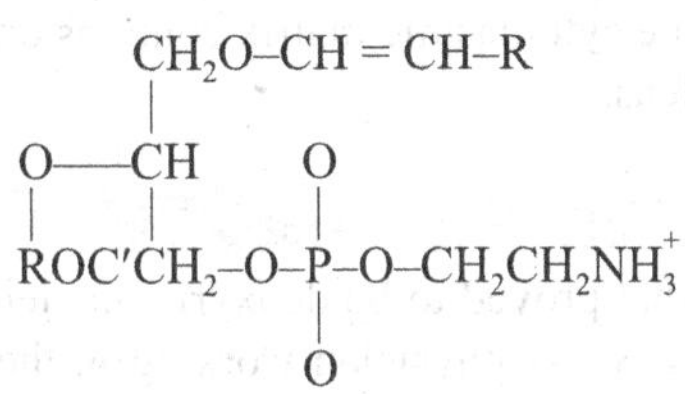

Plasmalogen

33. This disease is caused by the deficiency of *cholesterol acyl transferase* (*LCAT*).
34. Tay-Sach disease is caused due to accumulation of gangliosides in lysosomes of cells of the brain and nervous tissue. This lipid storage disease or lipidosis leads to progressive development of idiocy, blindness and paralysis. Ganglioside accumulation occurs because of absence of enzyme *N-acetyl-galactosaminidase* (= *hexosaminidase A*).
35. Accumulation of sphinogomyelins in brain, liver and spleen of children causes '**Niemann-Pick disease**'. The child suffering from this disease has enlarged liver, spleen, and enlarged abdomen and mental retardation.
36. Gaucher's disease is a hereditary disease characterised by the accumulation of glucocerebrosidase.

Multiple Choice Questions

1. (*c*)	**2.** (*a*)	**3.** (*c*)	**4.** (*c*)	**5.** (*d*)	**6.** (*c*)	**7.** (*a*)
8. (*a*)	**9.** (*d*)					

10 Nucleotides and Nucleic Acids

Nucleic acids represent the fourth major class of macromolecules. Like proteins and polysaccharides, they contain multiple similar monomeric units that are covalently joined to produce large polymers.

The nucleic acids are the complex macromolecular organic compounds of immense biological importance. They control the important biosynthetic activities of the cell and carry hereditary informations from generation to generation. There occur two types of nucleic acids in living organisms, *viz.*, **ribonucleic acid (RNA)** and **deoxyribonucleic acid (DNA)**. Both types of nucleic acids are the polymers of the nucleotides. A nucleotide is composed of nucleoside and phosphoric acid. Even the nucleoside is composed of the pentose sugars (Ribose or Deoxyribose) and nitrogen bases (Purines or Pyrimidines). The **purines** are **adenine** and **guanine** and the **pyrimidines** are the **cytosine**, **thymine** and **uracil**. The cytoplasmic matrix contains only RNA, while DNA exclusively remains concentrated in the nucleus.

10.1. HISTORICAL

The discovery of the substance that proved to be deoxyribonucleic acid (DNA) was made in 1869 by **Friedrich Miescher**, a young Swiss physician working in the laboratory (in Tubingen) of the German physiological chemist **Felix Hoppe-Seyler**. Miescher treated white blood cells (which came from the pus on discarded surgical bandages in the Franko-Prussian War) with hydrochloric acid to obtain nuclei for study. When the nuclei were subsequently treated with acid, a precipitate formed that contained carbon, hydrogen, oxygen, nitrogen and a high percentage of phosphorus. Miescher called the precipitate "nuclein", because it came from nuclei. Later, when it was found to be strongly acidic, its name was changed to nucleic acid. Although he did not know it, Miescher had discovered DNA. Soon afterward, **Hoppe-Seyler** isolated a similar substance from yeast cells; this substance is now known to be ribonucleic acid (RNA). Both DNA and RNA are polymers of nucleotides or polynucleotides.

In 1880, **Emil Fischer** identified pyrimidines and purines. The biochemist, **Albrecht Kossel** identified the constituent nitrogenous bases of nuclein as well as its 5-carbon sugar and phosphoric acid. It was **Altmann** who first suggested, in 1899, the use of the term **nucleic acid** to describe phosphorus-containing nuclein. **Kossel** was awarded the 1910 Nobel Prize for demonstrating the presence of two pyrimidines (cytosine and thymine) and two purines (adenine and guanine) in nucleic acids. Kossel's work and later investigations of **Ascoli**, **Levine** and **Jones** during the first quarter of the 1900s disclosed the two kinds of nucleic acids, deoxyribonucleic acid (DNA) and ribonucleic acid (RNA). Development of DNA-specific staining techniques by **Feulgen** and **Rossenbeck** in 1924 enabled Feulgen to demonstrate in 1937 that most of the DNA content of the cell is located in the nucleus. It was not until the 1950s that the inter-nucleotide bond was established by **A.R. Todd** (**Judson**, 1979).

10.2. NUCLEIC ACIDS

A living organism contains a set of instructions for every step required to construct a replica of itself. This information resides in the genetic material, or **genome**, of the organism. The genomes of all cells are composed of DNA. Some viral genomes are composed of RNA (called **genetic RNA**).

In general information that specifies the primary structure of a protein is encoded in the sequence of nucleotides in DNA (the genetic code). This information is enzymatically copied during the synthesis of RNA, a process known as **transcription**. Some of the information contained in the transcribed RNA molecule is **translated** during the synthesis of polypeptide chains, which are then folded and assembled to form protein molecules. Thus, we can generalize that the biological information stored in a cell's DNA flows from DNA to RNA to protein.

The DNA and RNA have almost similar chemical compositions except a few differences. Both have been compared in Table 10.1.

Table 10.1 Comparison of DNA and RNA.

	DNA		RNA
1.	It contains pentose sugar known as deoxyribose.	1.	It contains pentose sugar called the ribose.
2.	The molecule contains the phosphoric acid (phosphate) molecule which connects various sugars with one another.	2.	The molecule contains the phosphoric acid (phosphate) molecule which connects various sugars with one another.
3.	The nitrogen bases are:	3.	The molecule contains following nitrogen bases in its molecule:
	(*i*) **Purines**—adenine and guanine.		(*i*) **Purines**—adenine and guanine.
	(*ii*) **Pyrimidine**—cytosine and thymine.		(*ii*) **Pyrimidines**—cytosine and uracil.
4.	Molecules have four nucleotides as deoxyadenosine monophosphate, deoxyguanosine monophosphate, deoxycytidine monophosphate and thymidine monophosphate.	4.	Molecules have four nucleotides as adenosine monophosphate, guanosine monophosphate, cytidine monophosphate and uridine monophosphate.
5.	The molecule contains a double stranded helix structure in which many nucleotides remain arranged in pair.	5.	The molecules consist of single chain of polynucleotides.
6.	DNA is a genetic material and occurs in chromosomes, nucleoplasm and mitochondria, chloroplasts, etc.	6.	RNA is a carrier of genetic informations and it plays very significant role in the mechanism of protein synthesis. It mostly occurs in nucleolus, nucleoplasm and cytoplasm.

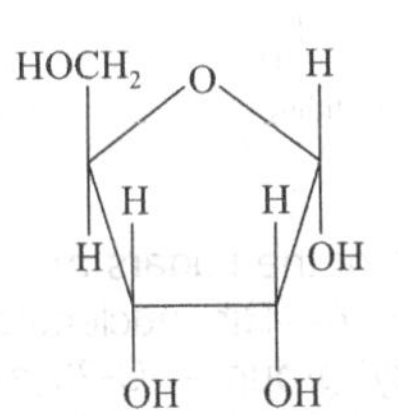

Fig. 10.1. Chemical formula of ribose sugar.

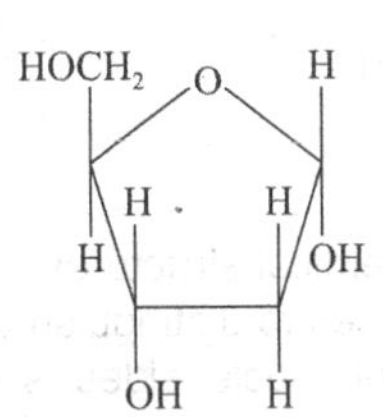

Fig. 10.2. Chemical formula of deoxyribose sugar.

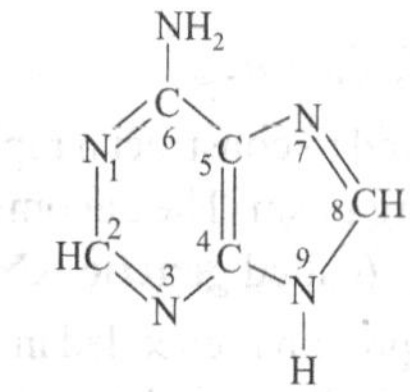

Fig. 10.3. Chemical formula of adenine.

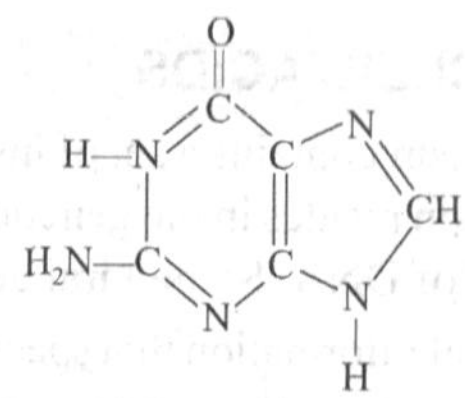

Fig. 10.4. Chemical formula of guanine.

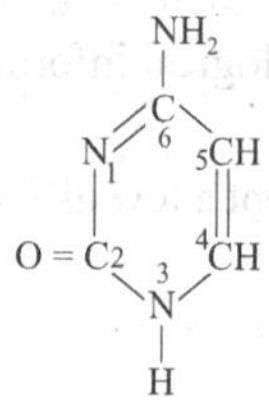

Fig. 10.5. Chemical formula of cytosine.

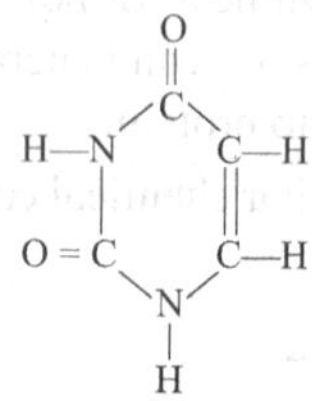

Fig. 10.6. Chemical formula of uracil.

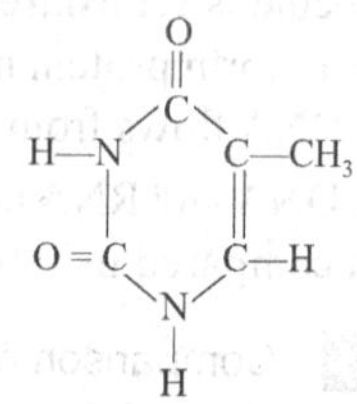

Fig. 10.7. Chemical formula of thymine.

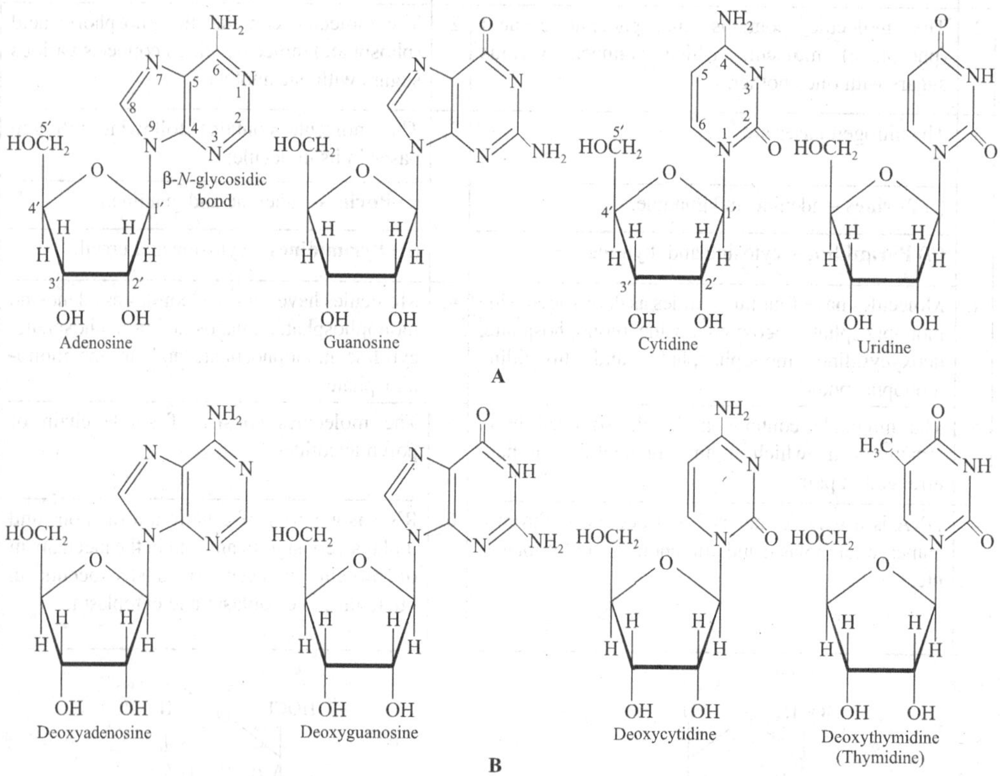

Fig. 10.8. Chemical structures of nucleosides. The carbon atoms of the sugars are numbered with primes to distinguish them from the atoms of the bases. A—Ribonucleosides. The sugar in ribonucleosides is ribose, which contains a hydroxyl group at C-2′, as shown here. The *β-N*-glycosidic bond of adenosine is also marked. B—Deoxyribonucleosides. In deoxyribonucleosides, there is an additional hydrogen atoms at C-2′ instead of a hydroxyl group (after Horton *et al.*, 2006).

10.3. NUCLEOSIDES

Nucleosides are composed of ribose or deoxyribose and a heterocyclic base. In each nucleoside, a β-N-glycosidic bond connects C-1 of the sugar to N-1 of the pyrimidine or N-9 of the purine. Nucleosides are therefore *N*-ribosyl or *N*-deoxyribosyl derivatives of pyrimidines or purines. The numbering convention for carbon and nitrogen atoms in nucleosides reflects the fact that they are composed of a base and a five-carbon sugar, each of which had its own numbering scheme. The designation of atoms in the purine or pyrimidine moieties takes precedence. Hence the atoms in the bases are numbered 1, 2, 3 and so on, while those in the funarose ring are distinguished by adding primes ('). Thus, the *β-N*-glycosidic bond connects the C-1′ or 1′, atom of the sugar moiety to the base. Ribose and deoxyribose differ at the C-2′, or 2′, position. The chemical structure of major ribonucleosides and deoxyribonucleosides are shown in Fig. 10.8.

The names of nucleosides are derived from the names of their bases. The ribonucleoside containing adenine is called **adenosine** (The systematic name, 9-β-D-ribofuranosyladenine, is seldom used), its deoxy counterpart is called **deoxyadenosine**. Similarly, the ribonucleosides of guanine, cytosine and uracil are **guanosine**, **cytidine**, and **uridine**, respectively. The deoxyribonucleosides of guanine, cytosine and thymine are **deoxyguanosine**, **deoxycytidine**, and **deoxythymidine**, respectively. Because thymine rarely occurs in ribonucleosides, deoxythymidine is often simply called **thymidine**. The single-letter abbreviations for pyrimidine and purine bases are also commonly used to designate ribonucleosides: A, G, C and U (for adenosine, guanosine, cytidine and uridine, respectively). The deoxyribonucleosides are abbreviated dA, dG, dC and dT when it is necessary to distinguish them from ribonucleosides.

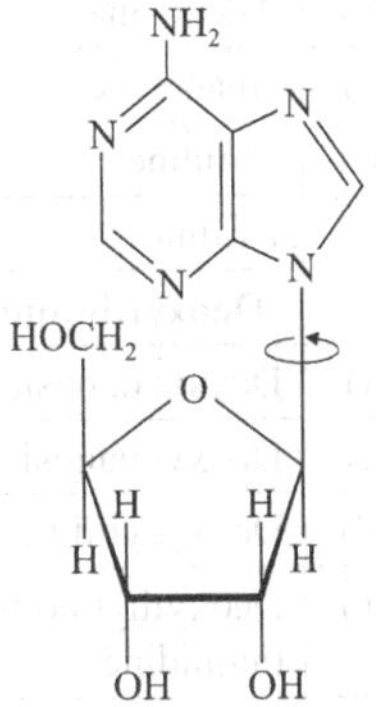

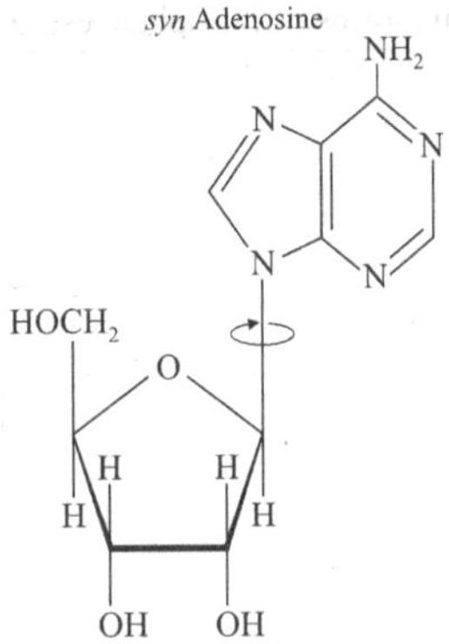

Fig. 10.9. The *syn* and *anti* conformations of adenosine. Some nucleosides assume either the *syn* or *anti* conformations. The *anti* form is usually more stable in pyrimidine nucleosides.

Rotation around the glycosidic bonds of nucleosides and nucleotides is sometimes hindered. In purine nucleotides, two conformations, *syn* and *anti*, are in rapid equilibrium (Fig. 10.9). In the common pyrimidine nucleosides, the *anti* conformation predominates. The *anti* conformations predominate in nucleic acids, the polymers of nucleotides.

10.4. NUCLEOTIDES

Nucleotides are phosphorylated derivatives of nucleosides. Ribonucleosides contain three hydroxyl groups that can be phosphorylated (2′, 3′ and 5′) and deoxyribonucleosides contain two such hydroxyl groups (3′ and 5′). In naturally occurring nucleotides, the phosphoryl groups are usually attached to oxygen atom of the 5′-hydroxyl group. By convention, a nucleotide is always assumed to be a 5′-phosphate ester unless otherwise designated.

The systematic names for nucleotides indicate the number of phosphate groups present. For example, the 5′-monophosphate ester of adenosine is known as **adenosine monophosphate (AMP)**. It is also simply called **adenylate**. Similarly, the 5′-monophosphate ester of deoxycytidine can be referred to as **deoxycytidine monophosphate (dCMP)** or **deoxycytidylate**. The 5′-monophosphate ester of the deoxyribonucleoside of thymine is usually known as **thymidylate** but is sometimes called **deoxythymidylate** to avoid ambiguity.

Table 10.2. Nomenclature of bases, nucleosides, and nucleotides.

Base	Ribonucleoside	Ribonucleotide (5′-monophosphate)
Adenine (A)	Adenosine	Adenosine 5′-monophosphate (AMP); adenylate*
Guanine (G)	Guanosine	Guanosine 5′-monophosphate (GMP); guanylate*
Cytosine (C)	Cytidine	Cytidine 5′-monophosphate (CMP); cytidylate*
Uracil (U)	Uridine	Uridine 5′-monophosphate (UMP); uridylate*
Base	**Deoxyribonucleoside**	**Deoxyribonucleotide (5′-monophosphate)**
Adenine (A)	Deoxyadenosine	Deoxyadenosine 5′-monophosphate (dAMP); deoxyadenylate*
Guanine (G)	Deoxyguanosine	Deoxyguanosine 5′-monophosphate (dGMP); deoxyguanylate*
Cytosine (C)	Deoxycytidine	Deoxycytidine 5′-monophosphate (dCMP); deoxycytidylate*
Thymine (T)	Deoxythymidine or thymidine	Deoxythymidine 5′-monophosphate (dTMP); deoxythymidylate* or thymidylate*

*Anionic forms of phosphate esters predominant at pH 7.4.

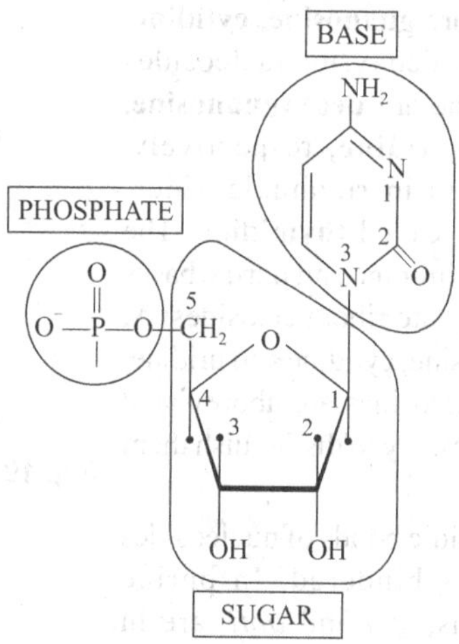

Fig. 10.10. Diagram showing composition of nucleotide.

10.5. DNA

By 1950 it was clear that DNA is a linear polymer of 2′-deoxyribonucleotide residues linked by 3′ – 5′ phosphodiesters. Such phosphodiester bonds exist between phosphate molecule of one nucleotide and 3rd carbon atom of sugar molecule of other nucleotide to form a polynucleotide chain (Fig. 10.11).

Moreover, **Erwin Chargaff** had deduced certain regularities in the nucleotide compositions of DNA samples obtained from a wide variety of prokaryotes and eukaryotes. Among other things, Chargaff observed that in the DNA of a given cell, A and T are present in equimolar amounts, as are G and C. In the DNA of all species, the ratio of purines to pyrimidines is always 1:1 (*i.e.*, Molar ratio of DNA is A + G = C + T).

The molecular and functional model of DNA molecule was proposed by **Watson** and **Crick** in 1953 was based on the known structures of the nucleotides, on X-ray diffraction patterns that **Rosalind Franklin** and **Maurice Wilkins** obtained from DNA fibers, and on the chemical equivalencies noted by Chargaff. The Watson-Crick model (see Chapter 34) accounted for the equal amounts of purines and primaridines by suggesting that DNA was double-stranded (*i.e.*, two polynucleotide chains arranged antiparallely) and that bases on one strand paired specifically with bases on the other strand: A with T and G with C. Watson and Crick's proposed structure is now referred to as the **B conformation of DNA** or simply **B-DNA**.

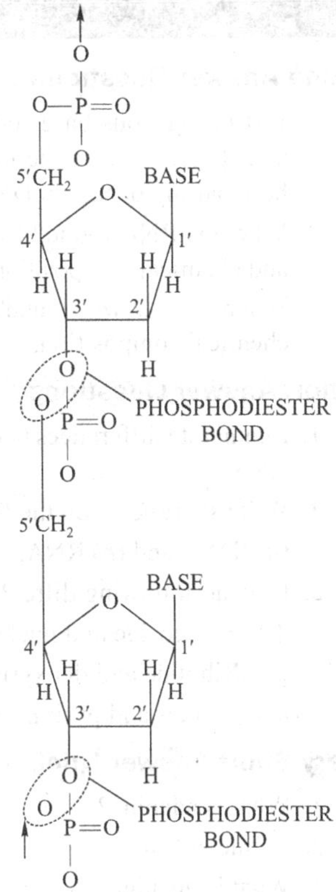

Fig.10.11. Phosphodiester bond between two nucleotides.

10.6. RNA

When RNA is the hereditary material, different types of RNA are **nongenetic**. RNAs are single-stranded molecules, but they often have complex secondary structure. RNA molecules participate in several processes associated with **gene expression**. RNA molecules are found in multiple copies and in several different forms within a given cell (see Chapter 39). There are four major classes of RNA in all living cells:

1. Ribosomal RNA (rRNA) molecules are integral part of ribosomes (intracellular ribonucleoproteins that are sites of protein synthesis). Ribosomal RNA is the most abundant class of ribonucleic acid (RNA), accounting for about 80% of the total cellular RNA.

2. Transfer RNA (tRNA) molecules carry activated amino acids to the ribosomes for incorporation into growing peptide chains during protein synthesis. The tRNA molecules are only 73 to 95 nucleotide residues long. They account for about 15% of the total RNA.

3. Messenger RNA (mRNA) molecules encode the sequences of amino acids in proteins. They are the "messengers" that carry information from DNA to the translation complex where proteins are synthesized. In general, mRNA account for only 3% of the total cellular RNA. These molecules are the least stable of the cellular ribonucleic acids.

4. Small RNA molecules are present in all cells. Some small RNA molecules have catalytic activity or contribute to catalytic activity in association with proteins. Many of these RNA molecules are associated with processing events that modify RNA after it has been synthesized.

Under physiological conditions, most single-stranded polynucleotides fold back on themselves to form stable regions of base-paired, double-stranded RNA. One type of secondary structure is a stem-loop that forms when short regions of complementary sequence form base pairs (*e.g.*, tRNA, rRNA and small RNAs).

QUESTIONS

Long Answer Questions

1. List nitrogenous bases occuring in DNA and RNA. What is the manner in which base pairing occurs in DNA molecule?
2. Where is RNA found? Describe its types and chemical composition.
3. Where is DNA found? Describe its chemical composition.

Short Answer Questions

1. Enumerate differences between DNA and RNA.
2. Write short notes on the following:
   (*a*) DNA and (*b*) RNA.
3. How do following differ?
   (*i*) Nucleoside and nucleotide
   (*ii*) Ribotide and deoxyribotide
   (*iii*) Purines and pyrimidines

Very Short Answer Quetions

1. What is adenine?
2. Define cytosine.
3. What is guanine?
4. Define thymine.
5. What is uracil?
6. Name various nitrogenous bases found in DNA.
7. What is Chargaff's rule about DNA molecular structure?
8. What are monomers of DNA?

Multiple Choice Questions

1. Nucleic acids are polymers of
   (*a*) nucleoproteins (*b*) nucleotides
   (*c*) nucleosides (*d*) nitrogenous bases
2. The pentose sugar present in DNA is
   (*a*) ribose (*b*) deoxyribose
   (*c*) sucrose (*d*) ribulose
3. Purines found in DNA are
   (*a*) uracil and guanine
   (*b*) guanine and adenine
   (*c*) adenine and cytosine
   (*d*) guanine and thymine
4. Adenine is
   (*a*) purine (*b*) pyrimidine
   (*c*) nucleoside (*d*) nucleotide
5. Purine base of RNA is
   (*a*) guanine (*b*) thymine
   (*c*) uracil (*d*) cytosine
6. The base found in RNA but not in DNA
   (*a*) adenine (*b*) uracil
   (*c*) thymine (*d*) guanine
7. Acidic character of DNA and RNA are due to
   (*a*) purine
   (*b*) pyrimidine
   (*c*) sugar
   (*d*) phosphate group
8. ATP molecule is
   (*a*) nucleosome
   (*b*) nucleoside
   (*c*) nucleotide
   (*d*) deoxyribose sugar
9. Which of the following found both in DNA and RNA?
   (*a*) adenine (*b*) ribose
   (*c*) uracil (*d*) thymine
10. RNA is
   (*a*) nucleoside (*b*) nucleotide
   (*c*) polynucleotide (*d*) protein
11. DNA and RNA differ by
   (*a*) N-base and phosphate group
   (*b*) number of C-atoms in sugars
   (*c*) N-bases and sugars
   (*d*) sugar and phosphate group
12. Nucleoside is made up of
   (*a*) phosphate, pentose sugar and base
   (*b*) pentose sugar and base
   (*c*) pentose sugar and phosphate
   (*d*) only pentose sugar
13. Cytidine is a
   (*a*) nucleoside
   (*b*) nucleotide
   (*c*) essential amino acid
   (*d*) nonessential amino acid
14. DNA differs from RNA in having
   (*a*) thymine (*b*) uracil
   (*c*) adenine (*d*) cytosine

15. A unit compound of a sugar and base linked by β-glycosidic bond is known as
    (*a*) nucleotide (*b*) nucleoside
    (*c*) glycoside (*d*) purine
16. Two rings are there in molecule of
    (*a*) guanine (*b*) thymine
    (*c*) uracil (*d*) cytosine
17. Which of the following has a single ring of atoms in their molecule
    (*a*) thymine (*b*) adenine
    (*c*) guanine (*d*) glycine
18. Which one is a nucleotide?
    (*a*) thymidine (*b*) cytosine
    (*c*) uridylic acid (*d*) glutamic acid

ANSWERS

Very Short Answer Questions

1. Adenine (A) is a nitrogen-containing aromatic base, chemically designated as a purine, that serves as an informational monomeric unit when present in nucleic acids with other bases in a specific sequence; forms a complementary bases pair with thymine (T) or uracil (U) by hydrogen bonding.
2. Cytosine (C) is a nitrogen-containing aromatic base, chemically designated as a pyrimidine, that serves as an informational monomeric unit when present in nucleic acids with other bases in a specific sequence; forms a complementary bases pair with guanine (G) by hydrogen bonding.
3. Guanine (G) is a nitrogen-containing aromatic base, chemically designated as a purine, which serves as an informational monomeric unit when present in nucleic acids with other bases in a specific sequence; forms a complementary bases pair with cytosine (C) by hydrogen bonding.
4. Thymine (T) is a nitrogen-containing aromatic base, chemically designated as a pyrimidine, which serves as an informational monomeric unit when present in DNA with other bases in a specific sequence; forms a complementary bases pair with adenine (A) by hydrogen bonding.
5. Uracil (U) is a nitrogen-containing aromatic base, chemically designated as a pyrimidine, that serves as an information monomeric unit when present in RNA with other bases in a specific sequence; forms a complementary base pair with adenine (A) by hydrogen bonding.
6. Adenine, Guanine, Cytosine and Thymine.
7. According to Chargaff's rule—(*i*) In a DNA molecule purines and pyrimidine occur in equal amounts. (*ii*) Amount of adenine is equivalent to thymine and of cytosine is equivalent to guanine.
8. Four types of nucleotides (deoxyribonucleotides).

Multiple Choice Questions

1. (*b*)	2. (*b*)	3. (*b*)	4. (*a*)	5. (*a*)	6. (*b*)	7. (*d*)
8. (*c*)	9. (*a*)	10. (*c*)	11. (*c*)	12. (*b*)	13. (*b*)	14. (*a*)
15. (*b*)	16. (*a*)	17. (*a*)	18. (*c*)			

11 Tubules and Filaments (Cytoskeleton)

11.1. CYTOSKELETON

The cytoskeleton, a fundamental component of all eukaryotic cells, comprises filamentous protein polymers of three types: **intermediate filaments (IFs)**, **acin filaments** or **microfilaments (MFs)** and **microtubules (Mts)** (Box 11.1). Intermediate filaments are the most "skeletal" in nature giving the cytoplasm mechanical strength. Actin filaments and microtubules not only are skeletal but also "muscular", they are responsible for the movements of cytoplasm and organelle. Actin filaments and microtubules are dynamic, turning over many times during their functional life times. These polymers also are polar, in the sense of different biochemical properties at each end. Their dynamic behaviour and polarity are enhanced by free energy from the hydrolysis of nucleoside triphosphates (**Buchanan** *et al.*, 2000).

Box 11.1

The existence of an organized fibrous array or cytoskeleton in the structure of the protoplasm was postulated in 1928 by **Koltzoff**. He conceived of a cytoskeleton that determines both the shape of the cell and the changes in its form.

The main proteins that are present in the cytoskeleton are **tubulin** (in the microtubules), **actin**, **myosin**, **tropomyosin** and other (in the microfilaments) and **keratins**, **vimentin**, **desmin**, **lamin** and others (in intermediate filaments). Tubulin and actin are globular proteins, while subunits of intermediate filaments are fibrous proteins.

The cytoskeleton evolved before plants diverged from animals, and the main features of the cytoskeleton have been conserved in both. However, prokaryotes, such as rod-shaped bacteria, have recently been shown to have proteins that function in a manner very similar to microfilaments (proteins from the *MreB* family), microtubules (the *FtsZ* protein) and intermediate filaments (one type is called *crescentin*) (see **Becker** *et al.*, 2007).

Cellular architecture depends on the unique properties of the different cytoskeletal components working together. Microtubules generally thought to resist bending when a cell in compressed, while microfilaments serve as contractile elements that generate tension. Intermediate filaments are elastic and can withstand tensile force. The mechanical integration of IFs, MFs and MTs is made possible by specific linker proteins that connect them. **Plectin**, for example, is a versatile linker protein that is found at sites where IFs are connected to MFs and MTs.

Functions of Cytoskeleton

The cytoskeleton is a network of inter-connected fibrous polymers that run throughout the cell within the cytosol. This network provides structural stability to cytoplasm, anchoring proteins and other molecules, and supporting organelles during and after their synthesis (Fig. 11.1). Besides structural stability, the cytoskeleton also gives cells the property of motility, both internal and external. Cellular components can move within the cell, such as during cytoplasmic streaming (cyclosis), and many cell types change shape and move within their environment. The cytoskeleton also participates in processing cellular information. For example, elements of the cytoskeleton are known to converge on the plasma membrane at sites associated with transmembrane receptors (Box 11.2). Finally, many elements of the cytoskeleton consist of asymmetric subunits so that the polymers themselves are like an arrow, providing directional cues (Fig. 11.1).

Box 11.2

The cytoskeleton plays important roles in cell movement and cell division, and its position and activity moves membrane-bounded organelles within the cytosol. It also plays a similar role for messenger RNA and other cellular components. Many enzymes in the cytosol are in fact probably not soluble at all, but are physically clustered and attached to the cytoskeleton in close proximity to other enzymes involved in the same pathways, thereby facilitating the channeling of intermediates within each pathway. The cytoskeleton is also involved in many forms of cell movement and is intimately related to other processes such as cell signalling and cell-cell adhesion. The cytoskeleton is altered by events at the cell surface and, at the same time, appears to participate in and modulate these events.

The cytoskeleton is attached to the plasma membrane via proteins. It has been suggested that **vinculin** found in many cells is a link between the plasma membrane and the cytoskeleton. This is particularly important in giving shape to the cell. Disruption of this attachment renders the cell lose its shape, as happens in case of cancerous cells.

11.2. MICROTUBULES

Microtubules (MTs) were first of all observed in the axoplasm of the myelinated nerve fibres by **Robertis** and **Franchi** (1953). They called them **neurotubules**. The exact nature of microtubules was brought into light when **Sabatini**, **Bensch** and **Barnett** (1963) made use of the glutarldehyde fixative in the electron microscopy. Microtubules of plant cells were first described in detail by **Ledbetter** and **Porter** (1963).

Occurrence

With rare exceptions such as the human erythrocytes, microtubules are found in all eukaryotic cells, either free in the cytoplasm or forming part of centrioles, cilia and flagella. The most abundant source of microtubules for the biochemical studies is vertebrate brain—high densities of microtubules exist in axons and dendrites of nerve cells. In the cytoplasm of animal and plant cells, microtubules occur at following seven sites: 1. Cilia and flagella, 2. Centrioles and basal bodies, 3. Nerve processes, 4. The mitotic apparatus, 5. The cortex of meristematic plant cells, 6. Elongating cells such as during the formation of the lens or during spermatogenesis of certain insects. 7. Selected structures in Protozoa such as the axostyle of parasitic flagellates, the axoneme of *Echinosphaerium*, the fibre systems of *Stentor*, and the cytopharyngeal basket of *Nassula*.

Types of Microtubules

Eukaryotes have following two types of microtubules:

1. Axonemal microtubules. These include the highly organized, stable microtubules found in specific subcellular structures associated with cellular movement, including cilia, flagella, basal bodies and centrioles to which these appendages are attached. The central shaft, or **axoneme**, of a cilium or flagellum consists of a highly ordered bundle of axonemal MTs and associated proteins.

2. Cytoplasmic microtubules. They include loosely organized, dynamic microtubules which pervade the cytosol of most eukaryotic cells. They were detected by the help of fluorescence microscopy around early 1960's.

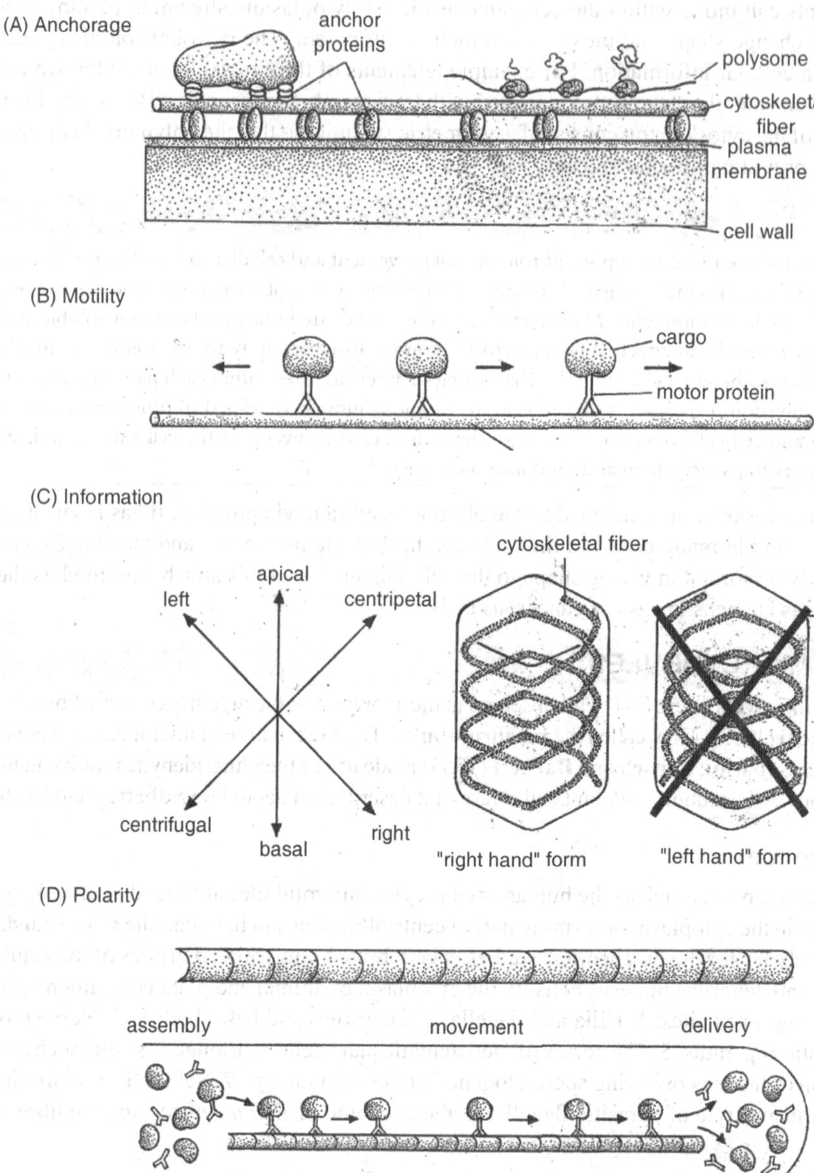

Fig. 11.1. Diagram showing different functions of the cytoskeleton. A—*Anchorage*. The cytoskeleton provides support to the plasma membrane and anchors organelles and other macromolecular assemblies (*e.g.*, polysomes). B—*Motility*. The cytoskeleton supports the directed intracellular movement of cellular components. C—*Information*. The cytoskeleton can provide spatial cues to cellular geometry. In this example, the cell builds cytoskeletal fibers into a right-handed helical array, not a left-handed one. The fibers define a third axis of symmetry in addition to the apical/basal and centripetal (toward the center)/centrifugal (away from the center) axes. D—*Polarity*. The informational content of cytoskeletal fibers depends on the polarity of the fibers, which are formed from asymmetric subunits that defines a direction along the polymer. In this example, cargo is assembled at one site, moved through the cell via the cytoskeleton, and delivered to a second location (after Buchanan *et al.*, 2000).

Structure

Microtubules constitute a class of morphologically and chemically related filamentous rods which are common to both plant and animal cells. A microtubule consists of long, unbranched, hollow tubules 24–25 nm in diameter, several micrometers long and with 6 nm thick wall having 13 subunits or **protofilaments**. Thus, the wall of the microtubule consists of 13 individual linear or spiralling filamentous structures about 5 nm in diameter, which in turn, are composed of tubulin proteins. These protofilaments have a centre-to-centre spacing of 4.5 nm. Application of negative staining techniques has shown that microtubules have a lumen 14 nm wide and a protofilament or subunit structure in the wall (Fig. 11.2).

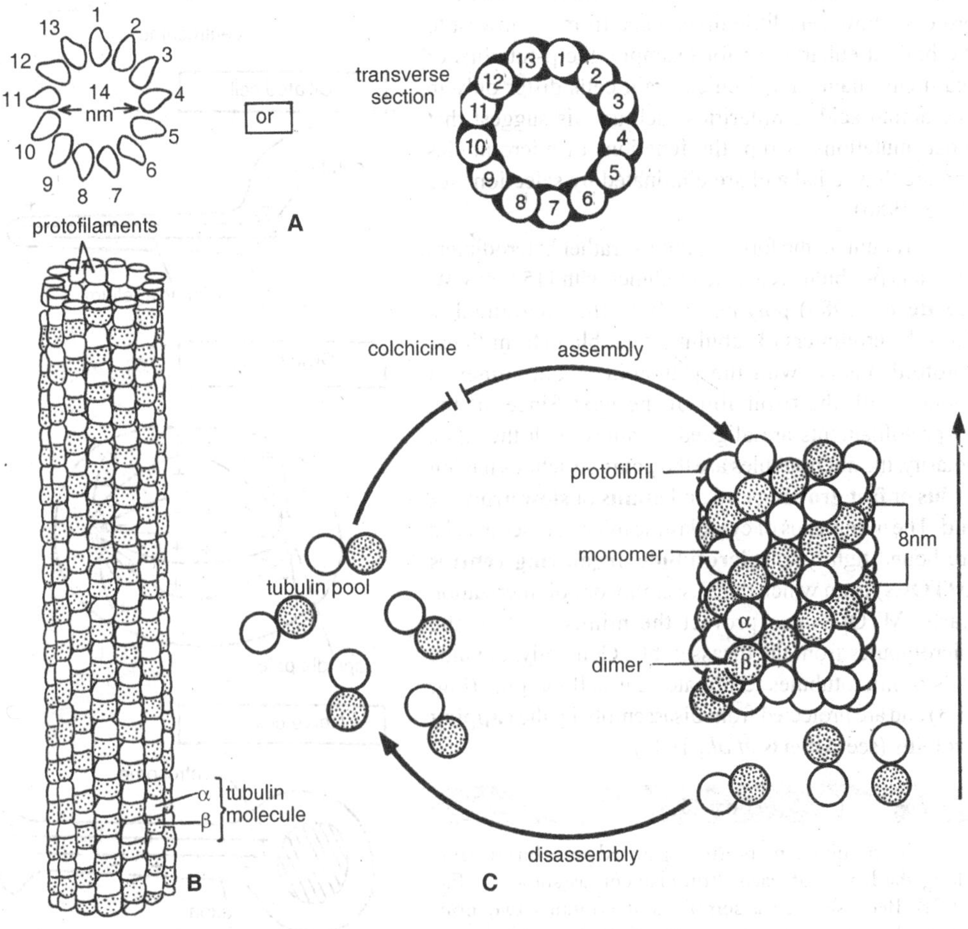

Fig. 11.2. Schematic diagrams of a microtubule, showing how the tubulin molecules pack together to form the cylindrical wall. A—13 tubulin molecules (subunits of protofilaments) in cross section; B—Side view of a short section of a microtubule, with the tubulin molecules aligned into rows, or protofilaments. Each of the 13 protofilaments is composed of a series of tubulin molecules, each with an α/β heterodimer; C—Assembly and disassembly of the microtubule. The microtubule is being disassembled at the bottom while being simultaneously assembled at the top. Colchicine, by blocking the assembly process, produces depolymerization of the microtubules (after De Robertis and De Robertis, Jr., 1987; Alberts *et al.*, 1989).

Chemical Composition

Biochemically, a protofilament of microtubule is made of a protein called **tubulin**. Tubulin is an acidic protein with a molecular weight of 115,000 and a sedimentation constant of 6S. It occurs in two different forms, called **α-tubulin** and **β-tubulin** (with molecular weights of 55,000 and 57,000) each containing about 500 amino acids. Both of these proteins have a distinct, though closely related, amino acid sequences and are thought to have evolved from a single ancestral protein. The two proteins show very little divergence from the lowest to the highest eukaryotes; for example, the β-tubulins of sea urchin flagella and chick brain cells differ only in one amino acid. Similarities such as this suggest that most mutations disrupt the functions of microtubules and are thus lethal and are eliminated by selection (see **King**, 1986).

Tubulin in the form of dimers (rather heterodimers of α- and β-tubulins; each heterodimer with 115,000 MW, see **Berns**, 1983) polymerizes into the microtubules. Thus, heterodimers of tubulins assemble to form linear "protofilaments" with the β-tubulin of one dimer in contact with the α-tubulin of the next. Since all the 13 protofilaments are aligned parallely with the same polarity, the microtubules are the **polar** structures having a **plus** or **fast growing end** and **minus** or **slow growing end**. The minus ends of cytoplasmic microtubules in cells are bound tightly to **microtubule organizing centres (MTOCs)** from which their assembly or polymerization starts. MTOCs also protect the minus ends of the microtubules from the disassembly. Generally, the plus ends of microtubules terminate near cell margins (Fig. 11.3) and are protected from disassembly by the **capping proteins** (see **Alberts** *et al.*, 1989).

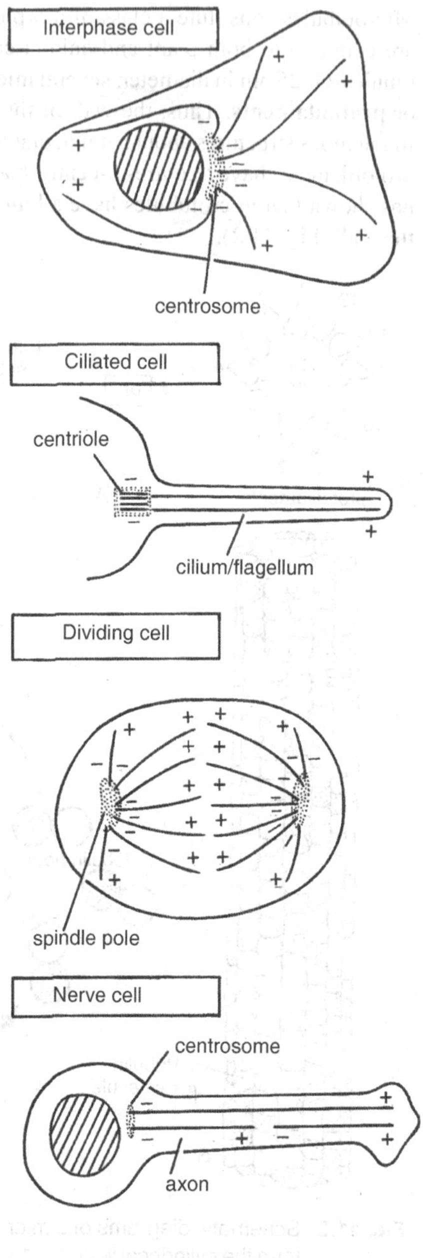

Fig. 11.3. The minus ends of microtubules in cells are generally embedded in a microtubule-organizing centre; while the plus ends are often near the plasma membrane (after Alberts *et al.*, 1989).

Box 11.3

The tubulin dimers are organised in a linear array along the length of each protofilament, as shown in Fig. 11.2B. Because each assembly unit contains two non-identical components (a heterodimer), the protofilament is asymmetric, with an α-tubulin at one end and a β-tubulin at the other end. All of the protofilaments of a microtubule have the same polarity. One end of a microtubule is known as **plus end** and is terminated by a row of β-tubulin subunits. The opposite end is the **minus end** and is terminated by a row of α-tubulin subunits. Such structural polarity of microtubules is an important factor in the growth of these structures and their ability to participate in **directed** mechanical activites (**Karp** 2010).

Microtubule-Associated Proteins (MAPs)

Recently, a number of proteins have been identified that associate with the surface of microtubules; these proteins are called **microtubule associated proteins** or **MAPs**. The following two major classes of MAPs have been isolated from brain in association with microtubules: 1. **HMW proteins** (= high molecular weight proteins) which have molecular weights of 200,000 to 300,000 or more; 2. **tau proteins**, with molecular weights of 40,000 to 60,000. Both classes of proteins have two domains, one of which binds to microtubules; because this domain binds to several unpolymerized tubulin molecules simultaneously, these MAPs tend to speed up the nucleation (= process of grouping around a central mass) step of tubulin polymerization *in vitro*. The other domain is believed to be involved in linking the microtubule to other cell components (Fig. 11.4). Antibodies to HMW and tau proteins show that both proteins; bind along the entire length of cytoplasmic microtubules.

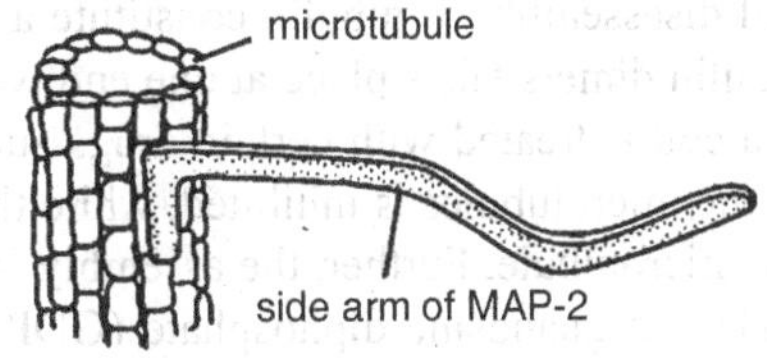

Fig. 11.4. A microtubule associated protein (known as MAP-2) showing its two domains (after Alberts *et al.*, 1989).

Mutations in the gene of tau protein are known to cause a brain disease in humans called **Alzheimer's** disease. Mutant *tau* protein is excessively phosphorylated and does not bind to microtubules (**Karp** 2010).

Microtubule Organizing Centres (MTOCs)

The microtubules are not found helter-skelter about the cell, but are organized in specific patterns designed to carry out specific function. Spontaneous nucleation, as seen *in vitro* (Fig. 11.2C), probably does not occur *in vivo*. Rather, initiation of assembly occurs at **microtubule organizing centres (MTOCs)**. Thus, MTOCs are nucleating centres that serve as templates for the polymerization of tubulin (see **Thorpe**, 1983). MTOCs exist in basal bodies (*e.g. Chlamydomonas*); in centrioles (*e.g.*, most animal cells); at the poles of mitotic spindles in dividing cells that do not have centrioles (*e.g.*, most plant cells); on chromosomes (*i.e.*, **kinetochore**); in membranes and probably many other places as well. Later studies have revealed that most cytoplasmic microtubules do not arise directly from the centrioles, but from a densely staining **pericentriolar material** that surrounds the centriole (see **King**, 1986).

Regardless of their diverse appearance, all MTOCs play similar roles in all cells: they control the number of microtubules, their polarity, the number of protofilaments that make up their walls, and the time and location of their assembly. In addition, all MTOCs share a common protein component—a type of tubulin discovered in the mid 1980s, the **γ-tubulin (Gamma tubulin)**. Unlike the α- and β-tubulins, which make up about 2.5 per cent of the protein of nonneural cell, γ-tubulin constitutes only about 0.005 per cent of the cell's total protein (**Karp** 2010).

Turning on and off of these organizing centres for microtubule assembly at different times in the cell's life are probably regulated by one or all the following factors: changes in nucleation centres, changes in Ca^{2+} concentration and involvement of MAPs.

Assembly and Disassembly of Microtubules

Cytoplasmic microtubules are highly dynamic structures, constantly forming and disappearing depending on cell activities. They, like the microfilaments, grow by the reversible addition of subunits, accompanied by nucleotide (GTP) hydrolysis and conformational change. The process of polymerization (assembly) and depolymerization (disassembly) of the microtubules appears to be a form of **self-assembly**. The assembly of microtubules from the tubulin dimers is a specifically oriented and programmed process. In the cell, the sites of orientation are MTOCs from which the polymerization is directed. The quantity of polymerized tubulin is high at interphase (cytoplasmic microtubules) and metaphase (spindle microtubules), but low at prophase and anaphase.

Within the cell, microtubules are in equilibrium with free tubulin. Phosphorylation of the tubulin monomers by a cyclic AMP-dependent kinase favours the polymerization. A definite relationship has been found between cell shape, the number and direction of microtubules and cAMP. The assembly and disassembly of tubulin constitute a polarized phenomenon. In a microtubule, the assembly of tubulin dimers takes place at one end, while disassembly is common at the other end (Fig. 11.2C). If a cell is treated with certain drugs such as **colchicine, vincristine** or **vinblastine**, the assembly of the microtubules is inhibited, while the disassembly continues, leading to the disorganization of the microtubule. Further, the assembly is accompanied by the hydrolysis of guanosine triphosphate (GTP) to guanosine diphosphate (GDP) and lack of GTP stops the assembly. *In vivo* control of assembly and disassembly of tubulin involves Ca^{2+} and the calcium-binding protein **calmodulin**. The addition of Ca^{2+} inhibits polymerization of tubulin; this effect is also enhanced by the addition of calmodulin.

The *in vivo* mechanism involved in self-assembly of the microtubules is still little understood, however, *in vitro* studies have revealed various interesting facts about it. Thus, in a classical study using isolated bovine brain tubulin, **Weingarten** *et al*., (1975) demonstrated that tubulin alone was not sufficient to bring about *in vitro* assembly into microtubules. Under normal conditions, if brain microtubules are isolated and caused to depolymerize into tubulin subunits, the tubulin molecules will reassemble into microtubules if Mg^{2+} and GTP (an energy source) are added to the mixture. However, according to **King** (1986), *in vitro* assembly of microtubules can occur in the presence of low calcium concentration, MAPs, GTP, and a level of free tubulin monomers above a threshold concentration.

In vitro polymerization evidently involves two distinct phases, one of **initiation** and the other of **elongation**. The initiation event seems to involve the formation of some multimeric "**nucleating**" centre, following which the addition of more subunits proceeds rapidly during elongation. Thus, during *in vitro* polymerization of microtubules, α- and β-tubulins combine to form heterodimers (Fig. 11.5). The heterodimers associate to form multimeric **rings**, **spirals** and other intermediate structures which eventually open up to form strands or protofilaments. Side-by-side assembly of the protofilaments creates sheet-like structures that curl to form a tube. Elongation of this short cylinder occurs by direct addition of new heterodimers at one end of the tubule (*i.e*., the plus end of tubule). It is believed that during anaphase, addition of dimers to one end of a microtubule is accompanied by the loss of dimers from the other end.

General Functions

Cytoplasmic MTs are responsible for a variety of functions. For example, in animal cells they are required to maintain axons, nerve cell extensions. Some migrating animal cells require cytoplasmic MTs to maintain their polarized shape. In plant cells, cytoplasmic MTs are thought to govern the orientation with which cellulose microfibrils are deposited during the growth of cell wall. Significantly, cytoplasmic MTs form the mitotic and meiotic spindles that are essential for the movement of chromosomes during mitosis and meiosis (see Chapter 46).

Specific Functions of Cytoplasmic Microtubules

Microtubules (MTs) have several functions in the eukaryotic cells such as follows:

1. Mechanical function. The shape of the cell (*e.g.*, red blood cells of non-mammalian vertebrates) and some cell processes or protuberances such as axons and dendrites of neurons, microvilli, etc., have been correlated to the orientation and distribution of microtubules.

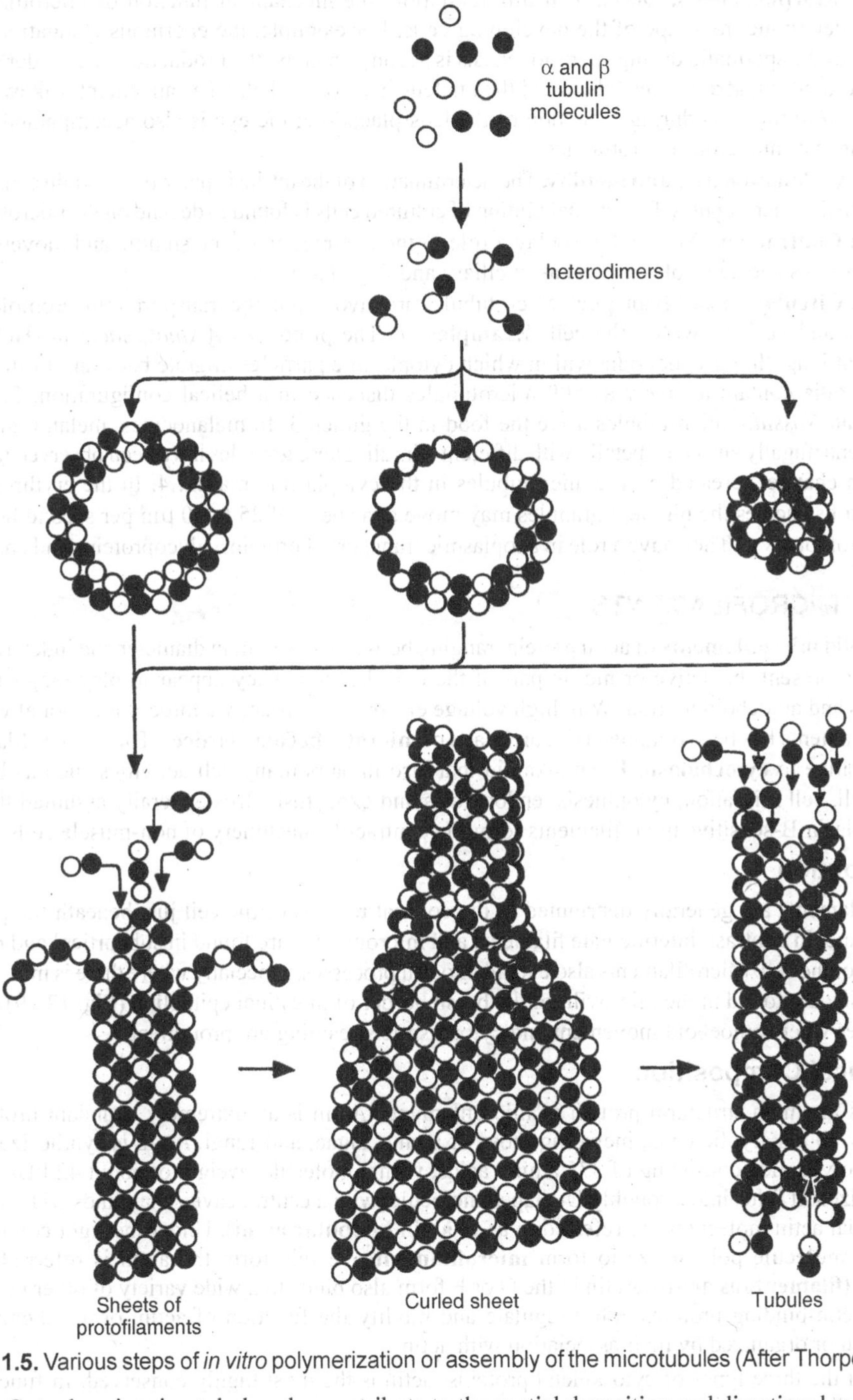

Fig. 11.5. Various steps of *in vitro* polymerization or assembly of the microtubules (After Thorpe, 1984).

Cytoplasmic microtubules also contribute to the spatial deposition and directional movement of vesicles and other organelles by providing an organized system of fibre to guide their movement. For example, cytoplasmic MTs help to govern the location of organelles, such as the Golgi apparatus and the endoplasmic reticulum, and are involved in active movement of vesicles.

2. Morphogenesis. During cell differentiation, the mechanical function of microtubules is used to determine the shape of the developing cells. For example, the enormous elongation in the nucleus of the spermatid during spermiogenesis is accompanied by the production of an orderly array of microtubules that are wrapped around the nucleus in a double helical arrangement. Likewise, the elongation of the cells during induction of the lens placode in the eye is also accompanied by the appearance of numerous microtubules.

3. Cellular polarity and motility. The determination of the intrinsic polarity of certain cells is also related to the microtubules. Directional gliding of cultured cells is found to depend on the microtubules.

4. Contraction. Microtubules play a role in the contraction of the spindle and movement of chromosomes and centrioles as well as in ciliary and flagellar motion.

5. Circulation and transport. Microtubules are involved in the transport of macromolecules, granules and vesicles within the cell. **Examples:** 1. The protozoan *Actinosphaerium* (Heliozoa) sends out long, thin pseudopodia within which cytoplasmic particles migrate back and forth. These pseudopodia contain as many as 500 microtubules disposed in a helical configuration. 2. In the protozoan *Nassula*, microtubules drive the food in the gullet. 3. In melanocytes, melanin granules move centrifugally and centripetally with different stimuli. These granules have been observed moving between channels created by the microtubules in the cytoplasmic matrix. 4. In the erythrophores found in fish scales the pigment granules may move at a speed of 25 to 30 μm per second between the microtubules. 5. They have a role in axoplasmic transport of proteins, glycoproteins and enzymes.

11.3. MICROFILAMENTS

Thin, solid **microfilaments** of actin protein, ranging between 5 to 7 nm in diameter and indeterminate length; represent the active or motile part of the cytoskeleton. They appear to play major role in cyclosis and amoeboid motion. With high voltage electron microscopy a three-dimensional view of microfilaments has been obtained (*i.e.*, an image of **microtrabecular lattice**). These microfilaments are sensitive to **cytochalasin-B**, an alkaloid that also impairs many cell activities such as beat of heart cell, cell migration, cytokinesis, endocytosis and exocytosis. It is generally assumed that the cytochalasin-B-sensitive microfilaments form the contractile machinery of non-muscle cells.

Distribution

Microfilaments are generally distributed in the cortical regions of the cell just beneath the plasma membrane. In contrast, intermediate filaments and microtubules are found in subcortical and deeper regions of the cell. Microfilaments also extend into cell processes, especially where there is movement. Thus, they are found in the microvilli of the brush border of intestinal epitheliun (Fig. 12.19) and in cell types where amoeboid movement and cytoplasmic streaming are prominent.

Chemical Composition

Actin is the main structural protein of microfilaments. Actin is an extremely abundant protein in virtually all eukaryotic cells, including those of plants, algae, and fungi. Actin is synthesized as a single polypeptide consisting of 375 amino acids, with a molecular weight of about 42 kDa. Once synthesized, it folds into a roughly U-shaped molecule, with a central cavity that binds ATP or ADP. Individual actin molecules are referred to as **G-actin (globular actin)**. Under the right conditions, G-actin molecule polymerize to form **microfilaments**; in this form the actin is referred to as **F-actin (filamentous actin)**. Actin in the G or F form also binds to a wide variety of other proteins. These actin-binding proteins either regulate and modify the function of actin, or are themselves regulated or organized by their association with actin.

Of the three types of cytoskeletal proteins, actin is the most highly conserved. In functional assays, all actins appear to be identical and actins from diverse organisms will copolymerize into filaments. Despite this high degree of sequence similarity, actins do differ among different organisms. Based on sequence similarity, actins can be broadly divided into two major groups: 1. The **muscle-specific actins** (*e.g.*, α-**actin**) and 2. **Non-muscle actins** (*e.g.*, β-**actin** and γ-**actin**).

In addition to the various types of actin, another class of protein known as **actin-related proteins (Arps)** show recognizable sequence similarity to actin Arp 2 and Arp 3 are found to be involved in nucleating assembly of new microfilaments in migrating cells.

Polymerization of G actins. Like tubulin dimers, monomers of G-actin polymerize into long filaments of F-actin with a diameter of about 7 nm (Fig. 11.6). In F-actin, a full turn of the helix occurs every 36–37 nm, with about 13.5 monomers required for a full turn. The addition of each G-actin monomer is followed by hydrolysis of ATP molecules so that it is tightly bound to the monomer, although the energy of ATP hydrolysis is not required to drive the polymerization reaction.

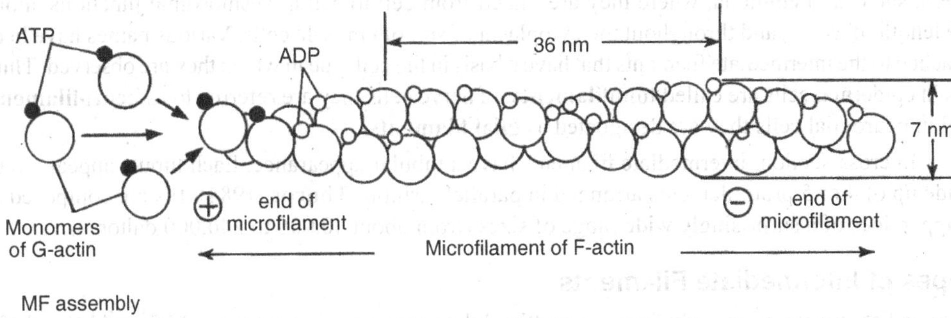

Fig. 11.6. A model for microfilament assembly in vitro (after Becker *et al.*, 2006).

Within a microfilament, all the actin monomers are oriented in the same direction, so that an MF, like a microtubule, has an inherent polarity, with one end differing chemically and structurally from the other end. This polarity can be readily demonstrated by incubating MFs with **heavy myosin (HMM)** or **myosin subfragment (S1)**, a proteolytic fragment of myosin (Fig. 11.7). HMM/S1 fragments bind to, or "decorate" the actin MFs to give a distinctive arrowhead pattern, the term **pointed end** and **barbed end** are commonly used to identify the **minus** and **plus ends** of a MF, respectively. The polarity of the microfilament is important, because it allows for independent regulation of actin assembly or disassembly of each end of the filament. Normally, plus end grows faster than the minus end.

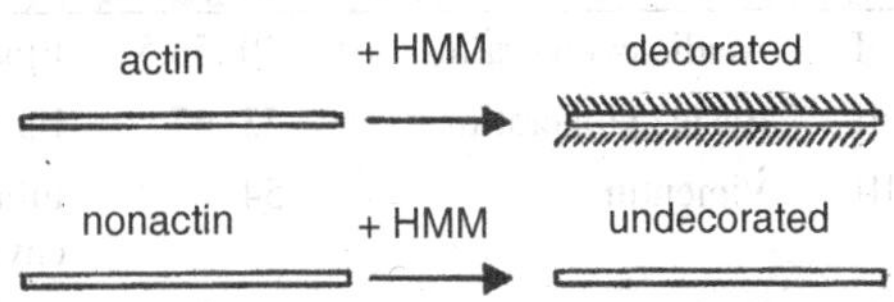

Fig. 11.7. Mode of "decoration" of actin filaments by heavy myosin or HMM. Intermediate filaments are not decorated by HMM. (after Thorpe, 1984).

Capping of MFs. Whether microfilament ends are available for further growth depends on whether or not the filament end is **capped**. Capping occurs when a **capping protein** binds the end of a filament and prevents further addition or loss of subunits, thereby stabilize it. One such protein that acts as a cap for the plus end of microfilament is appropriately named **Cap Z** (see **Becker** *et al.*, 2006).

Function

Involvement of actin and myosin in the contraction of microfilaments clearly points out to a similar mechanism as existing in muscle cells. Non-muscle cells carry out many functions such as changes in cells shape, locomotion (cell migration during embryonic development), cytokinesis, phagocytosis, and pseudopodial formation, which are attributed to the contractile function of microfilaments. One classic example of cell motility is **cyclosis** or cytoplasmic streaming involving actin-myosin interaction (*e.g.*, plant cell *Chara* and *Nitella*).

11.4. INTERMEDIATE FILAMENTS

Intermediate filaments (IFs) are tough and durable protein fibres in the cytoplasm of most higher eukaryotic cells. Constructed like woven ropes, they are typically between 8 nm to 10 nm in diameter, which is "intermediate" between the thin and thick filaments in muscle cells, where they were first described; their diameter is also between microfilaments (actin filaments) and microtubules. IFs are found resistant to colchicine and cytochalasin B and are sensitive to proteolysis.

In most animal cells IFs form a "basket" around the nucleus and extend out in gentle curving arrays to the cell periphery. IFs are particularly prominent where cells are subjected to mechanical stress, such as in epithelia, where they are linked from cell to cell at desmosomal junctions, along the length of axons, and throughout the cytoplasm of smooth muscle cells. Various names have been attached to the intermediate filaments that have a basis in the cell type in which they are observed. Thus, IFs in epidermal cells are called **tonofilaments**, in nerve cells they are referred to as **neurofilaments** and in neuroglial cells they are designated as **glial filaments**.

In cross-section, intermediate filaments have a tubular appearance. Each tubule appears to be made up of 4 or 5 protofilaments arranged in parallel fashion (**Thorpe**, 1984). IFs are composed of polypeptides of a surprisingly wide range of sizes (from about 40,000 to 240,000 daltons).

Types of Intermediate Filaments

Intermediate filaments are only found in multicellular organisms, in contrast to MTs and MFs, differ markedly in amino acid composition from tissue to tissue. Based on cell type in which they are found, IFs and their proteins can be grouped into five classes (Table 11.1).

Table 11.1 Classes of intermediate filaments (*Source:* Becker *et al.*, 2006 and Karp, 2010).

Class	IF Protein	Molecular Mass (kDa)	Tissue	Function
I	Acidic cytokeratins	40–56.5	Epithelial cells	Mechanical strength
II	Basic cytokeratins	53–67	Epithelial cells	Mechanical strength
III	Vimentin	54	Fibroblasts; cells of mesenchymal origin; lens of eye	Maintenance of cell shape
III	Desmin	53–54	Muscle cells, especially smooth muscle	Structure support for contractile machinery
III	Glial fibrillary acidic (GFA) protein	50	Glial cells and astrocytes	Maintenance of cell shape
IV	Neurofilament protein		Central and peripheral nerves	Axon strength; determines axon size
	NF-L (major)	62		
	NF-M (minor)	102		
	NF-H (minor)	110		
	Nestin	240	Neuroepithelial	
V	Nuclear lamins		All cell types	Form a nuclear scaffold to give shape to nucleus
	Lamin A	70		
	Lamin B	67		
	Lamin C	60		

Class I and II IFs comprise the **keratins**, proteins that make up the **tonofilaments** found in the epithelial cells that cover the body surfaces and line its cavities (the IFs visible beneath the terminal web in the intestinal mucosa cell consists of keratin). Class I keratins are **acidic keratins**, whereas class II are **basic** or **neutral keratins**.

Class III IFs include vimentin, desmin, synemin and glial fibrillarly acidic (GFA) protein. **Vimentin** is present in connective tissue and other cells derived from non-epithelial cells. Vimentin-containing filaments are often prominent feature, in cultured fibroblast cells. **Desmin** is found in muscle cells. **Synemin** is a protein of 230,000 daltons, which is also present in the intermediate filaments of muscle, together with desmin and vimentin. Vimentin and synemin containing IFs can be observed in the chicken erythrocytes. **Glial fibrillary acidic (GFA) protein** is characteristic of the glial cells that surround and insulate nerve cells.

Class IV IFs are the **neurofilaments** (NF) **proteins** found in the neurofilaments of nerve cells.

Class V IFs are the **nuclear lamins** A, B and C, which form a filamentous scaffold along the inner surface of the nuclear membrane of virtually all eukaryotic cells.

Neurofilaments found in cells in the embryonic nervous system are made of **nestin** protein, which constitute class IV IFs.

Box 11.4

The nuclear lamina is thought to stabilize the nuclear envelope, to maintain the shape of the organelle to ensure that chromatin can be successfully anchored and possibly to reinforce the structure of the large nuclear pore complexes. When the nuclear envelops breaks down during mitosis, the nuclear lamina dissolves after reversible protein **phosphorylation**. Lamin proteins contain several phosphorylation sites, subtrates for a cyclin-dependent **kinase** that is maximally active near the end of mitotic prophase. Conversely, after the transition from metaphase into anaphase this kinase activity declines and lamin-bound phosphates are removed. During telophase, the nuclear lamina and nuclear envelope reform around the decondensing chromosomes (**Buchanan** *et al.*, 2000).

Each of these IF proteins tends to assemble spontaneously *in vitro* to form homopolymers and will also co-assemble with the other Types III IF proteins to form **co-polymers** and **hetero-polymers**. In fact co-polymers of vimentin and desmin, or of vimentin and glial fibrillary acidic protein, are found in some type of cells. For example, desmin remains concentrated in the Z-lines and T-tubule system of striated or skeletal muscles, together with vimentin, synemin and α-actinin. Since desmin links actin to plasma membrane, from this fact the name of desmin has been derived by **Lazarides** and coworkers in 1976 (in Greek desmin means link or bond).

Plants are known to contain homologs of nuclear lamins, but for the other intermediate filament proteins, the evidence is no more than suggestive (**Buchanan** *et al.*, 2000).

General Structure of IFs

All six types of intermediate filaments (IFs) contain proteins which have a central rod like domain consisting of four **helical segments**. The central domain is highly conserved in size, secondary structure, and sequence, though sequence homologies are confined to the helical regions. In types I-IV, the helical segments contain a total of 276 amino acids and the linker segments are non-helical (Fig. 11.8). In type V, the helical segments contain 318 amino acids and their linker segments are also helical. The N- and C-terminal domains that flank the central section are non-helical and are much more variable in size and sequence.

Assembly of IFs

A current model of assembly of a intermediate filament includes the following steps: 1. Two identical **monomers** pair to form a **dimer** in which the conserved helical central regions are aligned in parallel and are wound together into a coiled coil (Fig. 11.9). 2. Two dimers then line up side-by-side to form a 48 nm by 3 nm **protofilament** containing four polypeptide chains. 3. These protofilaments then associate in a staggered manner to form successively larger structures. 4. The final 10 nm diameter of the intermediate filament is thought to be composed of 8-protofilaments (*i.e.*, 32 polypeptide chains) joined end on end to neighbours by staggered overlap to form the long rope-like filaments (Fig. 11.10). It is still not known whether IFs are polar structures (like actin and tubulin) or non-polar (like the DNA double helix).

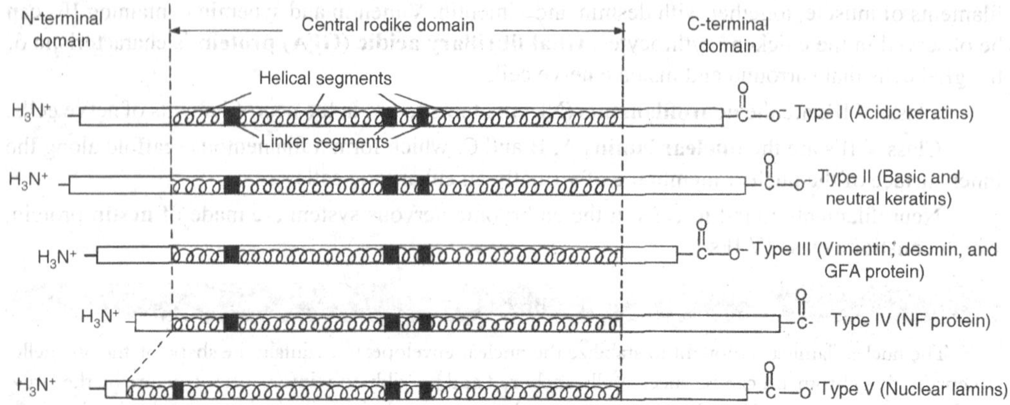

Fig. 11.8. Structural similarities of intermediate filament protein. Except VI or nestin, all the five IFs are shown (after Becker *et al.*, 2006).

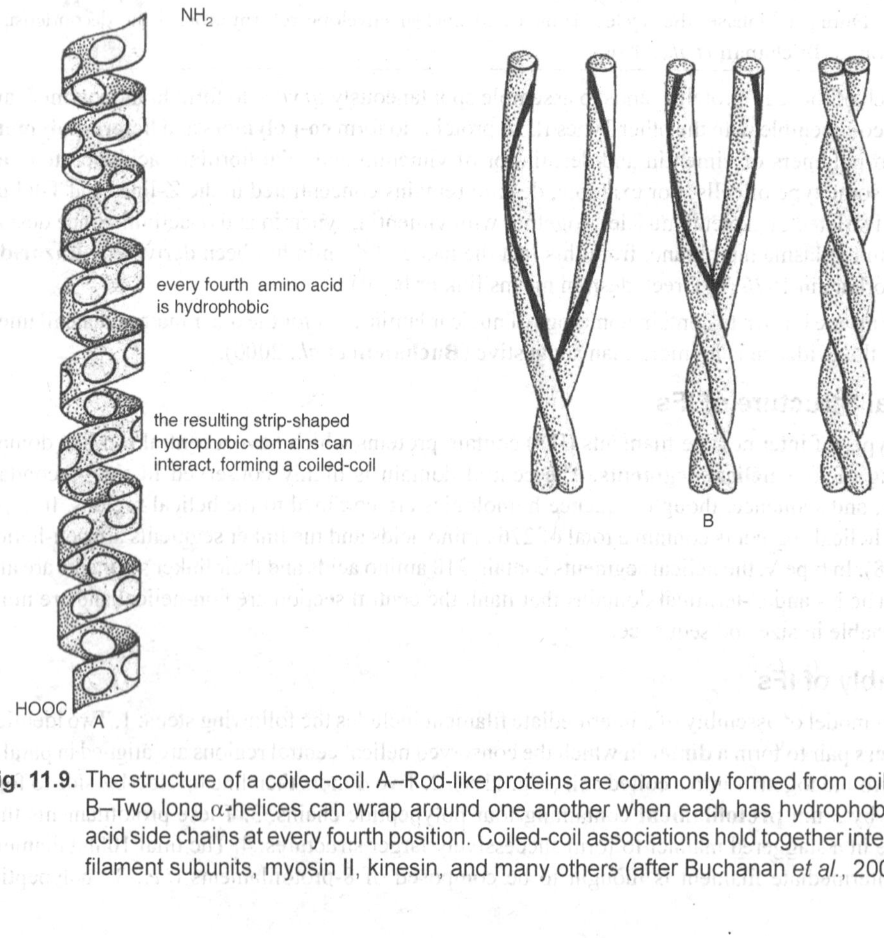

Fig. 11.9. The structure of a coiled-coil. A–Rod-like proteins are commonly formed from coiled-coils. B–Two long α-helices can wrap around one another when each has hydrophobic amino acid side chains at every fourth position. Coiled-coil associations hold together intermediate filament subunits, myosin II, kinesin, and many others (after Buchanan *et al.*, 2000).

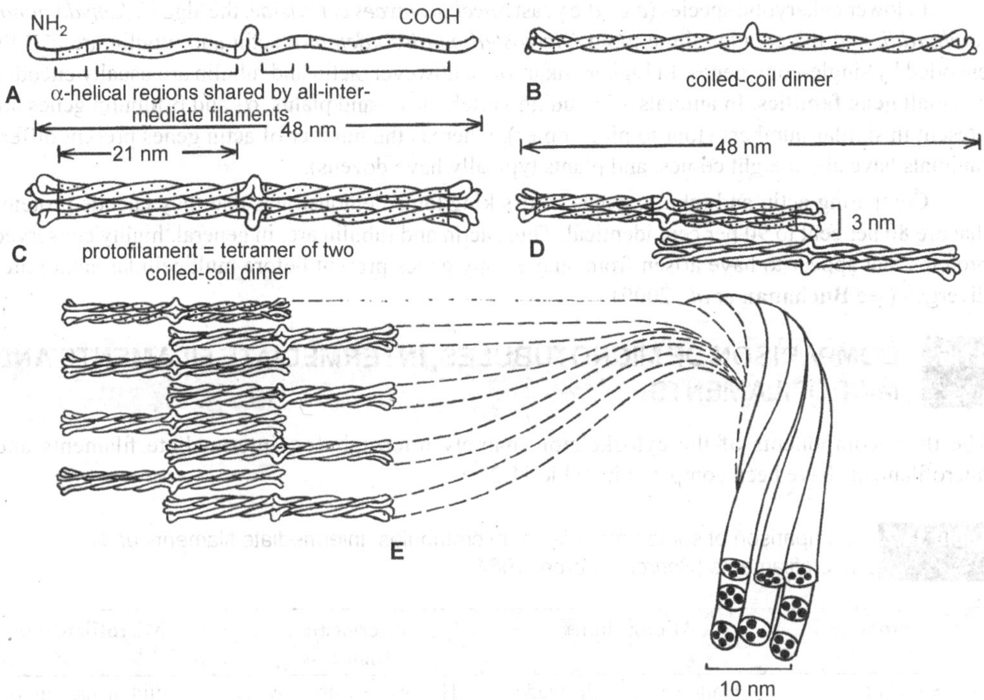

Fig. 11.10. A model to explain the mode of assembly of a 10-μm thick intermediate filament (after Albert *et al.*, 1989).

IFs During Mitosis

Mitosis of cultured epithelial cells shows striking changes in intermediate filaments of cytokeratin and vimentin. During prophase the 10 nm filaments unwind into threads of 2 to 4 nm and into spheroidal aggregates containing both types of proteins. At metaphase and anaphase most vimentin and cytokeratin appear as spheroid bodies, while at telophase the filamentous cytoskeleton become gradually reestablished. From these experimental studies, **Franke** (1982) has concluded that the living cells contain factors that promote the reversible disintegration and restoration of intermediate filaments during mitosis.

Functions of IFs

The main function of most intermediate filaments is to provide mechanical support to the cell and its nucleus. IFs in epithelia form a transcellular network that seems designed to resist external forces. The neurofilaments in the nerve cell axons probably resist stresses caused by the motion of the animal, which would otherwise break these long, thin cylinders of cytoplasm. Desmin filaments provide mechanical support for the sarcomeres in muscle cells, and vimentin filaments surround and probably support the large fat droplets in the fat cells.

11.5. ACTIN AND TUBULIN GENE FAMILIES

Microtubules are heterodimeric polymers of the globular proteins α- and β-tubulin; actin filaments (microfilaments) are polymers of the protein actin. Both actin and tubulin occur in all eukaryotes, and presumably they both evolved at the time eukaryotic cells became established. Recently, the product of the *E. coli FtsZ* gene was identified as a tubulin homolog, so the major cytoskeletal proteins may well have been present even before eukaryotes arose.

In lower eukaryotic species (*e.g.*, the yeast *Saccharomyces cerevisiae*, the alga *Chlamydomonas reinhardtii*, and the cellular slime mold *Dictyostelium discoidenum*), actin and tubulin are usually encoded by single-copy genes. In higher eukaryotes, however, actin and tubulin are usually encoded by small **gene families**. In animals—including vertebrates—and plants, α- and β-tubulin genes are present in similar numbers (four to nine copies), whereas the number of actin genes present differs (animals have about eight copies, and plants typically have dozens).

Comparing actin and tubulin genes across kingdoms demonstrates that they encode proteins that are 80 per cent to 90 per cent identical. Thus, actin and tubulin are, in general, highly conserved proteins that appear to have arisen from single copy genes present before multicellular eukaryotes diverged (see **Buchanan** *et al.*, 2000).

11.6. COMPARISON OF MICROTUBULES, INTERMEDIATE FILAMENTS AND MICROFILAMENTS

The three components of the cytoskeleton, namely microtubules, intermediate filaments and microfilaments have been compared in Table 11.2.

Table 11.2. Comparison of some properties of microtubules, intermediate filaments and microfilaments (*Source:* Thorpe, 1984).

Property	Microtubules	Intermediate filaments	Microfilaments
1. Structure	Hollow with walls made up of 13 protofilaments	Hollow with walls made up of 4 to 5 protofilaments	Solid made up of polymerized actin (F-actin)
2. Diameter (nm)	24–25	10	7–9
3. Monomer units	α- and β-tubulin	Six types of protein forming six major classes	G-actin
4. ATPase activity	Present in dynein arms	None	None
5. Functions	1. Motility of eukaryotes	1. Integrate contractile units in muscle	1. Muscle contraction
	2. Chromosome movement	2. Cytoskeletal structural function in cytoplasm	2. Cell shape changes
	3. Movements of intra-cellular materials		3. Protoplasmic streaming
	4. Contribute toward maintaining cell shape		4. Cytokinesis

11.7. CYTOSKELETAL ACCESSORY PROTEINS

If microfilaments and microtubules are the cell's scaffolding, then accessory proteins are the joints, motors and tools that link, move, and modify that scaffolding. The different functions performed by polymers of the highly conserved proteins actin and tubulin are determined largely by a collection of less highly conserved accessory proteins. Defined operationally as proteins that are co-purified with cytoskeletal polymers, accessory proteins can be further divided by function as follows:

1. Mechanochemical Enzymes or Motor Proteins

Mechanochemical enzymes are also called **motor proteins**. The three subfamilies of known motor proteins—myosin, dynein and kinesin—all use ATP as their energy source. Furthermore, all contain a globular, force-producing "head" domain that binds a cytoskeletal polymer and a rod-shaped "tail" domain that binds cargo. The families differ in which type of polymer they recognize, what type of cargo they will carry, and which biochemical pathways they use to convert chemical energy into work.

The first such protein to be characterized was **myosin**, the force-producing enzyme of muscle. We now know that members of the myosin family are present in all eukaryotic cell types and that they power many types of cellular motility involving actin filaments (Fig. 11.11). Of the 13 different classes of myosin described to date, only 2 have been demonstrated in plants.

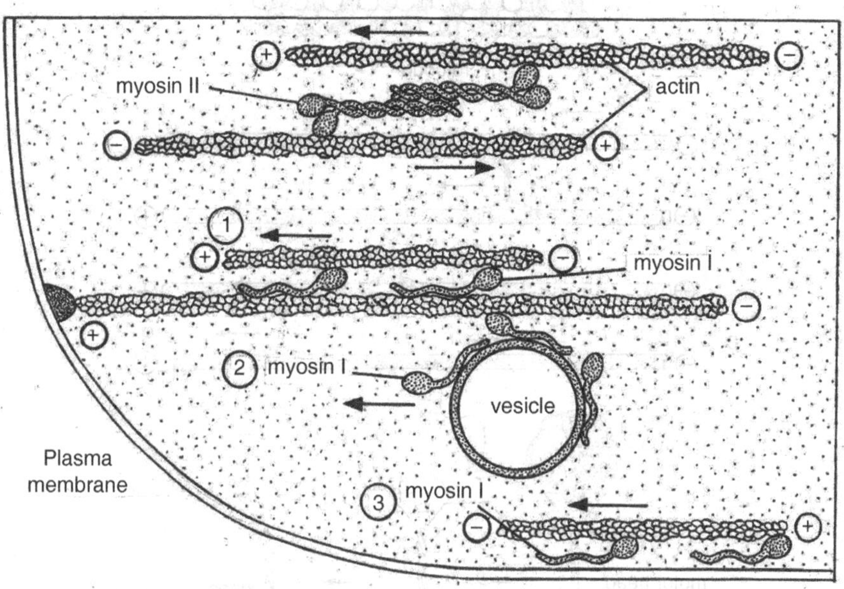

Fig. 11.11. Myosin proteins interact with actin filaments, moving from the minus (pointed) end to the plus (barbed) end of the polymer. Myosin II proteins, such as those in muscle, have long rod-like domains that promote assembly into bipolar (head-to-tail) filaments. Myosin II filaments pull actin filaments with opposite polarity past one another, mediating local contraction. Myosin I motors have a short tail domain that can bind an actin filament or a membrane. Force transduction by the myosin I head may then 1. move an actin filament relative to another; 2. move a vesicle along an actin filament; 3. move an actin filament along a membrane. Myosin proteins are found in plants, but the equivalent forms of myosin I and myosin II have not yet been described for plants (after Buchanan *et al.*, 2000).

The other two motor proteins, **dynein** and **kinesin**, bind microtubules; like myosin, they are represented by large gene families, with members present throughout the eukaryotes (Fig. 11.12). Dynein transports cargo to the minus end of the microtubule, whereas most, but not all, *kinesins* move towards the plus end. Dynein protein has two major forms, **axonemal** and **cytoplasmic**. The axonemal form, the force-producing protein of eukaryotic cilia, has probably been lost from seed plants, which lack ciliated sperm. The cytoplasmic form, which was recently identified in plants, supports a variety of microtubule-mediated motility in animal cells, including vesicular traffic toward the cell center and microtubule organization. Similarly, kinesins in both animal and plants are implicated in vesicle traffic and formation of mitotic spindles.

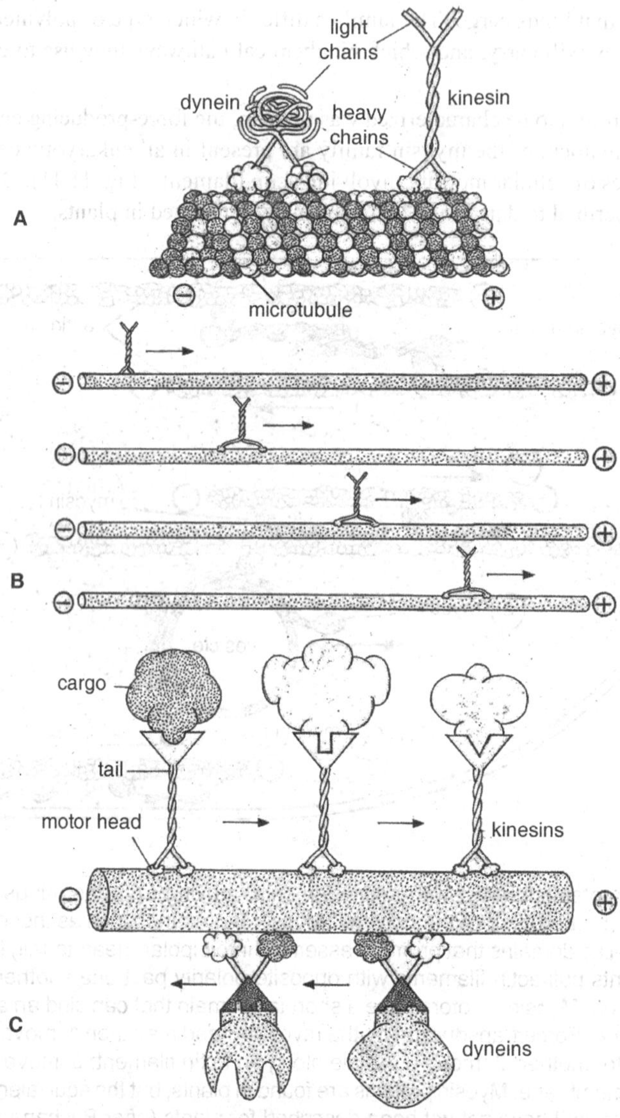

Fig. 11.12. Diagrams of two microtubule-associated motor proteins. A—Cytoplasmic dynein and kinesin. Each of these proteins contains a pair of heavy chains and several accessory light chains. The heavy chains form globular head domains and elongated tail domains. The head binds to the microtubule, contains the ATP-binding site, and produces most of the force. The tail domain is responsible for binding the specific cargo. The light chains may regulate the activity of the motor or may modify its binding properties. B—Schematic view of kinesin in action. Alternate binding by the two heads ensures that a head is almost always bound; hence, the motor with cargo can move along the microtubules for long distances before dissociating. Although this action often likened to a biped walking, the two heavy chains are the same, not mirror images (as left and right legs are), and so the head rotates by 180° with each step to be in proper orientation for binding the microtubule. C—Dyneins move cargo toward the minus end of a microtubule, whereas most kinesins move cargo toward the plus end (after Buchanan *et. al.*, 2000).

Other Accessory Proteins

In addition to motor proteins, several other types of accessory proteins affect cytoskeleton function (Fig. 11.13). Some called "cross-linking" or "building" proteins, form bonds between cytoskeletal polymers of the same type, *e.g.*, actin-binding protein such as fimbrin and α-actinin. **Fimbrin** links actin filaments (in a bundle) having the same polarity to form a parallel array in which all plus ends face the same direction. In contrast, α-actinin links actin filaments of opposite polarities, creating an antiparallel array in which half the plus ends face one direction and half the other. A parallel array can support movement in a uniform direction, as in cytoplasmic streaming, whereas an antiparallel array can be pushed apart or drawn together by a motor protein, as occurs in the **contractile ring** during animal cytokinesis (*i.e.*, furrow formation in cell division).

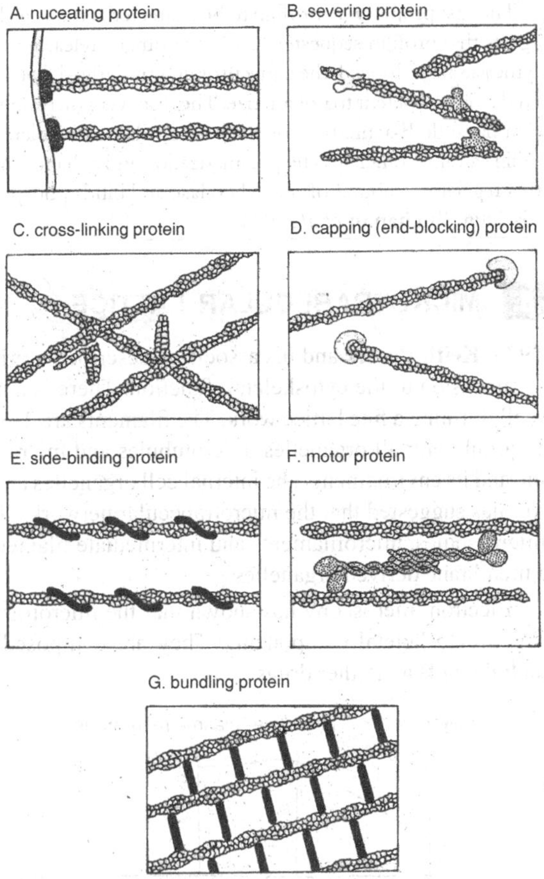

Fig. 11.13. Proteins that associate with the cytoskeleton have various functions, as illustrated here with actin filaments.

Other proteins (*e.g.*, filamin) cross-link actin filaments at an angle, promoting the formation of gel-like network that is known to influence motility and cytoplasmic structure of animal cells. Actin of plant cells are found to contain **profilin** which helps in the polymerization of actin and causes allergy in human beings (Box 11.5).

Box 11.5
ProfilinAn actin binding protein of plants and allergy

Some persons cry and sneeze around plants in flower, your allergy may be triggered by a fascinating, *actin-binding protein* present in pollen, called *profilin*. Medical research and plant cell biology, usually worlds apart, were united in 1991 by the discovery that many patients with pollen allergies synthesize antibodies against profilin. The antibodies recognized profilin from a variety of plants, a finding that offered the first explanation for the nonselectivity of typical pollen allergies. Even more intriguing, the patient's antibodies also recognized human profilin, suggesting the hypothesis that the allergy is aggravated through autosensitization.

The discovery of large quantities of profilin in pollen has proved a boon for students of plant skeleton. Profilin is a small (10 to 15 kDa) protein which occurs in all organs, although its abundance increases 10 to 100 fold as pollen matures. The conserved tertiary structure of profilin includes a core, composed of a six-stranded antiparallel β-sheet, and the *actin-binding domain*, composed predominantly of the C-terminal α-helix (H_3) and the bottom three strands of the central sheet (β4, β5 and β6).

The first property of profilin to be characterized was its high affinity for G-actin but not F-actin, which suggests that profilin sequesters actin monomers, releasing them locally when and where the cell needs to polymerize actin. Indeed, the name profilin was derived from "profilamentous" actin, a term for sequestered actin that is competent to polymerize. The discovery of profilin in mature pollen fits this conception: mature pollen has little F-actin, but abundant actin filaments (microfilaments) form within minutes of hydration. Profilin can also stimulate actin polymerization and enhance stability of actin filaments. Also, profilin interacts with a regulatory subunit of adenyl cyclase and binds phosphoinositidyl lipids with a greater affinity than it binds actin (**Buchanan** *et al.*, 2000).

11.8. MICROTRABECULAR LATTICE

In 1976, **Keith Porter** and his associates described a **microtrabecular system**, which has added a new dimension to the cytoskeleton function. There is a network of filaments extending throughout the cell, forming a fine lattice work. The filaments are 2–3 nm in diameter and 30 to 300 nm long and form a link with all organelles, microtubules and microfilaments (Fig. 11.14). There is no definite shape and its environment. The internal cell organelles are held in position with the help of filaments. Porter has suggested that the microtrabecular network, along with other cytoskeletal elements such as microtubules, microfilaments and intermediate filaments, gives support to the plasma membrane and membrane derived organelles.

Electron microscopy has shown that the microtrabecular lattice establish a true internal link between cytoskeletal components. They are composed of actin, since the lattice is affected by **cytochalasin B** and other drugs.

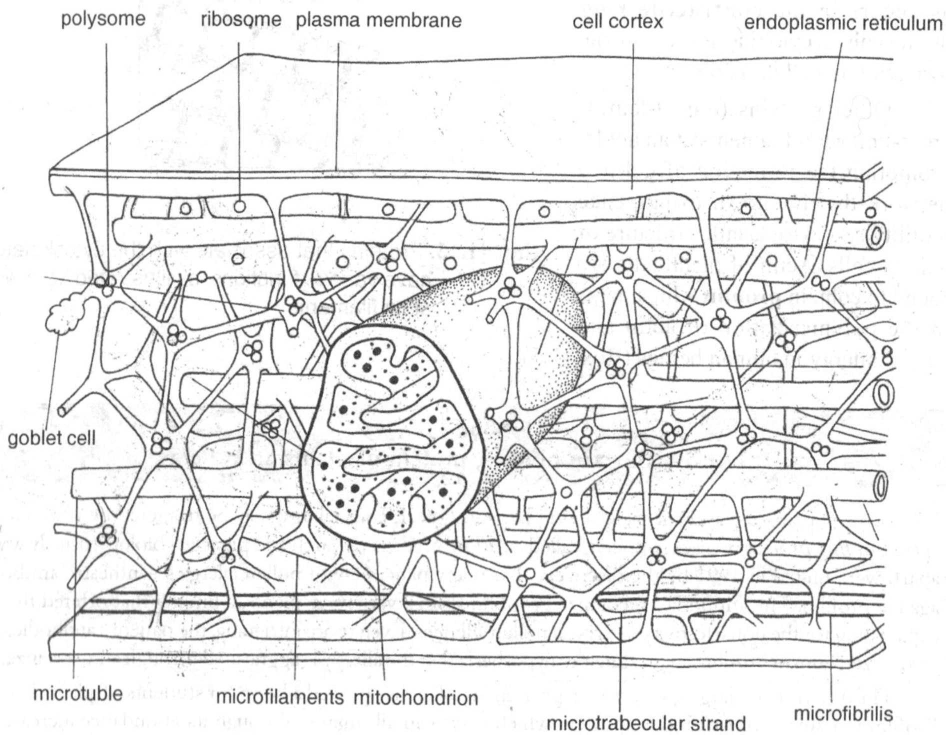

Fig. 11.14. A reconstruction of microtrabecular lattice showing various contractile components.

QUESTIONS

Long Answer Questions

1. What are microtubules? Describe their structure, assembly, disassembly and functions.
2. What are the intermediate filaments? Describe their types, structure and cellular functions.
3. Define the term microfilament. Describe the structure and function of microfilament in the cell.
4. How is it possible for the same vesicle to be transported along both microtubules and microfilaments?
5. Compare the structure and function of a fully assembled microtubule, microfilament (actin filament) and intermediate filament.

Short Answer Questions

1. Describe functions of microfilaments (actin filaments).
2. What is cytoskeleton? Write a short note on the cytoskeleton.
3. Make a comparison of three main components of the cytoskeleton: microtubules, intermediate filaments and microfilaments.
4. Describe structure and function of kinesins.
5. Give an account of structure and function of cytoplasmic dyneins.

Very Short Answer Questions

1. Write the names of main constituents of cytoskeleton.
2. How many subunits are found in a microtubule?
3. What is microtubule?
4. Define microfilament.
5. What is intermediate filament?
6. Define the MTOC.
7. What is microfilament-based movement?
8. Define motor MAP.
9. What is kinesin?
10. Define the cytoplasmic dynein.
11. Define myosin.

Multiple Choice Questions

1. Tubulin protein is absent in
   (*a*) cilia
   (*b*) flagella
   (*c*) plasma membrane
   (*d*) microtubules
2. Cytoskeleton is made up of
   (*a*) callose deposits
   (*b*) cellulosic microfibrils
   (*c*) proteinaceous filaments
   (*d*) calcium carbonate granules
3. Intermediate filaments are composed of
   (*a*) tubulin
   (*b*) actin
   (*c*) flagellin
   (*d*) keratin, vimentin, desmin or lamin

ANSWERS

Very Short Answer Questions

1. Microtubules, microfilaments and intermediate filaments.
2. 13 subunits.
3. Microtubule (MT) is a polymer of the protein tubulin, with a diameter of about 25 nm, that is an integral part of cytoskeleton and that contributes to the support, shape and motility of eukaryotic cells.
4. Microfilament (MF) is a polymer of actin, with a diameter of about 7 nm, that is an integral part of the cytoskeleton, contributing to the support, shape and mobility of eukaryotic cells.

5. Intermediate filament (IF) is a group of protein filaments that are the most stable components of the cytoskeleton of eukaryotic cells, exhibit a diameter of 8 to 12 nm, which is intermediate between the diameter of actin or microfilaments and microtubules.
6. MTOC stands for microtubule-organizing center. It is structure that initiates the assembly of microtubules, the primary example being the centrosome.
7. It is motility based on microfilaments composed of actin and their interaction with myosin; includes muscle contraction, amoeboid movement, cytoplasmic streaming, and cytokinesis in animal cells.
8. MAP stands for microtubule-associated motor protein; it includes proteins, such as kinesin or dynein, that uses energy derived from ATP to drive the transport of vesicles and organelles along microtubules or to generate sliding forces between microtubules.
9. It is family of motor proteins that generate movement along microtubules using energy derived from ATP hydrolysis.
10. Cytoplasmic dynein is a cytoplasmic motor protein that moves along the surface of microtubules in the plus-to-minus direction driven by energy derived from ATP hydrolysis; it is associated with dynactin which links cytoplasmic dynein to cargo vesicle.
11. Myosin is a family of motor proteins that create movements by exerting force on actin microfilaments using energy derived from ATP hydrolysis; makes up thick filaments that move the actin thin filament during muscle contraction.

Multiple Choice Questions

1. (*c*) **2.** (*c*) **3.** (*d*)

12 Plasma Membrane

A plasma membrane encloses every type of cell, both prokaryotic and eukaryotic cells. It physically separates the cytoplasm from the surrounding cellular environment. Plasma membrane is a ultrathin, elastic, living, dynamic and selective transport-barrier. 3It is a fluid-mosaic assembly of molecules of lipids (phospholipids), proteins and carbohydrates. Plasma membrane controls the entry of nutrients and exit of waste products, and generates differences in ion concentration between the interior and exterior of the cell. It also acts as a sensor of external signals (for example, hormonal, immunological, etc.) and allows the cell to react or change in response to environmental signals. The cells of bacteria and plants have the plasma membrane between the cell wall and the cytoplasm. For cells without cell walls (*e.g.*, mycoplasma and animal cells), plasma membrane forms the cell surface.

The plasma membrane is also called **cytoplasmic membrane, cell membrane,** or **plasmalemma.** The term cell membrane was coined by **C. Nageli** and **C. Cramer** in 1855 and the term plasmalemma has been given by **J.Q. Plowe** in 1931.

12.1. ISOLATION AND ANALYSIS

The plasma membrane is so thin that it cannot be observed by the light microscope. Structure of the plasma membrane of various cells has been studied by their isolation from the living systems and also by their artificial synthesis by using their constituent molecules (*e.g.*, liposome, see Chapter 9). The pure and isolated membranes are then studied by biochemical and biophysical methods. The purity of isolated membranes is controlled by **electron microscopy, enzyme analysis** and the **study of surface antigens**. A variety of cells such as mammalian red blood cell (erythrocytes), medullated nerve fibres, Ehrlich mouse ascites tumor cells, liver cells, striated muscle, *Amoeba proteus*, sea urchin eggs and bacteria, have been used in studying the ultrastructure of the plasma membrane. The mammalian erythrocytes and the myelin sheath of the nerve fibre, however, have provided the bulk of information regarding the structure and properties of the plasma membrane. For such experiments, human red blood cells or erythrocytes have been selected by **E. Gorter** and **F. Grendel** (1925) for following advantages: these cells are easy to obtain and are known to be extremely simple. Since these cells contain no intracellular organelles or membrane, so the only membrane structure to be considered is almost entirely that of the cell surface. Lastly, the plasma membrane of erythrocytes is relatively tough and does not readily fragment (See **Lucy**, 1975).

Plasma membranes are more easily isolated from erythrocytes subjected to haemolysis. The cells are treated with hypotonic solutions (to be discussed elsewhere in the chapter) that due to endosmosis produce swelling and then loss of the heamoglobin content (*i.e.*, haemolysis). The resulting membrane is called a **red cell ghost**. If haemolysis is mild, permeability functions of the membrane can be restored by certain treatment, such a ghost is called **resealed ghost**. But if haemolysis is more drastic (*i.e.*, there is complete removal of the haemoglobin) and there is no chance of its resealing, the resulting membrane is called **white ghost**. While the resealed ghosts can be used for the study of physiological as well as biochemical properties, white ghosts can only be used for the study of biochemical properties.

The cell wall of yeast, *Saccharomyces cerevisiae*, can be enzymatically removed by the help of a snail gut enzyme, and the resultant protoplast serves as a source of plasma membrane in a manner similar to that of mammalian erythrocytes.

12.2. CHEMICAL COMPOSITION

Chemically, plasma membrane and other membranes of different organelles are found to contain proteins, lipids and carbohydrates, but in different ratios (Table 12.1). For example, in the plasma membrane of human red blood cells proteins represent 52 per cent, lipids 40 per cent and carbohydrates 8 per cent.

Table 12.1. Chemical composition of some purified membranes (in percentages).

		Membrane	Protein	Lipid	Carbohydrate
1.	Myelin (Nerve cell)		18	79	3
2.	Plasma membrane:				
	(*i*)	Mouse liver	44	52	4
	(*ii*)	*Amoeba*	54	42	4
	(*iii*)	Human erythrocyte	52	40	8
3.	Spinach chloroplast lamellae		70	30	0
4.	Mitochondrial inner membrane		76	24	0

1. Lipids. Four major classes of lipids are commonly present in the plasma membrane and other membranes: **phospholipids** (most abundant), **sphingolipids, glycolipids** and **sterols** (*e.g.*, **cholesterol**). All of them are amphipathic molecules, possessing both hydrophilic and hydrophobic domains. The relative proportions of these lipids vary in different membranes. Phospholipids may be **acidic phospholipids** (20 per cent) such as **sphingomyelin** or **neutral phospholipids** (80 per cent) such as **phosphatidyl choline, phosphatidyl serine,** etc. Many membranes contain cholesterol. Cholesterol is especially abundant in the plasma membrane of mammalian cells and absent from prokaryotic cells. **Cardiolipin** (diphosphatidyl glycerol) is restricted to the inner mitochondrial membrane.

2. Proteins. The amount and types of proteins in the membranes are highly variable: in the myelin membranes which serve mainly to insulate nerve cell axons, less than 25 per cent of the membrane mass is protein, whereas, in the membranes involved in energy transduction (such as internal membranes of mitochondria and chloroplasts), approximately 75 per cent is protein. Plasma membrane contains about 50 per cent protein.

According to their position in the plasma membrane, the proteins fall into two main types: **integral** or **intrinsic proteins** and **peripheral** or **extrinsic proteins**, both of which may be either **ectoproteins**, lying or exposing to external or extracytoplasmic surface of the plasma membrane or **endoproteins**, lying or sticking out at the inner or cytoplasmic surface of the plasma membrane. The intrinsic proteins tend to associate firmly with the membrane, while the extrinsic proteins have a weaker association and are bound to lipids of membrane by electrostatic interaction. On the basis of their functions, proteins of plasma membrane can also be classified into three main types: structural proteins, enzymes and transport proteins (permeases or carriers). Some of them may act as **antigens**, **receptor proteins** (*e.g.*, insulin-binding sites of liver plasma membrane), **regulatory proteins** and so on. **Structural proteins** are extremely lipophilic and form the main bulk (*i.e.*, backbone) of the plasma membrane. **Enzymes** of plasma membrane are either **ectoenzymes** or **endoenzymes** and are of about 30 types (Table 12.2). **Transport proteins** transport specific substances across the plasma membrane and other cellular membranes.

Table 12.2. Some important enzymes present in the plasma membrane.

1. Acetyl phosphatase	11. Cholesterol esterase
2. Acetyl cholinesterase (Ectoenzyme of erythrocyte)	12. Guanylase cyclase
3. Acid phosphatase	13. Monoglyceride lipase
4. Adenosine triphosphatase	14. NAD-ase (Ectoenzyme of erythrocyte)
5. Mg^{2+} ATPase (Endoenzyme of erythrocyte)	15. Protein kinase (Endoenzyme of erythrocyte)
6. Na^+/K^+ ATPase (Ectoenzyme of erythrocyte)	16. Phospholipase A
7. Adenylate cyclase	17. Lactase
8. RNAase	18. Maltase
9. Alkaline phosphatase	19. Sialidase
10. Aminopeptidase	20. UDP glycosidase

3. Carbohydrates. Carbohydrates are present only in the plasma membrane. They are present as short, unbranched or branched chains of sugars (**oligosaccharides**) attached either to exterior ectoproteins (forming **glycoproteins**) or to the polar ends of phospholipids at the external surface of the plasma membrane (forming **glycolipids**). No carbohydrate is located at the cytoplasmic or inner surface of the plasma membrane. All types of oligosaccharides of the plasma membrane are formed by various combinations of six principal sugars (all of which are glucose-derivatives): **D-galactose, D-mannose, L-fucose, N-acetylneuraminic acid** (also called **sialic acid**), **N-acetyl-D-glucosamine** and **N-acetyl-D-galactosamine.**

12.3. STRUCTURE OF PLASMA MEMBRANE

1. Evolution of Fluid Mosaic Model of Membrane

The existence of the plasma membrane of the cell was difficult to prove by direct examination before 1930's (when electron microscopy was invented) because of technological limitations. The membrane is beyond the resolution of the light microscope, rendering a morphological approach of its study quite unfeasible with this instrument. Thus, most of the experimental approaches have been provided by only indirect evidences of the existence of such a membrane around the cells. Let us narrate in brief the saga of evolution of presently well accepted fluid-mosaic model of structure of the plasma membrane:

1. The plasmolysis of plant cells in hypertonic solutions suggests the existence of the plasma membrane in the plants.
2. The very fact that a cell, especially in animal cell which has no cell wall, can exist as a physically defined entity suggests that it must have some sort of boundary around it.
3. The presence of plasma membrane can be inferred because protoplasm leaks out of animal cells when cell surface is punctured.
4. After performing some 10,000 experiments with more than 500 different chemicals, in 1899, **Overton** concluded that the peculiar osmotic properties of living protoplasts are due to a **selective solubility mechanism**. Hydrophobic compounds entered cells more rapidly than hydrophilic ones. **Overton** believed this was because of an outer lipoid layer in which hydrophobic compounds were more soluble. He correctly speculated that this layer might contain cholesterol, lecithin and fatty oils.

5. **Hober** (1910) and **Fricke** (1925) found that the intact cell had low electrical conductivity, indicating the presence of a lipid layer around it.
6. If a lipid containing **hydrophilic groups** (such as the carboxyl groups of fatty acids or the phosphate groups of phospholipids) is dissolved in a highly volatile solvent (*e.g.*, benzene) and several drops of it are then carefully applied to the surface of the water, the lipid spreads out to form a thin, one-molecule-thick or **monomolecular film.** In this film, it is found that the hydrophilic parts of each molecule project into the water surface and the hydrophobic parts are directed up, away from the water.
7. In 1917, **Langmuir** (Nobel Laureate of 1932 in Chemistry) fabricated a **trough** or **film balance** (Fig. 12.1) for measuring the specific minimum surface area occupied by a monomolecular film of lipid and the force necessary to compress all the lipid molecules into this area. Langmuir trough consists of a shallow trough filled with water on which lipid substance can be spread to make a monomolecular film. A barrier can be pushed across the trough to compress the film.

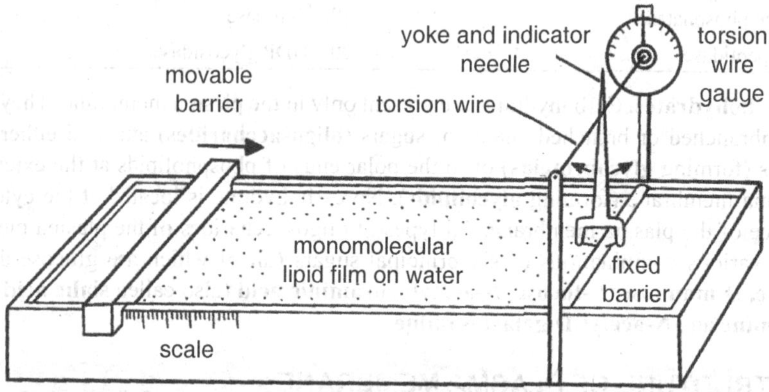

Fig. 12.1. Langmuir trough.

8. In 1925, **Gorter** and **Grendel** extracted the lipids from erythrocyte ghosts of a variety of mammals (such as dogs, sheep, rabbits, guinea pigs, goats and humans) and spread them out on monolayers in the Langmuir trough. These investigators discovered that the area covered by the lipid monomolecular layer film was twice than what was needed to cover the surface of the cells from which the lipid was extracted. Consequently, they safely concluded that *erythrocytes were covered by a layer of lipids two molecules thick* (**lipid bilayer** or **bimolecular lipid layer**) oriented with polar groups toward the inside and outside of the cell.
9. By studying the surface tension of cells **Harvey** (**Harvey** and **Cole**, 1931, **Danielle** and **Harvey**, 1935) suggested the presence of proteins in the plasma membrane, in addition to the lipids.
10. In 1935, **Danielli** and **Davson**, proposed a model, called **sandwich model**, for membrane structure in which a lipid bilayer was coated on its either side with hydrated proteins (globular proteins). Mutual attraction between the hydrocarbon chains of the lipids and electrostatic forces between the protein and the "head" of the lipid molecules, were thought to maintain the stability of the membrane. From the speed at which various molecules penetrate the membrane, they predicted the lipid bilayer to be about 6.0 nm in thickness, and each of the protein layer of about 1.0 nm thickness, giving a total thickness of about 8.0 nm.

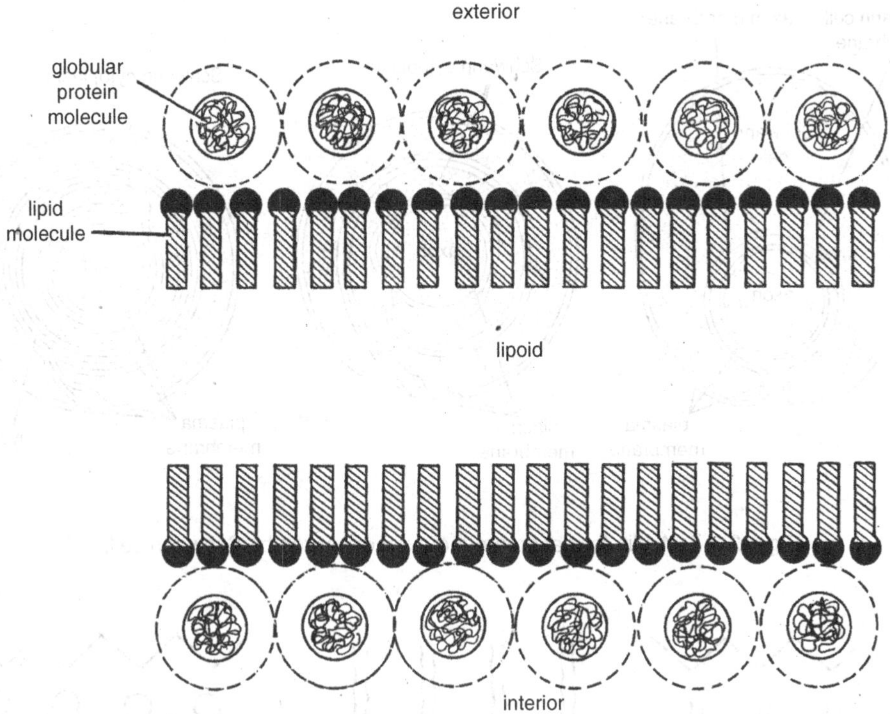

Fig. 12.2. The original Danielli-Davson model (1935) of membrane structure. The bimolecular layer of lipid molecules is of undefined thickness and is covered on each side by a continuous layer of globular proteins.

The Danielli-Davson model got support from electron microscopy. Electron micrographs of the plasma membrane showed that it consists of two dark layers (electron dense granular protein layers), both separated by a lighter area in between (the central clear area of lipid bilayer). The total thickness of the membranes is also turned out to be about 7.5 nm.

11. **Myelin as a model for unit membrane model.** To understand the relationship between the trilaminar appearance of plasma membrane and its underlying molecular structure, **Robertson** chose myelin as the model system for study. **Myelin** surrounds the axons of certain nerve cells and is composed of multiple layers of plasma membrane successively wound around one another. This stack of membrane is derived from the specialized **Schwann cells** that are intimately associated with the axon (Fig. 12.3). During the early stages of myelin formation, axons become surrounded by Schwann cell. The plasma membrane of Schwann cell start wrapping around the axon in a spiral manner. At the later stage of development, the plasma membrane of the Schwann cell become closely stacked upon each other and the cytoplasmic content of the cell becomes disorganized from the space between the membranes. This stack is closely packed membranes around the axon is called **myelin**.

Robertson used myelin, rather than a typical membrane as a model for study because, the repeating layers of membranes present in myelin creates a quasicrystalline structure agreeable to analysis by X-ray diffraction. The X-ray diffraction analyses, carried out by **F.O. Schmitt** and later on **J.B. Fineam** indicated that myelin consists of lipid bilayer alternating with layers of protein. These studies revealed that the general pattern of plasma membrane had one of three possible arrangements of protein and lipid (Fig. 12.4).

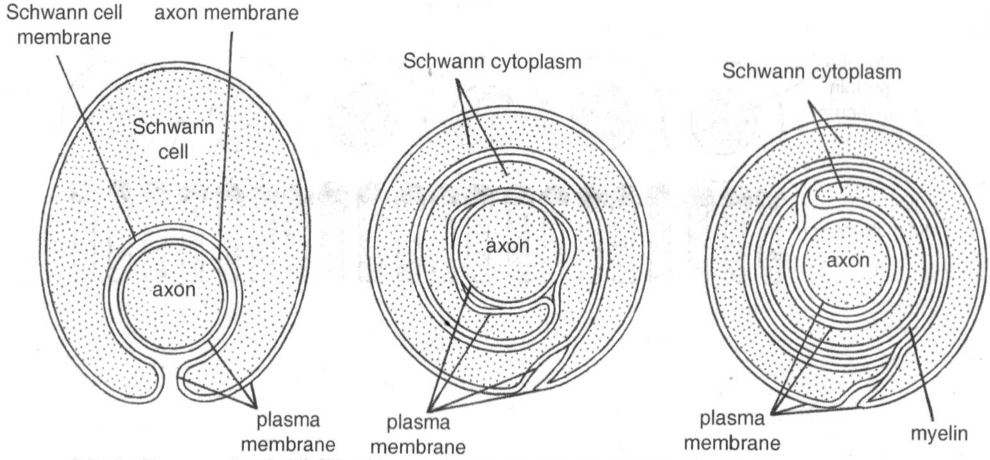

Fig. 12.3. Myelin formation around axon of neuron by Schwann cell.

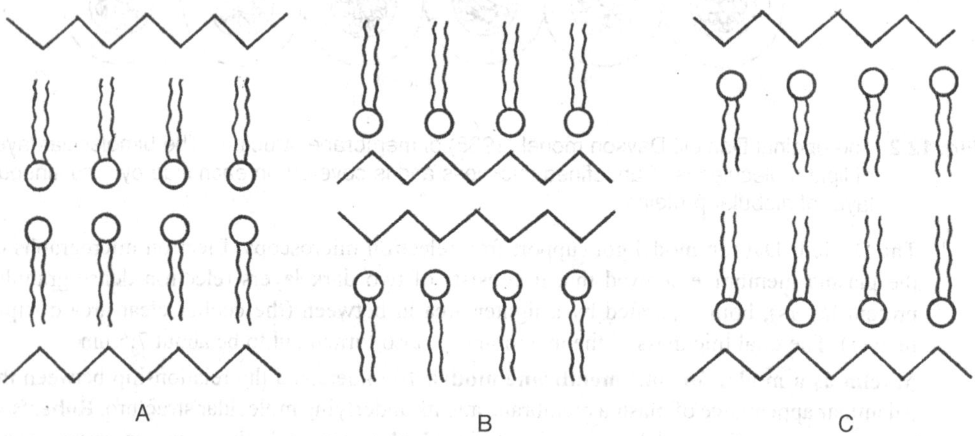

Fig. 12.4. Three possible arrangements of protein and lipid in plasma membrane.

To distinguish among these three alternative arrangements of lipids and proteins in plasma membrane, **Robertson** carried out investigations on electron microscope using different stains for lipids and protein molecules. These investigations led him to conclude that in electron micrographs of biological membranes, protein molecules as well as hydrophillic head groups of lipid molecules are stained. Thus, he found that lipids are present in two layers covered by protein with lipid head groups projecting outwards towards both membrane surfaces.

Later on by using evidence from various electron micrographs, **Robertson** in 1960, proposed the **unit membrane hypothesis** (Fig. 12.5). This hypothesis states that all cellular membranes have an identical **trilaminar** structure (or dark-light-dark or railway track pattern). However, thickness of the unit membrane has been found to be greater in plasma membrane (10 nm) than in the intracellular membranes of endoplasmic reticulum or Golgi apparatus (*i.e.*, 5 to 7 nm).

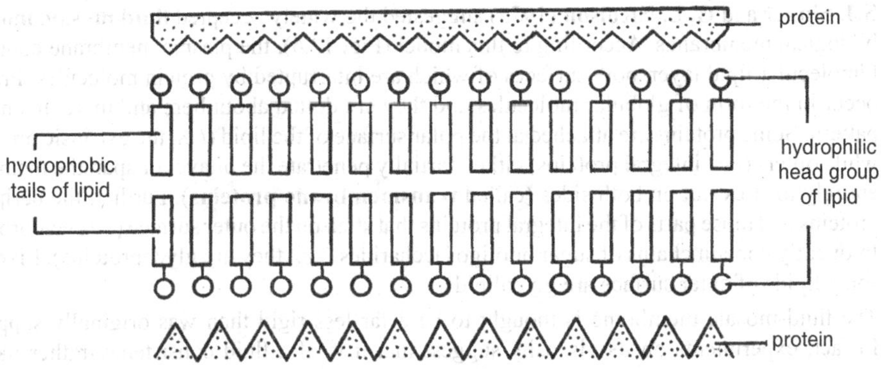

Fig. 12.5. Schematic diagram of the Robertson model of membrane structure. The lipid layer is defined as bimolecular, and the protein is extended but different on the two faces of the membrane.

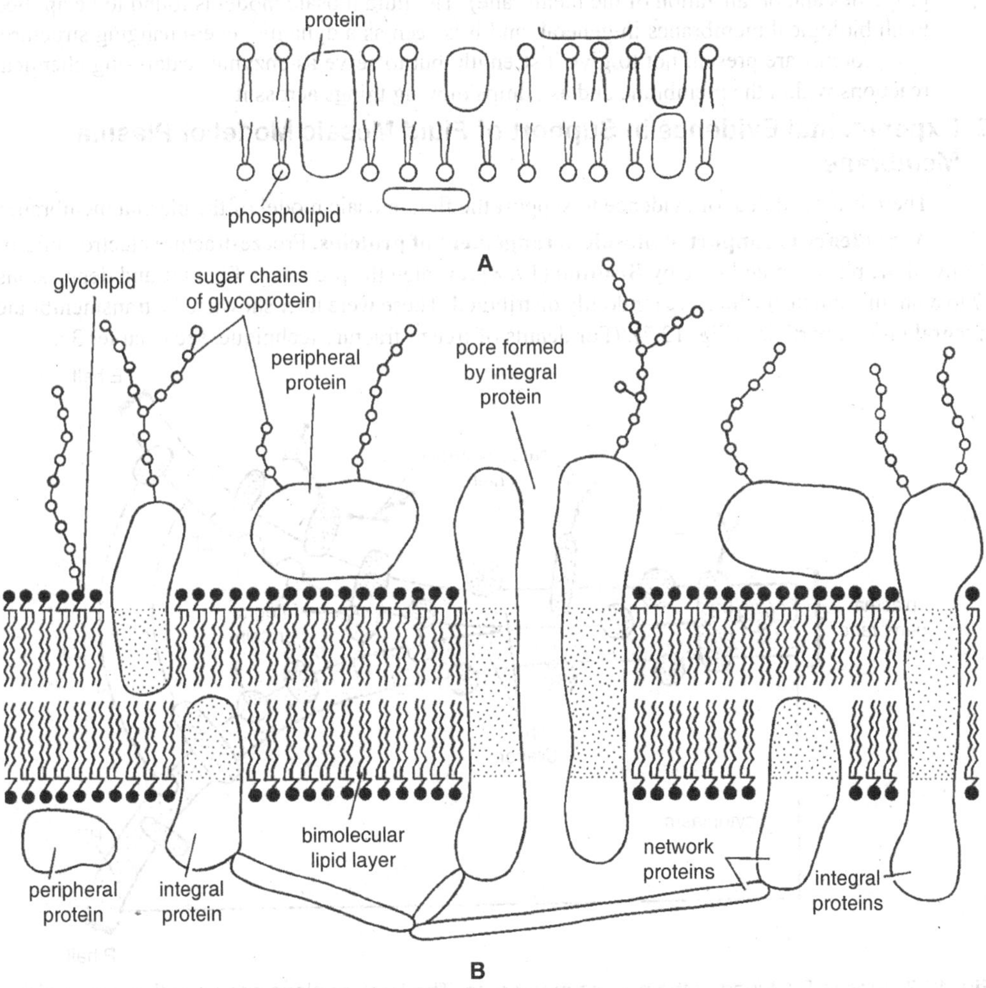

Fig. 12.6. Fluid mosaic model of the plasma membrane. A—Simplistic view; B—Complex view.

12. **S.J. Singer** and **G.L. Nicolson** (1972) suggested the widely accepted **fluid mosaic model** of biological membranes. According to this model (Fig. 12.6), the plasma membrane contains a bimolecular lipid layer, both surfaces of which are interrupted by protein molecules. Proteins occur in the form of globular molecules and they are dotted about here and there in a mosaic pattern. Some proteins are attached at the polar surface of the lipid (*i.e.*, the extrinsic proteins); while others (*i.e.*, integral proteins) either partially penetrate the bilayer or span the membrane entirely to stick out on both sides (called **transmembrane proteins**). Further, the peripheral proteins and those parts of the integral proteins that stick on the outer surface (*i.e.*, ectoproteins) frequently contain chains of sugar or oligosaccharides (*i.e.*, they are glycoproteins). Likewise, some lipids of outer surface are glycolipids.

    The fluid-mosaic membrane is thought to be a far less rigid than was originally supposed. In fact, experiments on its viscosity suggest that it is of a fluid consistency rather like the oil, and that there is a considerable sideways movement of the lipid and protein molecules within it. On account of its fluidity and the mosaic arrangement of protein molecules, this model of membrane structure is known as the "fluid mosaic model" (*i.e.*, it describes both properties and organization of the membrane). The fluid mosaic model is found to be applied to all biological membranes in general, and it is seen as a dynamic, ever-changing structure. The proteins are present not to give it strength, but to serve as enzymes catalysing chemical reactions within the membrane and as pumps moving things across it.

2. Experimental Evidence in Support of Fluid Mosaic Model of Plasma Membrane

There is a good deal of evidence to support the fluid mosaic model of the plasma membrane:

A. Evidence in support of mosaic arrangement of proteins. Freeze-fracture electron microscopy of the plasma membrane by **Branton** (1968) revealed the presence of bumps and depressions (7 to 8 nm in diameter) which are randomly distributed. These were later shown to be transmembrane integral protein particles (Fig. 12.7). (For details of freeze-fracture technique, see Chapter 3).

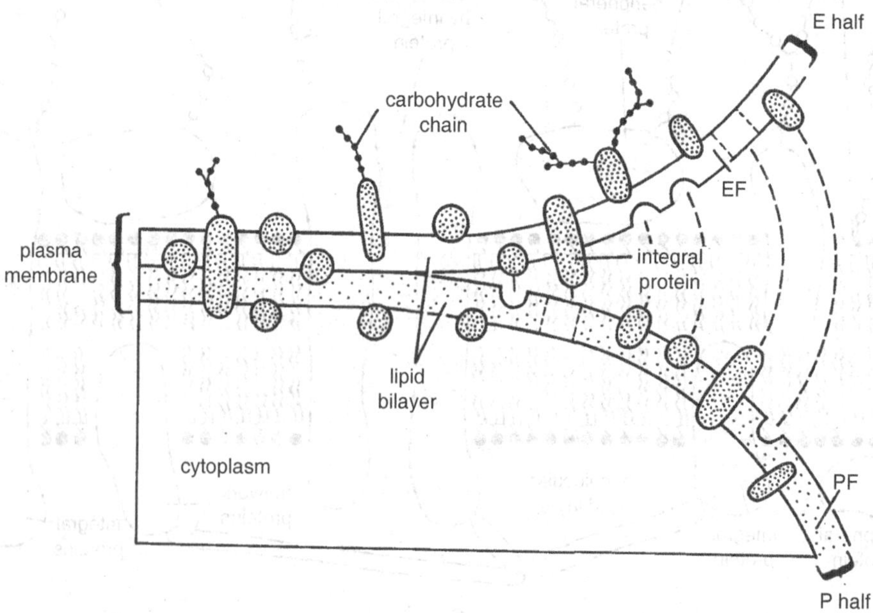

Fig. 12.7. Freeze fracturing of the plasma membrane. The fracture plane occurs at the centre of the lipid bilayer and passes over (or under) the integral membrane proteins. E = exterior, P = protoplasmic (cytosolic) side, F = fracture face.

B. Evidence in support of fluid property of lipid bilayer. Mobility of membrane proteins due to fluid property of lipid bilayer was demonstrated by a classical experiment of **D. Frye** and **M. Edidin** (1970). They fused two different types of cultured cells having different surface antigens (proteins). The **cell fusion** is achieved by the use of some fusogen such as an inactivated parainfluenza virus, called Sendai virus (named after a city of Japan). **A fusogen** is a membrane fusion promoting factor such as Sendai virus, lysophosphatides, oleic acid or an electric field. Sendai virus facilitates

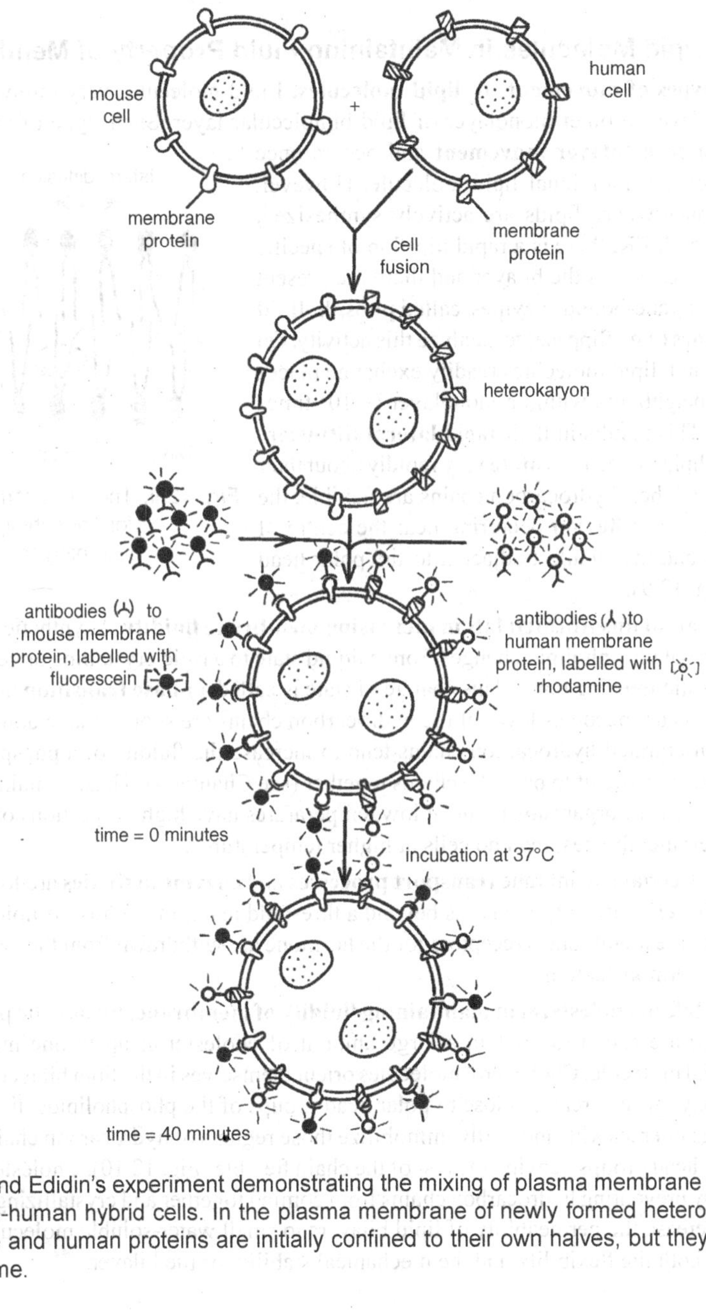

Fig. 12.8. Frye and Edidin's experiment demonstrating the mixing of plasma membrane proteins on mouse-human hybrid cells. In the plasma membrane of newly formed heterokaryon, the mouse and human proteins are initially confined to their own halves, but they intermixed with time.

fusion of the plasma membranes and cytoplasms of both cells to produce a **hybrid cell** or **heterokaryon** with two types of nuclei. If the two cells are originally labelled with fluorescent antibodies of different colours, such as fluorescein (green) and rhodamine (red), it is possible at the onset of fusion to recognize the parts of the plasma membrane corresponding to each cell. However intermixing occurs as the antigens are dispersed and the two colours become less and less detectable. After 40 minutes (at 37° C) the intermixing of two colours is complete and the two antigens can no longer be distinguished (Fig. 12.8).

Role of Lipid Molecules in Maintaining Fluid Property of Membrane

***(i)* Types of movements of lipid molecules.** Lipid molecules very rarely migrate from one lipid monolayer to other monolayer of lipid bimolecular layer. Such type of movement is called **flip-flop** or **transbilayer movement** and occurs once a month for any individual lipid molecule. However, in membranes where lipids are actively synthesized, such as smooth ER, there is a rapid flip-flop of specific lipid molecules across the bilayer and there are present certain membrane-bound enzymes, called **phospholipid translocators** (*e.g.*, flippase) to catalyze this activity. On the other hand, lipid molecules readily exchange places with their neighbours within a monolayer (~ 10^7 times a second). This results in their rapid **lateral diffusion.** Individual lipid molecules **rotate** very rapidly about their long axes and their hydrocarbon chains are flexible, the greatest degree of **flexion** occurring near the centre of the bilayer and the smallest adjacent to the polar head groups (Fig. 12.9).

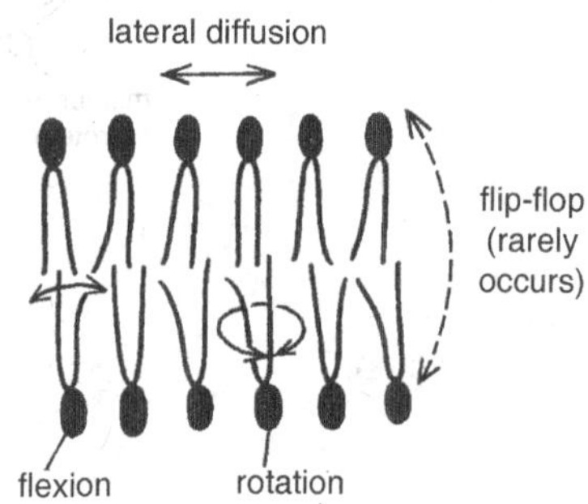

Fig. 12.9. The types of movement possible for phospholipid molecules in a lipid bilayer.

***(ii)* Role of unsaturated fats in increasing membrane fluidity.** A synthetic bilayer made from a single type of phospholipid changes from a liquid state to a rigid crystalline or gel (viscous) state at a characteristic freezing point. This change of state is called a **phase transition** and the temperature at which it occurs becomes lower if the hydrocarbon chains are short or have double bonds. Double bonds in unsaturated hydrocarbon chains tend to increase the fluidity of a phospholipid bilayer by making it more difficult to pack the chains together (see Chapter 9). Thus, to maintain fluidity of the membrane, cells of organisms living at low temperatures have high proportions of unsaturated fatty acids in their membranes, than do cells at higher temperatures.

In fact, certain membrane **transport processes** and **enzyme activities** are found to cease when the lipid bilayer's viscosity increases beyond a threshold level. In contrast, if lipid bilayer's fluidity is increased, the membrane's receptors for the hormone are withdrawn from the cell surface, thereby hampering hormone action.

***(iii)* Role of cholesterol in maintaining fluidity of membrane.** Eukaryotic plasma membranes (in animals) are found to contain a large amount of cholesterol; up to one molecule for every phospholipid molecule. Cholesterol molecules orient themselves in the lipid bilayer in such a way that their hydroxyl groups remain close to polar head groups of the phospholipids, their rigid plate-like steroid rings interact with and partly immobilize those regions of hydrocarbon chains that are closest to the polar head groups, leaving the rest of the chain flexible (Fig. 12.10). Cholesterol inhibits phase transition by preventing hydrocarbon chains from coming together and crystallizing. Cholesterol also tends to decrease the permeability of lipid bilayers to small water-soluble molecules and is thought to enhance both the flexibility and the mechanical stability of the bilayer.

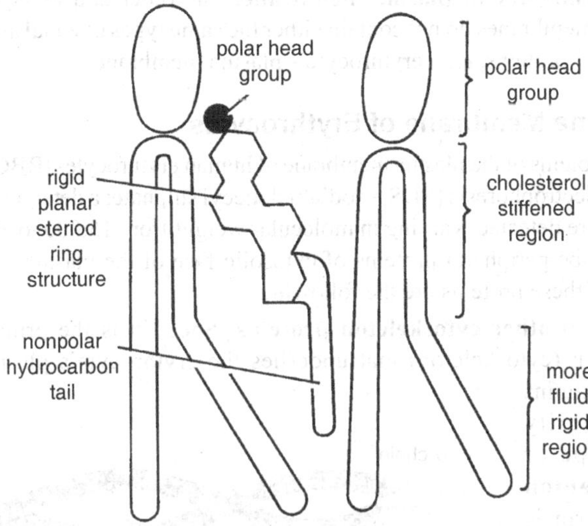

12.10. Cholesterol molecule (schematic) interacting with two phospholipid molecules in a monolayer.

Membrane Asymmetry

Both lipid and protein molecules have irregular distribution in both monolayers of the lipid bilayer, this is called **membrane asymmetry**.

A. Phospholipid asymmetry in plasma membrane. The lipid composition and state of fluidity of two halves of the lipid bilayer are found to be strikingly different. For example, in human erythrocyte's plasma membrane, outer half contains those phospholipids which have more saturated fatty acid chains, and inner half contains those phospholipids which contain terminal amino groups and less saturated fatty acid chains. As a result, inner monolayer is more fluid than the outer lipid monolayer. Such a phospholipid asymmetry is generated in smooth ER. The asymmetry of glycolipids such as galactocerebroside, ganglioside, etc., in myelin sheath of nerves (*i.e.,* they are found only in the outer half of lipid bilayer) is found to be originated in lumen of Golgi apparatus. The specific role of lipid asymmetry of the membrane is still not clear.

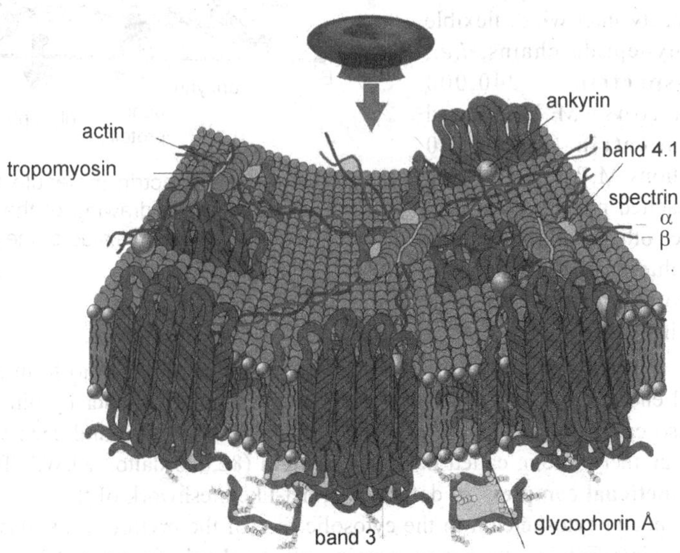

The plasma membrane of the human erythrocyte.

B. Protein asymmetry in plasma membrane. The outer and inner sides of the plasma membrane and other membranes do not contain either the same types or equal amounts of the various peripheral and integral proteins, *e.g.*, erythrocyte's plasma membrane.

Proteins of Plasma Membrane of Erythrocytes

When the extracted proteins of the plasma membrane of human erythrocytes (RBC) are studied by SDS polyacrylamide-gel electrophoresis (SDS = sodium dodecyl sulphate; a detergent), approximately 15 major protein bands are detected, varying in molecular weight from 15,000 to 25,000. Most of these proteins are found to be peripheral proteins of cytosolic face of the plasma membrane. Important properties of some of these proteins are the following:

***(i)* Spectrin and other cytoskeleton proteins.** Spectrin is the principal component of the protein meshwork (cytoskeleton) that underlies the erythrocyte's plasma membrane (Fig. 12.11). It, thus, maintains the structural integrity and biconcave shape of this membrane (**Branton** *et al.*, 1981). Spectrin is long, thin, flexible rod about 100 nm in length. It constitutes about 25 per cent of the membrane associated protein mass (about 2.5 × 10^5 copies per cell). Spectrin is a **heterodimer** and consists of two non-identical, antiparallel, loosely intertwind, flexible polypeptide chains, *i.e.*, **α-spectrin** (~ 240,000 daltons M.W.) and **β-spectrin** (~ 220,000 daltons M.W.), both being attached non-covalently to each other at multiple points including their ends (*i.e.*, phosphorylated 'head' and 'tail').

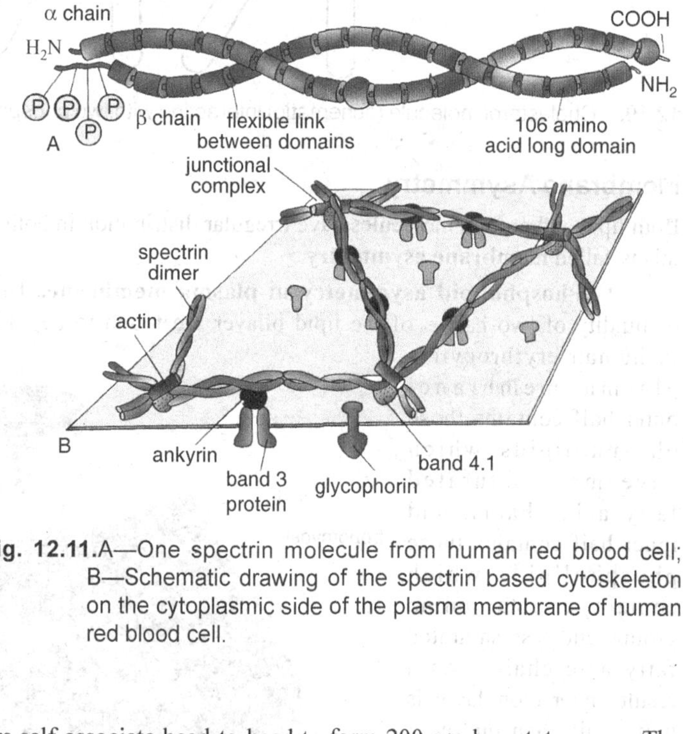

Fig. 12.11. A—One spectrin molecule from human red blood cell; B—Schematic drawing of the spectrin based cytoskeleton on the cytoplasmic side of the plasma membrane of human red blood cell.

The spectrin heterodimers self-associate head-to-head to form 200 nm long **tetramers**. The tail ends of five or six spectrin tetramers are linked together by binding to short **actin filaments** (also called **band 5 proteins**; with 43,000 dalton M.W.) and each with 15 actin monomers and to another protein, called **band 4.1 protein** (82,000 dalton M.W.). These three proteins form the "**junctional complex**" of deformable, net-like meshwork of the cytoskeleton. Further, the binding of spectin cytoskeleton to the cytosolic face of the erythrocyte's plasma membrane depends on a large intracellular attachment protein, called **ankyrin** (or **band 2.1 protein**; 210,000 dalton M.W.). Ankyrin tends to bind to both β-spectrin and to the cytoplasmic domain of a transmembrane protein, called **band 3 protein** (**Shen** *et al.*, 1986).

***(ii)* Glycophorin A.** It is a small transmembrane glycoprotein (single-pass membrane protein) having molecular weight of 55,000 daltons and 131 amino acid residues. This protein bears about 100 sugars on 16 separate oligosaccharide side chains (90 per cent of which is sialic acid). Erythrocytes have about 5 types of glycophorins, called A, B, C, D and E; glycophorin A being more common. There are more than 6×10^5 glycophorin molecules per cell (*i.e.*, erythrocyte). Glycophorins are found to contain certain antigenic determinants (carbohydrates) for the **MN blood groups**. Further, sialic acid confers a high negative charge to the cell surface of erythrocyte. Because of these charges, RBCs repel each other, which prevents the erythrocytes from clumping as they circulate through the body's tiny vessels (**Karp** 2010). Sialic and residue may be important in the life cycle of the erythrocytes as it has been shown that cells lose sialic acid as they age in the circulatory system. Correlated with this is the observation that *loss of sialic acid is a signal for removal and destruction of an erythrocyte by the spleen and liver*. In this way the life span of red blood cells may be regulated (see **King**, 1986).

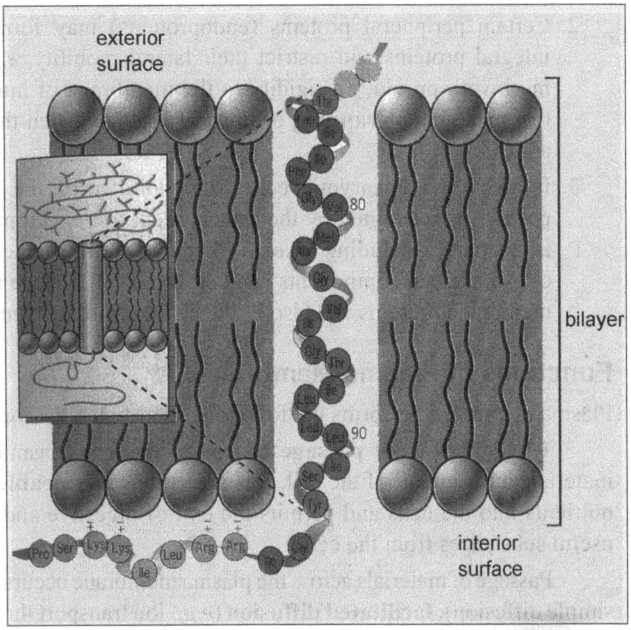

Glycophorin A, an integral protein with a single transmembrane domain.

Glycophorin A or B also happens to be the receptor utilised by the protozoa that causes malaria, providing a path for entry into the blood cell (**Karp** 2010).

***(iii)* Band 3 protein.** Like the glycophorin, **band 3 protein** (93,000 daltons M.W.) is a transmembrane protein, but it is a **multipass membrane protein**, *i.e.*, its highly folded polypeptide chain (about 930 amino acid long) extends across the lipid bilayer at least 10 times. Each human erythrocyte contains about 10^6 and band 3 proteins, each of which forms either a dimmer or tetramer in the membrane. Band 3 protein serves as a channel for the passive exchange of anions across the membrane, (see **Karp** 2010). As the erythrocytes pass through the lungs, they exchange bicarbonate (HCO^-) for chloride (Cl^-) through these hydrophilic channels during the process of CO_2 release (**chloride shift**).

Constraints on the Mobility of Membrane Molecules

In the fluid mosaic plasma membrane, there is not complete and independent freedom of movement for its different component molecules. The mobility of some part of lipid molecules is constrained since that remains tightly bound to some of the integral membrane proteins. For example, the mobility of lipid molecules surrounding **cytochrome oxidase** (an enzyme involved in the synthesis of ATP) are immobilized by the enzyme and makes boundary lipid layer. The immobilized boundary lipid makes 30 per cent of membrane lipid in the mitochondrial membrane.

In contrast to lipids, the mobility and distribution of protein molecules in the membrane is controlled by various ways:

1. Certain proteins of membrane are constrained by protein-protein interactions to form specialized ordered regions, representing 2 to 20 per cent of the membrane of a system, *e.g.*, gap junctions, synapsis of neurons and, plaques of halobacteria.

2. Certain peripheral proteins (endoproteins) may form a bridge-like lattice work between integral proteins and restrict their lateral mobility, *e.g.*, spectrin-ankyrin-actin cytoskeletal meshwork provides a rigidity to the membrane of human erythrocytes and does not permit the clustering or **capping** of integral proteins when the appropriate antibodies or lectins are added.
3. In nucleated eukaryotic cells, the mobility of the peripheral endoproteins and integral proteins is restrained by their attachment to the ectoplasmic cytoskeleton. The cytoskeleton is extensive, including **myosin** filaments, **actin** filaments and **microtubules**. Rearrangement of cytoskeletal components just below the cell surface manifests in the distribution of integral membrane proteins and also in the cellular motions, endocytosis and exocytosis.

Functions of Plasma Membrane

Plasma membrane performs the following vital functions for the cell:

1. Regulation of passage of materials. The plasma membrane regulates the passage of materials into and out of the cell. It is **selectively permeable**. It facilitates the entrance of required nutrients into the cells and permits the exit of digestive and nutritive wastes. It checks the exit of useful substances from the cell.

Passage of materials across the plasma membrane occurs by **passive transport** (such as osmosis, simple diffusion), **facilitated diffusion** (*e.g.,* ion transport through charged pores, Ca^{2+}-channels and D-hexose and anion exchange permeases of RBCs), **active transport** (by Na^+/K^+ ATPase, calcium ATPase, proton pump) and **bulk transport** (by exocytosis, phagocytosis, pinocytosis and receptor mediated endocytosis).

2. Maintenance of differential distribution of ions. Plasma membrane regulates and maintains the differential distribution of ions inside and outside the cell. For example, potassium ions are concentrated inside a living cell, whereas sodium and chloride ions are distributed outside. Such an unequal distribution of ions leads to a potential difference. This electric potential energy in all living cells is known as **membrane potential**.

3. Response to environment. The plasma membrane responds to changes in its environment with the help of **receptor proteins**. These receive chemical messages from other cells and thus help in cell recognition. The receptor proteins respond to **hormones, growth factors** and **neurotransmitters.**

4. Contact with neighbouring cells. The plasma membrane maintains structural and chemical relationships with the neighbouring cells by the help of certain glycoproteins.

5. Protection. Plasma membrane protects the cell and may be involved in cell movement, cell secretion and transmitting impulses.

12.4. ORIGIN OF PLASMA MEMBRANE

There is hardly any cell structure more important to the immediate health of the cell than the plasma membrane. If it is weakened or injured, the cell loses its ability to maintain gradients, to carry out the selective transport of nutrients, and to contain the pool of enzymes and organelles essential for the homeostasis. In consequence, new membranes may be added to existing membranes without altering the functions as a barrier and selective transporter. Also for maintaining the characteristic membrane asymmetry, the membrane must be assembled with precisely the correct molecular topography.

Thus, all cellular membranes grow from pre-existing membranes which act as **templates** for the addition of new precursors. All cells divide, daughter cells receive a full complement of membrane systems which undergo growth until the next division, to be passed on to subsequent progeny. Meanwhile the molecules within the membrane undergo continuous replacement.

The protein molecules of the plasma membrane are synthesized on both attached and free ribosomes. Proteins synthesized by free ribosomes may be inserted into the plasma membrane following their completion and release from the ribosomes. Proteins of plasma membrane synthesized on attached ribosomes of rough ER are **inserted** first into the membrane of RER and then **transferred** to the Golgi apparatus, **processed** there (*e.g.*, glycosylation) and ultimately are dispatched to the plasma membrane via the secretory vesicles. Likewise, the synthesis of phospholipid molecules of the plasma membrane takes place by the smooth ER (SER). Like the proteins, newly synthesized lipids are inserted into SER membranes, then they are passed to Golgi apparatus for the processing and ultimately are dispatched to the plasma membrane via small secretory vesicles. The cytosol also contains a number of **phospholipid transport proteins** that function to transfer phospholipid molecules from one cellular membrane to another (*e.g.*, from ER membranes to plasma membrane).

In fact, the process of glycosylation (or glycosidation, *i.e.*, addition of oligosaccharides containing the sugars such as galactose, fucose and/or sialic acid, to the molecules of proteins and phospholipids of the plasma membrane) is completed at the level of Golgi apparatus. However, some sugars are added to the proteins in the lumen of RER.

12.5. MEMBRANE TRANSPORT

The transport of substances into the cell is performed by the following methods:

A. Transport of Water

1. Osmosis (It is often included in Passive transport.)

B. Transport of Ions and Small Molecules

2. Passive transport
   (*i*) Simple diffusion
   (*ii*) Facilitated diffusion
3. Active transport

C. Transport of Large Molecules (Bulk Transport)

4. Phagocytosis
5. Endocytosis
   (*i*) Pinocytosis
   (*ii*) Receptor mediated endocytosis

(**Note.** For bulk transport see Chapter 23).

The plasma membrane acts as a semipermeable barrier between the cell and the extracellular environment. This permeability must be highly **selective** if it is to ensure that essential molecules such as glucose, amino acids and lipids can readily enter the cell, then these molecules and metabolic intermediates remain in the cell, and that waste compounds leave the cell. In short, the **selective permeability** of the plasma membrane allows the cell to maintain a constant internal environment (**homeostasis**). Transport across the membrane may be passive or active. It may occur via the phospholipid bilayer or by the help of specific integral membrane proteins, called **permeases** or **transport proteins**.

A. Passive transport. It is a type of **diffusion** in which an ion or molecule crossing a membrane moves down its electrochemical or concentration gradient. *No metabolic energy is consumed in passive transport.* Passive transport is of following three types:

1. Osmosis

The plasma membrane is permeable to water molecules. The to and fro movement of water molecules through the plasma membrane occurs due to the differences in the concentration of the solute on its either sides. The process by which the water molecules pass through a membrane from a region of higher water concentration to the region of lower water concentration is known as **osmosis** (*Gr.*, *osmos* = pushing). The process in which the water molecules enter into the cell is known as **endosmosis**, while the reverse process which involves the exit of the water molecules from the cell is known as **exosmosis**. In plant cells due to excessive exosmosis the cytoplasm along with the plasma membrane shrinks away from the cell wall. This process is known as **plasmolysis** (*Gr.*, *plasma* = molded, *lysis* = loosing) (Fig. 12.12).

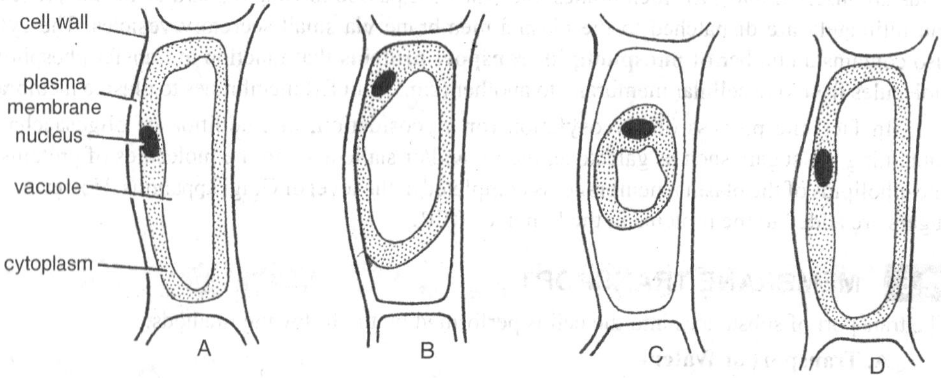

Fig. 12.12. Plasmolysis and deplasmolysis in plant cells. Plasmolysis occurs when a normal plant cell (A) is placed in a hypertonic solution. Water leaves the cell and the plasma membrane shrinks away from the cell wall (B, C). If solutes can penetrate the plasma membrane, the cell will eventually regain water—a process termed deplasmolysis (D).

Box 12.1 Plasma Membrane and Isotonic, Hypotonic and Hypertonic Solutions

A cell contains a variety of solutes in it, for instance, the mammalian erythrocytes contain the ions of potassium (K^+), calcium (Ca^+), phosphate (PO^-_4), dissolved haemoglobin and many other substances. If the erythrocyte is placed in a 0.9% solution of sodium chloride (NaCl), then it neither shrinks nor swells. In such case, because the intra-cellular and extra-celluar fluids contain same concentration, so no osmosis takes place. This type of extra-cellular solution or fluid is known as **isotonic solution** or **fluid**. If the concentration of NaCl solution is increased above 0.9% then the erythrocytes are shrinked due to excessive exosmosis. The solutions which have higher concentrations of solutes than the intracellular fluids are known as **hypertonic solutions**. Further, if the concentration of NaCl solution decreases below 0.9% the erythrocytes will swell up due to endosmosis. The extra-cellular solutions having less concentration of the solutes than the cytoplasm are known as **hypotonic solutions**.

Due to endosmosis or exosmosis the water molecules come in or go out of the cell. The amount of the water inside the cell causes a pressure known as **hydrostatic pressure**. The hydrostatic pressure which is caused by the osmosis is known as **osmotic pressure**. The plasma membrane maintains a balance between the osmotic pressure of the intra-cellular and inter-cellular fluids.

2. Simple Diffusion

In simple diffusion (Fig. 12.13), transport across the membrane takes place unaided, *i.e.*, molecules of gases such as oxygen and carbon dioxide and small molecules (*e.g.*, ethanol) enter the cell by crossing the plasma membrane without the help of any permease. During simple diffusion, a small molecule in aqueous solution dissolves into the phospholipid bilayer, crosses it and then dissolves into the aqueous solution on the opposite side. There is little specificity to the process. The relative rate of diffusion of the molecule across the phospholipid bilayer will be proportional to the concentration gradient across the membrane.

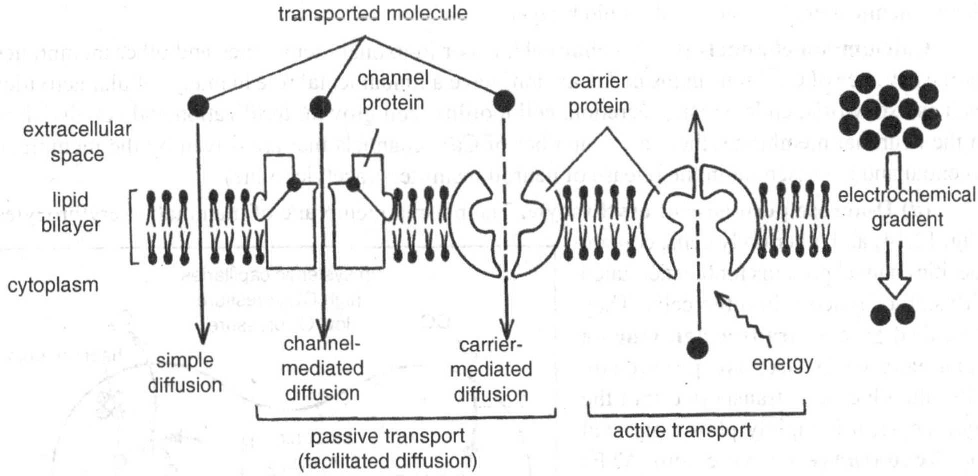

Fig. 12.13. Schematic diagram showing various types of transports across the membrane: simple diffusion, passive transport (down an electrochemical gradient) and active transport (against an electrochemical gradient).

3. Facilitated Diffusion

This is a special type of passive transport, in which ions or molecules cross the membrane rapidly because specific permeases in the membrane facilitate their crossing. Like the simple diffusion, facilitated diffusion does not require the metabolic energy and it occurs only in the direction of a concentration gradient. Facilitated diffusion is characterized by the following special features:

1. The rate of transport of the molecule across the membrane is far greater than would be expected from a simple diffusion.
2. This process is specific; each facilitated diffusion protein (called **protein channel**) transports only a single species of ion or molecule.
3. There is a maximum rate of transport, *i.e.,* when the concentration gradient of molecules across the membrane is low an increase in concentration gradient results in a corresponding increase in the rate of transport. Currently, it is believed that transport proteins form the **channels** through the membrane that permit certain ions or molecules to pass across the latter.

Examples of Facilitated Diffusion

***(i)* Ionic transport through charged pores.** Nerve conduction is propagated along the axonal membrane by **action potential** which regulates opening and closing of two main types of **ion channels** (*i.e.,* channel proteins with water filled pores): **Na^+ channels** (or **voltage-gated Na^+ channels**) and **K^+ channels** (or **K^+ leak channels**). At the point of stimulation there is a sudden and several hundred fold increase in permeability to Na+, which reaches its peak in 0.1 millisecond (*i.e.,* the membrane potential may depolarise from –90 mV and overshoot to +50 mV). At the end of the period, the membrane again becomes essentially impermeable to Na^+, but the K^+ permeability increases and this ion leaks out of the cell, repolarising the nerve fibre. In other words, during the rising phase of the spike, Na^+ enter through the Na^+ channels, and in descending phase K^+ is extruded through the K^+ channels.

Such ion channels also occur in other types of cells such as muscle, sperm and unfertilized ovum. They are not coupled to an energy source (ATP), so the transport they mediate is always passive ("down hill"), allowing specific ions mainly Na^+, K^+, Ca^{2+} and Cl^- to diffuse down their electrochemical gradient across the lipid bilayer.

Calcium ion channels (Ca^{2+} – channels) occur in axonal membranes and other membranes for the entrance of Ca^{2+} ions in the cell. Ca^{2+} ions have a fundamental role in many cellular activities such as exocytosis, endocytosis, secretion, cell motility, cell growth, fertilization and cell division. In the neuronal membrane, there are a number of Ca^{2+} channels that are driven by the membrane potential and are essential in the release of neurotransmitters (acetylcholine).

***(ii)* D-hexose permease of erythrocyte.** The plasma membrane of mammalian erythrocytes (Fig. 12.14) and other body cells, contain specific channel proteins for the facilitated diffusion of glucose into the cells. They are called **glucose transporter, glucose permease** or **D- hexose permease.** After the glucose is transported into the erythrocyte, it is rapidly phosphorylated (by **hexokinase** enzyme and ATP) to form **glucose-6-phosphate**. Once phosphorylated, the glucose no longer leaves the cell; moreover, the concentration of the simple glucose in the cell is lowered. As a result, the concentration gradient of glucose across the membrane is increased, allowing the facilitated diffusion to continue to import glucose. Since no cellular membrane (except the mitochondrial membranes) contains any permease for facilitated diffusion of phosphorylated compounds, so a cell can retain any type of molecule by phosphorylating them, *e.g.,* ATP and phosphorylated nucleosides are never released from the cells containing a normal intact plasma membrane. However, permeases for ATP and ADP do exist in a mitochondrial membrane to allow these molecules to move across it.

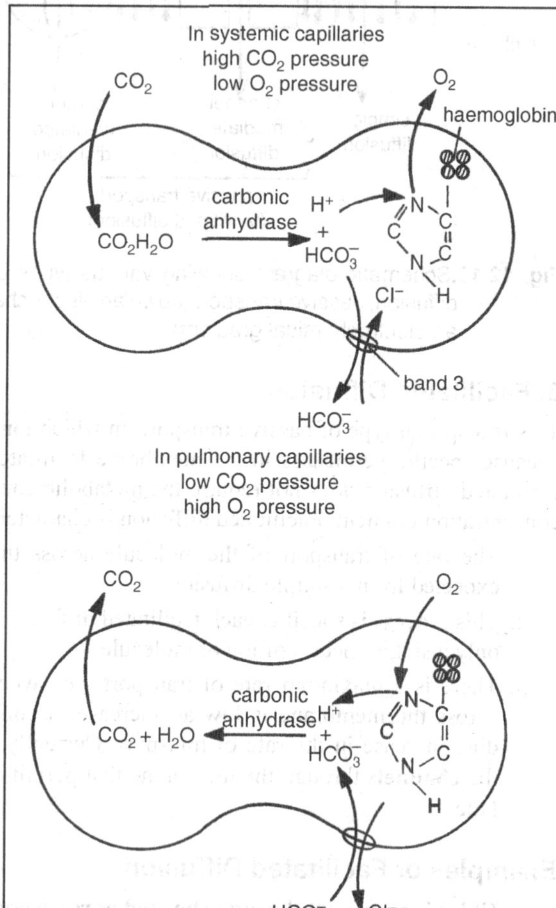

Fig. 12.14. Schematic drawing showing anion transport through the erythrocyte membrane in the capillaries and in the lungs. Band 3 protein (= anion exchange permease) catalyzes the exchange of the anions: Cl^- and HCO_3^- across the erythrocyte membrane.

***(iii)* Anion exchange permease of erythrocyte.** Band 3 polypeptide of plasma membrane of the erythrocytes and other cells is an **ion exchange permease protein** which catalyzes an one-for-one exchange of anions such as chloride (Cl^-) and bicarbonate (HCO_3^-) across the membrane (called **chloride shift;** erythrocyte has 100,000 times more permeability of Cl^- than other cells). The rapid flux of anions in the erythrocyte facilitates the transport in the blood of CO_2 from the tissues to the lungs. Waste CO_2 that is released from cell into the capillary blood, diffuses across the membrane of erythrocyte. In its gaseous form, CO_2 dissolves poorly in aqueous solutions such as blood plasma, but inside the erythrocyte the potent enzyme **carbonic anhydrase** converts it into a bicarbonate anion:

$$CO_2 + H_2O \underset{\text{anhydrase}}{\overset{\text{carbonic}}{\rightleftharpoons}} \underset{\text{proton}}{H^+} + \underset{\text{bicarbonate}}{HCO_3^-}$$

This process occurs while the haemoglobin in the erythrocyte is releasing its oxygen into the blood plasma. The removal of oxygen from haemoglobin induces a change in its conformation that enables a globin histidine (amino acid) side chain to bind to the proton produced by carbonic anhydrase enzyme. The bicarbonate anion formed by carbonic anhydrase is transported out of the erythrocyte in exchange for a chloride (Cl^-) anions:

$$\underset{\text{(in)}}{HCO_3^-} + \underset{\text{(out)}}{Cl^-} \rightleftharpoons \underset{\text{(out)}}{HCO_3^-} + \underset{\text{(in)}}{Cl^-}$$

B. Active transport. Active transport uses specific transport proteins, called **pumps,** which use metabolic energy (ATP) to move ions or molecules against their concentration gradient. For example, in both vertebrates and invertebrates, the concentration of sodium ion is about 10 to 20 times higher in the blood than within the cell. The concentration of the potassium ion is the reverse, generally 20 to 40 times higher inside the cell. Such a low sodium concentration inside the cell is maintained by the sodium-potassium pump or Na^+/K^+ – ATPase which was discovered in 1957 by **Jens Skou** in nerve cells of Krebs. There are different types of pumps for the different types of ions or molecules such as calcium pump, proton pump, etc.

Box 12.2

The sodium-potassium pump occurs only in animal cells. Plant cells have a **H^+-transporting, P-type,** plasma membrane pump. In plants, this proton pump plays a key role in the secondary transport of solutes, in the control of cytosolic pH and possibly in control of cell growth by means of acidification of the plant cell wall (**Karp** 2010).

Examples of Active Transport

1. **Na^+/K^+– ATPase.** It is an **ion pump** or **cation exchange pump** which is driven by energy of one ATP molecule to export three Na^+ ions outside the cell in exchange of the import of two K^+ ions inside the cell. Electrical organs of eels are found to be very rich in this enzyme or pump. N^+/K^+– ATPase is a transmembrane protein which is a dimer having two subunits: one smaller β unit which is a glycoprotein of 50,000 daltons M.W., which functions primarily in the maturation and assembly of the pump within the membrane (**Karp** 2010); and another larger α unit having 1,20,000 daltons M.W. The larger subunit of Na^+/K^+ – ATPase performs the actual function of cation transport. It has three sites on its extra cytoplasmic surface: two sites for K^+ ions and one site for the inhibitor **ouabain**. On its cytosolic side, the larger subunit contains three sites for three Na^+ ions and also has one catalytic site for a ATP molecule. It is believed that the hydrolysis of one ATP molecule somehow drives conformational changes in the Na^+/K^+– ATPase that allows the pump to transport three Na^+ ions out and two K^+ ions inside the cell (Fig. 12.15).

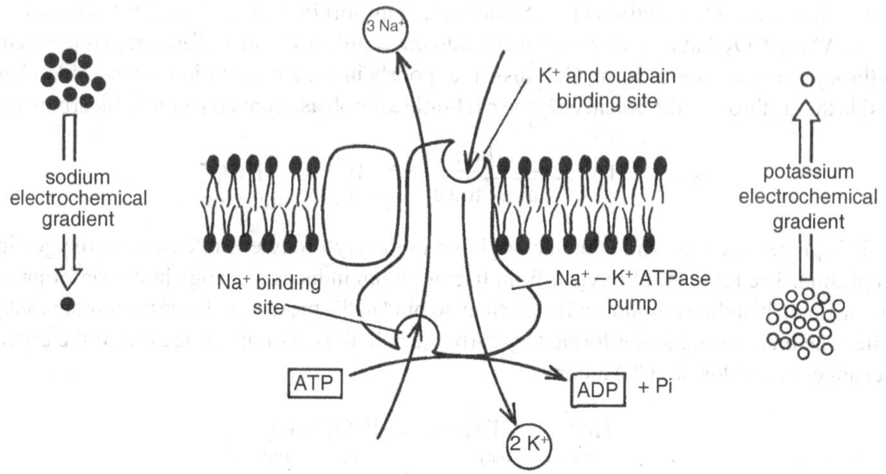

Fig. 12.15. The Na^+/K^+ – ATPase in the plasma membrane actively pumps Na^+ out and K^+ into a cell against their electrochemical gradients. For every molecule of ATP hydrolyzed inside the cell, 3 Na^+ ions are pumped out and $2K^+$ ions are pumped in.

Importance of Active Transport

Active transport moves the substances across the plasma membrane against their concentration gradient. It consumes a considerable amount of energy and is very essential for the cell since it enables the cell in:

1. The absorption of glucose against its concentration gradient in the uriniferous tubules.
2. The absorption of monosaccharides by the intestinal cells.
3. Maintenance of Na^+/K^+ – ATPase pump for establishing the unequal distribution of ions between inside and outside the plasma membrane. It helps in the conduction of nerve impulse in nerve cells and in maintaining a specific level of Na^+ in the blood.

Functions of Proteins of Membrane

Biological membranes (*e.g.*, plasma membrane and unit membrane) contain the following five different classes of proteins that carry out different functions:

1. Structural proteins. These form backbone of plasma membrane. These proteins provide elasticity and mechanical stability to plasma membrane. These have little catalytic activity.

2. Enzymes. These are catalytic proteins. They carry out different biological reactions in different parts of the cell and also in various types of cells.

3. Membrane transport proteins. These serve for the transport of polar molecules of various substances such as ions, monosaccharides, amino acids, nucleotides and certain metabolites across the plasma membrane in and out of the cell. These occur in many forms. Each transport protein is designed to transport a particular class of molecules. These are basically **multipass transmembrane proteins**, projecting on both the sides of lipid bilayer and traversing it many times. Membrane transport proteins are of following types:

(*i*) **Carrier proteins** bind the specific solute to be transported by active transport, *i.e.*, these undergo conformational change in order to transfer the solute across the membrane. These carrier proteins are of following three types:

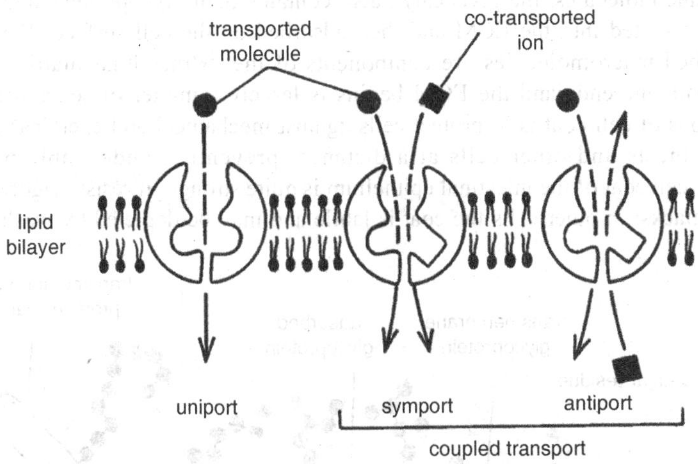

Fig. 12.16. Carrier proteins of membrane functioning as uniports, symports and antiports.

Uniport, symport and antiport. Those carrier proteins which simply transport a single solute from one side of the membrane to the other are called **uniports**. Others function as **coupled transporters**, in which the transfer of one solute depends on the simultaneous transfer of a second solute, either in the same direction (**symport**) or in the opposite direction (**antiport**). Both symport and antiport collectively form the **cotransport**. Most animal cells, for example, must take up glucose from the extracellular fluid, where the concentration of the sugar is relatively high, by passive transport through the glucose carriers (such as **D-hexose permease**) that operate as the uniports. By contrast, intestinal and kidney cells must take up glucose from the lumen of the intestine and kidney tubules, respectively, where the concentration of the sugar is low. These cells actively transport glucose by symport with Na^+ ions whose extracellular concentration is very high. The anion exchange permease of human erythrocytes operates as an antiport to the exchange of Cl^- for HCO_3^-.

(*ii*) **Channel proteins.** These are water filled pores that extend across the lipid bilayer. When these pores are open, they allow specific solutes to pass through them across the membrane.

4. Glycoproteins act as **cell receptors** (*e.g.*, **integrins**) and **cell antigens**.

5. Lipoproteins are **drug receptor proteins.**

12.6. CELL COAT OR GLYCOCALYX

Plasma membrane proteins, as a rule, do not protrude naked from the exterior of the cell. They are usually decorated by carbohydrates, which coat the surface of all eukaryotic cells. These carbohydrates occur as **oligosaccharide chains** covalently bound to membrane proteins (glycoproteins) and lipids (glycolipids). They also occur as the polysaccharide chains of integral membrane proteoglycan molecules. Proteoglycans, which consist of long polysaccharide chains linked covalently to a protein core, are found mainly outside the cell as part of the extra-cellular matrix or ECM (discussed in Chapter 14). But for some proteoglycans, the protein core either extends across the lipid bilayer or is attached to the bilayer by a GPI (= glycosyl phosphotidyl inositol) anchor.

The term **cell coat** or **glycocalyx** is often used to describe the carbohydrate-rich zone of the cell surface. This zone can be visualized by a variety of stains, such as **ruthenium red**, as well as by its affinity for carbohydrate-binding proteins called **lectins,** which can be labelled with a fluorescent dye or some other visible marker. The cell coat is a 10 to 20 nm thick layer and is in direct contact with the outer leaflet of the plasma membrane. In *Amoeba*, the cell coat is formed by fine filaments—5 to 8 nm thick and 100 to 200 nm long. Although most of the carbohydrate is attached to intrinsic plasma membrane molecules, the glycocalyx also contains both glycoproteins and proteoglycans that have been secreted into the ECM and then adsorbed on the cell surface (Fig. 12.17). Many of these adsorbed macromolecules are components of the extracellular matrix, so that where the plasma membrane ends and the ECM begins is largely a matter of semanitics. One of the possible functions of cell coat is to protect cells against mechanical and chemical damage and to keep foreign objects and other cells at a distance, preventing undesirable protein-protein interaction. The cell coat of the intestinal epithelium is quite strong—it resists vigorous mechanical and chemical attacks; in other cells the coat is labile and may be depleted by washing or enzyme exposure.

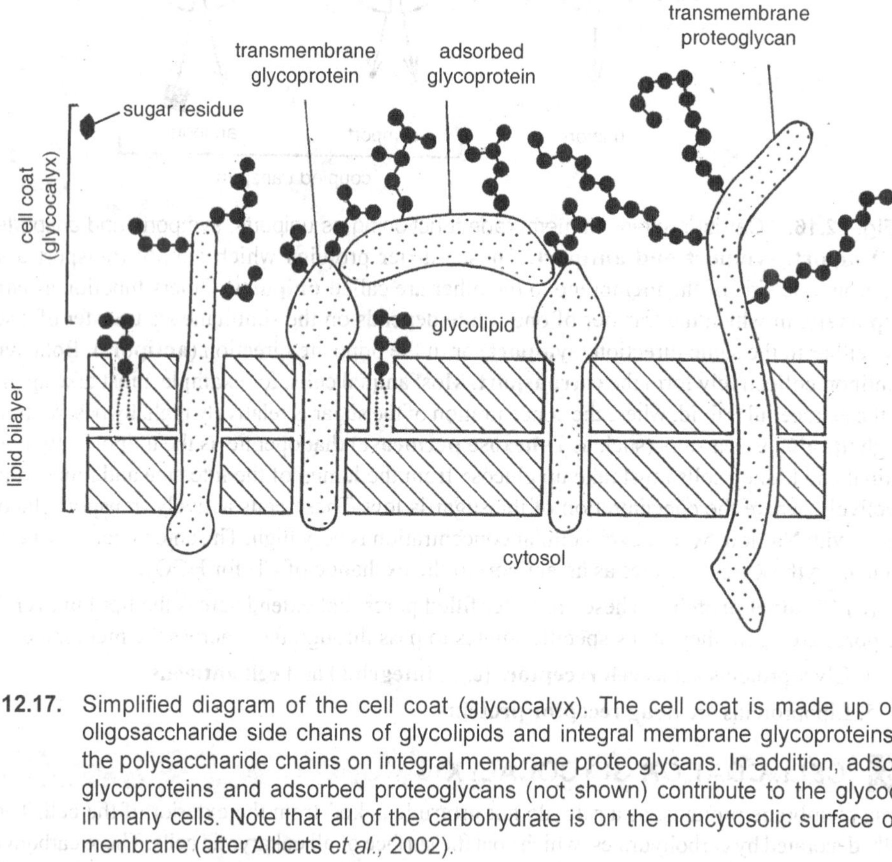

Fig. 12.17. Simplified diagram of the cell coat (glycocalyx). The cell coat is made up of the oligosaccharide side chains of glycolipids and integral membrane glycoproteins and the polysaccharide chains on integral membrane proteoglycans. In addition, adsorbed glycoproteins and adsorbed proteoglycans (not shown) contribute to the glycocalyx in many cells. Note that all of the carbohydrate is on the noncytosolic surface of the membrane (after Alberts *et al.*, 2002).

12.7. CELL SURFACE DIFFERENTIATIONS

In addition to specific molecular components plasma membrane consists of various cell surface differentiations such as microvilli, invaginations, etc.

1. Invaginations. The bases (inner ends) of certain cells, such as the cells of the kidney, perform active transportation and contain many **invaginations** or **infoldings** of the plasma membrane (Fig. 12.18). At the base of these folds, there develops a septa and, thus, narrow compartments of basal cytoplasm are formed. These infoldings contain many mitochondria. These mitochondria along with the enzymes of plasma membrane possibly provide energy rich compound, *viz.*, ATP to the plasma membrane for the active transportation of solutes.

2. Microvilli. Microvilli are finger-like, slender projections of plasma membrane (Fig. 12.19), which are found in mesothelial cells, hepatic cells, epithelial cells of intestine (**striated border**), uriniferous tubules (brush border), gall bladder, uterus, growing oocyte and yolk sac. Microvilli increase the effective surface of absorption. For example, a single epithelial cell of intestine may have as many as 3000 microvilli and in a square millimetre of intestine there may be 200,000,000. These microvilli are 0.6 to 0.8 μm long and 0.1 μm in diameter. The narrow spaces between the microvilli form a kind of sieve through which substances may pass during the process of absorption. Within the cytoplasmic core of a microvillus fine microfilaments are observed which in the underlying cytoplasm form a **terminal web.** The microfilaments contain actin and are attached to the tips of the microvilli by α-actinin; their function is to produce contraction of microvilli (Fig. 12.19).

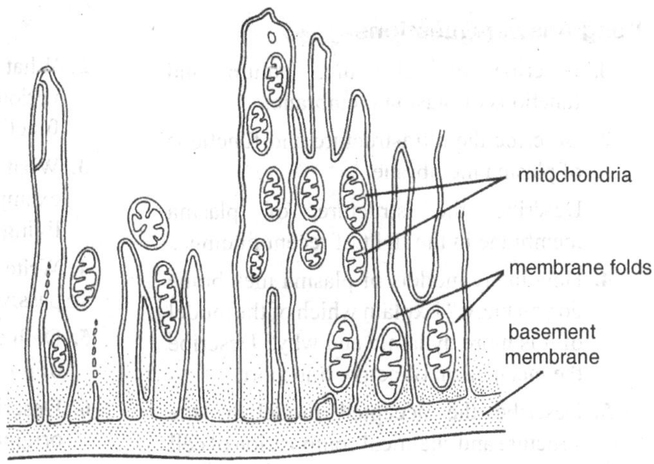

Fig. 12.18. Infoldings of plasma membrane.

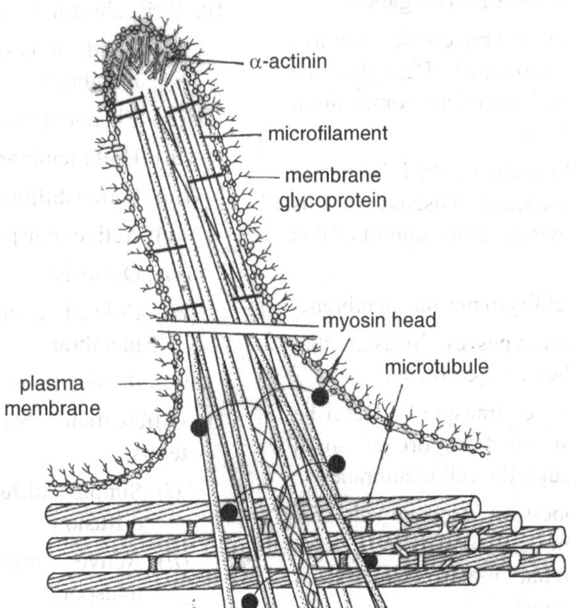

Fig. 12.19. Ultrastructure of a microvillus showing its different cytoskeleton components. Note glycoproteins embdded in the plasma membrane and connections between microfilaments and membrane proteins. Alpha actinin forms a non-muscle "Z line" to anchor the microfilaments. Myosin molecules are shown distributed along the base of microfilaments. Microtubules run parallel to the cell surface (after Dyson, 1978).

QUESTIONS

Long Answer Questions

1. Describe the structure, nature and functions of plasma membrane.
2. Describe the ultrastructure and functions of plasma membrane.
3. Describe the structure of plasma membrane in the light of recent findings.
4. How many models of plasma membrane do you know? Explain which of the model of it is more dynamic and why? Describe the mechanism of active transport.
5. Describe the accepted model of the structure and chemical composition of cell membrane. State briefly the functional significance of plasma membrane.
6. Describe briefly the ultrastructure of cell membrane based upon various theories.
7. Give an account of the fluid mosaic model of plasma membrane. Briefly describe the important functions of this organelle.
8. What are the characteristic features of a cell membrane? Describe the permeability and chemical composition of a cell membrane.
9. Describe fluid mosaic model of structure of plasma membrane. Discuss various experimental evidences in support of this model.
10. Describe the fluidity of plasma membrane.
11. Discuss various types of diffusion that occur across the cell membrane.
12. What is passive transport? Describe the mechanism of transport of small molecules through the cell membrane.
13. Why did Robertson used myelin for his unit membrane model of plasma membrane? What issues were raised against this model?

Short Answer Questions

1. 'Active transport moves solutes against their electrochemical gradients'– substantiate.
2. What is cell permeability? Describe the various types of permeability and their functions in living cells.
3. What is active transport? Give an example. Describe its importance in living organism.
4. Write what you know about active and passive transport.
5. Define osmosis and explain the movement of products through a plasma membrane.
6. Describe how is plasma membrane originated.
7. Describe various functions of proteins of biological membranes.
8. Describe the structure and importance of glycocalyx.
9. Describe the functions of plasma membrane.
10. Write short notes on the following:
    (*i*) Fluid mosaic model of plasma membrane
    (*ii*) Channel proteins
    (*iii*) Unit membrane
    (*iv*) Permeability of cell membrane
    (*v*) Active transport
    (*vi*) Osmosis
    (*vii*) Role of semipermeability of plasma membrane
    (*viii*) Passive transport
11. Differentiate between the following terms:
    (*i*) Simple diffusion and facilitated diffusion.
    (*ii*) Active transport and passive transport.
    (*iii*) Plasma membrane and endoplasmic reticulum.
12. Differentiate between uniport transport and passive diffusion.

13. Draw labelled diagram of the following (No description is required):
    (*i*) Fluid-mosaic model of plasma membrane.
    (*ii*) Microvillus.

Very Short Answer Questions

1. Who gave the concept of unit membrane?
2. Name the scientists who proposed bimolecular lipid layer model of plasma membrane.
3. Who proposed fluid mosaic model of plasma membrane?
4. Define passive transport.
5. Define diffusion.
6. What is flip-flop movement?
7. Name three substances passing the tunnels in the integral proteins of plasma membrane.
8. What causes diffusion of ions across the plasma membrane?
9. What is the name given to the proteins separated from plasma membranes of RBC?
10. Name any three sugars found in the glycocalyx.
11. Is glycocalyx found in prokaryotic plasma membrane?
12. How do large molecules of sugars and amino acids diffuse through plasma membrane quickly along the concentration gradient?
13. Name the integral components acting as a sodium-potassium pump.
14. Name the processes responsible for bulk transport of substances across the plasma membrane.
15. Name any two substances taken in by the plasma membrane through receptor-mediated endocytosis.
16. Name the organelle present in both prokaryotic and eukaryotic cells.

Mark the Statements True or False

1. Cell membrane is composed of lipoprotein.
2. Plasmolysis occurs in plant cells.
3. Plasma membrane is present in all cells.

Yes or No Questions

1. Thickness of plasma membrane ranges from 100 Å to 200 Å.
2. The concept of 'Unit membrane' was put forward by Robertson.

Multiple Choice Questions

1. Plasma membrane is made up of
   (*a*) glycoproteins and lipids
   (*b*) phospholipids and glycolipids
   (*c*) phospholipids and proteins
   (*d*) phosphoproteins and phospholipids
2. Phospholipid molecules are
   (*a*) acidic
   (*b*) alkaline
   (*c*) amphoteric
   (*d*) amphipathic
3. In lipid bilayer, the phospholipid molecules are arranged in such a fashion that
   (*a*) their hydrophobic ends are directed outwards
   (*b*) their hydrophobic ends are directed inwards
   (*c*) the hydrophobic ends of molecules of one layer make contact with the hydrophilic ends of the molecules of other layer
   (*d*) their hydrophobic ends are scattered indiscriminately
4. The hydrophobic ends of phospholipid molecules are
   (*a*) polar
   (*b*) nonpolar
   (*c*) neutral
   (*d*) bipolar

5. Spectrin, cytochrome-c and enzymes present on the plasma membrane are examples of
   (*a*) integral or intrinsic proteins
   (*b*) peripheral or extrinsic proteins
   (*c*) glycoproteins
   (*d*) lipoproteins
6. The membrane proteins that extend through both sides of lipid bilayer
   (*a*) acidic proteins
   (*b*) glycoproteins
   (*c*) intrinsic proteins
   (*d*) glycocolic acid
7. According to fluid mosaic model, the plasma membrane presents
   (*a*) a quasifluid structure
   (*b*) a micellar structure
   (*c*) a polar structure
   (*d*) a unit membrane structure
8. Fluidity of plasma membrane at body temperature is due to
   (*a*) fluidity of lipid bilayer
   (*b*) fluidity of glycolipids
   (*c*) presence of numerous unsaturated fatty acids in the lipid bilayer
   (*d*) presence of integral proteins in the form of glycophorins
9. Fluid mosaic model to explain structure of plasma membrane was proposed by
   (*a*) Hoffman
   (*b*) Singer and Nicolson
   (*c*) Robertson
   (*d*) Davson and Danielli
10. Unit membrane model of plasma membrane was proposed by
    (*a*) Nicolson
    (*b*) Danielli and Davson
    (*c*) Robertson
    (*d*) Mitchel
11. The enzyme essential for the transport of CO_2 as bicarbonate in blood is
    (*a*) carboxypeptidase
    (*b*) succinic dehydrogenase
    (*c*) carbonic anhydrase
    (*d*) lactase
12. Active transport is
    (*a*) movement of molecules against concentration gradient
    (*b*) movement of molecules along the concentration gradient
    (*c*) movement of molecules to and fro of plasma membrane
    (*d*) none of the above
13. Which protein of erythrocyte's plasma membrane prevents clumping of RBCs in the fine blood vessels?
    (*a*) spectrin
    (*b*) band 3 protein
    (*c*) ribophorin
    (*d*) ankyrin
14. Which integral protein of plasma membrane of human erythrocytes is exploited by *Plasmodium* for gaining entry inside RBCs?
    (*a*) band 3 protein
    (*b*) ribophorin
    (*c*) both (*a*) and (*b*)
    (*d*) none of these
15. MN blood group of humans is associated with which protein of plasma membrane of erythrocytes
    (*a*) spectrin
    (*b*) band 3 protein
    (*c*) ribophorin
    (*d*) ankyrin

ANSWERS

Very Short Answer Questions

1. Robertson.
2. Gorter and Grendel.
3. Singer and Nicholson.
4. Transport of substances through a membrane along their concentration gradients is called passive transport.
5. Movement of molecules and ions along an electrochemical gradient is diffusion.
6. Movement of lipid molecules from one layer to another of plasma membrane.
7. Water, oxygen and carbon dioxide (and also urea and glycerol).
8. Electrical difference between the inside and outside of a cell.
9. Tektins.
10. D–galactose, D–mannose, L–fucose, sialic acid, N–acetyl–D–glucosamine. (Note. students may choose any three of them.)
11. No.
12. Transportation of these substances across the plasma membrane takes place by specific carrier protein molecules of the plasma membrane.
13. Na^+/K^+–ATPase.
14. Pinocytosis, phagocytosis, endocytosis, receptor-mediated endocytosis and exocytosis.
15. Cholesterol and steroid hormones (and also antibodies and some proteins).
16. Ribosome or Plasma membrane.

True or False Statements

1. True **2.** True **3.** True.

Yes or No Questions

1. No **2.** Yes.

Multiple Choice Questions

1. (*c*) **2.** (*d*) **3.** (*b*) **4.** (*b*) **5.** (*b*) **6.** (*c*) **7.** (*a*)
8. (*c*) **9.** (*b*) **10.** (*c*) **11.** (*c*) **12.** (*a*) **13.** (*b*) **14.** (*b*)
15. (*c*).

13 The Cell Wall

The plant cell is always surrounded by a **cell wall** and this feature distinguishes them from animal cells. The cell wall is a non-living structure which is formed by the living protoplast (A plant cell without its cell wall is called a **protoplast**). In most of the plant cells, the cell wall is made up of cellulose, hemicellulose, pectin and protien. In many fungi, the cell wall is formed of chitin and in bacteria, the cell wall contains protein-lipid-polysaccharide complexes. Thus, the cell wall is a rigid and protective layer around the plasma membrane which provides the mechanical support to the cell. The cell wall also determines the shape of plant cells. Due to the shape of cell walls many types of plant cell as the parenchymatous, collenchymatous, etc., have been recognised.

The shape of the plant cell is dictated largely by its cell wall. When a living plant cell is treated with cell-wall-degrading enzymes to remove the wall, the resulting membrane-bound protoplast is invariably spherical.

The plant cell wall is a dynamic compartment that changes throughout the life of the cell. The new **primary cell wall** is born in the cell plate during cell expansion, in some cases it is more than a hundred-fold. The **middle lamella** forms the interface between the primary walls of neighboring cells. Finally, at differentiation, many cells elaborate within the primary cell wall a **secondary cell wall** (Fig. 13.1), building complex structure uniquely suited to the cell's function.

The plant cell wall is a highly organized composite of many different polysaccharides, proteins and aromatic substances. Some structural molecules act as fibers, others as a cross-linked matrix analogous to the glass-fibers- and plastic matrix in fiberglass. The molecular composition and arrangements of the wall polymers differ among species, among tissues of single species, among individual cells, and among regions of the wall around a single protoplast. Cell walls become specialized for the function of the approximately 40 cell types that plant comprise (see **Buchanan** *et al.*, 2000).

Not all specialized functions of cell walls are structural. Some cell walls contain molecules that affect patterns of development and mark a cell's position within the plant. Cell walls contain signalling molecules that participate in cell-cell and cell wall-nucleus communication. Fragments of cell wall polysaccharides may elicit the secretion of defense molecules, and the wall may become impregnated with protein and lignin to armor it against invading fungal and bacterial pathogens. In other instances, the cell walls participate in early recognition of symbiotic nitrogen fixing bacteria. Surface molecules on cell walls also allow plants to distinguish their own cells from foreign cells in pollen-style interactions.

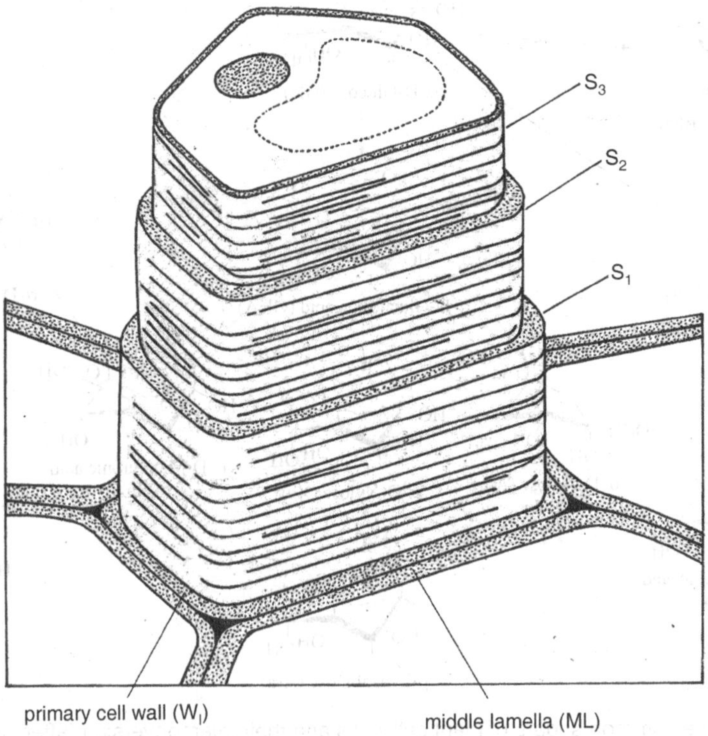

Fig. 13.1. When they have achieved their final size and shape, some cells elaborate a multilayered secondary cell wall within the primary wall. the lumen of cell is sometimes surrounded by several distinct kinds of secondary walls (here, S_1, S_2 and S_3) with the original primary cell wall (W_1) and the middle lamella (ML) constituting the outermost layers (after Buchanan *et al.*, 2000).

13.1. CHEMICAL COMPOSITION OF CELL WALL

1. Monosaccharides in cell wall. The monosaccharides in cell wall polymers are derived from glucose. **D-mannose (Man)** and **D-galactose (Gal)** are epimers (Box 13.1) of **D-glucose (Glc)** made by converting the C-2 and C-4 hydroxyl groups, respectively, from the equatorial position to axial positions (Fig. 13.2). The C-6 primary alcohols of these three sugars can be oxidized to a carboxylic acid group to form **D-mannuronic acid (ManA), D-galacturonic acid (GalA)** or **D-glucuronic acid (GlcA)**. Enzymatic removal of the carboxyl group from D-glucuronic acid forms the pentapyranose **D-xylose (Xyl)**, a sugar in which all of the carbons are part of the heterocyclic ring. A rare branched sugar, **D-apiose (Api)**, also is formed from D-GlcA. The C-4 epimer of D-xylose is **L-arabinose (Ara)**. The D to L conversion occurs, in this instance, the epimerization occurs at C-4, the last asymmetric carbon.

Box 13.1 Epimers

Two stereoisomers differing in configuration at one asymmetric center, in a compound having two or more asymmetric centers. In case of carbohydrates, any two sugars whcih differ from each other only in the configuration around a single asymmetric carbon atom other than the carbonyl carbon atom are called **epimers**. For example, glucose and galactose form an epimeric pair as they differ with respect to carbon 4 only. Similarly, glucose is also epimeric with mannose (see **Jain, J.L.**, 2012).

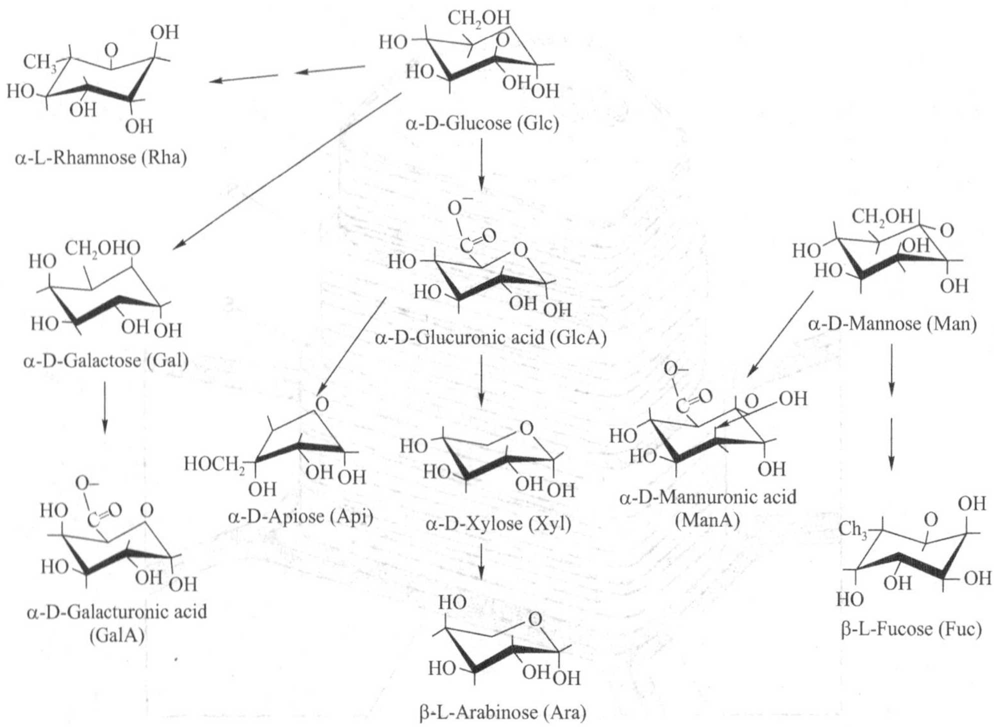

Fig. 13.2. The common sugars of plant cell walls and their interconversion (after Buchanan *et al.*, 2000).

Carbon 6 of some hexapyranoses also can be dehydrated to methyl groups, creating deoxysugars. In plants, the two major cell wall deoxysugars are 6-deoxy-L-mannose, called **L-rhamnose (Rha)** and 6 - deoxy-L-galactose, called **L-fucose (Fuc)**.

2. Polymers of specific sugars in cell wall. Sugars in polymers are always locked in pyranose or furanose rings. During sugar polymerization, the anomeric (Box 13.2) carbon of the sugar molecule is joined to the hydroxyl group of another sugar, a sugar alcohol, a hydroxyl-amino acid, or a **phenylpropanoid** compound in a **glycosidic-linkage.** A sugar can be attached to D-glucose at O-2, O-3, O-4, or O-6. Only the O-5 position is unavailable, because it constitutes part of the ring structure. Cell wall contains disaccharides such as cellobiose and lamenaribiose.

Box 13.2 Anomer

Anomer is a cyclic stereoisomer of a carbohydrate with isomerism involving only the arrangement of atoms or groups of aldehyde or ketone position. *Anomeric* is adjective of anomer.

A disaccharide can be described with respect to both linkage and anomeric configuration. For example, **cellobiose** is β-D-glucosyl-(1 $\rightarrow$ 4)-D-glucose. The anomeric linkage forms when the C-1 of one D-glucose residue is replaced by the equatorial hydroxyl at the C-4 position of the other D-glucose (Fig. 13.3). Only one D-glucose has locked the β-configuration; the other D-glucose is undesignated because its anomeric hydroxyl group is free to mutarotate in solution. Because aldehyde of this sugar is able to reduce copper under alkaline conditions, it is classically described as a **reducing sugar** and this end of even a very long polymer is called the **reducing end**. Branched polysaccharides will have a **nonreducing sugar** at end of each side chain and at the terminus of the backbone but only a single reducing end.

In cellobiose, the anormic carbon of one glucose is linked to the hydroxyl group farthest away from the anormic carbon of the other glucose. For the β-linkage to occur with the C-4 equatorial hydroxyl group, the sugars to be linked must be inverted almost 180° relative to each other; iteraction (*i.e.*, repetition) of this linkage produces a nearly linear molecule. In contrast, the units of laminaribiose or β-D-glycosyl (1 → 3)-D-glucose (Fig. 13.3) are linked somewhat askew; iteraction of this linkage produces a helical polymer. Polysaccharides are named after the principal sugars that constitute them. Most polysaccharides have a backbone structure, and the composition of this structure is indicated by the last sugar in the polymer's name. For example, *xyloglucan* has a backbone of C-4-linked glycosyl residues to which xylosyl units are attached, **glucuronoarabinoxylan**, has a backbone of C-4-linked xylosyl glucosyluronic acid and arabinosyl units are attached, and so forth.

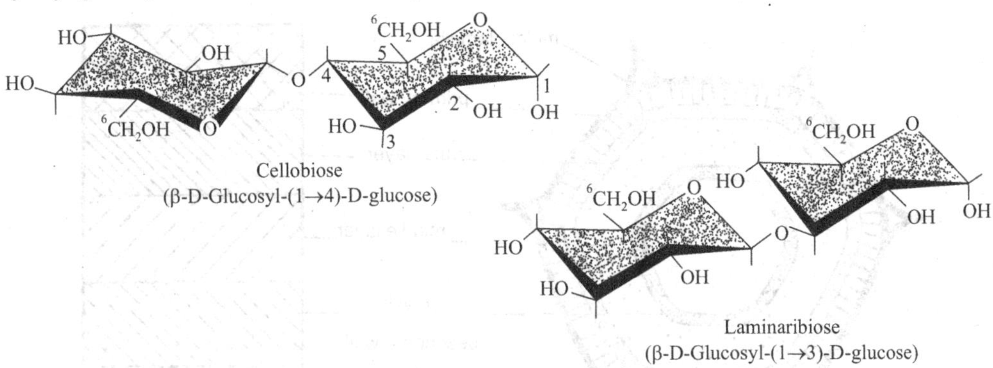

Fig. 13.3. Linkage structures of cellobiose and laminaribiose. The (1 → 4) β-D-linkage of cellobiose inverts the glycosyl unit about 180° with respect to each neighbor, whereas (1 → 3) β-D-linkage is only slightly askew (the shading of the glucose units is used to illustrate the 180° inversion) (Buchanan *et al.*, 2000).

3. Macromolecules in Cell Wall.

(*i*) Cellulose and Callose. Cellulose is the principal scaffolding component of all plant cell walls. It is the most abundant polysaccharide, accounting for 15 per cent to 30 per cent of the dry mass of all primary cell walls. Cellulose exists in the form of **microfibrils**, which are paracrystalline assemblies of several dozen (1 → 4) β-D-glucan chains hydrogen-bonded to one another along their length. In plants, on average each microfibril is 36 individual chains thick in cross-section, but microfibrils of algae can form either large, round cables or flattened ribbons of several hundred chains. Microfibrils of angiosperms have been measured to be betwen 5 and 12 nm wide in the electron microscope. Each (1 → 4) β-D-glucan chain may be just several thousand units (about 2 to 3 μm long), but individual chain begin and end at different places with microfibril to allow a microfibril to reach lengths of hundred of micrometers and to contain thousands of individual glucan chains. This structure is analogous to a spool of thread that consists of thousands of individual cotton fibers, each about 2 to 3 cm long.

Electron diffraction studies have shown that the (1 → 4) β-D-glucan chains of cellulose are arranged parallel to one another; that is, all of the reducing ends of the chains point in the same direction.

Callose. It differs from cellulose in consisting of (1 → 3) β-D-glucan chains, which can form helical duplexes and triplexes. Callose is made by a few cell types of specific stages of wall development, such as in growing pollen tubes and in the cell plates of dividing cells. It is also made in the response to wounding or to attempted penetration by invading fungal hyphae.

Box 13.3

Cellulose is a linear, unbranched polymer, consisting of straight polysaccharide chains made of glucose units linked by 1–4b–bonds (called **glycosidic bonds**). (**Note.** Complete hydrolysis of cellulose yields D-glucose and its partial hydrolysis yields disaccharide units, **cellobiose**). These are the glucan chains which by intra and intermolecular hydrogen bonding produce the structural units known as **microfibrils**, observable under electron microscopy and having toughness like the rubber. Each microfibril is ribbon-like flat fibre being 10 nm wide and 3 nm thick (or 25 to 30 nm in diameter) and is composed of about 2000 glucan chains in it. According to a classical estimate, each cellulose microfibril comprises three **micelles** or **elementary fibrils**: each elementary fibril contains about 100 cellulose molecules and each cellulose molecule is made up of 40 to 70 glucan chains (*i.e.*, one microfibril = 3 × 100 × 70 = 21000 glucan chains). Often numerous microfibrils get associated to form the **macrofibrils** having up to 0.5 μm diameter and observable under the light microscopy. Cellulose is synthesized by a wide variety of cells that include bacteria (*e.g.*, acetobacter, agrobacter and rhizobium), algae, fungi, cryptogams and seed plants.

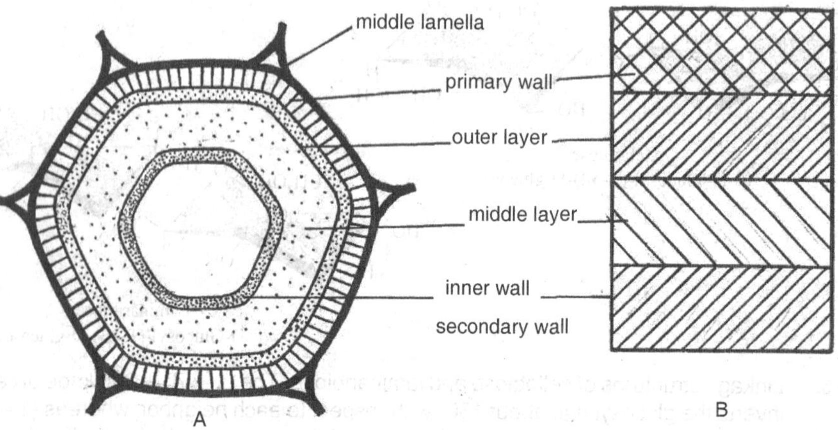

Fig. 13.4. Structural elements of cellulose.

***(ii)* Hemicellulose. Cross-linking glycans** are a class of polysaccharides that can hydrogen-bond to cellulose microfibrils: they may coat microfibrils but are also long enough to span the distance between microfibrils and link them together to form a network. Most cross-linking glycans are ofter called "hemicelluloses", a widely used but archaic term for all materials extracted from the cell wall with molar concentration of alkali, regardless of structure (see **Buchann** *et al.*, 2000).

The two major cross-linking glycans of all primary cell walls of flowering plants are **xyloglucans (XyGs)** and **glucuronoarabinose glycans (GAXs)**. The XyGs crosslink the cell walls of all dicots and about one-half of the monocots, but in the cell walls of "commelinoid" line of monocots, which include bromeliads, palms, gingers, cypresses and grasses, the major cross-linking glycan is GAX.

In the monocot order Poales, which contains the cereals and grasses, a third major cross-linking glycan, called "**mixed linkage**" (1 → 3) (1 → 4) β-D-glucans (β-glucans) distinguishes these species from the other commelinoid species. These unbranched polymers consist of 90% cellotriose and cellotetrose units in a ratio of about 2:1 and connected by (1 → 3) β-D-linkages. The cellotriosyl and cellotetrasyl units together make up cork-screw-shaped polymers about 50 residues long that are spaced by oligomers of four or more constiguous (1 → 4) Glc units.

Other much less abundant noncellulosic polysaccharides, such as **glucomannans**, **galactoglucomannans** and **galactomannas**, potentially interlock the microfibrils in some primary walls. These mannans are found in virtually all angiosperms examined.

***(iii)* Pectin matrix. Pectins**—a mixture of heterogenous, branched and highly hydrated polysaccharides rich in D-galacturonic acid—have been defined classically as material extracted from the cell wall by Ca^{2+}-chelators such as ammonium oxalate, FDTA, EGTA or cyclohexane diamine tetra-acetate. They are thought to perform many functions: determining wall porosity and providing charged surfaces that modulate wall pH and ion balance; regulating cell-cell adhesion at the middle lamella; and serving as recognition molecules that alert plant cells to the presence of symbiotic organisms, pathogens and insects. Particular cell wall enzymes may bind to the charged pectin netwrok, constraining their activities to local regions of the wall. By limiting wall porosity, pectins may affect cell growth, regulating the access of wall-loosening enzymes to their glycan substrates.

Pectin comprises of two fundamental constituents: **HGA** or **homogalacturonan** and **RGI** or **rhamnogalacturonan I**. HGAs are homopolymers of (1 → 4) α-D-Gal A that contain as many as 200 Gal A units and are about 100 nm long. There are two kinds of structurally modified HGAs, **xylogalacturonan** and **rhamnogalacturonan II (RG II)**. RG II has the richest diversity of sugars and linkage structures known, including apiose, aceric acid (3 – C′–carboxy-5-deoxy-L-xylose), 2-O-methyl fucose, 2-O-methyl xylose, kdo (3-deoxy D-manno-2-lyxo-2 heptulsaric acid). Its very highly conserved structure among flowering plants suggests an important function despite its low abundance in cell walls.

RG I is a rod-like heteropolymer of repeating (1 → 2) α-L-Rha-(1 → 4) α-D-Gal A disaccharide units.

Other polysaccharides, composed mostly of neutral sugars—such as **arabinans, galactans** and highly branched type 1 arabino-galactans (AGs).

4. Structural proteins of cell wall. There are four major classes of structural proteins, three of them named for their uniquely enriched amino acid: the **hydroxyl-proline-rich glycoproteins (HRGPs)**, the **proline-rich glycoproteins (PRPs)**, and the **glycine-rich protiens (GRPs)**. All of them are developmentally regulated, their relative amounts varying among tissues and species. Like other secretory proteins destined for the cell wall, the structural proteins are contranslationally inserted into the endoplasmic reticulum (ER). Thus, all mRNAs for cell wall proteins encode signal peptides that target the proteins to the secretory pathway.

The glycoprotein **extensin**, encoded by a multigene family, is one of the best-studied HRGPs of plants. Extensin consists of repeating Ser-$(Hyp)_4$ and Tyr-Lys-Tyr sequences that are important for secondary and tertiary structure. The repeating Hyp units project a "polyproline II" rod-like molecules.

The GRPs, some of which contain more than 70 per cent glycine, are predicted to be β-pleated sheets rather than rod-shaped molecules. GRPs are thought to form a plate-like structure at the plasma membrane – cell wall interface. The cell wall face of the pleated sheet contains an arrangement of aromatic amino acids of unknown function. GRPs constitute diverse group of glycoproteins that may function as structural elements inside the cell as well as in the cell wall.

The fourth major class of structural proteins, **AGPs (arabinogalactan proteins)** are more aptly name **proteoglycans** because they can be more than 95 per cent carbohydrate. AGPs constitute a broad class of molecules located in Golgi-derived vesicles, the plasma membrane, and the cell wall. The site of glycosylation of the AGPs remains unknown but is likely to occur in the Golgi apparatus because it involves the attachment of large, highly branched galactan chains and subsequent decoration with Ara units.

No clear-cut role has been described for any of the structural proteins. Some of these proteins may have more indirect architectural functions (*e.g.*, as nucleation sites for wall assembly, or may directly bind polymers together like the clamps that interlock scaffolding poles).

5. Aromatic substances of cell wall. The primary walls of the commelinoid orders of monocots and the chaenopodiaceae (such as sugar beet and spinach) contain significant amounts of aromatic substances in their nonlignified cell walls—a feature that makes them fluorescent under ultraviolet (UV) light. A large fraction of plant aromatics consists of **hydroxycinnamic acids**, such as **ferulic** and **p-coumaric acids**. Hydroxycinnamic acids are known to be reduced in the plant to **hydroxycinnamoyl alcohols** which form common precursors for lignin and lignan structures.

Other Molecules and Cell Wall

1. **Mannan** is a homopolysaccharide of mannose and is found in the cell wall of yeast, fungi and bacteria.
2. **Agar** is a polysaccharide, found in the cell wall of sea weeds and containing D- and L-galactose residues.
3. **Lignin** is a biological plastic and non-fibrous material. It occurs only in mature cell walls and is made of an insoluble hydrophobic aromatic polymer of phenolic alcohols (*e.g.*, hydroxyphenyl propane). During the later stages of growth of angiosperms large amounts of lignin are laid down in plant cells, and walls harden into wood.
4. The **chitin** is a polymer of glucosamine.
5. **Cutin** is also a biological plastic and is made of fatty acids (waxes). **Suberin** is a water-resistant substance, comprising of fatty acids and found in the cork and cell wall of many plants. **Sporopollenin** is a lipoidal polymer forming tough wall (with species-specific patterns) of pollen grains.
6. **Mineral** deposits occur in cuticle in the form of calcium and magnesium carbonates and silicates. Deposits of calcium compounds are found in the cell wall of cruciferous and cucurbitaceous plants. Silicate deposits are common in the cell wall of Graminae family.

13.2. STRUCTURE OF CELL WALL

The cell wall is complex in nature and is differentiated in the following layers (Fig. 13.5):

(*i*) Primary cell wall;

(*ii*) Secondary cell wall;

(*iii*) Tertiary cell wall.

(*i*) Primary cell wall. The first formed cell wall is known as primary cell wall. It is the outermost layer of the cell and in the immature meristematic and parenchymatous cells it forms the only cell wall. The primary cell wall is comparatively thin and permeable. Certain epidermal cells of the leaf and the stem also possess the cutin and cutin waxes which make the primary cell wall impermeable. The primary cell wall of the yeast and the fungi is composed of the chitin.

(*ii*) Secondary cell wall. The primary cell wall is followed by secondary cell wall. The secondary cell wall is thick, permeable and lies near the plasma membrane of the tertiary cell wall, if the latter occurs. It is composed of the three concentric layers (S_1, S_2 and S_3) which occur one after another. Chemically the secondary cell wall is composed of compactly arranged macrofibrils of the cellulose, in between which sometimes occurs lignin as a inter-fibrillar material.

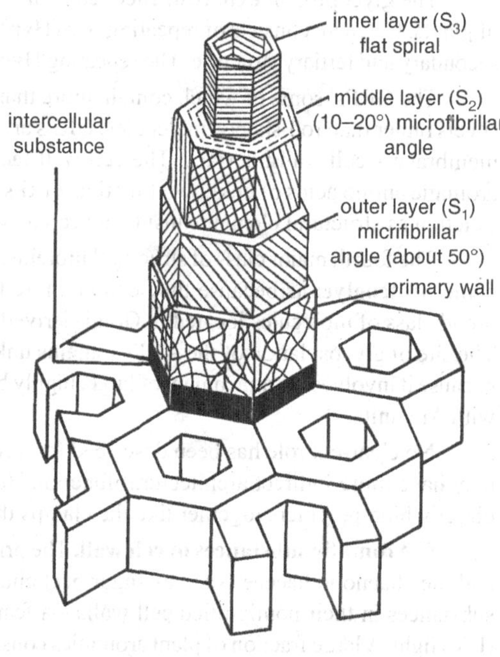

Fig. 13.5. Structure of a cell wall showing middle lamella, primary cell wall and three regions (S_1, S_2 and S_3) of the secondary cell wall.

(*iii*) Tertiary cell wall. In certain plant cells, there occurs another cell wall beneath the secondary cell wall which is known as tertiary cell wall. The tertiary cell wall differs from the primary and secondary cell wall in its morphology, chemistry and staining properties. Besides the cellulose, the tertiary cell wall consists of another chemical substance known as the xylan.

Middle lamella. The cells of plant tissues generally remain cemented together by an intercellular matrix known as the **middle lamella**. The middle lamella is mainly composed of the pectin, lignin and some proteins.

Ultrastructure. Electron microscopy has shown that the cell wall is constructed on the same architectural principle which applied well in the construction of animal bones and such common building materials as fibre glass (plastic + glass fibres) or reinforced concrete (concrete + metal framework), *i.e.*, strong fibres (*e.g.*, cellulose microfibrils) resistant to tension embedded in an amorphous matrix (comprising hemicellulose, pectin and proteins) resistant to compression. In the primary cell wall, the fibres and matrix molecules are cross-linked by a combination of covalent bonds and non-covalent bonds to form a highly complex structure whose composition is generally cell-specific (Fig. 13.6). In fact, hemicellulose molecules (*e.g.*, **xyloglucans**) are linked by hydrogen bonds to the surface of the cellulose microfibrils. Some of these hemicellulose molecules are cross-linked in turn to acidic pectin molecules (*e.g.*, rhamno-galacturonans) through short neutral pectin molecules (*e.g.*, arabinogalactans). Cell wall glycoproteins are tightly woven into the texture of the wall to complete the strucutre of matrix.

In the multilamellar secondary cell wall, cellulose microfibrils are laid down in layers, the microfibrils of each layer running roughly parallel with each other but at an angle to those in other layers (Fig. 13.5).

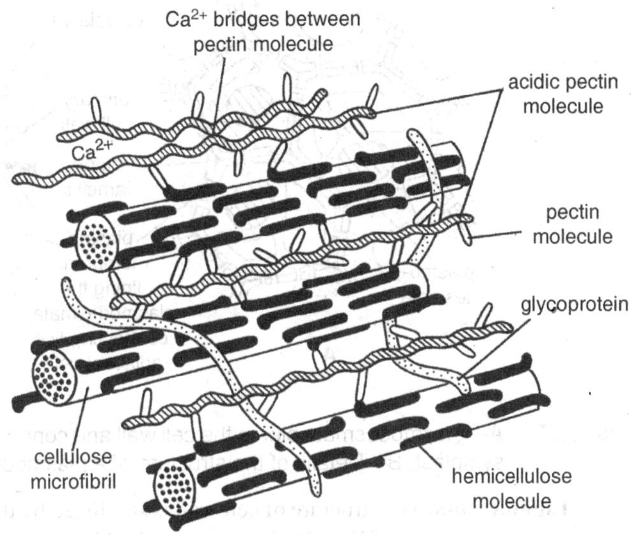

Fig. 13.6. Ultrastructure of primary cell wall showing interconnections between the two major components of the primary cell wall: the cellulose microfibrils and the matrix.

Plasmodesmata. Every living cell in a higher plant is connected to its living neighbours by fine cytoplasmic channels, each of which is called a **plasmodesma** (Gr., *desmos* = ribbon, ligament; plural, **plasmodesmata**) which pass through the intervening cell walls (Fig. 13.7). The plasma membrane of one cell is continuous with that of its neighbour at each plasmodesma. A plasmodesma is a roughly cylindrical, membrane-lined channel with a diameter of 20 to 40 nm. Running from cell to cell through the centre of most plasmodesmata is a narrower cylindrical structure, the **desmotubule**, which remains, continuous with elements of the SER membranes of each of the connected cells. Between the outside of the desmotubule and the inner face of the cylindrical plasma membrane is an **annulus of cytosol**, which often appears to be constricted at each end of the plasmodesmata. These constrictions may regulate the flux of molecules through the annulus that joins the two cytosols.

Plasmodesmata are formed around the elements of smooth endoplasmic reticulum (SER) that become trapped during cytokinesis (of mitotic cell division) within the new cell wall that will bisect the parental cell. Plasmodesmata function in intercellular communication, *i.e.*, they allow molecules to pass directly from cell to cell. For example, plasmodesmata are especially common and abundant in the walls of columns of cells that lead toward sites of intense secretion, such as in nectar-secreting glands (trichomes of *Abutilon* nectaries). In such cells there may be 15 or more plasmodesmata per square micrometer of wall surface, whereas there is often less than 1 per square micrometer in other cell wall.

In fact, experimental evidence has suggested that the plasmodesmata mediate transport between adjacent plant cells, much as gap junction of animal cells. They allow the passage of molecules with molecular weights of less than 800 daltons. Transport through the plasmodesmata is also found under complex regulations which may involve Ca^{2+} and protein phosphorylation. Thus, no dye movement is observed between the cells of the root cap and the root apex or between epidermal and cortical cells in either roots and shoots. However, certain plant viruses such as TMV can enlarge plasmodesmata in order to use this route to pass from cell to cell. Tobacco mosaic virus is known to synthesize a protein, called **P30** (30,000 dalton M.W.) that nullifies the normal regulatory mechanisms of plasmodesmata.

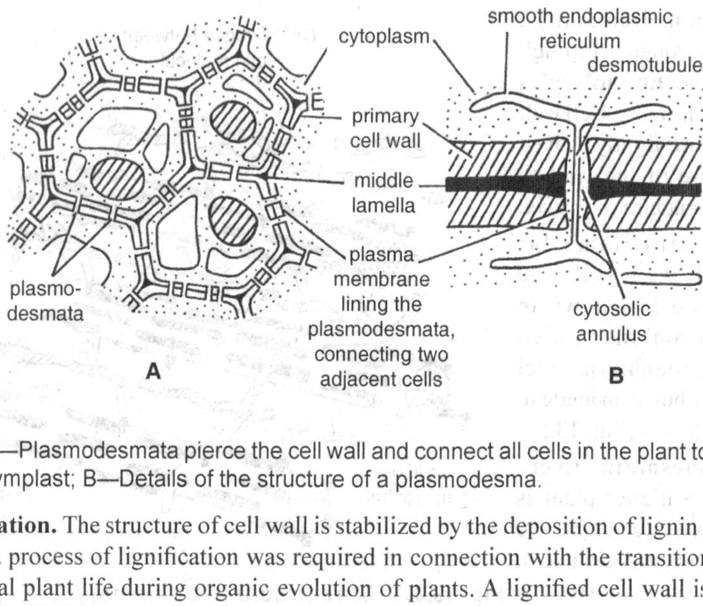

Fig. 13.7. A—Plasmodesmata pierce the cell wall and connect all cells in the plant together to form symplast; B—Details of the structure of a plasmodesma.

Lignification. The structure of cell wall is stabilized by the deposition of lignin in the cell wall matrix. Such a process of lignification was required in connection with the transition from aquatic to the terrestrial plant life during organic evolution of plants. A lignified cell wall is composed of microfibrils of cellulose embedded in the matrix containing large amount of lignin. Usually the primary cell wall becomes more lignified than secondary cell wall.

Primary Pit Fields and Pits

Primary pit fields are found in cells which contain only the primary cell wall. In certain areas the primary cell wall is thin and contains a group of pores. These areas are called **primary pit fields**. In some plants the secondary cell wall of sclerenchyma (fiber) and xylem cell (tracheids and vessels) has depressions or cavities called **pits** (Fig. 13.8; Box 13.4). Pits are of two types, simple pits (found in fiber) and bordered pits (xylem). Pits are usually found in nonliving cells such as tracheids and fibers whose function is conduction and support.

Box 13.4

1. **Pit.** A cavity in the secondary wall of a plant cell, formed where secondary deposition has failed to occur and primary cell wall remains uncovered; two main types are simple pits and bordered pits.

2. **Bordered pit.** A wood cell pit having the secondary cell wall arched over the cavity of the pit.
3. **How pit is formed?** No lignin is laid down where plasmadesmata were present in the original cell wall. These non-lignified areas are known as **pits** and they allow water to pass sideways between one xylem tracheid/vessel and the next. These lignified dead and hollow cells lack any cell content to restrict the flow of water.

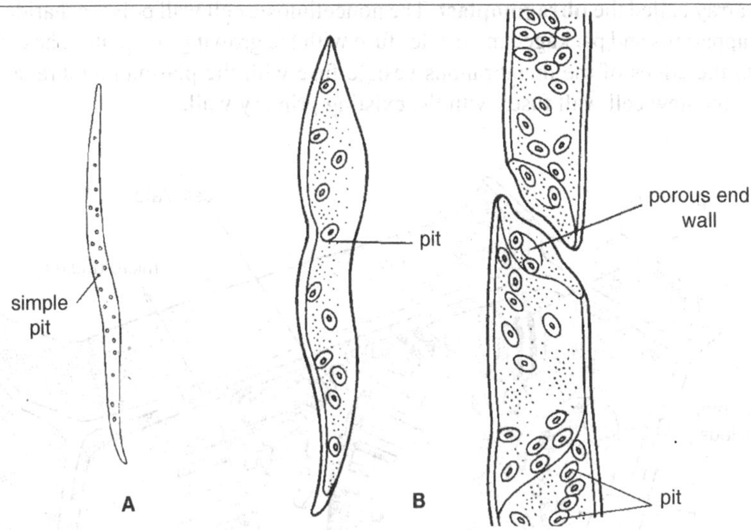

Fig. 13.8. A—Simple pits of sclerenchyma cell. B—Simple pits of xylem elements.

In a **bordered pit** (Fig. 13.9B) the secondary cell wall projects over the cavity of the pit, enclosing a **pit chamber** which opens outside through a **pit aperture**. In a **simple pit** (Fig. 13.9A), there is no projecting margin. Adjacent pits are separated by the middle lamella and the primary cell wall, which together form the **pit membrane**. The pit membrane may have a thickening called the **torus** which is formed by circular deposition of microfibrils. The part of the membrane surrounding the torus is called the **margo**. It is flexible and has radially arranged microfibrils which are anchored in the surrounding primary cell wall and hold the torus in position. The margo has small openings between the fibrils, through which aqueous substances flow from one cell to another. If the torus is pressed against the pit aperture due to pressure of aqeous substances in the cell, it stops the passage of substances. The pit and the pitted areas help in the passage of material from one cell to another.

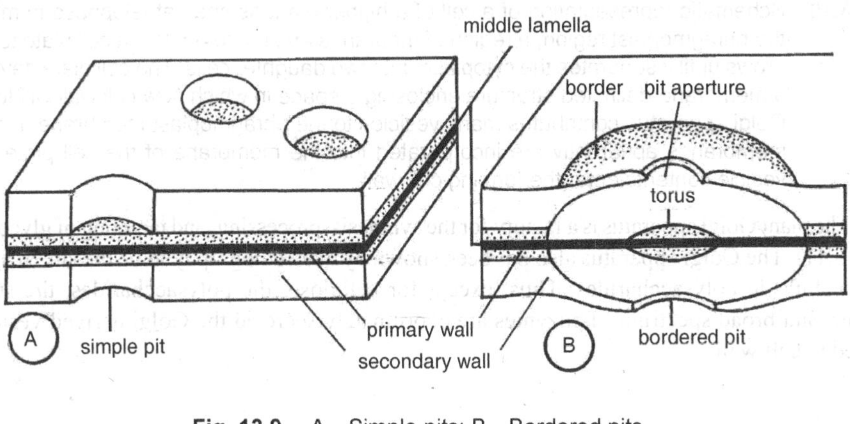

Fig. 13.9. A—Simple pits; B—Bordered pits.

13.3. CELL WALL BIOSYNTHESIS AND ASSEMBLY

1. Cell walls are born in the developing cell plate. Cell walls originate in the developing cell plate. As plant nuclei complete division during telophase of the mitotic cell cycle, the **phragmosome**, a flattened membranous vesicle containing cell wall components, forms across the cell within a cytoskeletal array called the **phragmoplast**. The noncellulosic cell wall polysaccharides synthesized in the Golgi apparatus and packaged in vesicles fuse with the growing cell plate. The cell plate grows outward until the edges of the membranous vesicle fuse with the plasma membrane, creating two cells. Finally, the new cell wall fuses with the existing primary wall.

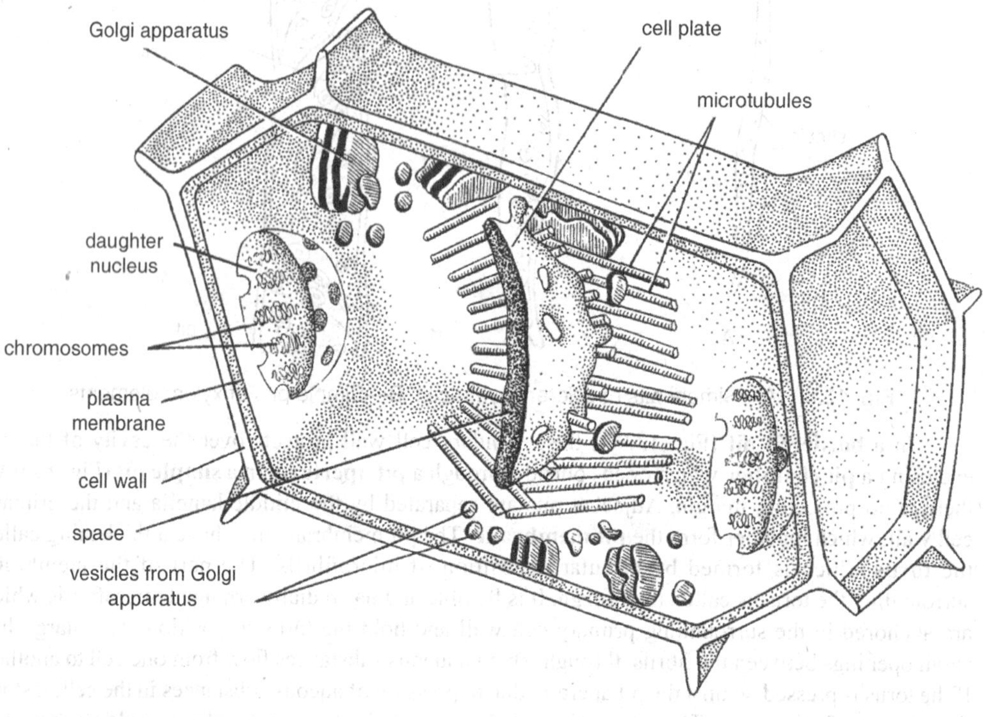

Fig. 13.10. Schematic representation of a cell of a higher plant as seen at telophase in mitosis. In the phragmoplast region, a region of mebranes and microtubules, a cell plate forms and grows until it separates the cytoplasm into two daughter cells. The cell plate develops as a membrane-delimited structure enclosing a space in which new cell wall will form. The Golgi apparatus contributes many vesicles to the phragmoplast membrane; the vesicle membranes apparently are incorporated into the membrane of the cell plate and the vesicle contents enter the forming cell wall.

The plant Golgi apparatus is a factory for the synthesis, processing, and targeting of glycoproteins (Fig. 13.11). The Golgi apparatus also has been shown by autoradiography to be the site of synthesis of noncellulosic polysaccharides. Thus, except for cellulose, the polysaccharides, the structural proteins, and broad spectrum of enzymes are coordinately secreted the Golgi derived vesicles and targeted to cell wall.

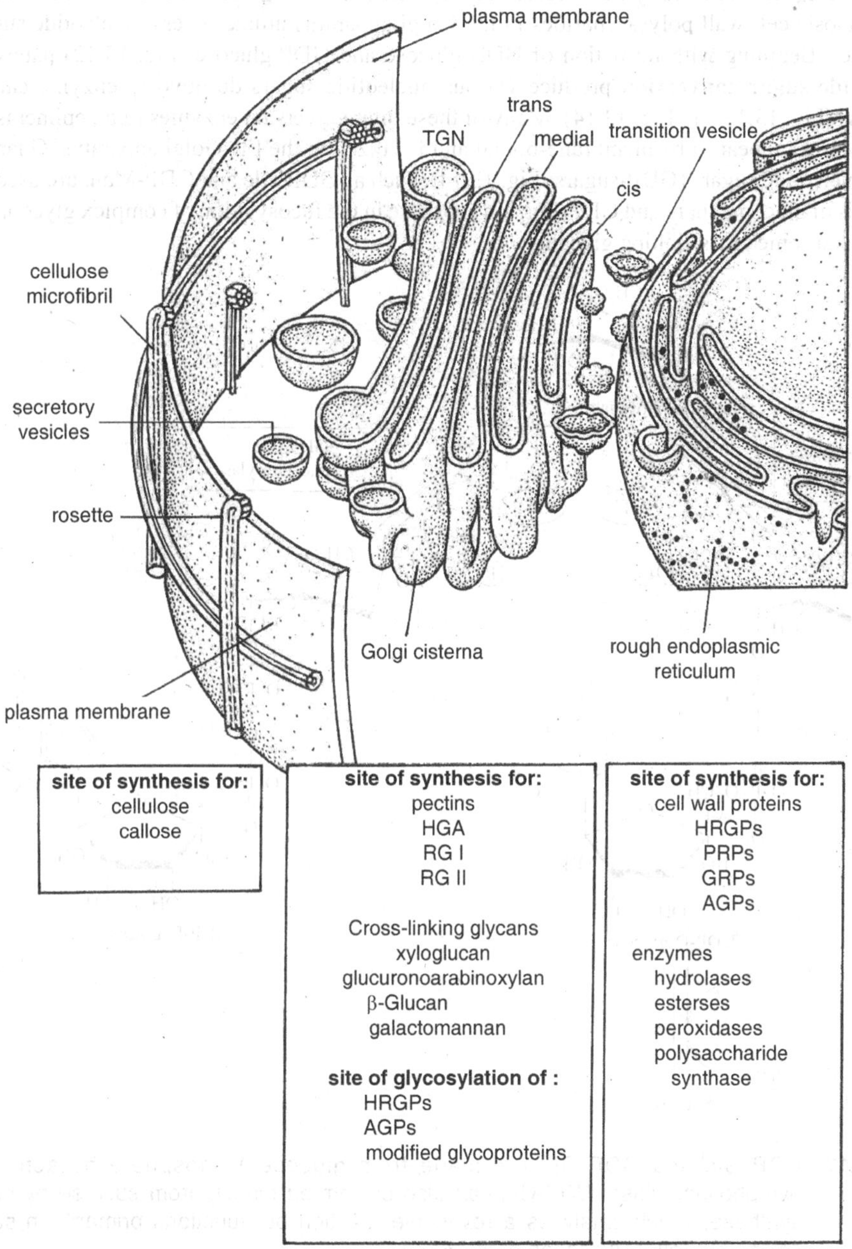

site of synthesis for:
cellulose
callose

site of synthesis for:
pectins
HGA
RG I
RG II

Cross-linking glycans
xyloglucan
glucuronoarabinoxylan
β-Glucan
galactomannan

site of glycosylation of :
HRGPs
AGPs
modified glycoproteins

site of synthesis for:
cell wall proteins
HRGPs
PRPs
GRPs
AGPs

enzymes
hydrolases
esterses
peroxidases
polysaccharide synthase

Fig. 13.11. Biosynthesis of the cell wall requires a coordination of the synthesis of cellulose microfibrils at the plasma membrane surface, with the synthesis and glycosylation of proteins and wall-modifying enzymes at the rough ER and the synthesis of all non cellulosic polysaccharides at the Golgi apparatus. Material destined for the cell wall is packaged into secretory vesicles, transported to the cell surface, and integrated with the newly synthesized microfibrils. Assembly of new wall stratum is estimated to begin no more than in 10 glucose residues of a cellulose chain are made (after Buchanan *et al.*, 2000).

2. Golgi-localized enzymes interconvert the nucleotide sugars. The reactions that synthesize noncellulosic cell wall polysaccharides in the Golgi apparatus utilize several nucleotide sugars as substrates. Begining with formation of UDB-glucose and GDP-glucose (Fig. 13.12) pathway for **nucleotide sugar conversion** produce various nucleotide sugars de novo in enzyme catalyzed reactions (Fig. 13.13 and Fig. 13.14). Many of these interconversion enzymes (*e.g.*, epimerases and dehydratases) appear to be membrane-bound and localized to the ER-Golgi apparatus. Guanosine-based nucleotide sugars (GDP-sugars; Fig. 13.14), such as GDP-Glc and GDP-Man, are used in the synthesis of glucomannan, and GDP-Fuc is a substrate in the fucosylation of complex glycoproteins, pectins and some cross-linking glycans.

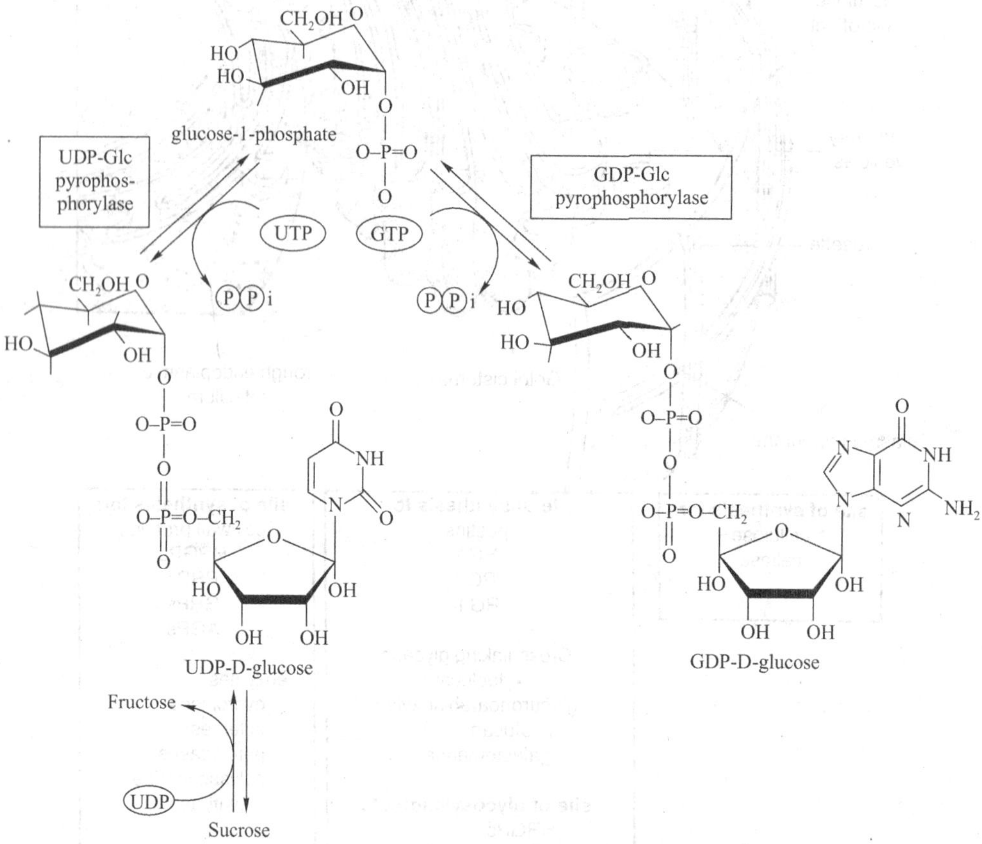

Fig. 13.12. UDP-Glc and GDP-Glc are made from glucose-1-phosphate by action of a pyrophosphorylase. UDP-Glc can also be formed directly from sucrose by sucrose synthase, which catalyzes a reversible reaction but functions primarily in sucrose catabolism (after Buchanan *et al.*, 2000).

L-arabinose is in the furanose ring conformation in most plant polymers containing this sugar, including GAX, 5-linked arabinans, AGP, and extensin, whereas UDP-Ara is exclusively in the pyranose form. An arabinosyl-transferase may differ from other glycosyl-transferases in its ability to permit ring rearrangement before the sugar is added to the polymer.

Distinct topographic location of the polysaccharide synthase enzymes on or within the Golgi membranes has not been established. For branched polysaccharides, the nucleotide sugar substrate used for synthesis of the backbone may be donated from either the cytosolic or the lumenal side of the Golgi, but the branch units probably are added only from the lumenal side. Hence, synthesis of complex polysaccharides must be coordinated with transport of some of the nucleotide sugars into the Golgi apparatus.

Formation of Nucleotide Sugars. It occurs via two distinct pathways:

(*i*) **The de novo synthesis pathways** initially produce the full array of nucleotide sugars, which are then used as substrate for the synthesis of polysaccharides, glycoproteins, and several other glycosylation reactions.

(*ii*) **Salvage pathways.** Several monosaccharides other than Glc may be incorporated into nucleotide sugars via salvage pathways involving C-1 kinases and nucleotide diphosphate (NDP)-sugar pyrophosphorylases (see Fig. 13.13 and Fig. 13.14). The salvage pathways are essential for reuse of these monosaccharides after their hydrolysis from polymers during cell wall assembly and during turnover of the cell wall. Some sugars, such as Rha and Xyl, do not have C-1 kinases (Fig. 13.13) and the carbons must be reused via other pathways. Xyl carbons are returned by way of the pentose phosphate pathway after isomerization to xylulose.

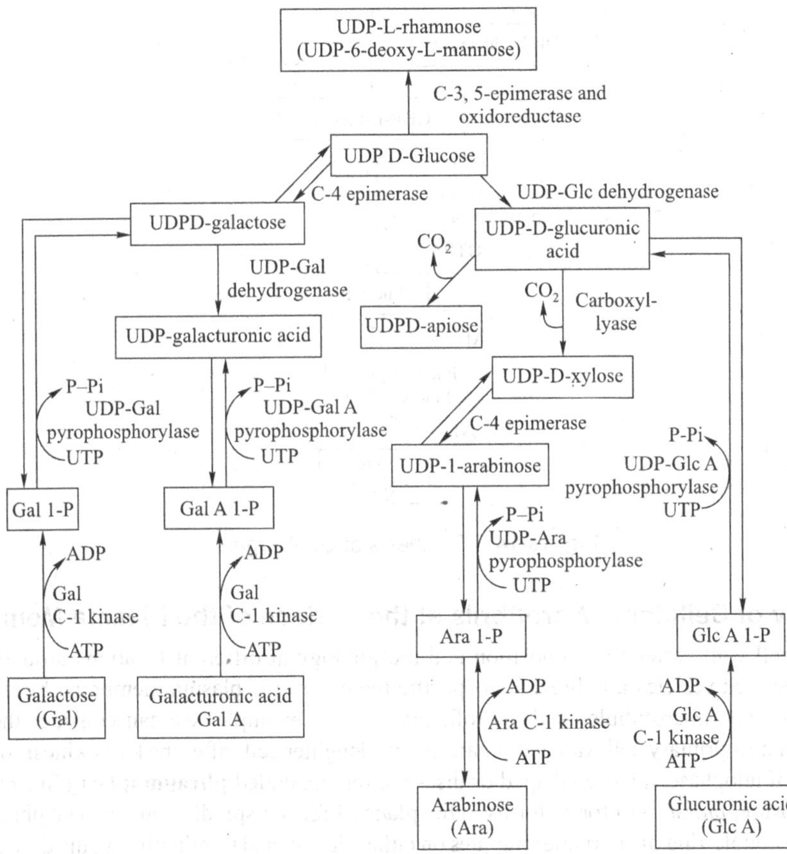

Fig. 13.13. Synthesis of UDP sugars.

In Vitro Synthesis of Noncellulosic Polysaccharides

Many reports of the synthesis of noncellulosic polysaccharides in vitro include use of membrane preparations enriched in Golgi membranes. In mixed-membrane preparations containing plasma membrane, Golgi, and UDP-Glc as a substrate, the predominant product is the (1 → 3) β-D-glucan, callose—a polymer thought to be made a default product by damaged cellulose synthase. This reaction is activated by calcium ions, the concentrations of which increase markedly in cells that have been damaged. Thus, the other noncellulosic polysaccharides of interest must be detected and quantified in the presence of huge background of callose.

By treating polysaccharides with sequence dependent **glycanases**, researchers can detect the characteristic unit structures despite the presence of wound-induced callose. This technique, which verifies that isolated membranes are capable of synthesizing unit structures identical to those made in vivo, has been particularly useful for studying in vitro synthesis of XyG.

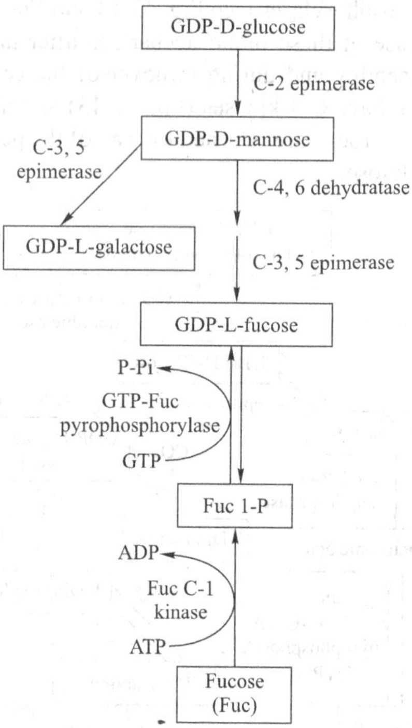

Fig. 13.14. Synthesis of GDP sugars.

Assembly of Cellulose Microfibrils at the Surface of the Plasma Membrane

Since, the cell wall varies in composition and morphology at different locations around the cell, the Golgi-derived vesicles are directed at specific regions of the plasma membrane by the help of cytoskeleton (*i.e.*, microtubules and microfilaments). For example, we can consider the case of formation of new primary cell wall that separates two daughter cells after the karyokinesis of mitosis. At the end of telophase, a barrel-shaped or disc-like region, called **phragmoplast** (Gr., *phragma* = hedge enclosure; *plasso* = to form) forms in the plane of former spindle equator. The phragmoplast comprises a double-ring of short microtubules on either side of, and terminating at, the division plane, and a set of microfilaments are coaligned with the microtubules. Golgi derived vesicles containing cell wall precursors, especially **pectin**, are guided inward along these oriented microtubules until they reach the phragmoplast, where they fuse with one another to form the cell plate (Fig. 13.10). Cell plate, at this stage, comprises following structures: (1) central middle lamella; (2) primary cell walls on both sides of middle lamella; and (3) plasma membrane lining cytoplasm of each daughter cell. Ultimately, the microtubular ring moves centrifugally outward as Golgi vesicles continue to add precursors to the growing cell plate. The cell plate fuses with the mother cell wall to create two separate daughter cells. It is not clear which component of the phragmoplast—the microtubules or the actin filaments (or both)—are responsible for the movement and guidance of the Golgi derived vesicles.

Mechanism of Cell Wall Formation

The mechanism of cell wall formation includes the following steps:

1. Formation of matrix. Most cell wall matrix components are transported via vesicles derived from the Golgi apparatus and secreted by exocytosis at the plasma membrane. Golgi apparatus of the plant cells is involved mainly in producing and secreting a very wide range of extracellular polysaccharides, rather than the glycoproteins typical of animal cells. This is perhaps due to following facts:

(*i*) Each of the polysaccharides of cell wall matrix is made of two or more sugars;

(*ii*) At least 12 types of monosaccharides are used in their polymerization;

(*iii*) Most of these polysaccharides are branched; and

(*iv*) Many covalent modifications are introduced in the polysaccharides after they are synthesized. It is estimated that several hundred different enzymes are engaged in the assembly of the polysaccharide component needed to form a typical primary cell wall. Most of these enzymes are found in the endoplasmic reticulum and Golgi apparatus. Some enzymes, which are concerned with later covalent modifications of the polysaccharides, are found in the cell wall itself.

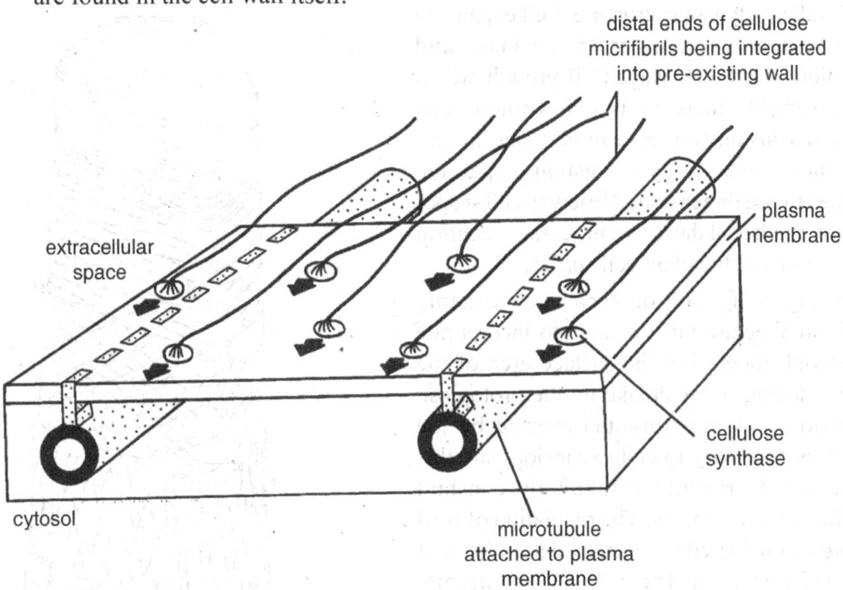

Fig. 13.15. A model explaining the mode of orientation of newly deposited cellulose microfibrils according to orientation of cortical microtubules.

2. Synthesis and orientation of cellulose microfibrils. In most plants, cellulose is synthesized on the external surface of the cell by a plasma membrane bound enzyme complex, called **cellulose synthetase** which uses a sugar nucleotide precursor supplied from the cytosol, probably UDP-glucose (Box 13.5). As they are being synthesized, the nascent cellulose chains spontaneously assemble into **microfibrils** that form a layer on the surface of the plasma membrane (a lamella) in which all the microfibrils have more or less the same alignment (**Note.** Each cellulose molecule has a polarity, having a 1′ and a 4′ end). Cellulose synthetase complexes are thought to be associated with the ends of growing microfibrils and the sugars present in the extracellular matrix are polymerized into cellulose at these "terminal complexes". Extension of a cellulose microfibrils is presumably achieved by lateral movement of the enzyme complex in the fluid phase of plasma membrane, with thê microfibril "spun out" on the outer surface of the membrane behind the moving enzyme complex (Fig. 13.15). The direction in which the complex moves and the orientation of the microfibril depend on some interactions between the membrane complex and the underlying cytoplasmic microtubules (*i.e.*, microtubules of cell cortex). Because the cellulose is synthesized at the plasma membrane, each new wall lamella forms internally to the last formed lamella. The cell wall, therefore, consists of concentrically arranged lamellae, with the oldest on the outside.

Box 13.5

The only polymers known to be made at the outer plasma membrane surface of plants are cellulose and callose. Cellulose synthesis is catalyzed by the multimeric enzyme complexes located at the termini of growing cellulose fibrils. These *terminal complexes* are visible in freeze-fracture replicas of plasma membrane of desmid *Micrasterias*. In desmids and angiosperms terminal complexes form *particle rosettes* (have 6 members in each rosette of *Micrasterias*). The orientation of cortical microtubule array is found to determine the orientation of cellulose microfibril deposition (see **Buchanan** *et al.*, 2000).

13.4. CELL'S GROWTH AND CELL WALLS

The cell wall is a dynamic structure. Cell expansion involves extensive changes in the mass and composition of the cell wall. Cell growth which is an irreversible increase in cell volume, can occur by expansion (increase in cell size in two or three dimensions) or by elongation (expansion constrained to one dimension). Variety in cell shapes may result if either of these two processes occurs at specific regions of the plant cell surface.

During elongation or expansion, existing cell wall architecture must change to incorporate new material, increasing the surface area of the cell and inducing water uptake by the protoplast. The osmotic pressure (turgour) exerted by the protoplast is necessary to cell expansion, and this pressure usually remains a relatively constant driving force for expansion. The regulation of **wall loosening** is considered the primary determinant of rates of cell expansion. The cell wall architecture must be extensible; that is, mechanisms must exist that allow discrete biochemical loosening of the cell wall matrix, permitting microfibril separation and insertion of newly synthesized polymers. Cell may extend to tens, hundreds, or even thousands of times their original length while maintaining a constant wall thickness. Thus, cell wall loosening and continued deposition of new material into the wall must be tightly integrated event.

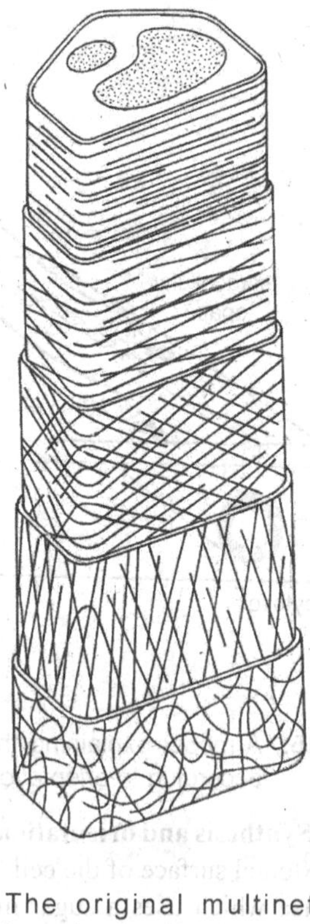

Fig. 13.16. The original multinet growth hypothesis explains that as walls stretch during growth, the microfibrils reorient passively from a transverse direction on the inner wall to a longitudinal direction at the outer wall (Buchanan *et al.*, 2000).

Following methods/hypotheses have been suggested regarding the growth of plant's cell wall:

1. **Multinet growth hypothesis.** From studies of developing cotton fibers, the **multinet growth hypothesis** was developed to explain how cellulose microfibrils that have been deposited in a transverse or slightly helical orientation are displaced axially as elongation proceeds (Fig. 13.16). New microfibrils deposited in strata on the inner surface of the wall in a generally transverse orientation functionally replace older microfibrils. The older microfibrils are pushed into the outer layers of the wall and are passively reoriented in a longitudinal direction as the cell elongates.
2. **Explanation of cell wall growth by biophysics.** Stress-relaxation is considered underlying basis of cell expansion (Fig. 13.17). When an elongating cell is streatched by turgor, the longitudinal stress (indicated by arrows in the figure) is borne more or less equally by the glycans tethering the cellulose microfibrils. If some of the tethers are disloged from the microfibrils, or hydrolyzed, they temporarily "relax" and the yield threshold is broken because the other tethers are strained. Water uptake results in expansion of the microfibrils to take up the slack of the relaxed glycans, which are once again placed under tensile stress.

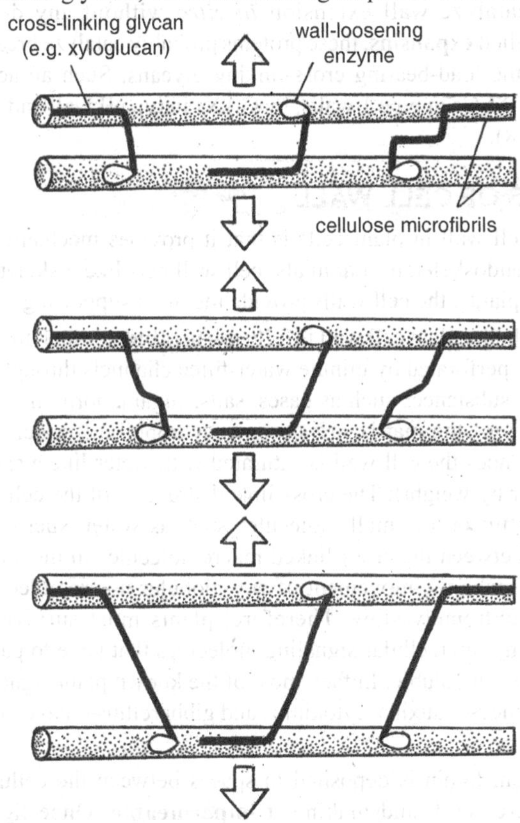

Fig. 13.17. Principles of biophysics are involved in stress-relaxation of microfibrils during cell wall growth (after Buchanan *et al.*, 2000).

3. **Acid-growth hypothesis.** The **acid-growth hypothesis** proposes that **auxin** activates a plasma-membrane proton pump, which acidifies the cell wall. The low pH, in turn, activates apoplast-localized growth-specific **hydrolases**, which cleave the load-bearing bonds that tether cellulose microfibrils to other polysaccharides. Cleavage of these bonds result in loosening of the cell wall, and the water potential difference causes uptake of water. Relaxation of the wall (*i.e.*, separation of the microfibrils) passively leads to an increase in cell size.

Box 13.6

Auxins are plant hormones that promote lengthwise growth (elongation of the cell) and control *abscission* and the plant's responses to light and gravity (*i.e.*, tropisms). Natural auxins are derivatives of indole (heterocyclic compounds). Synthetic auxins are used for crop control and as herbicides.

Abscission is the process whereby plant shed leaves, flowers and fruits controlled by plant hormones such as *auxins* and *abscisic acid*, leaf drop occurs in many plants through the formation of an abscission layer of corky cells in the leaf stem. This restricts sap flow to the leaf.

4. **Enzymatic theory.** At present, two kinds of enzymes are being evaluated as having possible wall-loosening activity. One of these, **xyloglucan endotransglycosylase (XET)**, carries out a transglycosylation of XyG in which one chain of XyG is cleaved and reattached to the nonreducing terminus of another XyG chain. Given such a mechanism, microfibrils could undergo a transient slippage but the overall tensile strength of the interlocking XyG matrix would not diminish (Fig. 13.18).

Other proteins catalyze wall extension *in vitro* without any detectable hydrolytic or transglycolytic events called **expansins**, these proteins probably catalyze breakage of hydrogen bonds between cellulose and the load-bearing cross-linking glycans. Such an activity could disrupt the tethering of cellulose by XyGs in type 1 walls, by GAXs in type II wall and by GAXs and β-glucans in grass walls (Fig. 13.18).

13.5. FUNCTIONS OF CELL WALL

The chief function of cell wall in plant cells is that it provides mechanical strength to the latter. Like the exoskelton or endoskeleton of animals, cell wall acts like a skeletal framework of plants. Particularly in vascular plants, the cell walls provide the main supporting framework.

Despite its strength, the plant cell wall is fully **permeable** to water and solutes. This is because the matrix of cell wall is perforated by minute water-filled channels through which free diffusion of water and water soluble substances such as gases, salts, sugars, hormones and like can take place. Moreover, the molecules of the matrix are strongly hydrophilic ("water-loving") with the result that in normal circumstances the cell wall is saturated with water like a sponge (*e.g.*, primary cell wall is 60 per cent water by weight): The cross-linked structure of the cell wall is, however, found to slightly block the diffusion of small molecules such as water, sucrose and K^+. The average diameter of the spaces between the cross-linked macromolecules in most cell wall is about 5 nm, this is small enough to make the movement of any globular macromolecules with a M.W. much above 20,000 daltons extremely slow. Therefore, plants must survive on molecules of low molecular weight, and any intercellular signaling molecules that have to pass through the cell wall must also be small and water soluble. In fact, most of the known plant signaling molecules, such as growth regulating substances—auxins, cytokinins and gibberellins—have molecular weights of less than 500 daltons.

During lignification, lignin is deposited in spaces between the cellulose molecule, making the cell wall much more rigid, and making it **impermeable**. Once lignification is complete the protoplasm can no longer absrob materials from outside the cell, which, therefore, dies. Hence, lignified tissue is always dead. Thus, a lignified tissue becomes well adapted for two types of functions:

1. It provides the mechanical strength due to its ligno-cellulose composition.
2. It transports water and salts, since, lignification involves loss of the protoplasm resulting in the formation of a hollow waterproof tube.

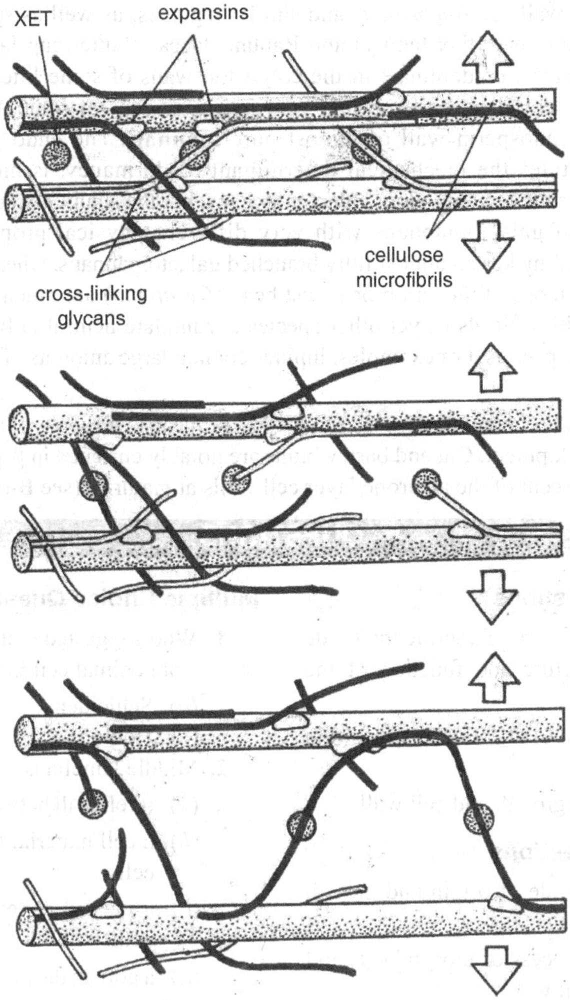

Fig. 13.18. Microfibril separation driven by osmotic pressure of the cell is facilitated by loosening of the cross-linking glycans that tether them. This may be accomplished by coordinate action of expansins, which break the steric interactions between the cross-linking glycans and reattaches one part of the chain to the nonreducing terminus of another. This action by XET may also function in forming new tethers because microfibrils from inner lamellae merge with microfibrils of the outer most lamellae as they are pulled apart during wall extension (after Buchanan *et al.*, 2000).

Storage Functions of Cell Wall

The secondary cell walls of the cotyledon and of the endosperm of developing seeds contain little or no cellulose but rather consist of a single noncellulosic polysaccharides typically found in the primary cell wall. These secondary walls serve two functions. First, they provide a strong wall to protect the embryo or impose mechanical dormancy. Second, they contain specialized storage carbohydrates that are digested during germination and converted to sucrose for transport to the growing seedling.

The cotyledon walls of *Tamarindus* and similar legumes, as well as species of Primulaceae (primrose family), Linaceae (flax family) and Ranunculaceae (butter cup family) abound in Gal-rich XyG. Glucomannans predominate in the cotyledon walls of some lilies and irises. Seeds of date, coconut and other palms; coffee bean, ivory nut; and seeds of some Apiaceae all contain a thick cotyledon or endosperm wall of almost pure **mannan**. The endosperm wall of lettuce seeds, which constitutes the mechanical determinant of dormancy, is more than 70 per cent mannan. All endospermic legumes store galactomannans, but the Man:Gal ratio can vary markedly, yielding a variety of galactomannans with very different physical properties. For example, fenugreek (*Trigonella*) makes an almost fully branched galactomannans, whereas guar (*Cyamopsis*) galactomannans and those of the carob or locust bean (*Ceratonia*) are much less branched, which changes their viscosities. Seeds of yet other species accumulate neutral polysaccharides typically found associated with pectins. For examples, lupines contain large amounts of $(1 \rightarrow 4)$ β-D-galactan and some arabinans.

All of the grasses accumulate $(1 \rightarrow 3)$, $(1 \rightarrow 4)$ β-D-glucan in the walls of the endosperm at some stage in embryo development. Oat and barley brans are notably enriched in β-glucans, which makes up as much as 30 per cent of the aleurone layer cell walls at maturity (see **Buchanan** *et al.*, 2000).

QUESTIONS

Long Answer Questions

1. What is the cell wall? Describe the mode of origin, structure and function of the cell wall.
2. Give an account of chemical composition of cell wall.
3. Describe cell's growth and cell wall.

Short Answer Questions

1. Describe the mode of origin and growth of cell wall.
2. Describe the process of biosynthesis and assembly of cell wall.
3. Write short note on following:
   (*i*) Plasmodesmata;
   (*ii*) Enzymatic theory of growth of cell wall.
4. Distinguish the following:
   (*i*) Plant cell wall from bacterial cell wall.
   (*ii*) Primary cell wall from secondary cell wall.

Very Short Answer Questions

1. What is hemicellulose?
2. What are pectins?

Multiple Choice Questions

1. Who suggested that plant cell is different from animal cell in having cell wall?
   (*a*) Schleiden (*b*) Schwann
   (*c*) Hooke (*d*) Robertson
2. Middle lamella is
   (*a*) a cell wall between two adjacent cells
   (*b*) a cell material between two adjacent cells
   (*c*) a plasma covering the two cells together
   (*d*) a pore in the plasma membrane of cell
3. Which is near to the plasma membrane?
   (*a*) tonoplast
   (*b*) primary cell wall
   (*c*) secondary cell wall
   (*d*) middle lamella
4. Most abundant water insoluble polysaccharide of plant cell wall is
   (*a*) cellulose
   (*b*) hemicellulose
   (*c*) pectin
   (*d*) lignin
5. The smallest unit of plant cell wall is
   (*a*) microfibril (*b*) micelle
   (*c*) middle lamella (*d*) tertiary wall

6. Pectin of the cell wall is
   (*a*) excretory product
   (*b*) waste product
   (*c*) secretory product
   (*d*) all above
7. The deposition of one of the following in plant cells make them impervious to gases and water
   (*a*) cellulose (*b*) hemicellulose
   (*c*) suberin (*d*) pectin
8. Cell wall is rigid because of
   (*a*) pecten
   (*b*) liginin
   (*c*) hemicellulose
   (*d*) cellulose
9. Outermost layer of cell wall is
   (*a*) primary wall
   (*b*) middle lamella
   (*c*) secondary wall
   (*d*) plasma membrane

ANSWERS

Very Short Answer Questions

1. Branched glycans or polysaccharides of the plant cell wall whose backbone consists of one sugar, such as glucose, and side chains of other sugars, such as xylose.
2. Pectins are a heterogenous class of negatively charged polysaccharides that make up the matrix of the plant cell wall. Pectins hold water and form a gel that fills in the spaces between the fibrous elements.

Multiple Choice Questions

1. (*b*) **2.** (*b*) **3.** (*c*) **4.** (*a*) **5.** (*b*) **6.** (*c*) **7.** (*c*)
8. (*b*) **9.** (*a*)

14 Cell Junctions, Cell Adhesion and the Extracellular Matrix (ECM)

The building technologies of multicellular animals and plants are different, and each type of organism is formed of many types of **tissues**, in which the cells are assembled and bound together in different ways. In both animals and plants, however, an essential part is played in most tissues by the **extracellular matrix (ECM)**. This complex network of secreted extracellular macromolecules has many functions, but the first and salient is that it forms a supporting framework. ECM helps hold cells and tissues together, and in animals, it provides an organized environment within which migratory cells can move and interact with one another in orderly ways. The extracellular matrix, however, is only half the story. In animals especially, the cells of most tissues are bound directly to one another by **cell-cell junctions**. These junctions too are of many types, serving many purposes in addition to mechanical attachment and without them, bodies of animals would disintegrate.

In vertebrates, the major tissue types are nerve, muscle, blood, lymphoid, epithelium, and connective tissues. Connective tissues and epithelial tissue represent two extremes of organization. In *connective tissue*, the extracellular matrix is plentiful, and cells are sparsely distributed within it. The matrix is rich in fibrous polymers, especially collagen, and it is the matrix-rather than the cells—that bears most of the mechanical stress to which the tissues is subjected. Direct attachment between one cell and another are relatively rare.

In **epithelial tissue**, by contrast, cells are tightly bound together into sheets called **epithelia**. The extracellular matrix is scanty, consisting mainly of a tiny mat called the **basal lamina**, which underlies the epithelium. The cells are attached to each other by cell-cell adhesions, which bear most of the mechanical stresses. For this purpose, strong intracellular protein filaments (component of cytoskeleton) cross the cytoplasm of each epithelial cell and attach to specialized junctions in the plasma membrane. The junctions, in turn, tie the surfaces of adjacent cells either to each other or to the underlying basal lamina.

Epithelial cell sheets line all the cavities and free surfaces of the animal body. The specialized junctions between the cells enable epithelia to form barriers that inhibit the movement of water, solutes and cells from one body compartment to another. As shown in (Fig. 14.1), epithelia almost always rest on a supporting bed of connective tissue. This supporting bed may in turn attach them to other tissues, such as the muscle shown in figure. In this way, tissues join together in various combinations to larger functional units called **organs**.

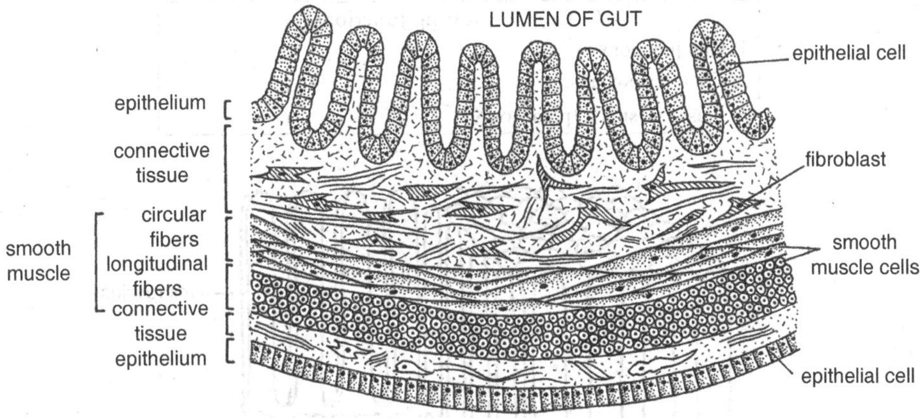

Fig. 14.1. A cross-sectional view of part of the wall of the intestine. This long, tube-like organ is constructed from epithelial tissue, connective tissue and muscle tissue. Each tissue is an organized assembly of cells held together by cell-cell adhesions, extracellular matrix or both (after Alberts *et al.*, 2002).

14.1. CELL JUNCTIONS

Specialized **cell junctions** occur at points of cell-cell and cell-matrix contact in all tissues, and they are particularly abundant in epithelia. Cell junctions are best visualised using either conventional or freeze-fracture electron microscopy, which reveals that the interacting plasma membranes (and often the underlying cytoplasm and the intervening intercellular space as well) are highly specialized in these regions.

Cell junctions (Fig. 14.2) are often divided into following three functional groups:

1. Occluding junctions. They seal cells together in an epithelium in a way that prevents even small molecules from leaking from one side of the sheet to the other. Indeed they form a selectively permeability barrier across epithelial cell sheet.

2. Anchoring junctions. They mechanically attach cells (and their cytoskeleton) to their neighbors or to the extracellular matrix. Thus, they connect the cytoskeleton of a cell to the cytoskeleton of it neighbors or to the ECM.

3. Communicating junctions. They mediate the passage of chemical or electrical signals from one interacting cell to its partner.

The major kinds of intercellular junctions within each group are listed in Table 14.1.

Table 14.1. The functional classification of cell junctions (after Alberts *et al.*, 2002).

A. Occluding Junctions
1. Tight junctions (vertebrates only)
2. Septate junctions (Invertebrates only)
B. Anchoring Junctions
I. Actin Filament Attachment Sites
1. Cell-cell junctions (adherens junctions)
2. Cell-matrix junctions (focal adhesions)
II. Intermediate Filament Attachment Sites
1. Cell-cell junctions (desmosomes)
2. Cell-matrix junctions (hemidesmosomes)

C. Communicating Junctions
1. Gap junctions
2. Chemical synapses
3. Plasmodesmata (plants only)

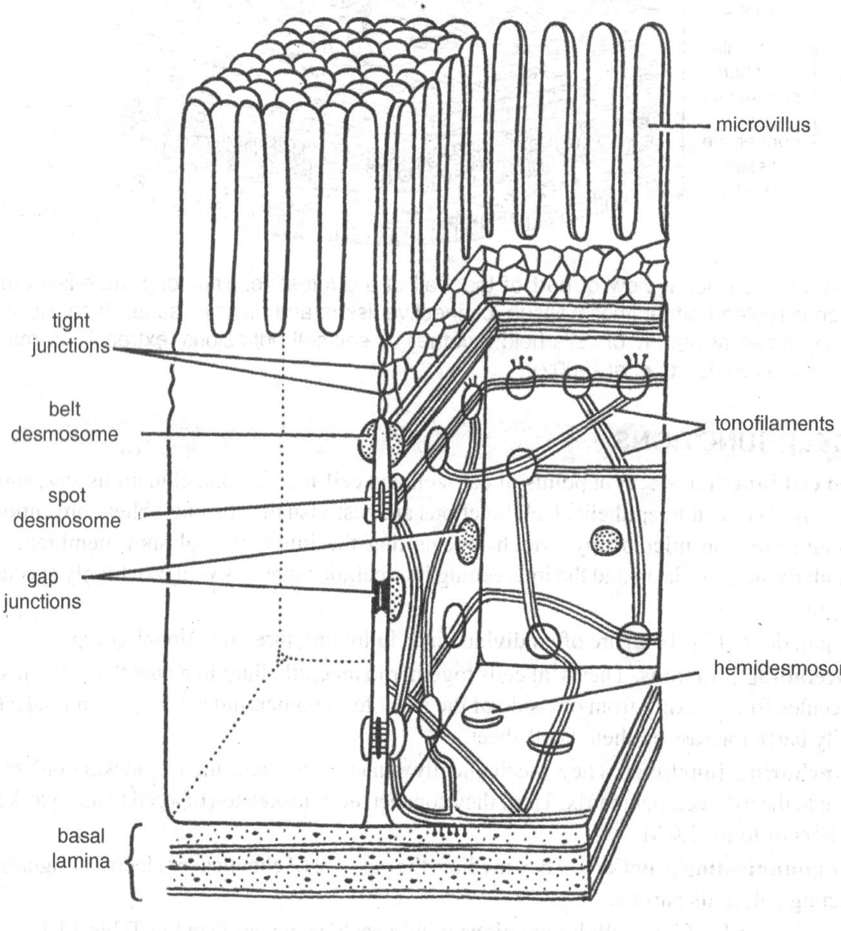

Fig. 14.2. Schematic diagram of the principal types of cell junctions, as found in the intestinal epithelial cell.

A. Occluding Junctions

All epithelia have at least one important function in common: they serve as selective permeability barriers, separating fluids on either side that have a different chemical composition. This function requires that the adjacent cells be sealed together by occluding junctions.

Occluding junctions are of following two types:

1. Tight junctions (TJs). They occur in vertebrates in the organs such as mammalian small intestine, or gut. Tight junctions are also abundant in the ducts and cavities of glands, such as liver and pancreas, that connect with the digestive tract, as well as in the urinary bladder, where they ensure that urine stored in the bladder does not seep out between cells. They form watertight epidermal TJs in skin and *blood-brain barriers* in brain (**Karp** 2010).

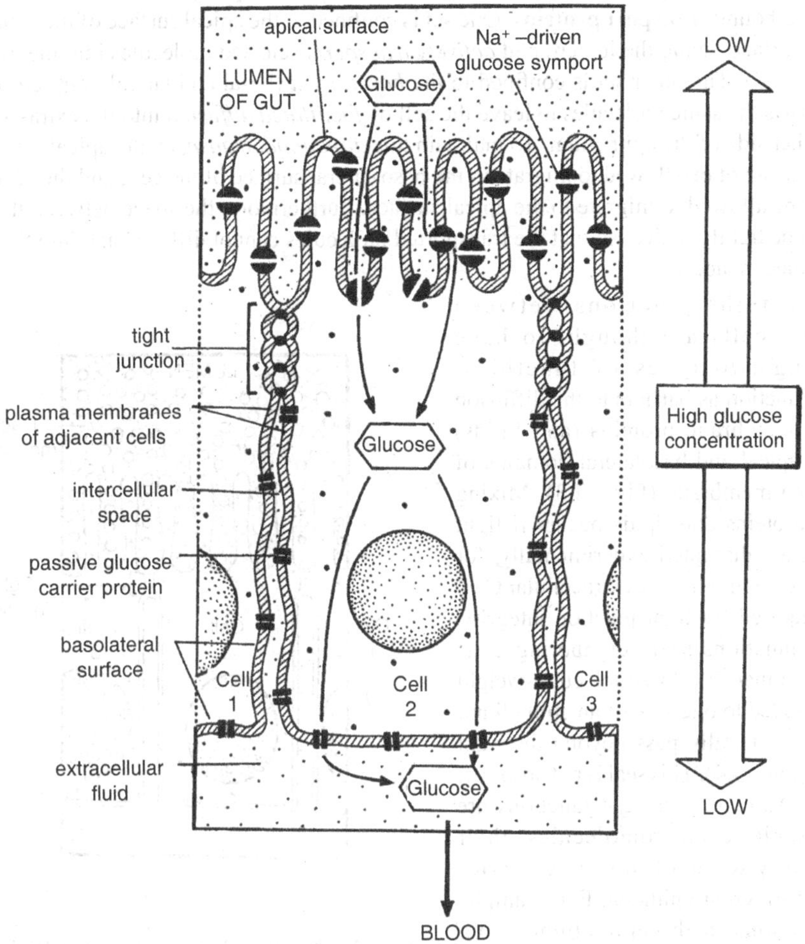

Fig. 14.3. The role of tight junctions in transcellular transport. The transport proteins are confined to different regions of the plasma membrane in epithelial cells of the small intestine. This separation permits a vectorial* transfer of nutrient across the epithelium from the gut lumen to the blood. In the example shown, glucose is actively transported into the cell by Na^+–driven glucose symports at the apical surface, and it diffuses out of the cell by facilitated diffusion mediated by glucose carriers in the basolateral membrane. Tight junctions are thought to confine transport proteins to their appropriate membrane domains by acting as diffusion barriers within the lipid bilayer of the plasma membrane; these junctions also block the backflow of glucose from the basal side of the epithelium into the gut lumen (after Alberts *et al.*, 2002).

***Vectorial.** Vector-like; vector represents a quantity that has magnitude and direction and that is usually represented by part of a straight line with the given direction and with a length representing a magnitude.

The epithelial cells lining the small intestine form a barrier that keeps the gut contents in the gut cavity, the **lumen**. At the same time, however, the cells must transport selected nutrients across the epithelium from the lumen into the extracellular fluid that permeates the connective tissue on the other side (Fig. 14.1). From there these nutrients diffuse into small blood vessels to provide nourishment to the organism. This is called **transcellular transport** and it depends on two sets of membrane-bound **transport proteins**. One set is confined to the apical surface of the epithelial cell (*i.e.*, the surface facing the lumen) and *actively transports* selected molecules into the cell from the gut (Fig. 14.3). The other set is confined to the *basolateral* (basal and lateral) *surfaces* of the cell, and it allows the same molecules to leave the cell by *facilitated diffusion* into the extra-cellular fluid on the other side of the epithelium. To maintain this *directional transport,* the apical set of transport proteins must not be allowed to migrate to the baso-lateral surface of the cell, and the basolateral set must not be allowed to migrate to the apical surface. Furthermore, the space between the epithelial cell must be tightly sealed, so that the transported molecules cannot diffuse back into the gut lumen through these spaces.

The tight junctions between epithelial cells are thought to have following two types of functions: 1. They function as barriers to the diffusion of some membrane proteins (and lipids) between apical and basolateral domains of the plasma membrane (Fig. 14.3). Mixing of such proteins and lipids occurs if tight junctions are disrupted experimentally, for example, by removing the extracellular Ca^{2+} that is required for tight junction integrity. 2. Tight junctions seal neighboring cells together so that, if a low-molecular-weight tracer is added to one side of an epithelium, it will not generally pass beyond the tight junction (Fig. 14.4). This seal is not absolute, however. Although, all tight junctions are impermeable to macromolecules, their permeability to small molecules varies greatly in different epithelia. For example, tight junctions in the epithelium lining the small intestine are 10,000 times more permeable to inorganic ions, such as Na^+, than the tight junctions in the epithelium lining the urinary bladder. These differences reflect differences in tight junction proteins that form the junctions.

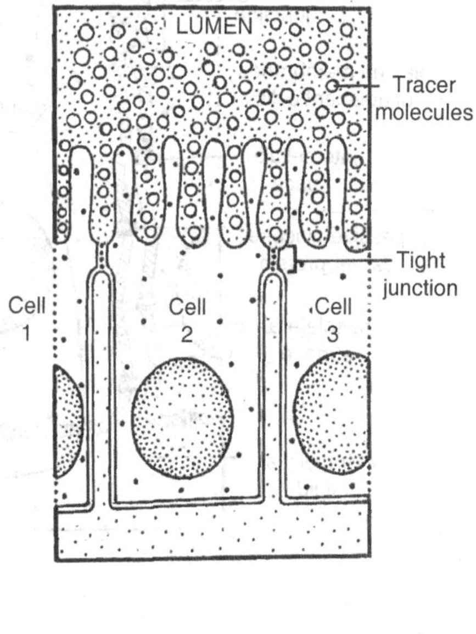

Fig. 14.4. The role of tight junctions in allowing epithelia to serve as barriers to solute diffusion. The drawing shows how a small extra-cellular tracer molecule added on one side of an epithelium cannot traverse the tight junctions that seal adjacent cell together (after Albert *et al.*, 2002).

Epithelial cells can temporarily alter their tight junctions to permit an increased flow of solutes and water through cracks in the junctional barriers. Such **paracellular transport** is especially important in the absorption of amino acids and monosaccharides from the lumen of the intestine, where their concentration can increase enough after a meal to drive passive transport in the desired direction. Thus, tight junctions serve as a regulatory center to help in coordination of multiple cell processes.

In pancreatic acinar tissue, they prevent the leakage of pancreatic secretary proteins, including digestive enzymes, into the blood.

Tight junctions are composed of thin bands that completely encircle a cell and are in contact with thin bands of adjacent cells. In a thin section, in the tight junction two adjacent plasma membranes appear to be fused at a series of points. However, in three-dimensional structure, revealed by freeze fracture technique the tight junctions appear as a network of ridges on the cytoplasmic half of the membrane, with complementary grooves in the outer half. The ridges appear to be composed of two rows of protein particles, as in zipper, each one belonging to the adjacent cells. The lines of these particles produce the sealing and for this reason have been named **sealing strands.** Often, sealing strands form a series of interconnected and anastamosing lines, like a row of stitches in a quilted surface (Fig. 14.5).

Each tight junction sealing strand is composed of long rows of transmembrane adhesion proteins embedded in each of the two interacting plasma membranes. The extracellular domains of these proteins join directly to one another to occlude the intercellular space (Fig. 14.5). The major transmembrane proteins in a tight junction are the **claudins**, which are essential for tight junction formation and function and differ in different tight junctions (Fig. 14.6). A specific claudin (*i.e.*, claudin-16) found in kidney epithelial cells, for example, is required for Mg^{2+} to be resorbed from the urine into the blood. A mutation in the gene encoding this claudin results in excessive loss of Mg^{2+} in the urine. A second major transmembrane protein in tight junctions is **occludin**, the function of which is uncertain. Claudins and occludins associate with intracellular peripheral membrane proteins called **ZO proteins** S (*Z* = zonula, O = occludens, another name of tight junctions) which anchor the strands to the actin cytoskeleton (Fig. 14.6).

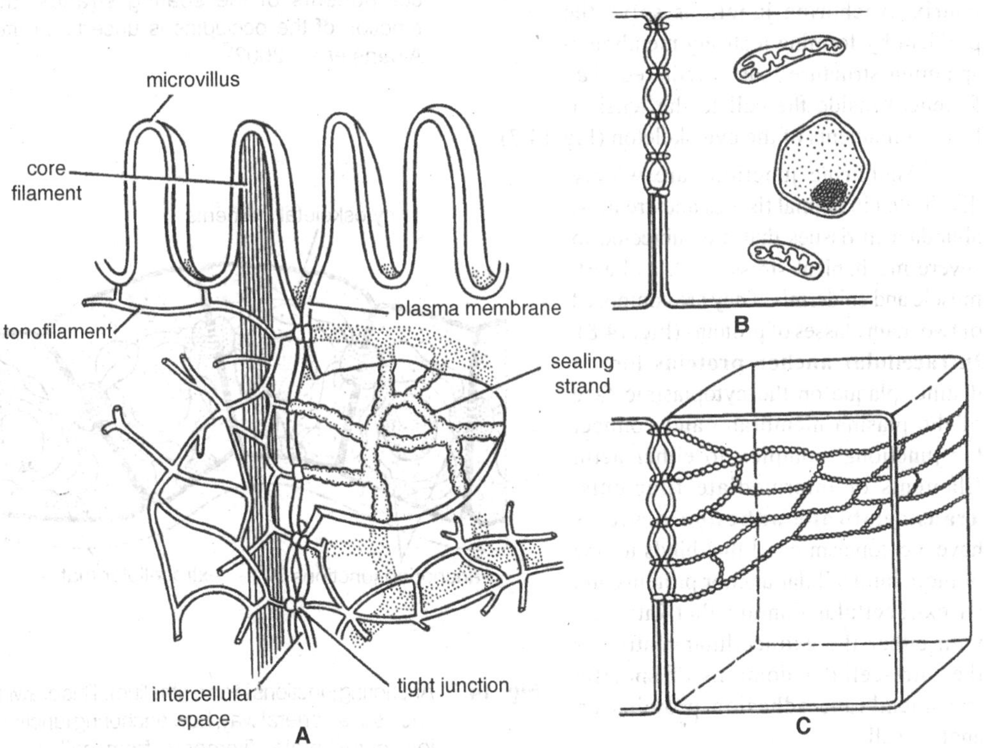

Fig. 14.5. Tight junctions: A—The adjacent plasma membranes are held firmly at the sealing strand which is composed of two rows of particles, as in zipper; B, C—Three-dimensional representations.

2. Septate junctions. These occluding junctions occur in invertebrates. Septate junctions are more regular in structure than the tight junctions and they form a continuous band around each epithelial cell. But their morphology is distinct because the interacting plasma membranes are joined by proteins that are arranged in parallel rows with a regular periodicity. A protein called **Discslarge**, which is required for the formation of septate junction in *Drosophila*, is structurally related to the ZO proteins found in vertebrate tight junction. Mutant flies that are deficient in this protein not only lack septate junctions but also develop epithelial tumors.

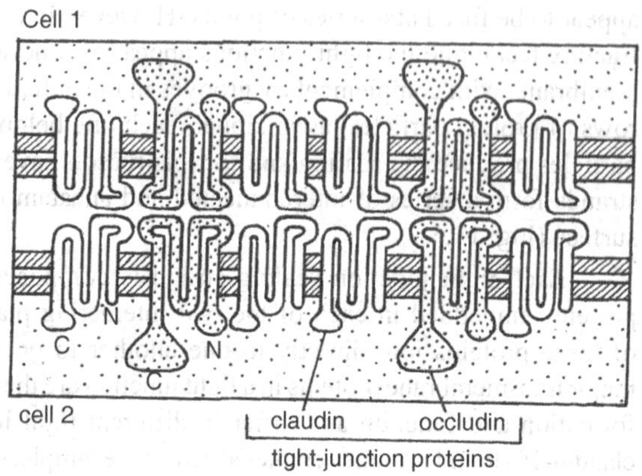

Fig. 14.6. A current model of a tight junction. Drawing showing the transmembrane claudin and occludin proteins in a tight junction. The claudins are the main components of the sealing strands; the function of the occludins is uncertain (after Alberts *et al.*, 2002).

B. Anchoring Junctions.

The lipid bilayer is flimsy and cannot by itself transmit large forces from cell to cell or from cell to extracellular matrix. Anchoring junctions solve the problem by forming a strong membrane-spanning structure that is tethered (*i.e.*, fastened) inside the cell to the tension-bearing filaments of the cytoskeleton (Fig. 14.7).

Anchoring junctions are widely distributed in animal tissues and are most abundant in tissues that are subjected to severe mechanical stress, such as heart, muscle and epidermis. They are composed of two main classes of proteins (Fig. 14.8). **Intracellular anchor proteins** form a distinct plaque on the cytoplasmic face of the plasma membrane and connect the junctional complex to either actin filaments or intermediate filaments. **Transmembrane adhesion proteins** have a cytoplasmic tail that binds to one or more intracellular anchor proteins and an extracellular domain that interacts with either the extracellular matrix or the intra-cellular domains of specific transmembrane adhesion proteins on another cell.

In addition to anchor proteins and adhesion proteins, many anchoring junction contain intracellular signalling proteins that enable the junctions to signal to the cell interior.

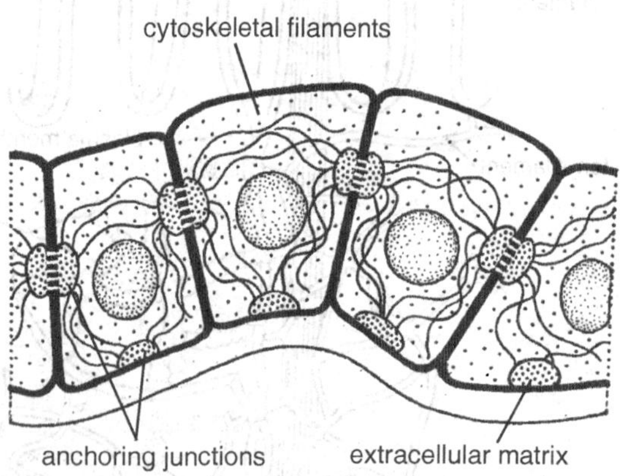

Fig. 14.7. Anchoring junctions in an epithelium. This drawing shows in a general way, how anchoring junctions join cytoskeletal filaments from cell to cell and from cell to the extracellular matrix (after Alberts *et al.*, 2002).

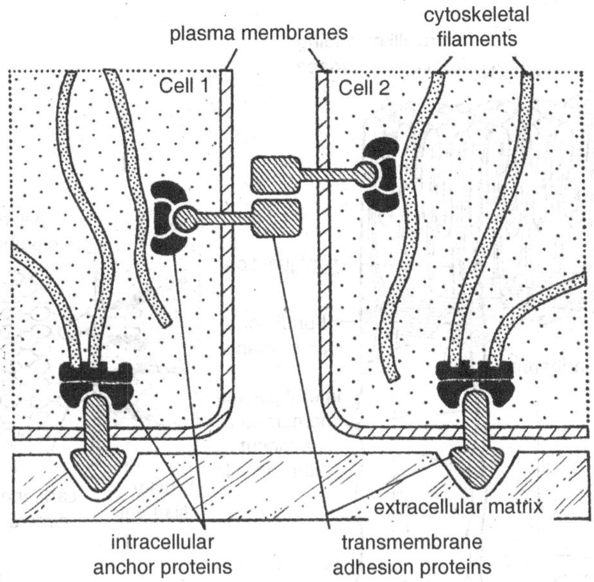

Fig. 14.8. The construction of an anchoring junction from two classes of proteins. This diagram shows how intracellular anchor proteins and transmembrane adhesion proteins form anchoring junctions (after Alberts *et al.*, 2002).

Anchoring junctions are of following types:

1. Adherens junctions and **desmosomes** hold cells together and are formed by transmembrane adhesion proteins that belongs to the **cadherin** family.

2. Focal adhesions and **hemidesmosomes** bind cells to the extracellular matrix and are formed by transmembrane adhesion proteins of the **integrin**.

On the intracellular side of the membrane, adherens junctions and focal adhesions serve as connection sites for actin filaments while desmosomes and hemidesmosomes serve as connection sites for intermediate filaments.

I. Actin Filament Attachment Sites

1. Adherens Junctions (Zonula adherens). Adherens junctions commonly occur in epithelia where they often form a continuous **adhesion belt** or **zonula adherens** just below the tight junctions, encircling each of the interacting cell in the sheet. The adhesion belt are directly apposed in adjacent epithelial cells, with the interacting plasma membranes held together by the **cadherins** that serve as transmembrane adhesion proteins.

Within each cell, a contractile bundle of actin filaments lies adjacent to the adhesion belt, oriented parallel to the plasma membrane. The actin is attached to this membrane through a set of intracellular anchor proteins, including **catenins, vinculin** and **a-actinin.** The actin bundles are thus linked, via the cadherins and anchor proteins, into an extensive transcellular network (Fig. 14.9). The network can contract with the help of **myosin** motor proteins and it is thought to help in mediating a fundamental process in animal morphogenesis—the folding of epithelial cell sheets into tubes and other related structure (Fig. 14.10).

The assembly of tight junctions between epithelial cells seems to require the prior formation of adherens junctions. Anti-cadherin antibodies that block the formation of adherens junctions, for example, also block the formation of tight junctions.

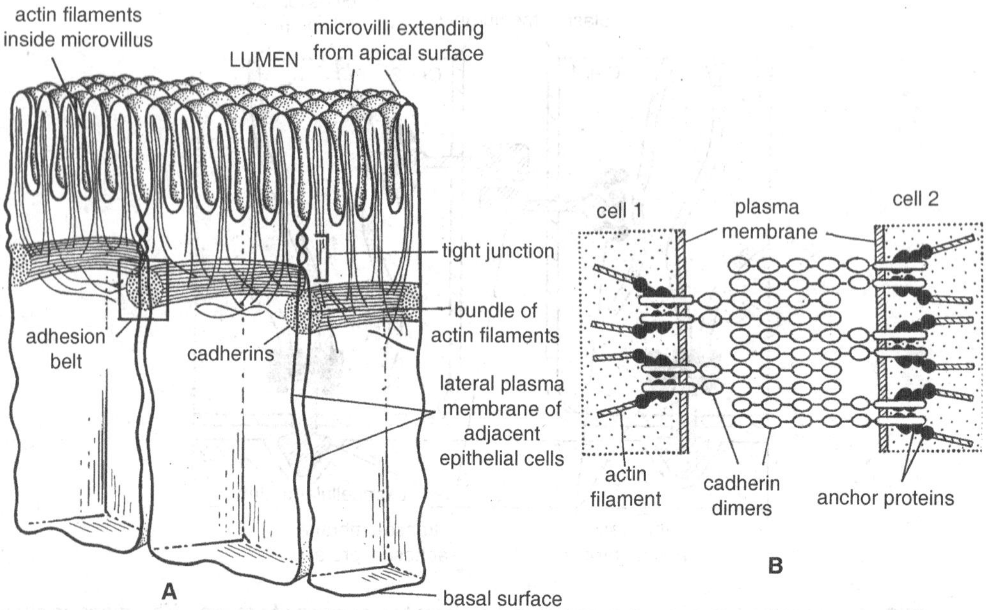

Fig. 14.9. Adherens junctions. A—Adherens junctions in the form of adhesion belts between epithelial cells in the small intestine. The belt like junction encircles each of the interacting cells. Its most obvious feature is a contractile bundle of actin filaments running along the cytoplasmic surface of the junctional plasma membrane. B—The actin filaments are joined from cell to cell by transmembrane adhesion proteins called **cadherins**. The cadherins form homodimers in the plasma membrane of each interacting cell. The extracellular domain of one cadherin dimer binds to the extracellular domain of an identical cadherin dimer on the adjacent cell. The intracellular tails of the cadherins bind to anchor proteins that tie them to actin filaments. These anchor proteins include α-catenin, β-catenin, γ-catenin (also called plakoglobin), α-actinin and vinculin (after Alberts *et al.*, 2002).

Selective Cell-Cell Adhesion and CAMs

Unlike adult vertebrate tissues, which are difficult to dissociate, embryonic vertebrate tissues are easily dissociated. This is usually done by treating the tissue with low concentrations of a proteolytic enzyme such as trypsin. Sometimes this technique is combined with the removal of extracellular Ca^{2+} and Mg^{2+} with a divalent-cation chelator (such as EDTA). These reagents disrupt the protein-protein interactions (many of which are divalent-cation-dependent) that hold cells together. Remarkably, the dissociated cell often reassemble *in vitro* into structures that resemble the original tissue. Findings of this nature reveal that tissue structure is not just a product of history (*i.e.*, origin and development); it is actively maintained and stabilized by the system of affinities that cells have for one another and for the extracellular matrix (ECM).

A salient example of this phenomenon is seen when dissociated cells from two embryonic vertebrate organs, such as liver and the retina, are mixed together and artificially formed into a pellet: the mixed aggregates gradually sort out according to their origin. More generally, disaggregated cells are found to adhere more readily to aggregates of their own organ than to aggregates of other

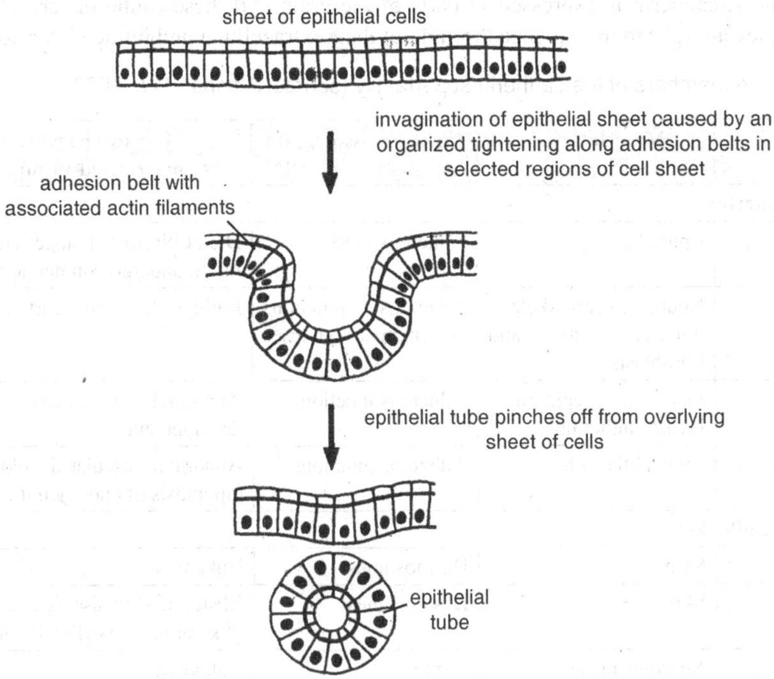

Fig. 14.10. The folding of an epithelial sheet due to adhesion belt to form an epithelial tube. The oriented contraction of the bundles of actin filaments running along adhesion belts causes the epithelial cells to narrow at their apex and helps the epithelial sheet to roll up into a tube. An example is the formation of the neural tube in early vertebrate development (during neurula stage). However, rearrangements of the cells within the epithelial sheet are also thought to have an important role in the process (after Alberts *et al.*, 2002).

organs. Evidently there exist **cell-cell recognition systems** that make cells of the same differentiated tissue preferentially adhere to one another; these adhesive preferences are presumably important in stabilizing tissue architecture.

Cells adhere to each other and to the extracellular matrix through cell surface proteins called **cell adhesion molecules (CAMs)**—a category that includes the transmembrane adhesion proteins such as cadherins and integrins. CAMs can be **cell-cell adhesion molecules** or **cell matrix adhesion molecules.** Some CAMs are Ca^{2+}-dependent (*e.g.*, cadherins), whereas others are Ca^{2+}-independent (*e.g.*, immunoglobulins). The Ca^{2+}-dependent CAMs seem to be primarily responsible for the tissue-specific cell-cell adhesion seen in early vertebrate embryos, explaining why these cells can be disintegrated with Ca^{2+}-chelating agents.

Examples of CAMs

1. Cadherins

The cadherins are the major CAMs which are responsible for Ca^{2+}-dependent cell-cell adhesion in vertebrate tissues. They are of following two main types (Table 14.2).

(*i*) Classical cadherins. The first three cadherins that were discovered were named according to the main tissues in which they were found: **E-cadherin** is present on many types of epithelial cells; **N-cadherin** on nerve, muscle and lens cells; and **P-cadherin** on cells in the placenta and epidermis. All are also found in various other tissues. For example, N-cadherin is expressed in fibroblasts and E-cadherin is expressed in parts of the brain. All these cadherins are **classical cadherins** and they are related in sequence throughout their extracellular and intracellular domains.

Table 14.2. Some members of the cadherin superfamily (source. Alberts *et al.*, 2002).

Name	Main location	Junction association	Phenotype when inactivated in mice
A. Classical cadherins			
1. E-cadherins	Epithelia	Adherens junctions	Die at blastocyst stage; embryos fail to undergo compaction
2. N-cadherins	Neurons, heart, skeletal muscle, lens and fibroblasts	Adherens junctions and chemical synapses	Embryos die from heart defects
3. P-cadherins	Placenta, epidermis, breast epithelium	Adherens junctions	Abnormal mammary gland development
4. VE-cadherins	Endothelial cells	Adherens junctions	Abnormal vascular development (apoptosis of endothelial cells)
B. Nonclassical cadherins			
1. Desmocollin	Skin	Desmosomes	Unknown
2. Desmoglein	Skin	Desmosomes	Blistering skin disease due to loss of keratinocyte cell-cell adhesion
3. Cadherin	Neurons, muscle	None	Unknown
4. Fat (in *Drosophila*)	Epithelial and CNS	None	Enlarged imaginal discs and tumors
5. Protocadherins	Neurons	Chemical synapses	Unknown

***(ii)* Nonclassical cadherins.** They are of many types; more than 50 nonclassical cadherins are expressed in the brain alone. The nonclassical cadherins include proteins with known adhesive function, such as the **desmosomal cadherins** and diverse **protocaderins** found in the brain. They also include proteins that appear to have nonadhesive functions, such as **T-cadherins**, which lack a transmembrane domain and is attached to the plasma membrane of nerve or muscle cells by a glycosulphosphatidylinositol (GPI) anchor, and the **Fat protein**, which was first identified as the product of a tumor-suppressor gene in *Drosophila*. Together, the classical and nonclassical cadherin proteins form the **cadherin superfamily**.

Occurence. Cadherins are found in both invertebrates and vertebrates. Virtually all vertebrates seem to contain one or more cadherins, according to the cell type. They are the main adhesion molecules holding cells together in early embryonic tissues. In culture, the removal of extracellular Ca^{2+} or treatment with anti-cadherin antibodies disrupts embryonic tissues. Mutations that inactivate the function of E-cadherin cause mouse embryos to fall apart and die early in development.

Molecular structure. Most cadherins are single-pass transmembrane glycoproteins about 700–750 amino acids long. Structural studies suggest that they associate in the plasma membrane to form **dimers** or larger **oligomers**. The large extracellular part of the polypeptide chain is usually folded into five or six **cadherin repeats**, which are structurally related to immunoglobulin (Ig) domains (Fig. 14.11A). The crystal structures of E-cadherin and N-cadherin have helped to explain the importance of Ca^{2+} binding for cadherin function. The Ca^{2+} ions positioned between each pair of cadherin repeats, locking the repeats together to form a stiff, rodlike structure: the more Ca^{2+} ions that are bound, the more rigid the structure is. If Ca^{2+} is removed, the extracellular part of the protein becomes floppy and is rapidly degraded by proteolytic enzymes (Fig. 14.11).

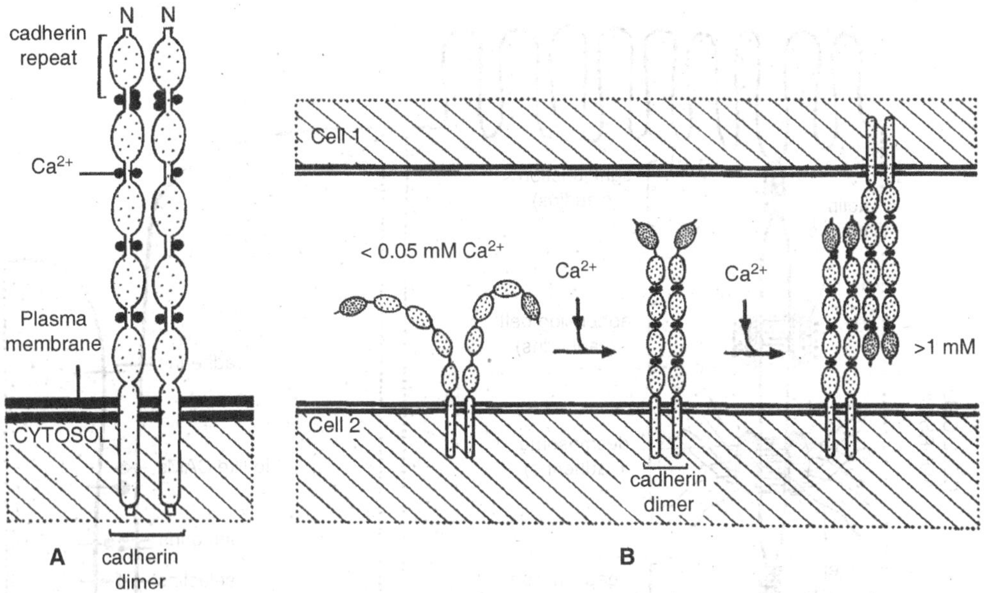

Fig. 14.11. The structure and function of cadherins. A—A classical cadherin molecule. The protein is a homodimer, with the extracellular part of each polypeptide folded into five cadherin repeats. There are Ca^{2+}-binding sites between each pair of repeats. B—The influence of extracellular Ca^{2+}. As the amount of Ca^{2+} increases, the extracellular parts of the cadherin chains become more rigid. When enough Ca^{2+} is bound, the cadherin dimer extends from the surface, where it can bind to a cadherin dimer on a neighboring cell. If Ca^{2+} is removed, the extracellular part of the protein becomes floppy and is degraded by proteolytic enzymes (after Alberts *et al.*, 2002).

Functions of Cadherins

(i) **Role of cadherins in development.** E-cadherin is the best characterized cadherin. It is usually concentrated in adherens junctions in mature epithelial cells, where it helps connect the cortical actin cytoskeletons of the cell it holds together (Fig. 14.9). E-cadherin is also the first cadherin expressed during mammalian development. It helps cause **compaction**, an important morphological change that occurs at the eight cell stage of mouse embryo development. During compaction, the loosely attached cells, called **blastomeres**, become tightly packed together and joined by intercellular junctions.

Further, as the neural tube forms during the tubulation embryonic stage and pinches off from the overlying ectoderm, neural tube lose E-cadherin and acquire other cadherins, including N-cadherin, while the cells in the overlying ectoderm continue to express E-cadherin.

(ii) **Role of cadherins in cell-cell adhesion.** How do cell-cell adhesion molecules such as the cadherins bind cells together? There can occur following three mechanisms (Fig. 14.13): 1. In **homophilic binding**, molecules on one cell bind to other molecules of the same kind on adjacent cells; 2. In **heterophilic binding**, the molecules on one cell bind to molecules of a different kind on adjacent cells; 3. In **linker-dependent binding**, cell surface receptors on adjacent cells are linked to one another by secreted multivalent linker molecules. Although all three mechanisms have been found to operate in animals, cadherins usually link cells by the homophilic mechanism.

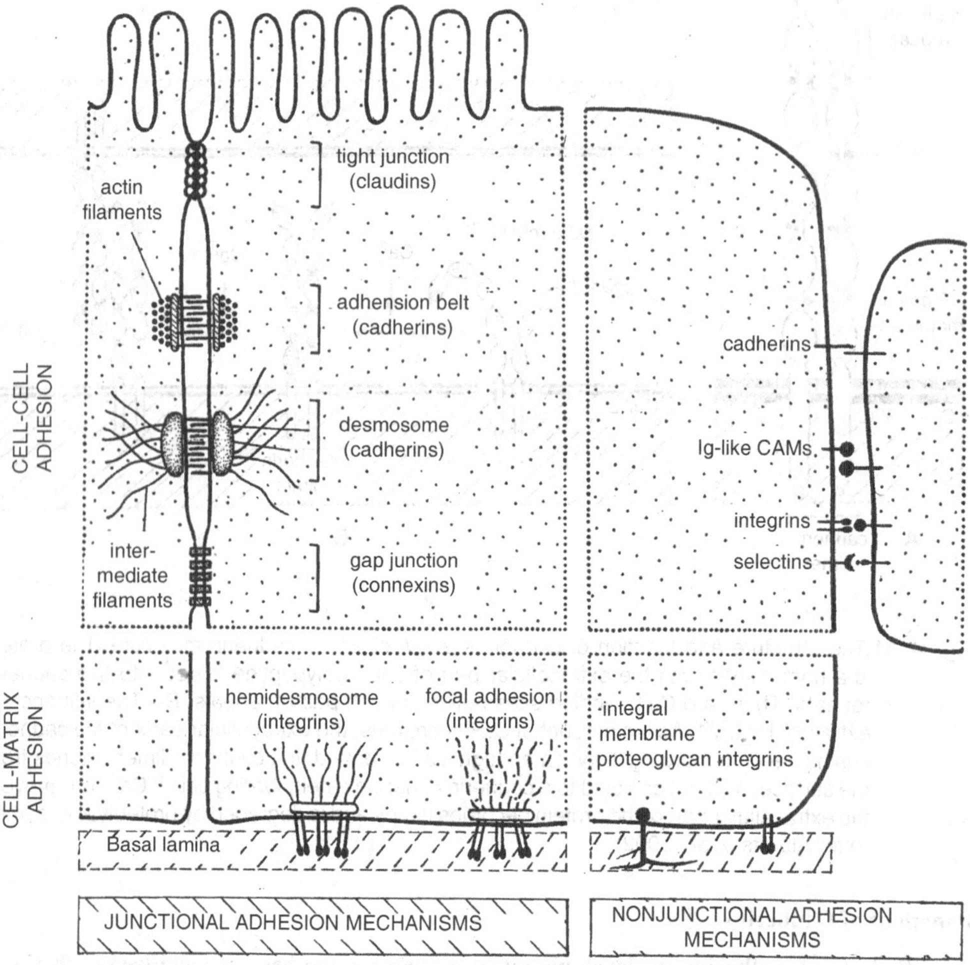

Fig. 14.12. A summary of the junctional and nonjunctional adhesive mechanisms used by animal cells in binding to one another and to the extracellular matrix (after Alberts *et al.*, 2002).

***(iii)* Role of cadherins in their linkage to actin filaments.** Most cadherins, including all classical and some nonclassical ones, function as transmembrane adhesion proteins that indirectly link the actin cytoskeletons of the cell they join together. This arrangement occurs in adherens junctions (Fig. 14.9B). The highly conserved cytoplasmic tail of these cadherins interacts indirectly with actin filaments by means of a group of intracellular anchor proteins called **catenins** (Fig. 14.14). This interaction is essential for efficient cell-cell adhesion, as classical cadherins that lack their cytoplasmic domain cannot hold cells strongly together.

Some embryonic cells (*e.g.*, neural tube formation) are known to regulate the adhesive activity of their cadherins. This is done by phosphorylation of anchor protein.

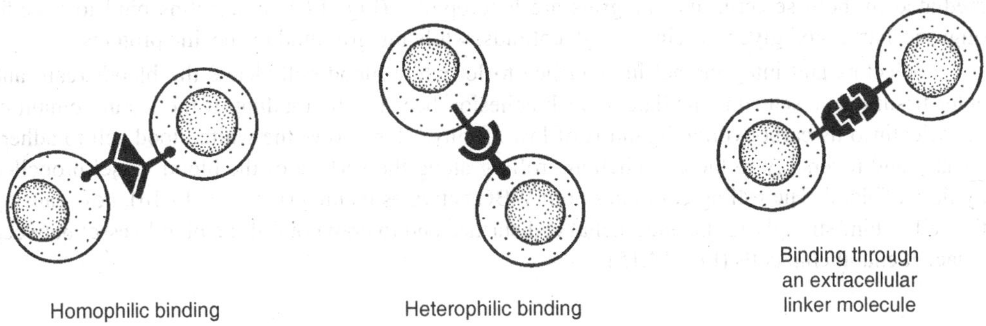

Fig. 14.13. Three possible mechanisms by which cell-surface molecules can mediate cell-cell adhesion (after Alberts *et al.*, 2002).

***(iv)* Transient cell-cell adhesions.** Leucocytes or white blood cells lead a nomadic life, moving to and fro between the bloodstream and the tissues, and this necessitates special adhesive properties. These properties depend on **selectins.** Selectins are cell surface carbohydrate-binding proteins (**lectins**) that mediate a variety of transient, Ca^{2+}-dependent, cell-cell adhesion interactions in the blood stream. These are of three types: 1. **L-selectin** on white blood cells, 2. **P-selectin** on blood platelets and on endothelial cells that have been locally activated by an inflammatory response, and 3. **E-selectin** on activated endothelial cells. Each selectin is a transmembrane protein with a highly conserved lectin domain that binds to a specific oligosaccharide on another cell (Fig. 14.15).

Selectins have an important role in binding white blood cells to endothelial cells lining blood vessels, thereby enabling the blood cells to migrate out of the bloodstream into a tissue. In a lymphoid organ, the endothelial cells express oligosaccharides that are recognized by L-selectin on lymphocytes, causing the lymphocytes to hang about and become trapped. Conversely, at sites of inflammation, the endothelial cells switch on expression of selectins, flagging the cells down to help deal with the local emergency. Selectins do not act alone, however; they colloborate with *integrins,* which strengthen the binding of the blood cells to the endothelium. The cell-cell adhesions

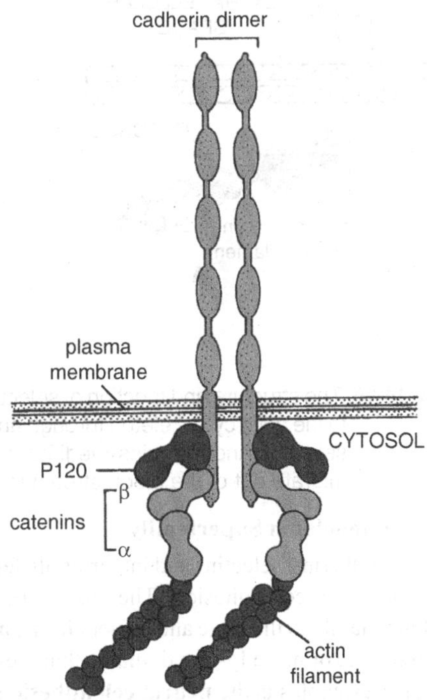

Fig.14.14. The linkage of classical cadherins to actin filaments. The cadherins are coupled indirectly to actin filaments by the anchor protein α-catenin and β-catenin. A third intracellular protein, called p120, also binds to the cadherin cytoplasmic tail and regulates cadherin function (after Alberts *et al.*, 2002).

mediated by both selectin and integrins are heterophilic (Fig. 14.12): selectins bind to specific oligosaccharides on glycoproteins and glycolipids, while integrin bind to specific proteins.

Selectins and integrins act in sequence to let white blood cells leave the bloodstream and enter tissues. The selectins mediate a weak adhesion because the binding of the lectin domain of the selectin to its carbohydrate ligand is of low affinity. This allows the white blood cell to adhere weakly and reversibly to the endothelium, rolling along the surface of the blood vessel propelled by flow of blood. The rolling continues until WBC activates its integrins (Fig. 14.16), now causing the cell to bind strongly to the endothelial cell surface and to crawl out of the blood vessel between adjacent endothelial cells (Fig. 14.15).

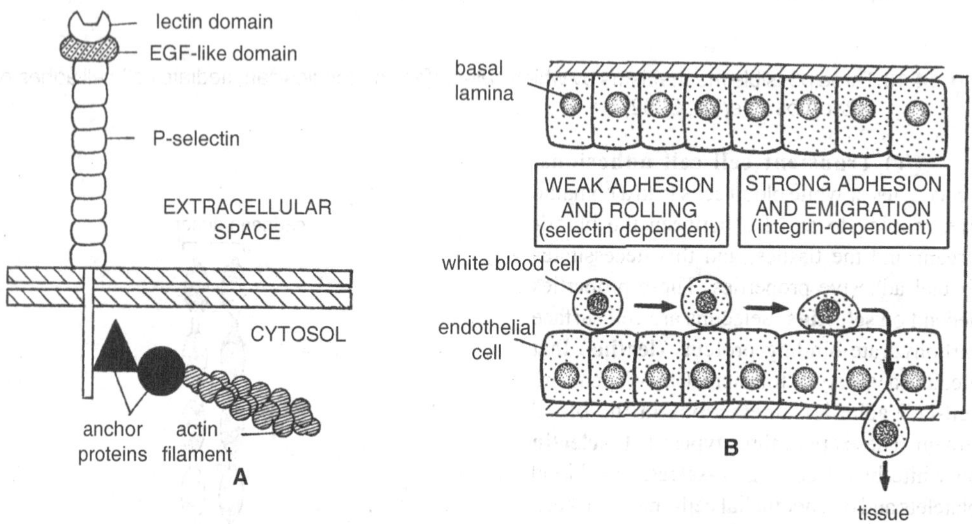

Fig. 14.15. The structure and function of selectins. A—The structure of P-selectin. The selectin attaches to the actin cytoskeleton through anchor proteins that are still poorly characterized. B—How selectins and integrins mediate the cell-cell adhesions required for a white blood cell to migrate out of the blood stream into a tissue (after Alberts *et al.*, 2002).

2. Immunoglobin Superfamily

Cadherins, selectins and integrins all depend on extra-cellular Ca^{2+} (or Mg^{2+} for some integrins) to function in cell adhesion. The molecules responsible for **Ca^{2+}-independent cell-cell adhesion** belong mainly to the large and ancient **immunoglobulin (Ig) superfamily** of proteins. These proteins contain one or more Ig-like domains that are characteristics of antibody molecules. One of the best-studied examples is the **neural cell adhesion molecule (N-CAM),** which is expressed by a variety of cell types, including most nerve cells. N-CAM is the most common of the Ca^{2+}-independent cell-cell adhesion molecule in vertebrates, and, like cadherin, it is thought to bind cells together by a homophilic mechanism (between N-CAM molecules on adjacent cells). Some Ig-like cell-cell adhesion proteins, however, use a heterophilic mechanism. For example, **intercellular adhesion molecules** or **ICAMs** on endothelial cells bind to integrins on white blood cells when WBCs migrate out of the blood stream.

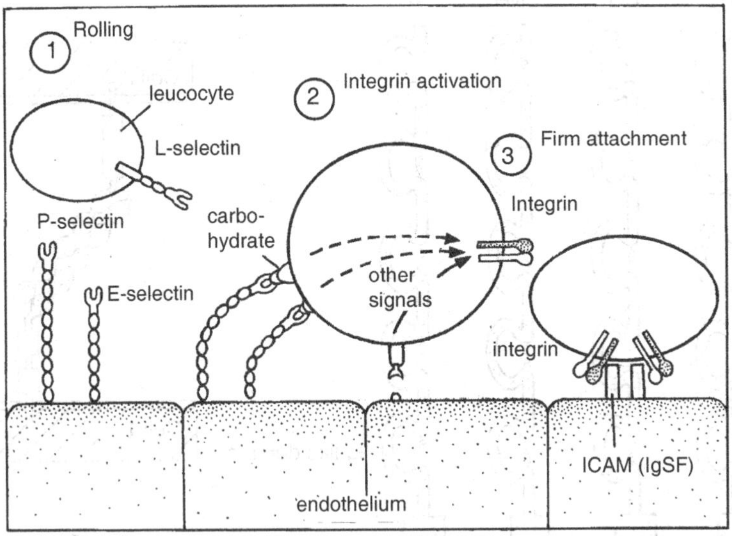

Fig. 14.16. Leucocyte adhesion and selectin. The initial attachment of leucocytes to endothelial cells that line blood vessels is mediated by selectins. (1) Attached leucocytes or white blood cells roll along the blood vessel wall, where they sense activating factors that are deposited on the surface of endothelial cells. (2) This leads to activation of leucocyte integrins, which (3) bind to IgSF proteins, (such as ICAMs), allowing leucocytes to adhere firmly to endothelial cells. Such adhesion allows leucocytes to stop and to pass through blood vessels to sites of inflammation (after Becker *et al.*, 2006).

Types of N-CAM. These are about 20 forms of N-CAM, all generated by alternative splicing of an RNA transcript produced from a single gene. In all forms, the large extracellular part of the polypeptide chain is folded into five Ig-like domains (Fig. 14.17). Some forms of N-CAM carry on unusually large quantity of sialic acid. By virtue of their negative charge, these long polysialic acid chains prevent cell adhesion, and there is increasing evidence that N-CAM heavily loaded with sialic acid serves to prevent adhesion, rather than cause it.

While cadherins and Ig family members are frequently expressed on the same cells, the adhesions mediated by cadherins are much stronger, and they are largely responsible for holding cells together, segregating cell collectives into discrete tissues, and maintaining tissue integrity. N-CAM and other members of the Ig family seem to contribute more to the **fine-tuning** of these adhesive interactions during development and regeneration. For example, in the developing rodent pancreas, the formation of the **islets of Langerhans** requires cell aggregation, followed by cell sorting. Whereas inhibition of cadherin function prevent cell aggregation and islet formation, loss of N-CAM only impairs the cell sorting process, so that disorganized islets forms.

Likewise, whereas mutant mice that lack N-cadherin die early in development, mutant mice that lack N-CAM develop normally, although they do have some defects in neural development. Mutations in other genes that encode Ig-like cell adhesion proteins, however, can cause more severe neural defects. *L1* gene mutations in human, for example, cause mental retardation and other neurological defects resulting from abnormalities in the migration of nerve cells and their axons.

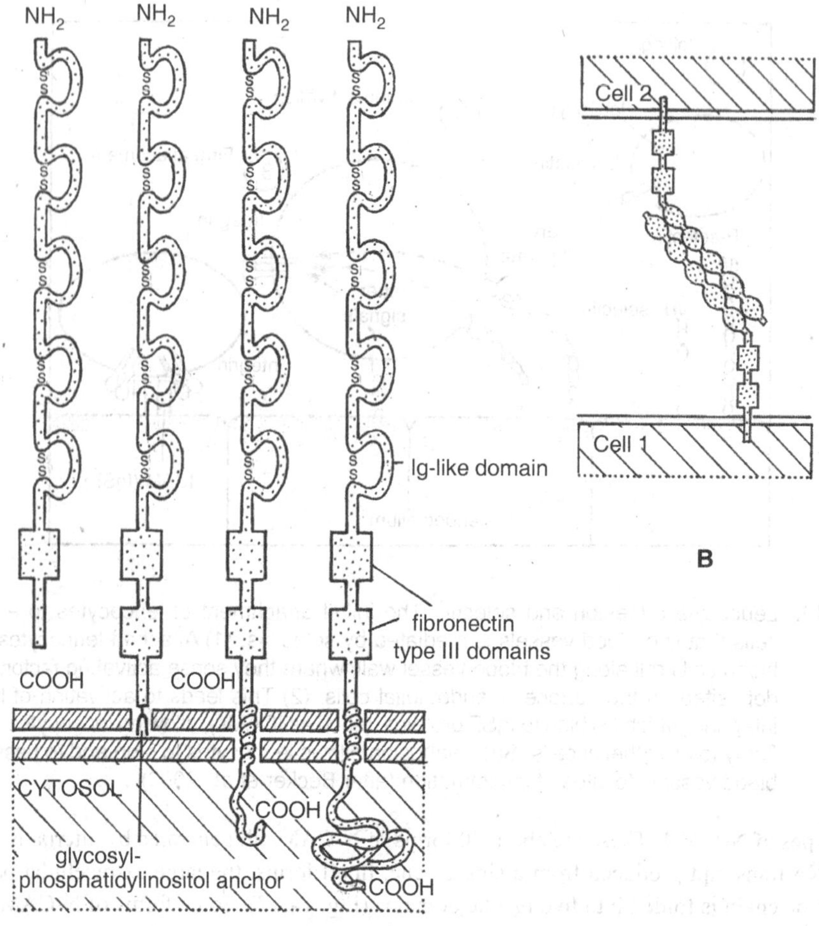

Fig. 14.17. The cell adhesion protein N-CAM. A—Four forms of N-CAM. The extracellular part of the polypeptide chain in each case is folded into five Ig-like domains (and one or two other domains called fibronectin type III repeats). Disulphide bonds (S–S) connect the ends of each loop that forms an Ig-like domain. B—A model for the homophilic interaction that allows N-CAM to mediate cell-cell adhesion (after Alberts *et al.*, 2002).

The significance of Ig-like cell adhesion proteins in connecting the neurons of the developing nervous system has been demonstrated dramatically in *Drosophila.* An N- CAM-like protein called **fasciclin III (FAS3)** is expressed temporarily on some motor neurons, as well as on the muscle cells they normally innervate. If FAS3 is genetically removed from these neurons, they fail to recognize their muscle targets and do not make synapses with them. Conversely, if motor neurons that normally do not express FAS3 are made to express this protein, they now synapse with FAS3-expressing muscle cells to which they normally do not connect. It seems that FAS3 mediates these synaptic connections by a **homophilic "matchmaking" mechanisms** (see **Alberts** *et al.*, 2002).

2. Cell-matrix junctions (Focal adhesions). Some type of anchoring junctions bind cells to the extracellular matrix (ECM) rather than to other cells. The transmembrane adhesion proteins in these cell-matrix junctions are **integrins**. Integrins form a large family of proteins distinct from the cadherins. **Focal adhesions** enable cells to get a hold on the ECM through integrins that link intracellularly to actin filaments. In this way muscle cells, for example, attach to their tendons at the **myotendinous junction**. Likewise, when cultured fibroblasts migrate on an artificial substratum coated with extracellular matrix (ECM) molecules, they also grip the substratum as focal adhesions, where bundles of actin filaments terminate. At all such adhesions, the extra-cellular domains of transmembrane integrin protein bind to a protein component of the extracellular matrix, while their intracellular domains bind indirectly to bundles of actin filaments via the intracellular anchor protein **talin, α-actinin, filamin** and **vinculin** (Fig. 14.18).

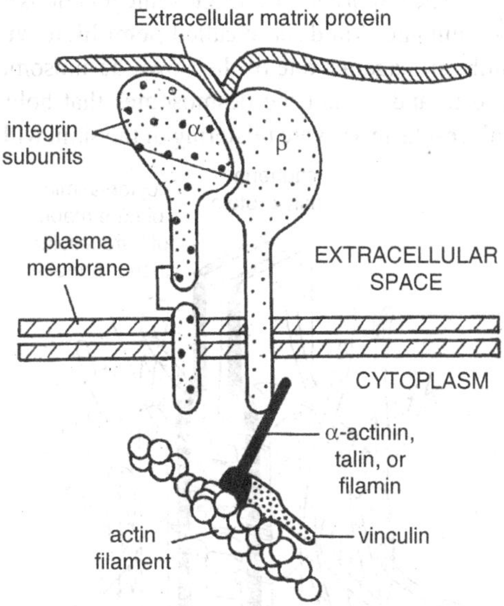

Fig. 14.18. Focal adhesion. Diagram showing some of the proteins that form focal adhesions. The transmembrane adhesion protein is an integrin heterodimer, composed of an α and a β subunit. Its extracellular domains bind to components of the extracellular matrix, while cytoplasmic tail or β subunit binds indirectly to actin filament via several intra-cellular anchor proteins such as vinculin and α-actinin, talin or filamin (after Alberts, *el al.*, 2002).

II. Intermediate Filament Attachment Sites of Anchoring Junctions

1. Desmosomes (or Maculae adherens). Desmosomes are disc-shaped adhesive junctions approximately 1 μm in diameter having a narrow (30 nm) extracellular gap. They are particularly numerous in tissues that are subjected to mechanical stress, such as cardiac muscle and the epithelial layers of the skin and uterine cervix (see **Karp**, 2002). In a way, desmosomes are buttonlike points of intercellular contact that rivet cells together (Fig. 14.19). Inside the cell, desmosomes serve as anchoring sites for ropelike **intermediate filaments (IFs)**, which form a structural framework of great tensile strength (Fig. 14.19A). Through desmosomes, the intermediate filaments of adjacent cells are linked into a net that extends throughout the many cells of a tissue. The particular type of intermediate filaments attached to the desmosomes depends on the cell type: for example, they are **keratin filaments** in most epithelial cells and **desmin filaments** in heart muscle cells.

The desmosome has a dense cytoplasmic plaque composed of a complex of intracellular anchor proteins (**plakoglobin** and **desmoplakin**) that are responsible for connecting the cytoskeleton to the transmembrane adhesion proteins (Fig. 14.19B). These adhesion proteins (**desmoglein** and **desmocollin**) like those at an adherens junction belong to the cadherin family. They interact through their extracellular domains to hold the adjacent plasma membrane together.

The importance of desmosome junctions is demonstrated by some forms of the potentially fatal autoimmune skin disease called **pemphigus vulgaris** of human beings. Affected individuals make antibodies against one of their own desmosomal cadherin protein (desmogleins). These antibodies bind to and disrupt the desmosomes that hold their skin epithelial cells (keratinocytes) together. This results in a severe blistering of the skin, with leakage of body fluids into the loosened epithelium.

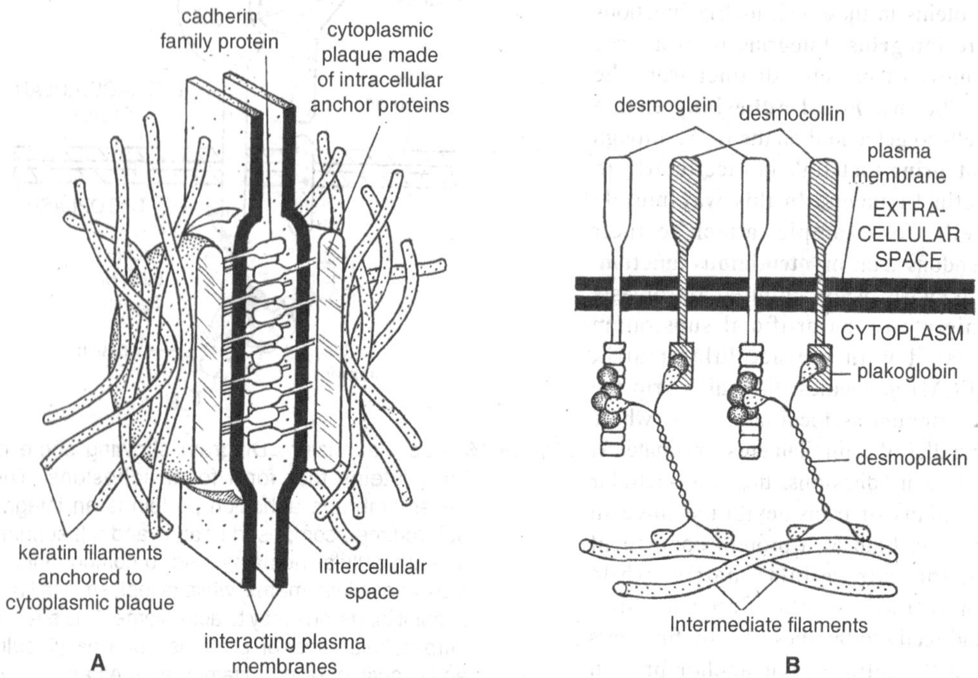

Fig. 14.19. Desmosomes. A—The structural component of a desmosome. On the cytoplasmic surface of each interacting plasma membrane is a dense plaque composed of a mixture of intracellular anchor proteins. A bundle of keratin/desmin intermediate filaments is attached to the surface of each plaque. Transmembrane adhesion proteins of the cadherin family bind to the plaques and interact through their extracellular domains to hold the adjacent membranes together by a Ca^{2+}-dependent mechanism. B—Some of the molecular components of a desmosome. Desmoglein and desmocollin are members of the cadherin family. Their cytoplasmic tails bind plakoglobin, which in turn binds to desmoplakin. Desmoplakin also binds to the sides of intermediate filaments, thereby tying the desmosome to these filaments (after Alberts *et al.*, 2002).

2. Hemidesmosomes. Hemidesmosomes or half-desmosomes resemble desmosomes morphologically and in connecting to intermediate filaments and, like desmosomes, they act as rivets to distribute tensile and shearing forces through an epithelium (Box 14.1). Instead of joining adjacent epithelial cells, however, hemidesmosomes connect the basal surface of an epithelial cell to the underlying basal lamina (Fig. 14.20). The extracellular domains of the $\alpha_6\beta_4$ **integrins** that mediate the adhesion bind to a **laminin** protein in the basal lamina, while an intracellular domain binds via an anchor protein (**plectin**) to keratin intermediate filaments. Whereas the keratin filaments associated with desmosomes make lateral attachments to the desmosomal plaques, many keratin filaments associated with hemidesmosomes have their ends buried in the plaque.

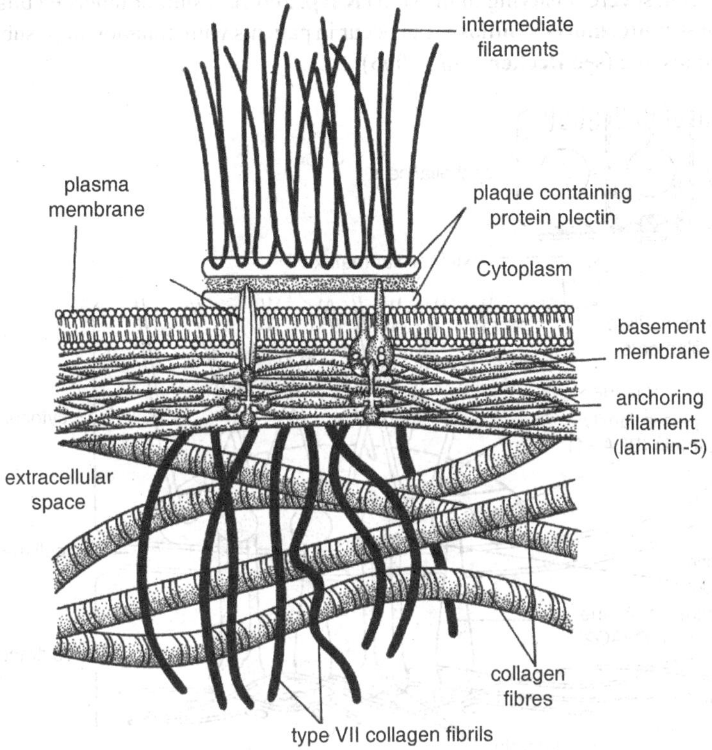

Fig. 14.20. Hemidesmosomes are differentiated sites at the basal surfaces of epithelial cells where the cells are attached to the underlying basement membrane. Schematic diagram showing the major components of a hemidesmosome connecting the epidermis to the underlying dermis. The $\alpha_6\beta_4$ integrin molecules of the epidermal cells are linked to cytoplasmic keratin intermediate filaments by a protein called plectin that is present in plaque and to the basement membrane by anchoring filaments of a particular type of laminin. A second transmembrane protein (BP180) is also present in hemidesmosomes. The collagen fibers are part of the underlying dermis (after Karp, 2003).

Box 14.1

Within body where occur **tightest attachment** between a cell and its extracellular matrix is seen at the basal surface of epithelial cells where the cells are anchored to the underlying basement membrane by a specialised adhesive structure called the **hemidesmosome**.

In addition, another transmembrane protein called BPAG2* and its associated plakin, called BPAG1 can serve as bridge between keratin and laminin (Fig. 14.21). (* BPAG stands for *b*ullous *p*emphigoid *a*ntigen).

The importance of hemidesmosomes is revealed by a rare disease of elderly people, called **bullous pemphigoid**, in which individuals produce antibodies that bind to proteins (the bullous pemphigoid antigens) present in these adhesive structures. Disease is caused by production of antibodies directed against one's own tissues (*i.e.*, autoantibodies), are called **autoimmune disorders** and are responsible for a wide variety of conditions. The presence of autoantibodies causes the lower layer of the epidermis to lose attachment to the underlying basement membrane (and thus to the underlying connective tissue layer of the dermis). The leakage of fluid into the space beneath the epidermis results in severe blistering of the skin (**Karp**, 2002). A similar inherited blistering disease, called **junctional epidermolysis bullosa**, can occur in patients with mutation in β_4 subunit of integrin of the hemidesmosome (see **Becker** *et al.*, 2006).

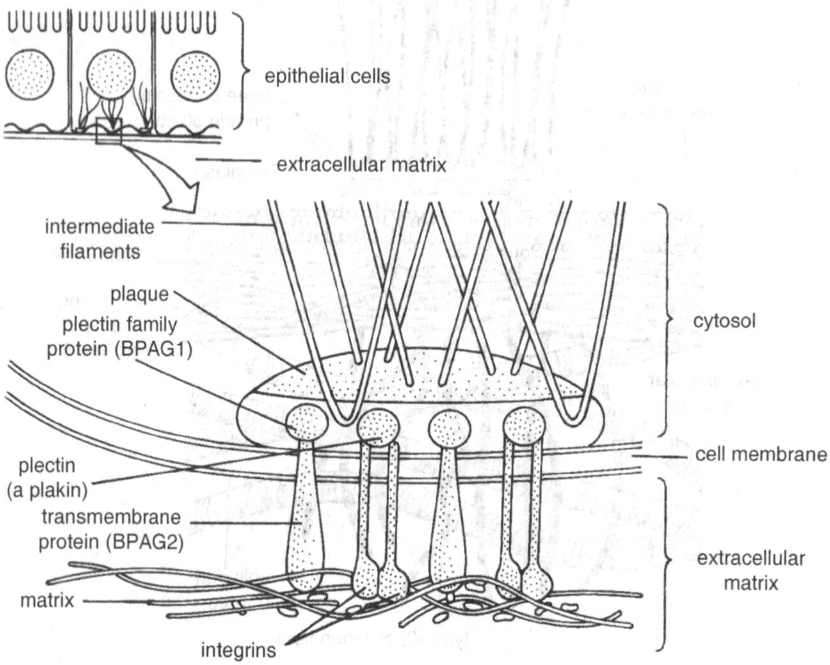

Fig. 14.21. Ultrastructure of a hemidesmosome (after Becker *et al.*, 2006).

C. Communicating Junctions

1. Gap Junctions. Many cells of the tissues of higher animals are coupled together by interconnecting **gap junctions, nexus** or **communicating junctions**. The presence of gap junctions explains the ionic or electronic connections between adjacent cells, *i.e.*, there are some cells which are **electrically coupled** and have regions of low resistance in the membrane through which there is a rather free flow of electrical current carried by ions. Such electrical coupling is found extensively in embryonic cells. In adult tissues it is usually found in epithelia, cardiac cells and liver cells. Skeletal muscles and most neurons do not show electrical coupling.

Gap junctions are found to permit molecules such as inorganic ions, sugars, amino acids, nucleotides and vitamins to pass with comparative freedom between one cell and another within a tissue, but they prevent larger molecules, such as proteins, nucleic acids and polysaccharides from being transferred. This observation also explains the phenomenon of **metabolic cooperation** or **metabolic coupling** between cells, *i.e.*, cells can transfer to neighbouring cells, the molecules which cannot be synthesized by the recipient cells. For example, in the tissue-culture experiments, the mutant cells which are deficient in the enzyme **thymidine kinase** can be shown by autoradiography to be capable of DNA synthesis only when grown in a culture vessel together with the wild type cells. This observation shows that required thymidine has been passed from a wild-type cell to a mutant cell, presumably via gap junctions. There are certain other molecules such as IP_3, AMP, ADP, ATP and cAMP that can pass through gap junctions.

A gap junction appears as a plaque-like contact in which the plasma membranes of adjacent cells are in close apposition, separated by a space of only 2 to 4 nm. Structurally, gap junctions consist of hollow channels around which a series of six protein subunits are located; a channel has a diameter of about 1.5 to 2 nm. A single major protein (a macromolecular unit, called **connexon**) of 27000 daltons has been isolated from rat liver preparations consisting of almost pure gap junction material. A connexon appears as an annulus of six **connexin** subunits surrounding the channel. It is believed that the sliding of the subunits caused the channel to open and close. The permeability of channel of gap junction is regulated by Ca^{2+} ions; if the intercellular Ca^{2+} ion level increases, the permeability is reduced or abolished. The gap junctions or connexons of adjacent cells are believed to line up to provide a continuous channel, made up of two connexons opposed end to end (Fig. 14.22).

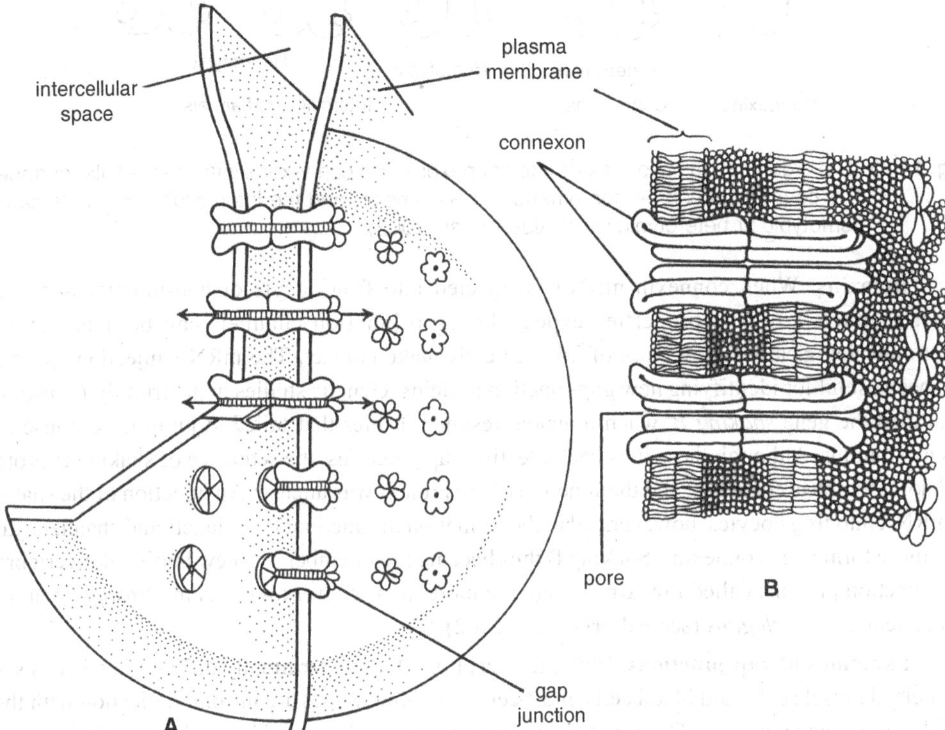

Fig. 14.22. Gap junctions. A—Location of gap junctions between two cells. The channels are made by particles in each membrane that traverse the intercellular space. The flow of fluid between the cells is indicated by arrows. B—Finer structure of unit structure or connexion of gap junctions. The channel has a pore about 2 nm and is formed by two hexamers (six subunits) traversing the lipid bilayers of two plasma membranes.

Gap junctions in different tissues can have different properties. The permeability of their individual channels can vary, reflecting differences in the connexins that form the junctions. In humans, for example, there are 14 distinct connexins, each encoded by a separate gene and each having a distinctive, but sometimes over-lapping, tissue distribution. Most cell types express more than one type of connexin, and two different connexin protein can assemble into a **heteromeric connexon**, the properties of which differ from those of a **homomeric connexon** (Fig. 14.23) constructed from a single type of connexin.

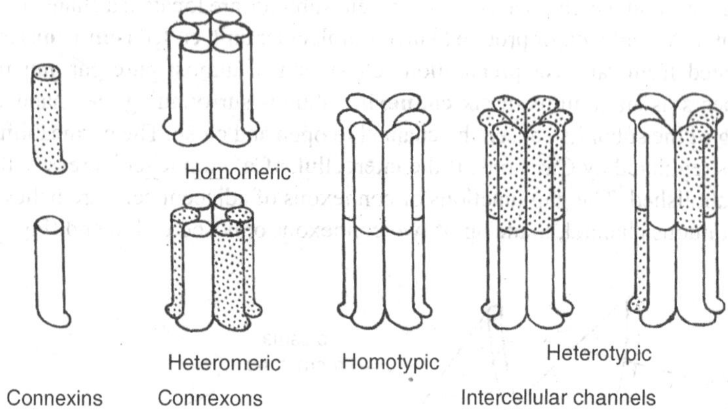

Fig. 14.23. The organization of connexins into connexons and connexons into inter-cellular channels. The connexons can be homomeric or heteromeric, and the intercellular channels can be homotypic or heterotypic (after Albert *et al.*, 2002).

Innexins. When connexin mRNA is injected into frog oocyte or gap-junctions-deficient cells, channels with the properties expected of gap-junction channels can be demonstrated electrophysiologically where pairs of injected cells make contact. The mRNA injection approach has been useful for identifying new gap-junction proteins. Genetic studies in the fruit fly *Drosophila* identified the gene *shaking B*, when mutated, resulted in flies that failed to jump in response to a visual stimulus. Although these flies had defective gap junctions, the sequence of shaking B protein did not resemble a connexin, and the function of the protein was unclear. An injection of the shaking B mRNA into frog oocytes, however, led to the formation of functional gap-junctional channels, just-like those formed by connexins. Shaking B thus became first member of a new family of invertebrate gap-junction proteins called **innexins.** There are more than 15 innexin genes in *Drosophila* and 25 in the nematode *C.elegans* (see **Alberts** *et al.*, 2002).

Functions of gap junctions. With the exception of a few terminally differentiated cells such as skeletal muscle cells and blood cells, most cells in animal tissues are in communication with their neighbour via gap junctions. Gap junctions occur in most vertebrate and invertebrate cell types. They are especially abundant in tissues such as muscle and nerve where extremely rapid communication between cells is required. In heart tissue, for example, gap junctions facilitate the flow of electrical current that causes the heart to beat; in the brain, they they are concentrated in the cerebellum, which is involved in coordinating rapid nuclear activities. These roles are confirmed by analyzing mutations in connexins and in nexins. For example, mice lacking in one type of connexin have defects in conducting electrical impulses in the heart.

Role of gap junctions in the liver. Gap junctions also occur in many tissues that do not contain electrically excitable cells. In principle, the sharing of small metabolites and ions provides a mechanism for coordinating the activities of individual cells in such tissues and for smoothing out random fluctuations in small molecule concentrations in different cells. For example, in the liver, the release of noradrenaline from sympathetic nerve endings in response to a fall in blood glucose levels stimulates hepatocytes to increase glycogen break-down and release of glucose into blood. Not all liver cells (hepatocytes) are innervated by sympathetic nerves, however. By means of the gap junctions that connect hepatocytes, the signal is transmitted from the innervated hepatocytes to the non-innervated ones. Thus, mice with a mutation in the major connexin gene expressed in the liver fail to mobilize glucose normally when blood glucose levels fall.

2. Role of gap junctions in mammalian ovary. The normal development of ovarian follicles also depends on gap-junction-mediated communication—in this case, between the oocyte and the surrounding granulosa cell. A mutation in the gene that encodes the connexin that normally couples these two cell types causes **infertility** (Fig. 14.24).

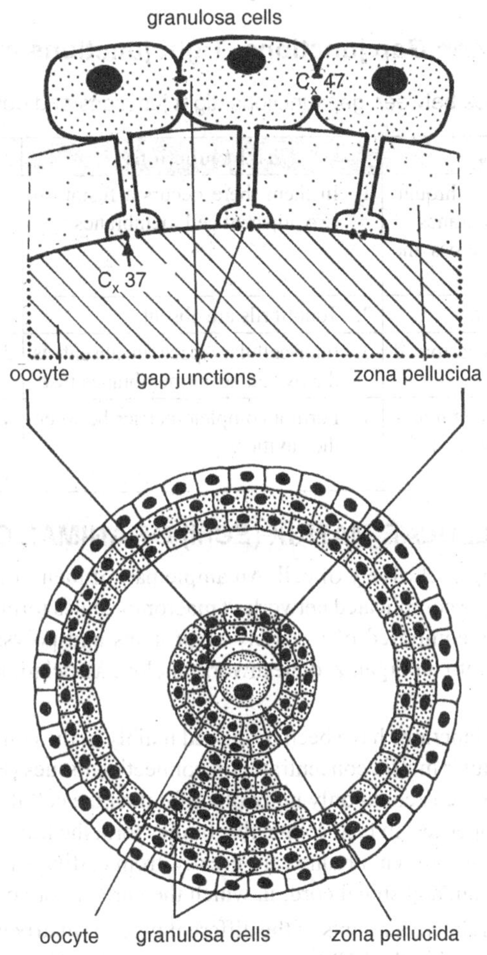

Fig. 14.24. Gap junction coupling in the ovarian follicle. The mammalian ovum is surrounded by a thick layer of extracellular or matrix called the zona pellucida. The surrounding granulose cells are coupled to each other by gap junctions formed by protein connexin 43 (C_x43). In addition, the granulosa cells extend processes through the zona pellucida and make gap junctions with the oocyte. These gap junctions contain a different connexin protein, called C_x 37. Mutation in the gene encoding C_x 37 causes infertility by disrupting the development of both the granulosa cells and the oocyte (after Alberts *et al.*, 2002).

3. Role of gap junctions in embryogenesis. Cell coupling via gap junctions also seem to be important in embryogenesis. In early vertebrate embryos, beginning with the late eight cell stage in mouse embryos, most cells are electrically coupled to one another. As specific groups of cells in the embryo develop their distinct identities and begin to *differtiate*, they commonly uncouple from surrounding tissue. As the neural plate folds up and pinches off to form the neural tube (Fig. 14.10) for example, its cells uncouple from the overlying epidermal ectoderm. Meanwhile, the cells within each group remain coupled with one another and therefore tend to behave a cooperative assembly all following a similar developmental pathway in a coordinated fashion.

Distinctions between Gap junctions, Tight junctions and Desmosomes

Table 14.3. Differences between gap junctions, tight junctions and desmosomes.

Gap junctions	Tight junctions	Desmosomes
1. Have channels or pores through the two plasma membranes of adjacent cells and across the inter-cellular space.	1. In them there occurs a fusion of two adjacent cell membranes.	1. They contain buttonlike electrondense regions in each of the opposing cell membranes.
2. Tonofibrils are absent.	2. Tonofibrils are absent.	2. Tonofibrils are present.
3. Inter-celluar space is narrowed to 5 nm.	3. Inter-cellular space in absent as the two adjacent membranes fuse.	3. Inter-cellular space 30 to 35 nm.
4. Provide electrical communi-cation between cells and for flow of ions.	4. Form a complete barrier between the cavities.	4. Hold cells tightly together.

14.2. EXTRACELLULAR MATRIX (ECM) OF ANIMAL CELLS

Tissues are not made up exclusively of cell. An ample part of their volume is **extracellular space,** which is mostly filled by a complicated network of macromolecules forming the **extracellular matrix (ECM)**. This matrix is composed of a variety of proteins and polysaccharides that are secreted locally and assembled into an organized meshwork in close association with the surface of the cell that produced them.

Whereas the cell junctions have been discussed mainly in the context of epithelial tissues, this account of the extracellular matrix concentrates on connective tissues (Fig. 14.25). The extracellular matrix in connective tissue is commonly more plentiful than the cell it surrounds, and it determines the tissue's physical properties. In fact, connective tissues form the framework of the vertebrate body, but the amounts found in different organs vary greatly-from cartilage to bone, in which they are the major component, to brain and spinal cord, in which they are minor component.

Variations in the relative amounts of the different types of matrix macromolecules and the way in which they are organised in the ECM give rise to a surprising diversity of forms, each adapted to the functional requirements of the particular tissue. For example, matrix can become calcified to form the rock-hard structures of bone or teeth, or it can form the transparent matrix of the cornea, or it can adopt the ropelike organization that gives tendons their massive tensile strength. At the interface between an epithelium and connective tissue, the matrix forms a **basal lamina,** which is important in controlling cell behaviour.

Once it was believed that ECM of vertebrates serve mainly as a relatively inert scaffold to stabilize the physical structure of tissue. But now, it is well established that the matrix has a far more active and complex role in regulating the behaviour of the cells that contact it, influencing their survival, development, migration, proliferation, shape and function.

The origins of the ECM are very ancient and virtually all multicellular organisms make it; examples include the cuticles of worms and insects, the shell of mollusks and the cell walls of plants.

1. Cells and molecules of ECM. The macromolecules that constitute the ECM are mainly produced locally by cells in the matrix. In most connective tissues, for example the matrix macromolecules are secreted largely by cells called **fibroblasts**. In certain specialized types of connective tissues, such as cartilage and bone, however, they are secreted by cells of the fibroblast family that have more specific names: **chondroblasts** form cartilage and **osteoblasts** form bone.

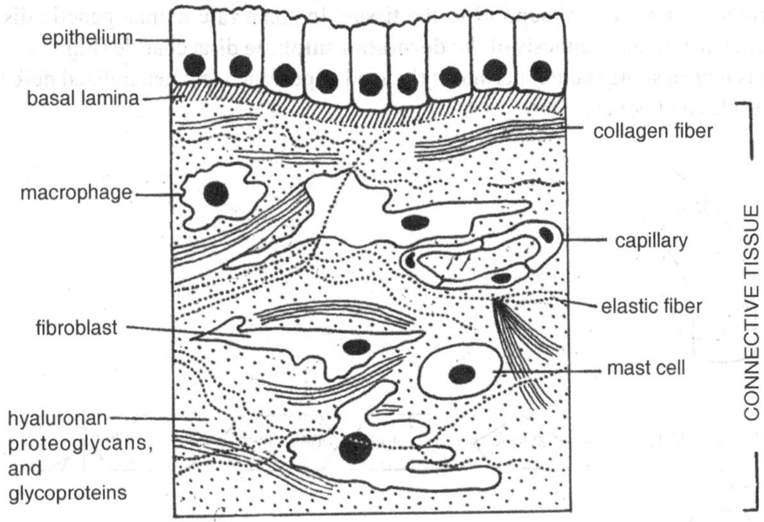

Fig. 14.25. The connective tissue underlying an epithelium. This tissue contains a variety of cells and ECM components. The predominant cell type is the fibroblasts, which secretes abundant extracellular matrix (after Alberts *et al.*, 2002).

Two main classes of extracellular macromolecules make up the matrix: 1. polysaccharide chains of the class called **glycosaminoglycans (GAGs)**, which are usually found covalently linked to protein in the form of **proteoglycans;** and 2. **fibrous proteins** including **collagen, elastin, fibronectin** and **laminin**, which have both structural and adhesive functions.

The proteoglycan molecules in connective tissue form a highly hydrated, gel-like **"ground substance"** in which fibrous proteins are embedded. The polysaccharide gel resists compressive forces on the matrix while permitting the rapid diffusion of nutrients, metabolites and hormones between the blood and the tissue cells. The collagen fibres both strengthen and help organize the matrix, and rubber like elastic fibers give it resilience. Finally, many matrix proteins help cells attach in the appropriate locations.

(*i*) GAGs (Glycosaminoglycans). These are unbranched polysaccharide chains composed of repeating disaccharide units. They are called GAGs because one of two sugars in the repeating disaccharide is always on **amino sugar** (*N*-acetylglucosamine or *N*-acetylgalactosamine), which in most cases is sulphated. The second sugar is usually a **uronic acid** (glucuronic or iduronic). Because there are sulphate or carboxyl groups on most of their sugars, GAGs are highly negative charged. Indeed, they are *the most anionic molecules produced by animal cells.* Four main groups of GAGs are distinguished according to their sugars, the type of linkage between the sugars and the number and location of sulphate groups: 1. **hyaluronan** (or hyaluronic acid); 2. **chondroitin sulphate** and **dermatan sulphate**; 3. **heparan sulphate** and 4. **keratin sulphate.**

GAGs are strongly hydrophilic. They tend to adopt highly extended conformations that occupy a huge volume relative to their mass, and they form **gels** even at very low concentrations. Their high density of negative charges attracts a cloud of cations, most notably Na^+, that are osmotically active causing large amounts of water to be sucked into the matrix. This creates a swelling pressure or **turgor**, that enables the matrix to withstand compressive forces (in contrast to collagen fibrils, which resist stretching forces). The cartilage matrix that lines the knee joint, for example, can support pressures of hundreds of atmospheres in this way.

The GAGs in connective tissue usually constitute less than 10% of the weight of the fibrous proteins. But, because they form porous hydrated gels, the GAG chains fill most of the extracellular space, providing mechanical support to the tissue. In some rare human genetic disease, there is a severe deficiency in the synthesis of the **dermatan sulphate** disaccharide (Fig. 14.26). The affected individuals have a short stature, permanently aged appearance and generalized defects in their skin, joints, muscles and bones.

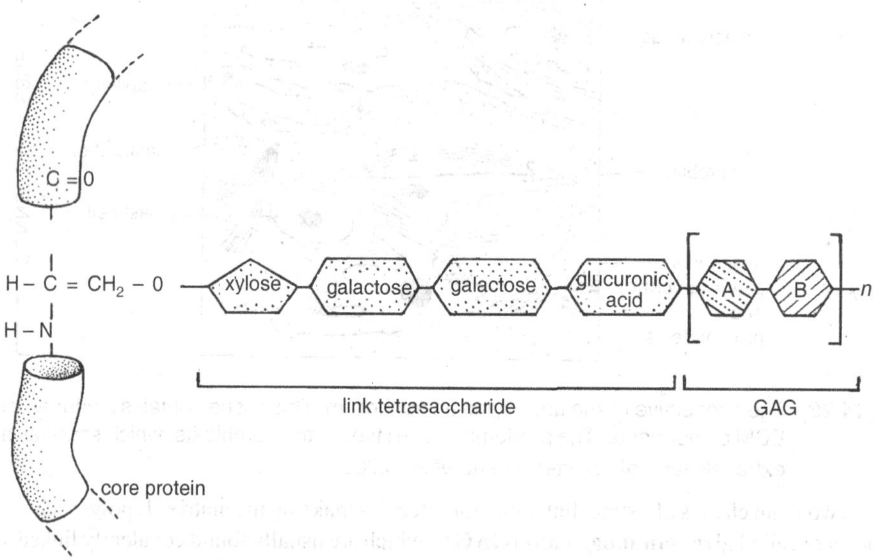

Fig. 14.26. The linkage between a GAG chain and its core protein in a proteoglycan molecule. A specific link tetrasaccharide is first assembled on a serine side chain. The rest of the GAG chain, consisting of mainly of a repeating disaccharide unit is then synthesized, with one sugar being added at a time (after Alberts *et al.*, 2002).

***(ii)* Proteoglycans.** Except for hyaluronan, all GAGs are found covalently attached to proteins in the form of **proteoglycans**, which are made by most animal cells. The polypeptide chain, called **core protein**, of a proteoglycan is made on membrane-bound ribosome and threaded in the lumen of the endoplasmic reticulum. The polysaccharide chains are mainly assembled on this core protein in the Golgi apparatus. First a special link **tetrasaccharide** is attached to a serine side chain on the core protein to serve as a primer for polysaccharide growth; then one sugar at a time is added by specific glycosyl transferase (enzyme) (Fig. 14.27). While still in the Golgi apparatus, many of the polymerized sugars are covalently modified by a sequential and coordinated series of reactions.

The GAG chains of proteoglycans can form gels of varying pore size and charge density; so, they serve as **selective sieves** to regulate the traffic of molecules and cells according to their size, charge, or both. Evidence suggests that a heparin sulphate proteoglycan called **perlecan** has this role in the basal lamina of the kidney glomerulus, which filters molecules passing into the urine from the blood stream.

Proteoglycans are thought to have a major role in chemical signalling between cell. They bind various secreted signal molecules, such as certain protein growth factors and can enhance or inhibit their signaling activity.

Proteoglycans as coreceptors. Not all proteoglycans are secreted components of the ECM. Some (*e.g.*, syndecans, betaglycans, etc.) are integral components of plasma membranes and have their core protein either inserted across the lipid bilayer or attached to the lipid bilayer by a GPI (*i.e.*, glycosylphosphatidylinositol) anchor. Some of these plasma membrane proteoglycans act as **coreceptors** that collaborate with conventional cell-surface receptor proteins, in both binding cells to the ECM and initiating the response of cells to some extracellular signal proteins. **Syndecans** are located on the surface of many types of cells, including fibroblasts and epithelial cells, where they serve as receptors for matrix proteins (fibronectin, integrins). Likewise, **betaglycan** binds TGF-β (transforming growth factor-β).

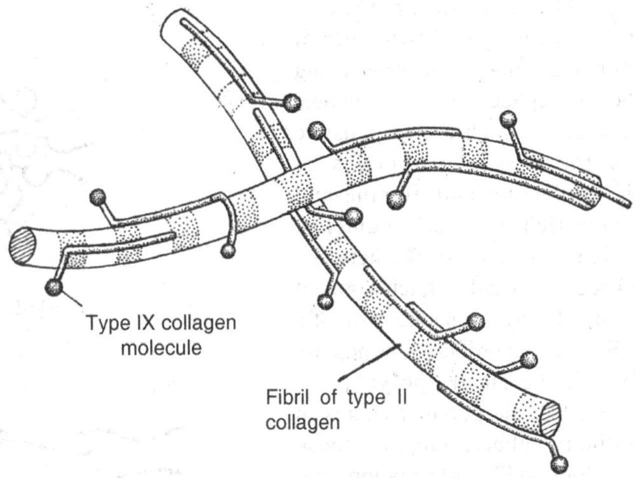

Fig. 14.27. Type IX collagen. Type IX collagen molecules binding in a periodic pattern to the surface of a fibril containing type II collagen (after Alberts *et al.*, 2002).

***(iii)* Collagen.** The **collagens** are a family of fibrous protein found in all multicellular animals. They are secreted by connective tissue cells as well as by a variety of other cell types. As a major component of skin and bone, they are the most abundant protein in mammals, constituting 25% of the total protein mass in these animals (see Chapter 7 for more details).

The **fibrillar collagens** (type I, II, III, V and XI collagens) are ropelike, triple-stranded helical molecules that aggregate into long fibrils in the extracellular space. The fibrils in turn can assemble into a variety of highly ordered arrays. **Fibril-associated collagen** molecules, such as type IX and XII, decorate the surface of collagen fibrils and influence the interactions of the fibrils with one another and with other matrix components (Fig. 14.28). In contrast, **type IV collagen** molecule assemble into a sheetlike meshwork that is a crucial component of all mature basal laminae.

***(iv)* Elastin.** Many vertebrate tissues, such as skin, blood vessels and lungs, need to be both strong and elastic in order to function. A network of **elastic fibers** in the extra-cellular matrix of these tissues gives them the required resilience so that they can recoil after transient stretch. Elastic fibers are at least five times more extensible than a rubber band of the same cross-sectional area. Long, inelastic collagen fibrils are interwoven with the elastic fibers to limit the extent of stretching and prevent the tissues from tearing.

The main component of elastic fibers is **elastin,** a highly hydrophobic protein (about 750 amino acid long), which like collagen, is unusually rich in proline and glycine, but unlike collagen, is not glycosylated and contain some hydroxy-proline but no hydroxylysine. The biosynthetic precursor of elastin is soluble **tropoelastin** which is secreted into the extracellular space and assembled into elastic fibers close to the plasma membrane, generally in cell-surface infoldings. After secretion, the tropoelastin molecules, generating an extensive network of elastin fibers and sheets. The cross-links are formed between lysines.

The elastin protein is composed mostly of two types of short segments that alternate along the polypeptide chain: hydrophobic segments, which are responsible for the elastic properties of the molecule; and alanine-and lysine rich α-helical segments, which form cross-links between adjacent molecules. Each segment is encoded by a separate exon of a split gene. According to one popular model, the elastin polypeptide chain, like the polymer chains in ordinary rubber, adopts a loose "random coil" conformation, and it is the random coil structure of the component molecules cross-linked into the elastic fiber network that allows the network to stretch and recoil like a rubber band (Fig. 14.28).

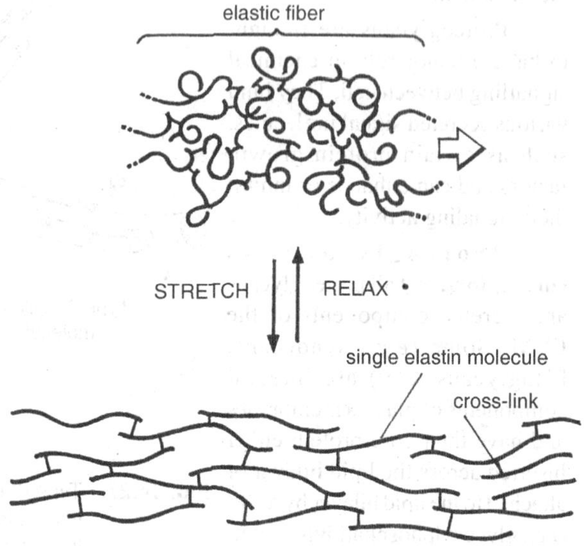

Fig. 14.28. Stretching a network of elastin molecule. The molecules are joined together by covalent bonds to generate a cross-linked network. Each elastin molecule in the network can expand and contract as a random coil.

(v) Fibronectin. Fibronectin and laminin are examples of large, multidomain matrix glycoproteins. By means of their multiple binding domains, such proteins help organize the matrix and help cells adhere to it.

Fibronectin protein is found in all vertebrates. It is a dimer composed of two very large subunits joined by disulphide bonds at one end. Each subunit is folded into a series of functionally distinct domains separated by regions of flexible polypeptide chain (Fig. 14.29). The domains in turn consist smaller modules, each of which is serially repeated and usually uncoded by a separate exon, suggesting that the fibronectin gene, like the collagen gene, evolved by multiple exon duplications. All forms of fibronectin are encoded by a single large gene that contains about 50 exons of similar size. Transcription produces a single large RNA molecule that can be

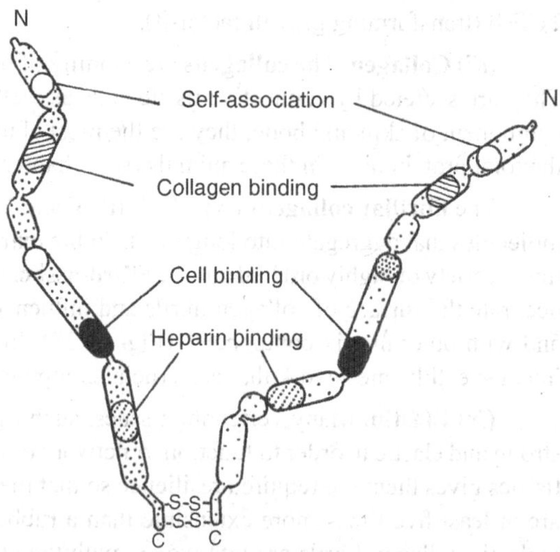

Fig. 14.29. The structure of a fibronectin dimer. The two polypeptide chains are similar but generally not identical (being made from the same gene but from different spliced mRNAs). They are joined by two disulphide bonds near the C-termini. Each chain is almost 2500 amino acids long and is folded into five or six domains connected by flexible polypeptide segments. Individual domains are specialized for binding to a particular molecule or to a cell as indicated for five of the domains (after Alberts *et al.*, 2002).

alternatively spliced to produce the various isoforms of fibronectin. The main type of module, called the **type III fibronectin repeat,** binds to integrins. It is about 90 amino acid long and occurs at least 15 times in each subunit. The type III fibronectin repeat is the most common of all protein domains.

The RGD sequence. One way to analyze a complex multifunctional protein molecule such as fibronectin is to chop it into pieces and determine the function of its individual domains. When fibronectin is treated with a low concentration of a proteolytic enzyme, the polypeptide chain is cut in the connecting regions between the domains, leaving the domains themselves intact. One can then show that one of its domain binds to collagen, another to heparin, another to specific receptors on the surface of various types of cells and so on (Fig. 14.29) Synthetic peptide corresponding to different segments of the cell-binding domain have been used to identify a specific **tripeptide sequence-Arg-Gly-Asp** or **RGD,** which is found in one of the type III repeats as a central feature of the binding site. Even very short peptides containing this **RGD sequence** can compete with fibronectin for the binding site on cells, thereby inhibiting the attachment of the cells to a fibronectin matrix. If these peptides are coupled to a solid surface, they cause cell to adhere to it.

The RGD sequence is not confined to fibronectin. It is found in a number of extracellular protein, including, for example, the blood-clotting factor *fibrinogen.* Fibrinogen peptides containing this RGD sequence have been useful in the development of anti-clotting drugs that mimic these peptides. Snake use a similar strategy to cause their victims to bleed: they secrete RGD-containing anti-clotting proteins called **disintegrins** into their venom.

Forms and functions of fibronectin. There are multiple isoforms of fibronectin. One, called **plasma fibronectin,** is soluble and circulates in the blood and other body fluids, where it is thought to enhance blood clotting, wound healing and phagocytosis. All of the other forms assemble on the surface of cells and are deposited in the extracellular matrix as highly insoluble **fibronectin fibrils.** In these cell-surface and matrix forms, fibronectin dimers are cross-linked to one another by additional disulphide bonds.

Fibronectin molecules assemble into fibrils only on the surface of certain cells. This is because additional proteins are needed for fibril formation, especially fibronectin-binding integrins. In the case of fibroblasts, fibronectin fibrils are associated with integrins at sites called **fibrillar adhesions.** These are distinct from focal adhesions, in that they are more elongated and contain different intracellular anchor proteins. The fibronectin fibrils on the cell surface are highly stretched and under tension. The tension is exerted by the cell and is essential for fibril formation. Some secreted proteins function to prevent fibronectin assembly in inappropriate places. **Uteroglobin,** for example, binds to fibronectin and prevent it from forming fibrils in the kidney. Mice that have a mutation in the uteroglobin gene accumulate insoluble fibronectin fibrils in their kidneys.

Fibronectins are also found to have a role in embryogenesis. Mutant mice that are unable to make fibronectin die early in embryogenesis because their endothelial cells fail to form proper blood vessels.

Fibronectin is important not only for cell adhesion to the matrix but also for guiding cell migrations in vertebrate embryos. Large amounts of fibronectin, for example, are found along the pathway followed by migrating prospective mesodermal cells during amphibian gastrulation (*i.e.*, involution of pharyngeal endoderm and chordamesoderm in frog's gastrulation).

Many matrix proteins are believed to have a role in guiding cell movements during development. The **tenascins** and **thrombospondins,** for example, are composed of several types of short amino acid sequences that are repeated many times and form functionally distinct domains. They can either promote or inhibit cell adhesion, depending on the cell type. Indeed, anti-adhesive interactions are as important as adhesive ones including cell migration (see **Alberts** *et al.*, 2002).

2. Basal Laminae and Laminin. Basal laminae are flexible, thin (40–120 nm thick) mats of specialized extracellular matrix that underlie all epithelial cell sheets and tubes. They also surround individual muscle cells, fat cells and Schwann cells. The basal lamina thus separates these cells and epithelia from the underlying or surrounding connective tissue. In other locations such as the kidney glomerulus, a basal lamina lies between two cell sheets and functions as a highly selective filter (Fig. 14.30). Basal laminae have more than simple structural and filtering functions, however, they are able to determine cell polarity, influence cell metabolism, organize the proteins in adjacent plasma membranes, promote cell survival, proliferation, or differentiation, and serve as specific highways for cell migration.

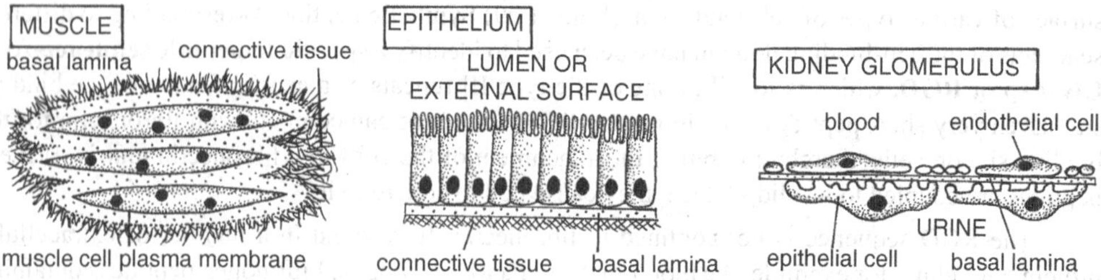

Fig. 14.30. Three ways in which basal laminae are organized. Basal laminae surround certain cells (such as skeletal muscle cells), underlie epithelia and are interposed between two cell sheets (as in the kidney glomerulus). Note that in the kidney glomerulus both cell sheets have gap in them, so that the basal lamina serves as the permeability barrier determining which molecules will pass into the urine from the blood (after Alberts *et al.*, 2002).

The basal lamina is synthesized largely by the cells that rest on it. In some multilayered epithelia, such as the stratified squamous epithelium that forms the epidermis of the skin, the basal lamina is tethered to the underlying connective tissue by specialized anchoring fibrils made of type VII collagen molecules. The term **basement membrane** is often used to describe the composite of the basal lamina and this layer of collagen fibrils (see **Alberts** *et al.*, 2002).

Although precise composition of the basal lamina varies from tissue to tissue and even from region to region, most mature basal laminae contain type IV collagen, the large heparin sulphate proteoglycan perlecan (*e.g.*, agrin) and the glycoproteins laminin and nidogen (or entactin).

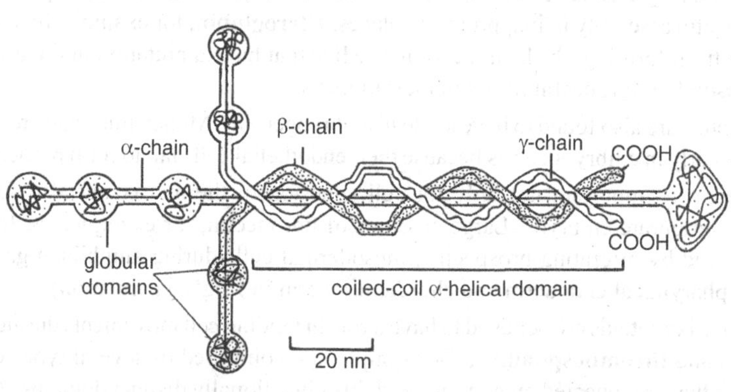

Fig. 14.31. The structure of laminin. The subunits of a laminin-I molecule. This multidomain glycoprotein is composed of three polypeptides (α, β, and γ) that are disulphide bonded into an asymmetric cross like structure. Each of the polypeptide chains, three types of β chains and three types of γ-chains are known; in principle, they can assemble to form 45 (5 × 3 × 3) laminin isoforms. Several such isoforms have been found each with a characteristic tissue distribution (after Alberts *et al.*, 2002).

Laminin. Early in development, basal laminae contain little or no type IV collagen and consist mainly of laminin molecules. **Laminin-I** (classical laminin) is a large, flexible protein composed of three vary long polypeptide chain (α, β, and γ) arranged in the shape of an asymmetric cross and held together by disulphide bonds (Fig. 14.31). Several isoforms of each type of chain can associate in different combinations to form a large family of laminins. The laminin γ-1 chain is a component of most laminin heterotrimers, and mice lacking it die during embryogenesis because they are unable to make a basal lamina. Like many other proteins in the ECM, the laminin in basement membranes consists of several functional domains: one binds to *perlecan*, one to *nidogen*, and two or more laminin receptor proteins on the surface of cells.

Integrins. The linkage of the extracellular matrix to the cell requires transmembrane cell adhesion protein that act as **matrix receptors** and tie the matrix to the cells cytoskeleton. Though some transmembrane proteoglycans function as coreceptors for matrix components, the main receptors on animal cells for binding most extracellular matrix proteins-including collagen, fibronectin and laminins-are the **integrins**. Integrins form a large family of homologous transmembrane, cell-matrix adhesion receptors. In blood cells, integrins also serve as cell-cell adhesion molecules, helping the cells bind to other cells, as well as the ECM.

Integrin are crucially important because they are the main receptor proteins that cells use to both bind to and respond to the extracellular matrix. An integrin molecule is composed of two noncovalently associated transmembrane glycoprotein subunits called α and β (Fig. 14.32). Because the same integrin molecule in different cell types can have different ligand-binding specificities, it

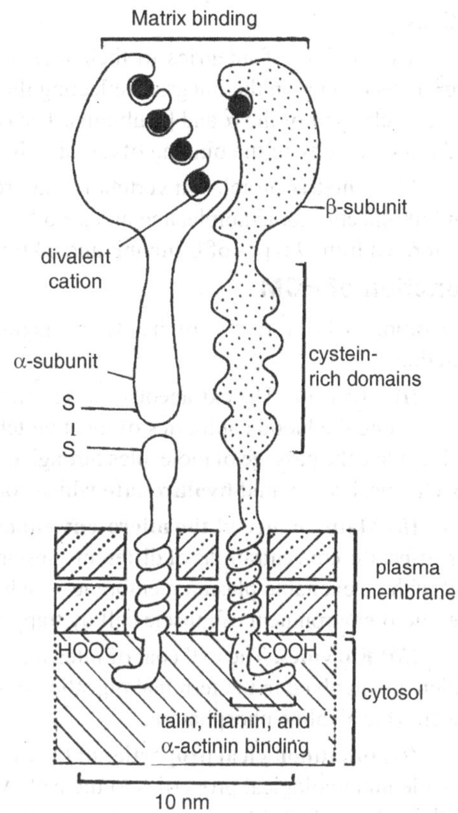

Fig. 14.32. The subunit structure of an integrin cell-surface matrix receptor. Electron micrographs of isolated receptors suggest that the molecule has approximately the shape shown here, with the globular head projecting more than 20 nm from the lipid bilayer. By binding to a matrix protein outside the cell and to the actin cytoskeleton (via the anchor proteins indicated) inside the cell, the protein serves as a transmembrane linker. The α and β subunits are held together by noncovalent bonds. In the fibronectin receptor shown, the a subunit is made initially as a single 1,40,000 dalton polypeptide chain, which is then cleaved into one small transmembrane domain and one large extracellular domain that contains four divalent-cation-binding sites; the two domains held together by a disulphide (S–S) bond. The extracellular part of the β subunit contains a single divalent-cation-binding site, as well as repeating cysteine-rich region where intrachain disulphide bonding occurs (after Alberts *et al.*, 2002).

seems that additional cell-type-specific factors can interact with integrins to modulate their binding activity.

The binding of integrins to their ligands depend on extracellular divalent cations (Ca^{2+} or Mg^{2+}), depending on the integrin, reflecting the presence of divalent-cation-binding domains in the extracellular part of the α and β subunits. The type of divalent cation can influence both the affinity and the specificity of the binding of an integrin to its ligands.

Many matrix proteins in vertebrates are recognized by multiple integrins. At least 8 integrins bind fibronectin, for example, and at least 5 bind laminin. A variety of human integrin heterodimers are formed from 9 types of β subunit and 24 types of α subunits.

Function of ECM

In addition to the protection of the plasma membrane, the cell coat performs the following important functions:

(*i*) **Filtration.** The extraneous coats sometimes act as filters. For instance, the extraneous coats surrounding the blood capillaries of most vertebrates, especially the kidney glomerulus act as filter and regulate the passage of molecules through it. The extracellular coats of connective tissues contain the chemical compound **hyaluronate** which controls the diffusion.

(*ii*) **Maintenance of the micro-environment of the cell.** The extraneous coats of animal cell can affect the concentrations of different substances at the surface of the cell. For example, a muscle cell with its excitable plasma membrane which is surrounded by a glycocalyx is found to maintain the micro-environment of muscle cell by trapping the sodium ions.

(*iii*) **Enzymes.** The cell coat of intestinal microvilli are found to contain a variety of enzymes which are involved in the terminal digestion of carbohydrates and proteins. For example it contains the enzyme alkaline phosphatase.

(*iv*) **Immunological properties of the extraneous coats.** Some substances of extraneous coats provide immunological properties to the cell. As for instance the plasma membrane of mammalian erythrocytes is found to contain some specific, genetically determined substances (carbohydrates and proteins) corresponding to the A, B and O blood groups. The major sialic acid containing glycoproteins of the red blood cell membrane carry the M and N antigens that appear infrequently in humans. The cell coat also contains the receptor sites for the influenza virus and for various lectins.

(*v*) **Histocompatibility.** The cell coats of some cells contain some antigens which provide **histocompatibility**, *i.e.*, they permit the recognition of the cells of one organism and rejection of other cells that are foreign to it (*e.g.*, the rejection of grafts from another organism).

14.3. FUNCTIONS OF CELL

The cell is the fundamental unit of life. It has all the chemical and physical factors necessary for its growth and maintenance so *it is a complete unit of metabolism*. Living organisms have various types of cells which have lot of differences in their shape and size. However, there exist some basic similarities in them at the functional level. Cells are found to contain the following four similar basic functions.

1. Functions as barrier. All cells maintain a barrier that protects the contents of cell from the external environment. Both prokaryotic and eukaryotic cells are enclosed by a thin membrane called **plasma membrane**. This membrane as a barrier maintains the concentration of the solutes in the cell by regulating the transport of materials from in and out of the cell. According to the requirement of the cell sometimes the plasma membrane is folded or invaginated to increase the surface area for more exchange of materials. The plasma membrane has the following roles in the cell:

(*i*) Membranes are selectively permeable allowing certain molecules to pass in and out of the cell.

(*ii*) Membranes can selectively allow certain specific molecules against a concentration gradient (active transport).

(*iii*) The cell receives substances from its environment by the process of **endocytosis.**

(*iv*) Undesired and excess amounts of substances present in the cell are thrown out with the formation of vesicles. These vesicles are pinched off and released from the cell. This type of outward movement is called **exocytosis.**

(*v*) The cell perform certain specific metabolic functions by the help of enzymes and proteins present on plasma membrane.

2. Functions of genetic material. Inheritance and transmission of genetic material from one generation to another are performed in the cell through cell division (mitosis and meiosis). In eukaryotic cells the genetic material is present in the nucleus which enclosed by nuclear envelope. The nuclear envelope regulates the transport of materials (chemical substances) from the nucleus to the cytoplasm and vice versa. It also serves as an anchoring point of chromosomes/choromatin fibers during interphase. In prokaryotic cells, organised nucleus is absent. The genetic material (DNA) is folded into a compact structure known as **nucleoid**.

The genetic material (DNA) both in prokaryotes and eukaryotes control the synthesis of proteins on small cytoplasmic structures called **ribosomes**. Three type of RNA molecules (*i.e.*, rRNA, tRNA and mRNA) are transcribed from DNA to form specific proteins in adequate amounts in correct form. The structure and function of ribosomes, in both prokaryotes and eukaryotes are similar.

3. Functions in cellular metabolism. All cells carry out series of chemical reactions for the synthesis of macromolecules, tapping energy, degradation of unused molecules, converting food substances into sugars, amino acids, etc. Most of the reactions are performed in the cytoplasm of the cell. The initial reactions of the cell start through the degradation of reserve foods (*e.g.*, glycogen) in the fluid phase of the cytoplasm (cytosol). The main biochemical reactions, which release energy such as ATP occur in the cytoplasmic organelle called **mitochondria**. The desired proteins are synthesized on the ribosomes according to genetic information of genes.

In photosynthetic eukaryotes, the energy from sunlight is used by converting the light energy into chemical energy. This conversion is done by some specialized organelles called **chloroplasts.** In prokaryotes such as photosynthetic bacteria and blue green algae (cyanobacteria) similar tapping of energy from sunlight is done by folding of the plasma membrane.

4. Functions in motility. Cells perform different types of motility, *i.e.*, from locomotion to the movement of some components of the cell. The movements may be either the movement of cell from one place to another (*e.g.*, amoeba, paramecium, white blood cells, sperm, etc.) or the movement of liquid over the surface of a cell by cilia and flagella (*e.g.*, ciliated surfaces of trachea, oviduct, etc.).

In eukaryotic cells, microtubules (Mts) are present which consist of a special protein called **tubulin**. The microtubules play an important role in forming spindle apparatus that helps in the movement of the chromosomes during the cell division. Microtubules also provide a rigid set of tracks for the transport of a variety of membrane-enclosed organelles and vesicles. The mechanical work needed for movement depends on **microtubule-associated motor proteins (motor MAPs)**, which attach to vesicles or organellels and then "walk" along the microtubule, using ATP to provide the needed energy. **Motor protein** or **mechanoenzyme** is a protein that uses energy derived from ATP to change shape in a way that exerts force and causes attached structures to move; includes three families of proteins (myosin, dynein, and kinesin) that interacts with cytoskeletal elements (microtubules and microfilaments) to produce movement (see **Becker** *et al.*, 2006).

In eukaryotic cells, other smaller filaments are found. These are made up of actin proteins and are called **microfilaments**. These microfilaments help in muscle contraction and cytoplasmic streaming (cyclosis).

In this way, the cells are capable of performing all the necessary metabolic activities essential for continuity of life.

QUESTIONS

Long Answer Questions

1. Describe the structure and functions of animal cell junctions. What is extracellular matrix?
2. Describe the main functions of cell.
3. What is cell adhesion? Describe the various components of extracellular matrix which are responsible for this process.
4. Describe various types of cell adhesion molecules which provide differential affinity.

Short Answer Questions

1. What are glycocalyx?
2. What are desmosomes? How many type of desmosomes are found? Describe their function.
3. Write a short note on extracellular matrix.
4. How does adhesion takes place?
5. Differentiate between gap junctions, tight junctions and desmosomes.
6. Write a short note on tight junctions or zona occludens.

Very Short Answer Questions

1. What is full form of ECM?
2. What are cadherins?
3. Write the names of main cell adhesion proteins.
4. What is fibronectin?
5. What is laminin?
6. Define the basal lamina?
7. What is collagen?
8. Define proteoglycans.
9. Define the plaque.
10. Define integrins.
11. In which tissue the gap junctions are mostly found?
12. What prevents leakage of pancreatic secretary proteins in blood?
13. Name the structure that provide mechanical adhesion between adjacent cells.

Multiple Choice Questions

1. The cell junctions called tight, adhering and gap junctions are found in
   (*a*) connective tissue
   (*b*) epithelial tissue
   (*c*) neural tissue
   (*d*) muscular tissue.
2. Desmosomes are concerned with
   (*a*) cell division
   (*b*) cellular excretion
   (*c*) cytolysis
   (*d*) cell adherance.
3. Cellular recognition and adhesion are facilitated by components of plasma membranes. These components are generally
   (*a*) lipid molecules
   (*b*) protein molecules
   (*c*) both lipid and protein molecules
   (*d*) glycolipids and glycoproteins.
4. Tonofibrils are anchored at this cell junction
   (*a*) desmosome
   (*b*) nexus
   (*c*) zonula occludens
   (*d*) zonula adherens.
5. Collagen contains
   (*a*) hydroxyproline (*b*) tryptophan
   (*c*) arginine (*d*) peptone.

ANSWERS

Very Short Answer Questions

1. Extracellular matrix.
2. Cadherins are Ca^{2+} dependent cell adhesion glycoproteins.
3. Adhesion proteins such as fibronectin, laminin and tenacin.
4. It is adhesive glycoprotein which is found in the extracellular matrix and loosely associated with the surface; it binds cells to the ECM and is important in determining the cell shape and guiding cell migration.
5. It is a glycoprotein of the extracellular matrix localized predominantly in the basal lamina of epithelial cells.
6. It is a thin sheet of specialized extra-cellular matrix material that separates epithelial cells from underlying connective tissue.
7. Collagen is a family of closely related proteins that form high-strength fibers found in high concentration in the ECM of animals.
8. They are complex between proteins and glycosaminoglycans found in the extracellular matrix.
9. It is a dense layer of fibrous material located on the cytoplasmic side of adhesive junctions such as desmosomes, hemidesmosomes, and adherens junctions; composed of intracellular attachment proteins that link the junction to the appropriate type of cytoskeletal filament.
10. Integrins are any of several plasma membrane receptors that bind to extracellular matrix components at the outer membrane surface and interact with cytoskeletal components at the inner membrane surface; includes receptors for fibronectin, laminin and collagen.
11. Embryonic tissue.
12. Zona occludens.
13. Desmosomes.

Multiple Choice Questions

1. (*b*) 2. (*d*) 3. (*d*) 4. (*a*) 5. (*a*)

15 Bioenergetics and Enzymes

15.1. BIOENERGETICS

A living cell bustles with activity. Macromolecules of all types are assembled from raw materials, waste products are produced and excreted, genetic instructions flow from the nucleus to the cytoplasm, vesicles are moved from the Golgi apparatus to the plasma membrane, ions are pumped across cell membranes, and so forth. To maintain such a high level of activity, a cell must acquire and expend energy. The study of various types of energy transformations that occur in living organisms is referred to as **bioenergetics**.

1. Energy Transductions

Energy is defined as the capacity to do work, that is, the capacity to change or move some thing. Energy exists in two alternate states: potential and kinetic. A rock perched on the edge of a cliff possesses **potential energy** because it has the potential to perform work. Rock has this potential because it exists in a field of force, in this case, a gravitational field. If the rock is pushed over the edge, then gravity can act on it and cause it to fall, while falling, the rock has **kinetic energy** and can accomplish work, for example, by lifting another object (*i.e.*, fish in Fig. 15.1). Similarly, a resting nerve cell, has potential energy when it maintains a high concentration of sodium ions on the outer side of its plasma membrane and a low concentration of sodium ions on the inner side. Like the flow of water through a dam, the opening of specific channels in the plasma membrane allows sodium ions to flow across the membrane into the cell. The directed movement of sodium ions into the cell is a form of kinetic energy that can be used to accomplish work, as occurs when a nerve impulse passes along the membrane to the nerve cell.

In both of these cases just described–the falling rock and the directed movement of sodium ions–as in all cases of measurement of energy at work, two factors must be considered: a potential factor and a capacity factor (Table 15.1). The **potential factor** is proportional to the intensity of the field of force, whereas the **capacity factor** provides some measure of the "size" of the subject being considered. In the case of falling rock, the **potential factor** is the distance it will fall, and the capacity factor is the rock's mass. For the movement of charged ions, the potential factor is the voltage, and the **capacity factor** is the combined charged of the particles. The work, or energy released during these events, is a multiple of these two factors; as either one increases, so too does the amount of energy.

Fig. 15.1. Accomplishing work. The rock perched on top of the cliff has potential energy. The energy available for work is proportional to the mass of the rock and the distance it can fall. Once it leaves the edge of the cliff, the rock has kinetic energy that can be used to accomplish work, in this by lifting both water and fish (after Karp, 2002).

Table 15.1. Potential and capacity factors in measuring energy (source: Karp 2002).

	Type of energy	Potential factor	Capacity factor
1.	**Mechanical**		
	Falling of rock	Height	Mass
	Compression	Pressure	Volume
	Stretching	Tension	Length
2.	**Electrical**		
	Movement of ions	Electric potential (voltage)	Charge
3.	**Osmotic**		
	Movement of water into a cell	Solute concentration gradient	Mass
4.	Energy available to do work	Temperature	Entropy

The Laws of Thermodynamics and the Concept of Entropy

Thermodynamics is the study of the changes in energy that accompany events in the universe.

The First Law of Thermodynamics. The first law of thermodynamics is the law of conservation of energy. It states that **energy** can neither be created nor destroyed. Energy can, however, be converted **(transduced)** from one form to another. Electric energy is converted to mechanical energy when we plug in a fan, and chemical energy is converted to thermal energy when fuel is burned in an oil heater. Cells are also capable of energy transduction. The chemical energy present in certain biological molecules, such as ATP, is converted to mechanical energy when organelles are moved from place to place in a cell, to thermal energy when heat is released during muscle cell contraction, or to electrical energy when ions flow across a membrane. *The most important energy transduction in the biological world is the conversion of sunlight into chemical energy—the process of photosynthesis*—which provides the fuel that directly or indirectly powers the activities of nearly all forms of life. A number of animals including fireflies and luminous fish are able to convert chemical energy back into light. Regardless of the transduction process, however, the total amount of energy in the universe remains constant.

To discuss energy transformations involving matter, we need to divide the universe into two parts: the **system** under study and the remainder of the universe, which we will refer to as the **surroundings**. A system can be defined in various ways: it may be a certain space in the universe or a certain amount of matter. For example, the system may be a living cell. The changes in a system's energy that occur during an event are manifested in two ways—as a change in heat content of a system and in the performance of work. Even though the system may gain or lose, energy, the first law of thermodynamics indicates that the loss or gain must be balanced by a corresponding gain or loss in the surroundings, so that the amount in the universe as a whole remains constant. The energy of the system is termed as **internal energy (E)**, and its change during a transformation in **ΔE**. One way to describe the first law of thermodynamics is that $\Delta E = Q - W$, where Q is the heat energy and W is the work energy.

Depending on the process, the internal energy of the system at the end can be greater than, equal to, or less than its internal energy at the start, depending on its relationship to its surroundings (Fig. 15.2). In other words ΔE (*i.e.*, change during the transformation) can be positive, zero, or negative.

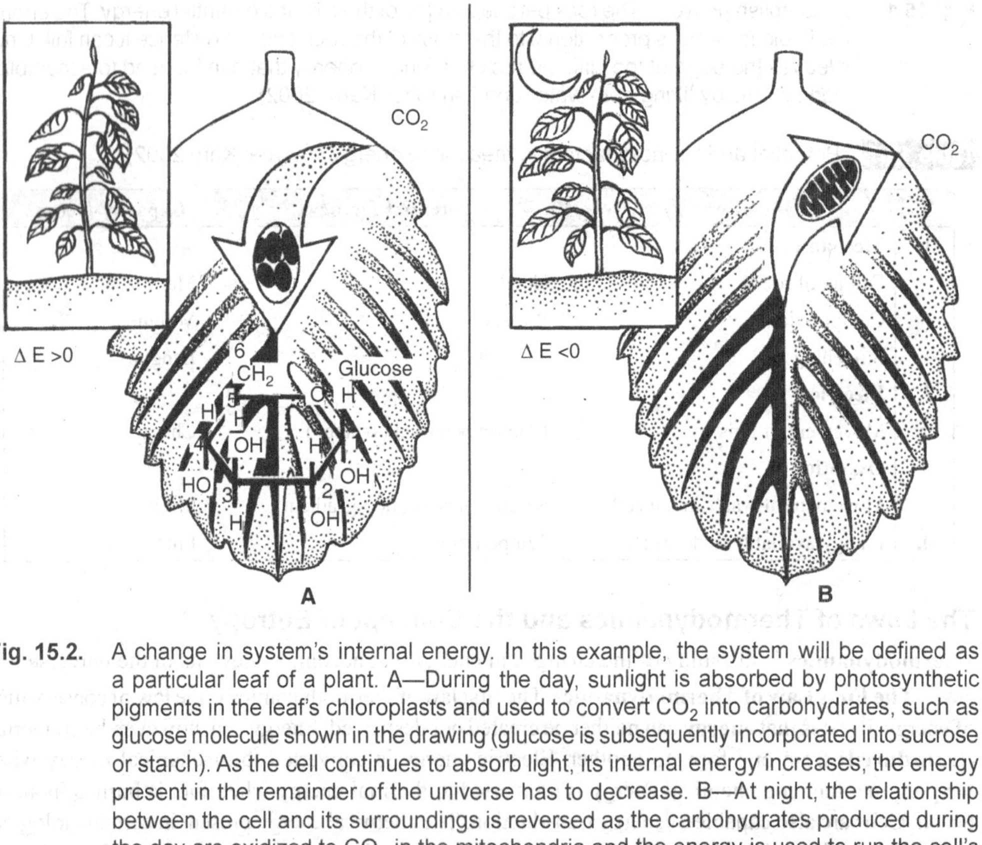

Fig. 15.2. A change in system's internal energy. In this example, the system will be defined as a particular leaf of a plant. A—During the day, sunlight is absorbed by photosynthetic pigments in the leaf's chloroplasts and used to convert CO_2 into carbohydrates, such as glucose molecule shown in the drawing (glucose is subsequently incorporated into sucrose or starch). As the cell continues to absorb light, its internal energy increases; the energy present in the remainder of the universe has to decrease. B—At night, the relationship between the cell and its surroundings is reversed as the carbohydrates produced during the day are oxidized to CO_2 in the mitochondria and the energy is used to run the cell's nocturnal activities (after Karp, 2002).

We can consider a system to be the contents of a reaction vessel. As long as there is no change in pressure or volume of the contents, there is no work being done by the system on its surroundings, or vice versa. In that case, the internal energy at the end of the transformation will be greater than at the beginning if heat is absorbed and less of heat is released. Reactions that lose heat are termed **exothermic**, and ones that gain heat are **endothermic**, and there are many reactions of both types.

The Second Law of Thermodynamics. The second law of thermodynamics expresses the concept that events in the universe have direction; they tend to proceed "downhill" from a state of higher energy to a state of lower energy. Thus, in any energy transformation, there is a decreasing availability of energy for doing additional work. Rocks fall off cliffs to the ground below, and once at the bottom, their ability to do additional work is reduced; it is very unlikely that they will lift themselves back to the top of the cliff. Similarly, opposite charges normally move together, not apart, and heat flows from a warmer to a cooler body not the reverse. Such events are said to be spontaneous, a term that indicates that they are thermodynamically favourable and *can occur without the input of external energy*.

The concept of the second law of thermodynamics was formulated originally for engines, and the law carried with the idea that it is thermodynamically impossible to construct a perpetual-motion machine. In other words, it is impossible for a machine to be 100 percent efficient, which would be required if it were to continue functioning without the input of external energy. Some of the energy is inevitably lost as the machine carries out its activity. A similar relationship holds true for living organisms. For example, when a giraffe browses on the leaves of a tree or a lion preys on the giraffe, most of the chemical energy in the food never becomes available to the animal having the meal. The energy that is unavailable for doing additional work after an event has occurred has a potential factor and a capacity factor (Table 15.1). The potential factor is temperature (in degrees) and the capacity factor is **entropy (S)**, which has the dimensions of energy per degree (*e.g.*, calories per degree). The unavailable energy is equal to $T\Delta S$ where ΔS is the change in entropy between the initial and final states.

The loss of available energy during a process is a result of a tendency for the randomness, or disorder, of the universe to increase every time there is a transfer of energy. Entropy provides a measure of this disorder; it is also associated with the random movements of the particles of matter, which, because they are random, cannot be made to accomplish a *directed* work process. According to the second law of thermodynamics, every event is accompanied by an increase in the entropy of the universe. When a sugar cube is dropped into a cup of hot water, for example, there is a spontaneous shift of the molecules from an ordered state in the crystal to a much more disordered condition when the sugar molecules are spread throughout the solution (Fig. 15.3A). As the molecules of the sugar cube dissolve into solution, their freedom of movement increases, as does the entropy of the system. The change from the concentrated to the dispersed state results from the random movements of the molecules. The sugar molecules eventually spread themselves equally through the available volume because the state of uniform distribution is the most probable state.

For example, the release of heat from the oxidation of glucose within a cell or from the friction generated as blood flows through a vessel, is another example of an increase in entropy. The release of thermal energy by living organisms increases the rate of random movements of atoms and molecules; it cannot be redirected to accomplish additional work. Since the energy of molecular and atomic movements increases with temperature, so too does the entropy. It is only at absolute zero (0°K), when all movements ceases, that the entropy is zero.

As with other spontaneous events, one must distinguish between the system and its surroundings. The second law of thermodynamics indicate only that the total entropy in the universe must increase; the disorder within one part of the universe (the system) can decrease at the greater expense of its surroundings. The dissolved sugar can decrease in entropy it can be recrystallized by evaporating the water (Fig. 15.3B). The consequence of this process, however, is an increase in the entropy of the surroundings. The increased freedom of movement of the water molecules in the gaseous phase more than balances the decrease in freedom of the molecules of the sugar crystals.

Life operates on in a similar principle. Living organisms are able to decrease their own entropy by increasing the entropy of the environment. Entropy is decreased in an organism when relatively simple molecules, such as amino acids, are ordered into more complex molecules, such as the protein myoglobin in a muscle cell. For this to occur, however, the entropy of the environment must increase, which is accomplished as complex, ordered molecules such as glycogen stored in liver or muscle tissue are converted into heat and smaller less ordered compounds (such as CO_2 and H_2O) are released to the environment. It is this feature of metabolism that allows living organisms to maintain such a highly ordered and improbable state.

Another measure of the energy state of a living organism is provided by the information content of its macromolecules. **Information** is a subject that is difficult to define but easy to recognise. Information can be measured in terms of the ordered arrangement of a structure's subunits. For example, proteins and nucleic acids, in which the specific linear sequence of the subunits is highly ordered, are low in entropy and high in information content. Maintaining a state of high information content (low entropy) requires the input of energy. We can consider here just one molecule of DNA located in one cell of our liver. That cell has dozens of different proteins whose sole job is to patrol the DNA, looking for damage and repairing it. Nucleotide damage in an active cell can be so great that, without this expenditure of energy, the information content of DNA would rapidly deteriorate (See **Karp** 2002).

Free energy. Together, the first and second laws of thermodynamics indicate that the energy of the universe is constant, but the entropy continues to increase toward a maximum. The concepts inherent in first two laws were put together by the American chemist, **Josiah Willard Gibbs** in 1878 into the expression $\Delta H = \Delta G + T\Delta S$, where ΔG is the change in **free energy**, that is, the change during a process in the energy available to do work; ΔH is the change in **enthalpy** or total energy content of the system (equivalent to ΔE), T is the absolute temperature (°K = °C + 273), and ΔS is the change in the entropy of the system. The equation states that the total energy change is equal to the sum of the changes in useful energy (ΔG) and energy that is unavailable to do further work ($T\Delta S$).

Processes that can occur spontaneously, that is, processes that are thermodynamically favoured (have a $-\Delta G$), are described as exergonic. In contrast, if the ΔG for a given process is positive, then it cannot occur spontaneously. Such processes are thermodynamically unfavourable and are described as **endergonic**.

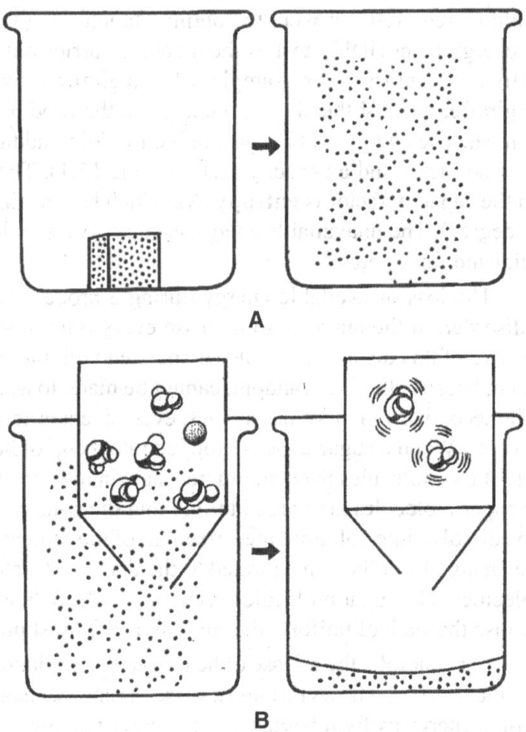

Fig. 15.3. Events are accompanied by an increase in the entropy of the universe. A—A sugar cube contains sucrose molecules in highly ordered arrangement in which the freedom of movement of the sucrose molecules is greatly increased, and their random movement causes them to become equally distributed throughout the available space. Once this occurs, there will be no further tendency for redistribution, and to entropy of the system is at a maximum. B—Sugar molecules spread randomly through a solution can be returned to a ordered state, but only if the entropy of the surroundings is increased, as occur when the more ordered water molecules of the liquid phase become disordered following evaporation (after Karp, 2002).

2. Metabolism

Metabolism is the collection of biochemical reactions that occur within a cell, which include a tremendous diversity of molecular conversions. Most of these reactions can be grouped into **metabolic pathways** containing a sequence of chemical reactions in which each reaction is catalyzed by a specific enzyme, and the product of one reaction is the substrate of the next. The enzymes constituting a metabolic pathway are usually confined to a specific region of the cell, such as mitochondria or the cytosol. Increasing evidence suggests that the enzymes of a metabolic pathway are often physically linked to one another, a feature that allows the product of one enzyme to be delivered directly as a substrate to the active sites of the next enzyme in the reaction sequence.

The compound formed each step along the pathway are **metabolic intermediate** (or **metabolites**) that lead ultimately to the formation of an **end product**. End products are the molecules with a particular role in the cell, such as an amino acid that can be incorporated into a polypeptide, or a sugar that can be consumed for its energy content. The metabolic pathways of a cell are interconnected at various points so that a compound generated by one pathway may be shuttled in a number of directions depending on the requirements of the cell at the time.

Metabolic pathways can be divided into two broad types. **Catabolic pathways** lead to the disassembly of complex molecules to form simple products. Catabolic pathways serve two functions: they make available the raw materials from which other molecules can be synthesized, and they provide chemical energy released by catabolic pathways is stored temporarily in two forms: a high energy phosphate (primarily ATP) and high-energy electrons (primarily in NADPH). **Anabolic pathways** lead to the synthesis of more complex compounds from simples starting materials. Anabolic pathways are energy requiring and utilize chemical energy released by the exergonic catabolic pathways.

Oxidation and Reduction

Both catabolic and anabolic pathways include key reactions in which electrons are transferred from one reactant to another. Reactions that involve a change in the electronic state of the reactants are called **oxidation-reduction** (or **redox**) **reactions**. Changes of this type involve the gain or loss of electrons, *e.g.*, the conversion of metallic iron (Fe^0) to the ferrous state (Fe^{2+}), by which iron atom loses a pair of electrons, thereby attaining a more positive state. When an atom loses one or more electrons, it is said to be **oxidized**. The reaction is reversible, meaning ferrous ions can be converted to metallic iron, a more negative state, by the acquisition of a pair of electrons. When an atom gains one or more electrons, it is said to be **reduced**.

For metallic iron to be oxidized, there must be some substance to accept the electrons that are released. Conversely, for ferrous ions to be reduced, there must be some substance to donate the necessary electrons. In other words, the oxidation of one reactant must be accompanied by the simultaneous reduction of some other reactant, and vice versa. One possible reaction involving iron might be

$$Fe^0 + Cu^{2+} \rightleftharpoons Fe^{2+} + Cu^0$$

The substance that is oxidized during an oxidation-reduction reaction, that is, the one that loses electrons, is called a **reducing agent**, and the one that is reduced, that is, the one that gains electrons, is called an **oxidizing agent**.

The oxidation or reduction of metals, such as iron or copper, involves the complete loss or gain of electrons. The same cannot occur with most organic compounds for the following reason: the oxidation and reduction of organic substrates during cellular metabolism involve carbon atoms that are covalently bonded to other atoms. When a pair of electrons are attracted more strongly to one of the two atoms of the polarized bond. In a C – H bond, the carbon atom has the strongest pull on the electrons; thus, it can be said that carbon atom is in a reduced state. In contrast, if a carbon atom is bonded to a more electronegative atom, as in a C – O or C – N bond, the electrons are pulled away from the carbon atom, which is thus is an oxidised state, since carbon has four outer-shell electrons it can share with other atoms, it can exist in a variety of oxidation states. This is illustrated by the carbon atom in a series of one carbon molecules (Fig. 15.4) ranging from the fully reduced state in methane (CH_4) to the fully oxidized state in carbon dioxide (CO_2). The relative oxidation state of an organic molecule can be roughly determined by counting the number of hydrogen versus oxygen and nitrogen atoms per carbon atom. The oxidation state of the carbon atoms in an organic molecule provides a measure of the molecules free-energy content.

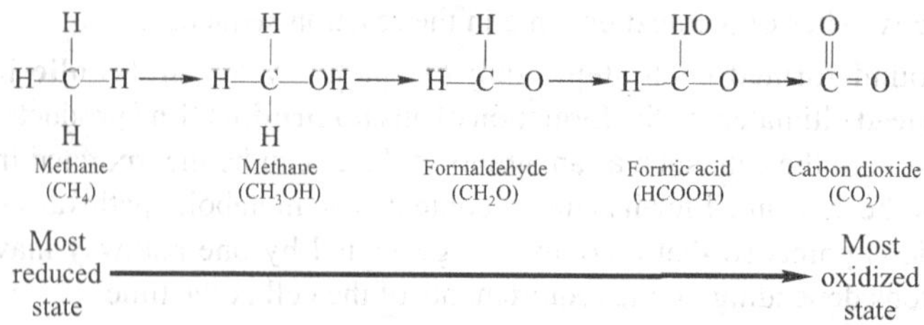

Fig. 15.4. The oxidation state of a carbon atom depends on the other atoms to which it is bonded. Each carbon atom can form a maximum of four bonds with other atoms. This series of simple, one-carbon molecules illustrates the various oxidation states in which the carbon atom can exist. In its most reduced state, the carbon is bonded to four hydrogens (forming methane); in its most oxidized state, the carbon atom is bonded to two oxygen (forming carbon dioxide) (after Karp 2002).

The Capture and Utilization of Energy

The compounds we use as chemical fuels to run automobiles are highly reduced organic compounds, such as natural gas (CH_4) and petroleum derivatives. Energy is released when these molecules are burned in the presence of oxygen, converting the carbon to more oxidized state, as in carbon dioxide and carbon monoxide gases. The degree of reduction of a compound is also a measure of its ability to perform chemical work within the cell. The more hydrogen atoms that can be stripped from a "fuel" molecule, the more ATP that ultimately can be produced. Carbohydrates are rich in chemical energy because they contain strings of (H–C–OH) units. Fats contain even greater energy per unit weight because they contain strings of more reduced (H–C–H) units.

As the sole building block of both starch and glycogen, glucose is a key molecule in the energy metabolism of plants and animals. The free energy released by the complete oxidation of glucose is very large:

$$C_6H_{12}O_6 + 6O_2 \rightarrow 6CO_2 + 6H_2O$$
$$\Delta G^{0\prime} = -686 \text{ kcal/mol}$$

By comparison, the free energy required to convert ADP to ATP is relatively small:

$$ADP + Pi \rightarrow ATP + H_2O$$
$$\Delta G^{0\prime} = 7.3 \text{ kcal/mol}$$

It is evident from these numbers that the complete oxidation of a molecule of glucose to CO_2 and H_2O can release enough energy to produce a large number of ATPs. As we will see in this chapter later on, upto 36 molecules of ATP are formed per molecule of glucose oxidized under conditions that exist in most cells. For this many ATP to be produced, the sugar molecule is disassembled in many small steps. Those steps in which the free-energy difference between the reactants and products is relatively large can be coupled to reactions that lead to the formation of ATP.

There are basically two stages in the catabolism of glucose, and they are virtually identical in all aerobic organisms. The first stage, **glycolysis**, occurs in the soluble phase of the cytoplasm (the cytosol) and leads to the formation of pyruvate. The second stage is the **tricarboxylic acid (TCA) cycle** or **Krebs cycle**, which occurs within the mitochondria of eukaryotic cells and the cytosol of prokaryotic cells and leads to the final oxidation of the carbon atoms to carbon dioxide. Most of the chemical energy of glucose is stored in the form of high-energy electrons, which are removed as substrate molecules are oxidized during both glycolysis and the TCA cycle. It is the energy of these electrons that is ultimately used to synthesize ATP.

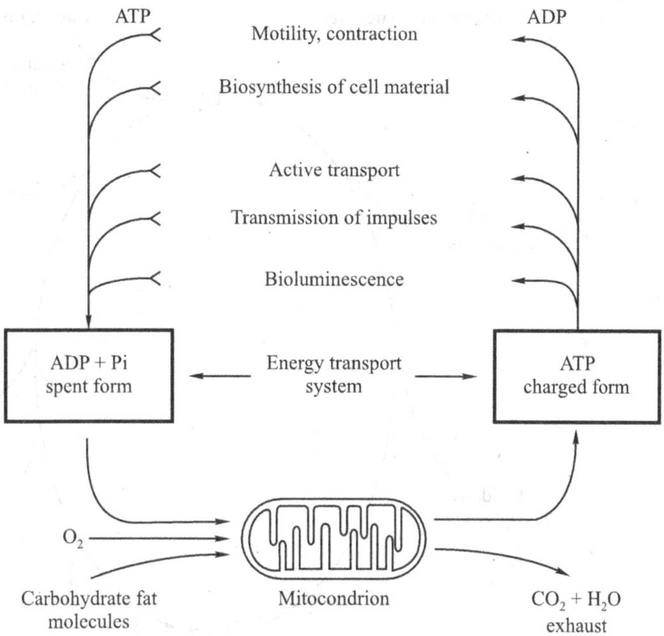

Fig. 15.5. Diagrammatic representation of the energy liberation in mitochondrion and its utilization in various cellular functions.

Adenosine Triphosphate or ATP

The ATP consists of a purine base **adenine**, a pentose sugar **ribose** and three molecules of the **phosphoric acids**. The adenine and ribose sugar collectively constitute the nucleoside **adenosine** which by having one, two or three phosphate groups forms the adenosine monophosphate (AMP), adenosine diphosphate (ADP) and adenosine triphosphate (ATP) respectively. In ATP the last phosphate group is linked with ADP by a special bond known as "**energy rich bond**" because when the last phosphate group of the ATP is broken the large amount of energy is released as shown by the following reaction:

$$\underset{\text{ATP}}{A{-}P{\sim}P{\sim}P} = \underset{\text{ADP}}{A{-}P{\sim}P} + \underset{\text{Phosphate group}}{Pi} + \underset{\text{Energy}}{7300 \text{ calories}}$$

In the above reaction, we have seen that by the breaking of the energy rich bond about 7300 calories of energy are released, while the common chemical bond releases only 300 calories of energy. The chemical reactions which synthesize the energy rich bond or ~P bond require great amount of energy which is supplied by the oxidation of the food-stuffs in the mitochondria. The utility of energy rich phosphate bond (~P) of the ATP is that great amount of energy which is kept stored in the ready state in a very limited space of the cell. The stored chemical energy is disposed of very quickly at the time of the need in various cellular functions such as respiratory cycle, protein and nucleic acid synthesis, nervous transmission, cell division, transportation and bioluminescence, etc. (Fig. 15.6).

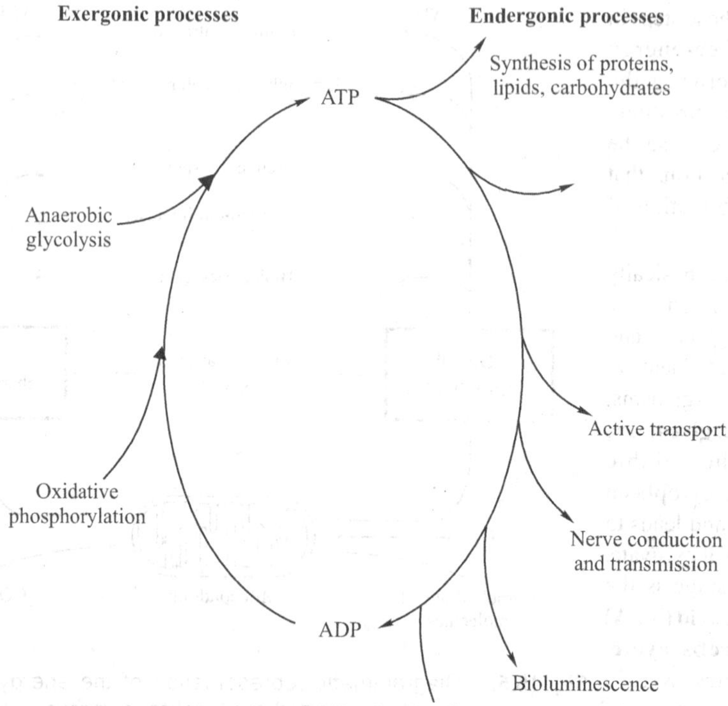

Fig. 15.6. Diagrammatic representation of uses and synthesis of ATP.

Because the terminal phosphate linkage in ATP is easily cleaved with release of free energy, ATP acts as an efficient phosphate donor in a large number of different phosphorylation reactions. In this way, ATP acts as a carrier molecule like the acetyl CoA and as coenzyme like the CoA or NAD.

Later on, besides ATP, certain other energy rich chemical compounds have been found to be active in the cellular metabolism. These are **cytosine triphosphate (CTP), uridine triphosphate (UTP)** and **guanosine triphosphate (GTP)**. These compounds, however, derive the energy from the ATP by nucleoside diphosphokinases. The energy for the production of ATP or other energy rich molecules is produced during the breakdown of food molecules including carbohydrates, fats and proteins (**catabolic** and **exergonic activities**).

Energy transducers. The cell organelles, such as chloroplasts and mitochondria, which convert one form of energy to another are known as **energy transducers**. The plants and animals have not only evolved highly effective energy transducers but have also developed very efficient control systems to regulate the energy transformations and to enable the cells to adjust to variations in environmental conditions.

15.2. ENZYMES

Enzyme is a biological catalyst almost always a protein (Ribozyme is an RNA molecule with enzymatic activity). Some enzymes may require additional cofactors for activity, virtually all biochemical reactions are catalyzed by specific enzymes.

A. Introduction

Enzymes are extraordinarily efficient selective biological catalysts. Every living cell has hundreds of different enzymes catalyzing the reactions essential for life. Even the simplest living organisms contain multiple copies of different enzymes. In multicellular organisms the complement of enzymes differentiates one cell type from another. Many enzymes catalyze the reactions of the central metabolic pathways necessary for the maintenance of life.

Most enzyme-catalyzed reactions do not proceed at significant rates under physiological conditions of enzymes in the absence of enzymes. The primary role of enzymes is to *enhance the rates of these reactions*. Enzyme-catalyzed reactions are typically 10^3 to 10^{20} times faster than the corresponding uncatalyzed reactions.

A catalyst is a substance that speeds up the attainment of equilibrium. A catalyst may be temporarily changed in the overall process, since it recycles to participate in multiple reactions. Reactants bind to a catalyst, and products dissociate from it. A catalyst does not change the position of reaction's equilibrium (*i.e.*, it does not make an unfavourable reaction favourable). Rather, it lower the amount of energy needed in order for the reaction to proceed. Catalysts speed up both the forward and reverse reactions by converting a one- or two-step process into several smaller steps, each needing less energy than the uncatalyzed reaction.

Enzymes are highly specific for the reactants, or **substrates**, they act on, and the degree of substrate specificity varies. Some enzymes act on a group of related substrates, and others on only a single compound. Many enzymes exhibit **stereospecifically**, meaning that they act on only a single stereoisomer of the substrate. Perhaps the most important aspect of enzyme specific is **reaction specificity** that is, the lack of formation of wasteful by-products. Reaction specificity is reflected in the exceptional purity of product (essentially 100%)—much higher that the purity of products of typical catalyzed reactions in organic chemistry. The specificity of enzymes not only saves energy for cells but also prevent the buildup of potentially toxic metabolic by-products.

Enzymes can do more than simply increase the rate of a single highly specific reaction. Some can also combine, or couple, two reactions that would normally occur separately. This property allows the energy gained from one reaction to be used in a second reaction. **Coupled reactions** are a common feature of many enzymes; the hydrolysis of ATP, for example, is often coupled to less favourable metabolic reactions.

Some enzymatic reactions function as a control points in metabolism. Metabolism is regulated in a variety of ways, including alterations with concentrations of enzymes, substrates, and enzyme inhibitors, and modulation of the activity levels of certain enzymes. Enzymes whose activity is regulated generally have a more complex structure than unregulated enzymes. With few exceptions, regulated enzymes are oligomeric molecules that have separate binding sites, for substrates and modulators, the compounds that act as regulatory signals. The fact that enzyme activity can be regulated is an important property that distinguishes biological catalysts from those encountered in a chemistry lab (*i.e.*, inorganic catalyst; Table 15.2).

Table 15.2. Differences between enzymes and chemical catalysts.

	Enzymes		Chemical catalysts
1.	Enzymes (biocatalysts) are always protein in nature [RNA enzymes (ribozymes) are notable exception].	1.	Inorganic or chemical catalysts are nonprotein in nature.
2.	They are highly specific and an enzyme catalyses a specific reaction.	2.	They may catalyze different reactions.
3.	After completion of reaction, the enzymes do not always return to their original state.	3.	Catalysts always returns to its original state.
4.	Catalysis can occur through the active site present in enzyme.	4.	The catalysis takes part as whole.
5.	Generally produced by living cells and acts independently of living cells.	5.	Generally, takes place outside the living cells.

B. History

The name enzyme is derived from a Greek and meaning "in yeast". It indicates that catalysts are present inside cells. In the late 1800s, scientists studied the fermentation of sugars by yeast cells. **Vitalists** (who maintained that organic compounds could be made only by living cells) said that intact cells were needed for fermentation. **Mechanists** claimed that enzymes in yeast cells catalyze the reactions of fermentation. The latter conclusion was supported by the observation that cell free extracts of yeast can catalyze fermentation. This finding was soon followed by the identification of individual reactions and the enzymes that catalyze them.

In 1926 **James B. Sumner** (1887–1955), a US biochemist, 1946 Nobel laureate and **"Father of Modern" Enzymology**, crystallised the first enzyme (**urease**) from jack bean (*Canavalia ensiformis*) and proved that it is a protein. In next decade, five more enzymes were purified and also found to be proteins: **pepsin**, **trypsin**, **chymotrypsin**, **carboxypeptidase** and **Old Yellow Enzyme** (a flavoprotein NAD oxidase). Since then, almost all enzymes have been shown to be proteins plus cofactors. Certain RNA molecules also exhibit catalytic activity, but they are not usually referred to as enzymes.

C. Sources and Sites of Action of Enzymes

On the basis of sites of their action, the enzymes are classified as follows:

1. Endoenzymes. As most enzymes are macromolecules (proteins), they can not normally get into the cell through the plasma membranes, and therefore every cell must synthesize its own enzymes for carrying out the reactions goining within the cell. The enzymes which operate within the cells are called **endoenzymes**. These enzymes catalyze the metabolic reactions of the cell. Hence they are also referred to as **metabolic enzymes**. For example, cytochrome oxidase is such endoenzyme.

Cytochrome oxidase was discovered by **Warburg** in 1926. It is responsible for the catalysis of the oxidation of cytochrome by oxygen. During this reaction in the presence of free oxygen and cytochrome oxidase, cytochrome gives up its hydrogen to the oxygen with the formation of water.

2. Exoenzymes. These are normally extracellular, *i.e.*, they work outside the cells after they have been secreted (*i.e.*, synthesized and released) by them, *e.g.*, digestive enzymes.

3. Antienzymes. These are proteins or protein-like substances produced in the body to inhibit the catalytic action of enzymes, *e.g.*, **antipepsin**, **antitrypsin**, **antiamylase**, **antirennin**, **antiurease**, etc. These enzymes are secreted by parasites of intestine such as *Ascaris*, *Taenia*, etc.

4. Coagulative enzymes. These convert soluble calcium caseinogen into insoluble calcium caseinate, *e.g.*, rennin (or chymosin) and rannet.

D. Properties or Characteristics of Enzymes

Enzymes have the following common characteristics or properties:

1. Physical nature. The enzymes are generally colorless, but some are colouerd (*i.e.*, yellow, brown, red or green). Most of the enzymes are soluble in water, but some such as those located in the mitochondria are insoluble.

2. Protein nature. Except ribozymes which contain RNA, all enzymes are proteins. The protein structure provides weak associations due to side groups of the amino acids within it. The enzyme-substrate complex involves the association of the enzyme and substrate by weak associations rather than covalent bondings. Enzymes exhibit all typical properties of proteins, *viz.* complex macromolecular organization, high molecular weight, colloidal behaviour, slow diffusion, inability to pass through most living membranes, and movement in response to an electric current.

3. Chemical activity. The enzymes may break a large molecule into two smaller molecules, or bring two small molecules together to form a large molecule.

4. Accelerate the rate of reaction. An enzymes can only change the rate of the biochemical reaction. Neither it can initiate a reaction nor it can change the direction of equilibrium. A reaction tends to proceed at a faster rate in presence of an enzyme, but the equilibrium between the substrate and product will remain constant.

5. Reversibility of enzyme action. Certain enzymes show reversibility of their action. This property is very helpful in metabolism. For example, *succinic dehydrogenase* catalyses reversibly the dehydrogenation of succinic acid. *Lipase*, a fat splitting enzyme, hydrolyses ethyl butyrate to butyric acid and ethyl alcohol. If this process continues ethyl butyrate is formed again.

$$\underset{\text{Ethyl butyrate}}{C_3H_7COOC_2H_5} \xrightleftharpoons{\text{Lipase}} \underset{\text{Butyric acid}}{C_3H_7COOH} + \underset{\text{Ethyl alcohol}}{C_2H_5OH}$$

However, reversion of reaction may be checked by removing the products of enzyme reaction as quickly as they are formed.

6. Remain unaltered in the end. The enzymes combine temporarily with the substrate but do not undergo any permanent change during reaction they catalyse. The same enzymes can enter another biochemical reaction with another molecule of the same substrates.

7. High efficiency. Enzymes are reused and are therefore, required in small quantity by biological systems. Most enzymes have high **turnover number**. A turnover number of molecules of a substance acted upon by one molecule of enzyme per minute. A molecule of enzyme catalase from cattle liver decomposes 5,000,000 molecules of hydrogen peroxide in one minute at 0°C. The turnover number of catalase is, thus, 5,000,000 at 0°C. High turnover numbers of enzymes explain their remarkable effectiveness even though they occur in a cell in minute quantities.

8. Action specificity. Enzymes catalyse only one specific reaction or act upon only one kind of substrate. The specificity may be of any kind, *e.g.*, reaction specificity, substrate specificity, group specificity or optical specificity. For example, the enzyme lactase catalyses the hydrolysis of lactose and no other disaccharide.

(*i*) Reaction specificity. Most enzymes can catalyse the same type of reactions (*e.g.*, phosphate transfer, oxidation-reduction, etc.) with several structurally related substrates.

(*ii*) Substrate specificity. A particular enzyme will act only on a certain substrate (Box 15.1) For example, the enzyme **urease** is a specific enzyme, that acts on urea only. It would not exert any influence on the catabolism of protein or carbohydrate.

Box 15.1

Substrate specificity of enzyme is based on the "information" the enzyme molecule carries about its catalytic role. The "information" lies in the enzyme molecule in the form of its shape. By virtue of its specific shape, the enzyme molecule can "recognise" the substrate molecules with a given form and holds them for its action to the place. Molecules of other substrates fail to bind to the enzyme molecule, which, therefore, does not act on them. In short, the matching surface of the enzyme and substrate molecules form the basis of specificity.

(*iii*) Group specificity. A particular enzyme act only on particular chemical groupings, *e.g.*, **glycosidase** on glycosides, **alcohol-dehydrogenase** on alcohols, **pepsin** on peptide bonds and **esterases** on ester linkages. **Chymotrypsin** hydrolyses peptide bonds in which the carboxyl group is contributed by the aromatic amino acids: phenylalanine, tyrosine or tryptophan. **Carboxypeptidases** and **aminopeptidases** split amino acids one at a time form the carboxyl or amino terminal end of polypeptide chains, respectively.

(*iv*) Optical specificity (stereospecifity). With the exception of *epimerases* which interconvert optical isomers, enzymes generally show absolute optical specificity. For example, **maltase** catalyses the hydrolysis of α-glucosidic bond, while enzymes of the Embden Meyerhof directs oxidative pathways and catalyse the interconversion of D-but not L-phosphosugars.

9. Denaturation. The protein enzymes become denatured by acids, high salt concentration, alkaloid reagent or UV-light. The denaturation causes loss of **enzymatic activity**.

10. Temperature sensitivity. The enzymes function best at an optimum temperature (25°C – 40°C). Their activity decreases with decrease as well as increase in temperature and stops at 0°C and above 80°C.

11. pH sensitivity. The enzymes show maximum activity at an optimum pH (6 – 7.5). Their activity slows with decrease and increase in pH till it stops.

12. Activation. Most enzymes can be activated by addition of specific agents. In the absence of such factors enzymes become inactive and sluggish.

13. Inhibition. Enzyme activity is inhibited by the addition of specific agents, called **enzyme inhibitors**. For example, physostigmine (which is a salicylate-containing alkaloid which is obtained from Calabar beans) is a competitive inhibitor of acetylcholine.

14. Team work. The enzymes generally work in teams in the cell, the product of one enzyme controlled reaction serving as the substrate for the next. In germinating seeds, starch is changed into glucose, by two enzymes: amylase and maltase. Amylase splits the starch into the double sugar (disaccharide) maltose, which is then broken by matlase into the single sugar (monosaccharide) glucose. Eleven different enzymes work sequentially to convert glucose to lactic acid in animals as well as in plant cells (during glycolysis).

15. Destruction by poisons. The enzymes are destroyed by poisons such as cyanide and iodoacetic acid. **Cyanide poisoning** occurs due to the destruction of the respiratory cytochrome enzymes by the cyanide.

16. Lack of side reaction. In enzyme catalysed processes, there are virtually no side reactions, *i.e.*, the only products formed are the desired ones.

E. Nomenclature and Classification of Enzymes

Enzymes are named and classified by a system developed by a committee **IUB** or **International Union of Biochemists** or **IUBMB or International Union of Biochemistry and Molecular Biology**.

Nomenclature of Enzymes

Following five different criteria are followed in naming the enzymes:

1. According to substrate. Some enzymes are named by adding suffix **'ase'** to the name of substrate, on which they work. Thus, enzymes acting on sucrose, lipids and proteins are named **sucrase**, **lipase** and **proteinase**, respectively.

2. According to the type of reaction. Some enzymes are named according to the type of reactions they catalyze. For example, **alcohol dehydrogenase** catalyzes the removal of hydrogen from alcohol (*i.e.*, oxidation of alcohols). Likewise the enzymes that removes CO_2 from the reactant are called **decarboxylases** and those add phosphate group to the reactant are **phosphatses**.

3. According to history. A few enzymes such as **trypsin** and **amylase** are known by their historic names.

4. According to genes or characters. Many newly discovered enzymes are named after their genes or for some nondescriptive characteristics. For example, **RecA** is named after *recA* and **HSP70** is a heat shock protein; both enzymes catalyse the hydrolysis of ATP.

5. IUB/IUBMB system of naming enzymes. This is the most recent method of naming enzymes. The system assigns a **systematic name** to each enzyme. The systematic name of an enzyme has two parts. The first part represents the substrate and the second part represents the type of reaction catalysed. For example, enzyme **DNA polymerase** catalyses polymerisation or synthesis of DNA. Enzyme **3-phosphoglyceraldehydrogenase** removes hydrogen from substrate 3-phosphoglyceraldehyde.

IUBMB classification scheme assigns a unique number called the **enzyme classification number**, or **EC number**, to each enzyme.

Classification of Enzymes

To date over 2000 different enzymes are known (see **Jain** *et al.*, 2012). According IUB/IUBMB system of classification, enzymes are divided into following six main groups, based on the type of chemical reactions, they catalyse:

1. Oxireductases. These enzymes catalyse the oxidation-reduction reactions by transfer of hydrogen or electrons from one compound to another. A compound is **reduced** on accepting H or e^-, and **oxidized** on losing H^+ or e^-. Hence the enzyme involved is called *oxireductase*. Most of these enzymes (*i.e.*, oxireductases) are commonly referred to as **dehydrogenases**. Other enzymes in this class are called *oxidases, peroxidases, oxygenases*, or *reductases*.

An important example of oxireductase is *lactate dehydrogenase* (EC 1.1.1.27) also called *lactate: NAD oxireductase*. This enzyme catalyses the reversible conversion of L-lactate to pyruvate. The oxidation of L-lactate is coupled to the reduction of the coenzyme NAD^+ (nictonamide dinucleotide).

$$\text{L-Lactate} + \text{NAD} \xrightleftharpoons[]{\text{Lactate dehydrogenase}} \text{Pyruvate} + \text{NAD} + H^+$$

Types and Examples of Oxyreductases

***(i)* Oxidase.** Oxidases bring about oxidation by addition of oxygen to a substrate or by removal of hydrogen or one or more electrons (Table 15.3). Oxidases or oxidising enzymes are intracellular enzymes (endoenzymes) which produce oxidative changes in the tissue cells during respiration and metabolism.

***(ii)* Reductases.** Reductases catalyse reduction by removing oxygen or by adding hydrogen or one or more electrons (Table 15.3).

Table 15.3. Summary of oxidation reduction reactions.

	Oxidation (with respect to A)		Reduction (with respect to A)
(*i*)	Addition of oxygen: $A + BO \rightarrow AO + B$	(*i*)	Removal of oxygen: $AO + B \rightarrow A + BO$
(*ii*)	Removal of hydrogen: $AH + B \rightarrow A + BH$	(*ii*)	Addition of hydrogen: $A + BH \rightarrow AH + B$
(*iii*)	Removal of electrons: $A \rightarrow A^+ + e^-$	(*iii*)	Addition of electrons: $A + e^- \rightarrow A^-$
	All the above reactions release energy.		All the above reactions store energy.

Examples:

(*i*) **Cytochrome oxidase** catalyses the transfer of hydrogen to oxygen, forming water in the last reaction of electron transport system in mitochondria:

$$\text{NADH} + H^+ + 1/2O_2 \xrightarrow{\text{Cytochrome oxidase}} \text{NAD}^+ + H_2O$$

(*ii*) **3-Phosphoglyceraldehyde dehydrogenase** catalyses transfer of hydrogen from

3-phosphoglyceraldehyde (PGAL) to NAD^+ in glycolysis:

$$\text{3-PGAL} + NAD^+ + Pi^2 \xrightarrow[\text{dehydrogenase}]{\text{3-phosphoglycer-aldehyde}} \text{1, 3-diphosphoglycerate} + NADH + H^+$$

2. Transferases. These enzymes catalyse group-transfer reactions, and many require the presence of coenzymes. In group transfer reactions, a portion of the substrate molecule usually binds covalently to the enzyme or its coenzyme. This group includes **kinases**, enzymes that catalyse the transfer of a phosphoryl group from ATP.

Examples:

(*i*) **Alanine transaminase** whose systematic name is *L-alanine: 2-oxyglutarate aminotransferase* (EC 2.6.12) is a typical example of transferase:

$$\text{L-Alanine} + \alpha - \text{Ketoglutarate} \xrightleftharpoons[]{\text{Alanine transaminase}} \text{Pyruvate} + \text{L-Glutamate}$$

(*ii*) **Hexokinase** transfers a phosphate group from ATP to glucose, forming glucose-6-phosphate in glycolysis.

(*iii*) **Glutamate pyruvate transaminase** transfers amino group from glutamate to pyruvate to form alanine.

$$\text{Glutamate} + \text{Pyruvate} \xrightarrow[\text{transaminase}]{\text{Glutamate pyruvate}} \text{Alanine} + \alpha\text{-Ketoglutarate}$$

3. Hydrolases. These enzymes catalyse hydrolysis. Hydrolases are special class of transferases, with water serving as the acceptor of the group transferred (**Horton** *et al.*, 2006).

Example: (*i*) Pyrophosphatase is a simple example of a hydrolase. The systematic name of pyrophosphatase enzyme is *diphosphate phosphohydrolase* (EC 3.6.1.1).

$$\text{Pyrophosphate} + H_2O \xrightarrow{\text{Pyrophosphatase}} \text{Phosphate}$$

(*ii*) All digestive enzymes are hydrolases (Box 15.2). For example, pancreatic amylase hydrolyses starch and glycogen to maltose.

$$\text{Starch/Glycogen} \xrightarrow{\text{Pancreatic amylase}} \text{Maltose} + \text{Isomaltose} + \text{"Limit" Dextrins}$$

Box 15.2

Hydrolases are mostly exoenzymes; they act outside the cell and mainly help in digestion. Example of digestive enzymes include **amylases, saccharases, proteinases, lipases**, etc. Hydrolases also include certain endoenzymes that operate inside the cells and involve the use of water, *e.g.*, **deaminases, arginases, carbonic anhydrases**, etc.

Types of Hydrolases. Hydrolases are divided into following 8 types of enzymes:

(*i*) **Proteolytic enzymes** which break down proteins into proteins and peptides, *e.g.*, pepsin, renin, trypsin and erepsin.

(*ii*) **Sucrolytic enzymes** that hydrolyse sugar molecules into simpler ones, *e.g.*, cellulase, amylase, lactase, fructase and galactase.

(*iii*) **Lipolytic enzymes** break down lipids or neutral fats into glycerol and fatty acids *e.g.*, lipase.

(*iv*) **Amylolytic enzymes** which bring about hydrolysis of starch into maltose.

(*v*) **Nucleases** break down nucleic acids (DNA and RNA) into nucleotides which are further hydrolysed by nucleotidases into nucleosides.

(*vi*) **Amidases** bring about breakages of urea, arginine, and purines, so accordingly are called urease, arginase, purinamidase, etc.

(*vii*) **Inverting enzymes** convert diasaccharides into monosaccharides, *e.g.*, invertase and maltase.

(*viii*) **Carbonic anhydrase** that splits up carbonic acid into CO_2 and water.

4. Lyases or Desmolases. Lyases catalyse breakage of specific covalent bonds and removal of some group without hydrolysis. Thus, lyases catalyse lysis of substrate, generating a double bond: these are **nonhydrolytic, nonoxidative, eliminative**, reactions. They play a role in respiration and fermentation.

Examples:

(*i*) **Pyruvate decarboxylase** splits pyruvate into acetaldehyde and carbon dioxide. The systematic name of pyruvate decarboxylase is *2-oxo-acid carboxylyase* (EC 4.1.1.1).

$$\text{Pyruvate} + H^+ \xrightarrow[\text{decarboxylase}]{\text{Pyruvate}} \text{Acetaldehyde} + CO_2$$

(*ii*) Enzyme **histidine decarboxylase** breaks histidine into CO_2 and histimine.

(*iii*) Enzyme **aldolase** splits fructose 1, 6-diposphate into dihydroxyacetone phosphate and 3-phosphoglyceraldehyde (3-PGAL) during glycolysis.

$$\text{Fructose 1, 6-diphosphate} \xrightarrow{\text{Aldolase}} \text{3-Phosphoglyceraldehyde (3-PGAL)} + \text{Dihydroxyacetone phosphate}$$

Synthase. In the reverse direction, lyases catalyse the addition of one substrate to a double bond of a second substrate. A lyase that catalyses an addition reaction in cells is often called a **synthase.**

5. Isomerases. Isomerases catalyse structural change within a single molecule (isomerization reactions). Thus, isomerases brings about the regrouping of atoms in the molecules (*i.e.*, intramolecular arrangement) and thus formation reactions have only one substrate and one product, they are among the simplest enzymatic reactions.

Examples:

(*i*) *Alanine racemase* (EC 5.1.1.1) is an isomerase that catalyses the interconversion of L-alanine and D-alanine. Its common name is the same as the systematic name.

$$\text{L-Alanine} \xrightarrow{\text{Alanine racemase}} \text{D-Alanine}$$

(*ii*) During glycolysis *phosphohexose isomerase enzyme* changes glucose-6-phosphate to fructose-6-phosphate.

$$\text{Glucose-6-phosphate} \xrightarrow{\text{Isomerase}} \text{Fructose-6-phosphate}$$

6. Ligases or Synthetases. These enzymes catalyse ligation or joining of two substrates. These reactions require the input of the chemical potential energy of a nucleotide triphosphate such as ATP.

Example:

(*i*) **Glutamine synthetase** or *L-glutamate*: *ammonia ligase* (ADP forming) (EC 6.3.1.2), uses the energy of ATP hydrolysis to join glutamate and ammonia to produce glutamine

$$\text{L-Glutamate} + \text{ATP} + NH_4^+ \xrightarrow[\text{synthetase}]{\text{Glutamine}} \text{L-Glutamine} + \text{ADP} + \underset{\text{(Inorganic phosphate)}}{\text{Pi}}$$

(*ii*) Enzyme **pyruvate carboxylase** catalyses joining of pyruvate and CO_2 to form oxaloacetate.

$$\text{Pyruvate} + CO_2 + \text{ATP} \xrightarrow[\text{carboxylase}]{\text{Pyruvate}} \text{Oxaloacetate} + \text{ADP} + \text{Pi}$$

(*iii*) Acetyl CoA synthesis is catalysed by enzyme **acetyl CoA synthetase**

$$\text{ATP} + \text{Acetate} + \text{CoA} \xrightarrow{\text{Acetyl CoA synthetase}} \text{Acetyl CoA} + \text{AMP} + \text{Pyrophosphate (PP)}$$

From the examples given above, we see that most enzymes have more than one substrate, although the second substrate may be only a molecule of water. Note also that although enzymes catalyse both forward and reverse reactions, one-way arrows are used when the equilibrium favours a great excess of product over substrate. At equilibrium, an enzyme catalyses both the forward and reverse reactions at the same rate.

F. Mechanism of Enzyme Action

An enzyme can speed up only those reactions which otherwise occur to some extent even in the absence of the enzyme. Several theories were put forward to explain how an enzyme speeds up the reaction.

1. Enzyme-substrate complex theory. This theory was put forward by **Heneri** and mathematically formulated by **Michaelis** (1875–1949) and **Menten** (1879–1960). According to this theory, an enzyme combines with its substrate to from an unstable **intermediate complex**, the **enzyme-substrate complex**. This intermediate complex immediately breaks down into the reaction products and the original enzyme. This can be represented in the following manner:

$$\underset{\text{Enzyme substrate}}{\text{E} + \text{S}} \underset{k_2}{\overset{k_1}{\rightleftharpoons}} \text{E} - \text{S complex} \xrightarrow{k_3} \underset{\text{Enzyme}}{\text{E}} + \underset{\text{Product}}{\text{P}}$$

Where k_1 = Rate constant of formation of ES complex

k_2 = Rate constant of dissociation of ES complex

k_3 = Rate constant of formation of products from ES complex

Since the enzyme-substrate complex is a very transitory compound, some difficulty was faced in actually proving its existence. Direct evidence of the existence of enzyme substrate complex was obtained by **David Keilin** and **Britton Chance**. Chance isolated a brown coloured peroxidase enzyme from horse raddishes (a cruciferous plant, *Armoracia rusticana*, with long lobed leaves) and when this was mixed with hydrogen peroxide, a green coloured enzyme substrate complex was formed. This was then changed to pale red enzyme substrate complex. The latter finally splitted up into enzyme and break down product.

The enzyme merely provides a "platform" or template on which certain molecules could react with each other. Such an enzyme platform brings reacting molecules in contact much faster than chance collisions at that temperature. The result is that the reactions are accelerated.

Enzymes reduce the energy of activation. The effect of enzymes is to lower the activation-energy requirements, thereby promoting appreciable reaction rates at lower temperatures than would be possible otherwise. For example, decomposition of hydrogen peroxide without a catalase has energy activation of about 18,000 cal/mol, when catalase is added this energy of activation is reduced to 2000 cal/mol.

Enzymes accelerate reactions by reducing the free energy of activation. This principle may be interpreted in terms of the **transition state theory**. According to this concept reactants in order to react to form the product C, must pass through a transition state (A – B) in which the activated complex is formed and certain amount of activation energy (Ea) is required. When the proper enzyme is present, less energy is required for activation (Ea'). In other words, the combination of substrate with enzyme creates a new reaction pathway that has a transition state of lower energy than in absence of enzyme.

2. Lock and Key Theory (or Rigid Model of Catalytic Site). The larger enzyme molecule comes in contact with smaller substrate molecules to form an intermediate complex. Enzymes bear distinct cavities or **active sites** or **catalytic sites** or active centers in which specific types of substrate molecules can form intimate association by their specific areas called **reactive sites.** Because these active sites are specifically designed, only specific substrate molecules with proper geometric shape can fit correctly. This combination occurs by electrostatic hydrogen and/or hydrophobic bonds and produces a strain on the bond which is to be broken so that lesser amount of energy is required to break it. This **intermediate complex** is called **enzyme substrate complex (Michaelis Complex).** It soon breaks to release the enzyme and product of biochemical reaction. It is believed that when substrate molecule binds to the active site, its parts are held together in such a way to cause distortion of chemical bonds, *i.e.*, the bonds are weakened. These weakened bonds speed up the rate of reaction. The overall process can be summarised by following reaction:

$$\text{Enzyme} + \text{Substrate} \rightarrow \text{Enzyme Substrate complex} \rightarrow \text{Enzyme} + \text{Product}$$

This simple model of enzyme activity indicates the **biphasic nature of enzyme mechanism.** First step involves binding of substrate molecule with the active site of enzyme while the second step involves the chemical modification of the substrate.

Nature of active sites. Active sites contain special groups, *e.g.*, –OH group of hydroxyamino acids such as serine; $-NH_2$ of basic amino acids such as lysine, –COOH of acidic amino acids; –SH of cysteine amino acid for establishing contact with substrate molecules. Sometimes more than one side chains may form the active site. As a particular lock can be opened by a particular type of key so particular enzyme can act on particular substrate due to their specificity, the name for the mechanism has been given as **Lock and Key Mechanism (Fischer** 1989). He noticed that glycosidases which act on glycosides are highly specific with respect to their substrate which is due to close similarity in the shape of enzyme and substrate molecules.

3. Induced-Fit Model. Normally a large enzyme molecule is required to catalyse a reaction at the site which is only a tiny portion on its surface. It may be debated that if the active site represents only a small part of the enzyme molecule, what is the function of the rest of the enzyme molecule? In 1959, **Koshland Jr.** suggested that the enzyme essentially consists of four categories of amino acids (Fig. 15.7):

(*a*) **Catalytic residues (catalytic site).** These amino acids make and break chemical bonds and are required for catalytic activity.

(*b*) **Binding residues (Binding sites).** These amino acids hold the substrates in place while the reaction (catalysis) is taking place.

(*c*) **Structural residues.** These amino acids hold the active site in the correct shape so that it can function properly.

(*d*) **Non-essential residues.** These amino acids have no specific function. They are often near the surface of the enzyme molecule and can be removed or replaced without loss of function.

The idea of an "*enzyme wrapping around*" a substrate to form a more stable structure is called **induced-fit hypothesis.** According to the induced-fit model, the enzyme molecule on coming in contact with the substrate, does not retain its original shape and structure. Rather, the contact of the substrate **induces** some configurational or geometrical changes in the active site of the enzyme molecule. Consequently, the enzyme molecule is **made to fit** completely the configuration and active centers of the substrates. At the same time, other amino acid residues may become buried in the interior of the molecule.

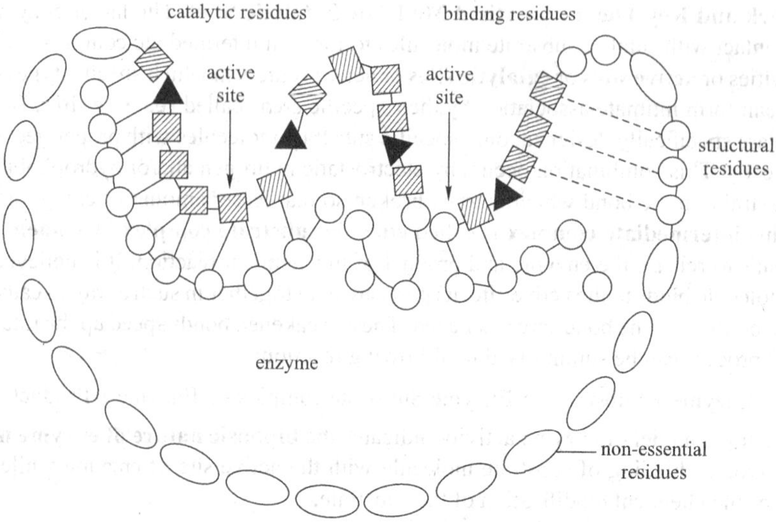

Fig. 15.7. Amino acid residues in an enzyme molecule. Active site is flexible to accommodate the substrate.

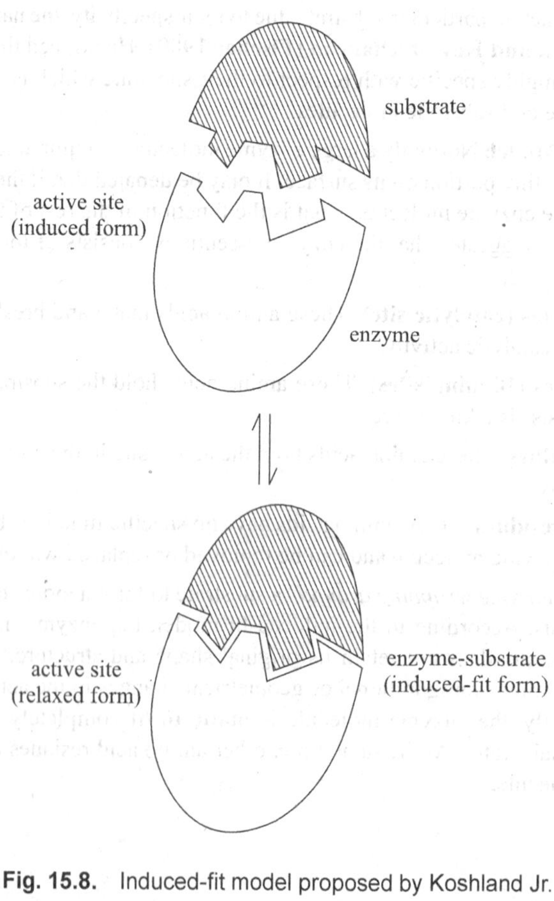

Fig. 15.8. Induced-fit model proposed by Koshland Jr.

G. Factors Affecting Enzyme Activity

Enzyme activity is affected by following nine important factors: 1. Enzyme concentration; 2. Substrate concentration; 3. Product concentration; 4. Temperature; 5. pH; 6. Poisons and radiations; 7. Activators; 8. Inhibitors; and 9. Enzyme-Substrate complex.

1. Enzyme concentration. The rate of a chemical reaction is influenced by particular enzyme. In most cases it increases with increasing concentration of the enzyme, provided that the substrate is present abundantly and is not a limiting factor (Fig. 15.9).

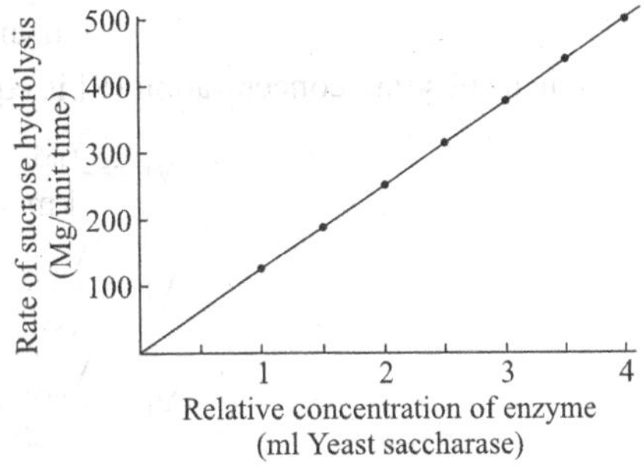

Fig. 15.9. Effect of enzyme concentration on the rate of a reaction when the substrate is available in excess.

2. Substrate concentration. If there are more enzyme molecules than substrate molecules, the two kinds of molecules do not collide frequently and the reaction proceeds slowly. In such a case, a progressive increase in the substrate molecule (S), increases the velocity (V) of their conversion to products because the collisions between the enzyme molecule and substrate molecules become more frequent. However, eventually the rate of reaction reaches a maximum (Fig. 15.10). At this state, the active sites of all the available enzyme molecules are occupied by the substrate molecule and the enzyme is said to be **saturated**. Therefore, further increase in concentration of substrate has no effect on the rate of reaction.

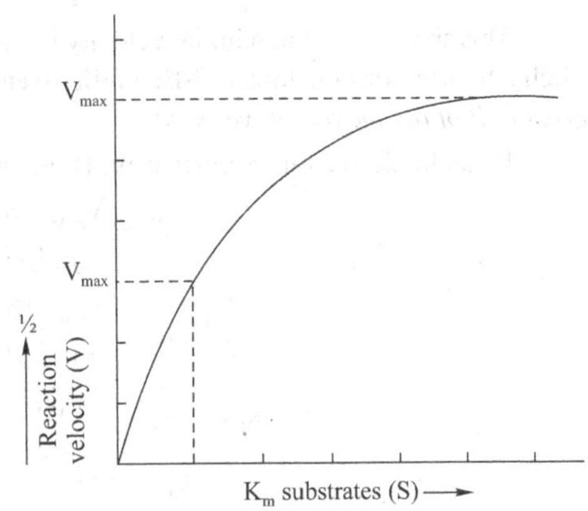

Fig. 15.10. Effect of substrate concentration on the velocity of enzyme action, S = substrates; V = reaction velocity, Vmax = Maximum velocity; Km = Michaelis-Menton constant.

Energy-kinetics of Enzymes (Michaelis-Menton Equation)

Michaelis and **Menton** (1913) proposed an equation which shows the inter-relationship between the enzyme activity and substrate concentration. This is called **Michaelis-Menton equation**. It is based on three basic assumptions:

1. ES complex is in a steady state, *i.e.*, concentration of ES complex remains constant.

2. Under saturating conditions, all enzyme molecules are involved to form ES complex and none is free.

3. Rate of formation of ES complex is maximum under saturating conditions.

Under these assumptions, equilibrium expression for the formation and breakdown of the ES complex is expressed as:

$$E + S \underset{k_2}{\overset{k_1}{\rightleftharpoons}} \text{ES complex} \xrightarrow{k_3} E + \text{Products}$$

So

$$km = \frac{k_2 + k_3}{k_1}$$

Rate of enzyme action (Vi) is expressed as:

$$Vi = \frac{V_{max} + [S]}{km + [S]}$$

S = Substrate concentration

Km = Michaelis-Menton constant

K_{max} = Maximum rate of enzyme activity under a given set of pH, temperature, ionic strength, etc., and is constant for specific enzyme

When substrate concentration [S] is equal to Km, then,

$$Vi = \frac{V_{max} + [S]}{km + [S]}$$

$$Vi = \frac{V_{max} + [S]}{[S] + [S]}$$

$$Vi = \frac{V_{max} + [S]}{2[S]}$$

$$Vi = \frac{V_{max}\ [S]}{2[S]}$$

$$Vi = \frac{V_{max}}{2} = \frac{1}{2}V_{max}$$

This shows that when initial velocity is half-maximal, then the substrate concentration indicates Michaelis-Menton constant, or **Michaelic-Menton constant** *is equal to substrate concentration that gives half of the maximum velocity*.

It can be derived in another way: If the initial velocity is equal to half maximal, then

$$Vi = \frac{V_{max}\ [S]}{km\ [S]}$$

$$\frac{1}{2}V_{max} = \frac{V_{max}\ [S]}{km\ [S]}$$

$$km + [S] = \frac{2V_{max}\ [S]}{V_{max}}$$

$$km = [S]$$

Thus, Km is equal to the substrate concentration at which the velocity of the reaction is half maximum.

The value of the Michaelis-Menton constant (Km) is inversely proportional to the enzyme activity. A large value of Km means that a high substrate concentration is needed to get half velocity of the maximum rate of reaction. In fact, it means that enzyme has lower affinity for the substrate.

When reciprocal values of enzyme activity and substrate concentration are plotted against each other, we get a straight line (Fig. 15.11). This double reciprocal plot is called the **Lineweaver-Burk plot**. From this plot, value of Km can be obtained by extending the line toward the abscissa. (Note. *Abcissa* is a mathematical term which means one of the coordinates of a two-dimensional coordinate system, usually the horizontal coordinate, denoted by *x*).

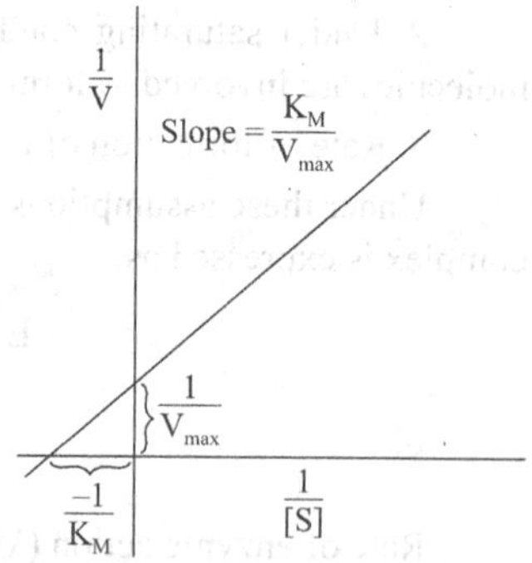

Fig. 15.11. Lineweaver-Burk plot.

3. Product concentration. Accumulation of the product of enzyme reaction lowers the enzyme activity. Enzyme molecules must be freed to combine with more substrate molecules. Normally the products are quickly removed from the site of formation and the reaction does not suffer.

4. Temperature. The optimum temperature for an enzyme usually corresponds to the body temperature of the organism. For example, the optimum temperature of enzyme of humans is 37°C which is also their normal body temperature. Most enzymes operate at their best in a temperature range of about 25°C to 40°C. Between a range of 0°C and 45°C the enzyme activity increases 2 – 3 times, with a rise of 10°C temperature. The increase in enzyme activity is due to two factors: (*i*) the influence of heat on the chemical reaction and (*ii*) the increased enzyme activity.

In the range of 45° to 60°C thermal **denturation** will destroy the catalytic activity of the enzyme (Box 15.3). This slows down the catalytic reaction and ultimately stops it. Enzyme's activity stops altogether at about 60° and also at 0°C.

Box 15.3 Denaturation

Rise in temperature increases the kinetic energy of the molecules. Therefore, at higher temperature an increasing number of molecules have the required activation energy and can take part in chemical reaction. At higher temperature, the kinetic activity of molecules in an enzyme becomes strong enough to break weak hydrogen bonds that maintain the tertiary structure of the enzyme. Modification in the physical form of the enzyme results in the loss of its catalytic activity. This change in structure is called **denaturation** of protein. This is a very permanent change, and the denatured enzyme protein remains inactive even if the temperature is then brought down.

However, the enzymes are not destroyed by freezing and regain their lost activity if the temperature is raised to normal.

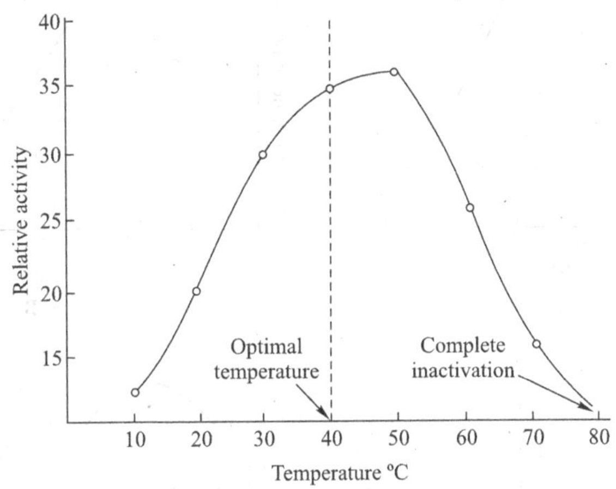

Fig. 15.12. Effect on increasing temperature on enzymatic activity.

Freezing preserve the foods for a long time because neither the microbial enzymes nor the enzymes in the food can act at low temperature to spoil them.

Enzymes differ in their sensitivity to temperature. Bacteria living in hot springs with temperature near 100°C have enzymes which can resist high temperature. On the other hand, the cellular proteins of the plants and cold blooded animals of the arctic denature readily at moderate temperature. Thus, the organisms have **biochemical adaptability** to environment.

5. Hydrogen ion concentration (pH). Some enzymes act best in an acid medium, others in an alkaline medium. For every enzyme, there is an optimum pH where its action is maximum. Most enzymes show maximum activity in a pH range of about 6.0 to 7.5 (Fig. 15.13A). A shift to alkaline or acid side rapidly decrease the enzyme activity and finally stop it altogether. This is due to denaturation of the enzyme molecule, *i.e.*, change in its physical structure. Some digestive enzymes have their optimum activity in the acidic or alkaline range, and drop in activity as the pH falls or rises (Fig. 15.13B, C). For example, **pepsin** of gastric juice has its optimum pH 2 (acidic) and **trypsin** of pancreatic juice show maximum activity of pH 8 (alkaline). Likewise, optimal range for activity of **amylases** is 5.2 and for **salivary amylase** between 6.7 and 6.8.

6. Poisons and radiations. Poisons, such as cyanide, and radiations destroy the tertiary structure of the enzymes, making them ineffective.

7. Activators. In living beings many proteins are manufactured and secreted in the form of inactive precursor proteins. These are called **proporteins**. Conversion of a proprotein to the mature protein involves one or more successive "**proteolytic clips**". As a result of such selective proteolysis the inactive proprotein is converted into active mature protein.

Enzymes are proteins. Certain enzymes are produced by the living cell in the inactive (nonfunctional) form. These inactive or precursor enzymes are called **proenzymes** or **zymogens**. For example, all the proteolytic enzymes secreted in inactive form, such as (*i*) propepsin is called **pepsinogen**; (*ii*) protrypsin is called **trypsinogen** and (*iii*) prochymotrypsin is called **chymotrypsinogen**.

One basic reason why proteolytic enzymes or proteases are secreted in inactive form is to avoid the **autodigestion** of origin, where these enzymes are synthesized, stored and secreted.

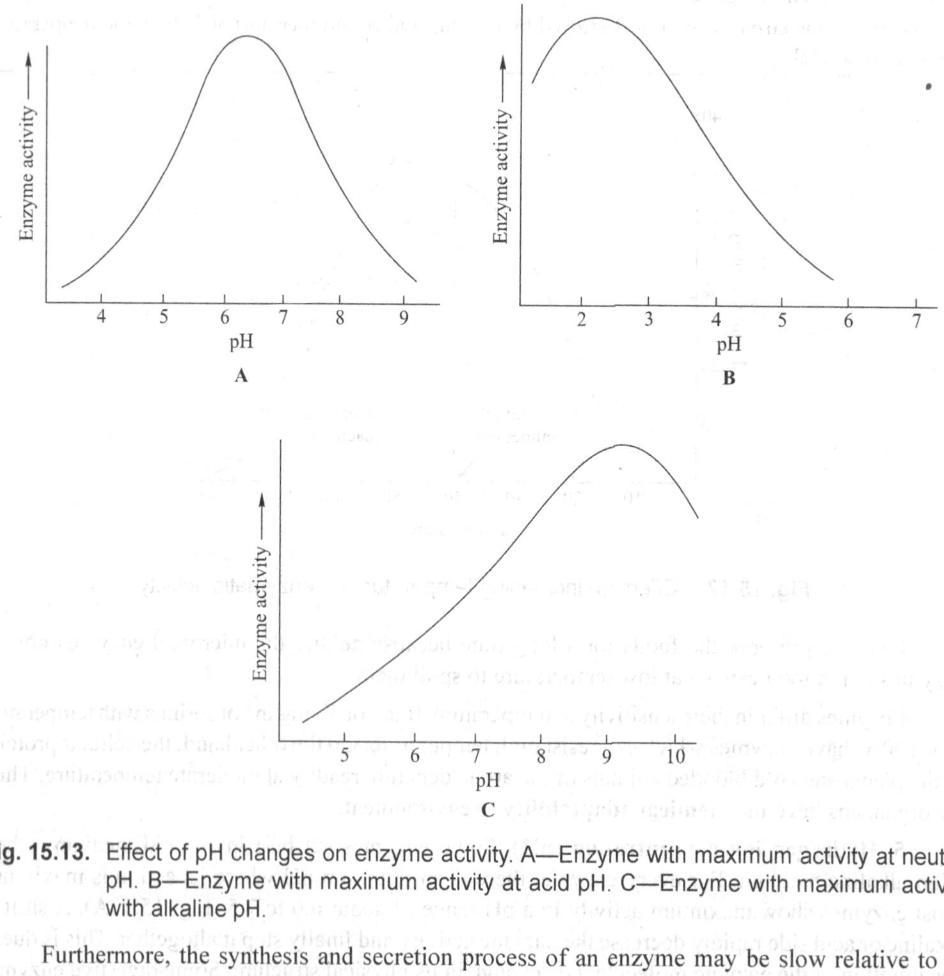

Fig. 15.13. Effect of pH changes on enzyme activity. A—Enzyme with maximum activity at neutral pH. B—Enzyme with maximum activity at acid pH. C—Enzyme with maximum activity with alkaline pH.

Furthermore, the synthesis and secretion process of an enzyme may be slow relative to its physiological demand. Hence an adequate amount of the required enzyme is stored in inactive or precursor form to be converted into active stage as and when required. So, secretion of proteases as zymogens is a protective device. It protects the mother cell from **autolysis** or **self-digestion**, **Mechanism of conversion of Proenzyme into active enzymes**. Maturation of a proenzyme into functional or physiologically active form involves selective proteolysis. This introduces essential conformational changes creating active or catalytic or substrate site and the allosteric site if any. For example, in the process of conversion of proenzyme prochymotrypsin (pro CT) to chymotrypsin (α-CT) involves three proteolytic clips and the formations of an inactive intermediate π-chymotrypsin (π-CT). In α-chymotrypsin A, B and C chains remain associated by 2 interchain disulphide bonds.

Enzyme activators. Activity of certain enzymes is activated by the presence of some substances other than the substrate. Such substances are called **activators**. These may be components of the enzyme, a loosely bound prosthetic group or the metabolite which binds to specific site of an enzyme and changes it into an active conformation. The substrate acts as an **allosteric effector**. Specific ions are required for the activity of certain enzymes. For example, phosphorylating enzymes require divalent cations, such as Mg^{2+} or Ca^{2+} for their activity.

Pepsinogen is changed into active pepsin by hydrogen ions in the stomach. Trypsinogen is activated to trypsin by an enzyme **enterokinase** in small intestine. Further, once small amount of pepsin or trypsin is formed, it itself catalyses the activation of remaining proenzyme. This process is called **autocatalytic reaction** or **autocatalysis**.

8. Inhibitors. Inhibitors are substances that inhibit the activity of enzymes. Enzymes inhibition can be caused in the following ways:

(*i*) Reversible inhibition. In reversible inhibition, the inhibitors prevent enzymes from combining with the substrates but the activity of enzymes is restored when inhibitor is removed. This is of following two types:

1. Competitive inhibition. Molecules of certain substances other than the substrate, bind to the active site of the enzyme stopping the substrate from binding with the enzyme. This makes the enzyme inactive. These substances are called **competitive inhibitors**. Their molecules are similar in shape to the substrate molecules (**substrate analogues**) but are different enough from them so that chemical reaction fails to occur.

The competitive inhibitor molecules compete with the substrate molecules. When enough of natural substrate is present, inhibition is less. This slows down the reaction. But when the inhibitor molecules are more, these displace the substrate molecules almost stopping the reaction.

Examples:

1. Enzyme succinate dehydrogenase catalyses the oxidation of succinate to fumarate in Krebs cycle. In the presence of malonate [a salt of malonic acid, $CH_2(COOH)_2$] the reaction rate is slowed down. Malonate has a molecular structure very similar to that of succinate. It occupies the active center of the enzyme. However, effect of malonate can be reversed if enough succinate is added.

Uses of Competitive Inhibitor

1. The phenomenon of competitive inhibition can be compared to a lock which is jammed by a key similar to the original one. Competitive inhibitors are used as drugs in the control of bacterial pathogens. Sulpha drugs such as **prontosil** are used to combat bacterial infection. In molecular structure prontosil is similar to p-aminobanzoic acid (PABA) which is used by bacteria to synthesize PABA, bacteria fails to synthesize folic acid and fail to grow and multiply.
2. Competitive inhibitors are used in treating some forms of cancer.

2. Non-competitive inhibition. The non-competitive inhibitor does not compete with the substrate for binding at the active or substrate-binding site of the enzyme, because it does not bear any structural similarity with the substrate. It gets attached to the enzyme at a place other than the active site rendering it inactive.

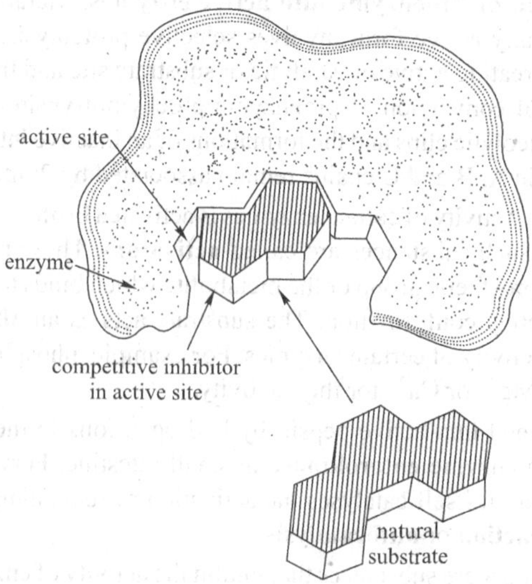

Fig. 15.14. Diagrammatic representation of competitive inhibition.

In this case, the degree of inhibition cannot be reduced by increasing the number of substrate molecules. However, the enzyme activity can be restored because non-competitive inhibitors can be removed by dialysis.

Examples:

1. Cyanide inhibits the mitochondrial enzyme cytochrome oxidase which is essential for cellular respiration. This kills the animals.
2. Ions of heavy metals such as mercury, silver and copper Ag^+, Hg^{2+}, and Cu^{2+} combine with thiol (–SH) groups in the enzyme breaking the disulphide bridges. These bridges are important in maintaining tertiary structure. When these bridges are broken, the enzyme become denatured and inactive.

(*ii*) Non-reversible inhibition. In non-reversible inhibition, the inhibitor molecule firmly (*i.e.*, permanently) attached to the R-group at the active site of enzyme. As a result the substrate molecule fails to bind to the enzyme molecule and the enzyme is rendered permanently inactive.

Example:

1. Many antibiotics are inhibitors of specific enzymes in bacteria. For instance, penicillin blocks the active site of an enzyme that many bacteria use to make their cell wall.
2. DDT and parathion are inhibitors of key enzymes in the nervous system.
3. Insecticide organophosphorus compound—**diisopropyl phosphofluoridate (DFP)** (marketed/trade name Malathion) reacts with a specific R-group of amino acid serine at the active site.

The same DFP is irreversible inhibitor to enzyme acetylcholinesterase which is essential for the orderly propagation of nerve impulse. Because of this characteristic DFP are classified as nerve gases and are used as insecticides.

3. Allosteric modulation. Allosteric regulation is the modulation of catalytic activity of certain regulatory enzymes by low molecular weight *allosteric effector* that have little or no structural similarity to the substrates or coenzymes for the regulated enzyme.

Enzymes having quaternary structure (*i.e.* formed of more than one polypeptide subunits) have a **regulatory site** or **allosteric site** in addition to the active or substrate site. **Allostery** means "different shape". An outside molecule, other than the substrate, binds to the regulatory site and can enhance or diminish the reactivity of the enzyme at the substrate site.

An allosteric hypothetical enzyme has minimum two units:

(*a*) **Catalytic subunit** which has full catalytic activity and bears active site or substrate site.

(*b*) **Regulatory subunit** which has the regulatory or allosteric site which binds to the inhibitor or activator.

The allosteric enzyme can exist in following two alternative forms:

(*i*) **The active form.** The active form of allosteric enzyme has full catalytic activity. It's active site on the catalytic subunit binds to the substrate and converts it to product or products. The regulatory or allosteric site on its regulatory subunit is deformed and cannot bind the inhibitor or activator.

(*ii*) **The inactive form.** The inactive form of allosteric enzyme is totally without enzyme activity. Its active site is distorted and unable to bind with the substrate. Its allosteric site can bind an effector (either inhibitor or activator).

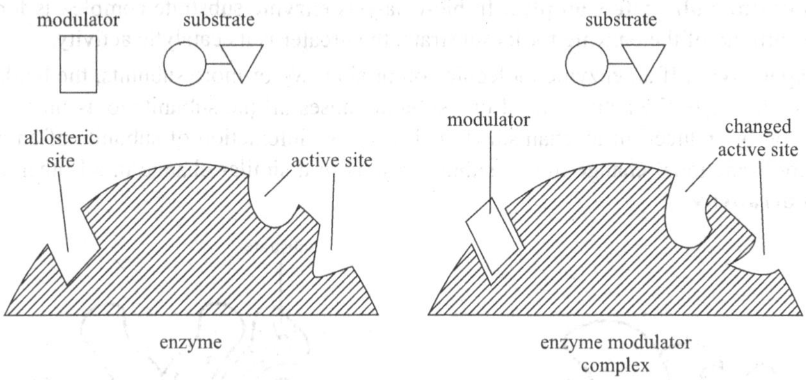

Fig. 15.15. Mechanism of allosteric regulation of enzyme action.

Mechanism of allosteric regulation. The active and inactive forms of allosteric enzyme occur in equilibrium.

When a substrate is added to the enzyme-substrate solution, some of it bind to the active sites of the active enzyme molecules. The active molecules of enzyme that have entered enzyme-substrate complex are incapable of being converted to inactive form. However, to maintain equilibrium, the inactive form of enzyme is converted to active form. So there is an increase in the concentration of active form of enzyme. This increases reactivity of enzyme.

On the contrary, addition of allosteric inhibitor decreases the concentration of free enzyme and inhibits the reaction by binding to the allosteric site of inactive form and preventing its conversion to active form.

A specific case of allosteric enzyme inhibition has been cited. Glucose is changed to glucose 6-phosphate in glycolysis with the help of the enzyme hexokinase. Glucose 6-phosphate causes allosteric inhibition of hexokinase (Fig. 15.16). This is called **feedback allosteric inhibition**.

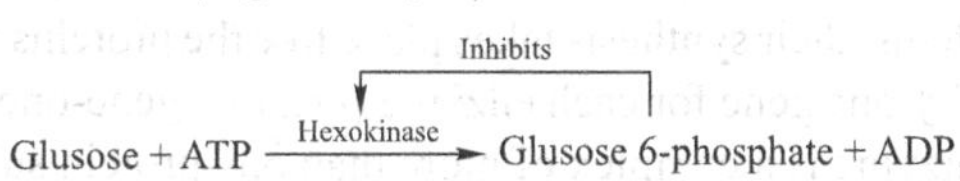

Fig. 15.16. Feedback allosteric inhibition.

The mechanism of allosteric inhibition and promotion was proposed by **Monod** and his colleagues.

Table 15.4. Differences between competitive inhibition and allosteric inhibition.

	Competitive inhibition		Allosteric inhibition
1.	Inhibitor binds to active sites of enzymes.	1.	Inhibitor binds to allosteric sites of enzymes.
2.	Substrate fails to join enzyme as its active site is occupied by inhibitors.	2.	Substrate fails to join enzyme as its active site is deformed by inhibitors.
3.	Inhibitor closely resembles substrates in structure.	3.	Inhibitor does not resembles substrates in structure.
4.	Inhibitors is not an intermediate or a product of metabolic pathway catalyzed by the enzyme.	4.	Inhibitors may be an intermediate or a product of metabolic pathway catalyzed by the enzyme.
5.	It has non-regulatory function.	5.	It regulates the metabolic pathway and checks the excessive formation of the product.

9. Enzyme Substrate Complex. In bio-catalysts enzyme substrate complex is formed. The greater the affinity of the enzyme for its substrate, the greater is its catalytic activity.

Cooperativity. If an enzyme molecule comprising two or more subunits, the binding of one substrate molecule to the active site of one subunit causes all the subunits to assume their active conformation by induced fit mechanism (Fig. 15.17). An interaction of subunits of an enzyme in which a conformational change in one subunit results in a similar change in all other subunits is called **cooperativity**.

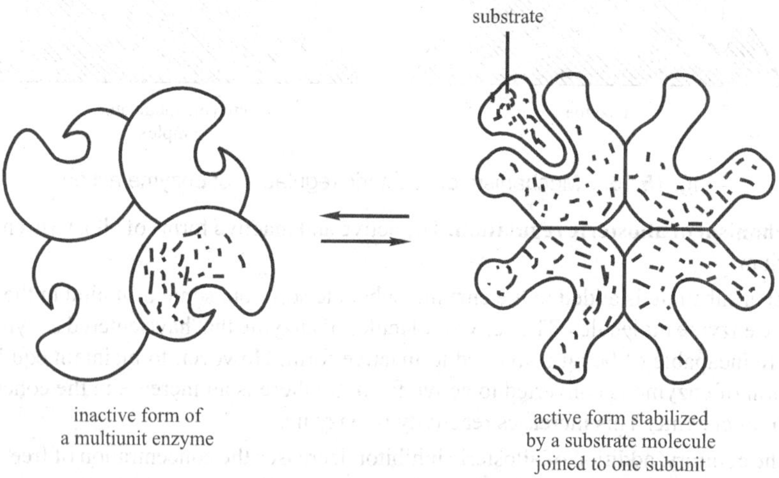

Fig. 15.17. Cooperativity. Joining of a substrate molecule to one subunit of an enzyme stabilizes the active form of all subunits.

H. Synthesis of Enzymes

Since all enzymes are proteins, their synthesis takes place like the proteins synthesis and is controlled by genes. There is generally one gene for each enzyme (*i.e.*, one gene-one enzyme theory of **Beadle** and **Tatum**). In case an enzyme is a complex of more than one protein subunit, more than one gene may control its synthesis.

Synthesis of specific enzymes in cells occurs at the right place and time and in right amount strictly according to the need. The presence of metabolite acts as a signal or stimulus for the synthesis of the desired enzyme or enzymes required for its metabolism.

I. Regulation of Gene Action

A large variety of chemical reactions occur in a cell simultaneously in a surprisingly coordinated manner, without interfering with one another. Each compound is produced when required and in the desired quantity. Such a precise regulation is achieved by two types of control mechanisms operating in the cells. One mechanism operates at the enzyme level and the other at the gene level.

1. Control at the enzyme level. In this type of gene regulation, the enzyme, substrate and product themselves regulate the chemical reaction. When the product of an enzyme reaction accumulates in the cell, it inhibits its own production by lowering the enzyme activity. This kind of control mechanism is called **feed back inhibition** or **negative feedback.**

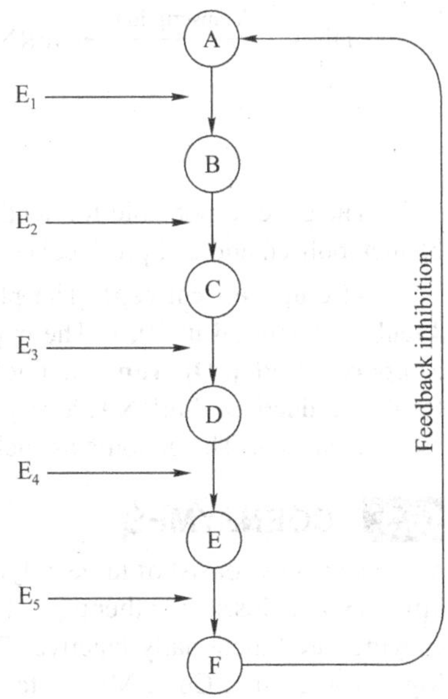

Fig. 15.18. Feedback inhibition.

Feedback Inhibition. Many multienzyme systems possess the capability of self-regulation of their overall reaction rate. In such a system, the end product of the reaction sequence can inhibit the first enzyme, with the result that the rate of entire sequence is determined by the steady-state concentration of the product. The first enzyme that is inhibited by the end-product is called the regulatory or **allosteric enzyme**, the first step is called the **committed step**.

The colon bacterium *Escherichia coli* synthesizes the amino acid **isoleucine** from a substrate **threonine** by a series of intermediate reactions. When isoleucine accumulates in amounts more than required, it stops its own production by inhibiting the activity of the enzyme **threonine deaminase** which catalyses the first reaction of the series (Fig. 15.19). The type of metabolic control, in which the first enzyme of a series is inhibited by the end product is known as end-product inhibition.

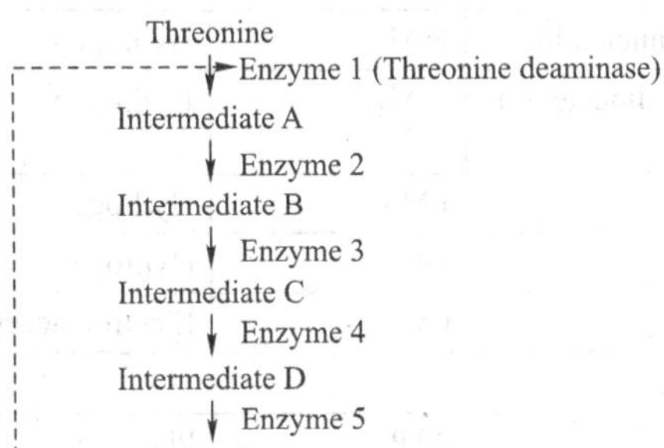

Fig. 15.19. Feedback inhibition of a pathway of *E.coli* involving five enzymes that catalyses successive steps. The end product inhibits the first enzyme.

2. Control at gene level. In this type of control mechanism, the gene regulates the production of enzymes. Proteins are the end product of gene (DNA) action. They are synthesized as summarised below:

$$\underset{1}{\text{DNA}} \xrightarrow{\text{Transcription}} \underset{2}{\text{mRNA}} \longrightarrow \underset{\underset{\text{Amino acid carrying tRNAs}}{\uparrow}}{\text{Ribosome}}\ 3 \xrightarrow{\text{Translation}} \underset{4}{\text{Protein (Enzyme)}}$$

The gene responsible for synthesis of enzyme is activated and inactivated by the substrate to be metabolised and end product accumulating in excess respectively.

The operon concept. The phenomenon of induction and repression were formulated by **Jacob** and **Monod** in 1961. The hypothesis propounded by them in context with *E.coli* is known as **operon model. Operon** is defined as a complete unit of gene expression, which includes genes for the production of mRNA, a gene which regulates the synthesis and the site at which the control mechanism acts. The genome as such is composed of clusters of genes.

15.3. COENZYMES

Many enzymes consist of large polypeptide or protein units (the protein portion of enzyme is called **apoenzyme**) plus some other type of substance called as **cofactor**. Without its cofactor, each of these enzymes are functionally inactive. The cofactor may be some simple divalent metalic ions such as Mg^{2+}, Ca^{2+}, Zn^{2+}, Co^{2+}, Mn^{2+}, etc., or sometimes a non-protein small molecular weight organic compound. If the organic cofactor is *loosely bound* with the protein (*i.e.*, apoenzyme), it is termed as **coenzyme** and if the cofactor is *firmly bound* with the apoenzyme it is called **prosthatic group**. The full intact enzyme sometimes referred to as **holoenzyme.**

$$\text{Cofactor} + \text{Apoenzyme} \rightarrow \text{Holoenzyme (or Enzyme)}$$

Nature of coenzyme. Coenzymes are nonprotein organic compounds. These usually have one of the vitamins as their active group.

1. Vitamins as coenzymes. The following vitamins and their derivatives act as coenzymes (Table 15.5):

Table 15.5. Derivatives of Vitamine B and AMP which act as coenzymes.

	Coenzyme	Abbreviation	Group transferred	Vitamin
	A. Oxido-reductase coenzymes (Hydrogen acceptor coenzymes)			
1.	Nicotin amide-adenine dinucleotide	NAD	Hydrogen	Nicotinamide
2.	Nicotin amide-adenine dinucleotide phosphate	NADP	Hydrogen	Nicotinamide
3.	Flavin mononucleotide	FMN	Hydrogen	Riboflavin
4.	Flavin dinucleotide	FAD	Hydrogen	Riboflavin
5.	Cytochromes	Cyt	Electron acceptor	—
	B. Transferase coenzymes			
1.	Adenosine triphosphate	ATP	Phosphate	—
2.	Adenosyl methionine	—	Methyl group of methionine	—
3.	Biotin	—	Carboxyl	Biotin
4.	Thiamine pyrophosphate	TPP	C_2-aldehyde	Thiamine (Vitamin B1)

Table 15.5. (*Contd...*)

	Coenzyme	Abbreviation	Group transferred	Vitamin
5.	Pyrioxal phosphate	PALP	α-amino acid	Pyridoxine (Vitamin B6)
6.	Coenzyme A	CoA	Acyl	Pantothenic acid
	C. Isomerase and Lyase coenzymes			
1.	B_{12} coenzyme	—	Carboxyl displacement	Cobalamine
2.	Pyridoxal phosphate	PALP	Decarboxylation (in amino acid metabolism)	Pyridoxine (Vitamin B6)
3.	Thiamine pyrophosphate	TPP	Decarboxylation	Thiamine (Vitamin B1)
4.	Uridine diphosphate	UDP	Monosaccharide isomerization	Thiamine (Vitamin B1)
	D. Decarboxylation coenzymes			
1.	Thiamine pyrophosphate	TPP	Removes CO_2	Thiamine (Vitamin B2)
2.	Pyroxal pyrophosphate	PALP	Removes CO_2	Pyridoxine (Vitamin B6)

(*i*) **Thiamine or Vitamin B_1** acts as coenzyme in the form of thiamine pyrophosphate.

(*ii*) **Riboflavin or Vitamin B_2** acts as coenzyme in the form of following two compounds:

(*a*) **Flavin mononucleotide (FMN)** and

(*b*) **Flavin adenine dinucleotide (FAD)**; (Fig. 15.20).

(*iii*) **Nicotinic acid** acts as coenzyme in the form of following two compounds:

(*a*) *Nicotinamide adenine dinucleotide* (*NAD*; Fig. 15.21) and

(*b*) *Nicotinamide adenine dinucleotide phosphate* (*NADP*) or *coenzyme-II.*

(*iv*) **Pantothenic acid or vitamin-B_3.** It acts as a component of coenzyme A.

(*v*) **Pyridoxine or Vitamin-B_6.** The phosphate derivative of this vitamin pyridoxal-phosphate acts as a coenzyme.

(*vi*) **Folic acid or pteroglutamic acid (PGA)** act as a coenzyme in DNA synthesis.

(*vii*) **Cyanocobalamin or Vitamin-B_{12}.** Cobamide and methyl B_{12} act as coenzymes as **methyl cobalamin** in methylation.

(*viii*) **Biotin or Vitamin-H.** This vitamin acts as coenzyme for various carboxylating enzymes in CO_2 transfer reactions.

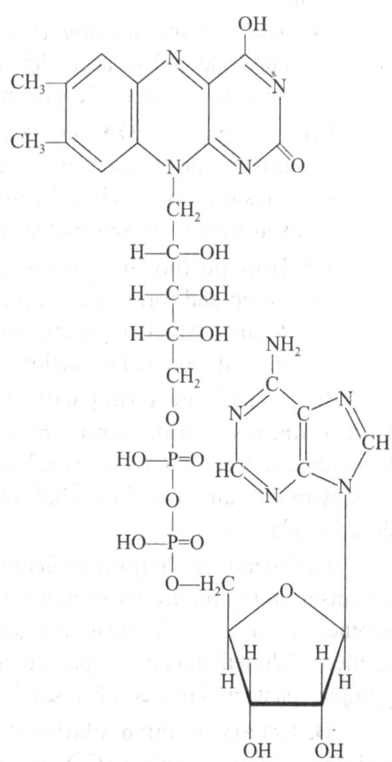

Fig. 15.20. Chemical formula of flavin adenine dinucleotide (FAD).

(*ix*) **Ascorbic acid or Vitamin-C.** This vitamin acts as a coenzyme with the enzyme that changes p-hydroxy-phenyl puruvic acid into 2, 5-dihydroxy-phenyl pyruvic acid.

2. Other Organic Compounds as Coenzymes

(*i*) **Lipoic acid** acts as coenzyme in the decarboxylation of α-ketoacid.

(*ii*) **Haem peroxidase** acts as coenzyme with catalase and cytochrome-oxidase enzymes.

(*iii*) **Uridine diphosphate glucose (UDPG)** acts as a coenzyme with uridine diphospho-galactose-5-epimerase.

(*iv*) **Ubiquinone Q_1, Q_2, Q_6, Q_7, Q_8 and Q_{10}.** These organic compounds act as coenzymes in electron transport system.

3. Metals as Cofactors/Coenzymes

Some metals also act as coenzymes or cofactors: (*i*) **Zinc (Zn)** in carbonic anhydrase; (*ii*) **Copper (Cu)** in tyrosinase, phenolase and ascorbic oxidase; (*iii*) **Iron (Fe)** in cytochrome enzymes, catalase and peroxidase; (*iv*) **Calcium (Ca)** in lecithinase, ATP ase and lipase; (*v*) **Cobalt (Co)** in peptidase; (*vi*) **Manganese (Mn)** in arginase and phosphoglucomutase and (*vii*) **Magnesium (Mg)** in phosphatases and hexokinases.

Mode of action of coenzymes. Coenzymes act as links between metabolic pathways. These can be considered as special intracellular enzyme-substrates. These help in oxido-reduction, group transfer and isomerization reactions and in reactions which form covalent bonds:

A. Coenzymes acting as hydrogen acceptor in oxido-reduction reactions. These are NAD and NADP. Both these coenzymes act as **hydrogen acceptors**. For example:

(*a*) The enzyme *alcohol dehydrogenase* oxidises ethyl alcohol to acetaldehyde by removing 2H only in presence of coenzyme NAD.

(*b*) Coenzymes **FMN** and **FAD** act as prosthetic group and accept hydrogen atoms from the substrate or giving hydrogen atoms to the substrate in presence of specific apoenzyme.

(*c*) **Iron porphyrin** compounds act as prosthetic group and form integral part of cytochromes *a*, *b*, and *c* which are respiratory enzymes involved in oxidation and reduction in the cell.

B. Coenzymes acting with decarboxylases and transaminases. Pyridoxamine phosphate or pyridoxal (derivative of vitamin B_6) act as prosthetic group for a number of enzymes of amino acid metabolism (transaminases and decarboxylases).

C. Coenzymes helping in acetylation. Coenzyme A accepts acetyl group from one metabolite and donates it to another. It thus helps in metabolism of fat. Coenzyme A is formed of thioethanolamine, pantothenic acid (vitamin B) pyrophosphate and nucleotide adenylic acid.

D. Coenzyme for oxidative decarboxylation. For splitting of pyruvic acid into CO_2 and acetaldehyde in yeast cell enzyme carboxylase requires thiamine pyrophosphate as a coenzyme. In animal cells for oxidative carboxylation

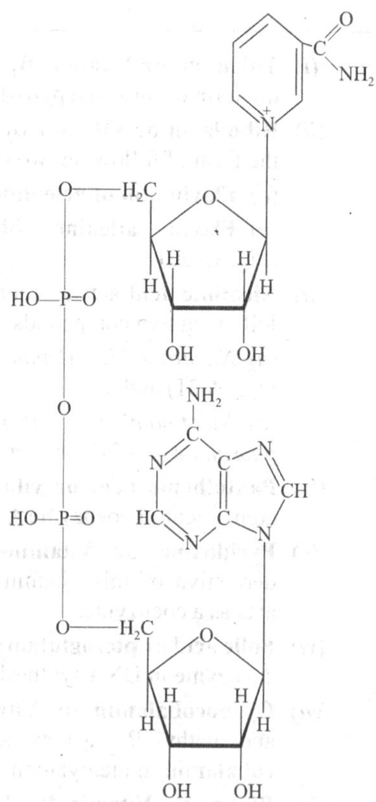

Fig. 15.21. Chemical formula of nicotinamide adenine dinucleotide (NAD^+).

of pyruvic acid to yield CO_2 and acetly CoA, a coenzyme is required which contains lipoid acid and thiamine pyrophosphate.

E. Other coenzymes. Glucose-1, 6-diphosphate, glyceric acid 2, 3-diphosphate and glucose-1-phosphate-uridine nucleotide, biotin, folic acid and vitamin B_{12} act as coenzymes.

Mechanism of Coenzyme Action

Coenzyme act as a second substrate or **cosubstrate** because the chemical changes in the coenzyme counterbalance the changes in the coenzyme occurring in substrate. The changed coenzyme reverses the reaction. Let us designate two metabolites as M_1 and M_2 and coenzyme A, its function can be shown as follows:

$$\text{Acetyl} - M_1 + \text{CoA} \xrightarrow{\text{Apoenzyme 1}} M_1 + \text{Acetyl CoA}$$

$$\text{Acetyl CoA} + M_2 \xrightarrow{\text{Apoenzyme 2}} \text{Acetyl} - M_2 + \text{CoA}$$

QUESTIONS

Long Answer Quetions

1. Give a list of some energy-rich compounds and explain why ATP is the ideal energy source in the cells.
2. Define high-energy phosphate bond. Why is phosphoenol pyruvate (PEP) a high energy compound?
3. What is enzyme specificity? Describe various specificities displayed by enzymes.
4. Explain enzyme inhibition. How would you know whether an inhibitor is competitive or non-competitive?
5. Describe feedback inhibition.
6. Why is it likely that the active site of an enzyme can fit and bind both the substrates and products?
7. Discuss the factors which affect enzyme activity.
8. Describe mechanims of enzyme action.

Short Answer Quetions

1. Define entropy and enthalpy.
2. Write short notes on the following:
   (*i*) Isoenzymes
   (*ii*) Allosteric enzymes
   (*iii*) Coenzymes
   (*iv*) Lock and key hypothesis
   (*v*) Induced fit theory
   (*vi*) Proenzyme
   (*vii*) Cofactors
   (*viii*) Zymogen
3. Write differences between following:
   (*i*) Endergonic and exergonic reactions
   (*ii*) Oxidation and reduction
   (*iii*) Endoenzymes and exoenzymes
4. List the characteristics of enzymes.
5. What effects do changes in pH have on enzyme-catalyzed reactions. Why?

Very Short Answer Questions

1. What is bioenergetics?
2. Define the metabolic pathway.
3. What is free energy?
4. Define exergonic.
5. What is endergonic?
6. How enzymes are named?
7. At what temperature enzymes act the best?
8. What is inactive form of trypsin?
9. What is the term used for modified binding site?
10. Give one example of a cofactor.
11. Why enzymes are required in very small amount?
12. Why enzymes are called biocatalysts?
13. What is the name given to that part of enzyme where catalytic activity is carried out?
14. Which structural level enables the proteins to become functional enzymes?

Multiple Choice Questions

1. How many classes of enzymes are in 'IUB' system?
   (*a*) two
   (*b*) six
   (*c*) eight
   (*d*) ten
2. The enzyme code of enzyme 2.7.1.1 refers to the following main group
   (*a*) lyase
   (*b*) ligase
   (*c*) hydrolase
   (*d*) transferase
3. ATP is formed in
   (*a*) mitochondria
   (*b*) Golgi apparatus
   (*c*) lysosome
   (*d*) ribosomes
4. Synthesis of enzymes takes place by
   (*a*) translation
   (*b*) deamination
   (*c*) transamination
   (*d*) none of these
5. Holoenzyme consists of
   (*a*) apoenzyme-cofactor complex
   (*b*) apoenzyme alone
   (*c*) cofactor
   (*d*) coenzyme
6. The word 'enzyme' was coined by
   (*a*) Kuhne
   (*b*) Summer
   (*c*) Paine
   (*d*) Berzelius
7. The term 'feed back' given by Nobel Laureate Umbarger refers to the effect of
   (*a*) substrate on the rate of enzymatic reactions
   (*b*) end products on the rate of enzymatic reactions
   (*c*) enzyme concentration on the rate of enzymatic reactions
   (*d*) an external compound on the rate of enzymatic reactions
8. Enzymes increase the rate of reactions
   (*a*) increasing the free energy of activation
   (*b*) decreasing the energy of activation
   (*c*) changing the equilibrium constant of the reaction
   (*d*) increasing the free energy change of the reaction
9. Enzymes with same function but with different molecular arrangement are
   (*a*) apoenzymes
   (*b*) coenzymes
   (*c*) isoenzymes
   (*d*) metalloenzymes
10. Enzymes acting within the cells are
   (*a*) endoenzymes (*b*) apoenzymes
   (*c*) exoenzymes (*d*) exopeptidases
11. LDH which catalyses pyruvate to lactate is an example of
   (*a*) antienzyme (*b*) isoenzyme
   (*c*) coenzyme (*d*) apoenzyme
12. Non-protein part of an enzyme is
   (*a*) isoenzyme (*b*) ribozyme
   (*c*) coenzyme (*d*) holoenzyme
13. Proteinous part of a conjugated enzyme is
   (*a*) holoenzyme (*b*) coenzyme
   (*c*) apoenzyme (*d*) prosthetic group
14. Prosthetic group of an enzyme is
   (*a*) loosely bound (*b*) tightly bound
   (*c*) can be organic or inorganic
   (*d*) only inorganic
15. The existence of enzymes inside the cells is in the form of
   (*a*) colloids (*b*) ions
   (*c*) solids (*d*) crystals
16. Lock and key hypothesis for enzyme action was given by
   (*a*) Emil Fischer (*b*) Koshland
   (*c*) Endomont (*d*) Sumner
17. Induced fit hypothesis was given by
   (*a*) Emil Fischer (*b*) Koshland
   (*c*) Endomont (*d*) Buchner
18. A mathematical explanation for enzyme action on substrate was formulated by
   (*a*) Leonor Michaelic Menten
   (*b*) Hans Gaffon
   (*c*) Melvin Calvin
   (*d*) Vant Hoff

ANSWERS

Very short Answer Quetions

1. It is an area of science that deals with the application of thermodynamic principles to reactions and processes in the biological world.
2. It is a series of cellular enzymatic reaction that convert one molecule to another via a series of intermediates.
3. Free energy (G) represents thermodynamic function that measures the extractable energy content of a molecule under conditions of constant temperature and pressure, the change in free energy is a measure of the ability of the system to do work.
4. It it pertaining to an energy-releasing reaction characterized by a negative free energy change ($\Delta G < O$).
5. It it pertaining to an energy-requiring reaction characterized by a positive free energy change ($\Delta G > O$).
6. By adding *ase* to the root word of substrate on which they act.
7. 35°C.
8. Trypsinogen.
9. Allosteric site.
10. Cu, Fe and Zn.
11. Because these are not used during reaction and enzymes have high turn over.
12. Because these control metabolic reactions inside living organisms.
13. Active site or substrate site.
14. Tertiary structure.

Multiple Choice Questions

1. (*b*)	**2.** (*d*)	**3.** (*a*)	**4.** (*a*)	**5.** (*a*)	**6.** (*a*)	**7.** (*b*)
8. (*b*)	**9.** (*c*)	**10.** (*a*)	**11.** (*b*)	**12.** (*c*)	**13.** (*c*)	**14.** (*c*)
15. (*a*)	**16.** (*a*)	**17.** (*b*)	**18.** (*a*)			

16 Mitochondria (Energy Metabolism)

The mitochondria (Gr., *mito* = thread, *chondrion* = granule) are filamentous or granular cytoplasmic organelles of all aerobic cells of higher animals, plants and also of certain microorganisms including algae, Protozoa and Fungi. These are absent in bacterial cells and red blood cells of mammals. The mitochondria have lipoprotein framework which contains many enzymes and coenzymes required for energy metabolism. They also contain a specific DNA for the cytoplasmic inheritance and ribosomes for the protein synthesis.

Mitochondria, chloroplasts and peroxisomes (microbodies) form **oxidative organelles** of eukaryotic cells. A typical eukaryotic cell depends on mitochondria or chloroplasts to supply most of its energy. Chloroplasts, found only in plant cells, capture light energy and transform it to chemical energy in the form of ATP and reduced coenzymes, a process called **photosynthesis**. Mitochondria, found in both animal and plant cells, oxidize reduced coenzymes and certain other organic molecules, using molecular oxygen from the air as an electron acceptor. **Oxidation** release energy, a portion of which is captured by mitochondria and used to form ATP. In spite of a few exceptions, mitochondria and chloroplasts are the "power houses" of most eukaryotic cells. Peroxisomes are another type of oxidative organelles of cells which use molecular oxygen, but produce no ATP molecule for the cell.

In this chapter, the most significant oxidative cellular organelle of plants and animals, the mitochondria, will be discussed. Remaining oxidative organelles have been discussed in forthcoming chapters.

Box 16.1 Historical

1. The mitochondria were first observed by **Kolliker** in 1850 as granular structures in the striated muscles. In 1888, he isolated them from insect muscles and named them sarcosomes.
2. **Richard Altmann** (1890) developed a specific stain that had useful specificity for the mitochondria. He named this organelle, the **bioblast**. He suggested that bioblasts were independent organelles.
3. The present name mitochondria was assigned by **Benda** (1897–98) to them. He stained mitochondria with alizarin and crystal violet.
4. **Sir Hans Adolph Krebs** (German biochemist in England; born 1900), in 1937 worked out various reactions of the **citric acid cycle** (or tricarboxylic acid or TCA cycle). Krebs received the Nobel Prize in 1953 along with **Lipmann** for his discovery of the citric acid cycle.
5. **Hogeboom** and coworkers (1948) first demonstrated that cellular respiration occurs in mitochondria.
6. **Kennedy** and **Lehninger** (1948–1950) showed that the citric acid cycle (Krebs cycle), oxidative phosphorylation and fatty acid oxidation took place in the mitochondria. In 1951, **Lehninger** proved that oxidative phosphorylation requires electron transport.

7. **Palade** (1954) described the ultra structure of cristae.
8. In 1961, **Mitchell** proposed the highly acclaimed "**chemiosmotic-coupling hypothesis**" for the ATP-production in mitochondria. He got the Nobel Prize in 1978 for the development of this model.
9. In 1963, **Nass** and **Nass** demonstrated the presence of **DNA fibres** in the matrix of mitochondria of embryonic cells.
10. **Attardi, Attardi** and **Aloni** (1971) reported the 70S type ribosomes inside the mitochondria.

Mitochondria are called "**power houses**" or '**atom bombs**' of the cell, since they produce and provide chemical energy (ATP) to the cell. Previously the mitochondria have been known by various names such as **fuchsinophilic granules, parabasal bodies, plasmosomes, plastosomes, fila, vermicules, bioblasts** and **chondriosomes.**

16.1. DISTRIBUTION OR LOCALIZATION

The mitochondria move autonomously in the cytoplasm, so they generally have uniform distribution in the cytoplasm, but in many cells their distribution is very restricted to an area of particularly heavy utilization of ATP. Thus, they are often found in rows adjacent to the contractile elements of muscle (Fig. 16.1), wrapped about the base of a sperm flagellum and adjacent to lipid droplet of fat cells. Likewise, in the rod and cone cells of the retina all the mitochondria are located in a portion of the inner segment, in certain muscles of the diaphragm the mitochondria are concentrated around the I band of the myofibril and forming a ring, and in the cells of renal distal tubules the mitochondria lie near and perpendicularly to the cell membranes. The concentration of the mitochondria in such cells has suggested their utility in providing more energy to these cells for active metabolic functions.

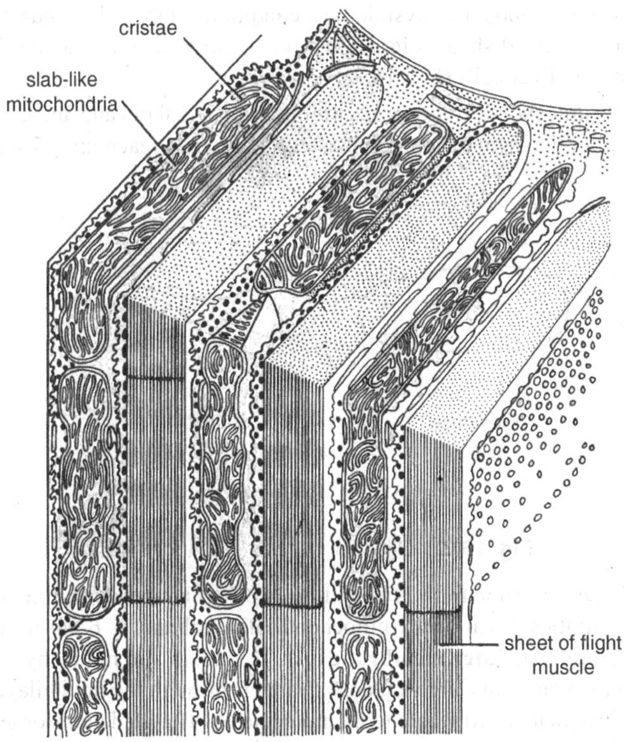

Fig. 16.1. Mitochondria of the flight muscle of a dragon fly, showing profuse cristae.

In some cells, the mitochondria are concentrated in the peripheral cytoplasm or cytoplasm just beneath the plasma membrane as in *Paramecium*, pancreatic cells, cells of the Malpighian tubules of the insects and young spermatids of the rat. While in certain cells the mitochondria remain concentrated around the nucleus. During the cell division (mitosis and meiosis) the mitochondria get concentrated around the spindle.

16.2. ORIENTATION

The mitochondria have definite orientation. For example, in cylindrical cells the mitochondria usually remain orientated in basal apical direction and lie parallel to the main axis. In leucocytes, the mitochondria remain arranged radially with respect to the centrioles. The orientation of the mitochondria is said to depend on the nature of the cytoplasmic matrix, vacuolar system and the direction of the diffusion currents of the cell.

16.3. MORPHOLOGY

Number. The number of mitochondria in a cell depends on the type and functional state of the cell. It varies from cell to cell and from species to species. Certain cells contain exceptionally large number of the mitochondria, *e.g.*, the *Amoeba*, *Chaos chaos*, contain 50,000; eggs of sea urchin contain 140,000 to 150,000 and oocytes of amphibians contain 300,000 mitochondria. Certain cells, *viz.*, liver cells of rat contain only 500 to 1600 mitochondria. The cells of green plants contain less number of mitochondria in comparison to animal cells because in plant cells the function of mitochondria is taken over by the chloroplasts. Some algal cells may contain only one mitochondrion.

Shape. The mitochondria may be filamentous or granular in shape and may change from one form to another depending upon the physiological conditions of the cells. Thus, they may be of club, racket, vesicular, ring or round-shape (Fig. 16.2). Mitochondria are granular in primary spermatocyte of rat, or club-shaped in liver cells (Fig. 16.3).

Size. Normally mitochondria vary in size from 0.5 μm to 2.0 μm and, therefore, are not distinctly visible under the light microscope. Sometimes their length may reach up to 7 μm.

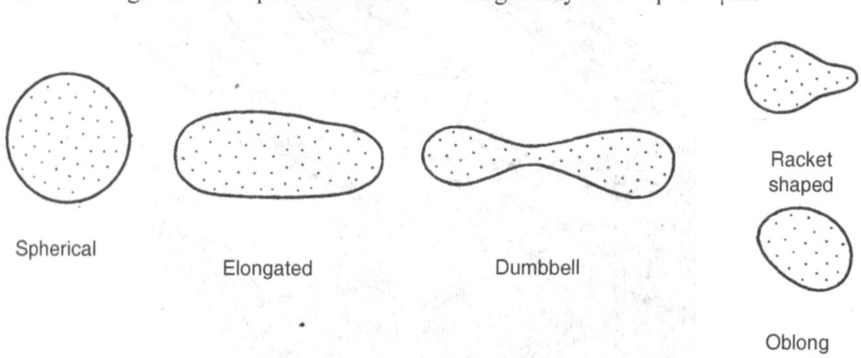

Fig. 16.2. Different shapes of mitochondria.

Structure. Each mitochondrion is bound by two highly specialized membranes that play a crucial part in its activities. Each of the mitochondrial membrane is 6 nm in thickness and fluid-mosaic in ultrastructure. The **outer membrane** is quite smooth and has many copies of a transport protein called **porin** which forms large aqueous channels through the lipid bilayer. This membrane, thus, resembles a sieve that is permeable to all molecules of 10,000 daltons or less, including small proteins. Inside and separated from the outer membrane by a 6–8 nm wide space is present the **inner membrane** (Fig. 16.4). The inner membrane is not smooth but is impermeable and highly convoluted, forming a series of infoldings, known as **cristae**, in the matrix space.

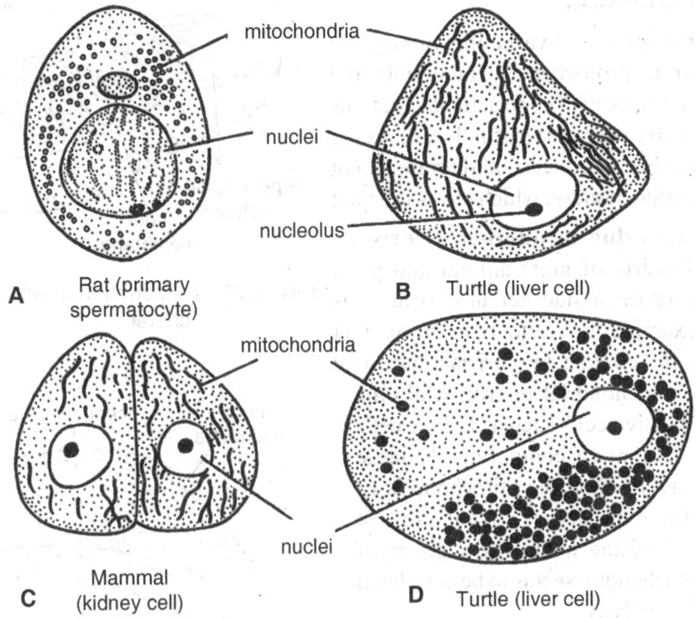

Fig. 16.3. Mitochondria of different types of animal cells.

Thus, mitochondria are double membrane envelopes in which the inner membrane divides the mitochondrial space into two distinct chambers:

1. The **outer compartment**, **perimitochondrial space** or the **inter-membrane space** between outer membrane and inner membrane. This space is continuous into the core of the crests or cristae.

2. The **inner compartment**, **inner chamber** or **matrix space**, which is filled with a dense, homogeneous, gel-like proteinaceous material, called **mitochondrial matrix**. The mitochondrial matrix contains lipids, proteins, circular DNA molecules, 55S ribosomes and certain **granules** which are related to the ability of mitochondria to accumulate ions. Granules are prominent in the mitochondria of cells concerned with the transport of ions and water, including kidney tubule cells, epithelial cells of the small intestine, and the osteoblasts of bone-forming cells. Further, the inner membrane has an outer **cytosol** or **C face** toward the perimitochondrial space and an inner **matrix** or **M face** toward matrix.

In general, the cristae of plant mitochondria are tubular, while those of animal mitochondria are lamellar or plate-like.

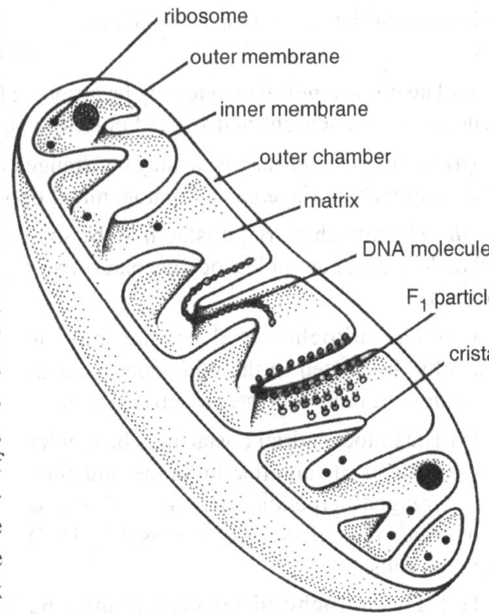

Fig. 16.4. A longitudinally cut mitochondrion showing its internal structure.

Structural Variations

There exists great structural variability among the mitochondria from protozoans to mammals and from algae to angiosperms. The structural variations in the mitochondria take place usually due to the difference in number, size, shape and arrangement of **cristae** in the mitochondria which are as follows:

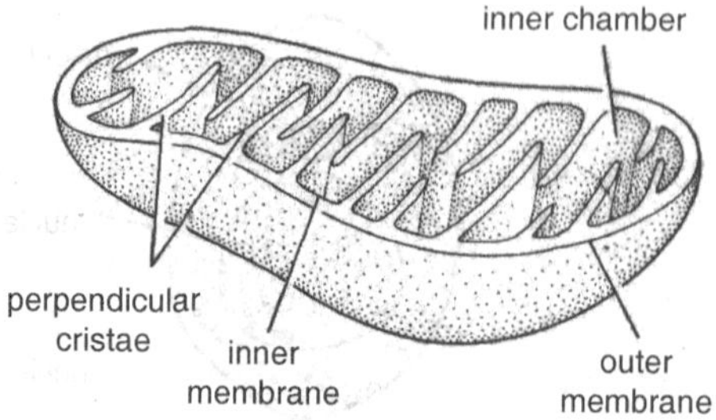

Fig. 16.5. Mitochondrion with perpendicular cristae.

1. Variation due to number of cristae. Among mitochondria of most animal and plant cells, the number of cristae per unit volume is variable. For example, the mitochondria of liver and germinal cells contain few cristae and great amount of the mitochondrial matrix, while the mitochondria of active cells such as kidney cells and muscle cells contain comparatively large number of cristae. Further, the mitochondria of the flight muscles of the insects contain greatest concentration of cristae in it. The number of cristae of the mitochondria seems to be correlated with its oxidative activity.

2. Variation due to the structure and arrangement of the cristae. Normally the cristae occur perpendicularly to the long axis of the mitochondria but various arrangements and shapes of cristae have been observed in the mitochondria of the different cells which are as follows:

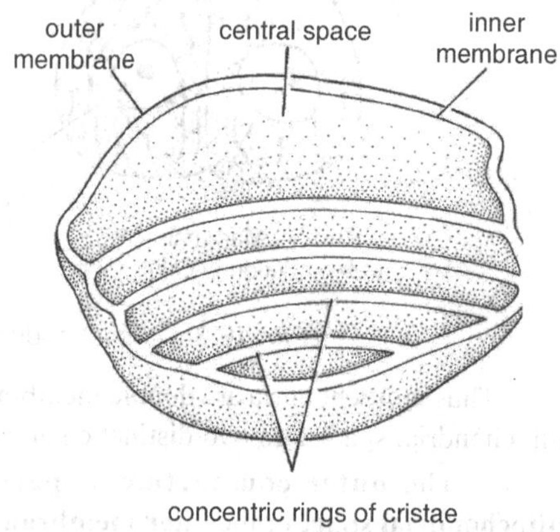

Fig. 16.6. Mitochondrion showing parallel cristae.

(*i*) The mitochondrial cristae may be arranged longitudinal and parallel to the main axis of the mitochondria, *e.g.*, neurons and striated muscles (Fig. 16.6).

(*ii*) The mitochondrial cristae may be arranged in concentric rings which lie parallel to the main axis of the mitochondria, *e.g.*, spermatogonia of oppossum and humans.

(*iii*) The mitochondrial cristae may be small, branched and anastomosingly arranged in the mitochondria, *e.g.*, human leucocytes, parathyroid gland cells.

(*iv*) The mitochondrial cristae may be haphazardly distributed in the mitochondria, *e.g.*, *Chaos chaos*, the multinucleate amoeba (Fig. 16.7).

(*v*) The mitochondrial cristae may be regular in shape instead of lamellar in shape and they may be arranged perpendicularly, *e.g.*, Protozoa (*Paramecium*), flight muscles of insects (Fig. 16.5) and adrenal cells.

(*vi*) The mitochondrial cristae may be spherical in shape, *e.g.*, spermatocytes (Fig. 16.8).

(*vii*) The mitochondrial cristae may be extremely reduced and irregular, *e.g.*, interstitial cells of oppossum testis (Fig. 16.9).

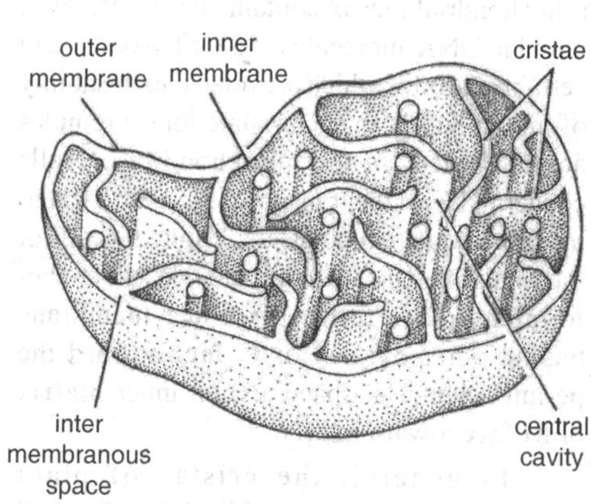

Fig. 16.7. Mitochondrion showing haphazardly distributed cristae.

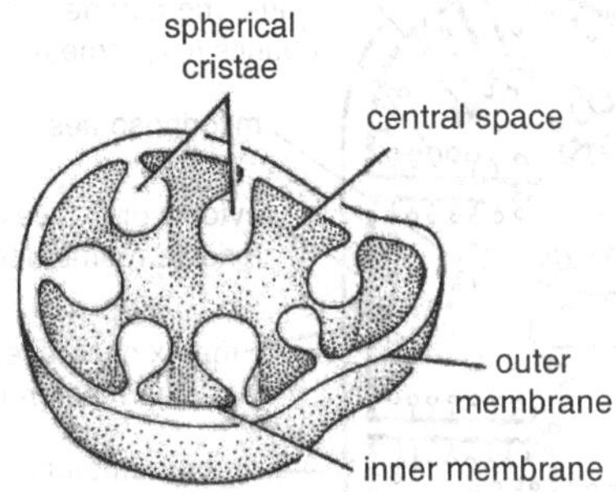

Fig. 16.8. Mitochondrion showing spherical cristae.

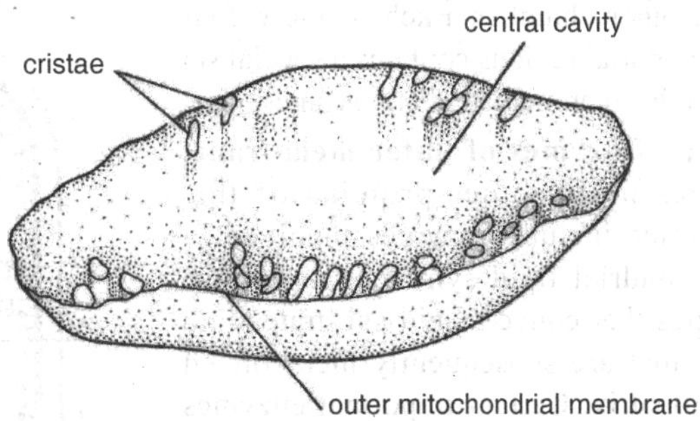

Fig. 16.9. Mitochondrion showing reduced cristae.

3. Variation due to the fusion. De Robertis (1957) has reported a peculiar type of behaviour in the mitochondria of spermatids of certain insects. In early spermatids all the mitochondria of the cytoplasm aggregate around the nucleus and all fuse together to form a single mitochondrial body.

Oxysomes. Attached to M face of inner mitochondrial membrane are repeated units of stalked particles, called **elementary particles, inner membrane subunits** or **oxysomes** (Fig. 16.11). They are also identified as $\mathbf{F_1}$ **particles** or $\mathbf{F_0-F_1}$ **particles** and are meant for ATP synthesis (phosphorylation) and also for ATP oxidation (*i.e.*, acting as ATP synthetase and ATPase). F_0-F_1 particles are regularly spaced at intervals of 10 nm on the inner surface of inner mitochondrial membrane. According to some estimates, there are 10^4 to 10^5 elementary particles per mitochondrion.

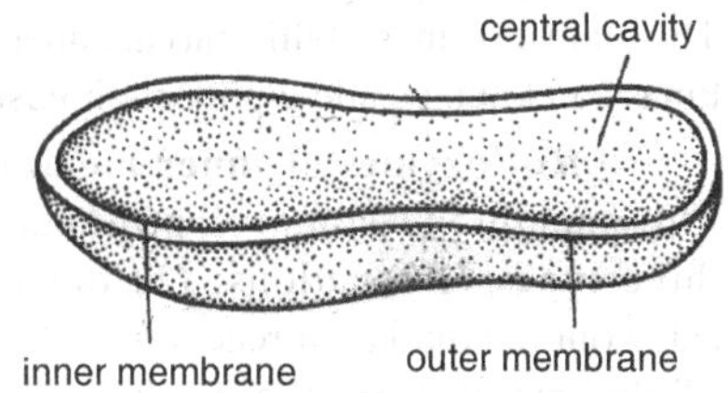

Fig. 16.10. Mitochondrion without cristae.

Mitoplast. When outer mitochondrial membrane is removed with digitonin (a detergent), the so-called **mitoplast** is formed. Mitoplast includes inner mitochondrial membrane with unfolded cristae and matrix. Mitoplast is found to carry out oxidative phosphorylation. The isolated outer membrane is revealed by negative staining and shows a "folded-bag" appearance (Fig. 16.12). Such isolation of two membranes and compartments has enabled localization of various enzyme systems of mitochondria.

16.4. CHEMICAL COMPOSITION

The gross chemical composition of the mitochondria varies in different animal and plant cells. However, the mitochondria are found to contain 65 to 70 per cent proteins, 25 to 30 per cent lipids, 0.5 per cent RNA and small amount of the DNA. The lipid contents of the mitochondria are composed of 90 per cent phospholipids (lecithin and cephalin), 5 per cent or less cholesterol and 5 per cent free fatty acids and triglycerides. The inner membrane is rich in one type of phospholipid, called **cardiolipin** which makes this membrane impermeable to a variety of ions and small molecules (*e.g.*, Na^+, K^+, Cl^-, NAD^+, AMP, GTP, CoA and so on).

The outer mitochondrial membrane has typical ratio of 50 per cent proteins and 50 per cent phospholipids of 'unit membrane'. However, it contains more unsaturated fatty acids and less cholesterol. It has been estimated that in the mitochondria of liver 67 percent of the total mitochondrial protein is located in the matrix, 21 per cent is located in the inner membrane, 6 per cent is situated in the outer membrane and 6 per cent is found in the outer chamber. Each of these four mitochondrial regions contains a special set of proteins that mediate distinct functions:

(*i*) Enzymes of outer membrane. Besides porin, other proteins of this membrane include enzymes involved in mitochondrial lipid synthesis and those enzymes that convert lipid substrates into forms that are subsequently metabolized in the matrix. Certain important enzymes of this membrane are monoamine oxidase, rotenone-insensitive NADH-cytochrome-C-reductase, kynurenine hydroxylase, and fatty acid CoA ligase.

(*ii*) Enzymes of intermembrane space. This space contains several enzymes that use the ATP molecules passing out of the matrix to phosphorylate other nucleotides. The main enzymes of this part are adenylate kinase and nucleoside diphosphokinase.

(*iii*) Enzymes of inner membrane. This membrane contains proteins with three types of functions: (1) those that carry out the oxidation reactions of the respiratory chain; (2) an enzyme complex, called **ATP synthetase** that makes ATP in matrix; and (3) specific transport proteins that regulate the passage of metabolites into and out of the matrix. Since an electrochemical gradient, that drives ATP synthetase, is established across this membrane by the respiratory chain, it is important that the membrane be impermeable to small ions. The significant enzymes of inner membrane are enzymes of electron transport pathways, *viz.*, nicotinamide adenine dinucleotide (NAD), flavin adenine dinucleotide (FAD), diphosphopyridine nucleotide (DPN), four cytochromes (Cyt. *b*, Cyt. *c*, Cyt. c_1, Cyt.*a* and Cyt. a_3), ubiquinone or coenzyme

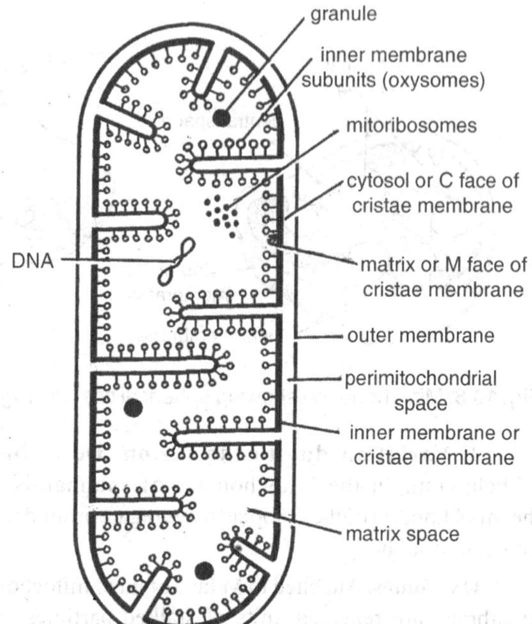

Fig. 16.11. A mitochondrion in sectional view to show its numerous oxysomes.

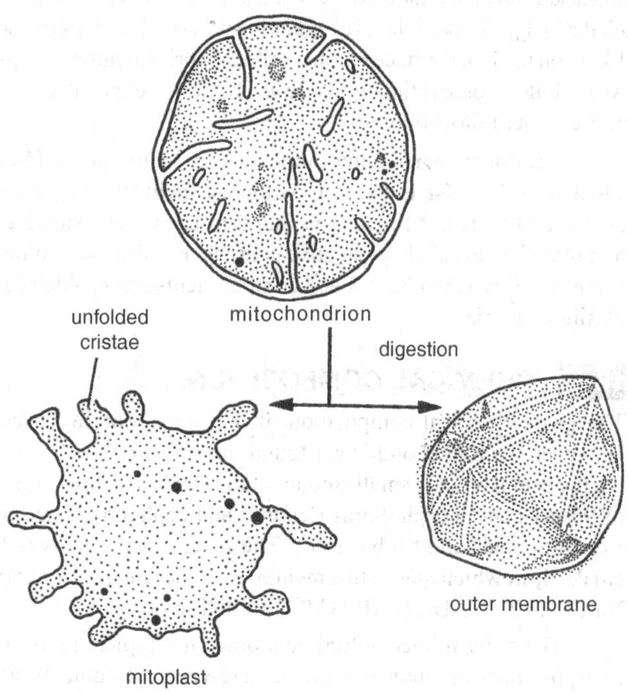

Fig. 16.12. Action of digitonin on liver mitochondrion to produce mitoplast and "folded-bag" of outer membrane.

Q_{10}, non-heme copper and iron, ATP synthetase, succinate dehydrogenase; β-hydroxy butyrate dehydrogenase; carnitine fatty acid acyl transferase (Fig. 16.13).

***(iv)* Enzymes of mitochondrial matrix.** The mitochondrial matrix contains a highly concentrated mixture of hundreds of enzymes, including those required for the oxidation of pyruvate and fatty acids and for the citric acid cycle or Krebs cycle. The matrix also contains several identical copies of the mitochondrial DNA, special 55S mitochondrial ribosomes, tRNAs and various enzymes required for the expression of mitochondrial genes. Thus, the mitochondrial matrix contains the following enzymes: malate dehydrogenase, isocitrate dehydrogenase, fumarase, aconitase, citrate synthetase, α-keto acid dehydrogenases, β-oxidation enzymes. Moreover, the mitochondrial matrix contains different nucleotides, nucleotide coenzymes and inorganic electrolytes—K^+, HPO_4^-, Mg^{2+}, Cl^- and SO_4^-.

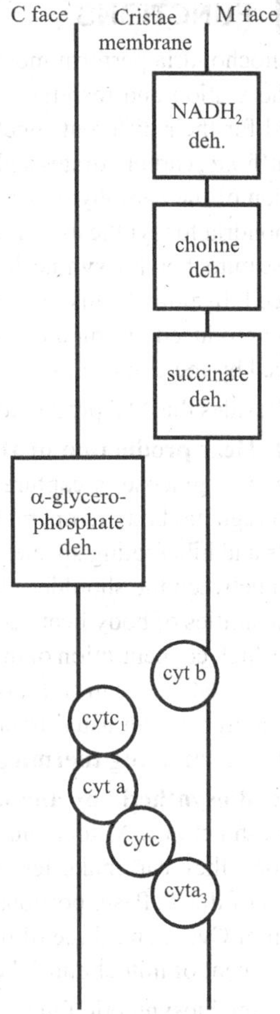

Fig. 16.13. A part of inner mitochondrial membrane (cristae) showing the distribution of different dehydrogenases and cyto-chromes on M face and C face.

Box 16.2 Coenzymes

Some coenzymes have a central role in mitochondrial function (see Chapter 15). **Coenzyme A** (CoA) is part of a group (Table 16.1) that is derived from a nucleoside (adenine-D-ribose) and contains pantothenic acid (a vitamin of B complex) linked to the ribose by pyrophosphoric acid. CoA can be easily transformed into an ester at the thiol end (–SH) by acetyl group making acetyl-CoA. Acetyl-CoA is a carrier molecule in which acetyl group is linked by reactive bonds so that they can be transferred efficiently to other molecules. The same carrier molecule will often participate in many different biosynthetic reactions in which its group (*i.e.*, acetyl group) is needed, *e.g.*, growing fatty acid.

Other mitochondrial coenzymes are **nicotinamide adenine dinucleotide (NAD^+)** which contains the vitamin nicotinic acid of B complex and **flavin mononucleotide (FMN)** and **flavin adenine dinucleotide (FAD)**, both of which contain riboflavin or vitamin B_2. NAD^+, FMN and FAD are important coenzymes not only in mitochondria but also in chloroplasts.

Table 16.1. Some important coenzymes which act as carrier molecules in transfer of a group.

	Coenzyme	Group transferred
1.	ATP	phosphate
2.	NADH, NADPH	hydrogen and electron (hydrogen ions)
3.	Coenzyme A	acetyl
4.	Biotin	carboxyl
5.	5′–Adenosyl-methionine	methyl

16.5. FUNCTIONS

The mitochondria perform most important functions such as oxidation, dehydrogenation, oxidative phosphorylation and respiratory chain of the cell. Their structure and enzymatic system are fully adapted for their different functions. They are the actual respiratory organs of the cells where the foodstuffs *i.e.*, carbohydrates and fats are completely oxidised into CO_2 and H_2O. During the biological oxidation of the carbohydrates and fats large amount of energy is released which is utilized by the mitochondria for synthesis of the energy rich compound known as **adenosine triphosphate** or **ATP**. Because mitochondria synthesize energy rich compound ATP, they are also known as "**power houses**" of the cell. In animal cells mitochondria produce 95 per cent of ATP molecules, remaining 5 per cent is being produced during anaerobic respiration outside the mitochondria. In plant cells, ATP is also produced by the chloroplasts.

Besides the ATP production, mitochondria serve the following important functions in animals:

1. Heat production or thermiogenesis. Only 45 per cent of the energy released during the oxidation of glucose is captured in the form of ATP, the rest 55 per cent is either lost as heat or used to regulate body temperature of warm-blooded animals. In some mammals, especially young animals and hibernating species, there is a specialized tissue called **brown fat**. This tissue, typically located between the shoulder blades, is especially important in temperature regulation; it produces large quantities of body heat necessary for arousal from hibernation. The colour of brown fat comes from its high concentration of mitochondria, which are sparse in ordinary fat cells. The mitochondria appear to catalyze electron transport in the usual way but are much less efficient at producing ATP. Hence, a higher than usual fraction of the oxidatively released energy is converted directly to heat (called **non-shivering thermiogenesis**).

2. Biosynthetic or anabolic activities. Mitochondria also perform certain biosynthetic or anabolic functions. Mitochondria contain DNA and the machinery needed for protein synthesis. Therefore, they can make less than a dozen different proteins. The proteins so far identified are subunits of the ATPase, portions of the reductase responsible for transfer of electrons from CoQ to the iron of Cyt c, and three of the seven subunits in cytochrome oxidase. Altogether, no more than 5–10 per cent of mitochondrial components can be attributed to mitochondrial genes.

Some biosynthetic functions of mitochondria are of primary benefit to the rest of the cell. For example, the synthesis of **haeme** (needed for cytochromes, myoglobin and haemoglobin) begins with a mitochondrial reaction catalyzed by the enzyme, delta or δ-aminolevulinic acid synthetase. Likewise, some of the early steps in the conversion of cholesterol to steroid hormones in the adrenal cortex are also catalyzed by mitochondrial enzymes.

3. Accumulation of Ca^{2+} and phosphate. In the mitochondria of **osteoblasts** present in tissues undergoing calcification large amount of Ca^{2+} and phosphate (PO_4^-) ions tend to accumulate. In them microcrystalline, electron-dense deposits may become visible. Sometimes, the mitochondria assume storage function, *e.g.*, the mitochondria of amphibian ovum store large amounts of yolk proteins and are transformed into yolk platelets.

16.6. MODE OF ENERGY PRODUCTION

The mitochondrion is the centre of oxidative metabolism in the cell, converting the products from carbohydrate, fat and protein catabolism into chemical energy stored in ATP.

I. Oxidation of Carbohydrates

The carbohydrates enter in the cell in the form of monosaccharides such as glucose or glycogen. The hexose sugars are first broken down into 3-carbon compound (pyruvic acid) by a series of chemical reactions by many known enzymes. The pyruvic acid enters in the mitochondria for its complete oxidation into CO_2 and water. The reactions which involve in the oxidation of glucose in to CO_2 and water are known to form the **metabolic pathways** and they can be grouped under the following heads:

1. Glycolysis or Embden-Meyerhof pathways (EMP) or Embden-Meyerhof Parnas pathways (EMPP);
2. Oxidative decarboxylation;
3. Krebs cycle; citric acid cycle or tricarboxylic acid cycle;
4. Respiratory chain and oxidative phosphorylation.

1. Glycolysis. Under anaerobic conditions (*i.e.*, in the absence of oxygen) glucose is degraded into lactic acid or lactate by a process called **glycolysis** (*i.e.*, lysis or splitting of glucose), *e.g.*, it commonly occurs in vertebrate muscles when the energy demand in heavy exercise exceeds the available oxygen.

If glycolysis is carried out under aerobic conditions the final products are pyruvic acid and coenzyme NADH. Glycolysis is achieved by a series of 10 enzymes all of which are located in the cytosol (cytoplasmic matrix). As shown in (Fig. 16.14), in this chain of reactions, the product of one enzyme serves as a substrate for the next reaction. To facilitate its analysis, the sequence of glycolysis can be subdivided into following three main steps: (*i*) activation (stage I); (*ii*) cleavage (stage II); and (*iii*) oxidation (stage III).

(*i*) Activation. In reactions 1 to 3 the glucose molecule is converted into **fructose-1-6-diphosphate**. This step uses two molecules of ATP and involves the following enzymes: hexokinase, phosphoglucose isomerase (or phosphohexoisomerase) and phosphofructokinase.

(*ii*) Cleavage. In reactions 4 and 5, fructose-1-6-diphosphate splits into two (3-carbon) end products, **glyceraldehyde-3-phosphate** molecules. During the step of cleavage only two enzymes are used: aldolase (fructoaldolase) and triose isomerase (triosephosphate isomerase).

(*iii*) Oxidation. In reactions 6 to 10, two molecules of glyceraldehyde-3-phosphate are oxidized and ultimately converted into two molecules of **pyruvic acid**. This step produces four molecules of ATP by **substrate-level-phosphorylation** and involves the following enzymes: phosphoglyceric dehydrogenase (glyceraldehyde phosphate dehydrogenase), phosphoglyceric kinase, phosphoglyceromutase, enolase and pyruvic kinase.

The net energy yield of chain reactions of glycolysis is the production of two ATP molecules from one molecule of glucose. Under aerobic conditions, the end products of glycolysis are pyruvic acid and reduced coenzyme NAD (*i.e.*, NADH). NADH carries two electrons, taken from glyceraldehyde-3-phosphate and contains little energy. However, under anaerobic conditions, pyruvic acid remains in the cytosol (cytoplasmic matrix) and is used as a hydrogen acceptor and converted into lactic acid:

$$\text{Pyruvic acid} \xrightarrow[\text{+NADH}]{\text{Lactic dehydrogenase}} \text{Lactic acid} + \text{NAD} + \text{Energy}$$

In above case, the following equation represents the overall reaction of glycolysis:

$$\underset{\text{Glucose}}{C_6H_{12}O_6} + Pi + 2ADP \rightarrow \underset{\text{Lactic acid}}{2C_3H_6O_3} + 2ATP + H_2O$$

2. Oxidative decarboxylation. In aerobic organisms, since pyruvic acid still contains a larger amount of energy, it must undergo further degradation, but this time inside the mitochondria. This is done in three consecutive steps: **oxidative decarboxylation** (removal of carboxyl or—COOH group), **Krebs cycle** and **oxidative phosphorylation**. Pyruvic acid directly enters the mitochondrial matrix and is converted into **acetyl-CoA** by the help of a huge enzyme, called **pyruvic acid dehydrogenase** (Fig. 16.15). The two NADH molecules (which are generated during glycolysis) cannot penetrate directly into the mitochondria, so their electrons are transferred to **dihydroxyacetone phosphate**, which shuttles them into the mitochondria. This process utilizes, one ATP molecule for each NADH; in all two ATP molecules are consumed for two NADH molecules. When both of these NADH pass through ETS, they tend to generate 6 ATP molecules.

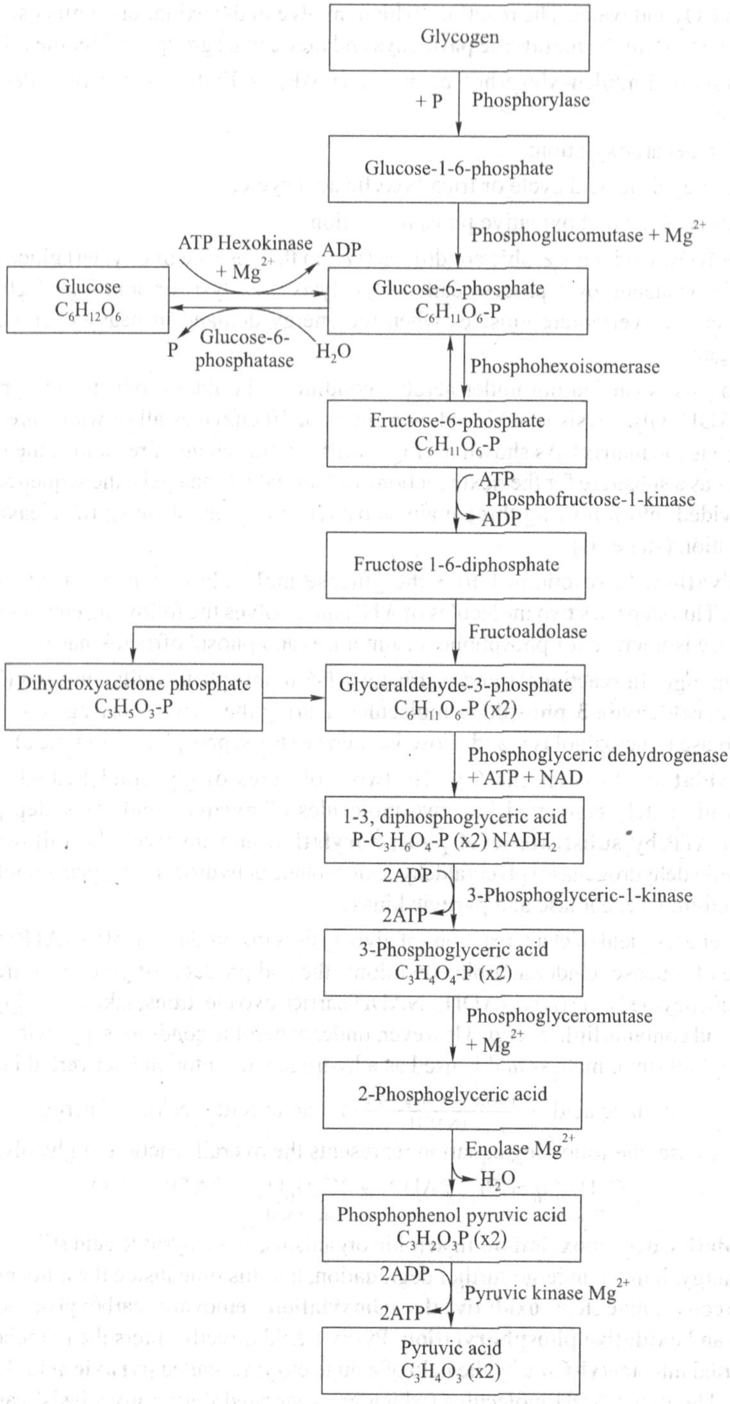

Fig. 16.14. Glycolysis or Embden-Meyerhof pathway showing the step-wise degradation of glucose to pyruvic acid.

Pyruvic acid dehydrogenase and its action. Sometimes two enzymes that catalyze sequential reactions form an **enzyme complex** and the product of the first enzyme does not have to diffuse through the cytoplasm to encounter the second enzyme. The second reaction begins as soon as the first is over. Some large enzyme (multienzyme) aggregates carry out whole series of reactions without losing contact with the substrate. For example, the conversion of pyruvic acid to acetyl CoA proceeds in *three* chemical steps, all of which take place on the same large multienzyme complex (*i.e.*, pyruvate dehydrogenase). Pyruvate dehydrogenase occurs in the mitochondrial matrix and is larger than a ribosome in size. It contains multiple copies of three enzymes namely **pyruvic acid dehydrogenase, dihydrolipoyl transacetylase** and **dihydrolipoyl dehydrogenase**. It also contains five coenzymes (*e.g.*, NAD, coenzyme A, etc.) and **two regulatory proteins** (*e.g.*, protein kinase and protein phosphatase; both regulating the activity of pyruvic acid dehydrogenase, turning it off whenever ATP levels are high).

Thus, during oxidative decarboxylation of one molecule of pyruvic acid, one mole of CO_2 is produced and one NAD is reduced to NADH. The end product of this reaction is a 2-carbon compound, the **acetyl group** which is attached to coenzyme A to produce the carrier molecule, called **acetyl CoA.**

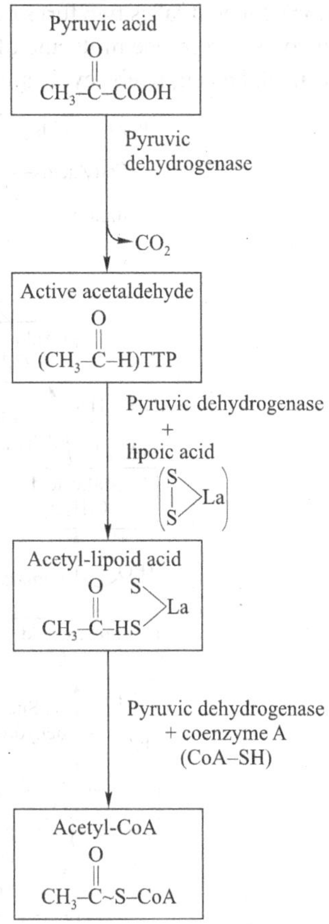

Fig. 16.15. The oxidative decarboxylation of pyrurvic acid to acetyl coenzyme A.

3. Krebs cycle. Two acetyl CoA molecules, produced above by oxidation of one molecule of glucose pass through a series of reactions of Krebs cycle to produce CO_2, H_2O and electrons. Enzymes and coenzymes of Krebs cycle are located in the mitochondrial matrix (some of them such as succinic dehydrogenase, are attached to M face of inner mitochondrial membrane, Fig. 16.13). As illustrated in (Fig. 16.16), the Krebs cycle involves the condensation of the acetyl group with oxaloacetic acid to make **citric acid** (6-carbon compound). This step is directed by the enzyme **citrate synthase.** From citric acid, H_2O is released twice by **aconitase** enzyme to produce **isocitric acid.** This is followed by a decarboxylation (loss of CO_2) by **isocitric dehydrogenase**, producing 5 carbon **α-ketoglutaric acid.** CO_2 is released by **α-ketoglutarate dehydrogenase** in the presence of CoA to produce **succinyl CoA** which changes by **succinyl kinase** enzyme (also called succinyl CoA synthetase) into a 4-carbon compound, the **succinic acid** (at this stage one GTP is generated by substrate level phosphorylation). The next enzyme of Krebs cycle, the **succinic dehydrogenase** converts succinic acid into **fumaric acid** and then **fumarase** enzyme produces **malic acid.** The mediation of **malate dehydrogenase** enzyme produces **oxaloacetic acid**, and thereby closes the Krebs cycle.

At each turn of the Krebs cycle, four pairs of hydrogen atoms are removed from the substrate intermediates by enzymatic dehydrogenation and two CO_2 molecules are released. These hydrogen atoms (or equivalent pairs of electrons) enter the respiratory chain, being accepted by either NAD^+ or FAD. Three pairs of hydrogen molecules are accepted by NAD^+, reducing it into NADH, and one pair by FAD, reducing it into $FADH_2$ (this pair of electrons come directly from the succinic dehydrogenase reaction). Since it takes two turns of the cycle to metabolize the two acetyl groups that are produced by glycolysis from one molecule of glucose, a total of six molecules of NADH and two of $FADH_2$ are formed. During Krebs cycle are also produced two ATP molecules (*i.e.*, via GTP molecules).

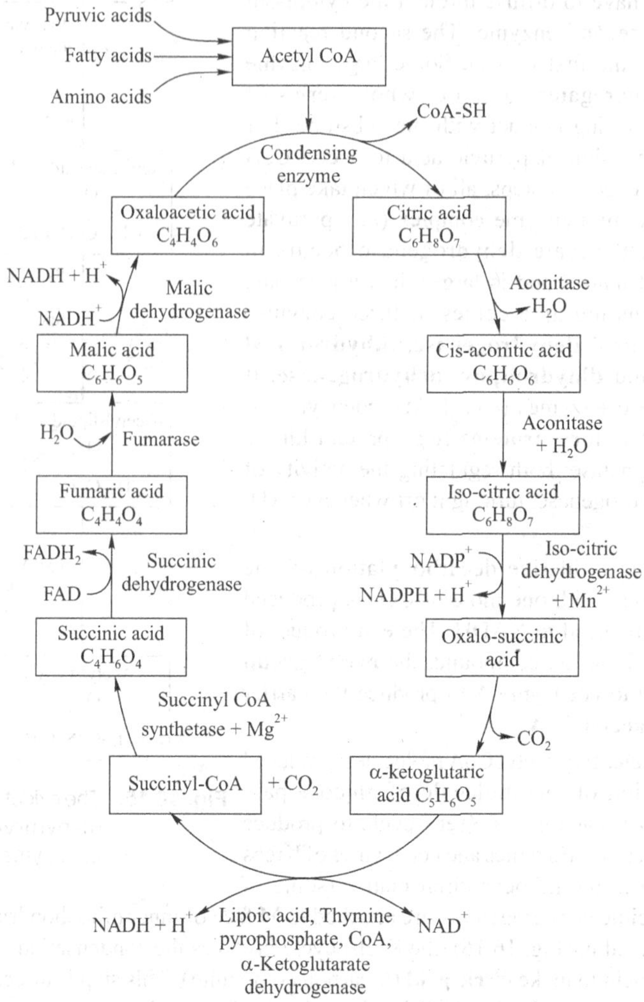

Fig. 16.16. Krebs cycle or tricarboxylic acid cycle.

The Krebs cycle itself releases no energy, but as the glucose breaks up, it also frees the hydrogen atoms attached to the carbons. Each hydrogen atom contains one proton (H^+) and one electron (e^-) The electrons which are released during complete oxidation of glucose (*i.e.*, glycolysis, oxidative decarboxylation and Krebs cycle) carry most of the energy of the glucose. In the final stage of cell respiration, the electron transport system, these electrons will at last release their energy to the cell.

4. Respiratory Chain and Oxidative Phosphorylation. Two molecules of $FADH_2$ and six molecules of NADH produced in Krebs cycle (from two molecules of acetyl-CoA) are oxidized by molecular O_2 in a **respiratory chain** or **electron transport system** or **ETS** involving a series of enzymes and coenzymes.

In the electron transport system, the successive electron acceptors are at lower and lower energy levels. With each transfer to a lower energy level, the electrons release some of their potential energy. That is why this series is called an **electron cascade,** like a cascade of falling water (Fig. 16.17). At each stage, the released energy is used to form ATP. Since electron transport involves oxidation as well as phosphorylation (*i.e.*, ADP + P = ATP) this process by which cell system traps chemical energy is called **oxidative phosphorylation** (Fig. 16.15). The passage of electrons from NAD to oxygen generates 3 ATP molecules, whereas the passage of electrons from FAD to oxygen generates only 2 ATP molecules.

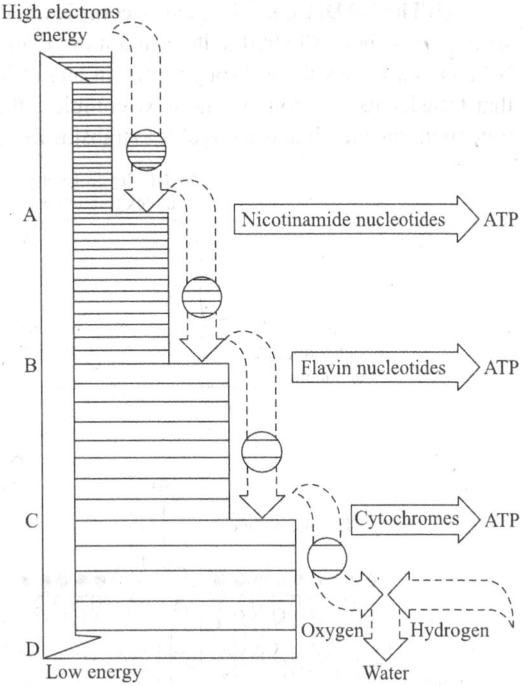

Fig. 16.17. Electron transfer in respiratory cascade or electron transport system.

(A) Compounds that occur in ETS. Following five types of compounds are associated with electron transport system of inner mitochondrial membrane:

(*i*) Pyridine-linked dehydrogenases require as their coenzyme either NAD^+ or $NADP^+$ both of which can accept two electrons at a time. There are about 200 dehydrogenases for which NAD^+–linked compounds such as pyruvic acid dehydrogenase, are involved in ETS.

(*ii*) Flavin-linked dehydrogenases (often called **flavoproteins** or **FPs**) require either FAD or FMN. Both are prosthetic groups whose isoalloxazine ring can accept two hydrogen atoms. Flavin-linked enzymes are commonly involved in a number of enzyme systems such as fatty acid oxidation, amino acid oxidation and Krebs cycle activity (*e.g.*, succinic dehydrogenase or SDH).

(*iii*) Ubiquinones were so named because of their occurrence in so many different organisms and their chemical resemblance to quinone. They are found in several different forms including the **plastoquinones** of chloroplasts. The form of ubiquinones present in mitochondria is often called **coenzyme Q_{10}** (CoQ_{10} or Q). CoQ_{10} is lipid soluble and accepts two hydrogen atoms (or two protons and two electrons) at a time.

(*iv*) Cytochromes are proteins containing iron-porphyrin (haem) groups. There are a large number of cytochrome in cells; most are found in mitochondria, although some also function in the ER and in chloroplasts. Mitochondria have five types of cytochromes which are arranged in the following order in an inner membrane: cyt. b, cyt. c_1, cyt. c, cyt. a and cyt. a_3. All of them transfer electrons by reversible valence changes of the iron atom (trivalent ferric or $Fe^{3+} \rightleftharpoons$ bivalent ferrous or Fe^{2+}).

(*v*) Iron-sulphur proteins (Fe_2S_2 and Fe_4S_4) are electron carriers of mitochondria containing iron and sulphur in equal amounts. The iron is reversibly oxidized during the electron transfer. Iron-sulphur proteins transfer one electron at a time.

All these components of ETS are arranged in the inner mitochondrial membrane in the following sequence: NAD-linked succinic dehydrogenase (SDH), flavoprotein (FAD), non-haem iron protein or iron-sulphur protein, flavoprotein (FAD), cytochrome b, ubiquinone or coenzyme Q_{10}, cytochrome c_1, cytochrome c, cytochrome a, cytochrome a_3 and three coupling sites, where phosphorylation coupled with oxidation leads to production of ATP.

(B) Three complexes. Evidently above described components of ETS occur in the mitochondria in the form of following three complexes:

***(i)* The NADH-dehydrogenase complex.** It is the largest of the respiratory enzyme complexes, with a mass about 800,000 daltons and more than 22 polypeptide chains. It accepts electrons from NADH and passes them through a flavin and at least five iron-sulphur centres to ubiquinone (Q) that transfer its electrons to the next complex, the b–c_1, complex. This complex spans the inner mitochondrial membrane and is able to translocate protons across it from M side to C side (Fig. 16.18).

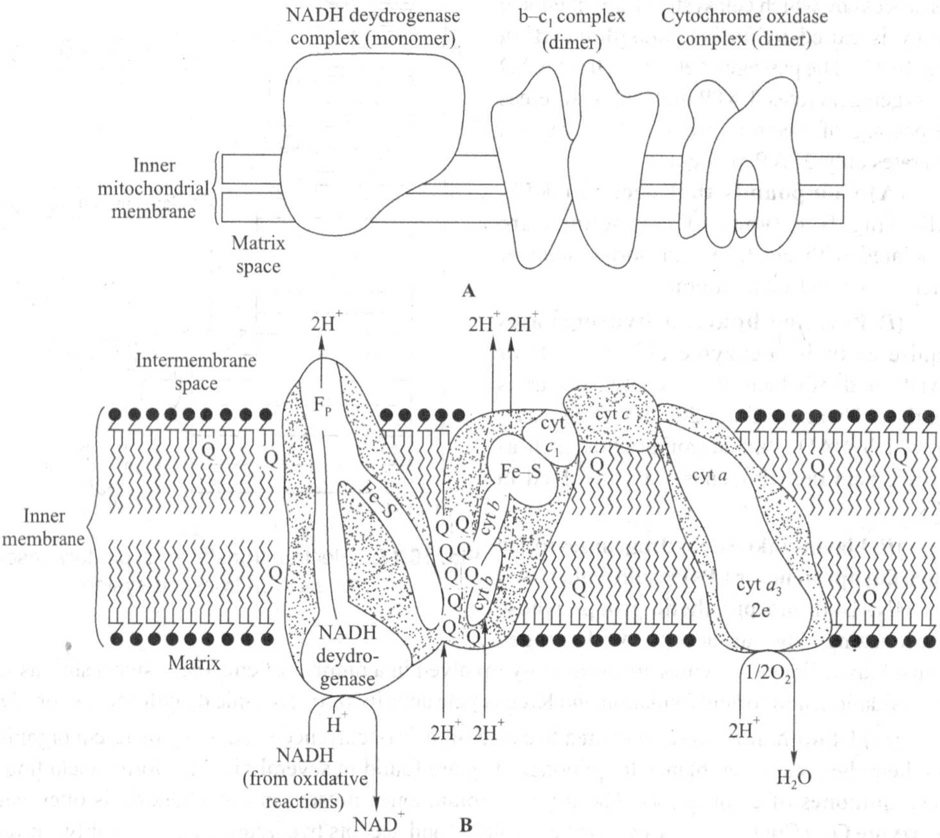

Fig. 16.18. Respiratory enzyme complexes. A—The relative sizes and shapes of the three respiratory enzyme complexes; B—Mode of transfer of H^+ through the three respiratory enzyme complexes of inner mitochondrial membrane.

***(ii)* The b–c_1 complex.** It contains at least 8 different polypeptide chains and is thought to function as a dimer of about 500,000 daltons. Each monomer contains three haemes bound to cytochromes and iron-sulphur protein. This complex accepts electrons from ubiquinone (Q) and passes them to cytochrome, a small peripheral membrane protein that carries its electrons to the cytochrome oxidase complex. In the topology of this complex the Q-site may be in the middle of the membrane in the hydrophobic area and the cytochrome c-site on the C side.

***(iii)* The cytochrome oxidase complex.** It comprises at least eight different polypeptide chains and is isolated as a dimer of about 300,000 daltons; each monomer contains two cytochromes (a, a_3) and two copper atoms. This complex accepts electrons from cytochrome c and passes them to oxygen and is thought to traverse the mitochondrial membrane, protruding on both surfaces. Such a trans-membrane orientation is associated with the vectorial transport of protons across the membrane.

The cytochrome oxidase reaction is estimated to account for 90 per cent of the total oxygen uptake in most cells. The toxicity of the poisons such as cyanide and azide is due to their ability to bind tightly to this complex and thereby block all electron transport.

(C) F_0–F_1 complex or coupling factors. One of the main proteins in the inner mitochondrial membrane is the multisubunit **coupling factor** (Fig. 16.19), the enzyme that actually synthesizes ATP and simultaneously acting as a proton pump (Box 16.3). A quite similar enzyme complex is located in the thylakoid membranes of chloroplasts and in the plasma membrane of bacterial cell. The coupling factor has two principal components:

(*a*) **F_0-complex.** It is an integral membrane complex, composed of very hydrophobic proteins—3 or 4 distinct polypeptides and one proteolipid—which together span the mitochondrial membrane. F_0-complex possesses the proton translocating mechanism. F_0-complex can be extracted only with strong detergents.

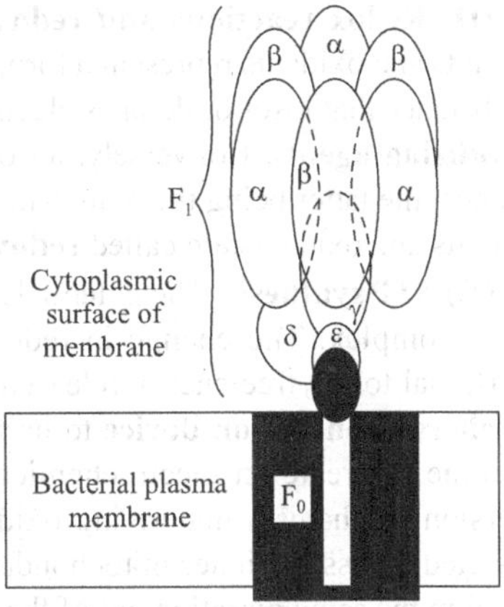

Fig. 16.19. Model of molecular structure of F_0 and F_1 particles in the bacterial plasma membrane. The core of F_1 particle is an $\alpha_3\beta_3$ complex. The δ polypeptide of F_1 and probably the γ are involved in binding F_1 to the membrane-embedded F_0-particle.

(*b*) **F_1-particle.** Attached to the F_0 complex is F_1 particle, a complex of five distinct polypeptides: alpha (α), beta (β), gamma (γ), delta (δ) and epsilon (ε), with the probable composition of $\alpha_3\ \beta_3\gamma\delta\varepsilon$. F_1 forms the knob or 'tadpole' that protrudes on the matrix side of the inner mitochondrial membrane. F_1 particle can be detached from the membrane by mechanical agitation and is water soluble. When physically separated from the membrane, F_1 particle is capable only of catalysing the hydrolysis of ATP into ADP and phosphate. Hence, it is often called the **F_1-ATPase.** However, its natural function is the synthesis of ATP.

Box 16.3 ATP Synthase

The ATP-synthesizing enzyme, which is called **ATP-synthase**, is a mushroom-shaped protein complex which is composed of two principal components: a spherical **F_1 head** (about 90A° diameter) and a basal section called **F_0**, embedded in the inner membrane. The two portions are connected by both a central and a peripheral stalk. A typical mammalian liver mitochondrion has roughly 15,000 copies of ATP synthase (**Karp** 2010).

The F_1 portions of the bacterial and mitochondrial ATP synthases are highly conserved, both contain five different polypeptides with a composition of α3β3δγε. The α and β subunits are arranged alternately within the F_1 head in a manner resembling the segments of an orange. Each F_1 contains three catalytic sites for ATP synthesis: One on each β subunit and the γ subunit which runs from the outer tip of the F_1 head through the central stalk and makes contact with the F_0 base piece. In the mitochondrial enzyme, all five polypeptides of F_1 are encoded by the nuclear DNA, synthesized in the cytosol and then imported.

The F_0 portion of ATP synthase resides within the membrane and consists of three different polypeptides with a stoichemistry of $a\ b_2\ c_{10-14}$. The number of subunits in the *c* ring is written as 10–14 because structural studies have revealed that this number can vary depending on the source of enzyme. Both the yeast mitochondrial and *E.coli* ATP synthase, for example have 10 *c* subunits, whereas a chloroplast enzyme has 14 *c* subunits. The F_0 base contains a channel through which protons are conducted from the intermembrane space to the matrix. According to **Boyers model** (supported by **Walker** *et al.*, 1994) ATP is synthesized by rotational catalysis in which one part (α and β subunits of F_1) of the ATP synthase rotates relative to another part (central stalk or γ subunit).

(D) Redox reactions and redox couples. The movements of electrons between cellular reductants and oxidants represent a form of energy transfer in cells. A reductant (or **reducing agent**) is a substance that loses or donates electrons to another substance; the latter substance is the oxidant (or **oxidizing agent**). Conversely, an oxidant is a substance that accepts electrons from another substance, the latter being the reductant. Reactions that involve the movement of electrons between reductants and oxidants are called **redox reactions**.

(E) ATP synthesis. The potentials drop in three large steps, one across each major respiratory enzyme complex. The change in redox potential between any two electron carriers is directly proportional to the free energy released by an electron transfer between them. Each complex acts as an **energy-conversion-device** to harness this free-energy change, pumping H^+ across the inner membrane to create an electrochemical proton gradient as electrons pass through. The energy conversion mechanism underlying oxidative phosphorylation requires that each protein complex be inserted across the inner mitochondrial membrane in a fixed orientation, so that all protons are pumped in the same direction out of the matrix space. Such a **vectorial organization** of membrane proteins has been experimentally proved.

Just as a flow of water from a higher to a lower level can be utilized to turn a water wheel or a hydroelectric turbine, the energy released by the flow of the protons down the gradient is utilized in the synthesis of ATP. Similarly, resultant electrochemical proton gradient is harnessed to make ATP by F_0–F_1 complex (acting as ATP synthase and proton pump), through which the protons flow back into the matrix (Fig. 16.20). The ATP synthase is a reversible coupling device that normally converts a back-flow of protons into ATP phosphate-bond energy, but it can also hydrolyze ATP (into ADP and phosphate) to pump protons in the opposite direction, if the electrochemical proton gradient is reduced.

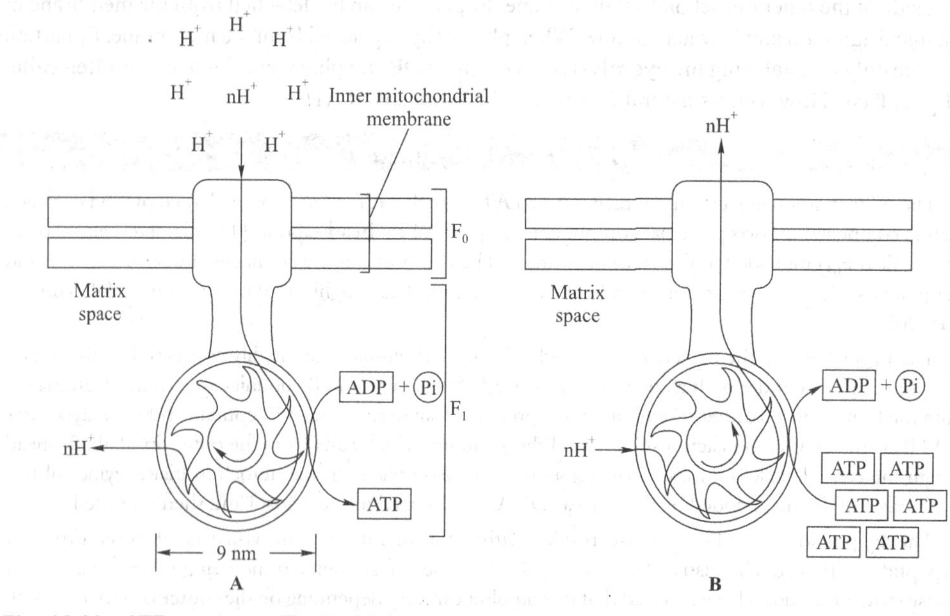

Fig. 16.20. ATP synthase (F_0–F_1 ATPase) is a reversible coupling device that interconverts the energies of the electrochemical proton gradient and chemical bonds. The ATP synthase can either synthesize ATP by harnessing the proton motive force (A) or pump protons against their electrochemical gradient by hydrolyzing ATP (B). The direction of operation at any given instant depends on the net free-energy change for the coupled processes of proton translocation across the membrane and the synthesis of ATP from ADP and Pi.

Chemiosmotic mechanism. The controlled movement of protons back across the membrane through ATP-synthesizing enzyme provides the energy required to phosphorylate ADP to ATP. The importance of proton movements on the formation of ATP was first proposed in 1961 by **Peter Mitchell** of the University of Edinburgh. This proposed mechanism is called **chemiosmotic mechanism**. For this contribution, Mitchell was awarded Nobel Prize in Chemistry in 1978.

According to the chemiosmotic model, the main energetic driving force for the synthesis of ATP is an ion gradient across a selectively permeable biological membrane; in chloroplasts and mitochondria, this gradient is known to be a **proton gradient** that is generated as a consequence of electron transfer. This gradient establishes a concentration difference of protons and can also result in difference in electric potential of the membrane. These sources of potential energy can be used for the phosphorylation of ADP by ATP synthase, an enzyme that energetically, favourable proton flow to the synthesis of ATP (**Buchanan** *et al.*, 2000).

(F) Energetics of glucose oxidation. Of the 686,000 calories contained in a mole of glucose, less than 10 per cent (*i.e.*, 58,000 calories) can be released by anaerobic glycolysis. The cell is able to store only 45 per cent of the chemical energy liberated by the combustion (oxidation) of glucose in the form of ATP (*i.e.*, only 36 ATP molecules). The rest of the energy is dissipated as heat or used for other cell functions.

At this stage, let us do stocktaking of ATP generation during aerobic respiration of one mole of glucose. We have seen that glycolysis and Krebs cycle can each generate 2 molecules of ATP per molecule of glucose by substrate level phosphorylation (total 4 ATP molecules). In addition 10 NADH (*i.e.*, 2NADH in glycolysis, 2NADH in oxidative phosphorylation and 6NADH in Krebs cycle) and $2FADH_2$ are produced which are equivalent to 34 ATP molecules. Thus, a total of 38 ATP molecules are produced per glucose molecule oxidized. However, in most eukaryotic cells 2 molecules of ATP are used in the transportation of 2 mole of NADH produced during glycolysis into the mitochondrion (*via* the malate shuttle) for their further oxidation (*via* ETS). Hence, the net gain of ATP is 38 – 2 = 36 molecules; since one high energy phosphate bond is equal to 36.8 kJ; so 36ATP = 1325 kJ or 360,000 calories:

$$\underset{\text{Glucose}}{C_6H_{12}O_6} + \underset{\substack{\text{Inorganic}\\ \text{phosphate}}}{36Pi} + 36ADP + 6O_2 + 6H_2O \rightarrow$$

$$6CO_2 + 36ATP\ (360{,}000 \text{ calories or } 1325 \text{ kJ}) + 12H_2O$$

II. Oxidation of Fats

Although carbohydrates are the major energy source for immediate use in the cell, fats are extremely important as an energy reserve. The energy in a gram of fat is more than twice that of a gram of starch; consequently, lipids are a convenient form of energy storage. Hibernating, migrating, or starving animals depend almost totally upon fats for energy. Fats can be stored in cells in a precipitated form, usually as fat droplets, and do not require water for their solution.

Fat molecules in cells are primarily neutral fats (triglycerides) or their products. A neutral fat consists of a glycerol molecule connected to three fatty acids. Lipase enzymes within the cell split the three fatty acids away from glycerol with the introduction of three water molecules, similar to the hydrolysis reaction occurring during digestion of fats in the small intestine. Glycerol and fatty acids are oxidized by different routes such as follows:

Glycerol is a 3-carbon compound and by addition of a phosphate from ATP and the removal of two hydrogens by NAD it is readily converted to dihydroxyacetone phosphate (DHAP):

$$\text{Glycerol} \xrightarrow[\text{NAD}\rightarrow\text{NADH}_2]{\text{ATP}\rightarrow\text{ADP(H)}} \text{Dihydroxyacetone phosphate (DHAP)}$$

Figure 16.21, shows that this DHAP is routinely converted into 3-phosphoglyceraldehyde. when it enters the mainstream of glycolytic pathway. Complete oxidation of one glycerol molecule produces a net gain of 22 ATP molecules.

Most fatty acids in cells have an even number of carbons (commonly 16 or 18) joined in a chain. The fatty acid molecules of various types are oxidized in a similar manner. In general, coenzyme A attaches to the carboxyl end of the fatty acid. This costs one ATP molecule. In a series of reactions where water is added, four hydrogen atoms are removed (two by NAD and two by FAD) from the two carbon atoms nearest to CoA. The 2-carbon fragment with CoA attachcd is split away from the rest of the fatty acid as another coenzyme A is introduced to the new end of the fatty acid. The rcsulting 2-carbon unit is acetyl-CoA. In a similar way other 2-carbon segments are chopped off the fatty acid, becoming acetyl-CoA (Fig. 16.21). The resultant acetyl-CoA molecules enter the Krebs cycle or TCA cycle for complete oxidation. The process of fatty acid oxidation is called **beta oxidation**, because the fatty acid is progressively severed at the second or beta carbon. Beta oxidation occurs within the cytoplasm of the cell and the acetyl-CoA molecules are transported across the mitochondrial membranes into the matrix where the oxidative process continues.

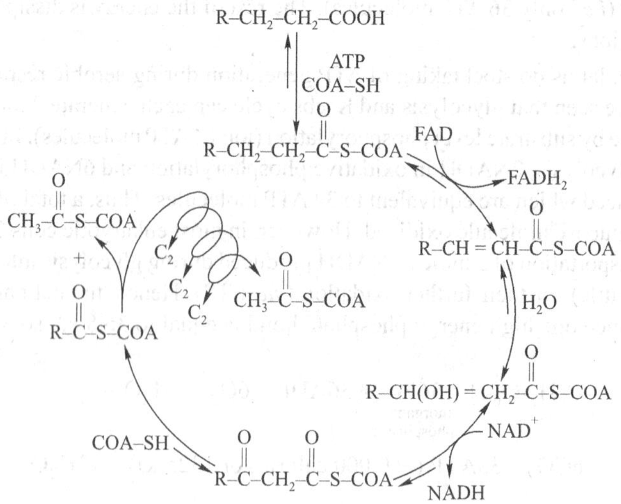

Fig. 16.21. The oxidation of fatty acids by the β-oxidation helicle scheme.

To calculate the energy derived from the total oxidation of a fatty acid, one must know the structure of the molecule. Suppose that the fatty acid had 18 carbons; the β-oxidation would cost an initial ATP molecule but would produce 9 acetyl-CoA molecules and 16 hydrogen pairs attached to coenzymes. The 9 acetyl-CoA molecules passing through the Krebs cycle would produce 12 ATP molecules each, for a total of 108 ATPs. Passage of 16 H_2 through the cytochrome system would yield 40 ATPs: 8 pairs would be handled via NAD and give 3 ATP molecules per hydrogen pair ($8 \times 3 = 24$ ATP) and 8 pairs would be handled via FAD and give 2 ATP molecules per hydrogen pair ($8 \times 2 = 16$ ATP). Thus adding 108 ATP from the oxidation of acetyl-CoA to 40 ATP generated by the passage of H_2 via the cytochrome system and substracting the initial ATP used in the first phosphorylation, we find that 147 ATP are produced for one 18-carbon fatty acid. Considering that each neutral fat has three such fatty acids along with one glycerol, we can estimate that 463 molecules of ATP are formed per triglyceride. This is over 12 times as much energy as that produced from one molecule of glucose. However, since the fat is a considerably larger molecule than glucose; energetic comparison is best based on the number of ATPs resulting per gram of nutrient "burned". On this basis, fat produces more than twice as many ATP molecules as the same weight of starch.

III. Oxidation of Proteins

Proteins are not stored within the cell in significant quantities. Those proteins present are either in the form of enzymes or an integral part of the cellular structure such as in the cell membrane. Therefore, proteins are not typically broken down and used as an energy source except in cases of starvation. Catabolism of proteins is also required for other cellular activities.

Before proteins can be introduced into the mainstream of metabolism (catabolism) they must be split into amino acids. The process is accomplished by protease enzymes similar to the process occurs in digestion. Each peptide bond is severed with the introduction of a water molecule, a hydrolytic reaction. Next, nitrogen is removed from amino acids by anyone following two processes: oxidative deamination and transamination.

During **oxidative deamination**, the amino group of the amino acid is split off from the rest of the molecule, forming ammonia (NH_3). The remainder of the amino acid then enters the main metabolic stream as a keto acid. Water is required for this process and two hydrogens are removed by coenzyme NAD. The ammonia formed during deamination may be immediately excreted or organized into another molecule before excretion; for example, in human beings the ammonia is converted into urea molecules by the liver cells before being sent via the blood to the kidney. The energy derived from oxidative deamination depends upon the amino acid involved. For example, oxidative deamination of glutamic acid involves its conversion into α-ketoglutaric acid, which is oxidized by the Krebs cycle:

$$\underset{\text{(amino acid)}}{\text{Glutamic acid}} \xrightarrow[\nearrow H_2O \quad \searrow NH_3]{NAD \rightarrow NADH_2} \underset{\text{(keto acid)}}{\alpha\text{-ketoglutaric acid}}$$

In this case, high-energy phosphates would be created by the transfer of hydrogen from $NADH_2$ (formed in deamination) through the cytochrome system.

Transamination reaction consists of an amino group being shifted from one molecule to another in exchange for an oxygen. For example, due to transamination amino acid glutamic acid being converted into α-ketoglutaric acid, as is shown by following reaction:

$$\underset{\text{(amino acid)}}{\text{Glutamic acid}} + \underset{\text{(keto acid)}}{\text{Oxaloacetic acid}} \rightarrow \underset{\text{(keto acid)}}{\alpha\text{-ketoglutaric acid}} + \underset{\text{(amino acid)}}{\text{aspartic acid}}$$

In this case, the amino group is not lost completely but is transferred to one of the substrate of Krebs cycle. Oxaloacetic acid loses its oxygen and picks up the NH_2 group and becomes the amino acid, aspartic acid. Thus one amino acid is converted into keto acid while another keto acid is transformed into an amino acid. The usual purpose of such reactions is to maintain a particular balance among amino acids and substrates rather than providing grist for the metabolic mill.

Further, amino acids have different structures, and they are fed into the main metabolic pathway at different points depending upon the number of carbons they possess. Amino acids with three carbons (*e.g.*, serine, cysteine, alanine, valine) are converted into 3-carbon pyruvic acid or pyruvate. The 2-carbon amino acid glycine is converted to acetyl-CoA; the 5-carbon glutamic acid is transformed into α-ketoglutaric acid; the 4-carbon aspartic acid is converted to oxaloacetic acid, and so on.

16.7. MITOCHONDRIA AS SEMI-AUTONOMOUS ORGANELLES

Recently the study of mitochondrial and chloroplast biogenesis became of great interest because it was demonstrated that these organelles contain DNA as well as ribosomes and are able to synthesize proteins. The term **semiautonomous organelles** was applied to the two structures in the recognition of these findings. This term also indicated that the biogenesis was highly dependent on the nuclear genome and the biosynthetic activity of the ground cytoplasm. It is well established now that the mitochondrial mass grows by the integrated activity of both genetic systems, which cooperate in time and space to synthesize the main components. The mitochondrial DNA codes for the mitochondrial ribosomes and transfer RNA and for a few proteins of the inner membrane. Most of the proteins of the mitochondrion, however, result from the activity of the nuclear genes and are synthesized on ribosomes of the cytosol (cytoplasmic matrix). The cooperation of two genomes has been greatly clarified by studies on the molecular assembly of cytochrome oxidase. This cytochrome, as studied in *Saccharomyces cerevisiae* is made up of seven polypeptide subunits for a combined molecular weight of 139,000 daltons. Three of the polypeptides are coded by mtDNA and assembled on mitochondrial ribosomes. They are very hydrophobic and high in molecular weight (23,000–40,000 daltons). The remaining four subunits are coded by nuclear DNA and made on cytoplasmic ribosomes. These are hydrophilic polypeptides of lower molecular weight (4500–14,000 daltons).

Mitochondrial DNA. Mitochondrial DNA (mtDNA) molecule is relatively small, simple, double-stranded and except for the DNA of some algae and protozoans, it is circular. The size of mitochondrial genome is very much large in plants than in animals. Thus, mtDNA varies in length from about 5 μm in most animal species to 30 μm or so in higher plants. The mtDNA is localized in the matrix and is probably attached to the inner membrane at the point where DNA duplication starts. This duplication is under nuclear control and the enzymes used (*i.e.*, polymerases) are imported from the cytosol.

Mitochondrial ribosomes. Mitochondria contain ribosomes (called **mitoribosomes**) and polyribosomes. In yeast and *Neurospora*, ribosomes have been ascribed to a 70S class similar to that of bacteria; in mammalian cells, however, mitoribosomes are smaller and have a total sedimentation coefficient of 55S, with subunits of 35S, and 25S. In mitochondria, ribosomes appear to be tightly associated with the inner membrane.

Mitochondrial protein synthesis. Mitochondira can synthesize about 12 different proteins, which are incorporated into the inner mitochondrial membrane. These proteins are very hydrophobic (*i.e.*, they are proteolipids). Thus, on the mitoribosomes are made the following proteins: three largest subunits of cytochrome oxidase, one protein subunit of the cytochrome b–c_1 complex, four subunits of ATPase and a few hydrophobic proteins. One of the best known differences between the two mechanisms of protein synthesis (*i.e.*, in the cytosol and in the mitochondrial matrix) is in the effect of some inhibitors. The mitochondrial protein synthesis is inhibited by **chloramphenicol**, while synthesis in the cytosol (cytoplasmic matrix) is not affected by this drug. In contrast, **cycloheximide** has the reverse effect.

Import mechanism of mitochondrial proteins. Most mitochondrial proteins are coded by nuclear genes and are synthesized on free ribosomes in the cytosol (cytoplasmic matrix). The import of these polypeptides involves similar mechanism both in mitochondria, and chloroplasts. The transport processes involved have been most extensively studied in mitochondria, especially in yeasts (For further details see chapter 22).

Mitochondrial lipid biosynthesis. The biogenesis of new mitochondria requires lipids in addition to nucleic acids and proteins. Chloroplasts tend to make the lipids they require. For example, in spinach leaves, all cellular fatty acid synthesis takes place in the chloroplast. The major glycolipids of the chloroplast are also synthesized locally. Mitochondria, on the other hand, import most of their lipids. In animal cells the phospholipids—**phosphatidyl-choline** and **phosphatidyl-serine**—are synthesized in the ER and then transferred to the outer membrane of mitochondria.

Symbiont Hypothesis or Prokaryotic Origin of Mitochondria. Early cytologists such as **Altmann** and **Schimper** (1890) have suggested the possibility of origin of the mitochondria from the prokaryotic cells. According to their hypothesis, the mitochondria and chloroplasts may be considered as intra-cellular parasites of the cells which have entered in the cytoplasm of eukaryotic cells in early evolutionary days, and have maintained the symbiotic relations with the eukaryotic cells (Fig. 16.22). The mitochondira are supposed to be derived from the bacterial cells (purple bacteria) while chloroplasts are supposed to be originated from the blue green algae (**Margulis**, 1981). Due to these reasons **Altmann** suggested the name "**bioblasts**" to the mitochondria and he also hinted about their self-duplicating nature.

Recent cytological findings have also suggested many homologies between the mitochondria and the bacterial cells. The similarities between the two can be summarised as follows:

1. **Similarity in inner mitochondrial membrane and bacterial plasma membrane.**
   (*i*) In the mitochondria the enzymes of the respiratory chain are localized on the inner mitochondrial membrane like the bacteria in which they remain localized in the plasma membrane. The bacterial plasma membrane resembles with the inner mitochondrial membrane in certain respects.
   (*ii*) The plasma membrane of certain bacterial cells gives out finger-like projections in the cytoplasm known as mesosomes. The mesosomes can be compared with mitochondrial crests. **Salton** (1962) has reported respiratory chain enzymes in the mesosomes.

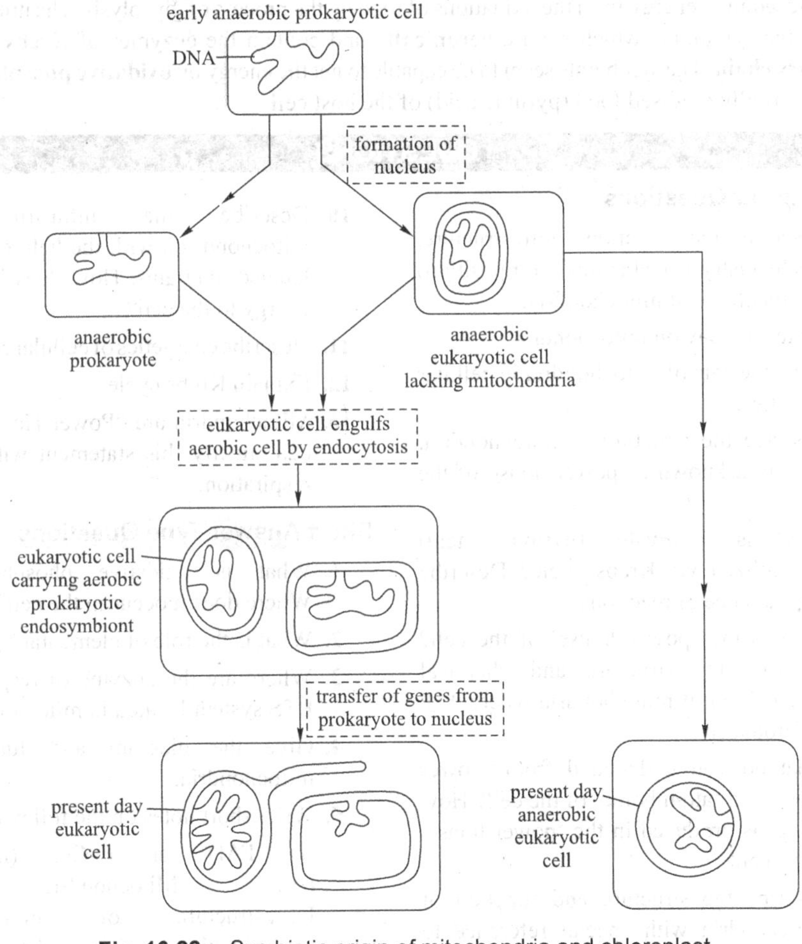

Fig. 16.22. Symbiotic origin of mitochondria and chloroplast.

   (*iii*) Because the outer mitochondrial membrane resembles with the plasma membrane, therefore, it may be assumed that the mitochondrial matrix and the inner mitochondrial membrane represent the symbiont which might be enclosed by the membrane of the cellular origin (outer mitochondrial membrane).

2. **Similarity in DNA molecule.** The DNA molecule of the mitochondria is circular like the DNA molecule of the bacterial cells. Further the replication process of the mitochondrial DNA is also similar to bacterial DNA.
3. **Similarity in ribosomes.** The mitochondrial ribosomes are small in size and resemble the ribosomes of the bacteria.
4. **Similarity in the process of protein synthesis.** The process of protein synthesis of both mitochondria and bacteria is fundamentally same because in both, the process of protein synthesis can be inhibited by same inhibitor known as chloramphenicol.

Further, the mitochondria for the process of protein synthesis depend partially on the mitochondrial matrix and DNA and partially on the nucleus and cytoplasm of the eukaryotic cells. This shows the symbiotic nature of the mitochondria.

Due to the above-mentioned similarities between the bacteria and mitochondria, the symbiont hypothesis postulated that the host cell (eukaryotic cell) represented an anaerobic organism which derives the required energy from the oxidations of food by the process of glycolysis. The mitochondria represent the symbionts which respire **aerobically** and contain the enzymes of **Krebs cycle** and **respiratory chain**. The symbionts seem to be capable to get the energy by **oxidative phosphorylation** from the partially oxidised food (pyruvic acid) of the host cell.

QUESTIONS

Long Answer Questions

1. Describe the location, ultrastructure, biochemistry (or chemical composition) and functions of mitochondria.
2. Write an essay on mitochondria.
3. Write the role of mitochondria in cellular activities.
4. Describe the structure of mitochondria. Why is it known as power house of the cell?
5. How is pyruvate (pyruvic acid) metabolized via Krebs cycle? Describe the sequence of reactions.
6. What is the 'power house' of the cell? Describe the structure and chemical composition of mitochondria where ATP are produced.
7. What do you understand from 'power house' and 'atom bomb' of the cell? How energy is produced in the 'power house' of the cell?
8. Describe the structure and function of mitochondria with special reference to electron transport system.
9. Describe steps involved in ATP synthesis in mitochondrion.
10. Describe the ultrastructure of mitochondrion with the help of neat and labeled diagrams. How does it provides energy to the cell?
11. Describe energetics of cellular respiration.
12. Explain Krebs cycle.
13. Mitochondria are "Power House" of the cell. Justify this statement with cellular respiration.

Short Answer Type Questions

1. What is oxidative phosphorylation? Where does it occur in the cell?
2. What is the role of elementary particles?
3. Where are the enzyme of respiratory or ETS system located in mitochondria?
4. Give the diagram and function of mitochondria.
5. Write short notes on the following:

   (*i*) FAD or FMN; (*ii*) ATP; (*iii*) Mitochondria; (*iv*) Ultrastructure of mitochondria; (*v*) Functions of mitochondria; (*vi*) Mitochondrial enzymes; (*vii*) Oxysomes; (*viii*) Glycolysis; (*ix*) Krebs cycle; and (*x*) Electron transport chain.

6. Draw a well labelled diagram of ultrastructure of mitochondria.
7. 'Mitochondrion as a semi-autonomous organelle'.

Very Short Answer Questions

1. Who coined the term mitochondria?
2. Name the cell type in which mitochondria are absent.
3. What will happen if all the mitochondria of a cell are destroyed?
4. What will happen if elementary particles of mitochondria are removed?
5. Name the organism which possess only one mitochondria.
6. Which cells possess highest number of mitochondria?
7. For which structure the name bioblast was used by Altman?
8. What is metabolism?
9. Where are enzymes of ETS located in mitochondria?
10. What is the location in enzymes involved in oxidation of pyruvic acid in the mitochondria?
11. Where is the site of oxidative phosphorylation?
12. Who gave the chemiosmotic hypothesis?
13. What is cellular respiration?
14. What is ATP.
15. Define coenzyme.
16. What is coenzyme A (CoA)?
17. Define coenzyme Q.
18. What is electron transport system (ETS)?
19. What are cytochromes?
20. Define FAD.
21. Define F_0–F_1 complex.
22. What is chemiosmotic coupling model?

True and False Statements

Indicate true or false statements by a tick mark:

1. ATP synthesis takes place in ribosome.
   (*a*) True (*b*) False
2. In a biological system an important source of energy is glucose.
   (*a*) True (*b*) False
3. Krebs cycle takes place in mitochondrial matrix.
   (*a*) True (*b*) False

YES OR NO QUESTIONS

Write Yes or No for the following statements:

1. Mitochondrion is made up of double membrane.
2. Cytochrome oxidase enzyme is found in mitochondria.
3. Citric acid cycle was first described by Bateson.

Fill in the Blanks

1. Metabolic reactions in the body are catalysed by ....
2. Citric acid cycle was worked out by ....

Multiple Choice Questions

Choose the correct answer from the four alternatives given:

1. One of the following is a coenzyme
   (*a*) glucose (*b*) protease
   (c) ATP (*d*) NAD
2. Unit of carbohydrate is
   (*a*) monosaccharide (*b*) amino acid
   (*c*) fatty acid (*d*) all
3. Cellular respiration, *i.e.*, transformation of chemical energy into usable bioenergy takes place
   (*a*) lysosomes
   (*b*) ribosomes
   (*c*) mitochondria
   (*d*) endoplasmic reticulum
4. Mitochondria are not found in
   (*a*) liver cells (*b*) nerve cells
   (*c*) maturing RBCs (*d*) mature RBCs
5. In mitochondria, cristae act as sites for
   (*a*) protein synthesis
   (*b*) oxidation-reduction reactions
   (*c*) breakdown of macromolecules
   (*d*) phosphorylation of flavoproteins

6. Cristae helps in
   (*a*) respiration (*b*) transpiration
   (*c*) photosynthesis (*d*) photo-oxidation
7. Cytochrome oxidases are found
   (*a*) on cristae of mitochondria
   (b) on outer wall of mitochondria
   (*c*) in the matrix of mitochondria
   (*d*) in the lysosomes
8. If the outer membrane is removed from a mitochondrion, the remaining structure is called
   (*a*) mitoribosome (*b*) mitoplast
   (*c*) microsome (*d*) peroxisome
9. The two types of cellular organelles that transform energy are:
   (*a*) chromoplasts and leucoplasts
   (*b*) mitochondria and chloroplasts
   (*c*) mitochondria and chromoplasts
   (*d*) chloroplasts and leucoplasts
10. At the end of the Krebs cycle, most of the energy removed from the glucose molecule has been transferred to the
    (*a*) NADH and $FADH_2$
    (*b*) ATP
    (*c*) oxaloacetic acid
    (*d*) citric acid
11. When a molecule is reduced it always
    (*a*) loses electrons
    (*b*) loses both proteins and electrons
    (*c*) loses neutrons
    (*d*) gains electrons
12. Last electron acceptor in ETS in
    (*a*) O_2 (*b*) water
    (*c*) cytochrome *c* (*d*) cytochrome a_3
13. Enzymes of glycolysis are present in
    (*a*) mitochondrial matrix
    (*b*) cristae
    (*c*) cytosol
    (*d*) all of the above
14. Oxidative phosphorylation means
    (*a*) photo-oxidation during photosynthesis
    (*b*) photolysis of water
    (*c*) ATP production in respiration
    (*d*) anaerobic respiration
15. End product of glycolysis is
    (*a*) acetyl Co-A
    (*b*) 1 molecule of pyruvic acid
    (*c*) 2 molecules of pyruvic acid
    (*d*) glucose 1-phosphate

ANSWERS

Very Short Answer Questions

1. C. Benda in 1897.
2. Mammalian red blood corpuscles (RBCs).
3. Cell will not get energy for performing various activities.
4. ATP molecules will not be produced.
5. *Micromonas*.
6. Muscle cells.
7. Mitochondria.
8. All chemical reactions occurring within a cell.
9. Inner membrane of mitochondria.
10. Mitochondrial matrix.
11. F_1 particles (of mitochondria).
12. Peter Mitchell.
13. It is oxidation-driven flow of electrons from reduced coenzymes to an electron acceptor, usually accompanied by the generation of ATP.

14. ATP or adenosine triphosphate is a adenosine with three phosphates linked to each other by phosphoanhydride bonds and to 5′ carbon of the ribose by a phosphoester bond; it is a principal energy storage compound of most cells, with energy stored in the high-energy phosphoanhydride bonds.
15. It is a small organic molecule that functions along with an enzyme by serving as a carrier of electrons or functional groups.
16. It is an organic molecule that serves as a carrier of acyl groups by forming a high-energy thioester bond with an organic acid.
17. Coenzyme Q (CoQ) is a nonprotein (quinone) component of the mitochondrial electron transport system that serves as a collection point for electrons from both FMN and FAD-linked dehydrogenases also called **ubiquinone.**
18. It is a group of membrane-bound electron carriers that transfer electrons from the coenzymes NADH and $FADH_2$ to oxygen.
19. These are heme-containing proteins of the ETS involved in the transfer of electrons from coenzyme Q to oxygen by the oxidation and reduction of the central iron atom of the heme group.
20. FAD or flavin adenine dinucleotide is a coenzyme that accepts two electrons and two photons from an oxidizable organic molecule to generate the reduced form, $FADH_2$; it is an important electron carrier in energy metabolism.
21. It is a protein complex in the mitochondrial inner membrane and the bacterial plasma membrane that consists of the F_1 complex bound to F_0 complex; the flow of proton through the F_0 component leads to the synthesis of ATP by the F_1 component.
22. It is a model postulating that electron transport pathways establish proton gradients across membranes and that the energy stored in such gradients can then be used to drive ATP synthesis.

True and False Statements

1. (*b*) **2.** (*a*) **3.** (*a*)

Yes or No Questions

1. Yes **2.** Yes **3.** No

Fill in the Blanks

1. Enzymes **2.** Hans A. Krebs

Multiple Choice Questions

1. (*d*) **2.** (*a*) **3.** (*c*) **4.** (*d*) **5.** (*b*) **6.** (*a*) **7.** (*a*)
8. (*b*) **9.** (*b*) **10.** (*a*) **11.** (*d*) **12.** (*a*) **13.** (*c*) **14.** (*c*)
15. (*c*)

17 Chloroplasts (Photosynthesis)

17.1. PLASTIDS

Plastids are present in all living plant cells and in *Euglena* (a protozoan). They are small bodies found free in the cytoplasm. Plastids are often more or less spherical or disc-shaped (1 μm to 1 mm in diameter), but may be elongated or lobed or show amoeboid characteristics. Their other identifying features are their double bounding membranes, the possession of **plastoglobuli** (spherical lipid droplets; store of lipids surplus to current requirements) and an internal membrane fretwork of many discrete internal vesicles. All plastids in a particular plant species contain multiple copies of same relatively small genome (DNA) and 70S-type ribosomes. They are self-replicating organelles containing a protein-synthesizing capacity comparable to that of mitochondria. They perform most important biological activities as the synthesis of food and storage of carbohydrates, lipids and proteins. Plastids are absent in the cells of fungi, bacteria, animals and male sperm cells of certain higher plants.

Types of Plastids. The term 'plastid' is derived from the Greek word "*plastikas*" (= formed or moulded) and was used by **A.F.W. Schimper** in 1885. **Schimper** classified the plastids into following types according to their structure, pigments and the functions (Fig. 17.1):

1. Leucoplasts. The leucoplasts (Gr., *leuco* = white; *plast* = living) are the colourless plastids which are found in embryonic and germ cells. They are also found in meristematic cells and in those regions of the plant which are not receiving light. Plastids located in the cotyledons and the primordium of the stem are colourless (leucoplasts) but eventually become filled with chorophyll and transform into chloroplasts. **True leucoplasts** occur in fully differentiated cells such as epidermal and internal plant tissues. They never become green and photosynthetic. True leucoplasts do not contain thylakoids and even ribosomes. They store the food materials as carbohydrates, lipids and proteins and accordingly are of following types:

***(i)* Amyloplasts.** The amyloplasts (L., *amyl* = starch; Gr., *plast* = living) are those leucoplasts which synthesize and store the starch. The amyloplasts occur in those cells which store the starch. The outer membrane of the amyloplast encloses the stroma and contains one to eight starch granules. In some plant tissues such as potato tuber, amyloplasts can grow to be as large as an average animal cell. In them starch granules may become so large that they rupture the encasing membrane. Starch granules of amyloplasts are typically composed of concentric layers of starch (Fig. 8.23).

***(ii)* Elaioplasts.** The elaioplasts store the lipids (oils) and occur in seeds of monocotyledons and dicotyledons. They also include sterol-rich **sterinochloroplast**.

***(iii)* Proteinoplasts.** The proteinoplasts are the protein storing plastids which mostly occur in seeds and contain few thylakoids.

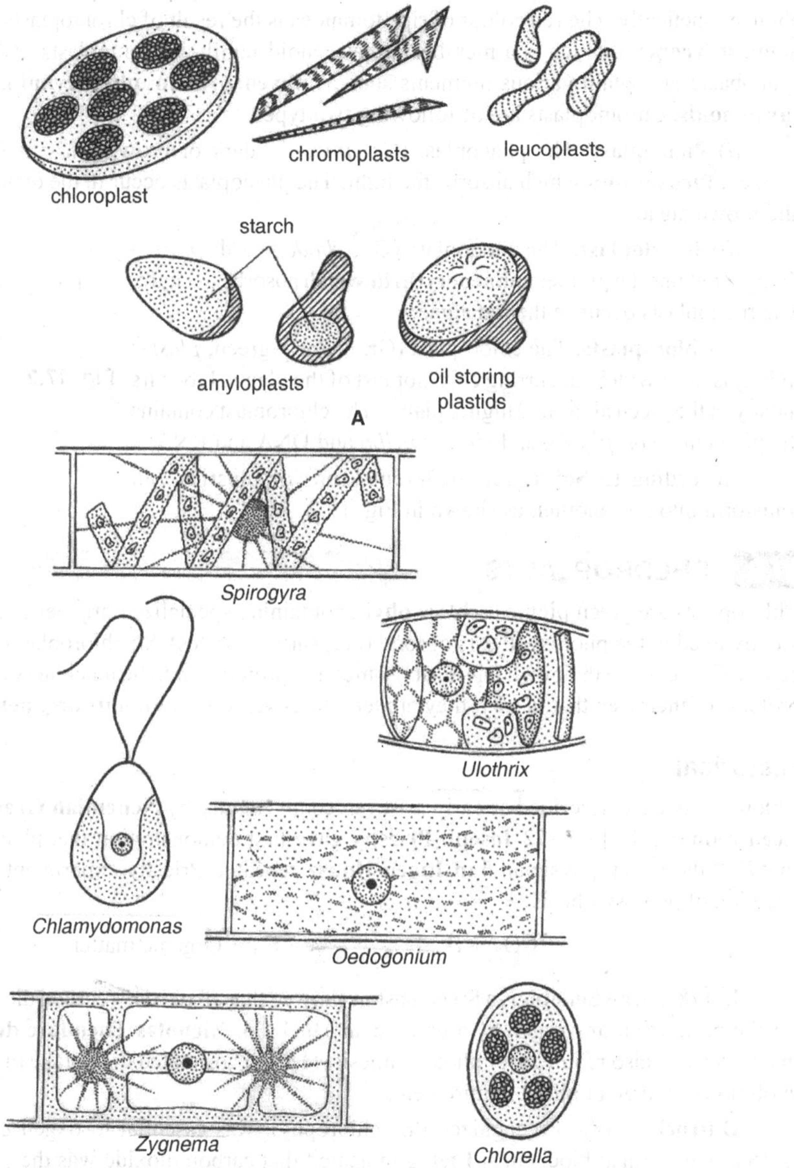

Fig. 17.1. A—Different kinds of plastids; B—Various types of choloroplasts found in Algae.

2. Chromoplasts. The chromoplasts (Gr., *chroma* = colour; *plast* = living) are the coloured plastids containing **carotenoids** and other pigments. They impart colour (*e.g.*, yellow, orange and red) to certain portions of plants such as flower petals (*e.g.*, daffodils, rose), fruits (*e.g.*, tomatoes) and some roots (*e.g.*, carrots). Chromoplast structure is quite diverse; they may be round, ellipsoidal, or even needle-shaped, and the carotenoids that they contain, may be localized in droplets or in crystalline structures. The function of chromoplasts is not clear but in many cases (*e.g.*, flowers and fruits) the colour they produce probably plays a role in attracting insects and other animals for pollination or seed dispersal.

In general, chromoplasts have a reduced chlorophyll content and are, thus, less active photosynthetically. The red colour of ripe tomatoes is the result of chromoplasts that contain the red pigment **lycopene** which is a member of carotenoid family. Chromoplasts of blue-green algae or cyanobacteria contain various pigments such as **phycoerythrin, phycocyanin, chlorophyll a** and **carotenoids**. Chromoplasts are of following two types:

(i) **Phaeoplast.** The phaeoplast (Gr., *phaeo* = dark or brown; *plast* = living) contains the pigment **fucoxanthin** which absorbs the light. The phaeoplasts occur in the diatoms, dinoflagellates and brown algae.

(ii) **Rhodoplast.** The rhodoplast (Gr., *rhode* = red; *plast* = living) contains the pigment **phaeoerythrin** which absorbs the light. The rhodoplasts occur in the red algae.

3. Chloroplasts. The chloroplast (Gr., *chlor* = green; *plast* = living) is most widely occurring chromoplast of the plants. It occurs mostly in the green algae and higher plants. The chloroplast contains the pigment *chlorophyll a* and *chlorophyll b* and DNA and RNA.

According to **Schimper**, different kinds of plastids can transform into one another, as shown in Fig. 17.2.

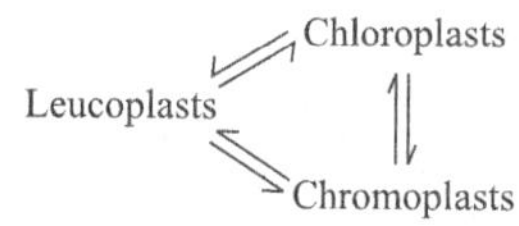

Fig. 17.2. Interconversion of three kinds of plastids.

17.2. CHLOROPLASTS

Chloroplasts are green pigment **chlorophyll**, containing specialized organelles of plant cells which are involved in the photosynthesis of food (*i.e.*, carbohydrates). So chloroplasts act as "**kitchens of cells**". They contain their own DNA, ribosomes and protein synthetic machinery to synthesize certain proteins of their use; that is why, they are termed as **semi-autonomous organelles**.

Historical

Chloroplasts were described as early as seventeenth century by **Nehemiah Grew** and **Antonie van Leeuwenhoek**. In 1771–72 **Joseph Priestly** gave first demonstration that plants produce oxygen. In 1779, the Dutch physician **Jan Ingen-Housz** repeated Priestly experiment and formulate first equation of photosynthesis:

$$CO_2 + H_2 \xrightarrow[\text{Green plants}]{\text{Light}} O_2 + \text{Organic matter}$$

In 1782, **Jean Senebier**, a Swiss pastor, showed that "fixed air" (carbon dioxide) was necessary for the production of oxygen by a green plant. In 1804, **Nicholas Theodore de Saussure** showed that water was also required for photosynthesis. He also correlated the uptake of carbon dioxide and evolution of oxygen to light requirement.

Dutrochet (1837) recognized that chlorophyll was essential to oxgen evolution by plants. In 1843 the German biochemist **Liebig** indicated that carbon dioxide was the source of all organic compounds synthesized by green plants. In 1862, **Sachs** proved that starch was synthesized by plants in a light-dependent reaction (photosynthesis). **Meyer** (1883) first described details of chloroplast structure. **Wilstatter** and **Stoll** (1913) isolated chlorophyll and later determined its structure. In 1923, **Thunberg** recognized that CO_2 was reduced and water oxidized during photosynthesis. In 1932, **Robert Emerson** and **William Arnold** of the California Institute of Technology carried out an experiment with green alga *Chlorella* suggesting that not all of the chlorophyll molecules in the chloroplast were directly involved in conversion of light energy into chemical energy. This finding paved the way for the discovery of **reaction centers** of chlorophyll.

The pathway of dark reaction is named after **Melvin Calvin**, who received a Nobel Prize in 1961 for the work he and his colleagues **Andrew Benson** and **James Bassham** did to elucidate the process. Taking advantage of the availability of radioactive isotopes following World War II, they were able to use $^{14}CO_2$ to show that the primary products of photosynthetic carbon fixation are **triose phosphates** (3-phosphoglycerate, G-3-P or glyceraldehyde-3-phosphate). The triose phosphates enter a variety of metabolic pathways, the most important of these being sucrose and starch biosynthesis.

Wood and **Werkman** (1936) distinguished between the "light reaction" and "dark reaction". **R. Hill** (1938) found that isolated chloroplasts evolved oxygen when illuminated, provided that appropriate electron acceptor was also made available. **K. Porter** and **S. Granick** (1947) described the ultrastructure of grana of chloroplasts. **Calvin** and **Benson** (1948) showed that phosphoglycerate was an early product of CO_2 fixation. **S. Ochoa** and **R.Vishniac** showed in 1950 that $NADP^+$ could substitute as the hydrogen acceptor in the Hill reaction. **R. Emerson** *et al.*, (1956) suggested the presence of two different traps in chloroplasts, each with a pigment system feeding energy into it. In 1966, **M. D. Hatch** and **S.R. Slack** suggested an alternate pathway for carbon fixation in corn.

Distribution

The chloroplasts remain distributed homogeneously in the cytoplasm of plant cells. But in certain cells, the chloroplasts become concentrated around the nucleus or just beneath the plasma membrane. The chloroplasts have a definite orientation in the cell cytoplasm. Since chloroplasts are motile organelles, they show passive and active movements.

Morphology

Shape. Higher plant chloroplasts are generally biconvex or plano-convex. However, in different plant cells, chloroplasts may have various shapes, *viz.*, filamentous, saucer-shaped, spheroid, ovoid, discoid or club-shaped. They are vesicular and have a colourless centre.

Size. The size of the chloroplasts varies from species to species. The chloroplasts generally measure 2–3 μm in thickness and 5–10 μm in diameter (*e.g.*, *Chlamydomonas*). The chloroplasts of polyploid plant cells are comparatively larger than the chloroplasts of the diploid plant cells. Generally, chloroplast of plants grown in the shade are larger and contain more chlorophyll than those of plants grown in sunlight.

Number. The number of the chloroplasts varies from cell to cell and from species to species and is related with the physiological state of the cell, but it usually remains constant for a particular plant cell. The algae usually have a single huge chloroplast. The cells of the higher plants have 20 to 40 chloroplasts. According to a calculation, the leaf of *Ricinus communis* contains about 400,000 chloroplasts per square millimeter of surface area. When the number of chloroplasts is inadequate, it is increased by division; when excessive, it is reduced by degeneration.

Chemical Composition

The chloroplasts are composed of the carbohydrates, lipids, proteins, chlorophyll, carotenoids (carotene and xanthophylls), DNA, RNA and certain enzymes and coenzymes. The chloroplasts also contain some metallic atoms as Fe, Cu, Mn and Zn.

The carbohydrates occur in very low percentage in the chloroplasts. The most common carbohydrates of the chloroplasts are the starch and sugar phosphates.

The chloroplasts contain 20–30 per cent lipids of its dry weight. The lipids are composed of 50 per cent fats, 20 per cent sterols, 16 per cent waxes and 7 to 20 per cent phospholipids. The most common alcohols of the lipids are the choline, inositol, glycerol, ethanolamine.

The proteins constitute 35 to 55 per cent of the chloroplast. About 80 per cent proteins are insoluble and forming the unit membranes of the chloroplasts along with the lipids. The 20 per cent proteins are soluble and occur in the form of the enzymes.

The chlorophyll is a green pigment of the chloroplasts. The chlorophyll contains an asymmetrical molecule which has hydrophilic head of four rings of the pyrols and hydrophobic tail of the phytol chain (Fig. 9.27). Chemically the chlorophyll is a porphyrin like the animal pigment haemoglobin and cytochromes except besides the iron (Fe), it contains Mg atom in between the rings of the pyrols which remain connected with each other by the methyl groups. The chlorophyll consists of 75 per cent **chlorophyll *a*** and 25 per cent **chlorophyll *b***.

The carotenoids are carotenes and xanthophylls, both of which are related to vitamin A. The carotenes have hydrophobic chains of unsaturated hydrocarbons in their molecules. The xanthophylls contain many hydroxy groups in their molecules.

DNA of chloroplast of *Chlamydomonas* represents non-chromosomal genetic system and has been found to be related with cytoplasmic heredity. **Ruth Sager**, who is pioneer on nonchromosomal genes, was able to prepare genetic map of *Chlamydomonas* chloroplast. She has shown that the genome of the chloroplast of *Chlamydomonas* is circular like that of bacteria.

Ultrastructure

A chloroplast comprises the following three main components (Fig. 17.3):

1. Envelope. The entire chloroplast is bounded by an **envelope** which is made up of double unit membranes. Across this double membrane envelope occurs exchange of molecules between chloroplast and cytosol (cytoplasmic matrix). Isolated membranes of envelope of chloroplast lack chlorophyll pigment and cytochromes but have a yellow colour due to the presence of small amounts of carotenoids. They contain only 1 to 2 per cent of the total protein of the chloroplast.

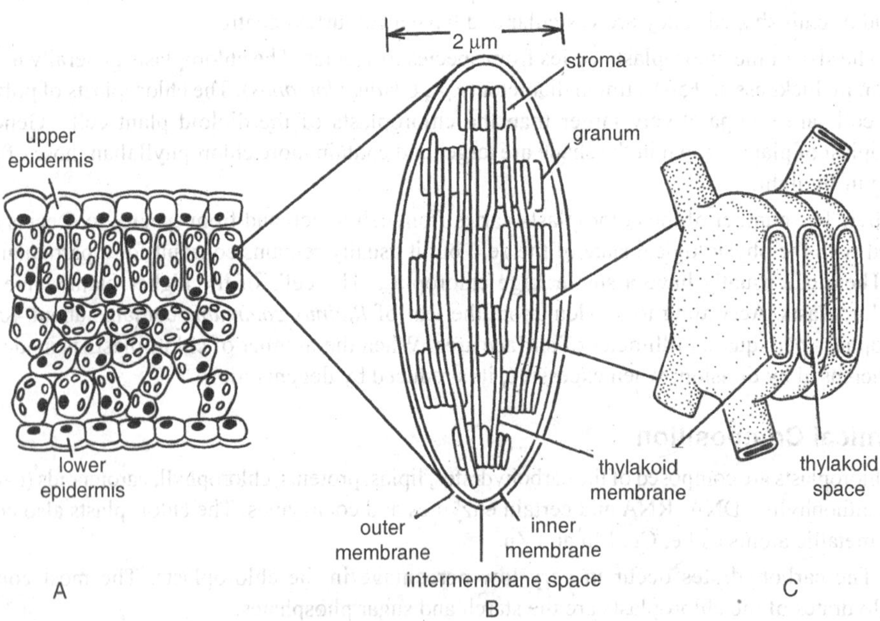

Fig. 17.3. A—Distribution of chloroplasts in mesophyll cells of a leaf; B—Ultrastructure of a chloroplast; C—Details of a granum.

2. Stroma. The matrix or stroma fills most of the volume of the chloroplasts and is a kind of gel-fluid phase that surrounds the thylakoids (grana). It contains about 50 per cent of the proteins of the chloroplast, most of which are soluble type. The stroma also contains ribosomes and DNA molecules (*i.e.*, 80 DNA molecules per chloroplast per cell of *Chlamydomonas*; 20 to 40 DNA molecules per chloroplast per cell of leaf of maize), both of which are involved in the synthesis of some of the structural proteins of the chloroplast. The stroma is the place where CO_2 fixation occurs and where the synthesis of sugars, starch, fatty acids and some proteins takes place.

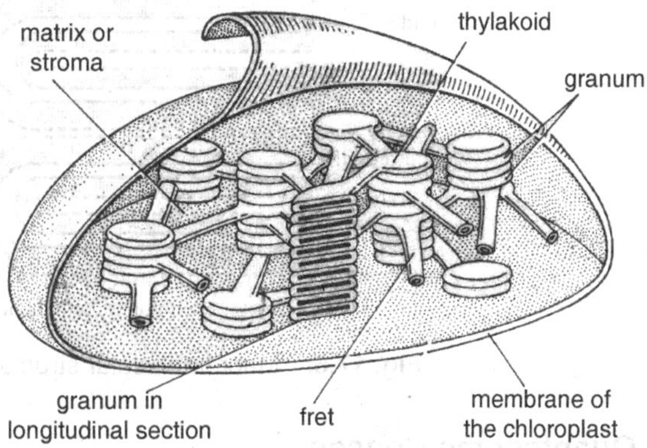

Fig. 17.4. Submicroscopic structure of the chloroplast.

3. Thylakoids. The thylakoids (thylakoid = sac-like) consists of flattened and closed vesicles arranged as a membranous network. The outer surface of the thylakoid is in contact with the stroma, and its inner surface encloses an **intrathylakoid space** (the third compartment). Thylakoids may be stacked like a neat pile of coins, forming grana or they may be unstacked, **intergranal**, or **stromal thylakoids**, forming a system of anastomosing tubules that are joined to the **grana thylakoids** (Fig. 17.4). There may be 40 to 80 grana in the matrix of a chloroplast. The number of thylakoids per granum may vary from 1 to 50 or more. For example, there may be single thylakoid (*e.g.*, red alga, Fig. 17.5), paired thylakoids (*e.g.*, Chrysophyta), triple thylakoids and multiple thylakoids (*e.g.*, green algae and higher plants).

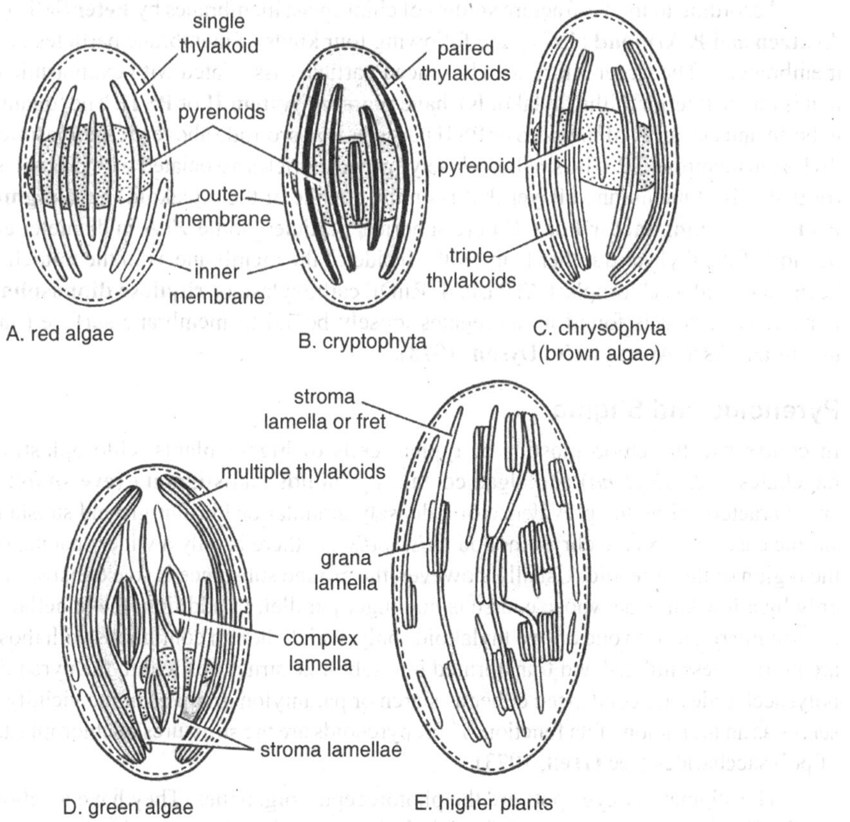

Fig. 17.5. Different kinds of chloroplasts containing variable number of thylakoids per granum.

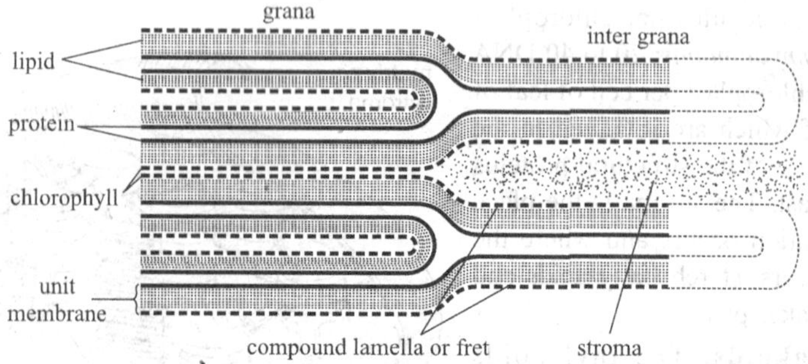

Fig. 17.6. Macromolecular structure of grana of a chloroplast.

Quantosome Concept

Electron microscopic studies of the inner surface of the thylakoid, or compartment forming the grana, has established the existence of a paracrystalline array of spheroid particles of 200 × 100 Å, called **quantosomes** (**Park** and **Pon**, 1963). When the fine structure of chloroplasts is studied by the freeze-etching technique by **Park** and associates (1964, 1967), quantosomes and some other particles were found to lie within the membrane of the thylakoid. Quantosomes were interpreted as photosynthetic units within thylakoid membrane each of which was capable of carrying out the photochemical reaction.

According to freeze-fracture studies of chloroplast membranes by **Peter Satir** (1970s) and **C.J. Arntzen** and **P. Armond** (1977), the following four kinds of membrane particles occur in thylakoid membrane: 1. The larger, more widely spaced particles associated with exoplasmic or E leaflets (or that is on the interior of the thylakoids), have **pigment system II** or **PS II**. The 'quantosomes' appear to be an unusual array of these ES or PS II particles that protrude above the exoplasmic surface (ES) of thylakoid membrane. 2. The smaller, closely spaced particles associated with protoplasmic (P) leaflets (or that half of membrane bilayer that is on the outside of the thylakoid) have **pigment system I** or **PS I**. 3. **Coupling factor I** or **CF I** are spherical particles, some 90 Å in diameter and occur on the outside of the thylakoid rather than on the inside of the membrane as in the mitochondria: CF1 has been identified as chloroplast ATPase. 4. RuDP carboxylase or **ribulose diphosphate carboxylase** is an enzyme that is found in aggregates loosely bound to membrane surface (outside thylakoid membrane) (see **Avers**, 1976; **Dyson**, 1978).

Pyrenoids and Stigma

In contrast to the chloroplastids of typical cells of higher plants, chloroplastids of protozoan flagellates (*e.g.*, *Euglena*) and algae contain **pyrenoids** and **stigmata (eye spots)**. The pyrenoids are characterized by the prevalence of a densely granular or fibrillar ground substance. In the most simple cases, *e.g.*, in the chrysomonod *Olisthodiscus*, there is only a wider spacing of the lamella in the region of the pyrenoid. Usually, however, the ground substance is so extensive that it is traversed only by a few lamellae whose course is no longer parallel. In addition, the lamellae of the pyrenoid are usually reduced to one or two thylakoids only, which though continuous with those of the plastid, are more or less inflated and transformed into tube-like structures within the pyrenoid. The fact that polysaccharides are condensed either as starch or paramylon granules in the vicinity of the pyrenoid serves as an indication of its function. Thus, pyrenoids are the structures of chloroplasts for the storage of polysaccharides (see **Grell**, 1973).

The stigmata or eye spots are the photoreceptor organelles. They have a yellow or red colour in life. This is due to an accumulation of pigment granules of carotenoid of equal size, arranged as a single layer or in several layers on top of each other. The layers may be bent like a dish and are separated by membranes evidently connected with thylakoids of chloroplast. Such chloroplastid stigmata are found in *Eudorina* (*Pleodorina*) *californica* (see **Grell**, 1973).

Photosynthetic Pigments of Chloroplast

There are three classes of photosynthetic pigments in the photosynthetic cells: **chlorophylls**, **carotenoids** and **phycobilins**. One or more kinds of chlorophylls and carotenoids occur in all photosynthetic species, but phycobilins are found in red algae and blue-green algae. Most photosynthetic species contain chlorophylls and carotenoids in tightly bound association with thylakoid membranes. However, phycobilins are loosely associated with these membranes and can be separated very easily from other photosensitive pigments. Phycobilins are concentrated in granules called **phycobilisomes** which look like amorphous particles in the electron micrographs of these algal cells (*e.g.*, red alga *Porphyridium cruentum*). Photosynthetic pigments are thus concentrated and localized in some fashion in all cells and are not dispersed at random in either stroma or cytosol.

1. Chlorophylls. Chlorophylls are most important and lipid-soluble photosynthetic pigments. A chlorophyll molecule consists of four rings of carbon and nitrogen bound together into what is called a **porphyrin ring**. A magnesium atom is nestled in the centre of this structure. A long chain of carbon and hydrogen atoms (**phytol chain**) trails off one side of the chlorophyll molecule.

There are several kinds of chlorophylls which vary in structure only slightly. Higher plants and algae, including prokaryotic blue-green algae, all contain **chlorophyll a**. Eukaryotes also contain a second chlorophyll, which varies according to the species group. The second chlorophyll may be **chlorophyll b** (higher plants and most greea algae), **chlorophyll c** (diatoms, dinoflagellates and brown algae), or **chlorophyll d** (red algae).

2. Carotenoids. Carotenoids also participate in photosynthesis in all organisms. These yellow or orange pigments are usually masked by the green colour of chlorophyll and are not commonly visible in leaves until autumn when the chlorophyll disintegrates. Carotenoids consist of a chain of carbon atoms connecting two six carbon rings. Two classes of carotenoids are important in photosynthesis: **carotenes** and **carotenols**.

3. Phycobilins. Phycobilins are accessory pigments found in eukaryotic red algae and prokaryotic blue-green algae. They are conjugated to specific protein. The protein conjugate of the pigment phycoerythrobilin in most red algae is called **phycoerythrin** and its analog in almost all blue-green algae is **phycocyanin**. In contrast to cyclic porphyrin component in chlorophyll, phycobilins are open chain porphyrin type.

Molecular Organization of Thylakoids

Molecular organization of the membrane of thylakoids is based on the fluid-mosaic model of the membrane which represents following main characteristics: **fluidity**, **asymmetry** and **economy** (*i.e.*, lack of movement in the third dimension). Lipids represent about 50 per cent of the thylakoid membrane; these include those directly involved in photosynthesis (called **functional lipids**) such as **chlorophylls**, **carotenoids** and **plastoquinones**. **Structural lipids** of thylakoids include glycolipids, sulpholipids and a few phospholipids. Most of these structural lipids are highly unsaturated which confer to the membrane of thylakoids a high degree of fluidity.

The protein components of thylakoid membrane are represented by 30 to 50 polypeptides which are disposed in the following five major supramolecular complexes (Fig. 17.7), which can be isolated with mild detergent:

1. Pigment system I or Photosystem I (PS I). This complex contains a reactive centre composed of P700 (Type of pigment which is bleached at the wavelength of 700 nm), several polypeptides, a lower chlorophyll *a*/*b* ratio and β-carotene. It acts as a light trap and is present in unstacked thylakoid membranes. In it light induced reduction of $NADP^+$ takes place.

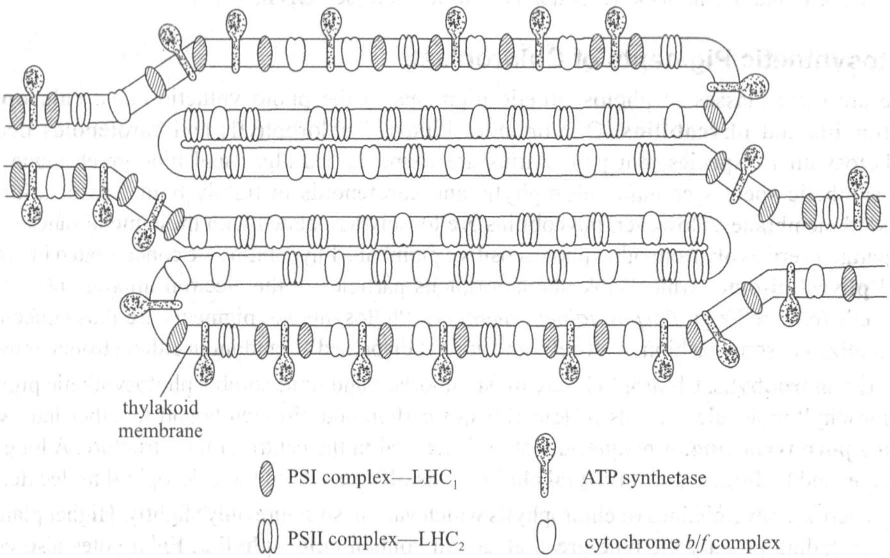

Fig. 17.7. Diagram showing the distribution of the main complexes within the thylakoid membranes both in the granal or stacked and stromal or unstacked regions.

2. Photosystem II (PS II). This complex comprises two intrinsic proteins that bind to the reaction centre of chlorophyll P680 (The pigment that bleaches when absorbing light at 680 nm). It contains a high ratio of chlorophyll *a*/*b* and β-carotene. Frequently, the PS IIs are associated with the light-harvesting complex and are involved in light induced release of O_2 from H_2O (*i.e.*, photolysis of water). PS II works as a light trap in photosynthesis and is mainly present in the stacked thylakoid membranes of grana.

3. Cytochrome b/f. This complex contains one cytochrome *f*, two cytochromes of *b*563, one FeS centre and a polypeptide. It is uniformly distributed in the grana and acts as the electron carrier.

These three complexes are related to the electron transport and are linked by **mobile electron carriers** (*i.e.*, plastoquinone, plastocyanin and ferredoxin). Electron transport through PS II and PS I finally results in the reduction of the coenzyme $NADP^+$. Simultaneously, the transfer of protons from the outside to the inside of the thylakoid membrane occurs.

4. ATP synthase. As in mitochondria, this complex consists of a $\mathbf{CF_0}$ hydrophobic portion, a proteolipid that makes a proton channel, and a $\mathbf{CF_1}$ (or coupling factor one) that synthesizes ATP from ADP and Pi, using the proton gradient provided by the electron transport. ATP synthetase complexes are located in stacked membrane (grana).

5. Light harvesting complex (LHC). The main function of LH complex is to capture solar energy. It contains two main polypeptides and both cholorophyll *a* and *b*. LH complex is mainly associated with PS II, but may also be associated with PS I. LHC is localized in stacked membranes and lacks photochemical activity.

Mutation and chloroplast structure. The organization of chloroplasts and other plastids is often modified due to mutation. **D.Von Wettstein** (1956) reported that the plastids of normal barley plant have a well organized system of grana and stroma. But the plastids of an albino mutant of barley, fail to develop beyond a particular stage and there occurs no differentiation of grana and stroma. Further, the plastids of a yellow-green mutant of barley develop somewhat further than plastids of an albino plant.

17.3. FUNCTIONS OF THE CHLOROPLAST: PHOTOSYNTHESIS

It is well evident now that the process of photosynthesis consists of the following two steps:

1. Light reaction. It is also called **Hill reaction, photosynthetic electron transfer reaction** or **photochemical reaction**. In light reaction solar energy is trapped in the form of chemical energy of ATP and as reducing power in NADPH. During it, oxygen is evolved by photolysis or splitting of water molecule. Light reaction occurs in thylakoid membranes.

2. Dark reaction. It is also called **Calvin cycle, photosynthetic carbon reduction cycle (PCR cycle), carbon-fixation reaction** or **thermochemical reaction**. In dark reaction, the reducing capacity of NADPH and the energy of ATP are utilized in the conversion of carbon dioxide to carbohydrate. Such a process of "**carbon fixation**" or "**CO_2-fixation**" occurs in the stroma of chloroplast (Fig. 17.8).

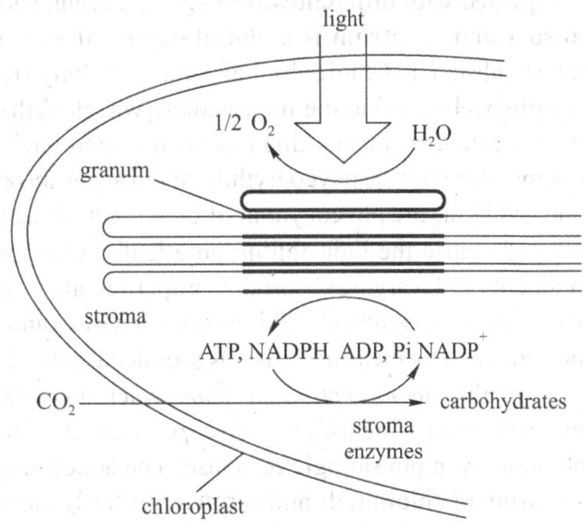

Fig. 17.8. Localization of the light and dark reaction of photosynthesis. The light reaction is catalyzed by chloroplast lamellae, espiecially in the grana. The dark reaction is catalyzed by enzymes of the stroma.

17.4. PHOTOSYNTHETIC UNITS AND REACTION CENTERS

Einstein suggested in 1905 that light and other electromagnetic radiations travel in discrete packets called **quanta** or **photons** and that when light interacts with matter it does so by annihilating complete photons, never a part of one. Further, according to Einstein's, **photoelectric theory**, it takes one photon to eject one electron. Increasing the intensity of light, or flux of photons, only increases the number of electrons ejected, not their velocities. On the other hand, changing the wavelength of light does change the velocity of ejected electrons, implying that the energy of a photon must be related to its wavelength.

When a molecule absorbs a photon of light, it is absorbing a **quantum** of energy. Several things can happen to this energy; (*i*) It Can be dissipated in molecular motion, manifest as heat. (*ii*) It can be reemitted as a new photon or light at a longer wavelength, with the shift representing losses to other processes—if remission occurs very quickly, it is called **fluorescence**; if there is a long lag (milli-seconds to seconds) between absorption and remission, the process is called **phosphorescence**. (*iii*) The energy of light can cause a chemical change in the compound that absorbs it. It is this latter possibility that can happen when a molecule of chlorophyll absorbs a photon. In fact, chloroplast acts as an energy transducer, *i.e.*, it can convert light energy to chemical energy, much as a solar battery uses light to run a transistor radio.

Visible light is only one small part of an immense spectrum of electromagnetic waves, which includes X-rays, ultraviolet radiations, visible light, infrared radiation and radio and television frequencies. The waves act as particles (quanta or photons) of variable energy content. Short wavelengths carry the most energy and can cause damage to living systems. Long wavelengths carry so little energy that they do not appear to affect cellular activities. Light waves in the visible range are most important in photosynthesis and have intermediate energy content.

Light energy must be absorbed by photosynthetic pigments in order to be useful for photosynthesis. Each pigment absorbs different wavelengths. The absorption characteristics of a pigment can be found by using a **spectrophotometer**. Light of a particular wavelength is passed through a sample of pigment and the relative absorption of light is determined. After this process is repeated with different wavelengths, a graph showing the absorption spectrum can be drawn. An **absorption spectrum** is a plot of degree of absorption of different wavelengths of a spectrum by a test substance. Chlorophyll absorbs both long (red and orange) and short (blue and violet) wavelengths well, leaving the intermediate wavelengths (yellow and green). The latter are not absorbed to any extent by plants; this light is reflected back to our eyes and we see the plants as green. In a comparable way the phycoerythrin of red algae absorbs blue and green wavelengths and reflects yellow and red light; the phycocyanin of blue green algae absorbs long wavelengths and reflects blue light.

Because the light falling on a leaf is composed of a variety of wavelengths, the presence of pigments with varying absorption properties ensures that a greater percentage of incoming photons will stimulate photosynthesis. This can be seen by examining an **action spectrum**. An action spectrum can be generated by measuring the O_2 produced by the leaves following exposure to various wavelengths.

Unlike an *absorption spectrum* which simply measures the wavelengths of light that are absorbed by particular pigments, an *action spectrum* identifies the wavelengths that are effective in bringing about a given physiologic response. The action spectrum for photosynthesis follows the absorption spectrum of chlorohylls and carotenoids fairly closely, reflecting the participation of these pigments in the photosynthetic process (see **Karp**, 2010).

Within a fraction of a second after light is absorbed by a photosynthetic pigment, the molecule is altered; some electrons associated with the pigment are raised to new energetic heights, changing their spin or modifying their position. If enough energy is absorbed, an electron may even be ejected and thus oxidation occurs. When these events happen, we say that the molecule is in an **excited state**. In fact, the presence of a conjugated bond structure (alternating double and single bonds) in the photosynthetic pigments (*e.g.*, chlorophyll) greatly increases the number of possible electronic configurations. Each double bond adds a pair of electrons that are shared by the whole pigment molecule. These are the electrons that are vulnerable to excitement.

Thus, chlorophyll *a* is altered within 10^{-15} second after being struck by a photon. The excitement lasts only for a fraction of a second before the energy is given up and the molecule returns to its initial unexcited state.

Photosynthetic units. The basic photosynthetic units seem to be groups of roughly 300 pigment molecules located in the chloroplast membranes (thylakoid disc). Although all the pigment molecules in the unit (carotenoids. chlorophyll *b*, etc,) are capable of capturing light energy, they must transfer it to a single key chlorophyll *a* molecule called the **reaction centre**. The latter then loses an electron to a series of electron carrier molecules. Thus the other 299 accessory pigment molecules are referred to as **antennal molecules** or **antenna pigments**, to designate their role in the capture of light energy. Energy is not passed randomly through these molecules but follows a definite pattern. In green plants excitation energy captured by the carotenoids is first passed to chlorophyll *b* and then through a sequence of chlorophyll *a* molecules with slightly different absorption characteristics, to ultimately rest in the reaction centre. At various stages of transfer among pigments, small amounts of energy are lost, so all the energy funneling into the reaction centre is ultimately at the same level.

We do not know the structural arrangement of the antenna pigments and the reaction-centre chlorophyll. Some investigators think that reaction-centres are positioned randomly throughout a mass of pigment molecules. A photon would be absorbed by any of the pigments and then the excitation energy would be transferred at random throughout the system until it reached a reaction-centre or reemitted as fluorescent light. This concept is known as the **"lake model"**. An alternative view, called the **"puddle model"** suggests that the various photosynthetic units are separated from one another as puddles dotted around the landscape; each unit has its own reaction centre surrounded by antenna pigments. Thus excitation energy cannot be transferred between units. Actually both models appear to have validity, and a hybrid model, called the **"connected puddle model"** has been proposed which suggests that although photosynthetic units are somewhat isolated functionally, some transference of energy does occur between them.

The transfer of excitation energy from one pigment molecule to another is very sensitive to the distance between the molecules. The chlorophyll molecules of an antenna are kept in close proximity (less than 1.5 nm separation) and proper orientation by noncovalent linkage to integral membrane polypeptides. A "rule" that operates among antenna pigments is that energy can only be transferred to an equal or less energy-requiring molecule. In other words, energy can only be passed to a pigment molecule that absorbs light of equal or longer wavelength (lower energy) than that absorbed by the donor molecule. As the energy "wander" through a photosynthetic unit (Fig. 17.9), it is repeatedly transferred to a pigment molecule that absorbs at a longer wavelength. The energy is ultimately transferred to a chlorophyll of the longer wavelength than any of its neighbors. Once the energy is received by the reaction center, the electron excited by light absorption can be transferred to its acceptor.

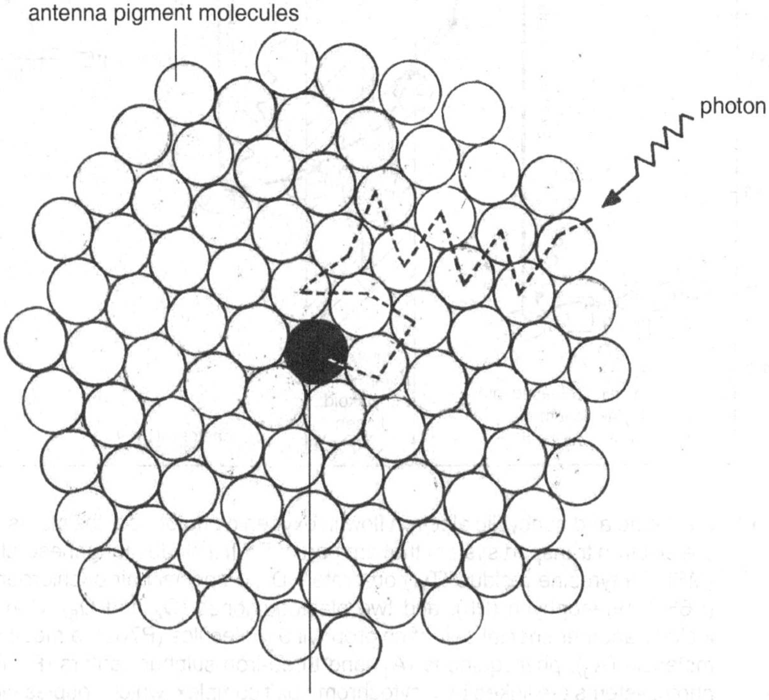

Fig. 17.9. The transfer of excitation energy. Energy is transferred randomly through a network of pigment moleculs that absorb light of increasing longer wavelength until the energy reaches a reaction-center chlorophyll, which transfers an excited electron to a primary acceptor (after Karp 2010).

Oxygen Formation

The light absorbing reactions of photosynthesis occur in large pigment-protein complex called **photosystems**. Two types of photosystems are required to catalyze the two light absorbing reactions utilized in oxygenic photosynthesis. One photosystem, **photosystem II (PS II)**, boosts electrons from an energy level below that of water to a midway point (Fig. 17.10). The other photosystem, **photosystem I (PSI)**, raises electrons from a midway point to an energy level well above that of $NADP^+$. The two photosystems act in series, that is one after the other.

The reaction center of photosystem II is a chlorophyll dimer referred to as P680, "P" standing for "pigment" and "680" standing for the wavelength of light that this particular pair of chlorophyll absorbs more strongly. The reaction center of the photosystem I is also a chlorophyll dimer and is referred to as

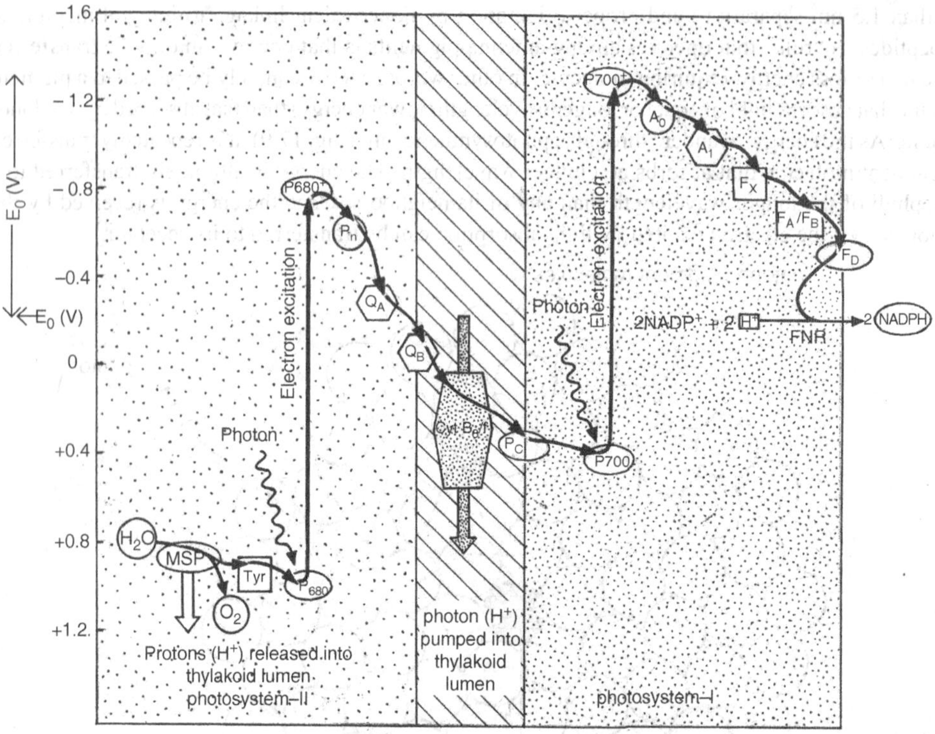

Fig. 17.10. Z scheme and noncyclic electron flow in oxygenic phototrops, the plants. Components of the electron transport system that are part of PS II include manganese-stabilizing protien (MSP), a tyrosine residue (Tyr) on protein D_1, a special pair of chlorophyll *a* molecules (P680), phenophytin (Ph), and two plastoquinones (Q_4 and Q_8). Components of PSI include another special pair of chlorophyll *a* molecules (P700), a modified chlorophyll *a* molecule (A_0), phylloquinone (A_1) and three-iron-sulphur centers (F_X, F_A and F_B). The photosystems are linked by a cytochrome b_0/f complex, which couples electron transport to pumping of protons into the thylakoid lumen, and by plastocyanin (Pc). P680 and P700 are unusual because absorption of photons causes a large decrease in their reduction potentials (vertical axis) establishing them to accept electrons from a donor with a highly positive reduction potential and donate electrons to an acceptor with a highly negative reduction potential, thereby conserving their captured energy. Ferredoxin-NADP$^+$ reductase (FNR) catalyzes the transfer of electrons from ferredoxin (Fd) to NADP$^+$ (after Baker *et al.*, 2006).

P700 for comparable reasons. When sunlight strikes a thylakoid membrane, energy is absorbed by antenna pigments of both PS II and PS I and passed to the reaction centers of both photosystems. Electrons of both reaction-center pigments are boosted to an outer orbital, and each electron is transferred to a **primary electron acceptor**. After losing their electrons, the reaction center chlorophylls of PS II and PS I become positively charged pigments referred to as P680$^+$ and P700$^+$, respectively. The electron acceptors, in turn, become negatively charged. In essence, this separation of charge within the photosystems is the *light reaction*—the conversion of light energy into chemical energy. Positively charged electron acceptors act as electron donors. Consequently, the separation of charge within each photosystem sets the stage for the flow of electrons along the chain of specific carriers.

The oxygenic photosynthesis (*i.e.*, in plants), where the two photosystems act in series, electrons flow occurs along three legs—from water to PS II, from PS II to PS I, and from PS I to $NADP^+$—an arrangement described as the **Z scheme** (Fig. 17.10) which was first proposed by **Robert Hill** and **Fay Bendall** of the University of Cambridge. Like the members of respiratory chain of mitochondria, most of the electron carriers of the Z scheme are found in part of large membrane protein complexes (Fig. 17.12). As in mitochondria, electron transfer releases energy, which is used to establish a proton gradient, which in turn drives the synthesis of ATP. ATP produced in the chloroplast is used primarily within the organelle in the synthesis of carbohydrates; ATP utilized outside the chloroplast is derived largely from that produced in plant mitochondria.

Working of PS II

Photosystem acts to absorb sunlight energy to obtain electrons by splitting of water. PS II used absorbed light energy for two interrelated activities: removing electrons from water and generating a proton gradient.

The PS II of plant cells is a complex of more than 20 different polypeptides, most of which are embedded in the thylakoid membrane. Two of these proteins, designated **D_1 protein** and **D_2 protein**, are particularly important because together they bind the P680 reaction center chlorophyll dimer and all the cofactors involved in electron transport through the photosystem II (Fig. 17.11).

The first step in PS II activation is the absorption of light by an antenna pigment. Most of the antenna pigments that collect solar energy for PS II reside within a separate pigment protein, called **light-harvesting complex II** or simply, **LHC II**. LHC II proteins bind both chlorophylls and carotenoids and are situated outside the core of the photosystem (Fig. 17.11). (*Note*. Carotenoids and chlorophyll-*b* are called **accessory pigments** because they hand over the energy absorbed by them to chlorophyll-*a*). LHC II is not always associated with PS II but, under appropriate conditions, can migrate along the thylakoid membrane and become associated with PS I, serving as a light-harvesting complex for the PS I reaction center as well.

The flow of electrons from PS II to plastoquinone (PQ). Excitation energy is passed from the outer antenna pigments to LHC II to a small number of inner-antenna chlorophyll molecules situated within the core of PS II. From there, the energy is ultimately passed to the PS II reaction center. The excited reaction center pigment (P680) responds by transferring a single photoexcited electron to a closely associated chlorophyll-like **phenophytin (Pheo)** molecule (step 1) which is the *primary electron acceptor*. This electron transfer generates a separation of charge and a negatively charged acceptor ($Pheo^-$). $P680^+$ is electron deficient and can accept electrons, making it an oxidizing agent. In contrast, *$Pheo^-$* has an extra electron that it will readily lose, making it a reducing agent. This event—the light-driven formation of an oxidizing agent and a reducing agent—takes less than one billionth of second and it is the essential first step in photosynthesis.

Because $P680^+$ and $Pheo^-$ charged, they exhibit an obvious reactivity with one another. Interaction between oppositely charged species is prevented by moving the separated charges further apart, ultimately to the opposite sides of the membrane, by passage through several different sites. $Pheo^-$ first transfers its electrons (step 2, Fig. 17.11) to a molecule of **plastoquinone (PQA)** bound near the outer (*i.e.*, stromal) side of the membrane. Plastoquinone (PQ) is a lipid-soluble molecule similar in structure to ubiquinone. The electron from PQ_A is transferred (step 3, Fig. 17.11) that remain firmly bound to the **D_1 protein** of the reaction center with each of these transfers, the electron moves closer to the stromal side of the thylakoid membrane.

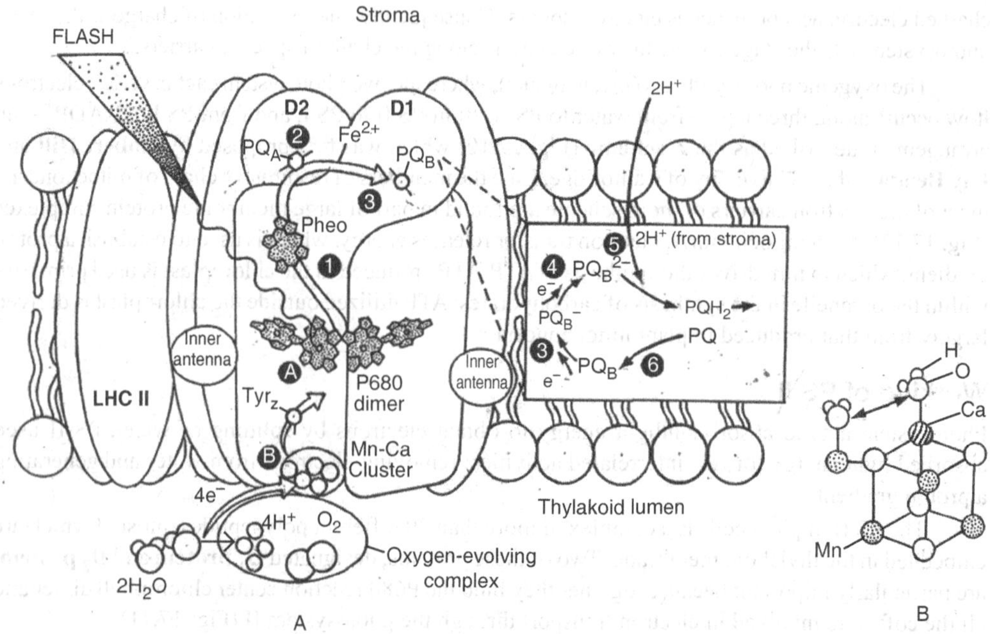

Fig. 17.11. The functional organization of photosystem II. A simplified model of the huge protein pigment complex, which catalyzes the light-driven oxidation of water and reduction of plastoquinone. Events begin with absorption of light by an antenna pigment in the outer light-harvesting complex (LHC II). Energy is transferred from LHC II through an inner-antenna pigment protein complex to a P680 reaction-center chlorophyll *a*, which is one of four closely spaced chlorophyll *a* molecule (the P680 dimer and two accessory chlorophyll *a* molecules). Absorption of this energy by P680 excites an electron, which is transferred to pheophytin (Pheo) (Step 1), the primary electron acceptor of PS II. (Pheophytin is a chlorophyll molecule that lacks the Mg^{2+} ion). The electron subsequently passes to a plastoquinone PQA (step 2) and then through a nonheme Fe^{2+} to PQB (step 3) to form a negatively charged free radical PQB^-. Absorption of a second photon sends a second electron along the same pathway, converting the acceptor to PQB^{2-} (step 4). Two protons then enter from stroma (step 5) generating PQH_2, which is released into the lipid bilayer and replaced by a new oxidized PQB molecule (step 6). As the above events are occurring, electrons are moving from H_2O by way of Tyrz to the positively charged reaction center pigment (step B and A) B–A model of crystal of metal ions and oxygen and water molecules (after Karp 2010).

The positively charged pigment ($P680^+$) is reduced back to P680, which charges the reaction center for the absorption of another photon. The absorption of a second photon sends a second energized electron along the path from P680 to pheophytin to PQA to (PQB^-), forming PQB^{2-} (step 4, Fig. 17.11) which combines with two photons to form **plastoquinol, PQH_2** (step 5, Fig. 17.11). The protons utilized in the formation of PQH_2 are derived from the stroma, causing the decrease in H^+ concentration of the stroma, which contributes to formation of the proton gradient. The reduced PQH_2 molecule dissociates from the D_1 protein and diffuses into the lipid bilayer. The displaced PQH_2 is replaced by a fully oxidized PQ molecule derived from a small "pool" of plastoquinone molecules in the bilayer (step 6, Fig. 17.11). This portion of PS II reaction can be summarized as follows:

$$2e^- + PQ + 2H^+_{stroma} \xrightarrow{\text{2 photons}} PQH_2$$

As we will see shortly, the oxidation of a molecule of water by PS II requires four photons, so we can revise this portion of the PS II reaction as follows:

$$4e^- + 2PQ + 4^{+}_{stroma} \xrightarrow{\text{4 photons}} 2PQH_2$$

Flow of Electrons From Water to PS II: Photolysis of water. Water is a very stable molecule made up of tightly held hydrogen and oxygen atoms. In fact, the splitting of water is the most thermodynamically challenging endergonic reaction known to occur in living organisms. Splitting water in a laboratory requires the use of a strong electric current or temperature approaching 2000°C. Yet a plant cell can accomplish this feat on a snowy mountain side using only the energy of visible light (see **Karp** 2010).

$P680^+$ (which is produced during step 1 in Fig. 17.11) is most powerful oxidizing agent yet to be discovered in a biological system. The redox potential of the oxidized form P680 is sufficiently storng to pull tightly held (low-energy) electrons from water (redox potential of +0.82 V), thus splitting the water molecule. The splitting of water during photosynthesis is called **photolysis**. The formation of one molecule of oxygen during photolysis is thought to require the simultaneous loss of four electrons from two molecules of water according to the following reaction:

$$2H_2O \xrightarrow{\text{4 photons}} 4H^+_{lumen} + O_2 + 4e^-$$

However, one PS II reaction center can only generate one positive charge ($P680^+$), or oxidizing equivalent, at a time. A solution to this problem was proposed around 1970 by **Pierre Joliot** and **Bessel Kok** as the **S-state hypothesis**, which allows the photosystem to accumulate the four oxidizing equivalents needed to oxidize water. Closely associated with the D_1 protein of PS II at its luminal surface is a cluster of five metal ions—four manganese (Mn^{2+}) ions and one calcium (Ca^{2+}) ion—that is stabilized and protected by a number of peripheral proteins that form the **oxygen-evolving complex (OEC)** (Fig. 17.11). (*Note*. According to **Baker** *et al.*, 2006, OEC contains three types of ions: Mn^{2+}, Ca^{2+} and Cl^- ions).

The Mn–Ca cluster accumulates four oxidizing equivalents by transferring four electrons, one at a time, to the nearby $P680^+$. The transfer of each electron from the Mn–Ca cluster to $P680^+$ (step B and A of Fig. 17.11) is accomplished by passage through an intermediate electron carrier, a tyrosine residue on the D_1 protein, termed **Tyrz.** After each electron is transferred to $P680^+$, regenerating P680, the pigment becomes reoxidized (back to $P680^+$) following the absorption of another photon by the photosystem. Thus, stepwise accumulation of four oxidizing equivalents by the Mn–Ca cluster is driven by the successive absorption of four photons of light by the PS II photosystem. Once this has occurred, the system can now catalyze the removal of $4e^-$ from two closely bound H_2O molecules as indicated in the following equation:

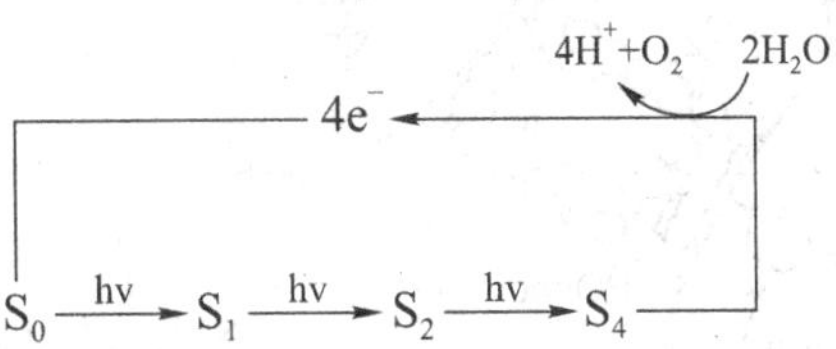

Here S indicates the number of oxidizing equivalents stored by the Mn–Ca cluster.

The protons produced in the photolysis reaction are retained in the thylakoid lumen (Fig. 17.11), where they contribute to the proton gradient. The four electrons produced in the photolysis reaction serve to regenerate the fully reduced Mn–Ca cluster (*i.e.*, So state), while the O_2 is released as a waste product into the environement.

From PS II to PS I. PQH_2 is a mobile electron carrier that diffuses through the lipid bilayer of the thylakoid membrane and binds to a large multiprotein complex called **cytochrome b6f**. Each PQH_2 molecule donates its two electrons to cytochrome *b6f*, while its two protons are released into the lumen. Thus, $2H^+$ (protons) are donated by PQH_2 and 2 additional H^+ are translocated through the complex from the stroma (Fig. 17.12). The electrons from cytochrome *b6f* are passed to another mobile electron carrier, a water soluble, copper containing, peripheral membrane protein, called **plastocyanin (PC)**, situated on the luminal side of the thylakoid membrane. Plastocyanin carries electrons to the luminal side of PS I, where they are transferred to $P700^+$, the positively charged reaction-center of pigment of PS I.

Box 17.1

Cytochrome *b6f* is related in structure and function to cytochrome bc_1 of electron transport chain of mitochondria. Both complexes have quinols as substrates and share similar redox groups, and both engage in a Q cycle that translocates $4H^+$ for each pair of electrons.

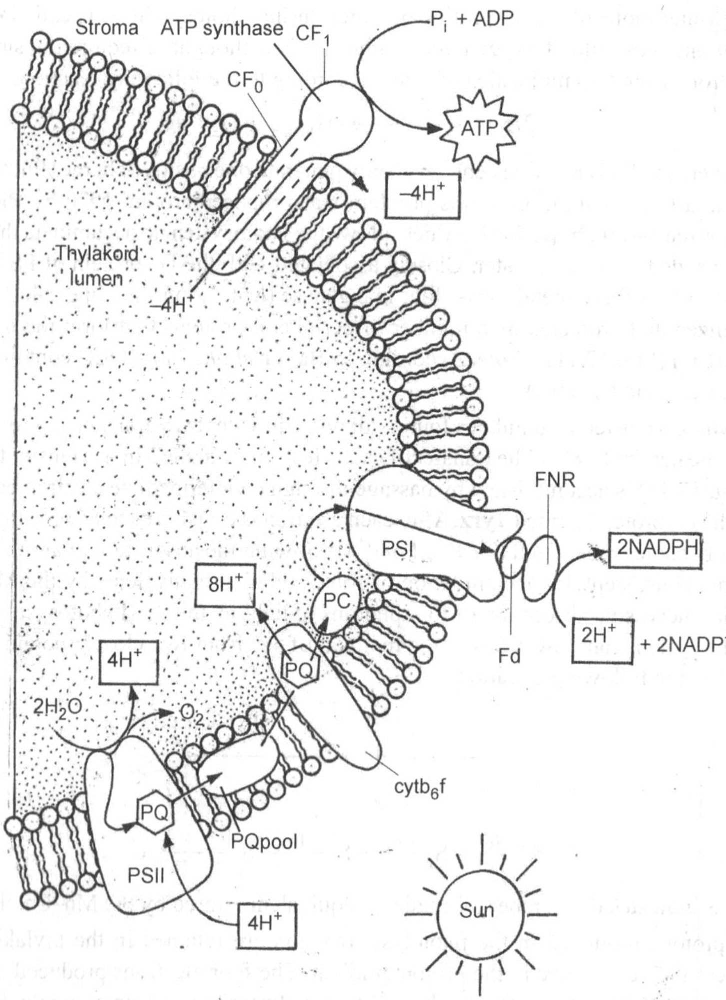

Fig. 17.12. Summary of the flow of electrons from water (H_2O) to NADPH through the three transmembrane complexes. This figure shows the estimated number of protons translocated throught the membrane as a result of the oxidation of two molecules of water, yielding two pairs of electrons. The ATP synthase of the thylakoid membranes is also shown. Approximately four protons are required for the synthesis of each molecule of ATP. (FNR = Ferredoxin NAD^+ reductase).

Working of PSI and production of NADPH. PSI of higher plants consists of a reaction-center core made up of 12–14 polypeptide subunits and a peripheral complex of protein-bound pigments called **light-harvesting complex I (LHCI)**. Light energy is absorbed by the antenna pigments of LHCI and passed to the PSI reaction-center pigment P700, which is a chlorophyll *a* dimer (Fig. 17.13). Following energy absorption, an excited reaction-center pigment ($P700^+$) transfers an electron to a separate monomeric chlorophyll a molecule (designated Ao), which acts as primary electron acceptor (step 1; Fig. 17.13). As in PS II, absorption of light leads to the production of two charged species: $P700^+$ and Ao^-. Ao^- is a very strong reducing agent with a redox potential of approximately –1.0 V. The positive charge of $P700^+$ pigment is neutralized by an incoming electron donated by plastocyanin (PC).

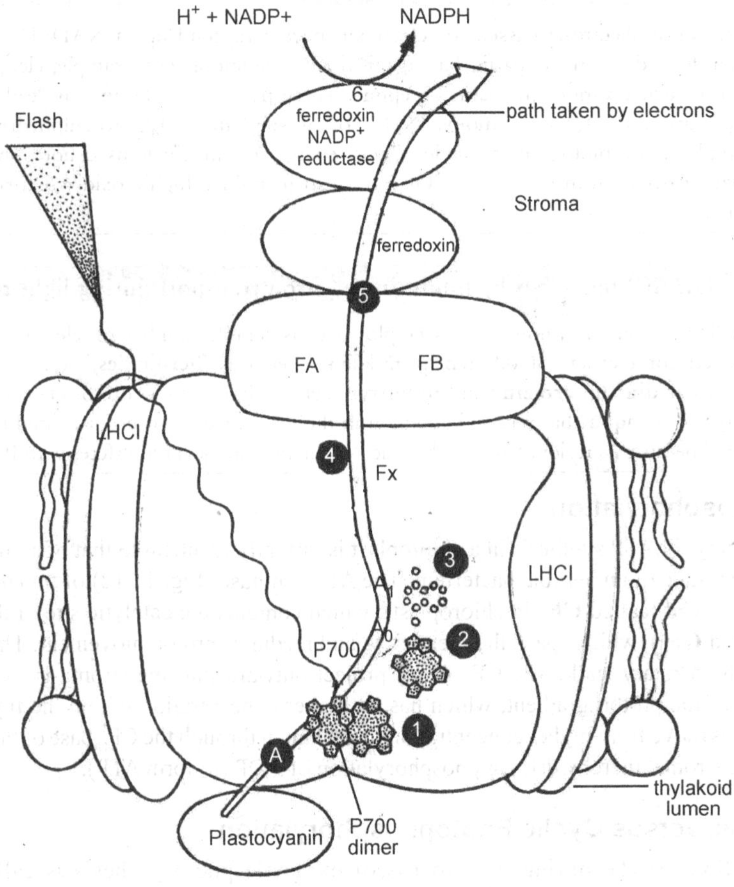

Fig. 17.13. The functional organization of photosystem I. F_A, F_B or Fx = Iron – sulphur center

The inaugural separation of charge in PS I is stabilized by passage of the electron from Ao^- through several cofactors beginning with a type of quinone called **phylloquinone** (designated **A1**) and then three iron-sulphur clusters (designated **Fx, Fb** and **FA**) (step 2 to 4, Fig. 17.13). The oxidation of P700 to $P700^+$ occurs at the luminal side of the membrane. The electron is subsequently transferred out of PSI to a small, water-soluble, iron-sulphur protein called **ferredoxin** (step 5, Fig. 17.13) associated with the stromal surface of the membrane. The reduction of $NADP^+$ to form NADPH (step 6, Fig. 17.13) is catalyzed by a large enzyme called **ferredoxin-NADP$^+$ reductase**, which contains an FAD prosthetic group capable of accepting and transferring two electrons. An individual ferredoxin molecule can donate only one electron so that two ferredoxins act together in the reduction.

$$2\text{Ferredoxin}_{red} + H^{+} + NADP^{+} \xrightarrow{\text{Ferridoxin-NADP}^{+}\text{reductase}} 2\text{Ferredoxin}_{ox} + NADPH$$

The removal of a proton from the stroma also adds to the proton gradient across the thylakoid membrane. Based on the absorption of four photons, the overall reaction for PS I can be written as follows:

$$4e^{-} + 2H^{+}_{stroma} + 2NADP^{+} \xrightarrow{\text{4 photons}} 2NADPH$$

Box 17.2

In PSI, not all electrons passed to ferredoxin necessarily end up in NADPH; alternate routes can be taken depending on the particular organism and conditions. For example, electrons from PS I can be used to reduce various inorganic acceptors. These paths for electrons can lead to the eventual reduction of nitrate (NO_3^-) to ammonia (NH_3), or of sulphate (SO_4^{2-}) to sulphydryl (–SH), which are key ingradients of biological molecules. Thus, energy in sunlight is used not only to reduce the most oxidized carbon atoms (those in CO_2), but also to reduce highly oxidized forms of nitrogen and sulphur atoms.

Box 17.3 Killing weeds by inhibiting electron transport during light reaction

The light reactions of photosystems employ a considerable number of electron carriers, which serve as **targets** for a variety of different plant-killing chemicals (**herbicides**). A number of common herbicides such as **diuron**, **arrazine** and **terbutryn**, act by binding to a core protein (PQB site) of PS II. The herbicide **paraquat** has received attention in the news media because it is used to kill marijuana (cannabis) and because its residues are highly toxic to humans. Paraquant interferes with PS I function.

Photophosphorylation

The machinery for ATP synthesis in a chloroplast is virtually identical to that of a mitochondrion or plasma membrane of an aerobic bacterium. The ATP synthase (Fig. 17.12) of thylakoid membrane consists of a head (called **CF_1** in chloroplast), which contains the catalytic site of the enzyme and a base (called **CF_0**), which spans the membrane and mediates proton movement. The two parts are connected by a rotary stalk. The CF_1 heads project outward into the stroma in keeping with the orientation of the proton gradient, which has its higher concentration within the thylakoid lumen. Thus, protons move from higher concentration in the lumen through the CF_0 base of the ATP synthase and into the stroma, thereby driving phosphorylation of ADP (to form ATP).

Non-cyclic versus Cyclic Photophosphorylation

The formation of ATP during the process of oxygenic photosynthesis is called **noncyclic photophosphorylation** because electrons move in a linear (*i.e.*, noncyclic) path from H_2O to $NADP^+$ (Fig. 17.12). During the 1950s, American botanist **Daniel Arnon** discovered that isolated chloroplasts were not only capable of synthesizing ATP from ADP but could do even in the absence of added CO_2 or $NADP^+$. These experiments indicated that chloroplasts had a means for ATP synthesis that did not require most of the photosynthetic reactions that would have led to oxygen production, CO_2 fixation, or $NADP^+$ reduction. All that was necessary was illumination, chloroplasts, ADP and Pi. The process **Arnon** had discovered was later called **cyclic photophosphorylation** and is a process that is carried out by PS I independent of PS II. Despite the fact that it was discovered over 50 years ago, cyclic photophosphorylation is not well understood. Recent studies suggest that there are two overrlapping pathways for cyclic electron tansport, one of which is outlined in Fig. 17.14.

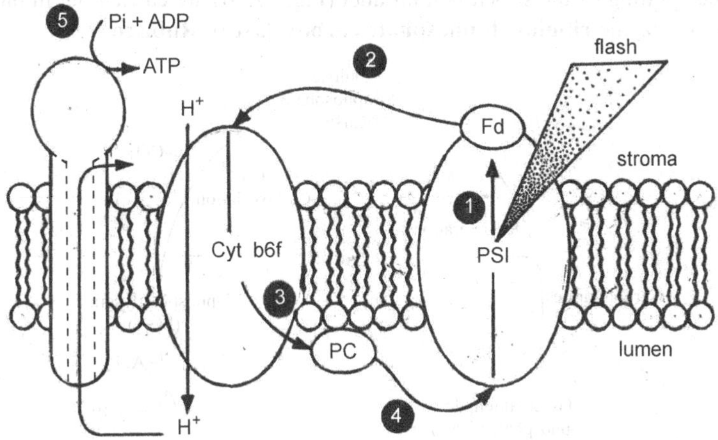

Fig. 17.14. Simplified scheme for cyclic photophosphorylation. Absorption of light by PS I excites an electron, which is transferred to ferredoxin (step 1) and on to cytochrome *b6f* (step 2), plastocyanin (step 3), and back to $P700^+$ (step 4). In the process, protons are translocated by cytochrome *b6f* to form a gradient utilized for ATP synthesis (step 5). Another cyclic pathway for electron transport that involves movement of electrons from PS I through NADPH to cytochrome *b6f* is not shown.

Cyclic electron transport begins with the absorption of a quantum of light by PS I and transfer of a high-energy electron to the primary acceptor. In the pathway depicted in Fig. 17.14, the electron is passed along to ferredoxin, as is always the case, but rather than being transferred to $NADP^+$, the electron is routed back to the electron-deficient reaction center to complete the cycle. During the flow of an electron around this course, sufficient free energy is released to translocate protons (estimated at two H^+/e^-) across the membrane by the cytochrome *b6f* complex and to build a proton gradient capable of driving ATP synthesis. Cyclic phosphorylation is thought to provide additional ATP required for carbohydrate synthesis (Fig. 17.16) and also for other ATP-requiring activities in the chloroplast (*e.g.*, molecular chaperone involvement in protein import). Inhibition of cyclic photophosphorylation leads to impaired development and growth of higher plants.

17.5. CARBON DIOXIDE FIXATION AND THE SYNTHESIS OF CARBOHYDRATE (CALVIN CYCLE)

1. Carbohydrate Synthesis in C_3 Plants

During experiment of **Melvin Calvin, Andrew Benson** and **James Bassham**, radiolabeled CO_2 ($^{14}CO_2$) was converted to reduced organic compounds very rapidly. If the incubation period was very short (upto a few seconds), one radioactive spot on the chromatogram predominated. The compound that formed the spot was determined to be **3-phosphoglycerate (PGA)**, one of the intermediates of glycolysis. Calvin group suspected that CO_2 was being covalently linked (or **fixed**) to a two-carbon compound to form the three-carbon PGA molecule. Because the first intermediate to be identified was a three-carbon molecules, plants that utilize this pathway to fix atmospheric CO_2 are referred to as **C_3 plants**.

After considerable investigation, it became apparent that the intial acceptor was not a two-carbon compound, but a five-carbon compound, **ribulose 1, 5-biphosphate (RuBP)**, which condensed with CO_2 to produce a six-carbon molecule. This six-carbon compound rapidly fragmented into two molecules of PGA, one of which contained the recently added carbon atom. Both the condensation of RuBP and the splitting of the six-carbon product (Fig. 17.16) are carried out in the stroma by a large multisubunit enzyme **ribulose biphosphate carboxylase** or **Rubisco**.

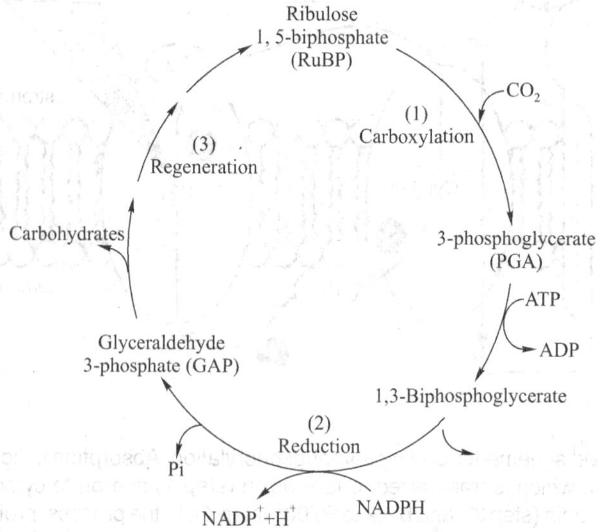

Fig. 17.15. Summary of the Calvin cycle (Dark reaction). The cycle has three stages; 1. carboxylation of ribulose 1, 5-biphosphate; 2. reduction of 3-phosphoglycerate to glyceraldehyde 3-phosphate, and 3. regeneration of ribulose 1, 5-biphosphate (after Horton et al., 2006).

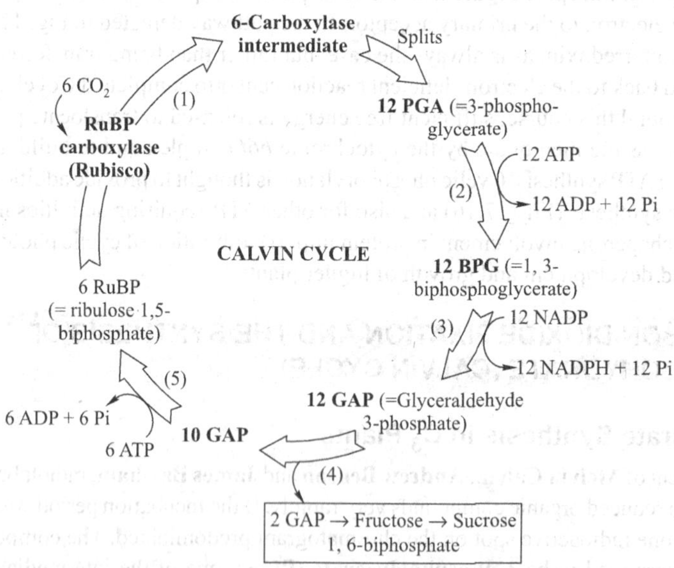

Fig. 17.16. Converting carbon dioxide (CO_2) into carbohydrate.

Calvin cycle. It is investigated out that the pathway for conversion of CO_2 into carbohydrate is **cyclic** and **complex**. This pathway is called **Calvin cycle** (or **Calvin-Benson cycle**) and it occurs in cyanobacteria and all eukaryotic photosynthetic cell. The Calvin cycle (Fig. 17.15) comprises three main parts: 1. Carboxylation of RuBP to form PGA; 2. Reduction of PGA to the level of a sugar (CH_2O) by formation of glyceraldehyde 3-phosphate (GAP) using the NADH and ATP produced in the light-dependent reactions; and 3. The regeneration of RuBP, which also requires ATP. It can be seen in Figure 17.16 that for every 6 molecules of CO_2 fixed, 12 molecules of GAP are produced. The atoms in 10 of these three-carbon GAP molecules are **rearranged** to regenerate 6 molecules of the five-carbon CO_2 acceptor, RuBP. The other 2 GAP molecules are regarded as **product**. These GAP molecules can be exported to the cytosol in exchange for phosphate ions (Fig. 17.17) and used to synthesize the disaccharide **sucrose**. Alternatively, GAP can remain in the chloroplast where it is converted to **starch**.

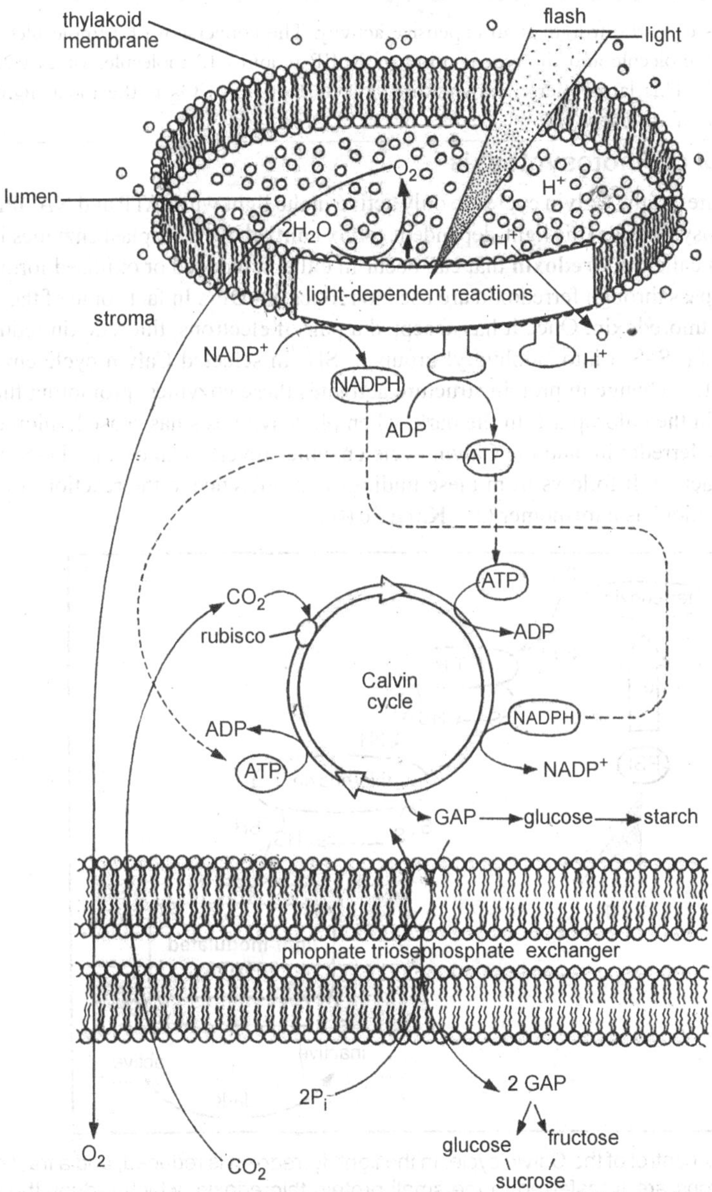

Fig. 17.17. An overview of the entire process of photosynthesis: light *reactions* (light absorption, oxidation of water, reduction of $NADP^+$, and translocation of protons), *phosphorylation of ADP*, the *Calvin cycle*, and the *synthesis of starch and sucrose*.

The sucrose molecules that are formed in the cytosol from the GAPs of the Calvin cycle are transported out of the leaf cells and into the phloem, where they are carried to the various nonphotosynthetic organs of the plant. Just as glucose serves as the source of energy and organic building blocks in most animals, sucrose serves an analogous role in most plants. Starch, on the other hand, is stored within chloroplasts as granules. Just as stored glycogen provides animals with readily available glucose in times of need, the starch stored in plant leaves provides it with sugars at night when the light reactions are not possible.

Box 17.4

The synthesis of carbohydrate is an expensive activity. The conversion of 6 molecules of CO_2 to 1 six-carbon sugar molecule and the regeneration of RuBP requires 12 molecules of NADPH and 18 molecules of ATP. This large energy expenditure reflects fact that CO_2 is the most highly oxidised form in which carbon can occur.

Redox Control of Photosynthesis

Several key enzymes of the Calvin cycle are only active in the light when ATP and NADPH are being produced by photosynthesis. This light-dependent **redox control** of chloroplast enzymes is exercised by a small protein called **thioredoxin** that can occur in either a reduced or oxidised form. Not all of the electrons that pass through ferredoxin are used to reduce $NADP^+$. In fact some of these electrons are transferred to thioredoxin. Once it has accepted a pair of electrons, thioredoxin reduces certain disulphide bridges (–S–S–) into sulphydryl groups (–SH) in selected Calvin cycle enzymes (Fig. 17.18). This covalent change in protein structure activates these enzymes, promoting the synthesis of carbohydrates in the chloroplast. In the dark, when photosynthesis has ceased, thioredoxin is no longer reduced by ferredoxin, and the Calvin cycle enzymes revert to an oxidized (–S–S–) state in which they are inactive. It follows from these findings that reference to the reactions of the Calvin cycle as "dark reaction" is a misnomer (see **Karp** 2010).

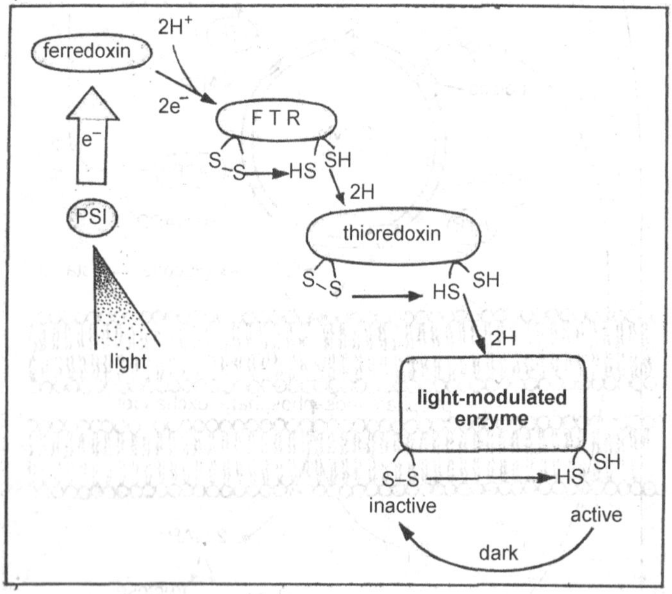

Fig. 17.18. Redox control of the Calvin cycle. In the light, ferredoxin is reduced, and a fraction of these electrons are transferred to the small protein thioredoxin, which redues the disulphide groups of certain Calvin cycle enzymes, maintaining them in an active state. In the dark, electron flow to thioredoxin ceases, the sulphydryl grops of the regulated enzymes become oxidized to the disulphide state, and the enzymes are inactivated.

17.6. OTHER PHOTOSYNTHETIC ASSIMILATION PATHWAY

Photosynthesis comprises more than carbon dioxide fixation and carbohydrate synthesis. In plants and algae, the ATP and NADPH generated by photosynthetic energy transduction reactions are consumed by a variety of other anabolic pathways found in chloroplasts. Carbohydrate synthesis is one example of carbon metabolism; the synthesis of fatty acids, chlorophyll, and carotenoids also occurs in chloroplasts. Moving beyond carbon metabolism, several key steps of nitrogen and sulphur assimilation are localized in chloroplasts. The reduction of nitrite (NO_2^-) to ammonia (NH_3) for example, is catalyzed by a reductase enzyme in the chloroplast stroma, with reduced ferredoxin serving as an electron donor. The ammonia is then channeled into amino acid and nucleotide synthesis, portions of which also occur in chloroplasts. Furthermore, much of the reduction of sulphate (SO_4^{2-}) to sulphide (S^{2-}) is catalyzed by enzymes in the chloroplast stroma. In this case, both ATP and reduced ferredoxin provide energy and reducing power. The sulphide, like ammonia, may then be used for amino acid synthesis.

1. Rubisco

The enzyme that catalyzes the capture of carbon dioxide (*i.e.*, inorganic carbon) and converts it into organic biological molecule, 3-phosphoglycerate is called **ribulose-1, 5-biphosphate carboxylase/oxygenase (RuBP carboxylase** or **rubisco)**. This relatively large enzyme (about 560 kDa) is unique to phototrophs and is found in all photosynthetic organisms (*i.e.*, plants, algae and cyanobacteria) except for a few photosynthetic bacteria. Considering its essential role in carbon dioxide fixation for virtually the entire biosphere, it is hardly surprising that rubisco is thought to be the most abundant protein on the planet Earth. About 10–25% of soluble leaf protein is rubisco (50% according to **Horton** *et al.*, 2006) and one estimate puts the total amount of rubisco on the Earth at 40 million tons, or about 7 kg for each living person.

Interestingly, rubisco's status as an abundant enzyme is due partly to the fact that it is not very efficient—the low turnover number $\sim 3_{s-1}$ (*i.e.*, fixes only about 3 molecules of CO_2 per second) means that large amounts of the enzyme are required to support CO_2 fixation! (see **Horton** *et al.*, 2006). Rubisco of plants, algae and cyanobacteria is composed of eight large (L) subunits and eight small (S) subunits. There are eight active sites located in the eight large subunits of rubisco.

2. Oxygenase Activity of Rubisco

The primary reaction catalyzed by rubisco acting as a **carboxylase** is the addition of carbon dioxide and water to ribulose-1, 5-biphosphate, forming two molecules of 3-phosphoglycerate (see Section 17.5). However, rubisco can also function as an **oxygenase**. Through this activity, rubisco catalyzes the addition of molecular oxygen, rather than carbon dioxide, to ribulose-1, 5-biphosphate:

$$\text{Ribulose-1, 5-biphosphate} \xrightarrow{O_2 \quad 2H^+} \text{3-phosphoglycerate + Phosphoglycolate}$$

The result of oxygenase activity of rubisco is a single three-carbon product—**3-phosphoglycerate** and one two-carbon product, **phosphoglycolate**. Because phosphoglycolate cannot be used in the next step of the Calvin cycle, it appears to be a wasteful diversion of material from carbon assimilation. Furthermore, alternative functions for rubisco's oxygenase activity have not been clearly demonstrated.

Indeed, not only does the production of phosphoglycolate appear to waste energy and carbon, the accumulation of phosphoglycolate may kill a plant by inhibiting the triose phosphate isomerase that maintains a balance of glyceraldehyde-3-phosphate and dihydroxyacetone phosphate in the chloroplast stroma.

Since rubisco has a much lower affinity for oxygen than for carbon dioxide and the oxygenase reaction proceeds much more slowly than the carboxylase reaction, so one should not expect the oxygenase reaction to be a serious problem. However, the low carbon dioxide and high oxygen concentration in the Earth's atmosphere—about 0.035 and 21% respectively—lead to significant competition between carbon dioxide and oxygen for binding to rubisco's catalytic site. Within a typical leaf or algal cell exposed to the atmosphere, the low CO_2:O_2 ratio leads to one oxygenation of ribulose-1, 5-biphosphate for every two or three carboxylation events, seriously reducing photosynthetic efficiency and generating large amounts of phosphoglycolate.

Photosynthetic porblems due to oxygenase activity of rubisco in desert plants. Plants living in hot arid environments under intense illumination are particularly affected by rubisco's oxygenase activity, since such conditions may further lower the CO_2:O_2 ratio in the chloroplast stroma. Although the solubilities of both carbon dioxide and oxygen decrease rapidly, thereby lowering the CO_2:O_2 ratio in solution. Another problem occurs when plants respond to drought by closing the stomata during the day to reduce water loss. When the stomata are closed, carbon dioxide cannot enter the leaf. Without the steady supply of carbon dioxide for assimilation, the concentration of carbon dioxide in leaf cells may decline. Moreover, water photolysis continues to generate oxygen, which accumulates because it cannot diffuse out of the leaf when the stomata are closed. Intense sunlight aggravates this problem by increasing the rate of water photolysis, which depends on the absorption of light that drives noncyclic electron flow and photoreduction. Why then does rubisco have this detrimental oxygenase activity?

According to one theory, the oxygenase activity is an evolutionary relic from a time when oxygen did not make up a large part of the Earth's atmosphere, and it cannot be eliminated without seriously compromising the carboxylase function. Not even natural selection appears to be up to task of altering this enzyme. Instead, phototrops that depend on rubisco have developed three alternative strategies for coping with the enzyme's apparently wasteful oxygenase activity:

A. Photorepiratory glycolate pathway;

B. C_4 photosynthesis; and

C. Crassulacean metabolism.

A. Photorespiratory glycolate pathway. In all photosynthetic plant cells, phosphoglycolate generated by rubisco's oxygenase activity is channeled into the **glycolate pathway**. This is regarded as a **salvage pathway** since it disposes of phosphoglycolate and returns about 75% of the reduced carbon present in phosphoglycolate to the Calvin cycle as 3-phosphoglycerate. Because the glycolate pathway is characterized by light-dependent uptake of oxygen and evolution of carbon dioxide, it is also referred to as **photorespiration**.

Several steps of the glycolate pathway are localized in a specific type of peroxisome called a **leaf peroxisome**. Typically, in photosynthetic plant cells such as mesophyll cell one can notice the close proximity of the leaf peroxisome to a chloroplast and a mitochondria (Fig. 17.19). Most likely, it reflects the participation of all three organelles in the glycolate pathway.

In the chloroplast, the phosphoglycolate generated by rubisco is rapidly dephosphorylated by a phosphatase in the stroma (reaction GP-1). The product is **glycolate**, which diffuses to a nearby leaf peroxisome, where an oxidase converts it to glyoxylate (reaction GP-2). Presumably, the close juxtaposition of chloroplasts and peroxisomes in photosynthetic cells-contributes to the efficient transfer of metabolites from one organelle to the other. The oxidation of glycolate is accompanied by the uptake of oxygen and the generation of *hydrogen peroxide*, which is immediately degraded to oxygen and water by catalase (reaction GP-C). During the next reaction in the peroxisome, an aminotransferase catalyzes the transfer of an amino group from glutamate to glyoxylate, forming **glycine** amino acid (reaction GP-3).

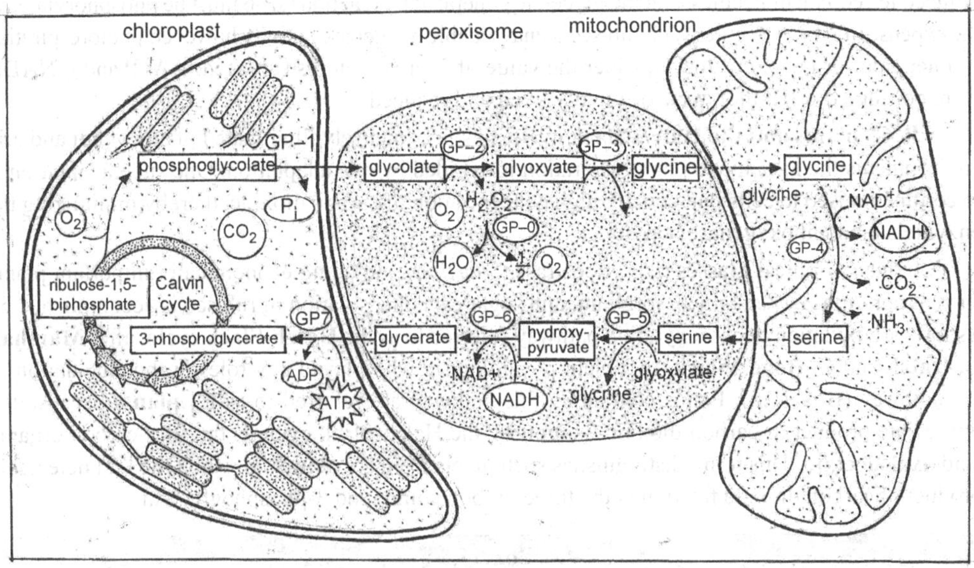

Fig. 17.19. The Glycolate pathway. Glycolate arises as a result of the oxygenase activity of rubisco. The immediate product is phosphoglycolate, which is converted to free glycolate by a phosphatase localized in the chloroplast membrane (reaction GP-1). Free glycolate diffuses out of the chloroplast stroma and is metabolized into glycerate by a five-step pathway (GP-2 through GP-6) that occurs partially in the peroxisome and partially in the mitochondrion. Glycerate then diffuses into the chloroplast and is phosphorylated to form 3-phosphoglycerate (reaction GP-7), which enters the Calvin cycle. The oxygen uptake and carbon dioxide evolution characteristic of photorespiration occur in the peroxisome (reaction GP-2) and mitochondrion (reaction GP-4), respectively (after Baker et al., 2006).

Glycine diffuses from the leaf peroxisome to a mitochondrion, where two enzyme activities working in series—a *decarboxylase* and an *hydroxylmethyl transferase*—convert glycine molecule to a single **serine** (amino acid), concomitant with the generation of NADH and the release of carbon dioxide and ammonia (reaction GP-4). Rubisco's oxygenase activity therefore leads not only to a loss of nitrogen. To prevent depletion of nitrogen reserves, the ammonia must be reassimilated at the expense of ATP and reductant.

The amino acid serine diffuses back to the peroxisome, where another aminotransferase removes the amino group, generating **hydroxypyruvate** (reaction GP-5). A reductase, using NADH as an electron donor, then reduces hydroxypyruvate to **glycerate** (reaction GP-6). Finally, glycerate diffuses to the chloroplast, where it is phosphorylated by glycerate kinase to generate **3-phosphoglycerate** (reaction GP-7), a key intermediate of the Calvin cycle.

Advantage of photorespiration. What is the advantage of this long salvage pathway, winding through several organelles? Three out of every four carbon atoms that exit the Calvin cycle as part of phosphoglycolate are recovered as 3-phosphoglycerate. Without this pathway, not only would phosphoglycolate accumulate to toxic levels, but also triose phosphates, essential for the regeneration of ribulose-1, 5-biphosphate and the continuation of the Calvin cycle, would be depleted. In terms of energy and reduced carbon, however, phosphoglycolate metabolism is expensive. For every three carbon salvaged, an ammonia molecule must be reassimilated at the expense of one ATP and two reduced ferredoxin molecules, and the glycerate generated by reaction GP-6 must be phosphorylated at the expense of ATP molecule. With rubisco's unavoidable oxygenase activity however, photorespiration is a net gain for the plant. Just consider the value of 3 carbon atoms salvaged: 9 ATP and 6 NADH were consumed when they were originally fixed and reduced.

B. C_4 plants and the Hatch-Slack cycle. In some intensely illuminated plants of hot and arid environments the loss due to photorespiration is avoided by one adaptation. One such adaptation is to confine rubisco to cells that contain high concentration of carbon dioxide, thereby minimizing the enzyme's inherent oxygenase activity.

In many tropical grasses (*e.g.*, rice grass, *Spartina*), including economically important plants such as corn/maize (*Zea mays*), sorghum, and sugar cane the isolation of rubisco is accomplished by a short carboxylation/decarboxylation pathway referred to as the **Hatch-Slack cycle**, after **Marshall D. Hatch** and **C. Roger Slack**, two plant physiologists who played key roles in the elucidation of the pathway (Box 17.5). Plants containing this pathway are referred to as **C_4 plants** because the immediate product of carbon dioxide fixation by the Hatch-Slack cycle is the four-carbon organic acid **oxaloacetate**. This term, distinguishes such plants from **C_3 plants**, in which the first detectable product of carbon dioxide fixation is the three-carbon compound 3-phosphoglycerate.

Box 17.5

The Russian scientist **I.A. Tarchevseki** and **Y.S. Karpilov** (1963) questioned the universal operation of the Calvin-Benson cycle in plants. In 1963 **Kortschask**, **Hart** and **Burr** reported that malate and aspartate were the major products formed in sugarcane leaves. This observation was confirmed by **Hatch** and **Slack** (1966). They showed that during short periods of photosynthesis in sugarcane leaves, four-carbon compounds such as **oxaloacetate**, **malate** and **aspartate** were formed. The pathway in which these compounds are formed has been called the **Hatch-Slack-Karschak** (**HSK**) **pathway**. Plants in which this cycle takes place are called C_4 plants.

Structural differences betwen leaves of C_3 and C_4 plants. The C_4 plants, unlike C_3 plants, have in their leaves two distinct types of photosynthetic cells—**mesophyll cells** and **bundle sheath cells**—that differ in their enzyme composition and hence their metabolic activities (Fig. 17.20). The first steps of carbon dioxide fixation within a C_4 plants are accomplished by the Hatch-Slack cycle in mesophyll cells, which are exposed to the carbon dioxide and oxygen that enter the leaf through its stomata. The carbon dioxide that is fixed in mesophyll cells is subsequently released in bundle sheath cells, which are relatively isolated from the atmosphere. The entire Calvin cycle, including rubisco, is confined to chloroplasts in the bundle sheath cells. Because of the activity of the Hatch-Slack cycle, the carbon dioxide concentration in bundle sheath cells may be as much as ten times the level in the atmosphere, strongly favouring rubisco's carboxylase activity and minimizing its oxygenase activity.

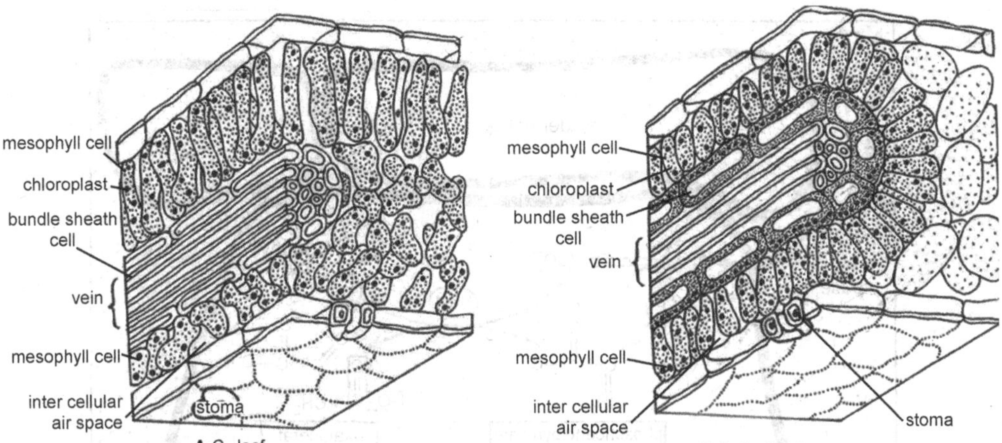

Fig. 17.20. Structure differences between leaves of C_3 and C_4 plants. A—In C_3 plants, the Calvin cycle occurs in mesophyll cells. B—In C_4 plants, the Calvin cycle is confined to bundle sheath cells, which are relatively isolated from atmospheric carbon dioxide and oxygen. C_4 plants utilize the Hatch-Slack cycle in concentrating it in bundle sheath cells. The bundle sheath cells surround the vascular bundles (veins) of the leaf, which carry carbohydrates to other parts of the plant. This concentric arrangements is called Kranz (German for "halo" or "wreath") anatomy and is essential to the photosynthetic efficiency of C_4 plants (after Baker et al., 2006).

Mechanism of Hatch and Slack cycle. The Hatch-Slack cycle begins with the carboxylation of **phosphoenolpyruvate (PEP)** to form **oxaloacetate** (reaction HS-1; Fig. 17.21). Carboxylation is catalyzed by a specific cytosolic form of PEP carboxylase, which is particularly abundant in mesophyll cells of C_4 plants. Not only does this carboxylase lacks rubisco's oxygenase activity, it is an excellent scavenger for carbon dioxide. In other words, it has a high affinity (a low *k*m) for its substrate **bicarbonate (HCO_3^-)**, and operates very efficiently even when the concentration of bicarbonate is quite low. (Bicarbonate forms when carbon dioxide dissolves in water; its concentration therefore reflect the availability of carbon dioxide gas).

In one version of the Hatch-Slack pathway, the oxaloacetate generated by PEP carboxylase is rapidly converted to malate by an NADPH-dependent malate dehydrogenase (reaction HS-2 in Fig. 17.21). Malate is a stable four-carbon acid that carries carbon from mesophyll cells to chloroplasts of bundle sheath cells, where decarboxylation by NAD^+-malic enzyme releases CO_2 (reaction HS-3). The liberated carbon dioxide is then refixed and reduced by Calvin cycle. (In other versions of Hatch-Slack pathway, decarboxylation occurs either in the mitochondrion, via NAD^+-malic enzyme, or in the cytoplasm, via PEP carboxylase. In all three versions, however, mesophyll cells contain the key enzymes PEP carboxylase and pyruvate, phosphate dikinase, which are important control points for light dependent regulation.) Because decarboxylation of malate is accompanied by the generation of NADPH, the Hatch-Slack cycle also conveys reducing power from mesophyll to bundle sheath cells. This might limit the demand for noncyclic flow from water to $NADP^+$ in the bundle sheath cells, thereby minimizing the formation of oxygen by PS II complexes and further favoring rubisco's carboxylase activity.

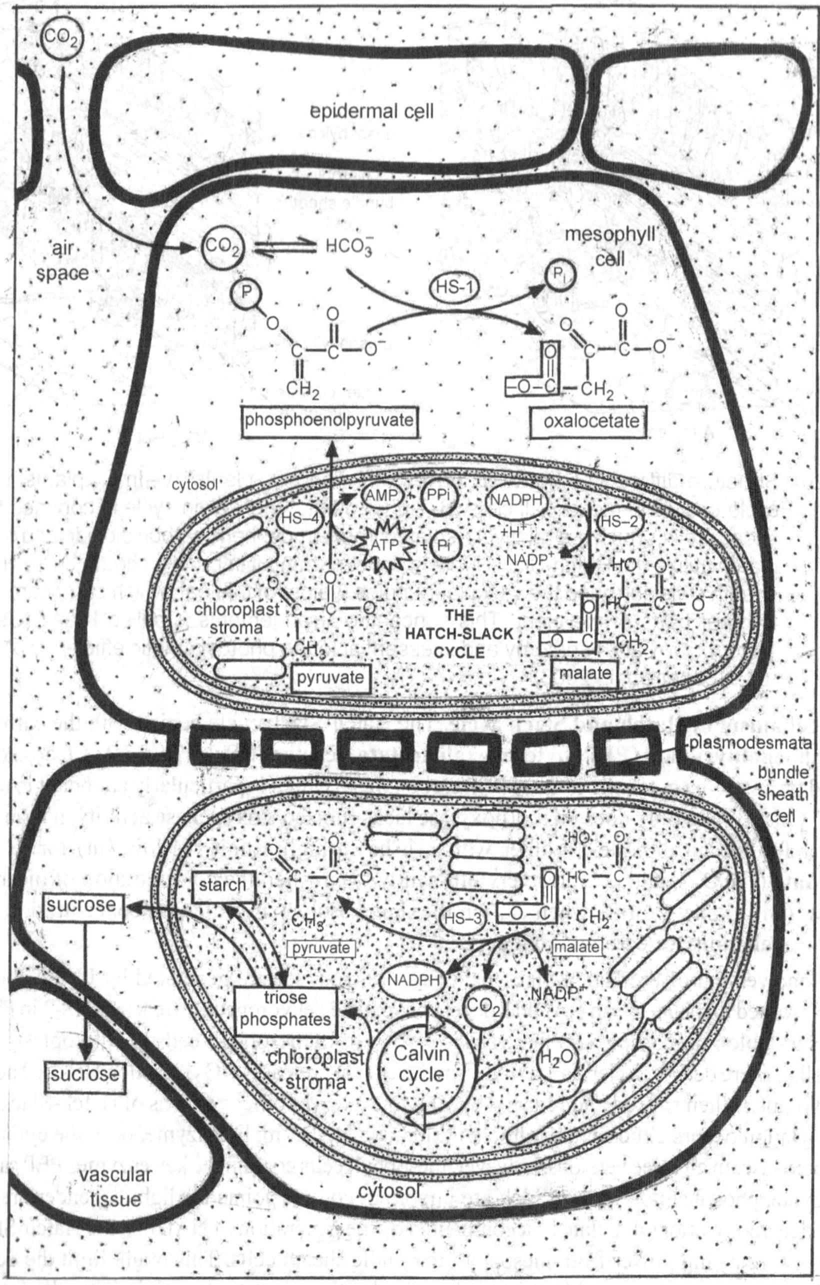

Fig. 17.21. Localization of the Hatch-Slack cycle within different cells of a C_4 leaf. Carbon dioxide fixation in C_4 plants initially occurs by the Hatch-Slack cycle within mesophyll cells, forming oxaloacetate. The malate formed by reduction of oxaloacetate is then passed inward to the bundle sheath cells, where it is decarboxylated. The carbon dioxide is refixed by the Calvin cycle, eventually yielding sucrose, which passes into the adjacent vascular tissue for transport to other parts of the plant. The enzymes that catalyze the Hatch-Slack reactions are listed in the box. The enzyme pyruvate, phosphate dikinase, which catalyzes phosphrylation of pyruvate is unique to Hatch-Slack cycle (HS–1 = Phosphoenolpyruvate (PEP) carboxylase, HS–2 = NADPH-dependent malate dehydrogenase, HS-3 = $NADP^+$ malic enzyme; HS-4 = Pyruvate, phosphate dikinase (after Baker et al., 2006).

The pyruvate generated by decarboxylation of malate diffuses into a mesophyll cell, where it is phosphorylated at the expense of ATP to regenerate PEP (reaction HS-4), the original carbon dioxide acceptor of the Hatch-Slack cycle. Thus, overall process is cyclic, and the net result is a feeder system that captures carbon dioxide in mesophyll cells and passes it to the Calvin cycle in bundle sheath cells. The Hatch-Slack cycle is not a substitute for the Calvin cycle, it is simply a preliminary carboxylation/decarboxylation sequence that concentrates CO_2 in the bundle sheath cells.

Because ATP is hydrolyzed to AMP in reaction HS-4 (Fig. 17.21), the actual cost of moving carbon from mesophyll to bundle sheath cells is equivalent to two ATP molecules per carbon dioxide molecule. Carbon assimilation within a C_4 plant therefore consumes a total of five ATP molecules per carbon atom, rather than the three required in C_3 plants. In an environment that enhances rubisco's oxygenase activity, however, the energy required to prevent the formation of phosphoglycolate may be far less than the energy that would otherwise lost through photorespiration.

When temperature exceeds about 30°C, the photosynthetic efficiency of a C_4 plant exposed to intense sunlight may be twice that of a C_3 plant. While the higher efficiency is largely due to reduced photorespiration in the C_4 plants and enhanced photorespiration in the C_3 plants, other factors are also important. In a C_3 plant, photosynthesis is often limited by the low atmosphereric concentration of carbon dioxide, not by the availability of sunlight. In a C_4 plant, however, the the Hatch-Slack cycle actively concentrate carbon dioxide in bundle sheath cells, where the Calvin cycle is localized, enabling the plant to take advantage of higher levels of illumination.

Water conservation by C_4 plants. Enrichment of carbon dioxide in the vicinity of rubisco by the Hatch-Slack cycle grants an additional advantage on C_4 plants. Because PEP carboxylase is an efficient scavenger of CO_2, gas exchange through the stomata of C_4 plants can be significantly reduced to conserve water without adversely affecting photosynthetic efficiency. As a result C_4 plants can be significantly reduced to conserve water without adversely affecting photosynthetic efficiency. As a result C_4 plants are able to assimilate over twice as much carbon as for each unit of water transpired. This adaptation makes C_4 plants suitable for regions of periodic drought, such as tropical savannas.

Nonefficient and efficient plants. The C_3 plants have been called "**nonefficient plants**" by some authors because they cannot grow fast at high temperatures and high light intensities. They carry out carbon fixation through the Calvin-Benson 3-carbon pathway. The C_4 plants are called "**efficient plants**" because they grow fast at high temperatures and light intensities. The predominant pathway for carbon fixation in C_4 plants in HSK pathway (or Hatch-Slack cycle) although the Calvin-Benson cycle also functions alongside. C_4 plants are mostly tropical plants growing in regions of high light intensities and temperature. They include monocots such as maize/corn, sorghum and sugarcane (*Note.* According to **Baker** *et al.*, 2006, some monocot plants producing cereal grains such as wheat and rice are C_3 plants). Examples of C_4 plants are also found among **dicots**, *e.g.*, members of families Amaranthaceae and Chaenopodiaceae have the C_4 dicorboxylic acid pathway of photosynthesis and carbon dioxide fixation (**Laetsch** 1968).

Chloroplast dimorphism in C_4 plants. In tropical grasses (C_4 plants), the chloroplasts of bundle sheath (see Kranz anatomy) cells differ from those in mesophyll cell. That monocots have two types of chloroplasts was pointed out as early as in 1904 by the German botanist **G. Haberlandt**. The light microscopy studies of **Rhoades** and **Carvalho** (1944) demonstrated **chloroplast diomorphism** in maize and sorghum. Later electron microscopy studies on maize (**Hodge**, **McLean** and **Mercer** 1955), many tropical grasses (**Johanson**, 1964) and sugarcane (**Laetsch**, **Stetler** and **Vlitos** 1955) have confirmed this dimorphism.

The bundle sheath cells around vascular tissue have large chloroplasts that are without grana and have many starch grains. The mesophyll cells contain smaller chloroplasts which possess well developed grana and have few starch grains. It has been found that bundle sheath cells apparently use the 3-carbon Calvin-Benson pathway while the mesophyll cells predominantly use the 4-carbon Hatch-Slack pathway.

C. Plants with Crassulacean Acid Metabolism (CAM). Finally, let us consider the third strategy used by some plants to cope with the wasteful oxygenase activity of rubisco. Certain plant species that live in deserts, salt marshes, and other environments where access to water is severly limited contain a preliminary carbon dioxide fixation pathway closely related to the Hatch-Slack pathway. The sequence of reactions is similar, but these plants segregate the carboxylation and decarboxylation reaction by **time** rather than by **space**. Because this pathway was first recognized in the family of succulent plants known as crassulacea, it is called **crassulacean acid metabolism (CAM)**, and plants that take advantage of CAM photosynthesis are called **CAM plants**. CAM photosynthesis has been found in about 4% of plant species investigated, including many succulents, cacti and orchids.

CAM plants, unlike most C_3 and C_4 plants, generally open their stomata only at night, when the atmosphere is relatively cool and moist. As carbon dioxide diffuses into mesophyll cells, it is assimilated by the first two steps of a pathway similar to the Hatch-Slack cycle, and accumulates as **malate**. Instead of being exported from mesophyll cells, however, the malate is stored in large vacuoles, which become very acidic. The process of moving malate into vacuoles consume ATP but is necessary to protect cytosolic enzymes from a large drop in pH at night.

During the day, CAM plants close their stomata to conserve water. The malate then diffuses from vacuoles to the cytosol, where the Hatch-Slack cycle continues, carbon dioxide released by decarboxylation of malate diffuses into the chloroplast stroma, where it is refixed and reduced by the Calvin cycle. The high carbon dioxide and low oxygen concentrations established when light is available for generating ATP and NADPH strongly favour rubisco's carboxylase activity and minimize the loss of carbon through phosphorylation. Both carboxylation of PEP and decarboxylation of malate tend to occur in the same compartment. Because of this, the activity of PEP carboxylase in CAM plants must be strictly inhibited during the day to prevent a futile cycle from developing.

With their remarkable ability to *conserve water*, CAM plants may assimilate over 25 times as much carbon as a C_3 plant for each unit of water transpired. Moreover, some CAM plants display a process called **CAM idling**, whereby the plant keeps its stomata closed night and day. Carbon dioxide is simply recycled between photosynthesis and respiration, with virtually no loss of water. Such plants will not display a net gain of carbohydrates. This ability, however, may enable them to survive droughts lasting up to several months.

The CAM metabolism shows various modifications. Well watered *Agave americana* shows normal daytime photosynthesis along with some CO_2 fixation at night. In *Agave deserti*, however, dark carboxylation stops and is replaced by normal C_3 daytime photosynthesis.

17.7. CHLOROPLAST AS SEMIAUTONOMOUS ORGANELLE

Like the mitochondria, the chloroplasts have their own DNA, RNAs and protein synthetic machinery and are semiautonomous in nature.

1. DNA of chloroplast. The chloroplasts of the algae and higher plants are found to contain DNA molecules. First of all **Ris** and **Plant** (1962) have reported DNA molecule in the chloroplast of the *Chlamydomonas*. Later on DNA molecule has been reported from the chloroplasts of other algae and higher plants. In general, chloroplasts have a double helical DNA circle with an average length of 45 μm (about 135,000 base pairs). The replication of chloroplast DNA has been followed with ^3H– thymidine. Maps of the location of genes (genetic maps) have been made in several chloroplast DNAs by the help of restriction enzymes. The gene for the large subunit of carboxydismutase enzyme has been fully sequenced and is found to contain 1425 nucleotides.

2. Ribosomes of chloroplasts. The chloroplasts contain the ribosomes which are smaller than the cytoplasmic ribosomes. The ribosomes of the chloroplast are of 70S type and resemble with the bacterial ribosomes. The ribosomes of the chloroplasts consist of two ribosomal RNAs, 23S rRNA and 16S rRNA. **Lyttleton** (1962) has separated polyribosomes or polysomes from the chloroplast. The chloroplasts also contain aminoacyl-tRNAs, aminoacyl-tRNA synthetases, methionyl-tRNA.

3. Protein synthesis. The DNA of chloroplast codes for chloroplast mRNA, rRNA, tRNA, and ribosomal proteins. It also codes for certain structural proteins of thylakoid membranes. The synthesis of other chloroplast components as chlorophyll, carotenoids, lipids and photosynthetic and starch synthesizing enzymes, is controlled by nuclear genes. The 70S ribosomes of *Euglena* chloroplast are found to require Mg^{++} for their stability and also have a requirement for N-formylmethionyl-tRNA in chain initiation protein synthesis like the bacteria. The protein synthetic mechanism of chloroplasts is inhibited by chloramphenicol like that of mitochondria and bacteria.

The mode of synthesis of proteins of chloroplasts indicates towards their **semiautonomous** or **symbiotic nature**. For example, of the 30 known thylakoid polypeptides that function in photosynthesis, so far 9 have been demonstrated to be synthesized on chloroplastic ribosomes and 9 are coded by nuclear genes and synthesized on cytoplasmic ribosomes. Synthesis of carboxydismutase (C Dase) presents a good case of cooperative action of two genetic systems (*i.e.*, chloroplastic and nuclear genetic systems). C Dase comprises 16 subunits: 8 subunits of high molecular weight (55,000 daltons) and 8 subunits of much smaller molecular weight (14,000 daltons). The large subunit is coded by genes present in chloroplastic DNA, while the small subunit is produced by nuclear genes. The small subunit (called P20) is synthesized as a precursor weighing 20,000 daltons on free ribosomes; it then enters **post-translationally** into the stroma to be cleaved to attain its final size. It is postulated that the chloroplastic envelope has receptor sites that recognize the proteins that are to be incorporated into the organelle. The extra sequence (acting as the **signal**) that is present in P20 is composed of acidic amino acids, in contrast to the hydrophobic ones in the signal sequence of secretory proteins. After entering the chloroplast the signal sequences are removed by a protease enzyme, which is present in the envelope of chloroplast, and the small subunit of C Dase is released into the stroma (**Ellis**, 1981). Thus, chloroplast proteins may be synthesized by three avenues:

1. by an exclusive chloroplastic mechanism,
2. by a mechanism involving nuclear genes and chloroplastic ribosomes, and
3. by nuclear genes and cytoplasmic ribosomes.

Protein transport into chloroplasts resembles transport into mitochondria in many respects: both occur post-translationally, both require energy, and both utilize hydrophilic amino-terminal signal peptides that are removed after use. However, there is at least one important difference that while mitochondria exploit the electrochemical gradient across their inner membrane to help drive the transport, chloroplasts (which have an electrochemical gradient across their thylakoid but not their inner membrane) appear to employ only ATP hydrolysis to import across their double-membrane outer envelope.

Translocation of proteins into the thylakoid space of chloroplasts requires two signal peptides and two translocation events. The precursor polypeptide contains an amino-terminal chloroplast signal peptide followed immediately by a thylakoid signal peptide. The **chloroplast signal peptide** initiates translocation into the stroma through a membrane contact site by a mechanism similar to that used for translocation into mitochondrial matrix. The signal peptide is then cleaved off, unmasking the thylakoid signal peptide, which initiates translocation across the thylakoid membrane (Fig. 17.22).

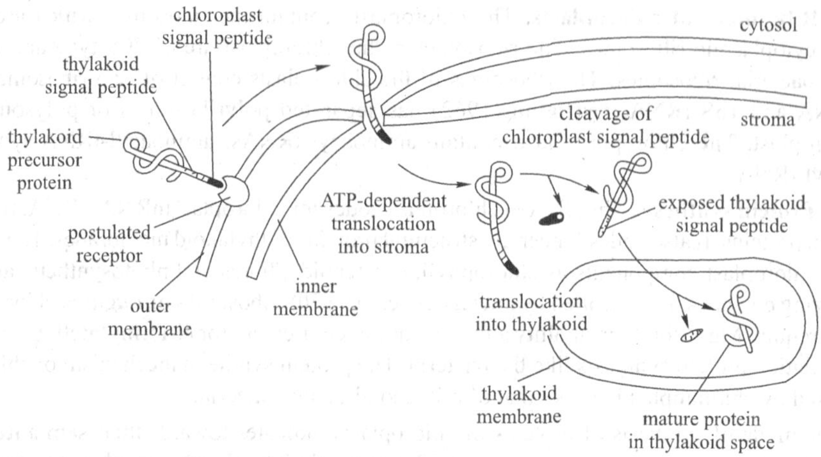

Fig. 17.22. Mode of translocation of proteins into stroma and thylakoid of chloroplast by the help of signal peptides.

17.8. BIOGENESIS OF CHLOROPLAST

The chloroplasts never originated *de novo*. Since the classic work of **Schimper** and **Meyer** (1883) it has been accepted that chloroplasts multiply by fission, a process that implies growth of the daughter organelles. This is easily observed in the alga *Nitella*, which contain a single huge chloroplast. In *Nitella* a division cycle of 18 hours has been cinematographically recorded for the chloroplast.

During the development of the chloroplast, the first structure to appear is the so-called **proplastid**, which has a double membrane. Development of proplastid into chloroplast takes place in the following steps:

1. In the presence of light, the inner membrane grows and gives off vesicles into the matrix that are transformed into discs (Fig. 17.23). These intrachloroplastic membranes are the thylakoids which, in certain regions, pile closely to form the grana. In the mature chloroplast the thylakoids are no longer connected to the inner membrane, but the grana remain united by inter-granal thylakoids.
2. In the absence of light, a reverse sequence of changes takes place. This is the process of **etiolation**, in which the leaves lose their green pigment and the chloroplast membranes become disorganized. The chloroplasts are transformed into **etioplasts**, in which there is a paracrystalline arrangement of tubules forming the so-called **prolamellar body**. Attached to these bodies are young thylakoid membranes that lack photosynthetic activity.

The regular crystal lattice of two prolamellar bodies surrounded by young thylakoid membranes is observed by **Osumi** *et al.*, (1984). If etiolated plants are re-exposed to light, thylakoids are reformed and the prolamellar material is used for assembly.

The symbiotic origin of the chloroplast. In certain characteristics, the chloroplasts are comparable with that of a semiautonomous or symbiotic organism living within the plant cells. They divide, grow and differentiate; they contain circular DNA, ribosomal RNA, messenger RNA and are able to conduct protein synthesis. By visualizing these similarities between chloroplast and micro-organism; it has been suggested that chloroplast might have resulted from a symbiotic relationship between an autotrophic microorganism, one which is able to transform radiant energy from sunlight and heterotrophic host cell. The symbiotic origin of the chloroplast appears very justified but **Kirk** (1966) has shown that certain important enzymes which are necessary for the development of the chlorophyll and for the photosynthetic mechanism are synthesized according to the codes of the nuclear DNA.

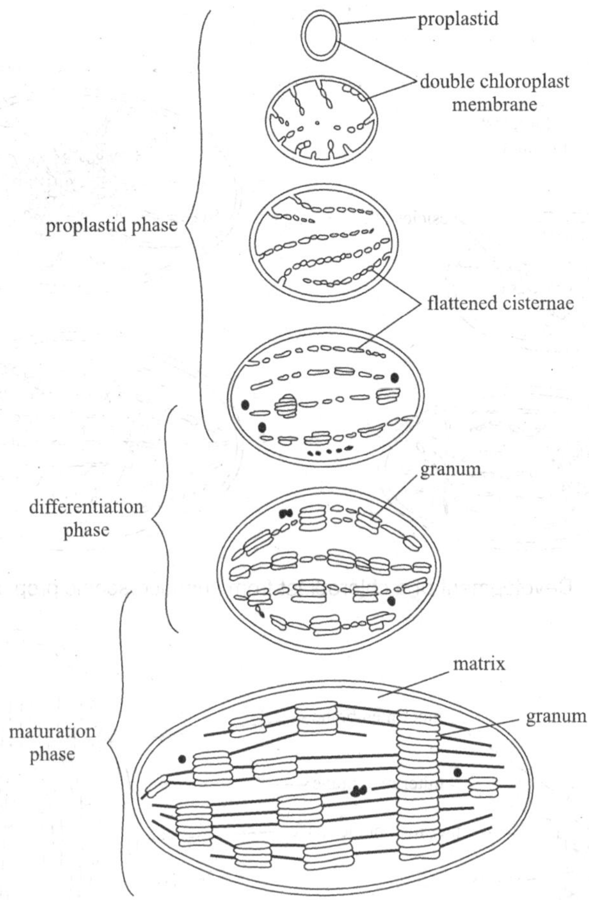

Fig. 17.23. Development of a chloroplast from a submicroscopic proplastid in the presence of light.

17.9. COMPARISON OF CHLOROPLASTS AND MITOCHONDRIA

Chloroplasts carry out their energy inter-conversions by chemiosmotic mechanisms in much the same way that mitochondria do and they are organized on the same principles. Thus, each chloroplast contains three distinct membranes which define three separate internal compartments—the intermembrane space, the stroma and the thylakoid space. The thylakoid membrane contains all the energy generating systems of chloroplasts.

Like the mitochondria, chloroplasts have a highly permeable outer membrane; a much less permeable inner membrane, in which special carrier or transport proteins are embedded; and a narrow intermembrane space in between. The inner membrane surrounds a large space called the **stroma**, which is analogous to the mitochondrial matrix and contains various enzymes, ribosomes, RNAs and DNA.

However, there is an important difference between the two: the inner membrane of the chloroplast is not folded into cristae and does not contain an electron-transport chain. Instead, the photosynthetic light-absorbing system, the electorn-transport chain and an ATP synthetase are all contained in a third distinct membrane that forms a set of flattened disc-like sacs, the **thylakoids**

(Fig. 17.25).

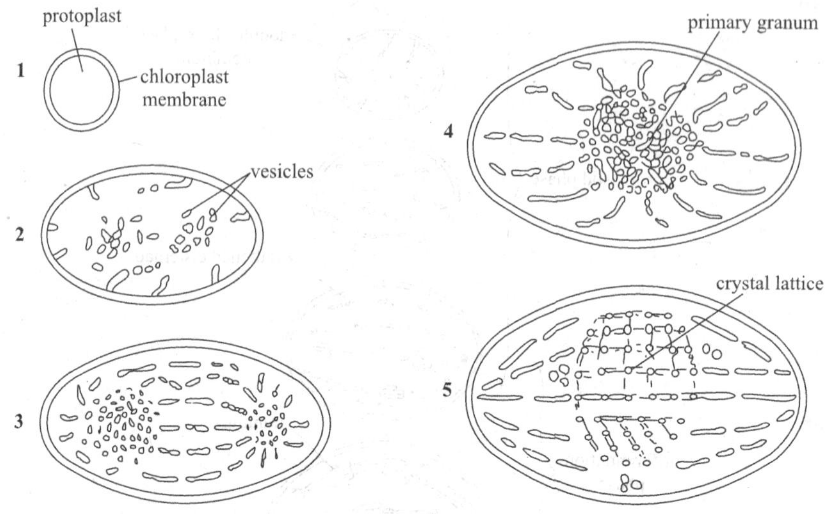

Fig. 17.24. Development of a chloroplast from submicroscopic proplastid in dark.

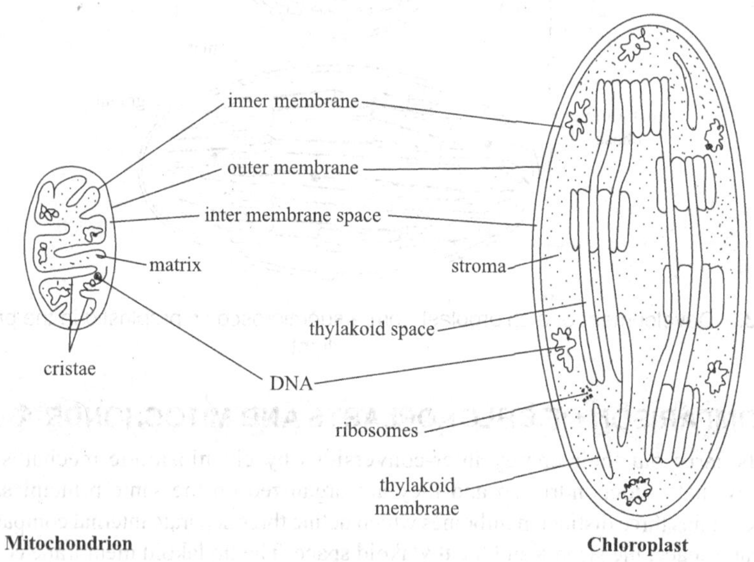

Fig. 17.25. Comparison of a mitochondrion and a chloroplast.

In a general way, one might view the chloroplast as a greatly enlarged mitochondria in which the cristae are converted into a series of interconnected submitochondrial particles in the matrix space. The knobbed ends of the chloroplast ATP synthetases ($F_0 - F_1$ coupling factors), where ATP is made, protrude from the thylakoid membrane into the stroma, just as they protrude into the matrix from the membrane of each mitochondrial crista.

QUESTIONS

Long Answer Type Questions

1. Define the chloroplast. Describe its structure, chemical composition and main functions.
2. Why chloroplast is considered as a semiautonomous organelle? Discuss it.
3. What are stroma and grana? Discuss their role in the process of photosynthesis.
4. Describe the process by which water is split during photolysis? How many photons must be absorbed by PS II for this to occur?
5. Describe the sequence of events following the absorption of a photon of light by the reaction center pigment of photosystem II. Describe the comparable events in photosystem I. How are the photosystems linked to one another?
6. How does the proton gradient leads to the formation of ATP in light reaction of chloroplasts?
7. Describe the basic plan of the Calvin cycle, indicating the reactions that require energy input. Why is it described as a cycle?
8. Describe the major structural and biochemical differences between C_3 and C_4 plants.

Short Answer Questions

1. Describe the origin or biogenesis of chloroplasts.
2. Compare the chloroplasts with mitochiondria.
3. Differentiate between leucoplasts and chromoplasts.
4. Compare absorption spectrum of Calvin cycle.
5. Write short notes on the following:
   (*a*) Chloroplasts
   (*b*) Photon
   (*c*) Z scheme.

Very Short Answer Questions

1. State the significance of thylakoids of chloroplasts.
2. Name the type of chlorophyll found in all green plants.
3. Name the site of occurrence of photochemical or light reaction.
4. What is the use of sunlight in light reaction?
5. Where does dark or synthetic reaction occur in chloroplasts?
6. What is antenna complex?
7. Define the chlorophyll.
8. What is Calvin cycle?
9. What is the biological significance of carbon atom?
10. Define the chloroplast.
11. What are eventual product of Calvin cycle?
12. Who described the ultrastructure of grana of chloroplast?

Multiple Choice Questions

1. Carbon dioxide is fixed in
   (*a*) light reaction
   (*b*) dark reaction
   (*c*) aerobic respiration
   (*d*) anaerobic respiration
2. Calvin cycle is
   (*a*) inhibited by light
   (*b*) dependent upon light
   (*c*) independent of light
   (*d*) supported by light
3. Which of the following occurs in dark reaction of photosynthesis?
   (*a*) formation of ATP
   (*b*) release of O_2
   (*c*) release of H_2
   (*d*) synthesis of PGAL
4. Photorespiration in C_4 plants starts from
   (*a*) phosphoglycerate
   (*b*) phosphoglycolate
   (*c*) glycerate
   (*d*) glycine
5. Tropical plants such as sugarcane show high efficiency of CO_2 fixation because of

(a) Calvin cycle
(b) Hatch-Slack cycle
(c) TCA-cycle
(d) Cyclic photophospharylation

6. In sugarcane plant $^{14}CO_2$ is fixed in the malic acid, in which the enzyme that fixes CO_2 is
(a) fructose phosphatase
(b) ribulose phosphate
(c) phosphoenol pyruvic acid carboxylase
(d) ribulose biphosphate carboxylase

7. Which is needed for photolysis of H_2O
(a) Mn^{2+} (b) Mg^{2+}
(c) Zn^{2+} (d) Ca^{2+}

8. The main site for the dark reaction of the photosynthesis is
(a) stroma (b) grana
(c) ribosomes (d) mitochondria

9. A cell that lacks chloroplasts does not
(a) evolve CO_2
(b) liberate oxygen
(c) require water
(d) utilise carbohydrates

ANSWERS

Very Short Answer Questions

1. Thylakoids have light absorbing system, electron transport system and ATP synthase enzyme.
2. Chlorophyll *a*.
3. Thylakoids of the chloroplasts.
4. Radiant energy present in sunlight is used by chloroplasts to form ATP and to oxidize water molecules.
5. Stroma of chloroplasts.
6. Clusters of hundreds of chlorophyll molecules which trap sunlight energy is called **antenna complex**.
7. Chlorophyll is a light-absorbing molecule that donates photoenergized electrons to organic molecules, initiating photochemical events that lead to the generation of the NADPH and ATP required for the Calvin cycle; because of its absorption properties, chlorophyll gives plants their characteristic green colour.
8. It is a cyclic series of reactions used by photosynthetic organisms for the fixation of carbon dioxide and its reduction to form carbohydrates.
9. Carbon is most important atom in biological molecules, capable of forming upto four covalent bonds.
10. It is a double membrane-enclosed cytoplasmic organelle of plants and algae that contains chlorophyll and enzymes necessary to carry out photosynthesis.
12. K. Porter and S. Granick in 1947.

Multiple Choice Questions

1. (b) 2. (d) 3. (d) 4. (b) 5. (b) 6. (c) 7. (a)
8. (a) 9. (b)

18 Biosynthesis of Lipids and Polysaccharides

In certain chapters we have discussed the biosynthesis of DNA (*viz.*, DNA replication), biosynthesis of RNA (*viz.*, transcription) and biosynthesis of proteins, all of which are macromolecules and are required by all cells for growth, metabolism and reproduction. But besides these macromolecules, there are many micromolecules (*e.g.*, nucleotides, monosaccharides, amino acids, fatty acids, etc.) and macromolecules (*e.g.*, lipids, polysaccharides, etc.) which are required by all cells as building blocks, energy source substances, etc. Therefore, a brief discussion of biosynthesis of lipids and polysaccharides by cells is earnestly required.

18.1 NUTRITIONAL PATTERNS

Of all the processes in nature, **photosynthesis** is the most fundamental. Transferring the free energy of sunlight, the green plant cells combine carbon dioxide and water, both of which are the end products of metabolism in all organisms, and from them builds free energy of all living beings. However, photosynthesis represents only one aspect of the synthetic activity of a chlorophyll-containing plant cell. The cell can use the products of photosynthesis to synthesize the complex spectrum of small molecules (monosaccharides, oligosaccharides, fatty acids, amino acids, vitamins) and macromolecules (lipids, polysaccharides, proteins, nucleic acids) which it needs for life and reproduction. Colourless plant cells, animal cells, bacteria and viruses ultimately depend upon the products of the green plant cell for their organic supplies—organic carbon, nitrogen and some vitamins. With these and inorganic salts, they too can synthesize the small molecules and macromolecules which are required for maintaining life, and for growth and cell division.

In other words, green plant cells (**photoautotrophic cells**) represent the pattern of nutrition with **maximal synthetic activity**, inasmuch as they synthesize all their organic requirement from the low energy compounds (carbon dioxide, water and salts) at the expense of light energy, all other types of cells (**heterotrophic cells**) may be considered cells in which the capacity for synthetic activities unique to the photoautotrophic cells has been progressively lost, the loss being greatest in heterotrophic cells requiring the most complex media for growth.

There are different degrees of dependence of heterotrophic cells upon supply of organic nutrients. Some organisms (*e.g.*, wild-type *Escherichia coli*), require only organic carbon (usually glucose) and various salts to synthesize all their other organic requirements for life. Others show progressive loss of capacity to synthesize various common molecules needed for functioning or growth and must obtain these directly from their nutrients. In this respect, animal cells generally are much more dependent upon their nutrients than other cells. It does not mean that animal cells are incapable of synthetic activity, as witnesses the genome of such cells which codes for a tremendous number of proteins, but rather that here synthesis start from the level of complex organic molecules which the cells obtain in their nutrients, and from these they synthesize their characteristic complex structures.

18.2. BIOSYNTHESIS OF LIPIDS

The synthesis of lipids is an essential part of cellular metabolism since lipids are crucial components of cell membranes and storage products (*e.g.*, adipose tissue, plant seeds).

Site of lipid biosynthesis Fatty acid biosynthesis in plants, for example, takes place within plastids (*e.g.*, chloroplasts). The enzymes involved in fatty acid synthesis are acetyl-CoA carboxylase (ACCase) and fatty acid synthase (FAS). FAS refers to a complex of various individual enzymes including a acyl-carrier protein (ACP) (see **Buchanan** *et al.*, 2000). Fatty acid synthesis begins with the carboxylation of acetyl-CoA to malonyl-CoA by ACCase (Fig. 18.1). Experimental results have shown that within plastids, pyruvate dehydrogenase (enzyme) can directly produce, **acetyl-CoA** from pyruvate generated during glycolysis. The central role of acetyl-CoA in cellular metabolism has been depicted in Fig. 18.2.

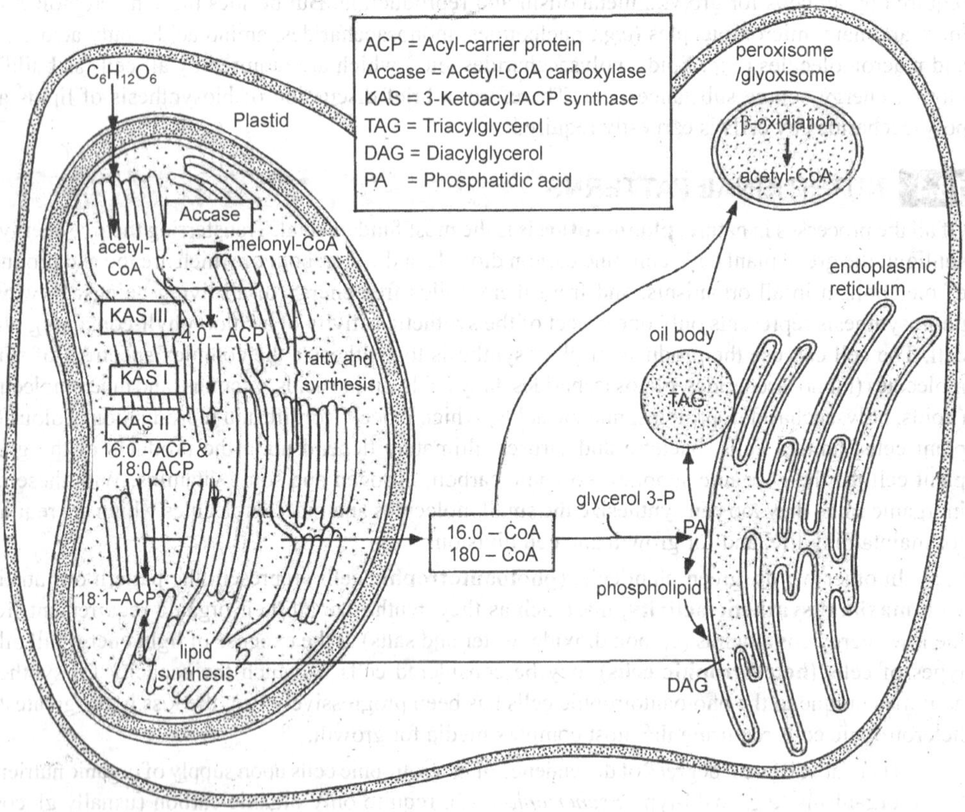

Fig. 18.1. Lipid synthesis and metabolism takes place in various organelles of a plant cell and in some cases involves movement of lipids from one cellular compartment to another. In the figure 16.0 stands for palmitic acid and 18.0 for stearic acid (after Buchanan *et al.*, 2000).

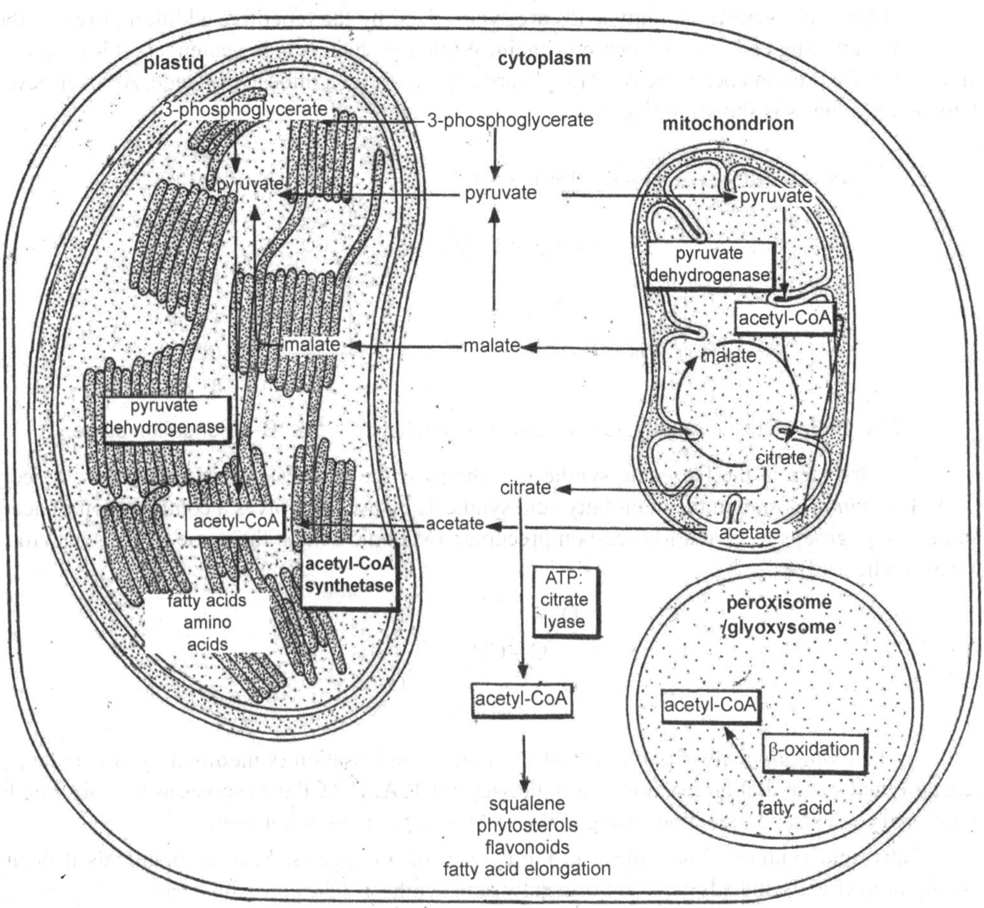

Fig. 18.2. The central role of acetyl-CoA in cellular metabolism. The major pathways involved in its production include glycolysis (via pyruvate hydrogenase and fatty acid oxidation). Acetyl-CoA is starting material for biosynthesis of fatty acids, several amino acids, flavonoids (via chalcone synthesis), sterols and many isoprenoid derivatives synthesized in the cytosol. During respiration, acetyl-CoA is the source of input into the citric acid cycle in the mitochondria. Despite this central role, acetyl-CoA is not believed to cross membranes and must be produced in the compartment in which it is synthesized (after Buchanan *et al.*, 2000).

In 1979, **Grattan Roughan** and **Roger Slack** first proposed that there are two distinct pathways for **membrane lipid biosynthesis** in higher plants and named these the prokaryotic pathway and the eukaryotic pathway. The **prokaryotic pathway** refers to the synthesis of lipids within the plastid. The **eukaryotic pathway** refers to the sequence of reactions involved in synthesis of lipids in ER, transfer of some lipids between the ER and the plastid, and further modification of the lipids within the plastid.

Sphingolipids synthesis is likely to take place in the ER. Similarly, elongation of fatty acids for wax synthesis and production of oxygenated fatty acids for cutin synthesis are also thought to take place in the ER.

Pathways of Biosynthesis of Lipids

The most important lipid synthesis pathways are 1-2. Fatty acid synthesis since they are required in triacylglycerols, 3. Cholesterol synthesis; 4. Eicosanoid synthesis and 5. Synthesis of spingolipids.

1. Fatty acid synthesis. Fatty acids are synthesized by the repetitive addition of two carbon units to the growing end of a hydrocarbon chain. While the chain is being extended, it is covalently attached to the protein coenzyme ACP (acyl carrier protein) by a thioester linkage. An overview of fatty acid synthesis is shown in Fig. 18.3.

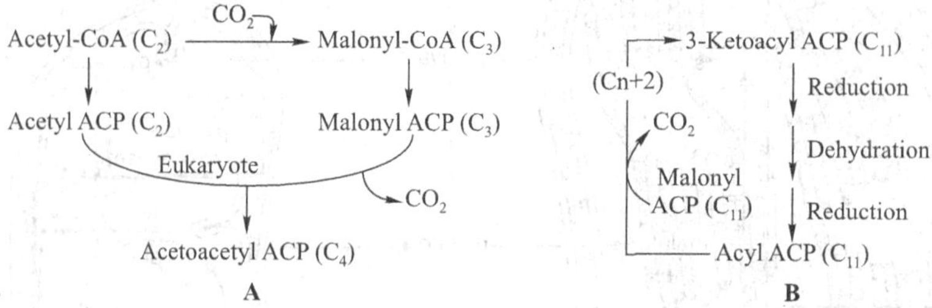

Fig. 18.3. Outline of fatty acid synthesis. A—Initiation stage; B—Elongation stage.

The first step in the fatty acid synthesis pathway is the synthesis of acetyl ACP from acetyl CoA. The *initiation step* in the main fatty acid synthesis pathway involves a condensation of acetyl and malonyl groups to give a four-carbon precursor. (Malonic acid is the name of the standard C_3 dicarboxylic acid).

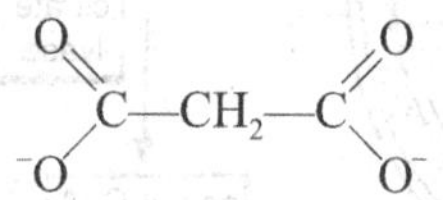

Malonate

In the *elongation step* the product of the initial condensation is modified by two reduction reactions and a dehydration reaction to produce acyl ACP. Acyl ACP then serves as the substrate for additional condensation reactions using malonyl ACP as the two-carbon donor.

Fatty acid synthesis takes place in the cytosol of all species. In adult mammals it occurs mainly in liver cells and adipocytes. Some fatty acid synthesis takes place in specialized cells such as mammary glands during lactation.

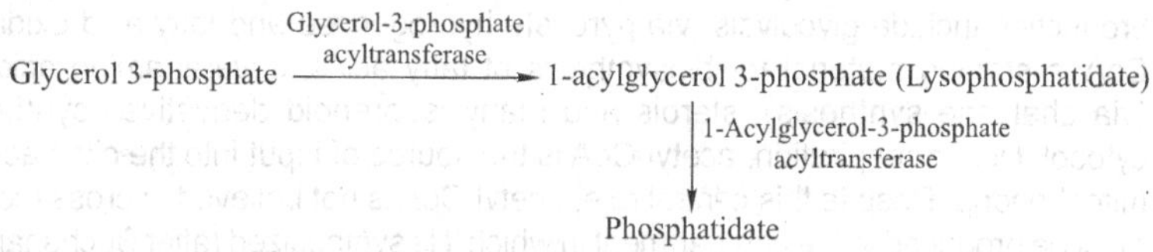

Fig. 18.4. Formation of phosphatidate. Glycerol-3-phosphate catalyzes esterification at C-1 of glycerol 3-phosphate. It has a preferance of saturated acyl chains. 1-acylglycerol-3-phosphate acyltransferase catalyzes esterification of C-2 and has a preference for unsaturated acyl chains.

2. Synthesis of triglycerols and glycerophospholipids. Most fatty acids are found in esterified forms as triacylglycerols or glycerophospholipids. Phosphatidate is an intermediate in the synthesis of triacylglycerols and glycerophospholipids. It is formed by transferring the acyl groups from fatty acyl CoA molecules to the C-1 and C-2 positions of glycerol 3-phosphate (Fig. 18.5). The reactions are catalyzed by two separate acyltransferases that use fatty acyl CoA molecules as the acyl-group donors (*e.g.*, Glycerol-3-phosphate acyltransferase and 1-acylglycerol-3-phosphate acyltransferase).

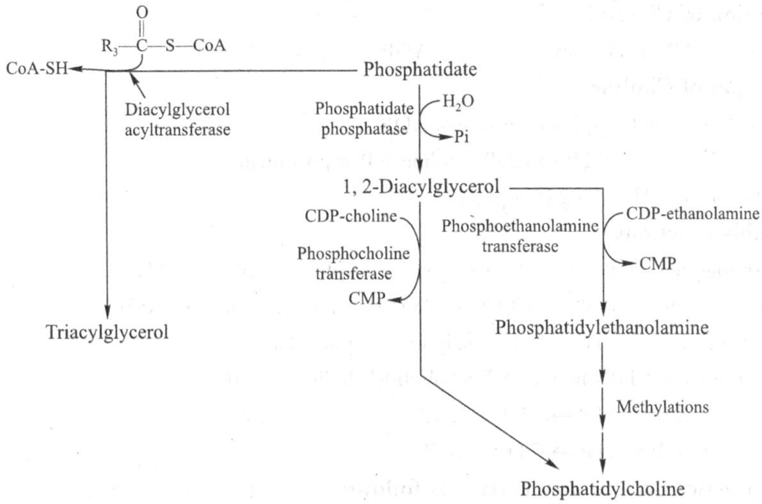

Fig. 18.5. Synthesis of triacylglycerols and neutral phospholipids. The formation of triacylglycerols, phosphatidylcholine, and phosphotidylethanolamine proceeds via a diacylglycerol intermediate. A cytosine-nucleotide derivative donates the polar head groups of the phospholipids. Three enzymatic methylation reactions, in which S-adenosyl methinonine is the methyl group donor, convert phosphatidyl-ethanolamine to phosphatidylcholine.

The formation of triacylglycerols and neutral phospholipids from phosphatidate begins with diphosphorylation, catalyzed by phosphatidate phosphatase (enzyme) (Fig. 18.5). The product of this reaction, a 1, 2-diacylglycerol, can be directly acylated to form a triacylglycerol. Alternatively, 1, 2-diacylglycerol can react with a nucleotide-alcohol derivative, such as CDP-choline (Box 18.1) or CDP-ethanolamine to form phosphatidylcholine or phosphatidylethanolamine, respectively. These derivatives are formed from CTP by the general reaction.

$$\text{CTP} + \text{alcohol phosphate} \rightarrow \text{CDP alcohol} + \text{PPi}$$

Phosphatidylcholine can also be synthesized by methylation of phosphatidylethanolamine by S-adenosylmethionine.

Phosphatidate is also the precursor of *acidic phospholipids*. In this pathway, phosphatidate is first activated by reacting with CTP to form CDP-diacylglycerol with the release of pyrophosphate (Fig. 18.6). In some bacteria, the displacement of CMP by serine produces phosphatidylserine. In both prokaryotes and eukaryotes, displacement of CMP by inositol produces phosphatidylinositol. Most eukaryotes use a different pathway for the synthesis of phosphatidyl-serine (**Horton** *et al.*, 2006).

Box 18.1. Summary of steps of biosynthesis of lecithins

Licithins are a group of phospholipids, containing glycerol, 2 moles of fatty acids (usually one saturated and the other unsaturated), phosphoric acid and a nitrogenous base choline (see Chapter 9). Their biosynthesis involves the following four steps (**Lehninger** 1965):

1. **Activation of Fatty Acids**

$$\underset{\text{Fatty acid}}{\text{RCOOH}} + \text{ATP} + \underset{\text{Coenzyme A}}{\text{CoASH}} \rightarrow \text{RCOSCoA} + \text{AMP} + \text{Pyrophosphate}$$

$$\text{R'COOH} + \text{ATP} + \text{CoASH} \rightarrow \text{R'COSCoA} + \text{AMP} + \text{Pyrophosphate}$$

$$2\ \text{AMP} + 2\ \text{ATP} \rightarrow 4\ \text{ADP}$$

$$2\ \text{Pyrophosphate} + 2\ H_2O \rightarrow 4\ \text{Phosphate}$$

2. **Activation of Glycerol**

   Glycerol + ATP → 3 Phosphoglycerol + ADP

3. **Activation of Choline**

   Choline + ATP → Phosphoryl choline + ADP

   Phosphoryl choline + CTP → CDP-Choline + Pyrophosphate

   Pyrophosphate + H_2O → 2 Phosphate

4. **Assembly Reactions**

   3-Phosphoglycerol + RCOSCoA → 1-Acyl 3-Phosphoglycerol + CoASH

   1-Acyl 3-phosphoglycerol + R′COSCoA → α-Phosphatidic acid + CoASH

   α-Phosphotidic acid + H_2O → 1, 2-diglyceride + phosphate

   1, 2-diglyceride + CDP-choline → Phosphotidyl choline + CMP

   CMP + ATP → CDP + ADP

   CDP + ATP → CTP + ADP

   These reactions can be summarised as follows:

   2 Fatty acids + Glycerol + Choline + 8 ATP → Phosphatidyl choline (lecithin) + 8 ADP + 8 phosphate

   Like the lecithin various other types of lipid molecules are synthesized by various complicated structures.

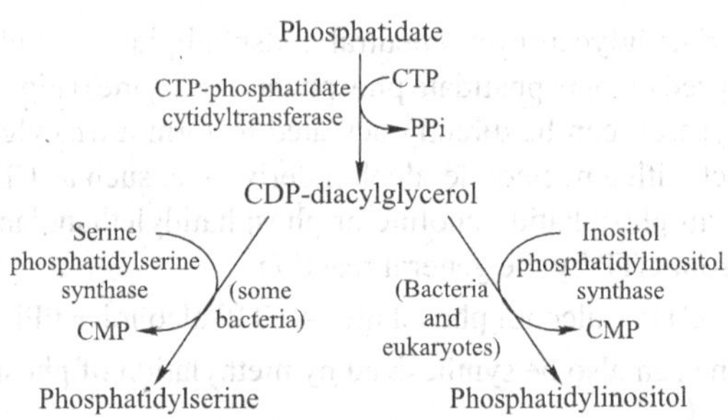

Fig. 18.6. Synthesis of acidic phospholipids. Phosphatidate accepts a cytidylyl group from CTP to form CDP-diacylglycerol. CMP is then displaced by a alcohol group of serine or inositol to form phosphatidylserine or phosphatidylinositol, respectively.

3. Synthesis of Cholesterol. The steroid cholesterol is an important component of many membranes and a precursor of steroid hormones and bile salts in mammals. All the carbon atoms in cholesterol come from acetyl CoA, a fact that emerged from early radioisotopic labeling experiments. **Squalene**, a C_{30} linear hydrocarbon is an intermediate in the biosynthesis of the 27-carbon cholesterol molecule. Squalene is formed from 5-carbon units related to isoprene. Thus, the stages in the cholesterol biosynthesis pathways are the following:

$$\text{Acetate } (C_2) \rightarrow \text{Isoprenoid } (C_5) \rightarrow \text{Squalene } (C_{30}) \rightarrow \text{Cholesterol } (C_{27})$$

A. Stage 1: Acetyl CoA to Isopentenyl Diphosphate. The first step in the cholesterol synthesis is the sequential condensation of three molecules of acetyl CoA. These condensation steps are catalyzed by acetoacetyl-CoA thiolase and HMG-CoA synthase. The product, HMG-CoA is then reduced to **mevalonate** in a reaction catalyzed by HMG-CoA reductase (Fig. 18.7). This is first committed step in cholesterol compound **isopentenyl diphosphate** by two phosphorylations followed by decarboxylation (Fig. 18.7). The conversion of two molecules of acetyl CoA to isopentenyl diphosphate requires energy in the form of three ATP and two NADPH.

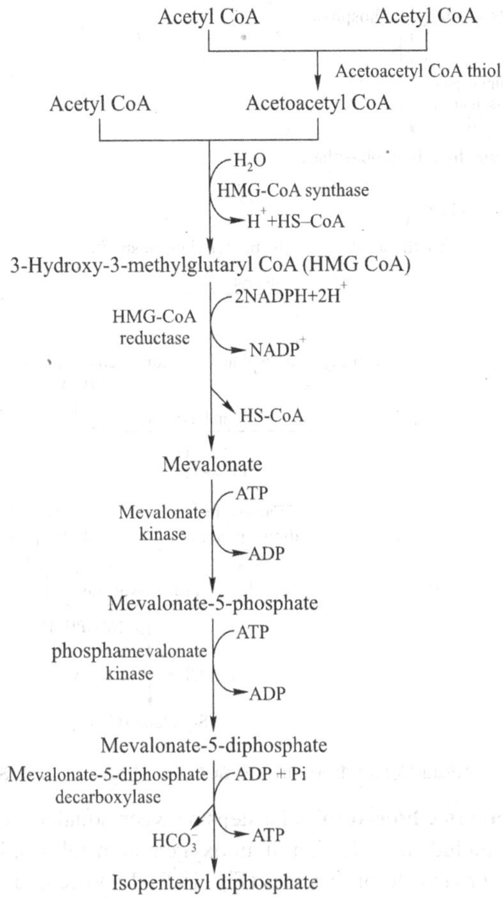

Fig. 18.7. Stage I of cholesterol synthesis: formation of isopentenyl diphosphate. The condensation of three acetyl CoA molecules leads to HMG-CoA, which is reduced to mevalonate. Mevalonate is then converted to the five-carbon molecule isopentenyl diphosphate via two phosphorylation and one decarboxylation.

B. Stage 2. Isopentenyl Diphosphate to Squalene. Isopentenyl diphosphate is converted to **dimethyl allyl diphosphate** by a specific isomerase (enzyme) called isopentenyl diphosphate isomerase (IDI). The two isomers are then joined in a *head-to-tail condensation* reaction catalyzed by prenyl transferase (Fig. 18.8). The product of this reaction is a C_{10} molecule (**geranyl diphosphate**) and pyrophosphate is released. A second condensation reaction of isoprenyl units produces a characteristic branched hydrocarbon with regularly spaced double bonds at the branch position. These isoprene units are present in a number of important cofactors.

Two molecules of farnesyl diphosphate are joined in a *head-to-head condensation* reaction to form the C_{30} molecule **squalene**. Pyrophosphate, whose hydrolysis drives reaction equilibria toward completion, is produced in three steps in the squalene synthesis pathway.

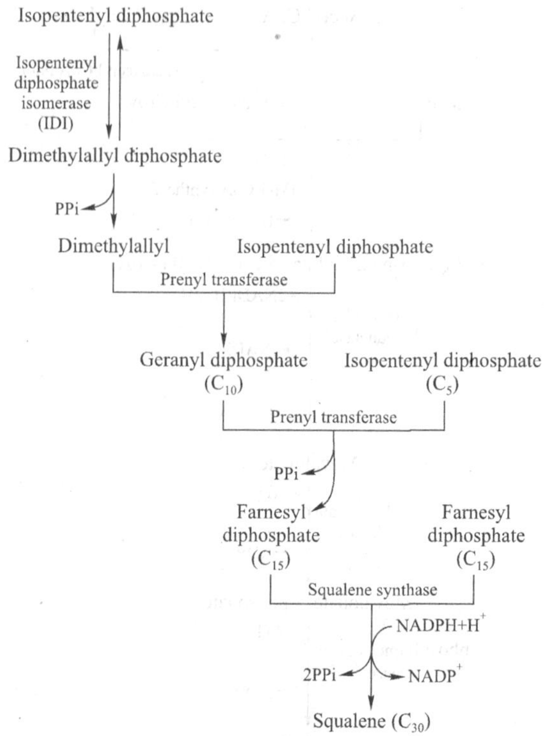

Fig. 18.8. Condensation reactions in the second stage of cholesterol synthesis.

C. Stage 3. Squalene to Cholesterol. The steps between squalene and the first fully cyclized intermediate, **lanosterol**, include the addition of an oxygen atom followed by a concerted series of cyclizations to form the four-ring steroid nucleus (Fig. 18.9). Lanosterol accumulates in appreciable quantities in cells actively synthesizing cholesterol. The conversion of lanosterol to cholesterol occurs via two pathways, both involving many steps.

D. Other Products of Isoprenoid Metabolism. A multitude of isoprenoids are synthesized from cholesterol or its precursors. Isopentenyl diphosphate, the C_5 precursor of squalene,is the precursor of a large number of other products, such as **quinones**, the lipid **vitamins A**, **E** and **K**, **carotenoids**, **terpenes**, the side chains of some cytochrome groups, and the **phytol** side chain of chlorophyll (Fig. 18.10).

Cholesterol is the precursor of *bile salts* that facilitate intestinal absorption of lipids; *vitamin D* that stimulate Ca^{2+} uptake from the intestine; *steroid hormones* such as **testosterone** and **β-estradiol** that control sex characteristics; and *steroids* that control salt balance. The principal products of steroid synthesis in mammals is cholesterol itself, which modulates membrane fluidity and is an essential component of the plasma membrane of animal cells.

4. Synthesis of Eicosanoids. There are two general classes of eicosanoids: prostaglandins + thromboxanes and leukotrienes. Arachidonate is the precursor of many eicosanoids. Arachidonate is synthesized from linoleoyl CoA.

Prostaglandins are synthesized by the cyclization of arachidonate in a reaction catalyzed by a bifunctional enzyme called *prostaglandin H synthase* (*PGHS*). The enzyme is bound to the inner surface of the endoplasmic reticulum. The cyclooxygenase (COX) activity of the enzyme catalyzes the formation of a hydroperoxide (**Prostaglandin G_2**). The PGHS enzyme contains a second active site for a hydroperoxidase activity that rapidly converts the unstable hydroperoxide to **prostglandin H_2**. This product is converted to various short-lived regulatory molecules including **prostacyclin**, **prostaglandins** and **thromboxane A_2**. Unlike hormones, which are produced by glands and travels in the blood to their sites of action, eicosanoids typically act in the immediate neighborhood of the cell in which they are produced. For example, **thromboxane A_2**, produced by blood platelets, leads to platelet aggregation and blood clots and constriction of the smooth muscles in arterial walls, causing localized changes in the blood flow. The uterus produces contraction—triggering prostaglandins during labor. Eicosanoids also mediate pain sensitivity, inflammation and swelling.

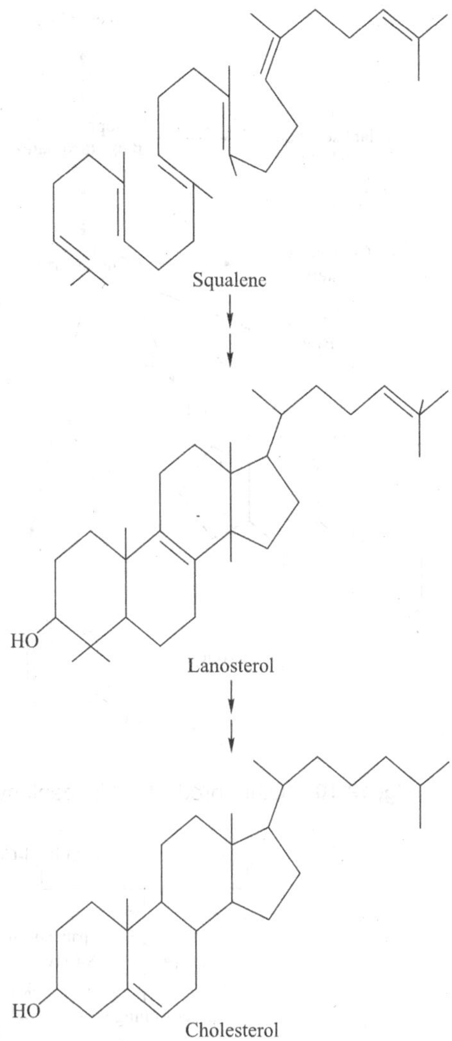

Fig. 18.9. Final stage of cholesterol synthesis: squalene to cholesterol requires up to 20 steps (after Horton *et al.*, 2006).

Box 18.2

Linoleate (*i.e.*, linoleic acid, an essential fatty acid found in oils of linseed, cottonseed, groundnut, soyabeans) must be supplied in the diet, usually from the plants, in order to support the synthesis of **arachidonate** and **eicosanoids**. One of the reasons why linoleate is essential is because it is required for synthesis of prostaglandins which are necessary for survival.

5. Synthesis of Sphingolipids. Sphingolipids are membrane lipids that have sphingosine (a C_{18} unsaturated amino alcohol) as their structural backbone. In the first step of sphingolipid biosynthesis, **serine** (a C_3 unit) condenses with **palmitoyl CoA**, producing 3-ketosphinganine and CO_2 (Fig. 18.11). Reduction of 3-ketosphinganine by NADPH produces **sphinganine**. Next, a fatty acyl group is transferred from acyl CoA to the amino group of sphinganine in an N-acylation reaction. The product of this reaction is **dihydroceramide** or ceramide without the characteristic double bond between C-4 and C-5 of a typical sphingosine. This double bond is introduced in a reaction catalyzed by dihydroceramide Δ^4-desaturase (enzyme). The final product is **ceramide** (*N*-acylsphingosine).

Ceramide is the source of all the other sphingolipids. It can react with phosphatidylcholine to form a **cerebroside**. Complex sugar-lipid conjugates, **gangliosides**, can be formed by reaction of a cerebroside with additional UDP-sugars and CMP-*N*-acetylneuraminic acid. Gangliosides are found in the outer leaflet of the plasma membrane, as are most glycolipids.

6. Synthesis of Ether Lipids. Plasmalogens are synthesized from other lipids by the formation of double bonds.

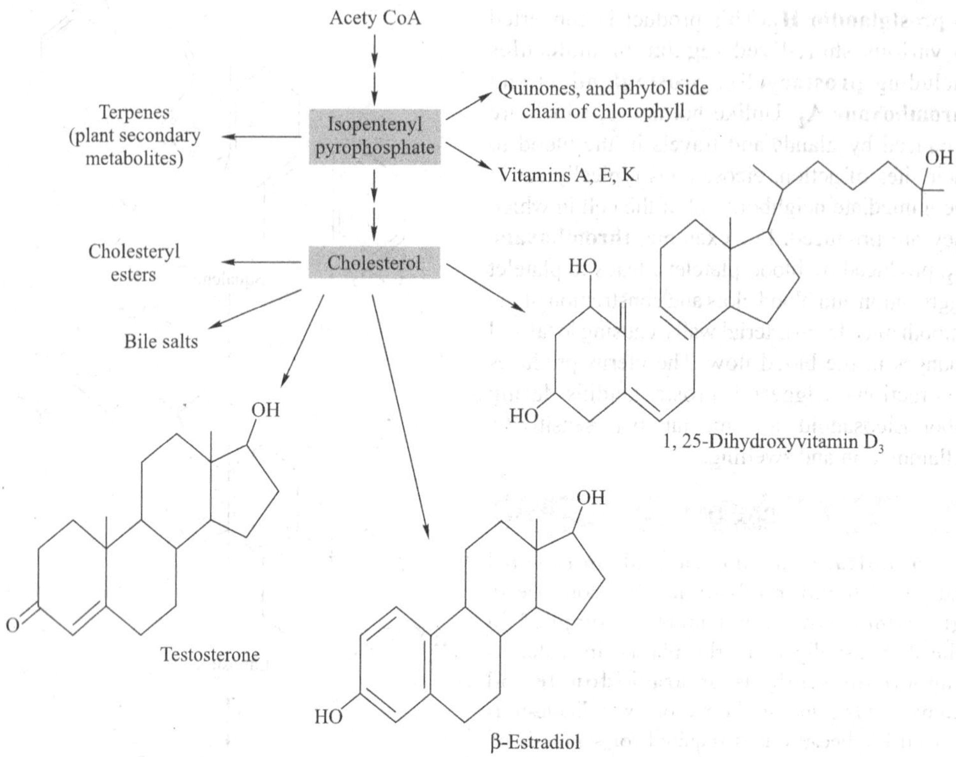

Fig. 18.10. Other products of isopentenyl pyrophosphate and cholesterol metabolism.

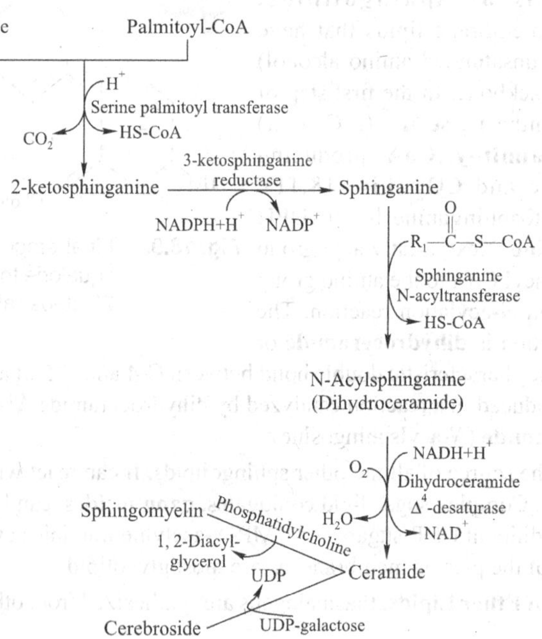

Fig. 18.11. Synthesis of sphingolipids.

18.3 CARBOHYDRATE SYNTHESIS

The Calvin cycle generates an interconvertible pool of glyceraldehyde-3-phosphate (G-3-P) and dihydroxyacetone phosphate (DHPP) (see Chapter 8), both of which can be transported out of the chloroplast by the phosphate translocator located in the chloroplast inner membrane. These triose phosphates are, in turn, the starting points for starch synthesis within the chloroplast and for sucrose synthesis in the cytosol (Fig. 18.12).

1. G-3-P and DHAP are Combined to form Glucose-1-Phosphate

The key hexose phosphate required for both starch and sucrose synthesis is glucose-1-phosphate, which is formed from triose phosphates as shown in Fig. 18.12, reactions S-1 through S-3. This set of reactions occur both in the cytosol (c) and in the chloroplast stroma (s). Hexoses and hexose-phosphates must be synthesized in each compartment because they cannot cross the chloroplast inner membrane. The first step in triose phosphate utilization is a condensation reaction catalyzed by aldolase that generates fructose-1, 6-biphosphate. This pathway is rendered exergonic by the hydrolytic removal of phosphate group from fructose-1, 6-biphosphate to form fructose-6-phosphate. The enzyme that catalyzes this reaction, fructose-1, 6-biphosphatase, exists in two forms, one in the cytosol and the other in the choroplast stroma. The multiple forms of such an enzyme are called **isoenzymes**-physically distinct proteins that catalyze the same reaction. The fructose-6-phosphate can then be converted, via glucose-6-phosphate to glucose-1-phosphate (reactions S-2 and S-3 in Fig. 18.12). Again separate isoenzymes catalyze each of these reactions in the cytosol and in the chloroplast stroma.

Very little free glucose accumulate in photosynthetic cells. Most glucose is converted into either transport carbohydrates such as sucrose or storage carbohydrates such as starch in plants or glycogen in photosynthetic bacteria.

2. Biosynthesis of Sucrose

Sucrose is a disaccharide consisting of one molecule each of glucose and fructose linked by a glycosidic bond (see Chapter 8). Sucrose is of interest because it is the major carbohydrate used for transporting stored energy and reduced carbon in most plant species. Moreover, in some species, such as sugar beets and sugar cane, sucrose also serves as a storage carbohydrate. As shown in Fig. 18.12A, sucrose synthesis is localized in the cytosol of a photosynthetic cell. Triose phosphates exported from the chloroplast stroma that are not consumed by other metabolic pathways are converted to fructose-6-phosphate and glucose-1-phosphate. Glucose from glucose-1-phosphate is then activated by the reaction of glucose-1-phosphate with **uridine triphosphate (UTP)**, generating **UDP-glucose** (reaction S-4c). Finally, the glucose is transferred to fructose-6-phosphate to form the phosphorylated disaccharide sucrose-6-phosphate, and hydrolytic removal of the phosphate group generates free sucrose (reaction S-5c and S-6c). (**Note:** In some plant species glucose may be transferred directly from UDP—glucose to free fructose). The sucrose is exported from leaves and conveys energy and reduced carbon to non-photosynthetic tissues of the organism.

Sucrose synthesis is precisely controlled to prevent conflict with degradation pathways. *Cytosolic* fructose-1, 6-biphosphatase, for example, is inhibited by fructose-2, 6-biphosphate (an important **regulator** of glycolysis and gluconeogenesis in liver cells). In plant cells, fructose-2, 6-biphosphate accumulates in response to high levels of fructose-6-phosphate and Pi (signals that indicate low sucrose demand) or low levels of 3-phosphoglycerate and dihydroxyacetone phosphate (signals that indicate high triose phosphate demand), another control point for sucrose synthesis is **sucrose phosphate synthase**, the enzyme that catalyzes the transfer of glucose from UDP-glucose to fructose-6-phosphate (reaction S-5c). This enzyme is stimulated by glucose-6-phosphate and inhibited by sucrose-6-phosphate, UDP and Pi.

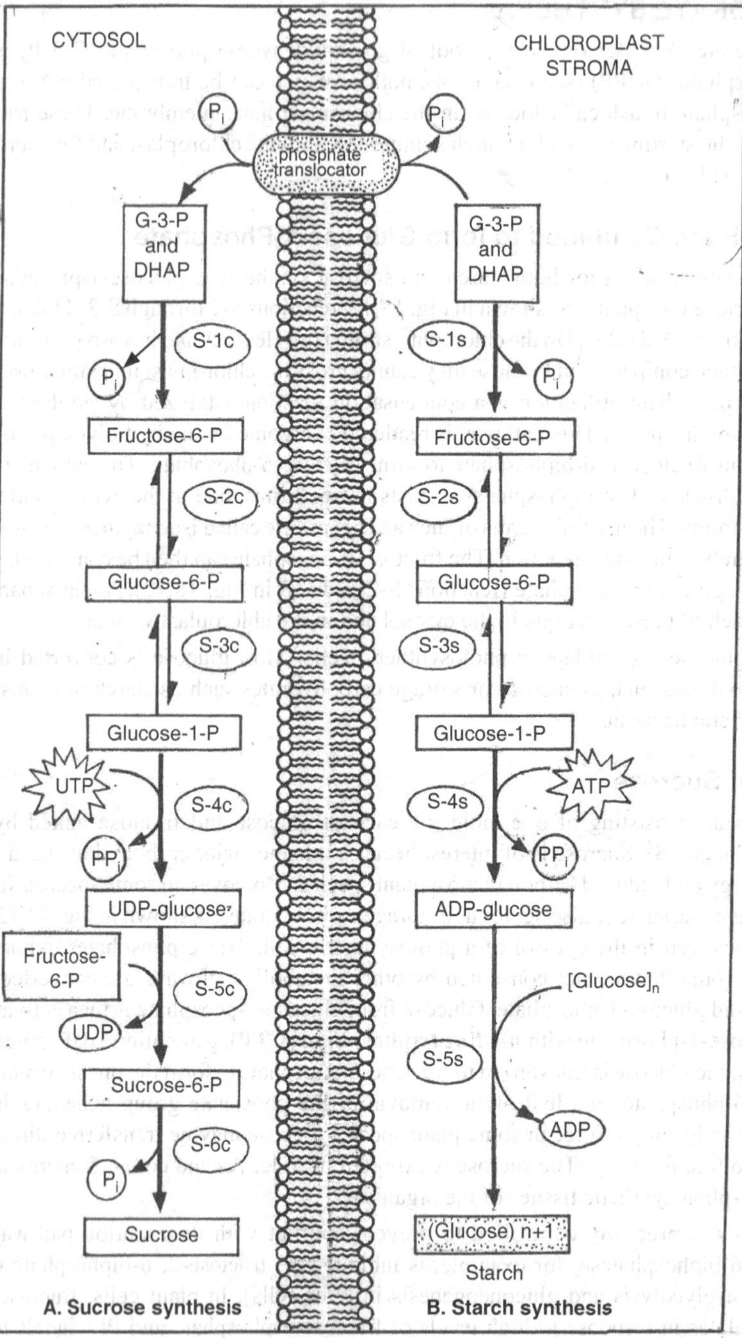

Enzymes that catalyze these reactions:

- S–1 : Aldolase and fructose – 1,6-bisphosphatase
- S–2 : Phosphoglucoisomerase
- S–3 : Phosphoglucomutase
- S–4c : UDP-glucose pyrophosphorylase
- S–5c : Sucrose phosphate synthase
- S–6c : Sucrose phosphatase
- S–4s : ADP–glucose pyrophosphorylase
- S–5s : Starch synthase

Fig. 18.12. The biosynthesis of sucrose and starch from products of the Calvin cycle. The triose phosphates glyceraldehyde-3-phosphate (G-3-P) and dihydroxyacetone phosphate (DHAP) from the chloroplast stroma are exchanged for inorganic phosphate from the cytosol via a phosphate translocator. A—Sucrose synthesis is confined to the cytosol, whereas B—Starch synthesis occurs only in the chloroplast stroma. The enzymes and isoenzymes that catalyze these reactions are restricted to either the cytosol (c) or stroma (s).

3. Biosynthesis of Starch

Starch synthesis in plant cells is confined to plastids. In photosynthetic plant cells, starch synthesis is generally restricted to chloroplasts, which are essentially photosynthetic plastids. When sufficient energy and carbon are available to meet the metabolic needs of a plant, the triose phosphates within the chloroplast stroma are converted to glucose-1-phosphate, which is then used for starch synthesis. As shown in Fig. 18.13B, glucose is activated by the reaction of glucose-1-phosphate with ATP, generating ADP-glucose (reaction S-4s). The activated glucose is then transferred directly to growing starch chain by **starch synthase**, leading to elongation of the polysaccharide (reaction S-5s). Starch may accumulate in large storage granules within the chloroplast stroma. When photosynthesis is limited by darkness or other factors, starch is degraded to triose phosphates, which can then enter glycolysis or be converted to sucrose in the cytosol and exported from the cell.

Starch synthesis is also precisely regulated. Chloroplast *fructose biphosphatase*, which channels fructose-1, 6-biphosphate into glucose and starch biosynthesis in the stroma, is activated by the same thioredoxin system that affects enzymes of the Calvin cycle. This regulation ensures that starch synthesis occurs only when there is sufficient illumination for photo-reduction. The key enzyme for regulation, though, is *ADP-glucose pyrophosphorylase*, which catalyzes the activation of glucose and commits to starch synthesis by forming ADP-glucose (reaction S-4s). ADP-glucose pyrophosphorylase is stimulated by glyceraldehyde-3-phosphate and inhibited by Pi. Thus, when triose phosphates are diverted to the cytosol and ATP is hydrolyzed to ADP and Pi (signals that indicate high energy demand), starch synthesis is blocked.

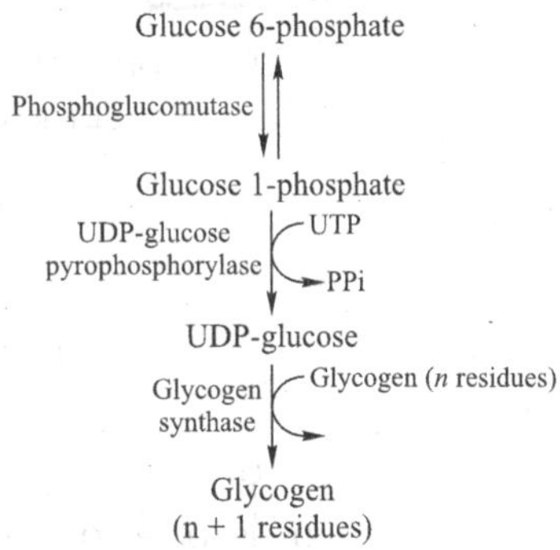

Fig. 18.13. Synthesis of glycogen in eukaryotes.

4. Glycogen Metabolism

Glucose is stored as the intracellular polysaccharides starch and glycogen. Glycogen is an important storage polysaccharides in bacteria, protists, fungi and animals. Most of the glycogen in vertebrates is found in muscle and liver cells. In electron micrographs, muscle glycogen appears as cytosolic granules with a diameter of 10 to 40 nm, about the size of ribosomes. Glycogen particles in liver cells are about three times, larger. The enzymes required for glycogen degradation are also found in muscle and liver cells.

Glycogen synthesis begins with glucose-6-phosphate, which can be converted to glucose-1-phosphate. Glycogen synthesis and degradation is mostly a way of storing glucose 6-phosphate until it is needed by the cell.

Glycogen synthesis in eukaryotes. Three separate enzyme-catalyzed reactions are required to incorporate a molecule of glucose 6-phosphate into glycogen (Fig. 18.13). First, *phosphoglucomatase* (enzyme) catalyzes the near-equilibrium conversion of glucose 6-phosphate to glucose 1-phosphate. In second step, glucose 1-phosphate is activated by reaction with UTP, forming UDP-glucose and pyrophosphate (PPi). In the third step, *glycogen synthase* (enzyme) catalyzes the addition of glucose residues from UDP-glucose to the nonreducing end of glycogen.

UDP-glucose can be a substrate for the synthesis of UDP-galactose which is used in the synthesis of gangliosides.

Glycogen synthesis is a polymerization reaction where glucose units are added one at a time to a growing polysaccharide chain. This reaction is catalyzed by glycogen synthase (Fig. 18.14).

UDP-glucose requiring glycogen synthases are present in protists, animals and fungi. Some bacteria synthesize glycogen using ADP-glucose. Starch synthesis in plants also requires ADP-glucose. The glycogen synthase reaction is major regulatory step of glycogen synthesis. In animals, there are hormones (*e.g.*, Insulin, glucagon and epinephrine) that control the rate of glycogen synthesis by altering the activity of glycogen synthase.

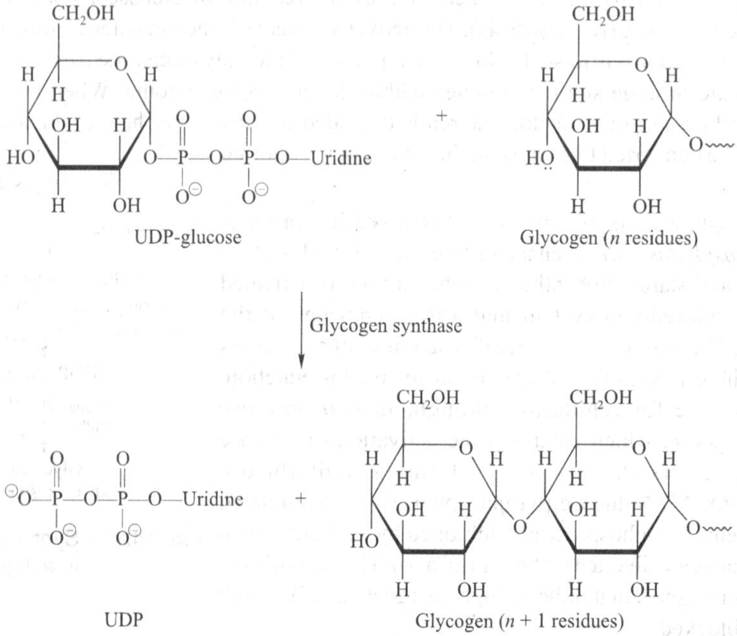

Fig. 18.14. Addition of a glucose residue to the non-reducing end of a glycogen molecule, catalyzed by glycogen synthase (after Horton *et al.*, 2006).

Glycogen synthase requires a preexisting **glycogen primer** of four to eight α-(1 → 4)-linked glucose residues. This primer is attached to a specific **tyrosine residue** of a protein called **glycogenin** via the 1-hydroxyl group of the reducing end of the short polysaccharide. The primer is formed in two steps. The first glucose residue is attached to glycogenin by action of a glucosyltransferase activity that require UDP-glucose as the glycosyl-group donor. Glycogenin itself catalyzes the extension of the primer by upto seven-more glucose residues. Thus, glycogen is both a protein scaffold for glycogen and an enzyme. Each glycogen complex contains a single glycogenin molecule and at least 50000 glucose residues. Glycogen synthase catalyzes further lengthening of glycogen as described above.

Another enzyme, **amylo-(1, 4 → 1, 6)-transglycosylase**, catalyzes branch formation in glycogen. This enzyme, also known as the **branching enzyme**, removes an oligosaccharide of at least six residues from the nonreducing end of an elongated chain and attaches it by an α-(1 → 6) linkage to a position at least four glucose residues from the nearest α-(1 → 6) branch point. These branches provide many sites for adding or removing glucose residues, thereby contributing to the speed with which glycogen can be synthesized or degraded.

QUESTIONS

Long Answer Questions

1. Give an account of metabolism of glycogen.
2. Describe the synthesis of sucrose in plant cells.
3. Give an account of biosynthesis of starch in green plants.
4. Describe the process of synthesis of cholesterol.
5. Give an account of synthesis of sphingolipids.

Short Answer Questions

1. Describe the mechanism of glycogenesis.

Very Short Answer Quetions

1. How many carbon atoms are present in squalene?
2. What is another name of the branching enzyme?
3. What is the site of biosynthesis of fatty acids?

ANSWERS

Very Short Answer Questions

1. 30.
2. Amylo-1, 4 → 1, 6-transglycosylase.
3. Cytosol.

19 Endoplasmic Reticulum (ER)

The cytoplasmic matrix is traversed by a complex network of inter-connecting membrane bound vacuoles or cavities. These vacuoles or cavities often remain concentrated in the endoplasmic portion of the cytoplasm; therefore, known as **endoplasmic reticulum**, a name derived from the fact that in the light microscope it looks like a "net in the cytoplasm." (Eighteenth-century European ladies carried purses of netting called **reticules**)The name "endoplasmic reticulum" was coined in 1953 by **Porter**, who in 1945 had observed it in electron micrographs of liver cells.

19.1. OCCURRENCE

The occurrence of the endoplasmic reticulum varies from cell to cell. The erythrocytes (RBC), egg and embryonic cells lack in endoplasmic reticulum. [**Note**. In the reticulocytes (immature red blood cells) which produce only proteins to be retained in the cytoplasmic matrix (cytosol) (*e.g.*, haemoglobin, the ER is poorly developed or non-existent, although the cell may contain many ribosomes]. The spermatocytes have poorly developed endoplasmic reticulum. The adipose tissues, brown fat cells and adrenocortical cells, interstitial cells of testes and cells of corpus luteum of ovaries, sebaceous cells and retinal pigment cells contain only **smooth endoplasmic reticulum (SER)**. The cells of those organs which are actively engaged in the synthesis of proteins such as acinar cells of pancreas, plasma cells, goblet cells and cells of some endocrine glands are found to contain **rough endoplasmic reticulum (RER)** which is highly developed. The presence of both SER and RER in the hepatocytes (liver cells) is reflective of the variety of the roles played by the liver in metabolism.

The classical literature distinguishes three types of ER membranes: **rough ER**, **smooth ER** and **nuclear envelope**. However, researchers now recognize many more morphologically distinct sub-domains that perform various different functions (Fig. 19.1). Despite this functional diversity, virtually all ER membranes are physically linked and enclose a single, continuous lumen that extends beyond the boundaries of individual cells via the plasmodesmata (**Buchanan** *et el.*, 2000).

19.2. ER AND ENDOMEMBRANE SYSTEM

The cytoplasm of eukaryotic cells contains a system of membranous organelles, including endoplasmic reticulum, Golgi apparatus and related to one another and to the plasma membrane. These various membranous organelles are part of a dynamic, integrated **endomembrane network/system** in which materials are shuttled back and forth as part of **transport vesicles** that form by budding from one compartment and fusing with another (Fig. 19.2). Mitochondria and chloroplasts are not part of this interconnected endomembrane system. Current evidence suggest that peroxisomes also are not components of endomembrane system. Peroxisomes have a dual origin. The basic elements of the boundary membrane of peroxisomes are thought to arise from the endoplasmic reticulum, but many of the membrane proteins and the soluble internal proteins are taken up from the cytosole (**Karp**, 2010).

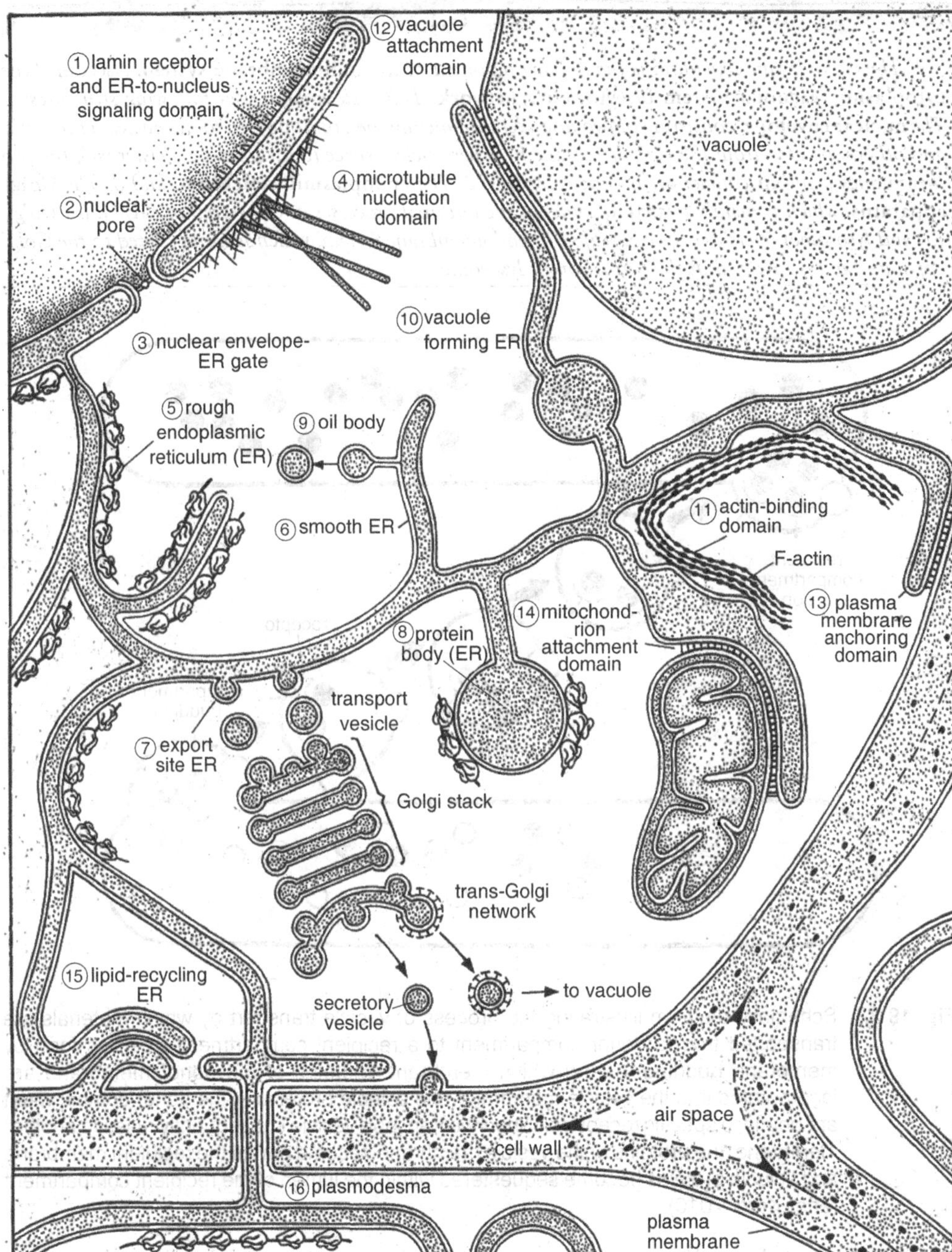

Fig. 19.1. Functional domains of the plant ER system (after Buchanan, *et al.*, 2000).

Box 19.1

The endoplasmic reticulum is the main component of the **endomembrane system,** *also called the* **cytoplasmic vacuolar system** *or* **cytocavity network**. *This system comprises following structures: 1. The* **nuclear envelope**, *consisting of two non-identical membranes, one opposed to the nuclear chromatin and other separated from the first membrane by a perinuclear space (both forming a cisternae), the two membranes being in contact at the nuclear pores; 2. The* **endoplasmic reticulum**; *and 3. The* **Golgi apparatus,** *which is mainly related to some of the terminal processes of cell secretion.* **GERL** *(or Golgi, ER and lysosome) refers to a special region of endomembrane system, which is more related to the Golgi apparatus and is involved in the formation of lysosomes.*

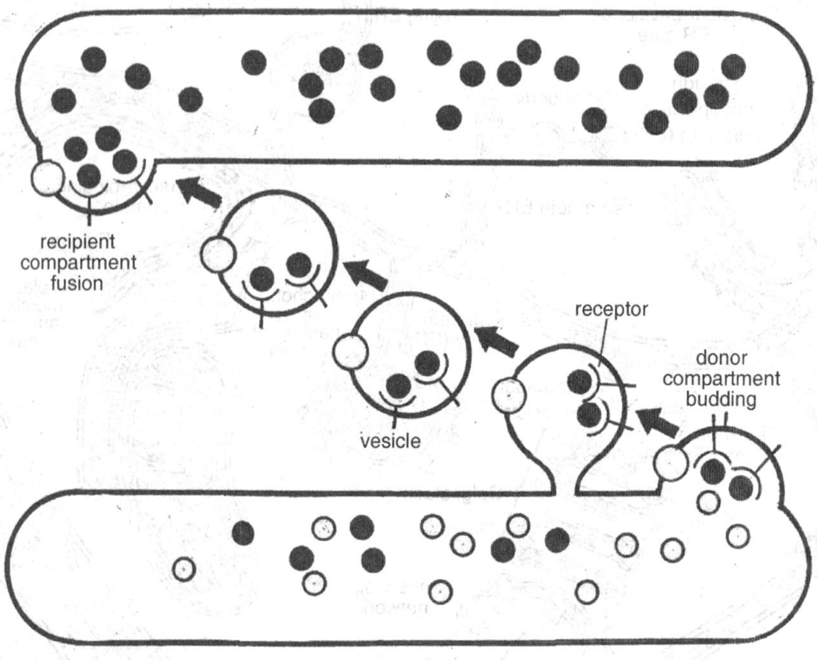

Fig. 19.2. Schematic diagram illustrating the process of vesicle transport by which materials are transported from a donor compartment to a recipient compartment. Vesicles form by membrane budding, during which membrane proteins of the donor membrane are incorporated into the vesicle membrane and soluble proteins in the donor compartment are bound to specific receptors. When the transport vesicle subsequently fuses to another membrane, the proteins of the vesicle membrane become part of the recipient membrane and soluble proteins become sequestered within the lumen of the recipient compartment (after Karp, 2010).

The entire endomembrane system represents a barrier separating cytoplasmic compartments. The membrane of each component of this system has two faces: (*i*) The **cytoplasmic** or **protoplasmic face** and (*ii*) The **luminal face** (Fig. 19.3). The luminal face borders the perinuclear cisternae, the cavities of ER and SER, and the Golgi elements. It also corresponds to the interior of the secretory granules, the lysosomes and peroxisomes and also to faces of mitochondrial membranes confronting to outer mitochondrial chamber.

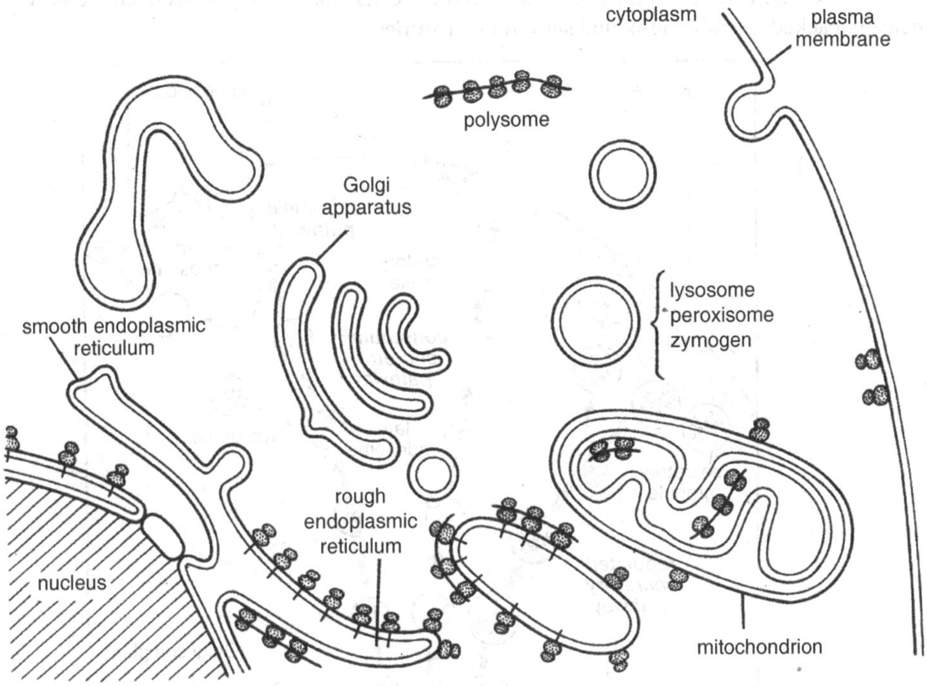

Fig. 19.3. Two faces of membranes of endomembrane system. In each membrane the luminal faces are shown in thick lines, while the cytoplasmic faces are depicted by thin lines. Ribosomes are always located on the cytoplasmic or matrix side.

Functions of Endomembrane System

For the most part, materials are shuttled between organelles of endomembrane system, for example, in small membrane-bounded **transport vesicles** that bud from a donor membrane compartment (Fig. 19.2). Transport vesicles move through the cytoplasm in a directed manner, often pulled by motor proteins that operate on tracks formed by microtubules and microfilaments of the cytoskeleton. When they reach their destination, the vesicles fuse with the membrane of the acceptor compartment, which receives the vesicles soluble cargo as well as its membranous wrapper. Repeated cycles of budding and fusion shuttle a diverse array of materials along numerous pathways that traverse the cell.

Several distinct pathways through the cytoplasm have been identified and are illustrated in Figure 19.4. A **biosynthetic pathway** can be recognized in which proteins are synthesized in endoplasmic reticulum, modified during passage through the Golgi apparatus, and transported from Golgi apparatus to various destinations, such as the plasma membrane, a lysosome, or a large vacuole of a plant cell. This route is also called the **secretory pathway**, as many of the proteins synthesized in the endoplasmic reticulum (as well as complex polysaccharides synthesized in the Golgi apparatus) are destined to be discharged (*i.e.*, **secreted**) from the cell. Secretory activities of cells can be divided into two types: constitutive and regulated (Fig. 19.4). During **constitutive secretion**, materials are transported in secretory vesicles from their sites of synthesis and discharged into the extracellular space in a continual manner. Most cells engage in constitutive secretion, a process that contributes not only to the formation of the extracellular matrix, but to the formation of the plasma membrane itself. During **regulated secretion**, materials are stored as membrane-bound packages and discharged only in response to an appropriate stimulus. Regulated secretion occurs, for example, in endocrine cells that release hormones, in pancreatic acinar cells that release digestive enzymes, and in nerve cells that release neurotransmitters. In some of these cells, materials to be secreted are stored in large, densely packed, membrane-bound **secretory granules**.

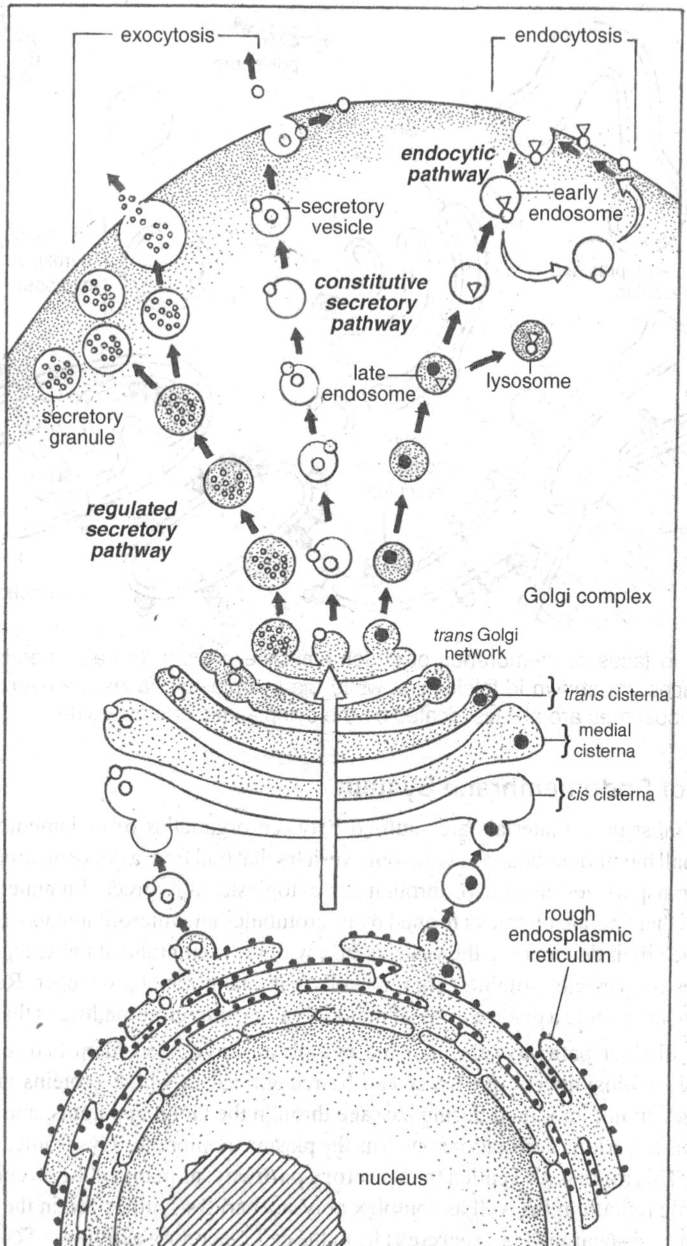

Fig. 19.4. Biosynthetic and endocytic pathways of endomembrane system. Materials follow the biosynthetic (or secretory) pathway from the endoplasmic reticulum, through the Golgi apparatus, and out to various locations including lysosomes, endosomes, secretory vesicles, secretory granules, vacuoles, and the plasma membrane (exocytosis). Materials follow the endocytic pathway from the cell surface to the interior by way of endosomes and lysosomes, where they are generally degraded by lysosomal enzymes (after Karp, 2010).

Proteins, lipids and complex polysaccharides are transported through the cell along the biosynthetic or secretory pathway. These proteins belong to several distinct classes: these include soluble proteins that are discharged from the cell, integral proteins of the various membranes and soluble proteins that reside within various compartments enclosed by the endomembranes (*e.g.*, lysosomal enzymes). Whereas materials move out of the cell by the secretory pathway (called **exocytosis**), the endocytic pathway operates in the opposite direction. By following the **endocytic pathway**, materials move from the outer surface of the cell to the compartments such as endosomes and lysosomes located within the cytoplasm.

The movement of vesicles and their contents along the various pathways of a cell is analogous to the movement of trucks carrying different types of cargo along the various **traffic patterns** to ensure that materials are accurately delivered to the appropriate sites. For example, protein trafficking within a salivary gland cell requires that the proteins of salivary mucus, which are synthesized by endoplasmic reticulum, are specifically targeted for secretory granules, while lysosomal enzymes, which are also manufactured in the endoplasmic reticulum, are specifically targeted to a lysosome. Different organelles also contain different integral membrane proteins. Consequently, membrane proteins must also be targeted to particular organelles, such as a lysosome or Golgi cisterna. These various types of cargo-secreted proteins, lysosomal enzymes, and membrane proteins—are routed to their appropriate cellular destinations by virtue of specific "address" or **sorting signals** that are encoded in the amino acid sequence of the proteins or in the attached oligosaccharides. The **sorting signals** are recognized by specific receptors that reside in the membranes or surface coats of budding vesicles, ensuring that the protein is transported to the appropriate destination. For the most part, the machinery responsible for driving this complex distribution system consists of soluble proteins that are **recruited** to specific membrane surfaces. In this chapter and other chapter, there will be an attempt to understand why one protein is recruited, for example, to the endoplasmic reticulum whereas another protein might be recruited to a particular region of the Golgi apparatus.

Great advances have been made over the past three decades in mapping the traffic patterns that exist in eukaryotic cells, identifying the specific address and receptors that govern the flow of traffic and dissecting the machinery that ensures that materials are delivered to the appropriate sites in the cell (**Karp**, 2010).

19.3. MORPHOLOGY (STRUCTURE) OF ER

Morphologically, the endoplasmic reticulum may occur in the following three forms: 1. Lamellar form or cisternae (A closed, fluid-filled sac, vesicle or cavity is called **cistern**); 2. Vesicular form or vesicle and 3. Tubular form or tubules.

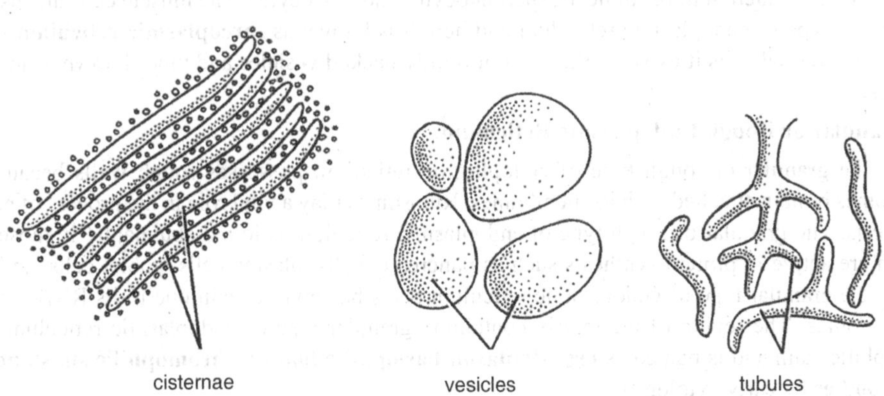

Fig. 19.5. Various components of the endoplasmic reticulum.

1. Cisternae. The cisternae are long, flattened, sac-like, unbranched tubules having the diameter of 40 to 50 μm. They remain arranged parallely in bundles or stakes. RER usually exists as cisternae which occur in those cells which have synthetic roles as the cells of pancreas, notochord and brain.

2. Vesicles. The vesicles are oval, membrane-bound vacuolar structures having the diameter of 25 to 500 μm. They often remain isolated in the cytoplasm and occur in most cells but especially abundant in the SER.

3. Tubules. The tubules are branched structures forming the reticular system along with the cisternae and vesicles. They usually have the diameter from 50 to 190 μm and occur almost in all the cells. Tubular form of ER is often found in SER and is dynamic in nature, *i.e.*, it is associated with membrane movements, fission and fusion between membranes of cytocavity network.

Ultrastructure

The cavities of cisternae, vesicles and tubules of the endoplasmic reticulum are bounded by a thin membrane of 50 to 60 Å thickness. The membrane of endoplasmic reticulum is fluid-mosaic like the unit membrane of the plasma membrane, nucleus, Golgi apparatus, etc. The membrane, thus, is composed of a bimolecular layer of phospholipids in which 'float' proteins of various sorts. The membrane of endoplasmic reticulum remains continuous with the membranes of plasma membrane, nuclear membrane and Golgi apparatus. The cavity of the endoplasmic reticulum is well developed and acts as a passage for the secretory products. **Palade** (1956) has observed secretory granules in the cavity of endoplasmic reticulum.

Sometimes, the cavity of RER is very narrow with two membranes closely apposed and is much distended in certain cells which are actively engaged in protein synthesis (*e.g.*, acinar cells, plasma cells and goblet cells). **Weibel** *et al.*, 1969, have calculated that the total surface of ER contained in 1 ml of liver tissue is about 11 square metres, two-third of which is of rough type (*i.e.*, RER).

19.4. TYPES OF ENDOPLASMIC RETICULUM

Two types of endoplasmic reticulum have been observed in same or different types of cells which are as follows:

1. Agranular or Smooth Endoplasmic Reticulum.

This type of endoplasmic reticulum possesses smooth walls because the ribosomes are not attached with its membranes. The smooth type of endoplasmic reticulum occurs mostly in those cells, which are involved in the metabolism of lipids (including steroids) and glycogen. The smooth endoplasmic reticulum is generally found in adipose cells, interstitial cells, glycogen storing cells of the liver, conduction fibres of heart, spermatocytes and leucocytes. The muscle cells are also rich in smooth type of endoplasmic reticulum and here it is known as **sarcoplasmic reticulum**. In the pigmented retinal cells it exists in the form of tightly packed vesicles and tubes known as **myeloid bodies**.

2. Granular or Rough Endoplasmic Reticulum.

The granular or rough type of endoplasmic reticulum possesses rough walls because the ribosomes remain attached with its membranes. Ribosomes play a vital role in the process of protein synthesis. The granular or rough type of endoplasmic reticulum is found abundantly in those cells which are active in protein synthesis such as pancreatic cells, plasma cells, goblet cells, and liver cells. The granular type of endoplasmic reticulum takes basiophilic stain due to its RNA contents of ribosomes. The region of the matrix containing granular type of endoplasmic reticulum takes basiophilic stain and is named as **ergastoplasm, basiophilic bodies, chromophilic substances** or **Nissl bodies** by early cytologists.

In RER, ribosomes are often present as polysomes held together by mRNA and are arranged in typical "rosettes" or spirals. RER contains two transmembrane glycoproteins (called **ribophorins I** and **II** of 65,000 and 64,000 dalton MW, respectively), to which are attached the ribosomes by their 60S subunits.

3. Annulate Lamellae.

Usually the endoplasmic reticulum has no pores or annuli in it but in certain cases the pores or annuli have been reported, *e.g.*, ER of invertebrates, oocytes and spermatocytes of the vertebrates. These annuli resemble with the pores or annuli of the nuclear membranes. Like the annuli of nuclear membranes it contains a diaphragm across it and possesses an octagonal symmetry. The annulate lamellae (pores) of the ER arise by the evagination from the nuclear envelope and have their association with the ribosomes.

4. Glycosomes.

Although the SER forms a continuous system with RER, it has different morphology. For example, in liver cells it consists of a tubular network that pervades major portion of the cytoplasmic matrix. These fine tubules are present in regions rich in glycogen and can be observed as dense particles, called **glycosomes**, in the matrix. Glycosomes measure 50 to 200 nm in diameter and contain glycogen along with enzymes involved in the synthesis of glycogen. Many glycosomes attached to the membranes of SER have been observed by electron microscopy in the liver and conduction fibre of heart.

19.5. ISOLATION

The membranes of the endoplasmic reticulum can be isolated by subjecting homogenized tissues to differential centrifugation. Electron microscopy of such ER preparations reveals that the membranes disrupt to form closed vesicles (~100 nm diameter) of either a rough or a smooth form. These membranous entities were named as "**microsomes**" by **Claude** in 1940, and the relationship between microsomes and the elements of endoplasmic reticulum in the intact cell was established by **Palade** and **Siekevitz** in 1956. Microsomes are not found in intact cells, *i.e.*, they are not natural cell organelles.

Microsomes derived from rough ER are studded with ribosomes and are called **rough** or **granular microsomes**. The ribosomes are always found on the outside surface, the interior being biochemically equivalent to the luminal space of the ER. Homogenate also contains **smooth** or **agranular microsomes** which lack attached ribosomes. They may be derived in part from smooth portion of the ER and in part from fragments of plasma membrane, Golgi apparatus, endosomes and mitochondria. Thus, while rough microsomes can be equated with rough portions of ER, the origin of smooth microsomes cannot be so easily assigned. However, since the hepatocytes of liver contain exceedingly large quantities of smooth ER, therefore, most of the smooth microsomes in liver homogenates are derived from smooth ER.

As rough microsomes can be readily purified in functional form, they are especially useful for studying many biochemical processes carried out by the ER, *e.g.*, protein synthesis, glycosylation and lipid synthesis.

In rat liver, the membranes of microsomes are 60 to 70 per cent protein and 30 to 40 per cent phospholipid by weight.

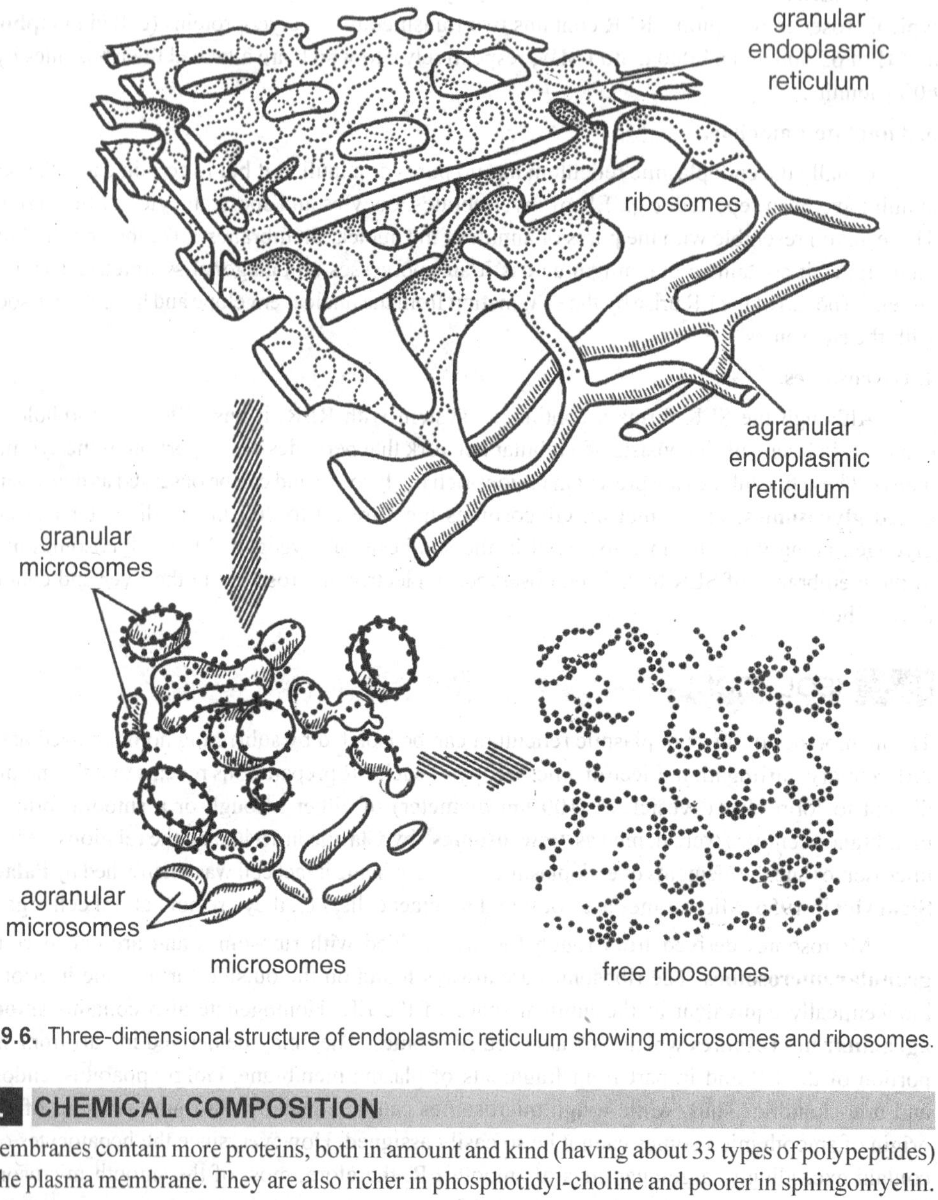

Fig. 19.6. Three-dimensional structure of endoplasmic reticulum showing microsomes and ribosomes.

19.6. CHEMICAL COMPOSITION

ER membranes contain more proteins, both in amount and kind (having about 33 types of polypeptides) than the plasma membrane. They are also richer in phosphotidyl-choline and poorer in sphingomyelin.

The membranes of the endoplasmic reticulum are found to contain many kinds of enzymes which are needed for various important synthetic activities. Some of the most common enzymes are found to have different transverse distribution in the ER membranes (Table 19.1). The most important enzymes are the esterases, NADH-cytochrome c reductase, NADH diphosphatase, glucose 6-phosphotase and Mg^{2+} activated ATPase. Certain enzymes of the endoplasmic reticulum such as nucleotide diphosphate are involved in the biosynthesis of phospholipid, ascorbic acid, glucuronide, steroids and hexose metabolism. The enzymes of the endoplasmic reticulum perform the following important functions:

1. Synthesis of glycerides, *e.g.*, triglycerides, phospholipids, glycolipids and plasmalogens.

Table 19.1. Transverse distribution of various enzymes in the membranes of endoplasmic reticulum.

Enzymes	Surface localization
1. Cytochrome b_5 (involved in synthesis of unsaturated fatty acids)	Cytoplasmic face
2. NADH-cytochrome b_5 reductase	Cytoplasmic face
3. NADP-cytochrome c reductase	Cytoplasmic face
4. Cytochrome P-450 (most abundant)	Both on cytoplasmic and luminal face
5. ATPase	Cytoplasmic face
6. 5-nucleotidase	Cytoplasmic face
7. Nucleoside pyrophosphatase	Cytoplasmic face
8. GDP-mannosyl transferase	Cytoplasmic face
9. Nucleoside diphosphatase	Luminal face
10. Glucose-6-phosphatase (histochemical marker enzyme)	Luminal face
11. Acetanilide-hydrolysing esterase	Luminal face
12. β-glucuronidase	Luminal face

2. Metabolism of plasmalogens.
3. Synthesis of fatty acids.
4. Biosynthesis of the steroids, *e.g.*, cholesterol biosynthesis, steroid hydrogenation of unsaturated bonds.
5. $NADPH_2 + O_2$—requiring steroid transformations: Aromatization and hydroxylation, side-chain oxidation, deamination, thio-ether oxidations, desulphuration.
6. L-ascorbic acid synthesis.
7. UDP-uronic acid metabolism.
8. UDP-glucose dephosphorylation.
9. Aryl-and steroid sulphatase.

19.7. ORIGIN OF ENDOPLASMIC RETICULUM

The exact process of the origin of endoplasmic reticulum is still unknown. But because membranes of ER resemble with the nuclear membrane and plasma membrane and also at the telophase stage the ER membranes are found to form the nuclear envelope. Therefore, it is normally assumed that the ER has originated by evagination of the nuclear membranes. **Seikevitz** and **Palade** (1960) have reported that the granular type of ER has originated first and later it synthesizes the agranular or smooth type of endoplasmic reticulum. According to **Leskes** (1971) and **Eytan** and **Ohad** (1972), the cell receives ER from its parent cell. Later on, ER grows by expansion.

The synthesis of membranes of ER is found to proceed in the following direction: RER → SER. In fact, membrane biogenesis is a multi-step process involving , first, the synthesis of a basic membrane of lipid and intrinsic proteins and thereafter the addition of other constituents such as enzymes, specific sugars, or lipids. The process by which a membrane is modified chemically and structurally is called **membrane differentiation**. The ER (especially SER) is the organelle containing the main phospholipid synthesizing and translocating enzymes (*i.e.*, there occurs an intense flip-flop of lipid components). The insertion of proteins into ER membranes occurs at the level of RER. Most of these proteins are formed on membrane-bound ribosomes. However, some of these are synthesized by free ribosomes in the cytosol (cytoplasmic matrix) and then are inserted into the membrane. For example, the enzyme **NAD-cytochrome-b_5-reductase** is synthesized in the cytosol (cytoplasmic matrix) and then becomes incorporated in various parts of the endomembrane system (*i.e.*, RER, SER and Golgi apparatus) and in the outer mitochondrial membrane.

19.8. FUNCTIONS OF ENDOPLASMIC RETICULUM

The endoplasmic reticulum acts as secretory, storage, circulatory and nervous system for the cell. It performs following important functions:

A. Common Functions of Granular and Agranular Endoplasmic Reticulum

1. The endoplasmic reticulum provides an ultrastructural skeletal framework to the cell and gives mechanical support to the colloidal cytoplasmic martix.
2. The exchange of molecules by the process of osmosis, diffusion and active transport occurs through the membranes of endoplasmic reticulum. Like plasma membrane, the ER membrane has permeases and carriers.
3. The endoplasmic membranes contain many enzymes which perform various synthetic and metabolic activities. Further the endoplasmic reticulum provides increased surface for various enzymatic reactions.
4. The endoplasmic reticulum acts as an intracellular circulatory or transporting system. Various secretory products of granular endoplasmic reticulum are transported to various organelles as follows: Granular ER → agranular ER → Golgi membrane → lysosomes, transport vesicles or secretory granules. Membrane flow may also be an important mechanism for carrying particles, molecules and ions into and out of the cells. Export of RNA and nucleoproteins from nucleus to cytoplasm may also occur by this type of flow.
5. The ER membranes are found to conduct intra-cellular impulses. For example, the sarcoplasmic reticulum transmits impulses from the surface membrane into the deep region of the muscle fibres.
6. The ER membranes form the new nuclear envelope after each nuclear division.
7. The sarcoplasmic reticulum plays a role in releasing calcium when the muscle is stimulated and actively transporting calcium back into the sarcoplasmic reticulum when the stimulation stops and the muscle must be relaxed.

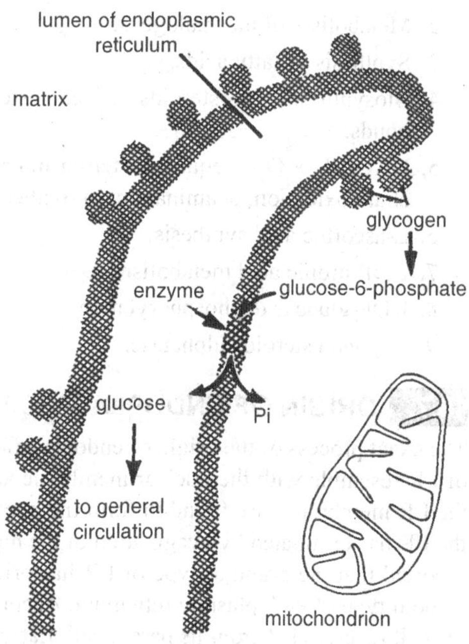

Fig. 19.7. Diagram of the intervention of the smooth endoplamic reticulum in glycogenolysis with the consequent release of glucose. The enzyme glucose-6-phosphatase, is present in the membrane and has a vectorial deposition by which it receives the glucose-6-phophate from the matrix surface. The product glucose penetrates the lumen of the endoplasmic reticulum.

B. Functions of Smooth Endoplasmic Reticulum.

Smooth ER performs the following functions of the cell:

1. Synthesis of lipids. SER performs synthesis of lipids (*e.g.*, phospholipids, cholesterol, etc.) and lipoproteins. Studies with radioactive precursors have indicated that the newly synthesized phospholipids are rapidly transferred to other cellular membranes by the help of specific cytosolic enzymes, called **phospholipid exchange proteins**.

2. Glycogenolysis and blood glucose homeostasis. The process of glycogen synthesis (glycogenesis) occurs in the cytosol (in glycosomes). The enzyme **UDPG-glycogen transferase**, which is directly involved in the synthesis of glycogen by addition of **uridine diphosphate glucose (UDPG)** to primer glycogen is bound to the glycogen particles or glycosomes (see Chapter 18).

SER is found related to **glycogenolysis** or breakdown of glycogen. An enzyme, called **glucose 6-phosphatase** (a marker enzyme) exists as an integral protein of the membrane of SER (*e.g.*, liver cell). Generally, this enzyme acts as a glucogenic phosphohydrolase that catalyzes the release of free glucose molecule in the lumen of SER from its phosphorylated form in liver (Fig. 19.7). Thus, this process operates to maintain homeostatic levels of glucose in the blood for the maintenance of functions of red blood cells and nerve tissues.

3. Sterol metabolism. The SER contains several key enzymes that catalyze the synthesis of **cholesterol** which is also a precursor substance for the biosynthesis of two types of compounds—the steroid hormones and bile acids:

(*i*) **Cholesterol biosynthesis.** The cholesterol is synthesized from the acetate and its entire biosynthetic pathway involves about 20 steps, each step catalyzed by an enzyme. Out of these twenty enzymes, eleven enzymes are bounded to SER membranes, rest nine enzymes are the soluble enzymes located in the cytosol and mitochondria. Examples of SER-bound enzyme include **HMG-CoA reductase** and **squalene synthase** (see Chapter 18).

(*ii*) **Bile acid synthesis.** The biosynthesis of the bile acids represents a very complex pattern of enzymes and products. Enzymes involved in the biosynthetic pathway of bile acids are hydroxylases, mono-oxygenases, dehydrogenases, isomerases and reductases. For example, by the help of the enzyme **cholesterol 7α-hydroxylase**, the cholesterol is first converted into 7α-hydroxyl cholesterol, which is then converted into bile acids by the help of hydroxylase enzymes. The latter reaction requires NADPH and molecular oxygen and depends on the enzymes of electron transport chains of SER such as **cytochrome P-450** and **NADPH-cytochrome-c-reductase**.

(*iii*) **Steroid hormone biosynthesis.** Steroid hormones are synthesized in the cells of various organs such as the cortex of adrenal gland, the ovaries, the testes and the placenta. For example, cholesterol is the precursor for both types of sex hormones—estrogen and testosterone—made in the reproductive tissues, and the adrenocorticoids (*e.g.*, corticosterone, aldosterone and cortisol) formed in the adrenal glands. Many enzymes (*e.g.*, dehydrogenases, isomerases and hydroxylases) are involved in the biosynthetic pathway of steroid hormones, some of which are located in SER membranes and some occur in the mitochondria. This biosynthetic pathway has the following steps:

Acetate (Cytosol) → Cholesterol (SER) → Pregneolone (Mitochondria) → Progesterone (SER) →
- 11-Deoxycortisol (SER) → Aldosterone (Mitochondria)
- 11-Deoxycortisterone (SER) → Cortisol (Mitochondria)

4. Detoxification. Protectively, the ER chemically modifies **xenobiotics** (toxic materials of both endogenous and exogenous origin), making them more hydrophilic, hence, more readily excreted. Among these materials are drugs, aspirin (acetyl-salicylic-acid), insecticides, anaesthetics, petroleum products, pollutants and carcinogens (*i.e.*, inducers of cancer; *e.g.*, **3-4-benzopyrene** and **3-methyl cholanthrene**).

The enzymes involved in the detoxification of aromatic hydrocarbons are **aryl hydroxylases**. It is now known that benzopyrene (found in charcoal-broiled meat) is not carcinogenic, but under the action of aryl hydroxylase enzyme in the liver, it is converted into **5, 6-epoxide**, which is a powerful carcinogen.

A wide variety of drugs (*e.g.*, phenobarbital), when administrated to animals, they bring about the proliferation of the ER membranes (first RER and then SER) and/or enhanced activity of enzymes related to detoxification.

5. Other synthetic functions. SER plays a role in the synthesis of triglycerides in intestinal absorptive cells and of visual pigments from vitamin A by pigmented epithelial cell of retina. In plant cells, SER forms the surface where cellulose cell walls are being formed.

C. Functions of Rough Endoplasmic Reticulum

1. Protein synthesis. The major function of the rough ER is the synthesis of protein. It has long been assumed that proteins destined for secretion (*i.e.*, export) from the cell or proteins to be used in the synthesis of cellular membranes are synthesized on rough ER-bound ribosomes, while cytoplasmic proteins are translated for the most part on free ribosomes. In fact, the array of the rough endoplasmic reticulum provides extensive surface area for the association of metabolically active enzymes, amino acids and ribosomes. There is more efficient functioning of these materials to synthesize proteins when oriented on a membrane surface than when they are simply in solution, mainly because chemical combinations between molecules can be accomplished in specific geometric patterns.

The membrane-bound ribosomes are attached with **specific binding sites** or **receptors** of rough ER membrane by their large 60S subunit, with small or 40S subunit sitting on top like a cap. These receptors are membrane proteins which extend well into and possibly through the lipid bilayer. The receptor proteins with bound ribosomes can float laterally like other membrane proteins and may facilitate formation of the polysome and probably translation which requires that mRNA and ribosome move with respect to each other.

Further, the secretory proteins, instead of passing into the cytoplasm, appear to pass instead into the cisternae of the rough ER and are, thus, protected from protease enzymes of cytoplasm. It is calculated that about 40 amino acid residues long segment at the—COOH end of the nascent protein remains protected inside the tunnel of 'free' or 'bound' ribosomes and rest of the chain, with—NH_2 end, is protected by the lumen of RER. The passage of nascent polypeptide chain into the ER cisterna takes place during translation leaving only a small segment exposed to the cytoplasm at any one time.

How the polypeptide chain gets through the lipid bilayer is not so clear, but it is quite reasonable to propose that the membrane proteins serving as ribosomal receptors also has a very fine channel through its core that opens into the cisterna of the rough ER. The chain may have great flexibility, permitting the amino acids to snake their way single file through the proposed pore. As soon as growing polypeptide chain reaches the cisterna, it folds into its secondary and tertiary structures and thus trapped in the cisterna of the rough ER.

2. Protein glycosylation. The covalent addition of sugars to the secretory proteins (*i.e.*, glycosylation) is one of the major biosynthetic functions of rough ER. Most of the proteins that are isolated in the lumen of RER before being transported to the Golgi apparatus, lysosomes, plasma membrane or extracellular space, become glycoproteins (a notable exception is albumin). In contrast, very few proteins in the cytosol (cytoplasmic matrix) are glycosylated and those that carry them have a different sugar modification.

The process of **protein glycosylation** in RER lumen is one of the most well understood cell biological phenomena. During this process, a single species of **oligosaccharide** (which comprises N-acetyl-glucosamine, mannose and glucose, containing a total of 14 sugar residues) is transferred to proteins in the ER. Because it is always transferred to the NH_2 group on the side chain of an asparagine residue of the protein, this oligosaccharide is said to be **N-linked** or **asparagine-linked** (Fig. 19.8A). The transfer is catalyzed by a membrane-bound enzyme (*i.e.*, **glycosyl transferase**) with its active site exposed on the luminal surface of the ER membrane. The preformed precursor oligosaccharide is transferred *en bloc* to the target asparagine residue in a single enzymatic step almost as soon as that residue emerges in the lumen of ER during protein translocation (Fig. 19.8B). Since most proteins are co-translationally imported into the ER, N-linked oligosaccharides are almost always added during protein synthesis, ensuring maximum access to the target asparagine residues, which are present in the sequences—*Asn-X-Ser* or *Asn-X-Thr* (where *X* is amino acid except proline). *These two sequences, thus, function as signals for N-linked glycosylation.*

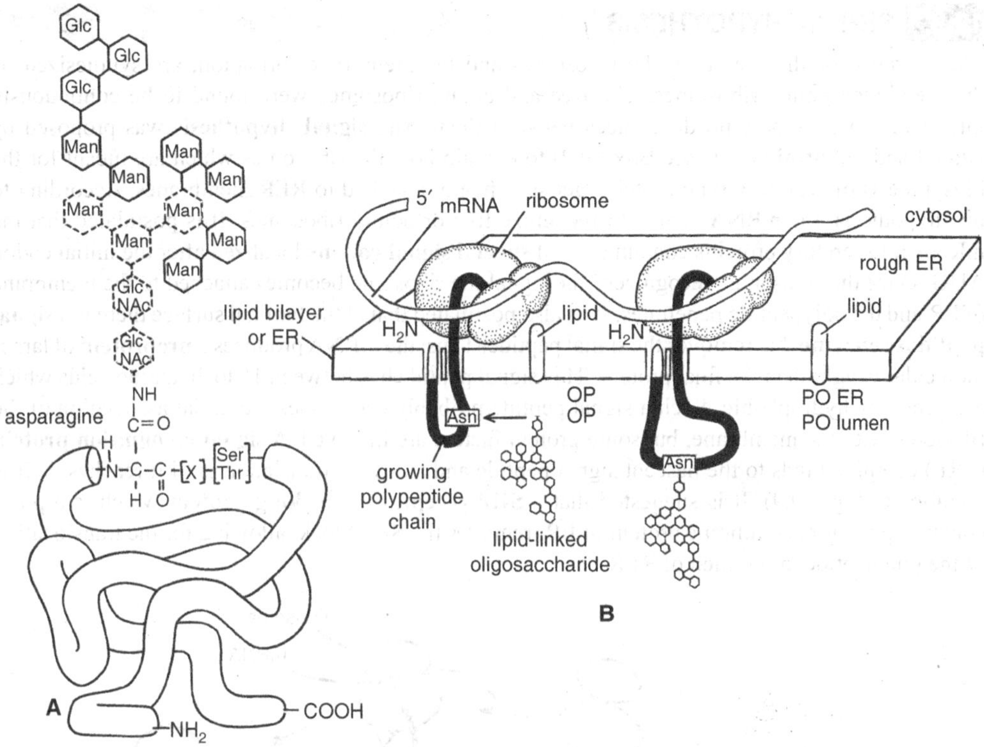

Fig. 19.8. N-linked glycosylation of protein in RER. A—The structure of asparagine-linked oligosaccharide. The sugars shown in shaded-form the 'core-region' of this oligosaccharide. For many glycoproteins, only the core sugars survive the extensive oligosaccharide trimming process in the Golgi apparatus; B—Mode of transfer of the oligosaccharide to the asparagine residues of the nascent protein inside RER lumen.

The precursor oligosaccharide is held in the ER membrane by a special lipid molecule, **dolicol** (the carrier). The oligosaccharide is linked to the dolicol by a high-energy **pyrophosphate bond** which activates the oligosaccharide for its transfer from the lipid to an asparagine side chain (*i.e.*, it provides activation energy for the glycosylation reaction). The oligosaccharide is built up sugar by sugar on the membrane-bound dolicol (towards the cytosolic side) prior to its transfer to a protein. Sugars are first activated in the cytosol (cytoplasmic matrix) by the formation of **nucleotide-sugar intermediates** (*e.g.*, UDP-glucose, UDP-N-acetylglucosamine, and GDP-mannose), which then donate their sugar (directly or indirectly) to the lipid in an orderly sequence. At some step of this process, the lipid-linked oligosaccharide is flipped from the cytosolic to the luminal side of the ER membranes. Dolicol is long and very hydrophobic: its 22 five-carbon units can span the thickness of lipid bilayer more than three times, so that the attached oligosaccharide is firmly anchored to the membrane.

While still in RER lumen, three glucose residues and one mannose residue are quickly removed from the oligosaccharides of most glycoproteins. Such oligosaccharide "trimming" or "processing" continues in the Golgi apparatus. If a glycoprotein is to contain a terminal glucose, fucose or sialic acid, then those sugars are added in the Golgi apparatus where the appropriate sugar transferase enzymes are localized.

3. Transportation of secretory proteins. From RER secretory proteins pass to Golgi apparatus, secretory vesicles and then to the outside of the cell (exocytosis).

19.9. SIGNAL HYPOTHESIS

The proteins for the secretion, the lysosomes and the membrane formation, are synthesized on the membrane bound ribosomes. The free and bound ribosomes were found to be continuously interchanging and show no differences between them. The **signal hypothesis** was proposed by **Blobel** and **Sabatini** (1971; see Box 19.2) to explain how the ribosomes which are meant for the biosynthesis of secretory type proteins get specifically attached to RER membranes. According to this hypothesis, the mRNA is able to recognize free or bound ribosomes. It is postulated that the mRNA for secretory proteins contain a set of **special signal codons** localized after the initial codon AUG. Once the ribosome "recognizes" the signal the ribosome becomes attached to the membrane of ER and the polypeptide penetrates. It is also postulated that at the luminal surface there is a **signal peptidase enzyme** that removes the signal peptide. Thus, the mRNA produces a **preprotein** of larger molecular weight than the final protein. This signal peptide has between 15 to 30 amino acids which are generally hydrophobic. Such a signal peptide probably establishes the initial association of the ribosome with the membrane, but some protein factors are involved. A **signal recognition protein** (SRP) complex binds to the nascent signal peptide and stops the translation until it reaches the ER membrane (Fig. 19.9). It is suggested that a **SRP receptor** or **docking protein** which is a pore-containing integral membrane protein of ER, removes the SRP block, allowing for the translocation of the polypeptide into lumen of RER.

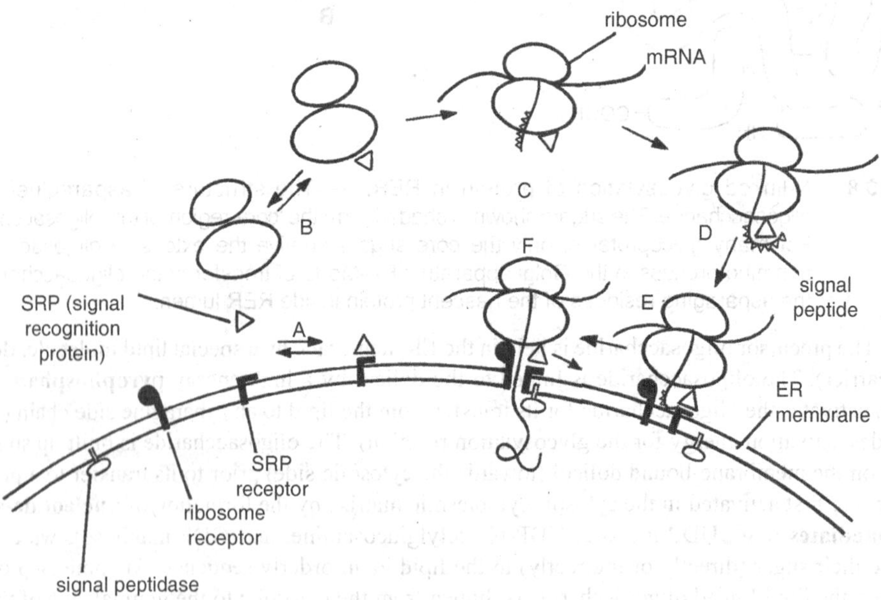

Fig. 19.9. Diagram explaining the signal hypothesis.

Box 19.2

Gunter Blobel of Rockefeller University, New York won the Nobel Prize for Medicine on 11.10.1999 for signal hypothesis. His research has shed new light on human disease including cystic fibrosis and early development of kidney stones.

19.10. ER EXPORT SIGNALS

Most proteins that enter the transitional ER move through the ER-Golgi intermediate compartment (ERGIC) and on the Golgi. These proteins are marked by sequences that signal either their export from or retention within the ER (Fig. 19.10). Many transmembrane proteins possess di-acidic or di-hydrophobic amino acid sequences in their cytosolic domains that function as ER export signals. Both GPI-anchored proteins and lumenal secretory proteins appear to be recognised and sequestered by these transmembrane receptor proteins (Box 19.3). Very few ER export signals have been detected on lumenal secretory proteins and their recognition may dpend on the shape of the correctly folded proteins. It is also possible that there is a default pathway where otherwise unmarked proteins in the ER lumen move to the Golgi and beyond.

Box 19.3

The GPI-anchored proteins are marked for export. GPI anchor stands for glycosylphosphophatidylinositol anchor. These are glycolipids containing phosphatidylinositol that anchor proteins to the external face of the plasma membrane (**Cooper** and **Hausman,** 2007).

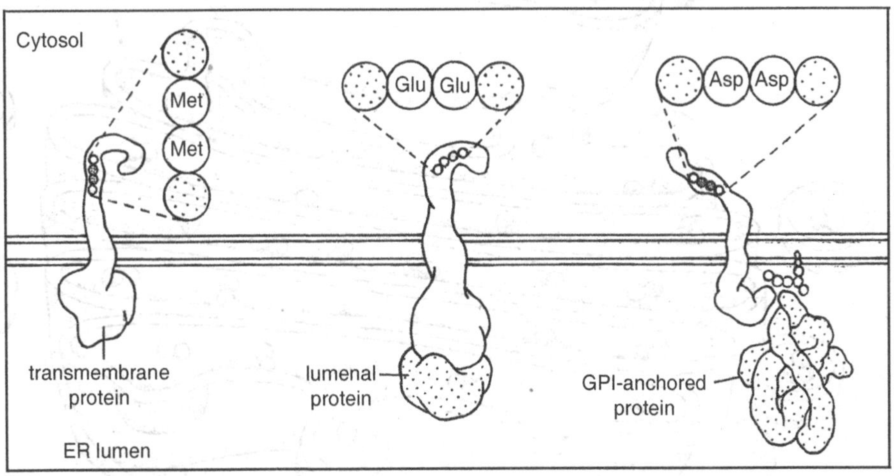

Fig. 19.10. ER export signals. Three types of secretory proteins are recruited into budding vesicles by cytosolic adaptor proteins. Transmembrane proteins are recognised by di-acidic (*e.g.*, Asp-Asp or Glu-Glu) or di-hydrophobic (*e.g.*, Met-Met) signal sequence in their cytosolic segments. Some of these are also receptor for lumenal and GPI-anchored membrane secretory proteins. The transmembrane receptors recognize the GPI-anchors and signal sequences or signal patches characteristic of the folded-lumenal proteins. (Asp=Aspartic acid; Glu=Glutamic acid; Met=Methionine) (after Cooper and Hausman, 2007).

If proteins that function within the ER (including BiP, signal peptidase, protein disulphide isomerase, and other enzyme discussed earlier) are allowed to proceed along the secretory pathway, they will be lost to the cell. Thus, many such proteins have a targeting sequence Lys-Asp-Glu-Leu (KDEL, in the single letter code) at their carboxy terminus that directs their retrieval back to the

ER. If this sequence is detected from a protein that normally function in the ER (*e.g.*, BiP or protein disulphide isomerase), the mutated protein is instead transported to the Golgi and secreted from the cell. Conversely, addition of KDEL sequence to the carboxy terminus of proteins that are normally secreted blocks their secretion. Some ER transmembrane proteins are similarly marked by short C-terminal sequences that contain two lysine residues (KKXX sequences).

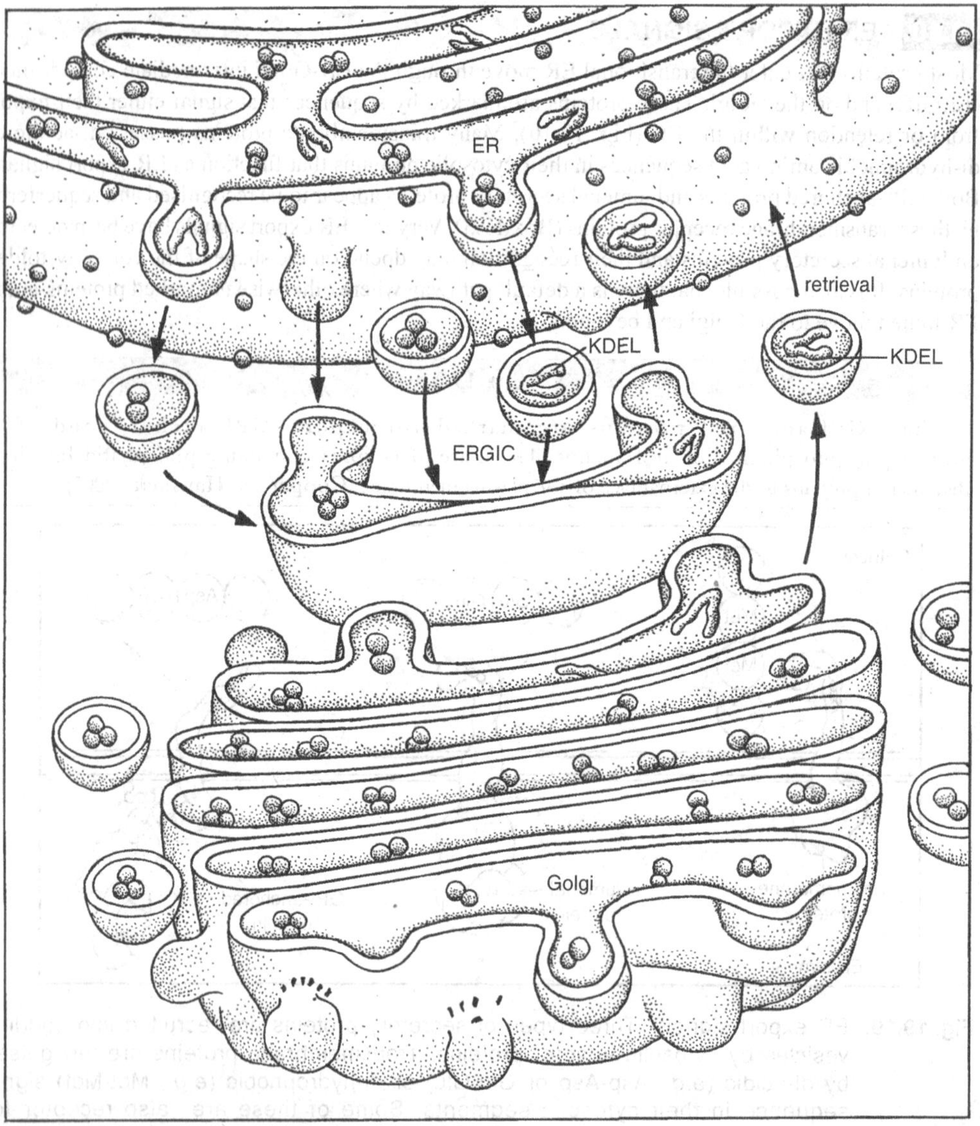

Fig. 19.11. Retrieval of resident ER proteins. Proteins destined to remain in the lumen of the ER are marked by the sequence Lys-Asp-Glu-Leu (KDEL) at their carboxy terminus. These proteins are exported from the ER to the Golgi, but they are recognised by a receptor in the ERGIC (ER Golgi intermediate compartment) or the Golgi apparatus and selectively returned to the ER (after Cooper and Hausman, 2007).

Curiously, the KDEL and KKXX signals do not prevent ER proteins from being packaged into vesicles and carried to the Golgi. Instead, these signals cause these ER resident proteins to be selectively retrieved from the ER-Golgi intermediate compartment or the Golgi-apparatus and returned to the ER via a recycling pathway (Fig. 19.11). Proteins bearing the KDEL and KKXX sequences bind to specific recycling receptors in the membranes of these compartments and are then selectively transported back to the ER. The KDEL and KKXX sequences are the best characterised retention/retrieval signals but there may be others. Other proteins are retrieved because they specifically bind to KDEL-bearing proteins such as BiP. Thus continued movement along the secretory pathway or retrieved back from the Golgi to the ER is the first branch point encountered by proteins being sorted to their correct destinations in the secretory pathway. Similar branch points arise at each subsequent stage of transport, such as retention in the Golgi versus export to lysosomes or the plasma membrane. In each case specific localization signals target proteins to their correct intracellular destinations.

Differences in Endoplasmic Reticulum and Golgi Complex/Apparatus

Table 19.2. Differences in endoplasmic reticulum and Golgi apparatus.

Endoplasmic reticulum	Golgi apparatus
1. It forms tubular network which traverses endoplasmic region of cytoplasm.	1. It contains stack of flattened tubules and has localized distribution in cytoplasm.
2. It is of two types: (*a*) Smooth ER (*b*) Rough ER	2. It is of one type.
3. It is involved in synthesis and transportation.	3. It is involved in processing and sorting of secretory products which are synthesized by smooth ER and rough ER.
4. Smooth ER is involved in metabolism of glycogen, lipids and steroids.	4. It is involved in carbohydrate synthesis, *e.g.*, cellulose, oligosaccharides.
5. Rough ER is involved in synthesis of protein for membrane, lysosomes and secretory vesicles. Process of glycosylation begins in it.	5. Process of glycosylation and sorting/targetting of proteins is completed in it.
6. It forms new nuclear envelope after each cell division.	6. It forms lysosomes.

QUESTIONS

Long Answer Questions

1. Describe the ultrastructure, types and functions of endoplasmic reticulum.
2. Give the location, structure and function of endoplasmic reticulum.
3. What are the functions of endoplasmic reticulum?
4. What do you mean by membrane system? Describe the same with reference to ER and write about its functions.
5. Describe the endoplasmic reticulum and ribosomes.
6. Give an account of the structure and functions of endoplasmic reticulum.
7. Give an account of endomembrane system.

Short Answer Questions

1. Give a brief account of microsomes.
2. Mention two important functions of RER.

3. What interrelation exists between SER and RER?
4. A cell has more of SER but just a few RER cisternae. What inference about the cell functions can be drawn?
5. Describe the mechanism of protein glycosylation in the lumen of RER.
6. Differentiate between rough ER and smooth ER.
7. Distinguish between endoplasmic reticulum and Golgi apparatus.
8. Write short notes on the following:
   (*i*) Endoplasmic reticulum or transporting system.
   (*ii*) Functions of endoplasmic reticulum.
   (*iii*) Functions of smooth ER.

Very Short Answer Questions

1. Who coined the term endoplasmic reticulum?
2. Which eukaryotic cell is devoid of ER?
3. Name the type of ER associated with glycogen metabolism.
4. What do you understand by sarcoplasmic reticulum?
5. Which cell organelle is responsible for the transport of secretory products to various other cell organelles?

Fill in the Blanks

(*i*) The type of ER is associated with glycogen metabolism is ....

(*ii*) It was possible to observe elements of ER with the help of ....

(*iii*) .... discovered endomembrane system.

(*iv*) .... coined the term endoplasmic reticulum.

(*v*) The eukaryotic cell which is without ER is ....

(*vi*) SER stands for ....

(*vii*) .... membranes develops from ER during cell division.

(*viii*) .... ER is associated with glycoprotein synthesis.

(*ix*) .... ER is developed in cell that secretes mucus.

(*x*) The appearance of ER is dependent on the .... of the cell.

Multiple Choice Questions

1. What part of the cell forms nuclear envelope during telophase?
   (*a*) cytoskeleton
   (*b*) centriole
   (*c*) Golgi complex
   (*d*) endoplasmic reticulum
2. Internal membranes or cytomembranes are extensions of cell membrane
   (*a*) the membranes have unit membrane structure
   (*b*) all the organelles are bounded by the same membrane
   (*c*) have same thickness as that of cell membrane
   (*d*) the membrane of ER are connected at certain points with the cell membrane
3. The enzyme glucose-6-phosphatase is present in
   (*a*) endoplasmic reticulum
   (*b*) lysosome
   (*c*) mitochondria
   (*d*) nucleus
4. RER is well developed in cells engaged in synthesis of
   (*a*) nucleotides
   (*b*) proteins
   (*c*) lipids
   (*d*) secretory products
5. The cells lacking the endoplasmic reticulum are
   (*a*) amphibian monocytes
   (*b*) mature erythrocytes of mammals
   (*c*) mature leucocytes
   (*d*) mammalian monocyte

ANSWERS

Very Short Answer Questions

1. Porter.
2. RBC or erythrocyte of mammals.
3. Smooth endoplasmic reticulum.
4. Smooth E.R. of muscle cells is called sarcoplasmic reticulum.
5. Endoplasmic reticulum.

Fill in the Blanks

(*i*) SER
(*ii*) Electron microscope
(*iii*) Porter
(*iv*) Porter
(*v*) Red blood cell of mammal
(*vi*) Smooth endoplasmic reticulum
(*vii*) Nuclear
(*viii*) Rough
(*ix*) Rough
(*x*) Secretion.

Multiple Choice Questions

1. (*d*) 2. (*a*) 3. (*a*) 4. (*b*) 5. (*b*)

20 Ribosomes

The ribosomes are small, dense, rounded and granular particles of the ribonucleoprotein. They occur either freely in the matrix of mitochondria, chloroplasts and cytoplasm (*i.e.*, cytoplasmic matrix) or remain attached with the membranes of the endoplasmic reticulum and nucleus. They occur in most prokaryotic and eukaryotic cells and are known to provide a scaffold for the ordered interaction of all the molecules involved in protein synthesis.

Box 20.1 Some Historical Facts.

1. **George E. Palade**, in 1952, described the ribosomes.
2. **Philip Siekevitz** and **Paul Zamecnik**, in 1952, showed that radioactive amino acids were incorporated into proteins on ribosomes.
3. In 1953, ribosomes were observed in plant cells (roots of bean) by **Robinson** and **Brown**.
4. Presence of ribosomes in both free and membrane attached form was confirmed in the animal cells by **Palade** and **Siekevitz** by electron microscopy.
5. In 1956, **Palade** detected RNA in ribosomes.
6. In 1958, the term **ribosome** was coined by **R.B. Roberts.**
7. **Tissieres** and **Watson** (1958) isolated *E.coli* ribosomes and showed that they consist of two subunits, 50S and 30S.
8. The first definitive evidence linking the nucleolus with eukaryotic ribosome formation was provided in 1960s by **Robert Perry.**
9. In 1970, **Oscar Miller, Jr.** developed techniques to visualise transcription and translation of silk protein (*i.e.*, polyribosomes) in silk gland cell.
10. **James A. Lake**, in 1981, suggested the well accepted **asymmetrical model** for 70S ribosome.

20.1. OCCURRENCE

The ribosomes occur in cells, both prokaryotic and eukaryotic cells. In prokaryotic cells the ribosomes often occur freely in the cytoplasm. In eukaryotic cells the ribosomes either occur freely in the cytoplasm or remain attached to the outer surface of the membrane of endoplasmic reticulum. The yeast cells, reticulocytes or lymphocytes, meristamatic plant tissues, embryonic nerve cells and cancerous cells contain large number of ribosomes which often occur freely in the cytoplasmic matrix. The cells in which active protein synthesis takes place, the ribosomes remain attached with the membranes of the endoplasmic reticulum. Such cells are the pancreatic cells, plasma cells, hepatic parenchymal cells, nissl bodies, osteoblasts, serous cells, or the submaxillary gland, chief cells of the glandular stomach, thyroid cells and mammary gland cells. The cells which synthesize specific proteins for the intracellular utilization and storage often contain large number of free ribosomes. Such cells are the erythroblasts, developing muscle cells, skin and hair.

20.2. TYPES OF RIBOSOMES

According to the size and the sedimentation coefficient (S) two types of ribosomes have been recognised (Fig. 20.1).

1. 70S ribosomes. The 70S ribosomes are comparatively smaller in size and have sedimentation coefficient 70S and the molecular weight 2.7 × 106 daltons. **(Dalton is the unit of molecular weight (MW)**; one dalton equals the weight of hydrogen atom. For example, a water molecule weighs 18 daltons). They occur in the prokaryotic cells of the blue green algae and bacteria and also in mitochondria and chloroplasts of eukaryotic cells.

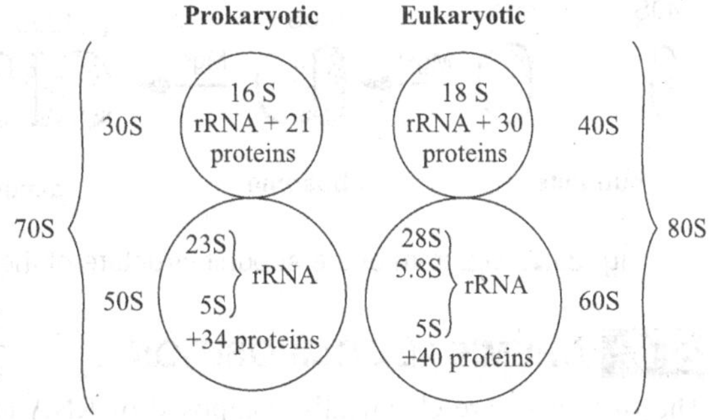

Fig. 20.1. Various components of prokaryotic (70S) and eukaryotic (80S) ribosomal subunits.

2. 80S ribosomes. The 80S ribosomes have the sedimentation coefficient of 80S and the molecular weight 40 × 10^6 daltons. The 80S ribosomes occur in eukaryotic cells of the plants and animals.

The ribosomes of mitochondria and chloroplasts are always smaller than 80S cytoplasmic ribosomes and are comparable to prokaryotic ribosomes in both size and sensitivity to antibiotics, although their sedimentation values vary in different phyla, *e.g.*, 77S in mitochondria of fungi, 55S in mitochondria of mammals and 60S in mitochondria of animals in general. The ribosomes of chloroplasts are 70S type.

20.3. NUMBER OF RIBOSOMES

An *E. coli* cell contains 10,000 ribosomes, forming 25 per cent of the total mass of the bacterial cell. In contrast, mammalian cultured cells contain 10 million ribosomes per cell, each of which is about twice as large as a prokaryotic ribosome.

20.4. STRUCTURE OF RIBOSOMES

The ribosomes are oblate spheroid structures of 150 to 250 Å in diameter. Each ribosome is porous, hydrated and composed of two subunits. One ribosomal subunit is large in size and has a dome-like shape, while the other ribosomal subunit is smaller in size and occurring above the larger subunit and forming a cap-like structure.

The 70S ribosome consists of two subunits, *viz.*, 50S and 30S. The 50S ribosomal subunit is larger in size and has the size of 160 Å to 180 Å. The 30S ribosomal subunit is smaller in size and occurs above the 50S subunit like a cap.

The 80S ribosome also consists of two subunits, *viz.*, 60S and 40S. The 60S ribosomal subunit is dome-shaped and larger in size. In the ribosomes which remain attached with the membranes of endoplasmic reticulum and nucleus, etc., the 60S subunit remains attached with the membranes. The 40S ribosomal subunit is smaller in size and occurs above the 60S subunit forming a cap-like structure. Both the subunits remain separated by a narrow cleft (Fig. 20.1).

The two ribosomal subunits remain united with each other due to high concentration of the Mg^{2+} (.001M) ions. When the concentration of Mg^{++} ions reduces in the matrix, both ribosomal subunits get separated. Actually in bacterial cells the two subunits are found to occur freely in the cytoplasm and they unite only during the process of protein synthesis. At high concentration of Mg^{++} ions in the matrix, the two ribosomes (called **monosomes**) become associated with each other and known as the **dimer**. Further, during protein synthesis many ribosomes are aggregated due to common messenger RNA and form the **polyribosomes** or **polysomes** (Fig. 20.2).

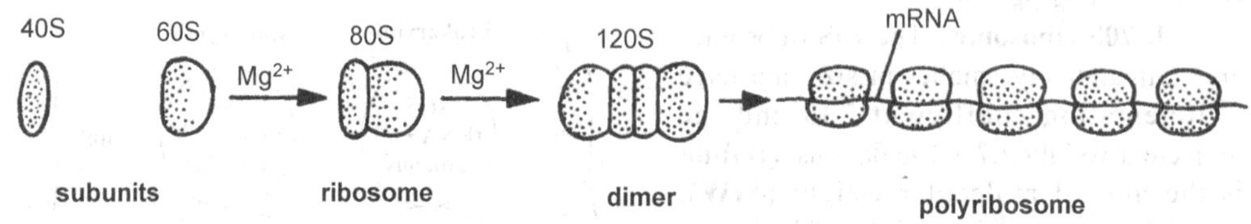

Fig. 20.2. Diagram of the subunit structure of the ribosome and the influence of Mg^{2+} ions.

20.5. CHEMICAL COMPOSITION

The ribosomes are chemically composed of RNA and proteins as their major constituents; both occurring approximately in equal proportions in smaller as well as larger subunit. However, the 70S ribosomes contain more RNA (60 to 40%) than the proteins (36 to 37%), *e.g.*, the ribosomes of *E. coli* contain 63% rRNA and 37% protein. While the 80S ribosomes contain less RNA (40 to 44%) than the proteins (60 to 56%), *e.g.*, yeast ribosomes have 40 to 44% RNA and 60 to 56% proteins; ribosomes of pea seedling contain 40% RNA and 60% proteins. There is no lipid content in ribosomes.

1. Ribosomal RNAs.

The 70S ribosomes contain three types of rRNA, *viz.*, **23S rRNA, 16S rRNA** and **5S rRNA**. The 23S and 5S rRNA occur in the larger 50S ribosomal subunit, while the 16S rRNA occurs in the smaller 30S ribosomal subunit. Assuming an average molecular weight for one nucleotide to be 330 daltons, one can calculate the total number of each type of rRNA. Thus, the 23S rRNA consists of 3300 nucleotides, 16S rRNA contains 1650 nucleotides and 5S rRNA includes 120 nucleotides in it.

The 80S ribosomes contain four types of rRNA, *viz.*, **28S rRNA** (or **25-26 rRNA** in plants, fungi and protozoa), **18S rRNA, 5S rRNA** and **5.8S rRNA**. The 28S, 5S and 5.8S rRNAs occur in the larger 60S ribosomal subunit, while the 18S rRNA occurs in the smaller 40S ribosomal subunit. About 60 per cent of the rRNA is helical (*i.e.*, double stranded) and contains paired bases. These double stranded regions are due to hairpin loops between complimentary regions of the linear molecule.

The 28S rRNA has the molecular weight 1.6×10^6 daltons and its molecule is double stranded and having nitrogen bases in pairs. The 18S rRNA has the molecular weight 0.6×10^6 daltons and consists of 2100 nucleotides. The 18S and 28S ribosomal RNAs contain a characteristic number of methyl groups, mostly as 2′-O-methyl ribose. The molecule of 5S rRNA has a clover leaf shape and a length equal to 120 nucleotides. The 5.8S rRNA is intimately associated with the 28S rRNA molecule and has, therefore, been referred to as **28S-associated ribosomal RNA (28S A rRNA).**

The 55S ribosomes of mammalian mitochondria lack 5S rRNA but contain **21S** and **12S rRNAs**. The 21S rRNA occurs in larger or 35S ribosomal subunits, while 12S rRNA occur in smaller or 25S ribosomal subunit.

It is thought that each ribosomal subunit contains a highly folded ribonucleic acid filament to which the various proteins adhere. But as the ribosomes easily bind the basic dyes so it is concluded that RNA is exposed at the surface of the ribosomal subunits, and the protein is assumed to be in the interior in relation to non-helical part of the RNA.

2. Ribosomal Proteins.

Each 70S ribosome of *E. coli* is composed of about 55 **ribosomal proteins**. Out of these 55 proteins, about 21 different molecules have been isolated from the 30S ribosomal subunit, and some 32 to 34 proteins from the 50S ribosomal subunit. The primary structure of several of these proteins has been elucidated. Most of the recent knowledge about the structure of ribosomal proteins has been achieved by dissociation of ribosomal subunits into their component rRNA and protein molecules. When both 50S and 30S ribosomal subunits are dissociated by centrifuging both of them in a gradient of 5 M cesium chloride, then there are two inactive core particles (40S and 23S, respectively) which contain the RNA and some proteins called **core proteins (CP)**. At the same time several other proteins—the so-called **split proteins (SP)** are released from each particle (Fig. 20.3). There are SP50 and SP30 proteins which may reconstitute the functional ribosomal subunit when added to their corresponding core. Some of the split proteins are apparently specific for each ribosomal subunit. The split proteins have been further fractionated and divided into acidic (A) and basic (B) proteins. **Nomura** *et al.*, (1968) fractionated at least six different groups of proteins in the ribosome (Fig. 20.3).

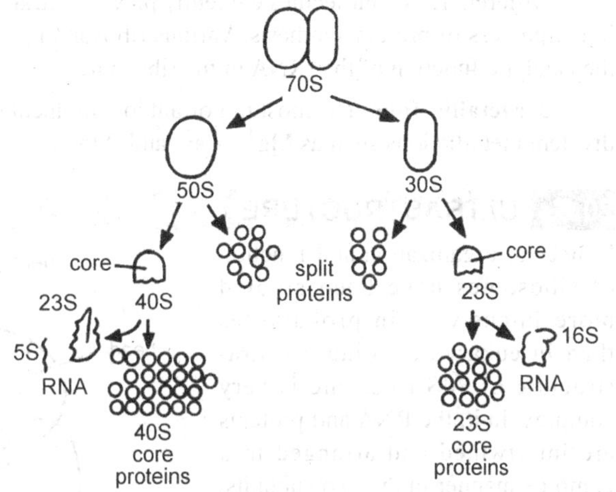

Fig. 20.3. Diagram showing the stepwise dismantling of the two subunits of 70S ribosome. Note that the proteins may be separated into split and core proteins. The 50S subunit contains 23S and 5S RNAs, and the 30S subunit has 16S RNA.

In all, 21 types of proteins have been isolated in smaller subunit (30S) of ribosome of *E. coli*. These are designated as S1 to S21. Similarly, in larger subunit (50S) 34 different proteins designated as L1 to L34, have been isolated. Thus, the 70S ribosome was thought to consist of 55 different proteins. However it was later shown that protein S20 is identical to L6, thus, the correct number of S proteins is 20. Likewise, L8 was shown to be an aggregate of proteins L7, L12 and L10; thus, the correct number of L proteins is 33. Thus, the prokaryotic 70S ribosome consists of 53 different proteins (20S + 33L = 53 proteins). Similar organization of ribosomal proteins and RNA is found in 80S ribosomes (Fig. 20.4). Small 40S subunit comprises 33 proteins and large 60S subunit contains more than 45 proteins (*i.e.*, 50 proteins; **Twyman** 1998).

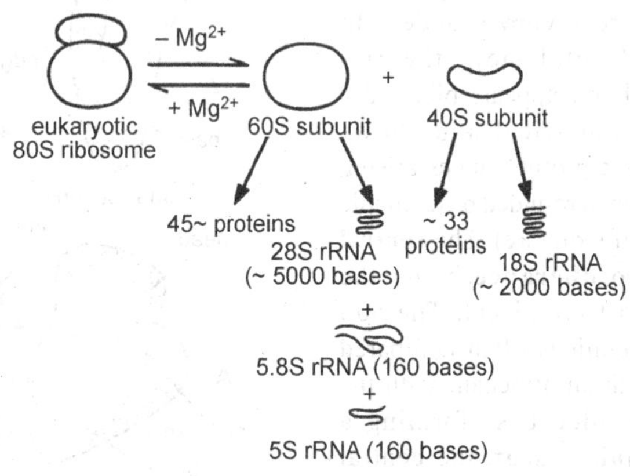

Fig. 20.4. Different RNA and protein components of eukaryotic ribosomes.

Different rRNA molecules evidently play a central role in the catalytic activities of ribosomes in the process of protein synthesis. Various ribosomal proteins have been found to mainly enhance the catalytic function of the rRNA in the ribosomes.

3. Metallic Ions. The most important low molecular weight components of ribosomes are the divalent metallic ions such as Mg^{2+}, Ca^{2+} and Mn^{2+}.

20.6. ULTRASTRUCTURE

Molecular organization and function of ribosomes have been studied more intensively in prokaryotes than in eukaryotes. Fine or ultra-structure of 70S ribosome is very complex. In it, the RNA and proteins are intertwined and arranged in a complex manner in the two subunits. Recently following two models have been suggested to explain the three-dimensional structure of prokaryotic or 70S ribosomes:

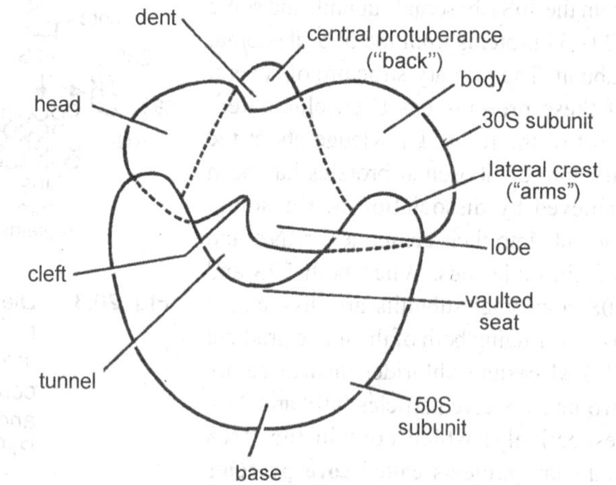

Fig. 20.5. Stoffler and Wittman's model of 70S ribosome.

1. Stoffler and Wittman's Model (Quasi-symmetrical model, 1977).

According to this model the 30S ribosomal subunit has an elongated, slightly bent prolate shape (Fig. 20.5). It is a bipartite structure. A transverse hollow or cleft divides the 30S subunit into two parts, a smaller **head** and larger **body**, giving it the appearance of a telephone receiver or embryo. In electron microscopy 50S ribosomal subunit showed various shapes depending on structure seen in different views such as frontal-maple leaf, lateral-kidney shaped or rear view-rounded. In a frontal view, the 50S subunit appears bilaterally symmetrical and shows three protuberances arising from a rounded base (maple leaf structure). The **central protuberance** being the most prominent. The 50S subunit is often compared with an armchair, with the rounded base forming a **vaulted seat**, the central protuberance forming the **back** and the lateral protuberances the **arms** of chair.

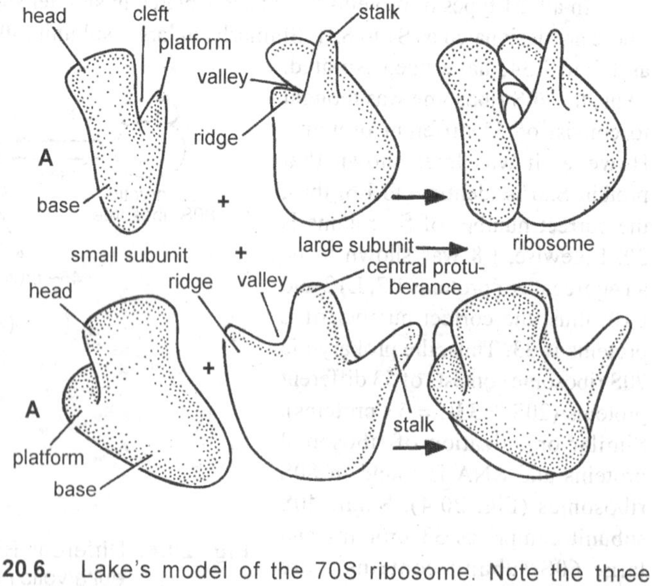

Fig. 20.6. Lake's model of the 70S ribosome. Note the three dimensional structure of the ribosome in two different orientations.

When 30S and the 50S subunits become associated to form the 70S ribosome, the frontal face of the 30S subunit with its hollow faces the vaulted seat of the 50S subunit. The long axis of 30S subunit is oriented transversely to the central protuberance of the 50S subunit. A **tunnel** is formed between the hollow of the small subunit and vaulted seat of the large subunit.

2. Lake's Model (Asymmetrical model, 1981).

This completely asymmetrical model of ribosome has been suggested by **James A. Lake** (1981). The smaller subunit has a **head**, a **base** and a **platform**. The platform separates the head from the base by the help of a **cleft**. This cleft is an important functional region; it is suggested to be the site of codon-anticodon interaction and as a part of binding site for initiation factors of protein synthesis.

The large subunit consists of a **ridge**, a **central protuberance** and a **stalk**. The ridge and the central protuberance are separated with the help of a valley (Fig. 20.6).

20.7. THREE DIMENSIONAL MODEL OF 80S RIBOSOME

In spite of the difference in overall sizes (as manifested in the greater molecular weights, sedimentation constants, sizes and numbers of rRNAs and proteins), the cytoplasmic ribosomes of eukaryotes (80S) are remarkably similar in morphology to those of prokaryotes. As in 30S subunits of prokaryote ribosomes, the 40S ribosomal subunit of eukaryotes is divided into **head** and **base** segments by a transverse groove (Fig. 20.7). The 60S ribosomal subunit is generally rounder in shape than the small subunit, although its one side is flattened; this is the side that becomes confluent with the small subunit during the formation of the monomer or monosome (*i.e.*, functional 80S ribosome).

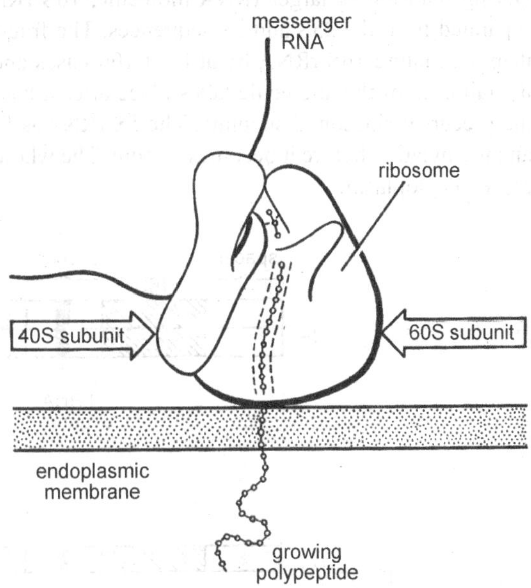

Fig. 20.7. A three dimensional model of eukaryotic ribosome.

Several **domains** of the ribosomes have particularly important functions during protein synthesis. The small subunit contains a binding site for tRNA. The **A-site (amino-acyl-tRNA site)** binds incoming charged tRNAs during elongation, the **P-site (peptidyl-tRNAs site)** binds the tRNA carrying the nascent polypeptide chain. Bacterial 70S ribosomes possess a third **E-site (exit site)** to which spent tRNA are dispatched prior to ejection. The large subunits possess a **peptidyl-transferase domain**, which provides the catalytic activity for peptide bond formation and a GTPase domain whose activities are required for translocation of the ribosome.

20.8. COMPARISON OF 70S AND 80S RIBOSOMES

Eukaryotic 80S ribosomes differ from prokaryotic 70S ribosomes in the following respects: 1. they are considerably larger; 2. they contain a large number of proteins (70–80 types of proteins instead of 53); 3. they have four types of RNA molecules instead of three types; 4. their proteins and nucleic acids are large-sized; 5. the RNA-protein ratio is near to 1 : 1 instead of 2 : 1 and 6. several antibiotics, such as **chloramphenicol**, inhibits bacterial but not eukaryotic ribosomes (this is the basis of the use of many antibiotics in medical treatment). Protein synthesis by eukaryotic ribosomes is inhibited by **cycloheximide**.

However, eukaryotic ribosomes do not differ functionally from those in prokaryotes in a fundamental way; they perform the same functions, by the same set of chemical reactions. The genetic code is the same for all living organisms, and eukaryotic ribosomes are able to translate bacterial mRNAs efficiently, provided that a "cap" is added enzymatically.

20.9. BIOGENESIS OF RIBOSOMES

Ribosomes are not self-replicating particles. Synthesis of various component of ribosomes such as rRNAs and proteins, are under genetic control, *i.e.*, rRNAs and mRNAs (for various ribosomal proteins) are transcribed by genes (DNA). Since the mechanism of biogenesis of 70S and 80S ribosomes differ greatly, so can be studied separately as follows:

1. Biogenesis of 70S ribosomes. In bacteria a single gene transcript containing the sequences of 16S, 23S and 5S rRNAs, is synthesized by a rRNA operon and this larger molecule undergoes both tailoring and chemical modifications before each rRNA molecule assumes its mature form. During tailoring of larger rRNA molecule, 16S rRNA sequence is first of all cleaved off and is separated from the 23S and 5S sequences. The fragment containing 16S information is still larger than the mature 16S rRNA by at least 100 bases and is not methylated: both the methylation and the tailoring of this molecule takes place after it has associated with a number of proteins to form the precursor ribosomal subunits. The 5S rRNA is found not to undergo the processes of tailoring and methylation before it becomes mature. The whole process of biogenesis of 70S ribosomes takes place in cytoplasm.

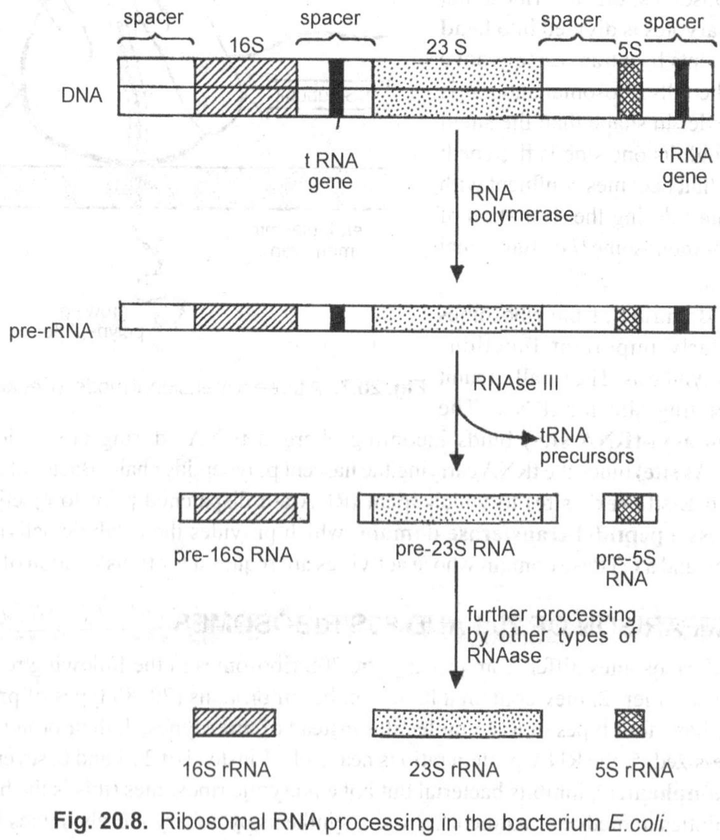

Fig. 20.8. Ribosomal RNA processing in the bacterium *E.coli*.

2. Biogenesis of 80S ribosomes. In eukaryotes, the biogenesis of ribosomes is much more complex and involves a long-lasting process in which several regions of cell are involved. The 5.8S, 18S and 28S rRNAs are transcribed as a much larger molecule in the **nucleolar organizer (NO)** which contains many copies 5.8S, 18S and 28S rRNA genes or **ribosomal DNA** (*i.e.*, there is gene redundancy or amplification). The DNA coding for the 5S rRNA is also highly repetitive, but the molecule is synthesized outside the nucleolus. It is in the nucleolus that newly synthesized rRNA accumulates and becomes associated, presumably by a self-assembly process, with 50 or more ribosomal proteins that have been synthesized in the cytoplasm by usual mechanism of protein synthesis and then migrate to the cytoplasm of cell, in the form of ribosomal subunits.

A. Ribosomal RNA synthesis inside nucleolus. All organisms have multiple rRNA genes. In case of *Xenopus*, each nucleolar organizer contains 450 rRNA genes. These genes are **tandemly repeated** or **reiterated** along the DNA molecule (in a head to tail arrangement) and are separated from each other by stretches of **spacer DNA**, which is not transcribed. These rRNA genes are being actively transcribed and the nascent RNA chains are spread perpendicularly to the DNA axis. Each gene is transcribed into a long RNA molecule (which varies in size from 40S to 45S according to species) which will eventually be processed to give rise to 18S, 28S and 5.8S rRNA (Fig. 20.8). Because each rRNA gene has a fixed initiation site (promoter) and a fixed termination site, the transcripts adopt the characteristic "christmas tree" or "fern leaf" configuration. Nucleolar rRNA genes are transcribed by **RNA polymerase I** (about 100 enzymes per gene). RNA polymerase I molecules are found to remain bound to the nucleolar organisers during mitotic metaphase and anaphase. During this period there is no RNA synthesis and so the enzyme molecules must remain in an inactive state.

Processing of rRNAs inside nucleolus. As already described, rRNA genes are transcribed into a long precursor RNA (which is 40S in *Xenopus* and 45S in HeLa and other human cells); this precursor must be cleaved into 18S, 28S and 5.8 rRNA. In the cleavage process about 50 per cent of the precursor RNA is degraded. In HeLa cell, the processing of rRNA involves the following steps:

(*i*) The first ribosomal RNA in HeLa cells is a large 45S molecule of 14,000 nucleotides. Within this precursor molecule the rRNAs are separated by stretches of spacer RNA and the order of transcription is: 5′ end—18S—5.8S—28S—3′ end. On a fully active gene about 100 RNA polymerase I enzymes (along with transcription factor I or TFI) are transcribing simultaneously on the rRNA gene.

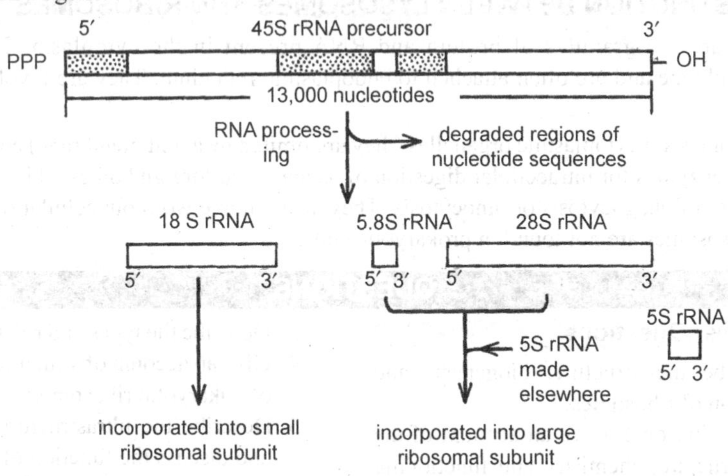

Fig. 20.9. The pattern of processing of 45S rRNA precursor molecule into three separate ribosomal RNAs.

(*ii*) In nucleolus, 45S RNA is rapidly **methylated**, even before transcription is completed. Methylations occur mostly on the ribose moiety (producing 2′-O- methyribose) and occur only in the 18S (46 methylations) and 28S (71 methylations) sequences that have to be conserved. Those segments of 45S which have to be degraded are not methylated.

(*iii*) 45S RNA has a lifetime of about 15 minutes and is then cleaved into smaller components as follows:

$$45S \rightarrow 41S \begin{cases} 32S \rightarrow 28S + 5.8S \\ 20S \rightarrow 18S \end{cases}$$

(*iv*) 20S RNA is rapidly processed in 18S rRNA and probably due to this reason the small ribosomal subunits appear in the cytoplasm earlier than the large ribosomal subunits. The large ribosomal subunits have a slower RNA processing.

(*v*) 32S RNA remains in the nucleolus for about 40 minutes and is then cleaved into 28S rRNA and 5.8S rRNA. Both of these rRNAs persist in the nucleolus for another 30 minutes before entering the cytoplasm as part of the large ribosomal subunit.

B. Biosynthesis of ribosomal proteins. The early embryo of frog is found to contain the mRNAs for all the 70 ribosomal proteins, and except four of these mRNAs, all are not translated until the midblastula stage when synthesis of rRNAs is switched on in the nucleolus. The remaining four mRNAs are translated at all times.

C. 5S RNA synthesis. The 5S rRNA is synthesized from 20,000 genes in the oocytes but only from 400 genes in the somatic cells, which differ slightly in sequence (six nucleotides out of 120). A gene for 5S rRNA contains an **internal control region (ICR)** in its middle region which is found essential for transcription. To this control region of gene remains attached a special protein, called **transcription factor IIIA** or **TF III A** which permits RNA polymerase III enzyme to recognize the promoter of a 5S rRNA gene (*i.e.*, TF III A initiates the synthesis of 5S rRNA).

20.10. FUNCTIONS

Ribosomes play a very significant role during biosynthesis of proteins. With the help of mRNA and rRNA ribosomes are able to form long polypeptide chain of amino acids from the cytoplasm.

20.11. DISTINCTION BETWEEN LYSOSOMES AND RIBOSOMES

Ribosomes are the granules of protein and RNA present in the cytoplasm. They lack any limiting membrane and are often attached to endoplasmic reticulum. They are involved in protein synthesis.

Lysosomes are cytoplasmic organelles (they are limited by a unit membrane) containing about 50 digestive enzymes for intracellular digestion of bacteria and foreign bodies which enter the cells by the process of phagocytosis or pinocytosis. They also remove worn out cellular organelles from the cells. Lysosomes are not found in prokaryotic cells.

QUESTIONS

Long Answer Questions

1. Describe the structure, biogenesis and function of ribosomes.
2. Define ribosome. Give an account of the ultrastructure, chemistry and functioning of ribosomes. Explain their role in protein synthesis.
3. Describe the types and roles of ribosomes.
4. Give an account of structure and functions of eukaryotic ribosomes.
5. Describe the ultrastructure of ribosomes and discuss the function of nucleolus as a ribosome factory.
6. Write an essay on ribosomes.

Short Answer Questions

1. Write short notes on the following:
   (*i*) Ribosomes.
   (*ii*) Polysomes.
   (*iii*) Ultrastructure of ribosome.
   (*iv*) Ribosome biogenesis.
   (*v*) Types of ribosomes.
   (*vi*) Different types of RNA.
   (*vii*) 80S subunit of ribosome.
2. Distinguish between ribosome and lysosome.
3. Define ribosome. Describe Lake's asymmetrical model of ribosome.
4. Prepare a well-labelled diagram of J.A. Lakes model of 70S ribosome.

Very Short Answer Questions

1. Define the ribosome.
2. What is polyribosome?
3. What is rRNA?
4. What is the main function of ribosome?
5. What are the main chemical constituents of ribosome?
6. Who gave the asymmetrical model of the ribosome?
7. Name the three parts of smaller subunit of ribosome.

YES OR NO ANSWER

(*i*) RNA plays an important role in protein synthesis.
Ans. Yes.

Fill in the Blanks

(*i*) Amino acids are linked together to form polypeptide chain by ....
(*ii*) Ribosome is the site of ....
(*iii*) Ribosome were first observed by ....
(*iv*) The term ribosome was used by ....
(*v*) The principal site of ribosomal RNA in the cell is the ....
(*vi*) During protein synthesis mRNA gets attached to .... subunit of ribosome.
(*vii*) Ribosome are formed in .... as .... and .... subunits.
(*viii*) The bacterial ribosomes are formed of .... and .... subunits.
(*ix*) Ribosomal RNA of large subunit is synthesized as .... RNA.

Multiple Choice Questions

Pick the correct option:

1. Ribosomes help in
   (*a*) lipogenesis
   (*b*) cellular digestion
   (*c*) protein synthesis
   (*d*) photosynthesis
2. The sedimentation constant of eukaryotic ribosome is generally 80S. What is the sedimentation constant of its subunits?
   (*a*) 40S and 30S (*b*) 50S and 30S
   (*c*) 60S and 30S (*d*) 60S and 40S
3. Proteins are synthesized by
   (*a*) ribosomes (*b*) mitochondria
   (*c*) Golgi apparatus (*d*) mesosomes
4. Ribosomes on hydrolysis produce
   (*a*) acidic proteins and rRNA
   (*b*) basic proteins and rRNA
   (*c*) split proteins and rRNA
   (*d*) proteins, rRNA and DNA
5. If all the ribosomes of a cell are destroyed
   (*a*) respiration will not take place
   (*b*) protein synthesis will not take place
   (*c*) photosynthesis will stop
   (*d*) it will have no effect
6. Cell organelle devoid of DNA is also lacking in
   (*a*) ribosomes (*b*) nucleus
   (*c*) plastids (*d*) mitochondria
7. Prokaryotic ribosomes are
   (*a*) 30S (*b*) 50S
   (*c*) 70S (*d*) 80S
8. Which of the following are non-membranous organelles?
   (*a*) microbodies (*b*) vacuoles
   (*c*) ribosomes (*d*) chloroplasts
9. Ribosomes are attached to ER through
   (*a*) ribophorins
   (*b*) rRNA
   (*c*) tRNA
   (*d*) hydrophobic interaction

ANSWERS

Very Short Answer Question

1. Ribosome is a small particle composed of rRNA and protein that function as the site of protein synthesis in the cytoplasm of prokaryotes and in the cytoplasm, mitochondria and chloroplasts of eukaryotes, it is composed of large and small subunits.
2. Polyribosome is a cluster of two or more ribosomes simultaneously translating a single mRNA molecule.
3. The rRNA is ribosomal RNA which is any of several types of RNA molecules used in the construction of ribosome.
4. Protein synthesis.
5. RNA and proteins.
6. J.A. Lake in 1981.
7. Head, base and platform.

Fill in the Blanks

(*i*) Peptide bond
(*ii*) Protein synthesis
(*iii*) Palade
(*iv*) R. B. Roberts
(*v*) Ribosome
(*vi*) Small
(*vii*) Nucleolus, ribonucleoproteins, ribosomal
(*viii*) 30S, 50S
(*ix*) 45S.

Multiple Choice Questions

1. (*c*) **2.** (*d*) **3.** (*a*) **4.** (*c*) **5.** (*b*) **6.** (*a*) **7.** (*c*)
8. (*c*) **9.** (*a*)

21 Golgi Apparatus

For the performance of certain important cellular functions such as biosynthesis of polysaccharides, packaging (compartmentalizing) of cellular synthetic products (proteins), production of exocytotic (secretory) vesicles and differentiation of cellular membranes, there occurs a complex organelle called **Golgi complex** or **Golgi apparatus** in the cytoplasm of animal and plant cells. The Golgi apparatus, like the endoplasmic reticulum, is a canalicular system with sacs, but unlike the endoplasmic reticulum it has parallely arranged, flattened, membrane-bounded vesicles which lack ribosomes and stainable by osmium tetraoxide and silver salts.

21.1. HISTORICAL

An Italian neurologist (*i.e.*, physician) **Camillo Golgi** in 1873 discovered and developed the **silver chromate method** (termed *la reazione nera*) for studying histological details of nerve cells. He, thus, opened a new field of scientific inquiry, called **neuromorphology**. In 1898, Golgi found that Purkinje cells (*i.e.*, nerve cells of cerebral cortex of brain) of barn owl contained an internal reticular network which stains black with the silver stain. He called this structure *apparato reticolare interno* (= internal reticular apparatus).

In the nerve cells of CNS, this reticular network (*i.e.*, Golgi apparatus) was located near the nucleus and was later identified in other cell types. Golgi got Nobel Prize in 1906 for this discovery. The Golgi apparatus remained a center of controversy for decades between those who believed that the organelle existed in living cells and those who believed that it was an **artifact**, that is, an artificial structure formed during preparation for microscopy. It was not until the Golgi apparatus was clearly identified in unfixed, freeze-fractured cells that its existence was verified beyond reasonable doubt (**Karp**, 2010).

Due to their presumed high lipid contents, Golgi apparatuses were called **lipochondria (Baker**, 1951, 1953). Since originally these were known to be networks, they were also called "**dictyosomes**" (Gr., *dictyes* = net). Currently, the term **Golgi apparatus** is more prevalent one, than many other names such as **Golgi complex, Golgiosome, Golgi bodies, Golgi material, Golgi membrane**, etc. The Golgi apparatus of the cells of plants and lower invertebrates is usually referred to as **Golgi body** or **dictyosome**.

21.2. OCCURRENCE

The Golgi apparatus occurs in all cells except the prokaryotic cells (*viz.*, mycoplasmas, bacteria and blue green algae) and eukaryotic cells of certain fungi, sperm cells of bryophytes and pteridiophytes, cells of mature sieve tubes of plants and mature sperm and red blood cells of animals. Their number per plant cell can vary from several hundred as in tissues of corn root and algal rhizoids (*i.e.*, more than 25,000 in algal rhizoids), to a single organelle in some algae. Certain algal cells such as *Pinularia* and *Microsterias*, contain largest and most complicated Golgi apparatuses. In higher plants, Golgi apparatuses are particularly common in secretory cells and in young rapidly growing cells. For example, small shoot apical cells of *Epilobium* (willow herb) contain approximately 20 Golgi stacks (i.e., dictyosomes), whereas the giant fiber cells of cotton contain more than 10,000 dictyosomes per cell. Onion root apical meristem cells contain approximately 400 dictyosomes or stacks (**Buchanan** *et al.*, 2000).

Plant Golgi stacks also remain intact and functionally active during mitosis to provide the plasma membrane and cell wall molecules needed for cell plate formation.

In animal cells, there usually occurs a single Golgi apparatus, however, its number may vary from animal to animal and from cell to cell. Thus, *Paramoeba* species has two Golgi apparatuses and nerve cells, liver cells and chordate oocytes have multiple Golgi apparatuses, there being about 50 of them in the liver cells.

21.3. DISTRIBUTION

In the cells of higher plants, the Golgi bodies or dictyosomes are usually found scattered throughout the cytoplasm and their distribution does not seem to be ordered or localized in any particular manner. However, in animal cells the Golgi apparatus is a localized organelle. For example, in the cells of ectodermal or endodermal origin, the Golgi apparatus remains polar and occurs in between the nucleus and the periphery (*e.g.*, thyroid cells, exocrine pancreatic cells and mucus-producing goblet cells of intestinal epithelium) and in the nerve cells it occupies a circum-nuclear position.

21.4. ULTRASTRUCTURE (ELECTRON MICROSCOPIC STRUCTURE)

The Golgi apparatus is morphologically very similar in both plant and animal cells. However, it is extremely **pleomorphic**: in some cell types it appears compact and limited, in others spread out and reticular (net-like). Its shape and form may vary depending on cell type. Typically, however, Golgi apparatus appears as a complex array of interconnecting *tubules*, *vesicles* and *cisternae*. There has been much debate concerning the terminology of the Golgi's parts. The classification given by **D.J. Morre** (1977) is most widely used. In this scheme, the simplest unit of the Golgi apparatus is the **cisterna**. This is a membrane-bound space in which various materials and secretions may accumulate. Numerous cisternae are associated with each other and appear in a stack-like (lamellar) aggregation. A group of these cisternae is called the **dictyosome**, and a group of dictyosomes makes up the cell's Golgi apparatus. All dictyosomes of a cell have a common function.

The detailed structure of three basic components of the Golgi apparatus can be studied as follows (Fig. 21.1 and Fig. 21.2):

1. Flattened Sac or Cisternae

The Golgi apparatus has a characteristic morphology consisting primarily of flattened, disclike, membranous cisternae with dilated rims and associated vesicles and tubules (Fig. 21.1). The cisternae, whose diameters are typically 0.5 to 1.0 μm, are arranged in an orderly stack (parallel bundle), much like a stack of pancakes. In each stack, cisternae are separated by a space of 20 to 30 nm which may contain rod-like elements or fibres. Each stack of cisternae forms a dictyosome which may contain 5 to 8 Golgi cisternae in animal cells or more cisternae in plant cells. Each cisterna is a closed compartment and is bounded by a smooth unit membrane (7.5 nm thick), having a lumen varying in width from about 500 to 1000 nm.

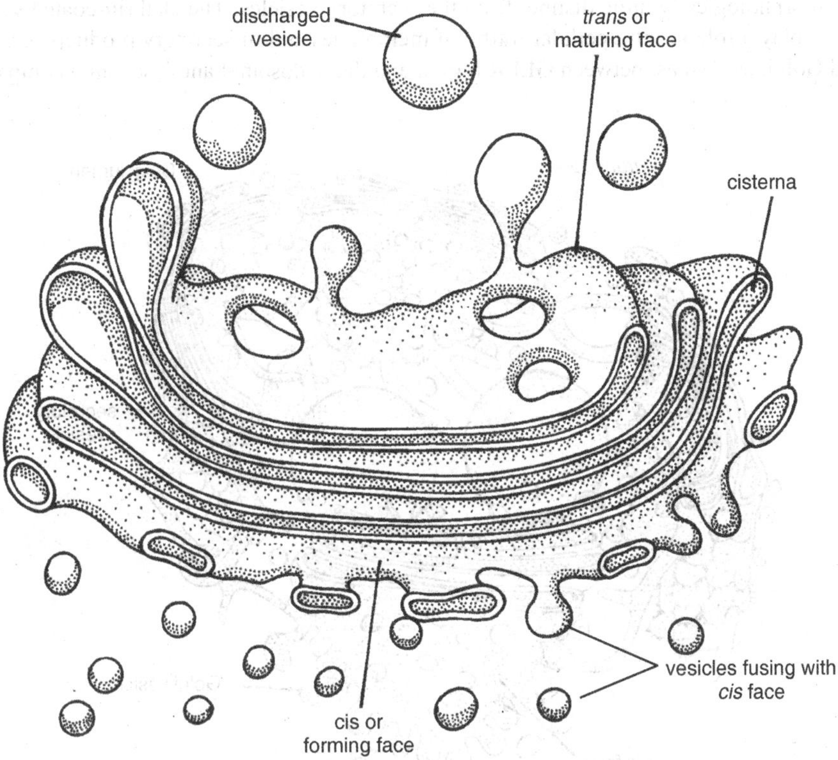

Fig. 21.1. Structure of Golgi apparatus.

Polarity. The margins of each cisterna are gently curved so that the entire dictyosome of Golgi apparatus takes on a shallow bowl or bow-like appearance. The cisternae at the convex end of the dictyosome comprise **proximal, forming** or **cis-face** and the cisternae at the concave end of the dictyosome comprise the **distal, maturing** or **trans-face**. The forming or cis-face of Golgi is located next to either the nucleus or a specialized portion of rough ER that lacks bound ribosomes and is called **"transitional" ER**. Trans-face of Golgi is located near the plasma membrane (Fig. 21.3). This polarization is called **cis-trans axis** of the Golgi apparatus.

2. Tubules

A complex array of associated **vesicles** and anastomosing **tubules** (30 to 50 nm diameter) surround the dictyosome and radiate from it. In fact, the peripheral area of dictyosome is fenestrated (lace-like) in structure.

3. Vesicles

The vesicles (60 nm in diameter) are of three types:

***(i)* Transitional vesicles** are small membrane limited vesicles which are thought to form as blebs from the transitional ER to migrate and converge to cis-face of Golgi, where they coalasce to form new cisternae.

***(ii)* Secretory vesicles** are varied-sized membrane-limited vesicles which discharge from margins of cisternae of Golgi. They, often, occur between the maturing face of Golgi and the plasma membrane.

***(iii)* Clathrin-coated vesicles** are spherical protuberances, about 50 μm in diameter and with a rough surface. They are found at the periphery of the organelle, usually at the ends of single tubules, and are morphologically quite distinct from the secretory vesicles. The clathrin-coated vesicles are known to play a role in intra-cellular traffic of membranes and of secretory products, *i.e.*, between ER and Golgi, as well as, between GELR region and the endosomal and lysosomal compartments.

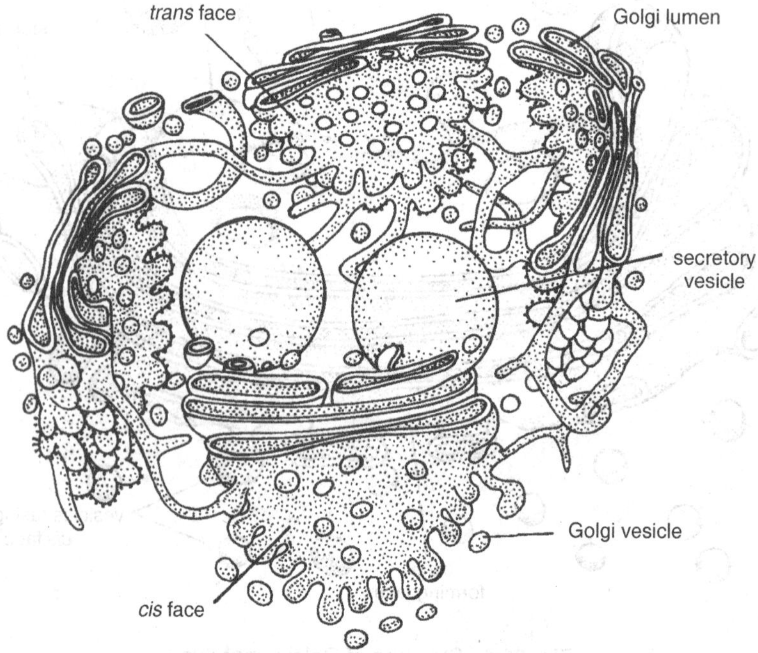

Fig. 21.2. Three-dimensional drawing of a Golgi apparatus; it is drawn from electron micrographs of a secretory animal cell.

CGN, TGN and Cytoskeletal Scaffold of Golgi

The Golgi apparatus is divided into several functionally distinct compartments arranged along an axis from the *cis* or entry face closest to the ER to the *trans* or exit face at the opposite end of the stack. The *cis*-most face of the organelle is composed of an interconnected network of tubules called ***cis* Golgi network (CGN)**. The CGN is thought to function primarily as a *sorting station* that distinguishes between proteins to be shipped back to the ER and those that are allowed to proceed to the next Golgi station.

The bulk of the Golgi apparatus consists of a series of large, flattened cisternae, which are divided into ***cis***, ***medial*** and ***trans*** **cisternae**. The *trans*-most face of the organelle contains a distict network of tubules and vesicles called ***trans* Golgi network (TGN)**. The TGN is a *sorting station* where proteins are segregated into different types of vesicles heading either to the plasma membrane or to various intracellular destinations (Fig. 21.6). The membranous elements of the Golgi apparatus are thought to be supported mechanically by a peripheral membrane skeleton or **scaffold** composed of a variety of proteins, including members of the **spectrin**, **ankyrin** and **actin** families—proteins that are also present as part of the plasma membrane skeleton. The Golgi scaffold may be physically linked with **motor proteins** that direct the movement of vesicles and tubules entering and exiting the Golgi apparatus. A separate group of fibrous proteins are thought to form Golgi "**matrix**", that plays a key role in the disassembly or reassembly of the Golig apparatus during mitosis.

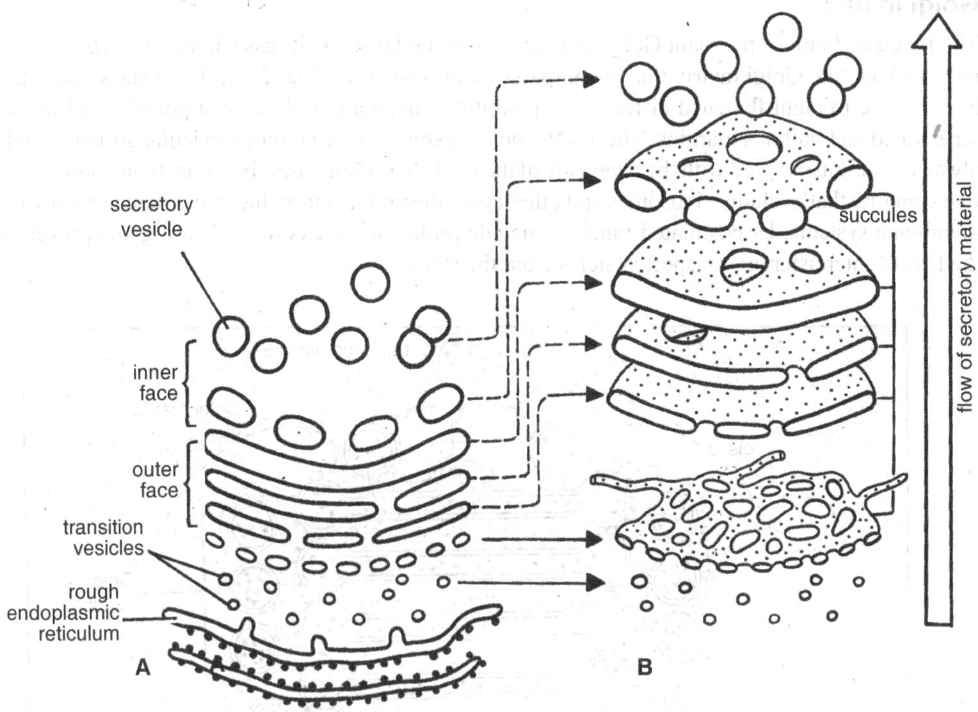

Fig. 21.3. The position and orientation of the Golgi apparatus in the secretory pathway. A—The components as they might be seen by electron microscopy of thin sections; B—A three-dimensional reconstruction of Golgi apparatus.

The Golgi apparatus is not uniform in composition from one end to the other. Differences in composition of the membrane compartments from the *cis* to the *trans* face reflect the fact that the Golgi apparatus is primarily a "**processing plant**". Newly synthesized membrane proteins, as well as secretory and lysosomal proteins, leave the ER and enter the Golgi apparatus at its *cis* face and then pass across the stack to the *trans* face. As they progress along the stack, proteins that were originally synthesized in the rough ER are sequentially modified in specific ways, *e.g.*, during Golgi activity, a protein's carbohydrates are modified by a series of stepwise enzymatic reactions (**Karp**, 2010).

Indeed, trans-Golgi cisternae are the most morphologically conspicuous because of the tight appression of their membranes and darkly staining contents. This is due to the fact that the *trans* cisternae and the TGN being the most acidic Golgi compartments [they are acidified by vacuolar (V-type) H^+-ATPase]. Aside from regulating enzyme activities, the low-pH environment appears to cause an *osmotic collapse* of the cisternae lumen, which squeezes the newly synthesized products into the budding vesicles (**Buchanan** *et al.*, 2000).

Intercisternal Elements

Intercisternal elements are another type of *trans*-Golgi cisterna-associated structure. These parallel protein fibers lie sandwiched between cisternae and may serve as anchors for the glycosyltransferases involved in the synthesis of large polysaccharide slime molecules, such as those secreted by the outer most cells (*i.e.*, columella cells) of the root caps (**Buchanan** *et al.*, 2000).

Golgi Matrix

The functional unit of the plant Golgi apparatus is the Golgi stack, its associated TGN (*trnas*-Golgi network) and the Golgi matrix that encompasses both structures (Fig. 21.4). Each stack consists of a set of five to eight flattened cisternae that exhibit a distinct morphological polarity and possess fenestrated and bulbous margins. The TGN tends to exhibit a more tubulo-vesicular structure and is always closely associated with the *trans* side of the stack. The **Golgi matrix** is fine, filamentous, cage-like structure that excludes ribosomes and other cytosolic proteins from the immediate vicinity of the membrane systems. Its postulated function include protecting stacks from shearing and preventing the loss of intercisternal transport vesicles from the stack.

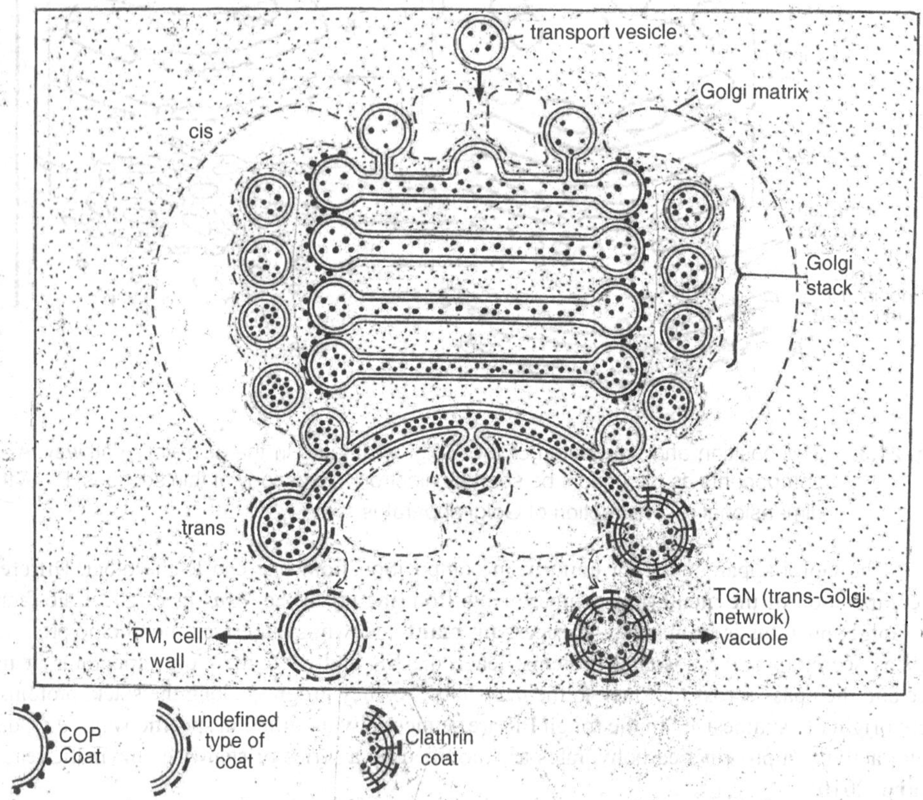

Fig. 21.4. Diagram illustrating the spatial relationship of plant Golgi stack to its associated *trans*-Golgi network (TGN) and the Golgi matrix. The distribution of COP- and clathrin-coated budding vesicles is also shown (after Buchanan *et al.*, 2000).

Zones of Exclusion

A Golgi body or Golgi apparatus is surrounded by a differentiated region of cytoplasm where ribosomes, glycogen, and organelles such as mitochondria and chloroplasts are scarce or absent. This is called **zone of exclusion** (**Morre** *et al.*, 1971). Endoplasmic reticulum within the zone of exclusion has a smooth surface (lacking ribosomes), and coated vesicles of the Golgi apparatus are restricted to this region. Similar zones of exclusion are associated with microtubules, centrioles, and regions of centriole formation.

21.5. CHEMICAL COMPOSITION

Chemically, Golgi apparatus of rat liver contains about 60 per cent lipid material. The Golgi apparatus of animal cells contains phospholipids in the form of **phosphatidyl choline**, whereas, that of plant cells contains **phosphatidic acid** and **phosphatidyl glycerol**. The Golgi apparatus also contains a variety of enzyme (Table 21.1), some of which have been used as cytochemical markers.

Table 21.1. Some important enzymes of the Golgi apparatus of animal cells.

Enzymes: class and types	Function
A. Glycosyl transferases: Glycoprotein biosynthesis	
1. Sialyl transferases	Transfers sialic acid from CMP-sialic acid
2. Galactosyl transferases	Transfer galactose to lipids or proteins
B. Sulpho- and glyco-transferases: Glycolipid biosynthesis	
3. Sulphotransferase	Transfer of sulphate from activated donor
4. Lysolecithin acetyltransferase	Transfer of acyl groups to phospholipid
5. Glycero-phosphate phosphatidyl transferase	Transfer of phosphatidyl group
C. Oxireductases: Oxidation and reduction	
6. NADH-cytochrome c-reductase	Removal or addition of hydrogen
7. NADPH-cytochrome c-reductase	Removal or addition of hydrogen
D. Phosphatases: Hydrolysis of phospholipids	
8. Glucose-6-phosphatase	Removal of phosphate
9. Thiamine pyrophosphatase (Nucleoside diphosphatase)	Hydrolysis of inorganic pyrophosphate
10. ATPase	Removal or addition of phosphate
11. Acid phosphatase	Removal of phosphate
E. Phospholipases: Hydrolysis of lipids	
12. Phospholipase A_1	Removal of non-specific fatty acid chains from phospholipids
13. Phospholipase A_2	Removal of fatty acid chains
F. Kinases: Phosphorylation	
14. Casein phosphokinases	Phosphorylation of casein
G. Mannosidases: Removal of mannose	
15. Mannosidase I and II	Removal of mannose residue from oligosaccharide

Cytochemical Properties of Golgi Apparatus

Different parts of Golgi apparatus have been histochemically identified by specific staining properties.

1. Osmium tetroxide (OsO_4) selectively impregnates the outer face (cis face) of the Golgi apparatus. This stain adheres well to lipids, especially phospholipids and unsaturated fats.

2. Phosphotungstic acid (H_3PO_4. 12 WO_3. 24 H_2O) selectively stain the maturing or trans face of Golgi stack. This stain is an anionic stain having special affinity for polysaccharides and proteins.

3. Glycosyl transferase and **thiamine pyrophosphatase** can be localized cytochemically in the trans cisternae of Golgi apparatus. Transferase enzymes are found to be located in the membranes of Golgi, not in the lumen of cisternae.

4. Acid phosphatase enzyme is cytochemically marked in the GERL region.

21.6. FORMATION OF GOLGI APPARATUS

Golgi apparatus is constantly being formed, changed, broken down and reformed. It has been variously described being formed from the plasma membrane, nuclear envelope, annulate lamellae and the endoplasmic reticulum (Fig. 21.5).

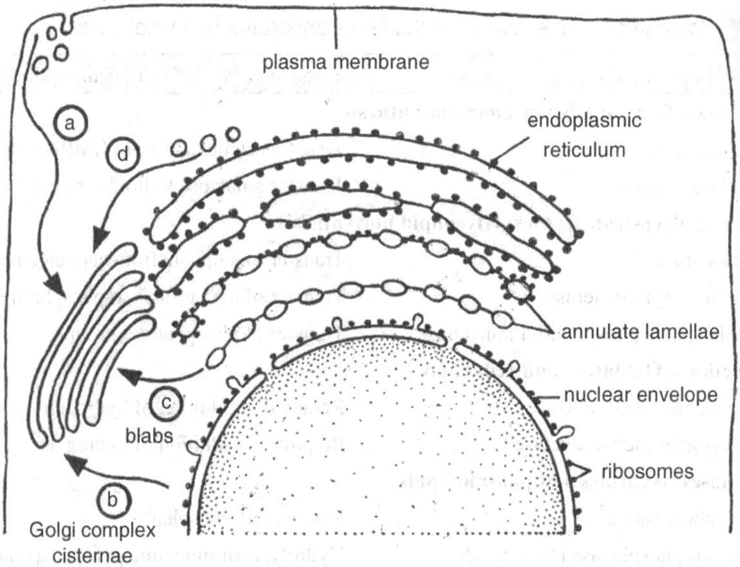

Fig. 21.5. Diagram showing different views of formation of the Golgi apparatus: (a) from the plasma membrane; (b) from the nuclear envelope; (c) from the annulate lamellae; and (d) from the endoplasmic reticulum.

1. From the plasma membrane. In the protozoan *Pelomyxa illionoisensis* the Golgi cisternae are described as originating from vesicles originating from the plasma membrane (**Daniels**, 1964). These vesicles are formed by pinocytosis and phagocytosis of the plasma membrane.

2. From the nuclear envelope. In the brown algae there is a close association between the nuclear envelope and the Golgi apparatus. **Bouch** (1965) has described vesicles, similar to those located on the formative face of the Golgi apparatus, arising from the nuclear envelope.

3. From the annulate lamellae. In the study of oocyte maturation in the frog, **Ward** (1965) has described vesicles derived from the annulate lamellae of using and giving rise to the Golgi membranes.

4. From the endoplasmic reticulum. Essner and **Novikoff** (1962) has described the formation of Golgi cisternae from the endoplasmic reticulum in certain hepatomas (*i.e.*, malignant tumor of the liver). **Beams** and **Kessel** (1968) have suggested that Golgi lamellae may be derived from the ER by loss of ribosomes.

In plants, new stacks (dictyosomes) arise by **fission** which can start either the *cis* or the *trans* side of the stack. Origin of new dictyosomes from the old ones usually occurs during G2 phase of the cell cycle (**Buchanan**, *et al.*, 2000).

Currently, it is believed that the Golgi apparatus arises from the rough endoplasmic reticulum which changes to smooth ER and then becomes the Golgi cisternae. The cisternae on the forming face (*i.e.*, *cis* face) are constantly being formed by fusion of vesicles derived from the ER. The cisternae on the maturing face (*i.e.*, *trans* face) are believed to form secretory vesicles.

Lastly, the presence of the nucleus is necessary to maintain the Golgi apparatus. In the absence of the nucleus the Golgi apparatus decreases in size and disappear.

21.7. FUNCTIONS

Golgi vesicles are often, referred to as the **"traffic police"** of the cell (**Darnell** *et al.*, 1986). They play a key role in **sorting** many of cell's proteins and membrane constituents, and in **directing** them to their proper destinations. To perform this function, the Golgi vesicles contain different sets of enzymes in different types of vesicles—**cis, middle** and **trans cisternae**—that react with and modify secretory proteins passing through the Golgi lumen or membrane proteins and glycoproteins that are transiently in the Golgi membranes as they are *en route* to their final destinations (Fig. 21.6). For example, a Golgi enzyme may add a "signal" or "tag" such as a carbohydrate or phosphate residues to certain proteins to direct them to their proper sites in the cell. Or, a proteolytic Golgi enzyme may cut a secretory or membrane protein into two or more specific segments (*e.g.*, molecular processing involved in the formation of pancreatic hormone insulin: preproinsulin → proinsulin → insulin).

Recently, in the function of Golgi apparatus, subcompartmentalization with a division of labour has been proposed between the *cis* region (in which proteins of RER are sorted and some of them are returned back possibly by coated vesicles), and the *trans* region in which the most refined proteins are further separated for their delivery to the various cell compartments (*e.g.*, plasma membrane, secretory granules and lysosomes).

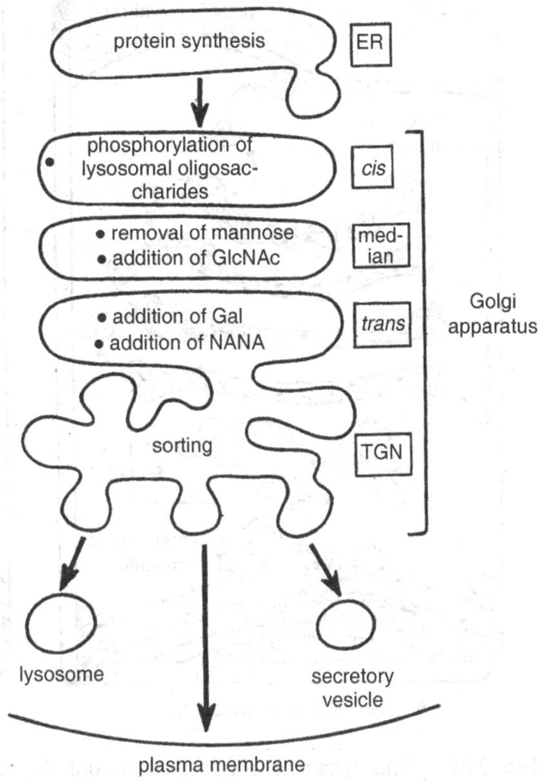

Fig. 21.6. The compartmentalization of the Golgi apparatus. GlcNAc = N-actyl-glucosamine galactose; NANA = N-acetyl neuraminic acid (sialic acid).

Movements of Materials through the Golgi Apparatus (Anterograde and Retrograde Movements of Transport Vesicles)

There exist following two views regarding the movement of the materials through the Golgi apparatus: (i) cisternal maturation model and (ii) vesicular transport model.

1. Cisternal maturation model. Up until the mid-1980's and again in early 2000's, it is generally accepted that Golgi cisternae are transient structures. It is supposed that Golgi cisternae formed at the *cis* face of the stack by fusion of membranous carriers from the ER and ERGIC (*i.e.*, endoplasmic reticulum Golgi intermediate compartment) and that each cisterna physically moved from the *cis* to the *trans* end of the stack, changing its composition as it progressed. This is known as the **cisternal maturation model** (Fig. 21.7B), because, according to the model, each cisterna "matures" into the next cisterna along the stack.

2. Vesicular transport model. From the mid-1980's to the mid-1990's, the maturation model of Golgi movement was largely abandoned and replaced by an alternate model, which was proposed that the cisternae of a Golgi stack remain in place as stable compartments. This model is called **vesicular transport model**. According to this model, the cargo (*i.e.*, secretory, lysosomal, and membrane proteins) is shuttled through the Golgi stack, from the CGN to the TGN, in vesicles that bud from one membrane compartment and fuse with a neighboring compartment farther along the stack.

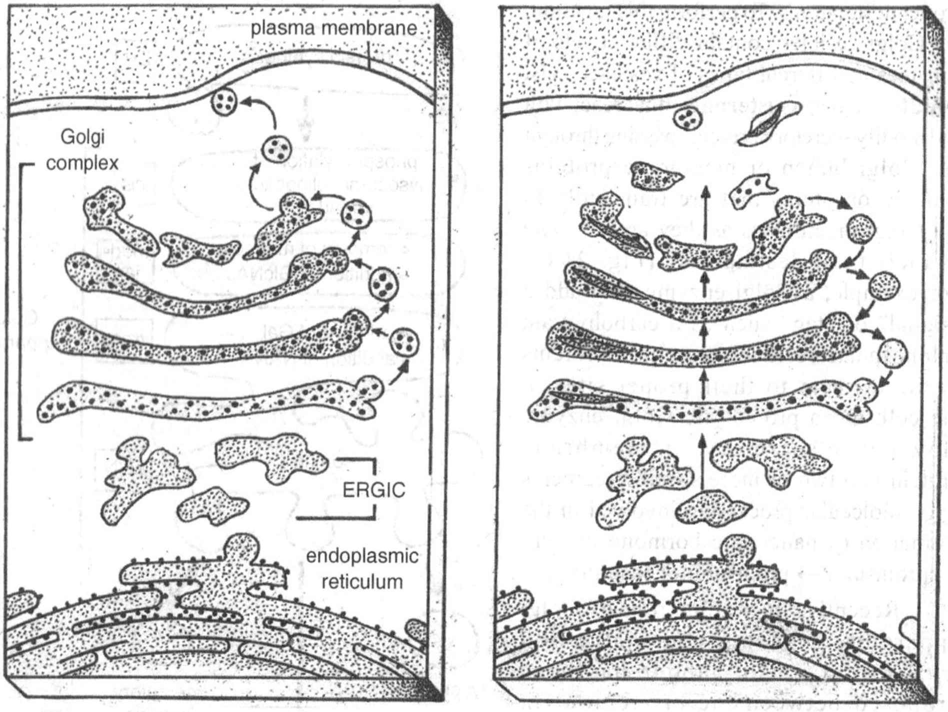

Fig. 21.7. The dynamics of transport through the Golgi apparatus. A—In the cisternal maturation model, the cisternae progress gradually from a *cis* to a *trans* position and then disperse at the TGN. Transport vesicles carry resident Golgi enzymes (indicated by the shaded vesicles) in a retrograde direction. The lens-shaped objects represent large cargo materials, such as procollagen complexes of fibroblasts. B—In the vesicular transport vesicles, while the cisternae themselves remain as stable elements (after Karp, 2010).

Example of Cisternal Maturation Model

1. The cisternal maturation model envisions a highly dynamic Golgi apparatus in which the major elements of the organelle, the cisternae are continually being formed at the *cis* face and dispersed at the *trans* face. According to this view, the very existence of the Golgi apparatus itself depends on the continual influx of transport carriers from the ER and ERGIC. As proposed by the cisternal maturation model when the formation of transport carriers from the ER is blocked either by treatment of cells with specific drugs or the use of temperature sensitive mutants, the Golgi apparatus simply disappears. When the drugs are removed or the mutant cells are returned to the permissive temperature, the Golgi apparatus rapidly reassembles as ER-to-Golgi transport is renewed.

2. Certain materials that are produced in the endoplasmic reticulum and travel through the Golgi apparatus can be shown to remain within the Golgi cisternae and never appear within Golgi-associated transport vesicles. For example, studies on fibroblasts indicate that large complexes of

procollagen molecules (the precursors of extracellular collagen) move from the *cis* cisternae to the *trans* cisternae without ever leaving the cisternal lumen.

3. It was assumed until the mid-1990s that transport vesicles always moved in a "forward" (**anterograde**) direction, that is, from a *cis* origin to a more *trans* destination. But a large body of evidence has indicated that vesicles can move in a "backward" (**retrograde**) direction, that is, from a *trans* donor membrane to a *cis* acceptor membrane.

4. Studies on live budding yeast cells containing fluorescently labelled Golgi proteins have shown directly that the composition of an individual Golgi cisterna can change over time-from one that contains early (*cis*) Golgi resident proteins to one that contains late (*trans*) Golgi resident proteins.

Examples of Vesicular Transport Model

1. Each of the various Golgi cisternae of a stack has a distinct population of resident enzymes [*e.g.*, reduced osmium tetroxide in *cis* cisternae, enzyme mannosidase II in medial cisternae and enzyme nucleoside diphosphatase in *trans* cisternae of Golgi (**Decker**, 1974; **Velasco** *et al.*, 1993)]. How could the various cisternae have such different properties if each cisternae gave rise to the next one in line, as suggested by the cisternal maturation model.

2. Large numbers of vesicles can be seen in electron micrographs to bud from the rim of Golgi cisternae. In 1983, **James Rothman** and his colleagues at Stanford University demonstrated using cell-free preparations of Golgi membranes, that transport vesicles were capable of budding from one Golgi cisterna and fusing with another Golgi cisterna *in vitro*. This landmark experiment formed the basis for a hypothesis suggesting that inside the cell, cargo-bearing vesicles budded from *cis* cisternae and fused with cisternae situated at a more *trans* position in the stack.

Currently, widely accepted cisternal maturation model acknowledge the role for transport vesicles, which have been clearly shown to bud from Golgi membranes. In this model, however, these transport vesicles do not shuttle cargo in an anterograde direction, but instead carry resident Golgi enzymes in a retrograde direction. This model of intra-Golgi transport is supported by electron micrographs (**Mironov** *et al.*, 2001; **Martinez-Menarguez** *et al.*, 2001).

Some Other Functions of Golgi Apparatus

The Golgi apparatus is primarily a **processing plant** of the cell. It is a centre of *reception, finishing, packaging*, and *dispatch* for a variety of materials in animal and plant cells. Thus, Golgi apparatus performs the following functions in the eukaryotic cells.

1. Glycosylation. Golgi apparatus facilitates linking of carbohydrates and proteins in the formation of **glycoproteins** or **proteoglycans**. Such addition of carbohydrates to proteins is called **glycosylation**. The process of glycosylation is started in the endoplasmic reticulum. Mainly one species of oligosaccharides is added to proteins in the ER. These proteins are transferred to Golgi by coated vesicles. In Golgi, the oligosaccharide chains of proteins are modified by removal (*e.g.*, mannoses) and addition of selected sugars in order to produce the final product. For this purpose the Golgi apparatus contains a variety of **glycosyl transferase enzymes** and sugar nucleotides. More glucose and other sugars such as fucose, galactose and sialic acid are added at the Golgi. Thus, glycosylation process begins in the ER and is completed in the Golgi apparatus.

2. Synthesis and secretion of polysaccharides. Studies on **goblet cells** by autoradiography, pulse-chase and electron microscopy have established the interrelationship between protein synthesis, carbohydrate addition (glycosylation) and sulphation. The goblet cells of the colon produce **mucigen**. This secretory material contains a large proportion of carbohydrate. The Golgi apparatus is found just above the nucleus of the goblet cell (Fig. 21.8). Towards the free surface of the cell are gradually enlarging mucigen granules. The proximal cisternae of the Golgi apparatus do not show any swelling, but at some distance across the stack the distal cisternae are quite suddenly converted into mucigen granules. The distal cisternae continually convert into mucigen granules every 2-4 minutes. New proximal cisternae are formed in compensation.

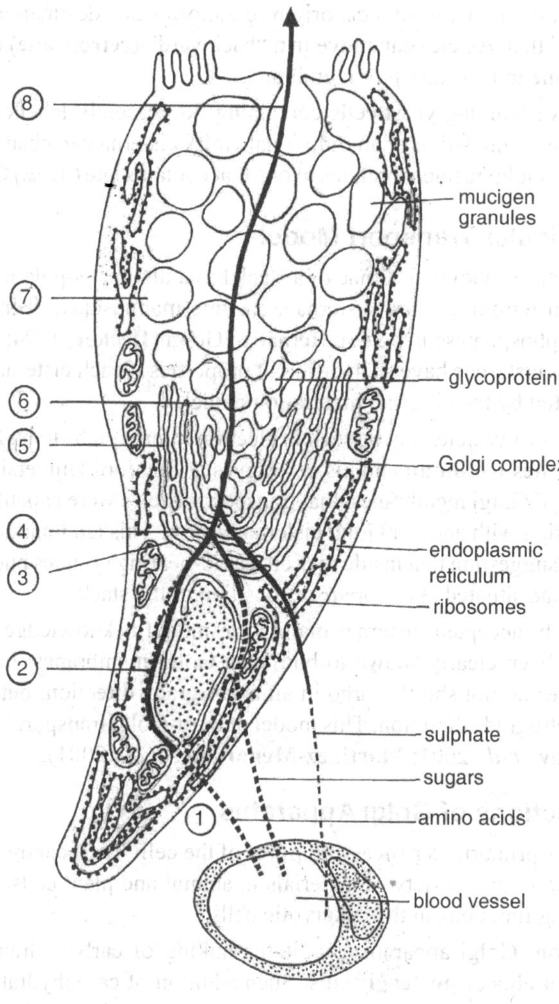

Fig. 21.8. Diagram of assembly of mucous in a goblet cell. The numbering follows that of the description in the text.

Autoradiographic studies by ^{3}H-glucose and ^{3}H-galactose labeling have provided direct evidence of polysaccharide synthesis in the Golgi apparatus (**Peterson** and **Leblond**, 1964; **Neutra** and **Leblond**, 1966). Five minutes after injection of ^{3}H-glucose the label is found over the Golgi apparatus. After 40 minutes the label crosses over to mucigen granules adjacent to the Golgi cisternae. Four hour after injection the label is found over the more apical mucigen granules. **Neutra** and **Leblond** also confirmed that sulphation has taken place in the Golgi apparatus.

On the basis of these observations the following scheme was proposed by **Neutra** and **Leblond** 1969 (Fig. 21.8): 1. Precursors enter the goblet cells from the capillaries of the vascular system. 2. The *amino acids* are synthesized into *proteins* on the ribosomes of the endoplasmic reticulum. 3. The proteins are then transferred to the cisternae of the Golgi apparatus. 4. The simple sugar molecules go directly from the blood stream to the Golgi apparatus cisternae, where they are complexed with protein to form *glycoprotein*. 5. The glycoprotein is then sulphated in the Golgi apparatus to form *mucigen*. 6. The distal cisternae become rounded off and are transformed into mucigen granules (= vesicles). 7. The mucigen granules are progressively displaced toward the apex of the cell. During this process the mucigen undergo further modification. 8. The mucigen granules are secreted into the intestine by exocytosis, during which the vesicle membrane fuses with the plasma membrane.

3. Sulphation. A number of secretory products also contain sulphates. The sulphation process requires the activity of **sulphotransferase** enzymes which are present in Golgi apparatus.

4. Concentration and storage of secretory products. The proteins, glycoproteins and lipoproteins are concentrated inside the **secretion vesicles** or in **condensing vacuoles** which bud off from the Golgi cisternae. Once concentrated, the products may be stored in the cytoplasm (inside the vesicle) or may be secreted to the outside of the cell through exocytosis.

5. Secretion. Golgi is mainly associated with the secretory activity of the cell. Various hormones, proenzymes or zymogens, skeletal components, mucus, etc., are secreted through Golgi apparatus. All these products are synthesized in different parts of the cell. Through endoplasmic reticulum these products reach Golgi apparatus where these are packed and concentrated for final secretion. Golgi apparatus of different organs produce following secretion (Table 21.2).

Table 21.2. Summary of secretions which are produced in Golgi apparatus.

Cell type	Golgi function
1. Exocrine cells of pancreas	Secretion of zymogens (proenzymes of digestive enzymes such as proteases, lipases, carbohydrases and nucleases)
2. Parotid (salivary) gland cells	Secretion of mucus and zymogen
3. Goblet cells of intestinal epithelium	Secretion of mucus and zymogen
4. Goblet cells of respiratory epithelium	Secretion of mucus (*i.e.*, mucin, a glycoprotein)
5. Paneth cells of intestine (*i.e.*, of crypts of Lieberkuhn)	Secretion of digestive enzymes (*e.g.*, chitinase)
6. Brunner's gland cells of intestine	Secretion of enzymes, mucus and hormones
7. Follicle cells of thyroid	Prethyroglobulin
8. Plasma cells of blood	Immunoglobulins
9. Hepatic cells of liver	Transformation and secretion of lipids
10. Cells of alveolar epithelium of mammary gland	Secretion of milk proteins (casein)
11. Nervous tissue (myelocytes, sympathetic ganglia, Schwann cells)	Sulphation reaction
12. Endothelial cells of blood vessels	Sulphation reaction
13. Cornea of birds	Secretion of tropocollagen and collagen
14. Most plant cells	Secretion of pectin and cellulose
15. Melanocytes	Secretion of melanin
16. Primary oocyte	Secretion of yolk and vitelline membrane
17. Root cap cells	Mucus

Types of Secretion

Molecular transport and diffusion across membranes account for the transfer of some materials from cytoplasm to extra-cellular space. Cellular secretion may be of the following three types: holocrine, apocrine and merocrine.

***(i)* Holocrine secretion.** In this kind of cellular secretion, the entire cell fills with secretory product and then is broken up and sacrificed. An example is the sebaceous gland, which secretes a fatty discharge onto adjacent hair and skin. It is this gland that, when overstimulated, gets plugged to form blackheads and ultimately the pimples of adolescent acne.

***(ii)* Apocrine secretion.** In this kind of secretion, the secretion vesicle buds through the plasma membrane carrying with it a portion of that membrane and hence a thin layer of cytoplasm. The secretion vesicle is thus covered by two layers of membrane, its own plus the plasma membrane. Milk fat droplets (MFD) of the cells of the lactating mammary gland are released as apocrine secretion (Fig. 21.9).

***(iii)* Merocrine or eccrine secretion.** It is a most common type of cellular secretion. This process is also called **exocytosis** because it is the function opposite of endocytosis. Thus, it involves the fusion of secretory vesicles with the plasma membrane and their opening to the exterior. It is the usual mode of secretion of most glands and of most other cells, both animal and plant. It is, for example, the mode of secretion of the **true sweat glands,** which are distributed over most of the human skin. In cells of lactating mammary gland, casein vesicles (CV) release their protein by merocrine secretion (Fig. 21.9).

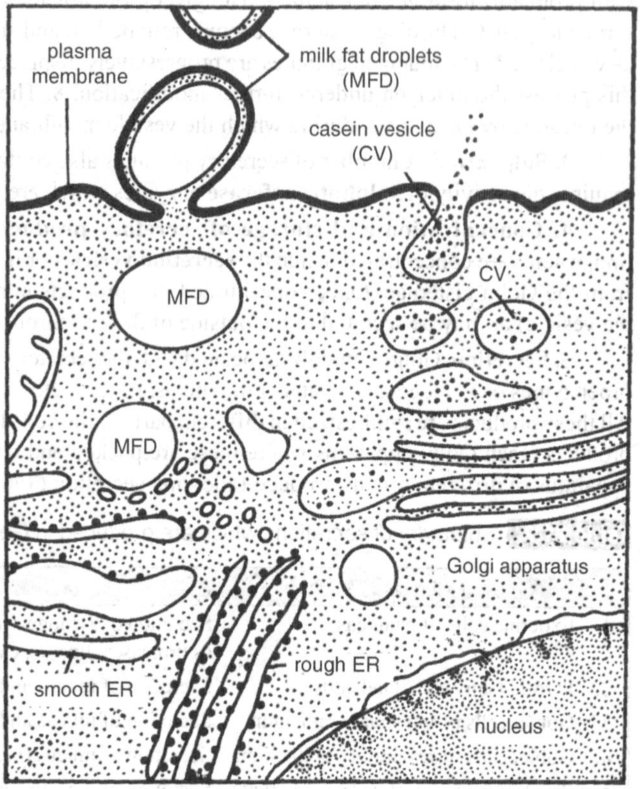

Fig. 21.9. Secretion in lactating mammary gland. Milk fat droplets are released by the apocrine mechanism, whereas casein vesicles normally release their protein by merocrine secretion. The former removes plasma membrane, the latter adds to it.

Pancreas which includes exocrine as well as endocrine cells has provided a best example of merocrine secretion.

6. Recycling of plasma membrane components. There occurs a continuous exchange of membrane between plasma membrane and Golgi apparatus due to exocytosis and endocytosis. The membrane material is reprocessed in the Golgi for reuse.

Mode of plasma membrane formation. Secretory vesicles originating from the Golgi apparatus fuse with the plasma membrane during exocytosis (Fig. 21.10). The membrane of secretory vesicles becomes incorporated into the plasma membrane, and thus contributes to the renewal of the plasma membrane constituents. The Golgi apparatus plays an important role in synthesis of the carbohydrate components of the plasma membrane. Autoradiographic studies of **Rambourg** and coworkers (1971) have shown that radioactive tritiated sugars (tracers) were first localized at the Golgi apparatus migrate to the cell surface and appear in the cell coat. The glycoproteins formed on the Golgi membrane remain attached to the secretory vesicle membrane after it fuses with the plasma membrane. Secretory glycoproteins on the other hand are released into the extracellular space after fusion of the secretory vesicle membrane with the plasma membrane.

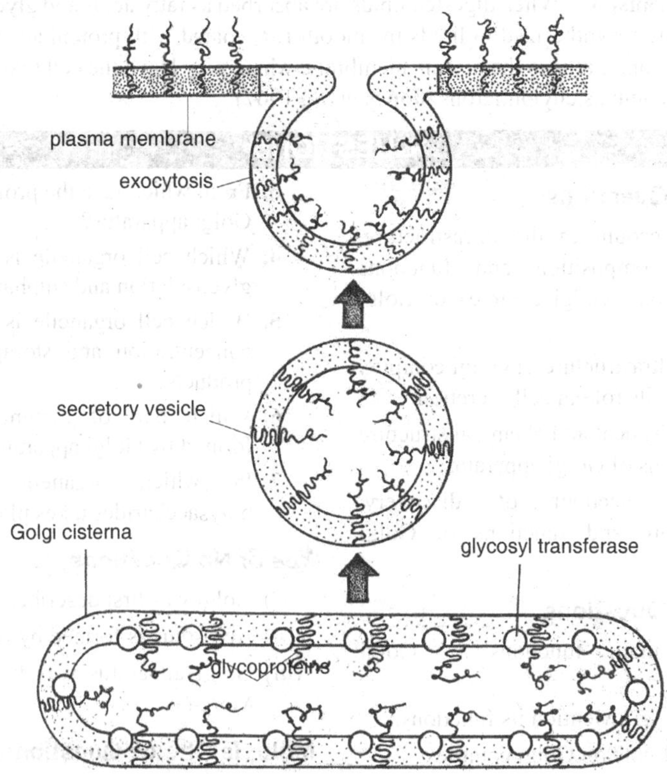

Fig. 21.10. Biogenesis of plasma membrane. Diagram showing formation of a secretory vesicle from the Golgi apparatus and its subsequent fusion with the plasma membrane (exocytosis). The glycoproteins remain attached to the membrane.

It has been shown that in plant cells the plasma membrane of the cells resulting from cell division are contributed by the Golgi apparatus. The Golgi apparatus contributed to the biogenesis of the plasma membrane by supplying it with glycosylated molecules.

7. Intracellular sorting of proteins. Proteins are exported from the Golgi apparatus to different intracellular and extracellular destinations. The accurate sorting of proteins for selective export is one of the major functions of Golgi. This sorting is done by **coated vesicles** which surround the cisternae. These contain specific receptors which recognize specific markers on the proteins to be exported.

8. Formation of cell plate and cell wall in plants. In plants, Golgi apparatus is mainly involved in the secretion of materials of primary and secondary cell walls (*e.g.*, formation and export of glycoproteins, lipids, pectins and monomers for hemicellulose, cellulose, lignin, etc.). During cytokinesis of mitosis or meiosis, the vesicles originating from the periphery of Golgi apparatus, coalesce in the phragmoplast area to form a semi-solid layer, called **cell plate**. The unit membrane of Golgi vesicles fuses during cell plate formation and becomes part of plasma membrane of daughter cells.

9. Development of acrosome in sperm. Acrosome is a membrane-bounded cap-like structure at the forwarding end of a spermatozoon (in animals). It contains proteolytic enzymes. The acrosome is formed from vesicles derived from the Golgi apparatus during sperm maturation.

10. Formation of primary lysosomes. Hydrolytic lysosomal enzymes produced in rough ER are sorted in TGN and targeted for lysosomes by virtue of phosphorylated mannose residues within their oligosaccharide chains.

11. Lipid transport. When digested lipids are absorbed as fatty acids and glycerol in the small intestine. They are resynthesized to lipids in smooth ER, coated with protein and then transported through the Golgi apparatus to the plasma membrane where they leave the cell (exocytosis), mainly to enter lymph system as chylomicrons (**Jones** *et al.*, 1967).

QUESTIONS

Long Answer Questions

1. Give an account of the ultrastructure, chemical composition and functions of Golgi body, Golgi complex or Golgi apparatus.
2. Describe ultrastructure of Golgi complex and discuss its role in cell secretion.
3. State the physical and chemical structure and functions of Golgi apparatus.
4. Give an account of discovery, ultrastructure and functions of Golgi complex.

Short Answer Questions

1. Describe main functions of Golgi complex.
2. What is Golgi? Mention its functions.
3. Write short note on Golgi complex.
4. Write short notes on the following:
   (*i*) Golgi apparatus
   (*ii*) Functions of Golgi body
   (*iii*) Acrosome formation
   (*iv*) Zones of exclusion

Very Short Answer Questions

1. Who first described the Golgi apparatus?
2. Which face of Golgi cisternae lie towards the plasma membrane?
3. From which face the proteins enter in the Golgi apparatus?
4. Which cell organelle is responsible for glycosylation and sulphation processes?
5. Which cell organelle is responsible for concentration and storage of secretory products?
6. Which part of mammalian sperm is formed by Golgi apparatus?
7. In which organelle synthesis of polysaccharides takes place?

Yes or No Questions

(*i*) Golgi was first described by Bateson.
(*ii*) Acrosome is formed by Golgi body.
(*iii*) Golgi apparatus was first described by Morgan.

Multiple Choice Questions

1. Golgi apparatus is concerned with
   (*a*) excretion (*b*) secretion
   (*c*) ATP synthesis (*d*) RNA synthesis
2. Golgi apparatus is absent in
   (*a*) yeast
   (*b*) liver cell
   (*c*) higher plant
   (*d*) blue green algae
3. Golgi apparatuses present in invertebrates and plant cells are called as
   (*a*) polysomes (*b*) golgisomes
   (*c*) dictyosomes (*d*) cisternae

ANSWERS

Very Short Answer Questions

1. Camillo Golgi. 2. Trans face. 3. Cis face.
4. Golgi apparatus. 5. Golgi apparatus. 6. Acrosome.
7. Golgi apparatus.

Yes or No Questions

1. No 2. Yes 3. No

Multiple Choice Questions

1. (*b*) 2. (*d*) 3. (*c*)

22 Intracellular Protein Sorting or Targeting and Chaperones

A typical plant cell contains about 5,000 to 10,000 different polypeptide sequences and billions of individual protein molecules (**Buchanan** *et al.*, 2000). If such a cell is to function properly, it must direct these proteins to specific metabolic compartments, cytoplasmic structures and membrane systems. Accurate protein sorting (or targeting) is required at all times, both when cellular structures are formed in dividing and differentiating cells, and when proteins in mature structures are degraded and replaced. Examples of the proteins that must be sorted include *soluble enzymes, intrinsic membrane proteins* in the cell wall matrix (Box 22.1).

Box 22.1 Destination of Proteins Synthesized by Free Ribosomes

Proteins synthesized by free ribosomes resides in the cytosol and devoid of signal sequence, hence these do not bind the ER. The completed polypeptide remains in the cytosol where **molecular chaperones** help these proteins to fold correctly so that they assume a final three dimensional conformation. However, all the proteins synthesized in the cytosol do not reside there, some of them exported to their appropriate destinations: nucleus, peroxisomes, mitochondria or chloroplasts.

Soluble enzymes are present in all subcellular compartments, including the cytosol, vacuole, cell wall, mitochondrial matrix, chloroplast stroma, thylakoid lumen, peroxisome/glyoxysome, lumen of the endoplasmic reticulum (ER), cisternae of Golgi apparatus and nuclear sap.

Membrane-bound proteins occur in more than a dozen different lipid layers that delimit these compartments (*e.g.*, the vacuolar membrane or tonoplast, the plasma membrane, ER, Golgi membranes, outer membrane of the chloroplast) including membranes that lie within organelles (*e.g.*, thylakoid).

Some proteins are unique to a particular structure, compartment or membrane. Alternatively, very similar proteins with comparable amino acid sequences, structures, and functions can occur in more than one compartment. For example, **acid invertases** occur in the vacuole and cell wall, and water-channel proteins (**aquaporins**) are found in the tonoplast and plasma membrane. Cells therefore require the necessary machinery to sort each protein and direct it to its proper destination.

22.1. MACHINERY OF PROTEIN SORTING

1. Protein sorting requires peptide address labels and protein-sorting machinery

How do thousands of proteins each find their way to the correct subcellular location? All proteins, except those that remain in the compartment where they are translated, include one or more **targeting domains** that act as an address label. Targeting domains are usually short peptides or amino acid motifs but can also be glycans (oligosaccharides), as in the case of mammalian lysosomal hydrolases.

Targeting domains are often located at the amino (N–) terminal end of a protein but may be present in the carboxyl (C) terminus. Two isoforms of invertase with different targeting domains are shown in Fig. 22.1. Specific cellular machinery interacts with this information to translocate the protein into, or retain, it in the proper compartment. Different names are applied to targeting domains, depending on the organelle to which a protein being targeted (Table 22.1).

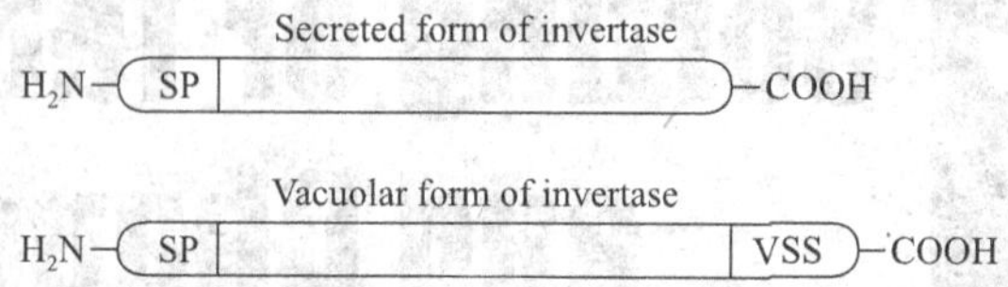

Fig. 22.1. Sorting domains direct proteins to specific compartments. Certain enzymes such as invertase, chitinase and glucanase exist in two forms: one resides in the vacuole and the other is secreted. The secreted form carries a single targeting domain at its N terminus, a signal peptide (SP) for entry in the ER. The vacuolar form has two targeting domains, a signal peptide at the N terminus (SP) and a vacuolar sorting signal (VS) at the C terminus (after Buchanan *et al.*, 2000).

Table 22.1. Peptide targeting domains for the transport to different organelles (Source: Buchanan *et al.*, 2000).

	Organelle	Address label (Targeting domain)
1.	ER	Signal peptide (SP)
2.	Chloroplast	Transit peptide
3.	Mitochondria	Presequence
4.	Nucleus	Nuclear localization signal (NLS)
5.	Peroxisome	Peroxisome targeting signal (PTS)
6.	Vacuole	Vacuolar sorting signal (VSS)

Each compartment and membrane system requires a different targeting domain and sorting machinery. In addition, some proteins contain domains required for interaction rather than localization. Certain cytosolic proteinous structures such as microtubules and microfilaments result from self-assembly of specific monomeric proteins. No protein sorting occurs in this process, but informational polypeptide domains that allow the monomers to form higher-order structures are required.

Although the targeting domain is essential for protein transport, it may not be part of the active protein, and proteases (enzymes) in the target location often remove the targeting domain to create a functional, mature polypeptide.

2. To reach its destination, a protein often has to cross at least one membrane (Chaperones)

Most proteins have hydrophilic surfaces and therefore do not readily pass through the hydrophobic core of the lipid bilayer. **Translocation** through a membrane involves a proteinaceous pore or channel through which a protein passes, not in its globular form but in an extended or unfolded configuration (Fig. 22.2). As a polypeptide passes through a pore, it is assisted by molecular **chaperones**, proteins that bind the polypeptide during the folding and assembly process. By inhibiting molecular interactions that may cause polypeptides to fold incorrectly and aggregate, chaperones increase the yield of correct tertiary structures but not the rate of protein folding. Some cytosolic chaperones interact with newly synthesized proteins, keeping them unfolded so they can pass through a protein pore to an appropriate compartment or membrane. Other chaperones bind to the amino acid chain as it emerges from the membrane and facilitate folding. Still other chaperones function as repair stations to correct minor misfolding. The many members of one family of chaperones—the 70 kDa

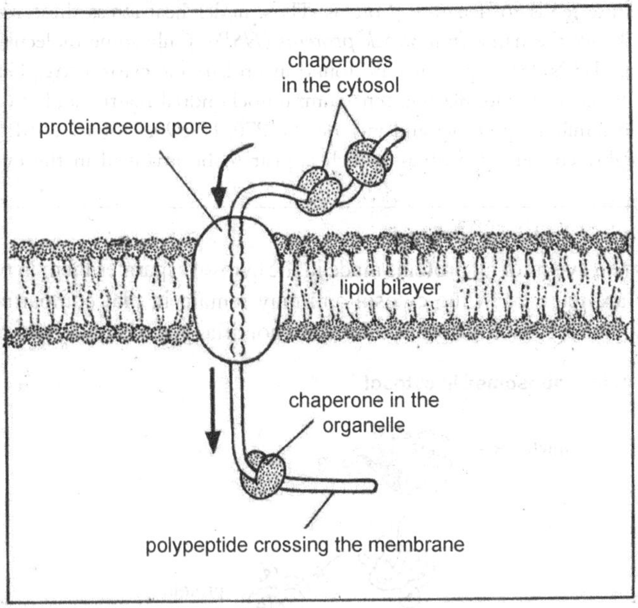

Fig. 22.2. Chaperones facilitate the passage of a protein through the lipid bilayer Chaperones bind a polypeptide as it makes its way through a proteinaceous pore in a lipid membrane. The chaperones keep the polypeptide in an unfolded state on one side of the membrane and help it fold correctly on the other side (after Buchanan *et al.*, 2000).

heat shock proteins (HSP70)—fill all these roles, interacting with a wide spectrum of proteins. HSP70 chaperones are present in all cells, and synthesis of many HSPs are unregulated in response to heat stress, perhaps to prevent misfolding and to repair misfolded proteins. HSP70 perform several different functions in protein transport across membranes, in addition to translocation and folding (Box 22.2).

Box 22.2 Heat Shock Proteins (HSPs) and Other Chaperones

There are a number of proteins whose role is to stabilize proteins that are unfolded. They assist in assembly and correct folding of proteins and also unfold them for translocation across the membranes or for degradation without being a part of them. These proteins are called **molecular chaperones**. Chaperones also play a major role to prevent partially folded proteins from aggregating. Many chaperones act in cooperation with ATP and other cofactors and facilitate in the repair or degradation of damaged proteins. Some of them also function as heat shock proteins (HSPs).

Rough endoplasmic reticulum (RER) is rich in molecular chaperones since the ER is involved in correct folding of proteins. Important examples of such proteins include 1. immunoglobulin heavy chain binding protein; 2. calnexin; 3. calreticulin; 4. protein disulphide isomerase (PDI); 5. peptidyl poly cis-trans isomerase (PPI), etc. (see **Rastogi** 2005). Calnexin and calreticulin are calcium binding proteins found in the ER membrane, while PDI and PPI help in shuffling of disulphide bonds and cis-trans isomerization of proline bonds in proline containing proteins respectively. Almost 80 per cent of newly synthesized proteins are stabilized by molecular chaperones that bind briefly to their surface until correctly folded and use ATP as energy source to achieve this task.

In plants, the most studied HSP60 protein is chaperon 60, a nuclear-encoded chloroplast protein that is involved in Rubisco assembly but do not increase in response to heat stress (see **Buchanan** *et al.*, 2000).

Functions of HSPs. Molecular chaperones are critical for the cells during their life time. During environmental stress, certain proteins are damaged or partly unfolded, losing their function. These shock proteins help in restoring their conformation or repair damage. Their role in numerous diseases such as cancer, ischaemia, etc., has been demonstrated.

High temperature tends to denature proteins. Thus, under heat stress these molecular chaperones swing into action, hence the name *heat shock proteins* (*HSPs*). Only some molecular chaperones act as heat shock proteins (HSPs). Such proteins are found in prokaryotic cytosol (*e.g.*, GroEL; **Buchanan** *et al.*, 2000), eukaryotic cytosol, endoplasmic reticulum, mitochondrial matrix, and stroma of chloroplasts. Heat shock protein family is quite big and includes HSP70, HSP90, HSP40, SSBI (in yeast), etc. They are a family of highly conserved proteins which appear to be involved in the evolution of modern enzymes.

3. Protein sorting is a multistep process

The first sorting event for all proteins made in the cytosol separates them in two groups. Proteins in the first group are released in the cytosol and may remain in that compartment or be targeted to a variety of destinations, including plastids, mitochondria, peroxisomes, and nuclei (Fig. 22.3A).

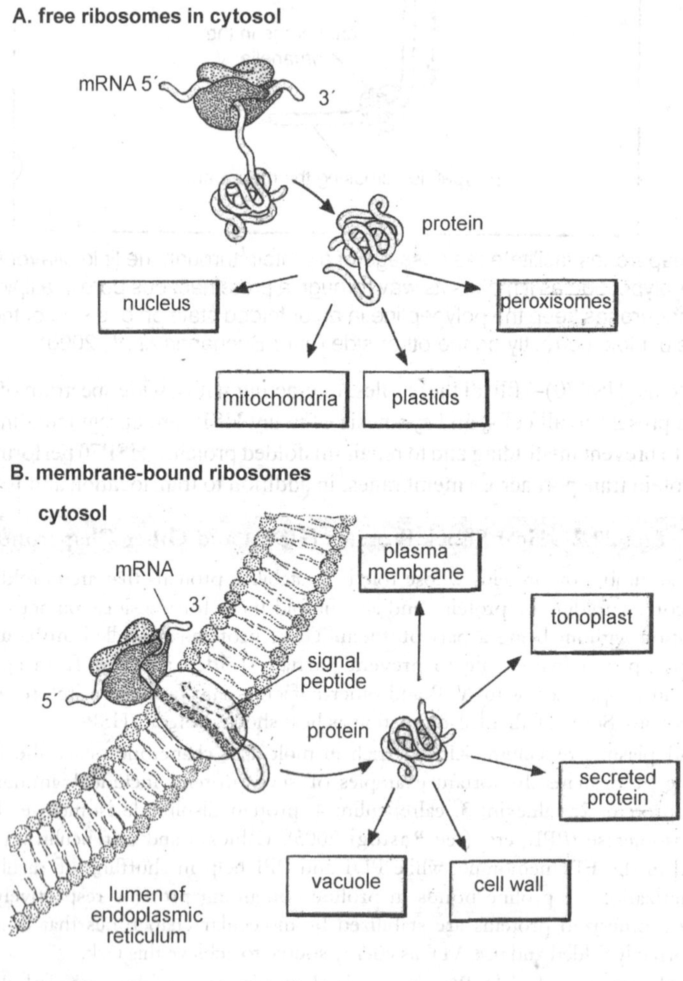

Fig. 22.3. A—proteins synthesized on free ribosomes either remain in the cytosol or are targeted to the nucleus, mitochondria, plastids or peroxisomes. B—proteins synthesized on membrane bound ribosomes are first translocated into the lumen of the ER and transported to the Golgi. These proteins may subsequently be targeted to the plasma membrane or the tonoplast, secreted or sent to the vacuole. Some proteins remain in the ER or the Golgi because they have special retention signals (Buchanan *et al.*, 2000).

By contrast, proteins in the second group are targeted to the ER by **signal peptides** located in the N terminus. Translation of a signal peptide causes the ribosome to bind to the ER during protein synthesis. ER studded with such ribosomes is known as **rough ER**. Proteins synthesized on the rough ER enter the **secretory pathway**, an intracellular system of vesicles and cisternae (flattened sacs) that includes the ER, Golgi apparatus, tonoplast, and plasma membrane. These proteins can then be secreted or targeted to the various compartments of the secretory system (Fig. 22.3B). Integral membrane proteins with signal peptides do not enter the ER lumen but are integrated into the ER membrane, from which they can translocate to other membranes of the secretory system. Translocation to a final destination or retention in a compartment usually requires a second targeting domain (Fig. 22.1).

The transport of both soluble and membrane proteins is mediated by vesicles that carry proteins from one compartment to the next. Thus, vesicles are constantly being formed, transported and fused with a compartment different from the one in which they originated. Such transport implies that vesicles can recognize their destinations according to the cargo they carry. This specificity of recognition is part of the sorting machinery. As polypeptide chains move through the secretory system, they first can be modified by the attachment of N-linked glycans, then folded correctly and assembled into oligomers—dimers, trimers, tetramers, and so forth. These events are catalysed by proteins that reside in the ER.

Ribosomes, when synthesizing proteins that lack a signal peptide, do not associate with the ER. These cytosol-localised ribosomes are described as "free", although they are often attached to the cytoskeleton. Both soluble proteins and proteins that later become incorporated into membranes can be translated by free ribosomes.

Although details vary, depending on the protein being transported, the general features of protein sorting appear to apply to any organelle. Most transported proteins carry a targeting domain that is recognized by a cytosolic factor, often a chaperone. The chaperone delivers the unfolded protein to a specific receptor on the target membrane. The targeting domain typically binds to receptor, opening a transmembrane channel within the membrane. Such channels usually consist of several polypeptides and are coupled to a peripheral protein translocation motor that hydrolyzes nucleoside triphosphates to drive protein transport. After passing through the membrane, the transported protein folds with the help of chaperones and ATP.

22.2. DESTINATION OF PROTEINS SYNTHESIZED BY FREE RIBOSOMES

1. Cytosol

A large number of proteins such as enzymes (*e.g.*, enzymes for glycolysis), tubulins (used for making the microtubules), actins (used for making the microfilaments), etc., are released from the ribosomes and start functioning in the cytosol itself.

2. Nucleus

A number of proteins are needed in the nucleus which have to be transported from the cytosol through the nuclear pores located in the nuclear membranes. Such proteins that are targeted to the nucleus contain a **nuclear localization sequence**, a sequence of 7–41 amino acids, rich in positively charged amino acids (lysine and arginine). The list of vital proteins includes histones, transcription factors, enzymes needed for metabolism of DNA and RNA, and ribosomal proteins that function in the nucleus.

3. Peroxisomes

Peroxisomes are very simple organelles having only two subcompartments in which an important protein can be placed: the boundary membrane and the internal matrix. Proteins destined for a peroxisome possess a **peroximal targeting signal (PTS)**, either PTS for a peroxisomal matrix protein or an mPTS for a peroxisomal membrane proteins. Several different PTSs, mPTSs and PTS receptors have been identified. PTS receptors bind to peroxisome-destined proteins in the cytosol and shuttle them to the surface of the peroxisome. The PTS receptor apparently accompanies the peroxisomal protein through the boundary membrane into the matrix and then recycles back to the cytosol to escort another protein. Unlike mitochondria and chloroplasts, whose imported proteins must assume an unfolded state, peroxisomes are somehow able to import peroxisomal matrix proteins in their native, folded conformation, even those that consist of several subunits. The mechanism by which peroxisomes are able to accomplish this daunting assignment is still not clear (**Karp** 2010).

4. Mitochondria

Mitochondria have four subcompartments into which proteins can be delivered: an outer mitochondrial membrane (OMM), inner mitochondrial membrane (IMM), intermembrane space and matrix. Although mitochondria synthesize a few of their own integral membrane polypeptides (13 in mammals), roughly 99 per cent of the organelle's proteins are encoded by the nuclear genome, synthesized in the cytosol, and imported post-translationally.

As with peroxisomal proteins, and proteins of other compartments, mitochondrial proteins contain signal sequences that target them to their home base. Most mitochondrial matrix proteins contain a removable targeting sequence (called the **presequence**) located at the N-terminus of the molecule (step 1, Fig. 22.4) that included a number of positively charged residues. In contrast, most proteins destined for the IMM contain internal targeting sequences that remain as part of the molecule.

Before a protein can enter a mitochondrion, several events are thought to take place. First, the protein must be presented to the mitochondrion in a relatively extended, or unfolded, state (steps 1 and I, Fig. 22.4). Several different molecular chaperones (*e.g.*, HSP70 and HSP90) have been implicated in preparing polypeptides for uptake into mitochondria, including ones that specifically direct mitochondrial proteins to the cytosolic surface of the OMM (Fig. 22.4). The OMM contains a protein-import complex, the **TOM-complex**, which includes 1. *receptors* that recognize and bind mitochondrial proteins and 2. protein-lined channels through which unfolded polypeptides are translocated across the outer membrane (steps 2 and II, Fig. 22.4). Proteins that are destined for the IMM or matrix must pass through the intermembrane space and engage a second protein-import complex located in the IMM, called a **TIM-complex**. The IMM contains two major TIM complexes: TIM22 and TIM23. **TIM22** binds integral proteins of the IMM that contain an internal targeting sequence and inserts them into the lipid bilayer (steps III, IV, Fig. 22.4). **TIM23**, in contrast, binds proteins with a N-terminal presequence, which includes all of the proteins of the matrix (as well as a number of proteins of the IMM). TIM23 recognizes and translocates the matrix proteins completely through the IMM and into the inner aqueous compartment (step 3, Fig. 22.4). Translocation occurs at sites where the outer and inner mitochondrial membranes come into close proximity so that the imported protein can cross both membrane simultaneously. Movement in the matrix is powered by the electric potential across the IMM acting on the positively charged targeting signal, if the potential is dissipated by addition of a drug such as DNP, translocation ceases.

Box 22.3

Quite interestingly, unlike the translocon of the ER or peroxisome, the pore-forming protein of the TOM complex (Tom 40) is a β-barrel protein, like the other integral proteins of the OMM, reflecting its evolution from the outer membrane of an ancestral bacterium. This has functional consequences, as the 13-barrel protein cannot open laterally to allow integral proteins to insert into the OMM. As a result, OMM proteins have to pass into the intermembrane space before entering the OMM bilayer (Karp 2010).

As it enters the matrix, a polypeptide interacts with mitochondrial chaperones, such as mt HSP70 (step 4, Fig. 22.4), that mediate entry into the aqueous compartment. Following two mechanisms have been proposed to explain the general action of molecular chaperones involved in the movement of protein across membranes, which is a wide spread phenomenon.

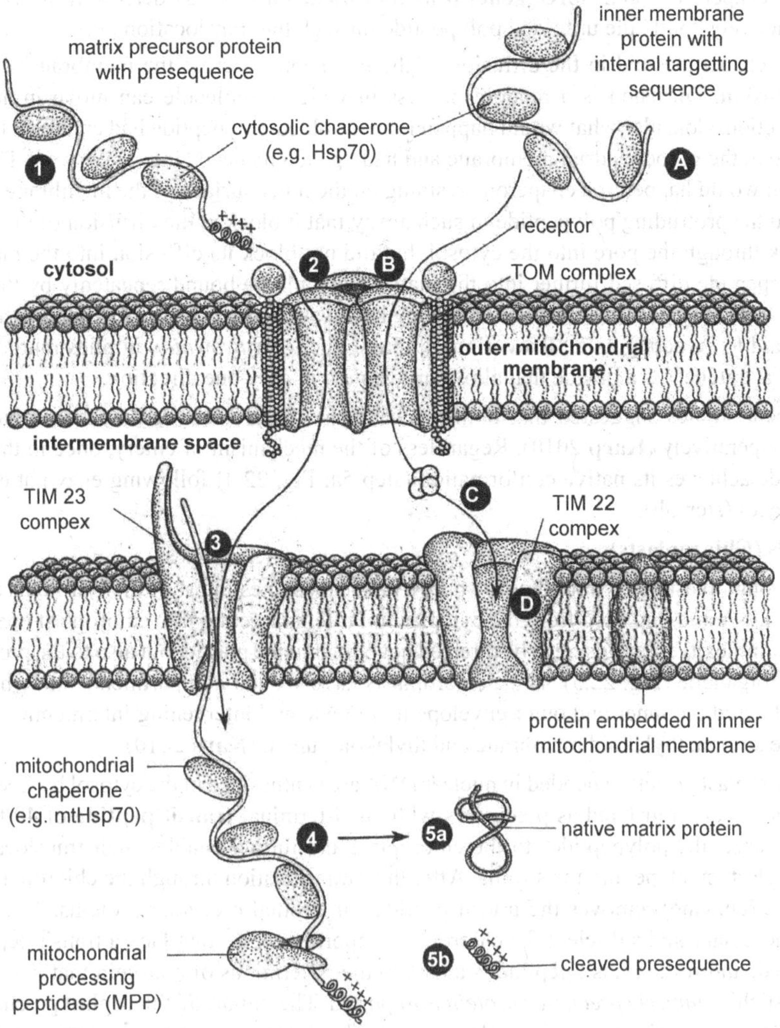

Fig. 22.4. Importing proteins into a mitochondrion. Proposed steps taken by protein imported post-translationally into either the mitochondrial matrix or inner mitochondrial membrane. The polypeptide is targeted to a mitochondrion by a targeting sequence, which is located at the N-terminus in the matrix protein (step 1) and is located internally in most inner membrane proteins (step I). Cytosolic HSP70 molecules unfold the polypeptides prior to their entry into the mitochondrion. The proteins are recognized by membrane receptors and translocated through the OMM by way of pores in the TOM complex of the OMM (step 2 or II). Most integral proteins of the IMM are directed to the TIM22 complex of the IMM (step III), which steers them into the lipid bilayer of the IMM (step IV). Mitochondrial matrix proteins are translocated through the TIM23 complex of the IMM (step 3). Once the protein enters the matrix, it is bound by a mitochondrion (step 4), which may either pull the polypeptide into the matrix or act like a Brownian ratchet to ensure that it diffuses into the matrix (these alternate chaperone mechanisms are discussed in the text). Once in the matrix, the unfolded protein assumes its native conformation (step 5a) with the help of HSP60 chaperone (not shown in figure). The presequence is removed enzymatically (step 5b) (after Karp 2010).

1. The chaperones act as **force-generating motors** that use energy derived from ATP hydrolysis to actively "pull" the unfolded polypeptide through the translocation pore.
2. The chaperones aid in the diffusion of the polypeptide across the membrane, called **biased diffusion**. Diffusion is a random process in which a molecule can move in any available direction. Consider what would happen if an unfolded polypeptide had entered a translocation pore in the mitochondrial membrane and had "poked its head" into the matrix. Then consider what would happen if a chaperone residing on the inner surface of the membrane were able to bind the protruding polypeptide in such a way that it blocked the diffusion of the polypeptide back through the pore into the cytosol, but did not block its diffusion into the matrix. As the polypeptide diffused further into the matrix, it would be bound repeatedly by the chaperone and at each stage prevented from diffusing backward. In such a biased diffusion the chaperone is said to be acting as a "**Brownian ratchet**"; the term *Brownian* implies random diffusion, and a "ratchet" is a device that allows movement in only one direction.

Recent studies suggested that both mechanisms of chaperone action are probably utilized and act cooperatively (**Karp** 2010). Regardless of the mechanism of entery, once in the matrix the polypeptide achieves its native conformation (step 5a; Fig. 22.4) following enzymatic removal of the sequences (step 5b).

5. Plastids (Chloroplasts)

Although plastids contain their own DNA and ribosomes, most of their proteins are encoded in nuclear DNA and imported from the cytoplasm. Protein targeting into chloroplast operates at two levels of complexity: proteins must be directed not only into the plastid, but also to the proper location within the organelle (Fig. 22.5). Thus, chloroplasts have six sub-compartments into which proteins can be delivered: an inner and outer envelope membrane and intervening intermembrane space, as well as the stroma, thylakoid membrane and thylakoid lumen (**Karp** 2010).

Chloroplast proteins encoded in nuclear DNA are synthesized in the cytosol by free ribosomes. These proteins are translated as precursors with an N-terminal **transit peptide** of 40 to 50 amino acids that target the polypeptides to the chloroplast and further enables their translocation across the chloroplast envelope into the stroma. After their translocation through the chloroplast envelope, a peptidase (enzyme) removes the transit peptides of stromal precursor proteins. Transit peptides are both necessary and sufficient for chloroplast import: proteins that lack a transit peptide cannot be imported, and if the transit peptide is added to the N-terminus of a protein that is foreign to the chloroplast, this *chimeric precursor protein* is imported. The import of chloroplast precursors occurs at contact sites (proteinaceous channels) between the outer and inner envelope membranes, and precursor proteins that are in transit span both envelope membranes.

Researchers have studied protein import into chloroplast in cell-free system and have proposed a model which entails two functional steps and requires several more proteins in addition to the transit peptide of the transported species. The process requires chaperones (proteins) on both sides of the chloroplast envelope and a group of proteins, collectively called the **protein import apparatus**, that span both the outer and inner membranes. Both steps require energy in the form of nucleotide triphosphates.

In *step one*, cytosolic chaperones (HSP70 homolog) hold proteins in an unfolded or partially folded state. Transit peptides on these proteins interact with the lipids of the outer membrane or with the proteins of the import apparatus, which form the proteinaceous channels through which imported polypeptide pass (Fig. 22.6). The import apparatus proteins include two main types of proteins—**Toc** (**translocon** of the outer membrane of the chloroplast) and **Tic** (tranlocon of the inner envelope membrane). Two of these proteins, **Toc34** and **Toc159**, are specific GTP-binding tightly anchored in the outer membrane with their GTP-binding domain exposed to the cytosol. A third protein, **Toc75** shows no obvious similarity to the proteins for which the functions have been identified; however, this component is deeply embedded in the outer membrane. A fourth protein, HSP70 exhibits the biochemical characteristics of an integral membrane protein. Only one component of the inner envelope translocation machinery, Tic 110, has also been identified.

In step 2, the protein enters the chloroplast stroma, where a protease enzyme removes the transit peptide (containing stroma targeting domain). Once inside the chloroplast, the polypeptide has several possible fates. If the protein is to remain in the stroma, a very large and complex chaperone made up of large (HSP60) and small (10 kDa or kilodaltons) subunits will help fold it in an ATP-dependent process. By contrast, soluble proteins that function in the thylakoid lumen, and membrane proteins that become part of the thylakoid, are held by chaperones in an unfolded configuration for insertion and passage to their next destination. Proteins targeted to the outer envelope membrane do not enter the transport channel but instead enter the membrane directly from cytosol.

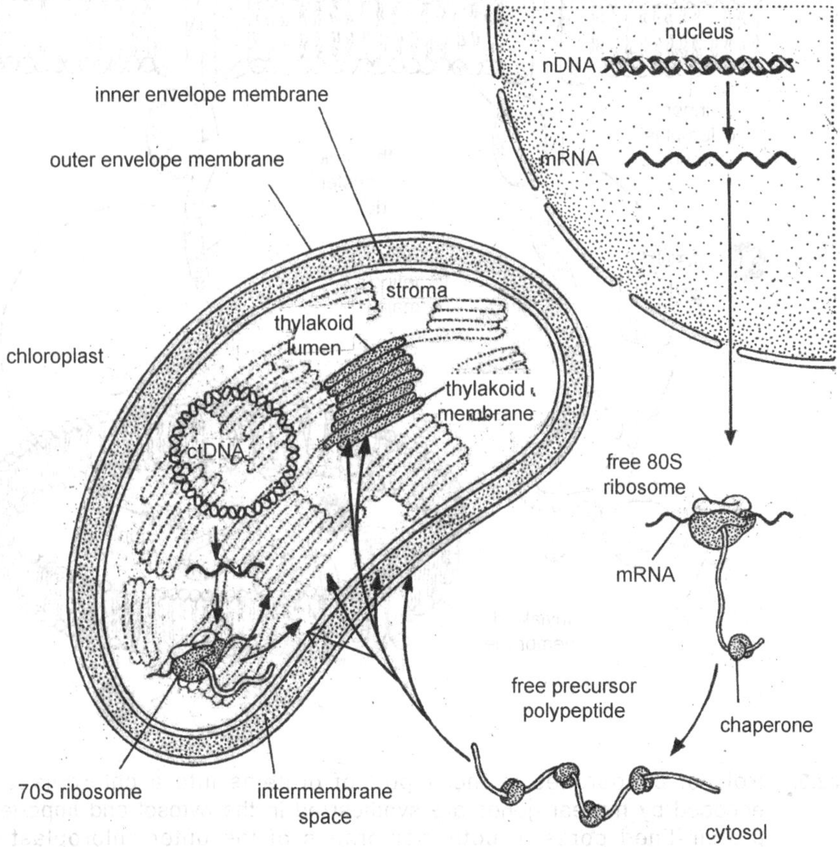

Fig. 22.5. Biosynthesis of chloroplast proteins and their targeting to six different compartments within the chloroplast. Chloroplast proteins may be encoded by nuclear DNA (nDNA) or chloroplast DNA (ct DNA); the respective mRNAs are translated by ribosomes in the cytosol (80S ribosomes) or in the chloroplast stroma (70S ribosomes). Proteins made as precursor polypeptides in the cytosol may be targeted to the outer envelope membrane or may enter the chloroplast stroma. Once across the envelope membranes, proteins may remain in the stroma compartment or may be targeted to the thylakoid membrane, thylakoid lumen, intermembrane space or inner envelope membrane (after Buchanan *et al.*, 2000).

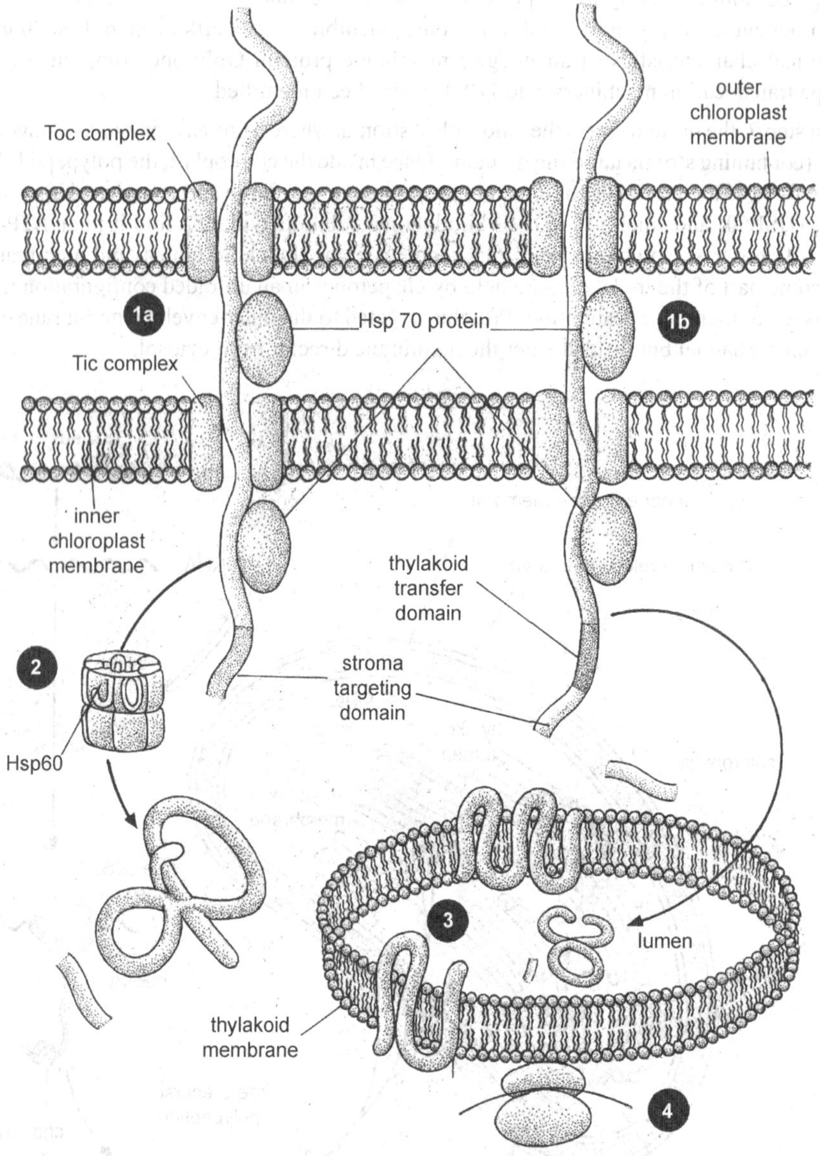

Fig. 22.6. Role of chaperones in the import of proteins into a chloroplast. Proteins encoded by nuclear genes are synthesized in the cytosol and imported through protein-lined pores in both membranes of the outer chloroplast envelope (step 1). Proteins directed for the stroma (step 1a) contain a stroma-targeting domain at their N-terminus, whereas proteins assigned for the thylakoid (step 1b) contain both a stroma-targeting domain and a thylakoid-transfer domain at their N-terminus. Stromal proteins remain in the stroma (step 2) following translocation through the outer envelope and removal of their single targeting sequence. The presence of the thylakoid transfer domain causes thylakoid proteins to be translocated either into or completely through the thylakoid membrane (step 3). A number of proteins of the thylakoid membrane are encoded by chloroplast genes and synthesized by chloroplast ribosomes that are bound to the outer surface of the thylakoid membrane (step 4) (after Karp 2010).

QUESTIONS

Long Answer Questions

1. How do cytosolic proteins enter the mitochondria and chloroplasts?
2. How are proteins, such as enzyme of the Krebs cycle, able to arrive in the mitochondrial matrix?
3. Describe the steps by which a polypeptide would move from the cytosol where it is synthesized to the thylakoid lumen.
4. What is the role of cytosolic and mitochondrial chaperones in the process of mitochondrial import?
5. Explain the roles of chaperones in the import of proteins in the chloroplast.

Short Answer Questions

1. Distinguish between two possible import mechanisms: biased diffusion and force-generating motors.
2. Briefly describe the following:
   (*i*) Molecular chaperones.
   (*ii*) Heat shock proteins.
   (*iii*) Biased diffusion.

Very Short Answer Quetions

1. What is chaperone?
2. What are chaperonins?
3. Define heat-shock proteins.
4. What is translocon?

ANSWERS

Very Short Answer Questions

1. Chaperone is a protein molecule that guides the folding of large segments of a newly formed protein and thus helps other proteins to avoid misfolding pathways that produce inactive or aggregated states.
2. Chaperonins is a class of molecular chaperones which are restricted to bacteria, mitochondria and chloroplasts.
3. Heat-shock proteins (or HSP, Hsp or hsp) are a group of proteins that are synthesized by living cells in response to increased temperature. They occur in both eukaryotes and prokaryotes and function mainly as molecular chaperones, protecting the cell's proteins as they become unfolded due to heating and enabling them to refold correctly.
4. It is a protein-lined channel which is embedded in the ER membrane; the nascent polypeptide is able to move through the translocon in its passage from the cytosol to the ER lumen.

23 Vesicular Traffic in the Secretory and Endocytic Pathway

Transport vesicles play a central role in the traffic or transport of molecules between different membrane-enclosed compartments of the secretory pathway. Vesicles are similarly involved in the transport of materials taken up at the cell surface (*i.e.,* at the level of plasma membrane through endocytosis). *Vesicular traffic* (= transport) is thus a major cellular activity, responsible for molecular traffic between a variety of specific membrane-enclosed compartments. The selectivity of such transport is therefore key in maintaining the functional organization of the cell. For example, lysosomal enzymes must be transported specifically from the Golgi apparatus to lysosomes not to the plasma membrane or to the ER. There exist some signals that target proteins to specific organelles, such as lysosomes. These proteins are transported within vesicles, so the specificity of transport is based on the selective packaging of the intended cargo into vesicles that recognize and fuse only with the appropriate target membrane. Because of the central importance of vesicular transport to the organization of eukaryotic cells, understanding the molecular mechanisms that control vesicle packaging, budding and fusion is a major area of research in cell biology.

Experimental Approaches for Understanding Vesicular Transport

Progress toward explaining the molecular mechanisms of vesicular transport has been made by three distinct experimental approaches: (*i*) Isolation of yeast mutants that are defective in protein transport and sorting; (*ii*) Reconstitution of vesicular transport in cell-free systems; and (*iii*) Biochemical analysis of synaptic vesicles, which are responsible for the regulated secretion of neurotransmitters by neurons (see Cooper and Hausman, 2007).

1. In 1980, **Randy Schekman** and his colleagues have pioneered the isolation of yeast mutants defective in vesicular transport. These include mutants that are defective at various stages of protein secretion (*sec* mutants), mutants that are unable to transport proteins to the vacuole, and mutants that are unable to retain resident ER proteins. The isolation of such mutants in yeasts led directly to the molecular cloning and analysis of corresponding genes, thereby identifying a number of proteins involved in various steps of the secretory pathway. For example, the protein Sec 61 forms a major component of the protein acting as translocation channel in RER.
2. Biochemical studies of vesicular transport using reconstituted systems have complemented these genetic studies and have enabled the direct isolation of proteins from mammalian cells. The first cell-free transport system was developed in 1984 by **James Rothman** and his colleagues, who analysed protein transport between compartments of the Golgi apparatus. Similar reconstituted systems have been developed to analyse transport between other

compartments, including transport from ER to Golgi apparatus and transport from the Golgi apparatus to secretory vesicles, vacuoles and the plasma membrane. The development of these *in vitro* systems has enabled biochemical studies of the transport process and functional analysis of proteins identified by mutations in years, as well as direct isolation of some of the proteins involved in vesicles budding and fusion.

3. Insights into the molecular mechanisms of vesicular support have come from studies of synaptic transmission in neurons, which represents a highly specialised form of regulated secretion. A **synapse** is the junction of a neuron with another cell, which may be either another neuron or an effector, such as muscle cell. Information is transmitted across the synapse by chemical neurotransmitters to be released into the synapse and stimulating the post-synaptic neuron of effector cell. Synaptic vesicles are extremely abundant in the brain, allowing them to be purified in large amounts for biochemical analysis. Some of the proteins isolated from synaptic vesicles are closely related to proteins that have been shown to play critical roles in vesicular transport by yeast genetics and reconstitution experiments, so biochemical analysis of these proteins has provided important insights into the molecular mechanisms of vesicle fusion.

Lastly, recent studies using GFP fusion proteins (Box 23.1) have allowed transport vesicles carrying specific proteins to be visualized by immunofluorescence as they move through the secretory pathway. In these experiments, cells are transfected with cDNA constructs encoding secretory proteins tagged with green fluorescent protein (GFP). The progress of the GFP-labelled proteins through the secretory pathway can then be followed in living cells, allowing characterisation of many aspects of the dynamics and molecular interactions involved in vesicular transport.

Box 23.1

GFP or green fluorescent protein is derived from the pacific jellyfish, *Aequoria victoria*.

23.1. SECRETORY PATHWAYS

Integral to the vesicular traffic shown in Figure 23.1 are *secretory pathways* by which proteins move from the ER through the Golgi apparatus to *secretory vesicles* and *secretory granules*, which then discharge their contents to the exterior of the cell. The collective roles of the ER and the Golgi apparatus in secretion were first demonstrated by *Georg Palade* and his colleagues, who used autoradiography to trace the fate of radioactively labelled protein through cells in the salivary gland.

In eukaryotic cells, mode of secretion may be of two types: 1. Constitutive secretion; and 2. Regulated secretion.

1. Constitutive secretion

After budding from the TGN (trans Golgi network), some secretory vesicles move directly to the cell surface where they immediately fuse with the plasma membrane and release their contents. This process, which is continuous and independent of specific extracellular signals, occurs in most eukaryotic cells and is called **constitutive secretion**. Examples include the continuous release of mucus by cells that line human intestine and the secretion of the glycoproteins of the extracellular matrix.

Default pathway model. Constitutive secretion was once assumed to be a **default pathway** for proteins synthesized by ribosomes attached to the rough ER (see **Becker** *et al.*, 2006). According to this model, all proteins destined for transport from the ER to the Golgi apparatus must have a tag that diverts them from constitutive secretion. Unless an amino acid sequence of oligosaccharide chain identifies a protein for retention in, or transport to, a specific organelle, the protein moves through the endomembrane system and is released outside the cell by default. Support for this model came from studies in which the **retrieval tags** on ER-specific proteins were removed and the fates of the proteins were traced. Removal of the KDEL sequence from resident protein of the ER generally led to secretion of the protein. More recent evidence, however, points to a variety of short amino acid sequence, that identify proteins for constitutive secretion by one of several distinct pathways for the process.

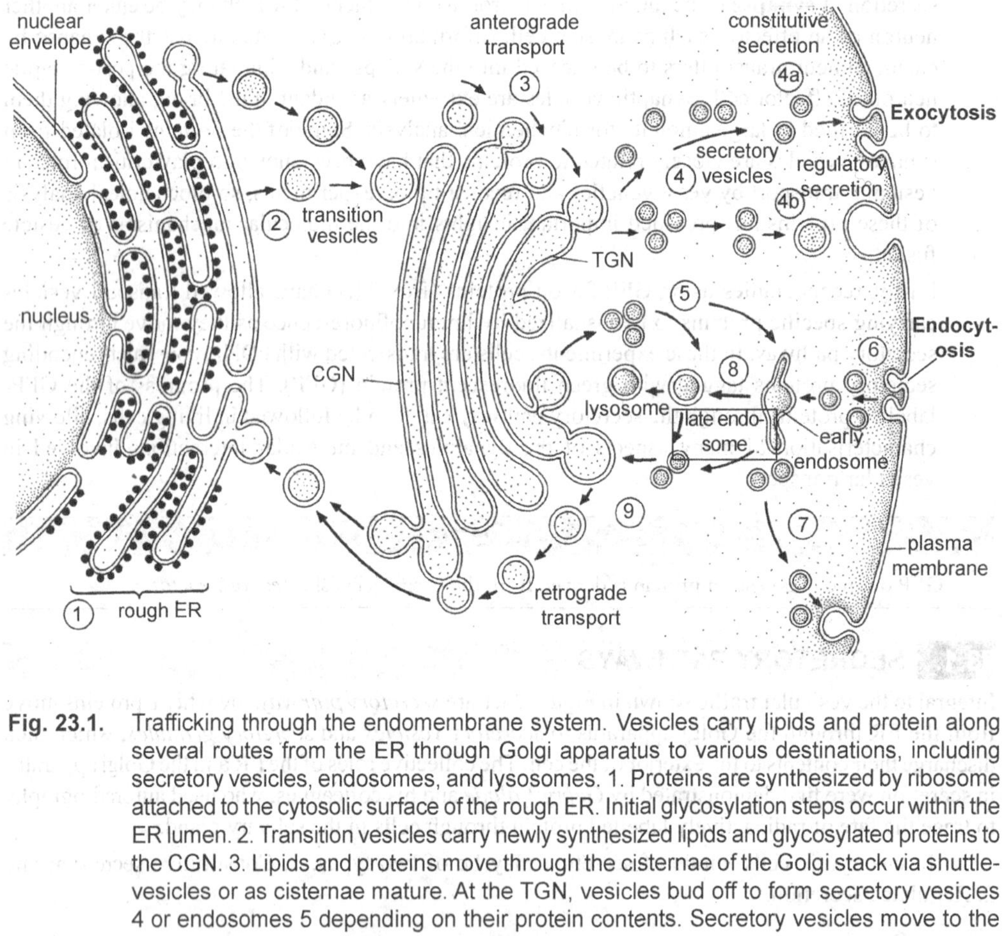

Fig. 23.1. Trafficking through the endomembrane system. Vesicles carry lipids and protein along several routes from the ER through Golgi apparatus to various destinations, including secretory vesicles, endosomes, and lysosomes. 1. Proteins are synthesized by ribosome attached to the cytosolic surface of the rough ER. Initial glycosylation steps occur within the ER lumen. 2. Transition vesicles carry newly synthesized lipids and glycosylated proteins to the CGN. 3. Lipids and proteins move through the cisternae of the Golgi stack via shuttle-vesicles or as cisternae mature. At the TGN, vesicles bud off to form secretory vesicles 4 or endosomes 5 depending on their protein contents. Secretory vesicles move to the plasma membrane, where they release their contents by exocytosis, either constitutively (4a) or in response to an appropriate signal (4b). 6. Protein and other materials are taken into the cell by endocytosis, forming endocytic vesicles that fuse with early endosomes. 7. Cellular components not destined for digestion following endocytosis are recycled to the plasma membrane. 8. Early endosomes containing material for digestion mature to form late endosomes and then lysosomes. 9. Retrograde traffic returns compartment-specific proteins to earlier compartments (after Becker *et al.*, 2006).

2. Regulated Secretion

While vesicles containing constitutively secreted proteins move continuously and directly from the TGN to the plasma membrane, secretory vesicles involved in **regulated secretion** accumulate in the cell and then fuse with the plasma membrane only in response to specific extracellular signals. An important example of regulated secretion is the release of neurotransmitters (Fig. 23.2). Two additional examples of regulated secretion are the release of insulin from the β cells of the pancreatic islets of Langerhans and the release of digestive enzymes from acinar cells of the pancreas.

Regulated secretory vesicles form by budding from the TGN as immature secretory vesicles, which lose their clathrin coats and undergo a maturation process. Maturation involves concentration of the proteins-referred to as **condensation** and frequently also the proteolytic processing of secretory proteins. The mature secretory vesicles then move close to the site of secretion and remain near the plasma membrane, until receiving the signal that triggers release of their contents by exocytosis.

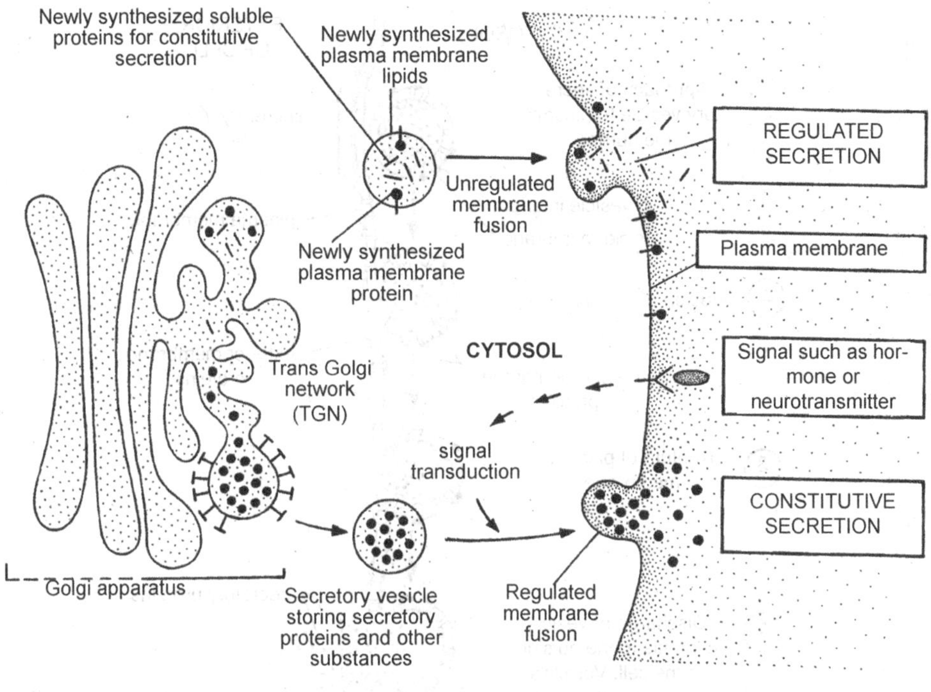

Fig. 23.2. Constitutive and regulates secretory pathways.

Mature regulated secretory vesicles are usually quite large and contain much more highly concentrated proteins than the constitutive secretory vesicles. Such large dense vesicles are often called **secretory** or **zymogen granules** to distinguish them from other secretory vesicles. For example, in an acinar (secretory) cell of exocrine pancreas of the rat zymogen granules (ZG) are concentrated in the region of the cell between the Golgi stacks from which they arise and the portion of the plasma membrane bordering the lumen into which contents of the granules are eventually discharged.

The information needed to direct a protein to a regulated secretory vesicle is presumably inherent in the amino acid sequence of the protein, though the precise signals and mechanisms for **sorting proteins** to secretory vesicles are not clear. Current evidence suggests that high concentrations of secretory proteins in secretory granules promote the formation of large **protein aggregates** that exclude non secretory proteins. This could occur in the TGN, where only aggregates would be packaged in vesicles destined for secretory granules, or it could occur in the secretory granules itself. The pH of the TGN and the secretory granule lumens may serve as a trigger favouring aggregation as material leaves the TGN. The soluble proteins that do not become part of an aggregate in the TGN or a secretory granules would be carried by transport vesicle to other locations.

23.2. TRANSPORTING MATERIAL ACROSS THE PLASMA MEMBRANE (BULK TRANSPORT BY THE PLASMA MEMBRANE)

Cells routinely import and export large molecules across the plasma membrane. Macromolecules are secreted out from the cell by *exocytosis* and are ingested into the cell from outside through *phagocytosis* and *endocytosis*.

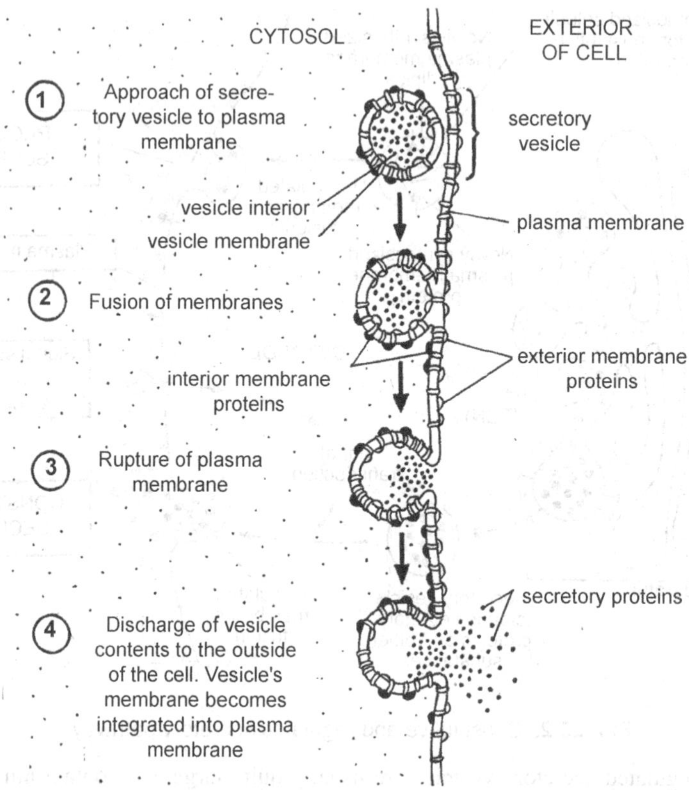

Fig. 23.3. Exocytosis. Release of the contents of a secretory vesicle or granule to the exterior of the cell (after Becker *et al.*, 2006).

1. Exocytosis

It is also called **emeiocytosis** and **cell vomiting**. In all eukaryotic cells, **secretory vesicles** are continually carrying new plasma membrane and cellular secretions such as proteins, lipids and carbohydrates (*e.g.*, cellulose) from the Golgi apparatus to the plasma membrane or to cell exterior by the process of exocytosis. The protein to be secreted are synthesized on the rough endoplasmic reticulum (RER). They pass into the lumen of the ER, glycosidated and are transported to the Golgi apparatus by ER-derived **transport vesicles.** In the Golgi apparatus the proteins are modified, concentrated, further glycosidated, sorted and finally packaged into vesicles that pinch off from trans Golgi tubules and migrate to plasma membrane to fuse with it and release the secretion to cell's exterior. In contrast, small molecules to be secreted (*e.g.*, histamine by the mast cells) are actively transported from the cytosol (where they are synthesized on the free ribosomes) into preformed vesicles, where they are complexed to specific macromolecules (*e.g.*, a network of proteoglycans, in case of histamine), so that, they can be stored at high concentration without generating an excessive osmotic gradient.

During exocytosis the vesicle membrane is incorporated into the plasma membrane. The amount of secretory vesicle membrane that is temporarily added to the plasma membrane can be enormous: in a pancreatic acinar cell discharging digestive enzymes, about 900 μm^2 of vesicle membrane is inserted into the apical plasma membrane (whose area is only 30 μm^3) when the cell is stimulated to secrete.

2. Phagocytosis

Sometimes the large-sized solid food or foreign particles are taken in by the cell through the plasma membrane. The process of ingestion of large-sized solid substances (*e.g.*, bacteria and parts of broken cells) by the cell is known as **phagocytosis** (Gr., *phagein*=to eat, *kytos*=cell or hollow vessel).

Occurrence of phagocytosis. The process of phagocytosis occurs in most protozoans and certain cells of multicellular organisms. In multicellular organisms such as mammals, the phagocytosis occurs very actively in granular leucocytes and in the cells of mesoblastic origin. The cells of the mesoblastic origin are collectively known as the cells of **macrophagic** or **reticuloendothelial system**. They can ingest bacteria, protozoa, cell debris or even colloidal particles by the process of phagocytosis.

Process of phagocytosis. In phagocytosis, first the target particle is bound, to the specific receptors on the cell's surface (process is called **adsorption**), then the plasma membrane expands along the surface of the particle and eventually engulfs it. Vesicle formed by phagocytosis is called **phagosome** and it is typically 1 to 2 μm or larger in diameter, much larger than those formed during pinocytosis and receptor-mediated endocytosis. The phagosomes migrate to the interior of the cell and fuse with the pre-existing lysosomes (to form phagolysosome). The food is digested by the hydrolytic enzymes (acid hydrolase) of the lysosomes and the digested food is ultimately diffused to the surrounding cytoplasm. In addition to the normal set of lysosomal hydrolases,

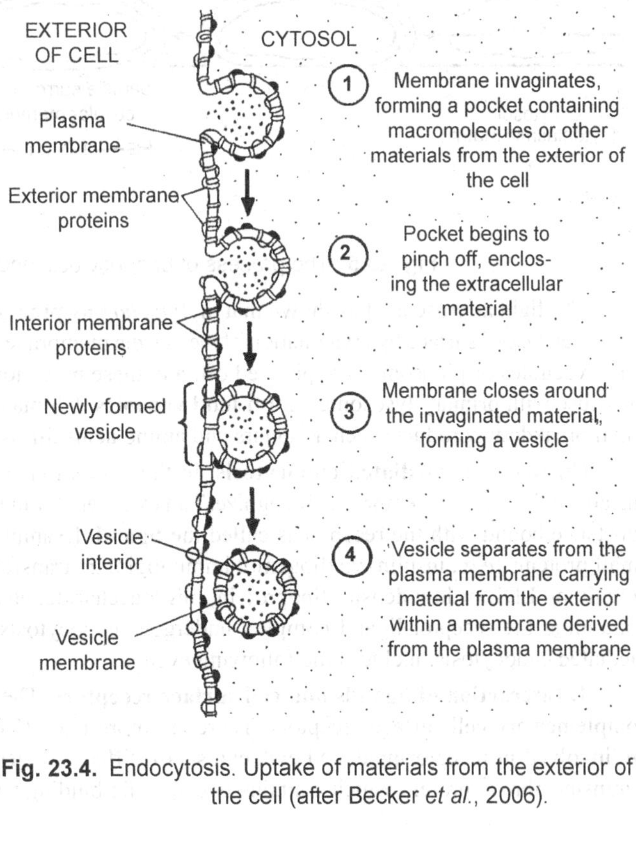

Fig. 23.4. Endocytosis. Uptake of materials from the exterior of the cell (after Becker *et al.*, 2006).

macrophage's lysosomes contain enzymes that generate hydrogen peroxide (H_2O_2) and other toxic chemicals that aid in the killing of the bacteria. The undigested food is expelled from the plasma membrane by the process of **ephagy** or **egestion**. In macrophages, the undigested parts of ingested material such as the cell walls of microorganisms, accumulate within lysosomes as residual bodies. Accumulation of residual bodies may be one reason why macrophages have a very short life time (*i.e.*, less than a few days).

3. Endocytosis

In endocytosis, small regions of the plasma membrane fold inwards or **invaginate** until it has formed new intracellular membrane limited vesicles. In eukaryotes, the following two types of endocytosis can occur: pinocytosis and receptor-mediated endocytosis.

(*i*) Pinocytosis. Pinocytosis (Gr., *pinein*=to drink; 'cell drinking') is the non-specific uptake of small droplets of extracellular fluid by **endocytic vesicles** or **pinosomes**, having diameter of about 0.1 µm to 0.2 µm. Any material dissolved in the extracellular fluid is internalized in proportion to its concentration in the fluid.

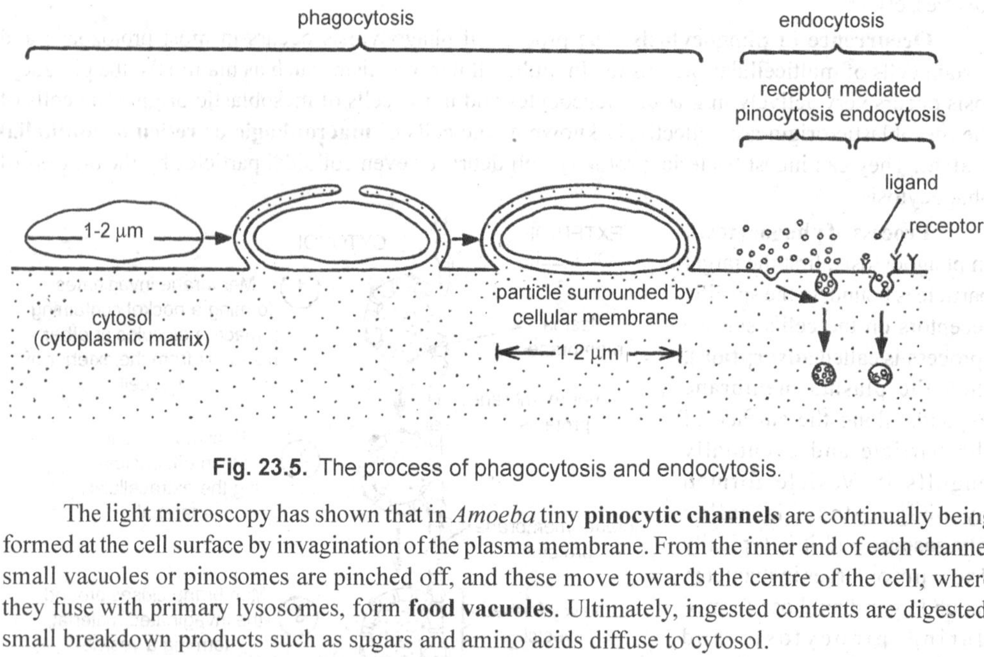

Fig. 23.5. The process of phagocytosis and endocytosis.

The light microscopy has shown that in *Amoeba* tiny **pinocytic channels** are continually being formed at the cell surface by invagination of the plasma membrane. From the inner end of each channel small vacuoles or pinosomes are pinched off, and these move towards the centre of the cell; where they fuse with primary lysosomes, form **food vacuoles**. Ultimately, ingested contents are digested, small breakdown products such as sugars and amino acids diffuse to cytosol.

(*ii*) Receptor-mediated endocytosis. In this type of endocytosis, a specific receptor on the surface of the plasma membrane "recognizes" an extracellular macromolecule and binds with it. The substance bound with the receptor is called the **ligand.** Examples of ligands may include viruses, small proteins (*e.g.*, insulin, vitellogenin, immunoglobin, transferrin, etc.), vitamin B_{12}, cholesterol containing LDL or low density lipoprotein, oligosaccharide, etc. The region of plasma membrane containing the receptor-ligand complex undergoes endocytosis. The whole process of receptor-mediated endocytosis, includes the following events:

1. Interaction of ligands and cell surface receptors. The macromolecules (ligands) bind to complementary cell-surface receptors. There are more than 25 different types of receptors which are involved in receptor-mediated endocytosis of different types of molecules. Such a receptor is a transmembrane protein which contains two specific binding sites: (1) **ligand-binding site** at the external surface of plasma membrane; and (2) **coated-pit binding site** at the inner or cytosolic face of the plasma membrane.

2. Formation of coated-pits and coated-vesicles. The endocytic cycle begins at specialized regions of the plasma membrane, called **coated-pits**. Coated-pits are depressions of plasma membrane having a coat of bristle-like structure towards their cytosolic side. The ligand-loaded receptors diffuse into these coated-pits. They invaginate into the cell and pinches off to form the **coated-vesicles.** The coat at coated vesicles is made up of protein, called **clathrin.**

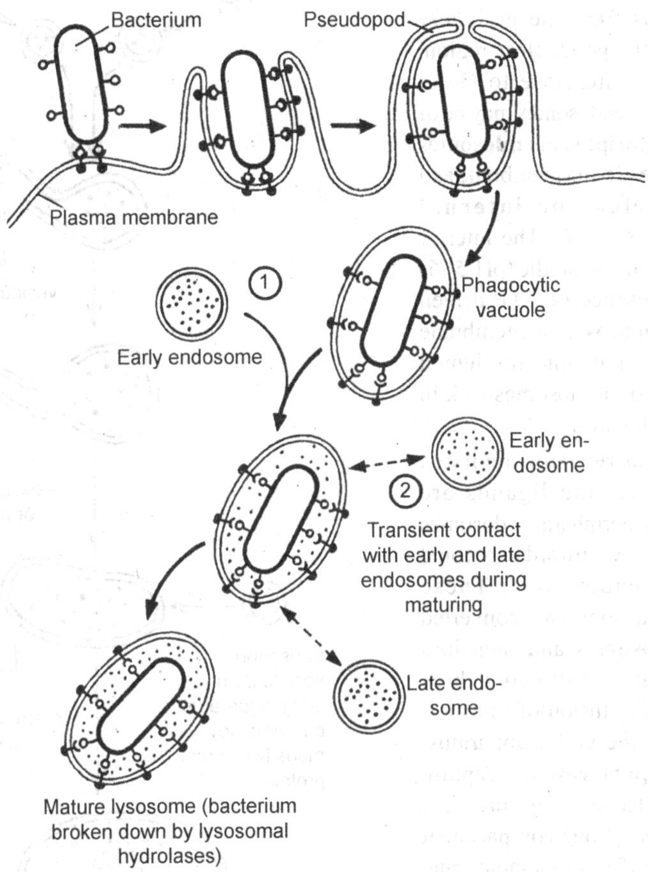

Fig. 23.6. Phagocytosis. Particles or microorganisms bind to receptors on the cell surface, triggering the onset of phagocytosis. In a process involving polymerization of actin, folds of membrane called pseudopods gradually surround the particle. Eventually, the pseudopods meet and engulf the particle, forming a phagocytic vacuole. The vacuole the (1) Fuses with an early endosome or (2) Forms transient connections (indicated by dashed line) with early and late endosomes and matures into a lysosome, in which digestion of the internalised material occurs. (after Becker *et al.*, 2006).

3. Fusion of endocytic vesicle and endosome. Once a coated vesicle is formed, the clathrin and associated proteins dissociate from the vesicle membrane and return to the plasma membrane to form a new coated-pit. The resultant **endocytic vesicle** gets fused with pre-existing endosomes and ultimately its contents are utilized by the cell.

Endosome or receptosome. Recently it has been found that in the cells exists a complex set of heterogeneous membrane-bound tubes and vesicles, called **endosome**, which extends from the periphery of the cell to the perinuclear region, where it lies quite close to Golgi apparatus. Thus, endosomes may be of two types: (*i*) **peripheral endosomes** just beneath the plasma membrane and (*ii*) **perinuclear** or **internal endosomes** (Fig. 23.7). The interior of the endosome is acidic (pH 5–6) due to the presence of ATP-driven **proton (H^+) pumps** in its membrane that pumps H^+ ions into the lumen from the cytosol. Endosomes lack in degradative enzymes.

Thus, via receptor-mediated coated-vesicles, the ligands are delivered to the peripheral endosomes which slowly move inward to become perinuclear endosomes. These perinuclear endosomes are converted into **endolysosomes** and then into **lysosomes** due to following three activities: 1. The fusion of transport vesicles from the Golgi apparatus. (**Note. Transport vesicles** capture a cargo of molecules. *e.g.*, proteins, from the lumen of one compartment as they pinch off from its membrane and then discharge that cargo into another compartment as they fuse with it. Thus, in such vesicular transport, the transported proteins do not cross any membrane and they are transferred from lumen to lumen). 2. Continuous membrane retrieval, and 3. Increased acidification. The endosomal compartment also acts as the main *sorting station* in the endocytic pathway. The acidic

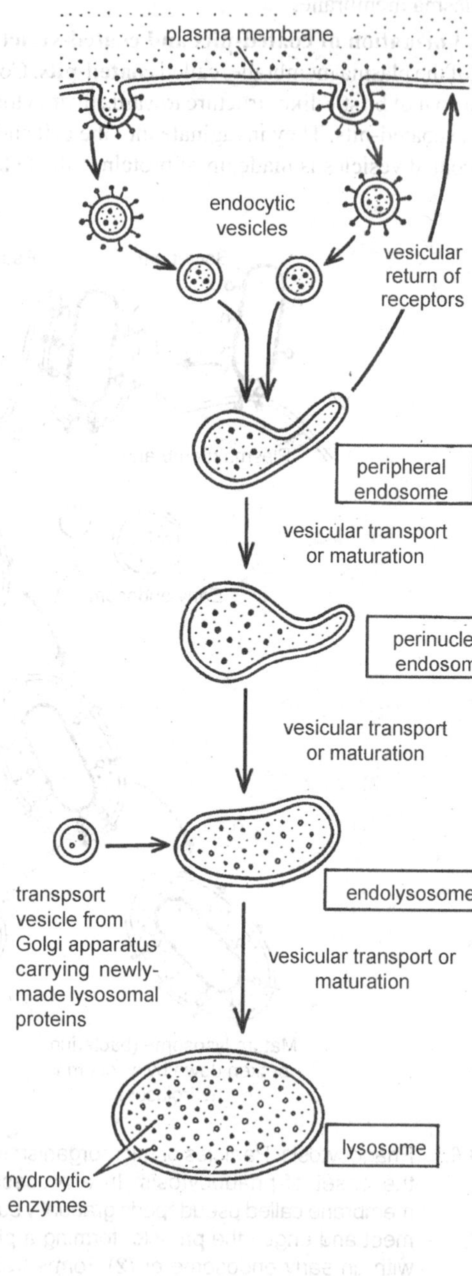

Fig. 23.7. An endocytic pathway: ligands which are endocytosed via coated pits, are delivered to the peripheral endosomes and then move sequentially to perinuclear endosomes, endolysosomes and lysosomes.

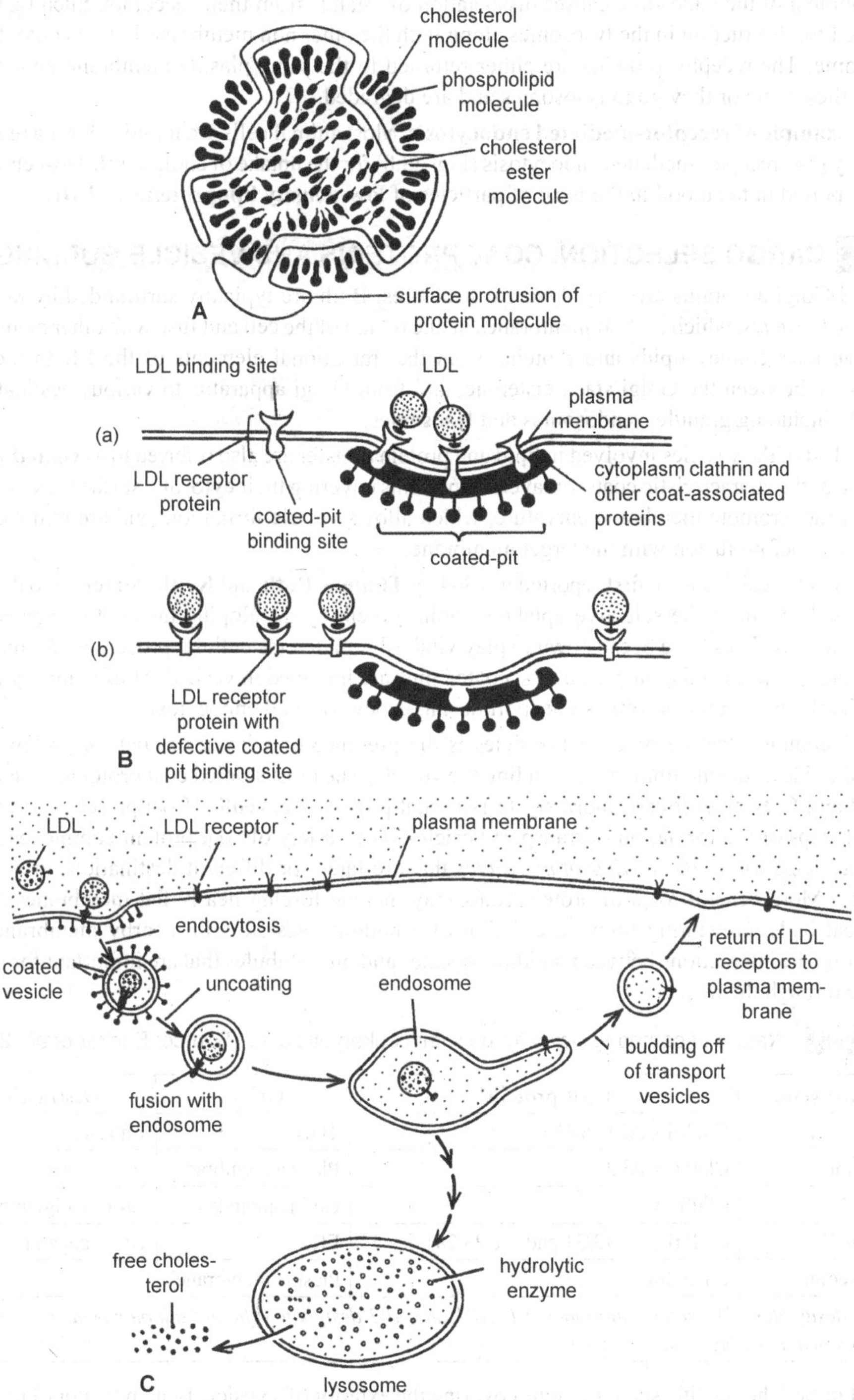

Fig. 23.8. Receptor-mediated endocytosis. A—A LDL particle in cross section; B—LDL receptor proteins binding to a coated-pit in the plasma membrane; C—Mechanism of receptor-mediated endocytosis of LDL particles.

environment of the endosome causes dissociation of ligands from their receptors. Such ligands are destined for destruction in the lysosomes along with the other non membrane-bound contents of the endosome: The receptor-proteins are either returned to the same plasma membrane domain from which they come or they go to lysosomes and are degraded.

Example of receptor-mediated endocytosis. Most animal cells are found to have a regulatory pathway (*i.e.*, receptor mediated endocytosis (Fig. 23.8) for the uptake of cholesterol. Most cholesterol is transported in the blood in the form of particles of **low-density lipoproteins** or **LDL**.

23.3. CARGO SELECTION, COAT PROTEINS AND VESICLE BUDDING

ER and Golgi apparatus are very dynamic structures. Both are typically surrounded by numerous *transport vesicles*, which bud off membranes in one region of the cell and fuse with other membranes. Such vesicles convey lipids and proteins from the transitional elements of the ER to the Golgi apparatus, between the Golgi stack cisternae, and from Golgi apparatus to various destinations in the cell, including granules, endosomes and lysosomes.

Most of the vesicles involved in lipid and protein transfer are also referred to as **coated vesicles** because of the characteristic coats, or layers of proteins covering their cytosolic surfaces as they form. These coats promote membrane curvature, which allows vesicle formation, and are removed from the vesicle before fusion with the target membrane.

Coated vesicles were first reported in 1964 by **Thomas Roth** and **Keith Porter**, who described their involvement in the selective uptake of yolk protein by developing mosquito oocytes. Since then, coated vesicles have been shown to play vital roles in diverse cellular processes. We have seen that coated vesicles and their precursors, coated pits, are involved in vesicular traffic throughout the endomembrane system as well as transport during exocytosis and endocytosis.

A common feature of coated vesicles is the presence of a layer, or coat, of protein on the cytosolic side of the membrane surrounding the vesicle. The most studied coat proteins are **clathrin, COP I** and **COP II**. A fourth, more mysterious coat protein is **caveolin**. Coat protein participate in several steps of the formation of transport vesicles. The variety of coat protein complexes reflects *their participation in the sorting of molecules* that are fated for different destinations into specific vesicles. More general roles of protein coats may include forcing nearly flat membranes to form spherical buds, preventing premature fusion of a budding vesicle with nearby membranes, and regulating the interactions between budding vesicles and microtubules that are important for moving vesicle through the cell.

Table 23.1 Nature of coated vesicles found within eukaryotic cells (Source: Becker *et al.*, 2006).

Coated vesicle	Coat protein*	Origin	Destination
1. Clathrin	Clathrin, AP1, ARF	TGN	Endosomes
2. Clathrin	Clathrin, AP2	Plasma membrane	Endosomes
3. COPI	COPI, ARF	Golgi apparatus	ER or Golgi apparatus
4. COP II	COP II (Sec 13/31 and sec 23/24), Sar 1	ER	Golgi apparatus
5. Caveolin	Caveolin	Plasma membrane	?

**ARF designates ADP ribosylation factor 1; AP1 and AP2 designate different adaptor protein complexes (also called assembly protein complexes).*

Further, the specific set of proteins covering the exterior of a vesicle is an indicator of the origin and destination of the vesicle within the cell (Table 23.1). For example, **clathrin-coated vesicles** are involved in the selective transport of protein from TGN to endosomes, and in the endocytosis of receptor-ligand complexes from the plasma membrane. **COPI-coated vesicles**, on the other hand, facilitate retrograde transport of proteins from the Golgi back to the ER, as well as between cisternae of the Golgi apparatus. **COP II-coated vesicles** are involved in the transport of material from the ER to the Golgi. The precise role of **caveolin-coated vesicles**, called **caveolae**, is still controversial. Transient caveolae form from the plasma membrane and internalized material appears in endosomes, but the pathway connecting the plasma membrane and endosome is not yet clear (see **Becker** *et at.*, 2006).

1. Clathrin-coated vesicles. Clathrin-coated vesicles are surrounded by coats composed of two multimeric proteins, **clathrin** and **adaptor protein (AP)**. The term *clathrin* comes from *clathratus* the Latin word for a "lattice" as clathrin and AP assemble to form polygonal protein lattices. *Flat clathrin lattices* are composed entirely of hexagons and pentagons.

Clathrin-coated vesicles readily dissociate into soluble clathrin complexes, adaptor protein complexes, and uncoated vesicles. These complexes and vesicles, in turn, can spontaneously reassemble under appropriate conditions. In a slightly acidic solution containing calcium ions, clathrin complexes will even assemble independent of adaptor protein and membrane-bounded vesicles, resulting in empty shells called **clathrin cages.**

***(i)* Components of clathrin lattices.** In 1981 **Ernst Ungewickell** and **Daniel Branton** visualised the basic structural units of clathrin lattices, three-legged structures called **triskelions** (Fig. 23.9A). Each triskelion is a multimeric protein composed of three large polypeptides and three small polypeptides radiating from a central vertex. The large polypeptides, each with a molecular weight of 192,000 are **clathrin heavy chains**; they form the legs of triskelion. Each leg is slightly curved near the middle, or knee, and has a globular domain at its outer tip. The small polypeptides, with a molecular weights in the range of 30,000 – 36,000, are **clathrin light chains**. The light chains are found to be associated with the inner half of each leg.

By combining information collected from electron microscopy and X–ray cyrstallography, researchers have assembled a model for the organization of triskelions into the characteristic hexagons and pentagons of clathrin-coated pits and vesicles (Fig. 23.9B). According to one popular model, one clathrin triskelion is located at each vertex of the polygonal lattice, with the knee of the clathrin heavy chains are located at adjacent vertices, each edge of the clathrin lattice is composed of overlapping legs from three triskelions. This arrangement of triskelions into overlapping networks ensures extensive longitudinal contact between clathrin polypeptides. Such contact may impart the mechanical strength needed when a coated vesicle forms from a membrane. The knees, however, do not interact with other polypeptides. This may provide flexibility for forming both hexagons and pentagons and also may accomodate vesicles of different sizes.

***(ii)* Adaptor protein.** The second major component of clathrin coats-the **adaptor protein (AP)**–was originally identified simply by its ability to promote the assembly of clathrin coats around vesicles; thus, AP is also called **assembly proteins**. Eukaryotic cells contain at least-four types of AP complexes, each composed of four polypeptides—two adaptin subunits, one medium chain, and one small chain. The four polypeptides, which are highly different in each type of AP complex, bind to different transmembrane receptor proteins and *impart specificity during vesicle budding and targeting*. In addition to ensuring that appropriate macromolecules will be concentrated in coated pits, AP complexes also mediate the attachment of clathrin to the plasma membrane. AP complexes are found to be sites for regulation of clathrin assembly and disassembly. For example, the ability of AP complexes to bind to clathrin is affected by pH, phosphorylation and dephosphorylation.

***(iii)* The assembly of clathrin coats drives the formation of vesicles from the plasma membrane and TGN.** The binding of AP complexes to the plasma membrane and the concentration of receptors or receptor-ligand complexes in coated pits require ATP and GTP—though perhaps only for regulation of the process. However, the assembly of clathrin coat around budding vesicles can occur without additional expenditure of ATP and GTP. An exception is the assembly of clathrin coats around vesicles forming from the TGN, which depends on the hydrolysis of nucleotide triphosphates.

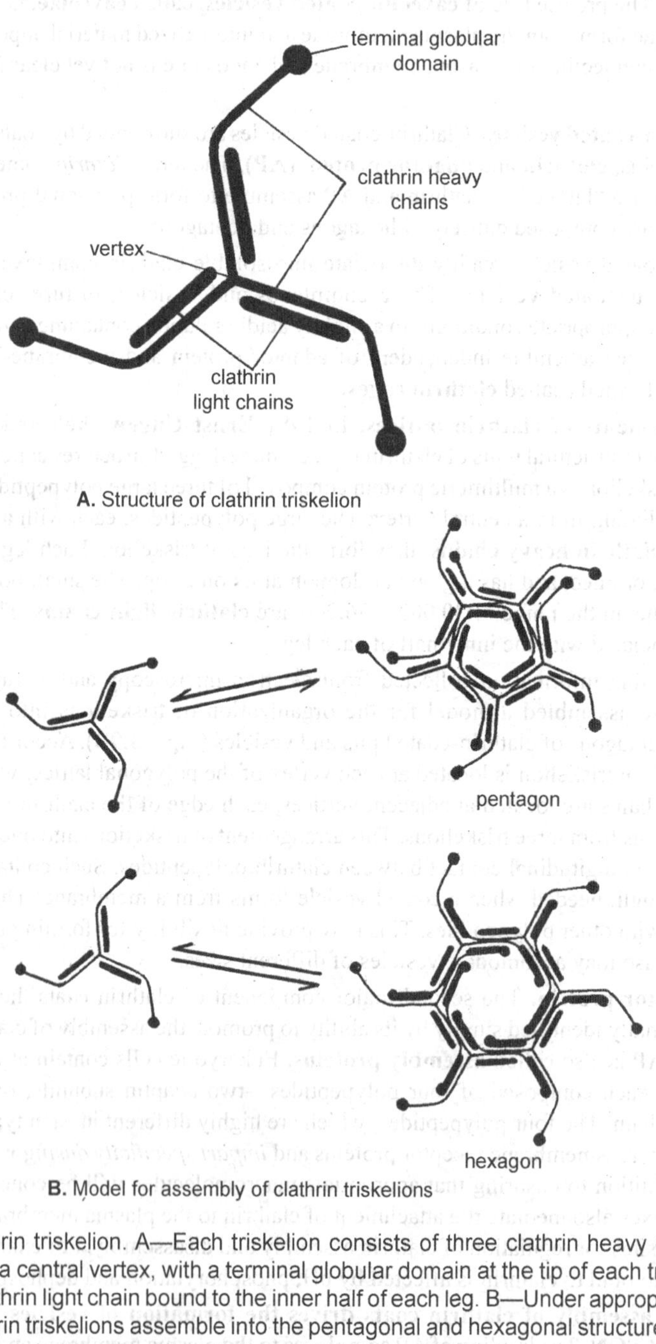

Fig. 23.9. Clathrin triskelion. A—Each triskelion consists of three clathrin heavy chains radiating from a central vertex, with a terminal globular domain at the tip of each triskelion leg and a clathrin light chain bound to the inner half of each leg. B—Under appropriate conditions, clathrin triskelions assemble into the pentagonal and hexagonal structures characteristic of coated pits and vesicles (after Becker *et al.*, 2006).

The accumulation of clathrin triskelions and the assembly of a clathrin coat on the cytosolic side of a membrane appear to provide part of the **driving force** for formation of a vesicle at the site. In the case of receptor-mediated endocytosis, the assembly of the clathrin coat on the inner side of the plasma membrane causes the membrane to fold inward. Initially, all clathrin units are hexagonal and form a planar, two-dimensional structure. As more clathrin triskelions are incorporated into the growing lattice, a combination of hexagonal and pentagonal units allows the new clathrin coat to curve around the budding vesicle.

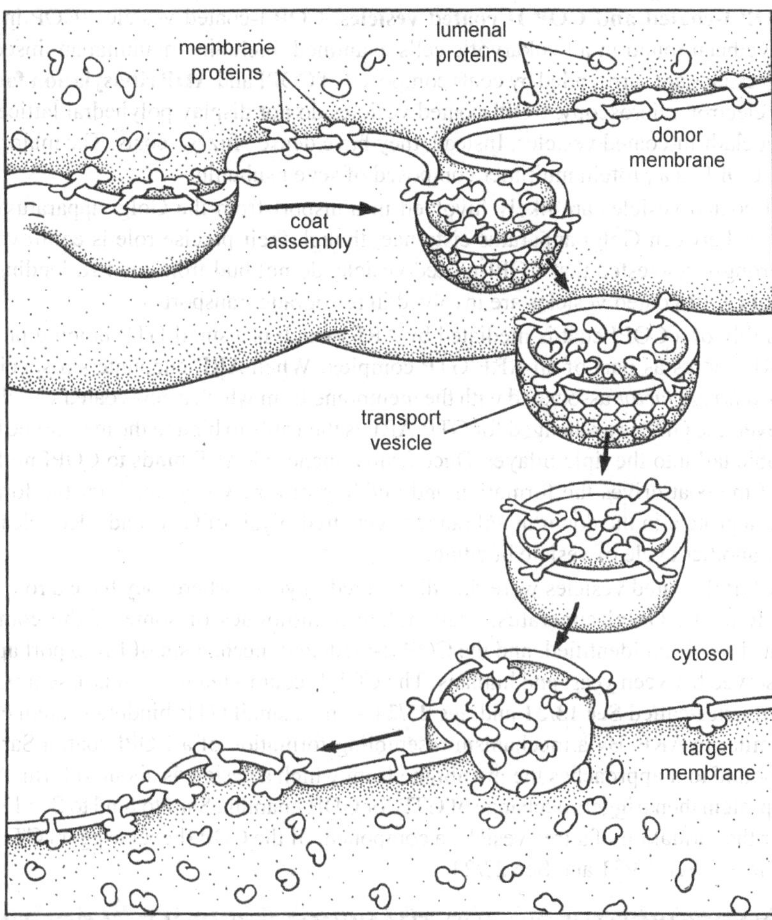

Fig. 23.10. Formation and fusion of a transport vesicle. Membrane and lumenal secretory proteins are collected into selected regions of a donor membrane where the formation of a cytosolic coat results in the budding off of a transport vesicle. During transport the coat is disassembled, and the transport vesicles docks and fuses with the target membrane (after Cooper and Hausman, 2007).

***(iv)* Sealing of the coated-vesicles.** As clathrin accumulates around the budding vesicle, at least one more protein-**dynamin**–participates in the process. Dynamin is a cytosolic GTPase (enzyme) required for coated pit *constriction and closing of the budding vesicle*. This essential protein was first identified in *Drosophila*. Flies expressing a temperature-sensitive form of dynamin were instantly paralyzed after a temperature shift. Further investigation revealed an accumulation of coated pits in the presynaptic membranes of neuromuscular junctions in affected flies. Binding of GTP probably allows dynamin to form helical rings around the neck of the coated pit. As GTP is hydrolyzed, the dynamin rings tighten and separate the fully sealed endocytic vesicle from the plasma membrane.

(*v*) **Uncoating of coated-vesicles.** Uncoating is done in a regulated manner because, in most cases, the clathrin coat remains intact as long as the membrane is part of a coated pit or budding vesicle but dissociate rapidly once the vesicle is fully formed. Like assembly, dissociation of the clathrin coat is an energy-consuming process, accompanied by the hydrolysis of about three ATP molecules per triskelion. At least one protein, an **uncoating ATPase** is essential for this process, though the uncoating ATPase releases only the clathrin triskelion from the APs; the factors responsible for releasing APs from the membrane have not yet been identified.

2. COP I-coated and COP II-coated vesicles. COP I-coated vesicles (COP indicates coat protein) have been found in all eukaryotic cells examined, including mammalian, insect and yeast cells. Such vesicles are surrounded by coats composed of COPI and **ADP ribosylation factor (ARF)**. Viewed by electron microscopy, COPI-coated vesicles do not display polyhedral lattices like those surrounding clathrin-coated vesicles. Instead, they have dense "fuzzy" coats. The major component of the coat, COPI, is a protein multimer composed of seven subunits.

COPI-coated vesicles are likely involved in transport from the Golgi apparatus back to the ER as well as between Golgi apparatus cisternae, though their precise role is controversial. Early evidence strongly suggested that COPI-coated vesicle, do not bud from the ER, lending support to the view that COPI-coated vesicles are involved in retrograde transport.

Assembly of a COPI coat is mediated by ARF, which is a small GTP-binding protein. In the cytosol, ARF occurs as part of an ARF-GTP complex. When ARF encounters a specific *guanine nucleotide exchange factor* associated with the membrane from which a new coated vesicle is about to form, however, the GDP is exchanged for GTP. ARF is then able to bind to the membrane by inserting a hydrophobic tail into the lipid bilayer. Once firmly anchored, ARF binds to COPI multimers, and assembly of the coat drives the formation and budding of a new vesicle. After the formation of a free vesicle, a protein in the donor membrane triggers hydrolysis of GTP, and ARF releases the coat proteins for another cycle of vesicle budding.

The COP II-coated vesicles were first discovered in yeast, where they have a role in transport from the ER to the Golgi apparatus. Mammalian homologues of some of the components of COP II coats have been identified, and the COP II-mediated mechanism of ER export appears to be highly conserved between yeast and humans. The COP II coat found in yeast is assembled from two protein complexes-called **Sec 13/31** and **Sec 23/24** – and a small GTP-binding protein called **Sar I**, which is similar to ARF. By a mechanism resembling formation of a COPI coat, a Sar I molecule with GDP bound to it approaches the membrane from which a vesicle is about to form. A peripheral membrane protein then triggers exchange of GTP for GDP, enabling Sar I to bind to Sec 13/31 and Sec 23/24. After the formation of a free vesicle, a component of the COP II coat triggers GTP hydrolysis and Sar I releases Sec 13/31 and Sec 23/24.

23.4. SNARE HYPOTHESIS FOR TRANSPORT VESICLE TARGETING AND FUSION

The SNARE hypothesis (proposed by Rothman *et al.*, 1993) provides a working model for sorting and targeting step in intracellular transport (Fig. 23.11). According to this hypothesis, the molecular components that facilitate sorting and targeting of vesicles in eukaryotic cells include two families of SNARE (SNAP receptor) proteins: the *v*-SNAREs (vesicle-SNAP receptors) found on transport vesicles and the ***t***-SNAREs (target-SNAP receptors) found on target membranes. The *v*-SNAREs and *t*-SNAREs are complementary molecules that, along with additional tethering proteins, allow a vesicle to recognise a target organelle. Both *v*-SNAREs and *t*-SNAREs were originally investigated because of their role in neuronal exocytosis. Since their discovery in brain tissue, both families of proteins have also been implicated in transport from the ER to Golgi apparatus in yeast and other organisms.

When a vesicle reaches its destination, a third family of proteins, the **Rab GTPases**, come into play. Rab GTPase are also specific (Table 23.2); vesicles fated for different destinations have distinct members of the Rab family associated with them. As shown in Fig. 23.11, the affinity of complementary *v*-SNAREs and *t*-SNAREs for one another forms a four-helix bundle. This ensures that when they collide they will remain in contact long enough for a Rab protein associated with the vesicle to lock the complementary *t*- and *v*-SNARE, together.

Following the vesicle fusion, another set of proteins, including **N-ethylmaleimide-sensitive factor (NSF)** and a group of **SNAPs (soluble NSF attachment proteins)**, mediates release of the *v*- and *t*-SNAREs of the donor and target membranes possibly accompanied by the hydrolysis of ATP. The NSF and SNAPs are involved in fusion between a variety of cellular membranes, indicating they are not responsible for specificity during targeting.

Box 23.2 Rab Proteins

Rab proteins, like the ARF family, participate in many of the vesicle budding and fusion reactions during vesicular transport. More than 60 different mammalian Rab proteins have been identified and shown to function in specific vesicle processes. They function in many steps of vesicle trafficking, including interacting with SNAREs to regulate and facilitate the formation of SNARE/SNARE complexes.

Table 23.2. Rab GTP-binding proteins and their sites in action (Source: Cooper and Hausman, 2007).

	Transport step	Rab protein involved
I.	**Exocytosis**	
1.	Transitional ER to Golgi	Rab 1, Rab 1b, Rab 2
2.	Golgi back to ER	Rab 6, Rab 6b
3.	Intra-Golgi	Rab 1, Rab 6, Rab 6b
4.	*Trans* Golgi network to plasma membrane	Rab 11a, Rab 11b
II.	**Endocytosis**	
1.	Plasma membrane to early endosome	Rab 5a, Rab 5b, Rab 5c
2.	Early endosome to plasma membrane	Rab 4, Rab 15, Rab 18
3.	Early endosome to late endosome	Rab 7
III.	**Special roles**	
1.	Exocytosis of secretory granules	Rab 8b
2.	Late endosome to *trans* Golgi network	Rab 9, Rab 11a, Rab 11b
3.	*Trans* Golgi network to basolateral membrane	Rab 8a
4.	*Trans* Golgi network to apical membrane	Rab 21

Recent research suggests that SNARE proteins alone cannot account for the specificity observed in vesicle targeting. Additional proteins known as **tethering proteins** act over longer distances and appear to provide specificity by connecting vesicles to their target membranes prior to *v*-SNARE/*t*-SNARE interaction (Fig. 23.11). For example, toxin-induced cleavage of SNARE proteins in vivo can block SNARE complex formation without vesicle association with the target membrane. Also, in an in vitro reconstituted system, ER-derived vesicles can attach to Golgi membranes without addition of SNARE proteins.

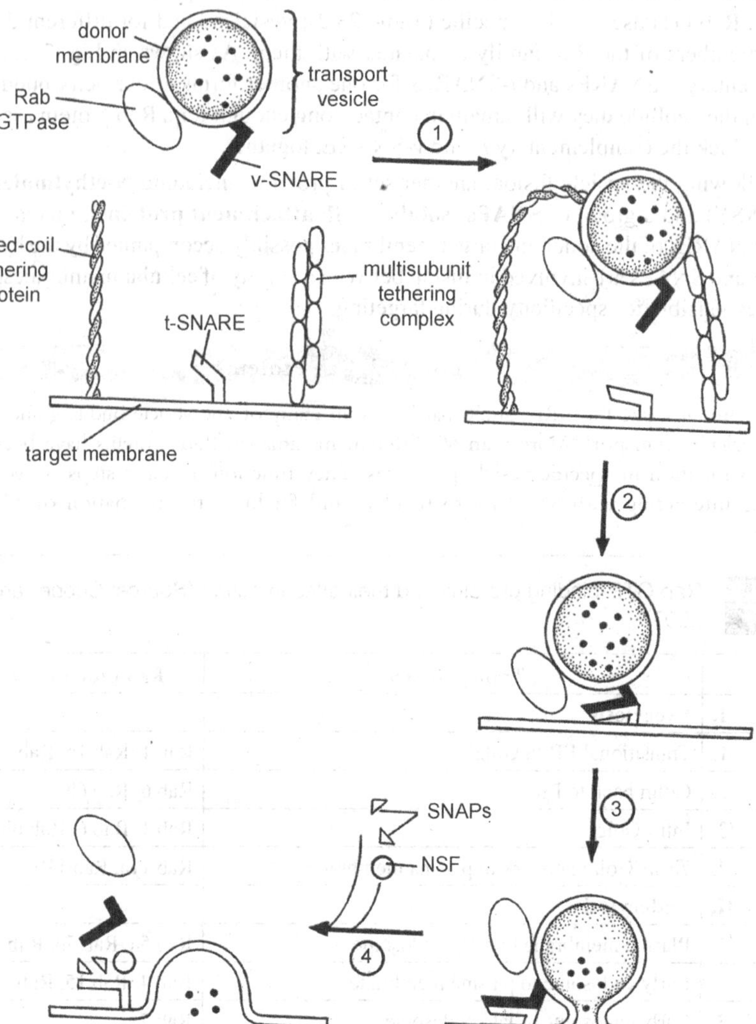

Fig. 23.11. The SNARE hypothesis for transport vesicle targeting and fusion The basic molecular components that mediate sorting and targeting of vesicles in eukaryotic cells include tethering proteins, *v*-SNAREs on the transport vesicles, t-SNAREs on target membranes, Rab GTPase, NSF, and several SNAPs. 1. The proper vesicle is recognised and bound by particular membrane-anchored tethering proteins: a coiled coil protein and/ or a multisubunit tethering complex. 2. A Rab GTPase bound to the incoming vesicle stimulates formation of a stable four-helix bundle composed of one *v*-SNARE helix and three *t*-SNARE helices (the tethering proteins are no longer shown). 3. Fusion of the vesicle membrane with the target membrane is promoted by the *v*-SNARE/ *t*-SNARE interaction. 4. Binding of NSF and SNAPs promotes dissociation of SNARE complexes. The exact timing of GTP or ATP hydrolysis is still unclear, but most likely occurs after vesicle fusion (after Becker *et al.*, 2006).

At present, two main groups of tethering proteins are known: **coiled coil proteins** and **multisubunit complexes.** Coiled-coil tethering proteins, such as the **golgins** are important in the initial recognition and binding of COP I- or COP II-coated vesicles to the Golgi. The golgins are anchored by one end to the Golgi membranes and use the other end to contact the appropriate passing vesicle. These proteins are also important in connecting Golgi cisternae to each other. Antibodies to certain golgins block the action of the golgins and disrupt the structure of the Golgi medial cisternae.

A second class of tethering proteins consists of several families of multisubunit protein complexes containing four to eight or more individual polypeptides. For example, the **exocyst** complex of yeast and mammals is important for protein secretion, binding both to the plasma membrane and to vesicles from the TGN whose contents are destined for export. Other types of multisubunit tethering complexes such as **COG (conserved oligomeric Golgi)** complex, the **GARP (Golgi associated retrograde protein)** complex and the **TRAPP (transport protein particle)** complex are implicated in the initial recognition and specificity of vesicle-target membrane interaction. Most of these proteins in these complexes are well conserved among organisms, as different as yeast and humans (see **Becker** *et al.*, 2006).

23.5. TRANSPORT OF NEUROTRANSMITTERS ACROSS NERVE SYNAPSES

Nerve cells (neurons) communicate with one another and with glands and muscles at *synapses,* which are of following two types:

1. Electrical synapse. Here, an axon of one neuron, called the **presynaptic neuron**, is connected to the dendrite (s) of another cell, the **postsynaptic neuron**, by gap junctions. As ions move back and forth between the two cells, the depolarisation in one cell spreads passively to the connected cell. Electrical synapses provide for transmission with virtually no delay and occur in places in the nervous system where speed of transmission is critical.

2. Chemical synapse. In this case, the presynaptic and postsynaptic neurons are close to each other but not directly connected (Fig. 23.12). Typically, the presynaptic membrane is separated from the postsynaptic membrane by a gap of about 20–25 nm, known as the **synaptic cleft**. A nerve signal arriving at the terminals of the presynaptic neuron cannot bridge the synaptic cleft as an electrical impulse. For synaptic transmission to take place, the electrical signal must be converted at the presynaptic neuron to a chemical signal carried by a neurotransmitter. Neurotransmitter molecules are stored in the **terminal bulbs** (also called **synaptic knobs**) of the presynaptic neuron. An action potential arriving at the terminal bulb causes the neurotransmitter to be secreted into and diffuse across the synaptic cleft. The neurotransmitter molecules then bind to specific proteins (called **receptors**) embedded within the plasma membrane of the postsynaptic neuron and are converted back to electrical signals, setting in motion a sequence of events that either stimulates or inhibits the production of an action potential in the postsynaptic neuron, depending on the kind of synapse.

***(i)* Neurotransmitters.** A **neurotransmitter** is a small molecule whose function is to bind to a cell surface protein called a **receptor** within the membrane of a postsynaptic neuron. When a neurotransmitter molecule binds to its specific receptor, the properties of receptor are altered and the postsynaptic neurons respond accordingly. An **excitatory neurotransmitter** causes depolarization of the postsynaptic neuron, whereas an **inhibitory neurotransmitter** typically causes the postsynaptic cell to hyperpolarise.

A neurotransmitter should have the following three properties: 1. It must elicit the appropriate response when introduced into the synaptic cleft; 2. it must occur naturally in the presynaptic neuron; and 3. it must be released at the right time when the presynaptic neuron is stimulated. At present, molecules known to meet these criteria include **acetylcholine**; a group of biogenic amines called **catecholamines**; certain amino acids (**histamine, serotonin**) and their derivatives (**GABA**) and some neuropeptides (**eukephalins**).

In vertebrates, acetylcholine is the most common neurotransmitter for synapse between neurons outside the central nervous system as well as for neuromuscular junctions. Acetylcholine is an excitatory neurotransmitter. Synapses that use acetylcholine as their neurotransmitter are called **cholinergic synapses.**

The catecholamines include **dopamine** and the hormone **norepinephrine** and **epinephrine**, all derivatives of the amino acid tyrosine. Because the catecholamines are also synthesized in the adrenal gland, synapses that use them as neurotransmitters are termed **adrenergic synapses**. Andrenergic synapses are found at the junctions between nerves and smooth muscles in internal organs such as intestine, as well as nerve-nerve junctions in the brain.

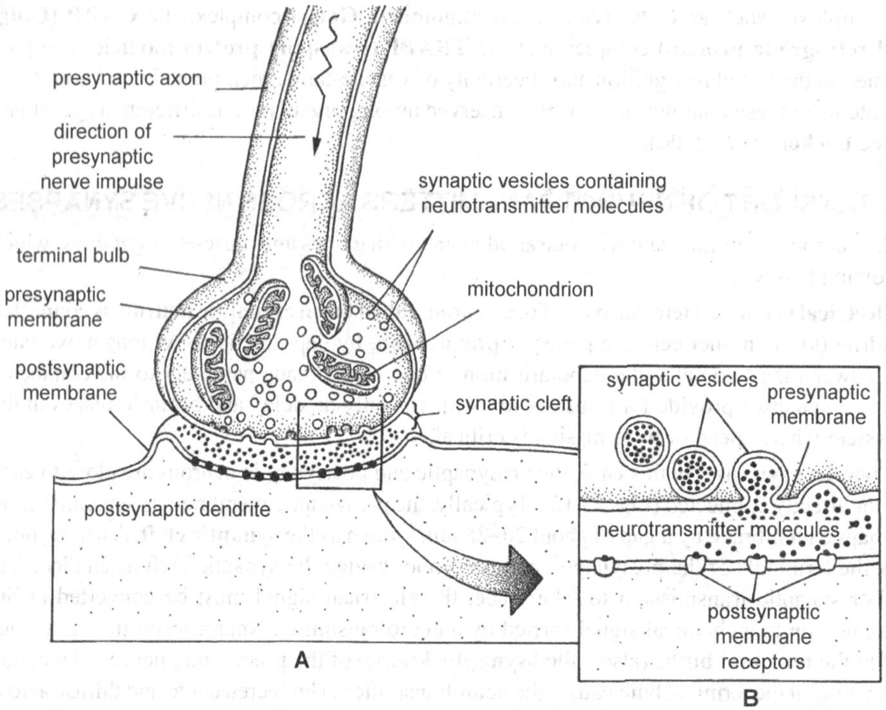

Fig. 23.12. Structure of a chemical synapse. A—When a nerve impulse from the presynaptic axon arrives at the synapse, it causes synaptic vesicle containing neurotransmitters in the terminal bulb to fuse with the presynaptic membrane, releasing their contents into the synaptic cleft. B—Neurotransmitter molecules diffuse across the cleft from the presynaptic (axonal) membrane to the postsynaptic (dendritic) membrane, where they bind to specific membrane receptors and change the polarisation of the membrane, either exciting or inhibiting the postsynaptic cell (after Becker *et al.*, 2006).

***(ii)* Role of calcium ions in secretion of neurotransmitters.** The secretion of neurotransmitters by the presynaptic cell is directly controlled by the concentration of calcium ions in the terminal bulb (Fig. 23.13). Each time an action potential arrives, the depolarization causes the calcium concentration in the terminal bulb to increase temporarily due to the opening of voltage-gated calcium channels in the terminal bulbs (**Note.** The calcium concentration outside the cell is about 10,000 times higher than that of the cytosol). Before they are released, neurotransmitter molecules are stored in

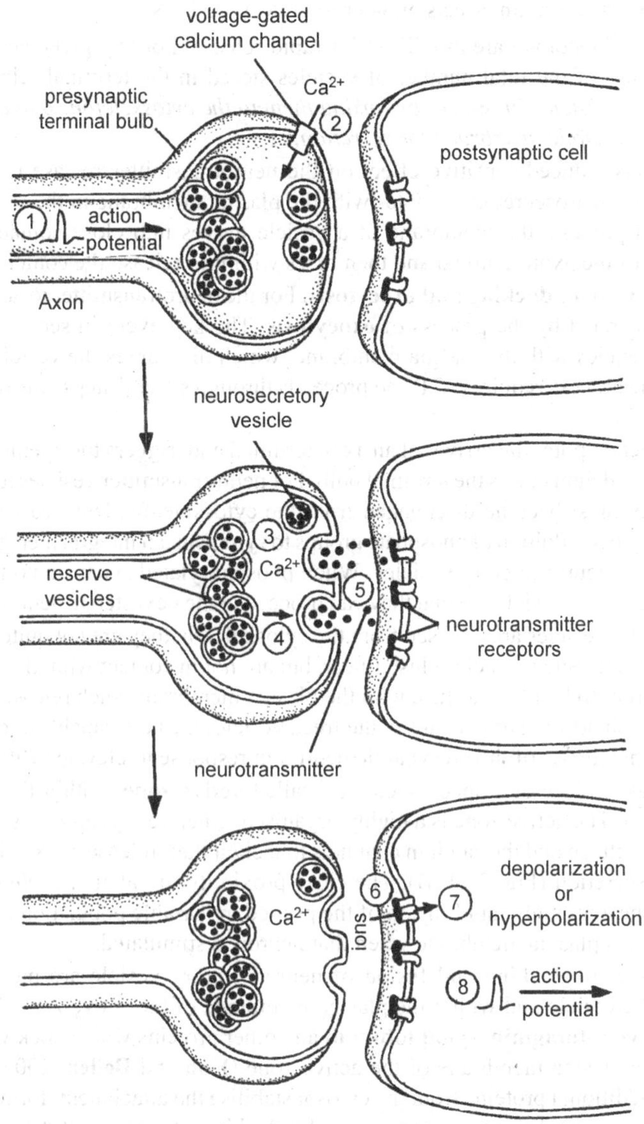

Fig. 23.13. The transmission of a signal across a synapse. 1. An action potential arrives at the presynaptic terminal bulb, resulting in a transient depolarization. 2. Depolarization opens voltage-gated calcium channels, allowing calcium ions to rush into the terminal. 3. This increase in the calcium concentration in the terminal bulb induces the secretion of a fraction of the neurosecretory vesicles. 4. Calcium also causes reserve vesicles to be released from the actin cytoskeleton so that they are ready for secretion. 5. Secreted neurotransmitter molecules diffuse across the synaptic clefts to receptors on the postsynaptic cell. 6. Binding of neurotransmitter to the receptor alters the receptor properties. 7. For receptors that are ligand-gated channels, the channel opening leads to either depolarization or hyperpolarization of the postsynaptic plasma membrane. 8. If depolarization results, a sufficient amount of excitatory neurotransmitter will result in an action potential in the postsynaptic cell (after Becker *et al.*, 2006).

small membrane-bounded neurosecretory vesicles in the terminal bulb. The release of calcium within the terminal bulb has two main effects on neurosecretory vesicles.

1. Vesicles held in storage are mobilised for rapid release. For any given action potential, only a tiny fraction of the total number of vesicles stored in the terminal release their contents. *Neurons hold vesicles in reserve by linking them to the cytoskeleton so that they cannot move close to the synaptic membrane for secretion.*
2. Calcium ions induced sensitive effect on the neurotransmitter release is the rapid docking and fusion of neurosecretory vesicles with the plasma membrane in the terminal bulb region. During this process, the membrane of a vesicle moves into close contact with the plasma membrane of the axon terminal and then fuses with it to release the content of the vesicle.

***(iii)* Mechanism of docking and exocytosis.** For the neurotransmitter to act on a postsynaptic cell, it must be secreted by the process of **exocytosis**. The key event in secretion is the fusion of neurosecretory vesicles with the plasma membrane, which discharges the vesicle contents into the synaptic cleft. The process requires ATP and proceeds through several steps, one of which is calcium dependent.

When the action potential arrives at an axon terminal and triggers the opening of voltage-gated calcium channels, calcium enters the terminal bulb. For neurotransmitter vesicles to become available for secretion, they must become disengaged from the cytoskeleton. This process is stimulated by calcium release, and the addition of phosphate groups to **synapsin**, an integral membrane protein found in the membrane of neurosecretory vesicles. When phosphorylated synapsin no longer binds to the cytoskeleton, freeing the vesicles so that they can proceed to the next step in neurotransmitter release.

Once synaptic vesicles are released from the cytoskeleton, they are not quite ready to fuse with the plasma membrane. Such vesicles cluster near, but are not in contact with, the plasma membrane (Fig. 23.14). In order to be ready to fuse with the plasma membrane, such released vesicles must be "primed". Once such priming has occurred, the freed vesicles are now capable of docking and fusing with the plasma membrane of the presynaptic neuron in response to elevated calcium.

Docking takes place at a specialized site, called **active zone**, within the membrane of the presynaptic neuron. The active zone is a highly organised structure. Synaptic vesicle, the proteins to which the vesicles attach and the calcium channels that elicit their release are all clustered in ordered rows poised for secretion (Fig. 23.14B). The close proximity of calcium channels to the vesicles helps to explain the extremely rapid fusion of the primed, releasable population of vesicles with the presynaptic neuron's plasma membrane when that neuron is stimulated.

The processes of docking and fusion of neurosecretory vesicle are mediated by docking proteins within the vesicle and within the plasma membrane of the active zone. Within the vesicle membrane exist **synaptotagmin**, **synaptobrevin** and other proteins which dock with proteins, such as **syntaxin** at the plasma membrane of the active zone (**Koh** and **Bellen**, 2003). Once the initial docking occurs, additional proteins from the cytosol stabilise the attachment, forming a multiprotein docking complex. Fusion depends on the successful docking of vesicles at the active zone and the formation of a complete docking complex.

Human diseases due to defects in docking events. Two common and prospective fatal illnesses of humans result from interference with the docking events. Both **tetanus** and **botulism** result from interference by **neurotoxins** with vesicle docking and release. **Tetanus toxin** prevents the release of neurotransmitter from inhibitory neurons in the spinal cord, resulting in uncontrolled muscle contraction (which is why tetanus has been familiarly referred to as "lockjaw"). **Botulinum toxin** prevents release of neurotransmitter from motor neurons, resulting in muscle weakness and paralysis (see **Becker** *et al.*, 2006).

***(iv)* The acetylcholine receptor.** Acetylcholine binds to a ligand-gated sodium channel called **acetylcholine receptor**. When two molecules of acetylcholine bind with receptor, the channel opens and lets sodium ions rush into the postsynaptic neuron causing a depolarization (Fig. 23.15).

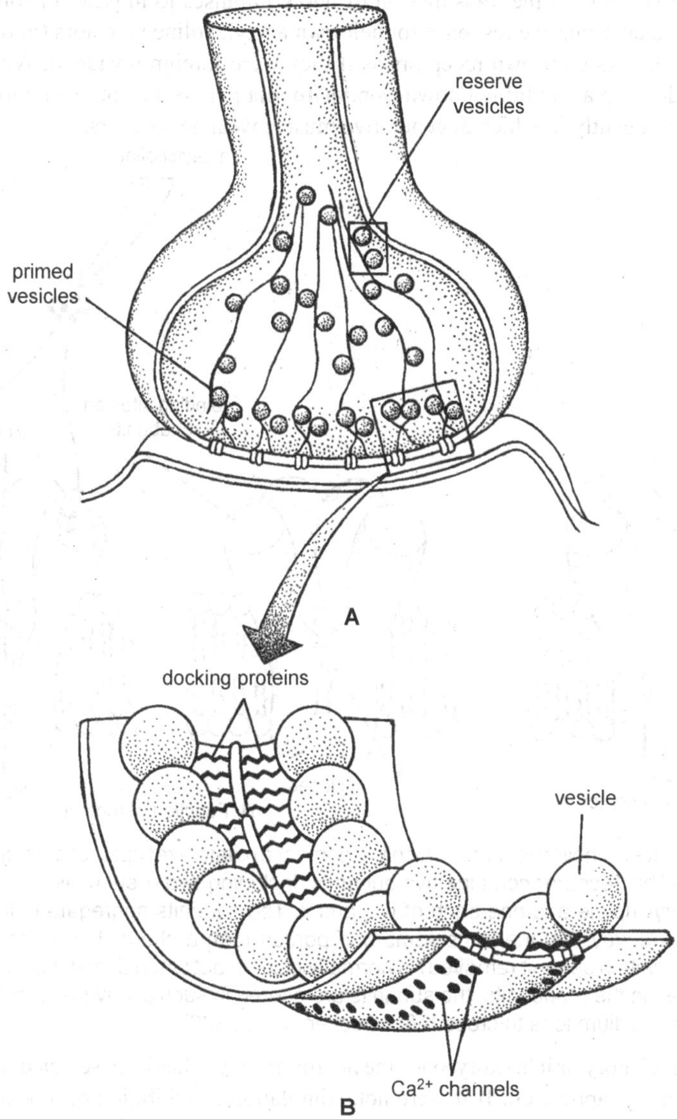

Fig. 23.14. Docking of synaptic vesicle with the plasma membrane of the presynaptic neuron. A—In response to local elevation of calcium in the presynaptic neuron, synaptic vesicles are released from the cytoskeleton, allowing them to fuse with the plasma membrane of the terminal bulb within the active zone. Vesicles bound to the cytoskeleton via synapsin form a reserve pool of vesicles. Once released from the cytoskeleton, vesicles move near the plasma membrane and become "primed" (*i.e.*, charged or prepared). Some vesicles become tightly associated with the plasma membrane, releasing their contents. B—A drawing based on an actual TEM reconstruction of the active zone of a motor neuron from a frog. Docked secretory vesicles are arranged in rows and connected by a complex of proteins associated with docking. Calcium channels lie immediately beneath the vesicles (after Becker *et al.*, 2006).

The purified acetylcholine receptor has molecular weight of about 300,000 and consists of four kinds of subunits-α, β, γ and δ-each containing about 500 amino acids (Fig. 23.15). Acetylcholine receptors play a key role in the transmission of nerve impulses to muscle. In some cases, human patients develop a autoimmune response to their own acetylcholine receptors (in other words, their immune system attacks their own receptors as if they were foreign invaders). When this happens, the patient can develop a condition known as **myasthenia gravis**, receptors on motor end plate are reduced and consequently in which degenerative muscle weakness occurs.

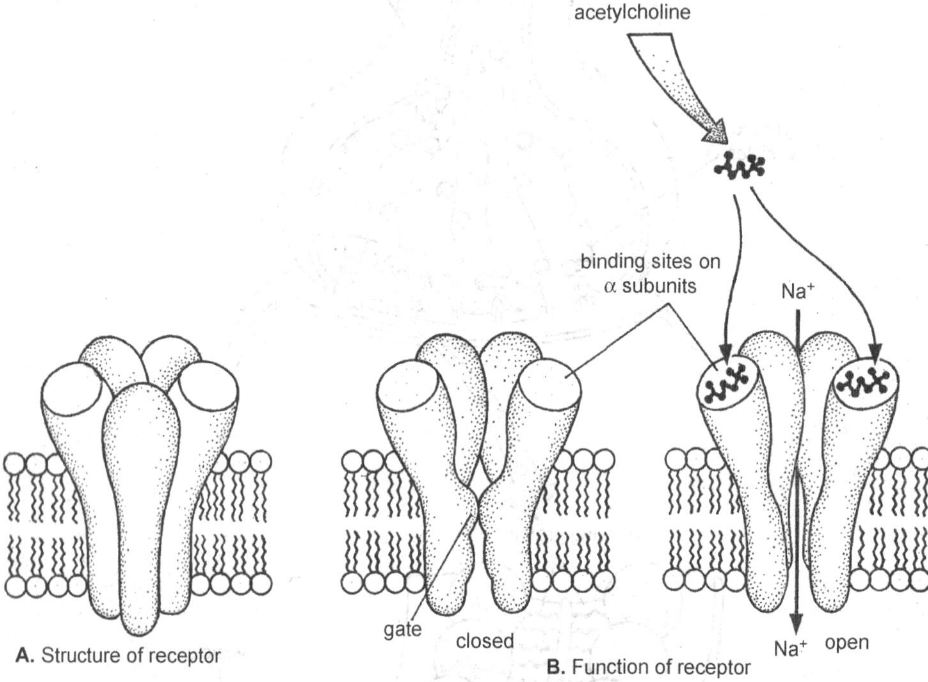

Fig. 23.15. The acetylcholine receptor which is a primary excitatory receptor of central nervous system. A—This receptor contains five subunits, including two α subunits with binding sites for acetylcholine and one each of β, γ and δ. The subunits aggregate in the lipid bilayer in such a way that the transmembrane portion form a channel. B—The channel (shown here with β subunit removed) is normally closed, but when acetylcholine binds to the two sites on the α subunits, the subunits are altered in such a way that the channel opens to allow sodium ions to cross (after Becker *et al.*, 2006).

Whether excitatory or inhibitory, once the neurotransmitter has been secreted, it must be rapidly removed from the synaptic cleft. If it were not, stimulation or inhibition of a postsynaptic neuron would be abnormally prolonged even in the absence of further signals from presynaptic neurons. In fact, the persistence of an excitatory neurotransmitter such as acetylcholine renders muscles unable to relax, ultimately leading to death.

Neurotransmitters are removed from the synaptic cleft by two specific mechanisms: degradation into inactive molecules or reuptake into the presynaptic terminals (*e.g.*, Serotonin). An example of the first mechanism is found in the case of the acetylcholine into acetic acid (or acetate ions) and choline, neither of which stimulatory to the acetylcholine receptor.

QUESTIONS

Long Answer Questions

1. What is bulk transport? Describe the process of endocytosis and exocytosis in brief.
2. What is vesicular transport? Give an outline of biosynthetic (secretory) pathway.

Short Answer Questions

1. What is the fate of receptor proteins that take part in receptor mediated endocytosis?
2. Write short notes on the following:
   (*i*) Phagocytosis
   (*ii*) Receptor-mediated-endocytosis
   (*iii*) Coated vesicles
   (*iv*) Exocytosis
   (*v*) Pinocytosis
3. Describe the mechanism of transport of neurotransmitters in nerve cells.
4. Differentiate between the following terms:
   (*i*) Phagocytosis and pinocytosis.

Very Short Answer Questions

1. What is early endosome?
2. What is late endosome?
3. Define secretory vesicle.
4. What is secretory granule?
5. Define secretory pathway.
6. What is exocytosis?
7. Define endocytosis.
8. What is endocytic vesicle?
9. What is SNARE hypothesis?
10. Define transition vesicles.
11. What is transport vesicle?
12. What is coated vesicle?
13. Define phagocytosis.
14. What is receptor-mediated endocytosis?
15. Define synapse.
16. What is neurosecretory vesicle?
17. Write the names of two types of cellular secretions?
18. Name of the cells in mammals that are engaged in phagocytosis.

Multiple Choice Questions

1. Unidirectional transmission of a nerve impulse through nerve fibre is due to the fact that
   (*a*) nerve fibre has nodes of Ranvier
   (*b*) nerve fibre is insulated by a medullary sheet
   (*c*) neurotransmitters are released by dendrites and not by axon endings
   (*d*) neurotransmitters are released by the axon endings and not by dendrites
2. Among which one of the following groups of chemicals, all are neurotransmitters?
   (*a*) somatostatin, serotonin, acetylcholine
   (*b*) nor-adrenaline, somatostatin, threonine
   (*c*) glycine, dopamine, melatonin
   (*d*) acetylcholine, nor-adrenaline, dopamine
3. The process by which a cell secretes macromolecules through fusion of secretory vesicles to the plasma membrane is called
   (*a*) pinocytosis (*b*) endocytosis
   (*c*) phagocytosis (*d*) exocytosis
4. Membrane receptors are
   (*a*) carbohydrates (*b*) nucleic acids
   (*c*) proteins (*d*) lipids
5. Pinocytosis means endocytosis of
   (*a*) droplet of extracellular fluid
   (*b*) sub-microscopic particles
   (*c*) solid particles
   (*d*) extracellular fluid with particles
6. The neurotransmitter produced at electric synapse is
   (*a*) GABA (*b*) acetylcholine
   (*c*) nor-epinephrine (*d*) none of these
7. A person with Parkinson's disease shows defective action of which neurotransmitter in brain
   (*a*) dopamine (*b*) serotonin
   (*c*) noradrenaline (*d*) enkalphia
8. In Myasthenia gravis acetylcholine
   (*a*) receptors on motor end plate are reduced
   (*b*) secretion from nerve terminals is reduced
   (*c*) esterase activity is inhibited
   (*d*) secretion from nerve terminals is enhanced

ANSWER

Very Short Answer Questions

1. Early endosomes are vesicles budding off the trans-Golgi network that are sites for the sorting and recycling of extracellular material brought into the cell by endocytosis.
2. Late endosome is a vesicle containing newly synthesized acid hydrolases plus material fated for digestion; activated either by lowering the pH of the late endosome or transferring its material to an existing lysosome.
3. Secretory vesicle is a membrane bounded compartment of a eukaryotic cell that carries secretory proteins from the Golgi apparatus to the plasma membrane for exocytosis and that may serve as a storage compartment for such proteins before they are released.
4. It is a large, dense secretory vesicle.
5. Secretory pathway is a pathway in which newly synthesized proteins move from ER through the Golgi apparatus to secretory vesicles and secretory granules, which then discharge their contents to the exterior of the cell.
6. Exocytosis is fusion of vesicle membranes with the plasma membrane so that contents of the vesicle can be expelled or secreted to the extracellular environment.
7. Endocytosis is uptake of extracellular materials by infolding of the plasma membrane, followed by pinching off of a membrane-bound vesicle containing extracellular fluid and materials.
8. It is a membrane vesicle formed by pinching off of a small segment of plasma membrane during the process of endocytosis.
9. It is a model explaining how membrane vesicles fuse with the proper target membrane; based on specific interactions between *v*-SNAREs (vesicle-SNAP receptor) and *t*-SNAREs (target-SNAP receptors) (SNAP=Soluble NSF attachment proteins).
10. It is a membrane vesicle that shuttles lipids and proteins from the endoplasmic reticulum to Golgi apparatus.
11. Vesicle that buds off from a membrane in one region of the cell and fuses with other membranes; includes vesicles that convey lipids and proteins from the ER to the Golgi apparatus, between the Golgi stack cisternae, and from the Golgi apparatus to various destinations in the cell, including secretory vesicles, endosomes and lysosomes.
12. Any of several types of membrane vesicles involved in vesicular traffic within the endo-membrane system; surrounded by a coat protein such as clathrin, COP I, COP II or caveolin.
13. It is a type of endocytosis in which particulate matter or even an entire cell is taken up from the environment and incorporated into vesicles for digestion.
14. It is a type of endocytosis initiated at coated pits and resulted in coated vesicles; believed to be a major mechanism for selective uptake of macromolecules and peptide hormones.
15. Synapse is a tiny gap between a neuron and another cell (neuron, muscle fibre, or gland cell), across which nerve impulse is transferred by direct electrical connections or by chemicals called neurotransmitters.
16. It is a small vesicle containing neurotransmitter molecules; located in the terminal bulb of an axon.
17. Constitutive secretion and regulated secretion.
18. Macrophages and neutrophils.

Multiple Choice Questions

1. (*d*) 2. (*d*) 3. (*d*) 4. (*c*) 5. (*d*) 6. (*b*) 7. (*a*)
8. (*a*)

24 Lysosomes

The lysosomes (*Gr., lyso* = digestive + *soma* = body) are tiny membrane-bound vesicles involved in intracellular digestion. They contain a variety of hydrolytic enzymes that remain active under acidic conditions. The lysosomal lumen is maintained at an acidic pH (around 5) by an ATP-driven proton pump in the membrane. Thus, these remarkable organelles are primarily meant for the digestion of a variety of biological materials and secondarily cause aging and death of animal cells and also a variety of human diseases such as cancer, gout, Pompe's disease, silicosis and I-cell disease.

Box 24.1

Lysosomes were observed and named by **Christian de Duve** in 1955 as cytoplasmic rounded dense bodies. They were termed **suicide bags** by **de Duve** in 1959.

Lysosomes were investigated according to following two schools: 1. **C. de Duve** and his co-workers (1963, 1964, 1974) worked in Belgium and their approach was biochemical one. 2. **Alex Novikoff** and his research group (1962, 1964) worked in United States and their approach was morphological and cytochemical. For the discovery of lysosomes and a brilliant series of experiments on them, **de Duve** shared the 1974 Nobel Prize for Physiology with **Palade** and **Claude**, both of them were pioneer cell biologists.

24.1. OCCURRENCE

The lysosomes occur in most animal and few plant cells. They are absent in bacteria and mature mammalian erythrocytes. Few lysosomes occur in muscle cells or in acinar cells of the pancreas. Leucocytes, especially granulocytes are a particularly rich source of lysosomes. Their lysosomes are so large-sized that they can be observed under the light microscope. Lysosomes are also numerous in epithelial cells of absorptive, secretory and excretory organs (*e.g.*, intestine, liver, kidney, etc.). They occur in abundance in the epithelial cells of lungs and uterus. Lastly, phagocytic cells and cells of reticuloendothelial system (*e.g.*, bone marrow, spleen and liver) are also rich in lysosomes.

24.2. STRUCTURE

The lysosomes are round vacuolar structures which remain filled with dense material and are bounded by single unit membrane. Their shape and density vary greatly. Lysosomes are 0.2 to 0.5 μm in size. Since, size and shape of lysosomes vary from cell to cell and time to time (*i.e.*, they are polymorphic), their identification becomes difficult. However, on the basis of the following three criteria, a cellular entity can be identified as a lysosome: 1. It should be bound by a limiting membrane; 2. It should contain two or more acid hydrolases; and 3. It should demonstrate the property of enzyme latency when treated in a way that adversely affects organelle's membrane structure.

24.3. CHEMICAL COMPOSITION

The **acid phosphatase** is the principal enzyme which is used as a marker for the lysosomes by the use of **Gomori' staining technique** (Gomori,1952). Specific stains are also used for other lysosomal enzymes such as **B-glucuronidase, aryl sulphatase, N-acetyl-B-glucosaminidase** and **5-bromo-4-chloroindolacetate esterase.**

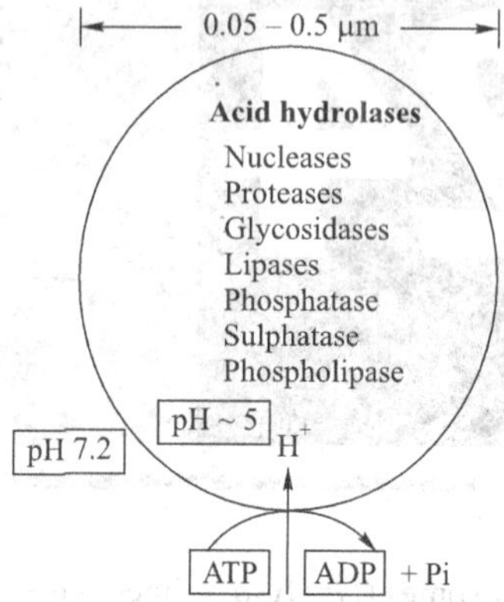

Fig. 24.1. A lysosome showing its various characteristics.

Lysosomal Enzymes

According to a recent estimate, a lysosome may contain up to 50 types of hydrolytic enzymes. They include **proteases** (*e.g.*, cathepsin for protein digestion), **nucleases, glycosidases** (for digestion of polysaccharides and glycosides), **lipases, phospholipases, phosphatases** and **sulphatases** (Table 24.1). All lysosomal enzymes are acid hydrolases, optimally active at the pH5 maintained within lysosomes. The membrane of the lysosome normally keeps the enzymes latent and out of the cytoplasmic matrix or cytosol (whose pH is about ~ 7.2), but *the acid dependency of lysosomal enzymes protects the contents of the cytosol (cytoplasmic matrix) against any damage even if leakage of lysosomal enzymes should occur.*

Table 24.1 Some lysosomal enzymes and their substrates.

	Enzyme	Substrate
A.	**Phosphatases**	
1.	Acid phosphatase	Phosphate monoesters
2.	Acid pyrophosphatase	ATP, FAD
3.	Acid phosphodiesterase	Oligonucleotides, phosphodiesters
B.	**Sulphatase**	
4.	Aryl sulphatase	Aryl sulphates
C.	**Proteases and peptidases**	
5.	Cathepsin A, B, C, D, and E	Various proteins and peptides
6.	Collagenase	Collagen
7.	Peptidases	Peptides
D.	**Nucleases**	
8.	Acid ribonuclease	RNA
9.	Acid deoxyribonuclease	DNA
E.	**Lipases**	
10.	Triglyceride lipase (Acid lipase)	Triacyglycerols
11.	Phospholipase	Lecithin, phosphatidyl ethanolamine
12.	Esterase	Fatty acid esters
13.	Sphingomyelinase	Sphingomyelin

	Enzyme	Substrate
F.	**Glycosidases**	
14.	α-Glucosidase	Glycogen
15.	β-Glucosidase	Glycoproteins (β-Glucosides)
16.	β-Galactosidase	β-Galactosides
17.	α-Mannosidase	Mannosyl oligosaccharides
18.	α-Fucosidase	Glycoprotein
19.	β-Xylosidase	Glycoprotein
20.	β-Glucocerebrosidase	Glycolipids
21.	α-N-Acetylhexosaminidase	Heparin
22.	β-N-Acetylhexosaminidase	Glycoproteins, glycolipids
23.	Sialidase	Sialic acid derivatives
24.	Lysozyme	Mucopolysaccharides, bacterial cell wall
25.	Hyaluronidase	Hyaluronic acid, chondroitin sulphates
26.	β-Glucuronidase	Polysaccharides, mucopolysaccharides

The so-called **latency** of the lysosomal enzymes is due to the presence of the membrane which is resistant to the enzymes that it encloses. Most probably this is due to the fact that most lysosomal hydrolases are membrane-bound, which may prevent the active centres of enzymes to gain access to susceptible groups in the membrane.

Lysosomal Membrane

The lysosomal membrane is slightly thicker than that of mitochondria. It contains substantial amounts of carbohydrate material, particularly sialic acid. In fact, lysosomal membranes contain two groups of acidic highly glycosylated integral proteins, called lgp-A and lgp-B, whose function may be to protect the membrane from attack of its enclosed enzymes.

The entire process of digestion is carried out within the lysosome. Most lysosomal enzymes act in an acid medium. Acidification of lysosomal contents depends on an ATP-dependent proton pump which is present in the membrane of the lysosome and accumulates H^+ inside the organelle. Lysosomal membrane also contains transport proteins that allow the final products of digestion of macromolecules to escape so that they can be either excreted or reutilized by the cell.

24.4. KINDS OF LYSOSOMES (POLYMORPHISM IN LYSOSOMES)

Lysosomes are extremely dynamic organelles, exhibiting polymorphism in their morphology. Following four types of lysosomes have been recognized in different types of cells or at different times in the same cell. Of these, only the first is the primary lysosome, the other three have been grouped together as secondary lysosomes.

1. Primary Lysosomes

These are also called **storage granules, protolysosomes** or **virgin lysosomes**. Primary lysosomes are newly formed organelles bounded by a single membrane and typically having a diameter of 100 nm. They contain the degradative enzymes which have not participated in any digestive process. Each primary lysosome contains one type of enzyme or another and it is only in the secondary lysosome that the full complement of acid hydrolases is present.

2. Heterophagosomes

They are also called **heterophagic vacuoles**, **heterolysosomes** or **phagolysosomes**. Heterophagosomes are formed by the fusion of primary lysosomes with cytoplasmic vacuoles containing **extracellular substances** brought into the cell by any of a variety of endocytic processes (*e.g.*, pinocytosis, phagocytosis or receptor-mediated endocytosis). The digestion of engulfed substances takes place by the enzymatic activities of the hydrolytic enzymes of the secondary lysosomes. The digested material has low molecular weight and readily passess through the membrane of the lysosomes to become the part of the matrix (Fig. 24.2).

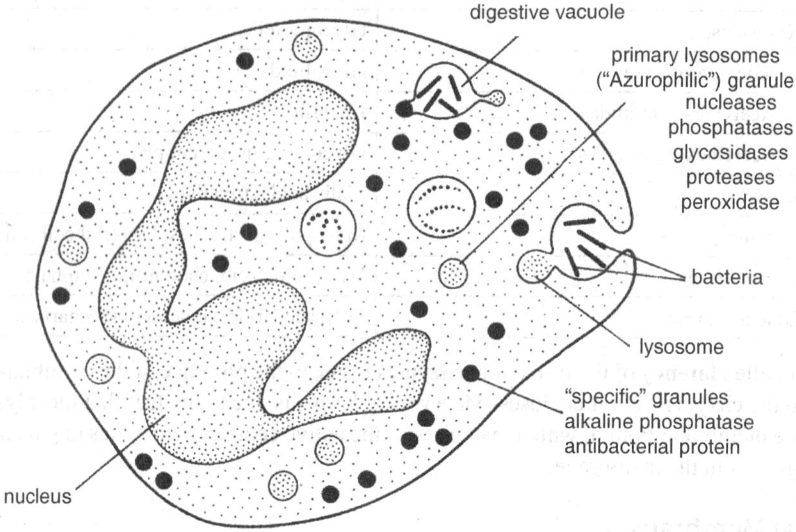

Fig. 24.2. Diagram of a white blood cell (neutrophil) ingesting bacteria. Two types of granules fuse with the phagocytotic vacuoles and contribute digestive enzymes and other components.

3. Autophagosomes

They are also called **autophagic vacuole, cytolysosomes** or **autolysosomes**. Primary lysosomes are able to digest **intracellular structures** including mitochondria, ribosomes, peroxisomes and glycogen granules. Such autodigestion (called **autophagy**) of cellular organelles is a normal event during cell growth and repair and is especially prevalent in differentiating and dedifferentiating tissues (*e.g.*, cells undergoing programmed death during metamorphosis or regeneration) and tissue under stress. Autophagy takes several forms. In some cases the lysosome appears to flow around the cell structure and fuse, enclosing it in a double membrane sac, the lysosomal enzymes being initially confined between the membranes. The inner membrane then breaks down and the enzymes are able to penetrate to the enclosed organelle.In other cases, the organelle to be digested is first encased by smooth ER, forming a vesicle that fuses with a primary lysosome (Fig. 24.4). Lysosomes also regularly engulf bits of cytosol (cytoplasmic matrix) which is degraded by a process, called **microautophagy**.

As digestion proceeds, it becomes increasingly difficult to identify the nature of the original secondary lysosome (*i.e.*, heterophagosome or autophagosome) and the more general term **digestive vacuole** is used to describe the organelle at this stage.

4. Residual Bodies

They are also called **telolysosomes** or **dense bodies**. Residual bodies are formed if the digestion inside the food vacuole is incomplete. Incomplete digestion may be due to absence of some lysosomal enzymes. The undigested food is present in the digestive vacuole as the residues and may take the form of whorls of membranes, grains, amorphous masses, ferritin-like or myelin fibres (Fig. 24.3).

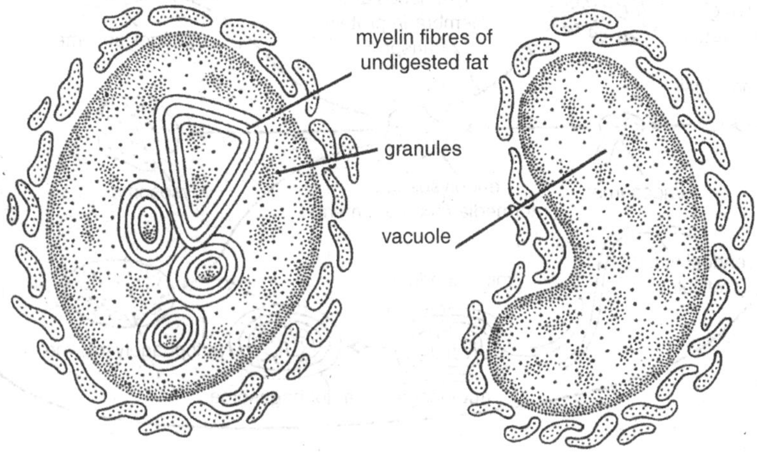

Fig. 24.3. Lysosomes of the kidney cells of rat, showing the presence of residues.

Residual bodies are large, irregular in shape and are usually quite electron-dense. In some cells, such as *Amoeba* and other potozoa, these residual bodies are eliminated by **defecation**. In other cells, residual bodies may remain for a long time and may load the cells to result in their aging. For example, pigment inclusions (age pigment or **lipofuscin granules**) found in nerve cells (also in liver cells, heart cells and muscle cells) of old animals may be due to the accumulation of residual bodies.

24.5. ORIGIN

The biogenesis (origin) of the lysosomes requires the synthesis of specialized lysosomal hydrolases and membrane proteins. Both classes of proteins are synthesized in the ER and transported through the Golgi apparatus, then transported from the trans Golgi network to an intermediate compartment (an endolysosome) by means of transport vesicles (which are coated by clathrin protein; Fig. 24.4). The lysosomal enzymes are glycoproteins, containing N-linked oligosaccharides that are processed in a unique way in the cis Golgi so that their mannose residues are phosphorylated. These mannose 6-phosphate (M6P) groups are recognized by M6P-receptors (which are transmembrane proteins) in the trans Golgi network that segregates the hydrolases and helps to package them into budding clathrin-coated vesicles which quickly lose their coats. These transport vesicles containing the M6P-receptors act as shuttles that move the receptors back and forth between the trans Golgi network and endolysosomes. The low pH in the endolysosome dissociates the lysosomal hydrolases from this receptor, making the transport of the hydrolases unidirectional.

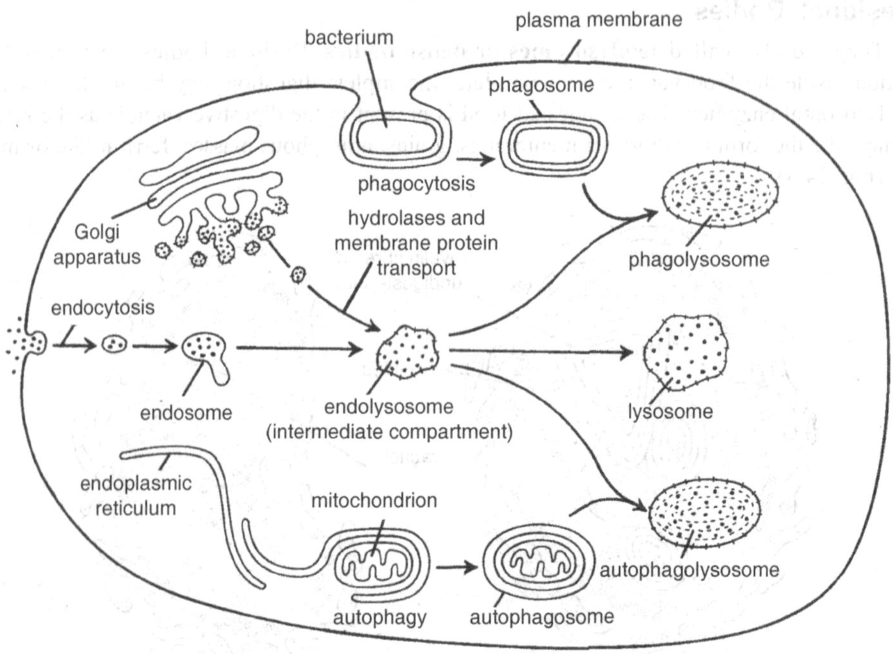

Fig. 24.4. Origin of three types of lysosomes : phagolysosome, lysosome (the classical secondary lysosome) and autophagolysosome. Transport vesicles (the classical primary lysosomes) originate from trans Golgi network to fuse with endolysosome which contains already endocytosed materials for digestion.

24.6. FUNCTIONS OF LYSOSOMES

The important functions of lysosomes are as follows:

1. Digestion of large extracellular particles. The lysosomes digest the food contents of the phagosomes or pinosomes. The lysosomes of leucocytes enable the latter to devour the foreign proteins, bacteria and viruses.

2. Digestion of intracellular substances. During the starvation, the lysosomes digest the stored food contents, *viz*., proteins, lipids and carbohydrates (glycogen) of the cytoplasm and supply to the cell necessary amount of energy.

3. Autolysis. In certain pathological conditions the lysosomes start to digest the various organelles of the cells and this process is known as **autolysis** or **cellular autophagy**. When a cell dies, the lysosome membrane ruptures and enzymes are liberated. These enzymes digest the dead cells. In the process of metamorphosis of amphibians and tunicates many embryonic tissues, *e.g*., gills, fins, tail, etc., are digested by the lysosomes and utilized by the other cells.

4. Extracellular digestion. The lysosomes of certain cells such as sperms discharge their enzymes outside the cell during the process of fertilization. The lysosomal enzymes digest the limiting membranes of the ovum and form penetration path in ovum for the sperms. Acid hydrolases are released from **osteoclasts** and break down bone for the reabsorption; these cells also secrete lactic acid which makes the local pH enough for optimal enzyme activity. Likewise, preceding ossification (bone formation), **fibroblasts** release cathepsin D enzyme to break down the connective tissue.

24.7. LYSOSOMAL STORAGE DISORDERS (LYSOSOMES AND DISEASE)

Malfunctioning of lysosomes often results in various pathological disorders affecting the life of the cell or an individual. Some of these are inborn diseases, caused by gene mutation (*e.g.*, I-cell disease, gout, Pompe's disease, Tay-Sach's disease, etc.) and others are induced by some environmental pollutants (*e.g.*, silicosis). Typically, the accumulated materials (*e.g.*, low-molecular weight materials, drugs, dyes, etc.) may cause malignant transformation of cells by bringing about leakage of lysosomal enzymes that attack the genetic material in the DNA.

Defects in normal functioning of lysosomes cause the following diseases (about 30 diseases):

1. Silicosis and asbestosis. Silicosis is a miner's disease which results from the uptake of silica fibres by macrophages (phagocytic cells) in the lungs. The fibres become enclosed within lysosomes but cannot be digested; instead they cause the lysosomal membrane to leak, spilling the contents of digestive enzymes into the cell and damaging the tissue of the lungs.

A similar result occurs when asbestos fibres are taken up by scavenging cells, resulting in the disease **asbestosis**. Both of these conditions, can be exhausting and may even be fatal.

2. Rheumatoid arthritis. This inflammatory disease results, in part, from the release of lysosomal enzymes from the immune cells into the extracellular space, causing damage to materials in the joints.

3. I-Cell disease. Patients of this inheritable disease has a deficiency of a lysosomal enzyme, called **N-acetylglucosmine phosphotransferase** (a lysosomal enzyme) which is required for phosphorylation of **mannose**. This lysosomal enzyme lacks mannose phosphate residues and due to this defect is not included in the lysosomes of fibroblasts of patients of I-cell disease. As a result, such lysosomes have bloated appearance due to undigested materials.

4. Pompe's disease. In the absence of another lysosomal enzyme, called **α-glucosidase**, undigested glycogen is accumulated in the lysosomes. This results in swelling of these organelles and irreversible damage to the cells and tissues.

5. Tay-Sachs disease. This disease is caused from the deficiency of the enzyme **β-N-hexosaminidase A**, an enzyme required to degrade the ganglioside G_{M2}, a common component of the membranes of brain cells. In its severest form, which strikes during infancy, the disease is characterized by progressive mental and motor retardation, as well as skeletal, cardiac and respiratory abnormalities. Tay-Sachs disease results in mental retardation, blindness and death by the age of 3.

24.8. VACUOLES IN PLANTS

Matile (1975) has divided vacuoles of plants into following three types:

1. Vacuoles

Vacuoles, fluid-filled compartments encompassed by a single membrane called the **tonoplast**, are conspicuous organelles of most plant cells. As much as 90 per cent of the volume of many plant cells is occupied by a central vacuole. Vacuoles store a large variety of molecules, including inorganic ions, organic acids, sugars, enzymes, storage proteins and many types of secondary metabolites. Proteins in the tonoplast transport all these molecules—except for storage proteins—into the vacuole. The resulting accumulation of solutes in the vacuole drives the osmotic uptake of water, producing the turgor pressure needed for cell enlargement. Many hydrolytic enzymes found in vacuoles resemble those present in the lysosome of animal cells suggesting a role for vacuoles in the turnover of cellular constituents in plant cells.

Functions of Vacuoles

Vacuoles play a wide range of essential roles in the life of a plant.

1. Growth. Plants use vacuoles to produce large cells cheaply. One of the major challenges faced by plants during evolution was to develop a structural design capable of producing large solar collectors at a metabolic price that could be recovered from the energy trapped and utilized by chloroplasts in growing season. This problem was solved by increasing the volume of the vacuolar compartment to drive cell enlargement while keeping the amount of nitrogen-rich cytoplasm constant. The latter point is particularly important for plants, whose growth is often limited by nitrogen availability. By filling a large volume of the cell with "inexpensive" vacuolar contents, mostly water and minerals, plants are able to reduce drastically the cost of making expanded structures such as leaves, which are essentially throw-away solar collectors.

Plant cell expansion is driven by a combination of osmotic uptake of water into the vacuoles and altered cell wall extensibility. The water taken into the vacuoles generates **turgor pressure**, which not only expands the primary cell wall, but also creates stiff, load-bearing structures in conjunction with the walls. This exploitation of internal *hydrostatic pressure* to stiffen thin primary cell walls resembles the use of air pressure in an inner tube to convert a pliable flat bicycle tire into a stiff circle capable of supporting heavy loads. **Wilting** and the associated softening of plant organs is caused by the loss of water from the vacuoles and surrounded cytosol.

To maintain the turgor pressure of continuously expanding cells, solute must be actively transported into the growing vacuole to maintain its osmolarity. An electrochemical gradient across the tonoplast provides the driving force for this uptake of solutes. The gradient, in turn, is produced and maintained by two electrogenic **proton pumps**: **V-type H^+–ATPase** and vacuolar **H^+–pyrophosphatase (H^+–PPase)**. The principal solutes in vacuoles include the ions K^+, Na^+, Ca^{2+}, Mg^{2+}, Cl^-, SO_4^{2-}, PO_4^{3-}, and NO^{3-}, and primary metabolites, such as amino acids, organic acids and sugars. The movement of water across the tonoplast is mediated by **aquaporin channels** consisting of **tonoplast intrinsic proteins (TIPs)**.

2. Storage. Aside from the ions, sugars, polysaccharides, pigments, amino acids, and organic acids mentioned above, plants also store large amounts of proteins in their vacuoles, especially in seeds. All of these primary metabolites can be retrieved from vacuoles and used in metabolic pathways to sustain growth. Interestingly, most of the flavors of fruits and vegetables can be traced to compounds that are stored in the vacuoles.

3. Digestion. Vacuoles contain the same types of acid hydrolases found in animal cell lysosomes. These enzymes, which include **proteases**, **nucleases**, **glycosidases** and **lipases**, together allow for the breakdown and recycling of nearly all cellular components. Such recycling is needed not only for the normal turnover of cellular structures but also for the retrieval of valuable nutrients during programmed cell death (apoptosis), associated with development and senescence.

4. pH and ionic homeostasis. Large vacuoles serve as reservoirs of protons and metabolically important ions such as calcium. Typically, plant vacuoles have a pH between 5.0 and 5.5, but the range extends from 2.5 or so (lemon fruit vacuoles) to greater than 7.0 in unactivated protein storage vacuoles. By controlling the release of protons and other ions into the cytosol, cells can regulate not only cytosolic pH but also the activity of enzymes, the assembly of cytoskeletal structures and membrane fusion events.

5. Defense against microbial pathogens and herbivores. Plant cells accumulate an amazing variety of toxic compounds in their vacuoles, both to reduce feeding by herbivores and to destroy microbial pathogens. These compounds include the following:

(*i*) Phenolic compounds, alkaloids, cyanide-containing glycosides and glucosinolates and protease inhibitors to discourage insect and animal herbivores.

(*ii*) Cell wall-degrading enzymes such as **chitnase** and **glucanase** and defense molecules, such as saponins, to destroy pathogenic fungi and bacteria.

(*iii*) Latexes, wound-clogging emulsions of hydrophobic polymers that possess insecticidal and fungicidal properties and also serve as antiherbivory agents.

(*iv*) Some of these excretory byproducts such as digitalis, have proven to have important chemical value. Digitalis is a drug which is prepared from the dried leaves of foxgloves and contain substances that stimulate the heart muscles.

6. Sequestration of toxic compounds. Plants cannot escape from toxic sites, nor can they efficiently eliminate by excretion toxic materials such as heavy metals and toxic metabolites such as **oxalate**. Instead, plants sequester these compounds into vacuoles. For example, to remove oxalate, specific cells develop vacuoles containing an organic matrix within which oxalate is allowed to react with calcium to form calcium oxalate crystals. In other plant cell types, members of the ABC family of transporters are used to transport **xenobiotics** (chemicals of human manufacture) from the cytoplasm into the vacuoles. Accumulation of toxic compounds in leaf vacuoles is one of the reasons leaves are shed on a regular basis.

7. Pigmentation. Vacuoles that contain anthocyanin pigments are found in many types of plant cells. Pigmented flower petals and fruits are used to attract pollinators and seed dispersers, respectively. Some leaf pigments screen out UV and visible light preventing photoxidative damage to the photosynthetic apparatus. This screening appears to be essential for survival of the leaves of evergreens that grow in climates where freezing conditions during winter months prevent the absorbed light energy from being used in photosynthesis.

Two Different Vacuole Systems

For years, plant researchers puzzled about how storage proteins and hydrolytic enzymes might coexist in vacuoles (see Buchanan, *et al.*, 2000). This problem has now finally been resolved by the discovery that many plant cells contain, at least during some stage of development, two functionally different kinds of vacuoles: neutral protein storage vacuoles and acidic, lytic vacuoles. During development, storage products may be mobilized either by fusing the two vacuole types or by delivering specific lytic enzymes to the storage vacuoles.

Evidence in support of discovery has come from investigations using several types of probes, including heavy metal stains, fluorescent molecules sensitive to pH or lytic enzyme activity and antibodies directed against two types of tonoplast proteins: α-TIP and TIP-Ma27. Anti-α-TIP antibodies label only vacuoles that have a pH near 7 and contain storage proteins such as lectins. In contrast, anti-TIP-Ma27 antibodies label only acidified vacuoles, which contain hydrolytic enzymes such as the cysteine protease aleurain (**Buchanan**, *et al.*, 2000).

The vacuole of a mature plant cell is formed from the enlargement and fusion of smaller vacuoles present in meristamatic cells; these **provacuoles**, which are believed to be derived from the ER and possibly the Golgi apparatus and contain acid hydrolases.

2. Spherosomes

The spherosomes are membrane-bounded, spherical particles of 0.5 to 2.5 μm diameter, occurring in most plant cells. They have a fine granular structure internally which is rich in lipids and proteins. They originate from the endoplasmic reticulum (ER). Oil accumulates at the end of a strand of ER and a small vesicle is then cut off by constriction to form particles, called **prospherosomes**. The prospherosomes grow in size to form spherosomes. Basically, the spherosomes are involved in lipid synthesis and storage. But, the spherosomes of maize root tips and spherosomes of tobacco endosperm tissue have been found rich in hydrolytic digestive enzymes and so have been considered as lysosomes. Like lysosomes they are not only responsible for the accumulation and mobilization of reserve lipids, but also for the digestion of other cytoplasmic components incorporated by phagocytosis.

3. Aleurone grain

The aleurone grains or protein bodies are spherical membrane-bounded storage particles occurring in the cells of endosperm and cotyledons of seeds. They are formed during the later stages of seed ripening and disappear in the early stages of germination. They store protein (*e.g.*, globulins) and phosphate in the form of phytin. **Matile** (1968) has demonstrated that aleurone grains from pea seed contain a wide range of hydrolytic enzymes including protease and phosphatase which are required for the mobilization of stored protein and phosphate, although the presence of other enzymes such as β-amylase and RNAase suggest that other cell constituents may also be digested. Thus, like spherosomes, aleurone grains store reserve materials, mobilize them during germination and in addition form a compartment for the digestion of other cell components. The aleurone grains are derived from the strands of the endoplasmic reticulum.

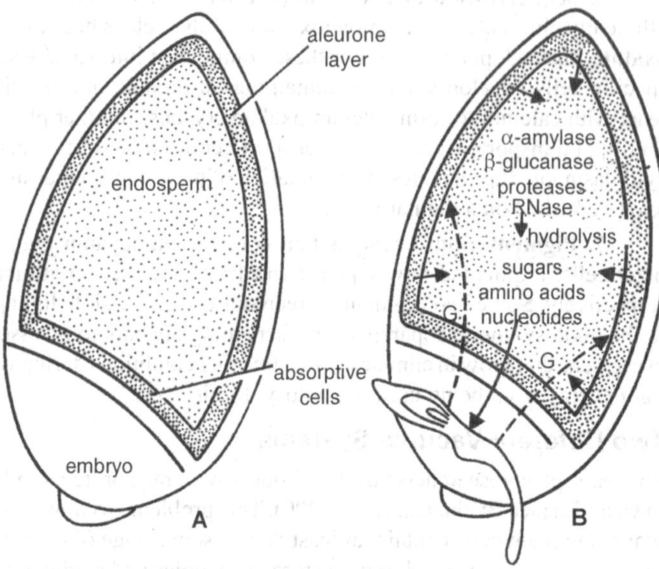

Fig. 24.5. Role of hormone and hydrolytic enzymes in seed germination. A—Ultrastructure of seed before germination; B–During germination, gibberellins (G) pass from the embryo to the aleurone layer, where the *de novo* synthesis of hydrolytic enzymes is induced. These enzymes break down the macromolecular stored reserves and the low molecular weight products are transported to the embryo, where they function as nutrients.

During germination of barley seed, the activity of hydrolases is found to be controlled by hormones such as **gibberellic acid** (Fig. 24.5). Gibberellic acid, a plant growth hormone, is released by the embryo to the aleurone layer where, in turn, the hydrolases are released to the endosperm. This hormone operates by derepressing appropriate genes in the aleurone cells, which then begin to crank out new hydrolytic proteins.

Extra-cellular digestion by plants. Plant cells are generally unable to engulf large particles, presumably because of the restrictions imposed on the cell by cell wall. The secretion of hydrolases to carry out extracellular digestion, therefore, becomes an important process. Hydrolases are commonly secreted by fungi, enabling the organism to degrade and grow on macromolecules it cannot transport into the cell. Higher plants also secrete hydrolases, a notable example being the insectivorous pitcher plants, which produce a proteinase-containing liquid in which victims are trapped and digested.

QUESTIONS

Long Answer Questions

1. Describe in detail the structure and functions of lysosomes.
2. What are lysosomes? Give the ultrastructure of lysosomes and their functional significance.
3. "Lysosomes are said to be the suicidal bags in the cells". Explain it. Give structure, chemical composition and functions of lysosomes.
4. Describe vacuoles in plants.
5. Write an essay on lysosome.

Short Answer Questions

1. Give an account of lysosomal polymorphism.
2. Explain the involvement of lysosomes in human disease.
3. What are suicide bags of cell? Describe their functions.
4. Write short notes on the following:
   (*i*) Lysosome.
   (*ii*) Lysosomes or Suicide bags.

(*iii*) Polymorphism in lysosome.
(*iv*) Enzymes of lysosomes.

5. Lysosomes are suicide bags of the cell. Explain.
6. Write a brief account of polymorphism in lysosomes.
7. Give a brief account of the functions of lysosomes.
8. How and where lysosomes are formed in the cell?
9. Give significance of autophagic vacuole.
10. Lysosomal enzymes are inactive as long as they remain inside the lysosomes but become active as soon as they come in contact with some other substance. Why?
11. Differentiate between the following:
    (*i*) Primary and secondary lysosomes.
    (*ii*) Autophagic vacuole and digestive vacuole.
    (*iii*) Peroxisomes and lysosomes.
    (*iv*) Lysosomes and zymogen granules.

Very Short Answer Questions

1. What is the role of acrosomal granule in sperm penetration?
2. Name the animal cells which do not contain lysosomes.
3. Where does lysosome formed in the cell?
4. Which secondary lysosome stores undigested wastes of cellular digestion?
5. What is autophagy?
6. What is pH of lumen of lysosomes?
7. Give the name of marker enzyme of the lysosome.

Fill in the Blanks

1. Lysosomal function is related to ....of substance.
2. Lysosomes with phagosomes form ....
3. Microbodies containing oxidative enzymes are called............
4. The microbodies with specific lipid contents in their matrix are known as............

Yes or No Questions

Write Yes or No as answers of the following:

1. Lysosomes help in protein synthesis.
2. Vaculoes of a cell are known as "suicide bags".

Multiple Choice Questions

1. Lysosomes are absent in
   (*a*) erythrocytes (*b*) plasma cells
   (*c*) nerve cells (*d*) muscle cells
2. Which one of the following is called as the suicidal sac of the cell?
   (*a*) centrosome (*b*) lysosomes
   (*c*) microsomes (*d*) desmosomes
3. At which pH, enzymes of lysosomes are usually active
   (*a*) pH 5 (*b*) pH 7
   (*c*) pH 8 (*d*) at any pH
4. Hydrolytic enzymes are stored in
   (*a*) Golgi bodies
   (*b*) lysosomes
   (*c*) endoplasmic reticulum
   (*d*) mitochondria

ANSWERS

Very Short Answer Questions

1. Lysis. **2.** Red blood corpuscles. **3.** Golgi apparatus. **4.** Residual bodies. **5.** Digestion of portions of a cell's own organelles by lysosomes. **6.** 5. **7.** Acid phosphatase.

Fill in the Blanks

1. Hydrolysis (*i.e.,* intracellular digestion). **2.** Heterolysosomes (or food vacuoles). **3.** Peroxisomes. **4.** Sphareosomes.

Yes or No Questions

1. No **2.** No

Multiple Choice Questions

1. (*a*) **2.** (*b*) **3.** (*c*) **4.** (*b*)

25 Peroxisomes

Peroxisomes are small, simple, single-membrane enclosed organelles with a diameter of 0.1 to 1.0 μm. They often contain a dense, crystalline core of oxidative enzymes. Peroxisomes are multifunctional organelles, so contain enzymes involved in a variety of metabolic reactions, including several aspects of energy metabolism. They are found in all eukaryotic cells; most human cells contain about 500 peroxisomes. Peroxisomes do not have their own genomes and ribosomes. All their proteins, called **peroxins** (Pex1, Pex2, etc.) are synthesized from the nuclear genome. Most peroxins are synthesized on free ribosomes and then imported into peroxisomes as completed polypeptide chains.

25.1. COMPARISON OF PEROXISOMES AND MITOCHONDRIA

Peroxisomes share several properties with mitochondria:

1. Both mitochondria and peroxisomes form by splitting from preexisting organelles. (However, unlike mitochondria, peroxisomes can also be regenerated even if entirely lost to the cell).
2. Both types of organelles import preformed proteins.
3. Both types of organelles engage in similar types of oxidative metabolism. In fact, at least one enzyme, alanine/glyoxylate aminotransferase, is found in the mitochondria of some mammals (*e.g.*, cats and dogs) and the peroxisomes of other mammals (*e.g.*, rabbit and humans; see **Cooper** and **Hausman**, 2007).
4. Peroxisomes use molecular oxygen like mitochondria, but instead of having cytochromes and capacity of ATP synthesis like them, they contain flavin-linked oxides and catalases for the hydrogen peroxide metabolism and also enzymes for fatty acid metabolism.

Lastly, while mitochondrial (and plastid) proteins resemble those of prokaryotes, reflecting their endosymbiotic origin, the peroxins resemble typical eukaryotic proteins.

Peroxisomes differ from mitochondria and chloroplasts in many ways. Most notably, these organelles are surrounded only by a single membrane, and they do not contain DNA (genome) or ribosomes. However, they resemble ER in being self-replicating membrane bound organelle.

25.2. HISTORICAL

Since the mid-1950s electron microscopists have observed small structures or bodies in cells that on morphological grounds have been aptly termed **microbodies**. **C.De Duve** and **P. Baudbuin** (1966) coined the term **peroxisome** for the microbodies of mammalian systems and studied their structure and function. **Leaf peroxisomes** were first isolated from spinach leaf homogenate (*i.e.*, from mesophyll cells) by **Tolbert's** group in Michigen in 1968. Glyoxylate (= Glycolate) cycle containing peroxisomes, called **glyoxysomes**, were discovered in 1969 by **Beevers** in the endosperm cells of germinating castor bean (*Ricinus*). Biogenesis of peroxisomes has been investigated by following cell biologists: **Purdue** and **Lazarow** (2001); **Mullen, Flynn** and **Trelease** (2001) **Dammai** and **Subramani** (2001) and **Lazarow** (2003).

25.3. OCCURRENCE

Peroxisomes are found in all eukaryotic cells but are especially prominent in mammalian kidney and liver cells, in algae and photosynthetic cells of plants, and in germinating seedlings of plant species that store fat in their seeds.

25.4. STRUCTURE

Peroxisomes are variable in size and shape, but usually appear circular in cross section having diameter between 0.2 and 1.5μm (0.15 to 0.25 μm diameter in most mammalian tissues; 0.5 μm in rat liver cells). They have a single limiting unit membrane of lipid and protein molecules, which encloses their granular matrix. In some cases (*e.g.*, in the festuciod grasses) the matrix contains numerous threads or fibrils, while in others they are observed to contain either an amorphous nucleoid or a dense inner core which in many species shows a regular crystalloid structure (*e.g.*, tobacco leaf cell, **Newcomb** and **Frederick**, 1971). Little is known about the function of the core, except that it is the site of the enzyme urate oxidase in rat liver peroxisomes and much of the catalase in some plants (see **Hall** *et.al.*, 1974).

25.5. ISOLATION AND CHEMICAL COMPOSITION

The technique of **isolation** of microbodies from animal and plant tissues includes the following steps: 1. Tissues are ground very carefully to save microbodies from disruption. 2. The homogenate is treated with differential centrifugation to obtain a fraction of the cell homogenate which is rich in microbodies. 3. The enriched fraction is subjected to isopycnic ultracentrifugation on discontinuous or continuous sucrose density gradient.

Animal peroxisomes often contain a distinct **crystalline core**, which is usually consists of a crystalline form of *urate oxidase.* Crystalline cores are also often present in the peroxisomes of plant leaves, but these usually consist of *catalase* instead. When such cores are present, it is easy to identify microbodies as peroxisomes, since urate oxidase and catalase are two of the enzymes by which peroxisomes are defined. Indeed, these organelles were named "peroxisomes" because they are the site of synthesis and degradation of hydrogen peroxide.

In the absence of crystalline core, it is not always easy to spot peroxisomes ultrastructurally. A useful technique in such cases is a **cytochemical test** for catalase called the **diaminobenzidine (DAB) reaction**. This assay depends on the ability of catalase to oxidize DAB to a polymeric form that causes deposition of electron dense osmium atoms when the tissue is treated with osmium tetroxide (OsO_4). The resulting electron-dense deposits can be readily seen in cells from stained tissue. In animal peroxisomes, the entire internal space often stains intensely with DAB, indicating that catalase exists as a soluble enzyme uniformly distributed throughout the matrix of the organelle. In plant leaf cells, DAB treatment preferentially stains the crystalline cores of the peroxisomes, thereby definitively identifying the cores as crystalline catalase. Because catalase is the single enzyme present in all peroxisomes and does not routinely occur in any other organelle, the DAB reaction is a very reliable and highly specific means of identifying organelles unambiguously as peroxisomes.

Peroxisomes contain at least 50 different enzymes, which are involved in a variety of biochemical pathways in different types of cells. However, peroxisomes are diverse organelles and even in various cell types of a single organism they may contain different sets of enzymes. They can also adapt remarkably to changing conditions. Yeast cells grown on sugar, for example, have small peroxisomes. But when some yeasts are grown on methanol, they develop large peroxisomes that oxidise methanol; and when grown on fatty acids, they develop large peroxisomes that break down fatty acids to acetyl CoA by β–oxidation (see **Alberts** *et.al.,* 2002).

25.6. FUNCTIONS

There are at least **five** general categories of peroxisomal functions:

1. Hydrogen peroxide metabolism;
2. Detoxification of harmful substances;
3. Oxidation of fatty acids;
4. Metabolism of nitrogen-containing compounds, and
5. Catabolism of unusual substances.

1. Hydrogen peroxide metabolism. The most obvious role of peroxisomes in eukaryotic cells is the detoxification of H_2O_2, which is accomplished by the *coexistence* in the same organelle of catalase and the oxidases that generate H_2O_2. The oxidases in peroxisomes vary considerably in the specific reactions they catalyze but they all share the property of **transferring** electrons from their respective substrates directly to molecular oxygen (O_2), forming H_2O_2. Using RH_2 to represent an oxidizable organic substrate, the general reaction catalyzed by oxidases can be written as follows:

$$RH_2 + O_2 \rightarrow R + H_2O_2 \qquad \text{...(Equation 25.1)}$$

The hydrogen peoxide formed in this manner is broken down by catalase in one of two ways. Usually catalase functions in what is called its **catalytic mode**, in which one molecule of H_2O_2 is oxidized to oxygen and a second is reduced to water:

$$2H_2O_2 \rightarrow O_2 + 2H_2O \qquad \text{...(Equation 25.2)}$$

Dividing reaction 25.2 and adding it to reaction 25.1 yields a summary reaction for the two-step process:

$$RH_2 + \frac{1}{2}O_2 \rightarrow R + H_2O \qquad \text{...(Equation 25.3)}$$

Alternatively, catalase can function in its **peroxidatic mode**, in which electrons derived from an organic donor are used to reduce H_2O_2 to water:

$$R'H_2 + H_2O_2 \rightarrow R' + 2H_2O \qquad \text{...(Equation 25.4)}$$

(The prime on R group (*i.e.*, R′) simply indicates that this substrate is likely to be different from the substrate in reaction 25.1) The corresponding summary reaction in this case is

$$RH_2 + R'H_2 + O_2 \rightarrow R + R' + 2H_2O \qquad \text{...(Equation 25.5)}$$

The result is the same in either case: Hydrogen peroxide is degraded without ever leaving the peroxisome. Given the toxicity of hydrogen peroxide (which is the main active ingredient in a variety of disinfectants), it makes good sense for the enzymes responsible for peroxide generation to be compartmentalized together with the catalase that catalyzes its degradation. Indeed, catalase is the most abundant protein in most peroxisomes, representing upto 15% of the total protein content of this organelle. In this way, virtually every molecule of H_2O_2 generated by oxidases will encounter a molecule of catalase almost immediately be promptly degraded.

H_2O_2 and aging. Most cytosolic H_2O_2 is produced by mitochondria and membranes of ER, although there are also H_2O_2 producing enzymes localized in the cytoplasmic matrix. Catalase acts as a "**safety valve**" for dealing with the large amounts of H_2O_2 generated by peroxisomes; however, other enzymes such as **glutathione peroxidase;** are capable of metabolizing organic hydroperoxides and also H_2O_2, in the cytosol (cytoplasmic matrix) and mitochondria. The production of superoxide anion (O_2^-) in mitochondria and cytosol (cytoplasmic matrix) is regulated mainly by the enzyme

superoxide dismutase. All of these protective enzymes are present in high levels in aerobic tissues.

Recently, a possible relationship has been stressed between peroxides and free radicals (such as superoxide anion–O_2^-) with the process of aging. These radicals may act on DNA molecule to produce **mutations** altering the transcription into mRNA and the translation into proteins. In addition, free radicals and peroxides can affect the membranes by causing peroxidation of lipids and proteins. For these reasons reducing compounds such as vitamin E or enzymes such as superoxide dismutase could play a role in keeping the healthy state of a cell.

2. Detoxification of harmful compounds. In its peroxidatic mode (reaction 25.4), catalase can use a variety of substances as electron donors, including methanol, ethanol, formic acid, formaldehyde, nitrites and phenols. Because all of these compounds are harmful to cells, *their oxidative detoxification by catalase* may be a vital peroxisomal function. The prominent peroxisomes of liver and kidney cells are thought to be important in such detoxification reactions.

3. Oxidation of fatty acids. Peroxisomes found in animal, plant and fungal cells contain enzymes necessary for oxidising fatty acids. This process, called **β-oxidation**, also occurs in the mitochondrion as was discussed in Chapter 16. About 25 to 50% of fatty acid oxidation in animal tissues occurs in peroxisomes, with the remainder localized in mitochondria. In plant and yeast cells, on the other hand, all β-oxidation is confined to peroxisomes.

In animal cells, peroxisomal β-oxidation appears to be especially important for the catabolism of long-chain (16 to 20 carbons), very long chain (24–26 carbons), and branched fatty acids. The primary product of β-oxidation, acetyl-CoA, is then exported to the cytosol and enters biosynthetic pathways. Once fatty acids are shortened to fewer than 16 carbon, further oxidation usually occurs in the mitochondria. Thus, in animal cells, the peroxisome is important for shortening fatty acids in preparation for subsequent metabolism in the mitochondrion rather than completely breaking them down to acetyl-CoA. In plants and yeast, on the other hand, peroxisomes are essential for the complete catabolism of fatty acids to acetyl-CoA.

4. Metabolism of nitrogen containing compounds. Except for primates, most animals require urate oxidase (also called uricase) to oxidize urate, a purine that is formed during the catabolism of nucleic acids and some proteins. Like other oxidases, urate oxidase catalyzes the direct transfer of electrons from the substrate to molecular oxygen, generating H_2O_2:

$$\text{Urate} + O_2 \rightarrow \text{Allantoin} + H_2O_2 \qquad \text{...(Equation 25.6)}$$

As noted earlier, the H_2O_2 is immediately degraded in the peroxisome by catalase. The allantoin is further metabolized and excreted by the organism, either as allantoic acid, or in case of crustaceans, fish, and amphibians, as urea.

Additional peroxisomal enzymes involved in nitrogen metabolism include aminotransferase. Members of this collection of enzymes catalyze the transfer of amino acids to α-keto acids.

$$\underset{\text{Amino acid}}{R-\overset{H_3N^+}{\overset{|}{C}}-\overset{O}{\overset{\|}{C}}-O^-} + \underset{\alpha\text{-keto acid}}{R'-\overset{O}{\overset{\|}{C}}-\overset{O}{\overset{\|}{C}}=O^-} \rightleftharpoons \underset{\alpha\text{-keto acid}}{R-\overset{O}{\overset{\|}{C}}-\overset{O}{\overset{\|}{C}}-O^-} + \underset{\text{Amino acid}}{R'-\overset{H_3N^+}{\overset{|}{C}}-\overset{O}{\overset{\|}{C}}=O^-} \qquad \text{...(Equation 25.7)}$$

Such enzymes play important roles in the biosynthesis and degradation of amino acids by moving amino groups from one molecule to another.

5. Catabolism of unusual substances (*e.g.*, xenobiotics). Some of the substrates for peroxisomal oxidases are rare compounds for which the cells has no other degradive pathways. Such compounds include D-amino acids, which are not recognized by enzymes capable of degrading the L-amino acids that constitute polypeptides. In some fungi, the peroxisomes also contain enzymes that break down unusual substances called **xenobiotics,** chemical compounds foreign to biological organisms. This category include **alkanes**, short-chain hydrocarbon compounds found in oil and other petroleum products. Fungi containing enzymes capable metabolizing such xenobiotics may turn out to be useful for cleaning up oil spills that would otherwise contaminate the environment (see **Becker** *et.al.*, 2006).

6. Synthetic activities. In addition to providing a compartment for oxidation reactions, peroxisomes are involved in biosynthesis of lipids and the amino acid, lysine. In animal cells, cholesterol and dolicol are synthesized in peroxisomes as well as in ER. In the liver, peroxisomes are also involved in the synthesis of bile acids, which are derived from cholesterol. In addition, peroxisomes contain enzymes required for the synthesis of **plasmalogens** a family of phospholipids in which one of the hydrocarbon chains is joined to glycerol by an **ether bond** rather than by ester bond (Fig. 25.1). Plasmalogens are important membrane components in some tissues. Particularly heart and brain, although they are absent in others. Deficiency of plasmalogens causes profound abnormalities in the **myelination** (formation of myelin-sheath) of nerves, which is one reason why many peroxisomal disorders lead to neurological disease (See **Alberts** *et.al.*, 2002) (Box 25.1).

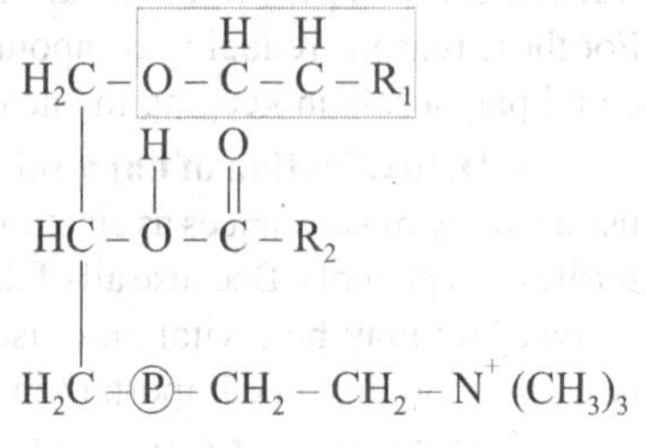

Fig. 25.1. Structure of a plasmalogen.

Box 25.1

The most common peroximal disorder is **X-linked adrenoleukodystrophy.** The defective protein causing this disorder is an integral membrane protein that may be responsible for transporting very long-chain fatty acids into the peroxisomes for β-oxidation. Accumulation of these long-chain fatty acids in human body fluids destroys myelin sheath in nervous system (See **Becker** *et. al.*, 2006).

Roles of Plant Specific Peroxisomes

1. Leaf peroxisomes. In green leaves, there are peroxisomes that carry out a process called **photorespiration** which is a light-stimulated production of CO_2 that is different from the generation of CO_2 by mitochondria in the dark. In photorespiration, **glycolic acid (glycolate)**, a two-carbon product of photosynthesis is released from chloroplasts and oxidized into **glyoxylate** and $\mathbf{H_2O_2}$ by a peroxisomal enzyme called **glycolic acid oxidase**. Later on, glyoxylate is oxidized into CO_2 and **formate**:

$$CH_2OH.COOH + O_2 \rightarrow CHO - COOH + H_2O_2$$

$$CHO - COOH + H_2O_2 \rightarrow HCOOH + CO_2 + H_2O$$

Photorespiration is so-called because light induces the synthesis of glycolic acid in chloroplasts. The entire process involves intervention of two basic organelles : chloroplasts and peroxisomes.

Lastly, photorespiration is driven by atmospheric conditions in which the O_2 tension is high and the CO_2 tension low. Apparently O_2 competes with CO_2 for the enzyme **ribulose diphosphate carboxylase** which normally is the key enzyme in CO_2 fixation during photosynthesis. When O_2 is used by the enzyme, an unstable intermediate is formed which breaks down into **3-phosphoglycerate** and **phosphoglycolate**. The latter tends to increase the glycolate concentration by removal of its phosphate group and, therefore, more glycolate is available for additional oxidation and CO_2 release.

Photorespiration is a wasteful process for the plant cell, since, it significantly reduces the efficiency of the process of photosynthesis (*i.e.*, it returns a portion of fixed CO_2 to the atmosphere).

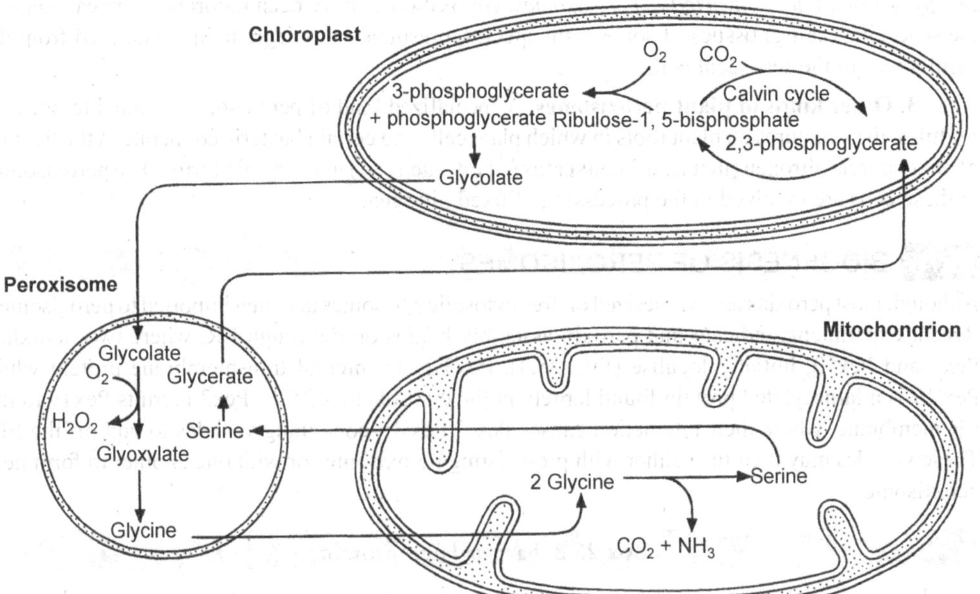

Fig. 25.2. Role of peroxisomes in photorespiration. During photosynthesis CO_2 is converted to carbohydrates by the Calvin cycle, which initiates with the addition of CO_2 to the five-carbon sugar ribulose–1, 5–diphosphate. However, the enzyme involved sometimes catalyzes the addition of O_2 instead, resulting in production of the two-carbon compound phosphoglycolate. Phosphoglycolate is converted to glycolate, which is then transferred to peroxisomes, where it is oxidized and converted to glycine. Glycine is then transferred to mitochondria and converted to serine. The serine is returned to peroxisomes and converted to glycerate, which is transferred back to chloroplasts (after Cooper and Hausman, 2007).

Glycolate cycle. Peroxisomes of plant leaves contain catalaze together with the enzymes of **glycolate pathway**, as glycolate oxidase, glutamate glyoxylate, serine-glyoxylate and asparate-α-ketoglutarate aminotransferases, hydroxy pyruvate reductase and malic dehydrogenase. They also contain FAD, NAD and NADP coenzymes. The glycolate cycle (Fig. 25.2) is thought to bring about the formation of the amino acids-glycine and serine-from the non-phosphorylated intermediates of photosynthetic carbon reduction cycle, *i.e.*, glycerate to serine, or glycolate to glycine and serine in a sequence of reactions which involve chloroplasts, peroxisomes, mitochondria and cytosol (**Tolbert,** 1971). The glycolate pathway also generates C_1 compounds and serves as the generator of precursors for nucleic acid biosynthesis.

2. Glyoxysomes. It occurs transiently in seedlings of plant species that store carbon and energy reserves in the seed of fat (primarily triacylglycerols). In such species, stored triglycerols are mobilized and converted to sucrose during early postgerminative development by a sequence of events that includes β-oxidation of fatty acids as well as a pathway known as the glyoxylate (= glycolate) cycle. All of the enzymes needed for these processes are localized to specialized peroxisomes called **glyoxysomes**. Glyoxysomes are found only in the tissues in which the fat is stored (endosperm or cotyledons, depending on the species) and are present only for the relatively short period of time required for the seedling to deplete its supply of stored fat. *Once they fulfill their role in the seedlings, the glyoxysomes are converted to peroxisomes*. Glyoxysomes have been reported to appear again in the senescing (aging) tissues of some plant species, presumably to degrade lipids derived from the membranes of the senescent cells.

3. Other kinds of plant peroxisomes. A specialized kind of peroxisome is found to occur in **nodules**, the structures on plant roots in which plant cells and certain bacteria cooperate in the fixation of atmospheric nitrogen (that is, the conversion of nitrogen (N_2) into organic form). The peroxisomes in these cells are involved in the processing of fixed nitrogen.

25.7. BIOGENESIS OF PEROXISOMES

Although most peroxins are synthesized on free cytosolic ribosomes and then imported to peroxisomes. Recent experiments indicate that peroxin assembly begins on the rough ER, where two peroxins, Pex3 and Pex19, initially localise (Fig. 25.3). Pex3 is an integral transmembrane protein while Pex 19 is a farnesylated protein found largely in the cytosol (Box 25.2). Pex3 recruits Pex19 to the ER membrane, where their interaction causes Pex 3/Pex 19-containing vesicles to bud off the ER. These vesicles may then fuse either with pre-existing peroxisomes or with one another to form new peroxisomes.

Box 25.2 Farnesylated protein

Lipid-anchored proteins are located on one of the surfaces of lipid bilayer but are covalently bond to lipid molecules embedded within the bilayer. Farnesylated proteins are the *prenylated membrane proteins* which are synthesized as soluble cytosol proteins before being modified by addition of a prenyl group, usually a 15 carbon **farnesnyl** group. After attachment, the farnesnyl group is inserted into the lipid bilayer of the membrane (see **Becker** *et.al*., 2006).

Pex3, Pex19, and other peroxisomal membrane proteins then act as receptors for import of the other peroxins, which are translated on free cytosolic ribosomes and then transported into peroxisomes as completed and folded polypeptides. They are targeted to the interior of peroxisomes by at least two pathways, which are conserved from yeasts to human.

Most peroxins are targeted by the simple amino acid sequence Ser-Lys-Leu at their carboxyl terminus (**peroxisome targeting signal 1 or PTS1**). A small number of peroxins are targeted by a sequence of nine amino acids at their amino terminus (**peroxisome targeting signal 2** or **PTS2**). PTS1 and PTS2 are recognized by distinct cytosolic receptors and then passed through a poorly understood channel in the peroxisomal membrane into the matrix. The receptors are subsequently retrieved from the peroxisome and recycled. Unlike the translocation of polypeptide chains across the membranes of the endoplasmic reticulum, mitochondria and chloroplasts, the targeting signals are usually not cleaved during the import of proteins into peroxisomes and the mechanism of translocation is not known (see **Cooper** and **Hausman** 2007).

Protein import together with the continuing addition of lipids from the rough ER, results in peroxisome growth, and new peroxisomes can be formed by division of old ones. In addition, peroxisomes undergo a complex **maturation process** that involves the import of different classes of proteins from the cytosol at different times. As a result, the enzyme content, and the metabolic activities of peroxisomes may change as they mature.

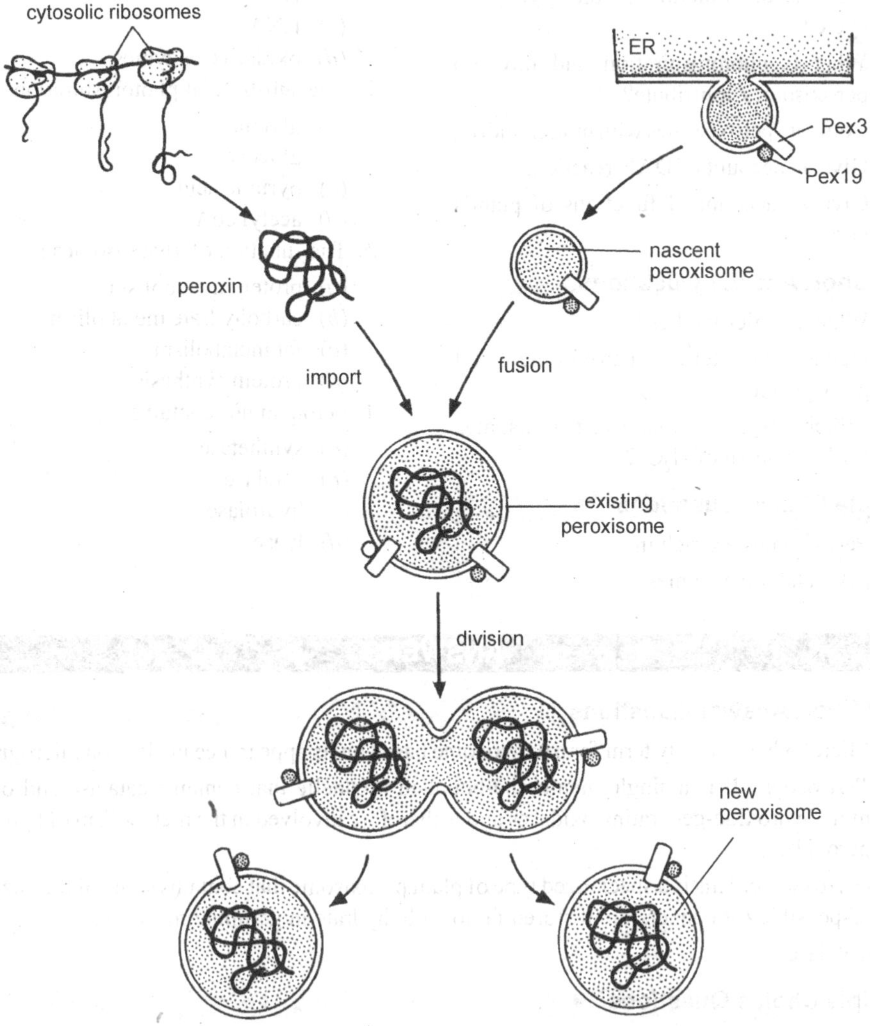

Fig. 25.3. Assembly of peroxisomes. Initiation of peroxisome assembly begins in the rough ER when the transmembrane protein peroxin (Pex3) recruits the soluble farnesylated protein peroxin (Pex19) and initiate budding of a nascent peroxisome. The nascent peroxisomes fuse either with each other or with existing peroxisomes. Additional peroxins are synthesized on free cytosolic ribosomes and imported as completed polypeptide chains to form functional peroxisomes, which grow larger and divide (after Cooper and Hausman, 2007).

QUESTIONS

Long Answer Questions

1. Write an account of structure, synthesis, function and histochemical localization of the catalase enzyme in eukaryotic cells.
2. Describe the mode of biogenesis and various functions of peroxisomes in animal cells.

Short Answer Questions

1. What would be the most likely role of peroxisome in cells lacking mitochondria?

2. What is the function of the glyoxylate cycle?
3. What is photorespiration and how do peroxisomes contribute?
4. Compare peroxisomes with mitochondria.
5. Give an account of DAB reaction.
6. Give an account of functions of plant's peroxisomes.

Very Short Answer Questions

1. What are microbodies?
2. Define the terms peroxisome and glyoxysome.
3. Which enzyme is found in peroxisomes for breakdown of H_2O_2?

Multiple Choice Questions

1. Peroxisomes are rich in
   (*a*) catalytic enzymes
   (*b*) DNA
   (*c*) RNA
   (*d*) oxidative enzymes
2. The substrate of photorespiration is
   (*a*) glycolate
   (*b*) glucose
   (*c*) pyruvic acid
   (*d*) acetyl coA
3. The function of glyoxysomes is
   (*a*) protein metabolism
   (*b*) carbohydrate metabolism
   (*c*) fat metabolism
   (*d*) protein synthesis
4. Peroxisomes contains
   (*a*) synthetase
   (*b*) catalase
   (*c*) hydrolase
   (*d*) lyase

ANSWERS

Very Short Answer Questions

1. Microbody is an early term for a peroxisome based on its appearance in electron micrographs.
2. *Peroxisome.* It is a single membrane-bounded organelle that contains catalase and one or more hydroxide-generating oxidases and is therefore involved in the metabolism of hydrogen peroxide.

   Glyoxysome. This is a specialized type of plant peroxisome that contains some of the enzymes responsible for conversion of stored fat to carbohydrate in germinating seeds.
3. Catalase.

Multiple Choice Questions

1. (*d*) **2.** (*a*) **3.** (*c*) **4.** (*b*)

26 Cell Signalling (Messages, Receptors and Signal Transduction)

No cell lives in isolation. In eukaryotic microorganisms such as yeast, slime molds and protozoans, secreted molecules called **pheromones** coordinate the aggregation of free-living cells for sexual mating or differentiation under certain environmental conditions (see **Lodish** *et al.*, 2004). The amoeboid cells of the slime mold *Dictyostelium* secrete a compound called cyclic AMP (cAMP) at one stage in its life cycle. The compound binds to receptors on the surfaces of neighbouring cells, triggering a process whereby thousands of individual amoeboid cells aggregate and differentiate into a multicellular organism (see **Becker** *et al.*, 2006).

Factors determining the mating-types of yeast are a well understood example of pheromone-mediated cell-to-cell signalling. More important in plants and animals are extracellular **signalling molecules** that function *within* an organism to control metabolic processes within cells, the growth and differentiation of tissues, the synthesis and secretion of proteins, and the composition of intracellular and extracellular fluids. Adjacent cells often communicate by direct cell-cell contact. For example, gap junctions in the plasma membrane of adjacent cells permit them to exchange small molecules and to coordinate metabolic responses.

Extracellular signalling molecules are synthesized and released by **signalling cells** and produce a specific response only in **target cells** that have receptors for the signalling molecules. In multicellular organisms, an enormous variety of chemicals including small molecules (*e.g.*, amino acid or lipid derivatives, acetylcholine, peptides) and proteins are used in this type of cell-to-cell communication. Some signalling molecules, especially hydrophobic molecules, such as steroids, retinoids, and thyroxine, spontaneously diffuse through the plasma membrane and bind to intracellular receptors.

Box 26.1 Importance of Signals

Intracellular signalling is essential to the survival of organisms providing an opportunity to adapt suitably. Signals perform the following important functions:

1. Maintenance of homeostasis.
2. Control of cell division and cell death.
3. Adaptation to environmental conditions.
4. Control of development and growth.
5. Release and production of hormones and other regulatory molecules.
6. Response elicited between organisms such as establishment of pathogenesis, activation of defenses, establishment of symbiosis, etc.

In this chapter we focus on signalling from a diverse group of receptor proteins located in the plasma membrane (*e.g.*, G protein-linked receptors, cytokine receptors, receptor tyrosine kinases, TGFB receptors, etc.). The signalling molecule acts as a **ligand**, which binds to a structurally complementary site on the extracellular or membrane-spanning domains of the receptor. Binding of a ligand to its receptor causes a conformational change in the cytosolic domain (or domains) of the receptor that ultimately induces specific cellular responses. The overall process of converting signals into cellular responses, is termed **signal transduction**. As we will see, signal transduction pathways may involve relatively few or many components.

Communication by extracellular signals usually involves the following 7 step: 1. synthesis and 2. release of the signalling molecule by the signalling cell; 3. transport of the signal to the target cell; 4. binding of the signal by a specific receptor protein leading to its activation; 5. initiation of one or more intracellular signal transduction pathways by the activated receptor; 6. specific changes in cellular function, metabolism or development; and 7. removal of the signal, which often terminates the celltular response.

The vast majority of receptors are activated by binding of secreted or membrane bound molecules (*e.g.*, hormones, growth factors, neurotransmitters and pheromones). Some receptors, however, are activated by changes in the concentration of a metabolite (*e.g.*, oxygen or nutrients) or by physical stimuli (*e.g.*, light, touch, heat). In bacterium *E. coli*, for example, receptors in the cell surface membrane trigger signalling pathways that help the cell respond to changes in the external level of phosphate and other nutrients.

Signalling Pathways in *E. coli*

Bacteria (*e.g.*, *E. coli*) have two-component regulatory systems, one protein acts as a **sensor**, monitoring the level of nutrients or other components in the environment. The other protein acts as **response regulator** which in phosphorylated state binds to DNA regulatory sequences, thereby stimulating or repressing transcription of specific genes (Fig. 26.1).

For example, in *E. coli* two membrane proteins **PhoR** and **PhoB** are found to regulate transcription in response to concentration of free phosphate. PhoR is a transmembrane protein (sensor), located in the inner (*i.e.*, plasma) membrane, whose periplasmic domain binds phosphate with moderate affinity and whose cytosolic domain has protein kinase activity; PhoB (*i.e.*, response regulator) is a cytosolic protein.

Large protein pores in the *E. coli* outer membrane allow ions to diffuse freely between the external environment and the periplasmic space. Consequently, when the phosphate concentration in the environment falls, it also falls in the periplasmic space, causing phosphate to dissociate from the PhoR periplasmic domain (Fig. 26.1). This causes a conformational change into the PhoR cytoplasmic domain that activates its protein kinase activity. The activated PhoR initially transfers a γ-phosphate from ATP to a histidine side chain in the PhoR kinase domain itself. The same phosphate is then transferred to a specific aspartic acid side chain in PhoB; converting it from an inactive to an active transcriptional activator. Phosphorylated, active PhoB then induces transcription from several genes (*e.g.*, *phoA*, *phoS*, *phoE* and *ugpB*) that help the cell cope with low phosphate condition.

26.1. TYPES OF CHEMICAL SIGNALS

A variety of compounds can function as chemical messengers that relay signals between cells (Fig. 26.2). Signalling molecules are often classified based on the distance between their site of production and the target tissue(s) upon which they act.

1. Endocrine signals. These messengers are called **hormones** and they act as **endocrine signals** (endocrine = from Greek words that mean to secrete into). They are produced at great distances from their target tissues, and are carried by the circulatory system to various sites of the body.

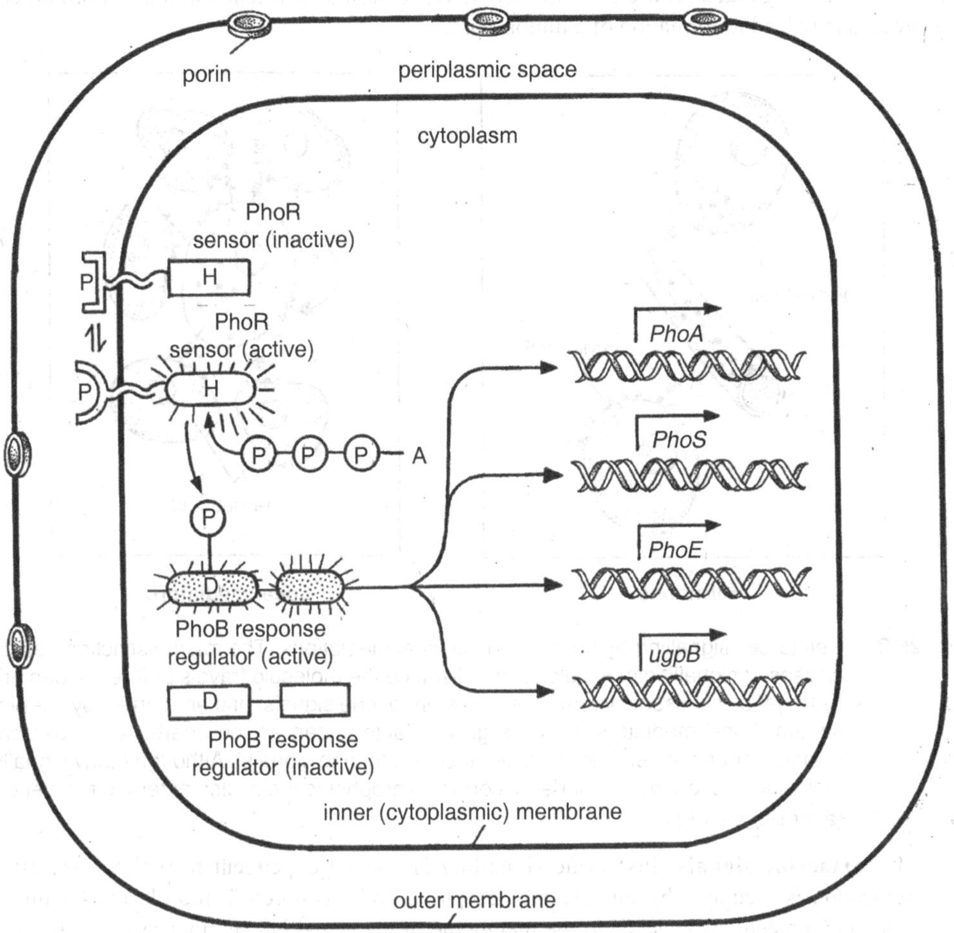

Fig. 26.1. The PhoR/PhoB two-component regulatory system in *E. coli* (see text for explanation of figure, after Lodish *et. al.*, 2004).

2. Paracrine signals. Some signals, such as growth factors, are released locally, where they diffuse to act at short range on nearby tissues. Such signals are referred to as **paracrine signals** (paracrine = form the Greek *para*, which means "beside"). The conduction by a neurotransmitter of a signal from one nerve cell to another or from a nerve cell to a muscle cell (inducing or inhibiting muscle contraction) occurs via paracrine signalling. Many growth factors regulating development in multicellular organisms also act at short range. Some of these molecules bind tightly to the extracellular matrix, unable to signal, but subsequently can be released in an active form. Many developmentally important signals diffuse away from the signalling cell, forming a concentration gradient and inducing various cellular responses depending on their concentration at a particular target cell.

3. Autocrine signals. Some local mediators act on the same cell that produces them; such signals are called **autocrine signals**. Some growth factors act in this fashion. Cultured cells often secrete growth factors that stimulate their own growth and proliferation. Autocrine signalling is particularly common in tumor cells, many of which overproduce and release growth factors that stimulate inappropriate, unregulated proliferation of themselves as well as adjacent nontumor (normal) cells; this process may lead to formation of a tumor mass.

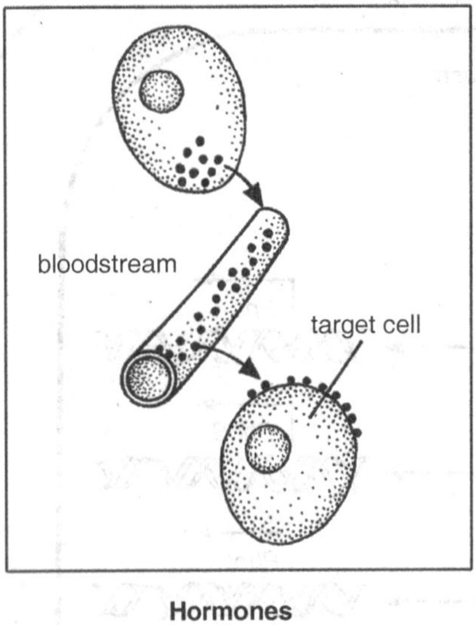

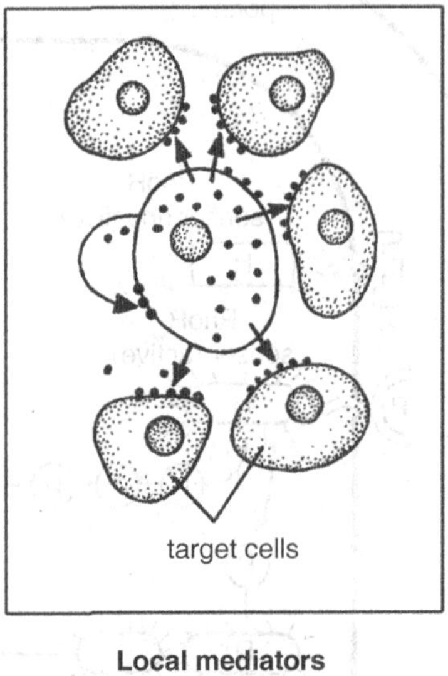

Fig. 26.2. Cell-to-cell signalling by hormones and local mediations. The main distinction between classes of signalling molecules is the distance the molecule travels before encountering its target cell or tissue. Hormones act as endocrine signals, and are carried by the blood stream. Local mediators, such as growth factors, can act on nearby cells (paracrine signals), or on the cell that produces them (autocrine signals). Although many signalling molecules act on the cell surface, certain hydrophobic molecules enter their target cells (after Becker *et al.*, 2006).

4. Juxtacrine signals. Juxtacrine signalling is contact-dependent in which intracellular communication is mediated by ligands and receptors, both anchored in the plasma membrane. The signalling molecule is attached to the membrane of the signalling cell that receives the signal (see **Rastogi**, 2005).

Some signalling molecules can act at both short range and long range. **Epinephrine,** for example, functions as a neurotransmitter (paracrine signalling) and as a systemic hormone (endocrine signalling). Another example is epidermal growth factor (EGF), which is synthesized as an integral plasma membrane protein. Membrane bound EGF can bind to and signal an adjacent cell by direct contact. Cleavage by an extracellular protease releases a soluble form of EGF, which can signal in either an autocrine or a paracrine manner.

Primary and Secondary Messengers

Ligands are the **first** or **primary messengers**. Once a messenger reaches its target tissue, it binds to receptors on the surface of the target cells, initiating the signalling process (Fig. 26.3). A molecule coming from either a long or a short distance functions as a ligand by binding to a receptor. A ligand often binds to a receptor embedded within the plasma membrane of the cell receiving the signal. In case of steroid hormones, the ligand binds to a receptor inside the cell.

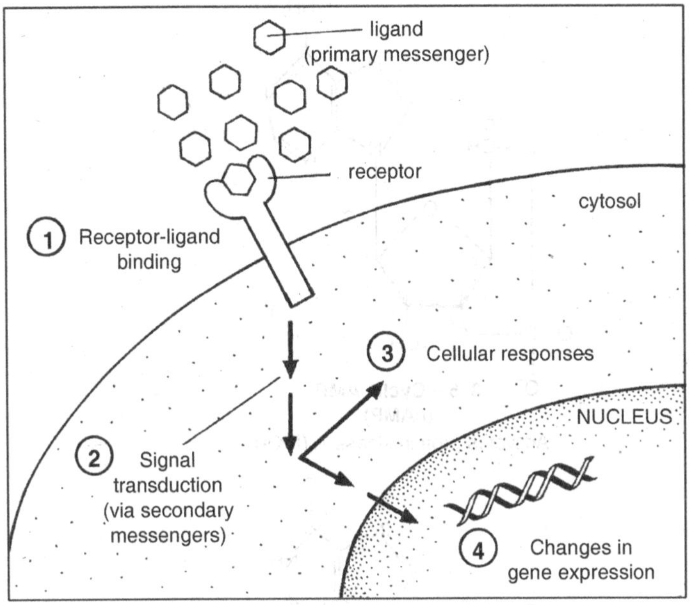

Fig. 26.3. The overall flow of information during cell signalling. Binding of ligand, the primary messenger, by a receptor activates a series of events known as signal transduction, which relays the signal to the interior of the cell, resulting in specific cellular responses and/or changes in gene expression (after Becker *et al.*, 2006).

The binding of ligand to receptor often results in a short-lived increase (or decrease) in the concentration of certain low-molecular weight intracellular signalling molecules termed **second** or **secondary messengers**. These molecules include cAMP, cGMP, DAG and IP_3 (Fig. 26.4). Other important second messengers are Ca^{2+} and various inositol phospholipids, also called **phosphoinositides**, which are embedded in cellular membranes. Second messengers relay the signals from one location in the cell, such as the plasma membrane, to the interior of the cell, initiating a cascade of chances within the receiving cell. Often these events affect the expression of specific genes within the receiving cell. The ability of a cell to translate a receptor-ligand interaction to changes in its behaviour or gene expression is the **signal transduction** pathways.

Signal transduction pathways. When a ligand (first messenger) binds to its specific (or cognate) receptor, the receptor is altered in a way that causes changes in cellular activities. In general, the binding of a ligand either induces a change in receptor conformation or causes receptors to cluster together. Once one of these changes takes place, the receptor initiates a pre-programmed sequence of events inside the cell. By *preprogammed* cell biologists mean that cells have a greater repertoire of functions than are in use at any particular time. Some of these cellular processes remain unused until particular signals are received that trigger them (see **Becker** *et al.*, 2006).

Signal amplification. Signal transduction pathways allow another important aspect of a cell's response to an external signal; it is called **signal amplification**. Often exceedingly small quantities of a ligand are sufficient to elicit a response from a target cell, yet the responding cell however responds in dramatic ways. Often the strong response of the target cell results from a signalling cascade with the responding cell. At each step in the cascade, a signalling intermediate persists long enough to stimulate the production of many molecules required for the next step in the cascade, thereby multiplying

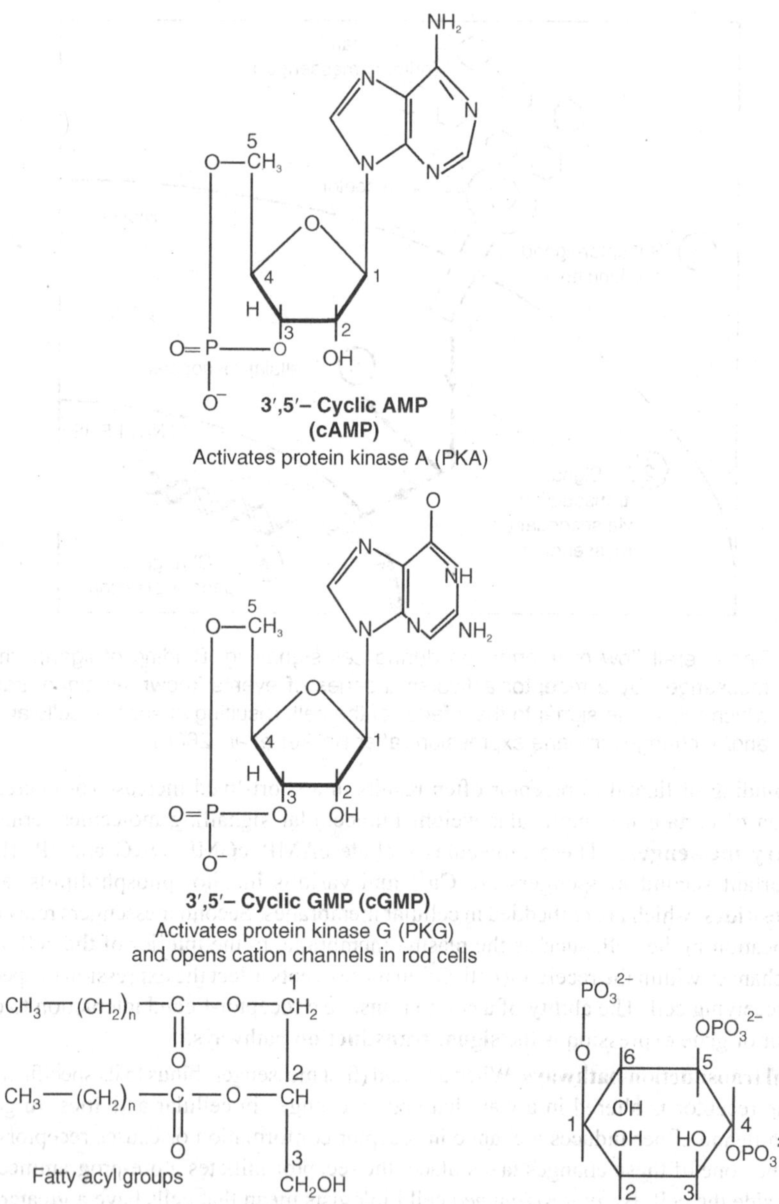

3′,5′– Cyclic AMP (cAMP)
Activates protein kinase A (PKA)

3′,5′– Cyclic GMP (cGMP)
Activates protein kinase G (PKG) and opens cation channels in rod cells

1,2-Diacylglycerol (DAG)
Activates protein kinase C (PKC)

Inositol 1,4,5-trisphosphate (IP_3)
Opens Ca^{2+} channels in the endoplasmic reticulum

Fig. 26.4. Four common intracellular second messengers. The major direct effect or effects of each compound are indicated below its structural formula (after Lodish *et al.*, 2004).

the effects of a single receptor/ligand interaction on the cell surface. A well-known example of such signal amplification involves the **breakdown of glycogen** in liver cells in response to the hormone *epinephrine* (Fig. 26.5). This process involves a G protein-linked receptor on the cell surface, which in turn initiates a signalling cascade within the liver cell. As a result of the cascade, a single epinephrine ligand can elicit the production of hundreds of millions of glucose molecules.

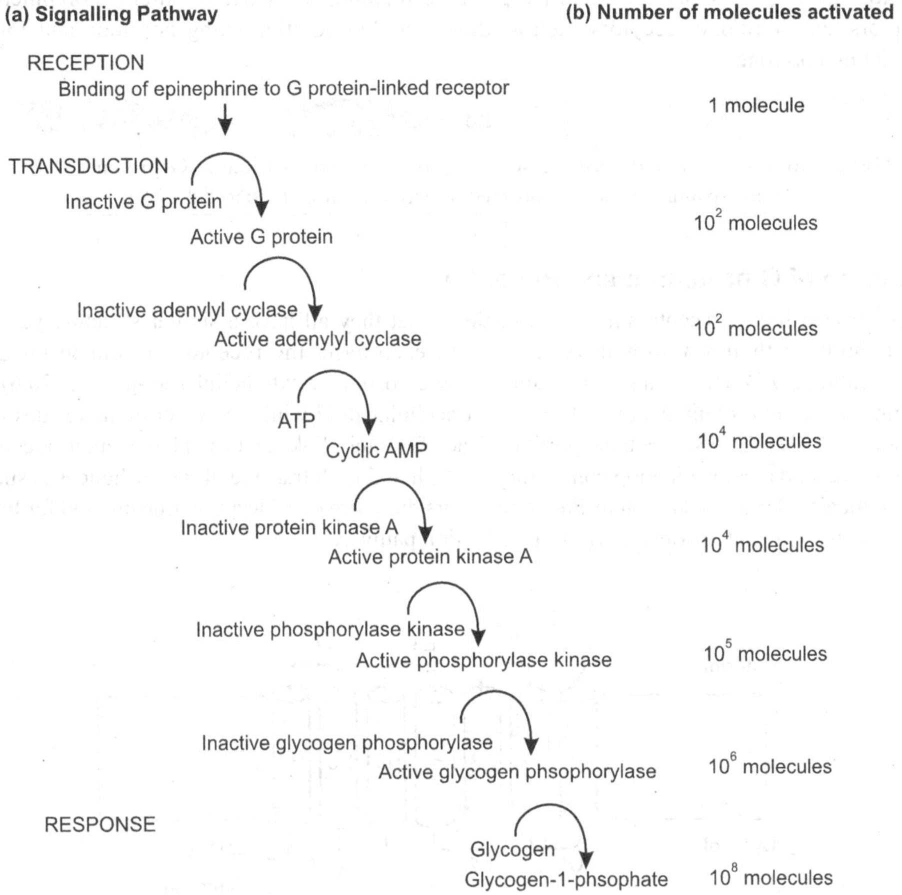

Fig. 26.5. Signal transduction pathways can amplify the cellular response to an external signal. Liver cells respond to hormone epinephrine by breaking down glycogen to liberate glucose-1-phosphate. (*a*) The epinephrine receptor is a G protein-linked receptor, which activates an enzyme known as adenylyl cyclase. Adenylyl cyclase catalyses formation of a second messenger, cAMP, which activates protein kinase (protein kinase A), which in turn activates another kinase (phosphorylase kinase). Ultimately, the enzyme glycogen phosphorylase is activated, which catalyses the breakdown of glycogen. (*b*) The approximate number of molecules produced at each step is shown on the right. One epinephrine molecule is capable of triggering the production of hundreds of millions of glucose-1-phosphate molecules (after Becker *et al.*, 2006).

26.2. G PROTEIN-LINKED RECEPTORS (OR G PROTEIN-COUPLED RECEPTORS OR GPCRs)

G protein-linked receptor family (Box 26.2) is so named because ligand binding causes a change in receptor conformation that activates a particular **G protein** (an abbreviation for **guanine-nucleotide binding protein**). The activated G protein in turn binds to a target protein such as an enzyme or a channel protein, thereby altering the target's activity. All G protein-linked receptors initiate signal transduction inside the cell in this way. Examples of G protein-linked receptors include olfactory receptors (responsible for human's sense of smell), norepinephrine receptors, and hormone receptors such as those for thyroid-stimulating hormone and follicle-stimulating hormone.

Box 26.2

GPCRs are transmembrane cell-surface receptors coupled to trimeric G proteins. All GPCRs have seven membrane spanning regions connected by extracellular and intracellular loops.

Structure of G protein-linked Receptor

The G protein-linked receptors are remarkable in that they all have a similar structure yet differ significantly in their amino acid sequences. In each case, the receptor protein forms seven transmembrane α helices connected by alternating cytosolic or extracellular loops (Fig. 26.6). The N-terminus of the protein is exposed to the extracellular fluid while the C-terminus resides in the cytosol (Fig. 26.7). The extra-cellular portion of each G protein-linked receptor has a unique messenger binding site, and a cytosolic loop connecting the fifth and sixth transmembrane α helices is specific for a particular G protein. G protein-linked receptors therefore provide a versatile method for linking different messengers to different signal transduction pathways.

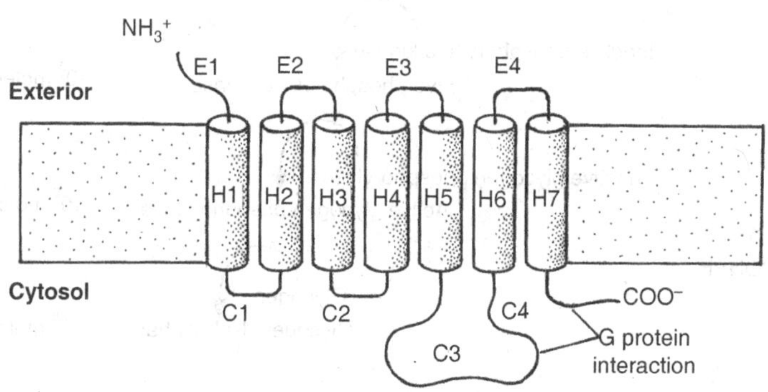

Fig. 26.6. Schematic diagram of the general structure of G protein-linked receptors. All receptors of this type have the same orientation in the membrane and contain seven transmembrane α-helical regions (H1 to H7), four extracellular segments (E1-E4), and four cytosolic segments (C1-C4). The carboxyl-terminal segment (C4), the C3 loop, and, in some receptors, also the C2 loop are involved in interactions with a linked/coupled trimeric G protein (after Lodish *et al.*, 2004).

Nature of G proteins. G proteins can be described as a type of molecular switch whose "on" or "off" state depends on whether the a protein is bound to GTP or GDP. There are two distinct classes of G proteins: the **large heterotrimeric G proteins** and the **small monomeric G proteins.** The large heterotrimeric G protein contains three different subunits, called G alpha (Gα), G beta (Gβ) and G gamma (Gγ). Heterotrimeric G protein mediates signal transduction through G protein-linked receptors. The small monomeric G proteins include **Ras** (protein) (Box 26.3).

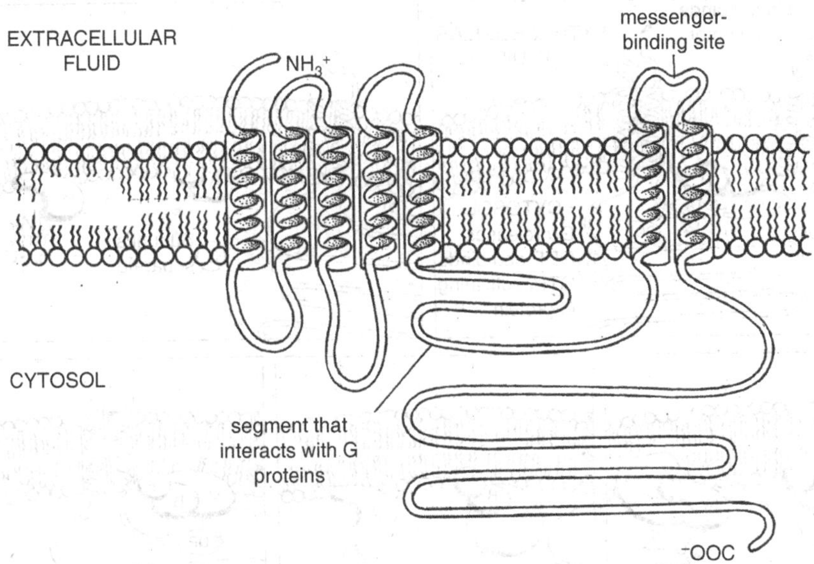

Fig. 26.7. The structure of G protein-linked receptor. Each G protein-linked receptor has seven transmembrane α helices. The primary messenger binds to the extracellular portion of the receptor. This binding causes an intracellular portion of the receptor to activate an adjacent G protein (after Becker *et al.*, 2006).

Box 26.3 Ras protein

The most famous member of a large family of GTP-binding proteins (called **monomeric GTPases**) that help relay signals from cell surface receptors to the nucleus. Named for the *ras* gene, first identified in viruses that cause rat sarcomas (**Alberts** *et al.*, 2002).

All G proteins have the same basic structure and mode of activation. Of the three subunits in the Gαβγ heterotrimer, Gα, the largest, binds to a guanine nucleotide (GDP or GTP). When Gα binds to GTP, it also detaches from the Gβγ complex. The Gβ and Gγ subunits, on the other hand, are permanently bound together. Some G proteins, such as **Gs,** act as stimulators of signal transduction (hence *s*, for stimulatory); others, such as G*i*, act to inhibit signal transduction (hence *i*, for inhibitory).

When a messenger (ligand) binds to a G protein-linked receptor on the surface of the cell, the change in conformation of the receptor causes the G protein to associate with the receptor, which in turn causes the *G*α subunit to release its bound GDP, acquire a GTP, and then detach from the complex (Fig. 26.8). Depending on the G-protein and the cell type, either the free GTP-Gγ subunit or the Gβγ complex can then initiate signal transduction events in the cell. Each portion of the G protein exerts its effect by binding to a particular enzyme or other protein in the cell. In some cases, both the GTP-Gγ and GTP-Gβγ, subunits simultaneously regulate different processes in the cytosol. However, the activity of the G protein persists only as long as the Gα is bound to GTP and the subunits remain separated. Because the Gα subunit hydrolyzes GTP, it will only remain active for a short time before reverting to the GDP-bound state and reassociated with Gβγ. This characteristic allows the signal transduction pathway to shut down rapidly when the messenger is removed.

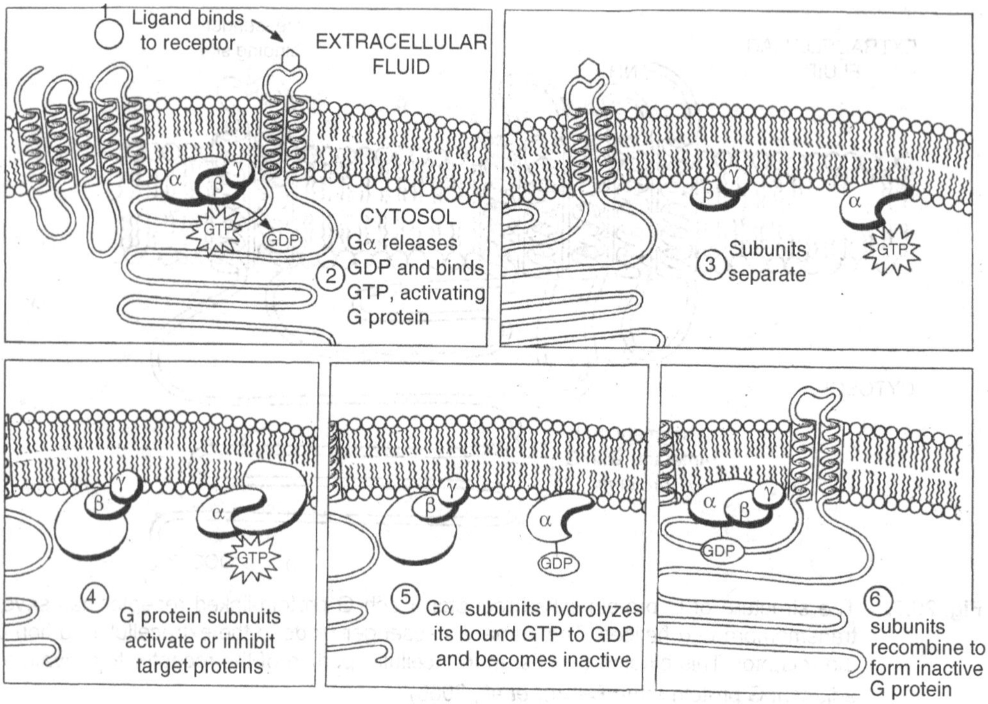

Fig. 26.8. The G protein activation and inactivation. 1. When the ligand binds, 2. The receptor activates a G protein by causing the Gα subunit to release GDP and acquire GTP, 3. The Gα and Gβγ subunits then separate and 4. Initiate signal transduction events. 5. The GTP-Gα subunit eventually hydrolyzes its bound GTP, converting the subunit back to its inactive GTP-Gα form. 6. The inactive GTP-Gα subunit then recombines with Gβγ to form the inactive heterotrimeric G protein (after Becker *et al.*, 2006).

Role of Cyclic AMP in Signal Transduction by G protein

The second messenger, the cyclic AMP (cAMP) is formed from cytosolic ATP by the enzyme **adenylyl cyclase** (Fig. 26.9) anchored in the plasma membrane with its catalytic portion protruding into cytosol. Normally, the enzyme is inactive until it binds to the activated α subunit of a specific G protein, Gs. When the G protein-linked receptor is coupled to Gs, the binding of ligand stimulates the Gsα subunit to release GDP and acquires GTP (Fig. 26.10). This in turn causes GTP-Gsα to detach from the Gsβγ subunits and bind to adenylyl cyclase, the enzyme becomes active and converts ATP to cAMP.

The G proteins respond quickly to changes in ligand concentration because they remain active for only a short period of time before the Gα subunit hydrolyzes its bound GTP and converts to the inactive state. Once the G protein becomes inactive, the adenylyl cyclase cease to make cAMP. However, cAMP levels would still remain elevated in the cell if not for the enzyme **phosphodiesterase**, which degrades cAMP. This process further ensures that the signal transduction pathway will shut down promptly when the concentration of the ligand outside the cell falls.

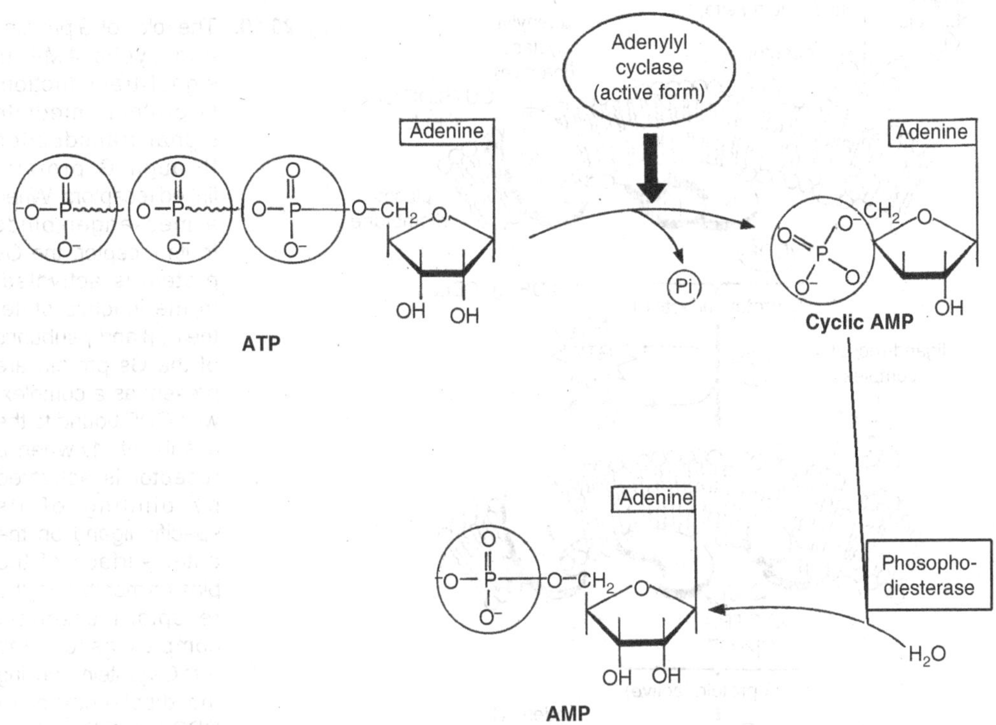

Fig. 26.9. The structure and metabolism of cAMP. Cyclic AMP (adenosine-3′, 5′-cyclic monophosphate) is generated from ATP in a reaction catalyzed by the active form of the enzyme adenylyl cyclase; it is inactivated by hydrolysis to AMP, a reaction catalyzed by the enzyme phosphodiesterase. Adenylyl cyclase is a membrane-bound enzyme, whereas phosphodiesterase is located in the cytosol (after Becker *et al.*, 2006).

Cyclic AMP has one main intracellular target, an enzyme known as cAMP-dependent kinase, or **protein kinase A (PKA)**. PKA enzyme phosphorylates a wide variety of cellular proteins by transferring phosphate from ATP to a serine or threonine found within the target protein. The cAMP tends to regulate the activity of PKA.

An increase in cAMP concentration can produce many different effects in different cell types. 1. When cAMP is elevated in skeletal and liver cells, the breakdown of glycogen is stimulated. 2. In cardiac muscles, the elevation of cAMP strengthens heart contraction; 3. Whereas in smooth muscle contraction is inhibited. 4. In blood platelets, the elevation of cAMP inhibits their mobilization during blood clotting, and 5. In intestinal epithelial cells, it causes the secretion of salts and water into the lumen of the gut. Each of these reactions is an example of the preprogrammed response discussed earlier.

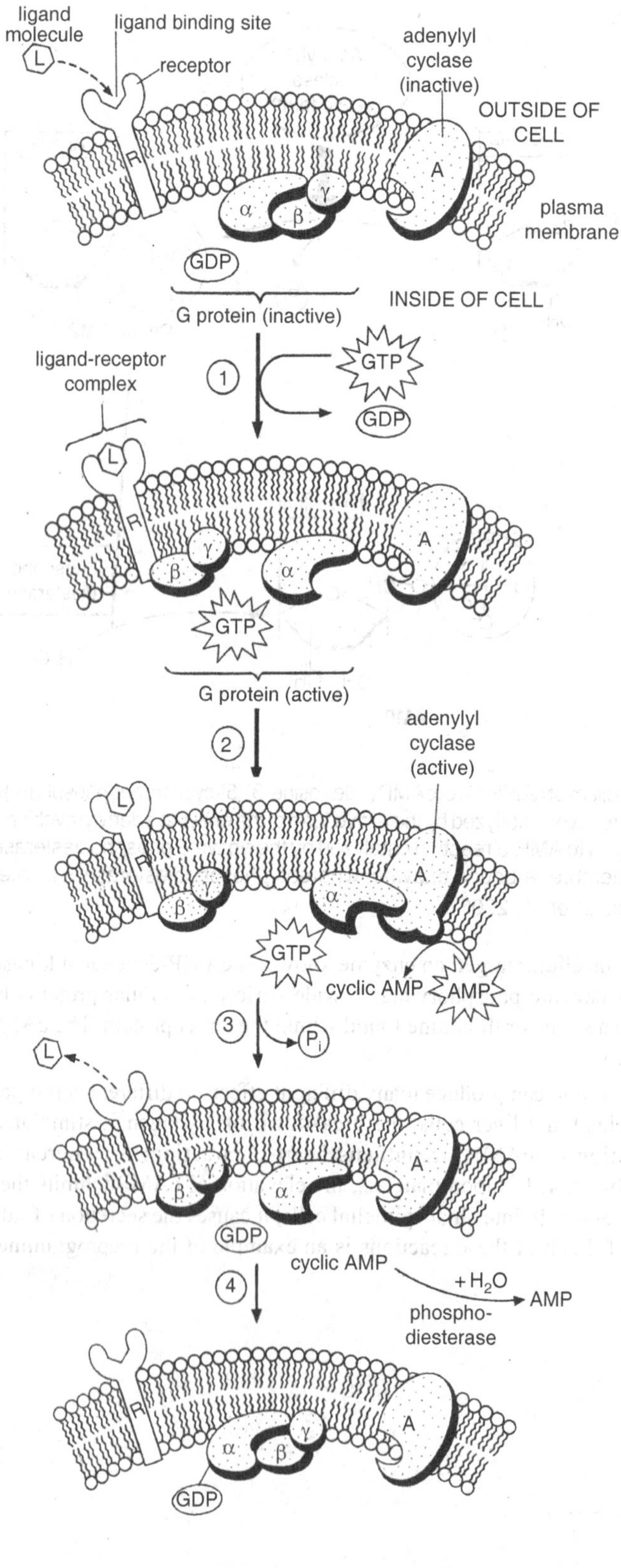

Fig. 26.10. The roles of G proteins and cyclic AMP in signal transduction. G proteins mediate signal transduction through G protein-linked receptors. When a messenger binds to its receptor, the Gs protein is activated. In the inactive state, the α, β and γ subunits of the Gs protein are present as a complex, with GDP bound to the α subunit. 1. when a receptor is activated by binding of its specific ligand on the outer surface of the plasma membrane, the receptor-messenger complex associates with Gs protein, causing the displacement of GDP by GTP and the dissociation of Gsα-GTP complex. 2. The GTP-Gsα complex then binds tightly to a molecule of membrane-bound adenylyl cyclase, activating it for synthesis of cAMP. 3. Activation ends when the ligand leaves the receptor, the GTP is hydrolyzed to GDP by the GTPase activity of the Gsα subunit, and the Gsa dissociates from the adenylyl cyclase. 4. Adenylyl cyclase then reverts to the inactive form, the Gsα reassociates with the Gsβγ complex, and cAMP molecules in the cytosol are hydrolyzed to AMP by enzyme phosphodiesterase (after Becker *et al.*, 2006).

Box 26.4

The intracellular level of cAMP can be raised without the ligands, for example, the enzyme phosphodiesterase is inhibited by **methylxanthine**, compounds such as *caffeine* and *theophylline*, found in coffee, tea and soft drinks. Theophylline is often used to treat **asthma** because it relaxes bronchial smooth muscles (see **Becker** *et al.*, 2006).

What would happen if the G protein-adenylyl cyclase system could not be shut off? Two human diseases caused by pathogenic bacteria are found to be due to certain defects in the heterotrimeric G protein (Box 26.5).

Box 26.5 Pathogenic toxins as cell biological tools

Both G*s* and G*i* proteins are targets for some medically important bacterial toxins. **Cholera toxin**, which is produced by the bacterium that causes cholera (*Vibrio cholere*) is an enzyme that catalyse the transfer of ADP ribose from intracellular NAD^+ to the α subunit of G*s* (*i.e.*, G*s*α). This ADP ribosylation alters the α subunit so that it can no longer hydrolyse its bound GTP, causing it to remain in an active state that stimulates adenylyl cyclase indefinitely. The resulting prolonged elevation in cAMP level within intestinal epithelial cells causes a large efflux of Cl^- and water into the gut, thereby causing the severe diarrhoea that characterizes cholera.

Pertussis toxin, which is made by the bacterium (*Bordetella pertussis*) that causes pertussis (whooping cough), catalyses the ADP ribosylation of the α subunit of G*i* (*i.e.*, G*i*α), preventing the subunit from interacting with receptors; as a result, this α subunit retains its bound GDP and is unable to regulate its target proteins. These two toxins are widely used as tools to determine whether a cell's response to a signal is mediated by the G*s* or by G*i* (see **Alberts** *et al.*, 2002).

Role of IP_3 and DAG as Second Messengers

The importance of inositol phospholipids in cell signalling was first brought to light in the pioneering studies of **Robert Michell** and **Michael Berridge**. In the early 1980s, **Berridge** noticed that when the salivary glands of certain insect larvae were stimulated to secrete, changes occurred in membrane inositol phospholipids. It is now known that **inositol 1, 4, 5-triphosphate (IP_3)**, one of the breakdown products of inositol phospholipids functions as a second messenger. IP_3 is generated from **phosphatidylinositol-4, 5-biphosphate (PIP_2)**, a relatively uncommon membrane phospholipids, when the enzyme **phospholipase C** is activated. Phospholipase C cleaves PIP_2 into two molecules: inositol triphosphate (IP_3) and **diacylglycerol (DAG)**. Later on, IP_3 and DAG were shown to be *second messengers* in a variety of regulated cell functions (Table 26.1).

Table 26.1 Examples of cell functions regulated by inositol triphosphates (IP_3) and diacylglycerol (DAG) (source: Becker *et al.*, 2006).

	Regulated function	Target tissue	Messenger
1.	Platelet activation	Blood platelets	Thrombin
2.	Muscle contraction	Smooth muscle	Acetylcholine
3.	Insulin secretion	Pancreas endocrine	Acetylcholine
4.	Amylase secretion	Pancreas exocrine	Acetylcholine
5.	Glycogen degradation	Liver	Antidiuretic hormone
6.	Antibody production	B lymphocytes	Foreign antigens

The functions of IP_3 and DAG as second messengers in the inositol-phospholipid-calcium pathway are shown in Fig. 26.11. The sequence begins with the binding of a ligand to its membrane receptor, leading to the activation of a specific G protein, called Gp (the "p" stands for activation of phospholipase C). Gp then activates a form of phospholipase C known as Gβ, thereby generating both IP_3 and DAG.

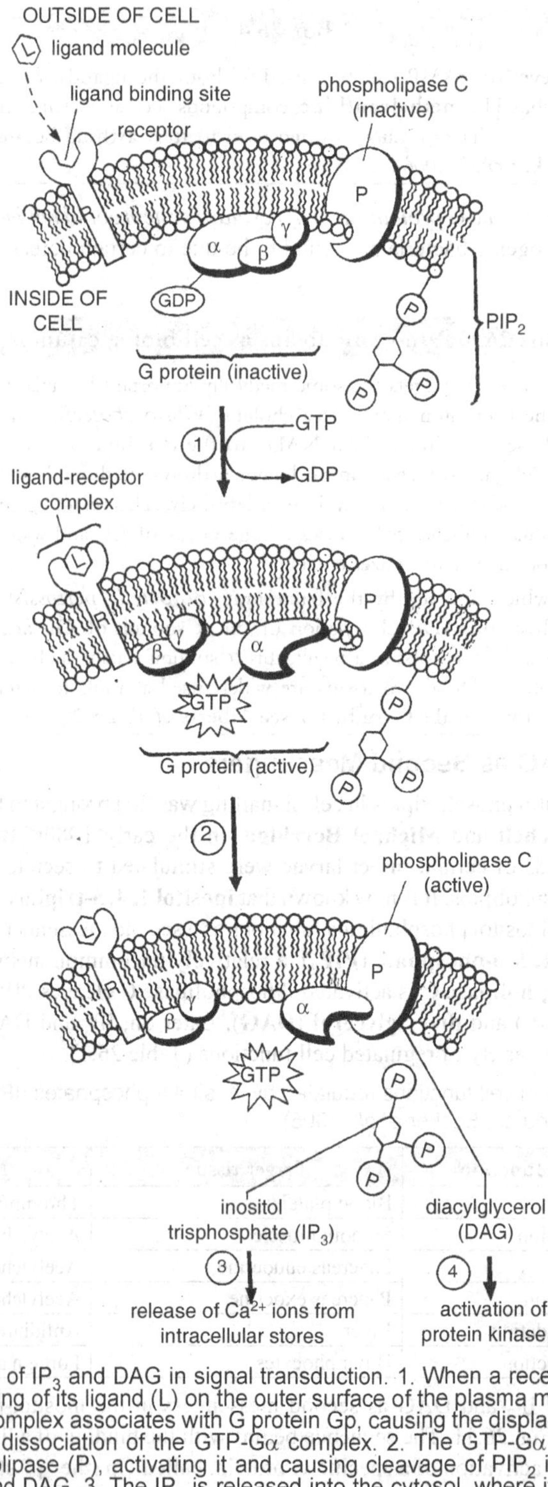

Fig. 26.11. The role of IP_3 and DAG in signal transduction. 1. When a receptor (R) is activated by the binding of its ligand (L) on the outer surface of the plasma membrane, the receptor-ligand complex associates with G protein Gp, causing the displacement of GDP by GTP and the dissociation of the GTP-Gα complex. 2. The GTP-Gα complex then binds to phospholipase (P), activating it and causing cleavage of PIP_2 into one molecule each of IP_3 and DAG. 3. The IP_3 is released into the cytosol, where it triggers the release of calcium. 4. The DAG remains in the membrane, where it activates protein kinase C (after Becker et al., 2006).

IP_3 is water-soluble, so quickly diffuses through the cytosol, binding to a ligand-gated calcium channel known as the **IP_3 receptor channel** in the endoplasmic reticulum. When IP_3 binds, the channel opens, releasing calcium ions into the cytosol. Calcium then binds to a protein known as **calmodulin**, and the calcium-calmodulin complex activates the desired physiological process.

The DAG remains in the membrane, where it activates members of the **protein kinase C (PKC)** family of enzymes. PKC can then phosphorylates specific serine and threonine groups on a variety of target proteins, depending on the cell type. A variety of cellular effects have been linked to the activation of protein kinase C, including 1. The stimulation of cell growth; 2. The regulation of ion channels; 3. Changes in the cytoskeleton; 4. Increase in cellular pH, and 5. Effects on secretion of proteins and other substances.

Regulation of Calcium Ion Concentration in Cytosol

The calcium ions (Ca^{2+}) play an essential role in regulating a variety of cellular functions. An overview of the various mechanisms of calcium is normally maintained at very low levels in the cytosol because of the presence of **calcium ATPase** in the plasma membrane and the endoplasmic reticulum. Calcium ATPases in the plasma membrane transport calcium out of the cell, where as the calcium ATPase in the ER sequester (Latin for "commit for safekeeping") calcium ions in the lumen of ER. In addition, some cells have sodium-calcium exchangers that further reduce the cytosolic calcium concentration. Finally, mitochondria can transport calcium into the mitochondrial matrix. For most cells in their resting state, the action of calcium ATPase maintains the concentration of the calcium in cytosol at about 1×10^{-4} mM.

Cytosolic calcium concentrations can be increased by various methods. One way is by opening of calcium channels in the plasma membrane of neurons. The calcium concentration in the extracellular fluid and the blood is about 1.2 mM, more than 10,000 times as high as that of the cytosol. As a result, when calcium channels open, calcium ions rush into the cell.

The levels of calcium ions in the cytosol can also elevated by the release of calcium from intracellular stores. Calcium ions sequestered in the ER can be released through the IP_3 receptor channel, discussed earlier, and through the **ryanodine receptor channel**. The ryanodine receptor channel, so named because it is sensitive to the plant alkaloid **ryanodine**, is particularly important for calcium release from the sarcoplasmic reticulum of cardiac and skeletal muscle, but nonmuscle cells such as neurons also have ryanodine calcium release channels.

Calcium induced calcium release. Quite interestingly, both the ryanodine receptor channel and IP_3 receptors are sensitive to calcium itself. When a neuron is depolarized, for example, calcium channels in the plasma membrane open and allow some calcium to enter the cytosol. Upon exposure to a rapid increase in calcium ions, the ryanodine receptors channel opens, allowing calcium to escape from the ER into the cytosol.

Function of calmodulin. Calcium can bind directly to some proteins and thereby alter their activity, but more often it works through the protein **calmodulin**. This protein has pivotal function in the regulation of a variety of cellular processes.

How does calmodulin mediate calcium activated events in the cell? The calmodulin molecule has been compared to a flexible "arm" with a "hand" at each end. Two calcium ions bind at each of two "hand" regions, causing the calmodulin to undergo a change in shape that form the active **calcium-calmodulin complex** (Fig. 26.12). When a protein is present that contains a calmodulin-binding site, the hands and arm bind to it by wrapping around the binding site (Fig. 26.12).

Calmodulin protein is found to have a remarkable affinity for calcium: calmodulin binds to calcium when the cytosolic calcium concentration increases to about 10^{-3} m*M*, but it releases calcium when cytosolic calcium levels decline back to the resting level of 10^{-4} m*M*. Thus calmodulin protein is uniquely suited to operate within the typical range of cytosolic calcium concentrations.

Most calmodulin binding proteins are enzymes such as *protein kinases* and *protein phosphatases*. The response of a target cell to an increase in calcium concentration depends on the particular calmodulin-binding proteins that are present in the cell. *This means that the same change in calcium concentration can produce markedly different effects in two target cells if each possesses different calmodulin-sensitive enzyme system.*

Function of calcium in slow block to polyspermy. The process of fertilization of animal eggs (*i.e.*, ova) is an outstanding example of the importance of calcium-mediated signal transduction following a receptor-ligand interaction. In many animals, when sperm that have undergone prior steps of activation bind to the surface of mature eggs and unite with them at fertilization, a striking sequence of events arises. One of the early responses of the egg-within 30 seconds to several minutes after fertilization is the release of calcium from internal stores. Calcium release occurs initially at the site where the sperm penetrates the egg surface, and then spreads across the egg, much as a ripple on the surface of a pond spreads away from the site where a pebble strikes the water. The wavelike propagation of calcium release can be visualized using calcium indicators (*e.g.*, **aequorin**, which is a dye that when injected in fertilizing fish egg, emits light when bound to free Ca^{2+} ions in the cytosol).

Release of calcium is necessary for two crucial events. First, it stimulates the fusion of vesicles, known as **cortical granules**, with the egg plasma membrane, resulting in the release of granule contents outside the cell (exocytosis; Fig. 26.13). Cortical granules contain several proteins and enzymes, the release of which results in alterations of the protein coat surrounding most eggs (typically known as **vitelline envelope**). These changes make the egg unable to bind additional sperm thereby preventing more than one sperm from fertilizing the egg. This process is known as **slow block to polyspermy** (An earlier process called **fast block to polyspermy** involves a transient depolarization of the egg plasma membrane).

The second major function of calcium is **egg activation**. Egg activation involves the restart of many metabolic processes, the reorganization of

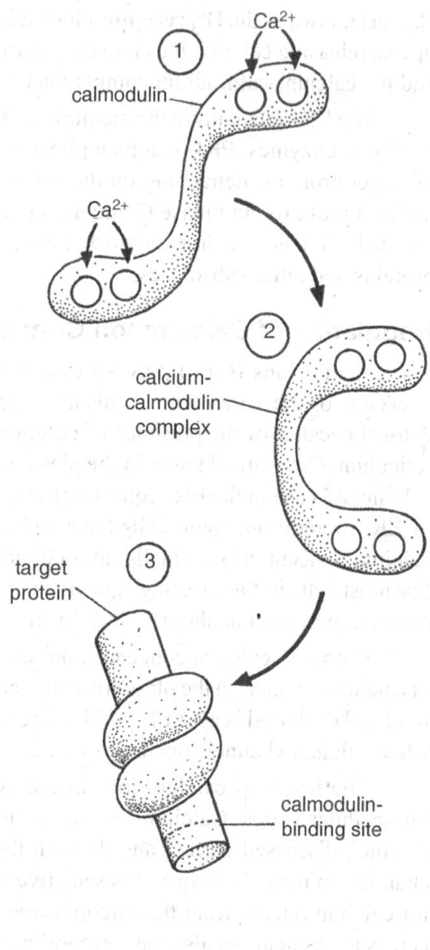

Fig. 26.12. The structure and function of the calcium-calmodulin complex. A—Calmodulin is a cytosolic calcium-binding protein. This model of molecular structure of calmodulin is based on data from X-ray crystallography. The molecule consists of two globular ends ("hands") joined by a helical region ("arm"). Each end has two calcium binding sites. B—To form a functional calcium-calmodulin complex, 1. Calmodulin binds four calcium ions. 2. Calmodulin changes conformation, resulting in an active complex. 3. The two globular "hands" of the complex wrap around a calmodulin binding site on a target protein (after Becker *et al.*, 2006).

the internal contents of the egg, and other events that initiate the process of embryonic development. Many features of the slow block to polyspermy and egg activation can be initiated by treating unfertilized eggs with calcium ionophore in the sperm, demonstrating the key role of elevated calcium levels within the egg play in its activation.

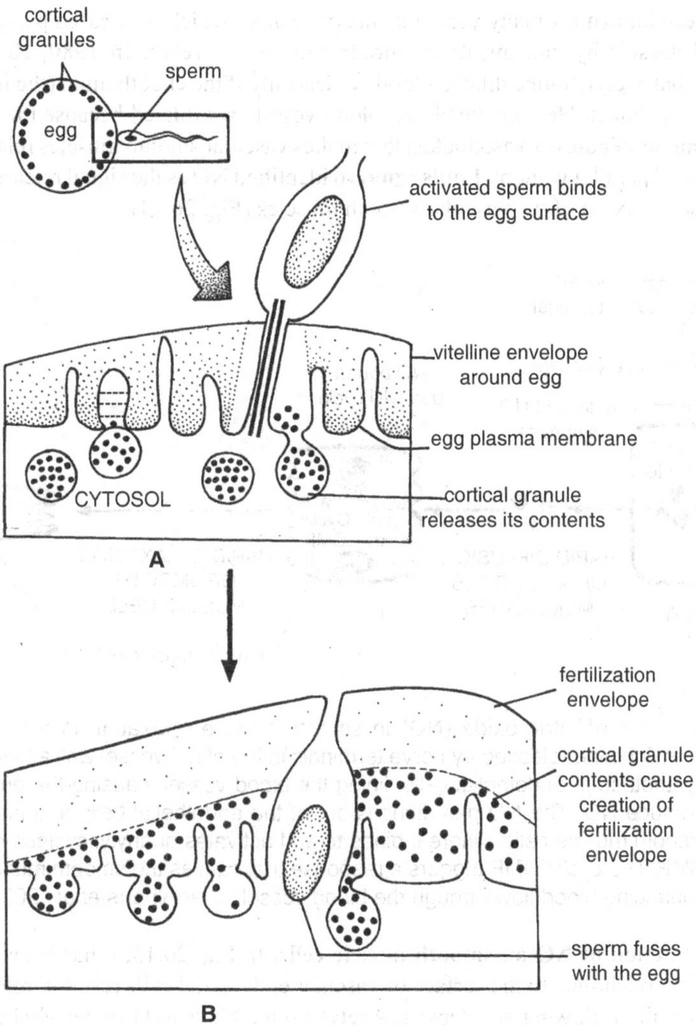

Fig. 26.13. The role of calcium in the slow block to polyspermy in sea urchins. A—A sperm cell that has been activated binds to the egg surface, resulting in local calcium release and exocytosis of cortical granule. B—The result is the creation of the fertilization envelope, which prevents additional sperm from penetrating the egg (after Becker *et al.*, 2006).

Oscillation of Ca^{2+} concentration. The oscillation of calcium concentration over time is found to elicit a cellular response in neurons and fertilized mammalian eggs. This contributes to stable changes in the state of these cells. Calcium oscillations are also important in regulating the opening and closing of stomata in plants.

Function of Nitric Oxide in G protein-linked Receptor

An important signalling molecule in the cardiovascular system is **nitric oxide (NO)**, a toxic short-lived (half life about 5–10 seconds) gas molecule produced by the enzyme **NO synthetase**, which converts the amino acid arginine to NO and citrulline. Nitric oxide's mode of action on blood vessels involves G protein. NO is hydrophobic and small enough to pass readily across target cell plasma membrane.

It has been known for many years that *acetylcholine,* which is released by autonomic nerves, dilates blood vessels by causing their smooth muscles to relax. In 1980, **Robert Furchgott** demonstrated that acetylcholine dilated blood vessels only if the endothelium (the inner lining of the blood vessel, was intact. He concluded that blood vessels are dilated because the endothelial cells produce a **signal molecule** (or vasodilator) that makes vascular smooth muscles relax. In 1986, work by **Furchgott** and parallel work by **Louis Ignarro** identified NO as the signal released by endothelial cells that causes relaxion of the vascular smooth muscles (Fig. 26.14).

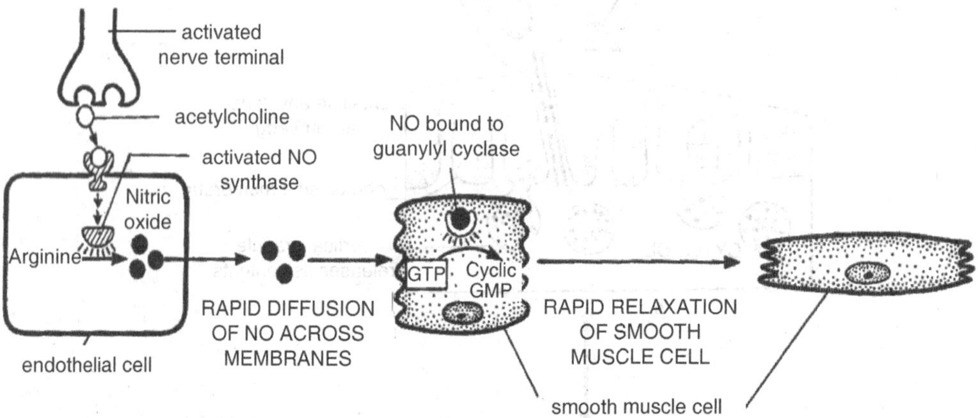

Fig. 26.14. The role of nitric oxide (NO) in smooth muscle relaxation in a blood vessel wall. Acetylcholine released by nerve terminals in the blood vessel wall activates enzyme NO synthetase in endothelial cells, lining the blood vessel, causing the endothelial cells to produce NO. The NO gas diffuses out of the endothelial cells and into the underlying smooth muscle cells, where it binds to and activates guanylyl cyclase to produce cyclic GMP. The cyclic GMP triggers a response that causes the smooth muscle cells to relax, enhancing blood flow through the blood vessel (after Alberts *et al.*, 2002).

Mode of action of NO on smooth muscle cells. In Fig. 26.15, it has been shown that how the binding of acetylcholine to the surface of vascular endothelial cells result in release of NO. This process involves the following six steps: 1. Acetylcholine binds to G protein-linked receptors that activate the phosphoinositide signalling pathway, causing IP_3 to be produced by the endothelial cells. 2. IP_3 causes the release of calcium from the endoplasmic reticulum. 3. The calcium ions bind to calmodulin, forming a complex that stimulates NO synthetase enzyme to produce nitric oxide. 4. Nitric oxide is a gas that readily diffuses through plasma membranes, allowing it to pass from the endothelial cell into the adjacent smooth muscle cell. 5. Once inside the smooth muscle cell, NO activates the enzyme **guanylyl cyclase**, which catalyzes the formation of cyclic GMP (cGMP). Cyclic GMP is derived from GTP in a manner similar to the production of cAMP from ATP, and like cAMP, cGMP concentration activates a protein known as **protein kinase G**, which induces muscle relaxation by catalyzing the phosphorylation of the appropriate muscle proteins.

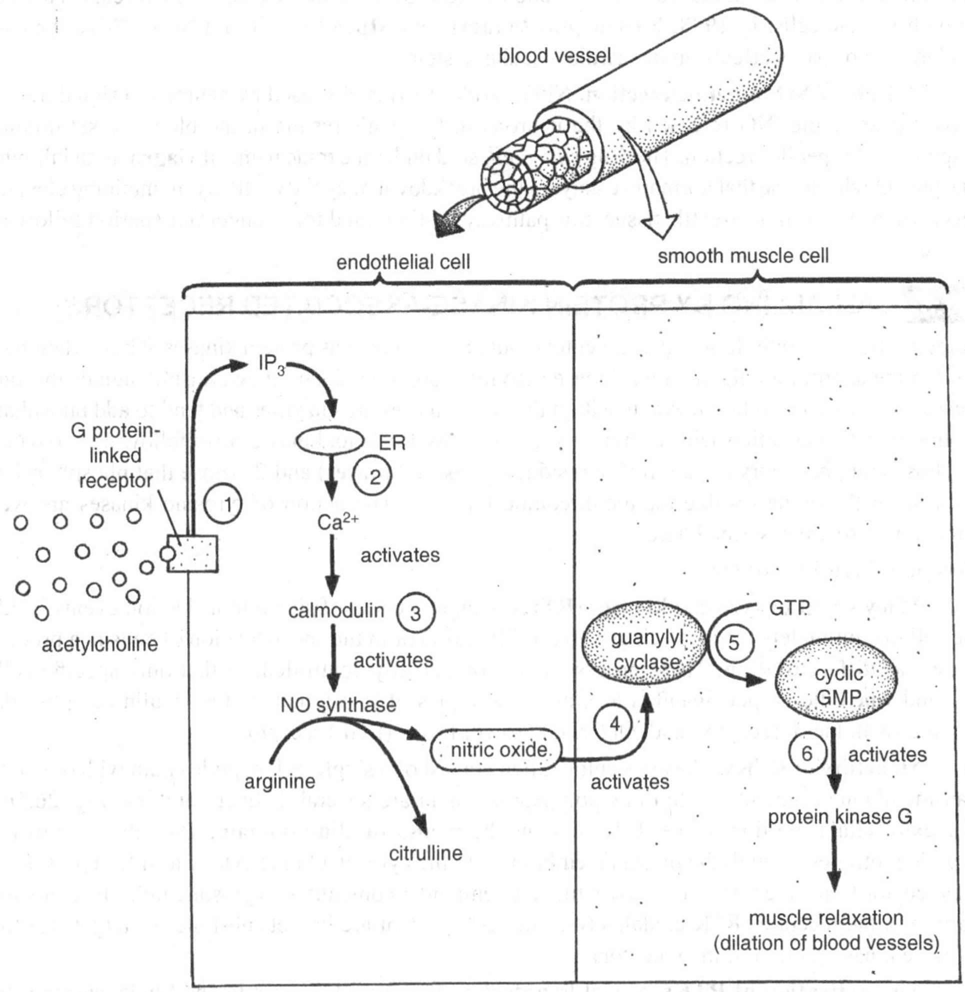

Fig. 26.15. The action of nitric oxide on blood vessels. The binding of acetylcholine to endothelial cells triggers the production of nitric oxide (NO), which diffuses into the adjacent smooth muscle cells and stimulates guanylyl cyclase, thereby leading to muscle relaxation (after Becker *et al.*, 2006).

Box 26.6

Gases such as NO (nitric oxide) and CO (carbon monoxide) are paracrine signaling molecules. They are used as intercellular signals in mammals and certain plants. For example, activated macrophages and neutrophils use NO in killing invading microorganisms. In plants, NO is involved in defensive responses to injury or infection.

Use of NO in treatment of some human ailments.

1. Role of NO in treatment of angina. The mechanism by which acetylcholine stimulation of the endothelial cells leads to smooth muscle relaxation also explains the mechanism of action of **nitroglycerin**. Nitroglycerin is often taken by patients with **angina** (chest pain due to inadequate blood flow to the heart) to relieve constriction of coronary arteries. In 1977, **Ferid Murad** found that nitroglycerin and similar vasodilators elicit release of nitric oxide (NO), which relaxes arterial smooth muscle cells. In 1998, **Furchogott, Ignarro** and **Murad** received a Nobel Prize for their explanation of NO's effects on the cardiovascular system.

2. Role of NO in penile erection. Nitric oxide (NO) is also used by neurons to signal nearby cells. For example, NO released by the neurons in the penis results in the blood vessel dilation responsible for penile erection. The drug *slidenafil,* sold under the trade name of **viagra**, is an inhibitor of a phosphodiesterase that normally catalyzes the breakdown of cyclic GMP. By maintaining elevated levels of cyclic GMP in erectile tissue, this pathway is stimulated for a longer time period following NO release.

26.3. SIGNALLING BY PROTEIN KINASE-ASSOCIATED RECEPTORS

Some proteins not only function as receptors, but are also serve as protein kinases. Thus, when they bind to the appropriate ligand, their kinase activity is stimulated, and they transmit signals through a cascade of phosphorylation events within the cell. Kinases are enzymes and tend to add phosphate groups to particular amino acids within substrate proteins. Receptor kinases are of following two types: 1. Those that phosphorylate a tyrosine residue **(tyrosine kinases)** and 2. Those that phosphorylate a serine or threonine residue (serine/threonine kinases). The action of tyrosine kinases are well understood, so are described here.

Receptor Tyrosine Kinases.

Many **receptor tyrosine kinases (RTKs)** trigger a chain of signal transduction events inside the cell that ultimately leads to cell growth, proliferation or in the specialization of cells in a process known as **differentiation**. These processes are competently controlled, so that only specific cells respond when the proper ligand is available. Examples of RTKs include the insulin receptor, the nerve growth factor receptor, and epidermal growth factor (EGF) receptor.

Structure of RTKs. These receptors often consist of a single polypeptide chain with only one transmembrane segment. Within this polypeptide chain are several distinct domains (Fig. 26.16). The extracellular portion of the RTK contains the **ligand-binding domain**. The other end of the peptide protrudes through the plasma membrane into the cytosol. On the cytosolic side, a portion of the receptor forms the tyrosine kinase while the remainder constitutes a cytosolic tail. The cytosolic portion of the receptor (RTK contains tyrosine residues that are in fact substrates or targets for the tyrosine kinase portion of the receptor).

The Activation of RTKs. Signal transduction is initiated when a ligand binds, causing the receptor tyrosine kinases (RTKs) to aggregate (Fig. 26.16B). Often two receptors cluster in this way, the tyrosine kinase associated with each receptor phosphorylates the tyrosines of neighbouring receptors (**cross-phosphorylation**). Since the receptors phosphorylate other receptors of the same type, this process is referred to as **autophosphorylation**.

Signal transduction through RTKs. Once autophosphorylation of tyrosine residues on the cytosolic portion of the receptor occurs in response to ligand binding [(1) and (2) in Fig. 26.17], it recruits a number of cytosolic proteins to interact with itself. Each of these proteins binds to the receptor at a phosphorylated tyrosine residue. To bind to the receptor, each cytosolic protein must contain a stretch of amino acids that recognizes the phosphotyrosine and a few neighbouring amino acids on the receptor. The portion of a protein that recognizes one of these phosphorylated tyrosines is called an **SH_2 domain** [SH_2=Src homology (domain)^2].

Enrolment of different SH_2 domain-containing proteins activates different signal transduction pathways. As a result, receptor tyrosine kinases can activate several different signal transduction pathways at the same time. These include the inositol-phospholipid-calcium second messenger pathway, which we have already discussed, and the **Ras pathway**, which ultimately activates the expression of genes involved in growth or development.

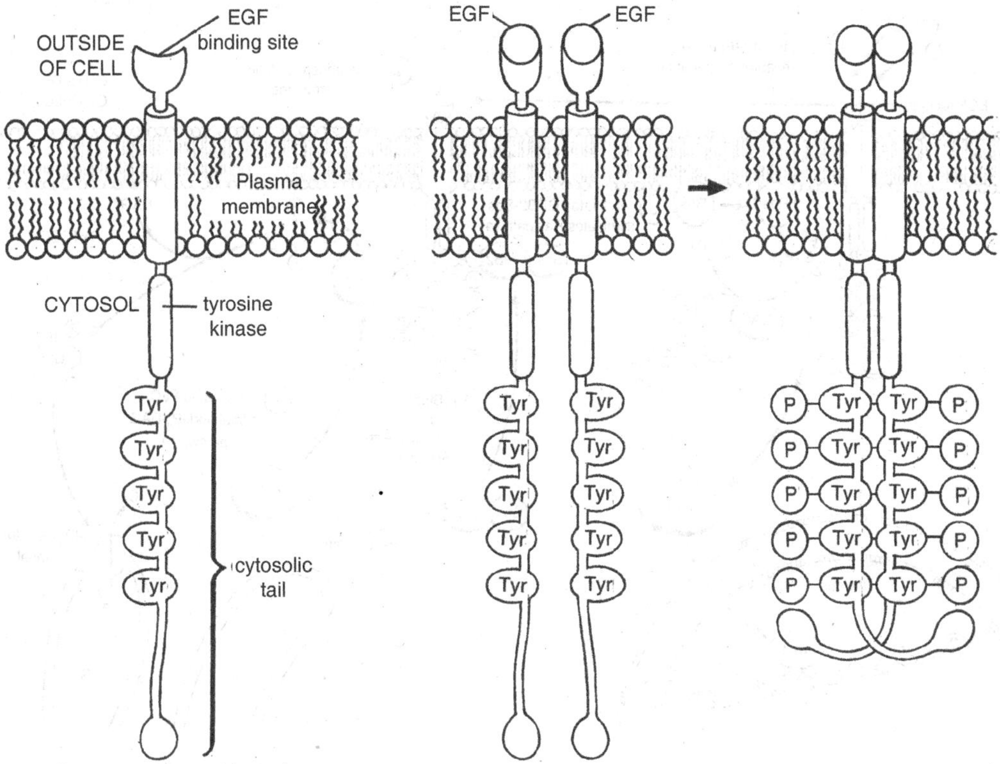

A. Structure of the epidermal growth factor (EGF) receptor

B. Activation of the EGF receptor

Fig. 26.16. The structure and activation of a receptor tyrosine kinase (RTK). A—The receptor for epidermal growth factor (EGF), shown here, is typical of many RTKs. These receptors often have only one transmembrane segment. The extracellular portion of the receptor binds to the ligand (EGF in this case). Inside the cell, a portion of the receptor has tyrosine kinase activity. The remainder of the receptor contains a series of tyrosine residues that are substrates for the tyrosine kinase. B—The activation of a receptor tyrosine kinase starts with the binding of a messenger (EGF), causing receptor aggregation or clustering. Once the receptors aggregate, they cross-phosphorylate each other at a number of tyrosine amino acid residues. The formation of tyrosine phosphate (Tyr-P) residues on the receptor creates binding sites for cytosolic proteins that contain SH_2 domains (after Becker *et al.*, 2006).

Characteristics of Ras pathway. Ras is important in regulating the growth of cells. It comprises a single subunit. Ras and other small monomeric G proteins are important signalling molecules. Like other types of G proteins, Ras can be bound to either GDP or GTP, but is active only when bound to GTP. In the absence of receptor stimulation, Ras is normally in the GDP-bound state. For Ras to become active, it must release GDP and acquire a molecule of GTP. For this to take place, Ras requires the help of another type of protein called a **guanine-nucleotide exchange factor (GEF)**.

The GEF that activates Ras is a Sos (so called because it was originally identified from a genetic mutation in fruit flies called *Son of sevenless* that results in the failure of cells in the compound eye to develop properly). For Sos to become active, it must bind indirectly to the receptor tyrosine kinase through another protein called **GRB2** (3. in Fig. 26.17), which contains a SH_2 domain. Thus, to activate Ras, the receptor becomes tyrosine phosphorylated, and GRB2 and Sos form a complex that binds to the receptor, activating Sos. Sos then stimulates Ras to release GDP and acquire GTP, which converts Ras to its active state.

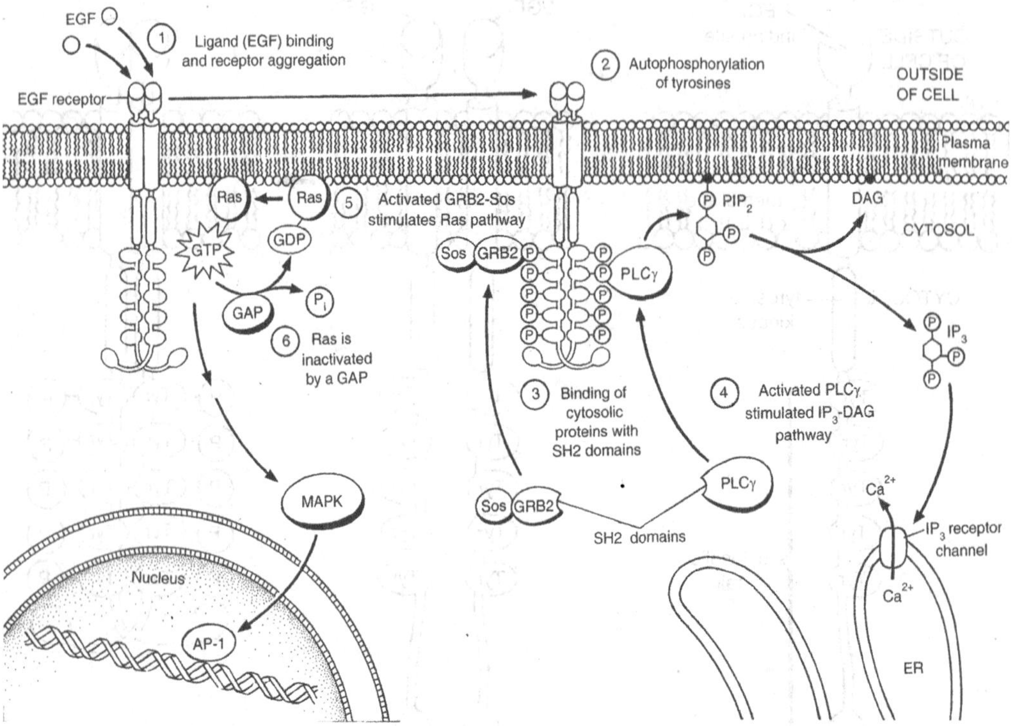

Fig. 26.17. Signal transduction through receptor tyrosine kinases. 1. Upon ligand binding, receptor tyrosine kinases, such as that shown here for the epidermal growth factor (EGF), aggregate and 2. Undergo autophosphorylation. Once a receptor is phosphorylated at residues in its cytosolic tail, 3. Proteins with SH_2 domains such as phospholipase C (PLC) and GRB2 bind to the receptor. 4. The binding of phospholipase C results in its activation and the cleavage of PIP_2 into IP_3 and DAG. 5. The binding of GRB2 causes the activation of Sos, to which it is bound. Sos (a GEF) then causes the activation of the Ras protein by helping it release GDP and acquire GTP. Activated Ras initiates a cascade of events that ultimately results in the formation of AP-1, a transcription factor in the nucleus that stimulates the expression of genes needed for growth of cell. 6. Ras is inactivated by hydrolysis of its bound GTP, a step facilitated by a GAP (after Becker *et al.*, 2006).

Once Ras is active, it triggers a cascade of cellular events collectively referred to as the **Ras pathway** (5. in Fig. 26.17). One significant event in the Ras pathway is the activation of **mitogen-activated protein kinases**, or **MAP kinases (MAPKs)**. MAPKs are activated when cells receive a stimulus to grow and divide. One of the functions of MAPKs is to phosphorylate a nuclear protein called **Jun**. When Jun is phosphorylated, it assembles along with other proteins in a transcription factor called **AP-1**, which appears to stimulate the production of proteins that are needed for cells to grow and divide.

Soon Ras has to be inactivated by hydrolysis of the GTP bound to it to avoid continued stimulation of the Ras pathway. GTP hydrolysis is facilitated by a **GTPase activating protein (GAP)**. GAPs can speed up inactivation of Ras a hundred fold.

26.4. GROWTH FACTORS AS MESSENGERS

Growth factors are secreted molecules that act at short range (*i.e.*, they are mostly paracrine signals) and have specific effects on cells possessing the appropriate receptor to sense the presence of the growth factor.

1. Discovery of growth factors. In order for a cell to grow, it must have all the nutrients needed for synthesis of its component parts, but the availability of nutrients is usually not itself sufficient for growth. Cells often also need messengers that act on specific receptors to stimulate cell growth. Biologists encountered the requirements for cell growth when they first tried to culture cells in nutrients, including the presence of blood plasma, cells would not grow. A turning point came when blood serum was used instead of plasma: *Serum was able to support the growth of cells, whereas plasma would not.* Many of the messengers present within the serum have now been purified, and they are members of various classes of proteins known as **growth factors**.

The difference between blood serum and blood plasma held an important clue about growth factors. *Plasma* is whole blood, including unreacted platelets (which contain clotting components) but without the red and white cells. *Serum* is the clear fluid remaining after blood has clotted. During clotting, platelets secrete growth factors into the blood that stimulate the growth of cells called **fibroblasts**, which form the new connective tissue that makes up a scar. After clotting, the resulting serum is full of **platelet-derived growth factor (PDGF)**. Plasma does not contain this factor because the clotting reaction has not taken place (see **Becker** *et al.*, 2006).

2. Examples of growth factors. Receptor for PDGF is found to be a receptor for tyrosine kinase. In fact, several growth factors act by stimulating receptor tyrosine kinases, including **insulin**, **insulin-like growth factor-I**, **fibroblast growth factor**, **epidermal growth factor**, and **nerve growth factor**. Many other types of growth factors have been isolated (Table 26.2).

Although collectively known as growth factors, however, these proteins have roles in many diverse events. These include not only growth and cell division, but also crucial events during the development of embryos, responses to tissues injury, and many other activities.

Table 26.2 Examples of growth factor families/superfamilies (source: Becker *et al.*, 2006).

	Growth factor	Target cells	Type of receptor complex
1.	Epidermal growth factor (EGF)	Wide variety of epithelial and mesenchymal cells	Tyrosine kinase
2.	Transforming growth factor-α (TGFα)	Same as EGF	Tyrosine kinase
3.	Platelet-derived growth factor (PDGF)	Mesenchyme, smooth muscle, trophoblast	Tyrosine kinase
4.	Transforming growth factor-β (TGFβ)	Fibroblastic cells	Serine-threonine kinase
5.	Fibroblast growth factor (FGF)	Mesenchyme, fibroblasts, many other cell types	Tyrosine kinase
6.	Interleukin-2 (IL-2)	Cytotoxic T lymphocytes	Complex of three units
7.	Colony stimulating factor-I (CSF-1)	Macrophage precursors	Tyrosine kinase
8.	Wnts	Many types of embryonic cells	Seven-pass protein

***(i)* Disruption of growth factor signalling through receptor tyrosine kinases.** Many receptors for growth factors are tyrosine kinases, falling into families based on which growth factor (s) they can bind. One well-studied class of growth factors are the **fibroblast growth factors (FGFs)** and **fibroblast growth factor receptors (FGFRs)**. FGFs and FGFRs are used in signalling events in both adult animals and embryos.

For example, FGFRs have been shown to play an important role in the development of cells derived from the embryonic cell layer or germinal layer of early embryo, known as **mesoderm**. The mesoderm forms many cell types including muscle, cartilage, bone and blood cells, as well as the forerunner of the vertebral column (the notochord). When specific FGFRs fail to function properly, the development of particular mesodermal tissue is affected. In one class of FGFR defects, a **mutation** in the receptor results in *dominant effects* on the developing embryo. In other words, even though the embryo makes a substantial quantity of normal functional receptor, the presence of the mutant receptor within the embryo's cells prevents the normal receptors from functioning properly. Normal function is **inhibited** because FGFRs must act together as dimers to bind FGFs. If a normal receptor dimerizes with a mutant receptor, then the phosphorylation event that normally occur within the tyrosine kinase portion of the receptor fail to occur, blocking signal transduction (Fig. 26.18). Such a mutation that cancels the function of the normal receptor is called a **dominant negative mutation**.

Examples of dominant negative mutations. Dominant negative mutations are found to have dramatic effects on cells in developing embryos. For example, when genetically engineered dominant negative FGFRs are expressed in frog eggs and the eggs are fertilized, the embryos fail to develop mesodermal tissues in the trunk and tail, resulting in tadpoles with heads, but no bodies. In humans, dominant mutations in the transmembrane portion of the FGFR-3 gene result in the most common form of dwarfism, called **achondroplasia**. Heterozygous individuals have abnormal bone growth, in which the long bones suffer from abnormal ossification (*i.e.*, the process by which cartilage is converted to bone during childhood). A related condition known as **thanatophoric dysplasia** (a congenital dwarfism) often results from a single amino acid change in the cytosolic portion of FGFR-3 protein. In this case, more severe bone abnormalities result, and affected individuals die soon after birth.

***(ii)* Disruption of growth factors signalling through receptor serine/threonine kinases.** Another major class of protein kinase-associated receptors uses a very different set of signal transduction pathways to elicit changes within the cell. These receptors phosphorylate serine and threonine residues rather than tyrosines. One major class of **serine/threonine kinase receptors** comprises a family of proteins that bind members of the **transforming growth factor-β** (TGF-β) superfamily of growth factors (Fig. 26.19). TGF-β superfamily consists of a large number of structurally related, secreted, dimeric proteins. They act either as hormones or, more commonly as local mediator, to regulate a wide range of biological functions in all animals. During development, they regulate pattern formation and influence various cell behaviours, including proliferation, differentiation, extracellular matrix production, and cell death. In adults, they are involved in tissue repair and in immune regulation, as well as in many other processes. The super family include the TGF-βs themselves, the *activins* and the **bone morphogenetic proteins** (BMPs).

Once activated, the receptor complex uses a strategy for rapidly relaying the signal to the nucleus. The **type I receptor** directly binds and phosphorylates a latent gene regulatory protein of the **Smad family** (named after the first two identified proteins, Sma in *C.elegans* and Mad in *Drosophila*). Activated TGF-β receptors and activin receptors phosphorylate Smad 2 or Smad 3, while activated BMP receptors phosphorylate Smad 1, Smad 5 and Smad 8. Once one of these Smads has been phosphorylated, it dissociates from the receptor and binds to Smad 4, which can form a complex with any of the above five *receptor-activated Smads*. The Smad complex then moves into the nucleus, where it associates with other gene regulatory proteins, binds to specific sites in DNA, and activates a particular set of target genes. Several mutations in Smad proteins in human cells appear to result in specific forms of cancer.

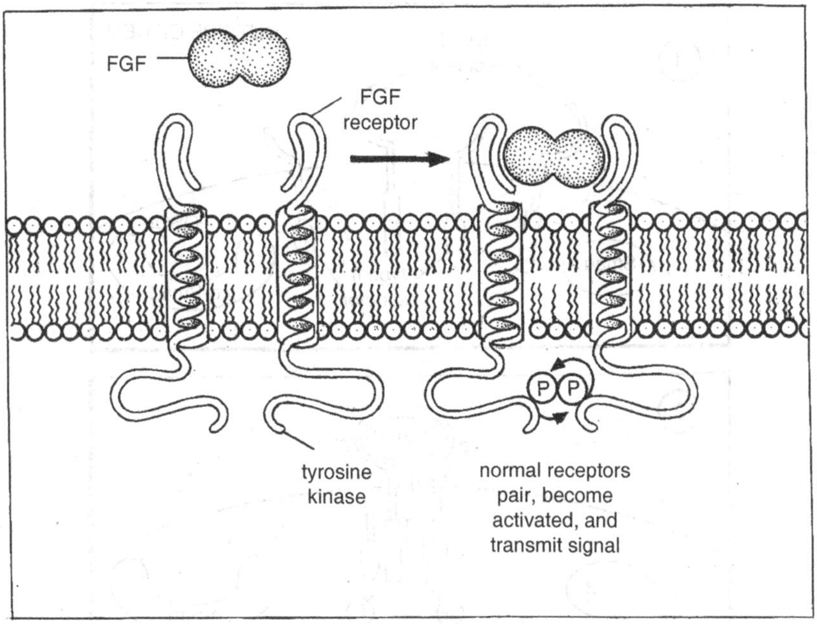

A. Normal FGFR: FGF binds; FGFRs form a dimer

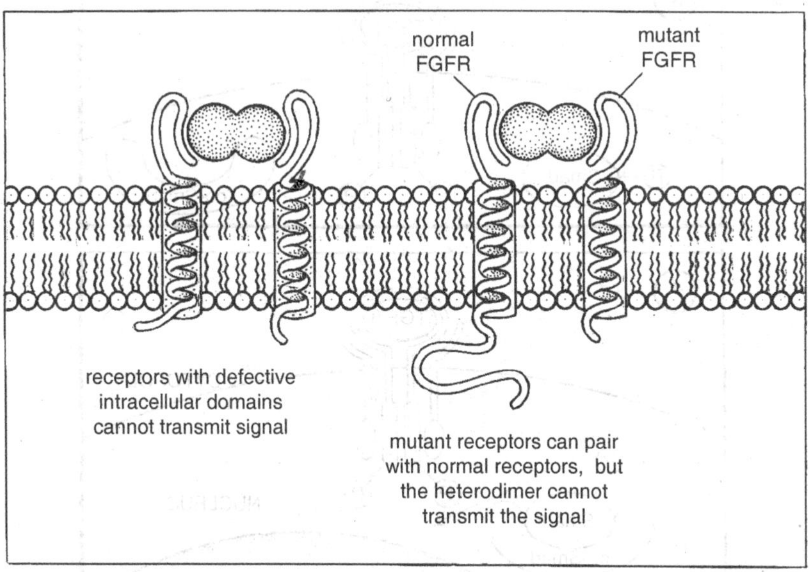

B. Dominant negative mutation in FGFR

Fig. 26.18. Dominant negative disruption (due to mutation) of FGF receptor (FGFR) function. A—Normal receptors form dimers after binding FGF, and transmit the appropriate signal via Ras and MAPK. B—When a cell makes mutant FGF receptors, normal receptors can dimerize as in A, or the defective receptors can bind to FGF and dimerize with normal receptors. In this case, no signal is transmitted (after Becker *et al.*, 2006).

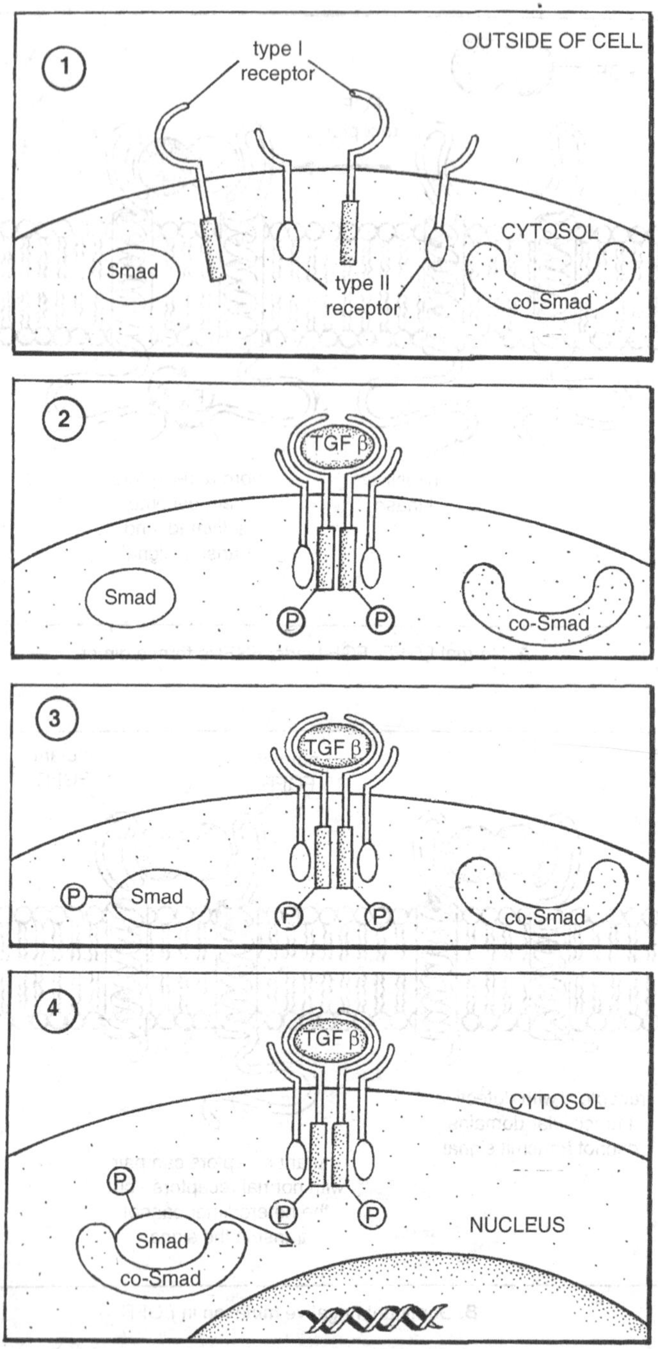

Fig. 26.19. Signal transduction by TGFβ receptor family proteins. 1. Type I and Type II receptors for TGFβ in a cell prior to binding of the growth factor. 2. Binding of growth factor results in clustering of type I and type II receptors. 3. The activated type I receptors then phosphorylate particular receptor-mediated Smads. 4. These Smads then bind to other Smads, and together they enter the nucleus (after Becker *et al.*, 2006).

Growth factor receptors and cancer. It is well known that growth factors regulate events such as cell proliferation, cell movement, and gene expression. Embryonic development requires growth factors to tightly regulate cell biological events. If development is a prime example of the processes that are correctly regulated by growth factors, many types of cancer are exactly the opposite. It is found that a number of cancers result from loss of regulation of growth factor signalling. For example, one of the first cancer-causing genes to be identified was a mutant form of Ras. Mutations in Ras are often associated with cancer. In the case of receptor tyrosine kinases, for example, mutations in the epidermal growth factor receptors can result in **breast cancer, glioblastoma** (a cancer of glial cells in the brain), and **fibrosarcoma** (a type of cancer of long bones). Similar discoveries have been made for receptor serine threonine kinases. Mutations in TGF-β type I receptor occur in one-third of **ovarian cancers**, and mutations in the type II receptor occur in many **colorectal cancers**. Mutations in one of the Smads (Smad 4) occur in one-half of all **pancreatic cancers**.

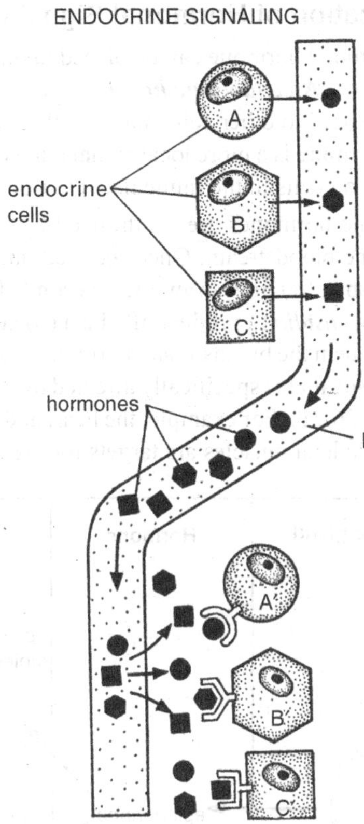

Fig. 26.20. Endocrine cells secrete hormones into the blood, which signal only the specific target cells that recognize them. These target cells have receptors for binding a specific hormone, which the cells "pull" from the extracellular fluid (after Albert *et al.*, 2002).

26.5. SIGNALLING BY HORMONES

Hormones are the chemical signals which are used by the animals to regulate the function of various cells and tissues. They are chemical messengers secreted by one tissue that regulate the function of other cells or tissues in the same organisms. In contrast to growth factors, hormones are often act over large distances. In animals, hormones are often transported via the vasculature. Hundreds of different hormones regulate a wide variety of functions, many critical for maintaining the physiological steady state of an organism.

Although hormones can be considered as a group based on their regulatory functions, however, they differ in many ways. Some hormones are steroids or other hydrophobic molecules that are targeted to intracellular receptors. Other hormones, such as adrenergic hormones are targeted to a wide variety of different G protein-linked receptors. Still others, such as insulin, are ligands, for receptor tyrosine kinase.

1. Classification of Hormonal Signals

Every animal hormone can be placed in one of two categories depending on the distance over which it operates. An ***endocrine hormone*** travels by means of the circulatory system from the cells where it is released to other cells (target cells), where it regulates one or more specific functions. A **paracrine hormone** is a more local signal that is taken up, destroyed, or immobilized so rapidly that it can act only on cells in the immediate environment.

Endocrine hormones are synthesized by the **endocrine tissue** of the body and are secreted directly into the bloodstream. Once secreted into the circulatory system endocrine hormones have a limited life span, ranging from a few seconds for *epinephrine* (a product of the adrenal gland) to many hours for *insulin* (a product of a beta (β) cells of islets of Langerhans of endocrine pancreas). As they circulate in the bloodstream, hormones come into contact with receptors in tissues throughout the body. A tissue that is specifically affected by a particular hormone is called a **target tissue** for that hormone (Fig. 26.21). For example, the heart and the liver are target tissues for epinephrine, whereas the liver and skeletal muscles are targets for insulin.

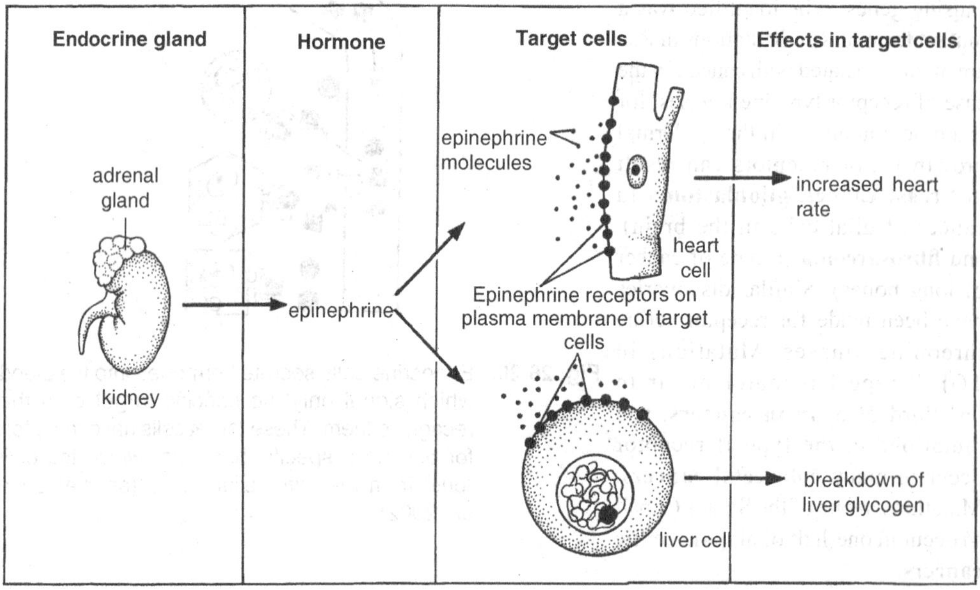

Fig. 26.21. Target tissues for endocrine hormones. Cells in a target tissue have hormone-specific receptors embedded in their plasma membrane (or in case of the steroid hormones, present in nucleus and cytosol). Heart and liver cells can respond to epinephrine synthesized by the adrenal glands because these cells have epinephrine specific receptors on their outer surfaces. A specific hormone may elicit different responses in different target cells. Epinephrine causes an increase in heart rate but stimulates glycogen breakdown in the liver (after Becker *et al.*, 2006).

2. Functions of Hormones

Hormones regulate a wide range of physiological functions, including growth and development, rate of body processes (metabolism), concentration of substances and responses to stress and injury (Table 26.3). For example, **somatotropin** is involved in the regulation of overall growth of the body, whereas **androgens** and **estrogens**, the sex hormones, control the differentiation of tissues and the consequent attainment of secondary sex characteristics. **Thyroxine** regulates the rate at which the body makes energy available and is therefore an example of a rate-controlling hormone. Hormones that control the concentration of substances include **insulin** (control of blood glucose level), **aldosterone** (control of blood sodium and potassium levels) and **parathyroid hormones** (control of blood calcium level). The body's response to stress is regulated by **epinephrine, norepinephrine** and **cortisol** and its response to local injury is regulated by the release of **histamine** and the production of **prostaglandins.**

Table 26.3. Physiological functions of hormones (Source: Becker *et al.*, 2006).

Function under hormonal control	Hormones	Source (Endocrine tissue)
A. Growth and Development		
1. Body size	Somatotropin (growth hormones)	Anterior pituitary
2. Sexual development	Androgens (males) Estrogen (females)	Testes Ovaries
3. Reproductive cycle	Luteinizing hormone Follicle-stimulating hormone Chorionic gonadotropin	Anterior pituitary Anterior pituitary Follicle
B. Rates of body processes		
1. Hormone secretion	Tropic hormones	Anterior pituitary
2. Basal metabolism	Thyroxine	Thyroid
3. Glucose uptake	Insulin	Pancreas
4. Kidney filteration	Antidiuretic hormone (vasopressin)	Posterior pituitary
5. Uterine contraction	Oxytocin	Posterior pituitary
C. Concentration of substances		
1. Blood glucose	Glucagon, insulin	Pancreas
2. Mineral balance	Corticosteroids	Adrenal cortex
3. Blood calcium	Parathyroid hormone	Parathyroid
D. Responses to stress and injury		
1. Heart rate	Epinephrine	Adrenal medulla
2. Blood pressure	Epinephrine	Adrenal medulla
3. Inflammation	Histidine Prostaglandins Corticosteroids (cortisol)	Mast cells All tissues Adrenal cortex

3. Chemical Type of Hormones

Chemically, the endocrine hormones are of four types: amino acid derivatives, peptides, proteins and lipid-like hormones such as steroids. An example of amino acid derivative is epinephrine, derived from tyrosine. **Antidiuretic hormone** (also called **vasopressin**) is an example of peptide hormone, whereas **insulin** is an example of protein hormone. **Testosterone** is an example of a steroid hormone. The steroid hormones are derivatives of cholesterol that are synthesized either in the gonads (the **sex hormones**) or in the adrenal cortex (the **corticosteroids**).

Examples of paracrine hormones are histamine and the prostaglandins. **Histamine** is produced by decarboxylation of the amino acid histidine and is responsible for local inflammatory responses. The **prostaglandins,** so named because they were first discovered in human semen as a secretion of prostate gland, are derived from arachidonic acid and are important in smooth muscle function.

Table 26.4 Chemical classification and functions of hormones (Source: Becker *et al.*, 2006).

Chemical classification	Examples	Regulated functions
A. Endocrine Hormones		
1. Amino acid derivatives	1. Epinephrine (adrenaline) and nor-epinephrine (both derived from tyrosine)	Stress response: regulation of heart rate and blood pressure; release of glucose and fatty acids from storage sites
	2. Thyroxine (derived from tyrosine)	Regulation of metabolic rate
2. Peptides	1. Antidiuretic hormones (vasopressin)	Regulation of body water and blood pressure
	2. Hypothalamic hormones (releasing factors)	Regulation of tropic hormone release from pituitary gland
3. Proteins	1. Anterior pituitary hormones	Regulation of other endocrine systems
	2. Sex hormones (androgens and estrogens)	Development and control of reproductive capacity
	3. Corticosteroids	Stress responses; control of blood electrolytes
B. Paracrine Hormones		
1. Amino acid derivatives	Histamine	Local responses to stress and injury
2. Arachidonic acid derivatives	Prostaglandins	Local responses to stress and injury

4. Signal Transduction by Hormones

Hormones are very efficient signal transducing molecules. They have been divided into two groups: lipid-soluble (lipophilic) hormones and water-soluble (hydrophilic) hormones (Table 26.5). Both groups of hormones show different modes of transducing properties.

Table 26.5 Grouping of hormones into lipophilic and hydrophilic types.

Lipophilic hormones	Hydrophilic hormones
1. Glucocorticoids	1. ACTH (adrenocorticotropic hormone)
2. Progestins	2. ADH (antidiuretic hormone)
3. Mineralocorticoids	3. Calcitonin
4. Estrogens	4. Somatostatin
5. Thyroid hormones	5. Glucagon
6. Androgens	6. FSH (follicle stimulating hormone)
7. Retinoic acid	7. Luteinizing hormone (LH)
8. Vitamin D	8. TSH (thyroid stimulating hormone)
9. cAMP	9. MSH (melanocyte stimulating hormone)
10. cGMP	10. PTH (parathormone)
	11. HCG (human chorionic gonadotropin)

1. Mode of action of lipophilic hormones. Lipid-soluble signalling molecules (*i.e.*, lipophilic hormones) diffuse through the plasma membrane cells and act on intracellular receptors. These receptors are specific for the molecule forming a hormone-receptor complex, thereby activating the receptors. Intracellular receptors are distributed in the cytoplasm as well as in the nucleus. At least two mechanisms have been found out for activation.

(*a*) **Mode of action of glucocorticoids.** In case of steroid hormones such as glucocorticoids or progestins the hormone diffuses into cytoplasm and reacts with the intracellular receptor specific of the hormone forming a hormone-receptor complex with a concomitant dissociation of heat-shock protein (*e.g.*, Hsp 90) from the receptor. The receptor is now activated undergoing a conformational change, which then travels to the nucleus where it binds the **hormone response element** (Fig. 26.22). The signal is transferred to DNA in which specific gene is activated for transcription.

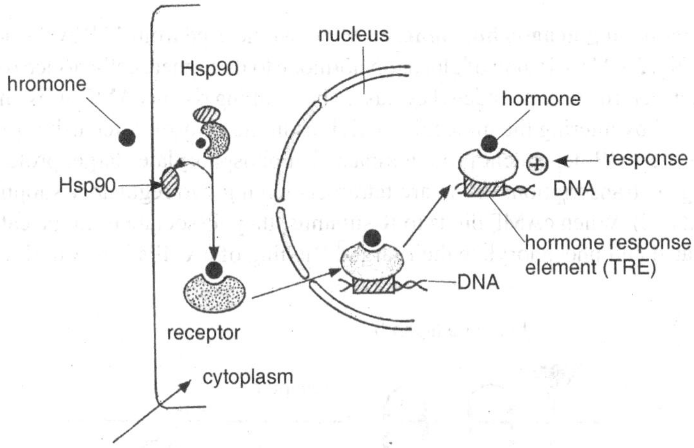

Fig. 26.22. Mechanism of signal transduction by lipophilic glucocorticoid hormone. The ligand (hormone) binds the receptor in the cytoplasm, releasing heat shock protein (Hsp 90). Ligand receptor complex then migrate to the nucleus and binds DNA that is associated with a hormone-response element, followed by gene transcription.

(*b*) **Mode of action of thyroid hormones.** In case of thyroid hormones, the thyroid hormone diffuses into the cytoplasm and directly move into the nucleus where it binds the specific receptor which is already tethered to a **thyroid response element** (**TRE**) associated with DNA. Transcription will proceed only after a corepressor bound to the receptor is dissociated (Fig. 26.23).

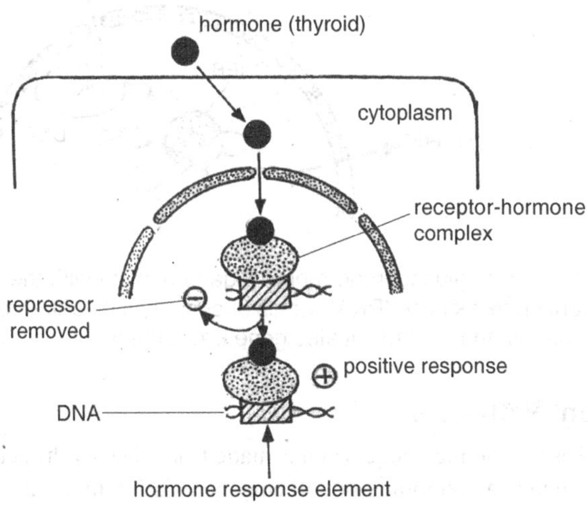

Fig. 26.23. Mechanism of signal transduction by thyroid hormones. The hormone enters the nucleus and binds the receptor associated with hormone response element and a repressor. Receptor-ligand binding triggers removal of repressor and gene transcription follows.

2. Mode of action of hydrophilic (or peptide) hormones. Hydrophilic hormones such as glucagon and antidiuretic hormone (ADH), directly bind to receptors embedded in the plasma membrane. The ligand (hormone) binds to cell surface receptor to generate a specific intracellular signal, *initiation signal transduction* process in which a signal is passed along the cell, often modifying one form of the signal to another. The intracellular signals include cyclic nucleotides (cAMP, cGMP), Ca^{2+} ions, IP_3 and DAG which are generally called **second messengers** affecting gene transcription. Some of the hormones that achieve their effects through cAMP are adrenaline, glucagon, luteinizing hormone, etc.

Mode of action of glucagon hormone. cAMP is synthesized from ATP by the action of enzyme adenyl cyclase (Fig. 26.24). Binding of glucagon hormone to its cognate cell-surface receptor activates a G protein, which in turn, activates adenyl cyclase. The resulting rise in cAMP turns on the appropriate response in the cell by altering the molecular activities in the cytosol, often using **protein kinase A (PKA)** which is a cAMP-dependent protein kinase that phosphorylates target proteins turning on a new pattern of gene transcription. PKAs are tetramers having two regulatory subunits (R) and two catalytic subunits (C). When cAMP binds to R-subunits, they dissociate from the catalytic subunits, which are activated and phosphorylate their target. Binding of cAMPs is required for release of the C-subunits.

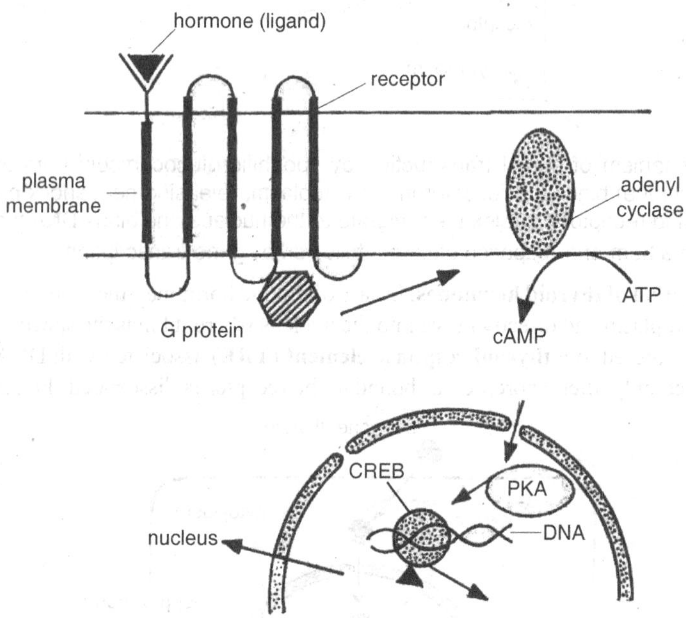

Fig. 26.24. Mode of action of glucagon hormone. Ligand binding with the receptor triggers cAMP-dependent protein kinase (PKA), which phosphorylates cAMP response element binding (CREB) protein that in turn initiates gene transcription.

cGMP-Dependent Pathways

Cylic GMP also acts as second messengers and is made from GTP by the action of an enzyme called **quanylyl cyclase**, which is a membrane-bound enzyme existing in isoenzyme forms. Cyclic GMP serves as the second messenger for **atrial natriuretic peptides** (atriopeptins or ANP), nitric oxide and the response of the rods of retina to light. The response of cGMP is mediated through **protein**

kinase G (PKG) which is a cGMP-dependent protein kinase (enzyme) that phosphorylates target proteins in the cell.

5. Modes of endocrine regulation of target cells. Since hormones modulate the function of particular target tissues, an important aspect of studying hormones is understanding the specific functions of these target tissues. To understand how endocrine hormones act, we will look more closely at the water-soluble (nonlipid) hormones such as **adrenergic hormones** epinephrine and norepinephrine. (Epinephrine is also called adrenaline, the two words are of Greek and Latin derivation, respectively, and mean "above or near, the kidney", referring to the location in the body of the adrenal glands, which synthesize this hormone). When secreted into the blood-stream, epinephrine and norepinephrine stimulate changes in many different tissues or organs, all aimed at preparing the body for dangerous or stressful situations (the so-called "fight-or-flight response"). On the whole, the adrenergic hormones trigger increased cardiac output, shunting blood from the visceral organs to the muscles and the heart, as well as dilation of arterioles to facilitate oxygenation of the blood. In addition, these hormones stimulate the breakdown of glycogen to supply glucose to the muscles.

Adrenergic hormones bind to a family of G protein-linked receptors known as **adrenergic receptors**. The individual members of this family differ mainly in their preference for epinephrine or norepinephrine and in which G protein is linked to the receptor. They are broadly classified into α- and **β-adrenergic receptors**. The α-adrenergic receptors bind both epinephrine and norepinephrine. These receptors are located on the smooth muscles that regulate blood flow to visceral organs. The β-adrenergic receptors bind epinephrine much better than norepinephrine. These receptors are found on smooth muscles associated with arterioles that feed the heart, smooth muscles of the bronchioles in the lungs, and skeletal muscles.

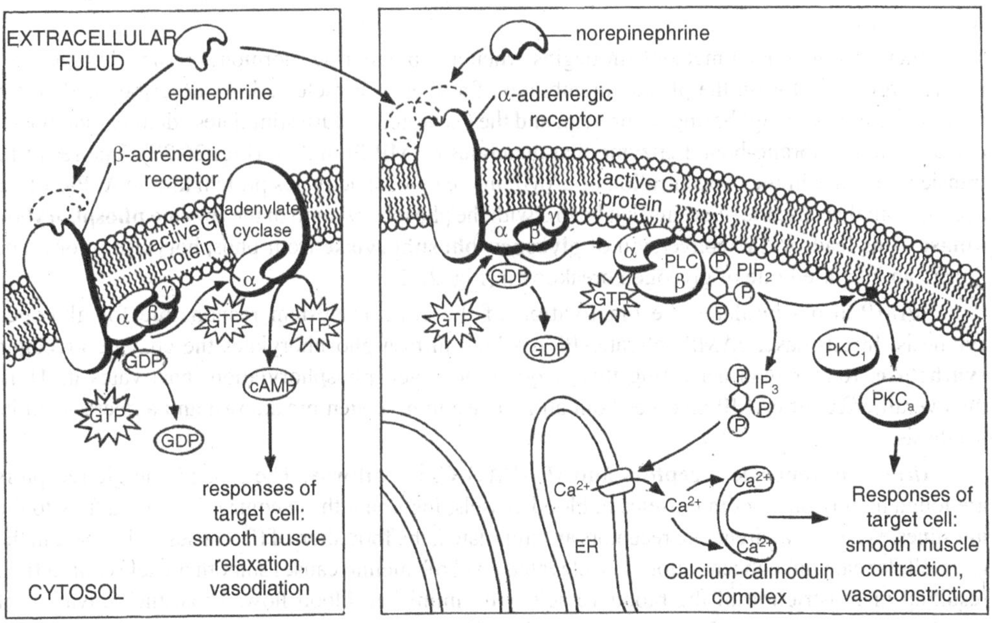

Fig. 26.25. The stimulation of G protein-linked signal transduction pathways by α- and β-adrenergic receptors (after Becker *et al.*, 2006).

The α- and β-adrenergic receptors stimulate different signal transduction pathways, because they are linked to different G proteins (Fig. 26.25). For example, the G proteins activated by one type of α-adrenergic receptor, the α_1*-adrenergic receptors*, are Gp proteins, whereas the β-adrenergic receptors activate Gs. As we discussed earlier, activation of G*s* stimulates the *cAMP signal transduction pathway*, leading to relaxation of certain smooth muscles. Activation of G*p* stimulates phospholipase C, leading to the production of IP_3 and DAG, which in turn elevates intracellular calcium levels. Table 26.6, summarises some of the main physiological functions that are regulated by cAMP. As a specific example of cAMP-mediated regulation, here will be considered the control of glycogen degradation by epinephrine in liver or muscle cells.

Table 26.6 Examples of cell functions regulated by cAMP (Source: Becker *et al.*, 2006).

Regulated function	Target tissue	Hormone
1. Glycogen degradation	Muscle, liver	Epinephrine
2. Fatty acid production	Adipose	Epinephrine
3. Heart rate, blood pressure	Cardiovascular	Epinephrine
4. Water reabsorption	Kidney	Antidiuretic hormone
5. Bone reabsorption	Bone	Parathyroid hormone

***(i)* Control of glycogen degradation by cAMP pathway.** The breakdown of glycogen to provide muscle cells with an adequate supply of glucose is stimulated by the adrenergic hormones. The breakdown of glycogen is catalysed by the enzyme **glycogen phosphorylase**, which cleaves glucose units from glycogen as glucose-1-phosphate by the addition of inorganic phosphate (Pi). The glycogen phosphorylase system was the first cAMP-mediated regulatory sequence to be explained. The original work was published in 1956 by **Earl Sutherland**, who received a Nobel Prize in 1971 for this discovery.

Increased glycogen metabolism begins when an epinephrine (hormone) molecule binds to a β-adrenergic receptor on the plasma membrane of a liver or muscle cell. As described earlier, the receptor activates a neighboring Gs protein, and the Gs protein in turn stimulates adenyly (adenylate) cyclase, the membrane-bound enzyme that generates cAMP from ATP (Fig. 26.27). The resulting transient increase in the concentration of cAMP in the cytosol activates protein kinase A. PKA then activates another cascade of events that begins with the phosphorylation of the enzyme **phosphorylase kinase**. This leads to the conversion of **glycogen phosphorylase** from phosphorylase b form and thus to an increased rate of glycogen breakdown (Fig. 26.26).

cAMP also stimulates the inactivation of the enzyme system responsible for glycogen synthesis. In this case, cAMP activates PKA which in turn phosphorylates the enzyme **glycogen synthetase**. Rather than activating this enzyme, however, phosphorylation inactivates it. Thus, the overall effect of cAMP involves both an increase in glycogen breakdown and a decrease in its synthesis.

***(ii)* α-Andrenergic receptors and IP_3-DAG-Ca^{2+} pathway.** The α_1-andrenergic receptors are found mainly on smooth muscles in blood vessels, including those controlling blood flow to the intestines. When α_1-andrenergic receptors are stimulated, the formation of IP_3 causes an increase in the intercellular calcium concentration. The elevated level of calcium causes smooth muscle contraction, resulting in constriction of the blood vessels and diminished blood flow. Thus, the activation of α_1-adrenergic receptors affects smooth muscle cells in a manner opposite to that of β-adrenergic activation, which causes smooth muscles to relax.

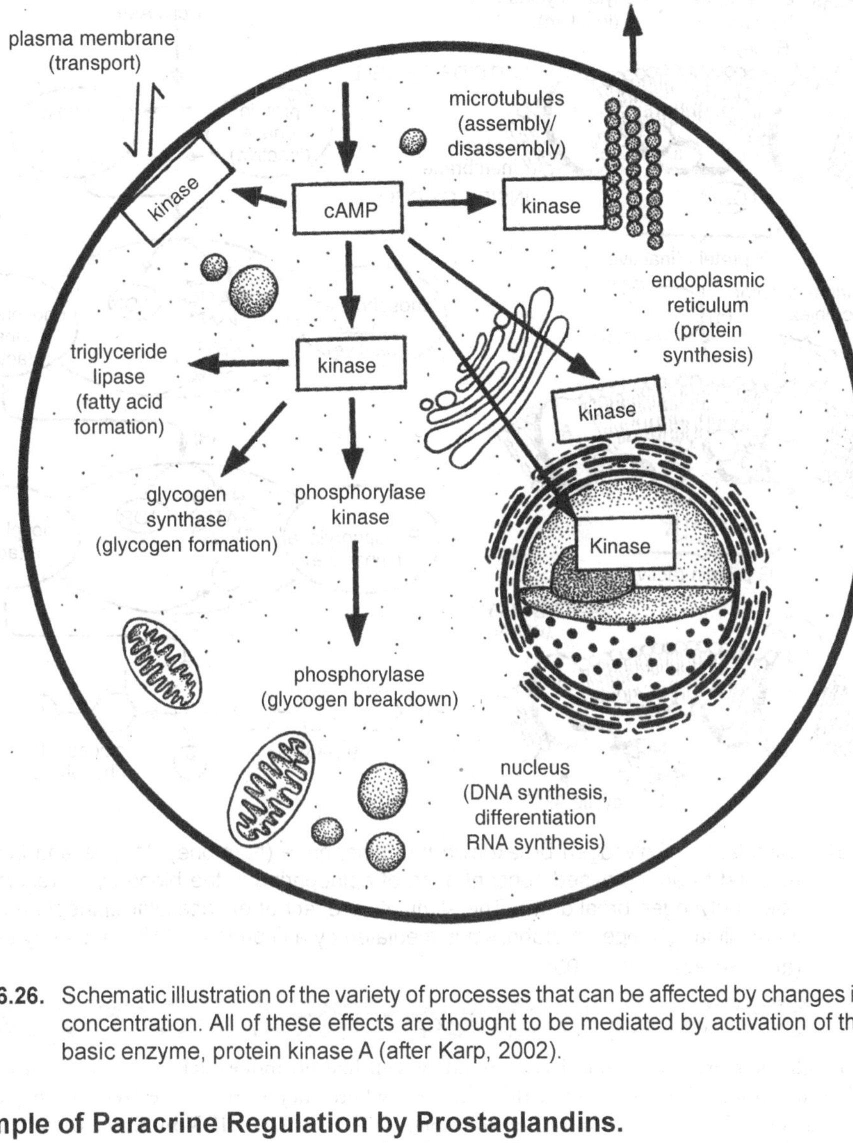

Fig. 26.26. Schematic illustration of the variety of processes that can be affected by changes in cAMP concentration. All of these effects are thought to be mediated by activation of the same basic enzyme, protein kinase A (after Karp, 2002).

Example of Paracrine Regulation by Prostaglandins.

Typically, the prostaglandins act on G protein-linked receptors to stimulate either the cAMP or IP_3-DAG-calcium (second messenger) pathway. Prostaglandins have a variety of effects, many of which involve smooth muscle. For example, prostaglandins contained in semen stimulate uterine smooth muscle contraction, which help transport sperm to egg. Prostaglandins also help to initiate smooth muscle contraction during labour and can be used to induce labour clinically. Some prostaglandins cause smooth muscle relaxation and can, for example, cause bronchiole dilation or lowering of blood pressure. Prostaglandins are also important in the activation of blood platelets, essential components of the blood-clotting mechanism that plug sites where blood vessels are ruptured.

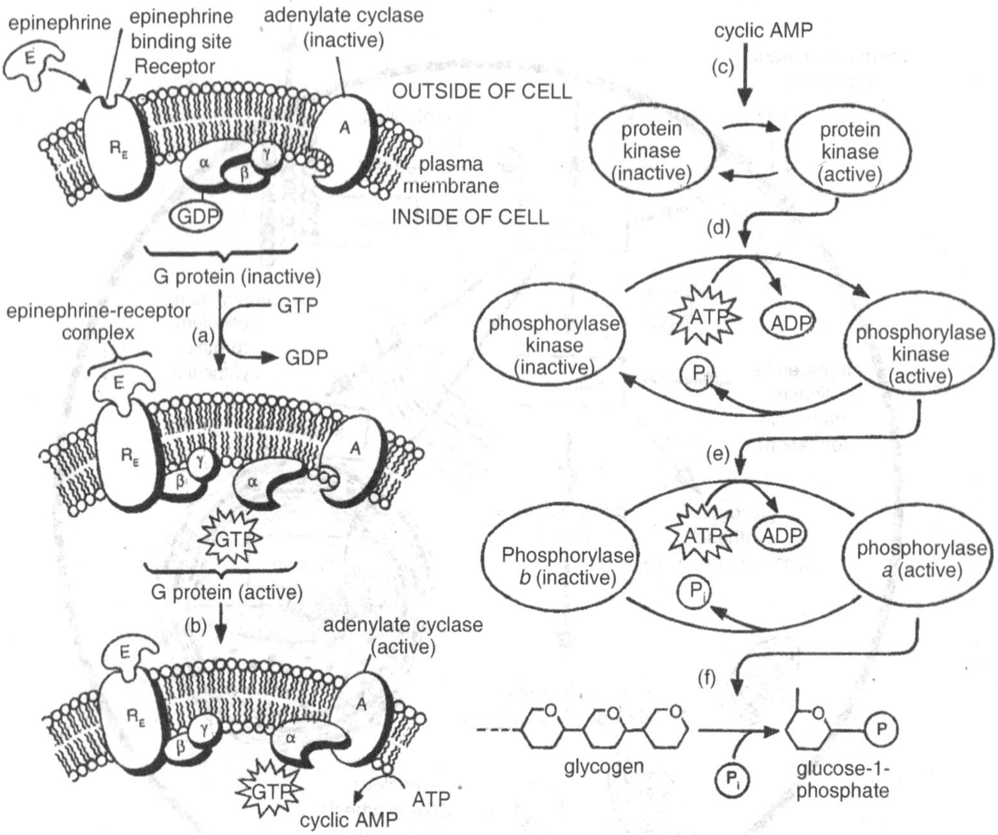

Fig. 26.27. Stimulation of glycogen breakdown by epinephrine (hormone). Muscle and liver cells respond to an increased concentration of epinephrine in the blood by increasing their rate of glycogen breakdown. The stimulatory effect of extracellular epinephrine on the intracellular glycogen metabolism is mediated by a G protein-cAMP regulatory cascade (after Becker *et al.*, 2006).

Box 26.7

Prostaglandins are the chemical compounds which like hormones act as chemical messengers but unlike hormones they act right within the cells where they were synthesized. Prostaglandins were first discovered and isolated from human semen in 1935 by **Dr. Ulf Von Euler** of Sweden. He thought that they had come from the prostate gland and he named them as prostaglandins. It has since been determined that they exist and are synthesized virtually in every cell of the body. Chemically prostaglandins are a class of compounds known as eicosanoids (see Chapter 9) which are fatty acid derivatives (see Sharma, 2010).

For this landmark discovery Dr. Euler got Nobel Prize in 1970.

26.6. CELL SIGNALS FOR APOPTOSIS

Cell signalling also regulates a kind of **programmed cell death (PCD)** or **apoptosis** (see Chapter 53). Apoptosis is a key event in many biological processes. In embryos, apoptosis occurs in a variety of circumstances. Examples include removal of the webbing between the digits (fingers and toes) during the development of hand and feet, the resorption of the tail of tadpoles when they undergo metamorphosis (Fig. 26.28) and the "pruning" of neurons that occur in human infants during first few months of life as connections mature within the developing brains. In adult humans, apoptosis occurs continually; when cells become infected by pathogens or when white blood cells reach the end of their life span, they are eliminated through apoptosis. As a result, millions of cells die every minute in the human body. When cells that should die via apoptosis do not, the consequences can be critical. It is currently known that mutations in some of the proteins that participate in apoptosis can lead to cancer; melanoma frequently results from a mutations in Apaf-1, a protein that participates in apoptosis.

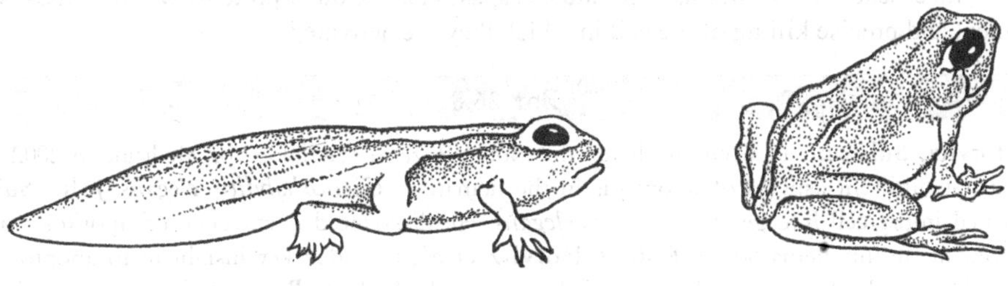

Fig. 26.28. Apoptosis during the metamorphosis of a tadpole into a frog. All the changes that occur during metamorphosis, including the induction of apoptosis of tail, are stimulated by an increase in thyroid hormone in the blood (after Albert *et al.*, 2002).

Apoptosis is very different from another type of cell death, known as **necrosis**, which sometimes follows massive tissue injury. Whereas necrosis involves a specific series of events that leads to the dismantling of the internal contents of the cell (Fig. 26.29). During the early phases of apoptosis, the cell's DNA segregates near the periphery of the nucleus and volume of the cytoplasm decreases. Next, the cell begins to produce small bubble-like cytoplasmic extensions ("blebs"), and the nucleus and organelles begin to fragment. The cell's DNA is cleaved by an apoptosis-specific **DNA endonuclease** or **DNase** (an enzyme that digests DNA), at regular intervals along the DNA. As a result, the DNA fragments, which are multiples of 200 base pairs in length, form a diagnostic "ladder" of fragments. Eventually the cell is demolished into small pieces called **apoptotic bodies.** Ultimately the remnants of the affected cell are engulfed by other nearby cells (typically macrophages) via phogocytosis. The macrophages act as scavengers to remove the resulting cellular debris.

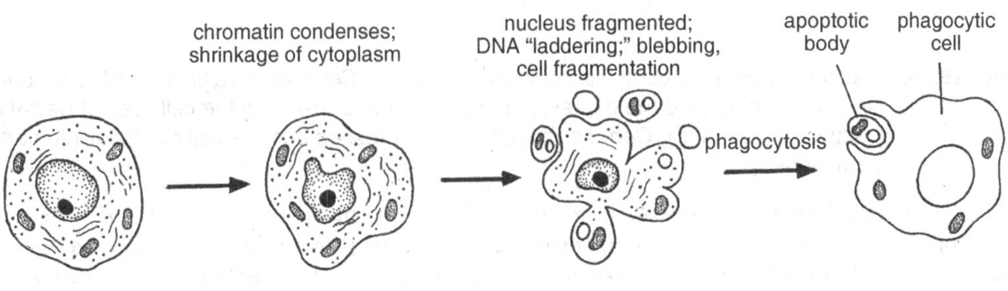

Fig. 26.29. Major steps in apoptosis. As a cell begins to undergo apoptosis, its chromatin condenses and the cytoplasm shrinks. Eventually the nucleus becomes fragmented, its DNA is digested at regular intervals ("laddering"), the cytoplasm becomes fragmented, and the cell extends numerous blebs. Ultimately the remnants of the dead cells (apoptotic bodies) are ingested by phagocytic cells (after Becker *et al.*, 2006).

Discovery of Apoptosis. The cells have a "death program" was first conclusively demonstrated in the nematode, *Caenorhabditis elegans*, where key genes that control apoptosis were first identified (Box 26.8). Subsequently research showed that many other organisms, including mammals, use similar proteins during apoptosis. As a result of these investigations many of the molecular events that underlie apoptosis are now known. For example, a key event in apoptosis of mammals is the activation of a series of enzymes called **caspases** (Caspases get their name because they contain a *c*ysteine at their active site, and they cleave proteins at sites that contain *aspartic* acid residue followed by four amino acids that are specific to each caspase). Caspases are produced as inactive precursors known as **procaspases**, which are subsequently cleaved to create active enzymes, often by other caspases, in a proteolytic cascade. Once they are activated, caspases cleave other proteins within cells, resulting in efficient and precise killing of the cell in which they are activated.

Box 26.8

Horvitz, Sulston and **Brenner** shared the Nobel Prize in Physiology or Medicine in 2002. They worked out different aspects of apoptosis in the nematode *Caenorhabditis elegans.* **John Sulston** performed initial cell lineage studies on *C.elegans.* He discovered occurrence of apoptosis during development of this nematode **H.Robert Horvitz** *et al.*, provided key insight in to apoptosis and cell signalling in the formation of vulva of this nematode. **Sydney Brenner** suggested genetic basis of apoptosis in *C.elegans.*

In *C.elegans*, development relies on programmed cell death. The number of cells that die during the worm's development is always the same: 131 cells of total 1090 cells in the hermaphrodites and 147 cells of total 1178 cells in males. In addition, the point in development at which a given cell dies and the identity of the cells that die are always the same.

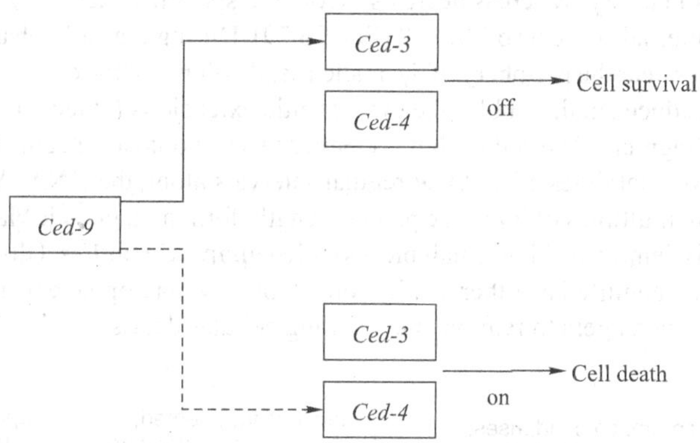

Fig. 26.30. The genetic pathway controlling cell death, the gene *Ced-9* act as a binary switch. If *Ced-9* is active, it repressed the expression of *Ced-3* and *Ced-4*, and the cell lives. If *Ced-9* is inactive (broken line), *Ced-3* and *Ced-4* are expressed and the cell die (after Klug and Cummings, 2001).

Genetic analysis of mutants indicates that although apoptosis occurs in cells of different developmental origins, all the cells die by using the same genetic pathway. In *C. elegans*, 15 genes are involved in cell death (apoptosis). These genes form four groups: 1. decision makers; 2. execution of the decision; 3. engulfment of dying cells, and 4. degradation of cell debris in the engulfing cells. For example, expression of *Ced-3* and *Ced-4* (= *c*ell *d*eath defective) are necessary for execution of the cell death programme; mutations that inactivates either of these genes result in survival of cells that normally die. Expression of *Ced-3* and *Ced-4* is controlled by *Ced-9.* Gain-of-function mutations that cause constitutive expression or overexpression of *Ced-9* prevent cell death. Conversely, loss-of-function mutations that inactivate *Ced-9* cause embryonic lethality. Based on these observations, it has been concluded that *Ced-9* gene works by preventing expression of *Ced-3* and *Ced-4* genes in cells that survive. This means that *Ced-9* is a **binary switch gene** for programmed cell death. Cells that express *Ced-9* survive and do not die.

Target proteins for caspases. Caspases are known to cleave the following target proteins of the cell (see **Karp**, 2002):

1. **More than a dozen protein kinases, including focal adhesion kinases (FAK), PKB, PKC and Raf 1**. Inactivation of FAK, for example, is presumed to disrupt cell adhesion, leading to detachment of the apoptotic cell from its neighbours.

2. **Lamins**, which makes up the inner lining of the nuclear envelope. Cleavage of lamins leads to the disassembly of the nuclear lamina and shrinkage of the nucleus.

3. **Proteins required for cell structure**, such as those of intermediate filaments, actin and gelsolin. Cleavage and consequent inactivation of these proteins lead to changes in cell shape.

4. **An endonuclease called capsin activated DNase (CAD)**, which is activated following caspase cleavage of an inhibitory protein. Once activated, CAD translocates from the cytoplasm to the nucleus where it attacks DNA, serving it into fragments.

5. **Enzymes involved in DNA repair**, which are inactivated by caspase cleavage. DNA repair is a homeostatic activity that is inappropriate in an apoptotic cell in which the DNA is being dismantled.

Triggering of Apoptosis. There are two main routes by which cells can activate caspases and enter the apoptic pathway (Fig. 26.31).

1. **Cell death signals.** When cells in the human body are infected by certain viruses, a population of **killer lymphocytes** are activated and induce the infected cells to initiate apoptosis. How do lymphocytes induce cells to commit suicide via apoptosis? Typically, such activation is triggered when cells receive **cell death signals**. Two well known death signals are *tumor necrosis factor* and *CD95/Fas*. For example, **CD95** is a protein which is present on the surface of infected cells. Lymphocytes have a protein on their surfaces that binds to CD95, causing the CD95 within the infected cell aggregate (Fig. 26.31). CD95 aggregation results in the attachment of adaptor proteins to the clustered CD95, which in turn recruits a procaspase (**procaspase-8**) to sites of receptor clustering, when the procaspase is activated, it acts as an initiator of the caspase cascade. A key action of such **initiator caspases**, known as **caspase-3**. Active caspase-3 is important for activating many steps in apoptosis.

2. **Survival factors.** In other cases, apoptosis is triggered indirectly. One of the best-studied cases of this type of apoptosis involves **survival factors**. Survival factors mediate one of the many events associated with growth factors: cell survival. When such factors are withdrawn, the cell enters apoptosis. Surprisingly, a key site of action of this second pathway is the mitochondrion. The connection between mitochondria and cell death may be surprising, but it is that in addition to their role in energy production, mitochondria are important in apoptosis. If withdrawal of trophic factors is the sentence of execution, then the executioners are the mitochondria.

How does mitochondrion speed up cell death? In a healthy cell that is not committed to apoptosis, there are several **anti-apoptotic proteins** in the outer mitochondrial membrane that prevent apoptosis, but only as long as a cell continues to be exposed to trophic factors. These proteins are structurally related to a protein known as **Bcl-2**, the best understood of these anti-apoptotic proteins. Bcl-2 and other anti-apoptotic proteins exert their effects by counteracting other proteins that are also structurally similar to Bcl-2. These proteins, however, *promote* apoptosis, and so they are collectively called **pro-apoptotic proteins**. These proteins (such as proteins called **Bax** and **Bad**), can insert into the mitochondrial outer membrane, but they only do so at high levels when trophic factors are no longer present. Thus, pro- and anti-apoptotic proteins carry on an ongoing battle in many cells; when the balance shifts toward pro-apoptotic proteins, a cell is more likely to undergo apoptosis.

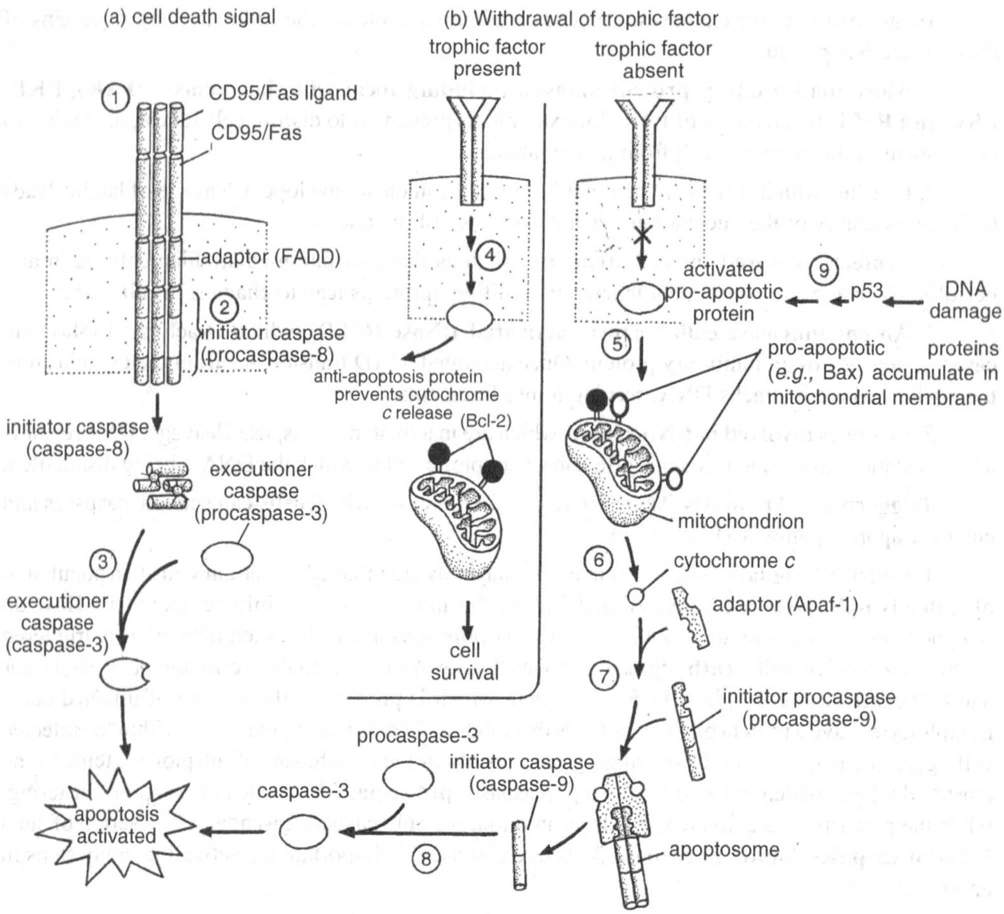

Fig. 26.31. Induction of apoptosis by cell death signals or by withdrawal of survival factors. A—Cell death signals, such as CD95/Fas ligand on the surface of a killer lymphocyte, can lead to apoptosis. 1. CD95/Fas ligand binds to CD95/Fas protein on the surface of a target cell. Binding causing clustering of receptors and recruitment of adaptor proteins in the target cell, resulting in clustering of interior procaspase (procaspase-8) protein. 2. Initiator caspases then become activated. After they are activated, 3. The initiator capases in turn activate the executioner caspase, caspase-3, a key initiator of apoptosis. B—4. When trophic factors are present, cell signalling results in inactivation of pro-apoptotic proteins and the cell does not enter apoptotis. 5. When trophic factors are withdrawn, pro-apoptotic proteins are activated, and some accumulate in the outer membrane of the mitochondrion. 6. The balance between pro-apoptotic and anti-apoptotic proteins (such as Bcl-2) at the mitochondrial outer membrane determines whether the mitochondrion releases cytochrome *c*. 7. Cytochrome *c* forms a complex with other proteins, resulting in activation of an initiator caspase (caspase-9). 8. Initiator caspase in turn activates the executioner caspase, caspase-3, triggering apoptosis. 9. DNA damage can also lead to apoptosis through the activity of the P53 protein (after Becker *et al.*, 2006).

Mitochondria trigger apoptosis by releasing **cytochrome c** into the cytosol. As we already know that cytochrome c is normally involved in electron transport, it also has other important functions in triggering apoptosis in at least two ways. 1. Cytochrome c stimulates calcium release from adjacent mitochondria and from the endoplasmic reticulum, where it binds IP_3 receptors. 2. Cytochrome can activate an initiator procaspase associated with mitochondria, known as **procaspase-9**. It does this by recruiting a cytosolic adaptor protein (known as **Apaf-1**) that assembles procaspase-9 into a complex sometimes called an **apoptosome**; the apoptosome promotes the production of active caspase-9. Like other initiator caspases, caspase-9 activates the executioner caspase, caspase-3. Thus, in the end, both cell death mechanism lead to activation of a common caspase that sets apoptosis in motion.

There is another situation that can trigger the mitochondrial pathway to apoptosis. When a cell suffers so much damage that it is unable to repair itself, it may trigger its own demise. In particular, when a cells DNA is damaged (for example, by radiation or ultraviolet light), it can enter apoptosis via the activity of a protein known as **P53** (P53 is an important regulator of the cell cycle). Like the withdrawal of trophic factors, the P53 pathway can also activate pro-apoptotic proteins to trigger apoptosis (see **Becker** *et al.*, 2006).

QUESTIONS

Long Answer Questions

1. What is cell signalling? Describe various type of receptors responsible for the function of cell signalling.
2. What are cell surface receptors? Discuss their types and roles in cell signalling.
3. 'Apoptosis is mediated by an intracellular proteolytic cascade'—substantiate.
4. What is signal transduction? Describe the steps that occur during its accomplishment.
5. Give the structure of GPCR. What is their significance? Explain the events that lead to sensitisation of a GPCR.
6. How does a peptide hormone act? Describe the second messenger hypothesis.
7. In what manner does calcium helps in the signalling mechanism? Explain it with the help of an example.

Short Answer Questions

1. Explain the mechanism of cell signalling.
2. What is ligand? What is its function?
3. What are second messengers? How do they participate in cell signal transduction?
4. Explain the role of mitochondria in apoptosis.
5. Write short notes on the following:
   (*i*) Signal transduction;
   (*ii*) Apoptosis;
   (*iii*) Protein kinases;
   (*iv*) Calmodulin;
   (*v*) Nuclear receptors.

Very Short Answer Questions

1. What is a ligand?
2. What is the function of ligand?
3. What is the function of cell surface receptors?
4. What is the chemical nature of receptors?
5. What is receptor?
6. Define cAMP.
7. What is full form of IP_3?
8. What is the function of IP_3?
9. Define calmodulin.
10. What is a paracrine signal?
11. Define the endocrine signal.
12. What is a autocrine signal?
13. Define signal transduction.
14. What is G protein?

Multiple Choice Questions

1. One of the following is a calcium-binding protein of the cell
   (*a*) actin
   (*b*) tubulin
   (*c*) calmodulin
   (*d*) dynein
2. A molecule acting as 'second messenger' in biological system is
   (*a*) cDNA
   (*b*) cAMP
   (*c*) tRNA
   (*d*) hn RNA

3. Epinephrine and nor-epinephrine together are known as
   (*a*) steroids
   (*b*) proteins
   (*c*) catecholamine
   (*d*) none of these
4. Substances which are formed at one place and express their effect at a distant place are called as
   (*a*) enzymes
   (*b*) hormones
   (*c*) pheromones
   (*d*) WBC
5. Insulin facilitates glucose uptake in
   (*a*) kidney tubules
   (*b*) red blood cells
   (*c*) brain
   (*d*) skeletal muscle
6. Cortical granules found in the egg cortex of many animals consist of
   (*a*) lipids
   (*b*) proteins
   (*c*) mucopolysaccharides
   (*d*) none of the above

ANSWERS

Very Short Answer Questions

1. Ligand is any molecule that binds to a specific site on a protein or other molecule (From Latin *ligare*, to bind).
2. Binding of ligand changes the property of the surface receptors and activates changes.
3. Communication with neighbouring cells, to bind matrix components with extracellular matrix and cell adhesion.
4. The receptors are protein molecules. These are proteoglycans.
5. A protein that contains a binding site for a specific signalling molecule.
6. cAMP is an intracellular molecule which acts as second messenger in eukaryotes and mediates the effects of various signalling molecules by activating protein.
7. Inositol-1, 4, 5-triphosphate.
8. IP_3 functions as a second messenger by triggering the release of calcium ions from storage sites within the endoplasmic reticulum.
9. Calmodulin is a calcium-binding protein involved in mediating many of the intracellular effects of calcium ions in eukaryotic cell.
10. It is a growth factor or other secreted molecule that acts on nearby cells.
11. It is a hormone or other molecule that is released into the circulatory system so that it can act on distant target cell.
12. It is a growth factor or other molecule secreted by a cell that acts on the cell that produces it.
13. Signal transduction is detection of specific signals at the cell surface and the mechanisms by which such signals are transmitted into the cell's interior, resulting in changes in cell behaviour and/or gene expression.
14. Any of numerous GTP-binding regulatory proteins located in the plasma membrane that mediate signal transduction pathways, usually by activating a specific target protein such as an enzyme or channel protein.

Multiple Choice Questions

1. (*c*) 2. (*b*) 3. (*c*) 4. (*b*) 5. (*d*) 6. (*c*)

27 Electrical Signals in Nerve Cells

All organisms respond to external stimulation, a property known as **irritability**. Even a single-celled amoeba, if poked with a fine glass needle, responds by withdrawing its pseudopodia, rounding up, and moving off in other direction. Irritability in an amoeba depends on the same basic properties of membranes that lead to formation and propagation of nerve impulses.

27.1. NEURONS

Almost all animals have a nervous system, in which electrical impulses are transmitted along the specialized plasma membranes of nerve cells. The nervous system performs *three* functions: It **collects** information from the environment, it **processes** that information and it **elicits responses** to that information by triggering specific **effectors**, usually muscle tissue or glands.

Cells that make up the nervous system can be broadly divided into two groups: neurons and glial cells. All **neurons** send or receive electrical impulses. Neurons can be subdivided into *three* basic types based on functions: sensory neurons, motor neurons and interneurons. **Sensory neurons** are a diverse group of cells specialized for the detection of various types of stimuli; they provide a continuous stream of information to the brain about the state of the body and its environment from various sensory receptors. Examples of sensory neurons include the photoreceptors of the retina (used in vision), olfactory neurons (smell), and the various touch, pressure, pain and temperature-sensitive neurons located in the skin and joints. **Motor neurons** transmit signals from the CNS (Central nervous system which includes brain, spinal cord and ganglia) to the muscles or glands they innervate—that is, the tissues to which they send signals. **Interneurons** processes signals received from other neurons and relay the information to other parts of the nervous system.

The term **glial cell** (Greek *glia* = glue) encompasses a variety of different cell types, including microglia, oligodendrocytes, Schwann cells and astrocytes. **Microglia** are phagocytic cells that fight infections and remove debris. **Oligodendrocytes** and **Schwann cells** form the insulating *myelin sheath* around neurons of CNS and those of peripheral nerves, respectively. **Astrocytes** control access of blood vessels to the extracellular fluid surrounding nerve cells, thereby forming the blood-brain barrier.

Structure of neurons. Nerve cells or neurons are specialized for the collection, conduction and transmission, which is coded in the form of fast-moving electrical impulses. The basic parts of a typical neuron are illustrated in Fig. 27.1.

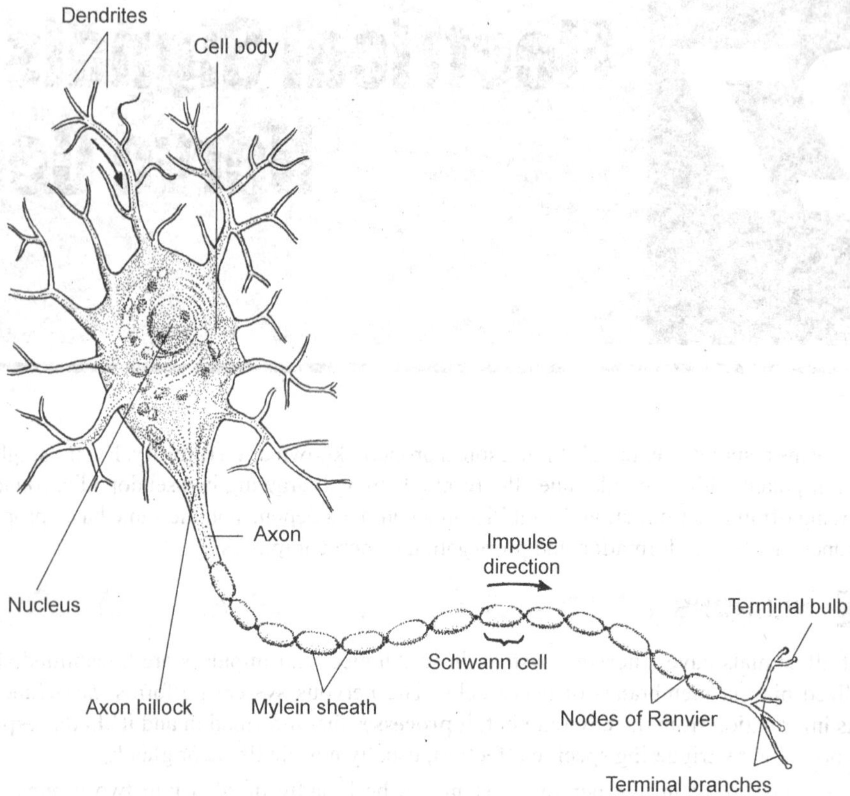

Fig. 27.1. The structure of a typical motor neuron. The cell body contains the nucleus and most of the usual organelles. Dendrites conduct signals passively inward to the cell body; starting near the axon hillock, the axon transmits signals outward. At the end of axon are numerous terminal bulbs (=knobs). Axons of some neurons have a discontinuous myelin sheath around their axons to insulate them electrically. Each segment of sheath consists of a concentric layer of membranes wrapped around the axon by a Schwann cell (or an oligodendrocyte in CNS). The breaks in the myelin sheath, called nodes of Ranvier, are concentrated regions of electrical activity (after Baker *et al.*, 2006).

Most expanded region of a neuron is called the **cell body**. The cell body of most neurons is similar to that of other cells; its cytoplasm contains most of the same organelles (*e.g.*, nucleus with nucleolus, mitochondria, Golgi apparatus, neurofibrils or microtubules, Nissl bodies, *i.e.*, RER). Cell body is the metabolic center of the nerve cell and the site where most of its material contents are manufactured. Neurons also contain extensions, or branches, called **processes**. There are two types of processes: those that receive signals and combine them with signals received from other neurons are called **dendrites**, and those that conduct signals, sometimes over long distances, are called **axons**. The cytoplasm within an axon is commonly referred to as **axoplasm**. Most vertebrate axons are surrounded by a discontinuous **myelin sheath**, which insulates the segments of axon separating the **nodes of Ranvier** (Fig. 27.2). Axons can be very long—upto several thousand times longer than the diameter of the cell body. For example, the motor neuron that innervates human foot has its cell body in the spinal cord (CNS) and its axon extends approximately a meter down the leg. A **nerve** is simply a tissue composed of bundles of axons.

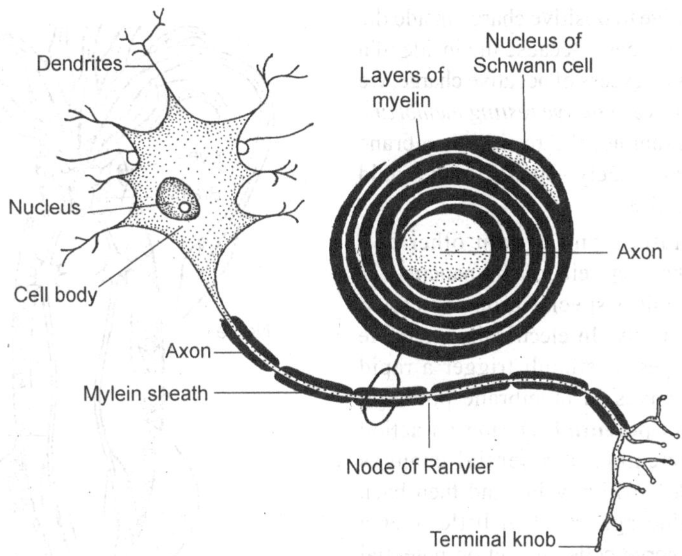

Fig. 27.2. Structure of a nerve cell showing details of myelin sheath and two synaptic terminals of presynaptic neurons (after Karp 2010).

A motor neuron has multiple, branched dendrites and a single axon leading away from the cell body. The axon of a typical neuron is much longer than the dendrites and forms multiple branches. Each branch terminates in structures called **terminal bulbs** or **terminal knobs** or **boutons**. The terminal knobs are responsible for transmitting the signals to the next cell, which may be another neuron or a muscle or gland cell. In each case, the junction is called a **synapse**. For neuron-to-neuron junction, synapses usually occur between an axon and a dendrite, but they can occur between two dendrites. Typically, neurons have synapses with many other neurons.

Functionally a nerve cell or neuron can be distinguished into three regions:

1. **Generator region** includes dendrites, cyton and colaterals. It is the part where initial polarisation or **generator potential** develops.
2. **Conductile region** includes the axon which carries the stimulus.
3. **Transmissional region** comprises of the nerve endings or terminal branches of the axon which end in synaptic vesicles.

Axoplasm plays no direct role in the generation of membrane potentials (see Section 27.2). An axon, having its axoplasm replaced with an artificial salt solution, behaves amazingly like a normal axon, it remains excitable for many hours and can give several hundred thousand action potentials. Only the axon membrane is essential for membrane potentials.

27.2. THE RESTING MEMBRANE POTENTIAL

Membrane potential is a fundamental property of all cells. It results from an excess of negative charge on one side of the plasma membrane and an excess of positive charge on the other side. Cells at rest normally have an excess of negative charge inside and an excess of positive charge outside the cell; the resulting electrical potential is called the **resting membrane potential**, denoted **Vm**. The membrane potential can be measured by placing one tiny electrode inside the cell and another outside the cell (Fig. 27.4); its value is generally given in **millivolts (mV)**. The electrode compare the ratio of negative to positive charge inside the cell and outside the cell. Because the inside of a cell typically has an excess of negative charge, we say that the cell has a *negative resting membrane potential*. For example, the resting membrane potential is approximately –60 mV for the squid giant axon (Fig. 27.3).

Nerve, muscle, and certain other cell types such as the islet cells of the pancreas of vertebrates exhibit a special property called **electrical excitability**. In electrically excitable cells, certain types of stimuli trigger a rapid sequence of changes in membrane potential known as **action potential**. During an action potential, the membrane potential changes from negative to positive value and then back to negative value again, all in little over a millisecond. In nerve cells, the action potential has the specific function of transmitting an electrical signal along the axon. To understand how nerve cells use action potentials to transmit signals, we must first examine how cells generate a resting membrane potential and how the membrane potential changes during an action potential.

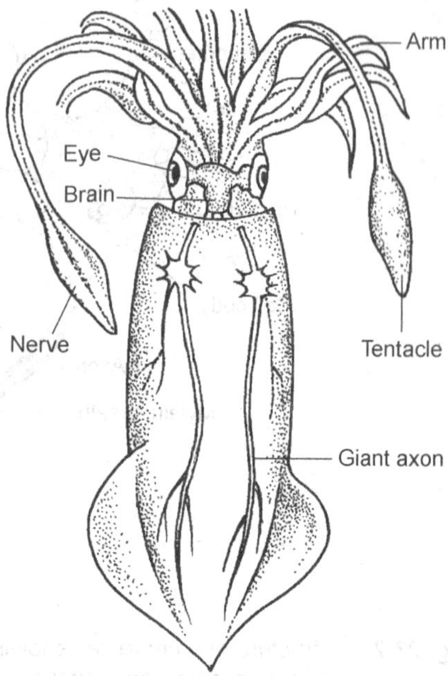

Fig. 27.3. Squid giant axons. The squid nervous system include motor nerves that control swimming movements. The nerves contain giant axons (fibers) with diameters ranging upto 1 mm, providing a convenient system for studying resting and action potentials in a biological membrane.

Generation of a Membrane Resting Potential

The resting membrane potential develops because the cytosol of the cell and the extracellular fluid contain different compositions of cations and anions. Extracellular fluid is a watery solution of salts, including sodium chloride and lesser amounts of potassium chloride. The cytosol contains potassium rather than sodium as its main cation because of the action of the Na^+/K^+ pump. The anions in the cytosol consist largely of macromolecules such as proteins, RNA and a variety of other molecules that are not present outside the cell. These negatively charged macromolecules cannot pass through the plasma membrane, and therefore, remain inside the cell. The concentrations of important ions for two types of well-studied neurons, the squid axon and one type of mammalian axon, are shown in Table 27.1.

Table 27.1 Ionic concentrations inside and outside of axons and neurons (source: Baker *et al.*, 2006).

		Squid axon		Mammalian neuron (Cat motor neuron)	
	Ion	Outside (mM)	Inside (mM)	Outside (mM)	Inside (mM)
1.	Na^+	440	50	145	10
2.	K^+	20	400	5	140
3.	Cl^-	560	50	125	10

For understanding the membrane potential forms, one has to recall a few basic physical principles. First, all substances tend to diffuse from an area where they are more highly concentrated to an area of their lower concentration. Cells normally have a high concentration of potassium ions inside and a low concentration of potassium ion outside. Such a uneven distribution of potassium ions is known as a **potassium ion gradient**. By convention, the potassium ion gradient is expressed as the molar concentration of potassium ions outside the cell $[(K^+)_{outside}]$ divided by the molar concentration of potassium in the cytoplasm $[(K^+)_{inside}]$, *i.e.*, $[(K^+)_{outside}/(K^+)_{inside}]$. Given the large potassium gradient, potassium ions will tend to diffuse out of the cell.

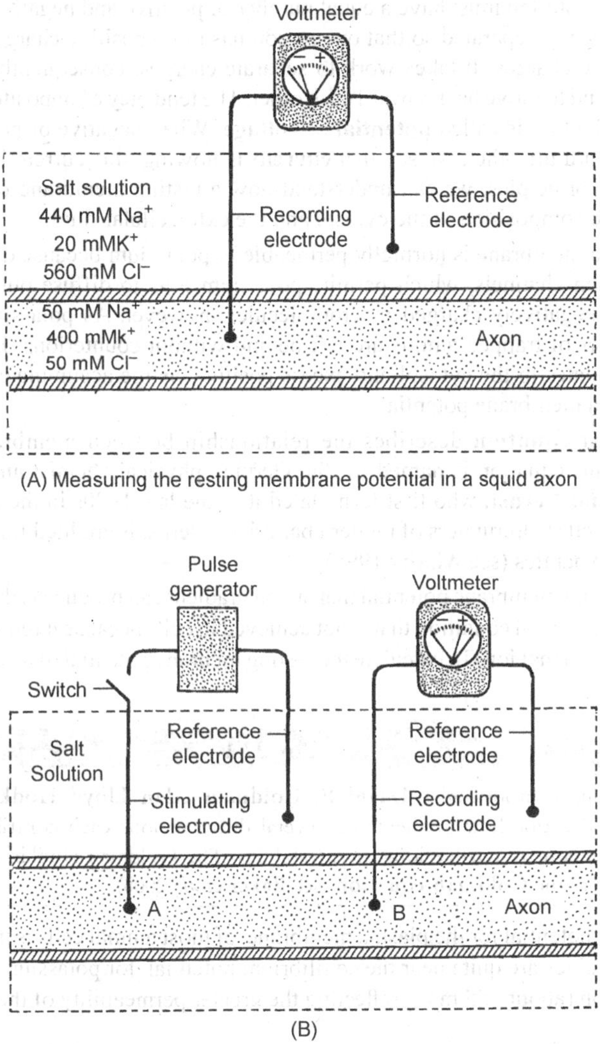

Fig. 27.4. An apparatus for measuring membrane potentials. A—Measurement of the resting membrane potential requires two electrodes, one inserted inside the axon (the recording electrode) and one placed in the fluid surrounding the cell (the reference electrode). Differeces in potential between the recording and reference electrodes are amplified by a voltage amplifier and displayed on a voltameter, an oscilloscope, or a computer monitor. B—Measurement of an action potential requires four electrodes, one in the axon for stimulation, another in the axon for recording, and two in the fluid surrounding the cell for reference. The stimulating electrode is connected to a pulse generator, which delivers a pulse of current to the axon when the switch is momentarily closed. The nerve impulse thus generates is propagated down the axon and can be detected a few milliseconds later by the recording electrode. The impulse is detected as a transient change in transmembrane potential, measured with respect to the reference electrode (after Baker et al., 2006).

The second basic principle is that of **electroneutrality**. When ions are in solution, they are always in pairs, one positive ion for negative ions, so that there is not net charge imbalance. For any given ion, which we will call A, there must be an oppositely charged ion B as the *counterion* for A. In the cytosol, potassium ions (K^+) serve as the counterions for the trapped anions. Outside the cell, sodium (Na^+) is the main cation and chloride (Cl^-) is its counterion.

Although a solution must have a equal number of positive and negative charges overall, these charges can be locally separated so that one region has more positive charges while another region has more negative charges. It takes work to separate charges; consequently, once they have been separated, they tend to move back toward each other. The tendency of oppositely charged ions to flow back toward each other is called **potential** or **voltage**. When negative or positive ions are actually moving, one toward the other, we say that **current** is flowing; this current is measured in amperes (A). Given these principles, we can understand how a resting membrane potential will form as a result of the ionic composition of the cytosol and the extracellular fluid.

The plasma membrane is normally permeable to potassium because of the leakiness of some types of potassium channels, which permits potassium ions to diffuse out of the cell. There are no channels for negatively charged macromolecules, however. As potassium leaves the cytosol, increasing number of trapped anions are left behind without counterions. Excess negative charge therefore accumulates in the cytosol and excess positive charge accumulates on the outside of the cell, resulting in a membrane potential.

The **Nernst equation** describes the relationship between membrane potential and ion concentration. This equation is named for the German physical chemist and 1920 Nobel laureate **Walther Hermann Nernst**, who first formulated it in the late 1880s in the context of his work on electrochemical cells (forerunners of modern batteries). Nernst is credited for the discovery of Third Law of Thermodynamics (see **Allaby** 1995).

The maximum membrane potential that an ion gradient can produce is the *equilibrium potential* for that ion—a theoretical condition that is not achieved in cells because it requires that the membrane be permeable only to that ion. To calculate the resting mebrane potential of a cell, **Goldman equation** is used (Box 27.1).

Box 27.1

The pioneering neurobiologist, **David E. Goldman**, **Alan Lloyd Hodkin** and **Bernard Katz** were the first to describe how gradients of several different ions each contribute to the membrane potential as a function of relative ionic permeabilities. The Goldman-Hodkin-Katz equation is more commonly known as the **Goldman equation** (see **Baker** *et al.*, 2006).

The resting potential for the plasma membrane of most animal cells is usually in the range –60 to –75 mV. The values are quite near the equilibrium potential for potassium ion, but vary for from that for sodium ion (about +55 mV), reflecting the greater permeability of the resting membrane for potassium.

27.3. VOLTAGE-GATED CATION CHANNELS

The plasma membrane of all electrically excitable cells such as neurons, muscle, endocrine and egg cells contain **ligand-gated channels** or **voltage-gated cation channels**, which are responsible for generating the action potentials. An action potential is triggered by a **depolarization** of the plasma membrane—that is by a shift in the membrane potential to a less negative value. In nerve and skeletal muscle cells, a stimulus that causes sufficient depolarization promptly causes **voltage-gated Na^+ channels** to open, allowing a small amount of Na^+ to enter the cell down its electrochemical gradient. The influx of positive charge depolarizes the membrane further, thereby opening more Na^+ channels, which admit more Na^+ ions, causing still further depolarization. This process continues in a self-amplifying fashion until, within a fraction of a millisecond, the electrical potential in the local region of membrane has shifted from its resting value of about –70 mV to almost as far as the Na^+ equilibrium potential of about +50 mV. At this point, when the net electrochemical driving force for the flow of Na^+ is almost zero, the cell would come to a new restive state, with all of its Na^+ channels permanently open, if the open conformation of the channel were stable. The cell is saved from such a permanent electrical spasm by *two* mechanisms that act in concert: inactivation of the Na^+ channels, and opening of voltage-gated K^+ (Kv) channels.

(*i*) Inactivation of sodium ion (Na^+) channels. The Na^+ channels have a automatic inactivating mechanism, which causes the channels to reclose rapidly even though the membrane is still depolarized. The Na^+ channels remain in this **inactivated state**, unable to reopen, until a few milliseconds after the membrane potential has returned to its initial negative value. The Na^+ channel can therefore exist in three distinct states: **closed, open** and **inactivated**. All this contribute to the rise and fall of the action potential as shown in Figure 27.5.

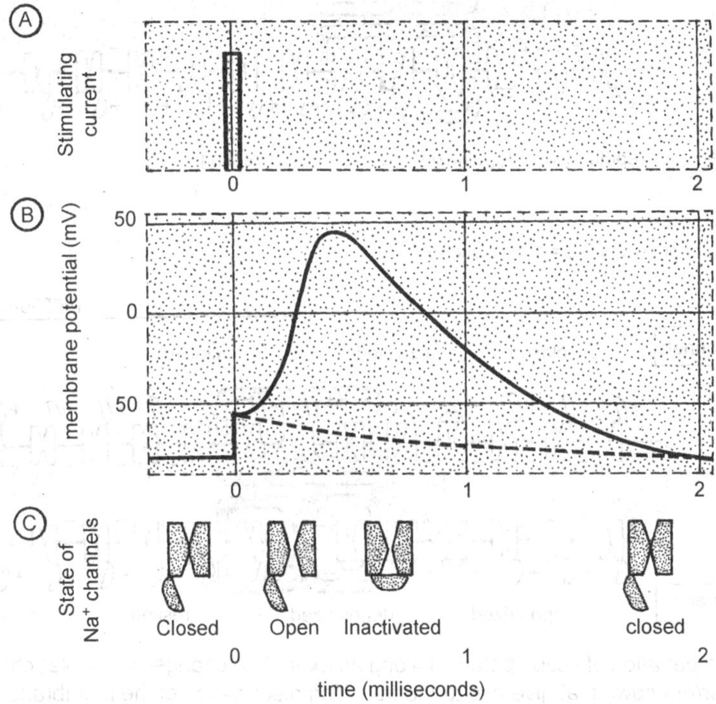

Fig. 27.5. An action potential. A—An action potential is triggered by a brief pulse of current, which partially depolarizes the membrane as shown in the plot of membrane potential versus time (B). B—The broken line curve shows how the membrane potential would have simply relaxed back to the resting value after the initial depolarising stimulus if there had been no voltage-gated ion channels in the membrane; this relatively slow return of the membrane potential to its initial value of –70 mV in the absence of open Na^+ channels occurs because of efflux of K^+ through K^+ channels, which open in response to membrane depolarization and drive the membrane back toward the K^+ equilibrium potential. The solid line curve shows the course of the action potential that is caused by the opening and subsequent inactivation of voltage-gated Na^+ channels, whose state is shown in (C). C—The membrane cannot fire a second action potential until the Na^+ channels have returned to the closed conformation; until then, the membrane is refractory to stimulation (after Alberts *et al.*, 2002).

***(ii)* Voltage-gated K^+ (Kv) channels.** These channels provide a second mechanism in most nerve cells to help bring the activated plasma membrane more rapidly back toward its original resting potential, ready to transmit a second impulse. Potassium ion (K^+) channels open, so that the transient influx of Na^+ is rapidly overcame by an efflux of K^+, which quickly drives the membrane back toward the K^+ equlibrium potential, even before the inactivation of the Na^+ channels is complete. These K^+ channels respond to changes in membrane potential in much the same way as the Na^+ channels do, but with slightly slower kinetics: for this reason; they are sometimes called **delayed K^+ channels** (**Alberts** *et al.*, 2002).

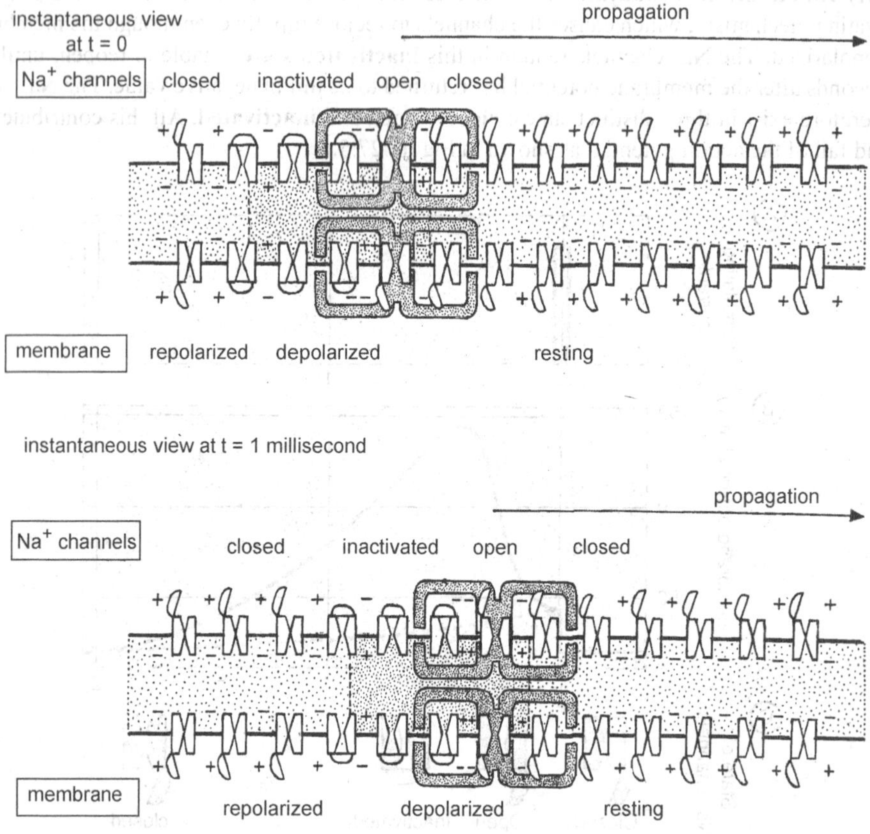

Fig. 27.6. Propagation of action potential along an exon. The changes in the Na^+ channels and the current flows that give rise to the travelling disturbance of the membrane potential. The region of the axon with a depolarized membrane is shaded. Note that an action potential can only travel away from the site of depolarization, because Na^+ channels inactivation prevents the depolarization from spreading backward.

Molecular structure of Kv channels and their working. The Kv channel or voltage-gated potassium ion channel of eukaryotes contain six membrane associated polypeptide helices, named **S1-S6** whcih are grouped into two functionally distinct domains: 1. Pore domain and 2. Voltage-sensing domain. Each Kv channel typically contains a large cytoplasmic structure (besides two transmembrane domains; Fig. 27.7A) whose composition varies among different channels. Inactivation of the channel is accomplished by movement of a small **inactivation peptide** that dangles from the cytoplasmic portion of the protein of Kv channel. The inactivation peptide is thought to gain access to the cytoplasmic mouth of the pore by snaking its way through one of four "side windows" indicated in the figure. When one of these dangling peptides moves up into the mouth of the pore (Fig. 27.7A), the passage of ions is blocked, and the channel is inactivated. And a subsequent stage of the cycle, the inactivation peptide is released and the gate to the channel is closed.

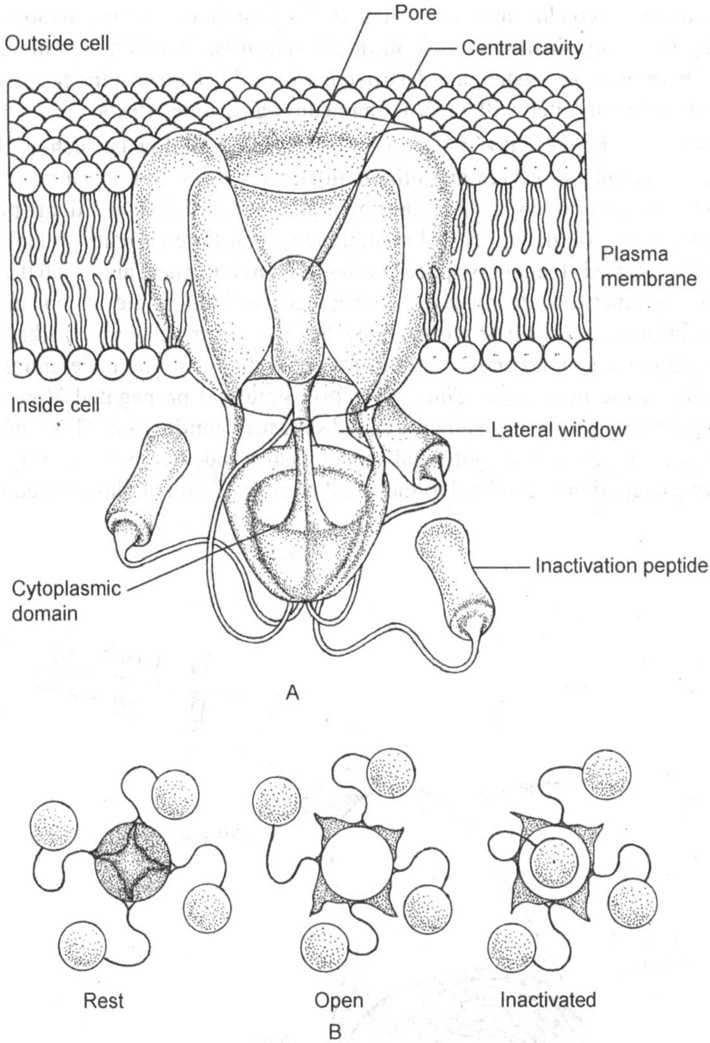

Fig. 27.7. Conformational states of a voltage gated potassium ion channel (Kv channel). A—Three-dimensional model of a eukaryonic voltage-gated K^+ ion channel. Inactivation of channel activity occurs as one of the inactivation peptides, which dangle from the cytoplasmic portion of the complex, fits into the cytoplasmic opening of the channel. B—schematic representation of a view into K^+–ion channel, perpendicular to the membrane from the cytoplasmic side, showing the channel in the closed (resting), open and inactivate state (after Karp 2010).

Once opened, more than ten million potassium (K^+) ions can pass through the Kv channel per second, which is nearly the rate that would occur by free diffusion in solution. Because of the large ion flux, the opening of a relatively small number of Kv channels has significant impact on the electrical properties of the membrane. After a channel is open for a few milliseconds, the movement of K^+ ions is "automatically" stopped by a process called **inactivation (Karp** 2010).

27.4. MYELINATION OF AXONS AND SALTATORY CONDUCTION

The axons of many vertebrate neurons are insulated by **myelin sheath**, which greatly increases the rate at which an axon can conduct an acton potential. The importance of myelination is dramatically demonstrated by the demyelinating disease **multiple sclerosis**, in which myelin sheaths in some regions of the central nervous system are destroyed; where this happens, the propagation of nerve impulses is greatly **slowed**, often with devastating neurological side effects.

Myelin is formed by specialized supporting cells called **glial cells**, such as **Schwann cells** myelinate axons in peripheral nerves and **oligodendrocytes** do so in the central nervous system. These glial cells wrap layer upon layer of their own plasma membrane in a tight spiral around the axon (Fig. 27.8), thereby insulating axonal membrane so that little current can leak across it. The myelin sheath is interrupted at regularly spaced **nodes of Ranvier**, where almost all the Na^+–channels in the axon are concentrated. Because the ensheathed portions of the axonal membrane have excellent cable properties (in other words, they behave electrically much like well-designed underwater telegraph cables), a depolarization of the membrane at one node almost immediately spreads passively to the next node. Thus, an action potential propagated along a myelinated axon by jumping from node to node, a process called **saltatory conduction**. This type of conduction has two main advantages: action potential travel faster and metabolic energy is conserved because the active excitation is confined to the small regions of axonal plasma membrane at nodes of Ranvier.

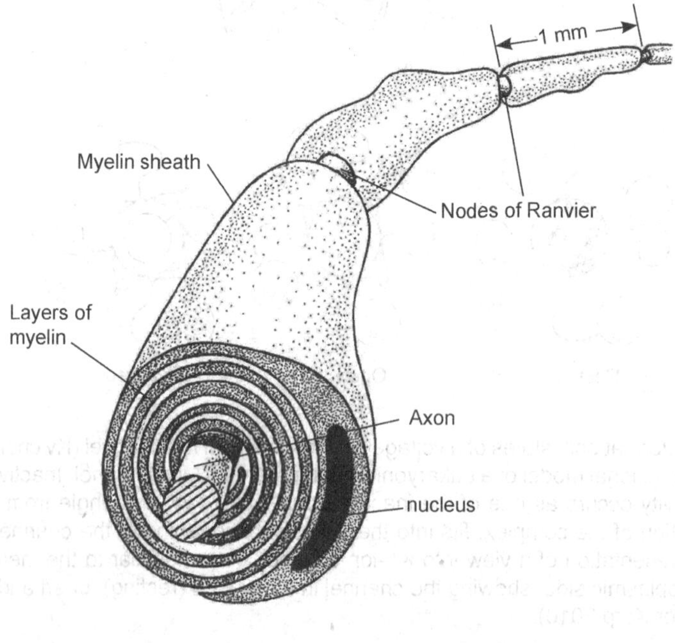

Fig. 27.8. Myelination. A myelinated axon from a peripheral nerve. Each Schwann cell wraps its plasma membrane concentrically around the axon to form a segment of myelin sheath about 1 mm long.

27.5. PATCH-CLAMP RECORDING

Neurons and skeletal muscle cell plasma membranes contain many thousands of voltage-gated Na^+ channels, and the current crossing the membrane is the sum of the currents flowing through all of these. This aggregate current can be recorded with an intracellular microelectrode. Remarkably, however, it is also possible to record current flowing through individual channels. This is achieved by means of **single-channel recording** or **patch-clamp recording** (=patch clamping), method that has revolutionized the study of ion channels by allowing researchers to examine transport through a single molecule of channel protein in a small patch of membrane covering the mouth of a micropipette (Fig. 27.9). (Technique of patch clamping was developed by **Erwin Neher** and **Bert Sackman**, who earned a Nobel Prize in 1991 for their discovery.) With this simple but powerful technique, the detailed properties of ion channels can be studied in all sorts of cell types. This work has led to the discovery that even cells that are not electrically excitable usually have a variety of gated ion channels in their plasma membrane.

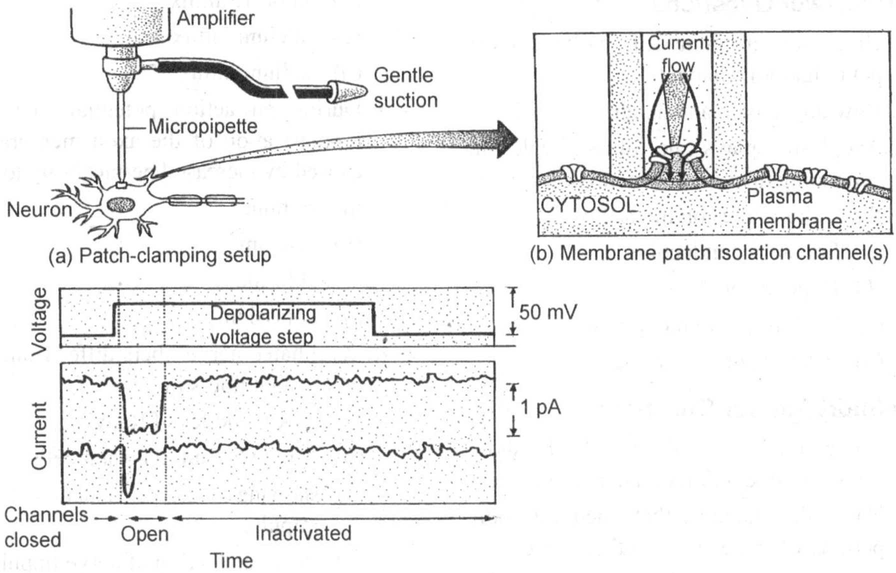

Fig. 27.9. Patch Clamping. A—A firepolished micropipette with a diameter of about 1 mm is carefully placed against a cell, such as the neuron, and B—gentle suction is applied to form a tight seal between the pipette and the plasma membrane. The density of channels is usually low enough that only one or a few channels will be in the membrane under the mouth of the pipette.

Patch-clamp recording indicates that individual voltage-gated Na^+ channels open in all-or-nothing fashion. The times of a channel's opening and closing are random, but when open, the channel always has the same large conductance, allowing more than 1000 ions to pass per millisecond. Therefore, the aggregate current crossing the membrane of an entire cell does not indicate the degree to which a typical individual channel is open but rather the total number of channels in the membrane that are open at any one time.

Lastly, Na^+ channels are not only kind of voltage-gated cation channel that can generate an action potential. The action potential in some muscle, egg and endocrine cells, for example, depend on **voltage-gated Ca^{2+} channels** rather than on Na^+ channels.

QUESTIONS

Long Answer Questions

1. What is action potential and how is it generated?
2. What is nerve impulse and how it is propagated along a nerve fiber?
3. Describe briefly the mechanism of conduction of nerve impulse along a nerve fiber.
4. How does the pattern of nerve conduction vary in myelinated and unmyeinated nerve fibers?

Short Answer Questions

1. Give an account of voltage-gated potassium ion channel.
2. How do an axon and dendrite differ?
3. Axoplasm plays no direct role in generating membrane potential. Explain.
4. Write short notes on the following:
   (*a*) Saltatory conduction
   (*b*) Depolarisation
   (*c*) Sodium-potassium pump
   (*d*) Patch-clamp recording

Very Short Answer Questions

1. Name the nerve fibers which show saltatory conduction of a nerve impulse.
2. Name the channels that open for short periods when needed and then close.
3. What is the effect of temperature on the conduction of nerve impulse?

Multiple Choice Questions

1. A nerve impulse leaves a neuron through the
   (*a*) dendrite
   (*b*) cyton
   (*c*) axon
   (*d*) Nissl bodies
2. The dominant current during the membrane depolarization phase in an axon is
   (*a*) potassium influx
   (*b*) chloride influx
   (*c*) calcium influx
   (*d*) sodium influx
3. During an action potential, the rapid repolarization of the axon membrane is caused by increased permeability to
   (*a*) sodium
   (*b*) calcium
   (*c*) chloride
   (*d*) potassium
4. Myelinated nerve fibers differ from non-myelinated fibers in
   (*a*) lacking nodes of Ranvier
   (*b*) being without Schwann cells
   (*c*) showing saltatory conduction of nerve impulse
   (*d*) slow conduction of nerve impulse

ANSWERS

Very Short Answer Questions

1. Medullated nerve fibers of vertebrates.
2. Sodium ion channels.
3. Rise in temperature accelerates rate of nerve impulse conduction.

Multiple Choice Questions

1. (*c*) 2. (*d*) 3. (*d*) 4. (*d*)

28 Centrioles and Basal Bodies

Cytoplasm of some eukaryotic cells contains two cylindrical, rod-shaped, microtubular structures, called **centrioles**, near the nucleus. The portion of the cytoplasm in which centriole formation takes place is called **centrosome**. Centrioles lack limiting membrane and DNA or RNA and form a spindle of microtubules, the mitotic apparatus during mitosis or meiosis and sometimes get arranged just beneath the plasma membrane to form and bear flagella or cilia in flagellated or ciliated cells (**Fulton**, 1971). When a centriole bears a flagellum or cilium, it is called **basal body**. There are many synonyms for basal body, including kinetosome, kinetoplast, blepharoplast, basal granule, basal corpuscle, and proximal centriole.

Due to the classic works of **Henneguy** and **Lenhossek** (1897), it has been proposed that basal bodies of cilia and flagella are homologous with the centrioles found in mitotic spindle.

28.1. OCCURRENCE

Centrioles are present in all animal groups, in lower plants (algae, moss and ferns) and in primitive flagellates such as dinoflagellates and *Chlamydomonas*. In flagellated animal cells centrioles play an important role in organising the mitotic apparatus, while in primitive flagellates they function, exclusively as basal bodies for the flagella. They are absent in prokaryotes, red algae, yeast, cone-bearing and flowering plants (conifers and angiosperms) and some non-flagellated or non-ciliated protozoans (such as amoebae). Some species of amoebae have a flagellated stage as well as an amoeboid stage; a centriole develops during the flagellated stage but disappears during the amoeboid stage. Apparently the original function of the centrioles is locomotion and their role in mitotic division is secondarily acquired.

28.2. STRUCTURE

Centrioles and basal bodies are cylindrical structures which are 0.15–0.25 μm in diameter usually 0.3–0.7 μm in length, though, some are as short as 0.16 μm and others are as long as 8 μm (see **Fulton**, 1971). Both have following ultrastructural components:

1. Cylinder wall. The most striking and regular ultrastructural feature of centrioles and basal bodies is the array of nine triplet microtubules equally spaced around the perimeter of an imaginary cylinder (Fig. 28.1). The space between and immediately around the triplet is filled with an amorphous, electron-dense material (see Chapter 46). In transverse section the triplets are arranged like vanes or blades of pinwheel or turbine. Each triplet or blade is tilted inward to the central axis at an angle of about 450 to the circumference; within each blade the tubules twist from one end to the other or describe a helical course. Since centrioles have no outer membrane, the triplets are considered to form the wall of the cylinder, and arbitrarily define the inside and outside of the centriole.

2. Triplets. The nine triplets that make up the wall are basically similar in centrioles and basal bodies. The three subunit microtubules have been designated A, B, and C, with the innermost tubule being A (Fig. 28.2). Individual tubules are 200–260 Å in diameter. Only the A tubule is round; the others are incomplete, C-shaped and share their wall with the preceding tubule. At both ends the C tubule often terminates before the A and B tubules.

The substructure of A, B and C tubules, is similar to the structure of other microtubules (**Stephens**, 1970). The A tubule has 13, 40–45 Å globular subunits around its perimeter. Three or four of these subunits are shared with the B tubules, which in turn share several of its subunits with the C tubules.

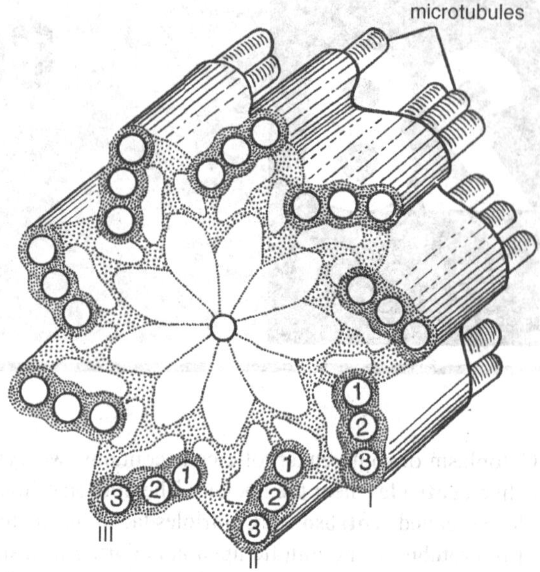

Fig. 28.1. Cross-sectional structure of a centriole. There are nine groups of three microtubules (after De Witt, 1977).

Often the triplets are thought to run parallel to one another and to the long axis of the cylinder, but this is not always the case. In the basal bodies of some organisms, the triplets get closer toward, the proximal end, so the diameter of the cylinder gets smaller. In some centrioles the triplets are parallel to one another but turn in a long-pitched helix with respect to the cylinder axis (**Fulton**, 1971).

3. Linkers. The A tubule of each triplet is linked with C tubule of neighbouring triplet by protein **linkers** at intervals along their entire length. These linkers hold the cylindrical array of the microtubules and maintain the typical radial tilt of the triplets.

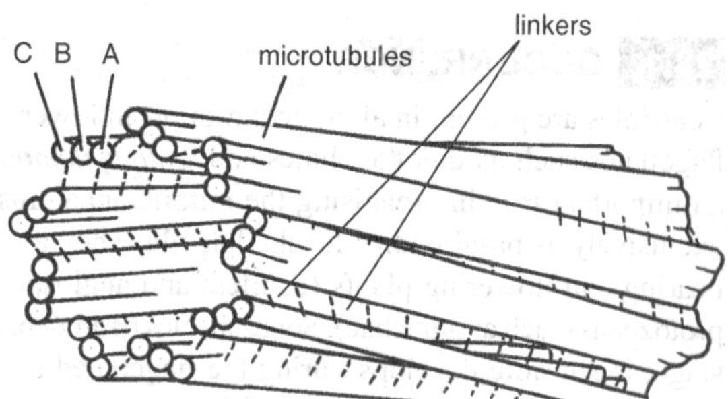

Fig. 28.2. Ultrastructure of a centriole (after Alberts *et al.*, 1989).

4. Cartwheel. There are no central microtubules in the centrioles and no special arms. However, often faint protein spokes are radiate out to each triplet from a central core, forming a pattern like a cartwheel. Such a cartwheel configuration determines the **proximal end** of a centriole and, thus, provides a structural and functional **polarity** to it. The growth of the centriole takes place from the distal end, and in the case of basal bodies, it is from this end that cilium is formed. Moreover, the procentrioles which are formed at right angles to the centriole, are located near the proximal end.

5. Ciliary Rootlets. In some cells, from the basal ends of the basal bodies originate the ciliary rootlets which are of following two typs:

(*i*) **Tubular root fibrils.** The tubular root fibrils have the diameter of 200Å.

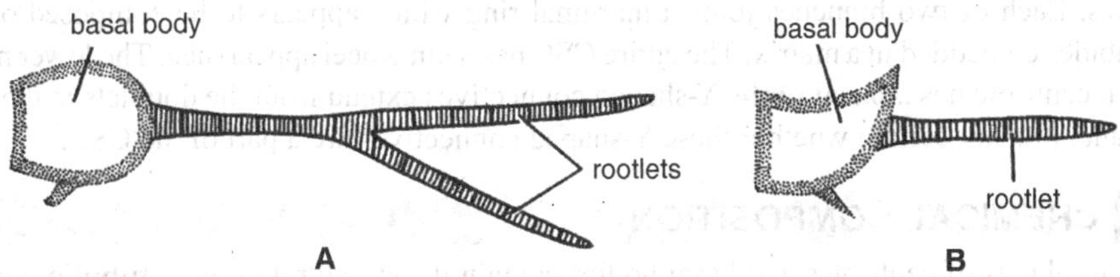

Fig. 28.3. Relation of rootlets with the basal body. A—Double rootlets (*e.g.*, molluscs); B—Single rootlet (*e.g.*, frog-*Rana*).

(*ii*) **Striated rootlets.** Most ciliary rootlets are striated, having a regular cross-banding with a repeating period of 55 to 70 nm. The striated fibres of rootlets are composed of parallel microfilaments; 3 to 7 nm in diameter, which in turn are formed of globular subunits. These fibres and filaments may have a structural role such as anchoring the basal body. Due to microfilaments, ciliary rootlets also have a contractile role. The rootlet may be double (*e.g.*, molluscs) or single (*e.g.*, the frog *Rana*).

6. Basal feet and satellites. The **basal feet** are dense processes that are arranged perpendicularly to the basal body. These processes impose a structural asymmetry on the basal body that is related with direction of the ciliary beat. A basal foot is composed of microfilaments that terminate in a dense bar. It may act as a focal point for the convergence of microtubules.

Satellites or **pericentriolar bodies** are electron-dense structures lying near the centriole that are probably nucleating sites for the microtubules.

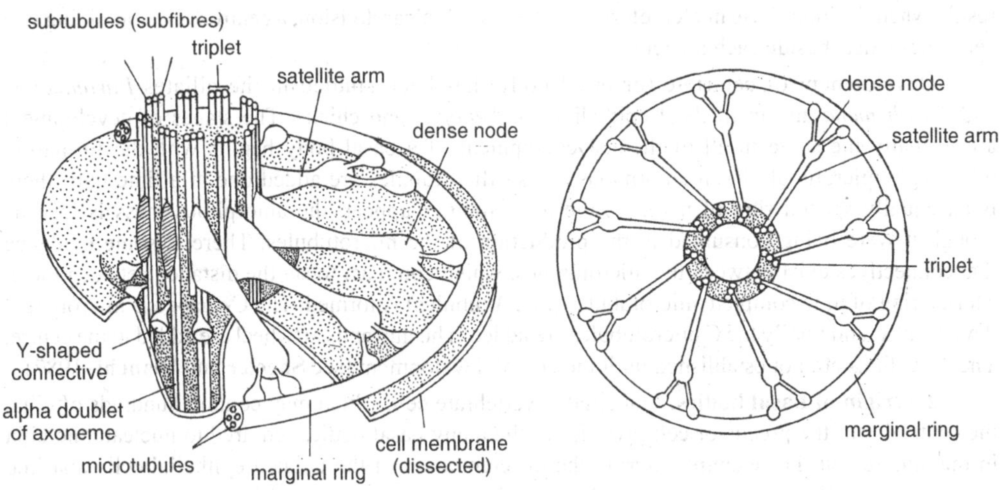

Fig. 28.4. A—Morphology of CSC (centriolar satellite complex) in the sea cucumber. B—Diagramatic transverse section through the CSC.

The Centriolar Satellite Complex (CSC)

The *centriolar satellite complex* (*CSC*) of sperms of sea cucumber *Cucumaria miniata* (an enchinoderm animal) has been described by **Fontaine** and **Lambert** (1976) (Fig. 28.4). The B-subfiber of each centriolar triplet gives off a satellite arm. The arm thickens distally to form a dense node and then bifurcates. Each of two branches joins a marginal ring which appears to be composed of several microtubules embedded in a matrix. The entire CSC has a pin wheel appearance. The lower margin of the distal centriole has a **basal plate**. Y-shaped connectives extend from the doublets to the flagellar membrane. It is not certain whether these Y-shaped connectives are a part of the CSC.

28.3. CHEMICAL COMPOSITION

The microtubules of centrioles and basal bodies contain the structural protein, **tubulin**, along with lipid molecules (**Fulton**, 1971). The centrioles and basal bodies contain a high concentration of ATPase enzyme. There exists a controversy that whether centrioles and basal bodies have DNA and RNA. **Fulton** (1971) has doubted the presence of nucleic acids in these organelles.

28.4. ORIGIN OF CENTRIOLES AND BASAL BODIES

The idea prevalent years ago that new centrioles arise by the division of existing centrioles is no longer accepted. Rather it appears that new centrioles are either produced *de novo* or are synthesized using an existing centriole as some form of template (semi-autonomous replication).

1. Origin of centrioles by duplication of pre-existing centrioles. In cultured fibroblasts, centriole doubling begins at around the time that DNA synthesis begins (interphase). First the two members of a pair of centrioles separate; then a daughter centriole, called **procentriole**, is formed perpendicular to each original centriole, the two organelles being separated from each other by a distance of 50 to 100 nm. An immature centriole contains a ninefold symmetric array of single microtubules; each microtubule then presumably acts as a template for the assembly of the triplet microtubules of mature centrioles. Each daughter centriole grows to mature size in late prophase maintaining their close proximity to and orientation at right angles to the mother centriole. As a result, when the interphase nuclei reform at the end of nuclear division, a centrosome containing two centrioles exists beside each nucleus.

Development of centriole (or basal body) has been studied in the ciliates *Paramecium* and *Tetrahymena* and in tracheal epithelium of *Xenopus* and chicks. The stages of development are virtually the same in all of them. Development of a basal body begins with the formation of a single microtubule in an amorphous mass. Microtubules are added one at a time until there is an equally spaced ring of nine. As the microtubules appear, the amorphous mass is lost, as though it were being consumed in the production of the microtubules. There is some evidence that connectives exist between the microtubules, which could act to set the distance between them. Thus, a ring of nine complete microtubules (*i.e.*, A tubules) is formed. The C-shaped B microtubule develop next and finally the C microtubules are added. The hub and cartwheel are added in the centre. The A–C links are not established until the end of development (see **Sheeler** and **Bianchi**, 1987).

2. Origin of basal bodies. In a ciliated vertebrate cell, which may contain hundreds of cilia, the centrioles of the precursor cell give rise to the many basal bodies required to nucleate the cilia in the mature cell. For example, during the differentiation of the ciliated epithelial cells that line the oviduct and the trachea, the centriole pair migrates from its normal location near the nucleus to the apical region of the cell where the cilia will form. There, instead forming a single daughter centriole in the typical manner, each centriole in the pair forms numerous electron-dense fibrogranular satellites. Many basal bodies then arise from these satellites and migrate to the membrane to initiate the formation of cilia.

3. The *de novo* origin of centrioles and basal bodies. There are certain cases where centrioles seem to arise *de novo*. For example, unfertilized eggs of many animals lack functional centrioles and use the sperm centriole for the first mitotic division (for cleavage), however, under certain experimental conditions—such as extreme ionic imbalance or electrical stimulation—the unfertilized egg can produce a variable number of centrioles. Each of these centrioles nucleates the formation of a small aster, one of which can be used by the egg for cleavage division, so that a haploid organism develops by a process called **parthenogenesis**. In fact, centriole precursors are stored in the cytoplasm of unfertilized eggs and can be activated to form a new centriole under special situation.

Like the centrioles, the basal bodies are found to possess some capacity for self-assembly and they appear suddenly in *Naegleria* as it changes from its amoeboid form to a typical ciliate (see **Reid** and **Leech**, 1980).

The unusual mode of duplication of centrioles and their continuity over many generations led to the earlier suggestion that centrioles might be fully **autonomous, self-replicating** organelles. Although it is now known that this is not the case and that under certain situation centrioles can arise *de novo* in the cytoplasm. So, it is possible that some information necessary for centriole formation is usually carried in the centriole itself (just as the replication of mitochondria and chloroplasts depends on extrachromosomal genes carried in the organelles). For example, in *Chlamydomonas* a set of genes that encode proteins involved in basal body structure and flagellar assembly is carried on a separate genetic element that segregate independently of the major chromosomes. The nature and location of this genetic element have still to be investigated (**Ramanis** and **Luck**, 1986).

28.5. FUNCTIONS

1. Formation of basal bodies and ultimately the cilia is the specialized function of the centrioles in the cell (Box 28.1).

Box 28.1 Ciliogenesis

The primitive function of the centrioles seems to have been the generation of cilia and flagella, the primitive locomotory organelles of the eukaryotes. Both members of a pair of centrioles can give rise to cilium. Usually, however, only one of them forms the cilium. Cilia and flagella always develop in the presence of basal bodies. The basal body acquires a basal plate, and the cilium then develops distal to the basal plate. Basal bodies produce the ciliary rootlets from their basal ends and the cilia from their apical ends.

2. The normal function of a pair of centrioles in most animal cells is to act as a focal point for the centrosome. The centrosome (also called the **cell centre**) organizes the array of cytoplasmic microtubules during interphase and duplicates at mitosis to nucleate the two poles of the **mitotic spindle**.
3. Sometimes centrioles can serve first one function and then another in turn: for example, prior to each division in *Chlamydomonas*, the two flagella resorb and the basal bodies leave their position to act as mitotic poles.
4. In spermatozoon one centriole gives rise to the tail fibre or flagellum.
5. Centrioles and basal bodies are also found to be involved in ciliary and flagellar beat.
6. Centrioles and basal bodies have a role in the reception of optical, acoustic and olfactory signals.
7. Recently, it has been suggested that centrioles could serve as devices for locating the directions of signal sources. Such a role for them has been conceived by comparing the geometric design of centrioles (with their disposition in pairs at the right angles and their ninefold symmetry) with human-made devices such as radar scanner, that detect directional signals (**Albrecht-Buehler**, 1981).

QUESTIONS

Long Answer Question

1. Describe the ultrastructure of basal body and centrioles.

Short Answer Questions

1. How are the centrioles and basal bodies (kinetosomes) are originated in the cell?
2. What are the main functions of the basal bodies and the centrioles?

Very Short Answer Questions

1. Define the terms basal bodies and centrioles.

Multiple Choice Questions

1. Organelle important in spindle formation during nuclear division is
   (*a*) centriole
   (*b*) Golgi apparatus
   (*c*) chloroplast
   (*d*) mitochondrion
2. Which of the following organelle is not present in plant cells?
   (*a*) peroxisomes
   (*b*) ribosomes
   (*c*) centrioles
   (*d*) mitochondria

ANSWERS

Very Short Answer Questions

1. *Basal body* is a microtubule containing structure located at the base of a eukaryotic flagellum or cilium that consists of nine sets of triplet microtubules; identical in appearance to a centriole. *Centriole* is microtubule containing structure embedded within the centrosome of animal cells, where centrioles lie at right angles to each other; identical in structure to the basal body of eukaryotic cilia and flagella.

Multiple Choice Questions

1. (*a*) 2. (*c*)

29 Cellular Movement: Muscle Contractility and Amoeboid Movement

Cellular movement involves motility and contractility. **Motility** occurs at the tissue, cellular and subcellular levels. The most conspicuous examples of motility, particularly in the animal world, take place at **the tissue level**. The muscle tissues common to most animals consist of cells specifically adapted to for *contraction*, and the movements produced are often obvious, whether manifested as the bending of a limb, the beating of a heart, or a uterine contraction during childbirth.

Equally important is the movement of intracellular components, which might be regarded as **motility at the subcellular level**. For example, highly ordered microtubules of the mitotic spindle play a key role in the separation of chromosomes during cell division (see Chapter 46). In addition, some cells display the phenomenon of cytoplasmic streaming, in which the cytoplasm undergoes rhythmic patterns of flow (*e.g.*, cyclosis in *Paramecium*). Other examples of mechanical work at the subcellular level include the characteristic movements of molecular structures that occur during cell growth and differentiation. An example of such a process is the transport of cellulose to the growing wall of a dividing or differentiating plant cell.

The microfilaments and microtubules of the cytoskeleton provide a basic scaffolding for specialized **motor proteins**, which interact with the cytoskeleton to produce motion at the molecular level (Table 29.1). The combined effects of these molecular motions produce movement at cellular level. In cases such as muscle contraction, the combined effects of many cells moving simultaneously produce motion at the tissue level.

Table 29.1 Selected motor proteins (or mechanoenzymes) of eukaryotic cells; they produce motion at the molecular level (Source: Becker *et al.*, 2006).

		Molecules	Typical function
A.	**Microtubule associated proteins**		
	1.	Cytoplasmic dynein	Motion toward minus end of microtubule
	2.	Axonemal dynein	Activation of sliding in flagellar microtubule
	3.	Kinesins	Motion toward plus end of microtubules
B.	**Microfilament-associated (actin-binding) proteins**		
	1.	Myosin I, monomer	Motion along actin filament
	2.	Myosin II, filament	Slides actin filaments in sarcomere of muscle

In eukaryotes, there are two motility systems. The first involves interactions between specialized motor molecules and microtubules. Such **microtubule-based movement** occurs in **fast axonal transport**, one of the processes by which a nerve cell transport materials between the central part of the cell and outlying regions. The second type of eukaryotic motility is based on interactions between the actin microfilaments and members of the myosin family of motor molecules. A familiar example of **microfilament-based movement** in muscle contraction.

29.1. INTRACELLULAR MICROTUBULE-BASED MOVEMENT

Microtubules (MTs) provide a rigid set of tracks for the transport of a variety of membrane-enclosed organelles and vesicles. The centrosome (including centrioles) provides organization and orientation to MTs, because the minus ends of most microtubules are embedded in the centrosome. The centrosome is generally located near the center of cell, so traffic towards the minus ends of MTs might be considered "inbound" traffic. Traffic directed toward the plus ends might likewise be considered "outbound", meaning that it is directed toward the periphery of the cell.

While microtubules provide an organized set of tracks along which organelles can move, they do not directly generate the force necessary for movement. The mechanical work needed for movement depend on **microtubule-associated motor proteins (motor MAPs)** which attach to vesicles or organelles and then "walk" along the microtubule using ATP to provide the needed energy. These motor MAPs recognize the polarity of the MT, with each motor MAP having a preferred direction of movement. At present, only two major families of motor MAPs namely **kinesins** and **dyneins** are known.

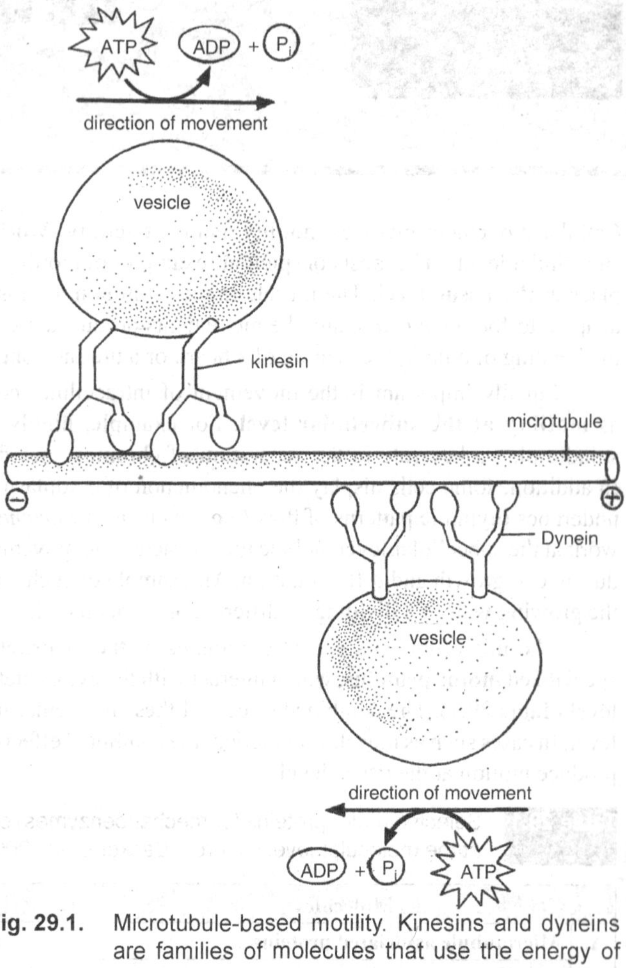

Fig. 29.1. Microtubule-based motility. Kinesins and dyneins are families of molecules that use the energy of ATP hydrolysis to "walk" along microtubules. In the process, they move intracellular structures (vesicles and organelles) along MTs. In general, members of the kinesin family move vesicles or organelles toward the plus ends of MTs-that is, from the center of the cell to the periphery. Dynein move in opposite direction toward the minus ends of MTs, and therefore toward the center of cell where the MTOC (microtubule-organizing center) is located (after Becker *et al.*, 2006).

Role of Kinesins and Dyneins in Movement of Organelles

A historically important cell type for studying microtubule-dependent intracellular movement is the neuron. In particular, the squid (*Loligo*, a molluscan animal) giant axon has been extremely useful for the biochemical purification of components that interact with MTs. A receptor protein or neurotransmitter synthesized in cell body of the neuron must be transported over distances upto a meter between the cell body and the nerve ending. The need for such transport arises because ribosomes are present only in the cell body, so no protein synthesis occurs in the axons or synaptic knobs. Instead proteins and membranous vesicles are synthesized in the cell body and transported along the axons to the synaptic knobs. Some form of energy-dependent transport is clearly required mechanism. The process is called **fast axonal transport** and appears to involve the movement of protein containing vesicles and other organelles.

The role of microtubules in axonal transport was initially suggested because the process is inhibited by **colchicine** and other drugs that impair MT function but is insensitive to drugs such as **cytochalasins** that affect microfilaments. Since then, MTs have been visualized along the axon by the deep-etch technique and shown to be prominent feature of the axonal cytoskeleton. Moreover, axonal MTs have small membranous vesicles and mitochondria associated with them.

Further extensive investigations have revealed that in a nerve cell, **kinesin** protein mediates transport from the cell body down the axon to the nerve ending; it is called **anterograde axonal transport**. Likewise, **cytoplasmic dynein** mediates transport in the opposite direction, towards the minus ends of the MTs; it is called **retrograde axonal transport**. These two motors transport the materials in opposite direction within the cytoplasm (Fig. 29.1).

Structure and Function of Kinesins

The first kinesins were originally identified in the cytoplasm of squid giant axons. Such "classic" kinesins consists of three parts (Fig. 29.2): 1. A globular head region that attaches to the microtubules

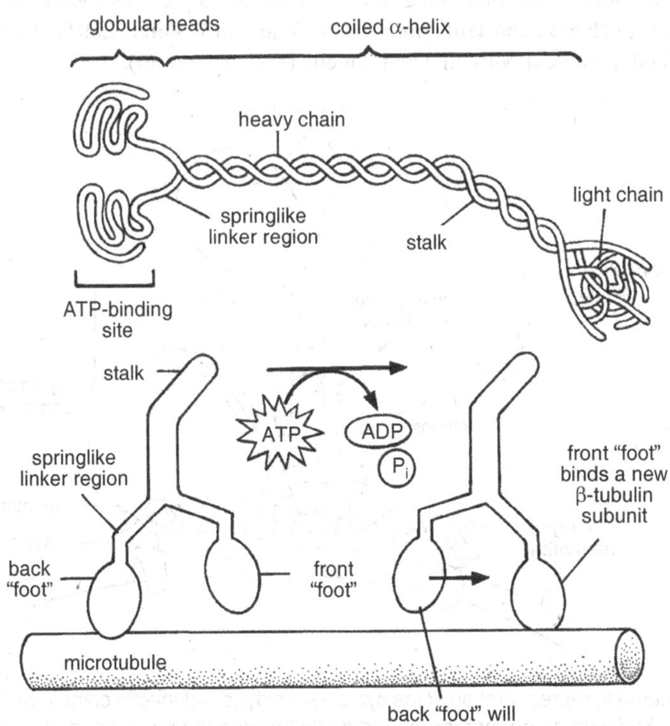

Fig. 29.2. Structure and movement of kinesin. A—The basic structure of a kinesin molecule. B—Kinesins "walk" along microtubules. The front "foot", one of the kinesin's two globular heads, detaches from a β-tubulin subunit by hydrolysing ATP, and moves forward. The rear "foot" then releases by hydrolysing ATP and springs forward, moving to the front in a "hand-over-hand fashion" (after Becker *et al.*, 2006).

and is involved in hydrolysis of ATP; 2. A coiled helical region; and 3. A light-chain region that is involved in attaching the kinesin to other proteins or organelles.

Classic kinesins move along microtubules in 8 nm steps: one of the two globular heads moves forward to make an attachment to a new β-tubulin subunit, followed by detachment of the trailing globular head which can now make an attachment to a new region of the MT (Fig. 29.2B). This movement occurs in a "hand-over-hand" fashion, as the two heads take turns serving as the front "foot". This movement is coupled to the exchange of ATP and ADP at specific sites within the heads. The result is that kinesin moves toward the plus end of a MT in an ATP-dependent fashion. A single kinesin molecule exhibits **processivity**. It can cover long distances before detaching from a MT. A single kinesin molecule can move as far as 1 μm which is a great distance relative to its size. Lastly, as a molecular motor, kinesin appears to be quite efficient, estimates of its efficiency in converting the energy of ATP hydrolysis to useful work are on the order of 60–70% (**Endow**, 2003).

Structure and Functions of Cytoplasmic Dynein

The dynein family of motor MAPs consists of two basic types: 1. Cytoplasmic dynein and 2. Axonemal dynein. **Cytoplasmic dynein** contains two heavy chains that interacts with microtubules, three intermediate chains and four light chains. In contrast to most kinesins, cytoplasmic dynein moves toward the minus ends of MTs. Also, unlike kinesins, cytoplasmic dynein cannot efficiently bind to cytoplasmic organelles on its own. Instead, dynein is associated with the protein complex known as **dynactin** (Fig. 29.3). The dynactin complex helps to link cytoplasmic dynein to the cargo (such as membraneous vesicle) it transports along MTs, by binding to proteins such as spectrin attached to the membrane cargo (**Karki** and **Holzbaur**, 1999; **Asai** and **Koonce**, 2001). The axonemal dynein has been described in context with cilia and flagella (see Chapter 30).

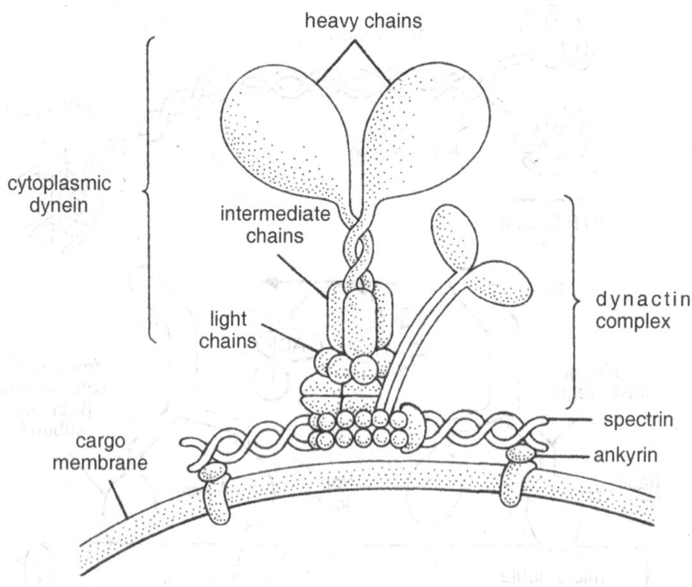

Fig. 29.3. Schematic representation of the cytoplasmic dynein-dynactin complex. Cytoplasmic dynein is linked with cargo membrane indirectly through the dynactin multiprotein complex. The dynactin complex binds to spectrin in the membrane of the cargo vesicle (after Becker *et al.*, 2006).

Figure 29.4 has shown how motor MAPs are involved in transport of intracellular vesicles (**Allan, Thompson** and **McNiven**, 2002).

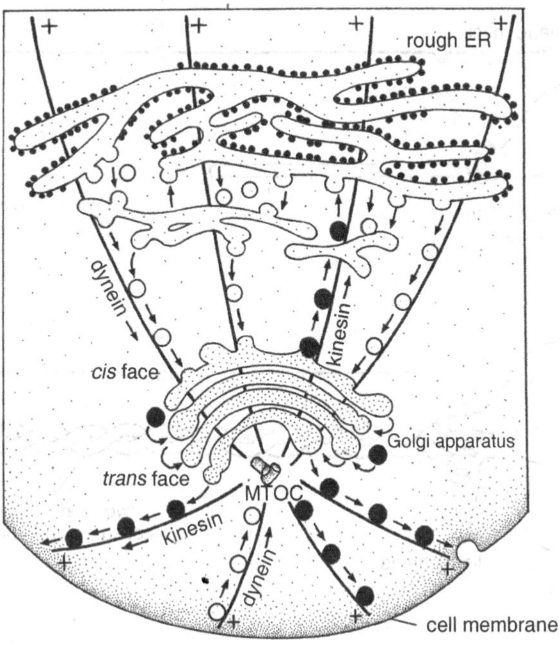

Fig. 29.4. Microtubules, Motor MAPs and the Golgi apparatus. According to this model, vesicles going to and from the Golgi apparatus are attached to microtubules and are thought to be carried by motor MAPs similar to dynein and kinesin. Dynein is a minus-end directed motor MAP, whereas kinesin is plus-end directed MAP. Thus, vesicles derived from either the ER or the plasma membrane are carried toward the Golgi and MTOC by dynein, whereas vesicle derived from the Golgi apparatus are carried toward either the ER or the cell periphery by kinesins (Becker *et al.*, 2006).

Microtubule-Based Motility

Microtubules are crucial not only for movement within cells, but also for the movement of **flagella** and **cilia**, the motile appendages of eukaryotic cells (cilia and flagella have been dealt in quite detail in the next chapter).

Actin-Based Cell Movements: The Myosins

Movement of molecules and other cellular components also occurs along another major filament system in the cell, the actin cytoskeleton. As with microtubules, mechanoenzymes act as ATP-dependent motors that exert force on actin microfilaments within cells. These mechanoenzymes are all members of a large superfamily of proteins known as **myosins**. Currently there are at least 18 known classes of myosins. All myosins have at least one polypeptide chain, called the **heavy chain**, with a globular head group at one end attached to a tail of varying length (Fig. 29.5). The globular head binds to actin and uses the energy of ATP hydrolysis to move along an actin filament. The structure of the tail region varies among the different kinds of myosin. Myosins typically contain small polypeptides bound to the globular head group. These polypeptides, referred to as the **light chains**, often play a role of regulating the activity of the myosin ATPase.

Several myosins have been shown to have specific functions in events as wide-ranging as muscle contraction, cell movement and phagocytosis. For example, mutations in the mouse indicate that a myosin V is required for transfer of pigment granules from melanocytes (cells that produce pigment) to keratinocytes (cells in the hair shaft that normally take up the pigment).

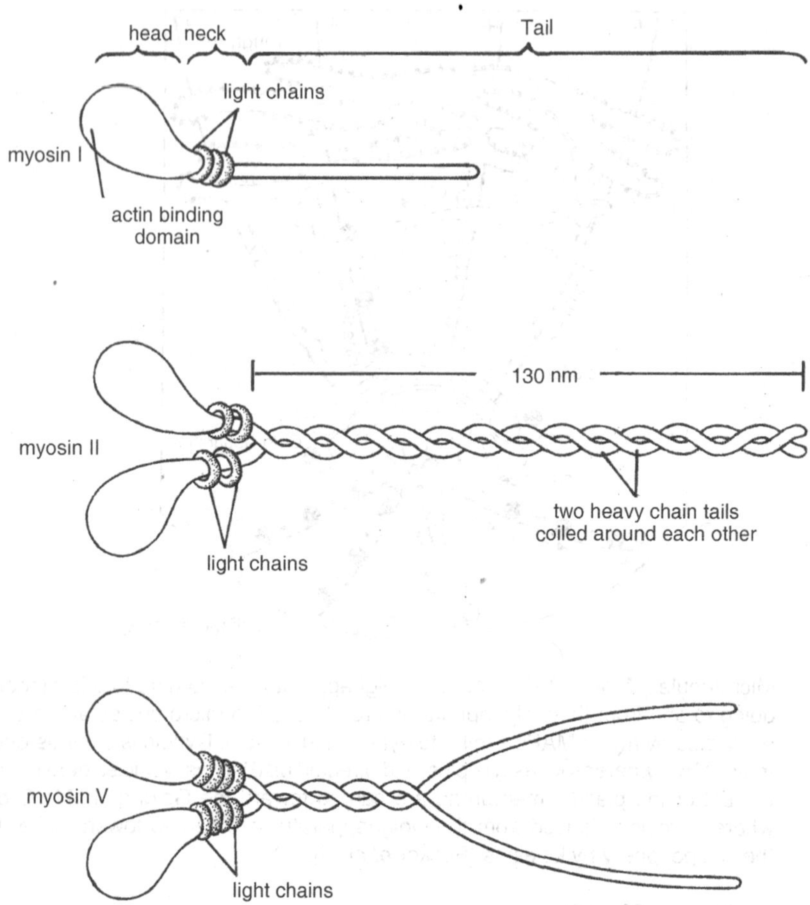

Fig. 29.5. Myosin family members. All myosins have an actin-and ATP-binding heavy chain "head" and typically have two or more regulatory light chains. Some myosins, such as myosin I, have one head. Others, such as myosin II and myosin V, associate via their tails into two headed proteins (after Becker *et al.*, 2006).

Cooperation Between Microtubules and Microfilaments

Two types of motors (*i.e.*, kinesin and dynein) may be physically linked to one another. The movement of vesicles within animal cells occurs on microtubules, as previously described. However, once they approach the end of the microtubule, these membraneous vesicles are often thought to switch over to microfilament tracks for the local movement through the actin-rich periphery of the cell (Fig. 29.6).

Cooperation between microtubules and microfilaments has been best studied in pigment cells (Fig. 29.6). In mammals pigment cells (*i.e.*, melanosomes) are transported into fine peripheral processes of the pigment cell by one of the myosin V isoforms called Va. Melanosomes are then transferred to hair follicles where they become incorporated into a developing hair. Mice lacking myosin Va activity are unable to transfer melanosomes into hair follicles, causing their coat to have a much lighter colour. Humans lacking a normal gene encoding myosin Va suffer from a rare disorder, called **Griscelli syndrome**; these individuals exhibit partial **albinism** (lack of skin colouration) and suffer other symptoms thought to be related to defect in vesicle transport. In 2000 it was discovered that a subset of Griscelli patients had a normal myosin Va gene, but lacked a functional gene encoding a peripheral membrane protein called **Rab27a**. The Rab family of protein molecules tend to tether vesicles to target membranes. Rabs are also involved in the attachment of myosin (and kinesin) motors to membrane surfaces (Fig. 29.7).

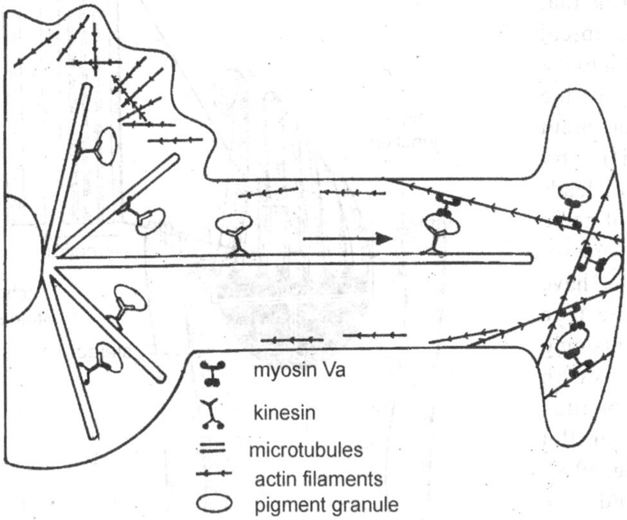

Fig. 29.6. Role of myosin Va in the movement of pigment granules. Most vesicle transport is thought to be mediated by members of kinesin and dynein families, which carry their cargo over long distances. It is thought that some vesicles also carry myosin motors, such as myosin Va, which transport their cargo over microfilaments, including those present in the peripheral (cortical) regions of the cell. The two types of motors may act in cooperative manner, as illustrated here in the case of a pigment cell in which pigment granules move back and forth within extended cellular processes (after Karp, 2010).

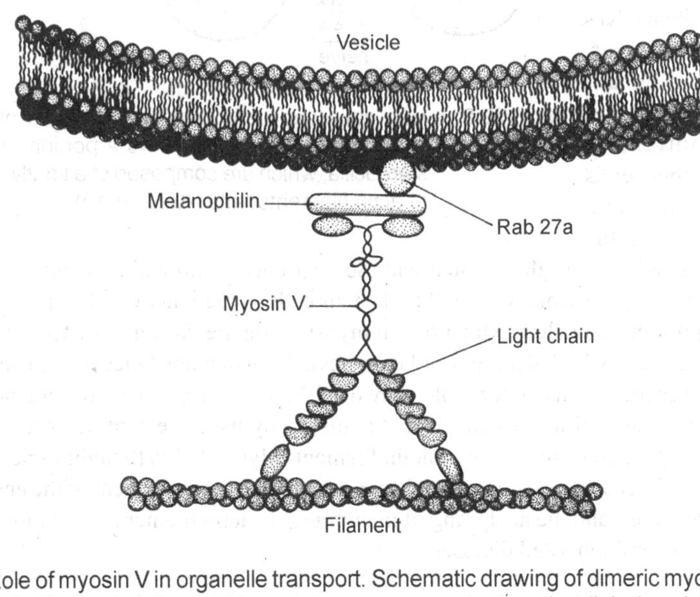

Fig. 29.7. Role of myosin V in organelle transport. Schematic drawing of dimeric myosin V molecule. Rab 27a and melanophilin proteins serve as adaptors that link the globular tail of the motor to the vesicular membrane (after Karp, 2010).

Hair Cells and Usher 1B Syndrome

Hair cells are named for the bundle of stiff, hairlike **stereocilia** that project from the apical surface of the cell into the fluid-filled cavity of the inner ear. Displacement of the sterocilia by mechanical stimuli leads to the generation of nerve impulses that we perceive as sound. Stereocilia have no relation to the true cilia. Instead of containing microtubules, each stereocilium is supported by a bundle of parallel actin filaments (Fig. 29.8), whose barbed ends are located at the outer tip of the projection and pointed ends at the base. Stereocilia have provided some of the most striking images of the dynamic nature of the actin cytoskeleton. While the stereocilia are themselves are permanent structures, the actin bundles are in constant flux. Actin monomers continually bind to the barbed ends of each filament treadmill through the body of the filament, and dissociate from the pointed end. Several unconventional myosins (*i.e.*, I, V, VI, VII and XV) are localized at various sites within the hair cells of the inner ear. The precise role of these various motor proteins is unclear. Mutations in myosin VIIa are the cause of **Usher 1B syndrome**, which is characterized by both deafness and blindness. As in humans, mice that are homozygous for the mutant allele encoding this motor protein are deaf. Myosin VI, a processive organelle transporter protein in the cytoplasm of many cells, is distinguished by its movement in a "*reverse direction*", that is toward the pointed (minus) end of actin filament. Myosin VI is thought to be involved in the formation of clathrin-coated vesicles at the plasma membrane, the movement of the uncoated vesicles to the early endosomes, and the anchoring of membranes to actin filaments. Mutations in myosin VI are the cause of several inherited diseases.

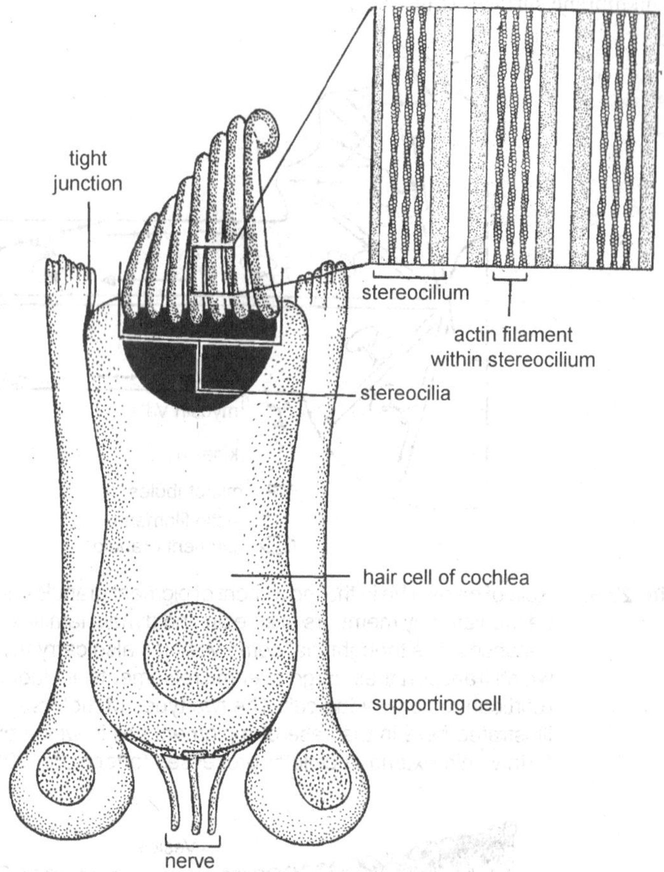

Fig. 29.8. Hair cells and actin bundles. Drawing of a hair cell of the cochlea. The inset shows a portion of several of the stereocilia, which are composed of a tightly grouped bundle of actin filaments (after Karp, 2010).

29.2. MUSCLE CONTRACTILITY

I. Contractility of Skeletal Muscles

Skeletal muscles derive their name from the fact that most of them are anchored to bones that they move. They are under voluntary control and can be consciously commanded to contract. Skeletal muscle cells have a highly irregular structure. A single, cylindrically shaped muscle cell is typically 10 to 100 µm thick, over 100 mm long, and contain hundreds of nuclei (*i.e.*, it is multinucleated). Because of these properties, a skeletal muscle cell is more appropriately called a **muscle fiber**. Muscle fibers have multiple nuclei because each fiber is a product of fusion of large numbers of mononucleated **myoblasts** (premuscle cells) in the embryo.

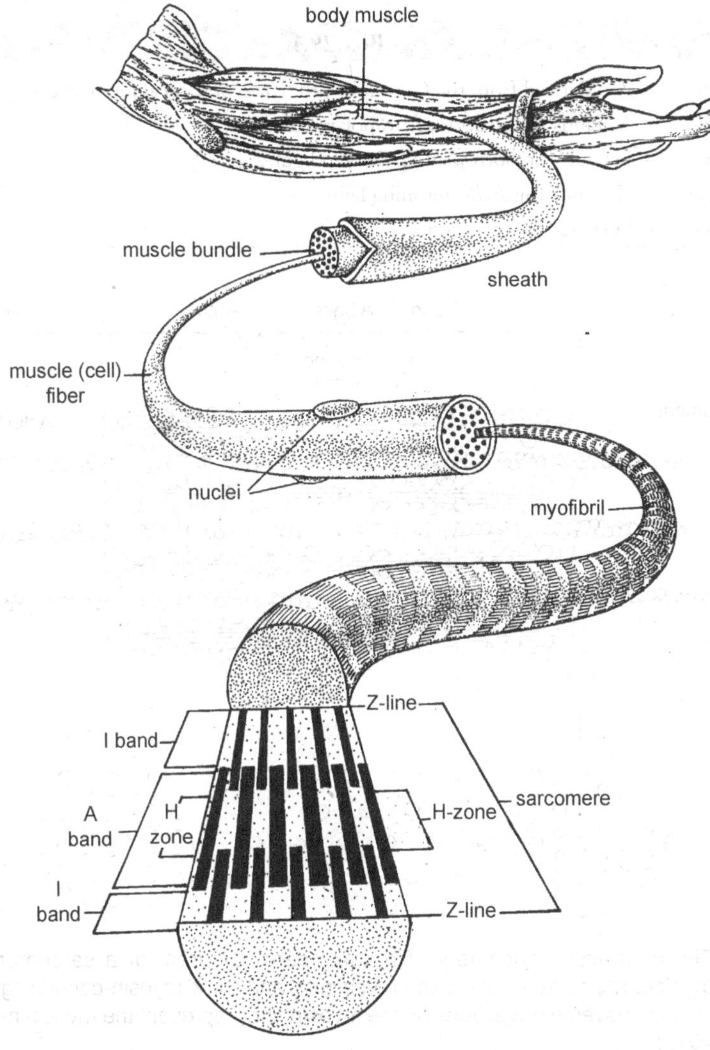

Fig. 29.9. The structure of skeletal muscle (after Karp, 2010).

A longitudinal section of muscle fiber, reveals a cable made up of hundreds of thinner, cylindrical strands, called **myofibrils**. Each myofibril consists of repeating linear array of contractile units, called **sarcomeres**. Each sarcomere in turn exhibits a characteristic banding pattern, which gives the muscle fiber a **striped** or **striated** appearance (Fig. 29.9). Examination of stained muscle fiber in the electron microscope shows the banding pattern to be the result of the partial overlap of two distinct types of filaments, **thin filaments** (also called **secondary filaments**) and **thick filaments** (also called **primary filaments**) (Fig. 29.10). Each myofibril is composed of about 1500 adjacent thick **myosin** filaments and 3000 thin **actin** filaments (**Guyton** and **Hall**, 2011). Each sarcomere extends from one Z line to the next Z line (Box 29.1) and contains several *dark bands* and *light* zones. A sarcomere has a pair of lightly staining **I bands** located at its outer edges, a more densely staining **A stand** located between the outer I bands, a lightly staining **H zone** located in the center of the A band. A densely staining **M line** lies in the center of the H zone. The I band contains only thin filaments, the H zone only thick filaments and that part of the A band on either side of H zone represents the region of overlap and contain both types of filaments.

Box 29.1

Z line	= Z is derived from the German word '*Zwischen*' meaning between.
A zone	= A stands for *anisotropic* to polarized light.
I band	= I stands for *isotropic* to polarized light.
H zone	= H stands for '*helle*' meaning light.
M line	= M stands for myomesin.

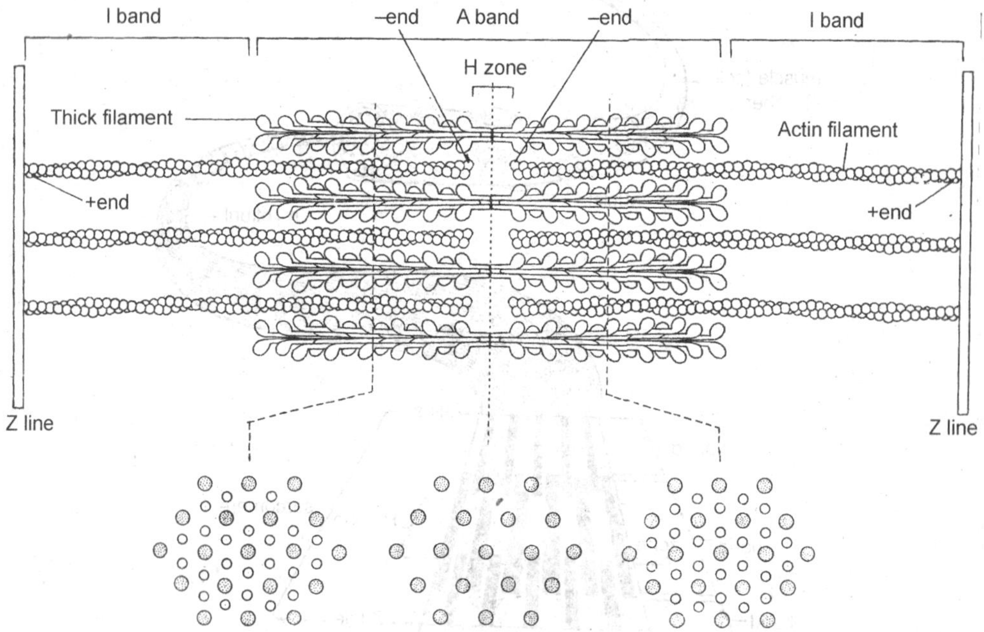

Fig. 29.10. The contractile machinery of a sarcomere. Diagram of a sarcomere showing the overlapping array of thin actin-containing and thick myosin-containing filaments. The small transverse projections on the myosin fiber represent the myosin heads (after Karp 2010).

Cross sections through the region of overlap show that the thin filaments are organized in a hexagonal array around each thick filament (Fig. 29.10). Longitudinal section show the presence of projections from the thick filaments at regularly placed intervals. The projections represent cross-bridges capable of forming attachments with neighboring thin filaments.

Sarcolemma consists of a true cell membrane, the plasma membrane, and an outer coat made of a thin layer of polysaccharide material that contains numerous thin collagen fibrils. At each end of the muscle fiber, this surface layer of the sarcolemma fuses with a tendon fiber. The tendon fibers in turn collect into bundles to form the muscle tendons that then insert into the bone.

The many myofibrils of each muscle fiber are suspended side by side in the muscle fiber. The spaces between the myofibrils are filled with intracellular fluid called **sarcoplasm** contaning large quantities of potassium, magnesium, and phosphate, plus multiple protein enzymes. Also present are tremendous numbers of mitochondria that lie parallel to the myofibrils. These supply the contracting myofibrils with large amounts of energy in the form of adenosine triphosphate (ATP) formed by the mitochondria.

Also in the sarcoplasm surrounding the myofibrils of each muscle fiber is a extensive reticulum, called the **sarcoplasmic reticulum**. The reticulum has a special organization that is extensively important in controlling muscle contraction.

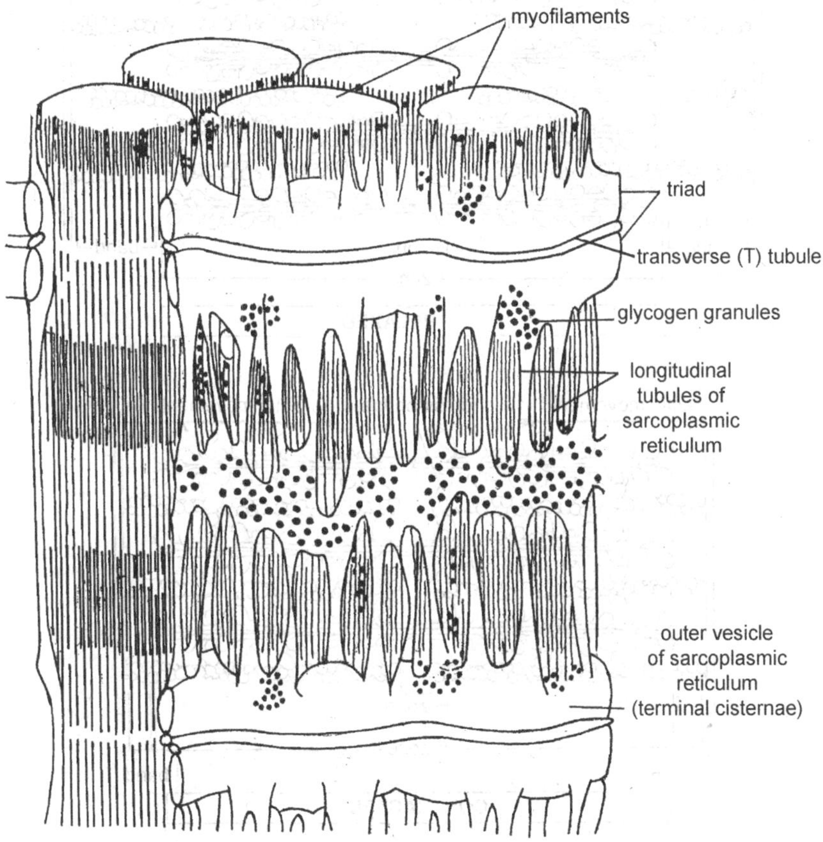

Fig. 29.11. T tubules and sarcoplasmic reticulum in striated muscle fibre.

The Sliding Filament Model of Muscle Contraction

All skeletal muscles operates by shortening; there is no other way they can perform work. The units of shortening are the sarcomeres, whose combined decrease in length accounts for the decrease in length of the entire muscle. The most important evidence to the mechanism underlying muscle contraction came from observation of the banding pattern of the sarcomeres at different stages in the contractile process. As a muscle fiber shortened, the A band in each sarcomere remained essentially constant in length, while the H band and I bands decreased in width and then disappeared altogether. As shortening progressed, the Z lines on both ends of the sarcomere moved inward until they contacted the outer edges of the A band (Fig. 29.12).

In 1957, two groups of British investigators, **Andrew Huxley** and **Rolf Niederke** and **Hugh Huxley** and **Jean Hanson** proposed a far-reaching model to explain muscle contraction. They proposed that the shortening of individual sarcomeres did not result from the shortening of filaments, but rather from their sliding over one another. The sliding of the thin filaments toward the center of the sarcomere would result in the observed increase in overlap between the filaments and the decreased width of the I and H bands (Fig. 29.12). The sliding filament model was rapidly accepted, and evidence in its favour continues to accumulate.

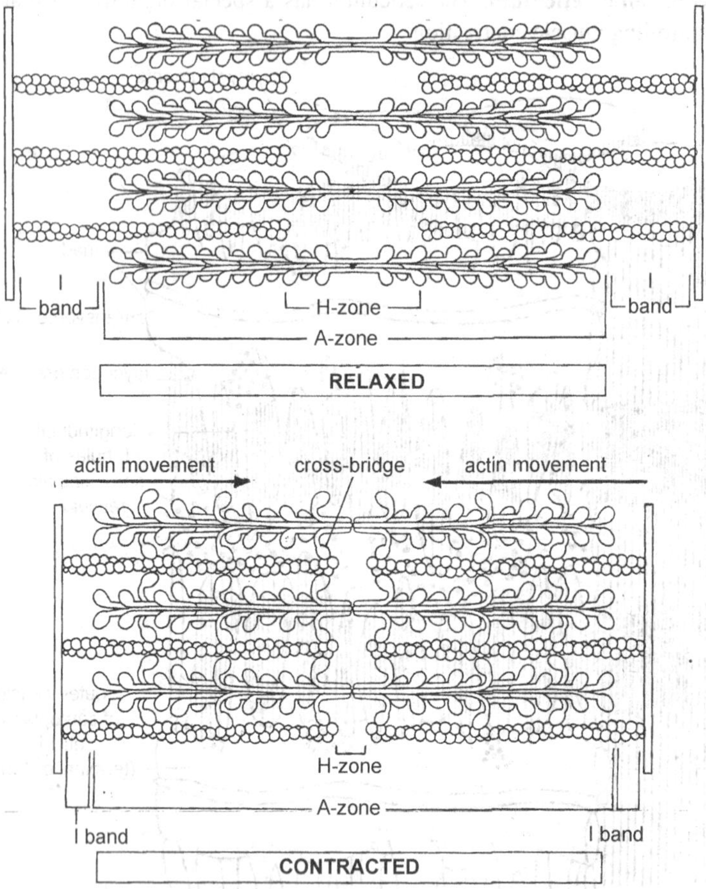

Fig. 29.12. The shortening of the sarcomere during muscle contraction. Schematic diagram showing the difference in the structure of the sarcomere in a relaxed and contracted muscle. During contraction, the myosin cross-bridge make contact with the surrounding thin filaments, and the thin filaments are forced to slide toward the center of the sarcomere. Cross-bridges work asynchronously, so that only a fraction are active at any given instant (after Karp 2010).

Evidences Supporting Sliding Filament Model

1. Composition and organization of thick and thin filaments. In addition to actin protein, the thin filaments of a skeletal muscle contain two other proteins, tropomyosin and troponin (Fig. 29.13). **Tropomyosin** is an elongated molecule (approximately 40 nm long) that fits securely in to the grooves within the thin filaments. Each rod-shaped tropomyosin molecule is associated with seven actin subunits along the thin filaments (Fig. 29.13).

Troponin is a globular protein complex composed of three subunits, each having an important and distinct role in the overall function of the molecule. Troponin molecules are spaced approximately 40 nm apart along the thin filament and contact both the actin and tropomyosin components of the filament. The actin filaments of each half sarcomere are aligned with their barbed ends linked to the Z line.

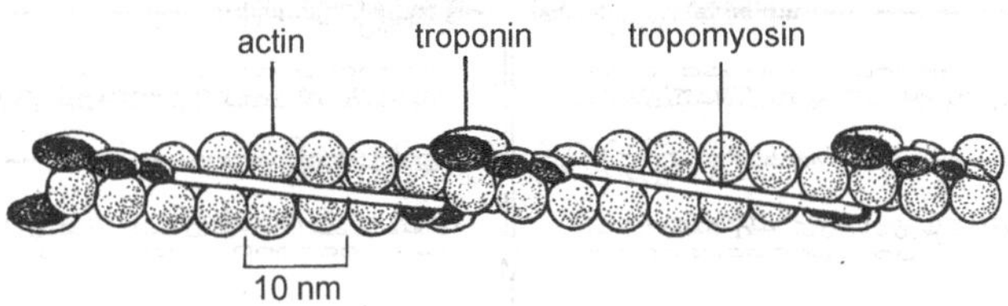

Fig. 29.13. The molecular organization of the thin filaments. Each thin filament consists of a helical array of actin subunits with rod-shaped tropomyosin molecules. The positional changes in these proteins that trigger contraction are shown in Figure 29.18 (after Karp 2010).

Each thick filament is composed of several hundred myosin II molecules together with small amounts of other proteins. Like the filaments that form *in vitro* (Fig. 29.14), the polarity of the thick filaments of muscle cells is reversed at the center of the sarcomere. The center of the filament is composed of the opposing tail regions of the myosin molecules and is devoid of heads. The myosin heads project from each thick filament along the remainder of its length due to the staggered position of the myosin molecules that make up the body of the filament.

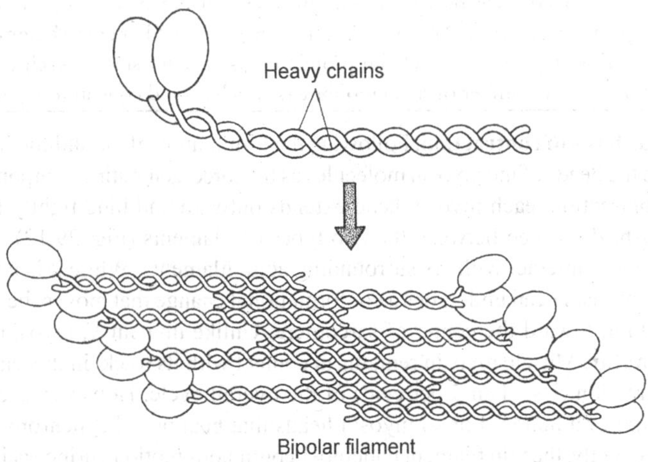

Fig. 29.14. Schematic diagram of the staggered arrangement of the individual myosin molecules in a bipolar myosin II filament (after Karp 2010).

The third most abundant protein of vertebrate skeletal muscle is **connectin** or **titin**, which is the largest protein yet to be discovered in any organism (see **Karp**, 2010). The entire titin gene (which can give rise to isoforms of different length) encodes a polypeptide more than 3.5 million daltons in molecular mass (**Karp**, 2010) and containing more than 38,000 amino acids. (According to **Guyton** and **Hall**, 2011, each titin molecule has a molecular weight about 3 million). Titin molecule originate at the M line in the center of each sarcomere and extend along the myosin filament, continuing past the A band and terminating at the Z line (Fig. 29.15). Titin is a highly elastic (springy) protein that stretches like a molecular spring as certain regions within the molecule become uncoiled. Titin is thought to prevent the sarcomere from becoming pulled apart during muscle stretching. Titin also maintains myosin filaments in their proper position within the center of the sarcomere during muscle contraction.

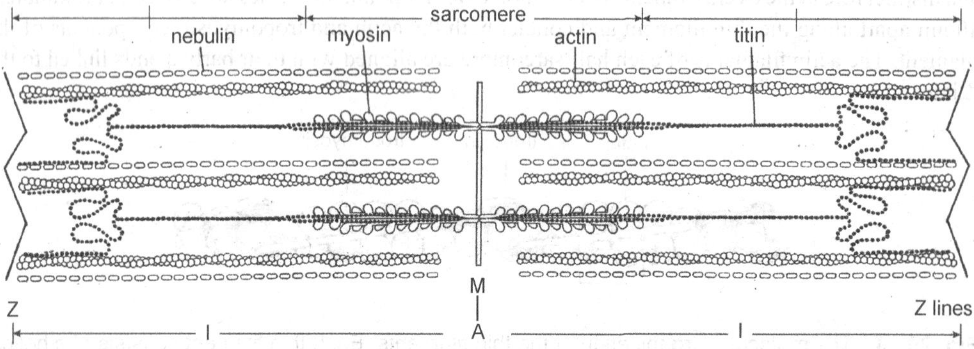

Fig. 29.15. The arrangement of titin molecules within the sarcomere. These huge elastic molecules stretch from the end of the sarcomere at the Z line to the M band in the sarcomere center. Titin molecules are thought to maintain the thick filaments in the center of sarcomere. The I-band portion of the titin molecule contains spring-like domains and is capable of great elasticity. The nebulin molecules thought to act like a "molecular ruler" by regulating the number of actin monomers that are allowed to assemble into a thin filament (after Karp 2010).

Box 29.2 Nebulin: The Molecular Ruler

Nebulin is an elongated inelastic cytoskeletal protein that extends along the length of the thin filament and may participate in regulation of the length of thin filaments (**Koeppen** and **Stanton** 2010). According to **Keller** (1995), one nebulin protein exists at outer side of each thin filament of the sarcomere and it controls the number of actin monomers which should constitute a single thin filament.

2. Molecular basis of contraction. Following the formulation of the sliding filament hypothesis, attention turned to the heads of the myosin molecules as the force-generating components of the muscle fiber. During a contraction, each myosin head extends outward and bind tightly to a thin filament, forming the cross-bridges seen between the two types of filaments (Fig. 29.12). The heads from a single myosin filament interact with six surrounding actin filaments. While it is bound tightly to the actin filament, the myosin head undergoes a conformational change that moves the thin actin filament approximately 10 nm toward the center of sarcomere. Unlike myosin V, myosin II of muscle is a **non-processive motor**. Muscle myosin remains in contact with its track, in this case a thin filament, for only a short fraction (less than 5 per cent) of the overall cycle. However, each thin filament is contacted by a team of a hundred or so myosin heads that beat out of synchrony with one another (Fig 29.12) Consequently, the thin filament undergoes continuous motion during each contractile cycle. It is estimated that a single thin filament in a muscle cell can be moved several hundred nanometers during a period as short as 50 milliseconds.

Muscle physiologists have long wanted to understand how a single myosin molecule can move an actin filament approximately 10 nm in a single power stroke. The publication of the first atomic structure of the S1 fragment of myosin II (Fig. 29.16) in 1993 by **Ivan Rayment**, **Hazel**

Holden, and their collegues at the University of Wisconsin focused attention on a model to explain its mechanism of action. In this model, the energy released by ATP hydrolysis induces a small (0.5 nm) conformational change within the head while the head is tightly bound to the actin filament. The small movement within the head is then amplified approximately 20-fold by the swinging movement of an adjoining α-helical neck (Fig. 29.17). According to this hypothesis, the elongated myosin II neck acts as a rigid "**lever arm**" causing the attached actin filament to slide a much greater distance than would otherwise be possible. The two light chains, which are wrapped around the neck of myosin II molecule are thought to provide rigidity for the lever.

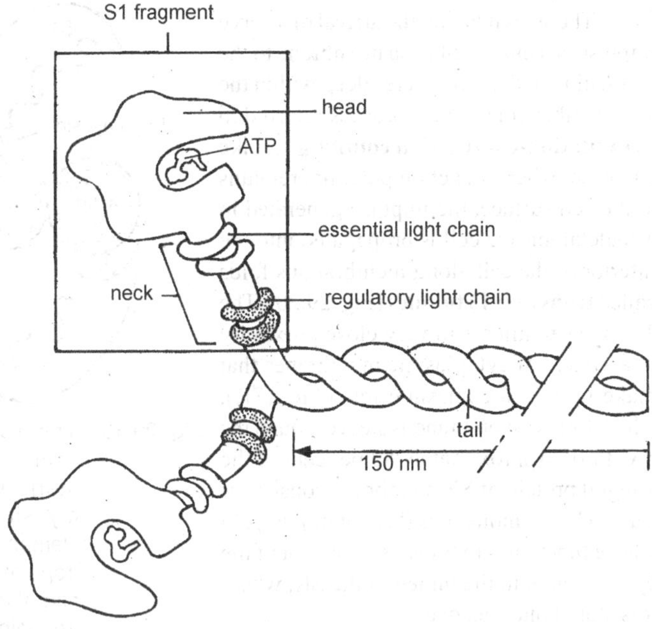

Fig. 29.16. S1 fragment. A highly schematic drawing of a myosin II molecule (molecular mass of 520,000 daltons). The molecule consists of one pair of heavy chains (dark shaded) and two pairs of light chains, which are named as indicated. The paired heavy chains consist of a rod-shaped tail in which portions of two polypeptide chains wrap around one another to form a coiled coil and a pair of globular heads. When treated with a protease under mild conditions, the molecule is cleaved at the junction between the neck and tail, which generates the S1 fragment (after Karp, 2010).

3. Excitation contraction coupling. Muscle fibers are organized into groups termed **motor units**. All the fibers of a motor unit are jointly innervated by branches from a single motor neuron and contract simultaneously when stimulated by an impulse transmitted along the neuron. The point of contact of a terminus of an axon with a muscle fiber is called **neuromuscular junction** (Fig. 29.20). The neuromuscular junction is a site of transmission of the nerve impulse from axon across a synaptic cleft to the muscle fiber, whose plasma membrane is also excitable and capable of conducting an action potential.

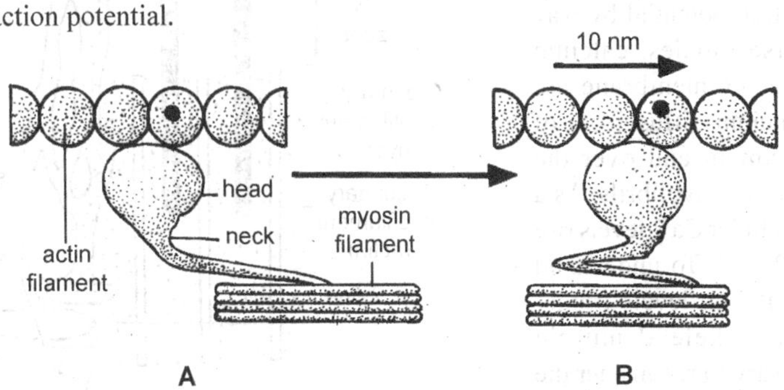

Fig. 29.17. Model of the swinging lever arm of a myosin II molecule. During the power stroke, the neck of myosine molecule moves through a rotation of approximately, 70°, which would produce a movement of the actin filament of approximately 10 nm (after Karp, 2010).

The steps that link the arrival of a nerve impulse at a muscle plasma membrane to the shortening of the sarcomeres deep within the muscle fiber constitute a process referred to as **excitation-contraction coupling**. Unlike a neuron, where an action potential remains at the cell surface, the impulse generated in a skeletal muscle cell is propagated into the interior of the cell along membranous folds called **transver (T) tubules** (Fig. 29.20). The T tubules terminate in very close proximity to a system of cytoplasmic membranes that make up the sarcoplasmic reticulum (SR), which forms a membranous sleeve around the myofibril. Approximately 80 per cent of the integral protein of SR membrane consists of Ca^{2+}-ATPase molecules ("calcium pumps") whose function is to transport Ca^{2+} out of the cytosol and into the lumen of the SR, where it is stored until needed.

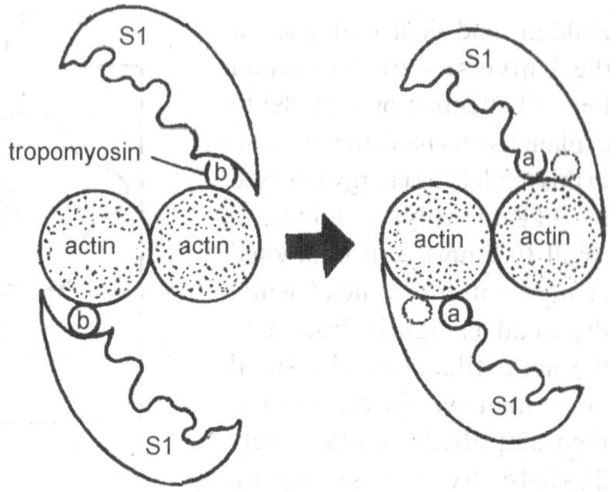

Fig. 29.18. The role of tropomyosin in muscle contraction. Schematic diagram of the steric hindrance model in which the myosin-binding site on the thin actin filaments is controlled by the position of the tropomyosin molecule. When calcium level rise, the interaction between calcium and troponin (not shown) leads to a movement of the tropomyosin from position *b* to position *a*, which exposes the myosin-binding site on the thin filaments to the myosin head (after Karp, 2010).

The importance of calcium in muscle contraction was first shown in 1882 by **Syndey Ringer**, an English physician. Ringer found that an isolated frog heart would contract in a saline solution made with London tap water, but failed to contract in a solution made with distilled water. Ringer determined that calcium ions present in tap water were an essential factor in muscle contraction. In the relaxed state, the Ca^{2+} levels within the cytoplasm of muscle fiber are very low (approximately 2×10^{-7} M)—below the threshold concentration required for contraction. With the arrival of an action potential by way of the transverse tubules, calcium channels in the SR membrane are opened, and calcium diffuses out of the SR compartment and over the short distance to the myofibrils. As a result, the intracellular Ca^{2+} levels rise to about 5×10^{-5} M. To understand how elevated calcium levels trigger contraction in a skeletal muscle fiber, it is necessary to reconsider the protein makeup of the thin filaments.

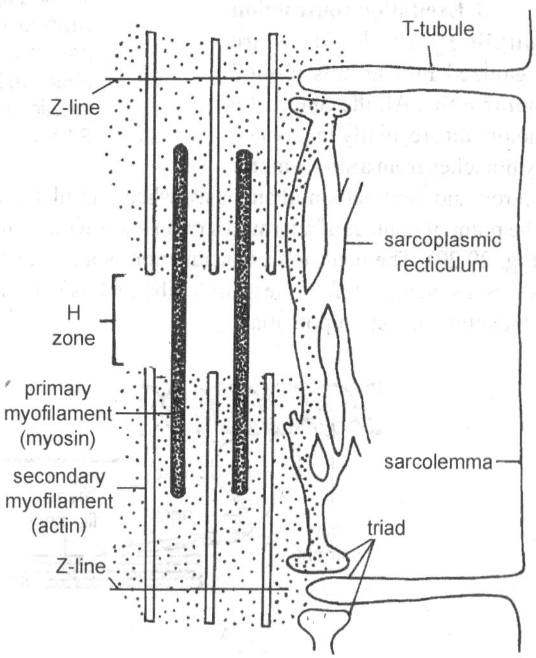

Fig. 29.19. Structure of a sarcomere showing T-tubule and sarcoplasmic reticulum.

When the sarcomere is relaxed, the tropomyosin molecules of the thin filaments (Fig. 29.13) block the myosin-binding sites on the actin filaments. The position of the tropomyosin within the groove is the under the control of the attached troponin molecule. When Ca^{2+} levels rise, these ions bind to one of the subunits of troponin (troponin C), causing a conformational change in another subunit of the troponin molecule. Like the collapse of a row of dominoes, the movement of troponin is transmitted to the adjacent tropmyosin, which moves approximately 1.5 nm closer to the center of the filaments groove (from position *b* to *a* in Fig. 29.18). This shift in position of the tropomyosin exposes the myosin binding sites on the adjacent actin molecules, allowing the myosin heads to attach to the thin filaments. Each troponin molecule controls the position of one tropomyosin molecule, which in turn controls the binding capacity of seven actin subunits in the thin filaments.

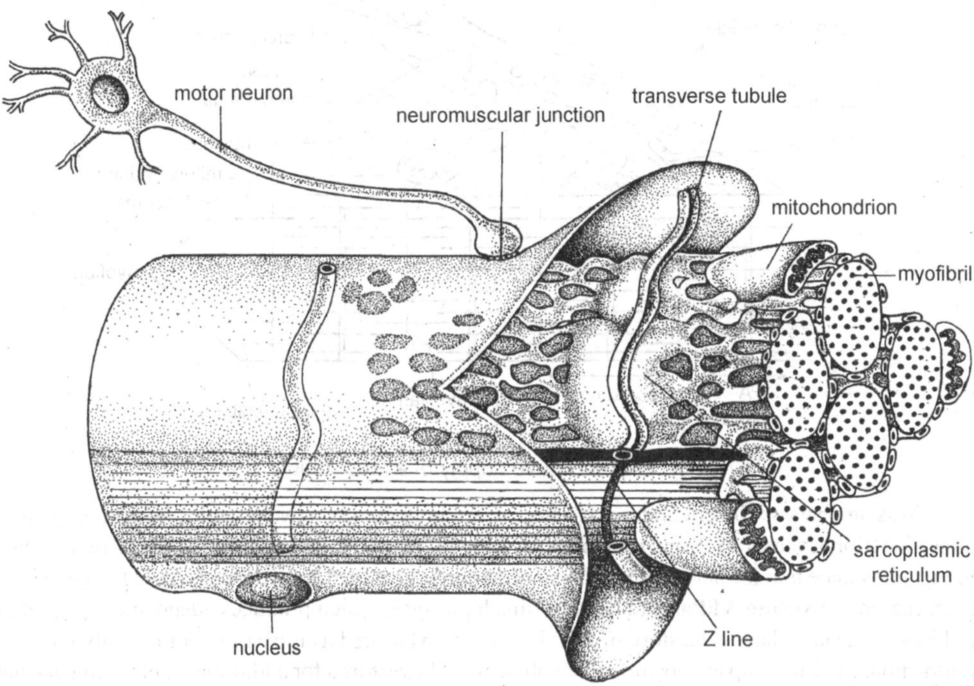

Fig. 29.20. The functional anatomy of a muscle fiber. Calcium is housed in the elaborate network of internal membranes that make up the sarcoplasmic reticulum (SR). When an impulse arrives by means of a motor neuron, it is carried into the interior of the fiber along the membrane of the transverse tubule to the SR. The calcium gates of the SR open, releasing calcium into the cytosol. The binding of calcium ions to troponin molecules of the thin filaments leads to the events described in Figure 29.18 (after Karp, 2010).

Once stimulation from the innervating motor nerve fiber ceases, the Ca^{2+} channels in the SR membrane close, and the Ca^{2+}-ATPase molecules in that membrane remove excess calcium from the cytosol. As the Ca^{2+} concentration decreases, these ions dissociate from their binding sites on troponin, which causes the tropomyosin molecules to move back to a position where they block the actin-myosin interaction. The process of relaxation can be thought of as a competition for calcium between the transport protein of the SR membrane and troponin. The transport protein has a greater affinity for the ion, so it preferentially removes it from the cytosol, leaving the troponin molecules without bound calcium.

4. Energy for the muscle contraction. The ultimate source of energy for muscle contraction is usually glycogen, but may be fatty acids. Metabolism of these substances during respiration generates ATP. Normally, the oxygen used in aerobic respiration is supplied by haemoglobin. But the muscles also have their own store of oxygen in the protein **myoglobin**, which releases oxygen only when haemoglobin fails to meet its demand as in strenuous exercise.

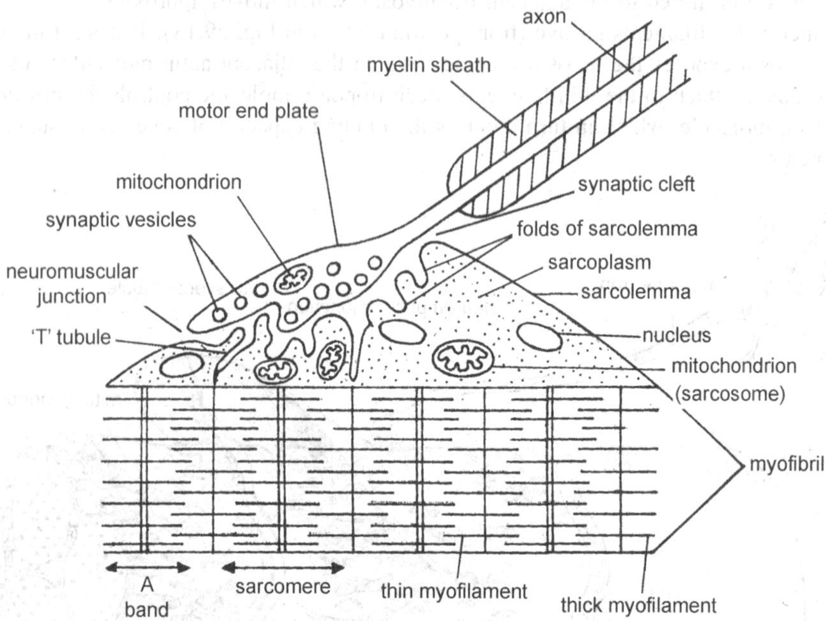

Fig. 29.21. Nerve supply of a muscle fiber.

Muscle fibers of all vertebrates and many invertebrates contain two organic phosphates, namely, adenosine triphosphate (ATP) and **phosphocreatine (PCr)**, also called **creatine phosphate**, with high energy phosphate bond. Energy for muscle contraction is provided by hydrolysis of ATP. Myosin acts as an enzyme, **myosine ATPase**, to catalyze this hydrolysis which produces adenosine diphosphate and inorganic phosphate. The store of ATP is quickly exhausted as it is sufficient for only ten rapid contractions. The used up adenosine triphosphate must be restored for additional contraction because ATP is the only energy compound a muscle can use directly. Regeneration of ATP is done by creatine phosphate which is present in large amounts. Phosphocreatine donates its high energy phosphate to ADP, producing ATP. This reaction is catalysed by an enzyme **creatine phosphotransferase (creatine kinase)**. A substance, such as phosphocreatine, that serves as a reservoir for high energy phosphate bonds ~P, is called a **phosphagen**. Phosphocreatine can fulfil total energy demand for about 5 to 10 seconds, and normally about 70 per cent of the store is used up in one minute of heavy exercise. When phosphocreatine is used up, new ATP is generated by aerobic respiration of carbohydrates (glycogen) or fatty acids in muscle cells. If ATP is used faster than the muscle cell can produce it aerobically, the muscle fibers start anaerobic respiration (glycolysis) to replenish ATP. This produces lactic acid as a byproduct. Lactic acid accumulates in the muscle cells and diffuses out into the blood. A small portion of it reenters the relaxing muscle cells and is changed into glycogen. Major part of lactic acid passes into the liver. Here, one-fifth of lactic acid is oxidized to CO_2 and H_2O. The energy (ATP molecules) derived from this oxidation is used in changing the remaining four-fifth of liver lactic acid into glucose. Some of the glucose is transported back to the muscle where it is changed into glycogen. The rest of the glucose is converted to glycogen and stored in liver. Phosphocreatine is regenerated in a relaxing muscle cell by using ATP produced by carbohydrate oxidation. The chemistry of muscle contraction can be summarized as follows:

(*i*) $ATP + H_2O \xrightarrow{\text{Myosine ATPase}} ADP + Pi + \text{Energy for muscle contraction}$

(*ii*) $\text{Creatine} - \text{phosphate} + ADP \xrightarrow{\text{Contracting muscle}} ATP + \text{Creatine}$

(*iii*) $\text{Glycogen} \rightarrow \text{Intermediates} \rightarrow \text{Lactic acid} + \text{Energy (used for ATP generation)}$

(*iv*) $1/5\,\text{Lactic acid} + O_2 \rightarrow CO_2 + H_2O + \text{Energy (ATP) in liver cells}$

(*v*) $4/5\,\text{Lactic acid} + \text{Water} \xrightarrow{ATP} \text{Glycogen in liver cells}$

(*vi*) $\text{Creatine} + ATP \xrightarrow{\text{Relaxing muscle}} \text{Phosphocreatine} + ADP$

All these reactions are catalyzed by specific enzymes. The first 3 steps do not involve oxygen and can occur under anaerobic conditions. Fourth step needs oxygen. In other words, contraction of a muscle and part of its recovery from contraction occur without oxygen. This is of great importance as it enables us to undertake great exertions. Oxygen is required only in the recovery from contraction.

Cori cycle. The lactic acid formed in reaction number (*iii*) is carried in the blood to the liver, where it is converted into glycogen. The glycogen releases glucose into the blood, which is reconverted into glycogen in the muscles. This lactic acid → glycogen → glucose → glycogen cycle is called the **Cori cycle** (Fig. 29.22). It shows the role of liver in the metabolism of muscle contraction.

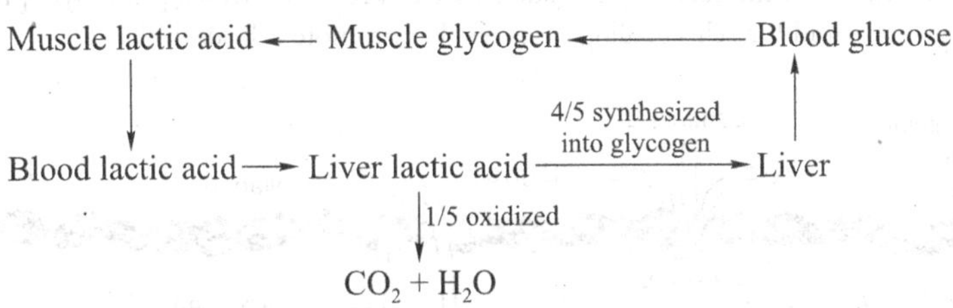

Fig. 29.22. Cori cycle in muscle and liver cells.

This cycle was discovered by **Carl Ferdinand Cori** and his wife **Gerty Theresa** in 1943 and for this discovery the Coris were awarded in 1947 the Nobel Prize for Physiology and Medicine.

General Mechanism of Muscle Contraction

The initiation and execution of muscle contraction occur in the following sequential steps:

1. An action potential travels along a motor nerve to its endings on muscle fibers.
2. At each endings, the nerve secretes a small amount of the neurotransmitter substance acetylcholine.
3. The acetycholine acts on a local area of the muscle fiber membrane to open multiple "acetylcholine-gated" cation channels through protein molecules floating in the membrane.
4. Opening of the acetylecholine-gated channels allows large quantities of sodium ions to diffuse to the interior of the muscle fiber membrane. This causes a local depolarization that in turn leads to opening of voltage-gated sodium channels. This initiates an action potential at the membrane.
5. The action potential travels along the muscle fiber membrane in the same way that action potential travel along the nerve fiber membrane.
6. The action potential depolarizes the muscle membrane, and much of the action potential electricity flows through the center of the muscle fiber. Here it causes the sarcoplasmic reticulum to release large quantities of calcium ions that have been stored within this SR.
7. The calcium ions initiates attractive forces between the actin and myosin filaments, causing them to slide along side each other, which is the contractile process.

8. After a fraction of a second, the calcium ions are pumped back into the sarcoplasmic reticulum by a Ca^{2+} membrane pump and remain stored in the reticulum until a new muscle action potential comes along; this removal of calcium ions from the myofibrils causes the muscle contraction to cease (see **Guyton** and **Hall**, 2011).

Box 29.3 Dystrophin and DMD

Dystrophin is a protein and a member of spectrin family that is found in the membrane skeleton of muscle cells. Dystrophin relays nerve's signals to the calcium storage sacs.

The muscular dystrophies constitute a group of genetically determined degenerative disorders. **Duchennes muscular dystrophy** (**DMD**; described by **G.B. Duchenne** in 1861) is the most common of the muscular dystrophies and affects 1 in 3500 boys (3 to 5 years of age). Severe muscle wasting occurs, with most patients being wheelchair bound by the age of 12 and many dying of respiratory failure in adulthood (30 to 40 years of age). DMD is an X-linked recessive disease that has been linked to a defect in the dystrophin gene that leads to a deficiency of the dystrophin protein in skeletal muscle, brain, retina and smooth muscle. Dystropin is a large (427 kDa) protein that is present in low abundance (0.025%) in skeletal muscle. It is localized on the intracellular surface of the sarcolemma in association with several integral membrane glycoproteins (forming a dystrophin-glycoprotein complex). This dystrophic-glycoprotein complex provides a structural link between the subsarcolemmal cytoskeleton of the muscle cell and the extracellular matrix (Fig. 29.23) and appears to stabilize the sarcolemma and hence prevent contraction-induced injury (such as rupture). The dystrophin-glycoprotein complex may also serve as a scaffold for cell signaling cascades that promote cell survival.

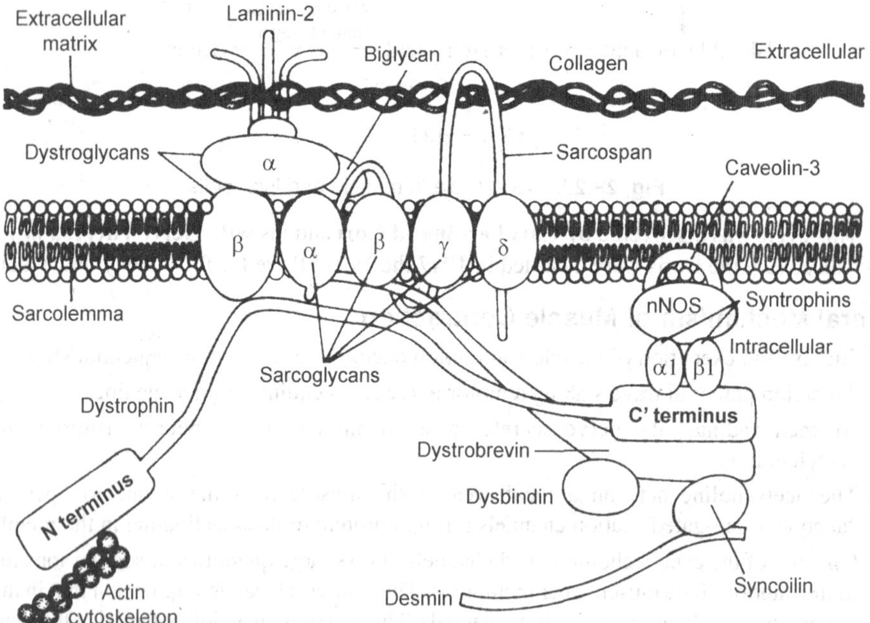

Fig. 29.23. Organization of the dystrophin-glycoprotein complex in skeletal muscle. The dystrophin-glycoprotein complex provides a structural link between the cytoskeleton of the muscle cell and the extracellular matrix, which stabilize the sarcolemma and hence prevents contraction—induced injury (rupture). Duchenne's muscular dystrophy (DMD) is associated with loss of dystrophin (after Koeppen and Stanton 2010).

II. Contractility of Cardiac Muscles

Cardiac (heart) muscle is responsible for the beating of the heart and the pumping of blood through the body's circulatory system. Cardiac muscle function continuously; in one year, human heart beats about 40 million times. Cardiac muscle is very similar to skeletal muscle in the organization of actin and myosin filaments and has the same striated appearance (Fig. 29.24). Cardiac muscle (cell) has very large mitochondria that are densely packed between the myofibrils (Fig. 29.25). These mitochondria have numerous closely packed cristae. In contrast to skeletal muscle, most of the energy required for the beating of heart under resting conditions is provided not by blood glucose, but by free fatty acids that are transported from adipose (fat storage) tissue to the heart by serum albumin, a blood protein.

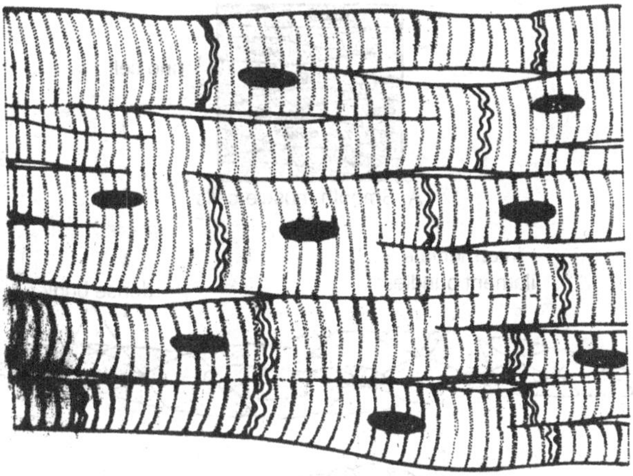

Fig. 29.24. Cardiac muscle cells. Cardiac muscle cells have a contractile mechanism and sarcomeric structure similar to those of skeletal muscle cells. However, unlike skeletal muscle cells, cardiac muscle cells are joined together end-to-end at intercalated discs, which allow ions and electric signals to pass from one cell to the next. This ionic permeability enables a contraction stimuli to spread evenly to all the cells of the heart (after Guyton and Hall, 2011).

The other difference between cardiac and skeletal muscle is that heart muscle cells are not multinucleate. Instead cells are joined end-to-end through structures called **intercalated discs**. The discs have gap junctions (*i.e.*, desmosome junctions, adherent type junctions and communicating gap junctions) that electrically couple neighboring cells, allowing depolarization wave to spread throughout the heart during its contraction cycle. The heart is not activated by nerve impulses, as skeletal muscle is, but contracts spontaneously once every second or so (*i.e.*, heart muscles are **myogenic**). The heart rate is controlled by a "pacemaker" region in an upper portion of the heart (right attrium). The depolarization wave is initiated by the pacemaker and then spreads to the rest of the heart to produce the heart beat.

III. Contractility of Smooth Muscle

Smooth muscle is responsible for involuntary contractions such as those of the stomach, intestines, respiratory tract, uterus, and blood vessels. In general, such contractions are slow, taking upto five seconds to reach maximum tension. Smooth muscle contractions are also of greater duration than those of skeletal or cardiac muscle. Though smooth muscle is not able to contract rapidly, it is well adapted to maintain tension for long periods of time, as is required in these organs and tissues.

Structure of smooth muscle. Smooth muscle cells are long and thin, with pointed ends. The long cigar-shaped nucleus of smooth muscle cells is located in the center of the cell (**Ross** *et al.*, 1995). Unlike skeletal or heart muscle, smooth muscle has no striations (Fig. 29.25A). Smooth muscle cells do not contain Z lines, which are responsible for the periodic organization of the sarcomeres found in skeletal and cardiac muscle cells. Instead, smooth muscle cells contain **dense bodies** (containing α-actinin protein), plaque-like structures in the cytoplasm and on the plasma membrane (Fig. 29.25B) that contain **desmin** containing intermediate filaments. Bundles of actin filaments are anchored at their ends to these dense bodies. As a result, actin filaments appear in a criss-cross pattern, aligned obliquely to the long axis of the cell. Cross-briges connect thick and thin filaments in smooth muscle, but not in the regular repeating pattern seen in skeletal muscle. Smooth muscle cells make contact with neighboring cells by **gap junctions**. The gap junctions between two smooth muscle cells is called **nexus** (Fig. 29.26).

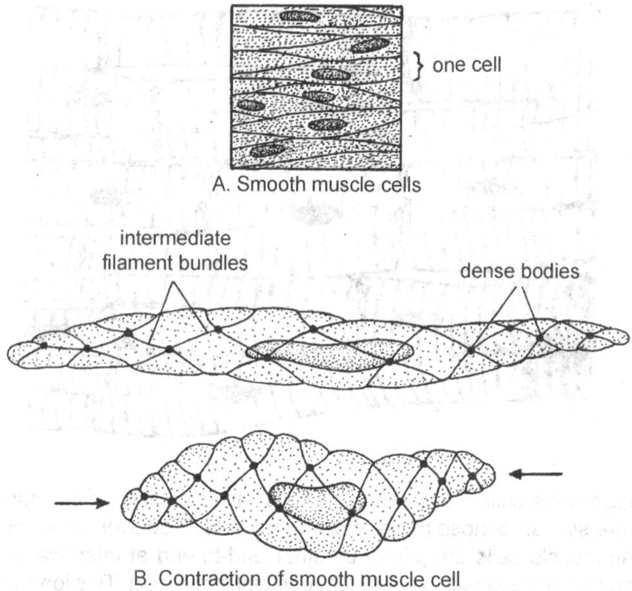

Fig. 29.25. Smooth muscle and its contraction. A—Individual smooth muscle cells are long and spindle-shaped with no Z lines or sarcoplasmic structures. B—In the smooth muscle cell, contractile bundles of actin and myosin appear to be anchored to plaque-like structures called dense bodies. The dense bodies are connected to each other by intermediate filaments, thereby orienting the actin and myosin bundles obliquely to the long axis of the cell. When the actin and myosin bundles contract, they pull on the dense bodies and IFs, producing the cellular contraction shown here (after Becker *et al.*, 2006).

Regulation of Contraction in Smooth Muscle Cells

Smooth muscle cell contraction and nonmuscle cell contraction are regulated in a manner distinct from that of skeletal muscle cells. Although skeletal and smooth muscle cells are both stimulated to contract by an increase in the sarcoplasmic concentration of calcium ions, the mechanisms involved are quite different. When sarcoplasmic calcium concentration increases in the smooth muscle and nonmuscle cells, a cascade of events takes place that includes the activation of **myosin light-chain kinase (MLCK)**. Activated MLCK then phosphorylates one type of myosin light chain known as a **regulatory light chain** (myosin II).

Myosin light-chain phosphorylation affects myosin in two ways. First, some myosin molecules are curled up so that they cannot assemble into filaments. When the myosin light chain is phosphorylated, the myosin tail uncurls and becomes capable of assembly. Second, the phosphorylation of the light chains activates myosin, enabling it to interact with actin filaments to undergo the cross bridge cycle.

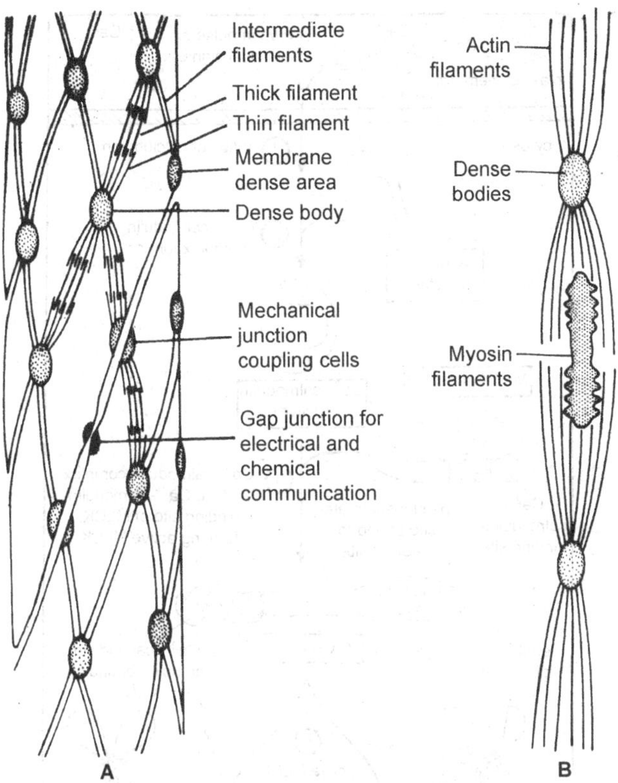

Fig. 29.26. A—cell-to-cell contacts, cytoskeleton, and myofilaments in smooth muscle cells. Small contractile elements, functionally equivalent to a sarcomere underlie the similarities in mechanics between smooth and skeletal muscles. Linkages consisting of specialized junctions or interstitial fibrillar material functionally couple the contractile apparatus of adjacent cells. B—Details of actin and myosin fibers (after Koeppen and Stanton, 2010).

The cascade of events involved in the activation of smooth muscle and non-muscle myosin is shown in Fig. 29.27. In response to a nerve impulse or hormonal signal reaching the smooth muscle cell, an influx of extra-cellular calcium concentration occurs causing contraction. The effect of the increased calcium concentration on contraction is mediated by the binding of calcium to **calmodulin**. The *calcium-calmodulin complex* can bind to myosin light chain kinase thereby activating the enzyme. As a result, myosin light-chains become phosphorylated, and myosin can interact with actin to cause contraction.

The regulation of MLCK by calcium and calmodulin illustrates a common theme in the regulation of protein kinases. MLCK contains a peptide sequence at one end called a **pseudosubstrate**, a sequence of amino acids similar to the enzyme's normal substrate. In this case, the real substrate is a sequence of amino acids on the myosin light chain that is recognized by MLCK. Within this substrate's sequence of amino acids is the serine that actually accepts the phosphate from ATP. In the pseudosubstrate, this serine is replaced by a different amino acid that cannot accept a phosphate group.

The pseudosubstrate region of MLCK regulates the activity of this enzyme by binding to the active site of the enzyme and thus inhibiting its activity. Inhibition of enzyme activity in this manner is referred to as **autoinhibition**, because, in effect enzyme inhibits itself. However, there is also a binding site for calcium-calmodulin binds MLCK, the pseudosubstrate region is prevented from inhibiting the enzyme, and the enzyme becomes active. Once activated, MLCK can then phosphorylate myosin. When the myosin light chain is phosphorylated, myosin becomes active and capable of interacting with actin, and the tail of myosin in the curled form straightens out and can assemble with their

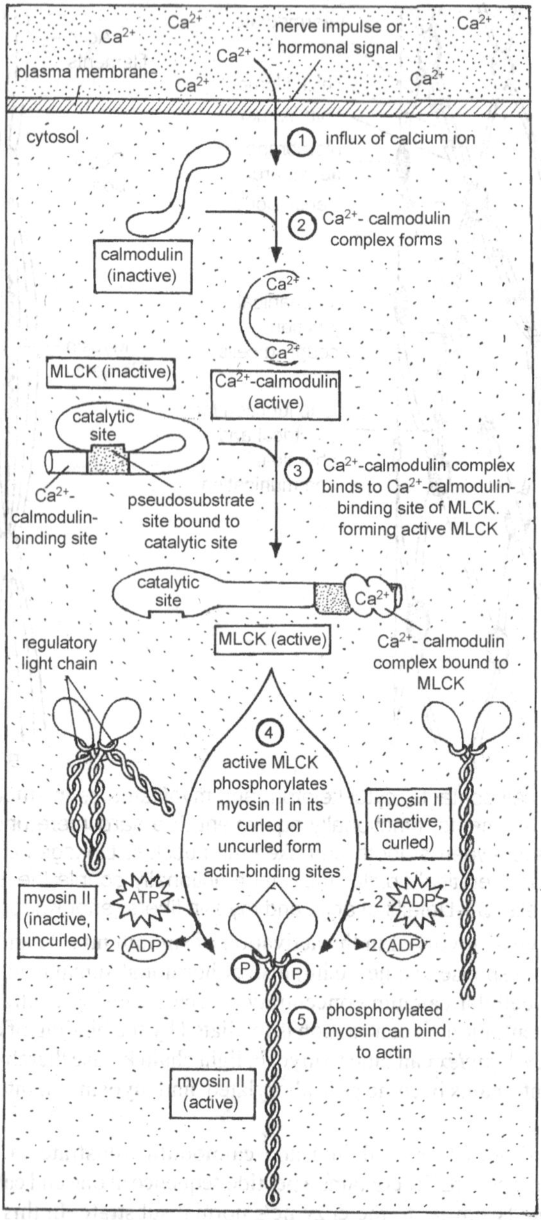

Fig. 29.27. Phosphorylation of smooth muscle and non-muscle myosin. The functions of both smooth muscle and non-muscles myosin II are regulated by phosphorylation of the regulatory light chains. 1. An influx of calcium ions into the cell is triggered by a nerve impulse or a hormonal signal. 2. When present at a sufficiently high concentration, calcium ions bind to calmodulin, forming an active calcium-calmodulin complex. 3. The calcium-calmodulin complex in turn binds to a region of myosin light-chain kinase (MLCK) that overlaps the pseudosubstrate site. When this happens, the pseudosubstrate strech of amino acids is pulled away and prevented from binding to the active site of MLCK, thus activating MLCK. 4. Activated MLCK phosphorylates the myosin light chains, whether the myosin is curled or uncurled. 5. The activated (and uncurled) myosin can then bind to actin and undergo the cross-bridge cycle (after Becker *et al.*, 2006).

myosin molecules into filaments. As the calcium levels within smooth muscle cells drop again, the MLCK is inactivated and a second enzyme, *myosin light-chain phosphatase*, removes the phosphate group from the myosin light chain. Since the dephosphorylated myosin can no longer bind to actin, the muscle cell relaxes.

Thus, both skeletal muscle and smooth muscle are activated to contract by calcium ions, but from different sources and by different mechanisms. In skeletal muscle, the calcium comes from the sarcoplasmic reticulum. Its effect on active myosin interaction is mediated by troponin and is very rapid because it depends on conformation changes only. In smooth muscle, the calcium comes from outside the cell, and its effect is mediated by calmodulin. The effect is much slower in this case because it involves a covalent modification (phosphorylation) of the myosin molecule.

29.3. ACTIN BASED MOTILITY IN NON-MUSCLE CELLS

I. Cell Crawling

Actin microfilaments (MFs) are required for the movement of most non-muscle cells in animals. Many non-muscle cells such as **fibroblasts**, the **growth cones** of neurons and many **embryonic cells** in animals, are capable of crawling over a substrate. In this section, we will consider a type of crawling involving **lamellipodia** and **filopodia**.

Process of cell crawling. Cell crawling involves several distinct events (Fig. 29.28):

1. Extension of a protrusion at the cell's leading edge;
2. Attachment of protrusion to the substrate;
3. Generation of tension, which pulls the cell forward;
4. Release of attachments and retraction of the "tail".

1. Extending protrusions. In order to crawl, crawling cells (*e.g.*, fibroblasts, macrophages, etc.) produce specialized extensions or **protrutions**, at their front or **leading edge.** One type of protrusion is a thin sheet of cytoplasm called a **lamellipodium** (plural: **lamellipodia**). Another type of protrusion is thin, pointed structure known as **filopodium** (plural: **filopodia**). Crawling cells often exhibit inter-conversion of these two types of protrusions as they migrate.

Retrograde flow. Fundamental to the dynamics of protrusions is the phenomenon of **retrograde flow** of F-actin. Retrograde flow has been extensively studied in the crawling

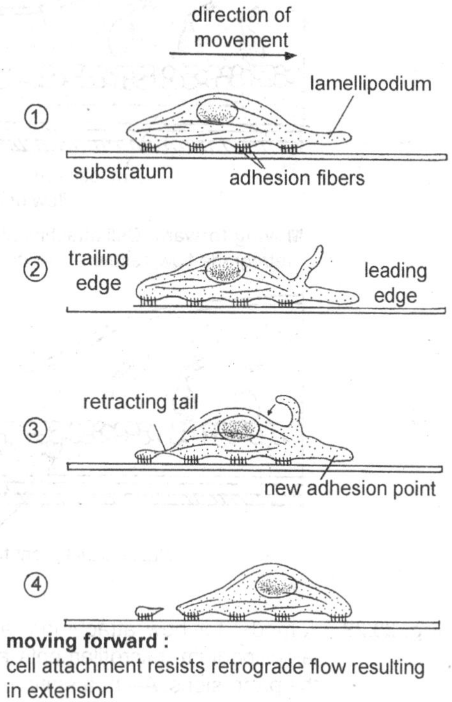

Fig. 29.28. Steps in cell crawling. Several different processes are involved in cell crawling, including cell protrusion, attachment, and contractile activities. Illustrate schematically here is a macrophage. 1. The leading edge of the cell forms protrusions (lamellipodia) that extend in the direction of travel. 2. These protrusions can adhere to the substrate, or they can be pulled upward and back toward the cell body. 3. When the protrusions adhere, they provide anchorage points for actin filaments. 4. Tension on the actin filaments can then cause the rest of the cell to pull forward. When the cell adheres very tightly to its substrate, a piece of the cell tail may break off and be left behind as the cell crawls forward (after Becker *et al.*, 2006).

appendages of cultured nerve cells known as **growth cones**, and it seems to be a fundamental process in many migrating cells. During normal retrograde flow, there is bulk movement of microfilaments toward the rear of the protrusion as it extends. These filamentous structures disappear gradually when cells are treated with *cytochalasin*, demonstrating that they are made predominantly of actin. Retrograde flow appears to result from two simultaneous process: **Actin assembly** at the tip of the growing lamellipodium or filopodium, and **rearward translocation** of actin filaments toward the base of the protrusion (Fig. 29.29). In a typical cell, forward assembly and rearward translocation balance one another and a protrusion can be extended or retracted.

- **Stationary:** Polymerization and retrograde flow balanced

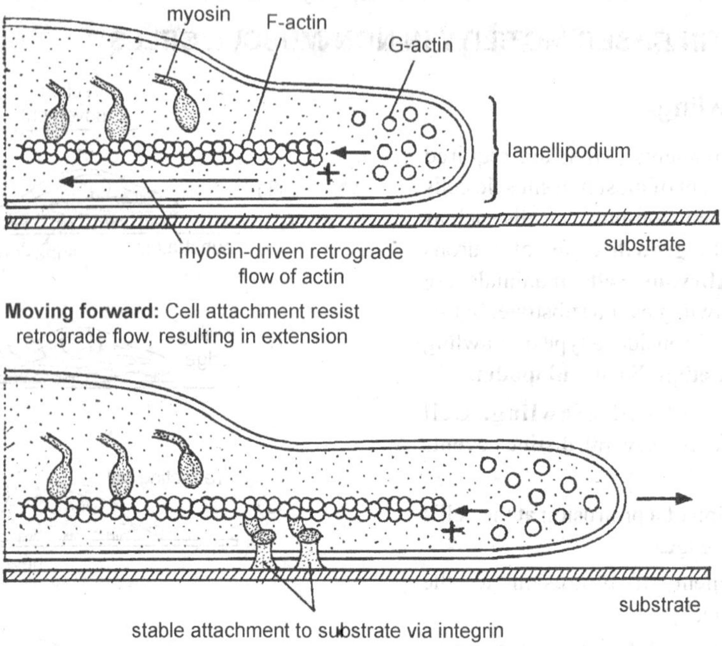

Fig. 29.29. A model for how attachment acts as a "clutch" that allows forward movement. In a lamellipodium, microfilaments are oriented with their plus (barbed) ends at the tip of the protrusions. A—In a stationary protrusions, retrograde flow driven by myosin moves microfilaments rearward as new G-actin monomers are added at the tip. B—When retrograde flow is resisted by attachment to the underlying substrate (*i.e.*, engaging the "clutch"), polymerization result in net elongation of the protrusion and forward movement (after Becker *et al.*, 2006).

At the same time that extension of the tip of the protrusion is occurring, the polymerized actin is drawn toward the base of the protrusion, where it is disassembled. Released actin monomers are then available for addition to plus ends of new or growing microfilaments as the cell continues to crawl forward. The rearward translocation of actin polymers is likely driven by myosins. Non-muscle myosin II is found to be abundant at the base of lamellipodia in several types of cells (*e.g.*, fibroblast) and biochemical agents that disrupt myosin motors inhibit retrograde flow.

2. Cell attachment. Adhesion, or attachment of the cell to its substrate, is also necessary for cell crawling. New sites of attachment must be formed at the front of a cell, and contacts at the rear must be broken. Attachment sites between the cell and the substrate are complex structures involving the attachment of proteins in the plasma membrane to other proteins both outside and inside the cell. One family of attachment proteins is the **integrins**. On the outside of the cell, integrins attach to extracellular matrix proteins. Inside the cell, integrins are connected to actin filaments through linker proteins.

How firmly a cell is attached to the underlying substrate probably helps to shift the balance between the processes of retrograde flow and plus-end addition of actin subunits.

3. Translocation and detachment. Cell crawling coordinates protrusion formation and attachment with forward movement of entire cell body. Contraction of the rear of cell squeezes the cell body forward and releases the cell from attachment at its rear. Evidence suggests that contraction is due to interactions between actin and myosins. Non-muscle myosin II is localized not only at the base of protrusions, but also further toward the rear of the cell. In mutant cells from the cellular slime mold *Dictyostelium* that lack myosin II, the ability of trailing edge of the cell to retract is reduced, supporting the idea that myosins are involved in contraction.

Contraction of the cell body must be linked to detachment of the trailing edge of the cell. Detachment requires breaking adhesive contacts. In general, how firmly a cell attaches to a substrate affects how quickly that cell can crawl (**Becker** *et al.*, 2006).

II. Amoeboid Movement

Amoeboid movement is movement of entire cell in relation to its surroundings, such as movement of white blood cells through tissues. It receives its name from the fact that amebae move in this manner and have provided an excellent tool for studying the phenomenon.

Typically, amoeboid locomotion begins with protrusion of a **pseudopodium** from one end of the cell. The pseudopodium projects far out, away from the body of the cell and partially secures itself in a new tissue area. Then the remainder of the cell is pulled toward the pseudopodium. Figure 29.30 shows this process, showing an elongated cell, the right-end of which is a protruding pseudopodium. The membrane of this end of the cell is continually moving forward, and the membrane at the left-hand of the cell is continually following along as the cell moves.

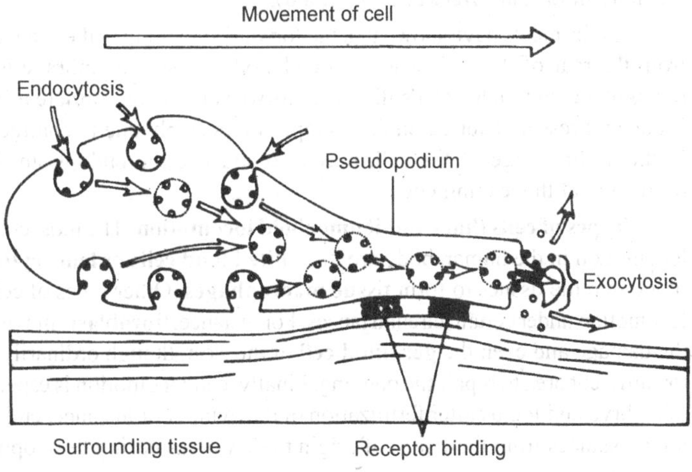

Fig. 29.30. Amoeboid motion by a cell.

Mechanism of amoeboid locomotion. Basically, amoeboid locomotion results from continual formation of new plasma membrane at the leading edge of the pseudopodium and continual absorption of the plasma membrane in mid and rear portions of the cell. Also, two other effects are essential for forward movement of the cell. The first effect is **attachment** of the pseudopodium to surrounding tissues so that it becomes fixed in its leading position, while the remainder of the cell body is pulled forward toward the point of attachment. This attachment is effected by **receptor proteins** that line the inside of the exocytotic vesicles. When the vesicles become part of the pseudopodium membrane, they open so that their insides **evert** to the outside, and the receptors now protrude to the outside and attach to ligands in the surrounding tissues.

At the opposite end of the cell, the receptors pull away from their ligands and form new endocytic vesicles. Then, inside the cell, these vesicles stream toward the pseudopodial end of the cell, where they are used to form still new membrane, for the pseudopodium.

The second essential effect for locomotion is to provide the energy required to pull the cell body in the direction of the pseudopodium.

Experiments suggest the following as an explanation: in the cytoplasm of all cells is a moderate to large amount of the protein **actin**. Much of the actin is in the form of single molecules that do not provide any motive power; however, these polymerize to form a filamentous network, and the network contracts when it binds with an actin-binding protein such as **myosin**. The whole process is energized by the high-energy compound ATP. This is what happens in the pseudopodium of a moving cell, where such a network of actin filaments forms anew inside the enlarging pseudopodium. **Contraction** also occurs in the ectoplasm of the cell body, where a pre-existing actin network is already present beneath the plasma membrane.

Gelation and solation during amoeboid movement. Cells that undergo amoeboid movement have an outer layer of thick, *gelatinous cytoplasm* called the **ectoplasm** and an inner layer of more *fluid cytoplasm* called the **endoplasm**. In an amoeba, as a pseudopod is extended from the cell, fluid endoplasm streams forward in the direction of extension and congeals (*i.e.*, becomes semisolid) into ectoplasm at the tip of the pseudopod. Meanwhile, at the rear of the moving cell, ectoplasm changes into more fluid endoplasm and streams toward the pseudopod. The alteration between these states of the actin-based cytoskeleton that occurs during such transitions is called **gelation-solation.** Proteins such as **gelsolin** that are present within these gels may be activated by calcium to convert the gel to the more fluid state (**Becker** *et al.*, 2006).

Experiments have shown that the forward streaming in the pseudopod does not require squeezing from the rear of the cell: when a pseudopod's plasma membrane is removed using detergent, the remaining components can still stream forward if the appropriate mixture of ions and other chmical is added. Pressure exerted on the endoplasm, possibly due to contraction of an actomyosin network in the trailing edge of the cell, may also squeeze the endoplasm forward, aiding formation of a protrusion at the leading edge.

Types of cells that exhibit amoeboid locomotion. The most common cells to exhibit amoeboid locomotion in the human body are the **white blood cells** or **leucocytes**. WBCs often move out of the blood into the tissues to form **tissue macrophages.** Other types of cells can also move by amoeboid locomotion under certain circumstances. For instance, **fibroblasts** move into damaged area to help repair the damage and even the **germinal cell** of the skin, though ordinarily completely sessile cells, move toward a cut area to repair the opening. Finally, cell locomotion is especially important in development of embryo and foetus after fertilization of the ovum. For instance, embryonic cells often must migrate long distances from their sites of origin to new areas during development of special structures.

Control of amoeboid locomotion—chemotaxis. The most important initiator of amoeboid locomotion/movement is the process called **chemotaxis**. This results from the appearance of certain chemical substances in the tissues. Any chemical substance that causes chemotaxis to occur is called a **chemotactic substance**. Most cells that exhibit amoeboid locomotion move toward the source of a chemotactic substance—that is, from an area of lower concentration toward an area of higher concentration—which is called **positive chemotaxis**. Some cells move away from the source, which is called **negative chemotaxis**.

But how does chemotaxis control the direction of amoeboid locomotion? Although the answer is not certain, it is known that the side of the cell most exposed to the chemotactic substance develops membrane changes that cause pseudopodial protrusion (see **Guyton** and **Hall**, 2011).

III. Cytoplasmic Streaming

Cytoplasmic streaming is a actomyosin-dependent movement of the cytosol within a cell and is seen in a variety of organisms that do not display amoeboid movement. In slime molds such as *Physarum polycephalum*, for example, cytosol streams back and forth in the branched network that constitutes the cell mass.

Many plant cells display a circular flow of cell contents around a central vacuole. This streaming process, called **cyclosis**, has been studied most extensively in the giant algal cell *Nitella*. In this case, the movement seems to circulate mix cell contents. Cytoplasmic streaming requires actin filaments, as it is inhibited in cells treated with **cytochalasin**. In *Nitella*, a dense set of aligned microfilaments are found near sites where cyclosis occurs. Cyclosis probably involves specific myosins that provide the force of movement of components within the cytoplasm. When latex beads coated with various types of myosin are added to *Nitella* cells that have been broken open, the beads move along the actin filaments in an ATP-dependent manner in the same direction as normal organelle movement.

Box 29.4 Brownian Movement

Robert Brown (1773–1858) discovered the motion that bears his name while pursuing his passion, botany. Even though Brown studied medicine at university and serving in the army, he had amassed so great knowledge of botany that **Sir Joseph Banks** hired him to sail as a ship's naturalist aboard the ship *The Investigator* on a voyage of discovery to Australia. The voyage lasted for years, circumnavigating Australia twice. During that time, **Brown** collected more than 4,000 species, most of them new to science and many in previously unknown genera. The difficult times he faced were physical and intellectual: The ship *Investigator* was cramped, damp and unseaworthy and the botany of his day had only just begun to classify plants "naturally" (that is evolutionarally). However, Brown's treatment of the Australian flora remains essentially intact today.

Brown's zeal for botany stimulated to become an excellent light microscopist. He realized that traits needed to support a natural classification of plants could be found in the early stages of development, which are microscopic. In Brown's day, the best microscopes had only a single lens, because theory for properly correcting compound lenses had yet to be developed. Despite their simplicity, single lens microscopes can reveal even subcellular detail. An early micrograph of peel of onion epidermis viewed through a microscope used by Brown showed the nuclei clearly. From his observations, Brown coined the term '*nucleus*' and before the advent of cell theory he correctly surmised that it was a feature of every plant cell. He also discovered cytoplasmic streaming in the stamen hairs of *Tradescantia* and delighted in showing this vivid cellular action to his friends, including such luminaries as **Charles Darwin** and **William Hyde Wollaston** (British chemist and invertor of prism used in Nomarski optics, which today provides best images of streaming in the stamen hairs). Brown's major microscopical interest was the fertilization of plants. He was the first to correctly outline the anatomy of seeds and their embryos, and discovered the naked ovule of gymnosperms, which allowed to rationalize the classification of this group.

However, Brown's enduring fame rests on an 1827 discovery, **Brownian motion**, made while examining pollen. He saw particles inside the pollen grain moving randomly—rapidly and without cessation. Brown was not the first person to observe this incessant movement; other had seen it before and hoped they were seeing the essence of life. But Brown observed the motion in pollen grains preserved for months in alcohol as well as in suspensions made by grinding a variety of rocks and minerals. He correctly concluded that the motion has a physical, not an organic, explanation. Later scientists quantified the motion to drive fundamental insight into the stochastic character of molecular world. For example, French physicist and Nobel Laureate **Jean-Baptiste Perrin** used **Albert Einsteins** formula for Brownian motion to conclude the size of the water molecule. It is apt that the incessant movement of particles bears the name of the indefatigable botanist Robert Brown (**Buchanan** *et al.*, 2000).

QUESTIONS

Long Answer Questions

1. Describe various types of motility systems of the cells and their utility in the organism.
2. Eukaryotic cells show intracellular motility. Describe the elements that are involved in maintaining this.
3. How do the skeletal muscles contract? Explain the role of each type of muscle protein.
4. List the main events in the contraction of a skeletal muscle.
5. How is energy supplied for muscle contraction?

Short Answer Questions

1. What is titin?

   Ans. Titin is largest, highly elastic, spring like protein of a sarcomere of vertebrate skeletal muscle. Its molecules originate at the M line in the center of each sarcomere and extend along the myosin filament, continuing past A band and terminating at Z line.
2. Define nebulin.

   Ans. Neulin is a molecular ruler of sarcomere. One nebulin protein exists at outer side of each thin filament of the sarcomere and regulates the number of actin filaments which should constitute a single thin filament.
3. What is the function of titin?

   Ans. Titin prevents the sarcomere from becoming pulled apart during muscle stretching. Titin also maintains myosin filaments in their proper position within the center of the sarcomere during muscle contraction.
4. Discuss sliding filament theory of muscle contraction.
5. What is energy source of contraction and how is it generated?
6. Explain the following:

   (*i*) Mechano-chemical coupling during muscle contraction.

   (*ii*) Role of calcium in muscle contraction.
7. What is biological significane of

   (*a*) Myoglobin and (*b*) Phosphocreatine?
8. What is sarcomere? Name the proteins found in it.
9. Describe the T system of sarcoplasmic reticulum.

Very Short Answer Questions

1. What is the contractile unit of a muscle celled skeletal muscle?
2. Name the two types of protein filaments (myofilaments) found in a sarcomere. In which muscle they are best known?
3. Which myofilaments slide in a sarcomere during muscle contraction?
4. Where are the intercalated discs found in the cardiac muscle fibers?
5. What is the fatigue of a muscle due to?
6. Which compounds provide energy for muscle contraction?
7. Name the muscle which causes peristalsis.
8. By which tissue are the muscles attached to the bones?
9. Name the components of a triad.
10. What does a F-actin chain consist of?
11. Give the location of M line in a sarcomere.
12. Vertebrate heart muscle is said to be myogenic. Why?
13. Which of the three types of muscle tissue is most specialized?

Fill in the Blanks

1. A sarcomere has two types of myofilaments: (*i*) thick ________ myofilaments and (*ii*) thin ________ myofilaments.
2. In a contracting myofibril, only the ________ band actually shorten.
3. Sarcolemma is stimulated (or depolarised) by ________ released from the axon at the neuromuscular juction.

4. Depolarisation of sarcolemma causes the release of ____________ from the ____________ of the muscle fiber.
5. Energy-rich reserve phosphatase in the muscle cell is ____________.
6. A group of muscle fibers innervated by an axon is called a ____________.
7. The genetic disease muscular dystrophy is due to lack of a protein ____________.
8. ____________ muscle has a long refractory period.

Multiple Choice Questions

1. The ions that play an important role in muscle contraction are
   (*a*) potassium (*b*) sodium
   (*c*) magnesium (*d*) calcium
2. The sarcoplasmic reticulum
   (*a*) releases calcium into sarcoplasm
   (*b*) conducts impulse inward
   (*c*) withdraws calcium from sarcoplasm
   (*d*) does (*a*) and (*c*)
3. The A band of skeletal muscle contains
   (*a*) all myosin, but no actin
   (*b*) all actin but no myosin
   (*c*) troponin and tropomyosin
   (*d*) all actin and some myosin
4. When skeletal muscle contracts, there is
   (*a*) shortening of sarcomere
   (*b*) shortening of thick and thin myofilaments
   (*c*) decrease in the width of a band
   (*d*) increase in the distance between Z line

ANSWERS

Very Short Answer Questions

1. Sarcomere.
2. Myosin (thick filament) and actin (thin filament); skeletal muscle.
3. Actin myofilament.
4. At the end of cardiac muscle fibers.
5. Accumulation of lactic acid.
6. ATP and phosphocreatine.
7. Smooth muscle (also called visceral, unstriated or unstriped muscles).
8. Tendons.
9. A tubule and a pair of SR cisternae.
10. G-actin molecules.
11. Middle of A band or H band.
12. Because rhythmic contraction originate in the heart muscle tissue itself.
13. Skeletal (striated) muscle.

Fill in the Blanks

1. myosin, actin; 2. I; 3. acetylcholine;
4. calcium, sarcoplasmic reticulum; 5. phosphocreatine; 6. motor unit;
7. distrophin; 8. cardiac muscle.

Multiple Choice Questions

1. (*d*) 2. (*d*) 3. (*c*) 4. (*a*)

30 Microtubule-Based Cell Movement: Cilia and Flagella

The cilia (*L.,cili*=eye lash) and flagella (*L.*, little whip) are microscopic, contractile and filamentous processes of the cytoplasm which create food currents, act as sensory organs and perform many mechanical functions of the cell. Morphologically and physiologically, the cilia and flagella are identical structures but even then both can be distinguished from each other by their number, size and functions. Their distinguishing features are as follows:

1. The flagella are less (1 or 2) in number than the cilia which may be numerous (3000 to 14000 or more) in number.
2. The flagella occur at one end of the cell, while the cilia may occur throughout the surface of the cell.
3. The flagella are longer (up to 150 μm) processes, while the cilia are short (5 to 10 μm) appendages of the cytoplasm.
4. The flagella usually beat independently, while the cilia tend to beat in a coordinated rhythm.
5. The flagella exhibit undulatory motion, while the cilia move in a sweeping or pendular stroke.

30.1. STEREOCILIA AND KINOCILIA

The cell sometimes gives out immobile cytoplasmic extensions known as **stereocilia**. The sterocilia differ from the true cilia which are known as **kinocilia**. The sterocilia occur in most epithelial cells of the epididymis and macula and crista of the internal ear. Sterocilia of the hair cells of the inner ear are responsible for the transduction of sound. These and other sterocilia do not contain microtubules (see Chapter 29). They contain, however, about 3000 microfilaments which are disposed longitudinally but have a definite polarity and a helical symmetry, with cross-bridges around the filaments (**De Rosier**, 1980).

30.2. DISTRIBUTION OF THE CILIA AND FLAGELLA

The flagella occur in the protozoans of the class Flagellata, choanocyte cells of the sponges, spermatozoa of the Metazoa and among plants in the algae and gamete cells. The cilia occur in the protozoans of the class Ciliata and members of other classes and ciliated epithelium of the Metazoa. The cilia may occur on external body surface and may help in the locomotion of such animals as the larvae of certain Platyhelminthes, Nemertines, Echinodermata, Mollusca and Annelida. The cilia may line the internal cavities or passages of the metazoan bodies as air passage of the respiratory system and reproductive tracts. The nematode worms and arthropods have no cilia.

Except for sperm, the cilia in mammalian systems are not organelles of locomotion. But their effect is the same, that is, to move the environment with respect to the cell surface.

30.2. STRUCTURE OF THE CILIA AND FLAGELLA

The ciliary apparatus is composed of following basic components: 1. The **shaft** or **cilium**, which is the slender cylindroid process that projects from the free surface of the cell; 2. The **basal body** or granule, a centriole like cellular organelle from which the cilium originates; and 3. In some cells fine fibrils—called **ciliary rootlets**. Basal body remains separated from cilium by a **ciliary** or **basal plate** which has two functions: termination of the C tubule of each triplet of basal body; and beginning of two central microtubules. The cilia and flagella are extremely delicate, permanently formed, thread-like extension of cytoplasm and their thickness is often at the limit of the resolving power of light microscope.

Isolation and Chemical Composition of Cilia and Flagella

The first detailed chemical analysis of the protein components of the cilia of *Tetrahymena pyriformis* was conducted by **I. R. Gibbons** (1963). Ciliary movements can be analyzed easily by scraping the pharyngeal epithelium of a frog or toad with a spatula and placing the scrapings in a drop of physiological salt solution between a slide and a coverglass. By certain recent techniques, a flagellum can be severed from a cell by a laser beam and ciliary membrane can be peeled off by detergent treatments.

Axoneme of cilia has a variety of proteins such as A and B **tubulins** in the microtubules, **dynein** (the microtubule ATPase), **nexin** and others.

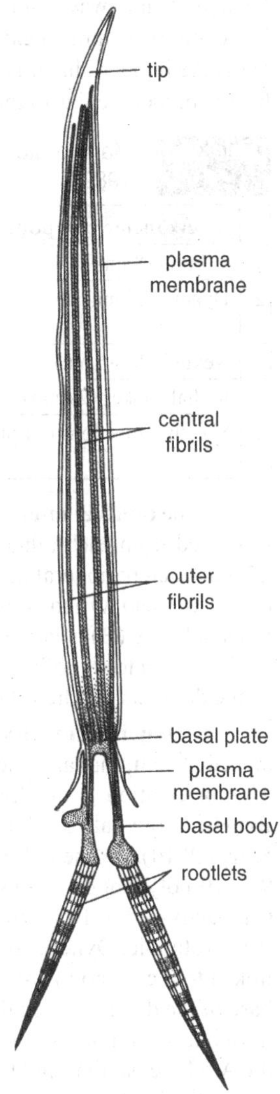

Fig. 30.1. A ciliary apparatus in L.S. showing fundamental structure of cilium or shaft, basal bodies and ciliary rootlets.

Ultrastructure of the Cilia and Flagella

An eukaryotic cilium or flagellum is composed of three major parts: a central axoneme or shaft, the surrounding plasma membrane and the interposed cytoplasmic matrix (Fig. 30.1).

1. Ciliary membrane. Though the ciliary membrane (9.5 nm thick) is physically continuous with plasma membrane of the cell, but it contains far less amount of proteins than the latter (*i.e.*, it is a typically protein poor; **Satir**, 1977). Further, some of the proteins present in the ciliary membrane are specific to it and have a role as the barrier against the loss of ATP and certain essential ion that are required at appropriate concentrations to provide the energy for the ciliary movement.

Ciliary necklace. An unusual feature of the membranes of all somatic cilia is the presence of multiple strands (2 to 6 and up to 11) of particles, called **ciliary necklace**, at the base of the organelle. These particles can be seen in the electron micrographs of freeze-fractured cilium (see **Fumi Suzuki**). The ciliary necklace is found at a region in the cilium where microtubules and basal bodies make contact with the membrane. According to **Thorpe** (1984), ciliary necklace may have following two functions: 1. It may position the underlying basal body from which the cilium is originated. 2. It may help in the differentiation of ciliary membrane; *i.e.*, the rings of particles may retain proteins that would otherwise diffuse out and be incorporated into ciliary membrane.

2. Matrix. The bounded space of the cilium contains a watery substance known as **matrix**. In the ciliary matrix are embedded eleven microtubules of axoneme and other interconnecting proteins.

3. Axoneme. The axial basic microtubular structure of cilia and flagella is called **axoneme.** It is the essential motile element of these organelles. The axoneme is about 0.2 to 10 μm in diameter and may range from a few microns to 1 to 2 mm in length. The cilia may be thicker at the base and may become thinner gradually along the length.

The axonemal elements of nearly all cilia and flagella (as well as the tails of sperm cells) contain the same 9 + 2 arrangement of microtubules. In the centre of the axoneme are two **singlet microtubules** or fibrils that run length of the cilium. Each of the central microtubules (25 nm in diameter) is composed of 13 protofilaments. The central fibrils, each has a wall of 6 nm thick and are located 35 nm away from each other. Both central fibrils are connected by a bridge and are enclosed in a common central **sheath**. A plane perpendicular to line joining the two central tubules divides the axoneme into a right and a left symmetrical half. It is generally accepted that the plane of the ciliary beat is perpendicular to this plane of symmetry.

Table 30.1 Major protein structures of the axoneme of the cilia and flagella (Source: Alberts *et al.*, 1989).

	Axoneme component	Function
1.	Tubulin (8 nm)	Principal component of microtubules.
2.	Dynein (24 nm)	Project from microtubule doublets and interact with adjacent doublets to produce bending.
3.	Nexin link (86 nm)	Hold adjacent microtubule doublets together.
4.	Radial spokes (29 nm)	Extend from each of the nine outer doublets inward to the central pair.
5.	Sheath projections (14 nm)	Project as a series of side arms from the central pair of microtubules; together with the radial spokes these regulate the form of the ciliary beat.

Nine **doublet microtubules** (each 36 nm in diameter) surround the central sheath; they remain separated from each other by a distance of 20 nm and from the ciliary membrane by a distance of 25 nm. Each peripheral doublet consists of two microtubules or subfibres (18 to 25 nm in diameter), one is smaller (A) and complete, having 13 protofilaments of tubulin and lying closer to the axis; the other subfibre (B) is larger and incomplete, having only 11 protofilaments. The B subfibre lacks the wall adjacent to A subfibre and is skewed at about 10° away from the axis. Other associated structures of the doublets are the following:

1. Dynein arms. Extending from each A subfibre are two **dynein arms**—an **outer arm** and an **inner arm**, that are oriented in the same direction in all microtubules (*i.e.*, peripheral doublets). This orientation is clockwise when the axoneme is viewed from base to tip. The arms contain **dynein** (Now called **ciliary** or **axonemal dynein** to distinguish it from cytoplasmic dynein; see **Karp**, 2010). Axonemal dynein is a large protein complex (nearly 2 million daltons) composed of 9 to 10 polypeptide chains, the largest of which are about 450,000 daltons. Dynein is a Mg^{2+} and Ca^{2+}-activated ATPase enzyme which after solubilization can recombine at the same position on the A microtubule. Dynein contains two or three elliptical or globular **heads** (depending on the source) linked to a common **root**, **foot** or **base** by the thin flexible **strands** or **stalks** (Fig. 30.3). Thus, the base of the dynein molecule attaches only to A subfibre, leaving the heads free to make contact with the adjacent B tubules of neighbouring doublet. It indicates that B tubule has different structure than the A tubule, so that, the base of dynein cannot attach to it. The resulting asymmetry is required to prevent a fruitless tug-of-war between the neighbouring microtubules, which presumably explains why each of nine outer microtubules is an A–B doublet (see **Alberts** *et al.*, 1989).

2. Nexin links. Adjacent doublets are joined or linked by **peripheral, interdoublet** or **nexin links**; the nexin links have a periodicity of 86 nm. Nexin links extend from A tubule of one doublet to B tubule of adjacent doublet. Nexin protein has a molecular weight of about 150,000 to 160,000 daltons. Nexin links are highly elastic: their normal length is 30 nm, but they can be stretched to 250 nm without breaking (see **Darnell,** 1986). They are thought to function like the rubber bands to resist the sliding between adjacent double microtubules (*i.e.*, they maintain integrity of the axoneme during the sliding motion.

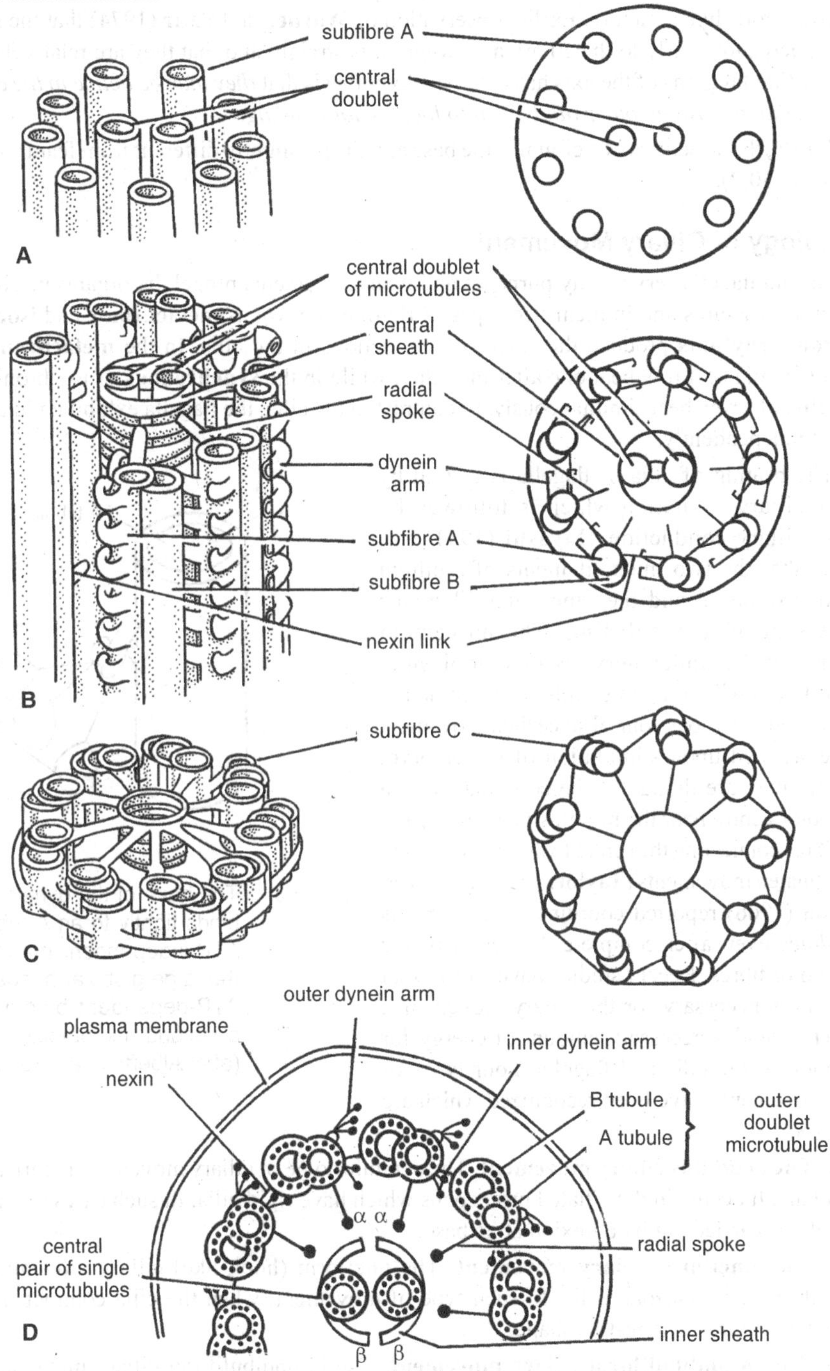

Fig. 30.2. Schematic diagram of a flagellar axoneme, showing relationships between its components at the flageller tip (A), in the middle of the flagellum (B), and at the basal body (C). More detailed picture of flagellum in the cross section is shown in D (after Berns, 1983; Albert *et al.*, 1989).

3. Radial spokes. There are 36 nm long radial bridges or links between the A subfibre and the sheath containing the central microtubules. These spokes terminate in a dense **knob** or **head**, which may have a fork-like structure. Earlier observation of **Warner** and **Satir** (1974) that the spokes are attached perpendicularly to the ciliary axis where it is straight and that they are relatively detached in bent or tilted regions of the axis has led to the hypothesis that *they may be active in the conversion of active sliding between outer doublets into local axial bending.*

Lastly, the structures of a cilium at the base and tip are slightly different from features described above (Fig. 30.2).

Physiology of Ciliary Movement

The cilia and flagella serve many purposes and their movements propel the organism. The cilia are contractile structures and in them two types of rhythms known as **metachronic** and **isochronic** or **synchronic rhythms** produce the wave of contractions in the cilia. In the **metachronic** type of rhythm the cilia of a row beat one after the other, while in the **synchronous** or isochronic rhythm, all the cilia of a row beat simultaneously. In contrast to the cilia, the flagella exhibit undulant motion and beat independently.

The beating of cilia or flagella is caused by the intraciliary excitation which is followed by the interciliary conduction **Hayashi** (1961) has reported that the two inner filaments of a cilium transmit excitation and the nine outer filaments are the seat of ATP splitting. The movement of cilia may be under nervous or cytoplasmic control. In a few invertebrate embryos the cilia are probably under nervous control since their movement may be stopped upon stimulation of the embryo. In ciliates they are thought to be coordinated by a neuromotor centre near the mouth since destruction of the fibres connecting the centre to the cilia results in uncoordinated movements (**Taylor**, 1920). However, **Okajima** (1966) reported coordinated movements in *Euplotes* even after complete dissection of the neuro-motor fibres. Recent studies have shown that cytoplasm is necessary for the ciliary movements. The ATP provides necessary amount of energy for the motion of the cilia and flagella. Four types of ciliary movements have been recognized which are as follows:

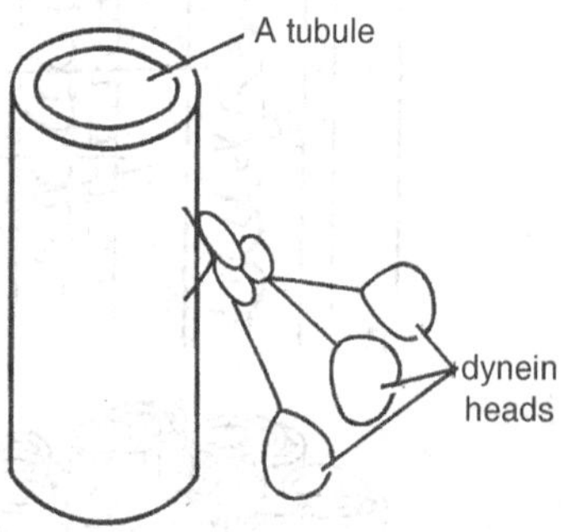

Fig. 30.3. The base of the dynein molecule binds tightly to an A tubule in an ATP-independent manner, while the large globular heads have an ATP-dependent binding site for a second microtubule (B tubule) (after Alberts *et al.*, 1989).

1. The pendulus ciliary movement. The pendulus type of ciliary movement is carried out in a single plane. It occurs in the ciliated protozoans which have rigid cilia. In such cases the movement of the cilia is carried out by a flexion at its base.

2. The unciform ciliary movement. The unciform (hook-like) ciliary movement occurs commonly in the metazoan cells. In such type of movement, when the cilia contract it becomes double and acquires a hook-like shape.

3. The infundibuliform ciliary movement. The infundibuliform ciliary movement occurs due to the rotary movement of the cilium and flagellum. In this case, the cilium or flagellum is passed through three mutually perpendicular planes in the space and makes conical or funnel-shaped shape.

4. The undulant movement. The undulant movement is the characteristic of the flagellum. In undulant movement the waves of the contraction proceed from the site of implantation and pass to the border.

Each beat of cilium or flagellum involves the same pattern of microtubule movement. Each cilium moves with a whip-like motion and its beat may be divided into two phases: 1. The fast **effective stroke** (or forward active stroke or power stroke) in which the cilium is fully extended and beating against the surrounding liquid (*i.e.*, it is like the action of an oar in a rowboat; Fig. 30.4B). 2. The slow **recovery stroke**, in which the cilium returns to its original position with an unrolling movement that minimizes viscous drag (Fig. 30.4C). The cycles of adjacent cilia are almost but not quite in synchrony, creating a wave-like patten.

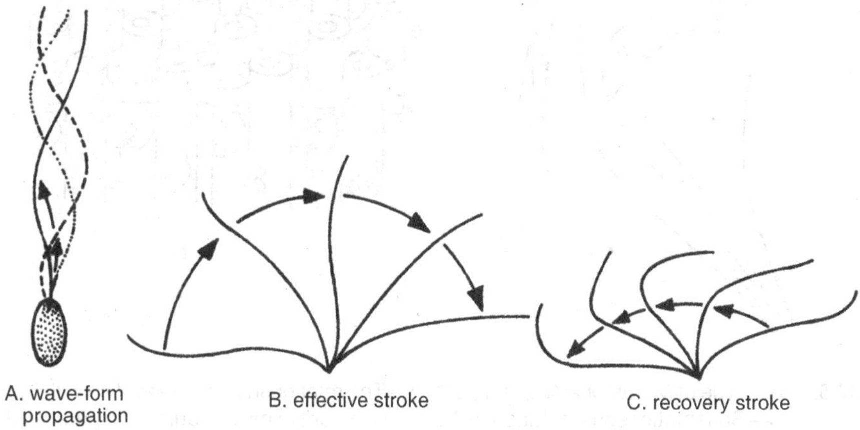

Fig. 30.4. Movement of cilia and flagella. A—Typical wave-form propagation of a flagellum, such as that found in sperm; B—Effective stroke common with cilia of all types; C—Recovery of cilium prior to the next effective stroke (after Dyson, 1978).

The flagellum instead of making whip-like movements, propagates quasi-sinusoidal waves (Fig. 30.4A), *i.e.*, successive waves move toward the tip of the flagellum, propelling the cell (*e.g.*, sperm) in the opposite direction.

The mechanism of force and movement (bending) by the flagellum has recently been studied extensively. It is well established now that the ciliary movement is generated by the microtubules and the associated structures of the flagellum. It was shown that the cell free flagella can be caused to move by adding an energy source such as ATP. Even broken pieces of cilia or isolated axoneme itself continue to beat, suggesting the role of microtubules in the movement. The contractile **axostyle** of some microorganisms such as *Metamonadida* (a dinoflagellate that lives in the gut of termite; Fig. 30.5) is another example of microtubule mediated motile process (see **Berns**, 1983). In fact, bending force is produced by the sliding of microtubules. This has been shown by exposing isolated axonemes to proteolytic enzymes, which disrupts both the nexin links and the radial spokes but leaving the dynein arms and the microtubules themselves intact. If such a partially digested structure is exposed to as little as 10 μM ATP, the axoneme elongates until it is up to nine times its original length, the component microtubules in the axoneme telescoping out of the loosened structure. It seems that the adjacent peripheral doublets can actively slide against each other once they have been freed of their lateral cross-links (such as those made of nexin). Apparently, in intact structure this sliding movement is converted to bending. Further, since the adjacent outer doublets actively slide against each other, a force must be generated between them. This force is apparently generated by dynein arms which 'walk' along the doublets, as has been suggested by **Peter Satir's** (1968) sliding filament hypothesis.

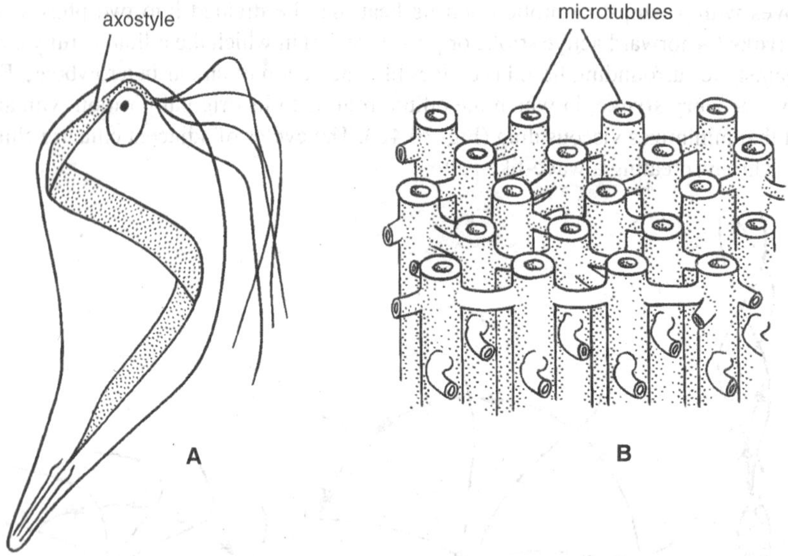

Fig. 30.5. Axostyle of *Metamonadida* flagellate. A—The relationship of the axostyle with the cell body; B—Schematic representation of the arrangement of microtubules into parallel sheets with protein cross bridges between them as seen in a T.S. of axostyle (after Berns, 1983).

30.3. SLIDING FILAMENT HYPOTHESIS

Recent experimental work on ciliary motion has shown notable similarities with the sliding mechanism involved in the interaction of actin and myosin in muscle. The dynein arms attached to subfibre A have been compared with the cross bridges of myosin and it has been postulated that they form intermittent attachments, by which one doublet (N_1) is able to push the adjacent one ($N_1 + 1$) toward the tip of the axoneme (Fig. 30.6). Under normal conditions, the attachment of subfibre A of N to subfibre B of N + 1 by dynein arms is not observed in an intact cilium. Only when the ciliary membrane is extracted with a detergent, the axoneme enters in a state of **rigor** in which the attachment is produced (Fig. 30.6A). Addition of ATP to axonemes in the state of rigor restores motility and causes release of the dynein arm (Fig. 30.6B). In this mechano-chemical cycle, the next step would be **re-extension** of the dynein arm (Fig. 30.6C) and its **rebinding** at an angle, with a new, more proximal site on subfibre B (Fig. 30.6D). This step involves the hydrolysis of ATP to ADP + Pi. In the last step, the arm returns to the rigor position and displacement of the doublets results (Fig. 30.6E).

Force is generated when dynein arms move. The movement of sliding is converted to bending by virture of radial spokes that bridge each other doublet to the inner pair of microtubules (**Warner** and **Satir**, 1974; **Huang** *et al.,* 1981). The wave that is generated by sliding is propagated down the organelle from base to tip, with the cell generally moving in a direction opposite from that of wave propagation.

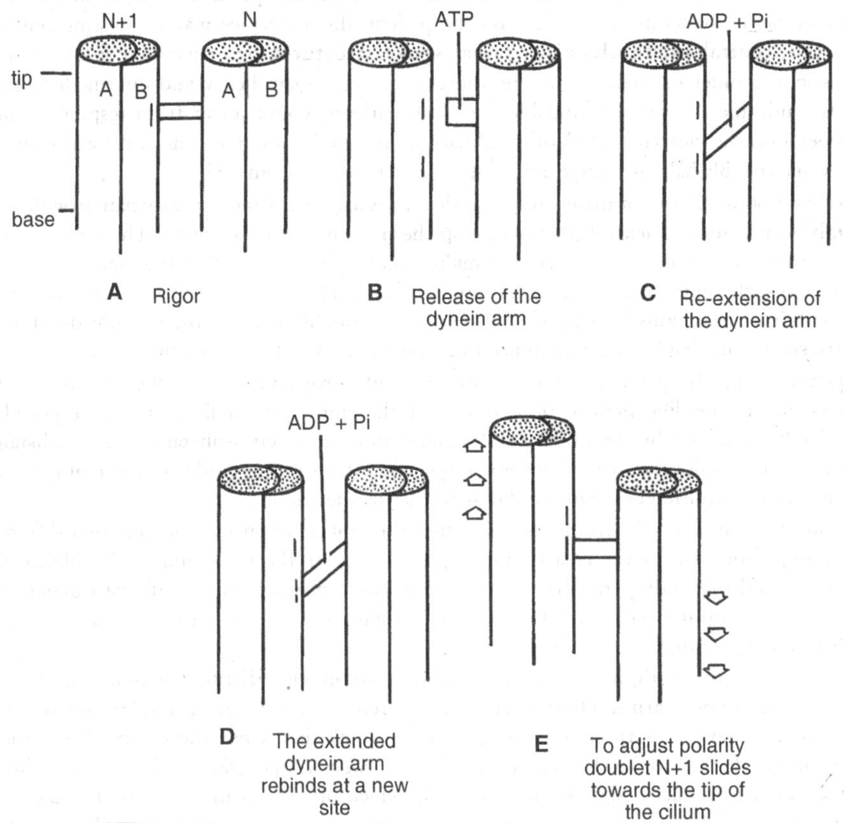

Fig. 30.6. Schematic representation of the mechano-chemical cycle involved in sliding of filament in ciliary movement (after De Robertis and De Robertis, Jr., 1987).

30.4. IMMOTILE CILIA SYNDROME (KARTAGENER'S SYNDROME)

Ciliary motion can be affected by many deficiencies in the protein composition of the organelle. For example, in immotile cilia syndrome, a condition characterized by severe respiratory difficulty (chronic bronchitis and sinusitis) and male sterility, the underlying genetic defect is the absence of inner and outer dynein arms on the peripheral doublets of both cilia and flagella. The symptoms of this syndrome result from the immobility of cilia in the respiratory tract and of the flagella in the sperm (Box 30.1).

Box 30.1 Kartagener Syndrome and Situs Inversus (Primary cilia and role of cilia in embryonic development and disease)

When we look into a mirror, we see a relatively symmetrical organism, one whose left half is essentially a mirror image of the right half. Surgeons, however, see a strikingly asymmetric organism when they open a person's thoracic and abdominal cavity. The stomach, heart, and spleen, for example, are shifted to the left side of the body, whereas the liver is largely on the right side. On occasion, a physician may see a patient in which the left-right asymmetry of the visceral organs is reversed (a condition called **situs invertus**).

Situs invertus is seen in persons with **Kartagener syndrome**, which is also characterized by recurrent sinus and respiratory infections and infertility in males. The first clues to the underlying cause of this disorder came in 1970s when it was discovered that the immotile sperm from these individuals had an abnormal axonemal structure. Depending on the patient, the axonemes may be missing outer or inner **dynein arms, central microtubules,** or **radial spoke structures.** Subsequent studies have shown that mutations in a number of genes, including those encoding dynein heavy and intermediate chains, can cause this syndrome. It is understandable that such patients would suffer from respiratory infections, which depend on the clearance of debris and bacteria by respiratory tract cilia and form male infertility, but why would roughly half of them exhibit a reversed left-right asymmetry?

The basic body plan of a mammal is laid down during gastrulation in association with a structure called **embryonic node**. Each cell that makes up the node has a single cilium. These cilia have unusual properties; they are missing the two central microtubules (they have a 9 + 0 axonemal structure) and exhibit an unusual rotary (*i.e.*, twirling) motion. If the motility of these cilia is impaired, as happens in mice harboring mutations in a flagellar dynein gene, roughly half of the animals develop reversed asymmetry, suggesting that left-right asymmetry in these mutants is determined by chance.

Experiments involving microscopic fluorescent beads have revealed that the rotation of the nodal cilia moves the surrounding fluid to the left side of the embryonic midline. It was proposed that the extracellular fluid moved by the nodal cilia contains morphogenetic substances (*i.e.*, substances that direct embryonic development) that become concentrated on the left side of the embryo, leading to eventual formation of different organs on different sides of the midline.

An alternate but related hypothesis holds that the embryonic node contains two different types of cilia, a population of motile cilia located in the center of the node and nonmotile primary cilia distributed around the node periphery. According to this hypothesis, the motile cilia generate the left ward flow and the nonmotile cilia act as sensory structures that detect the movement and transmit signals that lead to asymmetry.

Many cells of the body have a single nonmotile **primary cilium** that is lacking both central microtubules and dynein arms. These cilia were ignored by researchers for years but recent studies suggest that they have important functions as "**antennae**" that sense the chemical and mechanical properties of the fluids into which they project. We can consider the primary cilia on the epithelial cells that line the lumen of microscopic kidney tubules in which urine formation occurs. The importance of these cilia was revealed when it was discovered that a pair of membrane protein called polycystins are located on the surface of these kidney cilia where they form a Ca^{2+} ion channel. Mutations in *PKD1* and *PKD2*, the genes that encode polycystins, lead to polycystic kidney disease, in which the kidney develops multiple cysts that destroy kidney functions. PKD is thought to be a condition that result from breakdown in the regulation of cell division because the cysts are the result of an abnormally high level of proliferation of the epithelia cells that lines parts of kidney tubules. Mutations in the polycystins are thought to alter the response of the primary cilia to fluid flow, which leads to a disturbance in calcium flux across the ciliary membrane, which in turn impairs the transmission of signals to the body of the cells and the nucleus, resulting in abnormal proliferation.

Bardet-Biedl Syndrome (BBS)

The importance of cilia in human development became even more evident with the discovery that Bardet-Biedl syndrome (BBS) is caused by mutations in any one of a number of genes that affect the assembly of basal bodies and cilia. Persons afflicted with BBS exhibit a remarkable range of abnormalities, including polydactyly (extra fingers or toes), situs inversus, obesity, kidney disease, heart defects, mental retardation, genital abnormalities, retinal degeneration, loss of hearing and smell, diabetes and high blood pressure. The fact that all of these dysfunctions can be traced to abnormalities in the basal bodies and cilia shows the widespread importance of these structures in organ development and function. Many of the genes responsible for these various cilia-based disorders ("ciliopathies") were first identified in model organisms such as *Chlamydomonas* or *Caenorhabditis elegans* (nematode), which provide another example of the importance of basic research on nonchordates in furthering our understanding of human diseases (see **Karp**, 2010).

In *Chlamydomonas* several mutational defects have been studied in the axoneme of flagellum which may lead to paralysis of the flagellar function (**Luck**, 1984).

30.6. OTHER FUNCTIONS OF THE CILIA AND FLAGELLA

1. The ciliary or flagellar movement provides the locomotion to the cell or organism.
2. The cilia create food currents in lower aquatic animals.
3. In the respiratory tract, the ciliary movements help in the elimination of the solid particles from it.
4. The eggs of amphibians and mammals are driven out from the oviduct by the aid of vibratile cilia of the latter.

Thus, the cilia and flagella serve many physiological processes of the cell, such as locomotion, alimentation, circulation, respiration, excretion and perception of sense.

30.7. ORIGIN OF CILIA

The newly formed basal bodies become aligned in rows beneath the apical plasma membrane and each basal body may then produce satellites from the side, a root from its base and a cilium from its apex.

The formation of the cilia from the basal bodies is started by the formation of a vesicle-like structure of the cytoplasm towards the distal end of the basal bodies. The walls of the vesicle are invaginated due to rapid growth of ciliary shaft. The walls of the vesicle are temporary and are replaced by the new and permanent ciliary sheaths.

30.8. DERIVATIVES OF CILIA

The cilia are modified into a variety of structures such as the rods and cones of the retina, crown cell of saccus vasculosus of third ventricle of fishes, primitive sensory cells of the pineal eye and cnidocil of the nematocysts of the coelenterates.

QUESTIONS

Long Answer Questions

1. Explain sliding microtubule hypothesis of ciliary movement.
2. Flagella and cilia, though identical in structure, commonly exhibit a quite different pattern of movement. Describe the two patterns and conditions under which one or the other would be more appropriate.

Short Answer Questions

1. Differentiate between cilia and flagella. Describe the structure of the axoneme.
2. Write short notes on the following:
   (*a*) axoneme;
   (*b*) sterocilia;
   (*c*) ciliary necklace;
   (*d*) axostyle;
   (*e*) sliding filament hypothesis; and
   (*f*) Kartagenre's syndrome.

Very Short Answer Questions

1. What is cilium?
2. What is axoneme?
3. What is microtubule-based movement?
4. What is found in the core of cilia?

Multiple Choice Questions

1. Microtubular structure of cilium is called
   (*a*) basal body
   (*b*) ciliary rootlet
   (*c*) axoneme
   (*d*) kinetosome
2. ATPase activity is associated with structural component of cilia called
   (*a*) A tubule (*b*) B tubule
   (*c*) radial spoke (*d*) dynein
3. Stereocilia can be found in
   (*a*) trachea (*b*) epididymis
   (*c*) sperm (*d*) buccal cavity

ANSWERS

Very Short Answer Questions

1. Cilium (plural cilia) is a membrane-bounded appendage on the surface of a eukaryotic cell, composed of a specific arrangement of microtubules and responsible for motility of the cell. It is shorter and more numerous than closely related organelles called flagella.
2. Axoneme is a group of interconnected microtubules that form the backbone of a eukaryotic cilium or flagellum, usually arranged as nine outer doublet microtubules surrounding a pair of central microtubules.
3. It is motility based on microtubules; it includes motility involving cilia, flagella and sperm tails, as well as chromosomal movements mediated by spindle microtubule.
4. Matrix with axoneme having 9 + 2 arrangement of microtubules.

Multiple Choice Questions

1. (*c*) 2. (*d*) 3. (*b*)

31 Nucleus, Nucleopore and Nucleolus

The nucleus (*L.*, *nux* = nut) is the heart of the cell. It is here that almost all of the cell's DNA is confined, replicated and transcribed. The nucleus, thus, controls different metabolic as well as hereditary activities of the cell. A synonymous term for this organelle is the Greek word **karyon**. Nucleus serves as the main distinguishing feature of eukaryotic cells, *i.e.*, this is the true nucleus as opposed to the nuclear region, prokaryon or nucleoid of the prokaryotic cells.

Nuclei were first discovered and named by **Robert Brown** in 1833 in the plant cells and were quickly recognized as a constant feature of all animal and plant cells. Nucleoli were described by **M.J. Schleiden** in 1838, although first noted by **Fontana** (1781). The term nucleolus was coined by **Bowman** in 1840. In 1879, **W. Flemming** coined the term chromatin for chromosomal meshwork.

31.1. OCCURRENCE AND POSITION

The nucleus is found in all the eukaryotic cells of the plants and animals. However, certain eukaryotic cells such as the mature sieve tubes of higher plants and mammalian erythrocytes contain no nucleus. In such cells nuclei are present during the early stages of development. Since mature mammalian red blood cells are without any nuclei, they are called red blood "corpuscles" rather than cells (*L. corpus* = body, especially dead body or corpse).

Usually the nucleus remains located in the centre. But its position may change from time to time according to the metabolic state of the cell. For example, in the embryonic cells the nucleus generally occupies the geometric centre of the cell but as the cells start to differentiate and the rate of the metabolic activities increases, the displacement in the position of the nucleus takes place. In certain cells such as the glandular cells the nucleus remains located in the basal portion of the cell.

31.2. MORPHOLOGY

Number. Usually the cells contain single nucleus but the number of the nucleus may vary from cell to cell. According to the number of the nuclei following types of cells have been recognised:

1. **Mononucleate cells.** Most plant and animal cells contain single nucleus, such cells are known as **mononucleate cells.**
2. **Binucleate cells.** The cells which contain two nuclei are known as **binucleate cells**. Such cells occur in certain protozoans such as *Paramecium* and cells of cartilage and liver.
3. **Polynucleate cells.** The cells which contain many (from 3 to 100) nuclei are known as **polynucleate cells**. The polynucleate cells of the animals are termed as **syncytial cells**, while the polynucleate cells of the plants are known as **coenocytes**. The most common example of the syncytial cells are the osteoblasts (polykaryocytes of the bone marrow) which contain

about 100 nuclei percell and striated muscle fibres each of which contains many hundred nuclei. The siphonal alga *Vaucheria* contains hundreds of nuclei and certain fungi are the best example of the coenocytic cells.

Shape. The shape of the nucleus normally remains related with the shape of the cell, but certain nuclei are almost irregular in shape. The spheroid, cuboid or polyhedral cells (isodiametrical cells) contain the **spheroid** nuclei. The nuclei of the cylindrical, prismatic or fusiform cells are **ellipsoid** in shape. The cells of the squamous epithelium contain the **discoidal** nuclei. The leucocytes, certain infusoria, glandular cells of some insects and spermatozoa contain the irregular shaped nuclei. Nuclei of cells of silk glands of silk worm have finger-like extensions that greatly increase their surface area (Fig. 31.1).

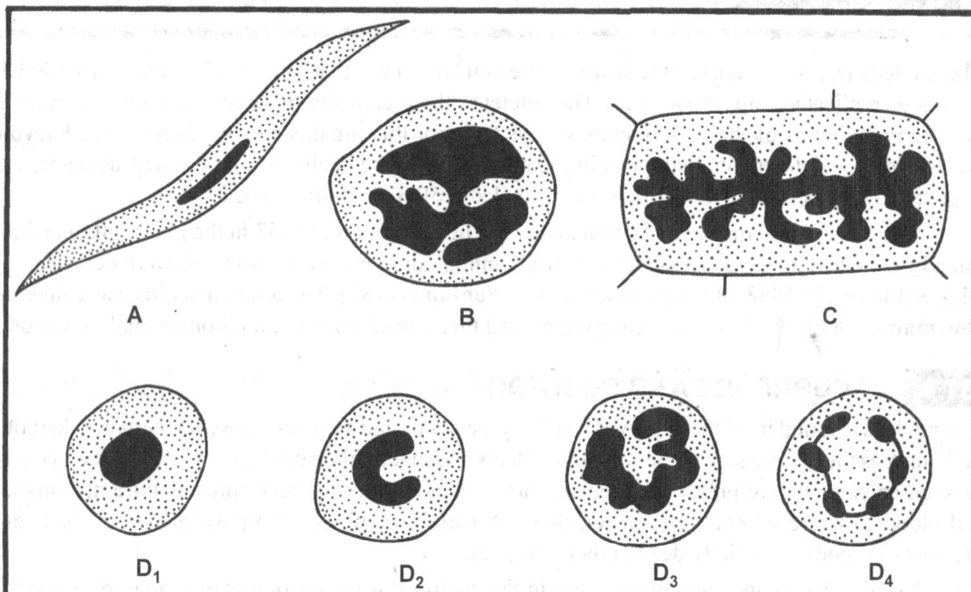

Fig. 31.1. Different shapes of the nucleus in animal cells. A—Elongated in muscle cell: B—Lobed in a human neutrophil cell: C—Branched in a silk spinning cell of an insect larva: D1 to D4—Variable shapes in leucocytes.

Size. Generally nucleus occupies about 10 per cent of the total cell volume. Nuclei vary in size from about 3 µm to 25 µm in diameter, depending on cell type and contain diploid set of chromosomes. The size of the nucleus is directly proportional to that of the cytoplasm. **R. Hertwig** has given the following formula for the deduction of the size of the nucleus of a particular cell:

$$NP = \frac{V_n}{V_c - V_n}$$

where NP is the nucleoprotein index, V_n is the volume of the nucleus and V_c is the volume of the cell.

Moreover, the size of the nucleus is related with the number of the chromosomes or ploidy. The haploid cells contain small-sized nuclei in comparison to the nuclei of the diploid cells. Likewise the polyploid cells contain larger nuclei than the diploid cells. Thus, the size of the nucleus of a cell depends on the volume of the cell, amount of the DNA and proteins and metabolic phase of the cell.

31.3. ULTRASTRUCTURE

The nucleus is composed of following structures: 1. The nuclear membrane or karyotheca or nuclear envelope; 2. The nuclear sap or nucleoplasm; 3. The chromatin fibres; and 4. The nucleolus (Fig. 31.2).

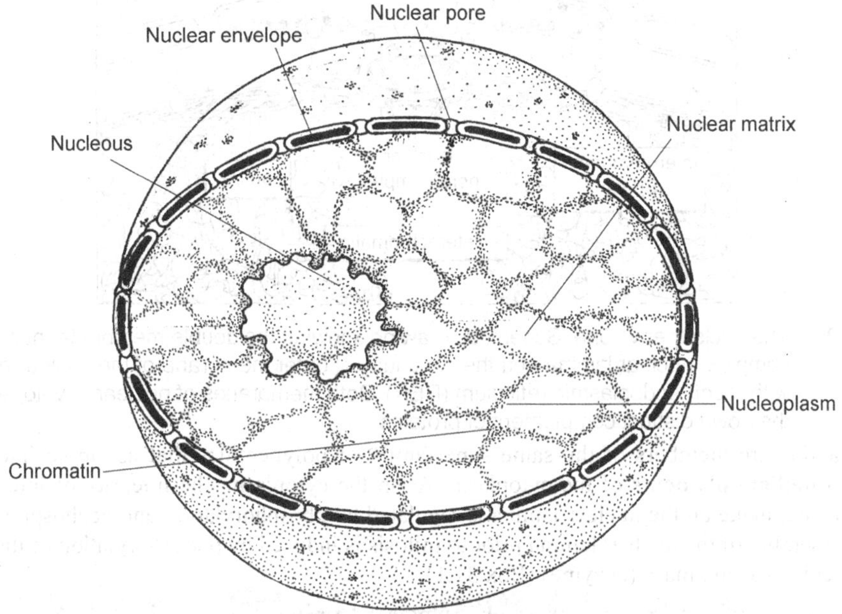

Fig. 31.2. The cell nucleus. Schematic drawing showing some of the major components of the nucleus.

I. Nuclear Envelope

The separation of a cell's genetic material from the surrounding cytoplasm may be the single most important feature that distinguishes eukaryotes from prokaryotes, which makes the appearance of a nuclear envelope a landmark in biological evolution (**Karp** 2010). The **nuclear envelope** consists of two cellular membranes arranged parallel to one another and separated by a 10 to 50 nm distance (Fig. 31.3). The membranes of nuclear envelope serve as a barrier that keeps ions, solutes and macromolecules from passing freely between the nucleus and cytoplasm. The two nuclear membranes of the envelope are fused at sites forming circular pores that contain complex assemblies of proteins. The average mammalian cells contain several thousand nuclear pores.

The outer nuclear membrane is generally studded with ribosomes and is continuous with the membrane of rough endoplasmic reticulum. The space between the nuclear membranes is called **perinuclear space** and it is continuous with the RER lumen (Fig. 31.3).

The inner surface of the nuclear envelope of animal cells is bound by integral membrane proteins to a thin filamentous meshwork, called the **nuclear lamina**. The nuclear lamina provides mechanical support to the nuclear envelope, serves as a site of attachment for chromatin fibers at the nuclear periphery. Nuclear lamina also has a poorly understood role in DNA replication and transcription. The filaments of nuclear lamina are approximately 10 nm in diameter and composed of polypeptides, called **lamins**.

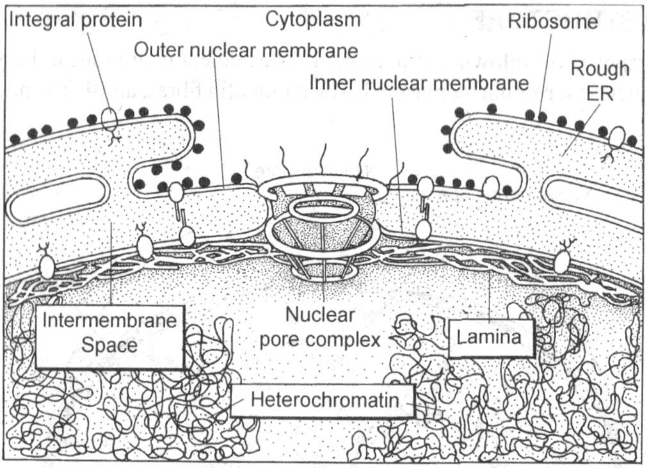

Fig. 31.3. The nuclear envelope. Schematic drawing showing the double membrane, nuclear pore complex, nuclear lamina and the continuity of outer membrane of the nuclear envelope with rough endoplasmic reticulum (RER). Both membranes of nuclear envelope contain their own distinct complement of proteins.

Lamins are members of the same superfamily of polypeptides that assemble into the 10 nm **intermediate filaments** of the cytoplasm. As in the cytoplasm, the integrity of intermediate filaments that make up the nuclear lamina is regulated by phosphorylation and dephosphorylation. The disassembly of the nuclear lamina prior to mitosis is induced by phosphorylation of the lamins by a specific protein kinase (enzyme) (Box 31.1).

Box 31.1 Nuclear Lamina

Nuclear lamina is also called **fibrous lamina, zonula nucleum limitans, internal dense lamella, nuclear cortex** and **lamina densa**.

The nuclear lamina is a protein meshwork which is 10 nm thick. It lines the inside surface of the inner nuclear membrane, except the areas of nucleopores, and consists of a square lattice of intermediate filaments. In mammals, these intermediate filaments are of three types: **lamins A, B** and **C** having M.W. 74,000, 72,000 and 62,000 daltons, respectively. The lamins form dimers that have a rod-like domain and two globular heads at one end. Under appropriate conditions of pH and ionic strength, the dimers spontaneously associate into filaments that have a diameter and repeating structure similar to those of cytoplasmic filaments.

The nuclear lamina is a very dynamic structure. In mammalian cells undergoing mitosis, the transient phosphorylation of several serine residues on the lamins causes the lamina to reversibly disassemble into tetramers of hypo-phosphorylated lamin A and lamin C and membrane associated lamin B. As a result, lamin A and C become entirely soluble during mitosis, and at telophase they become dephosphorylated again and polymerize around chromatin. Lamin B seems to remain associated with membrane vesicles during mitosis, and these vesicles in turn remain as a distinct subset of membrane components from which nuclear envelope is reassembled at telophase. Inside an interphase nucleus, chromatin binds strongly to the inner part of the nuclear lamina which is believed to interfere with chromosome condensation. In fact, during meiotic chromosome condensation, the nuclear lamina completely disappears by the pachytene stage of prophase and reappears later during diplotene in oocytes, but does not reappear at all in spermatocytes.

The lamins may play a crucial role in the assembly of interphase nuclei. For example, when cells are left for a long time in colchicine (drug which arrests cells in metaphase), the lamins assemble around individual chromosomes, which then surrounded by nuclear envelopes give rise to micronuclei containing only one chromosome. A similar phenomenon occurs during normal amphibian development. In the first few cleavages of amphibian development, the nuclear envelope initially forms around individual chromosomes, forming several vesicles that then fuse together to form a single nucleus. This suggests that chromatin is the nucleating centre for the deposition of a nuclear lamina and envelope.

The nuclear envelope is the barrier between the nucleus and the cytoplasm, and nuclear pores are the gateways across that barrier. Unlike the plasma membrane, which prevents passage of macromolecules between the cytoplasm and extracellular space, the nuclear envelope is a hub of activity for the movement of RNAs and proteins in both directions between the nucleus and cytoplasm (Box 31.2). The replication and transcription of genetic material within the nucleus requires the participation of large number of proteins that are synthesized in the cytoplasm and transported across the nuclear envelope. Conversely, the mRNAs, tRNAs and ribosomal subunits that are manufactured in the nucleus must be transported through the nuclear envelope in the opposite direction. Some components, such as snRNAs of the spliceosome move in both directions; they are synthesized in the nucleus, assembled into RNP particles in the cytoplasm and then shipped back to the nucleus where they function in mRNA processing. For appreciating the magnitude of traffic between the two major cellular compartments, **Karp** (2010) has cited the following example: a HeLa cell, which is estimated to contain about 10 million (one crore) ribosomes. To support its growth, a single HeLa cell nucleus must import approximately 560,000 ribosomal proteins and export approximately 14,000 subunits every minute.

Box 31.2 Rate of Transport Through the Nuclear Pores

The nuclear envelope of a typical mammalian cell contains 3000 to 4000 pores (about 11 pores/μm^2 of membrane area). If the cell is synthesizing DNA, it needs to import about 10^6 histone molecules from the cytoplasm every 3 minutes in order to package newly made DNA into chromatin, which means that on an average each pore needs to transport about 100 histone molecules per minute. Further, if the cell is growing rapidly, each nuclear pore needs to export about three newly assembled ribosomes per minute to the cytoplasm, since ribosomes are produced in nucleus but function in the cytoplasm. The export of new ribosomal subunits is particularly problematic since these particles are about 15 nm in diameter and are much too large to pass through the 9 nm channels of nuclear pores, it is believed that they are specifically exported through the nuclear pores by an active transport system. Similarly, mRNA molecules complexed with special proteins to form ribonucleoprotein particles, are thought to be actively exported from the nucleus.

Number of nuclear pores (Pore density). In nuclei of mammals it has been calculated that nuclear pores account for 5 to 15 per cent of the surface area of the nuclear membrane. In amphibian oocytes, certain plant cells and protozoa, the surface occupied by the nuclear pores may be as high as 20 to 36 per cent. The number of pores in the nuclear envelope or **pore density** seems to correlate with the transcriptional activity of the cell. Thus, pore densities as low as ~3 pores/μm^2 are seen in nucleated red blood cells and lymphocytes (which are inactive in transcription). These cells are highly differentiated but metabolically inactive and they are non- proliferating cells. The majority of proliferating cells have pore densities between 7 and 12 pores/μm^2. Among cells of a third type, differentiated but highly active, pore densities are often 15 to 20 pores/μm^2. Liver, kidney and brain cells fall into this category. Still higher pore densities are found in specialized cells, such as salivary gland cells (~40 pores/μm^2) and the oocytes from *Xenopus laevis* (~50 pores/μm^2), both of which are very active in transcription.

Arrangement of nuclear pores on nuclear envelope. In somatic cells, the nuclear pores are evenly or randomly distributed over the surface of nuclear envelope. However, pore arrangement in other cell types is not random but rather range from **rows** (*e.g.*, spores of *Eqisetum*) to **clusters** (*e.g.*, oocytes of *Xenopus laevis*) to **hexagonal** (*e.g.*, Malpighian tubules of leaf hoppers) packing order.

Nuclear pore complex (NPC). Nuclear pores contain an intricate structure, called **nuclear pore complex (NPC)** that appears to fill the pore like a stopper, projecting into both the cytoplasm (Fig. 31.3). The NPC is a huge, supermolecular complex –15 to 30 times the mass of a ribosome that exhibit octagonal symmetry due to the eightfold repitition of a number of structures (Fig. 31.4). Despite their considerable size and complexity, NPCs contain only 30 different proteins, called **nucleoporins**, which are largely conserved between yeast and vertebrates. Each nucleoporin is present in at least eight copies, in keeping with octagonal symmetry of the structure. The NPC is not a static structure, as evidenced by the finding that many of its component proteins are replaced with new copies over a time period of seconds to minutes.

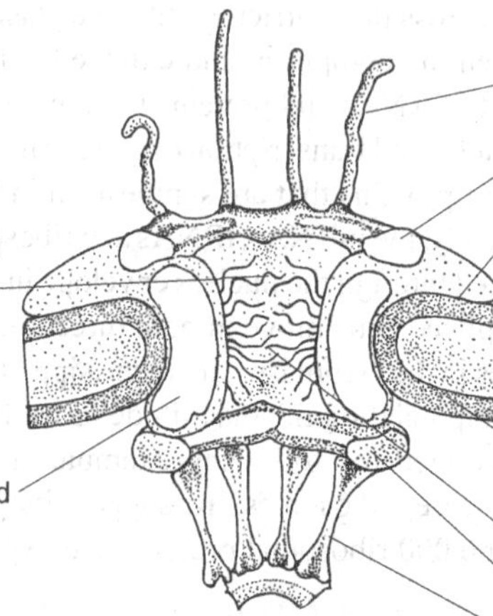

Fig. 31.4. Three dimensional representation of a vertebrate NPC as it is situated within the nuclear envelope. This elaborate structure consists of several parts, including a scaffold that anchors the complex to the nuclear envelope, a cytoplasmic and nuclear ring, a nuclear basket, and eight cytoplasmic filaments. The FG (phenylalanine-glycine)-containing nucleoporins line the channel with this disordered FG-containing domains extending into the opening and forming a hydrophobic meshwork (after Karp 2010).

The amino acid sequences of some nucleoporin proteins include a large number of phenylalanine-glycine repeat (*i.e.*, FG, by their single letter names). The FG repeats are clustered in a particular region of each molecule called the **FG domain**. Because of their unusual amino acid composition, the FG domains possess a disordered structure that gives them an extended and flexible organization. The FG repeat containing nucleoporins are thought to line the central channel of the NPC with their filamentous FG domains extending into the heart (center) of the 20 to 30 nm wide channel. The FG domains form a hydrophobic meshwork or **sieve** that blocks the diffusion of larger macromolecules (greater than about 40,000 Daltons) between the nucleus and cytoplasm.

Functions of Nuclear Pore: Nuclear Transport. Laskey *et al.*, (1982) has found that **nucleoplasmin**, one of the more abundant nuclear protein of amphibian oocytes, contains a stretch of amino acids near its C-terminus that functions as a **nuclear localization signal (NLS)**. This sequence enables a protein to pass through the nuclear pores and enter the nucleus. The best studied or "classical" NLSs, consist of one or two short stretches of positively charged amino acids. The T antigen encoded by the virus SV40, for example, contains an NLS identified as –Pro–Lys–Lys–Lys–Arg–Lys–Val. If one of the basic amino acids in this sequence is replaced by a nonpolar amino acid the protein fails to become localized in the nucleus. Contrary to this, if this NLS is fused to a nonnuclear protein, such as serum albumin and injected into the cytoplasm, the modified protein becomes concentrated in the nucleus. Thus, **targeting of proteins** to the nucleus is similar in principle to trafficking of other proteins that are destined for segregation within a particular organelle, such as a mitochondrion, a chloroplast or a peroxisome. In all of these cases, the proteins possess a specific "address" that is recognized by a specific receptor that mediates its transport into the organelle.

Importins and Exportins. Recently, researchers have identified a family of proteins that function as **mobile transport receptors**, ferrying macromolecules across the nuclear envelope. Within this family, **importins** move macromolecules from the cytoplasm into the nucleus and **exportins** move macromolecules in the opposite direction.

Nuclear imports. During nuclear import of a protein, such as nucleoplasmin, that contains a classical NLS, following steps are involved (Fig. 31.5). Step 1. Import of a protein begins as the NLS-containing cargo protein binds to a heterodimeric, soluble NLS receptor, called **importin** α/β, that resides in the cytoplasm. Step 2. The transport receptor is thought to escort the protein cargo to the outer surface of the nucleus where it likely docks with the cytoplasmic filaments that extend from the outer ring of the NPC. Step 3. The receptor-cargo complex then moves through the nuclear pore by engaging in a series of successive interactions with the FG domains of the FG-containing nucleopores. These interactions are thought to "dissolve" portions of FG-rich meshwork that fills the interior of the channel, allowing passage of the receptor-cargo complex through the NPC.

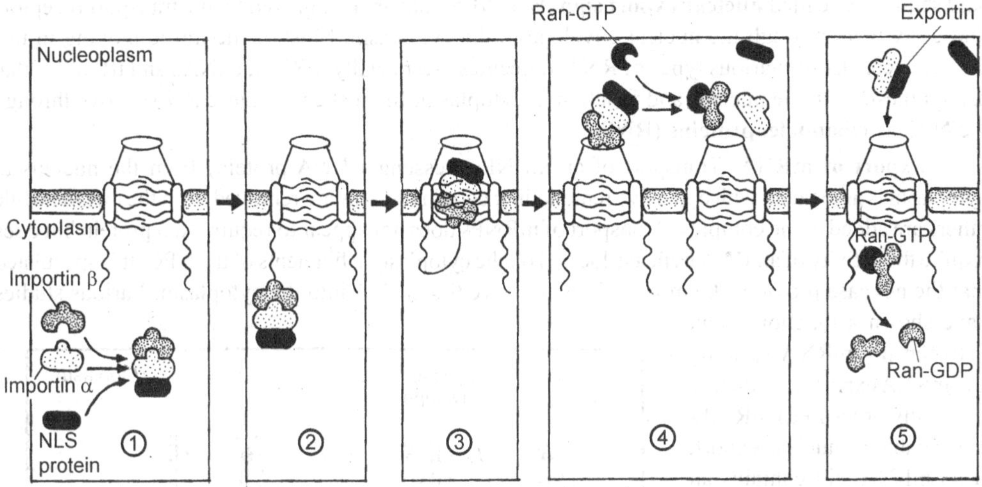

Fig. 31.5. Mechanism of import from the cytoplasm into the nucleus. The protein bearing a NLS (= nuclear localization signal) binds to the heterodimeric receptor (importin α/β) (step 1) forming a complex that associates with a cytoplasmic filament (step 2). The receptor-cargo complex moves through the nuclear pore (step 3) and into nucleoplasm where it interacts with Ran-GTP and dissociates (step 4). The importin β subunit, in association with Ran-GTP, is transported back to the cytoplasm where the Ran-GTP is hydrolyzed (step 5). Ran-GDP is subsequently transported back to the nucleus, where it is converted to Ran-GTP. Conversely, importin α is transported back to cytoplasm (after Karp 2010).

Role of Ran protein in import. In nucleus, there is a GTP-binding protein called **Ran** which has a role in next two remaining steps of import of proteins into nuclear compartment. Ran can exist in an *active* GTP-bound form or in *inactive* GDP-bound form. Ran's role in regulating nucleocytoplasmic transport is based on a mechanism in which the cell maintains a high concentration of Ran-GTP in the nucleus and a very low concentration of Ran-GDP in the cytoplasm. The steep gradient of Ran-GTP across the nuclear envelope depends on the compartmentalization of certain accessory proteins such as RCC1 and RanGAP1. **RCC1** is sequestered in the nucleus where it promotes the conversion of Ran-GDP to Ran-GTP, thus maintaning the high level of Ran-GTP inside nucleus. **RanGAP1** resides in the cytoplasm where it promotes hydrolysis of Ran-GTP to Ran-GDP, thus, maintaining the low cytoplasmic level of Ran-GTP. In this way, the energy released by GTP hydrolysis is used to maintain the Ran-GTP gradient across the nuclear envelope. The Ran-GTP gradient drives nuclear transport by a process that depends only on receptor-mediated diffusion; *no motor proteins or ATPases have been implicated.*

So, when the importin-cargo complex arrives in the nucleus during step 3, it is by a molecule of Ran-GTP, which binds to the complex and promotes disassembly of complex (step 4). The imported cargo is released into the nucleoplasm, and one portion of NLS receptor (the importin β subunit) is shuttled back to the cytoplasm together with the bound Ran-GTP (step 5). Once in the cytoplasm, the GTP molecule bound to Ran is hydrolyzed, releasing Ran-GDP from the importin β subunit. Ran-GDP is returned to the nucleus where it is converted back to the GTP-bound state for additional rounds of activity. Importin α is transported back to the cytoplasm by one of the exportins.

Nuclear export. Ran-GTP plays a key role in the escort of macromolecules from the nucleus, just as it does in their import from the cytoplasm. As it has been mentioned that Ran-GTP is essentially confined to the nucleus. Whereas the Ran-GTP induces the disassembly of imported complexes, it also promotes the assembly of exported complexes. Proteins exported from the nucleus contain amino acid sequences, called **nuclear export signals** or **NES**, that are recognized by the transport receptors that carry them through the nuclear envelope to the cytoplasm. Most of the traffic moving in this direction consists of various types of RNA molecules—especially mRNA, rRNAs and tRNAs—that are synthesized in the nucleus and function in cytoplasm. In most cases, these RNAs move through the NPC as **ribonucleoproteins (RNPs)**.

Export of mRNA. Transport of an mRNP (messenger RNA protein) from the nucleus to cytoplasm is associated with extensive remodeling; certain proteins are removed from the mRNP, while others are added to the complex. Transport of mRNPs does not appear to require Ran protein but does require the activity of an **RNA helicase** located on the cytoplasmic filaments of the NPC. It is speculated that the helicase provides the motive force to move the mRNA into the cytoplasm. Various studies have shown a functional link between pre-mRNA splicing and mRNA export; only mature (*i.e.*, fully processed) mRNAs are capable of nuclear export. If an mRNA still contains an unspliced intron, that RNA is retained in the nucleus.

Lastly, nuclear pores are not the only avenues for nucleocytoplasmic exchanges. For example, small molecules and ions readily permeate both nuclear membranes. Larger molecules and particles may pass through the membrane by formation of small pockets and vesicles that traverse the envelope and empty on the other side (Fig. 31.6).

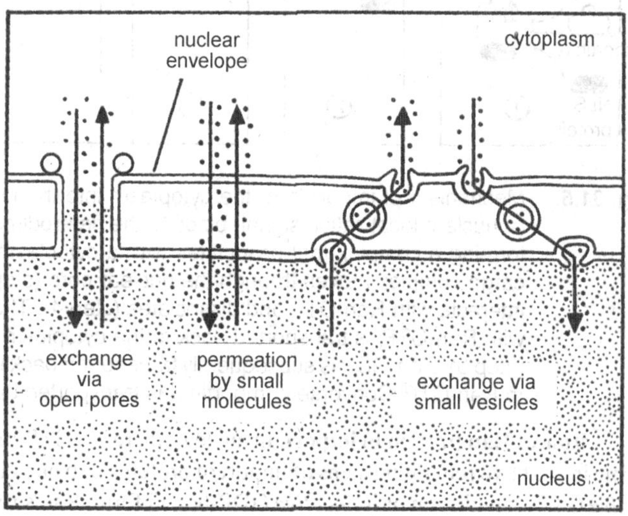

Fig. 31.6. Various avenues of transport of material from the nucleus to the cytosol (cytoplasmic matrix).

II. Nucleoplasm

The space between the nuclear envelope and the nucleolus is filled by a transparent, semi-solid, granular and slightly acidophilic ground substance or the matrix known as the **nuclear sap** or **nucleoplasm** or **karyolymph**. The nuclear components such as the chromatin threads and the nucleolus remain suspended in the nucleoplasm.

The nucleoplasm has a complex chemical composition. It is composed of mainly the nucleoproteins but it also contains other inorganic and organic substances, *viz.*, nucleic acids, proteins, enzymes and minerals.

1. Nucleic acids. The most common nucleic acids of the nucleoplasm are the DNA and RNA. Both may occur in the macromolecular state or in the form of their monomer nucleotides.

2. Proteins. The nucleoplasm contains many types of complex proteins. The nucleoproteins can be categorized into following two types:

***(i)* Basic proteins.** The proteins which take basic stain are known as the basic proteins. The most important basic proteins of the nucleus are **nucleoprotamines** and the **nucleohistones**.

The nucleoprotamines are simple and basic proteins having very low molecular weight (about 4000 daltons). The most abundant amino acid of these proteins is **arginine** (pH 10 to 11). The protamines usually remain bounded with the DNA molecules by the salt linkage. The protamines occur in the spermatozoa of the certain fishes. The nucleohistones have high molecular weight, *e.g.*, 10,000 to 18,000 daltons. The histones are composed of basic amino acids such as **arginine, lysine** and **histidine**. The histone proteins remain associated with the DNA by the ionic bonds and they occur in the nuclei of most organisms. According to the composition of the amino acids following types of histone proteins have been recognised, *e.g.*, histones rich in lysine, histones with arginine and histones with poor amount of the lysine.

***(ii)* Non-histone or Acidic proteins.** The acidic proteins either occur in the nucleoplasm or in the chromatin. The most abundant acidic proteins of the euchromatin (a type of chromatin) are the **phosphoproteins** (Box 31.3).

Box 31.3

Nucleoproteins are the main and the most important components of nucleus. These are compounds of nucleic acids and proteins. The proteins present in the nucleus are specific and are of following two types:

1. **Histones.** Histones (which are of five basic types such as H2A, H2B, H3, H4 and H1 or H5) are highly basic proteins rich in positively charged **lysine** and **arginine** residues. Histones H3 and H4 are most strongly conserved, while linker histones (H1 or H5) show most diversity.

   All the histones except H4 exist as multiple isoforms (isohistones) whose relative predominance in chromatin varies in a cell-type-specific manner. Histones represent the bulk of protein in chromatin and are relatively homogeneous in nature.

2. **Non-histone protein.** Non-histone proteins are acidic in nature and are rich in **tryptophan** and **tyrosine**. Non-histone proteins represent small but extremely heterogenous fraction. The non-histone proteins include **enzymes** involved in DNA and histone metabolism, replication, repair, recombination, and transcriptional regulation. They also include **scaffold proteins** which organize higher order chromatin structure and **high-mobility group proteins** (*HMG proteins*) which are highly charged proteins with various functions in gene regulation and structural organization. A further class of non histone proteins, termed **protamines**, facilitate the packaging of DNA into sperm head (**Twyman** 1998).

---

3. Enzymes. The nucleoplasm holds many enzymes which are necessary for the synthesis of the DNA and RNA. Most of the nuclear enzymes are composed of non-histone (acidic) proteins. The most important nuclear enzymes are the **DNA polymerase, RNA polymerase, NAD synthetase, nucleoside triphosphatase, adenosine diaminase, nucleoside phosphorylase, guanase, aldolase, enolase, 3-phosphoglyceraldehyde dehydrogenase** and **pyruvate kinase**. The nucleoplasm also contains certain cofactors and coenzymes such as **ATP** and **acetyl CoA**.

4. Lipids. The nucleoplasm contains small lipid content.

5. Minerals. The nucleoplasm comprises several inorganic compounds such as phosphorus, potassium, sodium, calcium and magnesium. The chromatin comparatively contains large amount of these minerals than the nucleoplasm.

III. Chromatin Fibres

The nucleoplasm contains many thread-like, coiled and much elongated structures which take readily the basic stains such as the basic fuchsin. These thread-like structures are known as the **chromatin** (*Gr.*, *chrome*=colour) **substance** or **chromatin fibres**. Such chromatin fibres are observed only in the interphase nucleus. During the cell division (mitosis and meiosis) chromatin fibres become thick ribbon-like structures which are known as the **chromosomes**.

Chemically, chromatin consists of DNA and proteins. Small quantity of RNA may also be present but the RNA rarely accounts for more than about 5 per cent of the total chromatin present. Most of the protein of chromatin is histone, but "non histone" proteins are also present. The protein: DNA weight ratio averages about 1:1. Histones are constituents of the chromatin of all eukaryotes except fungi, which, therefore, resemble prokaryotes in this respect.

The fibres of the chromatin are twisted, finely anastomosed and uniformly distributed in the nucleoplasm. Two types of chromatin material have been recognised, *e.g.*, heterochromatin and euchromatin.

A. Heterochromatin. The darkly stained, condensed region of the chromatin is known as heterochromatin. The condensed portions of the nucleus are known as **chromocenters** or **karyosomes** or **false nucleoli** (Box 31.4). The heterochromatin occurs around the nucleolus and at the periphery. It is supposed to be metabolically and genetically inert because it contains comparatively small amount of the DNA and large amount of the RNA.

Box 31.4

Chromocenters are darkly stained heterochromatic regions of chromatin of interphase nucleus. They may represent a heterochromatic region of several chromosomes or of all the chromosomes of the nucleus. Chromocenters are well marked in the nuclei of salivary gland cells of *Drosophila*. In this case, pericentromeric heterochromatin of all polytene chromosomes coalesces in a chromocenter.

B. Euchromatin. The light stained and diffused region of the chromatin is known as the euchromatin. The euchromatin contains comparatively large amount of DNA.

Heteropyknosis

Differential condensation of certain chromosomes, such as sex chromosomes, or chromosome parts is called **heteropyknosis** (*Gr.*, differential staining). During interphase and prophase stages of cell division most of the chromosomal substance is uncoiled and is in the form of light stained fine threads. But some chromosomes or their parts remain condensed and can be recognized by their dark stain. Thus, heteropyknosis can be defined as *the property of certain chromosomes or their parts to remain more condensed and stain more intensely than other chromosomes or their parts during the nuclear cycle*.

Heteropyknosis is the characteristic of sex chromosomes of many species, but may also be observed in other chromosomes as well. For example, one of two X chromosomes in females of all the mammals undergoes heteropyknosis. The heteropyknosis may involve entire chromosome, may be intercalated or localized at the extremities.

When the heteropyknotic chromosomes (*e.g.*, sex chromosomes) are darkly stained, it is called **positive heteropyknosis,** however, in the toad, *Bufo arenarus*, the chromosomes exhibit **negative heteropyknosis** (under-condensation).

IV. Nucleolus

Most cells contain in their nuclei one or more prominent spherical colloidal acidophilic bodies, called **nucleoli**. However, cells of bacteria and yeast lack nucleoli. The size of the nucleolus is found to be related with the synthetic activity of the cell. Therefore, the cells with little or no synthetic activities, *e.g.*, sperm cells, blastomeres, muscle cell, etc., are found to contain smaller or no nucleoli, while the oocytes, neurons and secretory cells which synthesize the proteins or other substances contain comparatively large-sized nucleoli. The number of the nucleoli in the nucleus depends on the species and the number of the chromosomes. The number of the nucleoli in the cells may be one, two or four. The position of the nucleolus in the nucleus is eccentric.

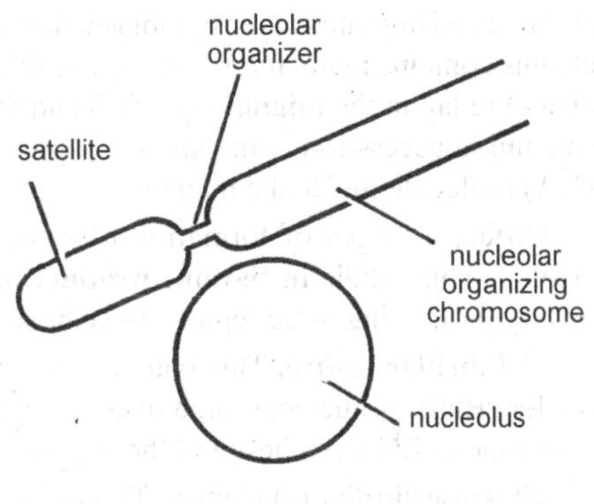

Fig. 31.7. A satellited chromosome and an attached nucleolus.

A nucleolus is often associated with the **nucleolar organizer (NO)** which represents the secondary constriction of the nucleolar organizing chromosomes, and are 10 in number in human beings (Fig. 31.7). In corn, *Zea mays* chromosome 9 and 6 contain 'darkly staining knobs' or nucleolar organizers. Nucleolar organizer consists of the genes for 18S, 5.8S and 28S rRNAs. The genes for fourth type of rRNA, *i.e.*, 5S rRNA occur outside the nucleolar organizer.

Chemical composition of nucleolus. Nucleolus is not bounded by any limiting membrane; calcium ions are supposed to maintain its intact organization. Chemically, nucleolus contains DNA of nucleolar organizer, four types of rRNAs, 70 types of ribosomal proteins, RNA binding proteins (*e.g.*, nucleolin) and RNA splicing nucleoproteins (U_1, U_2......U_{12}). It also contains phospholipids, orthophosphates and Ca^{2+} ions. Nucleolus also comprises of some enzymes such as acid phosphatase, nucleoside phosphorylase and NAD^+ synthesizing enzymes for the synthesis of some coenzymes, nucleotides and ribosomal RNA. **RNA methylase** enzyme which transfers methyl groups to the nitrogen bases, occurs in the nucleolus of some cells.

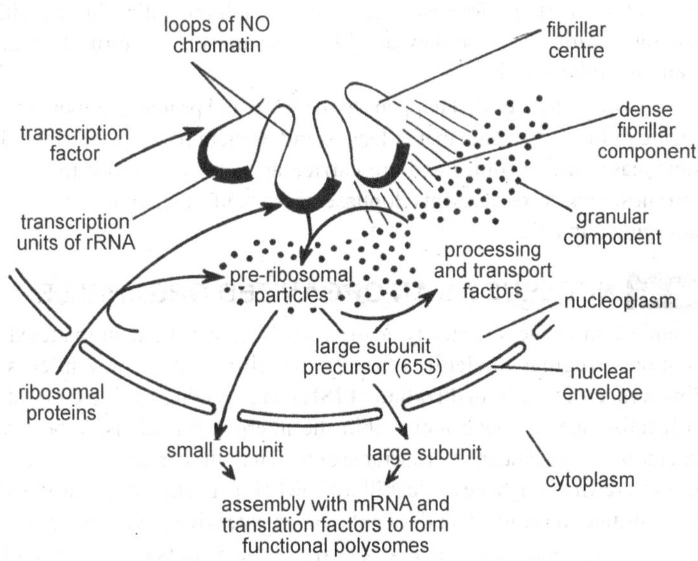

Fig. 31.8. Three different regions of the nucleolus and their involvement in ribosome assembly.

Ultrastructure and function of nucleolus. Nucleolus is the site where biogenesis of ribosomal subunits (*i.e.*, 40S and 60S) takes place. In it three types of rRNAs, namely 18S, 5.8S and 28S rRNAs, are transcribed as parts of a much longer precursor molecule (45S transcript) which undergoes processing (RNA splicing, for example) by the help of two types of proteins such as nucleolin and U3 snRNP (U3 is a 250 nucleotide containing RNA, snRNP represents small nuclear ribonucleoprotein). The 5S rRNA is transcribed on the chromosome existing outside the nucleolus and the 70 types of ribosomal proteins are synthesized in the cytoplasm. All of these components of the ribosomes migrate to the nucleolus, where they are assembled into two types of ribosomal subunits which are transported back to the cytoplasm. The smaller (40S) ribosomal subunits are formed and migrated to the cytoplasm much earlier than larger (60S) ribosomal subunits; therefore, nucleolus contains many more incomplete 60S ribosomal subunits than the 40S ribosomal subunits. Such a time lag in the migration of 60S and 40S ribosomal subunits, prevents functional ribosomes from gaining access to the incompletely processed heterogeneous RNA (hn RNA; the precursor of mRNA) molecule inside the nucleus.

Different stages of formation of ribosomes are completed in three distinct regions of the nucleolus. Thus, their **initiation**, **production** and **maturation** seem to progress from centre to periphery. Following three regions have been identified in the nucleolus (Fig. 31.8):

(*i*) **Fibrillar centre.** This pale-staining part represents the innermost region of nucleolus. The genes for rRNA of nucleolar organizer of chromosomes are located in this region. The transcription (*i.e.*, ribosomal RNA synthesis) of these genes is also initiated in this region.

(*ii*) **Dense fibrillar component.** This region surrounds the fibrillar centre and RNA synthesis progresses in this region. The 70 ribosomal proteins (rps) also bind to the transcripts in this region.

(*iii*) **Cortical granular components.** This is the outermost region of the nucleolus where processing and maturation of pre-ribosomal particles occur.

Nucleolar Cycle

The appearance of nucleolus changes dramatically during the cell cycle. During meiosis as well as during mitosis the nucleolus disappears during prophase. As the cell approaches mitosis, the nucleolus first decreases in size and then disappears as the chromosomes condense and all RNA synthesis stops, so that generally there is no nucleolus in a metaphase cell. When ribosomal RNA synthesis restarts at the end of mitosis (in telophase), tiny nucleoli reappear at the chromosomal locations of the ribosomal RNA genes (NOs). For example, in humans the rRNA genes are located near the tips of each of the 5 different chromosomes (*i.e.*, paired autosomes 13, 14, 15, 21 and 22). Accordingly, 10 small nucleoli are formed after mitosis in a human diploid cell, although they are rarely seen as separate entities because they quickly grow and fuse to form the single large nucleolus typical of many interphase cells.

Now let us see what happens to the RNA and protein components of the disintegrated nucleolus during mitosis? It seems that at least some of them become distributed over the surface of all of the metaphase chromosomes and are carried as cargo to each of the two daughter cell nuclei. As the chromosomes decondense at telophase, these "old" nucleolar components help re-establish the newly emerging nucleoli.

31.4. NUCLEUS AS AN ORGANIZED ORGANELLE

Examination of the nucleus, typically reveals little more than scattered clumps of chromatin and one or more irregular nucleoli. With the development of new microscopic techniques, including **florescence in situ hybribization** (**FISH**) and imaging of live GFP-labeled cells, it became possible to localize specific gene loci within the interphase nucleus. It became evident from these studies that the nucleus maintains considerable order. Thus, the chromatin fibers of a given interphase chromosome are not strewn through the nucleus like a bowl of spaghetti (*i.e.*, pasta in long strands), but are concentrated into a distinct **territory** that does not overlap extensively with the territories of other chromosomes.

Even though genes are typically transcribed while they reside within their territories, individual chromatin fibers can extend away from these territories for great distances. Further more, DNA sequences that participate in a common biological response but reside on differenet chromosomes can apparrently come together within the nucleus where they can influence gene transcription.

1. 3C technique. Interactions between different loci have been revealed by the invention of a variety of techniques that involve **3C**, *i.e.*, **chromosome conformation capture**. In this approach cells are killed (*i.e.*, fixed) by treatment with formaladehyde, which causes DNA sequences residing in close proximity to become covalently cross-linked to one another (they are said to be "captured"). After the fixation process the DNA is **isolated** and subjected to digestion by **restriction enzymes** and the digestion products are analyzed to determine which DNA sequences in the genome were interacting with a given DNA sequence (called the "bait" sequences) at the time of fixation. For example, a researcher might want to know which DNA sequences throughout the genome interact with the β-globin locus during the differentiation of erythrocytes in the bone marrow. DNA sequences that are found to interact using this technique are typically sequences that are present on the same chromosome, *e.g.*, interaction between an enhancer and a promoter for the same gene, which come into close proximity.

2. Interchromosomal interactions and transcription factories. A good example of interchromosomal interactions comes from a study in which human cultured breast cells (both normal and malignant versions) were treated with the horomone estrogen. **Estrogen** induces the transcription of a large number of target genes through its binding to an estrogen receptor (ERα). Two of estrogen's target genes in humans are *GREB1*, located on chromosome 2 and *TRFF1*, located on chromosome 21. Prior to treating the cells with estrogen, the chromosome 2 and 21 territories are at distant locations from one another, and the *GREB1* and *TRFF1* gene loci are tucked away within the interior of their respective chromosome territories. However, within minutes after cells are exposed to estrogen these two chromosome territories move into close physical proximity to one another and two gene loci become localized together (co-localized) on the periphery of their territories.

The studies support the ideas that genes are physically moved to sites within the nucleus called **transcription factories**, where the transcription machinery is concentrated and that genes are involved in the same response tend to become co-localized in the same factory (Fig. 31.9; **Hu** *et al.*, 2008).

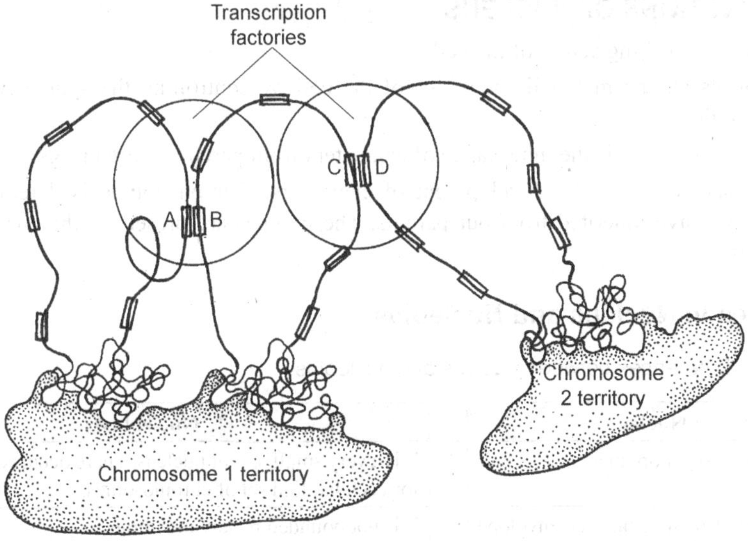

Fig. 31.9. Interactions can occur between distantly located genes in response to physiological stimuli. A drawing illustrating how genes on different regions of the same chromosome (A and B), or on different chromosomes (C and D), can come together within the nucleus. In some cases, DNA sequences from distant loci may influence one another's transcription activity, and in other cases these distant genes may simply share a common pool of proteins involved in the transcription process (after Karp 2010).

3. Speckles. Another example of the interrelationship between nuclear organization and gene expression is seen in the presence of **speckles** in the nucleus. When a cell is stained with a fluorescent antibody against one of the protein factors involved in pre-mRNA splicing, these speckles can be observed in electron micrographs. Rather than being spread uniformly throughout the nucleus, the processing machinery is concentrated within 20 to 50 irregular domains as the speckles (**Misteli** and **Spector** 1998). The various structures of the nucleus, including nucleoli, and speckles are dynamic, steady-state compartments whose existence depends on their continued activity. If that activity is blocked, the compartment simply disappears as its parts are dispersed into the nucleoplasm.

In addition to nucleoli and speckles, several other types of nuclear bodies, such as **cajal bodies, GEMs** and **PML bodies** are often seen under the microscope. Each of these nuclear bodies contain large number of proteins that move in and out of the structure in a dynamic manner. Because none of these nuclear bodies are bounded by a membrane, no special transport mechanism are required for these large-scale movements.

4. The nuclear matrix. When isolated nuclei are treated with nonionic detergents and high salt (*e.g.*, 2M NaCl), which remove lipids and nearly all of the histone and non-histone proteins of the chromatin, the DNA is seen as a halo surrounding a residual nuclear core. If the DNA fibers are subsequently digested with DNase (enzyme), the structure that remains possesses the same shape as the original nucleus but is composed of a network of thin protein-containing fibrils criss-crossing through the nuclear space (Fig. 31.2). This insoluble fibrillar network has been called **nuclear matrix** (**Jackson, McCready** and **Cook** 1981). Some workers have regarded this matrix as an artifact of preparation, but majority of them thought that nuclear matrix serves as more than a skeleton to maintain the shape of the nucleus or a scaffold on which to organize the loops of chromatin, it also serves to anchor much of the machinery that is involved in the various activities of the nucleus, including transcription, RNA processing and replication (**Karp** 2010).

31.5. FUNCTIONS OF NUCLEUS

Nucleus is the controlling centre of the cell.

(*i*) It controls all the metabolic activities of the cell by controlling the synthesis of enzymes required there in.

(*ii*) The nucleus controls the inheritance of characters from parents to offsprings.

(*iii*) It is responsible for the development of characters. Our phenotype is determined by the genes we have inherited from our parents. These genes are present in the chromosomes of nucleus.

Differences in Nucleus and Nucleolus

Table 31.1 Differences between nucleus and nucleolus.

	Nucleus		Nucleolus
1.	It is a large sized organelle.	1.	It is a very small part of nucleus. There occurs one nucleolus for each haploid set of chromosomes.
2.	It is bounded by a nuclear envelope of double unit membranes. Nuclear envelope bears nucleopores.	2.	It is not bounded by any membrane.
3.	It is site of replication, recombination and transcription of DNA.	3.	It is site of transcription of rDNA of nucleolar region of chromosomes. Tailoring and modifications of various types of rRNAs take place in nucleolus and they are assembled with ribosomal proteins to form ribosomal subunits.

31.6. HAMMERLING'S EXPERIMENT

The evidences for nucleo-cytoplasmic communication as a factor in cell maintenance and development have been known before the rediscovery of Mendel's "genes". In the late nineteenth century, **Verworm, Balbiani** and others showed that following microsurgery, nucleated halves of various protozoans survived and grew, whereas the enucleated halves degenerated and died. Later, in the 1930s it was shown that insertion of nuclei into enucleated amoebae restored pseudopodial activity, a feeding behaviour and growth. It was also shown that the nucleus was essential for the growth and regeneration of the morphologically complicated ciliate *Stentor*. In a classical series of experiments spanning between 1934 and 1954, on the unicellular alga *Acetabularia*, **Hammerling** demonstrated by means of interspecific nuclear transplants, that morphological features, notably the shape of cap, were determined by the nucleus. He also showed that even after removal of the nucleus, the cell was able to continue morphogenesis for a time and proposed that the cytoplasm contained a store of morphogenetic material (later on recognized as mRNA molecules) that had been produced by the nucleus. Let us closely examine the Hammerling's classical nuclear transplantation experiments:

Hammerling's experiment. The body of an alga *Acetabularia* is about six centimeters long and is differentiated into a foot, a stalk and a cap. The cap has a characteristic shape for each species and is easily regenerated if removed. The single nucleus is situated in the rhizoid portion. *Acetabularia crenulata* has a cap, with about 31 rays, the tips of which are pointed, but *Acetabularia mediterranea* has about 81 rays with rounded tips. If the cap, stalk or even

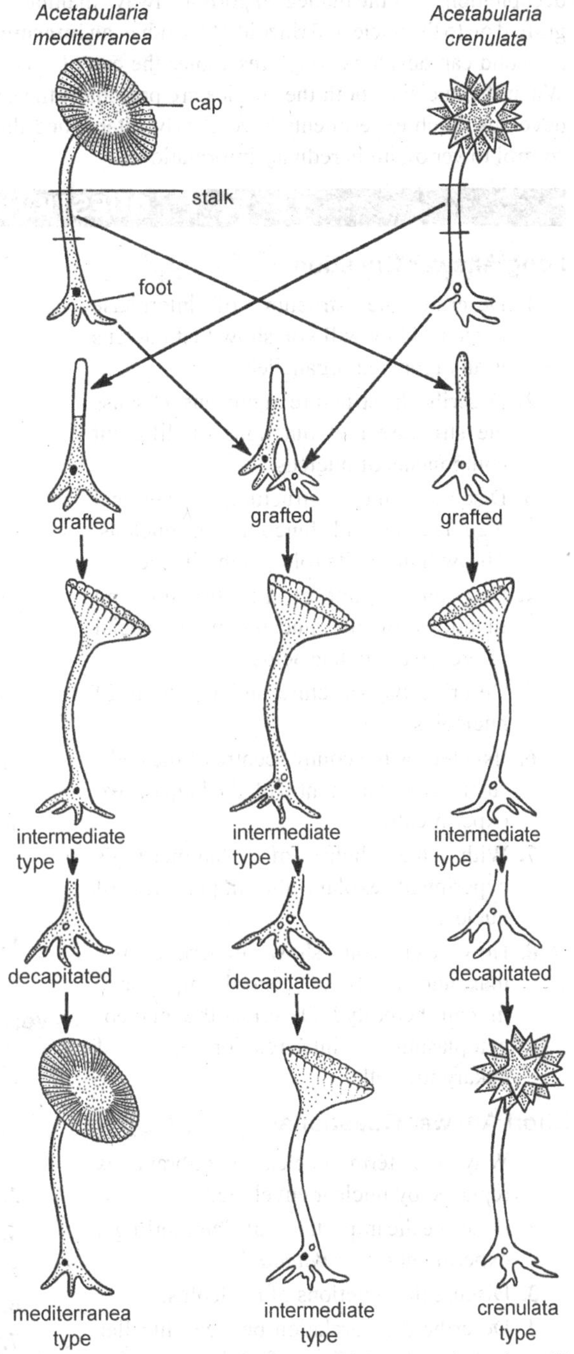

Fig. 31.10. Summary of the grafting experiments with *Acetabularia* to prove that the hereditary characters are determined by the nucleus and not by the cytoplasm.

the nucleated portion of the rhizoid is removed, the remaining portion of the alga has the capacity to regenerate into a whole plant. The enucleated part loses the regeneration capacity after a few decapitations, but the nucleated portion always maintains this ability. When the stalk of one species is grafted on to the nucleated rhizoid of the other, an **intermediate type** of cap is formed. On decapitation, a second cap develops which resembles the cap of species which provides the nucleus (Fig. 31.10). When the nuclei of both the species are present in the same cytoplasm, an intermediate type of cap develops. Such experiments have clearly established that the nucleus is the storehouse for and the control tower of, all hereditary information.

QUESTIONS

Long Answer Questions

1. Describe the structure of interphase nucleus. How will you show that nucleus is an organized organelle?
2. Describe the structure of nucleus. Discuss the structure and functions of different components of nucleus.
3. Discuss the structure, chemical organization and functions of nucleus. Throw light on its role in inheritance.
4. Give an account of the structure and functions of the nucleus with special reference to nucleic acids.
5. Describe the structure and functions of nucleolus.
6. 'Nucleus is the control centre of the cell'. Justify this statement with the help of two experiments.
7. With the help of Hammerling's experiment explain the importance of nucleus.
8. How can you show experimentally that nucleus is of central importance in cell heredity? Describe the nucleo-cytoplasmic inter-relationship of eukaryotic cell.

Short Answer Questions

1. Why the term nuclear membrane is replaced by nuclear envelope?
2. Describe the importance of Hammerling's experiment on *Acetabularia.*
3. Discuss the functions of nucleolus.
4. Describe the corelation between fibrillar and granular portions of nucleolus.
5. Describe in brief the nucleolar cycle.
6. Give reasons why nucleolus is small in muscle cell but large in liver cells.
7. Write short notes on the following:

(*i*)	Euchromatin	(*ix*)	Functions of nucleus
(*ii*)	Functions of nuclear envelope	(*x*)	Nuclear pores
(*iii*)	Functions of nucleolus	(*xi*)	Nuclear pore complex
(*iv*)	Heteropyknosis	(*xii*)	Nuclear lamina
(*v*)	Nucleoproteins	(*xiii*)	Importins and exportins
(*vi*)	Chromocentres	(*xiv*)	Ran proteins
(*vii*)	Nucleolar cycle	(*xv*)	Nuclear export
(*viii*)	Nucleolus	(*xvi*)	Transcription factories

8. Differentiate between the following:
   (*i*) Euchromatin and heterochromatin
   (*ii*) Nucleus and nucleolus.
9. Draw a neat and labelled diagram of nucleus.

Very Short Answer Questions

1. Why mammalian RBCs have a very short life span?
2. What will happen if nucleus in *Amoeba* is destroyed?
3. Who discovered the nucleus?
4. What is the nucleus?
5. Define the nucleolus.
6. What is nuclear envelope?
7. Define nuclear lamina.
8. What is the nuclear pore?
9. Define nuclear pore complex.
10. What is chromatin?

11. How many types of chromatin occur in the nucleus?
12. What is heterochromatin?
13. Define euchromatin.
14. What is NOR?
15. What is main function of nucleolus?

Multiple Choice Questions

1. Site of formation of immature ribosomal subunits in a eukaryotic cell is
   (*a*) cytoplasm (*b*) nucleolus
   (*c*) nucleus (*d*) nuclear pore complex
2. The role of nucleus in heredity was at first demonstrated by
   (*a*) Max Hammerling
   (*b*) Robert Brown
   (*c*) W. Fleming
   (*d*) Benda Hooker
3. Who coined the term nucleolus?
   (*a*) Fontana (*b*) Robert Brown
   (*c*) Camillo Golgi (*d*) Strasburger
4. Nucleolus is the site for the synthesis of
   (*a*) DNA (*b*) mRNA
   (*c*) tRNA (*d*) ribosomes
5. An example for the enucleated living plant cell is
   (*a*) xylem parenchyma
   (*b*) RBCs
   (*c*) sieve tubes of phloem
   (*d*) companion cells of phloem

ANSWERS

Very Short Answer Questions

1. Mammalian RBCs are without nucleus.
2. Such *Amoeba* will die.
3. Robert Brown in 1833.
4. Nucleus is a large, double membrane enclosed organelle that contains the chromosomal DNA of a eukaryotic cell.
5. Nucleolus is a large, spherical structure present in the nucleus of an eukaryotic cell; it is site of ribosomal RNA synthesis and processing and of the assembly of ribosomal subunits.
6. Nuclear envelope is a double membrane around the nucleus that is interrupted by numerous small pores.
7. Nuclear lamina is a thin, dense, meshwork of fibre that lines the inner surface of the inner nuclear membrane and helps support the nuclear envelope.
8. Nuclear pore is a small opening in the nuclear envelope through which molecules enter and exist the nucleus; it is lined by an intricate protein structure called the nuclear pore complex (NPC).
9. Nuclear pore complex is a complex, basket-like apparatus of proteins that fills the nuclear pore like a stopper, projecting outward into both the cytoplasm and the nucleoplasm.
10. Chromatin is DNA-protein fibre that make up chromosomes. It is constructed from nucleosomes spaced regularly along a DNA chain.
11. Euchromatin and heterochromatin.
12. Heterochromatin is highly compacted form of chromatin present during interphase; it contains DNA that is not being transcribed.
13. Euchromatin is loosely packed, uncondensed form of chromatin present during interphase; it contains DNA that is being actively transcribed.
14. NOR stands for nucleolus organizer region which is a stretch of DNA in certain chromosomes where multiple copies of the genes for ribosomal RNA are located and where nucleoli form.
15. Nucleolus is a ribosome producing organ.

Multiple Choice Questions

1. (*b*) 2. (*a*) 3. (*a*) 4. (*d*) 5. (*c*)

32 Chromosomes

The chromosomes (Gr., *chrom* = colour, *soma* = body) are the nuclear components of special organisation, individuality and function. They are capable of self-reproduction and play a vital role in heredity, mutation, variation and evolutionary development of the species.

Chromosomes are most distinct in metaphase stage of cell division. In an interphase nucleus these form an interwoven net of fine chromatin threads. The term **chromosome** was introduced by **Waldeyer** in 1888 because of their affinity towards dyes or stains.

32.1. TYPES OF CHROMOSOMES

1. Viral chromosomes. In viruses, a single, linear or circular molecule of DNA or RNA represent the chromosome. Nucleic acid of the viral chromosome may be single-stranded or double-stranded.

2. Prokaryotic chromosomes (or Bacterial chromosomes). In bacteria and cyanobacteria (blue green algae), the hereditary material (DNA) is organized into a single circular chromosome. Chromosomal DNA is double stranded and lacks histone proteins. Prokaryotic chromosomes are also found in certain cellular organelles such as mitochondria and chloroplasts.

3. Eukaryotic chromosomes. In eukaryotic cells (*e.g.*, protistans plants, fungi and animals), chromosomes are formed of DNA and proteins (histones and non-histones). DNA in linear and double stranded.

In the chapter the structure and function of eukaryotic chromosomes only will be described.

Box 32.1 Distinction between viral and bacterial chromosomes

In viruses (including bacteriophages) a single molecule of DNA or RNA forms the viral chromosome. It is about 20 Å wide corresponding to the width of DNA molecule. The viral chromosome may be linear or circular.

Bacterial chromosome is a single, large circular chromosome, formed of a circular DNA molecule.

Box 32.2 Distinction between prokaryotic and eukaryotic chromosomes

Prokaryotic or bacterial chromosome is a large circular DNA molecule. It is without histones but may contain some polyamines. It lacks centromere and nucleosomes, but is folded into a series of **looped domains**.

Eukaryotic chromosome is much complicated structure having centromere and nucleosomes. Each chromosome is formed by the condensation of a long chromatin thread. The chromatin thread is formed of supercoiling of a series of nuclesosomes. Each nucleosome is formed of histones and DNA.

32.2. CHEMICAL COMPOSITION OF CHROMOSOMES

Chromatin which has been isolated from rat liver contains DNA, RNA and protein. The protein of chromatin is of two types : the histones and the non-histones. Rat liver chromatin has been used as a model for chromatin. It possesses a histone to DNA ratio near 1 : 1, a non-histone protein to DNA ratio of 0.6 : 1 and a RNA/DNA ratio of 0.1 : 1.

1. DNA

DNA is the most important chemical component of chromatin, since it plays the central role of controlling heredity. The most convenient measurement of DNA is picogram (10^{-12} gm). DNA of chromatin represent the following two phenomena:

The C-value. The DNA in nuclei was stained by using the **Feulgen reactions** and the amount of stain in single nuclei was measured using a special microscope (called **cytophotometer**). Both of these techniques demonstrated that nuclei contain a constant amount of DNA. Thus, all the cells in an organism contain the same DNA content (2C) provided that they are diploid. Gametes are haploid and, therefore, have half the DNA content (1C). Some tissues such as liver, contain occasional cells that are polyploid and their nuclei have a correspondingly higher DNA content (4C or 8C) (see Table 32.1).

Table 32.1 DNA content and chromosome component.

	Cells	Mean DNA-Feulgen content	Presumed chromosome set
1.	Spermatid	1.68 (1C)	Haploid (n)
2.	Liver	3.16 (2C)	Diploid (2n)
3.	Liver	6.30 (4C)	Tetraploid (4n)
4.	Liver	12.80 (8C)	Octoploid (8n)

Thus, each species has a characteristic content of DNA which is **constant** in all the individuals of that species and has, thus, been called the C-value.

The C-value paradox. Eukaryotes vary greatly in DNA content but always contain much more DNA than prokaryotes. Lower eukaryotes in general have less DNA, such as nematode *Caenorhabditis elegans* which has only 20 times more DNA than *E.coli*, or the fruit fly *D. melanogaster* which has 40 times more DNA (*i.e*, 0.18 pg or picogram per haploid genome). Vertebrates have greater DNA content (about 3 pg), in general about 700 times more then *E. coli*. One of the highest DNA content is that of the salamander *Amphiuma* which has 84 pg of DNA. Each human being has about 3 pg of DNA per haploid genome, or 3×10^9 base pairs, *i.e.*, the human genome could accommodate about 3 million average sized proteins. If all the DNA were coding (or containing structural genes) and if this was true, salamanders would have 30 times more genes than human beings. This is called **C-value paradox** (**Gall**, 1981). It was detected quite early that there was little connection between the morphological complexity of eukaryotic organisms and their DNA content. For example, *E. coli* (containing 3,400,000 base pairs in its DNA) has about 3000 genes. Although it is difficult to estimate how many different genes exist in the human genome, there are probably not more than 20,000 to 30,000 genes (Note : According to a most recent estimate, there are 100,000 genes in human genome, see **Deviah**, 1994). There is no reason to believe that salamanders should have any more. From these facts, it can be easily concluded that most of the DNA in the eukaryotic genomes must be of a non-coding nature.

To explain the C-value paradox, researches examined the **repetitiveness of DNA**, and more recently probed and sequenced DNA to understand its properties.

(*i*) Unique DNA and Repetitive DNA. The **DNA-DNA hybridization** is the process of taking DNA from the same or different sources and heating and then cooling it, causing double helices to reform at homologous regions. This technique is useful for determining sequence similarities and degrees of repetitiveness among DNAs. **Britten** and **Kohne** (1968), using the technique of **DNA-DNA hybridization**, first systematically analyzed the repetitiveness of the DNA within eukaryotes. When DNA is heated, it cools, it **renatures**. The rate of renaturation depends on the DNA sequences. If the sample contains DNA with repeated sequences, it will hybridize faster than DNA that does not have repeated sequences. From these studies, **Britten** and **Kohne** (1968) found that eukaryotic chromosomes contain regions of unique, moderately repetitive and highly repetitive DNA. **Unique DNA** is, as its name implies, DNA with unrepeated sequences. **Repetitive DNA** is DNA whose sequences are repeated in the genome. During the last 39 years, hundreds of plants and animals have been analyzed for the relative proportion of repetitive DNA in their genome (Table 32.2).

Table 32.2 Relative proportions (per cent of haploid genome) of repetitive sequences.

	Organism	Proportion of repetitive sequences (%)
I. Animals	(*i*) Sea urchin	38
	(*ii*) Snail	60
	(*iii*) Toad	45
	(*iv*) Mouse	40
	(*v*) Calf	45
	(*vi*) Human	30
II. Plants	(*i*) Bread wheat	80
	(*ii*) Oats	83
	(*iii*) Maize	78
	(*iv*) Garden pea	75
	(*v*) Barley	76
	(*vi*) Rye	92

Satellite DNA, found around centromeres is highly repetitive DNA with a unique repeat length of about two hundred base pairs. Unique DNA makes up most of the transcribed genes of an organism. The rest of the DNA is repetitive DNA in a few to several hundred thousand copies. This repetitive DNA comprises three categories. One is "junk" DNA, DNA that is not useful to the organism, made up of untranscribed and parasitic sequences (selfish DNA). Another category is transcribed genes in many copies that have diverged from each other, such as antibody, collagen, and globin genes. The term diver **gene family** has been given to the genes that have arisen by duplication, with or without divergence, from an ancestral gene. Finally, the transcribed genes in many copies that are virtually identical such as ribosomal RNA genes and histone genes, make up a third category of repetitive DNA.

***(ii)* Junk DNA.** Transposons in prokaryotes are generally viewed as selfish or parasitic: they serve no purpose to the cell. The transposons replicate on their own, increasing in number. Eukaryotic transposons are mostly **retrotransposons**, *i.e.*, transposable elements that move by way of an RNA intermediate. In other words, the retrotransposon is transcribed into RNA and then, by reverse transcription, converted to a cDNA and then inserted into the genome. These elements can make up 50 per cent of the eukaryotic genome, existing in hundreds of thousands of copies. They generally fall into two categories: LINES and SINES.

Box 32.3

The term 'junk DNA' is falling out of favour of molecular biologists. This is partly because the number of surprises resulting from genome research over the last few years has meant that molecular biologists have become less confident in asserting that any part of the genome is unimportant simply because we not currently know what its function might be. One thing that is clear that the bulk of the intergenic DNA is made up of repeated sequences of one type or another (see **Brown**, 2002).

***(a)* LINES or Long Interspersed Elements.** They contain upto seven thousand pairs each and contain genes for **reverse transcription, RNA binding** and **endonuclease activity**. They thus have the ability to jump by way of an RNA intermediate. Human DNA is believed to be composed of about 15 per cent LINES.

***(b)* SINES or Short Interspersed Elements.** They are generally derivatives of transfer RNA (tRNA) genes and do not have the ability to retrotranspose on their own. That is, in the past, their transcripts were modified, converted to cDNA by reverse transcription and then reinserted into the host's genome. They depend on the reverse transcriptase enzyme provided by the genes of LINES or retroviruses. One group of SINES not derived from tRNA is derived from RNA of the signal recognition particle; members of this group occur in human beings in about five hundred thousand copies of a three hundred base pair sequences. Because these sequences are cleaved by the restriction endonuclease *alu I*, they are called the *Alu family*. The human genome is also pervaded by remnants of at least a dozen distinct families of ancient retroviruses scattered throughout our chromosomes.

At this juncture, some explanation can be forwarded for the C-value paradox. Much eukaryotic DNA is junk. Apparently doing no harm. In some cases, 97% of the host genome is composed of junk DNA. Recent work has explained that gross differences in DNA content between higher organisms may be due to the differing abilities of different species to rid themselves of this parasitic DNA. If it builds up without being removed, the DNA content of the species can rise. Thus, wide differences in DNA content among higher eukaryotes mentioned at the beginning of this section have little to do with the complexity of the organism, but rather with the ability of the organism to remove the junk DNA as it forms (see **Tamarin**, 2002).

2. Histones

Histones are very basic proteins, basic because they are enriched in the amino acids arginine and lysine to a level of about 24 mole present. Arginine and lysine at physiological pH are cationic and can interact electrostatically with anionic nucleic acids. Thus, being basic, histones bind tightly to DNA which is an acid. There are five types of histones in the eukaryotic chromosomes, namely H1, H2A, H2B, H3 and H4.

One of the important discoveries that has come from chemical studies is that the primary structures of histones have been highly conserved during evolutionary history. For example, histone H4 of calf and of garden pea contains only two amino acid differences in a protein of 102 residues (**DeLange**, 1969). Likewise, the sequence of histone H3 from rat differs only in two amino acids from that of peas, out of 102 total amino acid residues. These organisms are estimated to have an evolutionary history of at least 600 million years, during which they diverged structurally. This conservation of structure suggests that over the eras, histones have had a very similar and crucial role in maintaining the structural and functional integrity of chromatin. Such an evolutionary conservation suggests that the functions of these two histones involve nearly all of their amino acids so that a change in any position is deleterious to the cell.

Histone H1 is the least rigidly conserved histone protein. It contains 210 to 220 amino acids and may be represented by a variety of forms even within a single tissue. H1 is present only once per 200 base pairs of DNA (in contrast to rest of the four types of histones each of which is present twice) and is rather loosely associated with DNA. H1 histone is absent in yeast, *Saccharomyces cerevisiae*. Histones besides determining the structure of chromatin, play a regulatory role in the repression activity of genes.

3. Non-histones

In contrast to the modest population of histones in chromatin, non-histone proteins display more diversity. In various organisms, number of non-histones can vary from 12 to 20. Heterogeneity of these proteins is not conserved in evolution as the histones. These non-histones differ even between different tissues of the same organism suggesting that they regulate the activity of specific genes.

About 50 per cent non-histones of chromatin have been found to be structural proteins and include such proteins as **actin**, and α- and β-**tubulins** and **myosin**. Although for sometime these contractile proteins were thought to be contaminants, it is now believed that they are vital ingredients of the chromosome, functioning during chromosome condensation and in the movement of chromosomes during mitosis and meiosis (see **Thorpe**, 1984). Many of the remaining 50 per cent of non-histones include all the enzymes and factors that are involved in DNA replication, in transcription and in the regulation of transcription. These proteins are not as highly conserved among organisms, although they must carry out similar enzymatic activities. Apparently they are not as important as the histones in maintaining chromosome integrity.

32.2. STRUCTURE OF CHROMOSOMES

Shape

The shape of the chromosomes is changeable from phase to phase in the continuous process of the cell growth and cell division. In the resting phase or interphase of the cell, the chromosomes occur in the form of thin, coiled, elastic and contractile, thread-like stainable structures, the **chromatin** threads. In the metaphase and the anaphase, the chromosomes become thick and filamentous. Each chromosome contains a clear zone, known as **centromere**, along their length. The centromere divides the chromosomes into two parts, each part is called **chromosome arm**.

Depending upon the number of centromeres, an eukaryotic chromosome may be of following types:

1. **Monocentric chromosome** has one centromere. Most normal chromosomes are monocentric.
2. **Holocentric chromosome** has a diffuse centromere, so that spindle attachment occurs over the entire chromosome, *e.g.*, Lepidoptera.
3. **Dicentric chromosome** is with two centromeres.
4. **Polycentric chromosome** is with more than two centromeres.
5. **Acentric chromosome** is without any centromere. Such chromosomes represent freshly broken segments of chromosomes which do not survive for long.

In a monocentric chromosome, the position of centromere varies form chromosome to chromosome and it provides different shapes to the latter which are following (Fig. 32.1):

1. Telocentric (Telomictic). It is a rod-like monocentric chromosome with a terminal centromere. It is also known as **T-chromosome** or **monobrachial chromosome**. Such chromosomes are very rare in nature.

2. Atelocentric (Atelomictic). It is a monocentric chromosome with a nonterminal centromere. It is also called **dibrachial chromosome**. Atelocentric chromosomes are of following four types:

***(i)* Metacentric.** It is a monocentric V-shaped chromosome with a central or near central (Fig. 32.2) centromere. If the centromere is exactly at the median point, it is classified as a **M-chromosome**, whilst if the centromere is not central but lies within the median region, it is classified as a **m-chromosome**. M-chromosomes are rare in nature.

(ii) **Submetacentric.** It is a monocentric chromosome with a centromere in the **submedian region.** It is J-or L-shaped.

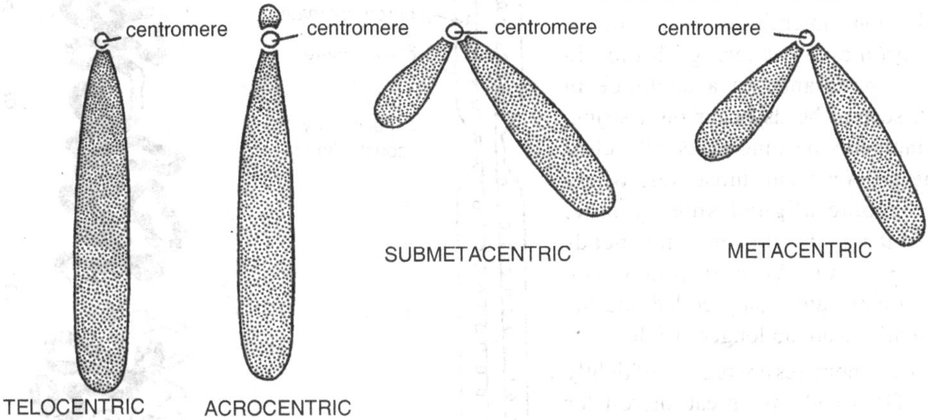

Fig. 32.1. The four morphological types of chromosomes according to the position of the centromere.

(iii) **Subacrocentric.** It is a monocentric chromosome with a centromere in the **subterminal region** (Fig. 32.2).

(iv) **Acrocentric.** It is a monocentric chromosome with its centromere in **terminal** region. It is also called **t-chromosome**, since most so-called telocentric chromosomes are actually acrocentric.

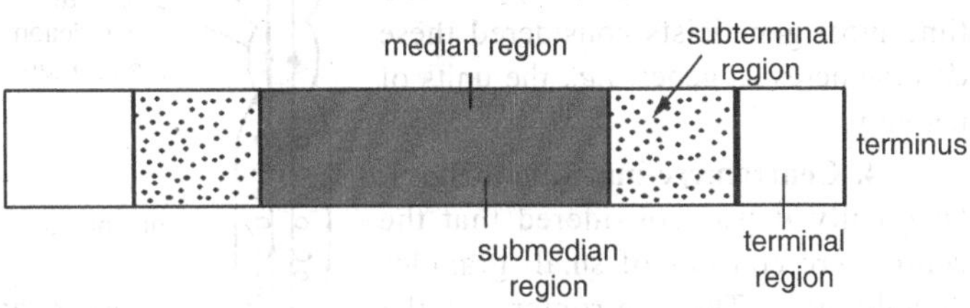

Fig. 32.2. Various regions of a chromosome used in classification of chromosomes.

Structure (Parts of Chromosomes)

In the metaphase chromosomes following structures or parts have been recognized:

1. Chromatid. At mitotic metaphase each chromosome consists of two symmetrical structures, called **chromatids**. Each chromatid contains a single DNA molecule. Both chromatids are attached to each other only by the centromere and become separated at the beginning of anaphase, when the sister chromatids of a chromosome migrate to the opposite poles.

2. Chromonema (ta). During mitotic prophase the chromosomal material becomes visible as very thin filaments, called **chromonemata** (a term coined by **Vejdovsky** in 1912). A chromonema represents a chromatid in the early stages of condensation. Therefore, 'chromatid' and 'chromonema' are two names for the same structure: a single linear DNA molecule with its associated proteins. The chromonemata form the gene-bearing portions of the chromosomes.

According to old view, a chromosome may have more than one chromonemata which are embedded in the achromatic and amorphous substance, called **matrix**. The matrix is enclosed in a sheath or **pellicle**. Both matrix and pellicle are non-genetic materials and appear only at metaphase when the nucleolus disappears. It is believed that nucleolar material and matrix are interchangeable, *i.e.*, when chromosomal matrix disappears, the nucleolus appears and vice versa. Electron microscopic observations, however, have questioned the occurrence of pellicle and matrix in them.

3. Chromomeres. The chromomeres are bead-like accumulations of chromatin material that are sometimes visible along interphase chromosomes. The chromomere-bearing chromatin has an appearance of a necklace in which several beads occur on a string. Chromomeres become especially clear in the polytene chromosomes, where they become aligned side by side, constituting the chromosome beads (Fig. 32.3A). At metaphase the chromosomes are tightly coiled and the chromomeres are no longer visible.

Chromomeres are regions of tightly folded DNA and have great interest for the cell biologists. They are believed to correspond to the units of genetic function in the chromosomes. In fact, for long time most geneticists considered these chromomeres as genes, *i.e.*, the units of heredity.

4. Centromere and kinetochore. Originally it was considered that the centromere consists of small granules or spherules. The centromere of the chromosome of the *Trillium* has the diameter of 3μm and the spherules have the diameter of 0.2 μm. The chromonema remains connected with the spherules of the centromere. Currently it is held that centromere is the region of the chromosome to which are attached the fibres of mitotic spindle. The centromere (a term much preferred by the geneticists) lies within a thinner segment of chromosome, the **primary constriction**. The regions of chromosome flanking the centromere contain highly repetitive DNA and may stain more intensely with the basic dyes. (*i.e.,* it is a constitutive heterochromatin, Fig. 32.4B).

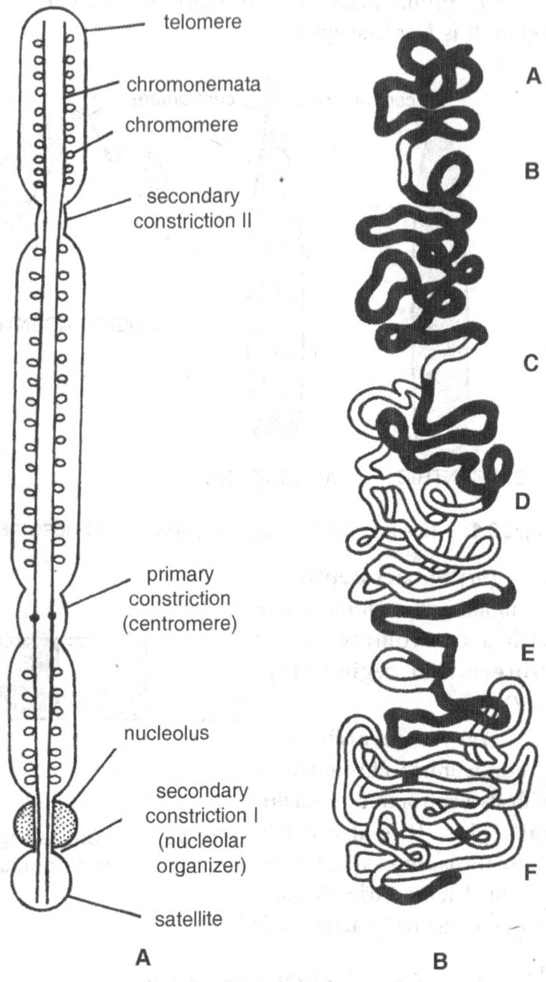

Fig. 32.3. A—Structure of a typical chromosome; B—Model of constitutive heterochromatin in a mammalian metaphase chromosome. [A—Constitutive heterochromatin; B—Secondary constriction; I or nucleolar organizer; C–Primary constriction or centromere; D—Euchromatin; E—Secondary constriction. II—possible site of 5S rRNA cistrons; F—Telomere].

Centromeres are found to contain specific DNA sequences with special proteins bound to them, forming a disc-shaped structure, called **kinetochore** (a term that is much preferred by the cytologists). Under the EM, the kinetochore appears as a plate-or cup-like disc, 0.20 to 0.25 nm in diameter situated upon the primary constriction or centromere. In thin electron microscopic sections, the kinetochore shows a trilaminar structure, *i.e.,* a 10 nm thick dense **outer proteinaceous** layer, a **middle** layer of low density and a dense **inner** layer tightly bound to the centromere

(Fig. 32.4). The DNA of centromere does not exist in the form of nucleosome.

5. Telomere. (Gr., *telo* = for; *meros* = part). Each extremity of the chromosome has a polarity and therefore, it prevents other chromosomal segments to be fused with it. The chromosomal ends are known as the **telomeres**. If a chromosome breaks, the broken ends can fuse with each other due to lack of telomeres.

6. Secondary constriction. The chromosomes besides having the primary constriction or the centromere possess secondary constriction at any point of the chromosome. Constant in their position and extent, these constrictions are useful in identifying particular chromosomes in a set. Secondary constrictions can be distinguished from primary constriction or centromere, because chromosome bends (or exhibits angular deviation) only at the position of centromere during anaphase.

7. Nucleolar organizers. These areas are certain secondary constrictions that contain the genes coding for 5.8S, 18S and 28S ribosomal RNA and that induce the formation of nucleoli. The secondary constriction may arise because the rRNA genes are transcribed very actively and, thus, interfering with chromosomal condensation. In human beings, the nucleolar organizers are located in the secondary constrictions of chromosomes 13, 14, 15, 21 and 22, all of which are acrocentric and have satellites.

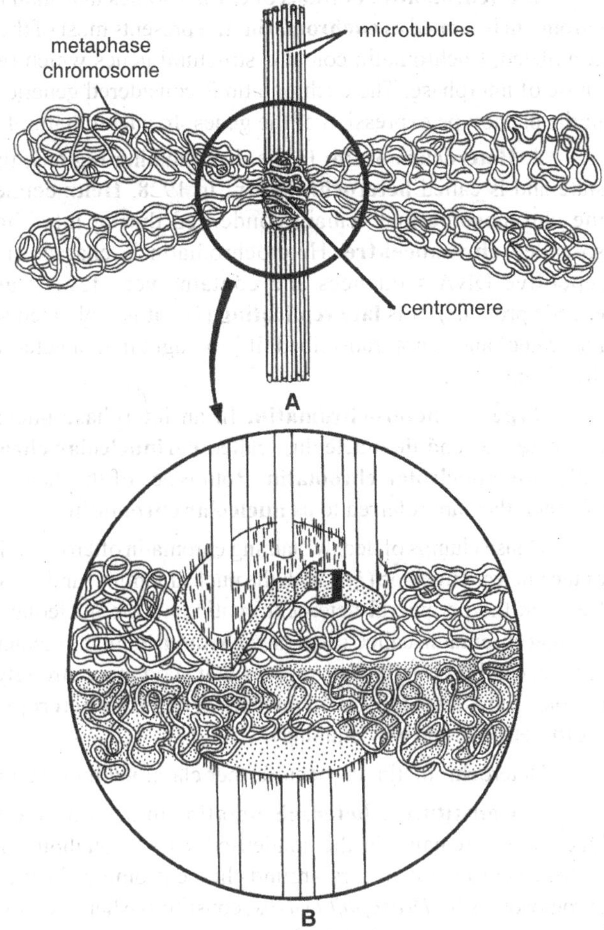

Fig. 32.4. A—A metaphase chromosome showing the folded fibre structure and the centromere with implanted microtubules; B—Inset: a higher magnification, showing the convex electron dense layer and the fibrillar material forming the "corona" of the centromere.

8. Satellite. Sometimes the chromosomes bear round elongated or knob-like appendages known as **satellites**. The satellite remains connected with the rest of the chromosome by a thin chromatin filament. The chromosomes with the satellite are designated as the **sat chromosomes**. The shape and size of the satellite remain constant.

Chromosome satellites are a morphological entity and should not be confused with satellite DNAs which are highly repeated DNA sequences.

Material of the Chromosomes

The material of the chromosomes is the chromatin. Depending on their staining properties, the following two types of chromatin may be distinguished in the interphase nucleus:

1. Euchromatin. Portions of chromosomes that stain lightly are only partially condensed; this chromatin is termed as **euchromatin**. It represents most of the chromatin that disperse after mitosis has completed. Euchromatin contains structural genes which replicate and transcribe during G_1 and S phase of interphase. The euchromatin is considered genetically active chromatin, since it has a role in the phenotype expression of the genes. In euchromatin, DNA is found packed in 3 to 8 nm fibre.

2. Heterochromatin. In the dark-staining regions, the chromatin remains in the condensed state and is called **heterochromatin**. In 1928, **Heitz** defined heterochromatin as those regions of the chromosome that remain condensed during interphase and early prophase and form the so-called **chromocentre**. Heterochromatin is characterized by its especially high content of repetitive DNA sequences and contains very few, if any, structural genes (*i.e.*, genes that encode proteins). It is **late replicating** (*i.e.*, it is replicated when the bulk of DNA has already been replicated) and is not transcribed. It is thought that in heterochromatin the DNA is tightly packed in the 30 nm fibre.

Types of heterochromatin. In an interphase nucleus, usually there is some condensed chromatin around the nucleolus, called **perinucleolar chromatin**, and some inside the nucleolus, called **intranucleolar chromatin**. Both types of this heterochromatin appear to be connected and together, they are referred to as **nucleolar chromatin**.

Dense clumps of deeply staining chromatin often occur in close contact with the inner membrane of the nuclear envelope (*i.e.*, with the nuclear lamina) and is called **condensed peripheral chromatin**. Between the peripheral heterochromatin and the nucleolar heterochromatin are regions of lightly staining chromatin, called **dispersed chromatin**. In the condensed chromosomes, the heterochromatic regions can be visualized as regions that stain more strongly or more weakly than the euchromatic regions, showing the so-called **positive** or **negative heteropyknosis** of the chromosomes (Gr., *hetero* = different + *pyknosis* = staining).

Heterochromatin has been further classified into the following types:

1. Constitutive heterochromatin. In such a heterochromatin the DNA is permanently inactive and remains in the condensed state throughout the cell cycle. This most common type of heterochromatin occurs around the centromere, in the telomeres and in the C-bands of the chromosomes. In *Drosophila virilis*, constitutive heterochromatin exists around the centromeres and such **pericentromeric heterochromatin** occupies 40 per cent of the chromosomes. In many species, entire chromosomes become heterochromatic and are called **B chromosome, satellite chromosomes** or **accessory chromosomes** and contain very minor biological roles. Such chromosomes comprising wholly constitutive heterochromatin occur in corn, many phytoparasitic insects and salamanders. In the fly *Sciara*, large metacentric heterochromatic chromosomes are found in the gonadal cells, but are absent in somatic cells. Entire Y chromosome of male *Drosophila* is heterochromatic, even though containing six gene loci which are necessary for male fertility.

Constitutive heterochromatin contains short repeated sequences of DNA, called **satellite DNA**. This DNA is called satellite DNA because upon ultracentrifugation, it separates from the main component of DNA. Satellite DNA may have a higher or lower G + C content than the main fraction. For example, the mouse satellite DNA is a 240 base pair sequence that is repeated about 1000,000 (10^6) times in the mouse genome, constituting 10 per cent of the total mouse DNA. The exact significance of constitutive heterochromatin is still unexplained.

2. Facultative heterochromatin. Such type of heterochromatin is not permanently maintained in the condensed state; instead it undergoes periodic dispersal and during these times is transcriptionally active. Frequently, in facultative heterochromatin one chromosome of the pair becomes either totally or partially heterochromatic. The best known case is that of the X-chromosomes in the mammalian female, one of which is active and remains euchromatic, whereas the other is inactive and forms at interphase, the **sex chromatin** or **Barr body** (Named after its discoverer, Canadian cytologist **Murray L. Barr**). Barr body contains DNA which is not transcribed and is not found in males. Indeed, the number of Barr bodies is always one less than the number of X chromosomes (*i.e.*, in humans, XXX female has two Barr bodies and XXXX female has three Barr bodies; **M.L. Barr**, 1959).

32.3. ULTRASTRUCTURE AND MOLECULAR ORGANIZATION

The structural unit of each chromosome is single DNA complexed with basic proteins (histones) in 1 : 1 ratio. This is called **single stranded hypothesis**.

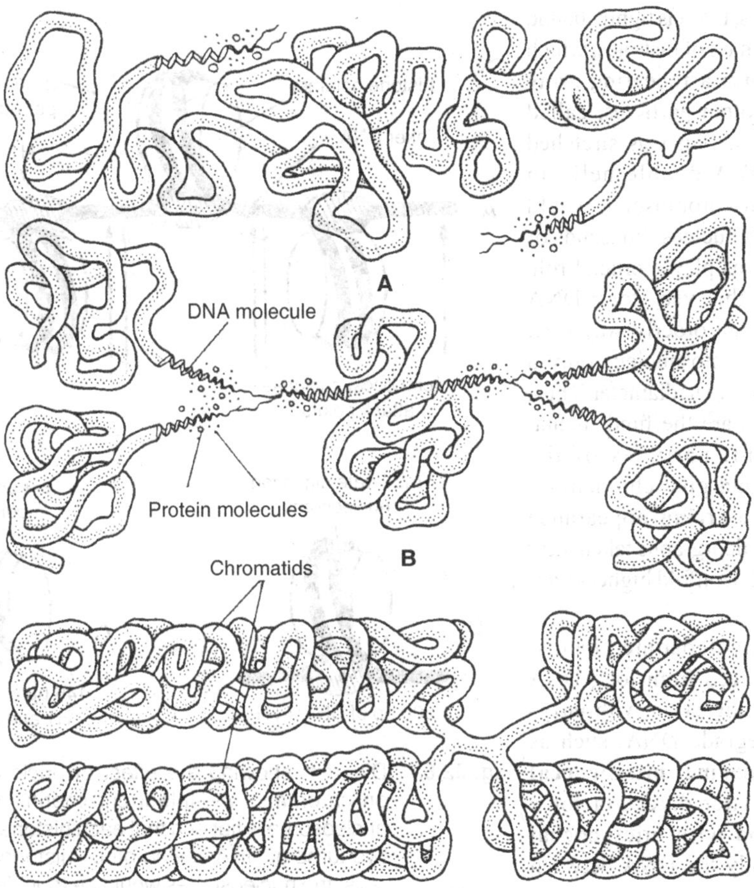

Fig. 32.5. Dupraw's folded fibre model of chromatin in interphase (A and B) and in metaphase (C).

The manner of coiling and folding of DNA was a matter of debate and dozens of models were available for this purpose; of them only two stand out and are important. A popular model is the **folded fibre model**, proposed by **E.J. Dupraw** in 1965. According to it, the bulk of the chromosome is visualized to be composed of a tightly folded fibre which has a rather homogeneous diameter of 200 to 300 A°. This folded fibre is supposed to contain the DNA histone helix (of 30 A° diameter) in a supercoiled condition (Fig. 32.5). Another model is most significant and universally accepted one and is called **nucleosome model** which was proposed by **R.D. Kornberg** (1974) (Fig. 32.6) and confirmed and christened by **P. Oudet** *et al.*, (1975). Thus, while in the folded-fibre model, it was proposed that the histones were bound on the outside of the DNA coils (*i.e.*, histone shell around DNA), the nucleosome model has proposed the opposite (*i.e.*, histone particle with DNA round it). In other words, the earlier theory that basic chromatin fibre had DNA core surrounded by histones was incorrect. In fact, from a genetic perspective, a significant feature of packing mechanism through the nucleosomes lies in its topology: *at no point is the DNA buried;* instead, it is freely exposed along the entire surface of the "spool", available for genetic expression. Nucleosomes seem to be universal devices for compacting the long DNA molecules of eukaryotic cells.

Nucleosomes and Solenoid Model of Chromatin

In eukaryotes, DNA is tightly bound to an equal mass of histones, which serve to form a repeating array of DNA-protein particles, called **nucleosomes**. If it was stretched out, the DNA double-helix in each human chromosome would span the cell nucleus thousands of time. Histones play a crucial role in packing this very long DNA molecule in an orderly way (*i.e.*, nucleosome) into nucleus only a few micrometres in diameter. Thus, nucleosomes are the fundamental packing unit particles of the chromatin and give chromatin a "**beads-on-a-string**" appearance in electron micorgraphs taken after treatments that unfold higher-order packing.

The nucleosome 'beads' can be removed from long DNA "string" by digestion with enzymes that degrade DNA, such as bacterial enzyme, **micrococcal nuclease**. After digestion for a short period with micrococcal nuclease, only the DNA between the nucleosome beads is degraded (Fig. 32.6). The rest is protected from digestion and remains as double-stranded DNA fragments 146 nucleotide pairs long bound to a specific complex of 8 nucleosome histones (the **histone octamer**). The nucleosome beads obtained in this way have been crystallized and analyzed by X-ray diffraction.

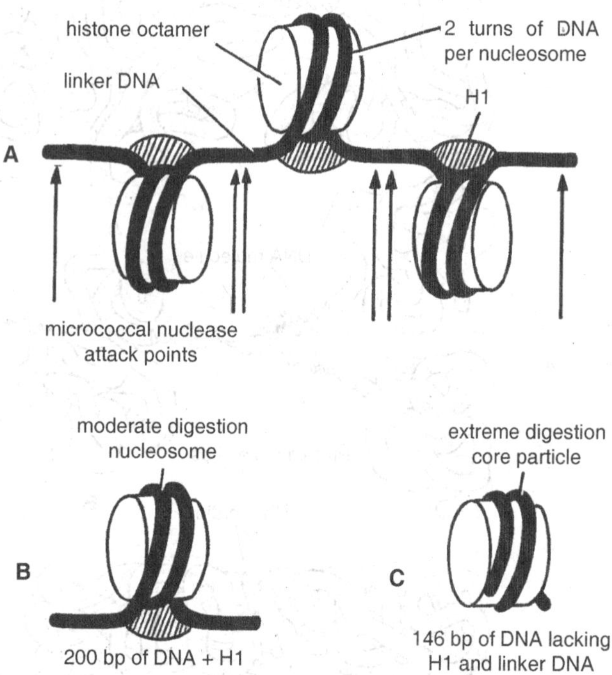

Fig. 32.6. Diagram showing the effect of nuclease enzymes on chromatin. A—Intact chromatin, note that the nucleosome is flat and that each histone octamer has two turns of DNA sealed off by histone H1. In living cells, the nucleosomes would be touching each other to form a 10 nm fibre and not stretched out (to form beads-on-a-string preparation); B—A nucleosome released by moderate digestion; C—Core particle of nucleosome obtained by extensive digestion.

Each nucleosome is a disc-shaped particle with a diameter of about 11 nm and 5.7 nm in height containing 2 copies of each 4 nucleosome histones—H2A, H2B, H3 and H4. This histone octamer forms a protein core [(*i.e.*, a core of histone tetramer $(H3, H4)_2$ and the apolar regions of 2(H2A and H2B)] around which the double-stranded DNA helix is wound 1¾ time containing 146 base pairs. In undigested chromatin the DNA extends as a continuous thread from nucleosome to nucleosome. Each nucleosome bead is separated from the next by a region of **linker DNA** which is generally 54 base pair long and contains single H1 histone protein molecule. Generally, DNA makes two complete turns around the histone octamers and these two turns (200 bp long) are sealed off by H1 molecules.

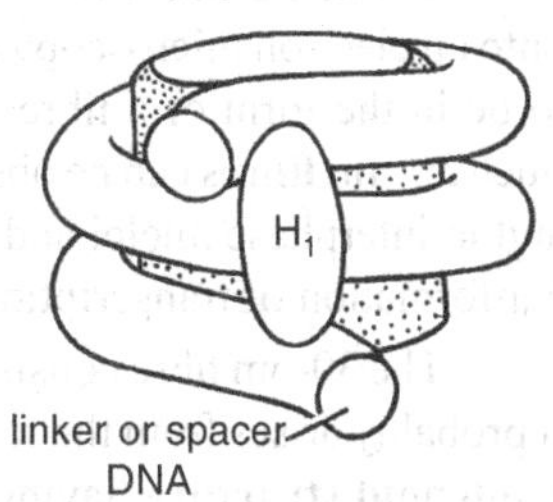

Fig. 32.7. A nucleosome.

[**Note** : In some organisms nucleosome DNA may vary from 162 base pairs (*e.g.*, rabbit cortical neurons) to 242 base pairs (*e.g.*, sea urchin sperm)]. Thus, on an average, nucleosomes repeat at intervals of about 200 nucleotides or base pairs. For example, an eukaryotic gene of 10,000 nucleotide pairs will be associated with 50 nucleosomes and each human cell with 6×10^9 DNA nucleotide pairs contains 3×10^7 nucleosomes.

Solenoid Models

H1 is reported to be phosphorylated just before mitotic and meiotic cell division to make possible the higher levels of coiling. During mitosis or meiosis, the prophase is the stage during which the chromosomes become shorter and thicker due to multiple coiling as proposed by **DuPraw** and others. The hypothesis of a solenoidal structure, with coils of coils had renewal since nucleosomal substructure has been discovered. Thus, due to solenoid coiling of nucleosome containing fibre, the following types of chromosomal structures can be observed during the cell cycle.

Box 32.4 Distinction between centrosome and nucleosome

Centrosome is an extranuclear structure which includes **centriole** with surrounding cytoplasm. It helps in spindle formation during cell division. Nucleosome is the unit of chromatin fibre in which 4 pairs of histones [2 (H2A, H2B, H3, H4)] jointly form a core particle and a DNA segment of about 200 base pairs long encircle the core particle. Another histone (H1 or H5) maintains this configuration.

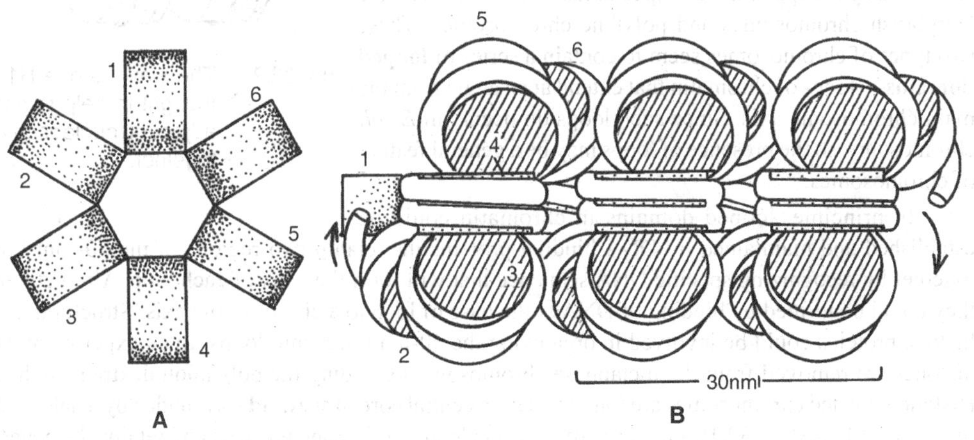

Fig. 32.8. A model suggested to explain how the "Beads-on-a-string" form of nucleosomes is packed to form the 30-nm fibre. A—Top view of 30-nm fibre; B—Side view of 30 nm fibre.

1. The 10-nm fibre. When nucleosomes are in close apposition, they form the **10-nm filaments**, in which packing of DNA is about five-to seven-fold, *i.e.*, five to seven times more compact than free DNA.

2. The 30-nm fibre. When nuclei are very gently lysed onto an electron microscopy grid, most of the chromatin is seen to be in the form of a fibre, with a diameter of about 30 nm. Such **30-nm fibres** can be observed in metaphase chromosomes and in interphase nuclei and it probably represents the natural conformation of transcriptionally inactive chromatin.

The 30-nm fibre consists of closely packed nucleosomes. It probably arises from the folding of the nucleosome chain into a **solenoid structure** having about six nucleosomes per turn. The DNA of 30-nm solenoid has a packing that is about 40-fold.

Histone H1 molecules are found responsible for packing nucleosomes into the 30-nm fibre. The H1 histone molecule has an evolutionarily conserved globular core or central region linked to extended amino-terminal and carboxyl-terminal "arms", whose amino acid sequence has evolved much more rapidly. Each H1 molecule binds through its globular portion to a unique site on a nucleosome and has arms that are thought to extend to contact with other sites on the histone cores of adjacent nucleosomes, so that the nucleosomes are pulled together into a regular repeating array (Fig. 32.9). The binding of H1 molecule to chromatin tends to create a local polarity that the chromatin otherwise lacks.

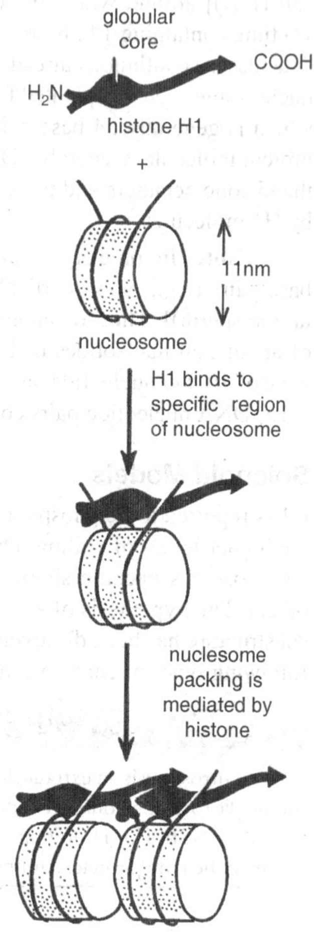

Fig. 32.9. The way histone H1 is thought to help to pack adjacent nucleosomes together.

3. Radial loops of 30-nm fibre and metaphase chromosome. The nucleus is typically about 5μm (5×10^{-4} cm) in diameter. The packaging of DNA into a 30-nm chromatin fibre leaves a human chromosome about 0.1 cm long, so there must be several higher orders of folding. The probable nature of one further level of folding was originally suggested by the appearance of specialized chromosomes—the lampbrush chromosomes and polytene chromosomes. These two types of chromosomes seem to contain a series of **looped domains**— loops of chromatin that extend at an angle from the main chromosome axis. Since such loops do occurs in *E.coli* chromosome, so the presence of loops may be a general feature of chromosomes.

In principle, looped domains in chromatin could be established and maintained by DNA binding proteins that clamp two regions of the 30- nm fibre together by recognizing specific DNA sequence that will form the neck of each loop. Alternatively, they could be formed by binding of DNA at the base of loop to a chromosome axis. Structural non-histone proteins could be involved in organizing the 30-nm fibres into loops. In an experiment, the histones are removed from the metaphase chromosome by adding the polyanion dextran sulphate. Histone-depleted chromosomes are found to have a central core of **scaffold** surrounded by a halo made of loops of DNA (Fig. 32.10A). The scaffold is made of non-histone proteins and retains the general shape of the metaphase chromosome. Each chromosome has two scaffolds, one for each chromatid, and they are connected together at the centromere region. When the histones are removed, the DNA which has packed about 40-fold in the 30-nm chromatin fibre, becomes extended and produces loops with an average length of 25 μm (75,000 base pairs). In each loop the DNA exits from the scaffold and returns to an adjacent point. On the basis of these observations a model of chromosome structure has been proposed by **Laemmli** and co-workers (1979, 1984). In Laemmli's **radial loop model** DNA is arranged in loops anchored to the non-histone scaffold. Because the lateral loops have 25 μm DNA, after contracting 40-fold in the 30-nm fibre, they would be only about 0.6 μm long, a length consistent with the diameter of metaphase chromosome (1 μm). Figure 32.10B shows how the chromatin is arranged in loops which during metaphase, become arranged so that the base of the loops forms a scaffold in the centre of the chromatid. The base of the loop might be arranged on a helical coiled path (*e.g., Trillium* and *Tradescantia*).

The chromatin in mitotic chromosomes is transcriptionally inert: all RNA synthesis ceases as the chromosomes condense. Presumably the condensation prevents RNA polymerase from gaining access to the DNA, although other controlling factors might also be involved.

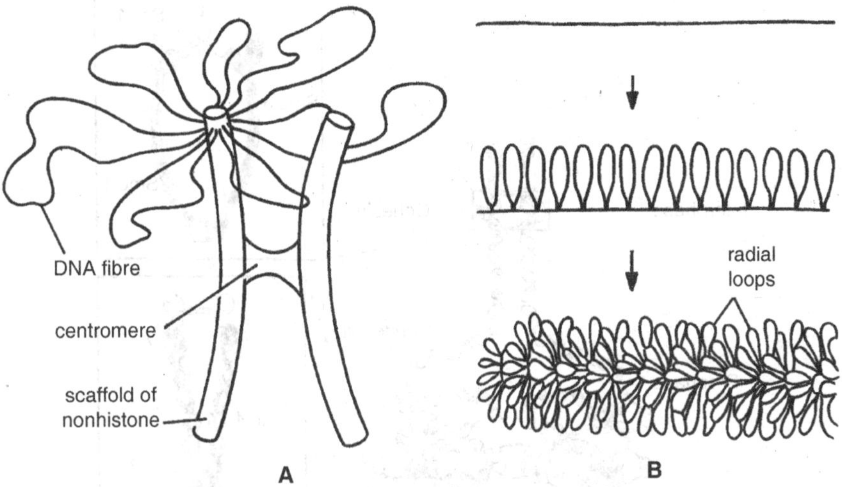

Fig. 32.10. Two methods by which DNA loops may form a metaphase chromosome. A—The non-histone proteins form two scaffolds, one per chromatid, while the naked DNA fibres form a halo around it; B—Laemmli's radial loop model of chromosome structure showin how 10-nm fibre form 30-nm fibre which further folds into radial loops.

Formation of Mitotic Chromosomes: Chromosome Compaction

The nucleus of an interphase cell contains tremendous lengths of chromatin fibers. The extended state of interphase chromatin is ideally suited for the processes of **transcription** and **replication** but not for segregation into two daughter cells. Before segregating its chromosomes, a cell converts them into much shorter, thicker structures by a remarkable process of **chromosome compaction** (or chromosome condensation) which occurs during early prophase.

In recent years, reserach on chromosome compaction has focused on an abundant multiprotein complex called **condensin**. The proteins of condensin were discovered by incubating nuclei in frog egg extracts and identifying those proteins that associated with the chromosomes as they underwent compaction. Removal of condensin from the extracts prevented normal chromosome compaction.

How is condensin involved in such dramatic changes in chromatin architecture? It has been found that in the presence of a topoisomerase and ATP, condensin is able to bind to DNA in vitro and curl the DNA into positively supercoiled loops. This finding fits nicely with the observation that chromosome compaction at prophase requires topoisomerase II, which along with condensin is present as part of the mitotic chromosome scaffold. A speculative model for condensin has been shown in Fig. 32.11. Condensin is activated at the onset of mitosis by phosphorylation of several of its subunits by the cyclin-Cdk (Cdk = cyclin – dependent kinases) responsible for driving cells from G_2 into mitosis. Thus condensin is presumably one of the targets through which Cds are able to trigger cell cycle activities. The subunit structure of a V-shaped condensin molecule is shown in the lower inset of Fig. 32.11.

As the result of compaction, the chromosomes of a mitotic cell appear as distinct, rod-like structures. Close examination of mitotic chromosomes reveals each of them to be composed of two mirror-image, "sister" **chromatids.** Sister chromatids are a result of replication in the previous interphase.

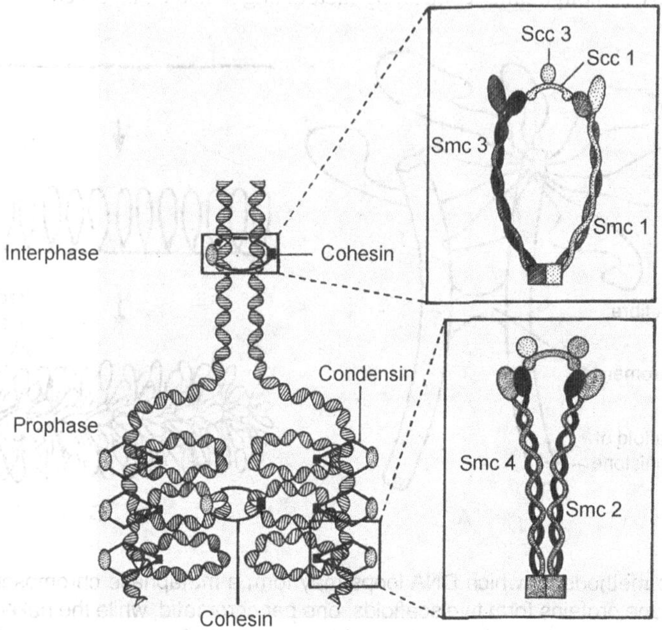

Fig. 32.11. Model for the roles of condensin and cohesin in chromosome compaction in mitosis. Just after replication, the DNA helices of a pair of sister chromatids would be held in association by cohesin molecules that encircled the sister DNA helices. As the cell entered mitosis, the compaction process would begin, aided by condensin molecules. In this model, condensin brings about chromosome compaction by forming a ring around supercoiled loops of DNA within chromatin. Cohesin molecules would continue to hold the DNA of sister chromatids together. It is proposed that cooperative interactions between condensin molecules would then organize the supercoiled loops into larger coils, which are then folded into mitotic chromosome fiber. The top and bottom insets show the subunit structure of an individual cohesin and condensin complex, respectively. Both complexes are built around a pair of SMC subunits (*i.e.*, Smc1 and Smc3; Smc2 and Smc4). Each of the SMC polypeptide folds back on itself to form a highly elongated antiparallel, coiled coil with an ATP-binding globular domain where the N- and C-termini come together. Cohesin and condensin also have two or three non-SMC subunits (*i.e.*, Scc1 and Scc3) that complete the ring like structure of these proteins (after Karp 2010).

Prior to replication, the DNA of each interphase chromosome becomes associated at sites along its length with a ring-like multiprotein complex called **cohesin** (Fig. 32.11). Following replication, cohesin holds the two sister chromatids together through G_2 and into mitosis when they are ultimately separated. As indicated in insets of Fig. 32.11, condensin and cohesin have a similar structural organization.

In vertebrates, cohesin is released from the chromosomes in two distinct stages. Most of the cohesin dissociates from the arms of the chromosomes as they become compacted during prophase. Dissociation is induced by *phosphorylation* of cohesin subunits by two important mitotic enzymes called **Polo-like kinase** and **Aurora B kinase**. In the wake of this event, the chromatids of each mitotic chromosome are held relatively loosely along their extended arms, but much more tightly at their centromeres. Cohesin is thought to remain at the centromeres because of the presence there of a **phosphatase** (enzyme) that removes any phosphate groups added to the protein by the kinases. The release of cohesin from the centromeres is normally delayed until anaphase.

Functions

The function of chromosomes is to carry the genetic information from one cell generation to another. DNA being the only permanent component of chromosome structure, is the sole genetic material of eukaryotes.

32.4. GIANT CHROMOSOMES

Some cells at certain particular stages contain large nuclei with giant or large-sized chromosomes. The giant chromosomes are the **polytene** and **lampbrush** chromosomes.

1. Polytene Chromosomes (Salivary Gland Chromosomes)

An Italian cytologist **E.G. Balbiani** (1881) had observed peculiar structures in the nuclei of certain secretory cells (*e.g.*, of salivary glands) of midge, *Chironomus* (Diptera). These structures were long and sausage-shaped and marked by swellings and cross striations (transverse bands). Unfortunately, he did not recognize them as chromosomes, and his report remained buried in the literature. It was not until 1933 that **Theophilus Painter**, **Ernst Heitz**, and **H. Bauer** rediscovered them in *Drosophila* and recognized them as the chromosomes. Since these chromosomes were discovered in the salivary gland cells, they were called **salivary gland chromosomes** (Fig. 32.12). The present name **polytene chromosomes** was suggested by **Kollar** due to the occurrence of many chromonemata (DNA) in them.

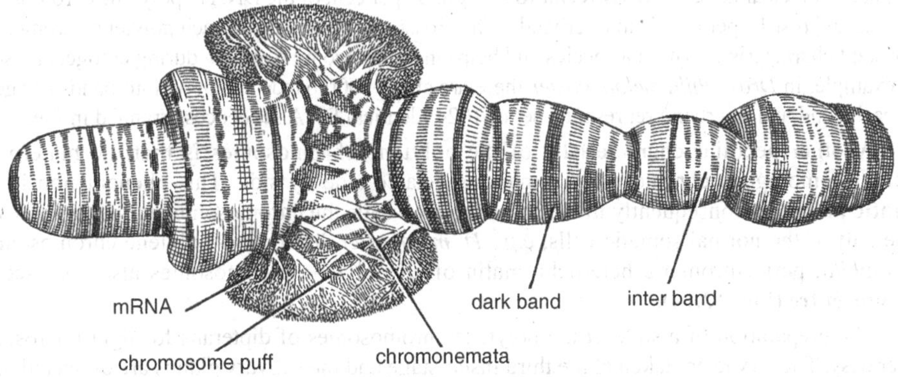

Fig. 32.12. Polytene chromosome of an insect, showing bands and interbands and a puff or Balbiani ring.

Thus, some cells of the larvae of the dipteran insects such as flies (*e.g., Drosophila*), mosquitoes and midges (*Chironomus*) become very large having high DNA content. These cells are unable to undergo mitosis and are destined to die during metamorphosis (Those cells of larva which are destined to produce the adult structures after metamorphosis, *i.e.*, imaginal discs remain diploid).

Such polytenic cells are located most prominently in the salivary gland, but also occur in Malpighian tubules, rectum, gut, foot pads, fat bodies, ovarian nurse cells, etc. Polyteny of giant chromosomes is achieved by replication of the chromosomal DNA several times without nuclear division (**endomitosis**); and the resulting daughter chromatids do not separate but remain aligned side by side. In the process of endomitosis the nuclear envelope does not rupture and no spindle formation takes place. In fact, polyteny differs from polyploidy, in which there is also an excess DNA per nucleus, but in which the new chromosomes are separated from each other.

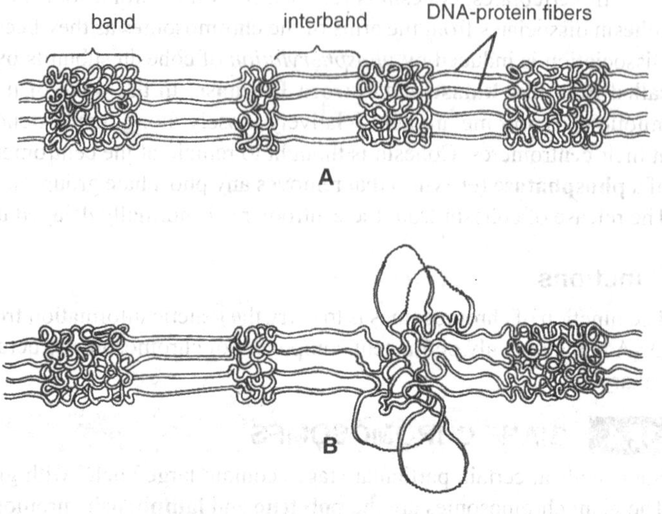

Fig. 32.13. A—Model of Dupraw and Rae showing band and interband structure in giant chromosome. B—The same model showing a possible method of "puffing".

A polytene chromosome of *Drosophila* salivary **gland** has about 1000 DNA molecules which are arranged side by side and which arise from 10 rounds of DNA replication ($2^{10} = 1024$). Other dipteran species have more DNA, for example, *Chironomus* has 16000 DNA molecules in their each polytene chromosomes. Further, the polytene chromosomes are visible during interphase and prophase of mitosis. In them the chromomere (regions in which the chromatin is more tightly coiled) alternate with regions where the DNA fibres are folded more loosely (Fig. 32.13). The alignment of many chromomeres gives polytene chromosomes their characteristic morphology, in which a series of dark transverse **bands** alternates with clear zones called **interbands**. About 85 per cent of the DNA in polytene chromosomes is in bands and rest 15 per cent is in interbands. The crossbanding pattern of each polytene chromosome is a constant characteristic within a species and helps in chromosome mapping during cytogenetic studies. For example, in *Drosophila melanogaster* there are about 5000 bands and 5000 interbands per genome and each band and interband represent a set of 1024 identical DNA sequences arranged in file.

Another peculiar characteristic of the polytene chromosomes is that the maternal and paternal homologous chromosomes remain associated side by side. This phenomenon is called **somatic pairing**. Consequently in the salivary gland cells the chromosome number always appear to be half of the normal somatic cells, *e.g., D. melogaster,* has only 4 polytene chromosomes. In *Drosophila*, pericentromeric heterochromatin of all polytene chromosomes also coalesces in a **chromocentre** (Fig. 32.14).

The preparation of a slide of the polytene chromosomes of dipterans for light microscopy is rather easy. The larvae are taken at the third instar stage and the salivary glands are dissected out and squashed in acetocarmine. In such preparations, these chromosomes in aggregate reach a length of as much as 2000 μm in *D. melanogaster*. In female *Drosophila,* the polytene chromosomes are found in the form of five long and one short strands radiating from a single more or less amorphous chromocentre (Fig. 32.14). One long strand corresponds to the X chromosome and remaining four long strands are the left and right arms of II and III chromosomes. The shortest strand represents the small dot-like IV chromosome. Each of these chromosomes contains maternal and paternal homologs in somatic pairing which lacks in the sex chromosomes of male fruit flies. Thus, in male *Drosophila,* X chromosome remains single and thin and Y chromosome exists indistinctly fused with the chromocentre.

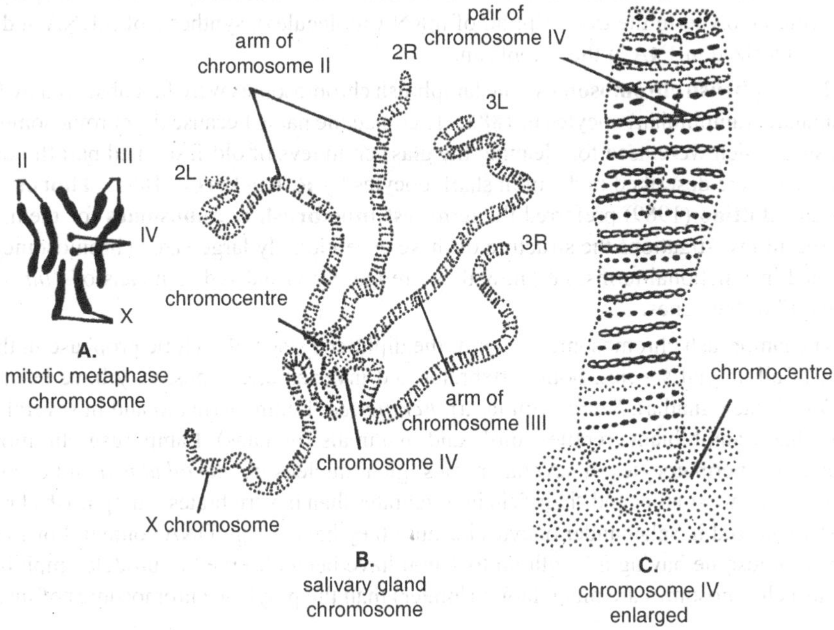

Fig. 32.14. Normal and polytene chromosomes of *Drosophila melanogaster.* A—Normal mitotic chromosomes; B—Polytene chromosomes of female; C—An enlarged IV chromosome.

One-band, one-gene hypothesis. The fixed pattern of bands and interbands in a *Drosophila* polytene chromosome suggested the early cytologists such as **Painter** (1933) and **Bridges** (1936) that each band might correspond to a single gene. Accordingly, they concluded that *Drosophila* might contain only 5000 essential genes. It was also believed that bands were the sites of genes (DNA) and interbands were relatively inert linker regions. Recent data, however, have contradicted this simple "one-band, one-gene" hypothesis, now it is held that bands as well as inter-bands contain active genes and a band may even contain more than one genes.

At this juncture a question arises, why is the single long strand of chromatin in each chromosome subdivided into so many distinct regions ? Exact explanation of this question is still not known. However, **Alberts** *et al.,* (1989) believed that this type of organization (*i.e.,* banding and interbanding of chromosomes in general) may help to: (1) Keep the DNA organized; (2) Isolate genes from their neighbours and thereby prevent biological "crosstalk", or (3) Regulate gene transcription for the cytodifferentiation, for example, constitutively expressed "**housekeeping**" **genes** could be located in interbands, whereas **cell-type-specific genes** could be confined to the bands.

Chromosome puffs or Balbiani rings. Chromosome puffs or **Balbiani rings** are the swellings of bands of the polytene chromosomes (Fig. 32.15) where DNA unfolds into open loops as a consequence of intense gene transcription (*e.g.,* mRNA formation). In 1954, **W. Beerman** compared the polytene chromosomes of different tissues of *Chironomus* larvae and showed that although the pattern of bands and interbands was similar in all tissues, the distribution of puffs differred from one tissue to another. These puffs represent regions where the tightly coiled chromosomal fibres open out to form many loops. In fact, puffing is a cyclic and reversible phenomenon: at definite times and in different tissues of the larvae, puffs may appear, grow and disappear. In salivary glands the appearance of some puffs has been corelated with the production of specific proteins which are secreted in large amounts in the larval saliva. The process of puffing involves several processes such as the accumulation of acidic proteins, despiralization of DNA, accumulation of **RNA polymerase II** (an enzyme involved in the transcription of mRNA molecules), synthesis of mRNA and release of newly synthesized mRNA in the cytoplasm.

2. Lampbrush chromosomes. The lampbrush chromosomes were first observed by **Flemming** in salamander (amphibian) oocytes in 1882. He coined the name because the chromosomes look like the brushes which were used for cleaning the glass chimneys of old-fashioned paraffin or kerosene lamps. They were described in detail in shark oocytes by **R. Ruckert** in 1892. **Thorpe** (1984) and **Burns** and **Bottino** (1989) preferred the term **test tube brush chromosomes** for them. However, due to recent investigations the structure of these exceptionally large-sized chromosomes has been interpreted in functional terms, *i.e.*, now they are merely visualized as means of *"turning on and turning off"* of the genes.

The lampbrush chromosomes occur at the diplotene stage of meiotic prophase in the primary oocytes of all animal species, both vertebrates and invertebrates. Thus, they have been described in *Sagitta* (Chaetognatha), *Sepia* (Mollusca), *Echinaster* (Echinodermata) and in several species of insects, shark, amphibians, reptiles, birds and mammals (humans). Lampbrush chromosomes are also found in spermatocytes of several species, giant nucleus of *Acetabularia* and even in plants. Generally, they are smaller and "hairy" in invertebrates than in vertebrates. Lamphrush chromosomes are best visualized in salamander oocytes because they have a high DNA content. For example, the largest chromosome having a length up to 1 mm have been observed in urodele amphibian. Thus, lampbrush chromosomes are much larger (longer) than the polytene chromosomes of insects.

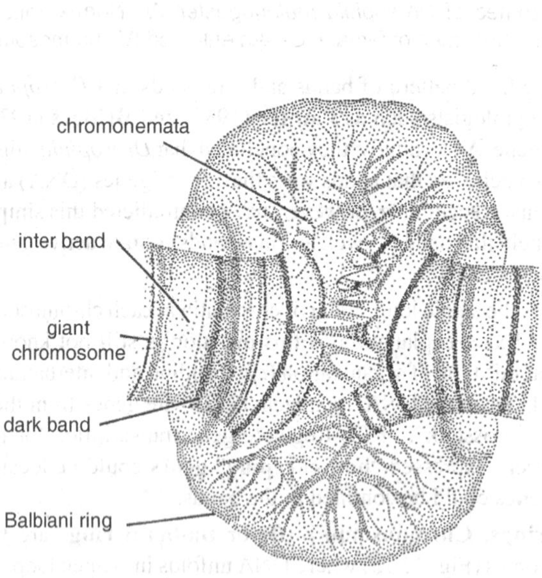

Fig. 32.15. A Balbiani ring of a polytene in chromosome.

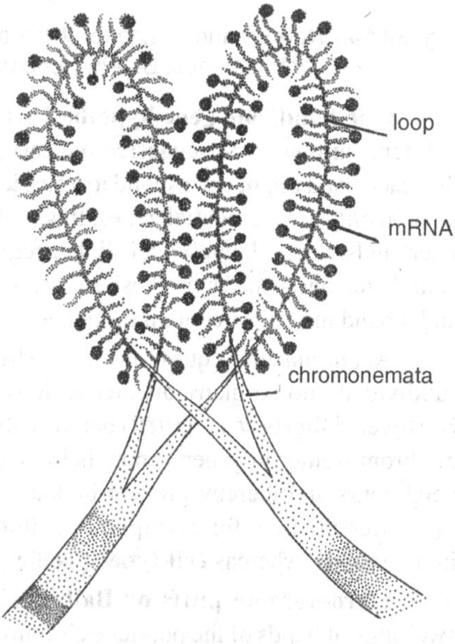

Fig. 32.16. Suggested form of the individual loops chromosome puffs. Note the tiny fibrils of RNA and proteins attached to the loop.

Since the lampbrush chromosomes are found in the prolonged diplotene stage of meiotic prophase I, they are present in the form of **bivalents** in which the maternal and paternal chromosomes are held together by **chiasmata**, at those sites where crossing over has previously occurred (Fig. 32.17). The paired homologs are not condensed as usual chromosomes would be; instead, they are very long and stretched out. Each bivalent has four chromatids, two in each homolog. The axis of each homologue consists of a row of granules or **chromomeres** from which **lateral loops** extend. The loops are always symmetrical, each chromosome having two of them, one for each chromatid. The loops can be categorized by size, thickness and other morphological characteristics. Each loop appears at a constant position in the chromosome; this fact helps in the chromosome mapping. There are about 10,000 loops per chromosome set or haploid set (*e.g.*, oocytes of slamander *Triturus*). Each loop has an **axis** which is made of single DNA molecule that is unfolded from the chromosome for the intense RNA synthesis. Thus, about 5 to 10 per cent of the DNA exists in the lateral loops, the rest being tightly condensed state in the chromomeres which are transcriptionally inactive. The centromeres of the chromosomes bear no loops.

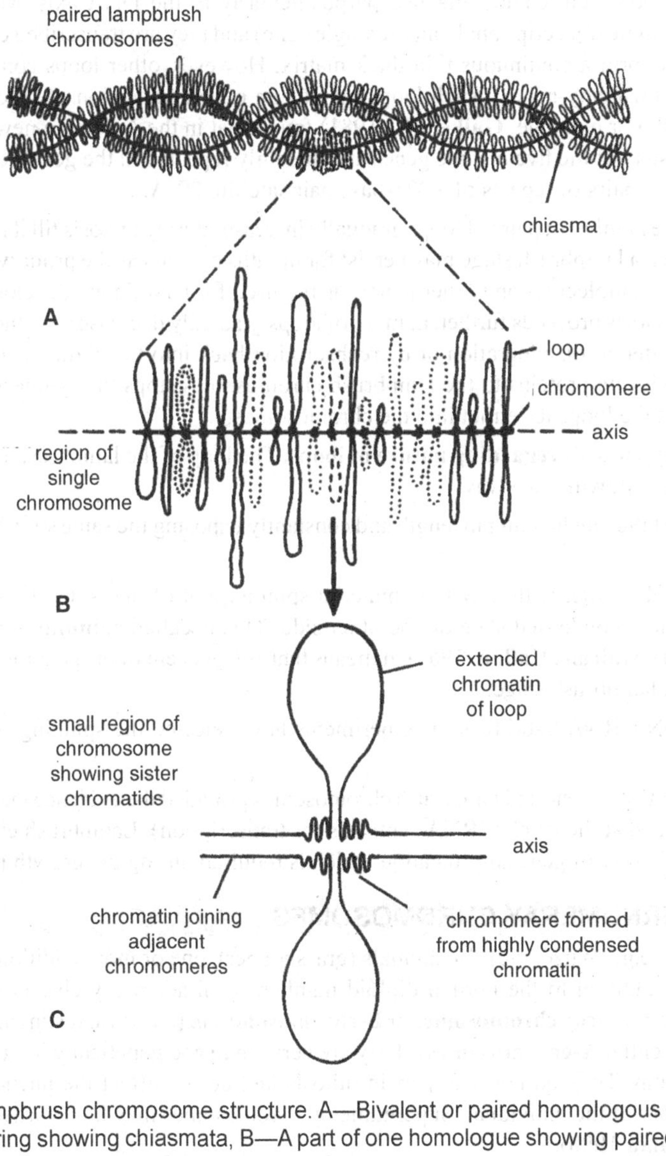

Fig. 32.17. Lampbrush chromosome structure. A—Bivalent or paired homologous chromosomes in pairing showing chiasmata, B—A part of one homologue showing paired loops given out by two chromatids; C—Single pair of loop.

Each loop of lampbrush chromosomes is found to perform intense transcription of **hnRNA** or **heterogeneous RNA** molecules. (*i.e.,* precursors of mRNA molecules for various ribosomal proteins and for five types of histone proteins). Electron microscopy of the loops has shown that **RNA polymerase** enzyme molecules are attached to the principal axis (DNA) of the loop from which RNA fibrils of increasing length extend. As transcription continues along the DNA strand of loop, the fibrils of RNA (*i.e.,* hnRNA) lengthen. Proteins get associated with these RNA fibrils as they are formed and ultimately ribonucleoprotein product is released.

Thus, each lateral loop is covered by a **matrix** that consists of RNA transcripts with hnRNA-binding proteins attached to them. Generally this matrix is asymmetrical, being thicker at one end of the loop than at the other. RNA synthesis starts at the thinner end and progresses toward the thicker end. Preparations spread for electron microscopy exhibit the typical '**christmas tree**" images with nascent ribonucleoprotein chains attached perpendicularly to the DNA axis. Many of the loops correspond to a single transcriptional unit (or single gene) and they are transcribed continuously from end to end; they form a continuous thin-thick matrix. However, other loops contain several units of transcription (or many genes); such loops include an extended section of chromatin that is not transcribed at all. For example, **Gall** *et al.,* (1981) found that in the American newt *Notophthalmus viridescens*, clusters of the five histone genes are tandemly repeated in the genome but separated by about 50,000 base pairs of repeats of a 225-base pair satellite DNA.

Further, the number of pairs of loops gradually increases during meiosis till it reaches maximum in diplotene. Such a lampbrush stage may persist for months or years as the primary oocyte builds up a supply of mRNA molecules and other materials required for its ultimate development into a new individual. As meiosis proceeds further, number of loops gradually decreases and the loops ultimately disappear either due to disintegration or by reabsorption back into the chromosome. For example, the addition of histone proteins to the lampbrush chromosomes stops the synthesis of RNA on the loops and causes the loops to retract into the chromosomes.

Certain hypothesis regarding nature of loops. The loop of the lampbrush chromosomes can be viewed in the following two ways:

1. It may be static, unchanging in length and constantly exposing the same stretch of chromosome fibre.
2. It may be dynamic, with new loop material spinning out of one side of a chromomere and returning to a condensed state on the other side. This is called **spinning out** and **retraction hypothesis** (**Gall** and **Callan**, 1962). It means that 100 per cent of the genome can be expressed during the lampbrush stage.

Recent, DNA-RNA hybridization experiments have rejected the spinning out and retraction hypothesis.

Study of both polytene and lampbrush chromosomes provided the evidence that eukaryotic gene activity is regulated at the level of RNA synthesis (or transcription). Lampbrush chromosomes also show the possible way of gene amplification which is required during the growth phase of oocytes.

32.5. SUPERNUMERAY CHROMOSOMES

In certain plants (e.g., maize, rye) and animals (grasshopper), one or more additional chromosomes are observed in addition to the normal diploid number. Such accessory chromosomes are called **satellite** or **supernumeray chromosomes** or **B-chromosomes in plants** (to distinguish them from the invariant and essential **A-chromosomes**). They are very small and genetically inert being composed of heterochromatin. They do not take part in mitosis and do not affect the phenotype but reduce fertility and vigour of the individual. Supernumeray chromosomes may be regarded as **giant linear plasmids** (**Twyman** 1998).

32.6. HISTONE TAILS

The formation of the 30 nm fiber (see section 32.3) probably involves the highly flexible tails of the core histones, which extend from the nucleosome. It is thought that these tails may help attach one nucleosome to another—thereby allowing a string of them, with the aid of histone H1 to condense into the 30 nm fiber (Fig. 32.18). This model has been supported by X-ray diffraction analyses and evidence that the histone tails interact with DNA.

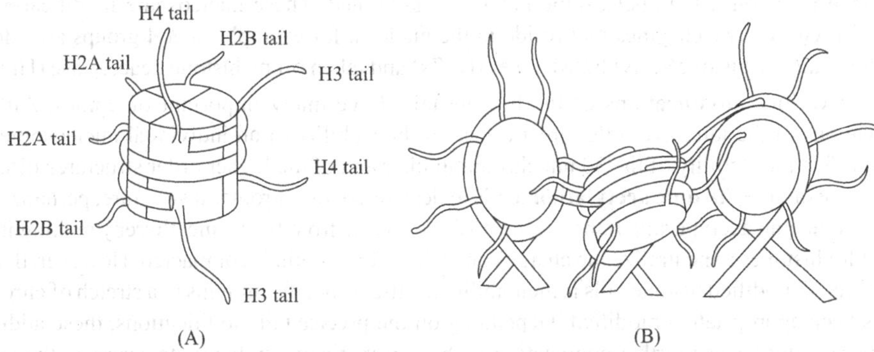

Fig. 32.18. A speculative model for histone tails in the formation of the 30 nm fiber. A—The approximate exit points of the eight histone tails (one from each histone subunit) from histone octamer. B—A speculative model showing how the histone tails may help to pack nucleosomes together (after Alberts *et al.*, 2002).

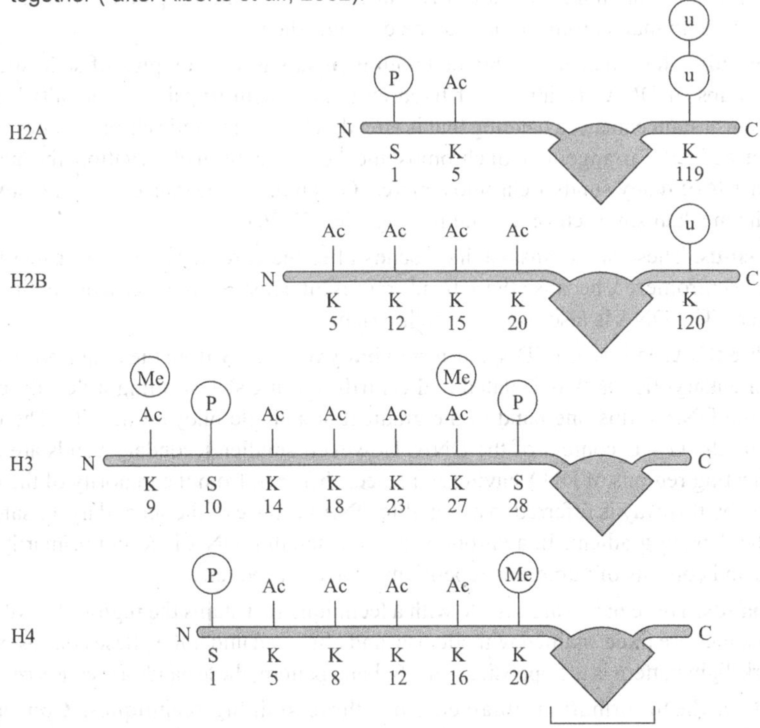

Fig. 32.19. Covalent modification of core histone tails. Known modifications of the four histone core proteins are indicated: K = lysine, S = serine, Me = methyl group, Ac = acetyl group, P = phosphate, u = ubiquitin. Note that some positions (*e.g.*, lysine of H3) can be modified in more than one way. Histone H2B can be modified by a single ubiquitin molecule; H2A can be modified by the addition of several ubiquitins (after Alberts *et al.*, 2002).

Roles of histone tails in gene regulation. The N-terminal tails of each of the four core histones (Fig. 32.18) are highly conserved in their sequence, and perform crucial functions in regulating chromatin structure. Each tail is subject to several types of covalent modifications, including acetylation of lysines, methylation of lysines, and phosphorylation of serines (Fig. 32.19). Histones are synthesized in the cytosol and then assembled into nucleosomes. Some of the modifications of histone tails occur just after their synthesis, but before their assembly. For us those modifications are significant however, which take place once the nucleosome has been assembled. These nucleosome modifications are added and removed by enzymes that reside in the nucleus; for example, acetyl groups are added to the histone tails by histone acetyl transferases (HATs) and taken off by histone deacetylase (HDACs).

The various modifications of the histone tails have many important outcomes. Although modifications of the tail have little direct effect on the stability of an individual nucleosome, they seem to affect the stability of the 30 nm chromatin fiber and of the higher order structures discussed below. For example, **histone acetylation** tends to destabilize the chromatin structure, perhaps in part because adding an acetyl group removes the positive charge from the lysine, thereby making it more difficult for histones to neutralize the charges on DNA as chromatin is compacted. However, the most acute effect of modified histone tails is their ability to attract specific proteins to a stretch of chromatin that has been appropriately modified. Depending on the precise tail modifications, these additional proteins can either cause further compaction of the chromatin or can facilitate access to the DNA.

32.7. CHROMOSOME BANDING

Various chromosomal staining techniques reveal consistent banding patterns. By means of these patterns, all of the human chromosomes can be distinguished.

1. G-bands. These bands are obtained with **giemsa stain**, a complex of stain specific for the phosphate groups of DNA. Treatment of fixed chromatin with trypsin or hot salts brings out the G-bands. Giemsa stain enhances banding that is already visible in mitotic chromosomes. The banding pattern is caused by the arrangement of chromosomes. Under careful observation, the major G-bands prove to consists of many smaller chromosomeres. G-banding appearance bas led **Comings** (1978) to suggest the mechanism of chromosome folding (Fig. 32.20).

2. C-bands. These are giemsa-stained bands after the chromosomes are treated with NaOH. The C is for "centromere", because these bands represent constitutive heterochromatin surrounding the centromers. The DNA is also usually satellite rich.

Satellite DNA. This type of DNA differs in **buoyan density** from the major portion of cellular DNA. When eukaryotic DNA is isolated and centrifused in CsCl, forming a density gradient, the majority of the DNA forms one band in the gradient at a single buoyant density. The buoyancy is determined by the G – C content of the DNA. However, smaller secondary bands are also usually present, indicating regions of DNA having sequences different from the majority of the cell's DNA. DNA isolated by this way is referred to as satellite DNA because of the secondary, or satellite, bands formed in the density gradient. In a chromosome, the satellite DNA is found primarily around the centromeres and consists of numerous repetitions of a short sequence.

3. R-bands. These bands are visible with a technique that stains the regions between G-bands. The chromosomes are fixed, stained with giemsa, and observed under the phase contrast microscope. Since the dark-light pattern is the opposite of the G-band pattern, these bands are called **reverse bands.**

Based on the information obtained from these staining techniques, **Comings** (1978), distinguished between three basic chromatin types: euchromatin, constitutive heterochromatin and intercalary heterochromatin (Table 32.3). The only chromatin involved in transcription is **euchromatin. Constitutive heterochromatin** surrounds the centromere and is rich in satellite DNA. **Intercalary heterochromatin** is found throughout the chromosome. Thus, it becomes apparent that the eukaryotic chromosome is a relatively complex structure.

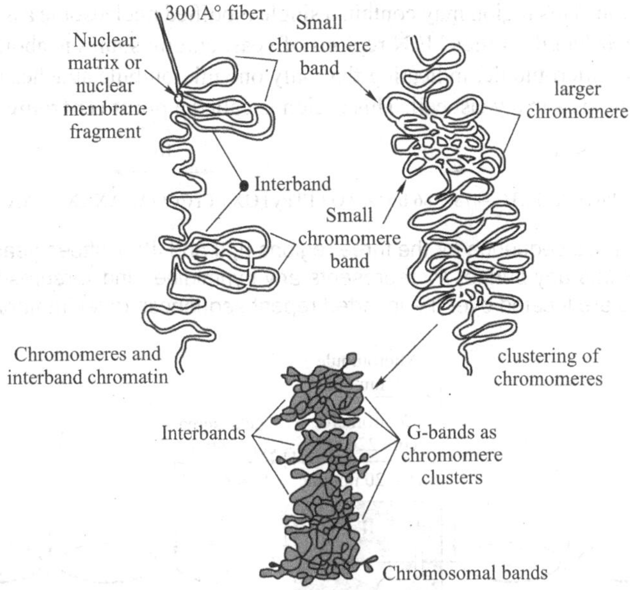

Fig. 32.20. Model of eukaryotic (mammalian) chromosomal banding. G—bands are chromomere clusters, which result from the contraction of smaller chromomeres. These, in turn, result from looping of the 300 A° fiber (after Tamarin, 2002).

Table 32.3 The three major types of chromatin in eukaryotic chromosome (Source: Tamarin, 2002).

	Feature	Euchromatin	Centromeric constitutive heterocromatin	Intercalary heterochromatin
1.	Relation to bands	In R-bands	In C-bands	In G-bands
2.	Location	Chromosome arms	Usually centromeric	Chromosome arms
3.	Condition during interphase	Usually dispersed	Condensed	Condensed
4.	Genetic activity	Usually active	Inactive	Probably inactive
5.	Relation to chromomeres	Interchromomeric	Centromeric chromomere	Intercalary chromomeres

32.8. DNA OF CENTROMERE

The centromere is a specific region of eukryotic chromosome. It is involved in movement of the chromosome during mitosis and meiosis.

Most of the genetical and molecular biological information of centromere has been obtained by studies in the fungus, yeast (*Saccharomyces cervisiae*) and the plant, *Arabidopsis thaliana*. Cells did not maintain most artificially created yeast plasmids because they were lost during mitosis. However, some plasmids were isolated that did replicate normally during cell division. Presumably, they contained centromeres, allowing them to replicate and move in synchrony with the host's chromosomes. Genetic engineering also made it possible to isolate smaller and smaller regions that could serve as centromeres. For example, after sequencing the centromeres of fifteen or the sixteen yeast chromosomes, it was possible to conclude that the centromere from yeast is about 250 base pairs long with three consensus regions (**Clarke** and **Carbon**, 1985; Fig. 32.21). In the light of current molecular biological data, the centromere can be defined as a sequence of DNA, called the *C E N* *locus* or C E N region. This region may contain a single modified nucleosome associated with region II. The 250 base-pair length of the C E N regions of yeast chromosomes is about 200 A°, the same as the diameter of a microtubule, indicating that only one microtubule attaches to each centromere during mitosis or meiosis in a yeast cell. This region is called a **point centromere** (Fig. 32.22).

I II III

PuTCACPuTG – 78 – 86 bp — TGTTTPyTGXTTTCCGAAAXXXXAAA

Fig. 32.21. Consensus sequence for the three regions (I, II and III) of fifteen yeast centromeres. *Pu* represents any purine, *Py* represents any pyrimidine, and *X* represents any base. The arrows are located over the inverted repeat sequences (after Tamarin, 2002).

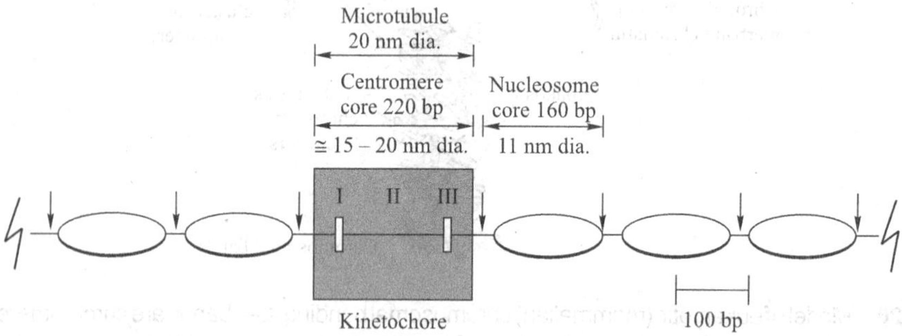

Fig. 32.22. Schematic view of a yeast centromeric region. The arrows indicate nuclease-hypersensitive sites. A microtubule is about the same width as the centromeric region (after Tamarin, 2002).

Higher eukaryotes have large centromeric regions each of which attaches more microtubules, *e.g.*, 4 to 7 in rat fetus and 70 to 150 in Plant *Haemanthus*. These regions are referred to as **regional centromeres**. Regional centromeres range from 19 to 100 kb, with unique and satellite (repeated sequence). DNA is heterochromatic and may include expressed genes.

Arabidopsis centromeres span 0.9 – 1.2 Mb of DNA and each one is made up largely of 180 bp repeat sequences. In humans, the equivalent sequences are 171 bp and are called **alphoid DNA**. DNA of centromere of *Arabidopsis* contains multiple copies of genome-wide repeats, along with a few genes, the latter at a density of 7 – 9 per 100 kb compared with 25 genes per 100 bp for the noncentromeric regions of *Arabidopsis* chromosomes (**Copenhaver** *et al.*, 1999). The discovery that centromere DNA contains genes was a big surprise because it was thought that these regions were genetically inactive.

The **special centromeric proteins** in humans include at least seven that are not found else where in the chromosome (**Warburton**, 2001). One of these proteins, **CENP-A** is very similar to histone H3 and is thought to replace this histone in the centromeric nucleosomes. Part of the human kinetochore is made up of alphoid DNA plus CENP-A and other proteins, but its structure has not been decribed in detail (**Vafa** and **Sullivan**, 1997).

32.9. DNA OF TELOMERES

It is already known that eukaryotic chromosomes are linear, each has two ends, referred to as **telomeres**. The telomeres not only mark the termination of the linear chromosome but also have several specific functions. For example, telomeres must prevent the chromosomal ends from acting in a "sticky" fashion the way the broken chromosomal ends act. In other words, chromosomal ends must also prevent the ends of chromosomes from being degraded by **exonucleases** and must allow chromosomal ends to be properly replicated.

Most telomeres isolated so far are repetitions of sequences of five to eight bases. In human beings, the telomeric sequence is T T A G G G, repeated 300 to 5000 times at the end of each chromosome. The human telomere was discovered by **Robert K. Moyzis** and his colleagues (1988) when they probed the *highly repetitive segment of human DNA*. When a probe for this sequence was applied to human chromosomes, the sequence was found at the tip of each chromosome in roughly the same quantity. This is a highly conserved sequence, found in all vertebrates studied as well as in unicellular trypanosomes. Similar sequences are found in various other eukaryotes (Table 32.4); the first sequence was isolated by **E. Blackburn** and **J. Gall** in 1978.

Table 32.4 Telomeric sequences in eukaryotes; the G-rich strand of the double helix is shown (Source: Tamarin, 2002).

	Organism	Telomeric repeat
1.	Human beings, other mammals, birds and reptiles	T T A G G G
2.	Trypanosomes	T T A G G G
3.	*Tetrahymena* (holotrichous ciliates)	G G G G T T
4.	*Stylonychia* (hypotrichous ciliates)	G G G G T T T T
5.	Yeast	G T, G G T and G G G T
6.	Plants	T T T A G G G

When a linear DNA molecule is replicated, the 3′ – 5′ strand can be replicated up to the end. The 5′ – 3′ strand, however, is replicated with DNA primers that are then degraded, leaving a short gap on the progeny strand (Fig. 32.23). It is always the G-rich strand of telomeric DNA that ends up single stranded, forming a 3′ overhang of twelve to sixteen nucleotides. Thus, *the normal replication process of a linear DNA molecule leaves an incomplete terminus*. Hence, molecular biologists suspected that there would be a unique mechanism for the replication of telomeres.

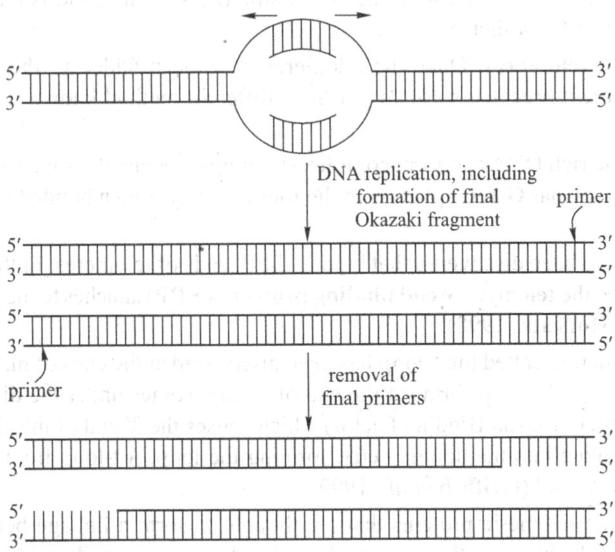

Fig. 32.23. Removal of final primers after the replication of linear DNA creates single stranded ends (after Tamarin, 2002).

The Telomerase Enzyme

Telomeric sequences appear to be added *de novo* without DNA template assistance by an enzyme called **telomerase**. This enzyme was discovered by **Blackburn** and her colleagues (1990). This was seen when telomeres from another species were engineered into yeast cells. After a cell cycle, the yeast telomeric sequence had been added on at the ends of the foreign chromosome, the result presumably of the telomerase enzyme.

Further, when **Blackburn** and her colleagues (1992) isolated telomerase, they discovered that a segment of RNA, about 160 nucleotides in length is an integral part of the enzyme. Telomerase RNA has a region that is complementary of the G-rich repeat of the telomeric DNA sequence of the species. After careful experimentation, including the modifying the gene for the telomerase RNA, **Blackburn** and her colleagues conducted that telomerase enzyme uses its RNA as a template for adding telomeric repeats to the ends of chromosomes. Telomerase enzyme is thus a reverse **transcriptase**, using RNA nucleotides as a template to polymerize DNA nucleotides.

Box 32.5

Telomerase is an unusual enzyme in that it consists of both protein and RNA. In the human enzyme the RNA component is 450 nucleotides in length and contains near its 5′ end the sequence 5′–C U A A C C C U A A C–3′, whose central region is the reverse complement of the human telomere repeat sequence 5′–T T A G G G–3′ (**Feng** *et al.*, 1995). This enables telomerase to extend the telomeric DNA at the 3′ end of a polynucleotide by the copying mechanism, in which the telomerase RNA is used as a template for each extension step, the DNA synthesis being carried out by the protein component of enzyme which is a reverse transcriptase (**Lingner** *et al.*, 1997).

Mode of working of telomerase. The first step in, telomerase extension is **hybridization** of the 3′ end of the telomere with the RNA component of telomerase (Fig. 32.24A). Then, with the telomerase RNA as a template, the 3′ end of the telomere is extended (Fig. 32.24B). Finally, a translocation step takes place that displaces the telomere in respect to the RNA, returning to the configuration at the beginning of the process (Fig. 32.24C). The single stranded C-rich strand is then synthesized with **DNA polymerase** and **DNA ligase**.

Protection of telomeres. Once the telomeres have been added to the ends of eukaryotic chromosomes, different organisms use any of *three* different method known to protect the ends of the chromosomes.

1. The guanine rich DNA can form complex structures. Biochemists have discovered that four guanines can form a planar **G-tetraplex**, with the four bases hydrogen bonded to each other (**Wang** and **Patel**, 1993).

2. Proteins have been discovered that bind to the 3′ ends of telomeres. In the ciliate, *Oxytricha nova*, a protein called the **telomerase end binding protein (TEBP)** attaches to the 3′ ends of telomeres and protects them (**Horvath**, 1998).

3. A novel structure called the **t-loop** has been discovered at the ends of mammalian telomeres (**Griffith** *et al.*, 1999). The loop forms at the end of chromosomes under the direction of a protein called **TRF2** (Telomere Repeat Binding Factor) which causes the 3′ end of the chromosome to loop around and interdigitate into the double helix, forming the loop. A t-loop from mouse liver cell is about 10,000 bases around (**Griffith** *et al.*, 1999).

There are some telomeric proteins that are thought to form a linkage between the telomere and the periphery of the nucleus, the area in which the chromosome ends are localized (**Tham** and **Zakian**, 2000).

How do cells keep track of the number of their telomeric repeats? Proteins have been isolated that bind to telomeres (Rap1 in yeast *Saccharomyces cerevisiae*, **Marcand** *et al.*, 1997; TRFl in human beings. **Van Steensel** *et al.*, 1997). By mutating these proteins or the telomeric sequences, molecular biologists have changed the equilibrium number of telomeric repeats. This led to the current model that the cell counts the number of these proteins bound to the telomeres, not the number of telomeres directly, to know whether telomeres should be added.

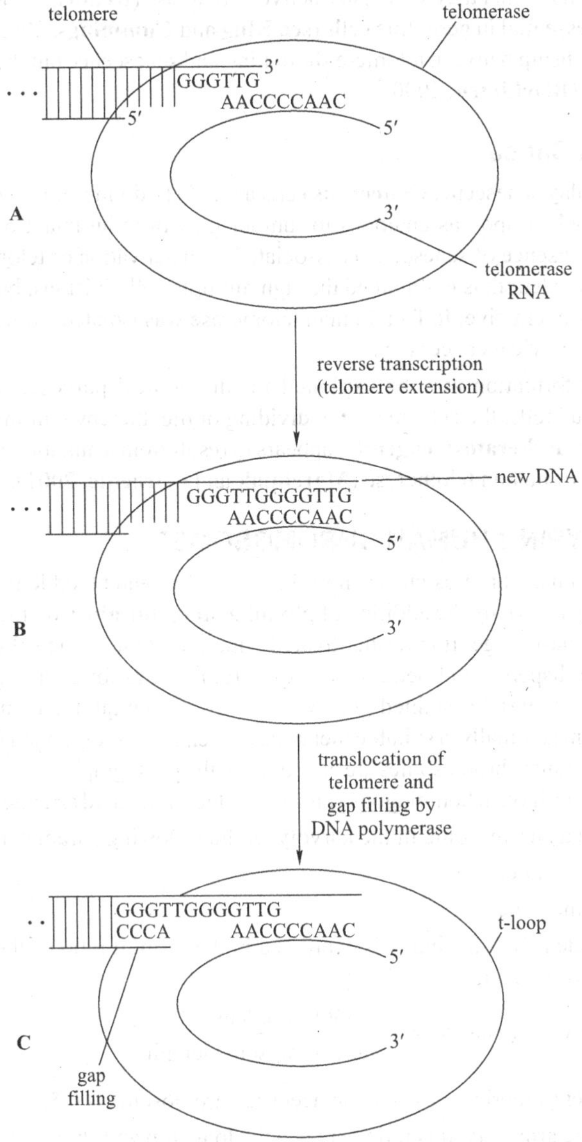

Fig. 32.24. Telomerase enzyme extends telomeres using telomerase RNA as a template. Gap filling by DNA polymerase and ligase complete the double helix (after Tamarin, 2002).

Telomerase and Aging

In yeast, protozoa and other single-celled organisms, telomerase enzyme is active, keeping the ends of the chromosomes at the appropriate length. These cells can divide potentially forever. In contrast, in higher organisms, telomere shortening leads to the termination of a cell lineage. For several years the biologists have attempted to link this process with **cell senescense**, a phenomenon originally observed in cell cultures. All normal cell cultures have a limited lifetime: after a certain number of divisions the cells enter a senescent state in which they remain alive but cannot divide. With some mammalian cell lines notably fibroblast cultures (connective tissue cells), senescence can be delayed by engineering the cells so that they synthesize active telomerase (**Reddel**, 1998). In mammals active telomerase is found essential in germ line cells (see **Klug** and **Cummings**, 2003). These experiments suggest a clear relationship between telomere shortening and senescence but the exactness of the link has been questioned (**Blackburn**, 2000).

Telomerase and Cancer

Not all cell lines display senescene. **Cancerous cells** are able to divide continuously in culture, their immortality being looked upon as analogus to tumour growth in an intact organism with several types of cancer. The absence of senescene is associated with activation of telomerase, sometimes to the extent that telomere length is maintained through multiple cell divisions, but often in such a way that the telomerase is overactive. In fact, human telomerase was isolated from an immortal cell line (HeLa) derived from cervical cancer cells.

This current information may be exploited for the clinical purpose: if, telomerase can be **deactivated** in tumour cells, the cells may stop dividing or die, thereby eliminating the cancer. One type of human cancer, **dyskeratosis cogenita**, appears to result from a mutation in the gene specifying the RNA component of human telomerase (**Marciniak** and **Curaente**, 2001).

32.10. KARYOTYPING HUMAN CHROMOSOMES

For karyotyping of human chromosomes venous blood is taken and blood leucocytes are stimulated to divide (by mitosis) *in vitro* by the addition of **phytohaemagglutinin**. Colchicine is added to arrest cell division at metaphase stage. It is further treated with hypotonic saline solution which results in swelling of cells and dispersal and better clarity of chromosomes for counting and morphological study. There after, the material is stained (*e.g.*, with Giemsa technique) to demonstrate the banding patterns of chromosomes. Finally, a suitable metaphase spread is photographed through a high power microscope. The individual chromosomes are cut out from the photograph. The chromosomes are then arranged in an orderly fashion in homologous pairs, to produce a standard arrangement, the **karyotype**.

To characterize a chromosome in the karyotype, the following parameters are used:

1. Shape of chromosome;
2. Length of chromosome;
3. Centromeric index, *i.e.*, this index is expressed in the form of ratio of the short arm length to the total chromosome length:

$$\text{Centromeric index} = \frac{\text{Short arm length}}{\text{Total chromosome length}}$$

   For example, centromeric index in a metacentric chromosome is 0.5.

4. Proportion of the arms, *i.e.*, it is ratio between the long arm and short arm of the chromosome. This ratio is 1:1 in a typical metacentric chromosome.

Classification. The human metaphase chromosomes were first of all classified by a conference of cytogeneticists at Denver, Colorado in 1960 and is known as the **Denver classification**. To follow this classification, each of the 22 pairs of autosomes has been numbered from 1 to 22 according to their decreasing size. **Patau** (1960) divided the human chromosomes into the following seven groups designated A to G:

1. A group : 1 to 3 pairs — Metacentric
2. B group : 4 to 5 pairs — Submetacentric
3. C group : 6 to 12 pairs — Submetacentric
4. D group : 13 to 15 pairs — Acrocentric
5. E group : 16 to 18 pairs — Submetacentric (16 is metacentric)
6. F group : 19 to 20 pairs — Metacentric
7. G group : 21 to 22 pairs — Acrocentric

Group A consists of longest metacentric chromosomes.

Group G consists of the shortest acrocentric chromosomes. These chromosomes have satellites that correspond to nucleolar organizers. Chromosomes of group D also contains satellites. In males, group G includes a variable Y chromosome which lacks the satellites.

The X chromosome is the member of group C and can be identified by special banding or staining methods.

QUESTIONS

Long Answer Questions

1. Describe the occurrence, ultrastructure and chemical composition of chromosome. What are its different types?
2. Describe the structure and function of chromosome.
3. Describe the structure, chemical composition and function of the eukaryotic chromosomes.
4. Describe the structure of a typical chromosome.
5. Describe the structure and physiology of chromosomes.
6. Give an account of the structure of chromosome. Distinguish between chromonema and chromatid. Write a short note on the different chromosomal shapes at anaphase.
7. Give a detailed account of chromatin.
8. Describe the Oudet and other nucleosome models of chromosome and explain as to how Oudet's model is considered better than Dupraw's folded fibre model? Describe as to how many structure groups human chromosomes belong?
9. Give the biochemical components of the chromatin and discuss the structural organization of chromatin emphasizing the nucleosome organization.
10. What are chromosomes? Describe the chemical composition and organization of metaphase chromosome.
11. Give an account of the chemical organization of chromosomes. What are nucleosomes?
12. What do you mean by giant chromosome? Describe the physico-chemical structure of polytene chromosome.
13. What is a giant chromosome? Describe the structure of a giant chromosome and its role in heredity.
14. Give a brief description of the significance of polytene chromosomes in understanding gene regulation.
15. Describe the structure of polytene chromosome and state their significance.
16. What are lampbrush chromosomes and Balbiani rings?
17. Discuss the structure and functions of lampbrush chromosomes.
18. State the difference between the lampbrush and the salivary gland chromosomes. Write the structural and functional significance of the salivary gland chromosomes.

Short Answer Questions

1. What is the role of basic proteins which are associated with the DNA molecule?

2. What are Feulgen positive regions?
3. Why secondary constriction is known as nucleolar organizer?
4. What are supernumerary chromosomes?
5. Describe the role of centromere.
6. What is the nucleolar organizer (NO) region in the chromosome? What is its role?
7. Describe unistranded folded fibre model of chromosome structure.
8. How protein molecules are associated with DNA molecule?
9. What is nucleosome concept?
10. Describe structure of a nucleosome.
11. What do you mean by telocentric and metacentric chromosome?
12. What do you mean by polyteny? Who introduced this concept?
13. Explain significance of puffing.
14. What do you mean by gene amplification? What is its need? Explain.
15. Define chromosome. Write the names of various types of special chromosomes.
16. Write short note on the following:
    (*i*) Chromosomes
    (*ii*) Chromosome shapes
    (*iii*) Types of chromosomes
    (*iv*) Centromere
    (*v*) Autosomes
    (*vi*) Barr bodies
    (*vii*) Nucleosome
    (*viii*) Heterochromatin and euchromatin
    (*ix*) Fine structure of chromosome
    (*x*) Polytene (or polytenic) chromosome
    (*xi*) Lampbrush chromosomes
    (*xii*) Balbiani chromosome
    (*xiii*) Histones
    (*xiv*) Alphoid DNA
17. Prepare a well labelled diagram of each of the following:
    (*i*) Lampbrush chromosomes
    (*ii*) Fine structure of chromosome
18. Differentiate between the following:
    (*i*) Viral and bacterial chromosomes
    (*ii*) Prokaryotic (bacterial) and eukaryotic chromosomes
    (*iii*) Primary constriction and secondary constriction
    (*iv*) Centrosome and nucleosome
    (*v*) Euchromatin and heterochromatin
    (*vi*) Lampbrush chromosome and salivary gland chromosome
    (*vii*) Heterochromatin and Barr bodies
    (*viii*) Centromere and chromocentres
    (*ix*) Histones and non-histones

Very Short Answer Questions

1. Name the scientists.
    (*i*) Who coined the term chromosome?
    (*ii*) Who described parallelism between chromosomes and transmission of hereditary factors?
2. Name the part of a chromosome where its two chromonemata are joined together.
3. Name the organism where RNA act as hereditary material.
4. What is nucleoid?
5. How many DNA double helical fibres are present in a chromosome?
6. 'Beads in the string' appearance of chromatin represents what?
7. What is the importance of telomere in chromosome?
8. Lampbrush chromosomes are found during which stage of cell division?

True/False Questions

1. Histone are basic proteins rich in lysine and arginine.

Fill in the Blanks

1. Polytene chromosomes are also known as ....
2. Number of chromosomes in human genome is ....

Multiple Choice Questions

1. Number of chromatids present in a metaphase chromosome is
   (*a*) four
   (*b*) three
   (*c*) two
   (*d*) one
2. Number of chromosomes in definite nucleus
   (*a*) haploid
   (*b*) diploid
   (*c*) triploid
   (*d*) polyploid
3. Diploid chromosome number in humans is
   (*a*) 42
   (*b*) 44
   (*c*) 46
   (*d*) 48
4. Two sister chromatids are attached with the
   (*a*) spindle fibre
   (*b*) chromatid
   (*c*) centromere
   (*d*) chromocentre
5. Autosomes in humans are
   (*a*) 11 pairs
   (*b*) 22 pairs
   (*c*) 23 pairs
   (*d*) 43 pairs
6. Eukaryotic chromosome is made of
   (*a*) DNA
   (*b*) DNA + lipid
   (*c*) DNA + proteins
   (*d*) RNA
7. The term 'nucleosome' was given by
   (*a*) Dupraw
   (*b*) Oudet
   (*c*) Flemming
   (*d*) Emil Heitz
8. The nucleosome
   (*a*) surrounds nuclear pores
   (*b*) has only DNA and non-histone proteins
   (*c*) is fully responsible for packing DNA into chromosomes
   (*d*) contains a core of histones with DNA wrapped around it
9. Balbiani rings occur in
   (*a*) polytene chromosomes
   (*b*) lampbrush chromosomes
   (*c*) polysomes
   (*d*) heterosomes
10. Chemical composition of chromomere is
   (*a*) DNA and proteins
   (*b*) DNA and lipids
   (*c*) DNA and carbohydrates
   (*d*) proteins and lipids
11. A chromosome in which the centromere is situated near one end is known as
   (*a*) metacentric
   (*b*) submetacentric
   (*c*) acrocentric
   (*d*) telocentric
12. The total number of genes present in a haploid cell is
   (*a*) genome
   (*b*) genotype
   (*c*) karyotype
   (*d*) gene pool
13. The centromere is found at the centre of the two arms in ....
   (*a*) telocentric
   (*b*) acrocentric
   (*c*) submetacentric
   (*d*) metacentric
14. Chromosomes can be stained with one of the following chemicals?
   (*a*) acetocarmine
   (*b*) safranine
   (*c*) light green
   (*d*) eosin

ANSWERS

Very Short Answer Questions

1. (*i*) Waldeyer.
   (*ii*) S.W. Sutton.
2. Centromere.
3. Some viruses, *e.g.*, TMV.
4. The nuclear regions in bacterial cells (= prokaryotic cells) where DNA molecule of bacterial chromosome lies in a coiled state. Bacterial cells lack a nucleus.
5. One in each chromosome.
6. The 10 nm chromatin fibre (*i.e.*, nucleosomes in close apposition).
7. The chromosomal ends are called **telomeres**. Each telomere provides polarity to the chromosome and prevents other chromosomal segments to fuse with it.
8. Diplotene of prophase I of meiosis.

True/False Questions

1. True.

Fill in the Blanks

1. Salivary gland chromosomes. 2. 23.

Multiple Choice Questions

1. (*c*)	2. (*b*)	3. (*c*)	4. (*c*)	5. (*b*)	6. (*c*)	7. (*b*)
8. (*d*)	9. (*a*)	10. (*a*)	11. (*c*)	12. (*a*)	13. (*d*)	14. (*a*)

33 Chromosomal Variations: 1. Morphological; 2. Numerical

The chromosomes of each species has a characteristic morphology (structure) and number. But, sometimes due to certain accidents or irregularities at the time of cell division, crossing over or fertilization some alterations in the morphology and number of chromosomes take place. The slightest variation in the organisation of chromosomes is manifested phenotypically and is of great genetical interest. The changes in the genome involving chromosome parts, whole chromosomes, or whole chromosome sets are called **chromosome aberrations**, **chromosome mutations** or **chromosomal variations**. Chromosome mutations have proved to be of great significance in applied biology—agriculture (including horticulture), animal husbandry and medicine.

Chromosome variations are inherited once they occur and are of the following types:

A. Structural changes in chromosomes:
   1. Changes in number of genes
      (*a*) Loss: deletion
      (*b*) Addition: Duplication
   2. Changes in gene arrangement:
      (*a*) Rotation of a group of genes 180° within one chromosome: inversion
      (*b*) Exchange of parts between chromosomes of different pairs: translocation.

B. Changes in number of chromosomes:
   1. Loss, or gain, of a part of the chromosome set (aneuploidy)
   2. Loss, or gain, of whole chromosome set (euploidy)
      (*a*) Loss of an entire set of chromosomes (haploidy).
      (*b*) Addition of one or more sets of chromosomes (polyploidy).

Both types of changes (structural and numerical) in chromosomes can be detected not only with a microscope (*i.e.*, cytologically) but also by standard genetic analysis. This gave birth to a hybrid science, called **cytogenetics** which attempts to correlate cellular events, especially those of chromosomes, with genetic phenomena.

33.1. STRUCTURAL CHANGES IN CHROMOSOMES

For better understanding of the abnormalities of chromosome structure, let us consider two important features of chromosome behaviour: (1) During prophase I of meiosis, homologous regions of chromosomes show a great affinity for pairing and they often go through considerable contortions in order to pair. This property results in many curious structures observed in cells containing one normal chromosome set plus an aberrant set. (2) structural changes usually involve chromosome breakage; the broken chromosome ends are highly "reactive" or "sticky", showing strong tendency to join with broken ends.

Types of Structural Changes in Chromosome

Structural changes in chromosome may be of the following types (Fig. 33.1) : 1. **deficiency** or **deletion** which involves loss of a broken part of a chromosome; 2. **duplication** involves addition of a part of chromosome (*i.e.*, broken segment becomes attached to a homolog which, thus, bears one block of genes in duplicate); 3. **inversion** in which broken segment reattached to original chromosome in reverse order, and 4. **translocation** in which the broken segment becomes attached to a non-homologous chromosome resulting in new linkage relations.

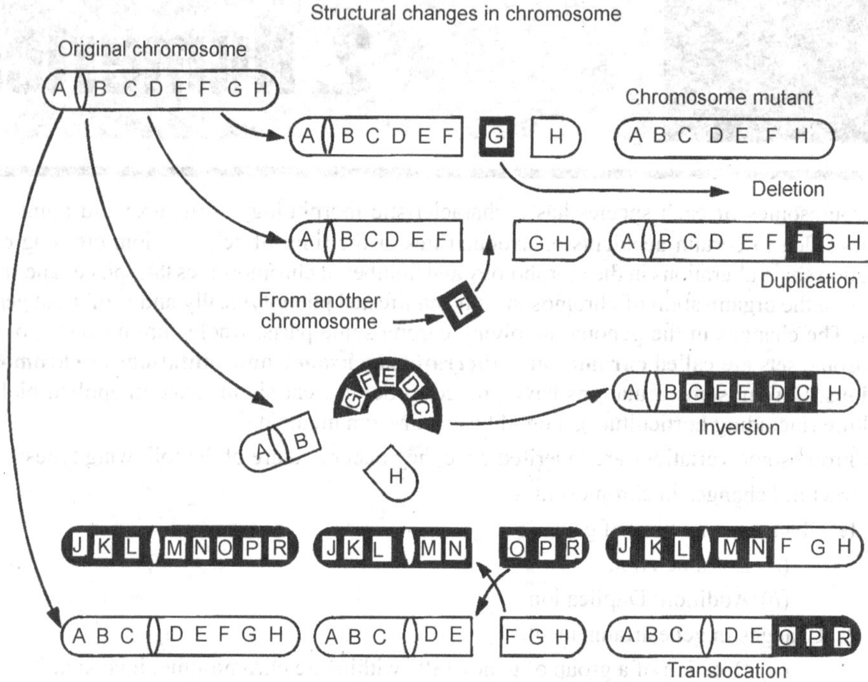

Fig. 33.1. Structural changes in chromosome (after Savage, 1969).

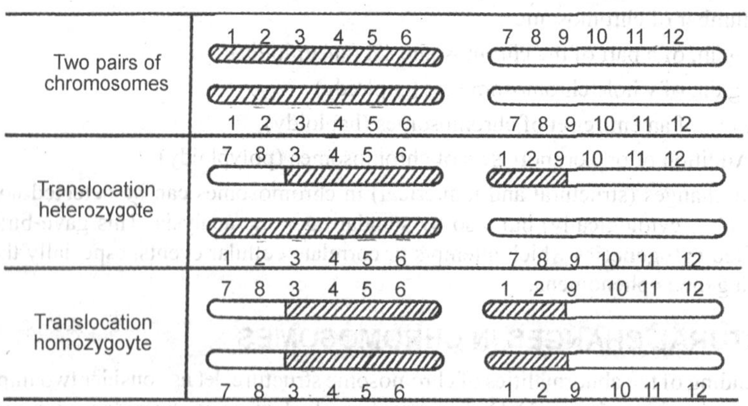

Fig. 33.2. Chromosome constitution of a translocation heterozygote and a translocation homozygote.

Further, structural abnormalities can occur in both homologous chromosomes of a pair or in only one of them. When both homologous chromosomes are involved, these are called **structural**

homozygotes, *e.g.*, deletion homozygote, duplication homozygote, etc. When only one homologous chromosome is involved, it is called **structural heterozygote**. In Fig. 33.2 have been shown translocation heterozygote and translocation homozygote.

1. Deletion (or Deficiency)

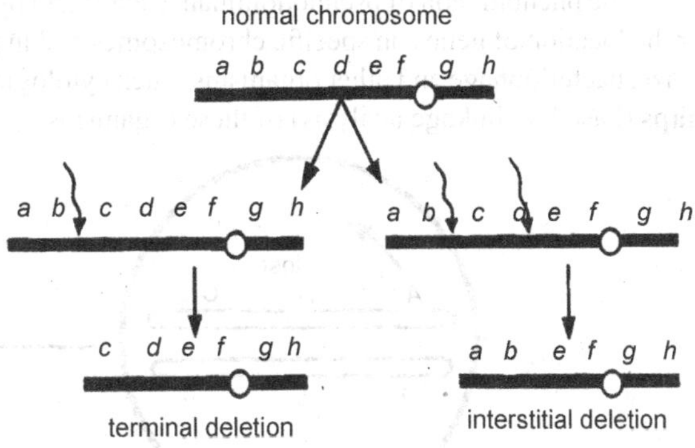

Fig. 33.3. Production of terminal and interstitial deletion. Chromosomes can be broken when struck by ionizing radiation (wavy arrows).

The simplest result of breakage is the loss of a part of a chromosome. Portions of chromosomes without a centromere (called **acentric fragments**) lag in anaphase movement and are lost from reorganizing nuclei or digested by nucleases. Such loss of a portion of a chromosome (and of some genes) is called **deletion.** The chromosomes with deletions can never revert to a normal condition. If gametes arise from the cells having a deleted chromosome, this deletion is transmitted to the next generation. Further, a deletion can be terminal or intercalary (interstitial). In **terminal deletion** a terminal section of a chromosome is absent and it is resulted by only one break (Fig. 33.3). While in the **intercalary deletion**, an intermediate section or portion of chromosome is lost and it is caused by two breaks—one on either end of the deleted region (Fig. 33.3). Thus, in the latter case, the chromosome is broken into three pieces, the middle one of which is lost and the remaining two pieces get jointed agan.

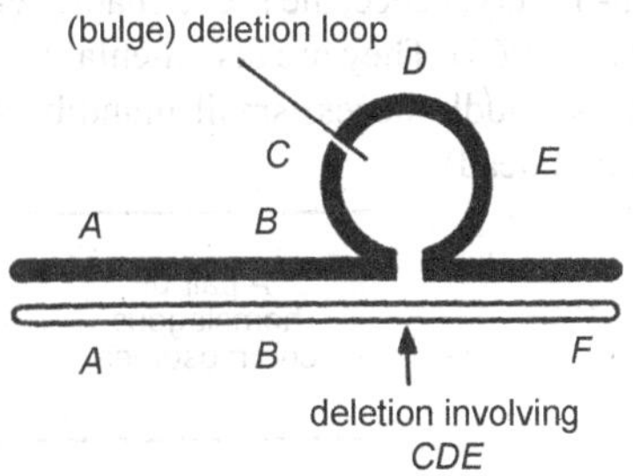

Fig. 33.4. Formation of a deletion loop during synapsis in a deletion heterozygote (after Stansfield, 1986).

Experimental proof for deletion was obtained by **Bridges** (1916–1919) who studied the inheritance pattern of sex-linked lethal characters which have arisen spontaneously in a population of *Drosophila.*

Stadler had produced breaks in the chromosomes of the plants by the help of ionizing radiation such as X-rays.

In general, if a homozygous deletion is made, it is lethal. Even individuals heterozygous for deletion (deletion in one of the homologous chromosomes) may not survive. However, smaller deletion in heterozygous condition can be tolerated by the organisms. If meiotic chromosomes in such heterozygotes are examined, the region of deletion can be detected by the failure of the corresponding segment on the normal chromosome to pair properly; so a "**deletion-loop**" results. Deletion loops are also detected in polytene chromosomes of *Drosophila*, where the homologs exist in permanent state of pairing (Fig. 33.4). The cytological studies of pairing between normal and deleted chromosomes have helped a lot in finding out the relative positions of genes in chromosomes.

Genetical effects of deletion. Deletion of some chromosome regions produce their own unique phenotypes. A good example of this is a dominant **notch-wing** mutation in *Drosophila*. In fact, this is a small deletion and acts as a recessive lethal in this regard. Further, in the presence of a deletion, a recessive allele of the normal homologous chromosome will behave like a dominant allele, *i.e.*, it will be phenotypically expressed, this phenomenon is called **pseudodominance** (Fig. 33.5).

The phenomenon of pseudodominance exhibited by deficiency heterozygotes has been utilized tor the location of genes on specific chromosomes and in preparing **cytological maps** in *Drosophila*, maize, bacteriophage and other organisms. Such cytological maps are often used to verify the **genetic maps** (based on linkage analysis) of these organisms.

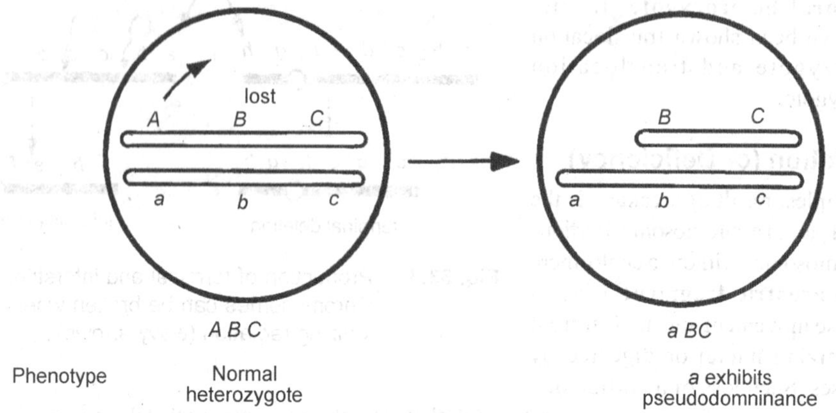

Fig. 33.5. Pseudodominance. A deficiency in the segment of chromosome bearing the dominant gene A allows the recessive allele a to become phenotypically expressed (after Stansfield, 1986).

Examples of pseudodominance (deletion).

1. Human babies missing a portion of the short arm of chromosome 5 (autosome) have a distinctive cat-like cry; hence, the French name "**cri du chat**" (cry of the cat) **syndrome** (first described by **Lejeune** *et al.*, 1963). They are also mentally retarded (IQ below 20), have malformation in the larynx, moon faces, saddle noses, small mandibles (micrognathia), malformed low-set ears and microcephally (small head).

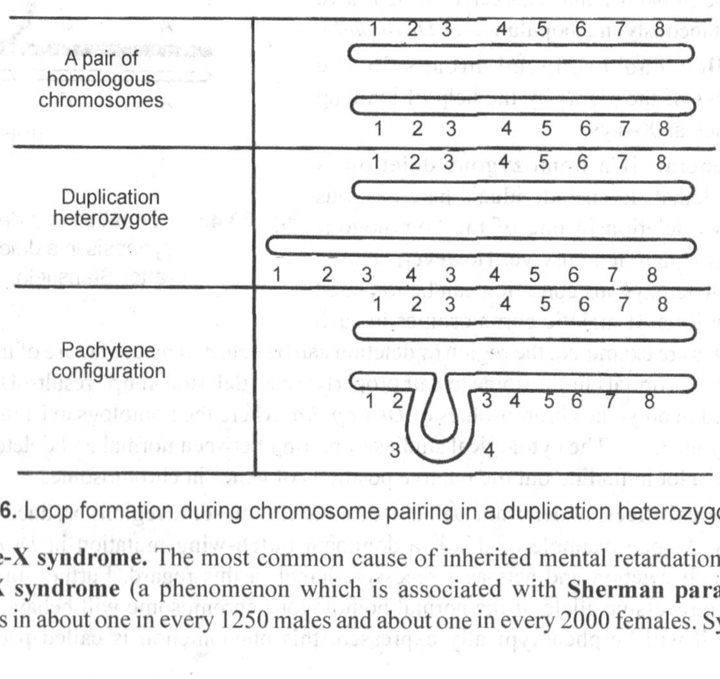

Fig. 33.6. Loop formation during chromosome pairing in a duplication heterozygote.

2. **Fragile-X syndrome.** The most common cause of inherited mental retardation in humans is the **fragile-X syndrome** (a phenomenon which is associated with **Sherman paradox**). This syndrome occurs in about one in every 1250 males and about one in every 2000 females. Symptoms of

fragile-X syndrome include mental retardation, altered speech patterns, and other physical attributes. The condition is called the fragile-X syndrome because it is related to a region at the X chromosome tip that break more frequently than other chromosomal regions. However, the break is not required for the syndrome to occur, and the fragile X-chromosome is identified by *the lack of chromosome condensation* at the site. In fact, under the microscope, it appears that the tip of the X chromosome is being held in place by a thread (**Craig** 1991). The gene responsible for the syndrome is called *FMR-1*, for fragile-X mental retardation. At the molecular level, *FMR-1* genes have between 230 to 2000 copies of a three-nucleotide repeat, CCG. Repeat number above 230 results in the inactive gene (*i.e.*, FMR-1) (**Tamarin** 2002).

2. Duplication

The presence of a part of a chromosome in excess of the normal complement is known as **duplication**. Thus, due to duplication some genes are present in a cell in more than two doses. If duplication is present only on one of two homologous chromosomes, at meiosis the chromosome bearing the duplicated segment forms a **loop** to maximize the juxtaposition (during pairing) of homologous regions (Fig. 33.7).

Extra segments in a chromosome may arise in a variety of ways such as follows:

1. Tandem duplication. In this case the duplicated region is situated just by the side of the normal corresponding section of the chromosome and the sequences of genes are the same in normal and duplicated region. For example, if the sequences of genes in a chromosome is ABC. DEFGH (The full stop depicts the centromere) and if the chromosomal segment containing the genes DEF is duplicated, the sequence of genes in tandem duplication will be ABC. DEFDEFGH.

2. Reverse tandem duplication. Here, the sequence of genes in the duplicated region of a chromosome is just the reverse of a normal sequence. In the above mentioned example, therefore, the sequence of genes due to reverse of a normal sequence. In the above mentioned example, therefore, the sequence of genes due to reverse tandem dupliation will be ABC. DEFFEDGH.

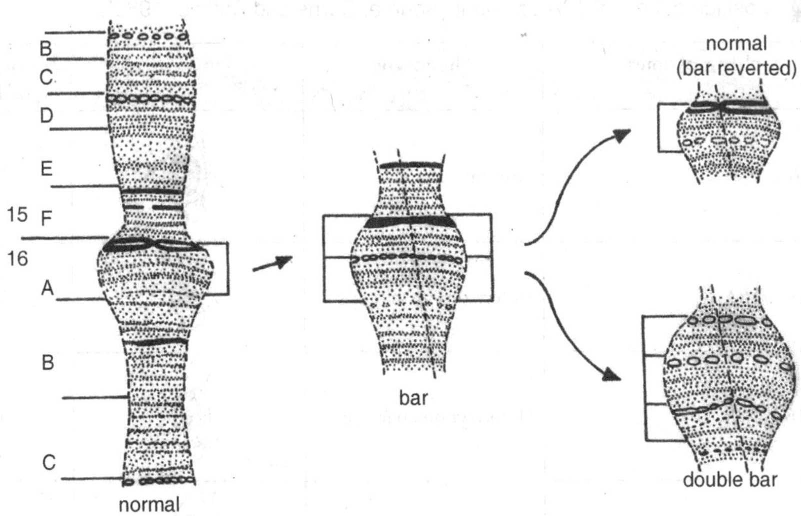

Fig. 33.7. Production of double-bar (or ultrabar) and bar-revertant (normal) chromosomes by asymmetric pairing and recombination in duplication homozygote. A—Diagrammatic representation of formation of unequal cross-over; B—Cytological representation (after Suzuki, *et al.*, 1986).

3. Displaced duplication. In this case the duplicated region is not situated adjacent to the normal section. Depending on whether the duplicated portion is on the same side of the centromere as the original section or on the other side, the displaced duplication can be termed either **homobranchial** or **heterobranchial.**

Example. Homobranchial duplication = ABC. DEFG DEFH

Heterobranchial duplication = A DEFB C. DEFGH

4. Transposed duplication. Here, the duplicated portion of chromosome becomes attached to a non-homologous chromosome. For example, if ABC.DEFGH and LMNOPQ. RST represent the gene sequences of two nonhomologous chromosomes, a transposed duplication will result into chromosomes with gene sequence ABC.GH and LMNDEF OPQ. RST. Such a transposed duplication may be either **interstitial** (*e.g.*, LMN DEF OPQ. RST) or **terminal** (*i.e.*, LMN OPQ. RSTDEF).

5. Extra-chromosomal duplication. In the presence of centromere the duplicated part of a chromosome act as independent chromosome.

Genetical effects of duplication. Due to duplication, there occur unequal crossing over which results in deletion and reduplication which produce distinct phenotypes as shown by the following examples:

1. **Bar eye in Drosophila.** The Bar phenotype of *Drosophila* is characterized by narrower, oblong, bar-shaped eye with few facets. It is determined by a X-linked recessive allele B. The classical studies of **Bridges** (1936) showed that the bar trait of *Drosophila* is associated with the duplication of a segment of the X-chromosome, called **section 16A**, as observed in salivary gland chromosomes (Fig. 33.7). Each added section of 16A intensifies the bar phenotype (*i.e.*, duplication behaves genetically as a dominant factor). However, the narrowing effect is greater if the duplicated segments are on the same chromosome (called **position effect**) (Table 33.1). Thus, cis and trans arrangements of the same number of 16A segments give different phenotypes (compare heterozygous ultrabar and homozygous bar eyes).

Table 33.1 Comparison of genotypes and phenotypes for bar eye in *Drosophila* females showing position effects of 16A segment (Source: Burns and Bottino, 1989).

	X chromosomes	Phenotype		Mean number of facets
1.	16A/16A	Normal		779
2.	16A, 16A/16A	Heterozygous bar eye		358
3.	16A, 16A/16A, 16A	Homozygous bar eye		68
4.	16A, 16A, 16A/16A	Heterozygous ultrabar (=double bar)		45
5.	16A, 16A, 16A/16A, 16A, 16A	Homozygous ultrabar		25

Some of the other well known duplications of *Drosophila* lead to following phenotypic effects: (*i*) a reverse repeat in chromosome 4 causes *eyeless* dominant (Ey); (*ii*) a tandem duplication in chromosome 3 causes *confluens* (Co) resulting in thickened veins, and (*iii*) another duplication causes *hairy wing* (Hw).

2. In humans, unequal crossing over between homologous chromosomes bearing σ (sigma) and β (beta) genes for σ and β subunits of adult haemoglobin (HbA), results in deletions and duplications of these genes. Deletions result in **Lepore** and **Kenya** variants of adult haemoglobin (HbA), both causing anaemia (*i.e.*, one type of thalassemia), while duplication result in Anti-Lepore and Anti-Kenya variants of haemoglobin A (see **Suzuki** *et al.*, 1986).

3. Inversion

Inversion involves a rotation of a part of a chromosome or a set of genes by 180° on its own axis. It essentially involves occurrence of *breakage* and *reunion.* The net result of inversion is neither a gain nor a loss in the genetic material but simply a rearrangement of the gene sequence. An inversion can occur in the following way: suppose that the normal order of segments within a chromosome is 1–2–3–4–5–6; breaks occur in regions 2–3 and 5–6 and broken piece is reinserted in reverse order. This results in an inverted chromosome having segments 1–2–5–4–3–6 (Fig. 33.8).

An **inversion heterozygote** has one chromosome in the inverted order and its homologue in the normal order. The location of the inverted segment can be detected cytologically in the meiotic nuclei of such heterozygotes by the presence of an **inversion loop** in the paired homologs. The location of the centromere relative to inverted segment determines the genetic behaviour of the chromosomes. If the centromere is not included in the inversion it is called **paracentric inversion** and when inversion includes the centromere it is called **pericentric inversion** (Fig. 33.9). Homologous chromosomes, with identical inversions in each member, pair and undergo normal distribution in meiosis. However, crossing over in inversion heterozygotes produce deletions, duplications and other curious configurations.

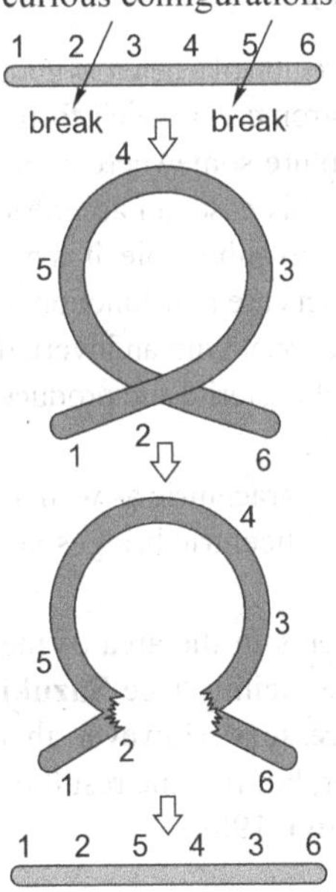

Fig. 33.8. Origin of an inversion in a chromosome.

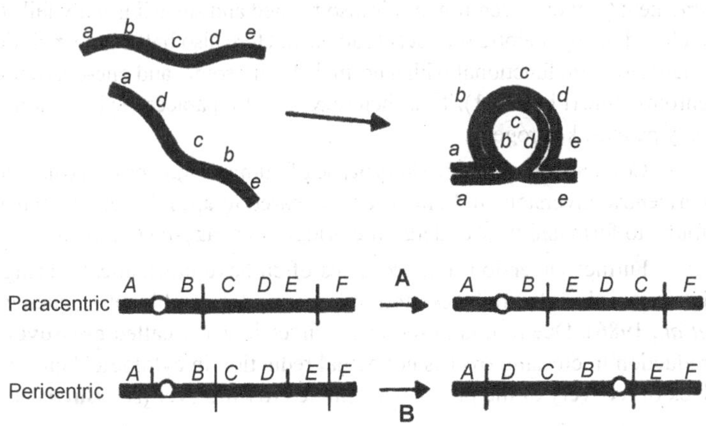

Fig. 33.9. A—An inversion loop in paired homologs of an inversion heterozygote; B—The location of the centromere relative to the inverted segment.

A. Crossing over in pericentric inversion. Crossing over in a heterozygous pericentric inversion result in deletions and duplications and also produce rod-shaped (acrocentric) chromosomes. The first meiotic anaphase figures appear normal, but the two chromatids of each chromosome usually have arms of unequal length depending upon where the crossing over occurred (Fig. 33.10). Half of the meiotic products (gametes/pollen grains) are non-functional and inviable due to the presence of duplications and deletions in them. The other half of the gametes are functional: one-quarter have the normal chromosome order, one-quarter have the inverted arrangement.

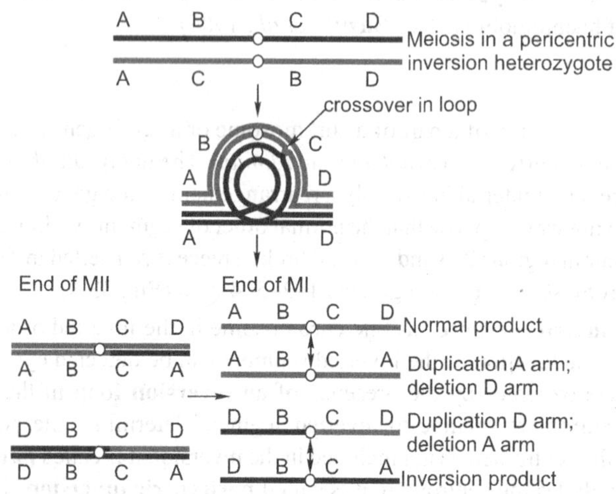

Fig. 33.10. Meiotic products resulting from a single crossover within a heterozygous pericentric inversion (MI = Meiosis I; MII = Meiosis II) (after Suzuki *et al.*, 1986)

B. Crossing over in paracentric inversion. A crossing over in the inverted region of a heterozygous paracentric inversion produces a **dicentric chromosome** (possessing two centromeres) which forms a bridge from one pole to the other during first anaphase. The bridge will rupture somewhere along its length and resulting fragments will contain duplication and/or deletion. In this case, an accentric fragment (without a centromere) is also formed and since it usually fails to move to either pole, it is not included in any meiotic products (gametes). Here also half of the meiotic products are non-functional, one-quarter are functional with a normal chromosome, and one-quarter are functional with an inverted chromosome (Fig. 33.11). Thus, heterozygotes for paracentric inversions are highly sterile and produce only parent-like progeny.

Comparison of banded karyotype of humans and apes reveals numerous paracentric as well as pericentric inversions in humans as compared to apes. In female maize plants dicentric bridges are found to form and they undergo the bridge-breakage-fusion-cycle.

Further, inversion heterozygotes often have mechanical pairing problems in the area of the inversion; this also reduces crossing over and recombinant frequency in the vicinity (see **Suzuki** *et al.*, 1986). Due to this observation, inversions are called **crossover-suppressors.** However, this reduction in crossing over is not actual reduction in cytological crossing over, but it is the result of lack of recovery of the products of single crossing over (see **Burns** and **Bottino**, 1989).

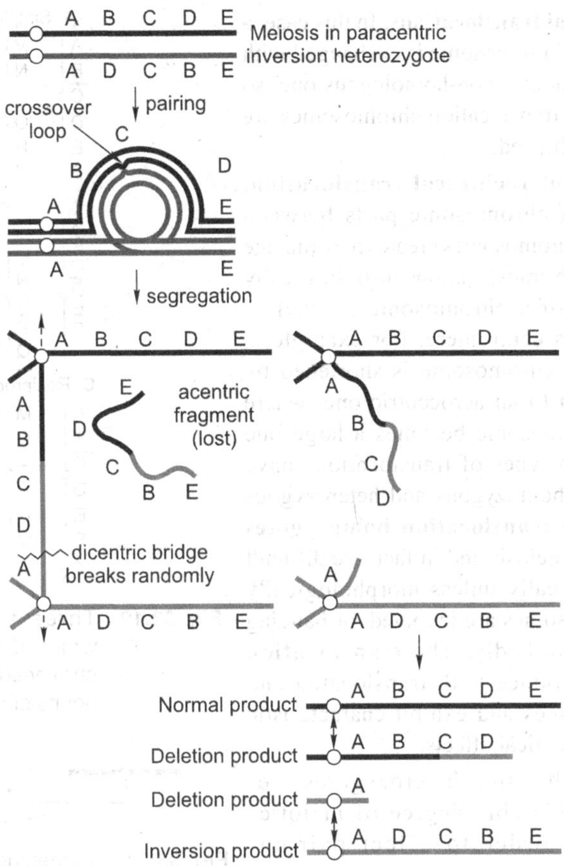

Fig. 33.11. Meiotic products resulting from a single crossing over within a heterozygous paracentric inversion loop (after Suzuki *et al.*, 1986).

Advantages of inversions. Fertility of inversion homozygotes and sterility of inversion heterozygotes lead to establishment of two groups (or varieties) which are mutually fertile but do not breed well with the rest of the species. Both varieties evolve in different directions and later become reproductively isolated species. There is plenty of cytological evidene to prove that such evolutionary mechanisms have and are operating in *Drosophila* and a number of other organisms.

4. Translocation

The shifting or transfer of a part of a chromosome or a set of genes to a non-homologous one, is called **translocation**. There is no addition or loss of genes during translocations, only a rearrangement (*i.e.*, change in the sequence and position of a gene). Translocations may be of following three types (Fig. 33.12):

1. Simple translocations. They involve a single break in a chromosome. The broken piece gets attached to one end of a non-homologous chromosome.

2. Shift translocations. In this type of translocation, the broken segment of one chromosome get inserted interstitially in a non-homologous chromosome.

3. Reciprocal translocations. In this case, a segment from one chromosome is exchanged with a segment from another non-homologous one, so that in reality two translocation chromosomes are simultaneously achieved.

Outcomes of reciprocal translocation. The exchange of chromosome parts between nonhomologous chromosomes creates new linkage relationships. Such translocations also drastically change the size of a chromosome as well as the position of its centromere. For example, a large metacentric chromosome is shortened by one-half in length to an acrocentric one, where as the small chromosome becomes a large one (Fig. 33.13). Two types of translocations have been recognized: homozygous and heterozygous (Fig. 33.14). The **translocation homozygotes** may have normal meiosis and in fact, are difficult to detect cytologically unless morphologically dissimilar chromosomes are involved, or banding patterns differ markedly. The **translocation heterozygotes** produce both translocated and normal chromosomes and exhibit characteristic cytological and genetical effects.

Thus, translocation heterozygotes are marked by considerable degree of meiotic irregularity. In order to affect pairing of all homologous segments, peculiar and characteristic formations occur during synapsis.

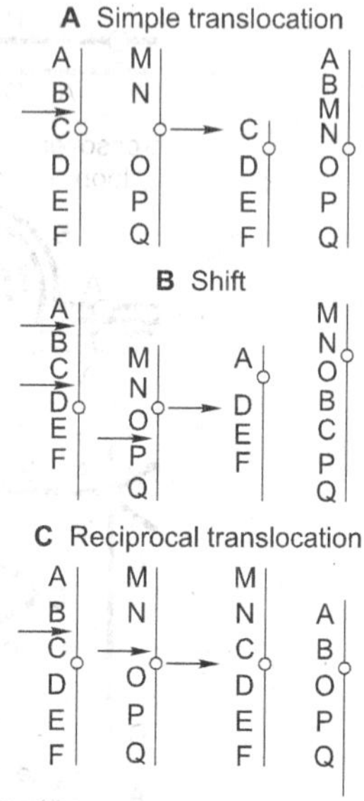

Fig. 33.12. Three types of translocation which occur in nonhomologous chromosomes. Arrows indicate the points of breaks.

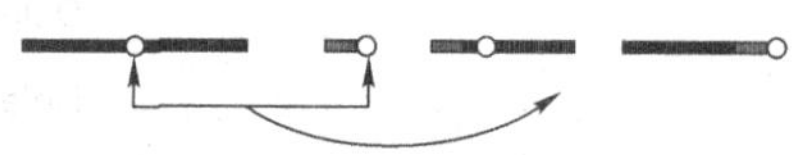

Fig. 33.13. Shortening of metacentric chromosome and lengthening of acrocentric chromosomes due to translocation.

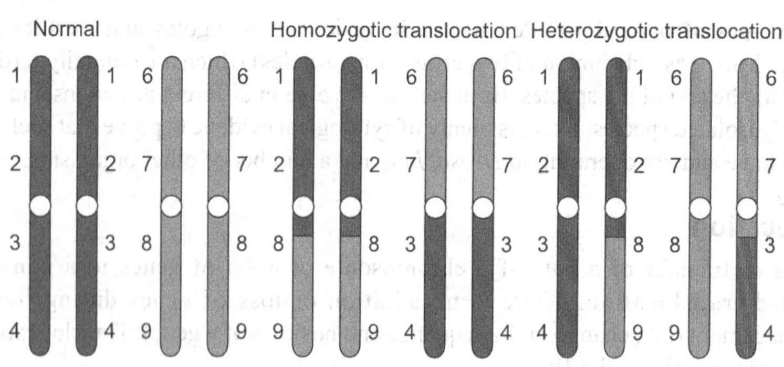

Fig. 33.14. Schematic represenation of homozygotic and heterozygotic reciprocal translocations compared with the normal arrangement.

Typically, a cross-shaped configuration is seen in prophase-I. This structure opens into either a ring or a figure of 8, both comprising four chromosomes (Fig. 33.15). In case of 8 figure formation, both normal chromosomes move to one pole and both translocation chromosomes move to the other pole at anaphase-I. This is called **alternate segregation** (or **disjunction**) and in it functional meiospores

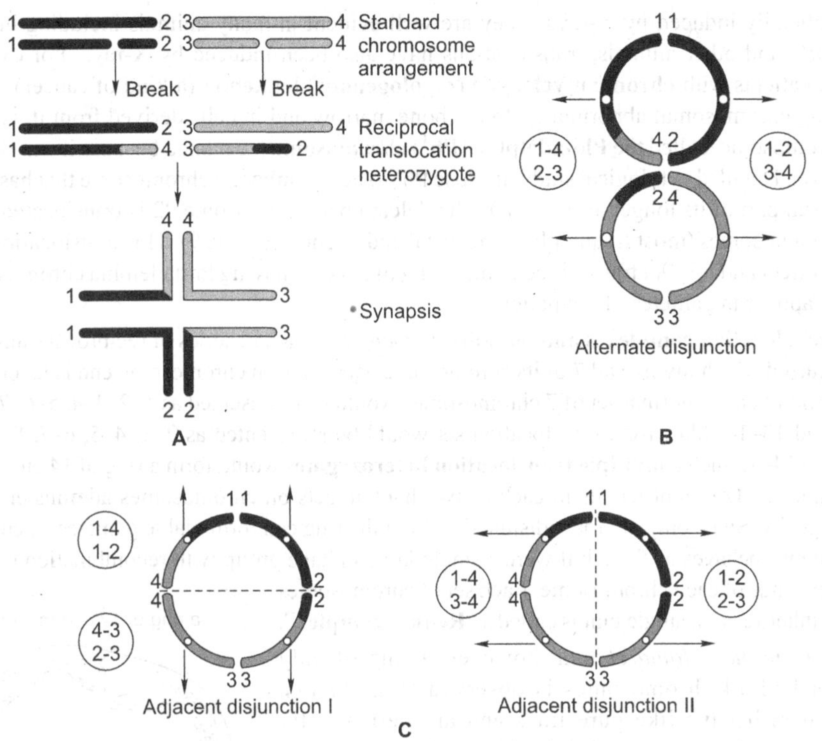

Fig. 33.15. A—Cross shaped figure formation during heterozygous reciprocal translocation; B—Alternate disjunction; C—Adjacent I and II disjunctions (after Stansfield, 1986).

(in higher plants) or gametes (animals) will be produced because the meiotic products will have a full gene complement. In case of ring formation, **adjacent type segregation** may occur and as a result each of translocation chromosomes and normal chromosomes move to opposite poles. In this case all of the gametes will contain some extra segments (duplications) and some pieces will be missing (deletion). Semisterility resulting from adjacent type segregation during reciprocal translocation is easily observed in such plants as corn (maize), wheat, pea and *Datura*. Ears of com lack about half the kernels, and these are arranged irregularly. Abortive pollen is reduced in size (Fig. 33.16).

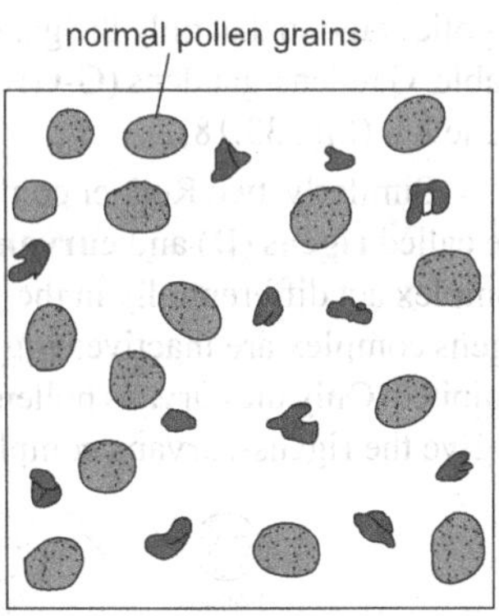

Fig. 33.16. Normal and aborted pollen in a semisterile corn plant. The small, shriveled pollen grains contain an euploid meiotic products of reciprocal translocation heterozygote (after Suzuki *et al.*, 1986).

Lastly, some genes which formerly were on nonhomologous chromosomes will no longer appear to be assorted independently. And the phenotypic expression of a gene may be modified when it is translocated to a new position in the genome. Such position effects are particularly evident when genes in euchromatin are shifted near heterochromatin region.

The first case of translocation was studied in the evening primrose (*Oenothera*) which was originally described as a mutation by **de Vries**. In *Oenothera*, *Tradescantia* and *Rhoeo* translocations in heterozygous condition are frequently found in nature. In many other crop plants translocations have been artificially induced by X-rays. They are well evident in many animals including humans. In *Drosophila* and other animals, translocations have also been induced by X-rays. For example, in humans, patients with **chronic myelocytic (myologenous) leukemia** (a kind of cancer) display an interesting chromosomal abnormality. In the bone marrow and in cells derived from it, is present a short chromosome, called the **Philadelphia (Ph^1) chromosome** (so named because it was discovered in that city). Detailed cytological study disclose Ph^1 to be a number 22 chromosome that has lost most of the distal part of its longer arm (22 q^-). The deleted part of autosome 22 is translocated to one of the larger autosomes (most frequently to the distal end of chromosome 9). This translocation exhibits position effect and it is not transmitted to offspring of persons having Philadelphia chromosome (Ph^1 does not appear in gametes of the patients).

Translocation complexes and lethality. In *Oenothera*, a rare series of reciprocal translocations have occurred which involve all 7 of its chromosome pairs. If each chromosome end is labelled with a different number, the normal set of 7 chromosomes would be represented as 1–2, 3–4, 5–6, 7–8, 9–10, 11–12, and 13–14; likewise a translocation set would be represented as 2–3, 4–5, 6–7, 8–9, 10–11, 12–13 and 14–1. Such a **multiple translocation heterozygotes** would form a ring of 14 chromosomes during meiosis. Different lethals in each of two haploid sets on chromosomes administer structural heterozygosity. Since only alternate disjunction from the ring can form viable gametes, each group on chromosomes behaves as though it were a single large linkage group with recombination confined to the pairing ends of each chromosome. Each set of chromosomes which is inherited as a single unit is called a "**Renner complex**".

In *Oenothera lamarckiana,* however, a ring of only 12 instead of 14 chromosomes is observed (Fig. 33.17). Its members behave like pure lines and are permanently heterozygotes. Permanent hybridity is maintained in some species of *Oenothera* due to operation of balanced **lethal system** which may function due to gametic and zygotic lethality. For example, in *O. lamarckiana* one of the Renner complexes is called **gaudens** and the other is called **velans.** This species is mainly self-pollinated. The lethals become effective in the zygotic stage so that only the gaudens-velans (G-V) zygotes are viable. Gaudens-gaudens (G-G) or velans-velans (V-V) zygotes are lethal (Fig. 33.18).

Similarly, two Renner complexes in *Oenothera muricata* are called **rigens (R)** and **curvans (C)**. Gametic lethals in each complex act differentially in the gametophytes. Pollen with the rigens complex are inactive; egg with the curvans complex are inhibited. Only the curvans pollen and rigens eggs are functional to give the rigens-curvans complex in the zygote (Fig. 33.19).

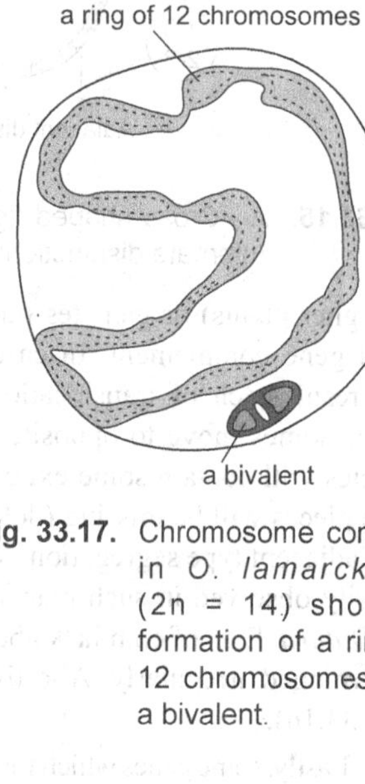

Fig. 33.17. Chromosome complex in *O. lamarckiana* (2n = 14) showing formation of a ring of 12 chromosomes and a bivalent.

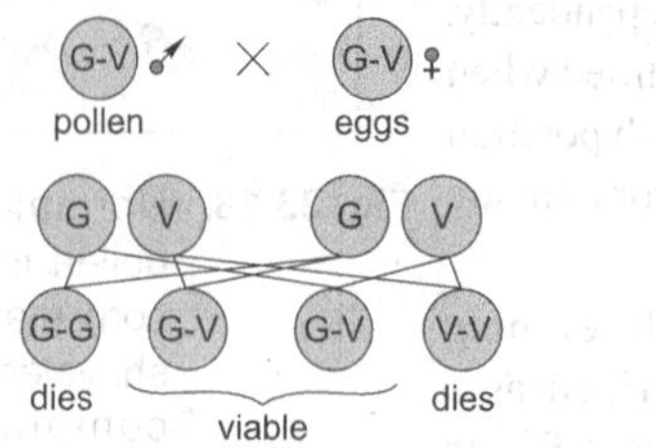

Fig. 33.18. Zygotic lethality due to translocation complexes.

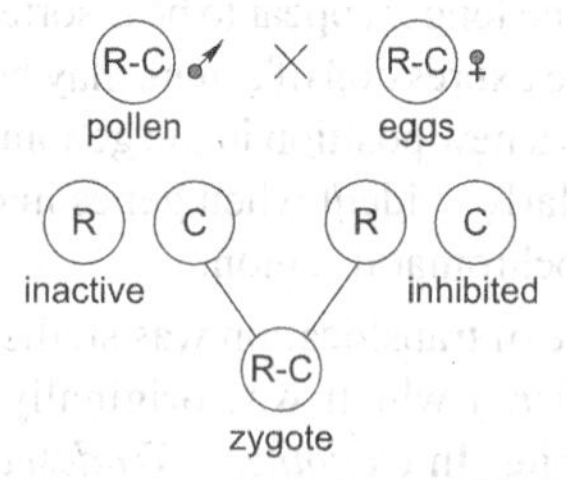

Fig. 33.19. Gametic lethality due to translocation complexes.

Variation in Chromosome Morphology

Various changes in chromosome structure often produce variation in chromosome morphology such as isochromosomes, ring chromosomes and Robertsonian translocation.

1. Isochromosomes. An isochromosome is a chromosome in which both arms are identical. It is thought to arise when a centromere divides in the wrong plane, yielding two daughter chromosomes, each of which carries the information of one arm only but present twice. For example, telocentric X chromosome of *Drosophila* may be changed into an "attached-X" which is formed due to misdivision of the centromere (Fig. 33.20).

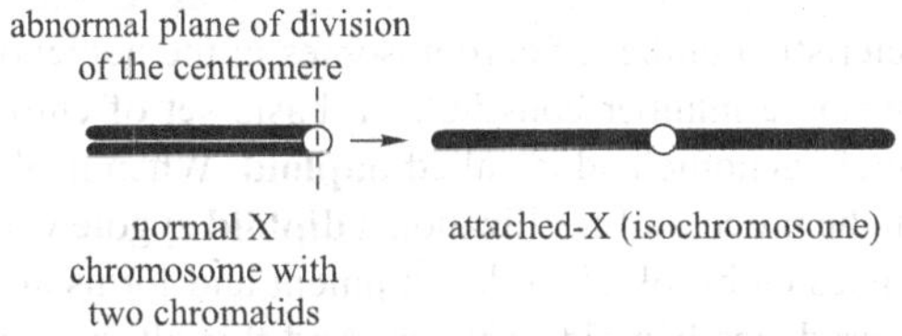

Fig. 33.20. Origin of an attached-X chromosome.

2. Ring chromosomes. Chromosomes are not always rod-shaped. Occasionally ring chromosomes are encountered in higher organisms. Sometimes breaks occur at each end of the chromosome and broken ends are joined to form a ring chromosome. Crossing over between ring chromosomes can lead to bizarre anaphase figures (Fig. 33.21).

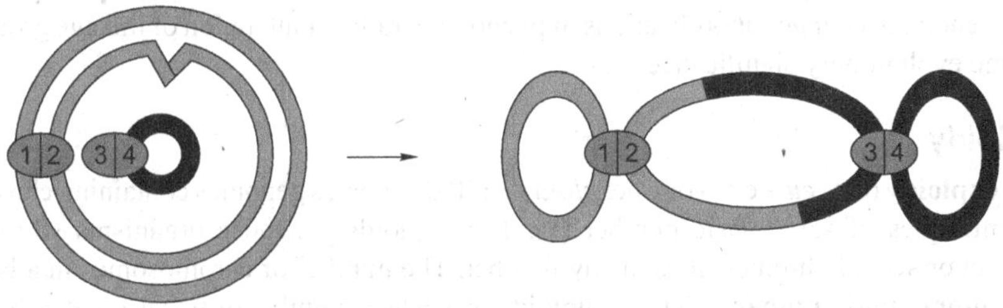

Fig. 33.21. Formation of double bridge during single crossing over in ring chromosomes (after Stansfield, 1986).

3. Robertsonian translocation. Sometimes whole arm fusions occur in the non-homologous chromosomes. It is called **Robertsonian translocation**. Thus, Robertsonian translocation is an eucentric reciprocal translocation where the break in one chromosome is near the front of the centromere and the break in the other chromosomes is immediately behind its centromere. The resultant smaller chromosome consists of largely inert heterochromatic material near the centromere; it normally contains no essential genes and tends to become lost. Thus, Robertsonian translocation results in a reduction of the chromosome number (Fig. 33.22).

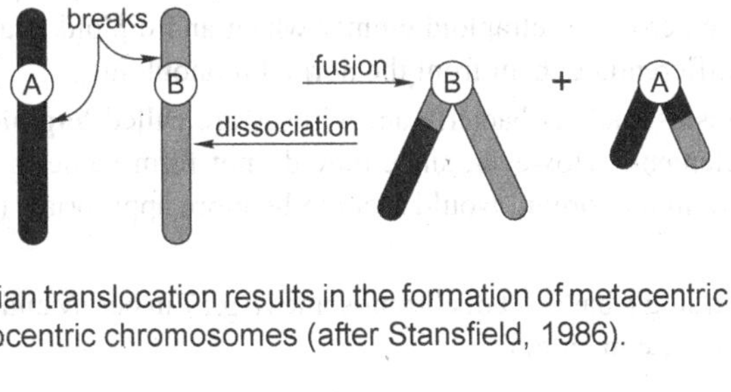

Fig. 33.22. Robertsonian translocation results in the formation of metacentric chromosome by fusion of two acrocentric chromosomes (after Stansfield, 1986).

The Robertsonian translocation is found to have a role in the evolution of human beings. Humans have 46 chromosomes whereas the great apes (Chimpanzees, Gorillas and Orangutans) have 48 chromosomes. Cytogeneticists believe that humans evolved from a common human/ape ancestor due to centric fusion of two acrocentric chromosomes to produce a single large chromosome containing the combined genetic content of two acrocentric chromosomes. It is suspected that structural rearrangements of chromosomes may lead to reproductive isolation and the formation of new species.

33.2. NUMERICAL CHANGES IN CHROMOSOMES

Each species has a characteristic number of chromosomes in the nuclei of its gametes and somatic cells. The gametic chromosome number constitutes a basic set of chromosomes called **genome**. A gamete cell contains single genome and is called **haploid**. When haploid gametes of both sexes (male and female) unite in the process of fertilization a **diploid** zygote with two genomes is formed. The diploid zygote undergoes embryological development and forms an adult animal which upon attaining sexual maturity produces haploid gametes. And this alternation of generation continues between haploidic and diploidic generation in most species. However, sometimes irregularities occur in nuclear division or "accidents" (as from radiations) may befall interphase chromosomes so that cells or entire organisms with aberrant genomes may be formed. Such chromosomal aberrations may include whole genomes and entire single chromosomes. Changes in number of whole chromosomes is called **heteroploidy** (see **Burns** and **Bottino**, 1989). Heteroploidy may involve entire sets of chromosomes (**euploidy**), or loss or addition of single whole chromosomes (**aneuploidy**). Each may produce phenotypic changes, modifications of phenotypic ratio, or alteration of linkage groups. Many are of some evolutionary significance.

A. Euploidy

The term **euploidy** (Gr., *eu* = even or true; *ploid* = unit) designates genomes containing chromosomes that are multiples of some basic number (x). The euploids are those organisms which contain balanced set or sets of chromosomes in any number. The number of chromosomes in a basic set is called the **monoploid number**, x. Those euploid types whose number of sets is greater than two are called **polyploid**. Thus, $1x$ is **monoploid**, $2x$ is **diploid**; and the polyploid types are $3x$ (**triploid**), $4x$ (**tetraploid**), $5x$ (**pentaploid**), $6x$ (**hexaploid**), $8x$ (**octaploid**), $10x$ (**decaploid**), $12x$ (**dodecaploid**) and so on. The **haploid (n)** refers strictly to the number of chromosomes in gametes. In most animals and many plants the haploid number and monoploid number are the same. Hence n or x (or $2n$ or $2x$) can be used interchangeably. However, in case of polyploids the usage of n may create confusion. For example, in modern wheat x and n are different. Wheat has 42 chromosomes, but careful study reveals that in hexaploid there are six rather similar but not identical sets of seven chromosomes. So, $6x = 42$, and $x = 7$. However, gametes of wheat contain 21 chromosomes, hence, $2n = 42$ and $n = 21$ (see **Suzuki** *et al.*, 1986). A triploid hybrid of wheat, from which a hexaploid *Triticum spelta* has been obtained due to colchicine treatment, contains $2n = 3x = 21$. Such haploids, since are obtained from the polyploids (*i.e.*, cross of tetraploid emmer wheat and diploid goat grass), they are called **polyhaploids**, just to differentiate them from the normal monoploids.

The lower organisms such as bacteria and viruses are called **haploids** because they have a single set of genetic elements. However, since they do not form gametes comparable to those of higher organisms, the term monoploid would seem to be more appropriate (see **Stansfield**, 1986).

(1) Monoploidy

Monoploid have a single basic set of chromosomes *e.g.*, 7 in barley and 10 in corn. Monoploidy is common in plants and rare in animals.

***(i)* Origin and production of monoploids.** Monoploids in some cases are found normally and are produced due to parthenogenesis, as in male (drone) hymenopteran insects such as bees, wasps and ants. In these insects, queen and workers are diploid females. In angiosperms (flowering plants) monoploids may also originate spontaneously due to parthenogenetic development of egg. Such rare monoploids have been obtained in tomatoes and cotton under cultivation. Rarely monoploid plants may originate from the pollen tube, synergids and antipodals of the embryo sac and are called **androgenic monoploids** (or androgenic haploids).

Monoploids can be produced by artificial means by the following methods: (1) X-ray treatments, (2) delayed pollination, (3) temperature shock (cold treatment), (4) colchicine treatment, (5) distant (interspecific or intergeneric hybridization), (6) anther or pollen culture. Among these techniques, the most important ones are distant hybridization and anther culture.

***(a)* Distant hybridization.** Interspecific crosses in genera of *Solanaceae* (*e.g.*, *Solanum* and *Nicotiana*) have been employed for the production of both parthenogenetic and androgenic monoploids. By this technique monoploids have been obtained in large number in potato. **Kasha** and **Kao** (1970) have used this technique for producing monoploids in large number in barley. They discovered that when diploid barley, *Hordeum vulgare*, is pollinated using a diploid wild relative called *Hordeum bulbosum*, fertilization occurs, but during the ensuing somatic cell divisions, the chromosomes of *H. bulbosum* are preferentially eliminated from the zygote, resulting in a haploid embryo (such a **haplodization process** appears to be caused by a genetic incompatiability between the chromosomes of the different species). The resulting haploids can be doubled with colchicine treatment. This technique has resulted in the rapid production of new varieties of barley and applied to other plant species also.

***(b)* Anther or pollen culture.** The production of monoploids in tobacco plant by anther and pollen culture was demonstrated for the first time in the laboratory of **Prof. S.C. Maheshwari** of Dept. of Botany of Delhi University (**Guha** and **Maheshwari**, 1967). In this technique, a cell which is destined to become a pollen grain, may be induced by cold treatment to grow instead into an **embryoid**, a small dividing mass of cells. The embryoid may be grown on agar to form a monoploid plantlet, which can then be potted in soil to mature. Subsequently, such monoploids were produced for various crop plants such as soyabean, rice, wheat, mustard, and tobbaco. Presently, this technique is regarded as a very potential source of monoploid production.

***(ii)* Morphology of monoploids.** Monoploid plants have reduced size of all vegetative and floral parts. In monoploid *Nicotiana* **Kostoff** reported that the leaves, flowers and over all plant size were smaller. The size of seed and stomata as well as diameter of pollen were found smaller in monoploids than in the diploids. Even the size of nucleus (or the nuclear volume) of a monoploid often was found to be just half than the nucleus of the diploid cell.

***(iii)* Cytology of monoploids.** In monoploids each chromosome is represented only once due to which there is no zygotene pairing and all the chromosomes appear as **univalents** on the metaphase plate at the time of meiosis. During anaphase each chromosome move independently of the other and goes to either of the two poles. According to law of probability the chance that a particular chromosome will go to a particular pole as half and the chance that all the chromosomes of a monoploid set will go to the same pole is $\frac{1}{2} \times \frac{1}{2} \times \frac{1}{2} \dots n$ times, where n = number of chromosomes in the monoploid set. So, the frequency of gametes with the haploid set or n number of chromosomes will be $(\frac{1}{2})^n$. This indicates that higher the number of chromosomes in a haploid set, lesser will be the frequency of all of them being included in the same gamete. Gametes containing less than the haploid number of chromosomes are normally not viable. Therefore, monoploid organisms are highly sterile. For instance, a monoploid in maize ($2n = 20$) will have 10 chromosomes and the number of chromosomes in a gamete can range from 0 – 10. Consequently, considerable sterility is found in such monoploid maize plants.

In contrast, in monoploid male honey bees during spermatogenesis the meiosis is bypassed by mitosis. As a result, their sperms are haploid and viable.

(iv) **Uses of monoploids.** In a monoploid, since there is only one copy of each chromosome and only one allele of each gene, so, in it each gene is expressed whether it is dominant or recessive. This facilitates genetic experiments and this is the reason why microorganisms have been helpful in genetic studies. For the same reason, scientists are trying hard to develop haploid strains of the flowering plants. Success has been achieved in developing monoploid strains of *Nicotiana, Datura* and *Triticum*. From these monoploid strains have also been developed pure breeding strain; which are resistant for the insecticides and also for toxic compounds normally produced by the parasites of these plants.

(2) Polyploidy

Any organism with more than two genomes ($2x$) is called a **polyploid**. Many plant genera include species whose chromosome numbers constitute a euploid series. For example, the rose genus *Rosa* includes species with the somatic numbers 14, 21, 28, 35, 42 and 56. These numbers are the multiples of 7. Therefore, this is a euploid series of the basic monoploid number 7, which gives diploid, triploid, tetraploid, pentaploid, hexaploid and octaploid species. Except diploids, rest of these belong to polyploid category. Ploidy levels higher than tetraploid are not commonly encountered in natural populations, but our most important crops and ornamental flowers are polyploids, *e.g.*, wheat (hexaploid $6x$), strawberries (octaploid, $8x$), many commercial fruits and ornamental plants. Generally, polyploidy is common in plants (more common in monocots) but rare in animals.

Types of polyploidy. There are following three different kinds of polyploids: *(i)* autopolyploid, *(ii)* allopolyploids and *(iii)* autoallopolyploids. Suppose that there are four different haploid sets of chromosomes A, B_1, B_2 and C, in which B_1 and B_2 genomes are related. By using these genomes, all three types of polyploids can be derived as have been shown in Fig. 33.23.

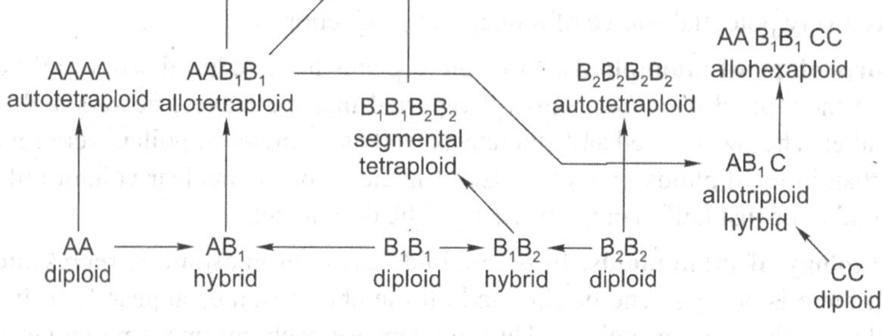

Fig. 33.23. Mode of formation of different kinds of polyploid.

(a) **Autopolyploids.** The autopolyploids are those polyploids, which consist of same basic set of chromosomes multiplied. For example, if a diploid species has two similar sets of chromosomes or genomes (AA), an autotriploid will have three similar genomes (AAA), and an autotetraploid, will have four such genomes (AAAA).

(i) **Origin and production of autopolyploids.** The autopolyploids may occur in nature or may be produced artificialy. When they are found in nature, their autopolyploidy nature is deduced by their meitotic behaviour. One of the common example of natural autopolyploidy is 'doob' grass (*Cynodon dactylon*) which is quite commonly cultivated in U.P. and Bihar. Its **autotriploid** status was established from its meiotic behaviour by **Prof. P.K. Gupta**, an eminent cytogeneticists of Northern

India, working in Dept. of Agriculture Botany of Meerut (C.C.S.) University (**Gupta** and **Srivastava**, 1970). *Cynodon* is quite successful in the cultivation mainly due to its efficient way of vegetative propagation (since being triploid, it is sterile and setting no seed). Polyploids may arise naturally by following means: (*i*) in natural populations polyploidy may arise as a result of interference with cytokinesis, once chromosome replication has occurred; (*ii*) it may occur either in somatic tissues which give rise to tetraploid branches or during meiosis which produces unreduced gametes. All these natural inductions of polyploidy may occur due to chilling.

Some of common examples of autotriploid crop plants, which are mainly produced by artificial methods, are seedless varieties of watermelons, sugar beet, tomato, grapes and banana. Similarly, many important crop plants include **autotetraploids** such as rye (*Secale cereale*), corn (*Zea mays*), red clover (*Trifolium pratense*), berseem (*Trifolium alexandrium*), marigolds (*Tagetes*), snapdragons (*Antirrhinum*), *Phlox*, grapes, apples, *Oenothera lamarkiana* (which was recognized as mutation by **Hugo de Vries**).

Induced autopolyploidy. The autopolyploidy have been induced in many plant and animal cells by artificial means such as chemical (*e.g.*, chloral hydrate, colchicine, acenapthene, indole acetic acid, sulphanil amide, mercury chloride, hexachlorcyclohexane, etc.), radioactive substances, *e.g.*, radium and X-ray), temperature shocks (*e.g.*, maize plant), and decapitation (after removing the bud, some tissues developing from scar tissue were found to be tetraploid). These inducers usually disturb the mitotic or meiotic spindle and cause non-segregation of already duplicated chromosomes, during cell divisions.

Colchicine is a drug (*i.e.*, an alkaloid obtained from the corms of plants—*Colchicum autmunale* and *C. luteum*) and its aqueous solution is found to prevent the formation and organization of spindle fibres, so the metaphase chromosomes of the affected cells (called **C-metaphase** or **colchicine metaphase**) do not move to a metaphase plate and remain scattered in the cytoplasm. Even the process of cytokinesis is prevented by colchicine and with duplications of chromosomes the number goes on increasing. As colchicine interferes with spindle formation, its effects are limited to dividing and meristematic cells (Box 33.1).

Box 33.1

Colchicine was first obtained by **Houde** (1887) and was first used for inducing polyploidy by **Dustin** (1934). **Blakeslee**, **Avery** and **Nebel** (1937) found that the alkaloid **colchicine** interferes with spindle formation. When cell of the root tip were placed in colchicine solution, spindle formation was inhibited. The chromosomes, however, divided normally. Chromosome division without cytoplasmic division results in the doubling of the chromosome number. A number of polyploids have been developed artificially using colchicine treatment. Onion cells kept in colchicine for four days had as many as a thousand chromosomes in a single nucleus.

(*ii*) Effects of autopolyploidy. Autopolyploidy results in **gigantism** of plant cells, *i.e.*, leaves, flowers and fruits of an autopolyploid are larger in size than a diploid plant. For example, the size of lower epidermis of leaf of a tetraploid *Saxifraga pensylvanica* was found greater than the diploids (Fig. 33.24). Some of significant effects of autopolyploidy are as follows: (1) With the increase in cell size, the water content increases which leads to a decrease in osmotic pressure. This results into loss of resistance against frost, etc. (2) Due to slower rate of cell division, the plant's growth rate decreases. This leads to a decrease in **auxin** supply and a decrease in respiration. (3) Due to slow growth rate, the time of blooming of an autopolyploid is delayed. (4) At higher ploidy level, such as autooctoploids, the adverse effects become highly pronounced and lead to the death of the plants.

Polyloid varieties with an even number of genomes (*e.g.*, tetraploids) are often fully fertile (Fig. 33.25), whereas those with an odd number (*e.g.*, triploids) are highly sterile (Fig. 33.26).

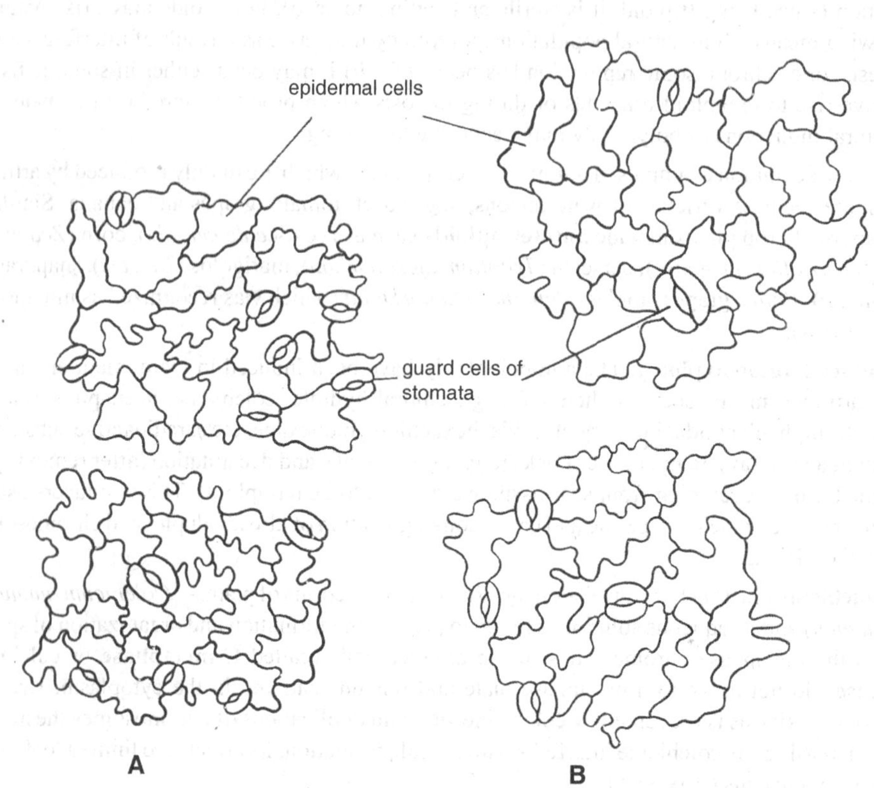

Fig. 33.24. Comparison of size of leaf epidermal cells of a diploid (A) and tetraploid (B) saxifrage (after Burns and Bottino, 1989).

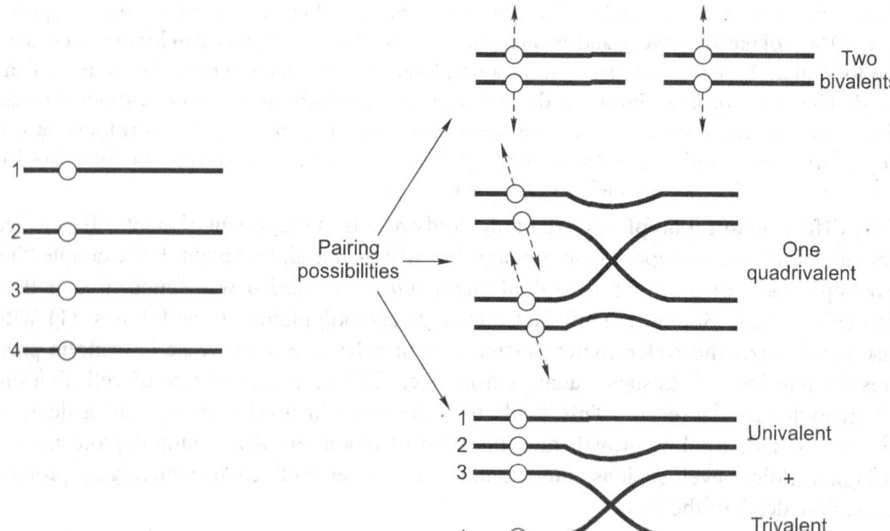

Fig. 33.25. Meiotic pairing possibilities in tetraploids (each chromosome is really two chromatids).

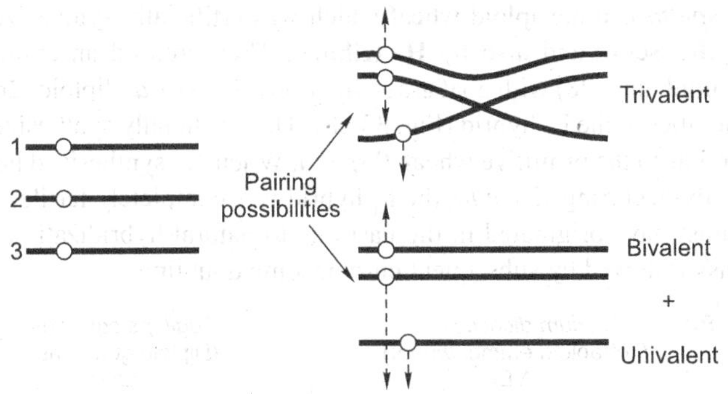

Fig. 33.26. Meiotic pairing possibilities in triploids (each chromosome is really two chromatids).

Uses of induced polyploidy. Since in the induced polyploids, the fertility level and seed set are low, so seedless fruits can be produced by using triploids as in case of seedless watermelons which were produced by a Japanese scientist, **Dr. Hitoshi Kihara**. These triploids are obtained from seeds raised by a cross of tetraploid and diploid plants. The tetraploids have been produced from the diploids by colchicine treatment. By adopting these methods a variety of triploids such as sugar beet, tomato, apples, tulips, lilies and grapes and tetraploids such as rye, barley, corn, kinnows, grapes, marigolds, snapdragons, lily, phlox etc., have been obtained. Among the forage crops, tetraploid barseem is a very popular crop in Northen India.

(*b*) Allopolyploids. When the polyploidy results due the doubling of chromosome number in F_1 hybrid which is derived from two distinctly different species, then, it is called **allopolyploidy (amphiploidy)** and the resultant species is called an allopolyploid. Let A represent a set of chromosomes (genome) in species X, and let B represent another genome In a species Y. The F_1 hybrids of these species then would have one A genome and another B genome. The doubling of chromosomes in the F_1 hybrids will give rise to allotetraploids with two A and two B genomes (see Fig. 33.27).

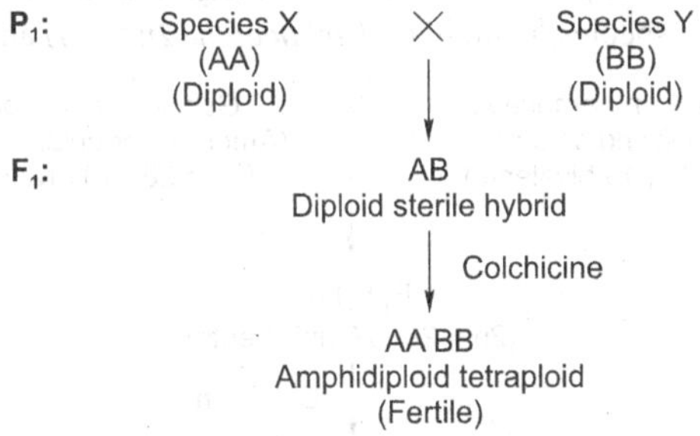

Fig. 33.27. Formation of an amphidiploid (allopolyploid) tetraploid.

Raphanobrassica is a classical example of allopolyploidy or amphipolyploidy. In 1927, Russian geneticist, **G.D. Karpechenko** performed a cross between radish (*Raphanus sativum*, 2n = 18 and cabbage (*Brassica oleracea*, 2n = 18) and in F_1 got sterile (diploid) hybrids. Among these F_1 hybrids, he found certain fertile plants which were found to contain 36 chromosomes. These fertile tetraploids were called *Raphanobrassica*.

Synthesized Allopolyploids

To find out the origin of naturally occurring allopolyploids some cytogeneticists produced certain allopolyploids in laboratory by employing artificial means. Common hexaploid wheat and tetraploid cotton furnish two such examples.

(*i*) ***Triticum spelta*** is a hexaploid wheat which was artificially synthesized in 1946 by **E.S. McFadden** and **E.R. Sears** and also by **H. Kihara**. They crossed an emmer wheat, triticum *dicoccoides*, (tetraploid: 2n = 28) with goat grass, *Aegilops squarrosa* (diploid; 2n = 14) and doubled the chromosome number in the F_1 hybrid (Fig. 33.28). This artificially synthesized hexaploid wheat was found to be similar to the primitive wheat *T. spelta*. When the synthesized hexaploid wheat was crossed with naturally occurring *T. spelta*, the F_1 hybrid was completely fertile. This suggested that hexaploid wheat must have originated in the past due to natural hybridization between tetraploid wheat and goat grass followed by subsequent chromosome doubling.

P_1: *Triticum dicoccoides* (Tetraploid emmer wheat) AA BB (2n = 28 ; 14 bivalents) × *Aegilops squarrosa* (Diploid goat grass) DD (2n = 14 ; 7 bivalents)

↓

F_1: ABD
Triploid hybrid
(2n = 21 ; 21 univalents)

↓ Colchicine

AA BB DD
Synthesized hexaploid wheat
(*Triticum spelta*)
(2n = 42 ; 21 bivalents)

Fig. 33.28. Artificial synthesis of hexaploid wheat.

(ii) Gossypium hirsutum, the New World cotton plant, is another interesting example of allopolyploidy. Old World cotton, *Gossypium herbaceum*, has 13 pairs of chromosomes, while, American or "upland cotton" also contains 13 pairs of chromosomes. **J.O. Beasley** crossed the Old World and American cottons and doubled the chromosome number in the F_1 hybrids. The allopolyploids, thus, produced resembled the cultivated New World cotton and when crossed with it gave fertile F_1 hybrids (Fig. 33.29). These results, thus, suggested that tetraploid *Gossypium hirsutum* originated from two diploid species, namely *G. herbaceum* (2n = 26) and *G. raimondii* (2n = 26).

Gossypium herbaceum (Old World cotton) (2n = 26 ; 13 bivalents) × *Gossypium raimondii* (American or upland cotton) (2n = 26 ; 13 bivalents)

↓

F_1 hybrid
(2n = 26 ; 26 univalents)

↓ Colchicine

New World cotton
(*Gossypium hirsutum*)
(2n = 52 ; 26 bivalents)

Fig. 33.29. Artificial synthesis of New World cotton.

***(iii)* Triticale (*Triticosecale wittmack*)** is the first human made cereal which has been developed in recent years and is cultivated on about one million hectares of land throughout the Globe for the commercial use. Triticale is an artificial allopolyploid which has been derived by crossing wheat (*Triticum*) and rye (*Secale*) (*i.e.*, intergenic hybridization). Depending upon whether *Triticum* is a tetraploid (2n = 4*x* = 28) or hexaploid (2n = 6*x* = 42), one would get hexaploid triticale (2n = 6*x* = 42; Fig. 33.30) or octaploid triticale (2n = 8*x* = 56; Fig. 33.31), respectively. In each case, only diploid rye (2n = 4*x* = 14) was used.

Triticum durum (Tetraploid wheat; 2n = 28) × *Secale cereale* (Diploid rye; 2n = 14)

↓

F_1 hybrid (sterile)
(Triploid ; 2n = 21)

↓ (Chromosome doubling)

Hexaploid triticale
(2n = 42)

Fig. 33.30. Artificial synthesis of a hexaploid triticle.

(*c*) **Segmental allopolyploids.** Different genomes of some allopolyploids are not quite different from each other. Consequently in these polyploids chromosomes belonging to different genomes do pair together to some extent. This indicates that segments of chromosomes and not the whole chromosomes are homologous. Therefore, such allopolyploids are called **segmental allopolyploids** (**Stebbins**, 1943, 1950). The segmental allopolyploids are intermediate between autopolyploids and allopolyploids and can be identified by their peculiar meiotic behaviour.

Triticum aestivum (Hexploid wheat; 2n = 42) × *Secale cereale* (Diploid rye; 2n = 14)

↓

F_1 hybrid (sterile)
(Triploid ; 2n = 28)

↓ (Chromosome doubling)

Octoploid triticale
(2n = 56)

Fig. 33.31. Artificial synthesis of a octoploid triticle.

It is generally believed that most naturally occurring polyploids are segmental allopolyploids. Our common hexaploid bread wheat (*Triticum aestivum*) too is found to be a natural segmental allopolyploid.

Polyploidy in animals. Polyploidy is rare in animals but occur in flatworms, leeches and brine shrimp. In mice, also, 40 per cent liver cells are tetraploids, and about 5 per cent are octoploids. Polyploidy in humans have been found in liver cells and cancer cells. In them polyploidy is whether complete or as a mosaic, it leads to gross abnormalities and death.

Phenotypic Effects of Polyploidy

The increase in the genome's size beyond the diploid level is often caused following detectable phenotypic characteristics in a polyploid organism:

(*i*) **Morphological effect of polyploidy.** The polyploidy is invariably related with gigantism. The polyploid plants have been found to contain large-sized pollen grains, cells, leaves, stomata, xylem, etc. The polyploid plants are more vigorous than diploids.

(*ii*) **Physiological effect of polyploidy.** The ascorbic acid content has been reported to be higher in tetraploid cabbages and tomatoes than in corresponding diploids. Likewise corn meal of a tetraploid maize seed contain 40 per cent more vitamin A than corn meal from a diploid plant.

(*iii*) **Effect on fertility of polyploidy.** The most important effect of polyploidy is that it reduces the fertility of polyploid plants in variable degrees.

(*iv*) **Evolution through polyploidy.** Interspecific hybridization combined with polyploidy offers a mechanism whereby new species may arise suddenly in natural populations.

B. Aneuploidy

Changes that involve parts of a chromosome set results in individuals, called **aneuploids** (Gr. *aneu* = uneven; *ploid* = unit). Aneuploidy can be either due to the loss of one or more chromosomes (**hypoploidy**, *e.g.*, monosomy and nullisomy) or due to addition of one or more chromosomes to the complete chromosomes set (**hyperploidy**, *e.g.*, trisomy, tetrasomy and pentasomy). Hypoploidy is mainly due to the substraction (or loss) of a single chromosome, called **monosomy** (2n – 1) or due to the the loss of one pair of chromosome called **nullisomy** (2n – 2; two lost chromosomes are homologs). Likewise, hyperploidy may involve addition of either a single chromosome, called **trisomy** (2n + 1) or a pair of chromosomes, called **tetrasomy** (2n + 2). In the monoploid organisms, addition of single chromosome produces **disomy** (n + 1). All of these aneuploids are probably produced by nondisjunction during mitosis or meiosis (Fig. 33.32).

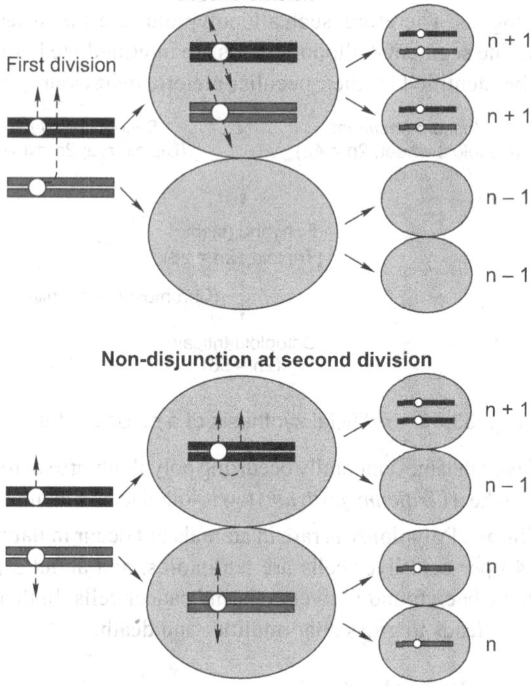

Fig. 33.32. The mode of origin of aneuploid gametes by non-disjunction at either the first or second meiotic division (after Suzuki *et al.*, 1986).

1. Monosomy

Diploid organisms which are missing one chromosome of a single pair are monosomic with the genomic formula 2n – 1. A monosomic individual forms gametes of two types, (n) and (n – 1). The n – 1 gametes do not survive in plants, but, in animals that may cause genetic imbalance, which is manifested by high mortality or reduced fertility of resulted organism (Fig. 33.33).

Monosomy in diploids is not tolerated, since it creates imbalance due to loss of one complete chromosome.

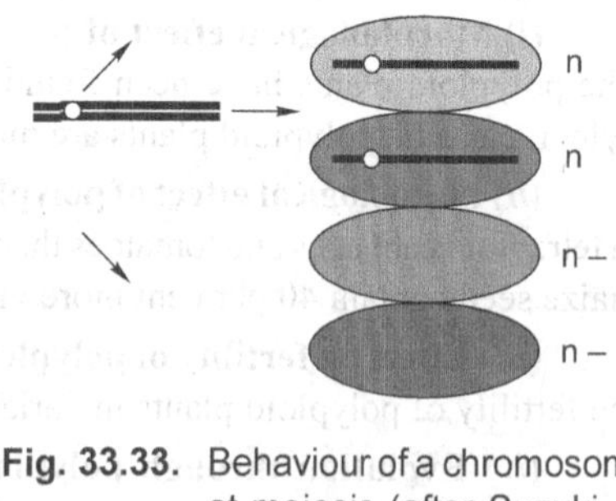

Fig. 33.33. Behaviour of a chromosome at meiosis (after Suzuki, *et al.*, 1986).

However, in polyploids, monosomy has no apparent effect, since they have several chromosomes of same type and loss of one chromosome can be easily tolerated. The number of possible monosomics in an organism will be equal to the haploid chromosome number. For example, in common wheat, since 21 pairs of chromosomes are present, 21 possible monosomics are known. These 21 monosomics in wheat were produced by **E.R. Sears** in variety called "**chinese spring**" and being used for genetic studies all over the world. Monosomics were also reported in cotton (2n = 52) by **J. Endrizzi** and his coworkers and in tobacco (2n = 48) by **E.R. Clausen** and **D.R. Cameron**. Monsomics have also been produced in maize and tomato (2n = 24) despite their being diploids. **Double monosomics** (2n – 1 – 1) or **triple monosomics** (2n – 1 – 1 – 1) could also be produced in polyploids such as wheat. Double monosomic means that the chromosome number is 2n – 2 like that in a nullisomic, but the missing chromosomes are nonhomologous. The same explanation is applied to the triple monosomics.

2. Nullisomy

An organism which has lost a chromosome pair is a nullosomic. The nullosomic organism has the genomic formula (2n – 2). A nullosomic diploid often does not survive, however, a nullosomic polyploid (*e.g.*, hexaploid wheat, 6*x* – 2) may survive but exhibit reduced vigour and fertility.

3. Trisomy

Trisomics are those diploid organisms which have an extra chromosome (2n + 1). Since the extra chromosome may belong to anyone of different chromosomes of a haploid complement, the number of possible trisomics will be equal to the haploid chromosome number. For example, haploid chromosome number of barley is 7, consequently in it seven trisomics are possible. Further, when the extra chromosome is identical to its homologs, such a trisomic is called **primary trisomic**. There are also secondary and tertiary trisomics. While the **secondary trisomic** means that the extra chromosome should be an isochromosome (*i.e.*, both chromosome arms genetically similar), a **tertiary trisomic** would mean that the extra chromosome should be the product of translocation (Fig. 33.34). Trisomics were obtained for the first time in jimson weed (*Datura stramonium*) by **A.F. Blakeslee** and **J. Belling** (1924). Since the haploid chromosome number of this species is n = 12, so here, 12 primary trisomics, 24 secondary trisomics and a large number of tertiary trisomies are possible. Most of the trisomics were identified by the size, shape and other morphological features of the fruit of jimson weed (Fig. 33.35). In barley, such a trisomic series was produced and extensively studied by **T. Tsuchiya**.

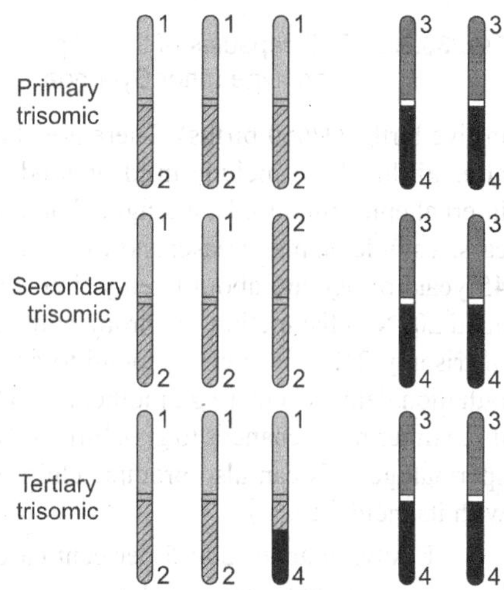

Fig. 33.34. Three kinds of trisomics.

Trisomy in humans. In human beings, the following three trisomic syndromes have been studied:

A. Down's syndrome (DS) or Trisomy-21. Down's syndrome is named after the physician **J. Langdon Down** who first described this genetic defect in 1866 and it was formally called **mongolism** or **mongolian idiocy**. It is usually associated with a trisomic condition for one of the smallest human autosomes (*i.e.*, chromosome 21). It is the most common chromosomal abnormality

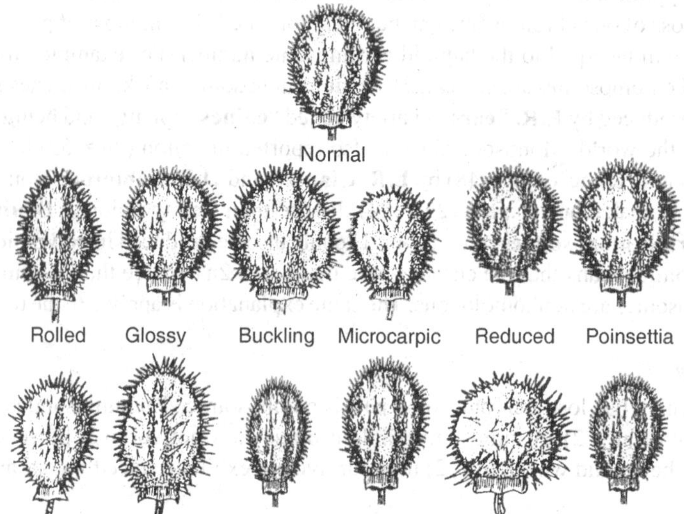

Fig. 33.35. Fruit capsules of the 12 primary trisomics of *Datura stramonium*, each with its particular phenotype (after Sybenga, 1972).

in live births (1/650 births). There are about 50 physical characteristics shown by DS infants soon after birth. These include mild or moderete mental retardation; eyes that slant up and out with internal epicanthal folds; a tongue that is large, swollen and protruding; small and under developed ears; a single palmar crease; short stature; stubby fingers; an enlarged liver and spleen. Women over 45 years of age are about twenty times more likely to give birth to a child with DS than women aged 20. Nondisjunction of chromosome pair 21 during oogenesis is the main cause of occurrence of trisomy-21. This event is found to be affected either by senescence of oocytes, virus infection, radiation damage, etc. (*e.g.*, mothers who have had infectious hepatitis prior to pregnancy may have three times more chances to give birth to DS infants). Nondisjunction of chromosome pair 21 during spermatogenesis can also produce child with DS, but paternal age does not seem to be associated with its incidence.

Lastly, in about 2 to 5 per cent cases, the normal chromosome number is present (2n = 46), but the extra chromosome 21 is attached (translocated) to one of the larger autosomes (usually chromosome 14).

B. Edward's syndrome or Trisomy-18. First described in 1960 by **John H. Edwards** and his colleagues, **trisomy-18** is found to contain an incidence of about 0.3 per 1000 births. It is characterized by multiple malformations, primarily low-set ears; small receding lower jaw; flexed and clenched fingers; cardiac malformations; and various deformaties of skull, face and feet. Harelip and cleft palate often occurs. Death takes place around 3 to 4 months of age. Trisomy-18 children show evidence of severe mental retardation, which is more pronounced in females (the reason is still not clear). Like the Down's syndrome, occurrence of Edward's syndrome is too related with maternal age (*i.e.*, 35 to 45 year old mothers have more chance of giving birth to trisomy-18 infant).

C. Patau syndrome or Trisomy-13. This syndrome was described in 1960 by **Klaus Patau** and coworkers. Its incidence is about 0.2 per 1000 births. Individuals with Patau syndrome appear to be markedly mentally retarded; have sloping forehead, harelip and cleft palate. Polydactyly (both hands and feet) is almost always present; the hands and feet are deformed. Cardiac and various internal defects (of kidney, colon, small intestine) are common. Death usually occurs within hours or days, but the foetus may abort spontaneously.

Trisomy in non-humans. Trisomy-22 has been reported in chimpanzees (**McClure** *et al.*, 1969); this shows Down syndrome-like phenotypic features. **Trisomy-21** has been reported in the gorilla.

Cytology of trisomics. The trisomics have an extra chromosome which is homologous to one chromosome of the diploid complement. Therefore, it forms **trivalent** which may take a variety of shapes in primary and secondary trisomics (Fig. 33.36). In a tertiary trisomic a characteristic pentavalent is observed.

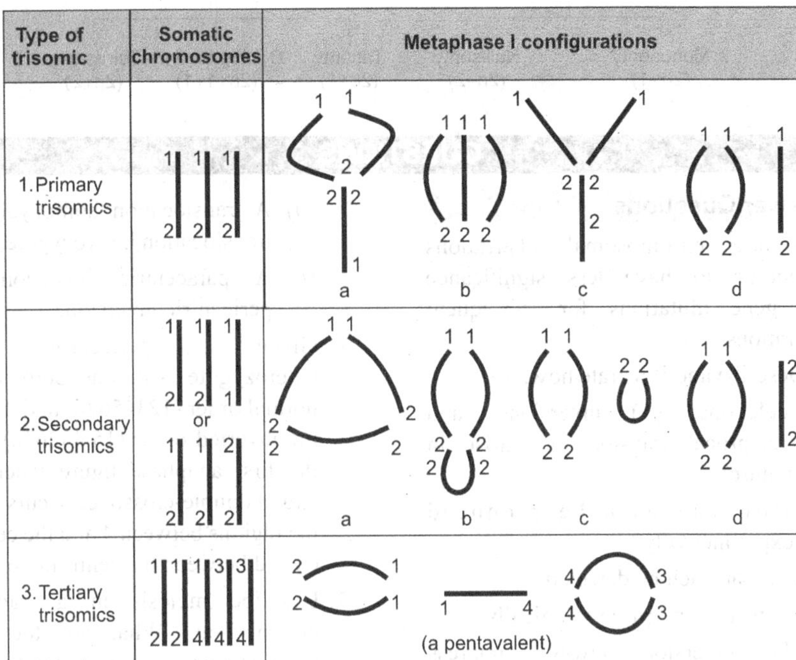

Fig. 33.36. Meiotic configurations formed at metaphase I in different types of trisomics.

4. Double Trisomy

In a diploid organism when two different chromosomes are represented in triplicate, the double trisomic is resulted. The double trisomic causes great genetic imbalance and has the genomic formula 2n + 1 + 1.

5. Tetrasomy

The diploid organisms having two extra chromosomes are known as **tetrasomic**. They have the genomic formula 2n + 2. All the 21 possible tetrasomics are available in wheat.

6. Pentasomy

The diploid organism (human female) with three extra X chromosomes (*i.e.*, 22 autosomes + XXXXX), called **XXXXX syndrome.**

Different types of ploidies can be summarized in the following chart:

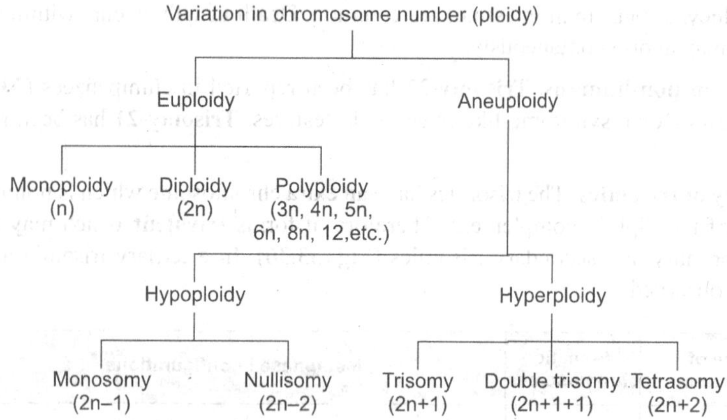

QUESTIONS

Long Answer Questions

1. Why are chromosomal aberrations considered to have less significance than gene mutations for subsequent generations?
2. (*a*) Describe and illustrate how:
   (*i*) deletions, (*ii*) inversions, and (*iii*) reciprocal translocations arise in nature?
   (*b*) How can each be produced experimentally?
   (*c*) How can each be detected:
   (*i*) genetically and (*ii*) cytologically?
3. (*a*) Do inversions always suppress crossing-over? Give reasons for your answer.
   (*b*) If your answer to:
   (*a*) is no, why are inversions referred to as crossover-suppressors?
   (*c*) Show how:
   (*i*) paracentric and
   (*ii*) pericentric inversions can act as crossover suppressors?
4. Why are salivary gland chromosomes of *Drosophila* commonly used for study of structural changes in chromosomes? Draw the configurations, you would observe in salivary gland chromosomes due to either a deficiency or a duplication.
5. How will you distinguish cytologically between the following?
   (*a*) A translocation homozygote and a translocation heterozygote;
   (*b*) A paracentric inversion and a pericentric inversion.
6. Given a pericentric inversion heterozygote with one chromosome in normal order (12345678) and the other in the inverted order (15.432678), diagram the first anaphase figure when a four-strand double-crossover occurs involving the regions between 4 and the centromere (.) and between the centromere and 5.
7. Describe meiosis in a translocation heterozygote. What are the different types of gametes expected to be formed?
8. How can you differentiate between the terms haploidy and monoploidy? How can monoploids be produced and utilized in plant breeding?
9. What is polyploidy? What are different kinds of polyploids? How will you distinguish between autopolyploids and allopolyploids.
10. How can triploidy leads to seedlessness? Discuss this by using examples of Kihara's seedless watermelons.
11. Some of the most desirable apples are triploids. If a desirable mutation occurred in a branch of triploid tree, how would you establish an orchard of trees with this mutation?

12. Briefly discuss the significance and role of polyploidy in evolution. Be sure to state whether autopolyploidy or allopolyploidy has been more important in speciation and why?
13. What is triticale? What is the difference between primary and secondary triticales?

Short Answer Questions

1. What may be an important role of chromosomal duplications in evolution?
2. Differentiate between paracentric and pericentric inversions.
3. Write short notes on the following:
   (*a*) Philadelphia chromosome;
   (*b*) Ring chromosomes;
   (*c*) Dicentric bridge;
   (*d*) Robertosonian translocation;
   (*e*) Pseudodominance;
   (*f*) Deficiency;
   (*g*) Renner complex;
   (*h*) *Oenothera.*
4. How will you distinguish cytologically:
   (*i*) between a double monosomic and a nullisomic;
   (*ii*) between a primary trisomic and a secondary trisomic.
5. Write short notes on the following:
   (*a*) Aneuploidy;
   (*b*) Euploidy;
   (*c*) Nullisomic;
   (*d*) Evolution of wheat;
   (*e*) *Raphanobrassica*;
   (*f*) Colchicine treatment;
   (*g*) Down's syndrome;
   (*h*) Trisomy-18; and
   (*i*) Patau syndrome.

Very Short Answer Questions

1. Assume a chromosome with the following gene sequence [the full stop (.) represents the centromere):

   ABCD. EFGH

   You find the following aberrations in this chromosomes; for each identify the specific kind of aberration:

   (*a*) ABCD. EFH; (*b*) ADCB. EFGH; and (*c*) ABCDCD. EFGH.
2. A chromosome with segments in the normal order is (abcdefgh). An inversion heterozygote abnormal order (abfedcgh). A three-strand double crossover occurs involving the regions a and b and between d and c. Diagram and label the first and second anaphase figures.
3. Abyssinian oat (*Avena abyssinica*) appears to be a tetraploid with 28 chromosomes. The common cultivated oat (*Avena sativa*) appears to be a hexaploid in the same series. How many chromosomes does the common oat possess?
4. The diploid number of an organism is 12. How many chromosomes would be expected in (*a*) a monosomic, (*b*) a trisomic, (*c*) a tetrasomic, (*d*) a double trisomic, (*e*) a nullisomic, (*f*) a monoploid, (*g*) a triploid, (*h*) an autotetraploid?
5. Application of colchicine to a vegetative bud of a homozygous tall diploid tomato plant (DD) caused development of a tetraploid branch. What is the genotype of the somatic cells of this branch?
6. Different species of rhododendron have somatic chromosome numbers of 26, 39, 52, 78, 104 and 156. By what means does evolution appear to be taking place in this genus?
7. How many sets are represented in the species with 156 chromosomes?
8. Both autoploidy and allopolyploidy can result in a species with doubling of the original chromosome number. If you had a plant with 4n complement, how would you determine cytologically if it were an autopolyploid or an allopolyploid?
9. The loci of genes A and B are on different chromosomes. A dihybrid autotetraploid plant of genotype AA aa BB bb is self-pollinated. Assume that only diploid gametes are formed and that the loci A and B are very close to their respective centromere (chromosome segregation).

Find the phenotypic expectations of the progeny.

10. The European raspberry (*Rubus idaeus*) has 14 chromosomes. The dewberry (*Rubus caesius*) is a tetraploid with 28 chromosomes. Hybrids between these two species are sterile F_1 individuals. Some unreduced gametes of the F_1 are functional in backcrosses. Determine the chromosome number and level of ploidy for each of the following: (*a*) F_1; (*b*) F_1 backcrossed to *R. idaeus*; (*c*) F_1 backcrossed to *R. caesius*; to (*d*) chromosome doubling of F_1 (*R. maximus*).

11. If a plant were trisomic for one of its chromosomes and these chromosomes carried the alleles A, A^1, and A^2, respectively, what would be the genotypes of the gametes produced with respect of these genes?

12. What types and proportions of eggs will be produced by a *Drosophila* female trisomic for chromosome 4 and of genotype +/+ +/ey ? (ey = eyeless is a recessive fourth chromosome gene).
    (*a*) If the female is crossed to an ey/ey male, what phenotypic ratio is expected in the offspring?
    (*b*) What proportion of the diploid offspring will be eyeless?

Multiple Choice Questions

1. Colchicine is an inhibitory chemical which
   (*a*) stops the functioning of centriole
   (*b*) prevents attaching of centromeres with rays
   (*c*) prevents the spindle formation in mitosis
   (*d*) prevents the formation of equatorial plane
2. Chromosomal aberration is due to
   (*a*) physical effects
   (*b*) change in structure or number of chromosomes
   (*c*) polyploidy
   (*d*) none of the above
3. Down's syndrome is characterized by
   (*a*) 21 trisomy
   (*b*) two X and one Y chromosome
   (*c*) 19 trisomy
   (*d*) only one X chromosome
4. Patau's syndrome occurs due to
   (*a*) trisomy of 21st chromosome
   (*b*) trisomy of 18th chromosome
   (*c*) trisomy of 22nd chromosome
   (*d*) trisomy of 13th chromosome
5. Which of the following cannot be detected in a developing fetus by aminocentesis?
   (*a*) Down syndrome
   (*b*) jaundice
   (*c*) Klinefelter syndrome
   (*d*) sex of the foetus
6. The loss of a chromosomal segment is due to
   (*a*) polyploidy
   (*b*) deletion
   (*c*) duplication
   (*d*) inversion
7. Respective haploid and monoploid numbers of chromosomes of hexaploid wheat are
   (*a*) 21 and 42
   (*b*) 7 and 21
   (*c*) 21 and 7
   (*d*) 42 and 21
8. Addition or deletion of one or few chromosomes from diploid set of chromosome results in
   (*a*) polyploidy
   (*b*) aneuploidy
   (*c*) euploidy
   (*d*) none of these
9. Loss or gain of one or more complete set of chromosomes alongwith the diploid complement is known as
   (*a*) aneuploidy
   (*b*) euploidy
   (*c*) reverse tandem duplication
   (*d*) substitution mutation

ANSWERS

Very Short Answer Questions

1. (*a*) Deletion; (*b*) Inversion; (*c*) Duplication.
2. First anaphase: a diad, a loop chromatid and an acentric fragment; second anaphase: the diad splits into two monads, the loop forms a bridge and the acentric fragments become lost.
3. 42.
4. (*a*) 11; (*b*) 13 ; (*c*) 14; (*d*) 14; (*e*) 10; (*f*) 6; (*g*) 18.
5. DDDD.
6. Euploidy; the different chromosome numbers are 2n, 3n, 4n, 6n, 8n, and 12n respectively.
7. 12 sets (*i.e.*, 12 × 13).
8. Examine meiotic cells cytologically. If an autopolyploid, complex synaptic associations involving chromosomes (eight chromatids) will be evident. If an allopolyploid, tetrads composed of two chromosomes (four chromatids) each should occur.
9. 1225 AB: 35 Ab: 35 aB: 1 ab.
10. (*a*) 21 triploid; (*b*) 28 tetraploid; (*c*) 35 pentaploid; (*d*) 42 hexaploid.
11. Six kinds of gametes could be produced: AA^1, A^2, AA^2, A^1, A^1A^2 and A.
12. (*a*) Eggs: 1 +/+ : 2 +/ey : 2 + : 1 ey
    Phenotypic ratio: 5 + : 1 ey (*i.e.*, 5 wild type: 1 eyeless);
    (*b*) 1/3 of the diploid progeny will be eyeless.

Multiple Choice Questions

1. (*c*) 2. (*b*) 3. (*a*) 4. (*d*) 5. (*b*) 6. (*b*) 7. (*c*)
8. (*b*) 9. (*b*)

34 DNA: Molecular Structure

When it became evident that the chromosomes were the organs of heredity, because (*i*) they formed the only link between two generations, (*ii*) they carried linearly arranged genes, (*iii*) they occurred in every organism in specific number and had specific morphology for a particular species, and (*iv*) any variation in their number or morphology affected the phenotype of the species, then, various attempts were made by early molecular geniticists to identify the physical and chemical nature of genes. But the genes were found to be so minute structures that their physical identity remained almost impossible. However, the extensive chemical analysis of chromosomes of different organisms have revealed that chromosomes contain proteins and nucleic acids (DNA and RNA), and it was thought that genes might have either proteins or nucleic acids as their component molecules. Early molecular geneticists have assigned the informational roles of genes to the chromosomal proteins because, they found nucleic acids too simple to carry genetic informations. (This is called **tetranucleotide hypothesis**).

The controversy about the assignment of genetic role either to chromosomal proteins or to chromosomal DNA, existed up to 1949 when **A. Mirsky** and **H. Ris** had found that all cells of an organism appeared to contain the same amount of DNA, whereas different cell types contained quite different amounts and kinds of protein. Its constancy, therefore, favoured DNA as the genetic material. Around 1953, it was universally accepted that DNA is the genetic substance (*i.e.*, chemical of which genes are composed) of most microorganisms and higher organisms. Later on, RNA was found to be the genetic material of some viruses. The concept that DNA or RNA is the genetic material of most organisms has been developed and supported by the various direct and indirect evidences.

34.1 HISTORICAL

In 1869, **Friedrich Miescher**, 22-year old Swiss physician (working in the laboratory of Felix Hoppe-Seyler in Tubingen) had isolated from pus cells obtained from discarded bandages in the Franko-Prussian War, and from salmon sperm, a previously identified macromolecular substance, to which he gave the name **nuclein**. Although he was unaware of the structure and function of nuclein, he submitted his findings for publication. The editor who received the paper was doubtful about some aspects of the Miescher's report and delayed publication for two years while he tried repeating some of the more questionable aspects of Miescher's work. Finally, in 1871, Miescher's report was published, but it made little immediate impact. He continued his careful work up to his death in 1895, recognizing (with the help of his student Richard Altmann in 1889) that nuclein was of higher molecular weight and was associated in some way with a basic protein, to which he gave the name **protamine**.

In 1880, **Emil Fischer** identified pyrimidines and purines. The biochemist, **Albrecht Kossel** identified the constituent nitrogenous bases of nuclein as well as its 5-carbon sugar and phosphoric acid. It was **Altmann** who first suggested, in 1899, the use of the term **nucleic acid** to describe phosphorus-containing nuclein. **Kossel** was awarded the 1910 Nobel Prize for demonstrating the presence of two pyrimidines (cytosine and thymine) and two purines (adenine and guanine) in nucleic acids. Kossel's work and later investigations of **Ascoli, Levine** and **Jones** during the first quarter of the 1900s disclosed the two kinds of nucleic acids, deoxyribonucleic acid (DNA) and ribonucleic acid (RNA). Development of DNA-specific staining techniques by **Feulgen** and **Rossenbeck** in 1924 enabled Feulgen to demonstrate in 1937 that most of the DNA content of the cell is located in the nucleus. It was not until the 1950s that the inter-nucleotide bond was established by **A.R. Todd (Judson,** 1979).

34.2. DOXYRIBONUCLEIC ACID OR DNA

Highly purified DNA, extracted from a wide variety of plants, animals, bacteria and viruses, has been found to be complex macromolecular or polymeric chemical compound which contains four kinds of smaller building blocks (monomers) called **deoxyribotids** or **deoxyribonucleotides**. Each deoxyribonucleotide is made up of three moieties: a **phosphoric acid molecule** (biologically called phosphate); a pentose sugar called **deoxyribose**; and **pyrimidine** and **purine** nitrogenous bases. Four major kinds of nitrogenous bases have been found in four kinds of deoxyribonucleotides of DNA: two are heterocyclic and two-ringed purines, **adenine (A)** and **guanine (G)**, and two are one ringed pyrimidines, **cytosine (C)** and **thymine (T)**. The pyrimidine ring can be numbered in two different ways (Fig 34.6). In this book, we will follow the old numbering system.

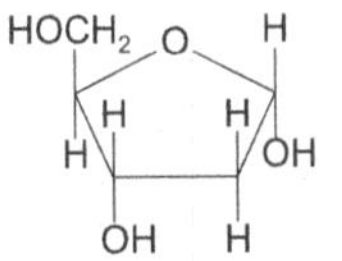

Fig. 34.1. Chemical formula of deoxyribose.

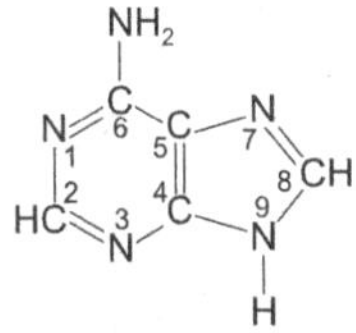

Fig. 34.2. Chemical formula of adenine (6-aminopurine mw = 151.13 daltons).

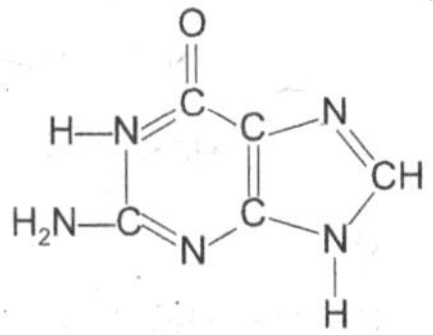

Fig. 34.3. Chemical formula of guanine (2-amino-6-hydroxipurine, mw = 135.13 daltons).

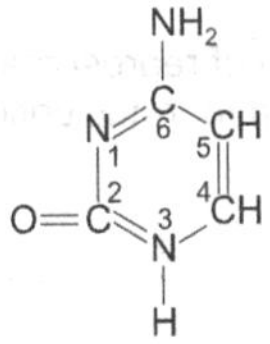

Fig. 34.4. Chemical formula of cytosine (2-hydroxy-6-amino pyrimidine, mw = 111.10 daltons).

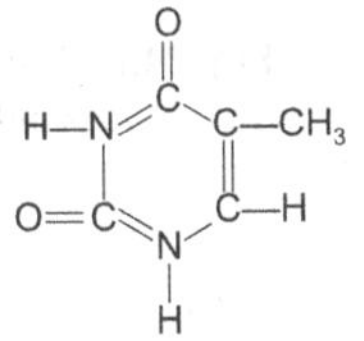

Fig. 34.5. Chemical formula of thymine (2,6-dihydroxy-5-methylpyrimidine, mw = 126.12 daltons).

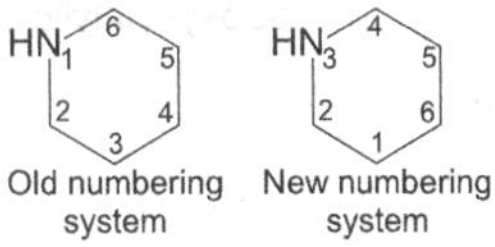

Fig. 34.6. Numbering of carbon atoms in a pyrimidine ring.

That part of each nucleotide which contains a nitrogenous base and deoxyribose is called **deoxyribonucleoside**. The four kinds of deoxyribonucleosides and deoxyribonucleotides can be tabulated as follows:

Table 34.1 Four nitrogen bases, nucleosides and nucleotides of DNA molecule.

Nitrogen base	Base + deoxyribose = deoxyribonucleoside	Deoxyribonucleoside + phosphoric acid = deoxyribonucleotide	Abbreviation for nucleotide
1. Adenine (A)	Deoxyadenosine	Deoxyadenylic acid (Deoxyadenosine monophosphate)	dAMP
2. Guanine (G)	Deoxyguanosine	Deoxyguanylic acid (Deoxyguanosine monophosphate)	dGMP
3. Cytosine (C)	Deoxycytidine	Deoxycytidylic acid (Deoxycytidine monophosphate)	dCMP
4. Thymine (T)	Thymidine	Thymidylic acid (Thymidine monophosphate)	TMP

The four deoxyribonucleotides besides occurring in DNA molecule, occur also in nucleoplasm and cytoplasm, but in their triphosphate forms such as deoxyadenosine triphosphate (dATP), deoxyguanosine triphosphate (dGTP), deoxycytidine triphosphate (dCTP) and thymidine triphosphate (TTP). The significance of occurrence of deoxyribonucleotides in triphosphate forms lies in the fact that during DNA replication, the DNA polymerase enzyme can act only on triphosphate of deoxyribonucleotides.

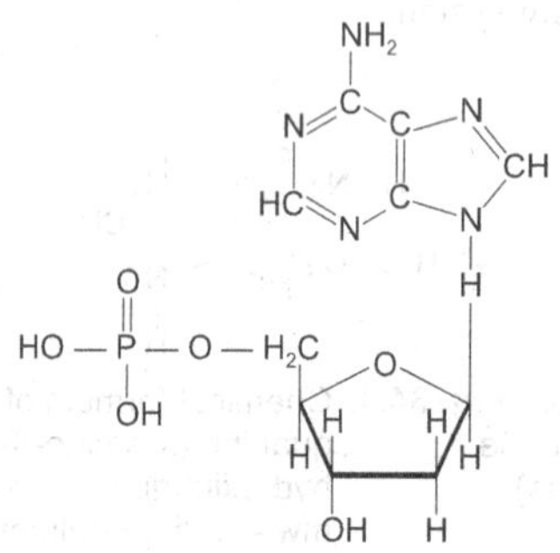

Fig. 34.7. Chemical formula of deoxycytidylic acid.

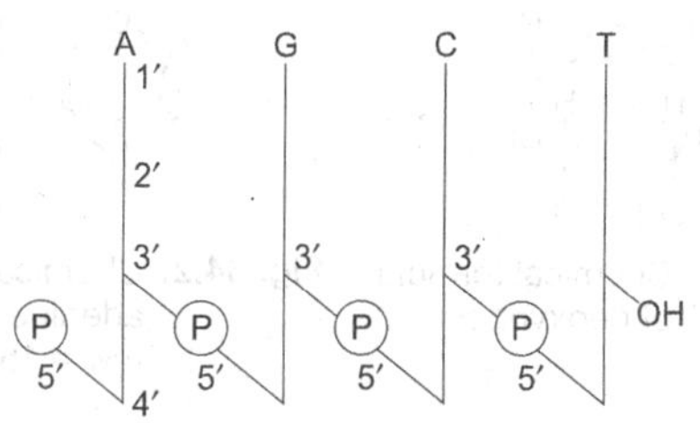

Fig. 34.8. A short hand system of representation of a polynucleotide with four nucleotides.

Molar Ratios of Nitrogen Bases in DNA Molecule

When many samples of DNA were isolated, purified and analysed by various techniques, such as paper chromatography, etc., it was found by **Hotchkiss** and **Chargaff** in 1948 that contrary to Levene's (1920's) tetranucleotide theory which considered DNA as a monotonous polymer having four DNA bases in approximately equal molar proportions, the four nucleotide bases are not necessarily present in DNA in exactly equal proportions. Thus, **Chargaff** reported that the DNA extracted from calf thymus nuclei contains the four bases in the following molar proportions: 28% adenine, 24% guanine, 20% cytosine and 28% thymine. When he analyzed the DNA samples of varied animals, he found that the exact base composition of DNA differs according to its biological source. The relative amounts of purines and pyrimidines in samples of DNA of different living organisms are tabulated as follows:

Table 34.2 Relative amounts of nitrogen bases in different samples of DNA (**Chargaff** and **David son**, 1955).

S.N.	Source	Adenine	Guanine	Cytosine	Thymine	$\frac{A+T}{G+C}$
1.	Beef sperm	28.7	22.2	22.0	27.2	1.26
2.	Human thymus	30.9	19.9	19.8	29.4	1.52
3.	Human liver	30.3	19.5	19.9	30.3	1.53
4.	Human sperm	30.7	19.3	18.8	31.2	1.62
5.	Hen red cells	28.8	20.5	21.5	29.2	1.38
6.	Rat bone marrow	28.6	21.4	21.5	28.4	1.33
7.	Herring sperm	27.8	22.2	22.6	27.5	1.23
8.	*Paracentrotus lividus* (sea urchin) sperm	32.8	17.7	18.4	32.1	1.85
9.	Salmon	29.7	20.8	20.4	29.1	1.43
10.	Wheat germ	26.5	23.5	23.0	27.0	1.19
11.	Yeast	31.3	18.7	17.1	32.9	1.79
12.	*Diplococcus pneumoniae*	29.8	20.5	18.0	31.6	1.59
13.	*K-12 Escherichia coli*	26.0	24.9	25.2	23.9	1.00
14.	*Mycobacterium tuberculosis*	15.1	34.9	35.4	14.6	0.42
15.	Bacteriophage T_2	32.5	18.2	16.7	32.6	1.86

These different values of A, G, C, and T in different samples of DNA are suggesting that DNA rather than being a monotonous polymer, carries genetic information in the form of specific nucleotide base sequences.

The equivalence rule. Chargaff, in 1950, formulated the equivalence rule which suggested that despite wide compositional variations exhibited by different types of DNA, the total amount of purines equaled the total amount of pyrimidines ($A + G = T + C$); the amount of adenine equaled the amount of thymine ($A = T$) and the amount of guanine equaled the amount of cytosine ($G = C$). **Chargaff's equivalence rule** has been found to apply almost universally in different organisms (viruses, bacteria, plants and animals). However, DNA isolated from higher plants and animals was in general rich in adenine and thymine (A : T) and relatively poor in guanine and cytosine (G : C) (*e.g.*, AT/GC ratio of DNA of human beings was 1.40 : 1); whereas DNA isolated from microorganisms (viruses, bacteria and lower plants and animals) was in general rich in guanine and cytosine and relatively poor in adenine and thymine (*e.g.*, AT/GC ratio of DNA of *Mycobacterium tuberculosis* was 0.60 : 1). These differences in AT/GC ratios of microorganisms and higher organisms undoubtedly reflect the difference in genetic information carried by these hereditary molecules and also phylogenetic, evolutionary and taxonomical significance of them.

Physical, Molecular or Geometrical Organization of DNA

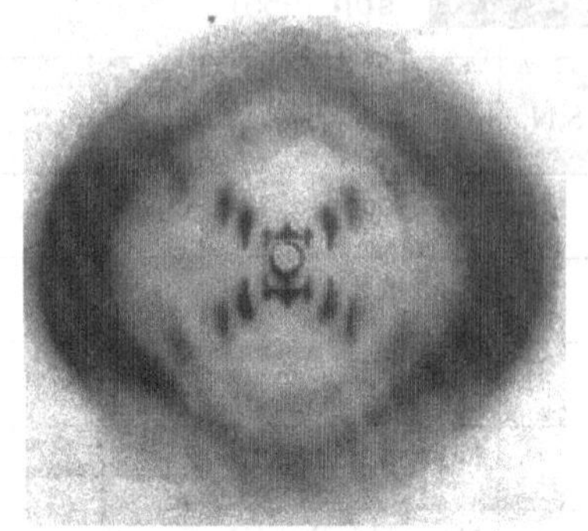

Franklin's original X-ray diffraction photograph.

The first person to give any thought to the three dimensional structure of DNA was **W.T. Astbury** who by his X-ray crystallographic studies of DNA molecule concluded in 1938 that because DNA has high density, so, its polynucleotide was a stack of flat nucleotides, each of which was oriented perpendicularly to the long axis of the molecule and was situated every 3.4 A° along the stack. The X-ray crystallographic studies of Astbury were continued by **Wilkins** and his associates who managed to prepare highly oriented DNA fibres that allowed them to obtain an X-ray diffraction photograph. One of his female associates **Rosalind Franklin** obtained a superior X-ray diffraction photograph of DNA which confirmed Astbury's earlier inference of 3.4 internucleotide distance. Such studies demonstrated that DNA was a helical structure with a diameter of 20 A° and a pitch (one round) of about 34 A°. **Watson** and **Crick** who were already engaged in constructing some suitable model for DNA structure, when observed Franklin's picture of DNA molecule, they immediately utilized that information in constructing a molecular model for DNA. In April 1953, **Watson** and **Crick** published their conclusions about the structure of the DNA in the same issue of '**Nature**', in which **Wilkins** and his colleagues presented the X-ray evidence for that structure.

Considerations of Watson and Crick in the construction of double helical structure of DNA molecule. Watson and **Crick** concluded directly from the X-ray diffraction photograph of DNA taken by **Franklin** that (1) the DNA polynucleotide chain has the form of a regular helix, (2) the helix has a diameter of about 20 A° and, (3) the helix makes one complete turn every 34 A° along its length, and hence, since the internucleotide distance is 3.4 A°, consists a stack of ten nucleotides per turn. Considering the known density of the DNA molecule, **Watson** and **Crick** next concluded that the helix must contain two polynucleotide chains, or two stacks of ten nucleotides each per turn, since the density of a cylinder 20 A° in diameter and 34 A° long would be too low if it contained but a single stack of ten, and too high if it contained three or more stacks of ten nucleotides each. Before trying to arrange these two polynucleotide chains into a regular helix of the required dimensions, however, **Watson** and **Crick** placed a further restriction on their model—a restriction that derived from their knowledge that DNA is, after all, the genetic material. If DNA is to contain hereditary information as they reasoned, and if that information is inscribed as a specific sequence of the four bases along the polynucleotide chain, then the molecular structure of DNA must be able to accommodate any arbitrary sequence of bases along its polynucleotide chains. Otherwise, the capacity of DNA as an information carrier would be too severely limited. Hence, they felt the need of construction of such a regular helix that though is composed of two polynucleotide chains containing an arbitrary sequence of nucleotide bases every 3.4 A° along their length, it would, nevertheless, have a constant diameter of 20 A°. Since the dimension of the purine ring is greater than that of the pyrimidine ring, **Watson** and **Crick** hit upon the idea that the two-chain helix could have a constant diameter if there existed a complementary relation between the two nucleotide stacks, so that at every level one stack harbours a purine base and the other a pyrimidine base. Finally, to endow the helix with thermodynamics stability, the structure would have ample opportunities for the formation of hydrogen bonds between amino or hydroxyl-hydrogens and keto-oxygens or immino-nitrogens of the purines and pyrimidine bases. These considerations led them to construct a double helix model for the molecular structure of DNA molecule.

Watson and Crick's Structural Model of DNA

In DNA molecule the adjacent deoxyribonucleotides are joined in a chain by phosphodiester bridges or bonds which link the 5′ carbon of the deoxyribose of one mononucleotide unit with the 3′ carbon of the deoxyribose of the next mononucleotide unit. According to **Watson** and **Crick** DNA molecule consists of two such polynucleotide chains wrapped helically around each other, with the sugar-phosphate chain on the outside (forming ribbon-like backbone of double helix) and purines and pyrimidines on the inside of the helix (projecting between two sugar phosphate backbones as transverse bars). The two polynucleotide strands are held together by hydrogen bonds between specific pairs of purines and pyrimidines.

The hydrogen bonds between purines and pyrimidines are such that adenine can bond only to thymine by two hydrogen bonds, and guanine can bond only to cytosine by three hydrogen bonds and no other alternative is possible between them. The specificity of the kind of hydrogen bonds that can be formed assures that for every adenine in one chain there will be thymine in the other. For every guanine in first chain there will be a cytosine in the other and so on. Thus, the two chains are complementary to each other; that is, the sequence of nucleotides in one chain dictates the sequence of nucleotides in the other. The two strands run anti-parallely, that is, have opposite directions. One strand has phosphodiester linkage in 3′ → 5′ direction, while other strand has phosphodiester linkage in just reverse or 5′ → 3′ direction. Further, both polynucleotides strands remain separated by 20 A° distance. The coiling of double helix is right handed and a complete turn occurs every 34 A°. Since each nucleotide occupies 3.4 A° distance along the length of a polynucleotide strand, ten mononucleotides occur per complete turn. (The base pairs are rotated 36° with respect to each adjacent pair). The DNA double helix has two grooves of unequal width because of the way the base pairs stack and the sugar phosphate backbones twist. These grooves are called the **major groove** and the **minor groove** (Fig. 34.9). Within each groove, functional group on the edges of the base pairs are exposed to water. Each base pair has a distinctive pattern of chemical groups in the grooves. Because the base pairs are accessible in the grooves, molecules that interact with particular base pairs can identify them without disrupting the helix. This is particularly important fo proteins that must bind to double stranded DNA and "read" a specific sequence.

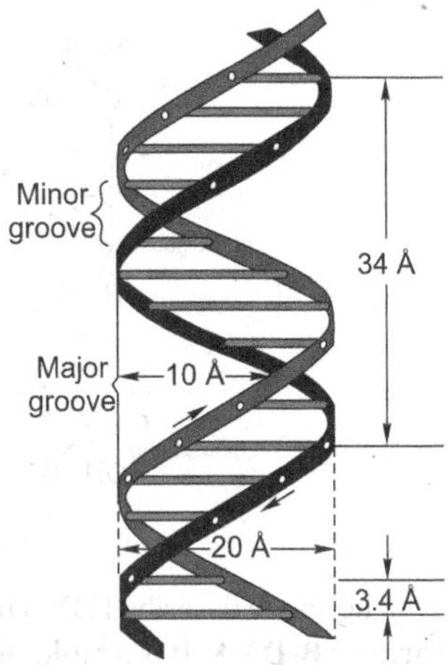

Fig. 34.9. Watson-Crick double helical DNA molecule.

B-DNA is a right-handed helix with a diameter of 2.37 nm. The **rise** of the helix (the distance between one base pairs and the next along the helical axis) averages 0.33 nm, and the **pitch** of the helix (the distance to complete one turn) is about 3.40 nm. These values vary to some extent, depending on the base composition. Because there are about 10.4 base pairs per turn of the helix, the angle of rotation between adjacent nucleotides within each strand is about 34.6 (360/10.4).

Polymorphism of DNA Helix (Or Alternative Forms of DNA Double Helix)

For about 20 years after the discovery of the DNA double helix in 1953, DNA was thought to have the same monotonous structure, with exactly 36° of helical twist between its adjacent base pairs (10 nucleotide pairs per helical turn) and a uniform helix geometry. Subsequent experiments have shown that DNA is much more **polymorphic** than expected. Thus, DNA can be of the following types:

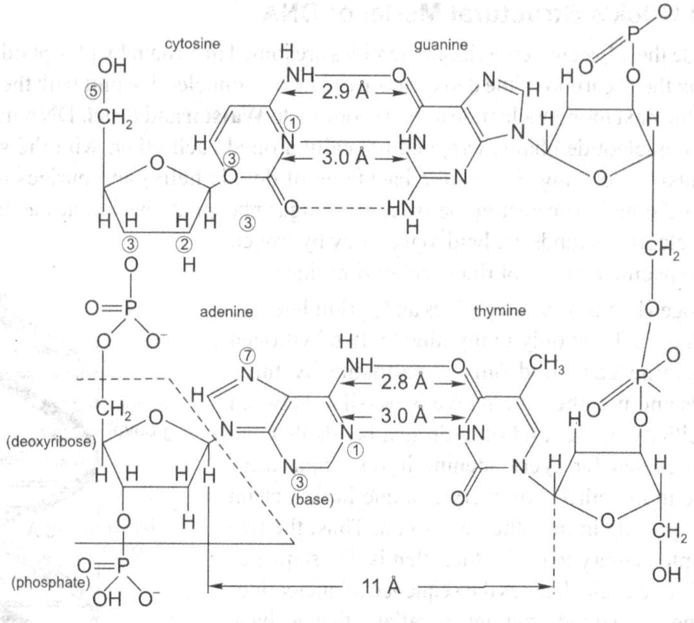

Fig. 34.10. Segment of a DNA molecule.

The above described DNA molecule contains the right handed helical coiling and has been called **B-form** or **B-DNA**. It is a biologically important form of DNA that is commonly and naturally found in most living systems. This double helical structure of DNA is found to exist in other alternative forms (such as A-form and C-form) which differ in features such as (*i*) number of residues (monomers) per turn ("n") or (*ii*) the spacing of residues along the helical axis ("h") (Table 34.3). For example, **A-form DNA** (A-DNA) is right handed but less hydrated than the B-form DNA. A-DNA is more compact with 11 base pairs per turn of the helix and it is 23 angstroms in diameter. The bases are tilted more in relation to the axis of the helix than in the B-DNA. The A-DNA may occur under experimental conditions.

Table 34.3 Important features of different forms of DNA double helical structures.

Helix type	Conditions	Base pair per turn	Rotation per bp	Vertical rise per bp	Helical diameter
A	75 per cent relative humidity; Na^+, K^+, Cs^+, ions	11	+ 32.7° (right handed)	2.56 A°	23 A°
B	92 per cent relative humidity, low ionic strength	10	+ 36.0° (right handed)	3.38 A°	19 A°
C	66 per cent relative humidity; Li^+ ions	9.33	+ 38.6° (right handed)	3.32 A°	19A°
Z	Very light salt concentration	12	– 30.0° (left handed)	3.71 A°	18 A°

The B-DNA is found in fibres of living cells at a very high 92 per cent relative humidity and low ionic strength. Likewise, A-form of DNA is found at 75 per cent humidity in the presence of high ionic strength of Na^+, Ka^+ or Cs^+ ions. C-form of DNA is found at 66 per cent relative humidity in the presence of lithium (Li^+) ions. These three forms are assumed to be found in all DNAs.

There are certain other forms of DNA such as **D-form** and **E-form**, both of which are found as rare extreme variants and contain only 8 and 7½ base pairs per turn respectively. These rare DNA variants are found only in some DNA molecules which lack guanine.

Z-DNA (or Left-handed DNA). Crystallographic studies on synthetic nucleotides consisting of alternating purines and pyrimidines such as GCGCGCGC have shown that left-handed DNA can also exist. This DNA is called **Z-DNA** because of its zigzag structure. Z-DNA can also be found in solutions of high-ionic strength, such as 2M NaCI. In Z-form of DNA, the molecule still consists of two anti-parallel chains, but, otherwise, it is quite different from the A or B form (Fig. 34.11). The helix of Z-DNA is 18 angstrom in diameter, containing 12 base pairs per turn. The differences and similarities of Z-DNA and B-DNA have been summarized in Table 34.3; some other features are the following:

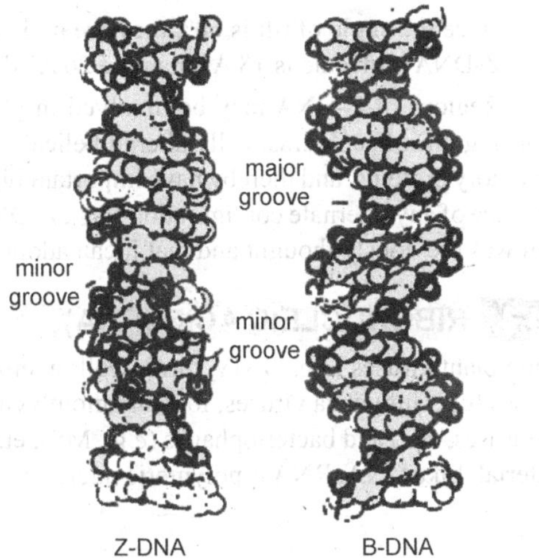

Fig. 34.11. Two molecular forms of DNA double helix, right-handed B-DNA and left-handed Z-DNA. The heavy black line in each molecule goes from phosphate group to phosphate group, indicating a smooth right-handed helix in B-DNA and irregular or zigzag left-handed helix in Z-DNA (after De Robertis and De Robertis Jr., 1987).

A. Similarities between Z-DNA and B-DNA

1. Both are double helical.
2. In both DNAs, two polynucleotide strands of double helix are antiparallel.
3. Both forms exhibit G ≡ C pairing.

B. Differences between Z-DNA and B-DNA

(*i*) Z-DNA has left-handed helical sense, while B-DNA has right-handed helical sense (Fig. 34.12).

(*ii*) The phosphate backbone of Z-DNA follows a zigzag course, while in B-DNA this backbone is regular.

(*iii*) In Z-DNA, the adjacent sugar residues have opposite orientation, while in B-DNA they have same orientation.

(*iv*) In Z-DNA, one complete helix (*i.e.*, a twist through 360°) has twelve base pairs, while in B-DNA one complete helix has only ten base pairs or 10 repeating units.

(*v*) The angle of twist (rotation) per repeating unit (dinucleotide) in Z-DNA is 60° than the 36° of mononucleotide in B-DNA.

(*vi*) In Z-DNA, one complete turn of helix is 45A° long, while in B-DNA it is 34 A° long.

(*vii*) Since bases get more length spread out in Z-DNA and

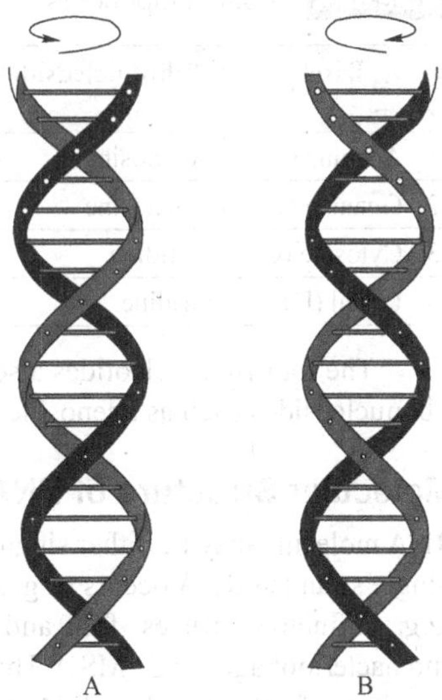

Fig. 34.12. The normal right handed helical sense of B-DNA (A) and rare left handed helical sense of Z-DNA (B).

since the angle of tilt is 60°, they are more closer to the axis. Due to this fact the diameter of Z-DNA molecule is 18 A° than the 20 A° diameter of B-DNA.

Regions of Z-DNA may be involved in gene regulation in the cells of "higher" organisms. Short regions of such drastically altered helical geometry could be specifically recognized by gene regulatory proteins, and thereby have important biological roles (see **Alberts** *et al.*, 1989). Lastly, the presence of any alternate configuration (*e.g.*, Z-DNA) suggests that DNA is a more flexible molecule than was previously thought and that it can adopt in the genome of a variety of forms (**Rich**, 1980).

34.3. RIBONUCLEIC ACID (RNA)

Some plant viruses (*e.g.*, TMV, turnip yellow mosaic viruses, wound tumour viruses, etc.), animal viruses (*e.g.*, influenza viruses, foot and mouth viruses; rous sarcoma viruses, poliomyelitis viruses, reoviruses, etc.) and bacteriophages (*e.g.*, MS_2, etc.) contain ribonucleic acid (RNA) as their genetic material. Like DNA, RNA is polymeric nucleic acid of four monomeric **ribotids** or **ribonucleotides**.

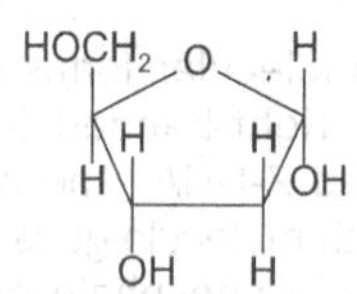

Fig. 34.13. Chemical formula of D-ribose.

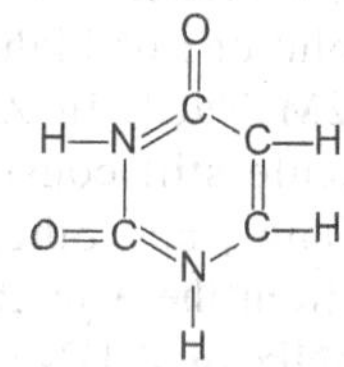

Fig. 34.14. Chemical formula of uracil.

Each ribonucleotide contains a pentose sugar (**D-ribose**); a molecule of phosphate group and nitrogen base. The nitrogen bases of RNA are two purines, **adenine** and **guanine** and two pyrimidines, **cytosine** and **uracil**. The four bases, ribonucleosides and ribonucleotides of RNA can be tabulated as follows:

Table 34.4 Four components of RNA.

Base	Ribonucleoside	Ribonucleotide	Abbreviation for ribonucleotide
1. Adenine (A)	Adenosine	Adenylic acid (Adenosine monophosphate)	AMP
2. Guanine (G)	Guanosine	Guanylic acid (Guanosine monophosphate)	GMP
3. Cytosine (C)	Cytidine	Cytidylic acid (Cytidine monophosphate)	CMP
4. Uracil (U)	Uridine	Uridylic acid (Uridine monophosphate)	UMP

The four ribonucleotides also occur freely in nucleoplasm but in the form of triphosphates of ribonucleosides such as adenosine triphosphate (ATP) and uridine triphosphate (UTP).

Molecular Structure of RNA

RNA molecule may be either single stranded or double stranded but not helical like DNA molecule. Single stranded RNA occurs as genetic material in plant viruses (*e.g.*, TMV, TYM), animal viruses (*e.g.*, influenza viruses, foot and mouth viruses, rous sarcoma viruses, poliomyelitis viruses) and bacteriophages (*e.g.*, MS_2). The non-genetic RNAs except tRNA of prokaryotes and eukaryotes, also have single stranded RNA molecules. The double stranded but non-helical RNA occurs as the genetic material in some plant viruses (*e.g.*, reoviruses). The transfer or soluble RNA (tRNA or sRNA) which is non-genetic RNA of prokaryotes and eukaryotes, is double stranded but non-helical structure.

Each strand of RNA is polynucleotidic, that is, made up of many ribonucleotides. In the polynucleotide strand of RNA, the ribose and phosphoric acids of nucleotides remain linked by phosphodiester bonds. The organisms which have only RNA, is called **genetic RNA**. While the organisms which have DNA along with RNA, they use the RNA in carrying the orders of DNA and in them because RNA has no genetic role, so called **non-genetic RNA**. The non-genetic RNA is heterogeneous and includes the following three genera: ribosomal RNA (rRNA), transfer RNA (tRNA) and messenger RNA (mRNA). Each genera of non-genetic RNA has **DNA-dependent replication** (transcription) of itself, that is, it is not self-replicating like DNA and is transcribed by DNA.

Replication of Genetic RNA

The genetic RNA of viruses is self-replicating, that is, it can produce its own replica by itself. So, its mode of replication is called **RNA-dependent RNA synthesis**. The genetical research on genetic RNA has revealed the following facts about it.

1. The viral RNA functions directly as a messenger RNA which, in association with the ribosomal apparatus of the host, directs the synthesis of both the **RNA polymerase enzyme** (required for RNA replication) and the proteins of the viral coat.
2. With the mediation of RNA polymerase and on the standard base-pairing principles, the viral RNA serves as a template in the synthesis of a complementary RNA chain, and thus a double stranded structure is produced.

Table 34.5 DNA-RNA similarities and differences. (Source: Sylvia S.Mader, 1998).

A. DNA-RNA similarities	
1. Both are nucleic acids.	
2. Both are composed of nucleotides.	
3. Both have a sugar-phosphate backbone.	
4. Both have four different types of bases.	
B. DNA-RNA differences	
DNA	**RNA**
1. Found in the nucleus.	1. Found in nucleus and cytoplasm.
2. The genetic material.	2. Helper to DNA.
3. Sugar is deoxyribose.	3. Sugar is ribose.
4. Bases are A,T,C,G.	4. Bases are A,U,C,G.
5. Double stranded.	5. Single stranded.
6. Is transcribed (to give mRNA).	6. Is translated (to give proteins).

QUESTIONS

Long Answer Questions

1. Discuss, in brief, the structure of deoxyribonucleic acid and compare it with that of ribonucleic acid.
2. Give an account of Watson and Crick's double-stranded model of DNA molecule.
3. Do the two strands of DNA helix carry the same genetic information? Explain.
4. What kind of evidence indicates (*i*) that DNA can reproduce itself and (*ii*) that nucleotides occur in matched pairs in DNA molecule?

Short Answer Questions

1. Write short notes on the following:
   (*i*) Z-DNA,
   (*ii*) B-DNA,

(*iii*) Tetranucleotide hypothesis,

(*iv*) Left-handed DNA versus right-handed DNA.

2. Assume the following base sequence was found in a 20 base DNA strand:

   3′ ATT CGA CCT TAT TAC TGC AC 5′

   (*a*) What would be the first 5 bases in the 3′ end of the complementary strand?

   (*b*) What would be the 10 bases of the 5′ end of the complementary strand?

   (*c*) Assuming the presence of complementary strands, what is the per cent composition of the polymer with respect to AT base pairs and with respect to GC base pairs?

3. Analysis of four double-stranded DNA samples yielded the following information:

   1.15% cytosine; 2.12% guanine; 3.35% thymine; 4.28% adenine.

   (*a*) What would be the percentage of the other bases in each sample?

   (*b*) Could any of these samples have been obtained from the same organism? If so, which ones?

4. Four samples of nucleic acid were analyzed for the proportion of the different bases present, with the following results:

   (*i*) A = 30%, C = 30%,
   G = 20%, T = 20%.

   (*ii*) A = 27.5%, C = 22.5%,
   G = 22.5%, T = 27.5%.

   (*iii*) A = 18%, C = 32%,
   U = 32%, G = 18%.

   (*iv*) A = 18%, C = 32%,
   U = 18%, G = 32%.

   Which of these samples were DNA and which were RNA? Which were double-stranded?

Very Short Answer Questions

1. Ratio of the bases present in different samples of nucleic acid yielded the following results:

   (*i*) (A + C)/(T + G) = 1,

   (*ii*) (A + C)/(U + G) = 0.8,

   (*iii*) (A + G)/(T + C) = 1.5.

   Which were RNA and which were DNA? Which were single and double-stranded?

2. If one DNA sample had a melting temperature of 85.5°C and another showed a melting temperature of 88°C, what might you conclude concerning the base composition of the two samples?

3. Assume an average-sized gene consisting of a linear sequence of 1000 bases and there were 1000 genes in a bacterial chromosome.

   (*a*) How many bases would such a chromosome contain in each strand of the double helix?

   (*b*) If 10 nucleotides = 34 A°, how long would this chromosome would be in millimeters?

4. (*a*) What background material did Watson and Crick have available for developing a model of DNA?

   (*b*) What was their contribution to the building of the model?

5. (*a*) Why was a double helix chosen for the basic pattern of the molecule?

   (*b*) Why were hydrogen bonds placed in model to connect the base?

6. What are differences between DNA and RNA?

7. RNA was extracted from TMV (tobacco mosaic virus) particles and found to contain 20 per cent cytosine. Using this information, is it possible to predict what percentage of the bases in TMV are adenine? If so, what percentage? If not, why not?

Multiple Choice Questions

Choose the correct answer from the four alternatives given.

1. In a sample, DNA is found to have the base composition (mole ratio) of adenine = 40, T = 22, G = 21 and cytosine = 17. It shows that

   (*a*) DNA is circular duplex

(*b*) DNA is linear duplex
(*c*) DNA is single stranded
(*d*) DNA has melting point

2. In the double helix model of DNA, how far is each base pair from the next base pair?
(*a*) 3.4 nm
(*b*) 0.34 nm
(*c*) 2.0 nm
(*d*) 34 nm

3. The scientist who developed cytochemical technique for identification DNA was
(*a*) McClintok
(*b*) Abbe
(*c*) Fuelgen and Rossenbeck
(*d*) Waldeyer

4. What is common point of similarity between DNA and RNA?
(*a*) both are double stranded
(*b*) both have identical sugar molecules
(*c*) both have identical pyrimidine bases
(*d*) both are polymers of nucleotides

5. DNA strands are antiparallel because of
(*a*) hydrogen bonds
(*b*) phosphodiester bonds
(*c*) disulphide bonds
(*d*) glycosidic bonds

6. Chargaff's rule is applicable to
(*a*) single stranded RNA
(*b*) single stranded DNA and RNA
(*c*) single stranded DNA
(*d*) double stranded DNA

7. The diameter of Z-DNA molecule is
(*a*) 18A°
(*b*) 22A°
(*c*) 45A°
(*d*) 34A°

8. What is the type of coiling in DNA?
(*a*) right handed
(*b*) left handed
(*c*) opposite
(*d*) zig-zag

9. Diameter of DNA helix is
(*a*) 20A°
(*b*) 10A°
(*c*) 30A°
(*d*) 40A°

10. Which ratio is constant for DNA?
(*a*) A + G / T + C
(*b*) A + T / G + C
(*c*) A + C / U + G
(*d*) A + U / C + G

ANSWERS

Very Short Answer Questions

1. (*i*) Double-stranded DNA, (*ii*) Single-stranded RNA, (*iii*) Single-stranded DNA.
2. The second sample has a higher G-C content.
3. (*a*) 1 million, (*b*) 3.4 mm.
4. (*a*) The ladderlike pattern was known from X-ray diffraction studies. Chemical analyses had shown that 1 : 1 relationship existed between the organic bases adenine and thymine and between cytosine and guanine. Physical data concerning the length of each spiral and the stacking of bases were also available.

   (*b*) Watson and Crick developed the model of a double helix, with the rigid strands of sugar and phosphorus forming spirals around an axis, and hydrogen bonds connecting the complementary bases in base-pairs.

5. (*a*) A multistranded, spiral structure was suggested by the X-ray diffraction patterns. A double-stranded helix with specific base-pairing fits the 1 : 1 stoichiometry observed for A : T and G : C in DNA (*stoichiometry* is a branch of chemistry that treats of the proportions of elements or compounds involved in reactions, and the methods of calculating them).
   (*b*) Use of the known hydrogen-bonding potential of the bases provided a means of holding the two complementary strands in a stable configuration in such a double helix.
6. DNA has one atom less of oxygen than RNA in the sugar part of the molecule. In DNA, thymine replaces the uracil that is present in RNA. (In certain bacteriophages, DNA containing a uracil is present). DNA is most frequently double-stranded, but bacteriophages such as $\phi \times 174$ contain single-stranded DNA. RNA is most frequently single-stranded. Some viruses, such as reoviruses, however, contain double-stranded RNA chromosomes.
7. No. TMV RNA is single stranded. Thus, the base-pair stoichiometry of DNA does not apply.

Multiple Choice Questions

1. (*c*)	**2.** (*b*)	**3.** (*c*)	**4.** (*d*)	**5.** (*b*)	**6.** (*d*)	**7.** (*a*)
8. (*a*)	**9.** (*a*)	**10.** (*a*)				

35 DNA: Replication, Repair and Recombination

As a carrier of genetic information DNA has the following two important functions: **1. Heterocatalytic function**—When DNA directs the synthesis of chemical molecules other than itself (*e.g.*, synthesis of RNA, proteins, etc.), then such functions of DNA are called heterocatalytic functions. **2. Autocatalytic functions**—The functions of DNA which directs the synthesis of DNA itself, are called autocatalytic functions. Here, we are concerned only with autocatalytic functions of DNA.

35.1. WATSON AND CRICK'S MODEL FOR DNA REPLICATION

The double helix model of DNA molecule of Watson and Crick is beautifully embodied a built-in template system for self-replication or autocatalytic function. Because of the specificity of base pairing, the sequence of bases along one chain automatically determines the base sequence along the other. Thus, each chain of the, double helix can serve as **template** for the synthesis of the other. For the replication of DNA molecule, **Watson** and **Crick** proposed that replication involved the disruption of hydrogen bonds followed by a rotation and separation of the two polynucleotide strands. Each purine and pyrimidine base of each polynucleotide strand is thought to attract a complementary free nucleotide available for polymerization in cell and to hold it in place by means of the specific hydrogen bonds.

Once held in place on the parent template chain, the free nucleotides were joined together by formation of the phosphate diester bonds that linked adjacent deoxyribose residues, forming a new polynucleotide molecule of predetermined base sequence. Thus, two double helical molecules, identical with each other, are formed (Fig. 35.1).

The Watson-Crick's model of DNA structure and replication suggested that once DNA replication is initiated, the two original polynucleotide strands of the duplex of helix will unwind, at least locally, so that each can serve as a template for a new strand.

An immediate prediction follows from this proposal: both duplexes that result from replication should be **hybrid** in nature, each containing an old strand derived from the original molecule and a new strand which has been formed during the replication process (Fig. 35.1). Since each of the two double helices or duplexes conserves only one of the parent polynucleotide strands, the process is said to be **semiconservative**. This prediction is diagrammed in Fig 35.2 in which old DNA is shown in black and new DNA in white. Figure 35.2 also outlines what would be predicted if these two hybrid duplexes went on to replicate themselves. Four duplexes would result, two of which would contain a single strand derived from the original chromosome and two of which would contain totally new DNA strands.

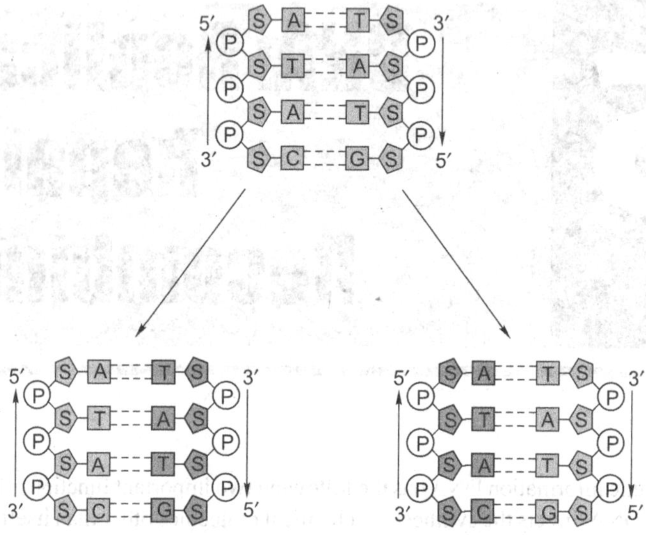

Fig. 35.1. Semi-conservative model of DNA replication.

Besides the semi-conservative mode of DNA replication, the following two methods of DNA replication were deemed equally feasible: (*i*) **conservative replication**, in which both strands of parent double helix would be conserved and the new DNA molecule would consist of two newly synthesized strands; and (*ii*) **dispersive replication**, in which replication would involve fragmentaion of the parent double helix and the intermixing of pieces of the parent strands with newly sythesized pieces. thereby forming the two new double helices.

1. Experimental Evidence for Semiconservative DNA Replication in *E.coli*

(*i*) Meselson and Stahl's experiment. M. Meselson and **F.W. Stahl** (1958) verified the semiconservative nature of DNA replication in a series of elegant experiments using isotopically labelled DNA and a form of isopycnic density gradient centrifugation. They cultured *Escherichia coli* cells in a medium in which the nitrogen was ^{15}N (a 'heavy' isotope of nitrogen, but not a radioisotope) instead of commonly occurring and lighter ^{14}N. In time, the purines and pyrimidines of DNA in new cells contained ^{15}N (where ^{14}N normally occurs) and, thus, the DNA molecules were denser. DNA in which the nitrogen atoms are heavy (^{15}N) can be distinguished from DNA containing light nitrogen (^{14}N), because during isopycnic centrifugation, the two different DNAs band at different density positions in the centrifuge tube.

Depending on its content of ^{15}N and ^{14}N, the DNA bands at a specific position in the density gradient. Because the DNA synthesized by *E.coli* cells grown in ^{15}N would be denser than ^{14}N containing DNA, it would band further down the tube.

E.coli cells grown for sometime in the presence of ^{15}N-medium were washed free of the medium and transferred to ^{14}N-containing medium and allowed to continue to grow for specific lengths of time (*i.e.*, for various numbers of generation time). DNA isolated from cells grown for one generation of time in the ^{14}N medium had a density intermediate to that of the DNA from cells grown only in ^{15}N-containing medium (identified as *generation O*; Fig. 35.3) and that of DNA from cells grown only in ^{14}N-containing medium (*the controls*). Such a result immediately ruled out the possibility that DNA replication was conservative, because the conservative replication would have yielded two DNA bands in the density gradient for **generation 1** (*i.e.*, F_1 cells). The single band of intermediate density (identified as "**hybrid DNA**") consisted of DNA molecules in which one strand contained

^{15}N and the other contained ^{14}N. When the incubation in the ^{14}N-medium was carried out for two generations of time (*i.e.*, **generation 2**), two DNA bands were formed—one at the same density position as the DNA from cells grown exclusively in ^{14}N medium (*i.e.*, light controls) and the other of intermediate density. Subsequent generations produced greater numbers of DNA molecules that banded at the "light" (^{14}N-containing DNA) position in the density gradient. These results are consistent only with the model of semiconservative replication. Studies using other prokaryotes as well as eukaryotes indicates that semiconservative replication of DNA is probably a universal mechanism.

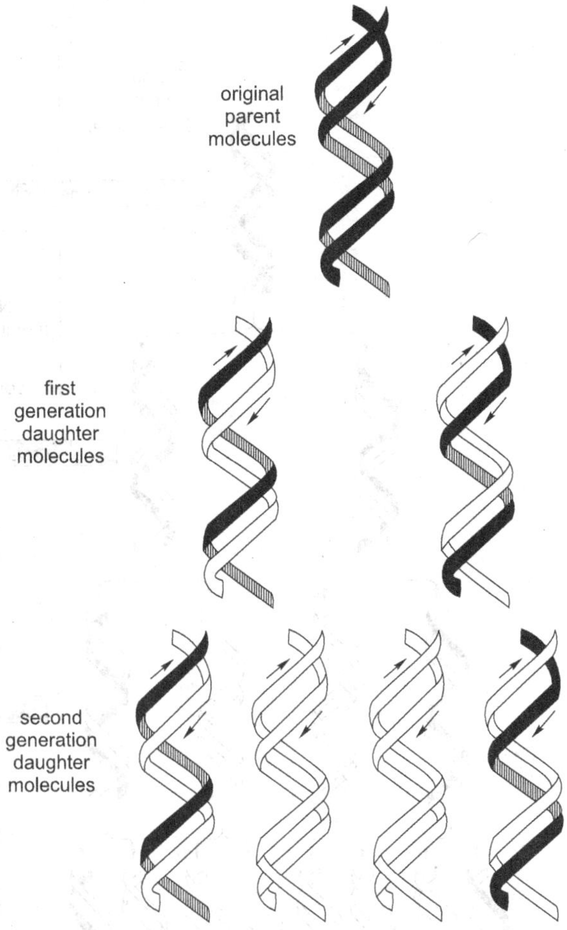

Fig. 35.2. Semiconservative DNA replication.

***(ii)* Visualization of replication in *E.coli*.** In 1963, **J.Cairns** developed a technique employing a combination of microscopy and autoradiography that made it possible to visualize the replication of the chromosome of *E.coli*. Cairns placed *E.coli* cells in a medium containing ^{3}H-thymidine (tritiated-thymidine) for various periods of time so that the radioactive thymidine was incorporated into the DNA as the chromosome was replicated in successive generations of cells. *E.coli* cells were removed from the medium after various periods of incubation and gently lysed to release the chromosome from the cell (since the shear forces created by harsh lysis break the chromosome into small pieces). The chromosomes were then transferred to glass slides and coated with a photographic emulsion sensitive to the low energy beta-particles emitted by the ^{3}H-thymidine. After exposing the emulsion to the beta rays, the emulsion was developed and examined by light microscopy. Wherever decay of labelled thymidine had occurred in a chromosome, the emulsion was exposed and created visible grains. A chromosome not engaged in replication appeared as a circular structure formed from a close succession of exposed spots. Chromosomes "caught in the act" of replication gave rise to what are called **theta configurations** because they have the appearance of the Greek letter theta (*i.e.*, θ, Fig. 35.20) The theta structures reveal the positions of the replication forks in the circular chromosomes. Cairns' observations also clearly supported the semiconservative nature of replication.

The rate at which the replication proceeds could also be worked out by measuring the length of DNA undergoing replication in a known interval of time. Cairns worked out generation time of *E.coli* as 30 minutes. The length of the chromosome was worked out to about 1 mm. The rate of replication, thus, would be approximately 30 μm – 40 μm per minute (1 mm = 1000 μm).

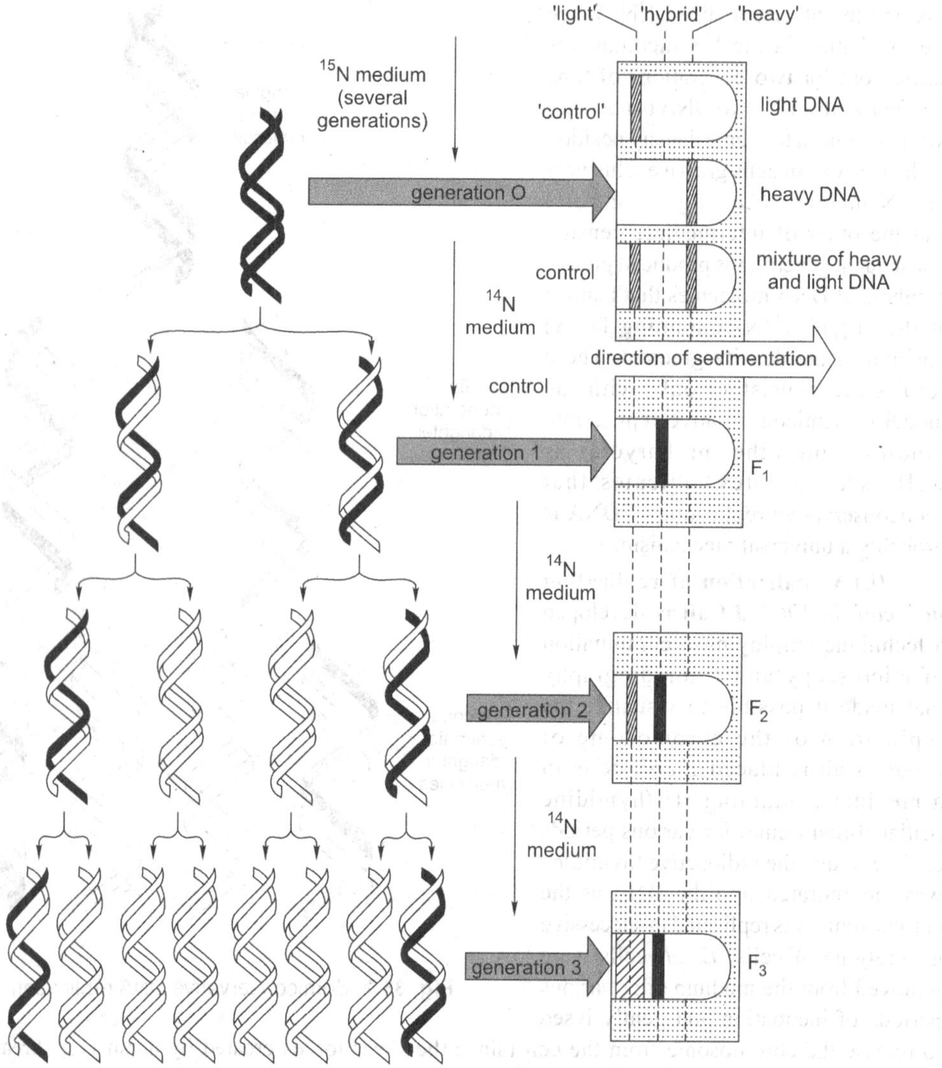

Fig. 35.3. Results of the Meselson-Stahl experiments (right) and their interpretation (left) (after Sheeler and Bianchi, 1987).

2. Evidence for Semiconservative Replication of Chromosomes (or DNA) in Eukaryotes

J.H. Taylor and **P. Woods** (1957) provided evidence in support of semiconservative mode of DNA replication in eukaryotes by using the technique of autoradiography and light microscopy in dividing root tip cells of the bean, *Vicia faba*. After incorporation of tritiated thymidine, when root tips were transferred to unlabelled culture medium (and colchicine was added to the medium to prevent anaphase separation of sister chromatids), in the first generation of duplication both chromatids were labelled (this is interpreted as one DNA double helix in each chromatid and only one of the two strands labelled). In the second cycle of duplication (in the unlabelled medium) in each chromosome, one of the two chromatids was found to be labelled (Fig. 35.4). This was interpreted , as showing semiconservative mode of duplication.

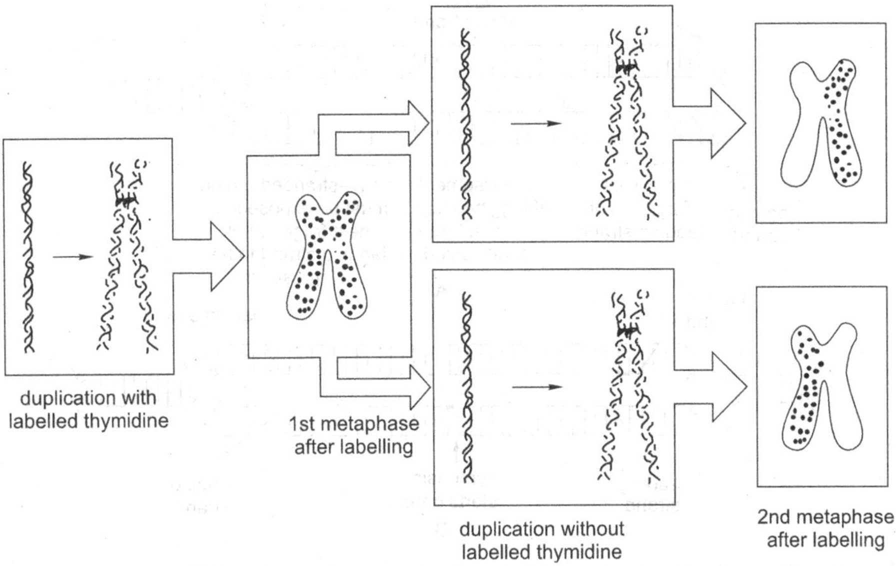

Fig. 35.4. Taylor and Wood's experiment on *Vicia faba* root tips using autoradiography technique (after Sheeler and Bianchi, 1987).

3. Semidiscontinuous DNA Replication

Various experimental evidences have suggested that DNA synthesis is continuous on one strand (3′ to 5′ strand), called **leading strand** and discontinuous on the other strand (5′ to 3′ strand), called **lagging strand**. Since DNA synthesis always proceeds in 5′ to 3′ direction, so, on the lagging strand, synthesis takes place discontinuously in pieces, called **Okazaki fragments** (after the name of discoverer **R. Okazaki**, 1968). Okazaki fragments average about 1,500 nucleotides in prokaryotes and 150 in eukaryotes (see **Tamarin**, 2002). Later on, these pieces are fused with the help of ligase enzyme to form an intact lagging strand. Such a DNA replication, where the leading strand is synthesized continuously and the lagging strand is synthesized discontinuously, is called **semidiscontinuous replication**. Once initiated, continuous DNA replication can proceed indefinitely. DNA polymerase III on the leading-strand template has what is called **processivity**: once it attaches, it does not release until the entire strand is replicated. Discontinuous replication, however, requires the repetition of four steps: primer synthesis, elongation, primer removal with gap filling and ligation.

4. Unidirectional and Bidirectional DNA Replication

Regarding the direction of replication, DNA replication may be of the following two types:

(*a*) **Bidirectional replication.** All known DNA molecules, with only few exceptions, replicate as circles (or bubbles/eyes) and, hence, initiate within the helix. In the electron microscope, eukaryotic chromosomes are found to contain multiple expanding replication eyes, in contrast to a single eye of the prokaryotic DNA. Further DNA synthesis within a given replication unit eye is initiated some where at or near the midpoint of the unit at a site termed as the origin (O); prokaryotes contain a solitary origin, while eukaryotes have multiple (up to several thousands) origins for DNA replication. Both ends of the eye are moving and serve as **replication forks** (Fig. 35.5).

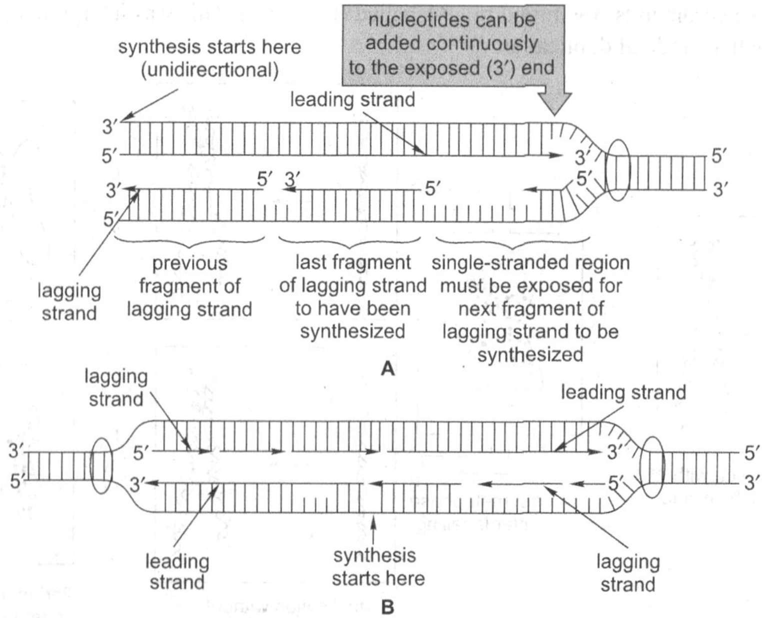

Fig. 35.5. A—Unidirectional DNA replication; B—Bidirectional DNA replication.

Two replicating forks are then believed to travel in opposite directions until they reach either end of that unit, the two end points being called **termini(T)** (Fig. 35.6). A given replication unit may or may not undergo bidirectional synthesis in synchrony with continuous units.

In either case the newly replicated strands in adjacent units will eventually meet. These strands are then linked, perhaps, by a DNA ligase to form long, continuous daughter DNA strands (Fig. 35.6).

***(b)* Unidirectional replication.** In case of this type of DNA replication, one of two ends of the replication eye remains stationary and the other end serves as the replication fork and moves with replication. An example of unidirectional replication is the replication of mitochondrial DNA (mt DNA) by D-loop (or displacement loop) in vertebrates.

35.2. ENZYMES AND PROTEINS OF DNA METABOLISM

Different prokaryotic and eukaryotic cells have been found to contain three kinds of nuclear enzymes or enzymatic activities that act on DNA, namely nucleases, polymerase and ligases.

A. Enzymes involved in DNA Replication

(I) Nuclease enzymes. The nuclease enzymes act to hydrolyze or break down a polynucleotide chain into its component nucleotides. A polynucleotide is held together by 3′, 5′ phosphodiester bonds and a nuclease enzyme will attack either the 3′ or the 5′ end of this linkage. The nuclease enymes may be of the following two kinds.

***(a)* Exonuclease enzyme.** A nuclease enzyme which begins its attack from a free end of a polynucleotide is called **exonuclease**. Therefore, depending on the specificity of the enzyme, an exonuclease will either begin at a free 3′-OH end of a polynucleotide and progressively cleave the bonds on the 3′-OH side of the phosphodiester backbone or it will begin at a free 5′-P end and digest the polynucleotide in a 5′ → 3′ direction (Fig. 35.7). In both cases the enzyme travels along the chain in a stepwise manner, liberating single nucleoside monophosphate molecules and eventually digesting entire polymer.

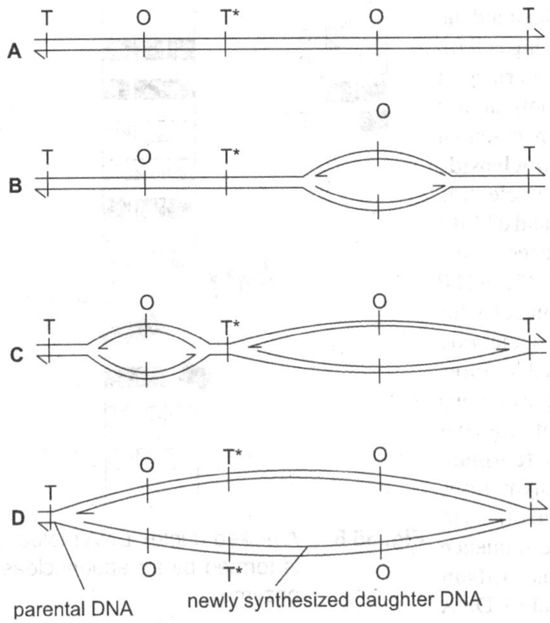

Fig. 35.6. Bidirectional mode for mammalian chromosomal DNA replication. Origins are indicated by O and termini by T. A—DNA prior to replication; B—replication started in right-hand replication unit; C—replication started in left-hand replication unit and completed at termini of right-hand unit; D—replication completed, sister duplexes joined at common terminus T.

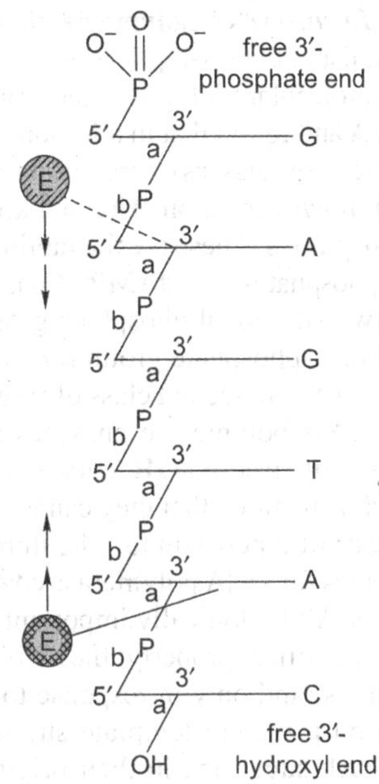

Fig. 35.7. Exonuclease action on a nucleotide chain. The 3′ side of each phosphodiester linkage is labelled a, the 5′ side is labelled b. An exonuclease that cleaves a linkage, starting from a free 3′-OH end (*e.g.*, snake venom phosphodiesterase) is shown in check. An exonuclease that cleaves b linkages, starting from a free 5′-p end (*e.g.*, bovine spleen phosphodiesterase) is shown in line shade (after Lehninger, 1970).

(*b*) Endonuclease enzyme. Endonuclease enzyme also attacks one of the two sides of phosphodiester linkages, but they react with those bonds that occur within the interior of a polynucleotide chain. If the polynucleotide chain is single-stranded (*e.g.*, viral DNA), such an attack will obviously cut the chain into two pieces. If, however, the polynucleotide strand is a member of a DNA helix (*e.g.*, prokaryotic and eukaryotic DNA), a single endonucleolytic cut will create a **nick** in the helix (Fig. 35.8); the helix remains in one piece but it now possesses a gap that contains two free ends, which can serve as substrates for exonucleases. A nicked double helix suffers a localized disruption of its secondary structure. The molecule becomes free to bend or rotate around its intact strand and the two broken ends are free to dangle. The increased molecular motion in the region of the nick will very likely interrupt hydrogen bonding between bases in the vicinity of the nick, thus, effecting a limited "untravelling" of the helix.

(II) Polymerase or replicase enzymes. A polymerase enzyme catalyses the formation of a polymer and cellular polymerase enzymes of genetic interest are those that bring about the synthsis of one polynucleotide chain that is a copy of another. A polymerase enzyme is called **replicase** enzyme when the copy of polynucleotide chain so produced is inherited by daughter cells or viruses, that is, when the enzyme brings about chromosome replication (see **Goodenough** and **Levine**, 1974).

***In vitro* DNA polymerization.** To understand the mechanism of DNA-replication inside the living cell (*in vivo*), molecular biologists tried *in vitro* polymerization of DNA and found that in addition to DNA polymerase enzyme, three classes of organic molecules are essential for an *in vitro* reaction. The first are **deoxynucleoside triphosphates.** These are the familiar deoxynucleotide monophosphates (*i.e.*, dAMP, dCMP, dGMP and dTMP) with two additional phosphate groups attached to the initial or α-phosphate group (*i.e.*, dATP, dCTP, dGTP and dTTP). The second class of essential molecules for many DNA polymerase enzymes are polynucleotide chains with free **3′-OH** ends (often called **primer strands**), meaning that they cannot initiate the *de novo* synthesis of a new strand. The third class of essential molecules for DNA polymerase enzymes are **template strands.** All biologically important DNA polymerases possess a critical property: they will add nucleotides to a primer strand only in response to the base sequence found on a second template strand. Just as **Watson** and **Crick** suggested in their original model of DNA replication, the polymerase enzymes observe the rule of complementary base pairing. Thus, if a free 3′-OH group on a primer strand lies opposite a thymine on a template strand (see Fig. 35.9), a polymerase enzyme will add only an adenine group to the primer, even when dCTP, dTTP, dGTP are also present in the reaction mixture.

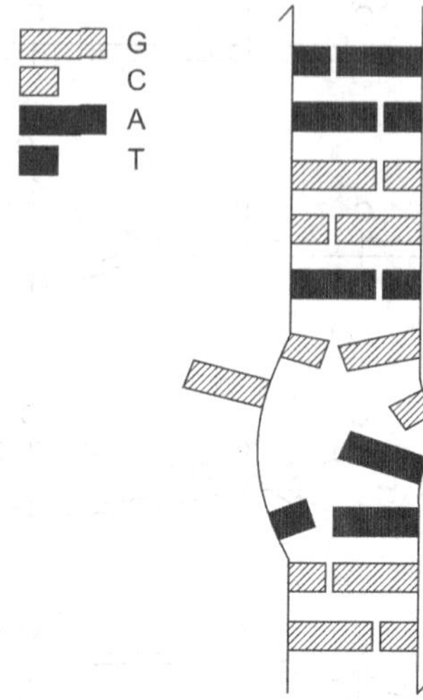

Fig. 35.8. A nicked duplex DNA molecule is formed by an endonuclease enzyme.

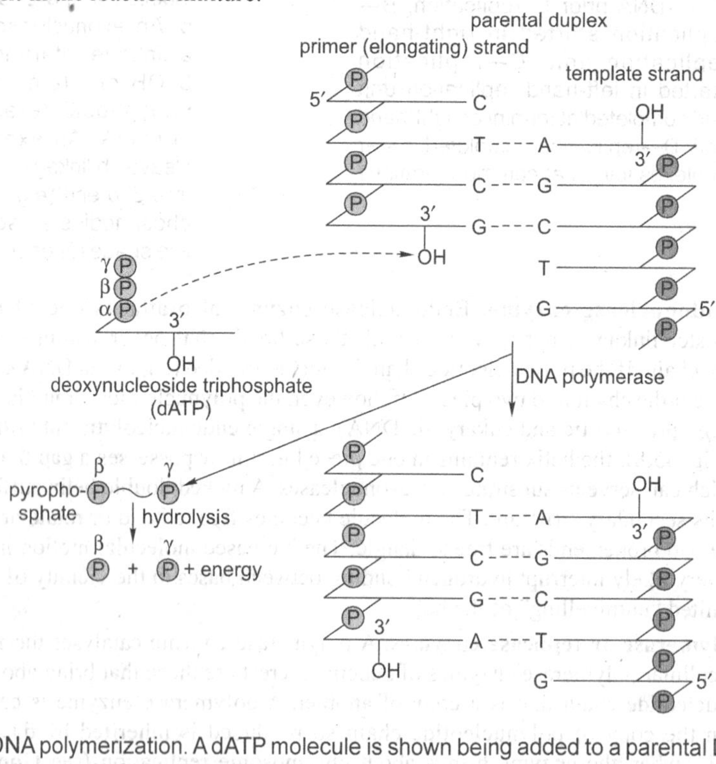

Fig. 35.9. DNA polymerization. A dATP molecule is shown being added to a parental DNA duplex or helix at its 3′-OH end. As DNA polymerase catalyzes the formation of the phosphodiester bond, pyrophosphate (P ~ P) is hydrolyzed into two phosphate molecules.

Once the appropriate molecules are present and the appropriate ionic conditions are maintained, *in vitro* polymerization reaction summarized in Fig. 35.9 will occur. The α-phosphate of a nucleoside triphosphate molecule forms a 3′, 5′ phosphodiester bond with a free 3′-OH in the growing polynucleotide chain, and a molecule of **pyrophosphate** (P ~ P) is simultaneously released. Pyrophosphate contains a "high-energy" or " ~ " bond, meaning that when the released pyrophosphate is hydrolyzed into two phosphate molecules, energy is liberated which drives the polymerization process forward.The resultant polymerization will always proceed in a net 5′ → 3′ direction, meaning that the nucleotide at the 3′ end is always the most recently added to the chain.

(1) Prokaryotic DNA polymerases. Three different DNA polymerases are known in *E.coli* and other prokaryotes (see Table 35.1), of which **DNA polymerase I** and **II** are meant for DNA repair and **DNA polymerase III** is meant for actual DNA replication.

Table 35.1 Comparison of different characters of three types of DNA polymerases of *E.coli*.

	Character	DNA polymerase		
		I	II	III
1.	Polymerization 5′ → 3′	Yes	Yes	Yes
2.	Exonuclease 3′ → 5′	Yes	Yes	Yes
3.	Exonuclease 5′ → 3′	Yes	No	No
4.	Use of primer single strands	Yes	No	No
5.	Use of nicked duplex or helix	Yes	No	No
6.	Molecular weight	109,000 (single chain)	120,000 (single chain)	>250,000 (heteromultimeric chain)
7.	Molecules per cell	400	Not known	10 – 12
8.	Gene	*pol A*	*pol B*	*pol C* (*dna E*), *dna N*, *dna X*, *dna Q*, *dna T*
9.	Nucleotides polymerized at 37°C molecule/minute	up to 1,000	up to 50	up to 15,000
10.	Affinity for nucleoside triphosphate (NTPs) precursors	low	low	high

1. DNA polymerase I. This enzyme was isolated around 1960 by **Arthur Kornberg** and was the first enzyme suggested to be involved in DNA replication. It is also called **Kornberg enzyme.** DNA polymerase I enzyme is now considered to be a DNA repair enzyme rather than a replication enzyme. This enzyme is known to have five active sites, namely **template site, primer site, 5′ → 3′ cleavage** or **exonuclease site, nucleoside triphosphate site** and **3′ → 5′ cleavage site** (or **3′ → 5′ exonuclease site**) (Fig. 35.10). DNA polymerase I is mainly involved in removing RNA primers from Okazaki or precursor fragments and filling the resultant gaps due to its 5′ → 3′ polymerizing capacity. DNA polymerase I enzyme can also remove thymine dimers produced due to UV-irradiation and fill the gap due to excision. Both polymerization (= chain elongation) and exonuclease activity of DNA polymerase I have been shown in Fig. 35.13. This is called **proof reading** or **editing function** of this enzyme.

Klenow fragment/polymerase. T. Steitz and coworkers have done much X-ray crystallography work that has given us a good look at the structure of a DNA polymerase. In fact, most of their work has been done on a fragment of DNA polymerase I called **Klenow fragment**.

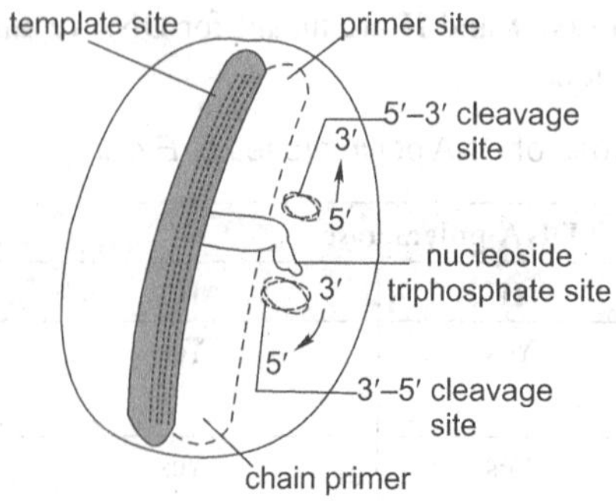

Fig. 35.10. A model of DNA polymeras I enzyme, showing its five-different active sites.

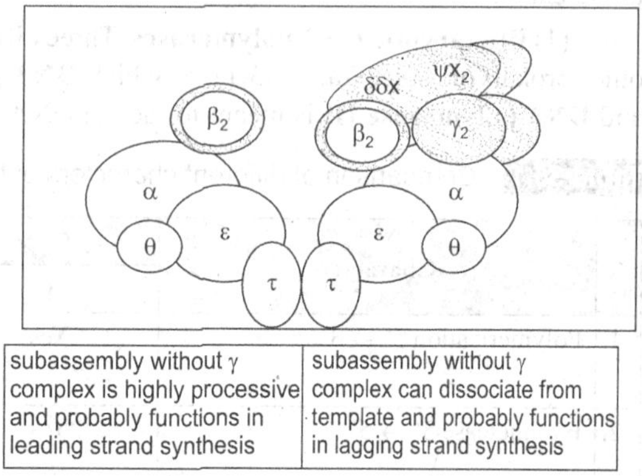

Fig. 35.11. Heterodimeric assembly of *E.coli* polymerase III holoenzyme to facilitate simultaneous leading and lagging strand synthesis.

Klenow fragment/polymerase was initially prepared by cutting the natural *E.coli* DNA polymerase I enzyme into two segments with a protease. One of these segments retained the polymerase and 3′ → 5′ exonuclease activity, but lacked the 5′ → 3′ exonuclease. Klenow enzyme is shaped like a cupped right hand with enzymatic activity taking place in two places, separated by a distance of about two or three nucleotides. It is proposed that when the polymerization site senses a mismatch, the DNA is moved so that the 3′ end enters the exonuclease site, where the incorrect nucleotide residue is then cleaved. Polymerization then continues (see **Tamarin**, 2002).

2. DNA polymerase II. This enzyme resembles DNA polymerase I in its activity, but is a DNA repair enzyme. It brings about the growth in 5′ → 3′ direction, using free 3′-OH groups.

3. DNA polymerase III. DNA polymerase III or Pol III enzyme (Fig. 35.11) plays an essential role in DNA replication. It is a multimeric enzyme or holoenzyme having ten subunits such as alpha (∝), beta (β), epsilon (ε), theta (θ), tau (τ), gamma (γ), delta (δ), delta dash (δ′), chi (χ) and psi (Ψ) (Table 35.2). All these ten subunits are needed for DNA replication *in vitro*; however, all having different functions. For example, three of the subunits α, ε and θ form the **polymerization core**, with both 5′ → 3′ polymerase activity and 3′ → 5′ exonuclease activity. One subunit, the β **subunit**, is a "**processivity clamp**". As a dimer (two identical copies attached head to tail), the protein forms "doughnut" around the DNA so it can move freely on the DNA. When it is attached to the core enzyme, the polymerase is held tightly to DNA and shows high processivity: the leading strand is usually synthesized entirely without the enzyme leaving the template (Fig. 35.12). The remaining subunits are involved in processivity control and replisome formation. They allow polymerase to move off and on DNA of lagging-strand template as Okazaki fragments are completed (a process known as **polymerase cycling**).

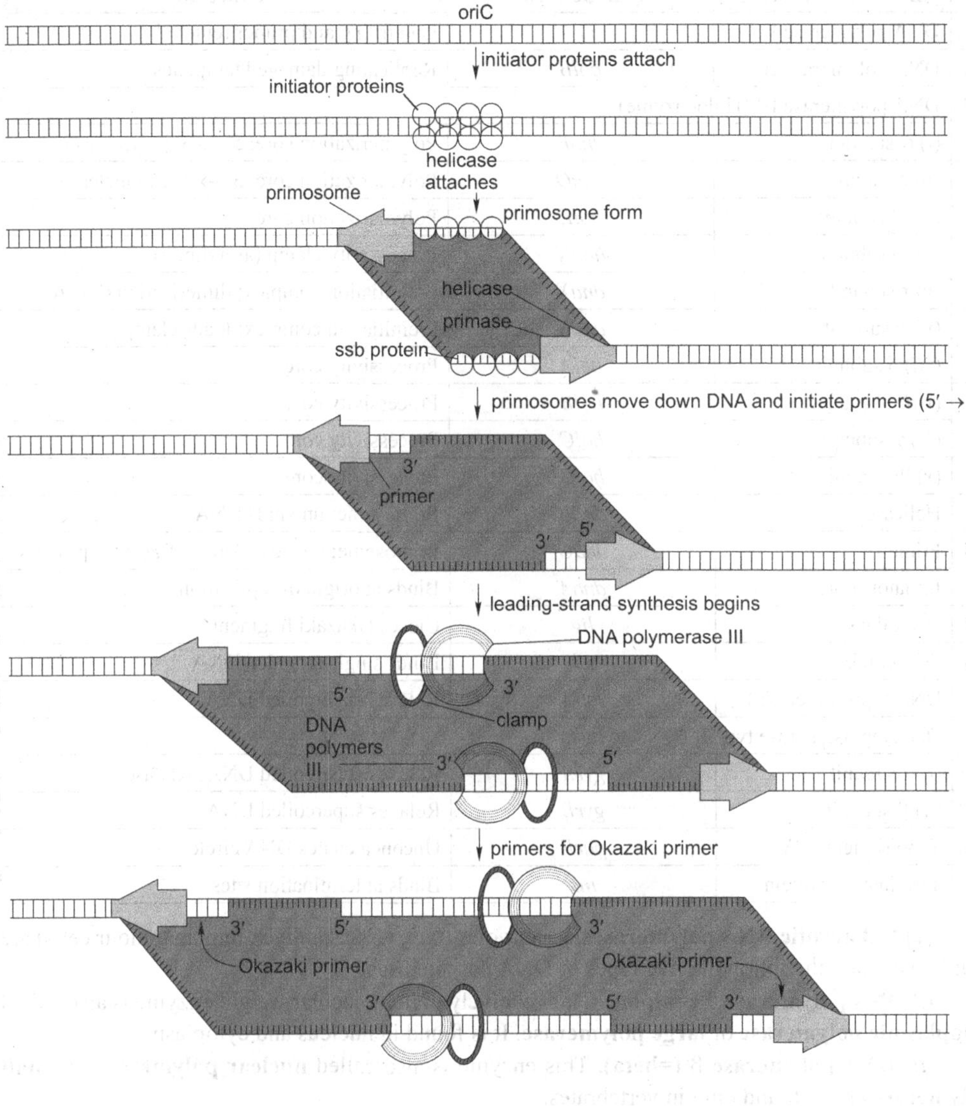

Fig. 35.12. Events at the origin of DNA replication in *E.coli*. The DNA opens up at *oriC* to create two moving Y-junctions (replication forks). Initiator proteins attach and then bind helicase. The helicase then binds primase, forming a primosome. After the primer forms and two copies of DNA polymerase III are bound, the polymerization begins (after Tamarin, 2002).

Table 35.2 Summary of the enzymes and proteins involved in DNA replication in *E.coli* (Source: Tamarin, 2002).

	Enzyme or Protein	Genetic locus	Function
1.	DNA polymerase I	*polA*	Gap filling and primer removal
2.	DNA polymerase II	*polB*	Replicating damaged templates
3.	DNA polymerase III (Holoenzyme)		
	(*i*) α subunit	*dnaE*	Polymerization core; 5′ → 3′ polymerase
	(*ii*) ε subunit	*dnaQ*	Polymerization core; 3′ → 5′ exonuclease
	(*iii*) θ subunit	*holE*	Polymerization core
	(*iv*) β subunit	*dnaN*	Processivity clamp (as a dimer)
	(*v*) τ subunit	*dnaX*	Preinitiation complex; dimerization of core
	(*vi*) γ subunit	*dnaX*	Preinitiation complex; loads clamp
	(*vii*) δ subunit	*holA*	Processivity core
	(*viii*) δ′ subunit	*holB*	Processivity core
	(*ix*) χ subunit	*holC*	Processivity core
	(*x*) Ψ subunit	*holD*	Processivity core
4.	Helicase	*dnaB*	Primosome; unwinds DNA
5.	Primase	*dnaG*	Primosome; creates Okazaki fragment primers
6.	Initiator protein	*dnaA*	Binds at origin of replication
7.	DNA ligase	*lig*	Closes Okazaki fragment
8.	Ssb protein	*ssb*	Binds single stranded DNA
9.	DNA topoisomerase I	*topA*	Relaxes supercoiled DNA
10.	DNA topoisomerase type II (DNA gyrase)		
	(*i*) α subunit	*gyrA*	Relaxes supercoiled DNA; ATPase
	(*ii*) β subunit	*gyrB*	Relaxes supercoiled DNA
11.	Topoisomerase IV	*parE*	Unconcatenates DNA circles
12.	Termination protein	*tns*	Binds at termination sites

(2) Eukaryotic DNA polymerases. Eukaryotes (*e.g.*, yeast, rat liver, human tumour cells) are found to contain the following five types of DNA polymerases:

(*i*) **DNA polymerase α (=alpha).** This relatively high molecular weight enzyme is also called **cytoplasmic polymerase** or **large polymerase**. It is found in nucleus and cytoplasm.

(*ii*) **DNA polymerase β (=beta).** This enzyme is also called **nuclear polymerase** or **small polymerase** and is found only in vertebrates.

(*iii*) **DNA polymerase γ (=gamma).** This enzyme is called **mitochondrial polymerase** and is encoded in the nucleus.

(*iv*) **DNA polymerase δ (=delta).** This enzyme is found in mammalian cells and is PCNA dependent for DNA-synthesis processivity (PCNA = proliferating cell nuclear antigen).

(*v*) **DNA polymerase ε (=epsilon).** It was previously known as **DNA polymerase II**. This enzyme is PCNA independent and occurs in mammalian HeLa cells and budding yeast.

The large DNA polymerase α is the predominant DNA polymerase enzyme in eukaryotic cells and was believed for long time to be only enzyme involved in DNA replication. But now one more polymerase, namely DNA polymerase δ is also found to be involved in eukaryotic DNA replication.

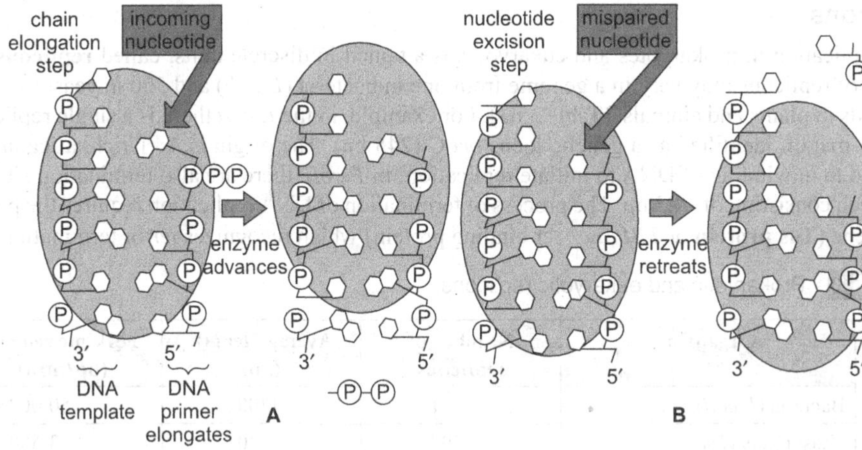

Fig. 35.13. Functions of DNA polymerase-I enzyme. A—Mode of polymerization of nucleotides, on the primer site of the enzyme, B—Removal of mispaired nucleotide activity of the enzyme.

(III) DNA ligases. DNA ligase enzymes are capable of catalyzing phosphodiester bond formation between free 3′-OH and free 5′-P groups of a nick of DNA which is created by endonuclease enzyme, thereby restoring an intact DNA duplex. Many DNA ligases have already been discovered. The ligase enzyme from *E.coli* requires the presence of oxidized nicotinamide adenine dinucleotide (NAD^+) a cofactor, whereas the ligase enzyme specified by T_4 bacteriophage requires ATP to bring about the joining reaction.

Roles of DNA Primers in DNA Replication

No known DNA polymerase can initiate synthesis of DNA without the availability of a **primer RNA strand**. So, before actual DNA replication starts, short RNA oligonucleotide segments, called **RNA primers** or simply the **primers**, have to be synthesized by **DNA primase** enzyme utilizing ribonucleoside triphosphates (Fig. 35.14). This RNA primer is synthesized by copying a particular base sequence from one DNA strand and differs from a typical RNA molecule in that after the synthesis the primer remains hydrogen-bonded to the DNA template (**Freifelder**, 1985). The primers are about 10 nucleotides long in eukaryotes and they are made at intervals on the lagging strand where they are elongated by the DNA polymerase enzyme to begin each Okazaki fragment. These RNA primers are later excised and filled with DNA with the help of DNA repair system in eukaryotes (or DNA polymerase I in *E.coli*).

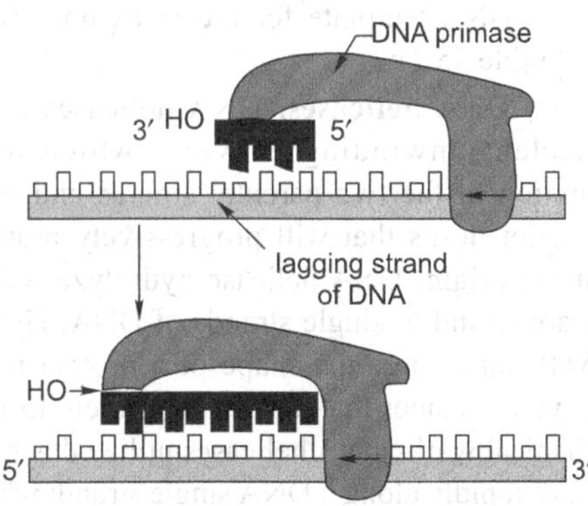

Fig. 35.14. Formation of RNA primer on lagging strand by DNA primase enzyme. Unlike DNA polymerase, this enzyme can start a new polynucleotide chain by joining two nucleotide triphosphates together (after Alberts *et al.*, 1989).

In bacteria, two different enzymes are known to synthesize primer RNA oligonucleotides—**RNA polymerase** (on the leading strand) and **DNA primase** (on the lagging strand).

Replicons

DNA replication in prokaryotes and eukaryotes is attained in discrete units, called **replicons**. The number of replicons may vary in a genome from one in bacteria (*E.coli*) and 500 in yeast to several thousands in plants and animals (Table 35.3). For example, in the *E.coli* there is a single replication with the **origin**, identified as a genetic locus *ori* C (245 bp). The origin is A: T rich, a feature that is related to unwinding of DNA to initiate replication. In *E.coli*, there are also termination sites (*ter A-F*), each consisting of ~ 23 bp. The process of termination of DNA replication requires the product of *tus* gene (**Tus protein** or *TBP*, *i.e.*, ter binding protein) which recognizes *ter* or termination sites.

Table 35.3. Prokaryotic and eukaryotic replicons.

	Organism	Number of replicons	Average length (kb)	Fork movement (bp/min)
1.	Bacteria (*E.coli*)	1	4200	50,000
2.	Yeast (*S.cervisiae*)	500	40	3,600
3.	Fruit fly (*D.melanogaster*)	3,500	40	2,600
4.	Toad (*Xenopus laevis*)	15,000	200	500
5.	Mouse (*Mus musculus*)	25,000	150	2,200
6.	Bean (*Vicia faba*)	35,000	300	Not known

B. Proteins Involved in Opening of DNA Helix

The following three types of proteins are needed to help the DNA double helix to open and to provide exposed DNA template for the DNA polymerase to copy (Table 35.2):

***(i)* DNA helicases.** DNA helicases are ATP-dependent unwinding enzymes which promote separation of the two parental strands and establish replication forks that will progressively move away from the origin. DNA helicase hydrolyze ATP when they are bound to single strands of DNA. Hydrolysis of ATP can change the shape of a protein molecule in a cyclic manner that allows the protein to perform mechanical work. DNA helicases utilize this principle to move rapidly along a DNA single strand; when they encounter a region of double helix, they continue to move along their strand, thereby unwinding the helix (Fig. 35.15). Unwinding of the template DNA helix at a replication fork could in principle be catalyzed by two DNA helicases, acting in concert, one running along the leading strand and the other along the lagging strand.

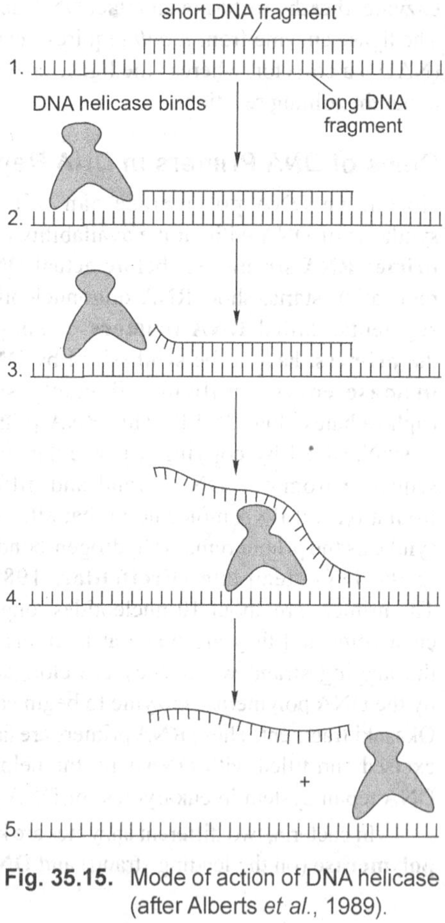

Fig. 35.15. Mode of action of DNA helicase (after Alberts *et al.*, 1989).

***(ii)* Helix-destabilizing proteins.** (also called **single strand DNA-binding proteins** or **SSBPs**). Behind the replication fork, the single DNA strands are prevented from rewinding about one another (or forming double-stranded hair-pin loops in each single strands) by the action of SSB proteins. SSB proteins bind to exposed DNA strands without covering the bases, which, therefore, remain available for the templating process.

***(iii)* Topoisomerases (DNA gyrases).** The action of a helicase introduces a positive supercoil into the duplex DNA ahead of the replication fork. Enzymes, called **topoisomerases**, relax the supercoil by attaching to the transiently supercoil duplex, nicking one of the strands and rotating it through the unbroken strand. The nick is then released. Thus, a DNA topoisomerase can be viewed as a "reversible nuclease" that adds itself covalently to a DNA phosphate, thereby breaking a phosphodiester bond in the DNA strand. because the covalent linkage that joins a topoisomerase to DNA phosphate retains the energy of the broken phosphodiester bond, the breakage reaction is reversible; resealing is rapid and does not require additional energy input. The rejoining mechanism is different from DNA ligase enzyme.

One type of topoisomerase (*i.e.*, **topoisomerase I**) causes a single-strand break or nick which allows the two sections of DNA helix on either side of nick to rotate freely relative, to each other, using the phosphodiester bond in the strand opposite the nick as a swivel point. A second type of topoisomerase (*i.e.*, **topoisomerase II**) forms a covalent bond to both strands of DNA helix at the same time, making transient **double-strand break** in the helix.

Replisome and Primosome

N.K. Sinha and **A. Kornberg** have suggested that the DNA polymerases, RNA primases and helicases may be associated with one another to form a multienzyme complex—the **replisome** that carries out the synthesis of leading and lagging strands in a **coordinated fashion** (Fig. 35.12). Such a complex would be highly processive and assure rapid replication of the DNA.

Likewise, the proteins at a replication fork cooperate to form replication machine, *i.e.*, the primase molecule is linked directly to the helicase to form a unit on the lagging strand called a **primosome** which moves with the fork, synthesizing RNA primers as it moves (Fig. 35.16).

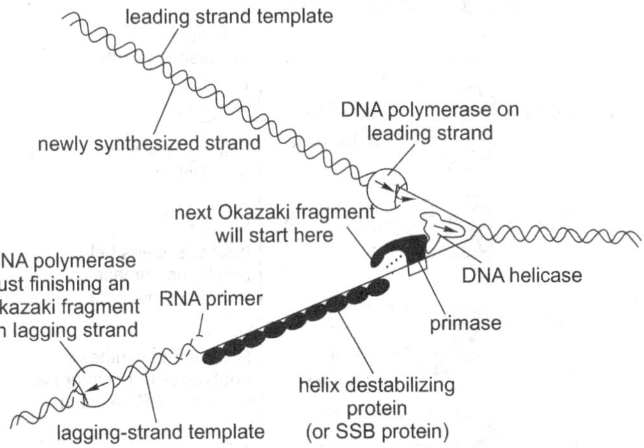

Fig. 35.16. A primosome in action (after Alberts *et al.*, 1989).

35.3. MECHANISM OF DNA REPLICATION IN PROKARYOTES

In vitro DNA replication has been extensively studied in *E.coli* and in the phages and plasmids of *E.coli*. In *E.coli*, the process of DNA replication involves the following three steps.

1. Initiation of DNA replication. This process comprises three steps: (*i*) recognition of the origin (O), (*ii*) opening of DNA duplex to generate a region of single stranded DNA, and (*iii*) capture of **Dna B protein** (*i.e.*, $5' \rightarrow 3'$ helicase; also acts as the activator of primase). Thus, Dna-A (or initiator protein)-ATP complex binds at 9 bp inverted repeat regions (R_1, R_2, R_3, R_4) of oriC of *E.coli* and promotes opening of the DNA duplex in a region of three direct repeats of 13-bp sequence (called **13-mers**). The opening occurs from right 13-mer leftwards and requires negatively supercoiled DNA and HU or IHF initiator proteins. Dna B (=helicase) is transferred to exposed single stranded DNA (Fig. 35.17) and causes unwinding of the DNA in the presence of ATP, SSB protein and **DNA gyrase** (a topoisomerase). This results in unwinding of DNA duplex and the replication from ori proceeds in both directions (bidirectional); SSB binding occurs on single stranded regions and two Dna B complexes (=primosomes) are loaded one on each strand.

2. Elongation of DNA chain. This step requires the presence of the following enzymes and factors: 1. Dna B or helicase (also called **mobile promoter**); 2. Primase (Dna G); 3. DNA polymerase holoenzyme (or DNA pol III HE); 4. SSB proteins; 5. RNAase H which removes RNA primers; 6. DNA polymerase I which is used for filling the gap created due to RNA primers and 7. DNA ligase (which converts primerless Okazaki fragments into continuous strand). During initiation to elongation transition, the following events occur:

1. As helicase (or Dna B) travels in $5' \rightarrow 3'$ direction, it generates a replication fork by opening the DNA duplex.

2. The DNA strand having helicase becomes the lagging strand. DNA primase associates with Dna B helicase, forming the primosome which synthesizes multiple primers for lagging strand and single RNA primer for the leading strand.

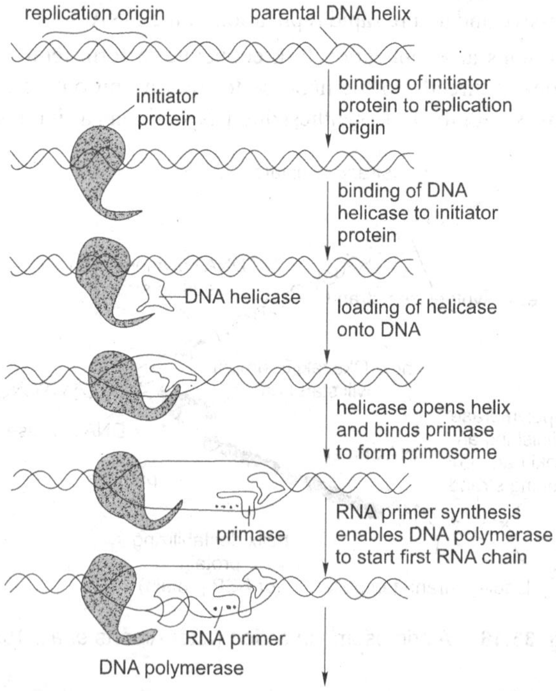

Fig. 35.17. Initial steps leading to formation of replication forks at the *E.coli* (after Alberts *et al.*, 1989).

3. For the synthesis of lagging strand, the DNA pol III HE has to work on the same strand to which Dna B helicase is bound, but it travels in opposite direction.

4. Dna B helicase, Dna G primase and DNA pol III HE work together in strand elongation. Helicase and DNA polymerase assembly remains *processive*, *i.e.*, they remain tightly bound to the fork and stay bound throughout the reaction.

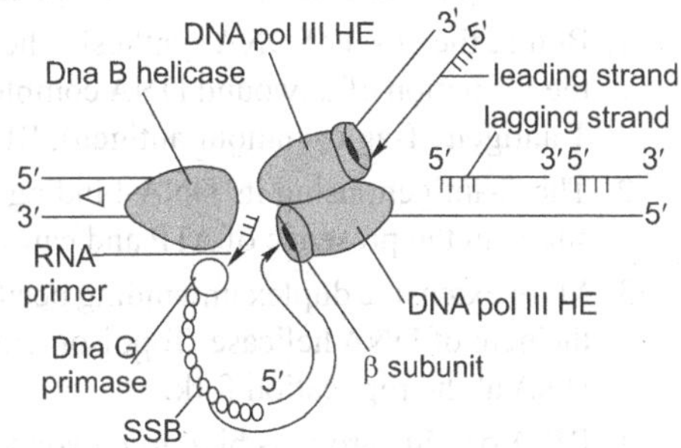

Fig. 35.18. Model of a DNA replication fork in prokaryotes during elongation.

Synthesis (=elongation) of lagging and leading strands takes place by somewhat different methods; it is far more complex for lagging strand than for the leading strand:

A. Discontinuous synthesis on lagging strand. 1. Primase is taken up from solution and is activated by helicase (Dna B) to synthesize a RNA primer 10 to 20 nt or nucleotides long) on the lagging strand. 2. The RNA primers are recognized by DNA pol III HE on the lagging strand and are utilized for synthesis of precursor or Okazaki fragments. In fact, each new RNA primer is recognized by the gamma (γ) subunit of DNA pol III HE and loaded with a subunit of the same polymerase. This preloaded β subunit may then capture the core of DNA poly III HE when it becomes available after finishing its synthetic job on the preceding Okazaki fragments. 3. On completion of the Okazaki fragments, the RNA primers are excised by DNA polymerase I, which then fills the resulting gaps with DNA. 4. After DNA **polymerase I** adds the final deoxyribonucleotides in the gap left by the excised primer, the enzyme **DNA ligase** forms the phosphodiester bond that links the free 3′ end of the primer replacement to the 5′ end of the Okazaki fragment.

B. Continuous synthesis on leading strand. (1) In bidirectional DNA replication, the leading strand is primed once on each of the parental strands. (2) The RNA primer of the leading strand is synthesized by RNA polymerase enzyme. (3) DNA pol III HE causes elongation of the leading strand and finally DNA pol I and ligase enzymes give final touch to the leading strand as in case of the lagging Strand.

35.4. DNA REPLICATION IN EUKARYOTES

Eukaryotic DNA replication requires two different DNA polymerase enzymes, namely **DNA polymerase** α and **DNA polymerase** δ (Fig. 35.19). DNA polymerase δ synthesizes the DNA on the leading strand (continuous synthesis), whereas DNA polymerase α synthesizes the DNA on the lagging strand (discontinuous DNA synthesis). Besides these two enzymes, six more factors are involved in eukaryotic DNA replication: (1) T antigen; (2) replication factor A or RF-A (also called RP-A or eukaryotic SSB); (3) topoisomerase I; (4) topoisomerase II; (5) proliferating-cell nuclear antigen (PCNA), also called **cyclin**), and (6) replication factor C or RF-C.

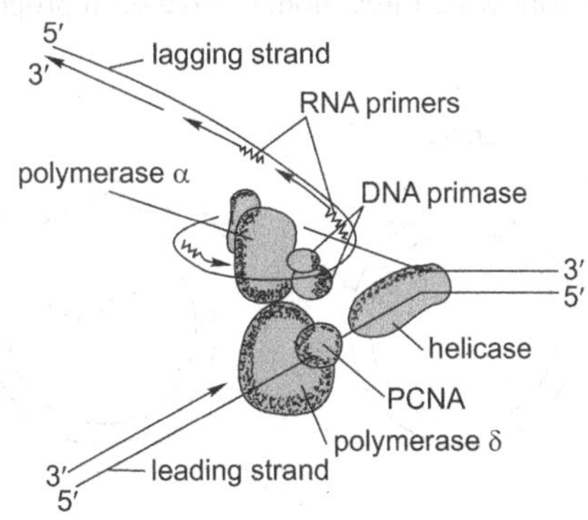

Fig. 35.19. Two different DNA polymerases, synthesizing leading and lagging strands at a eukaryotic DNA replication fork.

The process of eukaryotic DNA replication involves the following steps:

1. Before the onset of DNA synthesis, there is a presynthetic stage of 8 – 10 minutes duration for the formation of unwound DNA complex. This step needs only three purified proteins, namely T antigen (T-ag or tumour antigen), RF-A and topoisomerase I and II.
2. The T-antigen, using its DNA-binding domain, forms a multisubunit complex with site I and site II in the presence of ATP and caused local unwinding.
3. More extensive duplex unwinding occurs due to association of RF-A and a topoisomerase with the help of DNA helicase. Topoisomerases help in unwinding of DNA by altering topology of DNA at the replication fork.
4. RF-A or SSB proteins bind to unwound single stranded DNA.
5. The primer RNA synthesis is performed by primase which is tightly associated with DNA polymerase α.
6. DNA polymerase α helps in synthesis of an Okazaki fragment in 5′ to 3′ direction. In eukaryotes, Okazaki fragments are much smaller—only about 135 bases long, about the size of the DNA on a nucleosome (see **Mathews**, 2000).
7. Replication factor C (or RF-C) and PCNA (cyclin) help in switching of DNA polymerases so that pol α is replaced by pol δ which then continuously synthesized DNA on the leading strand.
8. Another Okazaki fragment is then synthesized from the replication fork on the lagging strand by pol α-primase complex and this step is repeated again and again, till the entire DNA molecule is covered.
9. The RNA primers are removed and the gaps are filled as in prokaryotic DNA replication.

Recently, role of DNA polymerase ε in DNA replication has been stressed upon, so that three DNA polymerases (α, δ and ε) are now known to be involved in eukaryotic DNA replication. **A. Sugino** and coworkers have proposed that DNA polymerase α might function at both the leading and lagging strands (since polymerase α has primase activity), whereas polymerase ε and polymerase δ are involved in elongation of the leading and lagging-strands respectively.

35.5. MODELS OF DNA REPLICATION

The following three models have been proposed for DNA replication in different organisms:

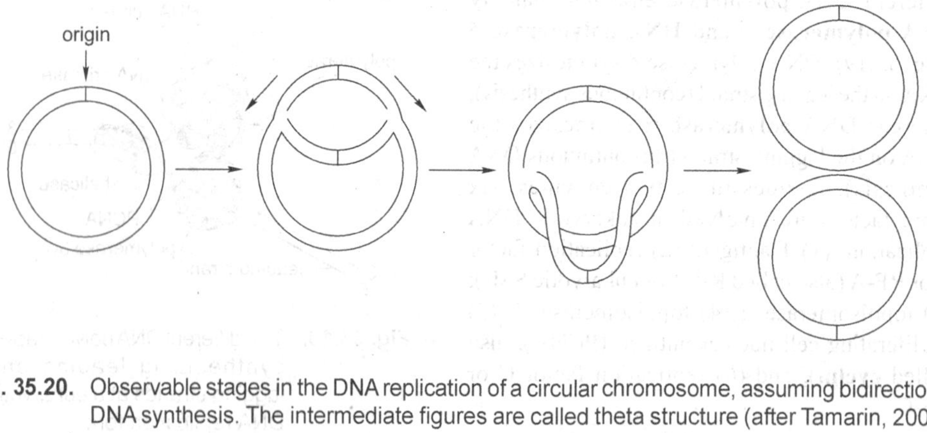

Fig. 35.20. Observable stages in the DNA replication of a circular chromosome, assuming bidirectional DNA synthesis. The intermediate figures are called theta structure (after Tamarin, 2002).

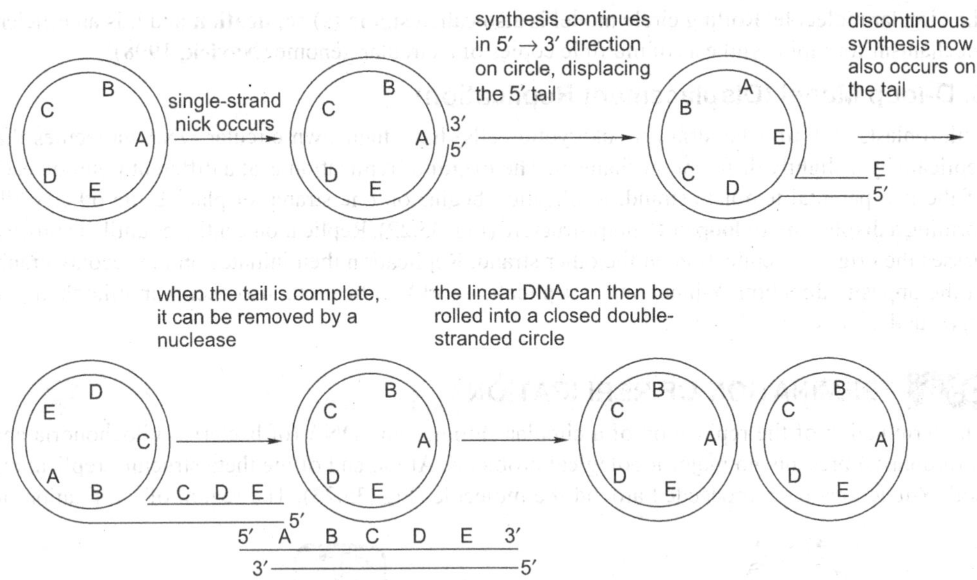

Fig. 35.21. Rolling circle model of DNA replication. The letters A to E provide landmarks on the chromosome (after Tamarin, 2002).

1. Replication Fork Model

It occurs both in linear and circular DNA molecules and involves the formation of **replication forks** which either move in one direction in unidirectional replication or both directions in bidirectional replication.

In *E.coli*, replication of circular DNA is of bidirectional type involving intermediate theta (θ) type structures.

2. Rolling Circle Model (Rolling Circle Replication)

In the rolling circle mode of DNA replication, a **nick** (a break in one of the phosphodiester bonds) is made in one of the strands of the circular DNA and this nick has 3′-OH and 5′-P termini. Due to a helicase and *SSB* protein α, Y-fork is generated. Ultimate result is replication of a circle and a tail (Fig. 35.21). This form of replication occurs in F plasmid or *E.coli* Hfr chromosome during conjugation. The F^+ or Hfr bacterial cell retains the circular daughter while passing the linear tail into the F^- cell, where replication of the tail takes place. Several phages (including 1 phage) also use this DNA replication method, filling their heads (protein coats) with linear DNA replicated from

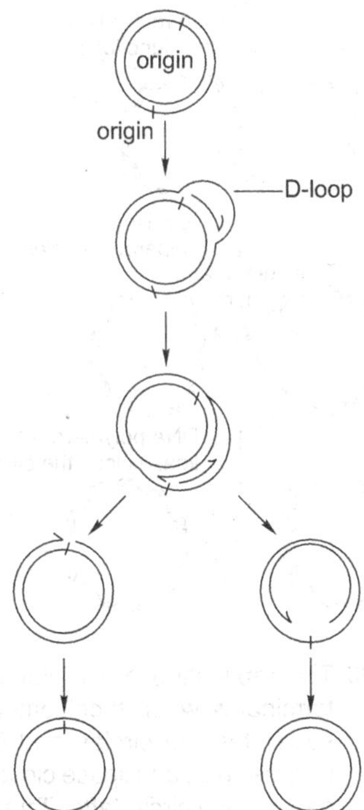

Fig. 35.22. D-loop form during mito-chondrial and chloroplast DNA replication because the origins of replication are located at different places on the two strands of the double helix (after Tamarin, 2002).

the circular molecule. Rolling circle model is also called **sigma (s) replication** and it is an efficient mechanism for rapid synthesis of multiple copies of a circular genome (**Novick**, 1998).

3. D-loop Model (Displacement Replication)

Chloroplasts and mitochondria (in eukaryotic cells) have their own circular DNA molecules that replicate by a slightly different mechanism. The **origin** of replication is at a different point on each of the two parental template strand. Replication begins on one strand, displacing the other while forming a **displacement loop** or **D-loop** structure (Fig. 35.22). Replication continues until the process passes the origin of replication on the other strand. Replication then initiates on the second strand, in the opposite direction. Y-junction (or replication fork) replication, also occurs in mitochondrial DNA under some growth condition.

35.6. TERMINATION OF REPLICATION

The termination of the replication of a circular chromosome DNA (of bacteria, mitochondria and chloroplasts) presents no major topological problems. At the end of the theta structure replication, both Y-junctions have proceeded around the molecule (Fig. 35.23). The region of termination on

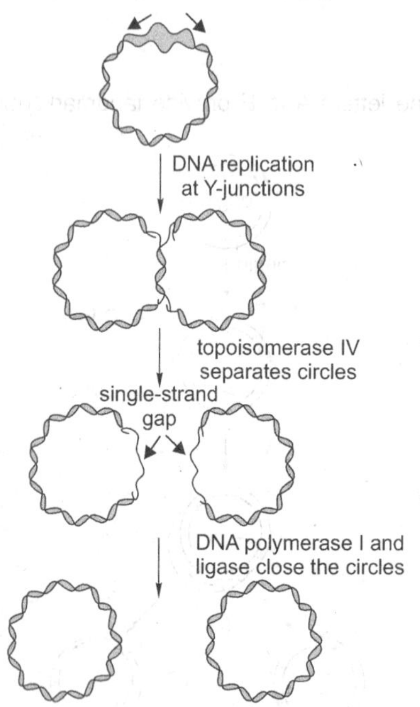

Fig. 35.23. The replication of circular DNA terminates when topoisomerase separates the circles and DNA polymerase I and ligase close the gap in each circle (after Tamarin, 2002).

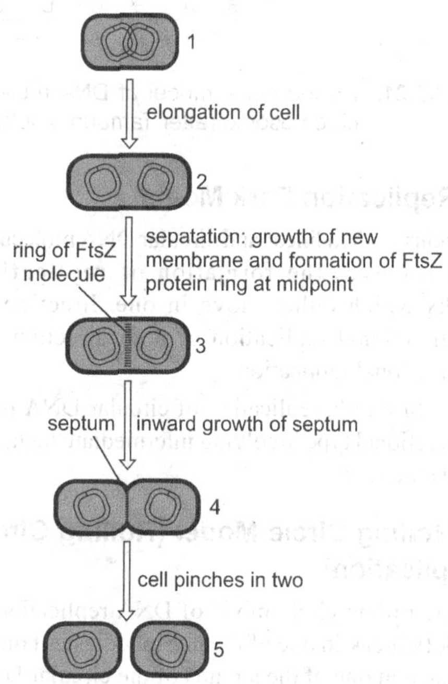

Fig. 35.24. Binary fission in *E.coli*. After the DNA is replicated, the cell elongates and the DNA is partitioned in the cell. Separation then begins, in which new cell membrane material begins to grow and form a septum at approximately the midpoint of the cell. A protein molecule, called FtsZ, (Which is a tubulin-like protein) facilitates this process, *i.e.*, FtzZ forms a ring at approximately the midpoint of the cell. When the septum is completed, the cell pinches in two (after Raven *et al.*, 2005).

the *E.coli* chromosome, the **terminus** (*Ter*) is 180 degrees from *OriC* on the circular chromosome, between minutes 28 and 36. There are six terminator sites; three arrest the Y-junction from the left, and three arrest the one from the right when bound by a termination protein, the product of the *tus* gene. (*Tus* stands for terminus utilization substance; each *ter* site is about twenty base pairs). One interesting aspect of the termination of *E.coli* DNA replication is that the cells are viable even if the whole terminator region is deleted. There are fewer viable cells and some growth problems, but in general, *E.coli* can successfully terminate DNA replication even without formal termination sites. A topoisomerase (**topoisomerase IV**) then releases the two circles and DNA polymerase I and ligase enzymes close them up (Fig. 35.23).

DNA Partitioning in *E.coli*

Until very recently, geneticists believed that the partitioning of the *E.coli* chromosome was a passive process, unlike that in eukaryotes. Now, however, we know that *E.coli* DNA partitioning is indeed a very complex process. When DNA replication begins, the newly replicated origins of replication are segregated to opposite ends of the bacterial cell, acting as centromeres do. A ring of FtsZ protein (Fig. 35.24), the products of *ftsZ* gene, form a ring at the middle of the already elongated cell and begins to create septum that will divide the elongated cell into two daughter cells.

35. 7. DNA REPAIR

Radiation, chemical mutagens, heat, enzymatic errors, and spontaneous decay constantly damage DNA. For example, it is estimated that several thousand DNA bases are lost each day in every mammalian cell due to spontaneous decay (**Tamarin**, 2002). Some types of damage interfere with DNA replication and transcription. In the long evolutionary challenge to minimize mutation, cell have evolved numerous mechanisms to repair incorrectly replicated DNA. Many enzymes, acting alone or in concert with other enzymes repair DNA. DNA repair systems are generally placed in four broad categories: 1. *damage reversal*, 2. *excise repair*, 3. *double strand breakage repair* and 4. *post replicative repair*. Enzymes that process DNA repair steps have been conserved during evolution. That is, enzymes found in *E.coli* have homologues in yeast, fruit flies, and human beings.

1. Damage Reversal

Ultraviolet (UV) rays cause linkage or **dimerization** of adjacent pyrimidines in DNA (Fig. 35.25). Although cytosine-cytosine and cytosine-thymine dimers are occasionally produced, the principal products of UV irradiation are thymine-thymine **dimers**. These can be repaired in several different ways. The simplest is to reverse the dimerization process and restore the original unlinked thymines.

In *E.coli*, an enzyme called **DNA photolyase**, the product of the *phr* gene (for photoreactivation), binds to dimerized thymines. When light shines on the cell, the enzyme breaks the dimer bonds with light energy. The enzyme then falls free of the DNA. Another example of an enzyme that performs direct DNA repair is **O^6-mGua DNA methyltransferase**, which removes the methyl groups from O^6-methylguanine, the major product of DNA-methylating agents. Photoreactivation does not occur in human beings.

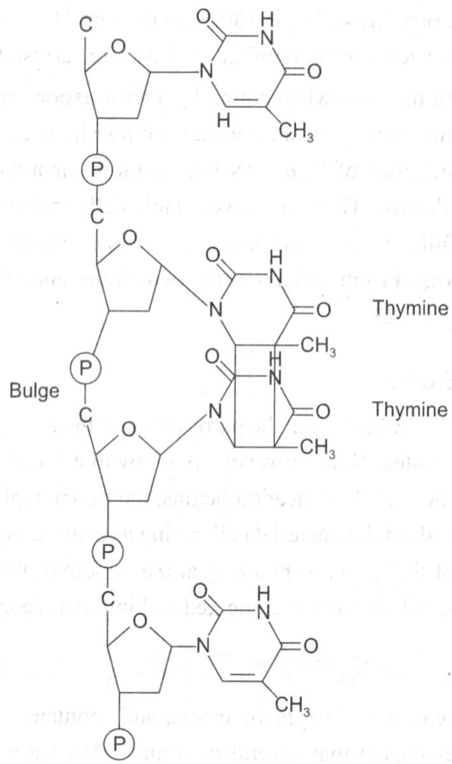

Fig. 35.25. UV-induced dimerization of adjacent thymines in DNA.

2. Excision Repair

Excision repair refers to the general mechanism of DNA repair that works by removing the damaged portion of a DNA molecule. Various enzymes can detect damage or distortion in the DNA double helix. During excision repair bases and nucleotides are removed from the damaged strand. The gap is then patched using complementarity with the remaining strand. These systems are of three types: base excision repair, nucleotide excision repair and mismatch repair.

(*i*) **Base excision repair.** A base can be removed from a nucleotide within DNA in several ways: by direct action of an agent such as radiation, by spontaneous hydrolysis, by an attack of oxygen free radicals, or by **DNA glycosylases** which are enzymes that detect damaged bases and removes them. According to **Tamarin** (2002), there are five glycoylases (enzymes) for excision repairs. For example, **uracil-DNA glycosylase**, the product of the *ung* gene in *E.coli*, recognizes uracil within DNA and cleaves it out at the base-sugar (glycosidic) bond. The resulting site is called an **AP** (= apurinic-apyrimidenic) **site**, because of the lack of a purine or pyrimidine at the site. An **AP endonuclease** then senses the minor distortion of the DNA double helix and initiates excision of the single AP nucleotide in a process known as **base excision repair**. The AP endonuclease nicks the DNA at the 5′ side of the base-free AP site. A **DNA polymerase** then inserts a nucleotide at the AP site; an **exonuclease, lyase** or **phosphodiesterate** enzyme then removes the base-free nucleotide. (Lyases are enzymes that can break C-C, C-O and C-N bonds.) **DNA ligase** then closes the nick (Fig. 35.26). The replacement of just one base occurs 80–90% of the time. In the remaining 10–20% of cases, several nucleotides may be removed, depending probably on which DNA polymerase (I or III) first repairs the site (**Lindahl** 1996).

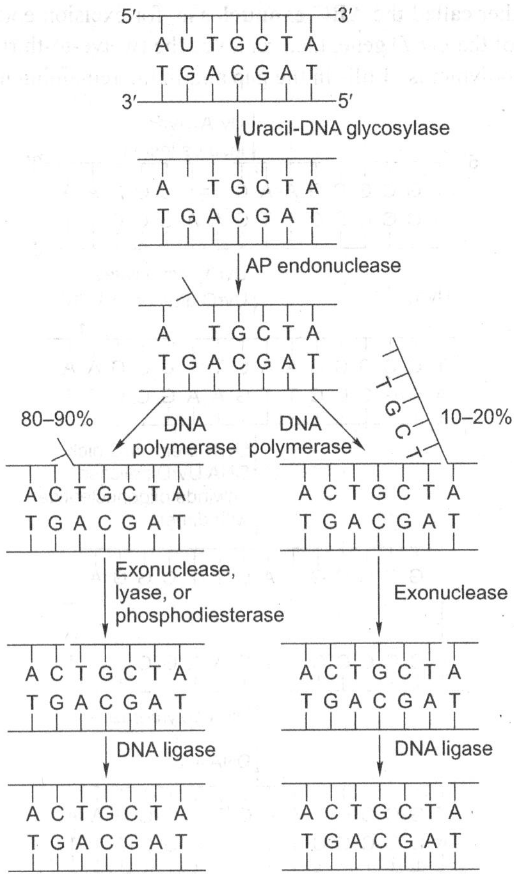

Fig. 35.26. Mechanism of base excision repair. In this case, a uracil-DNA glycosylase enzyme removes a uracil from DNA. An AP endonuclease nicks the DNA on the 5′ side of the base-free site. Between 80 and 90% of the time, a DNA polymerase will replace the single nucleotide, an exonuclease, lyase or phosphodiesterase will remove the base-free nucleotide. The final nick is sealed with DNA ligase. Between 10 to 20% of the time, the DNA polymerase will extend polymerization beyond the single nucleotide. In these cases, an exonuclease and DNA ligase finish the repair (after Tamarin 2002).

In mammals, **DNA polymerase** β performs two roles in base excision repair: it both inserts a new base where the AP site was and also eliminates the AP nucleotide residue by exonuclease activity.

***(ii)* Nucleotide excision repair (NER).** Nucleotide excision repair is initiated by enzymes that sense distortions in the DNA backbone and replace a short stretch of nucleotides. For example, six enzymes in *E.coli* excise a short stretch of DNA containing thymine dimers if the dimerization is not reversed by photoreactivation. Two copies of the protein product of the *uvr A* gene (for ultraviolet light UV-repair) combine with one copy of the product of the *uvr B* gene to from a **Uvr A_2 Uvr B complex** that moves along the DNA, looking for damage (Fig. 35.27). (The complex has 5′ helicase activity.) When the complex finds damage such as a thymine dimer, with moderate to large distortion of the DNA double helix, the Uvr A_2 dimer dissociates, leaving the Uvr B subunit alone. This causes the DNA to bend and attracts the protein product of the *uvr C* gene, *Uvr C*. The Uvr B subunit first nick (hydrolyzes), the DNA four to five nucleotides on the 3′ side of the lesion; next, the Uvr C subunit nicks the DNA eight nucleotides on the 5′ side of the lesion. (The three components, Uvr A, Uvr

B, and Uvr C, are together called the **ABC exonuclease**, for excision endonuclease.) The enzyme helicase II, the product of the *uvr D* gene, then removes the twelve-to-thirteen base oligonucleotide as well as Uvr C. DNA polymerase I fills in the gap and in the remaining nick (Fig. 35.27).

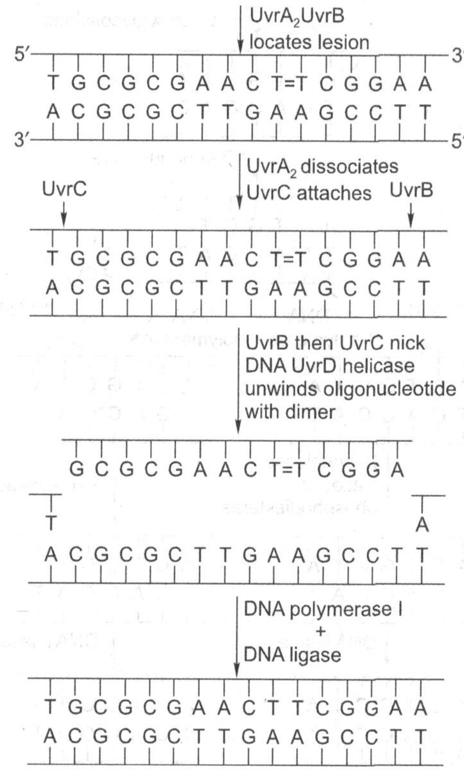

Fig. 35.27. Nucleotide excision repair. A lesion in DNA (a thymine dimer) is located by a protein made of two copies of Uvr A and one of Uvr B. Then, the Uvr A subunits detach, and Uvr C attaches on the 5′ side of the lesion. Uvr B nicks the DNA on the 3′ side and Uvr C on the 5′ side of the lesion; Uvr D helicase unwinds the oligonucleotide containing the lesion. DNA polymerase I and DNA ligase then repair the patch (after Tamarin 2002).

Like the excision repair, nucleotide excision repair is present in all organisms. In yeast, approximately twelve genes are involved, what is called the **RAD 3 group**. In human beings, twenty five proteins are involved; they remove twenty-seven to twenty-nine nucleotides, as compared to twelve to thirteen in *E.coli*.

Transcription and nucleotide excision repair are linked in eukaryotes. Transcription factor TFIIH is involved in repair of UV damage. In human beings, the autosomal recessive trait **xeroderma pigmentosum** is caused by an inability to repair thymine dimerization induced by UV light. Persons with this trait freckle heavily when exposed to UV rays of the sun, and they have a high incidence of skin cancer. Defect in XPD component of TFIIH is main cause of this ailment.

***(iii)* Mismatch repair.** Excision repair triggered by mismatches is referred to as **mismatch repair**, which encompasses about 99% of all DNA repairs. As DNA polymerase replicates DNA, some errors are made that the proofreading polymerase does not correct. For example, a template G can be paired with a T rather than a C in the progeny strand. The GT base pair does not fit correctly in the DNA duplex. The mismatch repair system, which follows behind the replicating fork, recognizes this problem. This system, whose members in *E.coli* are specified by the *mut H*, *mut L*, *mut S* and *mut U* genes, is responsible for the removal of the incorrect base by an excision repair process. (The genes are called *mut* for mutator because mutation of these genes cause high levels of spontaneous mutation in the cell. The *mut U* gene is also known as *uvr D*.) The mismatch repair enzymes initiate the removal of the incorrect base by nicking the DNA strand on one side of the mismatch.

Let's know how the mismatch repair system recognizes the progeny rather than the template, base as the wrong one. After all in a mismatch, there are defective bases—theoretically, either partner could be the "wrong" base. In *E.coli*, the answer lies in the methylation state of the DNA. **DNA methylase**, then product of the *dam* gene, methylates 5′-GATC-3′ sequences, which are relatively common in the DNA of *E.coli*, at the adenine residue. Since the mismatch repair enzymes follow the replication fork of the DNA, they usually reach the site of mismatch before the DNA methylase does. Template strands will be methylated, where as progeny strands, being newly synthesized, will not be.

Mechanism of mismatch repair (Fig. 35.28). The **Muts protein**, in the form of homodimer—two copies of the same protein—finds the mismatch. **Mut L protein**, also in the form of a homodimer, then binds, and together they find methylation signal. They also activate the endonuclease Mut H, which then nicks the unmethylated strand at the 3′-CTAG-5′ recognition site which can be one thousand

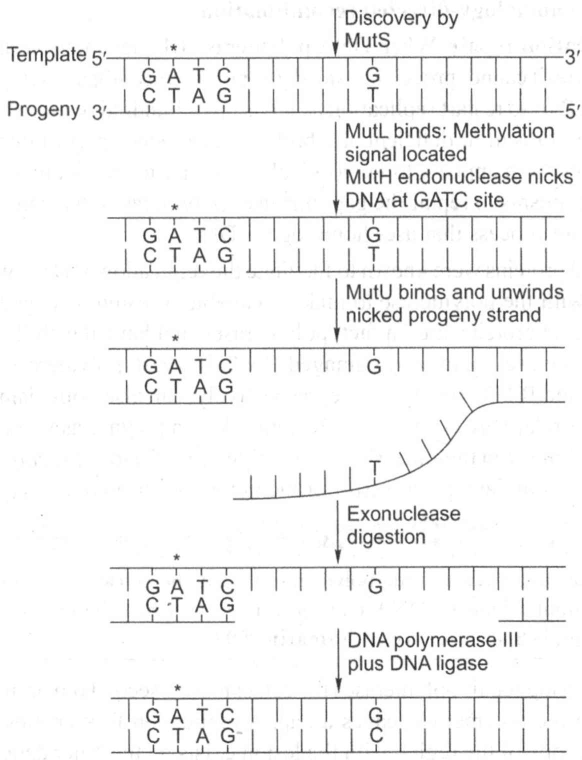

Fig. 35.28. Mismatch repair. The Mut S protein discovers mismatches; Mut L binds and Mut H endonuclease nicks the progeny strand at the 3′-CTAG-5′ sequence. Mut U helicase unwinds the nicked oligonucleotide with this mismatch. Exonuclease digestion, followed by DNA polymerase III and DNA ligase repair, completes the operation (after Tamarin 2002).

to two thousand bases away from the mismatch. At the recognition site, the Mut S – Mut L tetramer loads the helicase Mut U (Uvr D), which then unwinds the nicked strand. Any one of at least four different exonucleases then attacks the unwound oligonucleotide. DNA polymerase III then repairs the gap, and DNA ligase seals it. This sequence of events highlights a common theme in DNA repair: once a lesion is found, the damaged DNA has some protein bound to it until the repair is finished.

At this juncture we can answer an evolutionary question—*why does DNA have thymine while RNA has uracil*? If we live in an RNA world, in which RNA evolved first, why don't DNA and RNA both contain uracil? One answer is that a common damage to cytosine, spontaneous deamination, result in uracil. If uracil were a normal base in DNA, the conversion of cytosine to uracil by deamination would not leave any clue to a mismatch repair system that a mutation had occurred. Thus, thymine replaces uracil in DNA, since thymine is not confused with any other normal base in DNA by common spontaneous changes. Infact, cytosine, guanine, adenine and thymine are not converted simply to any other of the bases in DNA. Hence, changes of these bases leave clues for the repair systems.

***(iv)* Double-strand break repair.** Some damage to DNA, such as that caused by ionizing radiation, is capable of breaking both strands of the double helix. When that happens, the cell uses following two mechanisms to repair the broken ends: (1) It can simply bring the ends back together. This is called **nonhomologous end joining.** (2) It can use a mechanism that relies on the nucleotide sequences of a homologous piece of DNA, such as a sister chromatid or a homologous chromosome. This method is called **homology-directed recombination**.

***(v)* Post-replication repair.** When DNA polymerase III encounters certain damage in *E.coli*, such as thymine dimers, it cannot proceed. Instead, the polymerase stops DNA synthesis and leaving a gap, skips down the DNA to restart replication as far as eight hundred or more bases away. If allowed to remain, this gap will result in deficient and broken DNA. Since part of one strand is absent and the other has damage, there appears to be no viable template for replicating new DNA. However, the cell has two mechanisms to repair the gap: one uses polymerases that can replicate these lesions and the other is a repair process that uses homologous DNA.

Initially, several proteins were known to facilitate the replication of DNA with lesions; they were believed to interact with the polymerase to make it capable of using damaged DNA as a template, we now know that these proteins are, in fact, polymerases that have the ability to replicate damage DNA. In *E.coli*, polymerase V can copy damaged DNA. In yeast, polymerases η (eta) and ζ (zeta), also called REV 3/7 and RAD 30 polymerase, respectively, can also copy damaged DNA. Some of these polymerases are relatively error free: polymerase V and polymerase η put adenine-containing nucleotides opposite dimerized thymine. However, polymerase ζ and the *E.coli* polymerase IV, which also appears at the times of damage, are error prone in their replicative roles (Box 35.1).

Box 35.1

One possible reason for error-prone polymerase is that they create mutation at a time when the cell might need variability. That is, DNA damage can occur when the environment is stressful for the cell; variability might help the cell survive (**Tamarin** 2002).

In addition, by using repair polymerases the cell can use a second repair mechanism to replicate damaged DNA when the polymerase leaves a gap. A replication fork creates two DNA duplexes. Thus, an undamaged copy of the region with the lesion exists on the other daughter duplex. A group of enzymes, with one specified by the *rec A* locus having central importance, repairs the gap. Since the repair takes place at a gap created by the failure of DNA replication, the process is called **post-replicative repair**.

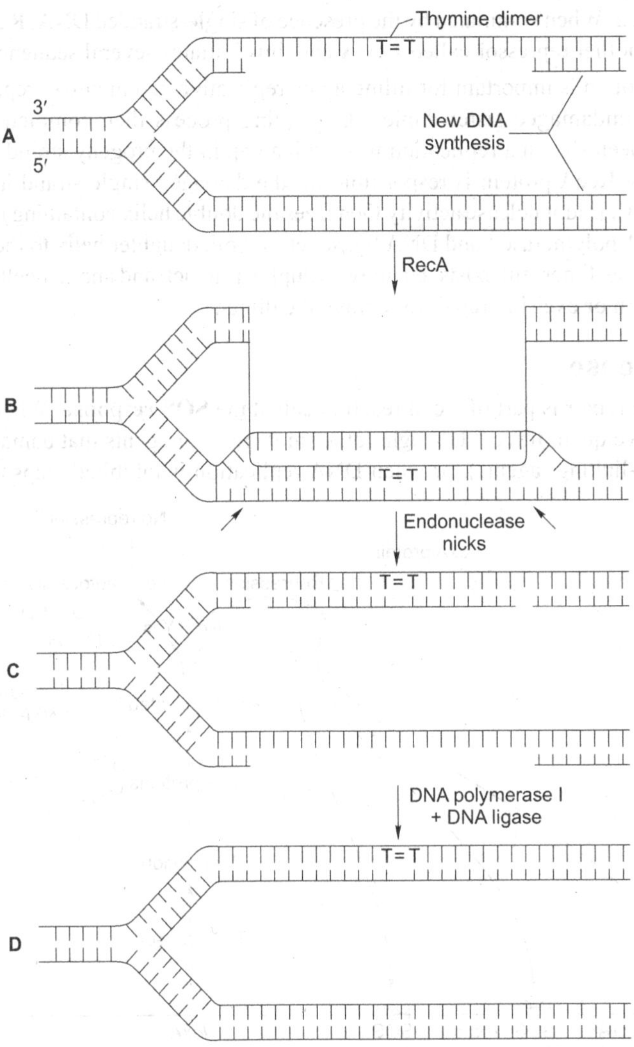

Fig. 35.29. RecA protein dependent postreplicative DNA repair. A—DNA polymerase III skips past a thymine dimer during DNA replication; it produces a normal duplex and abnormal duplex (with gap against the thymine-dimer); B—With the help of RecA protein, the single strand with the thymine dimer invades the normal sister duplex; C—An endonuclease nicks the new duplex at either side of the thymine dimer site, freeing the new duplex, with the thymine dimer and leaving the sister duplex single-stranded; D—Repair enzymes (DNA polymerase I and DNA ligase) then create two intact daughter duplexes (after Tamarin 2002).

The RecA protein. The RecA protein has two major properties: 1. It coats single-stranded DNA and causes that coated, single-stranded DNA to invade double-stranded DNA. By invasion, we mean that the single-stranded DNA attempts to form complementary base pairs with the antiparallel strand of the double stranded DNA while displacing the other strand of that double helix. RecA protein continues to move the single-stranded DNA along the double stranded DNA until a region of homology is found. 2. When stimulated by the presence of single-stranded DNA, RecA protein causes **autocatalysis** of another repressor called **Lex A** and thus initiates several sequences of reactions.

The RecA protein is important for filling a postreplicative gap in newly replicated DNA with a strand from the undamaged sister duplex. Gap-filling process then complete both strands. In Fig. 35.29A have been shown a replication fork with a gap in the progeny strand in the region of a thymine dimer. The RecA protein is responsible for the damaged single strand invading the sister duplex (Fig. 35.29B). Endonuclease activity then frees the double helix containing the thymine dimer (Fig. 35.29C). DNA polymerase I and DNA ligase return both daughter helix to the intact state (Fig. 35.29D). The thymine dimer still exists, but now its duplex is intact, and another cell cycle is available for photoreactivation or excision repair to remove the dimer.

The SOS Response

The postreplicative repair is part of a cell reaction called the **SOS response**. When an *E.coli* cell is exposed to excessive quantities of UV light, other mutagens, or agents that damage DNA (such as alkylating or cross-linking agents), or when DNA replication is inhibited, gaps are created in the

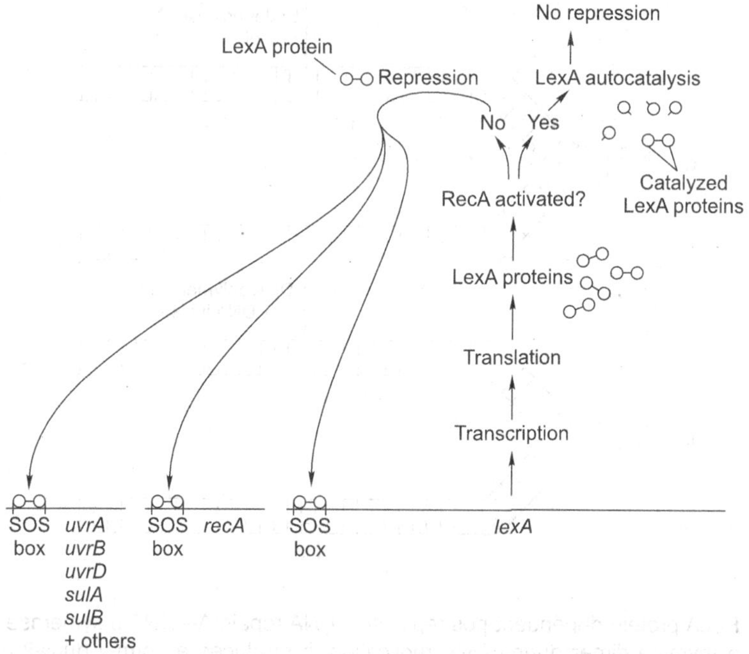

Fig. 35.30. The Lex A protein represses its own gene: *recA* and several other loci (*uvrA*, *uvrB*, *sulA* and *sulB*) by binding at the SOS box in each of the loci. Activated RecA protein causes autocatalysis of LexA, eliminating the repression of all loci, which are then transcribed and translated (after Tamarin 2002).

DNA. In the presence of this single-stranded DNA, the RecA protein interacts with the **LexA protein** normally represses about eighteen genes include *recA*, *uvrA*, *uvrB*, and *uvrD*; two genes that inhibit cell division, *sulA* and *sulB*; and several others. Each of these genes has a consensus sequence in its promoter called the **SOS box**; 5′-CTG X_{10} CAg (where X_{10} refers to any ten bases). The Lex A protein normally binds at the SOS box, limiting the transcription of these genes. When single stranded DNA activates RecA; RecA interacts with the Lex A protein to trigger the autocatalytic properties of Lex A

(Fig. 35.30). Transcription then follows from all the genes, presumably increase the amount of time the cell has to repair the damage before the next round of DNA replication.

35.8. RECOMBINATION

Although recombination, the nonparental arrangement of alleles in progeny, can come about both by independent assortment and crossing over, we are concerned here with recombination due to crossing over between homologous pieces of DNA, called **homologous recombination**.

Recombination is a **breakage-and-reunion** process. Homologous parts of chromosomes come into apposition and then reconnected in a crosswise fashion. This general model fits what we know about the concordance of recombination and repair: both involve breakage of the DNA and a small amount of repair synthesis and both involve some of the same enzyme. (Concordance means the amount of phenotypic similarity between individuals).

Double Strand Break Model of Recombination

In 1964, **R. Holliday** suggested a model of homologous recombination that involved simultaneous breaks in one strand each of the two double helices that were to cross over. In 1983, **J. Szostak** and colleagues put forth a different model, initiated by double-strand break in one of the double helices. At first, this model was not considered seriously because a double-strand break was thought too dangerous a DNA lesion for cellular enzymes to create. However, we now know that the double-strand break model is generally correct, and we refer to the **Holliday junction** for an intermediate stage in the process. The model depends on DNA complementarity between the recombining molecules and is thus a model of great precision (**Tamarin** 2002).

One such model is shown in Fig. 35.31. It begins with two paired DNA duplexes or homologs (Fig. 35.31A), in each of which, an **endonuclease** enzyme introduces a single-stranded **nick** at an identical position (breakage; Fig. 35.31B). The ends of the strands produced by these cuts are then **displaced** by helicases and subsequently **pair** with their complements on the oppsite DNA duplex by the help of **rec A protein** (Fig. 35.31C). A **ligase** enzyme then seals the loose ends (Fig. 35.31D), creating hybrid duplexes called **heteroduplex** DNA molecules. The exchange creates a **cross-bridge** structure. The position of this corss-bridge can then move down the chromosome by a process called **branch migration** (Fig. 35.31E), which occurs as a result of zipper-like action as hydrogen bonds are broken and then reformed between complementary bases of the displaced strands of each duplex. This migration yields an increased length of heteroduplex DNA on both homologs.

If the DNA duplexes now separate (Fig. 35.31F), and the bottom portions rotate 180° (Fig. 35.31G), an intermediate planar structure called a χ form (= *chi* form)—the characteristic **Holliday structure** or **Holliday junction**—is created. Separation, or **resolution**, of the Holliday structure back into individual double stranded molecules occurs by cleavage across the branch point. *This is key to the entire process* because the cut can be made in either of two orientations, as becomes apparent when the three-dimensional configuration or chi-form of the Holliday structure is examined. These two cuts produce very different results. If the cut is made left-right across the chi-form (*i.e.*, **horizontal resolution**), then all that happens is that a short segment of polynucleotides corresponding to the distance migrated by the branch of the Holliday structure, is transferred between the two molecules. On the other hand, an up-down (*i.e.*, **vertical resolution**) results in **reciprocal strand exchange**, double-stranded DNA being transferred between the two molecules so that the end of one molecule is exchanged for the end of the other molecule. This type of DNA transfer is seen as crossing over.

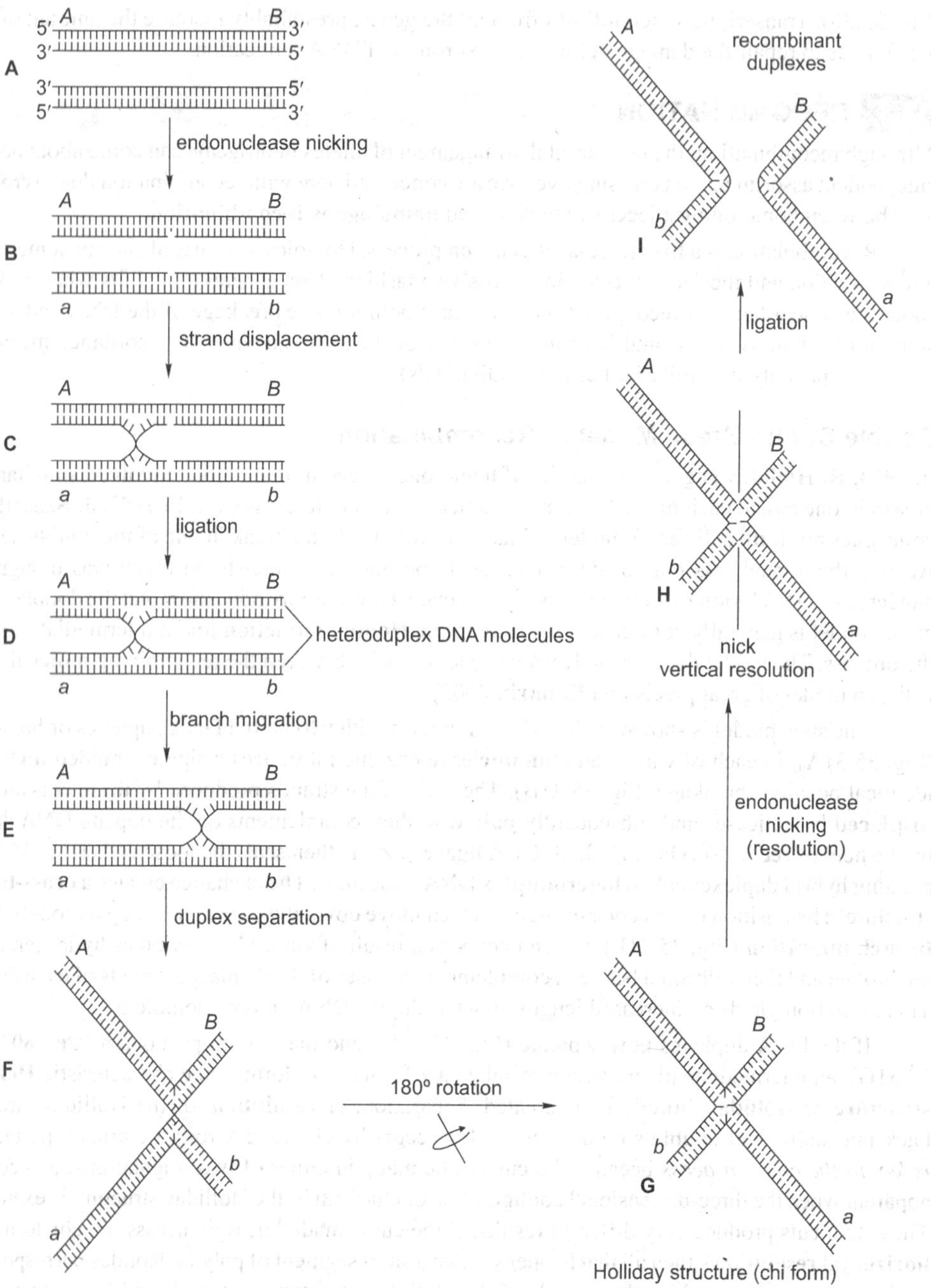

Fig. 35.31. Updated model of Holliday depicting how genetic/homologous recombination can occur as a result of the breakage and rejoining of heterologous DNA strands (after Klug and Cummings, 2003).

Evidences supporting this model includes the electron microscopic visualisation of χ-form (chi-form) planar molecules from bacteria in which four duplex arms are joined at a single point of exchange. Further important evidence comes from the discovery of the **RecA protein** in *E.coli*. The molecule promotes the exchange of reciprocal single-stranded DNA molecule as occurs in step C of

Fig. 35.31. RecA protein also enhances the hydrogen-bond formation during strand displacement, thus, initiating heteroduplex formation. Finally, many other enzymes essential to the nicking and ligation process have also been discovered and investigated. The products of the *rec B*, *rec C* and *rec D* genes are thought to be involved in the nicking and unwinding of DNA. Numerous mutations that prevent genetic recombination have been found in viruses and bacteria. These mutations represent genes whose products play an essential role in the process.

QUESTIONS

Long Answer Questions

1. Give an account of enzymes and proteins which are involved in prokaryotic DNA replication.
2. Describe the mechanism of DNA replication in prokaryotes.
3. Give an account of DNA replication in eukaryotes.
4. Give an detailed account of various ways of DNA repair.
5. Give an account of double strand break model of DNA recombination.

Short Answer Questions

1. Make a well labelled diagram of replication of DNA in prokaryotes.
2. Describe rolling circle model of DNA replication.
3. Describe the SOS response.
4. Write short notes on the following:
   (*i*) Meselson and Stahl experiment;
   (*ii*) Okazaki fragments;
   (*iii*) Unwinding proteins;
   (*iv*) Replisome;
   (*v*) Excision repair;
   (*vi*) NER;
   (*vii*) Mismatch repair.

Very Short Answer Questions

1. Define primer and template.
2. What is the role of RNA in DNA replication?
3. What is mismatch repair?
4. Define DNA ligase?
5. What is replication?
6. What is replication fork.

Multiple Choice Questions

1. The replisome of *E.coli* comprises
   (*a*) a polymerase and a primosome
   (*b*) a polymerase and a helicase
   (*c*) a polymerase, a primosome and a helicase
   (*d*) a primosome and a helicase
2. In proofreading during DNA replication
   (*a*) wrong nucleotides are inserted
   (*b*) correct nucleotides are taken out
   (*c*) wrong nucleotides are taken out
   (*d*) mutations are prevented
3. Which of the following is an autocatalytic function of DNA?
   (*a*) synthesis of RNA
   (*b*) synthesis of DNA
   (*c*) synthesis of protiens
   (*d*) all of these
4. Semiconservative replication of DNA was first demostrated in
   (*a*) *Escherichia coli*
   (*b*) *Streptococcus pneumoniae*
   (*c*) *Salmonella typhimurium*
   (*d*) *Drosphila melanogaster*
5. In eukaryotes, initiation of DNA replication is carried out by
   (*a*) DNA polymerase δ
   (*b*) DNA polymerase β
   (*c*) DNA polymerase α
   (*d*) DNA polymerase ε
6. Unwinding of DNA helix is carried out by
   (*a*) topoisomerase
   (*b*) ligase
   (*c*) replicase
   (*d*) helicase

7. Okazaki fragments are small segments of DNA
   (*a*) synthesized on the leading strand of DNA replication fork
   (*b*) synthesized on the lagging strand of DNA replication fork
   (*c*) produced during the digestion of DNA by restriction endonucleases
   (*d*) none of these

8. Meselson and Stahl (1958) conducted their experiments on
   (*a*) *E.coli*
   (*b*) *Vicia faba*
   (*c*) *Drosophila*
   (*d*) bacteriophage

ANSWERS

Very Short Answer Questions

1. Primer means nucleotide bound to DNA and having a 3′-OH group.
   Template means a polynucleotide strand whose base sequence can be copied.
2. RNA serves as a primer.
3. Mismatch repair is a type of DNA repair system that removes mismatched bases that are incorporated by the DNA polymerase and escape the enzyme's proof reading exonuclease.
4. The DNA ligase enzyme is reponsible for joining DNA fragments into a continuous strand.
5. Replication is duplication of the genetic material.
6. Replication fork is a Y shaped site on replicating DNA, where (*i*) the parental double helix is undergoing strand separation and (*ii*) nucleotides are being incorporated into the newly synthesized complementary strands.

Multiple Choice Questions

1. (*c*) **2.** (*c*) **3.** (*b*) **4.** (*a*) **5.** (*c*) **6.** (*d*) **7.** (*b*)
8. (*a*)

36 Genes

The hereditary units which are transmitted from one generation to the next generation are called **genes**. A gene is the fundamental biologic unit, like the atom which is the fundamental physical unit. **Mendel** while explaining the results of his monohybrid and dihybrid crosses, first of all conceived of the genes as particulate units and referred them by various names such as **hereditary factors** or **hereditary elements** (or **Merkmal**). In the beginning of present century Mendel's factor came to be known as **gene** (**Johannsen**, 1909). The presence of the gene is detected only when a mutation occurs in it. Initially genes were considered as beads and chromosomes as strings of beads (**Morgan**, 1911). Mutation was supposed to alter the bead structure and recombination (or crossing over) was regarded to, involve a breakage between two beads followed by their exchange between paired chromosomes. Each bead was thought of to control one character by controlling some biochemical step. Thus, a gene was considered to be a unit of mutation, recombination and function.

36.1. THE GENOME AND THE PLASMON

The total hereditary material of an organism may be divided into the **genome** and the **plasmon**. The *genome* is the hereditary material of the chromosomal complex. The *plasmon* is the total hereditary materials outside the chromosome. The term plasmon was introduced by **Fritz von Wettstein** in 1924. The smallest heritable unit of the plasmon is the **plasmagene**, a term introduced by **Darlington** in 1939. Plasmagenes are known by names according to their location in different cell structures. The plasmagenes of plastids are called **plastogenes** and those of mitochondria **chondriogenes**.

36.2. FINE STRUCTURE OF GENES

1. Recon, Muton and Cistron. From the terms *cis* and *trans*, **Seymour Benzer** in 1950's coined the term *cistron* for the smallest genetic unit (length of genetic material) that exhibits a *cis-trans* position effect (Box 36.1). We thus have a new word for the gene, one in which function is more explicit. We have, in principle, refined **Beadle** and **Tatum's** one-gene-one-enzyme hypothesis to a more accurate one-cistron-one-polypeptide concept. The cistron is the smallest unit that codes for a messenger RNA that is then translated into a single polypeptide or expressed directly (transfer RNA or ribosomal RNA).

Box 36.1

The functional gene has been called *cistron* by Benzer because it is a chromosomal segment within which the cis-trans effect operates. The cistron represents a segment of DNA molecule, and consists of a linear sequence of nucleotides which control some cellular functions. The cistron starts with an *initiation codon* and ends with a *terminating codon*.

Fine structure mapping. After Beadle and Tatum established in 1914 that a gene controls the production of an enzyme that then controls a step in biochemical pathway, **Benzer** used analytical techniques to dissect the fine structure of the gene. Fine structure mapping means examining the size and number of sites within a gene that are capable of mutation and recombination. In the late 1950s (*i.e.*, 1955, 1956), when biochemical techniques were not available for DNA sequencing, Benzer used classical recombinational and mutational techniques with bacterial viruses (*i.e.*, rII locus of T_4 bacteriophage) to provide reasonable estimates on the details of fine structure and to give insight into the nature of the gene. He coined the terms **muton** for the smallest mutable site and **recon** for the smallest unit of recombination. It is now known that both muton and recon are a single base pair (**Tamarin**, 2002).

2. Loci and alleles. A gene occupies a definite position within the chromosome. This position is called the **locus** (plural **loci**) (**Demerec** 1955). Chromosomes exist in homologous pairs, each cell contains two kinds of genes which are found in pairs. The two members of a pair of a gene are called **alleles**.

3. Gene regulation. Jacob and **Monod** (1961) proposed the **operon model** to explain the control of gene activity. It was assumed that there are three kinds of genes: *structural genes*, *operator genes* and *regulator genes*. **Structural gens** are those genes that produce mRNA. **Operator genes** control structural genes by acting as switches in controlling mRNA synthesis by the latter. **Regulator genes** produce substances, called **repressors** which block the operator genes, so that mRNA synthesis by structural genes is prevented. The structural and operator genes, both combine to form a **operon**.

36.3. SPLIT GENES

Split genes means that the (DNA) sequences containing actual information of the gene (called **exons**) are interrupted by other sequences (called **introns**) which are spliced out after transcription.

A. Discovery of Split Genes

With the discovery of DNA as the genetic material, a gene was regarded as a continuous segment of DNA. In 1964, it was also proved that there was co-linearity between sequence of nucleotides on DNA and sequence of amino acids in corresponding protein. Therefore, there was absolutely no doubt about the continuity of nucleotides in a gene represented by a DNA segment. This was proved to be true both for prokaryotes and eukaryotes. However, a big surprise came in 1977, when geneticists came to know that in some mammals, birds and amphibians, a gene may not be represented by a continuous sequence of nucleotides but may be interrupted by some intervening sequences which are not represented in mRNA transcribed from the gene and utilized for the synthesis of proteins. Such genes with intervening sequences were called **split genes** or **interrupted genes** (Fig. 36.1). Split genes are also called **discontinuous** or **mosaic genes** (see **Brown**, 1998).

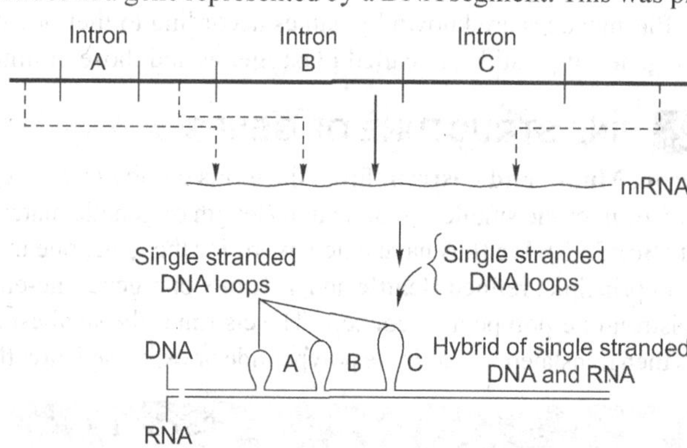

Fig. 36.1. The upper figure shows a DNA sequence representing an split or interrupted gene with their introns (A, B, C) and the synthesis of its mRNA; the bottom figure shows the result of hybridization of mRNA with single stranded DNA obtained after denaturation of native DNA (note the loops formed by three intron genes).

The discovery of split genes was made in 1977 by various groups of biologists in a variety of materials: (*i*) Two research groups separately headed by **Philip A. Sharp** and **Richard J. Roberts** studied genes of adenovirus 2. (*ii*) Research groups of **P. Chambon**, **P. Leader** and **R.A. Flavell** studied β globin genes, ovalbummin genes and tRNA genes. In all these cases the genes were found to be interrupted by intervening sequences. The credit for discovery of split genes, however, goes to **Philip Sharp** and **Richard Roberts**, who won in 1993 Nobel Prize for Medicine or Physiology for their independent work on split genes. They analyzed the hybrids of *late mRNA* of adenovirus 2 with the adenovirus genomic DNA. When these mRNA-DNA hybrids were examined under electron microscope, the adjoining sequences of mRNA were found to be hybridized with discontinuous stretches of genomic DNA of adenovirus. The intervening DNA sequences were observed as loops and the phenomenon was later described as **R-looping**.

P. Chambon's group (France) and **B.W. O'Malley's** group (USA) showed that the **ovalbumin gene** of chickens is not contiguous but is made up of pieces scattered in the chromosome. These genes are the *split genes*. Within a gene there may be silent regions which are not represented in the polypeptide chain. The natural ovalbumin gene transcribes precursor RNA which contains intervening sequences not found in mRNA. The precursor RNA loops out these intervening sequences, which are then excised, to form mRNA.

Chambon's group compiled sequences of the boundaries of introns from a large number of protein coding eukaryotic genes (not ribosomal RNA or tRNA genes), which revealed the presence of *consensus sequences* at the intron-exon junctions. Of these GT (guanine-thymine) was always found at the 5′ side of the intron (**left splice junction**) and AG (adenine-guanine) at the 3′ side (**right splice junction**). This became popularly known as **GT-AG rule** or **Chambon's rule** (*Note:* Rarely, AT-AC occur at intron-exon junctions). Some of the diseases (*e.g.*, thalessemia) are caused by mutations, which created or abolished these splice junctions.

Methods of Investigations of Split Genes

A detailed study was conducted on **ovalbumin gene** found in chicken. This ovalbumin gene is responsible for synthesis of a protein, called **ovalbumin** consisting of 386 amino acids and synthesized only by highly specialized tubular gland cells of the oviduct at the time when the hen is laying eggs. The expression of this ovalbumin gene is controlled by some female sex hormones. Chambon and his collegues synthesized artificial ovalbumin gene in order to study its regulation. Such an artificial gene of ovalbumin could be synthesized by using ovalbumin mRNA which could give rise to cDNA (complementary DNA) with the help of enzyme **reverse transcriptase**. This cDNA was inserted into a plasmid and cloned in *E.coli* for its multiplication. When this cDNA was compared with corresponding genomic DNA, it was discovered (through DNA hybridization) that the genomic DNA had additional intervening sequences.

Another ingenious technique used for the study of split gene was the use of **restriction enzymes** which have the property of cleaving DNA at unique sites. More than 100 such restriction endonuclease enzymes are now available. In view of this, restriction endonucleases could be used to find out the presence or absence of a certain sequence in a particular gene. For example, when *EcoR1* and *Hind III* enzymes were used with cDNA for ovalbumin, it was found that no cleavage occurred suggesting that the sequences of six base pairs each recognized by these two restriction enzymes were absent. On the basis of this it was expected that if the DNA extracted from oviduct was cleaved by utilizing these two enzymes, the ovalbumin gene will not be broken and the cleavage will occur at other places thus making it possible to isolate ovalbumin gene from the living cells. This DNA segment representing ovalbumin gene was expected to be separated with the help of hybridization with cDNA artificially synthesized. When such hybridization was done with cDNA hybridized with diffferent fragments of DNA rather than with a single fragment. Hybridization between the single stranded DNA having the gene for ovalbumin and its mRNA also showed the formation of distinct loops at specific sites as observed in electron microscope (Fig. 36.1). Such DNA which exists in the loops is obviously missing in mRNA. This kind of conclusion eventually led to the discovery of split genes in 1977.

In subsequent years, it could be proved that split genes are present at least in two more cases, *i.e.*, gene for β–**globin** (a component of haemoglobin molecule) in rabbit and mouse and **immunoglobin gene** (antibody gene). Later on, split genes were reported to occur in higher organisms as a common phenomenon. For understanding the structure of a split gene, we can consider an example of a hypothetical split gene which has a DNA sequence including three pieces, called **exon 1**, **exon 2** and **exon 3** (Fig. 36.2). These three exons are separated by two long intervening sequences, called **intron A** and **intron B**. The terms exon and intron were used by **Walter Gilbert** (1985) for the first time and are being followed ever since.

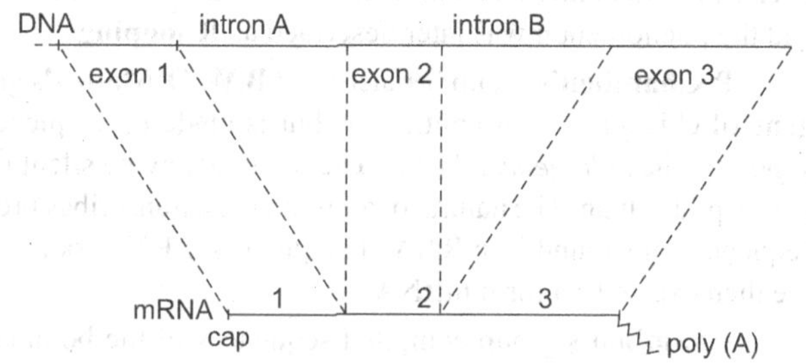

Fig. 36.2. Diagrammatic representation of a hypothetical split gene having three exons (exon 1, exon 2 and exon 3) and two introns (intron A and intron B) and its relationship with mRNA. There are no sequences in mRNA corresponding to those in two introns.

Figure 36.2 shows that in the transcription of DNA, the ultimate product, *i.e.*, mRNA had only those sequences which correspond to exons, the sequences representing introns being entirely absent. In subsequent years, it is discovered that both exons and introns are first transcribed and this primary transcript in then modified. The sequences corresponding to introns are removed from the transcript and the sequences corresponding to exons are joined together, in correct order to give rise to mRNA. A generalization has been made that the order of exons on DNA is the same as the order in which they are found in the processed mRNA.

Examples of split genes or interrupted genes occur in a variety of organisms: (*i*) nuclear genes for proteins; (*ii*) nuclear genes for rRNAs; (*iii*) nuclear genes for tRNAs; (*iv*) mitochondrial genes in yeast; (*v*) chloroplast genes in a wide variety of plants; (*vi*) genes in archaebaceria and (*vii*) genes in bacteriophages of *E.coli*. They were initially believed to be entirely absent in eubacterial genomes but in 1990s introns were discovered even in eubacteria.

B. Certain Examples of Split Genes

1. Structure of ovalbumin split gene of hen (chicken). In the split gene of ovalbumin there are eight exons and seven intron (Fig. 36.3 and Fig. 36.4). The size of ovalbumin gene is 7,700 base pairs long although the mature mRNA for ovalbumin is only 1,872 nucleotides. The entire ovalbumin gene with its 7700 base pairs is first transcribed to a precursor RNA to which a cap consisting of 7 mG is added at the 5′ end and poly A tail is added at the 3′ end. After the addition of cap and tail, five introns are excised in the first step and the remaining two in second step (Fig. 36.5). The exons produced due to splicing are then joined with the help of enzyme ligase to produce the mature mRNA of ovalbumin gene.

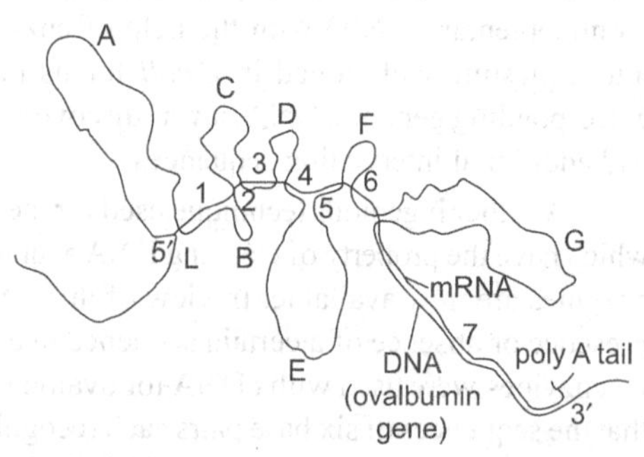

Fig. 36.3. The map of electron micrograph prepared after hybridizing mRNA (thin line) of the ovalbumin gene with its single stranded DNA (thick line). The loops A-G represent 7 different introns and segments L and 1-7 represent 8 exons.

2. Split genes in humans. In the human gene for the **cystic fibrosis** transmembrane regulator, the introns are much longer than the exons. This gene is 250 kb long and is split into 24 exons and 23 introns. The average length of the exons is 227 bp, so all the exons added together make up only 2.4% of the gene. These exons are scattered throughout the entrie length of gene, separated by introns that range in size from 2 to 35 kb (see **Brown**, 1998).

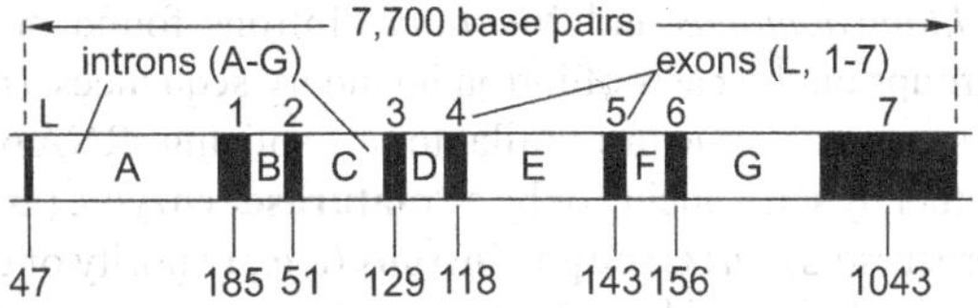

Fig. 36.4. Schematic representation of ovalbumin gene showing 7 introns (A-G) and 8 exons (l, 1-7); number of base pairs in each exon are also given.

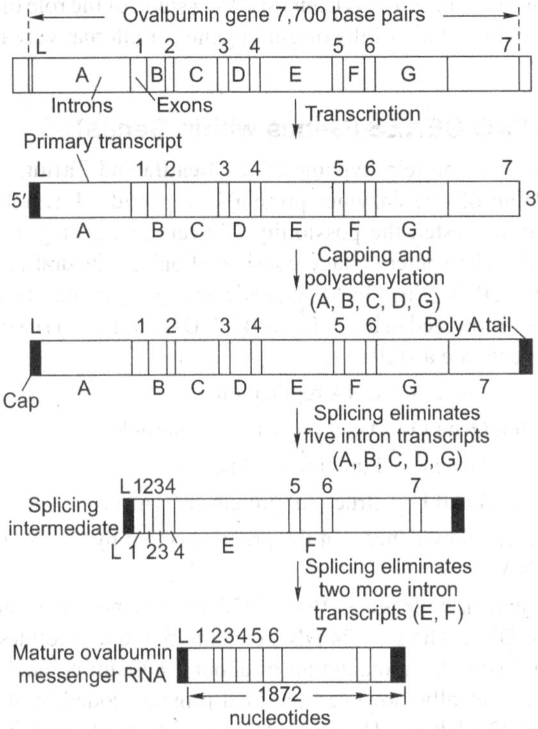

Fig. 36.5. Steps involved in the production of ovalbumin mRNA (1872 nucleotides long) from ovalbumin gene (7700 base pairs long); introns (A-G) and exons (L and 1-7).

3. Split genes in fungal mitochondria. Split genes are also found in the mitochondria. Introns of these split genes in fungal mitochondria are of two types: (*i*) **Group I introns** which are found in majority of the fungal mitochondria split genes, do not carry any conserved sequence, called **internal guide sequence**. By internal pairing, this guide sequence brings the two intron-exon junctions together and help in splicing out the introns. These group I introns are also found in the nuclear gene coding for rRNA in *Tetrahymena* (a ciliate) and *Physarum* (a slime mold). Features similar to those of group I introns are also reported from introns of phage T4 genes. (*ii*) **Group II introns** resemble nuclear genes and have consensus sequences (GT and APy) and a branch sequence that resembles the TACTAAC box. These introns are excised as lariats.

4. Split genes in chloroplasts. Split genes for ribosomal RNA (rRNA), transfer RNAs (tRNA) and some proteins have also been reported in the chloroplast genomes of several plants including

Chlamydomonas and *Nicotiana.* Introns found in chloroplast genes can be classified into three groups on the basis of intron boundary sequences. (*i*) *Group I introns* (*e.g.*, in *trnL*) can be folded in a secondary structure similar to self-splicing rRNA precursor of *Tetrahymena.* These can be removed either by self-splicing or by a "**maturase**" enzyme (as in cytochrome b and cytochrome oxidase mRNA precursors). (*ii*) **Group II introns** (*e.g.*, majority of genes including *trnA* and *trnI*) can be folded into a complex secondary structure (as in introns of mitochondrial genes for cytochrome oxidase in maize and yeast). (*iii*) **Group III introns** (*e.g.*, *trnG, trnK, trnV, rpl2, rps12, rps16*, etc.) have conserved sequences at their borders (GTGCGNY at 5′ end, and ATCNRYY(N)YYAY at 3′ end), similar to those in the eukaryotic nuclear genes (R = purine; Y = pyrimidine; N = any nucleotide).

Although introns are generally absent in protein-coding genes of bacteria, archaebacteria and some lower eukaryotes, class II type of self-splicing introns have recently been reported even in bacteria. Such a distribution of introns has been used for a study of the role of introns in the evolution of genes or alternatively for a study of the origin of genes or alternatively for a study of the origin of introns themselves.

36.4. OVERLAPPING GENES (Genes within Genes)

According to "one gene, one protein hypothesis" of **Beadle** and **Tatum** (1937–40) each gene is responsible for the coding of one enzyme (protein)/polypeptide. **Barrell** and coworkers (1976) first gave evidence that suggested the possibility of overlapping of genes in the bacterial virus (= bacteriophage) $\phi\times174$. The bacteriophage consists of an icosahedral protein capsid containing single-stranded circular DNA (Fig. 36.6). Genetic mapping techniques have shown that the bacteriophage has 9 genes (arranged in the order A–B–C–D–E–J–F–G–H) which codes for 9 proteins. The functions of these genes are as follows:

Gene A = DNA replication
Gene B and Gene D = Assembly of phase particle
Gene C = Function not known
Genes F, G and H = structural proteins of capsid

Barrell *et al.*, presented evidence that the proteins coded by genes D and E are specified by the same segment of DNA.

Fredrick Sanger and his colleagues (1976, 1977) have mapped the entire nucleotide sequence of $\phi \times 174$ and phage G_4 DNA. The $\phi \times 174$ DNA contains 5,386 nucleotides. On the basis of triplet code this number should code for a maximum of about 1,800 amino acids having a total weight of about 200,000 dalton. Actually, however, the total proteins coded by these nucleotides have a molecular weight of 250,000 daltons. The phage DNA which should have a coding capacity of 5 to 6 average sized protein molecules, in fact code for 9 proteins. The phage DNA is thus apparently 10 to 15% too short for the total protein coded.

A comparison of the DNA base sequence with the amino acid sequences of the protein coded revealed that in two cases the same gene coded for two different proteins. Thus two genes which should normally have coded two proteins actually coded four. Sanger and his coworkers found that gene B (136 nucleotides) was completely contained within gene A (1536 nucleotides). Simiarly, gene E (273 nucleotides) was contained in gene D (1456 nucleotides).

The protein synthesized by gene B is not simply a shorter version of that synthesized by gene A. The two proteins have completely different amino sequnces. Similar is the case with the proteins synthsised by genes D and E. It has been explained that the coding of two proteins by one gene is achieved by a **shift in frame reading**. For example, a reading frame of ...G, AAG, TTA, ACA ... nucleotides codes for the amino acids *lysine*, *leucine* and *threonine*. If the frame is read from one point earlier it becomes ... GAA, GTT, AAC, A ... which codes for *glutamine*, *valine* and *asparagine*. Thus, the same gene can code for two different sequences by a frame shift (overlapping code) and produce two totally different proteins. Since one gene codes for more than one protein is an amendment to the one-gene, one-protein dogma becomes necessary, at least in few cases such as $\phi\times174$ and G4 phages.

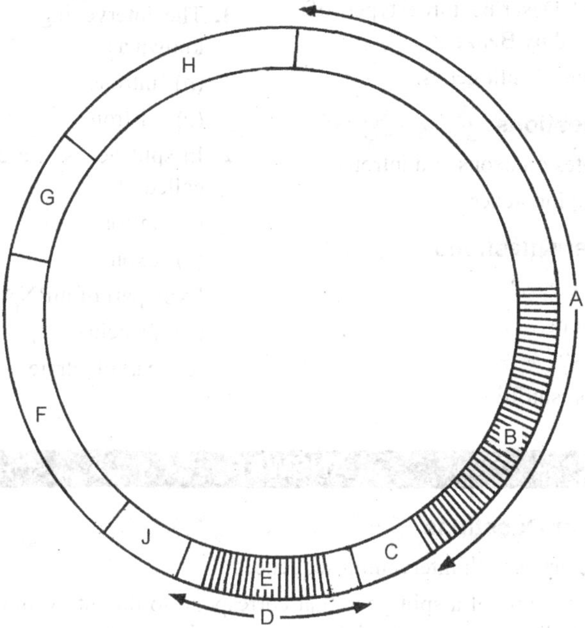

Fig. 36.6. Genetic map of the bacteriophage $\phi\times174$. Note the overlapping of gene E over D and of gene B over A.

Another interesting fact that the **termination codon** of gene D and the **initiation codon** of gene J overlap by one nucleotide:

$$---\underbrace{T-A-A}_{\text{Termination codon of gene D}}\!\!\!\!\!\!\overbrace{}^{}\underbrace{-T-G}_{\text{Initiation codon of gene J}}---$$

The bacteriophage G4 which infects *E.coli* also has overlapping genes. Gene B is completely contained within gene A and gene E within gene D (Fig. 36.7). Gene K is made up of the last 86 nucleotides of gene A and the first 89 nucleotides of gene C.

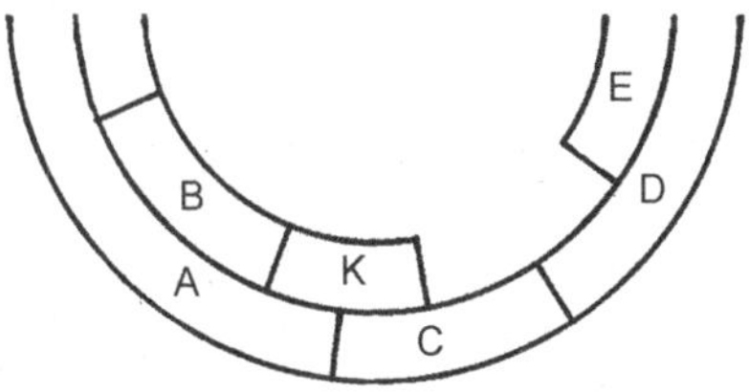

Fig. 36.7. A part of the genetic map of the virus G4 showing overlapping genes.

The condition of overlapping genes is not of common occurrence. It may have arisen in the smaller viruses because of the limitations of the quantity of DNA that can be enclosed in the small viral capsids. It seems to be an economic device to make better use of genetic information in less DNA.

QUESTIONS

Long Answer Questions

1. What is gene? Describe three types of genes recognised by Benzer.
2. Give an account of split genes.

Short Answer Questions

1. Write short notes on exons and introns.
2. Define overlapping genes.

Very Short Answer Questions

1. What are split genes?
2. Define the introns.
3. What are exons?
4. Define the exon shuffling.

Multiple Choice Questions

1. The intervening sequence of 'gene' are known as
   (*a*) introns (*b*) exons
   (*c*) cistron (*d*) codons
2. In split genes, the coding sequences are called
   (*a*) introns (*b*) operons
   (*c*) exons (*d*) cistrons
3. Exon part of mRNA have codes for
   (*a*) protein (*b*) lipid
   (*c*) carbohydrate (*d*) phospholipid

ANSWER

Very Short Answer Questions

1. Split genes are genes with intervening sequences.
2. Introns are those parts of a split gene that correspond to the intervening sequences. Or, an intron is a noncoding sequence that interrupts exons in a gene.
3. Exons are those parts of a split gene that contribute to a mature RNA product. Or, an exon is a segment of a gene that contains a coding sequence.
4. Exon shuffling is the movement of genetic "modules" among unrelated genes facilitated by the presence of introns; the introns act like inert spacer elements between exons.

Multiple Choice Questions

1. (*a*) 2. (*c*) 3. (*a*)

37 Point Mutation

Mutation (L. *mutare* = to change) is any heritable change in the nucleotide sequence of genomic DNA (or genomic RNA, in case of an RNA virus) (**Jain**, 2012). It is also called **point mutation** or **gene mutation**. A gene mutation is abrupt inheritable qualitative or quantitative change in the genetic material of an organism. Since in most organisms, genes are segments of DNA molecule, so a mutation can be regarded as a change in the DNA sequence which is reflected in the change of sequence of corresponding RNA or protein molecules. Such a change may involve only one base/ base pair or more than one base pair of DNA. Mutations occur in a **random** manner, *i.e.*, they are not directed according to the requirements of the organism. Most mutations occur **spontaneously** by the environmental effect, however, they can be induced in the laboratory either by radiations, physical factors or chemicals (called **mutagens**). A unicellular organism is more subjected to environmental attacks since it is at the same time a somatic or germ cell. In multicellular organisms, the germ cells are distinct cells, and are relatively protected from the environment. Mutation has a significant role to play in the origin of species or evolution.

37.1. HISTORICAL BACKGROUND

The earliest record of point mutations dates back to 1791, when **Seth Wright** noticed a lamb with exceptionally short legs in his flock of sheep. Visualising the economic significance of this short-legged sheep. *i.e.*, short-legged sheep could not cross the low stone fence and damage the crop fields in the vicinity, he produced a flock of sheep, each of which having short legs by employing artificial breeding techniques. The short-legged breed of sheep was known as **Ancon** breed. Later on, the trait of short legs was found to have resulted from a recessive mutation and the short-legged individuals were found to be homozygous recessive.

Hugo de Vries was the first hybridist who used the term "mutation" to describe the heritable phenotypic changes of the evening primrose, *Oenothera lamarckiana*. Many mutations described by de Vries in *O. lamarckiana*, are now known to be due to variation in chromosome number or ploidy and chromosomal aberrations (*viz.* **gross mutations**). The first scientific study of mutation was started in 1910, when **Morgan** started his work on fruitfly, *Drosophila melanogaster* and reported white-eyed male individuals among red-eyed male individuals. The discovery of white-eyed mutants in *Drosophila* was followed by an extensive search of other mutants of *Drosophila* by **Morgan** and his co-workers and other geneticists. Consequently, about 500 mutants of *Drosophila* have been reported by geneticists all over the world. Later on, several cases of mutations have been reported in a variety of microorganisms (*e.g.*, bacteriophages, bacteria (*Escherichia coli*), *Neurospora*, etc., plants (*i.e.*, pea, snapdragon, maize, etc.) and animals, (*i.e.*, nematodes, amphibians, rodents, fowls, human beings, etc.).

37.2. OCCURRENCE

Mutations occur frequently in nature and have been reported in many organisms, *e.g.*, *Drosophila*, mice and other rodents, rats, rabbits, guinea pigs and human beings. In *Drosophila*, mutation causes white and pink eyes, black and yellow body colours, and vestigial wings. In rodents the mutations are responsible for black, white and brown coats. In human beings, the mutations cause variation in hair colour, eye colour, skin pigmentation and several somatic malformations. Various genetical diseases of human beings such as haemophila, colour blindness, phenylketonuria, etc., form other examples of mutation in human beings.

How does a mutation act? Any change in sequence of nucleotides in the DNA will result in the corresponding change in the nucleotide sequence of mRNA. This may result in alignment of different tRNA molecules on mRNA (during protein synthesis). Thus, the amino acid sequence, and, hence, the structure and properties of the enzyme formed will be changed. This defective enzyme or structural protein may adversely affect the trait controlled by the protein. In consequence, a mutant phenotype makes its expression.

37.3. KINDS OF MUTATIONS

There exists a lot of controversy among geneticists about the possible kinds of mutations. Mutations have been classified variously according to different criteria as follows:

1. Classification of Mutation According to Type of Cells

According to their occurrence in somatic and germinal cells, following types of mutations have been classified:

A. Somatic mutations. The mutations occurring in non-reproductive body cells are known as **somatic mutations**. The genetical and evolutionary consequences of somatic mutations are insignificant, since only single cells and their daughter cells are involved. If, however, a somatic mutation occurs early during embryonic life, the mutant cells may constitute a large proportion of body cells and the animal body may be a mosaic of different types of cells. Somatic mutations have often been related with malignant (cancerous) growth. Examples of somatic mutation have been reported in *Oenothera lamarckiana* (by Hugo de Vries) and several other cases including human beings. In human beings, somatic mutation causes several fatal diseases such as *paraoxysmal nocturnal haemoglobinura, circumscribed neurofibroma, unilateral retinoblastoma and heterochromia of the iris.*

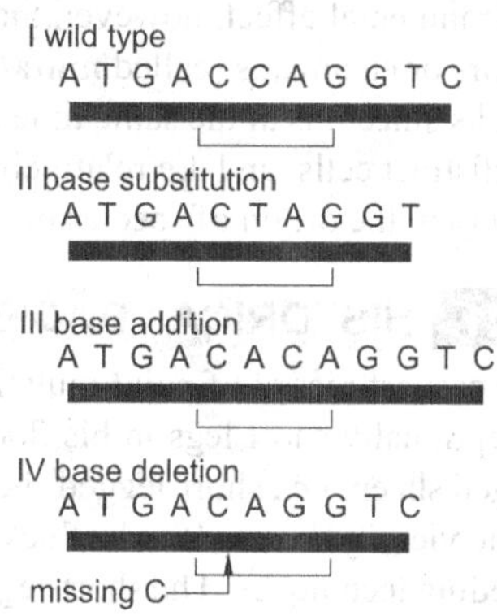

Fig. 37.1. Three types of point mutations. Only the base squence of one DNA strand is shown. Changes are shown in square; the horizontal brackets indicate the affected segment (after Freifelder, 1985).

B. Gametic mutations. The mutations occurring in gamete cells (*e.g.*, sperms and ova) are called **gametic mutations**. Such mutations are heritable and of great genetical significance. The gametic mutations only form the raw material for the natural selection.

2. Classification of Mutations According to the Size and Quality

According to size, following two types of mutations have been recognised:

A. Point mutation. When heritable alterations occur in a very small segment of DNA molecule, *i.e.*, a single nucleotide or nucleotide pair, then these types of **mutations** are called **"point mutations"**. The point mutations may occur due to following types of subnucleotide change in the DNA and RNA.

1. Deletion mutation. The point mutation which is caused due to loss or deletion of some portion (single nucleotide pair) in a triplet codon of a cistron or gene is called **deletion mutation** (Fig. 37.1). Deletion mutations have been frequently reported in some bacteriophages (Phage T_4).

2. Insertion or addition mutation. The point mutations which occur due to addition of one or more extra nucleotides to a gene or cistron are called **insertion mutations** (Fig. 37.1). The insertion mutations can be artificially induced by certain chemical substances called **mutagens** such as acridine dye and proflavin. A proflavin molecule, it is believed, insert between two successive bases of a DNA strand, thereby stretching the strand lengthwise. At replication, this situation would allow the insertion of an extra nucleotide in the complementary chain at the position occupied by the proflavin molecule.

The mutations which arise from the insertion or deletion of individual nucleotides and cause the rest of the message downstream of the mutation to be read out of phase, are called **frameshift mutations**. They result in the production of an incorrect, hence, inactive protein, due to which death of the cell may occur.

3. Substitution mutation. A point mutation in which a nucleotide of a triplet is replaced by another nucleotide, is called **substitution mutation** (Fig. 37.1). The substitution mutation affect only a particular triplet codon. Such an altered code word (triplet codon) may designate a different amino acid and may result in the production of a protein with a single amino acid substitution. The substitution mutations alter the phenotype of an organism variously and are of great genetical significance. They may be of following two types:

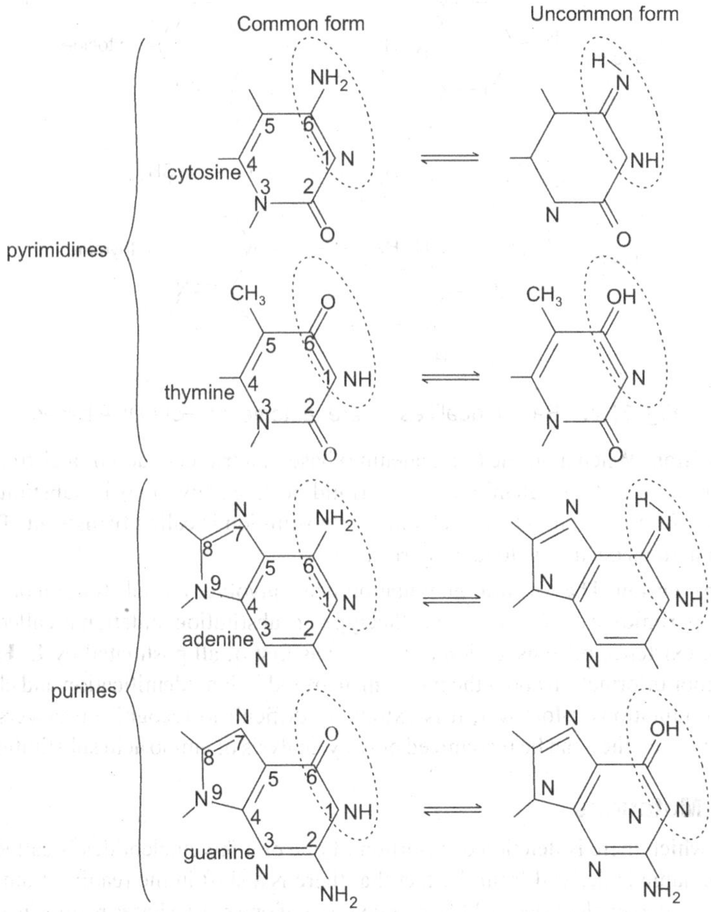

Fig. 37.2. The uncommon form of DNA bases.

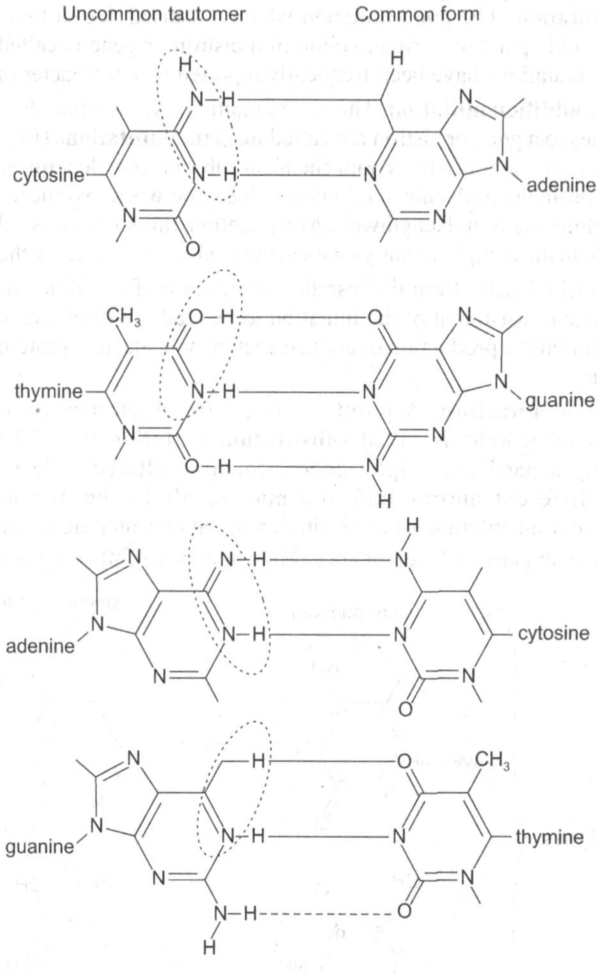

Fig. 37.3. Pairing qualities of rare tautomers of four DNA bases.

***(i)* Transition.** When a purine (*e.g.*, adenine) base of a triplet codon of a cistron is substituted by another purine base (*e.g.*, guanine) or a pyrimidine (*e.g.*, thymine) is substituted by another pyrimidine base, (*e.g.*, cytosine) then such kind of substitution is called **transition**. The transitional substitution mutations occur due to tautomerization.

***(ii)* Transversion.** The substitution mutation when involves the substitution or replacement of a purine with a pyrimidine or *vice versa* then that type of substitution mutation is called **transversion mutation**. The existence of transversion mutation was first of all postulated by **E. Freese** in 1959. We still have poor information about the mechanism of induction, identification and characterisation of transversion mutations. Moreover, it is extremely difficult to recognise transversion mutations genetically. However, they can be recognized only by analysis of amino acid substitutions in proteins.

Frameshift Mutations

A mutation in which there is deletion or insertion of one or a few nucleotides is called a **frameshift mutation**. The name is derived from the fact that there is a shift in the reading frame backward or forward by one or two nucleotides. Addition or deletion of one or two bases results in a new sequence of codons which may code for entirely different amino acids. This results in a drastic changes in the protein synthesized. The protein is usually nonfunctional. It should be noted that if the reading frame shifts by three nucleotides, the resulting protein is normal, except that it may lack one amino acid or may contain an extra amino acid.

Another effect of frameshift mutation is that stop-signal information will be misread (Fig. 37.4). One of the new codons may be a nonsense codon which causes translation to stop prematurally. Or, if the translation apparatus reaches the original nonsense codon, it is no longer recognized as such because, it is in a different reading frame, and therefore, the translation process continues beyond the end of the gene (**Tamarin** 2002).

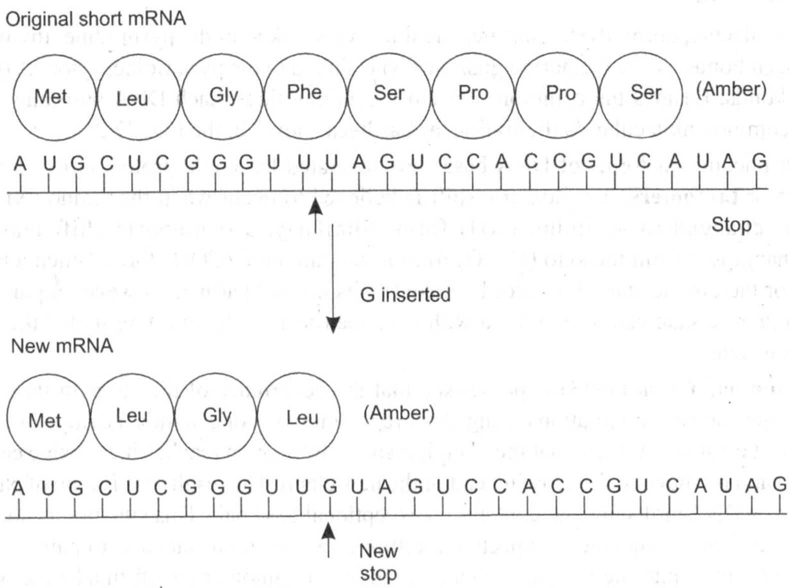

Fig. 37.4. Possible effect of a frameshift mutation. The insertion of a single base result in the creation of a new stop sequence (the amber). The result will be premature termination of translation.

Haemoglobinopathy due to substitution Mutation and Frame Shift Mutation

A change in only one amino acid can have a drastic effect on the phenotype. Globin protein of haemoglobin which is found in the RBCs consists of four polypeptide chains, *two alpha* (α) *chains* and *two beta* (β) *chains*. These chains consist of amino acids arranged in a definite sequence. Normal human RBCs are disc shaped having haemoglobin A (Hb^A). Change in haemoglobin structure result in a type of anaemia, called **sickle cell anaemia (SCA)**. In sickle cell anaemia, the RBCs become sickle shaped when oxygen tension is reduced and are much less effective in the transportation of oxygen. Death may occur in server cases of sickle cell anaemia.

Biochemical investigations have shown that globin protein of haemoglobin is a heteropolymer having two identical α polypeptide chains, each with 141 amino acid residues and two idential β polypeptide chains, each with 146 amino acid residues. In 1957, **Vernon Ingram** showed that β chain amino acid of 6th position is occupied by **glutamic acid** in haemoglobin A, when this amino acid is substituted by another amino acid, the **valine**, it forms the haemoglobin S:

	1		2		3		4		5		6		7		8	
Hb^A =	val	-	his	-	leu	-	thr	-	pro	-	glu	-	glu	-	lys	-
Hb^S =	val	-	his	-	leu	-	thr	-	pro	-	val	-	glu	-	lys	-

Fig. 37.5. Parts of β chains of globin molecules of haemoglobin showing differences in amino acids in the sixth position in Hb^A and Hb^S.

Human's SCA phenotype is controlled by a single autosomal gene on chromosome 11 with two alleles A and S (*i.e.*, Hb^A and Hb^S).

Tautomerization

In a DNA molecule, normally, the purine, adenine (A) is linked to the pyrimidine, thymine (T), by two hydrogen bonds, while the purine guanine (G) is linked to the pyrimidine, cytosine (C) by three hydrogen bonds. Besides the common molecular configurations, each DNA base may have some altered uncommon molecular configuration, as has been shown in the Fig. 37.2.

Such uncommon forms of DNA bases are generated by single proton shifts and are called **rare states** or **tautomers**. A tautomeric shift is believed to occur when the amino (NH_2) form of adenine is changed to an imino (NH) form. Similarly, a tautomeric shift may occur in thymine changing it from the keto (C = O) form to the rare enol (COH) form. When a base occurs in its rare or tautomeric state, it cannot be linked to its normal partner. However, a purine, such as adenine in its rare state can form a bond with cytosine (besides thymine), provided the cytosine is in its normal state.

Watson and **Crick** (1953) hypothesised that the occurrence of the bases in their rare states provides a mechanism for mutation during DNA replication. If, for example, adenine in an old chain is in its rare state at the moment that the complementary new chain reaches it, cytosine can pair with it (adenine) and be added to the growing end of the new chain. The result of this type of pairing is the formation of a DNA molecule that contains an exceptional base pair. This situation is not stable and at the next replication, adenine is expected to return to its common state and to pair with thymine. Cytosine introduced into the complementary strand due to tautomeric shift in adenine, would then pair with guanine. Thus, there would be formed two kinds of DNA molecules, one that is identical to the original DNA and another that has undergone a base pair substitution of G-C for A-T. This transitionally substituted DNA molecule has altered coding at a point and results in recognisable mutation. Such mutations which formed during DNA replication are called **copy error mutations**. Such copy error mutations have been shown by Fig. 37.6.

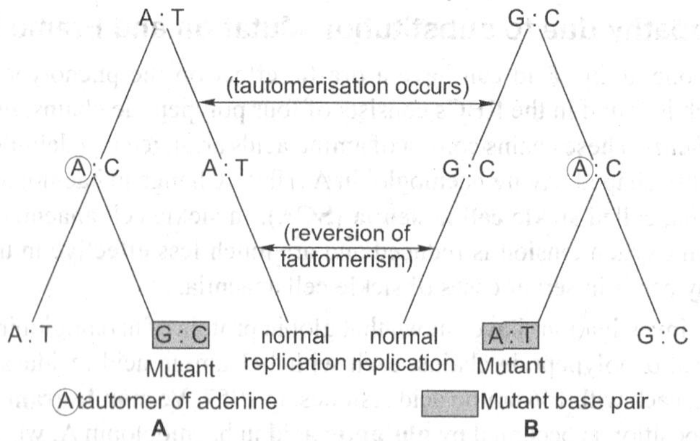

Fig. 37.6. Copy error mutation due to tautomerization of adenine.

The abnormal pairing due to transitional substitution may also occur due to **ionisation** of a base at the time of DNA replication. Ionisation involves the loss of the hydrogen from the number one nitrogen of a base. For example, in its ionised state, thymine can pair with guanine, if the guanine is in its common form. In a similar fashion, guanine in its ionised state can pair with thymine in its common form. From any such unstable base pair, a transition will result following the steps outlined in Fig. 37.6 for A-T to G-C and G-C to A-T.

Effects of Chemical Mutagens on Nucleotide Sequence

(*a*) Alteration in Resting Nucleic Acid

1. Deamination. Some chemical substance such as nitrous acid causes transitional mutation due to oxidative deamination of DNA bases. In the process of oxidative deamination, the amino group (NH_2) of a DNA base is replaced by hydroxyl (OH) group by the chemical mutagen. Thus, adenine is deaminated into **hypoxanthine** by nitrous acid as shown in the following Fig. 37.7:

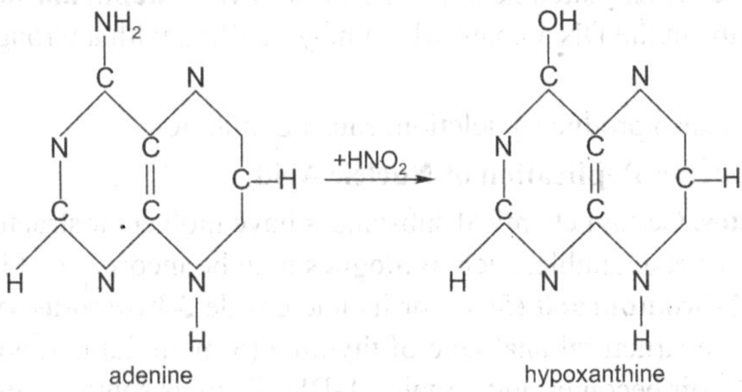

Fig. 37.7. Deamination of adenine into hypoxanthine.

By tautomeric shift the hypoxanthine (HX) is converted into more common keto-tautomer which pairs with cytosine. The A:T pair, thus, can be converted to a G:C pair.

Similarly, deamination converts cytosine to uracil, which has pairing properties similar to thymine and in such a case G : C pair would be changed into A : T pair.

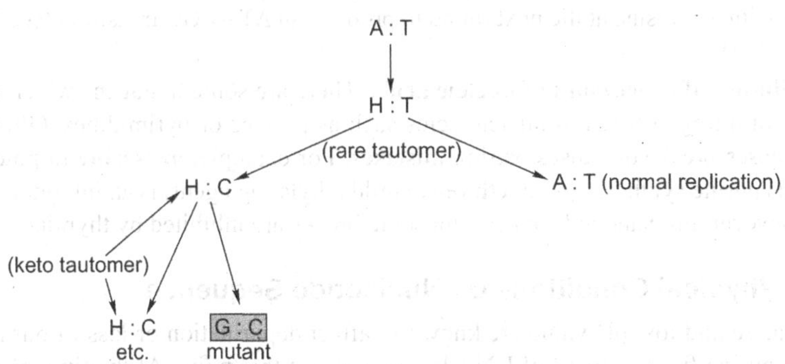

Fig. 37.8. Conversion of A : T pair into G : C pair due to keto-tautomerization of adenine.

2. Hydroxylamine (HA =NH_2.OH) and hydrazine (HZ =NH_2NH_2). When DNA is treated with hydroxylamine (HA), its cytosine base is the strongest reacting base. Hydroxylamine probably cause hydroxylation of cytosine at amino group giving rise to **hydroxylcytosine**, which then subsequently pair with adenine. Thus, hydroxylamine (HA) induces in DNA a GC → AT base pair transition (Fig. 37.9).

The hydrazine affects DNA by breaking of rings of uracil and cytosine giving rise to **pyrazolone** and **3-aminopyrasole,** respectively. The treatment of RNA or DNA with **anhydrous hydrazine** results in the destruction of their pyrimidines.

3. Alkylating agents. Some alkylating agents carry one, two, or more alkyl groups in a reactive form and act as strong mutagens. Examples of some most extensively studied alkylating agents include **diethyl sulphate (DES), dimethyl sulphate (DMS), methyl methane sulphonate (MMS), ethyl ethane sulphonate (EES)** and **ethyl methane sulphonate (EMS).** These mutagens produce mutations in the following ways:

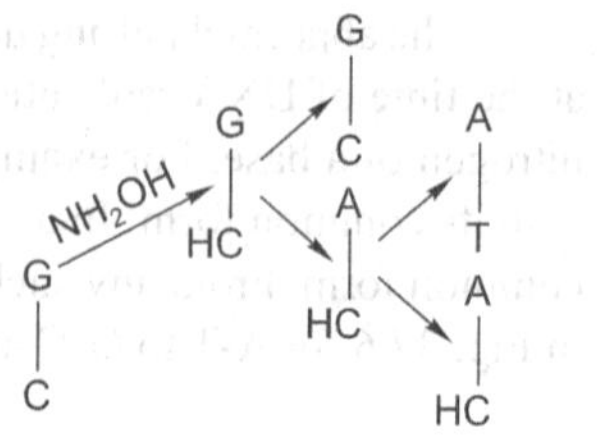

Fig. 37.9. Conversion of GC pair into AT pair due to conversion of cytosine (C) into hydroxylcytosine (HC).

(*i*) They add ethyl or methyl groups to guanine. This makes guanine the base analogue to adenine.

(*ii*) They remove the alkylated guanine. This is known as **depurination.** The loss of the base produces gaps in the DNA chain which may be filled with a wrong base, thus, producing mutation.

(*iii*) The gap may also produce a deletion, causing mutation.

(*b*) Alteration during Replication of Nucleic Acid

1. Base analogues. Certain chemical substances have molecular structure similar to the usual DNA bases that, if they are available, such **analogues** may be incorporated into a replicating DNA strand. For example, **5-bromouracil** (5BU) or its nucleoside **5-bromodeoxyuridine** (5-BUdR) in its usual (keto) form is a structural analogue of thymine (5-methyluracil) and it will substitute for thymine. Thus, an A-T pair becomes and remains A-BU. There is some *in vitro* evidence to indicate the BU immediately adjacent to an adenine in one of DNA strands causes the latter to pair with guanine. But, in its rare (enol) state, 5BU behaves similar to the tautomer of thymine and pairs with guanine. This converts A : T to G : C as shown in Fig. 37.10.

The **2-Aminopurine** (2-AP) is another base analogue which is a relatively undifferentiated purine that apparently can pair with cytosine and thymine. It is thought that 2-AP acts by "switching" pyrimidines: for example, it may be incorporated opposite thymine during one round of replication and then pair with a cytosine at the next round to produce an AT → GC transition (see **Goodenough** and **Levine**, 1974).

2. Inhibition of precursors of nucleic acids. There are some mutagens which interfere with the synthesis of nitrogen bases of nucleic acids such as purines or pyrimidines. Often lack of one base either causes breaks or causes pairing mistakes. For exampie, **azaserine** (a potent alkylating agent) inhibits purine synthesis and **urethane** (a mild alkylating agent) is an inhibitor of pyrimidine synthesis. However, urethane induced chromosome breaks are inhibited by thymine.

Effects of Physical Conditions on Nucleotide Sequence

High temperature and low pH value are known to affect depurination or loss of purine bases. The removal of a purine from a strand of DNA leaves a gap at that point. At the time of replication, it would be possible for any of the four bases to insert in the complementary newly formed strand. If the inserted nucleotide contained a purine, the complementary strand would contain a transversion.

B. Multiple mutations or gross mutations. When changes involving more than one nucleotide pair, or entire gene, then such mutations are called **gross mutations.** The gross mutations occur due to rearrangements of genes within the genome and may be of the following types:

1. The rearrangement of genes may occur within a gene. Two mutations within the same functional gene can produce different effects depending on gene whether they occur in the *cis* or *trans* position.

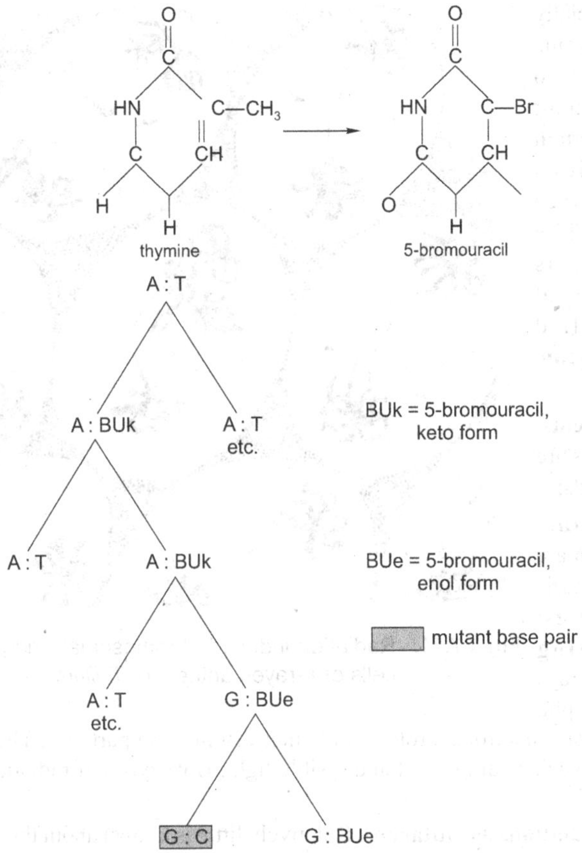

Fig. 37.10. Conversion of A : T base pair into G : C base pair due to keto and enol forms of bromouracil.

2. The rearrangement of gene may occur in number of genes per chromosome. If the number of gene replicas are non-equivalent on the homologous chromosomes, they may cause different types of phenotypic effects over the organisms.

3. Due to movement of a gene locus, new type of phenotypes may be created, especially when the gene is relocated near heterochromatin. The movement of gene loci may take place due to following method:

***(i)* Translocation.** Movement of a gene may take place to a non-homologous chromosome and this is known as **translocation**.

***(ii)* Inversion.** The movement of a gene within the same chromosome is called **inversion**.

3. Classification of Mutation According to the Origin

According to the mode of origin, following two kinds of mutations have been recognised:

(1) Spontaneous mutations. The spontaneous mutations occur suddenly in the nature and their origin is unknown. They are also called "**background mutation**" and have been reported in many organisms such as, *Oenothera*, maize, bread molds, microorganisms (bacteria and viruses), *Drosophila*, mice, human beings, etc.

(2) Induced mutations. Besides naturally occurring spontaneous mutations, the mutations can be induced artificially in the living organisms by exposing them to abnormal environment such as radiation, certain physical conditions (*i.e.*, temperature) and chemicals. The substances or agents which induce artificial mutations are called **mutagens** or **mutagenic agents**.

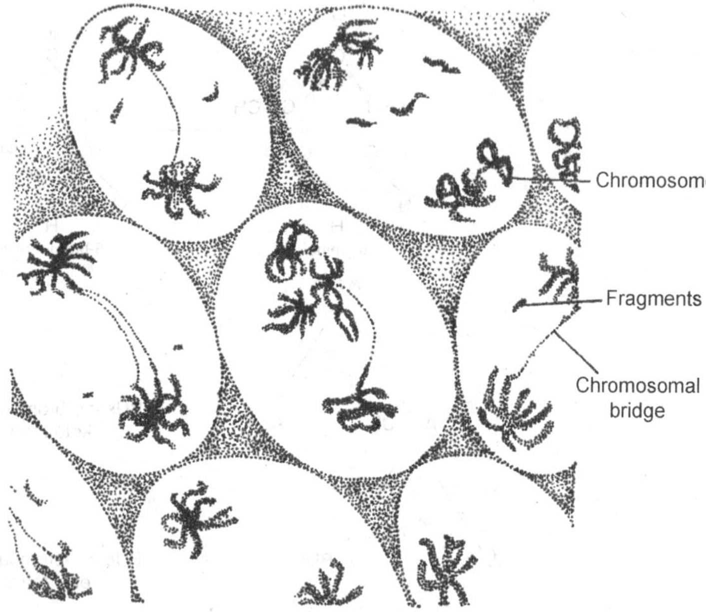

Fig. 37.11. Radiation-induced chromosomal bridges and fragments in cells of X-rayed anthers of *Trillium*.

Mutagenic agents. The mutagenic agents are of the following kinds:

A. Radiations. The radiations which are important in mutagenesis are of two categories: one type is **ionising radiations** such is X-rays and gamma rays; alpha and beta rays; electrons, neutrons, protons and other fast moving particles. The second type is **non-ionising radiations** such as ultraviolet and visible light. Both types of radiations induce mutations by following methods:

(*i*) **Ionising radiations as mutagens.** Relatively little is known about the mechanism by which ionising radiations cause mutation. As, we are already familiar that matter composed of atoms and atoms, in turn, are made up of a positively charged atomic nucleus (with neutrons, protons) and a surrounding constellation of negatively charged electrons. The charges of atomic particles remain so balanced that normal atoms are electrically neutral. When ionising radiations pass through matter, they dissipate their energy in part through the ejection of electrons from the outer shell of atoms and the loss of these balancing, negatively-charged particles (electrons) leaves atoms which are no longer neutral but are positively charged. The positively-charged atom is called **ion**. The ejected electrons move at high speed; knock other electrons free from their respective atoms and when their energy is dissipiated, become attach to other atoms and convert the atoms into negatively charged ions. To achieve their stable configuration (*i.e.*, neutral charge), ions undergo many chemical reactions and during these chemical reactions ionising radiation is thought to cause mutation.

Further, ionizing radiations cause breaks in poly-sugar phosphate backbone of DNA and, thus, causing chromosomal mutations such as break, deletion, addition, inversion and translocation. During breakage of DNA molecule due to ionising radiation, the active role of oxygen is predicted. Because, oxygen is important in the formation of H_2O_2 and H_2O in irradiated water and these products may induce breaks in DNA molecule.

(*ii*) **Non-ionising radiations as mutagens.** The ultraviolet (UV) light is a non-ionizing radiation which may cause mutation. The most effective wavelength of ultraviolet for inducing mutations is about 2,600 A°. This is a wavelength that is best absorbed by DNA and a wavelength at which proteins absorb little energy. When a substance absorbs sufficient energy from the ultraviolet light, some of their electrons are raised to higher energy levels, a state called **excitation**. The excited molecule becomes reactive and mutated and is called **photoproduct**.

Dimerization. The ultraviolet radiation produces several effects on DNA, one being the formation of chemical bonds between two adjacent pyrimidine molecules in a polynucleotide and particularly, between adjacent thymine residues as shown in Fig. 37.12. As two thymine residues associate, or **dimerize** to form a **dimer**, their position in the DNA helix becomes so displaced that they can no longer form hydrogen bonds with the opposing purines and thus regularity of the helix becomes distorted. Thus, dimerization interferes with the proper base pairing of thymine with adenine, and result in thymine's pairing with guanine. This will produce a T-A to C-G transition.

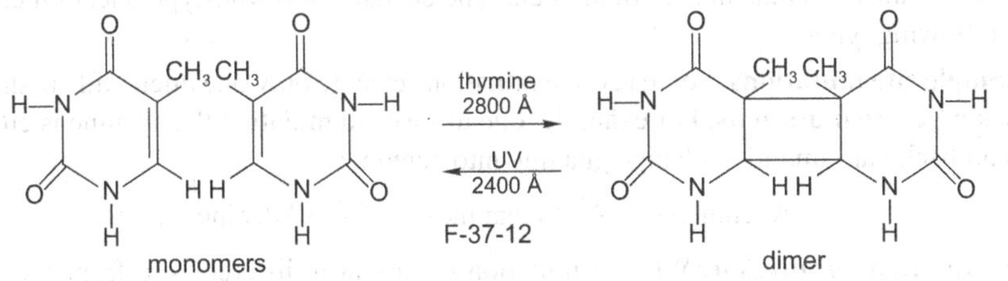

Fig. 37.12. Formation of a dimer of thymine.

B. Temperature as mutagen. The rate of all chemical reactions are influenced by temperature. It is not surprising that temperature can be mutagenic. It is reported that the rate of mutation is increased due to increase in temperature. For example, an increase of 10°C temperature increases the mutation rate two or three-fold. Temperature probably affects both thermal stability of DNA and the rate of reaction of other substances with DNA.

A study of Swedish nudist indicated that the scrotal temperature of human males in ordinary clothing is about 3°C higher than that of nude male. The higher temperature could well increase the mutation rate nearly two-fold, leading the investigators to suggest that the wearing of pants has possibly been much more unhygeinic than fall out from testing of nuclear devices threatens to be. They suggested the wearing of kilts as one solution.

C. Chemical mutagens. Many chemical substances have been responsible to increase the mutability of genes. The ability of chemicals to induce mutation was first of all demonstrated by **Auerbach** and **Robson** in 1947 using mustard gas, and related compounds as the nitrogen and sulphur mustards, mustard oil and chloracetony in experiments with male *Drosophila melanogaster*. Since then many chemical compounds which are ordinarily considered to be non-toxic have been found to be mutagenic in certain specific situations. Any chemical substance that affects the chemical environment of chromosomes is likely to influence, at least indirectly, the stability of DNA and its ability to replicate without error. A chemical mutagen can cause mutation only when it enters in the nucleus of the cell. It can affect the chromosomal DNA by following two ways:

(1) Direct gene change. Certain chemical mutagens affect DNA directly. They affect the constituents of DNA only when DNA is not replicating. For example, nitrous acid converts adenine into hypoxanthine and cytosine to uracil by deamination. Like the nitrous acid, nitrogen mustard, formaldehyde, epoxides, dimethyl and diethyl sulphonate, methyl and ethyl methanesulphonate (MMS and EMS) and nitrosoguanidine (NG) also have direct mutagenic effect on the DNA molecule.

(2) Copy error. Certain chemical compounds, called **base analogues** (*e.g.*, 5-bromouracil, 2-aminopurine, etc.) closely resemble with certain DNA bases and therefore, act as mutagens. During DNA replication, they are incorporated by DNA in place of the normal DNA bases. Certain other base analogues such as urethane triazine, caffeine (in coffee, tea and soft drinks), phenol and carcinogens, acridines (proflavin, etc.), have mutagenic effects. Certain inorganic substances such as manganese chloride is mutagenic for many organisms, as, they are the compounds which bind calcium and, thus, interfere with the integrity of the chromosome structure.

4. Classification of Mutation According to the Direction

According to their mode of direction, following types of mutations have been recognised:

(A) Forward mutations. In an organism when mutations create a change from wild-type to abnormal phenotype, then that type of mutations are known as **forward mutations**. Most mutations are forward type.

(B) Reverse or backward mutations. The forward mutations are often corrected by error correcting mechanism so that an abnormal phenotype changes into wild-type phenotype. They may be of the following types:

(*i*) Single site mutations. Some reverse mutations change only one nucleotide in the gene and are called **single site mutations**. For example, due to forward mutation the adenine is changed into guanine and backward mutation change guanine into adenine:

$$\text{Adenine} \xrightarrow{\text{forward}} \text{Guanine} \xrightarrow{\text{reverse}} \text{Adenine}$$

(*ii*) Mutation suppressor. When a mutation occurs at a different site from the site where already primary mutation occurred and that mutated gene reverse the effects of primarily mutated gene, then such (secondary) mutations are called **mutation suppressors**. They may be of following types:

(*a*) Extragenic suppressor. The extragenic suppressor mutation occurs in a different gene from that of the mutant gene. In *E.coli*, a gene mutation suppressor gene called **rec A** (*rec* for recombination) is known which is necessary for recombination and is found to repair ultraviolet induced thymine dimers of a gene by a process called **postreplication recombinational repair** (see **Goodenough** and **Levine**, 1974).

(*b*) Intragenic suppressor. The intragenic suppressor mutation occurs in a different nucleotide within the same gene and shift the reading frame back into register.

(*c*) Photoreactivation. In photoreactivation type reverse mutation, reversal of ultraviolet induced thymine dimers takes place by specific enzymes in the presence of visible light waves (see Chapter 35). During ultraviolet radiation, a particular enzyme is selectively bound to the bacterial DNA. During photoreactivation the enzyme is activated by visible light and that cleaves the pyrimidine or purine dimers into monomers and restores their original forms.

(*d*) Excision repair or Dark reactivation. In an ultraviolet (UV) induced mutation, the reverse mutation may also occur in the absence of light. According to **Howard Flanders** and **Boyce** (1964), dark reactivation includes following stages (see Chapter 35): (*i*) An enzyme possibly endonuclease makes a cut in the polynucleotide strand on either side of the dimer which may be formed due to ultraviolet radiation and excises a short, single-strand segment of the DNA. (*ii*) Another enzyme, possibly exonuclease widens the gap produced by the action of the endonuclease. (*iii*) DNA polymerase resynthesises the missing segment, using the remaining opposite strand as a template; and (*iv*) the final gap is closed by some enzymatic rejoining process, (*i.e.*, DNA ligase).

5. Classification of Mutation According to Magnitude of Phenotypic Effect

According to their phenotypic effects following kinds of mutations may occur:

1. Dominant mutations. The mutations which have dominant phenotypic expression are called dominant mutations. For example, in human beings the mutation disease **aniridia** (absence of iris of eyes) occurs due to a dominant mutant gene.

2. Recessive mutations. Most types of mutations are recessive in nature and so they are not expressed phenotypically immediately. The phenotypic effects of mutations of a recessive gene is seen only after one or more generations, when the mutant gene is able to recombine with another similar recessive gene.

3. Isoalleles. Some mutations alter the phenotype of an organism so slightly that they can be detected only by special techniques. Mutant genes that give slightly modified phenotypes are called **isoalleles.** They produce identical phenotypes in homozygous or heterozygous combinations.

4. Lethal mutations. According to their effects on the phenotype, mutations may be classified as lethals, subvitals and supervitals. Lethal mutations result in the death of the cells or organisms in which they occur. **Subvital mutations** reduces the chances of survival of the organism in which they occur. **Supervital mutations**, in contrast, cause the improvement of biological fitness under certain conditions.

6. Classification of Mutation According to Consequent Change in Amino Acid Sequence

1. Missense mutations. They change the meaning of a codon, changing one amino acid into another.

2. Temperature sensitive mutations or T_s mutations. If the substitution produces a protein that is active at one temperature (typically 30°C) and inactive at a higher temperature (usually 40–42°C).

3. Nonsense or chain termination mutations. They arise when a codon for an amino acid is mutated into a termination codon (UAG, UAA or UGA), resulting in the production of a shorter protein.

Since, temperature-sensitive and chain termination mutations exhibit the mutant phenotype only under certain conditions, they are called **conditional mutations**; they are the most versatile and useful mutations.

4. Silent mutations. They change a nucleotide but not the amino acid sequence because they affect the *third position* of the codon, which is usually less important in coding. This is a silent mutation because it leaves the protein sequence unchanged.

7. Classification of Mutation According to the Types of Chromosomes

According to the types of chromosomes, the mutations may be of following two kinds:

1. **Autosomal mutations.** This type of mutation occurs in autosomal chromosomes.
2. **Sex chromosomal mutations.** This type of mutation occurs in sex chromosomes.

37.4. MUTATION RATE

The frequency with which genes mutate spontaneously is called **mutation rate**. Most genes are relatively stable and mutation is a rare event. The great majority of genes have mutation rate of 1×10^{-5}, *viz.*, one gamete in 100,000 to one gamete in million would contain a mutation at a given locus. Mutations occur much more frequently in certain regions of the gene than in others. The favoured regions of mutations are called **hot spots**. The mutation rate is influenced by various factors which are as follows:

1. Genetic control of mutation rate. There are ample evidences which show that mutation rate is under genetic control, *viz.*, certain genes called **mutator genes** may increase the mutation rate in *Drosophila* (**Demere**, 1937), maize (**Rhoades**, 1938) and *E.coli* (**Goldstein**, 1955). However, certain suppressor genes may decrease the rate of mutation.

In bacteria, as well as in eukaryotes, spontaneous mutations most frequently are caused by transposons which are segments of DNA that have a tendency to jump around the genome.

2. Viral control of mutation rate. Virus reportedly affect the mutability of host's genes. **Sprague** (1963) experimented with maize suggested that virus may cause mutation. **Baumiliar** (1967) reported that viruses increase the mutation rate in *Drosophila melanogaster*.

3. Environmental control of mutation rate. There are three major environmental factors that affect mutation rates, *viz.*, temperature, certain radiations and chemicals.

37.5. METHOD OF DETECTION OF SEX-LINKED LETHAL MUTATION

H.J. Muller devised an easy method for detecting lethal mutations in the sex chromosomes of *Drosophila.* This is called a ClB method in which a special type of female fly is employed which carries a normal X chromosome and an abnormal X chromosome. The abnormal X chromosome contains an inversion mutation C (which prevents the chromosome to do crossing over with the normal X chromosome, therefore, called **crossover suppressor**), a recessive lethal mutational gene, l and a dominant gene B for bar-eye. In Muller's ClB technique, these ClB female flies are mated with males which are previously treated with some mutagenic agent (such as X-rays) to cause mutation in some of their sperms. The resulting zygotes are of four types and one of these, the ClB male, fail to survive because such embryos contain a recessive lethal which expresses itself when hemizygous. Thus, only one class of male (with ClB X chromosome) remains to fertilize the F_1 females. Each heterozygous ClB female results from the fertilization of a ClB egg and an irradiated X-bearing sperm and some of these sperms will contain mutated X-chromosomes. Mated F_1 heterozygous ClB females are distributed individually into culture tubes in which each lays fertile eggs and so produces a single F_2 culture. The culture produced by females bearing an induced lethal mutation contain only females; whereas female bearing irradiated X-chromosomes in which no recessive lethal has been induced yield cultures containing some wild-type males. Thus, if in a population of 1000 cultures, 990 contained some males and 10 contained only females, the induced rate of sex-linked, recessive lethal mutations would be 1 per cent.

Besides the ClB method, there are many more methods such as **Muller-5 method** (for the detection of sex-linked lethal mutations); **attached X-method** (for the detection of sex-linked visible mutations); **balanced lethal systems** (for the detection of autosomal mutations); and **Stadler's method** and **Singleton's method** (for the detection of specific loci in plants).

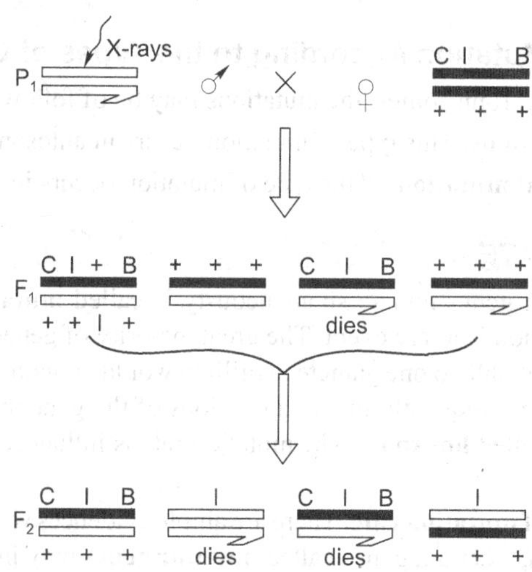

Fig. 37.13. Diagram of Muller's ClB technique for detecting sex-linked (X-linked), recessive lethal mutations.

37.6. SIGNIFICANCE OF MUTATION

The vast majority of mutations are deleterious to the organism and are kept at low frequency in the population by the action of natural selection. Mutant types are generally unable to compete equally with wild-type individuals. Even under optimal environmental conditions many mutants appear less frequently than expected.

QUESTIONS

Long Answer Questions

1. What are missense, nonsense, and frameshift mutations and what are the consequences of each?
2. What do you understand by spontaneous mutations and induced mutations? Discuss variations in mutation rates and frequencies at different loci within an organism.
3. (*a*) What critical evidence is needed to distinguish a gene mutation from a minute deletion, *i.e.*, one too small to be cytologically detectable?
   (*b*) Show how you would determine the rate of mutation from a recessive to a dominant allele in human beings.
   (*c*) Describe what is meant by a tautomeric shift?
   (*d*) Are the mutational consequences of tautomeric shifts base-pair transitions or base pair transversions? Explain, using a specific example.
4. Explain:
   (*i*) Why geneticists find most mutations to be deleterious?
   (*ii*) Why, nevertheless, the mutation process is considered to be the basis of evolutionary progress?
5. Answer each of the following questions as briefly as possible:
   (*a*) Which type of mutation, one induced by a base analog or one induced by proflavin, would you expect to be more deleterious to an organism and why?
   (*b*) What evidence is there that ionisation caused by X-rays need not occur in the gene itself to cause mutation?
   (*c*) What are the possible mechanisms by which a gene may change to many different allelic forms?
   (*d*) Why are sex-linked lethal mutations easier to detect than autosomal lethals?
6. Three repair mechanisms are known in *E.coli* for the repair of DNA damage (pyrimidine dimer formation) after exposure to ultraviolet light: (*i*) photoreactivation; (*ii*) excision (dark) repair; and (*iii*) postreplication repair. Compare and contrast these mechanisms, indicating how each achives repair and how the events occurring in each may lead to gene mutations.

Short Answer Questions

1. Describe different kinds of radiations and chemical mutagens utilised for induction of mutations.
2. Discuss the procedure used in the detection of sex-linked lethal mutations by ClB method.
3. Explain the difference between a transition and a transversion and give an example of each.
4. What possible explanations can you offer for the reversion of a mutant to the wild-type phenotype?
5. Compare the effects of nitrous acid, hydroxylamine and 5-bromouracil on DNA.
6. Write short notes on the following:
   (*a*) Base substitution;
   (*b*) Ionising radiation;
   (*c*) Balanced lethal;
   (*d*) Somatic mutation;
   (*e*) Acridine dye;
   (*f*) Lethal mutation.

Very Short Answer Questions

1. Why are most mutations in structural genes recessive to their wild-type alleles?
2. Which of the following would be likely to suffer the greatest, and which the least genetic damage from radiation exposure: (*a*) a haploid, (*b*) a diploid, and (*c*) a polyploid?
3. If the mutation rate of a certain gene is directly proportional to the radiation dosage and the mutation rate of

Drosophila is observed to increase from 3% at 1000 R to 6% at 2000 R. What percentage of mutations would be expected at 3500 R?

4. The X-linked recessive mutations are more easily studied in appropriate organisms than are autosomal ones. Why?
5. Some individuals have a patch of blonde hair in a head of brown hair. What types of mutation would this be?
6. If a drastic alteration occurred in the structure of one of the genes for 28S rRNA, do you think that the translation of mRNA into protein would cease? If not, why not?
7. What is the difference between intragenic and intergenic suppression? Give an illustration of each.
8. How many base pairs would have to be deleted in a mutational event to eliminate a single amino acid from a protein and not change the rest of the protein?
9. The "dotted" gene in maize (Dt) is a "mutator" gene influencing the rate at which the gene for colourless aleurone (*a*) mutates to its dominant allele (A) for coloured aleurone. An average of 7.2 coloured dots (mutations) per kernel was observed when the seed parent was dt/dt, a/a and the pollen parent was Dt/Dt, a/a. An average of 22.2 dots per kernel was observed in the reciprocal cross, How can these results be explained?
10. What is the difference between substrate and a template transition mutation?
11. 5-bromouracil, 2-arninopurine, proflavin, ethyl ethane sulphonate and nitrous acid are chemical mutagens. What does each do?

Multiple Choice Questions

Choose the correct answer from the four alternatives given.

1. Mutation is
   (*a*) change that is inherited
   (*b*) change in parent not inherited
   (*c*) plant growth controlling factor
   (*d*) change which affects the offsprings of F_2 generation
2. Gene mutation is
   (*a*) mutation in the genes of DNA
   (*b*) mutation in phosphodiester linkage
   (*c*) mutation in chromosomes
   (*d*) change in the sequence of nitrogenous bases
3. Mutations which normally happen randomly are considered one of the raw materials for evolution because they
   (*a*) are stable
   (*b*) contribute new variation in organism
   (*c*) cause death of organism
   (*d*) none of these
4. Proflavin and acridine orange induce
   (*a*) transitions
   (*b*) transversions
   (*c*) inversions
   (*d*) frameshift mutations
5. Induction of mutation by X-rays was discovered by
   (*a*) Morgan
   (*b*) Hugo de Vries
   (*c*) Muller
   (*d*) Luria
6. Who is associated with "Green revolution" in India?
   (*a*) B.P. Pal
   (*b*) M.S. Swaminathan
   (*c*) R.S. Paroda
   (*d*) E.J. Butler
7. Low temperature is mutagenic in
   (*a*) wheat
   (*b*) maize
   (*c*) rice
   (*d*) mustard
8. Frequency of mutation
   (*a*) varies with characters and organisms
   (*b*) can be increased by X-rays
   (*c*) is greatly affected by environmental factors
   (*d*) all of the above

9. Which of the following mutagens can be best used in inducing mutation in microorganisms?
   (*a*) X-rays
   (*b*) β-rays
   (*c*) UV-rays
   (*d*) γ-rays
10. Why are haploids are superior to diploids in study of mutations?
   (*a*) they have shorter life time
   (*b*) smaller number of chromosomes
   (*c*) they allow expression of recessive mutation immediately
   (*d*) obtained in large numbers
11. Haemoglobin S contains
   (*a*) glutamic acid in β-chain
   (*b*) valine in β-chain
   (*c*) glutamic acid in α-chain
   (*d*) valine in α-chain
12. Frame shift mutation occurs when
   (*a*) base is deleted
   (*b*) base is added
   (*c*) both (*a*) and (*b*)
   (*d*) anticodons are not present
13. Mutation is generally due to
   (*a*) recessive gene
   (*b*) lethal gene
   (*c*) dominant gene
   (*d*) complementary gene
14. A point mutation comprising the substitution of a purine by a pyrimidine is called
   (*a*) transition
   (*b*) translocation
   (*c*) deletion
   (*d*) transversion

ANSWERS

Very Short Answer Questions

1. Wild-type alleles usually code for complete, functional enzymes or other proteins. One active wild-type allele can often cause enough enzyme to be produced so that normal or nearly normal phenotypes result (dominance). Mutations of normally functioning genes are more likely to destroy the biological activities of proteins. Only in the complete absence of the wild-type gene product would the mutant phenotype be expressed recessiveness.
2. Haploid greatest, polyploid least. Most mutations are recessive, and recessives have only a very low probability of being expressed in polyploids with their multiple sets of chromosomes bearing normal (dominant) alleles. Both dominant and recessive mutations are expressed at once in haploids.
3. 10.5%.
4. Recessive mutations are more easily detected in hemizygous males.
5. Somatic mutation.
6. Translation would not cease since numerous genes for rRNA are present in the genome. The mutation of one of these would probably not interfere with protein synthesis.
7. *Intragenic mutation within the same codon*, either restoring the original amino acid or resulting in the presence of a compatible amino acid; *intragenic mutation within the same cistron*, such as one that restores the normal reading frame; *intergenic direct suppression*, such as alteration in some component directly involved in protein synthesis, for example, tRNA; *intergenic indirect suppression* by an alteration in the cellular milieu.
8. Three, as any other number of deletion (or addition) would cause a frameshift and other amino acid changes.

9. Seed parent contributes two sets of chromosomes to triploid endosperm; one Dt gene gives 7.2 mutations/kernel, two Dt genes increase mutations to 22.2/kernel.
10. In replicating DNA, a transition mutation can occur by tautomerization of a base in the template strand (template transition) or entering the progeny strand (substrate transition).
11. *5-bromouracil* (pyrimidine analogue) and *2-aminopurine* (purine analogue) are incorporated into DNA as thymine and adenine, respectively. However, each undergoes tautomeric shifts more frequently than the normal base. Both cause transitions. *Nitrous acid* also promotes transitions by converting cytosine into uracil, which acts like thymine, and adenine into hypoxanthine, which acts like guanine. *Proflavin* induces insertions and deletions by intercalating and buckling DNA. *Ethyl ethane sulphonate* removes purine rings and thus promotes transitions and transversions.

Multiple Choice Questions

1. (*a*)	**2.** (*d*)	**3.** (*b*)	**4.** (*d*)	**5.** (*c*)	**6.** (*b*)	**7.** (*c*)
8. (*d*)	**9.** (*c*)	**10.** (*c*)	**11.** (*b*)	**12.** (*c*)	**13.** (*a*)	**14.** (*d*)

38 Genetic Code

As DNA is a genetic material, it carries genetic information from cell to cell and from generation to generation. At this stage, an attempt will be made to determine that in what manner the genetic informations are existed in DNA molecule? Are they written in articulated or coded language on DNA molecule? If in the language of codes what is the nature of genetic code?

A DNA molecule is composed of three kinds of moieties: (*i*) phosphoric acid, (*ii*) deoxyribose sugar, and (*iii*) nitrogen bases. The genetic informations may be written in any one of the three moieties of DNA. But the poly-sugar-phosphate backbone is always the same, and it is, therefore, unlikely that these moiteies of DNA molecule carry the genetic informations. The nitrogen bases, however, vary from one segment of DNA to another, so the informations might well depend on their sequences. The sequence of nitrogen bases of a given segment of DNA molecule, actually, has been found to be identical to linear sequence of amino acids in a protein molecule. The proof of such a **colinearity** between DNA nitrogen base sequence and amino acid sequence in protein molecules has first obtained from an analysis of mutants of head protein of bacteriophage T_4 (**Sarabhai** *et al.*, 1964) and the A protein of tryptophan synthetase of *Escherichia coli* (**Yanofski** *et al.*, 1964). The colinearity of protein molecules and DNA polynucleotides has given the clue that the specific arrangement of four nitrogen bases (*e.g.*, A, T, C and G) in DNA polynucleotides chains, somehow, determines the sequence of amino acids in protein molecules. Therefore, these four DNA bases can be considered as four alphabets of DNA molecule. All the genetic informations, therefore, should be written by these four alphabets of DNA. Now the question arises that whether the genetic informations are written in articulated language or coded language? If genetic informations might have occurred in an articulated language, the DNA molecule might require various alphabets, a complex system of grammar and ample amount of space on it. All of which might be practically impossible and troublesome too for the DNA. Therefore, it was safe to conclude for molecular biologists that genetic informations were existed in DNA molecule in the form of certain special language of code words which might utilize the four nitrogen bases of DNA for its symbols. Any coded message is commonly called **cryptogram**.

38.1. BASIS OF CRYPTOANALYSIS

The basic problem of such a genetic code is to indicate how information written in a four letter language (four nucleotides or nitrogen bases of DNA) can be translated into a twenty letter language (twenty amino acids of proteins). The group of nucleotides that specifies one amino acid is a **code word** or **codon**. By the **genetic code** one means the collection of base sequences (codons) that correspond to each amino acid and to translation signals. We can consider here the classical but logical reasoning done by **George Gamov** (1954) about the possible size of a codon. The simplest possible code is a **singlet code** (a code of single letter) in which one nucleotide that codes for one amino acid could be specified. A **doublet code** (a code of two letters) is also inadequate, because it could specify only sixteen (4 × 4) amino acids, whereas a **triplet code** (a code of three letters) could specify sixty four (4 × 4 × 4) amino acids. Therefore, it is likely that there may be 64 triplet codes for 20 amino acids. The possible singlet, doublet and triplet codes, which are customarily represented in terms of "**mRNA language**" [mRNA is a complementary molecule which copies the genetic informations (cryptogram of DNA) during its transcription], have been illustrated in Table 38.1.

Table 38.1 Possible singlet, doublet and triplet codes of mRNA.

Singlet code (4 words)	Doublet code (16 words)	Triplet code (64 words)
		AAA AAG AAC AAU
		AGA AGG AGC AGU
		ACA ACG ACC ACU
		AUA AUG AUC AUU
		GAA GAG GAC GAU
		GGA GGG GGC GGU
A	AA AG AC AU	GCA GCG GCC GCU
G	GA GG GC GU	GUA GUG GUC GUU
C	CA CG CC CU	CAA CAG CAC CAU
U	UA UG UC UU	CGA CGG CGC CGU
		CCA CCG CCC CCU
		CUA CUG CUC CUU
		UAA UAG UAC UAU
		UGA UGG UGC UGU
		UCA UCG UCC UCU
		UUA UUG UUC UUU

The first experimental evidence in support to the concept of triplet code is provided by **Crick** and coworkers in 1961. During their experiments, when they added or deleted single or double base pairs in a particular region of DNA of T_4 bacteriophages of *E.coli*, they found that such bacteriophages ceased to perform their normal functions. However, bacteriophages with addition or deletion of three base pairs in DNA molecule, had performed normal functions. From this experiment, they concluded that a genetic code is in **triplet** form, because the addition of one or two nucleotides has put the reading of the code out of order, while the addition of third nucleotide resulted in a return to the proper reading of the message.

38.2. CODON ASSIGNMENT (Cracking the Code or Deciphering the Code)

The genetic code has been cracked or deciphered by the following kinds of approaches:

A. Theoretical Approach

The physicist **George Gamow** proposed the **diamond code** (1954) and the **triangle code** (1955) and suggested an exhaustive theoretical framework to the different aspect of the genetic code. **Gamow** suggested the following properties of the genetic code:

(*i*) **A triplet codon** corresponding to one amino acid of the polypeptide chain.

(*ii*) **Direct template translation** by codon-amino acid pairing.

(*iii*) Translation of the code in an **overlapping** manner.

(*iv*) **Degeneracy** of the code, *i.e.*, an amino acid being coded by more than one codon.

(*v*) **Colinearity** of nucleic acid and the primary protein synthesized.

(*vi*) **Universality** of the code, *i.e.*, the code being essentially the same for different organisms.

Some of these Gamow's proposals have been contradicted by the molecular biologists. For example, **Brenner** (1957) showed that the overlapping triplet code is an impossibility, and subsequent work has shown that the code is a **non-overlapping** one. Similarly, Gamow's idea of direct template relationship between nucleic acid and polypeptide chain was challenged when **Crick** proposed his **adaptor hypothesis**. According to this hypothesis, **adaptor molecules** intervene between nucleic acid and amino acids during translation. In fact, it is now known that tRNA molecules act as adaptors between codons of mRNA and amino acids of the resulting polypeptide chain.

B. The *in vitro* Codon Assignment

1. Discovery and use of polynucleotide phosphorylase enzyme. Marianne Grunberg-Manago and **Severo Ochoa** isolated an enzyme from the bacteria (*e.g.*, *Azobacter vinelandii* or *Micrococcus lysodeikticus*) that catalyzes the breakdown of RNA in bacterial cells. This enzyme is called **polynucleotide phosphorylase. Manago** and **Ochoa** found that outside of the cell (*in vitro*), with high concentrations of ribonucleotides, the reaction could be driven in reverse and an **RNA** molecule could be made (see **Burns** and **Bottino**, 1989). Incorporation of bases into the molecule is random and does not require a DNA template. Thus, in 1955 **Manago** and **Ochoa** made possible the artificial synthesis of polynucleotides = (mRNA) containing only a single type of nucleotides (U, A, C, or G respectively) repeated many times.

	Polynucleotide	Configuration
1.	Polyuridylic acid or poly (U)	UUUUUU
2.	Polyadenylic acid or poly (A)	AAAAAA
3.	Polycytidylic acid or poly (C)	CCCCCC
4.	Polyguanidylic acid or poly (G)	GGGGGG

Thus, the action of polynucleotide phosphorylase can be represented in the following way:

$$(\text{RNA})_n + \text{Ribonucleoside diphosphate} \underset{}{\overset{\text{polynucleotide phosphorylase}}{\rightleftharpoons}} (\text{RNA})_{n+1} + \text{Pi}$$

The polynucleotide phosphorylase enzyme differs from RNA polymerase used to transcribe mRNA from DNA in that: (*i*) it does not require a template or primer; (*ii*) the activated substrates are ribonucleoside diphosphates (*e.g.*, UDP, ADP, CDP and GDP) and not triphosphates; and (*iii*) orthophosphate (*Pi*) is produced instead of pyrophosphates (*PPi*).

The deciphering of the genetic code was made possible by the use of synthetic (or artificial) polynucleotides and trinucleotides. The different types of techniques used include the use of polymers containing a single type of nucleotide (called **homopolymers**), the use of mixed polymers (**copolymers**) containing more than one type of nucleotides (**heteropolymers**) in random or defined sequences and the use of trinucleotides (or "**minimessengers**") in ribosome-binding or filter-binding.

2. Codon assignment with unknown sequence. (*i*) Codon assignment by homopolymer. The first clue to codon assignment was provided by **Marshall Nirenberg** and **Heinrich Matthaei** (1961) when they used *in vitro* system for the synthesis of a polypeptide using an artificially synthesized mRNA molecule containing only one type of nucleotide (*i.e.*, homopolymer). Prior to performing the actual experiments, they tested the ability of a cell-free protein synthesizing system to incorporate radioactive amino acids into newly synthesized proteins. Their cell-free extracts of *E.coli* contained ribosomes, tRNAs, aminoacyl-tRNA synthetase enzymes, DNA and mRNA.The DNA of this extract was eradicated by the help of **deoxyribonuclease** enzyme, thus, the template which might synthesize new mRNA was destroyed. When twenty amino acids were added to this mixture along with ATP, GTP, K^+ and MG^{2+}, they were incorporated into proteins. This incorporation continued so long as mRNA was present in such a cell-free suspension. It also continued in the presence of synthetic polynucleotides (mRNAs) which could be made with the help of polynucleotide phosphorylase enzyme.

The first successful use of this technique was made by **Nirenberg** and **Matthaei** who synthesized a chain of uracil molecules (poly U) as their synthetic mRNA (homopolymer). Poly (U) seemed a good choice; because there could be no ambiguity in a message consisting of only one base. Poly (U) was a good choice for other reasons: it binds well to ribosomes and, as it turned out, the product protein was insoluble and easy to isolate. When poly (U) was presented as the message to the cell free system containing all the amino acids, one amino acid was exclusively selected from the mixture for incorporation into the polypeptide, called **polyphenylalanine**. This amino acid was phenylalanine and it could be concluded that some sequence of UUU coded for phenylalanine. Other homogeneous chains of nucleotides (Poly A, Poly C and Poly G) were inactive for phenylalanine incorporation. The mRNA code word for phenylalanine was, therefore, shown to be UUU. Thus, the first code word to be deciphered was UUU.

This discovery was extended in the laboratories of **Nirenberg** and **Ochoa**. The experiment was repeated using synthetic **poly (A)** and **poly (C)** chains, which gave **polylysine** and **polyproline** respectively. Thus, AAA was identified as the code for lysine and CCC as the code for proline. A poly (G) message was found non-functional *in vitro*, since it attains secondary structure and, thus, could not attach the ribosomes. In this way three of 64 codons were easily accounted for.

***(ii)* Khoran's copolymers. Dr. H.G. Khorana** devised an ingenious technique for codon assignment by using synthetic DNA. **Har Gobind Khorana** (b. 1922) is a India born US biochemist who shared with Holley and Nirenberg the 1968 Nobel Prize for Physiology or Medicine toward deciphering the Genetic code.

Khorana and his coworkers could prepare polyribonucleotides (RNA) with **known repeating sequences**. A repeating sequence means that, if CU are two bases, these will be repeatedly present throughout the length as follows:

CUCUCUCUCUCUCU

In a similar manner, if ACU are three bases, they will be present repeatedly as follows:

ACUACUACUACUACU

Such copolymers will direct the incorporation of amino acids in a manner which can be theoretically predicted. For instance, in (CU) coplymer: CUC/UCU/CUC/UCU only two codons are possible and these are CUC and UCU. These codon are present in alternating sequence. The result would be the polypeptide formed would have only two amino acids in alternating sequences, *i.e.*, leucine, serine (Table 38.2).

Table 38.2 Assignment of codons, having known sequences, with the help of copolymers having repetitive sequences of two bases.

	Copolymers used	Codons in copolymers	Amino acids incorporated	Codon assigned
1.	(CU)n	CUC/UCU/CUC	leucine/serine	CUC/UCU
2.	(UG)n	UGU/GUG/UGU	cysteine/valine	UGU/GUG
3.	(AC)n	ACA/CAC/ACA	threonine/histidine	ACA/CAC

We may similarly consider a repeating sequence of three bases, *e.g.*, (ACG)n. Depending upon where the reading is started, three kinds of homopolypeptides are expected (Table 38.3).

Table 38.3 Assignment of codons, having known sequences, with the help of copolymers having repetitive sequnces of three bases = (ACG)n.

	Codons	Homopolypeptide	Codon assignment
1.	ACG/ACG/ACG/ACG/ACG = Poly (ACG)	(Threonine)n	ACG = threonine
2.	A/CGA/CGA/CGA/CGA = Poly (CGA)	(Arginine)n	CGA = arginine
3.	AC/GAC/GAC/GAC/GAC = Poly (GAC)	(Aspartic acid)n	GAC = aspartic acid

***(iii)* Codon assignment by heteropolymers (Copolymers with random sequences).** Further exposition of the genetic code took place by using synthetic messenger RNAs containing two kinds of bases. This technique was used in the laboratories of **Ochoa** and **Nirenberg** and led the deduction of the composition of codons for the 20 amino acids. The synthetic messengers contained the bases at random (called **random copolymers**). For example, in a random copolymer using U and A nucleotides eight triplets are possible, such as UUU, UUA, UAA, UAU, AAA, AAU, AUU and AUA. Theoretically, eight amino acids could be coded by these eight codons. Actual experiments, however, yielded only six, *i.e.*, phenylalanine, leucine, tyrosine, lysine, asparagine and isoleucine. By varying the relative compositions of U and A in the random copolymer and determining the percentage of the different amino acids in the proteins formed, it was possible to deduce the composition of the code for different amino acids.

3. Assignment of codons with known sequences. (*i*) Use of trinucleotides or minimessengers in filter binding (Ribosome-binding technique). Ribosome technique of **Nirenberg** and **Leder** (1964) made use of the finding that aminoacyl-tRNA molecules specificially bind to ribosome-mRNA complex. The binding does not require the presence of a long mRNA molecule; in fact, the association of a **trinucleotide** or **minimessenger** with the ribosome is sufficient to cause aminoacyl-tRNA binding. When a mixture of such small mRNA molecules-ribosomes and amino acid-tRNA complexes are incubated for a short time and then filtered through a **nitrocellulose membrane**, then the mRNA-ribosome-tRNA-amino acid complex is retained back and rest of the mixture passes through the filter. By using a series of 20 different amino acid mixtures, each containing one radioactive amino acid at a time, it is possible to find out the amino acid corresponding to each triplet by analysing the radioactivity absorbed by the membrane, *e.g.*, the triplet GCC and GUU retain only alanyl-tRNA and valyl-tRNA respectively. All 64 possible triplets have been synthesized and tested in this way. Forty five of them have given clear-cut results. Later on, with the help of longer synthetic messages it has been possible to decipher 61 out of the possible 64 codons.

C. The *in vivo* Codon Assignment

The cell free protein synthetic systems, though have proved of great significance in deciphering of the genetic code, but they could not tell us whether the genetic code so deciphered is used in the living systems of all organisms also. Three kinds of techniques are used by different molecular biologists to determine whether the same code is also used *in vivo* (*a*) amino acid replacement studies (*e.g.*, tryptophan synthetase synthesis in *E.coli* (**Yanofsky** *et al.*, 1963) and haemoglobin synthesis in humans), (*b*) frameshift mutations (*e.g.*, investigations of **Terzaghi** *et al.*, 1966, on lysozyme enzyme of T_4 bacteriophages, and (*c*) comparison of a DNA or mRNA polynucleotide cryptogram with its corresponding polypeptide clear text (*e.g.*, comparison of amino acid sequence of the R_{17} bacteriophage coat protein with the nucleotide sequence of the R_{17} mRNA in the region of the molecule that dictates coat-protein synthesis by **S. Cory** *et al.*, 1970).

Thus, *in vitro* and *in vivo* studies, so far described, gave the way to formulate a code table for twenty amino acids (see Table 38.4 and Table 38.5).

Table 38.4 The genetic dictionary. The trinucleotide codons are written in the 5′ → 3′ direction.

First base	Second base: U	Second base: C	Second base: A	Second base: G	Third base
U	UUU Phe	UCU Ser	UAU Tyr	UGU Cys	U
	UUC Phe	UCC Ser	UAC Tyr	UGC Cys	C
	UUA Leu	UCA Ser	UAA Non-† sense codon	UGA Non-sense† codon	A
	UUG Leu	UCG Ser	UAG Non-† sense codon	UGG Trp	G
C	CUU Leu	CCU Pro	CAU His	CGU Arg	U
	CUC Leu	CCC Pro	CAC His	CGC Arg	C
	CUA Leu	CCA Pro	CAA Gln	CGA Arg	A
	CUG Leu	CCG Pro	CAG Gln	CGG Arg	G
A	AUU Ileu	ACU Thr	AAU Asn	AGU Ser	U
	AUC Ileu	ACC Thr	AAC Asn	AGC Ser	C
	AUA Ileu	ACA Thr	AAA Lys	AGA Arg	A
	AUG* Met	ACG Thr	AAG Lys	AGG Arg	G
G	GUU Val	GCU Ala	GAU Asp	GGU Gly	U
	GUC Val	GCC Ala	GAC Asp	GGC Gly	C
	GUA Val	GCA Ala	GAA Glu	GGA Gly	A
	GUG Val	GCG Ala	GAG Glu	GGG Gly	G

*AUG—Met or chain initiation codon; †UAA, UAG, UGA—Stop codon.

Table 38.5 Amino acids and their messenger RNA codons.

Amino acid	Codons	Amino acid	Codons	Amino acid	Codons	Amino acid	Codons
Alanine (Ala)	–GCA	Glycine (Gly)	– GGA	Leucine (Leu)	– CUA	Threonine (Thr)	– ACA
	GCG		GGG		CUG		ACG
	GCC		GGC		CUC		ACC
	GCU		GGU		CUU		ACU
Argine (Arg)	–AGA	Glutamine	– CAA		UUA		
	AGG	(Gln; Glun)	CAG		UUG	Tryptophan (Trp)	– UGG
	CGA	Glutamic acid		Methionine (Met)	– AUG		
	CGG	(Glu)	– GAA	(Starting codon)		Tyrosine (Tyr)	– UAC
	CGC		GAG				UAU
	CGU						
Aspartic acid	–GAC	Histidine (His)	– CAC	Phenylalanine	– UUC	Valine (Val)	– GUA
(Asp)	GAU		CAU	(Phe)	UUU		GUG*
Asparagine	–AAC	Isoleucine (Ile)	– AUC	Proline (Pro)	– CCA		GUC
(Asn; Aspn)	AAU		AUU		CCG		GUU
Cysteine	–UGC		AUA		CCC	Terminator	– UAA
(Cys)	UGU				CCU	(Nonsense	(ocher)
		Lysine (Lys)	– AAA	Serine (Ser)	– AGC	codons)	UAG
			AAG		AGU		(amber)
					UCA		UGA
					UCG		(opal)
					UCC		
					UCU		

*GUG is also used as a start codon for some proteins.

38.3. CHARACTERISTICS OF GENETIC CODE

The genetic code has the following general properties:

1. The code is a triplet codon. The nucleotides of mRNA are arranged as a linear sequence of codons, each codon consisting of three successive nitrogenous bases, *i.e.*, the code is a triplet codon. The concept of triplet codon has been supported by two types of point mutations: frameshift mutations and base substitution.

***(i)* Frameshift mutations.** Evidently, the genetic message once initiated at a fixed point is read in a definite frame in a series of three letter words. The framework would be disturbed as soon as there is a deletion or addition of one or more bases. When such frameshift mutations were intercrossed, then in certain combinations they produce wild type normal gene. It was concluded that one of them was deletion and the other an addition, so that the disturbed order of the frame due to mutation will be restored by the other (Fig. 38.1).

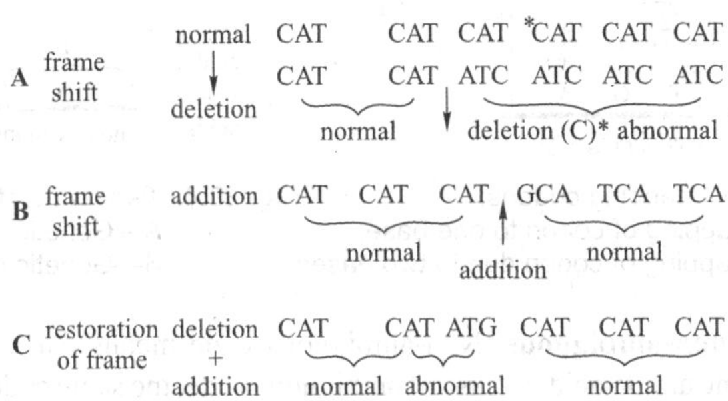

Fig. 38.1. Frame-shift mutations: A—Deletion; B—Addition; C—Restoration of frame.

***(ii)* Base substitution.** If in a mRNA molecule at a particular point, one base pair is replaced by another without any deletion or addition, the meaning of one codon containing such an altered base, will be changed. In consequence, in place of a particular amino acid at a particular position in a polypeptide, another amino acid will be incorporated. For example, due to substitution mutation, in the gene for tryptophan synthetase enzyme in *E.coli*, the GGA codon for glycine becomes a missence codon AGA which codes for arginine. **Missence codon** is a codon which undergoes an alteration to specify another amino acid.

A more direct evidence for a triplet code came from the finding that a piece of mRNA containing 90 nucleotides, corresponded to a polypeptide chain of 30 amino acids of a growing haemoglobin molecule. Similarly, 1200 nucleotides of "satellite" tobacco necrosis virus direct the synthesis of coat protein molecules which have 372 amino acids.

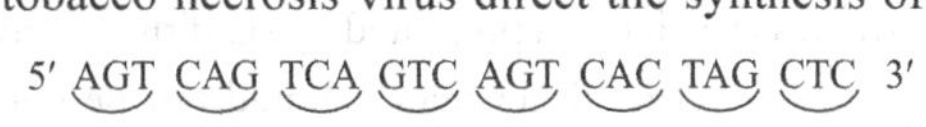

Fig. 38.2. Bases in mRNA are read sequentially in the 5′ to 3′ direction, in group of three bases.

2. The code is non-overlapping. In translating mRNA molecules the codons do not overlap but are "read" sequentially (Fig. 38.2). Thus, **a non-overlapping code** means that a base in a mRNA is not used for different codons. In Figure 38.3, it has been shown that an overlapping code can mean coding for four amino acids from six bases. However, in actual practice six bases code for not more than two amino acids. For example, in case of an overlapping code, a single change (of substitution type) in the base sequence will be reflected in substitutions of more than one amino acid in corresponding protein. Many examples have accumulated since 1956 in which a single base substitution results into a single amino acid change in insulin, tryptophan synthetase, TMV coat protein, alkaline phosphatase, haemoglobin, etc (see Chapter 37).

Recently, however, it has been shown that in the bacteriophage $\phi \times 174$ there is a possibility of overlapping of genes and codons (**Barrel** and coworkers, 1976; **Sanger**, *et al.*, 1977).

3. The code is commaless. The genetic code is commaless, which means that no codon is reserved for punctuations. It means that after one amino acid is coded, the second amino acid will be automatically coded by the next three letters and that no letters are wasted as the punctuation marks (Fig. 38.4).

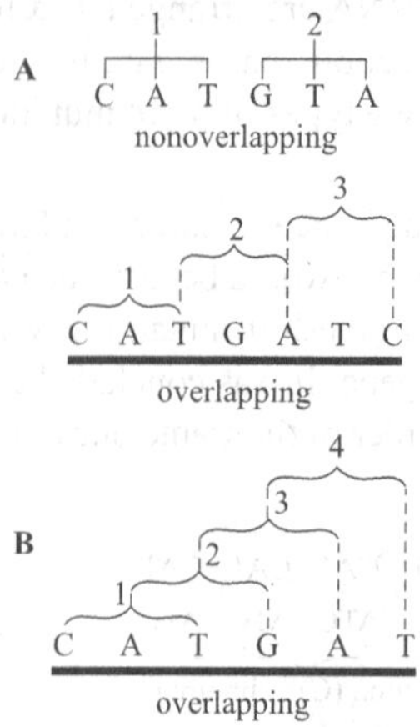

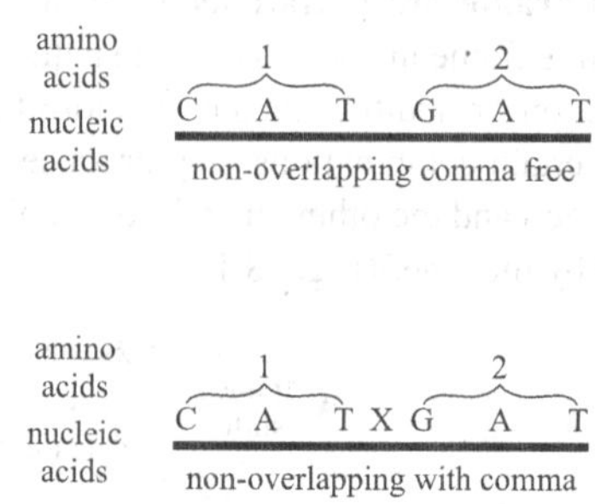

Fig. 38.3. A—Non-overlapping codons;
B—Overlapping of codon to one base;
C—Overlapping of codon due to two bases.

Fig. 38.4. Two forms of genetic code.
A—Genetic code without comma;
B—Genetic code with comma.

4. The code is non-ambiguous. Non-ambiguous code means that a particular codon will always code for the same amino acid. In case of ambiguous code, the same codon could have different meanings or in other words, the same codon could code two or more than two amino acids. Generally, as a rule, the same codon shall never code for two different amino acids. However, there are some reported exceptions to this rule: the codons AUG and GUG both may code for *methionine* and as initiating or starting codon, although GUG is meant for *valine*. Likewise, GGA codon codes for two amino acids *glycine* and *glutamic acid* (Box 38.1).

Box 38.1 Genetic code is degenerate not ambiguous

After the start and stop codons, the remaining 60 codons are far more than enough to code for the other 19 amino acids and indeed there are repeats. Thus we say that the genetic code is **degenerate**, that is, an amino acid may be represented by more than one codon. The degeneracy is not evenly divided among the amino acids. For example, methionine and tryptophan are represented by only one codon each, whereas leucine is represented by six different codons.

The term "degeneracy" should not be confused with "ambiguity". To say that the code was ambiguous would mean that a *single codon could specify either of two (or more) different amino acids*; there would be doubt whether to put in, say leucine or something else. The genetic code is not ambiguous. Degeneracy in the code means that *there is more than one clear way to say "put leucine here"*. In other words, a given amino acid may be encoded by more than one codon, but a codon can code for only one amino acid. But just as people in different places prefer different ways of saying the same thing—"Good-bye"! "See you" | "Ciao"! and "So long" ! have the same meaning different organisms prefer one or others of the degenerate codons (Ciao means informal hello or good bye). These preferences are important in genetic engineering.

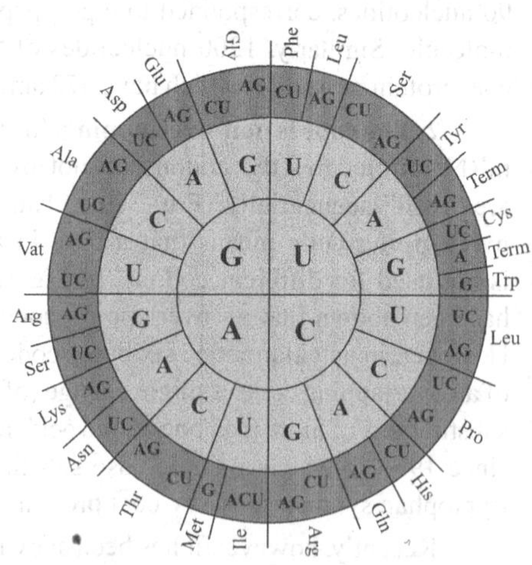

Genetic code degenerate.

5. The code has polarity. The code is always read in a fixed direction, *i.e.*, in the 5′ → 3′ direction. In other words, the codon has a **polarity**. It is apparent that if the code is read in opposite directions, it would specify two different proteins, since the codon would have reversed base sequence:

Codon:	UUG	AUC	GUC	UCG	CCA	ACA	AGG
Polypeptide: →	Leu	Ile	Val	Ser	Pro	Thr	Arg
	Val	Leu	Leu	Ala	Thr	Thr	Gly ←

6. The code is degenerate. More than one codon may specify the same amino acid; this is called **degeneracy** of the code. For example, except for *tryptophan* and *methionine*, which have a single codon each, all other 18 amino acids have more than one codon. Thus, nine amino acids, namely *phenylalanine*, *tyrosine*, *histidine*, *glutamine*, *asparagine*, *lysine*, *aspartic acid*, *glutamic acid* and *cysteine*, have two codons each. Isoleucine has three codons. Five amino acids, namely *valine*, *proline*, *threonine*, *alanine* and *glycine*, have four codons each. Three amino acids, namely *leucine*, *arginine* and *serine*, have six codons each (see Table 38.5).

The code degeneracy is basically of two types: partial and complete. **Partial degeneracy** occurs when first two nucleotides are identical but the third (*i.e.*, 3′ base) nucleotide of the degenerate codons differs, *e.g.*, CUU and CUC code for leucine. **Complete degeneracy** occurs when any of the four bases can take third position and still code for the same amino acid (*e.g.*, UCU, UCC, UCA and UCG code for serine).

Degeneracy of genetic code has certain biological advantages. For example, it permits essentially the same complement of enzymes and other proteins to be specified by microorganisms varying widely in their DNA base composition. Degeneracy also provides a mechanism of minimizing mutational lethality.

7. Some codes act as start codons. In most organisms, AUG codon is the **start** or **initiation** codon, *i.e.*, the polypeptide chain starts either with **methionine** (eukaryotes) or **N-formylmethionine** (prokaryotes). Methionyl or N-formylmethionyl-tRNA specifically binds to the **initiation site** of mRNA containing the AUG initiation codon. In rare cases, GUG also serves as the initiation codon, *e.g.*, bacterial protein synthesis. Normally GUG codes for valine, but when normal AUG codon is lost by deletion, only then GUG is used as initiation codon.

8. Some codes act as stop codons. Three codons UAG, UAA, UGA are the chain **stop** or **termination** codons. They do not code for any amino acids. These codons are not read by any tRNA molecules (*via* their anticodons), but are read by specific proteins, called **release factors** (*e.g.*, RF-1, RF-2, RF-3 in prokaryotes and RF in eukaryotes). These codons are also called as **nonsense codons**, since they do not specify any amino acid.

The UAG was the first termination codon to be discovered by **Sidney Brenner** (1965). It was named **amber** after a graduate student named **Bernstein** (= the German word for 'amber' and amber means brownish yellow) who helped in the discovery of a class of mutations. Apparently, to give uniformity the other two termination codons were also named after colours such as **ochre** for UAA and **opal** or **umber** for UGA. (*Ochre* means yellow red or pale yellow; *opal* means milky white and *umber* means brown). The existence of more than one stop codon might be a safety measure, in case the first codon fails to function.

9. The code is universal. Same genetic code is found valid for all organisms ranging from bacteria to human beings. Such universality of the code was demonstrated by **Marshall, Caskey** and **Nirenberg** (1967) who found that *E.coli* (bacterium), *Xenopus laevis* (amphibian) and guinea pig (mammal) amino acyl-tRNA use almost the same code. **Nirenberg** has also stated that the genetic code may have developed 3 billion years ago with the first bacteria, and it has changed very little throughout the evolution of living organisms.

Codon and Anticodon

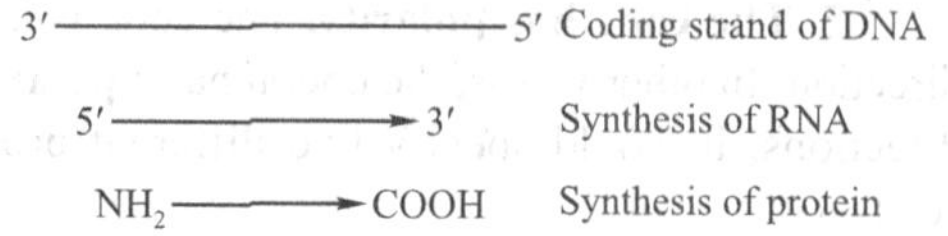

Fig. 38.5. Directions of synthesis of RNA and protein with respect to the coding strand of DNA (after Freifelder, 1985).

The codon words of DNA would be complementary to the mRNA code words (*i.e.*, DNA codes run in 3′ → 5′ direction and mRNA code words run in 5′ → 3′ direction) and so thereby the three bases forming the **anticodon** of tRNA (*i.e.*, bases of anticodons run in 3′ → 5′ direction). Three bases of anticodon pair with the mRNA on the ribosomes at time of aligning the amino acids during protein synthesis (translation of mRNA into proteins which proceeds in NH_2—COOH direction). For example, one of two mRNA and DNA code words for the amino acid phenylalanine is UUC and AAG respectively, and the corresponding anticodon of tRNA is CAA. This indicates that codon and anticodon pairing is antiparallel. In this case, C pairs with G and U pairs with A.

38.4. WOBBLE HYPOTHESIS

The genetic code is a degenerate code, meaning that a given amino acid may have more than one codon (Box 38.1). As you can see from Table 38.4, eight of the sixteen boxes contain just one amino acid per box (*e.g.*, serine, leucine, proline, arginine, threonine, valine, alanine and glycine). (A **box** is determined by the first and second positions, *e.g.*, the UUX box in which X is any of the four bases). Therefore, for these eight amino acids, the codon need only be read in the first two positions because the same amino acid will be represented regardless of the third base of the codon. These eight groups of codons are termed **unmixed families** of codons. An unmixed family is the four codons beginning with the same two bases that specify a single amino acid. For example the codon family GUX codes for valine.

Mixed families of codons code for two amino acids or for stop signals and one or two amino acids. Six of the mixed-family boxes are split in half so that the codons are differentiated by the presence of a purine or a pyrimidine in the third base. For example, CAU and CAC both code for histidine: in both, the third base, U (uracil) or C (cytosine) is a pyrimidine. Only two of the families of codons are split differently.

The lesser importance of the third position in the genetic code is linked with two facts about transfer RNAs. First, although there would seem to be a need for sixty-two transfer RNAs—since there are sixty-one (*i.e.*, one codons specifying amino acids and an additional codon for initiation) there are actually only about fifty different transfer RNAs in an *E.coli* cell. Second, a rare base such as inosine (I) can appear in the anticodon, usually in the position that is complementary to the third position of the codon. Visualising these two facts researchers believe that some kind of conservation of transfer RNAs is occurring and that rare base may be involved.

At this stage, one point should be very clear that both messenger RNA and transfer RNA bases are usually numbered from the 5′ side. Thus, the number one base of the codon is complementary to the number-three base of the anticodon (Fig. 38.6). Thus, the codon base of lesser importance is the number-three base, whereas its complement in the anticodon is the number-one base.

Since the first position of the anticodon (5′) is not as constrained as the other two positions, a given base at that position may be able to pair with any of several bases in the third position of the codon. **Crick** (1966) characterized this ability as **wobble** (wobble means to sway or move unsteadily) (Fig. 38.7). Table 38.6 shows the possible pairings that would produce a transfer RNA system compatible with the known code. For example, if an isoleucine transfer RNA has the anticodons 3′-UAI-5′, it is compatible with the three codons for that amino acid (see Table 38.4 and Table 38.5): 5′-AUU-3′, 5′-AUC-3′ and 5′-AUA-3′. That is inosine is the first (5′) position of the anticodon can recognise U, C, or A in the third (3′) position of the codon, and thus one transfer RNA complements with all three codon for isoleucine (Table 38.6).

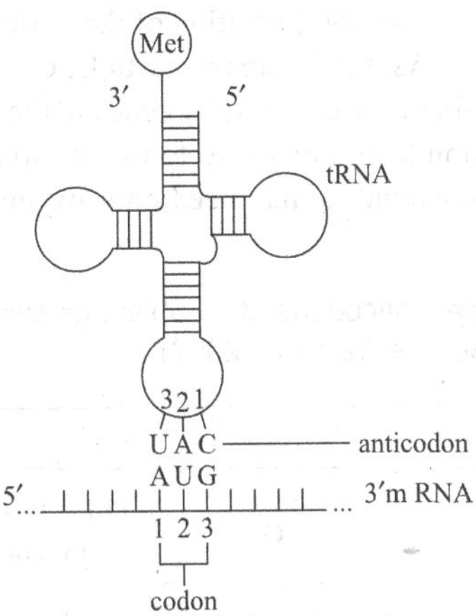

Fig. 38.6. Codon and anticodon base positions are numbered from the 5′ end. The 3′ position in the codon (5′ in the anticodon) is the wobble base (after Tamarin, 2002).

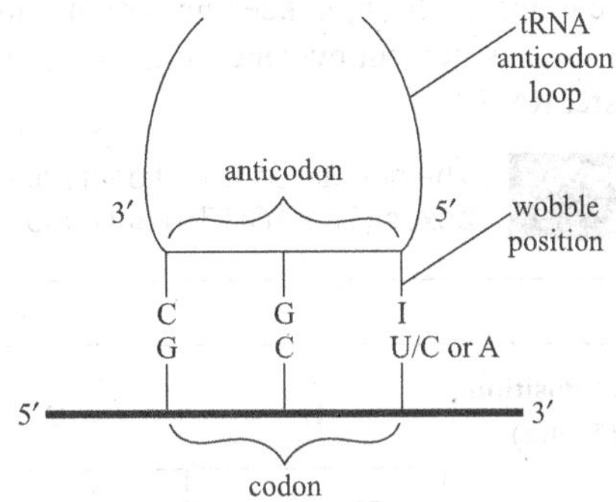

Fig. 38.7. Wobble hypothesis. In the third site (5′ end) of the anticodon I can take either of three wobble positions, thus being able to pair with U, C or A. This means that a single tRNA species carrying an amino acid can recognize three codons in mRNA: GCU, GCC and GCA (After Suzuki et al., 1986).

Table 38.6 Pairing combinations at the third codon position (Source: Tamarin, 2002).

Number-one base in tRNA (5′ end)	Number-three base in mRNA (3′ end)
G	U or C
C	G
A	U
U	A or G
I	A, U or C

38.5. DEVIATIONS FROM UNIVERSALITY OF GENETIC CODE

Molecular biologists concluded by 1979 that the genetic code was universal. That is the code dictionary (Table 38.4) was the same for *E.coli*, human beings and oak trees, as well as other species investigated up to that time. The universality of the genetic code was demonstrated, for example, by taking the ribosomes and messenger RNA from rabbit reticulocytes and mixing them with the aminoacyl-tRNA and other translational components of *E.coli*. Rabbit haemoglobin was synthesized.

Around 1980, however researchers noted divergence when sequencing mitochondrial genes for structural proteins. It was discovered that there were *two* kinds of deviations from universality in the way mitochondrial transfer RNAs read the code. First, fewer transfer RNAs were needed to read the genetic code, second, there were several cases in which the mitochondrial and cellular systems interpreted a codon differently.

According to Crick's wobble rules (Table 38.6), thirty-two transfer RNAs (including one for initiation) can complement all sixty-one non-terminating codons. For example, unmixed families of codons require two transfer RNAs, and mixed families of codons require one, two or three transfer RNAs, depending on the family. The yeast mitochondrial coding system apparently needs only one twenty-four transfer RNAs. The reduction in numbers is attained primarily by having only one transfer RNA recognize each unmixed family (Table 38.7). Because mitochondrial transfer

RNAs for unmixed families of codons have a U in the first (*i.e.*, wobble) position of the anticodon, apparently, given the structure of the mitochondrial transfer RNAs, the U can pair with U, C, A or G Most probably, evolutionary pressure has minimized the number of transfer RNA genes in the DNA of the mitochondrion, in keeping with its small size. Reduction from thirty-two to twenty-four is a 25% savings. Recent evidence suggests that mammalian mitochondria may need only twenty-two transfer RNAs.

Table 38.7 The genetic code dictionary of yeast mitochondria; anticodons (3′ → 5′) are given within boxes (The ACU Trp anticodon is predicted). (Source: Tamarin, 2002).

<table>
<tr><th colspan="6">Second Position</th></tr>
<tr><th>First Position (5′ end)</th><th>U</th><th>C</th><th>A</th><th>G</th><th>Third Position (3′ end)</th></tr>
<tr><td rowspan="2">U</td><td>Phe AAG</td><td rowspan="2">Ser AGU</td><td>Try AUG</td><td>Cys ACG</td><td>U
C</td></tr>
<tr><td>Leu AAU</td><td>stop</td><td>Trp ACU</td><td>A
G</td></tr>
<tr><td rowspan="2">C</td><td rowspan="2">Thr GAU</td><td rowspan="2">Pro GGU</td><td>His GUG</td><td rowspan="2">Arg GCA</td><td>U
C</td></tr>
<tr><td>Glon GUU</td><td>A
G</td></tr>
<tr><td rowspan="2">A</td><td>Ile UAG</td><td rowspan="2">Thr UGU</td><td>Asn UUG</td><td>Ser UCG</td><td>U
C</td></tr>
<tr><td>Met UAC</td><td>Lys UUU</td><td>Arg UCU</td><td>A
G</td></tr>
<tr><td rowspan="2">G</td><td rowspan="2">Val CAU</td><td rowspan="2">Ala CGU</td><td>Asp CUG</td><td rowspan="2">Gly CCU</td><td>U
C</td></tr>
<tr><td>Glu CUU</td><td>A
G</td></tr>
</table>

It has also been found that yeast mitochondria read the CUX family (*i.e.*, CUU, CUC, CUA, CUG) as threonine rather than as leucine (Table 38.4 and Table 38.7) and the terminator UCA (opal) as tryptophan rather than as termination. However, there have been found differences among mitochondria of different groups of organisms reading the CUX family. Human and *Neurospora* mitochondria appear to read the CUX codons as leucine just as cellular systems do. Of the groups so far analyzed, only yeast reads the CUX family as threonine. Similarly, human and *Drosophila* mitochondria read AGA and AGG as stop signals rather than as arginine (Table 38.8).

Table 38.8 Common and alternative meanings of codons (Source: Tamarin, 2002).

Codon	General meaning	Alternative meaning
1. CUX	Leu	Thr in yeast mitochondria
2. AUA	Ile	Met in mitochondria of yeast, *Drosophila* and Vertebrates
3. UGA	Stop	Trp in mycoplasmas and mitochondria other than higher plants
4. AGA/AGG	Arg	*Stop* in mitochondria of yeast and vertebrates; Ser in mitochondria of *Drosophila*
5. CGG	Arg	Trp in mitochondria of higher plants
6. UAA/UAG	*Stop*	Gln in ciliated protozoa
7. UAG	*Stop*	Ala or Leu in mitochondria of some higher plants

The ciliated protozoan *Paramecium* species read the UAA and UAG stop codons as glutamine within the cell. In addition, a prokaryote (*Mycoplasma capricolum*) reads UGA codon as tryptophan. We do not yet know how general this finding is : scientists have analysed the genetic code of very few species. We can thus conclude that *the genetic code seems to have universal tendencies among prokaryotes, eukaryotes* and *viruses*. Mitochondria however, read the code slightly differently : different wobble rules apply, and mitochondria and cells read at least one terminator and one unmixed family of codons differently.

Site specific variation. In this type of variation in codon reading, the interpretation of a codon depends on its specific location. For example, codons GUG and, rarely, UUG can serve as prokaryotic initiation codons. This means that they are recognized by $tRNA_f^{Met}$. However, they are not recognised by $tRNA_m^{Met}$. (*i.e.*, GUG and UUG are not misread internally in messenger RNAs). In some cases, two of the termination codons (UGA and UAG, but not UAA) are misinterpreted as codons for amino acids. That is, termination will not occur at the normal place, resulting in a longer-than usual protein. In some cases, these "read-through" proteins are vital—the organism depends on their existence. For example, in the phage θβ, the coat-protein gene is read through about 2% of the time. Without this small number of read through proteins, the phage coat (capsid) cannot be constructed properly.

Selenocysteine codon. Site-specific variation also involves the amino acid selenocysteine (*i.e.*, cysteine with a selenium atom replacing the sulphur). Although many proteins have unusual amino acids, almost all are due to post-translational modifications of normal amino acids. However, the amino acid selenocysteine is inserted directly into some proteins, such as formate dehydrogenase in *E.coli*, which has selenium in its active site. Selenocysteine is inserted into the protein by a new transfer RNA that recognises the termination codon, UGA, if that codon involved in a particular stem-loop secondary structure in the messenger RNA. The selenocysteine transfer RNA is originally charged with a serine that is then modified to a selenocysteine. In addition to the stem-loop structure 3′ (downstream) from the amber codon (UAG), a selenocysteine elongation factor (SELB) is also needed at the ribosome. This same mechanism may occur in eukaryotes, but not all of the components have yet been recognized.

38.6. EVOLUTION OF GENETIC CODE

Certain modern theories have suggested that the genetic code has the wobble in it because it originally arose from a code in which only the first two bases were needed for the small number of amino acids in use several billion years ago. As new amino acids with useful properties became available, they were incorporated into proteins by a code modified by the third base, albeit with less specificity. This view has been supported by the fact that codons starting with the same nucleotide come from the same biosynthetic pathway. This indicates that in early evolution, as biosynthetic pathways were extended to new amino acids, the newcomers were incorporated by use of the second and the third bases of the code.

Let us see whether there is a relationship between the codons and the amino acids they code for, or is the code just one of many random/chance possibilities. That is, whether the genetic code is highly evolved or just a "frozen accident". Recent computer simulations of random codes indicate that the current genetic code is far outside the range of random in its ability to protect the organism from mutation. This suggests that the genetic code is not a frozen accident, but rather is highly evolved. Numerous examples in the current code support this view.

For example, in the unmixed family 5′-CUX-3′, any mutation in the third position produces another codon for the same amino acid. Wobble in the third position and codon arrangement ensures that less than half of the mutations in the third codon position result in the specification of a different amino acid.

Some patterns also have been observed in the genetic code. For example, the mutation of one codon to another results in an amino acid of similar properties. A high probability exists that such a mutation will produce a functional protein. All the codons with U as the middle base, for example, are for amino acids that are hydrophobic (*e.g.*, phenylalanine, leucine, isoleucine, methionine and valine). Mutation in the first or third positions for any of these codons still codes a hydrophobic amino acids. Both of the two negatively charged amino acids, aspartic acid and glutamic acid, have codons that start with GA. All of the aromatic amino acids—phenylalanine, tyrosine and tryptophan—have codons that begin with uracil. Such patterns minimize the negative effects of mutation (see **Tamarin**, 2002).

QUESTIONS

Long Answer Questions

1. What do you understand by 'genetic code'? How will you show that the minimum size of a code word should be triplet?
2. A basic concept of molecular biology is the collinearity of gene, DNA, RNA and protein. Discuss the various lines of evidence supporting or providing this concept.
3. Gamow (1954) pointed out that since the genetic language contains only four letters, A, U(T), C and G, if all code words are of the same size, codons must be at least three bases long.
   (*a*) Show why codons cannot consist of one or of two bases.
   (*b*) Discuss two lines of research that indicate that codons are three base long.
   (*c*) Since codons are three base long, 64 different triplets can exist. Illustrate these using the branching method.
   (*d*) How many of the 64 triplets will contain (*i*) no adenine, (*ii*) at least one adenine?
4. (*a*) Explain what is meant by a degenerate code and illustrate your answer to show degeneracy in translation.
   (*b*) Which are the nonsense triplets and why are they so termed?
   (*c*) Do codons have the same meaning *in vitro* as *in vivo*?
5. Describe the general properties of the genetic code.
6. Give a brief account of the experiments which helped in deciphering the genetic code.
7. Discuss the experiments conducted by H.G. Khorana and his coworkers for cracking the genetic code.

Short Answer Questions

1. Write short notes on the following:
   (*i*) Homopolymers;
   (*ii*) Copolymers of repetitive sequences;
   (*iii*) Poly U;
   (*iv*) Frame shift mutation;
   (*v*) Overlapping genes;
   (*vi*) Wobble hypothesis.

Very Short Answer Questions

1. The size of haemoglobin gene in humans is estimated to consist of approximately 450 nucleotide pairs. The protein product of the gene is estimated to consist of about 150 amino acid residues. Calculate how many nucleotides are present in a genetic codon for each amino acid.
2. How many triplet codons can be made from the four ribonucleotides A, U, G and C containing (*a*) no uracils; (*b*) one or more uracils?
3. Assume a series of different one-base changes in the codon GGA, which produces following several new codons: (*a*) UGA; (*b*) GAA; (*c*) GGC, (*d*) CGA. Which of these represent(s) degeneracy, which missense, and which nonsense?
4. If single-base changes occur in DNA (and, therefore, in mRNA), which amino acid, tryptophan or arginine, is most likely to be replaced by another in protein synthesis?

5. (*a*) How was the genetic code first decoded? (*b*) What refinements have since been incorporated in the technique?
6. In what sense and to what extent is the genetic code (*a*) degenerate and (*b*) universal?
7. What evidence supports the hypothesis of colinearity between the nucleotide sequence in a gene and the amino acid sequence in a polypeptide?
8. Why is colinearity between codons and polypeptides significant?
9. A. Garen (1968) extensively studied a particular nonsense (Chain-termination) mutation in thealkalinephosphatasegeneof*E.coli*.This mutation resulted in the termination of the alkaline phosphatase polypeptide chain at a position where the amino acid tryptophan occurred in the wild-type polypeptide. Garen induced revertants (in this case, mutations altering the same codon) of this mutant with chemical mutagens that induced single base-pair substitutions and sequenced the mutant polypeptides. Seven different types of revertants were found, each with a different amino acid at the tryptophan position of the wild-type polypeptide (termination position of the mutant polypeptide fragment). The amino acids present at this position in the various revertants included tryptophan, serine, tyrosine, leucine, glutamic acid, glutamine and lysine. Was the nonsense mutation studied by Garen an *amber* (UAG), an *ochre* (UAA), or an *opal* (UGA) nonsense mutation? Explain the basis of your deduction.

Multiple Choice Questions

Choose the correct answer from the four alternatives given.

1. Genetic code is sequence of nitrogenous bases on
   (*a*) mRNA (*b*) tRNA
   (*c*) rRNA (*d*) DNA
2. The terminator codons are
   (*a*) UAA, UAG, UGA
   (*b*) AUG, UAG, UGA
   (*c*) UAC, AUG, UAG
   (*d*) AUG, ACG, GAG
3. Who discovered the entire genetic code?
   (*a*) Nirenberg (*b*) Ochoa
   (*c*) Khorana (*d*) Crick
4. A codon consists of how many nitrogen bases in it
   (*a*) 1 (*b*) 2
   (*c*) 3 (*d*) 4
5. The arrangement of three bases in DNA represents
   (*a*) protein (*b*) amino acid
   (*c*) plasmid (*d*) nucleic acid
6. Degeneration of a genetic code is attributed to the
   (*a*) first member of codon
   (*b*) second member of codon
   (*c*) entire codon
   (*d*) third member of codon
7. In the genetic code dictionary, how many codons are used to code for all the 20 essential amino acids?
   (*a*) 20 (*b*) 64
   (*c*) 61 (*d*) 60
8. What would happen if in a gene encoding a polypeptides of 50 amino acids, 25th codon (UAU) is mutated to UAA?
   (*a*) a polypeptides of 24 amino acids will be formed
   (*b*) two polypeptide of 24 and 25 amino acids will be formed
   (*c*) a polypeptide of 49 amino acids will be formed
   (*d*) a polypeptide of 25 amino acids will be formed
9. Which of the chemical characteristics is not common to all living beings?
   (*a*) ribosomes are sites of protein synthesis
   (*b*) types of protein present in the body
   (*c*) energy is stored by high phosphate bond
   (*d*) similar triplet codes for amino acid

ANSWERS

Very Short Answer Questions

1. Approximately 3 nucleotides code for each amino acid.

2. (*a*) Since uracil represents I among 4 nucleotides; the probability that uracil will be the first letter of the codon is 1/4 and the probability that U will not be the first letter is 3/4. The same reasoning holds true for the second and third letters of the codon. The probability that none of the three letters of the codon are uracils is $(3/4)^3 = 27/64$.

(*b*) The number of codons containing at least one uracil is $1 - 27/64 = 37/64$.

3. (*a*) Nonsense; UGA is a stop signal and does not translate.

(*b*) Missense; GGA codes for glycine, GAA for glutamic acid.

(*c*) Degeneracy; both GGA and GGC code for glycine.

(*d*) Missense; GGA codes for glycine, CGA for arginine.

4. Tryptophan, which has only one codon. Arginine has six codons.

5. (*a*) Synthetic RNA molecules (polyuridylic acid molecules) containing only the base uracil were prepared when these synthetic molecules were used to activate *in vitro* protein synthesis systems, small polypeptides containing only the amino acid phenylalanine (polyphenyl alanine molecules) were synthesized. Codons composed only of uracil were therefore shown to specify phenylalanine. Similar experiments were carried out using synthetic RNA molecules with different base compositions.

(*b*) Better *in vitro* systems activated with synthetic RNA molecules with known repeated base sequences were developed ultimately, *in vitro* systems in which specific aminoacyl-tRNAs were shown to bind to ribosomes activated with specific mini-mRNAs, which were trinucleotides of known base sequence, were developed and used in codon identification.

6. (*a*) The genetic code is degenerated in that all but 2 of the 20 amino acids are specified by two or more codons. Some amino acids are specified by six different codons. The degeneracy occurs largely at the third or 3′ base of the codons. "*Partial degeneracy*" occurs where the third base of the codon may be either of the two purines or either of the two pyrimidines and the codon still specifies the same amino acid. "*Complete degeneracy*" occurs where the third base of the codon may be anyone of the four bases and the codon still specifies the same amino acid.

(*b*) The code appears to be almost completely universal. Known exceptions to universality include strains carrying suppressor mutations that alter the reading of certain codons (with low efficiencies in most cases) and the use of UGA as a tryptophan codon in yeast and human mitochondria.

7. Colinearity received strong support from studies that showed a direct correlation between the linear sequence of mutational sites in a gene (established by genetic mapping experiments) and the linear sequence of mutational defects (amino acid substitutions or chain terminations) in the polypeptide gene-product (established by purification and amino acid sequencing, etc., of mutant polypeptides).

8. Colinearity is an important prediction of our present concepts of noncoding sequences ("introns" or "intervening sequences") within eukaryotic genes does not violate the concept of colinearity; their presence simply means that the colinear structures frequently contain interruptions.

9. *Amber* (UAG). This is the only nonsense codon that is related to tryptophan, serine, tyrosine, leucine, glutamic acid, glutamine and lysine codons by a single base pair substitution in each case.

Multiple Choice Questions

1. (*a*) **2.** (*a*) **3.** (*a*) **4.** (*c*) **5.** (*b*) **6.** (*d*) **7.** (*c*)
8. (*c*) **9.** (*d*)

39 Transcription and mRNAs, rRNAs and tRNAs

In the organisms (*viz.*, prokaryotes and eukaryotes), where coded genetic informations are contained in the DNA molecule, different genetically controlled functions of their cells are performed by a different kinds of nucleic acid, called **non-genetic ribonucleic acid (RNA)**. In the cells of such organisms, the DNA molecule occurs in the chromosomes of nucleus or nucleoids of eukaryotes or prokaryotes respectively, and the process of protein synthesis occurs in the cytoplasm. It is investigated that a DNA molecule does not leave the nucleus to participate directly in the process of protein synthesis but employs different types of non-genetic ribobucleic acid molecules for carrying its genetic informations from the nucleus to the site of protein synthesis, *i.e.*, ribosomes. Thus, gene expression is accomplished by the transfer of genetic information from DNA to RNA molecules and then from RNA to protein molecules. RNA molecules are synthesized by using the base sequence (triplet codons) of one strand of DNA as a template in a polymerization reaction that is catalyzed by enzymes called **DNA-dependent RNA polymerases** or simply **RNA polymerases**. The process by which RNA molecules are initiated, elongated, and terminated is called **transcription**.

39.1. CHEMICAL COMPOSITION OF NON-GENETIC RIBONUCLEIC ACID (RNA)

Chemically, the non-genetic RNA is closely related with DNA. However, the non-genetic RNA is a single stranded polymer of ribonucleotide units and is composed of phosphoric acid, ribose (pentose), sugar and nitrogen bases, which are purines (adenine and guanine) and pyrimidines (cytosine and uracil). The molecular weights of the various RNA molecules vary from about 25,000 to over one million. Its base composition does not follow the base composition of DNA, although, in general, the proportion of A+C roughly equals G+D, as has been shown in Table 39.1.

Table 39.1 Base ratios of RNA from various sources (as molar percentages) (From, Hall *et al.*, 1974).

	Source	Adenine	Guanine	Cytosine	Uracil
1.	*Allium cepa* (onion seed)	24.9	29.8	24.7	20.6
2.	*Phaseolus vulgaris* (bean seed)	24.9	31.4	24.1	19.6
3.	*Cucurbita pepo* (pumpkin seed)	25.2	30.6	24.8	19.4
4.	Ox liver	17.1	27.3	33.9	21.7
5.	Yeast	25.4	24.6	22.6	27.4
6.	*E.coli*	25.3	28.8	24.7	21.2

Further, some RNA molecules contain significant proportions of some methylated bases and an unusual nucleoside known as pseudouridine (ψU), in which the glycosidic bond is associated with position 5 of uracil rather than position 3.

39.2. COMPARISON BETWEEN DNA REPLICATION AND TRANSCRIPTION

The replication and transcription are two chief activities of DNA and they should be compared at the outset in the following manner:

1. In DNA replication an enzyme (or enzyme complex) matches up complementary deoxyribonucleoside triphosphates with a DNA template according to Watson-Crick pairing rules and the bases are polymerized to form a daughter DNA molecule. In DNA transcription a different enzyme, called **DNA-dependent RNA polymerase** mediates a similar process, except that complementary ribonucleoside-triphosphates are matched with the DNA template and RNA polymers are formed. The base uracil (U) replaces thymine in RNA but, otherwise, the DNA strand is faithfully copied. The RNA copy has the opposite polarity from the template, so that, for example, the sequence $\xrightarrow[\text{TACAAC}]{}$ in DNA is transcribed as $\xleftarrow[\text{AUGUUG}]{}$ in RNA.

2. The RNA transcripts do not ordinarily remain hydrogen-bonded with their DNA templates as DNA daughter strands do. Instead, they "peel off" the template as they are formed, thus, becoming available to participate in protein synthesis.

3. DNA of prokaryotes (*E.coli*) and eukaryotes has one or few initiation points for DNA replication, so that, at least, large portions of genome are copied into single, enormous daughter DNA molecules. In contrast, the synthesis of RNA molecules is initiated at close intervals along the DNA and, thus, relatively short RNA copies are produced.

4. Lastly, replication and transcription have to serve two different functions. The purpose of replication is to conserve the entire genome for next generation, whereas the purpose of transcription is to make RNA copies of individual genes that the cell can use in the biochemistry. The copies are themselves not endowed with the same kind of permanence as the genetic material; instead, they are typically degraded by cellular nucleases, once their functional usefulness has been spent.

39.3. MECHANISM OF PROKARYOTIC TRANSCRIPTION

Transcription involves the following three aspects: (1) The enzymatic synthesis of RNA; (2) The signals that determine at what points on a DNA molecule transcription starts and stops; and (3) The types of transcription products and how they are converted to the RNA molecules needed by the cell.

1. Enzymatic Synthesis of RNA

The essential chemical characteristics of the synthesis of RNA are the following:

1. The precursors in the synthesis of RNA are the four ribonucleotide 5′-triphosphates (rNTP)—ATP, GTP, CTP and UTP. On the ribose portion of each NTP, there are two OH groups—one each on the 2′- and 3′-carbon atoms.

2. In the polymerization reaction a 3′-OH group of one nucleotide reacts with the 5′-triphosphate of a second nucleotide; a pyrophosphate is removed and a phosphodiester bond results (Fig. 39.1).

3. The sequence of bases in RNA molecule is determined by the base sequence of the DNA. Each base, added to the growing end of the RNA chain, is chosen by its ability to base-pair with the DNA strand used as template; thus, the bases C, T, G and A in a DNA strand cause G, A, C and U respectively, to appear in the newly synthesized RNA molecule.

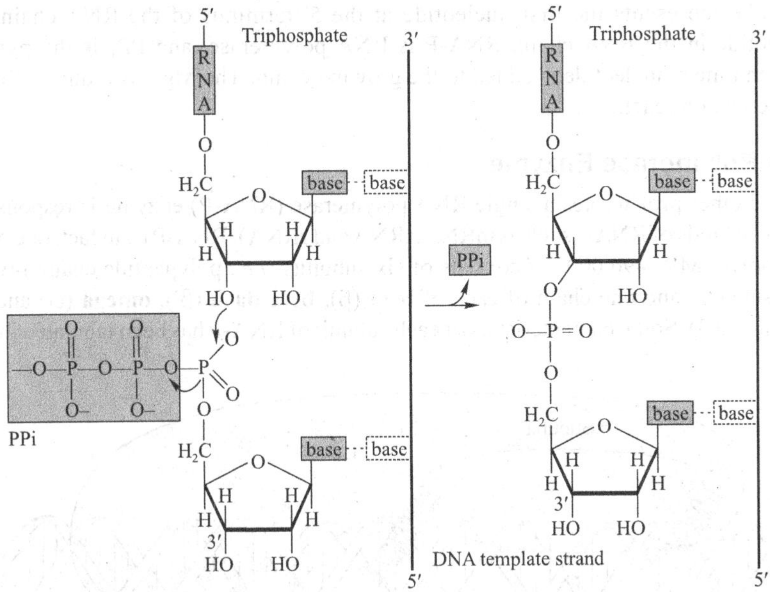

Fig. 39.1. Mechanism of the chain-elongation reaction catalyzed by RNA polymerase. The broken arrows join the reacting groups. The pyrophosphate group (PP) and the bold hydrogen do not appear in the RNA strand. The DNA template and the RNA strands are antiparallel as in double-stranded DNA (after Freifelder, 1985).

4. The DNA molecule being transcribed is double-stranded, yet in many particular regions only one strand serves as a template (Fig. 39.2).

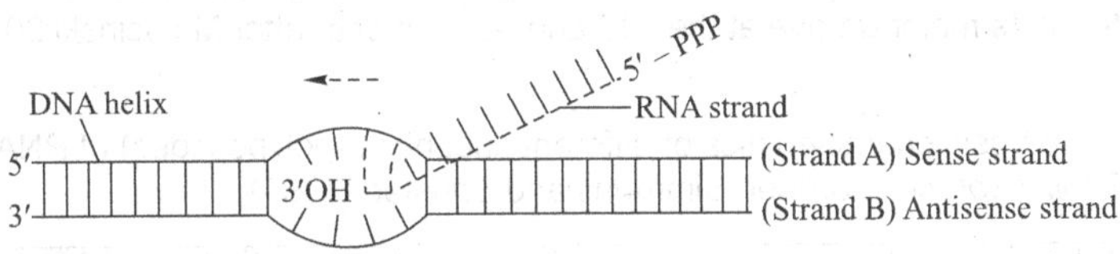

Fig. 39.2. An RNA strand is copied only from strand A (sense strand) of a segment of a DNA molecule. No RNA is copied from strand B (antisense strand) in the region of the DNA molecule. However, elsewhere, for example, in a different gene, strand B might be copied; in that case strand A would not be copied in that region of the DNA. The arrow shows the direction of RNA chain growth (after Freifelder, 1985).

5. The RNA chain grows in 5′ → 3′ direction; that is, nucleotides are added only to the 3′-OH end of the growing chain. (This is the same as the direction of chain growth in DNA synthesis). The RNA molecule is terminated by a 5′-triphosphate at the non-growing end. The RNA molecule is antiparallel to the DNA strand being copied. Once initated, RNA chains grow at a rapid rate—40 to 50 nucleotides per second in *E.coli* at 37°C.

6. RNA polymerases, in contrast with DNA polymerases, are able to initiate chain growth; that is, no primer is needed.

7. Only ribonucleoside 5′-triphosphates participate in RNA synthesis and the first base to be laid down in the initiation event is a triphosphate. Its 3′-OH group is the point of attachment of the subsequent nucleotide. Thus, the 5′ end of a growing RNA molecule terminates with a triphosphate.

The overall polymerization reaction can be written as follows:

$$_{n}NTP + XTP \xrightarrow[Mg^{2+}]{DNA,\ RNA\text{-}P} (NMP)_n\,XTP + nPP_1$$

in which XTP represents the first nucleotide at the 5′ terminus of the RNA chain, NMP is a mononucleotide in the RNA chain. RNA-P is RNA polymerase, and PP_1 is the pyrophosphate released each time a nucleotide is added to the growing chain. The Mg^{2+} is required for all nucleic acid polymerization reactions.

The RNA Polymerase Enzyme

In *E.coli* and other prokaryotes a single RNA polymerase (RNA-P) enzyme is responsible for the synthesis of all kinds of RNAs (such as mRNA, tRNA and rRNA). RNA-P is, in fact, one of the largest enzymes known (MW 490,000). It consists of six subunits (*i.e.*, polypeptide chains)-two identical **alpha** (α) subunits and one chain of each of **beta** (β), **beta dash** (β′), **omega** (ω) and **sigma** (σ) subunits (Fig. 39.3). Some characteristics of each subunit of RNA-P has been tabulated in Table 39.2.

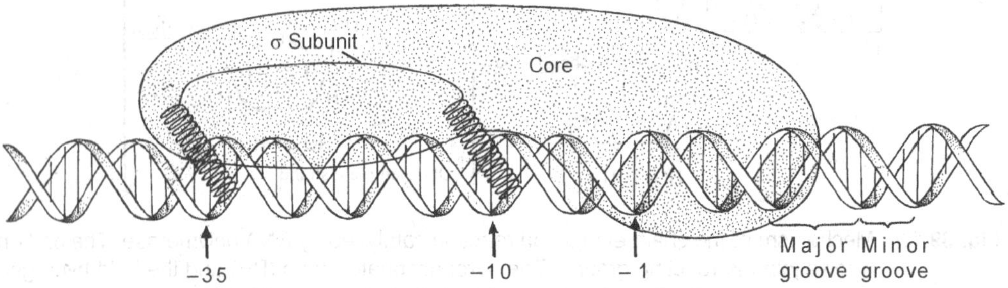

Fig. 39.3. RNA polymerase (Holoenzyme) bound to DNA at a promoter. The arrows indicate the position of the transcription start point (+1) and the –10 and –35 regions. The sigma factor is shown with two cylinders representing α-helical segments that make contact with the DNA in its major groove at the –10 and –35 regions (after Malacinski 2003).

Table 39.2 Some basic characteristics of different subunits (=polypeptides) of RNA polymerase of *E.coli.* (*Source*: Gardner, Simmons and Snustad, 1991)

	Subunit	Gene	Molecular weight (daltons)	Location	Function
1.	α_2	*rpo A*	41,000 each	Core enzyme	Promoter binding
2.	β	*rpo B*	155,000	Core enzyme	Nucleotide binding
3.	β′	*rpo C*	165,000	Core enzyme	DNA template binding
4.	ω	—	12,000	Core enzyme	—
5.	σ	*rpo D*	95,000	Sigma factor	Initiation

The complete RNA polymerase enzyme is termed **holoenzyme** and can be represented as $\alpha_2\beta\beta'\omega\sigma$ in which attachment of sigma (σ) subunit (or factor) is not very firm, but resultant **core enzyme** (α2ββ′) does not lose its catalytic activity of transcription. The active sites of core enzyme have been shown in Fig. 39.4. Functions of different polypeptide subunits of RNA polymerase are now known, but not in any detail. Thus, β and β′ subunits form the catalytic centre of RNA-P and help RNA polymerase in unwinding of DNA molecule for the purpose of transcription. The sigma (σ) factor helps in the recognition of start signals on DNA molecule and directs RNA polymerase in selecting the initiation sites (promoter). Once RNA synthesis is initiated and RNA molecule becomes 8–9 bases long, the σ factor dissociates from the holoenzyme and then the core enzyme brings about elongation of mRNA (or any other RNA).

2. Binding of RNA Polymerase to Promoter, Initiation, Elongation and Termination

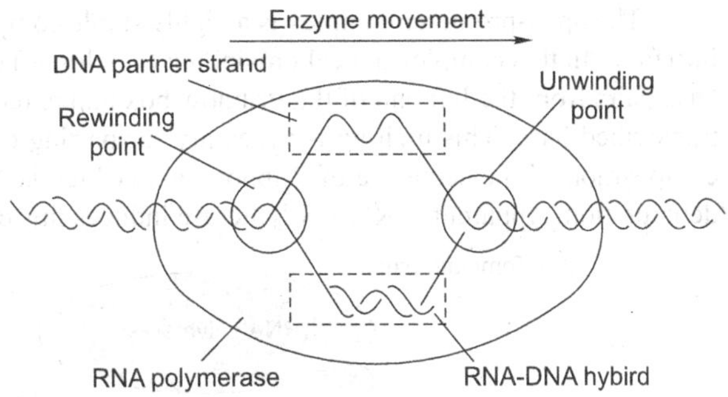

Fig. 39.4. Active centres in the core enzyme of bacterial RNA polymerase enzyme.

Unlike replication, transcription does not progress along the entire length of a chromosome. Instead, certain parts of the chromosome are transcribed. Only one of the two strands of a DNA duplex is transcribed; this strand is called the **sense strand**. The other strand of DNA which is not transcribed at the moment, is called **antisense strand**. Like DNA, RNA is synthesized in the 5′ → 3′ direction from the single-stranded region of the DNA template. This localized unwinding moves along the molecule followed by recoiling of the helix behind the newly synthesized RNA. The region of sense strand of DNA which is actually transcribed into RNA, is called the **coding region**.

The first step in transcription is *binding of RNA polymerase to a DNA molecule*. Binding occurs at particular sites, the **promoters**, which are specific sequences of 20 to 200 bases at which several interactions occur. Two special promoter regions have been identified that appear in all organisms. In a region of five to ten bases preceding the coding region is a sequence of seven bases that reads:

TATAATG

with minor variations. A sequence such as this is called a **consensus sequence**, because it is a sequence observed to occur with very little variation in many different organisms. In bacteria (*e.g.*, *E.coli*) this region is called the **Pribnow box** and in the eukaryotes the same region has the sequence:

TATAAAT

and is called the **Hogness box**, each after the person who originally described the region. This region is generally referred to as the **TATA box** and is believed to orient the RNA polymerase enzyme, so that synthesis proceeds from left to right. It is also the region at which the double helix opens to form the **open promoter complex**. Further, the centre of pribnow box lies usually 10 bp upstream of coding region, it contains T and is called—**10 sequence** which is recognized by RNA polymerase during the binding reaction.

Another important region, further upstream from TATA box, is located approximately thirty-five bases upstream from the coding region (= mRNA start). This is called **–35 sequence** (also called **recognition sequence**) and it consists of a nine-base consensus sequence, considered to be actual site of binding of the RNA polymerase. It seems likely that the sigma (σ) subunit first binds to the –35 sequence in a highly specific interaction and then, the appropriate region of this huge enzyme can come in contact with the –10 sequence of Pribnow box (Fig. 39.5).

```
CCAGGCTTTACACTTTATGCTTCCGGCTCGTATGATTGTGTGGAATTG
CTTTTTGATGCAATTCGCTTTGCTTCTGACTATAATAGACAGGGTAA
GGCGGTGTTGACATAAATACCACTGGCGGTGATACTGAGCACATCAG
GTGCGTGTTGACTATTTTACCTCTGGCFGGTGATAATGGTTGCATGTA
ATTGTTGTTGTTAACTTGTTTATTGCAGCTTATAATGGTTACAAATA
GCTAACACTTTACAGCGCCGCGTCATTTGATATGATGCGCCCCGCTT
```

–35 Sequence **mRNA start**

Fig. 39.5. Base sequence in the noncoding strand of six different *E.coli* promoters, showing the three important regions: mRNA start, Pribnow box (including –10 Sequence) and –35 sequence (after Freifelder, 1985).

The open-promoter complex is a highly stable complex and is the active intermediate in chain initiation. In this complex a local unwinding ("melting") of the DNA helix occurs starting about ten base pairs from the left end of the Pribnow box and extending to the end of the position of the first transcribed base. This melting is necessary for pairing of the incoming rebonucleotides. The base composition of the sequence of Pribnow box (which is A+T rich) makes the DNA strand open to denaturation. Apparently, RNA polymerase induces this conformational change.

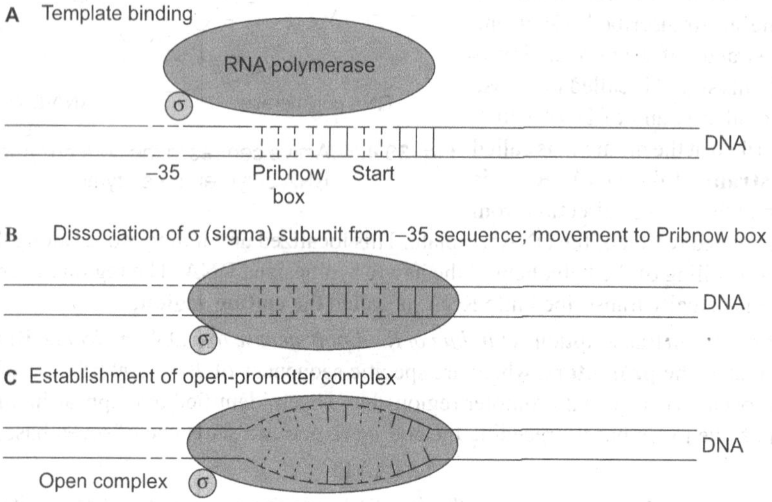

Fig. 39.6. A model for the binding of RNA polymerase to a promoter to form an open-promoter complex; PB=Pribnow box (after Freifelder, 1985).

Once an open-promoter complex has formed, RNA polymerase is ready to initiate RNA synthesis. RNA polymerase contains two nucleotides binding sites, called the **initiation site** and the **elongation site**. The initiation site binds only purine triphosphates, namely ATP and GTP, and one of these (usually ATP) is the first nucleotide in the growing RNA chain. Thus, the first DNA base that is transcribed is usually thymine (T). The initiating nucleoside triphosphate binds to the enzyme in the open-promoter complex and forms a hydrogen bond with the complementary DNA base (Fig. 39.7). The elongation site is then filled with a nucleoside triphosphate that is selected strictly by its ability to form a hydrogen bond with the next base in the DNA strand. The two nucleotides are then joined together, the first base is released from the initiation site, and initiation is completed. The dinucleotide remains hydrogen-bonded to the DNA. The **elongation phase** begins when the polymerase releases the base and then moves along the DNA chain.

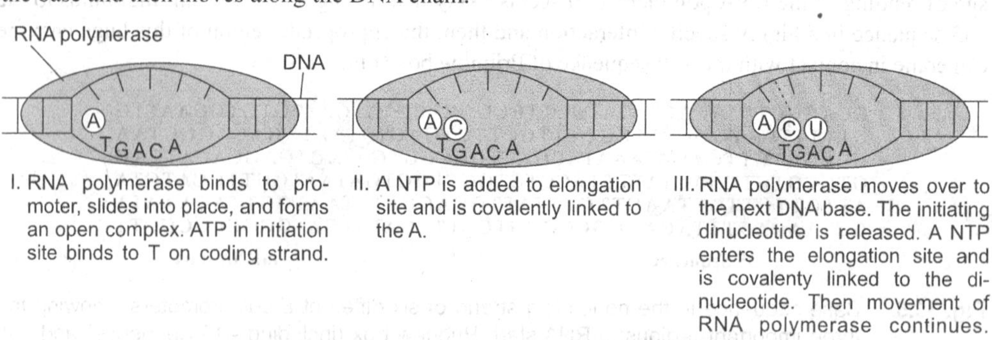

Fig. 39.7. Method of initiation of transcription. The enzyme is drawn without the sigma factor, NTP = Nucleoside triphosphate (after Freifelder, 1985).

After several nucleotides (approximately eight) are added to the growing chain, RNA polymerase changes its structure [forming stable **ternary** (= of three components) **elongation complex**] and loses the sigma factor. Thus, most elongation is carried out by the core enzyme (Fig. 39.8). The core enzyme moves along the DNA, binding a nucleoside triphosphate that can pair with the next DNA base and opening the DNA helix as it moves; thus, during elongation phase addition of 40 bases-per second at 37°C takes place. The open region extends only over a few base pairs; that is, the DNA helix recloses just behind the enzyme. The newly synthesized RNA is released from its hydrogen bonds with the DNA as the helix reforms; however, a few RNA bases remain paired with the DNA template during RNA synthesis. It should be noted that the promoter itself is not transcribed.

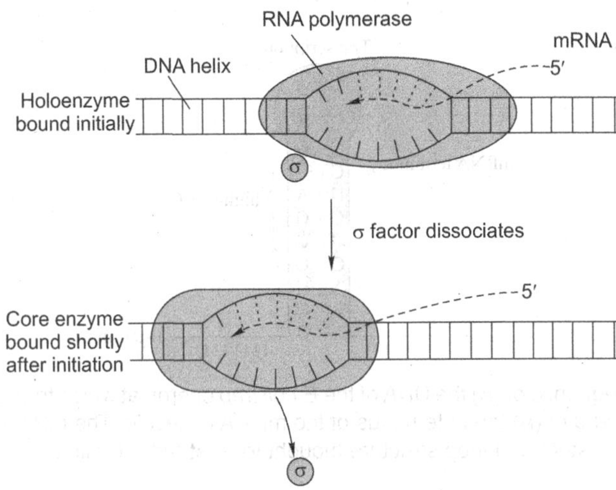

Fig. 39.8. Change in shape of RNA polymerase enzyme when the sigma factor gets dissociated from the core enzyme (after Freifelder, 1985).

Termination of RNA synthesis (or transcription) occurs at specific base sequences in the DNA molecule. Twenty termination sequences have so far been determined and each has the characteristics shown in Fig. 39.9. Termination region consists of the following three important regions: 1. First, there is an inverted repeat base sequence containing a central non-repeating segment: that is, the sequence in one DNA strand would read like—

ABCDEF—XYZ—F′E′D′C′B′A′

in which A and A′, B and B′ and so on are complementary bases. Thus, this sequence is capable of intrastrand base pairing, forming a "stem-and-loop" configuration in the transcript (RNA) and possibly in the DNA strands. 2. The second region is near the loop end of the presumed stem (sometimes totally within the stem) and is a sequence having a high G+C content. 3. A third region (sometimes absent) is a sequence of A.T pairs that yields in the RNA a sequence of six to eight uracils (U) often followed by adenine.

In fact, there are two types of termination events: those that depend only on the DNA base sequence and those that require the presence of termination protein, called *rho* (ρ; discovered by **J. Roberts**, 1969).

The final step in the termination process is dissociation of the core enzyme from the DNA. Following this event, the core enzyme interacts with a free σ (sigma) factor to reform the holoenzyme which becomes available for initiating RNA synthesis again.

3. Classes of RNA Molecules and Processing

This step of prokaryotic transcription will be explained later on along with that of eukaryotes.

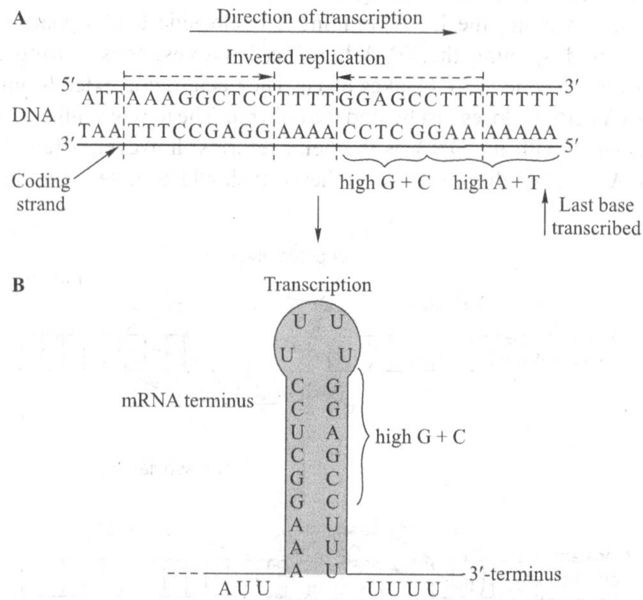

Fig. 39.9. Base sequence of (A) the DNA of the *E.coli trap* operon at which transcription termination occurs and of (B) the 3′ terminus of the mRNA molecule. The mRNA molecule is folded to form a stem-and-loop structure thought to exist (after Freifelder, 1985).

39.4. MECHANISM OF EUKARYOTIC TRANSCRIPTION

In eukaryotes, there are three major classes of RNA polymerases which are designated as I, II and III and are found in the nucleus. The three polymerases have different properties (see Table 39.3) and can be distinguished by the ions required for their activity, the optimal ion strength and their sensitivity to inhibition by various antibiotics (*e.g.*, α-amanitin). Each of these enzymes is a large protein (~50,000 daltons), with two large and several (8 to 10) smaller subunits.

Table 39.3. Properties of three different eukaryotic nuclear RNA polymerases.

	Enzymes	Location	Production and abundance	Sensitivity to α-amanitin
1.	RNA polymerase I	Nucleolus	rRNA (50–70%) (except 5S rRNA)	Not sensitive
2.	RNA polymerase II	Nucleoplasm	hnRNA (mRNA) (20–40%)	Sensitive
3.	RNA polymerase III	Nucleoplasm	tRNA (~10%) (and 5S rRNA)	Inhibited in animals at high levels; not in yeast and insects

Promoter, enhancer and silencers. The nature and function of RNA polymerase II (for hnRNA) is well studied than other two types of RNA polymerases. The promoters of genes of RNA polymerase II contain three distinct regions which are centred at sites lying between –25 bp and –100 P (Fig. 39.10). The least effective of these three regions is the TATA or Goldberg-Rogness box (7 bp long) located 20 bp upstream to the starting point. Further upstream is another sequence called **CAAT box** (or **CAT box**) which being necessary for initiation, is conserved in some promoters (*e.g.*, β-globin gene). CAAT box sequence lies between –70 and –80 base pairs and includes GGT/ACAATCT base composition. Another sequence called **GC box** (GGGCGG) is found in one or more copies at –60 or

–100 bp upstream in any orientation in many genes. It has been suggested that CAAT and GC boxes determine the efficiency of transcription, while TATA box aligns RNA polymerase at proper site with the help of proteins, called **transcription factors** or **TFs** (*e.g.*, TF II D).

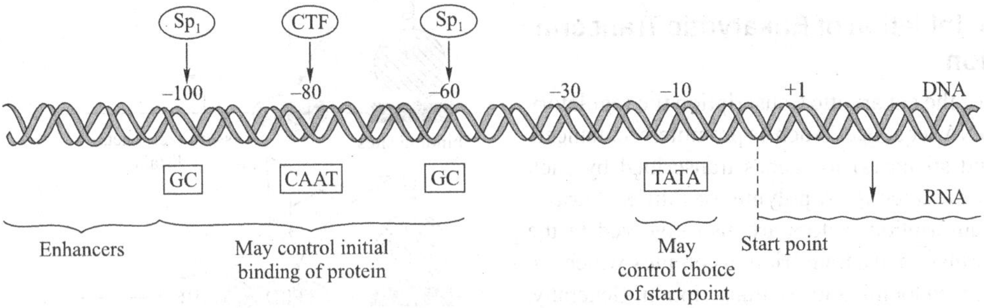

Fig. 39.10. An eukaryotic DNA segment showing promoter sites (TATA, CAT, GC boxes and enhancer sites).

Eukaryotic promoters also consist of sites located 100 to 200 base pairs upstream, which interact with proteins other than RNA polymerase and, thus, regulate the activity of promoter. These sites are called **enhancers**, since they lead upto 200-fold increase in the rate of transcription of an affected gene. Examples of enhancers are known in the genome of viruses (SV40) and eukaryotes (*i.e.*, in the genes for immunoglobulin, insulin, alpha amylase, etc.; **Picard**, 1985). In the spacer regions between *Xenopus* large ribosomal genes are multiple regions 60–80 bp long that confer a 20-fold increase in transcription rate compared with genes lacking them (see **Maclean** and **Hall**, 1987).

There are other regulatory sites known as **silencers** which repress gene expression. Both enhancers and silencers can function at great distance (often many kilobases) from the genes they enhance and repress respectively. Silencers are known to occur in yeast and repress expression of *HML* and *HMR* loci involved in switching of mating types.

On the other hand, in each of the 5S RNA genes which is transcribed by RNA polymerase III, the promoter lies in the middle of transcription unit, 50 b down-stream from the start point (Fig. 39.11). Such internal promoters also occur in the genes for different types of RNA polymerase III enzymes. The polymerase III enzyme is big enough to occupy start point (+1 bp) and the promoter region (+55 and +80) simultaneously and starts transcribing the start point without any apparent difficulty. The downstream promoters sequences of 5S rRNA and tRNA genes have subsequently been characterized more distinctly into **box A**, **box B** and **box C**. In 5S RNA genes box A and box C are found at +59 to +69 and +80 to +90 sequences, respectively. Likewise, in tRNA genes, box A and box B are located at +8 to +30 and +51 to +72 sequences, respectively. Box A of 5S RNA and tRNA genes contain similar conserved sequences and is recognized by the same transcription factor (TF III B).

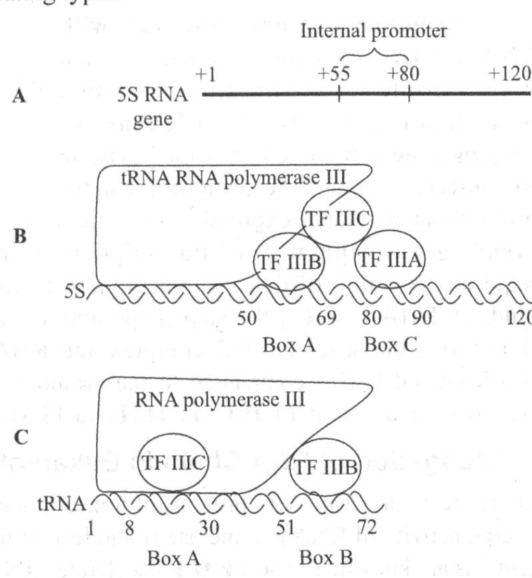

Fig. 39.11. A—Structure of internal promoter of a 5S RNA gene. B—Binding of TF III B to downstream promoters (box A), located at different positions in 5S and tRNA genes.

The eukaryotic transcription too involves the following three main steps: initiation, elongation and termination.

A. Initiation of Eukaryotic Transcription

For the eukaryotic transcription the regulatory DNA sequences (such as promoters, enhancers and silencers) for genes transcribed by each of the three RNA polymerases differ. Various transcription factors are also involved in the formation of a transcription complex which are needed for initiation of transcription. Generally, each of RNA polymerase is believed to have its own set of transcription factors, however, TF II D or a part of it (*e.g.*, TBP = TATA binding protein) is required for all the three RNA polymerases. The **transcription factors (TFs)** can be defined as proteins, which are needed for initiation of transcription, but are not part of the RNA polymerase. They help in DNA binding of a RNA polymerase to constitute the so-called **pre-initiation complex** or **transcription complex**. After the formation of this complex initiation of transcription occurs. All known transcription factors may recognize either DNA sequences, another factor or RNA polymerase.

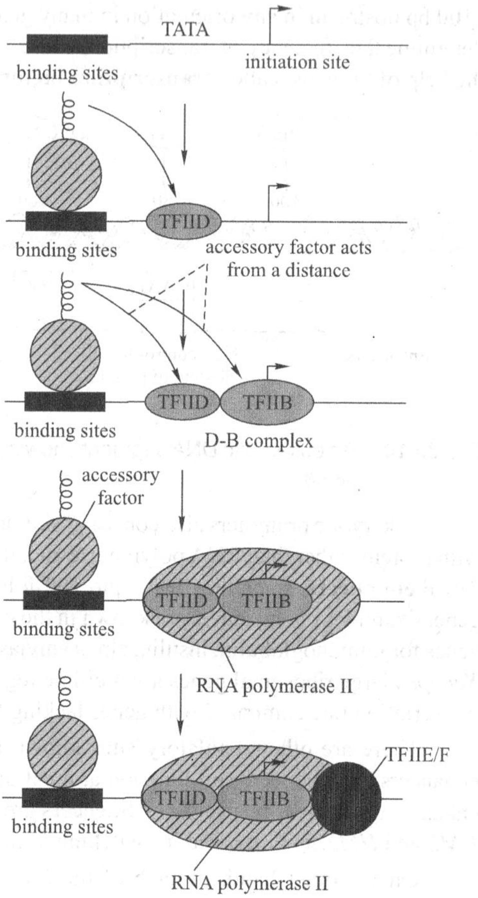

Fig. 39.12. The mode of formation of transcription complex at the promoter for RNA polymerase II.

Formation of transcriptosome with RNA pol II. A promoter sequence which is responsible for constitutive expression of common genes (also called **house keeping genes**) in all cells, is called **generic promoter**. The generic promoter cannot bring about regulated expression (*i.e.*, tissue or stimulus specific expression of genes, called **luxury genes**). Initiation of transcription on the generic promoter by RNA polymerase II requires the action of diverse transcription factors (TFs) in the following order: (*i*) **TF II D** binds at TATA box; (*ii*) the step (*i*) permits the association of **TF II A** and **TF II B**; (*iii*) TF II B forms the so-called DB complex and RNA polymerase II associates to promoter site; (*iv*) RNA pol II is accompanied to the promoter by **TF II F** to form a transcription complex; (*v*) orderly addition of **TF II E**, **TF II H** and **TF II J** helps the initiation process.

B. Elongation of RNA Chain in Eukaryotes

There are certain accessory proteins of transcription, called **elongation factors**, which enhance the overall activity of RNA polymerase II and lead to increase in the elongation rate. At least two such proteins are known: 1. The **TF II F** accelerates RNA chain growth relatively uniformly in concord with RNA polymerase II. 2. The **TF II S** (also called **S II**) helps in elongation of RNA chain by unburdening the obstruction in the path of such elongation. TF II S is known to act by first causing hydrolytic cleavage at 3′ end of RNA chain, thereby, helping in the forward movement of RNA polymerase through any block to elongation.

C. Termination of Eukaryotic Transcription

In eukaryotes, the actual termination of RNA polymerase II activity during transcription may take place through termination sites similar to those found in prokaryotes. However, the nature of individual sites is not known. Such termination sites are believed to be present away (sometimes up to one kilobase away from the site of the 3′ end of mRNA).

Chromatin Structure and Transcription

Decondensation of large chromosomal domains (coiled and supercoiled) is a prerequisite for transcripition. This process involves the following steps: 1. Some **activator proteins** bind and bring about the formation of nucleosome-free regions. These activator proteins gain access to specific DNA sequence in chromatin by the help of non-histone proteins in an unknown manner. The amino-terminal region of H3 and H4 histones may also undergo **acetylation**, which is often correlated with great amount of transcription. 2. In the so-generated nucleosome-free DNA, additional transcription factors and RNA polymerase bind. Transcription is, thus, initiated.

39.5. TYPES OF NON-GENETIC RNA AND PROCESSING

According to their specific functions during the process of protein synthesis, the following kinds of non-genetic RNA molecules have been recognized in prokaryotic and eukaryotic cells:

I. Ribosomal RNA (rRNA)

Ribosomal RNA (rRNA), stable or **insoluble RNA** constitutes the largest part (up to 80%) of the total cellular RNA. It is found primarily in the ribosomes although, since it is synthesized in the nucleus it is also detected in that organelle. It contains four major RNA bases with a slight degree of methylation and shows differences in the relative proportions of the bases between species. Its molecules appear to be single polynucleotide strands which are unbranched and flexible. At low ionic strength, rRNA behaves as a random coil, but with increasing ionic strength the molecule shows helical regions produced by base pairing between adenine and uracil and guanine and cytosine.

Types and synthesis of rRNA. The eukaryotic cells have four kinds of rRNA molecules, namely **28S rRNA** (the sedimentation constant varies between 25S and 30S depending on the species), **18S rRNA**, **5.8S** and **5S rRNA**. The 28S rRNA, 5.8S rRNA and 5S rRNA occur in 60S ribosomal subunit, while 18S rRNA occurs in 40S ribosomal subunit of 80S ribosomes of eukaryotes. The prokaryotic cells contain three kinds of rRNA molecules, namely **23S rRNA**, **16S rRNA** and **5S rRNA**. The 23S rRNA and 5S rRNA occur in 50S ribosomal subunit, while 16S rRNA occurs in 30S ribosomal subunit of 70S ribosomes of prokaryotes.

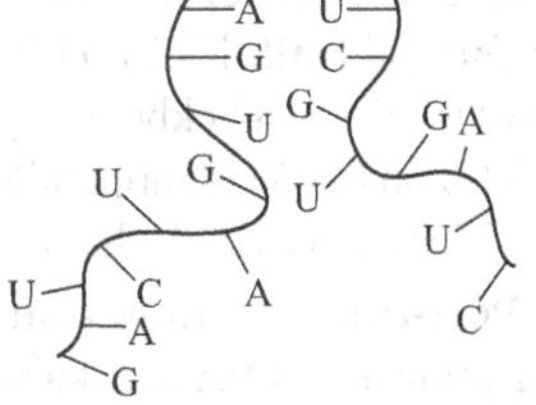

Fig. 39.13. The molecular (*i.e.*, secondary) structure of rRNA showing a helical region with complementary base pairing and a looped outer region of the helix at X (Davidson, 1972).

In bacteria, the genes for 5S, 23S and 16S rRNAs are clustered in one region to form a single operon working as a functional unit. The synthesis of rRNA molecules is initiated at promoter and completed at a terminator sequence.

In eukaryotes, rRNA genes (which belong to multigene families) occur in the region of nucleolar organizer (NO). The number of these genes may vary from 50 to 30,000 in a cell and this number may be unequally distributed on NOs, if more than such loci are present. The DNA comprising these genes is called **rDNA** (=ribosomal DNA) which is repetitive in nature. Each repeat unit has (*i*) a **coding region**, in which the genes for the 18S, 5.8S and 28S rRNA molecules exist next to each other in the order mentioned; (*ii*) a spacer region called **intergenic spacer (IGS)** and (*iii*) **internal transcribed spacers (ITS)**, one each between 18S and 5.8S genes and another between 5.8S and 28S genes. For the synthesis of above mentioned rRNAs, a transcription initiation factor (TIFI) is needed which brings RNA polymerase I to the promoter region. A primary transcript is made by all three rRNA genes, including spacer regions between the genes. The transcript is then processed into the functional rRNA molecules.

Further, the 5S RNA genes are located outside the nucleolar organizer. However, in prokaryotes and yeast, 5S RNA genes are present in close vicinity of rDNA. The 5S RNA genes are also organized in tandem repeats, each repeat consisting of a gene 120 bp long and a spacer region. The length of the complete repeat is 375 bp in *Drosophila*.

II. Messenger RNA (mRNA)

The RNA molecules which are transcribed from large number of genes of the total genome (*i.e.*, 99 per cent genes of the total genome of *E.coli*) and have base sequence complementary to DNA, carry DNA's genetic informations for the assembly of amino acids into the polypeptide chains (protein molecules), to the cytoplasmic sites of protein synthesis, the ribosomes, to which they become associated to participate in codon-anticodon interaction with tRNA, are called **informational** or **messenger RNAs (mRNA)**. The name messenger RNA has been proposed by **Jacob** and **Monod** (1961). The molecule of a mRNA is single-stranded like the rRNA molecule and it is DNA-like in its base composition so that GC contents of mRNA correspond to the GC contents of the genomes total DNA.

Messenger RNA synthesis in bacteria. Messenger RNA is **complementary** to chromosomal DNA; it forms RNA-DNA hybrids after separation of the two DNA strands. Synthesis of mRNA is accomplished with only one of the two strands of DNA, which is used as template. The enzyme RNA polymerase joins the ribonucleotides, thus, catalyzing the formation of 3′-5′-phosphodiester bonds that form the RNA backbone. In this synthesis the AU/GU ratio of RNA is similar to the AT/GC ratio of DNA. The mRNA synthesis is initiated at 5′ end and direction of growth is from the 5′ end to 3′ end. In bacteria, the RNA polymerase attaches to an initiator site of the structural gene, in the promoter and it catalyzes mRNA synthesis until termination site is reached.

In bacteria, the process of transcription of mRNA is simultaneous with translation, *i.e.*, as soon as the mRNA is being transcribed by RNA polymerases the ribosomes become attached to the mRNA to initiate protein synthesis.

Messenger RNA synthesis in eukaryotes. Transcription of eukaryotic DNA to produce mRNA begins with the synthesis of long precursor molecules by RNA polymerase II from the template strand of DNA. In an average cell nucleus, there is only one molecule of RNA polymerase II per 750 nucleosomes-worth of DNA, *i.e.*, one enzyme molecule exists per 150,000 base pairs of DNA (**Maclean** and **Hall**, 1987). This enzyme functions by catalyzing formation of $5' \rightarrow 3'$ phosphodiester bonds of the RNA "backbone" by "reading" the DNA template in the $3' \rightarrow 5'$ direction. The developing mRNA (or hn RNA) is antiparallel and its nucleotides are complementary to those of the DNA template strand. Messenger RNA chain growth is rapid—from 15 to 100 nucleotides per second *in vitro*.

Post-transcriptional modification or processing of mRNA. The immediate product of transcription of mRNA in eukaryotes is a molecule of many more ribonucleotides than that comprising the ultimate functional mRNA. This primary transcript may range from 500 to 50,000 nucleotides; it remains confined to the nucleus and is called **heterogeneous nuclear RNA (hnRNA)**. The fate of this hnRNA may be one of the followings: 1. RNA transcripts of some genes do not seem to give rise any cytoplasmic mRNA, but get degraded within the nucleus. 2. For each gene, only a small proportion (25 per cent) of RNA transcript takes part in RNA processing leading to formation of mRNA, the remaining 75 per cent undergoing degradation in the nucleus. Thus, only 5 per cent hnRNA (by mass) enters the cytoplasm. The hnRNA molecules which are destined to produce functional mRNA, undergo RNA processing which includes the following steps:

1. Addition of a cap of 7-MeG or m7G. During capping process, a cap of a methylated guanosine, called 7-methylguanosine (7-MeG or m7G), is added to 5′ end of primary transcript (*i.e.*, hnRNA) in a rare 5′-5′ linkage. Sometimes, this cap also includes methylation of additional sugars of both the 5′ nucleotides: (7-MeG)-5′PPP-5′ (G or A, with possibly methylated ribose)-3′-P—in which P and PPP refer to mono- and triphosphate groups, respectively.

Capping occurs shortly after initiation of synthesis of the mRNA, possibly before RNA polymerase II leaves the initiation site, and precedes all excision and splicing events. The biological significance of capping is that the cap may protect the mRNA from degradation by nucleases and may provide a feature for recognition by the protein-synthesizing machinery (*i.e.*, cap helps in recognition of ribosomes and thereby facilitates translation of mRNA). The 7-MeG caps are absent in mRNAs of histone proteins.

2. Addition of tail of poly-A. The 3′ end of mRNA is generated in two steps (Fig. 39.14): 1. Endonuclease enzyme cuts the primary transcript at an appropriate location. 2. Poly (A) is added to the newly generated end by an enzyme, called **poly (A) polymerase**, utilizing ATP as a substrate. This step is called **polyadenylation**. Studies have shown that ordinarily ~30 per cent of hnRNA and ~70 per cent of mRNA are poly-adenylated. In a region 11 to 30 nucleotides upstream of the site of poly(A) addition, there is a sequence AAUAAAA (in all higher eukaryotes except yeast) which perhaps provides a signal for nuclease cleavage.

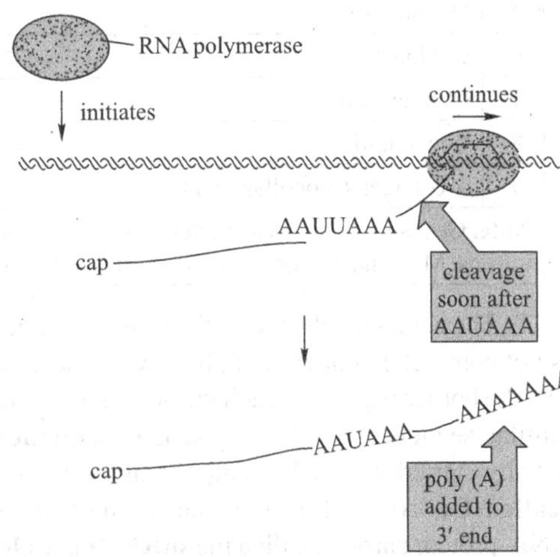

Fig. 39.14. Cleavage at AAUAAA sequence and poly adenylation at the cleaved 3′ end.

In mRNAs of most histone proteins, no polyadenylation occurs at 3′ ends, so that 3′ ends are processed without addition of poly (A). The **U7 snRNA** (56 bases long) is involved in this processing through extensive complementary base pairing with histone hn RNAs.

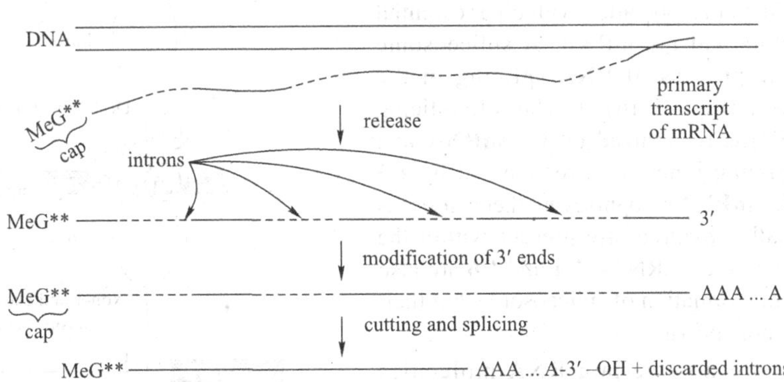

Fig. 39.15. Summary of various steps of production of mRNA. The primary transcript is capped before it is released. Then, its 3′-OH end is modified, and finally introns are excised and exons are rejoined to form functional mRNA. (MeG = 7-methyl guanosine; * = nucleotide whose ribose is methylated) (after Freifelder, 1985).

3. RNA splicing. This is the controlled excision of large **intervening sequences** or **introns** from the transcript and rejoining of the remaining fragments, called **coding sequences** or **exons**, together to produce the finished mRNA. The number of introns per gene varies greatly (Table 39.4) and for a given protein is not the same in all organisms.

Table 39.4 Translated eukaryotic genes in which introns have been demonstrated (Source: Freifelder, 1985).

	Gene	Number of introns
1.	α- Globin	2
2.	Immunoglobulin L chain	2
3.	Immunoglobulin H chain	4
4.	Yeast mitochondria cytochrome b	6
5.	Ovomucoid	6
6.	Ovalbumin	7
7.	Ovotransferrin	16
8.	Conalbumin	17
9.	α-collagen (procollagen α)	52

Note. Genes for histones and interferon are among the few protein-coding genes in eukaryotes that lack introns. Many human genes are comprised of more than 95% intron sequences (**Malacinski** 2003).

The actual mechanism of cutting and splicing is not completely understood; however, it is known that the border regions of each intron usually contain similar sequences (called **consensus sequences**), usually started with a GU and ending with an AG (called **GU-AG rule**); and that a small nuclear RNA-protein particle, called the **snRNP particle**, is involved in RNA splicing. Small nuclear RNAs (sn RNAs) present in the sn RNP particles are designed as U1, U2, U4, U5 and U6. The sn RNPs form a macromolecular complex (called **spliceosome**; 40 to 60 nm in size) in association with other essential protein factors and pre-mRNA. A spliceosome performs the process of RNA splicing in the following way (Fig. 39.16): 1. The left splicing junction (GU) is recognized by **U1 snRNA** and the right splicing junction is recognized by **U5 snRNA**. 2. **U2 snRNA** recognizes another consensus sequence, called **branch site** present within the intron. 3. Two other snRNAs, **U4** and **U6** are also involved in the formation of spliceosome, but their exact role is not known.

There are also certain RNA molecules which act as enzymes, called **ribozymes** which were discovered by **Thomas Cech** (1982) while working on RNA splicing in *Tetrahymena* (a ciliated protozoan). These RNA molecules (= ribozymes)

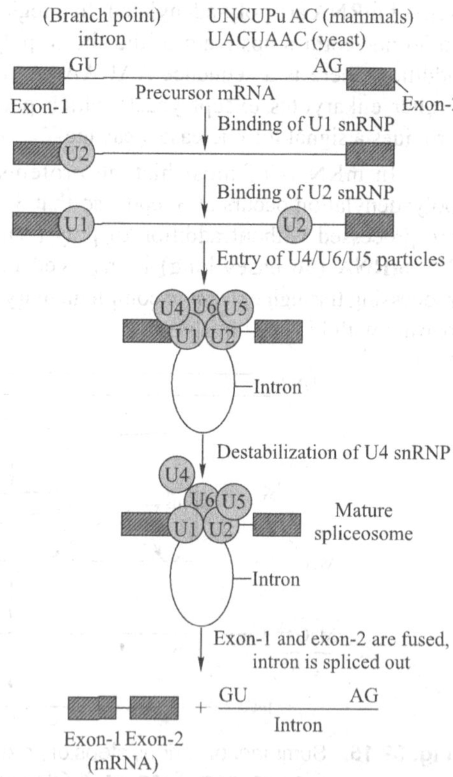

Fig. 39.16. Formation of a spliceosome and stages involved in RNA splicing.

act only themselves and not on other molecules. They cut, splice and assemble (= do processing of) precursor hn RNA into mRNA, or precursor rRNA into mature rRNA.

Heterogeneity and types of mRNA. When the total mRNA population of an organism is considered, it is found to be heterogeneous in size, showing a wide range of S values of 6 to 30. This property of mRNA reflects the fact that the size or length of the mRNA molecule is directly related with the size of the codons for different protein molecules, the sizes of which may be quite variable. According to the size, the following two types of mRNA molecules can be recognized.

***(a)* Monocistronic mRNA.** Mostly the mRNA carries the codons of single cistron (*i.e.*, codes for one complete protein molecule) of the DNA. Such mRNA molecule is called monocistronic mRNA. For example, for the synthesis of a polypeptide chain of 300 to 500 amino acid residues, a monocistronic mRNA of *E.coli* contains 900 to 1500 nucleotides in its molecule.

***(b)* Polygenic or polycistronic mRNA.** Sometimes a mRNA molecule carries the codes from several adjacent DNA cistrons and become much longer in size. This type of mRNA is called **polygenic** or **polycistronic mRNA**. For example, for the metabolism of the histidine protein the cell synthesizes about 10 specific enzymatic proteins and a mRNA in this case may carry codons for all the 10 enzymes.

Life-Span of mRNA

In most prokaryotic and eukaryotic cells, mRNA has short life time. For example, the average life of mRNA of *E.coli* is about 2 minutes, because it is attacked by the cytoplasmic ribonuclease enzyme. So that, at most times, mRNA makes up only 5% of the total cellular RNA. Likewise, in most eukaryotes the average life span of mRNA is one to four hours. However, in both bacteria and eukaryotes mRNAs are known that are apparently resistant to nucleases and survive for long period of time. For example, mRNA with life time of six hours has been detected in the bacterium *Bacillus cerus* at a time when the cells are induced to become long lived spores. Likewise, in differentiating eukaryotic cells mRNAs with a life time of days have been detected. For example, in the immature red blood cells (reticulocytes) of the mammals the mRNA is synthesised originally by the nucleus in early stages and expelled to the cytoplasm. In later stages, the nuclei of maturing reticulocytes degenerate but the mRNA exists up to 2 days for prolong utilization in the synthesis of globin protein of haemoglobin. Further, in extreme cases, such as in the state of dormancy adopted by many animal eggs and plant seeds, mRNA is maintained in a stable form for months or even years.

Informosomes. In the eukaryotic cells, the stability to the mRNA is provided by certain proteins. Several investigators, *e.g.*, **Spirin**, **Beltisina** and **Lerman** (1965), **Perry** and **Kelley** (1968) and **Henshaw** (1968) have reported that in certain eukaryotic cells the mRNA does not enter in the cytoplasm as a naked RNA strand but often remains ensheathed by certain proteins. **Spirin** has coined a new term **informosome** to this mRNA and protein complex. The informosome is used by the cell when there is a delay in the translation. For instance, in the embryo the genetic expression is manifested late during organogenesis. In such cases, mRNA occurs in the form of informosomes. The proteins of informosomes protect the mRNA from the degrading action of the enzyme, ribonuclease. These proteins may also control the synthesis at the level of the translation and, thus, may regulate or modulate the protein synthesis process.

III. Transfer RNA (tRNA)

The RNA which possesses the capacity to combine specifically with only one amino acid in a reaction mediated by a set of amino acid-specific enzymes, called **aminoacyl-tRNA synthetases**; transfers that amino acid from the "amino acid pool" to the site of protein synthesis and recognises the codons of the mRNA is known as the **soluble RNA (sRNA)** or **transfer RNA (tRNA)**. Thus, tRNA molecule has to perform several highly complex functions during protein synthesis—it interacts with a specific synthetase enzyme, possesses a site for binding an amino acid, possesses a second site for interacting with a ribosome, and contains an anticodon that must be exposed to the codons of mRNA.

Structure of tRNA. Robert Holley (1965) and his colleagues reported the complete nucleotide sequence of alanine tRNA of yeast (**Holley** received the Nobel Prize in 1968 for his work along with **Khorana** and **Nirenberg**). Nucleotide sequences are now known for more than 100 different "species" of tRNA.

Transfer RNA has several unique characteristics: 1. It is a relatively small molecule of 75 to 90 ribonucleotides and is, thus, smaller than either mRNA or any of the rRNAs, and has a sedimentation coefficient of 4S.

2. The ratios of A:U and G:C are near unity which suggests the formation of DNA-like double helical segments (secondary structure). In these double helical segments, G:C base pairs are more common than A:U as suggested by the ratio AU:GC = 0.7.

3. All tRNA molecules have a tertiary structure, the details for which are now known and Mg^{2+} ion concentration is important for its stabilization.

4. A number of "unusual" nucleotides are found in tRNA (*e.g.*, pseudouridine ψ or psi), inosine (I), dihydroxyuridine (DHU), etc.). Many of "unusual" nucleotides are methylated derivatives of common ones (*e.g.*, 1-methylguanylic acid, 1-methyladenylic acid, ribothymidylic acid and 5-methylcytosine).

The significance of these unusual bases of tRNA was understood well by molecular biologists during the construction of two-dimensional model from the primary-sequences of nucleotides of known tRNA. Thus, it was realized that most bases of tRNA pair according to Watson-Crick's pairing lule, but unusual bases fail to do so because they carry substitutions or alterations in those positions that usually participate in hydrogen bonding. Consequently, the presence of these bases forces the model builder to construct several non-base-paired loops in the tRNA molecule. By working on these lines, **R. Holley** (1965) first of all proposed a **clover leaf model** for yeast $tRNA^{ala}$. The clover leaf model of tRNA because accommodated several of the known functions of tRNA, so, it gained general acceptance. A typical clover-leaf model tRNA (Fig. 39.17) depicts the following structural peculiarities:

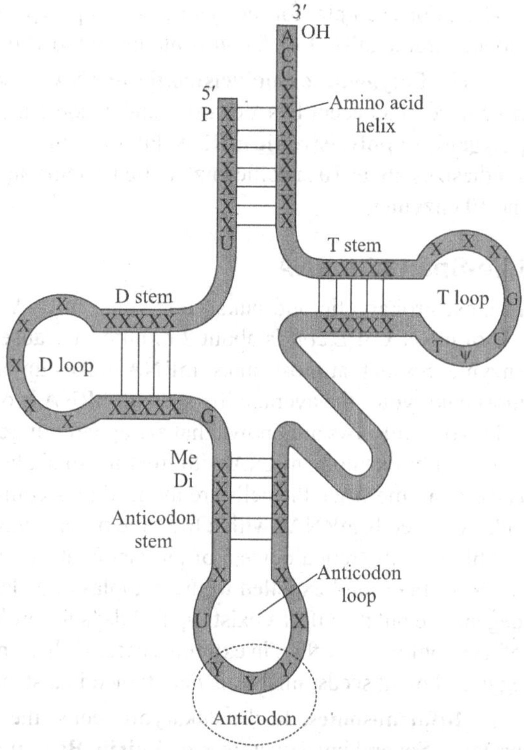

Fig. 39.17. Generalized two-dimensional clover-leaf model of tRNA, based on analyses of several yeast tRNA molecules by various investigators. Note the common 3′ terminal-CCA, the TCG in the T-loop, the anticodon-U (here YYYU) in the anticodon loop, the DiMeG between the anticodon loop and D loop and the U as the first unpaired base on the 5′ strand. As diagrammed here, the anticodon is read from right to left 3′ → 5′ (A = adenosine, C = cytidine, G = guanosine, T = ribothymidine, U = uridine, ψ = pseudouridine, DiMeG = dimethylguanosine, Y = any base of the anticodon (after Burns and Bottino, 1989).

(*i*) All tRNA molecules have guanine

residue G at the 5′ terminal end and unpaired (single stranded) C-C-A sequence at the 3′ end. This is called **amino acid attachment site**, because the amino acid becomes covalently attached to adenylic acid or A of CCA sequence during polypeptide synthesis.

(*ii*) The **amino acid stem** or helix consists of seven paired bases.

(*iii*) The **T-stem** is composed of five paired bases—the last (*i.e.*, nearest the **T-loop** or **TψC loop** is C-G. T-loop contains seven unpaired bases and is involved in the binding of tRNA molecules to the ribosomes.

(*iv*) The **anticodon stem** includes five paired bases. The **anticodon loop** consists of seven unpaired bases, the third, the fourth and the fifth of which (from the 3′ end of the molecule) constitute the **anticodon**. The anticodon permits temporary complementary pairing with three bases (triple codon) on mRNA.

(*v*) The base on the 3′ side of anticodon is a purine.

(*vi*) Immediately adjacent to the 5′ side of the anticodon, uracil and another pyrimidine occurs.

(*vii*) A purine, often dimethyl guanylic acid, is located in the "corner" between the anticodon stem and the D stem.

(*viii*) The **D-stem** is composed of three or four base pairs (depending on the "species" of tRNA). The **DHU-loop** or **D-loop** is also variable in size containing 8 to 12 unpaired bases. The D-loop help in binding of amino-acyl synthetase.

(*ix*) The extra arm is variable in nucleotide composition and is lacking entirely in some tRNA.

Three-dimensional structure of tRNA. In order to understand the structure-function relationship of tRNA, its three dimensional structure (TDS) was worked out by the help of X-ray crystallography study. **A. Klug**, the Nobel Laureate of 1982, has contributed much to the TDS of tRNAs. **S.H. Kim** (1973) proposed a most acceptable TDS model of tRNA (*i.e.*, phenyl-alanine tRNA of yeast cells). According to **Kim**, TDS of tRNA takes the shape of letter L with a thickness of 20A°. Each arm of the L doubled over by bonds holding complementary base together. Such an L-shape can also easily derived from two dimensional clover-leaf model.

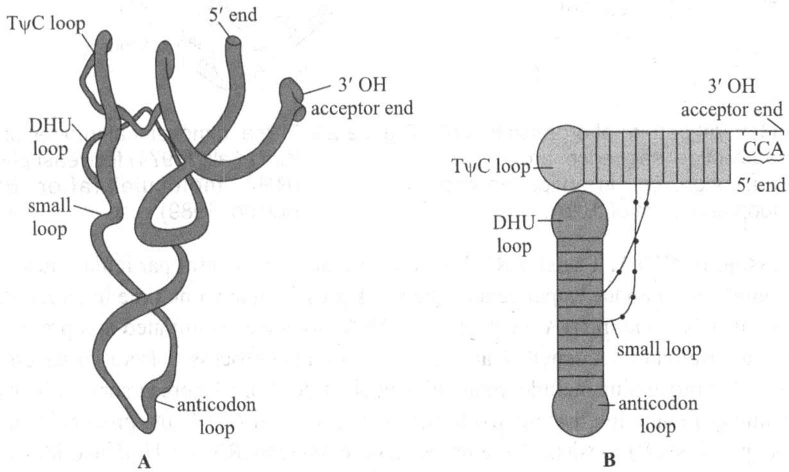

Fig. 39.18. Three dimensional structure of yeast phenylalanine tRNA. A—Actual appearance of the folding of the molecule. The polynucleotide chain is represented as a continuous coiled tube; B—The way that the clover leaf representation must be transformed in order to show the physical connections between various parts of the molecule. Two double-stranded helical regions are seen, each oriented at right angles to the other produces a L-shaped structure.

Extended anticodon hypothesis. Recently it is reported that the performance of anticodon, when isolated from tRNA, is weak and inaccurate. However, the performance of this anticodon triplet is enhanced if a matching sequence is present on the anticodon loop and on the stem on either sidee of anticodon triplet. As a result, an **extended anticodon hypothesis** has been proposed which suggests that the structures of anticodon loop and that of the proximal anticodon stem are related to the sequence of anticodon (**Michael Yarus**, 1982). Thus, the anticodon is extended into the nearby sequence and consists of in all 12 nucleotides arranged in the following way: (*i*) two nucleotides (*i.e.*, CU, ψU or UU) at the 5′ side of anticodon loop; (*ii*) three nucleotides of anticodon (3′ nucleotide of anticodon being very important and termed **cardinal nucleotide**); (*iii*) two nucleotides at the 3′ side of anticodon loop; and (*iv*) five pairs of nucleotides in the anticodon stem which can be conveniently written by giving only the bases on 3′ side of the stem (Fig. 39.19).

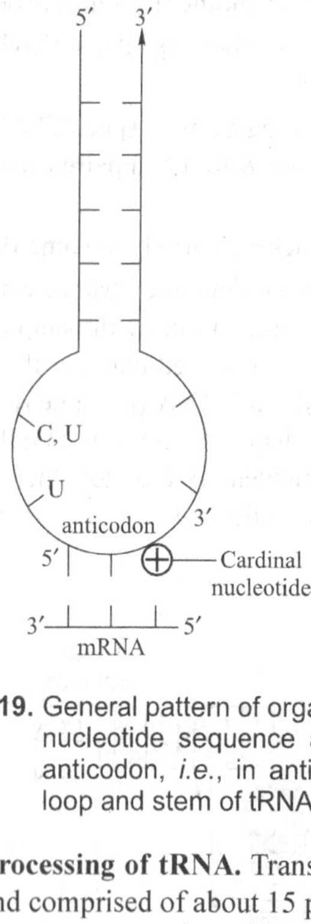

Fig. 39.19. General pattern of organization of nucleotide sequence around the anticodon, *i.e.*, in anticodon and loop and stem of tRNA.

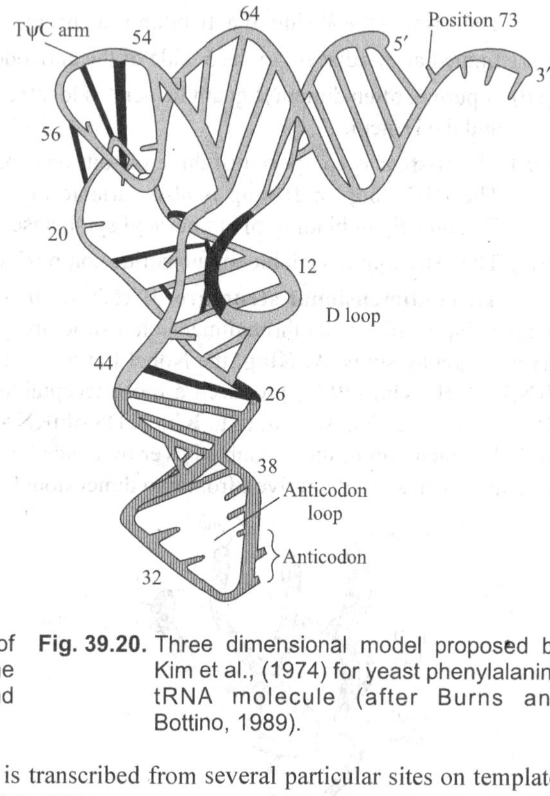

Fig. 39.20. Three dimensional model proposed by Kim et al., (1974) for yeast phenylalanine tRNA molecule (after Burns and Bottino, 1989).

Processing of tRNA. Transfer RNA is transcribed from several particular sites on template DNA and comprised of about 15 per cent of the RNA present at any one time in an *E.coli* cell. Like the synthesis of rRNA and mRNA, synthesis of tRNA molecules is initiated at a promoter site and completed at a terminator sequence. Transfer RNAs are also processed from larger precursors. In prokaryotes, tRNA processing includes removal of nucleotides from the precursor or primary transcript and modification of some internal nucleotides. For example, in *E.coli*, the precursor for the amino acid tyrosine (pre $tRNA^{tyr}$) consists of 128 bases; in processing to $tRNA^{tyr}$ 41 nucleotides are removed from the 5′ end and two are removed from the 3′ end (Fig. 39.21).

In addition to processing, methylation and inclusion of other "unusual" intercalary bases takes place after transcription, as does the addition of the 3′ terminal-C-C-A. The final functional tRNA molecule has 85 bases. Three-dimensional shape is achieved through hydrogen bonding. The processing of eukaryotic tRNAs resembles with that of prokaryotes. However, in them it is far more-complex.

In yeast, the process of tRNA splicing involves the following two steps: 1. Cutting reaction in which phosphodiester bond cleavage occurs by an endonuclease and this step does not require ATP. 2. Ligation reaction which requires ATP and involves bond formation with the help of **RNA ligase**.

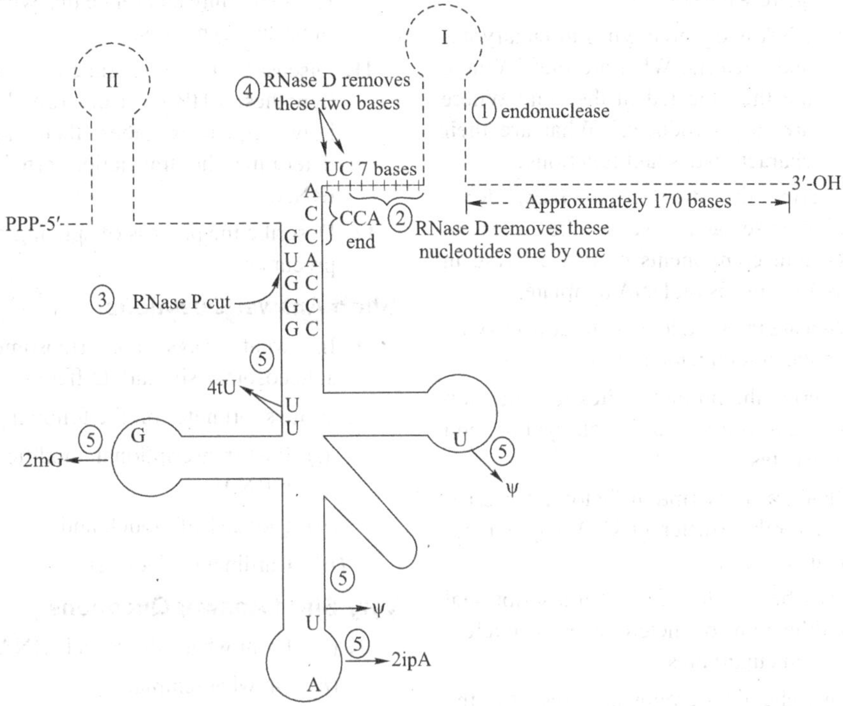

Fig. 39.21. Different stages in processing of the *E.coli* $tRNA^{Tyr}$ gene transcript. The five stages have been given Arabic numbers. Step 3 generates the 5′-P end. Step 4 generates the 3′-OH end (the CCA end). In step 5 six bases, all in or near the loops of tRNA molecule, are modified to form pseudouridine (ψ, psi), 2-isopentenyladenosine (2 ip A), 2-O-methylguanosine (2 mG), and 4-thiouridine (4tU). The continuous sequence that forms the final tRNA molecule is given in black (after Freifelder, 1985).

Genes for tRNA. There are, probably, at least 30 to 40 different tRNA genes and tRNA molecules in *E.coli*. Higher organisms are found to contain 60 tRNA molecules and 60 tRNA genes. Since the cell uses only 20 amino acids in protein synthesis (and probably only 20 synthetase enzymes), it follows that several tRNA will often have an affinity for the same amino acid. For example, *E.coli* cells contain five species of tRNA for leucine amino acid.

All the tRNA genes constitute far less than 1% of total genome in both *E.coli* and eukaryotic cells, yet some 10 to 15% of each cell's RNA may be in the form of tRNA. This discrepancy between the number of tRNA genes and gene transcripts occurs because of the following facts—(1) The tRNA molecules are relatively stable compared with many kinds of RNA. (2) The tRNA molecules are transcribed continuously and more quickly by tRNA genes than other RNAs because they are needed in plentiful amounts.

QUESTIONS

Long Answer Questions

1. (*a*) What genetic attributes does RNA share with DNA?
   (*b*) RNA is of three types in eukaryotes and bacteria. What are they? Where are they located in the cell? Where are they produced? What are their characteristics and functions?
2. Describe the structure of *E.coli* RNA polymerase and discuss the roles of different components of this enzyme in RNA synthesis on DNA template.
3. What are the functions of the core enzyme and the holoenzyme *in vivo*?
4. Describe the promoter sites for initiation of transcription in prokaryotes and eukaryotes.
5. What are transcription factors? Describe them for three different RNA polymerases in eukaryotes.
6. Describe the post-transcriptional modification of heterogeneous nuclear RNA in eukaryotes.
7. What chemical groups are present at the origin and terminus of a molecule of mRNA that has just been synthesized?
8. (*a*) What is mRNA ?
   (*b*) How does mRNA sometimes differ from a primary transcript?
   (*c*) Define coding strand and antisense strand.
   (*d*) Define cistron and polycistronic mRNA.
   (*e*) What parts of an mRNA molecule are not translated? What is meant by the terms "upstream" and "downstream"? What is Pribnow box? Describe its evolutionary and biochemical significance.
9. Describe the role of nucleolus in the processing of precursor of rRNAs and biogenesis of ribosomes.
10. Describe the primary, secondary and three dimensional structure (TDS) of transfer RNA structure and relate this with its role in protein synthesis.
11. Discuss the extended anticodon hypothesis of tRNA structure and desclibe how sequences other than anticodon determine the translation efficiency of tRNA.
12. Describe the process of splicing of tRNA precursor.

Short Answer Questions

1. In what ways are ribosomes and spliceosomes similar? Different?
2. Write short notes on the following:
   (*i*) Post-transcriptional modification of mRNA;
   (*ii*) Clover-leaf model; and
   (*iii*) Cardinal nucleotides.

Very Short Answer Questions

1. (*a*) From what substrates is RNA made?
   (*b*) On what template?
   (*c*) By which enzyme?
   (*d*) Is a primer required?
2. What is a transcription unit? Is it the same thing as a gene?
3. In what way, relating to polycistronic mRNA, do eukaryotic and prokaryotic protein synthesis differ?
4. Why do you think that most promoter regions are A-T rich?
5. (*a*) What are intervening sequences or introns?
   (*b*) What is mRNA splicing?
6. Many eukaryotic genes contain noncoding introns that separate the coding sequences or exons of these genes. At what stage during the expression of these mosaic genes are the noncoding intron sequences removed?

Multiple Choice Questions

Choose the correct answer from the four alternatives given.

1. Transcription is the transfer of genetic information from
   (*a*) chromosome to cytoplasm
   (*b*) tRNA to mRNA
   (*c*) DNA to mRNA
   (*d*) mRNA to rRNA
2. Nongenetic RNA is of
   (*a*) one type
   (*b*) two types
   (*c*) three types
   (*d*) non functional type
3. Regarding transcription which one of these is false?
   (*a*) uses RNA polymerase
   (*b*) one of the mother strands of DNA is the template
   (*c*) is bidirectional
   (*d*) reverse transcription used RNA dependent DNA polymerase
4. The enzyme required for transcription is
   (*a*) DNA polymerase
   (*b*) RNA polymerase
   (*c*) endonuclease
   (*d*) all the above
5. Pribnow box contribute to
   (*a*) protein synthesis
   (*b*) ATP synthesis
   (*c*) RNA synthesis
   (*d*) none of the above
6. The process involved in the RNA formation on DNA template is
   (*a*) translation (*b*) transduction
   (*c*) transcription (d) transformation
7. Transcription takes place in
   (*a*) matrix (*b*) nucleus
   (*c*) cytosol (*d*) cytoplasm
8. Essential components of eukaryotic cistron are
   (*a*) intron (*b*) exons
   (*c*) operons
   (*d*) operator and regulator genes
9. After initiation of transcription with core enzyme RNA polymerase the sigma factor is
   (*a*) functionless
   (*b*) released to take part again
   (*c*) used during closing of chain
   (*d*) retained and it performs special function
10. During transcription, the DNA site at which RNA polymerase binds is called
    (*a*) promoter (*b*) regulator
    (*c*) receptor (*d*) enhancer
11. Nuclear DNA sends information for protein synthesis through
    (*a*) tRNA (*b*) mRNA
    (*c*) rRNA (*d*) all the above
12. The RNA which carries the genetic message from the nucleus to the ribosome
    (*a*) hn mRNA (*b*) mRNA
    (*c*) tRNA (*d*) rRNA
13. The type of RNA specifically responsible for directing the proper sequence of amino acids in protein synthesis is
    (*a*) chromosomal RNA
    (*b*) rRNA
    (*c*) tRNA
    (*d*) mRNA
14. Maximum formation of mRNA occurs in
    (*a*) cytoplasm (*b*) nucleolus
    (*c*) ribosome (*d*) nucleoplasm
15. The ribozyme is a RNA
    (*a*) without sugar
    (*b*) with extra phosphate
    (*c*) with enzymatic property
    (*d*) without phosphate
16. mRNA is complimentary copy of
    (*a*) tRNA
    (*b*) rRNA
    (*c*) ribosomal DNA
    (*d*) a single strand of DNA
17. The cap structure of mRNA is of methyl GTP and tail of
    (*a*) methionine
    (*b*) formylmethionine
    (*c*) poly A
    (*d*) UAG codon

18. Synthesis of mRNA molecule is terminated by _____ factor

(*a*) alpha (*b*) beta
(*c*) sigma (*d*) rho

19. Which RNA is having least age?

(*a*) mRNA (*b*) tRNA
(*c*) rRNA (*d*) none of these

20. The function of nucleolus is the synthesis of

(*a*) DNA (*b*) mRNA
(*c*) rRNA (*d*) tRNA

21. Functional unit in synthesis of protein is

(*a*) peroxisome (*b*) dictyosome
(*c*) lysosome (*d*) polysome

22. An anticodon of tRNA represents

(*a*) Wobble hypothesis
(*b*) template hypothesis
(*c*) gene flow hypothesis
(*d*) Richmond and Long effect

23. Which of the following RNAs picks up specific amino acid from amino acid pool in the cytoplasm to ribosome during protein synthesis?

(*a*) tRNA (*b*) mRNA
(*c*) rRNA (*d*) all of them

24. Which site of the tRNA molecule hydrogen bonds to a mRNA molecule?

(*a*) codon (*b*) anticodon
(*c*) 5′ end of tRNA (*d*) 3′ end of tRNA

25. Clover leaf model belongs to

(*a*) tRNA (*b*) DNA
(*c*) centriole (*d*) flagella

26. An anticodon is complementary to the nucleotide triplet in

(*a*) tRNA (*b*) rRNA
(*c*) mRNA (*d*) cDNA

27. Which of the character is not applicable to tRNA?

(*a*) it is the smallest of the RNAs
(*b*) it acts as an adoptor for amino acids
(*c*) it has clover leaf structure
(*d*) it is the largest of the RNAs
(*e*) It bears anticodon

28. According to Wobble hypothesis

(*a*) first base is unstable
(*b*) second base is unstable
(*c*) third base is unstable
(*d*) the process of polypeptide chain elongation has been established

29. The codon for anticodon 3′ UUA 5′ is

(*a*) 5′ AAU 3′ (*b*) 3′ AAU 5′
(*c*) 5′ AAT 3′ (*d*) 3′ AAC 5′

30. tRNA recognizes aminoacyl synthetase enzyme by

(*a*) anticodon (*b*) DHU loop
(*c*) T Ψ C loop (*d*) AA-site

31. Shape of 3-D view of tRNA is

(*a*) Z-shaped (*b*) X-shaped
(*c*) Y-shaped (*d*) L-shaped

32. Amino acid binding site in tRNA is

(*a*) 5′ end (*b*) anticodon loop
(*c*) CCA 3′ end (*d*) DHU loop

33. Genetic information for the synthesis of ribosomal RNA is coded in

(*a*) DNA present in nucleus
(*b*) nucleolar associated chromatin
(*c*) granular zone of nucleolus
(*d*) amorphous zone of nucleolus

ANSWERS

Very Short Answer Questions

1. (*a*) Ribonucleoside 5′ triphosphates.
(*b*) Double-stranded DNA.
(*c*) RNA polymerase.
(*d*) No.

2. A transcription unit is a section of DNA extending from a promoter to an RNA polymerase termination site. It is usually not a gene, but typically includes many genes.

3. In eukaryotes all mRNA is monocistronic.

4. A double helix of DNA must unwind in order for transcription to occur. A-T pairs, because they have only two H-bonds, are more easily disrupted than G-C pairs.
5. (*a*) Untranslated sequences that interrupt the coding sequence of a transcript and that are removed before translation begins.

   (*b*) Removal of introns.
6. The entire nucleotide-pair sequences-including the introns-of the genes are transcribed by RNA polymerase to produce primary transcript that still contain the intron sequences. The intron sequences are then "spliced out" off the primary transcripts to produce the mature, functional RNA molecules. In the case of protein encoding nuclear genes of higher eukaryotes, the introns are "spliced out" by complex ribosome-like macro-molecular structures called **spliceosomes**.

Multiple Choice Questions

1. (*c*)	**2.** (*c*)	**3.** (*c*)	**4.** (*b*)	**5.** (*c*)	**6.** (*c*)	**7.** (*b*)
8. (*d*)	**9.** (*b*)	**10.** (*a*)	**11.** (*b*)	**12.** (*b*)	**13.** (*d*)	**14.** (*d*)
15. (*c*)	**16.** (*d*)	**17.** (*c*)	**18.** (*d*)	**19.** (*a*)	**20.** (*c*)	**21.** (*d*)
22. (*a*)	**23.** (*a*)	**24.** (*b*)	**25.** (*a*)	**26.** (*c*)	**27.** (*d*)	**28.** (*c*)
29. (*a*)	**30.** (*b*)	**31.** (*d*)	**32.** (*c*)	**33.** (*b*)		

40 Protein Synthesis

DNA, with its correct mechanism of replication, serves to carry genetic information from cell to cell and from generation to generation. This information is translated into proteins that determine the phenotype. Virtually, all the phenotypes examined so far are the result of biochemical reactions that occur in the cell. All of these reactions require enzymes and enzymes are proteins. In fact, more than 2000 types of enzymes have been identified in the living organisms. Each enzyme is a unique molecule catalyzing a specific chemical reaction. Other phenotypes are due primarily to the kinds and amounts of non-enzymatic proteins (including the structural proteins) present, for example, haemoglobin, myoglobin, gamma-globulin (*e.g.*, immunoglobulins or Igs), insulin, cytochrome C, fibroin (silk protein), or collagen. Proteins are composed of one or more, long linear polymers of amino acid residues (polypeptide chains) that are synthesized almost exclusively in the cytoplasm. The topic of this chapter is how the information present in the sequence of bases (= triplet codons) of the mRNA is translated into a sequence of amino acids in proteins. However, at the outset, one point should be clear that the genetic information of mRNA is undircctionally read beginning at the 5′-hydroxyl end by one or more ribosomes (polysomes) and 3′-end signify the C-terminal amino acid. The 5′-end of the mRNA correspond to the N-tenninal amino acid and ending at 3′-end in the completed protein.

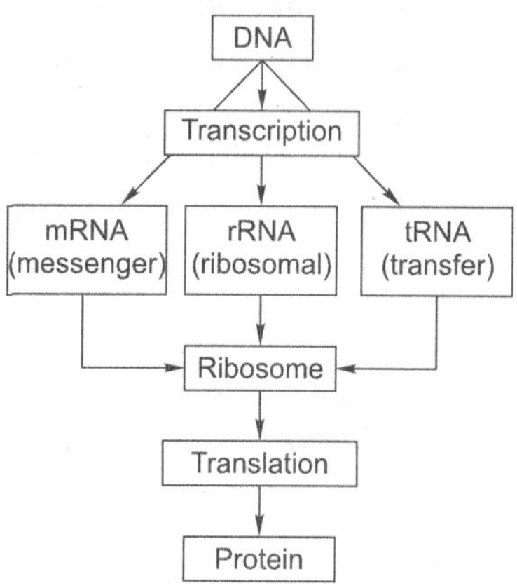

Fig. 40.1. Central dogma.

40.1. CENTRAL DOGMA AND CENTRAL DOGMA REVERSE

The process of synthesis of protein involves one of the central dogma of molecular biology (Fig. 40.1), which postulates that genetic information flows from nucleic acids to protein. (It was first forwarded by **Crick** in 1958). The first step of this central dogma is known as transcription and does not involve a change of code since DNA and mRNA are complementary. The second step involves a change of code from nucleotide sequences to amino acid sequences and is called translation. It can be illustrated as follows:

$$\text{DNA} \xrightarrow{\text{replication}} \text{DNA} \xrightarrow{\text{transcription}} \text{RNA} \xrightarrow{\text{translation}} \text{Proteins}$$

Thus, according to this central dogma, the flow of information is one way, *i.e.*, from DNA the information is transferred to RNA (mRNA) and from RNA to proteins. In 1968, **Barry Commoner** suggested a circular flow of information, *i.e.*, DNA transcribes RNA, RNA translates into proteins,

proteins synthesize RNA and RNA synthesizes DNA, as has been illustrated in following figure:

$$\begin{array}{ccc} \text{DNA} & \rightarrow & \text{RNA} \\ \uparrow & & \downarrow \\ \text{RNA} & \leftarrow & \text{Proteins} \end{array}$$

Later on, **Temin** (1970) reported the existence of an enzyme "**RNA-dependent DNA polymerase**" (**reverse transcriptase**) which could synthesize DNA from a single-stranded RNA template. **D. Baltimore** (1970) also reported the activity of this enzyme in certain tumour viruses. This exciting finding in molecular biology gave rise to the concept of "**central dogma reverse**" or **teminism**, suggesting that the sequence of information flow is not necessarily from DNA to RNA to protein but can also take place from RNA to DNA. The central dogma reverse can be illustrated as follows:

$$\text{DNA} \underset{\text{reverse transcription}}{\overset{\text{transcription}}{\rightleftarrows}} \text{RNA} \xrightarrow{\text{translation}} \text{Proteins}$$

40.2. MINIMUM NECESSARY MATERIALS FOR TRANSLATION

Success in polypeptide synthesis in *in vitro* cell free systems shows that the minimum necessary materials for translation are the following:

1. Amino acids (*i.e.*, 20 amino acids forming the pool of amino acids in the cytoplasm).
2. Ribosomes (each of which comprises two subunits which exist as separate subunits prior to the translation of mRNA and contain two tRNA binding sites: the **P site** or **peptidyl site** and an **A** or **aminoacyl site**. One more site called **E** or **exit site** has been recognized in the ribosomes by some workers.
3. mRNA
4. tRNA of several kinds
5. Enzymes
   (*a*) Amino-acid activating system (*e.g.*, aminoacyl-tRNA-synthetase).
   (*b*) Peptide polymerase system
6. Adenosine triphosphate (ATP) as an energy source
7. Guanosine triphosphate (GTP) for synthesis of peptide bonds
8. Soluble protein for initiation and transfer factors
9. Various inorganic cations (*e.g.*, K^+, NH^+ NH_4^+, Mg^{2+}).

Because, the proteins are the polymers of amino acids therefore, protein synthetic process requires the amino acids as the raw material. All the naturally occurring proteins of living organisms fundamentally are the polymers of about 20 amino acids. These amino acids occur in the matrix forming an "**amino acid pool**" and are readily available for the process of protein synthesis.

All amino acids present in amino acid pool are called α-*amino acids* because one carbon, the α-carbon has four specific groups attached to it: an amino group, a carboxyl (acidic) group, a hydrogen and one of the twenty different R groups (side chains), imparting the specific properties of that amino acid. (Technically, proline is called an **imino acid** because of its structure). Having these four groups attached to C imparts a property known as **chirality** on the amino acid: such as left and right-handed gloves, the mirror images cannot be superimposed. (*Chirality* means handedness or the property by which a form is distinguishable from its mirror image). Because of optical properties, the two forms of each amino acids are referred to as D and L, in which D comes from dextrorotatory (right turning) and L comes from levorotatory (left turning). All biologically active amino acids are of L form, and hence we need not refer to this designation. Proteins (polypeptides) are synthesized when peptide bonds form between any two amino acids called **residues** when incorporated into a protein—can join and all chains will have an amino (N-terminal) end and a carboxyl (C-terminal) end.

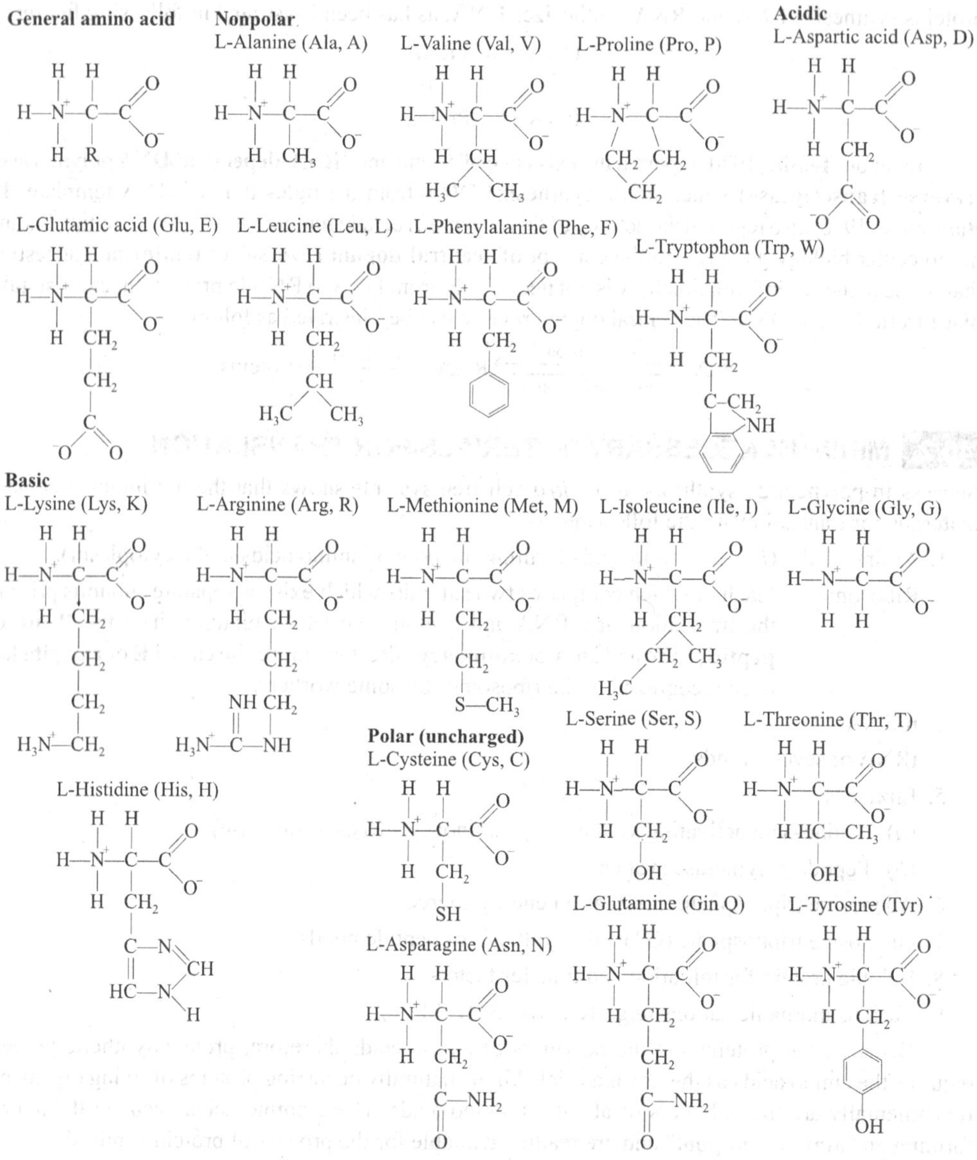

Fig. 40.2. The twenty amino acids found in proteins and their three- and one- letter abbreviations. At physiological pH, the amino acids usually exist as ions. Note the classification of the various R groups (after Tamarin, 2002).

Proteins (polypeptides) are synthesized by the formation of peptide bonds between two amino acids. The polypeptide chain synthesized in such a way will have an amino end (N-terminal) and a carboxyl (C-terminal) end.

Overview of Mechanism of Protein Synthesis

The RNA-directed assembly of a protein is called *translation* and *protein synthesis*. The actual mechanism of protein synthesis can be divided into three stages:

1. **Initiation.** The assembly of a ribosome on a mRNA molecule.
2. **Elongation.** Repeated cycle of amino acid addition to a growing peptide chain.
3. **Termination.** The release of the new protein chain.

In nutshell, the ribosome with its ribosomal RNA (rRNA) and proteins is the site of protein synthesis. The information from the gene is in the form of messenger RNA (mRNA) in which each group of three nucleotides-a *codon*-specifies an amino acid. The amino acids are carried to the ribosome attached to transfer RNAs (tRNAs) and these transfer RNAs have *anticodons*, three nucleotides complementary to a codon, located at the end opposite the amino acid attachment site. A peptide bond will form between the two amino acids present at the ribosome, freeing one transfer RNA (at codon 1 in Fig. 40.3) and lengthening the amino acid chain attached to the second transfer RNA (at codon 2 in Fig. 40.3). The messenger RNA will then move one codon with respect to the ribosome, a new transfer RNA will attach at codon 3. This cycle is then repeated, with the polypeptiqe lengthening by one amino acid each time. It is very significant that the translation process start precisely. Remember that the genetic code is translated in groups of three nuc l eotides (codons). If the reading of the messenger RNA begins one base too early or too late, the reading frame is shifted so that an entirely different set of codons is read (Fig. 40.4). The protein produced, if any, will probably bear no structural or functional similarity to the protein the gene is coded for.

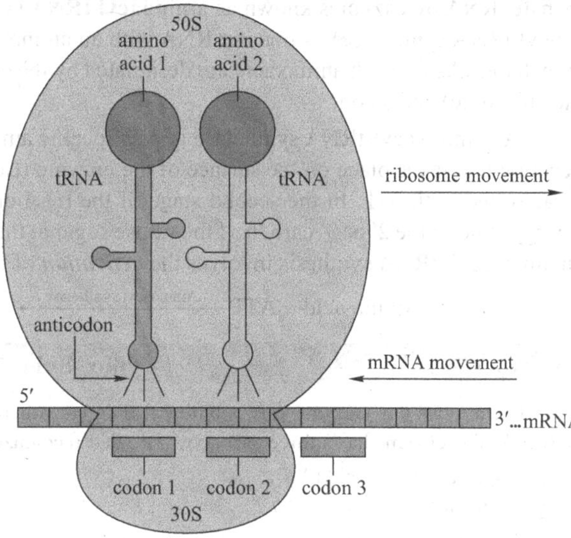

Fig. 40.3. A ribosome during the process of protein synthesis. Note the two charged mRNAs. They are in position to form the first peptide bond between the two amino acids attached to the transfer RNA (after Tamarin, 2002).

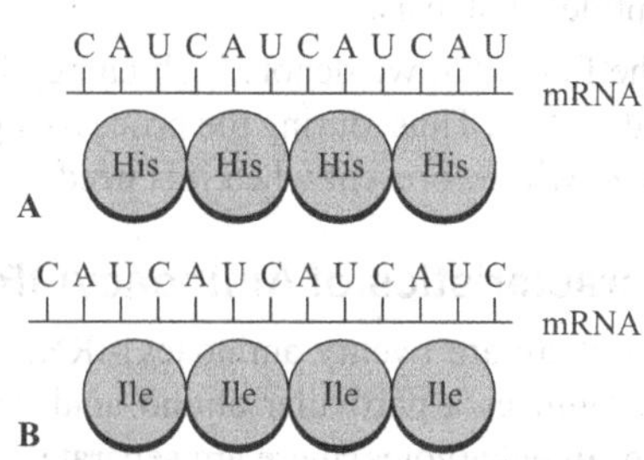

Fig. 40.4. A—In the normal reading of the messenger RNA, these codons are read as repeats of CAU, coding for histidine. B—A shift in the reading frame of the messenger RNA causes the codons to be read as AUC repeats coding for isoleucine (after Tamarin, 2002).

40.3. MECHANISM OF TRANSLATION OR PROTEIN SYNTHESIS

I. Aminoacylation of tRNA (Attachment of Amino Acid to Transfer RNA)

The function of tRNA (transfer RNA) is to ensure that each amino acid incorporated into a protein corresponds to a particular codon in the messenger RNA. The transfer RNA serves this function through its peculiar structure: it has an anticodon at one end and an amino acid attachment site at the other end. The "correct" amino acid, the amino acid corresponding to the anticodon, is attached to the transfer RNA by enzymes known as **aminoacyl tRNA synthetases** (*e.g.*, arginyl-tRNA synthetase, leucyl-tRNA synthetase). A transfer RNA with an amino acid attached is said to be "charged". The transfer RNAs for each amino acid are designated by the convention $tRNA^{Leu}$ (for leucine), $tRNA^{His}$ (for histidine) and so on.

An amino acyl-tRNA synthetase joins a specific amino acid to its transfer RNA in a two stage reaction that takes place on the surface of the enzyme (Box 40.1): In the first stage, the amino acid is activated with ATP. In the second stage of the reaction, the amino acid is attached with a high energy bond to the 2′ or 3′ carbon of the ribose sugar at the 3′ end of the transfer RNA: The first step in aminoacyl–tRNA synthesis involves the ***activation*** of the amino acid using energy from ATP:

$$\text{Amino acid} + \text{ATP} \xrightarrow{\text{Aminoacyl synthetase}} \text{amino acid} \sim \text{AMP} + \text{P} + \text{P}$$

Box 40.1

Each Activating enzyme (*i.e.*, aminoacyl-tRNA synthetase) is specific for one amino acid and for a tRNA. The enzyme has a three part active site that recognizes three smaller molecules:

1. a specific amino acid (AA);
2. ATP; and
3. a specific transfer RNA.

The aminoacyl ~ AMP intermediate is not normally released from the enzyme before undergoing the second step in aminoacyl tRNA synthesis, namely the reaction with appropriate tRNA:

$$\text{Amino acid} \sim \text{AMP} + \text{tRNA} \xrightarrow{\text{Aminoacyl-tRNA synthetase}} \text{amino acid} \sim \text{tRNA} + \text{AMP}$$

The resultant aminoacyl ~ tRNAs (amino acid ~ tRNAs) are immediate precursors of protein (= polypeptide) synthesis.

In the Fig. 40.5, we denote high energy bonds, bonds that liberate a lot of free energy when hydrolysed, as "~". Thus, during the process of protein synthesis, the energy for the formation of the peptide bond will be present when it is needed, at the point of peptide bond formation.

Few Characteristics of Aminoacyl-tRNA Synthetases

In bacteria, there are twenty aminoacyl-tRNA synthetases, one for each amino acid. A particular enzyme recognizes a particular amino acid, as well as, all the transfer RNAs that code for that amino acid. In eukaryotes, there are separate sets of twenty cytoplasmic and twenty mitochondrial aminoacyl-tRNA synthetases, all coded in the nucleus.

Aminoacyl-tRNA synthetases are a heterogeneous group of enzymes. In *E.coli*, these enzymes vary from monomeric proteins (one subunit) to tetrameric proteins, made up of two copies each of two subunits. The enzymes fall into two categories based on sequence similarity, structural features, and whether the amino acid is attached at the 2′-OH (in class I enzymes) or 3′-OH (in class II enzymes) of the 3′-terminal adenosine of the transfer RNA (Table 40.1; **Ibba** *et al.*, 2000).

Table 40.1. Features of aminoacyl-tRNA synthetases (Source: Brown, 2002).

	Feature	Class I enzymes	Class II enzymes
1.	Structure of the enzyme's active sites	Parallel β-sheet	Antiparallel β-sheet
2.	Interaction with tRNA	Minor groove of the acceptor stem	Major groove of the acceptor stem
3.	Orientation of the bound tRNA	V loop faces away from the enzyme	V loop faces the enzyme

	Feature	Class I enzymes	Class II enzymes
4.	Amino acid attachment	To the 2′-OH of the terminal nucleotide of the tRNA	To the 3′-OH of the terminal nucleotide of the tRNA
5.	Enzymes for*	Arg, Gys, Glu, Gln, Ile, Leu, LysI, Met, Trp, Tyr, Val	Ala, Asn, Asp, Gly, His, LysI, Phe, Pro, Thr, Ser

*The aminoacyl-tRNA synthetase for lysine in a class I enzyme in some archaea and bacteria and a class II enzyme in all other organisms.

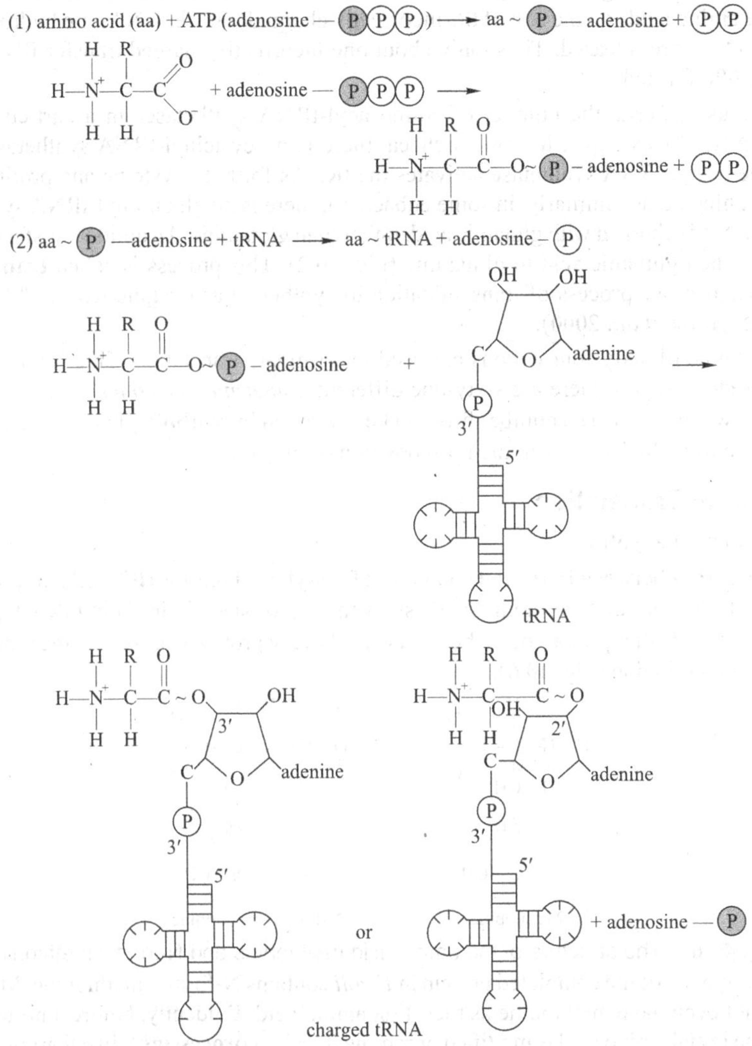

Fig. 40.5. Two steps of aminoacylation that is attachment of a specific amino acid to its transfer RNA by an aminoacyl synthetase enzyme. High-energy bonds are indicated by ~. In the first step, an amino acid is attached to AMP with a high-energy bond. In the second step, the high-energy bond is transferred to the tRNA, which is then referred to "charged". Depending on which class of aminoacyl-tRNA synthetase is involved, the amino acid will be attached to either the 2′ or 3′ carbon of the sugar of the 3′ terminal adenosine. The link is formed between the —COOH group of the amino acid and the —OH group attached to carbon on the sugar of the last nucleotide, which is always an A(= adenine).

To add its appropriate amino acid to the appropriate transfer RNA, a synthetase recognizes many parts of transfer RNA. This can be demonstrated by experiments that introduces specific changes in transfer RNAs by site-directed mutagenesis. In seventeen of the twenty synthetase enzymes of *E.coli*, recognition involves part of the anticodon itself.

Further, an aminoacyl-tRNA synthetase enzyme can initially make errors and attach the "wrong" amino acid to a tRNA. For example, isoleucine (Ile) synthetase will attach valine about one in 225 times. This type of error occurs because a similar, but smaller, amino acid can sometimes occupy the active site of the enzyme (compare chemical formulae of isoleucine and valine in Fig. 40.2). However, because of a proof-reading step, only 1 in 270 to 1 in 800 of the errors are released intact from the enzyme. The amino acids on the rest of the incorrectly charged transfer RNAs are hydrolyzed before the transfer RNAs are released. Thus, only about one incorrectly charged transfer RNA occurs per 60,000 to 80,000 formed.

In various bacteria, the number of amino acyl-tRNA synthetases in a particular organism is below twenty. For example, in some archaea, there is no cysteinyl-tRNA synthetase (enzyme). However, the prolyl-tRNA synthetase activates the tRNAs for both cysteine and proline with their appropriate amino acids. Similarly, in some eubacteria, there is no glutaminyl-tRNA synthetase; the glutaminyl-tRNA is charged with glutamic acid, rather than glutamine. An amido transferase (enzyme) then converts then glutamic acid to glutamine (Fig. 40.2). This process is called **transamidation.** Some archaea also use process of transamidation to synthesise asparagine-tRNA^{Asn} from aspartic acid-tRNA^{Asn} (**Ibba** *et al.*, 2000).

Lastly, three of sixty-four codons are used to terminate translation. Thus, sixty-one transfer RNAs are needed because there are sixty-one different *nonterminator codons*. About fifty transfer RNAs are known in *E.coli*. The number fifty can be explained by **Wobble phenomenon** (see chapter 10, Genetic Code); which occurs in the third position of the codon.

A. Initiation of Translation

I. Initiation in prokaryotes

1. Interaction between initiation codon and formyl methionine-tRNA. An important feature of initiation of polypeptide (= protein) synthesis is the use of specific initiating tRNA molecule. In *Escherichia coli* and other prokaryotes, the synthesis of every protein begins with the modified amino acid N-formyl methionine (Fig. 40.6).

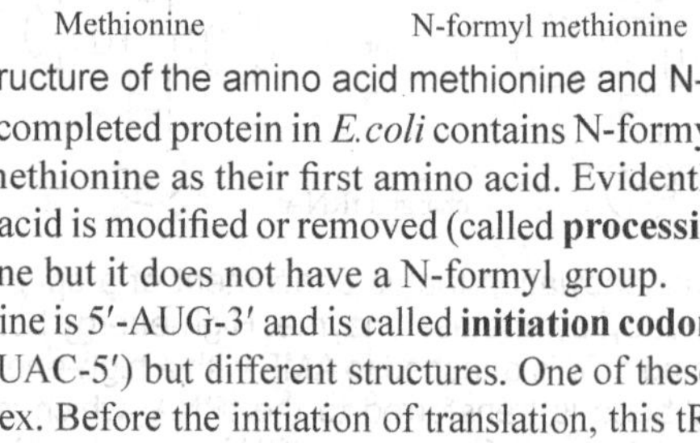

Fig. 40.6. The structure of the amino acid methionine and N-formyl methionine.

However, none of the completed protein in *E.coli* contains N-formyl methionine. Many of these proteins do not even have methionine as their first amino acid. Evidently, before a protein becomes functional, the initial amino acid is modified or removed (called **processing**). In eukaryotes, the initial amino acid is also methionine but it does not have a N-formyl group.

The codon for methionine is 5′-AUG-3′ and is called **initiation codon**. Methionine has two tRNAs with the same anticodon (3′-UAC-5′) but different structures. One of these tRNAs (tRNA$_f^{met}$) serves as a part of the initiation complex. Before the initiation of translation, this tRNA will have its methionine chemically modified to N-formyl methionine (fMet). The other transfer RNA will not have its methionine modified (tRNA$_m^{met}$). The translation process will use it to insert methionine into proteins, where it is required, in all but the first position. The cell thus has a device to make use of methionine in the normal way as well as to use a modified form of it (*i.e.*, formyl methionine) to initiate protein synthesis. Because of the structure of the prokaryotic initiation transfer RNA, it can recognize AUG, GUG and rarely UUG as initiation codons. In eukaryotes, CUG as well as AUG can serve as an initiation codon. Since the initiation methionine is not formylated in eukaryotes, the eukaryotic transfer RNA is designated $tRNA_i^{met}$; there is a separate internal methionine transfer RNA, termed $tRNA_m^{met}$, in eukaryotes as well as prokaryotes.

Since in protein synthesis the peptide chain always grows in a sequence from the free terminal amino (—NH_2) group towards the carboxyl (—COOH) end, so the function of formyl methionine-tRNA is to ensure that protein are synthesized in that direction. In the formyl methionine-tRNA, the amino (—NH_2) group is blocked by the formyl group leaving only the —COOH group available to react with the —NH_2 group of the second amino acid (AA_2). In this way, the synthesis of protein chain follows in the correct sequence.

2. Formation of initiation complex. The 30S and 50S ribosomal subunit of bacteria usually dissociate from each other when not involved in translation. To begin translation, an **initiation complex** forms, consisting of following components in prokaryotes (*i.e.*, *E.coli*): the 30S subunit of the ribosome, a messenger RNA, the charged N-formyl methionine tRNA (fmet-$tRNA_f^{Met}$), and three **initiation factors (IF1, IF2, IF3)**. Initiation factors (as well as elongation and termination factors) are proteins loosely associated with the ribosome (Table 40.2). They were discovered when ribosomes were isolated and then washed, losing the ability to perform translation (*i.e.*, polypeptide synthesis).

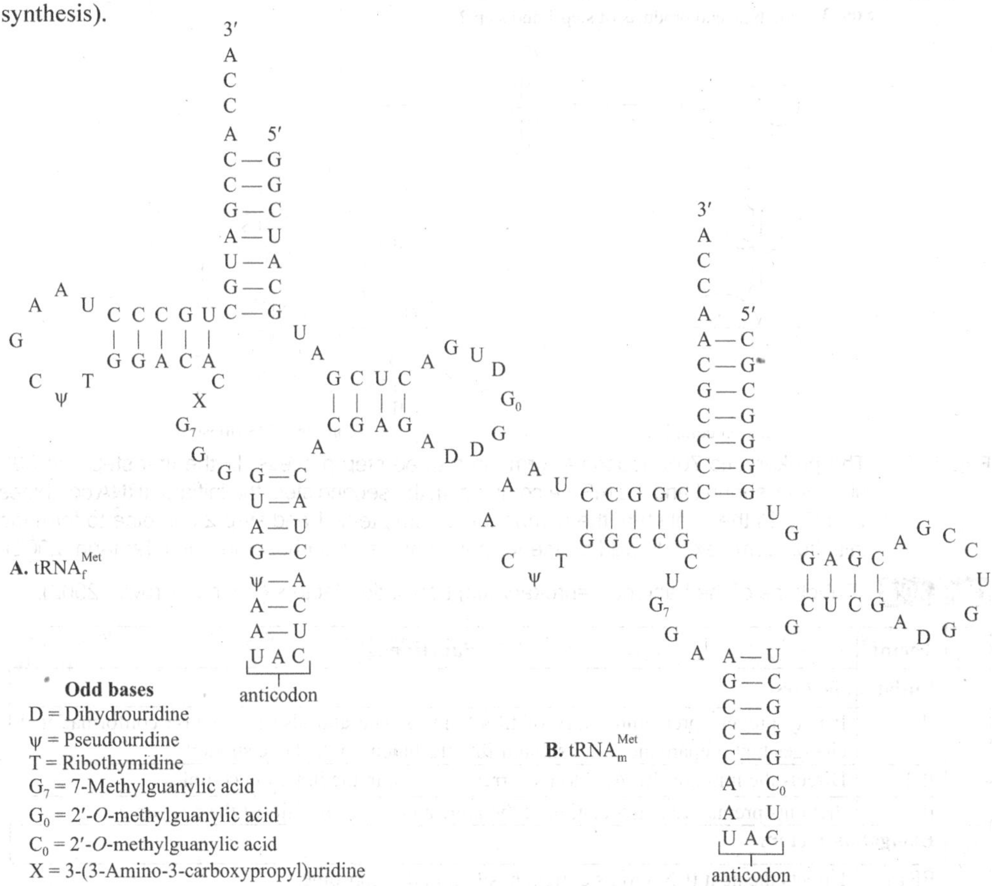

Fig. 40.7. Molecular structure of two tRNAs for methionine in *E.coli*. A—The initiator tRNA; B—The interior or internal tRNA (after Tamarin, 2002).

Various components of the initiation complex interact in a series of steps. It is known that IF3 binds to the 30S ribosomal subunit, allowing the 30S subunit to bind to messenger RNA (Fig. 40.8, step 1). Meanwhile, a complex forms with IF2, the charged N-formyl methionine tRNA (fmet–$tRNA_f^{Met}$) and GTP (guanosine triphosphate; Fig. 40.8, step 2).

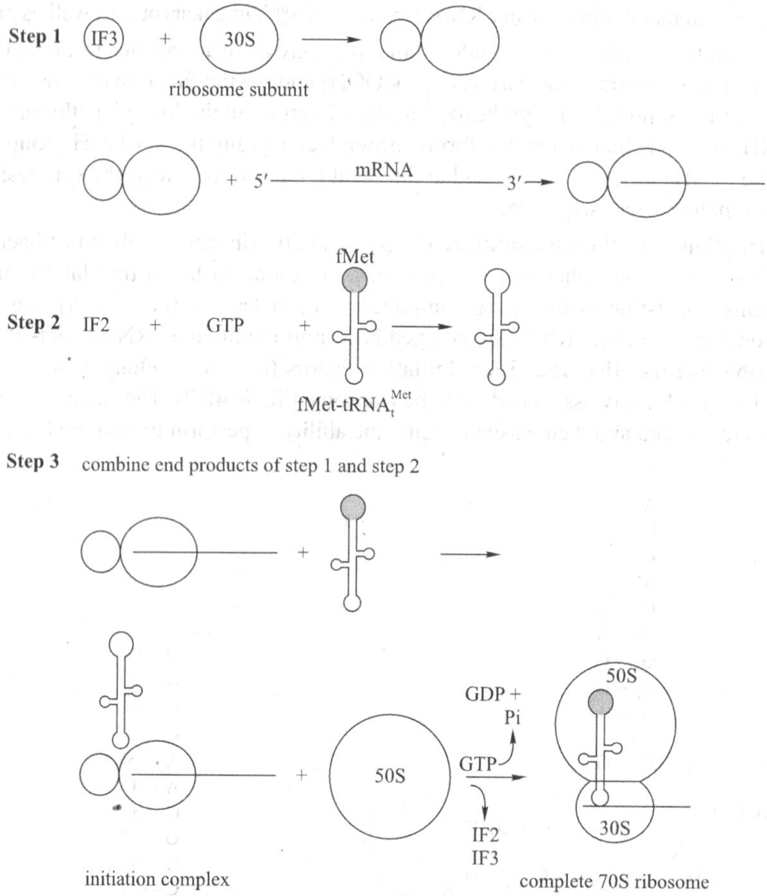

Fig. 40.8. The prokaryotic 70S ribosome forms in a three step process. In the first step, the 30S ribosomal subunit and the mRNA combine. In the second step, the initiator tRNA combines with IF2. In the final step, the components from steps 1 and step 2 combine to form the initiation complex, followed by the formation of the 70S ribosome (after Tamarin, 2002).

Table 40.2. Functions of the bacterial (=prokaryotic) translation factors (Source: Brown, 2002).

	Factor	Function
A.	**Initiation factors**	
	IF1	It may prevent premature entry of tRNA into A site and also may cause conformational changes that prepare the small subunit for attachment to the large subunit.
	IF2	Directs the initiator $tRNA_f^{Met}$ to its correct position in the initiation complex.
	IF3	Prevents premature reassociation of the large and small subunits of the ribosome.
B.	**Elongation factors**	
	EFTu	Directs the next tRNA to its correct position in the ribosome.
	EFTs	Regenerates EF–Tu after the latter yielded the energy contained in its attached GTP molecule.
	EFG	Mediates translocation.

	Factor	Function
C.	**Release factors**	
	RF1	Recognizes the termination codons 5′-UAA-3′ and 5′-UAG-3′.
	RF2	Recognizes 5′-UAA-3′ and 5′-UGA-3′.
	RF3	Stimulates dissociation of RF1 and RF2 from the ribosome after termination.
D.	**Ribosome recycling factor**	
	RRF	Responsible for dissociation of ribosomal subunits after translation has terminated.

It is IF2 that brings the initiator transfer RNA to the ribosome. IF2 binds only to the charged initiator transfer RNA, and without IF2, the initiator tRNA cannot bind to the ribosome. The final step in initiation-complex formation is bringing together the first two components (Fig. 40.8; step 3).

Here, GTP has unusual function. The hydrolysis of GTP to GDP + Pi (inorganic phosphate, the PO_4^{-3}) produces **conformational changes**; these changes allow the initiation complex to join the 50S ribosomal subunit to form the complete ribosome and then allow the initiation factors and GDP to be released. Generally, the hydrolysis of a nucleoside triphosphate (*i.e.*, ATP, GTP) in a cell occurs to release the energy in the phosphate bonds for use in a metabolic process. However, in the process of translation, the hydrolysis of GTP apparently changes the shape so that it and the initiation factors can be released from the ribosome after 70S particle has been formed. Thus, *hydrolysis of GTP in translation is for conformational change rather than covalent bond formation.* IF1 helps the other two initiation factors bind to the 30S initiation complex.

How does 70S ribosome recognizes mRNA? The ribosome recognizes the prokaryotic mRNA through complementarity of a region at the 3′ end of the 16S ribosomal RNA and a region slightly upstream from the initiation sequence (AUG) on the mRNA (Fig. 40.11A, 40.9). This consensus sequence is 5′-AGGAGGU-3′ and is called **Shine-Dalgarno sequence** (Fig. 40.9). It is located between 3 to 10 nucleotides upstream of the initiation codon.

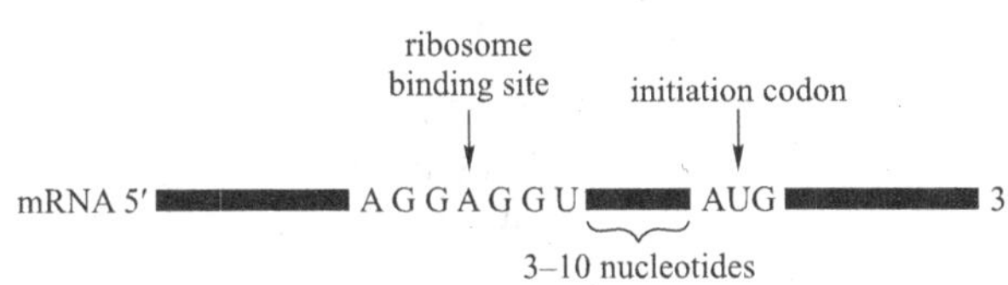

Fig. 40.9. The Shine-Dalgarno sequence or ribosome binding site for bacterial (*E.coli*) translation.

Moreover, in prokaryotes, most mRNAs contain the information for several genes. These RNAs are called **polycistronic mRNAs**. Each gene on the messenger RNA is translated independently: each gene has a Shine-Dalgarno sequence for ribosome reognition and an initiation codon (AUG) for fMet (Fig. 40.10).

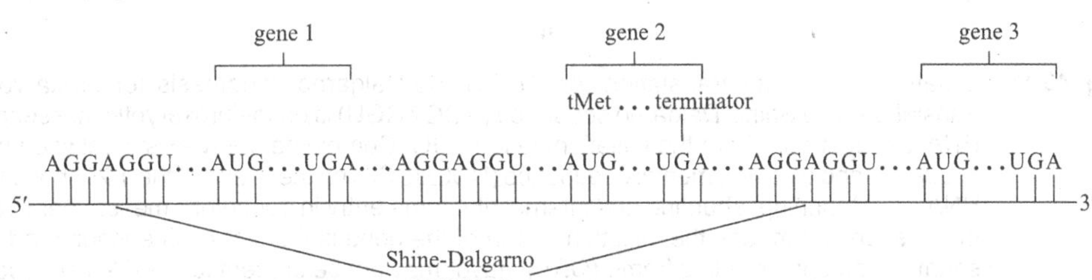

Fig. 40.10. A prokaryotic polycistronic mRNA. Note the several Shine-Dalgarno sequences for ribosomal attachment and the initiation and termination codons marking each gene (after Tamarin, 2002).

II Initiation in Eukaryotes

The process of initiation complex formation in eukaryotes is generally similar to prokaryotes but more complex. The eukaryotic initiation factor abbreviations are preceded by an "e" to denote that they are eukaryotic (eIF1, eIF2, etc.). At least eleven initiation factors are involved, including a specific cap-binding protein, eIF4E. The Shine-Dalgarno sequence is altogether absent in eukaryotes. The actual mechanism for recognizing the 5′ end of eukaryotic messenger RNA appears to be based on recognition of the **5′ cap** of the messenger RNA by the cap-binding protein with recruitment of other initiation factors (Table 40.3) and the small subunit of the ribosome. This is followed by the small subunit's movement down the messenger RNA. The ribosome scans the messenger RNA until it recognizes the initiation codon. This model is known as the **scanning hypothesis**.

In many eukaryotes, a process called **shunting** occurs, in which the first AUG does not serve as the initiation codon; rather, scanning begins, but it bypasses a region of the messenger RNA upstream of the initiation codon, called the **leader** or **5′ untranslated region** (5′ UTR), in favour of an AUG further down the messenger RNA. The cause of shunting seems to be secondary structure in the messenger RNA, upstream from the AUG codon that actually serves as the initiation codon.

In some eukaryotes, very small genes called **open reading frames (ORFs)** are present in this region of the messenger RNA and play some role in shunting. In the genes of some plant and animal viruses, ORFs are translated, and then the main gene is translated by the same ribosome in a process called **reinitiation** (Fig. 40.11B).

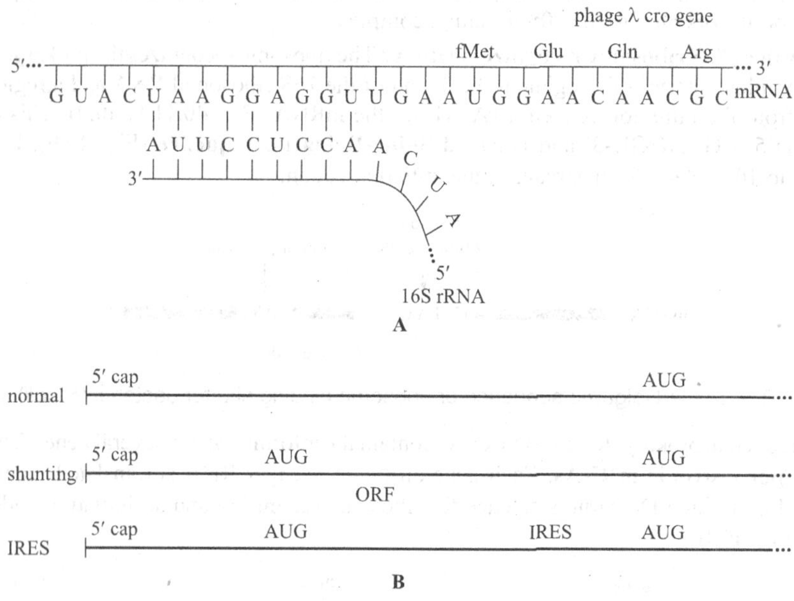

Fig. 40.11. Initiation signals for translation. A—The Shine-Dalgarno hypothesis for prokaryotic translation. The Shine-Dalgarno sequence (AGGAGGU) is on the prokaryotic messenger RNA just upstream from the initiation codon AUG. Complementarity exists between this sequence and a complementary sequence (UCCUCCA) on the 3′ end of the 16S ribosomal RNA. B—Scanning, shunting and internal ribosome entry in eukaryotic messenger RNA. In the scanning model, the initiation codon of the gene is the first AUG encountered. In shunting, an open reading frame (ORF) mayor may not be present to provide secondary structure in the messenger RNA to shunt scanning to the main gene. If the open reading frame (ORF) is translated, reinitiation of translation at the same ribosome may occur at the main gene. Finally, an internal ribosome entry site (IRES) allows translation to begin with the messenger RNA without scanning (after Tamarin, 2002).

In some eukaryotes, ribosomes can initiate protein synthesis within the mRNA if that mRNA contains a sequence called an **IRES** (*i.e.*, internal ribosome entry site). These sequences were discovered in RNA genome of the human poliovirus and rhinovirus (for cold) and in several cellular messenger RNAs. They are at least four hundred nucleotides long.

Table 40.3. Eukaryotic translation factors (Source: Brown, 2002).

A. Initiation factors	
eIF-1	Component of pre-initiation complex
eIF-1A	Component of pre-initiation complex
eIF-2	Binds to initiator tRNAMet within the ternary complex component of the pre-initiation complex; phosphorylation of eIF-2 results in a global repression of translation
eIF-3	Component of the pre-initiation complex; makes direct contact with eIF-4G and forms the link with the cap binding complex
eIF-4A	Component of the cap binding complex, a helicase that aids scanning by breaking intramolecular base pairs in the mRNA
eIF-4B	Aids scanning possible by acting as a helicase that breaks intramolecular base pairs in the mRNA
eIF-4E	Component of the cap-binding complex, possibly the component that makes direct contact with the cap structure at the 5′ end of mRNA
eIF-4F	The cap-binding complex, comprising eIF-4A, eIF-4E and eIF-4G, which makes the primary contact with the cap structure at the 5′ end of the mRNA
eIF-4G	Component of the cap-binding complex, forms a bridge between the cap binding complex and eIF-3 in the preinitiation complex; in at least some organisms, eIF-4G also forms an association with the poly (A) tail, via the polyadenylate-binding protein
eIF-5	Aids release of the other initiation factors at the completion of initiation
eIF-6	Associated with the large subunit of the ribosome, prevents large subunits from attaching to small subunits in the cytoplasm
B. Elongation factors	
eEF-1	Complex of four subunits (eEF-1a, eEF-1b, eEF-1d and eEF-1g); directs the next tRNA to its correct position in the ribosome
eEF-2	Mediates translocation
C. Release factors	
eRF-1	Recognizes the termination codon
eRF-2	Possibly stimulates dissociation of eRF-1 from the ribosome after termination; possibly causes the ribosome subunits to dissociate after termination of translation

(*i*) Initiation of eukaryotic translation with scanning. The first step involves assembly of the *pre-initiation complex*. This structure comprises of the 40S ribosomal subunit, a 'ternary complex' made up of the initiation factor eIF-2 bound to the initiator $tRNA^{Met}$ and a molecule of GTP, and three additional initiation factors, eIF-1, eIF-1A and eIF-3. As in bacteria, the initiator tRNA is distinct from the normal $tRNA^{Met}$ that recognizes internal 5′-AUG-3′ codons but, unlike bacteria, it is aminoacylated with normal methionine, not the formylated version.

After assembly, the pre-initiation complex associates with the 5′ end of the mRNA. This step requires the **cap binding complex** (sometimes called eIF-4F) which comprises the initiation factors eIF-4A, eIF-4E and eIF-4G. The contact with the cap might be made by eIF-4E alone (Fig. 40.12) or involve a more general interaction with the cap binding complex (**Pestova** and **Hellen**, 1999). The factor eIF-4G serves as a bridge between eIF-4E, bound to the cap, and eIF-3, attached to pre-initiation complex gets attached to the 5′ region of the mRNA. Attachment of the pre-initiation complex to the mRNA is also influenced by the poly (A) tail, at the distant 3′ end of the mRNA. This interaction is thought to be mediated by the polyadenylate-binding protein (PADP) which is attached to the poly (A) tail. In yeast and plants, PADP is found to form an association with eIF-4G and for such an association mRNA has to bend back on itself.

With its attachment to the 5′ end of the mRNA, the **initiation complex** has to scan along the mRNA molecule and find the initiation codon. The leader regions of eukaryotic mRNAs can be several tens or even hundreds of nucleotides in length and often contain regions that form hairpins and other base paired structures. These are probably removed by a combination of eIF-4A and eIF-4B. The eIF-4A, and possibly also eIF-4B, has a helicase activity and is able to break intramolecular base pairs in the mRNA, freeing the passage for the initiation complex (Fig. 40.12B). The initiation codon (AUG) is recognizable because it is contained in a short consensus sequence; 5′-ACCAUGG-3′, which is called **Kozak consensus** (see **Brown**, 2002).

As soon as, the initiation complex is positioned over the initiation codon, the large subunit (60S) of the ribosome attaches. As in bacteria, this requires hydrolysis of GTP and leads to release of the initiation factors. Two final initiation factors are involved at, this stage: eIF-5 which aids release of the other factors and eIF-6, which is associated with the unbound large subunit and prevents it from attaching to a small subunit in the cytoplasm.

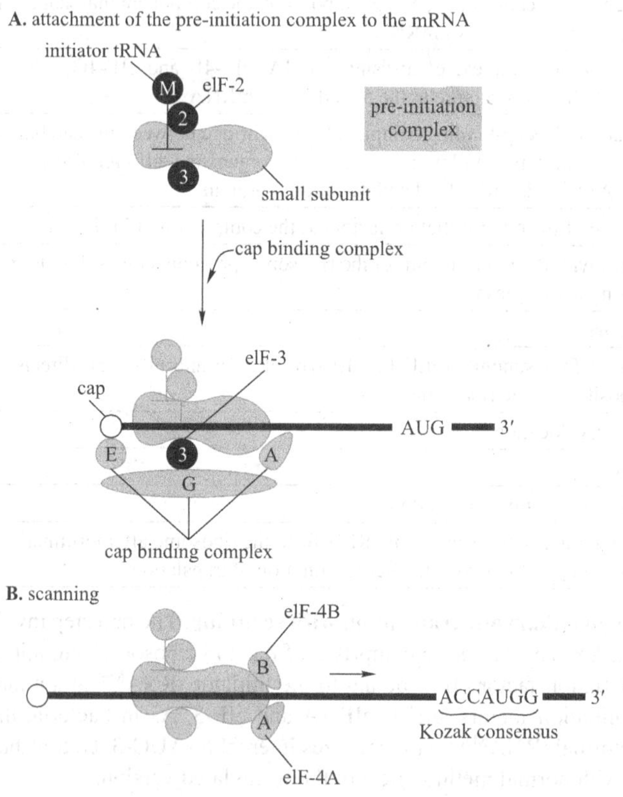

Fig. 40.12. Initiation of eukaryotic translation. A—Assembly of the pre-initiation complex and its attachment to the mRNAs (according to Hentze 1997). B—The pre-initiation complex scans along the mRNA until it reaches a initiation codon, which is recognisable because it is located within Kozak consensus sequence. Scanning is aided by eIF-4A, eIF-4B, which are thought to have helicase activity. Most probably, eIF-3 remains attached to the pre-initiation complex during scanning, as shown here. Scanning is an energy-dependent process that requires hydrolysis of ATP. M=methionine (after Brown, 2002).

***(ii)* Initiation of eukaryotic translation without scanning.** Transcripts (mRNA) of picorna viruses (*i.e.*, RNA viruses such as poliovirus and rhinovirus) are not capped but instead have an internal ribosome entry site (IRES). The presence of IRES on their transcripts means that picorna viruses can block protein synthesis in the host cell by inactivating the cap binding complex, without affecting translation of their own transcripts.

Quite interestingly, no virus proteins are required for the recognition of an IRES by a host ribosome. In other words, the normal eukaryotic cell possesses proteins and/or other factors that enable it to initiate translation by the IRES method (**Holick** *et al.*, 2000). Some nuclear genes are reported to have IRES. For example, IRES are present on mRNAs for the mammalian immunoglobulin heavy chain binding protein and the *Drosophila* antennapedia protein. IRES are also found on several mRNAs whose protein products are translated when the cell is put under stress, for example by exposure to heat, irradiation, on low oxygen conditions. Under these circumstances, cap dependent translation is globally suppressed. The presence of IRESs on the 'survival' mRNAs therefore enables these to undergo preferential translation at the time when their products are needed (see **Brown**, 2002).

Three Sites of Ribosomes

When the initiator transfer RNA joins the 30S ribosomal subunit of prokaryotes with its mRNA attached, it fits into one of three sites in the ribosome. These sites or cavities in the ribosome, are referred to as the **aminoacyl site (A site)**, the **peptidyl site (P site)** and the **exit site (E site**; Fig. 40.13). Here we concentrate on A and P sites, each of which contains a transfer RNA just before forming a peptide bond: the P site contains the transfer RNA with the growing peptide chain (peptidyl-tRNA); the A site contains a new transfer RNA with its single amino acid (aminoacyl-tRNA). The E site helps eject the depleted transfer RNAs after a peptide bond forms. When a complete 70S ribosome of Figure 40.9 has formed, the initiation fMet – $tRNA_f^{Met}$ is placed directly into the P site (Fig. 40.13), *the only charged tRNA that can be placed directly there*. The association of tRNA and ribosome is aided by a G-C base pairing between the 3′-CCA terminus of all transfer RNAs and a guanine in the 23S ribosomal RNA.

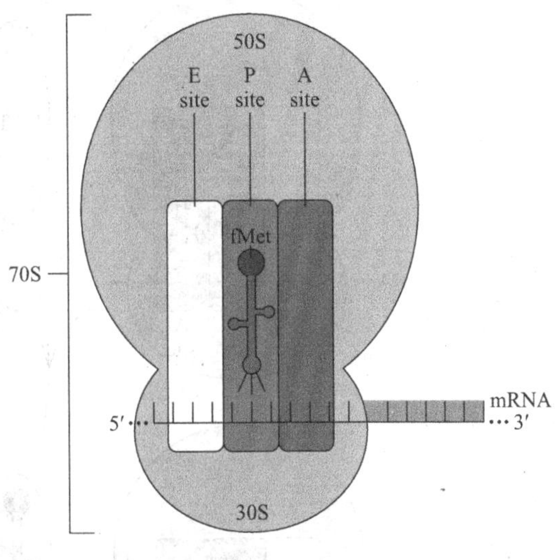

Fig. 40.13. The 70S ribosome contains an A site, a P site and an E site that can receive tRNAs. The mRNA runs through the bottom of the sites (after Tamarin, 2002).

B. Elongation

The main differences between translation in bacteria and eukaryotes occur during the initiation phase; the events after the large subunit of the ribosome becomes associated with the initiation complex are similar in both types of organisms.

1. Positioning a second transfer RNA. The second charged tRNA binds (due to a codon-directed binding) to the first ribosome at the latter's 'A' site with the help of the proteins, called elongation factors (*e.g.*, EF-Tu). EF-Tu carries a molecule of GTP. Correct hydrogen bonding with the mRNA template dictates the selection of a new tRNA, and the activity of the EF-Tu ensures the proper positions of the tRNA in the A site. Such a placement activity needs energy that is provided by the hydrolysis of the molecule of GTP to GDP and phosphate. After performing its function, the

EF-Tu protein dissociates from the ribosome, and in the cytoplasm is subsequently regenerated to its active form by another elongation factor, the **EF-Ts**. At this point, both sites of the ribosome are occupied by tRNA's each of which carries an amino acid, and each of which is hydrogen bonded to the template mRNA (Fig. 40.14).

The structures revealed by X-ray crystallography show that A, P and E sites are located in the cavity between the large and small subunits of the ribosome, the codon-anticodon interaction being associated with the small subunit and the aminoacyl end of the mRNA with the large subunit (**Yusupov** *et al.*, 2001; Fig. 40.14).

Further, it takes several milliseconds for the GTP to be hydrolysed, and another few milliseconds for the EF-Tu/GDP to actually leave the ribosome. During those two intervals of time, the codon-anticodon fit of the transfer RNA is scrutinized. If the correct tRNA is in place, a peptide bond forms. If not, the charged tRNA is released and a new cycle of EF-Tu/GTP-mediated testing of tRNAs begins. The error rate is only about one mistake in ten thousand amino acids incorporated into protein. The speed of amino acid incorporation is about fifteen amino acids per second in prokaryotes and about two to five per second in eukaryotes.

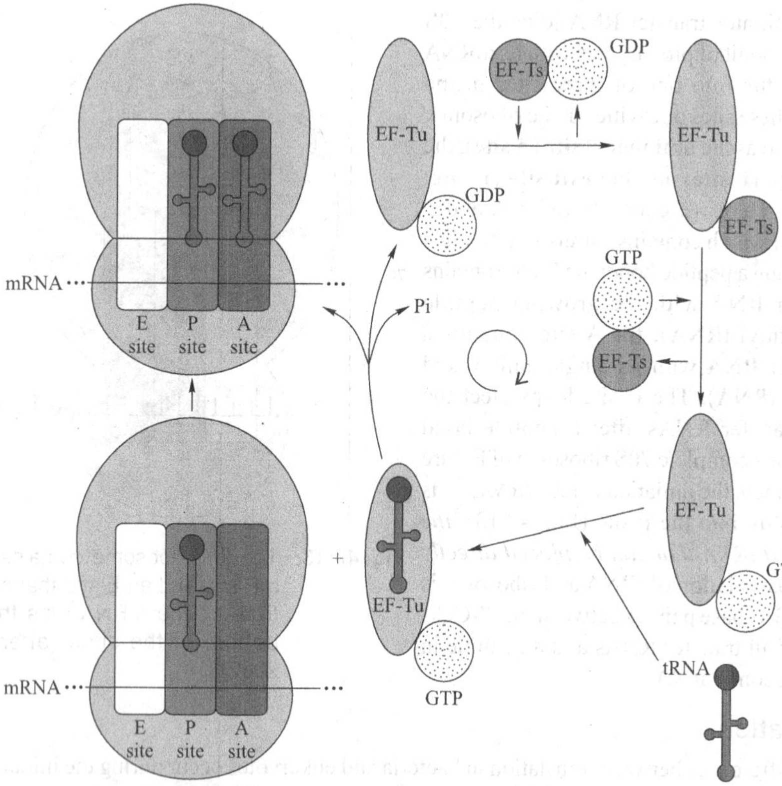

Fig. 40.14. The EF-Ts/EF-Tu cycle. EF-Ts and EF-Tu are required for a transfer RNA to attach to the A site of the ribosome. At *top centre*, there is EF-Tu attached to a GDP. The GDP is then displaced by EF-Ts, which in turn is displaced by GTP. A transfer RNA attaches and is brought to the ribosome. If the codon-anticodon fit is correct, the transfer RNA attaches at the A site with the help of the hydrolysis of GTP to GDP + Pi, allowing EF-Tu to release. The EF-Tu is now back where we started. Since EF-Tu has a strong affinity for GDP, the role of EF-Ts is to displace the GDP, and later to be replaced by GTP (after Tamarin, 2002).

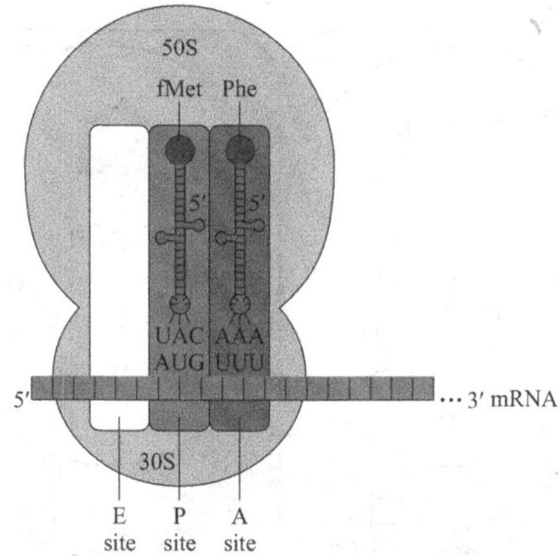

Fig. 40.15. A ribosome with two transfer RNAs attached. In this case, the second codon (UUU) is for the amino acid phenylalanine. The two amino acids are next to each other (after Tamarin, 2002).

2. Peptide bond formation. The next step is the formation of a peptide bond between the two amino acids. To accomplish this job, the first amino acid (N-formylmethionine) is removed from its attachment to its tRNA and transferred to the free-NH_2 terminus of the second amino acid. The first amino acid is, thus, placed "on top of" the second. The ensuing peptide bond, thus, joins the carboxyl group of the first amino acid with the amino group of the second amino acid (Fig. 40.16). The resulting compound is a dipeptide whose carboxyl end is still bonded to the second tRNA, but whose amino end is free. The reaction is catalyzed by an enzyme associated with 50S subunit and called **peptidyl transferase**. This enzymatic centre, an integral part of the 50S subunit, was originally believed to be composed of parts of several of the 50S proteins. Now, however, it is believed to have ribozymic activity (enzymatic activity) of 23S rRNA of 50S ribosomal subunit.

The enzymatic activity involves a bond transfer from the carboxyl end of N-formyl methionine to the amino end of the second amino acid (phenylalanine, Fig. 40.15). Every subsequent peptide bond is identical, regardless of the amino acids involved. The energy used is contained in the high-energy ester bond between the transfer RNA in the P site and its amino acid (Fig. 40.16). Immediately after the formation of the peptide bond, the transfer RNA with the dipeptide is in the A site, and a depleted tRNA is in the P site.

3. Translocation. The next stage in elongation is *translocation* of the ribosome in relation to the tRNA and the mRNA. Elongation factor **EF-G**, earlier called ***translocase***, catalyses the translocation process. The ribosome must be converted from *pretranslocational state* to the *post-translocational* state by the action of EF-G, which physically moves the mRNA and its associated transfer RNA (Fig. 40.17).

Thus, during translocation following three things happen at once:

1. The ribosome moves along the three nucleotides, so that the next codon enters the A site.
2. The dipeptide-tRNA in the A site moves to the P site.
3. The deacylated tRNA in the P site moves to a third position, the E or exit site in bacteria or in eukaryotes, is simply ejected from the ribosome.

Translocation requires hydrolysis of a molecule of GTP. Electron microscopy of ribosomes at different intermediate stages in translocation suggests that the two subunits rotate slightly in opposite directions, opening up the space between them and enabling the ribosome to slide along the mRNA (**Frank** and **Agarwal**, 2000).

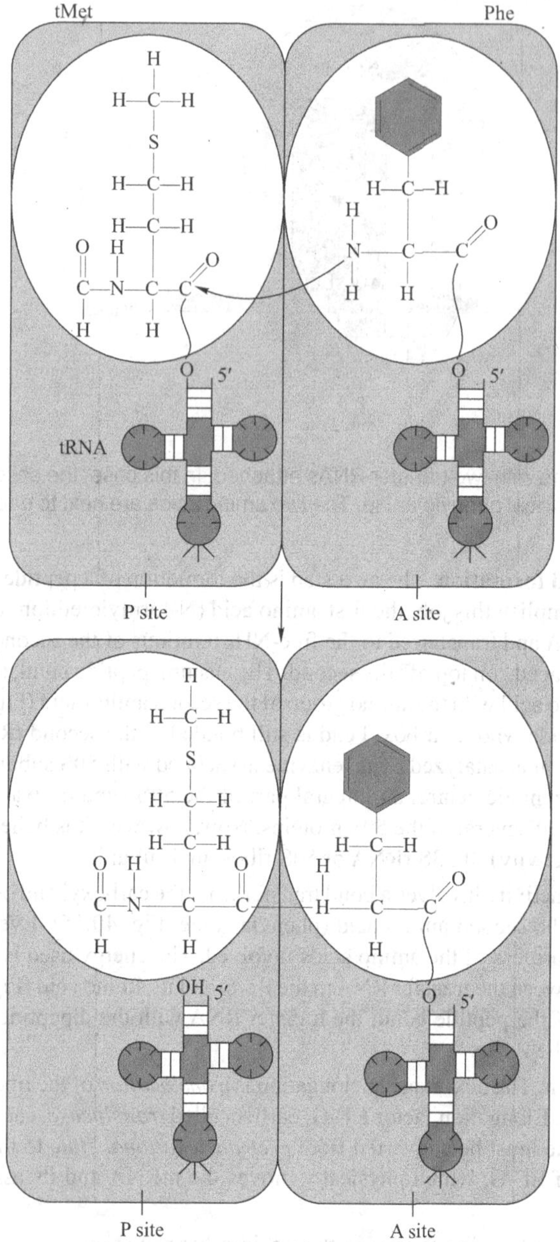

Fig. 40.16. Peptide bond formation on the ribosome (after Tamarin, 2002).

The sequential formation of a polypeptide continues in the manner described above. A tRNA in the 'P' site shifts its burden of growing polypeptide to the next succeeding tRNA, followed by translocation, exit of the discharged tRNA, and entrance of a new charged tRNA (having correct anticodon) to base pair with a new codon at 'A' site. Thus, the growing polypeptide is adopted in turn, by each tRNA, with each successive amino acid being added in effect, to the bottom of the stack. As process continues, the mRNA is progressively translated, codon by codon, from the 5′ end to the 3′ end.

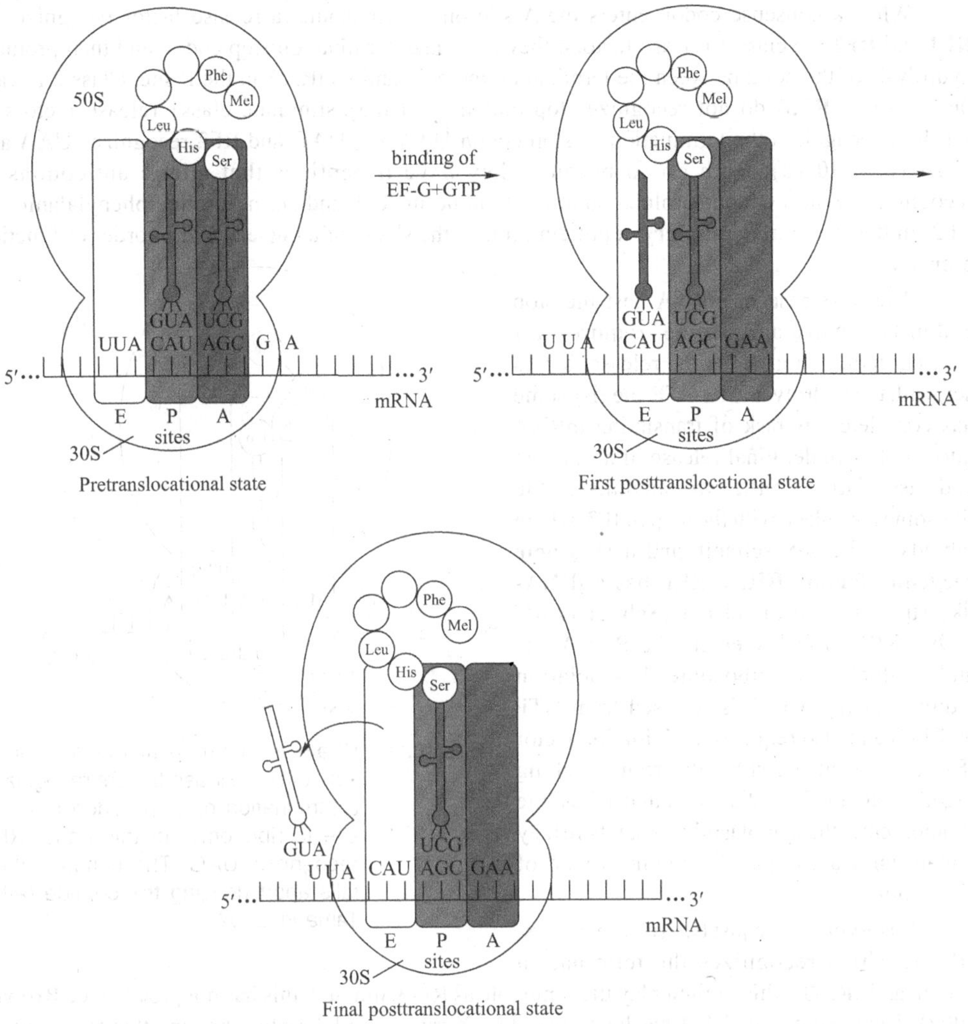

Fig. 40.17. Mode of translocation by the aid of EF-G. During translocation the ribosome converts itself from a pretranslocational state (P and A sites occupied) to a post-translocational state (E and P sites occupied). The uncharged tRNA in the E site is then ejected (after Tamarin, 2002).

In eukaryotes, three elongation factors perform the same tasks that EF-Tu, EF-Ts and EF-G perform in prokaryotes. The factor eEF1α replaces EF-Tu, factor eEF1βγ replaces EF-Ts and eEF2 replaces EF-G (see Table 40.2). According to Brown (2002), elongation factor eEF-1 is a complex of four subumts: eEF-1a, eEF-1b, eEF-1d and eEF-1g (Table 40.3). The first of these exists in at least two forms, eEF-1a1 and eEF-1a2, which are similar proteins that probably have equivalent functions in different tissues (**Hafezparast** and **Fisher**, 1998).

Termination

Termination of translation (or protein synthesis) in both prokaryotes and eukaryotes occur when one of three **nonsense codons** appears in the A site of the ribosome. These codons are UAG (amber), UAA (ochre) and UGA (opal). In prokaryotes, three proteins called **release factors** (**RF**) are involved in termination and a GTP is hydrolyzed to GDP + Pi.

When a nonsense codon enters the A site on the ribosome, a release factor recognizes it. **RF1** and **RF2** are class 1 release factors: they recognize termination/stop codon and then promote hydrolysis of the bond between the terminal amino acid and its tRNA in the P site. Class 2 release factors (*e.g.*, **RF3**) do not recognize stop codons, but they stimulate class1 release factors to act. RF1 recognizes the termination or stop codon UAA and UAG and RF2 recognizes UAA and UGA (Fig. 40.18). Both do so because they have tripeptides that mimic anticodons to recognize the stop codon: proline-alanine-threonine in RF1 and serine-proline-phenylalanine in RF2. In this **molecular mimicry**, a protein mimics the shape of a nucleic acid in order to function properly.

The next base in mRNA past the stop codon is usually an adenine, required for efficient termination. After the release factors act, with the hydrolysis of a GTP, the ribosome has completed its task of translating mRNA into a polypeptide. Final release of all factors and dissociation of the two subunits of the ribosome take place with the help of IF3, which rebinds to the 30S subunit, and a **ribosome recycling factor** (**RRF**). RRF has a tRNA-like structure like the eRF-1 (**Selmer** *et al.*, 1999). RRF probably enters the P or A site and "unlocks" the ribosome. Dissociation requires energy, which is released from GTP by EF-G and also requires the initiation factor IF-3 to prevent the subunits from attaching together again. The disassociated ribosome subunits enter the cytoplasmic pool, where they remain until used again in another round of translation.

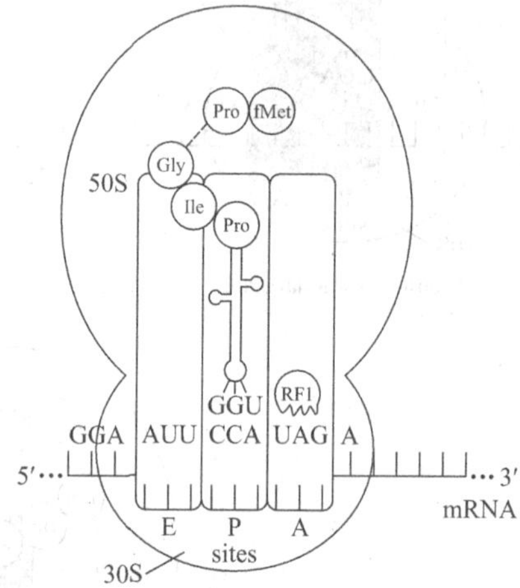

Fig. 40.18. Chain termination at the ribosome. One of two release factors recognizes a termination or stop codon in the A site in ribosome. In this case, RF1 recognizes UAG. The complex then falls apart rleasing the peptide (after Tamarin, 2002).

Eukaryotes have just two release factors: eRF-1, which recognizes the termination codon, and eRF-3, which might play the same role as RF-3 although this has not proven (see **Brown**, 2002). The structure of eRF-1 has been solved by X-ray crystallography, showing that the shape of this protein is very similar to that of a tRNA (**Kisselev** and **Buckingham**, 2000).

Further, in contrast to the often polycistronic mRNA of bacteria, mRNA of eukaryotes are monocistronic, containing the coding sequence only for one polypeptide. The initiation codon for this sequence is located near the 5′ end of the message. Thus, the association of mRNA takes place on 5′ end and not at the initiation codon AUG as in prokaryotes.

Comparison of Prokaryotic and Eukaryotic Translation

Table 40.4 Comparison between prokaryotic and eukaryotic translation (Source: Tamarin, 2000).

Factor	Prokaryotes	Eukaryotes
1. Initiation codon	AUG, occasionally GUG, UUG	AUG, occasionally GUG, CUG
2. Initiation amino acid	N-formyl methionine	Methionine
3. Initiation tRNA	$tRNA^{met}_{f}$	$tRNA^{Met}_{i}$
4. Internal methionine tRNA	$tRNA^{met}_{in}$	$tRNA^{Met}_{m}$
5. Initiation factors	IF1, IF2, IF3	eIF factors

Factor	Prokaryotes	Eukaryotes
6. Elongation factor	EF-Tu	eIF1α
7. Elongation factor	EF-Ts	eIF1βγ
8. Translocation factor	EF-G	eIF-2
9. Release factors	RF1, RF2, RF3, RRF	eRF1, eRF3

Polysomes and Coupled Transcription-Translation

The unit of translation is never simply a ribosome traversing a mRNA molecule, but is a more complex structure. After about 25 amino acids have been joined in a polypeptide chain, the AUG initiation site of the encoding mRNA molecule is completely free of the ribosome. A second initiation complex then forms. The overall configuration is of two 70S ribosomes moving along the mRNA at the same speed. When the second ribosome has moved along a distance similar to that traversed by the first, a third ribosome is able to attach. This process, *i.e.*, movement and reinitiation, continues until the mRNA is covered with ribosomes at a density of about one 70S ribosome per 80 nucleotides. This large translation unit is called a **polyribosome** or simply a **polysome**. This is the usual form of translation unit in all cells (Fig 40.19). The use of polysomes is advantageous to a cell, since the overall rate of protein synthesis is increased compared to the rate that would occur if there were no polysomes.

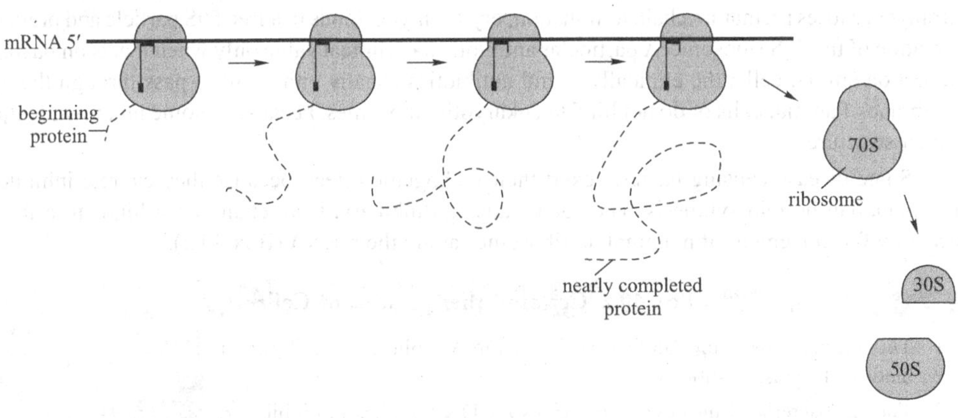

Fig. 40.19. Polysome formation: diagram shows the relative movement of the 70S ribosome and the mRNA, and growth of the protein chain (after Freifelder, 1985).

Further, a mRNA molecule being synthesized has a free 5′ terminus and translation also occurs in the 5′ → 3′ direction, so each cistron contained in the mRNA immediately starts its translation. As a result, the ribosome binding site is transcribed first, followed in order by the AUG codon, the region encoding the amino acid sequence, and finally the stop codon. Thus, in bacteria in which no nuclear membrane separates the DNA and the ribosomes, there is no obvious reason why the 70S initiation complex should not form before the mRNA is released from the DNA. With prokaryotes (*E.coli*) this does indeed occur; this process is called **coupled transcription-translation** (Fig. 40.20). This coupled activity does not occur in eukaryotes, because the mRNA is synthesized and processed in the nucleus and later on transported through the nuclear membrane to the cytoplasm where the ribosomes are located. Coupled transcription-translation, too, speeds up protein synthesis in the sense that translation does not have to await release of mRNA from the DNA.

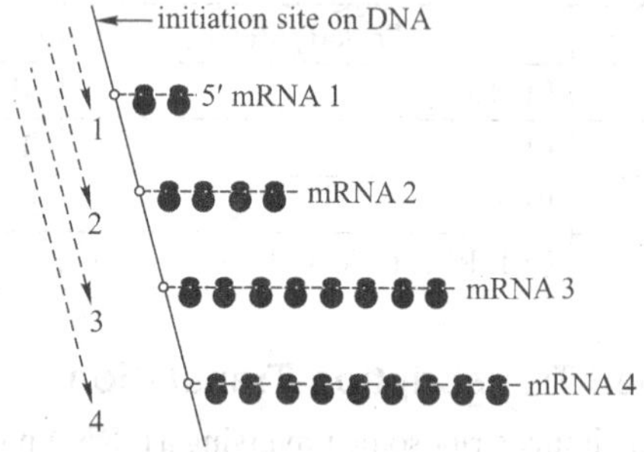

Fig. 40.20. Coupled transcription-translation. Transcription of a section of the DNA of *E.coli* and translation of the nascent mRNA. The dashed arrows show the distances of each RNA polymerase from the transcription initiation site, mRNA 4 probably becomes shorter due to partial digestion off its 5′ end by a RNase (after Freifelder, 1985).

40.4. ANTIBIOTICS AND PROTEIN SYNTHESIS

Many antibacterial agents (called **antibiotics**) have been isolated from fungi. Most of these are inhibitors of protein synthesis. For example, streptomycin and neomycin bind to a particular protein in the 30S particle and thereby prevent binding of tRNA^{fmet} to the 'P' site; the tetracyclines inhibit binding of charged tRNA; lincomycin and chloramphenicol inhibit the peptidyl transferase and puromycin causes premature chain termination; erythromycin binds to a free 50S particle and prevents formation of the 70S ribosome. A particular antibiotic has clinical value only when it acts on bacteria and not on animal cells; the clinically useful antibiotics usually either fail to pass through the cell membrane of animal cells or do not bind to eukaryotic ribosomes, because of some unknown feature of their structure.

Some disease-causing bacteria exert their pathogenic effect, because they excrete inhibitors of mammalian protein synthesis. The agent causing diphtheria is an example; it binds to a factor necessary for movement of mammalian ribosomes along the mRNA (Box 40.2).

Box 40.2 Certain other poisons of Cells

1. The antibacterial drug nalidixic acid is a DNA-replication-inhibitor; it binds with gyrase A subunit.
2. The antibacterial drug novobiocin is also a DNA-replication-inhibitor; it binds to gyrase B subunit and inhibits ATP cleavase.
3. The centrosome duplication is arrested in the mammalian cells by antibiotic amphidicolin. DNA polymerases L, S and E are sensitive to inhibition by amphidicolin (see **Mathews**, 2000).
4. Microtubular poisons are benomyl and nocodazole.

Nalidixic acid.

QUESTIONS

Long Answer Questions

1. Describe in brief the complete process of protein synthesis.
2. Summarize the events of initiation and termination during protein synthesis.

Short Answer Questions

1. How do the functions of the rRNA, mRNA and tRNA differ?
2. What are the functions of the A, P and E sites of the ribosome during protein synthesis?

3. Write short notes on the following:
   (*i*) Central dogma
   (*ii*) Teminism
   (*iii*) Kozak's scanning hypothesis

Very Short Answer Questions

1. Which process in protein synthesis requires hydrolysis of GTP?
2. Write down the Kozak consensus.
3. What is Shine-Dalgarno sequence?

Fill in the Blanks

1. The hydrolysis of GTP to GDP + Pi produces ________ changes in the initiation complex.
2. Another name of elongation factor, EF-G is ________.
3. 5′ UTR stands for ________.
4. In the termination step of protein synthesis, the protein factors RFs stand for ________.

Multiple Choice Questions

1. Initiation of polypeptide chain formation is always brought about in the site of a codon coding for an amino acid called
   (*a*) isoleucine (*b*) cysteine
   (*c*) phenylalanine (*d*) methionine
2. During protein synthesis anticodon of tRNA binds with
   (*a*) codon of mRNA
   (*b*) codon of tRNA
   (*c*) deoxyribonucleotide sequence of DNA
   (*d*) rRNA
3. tRNA recognises aminoacyl synthetase enzyme by
   (*a*) anticodon (*b*) DHU loop
   (*c*) TψC loop (*d*) AA-site
4. The starting tRNA of prokaryotes is loaded with
   (*a*) valine
   (*b*) methionine
   (*c*) formylated methionine
   (*d*) tryptophan
5. The site of protein synthesis in plants is the
   (*a*) mitochondria (*b*) pyrenoids
   (*c*) ribosomes (*d*) chloroplasts
6. The name of Temin and Baltimore is associated with
   (*a*) photorespiration
   (*b*) RNA synthesis
   (*c*) reverse transcription
   (*d*) all of these
7. Which one is not involved in protein synthesis?
   (*a*) transcription (*b*) initiation
   (*c*) elongation (*d*) termination

ANSWERS

Very Short Answer Questions

1. Formation of the 70S initiation complex and translocation.
2. In eukaryotes, it is 5′-ACCAUGG-3′. It contains initiation codon (AUG).
3. 5′-AGGAGGU-3′.

Fill in the Blanks

1. Conformational 2. Translocase 3. 5′ untranslated region
4. Release factors

Multiple Choice Questions

1. (*d*) 2. (*a*) 3. (*b*) 4. (*c*) 5. (*c*) 6. (*c*) 7. (*a*)

41 Regulation of Gene Action

41.1. REGULATION OF GENE ACTION IN PROKARYOTES

Bacteria are exposed to a wide range of environmental conditions. For example, *E.coli* may encounter rapidly changing growth conditions as they pass from mammalian intestinal tract to sewer systems to polluted rivers, lakes, ponds and so on. Each of these ecological niches will provide different organic molecules for use as energy sources. From this fact, it can be concluded that natural selection will have preserved those organisms that have evolved ways of adapting to the wide range of environmental conditions encountered during their evolution. Indeed, the available information indicates that most prokaryotes such as *E.coli* exhibit remarkable capacities to adapt to diverse environmental conditions.

To a great extent, the adaptability of bacteria and other prokaryotes depends on their ability to "turn on" and "turn off" the expression of specific sets of genes in response to the specific demands of the environmental milieu (*i.e.*, surroundings). In other words, prokaryotes exhibit a outstanding ability to **regulate the expression of specific genes in response to environmental signals** (Box 41.1). The expression of particular genes is "turned on" when the products of these genes are needed for growth in a given environment. Their expression is "turned off" when their products are no longer needed for growth in the existing environmental surroundings. Clearly, the ability of an organism to regulate gene expression in this way will increase its overall "fitness" (*i.e.*, its ability to grow and leave progeny under a variety of environmental conditions). The synthesis of gene transcripts (*i.e.*, various types of RNA molecules) and translation products (*i.e.*, proteins) require the expenditure of great amount of energy. By "turning off" the expression of genes when their products are not needed, an organism can avoid wasting energy to synthesise products that maximize the growth rate in the existing environmental surroundings.

Box 41.1

The chromosome of the bacterium *Escherischia coli*, a single celled organism, consists of a single circular DNA molecule of about 4.6 × 10^6 nucleotide pairs. This DNA encodes approximately 4300 proteins, although only a fraction of these are made at any one time. The expression of many of them is regulated according to the available food in the environment (see Alberts *et al.*, 2002).

Here, one can ask three sort of questions:

1. What are the mechanisms by which these organisms regulate gene expression in response to changes in the environment?

2. Is there a single mechanism by which the expression of different genes or sets of genes are regulated?
3. Or are different genes controlled by different mechanisms?

Some genes, for example, the genes specifying ribosomal RNAs, ribosomal proteins and transfer RNAs, are certainly expressed at some time in virtually all cells regardless of the environment conditions. The products of these genes are required for growth of all cells in all environments. However, the products of many other genes are required for growth only in certain environments, and *the expression of these genes is regulated such that the products are synthesized only when they are needed.* As a result, the expression of these genes is continually being "turned on" and "turned off" in response to changes in the environment. From this discussion, one can conclude that gene expression can be (and is) regulated at several different levels: for example, **transcription, mRNA processing, mRNA turnover, translation** and **enzyme function.** However, extensive data indicate that the regulation of transcription is the most important mode of control of gene expression, at least in prokaryotes.

The regulatory mechanisms of transcription in both prokaryotes and eukaryotes are of two basic types. The first, and best understood, category includes mechanisms involved in rapid turn-on and turn-off of gene expression in response to environmental changes. Such sort of mechanisms are very significant in microorganisms (bacteria) because these organisms are exposed to sudden changes in environment. They provide microorganisms with a great deal of "plasticity", an ability to rapidly adjust their metabolic processes in order to achieve maximal growth and reproduction under highly variable environmental conditions. These quick responding on-off switches seem to be less important in higher eukaryotes. This might be expected since the circulatory systems of higher eukaryotes buffer their cells against many sudden environmental changes.

The second major category of regulatory mechanisms includes the so called **preprogrammed circuits of gene expression**. In these cases, some event (*e.g.*, infection by a virus) triggers the expression of one set of genes. The product (or products) of one (or more) of these genes function by turning off the transcription of the first set of genes and/or turning on the transcription of the second set of genes. In turn, one or more of the products of the second set acts by turning on a third set, and so on. In these cases, *the sequential expression of genes is genetically preprogrammed*, and the genes usually cannot be turned on out of sequence. Such preprogrammed sequences of gene expression in viral infections are well worked out. In most of these preprogrammed sequences, it seems the circuit is cyclical. For example, in viral infections some event associated with the packaging of the viral DNA or RNA inside the protein coat somehow seems to reset the program so that the first set of genes will again be expressed when a progeny virus subsequently infects another host cell.

Discovery of Prokaryotic Gene Regulatory Proteins

Genetic analyses in bacteria carried out in the 1950s provided the first evidence for the existence of **gene regulatory proteins** that turn specific sets of genes on or off. One of these regulators, the **lambda repressor**, is encoded by a bacterial virus, the bacteriophage lambda. The repressor shuts off the viral genes that code for the protein components of new virus particles and thereby enables the viral genome to remain a silent passenger in the bacterial chromosome, multiplying with the bacterium when conditions are favourable for bacterial growth. The lambda repressor was among the first gene regulatory proteins to be characterised. Other bacterial regulators respond to nutritional conditions by shutting off genes encoding specific sets of metabolic enzymes when they are not needed. For example, **lac repressor**, the first of these bacterial proteins to be recognised, turns off the production of the proteins responsible for lactose metabolism when this sugar is absent from the medium.

The first step toward understanding gene regulation was the isolation of mutant strains of bacteria and bacteriophage lambda that were unable to shut off specific sets of genes. It was proposed at the time, and later proven, that most of these mutants were deficient in proteins acting as specific repressors for these sets of genes. Because, these proteins, like most gene regulator proteins, are present in small quantities, it was difficult and time consuming to isolate them. They were eventually purified by fractionating cell extracts. Once isolated, the proteins were shown to bind to specific DNA sequences close to the genes that they regulate. The precise DNA sequences that they recognized were then determined by a combination of classical genetics, DNA sequencing and DNA footprinting experiments.

Types of Gene Regulations in Prokaryotes

1. Constitutive Genes and Inducible Genes

Certain gene products, such as tRNA molecules, rRNA molecules, ribosomal proteins, RNA polymerase components (polypeptides), and other enzymes catalyzing metabolic processes that are commonly known for cellular "house keeping" functions, are essential components of almost all living cells. Genes that specify products of this type are continually being expressed in most cells. Such genes are said to be expressed **constitutively** and are called **constitutive genes**.

Rest of the gene products are needed for cell growth only under certain environmental conditions. Constitutive synthesis of such gene products would clearly be wasteful, using energy that could otherwise be utilized for more rapid growth and reproduction under the existing environmental conditions. The *evolution* of regulatory mechanisms that would provide for the synthesis of such gene products only when and where they were needed would clearly provide organisms with a selective advantage over organisms lacking these mechanisms. This undoubtedly explains why presently existing organisms, including the "primitive" bacteria and viruses, exhibit highly developed and very efficient mechanisms for the control of gene expression.

E.coli and most other bacteria are capable of growth using any one of several carbohydrates (*e.g.*, glucose, sucrose, galactose, arabinose, lactose) as an energy source. If glucose is present in the environment, it will be preferentially metabolised by *E.coli* cells. In the absence of glucose, however, *E.coli* cells can grow very well on other carbohydrates. Cells of *E.coli* growing in medium containing the sugar lactose, for example, as the sole carbon source synthesise two enzymes, **β-glalactosidase** and **β-galactoside permease**, that they uniquely synthesized for the catabolism of lactose (Note: A third enzyme, **α-galactoside transacetylase** is also synthesized). The enzyme β-galactosidase cleaves lactose into glucose and galactose, and enzyme β-galactoside permease pumps α-galactosides into the cell. Neither of these enzymes is of any use to *E.coli* cells when present in an environment not containing lactose. The synthesis of these two enzymes, ofcourse, requires the utilization of great amount of energy (in the form of ATP and GTP). Thus, *E.coli* cells have evolved a regulatory mechanism by which the synthesis of these lactose catabolic enzymes is turned on in the presence of lactose and turned off in its absence.

In natural environments (such as intestinal tracts and sewers), *E.coli* cells probably encounter an absence of glucose and the presence of lactose relatively rarely. Most of the time, therefore, *E.coli* cells growing on a carbohydrate other than lactose are transferred to medium containing lactose as the only carbon source, they rapidly begin sythesising the enzymes required for lactose utilization. This process, by which the expression of genes is turned on in response to a substance in the environment, is called **induction**. Genes whose expression are so regulated are called **inducible genes**. The substances or molecules responsible for induction are known as **inducers** (*e.g.*, in this case lactose is inducer).

Enzymes that are involved in **catabolic** (*i.e.*, degradative) **pathway**, such as in lactose, galactose or arabinose utilization, are characteristically inducible. Researches made it evident that *induction occurs at the level of transcription.* Induction alters the rate of synthesis of enzymes, not the activity of existing enzyme molecules. Induction should not be confused with **enzyme activation**, in which the binding of a small molecule to an enzyme increases the **activity** of the enzyme (but does not affect its rate of synthesis).

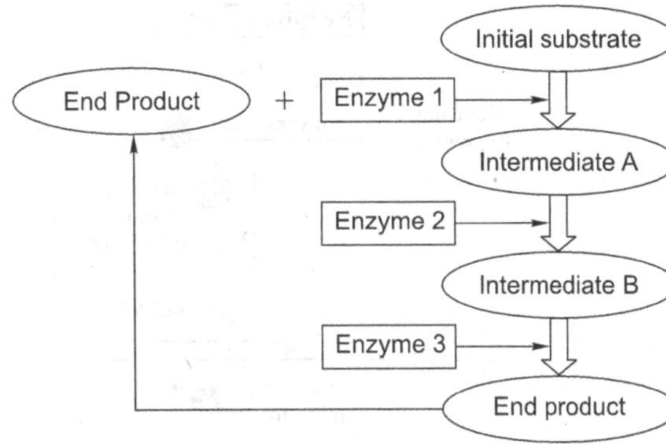

Feedback inhibition.

Further, *E.coli* and other bacteria possess the metabolic capacity to synthesise most of the organic molecules (such as amino acids, purines and vitamins) required for their growth. For example, *E.coli* has five genes coding for enzymes that are required in the synthesis of tryptophan (an amino acid). These five genes must be expressed in *E.coli* cells growing in an environment devoid of tryptophan in order to provide adequate amounts of this amino acid for ongoing protein synthesis. When *E.coli* cells are present in an environment containing concentrations of tryptophan sufficient to support optimal growth, the continued synthesis of the tryptophan biosynthetic enzymes would be a waste of energy, because these bacteria have the capacity to take in external tryptophan. Thus, a regulatory mechanism has evolved in *E.coli* by which the synthesis of the tryptophan is stopped until tryptophan is present in the external milieu (surrounding environment). This process of "turning off" the expression of sets of genes is called **repression**. A gene whose expression has been turned off in this way is said to be repressed; when its expression is turned on, a gene of this type is said to be **depressed**.

Enzymes that are components of **anabolic** (biosynthetic) **pathways** are frequently subject to **repression**. *Repression, like induction occurs at the level of transcription.* However, repression should not be confused with **feedback inhibition**, in which the binding of an end product to the first enzyme in a biosynthetic pathway **inhibits the activity** of the enzyme (but does not affect its synthesis).

2. Transcriptional Control Mechanisms

In bacteria, there occur several mechanisms of gene regulation at the level of transcription. A notable method depends on whether the enzymes being regulated act in catabolic (degradative) or anabolic (synthetic) metabolic pathways. For example, in a multistep catabolic system, the availability of the molecule to be degraded commonly determines whether the enzymes in the pathway will be synthesised. In contrast, in a biosynthetic pathway the final product is often the regulatory molecule. Even when a single protein molecule is translated from a monocistronic mRNA molecule, the protein may be **autoregulated**, *i.e.*, the protein itself may inhibit initiation of transcription and high concentrations of the protein may cause less transcription of the mRNA that encodes the protein. The molecular mechanisms for each of the regulatory patterns differ greatly and are of the following two types—negative regulation and positive regulation (Fig. 41.1). In a **negative regulated system**, an inhibitor is present in the cell and prevents transcription. An antagonist of the inhibitor, called an **inducer**, is needed to allow initiation of transcription. In a **positively regulated system**, an **effector** molecule (which may be protein, a small molecule, or a molecular complex) activates a promoter; no inhibitor must be abolished. Negative and positive regulation are not mutually exclusive, and some systems are both positively and negatively regulated, utilizing two regulators to respond to different conditions in the cell. Thus, a catabolic system may be regulated positively or negatively. In a biosynthetic (anabolic) pathway, the final product usually regulates negatively its own synthesis; in the simplest type of negative regulation, absence of the product increases its synthesis and presence of the product decreases its synthesis.

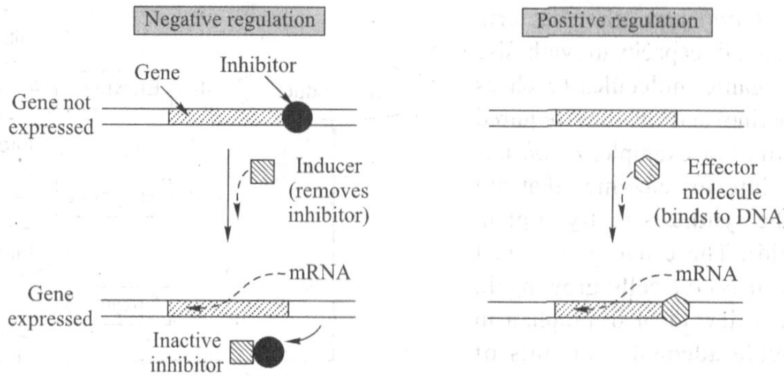

Fig. 41.1. The distinction between negative and positive regulation. In negative regulation an inhibitor, bound to the DNA, must be removed before transcription can occur. In positive regulation, an effector molecule must bind to the DNA (after Freifelder, 1985).

The Operon Model

Induction and repression of gene expression can be accomplished by essentially the same mechanism. This mechanism was first accurately described in 1961 when **Francois Jacob** and **Jacques Monod**, both 1965 Nobel Prize recipients, proposed the **operon model** to explain the regulation of genes encoding the enzymes required for lactose utilization in *E.coli*. **Jacob** and **Monod** proposed that the transcription of one or a set of contiguous structural genes (*i.e.*, genes coding for polypeptides) is regulated by two controlling elements (Fig 41.2). One of these controlling elements, called the **regulator gene** (or **repressor gene**), codes for a protein called the **repressor**; under the appropriate conditions, the repressor binds to the second element, the **operator**. The operator is always located contiguous to the structural gene or genes whose expression it regulates. *When the repressor is bound to the operator, transcription of the structural genes cannot occur*. We now know that this results because the binding of the repressor to the operator stearically prevents RNA polymerase (enzyme)

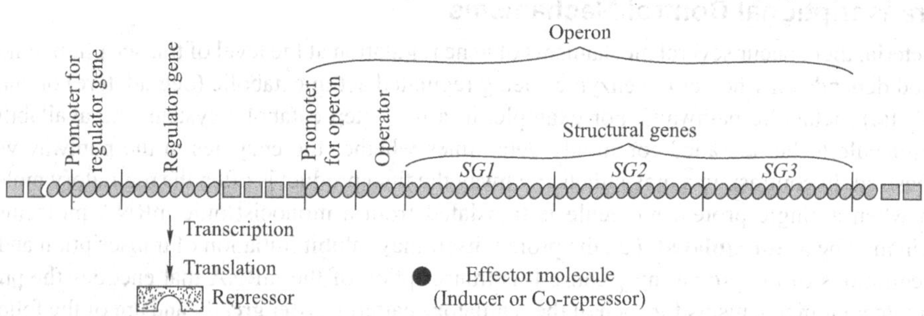

Fig. 41.2. The operon model for regulation of gene expression. The operon consists of one or more structural genes (three—*SG1*, *SG2* and *SG3*—are arbitrarily shown) and the adjoining operator and promoter sequences. The promoter for the operon (*PO*) is the site at which RNA polymerase enzyme must bind to initiate transcription of structural genes. The operator (*O*) is the site at which protein repressor (the products of the regulator gene or repressor gene) binds. The regulator gene need not be closely linked to the operon; in fact, it can be located at any position in the genome. The transcription of the regulator gene is initiated by RNA polymerase, which binds to its promoter (labeled *PR*, for promoter for regulator gene). When the repressor is bound to the operator, it stearically prevents RNA polymerase from binding to adjoining promoter (*PO*) and from initiating transcription of the structural genes. Whether the repressor binds to the operator or not, depends on the presence or absence of a metabolite called an effector molecule (after Gardner *et al.*, 2002).

from binding at the **promoter site** (the RNA polymerase binding site), which is always located contiguous with the operator sequence. The **operator** is usually located between the promoter and structural genes. (Note: The promoter was not recognized at the time of Jacob and Monod's proposal, but has since been shown to be as essential component of an operon). Thus, the complete contiguous unit, including the structural genes, the operator and the promoter is called the operon (**Gardner** *et al.*, 2002). In other words, an operon is a unit of prokaryotic gene expression which includes co-ordinately regulated (structural) genes and control elements (promoter and operator) which are recognized by regulatory gene products (**Turner** *et al.*, 2000).

Whether the repressor will bind to the operator and turn off the transcription of the structural genes in an operon is determined by the presence or absence of **effector molecules** (small molecules such as amino acids and sugars) in the environment. In the case of **inducible operons** (*e.g.*, *lac* operon), these effector molecules are called **inducers**. Those active on repressible operons (*e.g.*, *trp* operon) are called **co-repressors** act by binding to (or forming a complex with) the repressors.

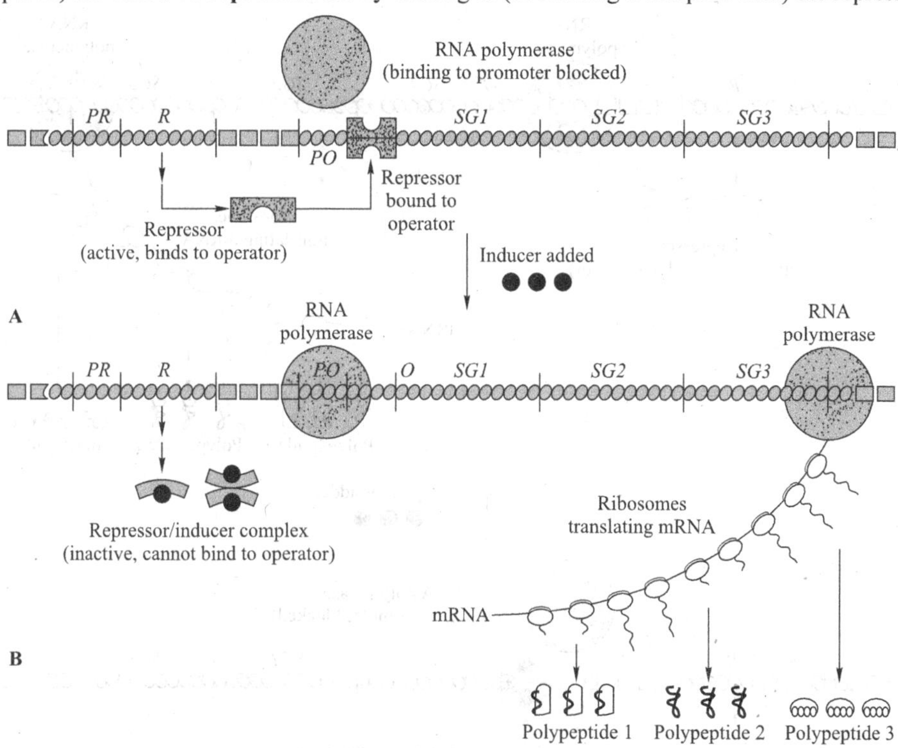

Fig. 41.3. Mode of regulation of gene expression for an inducible operon (*e.g.*, lac operon). A—The product of the regulator gene (*R*), the repressor, in the absence of the effector molecule (called an inducer, for inducible operon, *e.g.*, allolactose in lac operon), binds to the operator, preventing RNA polymerase from binding to the promoter for the operon (*PO*). Thus, transcription of the structural genes cannot occur. B—When inducer is added, it binds to the repressor, causing it to be released from the operator (*O*). This, in turn, allows RNA polymerase to bind to the promoter (*PO*) and initiate transcription of the structural genes. The resulting multigenic (= polycistronic) mRNA is rapidly translated by ribosomes, producing the three polypeptide products of the structural genes (after Gardner *et al.*, 2002).

The only essential difference between inducible operons and repressible operons is whether the naked repressor or the repressor effector molecule complex is active in binding to the operator. 1. In case of an inducible operon, the free repressor binds to the operator, turning off (Fig. 41.3). When the effector molecule or the inducer (*e.g.*, allolactose in *lac* operon) is present, it binds to the repressor, that is, the **repressor-inducer complex cannot bind to the operator**. Thus, the addition of inducer turns on (or induces) the transcription of the structural genes in the operon (Fig. 41.3). 2. In case of a repressible operon (*e.g.*, *trp* operon), the situation is just reversed. The free repressor-cannot bind to the operator. Only the repressor-effector molecule (co-repressor) complex is active in binding to the operator (Fig. 41.4). Thus, transcription of the structural genes in repressible operon is turned on in the absence of and turned off in the presence of the effector molecule (co-repressor). Except for this difference in the operator repressible operons are comparable.

A single mRNA transcript carries the coding information of an entire operon. Thus, the mRNAs of operons consisting of more than one structural gene are polygenic or polycistronic. For example, the tryptophan operon mRNA of *E.coli* is a huge macromolecule carrying the coding sequences that specify five different polypeptides because of their cotranscription, all the structural genes in an operon are coordinately expressed.

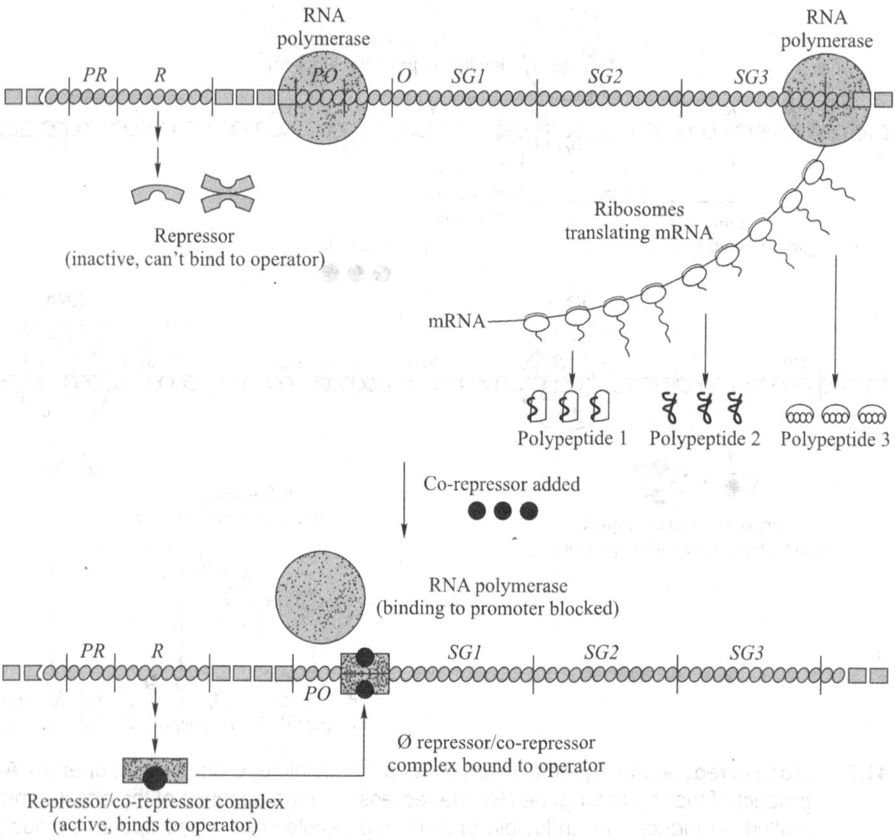

Fig. 41.4. Mode of regulation of gene expression for a repressible operon (*e.g.*, *trp* operon). In this case, the repressor can only bind to the operator in the presence of effector molecule (called a co-repressor, for repressible operons; *e.g.*, tryptophan amino acid in *trp* operon). In its absence, the operator is free, permitting RNA polymerase to bind at the adjoining promoter (*PO*) and to initiate transcription of the structural genes. When co-repressor is added, it forms a complex with the repressor. This repressor—co-repressor complex then binds to the operator (*O*). This, in turn, prevents RNA polymerase from binding at *PO* and stops transcription of the three structural genes (after Gardner *et al.*, 2002).

Examples of Operons

1. *Lac* Operons (Inducible System)

Lactose (milk sugar, a disaccharide) is a β-galactoside that *E.coli* can use for energy and as a carbon source after it is broken down into glucose and galactose. The enzyme that performs the breakdown is **β-galactosidase:**

$$\underset{\text{(a disaccharide)}}{\text{Lactose}} \xrightarrow{\beta\text{-galactosidase}+H_2O} \text{Galactose} + \text{Glucose}$$

The β-galactosidase enzyme can additionally convert lactose into allolactose, the inducer of *lac* opron (Fig. 41.5).

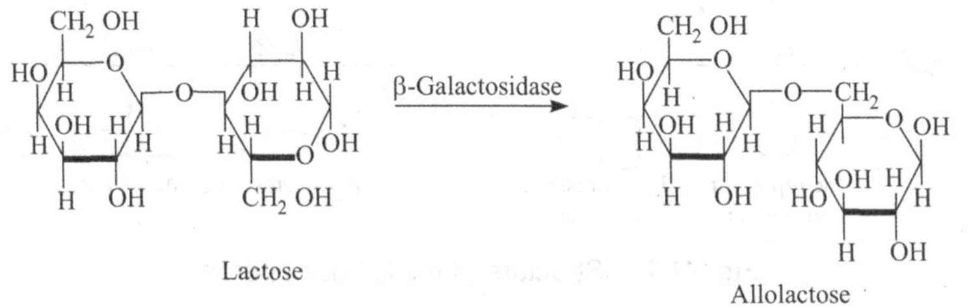

Fig. 41.5. The enzyme β-galactosidase can also convert lactose to allolactose (after Tamarin, 2002).

There are very few molecules of β-galactosidase enzyme in a wild type *E.coli* cell grown in the absence of lactose. Within minutes after adding lactose to the medium, however, this enzyme appears in large quantity within the bacterial cell. When the synthesis of β-galactosidase (encoded by the *lacZ* or *z gene*) is induced, the production of two additional enzymes is also induced, **β-galactoside permease** (encoded by the *lacY* or *y gene*) and **β-galactoside acetyltransferase** (encoded by the *lacA* or *a* gene). The permease (enzyme) is involved in transporting lactose into the cell. The transferase is believed to protect the cell from the build up of toxic products created by β-galactosidase (enzyme) acting on other sugars than lactose, the transferase prevents β-galactosidase from cleaving them (see **Tamarin**, 2002).

The regulator gene. Not only are three lac genes (*z*, *y* and *a*) induced together, but they are adjacent to one another in the *E.coli* chromosome; they are, in fact, transcribed on a single polycistronic messenger RNA (Fig. 41.6). Induction involves the protein product of another gene, the *regulator* gene, or *i* gene (*lac I*). Although the regulator gene is located adjacent to the three *lac* genes (*i.e.*, three structural genes), it is totally independent transcriptional entity. The regulator gene specifies a protein, the **repressor**, that interferes with the transcription of the genes involved in lactose metabolism.

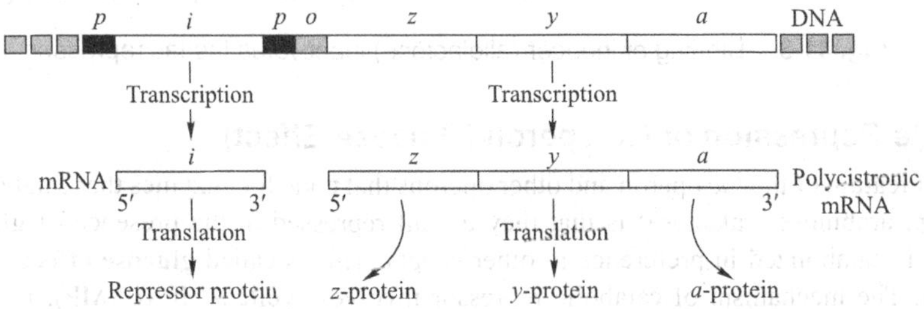

Fig. 41.6. The *lac* operon is transcribed as a multigenic (polycistronic) mRNA. The *z*, *y* and *a* indicate the *lac Z*, *lac Y* and *lac A* loci. The mRNA transcript is then translated as individual proteins. The regulator gene of *lac* operon is denoted as *i*; the *o* stands for operator and *p* for promoter. Both the operon and the regulator gene have their own promoters (*p*) (after Tamarin, 2002).

The operator. For the repressor protein to exert its influence over transcription, there must be a control element (receptor site) located near the beginning of the β-galactosidase (*lacZ*) gene. This control element is a region referred to as the **operator**, or operator site (Fig. 41.6). The operator site is a sequence of DNA that the product of the regulator gene, the repressor, recognizes. When the repressor is bound to the operator, it interferes with RNA polymerase binding or prevents the RNA polymerase from achieving the open complex. In either case, transcription of the operon is prevented (Fig. 41.8). The repressor is released when it combines with an **inducer**, a derivative of lactose called **allolactose** (Fig. 41.7).

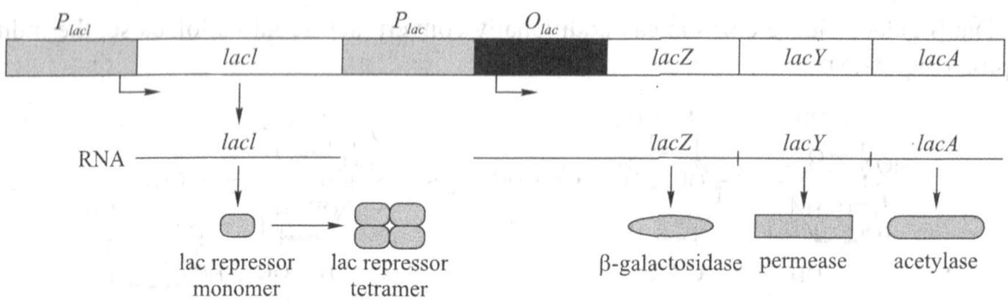

Fig. 41.7. Structure of the lactose operon.

Note that the promoter is not only recognized by RNA polymerase but also has elements in the immediate vicinity of the initiation site of transcription. At this stage, operon can be redefined as *a sequence of adjacent genes all under the transcriptional control of the same promoter and operator.*

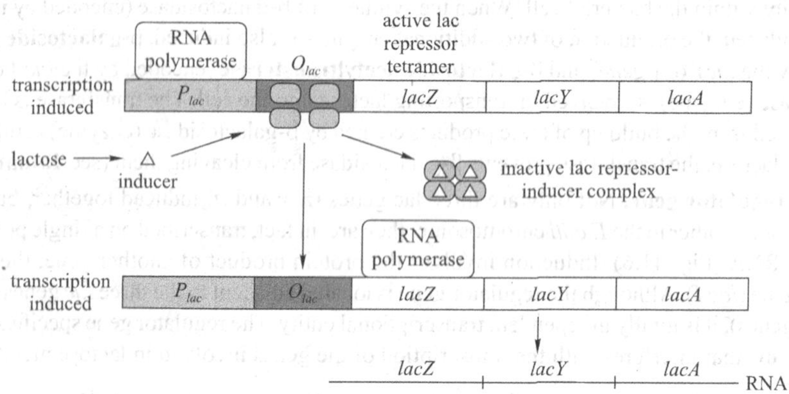

Fig. 41.8. Binding of inducer (allolactose) inactivates the lac repressor.

Catabolite Repression of *lac* operon (Glucose Effect)

An unusual feature of the *lac* operon and other operons that code for enzymes that catabolize certain sugars (*e.g.*, arabinose, galactose) is that they are all repressed in the presence of glucose. That is, glucose is catabolized in preference to other sugars. This is called **glucose effect** or **catabolic repression**. The mechanism of catabolic repressor involves cyclic AMP (cAMP). In eukaryotes, cAMP acts as a second messenger, an intracellular messenger regulated by certain extracellular hormones. Geneticists were surprised to discover cAMP in *E.coli*, where it works in association with another regulatory protein, the **catabolite activator protein (CAP),** to control the transcription of certain operons.

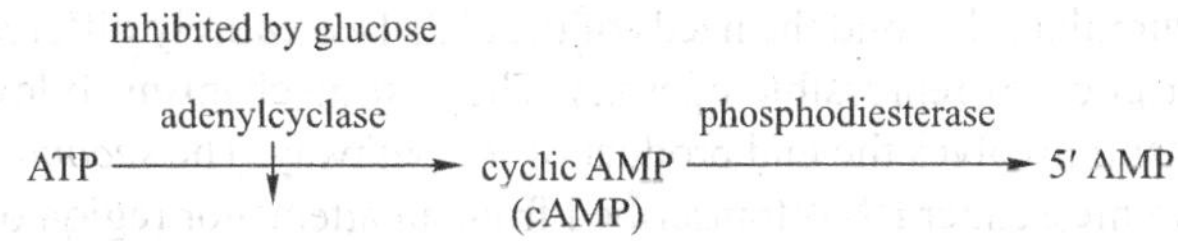

Fig. 41.9. Glucose effect. Glucose uptake lowers the quantity of cAMP (second messenger) in the cell by inhibiting the enzyme adenylcyclase, which converts ATP to cAMP.

When glucose is not present, cAMP combines with CAP, and the CAP-cAMP complex binds to a distal part of the promoter of operons with CAP sites (*e.g.*, the lac operon; see Fig. 41.10). This binding clearly increases the affinity of RNA polymerase for the promoter, because without the binding of the CAP-cAMP complex to the promoter, the transcription rate is very low. The uptake of glucose by *E.coli* cells causes the loss of cAMP from the cell, probably by inhibiting adenylcyclase (Fig. 41.9) and thus lowers the CAP-cAMP level. The transcription rate of operons with CAP sites will be reduced (Fig. 41.10). The same reduction of transcription rates has been observed in mutant strains of *E.coli* when this part of the distal end of the promoter is deleted. The binding of CAP-cAMP to the CAP site causes the DNA to bend more than 90 degrees. This bending by itself, may enhance transcription, making the DNA more available to RNA polymerase enzyme.

During process of initiation of transcription, the CAP is in direct contact with RNA polymerase. This was shown by photocross linking studies in which CAP was treated with a cross-linking agent that bound the subunit of RNA polymerase when irradiated with UV light. (**Photocrosslinking** is a technique used to determine which moieties (proteins, DNA) are in close proximity during a particular process). For the two proteins to cross-link, they must be in direct contact during the initiation of transcription.

Catabolite repression is an example of positive regulation: Binding of the CAP-cAMP complex at the CAP site enhances the transcription rate of the transcriptional unit. Thus, the *lac* operon is both positively and negatively regulated; the repressor exerts negative control, and the CAP-cAMP complex exerts positive control of transcription.

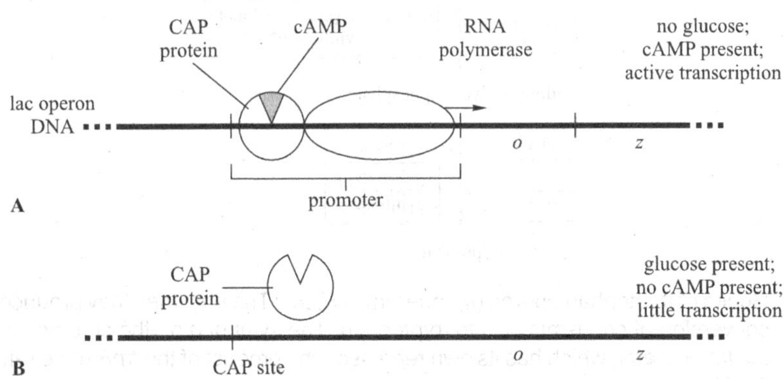

Fig. 41.10. Catabolite repression. When cAMP is present in the cell (no glucose is present), it binds with CAP protein and together they bind to the CAP site in various sugar metabolizing operons, such a *lac* operon shown here. The CAP-cAMP complex increases the transcription of the operon. When glucose is present, it inhibits the formation of cAMP. Thus no CAP-cAMP complex forms, and transcription of the same operon is reduced (after Tamarin, 2002).

2. TRP Operon

The **catabolic operons** or inducible operons are activated when the substrate that is to be catabolized enter the cell. **Anabolic operons** function in the reverse manner. They are turned off (repressed) when their end product accumulates beyond the needs of the cell. Two entirely different mechanisms seem to control the transcription of repressible operons. The *first* mechanism follows the basic scheme of inducible operons and involves the end product of the pathway. The *second* mechanism involves secondary structure in messenger RNA transcribed from an attenuator region of the operon.

(I) Tryptophan synthesis. Tryptophan (trp) operon is an example of a repressible system. It contains the five genes that code for the synthesis of the enzymes that build tryptophan amino acid, starting from **chorismic acid** (Fig. 41.11). It has a promoter operator sequence (*p*, *o*) as well as its own regulator gene (*trpR*).

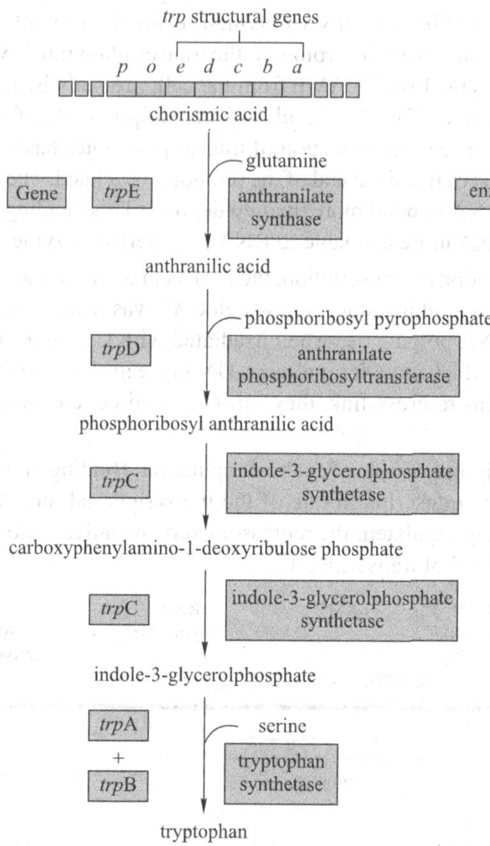

Fig. 41.11. Genes of tryptophan operon (*trp* operon) in *E.coli*. The enzymes they produce control the conversion of chorismic acid to tryptophan. The symbol *o* on the chromosome refers to the *trp* operator, which has its own repressor, the product of the *trpR* gene (after Tamarin, 2002).

(II) Operator control. In *trp* operon, the product of *trpR* gene is called **repressor**. It is inactive by itself and it does not recognize the operator sequence of *trp* operon. The repressor only become active when it combines with tryptophan. Thus, when tryptophan builds up, enough is available to bind with and activate the repressor. Tryptophan is, thus, called **corepressor**. The corepressor-repressor complex then recognizes the operator, binds to it, and prevents transcription by RNA polymerase.

After the available tryptophan in the cell is used up, the diffusion process causes tryptophan to leave the repressor, which then detaches from the trp operator. Since, the transcription process no longer is blocked, so proceeds normally (*i.e.*, the operon is now derepressed). Transcription continues until enough of the various enzymes are synthesized to again produce an excess of tryptophan. Some tryptophan molecules become available to bind to the repressor and make a functional complex, and the operon is again shut off. This process is repeated to ensure that tryptophan is being synthesized as needed (Fig. 41.12). This regulation is modified, however, by the existence of the second mechanism for regulating repressible operons—the attenuation.

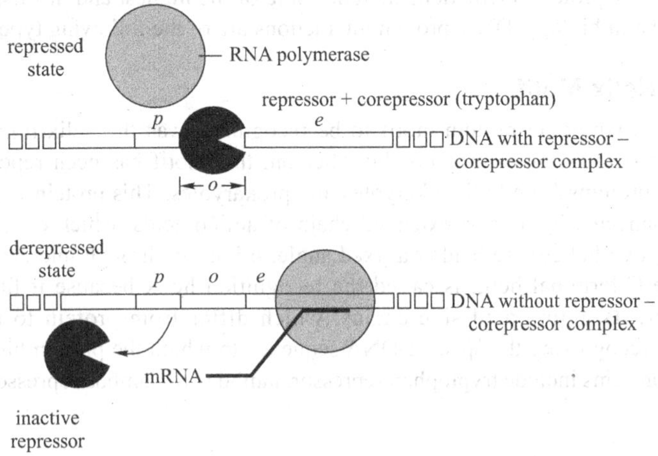

Fig. 41.12. The repressor-corepressor complex binds at the operator and prevents the transcription of the *trp* operon in *E.coli*. Without the co-repressor, the repressor cannot bind, and therefore transcription is not prevented (after Tamarin, 2002).

Mode of Working of Gene Regulatory Proteins

Gene regulatory proteins must recognize specific nucleotide sequences embedded within its structure. It was originally thought that these proteins might require direct access to the hydrogen bonds between base pairs in the interior of the double helix to distinguish between one DNA sequence and another. It is now clear, however, that the outside of the double helix is studded with DNA sequence information that gene regulatory proteins can recognize without having to open the double helix.

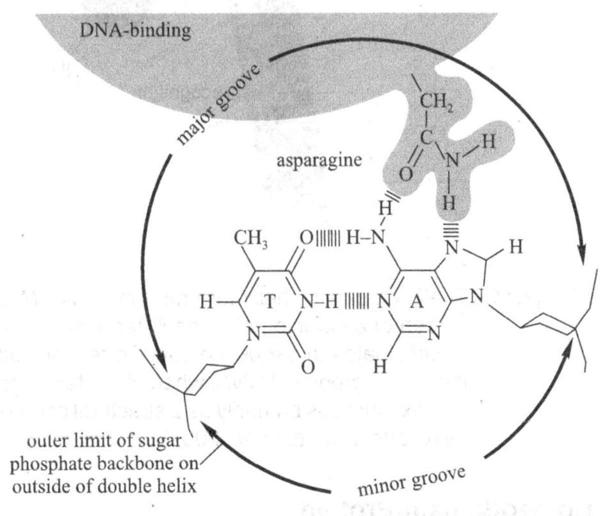

Fig. 41.13. The binding of a gene regulatory protein to the major groove of DNA. Only a single contact is shown. Typically, the protein DNA interface would consist of 10 to 20 such contacts, involving different amino acids, each contributing to the strength of the protein DNA interaction (after Alberts *et al.*, 2002).

Molecular Recognition

Gene regulatory proteins are found to contain structural motifs that read DNA sequences. Molecular recognition in biology generally relies on an exact fit between the surfaces of two molecules, and the study of gene regulatory proteins has provided some of the clearest example of this principle (see **Alberts** *et al.*, 2002). A gene regulatory protein recognizes a specific DNA sequence because the surface of the protein is extensively complementary to the specific surface features of the double helix in that region. In most cases, the protein makes a large number of contacts with the DNA, involving hydrogen bonds, ionic bonds and hydrophobic interactions. Although individual contact is weak, the 20 or so contacts that are typically formed at the protein-helix interface add together to ensure that the interaction is both highly specific and very strong (Fig. 41.13). In fact, DNA-protein interactions include some of the highest and most specific molecular interactions known in biology. DNA-protein interactions are of the following types.

1. Helix-turn-Helix Motif

The first DNA-binding protein motif to be recognized was the helix-turn-helix. This was originally identified in bacterial proteins; but later on, this motif has been reported in hundreds of DNA-binding proteins from both eukaryotes and prokaryotes. This protein is constructed from two α-helices connected by a short extended chain of amino acids, which constitutes the "turn" (Fig. 41.14). The two helices are held at a fixed angle, primarily through interactions between the two helices. The C-terminal helix is called the recognition helix because it fits into the major grooves of DNA; its amino acid side chains, which differ from protein to protein, play an important role in recognizing the specific DNA sequence to which the protein binds. Examples of helix-turn-helix proteins include tryptophan repressor, lambda cro, lambda repressor fragment, CAP fragment, etc.

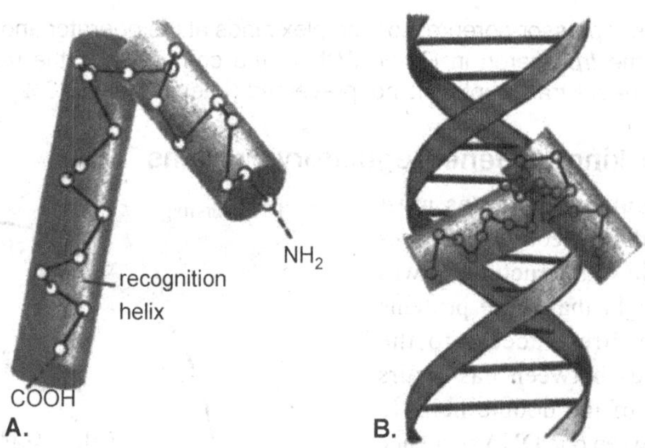

Fig. 41.14. The DNA-binding helix-turn-helix motif. A—Motif with circle which represents the central carbon of an amino acid. The C-terminal α-helix is called the recognition helix because it participates in sequence-specific recognition of DNA. B—The helix of motif fits into the major. groove of DNA, where it contacts the edges of the base pairs. The N-terminal α-helix functions primarily as a structural component that helps to position the recognition helix (after Alberts *et al.*, 2002).

Homeodomain Protein

Soon after the discovery of gene regulatory proteins in bacteria, genetic analyses in the fruitfly *Drosophila* led to the characterisation of an important class of genes, the **homeotic selector genes,** that play a critical part in coordinating fly development. Mutations in these genes cause one body part in the fly to be converted into another, showing that the proteins they encode control critical developmental decisions.

When the nucleotide sequence of several homeotic selector genes of *Drosophila* and higher animals was determined in the early 1980s, each proved to contain an almost identical stretch of 60 amino acids that defines this class of proteins and is termed as the **homeodomain**. When the three-dimensional structure of homeodomain was determined, it was shown to contain a helix-turn-helix motif related to that of the bacterial gene regulatory proteins, providing one of the first indications that the principle of gene regulation established in bacteria are relevant to higher organisms as well. More than 60 homeodomain proteins now have been discovered in *Drosophila* alone, and homeodomain proteins have been identified in virtually all eukaryotic organisms that have been studied, from yeasts to plants to humans.

The structure of a homeodomain that is bound to its specific DNA sequence, has been shown in Figure 41.15. Where as the helix-turn-helix motif of bacterial gene regulatory proteins is often embedded in different structural contexts, the helix-turn-helix motif of homeodomains is always surrounded by the same structure, suggesting that the motif is always presented in the same way.

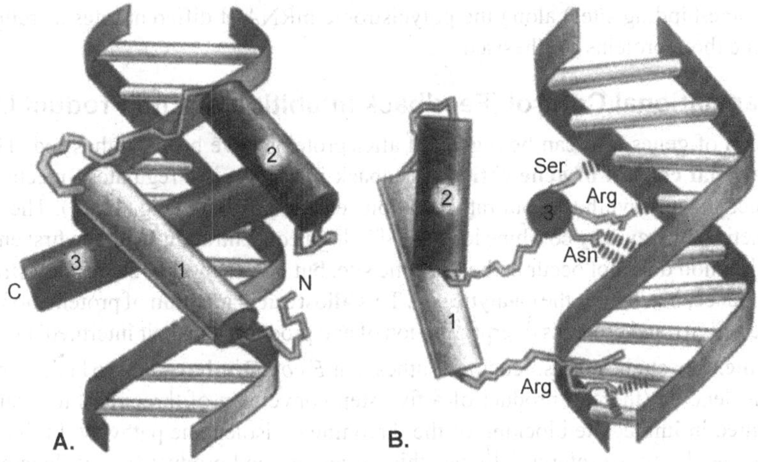

Fig. 41.15. A homeodomain (protein) of yeast bound to its specific DNA sequence. Two different views of the same structure are shown. A—The homeodomain is folded into three α-helices which are packed tightly together by hydrophobic interactions. The part containing helix 2 and 3 closely resembles the helix-turn-helix motif. B—The recognition helix (helix 3) makes important contacts with the major groove of DNA. The asparagine (Asn) of helix 3, for example, contacts an adenine. Nucleotide pairs are also contacted in the minor groove by a flexible arm attached to helix I (after Alberts *et al.*, 2002).

Other types of motifs of gene regulatory proteins includes the following (see **Alberts** *et al.*, 2002).

1. DNA—binding zinc finger motifs;
2. β-sheet proteins;
3. Leucine zipper motif; and
4. Heterodimerization.

3. Translational Control

In prokaryotic gene regulation at the translation level, the lifetime of a mRNA molecule may be genetically determined. Enzymatic degradation of rnRNA is from the 5′ to the 3′ end, *i.e.*, the end of the RNA that is first synthesised is also the end that is first degraded. The average lifetime of many mRNA molecules of *E.coli* is only two minutes at 37°C. The specific nucleotide sequences at the 5′ end may influence its susceptibility to enzymatic digestion. Further, catabolic enzymes are denied access to the mRNA when the ribosome are coated at their 5′ ends (*i.e.*, in case of polyribosomes). Hence, the lifetime of mRNAs may also be correlated with the number of free ribosomes available at any given moment to translate mRNA molecules. Bacteria vary their rates of proteins synthesis by varying their ribosomal content rather than by varying the translational rate.

Example. In the lactose system of *E.coli*, there are three structural genes under control of a common operator locus determining production of (1) β-galactosidase, (2) galactoside permease and (3) galactoside acetylase. These three proteins are produced in the respective ratios 1:1/2:1/5, reflecting their respective locations relative to the 5′ (operator) end of the polycistronic mRNA in which they are coded (these differences are the examples of translation regulations). Thus, there is a *polarity gradient* within the polycistronic mRNA that reduces the probability of cistron translation as a function of its distance from the 5′ end. It is hypothesised that ribosomes attach to different starting points (ribosome-binding sites) along the polycistronic mRNA at different rates as reflected by the amounts of the three proteins synthesised.

4. Post-translational Control (Feedback Inhibition or End Product Inhibition)

The expression of genes also can be regulated after proteins have been synthesised. This is called **post-translational control of gene action**. Feedback inhibition is a regulatory mechanism which does not affect enzyme synthesis, but rather inhibits enzyme activity (Fig. 41.16). The end product of a biosynthetic pathway may combine loosely (if in high concentration) with the first enzyme in the pathway. This union does not occur at the catalytic site, but it does modify the tertiary structure of an enzyme and, hence, inactivates the catalytic site. This **allosteric transition** of protein molecule blocks its enzymatic activity and prevents overproduction of end products and their intermediate metabolites.

Example. The studies on isoleucine synthesis in *E.coli* (**Umbarger**, 1961) demonstrated that addition of isoleucine (the end product of a five step conversion of threonine) to a culture of the bacteria resulted in immediate blocking of the threonine → isoleucine pathway. In the presence of added isoleucine, the cells preferentially use this exogenous end product (*i.e.*, isoleucine) and their own isoleucine synthesis becomes ceased. Moreover, the production of each of the five enzymes is not interfered with, but action of an enzyme responsible for deamination of threonine to α-ketobutyrate is inhibited by the end product, isoleucine.

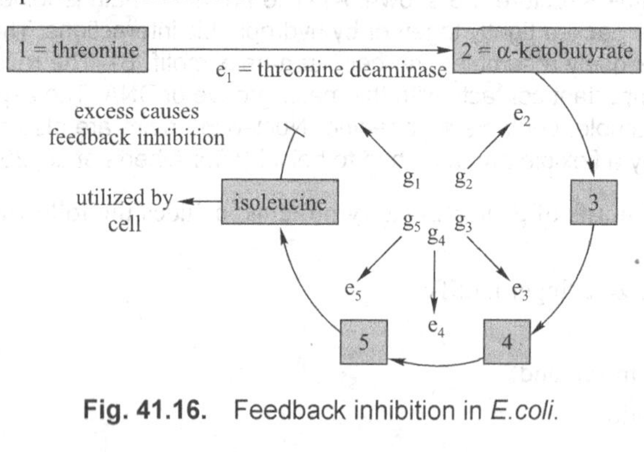

Fig. 41.16. Feedback inhibition in *E.coli*.

41.2. OPERONS OF PHAGE LAMBDA (Regulation of Phage λ Life Cycles)

When a bacteriophage infects a bacterial cell, it has to express its genes in an orderly fashion; some gene products are needed early in infection, and other products are required during later part of infection. **Early genes** usually control phage DNA replication; **late genes** usually determine phage coat proteins and the lysis of bacterial cell. A bacteriophage is considered most efficient if it expresses early genes first and the late genes last in the infection process. Also, temperate phages have the option of entering into lysogeny with the cell (see Box 41.2), here, also, control processes determine which path is taken. One generalisation has been made regarding the phages that their genes are clustered into early and late operons, with separate transcriptional control mechanisms for each.

Box 41.2

Lysis: Bursting of a cell by the destruction of plasma membrane following an infection by a virus.

Lysogenic: The state of a bacterial cell that has an integrated phage (prophage) in its chromosome.

Lysogenic bacteria: Bacteria harbouring temperate bacteriophage.

Temperate phage: A phage (virus) that invades but may not destroy (lyse) the host (bacterial cell); however, it may continue into the lytic cycle.

Virulent phage: A phage (virus) that destroys the host (bacterial) cell.

Phage λ (= lambda) is one of the best studied bacteriophage. It has a double stranded DNA (chromosome) of about 48,500 base pairs. Since it is a temperate phage, it can exist either vegetatively or as a prophage, integrated into the host chromosome. Life cycle of this phage is suitable for our analysis since its choice of pattern of life cycle is determined by an interesting and complex way. It provides a model system of operon controls. Here, complexity arises from having two conflicting life-cycle choices: one choice favours lytic cycle and other choice prefers nonlytic (*i.e.*, lysogenic) phage of life cycle.

The expression of one of the two life-cycle alternatives, lysogenic or lytic cycles, depends on whether two repressors, *CI* and *cro*, have access to operator site. The **CI repressor** acts to favour lysogeny: it expresses lytic cycle and represses lysogeny. The operator sites, when bound by either *CI* or *cro*, can either enhance or repress transcription. There are certain other control mechanisms which are involved in determining aspects of λ life cycle such as **antitermination** and **multiple promoters** for the same genes (see **Ptashne** and aoworkers 1982; **Ptashne**, 1987, 1989).

41.3. REGULATION OF GENE ACTION IN EUKARYOTES

The different types of cells in the body of a multicellular organism differ in structure and function, but their genes are identical, since all cells are ultimately derived from the zygote by cleavage (mitosis). The problem therefore is: how do cells with identical genetic complements differ so much in structure and function? The answer is that not all genes are active at one time. As development proceeds certain genes become active while others become inactive; *i.e.*, the genes are "switched on" and "switched off" at different times. This process is called **differential gene action**. When genes are active they direct the formation of enzymes which affect certain phenotypic traits.

Further, though a cell has the genes to produce hundreds of enzymes, only the enzymes required at a particular time are produced. Such a control mechanism ensures that the cell is not flooded with unnecessary enzymes. In eukaryotes, genes are active only when their products are required by the cell. The rest of the time they are "switched off". For example, in mouse liver cells, only about 3% of the genes are active (*i.e.*, transcribe RNA), and in brain cells about 9% of the genes are turned on.

In eukaryotes, the following two kinds of controls or regulations of gene expression occur: 1. **Short-term** or **reversible regulation** corresponds to the kind of regulation we studied in bacteria and it represents a cell's response to fluctuations in the environment, especially, it involves changes in activities of concentrations of enzymes as particular substrates and or hormone levels rise and fall. The changes a cell experiences during a cell cycle, particularly the fluctuations in rates of DNA, RNA, and protein synthesis that regularly occur. with respect to the time of mitosis can also be placed in this category. 2. **Long-term** or **irreversible regulation** includes the phenomena associated with **determination, differentiation**, or more generally development: it is involved in the numerous steps by which a fertilized egg becomes an organism of, perhaps, trillions, of cells with diverse and ultimately quite permanent roles to play in the maintenance of the whole. Short-term regulations also occur in developing and differentiating eukaryotic cells side by side of long-term regulation.

Both of these types of regulations of gene activates in eukaryotes, now, are considered to occur at the following levels involving diverse mechanisms: 1. Regulation at the level of DNA; 2. Regulation at the level of transcription; 3. Regulation at the level of translation; and 4. Regulation at the level of post-translation.

1. Regulation of Gene Action at the Level of Genome

In eukaryotic cells, it seems that certain classes of genes are transcribed more or less continuously, and only in extreme situations their activities are repressed. For example, genes coding for larger ribosomal RNA (28S or 18S rRNA)) or transfer RNA (tRNA) are present as multiple copies forming **simple multigene family**. These genes are transcribed uniquely by RNA polymerase I for the larger ribosomal RNA or by RNA polymerase III for tRNA and 5S RNA. Although the products of some of these genes, the ribosomes, are used continuously in all cells, it does not confirm that all of these multiple copies are continuously transcribed at maximum rate. Electron micrographs of spread chromatin from nucleoli often show that some of the repetitious rRNA genes are inactive. It is also true that in the nucleated erythrocytes of lower vertebrates such as *Xenopus*, all genes may be turned off (**Maclean** *et al.*, 1972), including those for ribosomal RNA and tRNA. Therefore, it is clear that mechanisms do exist for inactivating sequences even those regarded to be constitutive in normal cells.

Some of the clearest demonstrations that some specific genes are at least available for transcription in different kinds of differentiated cells are provided by *Drosophila* and other organisms (Fig 41.17). For example, the pattern of bands and interbands of **polytene chromosomes** of *Drosophila* does not vary between different larval tissues, yet it is now concluded that the interband regions probably represent **housekeeping genes** which code for essential proteins, that they are expressed in every cell, and that they are retained in a state of permanent decondensation (**Bautz** and **Kabisch**, 1983). Thus, 'housekeeping' genes may be 'left on' for much of the life of the cell when transcription of even the most essential housekeeping genes ceases (*i.e.*, during mitosis).

Further, when we consider the case of the **"cell-specific" genes**, **luxury genes** or **smart genes**, which code for the products only found in specialised tissues, it becomes immediately clear that differential expression is the rule. Whether expression of gene is measured at the level of the messenger RNA or the protein, genes coding for products such as globin, crystallin, fibroin, ovalbumin, casein and immunoglobulin give every indication of complete repression in all but the specialized tissue characterised by their presence.

Thus, at the level of genome (*i.e.*, DNA), the following five modes of regulation are operative:

(*i*) **Situation of total genetic shutdown.** (*a*) During mitotic phase of the cell cycle, chromatin is highly condensed to form chromosomes, and transcriptional activity of all genes is suspended.

(*b*) During meiotic division of germ cells a somewhat similar situation to (*a*) is evident, although in some rare cases, such as **lampbrush chromosomes** of meiotic diplotene in vertebrates (**Vlad**, 1983), transcription proceeds very actively.

(*c*) The nucleus of mature nucleated erythrocytes of amphibians is transcriptionally inactive. Chromatin in these cells is highly condensed but not organised into discrete chromosomes (**Chegini** *et al.*, 1981). However, transcription can be partially reactivated in these nuclei by transferring them into new cytoplasm or exposing them *in vitro* to altered environmental conditions.

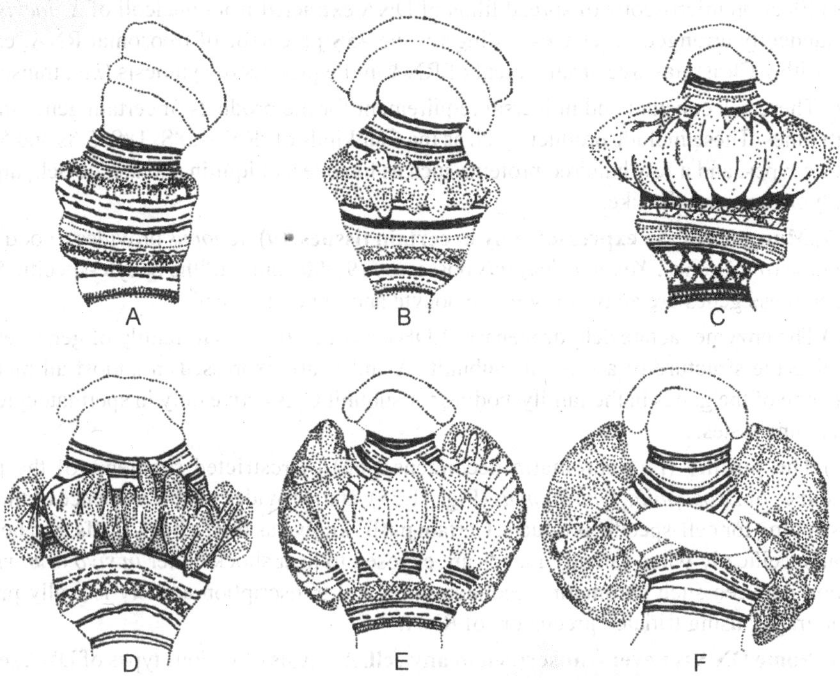

Fig. 41.17. The development of a chromosomal puff in a larval salivary gland cell nucleus of *Chironomus tentans*.

(*d*) In mammalian females, one of the two X chromosomes present in somatic cells undergoes condensation in early embryonic stages to become heterochromatic **sex chromatin** or **Barr body** (Dosage compensation). A variety of experiments indicate that most, though not all, genes of the condensed X chromosome are turned off. In developing oocytes, as opposed to somatic cells, Barr bodies are not present, the activities of both X chromosomes being required for normal oogenesis.

In somatic cells of normal XY males, genes of the single X chromosome remain active and Barr bodies are not found. However, in germ cells, the X chromosome is inactivated prior to spermatogenesis, otherwise, it may prevent sperm maturation and lead to sterility. In one extreme case, that of the creeping vole, *Microtus oregoni*, the X chromosome is eliminated from the germ cells of males by a special process of nondisjunction (see **Farnsworth**, 1988).

(*e*) Sperm cells clearly contain a complete genetic endowment but no transcription occurs until the sperm nucleus is activated within the egg cytoplasm.

(*f*) Complete suspension of transcriptional activity is also know in the following cases: cells of some plant seeds; cells within diapausing *Artemia* gastrulae; cells within inactive organisms such as desiccated *Tardigrada*; nuclei within bacterial and fungal spores; and nuclei within desiccated amoeba cells, as for example in the slime mould *Dictyostelium*.

***(ii)* Evidence for constitutive expression of some genes.** (*a*) If the interbands of *Drosophila* polytene chromosomes are correctly interpreted as being loci for "housekeeping" genes, then the evidence is that such chromatin is permanently decondensed and is transcribed at a low but constant rate (**Semehin** *et al.*, 1979).

(*b*) Electron microscopy of spread films of DNA extracted from nucleoli of *X. laevis* oocytes reveals tandemly arranged sequences coding for the 45S precursor of ribosomal RNA, each gene adorned with a Christmas-tree arrangement of RNA in the process of synthesis (*i.e.*, transcription).

(*c*) There is a constant and universal requirement for the products of certain genes in all cells and at all times. These include products such as the four kinds of rRNA-28S, 18S, 5.8s and 5S; tRNA of 20 basic types, and a few hundred proteins such as histones, ubiquitin and lactate dehydrogenase, RNA polymerase, and the like.

(*iii*) Many genes are expressed only in certain tissues. (*a*) *Xenopus* provides a good example of regulation of 5S genes. *Xenopus borealis* possesses 19,000 copies of the oocyte-specific 5S rRNA genes, and these genes are active only in the oocyte and in no other cell.

(*b*) The enzyme lactate dehydrogenase (LDH) is coded by a small family of genes, each gene determining the structure of a subunit. Subunits A and B are expressed in almost all mammalian cells, but one of the genes in the family, coding for subunit C, is active only in spermatocytes within the developing testes.

(*c*) The **puffing** (*i.e.*, chromatin decondensation) of restricted segments of the polytene chromosomes of *Drosophila, Chironomus* (Fig. 41.25), etc., provides visible evidence of the activity of genes coding for cell-specific products. Certain puffs, known as **heat-shock puffs**, can be induced to appear specifically when salivary glands are exposed to heat shock either *in vivo* or *in vitro*. The correlation between such chromatin decondensation and transcription activity is really proved by autoradiography using tritiated precursors of RNA.

(*iv*) Some DNA is never transcribed in any cell. Analysis of various types of DNA sequences existing in eukaryotic cells reveals that some DNA is comprised of tandemly repeated short sequences that are concentrated in **heterochromatin** such as centromeres of chromosomes and the Y chromosome. Current evidence indicates that much of this DNA is never transcribed in any cell. Some spacer sequences occur between genes, for example, between multiple copies of genes for ribosomal RNA. Such spacer sequences are often taken to be untranscribed, but now it is found that some of the spacer DNA may transcribe nuclear RNA molecules.

The very large size of the genomes of the higher eukaryotes certainly indicates that much of the DNA is redundant (=repetitious) and probably not utilized as coding or regulatory sequence. The **pseudogenes** found in many gene families, often presumed to have arisen as cDNA (=complementary DNA) copies of reverse transcribed message. They lack introns and contain many stop signals so that RNA polymerase molecules fail to move very far along them.

(*v*) Some DNA is spliced to cause gene rearrangement. Such a mechanism occurs during expression of immunoglobulin (Ig) genes.

2. Regulation of Gene Action at the Level of Transcription

(*a*) Chromatin reconstitution experiments. Chromatin has main components—DNA, histones and non-histones. While it is known since early 1960s that histones may be involved in repressing gene activity, the specific regulation by non-histones was shown only during 1970s. **Gilmour** and **Paul** (1973) performed a chromatin reconstitution experiment to demonstrate the positive role of non-histones in regulation of gene activity (Fig. 41.18). They isolated the chromatin from different tissues separately and then dissociated into DNA, histones and non-histones. This is followed by **chromatin reconstitution** using either the three components derived from the chromatin of the same tissue or by combining the non-histones of one tissue with the DNA and histones of another tissue (Fig. 41.18). From such experiments, it was demonstrated that the mRNA which is synthesized *in vitro* from reconstituted chromatin, mainly depended on the source of non-histone proteins (see **O'Malley** *et al.*, 1977).

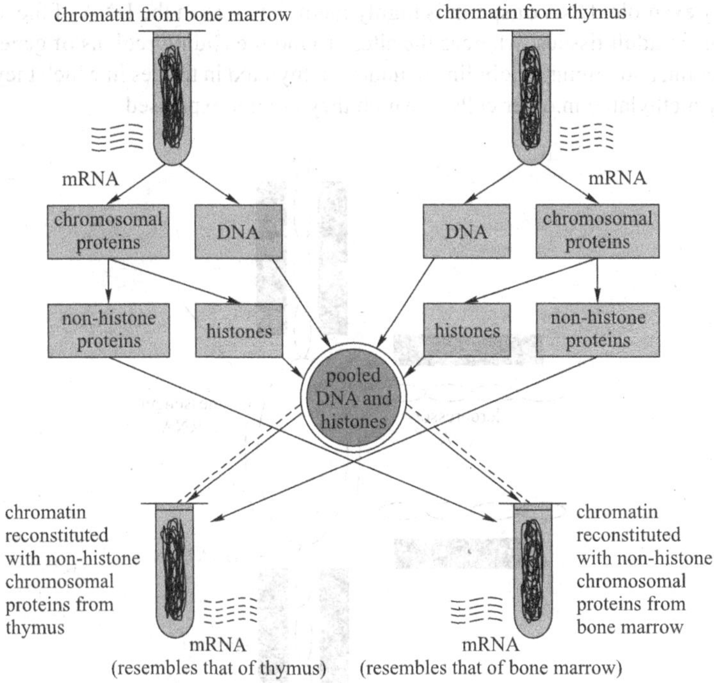

Fig. 41.18. Chromatin reconstitution experiment demonstration in positive role of non-histone proteins in transcription.

(*b*) Frenster's (1965) model of gene-specific depressor RNA (Fig. 41.19). Histones stabilise the DNA double helix by interacting with the negative phosphate groups of DNA. This prevents the separation of strands, and consequently transcription. The histones, thus, act as **general repressor** of protein synthesis. Histones are displaced from DNA by nuclear polyions which form complexes with histones. This permits separation of the DNA strands in a random manner. Gene-specific **derepressor RNA** hybridizes with the non-transcribing DNA strand at a particular locus and stabilizes the loop in the open position. This frees the other DNA strand and permits it to transcribe mRNA.

(*c*) Stein's model. According to **G.S. Stein**, **J.S. Stein** and **Kleinsmith** (1975), individual NHC (nonhistone chromosomal) proteins recognize certarn sites on DNA and bind to it there. They then pulls off the histones repressor from the site, thus, derepressing DNA. Thus, NHC proteins act by derepressing (Fig. 41.20).

(*d*) Change in chromatin conformation. Evidently nucleosomes continue to be present on most transcriptionally active DNA sequences, but they are probably reduced in number. Thus, although evidence from some laboratories suggests that the ribosomal genes of the amphibian nucleolus lack nucleosomes, active gene loci in *Drosophila* give a positive reaction to antibodies against H3 and H4, indicating that at least these subunits of the nucleosome persist on such DNA. In fact, in some active genes the nucleosomes are displaced or "phased" in these regions (**Samal** *et al.*, 1981).

(*e*) Modification of DNA sequences: DNA methylation. The genomic DNA of higher eukaryotes is modified following replication so that a large proportion of the cystosine (C) residues are present as **5-methylcytosine (SmC)**. However, such methylation has not ,been detected in the DNA of lower eukaryotes such as yeast and *Dictyostelium*, nor in *Drosophila*. The percentage of methylated C residues in DNA relative to unmethylated C residues is highly variable from less than 1 percent in some insects to over 50 percent in some higher plants and vertebrates. A much greater correlation exists between the methylation or under-methylation of sequence in the vicinity of gene promoters. For example, DNA of sperm is highly methylated, as in the DNA of the, oocyte-specific 5S rRNA genes in adult tissues, whereas the sites around the coding regions of genes such as adult globin, ovalbumin, and immunoglobulin are under-methylated in tissues in which they are expressed but are largely methylated in other cells in which they are not expressed.

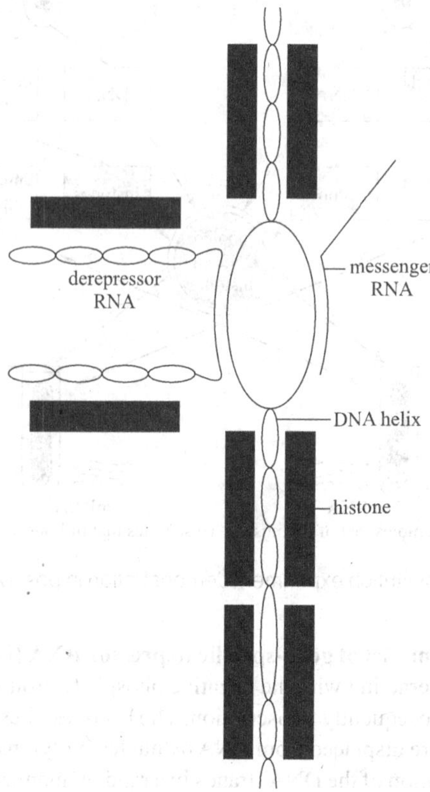

Fig. 41.19. Frenster's model of gene regulation.

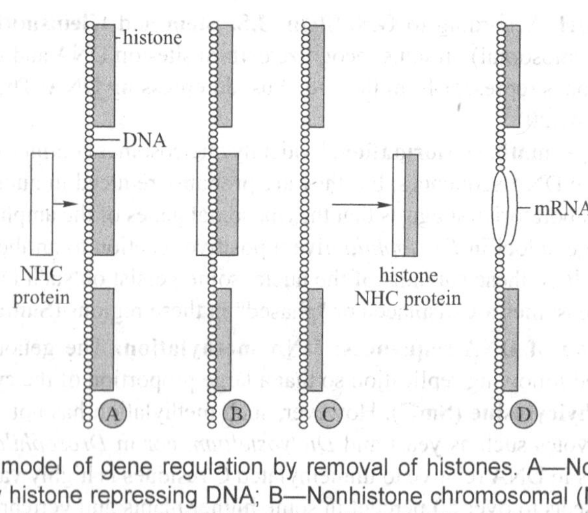

Fig. 41.20. Stein's model of gene regulation by removal of histones. A—Nonspecific masking of DNA by histone repressing DNA; B—Nonhistone chromosomal (NHC) protein binds to specific sites on DNA; C–NHC protein pulls out histone repressor; D—Unmasked DNA is derepressed and transcribes mRNA.

There is evidence that the cytosine methylation in DNA alters the structure of the double helix in a fundamental way and favours the transistion from B-form to Z-form DNA (**Bele** and **Felsenfed**, 1981). It is possible that B $\leftrightarrow$ Z transition is itself involved in gene regulation and this may be the way in which DNA methylation has its effects on transcription.

***(f)* Modifications of histones.** Histone component of chromatin is subject to three different post-synthetic modifications which have either direct or indirect effect on eukaryotic gene regulation; 1. **Histone methylation** affects only histones H3 and H4 and involves the irreversible methylation of a few lysine residues which alters the hydrophobic nature of the side chain of these histones. 2. **Histone phosphorylation** involves histone H1, phosphorylation affects **serines** and **threonines**, changing them from a state of neutral charge to one of negative charge and is a reversible reaction. The state of phosphorylation of H1 protein varies through the eukaryotic cell cycle, and after H1 phosphorylation, chromatin becomes much more strongly condensed, as it does in mitotic chromosomes. Evidently, activation of the **histone kinase** enzyme that is responsible for H1 phosphorylation may be the first step in the chain of events that leads to eventual chromatin condensation prior to mitotic cell division. 3. **Histone acetylation** is of two types. The first is the irreversible acetylation of the amino terminal **serines** of histones H1, H2A and H4. These modifications seem to be associated with histone synthesis. The second is reversible acetylation of **lysine** residues in the amino-terminal regions of histones H2A, H2B, H3 and H4. Acetylation converts the normally basic lysine side chain to a neutral acetyl lysine, and, thus, reduces the net basic charge of the amino-terminal ends of the affected histones. Both H3 and H4 can have up to four lysines in the acetyl form, and there is a strong correlation between this type of histone acetylation, especially **tetracetylation** of all available lysines, and transcriptionally active chromatin. Evidently, the acetylation of the core histone lysines would tend to loosen the nucleosomal structure which brings about the transition from a silent condensed gene to a transcriptionally active and extended one.

***(g)* Transcriptional regulation by protein A24.** The **A24** is an unusual hybrid protein, being a complex of histone **H2A** and the non-basic protein **ubiquitin**. The ubiquitin is covalently bound via the side chain amino group of lysine 19 of the histone. Some 10 per cent of H2A molecules are in the form of A24 and these specialised histones seem to be confined to interphase chromatin, disappearing as the chromosomes condense. A24 is found highly abundant in the chromatin of active genes.

(*h*) Gene regulatory molecules. Transcription of the eukaryotic genome is believed to be regulated by a variety of specific gene regulatory molecules which are produced by specific regulatory genes or by cytoplasm/cell surface. Examples of such gene regulatory molecules are the following:

(i) RNA polymerases. These enzymes are necessary for transcription and if they are short in supply, they tend to affect it. For example, there is a possible competition for type II polymerase (which is meant for hnRNA and mRNA) by the various promoter sequences that lie upstream of protein coding sequence.

(ii) Endonucleases. These enzymes are likely to affect the transcription, especially *in vitro* cell free system, by introducing nicks into DNA that may serve as initiation sites for some polymerases.

(iii) Topoisomerases, helicases and other DNA helix-destabilizing proteins. Various proteins are known that alter the three-dimensional structure of DNA and render it more available for processing which may affect transcription.

(iv) DNA methylase. The enzyme is likely to make DNA less available for transcription, and factors that antagonise methylation would enhance transcription.

(v) Histone acetylase and deacetylases. Such enzymes influence the rate of transcription by modulating acetylation.

(vi) Factors such as ATP. Such molecules may influence the transcription rate by changing the available energy.

(vii) Ions and small molecules. Many ions such as those of **calcium**, **magnesium** and **manganese** directly affect chromatin conformation, which modulates gene activity.

Britten-Davidson model or gene-battery model of transcription regulation. This model of eukaryotic gene regulation at the level of transcription was proposed by **Britten** and **Davidson** in 1969 and later on elaborated by them in 1973. The gene-battery model (Fig. 41.21) assumes the presence of four classes of sequences: 1. **producer gene** which is comparable to a structural gene of prokaryotic operon; 2. **receptor site** is located adjacent to each producer gene and is comparable to operator gene of prokaryotic operon; 3. **integrator gene** which is comparable to regulator gene and is responsible for synthesis of an **activator RNA** that may or may not give rise to proteins before it activates the receptor site; 4. **sensor site** regulates the activity of integrator gene, which can be transcribed only when the sensor site is activated. The sensor sites are recognised by agents which, like hormones and proteins, change the pattern of gene expression. For instance, hormone-protein complex or a transcription factor may bind to a sensor site and cause the transcription of integrator.

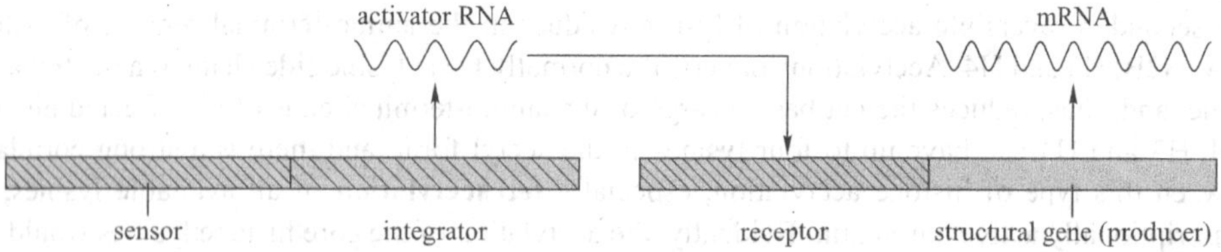

Fig. 41.21. Various components of Britten-Davidson's model for transcription regulation.

In this model, the genes (producer gene and integrator gene) are those sequences which are involved in RNA synthesis, while the receptor and sensor sites help only in recognition without taking part in RNA synthesis. Lastly, in Britten-Davidson's model, a set of structural genes controlled by one sensor site is called the **battery**.

3. Post-Transcriptional Regulation

Steps that come between transcription and translation are described as post-transcriptional and are the following:

1. Some RNA is capped and tailed. The precise functions of capping and tailing of mRNA are not known, but they seem to serve to identify a message or potential message, and tailing may also help in the final export of this message from the nucleus. The mRNA is capped at its 5′ end by a modified guanosine residue. Messenger RNAs that have unmodified caps are not translated. For example, stored mRNA in the oocyte of the tobacco hornworm moth has the guanosine added to its 5′ end, but the G is not modified. Hence, this stored mRNA is not translated. However after fertilization, the cap is modified, allowing the mRNA to be translated to produce proteins needed for early embryogenesis (see **Purves** *et al.*, 1998).

2. RNA is processed to remove intron sequences. Introns removal and **splicing** together of the remaining exons during processing of hnRNA must be absolutely precise. This is in part engineered by a distinct group of nuclear particles (SnRNPs containing U1, U2, U3, U4, U5 and U6 snRNAs). For example, **differential splicing** is used in different lymphocyte cell to produce different proteins from the same hnRNA molecule. As originally discovered by **Early** *et al.*, (1980), the two types of mRNA molecules are produced by part of an intron being omitted from one of the mRNAs but included in the exon splice used to produce the other mRNA. This allows production of two distinct protein both immunoglobulins (Ig), but *one* with a long strand of hydrophobic amino acids at its carboxyl terminus, and the *other* with only a short length of relatively hydrophilic amino acids. The Ig molecule with the long hydrophobic peptide is secreted from the cell. This change in splicing takes place within the life of a single lymphocyte cell and clearly explains the following observation. Immature lymphocytes retain antibody and simply insert the Ig molecules into their plasma membranes, whereas following stimulation with antigen, the same lymphocyte becomes secretory, releasing antibody molecules into circulation.

3. Most RNA is never exported from the nucleus. About 5 per cent of total transcribed RNA never leaves the nucleus. This is explained partly by removal of intron RNA and also by many RNA molecules which break up within the nucleus. The significance of this process is not clear, but some clues about the identification of RNA for export are coming to light. Although not all genes contain introns, most do, and it seems that the presence of some of these introns is essential for RNA export. In other words, introns are used as a means identifying or ticketing the molecules that are to be passed out of the nucleus (see **Maclean** and **Hall**, 1987).

4. Message degradation rates are significant. The rate at which eukaryotic mRNA is degraded in the cytoplasm is highly variable. This implies that differential message breakdown is an important method of regulating not only the rate of gene expression, but also the lag between transcriptional shutdown and the cessation of specific translation. For example, the survival of histone mRNA during the cell cycle explains this fact very clearly. New histone is required in massive amounts immediately at the start of the S period of DNA synthesis to provide the new DNA with nucleosomes. Recently, it is discovered that the restricted availability of histone message is not achieved as a result of transcriptional control alone but by differential breakdown rates for histone message.

Tubulin is a protein that polymerizes to form microtubules, a component of the cytoskeleton. When a large pool of free tubulin is available in the cytoplasm, there is no particular need for the cell to make more of it. Under these conditions, some tubulin molecules bind to tubulin mRNA, and this binding makes such mRNA molecules especially susceptible to breakdown, and less tubulin is made. Other examples illustrate the same mechanism—that the less time an mRNA stays in the cytoplasm, the less it can be translated into protein (see **Purves** *et al.*, 1998).

5. RNA silencing. Andrew Fire and **Craiz Melto** in 1998 demonstrated that **double-stranded RNAs (dsRNAs)** were taken up by cells where they induced a response leading to destruction of mRNAs having the same sequence as the added dsRNA. The phenomenon of dsRNA- mediated **RNA-interference (RNAi)** is an example of the broader phenomenon of RNA silencing, the which small RNAs, typically working in conjuction with protein machinery, act to inhibit gene expression in various ways. RNAi is thought to have involved as a type of **"genetic immune system"** to protect organisms from the presence of foreign or unwanted genetic material . RNAi probably has involved as a mechanism to block the replication of viruses and/or to suppress the movements of transposons within the genome because both of these potentially dangerous processes can involved the formation of double stranded RNAs. Cells can recognize dsRNAs as **"undesirable"** because such molecules can not be produced by the cell's normal genetic activities. Further , the double-stranded RNA that initiates the response is first cleaved into small (21-23 nucleotide), double-stranded fragments, called **small interfering RNAs (si RNAs)** by a particular type of ribonuclease, called **Dicer**.

Recent studies have indicated that germ cells of animals express a distinct class of small RNAs, called **piwi-interacting RNAs (piRNAs)** that suppress the movement of transposable elements in germline. The pi RNAs get their name from the proteins with which they associate. These proteins are called **PIWIS**. In fruit flies, deletion of PIWI proteins leads to defects in suppression of transposon movement in germ cells and ultimate failure of gamete formation. The piwiRNAs and PIWI proteins are also required for successful gamete formation in mammals, but their roles are poorly understood (**Karp** 2010).

4. Translational Control

In bacteria, most mRNA molecules are translated about the same number of times with only fairly small variation from gene to gene. In eukaryotes translational regulation occurs in which a mRNA molecule is not translated at all until a signal is received. Translational control may involve the following mechanisms:

1. Extension of lifetime of the mRNA. An important example of translational regulation is that of **informosomes** or **masked mRNA**. Unfertilized eggs are biologically static, but shortly after fertilization many new proteins must be synthesized, for example, the proteins of the mitotic apparatus, the cell membranes, histones for nucleosome formation as well as others. Unfertilized sea urchin eggs store large quantities of mRNA for many months in the form of mRNA-protein particles (=masked mRNA) made during formation of the egg. This mRNA is translationally inactive, but within minutes after fertilization, translation of these molecules begins. Here, the timing of translation is regulated; the mechanism for stabilizing the mRNA, for protecting it against RNases, and for activation are still unknown.

2. Regulation of rate of protein synthesis. This type of regulation also occurs in mature unfertilized eggs. These cells need to maintain themselves but do not have to grow or undergo a change of state. Thus, the rate of protein synthesis in eggs is generally low. This is not due to inadequate supply of mRNA but of a limitation of an as-yet-undefined element, called the **recruitment factor** which apparently interferes with formation of the ribosome-mRNA complex. A good example of translational control is the extension of the lifetime of silk fibroin mRNA in the silkworm *Bombyx mori*. During cocoon formation on the silk gland of the silk worm predominantly synthesises a single type of protein, **silk fibroin**. Since the worm takes several days to construct its cocoon, it is the total amount and not the rate of fibroin synthesis that must be great; the silk worm achieves this by synthesising a fibroin mRNA molecule that is very long lived.

Transcription of the fibroin gene is initiated at a strong promoter by an unknown signal and about 10^4 fibroin mRNA molecules are made in a period of several days (such a synthesis forms an example of transcriptional regulation). A typical eukaryotic mRNA molecule has a lifetime of about three hours before it is degraded. However, the fibroin mRNA survives for several days during which each, mRNA is translated repeatedly to yield 10^5 fibroin molecules. Thus, each gene is responsible for the synthesis of 10^9 protein molecules in four days. Altogether the silk gland makes 300 μg or 10^{15} molecules of fibroin during this period. If the lifetime of mRNA were extended, either 25 times as many genes would be needed or synthesis of the required fibroin would take about 100 days.

Further, free iron ions (Fe^{2+}) within a mammalian cell are bounded by a storage protein, **ferritin**. When iron is in excess, ferritin synthesis rises dramatically. Yet the amount of ferritin mRNA remains constant. This increase in ferritin synthesis is due to an increased rate of mRNA translation. *When the iron level in the cell is low*, a **translational repressor** protein binds to ferritin mRNA and prevents its translation by blocking its attachment to the ribosome. *When iron level rise*, the excess iron binds to the repressor and alters the three-dimensional structure of the translational repressor, causing it to detach from the mRNA and translation proceeds (**Purves** *et al.*,1998).

5. Post-Translation Modification of Proteins to Make them Active Ones

Some proteins are altered after synthesis, usually by partial degradation or trimming, as for example, by the enzymatic removal of the central section of the **proinsulin** molecule to yield the active protein, **insulin**. For their activity, many proteins also depend on being complexed into compound proteins together with other subunits, either the same or different in nature. Such post-translational control mechanisms do play a significant role in determining the activities of differential cells. For example, **haemoglobin** production is highly dependent on the availability of haem to complex with globin protein subunits which may be deficient in cases of iron-dependent anaemia.

6. Hormonal Control of Gene Expression

In higher plants and animals, intercellular communication is a very important phenomenon. Signals originating in various glands and/or secretory cells somehow stimulate **target tissue** or **target cells** to undergo dramatic changes in their metabolic patterns. These changes frequently include altered pattern of differentiation that are generally dependent on altered patterns of gene expression. **Peptide hormones** such as insulin, epinephrine, etc., and **steroid hormones** such as estrogen, progesterone, testosterone (in higher animals, *e.g.*, mammals) and ecdysone (in insects) control gene expression. In higher animals, hormones are synthesized in various specialized secretory cells (*i.e.*, endocrine cells) and are released into the blood stream. The peptide hormones do not normally enter cells because of their relative large size. Their effects appear to be mediated by receptor proteins located in target - cell membranes and by the intracellular levels of **cyclic AMP (cAMP)** (called **secondary messenger**). The cAMP activates a protein kinase (*e.g.*, A-kinase) which phosphorylates (activates) many specific enzymes. The steroid hormones, on the other hand, are small molecules that readily enter cells through the plasma membrane. Once inside the appropriate target cells, the steroid hormones become tightly bound to specific receptor proteins which are present only in the cytoplasm of target cells. The hormone-receptor protein complexes activate the transcription of specific genes or sets of genes according to following two methods (=hypotheses): 1. The hormone receptor protein complexes interact with specific non-histone chromosomal protein and this interaction stimulates the transcription of the correct genes (**J. Stein**, **G. Stein** and **L.Kleinsmith**, 1975). 2. The hormone receptor protein complexes activate transcription of target genes by binding to specific DNA sequences present in the cis-acting regulatory regions (the enhancers and promoter regions) of the genes (**R.M. Evans**, 1988). In both of these cases, these hormone-receptor protein complexes would function as positive regulators (or "activators") of transcription much like the CAP-cAMP complexes in prokaryotes.

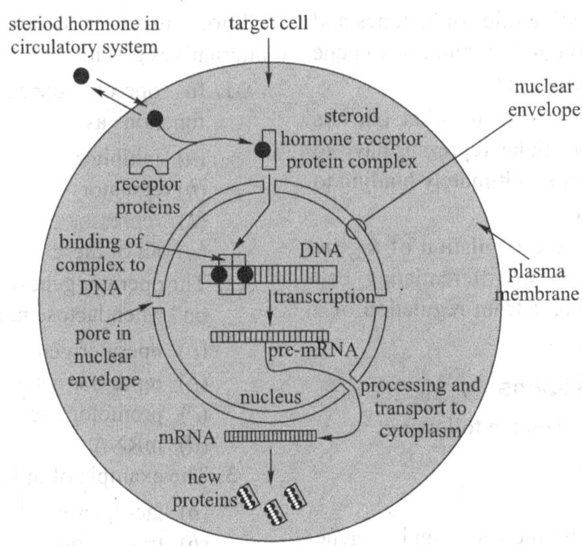

Fig. 41.22. Mode of action of steroid hormones on the gene expression (*i.e.*, transcription) (after Gardner, *et al.*, 1991).

During development of dipteran flies (*Drosophila melanogaster* and *Chironomus tentans*) the steroid hormone **ecdysone** is released and triggers moulting. If larvae of these insects are treated with ecdysone at stages of development prior to or between moultings, patterns of chromosome puffing occur that are identical to those occurring during natural moultings. Ecdysone tends to affect the gene expression at the level of transcription.

QUESTIONS

Long Answer Questions

1. How can inducible and repressible enzymes of microorganisms be distinguished?
2. Distinguish between (*a*) repression and (*b*) feedback inhibition caused by the end product of biosynthesis pathway. How do these two regulatory mechanisms complement each other to provide for the efficient regulation of metabolism?
3. Describe the 'lac operon' of *E.coli*. How does it operate for the regulation of lactose utilization.
4. Of what biological significance is the phenomenon of catabolite repression?
5. Describe the Britten-Davidson's model of regulation of gene activity in eukaryotes.
6. Describe regulation of gene action in eukaryotes at the level of DNA or genetic code.
7. Discuss the relative roles of histones and non-histone proteins in regulation of gene activity in the eukaryotes.
8. In an eukaryotic system, what barriers would an extracellular repressor have to pass through before ultimately binding to DNA?
9. In what way is the regulation of a gene that is active in a differentiating cell inherently different from regulation of a bacterial gene?

Short Answer Questions

1. Write short notes on the following:
   (*i*) lac operon
   (*ii*) repression
   (*iii*) steroid hormones and gene expression
   (*iv*) role of interferon in gene regualtion
   (*v*) heterochromatin

Very Short Answer Questions

1. In the *lac* operon of *E.coli*, what is the function of each of the following genes or sites:
   (*a*) regulator,
   (*b*) operator,
   (*c*) promoter,
   (*d*) structural gene *z*, and
   (*e*) structural gene *y*?
2. What would be the result of inactivation by mutation of the following genes or sites in the *E.coli lac* operon:
   (*a*) regulator;
   (*b*) operator;
   (*c*) promoter;
   (*d*) structural gene *z* and
   (*e*) structural gene *y*?
3. What kinds of mutations would completely eliminate translation of the entire sequence encoding a polypeptide?

Multiple Choice Questions

Choose the correct answer from the four alternatives given.

1. In operon concept, regulator gene functions as
   (*a*) inhibitor
   (*b*) regulator
   (*c*) repressor
   (*d*) all of these
2. The operator gene of lac operon is "turned on" when lactose molecule binds to
   (*a*) operator gene
   (*b*) repressor gene
   (*c*) promotor site
   (*d*) mRNA
3. The example of inducible operon is
   (*a*) lac-operon
   (*b*) try-operon
   (*c*) arginine operon
   (*d*) both (*a*) and (*b*)
4. How many promoters control the transcription of *E.coli* lac operon?
   (*a*) one
   (*b*) two
   (*c*) three
   (*d*) four

5. Which one of the following is not a component of lac-operon model?
   (*a*) promoter gene
   (*b*) structural gene
   (*c*) primer gene
   (*d*) regulator gene
6. In *E.coli*, presence of tryptophan amino acid causes
   (*a*) activation of operon
   (*b*) closure of operon
   (*c*) both of above
   (*d*) none
7. Methylation of DNA most commonly occurs in the sequence of
   (*a*) Cp G (*b*) Cp A
   (*c*) Cp T (*d*) Cp C
8. "Gene-battery model" of gene regulation in eukaryotes was proposed by
   (*a*) Jacob and Monod
   (*b*) Britten and Davidson
   (*c*) Beadle and Tatum
   (*d*) Kornberg and Ochoa
9. Fisher and Krebs got Nobel Prize in 1992 for their discovery of
   (*a*) organ and cell transplantation
   (*b*) reversible protein phosphorylation as biological regulatory mechanism
   (*c*) single ion channels
   (*d*) all of the above

ANSWERS

Very Short Answer Questions

1.

Gene or Regulatory element	Function
(*a*) Regulator gene	Codes for repressor
(*b*) Operator	Binding site repressor
(*c*) Promoter	Binding site of RNA polymerase and CAP–cAMP complex
(*d*) Structural gene *z*	Encodes β-galactosidase
(*e*) Structural gene *y*	Encodes β-galactoside permease

2. (*a*) Constitutive synthesis of the *lac* enzymes;
   (*b*) Constitutive synthesis of the *lac* enzymes;
   (*c*) Uninducibility of the *lac* enzymes;
   (*d*) No β-galactosidase activity; and
   (*e*) No β-galactoside permease activity.
3. A promoter mutation, which would prevent transcription; or mutation in the start codon, which would prevent initiation.

Multiple Choice Questions

1. (*c*) 2. (*b*) 3. (*a*) 4. (*b*) 5. (*c*) 6. (*b*) 7. (*a*)
8. (*b*) 9. (*b*)

42 Cell Commitment

Some cells of animal's body, such as neural cells, are highly specialised cells. They become specified through their interactions with other cells. There are following *four* stages through which the pluripotent cells of the epiblast or blastula become neural precursor cells, called **neuroblasts** (Fig. 42.1; **Wilson** and **Edlund** 2001).

1. **Competence**, wherein multipotent cells can become neuroblasts if they are exposed to the appropriate combination of signals.
2. **Specification**, wherein cells have received the appropriate signals to become neuroblasts, but progression along the neural differentiation pathway can still be repressed by other signals.
3. **Commitment (determination)**, wherein, the neuroblasts enter the neural differentiation pathway and will become neurons even in the presence of inhibitory signals.
4. **Differentiation**, wherein the neuroblasts leave the mitotic cycle and express those genes characteristic of neurons.

42.1. CELL COMMITMENT

American embryologist **William Keith Brooks**, wrote in 1883 of "the greatest of all wonders of the material universe: the existence, in a simple, unorganized egg, of a power to produce a definite adult animal." Brooks showed that this property is so complex that "we may fairly ask what hope there is of discovering its solution, of reaching its true meaning, its hidden laws and causes." These hidden laws and causes are now being sought in the way that *the cells interpret the genome that is the same in every embryonic cells*. How are certain genes activated and repressed in one group of cells to turn them into mesoderm, while a different set of genes is regulated to instruct to become endoderm? Moreover, are there any principles or strategies that characterize the origins of different cell types?

Levels of Commitment

The generation of specialized cell types is called **differentiation**. But differentiation is only the last, observable stage of a series of events that commit a particular blastomere (cleavage cell) to become a particular cell type (Table 42.1). A red blood cell obviously differs in its protein composition and cell structure from a lens cell or a nerve cell. But these observable changes in cellular biochemistry and function are preceded by a process resulting in the **commitment** of the cell to a certain fate. During these stages of commitment, the cell might not look differentiated, even though its developmental fate has become restricted.

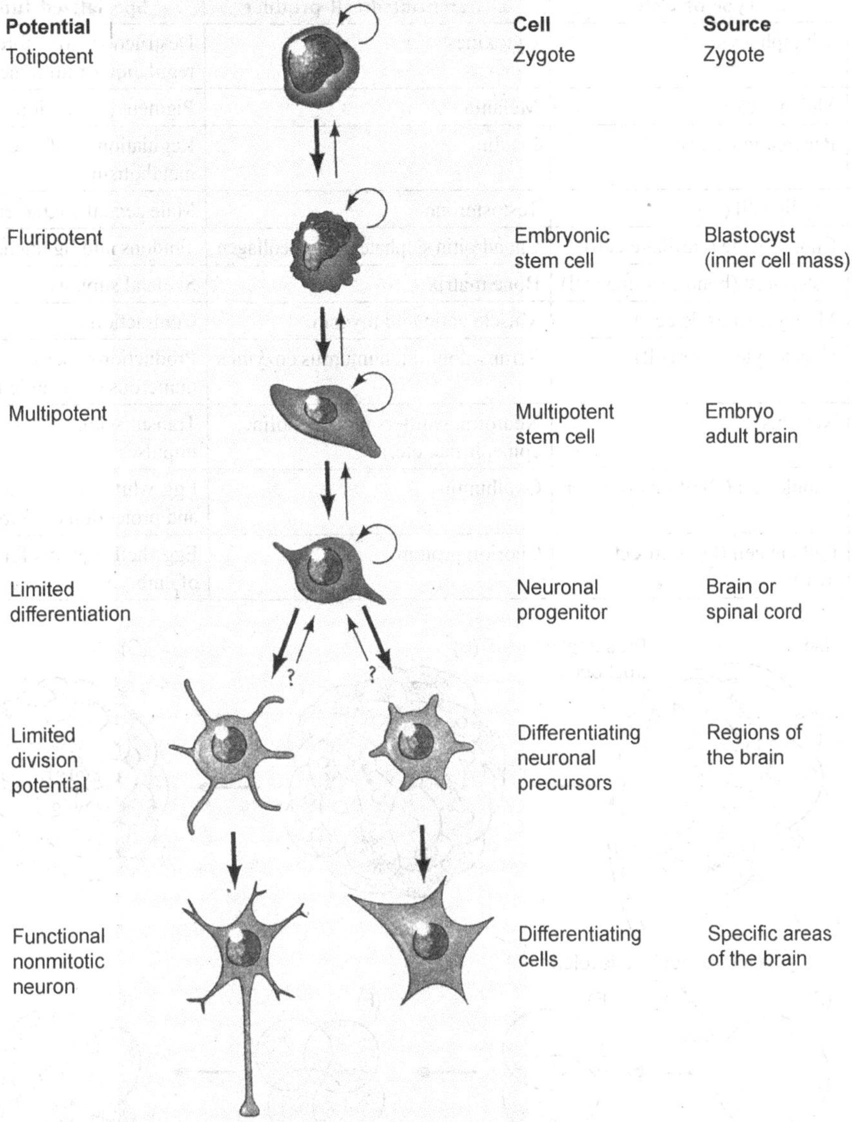

Fig. 42.1. Example of the maturational series of stem cells, here applied to the differentiation of neurons.

Table 42.1. Some differentiated cell types and their major products (*Source*: Gilbert 2010).

	Type of Cell	Differentiated cell product	Specialized function
1.	Keratinocyte (epidermal cell)	Keratin	Protection against abrasion, desiccation
2.	Erythrocyte (Red blood cell)	Haemoglobin	Transport of oxygen
3.	Lens cells	Crystallins	Transmission of light
4.	B lymphocyte	Immunoglobulins	Antibody synthesis

	Type of Cell	Differentiated cell product	Specialized function
5.	T lymphocyte	Cytokines	Destruction of foreign cells, regulation of immune response
6.	Melanocyte	Melanin	Pigment production
7.	Pancreatic islet cell	Insulin	Regulation of carbohydrate metabolism
8.	Leydig cell (♂)	Testosterone	Male sexual characteristics
9.	Chondrocyte (cartilage cell)	Chondroitin sulphate; type II collagen	Tendons and ligaments
10.	Osteoblast (bone-forming cell)	Bone matrix	Skeletal support
11.	Myocyte (muscle cell)	Muscle actin and myosin	Contraction
12.	Hepatocyte (liver cell)	Serum albumin; numerous enzymes	Production of serum proteins and numerous enzymatic functions
13.	Neurons	Neurotransmitters (acetylcholine, epinephrine, etc.)	Transmission of electric impulses
14.	Tubule cell (♀) of hen oviduct	Ovalbumin	Egg white proteins for nutrition and protection of embryo
15.	Follicle cell (♀) of insect ovary	Chorion protein	Egg shell proteins for protection of embryo

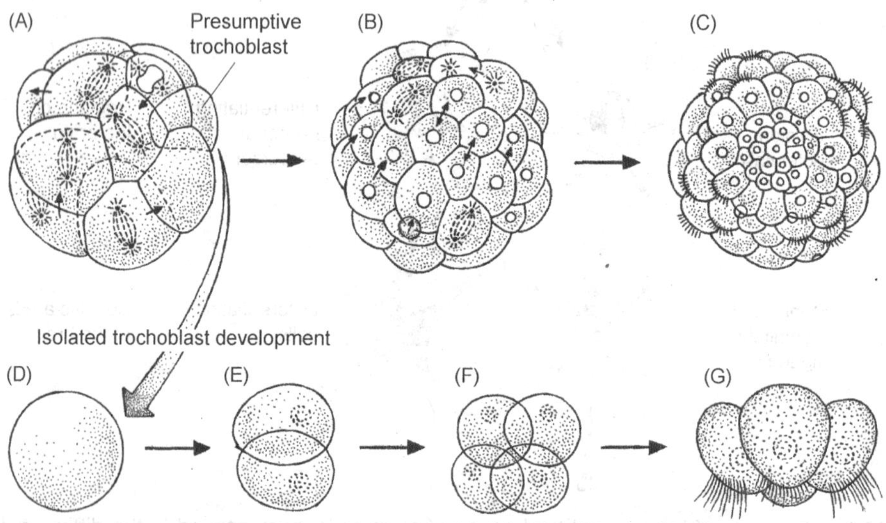

Fig. 42.2. Autonomous specification. (A–C) Differentiation of trochoblast (ciliated) cells of the mollusc *Patella*. A—16-cell stage seen from the side; the presumptive trochoblast cells are shaded. B—48-cell stage; C—Ciliated larval stage, seen from the animal pole (D–G) Differentiation of a *Patella* trochoblast cell isolated from the 16-cell stage embryo and cultured *in vitro*. E, F—Results of the first and second divisions in culture. G—Ciliated products of F. Even in isolated culture, these cells divide and become ciliated at the correct time.

42.2. SPECIFICATION

The process of commitment includes two stages (**Harrison** 1933; **Slack** 1991). The first stage is a labile phase called **speicification**. The fate of the cell or a tissue is said to be *specified* when it is capable of differentiating autonomously (*i.e.*, by itself) when placed into a petri dish or test tube that is, into an environment that is neutral with respect to the development pathway. At the stage of specification, cell commitment is still capable of being reversed.

The second stage of commitment is **determination**. A cell or tissue is said to be *determined* when it is capable of differentiating autonomously even when placed into another region of the embryo—a distinctly non-neutral environment. If a cell or tissue type is able to differentiate according to its specified fate even under these circumstances, it is assumed that commitment is irreversible.

Types of speicification. There are three major plans of commitment, and no one embryo uses only one of them. All three plans are based on mechanisms that apportion certain sets of *transcription factors* to different cells in the early embryo.

1. Autonomous specification. The first mode of commitment is autonomous specification. In this case, the blastomere inherits a set of transcription factors from the egg cytoplasm, and these transcription factors regulate gene expression, directing the cell into a particular path of development. In other words, the egg cytoplasm is not homogeneous, but rather contains different **morphogenetic determinants** (transcription factors or their mRNAs), which will influence the cell's development. In this type of specification, the cell "knows" what it is to become very early and without interacting with other cells (Fig. 42.2).

For instance, in embryos of tunicate (sea squirt), each blastomere will form most of its respective cell types even when separated from the remainder of the embryo (Fig. 42.3). In the 8-cell embryo, the two blastomeres that are going to generate tail muscles contain a yellow-pigmented cytoplasm that has within it a muscle-specific transcription factor called **Macho**. This transcription factor comes from the egg cytoplasm, and any blastomere that has this factor will become muscle cells, even if that blstomeres were to be isolated from the rest of the embryo. In fact, if Macho-containing cytoplasm is placed

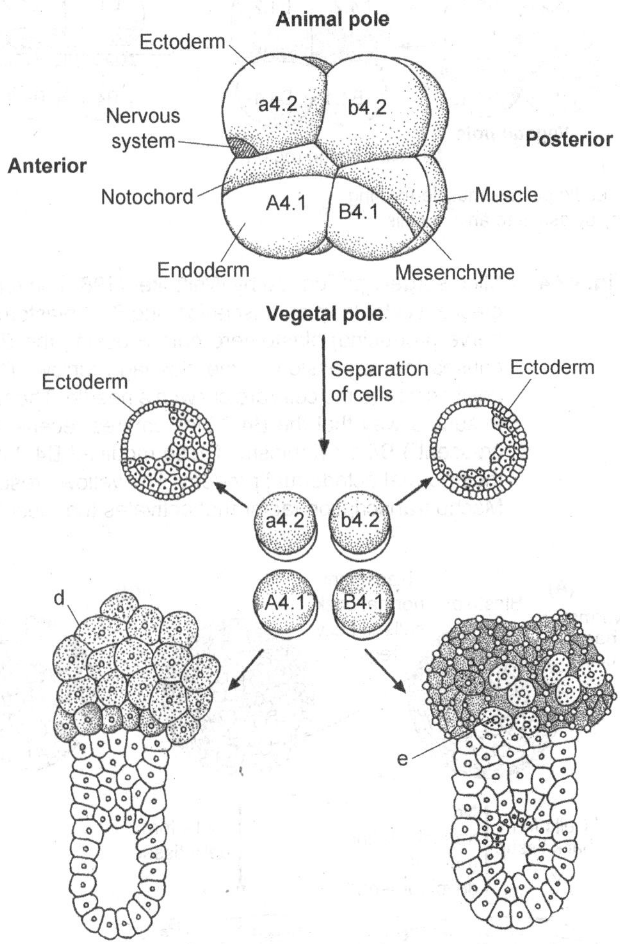

Fig. 42.3. Wilson's experiment in 1904 to show the autonomous specification in the early tunicate embryo. When the four blastomere pair of the 18-cell embryo are dissociated, each forms the structures it would have formed had it remained in the embryo. The nervous system, however, is conditionally specified. The fate map shows that the left and right sides of the tunicate embryo produce identical cell lineages.

into other cells, those cells will form tail muscles (Fig. 42.3; **Whittaker** 1973; **Nishida Sawada** 2001). If the cells containing this cytoplasm (B4.1 blastomeres) are removed from the embryo, the embryo will not form tail muscles (Fig. 42.3). Thus, the tail muscles of tunicates are formed autonomously by acquiring a transcription factor from the egg cytoplasm.

When most of the cells of an early embryo are determined by autonomous specification, it gives the appearance that the animal is fully specified this way. This is not the case, and even in tunicate embryos the nervous system arises conditionally by cell interactions. However, embryologists have traditionally called such embryos **mosaic embryos**, since they develop like a mosaic of individually laid tiles with each cell receiving its instructions independently, without cell-cell interactions.

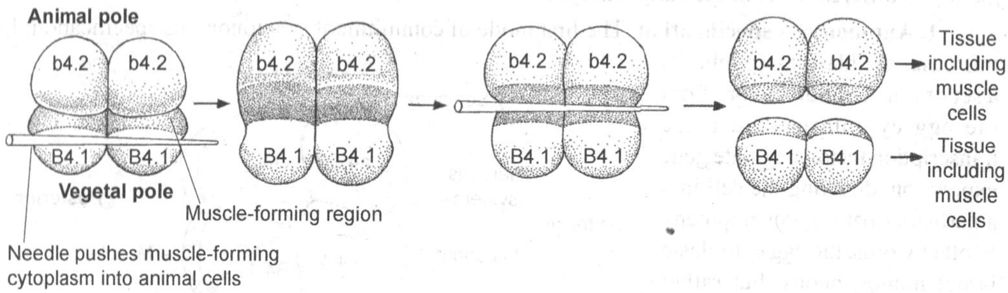

Fig. 42.4. Microsurgery performed by Whittaker (1982) on tunicate eggs forces some of the yellow crescent cytoplasm of muscle-forming B4.1 blastomeres to enter the B4.2 (epidermis- and nerve producing) blastomere pair. Pressing the B4.1 blastomeres with a glass needle causes the regression of the cleavage furrow. The furrow reforms at a more vegetal position where the cells are cut with a needle. The new furrow thereby separates the cells in such a way that the B4.2 blastomeres receive some of the muscle-forming ("yellow crescent") B4.1 cytoplasm. These modified B4.2 cells produce muscle cells as well as their normal ectodermal progeny. The yellow crescent cytoplasm is found to contain the Macho transcription factor that activates the muscle-specific genes.

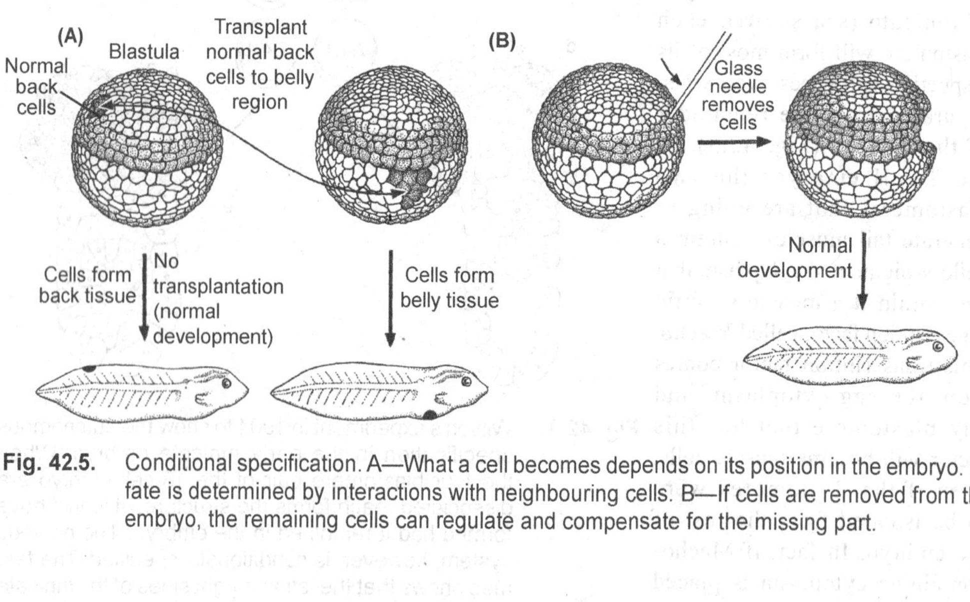

Fig. 42.5. Conditional specification. A—What a cell becomes depends on its position in the embryo. Its fate is determined by interactions with neighbouring cells. B—If cells are removed from the embryo, the remaining cells can regulate and compensate for the missing part.

2. Conditional specification. It is the ability of cells to achieve their respective fates by interactions with other cells (Fig. 42.5). In this case, what a cell becomes is in large measure specified by paracrine factors secreted by its neighbours. In 1888, **August Weismann** proposed the first testable model of cell specification, the **germ plasm theory**, in which each cell of the embryo would develop autonomously. He boldly proposed that the sperm and egg provided equal chromosomal contributions, both quantitatively and qualitatively, to the new organism. Moreover, he postulated that the chromosomes carried the inherited potentials of this new organism (Box 42.1). However, not all the determinants on the chromosomes were thought to enter every cell of the embryo. Instead of dividing equally, the chromosomes were hypothesized to divide in such a way that different chromosomal determinants entered different cells. Whereas the fertilized egg would carry the full complement of determinants, certain somatic cells would retain the "blood-forming" determinants while others retained the "muscle forming" determinants, and so forth (Fig. 42.6). Only the nuclei in those cells destined to become germ cells (gametes) were postulated to contain all the different types of determinants. The nuclei of all other cells would have only a subset of the original determinants.

Box 42.1

One should note that early embryologists were thinking in terms of chromosomal mechanisms of inheritance some 15 years before the rediscovery of Mendel s work. **Weismann** *(1892, 1893) also speculated that these nuclear determinants of inheritance functioned by elaborating substances that became active in the cytoplasm (***Gilbert** *2010).*

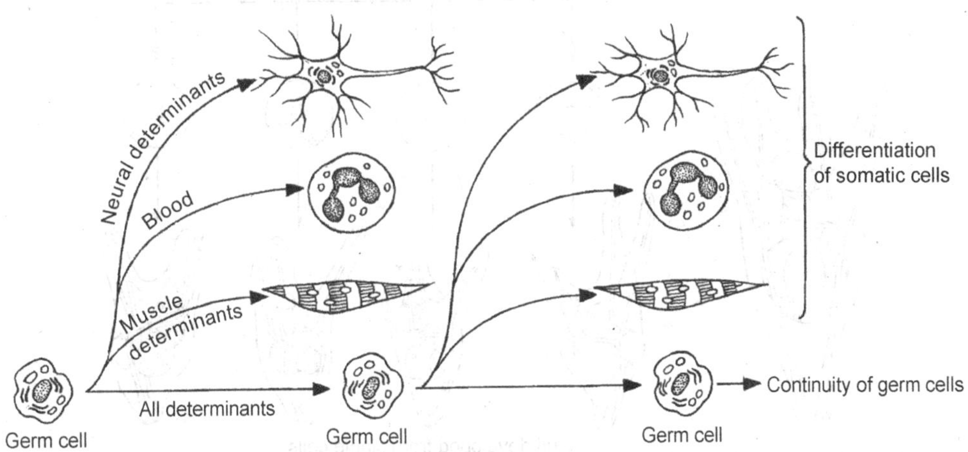

Fig. 42.6. Weismann's germ plasm theory of inheritance. The germ cell gives rise to the differentiating somatic cells of the body, as well as to new germ cells. Weismann hypothesized that only the germ cells contained all the inherited determinants, and types of determinants found in a somatic cell's nucleus would determine its differentiated types.

In postulating his germ plasm model, Weismann proposed a hypothesis of development that could be tested immediately. Based on the fate map of the frog embryo, Weismann claimed that when the first cleavage division separated the future right half of the embryo from the future left half, there would be a separation of "right" determinants from "left" determinants in the resulting blastomeres. **Wilhelm Roux** tested that hypothesis by using a hot needle to kill one of the cells in a two-cell frog embryo—with the result that only the right or left half of a larva developed (Fig. 42.6). Based on this result, Roux claimed that specification was mosaic, and that all the instructions for normal development were already inside each cell.

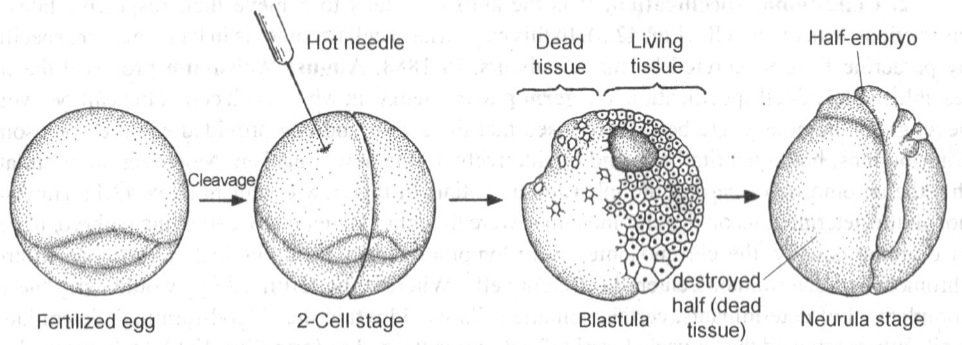

Fig. 42.7. Roux's defect experiment to show autonomous specification. Destroying (but not removing) one cell of a 2-cell frog embryo resulted in the development of only one half of the embryo.

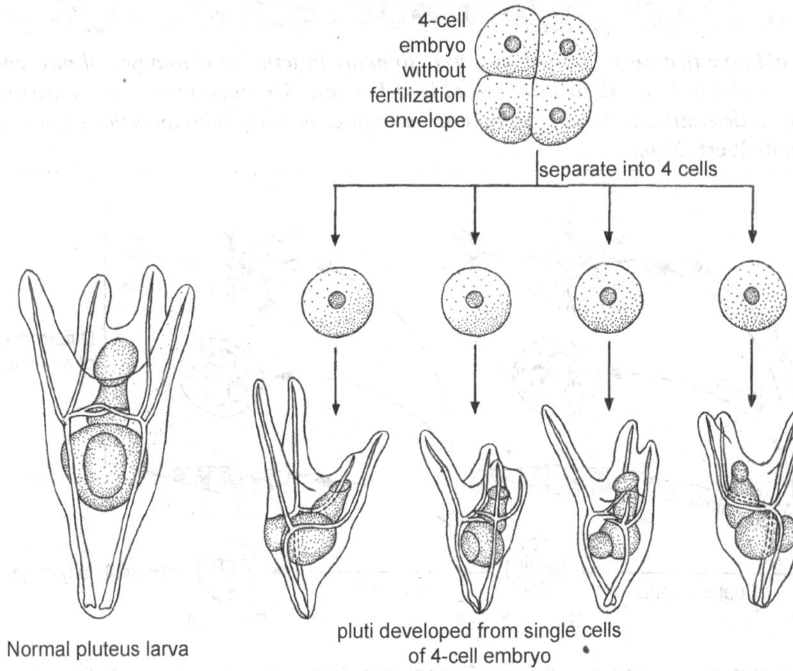

Fig. 42.8. Driesch's isolating experiment demonstrating conditional (regulative) specification. A–An intact 4-cell sea urchin embryo generates a normal pluteus larva. B–When one removes the 4-cell embryo from its fertilization envelope and isolates each of the four cells, each cell can form a smaller, but normal, pluteus larva.

Roux's colleague **Hans Driesch** (1892), however, obtained opposite results. While Roux's studies were **defect experiments** that answered the questions of how the remaining blastomeres of an embryo would develop when a subset of blastomeres was destroyed, **Driesch** (1892) sought to extend this research by performing **isolation experiments**. He separated sea urchin blastomeres from each other by vigorous shaking (or later, by placing them in calcium-free sea water). To Driesch's surprise, each of the blastomeres from a 2-cell embryo developed into a complete larva. Similarly, when Driesch separated the blastomeres of 4- and 8-cell embryos, some of the isolated cells produced entire pluteus larva (Fig. 42.8). In this case, the result was drastically different from the predictions of Weismann and

Roux. Rather self-differentiating into its future embryonic part, each isolated blastomere **regulated** its development to produce a complete organism. These isolation experiments provide the first experimentally observed evidence of conditional specification. Interactions between cells determine their fates, rather than some cytoplasmic factors that is particular to the type of cell.

Conditional development in sea urchin embryos was confirmed by Driesch by performing an complicated **recombination experiment.** In sea urchin eggs, the first two cleavage planes are normally meridional, passing through both the animal and vegetal poles, whereas third division is equatorial, dividing the embryo into four upper and four lower cells (Fig. 42.9 A). Driesch (1893) changed the direction of third cleavage division to be meridional like the preceding two. After he released the pressure, the fourth division was equatorial. This procedure reshuffled the nuclei, placing nuclei that normally would have been in the region destined to form endoderm into the presumptive ectoderm region. In other words, some nuclei that would normally have produced ventral structures were now found in the dorsal cells (Fig. 42.9 B). If segregation of nuclear determinants had occurred (as had been proposed by **Weismann** and **Roux**), the resulting embryo should have been strangely disordered. However, Driesch obtained normal larvae from these embryos. He thus concluded that "the relative position of a blastomere within the whole will probably in a general way determine what shall come from it."

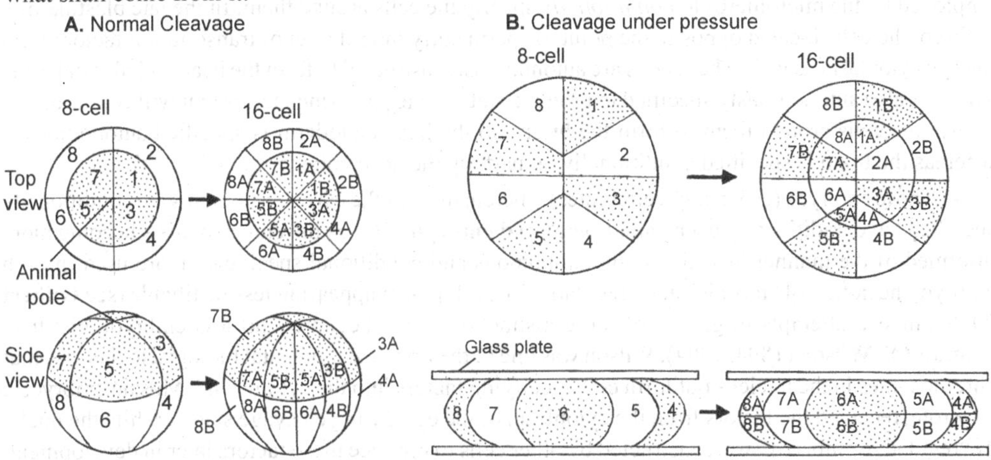

Fig. 42.9. Driesch's pressure-plate experiment for altering the distribution of nuclei. A–Normal cleavage in 8- to 16-cell sea urchin embryos, seen from the animal pole (upper sequence) and from the side (lower sequence). The nuclei are numbered. B–Abnormal cleavage planes formed under pressure, as seen from the animal pole and from the side. Some nuclei (such as 6A and 8A) are placed in different regions of the embryo.

The results of these experiments were very important, both for embryology and for Driesch personally (Box 42.2). First, Driesch had demonstrated that the prospective potency of an isolated blastomere (*i.e.*, those cell types it was possible for it to form) is greater than the blastomere's prospective fate (those cell types it would normally give rise to over the unaltered course of its development). According to Weismann and Roux, the prospective potency and the prospective fate of a blastomere should have been identical. Second, Driesch concluded that the sea urchin embryo is a "harmonious equipotential system" because all of its potentially independent parts interacted together to form a single organism. Driesch's experiment implies that cell interaction is critical for normal development, the community of cells must prevent it from doing so (**Hamburger** 1997).

Box 42.2

This idea of nuclear equivalence and the ability of cells to interact eventually caused Driesch to abandon science. Driesch, who thought the embryo was like a machine, could not explain how the embryo could make its missing parts or how a cell could change its fate to become another cell type. Harking back to Aristotle, he invoked a vital force, **entelechy** ("internal goal-directed force") to explain how development proceeds. Essentially Driesch believed that the embryo was imbued with an internal psyche and the wisdom to accomplish its goals despite the obstacles embryologists placed in its path. However, others, especially **Oscar Hertwig** (1894), were able to incorporate Driesch's experiments into a more sophisticated experimental embryology (see Gilbert 2010).

The third conclusion of Driesch was that the fate of a cell depended solely on its location in the embryo. The interactions between cells determined their fates. We now know that sea urchin embryos and frog embryos use both autonomous and conditional ways of specifying their early embryonic cells. Moreover, they use a similar plan and even similar molecules. In the 16-cell sea urchin embryo, a group of cells called the **micromeres** inherit a set of transcription factors from the egg cytoplasm. These transcription factors cause the micromeres to develop *autonomously* into the larval skeleton. These transcription factors also activate the genes for paracrine factors and **juxtacrine factors** that are employed by the micromeres to *conditionally* specify the cells around them. In the late blastula frog embryo, the cells located opposite the point of sperm entry inherit a set of transcription factors from the egg cytoplasm as well. These cells are autonomously instructed to form the head endodermal cells. They are also autonomously specified to produce and secrete paracrine factors that will conditionally induce the cells around them to form the brain. So the head endoderm is specified autonomously, whereas the brain is specified conditionally, in part, by the head endoderm.

Those embryos (especially vertebrates) wherein most of the early blastomeres are conditionally specified have traditionally been called **regulated embryos**. But as the embryologists become more informed of the manner in which both autonomous and conditional specification are used in each embryo, the notion of "mosaic" and "regulative" development appearing less justifiable (see **Gilbert** 2010). Indeed attempts to get rid of these distinctions were begun by no less embryologist than **Edmund B. Wilson** (1894, 1904). Wilson was one of the first developmental biologist to theorise that chromosomes in the nucleus put forth cell-specifying factors into the cytoplasm." Mosaic embryos", he wrote, received the factors from the cytoplasm of the egg during cleavage stages, while the nuclei of "regulative embryos" were instructed by other cells to produce these factors later in development.

Morphogen gradients and cell specification. A **morphogen** (Greek, "form-giver") is a diffusable biochemical molecule that can determine the fate of a cell by its concentration (Box 42.3). Morphogens can be transcription factors produced within cells (as in *Drosophila* embryos). They can also be paracrine factors that are produced in one group of cells and then travel to another population of cells, specifying the target cells differentially according to the concentration of morphogen. Uncommitted cells exposed to high concentrations of the morphogen (nearest its source of production) are specified as one cell type; when the morphogen's concentration drops below a certain threshold, the cells are determined to another fate. When the concentration falls even lower, a cell of the same initial uncommitted type is specified in yet a third manner.

Morphogen gradients provide a very important mechanism for conditional specification. The existence of morphogen gradients as a force in development and regeneration was predicted by **Thomas Hunt Morgan** (1905, 1906—before he become a geneticist), but it was many years before these gradient models were extended to explain how cells might be placed in specific positions along an embryonic axis (**Horstadius** 1939; **Stumpf** 1966; **Wolpert** 1968, 1969).

Box 42.3

Although there is overlap in the terminology, a **morphogen** *specifies cells in quantitative manner, while a* **morphogenetic determinant** *specifies cells in a qualitative manner. Morphogens are analog*[1], *morphogenetic determinants are digital*[2] *(see* **Gilbert** *2010).*

1. *Analog means having to do with electronic information or signals represented by a varying physical effects (e.g., voltage, the position of a pointer, etc.) rather than by a digital display*
2. *Digital means having to do with information represented as a series of binary digits, as in a computer.*

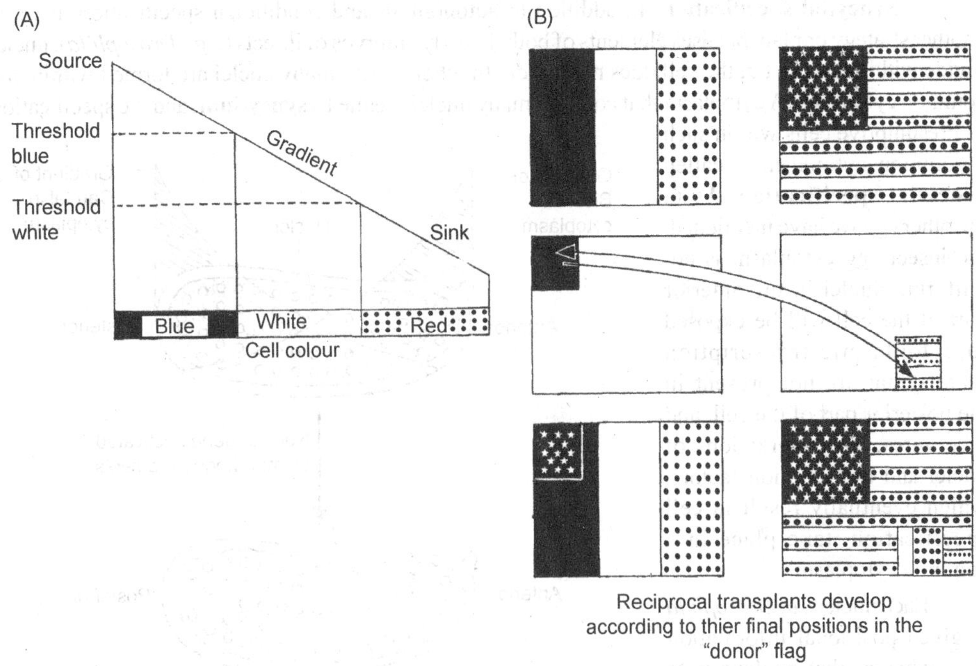

Fig. 42.10. Wolpert's "French flag" analogy for conditional specification. A–Positional information extending from a source to a sink. The thresholds indicated on the left are cellular properties that enable the gradient to be interpreted. In this analogy, cells become blue at one concentration of the morphogen. As the concentration declines below a certain threshold, cells become white, and where it falls below another threshold, cells become red. The resulting pattern is that of the French flag. B–An important feature of this model is that a piece of tissue transplanted from one region of an embryo to another retains its identity of origin, but differentiates according to its new positional instructions. This phenomenon is indicated schematically by reciprocal "grafts" between the flags of the United States and France.

Lewis Wolpert explained such a gradient of positional information using the "French flag" analogy. Imagine a row of "flag cells", each of which is "multipotential"—capable of differentiating into either a red, a white, or a blue cell. Then imagine a morphogen whose source is on the left-hand edge of the blue stripe and whose sink is at the other end of the flag, on the right-hand edge of the red stripe. A concentration gradient is thus formed, highest at one end of the "flag tissue" and lowest at the other (Fig. 42.10). The specification of the multipotential cells in this tissue is accomplished by threshold concentrations of the morphogen. Cells sensing the highest concentration of morphogen become blue. Then there is a threshold of morphogen concetration below which cells become white. As the declining concentration of morphogen falls below another threshold, the cells become red. According to such models (**Crick** 1970), the morphogen diffuses from its site of synthesis (called **source**) to its site of degradation (called **sink**), its concentration dropping along the way. This drop in concentration can be due to simple diffusion; to the cell's binding the morphogen and thus "using it up"; or to a combination of a source synthesizing the morphogen and an environment containing an enzyme that degrades it.

Later on the concept of morphogen gradients has been used to model how regions of the vertebrate body axis are established by **retinoic acid**, **Fgf8** and **Wnts**; how the different regions of the mesoderm are specified by **bone morphogenetic protein (BMP)** from the lateral plate mesoderm; and how vertebrate limbs and digits are specified by **Sonic hedgehog** (**Gilbert** 2010).

3. Syncytial specification. In addition to autonomous and conditional specification, there is another strategy or plan that uses elements of both. In early embryos of insects (*e.g.*, *Drosophila*), nuclei divide within the egg but the cell does not divide. In other words, many nuclei are formed within one common cytoplasm. A cytoplasm that contains many nuclei is called a **syncytium**, and the specification of presumptive cells within such a common cytoplasm is called **syncytial specification**. As in the other egg we have mentioned, the insect egg cytoplasm is not uniform. Nuclei in the anterior part of the cell will be exposed to cytoplasmic transcription factors that are not present in the posterior part of the cell, and vice versa. The interactions of nuclei and transcription factors, which eventually result in cell specification, take place in a common cytoplasm.

Each nucleus in *Drosophila* is given positional information (*i.e.*, whether that nucleus is to become part of anterior, posterior or midsection of the body) by transcription factors acting as morphogens. These transcription factors are made in specific sites in the embryo, diffuse over long distances, and form concentration gradients where the highest concentration is at the point of synthesis and gets lower as the morphogen diffuses away from its source and degrades over time. The concentration of specific morphogen at any particular site tells the nuclei where they are in relation to the source of the morphogens. The anterior most portion of *Drosophila* embryo produces a morphogen called

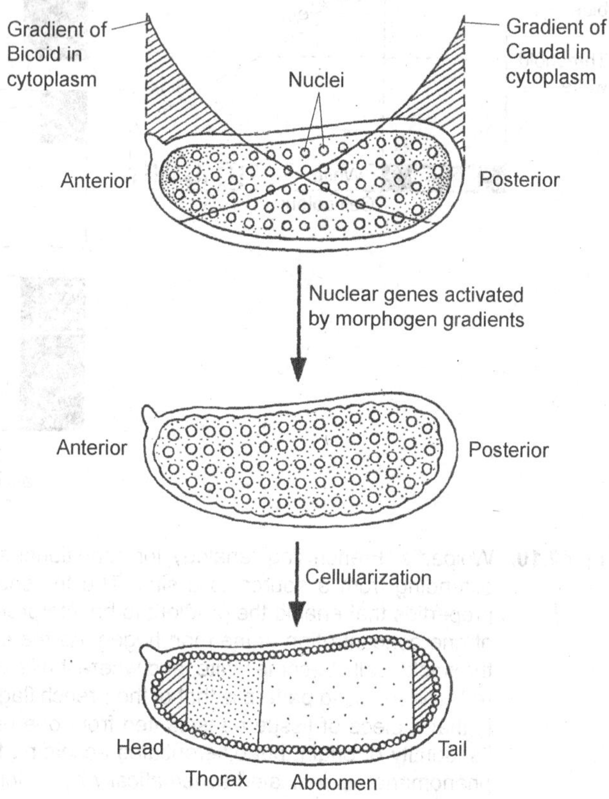

Fig. 42.11. Syncytial specification in *Drosophila melanogaster*. Anterior–posterior specification originates from morphogen gradients in the egg cytoplasm, specifically the morphogenetic transcription factors Biocoid and Caudal. The concentrations and ratios of these two proteins distinguish each position along the axis from any other position. When nuclear division occurs, the amounts and ratios of each morphogen differentially activate transcription of the various nuclear genes that specify the segment identities of the larval and the adult fly.

Bicoid with a concentration that is highest in the anterior and declines toward the posterior. The posterior-most portion of the egg forms a posterior-to-anterior gradient of the morphogen **Caudal**. Thus, the long axis of the *Drosophila* egg is spanned by opposing morphogen gradients—Bicoid coming from the anterior, and Caudal from the posterior (Fig. 42.11).

Bicoid and Caudal are both transcription factors, and different concentrations and ratios of Bicoid and Caudal proteins activate different sets of genes in the syncytial nuclei. Those nuclei in regions containing high amounts of Bicoid and little Caudal are instructed to activate those genes necessary for producing the head. Nuclei in regions with slightly less Bicoid and small amount of Caudal are instructed to activate genes that generate the thorax. Nuclei in regions that have little or no Bicoid but plenty of Caudal are instructed to form the abdominal structures (**Nusslein-Volhard** *et al.*, 1987). Thus when the syncytial nuclei are eventually incorporated into cells, these cells will have their *general* fate specified. Afterward, the specific fate of each cell will become determined both autonomously (from the transcription factors) acquired by the cell's nucleus from the egg cytoplasm) and conditionally by interactions between the cell and its neighbours.

Table 42.2. Modes of cell type specification and their characteristics (*Source*: Gilbert 2010).

A. Autonomous specification:
1. Predominates in most invertebrates.
2. Specification by different acquisition of certain cytoplasmic molecules present in the egg.
3. Invariant cleavages produce the same lineages in each embryo of the species.
4. Cell-type specification precedes any large-scale embryonic cell migration.
5. Produce "mosaic" development: cells cannot change fate if a blastomere is lost.
B. Conditional specification:
1. Predominates in vertebrates and a few invertebrates.
2. Specification by interactions between cell. Relative positions are important.
3. Variable cleavages produce no invariant fate assignments to cells.
4. Massive cell rearrangements and migrations precede or accompany specification.
5. Capacity for "regulative" development, allows cells to acquire different functions.
C. Syncytial specification:
1. Predominates in most insect classes.
2. Specification of body regions by interactions between cytoplasmic regions prior to cellularization of the blastoderm.
3. Variable cleavage produces no rigid cell fates for particular nuclei.
4. After cellularization, both autonomous and conditional specification are seen.

QUESTIONS

Long Answer Questions

1. Write an essay on determination.
2. Give a detailed account of specification.
3. Describe various types of experiments of Driesch for explaining conditional specification.

Short Answer Questions

1. Write a short note on the following:
   (*i*) Determinants.
   (*ii*) Syncytial specification.
2. Describe cell commitment.

Very Short Answer Questions

1. Define the competence.
2. What is determination?
3. Define the bicoid.
4. What are morphogens?

ANSWERS

Very Short Answer Questions

1. Competence is the ability of a cell to respond to specific inductive signal.
2. Determination is the second and irreversible stage of cell or tissue commitment in which the cell or tissue is capable of differentiating autonomously even when placed into a non-neutral environment.
3. Bicoid is anterior morphogen which is critical for establishing anterior posterior polarity in the *Drosophila* embryo. It functions as a transcription factor to activate anterior specific gap genes and as a translational repressor to suppress posterior specific gap genes.
4. Morphogens are substances that by their differing concentrations, differentially specify cell fates. Morphogens are made in specific sites in the embryo, diffuse over long distances and form concentraion gradients where the highest concentration is at the point of synthesis becoming lower as the morphogen diffuses away from its source and degrades over time.

43 Induction

When a cell undergoes mitosis, both of the resulting daughter cells receive a precise copy of the mother cell's genome. Yet those daughters will often have different specialized fates, and at some point, they or their progeny must acquire different characters. That may happen due to following *two* main reasons:

1. Asymmetric division. In some cases, the two sister cells are born different as a result of an **asymmetric cell division**, in which some significant set of molecules is divided unequally between the two daughter cells at the time of division. This asymmetrically segregated molecule (or set of molecules) then acts as a **determinant** for one of the cell fates by directly or indirectly altering the patterns of gene expression within the daughter cell that receives it (Fig. 43.1).

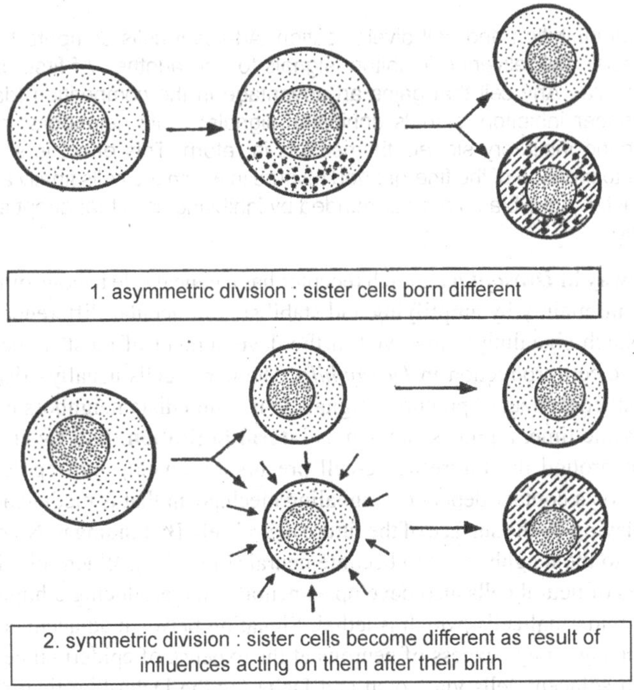

Fig. 43.1. Two ways of making sister cells different.

Asymmetric divisions are particularly common at the beginning of development, when the fertilized egg divides to give daughter cells with different fates, but they also occur at later stages, for example, in the genesis of nerve cells.

2. Inductive interaction. An alternative and by far commonest way to make cells different is by exposing them to different environments, and the most important environmental signals acting on cells in an embryo are signals from neighboring cells.

43.1. LATERAL INHIBITION

In some cases, adjacent similar cells exchange signals that drive them to become different from one another, as in a competition between identical twins. A sort of shouting match occurs, from which one cell or group of cells emerges as winner—not only specializing in a particular way but also delivering a signal to neighboring cells that inhibits them from doing likewise—a phenomenon called **lateral inhibition** (Fig. 43.2). Very often, this process is based on an exchange of signals at cell-cell contacts via the Notch pathway.

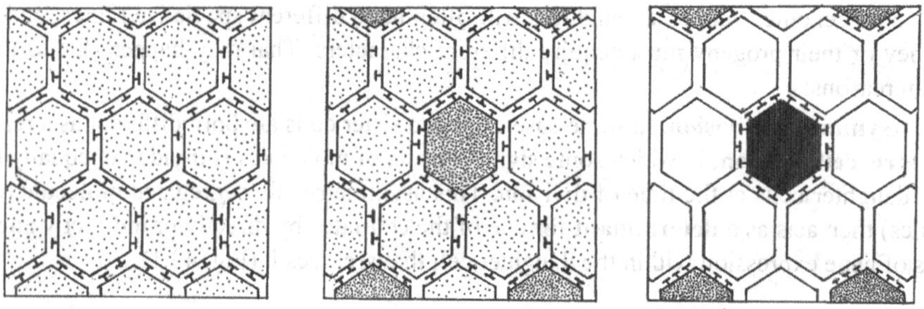

Fig. 43.2. Lateral inhibition and cell diversification. Adjacent cells compete to adopt the primary character, by delivering inhibitory signals to one another. Af first, all cells in the patch are similar. Any cell that gains an advantage in the competition (dark shade) delivers a stronger inhibitory signals (inhibition signals) to its neighbors, inhibiting them from delivering inhibitory signals themselves in return. This effect is self-reinforcing and it leads to creation of the fine-grained mixture in which the cells finally adopting the primary character (solid black) are surrounded by inhibiting cells that adopt a different character (blank).

Notch pathway in *Drosophila*. Notch receptor has a geneal role in controlling cell fate choices during development, mainly by amplifying and stabilising molecular differences between adjacent cells. Although Notch signaling is involved in the development of most tissues, it is best known for its role in nerve cell production in *Drosophila*. The nerve cells usually arise as isolated signal cells within an epithetial sheet of precursor signals to its immediate neighbors not to develop in the same way at the same time, a process known as **lateral inhibition**. In a fly embryo, for example, the inhibited cells around the future nerve-cell precursors develop into epidermal cells. Lateral inhibition depends on a contact-dependent signaling mechanism that is mediated by a signal protein called **Delta**, displaying on the surface of the future neural cell. By binding to Notch on a neighboring cell, Delta signals to the neighbor not to become neural (Fig. 43.3). When this signaling process in flies, the neighbors of neural cells also develop as neural cells, producing a huge excess of neurons at the expense of epidermal cells, which is lethal. Signaling between adjacent cells also develop as neural cells, producing a huge excess of neurons at the expense of epidermal cells, which is lethal. Signaling between adjacent cells via Notch and Delta (or the Delta-like ligand Serrate) regulates cell fate choices in a wide variety of tissues and animals, helping to cease fine-grained patterns of distinct cell types.

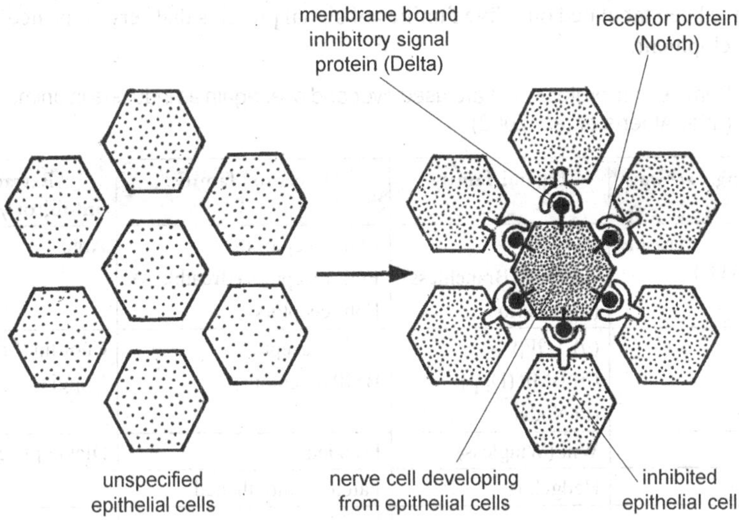

Fig. 43.3. Lateral inhibition mediated by Notch and Delta during nerve cell development in *Drosophila* when individual cells in the epithelium begin to develop as neural cells, they signal to their neighbors not to do the same. This inhibitory contact-dependent signaling is mediated by the ligand Delta that appears on the surface of the future nerve cell and binds to Notch proteins on the neighboring cells.

43.2. INDUCTIVE INTERACTION

In another strategy, perhaps the widely used of all, a group of cells start out all having the same development potential, and a signal from cells outside the group then drives one or more of the members of the group into a different developmental pathway, leading to a changed character. This process is called an **inductive interaction**. Generally, the signal is limited in time and space so that only a subset of the component cells—those closest to the source of the signal—take on the induced character (Fig. 43.4).

Some inductive signals are short range-notably those transmitted via cell-cell contacts; others are long range, mediated by molecules that can diffuse through the extracellular medium. The group of initially similar cells competent to respond to the signals is sometime called an **equivalence group** or a **morphogenetic field**. It can consist of as few as two cells or as many as thousands, and any number of the total can be induced depending on the amount and distribution of the signal.

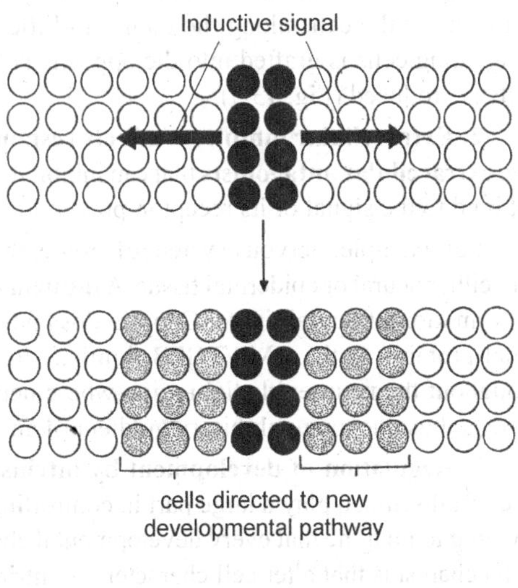

Fig. 43.4. Inductive signaling.

In principle, any kind of signal molecule could serve as an **inducer**. In practice, most of the known inductive events in animal development are governed by just a handful of highly conserved families of signal proteins, which are used over and over again in different contexts. In

Table 43.1 have been tabulated only five families of signal proteins that serve repeatedly as inducers in animal development.

Table 43.1. Some signal proteins that are used over and over again as inducers in animal development (after Alberts *et. al.*, 2002).

	Signaling pathway	Ligand family	Receptor family	Extracellular inhibitors/modulators
1.	Receptor tyrosine kinase (RTK)	(*i*) EGF (*ii*) EGF (Branchless) (*iii*) Ephrins	EGF receptors EGF receptors (Breathless) Eph receptors	Argos
2.	TGFβ	(*a*) TGFβ (*b*) BMP (Dpp) (*c*) Nodal	TGFβ receptor BMP receptors	Chordin (Sog) Noggin
3.	Wnt	Wnt (Wingless)	Frizzled	Dickkopf cerebrus
4.	Hedgehog	Hedgehog	Patched smoothened	
5.	Notch	Delta	Notch	Fringe

Note. Names peculiar to *Drosophila* are shown in parentheses.

The ultimate result of most inductive events is a change in **DNA transcription** in the responding cell: some genes are turned on and others are turned off. Differenet signaling molecules activate different kinds of gene regulatory proteins.

43.3. MODE OF ACTION OF INDUCERS

1. Graded effects of inducers. Responses of induced tissue are more finely graded: a **high concentration** of morphogen may direct target cells into one developmental pathways, an **intermediate concentration** of morphogen into another and a **low concentration** of morphogen into yet another. Development of vertebrate limb provides a striking example of graded effects of inducers. A specialized group of cells at one side of the embryonic limb bud secrete *Sonic hedgehog protein,* which is a member of the Hedgehog family of signal molecules or **morphogens**—and this protein spreads out from its source (inducer), forming a **morphogen gradient** that controls the characters of the cells along the thumb-to-little-finger axis of the limb bud. If an additional group of signaling cells is grafted into the opposite side of the bud, a mirror duplication of the pattern of digits is produced (Fig. 43.5).

2. Extracellular inhibitors affects response of inducer. Most developmental signal proteins have **extracellular antagonists** that can inhibit their function. These antagonists are generally proteins that bind to the signal or its receptor, preventing a productive interaction from taking place.

For example, nervous system of frog embryo arises from a field of cells that is competent to form either neural or epidermal tissue. A **inducing tissue** releases the protein **Chordin,** which favours the formation of neural tissue. Chordin does not have its own receptor. Instead it is an **inhibitor** of signal proteins of the BMP/TGFB family, which induce epidermal development and are present throughout the neuroepithelial region where neurons and epidermis form. The induction of neural tissue is thus due to an inhibitory gradient of an antagonistic signal.

3. Regulation of development by intrinsic programs of the cell. Signals such as those we have just discussed play a large part in controlling the timing of events in development, but it would be wrong to imagine that every developmental change needs an inductive signal to trigger it. Many of the mechanisms that alter cell character are intrinsic to the cell and require no signal from the cell's surroundings: the cell will step through its developmental program even when kept in a constant environment. There are numerous cases where one might suspect that something of this sort is occurring to control the duration of a developmental process. For example, in a mouse, the neural progenitor cells in the spinal cord carry on dividing and generating neurons for just 11 cell cycles, and those in the cerebral contex of the brain for just 23 cycle, after which they stop. Moreover, different kinds of neurons generated at different stages in this program, suggesting that as the progenitor cell ages, it changes the specifications that it supplies to the differentiating progeny cells.

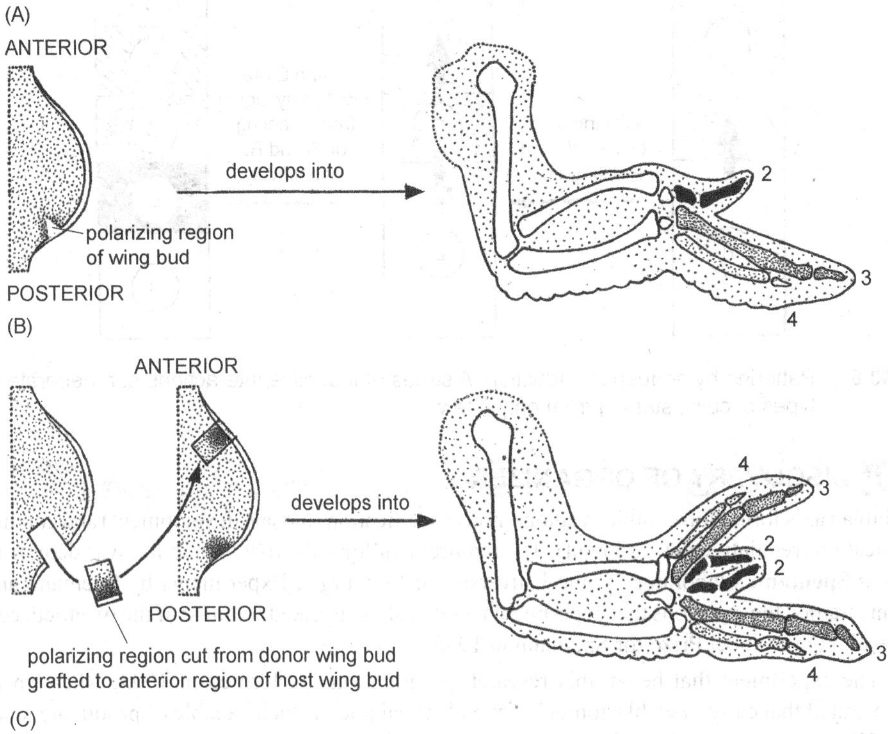

Fig. 43.5. Sonic hedgehog as a morphogen (inducer) in chick limb development. A—Normal wing development. B—A graft of tissue from the polarizing region causes a mirror-image duplication of the pattern of the host wing. The type of digit that develops is thought to be dictated by the local concentration of Sonic hedgehog protein; different types of digits (labeled 2, 3 and 4) therefore form according to their distance from a source of Sonic hedgehog (after Alberts *et al.*, 2002).

4. Sequential induction. The signals that organize the spatial pattern of an embryo generally act over short distances and govern relatively simple choices. A morphogen, for example, typically acts over a distance of less than 1 mm—an effective range for diffusion—and directs choices between no more than a handful of developmental options for the cells on which it acts. But the organs that eventually develop are much larger and more complex than this.

The cell proliferation that follows the initial specification accounts for the size increase, while the refinement of the initial pattern is explained by a series of *local inductions* that embroider successive level of detail on an initially simple sketch. As soon as two sorts of cells are present, one of them can produce a factor that **induces** a subset of the neighboring cells to specialize in a third way. The third cell type can in turn signal back to the other two cell types nearby, generating a fourth and a fifth cell type, and so on (Fig. 43.6).

This strategy for generating a progressive more complicated pattern is called **sequential**

induction. It is chiefly through sequential inductions that the body plan of a developing animal, after being roughed out in miniature, becomes elaborated with finer and finer details as development proceeds.

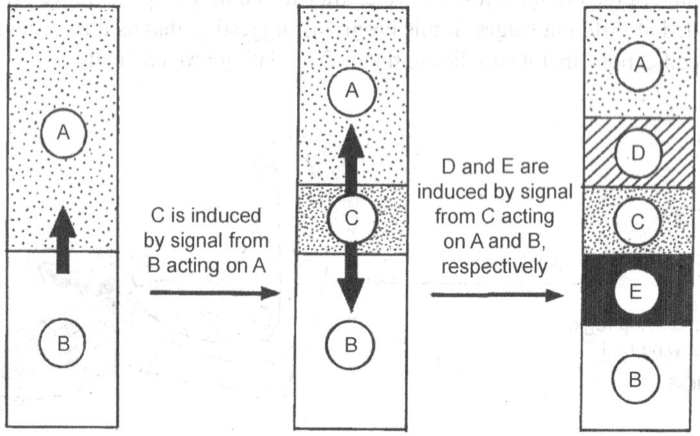

Fig. 43.6. Pattering by sequential induction. A series of inductive interactions can generate many types of cells, starting from only a few.

43.4. DISCOVERY OF ORGANIZER

Amphibian axis formation combines autonomous specification (mosaic development) and conditional specification (regulative development). The requirement for inductive interaction was demonstrated in **Hans Spemann's** laboratory at the University of Freiburger. Experiments by Spemann and his students framed the questions that experimental embryologists asked for most of the twentieth century and resulted in a Nobel Prize for Spemann in 1935.

The experiment that began this research program was performed in 1903, when Spemann demonstrated that early newt blastomeres have identical nuclei, each capable of producing an entire larva. His procedure was ingenious: shortly after fertilizing a newt egg, Spemann used a baby's hair (taken from his infant daughter; see **Gilbert** 2010), to "lasso" the zygote in the plane of first cleavage (*i.e.*, meridional plane). (**Note** *Lasso* is a rope with a noose at one end, used for catching cattle). He then partially constricted the egg, causing all the nuclear divisions to remain on one side of the constriction. Eventually—often as late as the 16 cell stage—a nucleus would escape across the constriction into the non-nucleated side. Cleavage then began on this side too, where upon Spemann tightened the lasso until the two halves were completely separated. Twin larvae developed, one slightly more advanced than the other (Fig. 43.7). Spemann concluded from this experiment that early amphibian nuclei were genetically identical and that each cell was capable of giving rise to an entire organism.

However, when Spemann performed a similar experiment with the constriction still longitudinal, but perpendicular to the plane of the first cleavage (*i.e.*, separating the future dorsal and ventral regions rather than the right and left sides), he obtained a different result altogether. The nuclei continued to divide on both sides of the constriction, but only one side—the future dorsal side of the embryo—gave rise to a normal larva. The other side produced an unorganised tissue mass of ventral cells, which Spemann called *Bauchstiic*—the belly piece. This tissue mass was a ball of epidermal cells (ectoderm) containing blood and mesenchyme (mesoderm) and gut cells (endoderm) but it contained no dorsal structures such as nervous system, notochord or somites.

Why should these two experiments give different results? One possibility was that when the egg was divided perpendicular to the first cleavage plane, some **cytoplasmic substance** was not equally distributed into the two halves. Fortunately, the egg of salamander was a good place to test that hypothesis. There are dramatic movements in the cytoplasm following the fertilization of amphibian eggs, and in some amphibians these movements expose a gray, crescent-shaped area of cytoplasm in the region directly opposite the point of sperm entry. The first cleavage plane normally splits this gray crescent equally between the two blastomeres. If these cells are separated, two complete larvae develop [Fig. 43.8(A)]. However, should this cleavage plane be aberrant (either in the natural event or in experiment), the gray crescent material passes into only one of the two blastomeres. Spemann's work revealed that when two blastomeres are separated such that only one of the two cells contains the crescent, only the blastomere containing the gray crescent develop normally [Fig. 43.8(B)].

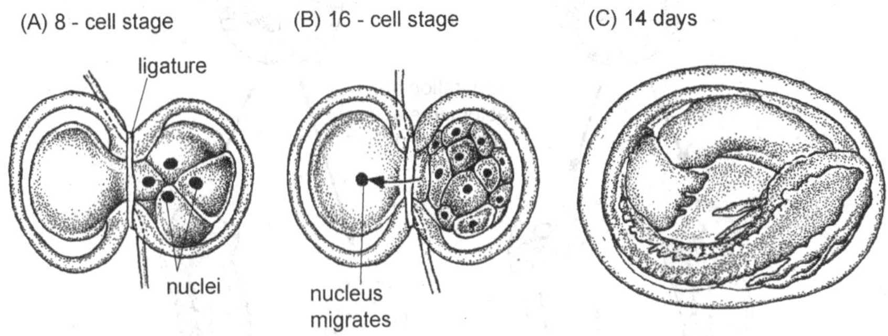

Fig. 43.7. In 1938, Spemann's demonstration of nuclear equivalence in newt cleavage. A—When the fertilized egg of the newt *Triturus taeniatus* was constricted by a ligature, the nucleus was restricted to one half of the embryo. The cleavage on that side of embryo reached the 8-cell stage, while the other side remained undivided. B—At the 16-cell stage, a single nucleus entered the undivided half and the ligature was further constricted to complete the separation of two halves. C—After 14 days, each side had developed into a normal embryo (after Gilbert 2010).

It appeared then, that something in the region of *the gray crescent was essential for proper embryonic development*. But how did it function? What role did it play in normal development? The most important clue came from the fate maps, which showed that the gray crescent region gives rise to those cells that form the dorsal lip of the blastopore. *These dorsal lip cells are committed to invaginate into the blastula, initiating gastrulation and formation of the head endomesoderm and notochord*. Because all future amphibian development depends on the interaction of cells that are rearranged during gastrulation, Spemann speculated that the importance of the gray crescent material lies in its ability to initiate gastrulation, and that *crucial changes in cell potency occur during gastrulation*. In 1918, he performed experiments that showed both statements to be true. He found that cells of the early gastrula were uncommitted, but that the fates of late gastrula cells were determined.

Spemann's experiment involved exchanging tissues between the gastrulae of two species of newt whose embryos were differently pigmented—the darkly pigmented *Triturus taeniatus* and the non-pigmented *Triturus cristatus*. When the region of prospective epidermal cells from an *early gastrula* of one species was transplanted into an area in an early gastrula of other species and placed in a region where neural tissue normally formed, the transplanted cells gave rise to neural tissue. When prospective neural tissue from early gastrulae was transplanted to the region fated to become belly skin, the neural tissue become epidermal [Fig. 43.9(A); Table 43.2]. Thus, cells of early newt gastrula exhibit conditional (*i.e.*, regulative or dependent) development, because their ultimate fate depends on their location in the embryo.

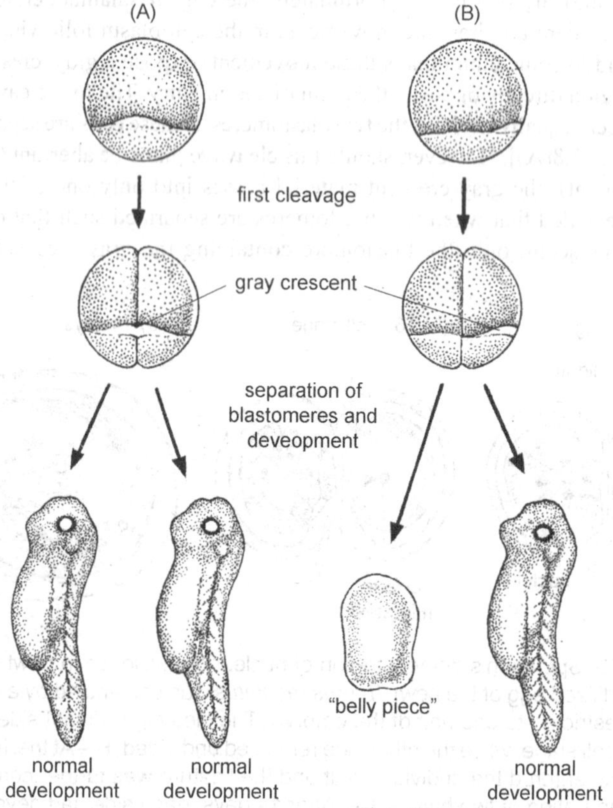

Fig. 43.8. Another experiment of Spemann in 1938 demonstrated asymmetry in the amphibian egg. A—When the egg is divided along the two blastomeres, each of which gets half of the gray crescent, each experimentally separated cell develops into a normal embryo. B—When only one of the two blastomeres receives the entire gray crescent, it alone forms a normal embryo. The other blastomere produces a mass of unorganized tissue (Belly piece) lacking dorsal structure (after Gilbert 2010).

Table 43.2. Results of tissue transplantation during early- and late-gastrula stages in the newt.

	Donor region	Host region	Differentiation of donor tissue	Conclusion
Early Gastrula				
1.	Prospective neurons	Prospective epidermis	Epidermis	Conditional development
2.	Prospective epidermis	Prospective neurons	Neurons	Conditional development
Late Gasrula				
1.	Prospective neurons	Prospective epidermis	Neurons	Autonomous development (determined)
2.	Prospective epidermis	Prospective neurons	Epidermis	Automous development (determined)

However, when the same interspecies transplantation experiment were performed on late gastrulae, Spemann obtained completely different results. Rather than differentiating in accordance with their new location, the transplanted cells exhibited autonomous (independent, or mosaic) development. Their prospective fate was **determined**, and the cells developed independently of their new embryonic location. Specifically, prospective neural cells now developed into brain tissue even when placed in the region of prospective epidermis [Fig. 43.9(B)], and prospective epidermis formed skin even in the region of the prospective neural tube. Within the time separating early and late gastrulation, the potencies of these groups of cells had become restricted to their eventual paths of differentiation. Something was causing them to become committed to epidermal and neural fates. In next set of experiments Spemann tried to know what was happening in the gastrulae.

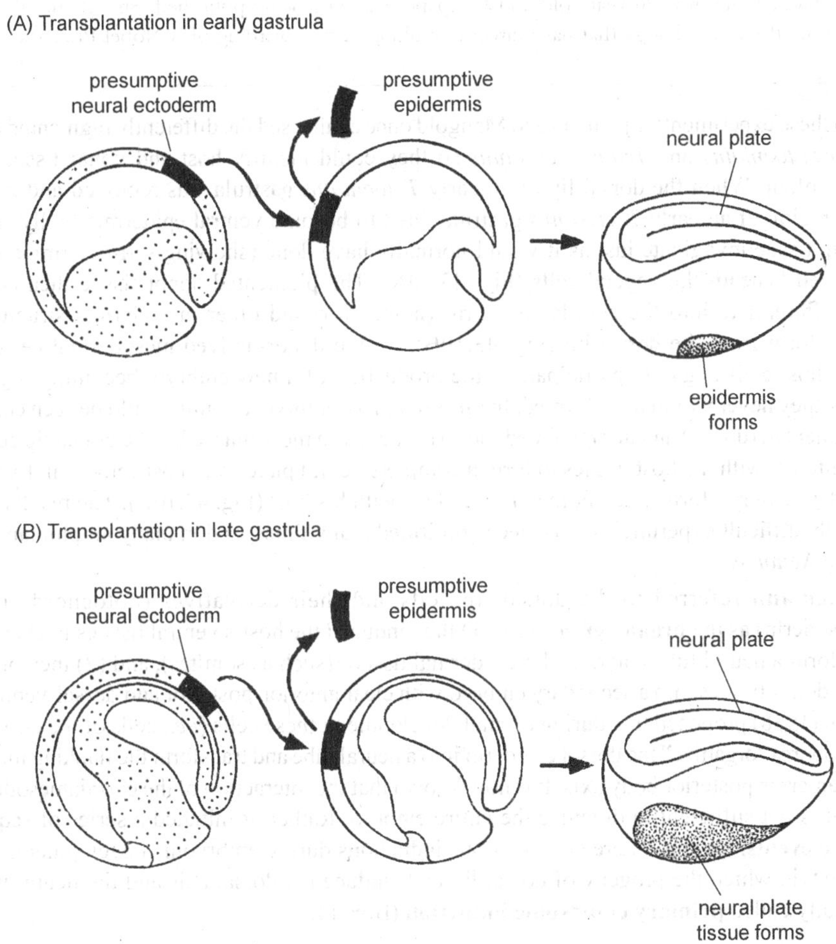

Fig. 43.9. Determination of ectoderm during newt gastrulation. Presumptive neural ectoderm from one newt embryo is transplanted into a region in another embryo that normally becomes epidermis. A—When the tissues are transferred between early gastrulae, the presumptive neural tissue develops into epidermis and only one neural plate is seen. B—When the same experiment is performed using late gastrula tissues, the presumptive neural cells form neural tissue, thereby causing two neural plates to form on the host (after Gilbert 2010).

Discovery of Primary Embryonic Induction

The most spectacular transplantation experiments were published by **Spemann** and **Mangold** in 1924 (Box 43.1). They showed that, of all the tissues in the early gastrula, only one has its fate autonomously determined. This self-determining tissue is the **dorsal lip** of the blastopore—the tissue derived from the gray crescent cytoplasm. When this tissue was transplanted into the presumptive belly skin region of another gastrula, it not only continued to be dorsal blastopore lip but also initiated gastrulation and embryogenesis in the surrounded tissue.

Box 43.1

Hilde Proescholdt Mangold died in a tragic accident in 1924, when her kitchen's gasoline heater exploded. She was 26 years old and her paper was just being published. She is one of the very few doctoral theses in biology that have directly resulting in the awarding of a Nobel Prize (See **Gilbert** 2010).

In these experiments, Spemann and Mangold once again used the differently pigmented embryos of *Triturus taeniatus* and *Triturus cristatus* so they could identify host and donar tissues on the basis of colour. When the dorsal lip of an early *T. taeniatus* gastrula was removed and implanted into the region of an early *T. cristatus* gastrula fated to become ventral epidermis (belly skin), the dorsal lip tissue invaginate just as it would normally have done (showing self-determination) and disappeared beneath the vegetal cells (Fig. 43.10A). The pigmented donor tissue then continued to self-differentiate into the chordamesoderm (notochord) and other mesodermal structures that normally form from the dorsal lip (Fig. 43.10B). As the donor-derived mesodermal cells moved forward, host cells began to participate in the production of a new embryo, becoming organs that normally they never would have formed. In this secondary embryo, a somite could be seen containing both pigmented (donor) and unpigmented (host) tissue. Even more markedly, the dorsal lip cells were able to interact with the host tissues to form a complete neural plate from host ectoderm. Eventually, a secondary embryo formed, conjoined face to face with his host (Fig. 43.10C). The results of these technically difficult experiments have been confirmed many times, and in many amphibian species, including *Xenopus*.

Spemann referred to the dorsal lip cells and their derivatives (notochord and head endomesoderm) as the **organizer** because (1) they induced the host's ventral tissues to change their fates to form a neural tube and dorsal mesodermal tissues (such as somites), and (2) they organized host and donor tissues into a secondary embryo with clear anterior-posterior and dorsal-ventral axes. Spemann (1938) proposed that during normal development, these cells (*i.e.*, cells of the dorsal lip of the blastopore) "organize" the dorsal ectoderm into a neural tube and transform the flanking mesoderm into the anterior-posterior body axis. It is now known that the interaction of the chordamesoderm and ectoderm is not sufficient to organize the entire embryo. Rather, it initiated a series of **sequential inductive events**. Because there are numerous inductions during embryonic development, this key induction—in which the progeny of dorsal lip cells induce the dorsal axis and the neural tube—is traditionally called **primary embryonic induction** (Box 43.2).

Box 43.2

This classical term (*i.e.*, primary embryonic induction) has been a source of confusion because the induction of the neural tube by the notochord is no longer considered the first inductive process in the embryo (**Gilbert** 2010). We will soon discuss some inductive events that precede the "primary induction".

How does the Organizer Form?

The organizer consists of pharyngeal endoderm, head mesoderm, notochord, and dorsal blastopore lip tissues. The organizer functions by secreting proteins (**Noggin**, **Chondrin** and **Follistatin**) that blocks the **BMP** signal that would otherwise ventralize the mesoderm and activate the epidermal genes in the ectoderm. BMPs stand for Bone Morphogenetic Proteins which induces ectoderm to become epidermis. The organizer tissue acts by secreting molecules that blocks BMPs thereby allowing the ectoderm to become neural tissue.

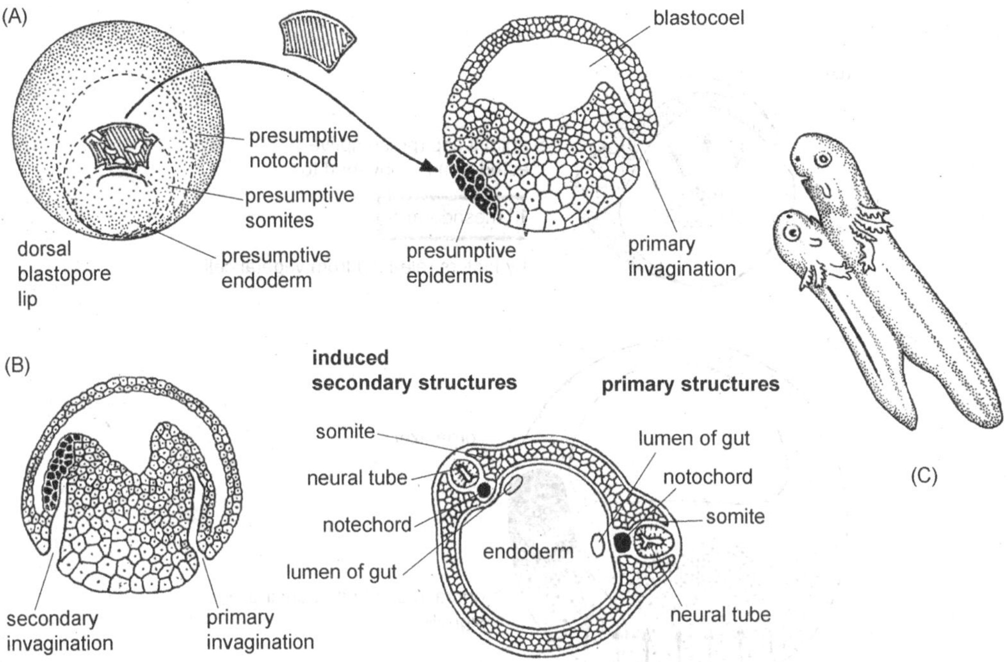

Fig. 43.10. Organization of a secondary axis by dorsal blastopore lip tissue A–C—Spemann and Mangold's 1924 experiments visualized the process by using differently pigmented newt embryo. A—Dorsal lip tissue from an early *T. taeniatus* gastrula is transplanted into a *T. cristatus* gastrula in the region that normally becomes ventral epidermis. B—The donor tissue invaginates and forms a second archenteron and then a second embryonic axis. Both donor and host tissues are seen in new neural tube, notochord and somites. C—Eventually, a second embryo forms joined to the host (after Gilbert 2010).

The organizer is itself induced by the **Nieuwkoop center**, located in the dorsal most vegetal cells. This center is formed by the translocation of the **Disheveled protein (Dsh)** and **Wnt 11** protein to the dorsal side of the egg to stabilize β-catenin in the dorsal cells of the embryo (Fig. 43.11).

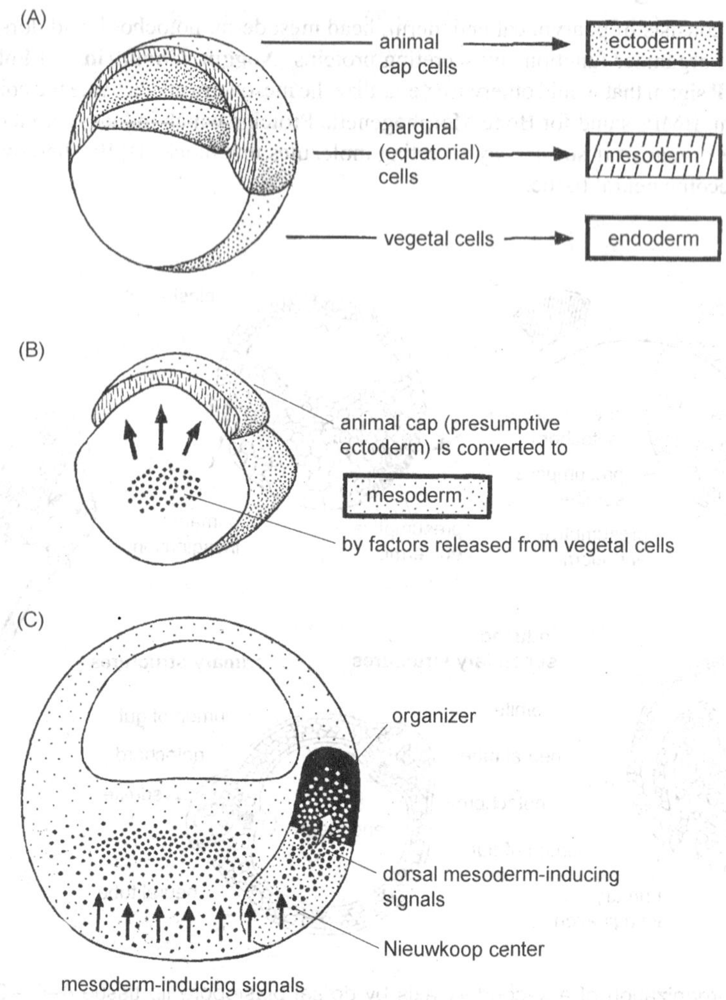

Fig. 43.11. Summary of experiments performed in 1970s in *Xenopus* embryos by Nieuwkoop and by Nakamura and Takasaki, showing mesodermal induction by vegetal endoderm. A—Isolated animal cap cells become a mass of ciliated ectoderm, isolated equatorial (marginal zone) cells become mesoderm, and isolated vegetal cells generate gut like tissue. B—If animal cap cells are combined with vegetal cells, many of the animal cells generate mesodermal tissue. C—Simplified model for mesoderm induction in *Xenopus*. A ventral signal (probably a complex set of signals from activin-like TGF-β factors and FGFs) is released throughtout vegetal region of the embryo. This signal induces the marginal cells to become mesoderm. On the dorsal side (away from the point of sperm entry), a signal is released by the vegetal cells of Nieuwkoop center. This dorsal signal induces the formation of the Spemann organizer in the overlying marginal zone (after Gilbert 2010).

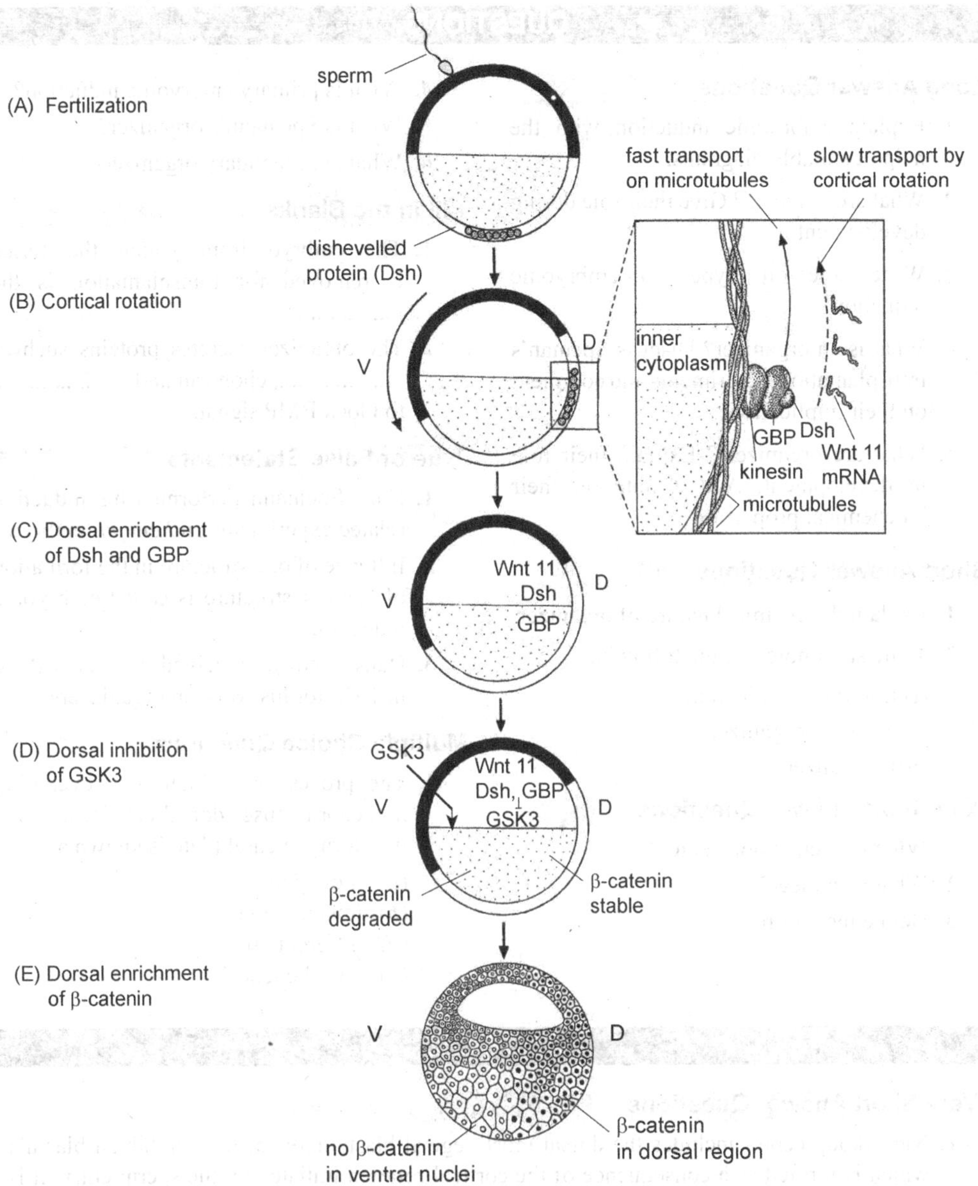

Fig. 43.12. Model of the mechanism by which the Disheveled protein (Dsh) stabilizes β-catenin in the dorsal portion of the amphibian egg. A—Disheveled (Dsh) and GBP (GSK3-binding protein, where GSK3 stands for glycogen synthase kinase 3) associate with kinesin at the vegetal pole of the unfertilized egg. Wnt11 is also in vesicles at the vegetal portion of the egg. B—After fertilization, these vegetal vesicles are translocated dorsally along subcortical microtubule tracks. Cortical rotation adds a "slow" form of transportation to the fast-track microtubule ride. C—Wnt11, Dsh, and GBP are then released from the microtubules and are distributed in the future dorsal third of the 1-cell embryo (*i.e.*, zygote). D—Dsh and GBP bind to and block the action of GSK3, thereby preventing the degradation of β-catenin on the dorsal side of the embryo. Wnt11 probably is needed to stabilize this reaction, keeping an active source of Dsh. E—The nuclei of the blastomeres in the dorsal region of the embryo receive β-catenin while the nuclei of those in the ventral region do not (after Gilbert 2010).

QUESTIONS

Long Answer Questions

1. Explain embryonic induction with the help of suitable diagrams.
2. What are inducers? Give their role during development.
3. Write note on 'types of embryonic induction'.
4. What is an organizer? Discuss Speman's transplantation experiments and comment on their implications.
5. What are organizers? Explain their role in development. Add a note on their biochemical properties.

Short Answer Questions

1. Explain the chemical nature of organizer.
2. Write short notes on the following:
   (i) Embryonic induction.
   (*ii*) Primary organizer.
   (*iii*) Organizer.

Very Short Answer Questions

1. What is Nieuwkoop center?
2. What is inducer?
3. Define induction.
4. What is primary embryonic induction?
5. What is Spemann's organizer?
6. What are secondary organizers?

Fill in the Blanks

1. The embryo from which the tissue is removed for transplantation is the ....
2. The organizer secretes proteins such as ...., chondrin and .... to block BMP signal.

True or False Statements

1. Hans Spemann performed his induction related experiments on *Xenopus* embryos.
2. Influnce of one structure in the formation of another structure is called embryonic induction.
3. Hans Spemann received the Nobel Prize in 1935 for his work on organizaers.

Multiple Choice Questions

1. The process by which the developing notochord causes dorsal ectoderm above it to form a neural plate is known as
   (*a*) induction
   (*b*) invagination
   (*c*) differentiation
   (*d*) morphogenesis

ANSWERS

Very Short Answer Questions

1. Nieuwkoop center includes the dorsal most vegetal blastomeres of the amphibian blastula which is formed as a consequence of the cortical rotation initiated by the sperm entry. It is an important signalling center on the dorsal side of the embryo. One of Nieuwkoop center's main functions is to induce Spemann's organizer.
2. Inducer is a tissue that produces a signal (or signals) that induces a cellular behaviour in some other tissue.
3. Induction is the process by which one cell population influences the development of neighbouring cells via interactions at close range.
4. The process whereby the dorsal axis and central nervous system forms through interactions with the underlying mesoderm, derived from the dorsal lip of the blastopore.

5. Spemann's organizer is more correctly known as the Spemann-Mangold Organizer. It represents, in amphibians the dorsal lip cells of the blastopore and their derivaties (notochord and head mesoderm). It is functionally equivalent to Hensen's node in chick, the node in mammals and the shield in fish. Organizer action establishes the basic body plan of the early embryo.
6. In development of frog, chordamesoderm, the primary organizer induces the formation of fore-brain and optic area in the anterior part of the embryo. The optic area evaginates forming the **optic vesicle**. By invagination it changes into a double-walled cup like structure, the **optic cup** which acts as a **secondary organizer** to induce the epidermis to form lens placode (lens) which acts as **tertiary organizer** to form cornea.

Fill in the Blanks

1. donor; **2.** noggin, follistatin

True or False Statements

1. False; **2.** True; **3.** True

Multiple Choice Questions

1. (*a*)

44 Cell Differentiation

A special characteristic of cell growth and cell division is *cell differentiation*, which refers to changes in physical and functional properties of cells as they proliferate in the embryo to form the different bodily structures and organs (**Hall** 2011). **Differentiation** is the process by which an unspecialized cell becomes specialized into one of the many cell types that make up the body of a multicellular organism.

In general, the cells in an organism can be grouped under three major categories on the basis of the levels of differentiation:

1. Undifferentiated cells. These cells are capable of undergoing division and development, for example, the **stem cells** (animals) and **meristematic cells** (plants).

2. Differentiated cells. These are post-mitotic cells which have undergone specialization or/ and exhibit the division of labour. Therefore, these cells acquire a distinct character and perform a definite function. For example, RBCs carry out transportation of oxygen and carbon dioxide, the muscle cells perform kinetic functions or movement and mesophyll cell (of leaf) carry out photosynthesis.

3. Dedifferentiated cells. Some differentiated cells are capable of reverting back to the undifferentiated state, when required. These cells are important for wound healing, regeneration and secondary growth. The process by which they lose their specialization is referred to as dedifferentiation.

Differentiation imparts tremendous benefits on the multicellular organisms. These are: (1) increased survivial; (2) increased specialization; (3) ensured uninterrupted life activity, and (4) a proper balance between the cell surface and cell volume for receiving external stimuli, exchange of materials, transport, secretion, etc.

The differentiation is based on **differential expression** of the genes which is a basic principle of developmental genetics: In spite of the fact that all the cells of an individual body contains the same genome, the specific proteins expressed by the different cell types are widely diverse. The following processes work to allow to extensive differentiation of cell types:

1. Differential gene expression;
2. Differential nRNA processing;
3. Differential mRNA translation; and
4. Differential protein modification.

44.1. DIFFERENTIAL GENE EXPRESSION

Eukaryotic genes are contained within a complex of DNA and protein called **chromatin**. The protein component constitutes about half the weight of chromatin and is composed largely of **histones**. The **nucleosome** is the basic unit of chromatin structure (Fig. 44.1). It is composed of an octamer of histone proteins (two molecules each of histones H2A, H2B, H3 and H4) wrapped with two loops containing approximately 140 base pairs of DNA. Histone H1 is bound to the 60 or so base pairs of "linker" DNA between the nucleosomes. There are 14 points of contact between the DNA and the histones.

Anatomy of the Gene

1. Active and repressed chromatin. Whereas classical geneticists have likened genes to "beads on a string", molecular geneticists liken genes to "string on the beads", an image in which the beads are nucleosomes. Most of time, the nucleosomes are wound into tight "solenoids" that are stabilized by histone H1 (Fig. 44.1B). This H1-dependent conformation of nucleosomes inhibits the transcription of genes in somatic cells by packing adjacent nucleosomes together into tight arrays that prevent **transcription factors** and **RNA polymerases** from gaining access to the gene. It is generally thought, then, that the "default" condition of chromatin is a repressed state, and that tissue-specific genes become activated by local interruption of this repression.

Histones as an activation switch. The histones are critical because they are responsible for maintaining the repression of gene expression. This repression can be locally strengthened (so that it becomes very difficult to transcribe those genes in the nucleosomes) or relieved (so that transcribing them becomes relatively easy) by modifying the histones (Fig. 44.1D). Repression and activation are controlled to a large extent by modifying the tails of histones H3 and H4 with two small organic groups: **methyl** (CH_3) and **acetyl** ($COCH_3$) residues. In general, **histone acetylation**—the addition of negatively charged acetyl groups to histones—neutralizes the basic charge of lysine and loosens the histones. *This activates transcription.* Enzyme known as **histone acetyltransferases** place acetyl groups on histones (especially on lysines in H3 and H4), destabilizing the nucleosomes so that they come apart easily. As might be expected, then enzymes that remove acetyl groups—**histone deacetylases**—stabilize the nucleosomes and *prevent transcription.*

Histone methylation, the addition of methyl groups to histones by **histone methyltransferase** enzyme, can either *activate or further repress transcription*, depending on the amino acid being methylated and the presence of other methyl or acetyl groups in the vicinity. For instance, acetylation of the "tails" of H3 and H4 along with methylation of the lysine at position 4 of H3 (*i.e.*, H3K4; remember that K is the abbreviation for lysine) is usually associated with highly repressed chromatin. In contrast, a combined lack of acetylation of the H3 and H4 tails and methylation of the lysine in 9th position of H3 (*i.e.*, H3K9) is usually associated with highly repressed chromatin. Indeed, lysine methylations at H3K9, H3K27, and H4K20 are often associated with highly repressed chromatin. Figure 44.2 shows a schematic drawing of a nucleosome, with the histone H3 tail having on it some residues whose modification can regulate transcription.

As might be expected, if methyl groups at specific places of the histones repress transcription, the getting rid of these methyl moieties should permit transcription. This has been shown in the activation of those genes responsible for specifying the posterior halves of vertebrate bodies. These genes, called **Hox genes**, encode transcription factors that are critical in giving cells their identities along the anterior-posterior axis. In early development, Hox genes are repressed by **H3K27 trimethylation** (the lysine at position 27 having three methyl groups). However, in differentiated cells, a demethylase (enzyme) that is specific for H3K27me3 is recruited to these promoters and enables the gene to be transcribed.

Histone regulation of transcriptional elongation. In addition to regulating the initiation of the transcriptional complex (*i.e.*, getting RNA polymerase on the promoter), nucleosome also appear to regulate the progression of RNA polymerase and the elongation of mRNA. Indeed recent evidence suggests that it is relatively common for RNA polymerase to be waiting at the promoters, ready to go.

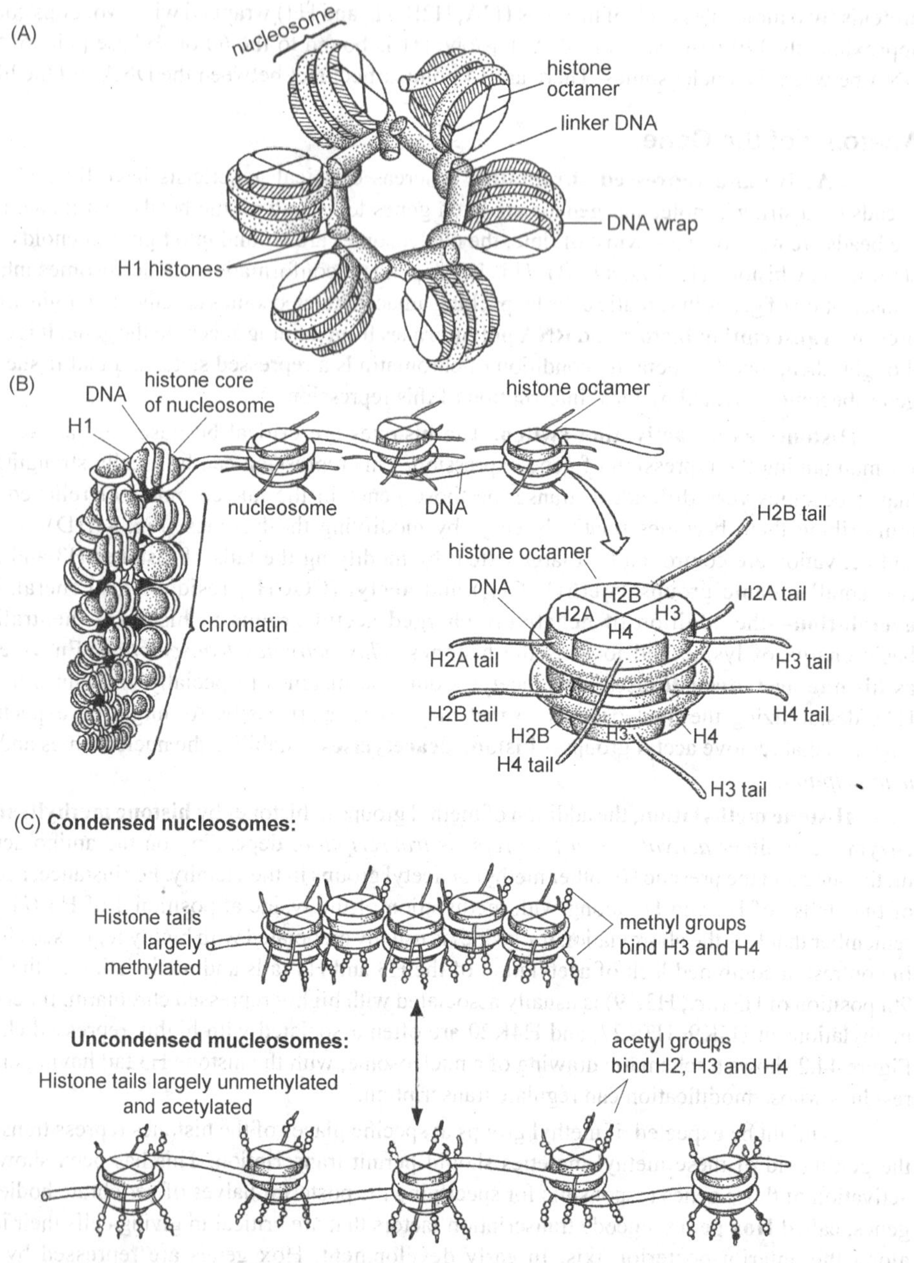

Fig. 44.1. Role of methylation and ethylation of histones on the packing of nucleosomes in chromatin. A–Histone H1 can draw nucleosomes together into compact forms. About 140 base pairs of DNA encircle each histone octamer, and about 60 base pairs of DNA link the nucleosomes together. B–Model for arrangement of nucleosomes in the highly compacted solenoidal chromatin structure. Histone "tails" protruding from the nucleosome subunits allow for the attachment of chemical groups. C–Methyl groups condense nucleosomes more tightly, preventing access to promoter sites and thus preventing gene transcription. Acetylation loosens nucleosome packing, exposing the DNA to RNA polymerase and transcription factors that will activate the genes (after Gilbert 2010).

For transcription to occur, these nucleosomes need to be modified, and it is possible that the **acetylation** of histone H3 at position 9 and 14, coupled with **trimethylation** of that histone at position 4, is critical for allowing elongation of the message.

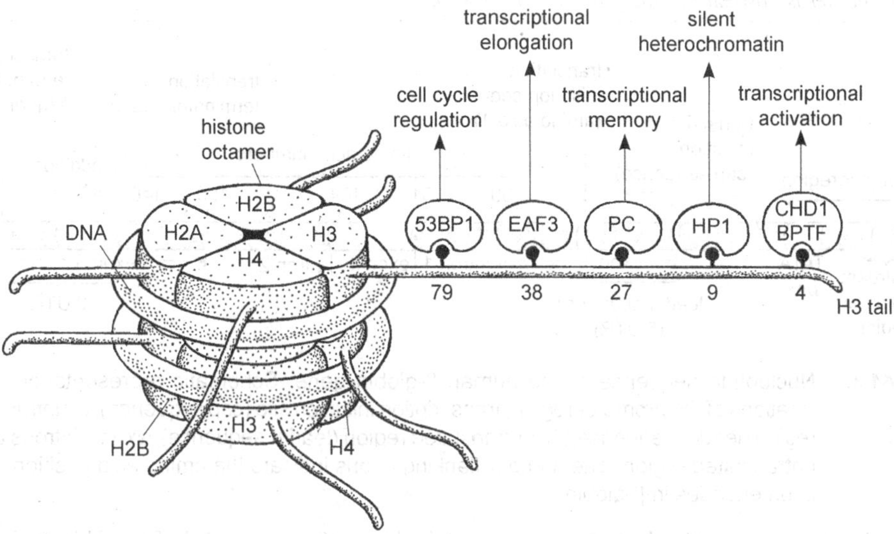

Fig. 44.2. Histone methylations on histone H3. The tail of histone H3 (its aminomost sequence at the beginning of the protein) sticks out from the nucleosome and is capable of being methylated or acetylated. Here, lysines can be methylated and recognized by particular proteins. Methylated lysine residues at positions 4, 38, and 79 are associated with gene activation, whereas methylated lysines at positions 9 and 27 are associated with repression. The proteins binding these sites are represented above the methyl groups (after Gilbert 2010).

2. Structure of an eukaryotic gene. Eukaryotic genes are different from prokaryotic genes. Eukaryotic genes are not co-linear with their peptide products. Rather, the single nucleic acid strand of eukaryotic mRNA comes from non-contiguous regions on the chromosome. Between **exons**—the regions of DNA that code for a protein—are intervening sequences called **introns** that have nothing to do with the amino acid sequence of the protein. The structure of a typical eukaryotic gene can be illustrated by the human **β-globin gene** (Fig. 44.3). The β-globin gene encodes part of the haemoglobin protein of the red blood cells, consists of following elements:

Box 44.1

Exon. It is nucleotide sequence of an eukaryotic gene that is represented in the mRNA (**Malacinski** 2003).

Intron. It is nucleotide sequence in a gene that is transcribed, but eventually removed from the mRNA by specific splicing enzymes (**Malacinski** 2003).

The term exon refers to a nucleotide sequence whose RNA "exits" the nucleus. It has taken on the functional definition of a protein-encoding nucleotide sequence. Leader sequences and 3'UTR sequences (of a mRNA) are also derived from exons, even though they are not translated into protein (**Gilbert** 2010).

(*i*) A promoter region, which is responsible for the binding of RNA polymerase and for the subsequent initiation of transcription. The promoter region of the human β-globin gene has three distinct units and extends from 95 to 26 pairs before ("upstream from" the transcription initiation site (*i.e.*, from –95 to –26).

Box 44.2

By convention, upstream, downstream, 5′ and 3′ directions are specified in relation to the RNA. Thus, the promoter is upstream of the gene, near its 5′ end.

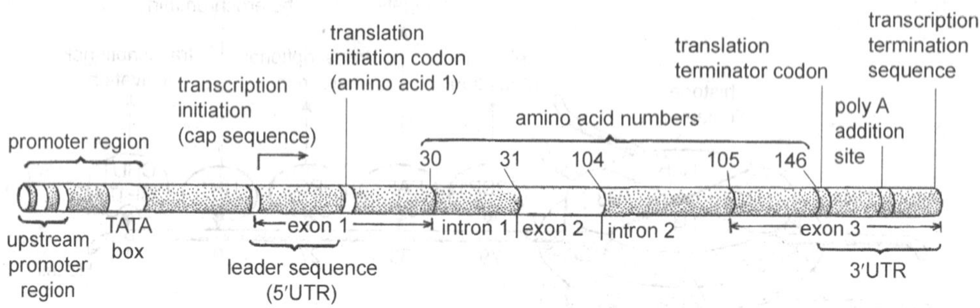

Fig. 44.3. Nucleotide sequence of the human β-globin gene. Schematic representation of the locations of the promoter region, transcription initiation site (cap sequence), 5′ untranslated region (leader sequence), 5′ untranslated region (leader sequence), exons, introns and 3′ untranslated region. The number flanking exons indicate the amino acid positions each exon encodes in β-globin.

(*ii*) **The transcription initiation site**, which for human β-globin is ACATTG. This site is often called the **cap sequence** because it represents the 5′ end of the RNA, which will receive a "cap" of modified nucleotides soon after it is transcribed. The specific cap sequence varies among genes.

(*iii*) **The translation initiation site, ATG.** This codon (which becomes AUG in mRNA) is located 50 base pairs after the transcription initiation site in the human β-globin gene (this distance differs greatly among different genes). The sequence of 50 base pairs interposing between the initiation points of transcription and translation is the **5′ untranslated region**, often called the **5′ UTR** or **leader sequence**. The 5′ UTR can determine the rate at which translation is initiated.

(*iv*) The first **exon**, which contains 90 base pairs coding for amino acids 1–30 of human β-globin protein.

(*v*) An **intron** containing 130 base pairs with no coding sequences for β-globin. However, the structure of this intron is important in enabling the RNA to be processed into mRNA and exit from the nucleus.

(*vi*) An **exon** containing 222 base pairs coding for amino acids 31–104.

(*vii*) A large *intron* –850 base pairs—having nothing to do with globin protein structure.

(*viii*) An *exon* containing 126 base pairs coding for amino acids 105–146 of the protein.

(*ix*) A **translation termination site, TAA.** This codon becomes UAA in the mRNA. The ribosome dissociates at this codon, and protein is released. (By convention, only the RNA-like strand of double helix is shown).

(*x*) A **3′ untranslated region (3′ UTR)** that although transcribed, is not translated into protein. This region includes the sequence AATAAA, which is needed for **polyadenylation**, the insertion of a "tail" of some 200–300 adenylate residues on the RNA transcript, about 20 base downstream of AAUAAA sequence. This poly A tail (1) grants stability to the mRNA, (2) allows the mRNA to exit the nucleus, and (3) permits the mRNA to be translated into protein.

(*xi*) **A transcription termination sequence.** Transcription continues beyond the AATAAA site for about 1000 nucleotide before being terminated.

The original transcription product is called **nuclear RNA (nRNA)**, sometimes called **heterogeneous nuclear RNA (hnRNA)** or **pre-messenger RNA (pre-mRNA).** Nuclear RNA contains the cap sequence, the 5′ UTR (leader sequence), exons, introns, and the 3′ UTR (Fig. 44.4). Both ends of these transcripts are modified before these RNAs leave the nucleus. A cap consisting of methylated guanosine is placed on the 5′ end of the RNA in opposite polarity to the RNA itself. This means there is no free 5′ phosphate group on the nRNA. The 5′ cap is necessary for the binding of mRNA to the ribosome and for subsequent translation. The 3′ terminus is usually modified in the nucleus by the addition of a poly A tail. The adenylate residues in this tail are put together enzymatically and are added to the transcript; they are not part of gene sequence. Both the 5′ and 3′ modifications may protect the mRNA from exonuclease enzymes that would otherwise digest it. The modifications thus stabilize the message and its precursor.

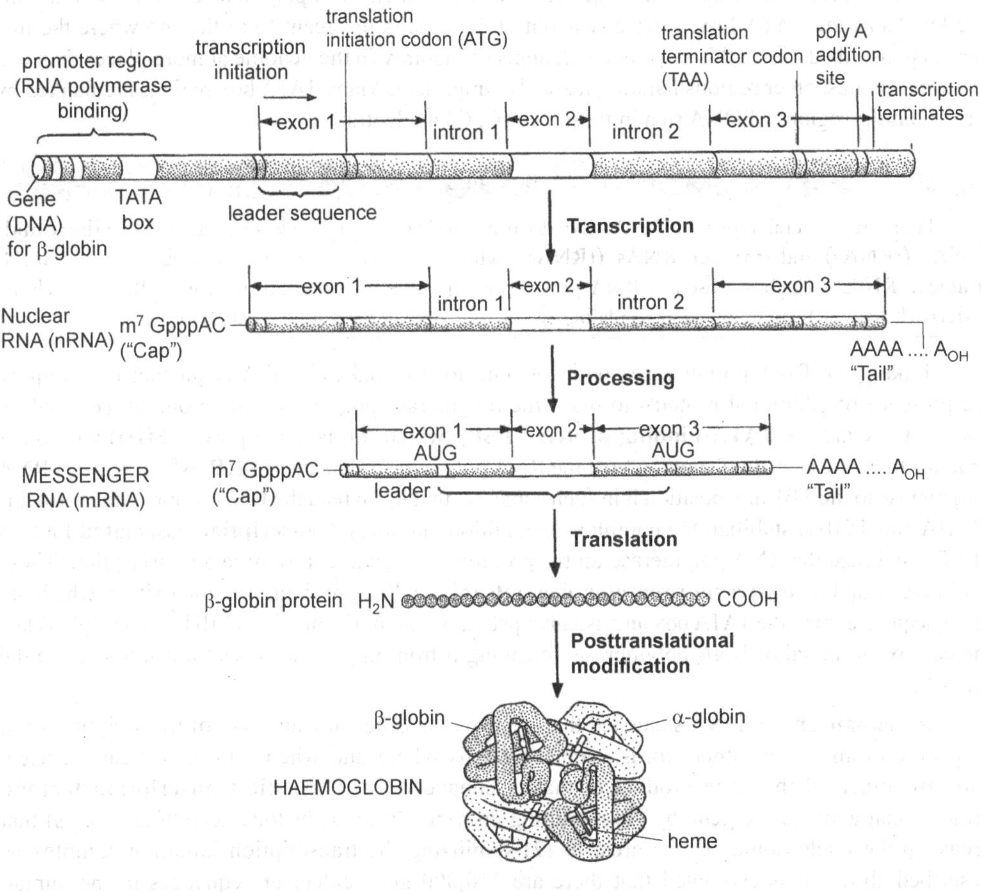

Fig. 44.4. Summary of steps involved in the production of β-globin and haemoglobin. Transcription of the gene creates a nuclear RNA (nRNA) containing exons and introns, as well as the cap, tail and 3′ end and 5′ UTRs. Processing the nuclear RNA (nRNA) into messenger RNA (mRNA) removes the introns. Translation on ribosomes uses the mRNA to encode a protein. The protein is inactive until it is modified and complexed with α-globin and haem to become active haemoglobin (after Gilbert 2010).

As the nRNA leaves the nucleus, its introns are removed and the remaining exons spliced together. In this way the coding regions of the mRNA—*i.e.*, the exons—are brought together to form a single transcript and this transcript is translated into a protein. The protein can be further modified to make it functional (Fig. 44.4).

3. Promoters and enhancers. In addition to the protein-encoding regions of the gene, there are **regulatory sequences** that can be located on either end of the gene (or even within it). These sequences are called the **promoters** and **enhancers** and they are necessary for controlling where and when a particular gene is transcribed.

Promoters are the sites where RNA polymerase enzyme binds to the DNA to initiate transcription. Promoters of genes that synthesize messenger RNAs (*i.e.*, genes that encode proteins; Fig. 44.3) are typically located immediately upstream from the site where the RNA polymerase initiates transcription. Most of these promoters contain the sequence TATA, to which RNA polymerase will be bound. This site known as the **TATA box**, is usually about 30 base pairs upstream from the site where the first base is transcribed. Since this sequence will appear randomly in the genome at more places than just at promoter sites, other regions flanking it are also important. Many TATA box regions are flanked by **CpG islands**, regions of DNA rich in those two (C, G) nucleotides.

Box 44.3

There are several types of RNA that do not encode proteins. These include the **ribosomal RNAs (rRNAs)** and **transfer RNAs (tRNAs)** (which are used in protein synthesis) and the **small nuclear RNAs** (which are used in RNA processing). In addition, there are regulatory RNAs (such as **microRNAs** which are involved in regulating gene expression and are not translated into peptides.

Eukaryotic RNA polymerases will not bind to the "naked" TATA sequence; they require the presence of additional proteins to place the polymerase properly on the promoter (Fig. 44.5). Two of these are the **TATA-binding protein (TBP)**, which forms a complex **(TFIID)** with other proteins to create a "saddle" upon which the RNA polymerase sits; and **TFIIB**, which recruits RNA polymerase to the TBP and position it in such a manner that it can read the DNA codon. Other proteins (TFIIA and TFIIH) stabilize the complex. In addition, auxiliary **transcription-associated factors (TAFs)** stabilize the RNA polymerase on the promoter and enable it to initiate transcription. These TAFs are bound by **upstream promoter elements** (also called proximal promoter sites), which are DNA sequences near the TATA box and usually upstream from it. Eventually, TFIIH will phosphorylate the carboxy terminal of RNA polymerase, releaving it from the saddle so that it can transcribe the mRNA.

An **enhancer** is a DNA sequence that controls the efficiency and rate of transcription from a specific promoter. In other words, enhancers tells where and when a promoter can be used, and how much of the gene product to make. Enhancers bind specific **transcription factors**, proteins that activate the gene by (1) recruiting enzymes (such as histone acetyltransferases) that break up the nucleosomes in the area, or (2) stabilizing the transcription initiation complex as described above. It is estimated that there are 110,000 gene enhancer sequences in the human genome.

Enhancers can activate only *cis*-linked promoters (*i.e.*, promoters on the same chromosome; Box 44.4). However, because of DNA folding, enhancers can regulate genes at great distances (some as great as a million bases away) from the promoter. Moreover, enhancers do not need to be on the 5′ (upstream) side of the gene; they can be at 3′ end, or even in the introns. The human β-globin gene has an enhancer in its 3′ UTR. This enhancer sequence is necessary for the temporal- and tissue-specific expressions of the β-globin gene in adult red blood cell precursors.

Box 44.4

Cis- and *trans-*regulatory elements are so named by analogy with *E. coli* genetics and organic chemistry. The *cis-* elements are regulatory elements that reside on the same strand of DNA (*cis* means "on the same side as") while *trans-* elements are those that could be supplied from another chromosome (*trans* means "on the other side of "). The term *cis-* regulatory elements now refers to those DNA sequences that regulate a gene on the same stretch of DNA (*i.e.*, the promoters and enhancers). *Trans-*regulatory factors are soluble molecules whose genes are located elsewhere in the genome and which bind to the *cis-*regulatory elements. They are usually transcription factors or microRNAs.

Functions of Transcription Factors

Natalie Angier (1992) has written about DNA that a series of new discoveries suggest that DNA is more like a certain type of politician, surrounded by a flock of protein handlers and advisers that must vigorously massage it, twist it, and on occasion, reinvent it before the grand blueprint of the body can make any sense at all. These "handlers and advisers" are the transcription factors. These factors can be grouped together in families based on similarities in structure (Table 44.1). The transcription factors within such a family share a common framework in their DNA-binding sites, and slight differences in the amino acids at the binding site to recognise different DNA sequences.

Table 44.1. Some major transcription factor families and subfamilies (*Source*: Gilbert 2010).

Family	Representative transcription factors	Some function
1. Homeodomain:		
(*i*) Hox	Hoxa1, Hoxb2, etc.	Axis formation
(*ii*) POU	Pit1, Unc-86, Oct-2	Pituitary development; neural fate
(*iii*) LIM	Lim1, Forkhead	Head development
(*iv*) Pax	Pax1, 2, 3, 6, etc.	Neural specification; eye development
2. Basic helix-loop	MyoD, MITF, daughterless	Muscle and nerve specification; *Drosophila* sex determination; pigmentation
3. Basic leucine zipper (bZip)	C/EBP, AP1	Liver differentiation; fat cell specification
4. Zinc finger:		
(*i*) Standard	WT1, Kruppel, Engrailed	Kidney, gonad, and macrophage development; *Drosophila* segmentation
(*ii*) Nuclear hormone receptor	Glucocorticoid receptor, estrogen receptor, testosterone receptor, retinoic acid receptors	Secondary sex determination; craniofacial development; limb development
(*iii*) Sry-Sox	Sry, Sox D, Sox 2	Bend DNA; mammalian primary sex determination; ectoderm differentiation

(*a*) **Enhancers** function by binding transcription factors, and each enhancer can have binding sites for several transcription factors. Transcription factors bind to the enhancer DNA with one part of the protein and use other sites on the protein to interact with one another to recruit histone-modifying enzymes. For example, the association of the Pax 6, Sox 2 and L-Maf transcription factors in lens cells recruits a histone acetyltransferase that can transfer acetyl groups to the histones and dissociate the nucleosomes in that area.

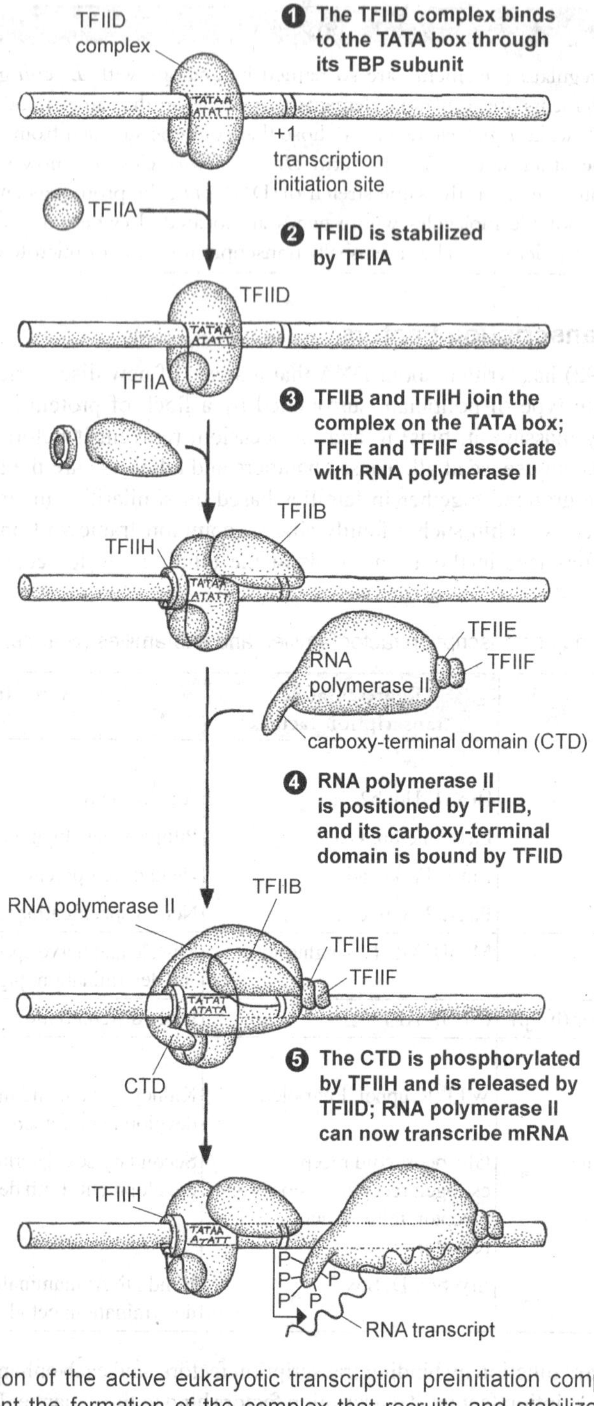

Fig. 44.5. Formation of the active eukaryotic transcription preinitiation complex. The diagrams represent the formation of the complex that recruits and stabilizes RNA polymerase onto the promoter. TF stands for transcription factors; II indicates that the factor was first identified as being needed for RNA polymerase II (the RNA polymerase that transcribes protein-encoding genes); and the letters designate the particular active fraction from the phospho-cellulose columns used to purify it (after Gilbert 2010).

In addition to recruiting nucleosome modifying enzymes, transcription factors can also work by stabilizing the transcription preinitiation complex that enables RNA polymerase bind to the promoter (Fig. 44.6).

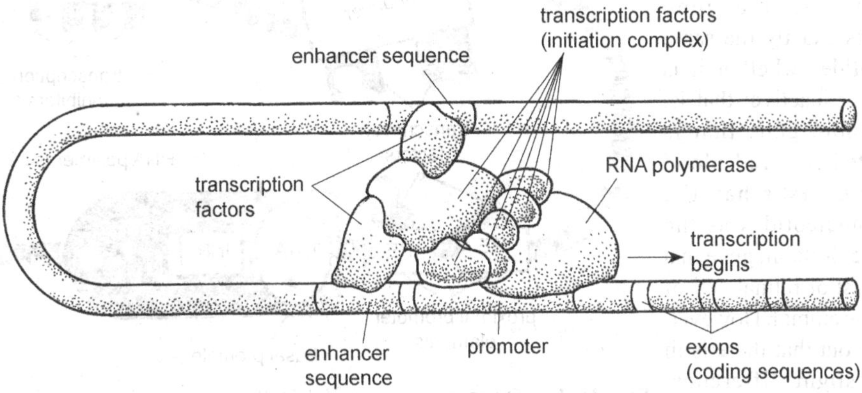

Fig. 44.6. RNA polymerase is stabilized on the promoter site of the DNA by transcription factors recruited by the enhancers. The TATA sequence at the promoter binds a protein that serves as a "saddle" for RNA polymerase. However, RNA polymerase would not remain bound long enough to initiate transcription were it not for the stabilization by the transcription factors (after Gilbert 2010).

***(b)* Pioneer transcription factors (TFs).** Finding a promoter is not easy, because the DNA is usually so wound up that the promoter sites are not accessible. Indeed, more than 6 feet of DNA is packaged into chromosomes of each human cell nucleus.

How can a transcription factor find its binding site, given that the enhancer might be covered by nucleosomes? Several studies have identified certain transcription factors that penetrate repressed chromatin and bind to their enhancer DNA sequences. They appear to be critical in establishing certain cell lineages.

Examples of pioneer TF. One of these transcription factors (TFs) is **FoxA1**, which binds to certain enhancers and opens up the chromatin to allow other transcription factors access to the promoter. Fox 8 proteins remain bound to the DNA during mitosis, providing a mechanism to re-establish normal transcription in presumptive liver cells. Another pioneer transcription factor is the **Pax7 protein**. It activates muscle-specific gene transcription in a population of muscle stem cells by binding to its DNA recognition sequence and being stabilized there by dimethylated H3K4 on the nucleosomes. Pax7 then recruits the histone methyltransferase (enzyme) that converts the dimethylated H3K4 into the trimethylated H3K4 associated with active transcription.

***(c)* Silencers.** Silencers are DNA regulatory elements that actively repress the transcription of a particular gene. They can be viewed as **"negative enhancers"** (see **Gilbert** 2010). For instance, in the mouse, there is a DNA sequence that prevents a promoter's activation in any tissue except neurons. This sequence is called **neural restrictive silencer element (NRSE)** and it has been found in several mouse genes whose expression is limited to the nervous system: those cells encoding *synapsin1*, *sodium channel type II*, *brain-derived neurotrophic factor*, *Ng-CAM*, and *L1*. The protein that binds to NRSE in a zinc finger transcription factor called **neural restrictive silencer factor (NRSF)**. NRSF appears to be expressed in every cell that is not a mature neuron.

***(d)* Insulators.** In the organizations of the chromosome, it is important to separate active genes that are being transcribed from genes that are repressed. This is achieved by chromosomal **insulators** (Fig. 44.7). These insulators are gene sequences that provide a barrier so that a specific gene is isolated against transcriptional influences from surrounding genes. Insulators can vary greatly in their DNA sequence and the proteins that bind to them. One way an insulator activity can be modulated is by **DNA methylation**.

DNA methylation and the control of transcription. It is often assumed that a gene contains exactly the same nucleotides whether it is active or inactive; that is, a β-globin gene that is activated in a red blood cell precursor has the same nucleotides as the inactive β-globin gene in a fibroblast or retinal cell of the same animal. However, it turns out that there is in fact a slight difference. In 1948, **R.D. Hotkiss** discovered a "fifth base" in DNA, **5-methylcytosine**. In vertebrates, this base is made enzymatically after DNA is replicated. At this time, about 5% of the cytosines in mammalian DNA are converted to 5-methylcytosine (Fig. 44.8A). This conversion can occur only when the cytosine residue is followed by a guanosine—in other words, at a CpG sequence. Numerous studies have shown that the degree to which the cytosines of a gene are methylated can control the level of the gene's transcription. Cytosine methylation appears to be a major mechanism of transcriptional regulation among vertebrates.

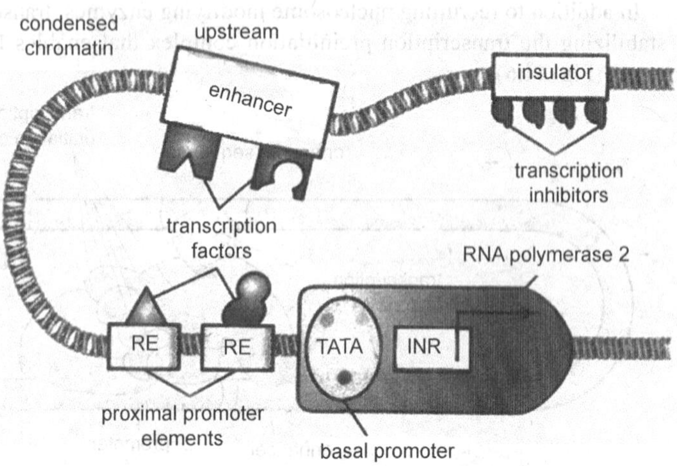

Fig. 44.7. Gene transcriptional in the eukaryotic cells. A complex arrangement of multiple clustered enhancer modules interspread with insulator elements, which can be located either upstream or downstream of a basal promoter containing TATA box (TATA), proximal promoter elements (response elements, RE), and initiator sequences (INR) (after Hall 2011).

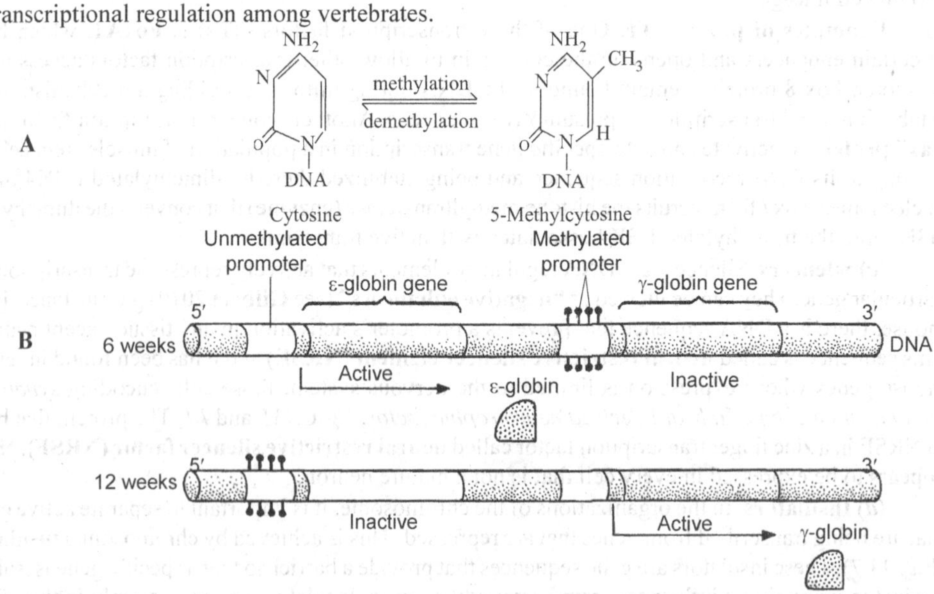

Fig. 44.8. Methylation of globin genes in human embryonic blood cells. A–Structure of 5-methylcytosine. B–The activity of the human β-globin genes correlates inversely with the methylation of their promoters (after Gilbert 2010).

In vertebrates, the presence of methylated cytosine in a gene's promoter correlates with the repression of transcription from that gene. In developing human and chick red blood cells, for example, the DNA of the globin gene promoters is almost completely unmethylated, whereas the same promoters are highly methylated in cells that do not produce globin. Moreover, the methylation pattern changes during development (Fig. 44.8B). The cells that produce haemoglobin in the human embryo have unmethylated promoters in the genes encoding the ε-globins ("embryonic globin chains") of embryonic haemoglobin. These promoters became methylated in the fetal tissue, as the genes for fetal-specific γ-globin (rather than the embryonic chains) become activated. Similarly, when fetal globin gives way to adult (β) globin, promoters of the fetal (γ) globin genes become methylated.

Methylation differences can also account for exmaples of **genomic imprinting**, wherein a gene transmitted through the sperm is expressed differently than the same gene transmitted through the egg.

DNA methylation in mammalian insulin-like growth factor 2 (IGF-2) gene. The mother's allele has an insulator between enhancer and promoter of the gene that allows for the binding of a transcriptional repressor. However, the paternal DNA sequence is methylated such that the transcriptional repressor cannot bind to the insulator and the IGF-2 gene is expressed from the paternal copy of the gene (**Hall** 2011).

Dosage compensation. Dosage compensation enables the X chromosome-derived products of males (which have one X chromosome per cell in fruit flies and mammals) to equal the X chromosome-derived products of females (which have two X chromosomes per cell). This compensation is accomplished at the level of transcription, either by accelerating transcription from the lone X chromosome in males (*Drosophila*) decreasing the level of transcription from each X chromosome by 50% (nematode *Caenorhabditis elegans*), or by inactivating a large portion of one of the two X chromosomes in females (mammals).

44.2. DIFFERENTIAL nRNA PROCESSING

The regulation of gene expression is not confined to the differential transcription of DNA. Even if a particular RNA transcript is synthesized, there is no guarantee that it will create a functional protein in a cell. To become an active protein, the RNA must be (1) processed into a messenger RNA by removal of introns; (2) translocated from the nucleus to the cytoplasm; and (3) translated in the protein—synthesized protein is not in its mature form and must be (4) post-translationally modifed to become active. Regulation of development can occur at any of these steps.

There are two major ways in which differential RNA processing can regulate development. The first involves "censorship"—selecting which nuclear transcripts are processed into cytoplasmic messages (called **nuclear RNA selection**). Different cells select different nuclear transcripts to be processed and sent to the cytoplasm as messenger RNA. Thus, the same pool of nuclear transcripts can give rise to different populations of cytoplasmic mRNAs in different cell types (Fig. 44.9A).

The second mode of differential RNA processing is the **splicing** of mRNA precursors into messages that specify different proteins by using different combinations of potential exons (it is called **differential splicing**). Thus, one cell type might use exons 1, 2, 4 and 5; a different cell type might use exons 1, 2, and 3; yet another cell type might use all five (Fig. 44.9B). Thus, a single gene can produce an entire family of proteins.

Example of Nuclear RNA Selection

More genes are transcribed in the nucleus than are allowed to become mRNAs in the cytoplasm. This "censoring" of RNA transcripts has been confirmed by probabing for introns and exons of specific genes. **Gagnon** and his colleagues (1992) performed such an analysis on the transcripts from the *Spec II* and *CyIIIa* genes of the sea urchin *Strongylocentrotus purpuratus*. These genes encode calcium-binding and actin proteins, respectively, which are expressed only in a particular part of the ectoderm of the sea urchin larva. Using probes that bound to an exon (which is included in the mRNA) and to an intron (which is not included in the mRNA), they found that these genes were being transcribed

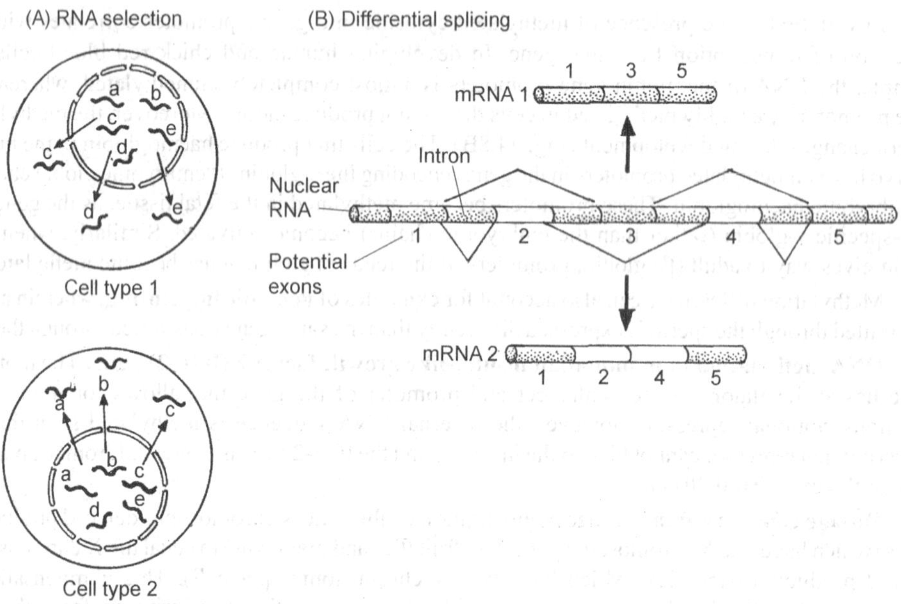

Fig. 44.9. Roles of differential RNA processing during development. By convention splicing paths are shown by fine V-shaped lines. A–RNA selection, whereby the same nuclear transcripts are made in two cell types, but the set that becomes cytoplasmic mRNA is different. B–Differential splicing, whereby the same nuclear RNA (nRNA) is spliced into different mRNAs by selectively using different exons (after Gilbert 2010).

not only in the ectodermal cells, but also in the mesoderm and endoderm. The analysis of the *CyIIIa* gene showed that the concentration of introns was the same in both the gastrula ectoderm and the mesoderm/endoderm samples, suggesting that this gene was being transcribed at the same rate in the nuclei of all cell types, but was made into cytoplasmic mRNA only in ectodermal cells. The unprocessed nRNA for *CyIIIa* is degraded while still in the nuclei of the endodermal and mesodermal cells.

Examples of Differential nRNA Splicing

Alternative nRNA splicing is a means of producing a wide variety of proteins from the same gene (Different proteins encoded by the same gene are called **splicing isoforms** of the protein). Researchers estimate that approximately 92% of human genes are alternatively spliced, and that such alternative splicing is a major way by which the rather limited number of genes can create a much larger array of proteins.

Thus, a gene with dozens of introns can create literally thousands of different related proteins through differential splicing. The current champion at making multiple proteins from the same gene is the *Drosophila Dscam1* gene. This gene encodes a membrane receptor protein involved in preventing dendrites from the same neuron from binding to one another. *Dscam1* gene contains 24 exons. However, a dozen different adjacent DNA sequences can be selected to be exon 4. Similarly, more than 30 mutually exclusive adjacent DNA sequences can become exons 6 to 9, respectively (Fig. 44.10). If all possible combinations of exons are used, this one gene can produce 38,016 different proteins, and random searches for these combinations indicate that a large fraction of them are in fact made. The nRNA of *Dscam1* has been found to be alternatively spliced in different axons, and when two dendrites from the same axon touch each other, they are repelled. This causes the extensive branching of the dendrites. It appears that the thousands of splicing isoforms are needed to ensure that each neuron acquires a unique identity (Fig. 44.11; **Schmucker** 2007). The *Drosophila* genome is thought to contain only 14,000 genes; but here is a single gene that encodes three time that number of proteins (see **Gilbert** 2010).

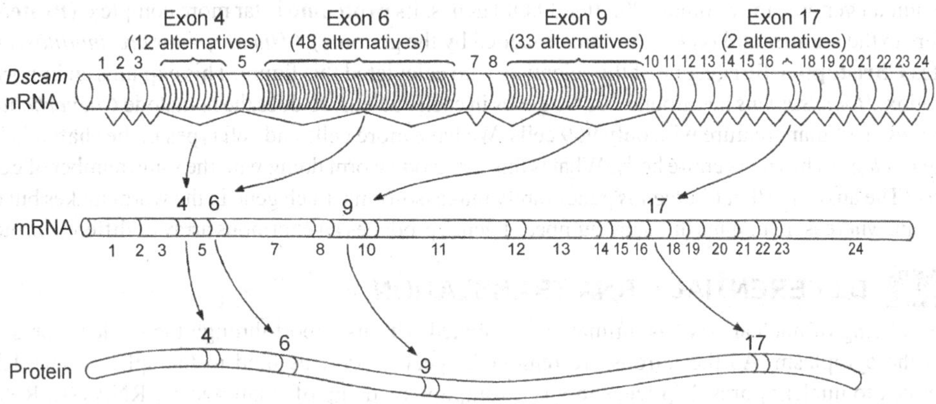

Fig. 44.10. The *Dscam* gene of *Drosophila* can produce 38,016 different types of proteins by alternative nRNA splicing. The gene contains 24 exons. Exons 4, 6, 9 and 17 are encoded by sets of mutually exclusive possible sequences. Each messenger RNA will contain one of the 12 possible exon 4 sequences, one of the 48 possible exon 6 alternatives, one of the 33 possible exon 9 alternatives, and one of the 2 possible exon 17 sequences. The *Drosophila Dscam* gene is homologous to a DNA sequence on human chromosome 21 that is expressed in the nervous system. Disturbances of this gene in humans may lead to the neurological defects of Down syndrome (after Gilbert 2010).

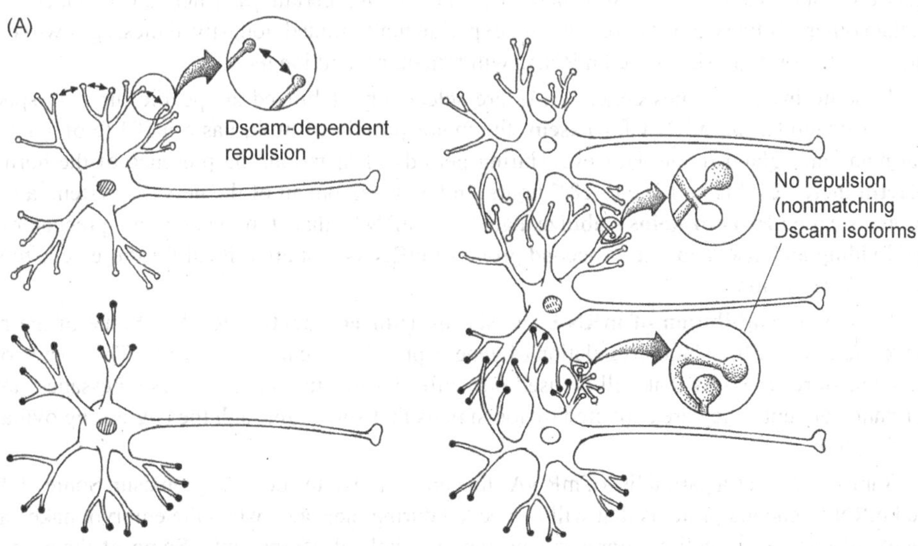

Fig. 44.11. Dscam protein is specifically required to keep dendrites from the same neuron from adhering to each other. A—When sister dendrites expressing the same splicing isoform of Dscam touch, the Dscam-Dscam interactions repel them and cause the dendrites to separate. B—Different neurons express different splicing isoforms that do not interact with one another (and therefore do not trigger repulsion), which allows neurons to interact (after Gilbert 2010).

Advantage of nRNA Splicing in Human Beings

About 92% of human genes are thought to produce multiple types of mRNA. Therefore, even though the human genome may contain 20,000–30,000 genes, its **proteome** is far more complex. (**Proteome** refers to the number and type of proteins encoded by the genome). "*Human genes are multitaskers*", notes **Christopher Burge**, one of the scientists who calculated this figure. This explains an important paradox. *Homo sapiens* has around 20,000 genes in each nucleus, so does the nematode *Caenorhabditis elegans*, a tubular creature with only 969 cells. We have more cells and cell types in the shaft of a hair than *C. elegans* has in its entire body. What is this nematode worm doing with the same number of genes as us? The answer is that *C. elegans* genes rarely make isoforms. Each gene in the worm makes but one protein, whereas in humans the same number of genes produces an enormous array of different protein.

44.3. DIFFERENTIAL mRNA TRANSLATION

The splicing of nuclear RNA is intimately associated with its export through the nuclear pores and into the cytoplasm. As the introns are removed, specific proteins bind to the spliceosome–RNA complex to nuclear pores. [**Spliceosome** is a complex, made up of small nuclear RNAs (sn RNAs) and splicing factors, that bind to splice sites and mediates the splicing of nRNA]. However, even the mRNA has reached to cytoplasm, there is still no guarantee that it would be translated. The control of gene expression at the level of translation can occur by following means:

1. Differential mRNA longevity. The longer an mRNA persists, the more proteins can be translated from it. If a message with relatively short half-life were selectively stabilized in certain cells at certain times, it would make large amounts of its particular protein only at those times and places.

The stability of a massage often depends on the length of its poly A tail. This, in turn, depends largely on sequences in the 3′ untranslated region (3′ UTR), certain of which allow longer poly A tails than others. If these 3′ UTR-regions are experimentally traded: long-lived messages will decay rapidly, while normally short-lived mRNAs will remain around longer.

In some instances, messenger RNAs are selectively stabilized at specific times in specific cells. For example, the mRNA for **casein**, the major protein of milk, has a half-life of 1.1 hours in rat mammary gland tissue. However during periods of lactation, the presence of the hormone **prolactin** increases this half-life to 28.5 hours. In the development of the nervous system, a group of proteins called **HuD proteins** stabilize a group of mRNAs that stop the neuronal precursor cells from dividing and also stabilizes a second group of mRNAs that are critical for these cells to start neuron differentiation.

2. Selective inhibition of mRNA translation (Stored oocyte mRNAs). Some of the most remarkable cases of translation regulation of gene expression occur in the oocyte. The oocyte often makes and stores mRNAs that will be used only after fertilization occurs. These messages stay in a dormant state until they are activated by ion signals that spread through the egg during ovulation or fertilization.

Table 44.2 gives a partial list of mRNAs that are stored in the oocyte cytoplasm. Some of these stored mRNAs encode proteins that will be needed during cleavage, when the embryo makes huge amounts of chromatin, cell membranes, and the cytoskeletal components. Some of them encode cyclin protein that regulate the timing of early cell division. Indeed, in many species (including sea urchins and *Drosophila*), maintenance of the normal rate and pattern of early cell divisions does not require a nucleus; rather, it requires continued protein synthesis from stored maternal mRNAs (Box 44.5). Other stored messages encode proteins that determine the fates of cells. These include *bicoid*, *caudal* and *nanos* messages that provide information in *Drosophila* embryo for the production of its head, thorax and abdomen.

Table 44.2. Some mRNAs stored in oocyte cytoplasm and translated at or near fertilization (*Source*: Gilbert 2010).

mRNA Encoding	Function(s)	Organism(s)
1. Cyclins	Cell division regulation	Sea urchin, clam, star fish, frog
2. Actin	Cell movement and contraction	Mouse, star fish
3. Tubulin	Formation of mitotic spindles, cilia, flagella	Clam, mouse
4. Small subunit of ribonucleotide reductase	DNA synthesis	Sea urchin, clam, star fish
5. Hypoxanthite phosphoribosyl-transferase	Purine synthesis	Mouse
6. Vgl	Mesoderm determination	Frog
7. Histones	Chromatin formation	Sea urchin, frog, clam
8. Cadherins	Blastomere adhesion	Frog
9. Metalloproteinases	Implantation in uterus	Mouse
10. Growth factors	Cell growth, uterine growth	Mouse
11. Sex determination factor FEM-3	Sperm formation	*C. elegans*
12. PAR gene products	Segregate morphogenetic determinants	*C. elegans*
13. SKN-1 morphogen	Blastomere fate determination	*C. elegans*
14. Hunchback morphogen	Anterior fate determination	*Drosophila*
15. Caudal morphogen	Posterior fate determination	*Drosophila*
16. Bicoid morphogen	Anterior fate determination	*Drosophila*
17. Nanos morphogen	Posterior fate determination	*Drosophila*
18. GLP-1 morphogen	Anterior fate determination	*C. elegans*
19. Germ cell-less protein	Germ cell determination	*Drosophila*
20. Oskar protein	Germ cell localization	*Drosophila*
21. Ornithine transcarbamylase	Urea cycle	Frog
22. Elongation factor 1a	Protein synthesis	Frog
23. Ribosomal proteins	Protein synthesis	Frog, *Drosophila*

Box 44.5

A good example of translational control is the extension of the lifetime of silk fibroin mRNA in the silkworm Bombyx mori. During cocoon formation an silk gland of the silk worm predominantly synthesises a single type of protein, **silk fibroin**. Since the worm takes several days to construct its cocoon, it is the total amount and not the rate of fibroin synthesis that must be great; the silk worm achieves this by synthesizing a fibroin mRNA molecule that is very long lived.

Transcription of the fibroin gene is initiated at a strong promoter by a unknown signal and about 10^4 fibroin mRNA molecules are made in a period of several days (such a synthesis forms an example of transcriptional regulation). A typical eukaryotic mRNA molecule has a lifetime of about three hours before it is degraded. However, the fibroin mRNA survives for several days during which each mRNA is translated repeatedly to yield 10^5 fibroin molecules. Thus, each gene is responsible for the synthesis of 10^9 protein molecules in four days. Altogether the silk gland makes 300μg or 10^{15} molecules of fibroin during this period. If the lifetime of mRNA were extended, either 25 times as many genes would be needed or synthesis of the required fibroin would take about 100 days.

Further, free iron ions (Fe^{2+}) within a mammalian cell are bounded by a storage protein, **ferritin**. When iron is in excess, ferritin synthesis rises dramatically. Yet the amout of ferritin mRNA remains constant. This increase in ferritin synthesis is due to an increased rate of mRNA translation. When the iron level in the cell is low, a **translational repressor** protein binds to ferritin mRNA and prevents its translation by blocking its attachment to the ribosome. When iron level rise, the excess iron binds to the repressor and alters the three-dimensional structure of the translational repressor, causing it to detach from the mRNA and translation proceeds (**Purves** *et al.,* 1998).

Most translational regulation in oocyte is negative, as the "default state" of the mRNA is to be available for translation. Therefore there must be inhibitors preventing the translation of these mRNA in the oocyte, and these inhibitors must somehow be removed at appropriate times around fertilization. The 5′ cap and 3′ untranslated region seem especially important in regulating the accessibility of mRNA to ribosome. If the 5′ cap is not made or if the 3′ UTR lacks a poly A tail, the message probably will not be translated. The oocytes of many species have "used these ends as means" to regulate the translation of their mRNAs.

Most mRNA of oocyte probably form circles, with their 3′ end being brought to their 5′ end. The 5′ cap is bound by **eukaryotic initiation factor-4E (eIF4E)**, a protein that is also bound to **eIF4A** (a helicase that unwinds double-stranded regions of RNA) and **eIF4G**, a scaffold protein that allows the mRNA to bind to the ribosomes through its interaction with eIF4E. The poly A binding protein, which sits on the poly A tail of the mRNA, also binds to the eIF4G protein. This brings 3′ end of the message next to the 5′ end and allows the messenger RNA to be recognized by the ribosome. Thus, the 5′ cap is critical for translation, and some animal's oocytes have used this as a direct means of translation control. For example, the oocytes of the tobacco hornworm moth makes some of the mRNAs without their methylated 5′ caps. In this state, they cannot be efficiently translated. However, at fertilization, a methyltransferase completes the formation of the caps, and these mRNAs can be translated.

In amphibian oocytes, the 5′ and 3′ ends of many mRNAs are brought together by a protein called **maskin**. Maskin links the 5′ and 3′ ends into a circle by binding to two other proteins, each at opposite ends of the message. First, it binds to the **cytoplasmic poly adenylation-element binding protein (CPEB)** attached to the UUUUAU sequence in the 3′ UTR, second, maskin also bind to the eIF4E factor that is attached to the cap sequence. In this configuration, the mRNA cannot be translated (Fig. 44.12A). **Mendez** and **Richter** (2001) have proposed an complex scenario to explain how mRNAs bound together by maskin become translated at about the time of fertilization (Fig. 44.12B).

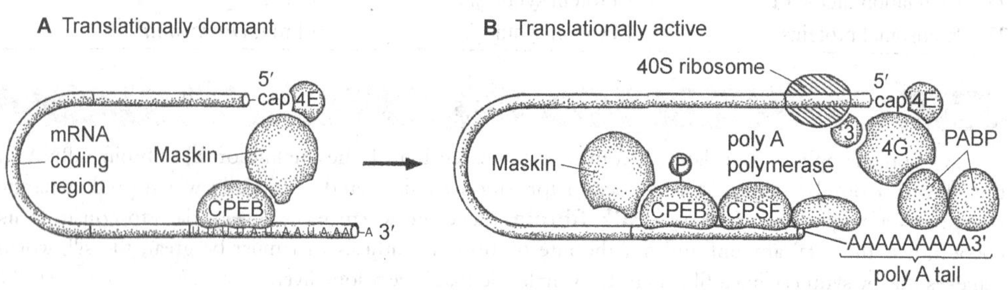

Fig. 44.12. Translational regulation in oocytes (**Mendez** and **Richter** 2001). A–In *Xenopus* oocytes, the 3′ and 5′ ends of the mRNA are brought together by maskin, a protein that binds to CPEB on the 3′ end and translation initiation factor 4E (eIF4E) on the 5′ end. Maskin blocks the initiation of translation by preventing eIF4E from binding eIF4G. B—When stimulated by progesterone during ovulation, a kinase phosphorylates CPEB, which then can bind CPSF. CPSF can bind poly A polymerase and initiate growth of the poly A tail. Poly A binding protein (PABP) can bind to this tail and then bind eIF4G in a stable manner. This initiation factor can then bind eIF4E and through its association with eIF3, position a 40S ribosomal subunit on the mRNA (after Gilbert 2010).

Role of microRNA in translation of mRNA. If proteins can bind specific nucleic acid sequences to block transcription or translation, one might think that RNA would do the job even better. After all, RNA can be made specifically to bind a particular sequence. Indeed, one of the most efficient means of regulating the translation of a specific message is to make a small RNA complementary to a portion of a particular mRNA. Such a naturally occurring antisense RNA was first seen in nematode *Caenorhabditis elegans*. Here, the lin-4 gene was found to encode a 21-nucleotide RNA called **microRNA (miRNA)**, that bound to multiple sites in the 3′ UTR of the *lin-14* mRNA. The *lin-14* gene encodes a transcription factor, **LIN-14**, that is important during the first larval phase of *C. elegans* development. It is not needed afterward, and *C. elegans* is able to inhibit synthesis of LIN-14 from these massages by activating the lin-4 gene. The binding of *lin-4* transcripts to the *lin-14* mRNA 3′ UTR causes degradation of the *lin-14* message.

The *lin-4* RNA is now thought to be the "founding member" of a very large group of microRNA (*miRNA*). These RNAs of about 22 nucleotides are made from longer precursors. These precursors can be in independent transcription units (the *lin-4* gene is far apart from the lin-14 gene), or they can reside in the introns of other genes. Many of the newly discovered microRNAs have been found in the regions between genes (regions previously considered to contain *junk DNA*). The initial RNA transcript (which may contain several repeats of the miRNA sequence) forms **hair-pin loops** wherein RNA finds complementary structures within its strand. These stem loop structures are processed by a set of RNases (*Drosha* and *Dicer*) to make single stranded microRNA (Fig. 44.13). The microRNA is then packaged with a series of proteins to make an **RNA induced silencing complex (RISC)**. Such small regulatory RNAs can bind to the 3′ UTR of messages and inhibit their translation. In some cases (especially when the binding of the miRNA to the 3′ UTR is tight), the site is cleaved. More usually, however, several RISCs attach to sites on the 3′ UTR and prevent the message from being translated.

Recent studies have shown that microRNAs are involved in differentiation of mammalian heart and blood cell. During mouse heart development, the microRNA *miR1* can repress the messages encoding the **Hand2** transcription factor. This transcription factor is critical in the proliferation of ventricle heart muscle cells, and **miR1** may control the balance between ventricle growth and differentiation. The *miR181* miRNA is essential for committing progenitor cells to differentiate into B lymphocytes, and ectopic expression of *miR181* in mice cause a dominance of B lymphocytes.

Role of microRNAs in Cleaning of used Maternal mRNAs from the Early Embryos

MicroRNAs are also used to "clean up" and fine-tune the level of gene products. We mentioned those maternal RNAs that allow early development to occur. How does the embryo get rid of maternal RNAs once they have been used and the embryonic cells are making their own mRNAs? In zebrafish this clean up operation is assigned to microRNAs such as *miR430*. This is one of the first genes transcribed by the fish embryonic cells, and there are about 90 copies of this gene in the zebrafish genome. So the level of *miR430* goes up very rapidly. This microRNA has hundreds of targets (about 40% of the maternal RNA types), and when it binds to the 3′ UTR of these target mRNAs, these mRNAs lose their poly A tails and are degraded. Slightly later in development, this same microRNA is used in the fish embryo to fine tune the expression of *Nodal* mRNA. The result of this latter use of *miR430* is the determination of how many cells become committed to be endoderm and how many become committed to be mesoderm.

Mode of Function of microRNAs

Although the microRNA is usually 22 bases long, it recognizes its target primarily through a "seed" region of about 5 bases in the 5′ end of microRNA (usually at positions 2–7). This seed region recognizes targets in the 3′ UTR of the message (Box 44.6).

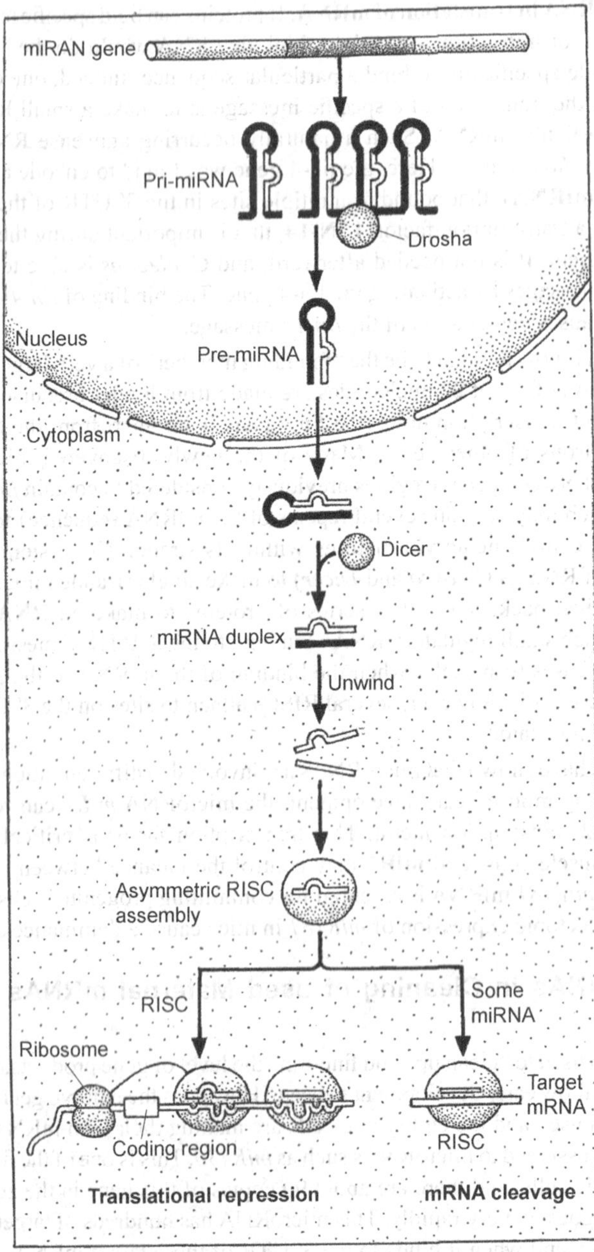

Fig. 44.13. Current model proposed by (**He** and **Hannon**, 2004) for the formation and use of microRNAs. The miRNA gene encodes a pri-miRNA that often has several hairpin regions, where the RNA finds nearby complementary bases with which to pair. The pri-miRNA is processed into individual pre-miRNA "hair pins" by the Drosha RNAase and these are exported from the nucleus. Once in the cytoplasm, another RNAase, Dicer, eliminates the non-base-paired loop. Dicer also acts as a helicase to separate the strands of the double-stranded miRNA. One strand (probably recognized by placement of Dicer) is packaged with proteins into the RNA induced silencing complex (RISC), which subsequently binds to the 3′ UTRs to affect translational suppression or cleavage, depending (at least in part) on the strength of the complementarity between the miRNA and its target (after Gilbert 2010).

Box 44.6 MicroRNAs and Texel Sheep

What happens, then, if an mRNA has a mutated 3′ UTR? Such a mutation appears to have given rise to the **Texel sheep**, a breed with a large and well-defined musculature that is the dominant meat-producing sheep in Europe. A mutation in the myostatin gene that prevent the proper splicing of the nRNA can produce a large muscled phenotype. Another way of reducing the levels of myostatin involves a mutation in its 3′ UTR sequence. Genetic techniques mapped the basis of the sheep's meaty phenotype to the myostatic gene. In the Texel breed, there has been a G-to-A transition in the 3′ UTR of the gene for myostatin, creating a target for the mir1 and mir206 microRNAs that are abundant in skeletal muscle (**Clop** *et al.*, 2006). This mutation causes the depletion of myostatin messages and the increases in muscle mass characteristic of these sheep.

The binding of microRNAs to the 3′ UTR can regulate translation in several ways (Fig. 44.14). First, they can block initiation of translation, preventing the binding of initiation factors or ribosomes. Second, they can recruit endonucleases that digest the mRNA, usually starting with the poly A tail. In a third mechanism, they allow translation to be initiated, but recruit proteolytic enzymes that digest the protein as it is being made. It is also possible that some microRNAs use more than one method, and it has been proposed that the microRNAs may first inhibit translation initiation and then consolidate mRNA silencing by causing the digestion of the message.

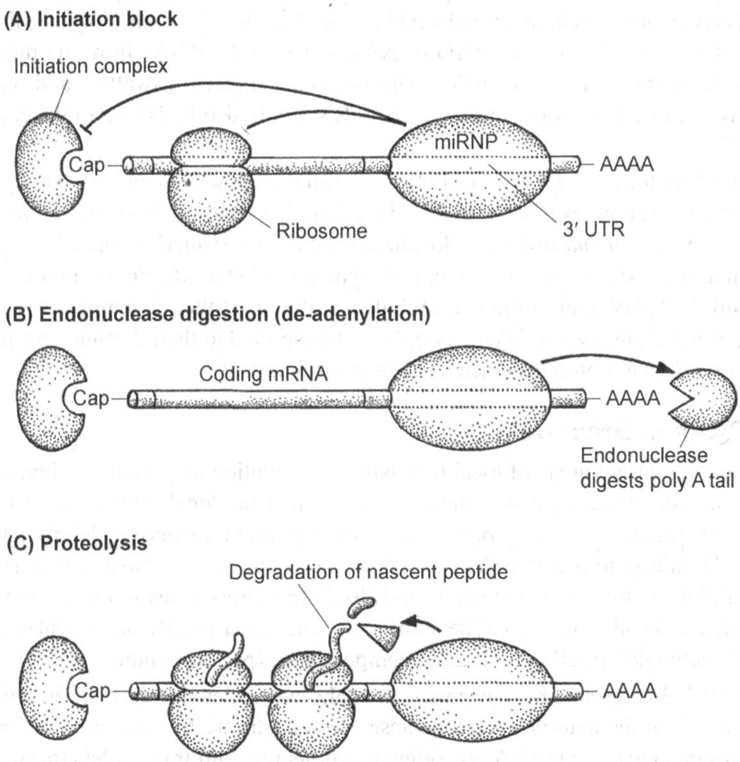

Fig. 44.14. The miRNA complex, including numerous proteins that bind to the miRNA (mi RNP), can block translation in several ways. These include A—Blocking the binding of the mRNA to initiation factors or ribosomes; B—Recruiting endonucleases to chew away the poly A tail of the mRNA, thereby causing its destruction; and C—Recruiting protein digesting enzymes that destroy the nascent protein (after Gilbert 2010).

Control of RNA Expression by Cytoplasmic Localization

Not only is the time of mRNA translocation regulated, but so is the place of RNA expression. A majority of mRNAs (about 70% in *Drosophila* embryos) are localized to specific places in the cell. Just like the selective repression of mRNA translation the **selective localization** of messages is often accomplished through their 3′ UTRs. There are following three main mechanisms for the localization of an mRNA:

1. Diffusion and local anchoring. Messenger RNAs such as *nanos* diffuse freely in the cytoplasm. However, when they diffuse to the posterior pole of the *Drosophila* oocyte, they are trapped there by proteins that reside particularly in these regions (Fig. 44.15A). These proteins also activate the mRNA, allowing it to be translated.

2. Localized protection. Messenger RNAs such as those encoding the *Drosophila* heat shock protein hsp83 (which helps protect the embryos from thermal extremes) also float freely in the cytoplasm. Like *nanos* mRNA, *hsp83* mRNA accumulates at the posterior pole, but its mechanism for getting there is different. Throughout the embryo, the protein is degraded. However, proteins at the posterior pole protect the hsp83 mRNA from being destroyed.

3. Active transport along the cytoskeleton. This is probably the most widely used mechanism for mRNA localization. Here, the 3′ UTR of the mRNA is recognized by proteins that can bind these messages to "motor proteins" that travel along the cytoskeleton to their final destination. These motor proteins are usually ATPases such as dynein or kinesin that split ATP for their motive force. For example, in *Drosophila* oocytes, the *bicoid* mRNA allows its message to bind to the microtubules through its association with two other proteins (Swallow and Staufen). If the bicoid 3′ UTR is attached to some message that mRNA will also be bound to the anterior pole of the oocyte.

The 3′ UTR of the *bicoid* message binds the Staufen protein that connects it to dynein. Dynein travels along the microtubules in the "minus" direction, that is, toward the site where microtubules begin. In this way, the *bicoid* mRNA is localized to the future anterior part of the oocyte. Other mRNAs, such as the *Oskar* message, in contrast, appear to bind to the kinesin motor protein, and it is taken toward the "plus" end of the microtubules, at the tip of their assembly. It is thereby taken to the posterior end of the *Drosophila* oocyte. Once transported to their destinations, mRNAs often bind to other cytoskeleton proteins (such as actin microfilaments).

Stored mRNAs in Brain Cells

One of the most important areas of local translational regulation may be in the brain. The storage of long term memory requires new protein synthesis, and the local translation of mRNA in the dendrites of brain neurons has been proposed as a control point for increasing the strength of synaptic connections. The ability to increase the strength of the connections between neurons is critical in forming the original architecture of the brain and also in the ability to learn. Indeed, in recent studies of mice, **Kelleher** and colleagues (2004) have shown that neuronal activity-dependent memory storage depends on the activation of eIF4E and other components of protein synthesis.

Several mRNAs appear to be transported along the cytoskeleton to the dendrites of neurons (the "receiving portion" of the neuron, where synapse connections are formed with the other neurons). These messages include those mRNAs encoding receptors for neurotransmitters (needed to transmit the signals from one neuron to another); activity-regulated enzymes; and the cytoskeletel components needed to build a synapse. One of the proteins responsible for constructing specific synapses in **brain-derived neurotrophic factor**, or **BDNF.** BDNF regulates neural activity and appears to be critical for new synapse formation. **Takei** and colleagues (2004) have shown that BDNF induce local translation of these neural messages in the dendrites.

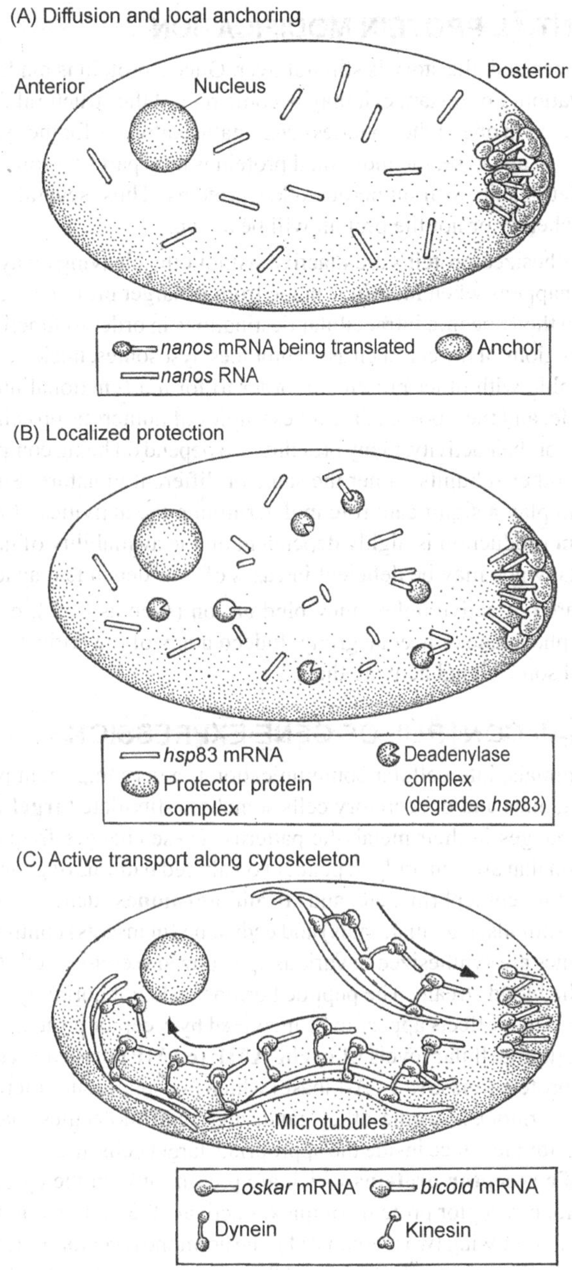

Fig. 44.15. Localization of mRNAs (**Palacois** 2007). A—Diffusion and local anchoring. *Nanos* mRNA diffuses through the *Drosophila* egg and is bound (in part by the Oskar protein) at the posterior end of oocyte. This anchoring allows the *nanos* mRNA to be translated. B—Localized protection. The mRNA for *Drosophila* heat shock protein (hsp83) will be degraded unless it binds to a protector protein (in this case, also at the posterior terminal of the oocyte). C—Active transport on the cytoskeleton, causing the accumulation of mRNA at a particular site. Here, *bicoid* mRNA is transported to the anterior pole of the oocyte by dynein and kinesin motor proteins. Meanwhile, *Oskar* mRNA is brought to the posterior pole by transport along microtubules by Kinesin ATPases (after Gilbert 2010).

44.4. DIFFERENTIAL PROTEIN MODIFICATION

When a protein is synthesized, the story is still not over. Once a protein is made, it becomes part of a larger level of organization. For instance, it may become part of the structural framework of the cell, or it may become involved in one of the countless enzymatic pathways for the synthesis or breakdown of cellular metabolites. In any case, the individual protein is now part of a complex "ecosystem" that integrates it into a relationship with numerous other proteins. Thus, several changes can still take place that determine whether or not the protein will be active.

Some newly synthesized proteins are inactive without the cleaving away of certain *inhibitory* sections. This is what happens when **insulin** is made from its larger protein precursor. Some proteins must be "addressed" to their specific intracellular destinations in order to function. Proteins are often sequestered in certain regions of the cell such as membranes, lysosomes, nuclei, or mitochondria. Some proteins need to assemble with other proteins in order to form a functional unit. The haemoglobin protein, the microtubule, and the ribosome are all examples of numerous proteins joining together to form a functional unit. For their activity, many proteins also depend on being complexed into compound proteins together with other subunits, either the same or different in nature. Such post-translational control mechanisms do play a significant role in determining the activities of differential cells. For example, **haemoglobin** production is highly dependent on the availability of haem to complex with globin protein subunits which may be deficient in cases of iron-dependent anaemia.

Some proteins are not active unless they bind an ion (such as Ca^{2+}), or are modified by the covalent addition of a phosphate or acetate group. Indeed many of the critical proteins in embryonic cells just sit there until some signal activate them.

44.5. HORMONAL CONTROL OF GENE EXPRESSION

In higher plants and animals, intercellular communication is a very important phenomenon. Signals originating in various glands and/or secretory cells somehow stimulate **target tissue** or **target cells** to undergo dramatic changes in their metabolic patterns. These changes frequently include altered pattern of differentiation that are generally dependent on altered patterns of gene expression. **Peptide hormones** such as insulin, epinephrine, etc. and **steroid hormones** such as estrogen, progesterone, testosterone (in higher animals, *e.g.*, mammals) and ecdysone (in insects) control gene expression. In higher animals, hormones are synthesized in various specialized secretory cells (*i.e.*, endocrine cells) and are released into the blood stream. The peptide hormones do not normally enter cells because of their relative large size. Their effects appear to be mediated by receptor proteins located in target-cell membranes and by the intracellular levels of **cyclic AMP (cAMP)** (called **secondary messenger**). The cAMP activates a protein kinase (*e.g.*, A-kinase) which phosphorylates (activates) many specific enzymes. The steroid hormones, on the other hand, are small molecules that readily enter cells through the plasma membrane. Once inside the appropriate target cells, the steroid hormones become tightly bound to **specific receptor proteins** which are present only in the cytoplasm of target cells (Fig. 41.22). The hormone-receptor protein complexes activate the transcription of specific genes or sets of genes according to following two methods: (1) The hormone receptor protein complexes interact with specific non-histone chromosomal protein and this interaction stimulates the transcription of the correct genes. (2) The hormone receptor protein complexes activate transcription of target genes by binding to specific DNA sequences present in the *cis*-acting regulatory regions (the enhancers and promoter regions of the genes. In both of these cases, these hormone-receptor protein complexes would function as positive regulators (or "activators") of transcription much like the CAP-cAMP complexes in prokaryotes.

During development of dipteran flies (*Drosophila melanogaster* and *Chironomus tentans*) the steroid hormone **ecdysone** is released and triggers moulting. If larvae of these insects are treated with ecdysone at stages of development prior to or between moultings, patterns of chromosome puffing occur that are identical to those occurring during natural moultings. Ecdysone tends to affect the gene expression at the level of transcription.

QUESTIONS

Long Answer Questions

1. What is cell differentiation? Describe the mechanism of differential gene expression.
2. Give an account of differential nRNA (=nuclear RNA) processing.
3. Describe the process of differential mRNA translation.
4. Describe in detail the structure of globin gene.

Short Answer Questions

1. Give an account of differential protein modification.
2. Write functions of transcription factors.
3. Enlist stored oocyte mRNAs.
4. Write short notes on
   (*a*) Cell differentiation
   (*b*) Enhancers
   (*c*) DNA methylation
   (*d*) DNA ethylation
   (*e*) Micro RNA.

Very Short Answer Questions

1. What is enhancer?
2. What is function of micro RNA?
3. Define cell differentiation.

45 Stem Cell Concept and Cell Replacement Therapy

Following gastrulation, three germinal layers interact with each other to begin *organogenesis*—the process of organ formation. Early organogenesis is a symphony of interactions between different parts of the embryo, and some of these interactions create privileged sites called **stem cell niches**. These niches provide a milieu of extracellular matrices and paracrine factors that allow cells residing them to remain relatively undifferentiated. These relatively undifferentiated cells are called **stem cells**, and their presence has become central to our vision of organogenesis and critical to the field of modern medicine.

45.1. THE STEM CELL CONCEPT

Karp (2010) has defined stem cells as undifferentiated cells that (1) are capable of self-renewal, that is production of more cells like themselves, and (2) are multipotent, that is, are capable of differentiating into two or more mature cell types, *e.g.*, haemopoietic stem cells (HSCs). Thus, a stem cell is a relatively undifferentiated cell that when divides produces (*i*) one that retains its undifferentiated character; (*ii*) a second cell that can undergo one or more paths of differentiation (Fig. 45.1). Thus, a stem cell has the potential to renew itself at each division (so that there is always a supply of stem cells) while also producing a daughter cell capable of responding to its environment by differentiating in a particular manner. (This potential is not always realizes; in some instances, stem cell divides symmetrically so that both daughter cells remain stem cells). In many cases, the stem cell remains in the niche while its sister leaves the niche and differentiate.

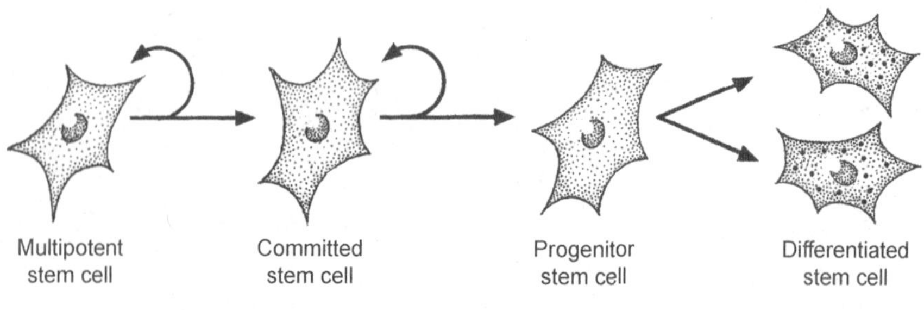

Fig. 45.1. Stem cells have two distinct progeny at division: another stem cell and a more committed cell that can differentiate.

In some human organs, such as the gut, epidermis, and bone marrow, stem cells regularly divide to replace worn-out cells and repair damaged tissues. In other human organs, such as the prostrate and heart, stem cells divide only under special physiological conditions, usually in response to stress or the need to repair the organ.

History

In the mid-twentieth century, biologists thought that cell specification occurred in the early embryo and that after this stage there was only the growth of the existing parts. This idea is preserved in the classification of human development into **embryonic stages** where specification takes place (*i.e.*, before week 9) and later **fetal stages** that are characterized by growth. However, beginning in 1960s, biologists studying blood cell development began to analyse a remarkable phenomenon. All of the blood cells—red blood cells (erythrocytes), white blood cells (granulocytes), platelets and even lymphocytes—are constantly being produced in bone marrow. Billions of blood cells are destroyed by the spleen each hour, but an equal number of blood cells are generated to replace them.

In an elegant series of experiments, **Ernest McCulloch** and **James Till** demonstrated that there was a common generative cell, the "haemopoitic (blood-forming) stem cell", that gave rise to all the different types of blood cells (Fig. 45.2). In a 1961 experiment, **Till** and **McCulloch** injected bone marrow cell from a donor mouse into lethally irradiated mice of the same genetic strain (Box 45.1). Some of the individual donor cells produced discrete nodules on the spleens of the host animals and these nodules contained erythrocytes, granulocytes and platelet precursor cells. Later experiments in which the donor marrow cells were irradiated to genetically mark each cell with random chromosome breaks confirmed that each of these different cell types in a nodule arose from a single cell, and that there were lymphocytes present in some of these nodules (**Becker** *et al*., 1963).

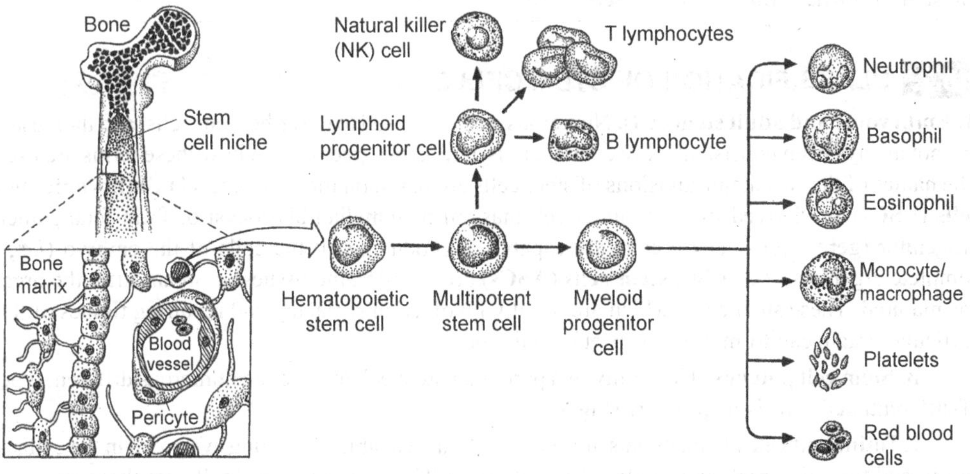

Fig. 45.2. Haemopoietic (blood-forming) stem cells and the bone marrow niche. The haemopoietic stem cell is located in the bone marrow and generates a second stem cell that is capable of becoming either a lymphocyte progenitor cell (which divides to form the cells of the immune system) or a myeloid stem cell (which forms the blood cell precursors). The type of lineage that the cells take is regulated by the niche, which involves contact between the stem cells and the matrices of bone cells, paracrine factors from stromal cells and systemic hormones and neural signals (*systemic* means supplying those parts of the body that receives blood through aorta). Mesenchymal stem cells also use the bone marrow as a niche.

Box 45.1

The mice were "lethally irradiated" in that X-rays were used to destroy their immune and blood cell precursors. The mice survived only if there were stem cells in the transplant that could replace these cells and regenerate blood cells and lymphocytes.

For this "colony-forming cell" to be true stem cell, however, it had to produce not only the differentiated blood cells but also more colony-forming cells. This was shown to be the case by taking the nodule derived from a single genetically marked colony-forming cell and injecting cells of the nodule into another irradiated mouse. Many spleen colonies emerged, each of them having the same chromosomal arrangement as the original colony. Thus, we see that a single marrow cell can form numerous different cell types and can also undergo self-renewal. This research on haemopoitic stem cells led to the establishment of the field of bone marrow transplantation (Box 45.2).

Box 45.2

The concept that the different blood cell types were continuously generated by haemopoietic stem cell was first proposed in 1909 by the Russian histologist **Alexander Maximov**. He is credited with coining the word *Stammzelle* to refer to the regenerative capacities of the cells. Maximov had been a student of **Oskar Hertwig**, one of the leading German embryologists and one of the originators of current *theory of epigenesis*. Returning to St. Petersburg to be a professor of embryology and histology, Maximov's work was cut short by the Russian Revolution. He managed to flee the Soviet Union in 1922, having bribed a guard at the Fennish border with a bottle of laboratory ethanol. Eventually Maximov settled in the United States, where he co-authored the leading text book of histology with his student **William Bloom** (see **Gilbert** 2010).

45.2. CLASSIFICATION OF STEM CELLS

A. Embryonic and adult stem cells. Numerous terms are used to describe stem cells, and their usage has not always been consistent. However, there is a general agreement on how these terms are used. The names of the two major divisions of stem cells are based on their sources. **(1) Embryonic stem cells (ESCs)** are derived from the inner cell mass of mammalian blastocyst or from fetal gamete progenitor (germ) cells. These cells are capable of producing all the cells of the embryo (*i.e.*, a complete organism). **(2) Adult stem cells (ASCs)** are found in the tissues of organs after the organ has matured. These stem cells, which are usually involved in replacing and repairing tissues of that particular organ, can form only a subset of cell types.

B. Stem cell potency. The ability of a particular stem cell to generate numerous different types of differentiated cells is its **potency** (Fig. 42.1).

1. Totipotent cells. In mammals, totipotent cells are capable of forming every cell in the embryo and in addition, the trophoblast cells of the placenta. The only totipotent cells are the zygote and (probably) the first 4–8 blastomeres to form prior to compaction.

2. Pluripotent stem cells. These cells have the ability to become all the cell types of the embryo except trophoblast. Usually these embryonic stem cells are derived from the inner cell mass of the mammalian blastocyst. However, germ cells and germ cell tumors (*e.g.*, teratocarcinomas) can also form pluripotent stem cells.

3. Multipotent stem cells. Commitment of these stem cells is limited to a relatively small subset of all the possible cells of the body. These are usually adult stem cells. The haematopoietic stem cell, for instance, can form the granulocyte, platelet, and red blood cells lineages. Similarly, the mammary stem cell can form all the different cell types of the mammary gland.

4. Unipotent stem cells. These adult stem cells are found in particular tissues and are involved in regenerating a particular type of cell. Spermatogonia, for example, are stem cells that give rise only to sperm.

Whereas pluripotent stem cells can produce cells of all three germ layers (as well as producing germ cells), the multipotent and unipotent stem cells are often grouped together as **committed stem cells**, since they have the potential to become relatively few cell types.

C. Progenitor cells. Although they are related to stem cells, but they are not capable of unlimited self-renewal; they have the capacity to divide only a few times before differentiating. Progenitor cells are sometimes called **transit-amplifying cells**, since they usually divide while migrating away from the stem cells niche. Both unipotent stem cells and progenitor cells have been called **lineage restricted cells**, but the stem cells have the capacity for self-renewal, while the progenitor cells do not. Progenitor cells are usually more differentiated than stem cells and have become committed to become a particular type of cell. In many instances, stem cell division generates progeny that become progenitor cells, as is seen in the formation of the blood cells, sperm cells and the nervous system.

45.3. ADULT STEM CELLS

1. Types of adult stem cells. Numerous adult organs contain committed stem cells that can give rise to a limited set of cells and tissue types. Developmental biologists have discovered following types of adult stem cells:

1. Haematopoietic stem cells (or HSCs);
2. Epidermal stem cells;
3. Neural stem cells;
4. Hair stem cells;
5. Melenocyte stem cells;
6. Muscle stem cells;
7. Tooth stem cells;
8. Gut stem cells; and
9. Germline stem cells.

These cells are not as easy to use as pluripotent embryonic stem cells; they are difficult to isolate, since they often represent fewer than 1 out of every 1000 cells in an organ. In addition, they appear to have a relatively low rate of cell division and do not proliferate readily. However, neither of these facts rule out their usefulness. Each year some 40,000 bone marrow transplant procedures are performed in which haematopoietic stem cells are transferred from one person to another (see **Gilbert** 2010). These multipotential stem cells are rare (about 1 in every 15,000 bone marrow cells), but such transplantation treatments work well for people suffering from red blood cell deficiencies or leukemias.

Techniques to selectively allow the growth and isolation of multipotent stem cells may allow some organ deficiencies to be treated in the same way as these blood cell deficiencies—by administering a source of committed stem cells. For example, in mice, very few (perhaps even one) blood stem cell will reconstitute the mouse's blood and immune systems (**Osawa** *et al.*, 1996); a single mammary stem cell will generate an entire mammary gland (including epithelium, muscles, and stroma; **Shackleton** *et al.*, 2006) and a single transplanted prostatic stem cell will produce an entire prostate gland (**Leong** *et al.*, 2008). **Carvey** *et al.*, (2001) have shown that when neural stem cells from the midbrain of adult rats are cultured in a mixture of paracrine factors, they will differentiate into dopaminergic neurons that can make better the rodent version of Parkinson disease.

2. Adult stem cell niches (or Regulatory microenvironments). Many tissues and organs contain stem cells that undergo continual renewal; these include the mammalian *epidermis, hair follicles, intestinal villi, blood cells* and *sperm cells*, as well as *Drosophila* intestine, sperm, and egg cells. Such stem cells must maintain the long-term ability to divide, producing some daughter cells that remains stem cells. The ability of a cell to become an adult cell is determined in large part by where it resides. The continuously proliferating stem cells are housed in compartments called **stem cell niches** (**Schofield** 1978; sometimes called **regulatory microenvironments**). These are particular places in the embryo that allow the controlled proliferation of the stem cells within the niche and the controlled differentiation of the cell progeny that leaves the niche.

Stem cell niches regulate stem cell proliferation and differentiation, usually by **paracrine factors** that are produced by the niche cells. These factors retain the cells in an uncommitted state. Once the cells leave the niche, the paracrine factors cannot reach them, and the cells begin differentiating.

1. Stem cell niche in mouse incisors. Mouse incisors (teeth) differ from human incisors in that they continue to grow throughout the life-time of the animal. Each mouse incisor has two stem cell niches; one is on the "inside", facing into the mouth, and one is on the "outside", facing the lips (Fig. 45.3). The stem cells that reside therein are kept in a proliferative and non-differentiating state by integrated network of paracrine factors, including **Fgf3** and **activin** (which increase the proliferation of stem cells), and their respective inhibitors, **BMP4** and **follistatin**. Teeth in humans and most other mammals lack stem cell niches and thus do not regenerate.

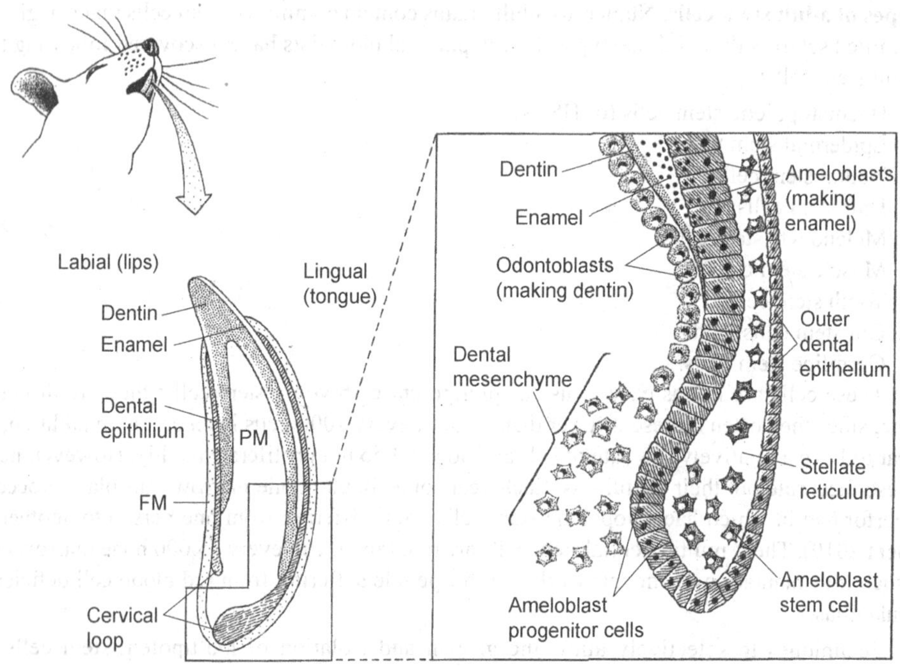

Fig. 45.3. The cervical loop of the mouse incisor is a stem cell niche for the enamel-secreting ameloblast cells. These cells migrate from the base of the stellate reticulum into the enamel layer, allowing the teeth to keep growing.

2. Hair follicle stem cell migration and differentiation. Mammalian hair follicle contains a "bulge" that houses the **melanocyte stem cells** which provide pigment to the hair (Fig. 45.4). The division of the melanocyte stem cells within this niche is coordinated with hair growth. Moreover, upon cell division, one daughter cell migrates toward the developing hair shaft. The migrating cell is committed melanocyte progenitor cell, and it will reside in the matrix of the bottom of the hair shaft, producing pigment forming melanocytes.

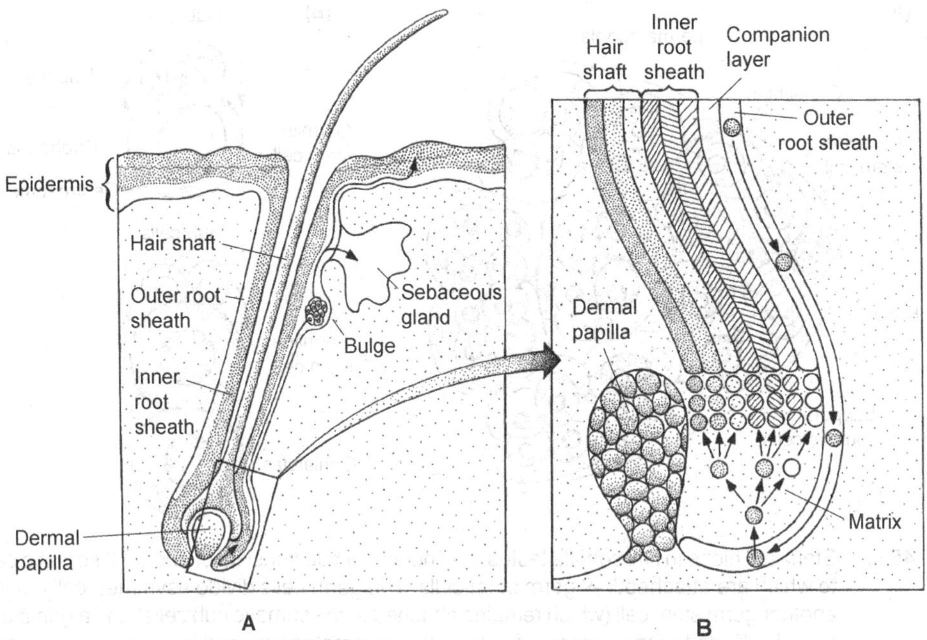

Fig. 45.4. Model of follicle stem cell migration and differentiation (proposed by **Fuchs** 2007 and **Blanpain** and **Fuchs** 2009). A–During the normal hair cycle, bulge stem cells are periodically activated to form a new hair follicle. As the hair follicle grows, the more committed progeny of the bulge cells migrate through the outer root sheath, becoming a highly proliferative population of cells that will form a new hair. B–As the bulge stem cell's progeny colonize the base of follicle, they subsequently differentiate along one to seven hair lineages. Differentiation occur at the base of the follicle.

3. Haematopoiesis. The mammalian haematopoietic niche is found in the hollow cavities of trabecular bones (such as sternum) where the bone marrow resides (Fig. 45.2). Here, the stem cells are in close proximity to the bone cell (osteocytes) and the endothelial cells that line the blood vessels. A complex cocktail of paracrine factors, including **Wnt**, **angiopoietin** and **stem cell factor**, combines with cell surface signals from **Notch** and **integrin** to regulate stem cell proliferation and differentiation. Hormonal signals and pressure from the blood vessels, as well as neurotransmitters from adjacent axons, also help regulate haematopoiesis (**Spiegel** *et al*., 2008; **Malhotra** and **Kincade**, 2009).

The hub. In many instances, germ cells are continuously produced in stem cell niches. In *Drosophila* testes, the stem cells for sperm reside in a regulatory microenvironment called the **hub** (Fig. 45.5). The hub consists of about a dozen somatic testes cells and is surrounded by 5–9 **germ stem cells**. The division of the sperm stem cell is asymmetric, always producing one cell that remains attached to the hub and one unattached cell. The daughter cell attached to the hub is maintained as a stem cell, while the cell that is not touching the hub becomes a **gonialblast**—a committed cell that will divide to become the precursor of the sperm cells. The somatic cells of the hub create this asymmetric proliferation by secreting the paracrine factor **Unpaired** (protein) onto the cells attached to them. Unpaired protein activates the JAK-STAT pathway in the adjacent germ stem cells to specify their self-renewal. Those cells that are distant from the paracrine factor cannot receive this signal, so their differentiation into the sperm cell lineage.

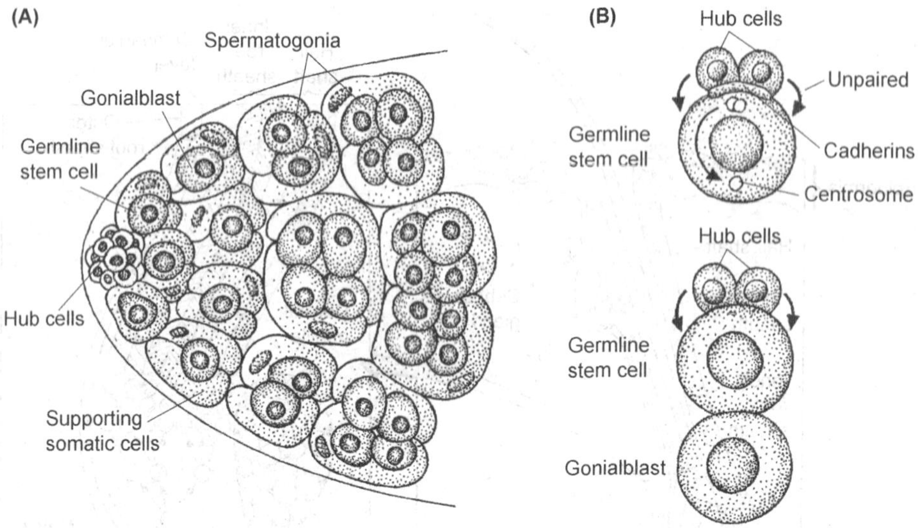

Fig. 45.5. Stem cell niche in *Drosophila* testes. A–The apical hub consists of about 12 somatic cells, to which are attached 5–9 germ stem cells. The germ cells divide asymmetrically to form another germ stem cell (which remains attached to the somatic hub cells) and a gonialblast that will divide to form sperm precursor (the spermatogonia and the spermatocyte cysts where meiosis is initiated). B–Cell division pattern of the germline stem cells, wherein one of the centrosomes remains in the cortical cytoplasm near the site of hub cell adhesion, while the other centrosome migrates to the opposite pole of the germline stem cell. This results in one cell remaining attached to the hub and the other cell detaching from it and differentiating.

Physically, this asymmetric division involves the interactions between the sperm stem cells and the somatic cells. In the division of stem cell, one centrosome remains attached to cortex at the contact site between the stem cell and the somatic cells. The other centrosome moves to the opposite side, thus establishing a mitotic spindle that will produce one daughter cell attached to the hub and one daughter cell away from it. (Similar inheritance of centrosomes occurs in the division of mammalian neural stem cells). The cell adhesion molecules linking the hub and stem cells together are probably involved in retaining one of the centrosomes in the region where the two cells touch.

The stem cell niche is a critical component of our phenotype, and it does nothing less than regulate the ratio of cell division to cell differentiation. This means that maintenance of such niches is critical for our health. Too much stem cell differentiation depletes the stem cells and promote the **phenotypes of aging** or **decay**. Too much stem cell division can cause **cancers** to arise. The "graying" **of mammalian hair** (in mice as well as humans) can involve a break-down in the maintenance of the stem cell niche such that both daughters of dividing stem cells differentiate, thereby leaving a smaller population of stem cells that cannot continue to pigment forming **melanocytes** (**Nishimura** *et al.*, 2005). Conversely, myeloproliferative disease—a cancer of blood stem cells and their (non-lymphocytic) derivatives—results when the stem cell niche is unable to provide the signals needed for proper blood cell differentiation (**Walkley** *et al.*, 2007).

Thus, stem cell niches provide micro-environments that regulate stem cell renewal, survival, and differentiation. Their paracrine factors, cell adhesion molecules, and architecture allow asymmetric cell divisions such that a stem cell divides in a manner that allows one of its daughter cells to have a high probability of leaving the niche and beginning to differentiate according to the new signals it encounters.

3. Mesenchymal stem cells: multipotent adult stem cells. Adult stem cells are used in the adult body to replace worn-out somatic cells on a regular basis. Our epidermis, our intestine epithelium, and our blood cells are continually being replaced with cells generated by dividing adult stem cells. Most adult stem cells are restricted to form only a few cells types (**Wagers** *et al.*, 2002). When haematopoietic stem cells marked with green fluorescent protein were placed in mice, their labelled descendants were found throughout the animal's blood but not in any other tissue. Some adult stem cells, however, appear to have a surprisingly large degree of plasticity. These multipotent cells are called **mesenchymal stem cells** or **MSCs** (sometimes called **bone-marrow derived stem cells** or **BMDSCs**), and their potency remains a controversial subject.

Medical usage of MSCs. Originally found in bone marrow, multipotent MSCs have also been found in adult tissues such as fat, muscle, thymus, and dental pulp, as well as in the umbilical cord (**Kuhn** and **Tuan** 2010). Indeed the finding that human umbilical cord and deciduous ("baby") teeth contain MSCs has led some physicians to propose that parents freeze cells from their child's umbilical cord or "milk" teeth so that these cells will be available for transplantation later in life. However, the crucial test for pluripotency—the ability of a mouse stem cell to generate cells of all germ layers when inserted into blastocyst—has not yet been achieved.

Mesenchymal stem cells are able to give rise to numerous bone, cartilage, muscle and fat lineages. The differentiation of MSCs is postulated on both paracrine factors and cell matrix molecules in the stem cell niche. Certain cell matrix components, especially **laminin**, have been implicated in keeping MSCs in a state of undifferentiated "stemness" (**Kuh** and **Tuan** 2010). Certain paracrine factors appear to direct development into specific lineages. In one study (**Ng** *et al.*, 2008), platelet derived growth factors was critical for chondrogenesis and fat formation, TGF-β signalling was important for chondrogenesis, and FGF signalling was crucial for differentiation of MSCs into bone cells.

In addition to paracrine factors, the repertoire (*i.e.*, the works known or regularly performed by a performer, here NSCs) of cell types from mesenchymal stem cells may also be enhanced by the surfaces on which the stem cells reside. Human MSCs can differentiate according to the elasticity of the surface on which they are placed. If placed on soft matrices of collagen-coated polyacrilamide, these stem cells differentiate into neurons. A moderately elastic matrix of the some materials causes the same stem cells to become muscle cells, while harder matrices causes the MSCs to produce bone cells (**Engler** *et al.*, 2006). It is not yet known whether this range of potency is found normally in the body.

Mesenchymal stem cells have recently been linked to normal growth and repair conditions in the human body. Indeed, one premature aging syndrome, called **Hutchinson-Gilford progeria**, appears to be caused by the inability of MSCs to differentiate into certain cell types, such as fat cells. These findings lead to speculation that the loss of MSC ability to differentiate may be a component of the normal aging syndrome. Moreover MSCs may work in ways other than differentiating needed cell types. They may produce paracrine factors that aid other, more specific stem cells to divide and repair tissues.

Stem cells and regenerative medicine. The ability to create, isolate and manipulate stem cells offers a vision of regenerative medicine, wherein a patient can have his or her diseased organs regrown and replaced by one's own stem cells. But beyond their potential medical uses, stem cells tell us a great many facts about how the body is constructed and how it maintains its structure. Organs often form by the regulation of stem cells, and we will see that the skin, hair, blood and parts of the nervous system routinely use stem cells in their construction. Stem cells certainly give support to the view that "**development never end**", and offer fascinating potential ways to modify development (see **Gilbert** 2010).

45.4. CELL REPLACEMENT THERAPY

To a person whose heart or liver is failing, an **organ transplant** is the best hope for survival and return to a normal life. Organ transplantation is one of the great success of modern medicine, but its scope is greatly limited by the availability of donor organs and the high risk of immunological rejection. Imagine the possibilities if we could grow cells and organs in the laboratory and use them to replace damaged or non-functional parts within our bodies. Recent studies have given researchers hope that one day this type of therapy will be a common place. To better understand the concept of cell replacement therapy, we can consider a present day procedure known as **bone marrow transplantation** in which cells are extracted from the interior of the pelvic bones of a donor and infused into the body of a recipient.

Bone marrow transplantation is used most often to treat **lymphomas** and **leukemias**, which are **cancers** that affect the nature and number of white blood cells. To carry out the procedure, the patient is exposed to a high level of radiation and/or toxic chemicals, which kills the cancerous cells, but also kills all the cells involved in the formation of red and white blood cells. This treatment has this effect because *blood-forming cells are particularly sensitive to radiation and toxic chemicals*. Once a person's blood-forming cells have been destroyed, they are replaced by bone marrow cells transplanted from a healthy donor. Bone marrow can **regenerate** the blood tissue of the transplant recipient because it contains small percentage of cells that can proliferate and restock the patient's blood-forming bone marrow tissue. [**Note:** According to **Karp** (2010), bone marrow transplantation can be contrasted to a simple blood transfusion where the recipient receives *differentiated* blood cells (especially red blood cells and platelets) present in the circulation]. These blood-forming cells in the bone marrow are termed **haemopoietic stem cells** (or **HSCs**), and they are normally responsible replacing the millions of red and white blood cells that age and die every minute in our bodies. Amazingly, a *single* HSC is capable of reconstituting the entire haematopoietic (blood-forming) system of an irradiated mouse.

Most of the organs in a human adult contain stem cells that are capable of replacing the particular cells of the tissue in which they are found. Even the adult brain, which is not known for its ability to regenerate, contains stem cells that can generate new neurons and glial cells (the supportive cells of the brain). Stem cells of adult skeletal muscles are called **satellite cells**. They are thought to divide and differentiate as needed for the repair of injured muscle tissue. Adipose cells too have stem cells.

A curious series of studies has recently been performed on a strain of golden retrievers (*Retriever* is a dog of a breed used for finding and bringing back game that has been shot) that suffer from an inherited disease very similar to the human skeletal-muscle disorder **muscular dystrophy**. Researchers have isolated stem cells from the muscles of these dogs, corrected the genetic disorder in the isolated cells and expanded the number of genetically modified cells they have to work with by growing them in culture. When these stem cells are injected back into the diseased animals, many of them return to a skeletal muscle where they take up residence. Once back in muscle tissue, the corrected **satellite cell** divide and differentiate into new muscle cells and in doing so, contribute to a marked improvement in the mobility and gait of the diseased animals.

There is great optimism that similar types of corrective approaches could be used for humans. The human heart, for example, contains **cardiac stem cells** that are capable of differentiating into the cells that form both the muscle tissue of the heart (the cardiomyocytes of the myocardium) and the heart's blood vessels. These stem cells might have the potential to regenerate healthy heart tissue in a patient who has experienced a serious heart attack or is suffering from congestive heart failure. Adult stem cells are an ideal system for cell replacement therapies because they can be isolated directly from the patient, grown in culture and reintroduced back into the same patient.

Although adult stem cells may ultimately prove to be an invaluable resource in cell replacement therapy, clinical studies carried out to date have been disappointing (with some current successes, Box 45.1). Much of the excitement that has been generated in the field over the past decade 2000 to 2010) has come from studies on **embryonic stem cells (ESCs)**, which are a type of stem cell isolated from very young mammalian embryos (Fig. 45.6). These are the cells in the early embryo that give rise to all of the various structures of the mammalian fetus. Unlike adult stem cells, ES cells are pluripotent; that is, they are capable of differentiating into every type of cell in the body. In most cases, human ES cells have been isolated from embryo provided by in vitro fertilization clinic. Worldwide, dozens of genetically distinct human ES cell lines, each derived from a single embryo, are available for experimental investigation.

Box 45.1

The World's first clinical trial of **brain stem cells to treat strokes** has passed the first safety test and is set to move to its next phase, says **Prof. Keith Muir** of Glasgow University who led the trial in 2010 and 2011 (Times of India, 06.09.2011).

Cell Replacement Therapy with Oligodendrocytes

The long-range goal of clinical researchers is to learn how to persuade ES cells to differentiate in culture into each of the many cell types that might be used for cell replacement therapy. Considerable progress has been made in this pursuit, and numerous studies have shown that transplants of differentiated, ES cell-derived cells can improve the condition of animals with diseased or damaged organs. The first trial in humans are likely to utilize cells, called **oligodendrocytes**, that produce the myelin sheaths that become wrapped around nerve cells. It was found by trial and error, that pure colonies of oligodendrocytes would differentiate from human ES cells that were cultured in a medium containing insulin and thyroid hormone. When implants of these human oligodendrocytes were transplanted into rats with paralyzing spinal cord injuries, the animals regained considerable motility. In 2009, the FDA (USA) approved the first clinical trial in which these ES cell-derived oligodendrocytes would be implanted into patients who have sustained recent damage to their spinal cord. Clinical trials are also planned for the treatment of type1 diabetes and the eye disease muscular degeneration. The **primary risk** with this type of procedure is the unnoticed presence of undifferentiated ES cells among the differentiated cell population. Undifferentiated ES cells are capable of forming a type of benign tumor, called a **teratoma**, which may contain a bizarre mass of various differentiated tissues, including hair and teeth. The formation of teratoma within the central nervous system could have severe consequences. In addition, the culture of ES cells at the present time involves the use of non-human biological materials, which also poses potential risks of causing disease.

Somatic Cell Nuclear Transfer (SCNT)

Although adult stem cells lack the unlimited differentiation capacity that is characteristic of ES cells, they do have an advantage over ES cells in that they can be isolated from the individual who is being treated and, thus will not face immunological rejection when used in subsequent cell replacement. However, it may be possible to "customise" ES cells so that they possess the same genetic makeup of the individual who is being treated, and thus would not be subject to attack by the recipients immune system. This can likely be accomplished by an indirect procedure called **somatic cell nuclear transfer (SCNT)** shown in Fig. 45.6, that begin with an unfertilized egg—a cell that is obtained from the ovaries of an unrelated woman donor. In this approach, the nucleus of a cell from the patient to be treated, which would cause the egg to have the same chromosome composition as that of the patient. The egg would then be allowed to develop to an early embryonic stage and the ES cells would be removed, cultured and induced to differentiate into the type of cells needed by the patient. Because this procedure involves the formation of a human embryo that is used only as a source of ES cells, there are major ethical questions that must be settled before it could be routinely practiced. In addition, the process of SCNT is so expensive and technically demanding that it is highly improbable that it could ever be practiced as part of any routine medical treatment. It is more likely that, if ES cell-based therapy is ever practiced, it would depend on the use of a bank of hundreds or thousands of different

ES cells. Such a bank could contain cells that are close enough as a tissue match to be suitable for use in the majority of patients.

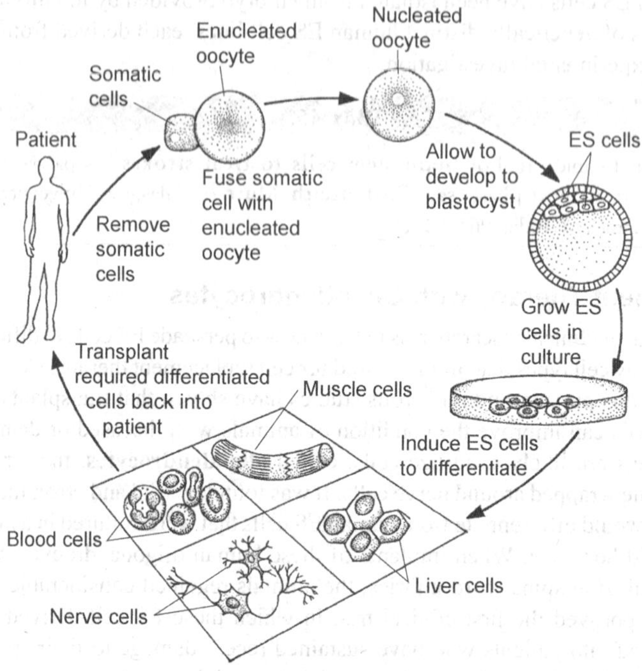

Fig. 45.6. A procedure for obtaining differentiated cells for use in cell replacement therapy. A small piece of tissue is taken from the patient, and one of the somatic cell is fused with a donor oocyte whose own nucleus had been previously removed. The resulting oocyte (egg), with the patient's cell nucleus, is allowed to develop into an early embryo, and the ES cells are harvested and grown in culture. A population of ES cells are induced to differentiate into the required cells, which are subsequently transplanted into the patient to restore organ function (Karp 2010).

It had long been thought that the process of cell differentiation in mammals was irreversible; once a cell had become a fibroblast, or white blood cell, or cartilage cell, it could never again revert to any other cell type. This concept was undermined in 2006 when **Shinya Yamanaka** and coworkers of Kyoto University announced a stunning discovery; his lab had succeed in *reprogramming* a fully differentiated mouse cell—in this case a type of connective tissue fibroblast—into a pluripotent stem cell. They accomplished the feat by introducing, into the mouse fibroblast, the genes that encoded four key proteins that are characteristic of ES cells. These genes (*Oct4, Sox2, Klf4* and *Myc*) are thought to play a key role in maintaining the cells in an undifferentiated state and allowing them to continue to self-renew. The genes were introduced into cultured fibroblasts using gene-carrying viruses, and those rare cells that became reprogrammed were selected from the others in the culture by specialized techniques. **Yamanaka** *et al.*, called this new type of cells **induced pluripotent cells (iPS cells)** and demonstrated that they were indeed pluripotent by injecting them into a mouse blastocyst and finding that they participated in the differentiation of all the cells of the body, including eggs and sperm. Within the next year or so the same reprogramming feat had been accomplished in several labs with human cells, and it was demonstrated that the humans iPS cells are virtually indistinguishable from authentic humans ES cells by numerous criteria. What this means is that researchers now have available to them an unlimited supply of pluripotent cells that can be directed to differentiate into various types of body cells using similar experimental protocols to those already developed for ES cells. Indeed, iPS cells have already been used to correct certain disease conditions in experimental animals, including sickle cell anaemia in mice (Fig. 45.7). The iPS cells have also been prepared from adult cells taken from patients with genetic disorders, such as Huntington's disease and type1 diabetes. Researchers will be able to follow the differentiation of these iPS cells into their diseased cell types in culture. It is hoped that such studies will reveal the mechanisms of disease formation as it unfolds in a culture dish just as it would normally occur in an unobservable way deep within the body. Such "diseased cells" might also serve as targets for testing the effects of newly developed drugs in halting disease progression.

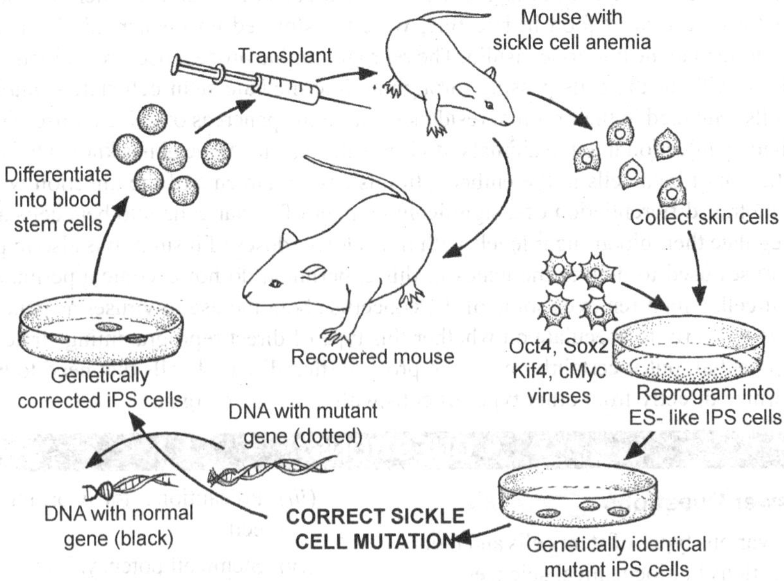

Fig. 45.7. Steps taken to generate induced pluripotent stem (iPS) cells for use in correcting the inherited disease sickle cell anemia in mice. Skin cells are collected from the diseased animal, reprogrammed in culture by introducing the four required genes that are ferried into the cells by viruses, and allowed to develop into undifferentiated pluripotent iPS cells. The iPS cells are then treated so as to replace the defective (globin) gene with a normal copy, and the corrected iPS cells are caused to differentiate into normal blood stem cells in culture. These blood stem cells are then injected back into the diseased mouse, where they proliferate and differentiate into normal blood cells, thereby curing the disorder (after Karp 2010).

Precautions for Working with iPS Cells

Unlike ES cells, the generation of iPS cells does not require the use of an embryo. This feature removes all of the ethical reservations that accompany work with ES cells and also makes it much easier to generate these cells in the lab. In fact, almost any laboratory that works with cultured mammalian cells should be able to jump into this exciting field—and many have. Just as well as cells, there are difficulties to overcome before iPS cells can be used as a source of cells for human therapy. It will be important to develop efficient cell reprogramming techniques that do not use gene-carrying viruses because such cells carry the potential of developing into cancers. Recent studies suggest this goal can be achieved. Like ES cells, undifferentiated iPS cells also give rise to teratomas, so it is essential that only fully differentiated cells are transplanted into human subjects. Also like ES cells, the iPS cells in current use have the same tissue antigens as the donors who originally provided them, so they would stimulate an immune attack if they were to be transplanted into other human recipients. Unlike the formation of ES cells, however, it will be much easier to generate a population of iPS cells from a specific donor. Consequently, if iPS cells are ever developed for corrective use, they would likely come from a large cell bank that could provide cells that are close tissue matches to most potential recipients. If this could be achieved, it might be possible to the recipient's immune system. Cells of this type could be used as the basis for **cell replacement** in everyone.

Transdifferentiation

In 2008 the field of cellular reprogramming took another unexpected turn with the publication that one type of differentiated cell had been converted directly into another type of differentiated cell, a case of "transdifferentiation". In this report, the acinar cells of pancreas, which produce enzymes responsible for digestion of food in intestine, were transformed into pancreatic beta cells, which synthesize and secrete the hormone insulin. The reprogramming process occurred directly, in a matter of a few days, without the cells passing through an intermediate stem cell state—and it occurred while the cells remained in their normal residence within the pancreas of a live mouse. This feat was accomplished by injection into the animals of viruses that carried three genes known to be important in differentiation of beta cells in the embryo. In this case, recipients of the injection were diabetic mice, and the transdifferentiation of a significant number of acinar cells into beta cells allowed the animal to regulate their blood sugar levels with much lower doses of insulin. It is also important that the adenoviruses used to deliver the genes in this experiment do not become a permanent part of the recipient cell, which removes some of the concerns about the use to viruses as gene carriers in humans. It is too early to speculate on whether this type of direct reprogramming strategy has real corrective potential, but it certainly raises the prospect that diseased cells that need to be replaced might be formed directly from other types of cells within the same organ.

QUESTIONS

Long Answer Questions

1. Write various types of stem cells and their applications in biochemical sciences.
2. Give an account of cell replacement therapy.

Short Answer Questions

1. Write short notes on the following:
   (*i*) Satellite cells
   (*ii*) Trans differentiation
   (*iii*) Precautions for working with iPS cell
   (*iv*) Stem cell potency.

Very Short Answer Questions

1. What are embryonic stem cells?
2. What are stem cells?
3. Write full form of each of following:
   (*i*) HSCs (*ii*) iPS
   (*iii*) ESCs (*iv*) ASCs
   (*v*) SCNT
4. Name key genes of ES cells?

ANSWERS

Very Short Answer Questions

1. Embronic stem cells (ESCs) are derived from the inner cell mass of mammalian blastocyst or from fetal gamete progenitor (*i.e.*, germ) cells. These are pluripotent stem cells, *i.e.*, are capable of producing all the cells of the embryo.
2. Stem cells are undifferentiated cells that (*i*) are capable of self-renewal and (*ii*) are multipotent, *e.g.*, Haemopoietic stem cells of bone marrow.
3. (*i*) Haemopoietic stem cells. (*ii*) Induced pluripotent cells.
   (*iii*) Embryonic stem cells. (*iv*) Adult stem cells.
   (*v*) Somatic cell nuclear transfer.
4. Oct4, Sox2, Klf4 and Myc.

46 Cell Growth, Cell Cycle and Mitosis

One of the most important characteristics of the living cells is their power to grow and divide. When a cell reaches the limit of its size peculiar to its species, it divides into two parts. These parts grow to their full size and they divide again. Thus, new cells always arise by the division of the pre-existing cell.

The **cell division** is a process by which the cell duplicates itself for growth and reproduction of organism. In cell division chromosomes occupy a central position. As the vehicles of heredity they determine the characteristics of the cell and its progeny, and it is essential that they should be correctly distributed between the daughter cells. Two basic types of cell divisions are recognized according to the behaviour of the chromosomes. In the first of these, the daughter cells finish up containing exactly the same number of chromosomes as the parent cell. This is called **mitotic cell division** (or **mitosis**) and is the type of cell division which takes place during an organisms growth. In the other type of the cell division, known as **meiotic cell division** (or **meiosis**) the daughter cells finish up with half the total number of chromosomes present in the parent cell. This kind of division generally takes place in the formation of gametes or spores which are required for sexual reproduction.

46.1. HISTORICAL

The first account of mitosis was provided by **A. Schneider** in 1873. **Walter Flemming** observed cell division in the salamander *Titurus maculosus* and in 1879, showed that nuclear division involves a longitudinal splitting of the chromosomes and a migration of the daughter halves to the daughter nuclei. In 1882, he coined the term **mitosis. E. Van Benden** studied cell division in the roundworm *Parascaris* in 1883 and showed that the gametes of round worm contain half as many chromosomes as the body cell. In 1892, **T. Boveri** described synapsis and meiosis in *Ascaris*. Further detailed information about cell division came from following cytologists: **Darlington**; **La Cour**; **Revell**; **Comandon** and **Jolly** (1913); **Belar** (1920; **Lewis** and **Lewis**; **Huges**; **A. Bajer** and **J. Bajer** (1950, 1969); and **Abercrombie, Ambrose** and **Easty** (1959). The division of interphase into three separate phases, namely G_1, S and G_2, based on the timing of DNA synthesis was first proposed in 1953 by **Alma Howard** and **Stephen Pelc**. Their work based on plant meristem cells. In recent years, electron microscopy, polarizing microscopy and other molecular biological studies of the following cytologists have provided modern understanding of cell division: **Swann** and **Mitchison**; **D. Mazia** (1961, 1974); **S. Inoue**; **A. Forer** (1965); **R.S. Brinkley** and **S. Stubblefield** (1970); **J.R. McIntosh** (1971, 1975); **R. Rappaport** (1971); **R. Wettstein** and **J. Sotelo** (1971); **P.B. Moens** (1973); **J. Rattner** and **S.G. Phillips** (1973, 1976); **H. Mohri** (1976) and many others.

46.2. TYPES OF CELL DIVISION

In prokaryotes and eukaryotes the following three types of cell divisions have been distinguished:

1. Direct cell division or amitosis.
2. Indirect cell division or mitosis.
3. Reduction division or meiosis.

However, in protozoans five kinds of cell divisions have been recognized (Table 46.1).

Table 46.1. Types of nuclear division in Protozoa (Meglitsch, 1972).

1.	Amitosis	Nuclear divisions with no evidence of chromosomes.
2.	Cryptomitosis	Nuclear divisions with a simple or complex spindle but no lining up of chronosomes or chromomeres on an equatorial plate.
3.	Paramitosis	Nuclear divisions with definite chromosome-like bodies on the spindle, if it is present, but with no shortening of the chromosomes during prophase.
4.	Promitosis	Nuclear divisions with a spindle and chromosomes, differing from mitosis in the persistence of the nuclear membrane throughout division.
5.	Eumitosis	Nuclear divisions with a spindle and chromosomes which shorten during prophase, and in which the nuclear membrane disintegrates during division.

Amitosis

The amitosis or direct cell division is the means of asexual reproduction in unicellular organisms such as bacteria and protozoans and also a method of multiplication or growth in foetal membranes of some vertebrates. In amitosis type of cell division the splitting of nucleus is followed by cytoplasmic constriction.

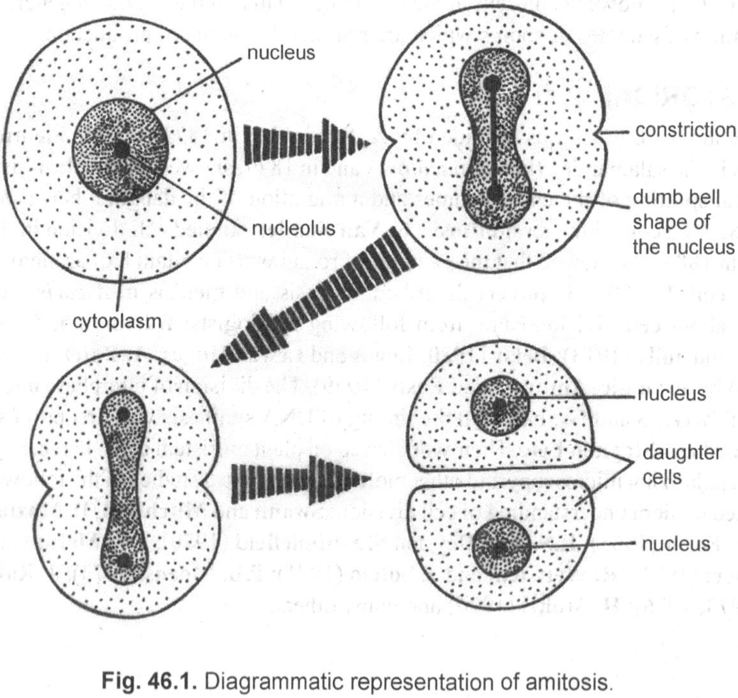

Fig. 46.1. Diagrammatic representation of amitosis.

During amitosis the nucleus elongates first and then assumes dumb-bell-shaped appearance. The depression or constriction increases in size and ultimately divides the nucleus into two nuclei. The division of nucleus is followed by the constriction of cytoplasm which divides the cell into two equal or approximately similar halves. Therefore, without the occurrence of any nuclear event two daughter cells are formed.

In bacteria, the distribution of replicated DNA to daughter cells is quite simple. The molecules are attached to the cell membrane and can be separated and distributed to daughter cells by growth of the membrane between them.

Mitosis and Cell Cycle

The mitosis (Gr., *mitos*-thread) occurs in the somatic cells and it is meant for the multiplication of cell number during embryogenesis and blastogenesis of plants and animals. Fundamentally, it remains related with the growth of an individual from zygote to adult stage or with the replacement of damaged or destroyed body cells.

One of the basic characteristic of mitotic cell division which is meant for growth due to multiplication is that it gives rise to two daughter cells, which resembles each other and also the parent cell qualitatively and quantitatively (*i.e.*, the chromosome number of mitotic product cells remains the same like the parent cell). The mitosis composed of two apparatuses, *viz.*, **chromatic apparatus** which includes the chromosomes and the nucleolus and the **achromatic apparatus** which in its turn includes the centrioles and spindle. The basic outline of mitosis remains the same in all living organisms. Mitosis is part of the overall **cell cycle** that includes replication of DNA, division of the nucleus (**karyokinesis**), and division of the cell itself (**cytokinesis**).

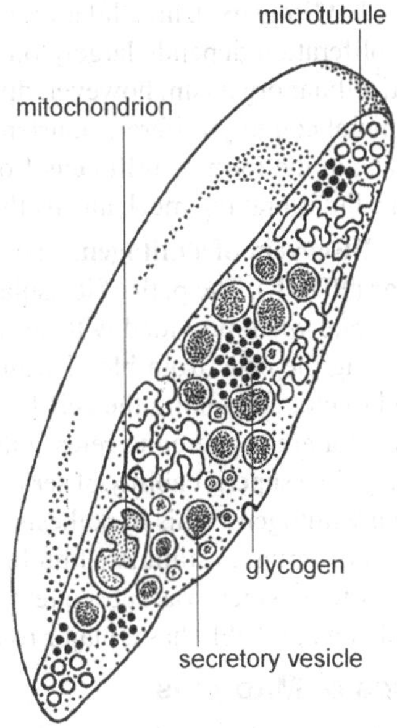

Fig. 46.2. A platelet. Platelets are miniature cells without a nucleus. They circulate in the blood and help stimulate blood clotting at sites of tissue damage, thereby preventing excessive bleeding. They also release various factors that stimulate healing. The platelet shown here has been cut in half to show its secretory vesicles, some of which contain platelet-derived growth factor (PDGF).

46.3. CELL GROWTH

The **size** of an organ or organism depends mainly on its total cell mass, which depends on both the total number of cells and the size of the cells. **Cell number**, in turn, depends on the amounts of cell division and cell death. Organ and body size ae therefore determined by *three* fundamental processes: **cell growth, cell division** and **cell death**. Each process is regulated both by intracellular programs and by extracellular signal molecules that control these programs.

Factors Regulating Cell Growth

The extracellular signal molecules that regulates the cell size and cell number are generally either *soluble secreted proteins, proteins bound to the surface of cells* or *components of the extracellular matrix*. The factors regulating cell growth are of following three types:

1. **Mitogens** which stimulate cell division, primarily by relieving intracellular negative controls that otherwise block progress through the cell cycle.
2. **Growth factors** which stimulate cell growth (an increase in cell mass) by promoting the synthesis of proteins and other macromolecules and by inhibiting their degradation.
3. **Survival factors** which promote cell survival by suppressing apoptosis.

1. Mitogens. Unicellular organisms tend to grow and divide as fast as they can, and their rate of proliferation depends largely on the availability of nutrients in the environment. The cells of a multicellular organism, however, divide only when more cells are needed by the organism. Thus, for an animal cell to proliferate, nutrients are not enough. It must also receive stimulatory extracellular signals, in the form of **mitogens** from other cells, usually its neighbors. Mitogens act to overcome intracellular braking mechanisms that block progress through the cell cycle.

Discovery of a mitogen. One of the first mitogens to be identified was **platelet-derived growth factor (PDGF)**. The path of its isolation began with the observation that fibroblasts in a culture dish proliferate when provided with serum but not when provided with plasma. **Plasma** is prepared by removing the cells from blood without allowing clotting to occur; **serum** is prepared by allowing blood to clot and taking the cell-free liquid that remains. When blood clots, **platelets** incorporated in the clot are triggered to release the contents of their secretory vesicles (Fig. 46.2) that stimulate healing. The superior ability of serum to support cell proliferation suggested that platelets contain one or many mitogens. This hypothesis was confirmed by showing that extracts of platelets could serve instead of serum to stimulate fibroblast proliferation. The crucial factor in the extracts was shown to be a protein, which was subsequently purified and named PDGF. In the body, PDGF liberated from blood clots probably has a major role in stimulating cell division during wound healing.

Types of Mitogens

About 50 proteins are known which act as mitogens. They are of following *three* main types: 1. Platelet-derived growth factor (PDGF); 2. Epidermal growth factor (EGF); and 3. Transforming growth factor-β (TGF-β).

Mitogens in Plants

There are certain chemical agents which stimulate mitosis in plant cells. Certain well known plant mitogens are the plant hormones such as **gibberellins**, **ethylene**, **indoleacetic acid** and **kinetin** (**D. Gross**, 1975; H.N. **Krishnamoorthy**, 1975; S. **Osborne**, 1977). For example, kinetin (6-furfurylaminopurine) increases the mitotic rate in meristems of *Allium* and at low concentrations, generally reduces the duration of interphase and increases the mitotic rate.

Functions of Mitogens

I. Effects of mitogens on cell division. Most mitogens can stimulate many type of cells to divide. PDGF acts on a range of cell types, including fibroblasts, smooth muscle cells, and neuroglial cells. Similarly, EGF acts not only on epidermal cells but also on many other cell types, including both epithelial and nonepithelial cells. At the opposite extreme lie narrow-specificity factors such as **erythropoietin**, which induces the proliferation of red blood cell precursors only.

Members of the TGF-β family act on some cells to stimulate cell proliferation and others to inhibit it, or stimulate at one concentration and inhibit at another. Indeed, like PDGF, many mitogens have other actions besides the stimulation of cell division: they can stimulate cell growth, survival, differentiation or migration, depending on the circumstances and the cell type.

2. Go or nondividing state of the cells. In the absence of a mitogenic signal to proliferate, Cdk inhibition in G_1 is maintained, and the cell cycle arrests. In some cases, cells partly disassemble their cell-cycle control system and exit from the cycle to a specialized, nondividing state, called G_0. Based on their G_0 state, cells are divided into following three categories:

(*i*) Most cells in human body are in G_0, but the molecular basis and reversibility of this state vary in different cell types. Neurons and skeletal muscle cells, for example, are in a terminally differentiated G_0 state, in which their cell cycle control system is completely dismantled: the expression of the genes encoding various Cdks and cyclins are permanently turned off, and cell division never occurs.

(*ii*) Other cell types withdraw from the cell cycle only temporarily and retain the ability to reassemble the cell cycle control system quickly and reenter the cycle. Most liver cells, for example, are in G_0, but they can be stimulated to divide if the liver is damaged.

(*iii*) Still other types of cells, including some lymphocytes, withdraw from and reenter the cell cycle repeatedly throughout the life time.

3. Stimulation by mitogens. For great majority of animal cells, mitogens control the rate of cell division by acting in the G_1 phase of cell cycle. Mitogens act to release the breakes on Cdk activity, thereby allowing S phase to begin.

4. Intracellular mechanisms that limit cell division. Many animal precursor cells divide a limited number of times before they stop and terminally differentiated into permanently arrested, specialized cells. Some intracellular mechanisms such as a progressive increase in **CkI proteins**, can limit the cell division.

5. Replicative cell senescence in human fibroblasts. The best-understood intracellular mechanism that limits cell proliferation occurs in fibroblasts. Fibroblasts taken from a normal human tissue go through only about 25–50 population doubling when cultured in a standard mitogenic medium. Toward the end of this time, proliferation slows down and finally halts, and the cells enter a nondividing state from which they never recover. This phenomenon is called **replicative cell senescence**, although it is unlikely to be responsible for the aging of the organism.

Replicative cell senescence in human fibroblasts seems to be caused by changes in the structure of the **telomeres**, the repititive DNA sequences and associated proteins at the ends of chromosomes. When a cell divides, telomeric DNA sequences are not replicated in the same manner as the rest of the genome but instead are synthesized by the enzyme **telomerase**. Telomerase is also found to promote the formation of protein cap structures that protect the chromosome ends. Because human fibroblasts, and many other human somatic cells, are deficient in telomerase, *their telomeres become shorter with every cell division*, and their protective protein caps progressively deteriorate. Eventually, DNA damage occurs at chromosome ends. The damage activates a p53-dependent cell-cycle arrest that resembles the arrest caused by other types of DNA damage.

II. Extracellular Growth Factors Stimulate Cell growth

The growth of an organism or organ depends on cell growth: cell division alone cannot release total cell mass without cell growth. In single celled organisms such as yeasts, cell growth (like the cell division) requires only nutrients. In animal, by contrast, cell growth and cell division both depends on signal from other cells.

The extracellular **growth factors** that stimulate cell growth bind to receptors on the cell surface and activate intracellular signaling pathway.These pathways stimulate the accumulation of proteins and other macromolecules, and they do so by both increasing their rate of synthesis and decreasing their rate of degradation.

One of the most important intracellular signaling pathways activated by growth factor receptors involves the enzyme **PI3-kinase**, which adds a phosphate from ATP to the 3 position of inositol phospholipids in the plasma membrane. The activation of PI3-kinase leads to the activation of several protein kinases, including **S6 kinase**. The S6 kinase phosphorylates ribosomal protein S6, increasing the ability of ribosomes to translate a subset of mRNAs, most of which encode ribosomal components. Protein synthesis therefore increases, when the gene encoding S6 kinase is inactivated in *Drosophila*, the mutant flies are small; whereas cell numbers are normal, cell size is abnormally small. Growth factors also activate a translation initiation factor called *eIF4E*, further increasing protein synthesis and cell growth.

Growth factor stimulation also leads to increased production of the gene regulatory protein **MyC**, which also plays an important part in signaling by mitogens. MyC increases the transcription of a number of genes that encode proteins involved in cell metabolism and macromolecular synthèsis. In this way, it stimulates both cell metabolism and cell growth.

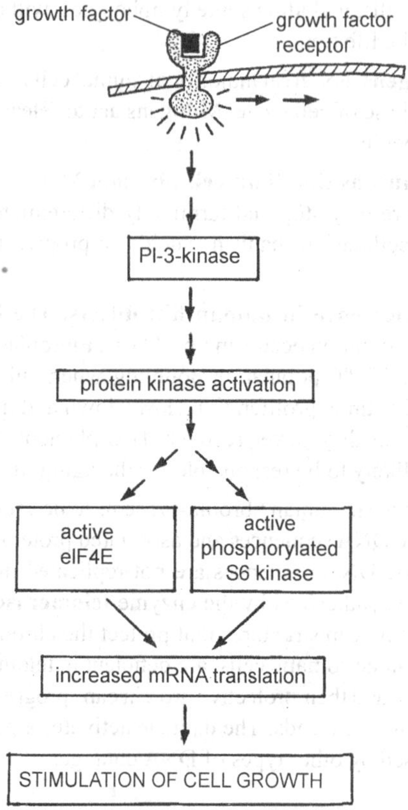

Fig. 46.3. One way in which growth factors promote cell growth. Activation of cell-surface receptors leads to the activation of PI3- kinase, which promotes protein synthesis, at least partly through the activation of eIF4E and S6 kinase (after Alberts *et al.*, 2002).

III. Extracellular Survival Factors

Animal cells need signals from other cells not only to grow and proliferate, but also to survive. If deprived of such **survival factors**, cells activate their intracellular death program and die by apoptosis. This arrangement ensures that cells survive only when and where they are needed. Nerve cells, for example, are produced in excess in the developing nervous system and then compete for limited amounts of survival factors that are secreted by target cells they contact. Nerve cells that receive enough survival factors live, while the others die by apoptosis. A similar dependence on survival signals from neighboring cells is thought to control cell numbers in other tissues, both during development and in adulthood.

Anchorage Dependence and Cell Shape

The shape of an animal cell changes as it spreads and crawls out over a substratum to occupy vacant space, as this can have a major impact on cell growth, cell division and cell survival.

When normal fibroblasts or epithelial cells, for example, are cultured in suspension, unattached to any solid surface and therefore rounded up, they almost never divide—a phenomenon known as **anchorage dependence of cell division**. But when these cells are allowed to settle and adhere to a sticky substrate, they rapidly form focal adhensions at sites of attachment, and then begin to grow and proliferate.

Focal adhesions are places where extracellular matrix molecules, such as laminin or **fibronectin** interact with cell surface matrix receptors, called **integrins**, which are linked to the actin cytoskeleton. The binding of extracellular matrix molecules to integrins leads to the local activation of protein kinases, including **focal adhesion kinase (FAK)**, which in turn leads to the activation of intracellular signaling pathways that can promote the survival, growth and division of cells.

The anchorage control operates in G_1 of interphase. Cells require anchorage to progress through G_1 and S phases but anchorage is not required for completing the cell cycle. In fact, cells commonly loosen their attachments and **round up** as they pass through M phase. This cycle of attachment and detachment presumably allows cells in tissues to rearrange their contacts with other cells and with the extracellular matrix. In this way, tissues can accommodate the daughter cells produced by cell division and then bind them securely into the tissue before they are allowed to begin the next division cycle.

The extracellular signal proteins discussed so far, *i.e.*, mitogens, growth factors and survival factors, are **positive regulators** of cell cycle progression, cell growth and cell survival, respectively. They therefore tend to increase the size of organs and organisms. In some tissues, however, cell and tissue size also is influenced by **inhibitory extracellular signal proteins** that oppose the positive regulators and thereby inhibit organ growth (**Alberts** *et. al.*, 2002).

46.4. CELL CYCLE

All cells are produced by divisions of pre-existing cell. Continuity of life depends on cell division. A cell born after a division, proceeds to grow by macromolecular synthesis, reaches a species-determined division size and divides. This cycle acts as a unit of biological time and defines life history of a cell. **Cell cycle** can be defined as the entire sequence of events happening from the end of one nuclear division to the beginning of the next. The cell cycle involves the following three cycles.

1. Chromosome cycle. In it **DNA synthesis** alternates with **mitosis** (or karyokinesis or nuclear division). During DNA synthesis, each double-helical DNA molecule is replicated into two identical daughter DNA molecules and during mitosis the duplicated copies of the genome are ultimately separated.

2. Cytoplasmic cycle. In it **cell growth** alternates with **cytokinesis** (or cytoplasmic division). During cell growth many other components of the cell (RNA, proteins and membranes) become double in quantity and during cytokinesis cell as a whole divides into two. Usually the karyokinesis is followed by the cytokinesis but sometimes the cytokinesis does not follow the karyokinesis and results into the multinucleate cell, *e.g.*, cleavage of egg in *Drosophila*.

3. Centrosome cycle. Both of the above cycles require that the **centrosome** be inherited reliably and duplicated precisely in order to form the two poles of the mitotic spindle; thus, centrosome cycle forms the third component of cell cycle.

Howard and **Pelc** (1953) have divided cell cycle into four phases or stages: G_1, S, G_2 and M phase. The G_1 phase, S phase and G_2 phase are combined to from the classical interphase.

1. G_1 Phase. After the M phase of previous cell cycle, the daughter cells begin G_1 of interphase of new cell cycle. The G_1 is a resting phase. It is called **first gap phase**, since no DNA synthesis takes place during this stage; currently, G_1 is also called **first growth phase**, since it involves synthesis of

RNA, proteins and membranes which leads to the growth of nucleus and cytoplasm of each daughter cell towards their mature size.

During G_1 phase, chromatin is fully extended and not distinguishable as discrete chromosomes with the light microscope. This is a time of resumption of normal cell metabolism which has slowed down during the previous cell division. Thus, G_1 involves transcription of three types of RNAs, namely rRNA, tRNA and mRNA; rRNA synthesis is indicated by the appearance of nucleolus in the interphase (G_1 phase) nucleus. Proteins synthesized during G_1 phase (1) Regulatory proteins which control various events of mitosis; (2) Enzymes (*e.g.*, DNA polymerase) necessary for DNA synthesis of the next stage; and (3) Tubulin and other mitotic apparatus proteins.

G_1 phase is most variable as to duration (Table 46.2); it either occupies 30 to 50 per cent of the total time of the cell cycle or lacks entirely in rapidly dividing cells (*e.g.*, blastomeres of early embryo of frog and mammals). Terminally differentiated somatic cells (*i.e.*, end cells such as neurons and striated muscle cells) that no longer divide, are arrested usually in the G_1 stage; such a type of G_1 phase is called **G_0 phase.**

Table 46.2 Different stages of a mitotic cell cycle and their duration in hours.

Parts of cell cycle	Phases	Description of phases	Duration in hours		
			Vicia faba	Mouse L cells	Human HeLa cells
Interphase	G_1	Pre-DNA-synthesis phase	12	12	12
	S	DNA-synthesis phase	6	6–8	10
	G_2	Post-DNA synthesis phase	12	3–4	3
Mitosis	M	Mitotic phase	1	1	1

2. S phase. During the S phase or **synthetic phase** of interphase, replication of DNA and synthesis of histone proteins occur. New histones are required in massive amounts immediately at the beginning of the S period of DNA synthesis to provide the new DNA with nucleosomes. Thus, at the end of S phase, each chromosome has two DNA molecules and a duplicate set of genes. S phase occupies roughly 35 to 45 per cent of cell cycle.

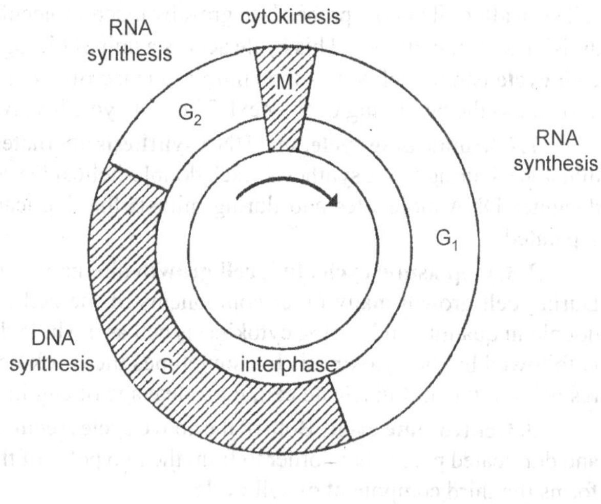

Fig. 46.4. The cell cycle or mitotic cycle, showing relative duration of phases (*e.g.*, interphase and mitotic phase) in a growing cell. S, synthesis of DNA; G_1 the first gap or growth phase; G_2, the second gap or growth phase; and M, mitotic phase.

3. G_2 phase. This is a **second gap** or **growth phase** or resting phase of interphase. During G_2 phase, synthesis of RNA and proteins continues which is required for cell growth. It may occupy 10 to 20 per cent time of cell cycle. As the G_2 phase draws to a close, the cell enters the M phase.

General Events of Interphase. The interphase is characterized by the following features:

The nuclear envelope remains intact. The chromosomes occur in the form of diffused, long, coiled and indistinctly visible chromatin fibres. The DNA amount becomes double. Due to accumulation of ribosomal RNA (rRNA) and ribosomal proteins in the nucleolus, the size of the latter is greatly increased. In animal cells, a daughter pair of centrioles originates near the already existing centrioles and, thus, an interphase cell has two pairs of centrioles.

In animal cells, net membrane biosynthesis increases just before cell division (mitosis). This extra membrane seems to be stored as **blebs** on the surface of the cells about to divide.

4. M phase or Mitotic phase. The mitosis (Gr., *mitos* = thread) occurs in the somatic cells and it is meant for the multiplication of cell number during embryogenesis and blastogenesis of plants and animals. Fundamentally, it remains related with the growth of an individual from zygote to adult stage. Mitosis starts at the culmination point of interphase (*i.e.*, G_2 phase). It is a short period of chromosome condensation, segregation and cytoplasmic division. Mitosis is important for replacement of cells lost to natural friction (**attrition**), wear and tear and for wound healing.

46.5. MITOSIS

Mitosis is the process that distributes the two sets of duplicated chromosomes into two daughter nuclei. As a process, mitosis is remarkably similar in all animals and plants. It is a smoothly continuous process and is divided arbitrarily into following stages or phases for convenient reference (Fig. 46.5 and Fig. 46.6):

1. Prophase. The appearance of thin-thread like condensing chromosomes marks the first phase of mitosis, called **prophase** (Gr., *pro* = before; *phasis* = appearance). The cell becomes spheroid, more refractile and viscous.

Each prophase chromosome is composed of two coiled filaments, the **chromatids**, which are the result of the replication of DNA during the S phase. As prophase progresses, the chromatids become shorter and thicker and two sister chromatids of each chromosome are held together by a special DNA-containing region, called the **centromere** or **primary constriction**. During prophase, proteins of the trilaminar **kinetochores** (one for each chromatid) start depositing or organizing on the centromere of each chromosome. Further, during early prophase, the chromosomes are evenly distributed in the nuclear cavity; as prophase progresses, the chromosomes approach the nuclear envelope, causing the central space of the nucleus to become empty.

In the cytoplasm, the most conspicuous change is the formation of the spindle or **mitotic apparatus**. In the early prophase, there are two pairs of centrioles, each one surrounded by the so-called **aster** which is composed of microtubules radiating in all directions. The two pairs of centrioles migrate to opposite poles of the cell along with the asters and become situated in antipodal positions. Between the separating centrioles forms a spindle.

Lastly, during prophase, the nucleolus gradually disintegrates. Degeneration and disappearance of the nuclear envelope marks the end of prophase. This process is incompletely understood. However, following two factors may be involved in this process: 1. **Enzymatic action** either by some mitochondrial enzymes, cytosolic MPF kinase or nuclear RNA (or ribozyme). 2. **Physical action**, *i.e.*, physical stress exerted by microtubules which become attached to the nuclear envelope.

2. Prometaphase. At the begining of prometaphase, compacted chromosomes are scattered throughout the space that was the nuclear region. As the microtubules of the spindle penetrate into the central region of the cell, the free (plus) ends of the microtubules are seen to grow and shrink in a dynamic fashion, as if they were "searching" for a chromosome. It is not certain whether searching is entirely random, as evidence suggest that microtubules may grow preferentially toward a site containing chromatin. Those microtubules that contact a kinetochore are "captured" and stabilized.

A kinetochore typically makes initial contact with the sidewall of a microtubule rather than its end. Once initial contact is made some chromosomes move activily along the wall of the microtubule, poleward by **motor proteins** located in the kinetochore. Soon, however, the kinetochore tends to become stably associated with the plus end of one or more spindle microtubules from one of the spindle poles. Eventually, the unattached kinetochore on the sister chromatid captures its own microtubules from the opposite spindle pole. It has also been reported that unattached kinetochores may serve as tubules that grow toward the opposite spindle pole. Regardles of how it occurs, the two sister chromatids of each mitotic chromosome ultimately connected by their kinetochores to microtubules that extend from opposite poles.

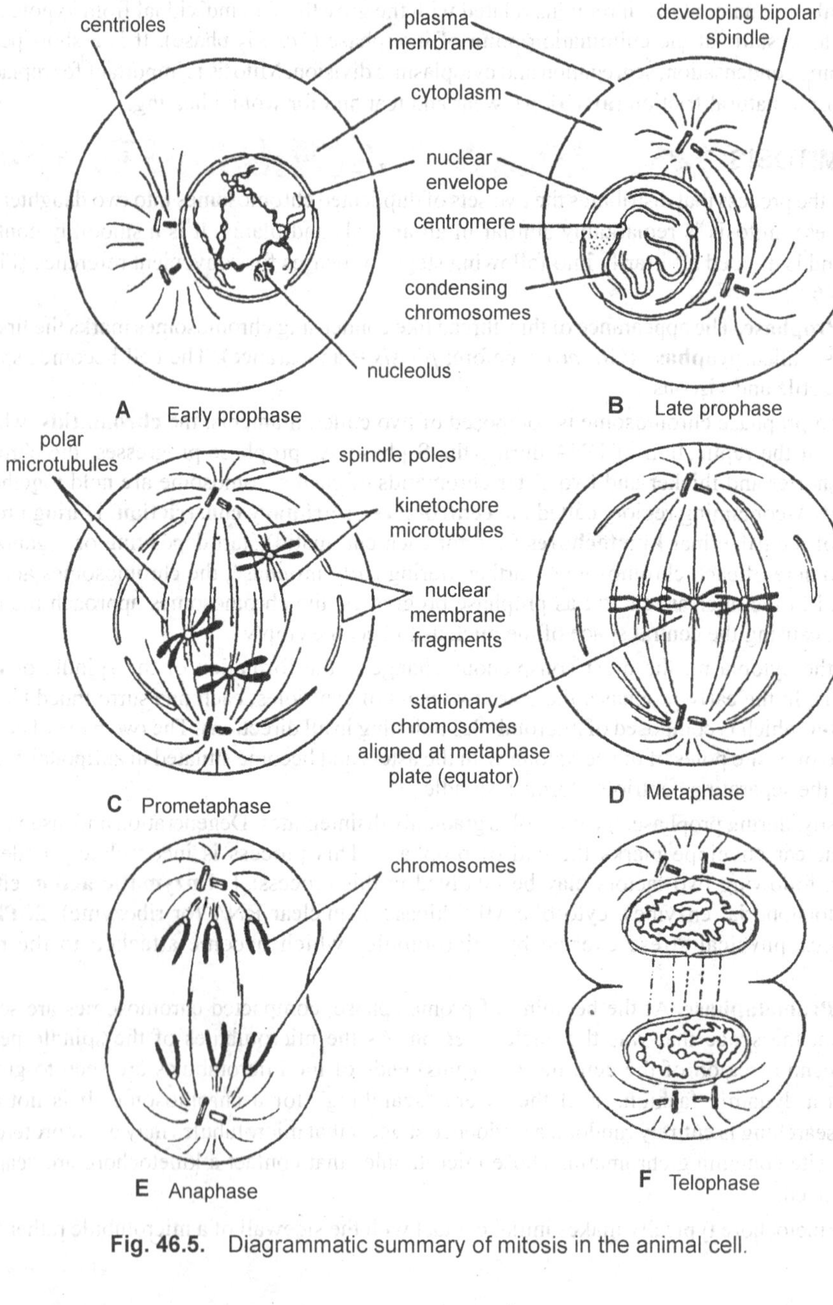

Fig. 46.5. Diagrammatic summary of mitosis in the animal cell.

Congression. Studies of living cells indicate that prometaphase chromosomes associated with spindle microtubules are not moved directly to the center of the spindle but rather oscillate back and forth in both a poleward and antipoleward direction. Ultimately, the chromosomes of a prometaphase cell are moved by a process called **congression** toward the center of the mitotic spindle, midway between the poles. The forces required for chromosome movements during prometaphase are generated by motor proteins associated with both the kinetochores and arms of the chromosomes (Fig. 46.15).

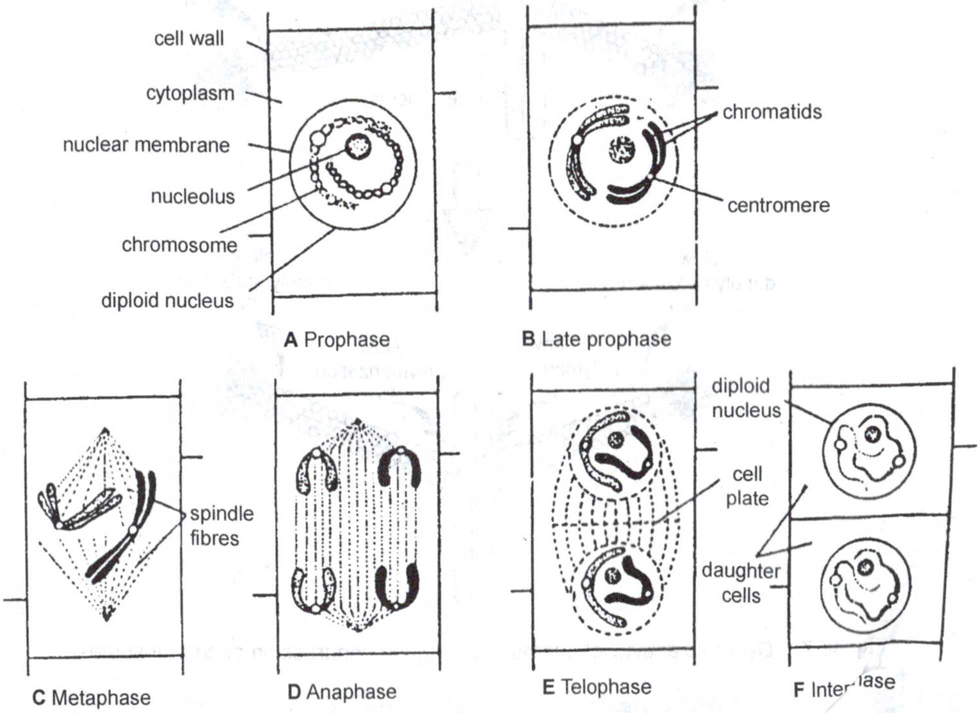

Fig. 46.6. Diagrammatic summary of mitosis in the higher plant is.

During prometaphse, microtubule **dynamics** also play a key role in aciating chromosome movements. As the chromosomes congress toward the center of the mit spindle, the longer microtubules attached to one kinetochroe are shortened, while the shor crotubules attached to the sister kinetochore are elongated. Shortening and elongation of mi ules occur primarily by loss or gain of subunits at the plus end of the microtubule (Fig. emarkably, this dynamic activity occurs while the plus end of each microtubule remain a to a kinetochore.

Centrosome Cycle

When an animal cell exits mitosis, the cytoplasm gle centrosome containing two centrioles situated at right angle to one another. Even tokinesis has been completed, the two centrioles of each daughter cell lose their close to one another (they are said to be "**disengaged**"). This event is triggered by the en ase, which becomes activated late in mitosis. Later, as DNA replication begins in the e onset of S phase, each centriole of the centrosome initiates its "replication" in the c progress begins with the appearance of a small **procentriole** next to each preexistin al) centrioles and oriented at right angles to it. Subsequent microtubule elongation c ocentriole into a full-length daughter centriole. At the beginning of mitosis, the centr to two adjacent centrosomes, each containing a pair of mother-daughter centrioles. The initiation of centrosome duplication at the $G_1 - S$ transition is normally triggered by phosphorylation of a centrosomal protein by **Cdk2**, the same agent responsible for the onset of DNA replication.

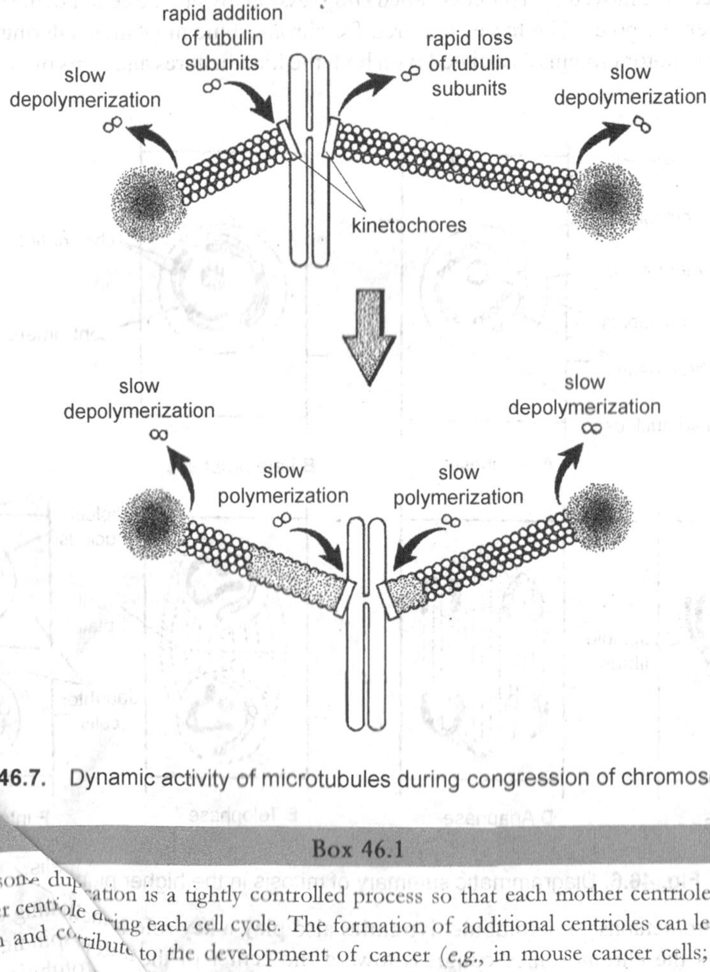

Fig. 46.7. Dynamic activity of microtubules during congression of chromosomes.

Box 46.1

Centrosome duplication is a tightly controlled process so that each mother centriole produces only one duaghter centiole during each cell cycle. The formation of additional centrioles can lead to abnormal cell division and contribute to the development of cancer (*e.g.*, in mouse cancer cells; **Goepfret** and **Brinkley**).

The *first stage* in formation of the mitotic spindle in a typical animal cell is the appearance of microtubules in a "su" arrangement or **aster**, around each centrosome during early prophase. As discussed in chapter 1" otubules grow by addition of subunits to their plus ends, while their minus ends remain associ h the centrosome. The process of aster formation is followed by separation of the centrosom one another and their subsequent movement around the nucleus toward opposite ends of the trosome separation is driven by **motor proteins** associated with the adjacent microtubules centrosome separate, the microtubules streching between them increase in number and elo entually, the two centrosomes reach points opposite one another, thus establishing the two p ipolar mitotic spindle (Fig. 46.8). Following mitosis, one centrosome will be distributed t hter cell.

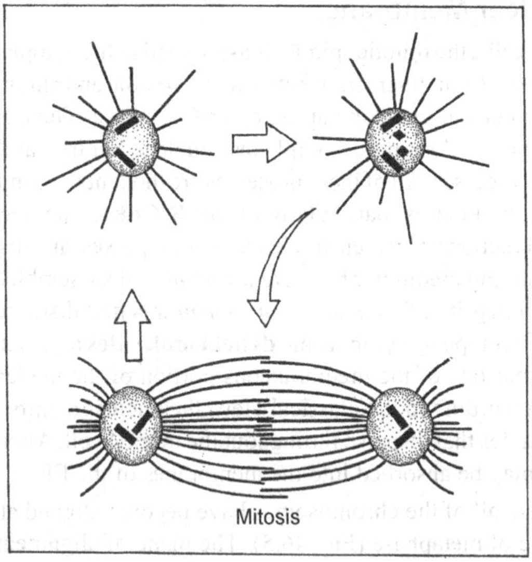

Fig. 46.8. The centrosome cycle of an animal cell. During G_1, the centrosome contains a single pair of centrioles that are no longer as tightly associated as they were during mitosis. During S phase, daughter procentrioles form adjacent to maternal centrioles so that two pairs of centrioles become visible within the centrosome. The daughter procentrioles continue to elongate during G_2 phase, and at the beginning of mitosis, the centrosome splits, with each pair of centrioles becoming part of its own centrosome. As they separate, the centrosomes organise the microtubule fibers that makeup the mitotic spindle.

Spindle formation in cells lacking centrosomes. A number of different types of animal cells (including those of the early mouse embryos), lack centrosomes, as do the cells of higher plants, yet all of these cells form a bipolar mitotic spindle and undergo a relatively typical mitosis. Functional mitotic spindles can even form in mutant *Drosophila* cells that lack centrosomes or in mammalian cells in which centrosome has been experimentally removed. In all of these cases, the microtubules of the mitotic spindle are nucleated near the chromosomes rather than at the poles where centrosome would normally reside. Then, once they have polymerized, the minus ends of the microtubules are brought together (*i.e.*, focused) at each spindle pole through the activity of motor proteins (Fig. 46.9). However, cells whether contain or lack the functional centrosome, some spindle microtubules are nucleated at the chromosome.

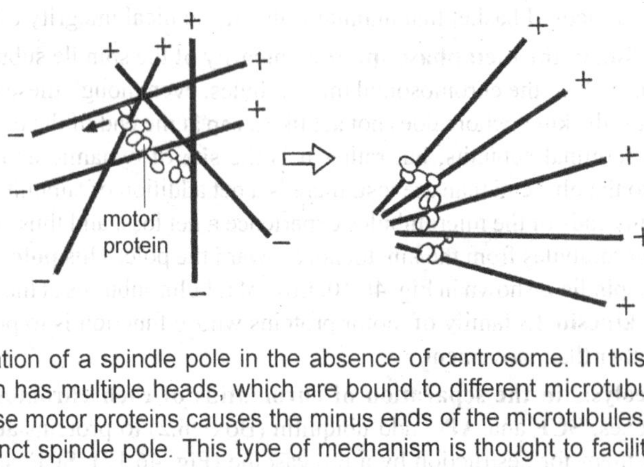

Fig. 46.9. Formation of a spindle pole in the absence of centrosome. In this model, each motor protein has multiple heads, which are bound to different microtubules. The movement of these motor proteins causes the minus ends of the microtubules to converge to form a distinct spindle pole. This type of mechanism is thought to facilitate the formation of spindle poles in the absence of centrosome but may also play a role when chromosomes are present.

Dissolution of Nuclear Membrane

In most eukaryotic cells, the mitotic spindle is assembled in the cytoplasm and the chromosomes are compacted in the nucleoplasm. Interaction between the spindle and chromosomes is made possible by the breakdown of the nuclear envelope at the end of prophase. The *three* major components of the nuclear envelope—the nuclear pore complexes, nuclear lamina, and nuclear envelope—are disassembled in separate processes. All of these processes are thought to be initiated by phosphorylation of key substrates by mitotic kinases, particularly **cyclin B-Cdk1**. The nuclear pore complexes are disassembled as the interactions between nucleoporin complexes are disrupted and the proteins dissociate into the surrounding medium. The *nuclear lamina* is disassembled by depolymerization of the lamin filaments. The integrity of the *nuclear membranes* is first disrupted mechanically as holes are torn into the nuclear envelope by cytoplasmic **dynein molecules** associated with the outer nuclear membrane. The subsequent fate of the membraneous portion of the nuclear membrane has been a subject of controversy. According to the classical view, the nuclear membranes are fragmented into a population of small vesicles that disperse throughout the mitotic cell. Alternatively, the membranes of the nuclear envelope may be absorbed into the membranes of the ER.

3. Metaphase. Once all of the chromosomes have become aligned at the spindle equator, the cell has reached the stage of metaphase (Fig. 46.5). The plane of alignment of the chromosomes at metaphase is referred to as the **metaphase plate**. The mitotic spindle of the metaphase cell contains a highly organized array of microtubules that is ideally suited for the task of separating the duplicated chromatids positioned at the center of the cell. Metaphase spindle comprises of following three types of microtubules:

***(i)* Astral microtubules.** They radiate outward from the centrosome into the region outside the body of the spindle. They help position the spindle apparatus in the cell and may help determine the plane of cytokinesis.

***(ii)* Chromosomal (or kinetochore) microtubules.** They extend between the centrosome and the kinetochore of the chromosomes. In mammalian cells, each kinetochore is attached to a bundle of 20 to 30 microtubules which form a spindle fiber. During metaphase, the chromosomal microtubules exert a pulling force on the kinetochores. As a result, the chromosomes are maintained in the equatorial plane by a "tug-of-war" between balanced pulling forces exerted by chromosomal spindle fibers from opposite poles. During anaphase, chromosomal microtubules are required for the movement of the chromosomes towards the poles.

***(iii)* Polar (or interpolar) microtubules.** They extend from the centrosomes. Polar microtubules from one centrosome overlap with their counterpart from the opposite centrosome. The polar microtubules form a structural basket that maintains the mechanical integrity of the spindle.

Microtubule flux in the metaphase spindle. Integrity of the spindle subunits are rapidly lost and added at the plus ends of the chromosomal microtubules, even though these ends are attached to the kinetochore. Thus, the kinetochore does not act like a cap at the end of the microtubule, blocking the entry or exit of terminal subunits, but rather it is the site of dynamic activity. Because more subunits are added to the plus end than are lost, there is a net addition of subunits at the kinetochore. Meanwhile, the minus ends of the microtubules experience a net loss, and thus subunits move along the chromosomal microtubules from the kinetochore toward the pole. This **poleward flux** of tubulin subunits in a mitotic spindle is shown in Fig. 46.10. Loss of tubulin subunits at the poles is likely aided by a member of the **kinesin-13** family of motor proteins whose function is to promote microtubule depolymerization rather than movement.

Role of proteolysis in the separation of chromatids of each chromosome. Two distinct multiprotein complexes, **SCF** and **APC** add ubiquitin (Box 46.2) to proteins at different stages of cell cycle, targeting them for destruction by a proteasome (Fig. 46.12). SCF acts primarily during interphase.

Box 46.2

Proteasome. Barrel-shaped, multiprotein complex in which cytoplasmic proteins are degraded. Proteins selected for destruction are linked to ubiquitin molecules and threaded into the central chamber of the proteasome.

Ubiquitin. A small, highly conserved protein that is linked to proteins targeted for internalization by endocytosis or degradation in proteasome.

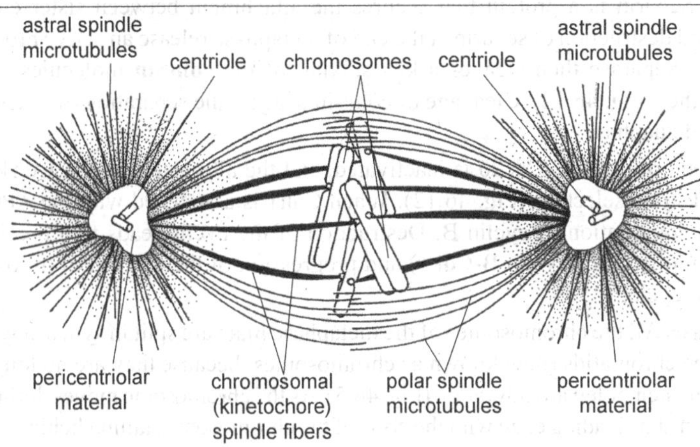

Fig. 46.10. Tubulin flux through the microtubules of the mitotic spindle at metaphase. Experimentally fluorescently labeled tubulin subunits are found to move through the microtubules of a metaphase spindle at a rate of about 1 μm/min.

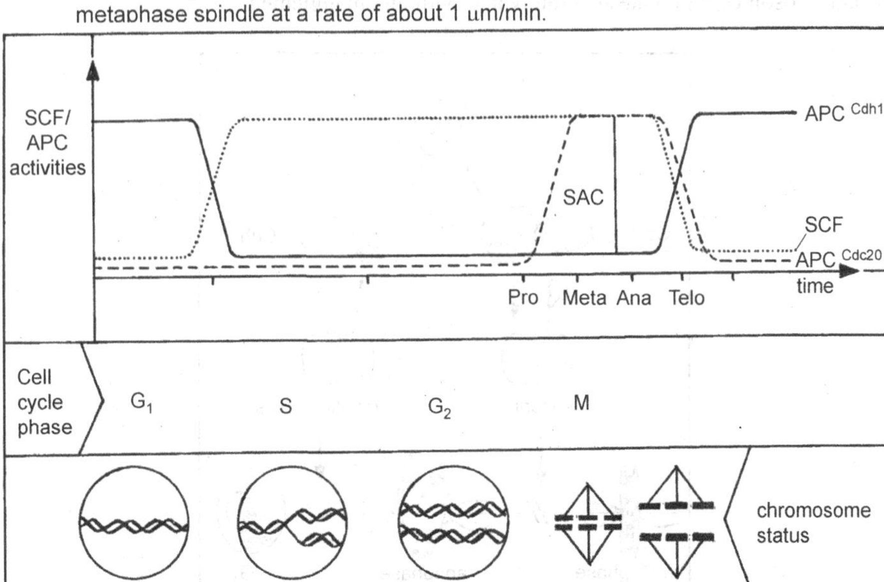

Fig. 46.11. SCF and APC activities during the cell cycle. SCF and APC are multisubunit complexes that ubiquitinate substrates, leading to their destruction by proteasomes. SCF is active during interphase, whereas APC is active during mitosis and G_1. Two different versions of APC are indicated. These two APCs differ in either containing either a Cdc20 or Cdh1 adaptor protein, which alters the substrates recognized by the APC. APC^{Cdc20} is active earlier in mitosis than is APC^{Cdh1}. The SAC (*i.e.*, spindle assembly check-point) prevents APC^{Cdc20} from triggering anaphase until all the chromosomes are properly aligned at the metaphase plate (after Karp 2010).

In contrast, APC (anaphase promoting complex) plays a key role in regulating events that occur during mitosis. The APC contains about a dozen core subunits, in addition to an "adaptor protein" that plays a key role in determining which proteins serve as the APC substrate. Two alternate versions of the adoptor protein—Cdc20 and Cdh1—play an important role in substrate selection during mitosis. APC complexes containing one or the other of these adaptors are known as APC^{Cdc20} or APC^{Cdh1} (Fig. 46.11).

APC^{Cdc20} becomes activated prior to metaphase and ubiquitinates a major anaphase inhibitor called **securin**. Securin is a protein that secures the attachment between sister chromatids. The ubiquitination and destruction of securin at the end of metaphase release an active protease (enzyme) called **separase**. Separase then cleaves a key subunit of the **cohesin** molecules that hold sister chromatids together (Fig. 46.12). Cleavage of cohesin triggers the separation of sister chromatids to mark the onset of anaphase.

Near the end of mitosis, Cdc20 is inactivated, and the alternate adaptor, Cdh1, takes control of the APC's substrate selection (Fig. 46.12). When Cdh1 is associated with the APC, the enzyme completes the ubiquitination of cyclin B. Destruction of the cyclin leads to a precipitous drop in activity of the mitotic Cdk (Cyclin B-Cdk1) and progression of the cell out of mitosis and into G_1 phase of the next cycle.

4. Anaphase. All the chromosomes of the metaphase plate are split in synchrony at the onset of anaphase, and the chromatids (now known as chromosomes, because they are no longer attached to their sisters) begin their poleward migration (Fig. 46.5). As the chromosome moves during anaphase, its centromere is seen at its leading edge with the arms of the chromosomes tailing behind. The movement of chromosomes toward opposite poles is very slow relative to other types of cellular movements, proceeding at approximately 1 μm per minute. The slow rate of movement of chromosomes ensures that the chromosomes segregate accurately and without entanglement.

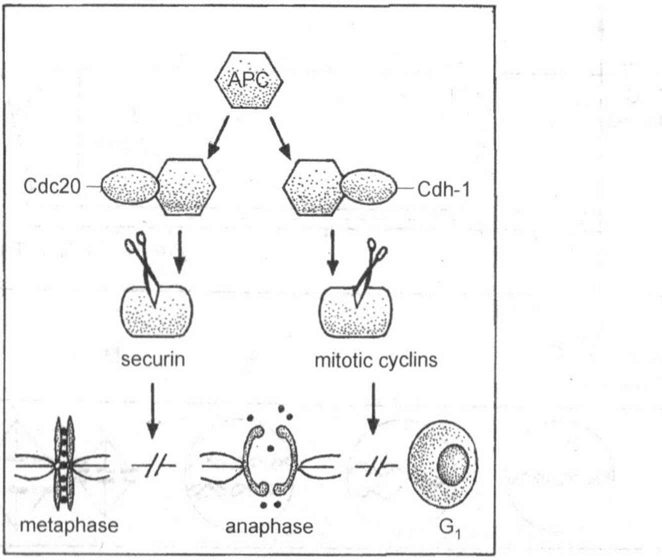

Fig. 46.12. Role of APC during cell cycle. APC^{Cdc20} is responsible for destroying proteins, such as securin, that inhibits a anaphase. Destruction of these substrates promotes the metaphase-anaphase transition. APC^{Cdh1} is responsible for ubiquitinating proteins, such as mitotic cyclins, that inhibit exit from mitosis. Destruction of these substrates promotes the mitosis-G_1 transition. APC^{Cdh1} activity during early G_1 helps maintain the low cyclin-Cdk activity required to assemble prereplication complexes at the origin of replication (after Karp 2010).

The poleward movement of chromosomes is accompanied by obvious shortening of chromosomal microtubules. It has long been appreciated that tubulin subunits are lost from the plus (kinetochore-based) ends of chromosomal microtubules during anaphase (Fig. 46.13). Subunits are lost from the minus ends of these microtubules as a result of the continued poleward flux of tubulin subunits that occurs during prometaphase and metaphase (Fig. 46.7 and Fig. 46.10).

The primary difference in microtubule dynamics between metaphase and anaphase is that subunits are added to the plus ends of microtubules during metaphase, keeping the length of the chromosomal fibers constant (Fig. 46.10), whereas subunits are lost from the plus ends during anaphase, resulting in shortening of chromosomal fibers (Fig. 46.11). This change in behaviour at the microtubule plus end is thought to be triggered by a change in tension on the kinetochores following separation of the sister chromatids.

The movement of the chromosomes toward the poles is referred to as **anaphase A** to distinguish it from a separate but simultaneous movement, called **anaphase B**, in which the two spindle poles move farther apart. The elongation of the mitotic spindles during anaphase B is accompanied by the net addition of tubulin subunits to the plus ends of the polar microtubules. Thus, subunits can be preferentially added to polar microtubules and removed from chromosomal microtubules at the same time in different regions of the same mitotic spindle (Fig. 46.13).

Forces Required for Chromosome Movements at Anaphase

Inoue (1960) suggested that depolymerization of the microtubules that comprise a spindle fiber could generate sufficient mechanical force to pull a chromosome forward. Following two models are quite popular for anaphasic movements of the chromosomes.

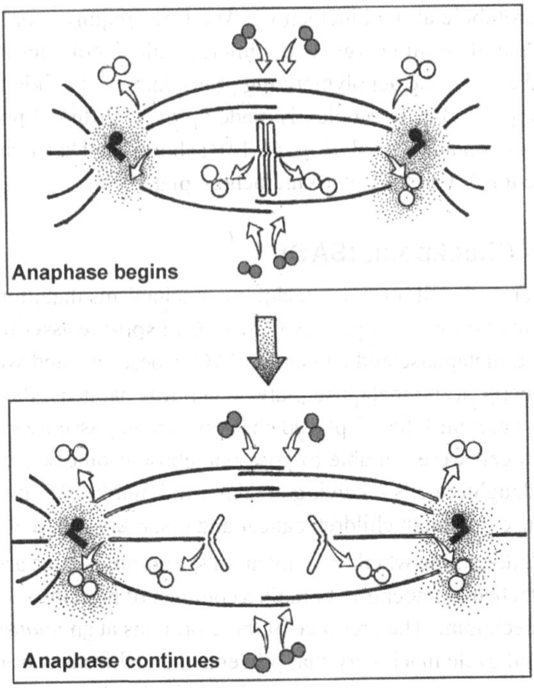

Fig. 46.13. Microtubule dynamics during anaphase. Tubulin subunits are lost from both ends of the chromosomal microtubules, resulting in shortening of chromosomal fibers and movement of the chromosomes toward the poles during anaphase A. Meanwhile, tubulin subunits are added to the plus ends of polar microtubules, which also slide across one another, leading to separation of the poles during anaphase B.

Anaphasic chromosome movements in budding yeast cell. During anaphase of a budding yeast cell or animal cell, the microtubules that comprise the chromosomal spindle fibers undergo depolymerization at both their minus and plus ends (Fig. 46.14). These combined activities lead to the movement of chromosomes toward the pole. Depolymerization at the microtubule minus ends serves to transport the chromosomes toward the poles due to poleward flux, reminiscent of a person standing on a **"moving walkway"** in an airport. In contrast, depolymerization at the microtubule plus ends serves to **"chew up"** the fiber that is towing the chromosomes. Some cells rely more on poleward flux, others on plus-end depolymerization. Studies of animal cells in anaphase have revealed that both the plus and minus ends of chromosomal fibers are sites where depolymerizing **kinesins** (members of kinesis-13 family) are localized. These polymerases (enzymes) are present at the opposite ends of the microtubule "depolymerases" is specifically inhibited, chromosome segregation during anaphase is at least partially disrupted. These findings suggest that ATP-dependent, kinesin-mediated depolymerization forms the basis for chromosomes segregation during mitosis.

Mechanism of holding of chromosomes by kinetochores. In the model depicted in Fig. 46.14, chromosome movement toward the poles is accomplished by a combination of **poleward flux**, which moves the body of each microtubule toward one of the poles, and simultaneous **depolymerization** of the microtubule at both ends. Depolymerizing **kinesins** of kinesin-13 family have been localized at both the plus (kinetochore) and minus (polar) ends of the chromosomal microtubules and are postulated to be responsible for depolymerization at their respective sites. 1. In Fig. 46.14A has been depicted model for budding yeast. In it, the chromosome is able to remain associated with the plus end of the microtubule as it depolymerizes by the presence of the **Dam1 ring**, which encircles the plus end of the microtubule at the kinetochore. The force required for chromosome movement is provided by the release of strain energy as the microtubule depolymerizes. The released energy is utilized by the curled ends of the depolymerizing protofilaments to slide the Dam1 ring along the microtubule toward the pole. Dam1 complex is made up of 10 different polypeptides. 2. Model of plus end of microtubule of an animal cell (Fig. 46.14B) shows the Dam1 ring like coupling device, called **Ndc80 protein complex** of the outer kinetochore plate.

Spindle Assembly Checkpoint (SAC)

As elsewhere discussed, cell possess checkpoint mechanisms that monitor the status of events during the cell cycle.One of these chackpoint, called **SAC** or **spindle assembly checkpoint** operates at the transition between metaphase and anaphase. SAC is best revealed when a chromosome fails to become aligned properly at the metaphse plate. When this happens, the checkpoint mechanism delays the onset of anaphase until the misplaced chromosome has assumed its proper position along the spindle equator. If a cell were not able to postpone chromosome segregation, it would greatly elevate the risk of the daughter cells receiving an abnormal number of chromosomes (aneuploidy) which are known to cause in human children cancer and disorders called MVA.

How does the cell determine whether or not all of the chromosomes are properly aligned at the metaphase plate? Unattached kinetochores contain a complex of proteins such as **Mad2** that mediate the spindle assembly checkpoint. The presence of these proteins at an *unattached kinetochore* sends a "wait" signal to the cell cycle machinery that prevents the cell from continuing on into anaphase. Once the wayward chromosome becomes attached to spindle fibers from both spindle poles and becomes properly aligned at the metaphase plate, the signaling complex leave the kinetochore, which turns off the "wait" signal and allows the cell to progress into anaphase.

As long as the cell contains unaligned chromosomes, Mad2 molecules are able to inhibit cell cycle progress. Such inhibition is found to be achieved through direct interaction between Mad2 and the APC activator Cdc20. During the period that Cdc20 is bound to Mad2, APC complexes would be unable to ubiquitinate the anaphase inhibitor securin thus keeping all of the sister chromatids attached to one another by their cohesion "glue".

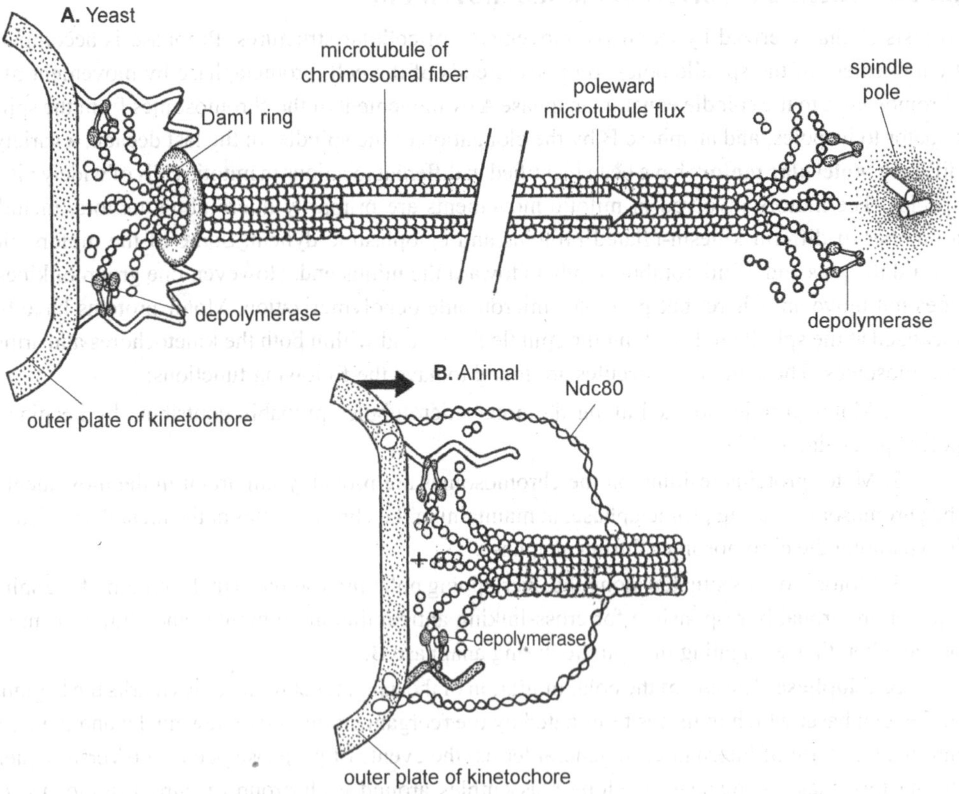

Fig. 46.14. Proposed mechanisms (or models) for the movement of chromosomes during anaphase in budding yeast (A) and animal cells (B).

Syntelic attachment and role of Aurora B kinase. It is well established that the spindle assembly checkpoint is activated by the presence of an unattached kinetochore, but there are other *chromosomal abnormalities* that arise during the progression to metaphase that require corrective measures. For example, on occasion, the two kinetochores of sister chromatids will become attached to microtubules from the same spindle pole, a condition called **syntelic attachment**. If not corrected, a syntelic attachment is very likely to lead to the movement of both sister chromatids to one of the daughter cells, leaving other daughter devoid of this chromosome.

Cells are able to correct syntelic attachments (and other types of abnormal microtubule connections) through the action of an enzyme called **Aurora B kinase**, which is part of a mobile protein complex that resides at the kinetochores during prometaphase and metaphase. Among the substrates of Aurora B kinase are several of the proteins thought to be involved in kinetochore microtubules attachment, including members of both Dam1 complex and the Ndc80 complex and the kinesin depolymerase. Studies suggest that Aurora B kinase molecules of an incorrectly attached chromosome phosphorylate these protein substrates, which destabilizes microtubule attachment to both kinetochores. Once freed of their bonds, the kinetochores of each sister chromatid have a fresh opportunity to become attached to microtubules from opposite spindle poles. Inhibition of Aurora B kinase in cells leads to misalignment and missegregation of chromosomes.

Motor Proteins Involved in Mitotic Movement

Mitosis is characterized by extensive movements of cellular structures. Prophase is accompanied by movement of the spindle poles to opposite ends of the cell, prometaphase by movement of the chromosomes to the spindle equator, anaphase A by movement of the chromosomes from the spindle equator to its poles, and anaphase B by the elongation of the spindle. In the last decade, a variety of different **molecular motors** have been identified in different locations in mitotic cells of widely diverse species. The motors involved in mitotic movements are primarily microtubule motors, including a number of differnt kinesin-related proteins and cytoplasmic dynein. Some of the motors move toward the plus end of microtubules, others toward the minus end. However, one group of kinesins does not move anywhere, but promotes microtubule depolymerization. Motor proteins have been localized at the spindle poles, along the spindle fibers, and within both the kinetochores and arms of chromosomes. These motor molecules are found to have the following functions:

1. Motor proteins located along the polar microtubules probably contribute by keeping the poles apart (Fig. 46.15).

2. Motor proteins residing on the chromosomes are probably important in the movements of the chromosomes during prometaphase, in maintaining the chromosomes at the metaphase plate and in separating the chromosomes during anaphase.

3. Motor proteins situated along the overlapping polar microtubules in the region of the spindle equator are probably responsible for cross-linking antiparallel microtubules and sliding them over one another, thus elongating the spindle during anaphase B.

5. Telophase. The end of the polar migration of the daughter chromosomes marks the beginning of the telophase; which in turn is terminated by the reorganization of two new nuclei and their entry into the G_1 phase of interphase. In general terms, the events of prophase occur in reverse sequence during this phase. A nuclear envelope reassembles around each group of chromosomes to form two daughter nuclei. The mitotic apparatus except the centrioles disappears; high viscosity of the cytoplasm decreases; the chromosomes resume their long, slender, extended form as their coils relax; and RNA- synthesis restarts causing the nucleolus to reappear.

46.6. CYTOKINESIS

Both DNA synthesis and mitosis are coupled to cytoplasmic division, or cytokinesis—the constriction of cytoplasm into two separate cells. During cytokinesis, the cytoplasm divides by a process, called **cleavage**. The mitotic spindle plays an important role in determining where and when cleavage occurs. Cytokinesis usually begins in anaphase and continues through telophase and into interphase. The first sign of cleavage in animal cells is **puckering** and **furrowing** of the plasma membrane during anaphase. The furrowing invariably occurs in the plane of the metaphase plate, at right angles to the long axis of the mitotic spindle. A cleavage furrow tends to form midway between asters originating from two centrosomes in fertilized sand dollar eggs.

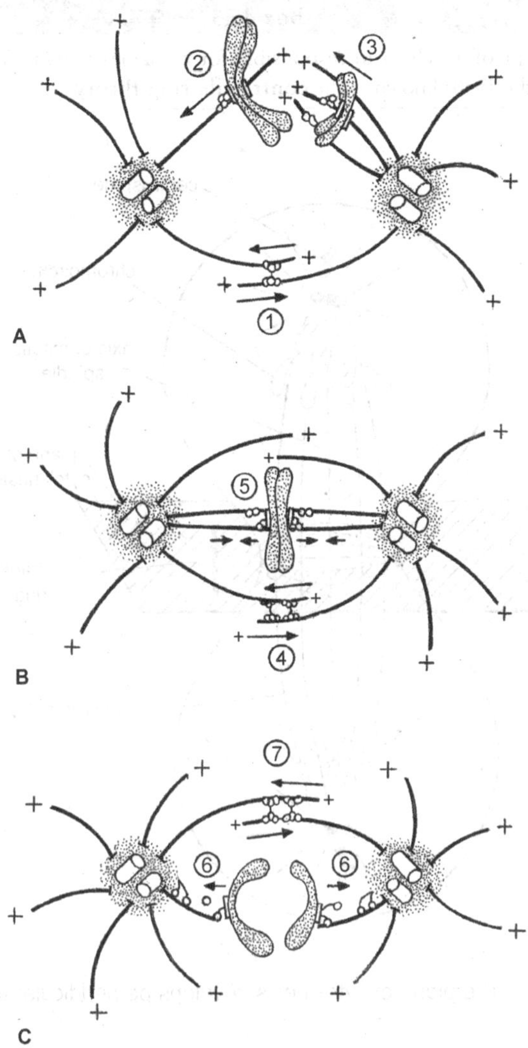

Fig. 46.15. Proposed activity of motor proteins during mitosis. A—Prometaphase. The two halves of the mitotic spindle are moving apart from one another to opposite poles, which is thought to result from the action of plus-end-directed motors that cause polar microtubules from opposite poles to slide relative to one another (step 1). Meanwhile, the chromosomes have become attached to the chromosomal microtubules and can be seen oscillating back and forth along the microtubules. Ultimately, the chromosomes are moved to the center of the spindle, midway between the poles. Poleward chromosome movements are mediated by minus-end-directed motors (*i.e.*, cytoplasmic dynein, residing at the kinetochore (step 2). Chromosome movements away from the poles are mediated by plus-end-directed motors (*i.e.*, kinesein-like proteins) residing at the kinetochore and especially along the chromosome arms (step 3). B—Metaphase. The two halves of the spindle maintain their separation as the result of plus-end-directed motor activity associated with the polar microtubules (step 4). The chromosomes are thought to be maintained at the equatorial plane by the balanced activity of motor proteins residing at the kinetochore (step 5). C—Anaphase. The movement of the chromosomes toward the poles is thought to require the activity of kinesin depolymerases that catalyze depolymerization at both the plus and minus ends of microtubules (step 6). The separation of the poles (anaphase B) is thought to result from the continuing activity of the plus-end-directed motors of the polar microtubules (step 7) (after Karp 2010).

Box 46.3

Our present concept of mechanism responsible for cytokinesis stems from a proposal made by **Douglas Marsland** in the 1950s known as the **contractile ring theory**.

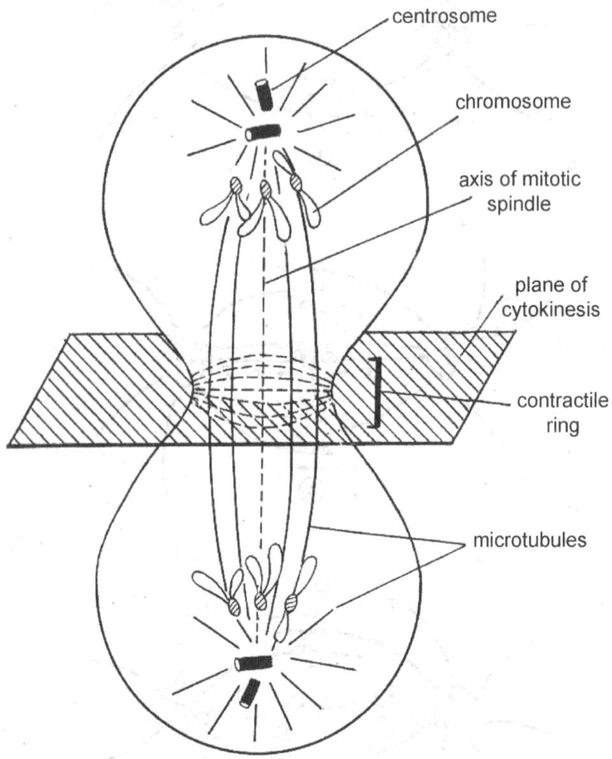

Fig. 46.16. Cytokinesis. The plane of cytokinesis develops perpendicular to the axis of the mitotic spindle.

Cleavage is accomplished by the contraction of a ring composed mainly of actin filaments. This bundle of filaments, called **contractile ring**, is bound to the cytoplasmic face of the plasma membrane by unidentified attachment proteins (Box 46.3). The contractile ring assembles in early anaphase, once assembled, it develops a force large enough to bend a fine glass needle inserted into the cell. Evidently this force is generated due to muscle-like sliding of actin and myosin filaments in the contractile ring. The actin-myosin interaction pulls the plasma membrane down into a furrow. During a normal cytokinesis, the contractile ring does not get thicker as the furrow invaginates, suggesting that it continuously reduces its volume by losing filaments (Fig. 46.17). When cleavage ends, the contractile ring is finally dispensed with altogether and the plasma membrane of the cleavage furrow narrows to form the **midbody**, which remains as a tether between two daughter cells (Tether means a rope for confining a beast within certain limits). The midbody contains the remains of the two sets of polar microtubules, packed tightly together with dense matrix material.

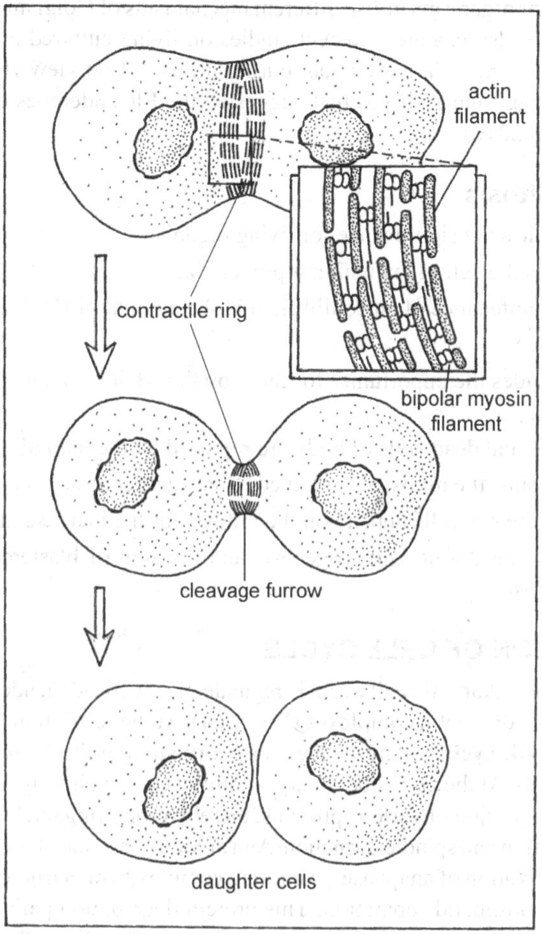

Fig. 46.17. Formation and operation of the contractile ring during cytokinesis. Actin filaments become assembled in a ring at the cell equator. Contraction of the ring requires the action of myosin and causes the formation of a furrow that splits the cell in two (after Karp 2010).

Cytokinesis greatly increases the total cell-surface area as two cells form from one. Therefore, the two daughter cells resulting from cytokinesis require more plasma membrane than in the plant cell.

Partitioning of Cytoplasmic Organelles

Some of the membraneous organelles of the cytoplasm remain relatively intact through mitosis, these include mitochondria, lysosomes, and peroxisomes, as well as the chloroplasts of a plant cell. Considerable debate has been generated in recent years over the mechanism by which the Golgi complex and endoplasmic reticulum are partitioned during mitosis. Accoding to one view, the contents of the Golgi complex become incorporated into the ER during prophase, and the Golgi complex cease to exist briefly as distinct organelle. According to an alternate view, the Golgi membranes become fragmented to form a distinct population of small vesicles that are partitioned between daughter cells. A third view based primarily on studies in algae and protists has the entire Golgi complex splitting in two, with each daughter cell receiving half of the original structure. Ultimately, we may learn that different types of cells or organisms utilise different mechanisms of Golgi inheritance. Our idea about the fate of the ER have also changed. Recent studies on living cultured mammalian cells suggest that the ER network remains relatively intact during mitosis. This view challenges earlier studies performed largely on eggs and embryos that suggested the ER undergoes extensive fragmentation during prophase (**Karp** 2010).

Significance of Mitosis

The mitosis has the following significance for living organisms:

1. The mitosis helps the cell in maintaining proper size.
2. It helps in the maintenance of an equilibrium in the amount of DNA and RNA in the cell.
3. The mitosis provides the opportunity for the growth and development to organs and the body of the organisms.
4. The old decaying and dead cells of body are replaced by the help of mitosis.
5. In certain organisms, the mitosis is involved in asexual reproduction.
6. The gonads and the sex cells depend on the mitosis for the increase in their number.
7. The cleavage of egg during embryogenesis and division of blastema during blastogenesis, both involve mitosis.

46.7. REGULATION OF CELL CYCLE

Progression through the eukaryotic cell cycle is regulated by cyclin-dependent kinases (cdks) which bind to cyclins to form Cdk-cyclin complexes (Box 46.4). At the restriction point (in yeast this point is called start), a **G1-Cdk-cyclin** catalyzes the phosphorylation of the *Rb protein* to trigger passage into S phase (Fig. 46.18). At the G_2–M boundary, a mitotic Cdk-cyclin triggers entry into mitosis by catalyzing the phosphorylation of various proteins, thereby promoting nuclear envelope breakdown, chromosome condensation and spindle formation. And at Metaphase-anaphase boundary, mitotic-Cdk-cyclin contribute to activation of anaphase-promoting complex, which triggers a protein degradation pathway that initiates chromatid separation. This protein degradation pathway also targets mitotic cyclin for breakdown, leading to an inactivation of mitotic Cdk activity that in turn triggers events associated with the exit from mitosis, including chromatin decondensation and assembly of nuclear envelope.

Box 46.4

Mitosis depends upon the activation of a pre-existing protein, the **M-phase kinase**. This protein has two subunits: 1. **Cyclin dependent kinase** (**Cdk**) which is activated by modification at the start of M-phase and 2. **Cyclin** which is accumulated during interphase but is destroyed during mitosis.

Cdk (kinase subunit) bring about *phosphorylation*. It associates with cyclin (regulatory subunit) to form a **Cdk-cyclin complex**. This complex is also known as **cell-cycles engine**. M-phase kinase acts as maturation promoting factor (MPF) to cause somatic cells to enter M-phase. M-phase protein (dimer) can phosphorylate a variety of protein substrates.

Rb protein is a protein whose phosphorylation controls passage through the restriction point in the cell cycle.

Lastly, cell utilises a series of **checkpoint mechanism** to monitor conditions within the cell and transiently halt the cell cycle if conditions are not suitable for continuing. The DNA replication checkpoint monitors the state of DNA replication to ensure that DNA synthesis is completed prior to permitting the cell to exit from G_2 and begin to mitosis. The **p53 protein** plays a central role in a series of DNA damage check-points that halt the cell cycle at various points of DNA damage. Finally, the *anaphase promoting complex* is involved in the spindle checkpoint, which prevents anaphase chromosome movements from the beginning before the chromosomes are attached to the spindle.

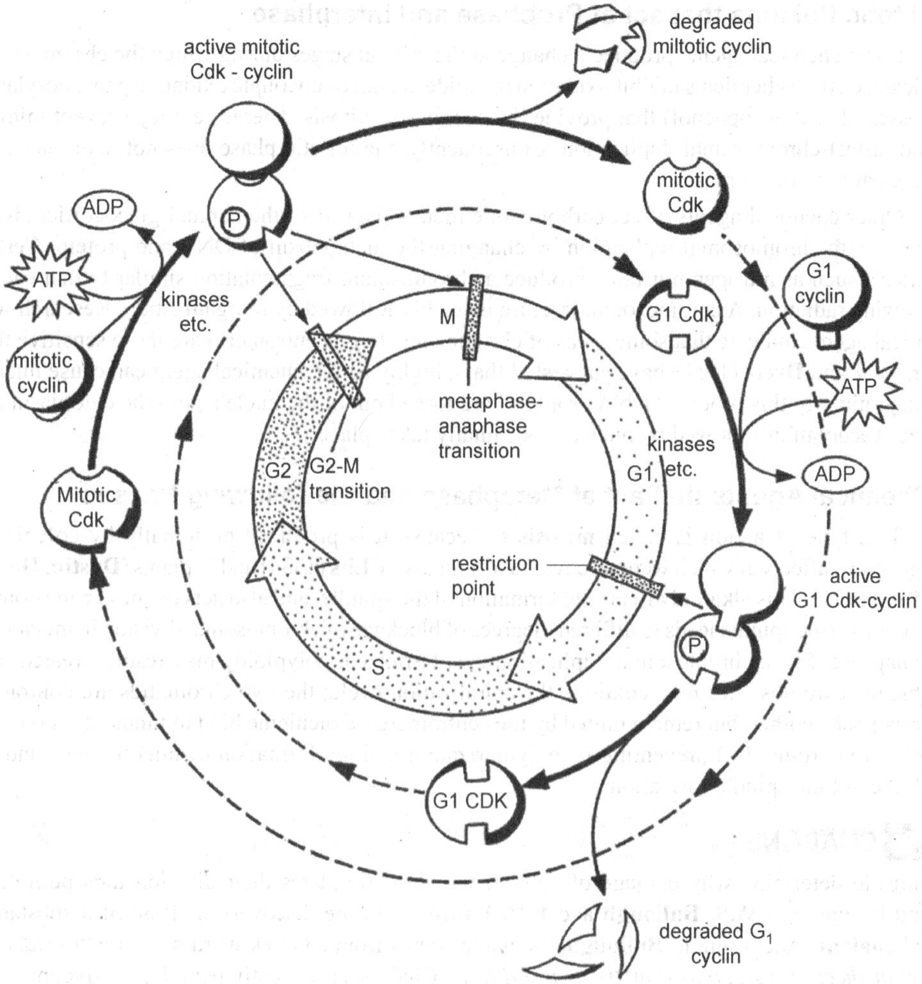

Fig. 46.18. A general model for cell cycle regulation. According to this model, passage through the three main transition points in the cell cycle is triggered by protein complexes made of cyclin and Cdk, whose phosphorylation of other proteins induces progression through the cell cycle (after Becker *et. al.*, 2006).

46.8. MITOTIC POISONS (Mitotic Inhibitors)

Certain chemical compounds act like inhibitors of the mitotic and meiotic cell divisions and they are generally called "**mitotic poisons**" (**Biesele**, 1958). These active agents can be endogenous (naturally occurring in tissues) or exogenous (developed or originating outside the organism). A mitotic poison may inhibit different phases of mitosis or all phases of mitosis with different intensity. One or several mitotic mechanisms may be involved, *e.g.*, nuclear membrane cycle, chromosomal condensation, behaviour of the centromere, formation of the spindle and chromosome movement, chromosome duplication, nuclear cycle and metabolism of nuclear DNA, nuclear and cytoplasmic RNA, proteins and the energy-producing processes for mitosis and meiosis. The mitotic poisons and their effect on cell division can be studied under the following headings:

1. Mitotic Poisons that act at Prophase and Interphase

Some chemical agents produce a change at the critical stages during which the chromosomes duplicate. Certain chemicals inhibit oxidation (cyanide, azide) or uncouple oxidative phosphorylation processes (2, 4-dinitrophenol) that provide the energy for mitosis. Therefore, they prevent mitosis, but not affect chromosomal duplication. Consequently, the mitotic phase does not take place, but the nuclear volume increases.

Other chemical agents affect carbohydrate metabolism (*e.g.*, the adrenal glucocorticoids) or interfere with chromosomal replication by changing the metabolism of DNA and protein. Certain chemicals such as nitrogen mustards produce a chromosomal fragmentation similar to that caused by ionizing radiation. A chromosomal fracture is readily followed by reorganization. Research with chemical agents indicate that some areas of chromosome (*e.g.*, centromere) are more sensitive than other. **Saez** and **Drets** (1958) have suggested that a highly active chemical agent can cause nuclear disintegration. In this process "DNA droplets" are forced out of the nucleus into the cytoplasm and nuclear vacuolation followed by nuclear lysis finally takes place.

2. Chemical Agents that act at Metaphase and the following Phases

This type of action is called **mitosis C** because it is produced principally by colchicine, cytogenetical affects of which were discovered in animals by **Lits** (1934) and in plants (**Dustin**, **Havas**, and **Lits**, 1937). This alkaloid affects the formation of the spindle, and also acts on the chromosomes. The action on the spindle leads to different degrees of blockage of chromosomal division in metaphase and anaphase. Since chromosomal duplication is not affected, polyploidy may result. Treated with colchicine, chromosomes may continue the spiralization cycle; the two chromatids are contracted and repel one another, but remain united by the centromeres. Colchicine bind to tubulin (a monomer protein of microtubules) preventing its polymerization and the formation of microtubules and so, block the mitotic spindle formation.

46.9. CHALONE

Attempt to determine why damage of cells in the skin stimulates their division thus permitting wound healing, led **W.S. Bullough** and **E.B. Laurence** to the discovery in 1960 of a substance, called **chalone**. According to **Bullough**, the name comes from a Greek word meaning "*to slack off the main sheet of a sloop to slow the vessel down*". Chalones are mostly peptides or glycoproteins and share the followmg characteristics:

1. They inhibit mitosis both *in vivo* and *in vitro*.
2. They are tissue-specific, but they are remarkably the same from one species to another.
3. The tissue affected by a particular chalone is also a tissue that makes the same chalone.
4. Mitotic suppression by chalones is reversible, apparently without harm to the cell.

Substances fitting this description have now been isolated from a number of animal tissues. Cells are stimulated to divide when chalone production by nearby cells is compromised, as when tissue is damaged. Inhibition is reestablished when an adequate concentration of healthy, chalone-producing cells is again present (**A. Thornley** and **E. Laurence**, 1975; **J. Houck**, 1976).

The subject of mitotic inhibitors is receiving much attention from medically oriented scientists because of its obvious application to the treatment of allergies and tumours. For example, the adrenal glucocorticoids such as cortisol or cortisone are used to treat severe allergies and inflammations, prevent rejection of transplanted organs and treat lymphocytic leukemia, since, lymphocytes are especially sensitive to these mitotic inhibitors. Malignant cells tend to be much less sensitive to mitotic inhibition by chalones and other natural inhibitors, but more sensitive to drugs that tend to have greater effects on rapidly dividing cells.

46.10. INDUCED DISASSEMBLY OF CYTOSKELETAL MICROTUBULES

Disassembly of cytoskeletal microtubules can be induced in living cell by **cold temperature, hydrostatic pressure, elevated Ca^{2+} concentration** and a variety **of chemicals** including *colchicine, vinblastine, vincristine, nocodazole* and *podophyllotoxin*. The drug **taxol** stops the dynamic activities of microtubule polymer, inhibiting its disassembly, thereby preventing the cell from assembling new microtubule structure as required.

How drugs kill the Cancer cell? Taxol and other drugs (compounds) mentioned here are used in **chemotherapy against cancer** because they preferentially kill tumor cells. It had been assumed for years that tumor cell were particularly sensitive to these drugs beacuse of their high rate of cell division. But recent researches has revealed there is more to the story. Normal cells are found to have a mechanism (or checkpoint) that stops them from dividing in the presence of drug, such as vinblastine or taxol, that alter the mitotic spindle. As a result,normal cells usually arrest their division activities until the drug has been eliminated from the body. In contrast, many cancer cells lack this mitotic checkpoint and attempt to complete their division even in the absence of functional mitotic spindle. This usually result in death of the tumor cell (**Karp** 2010).

46.11. DYNAMISM OF MICROTUBULES IN PLANTS

The microtubules of the cytoskeleton are normally subject to depolymerization and repolymerization as the requirements of the cell change from one time to another. The **dynamic character** of the microtubular cytoskeleton is well illustrated by plant cell. If a typical plant cell is followed from one mitotic division to the next, four distinct arrays of microtubules appear one after another (Fig. 46.19).

1. During most of interphase, the microtubules of a plant cell are distributed widely throughout the cortex (Fig. 46.19; stage 1). A search for γ-tubulin shows this *nucleation factor* to be localized along the lengh of cortical microtubules, suggesting that new microtubules might form directly on the surface of existing microtubules (Fig. 46.20). Once formed, the daughter microtubules are severed from the parent microtubule and incorporated into the parallel bundles that encircle the cell.

2. As the cell approches mitosis, the microtubules disappear from most of the cortex, leaving only a single transverse band, called the **preprophase band**, that encircles the cell like a belt (Fig.46.19, stage 2). The preprophase band marks the site of the future division plane.

3. As the cell progresses into mitosis, the preprophase band is lost and microtubules reappear in the form of the mitotic spindle (Fig. 46.19, stage 3).

4. After the chomosomes have been separated, the mitotic spindle disappears and is replaced by a bundle of microtubules called the **phragmoplast** (Fig. 46.19, stage 4), which plays a role in the formation of the cell wall that separates two daughter cells (Fig. 46.21).

Formation of cell plate during cytokinesis of plants. Cell wall formation starts in the center of the cell and grows outward to meet the existing lateral walls. The formation of a new cell wall begins with the construction of a simple precursor, which is called the **cell plate**.

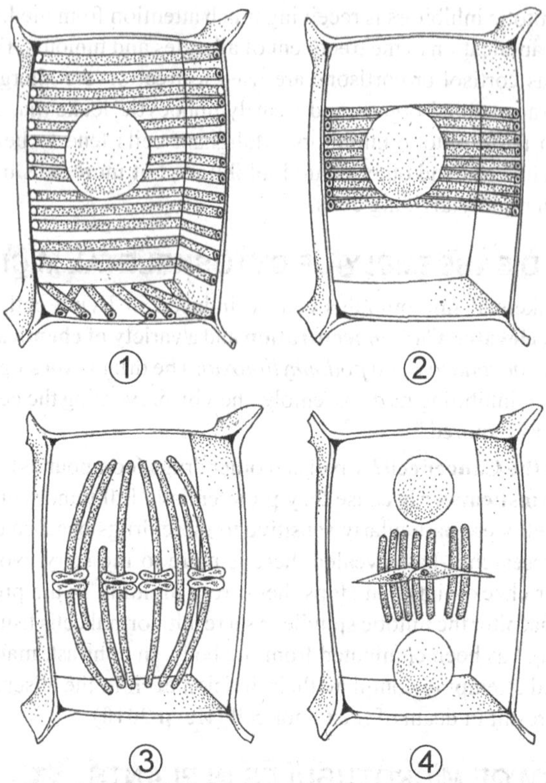

Fig. 46.19. Four major arrays of microtubules present during the cell cycle of plant cell.

The plane in which the cell plate forms is perpendicular to the axis of the mitotic spindle but, unlike the case for animal cells, is not determined by the position of the spindle. Rather, the orientation of both the mitotic spindle and cell plate are determined by a belt of cortical microtubules—*the preprophase* band—that forms on late G_2 (Fig.46.19). Even though the preprophase band has disassembled by premetaphase, it leaves an *invisible imprint* that determines the future division site. The first sign of cell plate formation is seen in late anaphase with the apearance of the **phragmoplast** consists of clusters of the interdigitation microtubules oriented perpendicular to the future plate together with actin filaments, membranous vesicles, and electron-dense material. The microtubules of the phragmoplasts, which arise from remnants of the mitotic spindle serve as tracks for the movement of small Golgi-derived secretory vesicles into the region. The vesicles become aligned along a plane between the daughter nuclei. Fig. 46.21 shows three steps of cytokinesis in tobacco plant cell (**Samuels, Giddings** and **Staehelin** 1995). In step 1, Golgi-derived vesicles send out finger-like tubules that contact and fuse with neighboring tubules to form an interwoven tubular network in the center of the cell (step 2). Additional

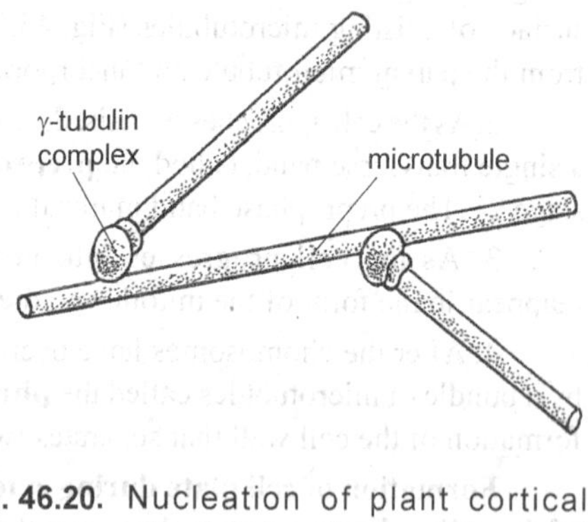

Fig. 46.20. Nucleation of plant cortical microtubules. A schematic model showing how new microtubules are nuleated at the sites of γ-tubulin present on the surface of an existing microtubule.

vesicles are then directed along microtubules to the lateral edges of the network. The newly arived vesicles continue the process of tubule formation and fusion, which extends the network in an outward direction (step 2). Eventually, the leading edge of the growing network contacts the parent plasma membrane at the boundary of the cell (Fig. 46.21, step 3). Ultimately, the tubular network loses its cytoplasmic gaps and matures into a continuous, flattened partition. The membranes of the tubular network become the plasma membrane of the two adjacent daughter cells, whereas the secretory products that had been carried within the vesicles contribute to the intervening cell plate. Once the cell plate is completed, cellulose and other materials are added to produce the mature cell wall.

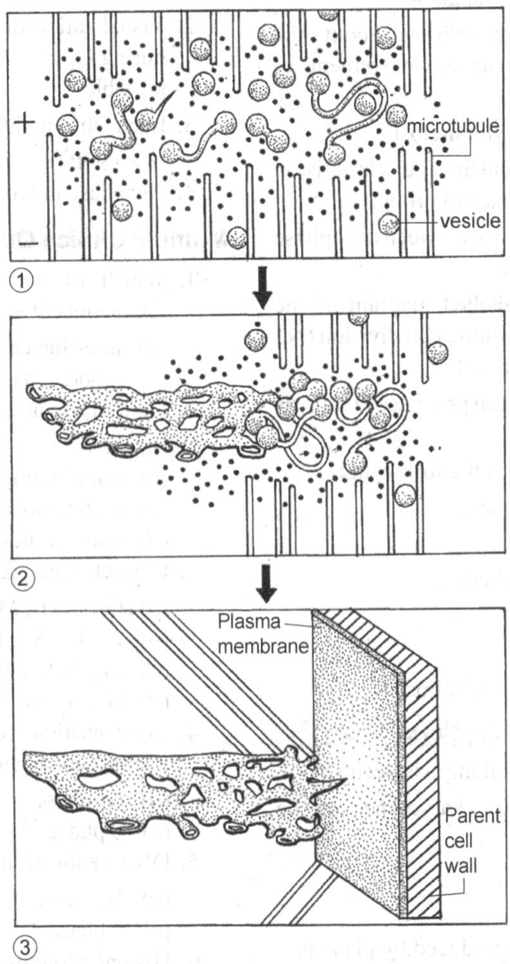

Fig. 46.21. Three steps in the formation of cell plate between two daughter plant nuclei during cytokinesis.

QUESTIONS

Long Answer Questions

1. What do you understand by cell cycle? Give an account of the salient features of various phases of cell cycle.
2. Give an illustrated account of mitosis.
3. Give a detailed account of mitotic cell division and add a note on the significance of mitosis.

4. Give an account of cell growth and cell cycle.
5. Describe mechanics of mitotic spindle.
6. Give an account of cytokinesis in animals and plants.

Short Answer Questions

1. Describe the molecular organization and functions of mitotic apparatus.
2. What are the hypotheses about the mechanism of anaphase movement of chromosomes?
3. Compare mitosis with meiosis.
4. Compare mitosis and first meiotic division with special reference to prophase.
5. Explain the difference between metaphase of mitosis and meiosis.
6. Draw neat and labelled diagram of the various stages of mitotic cell division (No description is required).
7. Differentiate between prophase of mitosis and meiosis.
8. Write short notes on the following:
   (*i*) Mitotic metaphase
   (*ii*) Cytokinesis
   (*iii*) S-phase of cell cycle
   (*iv*) Phragmoplast
   (*v*) Mitogens
   (*vi*) Motor proteins of mitosis.

Very Short Answer Questions

1. In which type of cell mitosis occurs?
2. In which phase the karyotype can be studied?
3. What is cyclin?
4. Define Cdk.
5. Which mitogen is produced by plantets?
6. Name three main types of growth factors.

Fill in the Blanks

1. The division of cytoplasm during cell division is called ....
2. The chromosomal microtubules extend between .... of chromosome and centrosome.
3. Give one word or few words for following sentences:
   Mode of cell division involved in the formation of sperm cells in haploid male bee ....

True and False Statements

1. Polytene chromosomes are in permanent interphase state.
2. Visualization of condensed chromosome during G_1 or G_2 phase of the cell cycle is possible.
3. Each chromatid is composed of single DNA duplex.
4. Anaphase movement is diphasic.

Multiple Choice Questions

1. Which of the following cells do not divide once it is differentiated?
   (*a*) intestine cells (*b*) nerve cells
   (*c*) blood cells (*d*) glial cells
2. Diploid somatic cells divide by
   (*a*) meiosis
   (*b*) mitosis only
   (*c*) both meiosis and mitosis
   (*d*) none of these
3. Correct sequence of stages in cell cycle is
   (*a*) G_1, S, G_2, M
   (*b*) G_1, G_2, S, M
   (*c*) M_1, S, G_1, G_2
   (*d*) G_2, G_1, M, S
4. Condensation of chromosome with visible centromere occurs during
   (*a*) G_1 phase (*b*) G_2 phase
   (*c*) S phase (*d*) M phase
5. DNA-replication takes place during
   (*a*) G_1 phase (*b*) G_2 phase
   (*c*) S phase (*d*) prophase
6. Histone proteins are synthesized in
   (*a*) S phase
   (*b*) G_1 phase
   (*c*) prophase
   (*d*) G_2 phase
7. The phase of cell cycle which lasts for longer duration
   (*a*) G_1 (*b*) G_2
   (*c*) S (*d*) M

8. Cells in Go phase of cell cycle
(*a*) exit cell cycle
(*b*) enter cell cycle
(*c*) suspend cell cycle
(*d*) terminate cell cycle

9. Spindle usually persists in the form of _____ during _____ method of cytokinesis
(*a*) phragmoplast, cleavage
(*b*) phragmoplant, cell plate
(*c*) cell plate, cell plate
(*d*) cell plate, cleavage

ANSWERS

Very Short Answer Questions

1. Somatic cell.
2. Metaphase.
3. Cyclin is any of proteins that activate the cyclin dependent kinases (Cdks) involved in regulating progression through the eukaryotic cell cycle.
4. Cdk stands for cyclin-dependent kinase, it is any of several protein kinases that are activated by different cyclins and that control progression through the eukaryotic cell cycle by phosphorylating various proteins.
5. PDGE (Platelet-derived growth factor).
6. (*i*) PDGF; (*ii*) EGF and (*iii*) TGF-β.

Fill in the Blanks

1. Cytokinesis **2.** Kinetochore **3.** Mitosis

True and False Statements

1. Yes **2.** No **3.** Yes **4.** True

Multiple Choice Questions

1. (*b*) **2.** (*b*) **3.** (*a*) **4.** (*d*) **5.** (*c*) **6.** (*b*) **7.** (*a*)
8. (*c*) **9.** (*b*)

47 Meiosis and Reproductive Cycle

The term meiosis (Gr., *meioum* = to reduce or to diminish) was coined by **J.B. Farmer** in 1905. Meiosis produces a total of four haploid cells from each original diploid cell. These haploid cells either become or give rise to gametes, which through union (fertilization) support sexual reproduction and a new generation of diploid organisms. Thus, meiosis is required to run the **reproductive cycle** of eukaryotes such as microorganisms *Chlamydomonas*, *Neurospora*; bryophytes; plants and animals.

For example, the reproductive cycle of *Chlamydomonas* includes a long haploid generation and a short diploid genetation which involves the zygote formation. The zygote undergoes reduction division (*i.e.*, meiosis) resulting in the formation of haploid spores (Fig. 47.1). In higher plants, however, the reproductive cycle includes a long dominant diploid and multicellular generation (called **sporophyte**) and a short, multicellular haploid generation, called **gametophyte** generation. The tiny gametophyte is nurtured in specialized tissues of sporophyte. Male and female haploid cells called **spores**, are produced by meiosis in the diploid (sporophyte) organism. Spores grow into multicellular male and female haploid (gametophyte) structures, which though meiosis produce haploid cells corresponding to the actual gamete.

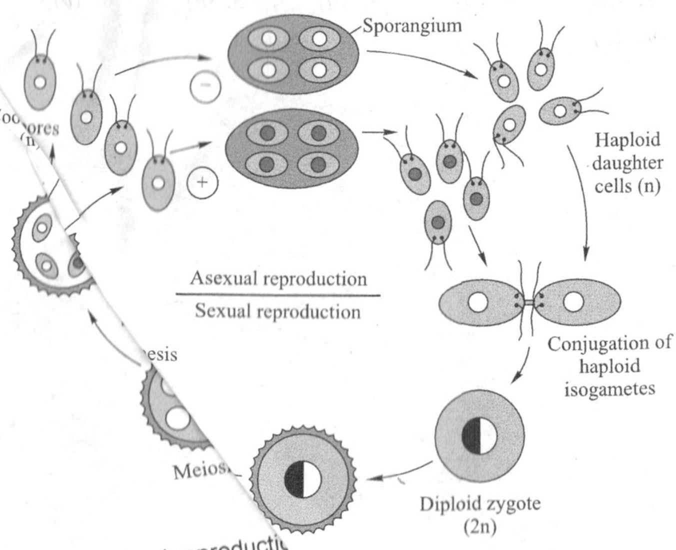

Fig. 47.1. Sexual reproductic ... hlamydomonas sp. (after Burns, 1969).

In both animals and plants, male and female gametes unite during fertilization to produce a **zygote** in which the diploid chromosome number is restored. In animals and simpler plants, the zygote matures to a new diploid organism. In the seed-producing plants, development is arrested at an early multicellular stage as a seed, which may remain stable for long time before germination permits a continuation of growth. Thus, reproductive cycle includes alternation of two generations: haploid and diploid and involves meiosis (Fig. 47.2, Fig. 47.3 and Fig. 47.4).

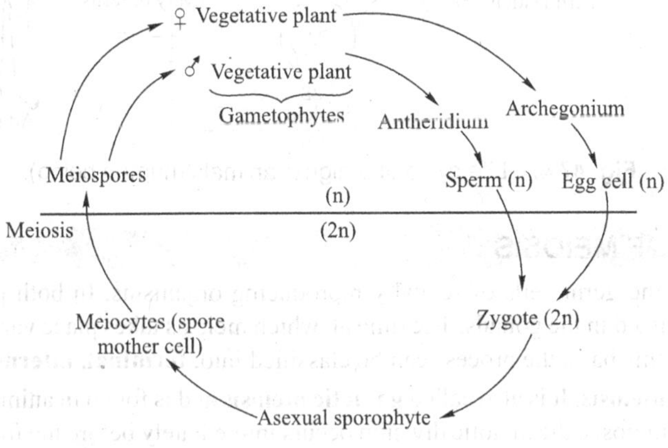

Fig. 47.2. Alternation of generation in Bryophyta (after Burns, 1969).

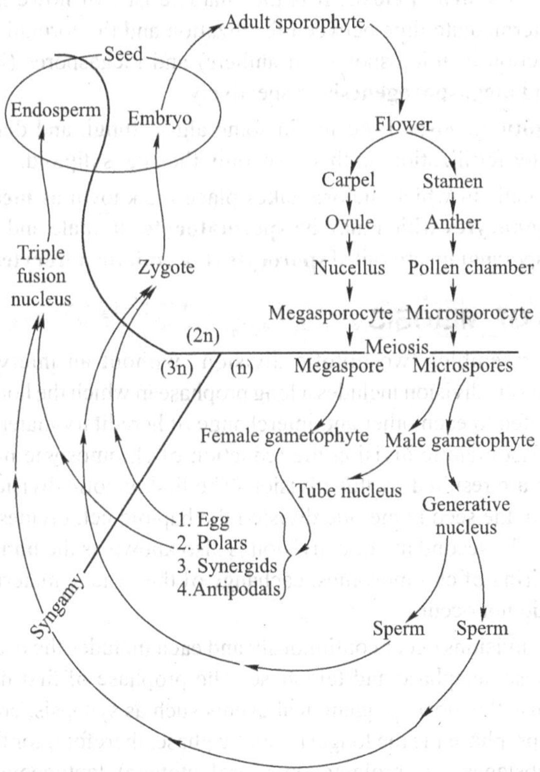

Fig. 47.3. Schematic diagram of the life cycle of an angiosperm (after Burns, 1969).

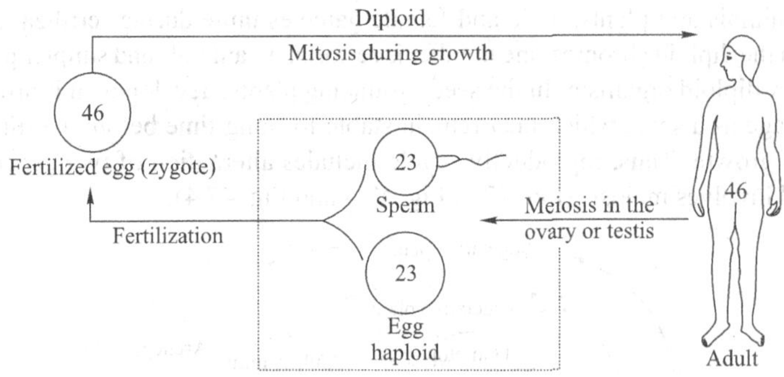

Fig. 47.4. Life cycle of a higher animals (human being).

47.1. KINDS OF MEIOSIS

Meiosis occurs in the germ cells of sexually reproducing organisms. In both plants and animals, germ cells are localized in the gonads. The time at which meiosis takes place varies among different organisms, and on this basis the process can be classified into: **terminal**, **intermediate** or **initial**.

1. Terminal meiosis. It is also called **gametic meiosis** and is found in animals and a few lower plants. In terminal meiosis, the meiotic division occurs immediately before the formation of gametes or **gametogenesis**.

2. Intermediate or **sporic meiosis**. It is the characteristic of flowering plants. This meiosis takes place at some intermediate time between fertilization and the formation of gametes. It is also involved in the production of microspores (in anthers) and megaspores (in ovary or pistil) or in **microsporogenesis** and **megasporogenesis**, respectively.

3. Initial or **zygotic meiosis**. It occurs in some algae, fungi, and diatoms. Meiotic division occurs immediately after fertilization; in this case, only the egg is diploid.

Meiocytes. The cells in which meiosis takes place are known as **meiocytes**. The meiocytes of gonads are called **gonocytes** which may be **spermatocytes** in male and **oocytes** in female. The meiocytes of the plant sporangium are called **sporocytes** (*i.e.*, **microsporocytes** and **megasporocytes**).

47.2. PROCESS OF MEIOSIS

Meiosis superficially resembles two mitotic divisions without an intervening period of DNA replication. The first meiotic division includes a long prophase in which the homologous chromosomes become closely associated to each other and interchange of hereditary material takes place between them. Further, in the frist meiotic division the reduction of chromosome number takes place and, thus, two haploid cells are resulted by this division. The first meiotic division is also known as the **heterotypic division**. In the second meiotic division the haploid cell divides mitotically and results into four haploid cells. The second meiotic division is also known as the **homotypic division.** In the homotypic division pairing of chromosomes, eachange of the genetic material and reduction of the chromosome number do not occur.

Both the meiotic divisions occur continuously and each includes the usual stage of the meiosis, *viz.*, prophase, metaphase, anaphase and telophase. The prophase of first meiotic division is very significant phase because the most cytogenetical events such as synapsis, crossing over, etc., occur during this phase. The prophase I is the longest meiotic phase, therefore, for the sake of convenience it is divided into six substages, *viz.*, proleptonema (proleptotene), leptonema (leptotene), zygonema (zygotene), pachynema (pachytene), diplonema (diplotene) and diakinesis. The successive meiotic substages can be represented as follows:

- **Meiosis**
  - **Division I** or **Heterotypic division**
    - Prophase I–
      - Proleptotene
      - Leptotene
      - Zygotene
      - Pachytene
      - Diplotene
      - Diakinesis
    - Prometaphase I
    - Metaphase I
    - Anaphase I
    - Telophase I
  - **Interphase**
  - **Division II** or **Homotypic division**
    - Prophase II
    - Metaphase II
    - Anaphase II
    - Telophase II

Heterotypic Division or First Meiotic Division

Meiosis starts after an **interphase** which is not very different from that of an intermitotic interphase. During the premeiotic interphase DNA duplication has occurred at the S phase. In the G_2 phase of interphase apparently there is a decisive change that directs the cell toward meiosis, instead of toward mitosis (**Stern** and **Hotta**, 1969). Further, in the beginning of the first meiotic division the nucleus of the meiocyte starts to swell up by absorbing the water from the cytoplasm and the nuclear volume increases about three folds. After these changes the cell passes to the first stage of first meiotic division which is known as prophase.

Prophase I

The first prophase is the longest stage of the meiotic division (Fig. 47.5). It includes following substages:

1. Proleptotene or Proleptonema. (Gr., *pro* = before; *leptas* = thin; *nema* = thread). The proleptotene stage closely resembles with the early mitotic prophase. In this stage the chromosomes are extremely thin, long, uncoiled, longitudianlly single and slender thread-like structures.

2. Leptotene or **Leptonema.** In the leptotene stage the chromosomes become more uncoiled and assume a long thread-like shape. The chromosomes at this stage take up a specific orientation inside the nucleus; the ends of the chromosomes converge toward one side of the nucleus, that side where the centrosome lies (the **bouquet stage**). The centriole duplicates and each daughter centriole migrates towards the opposite poles of the cell. On reaching at the poles, each centriole duplicates and, thus, each pole of cell possesses two centriles of a single diplosome.

On what basis do the homologous chromosomes recognize one another? Studies on yeast cells by **Nancy Kleckner** and her colleagues at Harward University demonstrated that homologous regions of DNA from homologous chromosomes are already associated with one another during leptotene. **Chromosome compaction** and **synapsis** during zygotene simply make this arrangement visible under the microscope. Studies in both yeast and mice suggest the DNA breaks occur in leptotene, well before chromosomes are visibly paired (**Karp** 2010).

Telomere clustering. In yeast, homologous chromosomes are paired to some extent before meiotic prophase begins. The telomeres of leptotene chromosomes are distributed throughout the nucleus. Then near the end of leptotene, there is a dramatic reorganization of chromosomes in many species so that the telomeres become localized at the inner surface of nuclear envelope at one side of the nucleus (Fig. 47.6). Clustering of telomeres at one end of the nuclear envelope occurs in a wide variety of eukaryotic cells (*e.g.*, male grasshoppers) and causes the chromosomes to resemble the clustured stems of a bouquet of flowers.

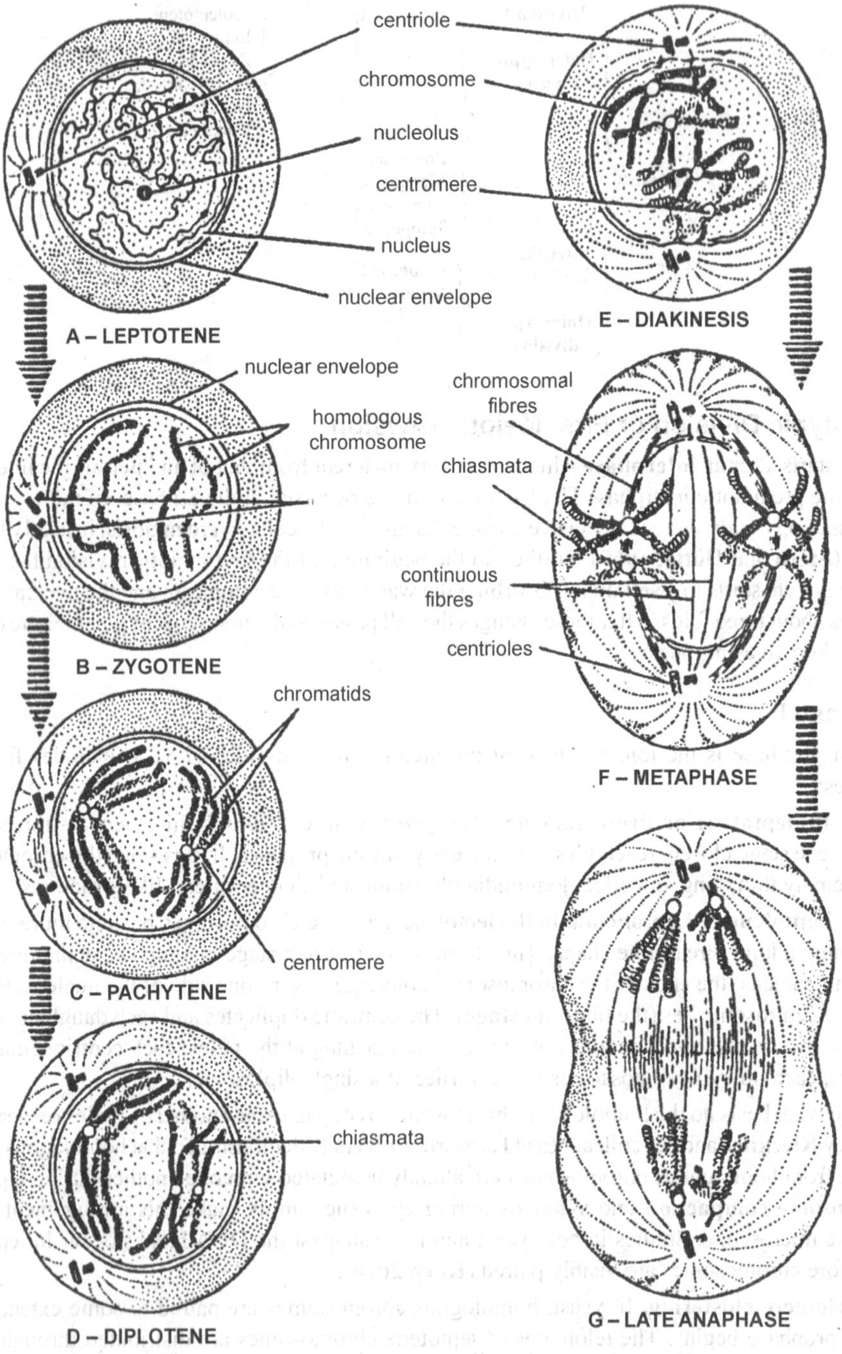

Fig. 47.5. Different stages of first meiotic division.

3. Zygotene or Zygonema. (Gr., *zygon* = adjoining). In the zygotene stage, the pairing of homologous chromosomes takes place. The homologous chromosomes which come from the mother (by ova) and father (by sperm) are attracted towards each other and their pairing takes place. The pairing of the homologous chromosomes is known as synapsis (Gr., *synapsis* = union). The synapsis begins at one or more points along the length of the homologous chromosomes. Three types of synapsis have been recognised.

***(i)* Proterminal synapsis.** In proterminal type of synapsis the pairing in homologous chromosomes starts from the end and continues towards their centromeres.

***(ii)* Procentric synapsis.** In procentric synapsis the homologous chromosomes start pairing from their centromeres and the pairing progresses towards the ends of the homologs.

***(iii)* Localized pairing or random synapsis.** The random type of synapsis occurs at various points of the homologous chromosomes.

The pairing of the homologous chromosomes is very exact and specific (*i.e.*, alignment of chromosomes is exactly gene-for-gene). The paired homologous chromosomes are joined by a roughly 0.2 μm thick, protein-containing framework called a **synaptonemal complex (SC)** (Fig. 47.6). This complex extends along the whole length of the paired chromosomes and is usually anchored at either end to the nuclear envelope.

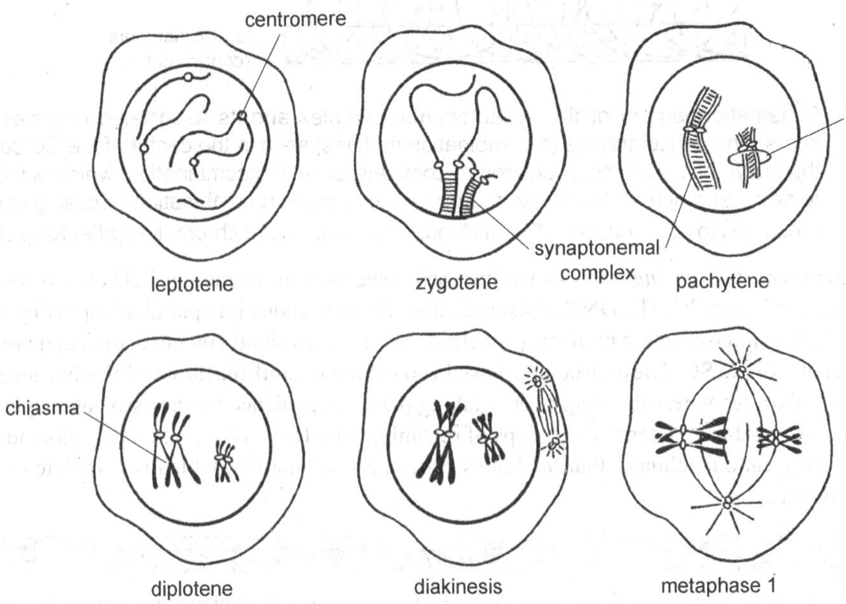

Fig. 47.6. Stages of prophase I.

Synaptonemal complex (SC) is a ladder-like structure with transverse protein filaments connecting the two lateral elements (Fig. 47.7). The chromatin of each homologue is organized into loops that extend from one of the lateral element of the SC. The lateral elements are composed primarily of **cohesin** protein, which presumably binds together the chromatin of sister chromatids.

Function of synaptonemal complex. For many years, the SC was thought to hold each pair of chromosomes in the proper position to initiate genetic recombination between strands of homologous DNA. It is now evident that the SC is not required for genetic recombination (**Karp** 2010). Not only does the SC form after genetic recombination has been initiated, but mutant yeast cells unable to assemble at SC can still engage in the exchange of genetic information between homologues. It is currently thought that the SC functions primarily as a *scaffold* to allow interacting chromatids to complete their crossover.

The complex formed by a pair of synapsed homologous chromosomes is called a **dyad** or **bivalent** or a **tetrad** or **quadrivalent**. The former term (*i.e.*, bivalent) indicates that the complex contains two homologues, whereas the latter term (*i.e.*, tetrad) explains the presence of four chromatids in the complex. The end of synapsis marks the end of zygotene and the beginning of the next stage of prophase I, called *pachytene* which is characterized by a fully formed synaptonemal complex.

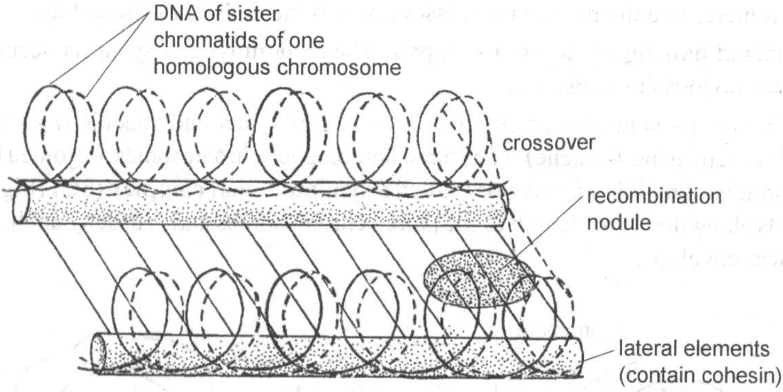

Fig. 47.7. Schematic diagram of the synaptonemal complex and its associated chromosomal fibers. The dense granule (recombination nodules) seen in the center of the SC contain the enzymatic machinery required to complete genetic recombination, which is thought to being at much earlier stage in prophase I. Genetic recombination (crossing over) is presumed to occur between the DNA loops from non–sister chromatids (after Karp 2010).

4. Pachytene (Gr., *pachus* = thick). During pachytene, the homologues are held closely together along their length by the SC. The DNA of sister chromatids is extended into parallel loops (Fig. 47.7). Under the electron microscope, a number of electron-dense bodies about 100 nm in diameter are seen within the center of the SC. These structures have been named **recombination nodules** because they correspond to the sites where crossing over is taking place, as evidence by the associated synthesis of DNA that occurs during intermediate steps of recombination (Box 47.1). Recombination nodules contain the enzymatic machinery that facilitates genetic recombination, which is completed by the end of pachytene.

Box 47.1

During pachytene stage an important genetic phenomenon called **crossing over** takes place. The crossing over involves reshuffling, redistribution and mutual exchange of hereditary material of two parents between two homologous chromosomes. According to recent views, one chromatid of each homologous chromosome of a bivalent may divide transversely by the help of an enzyme the **endonuclease** which is reported to increase in the nucleus during this stage by **Stern** and **Hotta** (1969). After the division of chromatids, the interchange of chromatid segments takes place between the non-sister chromatids of the homologous chromosomes. The broken chromatid segments are united with the chromatids due to the presence of an enzyme, **ligase** (**Stern** and **Hotta**, 1969). The process of interchange of chromatin material between one non-sister chromatid of each homologous chromosome is known as the **crossing over** which is accompanied by the **chiasmata formation**.

Stern and **Hotta** (1969) have reported that during the pachytene and zygotene stage, sythesis of small amount of DNA take place. This DNA amount is utilized in the repairing of broken DNA molecule of the chromatids during the chiasmata formation and crossing over.

The nucleolus remain prominent up to this stage and it is found to be associated with the nucleolar organizer region of the chromosome.

5. Diplotene. This stage of prophase I is recognized by the dissolution of the SC which leaves the chromosomes attached to one another at specific points by X-shaped structures, termed **chiasmata** (singular **chiasma**) (Gr. *chiasma* =cross piece). Chiasmata are located at sites on the chromosomes where crossing-over between DNA molecules from the two chromosomes had previously occurred (Fig. 47.8). Chiasmata are formed by covalent junctions between a chromatid from one homologue and a non-sister chromatid from the other homologue. These points of attachment provide a striking visual presentation of the extent of genetic recombination. The chiasmata are made more visible by the tendency for the homologues to separate from one another at the diplotene stage.

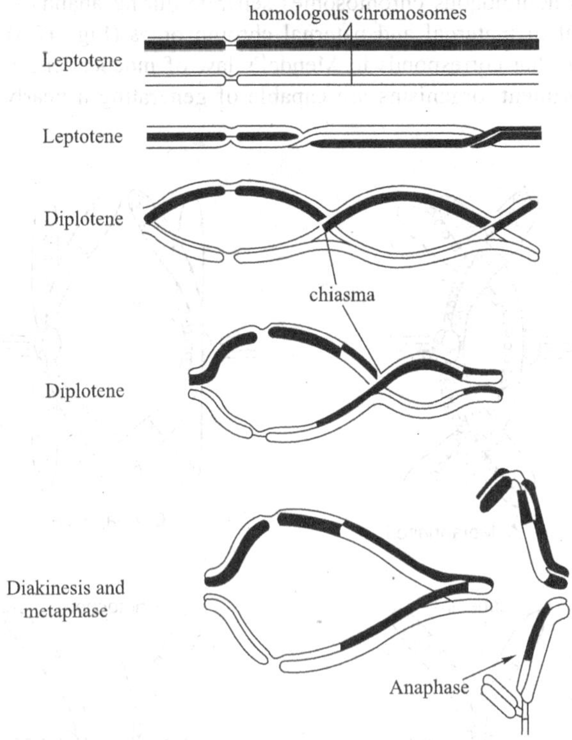

Fig. 47.8. Diagram of chromosomal exchange in the four-strand stage and of terminalization during first meiotic division.

In vertebrates (*e.g.*, frog), diplotene can be an extremely extended phase of oogenesis during which the bulk of oocyte growth occurs. Thus, diplotene can be a period of intense metabolic activity. **Transcription** during diplotene in the oocyte provides the RNA utilized for protein synthesis during both oogenesis and early embryonic development (cleavage and blastulation) following fertilization.

6. Diakinesis. During final stage of meiotic prophase I, called **diakinesis**, the meiotic spindle is assembled and the chromosomes are prepared for separation. In those species (*i.e.*, vertebrates), in which the chromosomes become highly dispersed during diplotene, the chromosomes become recompacted during diakinesis.

The nucleolus detaches from the nucleolar organizer and ultimately disappears. The nuclear envelope breaks down. During diakinesis the chiasma moves from the centromere towards the end of the chromosomes and the intermediate chiasmata diminish. This type of movement of the chiasmata is known as **terminalization**. The chromatids still remain connected by the terminal chiasmata and these exist up to the metaphase. The tetrads move to the metaphase plate.

In vertebrate oocytes, these events are trigered by an increase in the level of protein kinase activity of MPF (maturation protein factor).

METAPHASE I

At metaphase I, the two homologous chromosomes of each bivalent are connected to the spindle fibers from opposite poles (Fig. 47.9A). In contrast, sister chromatids are connected to microtubules from the same spindle pole, which is made possible by the side-by-side arrangement of their kinetochores (Lee *et al.*, 2000). The orientation of the maternal and paternal chromosomes of each bivalent member of a particular bivalent has an equal likelihood of facing either pole. Consequently, when homologous chromosomes separate during anaphase I, each pole receives a random assortment of maternal and paternal chromosomes (Fig. 47.5). Thus, anaphase I is the cytological event that corresponds to Mendel's law of independent assortment. As a result of independent assortment, organisms are capable of generating a nearly unlimited variety of gametes.

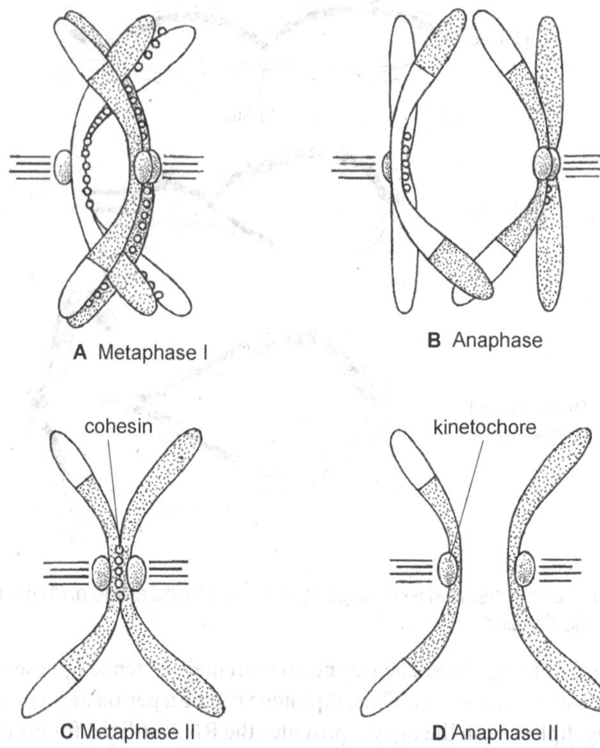

Fig. 47.9. Separation of homologous chromosomes during meiosis I and separation of chromatids during meiosis II. A—Schematic diagram of a pair of homologous chromosomes at metaphase I. The chromatids are held together along both their arms and centromeres by cohesin. The pair of homologous are maintained as a bivalent by the chiasmata. B—At anaphase I, the cohesin holding the arms of the chromatids is cleaved, allowing the homologues to separate from one another. Cohesin remains at the centromere, holding the chromatids together. C—At metaphase II, the chromatids are held together at the centromere, with microtubules from opposite poles attached to the two kinetochores. D—At anaphase II, the cohesin holding the chromatids together has been cleaved, allowing the chromosomes to move to opposite poles (after Karp 2010).

ANAPHASE I

Separation of homologous chromosomes at anaphase I requires dissolution of the chiasmata that hold the bivalents together. The chiasmata are maintained by cohesion between sister chromatids in region that flank these sites of recombination (Fig. 47.9A). The chiasmata disappear at the metaphase I—anaphase I transition, as the arms of the chromatids of each bivalent lose cohesion (Fig. 47.9B). Loss of cohesion between arms is accomplished by proteolytic cleavage of the **cohesion** molecules in those regions of the chromosome. In contrast, cohesin between the joined centromeres of sister chromatids remain strong, because the cohesin situated there is protected from proteolytic attack (Fig. 47.9B). As a result, sister chromatids remain firmly attached to one another as they move together toward a spindle pole during anaphase I. Thus, actual reduction and disjunction of chromosomes occur at anaphase I.

TELOPHASE I

Telophase I of meiosis I produces less dramatic changes than telophase of mitosis. Although chromosomes often undergo some dispersion, they do not reach the extremely extended state of the interphase nucleus. The nuclear envelope may or may not reform during telophase.

After karyokinesis, cytokinesis occurs and two haploid cells are formed.

Interkinesis

The stage between the two meiotic divisions (*i.e.*, meiosis I and meiosis II) is called **interkinesis** and is generally short-lived. In animals, cells in this fleeting stage are referred to as the **secondary spermatocytes** or **secondary oocytes**. These cells are characterized as being haploid because they contain only one member of each pair of homologous chromosomes (Box 47.2).

Box 47.2

Even though products of meiosis I are haploid, they have twice as much DNA as a haploid gamete because each chromosome is still represented by a pair of attached chromatids. Secondary spermatocytes are said to have a 2C amount of DNA, half as much as a primary spermatocyte which has a 4C DNA content, and twice as much as a sperm cell, which has a 1C DNA content.

Homotypic or Second Meiotic Division

The homotypic or second meiotic division is actually division which divides each haploid meiotic cell into two haploid cells. The second meiotic division includes following four stages.

Prophase II

In the prophase second, each centriole divides into two and, thus, two pairs of centrioles are formed. Each pair of centrioles migrates to the opposite pole. The microtubules get arranged in the form of spindle at the right angle of the spindle of first meiosis. The nuclear membrane and the nucleolus disappear. The chromosomes with two chromatids become short and thick.

Metaphase II

During metaphase II, the chromosomes get arranged on the equator of the spindle. The centromere divides into two and, thus, each chromosome produces two monads or daughter chromosomes. The microtubules of the spindle are attached with the centromere of the chromosomes.

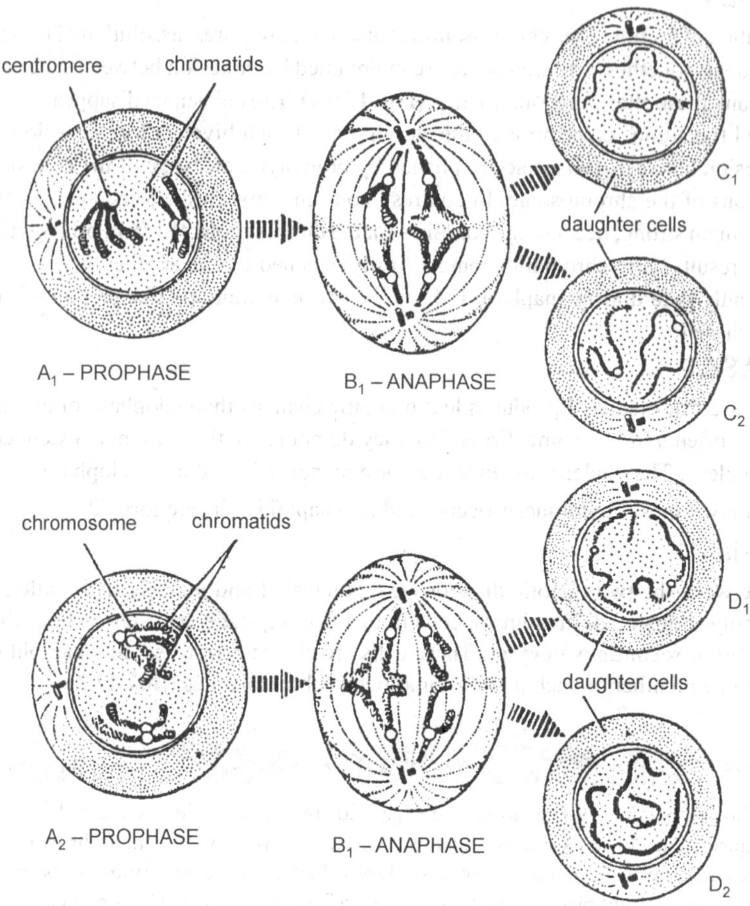

Fig. 47.10. Different stages of second meiotic division.

Anaphase II

The daughter chromosomes move towards the opposite poles due to the shortening of chromosomal microtubules and stretching of interzonal (polar) microtubules of the spindle.

Telophase II

The chromatids migrate to the opposite poles and now known as chromosomes. The endoplasmic reticulum forms the nuclear envelope around the chromosomes and the nucleolus reapears due to synthesis of ribosomal RNA (rRNA) by rDNA and also due to accumulation of ribosomal proteins.

After the karyokinesis, in each haploid meiotic cell, the cytokinesis occurs and, thus, four haploid cells are resulted. These cells have different types of chromosomes due to the crossing over in the prophase I.

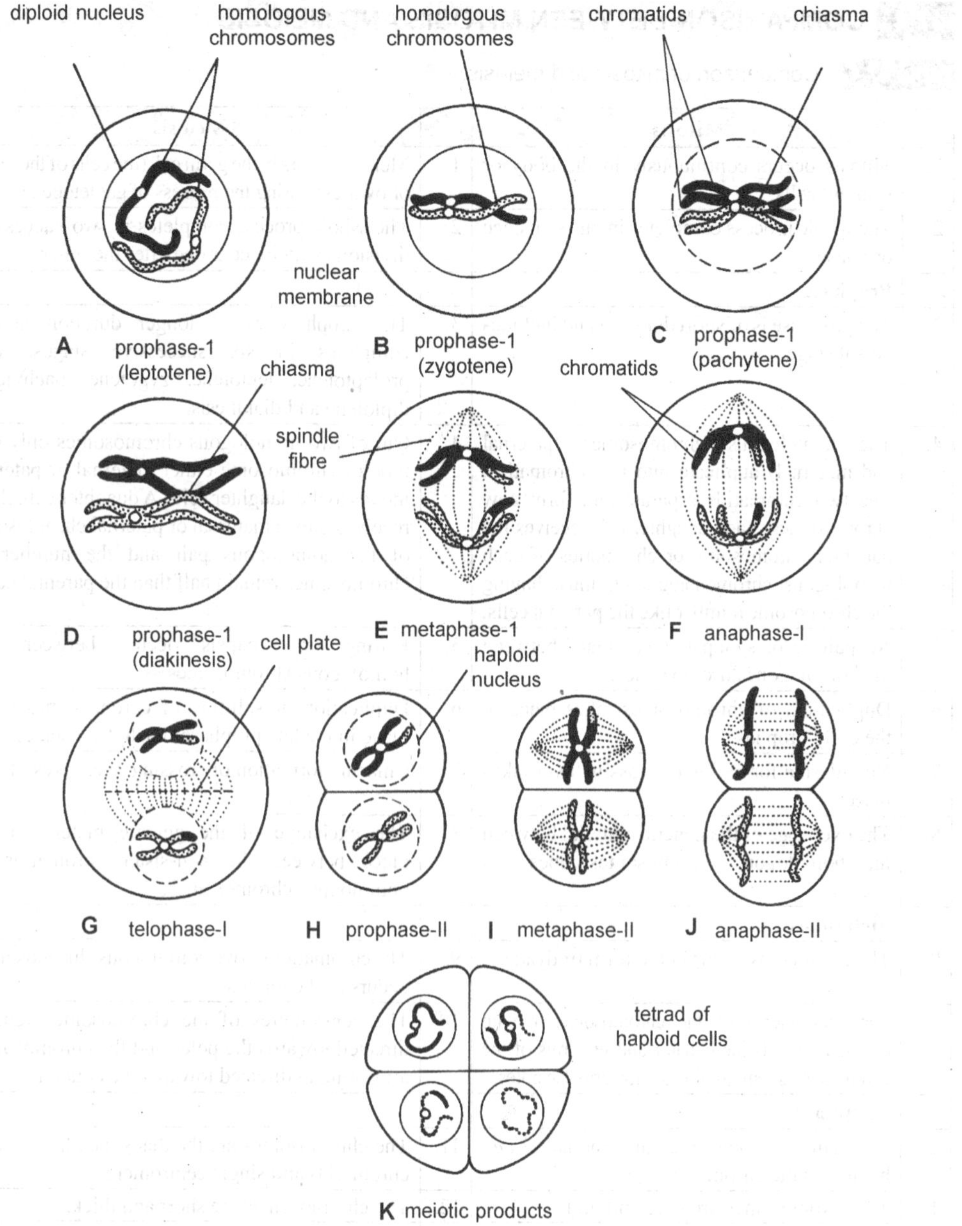

Fig. 47.11. Diagrammatic representation of meiosis in a plant cell showing one pair of homologous chromosomes (2n = 2).

47.3. SIGNIFICANCE OF MEIOSIS

The meiosis has the greatest significance for the biological world because of its following uses:

1. The meiosis maintains a definite and constant number of the chromosomes in the organisms.
2. By crossing over, the meiosis provides an opportunity for the exchange of the genes and, thus, causes the genetical variations among the species. The variations are the raw materials of the evolutionary process.

   Thus the meiosis is a peculiar taxonomic, genetical and evolutionary process.

47.4. COMPARISON BETWEEN MITOSIS AND MEIOSIS

Table 47.1 Comparison of mitosis and meiosis.

	Mitosis		Meiosis
1.	Mitosis occurs continuously in the body or somatic cells.	1.	Meiosis occurs in the germ cell (the cells of the testes or ovaries) during the process of gametogenesis.
2.	The whole process completes in one sequence or phase.	2.	The whole process completes in two successive divisions which occur one after the other.
	Prophase		
3.	The prophase is of short duration and includes no substage.	3.	The prophase is of longer duration and it completes in six successive stages, *viz.*, proleptotene, leptotene, zygotene, pachytene, diplotene and diakinesis.
4.	The homologous chromosomes (paternal and maternal) duplicate into two chromatids. The two chromatids separate and form new chromosomes. Each daughter cell receives the daughter chromosome or chromatids of each homologous chromosome and, thus, having the chromosome number like the parental cells.	4.	Out of two homologous chromosomes only one type of chromosome either maternal or paternal moves to the daughter cells. A duaghter cell, thus, receives only a maternal or paternal chromosome of the homologous pair and the number of chromosomes remain half than the parental cells.
5.	No pairing or synapsis takes place between the homologous chromosomes.	5.	Pairing or synapsis occurs between the homologous chromosomes.
6.	Duplication of chromosomes takes place in the early prophase.	6.	Duplication or splitting of chromosomes takes place in the late prophase (pachytene stage).
7.	No chiasma formation or crossing over takes place.	7.	Chiasma formation or crossing over takes place.
8.	The exchange of the genetic material between the homologous chromosomes does not occur.	8.	The exchange of the genetic material takes place between the non-sister chromatids of homologous chromosomes.
	Metaphase		
9.	The chromatids occur in the form of dyads.	9.	The chromatids of two homologous chromosomes occurs as the tetrads.
10.	The centromeres of the chromosomes remain directed towards the equator and the arms of the chromosomes remain directed towards the poles.	10.	The centromeres of the chromosomes remain directed towards the poles and the chromosomal arms remain directed towards the equator.
	Anaphase		
11.	The chromosomes are the monads, *i.e.*, having single chromatid.	11.	The chromosomes are the diads, *i.e.*, having two chromatids and single centromere.
12.	The chromosomes are long and thin.	12.	The chromosomes are short and thick.
	Telophase		
13.	The telophase always occurs.	13.	The fist telophase is sometimes omitted.
	Significance		
14.	The chromosome number in each daughter cell remains the same like the parent cell.	14.	In meiotic division the chromosome number is reduced to half in the daugher cells than the parental cells.
15.	A diploid cell produces two diploid cells by a mitotic division.	15.	A diploid cell produces four haploid cells by a meiotic division.

QUESTIONS

Long Answer Qeustions

1. Describe the meiosis.
2. Describe meiosis in detail and explain its significance.
3. Describe meiotic cell division and state the significance of crossing over.
4. Describe the prophase of meiosis. Give its significance.
5. What do you mean by cell division? Describe the different stages of meiotic cell division and its significance.
6. Explain the process of meiosis only by various diagrams.

Short Answer Questions

1. Write short note the significance of meiosis.
2. What are the main differences between mitosis and meiosis?
3. Differentiate between karyokinesis and cytokinesis.
4. Explain the meiotic cell division and its significance.
5. Write short notes on the following:
   (*i*) Leptotene
   (*ii*) Zygotene
   (*iii*) Pachytene
   (*iv*) Diplotene
   (*v*) Diakinesis
   (*vi*) Anaphase of meiosis
6. Draw neat and labelled diagram of the various stages of meiotic cell division (No description is required).

Very Short Answer Questions

1. When does synapsis occur during meiosis?
2. In which phase of meiosis chromosome number reduces to half?

Multiple Choice Questions

1. The term 'Meiosis' was coined by
   (*a*) Knoll and Ruska
   (*b*) Farmer and Moore
   (*c*) Beadle and Tatum
   (*d*) A. Flemming
2. Meiosis occurs in
   (*a*) liver
   (*b*) kidneys
   (*c*) gonads
   (*d*) brain
3. Pairing of homologous chromosomes occur during
   (*a*) leptotene
   (*b*) zygotene
   (*c*) pachytene
   (*d*) diplotene
4. Crossing-over occurs during
   (*a*) leptotene
   (*b*) pachytene
   (*c*) diakinesis
   (*d*) diplotene
5. Interkinesis is the resting stage between
   (*a*) two mitotic divisions
   (*b*) two meiotic divisions
   (*c*) mitotic and meiotic divisions
   (*d*) none of these
6. The structural basis for pairing of homologous chromosomes during prophase I.
   (*a*) kinetochore
   (*b*) synaptonemal complex
   (*c*) chromomere
   (*d*) centromere
7. The spindle microtubules attach to the
   (*a*) telomere
   (*b*) nucleosome
   (*c*) kinetochore
   (*d*) none
8. Synapsis between homologous chromosome occurs in
   (*a*) leptotene stage
   (*b*) pachytene
   (*c*) zygotene
   (*d*) diplotene

9. Chromosomes become tetravalent during
   (*a*) leptotene
   (*b*) zygotene
   (*c*) pachytene
   (*d*) diplotene
10. In which phase of meiosis homologous chromosomes separate?
   (*a*) metaphase I
   (*b*) metaphase II
   (*c*) anaphase I
   (*d*) anaphase II
11. The protein/factor which is involved in the separation of homologous chromosomes during meiosis I is
   (*a*) histone
   (*b*) cohesin
   (*c*) ATP
   (*d*) none of these
12. Interkinesis is followed by
   (*a*) prophase II
   (*b*) prophase I
   (*c*) interphase
   (*d*) telophase II
13. Best material to study meiosis is
   (*a*) root tip
   (*b*) ovary
   (*c*) young anther
   (*d*) pollen grain

ANSWERS

Very Short Answer Questions

1. Zygotene
2. Anaphase I.

Multiple Choice Questions

1. (*b*)	2. (*c*)	3. (*b*)	4. (*b*)	5. (*b*)	6. (*b*)	7. (*c*)
8. (*c*)	9. (*c*)	10. (*c*)	11. (*b*)	12. (*a*)	13. (*c*)	

48 Reproduction

This is the inherent property of the living organisms to continue their race by the mechanism of reproduction. The reproduction is a process by which the living beings propagate or duplicate their own kinds. The reproduction may be of following two types:

1. Asexual reproduction; 2. Sexual reproduction.

The word 'reproduction' implies replication, and it is true that biologic reproduction almost always yields a reasonable facsimile of the parent unit. However, sexual reproduction, performed by the majority of living organisms, produces the diversity which is required for survival in a world of constant change. The process, whether sexual or asexual, comprises a basic pattern: (1) the conversion of raw materials from the environment into the offspring, or sex cells that develop into offspring of a similar constitution, and (2) the transmission of a hereditary pattern or code (DNA of the genes) from the parent.

48.1. ASEXUAL REPRODUCTION

The development of new individuals without the fusion of the male and female gametes is known as **asexual reproduction**. The asexual reproduction usually includes amitotic or mitotic division of the body (somatic) cells, therefore, it is also known as **somatogenic** or **blastogenic reproduction**. The asexual reproduction is common in lower amimals and it may be of following types:

1. Fission; 2. Budding; 3. Gemmule formation; and 4. Regeneration.

1. Asexual reproduction by the fission. The fission is the most widely occurring type of asexual reproduction of the protozoans and various metazoans. In this method of reproduction the nuclear and cytoplasmic contents of the cell divide or split completely into smaller-sized daughter individuals. The fission itself may be of following types:

A. Binary fission. In the binary fission the animal body splits or divides in such a plane that two equal and identical halves are produced, It is most common in protozoans but it also occurs in certain lower metazoans. First of all the nucleus divides by amitotic or mitotic division and the division of the nucleus is followed by the division of the cytoplasm. According to the plane of fission following types of binary fission have been recognized in the organisms:

(*i*) Simple or orthodox type of binary fission. The simple or orthodox type of binary fission occurs in the irregular-shaped organims, *e.g.*, *Amoeba* in which the plane of division is difficult to observe.

(*ii*) Transverse binary fission. The transverse binary fission occurs in some protozoans, *e.g.*, *Paramecium* (Fig. 48.1) and some metazoans such as certain coelenterates, turbellarians and annelids.

In transverse binary fission the plane of the division is always transverse to the longitudinal axis of the body of the organisms.

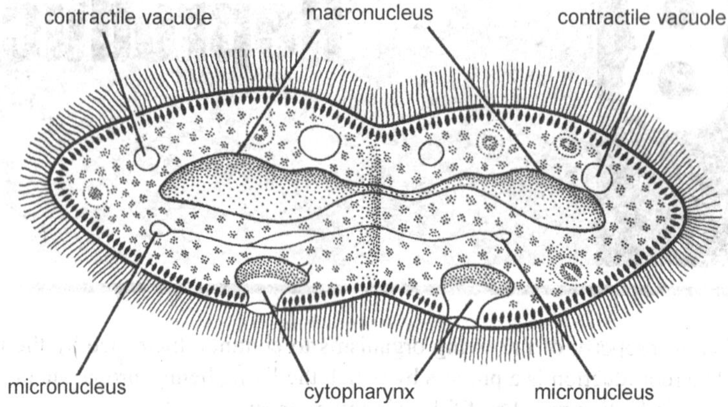

Fig. 48.1. Transverse binary fission in *Paramecium*.

***(iii)* Longitudinal binary fission.** The longitudinal binary fission occurs in certain ciliates and flagellates, *e.g.*, *Vorticella* and *Euglena* (Protozoa; Fig. 48.2) and some corals (Anthozoa). In longitudinal binary fission the nucleus and the cytoplasm divide in the longitudinal plane.

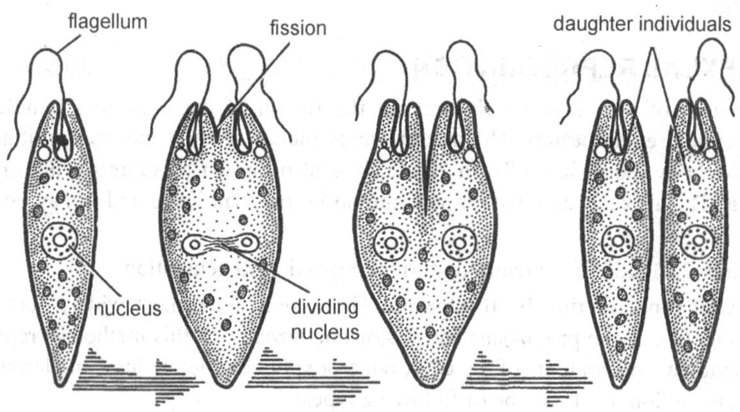

Fig. 48.2. Longitudinal binary fission in *Euglena*.

***(iv)* Oblique binary fission.** The oblique binary fission occurs in most dinoflagellates. In this type of fission the cell or body of the organism divides by the oblique division.

***(v)* Strobilation.** In certain metazoan animals a special type of transverse fission known as the strobilation occurs. In the process of the strobilation several transverse fissions occur simultaneously and giving rise to a number of individuals which often do not separate from each other immediately. The strobilation occurs in the scyphozoan (*Aurelia*; Fig. 48.3), certain polychaets and ascidians. In *Aurelia*, for instance, strobilation occurs during the formation of ephyra larva.

B. Asexual reproduction by multiple fission. In the multiple fission, the nucleus of the cell divides very rapidly into many nuclei. Each daughter nucleus in later stage is surrounded by the little mass of the cytoplasm and forms the asexually reproducing body such as schizodont, gamont, spore, etc. The multiple fission occurs in most algae, fungi and some protozoans, *e.g.*, *Amoeba*, *Plasmodium* and *Monocystis*, etc.

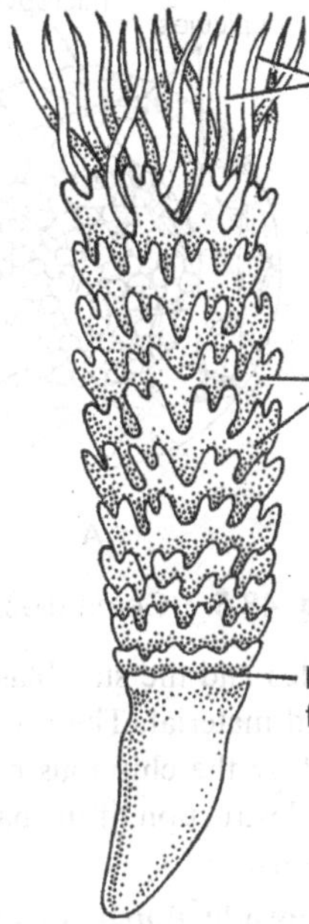

Fig. 48.3. Strobilated *Aurelia*.

2. Asexual reproduction by budding or gemmation. In certain multicellular animals such as *Hydra* (coelenterates) and certain tunicates, the body gives out a small outgrowth known as the **bud** (Fig. 48.4). The bud is supported by parent body and it ultimately develops into a new individual. The process of development of a bud into an adult animal is called **blastogenesis**. The developing individual gets its food from the body of the parent and when it becomes fully mature it is detached from the body of the parent and leads an individual existence.

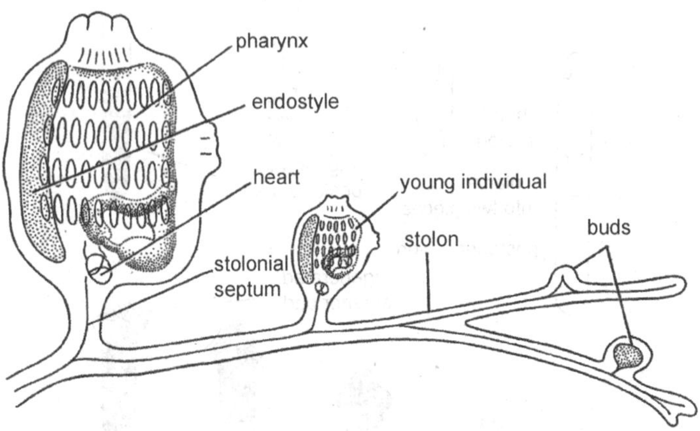

Fig. 48.4. Budding in a tunicate.

3. Asexual reproduction by gemmule formation. In certain metazoan animals the asexual reproduction is carried on by certain peculiar asexual bodies known as the **gemmules** (Fig. 48.5) and **statoblasts**. The gemmules occur in freshwater sponges (family Spongilidae) and the statoblasts occur in the bryozoans.

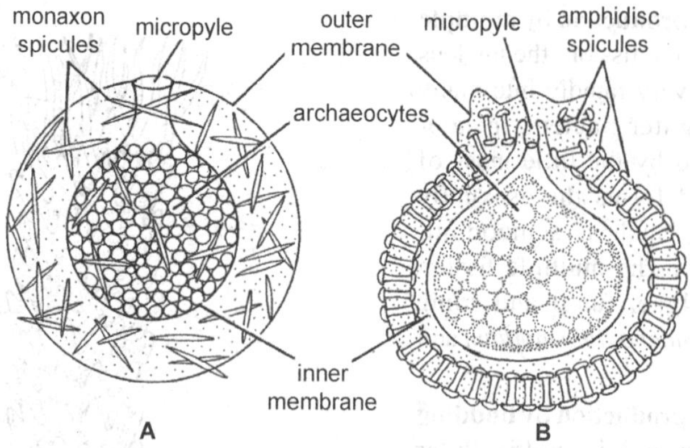

Fig. 48.5. Gemmules. A—*Spongilla* and B—*Ephydatia* (in section).

The gemmules and the statoblasts are composed of a group of undifferentiated cells which contain stored food material. These cells are enclosed and protected by the monaxon spicules in the gemmules and by the chitinous covering in the statoblasts. Both (gemmules and statoblasts) are set free by the destruction of the parental body and they develop into the new individuals in the favourable conditions.

4. Asexual reproduction by regeneration. The regeneration is a process by which the organisms develop or regenerate their lost or worn out parts. The regenenition is the best means of asexual reproduction in certain protozoans, sponges, coelenterates, planarians (Fig. 48.6) and echinoderms.

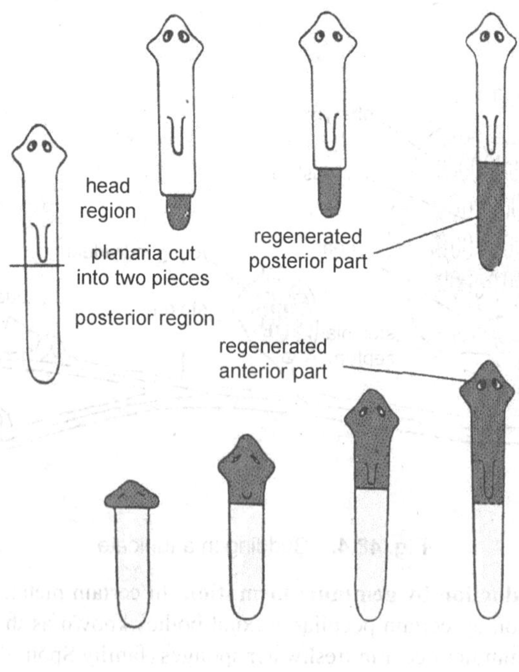

Fig. 48.6. Regeneration in *Planaria*.

48.1. SEXUAL REPRODUCTION

In the sexual reproduction, the development of the new indivdual takes place by the fusion of the sex cells or male and female gametes (Fig. 48.7). The sexual reproduction is the most common type of reproduction among animals. It may be of following types:

1. Syngamy. The syngamy is the most common type of sexual reproduction in the plants and animals. In syngamy (Gr., *syn* = together; *gam* = marriage) the fusion of two gametes takes palce completely and permanently. Following kinds of syngamy are prevalent among the living orgamsms:

***(i)* Autogamy.** In autogamy (Gr., *auto* = self; *gam* = mariage) the male and female gametes are produced by the same cell or organisms and both gametes fuse together to form a zygote, *e.g.*, *Actinosphaerium* and *Paramecium*.

***(ii)* Exogamy.** In exogamy (Gr. *exo* = external; *gam* = marriage) the male and female gametes are produced by different parents and both unite to form a zygote.

***(iii)* Hologamy.** In the lower organisms, sometimes the entire mature organisms start to act as gametes and the fusion of such mature individuals is known as the **hologamy**.

***(iv)* Paedogamy.** Paedogamy is the sexual union of young individuals produced immediately after the division of the adult parent cell by mitosis.

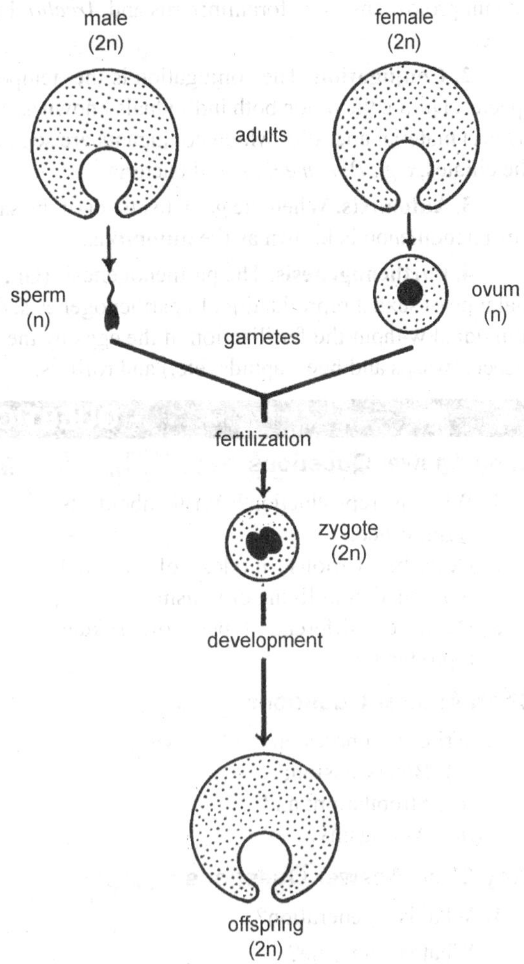

Fig. 48.7. A schematic representation of sexual reproduction.

***(v)* Merogamy.** In the merogamy (Gr., *meros* = part; *gam* = marriage), the fusion of smaller-sized and morphologically different gametes (**merogametes**) takes place.

***(vi)* Isogamy.** In isogamy (Gr., *is* = equal; *gam* = marriage) the fusion of morphologically and physiologically identical gametes (isogametes) takes place.

***(vii)* Anisogamy.** Some organisms produce two types of gametes. Both types of gametes differ from each other in their shape, size and behaviour and are collectively known as the **anisogametes** or **heterogametes**. The male gametes are motile and small in size and known as the **microgametes**. The female gametes are passive and have comparatively large size and known as the **macro-** or **mega-gametes**. The union of micro-and macrogametes is known as the anisogamy (Gr. *an* = without; *is* = equal; *gam* = marriage). The anisogamy occurs in higher animals and plants but it is customary to use the term **fertilization** in them instead of the anisogamy or syngamy.

***(viii)* Macrogamy.** The syngamy or fusion of the macrogametes is known as **macrogamy** (Gr., *macro* = large; *gam* = marriage).

***(ix)* Microgamy.** The microgamy (Gr., *micro* = small; *gam* = marriage) is common in certain protozoans, *e.g.*, foraminiferns and *Arcella*. In microgamy the fusion of microgametes takes place.

2. Conjugation. The conjugation is the temporary union of the two individuals of the same species. During the union both individuals known as **conjugants** exchange certain amount of nuclear (DNA) material and after which conjugants are separated. The conjugation is most common among the ciliates, *e.g.*, *Paramecium* and bacteria.

3. Automixis. When the gamete nuclei of the same cell unite together to form new individuals this phenomenon is known as the **automixis**.

4. Parthenogenesis. The parthenogenesis (Gr., *parthenos* = virgin, *genesis* = birth) is the special type of sexual reproduction. In parthenogenesis, the eggs of an organism develop into the young individual without the fertilization of the eggs by the sperms. The parthenogenesis occurs in certain insects (wasps and bees, aphids, etc.) and rotifers.

QUESTIONS

Long Answer Questions

1. What is reproduction? Write about its significance.
2. Describe various modes of asexual reproduction in living organisms.
3. Describe different types of sexual reproduction.

Short Answer Questions

1. Write short notes on the following:
   (*i*) Binary fission;
   (*ii*) Strobilation; and
   (*iii*) Anisogamy.

Very Short Answer Questions

1. What is regeneration?
2. What is gemmule?
3. Define budding.

Multiple Choice Questions

1. What is gemmule of sponge?
   (*a*) an external bud
   (*b*) an internal bud
   (*c*) an egg
   (*d*) collection of choanocytes
2. Most common modes of reproduction in *Hydra* is
   (*a*) sexual
   (*b*) budding
   (*c*) regeneration
   (*d*) all of these
3. Autotomy is recorded in
   (*a*) legs of crabs
   (*b*) tail of lizards
   (*c*) holothurian echinoderms
   (*d*) all of these

ANSWERS

Very Short Answer Questions

1. Regeration is regrowing of a lost or damaged part (tissue or organ) of an organism from a single cell or few cells, *e.g.*, *Hydra*, planarians, *Chaetopterus*, echinoderms, urodeles (amphibians), etc.
2. Gemmule is a cyst-like asexual reproductive strcture of many porifera that germinates when proper environmental conditions exist.
3. Budding is a form of asexual reproduction in which a new individual arises as an outgrowth of an older individual, *e.g.*, Hydra, tunicates.

Multiple Choice Questions

1. (*b*) 2. (*b*) 3. (*d*)

49 Gametogenesis

Two kinds of cells are recognised in the multicellular organisms:

1. **Somatic cells** which contain **somatoplasm** and do not take part in reproductive function; these cells are mainly concerned with the metabolic activities.
2. **Germ cells** which contain **germplasm** and produce reproductive sex cells or **gametes**.

Gametogenesis is a broad term that refers to the process by which germplasm is converted into highly specialised sex cells or gametes that are capable of uniting at fertilization and producing a new being. In simple terms, development or maturation of gametes from the primordial germ cells is the gametogenesis.

Primordial germ cell (PGCs) have germplasm and are thought not to be arising from any of the germ layers—neither from the ectoderm or mesoderm nor from endoderm (see **Gilbert**, 2000). PGCs are usually readily recognizable because of their large size and clear cytoplasm. They have certain distinct histochemical characteristics, such as high alkaline phosphatase activity in mammals and a high glycogen content in birds.

The maturation of germ cells or gametes is the direct result of the **meiosis** which reduces the chromosome numbers from **diploid (2n)** to **haploid (n)**. The formation of male gametes, the **sperm** or **spermatozoa**, is called **spermatogenesis** and occurs in testis (male gonad). The development of the female gamete, the **eggs** or **ova** is called **oogenesis** and occurs in ovary (female gonad). Both spermatogenesis and oogenesis are performed under the dictates of hormones produced by pituitary and gonads.

49.1. SPERMATOGENESIS

The male gametes or the spermatozoa are produced within the seminiferous tubules of the testis. The seminiferous tubules are separated by the endocrine cells, called **interstitial cells** or **cells of Leydig**.

Once the vertebrate PGCs arrive at the genital ridge of a male embryo, they become incorporated into the **sex cords**. They remain there until sexual maturity of the male, at which time the sex cords hollow out to form the seminiferous tubules of testis. The epithelium of the seminiferous tubules differentiates into the **Sertoli cells**.

The **initiation** of spermatogenesis during puberty is probably regulated by the synthesis of **BMP8B** (bone morphogenetic protein 8B) by the spermatogenic germs cells, the **spermatogonia**. When BMP8B reaches a critical concentration, the germ cells (sperm) begin to differentiate. The differentiating cells produce high levels of BMP8B which can then further stimulate their differentiation. Mice lacking this morphogenetic protein (*i.e.*, BMP8B) do not initiate spermatogenesis at puberty (**G. Zhao** *et al.*, 1996).

The Sertoli cells are radially distributed tall columnar cells. Their broad bases rests on the basement membrane and narrow tips extend into the lumen of the tubules. Each Sertoli cell contains a large nucleus with a prominent nucleolus. Sertoli cells are somatic cells and phagocytic in nature. They provide deep pockets or **recesses** into which spermatocytes and spermatids get anchored and receive necessary help from them, such as supply of nutrients to developing sperms and phagocytosis of residual bodies from metamorphosing spermatozoa. Sertoli cells also have some secretory (*i.e.*, endocrine) roles.

The spermatogenic, germ cells are bound to the Sertoli cells by **N-cadherin** molecules on both cell surfaces and by **galactosyltransferase** molecules on the spermatogenic cells that bind a carbohydrate receptor on the Sertoli cells (**Newton** *et al.*, 1993; **Pratt** *et al.*, 1993). Development pathway from germ cells to mature sperm occurs in the recesses of the Sertoli cells (Fig. 49.1).

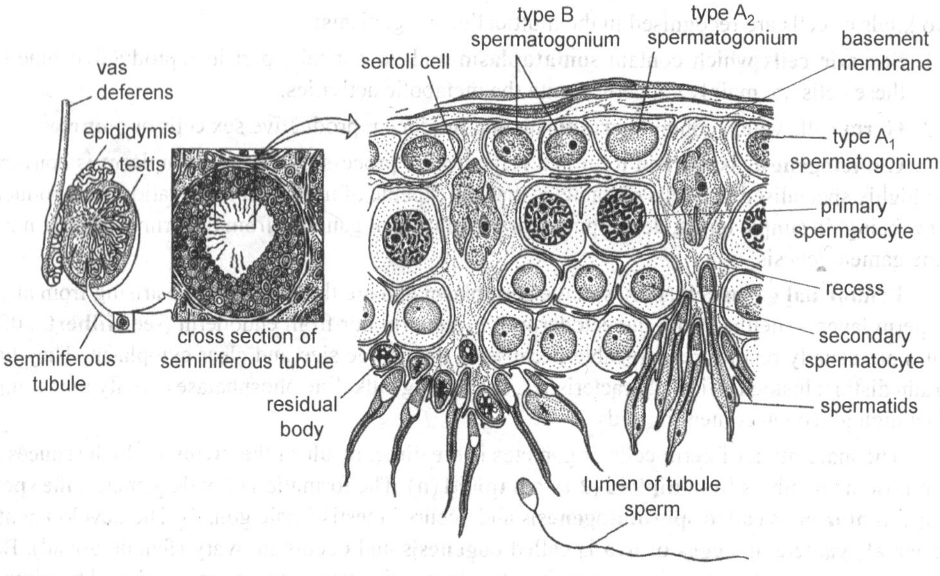

Fig. 49.1. Drawing of a section of the seminiferous tubule (*i.e.*, testis), showing the relationship between Sertoli cells and the developing sperms. As germ cells mature, they progress toward the lumen of the seminiferous tubule (after Gilbert, 2000).

Processes of Spermatogenesis

Spermatogenesis is a means to maintain a species-specific constant chromosome number. During spermatogenesis, the cells of germinal epithelium of seminiferous tubule of testis divide by **mitosis** and produce a large number of cells, each of which is capable of increasing in size (growth) and then dividing by **meiosis** to produce four spermatozoa (Fig. 49.2). The entire process of spermatogenesis can be divided into following four stages:

1. Multiplication phase
2. Growth phase
3. Maturation phase
4. Spermiogenesis

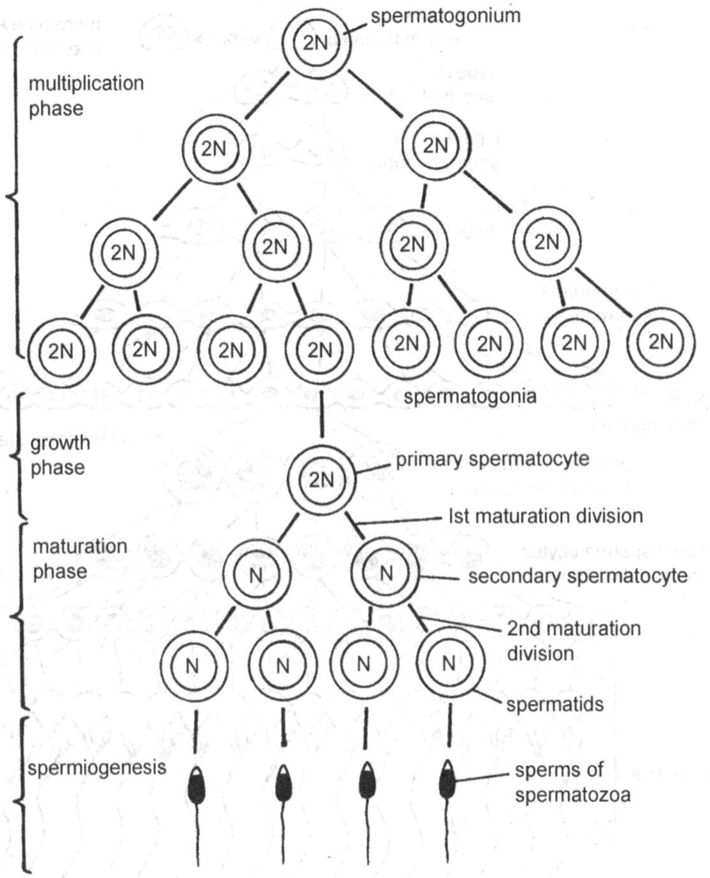

Fig. 49.2. Four steps of spermatogenesis.

1. Multiplication phase. By repeated mitotic multiplication, primary spermatogonial cells give rise to a **definitive number** of secondary spermatoginal cells which transform into primary spermatocytes. Thus, after reaching the testis, the PGCs divide to form **type A_1 spermatogonia**: These cells are smaller than the PGCs and are characterized by a ovoid nucleus that contains chromatin associated with the nuclear membrane (Fig. 49.1). The A_1 spermatogonia are found adjacent to the outer basement membrane of the sex cords or seminiferous tubules. They are **stem cells** (Box 49.1), and at maturity, they are thought to divide so as to make another A_1 spermatogonium as well as a second, polar type cell, the **type A_2 spermatogonium**. Thus, each type A_1 spermatogonium is a stem cell capable of regenerating itself as well as producing a new cell type. The A_2 spermatogonia divide to produce the **type A_3 spermatogonia**, which then be get the type A_4 spermatogonia. It is possible that each of type A spermatogonia are stem cells, capable of self renewal. The A_4 spermatogium has *three* options:

(*i*) it can form another A_4 spermatogonium (self-renewal);

(*ii*) it can undergo cell death (apoptosis); or

(*iii*) it can differentiate into first committed stem cell type, the **intermediate spermatogonium**. The intermediate spermotogonia are committed of becoming spermatozoa, and they divide mitotically one to form the **type B spermatogonia**. These cells are the precursors of the spermatocytes and are the base cells of the line that undergo mitotic division. They divide once to generate the **primary spermatocytes**—the cells that enter meiosis.

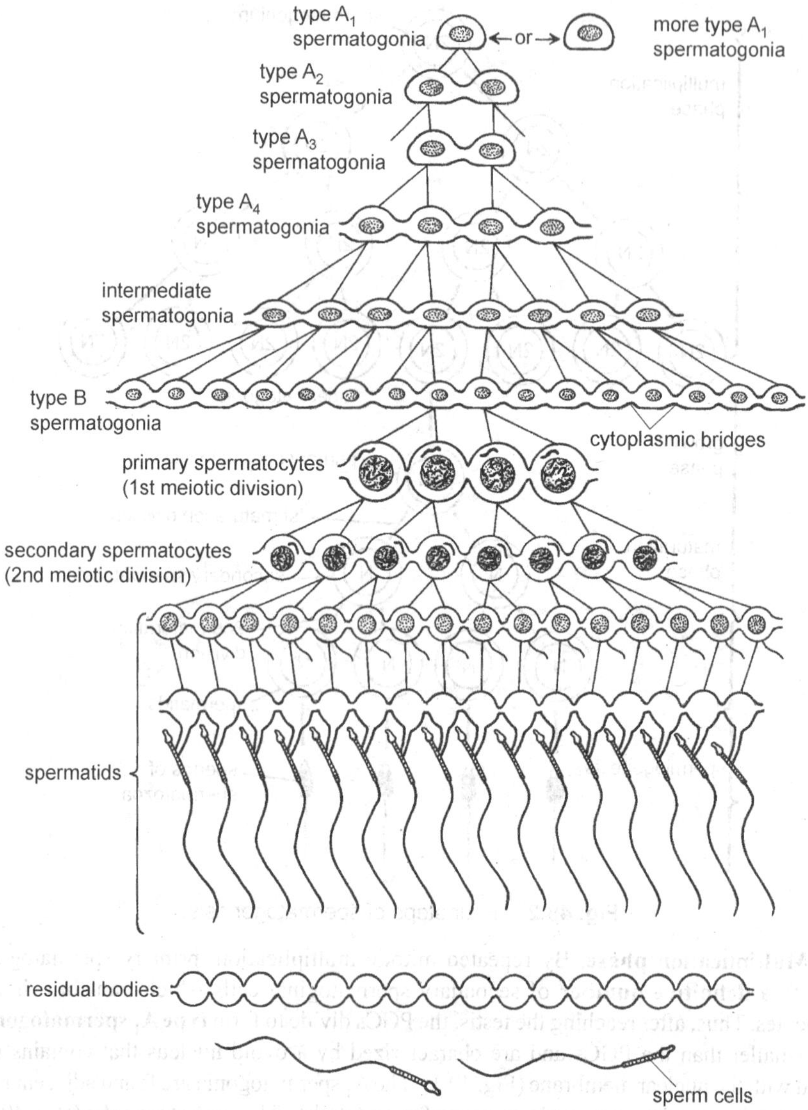

Fig. 49.3. Development of mammalian sperm as clones of interconnected cells. The cytoplasmic bridges between developing sperm cells are formed due to incomplete cytokinesis after mitotic as well as meiotic divisions (after Gilbert, 2000).

Box 49.1

A **stem cell** is a self-renewing cell that can divide *symmetrically* to give rise to two daughter cells with a developmental potential idential to the parent stem cell or *asymmetrically* to generate one daughter stem cell and one with a more restricted developmental potential than the parent cell (**Lodish** *et al.*, 2004).

A **clone** is a group of genetically identical cells/organisms all produced from the same parent individual by mitosis or asexual reproduction (**Bill Indge**, 1997).

Significance of Spermatogonial Syncytium

During the spermatogonial divisions, cytokinesis is not complete. Rather the cells form a **syncytium** whereby each cell communicates with the others via cytoplasmic bridges about 1 μm in diameter. The successive divisions produce **clones** of interconnected cells, and because ions and molecules readily pass through these intercellular bridges, each cohort (a group of cells with a shared feature), matures synchronously. Thus, due to formation of syncytium development of all sperm cells can be directed by gene products arising from both sets of parental chromosomes. This is particularly important in case of genes which are located on the sex chromosomes (X and Y chromosomes) which for the most part are nonhomologous. In the absence of cytoplasmic bridges, half of the secondary spermatocytes and their descendants would completely lack gene products from the Y chromosome and the other half would lack gene products from the X chromosome.

2. Growth phase. Primary spermatocytes increase in size by the accumulation of nourishing materials, *i.e.*, they undergo growth for preparation of maturation division or meiosis. During this time, the spermatocyte nucleus often transcribes genes whose products will be used later to form axoneme (sperm tail) and acrosome.

3. Maturation phase. This phase includes two vital steps: formation of four haploid spermatids from each primary spermatocyte and differentiation of these spermatids into motile sperm cells. The latter step is called **spermiogenesis**.

(*a*) Formation of spermatids. Each primary spermatocyte undergoes the first meiotic division (*i.e.*, reduction division) to yield a pair of **secondary spermatocytes**, which complete the second division of meiosis. The haploid cells thus formed are called **spermatids**. Spermatids are still connected to one another through their cytoplasmic bridges. The spermatids that are connected in this manner *have haploid nuclei, but are functionally diploid*, since a gene product made in one cell. readily diffuse into the cytoplasm of its neighbors (**Braun** *et al.*, 1989).

During the divisions from type A_1 spermatogonium to spermatid, the cells move farther and farther away from the basement membrane of the seminiferous tubule and closer to its lumen (Fig. 49.1.). Thus, each type of cell can be found in a particular layer of the tubule. The spermatids are located at the border of the lumen, and here they lose their cytoplasmic connections and differentiate into sperm cells. In humans, the progression from spermatogonial stem cell to mature sperm takes 65 days (**Dyin**, 1994).

(*b*) Spermiogenesis. A mammalian haploid spermatid is a round, unflagellated cell. For fertilization to occur, the sperm has to meet and bind with the egg, and spermiogenesis prepares the sperm for these functions of motility and interaction as shown in Fig. 49.4.

The first step of spermiogenesis involves the construction of the **acrosomal vesicle** from the Golgi apparatus. The characteristic feature of the acrosome is its specific staining with the carbohydrate stain PAS (periodic acid Schiffs) reagent. The acrosome forms a cap that covers the sperm nucleus. As the acrosome is formed, the nucleus rotates so that the cap will be facing the basal membrane of the seminiferous tubule. This rotation is necessary because the flagellum is beginning to form from the centriole on the other side of the nucleus and this flagellum will extend into the lumen.

During the last stages of spermiogenesis remaining part of head, midpiece and flagellum of a sperm are formed. The nucleus flattens, condenses and is streamlined, the remaining cytoplasm (the "cytoplasmic droplet" or "residual bodies") is jettisoned and the mitochondria form a ring around the base of the flagellum. Indeed, during spermiogenesis nucleus of sperm loses its entire fluid content, all its RNA, nucleolus and most of its proteins. Only one macromolecule remains untouched, *i.e.*, haploid amount of DNA. One of the major changes in the nucleus is the replacement of the histones by protamines. Transcription of the gene for protamines is seen in the early haploid cells (spermatids), although translation is delayed for several days (**Peschon** *et al.*, 1987). Protamines are relatively small basic proteins that are over 60 per cent arginine. During spermiogenesis, the nucleosomes dissociate, and the histones of the haploid nucleus are eventually are replaced by protamines. This causes complete shut-down of transcription in the nucleus and facilitates its assuming an almost crystalline structure. The resulting sperm then enter the lumen of seminiferous tubule.

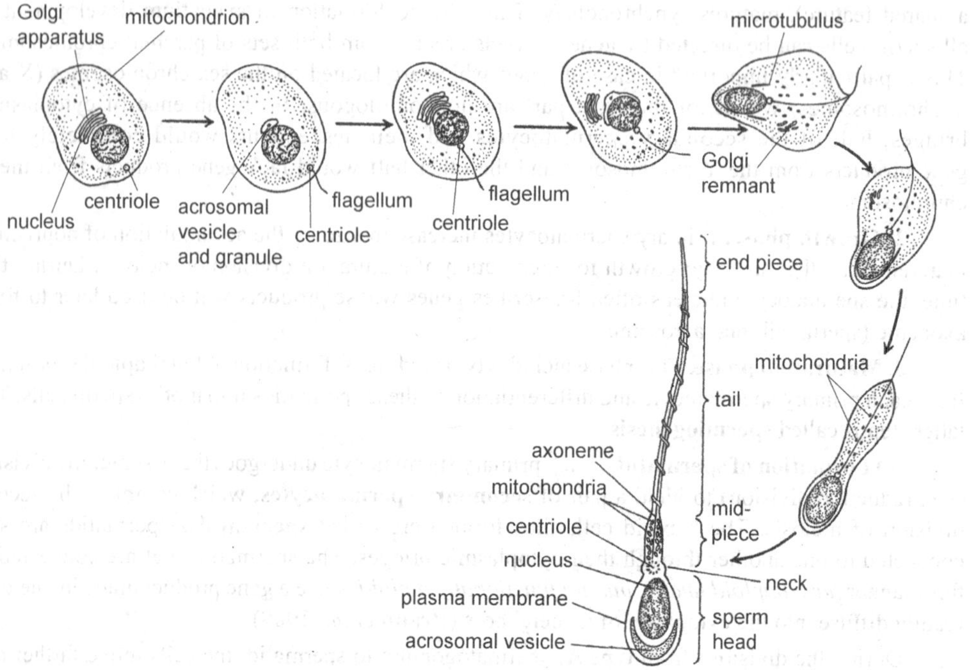

Fig. 49.4. Different stages in the spermiogenesis of a mammalian sperm (after Gilbert, 2000).

In the mouse, the entire process of spermatogenesis (*i.e.* from stem cell to spermatozoan) takes 34.5 days. The spermatogonial stages lasts 8 days, meiosis lasts 13 days and spermiogenesis takes up another 13.5 days. In humans, spermatogenesis takes nearly twice as long to complete. Because the type A_1 spermatogonia are stem cells, spermatogenesis can occur continuously. Each day, about 100 million sperms are made in each human testis. Each ejaculation release about 200 million sperm. Unused sperms are either resorbed or passed out of the body in urine. During his lifetime, a human male can produce 10^{12} to 10^{13} sperm (**Reijo** *et al.*, 1995).

Endocrine Regulation of Spermatogenesis

The production of spermatozoa in the vertebrate testis is regulated by two gonadotropic hormones, namely, **follicle stimulating hormone (FSH)** and **luteinizing hormone (LH)**, both produced from the anterior lobe of pituitary gland (*i.e.*, adenohypophysis). The adenohypophysis in turn is controlled by gonadotropin-releasing hormone from the hypothalamus. Within the testis, the initiation, maintenance and restoration of spermatogenesis is achieved by the testicular hormone, **testosterone**. The Leydig cells found in the interstitial space of the seminiferous tubules of the testis secrete testosterone in response to leuteinizing hormone (LH). FSH is also involved in the regulation of spermatogenesis by acting on Sertoli cells which produce a peptide, called **androgen binding protein (ABP)**. This protein binds to testosterone thereby sustaining its effects on spermiogenesis. Sertoli cells also secrete the hormone **inhibin** which inhibits the secretion of FSH by the pituitary gland and thereby plays an important feedback role in controlling the rate of spermiogenesis.

Molecular Events During Spermatogenesis

During spermatogenesis, specific genes necessary for sperm motility or binding to the egg are expressed. Following genes are expressed during meiotic prophase of spermatogenesis: (*i*) Y-linked genes (or holandric genes); (*ii*) β_2 *tubulin* gene (in *Drosophila*), necessary for meiotic spindles and the axoneme; (*iii*) the gene for the species-specific bindin protein, **bindin** (in sea urchin, *Strogylocentrotus purpuratus*). The following additional genes are expressed in spermatids: (1) genes for sperm protamine proteins that compact the sperm chromatin; (2) the gene for β_1, **4 galactosyltransferase** which helps binding of sperm with zona pellucida of the mammalian egg; (3) expression of **paternal effect genes** such as *spe*-11 in the nematode, *Caenorhabditis* (wild type allele of this gene from the male parent only will give normal embryo).

Imprinting of Paternal and Maternal Genomes

Paternal and maternal genomes are modified, or **imprinted**, during germ-cells differentiation. The paternal and maternal genomes contain the same set of genes, but the imprinting process turns off certain genes in either the sperm or egg, so that they are not expressed during development. For example, some of the genes necessary for yolk sac and placenta development are inactivated in the maternal genome, whereas some of those required for development of the embryo are turned off in the paternal genome. Imprinting implies that the affected genes carry a 'memory' of being in a sperm or an egg (**Reik** and **Walter**, 2001).

Structure of Spermatozoa

A typical mammalian spermatozoon is an intricate and highly polarised cell which is organised into following *three* distinct regions: head, midpiece and tail.

1. Head. Head is the anterior most part of the sperm. This is organ with which the sperm establishes contact with the egg during fertilization. The head of sperm consists of nucleus with an overlying cap-like acrosome.

(*a*) Nucleus. The haploid nucleus is housed within the head. It is highly streamlined and elongated due to extreme condensation of the chromatin (*i.e.*, DNA becomes tightly compressed by the help of protamine which replaces the histones of the nucleosomes). The tight packing of the DNA makes it less susceptible to physical damage or mutation during storage and transport to the site of fertilization. The chromosomal condensation and the resulting changes in the shape of the nucleus is also brought about by the pressure applied by microtubules present outside the sperm nucleus.

(*b*) Acrosome. In front of sperm nucleus lies a membrane bound **acrosome** which is derived from the Golgi apparatus. It has a functional homology with the lysosome in as much as it contains a variety of hydrolytic enzymes that digests proteins and complex sugars. These stored enzymes of acrosomal vesicle are used to lyse the outer coverings of the egg. Acrosome also contains non-enzymatic proteins, which serve mainly as egg recognition proteins. For example, in the sea urchin sperm, the acrosome releases a protein called **bindin** which is species-specific in its binding with vitelline envelope of eggs during fertilization.

Further, in many species, such as sea urchin a region (called **axial body**) of globular actin molecules lies between the nucleus and the acrosomal vesicle. These proteins are used to extend a finger-like **acrosomal process** from the sperm during the early stages of fertilization. Bindin is located on the tip of the acrosomal process.

2. Mid piece. Immediately posterior to the sperm head is a short **neck** which connects the sperm head distally to the thickened **midpiece**. Inside the neck two centrioles typically pressed against the posterior edge of nucleus. The **proximal centriole** of the sperm typically is donated to the egg and helps in forming the first mitotic spindle after fertilization. The **distal centriole** acts as a basal body for the flagellum of sperm tail.

The midpiece of sperm is characterized by the presence of helically arranged mitochondria, surrounding the inner axoneme as found in mammalian sperm. Sperm mitochondria have typical metabolic capabilities. Thus, sperm contains enzymes capable of aerobically metabolizing glycolysable sugars, glycerol, sorbitol, lactate, pyruvate, acetate, fatty acids and several amino acids. The sperm mitochondria also contain lactate dehydrogenase (LDH-C_4 isozymes) which can oxidise lactate to pyruvate, thereby allowing further energy production via the Krebs cycle.

The structure of midpiece is highly variable, since the mitochondria in different species assume different shape and structure. For example, in the sea urchin sperm, the midpiece is very short and simple, enclosing clusters of a few enlarged mitochondria. In many insects and mammals, the mitochondria in the midpiece are highly modified to give rise to a spherical structure called **nebenkern**. The mitochondrial cristae in the nebenkern are completely lost. The nebenkern is closely associated with the axonemal complex, suggesting their role in energy transfer and mobility of sperm.

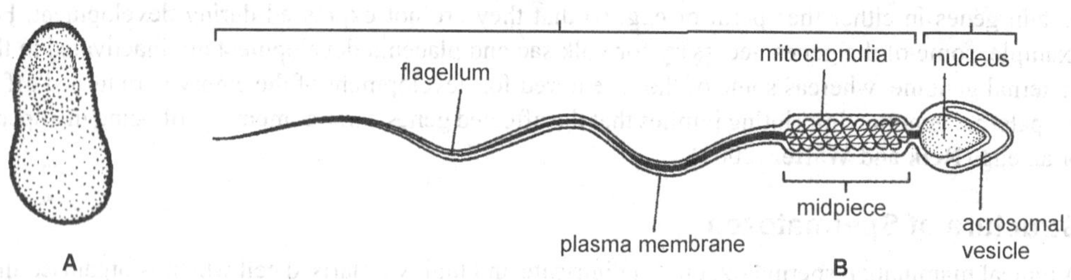

Fig. 49.5. A—Non-flagellated sperm of *Ascaris*; B—Flagellated sperm of humans (after Wolpert *et al.*, 2002).

3. Sperm tail. The means by which sperm are propelled vary according to how species has adapted to environmental conditions. In some species, such as the parasitic roundworm *Ascaris*, the sperm travel by the **amoeboid motion** of lamellipodia (Fig. 49.5A). In most species, however, each sperm is able to travel long distances by whipping its **flagellum**. The major motor portion of the flagellum is called the **axoneme**. Axoneme gives the cytoskeletal support to the sperm tail. Axoneme is formed by microtubules emanating from the distal centriole, found at the base of the sperm nucleus. Axoneme comprises the typical 9 + 2 array of microtubules. Thus, the core of the axoneme consists of two central microtubules surrounded by a row of nine doublet microtubules (Fig. 49.6).

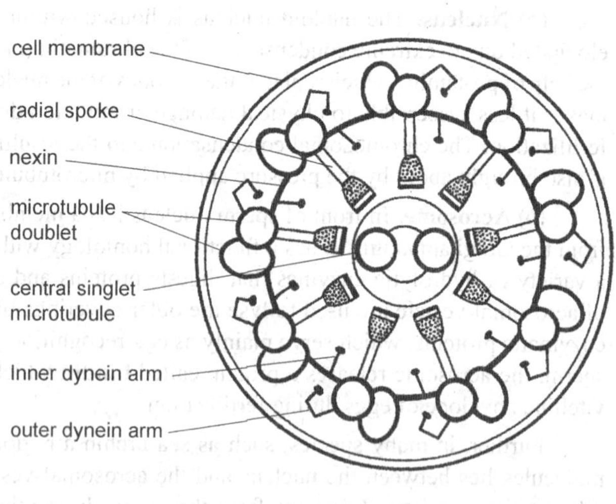

Fig. 49.6. Schematic diagram of the T.S. axoneme of sperm tail or flagellum (after Gilbert, 2000).

In fact, only one microtubule of each doublet is complete, having 13 protofilaments; the other is C-shaped and has only 10 or 11 protofilaments (see **Becker** *et al.*, 2006). Each protofilament is made by the contractile proteins, α- and β-tubulins. The peripheral doublets are interconnected by **nexin** bridges and are joined to the central sheath by nine spokes.

Although tubulin protein is the basis of the structure of the flagellum, other proteins are also critical for flagellar function (*i.e.*, lashing movement). The force for sperm propulsion is provided by **dynein**, a protein attached to the peripheral microtubules in the form of two arms, which tend to form transient cross bridges between adjacent microtubule doublets (Dynein is now referred to as **ciliary** or **axonemal dynein**). Dynein hydrolyses molecules of ATP and can convert the released chemical energy into the mechanical energy that propels the sperm. This energy allows the active sliding of the outer doublet microtubules, causing the flagellum to bend (**Ogawa** *et al.*, 1977; **Shinyo ji** *et al.*, 1998).

The mammalian sperm, which is required to swim in viscous genital fluid, gets additional structural support for the axoneme in the form of peripheral dense **fibrous sheath**. There are nine outer dense fibres located peripheral to each peripheral doublets of the axoneme, making the axoneme formula of 9 + 9 + 2 characterizing the mammalian sperm (Fig. 49.7). The fibres of fibrous sheath are thickest in the proximal half of the tail and progressively decrease in diameter towards the tip of sperm tail. This arrangement probably prevent the sperm head from being whipped around too suddenly (see **Gilbert**, 2000).

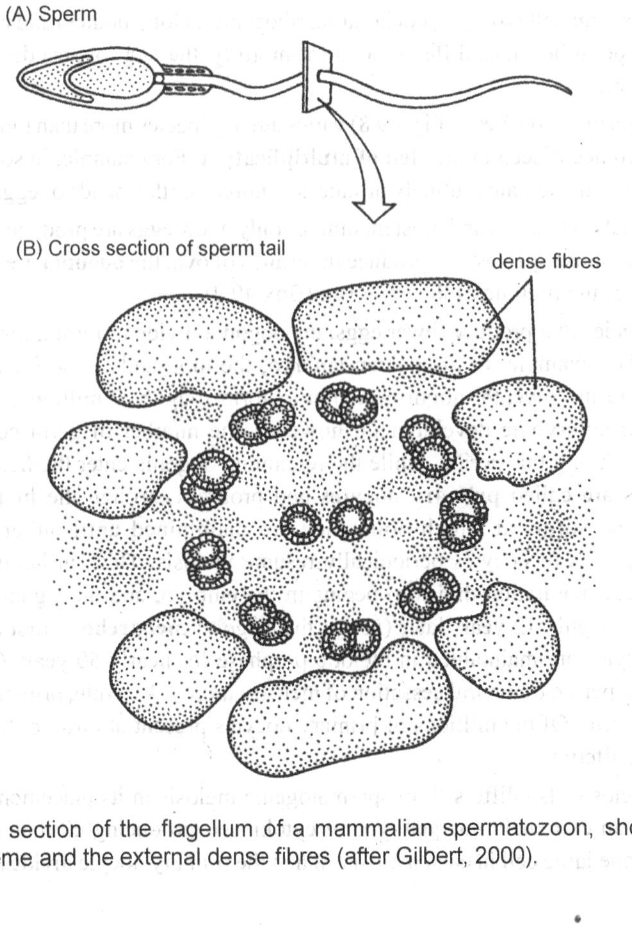

Fig. 49.7. Cross section of the flagellum of a mammalian spermatozoon, showing the central axoneme and the external dense fibres (after Gilbert, 2000).

Capacitation. The differentiation of mammalian sperm is not completed in the testes. After being expelled into the lumen of seminiferous tubules, the sperms are stored in the epididymis, where they acquire the ability to move. Motility of sperm is achieved through changes in the ATP-generating system (possibly through modification of dynein) as well as changes in the plasma membrane that make it more fluid (**Yanagimachi**, 1994). Lastly, the sperm released during ejaculation are able to move, yet they do not yet have the capacity to bind to and fertilize an egg. These final stages of sperm maturation (called **capacitation**) do not occur until the sperm has been inside the female reproductive tract for a certain period of time.

49.2. OOGENESIS

The process of formation of ovum (egg) in the female gonad (ovary) is called **oogenesis**. Oogenesis comprises the origin, multiplication, growth and differentiation, and maturation of female gametes. Before going through the details of mechanism of oogenesis, let us know the basic differences that exist in steps of multiplication and maturation of spermatogenesis and oogenesis.

Oogenesis Differs Drastically from Spermatogenesis

Whereas the gamete formed by spermatogenesis is essentially a motile nucleus, the gamete formed by oogenesis has all the materials needed to initiate and maintain metabolism and development.

Therefore, in addition to form a haploid nucleus, oogenesis also builds up a store of cytoplasmic enzymes, mRNAs, organelles (*e.g.*, mitochondria, ribosomes, Golgi bodies) and metabolic substrates. Thus, while the sperm becomes differentiated for motility, the egg (ovum) develops a remarkably complex cytoplasm.

The mechanism of oogenesis (Fig. 49.8) varies among species more than those spermatogenesis. One distinct difference is seen in the **step of multiplication**. For example, in some species, such as sea urchin and frogs, the female routinely produces hundreds or thousands of eggs at a time, where as in other species such as humans and most mammals, only a few eggs are produced during the lifetime of an individual. In those species that produce thousands of ova, the **oogonia** are self-renewing **stem cells** that persist for the lifetime of the organism (Box 49.2).

In those species that produce fewer eggs, the oogonia divide to form a limited number of egg precursor cells. For example, in the human embryo, the thousands or so oogonia divide rapidly from the second to the seventh month gestation to form roughly 7 million germ cells. After the seventh month of embryonic development, however, the number of germ cells drops abruptly. Most oogonia die during this period, while the remaining oogonia enter the **first meiotic division**. These latter cells are called **primary oocytes** and progress through the first meiotic prophase until the **diplotene stage**, at which point they are maintained until puberty. With the onset of adolescence, groups of oocytes periodically resume meiosis. Thus, in human female, the first part of meiosis (the maturation division) begins in the embryo, and the signal to resume meiosis is not given until roughly 12 years later (*i.e.*, at the stage of **menarche**—'first menstrual period'). In fact, some oocytes are maintained in meiotic prophase for nearly 50 years (*i.e.*, up to the time **menopause**—'the period of natural cessation of menstruation'). As a rule, primary oocytes continue to die even after birth. Of the millions of primary oocytes present at birth, only about 400 mature during a woman's lifetime.

Oogenic meiosis also differs from spermatogenic meiosis in its placement of the metaphase plate, this causes an extremely unequal type of cytokinesis producing ultimately three very small polar bodies and one large cell in each meiosis. When the primary oocyte divides, its nucleus, called the **germinal vesicle**, breaks down and the metaphase spindle migrates to the periphery of the cell. At telophase, one of the two daughter cells contains hardly any cytoplasm, whereas the other cell has nearly the entire volume of cellular constituents (Fig. 49.10). The smaller cell is called the **first polar body**, and the larger cell is referred to as **secondary oocyte**. During the second division of meiosis, a similar unequal cytokinesis takes place. Most of the cytoplasm is retained by the mature **egg (ovum)**, and a second polar body received little more than a haploid nucleus. Thus, oogenic meiosis conserves the volume of oocyte cytoplasm in a single cell rather than splitting it equally among four progenies.

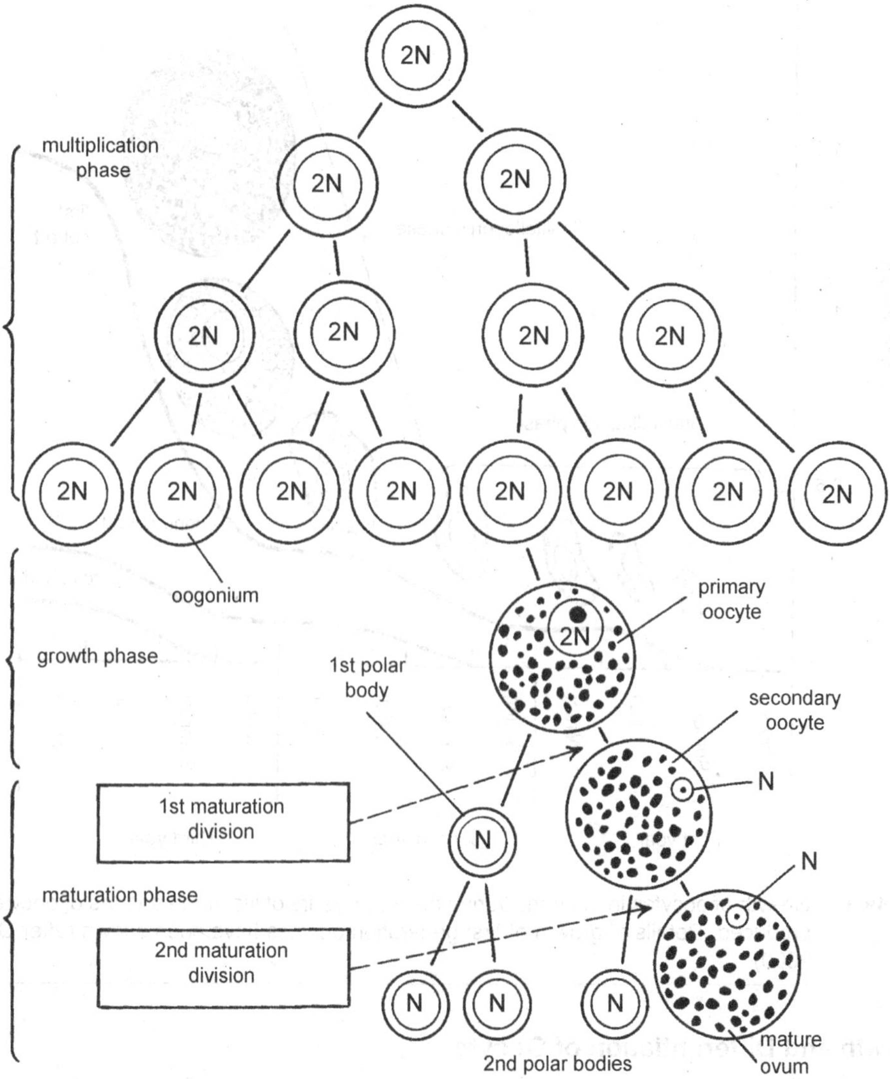

Fig. 49.8. Three steps of oogenesis.

Box 49.2

The eggs of fishes and amphibians are derived from an oogonial stem cell population that can generate a new cohort of oocytes each year. In the frog *Rana pipiens*, oogenesis takes 3 years. During the first two years, the oocyte increases its size very gradually. During the third year, however, rapid accumulation of yolk in the oocyte causes the egg to swell to its characteristically large size (Fig. 49.9). Egg matures in yearly batches, with the first cohort maturing shortly after metamorphosis; the next group matures a year later.

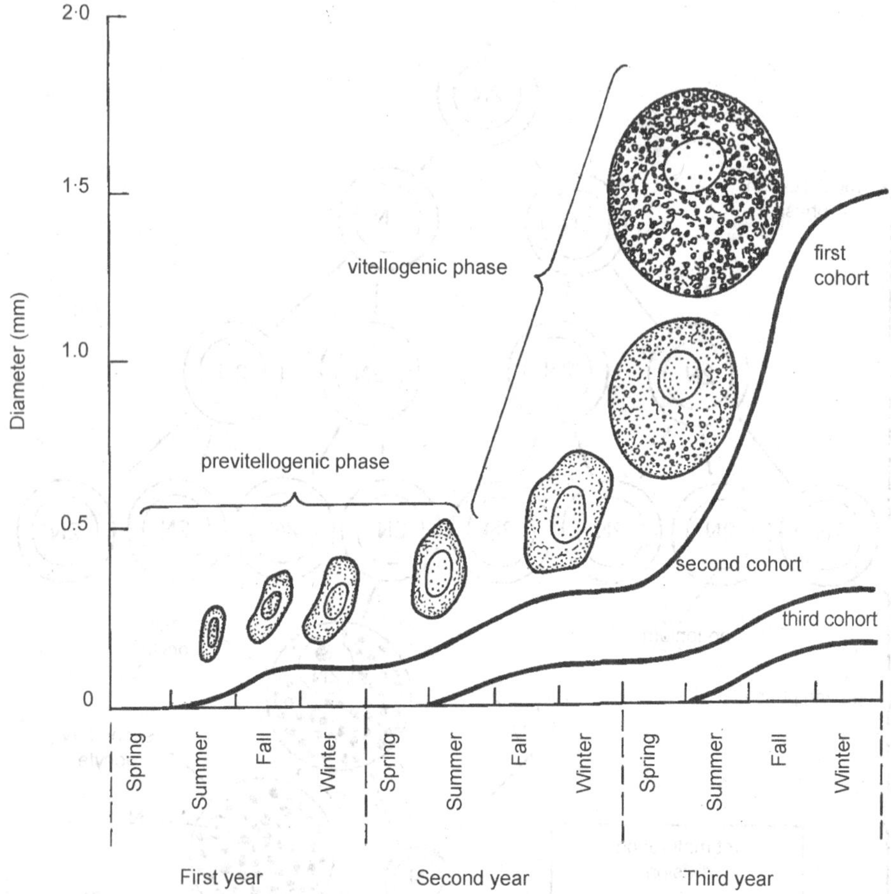

Fig. 49.9. Growth of oocytes in the frog. During the first 3 years of life, three cohorts of oocytes are produced. Details of growth of first generation oocytes have been shown (after Gilbert, 2000).

Growth and Differentiation of Oocyte

Growth and differentiation of the primary oocyte occur simultaneously. The accumulated material in the oocyte cytoplasm (*i.e.*, ooplasm) includes energy sources and energy-producing organelles (*i.e.*, the yolk and mitochondria); the enzymes and precursors for DNA, RNA and protein synthesis; stored messenger RNAs (Box 49.3); structural proteins and morphogenetic regulatory factors that control early embryogenesis (Table 49.1).

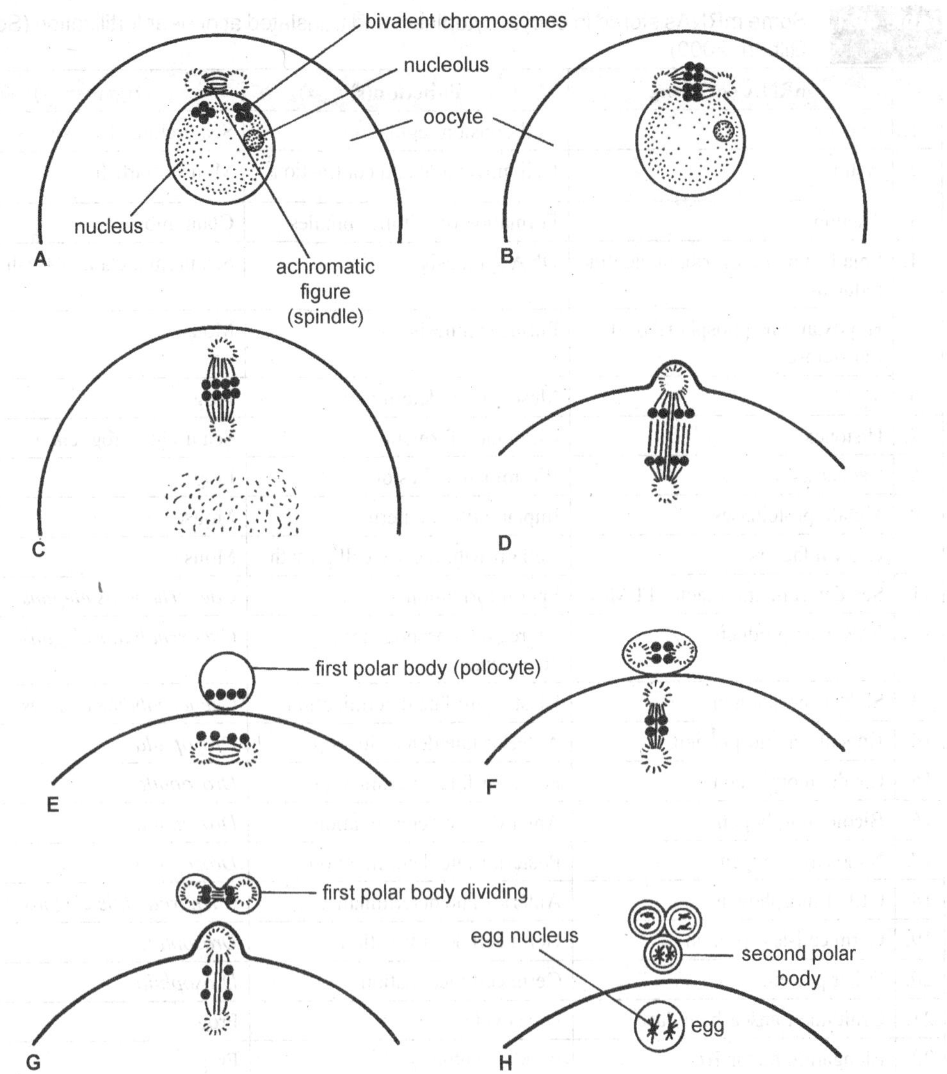

Fig. 49.10. Completion of oogenic meiosis and occurrence of unequal cytokinesis to produce three polar bodies and one egg cell.

Box 49.3

The oocytes of several species make two classes of mRNAs—those for immediate use in the oocyte and those that are stored for use during development (cleavage). In sea urchins, the translation of stored maternal mRNAs is initiated by fertilization, while in frogs, the signal for such translation is initiated by **progesterone** as the egg is about to be ovulated. Progesterone hormone is secreted by follicle cells in response to gonadotropic hormones (FSH, LH) secreted by the pituitary. One of the results of the MPF (mitosis promoting factor) activity induced by progesterone may be the phosphorylation of the proteins on the 3′UTR (= 3′ untranslated regions of mRNA, at poly A ends) of stored oocyte mRNAs. The phosphorylation of these factors is associated with the lengthening of the poly (A) tails in the stored messages (mRNAs) and the translation of the stored mRNAs (**Paris** *et al.*, 1991).

Table 49.1. Some mRNAs stored in oocyte cytoplasm and translated at or near fertilization (Source; Gilbert, 2000).

	mRNA encoding	Function(s)	Organism(s)
1.	Cyclins	Cell division regulation	Sea urchin, clam, starfish, frog
2.	Actin	Cell movement and contraction	Mouse, starfish
3.	Tubulin	Formation of mitotic spindles	Clam, mouse
4.	Small subunit of ribonucleotide reductase	DNA synthesis	Sea urchin, clam, starfish
5.	Hypoxanthine phosphoribosyl transferase	Purine synthesis	Mouse
6.	Vg1	Mesodermal determination	Frog
7.	Histones	Chromatin formation	Sea urchin, frog, clam
8.	Cadherins	Blastomere adhesion	Frog
9.	Metalloproteinases	Implantation in uterus	Mouse
10.	Growth factors	Cell growth; uterine cell growth	Mouse
11.	Sex determination factor FEM-3	Sperm formation	*Caenorhabditis elegans*
12.	PAR gene products	Segregate morphogenetic determinants	*Caenorhabditis elegans*
13.	SKN-1 morphogen	Blastomere fate determination	*Caenorhabditis elegans*
14.	Hunchback morphogen	Anterior fate determination	*Drosophila*
15.	Caudal morphogen	Posterior fate determination	*Drosophila*
16.	Bicoid morphogen	Anterior fate determination	*Drosophila*
17.	Nanos morphogen	Posterior fate determination	*Drosophila*
18.	GLP-1 morphogen	Anterior fate determination	*Caenorhabditis elegans*
19.	Germ cell-less protein	Germ cell determination	*Drosophila*
20.	Oskar protein	Germ cell localization	*Drosophila*
21.	Ornithine transcarbamylase	Urea cycle	Frog
22.	Elongation factor 1α	Protein synthesis	Frog
23.	Ribosomal proteins	Protein synthesis	Frog, *Drosophila*

Table 49.2. Cellular components stored in the mature oocyte of *Xenopus laevis*.

	Components	Approximate excess over amount in larval cells
1.	Mitochondria	100,000
2.	RNA polymerases	60,000 to 100,000
3.	DNA polymerases	100,000
4.	Ribosomes	200,000
5.	tRNA	10,000
6.	Histones	15,000
7.	Deoxyribonucleoside triphosphates	2,500

The period of growth and differentiation of primary oocyte is divided into *three* distinct phases: premeiotic, previtellogenic and vitellogenic.

1. The premeiotic phase. This phase is characterized by a regular sequence of nuclear events during prophase (*i.e.*, synapsis, crossing over; Fig. 49.11) of the first meiotic division.

2. Previtellogenic phase. During previtellogenic (Greek "before yolk formation") phase, the nuclear events proceed upto the stage of diplotene and then subsequent stages of meiosis are arrested. This event is called **meiotic I arrest** or **first meiotic block**. During first meiotic division, the oocyte nucleus swells up tremendously (by a factor of 8) and is called **germinal vesicle**. The chromosomes of germinal vesicle become very active in RNA synthesis or transcription. This stage is well suited for RNA synthesis because four chromatids and thus four sets of genes are available for transcription.

Premeiotic phase includes following main events in the germinal vesicle of primary oocyte:

(*i*) The genes coding for ribosomal RNAs (*i.e.*, 28S, 18S and 8.5 rRNAs), which form **nucleolar organizers**, are multiplied. Such sort of multiplication of genes without mitosis is called selective **gene amplification**. The increase in number of nucleolar organizers is reflected in an increase in the number of nucleoli. These nucleoli are formed on the periphery of the germinal vesicle (Fig. 49.11) and are involved in biogenesis of ribosomal subunits. Primary oocyte of *Xenopus* may contain upto 3000 nucleoli at this stage.

(*ii*) Hectic transcription of tRNAs and 5S rRNA takes place.

(*iii*) The chromosomes become dispersed in the nucleoplasm of germinal vesicle. In amphibian eggs and large yolky eggs of fishes, reptiles and birds, during the diplotene stage, certain chromosomes stretch out large lateral loops of DNA, causing the chromosomes to resemble a lamp brush. They are called **lampbrush chromosomes**. These loops synthesize different mRNAs that govern the synthesis of proteins that occurs in ooplasm. For example, the 100m long loops of American newt (*Notophthalamus*) are known to transcribe large mRNA molecules for five types of histones namely H1, H2A, H2B, H3 and H4.

Previtellogenic stage is often extended and the oocyte grows slowly with increase in ooplasm and in its organelles such as mitochondria, ribosomes ("differential products" of oocytes; Gilbert 2000), endoplasmic reticulum, Golgi apparatus, melanosomes and cortical granules.

3. Vitellogenic phase. Previtellogenic stage is followed by a vitellogenic phase during which yolk is elaborated. In general, yolk can be obtained by the primary oocyte by following two methods:

(*i*) Autosynthesis of yolk. In this case, the growing oocyte takes up the raw materials (*i.e.*, amino acids, glucose, fatty acids) from female parent's blood and synthesize the yolk by its membrane system of the rough ER, smooth ER and the Golgi apparatus. Autosynthesis of yolk generally occurs in animals having microlecithal eggs, *e.g.*, coelenterates, bivalve molluscs and echinoderms.

(*ii*) Heterosynthesis of yolk. In this case, yolk synthesis (vitellogenesis) is not done by the growing oocyte itself. In heterosynthesis, yolk precursor materials are synthesized in the extraovarian somatic organs such as liver of oviparous vertebrates and fat bodies of insects and are transported through blood for final incorporation (accumulation) into the oocytes. Specific uptake of the yolk precursor protein from blood is mediated by the follicle cells and micropinocytosis by oocyte, *e.g.*, insects, amphibians and birds.

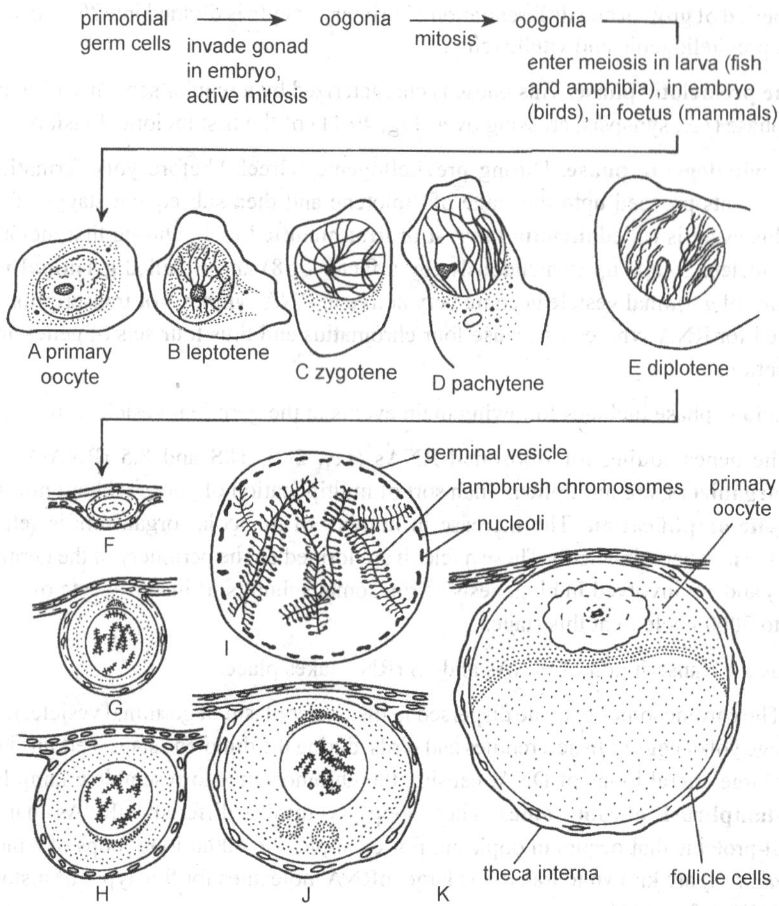

Fig. 49.11. Premeiotic and previtellogenic phases of oocyte's growth and differentiation phase. A—Primary oocyte (onset of meiosis); B to E—Meiotic prophase in cat, the leptotene to diplotene (at this stage meiosis I is arrested for many months or years); F to K—Vegetative growth phase in the amphibian oocyte.

Yolk and Types of Eggs

Yolk is not a single substance but a mixture of materials used for embryonic nutrition. It is the main nutritive material, accumulated in substantial quantities in the egg cytoplasm to meet the basic requirements of embryonic development in all oviparous animals. The composition of yolk could vary from species depending on the diet and their requirements during embryogenesis. Yolk is comprised of three chemical components, *viz.*, proteins, lipids and carbohydrates. *Protein yolk* is mainly found in the form of lipoproteins (lipovitellins) or phosphoproteins (phosvitins). Lipid reserves include fatty yolk granules, phospholipids and triglycerides. Carbohydrate yolk reserves include glycogen, and various polysaccharide protein complexes. Yolk protein is either present in the form of droplets (granules) or platelets and forms a major constituent of the ooplasm of many eggs. The precursor yolk protein after their synthesis in the extraovarian organs undergoes post-translational changes such as phosphorylation, glycosylation and sulphation in addition to lipid binding. The primary function of yolk is that it is used as a source of free amino acids for protein synthesis during embryogenesis. It also acts as a carrier protein during vitellogenesis for ions, lipids, sugars and vitamins in addition to transporting several steroid hormones for storage along with the yolk. The hormones thus transported through yolk protein serve as **morphogenetic hormones** during embryogenesis.

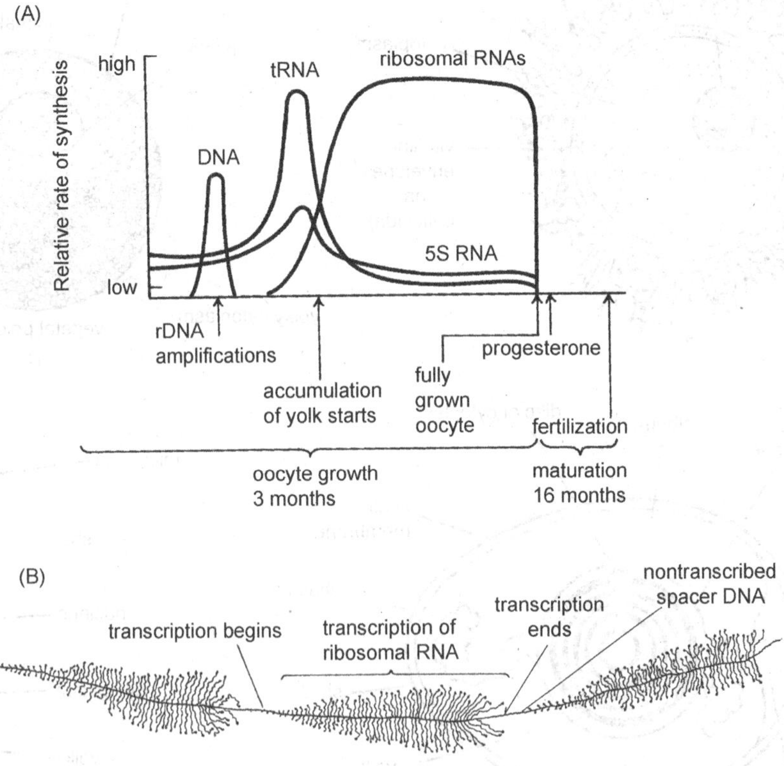

Fig. 49.12. Ribosomal RNA production in *Xenopus* oocytes. A—Rates of DNA, tRNA and rRNA synthesis in amphibian oogenesis during the last 3 months before ovulation. B—The transcription of the RNA precursors of 28S, 18S and 5.8S rRNAs. These units are tandemly linked together, some 450 per haploid genome (after Gilbert, 2000).

Further, the amount of yolk in the ooplasm has a definitive relationship with the type of fertilization and embryogenesis that the egg is going to have. For example, in all true viviparous mammals (eutherian mammals), where the embryogenesis occurs within the female reproductive system, the egg carries very little or almost no yolk for the simple reason that the embryonic development is not sustained by the stored yolk materials but by the supply of nutrient material from the mother through the placental structures. On the contrary, in oviparous animals (animals in which eggs are incubated externally by the parent, *e.g.*, hen) the egg contains enormous quantity of yolk. Here the entire embryonic development is independent of the mother and hence the yolk should supply nutrient material for the entire embryonic development (*e.g.*, hen's egg). Anyhow, the type of embryonic development in releasing a larva or a miniature adult (juvenile) will also have a bearing on the yolk distribution within the egg. For example, the eggs of sea urchin contain very little yolk and hence, the embryonic development is very quick and the larvae hatch out of the egg within few hours.

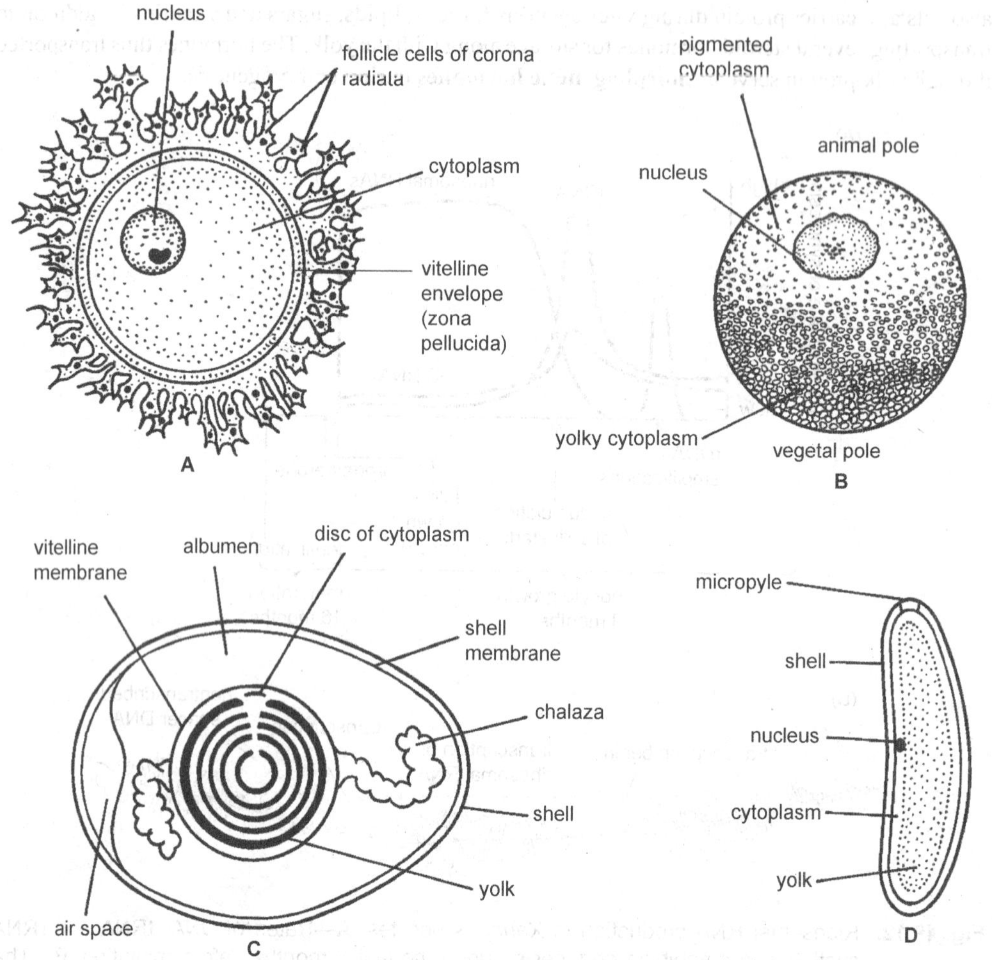

Fig. 49.13. Various types of ova (eggs). A—Alecithal ovum of humans; B—Mesolecithal and telolecithal ovum of frog; C—Macrolecithal and telolecithal egg of the hen; D—Macrolecithal and centrolecithal egg of fly (insect).

According to amount of yolk and distribution of yolk in ooplasm, following types of eggs have been recognised:

(*i*) **Alecithal eggs.** These eggs have no yolk, *e.g.*, eutherian mammals.

(*ii*) **Microlecithal and isolecithal eggs.** These eggs have very small amount of yolk and this yolk is evenly distributed in ooplasm, *e.g.*, sea urchin, *Amphioxus* and tunicates.

(*iii*) **Macrolecithal and telolecithal eggs.** Such eggs contain an enormous quantity of yolk material. Such a yolk mass almost fills the entire egg displacing the egg cytoplasm to the animal pole to form a cytoplasmic cap that contains the nucleus. Such eggs are found in bony fishes, birds, reptiles and egg-laying mammals.

Since in these eggs, yolk occurs at the vegetal pole, such eggs are also called **telolecithal eggs.**

(*iv*) **Mesolecithal and telolecithal eggs.** In amphibian eggs, there is a moderate amount of yolk. There is also a gradient of yolk distribution with the vegetal hemisphere having more of yolk than the animal hemisphere.

(*v*) **Centrolecithal eggs.** In insects (*e.g.*, *Drosophila*), the yolk assumes a central position and the cytoplasm is surrounding the centrally placed yolk as a thin coat. In the centre of yolk mass exists an island of cytoplasm which contains the nucleus. Insect eggs are macrolecithal eggs.

The Ovum (The Product of Oogenesis)

All the materials necessary for the beginning of growth and embryonic development must be stored in the mature egg (the **ovum**). Whereas the sperm has eliminated most of its cytoplasm, the developing egg (called the **primary oocyte** before it reaches the stage of meiosis at which it is fertilised) not only conserves its material, but is actively involved in accumulating more.The meiotic divisions that form the ovum conserve its cytoplasm (rather than equally distributing it to four meiotic products—ovum and three polar bodies). The primary oocyte either synthesizes or absorbs the yolk, that acts as food reservoirs for the developing embryo. Thus, birds eggs are enormous single cells, swollen with their accumulated yolk (Fig. 49.13). Even eggs with relatively sparse yolk are comparatively large. For example, the volume of a sea urchin egg is about 200 picoliters (more than 10,000 times the volume of the sperm). So, while sperm and egg have equal haploid nuclear components, the egg has a remarkable cytoplasmic storehouse that it has accumulated during its maturation.

A. Storage molecules of ooplasm. The cytoplasm of ovum (*i.e.*, ooplasm) store the following vital molecules:

1. Proteins. It will be after a long time when developing embryo is able to feed itself or obtain food from its mother. The early embryonic cells need a supply of energy and amino acids. In many species, this is achieved by accumulating yolk proteins in the egg. Many of the yolk proteins are made in other organs (*e.g.*, liver, fat body) and travel through the maternal blood to the egg.

2. Ribosomes and tRNA. The early embryo has to make many of its own proteins, and in some species, there is a burst of protein synthesis soon after fertilization. Protein synthesis is accomplished by ribosomes and tRNA, which exist in the egg. The growing oocyte has special mechanisms to synthesize ribosomes, and certain amphibian oocytes produce as many as 10^{12} ribosomes during their meiotic prophase.

3. Messenger RNA. In most animals, the instructions for proteins during early development are already packaged in the oocyte. It is estimated that the egg of sea urchins contain 25,000 to 50,000 different types of mRNA. This mRNA, however, remains dormant until after fertilization.

4. Morphogenetic factors. Molecules that direct the differentiation of cells into certain cell types are present in the egg. They appear to be localized in different regions of the egg and become segregated into different cells during cleavage.

5. Protective chemicals. The egg or embryo cannot run away from predators or move to a safer environment, so it must come equipped to deal with threats. Many eggs contain **ultraviolet filters** (*i.e.* mycosporin amino acid pigments of tunicates and green sea urchin) and **DNA repair enzymes** that protect them from sunlight. Some eggs contain molecules that potential predators find distasteful; and the yolk of bird eggs even contain antibodies.

B. Egg nucleus. Within the enormous volume of cytoplasm of egg resides a large nucleus. In some species (*e.g.*, sea urchins), the nucleus is already haploid at the time of fertilization (Fig. 49.14). In other species (including many worms and most mammals), the egg is still diploid, and the sperm enters before the meiotic divisions are completed. The stage of the egg nucleus at the time of sperm entry in different species is illustrated in Fig. 49.14.

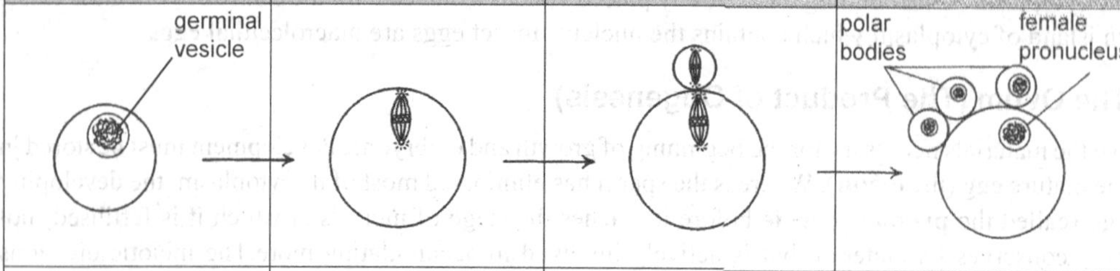

Primary oocyte	First metaphase	Second metaphase	Meiosis complete
1. The roundworm *Ascaris*	1. The nemertean worm *Cerebratulus*	1. The lancelet *Branchiostoma* (*Amphioxus*)	1. Cnidarian (e.g., anemones)
2. The mesozoan *Dicyema*	2. The polychaete worm *Chaetopterus*	2. Amphibians	2. Sea urchin
3. The sponge *Grantia*	3. The mollusc *Dentalium*	3. Most mammals	
4. The polychaete worm *Myzostoma*	4. The cork worm *Pectinaria*	4. Fish	
5. The clam worm *Nereis*	5. Many insects		
6. The calm *Spisula*	6. Starfish		
7. The echiuroid worm *Urechis*			
8. Dogs and foxes			

Fig. 49.14. Stage of egg maturation at the time of sperm entry in different animal species. The germinal vesicle is the name given to the large diploid nucleus of the primary oocyte. The polar bodies are seen as smaller cells (after Gilbert, 2000).

C. Egg membranes. Enclosing the cytoplasm is the egg's **plasma membrane.** This membrane must regulate the flow of certain ions during fertilization and must be capable of fusing with sperm plasma membrane (during fertilization). Outside the plasma membrane of sea urchin is the **vitelline envelope** (Fig. 49.15), which forms a fibrous mat around the egg. This envelope contains at least eight glycoproteins and is often involved in sperm egg recognition (**Correia** and **Carroll,** 1997). Vitelline envelope is supplemented by extensions of membrane glycoproteins from the plasma membrane and by proteinaceous vitelline posts that adhere the vitelline envelope to the membrane (**Mozingo** and **Chandler**, 1991). This envelope is essential for the species-specific binding of sperm. In mammals, the vitelline envelope is a separate and thick extracellular matrix, called the **zona pellucida.** The mammalian egg is also surrounded by a layer of cells called the **cumulus** (Fig, 49.16). Cumulus is made up of the ovarian follicular cells that were nurturing the egg at the time of its release from the ovary. Mammalian sperm have to get past these cells to fertilize egg. The innermost layer of cumulus cells, immediately next to zone pellucida, is called the **corona radiata** (Fig. 49.13A).

D. Egg cytoplasm. Lying directly beneath the plasma membrane of the egg is a thin shell (about 5 μm) of gel-like cytoplasm called the **ectoplasm** or **cortex**. The cytoplasm in this region is stiffer than the internal cytoplasm.The egg cortex is not displaced by centrifugation and cyclosis does not occur in it. Cortex contains high concentration of globular actin molecules and microtubules. The actin molecules polymerize to form long cables of actin known as **microfilaments**. Cytoskeletal elements (microtubules and microfilaments) are necessary for cell division, and they also are used to extend the

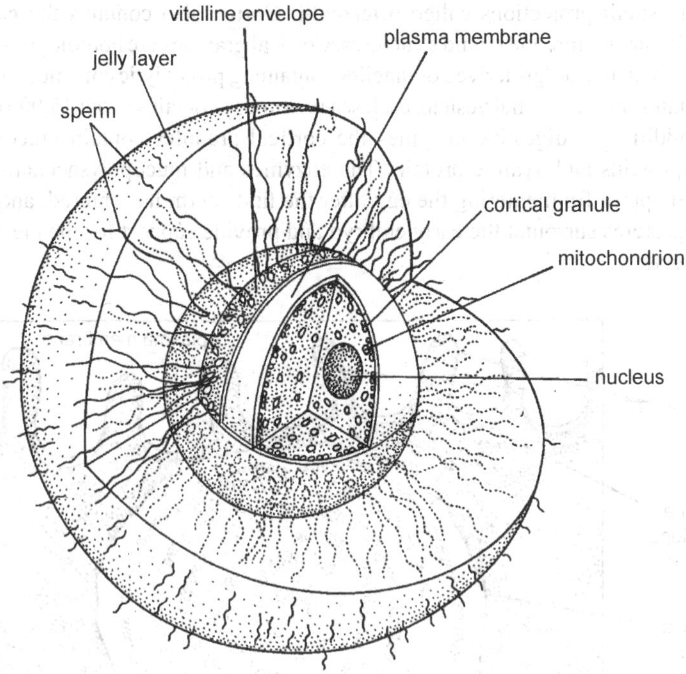

Fig. 49.15. Structure of the sea urchin egg during fertilization (after Gilbert, 2000).

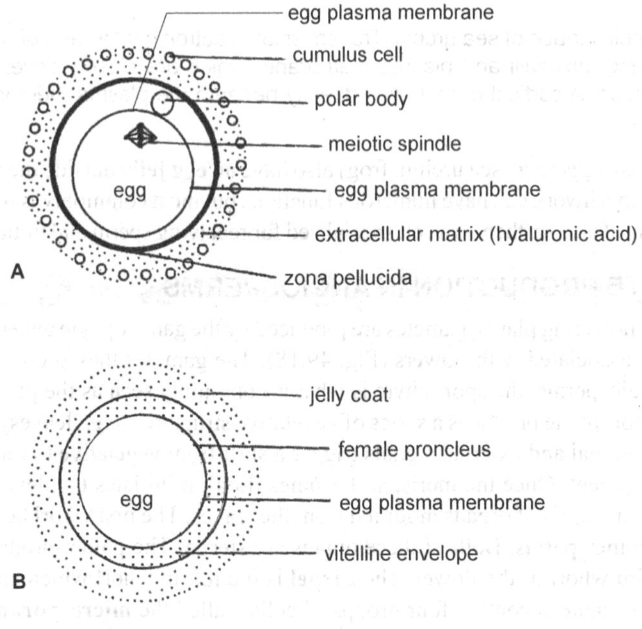

Fig.49.16. Schematic diagram of eggs and their extracellular layers. A—Mouse egg; B—Sea urchin egg (after Subramonian, 2002).

egg surface into small projections called **microvilli**. Cortex also contains the **cortical granules** (Fig. 49.17) which are membrane bound structures. Cortical granules are homologous to the acrosomal vesicle of sperm; both the Golgi-derived organelles containing proteolytic enzymes. However, whereas each sperm contains one acrosomal vesicle, each sea urchin egg contains about 15,000 cortical granules. Moreover, in addition to digestive enzymes the cortical granules contain mucopolysaccharides, adhesive glycoproteins and hyaline protein. The enzymes and mucopolysaccharides are active in preventing other sperm from entering the egg after the first sperm has entered, and the hyaline and adhesive glycoproteins surround the early embryo and provide support for the cleavage stage cells, called **blastomeres**.

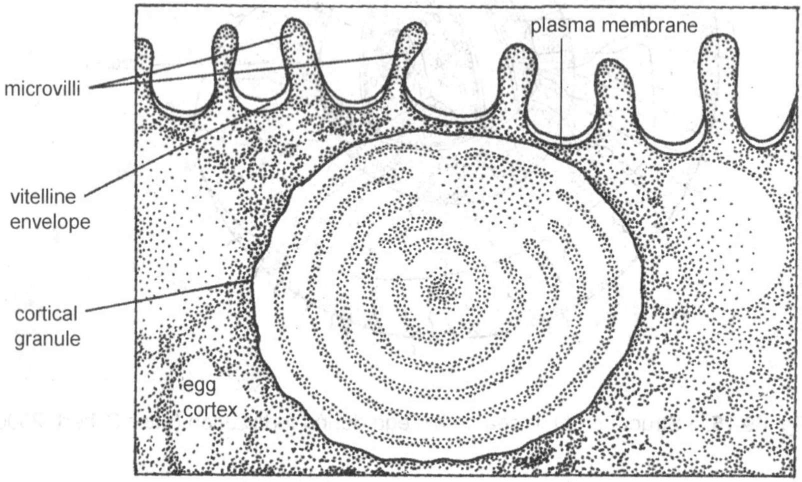

Fig. 49.17. Egg cell surface of sea urchin. Transmission electron micrograph of an unfertilized egg., showing microvilli and plasma membrane, which are closely covered by the vitelline envelope. A cortical granule lies directly beneath the plasma membrane (after Gilbert, 2000).

Many types of eggs (*e.g.*, sea urchin, frog) also have an **egg jelly** outside the vitelline envelope. This glycoprotein meshwork can have numerous functions but most commonly is used either to attract or to activate sperm. The egg, then, is a cell specialised for receiving sperm and initiating development.

49.3. GAMETE PRODUCTION IN ANGIOSPERMS

In angiosperms or flowering plants, gametes are produced by the gametophyte generation. Angiosperm gametophytes are associated with flowers (Fig. 49.18). The gametes they produce join to form the sporophyte. In angiosperms, the sporophyte is what is commonly seen as the plant body. The shoot meristem of the sporophyte produces a series of vegetative structures (*e.g.*, leaves). At a certain point in development, internal and external signals trigger a shift from vegetative to reproductive (flower producing) development. Once the meristem becomes floral, it initiates the development of flower parts sequentially in whorls of organs modified from the leaves. The first whorl becomes **sepals**. The second whorl becomes **petals**. Both of these organs are sterile. The pollen-producing **stamens** are initiated in the third whorl of the flower. The **carpel** is the fourth whorl which contains the **female gametophyte**. The stamens contain four groups of cells, called the **microsporangia** (pollen sacs), within an anther. The microsporangia undergo meiosis to produce microspores. Since, angiosperms are **heterosporous**, so the prefix *micro* is used to identify the spores that mitotically yield the male gametophyte—**pollen grains**. The inner wall of the pollen sac is called the **tapetum** and it provides nourishment for the developing pollen.

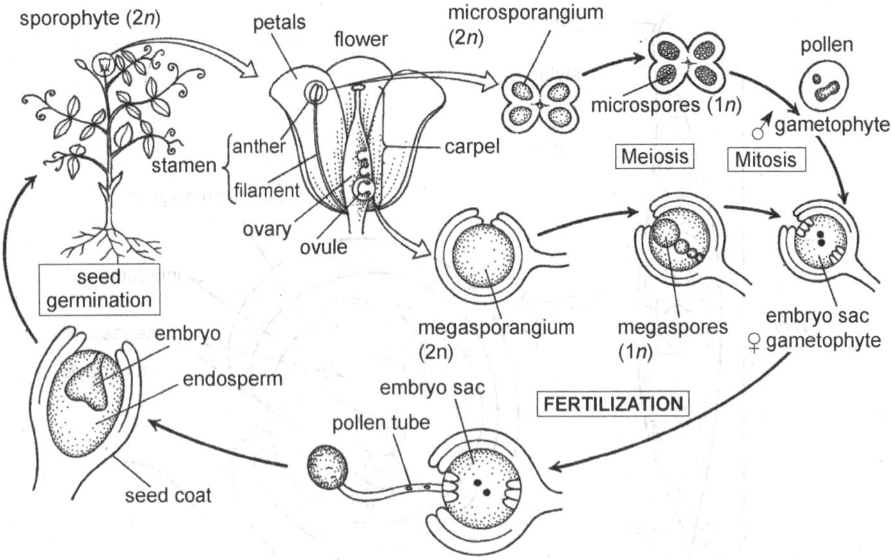

Fig. 49.18. Life cycle of an angiosperm, represented here by a pea plant (genus *Pisum*). The sporophyte is the dominant generation but mutlicellular male and female gametophytes are produced within the flowers of sporophyte (afrter Gilbert 2000).

Pollen

The pollen grain is a very simple multicellular structure. The outer wall of the pollen grain is called the **exine**; it is composed of resistant material provided by both the tapetum (sporophyte generation) and microspore (gametophyte generation). Exine of pollen grains contain the **sporopollenin**, a substance related to suberin and cutin but more resistant to decay. The inner wall, the intine, is produced by the microspore. A mature pollen grain consists of two cells, one within the other (Fig. 49.19). The **tube cell** contains a **generative cell** within it. The generative cell divides to produce two **sperm**. The tube cell nucleus guides pollen germination and the growth of the pollen tube after the pollen lands on the stigma of a female gametophyte. One of the two sperm will fuse with the egg cell to produce next sporophyte generation. The second sperm will participate in the formation of **endosperm**, a structure that provides nourishment for the embryo.

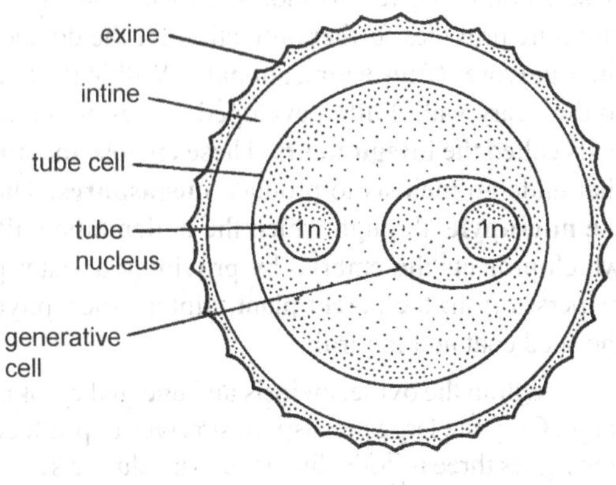

Fig. 49.19. Structure of a pollen grain. A pollen grain consists of a cell within a cell.The generative cell will undergo division to produce two sperm cells. One will fertilize the egg, and the other will join with the polar nuclei yielding the endosperm (after Gilbert, 2000).

The Ovary

The fourth whorl of organs within the flower forms the **carpel**, which gives rise to the female gametophyte (Fig. 49.20). The carpel consists of the **stigma** (where the pollen lands), the **style**, and, the

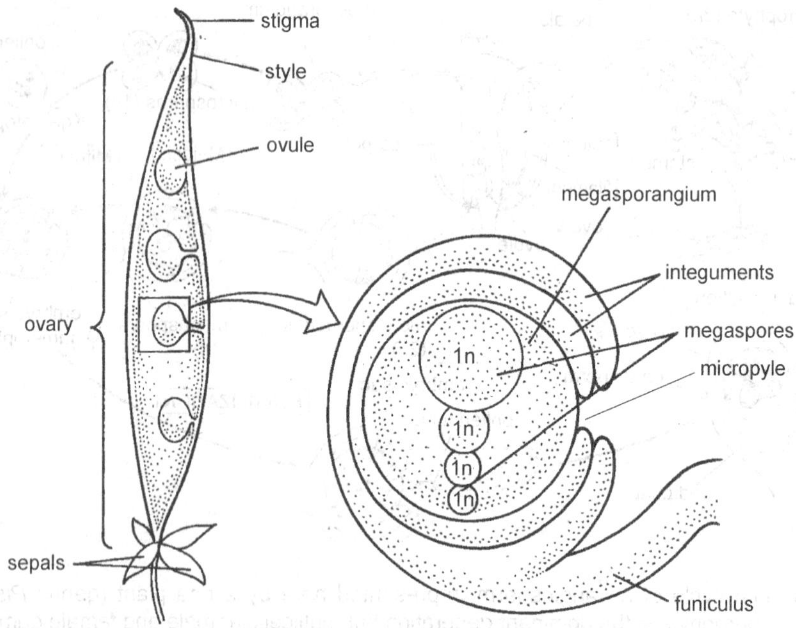

Fig. 49.20. The carpel contains one or more ovules; these contains megasporangia protected by two layers of integument cells. The megasporangia deivide meiotically to produce haploid megaspores. All of the carpel is diploid except for the megaspores, which divide mitotically to produce the embryo sac (the female gametophyte) (after Gilbert, 2000).

ovary. Following fertilization the ovary wall will develop into the **fruit**. This unique angiosperm structure provides further protection for the developing embryo and also enhances seed dispersal by the frugivores (fruit-eating animals). Within the ovary are one or more **ovules** attached by a **placenta** to the ovary wall. Fully developed ovules are called **seeds**. The ovule has one or two other layer of cells called the **integuments**. These enclose the **megasporangium**, which contains sporophyte cells that undergo meiosis to produce **megaspores**. There is a small opening in the integuments, called the **micropyle**, through which the pollen tube will grow. The integuments develop into seed coats, which protects the embryo by providing a water proof physical barrier. When the mature embryo disperses from the parent plant, diploid sporophyte tissue accompanies the embryo in the form of the seed coat and the fruit.

Within the ovule, meiosis and unequal cytokinesis yield four **megaspores**—three small and one large. Only the large megaspore survives to produce the embryo sac. The largest of these megaspores undergoes three mitotic divisions to produce a seven-celled embryo sac with eight nuclei (Fig. 49.21). One of these cells is the **egg**. The two **synergid cells** occur on either side of the egg; they seem to be evolutionary remnants of the archegonium (the female sex organs seen in mosses and ferns). The **central cell** consists of two or more polar nuclei, which will fuse with the second sperm nucleus and develop into the polyploid endosperm. Three **antipodal cells** form at opposite end of the embryo sac from the synergids and degenerate before or during embryonic development. There is no known function for the antipodals. Genetic analysis of female gametophyte development in maize and *Arabidopsis* has provided an insight into the regulation of the specific steps in this process (**Drews** *et at.*, 1998).

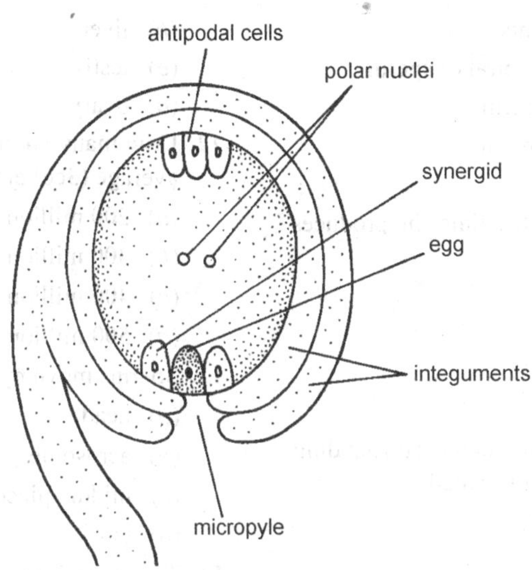

Fig. 49.21. The embryo sac is the product of three mitotic divisions of the haploid megaspores. Two of the nuclei are contained within the central cell; the other six cells contain one haploid nucleus each (after Gilbert, 2000).

QUESTIONS

Long Answer Questions

1. Describe the mechanism of spermatogenesis. Highlight the significance of spermatogonial syncytium.
2. Describe the process of spermiogenesis and also the structure of a mature mammalian spermatozoon.
3. Give an account of endocrine regulation of spermatogenesis in vertebrates.
4. Explain main points of differences between oogenesis and spermatogenesis.
5. Describe the structure of a mature ovum.
6. Describe development of an angiosperm upto the stage of embryo sac formation.

Short Answer Questions

1. Write short notes on the following:
   (*i*) Imprinting of paternal and maternal genomes.
   (*ii*) Molecular events during growth of primary oocyte.
   (*iii*) Sertoli cells.

Very Short Answer Questions

1. What is yolk?
2. What is gamete?
3. Define gametogenesis.
4. What are Sertoli cells?

Multiple Choice Questions

Choose the correct answer from the four alternatives given.

1. The function of Sertoli cell is
   (*a*) synthesis of testosterone
   (*b*) nourishment of sperm
   (*c*) storage of sperm
   (*d*) maturation of sperm
2. The preparation of sperm before penetration of ovum is called
   (*a*) coitous (*b*) capacitation
   (*c*) insemination (*d*) spermiation
3. The role of Leydig cells is
   (*a*) nourishment of sperms
   (*b*) give mortality of sperms
   (*c*) synthesize testosterone hormone
   (*d*) all above

4. Nebenkern represents
   (*a*) mitochondrial spiral of sperm
   (*b*) acrosome of sperm
   (*c*) centrioles of sperm
   (*d*) tail of sperm
5. During a woman's life time she produces about
   (*a*) 400–500 eggs
   (*b*) 4000 eggs
   (*c*) 365–800 eggs
   (*d*) 40 eggs
6. The thick clear membrane surrounding the mammalian egg is called
   (*a*) zona pellucida
   (*b*) zona indica
   (*c*) corpus luteum
   (*d*) zona cervica
7. Polar bodies contain
   (*a*) 23 autosomes
   (*b*) 22 autosomes and X chromosome
   (*c*) 22 autosomes and a Y chromosome
   (*d*) 22 autosomes and X or Y chromosome
8. Bird's eggs containing a large amount of yolk are called
   (*a*) microlecithal eggs
   (*b*) macrolecithal eggs
   (*c*) oligolecithal eggs
   (*d*) mesolecithal eggs
9. Pollen grains are able to withstand extremes of temperature and dessication because their exine is composed of
   (*a*) cutin
   (*b*) suberin
   (*c*) sporopollenin
   (*d*) callose
10. Pollen tube discharges its gametes into the
    (*a*) egg
    (*b*) central cell
    (*c*) healthy synergid
    (*d*) degenerating synergid
11. Seminiferous tubules occur in
    (*a*) kidney
    (*b*) liver
    (*c*) testis
    (*d*) ovary
12. How many sperm cells are present in an average (3cc) ejaculation?
    (*a*) 200 million
    (*b*) 300 million
    (*c*) 400 million
    (*d*) 500 million
13. Sperms move by
    (*a*) head
    (*b*) acrosome
    (*c*) middle piece
    (*d*) tail
14. The mobility of a mature sperm is controlled by the mitochondria located in the
    (*a*) head
    (*b*) middle piece
    (*c*) tail
    (*d*) in all of them
15. Supporting cells found in between the germinal epithelium of testis are called
    (*a*) interstitial cells of leydig
    (*b*) Sertoli cells
    (*c*) granular cells
    (*d*) phagocytes
16. The role of Leydig or interstitial cells is to
    (*a*) nourish the sperms
    (*b*) give motility to sperms
    (*c*) synthesize testosterone hormone
    (*d*) all above
17. During oogenesis, each diploid cell produces
    (*a*) four functional eggs
    (*b*) two functional eggs and two polar bodies
    (*c*) one functional egg and three polar bodies
    (*d*) four functional polar bodies
18. The preparation of sperm before penetration of ovum is called
    (*a*) coition

(*b*) capacitation
(*c*) insemination
(*d*) spermiation

19. Formation of yolk is known as
(*a*) vitellogenesis
(*b*) oogenesis
(*c*) histogenesis
(*d*) gametogenesis

20. Polar bodies are formed during
(*a*) somatic hybridization
(*b*) spermmatogenesis
(*c*) oogenesis
(*d*) spermiogenesis

21. The thick clear membrane surrounding the mammalian egg is called
(*a*) zona pellucida (*b*) zona indica
(*c*) corpus luteum (*d*) zona cervica

22. The Mullerian duct in the the female amniotes develop into
(*a*) oviduct
(*b*) ureter
(*c*) seminal receptacle
(*d*) uterus

23. Regressed Graafian follicle is called
(*a*) corpus albicans
(*b*) corpus callosum
(*c*) corpus luteum
(*d*) placenta

24. In humans, ovulation occurs during ________ phase of menstrual cycle
(*a*) proliferative
(*b*) luteal
(*c*) menstrual
(*d*) none of these

ANSWERS

Very Short Answer Questions

1. Yolk is stored food in the eggs of the majority of animals for the use of the developing embryo. It may consist of maily of protein (protein yolk) or of phospholipids and fats (fatty yolk).
2. Gamete is a specialized reproductive cell (a sperm or an egg) through which sexually reproducing parents pass chromosomes to their offspring.
3. Gametogenesis is production of gametes.
4. Sertoli cells are large secretory support cells in the seminiferous tubules of the testes involved in spermatogenesis in the adult through their role in nourishing and maintaining the developing sperm cells. They secrete AMF in the fetus and provide a niche for the incoming germ cells (AMF = Anti-Mullerian Factor that blocks development of the female ducts). They are derived from somatic cells, which are in turn derived from the genital ridge epithelium.

Multiple Choice Questions

1. (*b*)	**2.** (*b*)	**3.** (*c*)	**4.** (*a*)	**5.** (*a*)	**6.** (*a*)	**7.** (*b*)
8. (*b*)	**9.** (*c*)	**10.** (*d*)	**11.** (*c*)	**12.** (*b*)	**13.** (*d*)	**14.** (*b*)
15. (*b*)	**16.** (*c*)	**17.** (*c*)	**18.** (*b*)	**19.** (*a*)	**20.** (*c*)	**21.** (*a*)
22. (*a*)	**23.** (*c*)	**24.** (*b*)				

50 Fertilization

Fertilization is the central event in sexual reproduction to perpetuate the species. It is the process whereby two sex cells (gametes) fuse together to create a new individual with genetic potentials derived from both parents. Fertilization brings about two separate activities: 1. **Sex** (the combining of genes derived from the two parents); 2. **Reproduction** (the creation of new organisms). Thus, the first function of fertilization is to transmit genes from parent to offspring, and *second* function is to begin in the egg cytoplasm those reactions that permit development to proceed. Fertilization process occurs both in animals and plants.

50.1. FERTILIZATION IN ANIMALS

Fertilization in animals generally consists of four major events (Fig. 50.1).

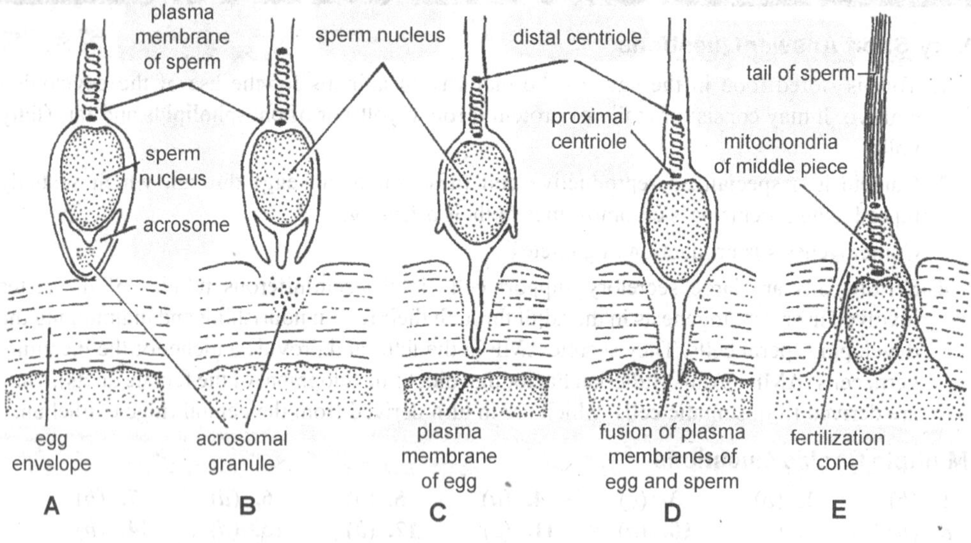

Fig. 50.1. Process of fertilization in animals (after Gilchrist, 1968).

1. **Contact and recognition between sperm and egg.** In most cases, this ensures that the sperm and egg are of the same species.
2. **Regulation of sperm entry into the egg.** Only one sperm can ultimately fertilize the egg. This is usually accomplished by allowing only one sperm to enter the egg and inhibiting any other from entering.
3. **Fusion of genetic material of sperm and egg.**
4. **Activation of egg metabolism to start development.**

I. Recognition of Egg and Sperm

Fertilization can be either **external**, as in sea urchin and frog, or **internal**, as in *Drosophila*, mammals and birds. In many species having external fertilization, the meeting of sperm and egg is not a simple matter. Many marine organisms release their gametes into the environment. That environment may be as small as a tide pool or as large as an ocean. Moreover, this environment is shared with other species that may shed their sex cells at the same time. These marine animals are faced with two problems: 1. How can sperm and egg meet in such a dilute concentration? and 2. How can sperm be prevented from trying to fertilize eggs of another species? Both of these problems have been solved by the following *two* mechanisms: 1. species-specific attraction of sperm and 2. species-specific activation.

1. Species-specific attraction of sperm. Such mechanism occurs in numerous species, including cnidarians, molluscs, echinoderms and urochordates. In many species, sperms are attracted toward eggs of their own species by **chemotaxis**, that is, by following a gradient of a chemical secreted by the egg (*e.g.*, cnidarian *Orthopyxis caliculata*; **Miller** 1978). Some chemotactic molecules, such as **resact** has been isolated from the egg jelly of the sea urchin *Arbacia punctulata* (**Ward** *et al.*, 1985). Resact exhibits species-specific sperm attraction and activation. This 14-amino acid peptide, on contact with the sperm cells, causes dramatic increase in sperm motility and oxygen consumption. Resact is found to increase in sperm cAMP and cGMP that appear to activate the dynein ATPase to stimulate tail beating in the sperm.

Fertilizin-Antifertilizin interactions. In 1914, **F.R. Lillie** proposed the fertilization theory. He observed that **egg water** (sea water surrounding sea urchins eggs) agglutinated the sperms and activated their motility (**Balinsky** 1970). The reaction was specific since some sperms from related species were unaffected. This factor (chemical molecule) was called **fertilizin** and found to be constituent of both jelly coats and egg envelopes such as vitelline envelope and plasma membrane of eggs of sea urchin, frog, etc., (**Grant** 1978).

Agglutinated sperms were found to contain receptor sites or **antifertilizin** molecules which interacted with egg's fertilizin molecules. Chemically fertilizin is mucopolysaccharide or proteoglycan (**Trinkaus** 1984). In contrast, antifertilizins are acidic proteins having molecular weight about 10,000 Daltons (**Verma** and **Agarwal** 2010).

In mammals, too, as the sperms are released from the testis, they acquire motility when they pass through the epididymis. However, the mammalian sperms require an additional phase of maturation, occurring within the female reproductive tract to prepare them for the acrosome reaction and fertilization. This maturation process is called **capacitation** and involves the following changes in sperm: alterations in plasma membrane such as rearrangements of intra-membranous particles, removal of sperm surface components and a decrease in net negative charge.

2. Acrosome reaction and species-specific recognition. In most marine invertebrates, the acrosomal reaction has *two* parts:

(*i*) Exocytosis of acrosomal vesicle. The fusion of the acrosomal vesicle with the sperm plasma membrane (*i.e.*, exocytosis that results in the release of the contents of the acrosomal vesicle). The acrosomal reaction in sea urchin is initiated by contact of the sperm with the egg jelly. Contact with egg jelly causes exocytosis of the sperm's acrosomal vesicle and the release of proteolytic enzymes that can digest a path across through the jelly coat to the egg surface (Fig. 50.2).

The acrosomal reaction in sea urchin is considered to be initiated by a fucose-containing polysaccharide present in the egg jelly that binds to the sperm and allows calcium to enter into the sperm head. The exocytosis of the acrosomal vesicle is caused by the calcium-mediated fusion of the acrosomal membrane with the adjacent sperm plasma membrane. The egg jelly factors that initiate the acrosomal reaction in sea urchins are often highly species-specific.

(*ii*) Extension of acrosomal process. The second part of the acrosomal reaction involves the extension of the acrosomal process. This protrusion arises through the polymerization of globular actin molecules into actin filaments. As soon as the acrosomal process of the sperm contacts the surface of the egg, a major **species-specific recognition step** occurs. The acrosomal protein mediating this recognition is called **bindin** (an insoluble 30,500 Da protein). Bindin occurs at the tip of acrosomal filament and adhere to species-specific receptors occurring on the surface of vitelline envelope of egg.

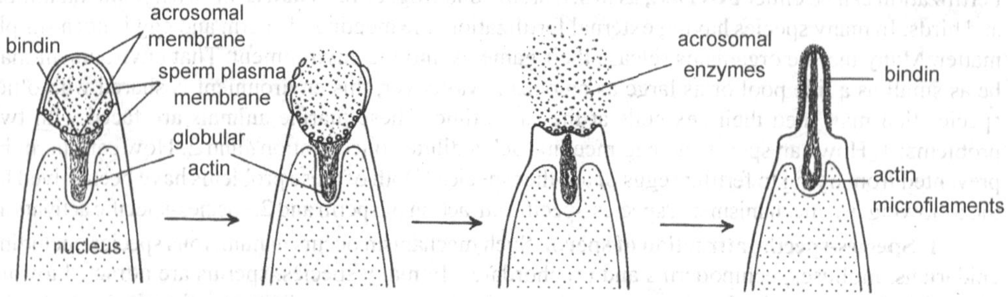

Fig. 50.2. Acrosomal reaction in sea urchin sperm (after Gilbert, 2000).

Gamete binding and recognition in mammals. In case of mammals, the sperm has to penetrate several physical barriers to enter the egg (Fig. 50.3). In mammalian egg, the first barrier is a layer of cumulus cells, embedded in a sticky mass of hyaluronic acid. Hyaluronidase activity on the surface of the sperm head helps it to penetrate this layer. The sperm next encounters the **zona pellucida**, a layer of glycoproteins surrounding the egg. This also acts as a physical barrier, but sperms are helped to penetrate it by acrosomal reaction—the release of the contents of the acrosomal vesicle located in the sperm head.

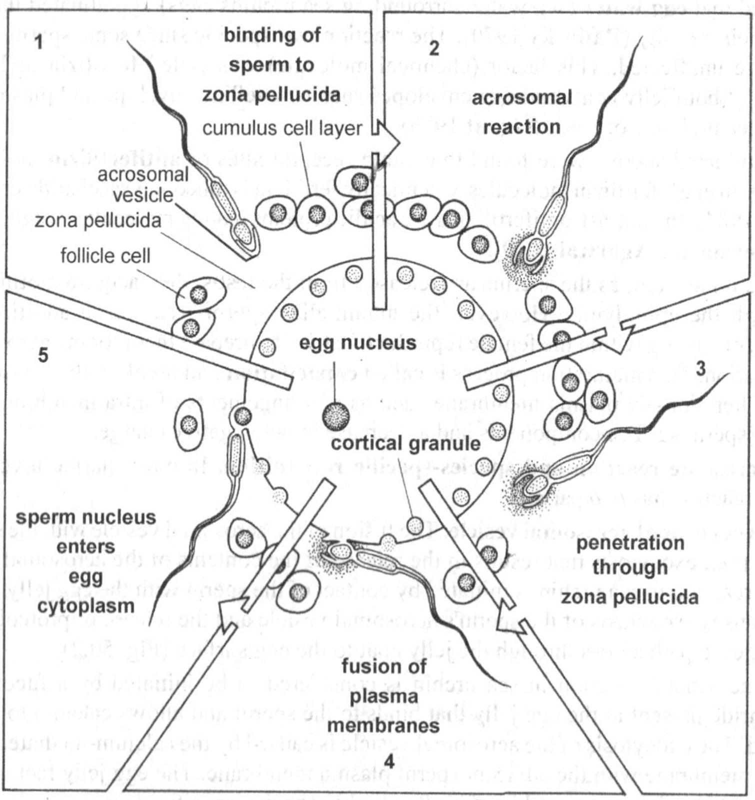

Fig. 50.3. Fertilization of a mammalian egg. A—After penetrating the follicle-derived cumulus cell layer, the sperm binds to the zona pellucida. B—Acrosomal reaction. C—Penetration of sperm through zona pellucida. D—Fusion of plasma membranes of sperm and ovum. E—Exocytosis of cortical granules due to egg activation and entry of sperm nucleus inside the ovum (after Wolpert, 2002).

Unlike acrosomal reaction of sea urchin, the acrosomal reaction in mammals occurs only after sperm has bound to the zona pellucida. Sperm cell surface of mammals has receptor proteins (β1, 4-galactosyltransferase) which specifically bind to the ZP3 (a glycoprotein) of the zona pellucida. The ZP3 protein initiates acrosomal reaction, *i.e.*, exocytosis of acrosome. The mouse sperm acrosomal reaction is induced by the cross-linking of ZP3 with receptors for it on the sperm membrane. This cross-linking opens calcium channels to increase the concentration of calcium in sperm (**Florman** *et al.*, 1998).

Box 50.1

Fertilization of mammalian eggs plays two major roles: 1. It binds the sperm; 2. It initiates the acrosomal reaction. The binding of sperm to the zona pellucida is relatively, but not absolutely, species-specific. According to **Gilbert** (2000), species-specifizc gamete recognition is not a major problem when fertilization occurs internally.

During the acrosomal reaction, the anterior portion of the sperm plasma membrane is shed from the sperm (Fig. 50.4). This region is where the ZP3-binding proteins are located, and yet the sperm must still remain bound to the zona pellucida in order to lyse a path through it. In mice, it appears that **secondary binding** to the zona is attained by proteins in the acrosomal membrane that bind specifically to ZP2 (**Bleil** *et al.*, 1988).

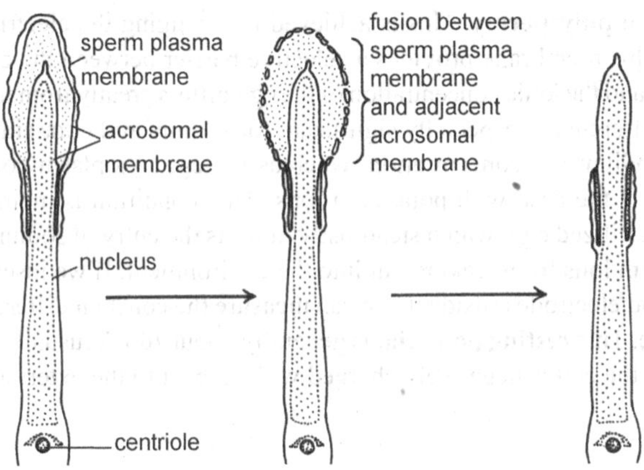

Fig. 50.4. Acrosomal reaction in hamster sperm (after Gilbert, 2000).

The enzymes released by exocytosis of acrosome include **β-N-acetyl glucosaminadase**, which breaks down the oligosaccharide side chains on the glycoproteins of zona pellucida and a protease enzyme, called **acrosin**. These enzymes allow the sperm to approach the egg plasma membrane.

II. Gamete Fusion and the Prevention of Polyspermy

This process involves the following two events:

1. Fusion of the egg and sperm plasma membranes. Sperm-egg binding in sea urchin causes the extension of several microvilli to form the **fertilization cone** by the egg surface. Homology between the egg and the sperm is again demonstrated, because the transitory fertilization cone, like the acrosomal process, appears to be extended by the polymerization of actin. The sperm and egg plasma membranes then join together and material from sperm membrane can later be found on the egg membrane. The sperm nucleus, mitochondria, centriole and tail pass through the resulting cytoplasmic bridge, which is widened by the action of actin polymerization. A similar process occurs during fusion of mammalian gametes.

Membrane fusion is an active process, often mediated by specific "**fusogenic**" **proteins.** In sea urchin, bindin is found to play a second role as a fusogenic protein. In mammals, the **fertilin** proteins of the sperm plasma membrane are essential for sperm membrane and egg membrane fusion. Mouse fertilin is localized to the posterior plasma membrane of the sperm head. A key egg receptor for the sperm's fertilin is the integrin-like protein **CD9**; interaction of these molecules may initiate sperm and egg fusion (**Wasserman** 1999).

2. Block to polyspermy. As soon as one sperm has entered the egg, the fusibility of the egg membrane, which was so necessary to get the sperm inside the egg, becomes a dangerous liability. In sea urchins and other animals, any sperm that enters the egg can provide a haploid nucleus and a centriole to the egg. In normal **monospermy**, in which only one sperm enters the egg, a haploid sperm nucleus and a haploid egg nucleus combine to form the diploid nucleus of the fertilized egg (zygote). This tends to restore the chromosome number appropriate for the species. The centriole, which is provided by the sperm, will divide to form the two poles of the mitotic spindle during cleavage.

The entrance of multiple sperm—**polyspermy**—leads to disasterous outcomes in most organisms. In some animals, such as birds, many sperms penetrate the egg but all but one destroyed in the cytoplasm. This is called **physiological polyspermy**.

Species have evolved ways to prevent the union of more than two haploid nuclei. The most common way is to prevent the entry of more than one sperm in the egg. Sea urchin egg has following *two* mechanisms to avoid polyspermy: one is a fast reaction and other is a slower reaction.

***(i)* Fast block to polyspermy.** This is achieved by changing the electric potential of the egg plasma membrane. This membrane provides a selective barrier between the egg cytoplasm and the outside environment, and the ionic concentration of the egg differs greatly from that of its surroundings. This concentration difference is especially significant for sodium and potassium ions. Sea water has a particularly high sodium ion concentration, whereas the egg cytoplasm contains relatively little sodium. The reverse is the case with potassium ions. This condition is maintained by the plasma membrane of the unfertilized egg, which steadfastly inhibits the entry of sodium ions into the oocyte and prevents potassium ions from leaking out into the environment. If we insert an electrode into an egg and place a second electrode outside it, we can measure the constant differences in charge across the plasma membrane. This **resting potential** is generally about 70m V, usually expressed as—70mV because the inside of the cell is negatively charged with respect to the exterior.

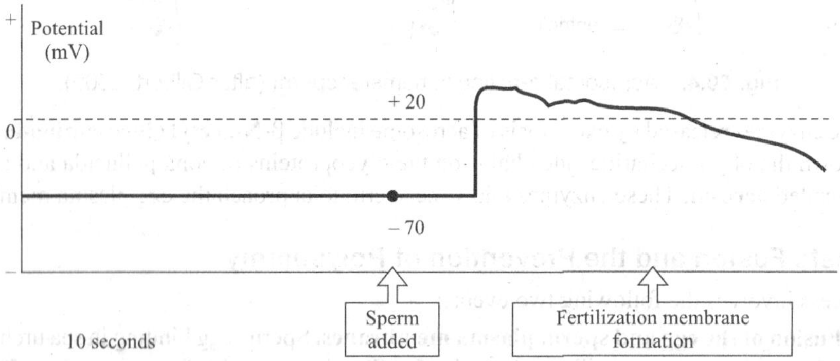

Fig. 50.5. Depolarization of the sea urchin egg plasma membrane at fertilization. The resting membrane potential of the unfertilized sea urchin egg is—70mV. At fertilization, it changes rapidly to +20mV, and then slowly returns to the original value. This depolarization may provide a fast block to polyspermy (after Wolpert, 2002).

Within 1–3 seconds after the binding of the first sperm, the membrane potential shifts to a positive level, about +20mV. This change is caused by a small influx of sodium ions into the egg. Although sperm can fuse with membranes having a resting potential of—70mV, they cannot fuse with membranes having a positive resting potential, so no more sperm can fuse to the egg.

***(ii)* Slow block to polyspermy.** Brief potential shift, however, is not sufficient to prevent polyspermy, which can still occur if the sperm bound to the vitelline envelope are not somehow removed. This removal is accomplished by the **cortical granule reaction** a slower, mechanical block to polyspermy that becomes active about a minute after the first successful sperm egg attachment.

In sea urchin, upon sperm entry, the cortical granules fuse with the egg plasma membrane and release their contents into the space between the plasma membrane and the fibrous mat of vitelline envelope. Several proteins are released during exocytosis of about 15000 cortical granules of sea urchin egg. 1. The *first* are **protease enzymes**. These enzymes dissolve the protein posts that connects the vitelline envelope proteins to the plasma membrane. Proteases also clip off the bindin receptors and any sperm attached to it. 2. **Mucopolysaccharides** released by the cortical granules produce an osmotic gradient that causes water to rush into the space between the plasma membrane and the vitelline envelope. This causes the envelop to expand and to become the **fertilization envelope**. 3. A third protein released by the cortical granules, a **peroxidase enzyme**, hardens the fertilization envelope by crosslinking *tyrosine* residues on adjacent proteins. Thus, the fertilization envelope starts to form at the site of sperm entry and continues its expansion around the egg (Fig. 50.6). As it forms, bound sperm are released from the envelope. 4. Finally, a fourth cortical granule protein, **hyaline**, forms a coating around the egg. The egg extends elongated microvilli whose tips attach to this **hyaline layer**. This layer provides support for the blastomeres during cleavage.

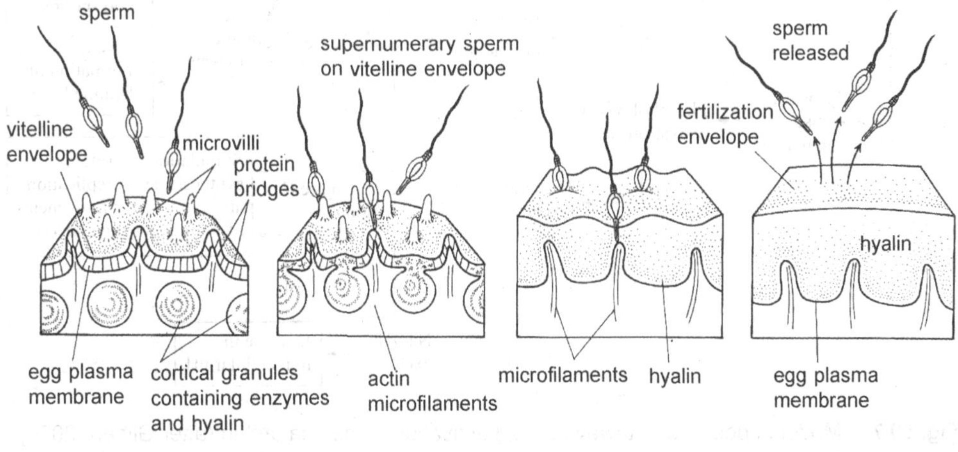

Fig. 50.6. Cortical granules exocytosis for slow block to polyspermy in sea urchin. Schematic diagram showing the events leading to the formation of the fertilization envelope and the hyaline layer (Gilbert, 2000).

Zona reaction of mammals. In mammals, the cortical reaction does not elevate the fertilization envelope, but the ultimate effect is the same: Released enzymes due to exocytosis of cortical granules modify the zona pellucida's sperm receptors in such a way that they can no longer bind the sperm. During this process, called **zona reaction**, both ZP3 and ZP2 glycoproteins of zona pellucida become modified.

The mechanism of the cortical granule reaction is similar to that of the acrosomal reaction. Upon fertilization, the intracellular calcium ion concentration of the egg increases greatly. In this high-calcium environment, the cortical granule membranes fuse with the egg plasma membrane, releasing their contents. Once the fusion of cortical granules begins near the point of sperm entry, a wave of cortical granule exocytosis propagates around the cortex to the opposite side of the egg. The calcium ions responsible for the cortical granule reaction are stored in the endoplasmic reticulum which occurs abundantly in the egg cortex.

III. Activation of Egg Metabolism

Generally fertilization is defined as merely the means to merge two haploid nuclei, it has an equally important role in initiating the process that begin development. These events happen in the cytoplasm of fertilized ovum and occur without the involvement of the nuclei. The mature sea urchin egg is a metabolically sluggish cell that is activated by the sperm. This activation is merely a stimulus, however; it sets into action a preprogrammed set of metabolic events. The responses of egg to the sperm can be divided into 'early' responses, which occur within seconds of the cortical reaction, and "late" responses, which take place several minutes after fertilization (Fig. 50.7).

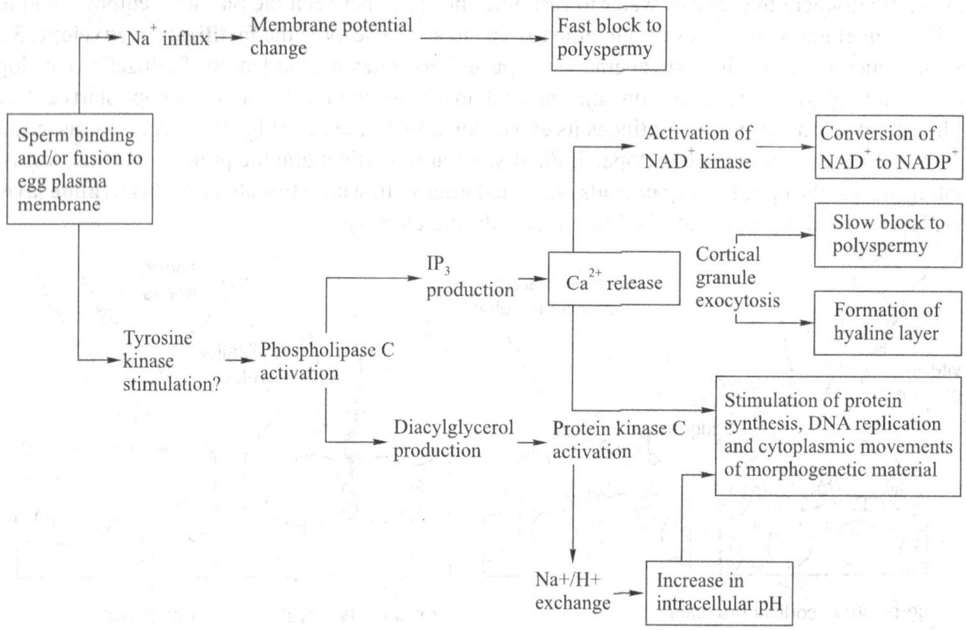

Fig. 50.7. Model of possible pathways of egg activation in the sea urchin (after Gilbert, 2000).

Early response of egg activation. It has been already discussed in previous section that how contact between sea urchin sperm and egg activates the two sorts of blocks to polyspermy: the fast block, initiated by sodium influx into the cell, and the slow block, initiated by the intracellular release of calcium ions. *The activation of all eggs appear to depend on an increase in the concentration of free calcium ions within the egg.* Such an increase of Ca^{2+} ions can occur in two ways : calcium ions can enter the egg from outside or calcium ions can be released from the endoplasmic reticulum within the egg. Both of these mechanisms are used to different degrees in different species. In snails and worms, much of the calcium probably enters the egg from outside, while in fishes, frogs, sea urchins and mammals, most of the calcium ions probably come from the endoplasmic reticulum. In both cases, a wave of calcium ions sweeps across the egg, beginning at the site of sperm-egg fusion (**Jaffe**, 1983).

Release of calcium activates a series of metabolic reactions (Fig. 50.7). One of these is the activation of the enzyme NAD^+ kinase, which converts NAD^+ to $NADP^+$. This change may have important consequences for lipid metabolism, since $NADP^+$ (but not NAD^+) can be used as a coenzyme for lipid biosynthesis. Thus, the conversion of NAD^+ to $NADP^+$ may be important in the synthesis of the many new cell membranes required during cleavage. Another effect of calcium release involves oxygen consumption. A burst of oxygen reduction (to hydrogen peroxide) is seen during fertilization, and much of this "respiratory burst" is used to crosslink the fertilization envelope. The enzyme responsible for this reduction of oxygen is also NADPH dependent. Lastly, NADPH helps regenerating **glutathione** and **ovothiols**, which may be crucial for scavenging free radicals that could otherwise damage the DNA of the egg and early embryo (**Mead** and **Epel**, 1995).

Box 50.2

Ovothiol. It is a recently described sulphur amino acid which is found in fertilized eggs, where it plays a role comparable to that glutathione. Ovothiol, which is present at 5 mM concentration in sea urchin eggs, protects the eggs against oxidative damage by peroxides at the egg surface early in fertilization. Ovothiol is in turn reduced by glutatione.

Glutathione. It is a cysteine-containing tripeptide, which is abundant in all cells and protects them against metabolic stresses. It can nonenzymatically reduce a number of substances, such as peroxides or free radicals, which accumulate in cells under oxidizing conditions. Glutathione maintains an intracellular reducing environment. Glutathione also detoxifies a variety of substances of cell (**Mathews** *et al.*, 2000).

Late responses of egg activation. Shortly after the calcium ion levels rise in a sea urchin eggs, its intracellular pH also increases. The rise in intracellular pH begins with a second influx of sodium ions, which causes a 1 : 1 exchange between sodium ions from the sea water and hydrogen ions from the egg. This loss of hydrogen ions causes pH to rise. It is thought that the pH increase and calcium ion elevation act together to stimulate new protein synthesis and DNA synthesis (**Rees** *et al.*, 1995). This protein synthesis does not depend on the synthesis of new mRNAs; rather, it utilises mRNAs already present in the oocyte cytoplasm. The messages include mRNAs encoding proteins such as histones, tubulins, actins and morphogenetic factors that are utilized during early development.

Fusion of Genetic Material

***(i)* In sea urchin.** In sea urchins, the sperm nucleus enters the egg perpendicular to the egg surface. After fusion of the sperm and egg plasma membrane, the sperm nucleus and its centriole separate from the mitochondria and the flagellum. The mitochondria and the flagellum disintegrate inside the egg, so very few, if any, sperm-derived mitochondria are found in developing or adult organisms. In mice, it is estimated that only 1 out of every 10,000 mitochondria is sperm derived. Thus, although each gamete contributes a haploid genome to the zygote, the mitochondrial genome is transmitted primarily by the maternal parent. Conversely, in almost all animals studied (the mouse being the major exception), the centrosome needed to produce the mitotic spindle of the subsequent divisions is derived from the sperm centriole.

Further, the egg nucleus, once it is haploid, is called the **female pronucleus**. Once inside the egg, the sperm nucleus decondenses to form the **male pronucleus**. The sperm nucleus undergoes a dramatic transformation. The nuclear envelope vesiculates into small packets, thereby exposing the compact sperm chromatin to the egg cytoplasm. The proteins holding the sperm chromatin in its condensed, inactive state are exchanged for other proteins derived from the egg cytoplasm. This exchange permits the decondensation of sperm chromatin. Once decondensed the DNA can begin transcription and replication.

Following the entry of sperm of the sea urchin inside the egg cytoplasm, the male pronucleus rotates 180°, so that the sperm centriole is placed between the sperm pronucleus and the egg pronucleus. The sperm centriole then acts as a microtubule organizing centre, extending its own microtubules and integrating them with egg microtubules to form an aster. These microtubules extend throughout the egg and contact the female pronucleus, and the two pronuclei migrate toward each other. Their fusion form the **diploid zygote nucleus.**

***(ii)* In mammals.** In mammals, the pronuclei disintegrate as they approach each other, and their chromosomes gather around a common metaphase plate. In zygote, microtubular changes cause cytoplasmic movements. These rearrangements of cytoplasm can be critical in specifying which portions of the egg are going to develop into which organ.

50.2. POLLINATION AND FORMATION OF SEED

In plants, the prerequisite of fertilization is pollination.

1. Pollination

Pollination refers to the landing and subsequent germination of the pollen on the stigma. Hence, it involves an interaction between the gametophytic generation of the male (the pollen) and the sporophytic generation of the female (the stigmatic surface of carpel). The arrival of viable pollen grains on a receptive stigma does not guarantee fertilization (Box 50.3).

Box 50.3 Incompatibility of gametes

In plants, interspecific incompatibility refers to the failure of pollen from one species to germinate and/or grow on the stigma of another species. Intraspecific incompatibility is incompatibility that occurs within a species. Self incompatibility blocks fertilization between two genetically similar gametes, increasing the probability of new combinations by promoting outcrossing (pollination by a different individual of same species). Groups of closely related plants can contain a mix of self-compatible and self-incompatible species.

If the pollen and the stigma are compatible, the pollen takes up water (hydrates) and pollen tube emerges. The pollen tube grows down the style of the carpel toward the micropyle. The **tube nucleus** and the **sperm cells** are kept at the growing tip by bands of **callose** (a complex carbohydrate). It is possible that this may be an exception to the "*plant cells do not move*" *rule*, as the generative cell(s) appear to move ahead via adhesive molecules (**Lord** *et al.* 1996). Pollen tube growth is quite slow in gymnosperms (up to a year), while in some angiosperms the tube can grow as rapidly as 1 cm/hour.

In the growth of the pollen tube calcium is found to play an essential role. Calcium accumulates in the tip of the pollen tubes, where open calcium channels are concentrated. There is a direct evidence that pollen tube growth in the field poppy is regulated by a slow-moving calcium wave controlled by the phosphoinositide signalling pathway (Fig. 50.8). The phosphoinositides play important roles as precursors to second messengers in **transmembrane signalling** (*i.e.*, the transmission of an extracellular signal to some element of the intracellular metabolic apparatus) (**Mathews** 2000). Cytoskeletal investigations show that organelle positioning during pollen tube growth depends on interactions with cytoskeletal components. This must link to signalling.

Genetic approaches have been useful in investigating how the growing pollen tube is guided toward unfertilized ovules. In *Arabidopsis*, the pollen tube appears to be guided by a long-distance signal from the ovule (**Hulskamp** *et al.*, 1995; **Wilhelmi** and **Preuss**, 1999). Analysis of pollen tube growth in ovule mutants of *Arabidopsis* indicates that the haploid embryo sac is particularly important in the long-range guidance of pollen tube growth. Mutants with defective sporophyte tissue in the ovule but a normal haploid embryo sac appear to stimulate normal pollen tube development.

Two *Arabidopsis* genes, *POP2* and *POP3*, have been identified that specifically guide pollen tubes to the ovule with no apparent effect on the plant (**Wilhelmi** and **Preuss**, 1996, 1999). These genes function in both the pollen and the pistil, thus implicating the sporophyte generation in the guidance system.

2. Fertilization

The growing pollen tube enters the embryo sac through the micropyle and grows through one of the synergids (Fig. 50.9). The two sperm cells are released and a **double fertilization** event occurs. One sperm cell fuses with the egg, producing the zygote that will develop into the sporophyte. The second sperm cell fuses with the bi- or multi-nucleate central cells, giving rise to **endosperm**, which nourishes the developing embryo. This second event is not true fertilization in the sense of male and female gametes undergoing syngamy (fusion). That is, it does not result in a zygote but in nutritionally supportive tissue. The other accessory cells in the embryo sac degenerate after fertilization.

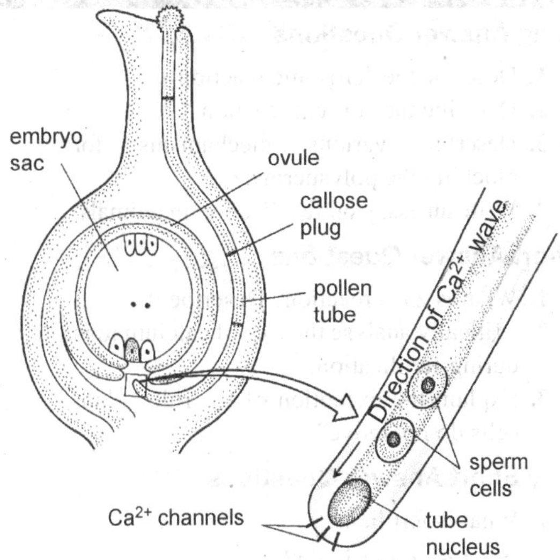

Fig. 50.8. Calcium and pollen tube tip growth. After compatible pollen germinates, the pollen tube grows toward the micropyle. Calcium plays a key role in the growth of the tube (after Gilbert, 2000).

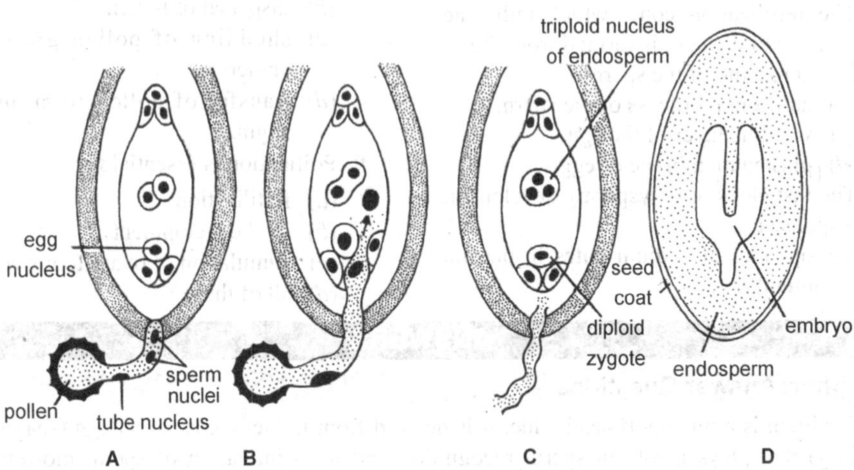

Fig. 50.9. Fertilization and development of a seed in an angiospermic plant (after Winchester, 1966; Stansfield, 1969).

QUESTIONS

Long Answer Questions

1. Describe the acrosome reaction.
2. Describe the cortical reaction.
3. Describe various mechanisms for blocking the polyspermy.
4. Write an essay on fertilization in animals.

Short Answer Questions

1. What is zona reaction? Describe it.
2. Critically analyse the role of calcium ions during fertilization.
3. Explain the exception of the rule "plant cells do not move".

Very Short Answer Questions

1. What is fertilizin?
2. What is antifertilizin?

Multiple Choice Questions

Choose the correct answer from the four alternatives given.

1. Fertilization of egg by the sperm in the human female genital tract takes place in
   (*a*) uterus
   (*b*) ovary
   (*c*) vagina
   (*d*) oviduct (Fallopian tube)
2. The fertilization cone, which pulls the sperm into the egg, is formed from the
   (*a*) acrosome of the sperm
   (*b*) acrosomal process of the sperm
   (*c*) vitelline layer of the egg
   (*d*) plasma membrane of egg
3. The fast block to polyspermy develops in response to
   (*a*) opening of sodium gates in plasma membrane
   (*b*) release of bindin
   (*c*) spreading of fertilizin around the egg
   (*d*) formation of fertilization membrane
4. The slow block to polyspermy develops in response to the
   (*a*) opening of sodium gates in the plasma membrane
   (*b*) release of bindin
   (*c*) spreading of fertilizin around the egg
   (*d*) formation of the fertilization membrane
5. Endosperm in an angiosperm is .
   (*a*) haploid (*b*) diploid
   (*c*) triploid (*d*) tetraploid
6. Function of synergids is to
   (*a*) attract and guide the pollen tube
   (*b*) protect the egg from pathogen
   (*c*) produce additional embryos
   (*d*) fuse with extra male gametes and form endosperm
7. The lytic enzyme released by sperm is
   (*a*) ligase (*b*) acrosome
   (*c*) androgamone (*d*) hyaluronidase
8. Pollination is
   (*a*) fertilization of plants
   (*b*) dispersal of pollen
   (*c*) shedding of pollen grains from anthers
   (*d*) transfer of pollen from anther to stigma
9. Pollination is essential for
   (*a*) fertilization
   (*b*) seed development
   (*c*) stimulation of ovary to grow
   (*d*) all of these

ANSWERS

Very Short Answer Questions

1. Fertilizin is a mucopolysaccharide. It is derived from the jelly coat of an egg (sea urchin, frog) that plays a role in sperm recognition and the stimulation of sperm motility and metabolic activity.
2. Antifertilizin is an immunologically specific substance produced by animal sperm to implement attraction by the egg before fertilization.

Multiple Choice Questions

1. (*d*) 2. (*d*) 3. (*a*) 4. (*d*) 5. (*c*) 6. (*a*) 7. (*d*)
8. (*d*) 9. (*d*)

51 Parthenogenesis

Usually an unfertilized ovum develops into a new individual only after the fertilization but in certain cases the development of the egg takes place without the fertilization. This peculiar mode of sexual reproduction in which egg development occurs without the fertilization is known as the **parthenogenesis** (Gr., *parthenos* = virgin; *genesis* = origin). An organism that has developed parthenogenetically is called a **parthenogenone** or **parthenote**. The phenomenon of the parthenogenesis occurs in different groups of the animals as in certain insects (Hymenoptera, Homoptera, Coleoptera), crustaceans, rotifers and also in some vertebrates such as several desert lizards, turkeys and some mammals.

There are certain conditions which are intermediate between parthenogenesis and fertilization, *e.g.*, partial fertilization, gynogenesis, androgenesis and merogony. 1. In **partial parthenogenesis**, the egg may be fertilized by only a part of sperm. For example, according to **Boveri** in sea urchin egg the fertilization of the egg (activation) takes place by the sperm aster. The sperm nucleus gets fused with the egg nucleus only in two cell stage. 2. In **gynogenesis**, the sperm penetrates the egg but takes no part in development. It degenerates in the egg without fusion with the egg nucleus, *e.g.*, *Rhabditis aberrans*. 3. In **androgenesis**, the egg is activated by the sperm and development takes place without the participation of the egg nucleus. For example, if the **ova** of frogs and toads are treated with radium and then fertilized by normal sperms, the egg nucleus does not take part in development, but sperm (paternal) nucleus participates in normal development. 4. In **merogony**, egg fragments devoid of nucleus develop when fertilized by a normal sperm. If sea urchin eggs are shaken to produce small pieces, the fragments round up to form spheres. Some of these spheres are without nuclei. If such enucleated spheres are normally fertilized, they may develop into dwarf larvae.

The parthenogenesis may be of two types:

1. Natural parthenogenesis; 2. Artificial parthenogenesis.

51.1. NATURAL PARTHENOGENESIS

In certain animals the parthenogenesis occurs regularly, constantly and naturally in their life cycles and is known as the **natural parthenogensis**. The natural parthenogenesis may be of two types, *viz.*, complete or incomplete.

***(i)* Complete parthenogenesis.** Certain insects have no sexual phase and no males. They depend exclusively on the parthenogenesis for the self-reproduction. This type of parthenogenesis is known as the **complete parthenogenesis** or **obligatory parthenogenesis.** It is found in some species of earthworms, badelloid rotifers, grasshoppers, roaches, phasmids, moths, gall flies, fishes, salamanders and lizards.

***(ii)* Incomplete parthenogensis.** The life cycle of certain insects includes two generations, the sexual generation and parthenogenetic generation, both of which alternate to each other. In such cases, the diploid eggs produce females and the unfertilized egg produce males. This type of parthenogenesis is known as the **partial** or **incomplete** or **cyclic parthenogenesis**.

Cyclic parthenogenesis shows several variations in the alternation of sexual (S) and parthenogenetic (P) generations : (1) In gall flies (*e.g.*, *Neuroterus*) there is an alternation of one sexual and one parthenogenetic generation per year (P, S, ... P, S, ... P, S). (2) In aphids (plant lice), daphnids and rotifers, the sexual generation may come after many parthenogenetic generations during the summer of the year (P, P, P, P, P, P, S, ... P, P, P, P, P, P, S ...). (3) In gall midge (*Miaster*) the larvae reproduce indefinitely by **paedogenetic parthenogenesis**. In this case, germ cells within the larvae develop parthenogenetically into parasitic larvae which feed on the mother larvae. These larvae tend to live under the bark of rotting logs and feed on fungus. Under favourable conditions winged males and females are produced. These stages reproduce sexually and help in dispersion. (4) In some groups there is no regularity between parthenogenetic and sexual generations.

The complete and incomplete type of natural parthenogenesis may be of following two types:

(*a*) Haploid or arrhenotokous parthenogenesis; (*b*) Diploid or thelytokous parthenogenesis.

(*a*) Haploid or arrhenotokous parthenogenesis. In the arrhenotokous parthenogenesis, the haploid eggs are not fertilized by the sperms and develop into the haploid individuals (Fig. 51.1). In these cases the haploid individuals are always males and the diploid individuals are the females *e.g.*, 1. Insects: (*i*) Hymenoptera (bees and wasps), (*ii*) Homoptera, (*iii*) Coleoptera (*Micromalthus debilis*), (*iv*) Thysanoptera (*Anthothrips verbasi*). 2. Arachnids, *e.g.*, ticks, mites and certain spiders (*Pediculoids ventricusm*). 3. Rotifers, *e.g.*, *Asplanchne amphora.*

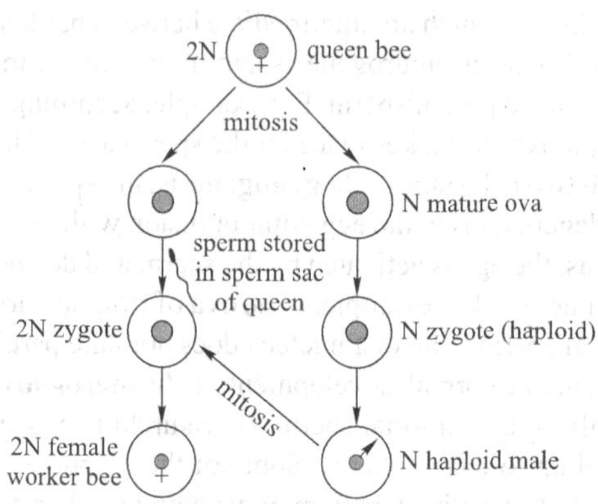

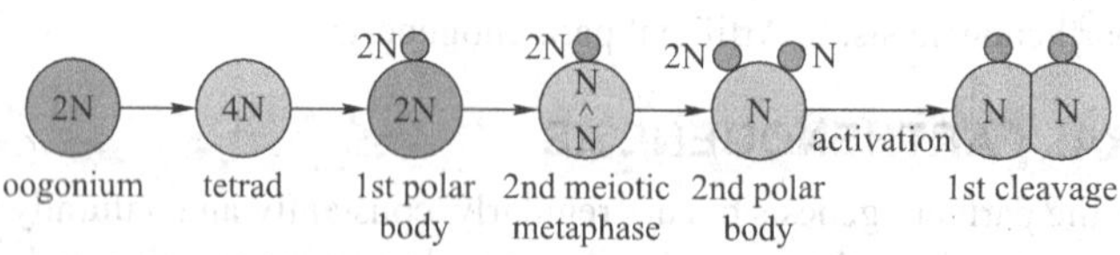

Fig. 51.1. Schematic representation of haploid parthenogenesis in bees (after Grant, 1978).

Thus, the queen bee is fertilized only once by one or many males (drones). She stores the sperm in her seminal receptacles and as she lays her eggs, she can either fertilize the eggs or allow them to pass unfertilized. The fertilized eggs become females (fertile queens or sterile workers depending upon the amount of royal jelly the developing young receives); the unfertilized eggs become fertile males or drones.

(*b*) Diploid or thelytokous parthenogenesis. In the diploid parthenogenesis, the young individuals develop from the unfertilized diploid eggs. The offspring of thelytoky could theoretically be either male or female; but normally it produces only diploid females (Fig. 51.2). For example, in aphids, females emerging in the spring produce several generations of females by diploid parthenogenesis resulting from suppression of first or second polar body. At summer's end some females produce sexual males and females by diploid parthenogenesis, males differing from females in lacking one sex chromosome. Males produce haploid gametes through normal meiosis which fuse to form diploid zygotes that emerge again in the spring as parthenogenetic females. Further, since thelytoky is also found in polyploid forms, it is also called **somatic parthenogenesis.**

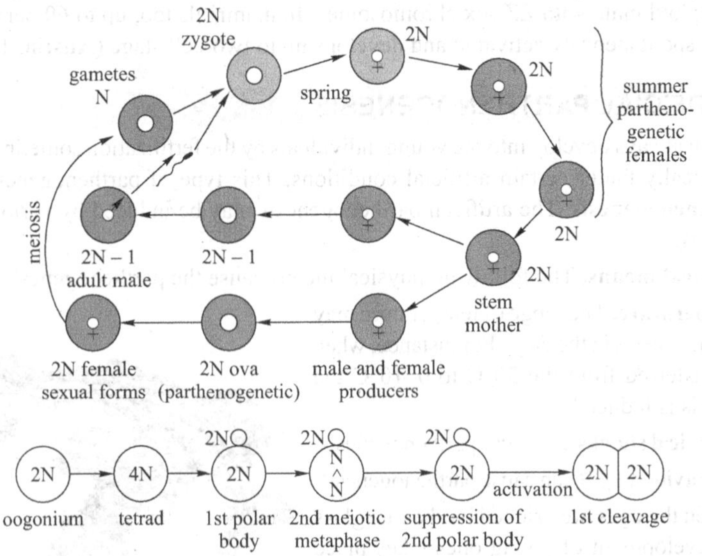

Fig. 51.2. Schematic representation of diploid parthenogenesis in aphids (after Grant, 1978).

Following types of the thelytoky have been recognised:

(*i*) Ameiotic parthenogenesis. Sometimes during the oogenesis, first meiotic or reduction divsion does not occur but second meiotic division occurs as usual. Such eggs contain diploid number of chromosomes and develop into new individuals without the fertilization. This type of parthenogenesis is known as **apomiotic** or **ameiotic parthenogenesis** and occurs in *Trichoniscus* (Isopoda), *Daphnia pulex* (Crustacea), *Compelona rufum* (Mollusca), weevils and long-horned grasshoppers.

(*ii*) Meiotic parthenogenesis. Certain eggs develop by the usual process of oogenesis but at certain stages **diplosis** or doubling of chromosome number and production of diploid eggs occur. Such eggs develop into the diploid individuals and this phenomenon is known as the **meiotic parthenogenesis**.

The **diplosis** of the diploid thelytoky may occur by the following methods:

(*a*) By autofertilization. In certain cases the oocyte divides meiotically up to the formation of ootid and secondary polocyte. But the ootid and the secondary polocyte unite together to form a diploid egg which develops into a new individual, *e.g.*, *Artemia salina* (Crustacea) and various other organisms.

(*b*) By restitution. Sometimes in primary oocyte karyokinesis forms a nucleus of the secondary oocyte and a nucleus of the first polocyte. But the karyokinesis is not followed by the cytokinesis. The chromosomes of both daughter nuclei are arranged on the equator and undergo second meiotic division to form a diploid ootid and a diploid polocyte. The dipioid ootid or ovum develops into a parthenogenetic diploid individual. This type of diplosis is known as the **restitution**, *e.g.*, insects of order Hymenoptera (*Nemertis conesceus*) and Lepidoptera.

Natural Parthenogenesis in Vertebrates

A few cases of natural parthenogenesis have also been reported in the vertebrates. The fish *Carasius auratus gibelio* is reported to consist of females only (**Lieder**, 1955). Likewise, males are found totally lacking in the lizard *Lacerta sexicola armeniaca* (**Lantz** and **Cyren**, 1936). In it females are reported to be originated by parthenogenesis. In turkeys 80 per cent of incubated eggs show early cleavage stages. Such parthenogentic forms have hatched and grown to reproducing adults which are found to be diploid male with ZZ sex chromosomes. In mammals too, up to 60 per cent of hamster eggs becomes spontaneously activated and develops up to two-cell stage (**Austin**, 1956).

51.2. ARTIFICIAL PARTHENOGENESIS

The eggs which always develop into the young individuals by the fertilization sometimes may develop parthenogenetically under certain artificial conditions. This type of parthenogenesis is known as **artificial parthenogenesis**. The artificial parthenogenesis may be induced by various physical and chemical means.

A. Physical means. The following physical means cause the parthenogenesis:

(*i*) **Temperature.** The range of temperature may induce parthenogenesis in the eggs. For instance, when the egg is transferred from the 30°C to 0–10°C the parthenogenesis is induced.

(*ii*) **Electrical shocks** can cause parthenogenesis.

(*iii*) **Ultraviolet light** can cause parthenogenesis.

(*iv*) When the eggs are pricked by the fine glass needles the development of young ones takes place parthenogenetically.

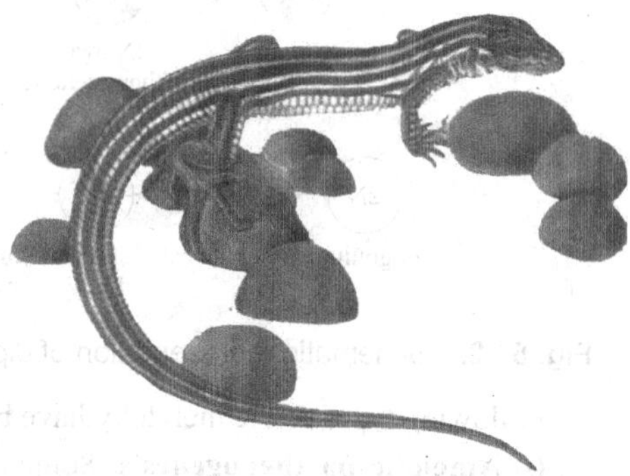

This whiptail lizard reproduces by parthenogenesis.

B. Chemical means. The following chemicals have been found to cause parthenogenesis in the normal eggs:

1. Chloroform. 2. Strychnine. 3. Hypertonic and Hypotonic sea waters. 4. Chlorides of K^+, Ca^{++}, Na^{++}, Mg^{++}, etc. 5. Acids as butyric acid, lactic acid, oleic acid and other fatty acids. 6. Fat solvents, *e.g.*, toulene, ether, alcohol, benzene and acetone. 7. Urea and sucrose.

The artificial parthenogenesis has been induced by above mentioned physical and chemical means by various workers in the eggs of most echinoderms, molluscs, annelids, amphibians, birds and mammals.

51.3. SIGNIFICANCE OF PARTHENOGENESIS

1. The parthenogenesis serves as the means for the determination of sex in the honey bees, wasps etc., and it supports the chromosome theory of inheritance.
2. The parthenogenesis is the most simple, stable and easy process of the reproduction, *e.g.*, aphids (insects).
3. The parthenogenesis eliminates the variation from the population, but encourages development of the advantageous mutant characters.
4. The parthenogenesis causes the polyploidy in the organisms.

5. Due to the parthenogenesis, there is no need for the organisms to waste their energy in the process of mating but it allows them to utilize that amount of energy in the feeding and reproduction.
6. Honey bee and other social insects also control their sex ratio by parthenogenesis.
7. In aphids, parthenogenesis is a means of rapid breeding; the females reproduce by diploid parthenogenesis during summer.

QUESIONS

Long Answer Question

1. What is parthenogenesis? Describe different types of natural parthenogenesis in animals.

Short Answer Questions

1. Give an account of artificial parthenogenesis.
2. What is the significance of parthenogenesis?

Fill in the Blanks

1. In honey bee ________ serves as the means for ________.

ANSWER

Fill in the Blanks

1. parthenogenesis, sex determination

52 Aging: The Biology of Senescence

Development does not cease once birth has occurred but continues throughout the stages of life: infancy, childhood, adolescence and adulthood. **Aging** or **senescence** encompasses these progressive changes that contribute to an increased risk of infirmity, disease and death (Box 52.1). The study of biology of senescence or aging is called **gerontology** (**Mader** 1998).

Box 52.1

When entropy wins, aging or senescence sets in (**Gilbert** 2010). Entropy is a measure of the unavailability of a system's thermal energy for conversion into mechanical work. In some contexts, entropy is interpreted as a measure of the degree of disorder or randomness in the system.

Aging is a slow process during which the body undergoes changes that eventually bring about death, even if no marked disease or disorder is present. Medical science is trying to extend the human life span and the health span, the length of time the body functions normally.

A multicellular organism is able to develop and maintain its identity for only so long before deterioration prevails over synthesis, and the organism ages. Aging can be defined as *the time-related deterioration of the physiological functions necessary for survival and fertility* (**Gilbert** 2010). The characteristics of aging—as distinguished from diseases of aging, such as cancer and heart disease—affect all individuals of a species. The aging process has two major facets. The first is simply how long an organism lives (maximum life span); the second concerns the physiological deterioration, or senescence, that characterizes old age. These topics are often viewed as being interrelated.

52.1. MAXIMUM LIFE SPAN (GENES AND AGING)

Genetic factors play roles both between species and within species. The **maximum life span**, which is the maximum number of years any member of a given species has been known to survive, is characteristic of a species. The maximum human life is estimated to be 121 years (**Arking** 1998). The life spans of some tortoises and lake trout are both unknown but are estimated to be more than 150 years. The maximum life span of a domestic dog is about 20 years and that of a laboratory mouse is 4.5 years. If a *Drosophila* fruit fly survives to eclose (in the wild, more than 90% die as larva), it has a maximum life span of 3 months. (**Note.** *Eclosion* means emergence of a fruit fly or any other holometabolous insect from its puparium; see **Agarwal** 2009).

The species-specific life span appears to be determined by genes that affect a compromise between early growth and reproduction and somatic maintenance (**Kirkwood** 1977). In other words, aging results from natural selection operating more on early survival and reproduction than on having a vigorous post-reproductive life. If longevity is a selectable trait, one should expect to find heritable variation within populations. Recently, long-term studies of wild populations of animals have provided convincing data that there is hereditable variation within a species for aging (**Wilson** *et al.*, 2007).

Molecular evidence indicates that certain genetic components of longevity are conserved between species: flies, worms, mammals, and even yeast cells all appear to use the same set of genes to promote survival and longevity (**Kenyon** 2001; **Vijg** and **Campisi** 2008). There are two sets of genes that are well known to be involved in aging and its prevention, and both sets appear to be conserved between phyla and even kingdoms of organisms. These are genes encoding DNA repair enzymes (proteins) and the genes encoding proteins involved in the insulin signalling pathway.

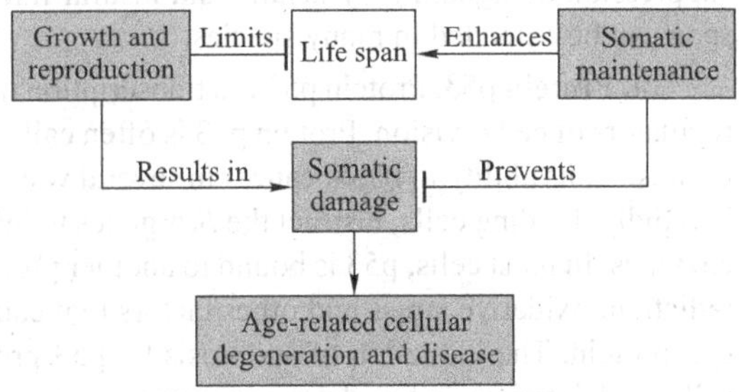

Fig. 52.1. Kirkwood's proposal that organisms have to affect a compromise between the energy allocated to reproduction and growth and the energy allocated to the maintenance and repair of bodily tissues.

I. Genes Encoding DNA Repair Proteins

1. Efficient DNA repair enzymes. DNA repair and synthesis may be important in preventing senescence. Individuals of species whose cells have more efficient DNA repair enzymes live longer (**Heart** and **Setlow** 1974; Fig. 52.2).

2. Premature aging syndromes. Certain premature aging syndromes, called **progerias,** in humans appear to be caused by mutations in certain DNA repair enzymes (**Sun** *et al.*, 1998; **Shen** and **Loeb** 2001). In humans, **Hutchinson-Gilford progeria** is a rapid-aging syndrome; children born with this condition age rapidly, dying (usually of heart failure) as early as 12 years of age. Symptoms of this type of progeria include thin skin with age spots, resorbed bone mass, hair loss, and arteriosclerosis—all characteristics of human senescent phenotype. Hutchinson-Gilford progeria is the result of a dominant mutation in the gene that encodes **lamin A**, a nuclear membrane protein, and these same mutations can be seen in age-related senescence (**Scaffidi** and **Misteli** 2006).

3. Klotho gene. Another type of progeria is reported in mice. It is caused by loss-of-function mutations of the *Klotho* gene (**Kuro** *et al.*, 1997). Conversely, the same gene's gain-of-function phenotype (causing its overexpression) has been known to prolong a mouse's life by 30% (**Kurosu** *et al.*, 2005). Klotho gene appears to encode a hormone that down regulates insulin signaling. The

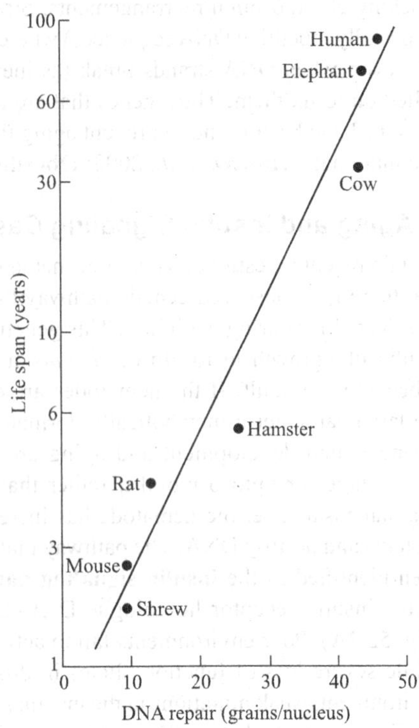

Fig. 52.2. Life span and the aging phenotype. Correlation between life span and the ability of fibroblasts to repair DNA in various mammalian species. Repair capacity is represented in auto-radiography by the number of grains from radioactive thymidine per cell nucleus. Note that y-axis (life span) is logarithmic scale (after Gilbert 2010).

suppression of signaling by insulin and insulin-like growth factor 1 (IGF-1) is one of the ways life span can be extended in many species.

4. Protein p53. Protein p53 is a transcription factor and is regarded as one of the most important regulators of cell division. Protein p53 is often called **"guardian of the genome"** (see **Gilbert** 2010) because of its ability to block cancer in several ways. It can stop cell cycle, cause cellular senescence in rapidly dividing cells, instruct the *Bax* genes to initiate cellular apoptosis, and activate DNA repair enzymes. In most cells, p53 is bound to another protein that keeps p53 inactive. However, ultraviolet radiation, oxidative stress and other factors that cause DNA damage will also separate and activate p53 protein. The induction of apoptosis by p53 protein can be beneficial (when destroying cancer cells) or deleterious (when destroying neurons or other vital cells). It is possible that animals with high levels of p53 have increased protection against cancer, but they may also age more rapidly (**Tyner** *et al.*, 2002). Indeed, p53 can be activated by the absence of lamin A (**Varela** *et al.*, 2005), thereby suggesting a mechanism for Hutchinson-Gifford progeria.

5. Sirtuin genes. Sirtuin genes encode histone deacetylation (chromatin-silencing) enzymes. They guard the genome, preventing genes from being expressed at the wrong times and places, and blocking chromosomal rearrangements. Sirtuin proteins are usually found in regions of chromatin (especially repetitive DNA sequences) where such mistaken chromosomal rearrangements can occur. However, when DNA strands break (as inevitably happens as the body ages), sirtuin proteins are called on to fix them. Thus, genes that are usually silenced become active as the cells age. Sirtuin proteins have been found to prevent aging throughout the eukaryotic kingdoms, including in yeasts and mammals (**Howitz** *et al.*, 2003; **Oberdoerffer** *et al.*, 2008).

II. Aging and Insulin Signaling Cascade

Certain recent investigations in mice, nematode *Caenorhabditis elegans*, and *Drosophila* suggested that there is a conserved genetic pathway that regulates aging. This genetic pathway was found to be selected for during evolution. This genetic pathway involves the response to insulin hormone or insulin like growth factor. In *C. elegans*, a larva proceeds through four larval stages, after which it becomes an adult. If the nematodes are overcrowded or if there is insufficient food, however, the larva can enter a metabolically dormant **dauer larval stage,** a non-feeding state of **diapause** during which development and aging are suspended. The nematode can remain in the dauer larval stage for upto 6 months, rather than becoming an adult that lives only a few weeks. In this diapausal state, the nematode has increased resistance to oxygen radicals that can cross-link proteins and destroy DNA. The pathway that regulates both dauer larva formation and longevity has been identified as the **insulin signaling pathway**. Favourable environments signal the activation of the insulin receptor homologue DAF-2, and this receptor stimulates the onset of adulthood (Fig. 52.3A). Poor environments fail to activate the DAF-2 receptor, and dauer formation happens. While severe loss-of-function alleles in this pathway cause the formation of dauer larvae in any environment, weak mutations in the insulin signaling pathway enable the animals to reach adulthood and live four times longer than wild-type animals.

There are many other functions of **down regulation** of insulin signaling pathway. First, it appears to influence metabolism, decreasing mitochondrial electron transport. When the DAF-2 receptor is not active, organisms have decreased sensitivity to **reactive oxygen species (ROS),** metabolic by-products that can damage cell membranes and proteins and even destroy DNA. Second, downregulating the insulin pathway increases the production of enzymes that prevent oxidative damage, as well as DNA repair enzymes. Third, this lack of insulin signaling decreases fertility. This increase in DNA synthetic enzymes and in enzymes that protect against ROS is due to the DAF-16 transcription factor. This forkhead-type transcription factor is inhibited by the insulin receptor (DAF-2) signal. When that signal is absent, DAF-16 can function, and this factor appears to activate the genes encoding several enzymes (such as catalase and superoxide dismutase) that are involved in reducing ROS, several enzymes that increase protein and lipid turnover, and several stress proteins.

Affect of insulin-signaling pathway in mammal's longevity. In mammals, insulin and insulin-like growth factor pathways are so deeply integrated with embryonic development and adult metabolism that mutations often have numerous and deleterious effects (such as diabetes or Donahue syndrome). However, there is some evidence that the insulin signaling pathway does affect life-span in mammals (Fig. 52.3B). Dog breeds with low levels of insulin-like growth factor 1 (IGF-1) live longer than breeds with higher levels of this factor. Mice with loss-of-function mutations of the insulin signaling pathway live longer than their wild-type litermates. Mice heterozygous for the insulin-like growth factor 1 receptor (IGF-1R) not only lived about 30% longer than their wild-type littermates, they also had greater resistance to oxidative stress. In addition, mice lacking one copy of their IGF-IR gene lived about 25% longer than wild-type mice (and had higher ROS resistance, but otherwise normal physiology and fertility).

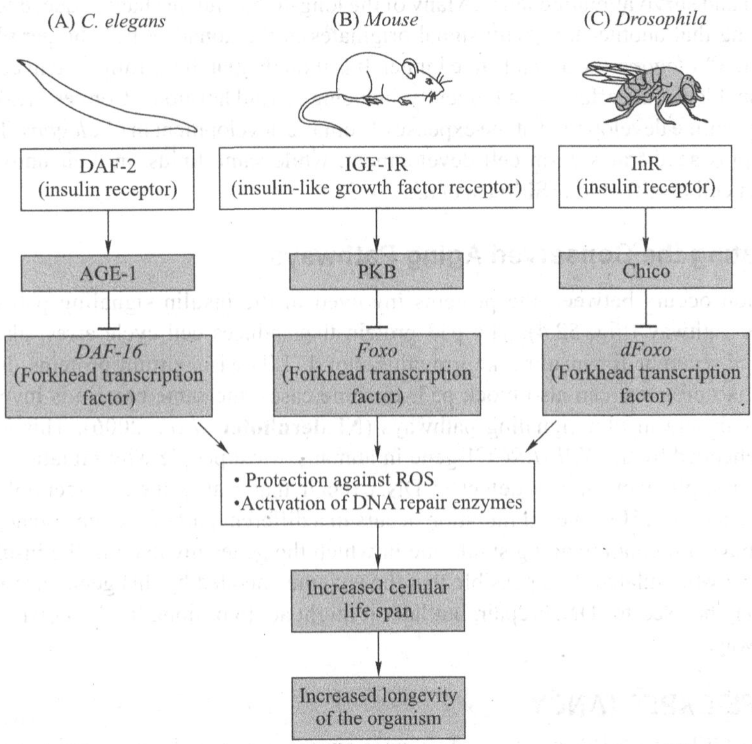

Fig. 52.3. A pathway for regulating longevity. In each case, the insulin signaling pathway inhibits the synthesis of proteins that would otherwise protect cells against oxidative damage caused by reactive oxygen species (ROS) that crosslink proteins and can damage DNA. These protective proteins may be particularly important in mitochondria. When insulin signaling is downregulated, Forkhead transcription factors may activate DNA repair enzymes that may protect against mutations caused randomly by ROS or other agents. Such protection against ROS and mutation may increase the functional life span of the cells and the longevity of the organism (after Gilbert 2010).

Affect of insulin-signaling pathway on *Drosophila's* longevity. The insulin signaling pathway also regulates life span in *Drosophila* (Fig. 52.3C). In 2001, research teams of **Clancy** and of **Tatar** have reported that flies with loss-of-function mutations of the insulin receptor gene or genes in the insulin pathway (such as *Chico*) live nearly 85% longer than wild-type flies. These long-lived mutants are sterile, and their metabolism resembles that of flies that are in diapause. In *Drosophila*, the insulin receptor is thought to regulate a Forkhead transcription factor (dFoxo), similar to the DAF-16 protein of *Caenorhabditis elegans*. When the *dFoxo* gene of *Drosophila* is activated in the fat body; it lengthen the fly's life span. Other studies have shown that when these enzymes (such as superoxide dismutase) are downregulated by mutation or by RNA interference, the resulting flies die early, have increased oxidative stress, and display higher levels of DNA damage. Conversely, over expression of superoxide dismutase genes can lengthen the *Drosophila* life span. While some evidence points to a correlation between longer life span, lower insulin signaling and elevated ROS protection in *Drosophila* (**Broughton** *et al.*, 2005), other studies suggest that some flies and other insects can obtain longer life spans without increasing the enzymes known to protect against oxidative stress (**Le Bourg** and **Fournier** 2004; **Parker** *et al.*, 2004).

From an evolutionary point of view, the insulin pathway may mediate a compromise between reproduction and survival/maintenance. Many of the long-lived mutants have reduced fertility. Thus, it is interesting that another longevity signal originates in the gonad. When the germline cells are removed from *C. elegans*, the animals live longer. It is thought that the germline stem cells produce a substance that blocks the effects of a longevity-inducing steroid hormone. Conversely, ROS appears to promote germline development at the expense of somatic development in *C. elegans.* The oxidation of certain lipids accelerates germ cell development, while same lipids, in their unoxidized form, prevent germ cell proliferation (**Shibata** *et al.*, 2003).

III. Integrating the Conserved Aging Pathways

An interaction occurs between the proteins involved in the insulin signaling pathway and the DNA repairs pathway (Fig. 52.4). The p53 protein that induces cell cycle arrest also blocks the activity of the receptor for insulin-like growth factor 1. Likewise, sirtuin proteins, in addition to activating Foxo proteins, can also block p53. In some cases, the same protein is involved in both the DNA repair and insulin signaling pathways (**Niedernhofer** *et al.*, 2006). This is the case in the protein encoded by the *XPF-ERCC*1 gene in humans. Most people with mutations of this gene have xeroderma pigmentosum, a defect in DNA repair that makes them susceptible to cancers, especially melanomas. However, if mutation occurs in a different part of the same gene, the affected individuals have a premature aging syndrome in which the genes involved in the insulin signaling pathway are downregulated. It is possible that the enzyme encoded by this gene has two functions. Initially it may be used for DNA repair, but later it might act to prolong life by downregulating the insulin pathway.

52.2. LIFE EXPECTANCY

According to **Gilbert** (2010), most people cannot expect to live 121 years, and most mice in the wild do not live to celebrate even their first birthday. **Life expectancy** means the length of time an average individual of a given species can expect to live. It is not a characteristic of species, but of populations. Life expectancy is usually defined as *the age at which half the population still survives.* A baby born in England during the 1780s could expect to live to be 35 years old. In Massachusetts during that same time, life expectancy was 28 years. These ages represent normal range of human life expectancy for most of the human race throughout recorded history (**Arking** 1998). Even today, in some areas of the world (such as Cambodia, Togo, Afghanistan, and several other countries) life expectancy is less than 40 years. In the United States, a male born in 1986 can expect to live 74 years, while females have a life expectancy of around 80 years.

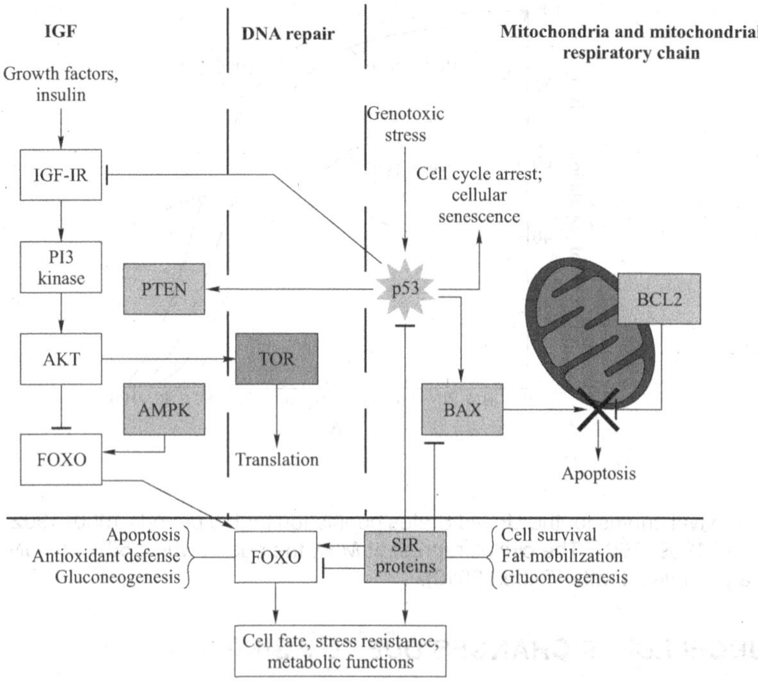

Fig. 52.4. Interactions of different evolutionarily conserved aging pathways. The insulin signaling pathway and mitochondrial pathway for apoptosis (cell death) are indicated. The pro-aging activities of these pathways are conserved across phyla, and energy sensors, such as AMPK, are potentially important integrators of these pathways. Many longevity signals converge on members of the FOXO and sirtuin protein families, which can interact. Note that sirtuin proteins can both activate and repress, FOXO, depending on the context. TOR is a positive regulator of general translational activity. Here, insulin-like factors and DNA-disrupting ("genotoxic") chemicals initiate these pathways (after Gilbert 2010).

Thus, in 1900, 50% of Americans were dead before the age of 60; in 1950, the comparable age was 72; by 2000, this "median survival" age had climbed to 80 years (Fig. 52.5). A 70 year old person was exceptional in 1900 but is commonplace today. People in 1900 did not have the "luxury" of dying from heart attacks or cancers, because these conditions are most likely to affect people over 50. Rather people died (as they are still dying in many parts of the world) from microbial and viral infections. In fact, before the discovery of antibiotics, the death rate of young women due to infections associated with childbirth was high throughout the world. Thus, the phenomena of senescence and diseases of aging are much more common today that they were a century ago. Until recently, relatively few people exhibited the general human senescent phenotype: gray hair, sagging and wrinkling skin, stiff joints, osteoporosis (loss of bone calcium), loss of muscle fibers and muscular strength, memory loss, eyesight deterioration and slowed sexual responsiveness.

The general senescent phenotype is characteristic of each species. But what causes it? This question can be asked at many levels. Here we will look primarily at the cellular level of organization.

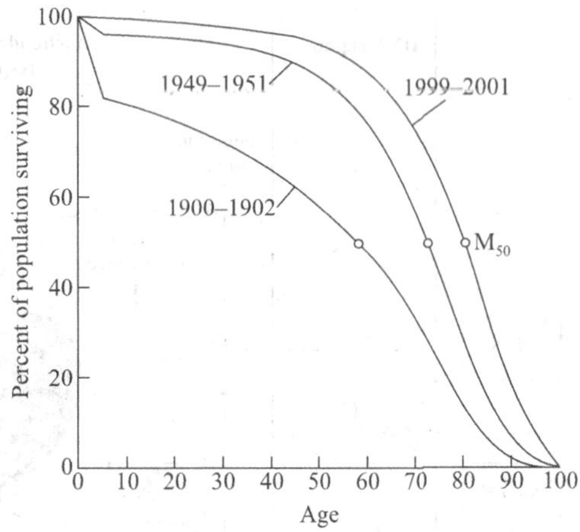

Fig. 52.5. Survival curves for the United States population for the periods 1900–1902, 1949–1951, and 1999–2001. The circles represent *M50*, the age at which 50% of individuals of that age survived (after Gilbert 2010).

52.3. SUBCELLULAR CHANGES DUE TO AGING

Due to the process of aging following changes occur in the different cellular organelles:

1. Mitochondria. Reiner (1947) has reported that due to aging the rate of carbohydrate metabolism (Krebs cycle) is decreased. The aging also affects the glycolysis. Such results have also been reported by **Rafsky** (1952) and **Barrows** (1958). **Dampsey** (1956) has shown that the mitochondria are related with the aging process of the animals and in old tissues the mitochondria become degenerated. **Zeuthen** (1947, 1953) has reported that the rate of respiration in multicellular organisms becomes slow and they become somewhat dehydrated.

2. Endoplasmic reticulum. Due to aging the amount of granular endoplasmic reticulum decreases in the cytoplasm of old cells. In the nerves of older animals and human the decrease of Nissl substance (rRNA) have been reported by various workers.

3. Pigment accumulation and lysosomes. Due to aging various pigmented inclusions such as lipofuscin, yellow pigment, brown degenerations, etc., become accumulated in the cells. The lipofuscin probably is the result of accumulation and autooxidation of lipid components of lysosomes. **Shock** (1962) and **Strehler** (1962, 1963) have reported the accumulation of calcium, various pigments and other inert matter in the aged cell. The inert matter are the accumulated metabolic excretory wastes which might have failed to exit across the plasma membrane.

4. Nucleus. Minot (1970) has suggested that the natural death of the cell is a consequence of cellular differentiation of which a change in nucleocytoplasmic ratio is an important index. **Smallwood** and **Philips** (1916) have shown that fatigue and age do not produce significant changes in the nuclear size. **Falzone** (1959) has studied the effect of the age on the ploidy or DNA contents of the nuclei of rat liver cells and reported that no significant effect of aging on the nuclei. However, various modern cytologists have shown that the age affects the nucleus variously.

5. Nuclear and plasma membrane. Lansing (1942, 1952) has reported that due to aging process the calcium becomes accumulated in the cellular membranes and causes various physiological changes in them.

52.4. HOW AGING AFFECTS BODY SYSTEMS OF HUMANS ?

1. Effect of Aging on Skin

As aging occurs, skin becomes thinner and less elastic because the number of elastic fibers decreases and collagen fibers undergo cross-linking. Also, there is less adipose tissue in the subcutaneous layer; therefore, older people are more likely to feel cold. The loss of thickness of skin partially accounts for sagging and wrinkling of the skin.

Homeostatic adjustment to heat is also limited because there are fewer sweat glands for sweating to occur. There are fewer hair follicles, so the hair on the scalp and the extremities thin out. The number of oil (sebaceous) glands is reduced, and the skin tends to crack. Older people also experience a decrease in the number of melanocytes, making hair gray and skin pale. In contrast, some of the remaining pigment cells are larger, and pigmented blotches appear in skin.

2. Processing and Transporting

Cardiovascular disorders are the leading cause of death among the elderly. The heart shrinks because there is a reduction in cardiac muscle size. This leads to loss of cardiac muscle strength and reduced cardiac output.

Because the middle coat of arteries contains elastic fibers, which most likely are subject to cross-linking, the arteries become more rigid with time, and their size is further reduced by plaque, a buildup of fatty material. Therefore, blood pressure reading gradually rise.

There is reduced blood flow to the liver, and this organ does not metabolize drugs as efficiently as before.

Circulatory problems often are accompanied by respiratory disorders, and vice-versa. Growing inelasticity of lung tissues means that ventilation is reduced. Because we rarely use the entire vital capacity, these effects are not noticed unless there is increased demand for oxygen.

There is also reduced blood supply to the kidneys. The kidneys become smaller and less efficient at filtering wastes. Salt and water balance are difficult to maintain, and the elderly dehydrate faster than young people. Difficulties involving urination include incontinence (lack of bladder control) and the inability to urinate. In men, the prostate gland may, enlarge and reduce the diameter of the urethera, making urination so difficult that surgery is often needed.

The loss of teeth, which is frequently seen in elderly people, is more apt to be the result of long-term neglect than aging. The digestive tract loses tone, and secretion of saliva and gastric juice is reduced, but there is no indication of reduced absorption. Therefore, an adequate diet, rather than vitamin and mineral supplements, is recommended (see **Mader** 1998). There are common complaints of constipation, increased amount of gas, and heartburn, but gastritis, ulcers and cancer can also occur.

3. Effect of Aging on Integration and Coordination

It is often mentioned that while most tissues of the body regularly replace their cells, some at faster rate than others, the brain and muscles ordinarily do not. However, contrary to previous opinion, recent studies show that few neural cells of the cerebral cortex are lost during the normal aging process. This means that cognitive skills remain unchanged even though there is characteristically a loss in short-term memory. Although the elderly persons learn more slowly than the young, they can acquire and remember new material. It is noted that when more time is given for the subject to respond, age differences in learning decrease.

Neurons are extremely sensitive to oxygen deficiency, and if neuron death does occur, it may not be due to aging itself but to reduced blood flow in narrowed blood vessels. Specific disorders, such as **depression, Parkinson disease,** and **Alzheimer disease**, are sometimes seen, but they are not common. Reaction time, however, does slow, and more stimulation is needed for hearing, taste, and smell receptors to function as before. After age 50, there is a gradual reduction in the ability to hear tones at higher frequencies and this can make it difficult to identify individual voices and to understand conversation in a group. The lens of the eye does not accommodate as well and may also develop a **cataract. Glaucoma**, the buildup of pressure due to increased fluid, is more likely to develop because of a reduction in the size of the anterior cavity of the eye.

Loss of skeletal muscle mass also occurs. There is a reduced capacity to do heavy labour, but routine physical work should be no problem. A decrease in the strength of the respiratory muscles and inflexibility of the rib cage contribute to the inability of the lungs to expand as before, and reduced muscularity of the urinary bladder contributes to difficulties with urination.

Aging is also accompanied by a decline in bone density. Osteoporosis, characterized by a loss of calcium and mineral from bone, is not uncommon, but there is evidence that proper health habits can prevent its occurrence. **Arthritis** which causes pain upon movement of the joint, is also seen.

Weight gain occurs because the basal metabolism decreases and inactivity increases. Muscle mass is replaced by stored fat and retained water.

4. Effects of Aging on the Reproductive System

Females undergo menopause, and thereafter the level of female sex hormones in blood falls markedly. The uterus and the cervix are reduced in size, and there is a thinning of the walls of the oviducts and the vagina. The external genitals become less pronounced. In males, the level of androgens fall gradually over the age span of 50–90, but sperm production continues until death.

It is of interest that as a group, females live longer than males. Although their health may be better, it is also possible that the female sex hormone estrogen offers women some protection against circulatory disorders when they are younger. Males suffer a marked increase in heart diseases in their forties, but an increase is not noted in females until after menopause. Then women lead men in the incidence of stroke. Men are still more likely than women to have a heart attack, however (see **Mader** 1998).

52.5. THEORIES OF AGING

No single theory of aging has gained general acceptance, but the following ones are being intensively investigated (see **Raven** *et al.*, 2005).

1. Accumulated Mutation Theory

The oldest general theory of aging is that cells accumulate mutations as they age, leading eventually to lethal damage. Careful studies have shown that somatic mutations do indeed accumulate during aging. As cells age, for example, they tend to accumulate the modified base 8-hydroxyguanine in which an OH group is added to the base guanine. There is little direct evidence, however, that these mutations cause aging. No acceleration in aging occurred among survivors of Hiroshima and Nagasaki despite their enormous added mutation load, arguing any general relationship between mutation and aging (see **Raven** *et al.*, 2005).

2. Telomere Depletion Theory

In an important experiment carried out in 1961, **Leonard Hayflick** demonstrated that fibroblast cells growing in tissue culture will divide only a certain number of time (Fig. 52.6). After about

50 population doublings, cell division stops—the cell cycle is blocked just before DNA replication. If a cell sample is taken after 20 doublings and frozen, when thawed it resumes growth for 30 more doublings, and then stops.

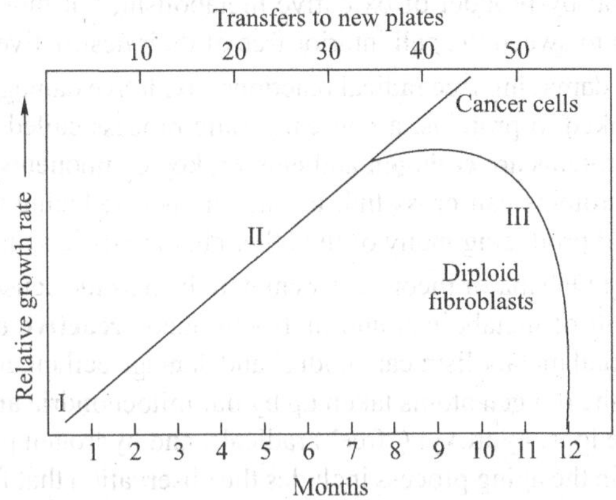

Fig. 52.6. Hayflick's experiment. Fibroblast cells stop growing after about 50 doublings. Growth is rapid in phase I and II but slows in phase III, as the cultures become senescent, until the final doubling. Cancer cells, by contrast, do not age (after Raven *el al.*, 2005).

An explanation of the "Hayflick limit" was suggested in 1986 when **Howard Cooke** first glimpsed an extra length of DNA at the ends of chromosomes. These **telomeres**, repeats of the sequence TTAGGG, were found to be substantially shorter in older somatic tissue, and Cooke speculated that a 100-base-pair portion of the telomere cap was lost by a chromosome during each cycle of DNA replication. Eventually, after about 50 replication cycles, the protective telomeric cap would be used up, and the cell line would then enter senescence, no longer able to proliferate. Cancer cells appear to avoid telomere shortening.

Research reported in 1998 has confirmed Cook's hypothesis, providing direct evidence for a causal relationship between telomeric shortening and cell senescence. Using genetic engineering researches transferred into human primary cell cultures a gene that leads to expression of **telomerase**, an enzyme that builds TTAGGG telomeric caps.

The result was clear-cut. New telomeric caps were added to the chromosomes of the cells, and the cells with the artificially elongated telomeres did not age at the Hayflick limit, continue to divide in a healthy and vigorous manner for more than 20 additional generations.

Evidences against telomere depletion theory. In some cases, no correlation between telomere length and the life span of an animal (humans have much shorter telomeres than mice) has been found, nor is there a correlation between human telomere length and a person's age (**Cristofalo** *et al.*, 1998, **Rudalph** *et al.*, 1999; **Karlseder** *et al.*, 2002). Nematodes can have mutations that extend or shorten longevity, and the length to the telomere does not correlate with the age in the roundworms (**Raices** *et al.*, 2005). Telomeres appear to be critical in stem cell maintenance, and the telomere-dependent inhibition of cell division might serve primarily as a defense against cancer (**Blasco** 2005; **Flores** *et al.*, 2005).

3. Wear-and-Tear Theories

Numerous theories of aging focus in one way or another on the general idea that cells wear out over time, accumulating damage until they are no longer able to function. Loosely dubbed the "wear-and-tear" hypothesis, this idea implies that there is no inherent designed-in limit to aging, just a statistical one—that is, disruption, wear, and damage over time erode a cell's ability to function properly.

Considerable evidence indicates that aging cells do accumulate damage. Some of the most interesting evidence concerns free radicals, fragments of molecules or atoms that contain an unpaired electron. Free radicals are very reactive chemically and can be quite destructive in a cell. Free radicals are produced as natural by-product of oxidative metabolism, but most are mopped up by special enzymes that function to sweep the cell interior free of their destructive effects.

One of the most damaging free radical reactions (oxidative damage) that occurs in cells causes glucose to become linked to proteins, a non-enzymatic process called **glycation**. Two of the most commonly glycated proteins are **collagen** and **elastin,** key components of the connective tissues in our joints. Glycated proteins can cross-link to one-another, reducing the flexibility of connective tissues in the joints and producing many of the other characteristic symptoms of aging.

ROS and aging: One major theory views metabolism as the cause of aging. According to this theory, aging is a result of metabolism and its by-products, **reactive oxygen species (ROS).** The ROS produced by normal metabolism can oxidize and damage cell membranes, proteins and nucleic acids. Some 2–3% of the oxygen atoms taken up by our mitochondria are reduced insufficiently and form ROS: superoxide ions, hydroxyl ("free") radicals, and hydrogen peroxide. Evidence that ROS molecules are critical in the aging process includes the observation that fruit flies over expressing the enzymes that destroys ROS (Catalase and superoxide dismutase) live 30–40% longer than do control flies. Moreover, flies with mutations in the *methuselah* gene (named after the biblical fellow said to have lived 969 years) live 35% longer than wild-type flies. These mutants have enhanced resistance to ROS (**Lin** *et al.*, 1998). In nematode *C. elegans*, too, individuals with mutations that result in either the degradation of ROS or the prevention of ROS formation live much longer than wild-type nematodes (**Feng** *et al.*, 2001). These findings not only suggests that aging is under genetic control, but also provide evidence for the role of ROS in the aging process.

4. Random Genetic Drift

A new variant of "wear-and-tear" theories is the hypothesis of **random genetic drift**. Given that appropriate **methylation** is essential for normal development, one can immediately see that diseases would result as a consequence of inappropriate epigenetic methylation. Recent evidences have confirmed that inappropriate methylation can be the critical factor in aging and cancers. Some of the evidences for this hypothesis comes from identical twins. Most "identical" twins start life with very few differences in appearance with age. Experiences counts, and both random events and life styles may be reflected in phenotypes. **Fraga** and colleagues (2005) found that twin pairs were nearly indistinguishable in methylation patterns when young but older monozygotic twins exhibited very different patterns of methylation. This affected their gene expression patterns, such that older twin pairs had different patterns of DNA expression, while younger twin pairs had

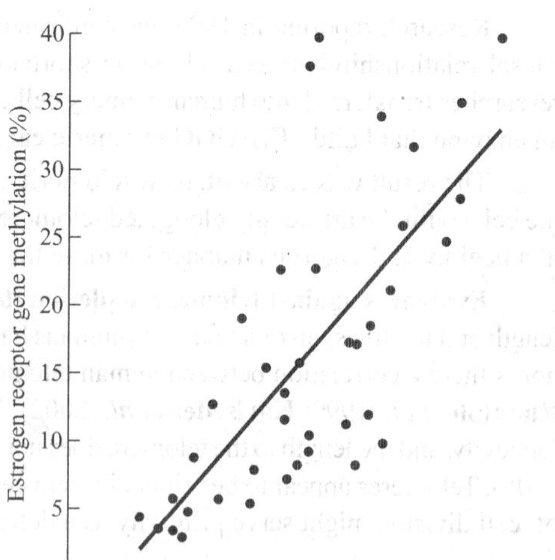

Fig. 52.7. Methylation of the estrogen receptor gene occurs as a function of a normal aging (after Gilbert 2010).

very similar expression patterns. **Fraga** *et al.*, (2005) have shown that monozygotic twin pairs start off with identical amounts of methylated DNA and acetylation of histones H3 and H4 (these are three **epigenetic markers**). As twins age, however, both methylation and acetylation increase, but to different extents and at different chromosomal locations in each twin.

The idea that random genetic drift inactivates important genes without any particular environmental cue gives rise to an entirely new hypothesis of aging. Instead of randomly accumulated mutations—which might be due to specific mutagens—we are at the mercy of *chance accumulations of errors* made by the DNA methylating and demethylating enzymes. Indeed, our DNA methylating enzymes, unlike the DNA polymerases, are prone to errors. The enzymes DNA methyltransferases are not the most fastidious of enzymes. At each round of DNA replication, they must methylate the appropriate cytosine residues and leave the others unmethylated. This is not always done properly, and such errors accumulate as we age.

For instance, the methylation of the promoter region of estrogen receptor is known to increase with age (**Issa** *et al.*, 1994). Figure 52.7 shows a linear relationship between the methylation of a promoter region in an estrogen receptor gene and increased age. The methylation of the promoters of the genes for the alpha and beta estrogen receptors increases with age, resulting in the inactivation of this gene in the smooth muscles of the circulatory system. Moreover, methylation of the estrogen receptor genes is even more prominent in the atherosclerotic plaques (thickened artery walls) that occlude the blood vessels. The atherosclerotic plaques show more methylation of estrogen receptor genes than does the tissue around it (**Post** *et al.*, 1999; **Kim** *et al.*, 2007). Thus, DNA methylation-associated inactivation of the estrogen receptor genes in vascular tissue may play a role in athrogenesis and aging of the vascular system. This potentially reversible defect may provide a new target for intervention in heart disease (**Gilbert** 2010).

5. Gene Clock Theory

There is little doubt that atleast some aspects of aging are under the direct control of genes. Just as genes regulate its rate of aging, some genes appear to promote longevity. For example, people over 100 years old are five times as likely to carry a mutation in mitochondrial DNA called **C150T**.

Other genes produce premature aging. For example, in 1996, a gene was identified that is responsible for **Werner's syndrome**, which produces premature aging and affects some 10 people per million worldwide. The syndrome is named after **Otto Werner,** who in 1904 reported in a German family affected by premature aging and believed a genetic component was at work. Werner's syndrome appears in adolescence, usually producing death before age 50 of heart attack or one of a variety of rare connective tissue cancers. Located on the short arm of chromosome 8, the gene seem to affect a helicase enzyme involved in repair of DNA. The genes which codes for 1432-amino acid protein, has been fully sequenced, and four mutant alleles identified. Helicase enzymes are needed to unwind the DNA double helix whenever DNA has to be replicated, repaired or transcribed. The high incidence of certain cancers among Werner's syndrome patients leads investigators to speculate that the mutant helicase may fail to activate critical tumour suppressor genes. The potential role of helicases in aging is the subject of heated research.

Research on aging in other animals strongly supports the hypothesis that genes regulate the rate of aging. A *Drosophila melanogaster* gene mutation called *Indy* ('I am not dead yet') doubles the fruit fly life span from the usual 37 days to an average of 70 days. When researchers isolated the DNA of the *Indy* gene and compared its DNA sequence with the Human Genome Project sequences, they found that the *Indy* gene is 50% similar to a human gene called *dicarboxylic cotransporter.* In humans, dicarboxylic cotransporter proteins move preliminary products of food metabolism (dicarbxylic acids of the Krebs cycle) across membranes to where the food processing takes place. In mutant *Indy* flies, poor dicarboxylic acid pumping means that less metabolic energy can be gleaned from the fly's food. In essence, the *Indy* mutation is the genetic equivalent of caloric restriction (*i.e.*, starving). Starving is known to prolong life in the nematode *Caenorhabditis elegans*, but *Indy's* caloric restriction does not involve the unpleasantness of starving. The *Indy* mutation in effect puts flies on a severe diet, while flies eat as much as normal and lead a normal vigorous life for far longer.

Box 52.2 Promoting Longevity

Several interacting agents may promote longevity. These include calorie restriction, protection against oxidative stress, and the factors activated by a suppressed insulin pathway. It is not yet known how these factors interact—whether they are part of a single "longevity pathway", or if they act separately. Moreover, genetics and diet do not appear to be the full answer to aging. Chance it seems, still play a role. When clonally indentical C. *elegans* (nematodes) are fed an identical diet, some organisms still live longer than others, and different organs deteriorate more rapidly in different individuals (**Herndon** *et al.*, 2002). Mutations are randomly occurring events, and they may play a role in the aging process.

6. Diet Caloric Restriction Theory

It is one of the few known ways of extending mammalian longevity (in fact at the expense of fertility) and it may do so through several routes. First, restricting caloric intake may reduce levels of IGF-1 and of circulating insulin (**Kenyon** 2001; **Roth** *et al.*, 2002; **Holzenberger** *et al.*, 2003). This association of increased longevity with the down regulation of the insulin pathway through diet is seen in yeast, flies, nematodes, and mice. Dietary restrictions may also work through the sirtuin proteins (**Lamming** *et al.*, 2005), thereby uniting the insulin metabolic pathway with the genomic protection hypothesis. The insulin pathway in mammals also negatively regulates *Foxo*, the gene for a transcription factor that activates ROS-protective enzymes (Fig. 52.3B; **Essers** *et al.*, 2004). Calorie restriction also represses a ribosomal activator whose absence is associated with increased longevity (**Selman** *et al.*, 2009).

7. Parabiosis Theory (Young Blood: Serum Factors and Progenitor Cells)

One of the distinctive features of aging is the declining ability of stem cells and progenitor cells to restore damaged or non-functioning tissues. A decline in muscle progenitor (satellite) cell activity when Notch signaling is lost results in a significant decrease in the ability to maintain muscle function. Similarly, an age-dependent decline in liver progenitor cell division impairs liver regeneration due to a decline in transcription factor CEBP-α.

The problem seems to exist in the environment of the stem cells. If an aged and young mouse are parabiosed (*i.e.*, their circulatory systems are surgically joined so that the two mice share one blood supply), the stem cells of the old mouse are exposed to factors in young blood serum (and vice-versa). This **heteronic parabiosis** has been seen to restore the activity of old stem cells. Notch signalling of the muscle stem cells regained its youthful levels, and muscle cell regeneration was restored. Similarly, liver progenitor cells regained "young" levels of CEBP-α and their ability to regenerate (**Conboy** *et al.*, 2005). When young haematopoietic stem cells were placed into "old" stem cell niches that had been exposed to old blood, they did not make many new cells when injected into lethally irradiated mice. However, when these young stem cells were placed into "old" niches that had been exposed by parabiosis to "young" blood cells as the young stem cells that had seen young niches exposed only to young blood (Fig. 52.8).

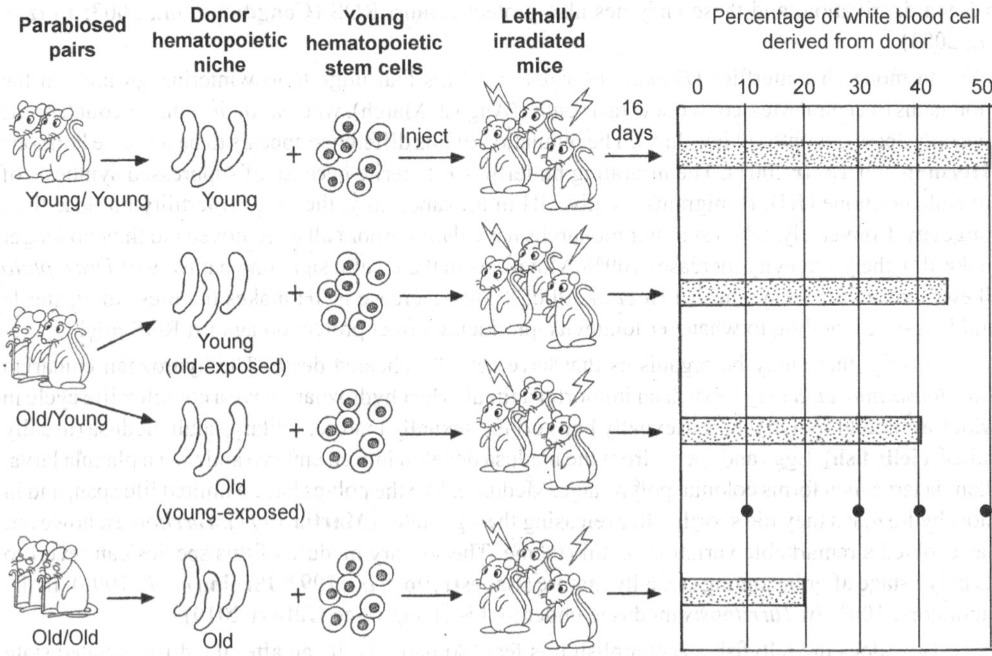

Fig. 52.8. Circulating factors "rejuvenate" haematopoietic niche cells. Haematopoietic niche cells were isolated from parabiosed (surgically conjoined) mouse pairs. The pairs were of either same age/young (2 months); same age/old (21 months); or one young and one old mouse. Niche cells were cultured with young haematopoietic stem cells and injected into lethally irradiated mice (*i.e.*, mice whose own stem cells had been destroyed by radiation). When the white blood cells produced by the injected stem cells were analyzed 16 weeks later, stem cells residing in young niche with young blood produced the most white blood cells; those residing in old niches bathed in old blood had the worst reconstitution. However, old-niche cells that developed in contact with young blood (*i.e.*, those from the "mixed" parabiosed pair) reconstituted the host while cell population almost as well as the "ever-young" cells (after Gilbert 2010).

According to **Mayack** and coworkers (2010), the factor involved in aging of the niche may be IGF-1 which is produced locally in the niche and is regulated by factors in the blood.

8. Pace maker theories of aging

Evidence is growing that aging is caused, in part, by progressive breakdown in the immunological system. **Burnet** has suggested that the **thymus gland** is the **biological clock** that determines how fast we age. It is the pacemaker for the whole body and its atrophy would be programmed event that leads to aging of the animal. It is well known that the thymus begins to atrophy shortly after puberty in humans.

52.6. EXCEPTIONS TO THE AGING RULE

There are a few species such as turtles, monarch butterflies and a hydrozoan in which aging seems to be optional, and these may hold some important clues to how animals can live longer and retain their health. Many turtle species not only live a long time, but they do not undergo the typical aging syndrome. In these species, for example, older females lay as many eggs as their younger counterparts. **Miller** (2001) showed that a 60-year old female three toed box turtle (*Terrapene carolina triunguis*) lays as many eggs annually as she ever did. Interestingly, turtles have special adaptations against oxygen deprivation, and these enzymes also protect against ROS (**Congdon** *et al*., 2003; **Lutz** *et al*., 2003).

In monarch butterflies (*Danaus plexippus*), adults that migrate to wintering grounds in the mountains of central Mexico live several months (August–March), whereas their summer counterparts live only about 2 months (May–July). The regulation of this difference appears to be juvenile hormone (**Herman** and **Tatar** 2001). The migrating butterflies are sterile because of suppressed synthesis of juvenile hormone (JH). If migrants are give JH in the laboratory, they regain fertility but lose their longevity. Conversely, when summer monarchs have their corpora allata removed (so they no longer make JH), their longevity increases 100%. Mutations in the insulin signaling pathway of *Drosophila* likewise decrease JH synthesis (**Tu** *et al*., 2005). This decrease in JH makes the flies small, sterile and long-lived, adding to whatever longevity-producing effect protection against ROS might have.

Finally, there may be organisms that have actually cheated death. The hydrozoan cnidarian *Turritopsis nutricula* may be such an immortal animal. Most hydrozoans have a complex life cycle in which a colonial (polyp) stage asexually buds off the sexually mature, solitary, adult medusa (usually called a **jelly fish**). Eggs and sperm from the medusa develop into an embryo and then a planula larva. Planula larva then forms colonial polyp stage. Medusae, like the polyps have a limited life span, and in most hydrozoans they die shortly after releasing their gametes (**Martin** 1997). *Turritopsis*, however, has evolved a remarkable variation on this theme. The solitary medusa of this species can revert to its polyp stage after becoming sexually mature (**Bavestrello** *et al*., 1992; **Piraino** *et al*., 1996). In the laboratory, 100% of *Turritopsis* medusae undergo this change (see **Gilbert** 2010).

How does the jellyfish a accomplish this feat? Apparently, it can after the differentiated state of a cell, transforming it into another cell type. Such a phenomenon is called **transdifferentiation** and is usually seen only when parts of an organ regenerate. However, it appears to occur normally in the *Turritopsis* life cycle. In the transdifferentiation process, the medusa is transformed into the stolons and polyps of a hydroid colony. These polyps feed on zooplankton and soon are budding off new medusae. Thus, it is possible that a organismic death does not occur in this species.

QUESTIONS

Long Answer Questions

1. Enumerate different theories of aging. Describe the telomerase depletion theory.
2. Write an essay on genes and aging.

Short Answer Questions

1. Write short notes on the following:
   (*i*) Wear and tear theories of aging
   (*ii*) Gene clock theory
   (*iii*) Parabiosis theory
   (*iv*) Life expectancy
   (*v*) Glycation

Very Short Answer Questions

1. Which protein is called guardian of the genome?
2. What is the function of sirtuin genes?
3. What is the name of non-feeding larva of *Caenorhabditis elegans*?
4. What is ROS?

Multiple Choice Questions

1. What is the study of senescence and senility called
   (*a*) dendrology
   (*b*) gerontology
   (*c*) palaeontology
   (*d*) craniology
2. During aging, collagen present in intercellular spaces becomes
   (*a*) destroyed
   (*b*) impermeable and rigid
   (*c*) more elastic
   (*d*) all of these

3. Decline in hearing begins after age of _________ years
   (*a*) 50
   (*b*) 35
   (*c*) 45
   (*d*) 12
4. The modern idea about aging is that our body slowly loses the power of defence pathogens. This process starts by disappearance of what organ?
   (*a*) thymus gland
   (*b*) parathyroid gland
   (*c*) pituitary gland
   (*d*) spleen
5. As the age advances the hairs in males start thining due to
   (*a*) reduced rate of protein synthesis
   (*b*) falling of hairs
   (*c*) low ATP formation
   (*d*) none of these
6. The number of dead brain cells at the age of 70 years constitute
   (*a*) 10%
   (*b*) 20%
   (*c*) 30%
   (*d*) 40%
7. Absence of one of these results in aging
   (*a*) young collagen
   (*b*) keratin
   (*c*) elastin
   (*d*) none of the above

ANSWERS

Very Short Answer Questions

1. Protein p53.
2. Sirutin genes encode deacetylation (chromatin-silencing) enzyme. They guard the genome, preventing genes from being expressed at wrong times and places and blocking chromosomal rearrangements.
3. Dauer larval stage.
4. ROS stands for reaction oxygen species which are metabolic byproducts that can damage cell membranes and proteins and even destroy DNA.

Multiple Choice Questions

1. (*b*) 2. (*b*) 3. (*a*) 4. (*a*) 5. (*a*) 6. (*b*) 7. (*a*)

53 Apoptosis (Programmed Cell Death)

Cells of a multicellular organism are member of a highly organized community. The number of cells in this community is tightly regulated—not simply by controlling the rate of cell division, but also by controlling the rate of cells death. If cells are no longer needed, they commit suicide by activating an intracellular death program. This process is therefore called **programmed cell death**. It is more commonly called **apoptosis** (from a Greek word meaning "falling off ", as leaves from a tree; Gr. *Apo* = away + *ptosis* = fall).

53.1. EXAMPLES OF APOPTOSIS

The amount of apoptosis that occurs in developing and adult tissues are exceptional. In developing vertebrate nervous system, for example, up to half or more of the nerve cells normally die soon after they are formed. In a healthy adult human beings, billions of cells die in the bone marrow and intestine every hour. It seems remarkable wasteful for so many cells to die, especially as the vast majority are perfectly healthy at the time they kill themselves.

53.2. PURPOSE OF APOPTOSIS

What purpose does this massive programmed cell death serve? In some cases, the answers are clear. Mouse paws, for example, are sculpted by cell death during embryonic development: they start out as spade-like structure, and the individual digits separate only as the cells between them die (Fig. 53.1).

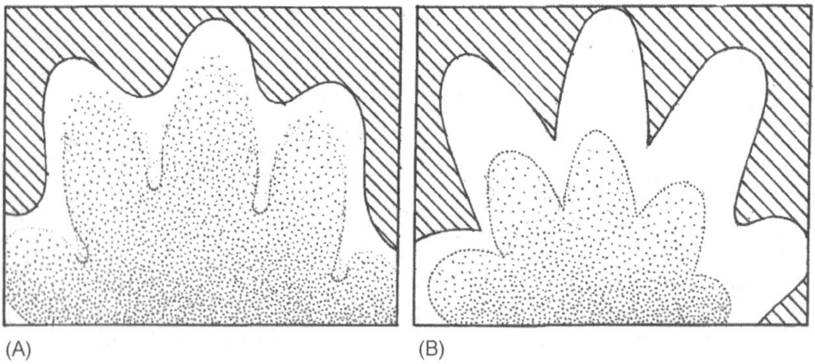

Fig. 53.1. Sculpting the digits in the developing mouse paw by apotosis.

In other cases, cells die when the structure they form is no longer needed. When a tadpole changes into a frog, the cells in the tail die, and the tail, which is not needed in the frog, disappears (Fig. 53.2).

Fig. 53.2. Apoptosis during the metamorphosis of a tadpole into a frog. As a tadpole changes into a frog, the cells in the tadpole tail are induced to undergo apoptosis: as a consequence, the tail is lost. All the changes that occur during metamorphosis, including the induction of apoptosis in the tail, are stimulated by an increase in thyroid hormone in the blood (after Albert *et al.*, 2002).

In many other cases, cell death helps regulate cell numbers. In the developing nervous system, for example, cell death adjusts the number of nerve cells to match the number of target cells that require innervation. In all these cases, the cells die by apoptosis.

In adult tissues, cell death exactly balances cell division. If this not were so, the tissue would grow to shrink. If part of the liver is removed in an adult rat, for example, liver cell proliferation increases to make up loss. Conversely, if a rat is treated with drug **phenobarbital**—which stimulates liver cell division (and thereby liver enlargement)—and then the phenobarbital treatment is stopped, apoptosis in the liver greatly increases until the liver has returned to its original size, usually within a week or so. Thus, the liver is kept at a constant size through the regulation of both the cell death rate and the cell birth rate.

53.3. MECHANISM OF APOPTOSIS

Cell that die as a result of acute injury typically swell and burst. They spill their contents all over their neighbors—a process called **cell necrosis**—causing a potentially damaging inflammatory response. By contrast, a cell that undergoes apoptosis dies neatly, without damaging its neighbors. The cell shrinks and condenses. The cytoskeleton collapses, the nuclear envelope disassembles, and the nuclear DNA breaks up into fragments. Most importantly, the cell surface is altered, displaying properties that cause the dying cell to be rapidly phagocytosed, either by a neighboring cell or by a macrophage (a specialized phagocytic cell), before any leakage of its contents occurs. This not only avoids the damaging consequences of cell necrosis but also allows the organic components of the dead cell to be recycled by the cell that ingests it.

The Caspase Cascade of Apoptosis

The intracellular machinery responsible for apoptosis seems to be similar in all animal cells. This machinery depends on a family of **proteases** (enzymes) that have a *cysteine* at their active site and cleave their target protains at specific aspartic acids. They are therefore called **caspases**. Caspases are synthesized in the cell as inactive precursors, or **procaspases**, which are usually activated by cleavage at aspartic acids by other caspases. Once activated, caspases cleave, and thereby activate, other procaspases, resulting in an amplifying proteolytic cascade (Fig. 53.3). Some of the activated caspases then cleave other key proteins in the cell. Some cleave the nuclear lamins, for example, causing the irreversible breakdown of the nuclear lamina, another cleaves a protein that normally holds a DNA-degrading enzyme (a DNAse) in an inactive form, freeing the DNAse to cut up the

DNA in the cell nucleus. In this way, the cell dismantles itself quickly and neatly, and its corpse is rapidly taken up and degraded by another cell.

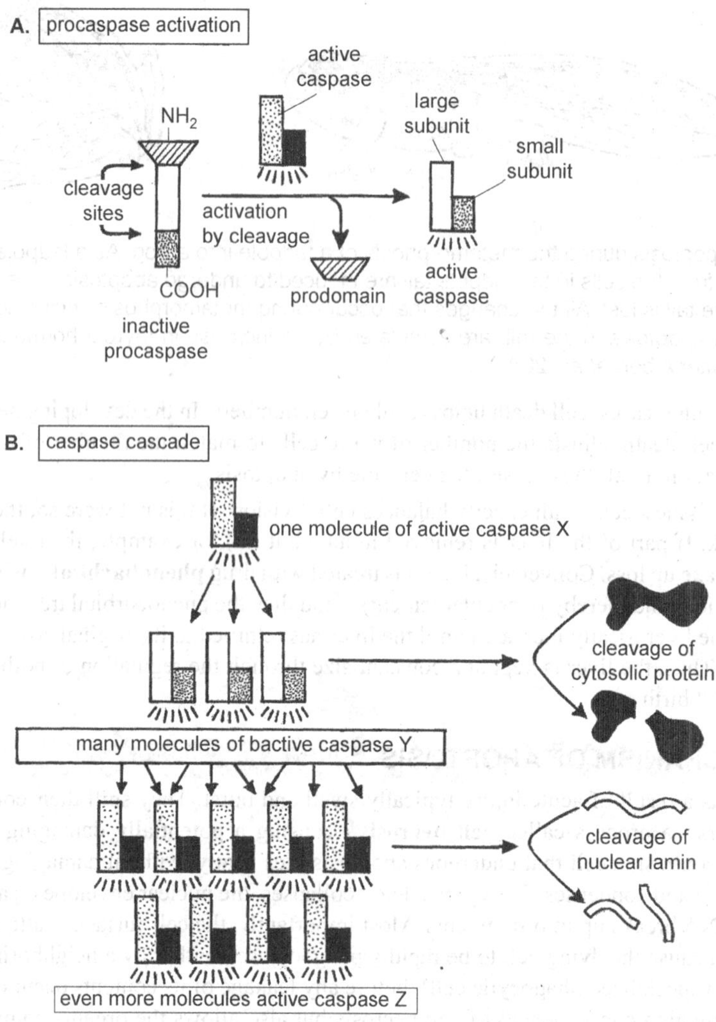

Fig. 53.3. The caspase cascade involved in apoptosis. A—Each suicide protease is made as an inactive proenzyme (procaspase), which is usually activated by proteolytic cleavage by another member of the caspase family. As indicated, two of the cleaved fragments associate to from the active site of the caspase. The active enzyme is thought to be a tetramer of two of these units. B—Each activated caspase molecule can cleave many procaspase molecules, thereby activating them, and these can then activate even more procaspase molecules. In this way, an initial activation of a small number of procaspase molecules (called **intiator caspases**) can lead, via an amplifying chain reaction (a cascade), to the explosive activation of a large number of procaspase molecules. Some of the activated caspases (called **effector caspases**) then cleave a number of key proteins in the cell, including specific cytosolic proteins and nuclear lamins, leading to controlled death of the cell (after Albert *et al.*, 2002).

Activation of the intracellular cell death pathway is like entry into a new stage of the cell cycle. It is usually triggered in a complete, all-or-none fashion. The protease cascade is not only destructive and self-amplifying but also irreversible, so that once a cell reaches a critical point along the path to destruction, it cannot turn back.

Activation of Procaspases

All nucleated animal cells contain the seeds of their own destruction, in the from of various inactive procaspases that lie waiting for a signal to destroy the cell. It is therefore not surprising that caspase activity is tightly regulated inside the cell to ensure that the death program is held in check until needed.

Activation of procaspases is triggered by **adaptor proteins** that bring multiple copies of specific procaspases, known as **initiator procaspases**, close together in a complex or aggregate. In some cases, the initiator procaspases have a small amount of protease activity, and forcing them together into a complex causes them to cleave each other, triggering their mutual activation. In other cases, the aggregation is thought to cause a conformational change that activates the procaspase. Within moments, the activated caspase at the top of the cascade cleave downstream procaspases to amplify the death signal and spread it throughout the cell (Fig. 53.3B).

Activation of procaspase can be triggered from outside the cell by the activation of **death receptors** on the cell surface. Killer lymphocytes, for example, can induce apoptosis by producing a protein, called **Fas ligand**, which binds to the death receptor protein **Fas** on the surface of the target cell (Fig. 53.4). The clustered Fas protein then recruit intracellular adaptor proteins that bind and aggregate procaspase-8 molecules, which cleave and activate one another. The activated caspase-8 molecules then activate downstream procaspase to induce apoptosis (Fig. 53.4A). Some stressed or damaged cell kill themselves by producing both the Fas ligand and the Fas protein, thereby, triggering an intracellular caspase cascade.

When cells are damaged or stressed, they can kill themselves by triggering procaspase aggregation and activation from within the cell. In the best understood pathway, mitochondria are induced to release the electron carrier protein **cytochrome c** into the cytosol, where it binds and activates an adaptor protein called **Apaf-1** (Fig. 53.4B). This mitochondrial pathway of procaspase activation is recruited in most form of apoptosis to initiate or to accelerate and amplify the caspase cascade. DNA damage, for example, can trigger apoptosis. This response usually requires p53, which can activate the transcription of genes that encode proteins that promote the release of cytochrome C from mitochondria. These proteins belong to the Bcl-2 family.

53.4. REGULATORS OF APOPTOSIS

The Bcl-2 family proteins and IAP proteins are main intracellular regulators of the cell death program. The **Bcl-2 family** of intracellular proteins helps regulate the activation of procaspases. Some members of this family, like **Bcl-2** itself or **Bcl-X**, inhibit apoptosis, at least partly by blocking the release of cytochrome c from mitochondria. Other members of the Bcl-2 family are not death inhibitors, but instead promote procaspase activation and cell death. Some of these apoptosis promoters, such as **Bad**, function by binding to and inactivating the death inhibiting members of the family, whereas others, like **Bax** and **Bak** stimulate the release of cytochrome c from mitochondria. If genes encoding Bax and Bak are both inactivated, cells are remarkably resistant to most apoptosis-inducing stimuli, indicating the crucial importance of these proteins in apoptosis induction. Bax and Bak are themselves activated by other apoptosis-promoting members of the Bcl-2 family such as Bid.

IAP family. Another important family of intracellular apoptosis regulators is the **IAP** (inhibitor of apoptosis) **family**. These proteins are thought to inhibit apoptosis in two ways: they bind to some procaspases to prevent their activation, and they bind to caspases to inhibit their activity. IAP proteins were originally discovered as proteins produced by certain insect viruses, which use them to prevent the infected cell from killing itself, before the virus has had time to replicate. When mitochondria release cytochrome c to activate Apaf-1, they also release a protein that block IAPs, thereby greatly increasing the efficiency of the death activation process.

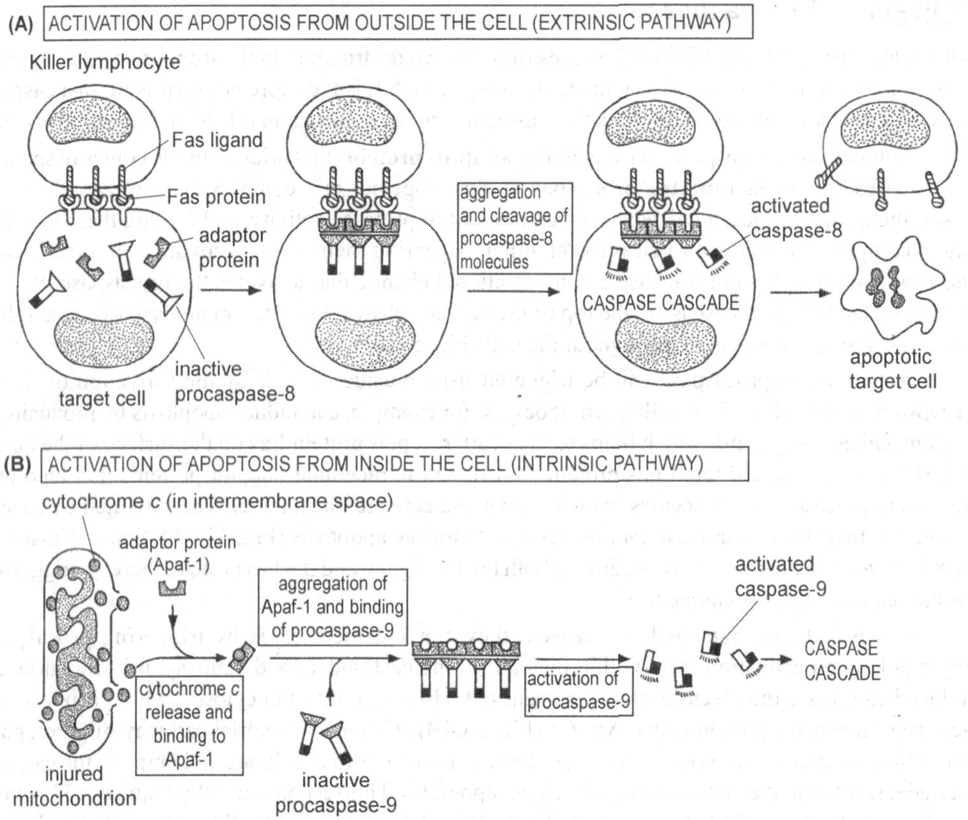

Fig. 53.4. Induction of apoptosis by either extracellular or intracellular activation. A—Extracellular activation. A killer lymphocyte carrying the Fas ligand binds and activates Fas proteins on the surface of the target cell. Adaptor proteins bind to the intracellular region of aggregated Fas protein, causing the aggregation of procaspase-8 molecules. These then cleave one another to initiate the caspase cascade. B—Intracellular activation. Mitochondria release cytochrome c, which binds to and causes the aggregation of the adoptor protein Apaf-1. Apaf-1 binds and aggregates procaspases-9 molecules, which leads to the cleavage of these molecules and the triggering of a caspase cascade. Other proteins that contribute to apoptosis are also released from the mitochondrial intermembrane space (not shown). (after Alberts *et al.*, 2002).

The intracellular cell death program is also regulated by extracellular signals, which can either activate apoptosis or inhibit it. These signal molecules mainly act by regulating the levels or activity of members of the Bcl-2 and IAP families.

53.5. PHYSIOLOGICAL SIGNIFICANCE OF APOPTOSIS

Apoptosis plays an important role during embryological development and also in adulthood:

1. It is responsible for regression of duct system during sex differentiation.
2. It is responsible for degeneration of neurons within CNS and for the formation of synapse.
3. It is responsible for removal of inappropriate clones of immune cells.

4. It is responsible for cyclic shedding of endometrium at the time of menstruation (**Jain**, 2012).
5. It is responsible for cell shed from the tip of villi in the small intestine.
6. It plays a role during metamorphosis of tadpole into frog.

QUESTIONS

Long Answer Questions

1. Write an essay on apoptosis.
2. Describe the mechanism of apoptosis.

Short Answer Questions

1. Write short notes on the following:
   (*i*) Caspase cascade.
   (*ii*) Regulators of apoptosis.
   (*iii*) Physiological significance of apoptosis.

Very Short Answer Questions

1. Define apoptosis.
2. What are the caspases?

ANSWERS

Very Short Answer Questions

1. Apoptosis is programmed cell death. It is an active process that prunes unneeded structures (*e.g.*, frog tails, male mammary tissue), controls the number of cells in particular tissues, and sculpts complex organs (*e.g.*, palate, retina, digits and heart). It should not be confused with *necrosis*, pathological cell death, caused by external factors such as inflammation or toxic injury.
2. Caspases is a family of cysteine proteases (enzymes) that are activated at an early stage of apoptosis and are responsible for the degradative events observed during cell death.

54 Cell Culture and Cryopreservation

The growth of cells or tissues of living organisms outside the body in a suitable sterile culture (or nutrient) medium under controlled environmental conditions is called **cell culture** or **tissue culture**. Depending on the nature of the explanted material, the techniques are called **cell**, **tissue**, **organ** or even **whole embryo culture**. The culture medium may be natural or artificial and either in solid form (*e.g.*, agar) or in liquid form (*e.g.*, physiological saline).

Box 54.1

Culture is a controlled growth of living cells in an artificial medium. The cells may be microorganisms isolated and studied in a pure culture, or they may be cells from animal or plant tissue. The culture medium usually contains water, gelatin or agar, salts and various nutrients (**Allaby** 1995).

Culture of bacterial cells. The culture medium of bacteria contains an **energy source** (some simple carbohydrates or some amino acids); **mineral salts** of calcium, potassium, sulphur and phosphorus; **traces of metals** (zinc, copper, magnesium, manganese, etc.) and **growth factors** (biochemical intermediates). The culture medium is maintained between pH 6 to 8 and at a temperature varying from 15–20°C or 30–37°C according to the species of bacteria.

The culture medium is prepared in *agar* in a petridish. Agar is a carbohydrate material derived from seaweed that is dissolved in a nutritive medium to make a semisolid gel on which bacteria can grow. When it sets, it forms a sheet on the bottom of the petridish. A very dilute suspension of bacteria is cultured in the petridish. As a result of growth and multiplication several colonies of bacteria are formed on the surface of agar. Each colony is derived from a single cell.

54.1. PLANT CELL CULTURE

The plant cell culture is based on an unique property of the cell—**totipotency** that may be defined as the ability of a plant cell to regenerate into whole plant.

1. Historical

Gettlieb Haberlandt, first attempted in 1902 to cultivate the mechanically isolated plant leaf cells on a simple nutrient medium. Though unsuccessful in achieving the growth and differentiation of the cultured cells, he predicted the concept of growth hormones, the use of embryo sac fluids, the cultivation of artificial embryos from somatic cells, etc. During the period 1902 to 1930 attempts were made to culture the isolated plant organs such as roots and shoot apices (called **organ culture**). But the fully differentiated tissues such as pith, secondary phloem, mesophyll and endosperm, etc., in basic culture medium grow into an unorganized and undifferentiated mass of tissues, called **callus**. **Gauherst** *et al.*, (1939) succeeded in growing callus *in vitro* from isolated plant parts. During the period 1940 to 1970, suitable nutrient media to culture plant tissues, embryos, anthers, pollen, cells and protoplasts, and the regeneration of complete plants (called *in vitro* **morphogenesis**) from cultured tissues and cells were developed.

2. Basic Technique of Plant Cell Culture

The whole plants can be regenerated virtually from any plant (referred to **explant**) or cells. The basic technique of plant cell culture (Fig. 54.1) involves the following steps:

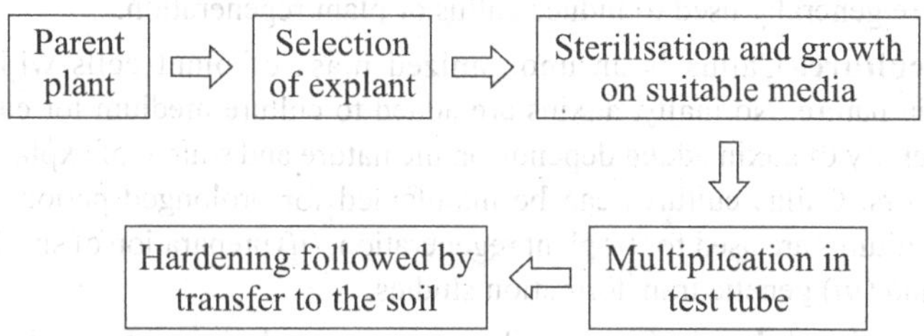

Fig. 54.1. Regeneration of whole plants using tissue culture techniques.

(*i*) **Selection of explants** such as shoot tip.

(*ii*) **Surface sterilization** of the explants by disinfectants (*e.g.*, sodium hypochlorite) and the **washing** the explant with sterile distilled water.

(*iii*) **Inoculation** (transfer) of the explants on to the suitable nutrient medium (which is sterilized by **autoclaving** or **filter-sterilized** to avoid microbial contamination) in the culture vessels under sterile conditions (*i.e.*, in laminar flow cabinet).

(*iv*) Growing the cultures in the **growth chamber** or **plant tissue culture rooms**, having the appropriate physical conditions *i.e.*, artificial light (16h photoperiod), temperature (~26°C) and relative humidity (50–60%).

(*v*) **Regeneration** of plants from cultured plants.

(*vi*) **Transfer** of the plants to the greenhouse or field conditions following the acclimatization (hardening) of the regeneration plants.

3. Culture Media

The *in vitro* culture of plant parts or cells requires a variety of nutrients and suitable physical conditions unlike the intact plants, which can synthesize their own food and many other essential compounds. The composition of plant tissue culture medium can vary, depending upon the type of plant tissues or cells that are used for culture.

A typical nutrient medium consists of inorganic salts (both micronutrients and macronutrients), a carbon source (usually sucrose), vitamins (*e.g.*, nicotinic acid, thymine, pyridoxine and myoinositol), amino acids (*e.g.*, arginine) and growth regulators (*e.g.*, auxins, cytokinins and gibberellins). Other compounds such as casein hydrolysate, coconut milk, malt extracts, yeast extract, tomato juice, etc., may be added for specific purposes. Plant hormones play a crucial role in growth and differentiation of cultured cells and tissues. An optimum pH (usually 5.7) is also very important. The choice of media is directed by the plant species, variety or plant part. The most extensively used nutrient medium is **MS medium** which was developed by **Murashige** and **Skoog** in 1962. Generally a gelling agent agar (a polysaccharide obtained from a red algae, *Gelidium amansii*) is added to the liquid medium for its solidification.

4. Types of Plant Cultures

1. Organ culture. It deals with the culture of the isolated organs (such as roots) under laboratory conditions (*in vitro*), and different names are given depending upon the organ utilized for the culture. For example, the culture of roots, endosperm, ovary, ovule and anther are termed as (*i*) *root culture*, (*ii*) *endosperm culture*, (*iii*) *ovary culture*, (*iv*) *ovule culture*, and (*v*) *anther culture*, respectively.

2. Explant culture. The culture of plant parts, the explants, is known as **explant culture**. The explants can be any part of the plant such as the piece of stem, leaf, cotyledon, hypocotyls, etc. The explant cultures are generally used to induce callus or plant regeneration.

3. Callus culture. Callus is an unorganized mass of plant cells which are generally parenchymatous in nature. Normally, auxins are added to culture medium for callus induction but the nature and quantity of auxin added depends on the nature and source of explant and its genotype besides other factors. Callus cultures can be maintained for prolonged periods by repeated sub-culturing. Callus cultures are used for (*i*) plant regeneration, (*ii*) preparation of single cell suspensions and protoplasts, and (*iii*) genetic transformation studies.

4. Cell suspension cultures. Single cells can be isolated from either callus or any other part of the plant and cultured in liquid medium. Both *mechanical* and *enzymatic* methods can be used for isolation of plant cells. The **mechanical methods** involve grinding of the tissue to a fine suspension in a buffered medium followed by filteration/centrifugation to get rid of cell debris. The **enzymatic method** is based on the usage of enzymes (such as pectinase, macroenzyme) which dissolve the middle lamella between the cells, *i.e.*, intercellular cement, to release single cells. Once the cells have been isolated, they may be cultured by *bath cultures* or *continuous cultures* (both of these cultures have been used for culture of bacteria). The cell suspension cultures can be used for (*i*) induction of somatic embryos or shoots, (*ii*) *in vitro* mutagenesis and mutant selection, (*iii*) genetic transformation, and (*iv*) production of secondary metabolites.

5. Batch culture. Batch culture is a closed culture system, which contains limited amount of nutrients. After the medium is inoculated with the cell or bacterial inoculum, the organism grows and shows normal growth phases *viz.*, lag, log (exceptional), stationary and decline. Growth results in the consumption of nutrients and excretion of microbial products. At stationary phase, the growth declines to zero. This means, that in such a culture, growing cells are exposed to continually changing environment due to gradual consumption of nutrients and accumulation of metabolites. Culturing microorganisms in the laboratory, in an ordinary flask, is basically a batch culture.

6. Fed-batch culture. If a batch culture is continuously or sequentially fed with fresh medium without removing the growing culture, it is called **fed-batch culture**. In such a culture, over a period of time, the volume in culture vessel goes on increasing. Fed-batch culture is preferred when high substrate concentration causes growth inhibition. To avoid this, substrate is fed at concentrations below the toxic level to achieve cell growth. By using fed-batch culture, high cell densities can be achieved in the same reactor volume in comparison to the normal batch culture.

7. Continuous culture. This method of culture is a way of getting a continuous supply of microbial growth and/or products. The growth medium is designed in such a way that one of the nutrients is in limited quantity. Therefore, during the exponential growth, as this nutrient is exhausted the growth will stop. However, just before the nutrient is fully exhausted, fresh medium containing the limited nutrient is added. In continuous culture, raw materials such as nutrients are supplied at a rate volumetrically equal to that at which the cells and products are removed.

In continuous culture, cells can be grown at a particular growth rate for an extended period of time. Most of the time the chemical environment inside a continuous culture is constant. In a **chemostat**, constant chemical environment is maintained whereas in a **turbidostat** constant cell concentration is maintained.

8. Mass cell culture. Plant cells can be cultured in specially designed plant bioreactors' which essentially do not have a stirrer (as plant cells are shear sensitive). In place of stirrer, gas is gently bubbled which provides stirring as well as meet the demand of a higher oxygen supply.

9. Protoplast culture. Protoplasts are plant cells without cell wall and can be isolated by **enzymatic method** (cellulases, hemicellulases and pectinases) from leaf, seedling, calli, pollen grains, embryo sacs, etc. As the protoplasts lack cell wall they can be utilized for many purposes such as (*i*) various biochemical and metabolic studies, (*ii*) fusion of two somatic cells to create somatic hybrids, (*iii*) fusion of enucleated and nucleated protoplasts to create cybrids (cytoplasmic hybrids), and (*iv*) genetic manipulation.

10. Meristem culture. This is growing of complete plants from explant tissue or shoot meristems on nutrient medium.

Of these culture methods, only two—namely protoplast culture and callus culture are largely used in laboratory and on commercial scale to raise whole plants, hence will be discussed in detail.

5. Technique of Protoplast Culture

Protoplast is the cell mass obtained by removing the cell wall. Cultured protoplasts have gained tremendous importance for they are used for somatic cell fusions and for introducing foreign DNA, cell organelles, bacteria and virus particles. The *totipotent* nature of plant cells has distinct advantage since any cell can be stripped of its cell wall and grown into a complete plant in suitable culture medium. The technique of protoplast culture includes the following four essential steps: 1. Isolation of protoplasts; 2. Purification; 3. Culturing; and 4. Regeneration of whole plant.

1. Isolation of protoplast. The protoplast is isolated either by mechanical method or by treating the cells with cell wall degrading enzymes. In mechanical method, plant cells are *plasmolysed* so that protoplast shrinks away from the cell wall, and then cut with a fine knife to release the protoplast. This is a crude method and provides a poor yield of protoplast. The enzymatic method is very efficient in which the tissue is treated with a mixture of enzyme such as cellulase and pectinase in a solution along with osmotic stabilizers (*e.g.*, sucrose solution).

The protoplasts can be isolated from leaves, roots, callus and any other part of a young plant. The most commonly used materials are the leaves which are sterilized and peeled off to remove epidermis before enzymatic treatment. The peeled leaves are kept in the enzyme solution for about 15–18 hours at 25°C and teased gently to liberate protoplasts. The mixture is filtered through a fine wire gauze to remove leaf debris, and transferred to screw capped tubes and centrifuged at 100g for about 2 minutes. The protoplasts form a pellet and the supernatant is discarded.

2. Washing. The pellet containing protoplast is washed with sorbitol, and subsequently centrifuged in 20% sucrose at 200 g for 1 minute to obtain isolated and purified protoplasts. All these steps are carried out in an aseptic chamber.

3. Protoplast culture and regeneration of whole plants. Isolated protoplast can be cultured to grow into individual plant by planting on semisolid or liquid media. Most popular method is **Bergmann's cell plating technique**, which utilizes liquid medium for culture. The free protoplasts are suspended in a liquid medium where the protoplasts start developing cell walls within 2–4 days. Later the cells along with the liquid medium are mixed with parafilm and then incubated at 25°C in dark for 3–4 weeks. One can ultimately observe small colonies.

6. Protoplast Fusion

The technique of protoplast fusion is also called **somatic cell hybridization**. This technique is significant since fusion of two protoplasts from two different varieties or species of plant would give rise to whole plants with improved characters. This is infact somatic hybridization, which was achieved first in animals and later in plants. Somatic hybridization has been a great success in producing hybrids between species of *Brassica*, *Nicotiana*, *Petunia* and *Solanum*.

In somatic hybridization, two protoplasts from different species of plants or from diverse sources are made to fuse in the presence of an inducer. Induced fusion is brought about but treatments with agents, such as $NaNO_3$, high pH with high concentration of Ca^{2+} ions, and polyethylene glycol (PEG). Use of PEG as a **fusogen** is most preferred since it is highly successful in many crop plants with a high frequency of fusion. Isolated protoplasts are mixed in 1 : 1 ratio and suspended in 1 ml of culture medium to which 1 ml of 56% solution of PEG is also added. The tube is shaken for 5 second and the protoplasts are allowed to settle down. The supernatant is removed and the protoplasts are washed with washing solution. The protoplasts are resuspended in the washing medium. Now two types of protoplasts are mixed in equal quantities and 4–6 microdrops are placed in a small petridish and allowed to settle at room temperature. Two or three drops (about 300 μl) of PEG are added in the petridish and incubated for 10–25 minutes at room temperature (24°C). This results in **agglutination** of the protoplasts. Subsequently, the protoplasts are washed with the culture medium at least 5 times and during this process most of the fusion is achieved.

The fused protoplasts are then cultured in a thin layer of culture medium to allow growth of plants. After the fusion process, the protoplast population may consist of **homokaryons** or **heterokaryons**. The population of heterokaryons can be selected by allowing them to grow on a medium in which homokaryons will be fail to grow.

7. Technique of Callus Culture

Callus is a mass of cells separated from a plant source or a seedling and then regenerated into a complete plant. For this purpose, a suitable plant species is selected in which morphogenesis can be readily induced by hormones. The plants that serve as good materials for the callus culture are carrot, tobacco, coffee, etc. In fact, it is also possible to grow isolated single cells into a mass of cells or callus. If callus is raised from single cells, the mesophyll are isolated from leaf by mechanical or enzymatic methods (*e.g.*, tobacco leaves) and free cells suspended in the Murashige skoog (MS) medium. The suspended cells are mixed with MS agar medium, shaken and transferred to petri dishes. The petri dishes are covered with parafilm and incubated in dark at 25°C for 3–4 weeks, one can observe colonies of cells on the agar surface.

Callus regeneration. The colonies of cells or calluses are now ready for regeneration. Alternatively, one can procure pith tissue or callus from a tobacco plant or carrot for the purpose of regeneration. The callus is taken into flasks containing MS liquid medium and shaken at about 200 rpm so that the callus is disintegrated into single cells mechanically. The cells are now transferred to a microchamber kept in a sterile petridish and incubated in a culture room in the presence of light at 25°C till such time the calluses are developed. Subsequently, the calluses are transferred in MS medium containing a hormone (2, 4-D) and incubated in light at 25°C for further growth of the callus into shoot, the mini plants are transferred to hormone-free MS medium to develop roots. The rooted plants are now ready for transfer to the pots containing soil, manure, nutrients, etc., and kept in a suitable environment where humidity, light and temperature are painstakingly maintained. These potted plants grow into full flowering plants. Many flowering plants are commercially being produced in advanced countries.

54.2. ANIMAL CELL CULTURE

Unlike plant cell culture, animal cell culture is much more complicated and demands sophisticated laboratory instrumentation and handling skills. Now the cultivation (or culture) of animal cells is widely used technique in many different disciplines from basic science of cell and molecular biology to biotechnology. This is being used to answer biochemical, physiological and morphological questions.

Individual cells released from complex tissues or organs can be maintained in artificial conditions and treated as discrete organisms *in vitro*. With the development of animal cell culture technique, the annual demand of more than 280 million of experimental animals worldwide has reduced (**Sobti** 2008).

The range of different cell types which can now be grown in culture is quite extensive and include connective tissue elements, such as fibroblasts, skeletal tissues (bone and cartilage, skeletal, cardiac and smooth muscles), epithelial tissue (*e.g.*, liver, lung, breast, skin, bladder and kidney cells), neural cells (glial cells and neurons), endocrine cells (adrenal, pituitary, pancreatic cells), melanocytes and many different types of tumours. Mammalian cells are now being cultured to produce a variety of pharmaceutically important macromolecules. Many animal cell cultures are made to transfer animal cells from laboratory to the production level.

Recently, animal cell culture has assumed tremendous importance as it is being exploited to produce a number of commercial and medicinal products such as enzymes, proteins, hormones, antibodies and vaccines. Evaluation of new drugs on culture cells is being done which is not only cost effective but also less time consuming. Stem cell technology is a boon to the medical sciences for the production of tissue grafts and gene replacement therapy.

1. History

Animal cell culture has a history of over 100 years when the first attempts to grow cells were made by **Ross Harrison** (1907) who was able to show the development of nerve fibres from frog embryo tissue cultured in a blood clot. The famous physiologist **Alexis Carrel** was able to keep fragments of chick embryo heart alive and breathing for three months *in vitro*. These studies demonstrated the utility of *in vitro* cultures as an experimental system. In late 1940's the discovery of a way to grow poliomyelitis virus in culture by **Enders**, **Weller** and **Robbins** gave a tremendous impetus to the study of this disease. It also brought an end to the **monkey era** (no longer did one depend on infection in the monkeys; **Kannan** 2003). Animal tissue culture opened up an easy and rapid method of growing polio virus for vaccine production and also led to the discovery and better understanding of new and unknown viruses. Tissue culture methods have provided virologists with a simple *in vitro* method for testing a multitude of chemical and antibiotic agents for their effect on the multiplication of viruses in living cells.

Alec Issacs (1966) discovered interferon by infecting cells in tissue culture with viruses. He took filterates from virus infected cells and grew fresh cells in the filtered medium. When challenged with the virus, the cells did not get infected. He went on to predict that cells infected with the virus secreted a molecule which coated onto uninfected cells and interfered with the viral entry. This molecule was called **interferon**. This observation was not very readily accepted and it was even said that Alec Issacs was mad. But the advent of genetic engineering led to the cloning and expression of interferon and proved the validity of the observation.

2. Features of Animal Cell Growth in Culture

1. Mortality. Animal cells can grow in simple glass or plastic containers. When we grow microbial and plant cells in culture and if nutritive media is constantly replenished the cell will grow indefinitely. However, animal cells, depending on the tissue they have been isolated from can grow even in the best nutritive media to only limited generations. Some cells such as *nerve cells* cannot divide in culture, cells from the connective tissue (skin), called *fibroblasts*, can grow somewhat longer but eventually all will die after several generations. Hence there is **mortality** associated with all normal animal cells.

2. Contact inhibition. In culture, animal cells tend to divide and fill the surface of the container they are growing in and then stop growing. We can relate this feature with what happens in the animal body. The infant animal grows only to adulthood and not any further. The cells comprising tissues and organs as the liver grow only to a certain size after which they cease to grow. This phenomenon which occurs in the normal body is observed also in cell culture and is termed "**contact inhibition**". This means that when cells grow and reach the walls of the container (*i.e.*, reach confluency) they are prevented further growth.

3. Cells in culture have different environment. In a cell culture, the environment of the cells is different from that *in vivo*. The major differences include the absence of cell-cell interaction, cell-matrix interaction, lack of three dimensional architecture and change in hormonal and nutritional environment. These differences lead to alterations in the way the cells adhere to the glass or plastic container they are grown in, the way the cells proliferate as well as the shape of the cells. It is worthnoting that in culture, cancer cells appear very different from normal cells. Cancer cell cultures lose contact inhibition and pile on each other due to uncontrolled growth and among other features appear more rounded in shape. These differences in growth patterns in normal versus cancer cells is ulilised by oncologists (cancer specialists) to determine whether tumours are cancerous (malignant) or not.

3. Types of Animal Cell Culture

1. Primary cell culture. Cells are dissociated from the parental tissue (such as kidney, liver) by mechanical or enzymatic methods and maintained in culture medium in suitable glass or plastic containers. The maintenance of growth of these cells under these conditions is known as **primary cell culture**. The enzymes used most frequently for separating cells in a given tissue (dispersion) are crude preparations of trypsin and collagenase. These enzymes are proteolytic in nature and cleave the cementing proteinaceous material between cells in a tissue. The characteristics of cells in culture usually depend on their original source within the animal.

In culture, cells can grow as **adherent cells** (anchorage-dependent) or as **suspension cultures** (anchorage-independent). Adherent cells are usually derived from tissues of organs such as kidney where they are not mobile and are embedded in connective tissue. They grow adhering to the cell culture vessel. On the other hand, suspension cells do not attach to the surface of the culture vessel. Virtually, all suspension cultures are derived from cells of the blood system. This is because these cells (*e.g.*, lymphocytes) are also suspended in plasma *in vivo*. The *drawbacks* of primary culture are that they are time consuming and require the use of live animals or fresh tissue. There can be considerable variation from one preparation to another particularly if prepared by different people. These difficulties can be overcome by the use of **secondary cell cultures** or **cell lines**.

2. Secondary cell cultures or cell lines. Once the primary cell culture is subcultured, it becomes known as **secondary culture** or **cell line**. Sub-culturing or "splitting cells" is required to periodically provide fresh nutrients and growing space for continuously growing cell lines. The frequency of subculture or density of cells plated, will depend on the characteristics of each cell type. If cells are split too frequently or at too low a density, the line may be lost. If cells are not split frequently enough, the cells may exhaust the medium and die.

Subculturing involves removing the growth media, washing the plate, disassociating the adhered cells, usually enzymatically (*e.g.*, with trypsin enzyme, although some cells may be removed by repeated pipetting or gentle scraping), and diluting the cell suspension into fresh media. Such cultures are the secondary cultures. Sometimes certain cells of these secondary cell cultures can spontaneously become transformed (Box 54.1) and give rise to *continuous cell* lines. These cell lines show immortality as they can grow indefinitely without dying in culture. The cultures can contain mixed cell types or can consist predominantly of a single cell type.

Types of cell lines. The various types of cell lines are categorised into two types, *i.e.*, finite cell line and continuous cell lines.

(*i*) **Finite cell lines.** They have a limited life span and they grow through a limited number of cell generations. Finite cell lines show the property of *contact inhibition, density limitation* and *anchorage dependence*. The mode of growth is in the monolayer form. The growth rate is slow and doubling time is around 24 to 96 hours.

(*ii*) **Continuous cell lines.** Cell lines transformed under *in vitro* conditions give rise to continuous cell-lines, *e.g.*, CHO continuous cell line from hamster ovary, Cos-7 cell line from monkey kidney, HeLa cells from human cervical carcinoma. The various properties associated with continuous cell lines are : the ploidy is either aneuploid or heteroploid, there is no contact inhibition and no anchorage dependence. The mode of growth is either monolayer or suspension form. The growth rate is rapid and doubling time is 12 to 24 hours. The density limitation is reduced or lost.

3. Suspension culture. In this culture, cells grow in suspension and do not attach to the substratum. It is also primary culture but anchorage dependent, *e.g.*, cell culture derived from haemopoietic system.

4. Monolayer culture. In this type of culture, a large number of cells are added to a culture dish; these cells settle and attach to the bottom and form a relatively uniform layer of cells. Those cells that survive will grow and divide and after a number of generations, form a monolayer of cells that cover the bottom of dish.

5. Clonal culture. In this type of culture, a relatively small number of cells is added to the dish, each of which, after settling and attaching to the surface is at some distance from its neighbours. In this case, the proliferation of cells generate individual colonies or **clones** of cells.

4. Material Preparation

A. Isolation of cells. Cells to be cultured are isolated from various tissues, which could be embryonic or normal parental tissue such as kidney, liver, skin, etc. Isolation of cells is done by following four methods:

(*a*) **Explant method.** In the explant method of tissue culturing, a small fragement of tissue is placed on a coverslip and covered with a drop of chick plasma and growth medium. The plasma sets to form a clot. The coverslip is mounted in a cavity slide with the clot attached to the under surface. The space in the cavity provides source of oxygen and allows for exchange of carbon dioxide.

This method is also called **Maximov double coverslip method** (Fig. 54.2).

(*b*) **Mechanical method.** In this method, the tissue is forced through cheese cloth or mechanically shaken in an appropriate buffer.

(*c*) **Enzymatic method.** In this method, tissues are dissociated by the use of hydrolytic enzymes that breakdown the extracellular matrix (*e.g.*, basement

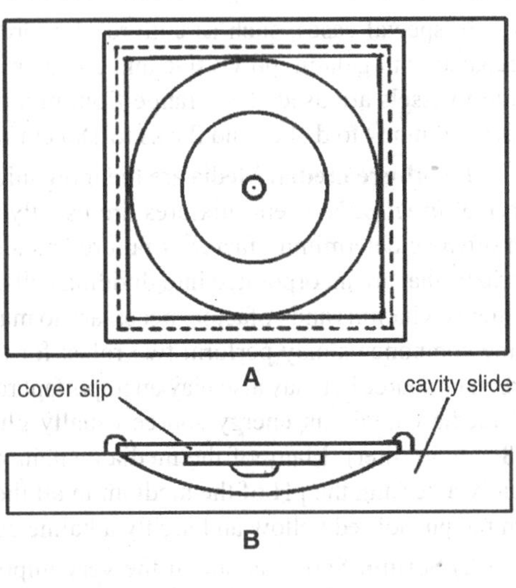

Fig. 54.2. Maximov double coverslip method of cell culture.

membrane). This method though efficient can damage the plasma membrane by removing its components. The exposure time to these chemicals or enzymes is very critical. The enzyme used for this purpose are *trypsin* (0.1–0.25%), *pronase* (0.05%), *collagenase* (0.01–0.15%), *elastase* (0.05%), *hyaluronidase* (0.1%), etc.

(*d*) **Chemical method.** In this method certain chemical agents such as EDTA or EGTA are used for the isolation of cells.

B. Sorting of cells. The separated cells are put in primary culture and they remain so until are passaged or subcultured. They are usually heterogenous and have a low growth fraction, but have more representatives of different cell types in the tissue from which they were derived and in the expression of tissue specific properties.

Once the cells have been dissociated into a single-cell suspension, they can either be cultured directly or separated according to cell type and placed in culture. A number of techniques including differential centrifugation or the fluorescence activated **cell sorter** can be used to separate cells. In the latter technique, the cell suspension is treated with a fluorescent antibody that specifically binds to the surface of the cell type to be cultured, and the suspension is passed through an electronic instrument that is able to deflect the cells lacking the fluorescent label.

5. Physical Environment for Culture of Animal Cells

The purpose of culturing of animal cells *in vitro* condition is to recreate physical, nutritional and hormonal environments in which the cells can grow. The physical environment includes controlling the temperature, pH, osmolarity and gaseous environment by providing a supporting surface (substrate and culture medium) and protecting the cells from chemical, physical and mechanical stresses.

1. Substrate. Some cells require a definite substrate for their growth. The nature of substrate depend on the type of cells and the use to which they will be put.

Diploid cells require for their growth a solid substratum with a defined surface charge. These requirements are fulfilled by **glass**. However, in the cleaning of the glass by detergents must be avoided and be replaced, *e.g.*, by phosphates. *Polystyrene* (*i.e.*, a synthetic resin) equipment which has been subjected to treatments that adjust the desired negative charged density is increasingly being used. Commercial tissue-culture grade flasks and dishes are made on these principles.

In special cases, such as culture of neurons, muscle cells and epithelial cells, the plastic is precoated with gelatin, polylysine and collagen. For providing a net positive charge, variable sized culture vessels are used. They range from multiwell plates (1 mm^2 square area) and microtitration plates (30 mm^2) to dishes and flasks to 180 cm^2 and multisurface propagators.

2. Culture media. Media are the inorganic salts and other nutrients capable of sustaining cell survival *in vitro*. Nutrient mixtures are usually called *media*. Having the correct nutrient mixture can often be determining factor as failure or success in cell culture. The medium provides essential nutrients that are incorporated into dividing cells, such as amino acids, fatty acids, sugars, ions, trace elements, vitamins and cofactors necessary to maintain the proper chemical environment for the cell. Some components may perform two roles; for example, the sodium bicarbonate may be used as a carbonate source but may also play an important role in maintaining the appropriate pH and osmolality. All media contain an energy source usually glucose, although its concentration can vary widely (0.8 – > 5 g/liter). Many of the media contain phenol red as pH indicator. This is very helpful in rapidly assessing the pH of the medium in all the cultures in an incubator. Highly acidic conditions turn the phenol red yellow and highly alkaline conditions turn the phenol red into pink colour.

(*i*) **Serum.** Serum is one of the very important components of animal cell culture. It should be supplemented along with the medium because it contains growth factors which promote cell proliferation, cell attachment and adhesion factors. Serum is also source of various amino acids, hormones, lipids, vitamins, polyamines and salts containing ions such as calcium, chloride, ferrous, ferric, potassium, etc.

***(ii)* Medium design.** There is a direct effect of culture system on medium design as all the three types of cultures (*i.e.*, suspension, monolayer and clonal) have their own requirements and limitations.

A. Growth media. These culture media are used for the growth of the cell.

1. The medium for suspension cultures should be able to support the relatively high yield, prevent cell aggregation and chemical damage to the cells and also the precipitation of serum proteins.

2. The medium for monolayer cultures should be enriched with calcium and magnesium ions.

3. The clonal cultures have special needs. In these, feeder layers are frequently used. Use of conditioned medium is a medium that has been removed from active growing or confluent cultures and clarified by centrifugation. It may be used after typically 3 : 1 or 1 : 1 dilution with fresh medium for cells that are very difficult to grow in culture or where very low cell densities are involved.

B. Maintenance media. These types of culture media are used for maintaining the cells in non-growing state. Such media have low levels of serum.

C. Sterilization of media. Almost all the nutrient culture media are used for animal culture contain thermolabile components, so it is impossible to sterilize them by autoclaving. The media are, therefore, sterilized by high pressure filteration through membrane filters with pore widths of about 0.2 mm. Such procedures can be carried out on clean benches with **horizontal laminar air flow** which offers the best protection against contamination from outside.

3. Temperature. Most mammalian cell culture are grown in incubators maintained at 37°C. This was chosen because it is the core body temperature of human beings (*Homo sapiens*). It has been observed that most cells from warm blooded animals will grow at this temperature.

4. pH. The regulation of extra-cellular and intra-cellular pH is essential for survival of individual mammalian cells. The pH is not only important for maintaining the appropriate ion balance but also for maintaining optimal function of cellular enzymes and for optimal binding of hormones and growth factors to cell surface receptors. Even transient changes in pH can alter cell metabolism and can lead to cell death. Most media strive to achieve and maintain pH between 7 and 7.4. Different cell types may have an optimum pH slightly outside this range. The regulation of pH is done through a variety of buffering systems. Most media use a bicarbonate CO_2 system as its major component. The interaction of CO_2 derived from cells or the atmosphere with water leads to drop in pH described by the equation below:

$$H_2O + CO_2 \rightleftharpoons H_2CO_3 \rightleftharpoons H^+ + HCO_3^-$$

Increasing the bicarbonate concentration neutralizes the effect of increased CO_2 due to the following: $NaHCO_3 \rightarrow Na^+ + HCO_3^-$. The increased HCO_3^- derives the equation above to the left until equilibrium is reached at pH 7.4. This kind of system is called **open system**.

The choice of buffering system will clearly affect both the pH and the osmolality of the final medium as well as other aspects of cell physiology controlled by the ionic environment.

5. Osmolality. The *osmolality* can be defined as the concentration of an osmotic solution especially when measured in osmols or milliosmols per 1000 grams of solvent. The osmolality of the medium used is determined by the media formulation. Salt and glucose are the major contributors to the osmolality of the medium, although amino acids may also contribute significantly. Altering the osmolality significantly can affect cell growth and function. For instance it may lead to loss of membrane integrity, as the outside osmotic pressure becomes higher or lower than that which must be maintained inside the cell and the cells shrink or burst respectively. Almost all commercial media are formulated to have a final osmolarity of around 300 mOsm. Osmolarity can be checked directly with an osmometer.

6. Equipments Required for Animal Cell Culture

1. Tissue culture hoods [lamilar air flow (LAF) hoods]. All tissue/cell culture manipulations are performed **aseptically**, *i.e.*, with the prevention of any bacterial or fungal contamination. Otherwise animal cell culture media can easily get contaminated with bacteria or fungi which grow more rapidly than animal cells. LAF hoods allow the work area to be free of contamination. A LAF hood essentially performs two functions: (*i*) It protects the tissue culture from the operator (*i.e.*, it maintains a sterile environment). (*ii*) It protects the operator from the tissue culture (*i.e.*, it check potential risks of infection).

Types of LAF hoods. Depending on the nature of the cells and organisms being handled (especially infective agents) tissue culture hoods are of following three types:

(*i*) Class I LAF hoods. They give good protection to the operator and to a lesser degree, the cell culture. Air is drawn from the open front (past the operator) over the cell culture, and out through the top of the hood. These hoods are found within specially designed sterile work areas (*i.e.*, sterile air is sucked into the hood) where user wear special protective clothing.

(*ii*) Class II LAF hoods. They offer protection to both the operator and the cell culture. Filtered air is drawn in through the top of the hood, down over the tissue culture and through the bottom of work area. In addition, air is drawn from the half open front past the operator and down through the grill in front of the work area. In this way the cell culture is protected in a stream of sterile air and the operator is protected from contamination by the inflow of air into the base of the work area. The class II hood is the most common type found in a tissue culture laboratory.

(*iii*) Class III LAF hoods. They are used for work with highly pathogenic organisms. In these, the worker is screened from the work by a full physical barrier. This is normally achieved by the replacement of the open front with glass or prespex, with a pair of heavy-duty protective gloves attached, through which the work is accessed. The use of a class III hood requires knowledge of special safety considerations—these should be assembled with the help of relevant local biological safety officer.

2. CO_2 incubator. The CO_2 incubator is designed to reproduce as closely as possible the environmental conditions of the living cells. The essential functions of the incubator are to maintain the sterility of the chamber, a constant temperature, an atmosphere with a fixed level of CO_2 and an atmosphere with a high relative humidity. The incubator chamber is made airtight by a silicon gasket on the inner door and is thus isolated from the external environment where there is contamination. The gas injected inside the chamber is High Efficiency Particulate Air (HEPA) filtered to maintain the sterility of the chamber. High relative humidity prevents desicattion of the medium and maintains the correct osmolality.

3. Inverted microscope. An inverted microscope is invaluable for visualizing cell cultures *in situ*. In tissue culture vessels, for example a petri plate, the cells are present on the bottom of the plate with the culture medium above. If the petri plate is placed on a conventional microscope one cannot clearly visualize the cells. The inverted microscope allows the cells at the bottom to be visualized because the optical system is at the bottom with the light source at the top. Looking at cultures in this way will give an immediate idea of their health and growth. A standard microscope with a movable slide holder is needed for counting cells using a counting chamber.

4. Centrifuges. For most cell cultures only low-speed centrifuges are required. A gently breaking action helps in preventing disruption of the separated bands of the cells. In most cases cells should be centrifuged at 20°C; nevertheless low operation temperature is useful to avoid exposing cells to uncontrolled higher temperatures due to heat from the motor.

7. Characterisation of Cell Lines

In order to analyze the growth characteristics of a particular cell type or cell line, a growth curve can be established from which one can obtain a population doubling time, a lag time, and a saturation density. A growth curve generally will show the cell populations **lag phase**, that is, the time it takes for the cells to recover from subculture, attach and spread; the **log phase**, in which the cell number begins to increase exponentially and a **plateau phase**, in which the growth rate slows or stops due to exhaustion of growth medium or confluency. An increase in cell number is also a frequently used method of assessing the effect of hormones, nutrients, and so forth on a specific cell type. The culture doubling time, allows prediction of the likely cell concentration at any time in the future.

In India, there are some laboratories and institutions which maintain various cell lines for scientists to use (for example, National Center for Cell Sciences, Pune, India). The advantage of using established cell lines is that their growth characteristics, media requirements and responses to selected reagents are established and therefore convenient for the scientists to use.

Once a cell line has been established, its karyotype has to be determined. This will confirm the species of origin, at least for those cells karyotyped, and determine the extent of gross chromosomal changes in the line. Karyotypes may vary from being near normal (*i.e.*, the vast majority of cells in the culture have normal karyotypes) to being aneuploid. The stability of karyotype depends on the species from which cell lines was derived, the growth conditions used, the way in which the cells are subcultured and wheter or not the cells are frozen.

8. Scale-up of Animal Culture Process

Scale up means increase in size, number or extent. The various scale up methods include roller bottles with microcarrier beads for adherent cell cultures and spinner flasks for suspension cultures.

1. Roller bottles. The roller bottles (Fig. 54.3) contain microcarrier beads. In them the cells adhere to the total curved surface area of the microcarrier beads, thereby markedly increasing the available space for growth. These tissue culture bottles can be used in specialized CO_2 incubators with attachments that rotate the bottles along the long axis. After each complete rotation of the bottle, the entire cell monolayer has transiently been exposed to the medium. The volume of the medium need only be sufficient to provide a shallow covering over the monolayer.

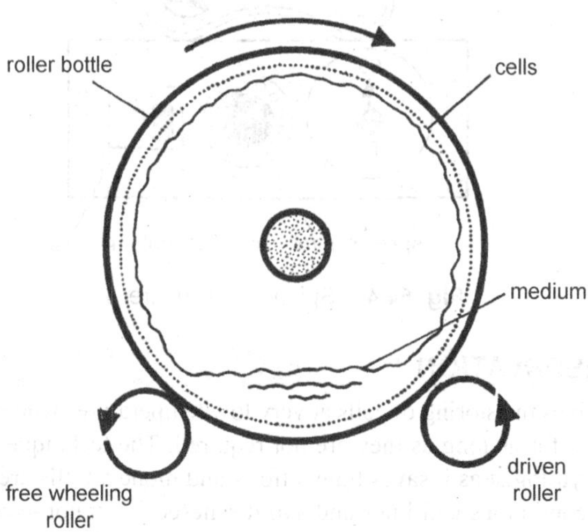

Fig. 54.3. Roller bottle cell culture.

Microcarrier beads. These beads are used to increase the number of adherent cells per flask and are either dextran or glass-based and come in a range of densities and sizes. The beads are buoyant and therefore can be used with spinner culture flasks. The surface area available for cell growth on these beads is huge. Microcarrier beads when resuspended at the recommended concentration provide 0.24 m^2 for every 100 ml of culture flasks. Under these circumstances adherent cells can be grown to very high densities before crowding becomes a problem. Cells growing at such high densities will rapidly exhaust the medium which may need replacing during culture.

2. Spinner culture. Spinner cultures are used for scaling up the production of suspension cells. They consist of a flat surface glass flask with a suspended central taflon paddle that turns and agitates the medium when placed on a magnetic stirrer (Fig. 54.4). Commercial versions incorporate one or more side arms for sampling and/or decantation. The cells are not allowed to settle to the bottom of the flask and thus cell crowding occurs only at very high densities. Stirring the medium improves gas exchange.

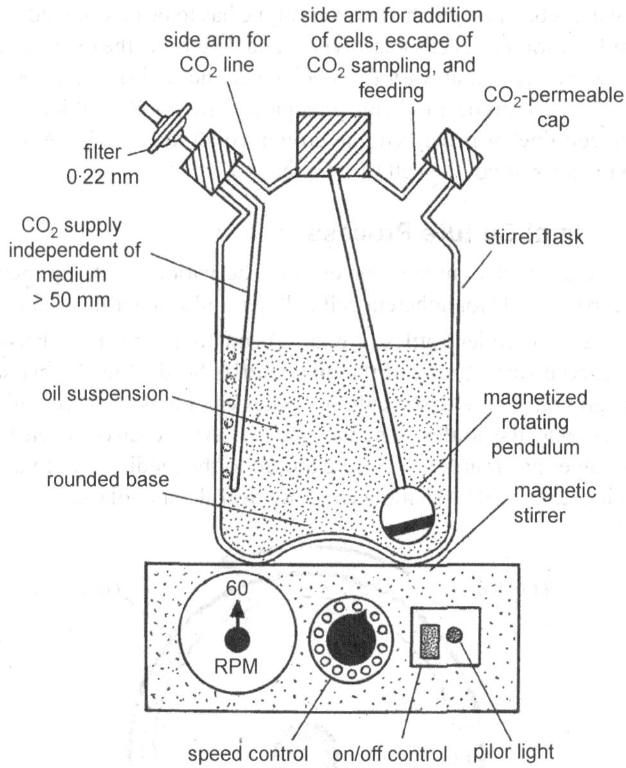

Fig. 54.4. Spinner cell culture.

54.3. CRYOPRESERVATION

Cryopreservation involves the storing of cells at very low temperatures where they remain in a state of suspended animation for as long as they are not required. The technique of cryopreservation is very significant for a cytologist as it saves time, efforts and money. Cells are also cryopreserved to avoid genetic drift in continuous cell lines and avoid senescence in non-continuous cell lines and guard against contamination. The ability of cryopreservation of cells is very vital. For the critical temperature interval between +20 and –70°C, programmable apparatuses have been developed, but the graduated use of refrigerators in combination with insulating materials fulfil the same purpose. For long term preservation, temperatures below –70°C are required, and for this purpose, **Dewar vessels with liquid nitrogen** or more sophisticated **"liquid nitrogen refrigerators"** are used. The procedure requires a **cryoprotective reagent**. The preferred additive is 10% glycerol except for those cells that are permeated too slowly and may require a cytotoxic agent such as 7.5% dimethyl sulphoxide. The critical lymphoid cells are preserved using 10% polyethyleneglycol (PEG 1000–2000).

1. Importance of Cryopreservation

Under culturing conditions, cells have an ability to change their properties from time to time. Differences are found to occur between cells from the same source when handled in different laboratories. To overcome this problem, cryopreservation is essential for the following reasons:

1. Research laboratories where consistent data are required over a period of time.
2. Cryopreservation is applicable for genome conservation in genetic engineering.
3. For preserving seed stock.
4. Low-passage stock of finite-life cells, *e.g.*, human diploid lines can be available for prolonged period.
5. To preserve parental stock for genetic studies.
6. Cryopreservation is also helpful in acting as a reserve against any loss due to contamination.
7. It is not necessary to repeatedly subculture cells that are not required immediately.
8. Cryopreservation reduces labour, risk of contamination and media costs, etc. Thus clones of fusion experiments can be investigated at ease.

2. Cooling and Warming Processes

Cells meant for cryopreservation are first grown in a proliferating medium and kept dormant. They are then put in a preserving medium. For this purpose serum is used, however, cells in the preserving medium are not growing. The procedure is very critical because there is always a fear of cell death. The cooling process must, therefore, start from inside of the cell.

Depending on the cooling rate, three events are supposed to occur during the process of cryopreservation: (*i*) formation of ice crystals, (*ii*) the removal of water, and (*iii*) increase in solute concentration.

1. Formation of ice-crystals. The intracellular ice formation occurs at fast cooling rates, *e.g.*, 100°C/min and high salt concentration which occurs at slower cooling rates, *e.g.*, < 1°C/min. This causes cell damage and loss of viability. So optimum cooling is a bridge between salt damage on one side and intracellular ice damage on the other.

2. The removal of water. During the process of cooling a cell, the temperature falls to freezing point of water, but due to the presence of salts, there is depression of this freezing point and the system is supercooled. During the formation of ice crystals, due to reduction in energy of water molecules, there is a release of latent heat of fusion. The movement of water molecules is inhibited. This results in rise of temperature to above the melting point which is the ice crystallization temperature. This confirms that there are no more nuclei for initiating crystallization and those that have already formed, will continue to grow until all free water is frozen. The temperature then starts declining continuously—period during which time the cell is subjected to functions of temperatures above and below the **eutectic point**. (*Note*. Eutectic means having a lowest freezing point of any possible proportions of a mixture's constituents). In remaining water, the salt concentration continuously increases. Thus, there is osmotically induced loss of water and shrinkage of cells.

3. Increase in solute concentration. During cryopreservation, the plasma membrane remains intact. It also acts as a barrier for ice crystals. This process is regarded efficient if it is proceeded at correct rate with high recovery and minimum damage. If the rate of cooling is too slow, excessive exposure to hypertonic conditions will cause damage to the cells. If it is too fast, then the latent heat of fusion is absorbed and the temperature will continue to fall, thus exposing many more nuclei for crystallization. Thus, a large number of ice-crystals are formed and convert all the free water to ice and there will not be enough time for controlled dehydration of the cell to occur.

At this critical temperature, the water still present within the cell will freeze and intracellular ice crystals will be formed. As intracellular ice is far more damaging than solute concentration, it is desirable to achieve fastest cooling rate that avoids any risk of such happening. The rate varies between cell types, but as a generalization, the following are generally recommended when cryoprotectants are present 0.5–2°C/min for lymphocyte cells, 1°C – 3°C/min for fibroblasts and 2°C – 10°C/min for epithelial cells. The damage to cell during cooling is manifested only until the cell is **thawed**. During rewarming, the cells are subjected to hypertonic stress, followed by rehydration. The rehydration of cells occurs when they take in water to return to their normal size. Whatever damage occurs during cooling is compensated during rewarming process. It determines whether the cell will survive or the injury is sufficient to lead to dilution/shock damage caused by excessive entry of water.

The above mentioned ill effects can, however, be overcome by using warming rate as fast as possible in order to minimize the time spent in hypertonic condition and to prevent growth of small ice-crystals into larger ones.

3. Tools and Technique of Cryopreservation

1. Volume and container used. Large volume of freezing mixture causes variable freezing and thawing rates. So, the volume used should be as small as possible to avoid a significant temperature gradient through the freezing mixture. The glass ampules are most suitable. The volume used should not be so large that the cooling rate is slower in the centre than that in the periphery. Screw cap plastic vials are not used, because leakage can occur, which may cause loss of materials, a risk of contamination and could be potential biological hazard.

2. Cryoprotectants. For general use, the cells are suspended in a mixture of growth medium and cryoprotectants at a concentration between 3×10^6 and 1×10^7 ml. Cryoprotective agents remain in solution at temperature well below the freezing point of water.

Cryoprotectants are of many types but all of them have a high solubility in water. They are of following two main types:

(*i*) **Penetrating.** Dimethyl sulphoxide (DMSO) and glycerol.

(*ii*) **Non-penetrating.** Polyvinyl pyrolidine and hydroxyethyl starch.

The *penetrating type of cryoprotectants* have the ability to reduce the temperature at which ice is formed postponing active leakage to higher osmolarities so that ice when formed is glass-like rather than crystalline and preventing over shrinkage of the cells. Such cryoprotectants tend to reduce the proportion of system (*i.e.*, water) that converts to ice.

The mode of action of *non-penetrating cryoprotectants* is not so clear. DMSO being 10%, a small lipid soluble molecule, enters the cell quickly by diffusion across the lipid bilayer of the plasma membrane where it is believed to alter the permeability characteristics of membranes. In the presence of DMSO, ice crystals which would otherwise rupture plasma membrane, causing cells to lyse, do not form. *Glycerol* has a similar effect. *Polyvinyl pyrolidine* has been recommended for lymphocytic cell lines. If DMSO is used, however, a precaution has to be taken—the mixture must be diluted at least 40 fold immediately on thawing to reduce its toxicity and the medium is changed as soon as the cells are attached to the substrate.

3. Freezing mixture. A double concentrated freezing mixture comprising of 40% v/v growth medium (containing 10% serum, 40% v/v FCS and 20% DMSO or glycerol) is used. It should be mixed well and made sterile by passing through a 0.2 mm sterile filter.

The high serum concentration probably contributes to cell integrity by maintaining the intercellular protein concentration of cells, made permeable by DMSO or glycerol.

The freezing mixture is mixed with an equal volume of cell suspension in complete medium which can then be used for cryopreservation.

4. Freezing down. Cells which are healthy and growing in exponential phase should be cryopreserved. Contaminated confluent or over grown cells should not be used. This is followed by following five steps:

(*i*) After counting, the cells should be centrifuged at 150–200 g at 4°C for 5 min, the supernatant discarded and the pallet resuspended in the residual medium.

(*ii*) The concentration is adjusted to double than that required finally with fresh ice-cold growth medium containing 10% FCS.

(*iii*) Cell suspension should be placed on ice and an equal volume of freezing mixture is added.

(*iv*) 1 ml aliquot of this should be put into cold prepared cryotubes. (*Aliquot* is a part or portion contained by the whole, an integral or whole number of times, *e.g.*, 4 is aliquot part of 12).

(*v*) The final concentration is 2×10^6 for adherent cells and 5×10^6 for suspension per vial. Care should be taken as not to over-tighten the screw caps of cryotubes. No antibiotic is added to these preparations.

5. Cooling rates. To obtain good results, a commercial biological freezer should be used. The cooling rates in these are well set which are maintained throughout phase transition without permitted prolonged electing point fluctuations to occur. The cooling rate should be controlled to at least 5°C, at which it is permissible to gradually increase the cooling rate. At –120°C to –140°C, the cells can be taken from the cooling chamber and put into liquid nitrogen.

The **biological freezers** work on the principle that the temperature difference between freezing chamber and inside of the ampoule can be programmed. The larger the difference, then greater the input of N_2 vapours and faster the fall in temperature. The cells can be driven straight through to the phase-transition point by increasing the temperature difference, thus absorbing the latent heat of fusion.

The alternatives to biological freezers is to lower the vials/ampoules slowly through nitrogen vapours into the liquid nitrogen.

A **freezing plug** allows cells to be cooled at a diffused rate in the vapour above the liquid nitrogen in the storage tank. The rate of cooling depends upon the number of vials and the type of plug used. A polystyrene box with a capacity of 10–20 vials and 5–10 min is used. The dead space in the box can be packed with tissue paper. This sealed box should be placed in a –70°C freezer. Cells are cooled at 1°C/min. The cells in 3 hours reach the temperature of liquid nitrogen and can then be placed in the liquid nitrogen tank.

6. Resuscitation (or reanimation). As mentioned earlier, cultures should be rewarmed as fast as possible. Transferring a one ml ampule in a container should be used.

The vial is then wiped with a tissue paper soaked in 70% alcohol and taken to a hood. The contents of the vials are emptied into a 10 ml sterile centrifuge tube containing 9 ml complete medium which has been prewarmed to 37°C and centrifuged at 150–200 g for 5 min. The freezing mixture is then immediately removed and the cells resuspended in 10 ml complete culture medium. Even the most optimally freeze-thawed cells are extremely vulnerable at this time due to membrane damage which as to be repaired. Thus the use of supplemented (rather than minimal) medium and even conditioned culture medium, results in higher recovery and faster establishment in culture.

A sample should be taken for counting and also to check its viability and then incubated at a subconfluent concentration for 24 hrs. The culture should be observed and the cells counted again for confirmation.

Special Methods of Cryopreservation

For cells for special importance, however, alternative cryopreservation method can be used to increase the change of high recovery rate. For example, two-step freezing method of **Farrant** is sometimes used.

In this method, the cell suspension is frozen as rapidly as possible to a temperature upto –40°C (the exact temperature has to be found out empirically for a given cell and system) held at this temperature for 10 minutes and then frozen rapidly to –196°C. The basis of this technique is that the cell is rapidly pushed through the hypertonic phase, stopped before intracellular ice is formed, dehydrated, then frozen rapidly again with the knowledge that no intracellular ice can be formed in a dehydrated cell.

Cryopreservation of Embryos

There are many advantages of freezing of embryos. It is possible to preserve all viable embryos till recipients at a matched physiological state are available. These embryos can be transported any time or to any place where recipients are available. To freeze embryos, the following steps are applied.

1. The embryo is placed in a culture medium containing a cryopreservative such as glycerol.
2. The embryo is cooled to –7°C, below the freezing point of the solution.
3. The vial contaning the embryo is seeded with ice crystals using a very cold rod held away from the embryo.
4. This cooled embryo is then transferred to liquid nitrogen.

QUESTIONS

Long Answer Questions

1. Describe the method for isolating and culturing protoplast.
2. What are the nutritional needs of cells in culture? How are they fulfilled? Describe in details.

Short Answer Questions

1. Explain the following:
   (*a*) Method of callus regeneration; and
   (*b*) Why do we culture the animal cells? Explain.
2. Write short notes on the following:
   (*i*) Cryopreservation;
   (*ii*) Callus;
   (*iii*) Cryoprotectants;
   (*iv*) Culture media; and
   (*v*) Cryopreservation of embryo.

Multiple Choice Question

1. Plant cells can be converted to protoplasts by treating them with
   (*a*) cellulase
   (*b*) cellulase + pectinase
   (*c*) cellulase + pectinase + lipase
   (*d*) cellulase + pectinase + lipase + protease

ANSWER

Multiple Choice Question

1. (*b*)

55 Cell Transformation and Cancer (Biology of Cancer)

The proliferation, differentiation and survival of individual cells in multicellular organisms are carefully regulated to meet the needs of the organism as a whole. This regulation is lost in cancer cells, which grow and divide in an uncontrolled manner, ultimately spreading throughout the body and interfering with the function of normal tissues and organs (**Cooper** and **Hausman**, 2007).

Cells which undergo rapid, abnormal and uncontrolled growth at the cost of remaining cells are called **neoplastic cells**. The growths resulting from the division of such cells are called **neoplastic growths** or **tumours**. Thus, any abnormal proliferation of cells is the tumour. Almost all type of differentiated cells of animals can become neoplastic or cancerous. The process of cell change in which a cell loses its ability to control its rate of division, and thus becomes a tumour cell, is called **cell transformation**.

Cancer is a genetic disease because it can be traced to alterations within specific genes, but in most cases, it is not an inherited disease. In an inherited disease, the genetic defect is present in the chromosomes of a parent and is transmitted to the zygote. In contrast, the genetic alterations that lead to most cancers arise in the DNA of a somatic cell during the lifetime of the affected individual. Because of these genetic changes, cancer cells proliferate uncontrollably, producing malignant tumours that invade surrounding healthy tissue (see **Karp**, 2002).

55.1. TYPES OF CANCER

Cancer can result from abnormal proliferation of any of the different kinds of cells in the body, so there are more than a hundred distinct types of cancers. These cancers can vary substantially in their behaviour and response to treatment. Based on differences in their growth patterns, tumours are classified as either benign or malignant. A **benign tumour** or **primary tumour**, such as a common skin wart, remains confined to its original location. A **malignant tumour** or **secondary tumour**, however, is capable of both invading surrounding normal tissue and spreading throughout the body via the circulatory or lymphatic systems (**metastasis**). Only malignant tumours are properly referred to as cancers and it is their ability to invade and metastasize that makes cancer so dangerous. The term **cancer** refers to any malignant tumour—that is, any tumour capable of spreading from its original location to other sites (Box 55.1). Whereas benign tumours can usually be removed surgically, the spread of malignant tumours to distant body sites frequently makes them resistant to such localized treatment.

Both benign and malignant tumours are classified according to the type of cell from which they arise. About two hundred distinct types of cancer have been recognized. These can be grouped into four main types: carcinomas, sarcomas, lymphomas and leukemias.

Box 55.1

The earliest recorded description of cancer is found in an Egyptian papyrus dating back to around 3000 B.C. (Papurus refers to a manuscript written on paper prepared from papyrus plant *Cyperus papyrus*). The term cancer, which means "crab" in Latin, was coined by Hippocrates in the fifth century B.C., to describe diseases in which tissues grow and spread unrestrained throughout the body, eventually choking off life (**Becker** *et al.*, 2006). Hippocrates used the words *carcinus, carcinoma* and *cancer*, which refer to crabs in Latin, to describe the tumours (**Cooper** and **Hausman**, 2007).

1. Carcinomas. These cancers account for 90 per cent of all human cancers. Carcinomas arise from the epithelial cells that cover external and internal body surfaces. These epithelial cells may be of ectodermal or endodermal origin. Carcinomas include cervical, breast, skin, brain, lung, colon prostate, kidney, bladder cancers. Carcinomas affect cells of ectodermal or endodermal origin.

2. Sarcomas. These cancers account for 4 per cent of all human cancers. Sarcomas are solid tumours of connective tissues such as cancers of muscle, bone, cartilage, fat cells and fibrous tissue. Sarcomas affect the cells of mesodermal origin.

3. Lymphomas. These cancers constitute about 4 per cent of human cancers. In lymphomas, there is an excessive production of lymphocytes by the lymph nodes and spleen. Hodgkin's disease is an example of lymphoma (Hodgkin's disease is a neoplastic disease that is characterized by progressive enlargement of lymph nodes, spleen, and liver and by progressive anaemia).

4. Leukemias. These cancers form about 4 per cent of human cancers. Leukemias are neoplastic growth of leucocytes (W.B.C.) and are characterized by excessive production of white blood cells (Box 55.2). Leukemias affect cells of mesodermal origin and are a class of sarcomas (**Lodish** *et al.* 2004). In addition to the four types of cancer mentioned above there may be **mixed malignant** tumours arising from ectodermal and mesodermal tissues, *e.g.*, tumours arising from ectodermal and mesodermal tissues.

Box 55.2

The leukemias grow as individual cells in the blood. The name **leukemia** is derived from the Latin for "white blood": the massive proliferation of leukemic cells can cause a patients blood to appear milky.

Box 55.3

Prior to 1900, most human deaths were due to infectious diseases such as pneumonia and tuberculosis, and life expectancy was less than 50 years. Cancer was rare disease that accounted for only a small percentage of deaths.

According to a recent survey conducted by American cancer society in 2005, more than 5 lac Americans die of cancer each year. The four most common cancers accounting for more than half of all cancer cases are those of the prostate, breast, lung, and colon/rectum. Lung cancer, by far the most lethal, is responsible for nearly 30% of all cancer death.

55.2. GROWTH PROPERTIES OF NORMAL AND CANCEROUS CELLS

1. Growth of cancer cells produce multilayered clumps. Information on the behaviour of human cancer cells has come largely from studies on cells grown *in vitro*. At the cellular level, the most important characteristic of a cancer cell—whether residing in the body or on a culture dish—is its loss of growth control. The **capacity** for growth and division is not drastically different between a cancer cell and most normal cells. When normal cells are grown in tissue culture, under conditions that promote cell proliferation, they grow and divide at a rate similar to that of their malignant counterparts. However, when the normal cells proliferate to the point where they cover the bottom of culture dish, their growth rate decreases markedly, and they tend to remain as a single layer (monolayer) of cells (Fig. 55.1A). Growth rates drop as normal cells respond to inhibitory influences from their environment. Growth inhibiting influences may arise as the result of depletion of growth factors in the culture medium or from contact with surrounding cells on the dish (called **contact inhibition**). In contrast, when malignant cells are cultured under the same conditions, they continue to grow, piling on top of one another to form clumps or foci (Fig. 55.1B). It is evident that malignant cells are not responsive to the types of regulatory signals that causes their normal counterparts to cease growth and division.

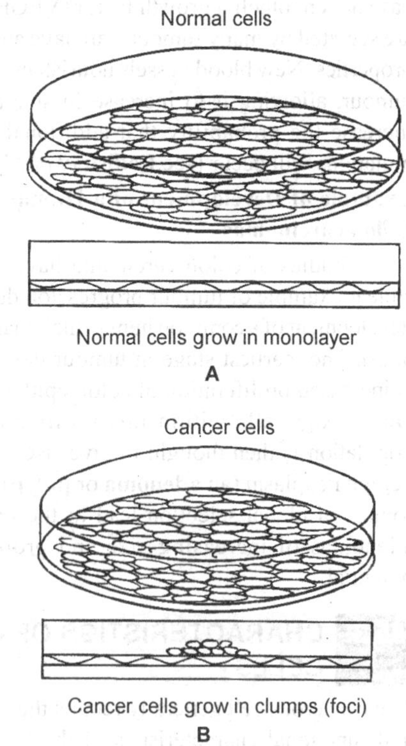

Fig. 55.1. Growth properties of normal and can-cerous cells. A—Normal cells typically grow in a culture dish until they cover the surface as a monolayer. B—In contrast, cells that have been transformed by viruses or carcinogenic chemicals (or malignant cells that have been cultured from tumours) typically grow in multilayered clumps or foci (after Karp, 2002).

Stages of Tumour Development.

At the cellular level, the development of cancer is viewed as multistep process involving **mutation** and **selection** for cells with progressively increasing capacity for proliferation, survival, invasion, and metastasis (Fig. 55.2).

1. Initiation. The step of tumour initiation is thought to be the result of a genetic alteration due to mutation leading to abnormal proliferation of a single cell. Cell proliferation then leads to the outgrowth of population of clonally derived tumour cells.

2. Tumour progression. This process continues as additional mutations occur within cells of the tumour population. Some of these mutations confer a selective advantage to the cell, such as more rapid growth, and the descendants of a cell bearing such mutations will consequently become dominant within the tumour population. The process is called **clonal selection**, since a new clone of tumour cells has evolved on the basis of its increased growth rate or other properties (such as survival, invasion or metastasis) that confer a selective advantage. Clonal selection continues throughout tumour development, so tumours continuously become more rapid-growing and increasingly malignant.

Angiogenesis

Both primary and secondary tumours require *angiogenesis*, the formation of new blood vessels, in order to grow to a large mass. Most tumours induce the formation of new blood vessels that invade the tumour and nourish it, a process called **angiogenesis**. This complex process requires several distinct steps:

1. Degradation of the basal lamina that surrounds a nearby capillary,
2. Migration of endothelial cells lining the capillary into the tumour,
3. Division of these endothelial cells and
4. Formation of a new basement membrane around the newly elongated capillary.

Many tumours produce growth factors that stimulate angiogenesis; other tumours somehow induce surround-ing normal cells to synthesize and secrete such factors. Basic fibroblast growth factors (bFGF), transforming growth factor α (TGFα) and vascular endothelial growth factor (VEGF), which are secreted by many tumours, all have angiogenic properties. New blood vessels nourish the growing tumour, allowing it to increase in size and thus increase the probability that additional harmful mutations will occur (see **Lodish** *et al.*, 2004).

Example of Development of a Human cancer: Colon carcinomas

Studies of colon carcinoma have provided a clear example of tumour progression during the development of a common human malignancy (Fig. 55.3). The earliest stage in tumour development is increased proliferation of colon epithelial cells. One of the cells within this proliferative cell population is then thought to give rise to a small benign neoplasm (an **adenoma** or **polyp**). Further rounds of clonal selection lead to the growth of adenomas of increasing size and proliferative potential.

55.3. CHARACTERISTICS OF CANCER CELLS

The cancerous cell generally retains the structural and functional characteristics of the normal cell type from which it is derived. For example, the cancerous cells of thyroid gland continue to secrete thyroxin. Neoplastic cells, however, differ from their normal counterparts in several ways.

1. Immortalization. Normal cell cultures do not survive indefinitely. For example, human cell cultures die after about 50 generations, and chicken cell cultures have a much shorter life expectancy. In contrast, transformed cell cultures are immortal and can grow indefinitely. A striking example is provided by HeLA cells, which were obtained in 1953 from a uterine cancer diagnosed in a women named **Henreitta Lacks** (hence the term "HeLa" cells). After the tumour was removed by surgeons, some of its cells were placed in culture. The cultured cells quickly began to proliferate and have continued to do so for more than 50 years, dividing more than 18,000 times with no sign of stopping (see **Becker** *et al.*, 2006). Similarly, cell cultures infected with mouse sarcoma virus can be maintained as long as nutrition is provided and overcrowding avoided.

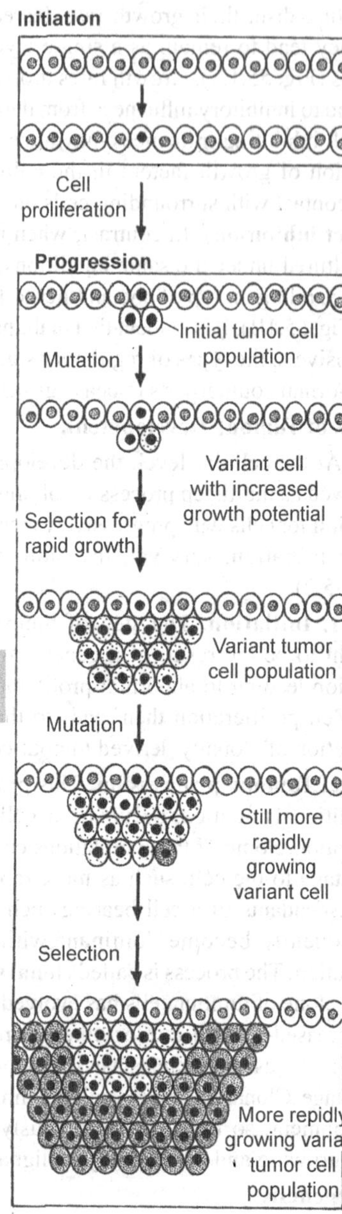

Fig. 55.2. Stages of tumour development. The development of cancer initiates when a single mutated cell begins to proliferate abnormally. Additional mutations followed by the selection for more rapid growing cells within the population then result in progressioin of the tumour to increasingly rapid growth and malignancy (after Cooper and Hausman, 2007).

Box 55.4

Cancer cells, like germ cells and stem cells but unlike most differentiated cells, produce **telomerase** enzyme, which prevent shortening of chromosomes during DNA replication and may contribute to their **immortalization**. The absence of telomerase is associated with resistance to generation of certain types of human tumours (*e.g.*, papillomas and colon tumours of mice) (see **Lodish** *et al.*, 2004).

2. Loss of contact inhibition. Normal cells in a culture stop growing when their plasma membranes come into contact with one another. This inhibition of growth after contact is called **contact inhibition**. Transformed cells are unable to go into a quiescent stage after division. They will grow (*i.e.,* devide) continuously until they kill themselves. Transformed cells apparently undergo a change in the property of their plasma membranes, which become less adhesive. This change enables the cells to dissociate from neighbouring cells and to infiltrate other organs, where they form metastatic tumours. Cancer cells thus, lack proper recognition and communication.

3. Reduced cellular adhesion. When normal cells become cancerous there is change in the **stickiness** or **adhesiveness** of their plasma membranes. If grown in a nutrient medium kept in a glass vessel, the normal cells stick to the glass rather than float in the medium. Transformed cells show a **decreased adhesiveness** and if grown in solid media, they stick to each other less than do normal cells.

Adhesiveness shows considerable specificity. For example, liver cells tend to stick to other liver cells and not to other cell types, *e.g.*, kidney cells. If the cells of the liver and the kidney are separated by the enzyme trypsin and incubated together, they aggregate to form small pieces of liver tissue and kidney tissue. Thus, kidney cells stick to kidney cells and liver cells to liver cells. Cancerous cells do show this property (*i.e.*, specificity of adhesiveness). Thus, if malignant skin cancer cells are mixed with normal kidney cells then aggregates formed contain both kidney and skin cells mixed together. This probably explains why malignant cell can invade several normal organs.

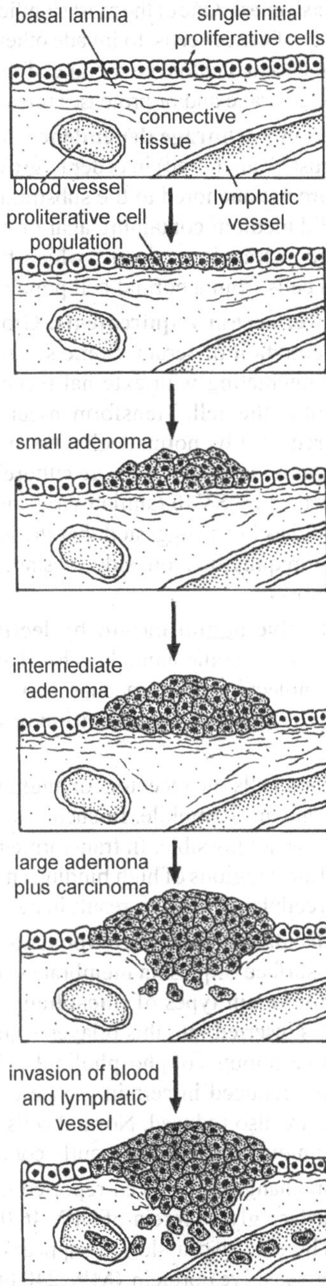

Fig. 55.3. Development of colon carcinomas. A single initially altered cell gives rise to a proliferative cell population, which progresses first to benign adenomas of increasing size and then to malignant carcinoma. The cancer cells invade the underlying connective tissue and penetrate blood and lymphatic vessels, thereby spreading throughout the body (after Cooper and Hausman, 2007).

4. Invasiveness. One of the most significant characteristics of transformed cells is their invasiveness, *i.e.*, the ability of cancer cells, to invade other tissues. In contrast to normal cells, transformed cells can penetrate the chorioallantoic membrane of the hen's egg. This invasiveness could be due to changes in the plasma membrane and/or proteases released by the cancerous cells.

5. Loss of anchorage dependence. Most normal cells must be attached to a rigid substratum (*i.e.*, they must be anchored) in order to grow in a culture medium. Transformed cells can grow even when they are not anchored to the substratum. For example, when transformed cells are suspended in a semisolid medium containing agar or methyl cellulose. This loss of anchorage is most striking characteristic of transformed cells which form malignant tumours. This property is used to select transformed cells from a normal cell population.

6. Lower serum requirements. Growth of normal cells in tissue culture medium requires a high concentration of serum. Some serum growth factors (*e.g.*, somatomedins) resemble insulin hormone in interacting with external receptors on the plasma membrane to regulate biochemical activities within the cell. Transformed cells can grow in a culture medium containing much less serum than required by normal cells. For example, normal 3T3 cells (established fibroblasts of the mouse line commonly used is tissue culture) grow optimally in 10 per cent foetal calf-serum, while cells transformed by SV40 (simian virus number 40) can grow equally well in 1 per cent of 10 per cent serum. It has been suggested that the lower serum requirement of transformed cells is because of their lesser requirement out side substances to lower their intracellular cAMP (cyclic AMP) level to trigger mitosis.

7. Selective agglutination by lectins. Lectins are proteins widely distributed in plants (*e.g.*, legumes) and some animals. They have the ability to bind to receptors, which are branched chain sugar molecules (oligosaccharides), on the surface of the plasma membrane. As a result of such binding lectin cause **agglutination** or **clumping** of cells. Due to this property, lectins work as agglutinins.

In normal cells the receptors or agglutin binding sites for lectins lie in a diffuse manner on the cell surface and are immobile. In such cases, lectins make few intercellular bridges, and therefore agglutination is not possible. In transformed cells, the receptors are more mobile within the plasma membrane. Local regions of high binding site concentration are formed. Lectins are thus able to form enough intercellular bridges to result in agglutination in the transformed cells.

8. Molecular changes in components of the plasma membrane. There are several differences between the surface of plasma membranes of normal and transformed cells. The plasma membrane consists of four main types of **phospholipids**, which form the lipid bilayer, with **glycolipids** and **glycoproteins** inserted into this bilayer. Cancerous cells apparently do not differ from normal cells in their relative amounts of phospholipids. However, **gangliosides** (glycolipids which contain sialic acids) become reduced in certain mouse cancer cells. Enzymes involved in biosynthesis of these gangliosides are also reduced. Normal cells possess four types of gangliosides, GM1a, GM1, GM2 and GM3. Tumour cells predominantly contain the simplest type, GM3.

Certain changes have been reported in the **glycoproteins** of plasma membrane of cancerous cells. The surface glycoprotein of MW 46,000 disappears early in transformation to the cancerous condition. There is also a slow disappearance of a major protein called **LETS** (large, external, transforming sensitive) protein (MW 240,000). Probably the most important protein to disappear after transformation is the one having MW 200,000.

9. Disorganisation of cytoskeleton. Normal cells have a well organised cytoskeleton which consists of **microtubules** (of tubulins), **microfilaments** (of actins) and **intermediate filaments** (of collagen for example). The fibre-like proteins have a regular arrangement and bring about coordinated cell movement. In transformed or cancer cells, the fibres are much fewer in number and usually much thinner. In transformed cells the cytoskeleton is found to undergo **depolymerisation**. The microtubules disaggregate. The microfilaments undergo depolymerisation and disappear, but diffuse actin persists.

The myosin-like filaments also disappear. Thus, in transformed cells the cytoskeleton proteins become less organised than in normal cells. It has been suggested that due to this disorganisation of the cytoskeleton, there is an increased mobility of plasma membrane proteins.

The disorganisation of the cytoskeleton also affects the cell surface in another way. When cancer cells are touched, there is a constant and uncoordinated throwing out and retraction of **blebs, microvilli** and **ruffles** from the cell surface. Tumour cells have a more ruffled surface than normal cells, with many more surface processes.

10. Increase in negative surface charge. Comparisons of surface membrane charge by microelectrophoresis have been made between normal and malignant cells. In malignant cells anodic mobility is usually higher, indicating increase in negative surface charge.

11. Defective electrical communication. Electrical connections normally occur between individual normal cells. In some cancer cells, however, it has been reported that such electrical connections are defective.

12. Increased sugar transport. Tumour cells consume much more glucose than normal cells because they have to grow and multiply. There is a great increase in the rate of sugar transport across the surface membrane (plasma membrane) after transformation of the cell. This tends to increase in sugar intake by malignant cells.

13. Increased rate of glycolysis. In the 1920's **Warburg** pointed out that oxidative (aerobic) respiration is depressed in tumour cells and that glycolysis (anaerobic respiration) increases. This has been demonstrated by an increase in lactic acid production in cells of solid tumour. There is a corresponding increase in the uptake of glucose.

14. Appearance of virus-specific transplantation rejection antigens. Plasma membranes of most transformed cells contain **antigens** which are not present in normal cells. Thus in cells transformed by adenoviruses and papovaviruses the **T-antigen** is always present. Similarly all cells transformed by the Epstein-Barr virus (EB virus) contain an antigen called the **EB nuclear antigen (EBNA)**. Tumour antigens can bring about an **immunity response** against themselves in genetically similar hosts. This is in contrast to normal histocompatibility antigens. The immunity response brought about by the antigens results in recognition and destruction of newly formed cancer cells and their descendants. Such a defence mechanism is called **immunological surveillance** and can lead to the elimination of transformed or cancer cells under favourable conditions. It has been suggested that only in the rare cases when this defence mechanism fails that tumour is formed.

15. Increased secretion of proteolytic enzymes. Large amount of proteolytic enzymes are secreted by all types of cancer cells, except those of blood forming tissues. The cancer cell secretes a protease called the **cell factor** (MW ~40,000). The cell factor acts on inert serum protein **plasminogen** (MW 85,000) to form a **plasmin**, a **proteolytic enzyme** (MW 76,000). It has been suggested that plasmin removes many proteins projecting from the cell surface by enzymatic digestion and signals the cell into division.

If normal cells are treated with proteases they show many characteristics of transformed cells. It has been speculated that viral proteins cause the release of extracellular protease. However, there is no direct evidence for this.

16. Aldolases. In most mammalian tissues the **aldolase** exists in the form of three isozymes A, B and C. Isozymes A and C predominate in embryonic tissues while in adult differentiated tissues the B isozyme is predominant. In some tumours, especially in poorly differentiated and rapidly growing cancers such as hepatomas, isozyme B is replaced by isozyme A, the embryonic form.

17. Most cancers develop later in life. The multiple mutations that lead to formation of a tumour may require many years to accumulate. That is why most cancers develop later in life. The occurrence of cancer after the age of reproduction may be one reason that evolutionary restraints have not done more to supress cancer. The requirement for multiple mutations also lowers the frequency of cancer compared with what it would be if tumerogenesis were triggered by a single mutation. However, huge numbers of cells are, in essence, mutagenized and tested for altered growth during our lifetime, a sort of evolutionary selection for cells that proliferate. *Fortunately the tumour itself is not inherited* (see **Lodish** *et al.*, 2004).

55.4. WHAT CAUSES CANCER?

Cancers are caused mainly by environmental agents and lifestyle factors, most of which act by triggering DNA mutations.

1. Epidemiological data. The first indication that a particular agent may cause cancer is usually provided by an approach called **epidemiology,** the branch of medical science that investigates the frequency and distribution of diseases in human populations. Epidemiological studies have revealed that cancers arise with differing frequencies in different parts of the world. For example, stomach cancer is frequent in Japan, breast cancer is prominent in the United States and liver cancer is common in Africa and Southeast Asia. To determine whether differences in heredity or environment are responsible for such differences, scientists have examined cancer rates in people who move from one country to another. For example, in Japan the incidence of stomach cancer is greater and the incidence of colon cancer is lower than in United States. When Japanese families move to the United States, their cancer rates come to resemble the rates in the United States, indicating that cancer rates are determined more by environment and lifestyle factors than by heredity.

Epidemiological data have played an important role in identifying environmental factors that can cause cancer. The most drastic facts involve lung cancer, a disease that has increased over tenfold in frequency in the United states since 1900. When the possible causes for this epidemic of lung cancer were investigated, it was discovered that virtually all lung cancer patients share one trait: a history of smoking cigarettes. As might be expected if cigarettes were responsible, heavy smokers develop lung cancer more frequently than light smokers, and long-term smokers develop lung cancer more frequently than do short-term smokers.

While the association between cigarette smoking and cancer was first detected through epidemiological studies, definitive proof of a cause-and-effect relationship requires direct experimental evidence. In the case of cigarettes and lung cancer, such evidence has come from studies showing that cigarette smoke contains several dozen chemicals that cause cancer when administered to animals. Such substances that cause cancer are called **carcinogens**.

2. Carcinogens. The idea that certain chemicals, such as those found in tobacco smoke, can cause cancer was first proposed more than 200 year's ago. In 1761 a London doctor, **John Hill**, reported that people who routinely use snuff (a powdered form of tobacco that is inhaled) experience an abnormally high incidence of nas[illegible]ancer, suggesting the presence of carcinogen (cancer-causing) chemicals in tobacco. A few years la[illegible] another British physician, **Percival Pott**, observed an elevated incidence of scrotum cancer among [illegible]n who had served as chimney sweepers in their youth. It was common practice at the time to emp[illegible] young boys to clean chimney flues because they fit into narrow spaces more readily than adults. P[illegible]culated that the chimney soot became dissolved in the natural oils of the scrotum, irritated the [illegible] and eventually triggered the development of cancer. This theory led to the discovery that scrot[illegible] cancer could be prevented among chimney sweepers through the use of protective clothing and [illegible] bathing practices.

In forthcoming years, the [illegible] known and suspected carcinogens has grown to include hundreds of different chemicals. [illegible]ls are usually labelled as carcinogens because humans or animals develop cancer when e[illegible] to them. This does not mean, however, that each of these substances causes cancer throug[illegible] direct action. For example, consider the behaviour of **2-naphthylamine**, a potent carcinogen [illegible]s bladder cancer in industrial workers and is present in tobacco smoke. As might be expected, feeding 2-naphthylamine to laboratory animals induces a high incidence of **bladder cancer**. But if 2-naphthylamine is implanted directly into an animal's bladder, cancer rarely develops. The explanation for the apparent difference is that 2-naphthylamine is ingested (by animals) or inhaled (by humans), it passes through the liver and is metabolically converted into chemical compounds that are actual causes of cancer. Placing 2-naphthylamine directly in an animal's bladder bypass this metabolic activation and consequently cancer does not arise.

Precarcinogens. Many carcinogens share the above need for metabolic activation before they can cause cancer. Substances exhibiting such behaviour are more accurately referred to as **precarcinogens**, a term applied to any chemical that is capable of causing cancer only after it has been **metabolically activated**. The activation of most precarcinogens is carried out by liver proteins that are members of cytchrome P450 enzyme family. Members of this enzyme family catalyze the oxidation of ingested foreign chemicals, such as drugs and pollutants, to make the molecules less toxic and easier to excrete from the body. However, in some cases these oxidation reactions accidentally convert foreign chemicals into carcinogens—a phenomenon called **carcinogen activation**.

Mode of action of carcinogens. Once it has been determined that chemicals can cause cancer, the question arose as to how they work. The idea that carcinogenic chemicals act by triggering **DNA mutations** was first proposed around 1950, but there was little supporting evidence at that time because nobody had systemically compared the mutagenic potency of different chemicals with their ability to cause cancer.

Ames test. The need for identifying potential carcinogens, inspired **Bruce Ames** (1973) to develop a simple rapid laboratory test for measuring a chemical's mutagenic activity. The procedure he developed, called the **Ames test**, utilise bacteria as a test organisms because they can be quickly grown in enormous numbers in culture. The bacteria used for the Ames test are a special strain that lack the ability to synthesize the amino acid **histidine**. As shown in Fig. 55.4, the bacteria are placed in a culture dish containing a growth medium without histidine, along with the chemicals being tested for mutagenic activity. Normally, the bacteria would not grow in the absence of histidine. However, it will trigger random mutations, some of which might restore the ability to synthesize histidine. Each bacterium acquiring such a mutation will grow into a visible colony, so that the total number of colonies is a measure of the **mutagenic potency** of the substance being investigated.

Since many chemicals that cause cancer only become carcinogenic after they have been modified by liver enzymes, the Ames test includes a step in which the chemical being tested is first incubated with an extract of liver cells to mimic the reactions that normally occur in the liver. The resulting chemcial mixture is then tested for its ability to cause bacterial mutations. When the Ames test is performed in this way, a strong correlation is observed between a chemical's ability to cause mutations and its ability to cause cancer.

Cancer arise through a multistep process. Chemicals induce the development of cancer through a multistep process involving **initiation, promotion**, and **tumour progression** (Fig. 55.2). **Initiation** is based on DNA mutation, whereas **promotion** involves proliferation of the initiated cells for a prolonged period of time and accompanied by a gradual **selection** of cells exhibiting enhanced growth properties. During **tumour progression**, cells acquire additional mutations and undergo changes in gene expression that produce cells with increasingly observant traits.

3. Role of ionizing and ultraviolet radiations in carcinogenesis. Shortly after the discovery of **X-rays** by **Wilhelm Roentgen** in 1895, it was noticed that people working with this type of radiation developed cancer at abnormally high rates. Animal studies subsequently confirmed that X-rays create DNA mutations and cause cancer in direct proportion to the dose administered.

X-rays type of radiations is emitted by many **radioactive elements.** An early example of the carcinogenic hazards posed by radioactivity occurred in 1920's in a New Jersey factory that produced glow-in-the-dark watch dials. A luminescent paint containing the radioactive element **radium** was used for painting the dials, and this paint was applied with a fine tipped brush that the workers frequently wetted with their tongues. As a result, tiny quantities of radium were carelessly ingested and became concentrated in their bones, leading to the development of **bone cancer**.

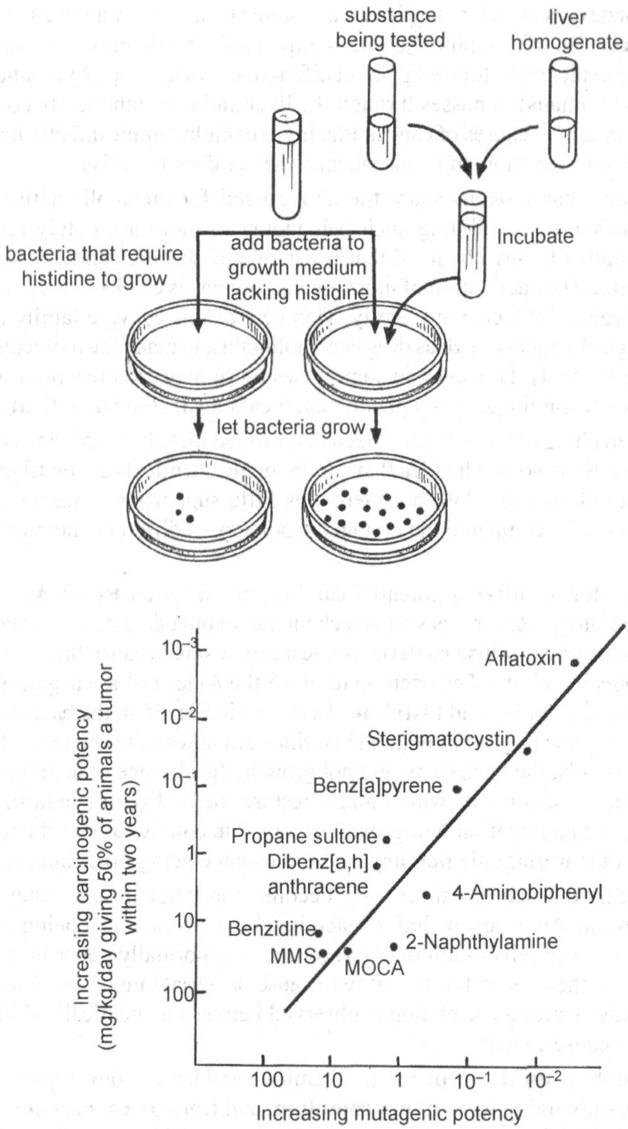

Fig. 55.4. Procedure of Ames test. Ames test is based on the grounds that most carcinogens are mutagens. A—The ability of chemicals to induce mutations is measured in bacteria that lacks the ability to synthesize the amino acid histidine. When placed in a growth medium that lacks histidine, the only bacteria that can grow are those that have acquired a mutation that allows them to make histidine. The number of bacterial colonies that grow is therefore related to the mutagenic potency of the substance being tested. Chemicals being investigated with the Ames test are first incubated with a liver homogenate because many of the chemicals to which humans are exposed only become carcinogenic after they have undergone biochemical modification in the liver. B—The data in the graph reveal that substances that exhibit strong mutagenic activity in the Ames test also tend to be strong carcinogens. Here, aflatoxin is the most potent mutagen and the most potent carcinogen. (*Abbreviations:* MOCA = 4–4′ methylene-bis2-chloroanaline, MMS = methyl methanesulphonate; after Becker *et. al.*, 2006).

High rates of cancer caused by exposure to radioactivity have also been observed in people exposed to radioactive fallout from nuclear explosions. The most dramatic incidents occurred in the Japanese cities of Hiroshima and Nagasaki after atomic bombs were dropped there in 1945. Another incident occurred in the area surrounding the Chernobyl nuclear power plant in the former Soviet Union (now Ukraine), which exploded in 1986.

X-rays and related forms of radiation emitted by radioactive elements, are called **ionizing radiation** because they remove electrons from molecules, thereby generating highly reactive ions that create DNA damage. **Ultraviolet radiation (UV)** is another type of radiation that causes cancer by damaging DNA. The ability of the UV radiation in sunlight to cause cancer was first deduced from the observation that skin cancer is most prevalent in people who spend long hours in the sun, especially in tropical regions where the sunlight is very intense. UV radiation is absorbed mainly by the skin, where it imparts enough energy to trigger **pyrimidine dimer** formation—that is, the formation of covalent bonds between adjacent pyrimidine bases in DNA. If the damage is not repaired, distortion of the double helix causes improper base-pairing during DNA replication that leads to distinctive mutation patterns. For example, a CC → TT mutation (conversion of two adjacent cytosines to thymines) is a unique product of UV exposure and can therefore be used as a distinctive "signature" to identify mutations caused by sunlight.

The existence of such **signature mutations** provided a way to prove that UV-induced mutations cause skin cancer. One of the first genes studied was the *p53* gene, which is known

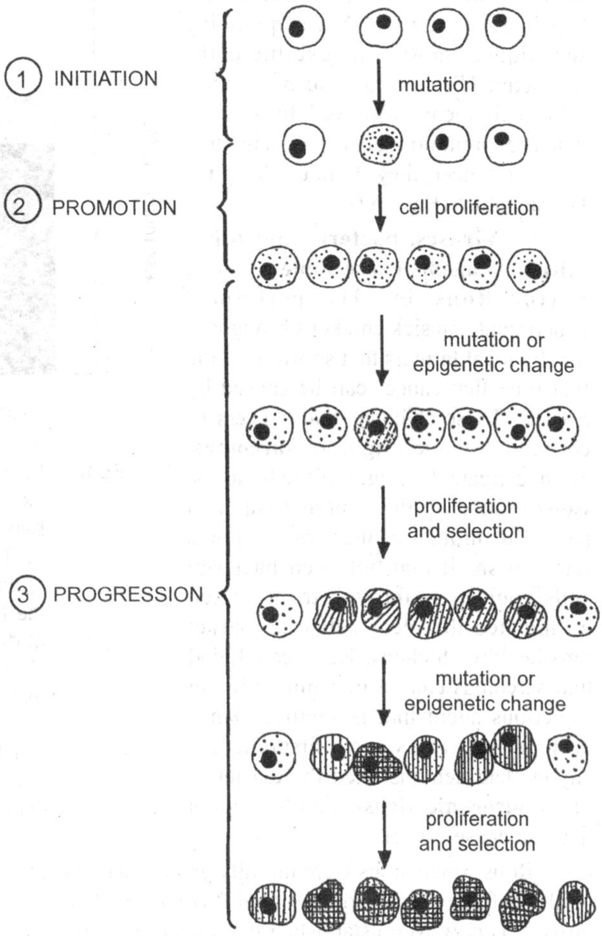

Fig. 55.5. Main stages in the development of cancer. Cancer arises by a complex process involving three main stages. 1. Initiation. It is based on DNA mutation. 2. Promotion. During this stage the initiated cell is stimulated to proliferate. 3. Progression. During tumour progression, further mutations and changes in gene expression create variant cells exhibiting enhanced growth rates or other aggressive properties that give certain cells a selective advantage. Such cells tend to outgrow their companions and become the predominant cell population in the tumour. Repeated cycles of this selection process create a population of cells whose properties gradually change over time (after Becker *et al.*, 2006).

to be mutated in many human cancers. When the *p53* gene of skin cancer cells, is examined using DNA sequencing techniques, mutations exhibited the distinctive UV signature (such as CC → TT) are frequently observed. In contrast, when *p53* mutations are detected in other types of cancer, they do not exhibit the UV signature (Fig. 55.6).

4. Viruses, bacteria and other infectious agents cause some cancers. **Peyton Rous**, in 1911, performed experiments on sick chickens brought to him by local farmers that showed for the first time that cancer can be caused by a virus. These chickens had cancers of connective tissue origin, or **sarcomas**. To investigate the origin of the tumours, **Rous** ground up the tumour tissue and passed it through a filter, whose pores were so small that not even bacterial cells could pass through them. When he injected the clear cell-free extract into healthy chickens, Rous concluded that sarcomas can be transmitted by an infectious agent that is smaller than a bacterial cell. This was the first time that anyone had demonstrated the existence of an **oncogenic virus**—that is, a virus that can cause cancer.

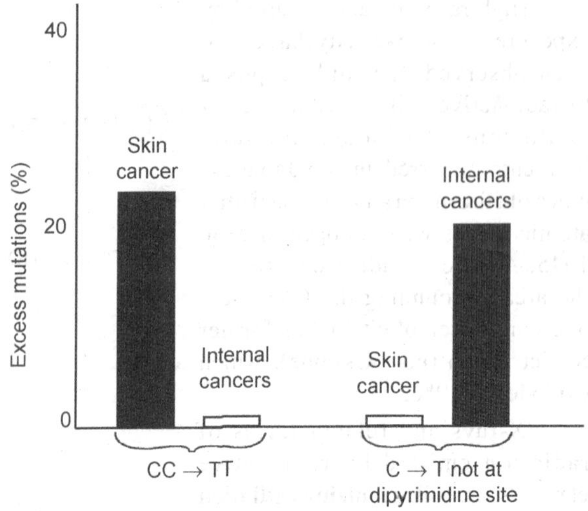

Fig. 55.6. Incidence of two types of *p53* mutations in skin cancer and internal cancers. The two bars on the left represent the frequency of CC → TT mutations, which are triggered by UV radiation. The two bars on the right represent the frequency of C → T mutations not located at dipyrimidine sites, which are not caused by UV radiation. Note that the UV-triggered type of mutation is found in the *p53* gene of skin cancers (squamous cell carcinomas), but not in cancers of internal organs. Mutations frequencies are plotted relative to what would be expected to occur randomly (after Becker *et al.*, 2006).

Rous conclusions were initially greeted with skepticism. It was not until 1966 that an 87-year old Rous finally received the Nobel Prize, more than 50 years after his discovery of the first cancer virus. It is now well established that dozens of viruses cause cancer in animals and that a smaller number cause cancer in humans. The first human example was discovered by **Danis Burkitt**, a British surgeon working in Africa in the late 1950's. At certain times of the year, Burkitt noted that a large number of his young patients developed massive **lymphocytic cancers** of the neck and jaw. Because this kind of cancer, now known as **Burkitt's lymphoma** occurred in periodic epidemics localized to specific geographical regions, Burkitt proposed that it was transmitted by an infections agent.

Burkitt's ideas soon attracted the attention of two virologists, named **Epstein** and **Barr**, whose electron microscopic studies revealed virus particles in Burkitt's lymphoma cells, the virus is now called the **Epstein-Barr virus** or **EBV** in recognition of these virologists (Box 55.5).

Box 55.5 EBV and Carcinogenesis

Unlike viruses that cause cancer in animals, it is difficult to prove that a virus such as EBV causes cancers in humans because ethical considerations prevent directly testing the hypothesis by injecting the virus into healthy individuals. Nonetheless, some indirect evidence supports the conclusion that EBV causes Burkitt's lymphomas: 1. DNA sequences and proteins encoded by EBV are detected in tumour cells obtained from patients with Burkitt's lymphoma, but not in normal cells from the same individuals; 2. Adding purified EBV to normal human lymphocyte cultures cause the cells to acquire the properties of cancer cells; and 3. Injecting EBV into monkeys induces the formation of lymphomas.

Following the discovery of EBV, several additional viruses have been identified as causes of human cancer. For example, liver cancer is caused by **hepatitis B** and **hepatitis C** viruses and cervical cancer is caused by **human papilloma virus (HPV)**. In addition to viruses, the bacterium *Helicobacter pylori* (*H. pylori*) causes stomach cancer, and parasitic flatworm infections have been linked to a small number of bladder and bile duct cancers.

Infectious agents trigger the development of cancer in two fundamentally different ways. One involves those agents such as the hepatitis B and C viruses, *H. pylori*, and parasitic flatworms that cause tissue destruction and chronic inflammation. Such destruction creates chronic inflammatory conditions in which cells of the immune system infiltrate the tissue and attempt to kill the infectious agent and repair the tissue damage. Unfortunately, the mechanisms used by immune cells to fight infections often produce mutagenic chemicals, such as **oxygen free radicles** (highly reactive forms of oxygen containing an unpaired electrons). This means that proliferation of replacement cells for the injured tissue takes place under conditions in which DNA damage is likely, thereby increasing the likelihood that cancer-causing mutations will arise.

The other way in which infectious agents cause cancer is related to the ability of certain viruses to directly stimulate the proliferation of infected cells. The mechanism by which this is accomplished varies among viruses. In some cases, it involves cancer-causing viral genes, whereas in other cases a virus alters the behaviour of host cell genes. The types of genes involved in such events play a role not just in viral cancers, but in cancers caused by chemicals and radiation as well.

55.5. GENES INVOLVED IN CANCERS

Oncogenes are genes whose presence can cause cancer. While they are sometimes introduced into cells by viruses, more often they arise from normal cellular genes called **proto-oncogenes** by *point mutation, gene amplification, chromosomal translocation, local DNA rearrangements* or *insertional mutagenesis.* Most proteins produced by oncogenes are signalling pathway components, such as growth factors, receptors, plasma membrane GTP-binding proteins, non-receptor protein kinases, transcription factors, and cell-cycle or cell-death regulators. Oncogenes code for abnormal forms or excessive quantities of such proteins, thereby leading to excessive stimulation of cell proliferation.

Tumour suppressor genes are genes whose loss or inactivation can lead to cancer. Susceptibility to develop cancer is increased in people who inherit defective tumour suppressor genes. Three important tumour suppressor genes are:

1. The *RB* gene, which produces a protein that restrains passage from G_1 into S phase.
2. The *p53* gene, which produces a protein that prevents cells with damaged DNA from proliferating.
3. The *APC* gene, which produces a protein that inhibits the Wnt pathway. The **Wnt pathway** plays a prominent role in controlling cell proliferation and differentiation during embryonic development. The central component of this wnt pathway is a protein called **β-catenin**.

55.6 DIAGNOSIS, SCREENING AND TREATMENT OF CANCER

Much progress has been made in recent years in explaining the genetic and biochemical abnormalities that underlie cancer development. One of the hopes for such research is that our growing understanding of the molecular alterations exhibited by cancer cells will eventually lead to improved strategies for cancer diagnosis and treatment.

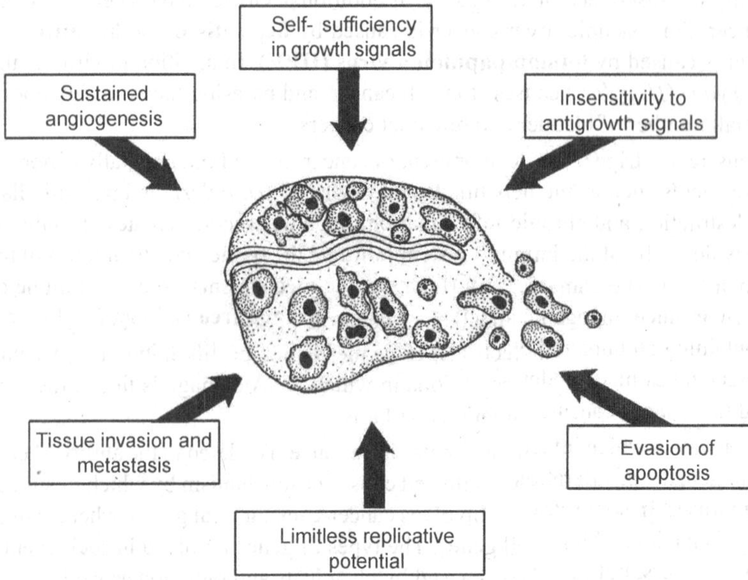

Fig. 55.7. Overview of changes in cells that cause cancer. During carcinogenesis, six fundamental cellular properties are altered, as shown here, to give rise to the complete, most destructive cancer phenotype. Less dangerous tumours arise when only some of these changes occur (after Lodish *et al.*, 2004).

1. Microscopic analysis of cancer.

Because cancer can arise in almost any tissue, few regular generalizations are possible regarding disease symptoms. A definitive diagnosis typically requires a **biopsy,** which involves surgical removal of a tiny tissue sample for microscopical examination. Under the microscope, cancer cells show a number of features that together indicate the presence of cancer (Table 55.1). For example, cancer cells

Table 55.1 Some differences in the microscopic appearance of benign and malignant tumours (source: Becker *et al.*, 2006)

	Trait	Benign	Malignant
1.	Nuclear size	Small	Large
2.	N/C ratio (Ratio of nuclear to cytoplasmic volume)	Low	High
3.	Nuclear shape	Regular	Pleomorphic (irregular shape)
4.	Mitotic index	Low	High
5.	Tissue organisation	Normal	Disorganised
6.	Differentiation	Well-differentiated	Poorly differentiated (anaplastic)
7.	Tumour boundry	Well-defined	Poorly-defined
			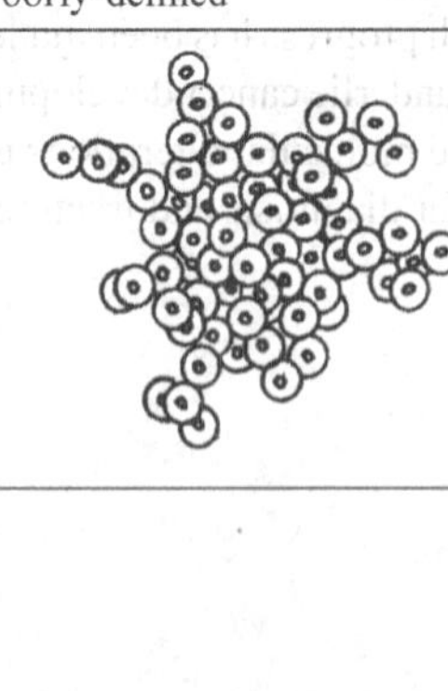

often have large, irregularly shaped nuclei, prominent nucleoli, and a high ratio of nuclear-to-cytoplasmic volume. Cancers also tend to exhibit significant variability in cell size and shape, as well as a loss of normal tissue organisation. To varying extents, cancer cells lose the specialized structural and biochemical properties of the cells normally residing in the tissue of origin. Cancers usually have more dividing cells than normal, which means that the mitotic index (Box 55.6) will be elevated.

Box 55.6 Mitotic index

Percentage of cells in a population that are in any stage of mitosis at a certain point in time; used to estimate the relative length of the M phase of cell cycle. For example, the mitotic index for cultured mammalian cells is often about 3–5 per cent, which means that M phase lasts less than an hour (usually 30–45 minutes).

Tumour grading. If a sufficient number of these seven traits (Table 55.1) are observed upon microscopic examination of sample, it can be concluded that cancer is present. In other words, the presence of these traits indicates a tumour that, if left untreated will eventually spread by invasion and metastasis. The severity of these observed microscopic, abnormalities varies significantly among cancers, even when they arise from the same cell type and in the same organ. These variability form the basis for **tumour grading**, which is the assignment of numerical grades to tumours based on differences in their microscopic appearance.

Lower numerical grades (*e.g.*, grade 1) are assigned to tumours whose cells exhibit normal differentiated features, divide slowly, and display only modest abnormalities in the traits listed in Table 55.1. Higher numbers (*e.g.*, grade 4) are assigned to tumours containing rapidly dividing, poorly differentiated cells that bear less resemblance to normal cells and exhibit severe abnormalities in the trait, listed in Table 55.1 The highest grade cancers contain cells that are **anaplastic**, which means that they are so poorly differentiated and abnormal in appearance and organization that they bear no resemblance to the cells of the tissue in which the tumour arose. Such anaplastic, high-grade cancers tend to grow and spread more aggressively and be less responsive to therapy than lower grade cancers (see **Becker** *et al.*, 2006).

2. Screening Techniques of Cancer

When cancer is detected before it has spread, cure rates tend to be very high. Therefore, a great need exists for screening techniques that can routinely detect cancers at an early stage. Some of these are following:

***(i)* Pap smear.** It is one of the most successful screening procedures that was developed in the early 1930s by **George Papanicolaou** (for whom it is named). The logic underlying this procedure is that the microscopic appearance of cancer cells is so distinctive that it is possible to detect the likely presence of cancer by simply examining a few isolated cells. A Pap smear is performed by taking a tiny sample of a woman's vaginal secretions and examining it with a microscope. If the cells in the fluid exhibit unusual features, such as large irregular nuclei or prominent variations in cell size and shape, it is a sign that cancer may be present and further tests need to be done. Because a Pap smear allows **cervical cancer** to be detected in its early stages before metastasis has occurred, this procedure has prevented hundreds of thousands of cancer deaths.

***(ii)* Mammography and other techniques.** The success of the Pap smear has led to development of screening techniques for other cancers. For example, **mammography** utilizes a special X-ray technique to look for early signs of **breast cancer** and **colonoscopy** uses a slender fiber-optic instrument to examinc the colon (*i.e.*, the part of the large intestine that extends from the cecum to the rectum) for early signs of **colon cancer**. The ideal screening test would allow doctors to detect cancers anywhere in the body with one simple procedure, such as a blood test. **Prostate cancer** is an example of cancer that can sometimes be detected this way. Men over the age of 50 are often advised to get a **PSA test,** which measures how much **prostate-specific antigen (PSA)** is present in the bloodstream. PSA, which is a protein produced by cells of the prostate gland, normally appears in only tiny concentrations in the blood. If a PSA test reveals a high concentration of PSA, it indicates the existence of a prostate problem and further tests are performed to determine whether or not cancer is actually present.

(iii) **Proteomic analysis.** Other cancers also release small amounts of specific proteins into the bloodstream, where their presence might be used to signal the existence of early disease. To investigate such tiny changes in blood proteins, scientists are experimenting with a general approach called **proteomic analysis** to analyse proteins present in the blood (the term proteomic refers to the complete set of proteins produced by an organism's genome). The key to most proteomic techniques is **mass spectrometry**, a high speed, extremely sensitive method for identifying proteins based on differences in mass and electrical charge. Because a blood sample contains thousands of different proteins, the data generated by proteomic analysis can be extremely complex. To deal with this problem, artificial intelligence software programs are used to compare the complex protein patterns seen in blood samples from individuals with or without cancer.

One of the first cancer to be investigated by proteomic analysis was **ovarian cancer**. When ovarian cancer is detected before it spreads, the five year survival rate exceeds 95 per cent. Using proteomic analysis, scientists have recently identified a pattern of five proteins in the blood of women with ovarian cancer that is not seen in the blood of other women.

3. Treatments of Cancer.

People diagnosed with cancer have various treatment options that depend both on the type of cancer involved and how far it has spread.

(i) **Surgery.** The most common treatment of cancer involves surgery to remove the primary tumour followed (if necessary) by radiation therapy and/or chemotherapy to destroy any remaining cancer cells.

(ii) **Radiation therapy.** This approach employ's high energy X-rays or other forms of ionizing radiation to kill cancer cells. We already know that DNA damage created by ionizing radiation can cause cancer, but paradoxically, the same type of radiation is also used in higher doses to destroy cancer cells in people who already have the disease. Ionizing radiation kills cells in two different ways. First, DNA damage caused by radiation activates the p53 signaling pathway, which then triggers cell death. However, many cancers have mutations that disable the p53 pathway, so p53-induced apoptosis plays only a modest role in the response of most cancers to radiation treatment. In the second mechanism, radiation kills cells by causing chromosomal damage that is so severe that it prevents cells from progressing through mitosis, and the cells therefore die while trying to divide.

(iii) **Chemotherapy**. Most forms of chemotherapy use drugs that, like radiation, are intended to kill dividing cells. Such drugs can be subdividing into four major categories:

(a) **Antimetabolites.** They inhibit metabolic pathways required for DNA synthesis by acting as competitive inhibitors that bind to enzyme active sites in place of normal substrate molecules. Examples of such antimetabolites include methotrexate, fluorouracil and mercaptopurine.

(b) **Alkylating agents.** They inhibit DNA function by chemically cross-linking the DNA double helix. Examples of such drugs include cyclophosphamide, chlorambucil and cisplatin.

(c) **Antibiotics.** These are substances made by microorganisms that inhibit DNA function by either binding to DNA or inhibiting topoisomerases required for DNA replication. Examples of such antibiotics include bleomycin and doxorubicin.

(d) **Plant-derived drugs.** These drugs either inhibit topoisomerases or disrupt the microtubules of the mitotic spindle. Examples of such drugs include the topoisomerase-inhibitor **etoposide** and the microtubule disrupting drug, **taxol**.

One problem with such drugs (and radiation therapy) is that they are toxic to normal dividing cells as well as to cancer cells. When cancer arises in a tissue whose growth requires a specific hormone, it may be treated in a less toxic manner using drugs that block the action of that particular hormone. For example, many breast cancers require estrogen for their growth. Estrogens tend to exert their effects by binding to nuclear receptor proteins that activate the expression of specific genes. The drug **tamoxifen**, a common treatment for breast cancer, binds to estrogen receptors in place of estrogen and prevents the receptors from being activated.

Newer treatment approaches include **immunotherapies** that exploit the ability of the immune system to attack cancer cells, **molecular** targeting drugs aimed at proteins that are critical to the cancer cells, and **antiangiogenic agents** that attacks a blood supply of the tumour.

QUESTIONS

Long Answer Questions

1. Define the cancer. Describe different types of cancer. Explain various characteristics of cancer cells.
2. Explain various causes of cancer.
3. Describe diagnosis, screening and treatment of cancer.
4. Write an essay on cancer.

Short Answer Questions

1. What are preventive and control measures of cancer?
2. What is cancer? Mention any four carcinogens.
3. Write a short note on cancer.
4. Explain cancer and cell transformation.

Very Short Answer Questions

1. Define cancer.
2. What is carcinogen?
3. What is Ames test?
4. What is a carcinoma?
5. Define the leukemia.
6. What is lymphoma?
7. Define the sarcoma.

Fill in the Blanks

1. One of the first cancers to be investigated by proteomic analysis was ....

Multiple Choice Questions

1. Carcinoma refers to
   (*a*) malignant tumours of the connective tissue
   (*b*) malignant tumours of the skin or mucous membrane
   (*c*) benign tumours of the connective tissue
   (*d*) none of the above
2. Cancer cells are more easily damaged by radiation than normal cells because they are
   (*a*) non-dividing
   (*b*) starved of food
   (*c*) undergoing rapid division
   (*d*) different in structure
3. Which one of the following gives rise to cancer?
   (*a*) unrestricted and fast multiplication of cells by meiosis
   (*b*) breakdown of mechanism regulating meiosis
   (*c*) unlimited enlargement of cells by failure of formation of cleavage furrow after telophase to separate the daughter cells
   (*d*) the two daughter cells after mitosis carrying different amount of DNA
4. Cancer can be defined as
   (*a*) hyperplastic growth, infiltration and destruction of tissues
   (*b*) non-malignant epithelial growth
   (*c*) repair of damaged parts
   (*d*) metabolic disorder
5. Blood cancer is called
   (*a*) leukaemia (*b*) sarcoma
   (*c*) melanosoma (*d*) carcinoma

6. The cells affected by leukaemia are
   (a) plasma cells
   (b) erythrocytes
   (c) thrombocytes
   (d) leucocytes
7. One of the faster and less expensive tests for preliminary screening of potential carcinogens include
   (a) Biuret
   (b) Dick
   (c) Ames
   (d) ELISA
8. Sarcoma is the cancer of
   (a) epithelial tissue
   (b) connective tissue
   (c) blood
   (d) endocrine tissue
9. Cancer treatment includes
   (a) surgery
   (b) radiotherapy
   (c) treatment with anticancer drug
   (d) all of these
10. Which type of cancer is found in lymph nodes, gland and spleen?
   (a) sarcoma
   (b) carcinoma
   (c) adenoma
   (d) leukaemia
11. The spread of cancerous cells to distant sites is termed
   (a) metamorphosis
   (b) metagenesis
   (c) metastasis
   (d) metachrosis
12. A cancer causing agent is known as
   (a) carcinoma
   (b) carcinogen
   (c) metastasis
   (d) sarcoma
13. Hodgkin's disease is
   (a) cancer of lymphoid tissue
   (b) cancer of liver
   (c) cancer of mammary gland
   (d) cancer of WBC
14. The most commonly used marker enzyme in clinical diagnosis of prostate cancer is
   (a) alkaline phosphatase
   (b) amylase
   (c) acid phosphatase
   (d) γ-GTP
15. Cancer cells are
   (a) HL-B
   (b) HeLa cells
   (c) Vero
   (d) BHK
16. Which of the following is used in treatment of thyroid cancer?
   (a) U-238
   (b) Ra-240
   (c) I-131
   (d) C-14
17. Which one of the following techniques is safest for the detection of cancers?
   (a) radiography (X-ray)
   (b) magnetic resonance imaging (MRI)
   (c) computed tomography (CT)
   (d) histopathological studies
18. Cancer is generally caused due to activation of ________ to ________ and/or inactivation of ________
   (a) oncogene, tumor suppressor gene, proto-oncogene
   (b) tumor suppressor gene, oncogene, proto-oncogene
   (c) proto-oncogene, oncogene, tumor suppressor gene
   (d) oncogene, proto-oncogene, tumor suppressor gene
19. A patient is suspecious of having breast cancer. What type of test will a physician conduct to diagnose the cancer?
   (a) Pap test
   (b) CT scan
   (c) blood test
   (d) mammography
20. The main or primary target in carcinogenesis is
   (a) RNA
   (b) rRNA
   (c) DNA
   (d) none of these

ANSWERS

Very Short Answer Questions

1. Cancer is uncontrolled, growing mass of cells that is capable of invading neighboring tissues and spreading via body fluids, especially the bloodstream, to other parts of body. Cancer is also called **malignant tumour**.
2. Carcinogen is any cancer-causing agent.
3. It is screening test for potential carcinogens that assesses whether a substance causes mutations in bacteria.
4. It is a malignant tumour (*i.e.*, cancer) arising from the epithelial cells that cover external and internal body surfaces.
5. It is cancer of blood or lymphatic origin in which the cancer cells proliferate and reside mainly in the blood stream rather than growing as a solid masses of tissue.
6. It is cancer of lymphatic origin in which the cancer cells grow as solid masses of tissue.
7. Sarcoma is any cancer arising from a supporting tissue, such as bone, cartilage, fat, connective tissue and muscle.

Fill in the Blanks

1. Ovarian cancer

Multiple Choice Questions

1. (*b*)	**2.** (*c*)	**3.** (*d*)	**4.** (*a*)	**5.** (*a*)	**6.** (*d*)	**7.** (*c*)
8. (*b*)	**9.** (*d*)	**10.** (*c*)	**11.** (*c*)	**12.** (*b*)	**13.** (*a*)	**14.** (*c*)
15. (*b*)	**16.** (*c*)	**17.** (*b*)	**18.** (*c*)	**19.** (*d*)	**20.** (*c*)	

56 Effect of Radiation on the Cell

This is a evident fact that all life of earth is ultimately dependent upon the sun's radiant energy that is stored by green plant cells during photosynthesis. In addition to this primary role of light in the living economy, a persistent environment of various radiations produces other effects, reactions and adaptations, which may influence the life activities of the organisms living within this environment.

Muller (1927, 1928), **Stadler** (1928) and **Altenburg** (1928) independently discovered the mutagenic effect of radiation. Experimenting with X-rays on *Drosophila melanogaster*, barley and maize, they found a considerable increase in the frequency of mutation. Radiation—X-rays, γ-rays, β-rays, fast neutrons, slow neutrons and ultraviolet rays can induce point mutations at the DNA level or chromosomal aberrations and thus, has a cytogenetical effect.

56.1. SOURCES OF RADIATIONS

In nature there are many sources of radiant energy. The natural radiations include the radiation of the sun's spectrum, radio waves from the sun and the stars, ionizing radiations from the breakdown of the radioactive elements in the earth's crust and the cosmic rays coming from the sun and from depth of the space. Solar flares and about 120 celestial bodies are the sources of X-rays which on absorption produce ionization 60 miles high in the earth's atmosphere (**Frier** and **Webber**, 1963; **Schnopper** and **Delvaille** 1972).

1. Radiations of sunlight. Sunlight is the major source of radiations on earth. Its spectrum includes ultraviolet radiation,visible light, infrared rays and radiowaves. The sun is also the major source of earth heat without which life as we know would be impossible on earth.

Here we have great concern with ultraviolet radiations only because that have great amount of effect on the biological world.

2. Ultraviolet radiations. The sun radiates three types of ultraviolet radiations, *e.g.*, very short, short and long, from it. The ultraviolet rays with very short wavelengths are absorbed in earth's atmosphere before they reach the surface of the earth. High in the atmosphere the oxygen absorbs ultraviolet radiations and forming a layer of ozone. The ozone in turn absorbs somewhat longer ultraviolet rays and reforming oxygen. The layer of ozone forming the so called **ozone umbrella** at the height of 25 to 30 miles in the atmosphere. The ozone umbrella is the agent which removes the ultraviolet radiations of very short wavelength (**Cadle** and **Allen**, 1971). The ultraviolet radiations of long wavelengths, however, reach the earth surface and effect biological world variously. The ultraviolet rays transmit their energy almost entirely by excitation and not by ionization.

Effect of ultraviolet radiations on the cells. The ultraviolet radiations cause the most commonly known abnormality such as sun burn in the human beings. In such a case ultraviolet radiations kill the epidermal cells of skin and the injured cells (prickle cell layer) of a sunburned epidermis liberate chemicals which diffuse out and cause a relaxation of the wall of blood vessels in the dermis, resulting in the reddening of the skin (erythema). If the injury is slight, the red or pink colour soon disappears and in most individuals after a few hour's delay a small amount of pigment (melanin) develops (**tanning**). If the injury is very severe, the prickle cells may die, whereupon the layer is invaded by white cells and serum accumulates, causing a blister which later dries and the skin "peels" off and a thicker new epidermis takes the place of the old (**Johnson** *et al.*, 1968). **Johnson** and his associates have also reported that UV radiation affect lysosomes of cell epidermis which lyse and release their hydrolytic enzymes and so digesting the internal contents.

Sometimes, UV radiation of sunlight induces changes in the dermis which appear later as premature aging of the skin-wrinkling, moulting, change in suppleness of the skin, dryness and alterations in the blood vessels (**Klingman**, 1969).

Another injurious effect of UV radiation is the **eye-burn** or **snow blindness**. UV radiation directly striking the eye, reinforced by similar radiation reflected from snow (or the surface of water), may kill the superficial layers of cells covering the cornea. These cells become opaque, blinding the individual until the layer of dead cells is shed after few days.

Prolong exposure of UV radiation is found to induce superficial skin cancer (**carcinomas**) in rats, mice and human beings. The relative frequency of carcinomas in human beings is statistically correlated with the latitude, the frequency for people with the same heredity increasing with proximity to the equator. The fair individuals who are burned by sun but do not tan are most susceptible. Such individuals are most frequently Irish, Scotch, Welsh and Swedes. Conversely, people with rich skin pigment, yellow or black, are essentially immune to such tumours (**Epstein**, 1970).

The effect of ultraviolet radiations of sunlight on cell and unicellular organisms is much more drastic. **Downes** and **Blunt** (1877) discovered that ultraviolet radiations kill bacteria, all types of cells, such as microbes, protozoans, eggs of marine animals, algae, fungi and cells in tissue culture. **Giese** (1968) has suggested that all drastic effects of UV radiation on cell such as retardation of cell division, induction of mutation, killing of bacteria, etc., are caused by the effects of this radiation on nucleic acids in cells. (For other effect of UV radiation on DNA see chapter 35). Experiments have shown that inhibition of DNA synthesis by ultraviolet radiations stops cell division. They also inhibit the synthesis of RNA, proteins and other chemical molecules (**Giese**, 1964).

3. Ionizing radiations. The radiant energy which is released due to breakdown of radioactive elements such as polonium, uranium, radium, etc., are known as **ionizing radiations**. For better understanding of the ionizing radiation, one should be clear about the structure of atom.

All the matter (solid, liquid and gases) is composed of different elemental, substances and identical atoms comprise these elements. The atom is composed of a positively charged nucleus and one or more negatively charged electrons orbiting in an elliptical or circular pathway around the nucleus. Although at least 32 elementary particles and antiparticles have been described in the nucleus, it may be viewed as essentially composed of uncharged neutrons and protons each of which contains a single positive charge. The number of protons in an atomic nucleus equals the number of orbital electrons which usually move in definite orbits around the nucleus. But, the electrons may be moved from one orbit to another unoccupicd orbit. To jump to a higher energy level an electron must absorb energy. When an electron moves to a lower energy level, there is a release or energy equal to the difference in energy between the two levels. Thc energy from the atom (atomic energy) is released in the form of electromagnetic radiations.

The atomic radiations are usually known as **ionizing radiations**. An atom can radiate out the ionising radiations in the following two forms:

(*i*) **Particulate radiations.** The ionizing radiations which are caused by subatomic particles such as alpha particles, beta particles, fast and slow neutrons, protons and heavy nuclei of the atoms are known as particulate radiations.

(*ii*) **Electromagnetic radiations.** The radiations which beside having particulate composition have electromagnetic vibrations or waves are known as electromagnetic radiations. They include gamma rays, hard or soft X-rays, ultraviolet light, visible light, infrared rays and radiowaves.

4. Characteristics of ionizing radiations. Alpha particles. The alpha particles are positively charged helium particles or helium nucleus produced by radioactive disintegration of natural or artificial radioactive elements. Since alpha particles are heavy and have a double charge, so they travel slowly and loss high amount of energy. A positively charged alpha particle loses energy by exciting and ionizing atoms of the substance through which it passes. The alpha particles has a strong attraction for a negatively charged orbiting electron. The electron may be excited, *i.e.*, pulled into a higher energy orbit, or pulled completely away to ionize the atoms. An alpha particle can penetrate the cell upto 100 μm deep. Alpha particles travelling slowly through the cell, excite and ionize atoms in a cell. The alpha particles cause breaks in the chromosomes and when their energy is completely lost, they capture two electrons in their outer orbit and become helium atoms.

Beta particles. The beta particles are the electrons emitted with high velocity from the nucleus of a radioactive atom. They have negative charge and known as beta-minus particles. The beta-positive particles (positrons) have a similar mass to electrons, but a different charge (positive). Both electrons and positrons transmit energy to matter in a similar way. They vary in speed and energy depending on the voltage by which they are accelerated. They can ionize the matter and are generally absorbed by superficial tissues.

Fast neutrons. They are uncharged, highly energetic particles produced by bombarding low molecular weight substances with protons and deutrons. They may give rise to gamma rays on an inelastic collision with some matter, but in tissue of organisms which consists of nuclei of atoms of low atomic weight, collision give rise to protons. The protons are positively charged particles which may pass through tissue and cause ionization.

Slow neutrons. They are also called **thermal neutrons** and are much less energetic. Consequently they do not cause ionization or dislodge atomic nuclei but react atomic nuclei to form a new isotope which may disintegrate to give rise to ionizing particles.

X-rays and gamma rays. The gamma rays and X-rays produce both excitation and ionization of the substance through which they pass. Both types of rays are photons and have wave lengths from 10^{-3} to 10 A°. Gamma rays are electromagnetic radiations emitted by atomic nuclei during certain nuclear transformations. They are produced when an unstable nucleus gains stability by releasing energy. X-rays are produced when any substance, such as tungsten is irradiated by high energy electrons. The electrons in the target substance are deflected from their path by the approaching high energy electrons in the form of electromagnetic radiations. X-rays energy, therefore, depends on the kinetic energy of the impinging electrons which produce X-rays of various energies. The X-rays and gamma radiations behave as if they were small bundles of energy travelling close to the speed of light. A bundle of this energy is called a **quantum** or **photon**. The amount of energy carried by the proton is directly proportional to the frequency of the wave form.

Ionizing radiations caused by X-rays and gamma rays transfer their energy to the media primarily by excitation and formation of ion pairs. When a gamma rays photon passes close to a nucleus it suddenly disappears and an electron and positron appears in its place. Energy (e) and mass (m) are related to each other according to the **Einstein formula** $E = mc^2$, where c is the velocity of light. Some energy is used in pair production, the rest is transferred to the medium by excitation and ionization.

56.2. THEORIES OF ACTION OF IONIZING RADIATIONS

The biological effects produced in cells by all types of ionizing radiations are qualitatively the same. The physical events causing these biological effects do so by exciting and ionizing atoms. Thus, excitation and ionization are the two major mechanisms whereby energy is transferred in radiation. In excitation, an electron in an atom is raised to a higher energy level. In ionization, an orbital electron is ejected from an atom. An atom can receive energy directly from the incident radiation or an atom may, however, receive energy indirectly by transfer from another atom. It is indirect transfer effect which is important in aqueous solutions in that free radical formation in water occurs. A frre radical is extremely reactive. It is an atom or group of atoms containing an unpaired electrons. Usually free radicals are formed in radiation as intermediates between final chemical products and ion pairs. During an ionizing radiation an electron is ejected from a water molecule as follows:

$$H_2O \rightarrow H_2O^+ + e^-$$

This high energy electron may be picked up by another water molecule as shown in the following reaction.

$$e^- + H_2O \rightarrow H_2O^-$$

In this way an ion pair, H_2O^+ and H_2O^-, are formed. Each ion thus may, in the presence of another water molecule, form a hydrogen ion and a free radical as shown in the following reaction.

$$H_2O^+ \xrightarrow{C-H_2O} H^+ + OH^\circ$$

or as

$$H_2O^- \xrightarrow{C-H_2O} H^\circ + OH^-$$

The H^+ and OH^- will combine to form water. The H° and OH° free radicals are very reactive. In fact many of them can react to form H_2O_2. In cells containing catalase and peroxidases (enzymes) the hydrogen peroxide formation may not be significant but in other cells without these nuclei the H_2O_2 may have biological importance.

Effect of Ionizing Radiations on Cells

Brief exposure of the cells of the skin to ionizing radiation leads to a reddening or erythema which is followed by slight tanning. However, the effect of ionizing radiation is not confined to the surface (except in the action of alpha rays or very soft X-rays), and cells deep in the body may be injured. Certain kinds of cells such as lymphocytes, synthetically active cells and proliferating cells (*e.g.*, gametogenic cells, cells of germinative layer in the skin, blood forming cells, etc.) are specifically sensitive to ionizing radiation; they are destroyed and killed by it (**Haber** and **Rothstein**, 1969).

The radiosensitivity of cells in an organism depends on various biological and chemical factors. The biological factors are age, sex, genetic constitution, body weight, health, diurnal variation, stress, and diet. Generally, radiosensitivity decreases as an organism gets older. When old age, however, is attained the organism again becomes more radiosensitive. The females are slightly less sensitive to radiation than males. The organism of lesser body weight seems more susceptible to radiation than an organism of greatet body weight. Unhealthy animals are less resistant to radiation' than healthy ones.

Recently, the effect of radiation on cellular organelles have been extensively studied. The ionizing radiation have following types of effects on different cellular organelles:

1. Cytoplasm. The ionizing radiations have small effect on cytoplasm. However, the endoplasmic reticulum vesicles, cellular membranes and mitochondria are affected by the radiations. The endoplasmic reticulum are dilated due to radiations. Membranes of lysosomes and plasma membrane are found much sensitive for ionizing radiations. Lysosome membrane are probably damaged by irradiation induced oxidation of the sulphydral groups of amino acid residues in proteins; very large doses induce peroxidation of membrane lipids. Consequently lysosomes are broken down and hydrolytic enzymes are set free, causing cellular damage (**Harris**, 1970).

The ionizing radiation injures the mitochondrial membranes. However, the oxidative phosphorylating system in mitochondria does not appear to be involved in primary radiation damage. Due to radiation the mitochondria swell and their cristae rupture.

2. Nucleus. The nucleus is more sensitive to ionising radiations than the cellular cytoplasm. The effect of ionizing radiations on chromosomes and DNA is drastic. The ionizing radiations cause mutation and chromosomal aberration in the chromosomes. The DNA is greatly effected by ionizing radiations. They change the genetic codes of DNA by affecting the nitrogen bases. The purines of DNA are less sensitive to radiation than the pyrimidines. Of these, thymine is the most sensitive pyrimidine. In fact, large doses of ionizing radiation destroy thymine, cytosine and uracil in aqueous solutions.

Ionizing radiations cause depolymerization of DNA. In doing so, it may prevent DNA replication and halt genetic transcription. Mutagenesis or death or both, may be the biological consequence of such damage to DNA. Depending on the nature and amount of DNA damage, complete reconstruction of the genome can occur. In some cells there exist a multienzyme mechanism which repairs DNA lesions caused not only by ionizing radiation, but also by ultraviolet irradiation, chemical mutagens and carcinogens.

The radiations destroy DNA either directly or indirectly. There are various types of direct destruction by radiation to DNA molecules. Hydrogen bonds break between chains. A base may be deleted, or changed in some way as in deamination. Single or double chain fracture may occur. And crosslinking may occur within the DNA double helix, or with protein molecules. Indirect damage can be done from peroxides, or free radicals, formed from water or organic compounds by ionizing radiation to act on nucleic acids or other precursors. Inhibition of DNA synthesis results from a direct effect. If the cell is irradiated in the 'S' phase (DNA synthesis phase of interphase) there is an inhibition of DNA replication which prevents mitosis and results in cell death.

3. Mitotic apparatus. The radiations may delay mitosis by inhibiting spindle formation and DNA synthesis during cell divisions.

4. Other influencing factors. No living cell is completely resistant to radiation. But several factors seem to be related to radiation effects in a cell. A clear correlation is apparent between lethal exposure and chromosome volume. Because the cells with the largest chromosome volumes are the most radiosensitive. Further, the haploid cells are more radiosensitive than diploid cells and polyploid cells are most radioresistant. There are fewer chromosomal aberrations in those cells having extra nucleoli. It appears that the nuclear cytoplasmic volume is an influencing factor. These cells with a large cytoplasmic volume as compared to their nuclear volume are more radioresistant than those with a small cytoplasmic volume. The cells with large number of mitochondria are also more radioresistant. Younger cells of particular tissue are more radioresistant than mature cells. The cells with a rapid mitotic rate are less radioresistant than those cells having a low rate of mitosis. The less differentiated cells seem more radiosensitive than highly specialized cells, but many exceptions occur.

56.3. GROSS EFFECTS OF RADIATIONS

The effects of radiations on the various organisms are as follows:

1. Viruses. The ultraviolet radiations inactivate viruses and destroy their infectivity and immunizing capacity. Both ultraviolet and ionizing radiations produce mutations in viruses. Viruses with DNA are more radiosensitive than viruses having RNA.

2. Microorganisms. Nearly all microorganisms are radioresistant. However, the radiations may cause inactivation and death in the microorganisms.

3. Plants. Generally plants arc more radioresistant than animals. They remain more radioresistant in dormant stage, but gradually become more radiosensitive in the early development and growth stages. When growing plants are irradiated, they may be killed, their growth inhibited or mutations may result. Modifications of roots, stems, leaves and flowers in regard to both gross morphology and histology have been seen to occur in irradiated plants. Ionizing radiations also increases the incidence of tumour formation in plants.

4. Animals. In animals the effect of radiation is not only mutagenic and chemogenic but is also carcinogenic. A quantitative relation exists between the amount of radiation tolerated by different groups of organisms and their nucleic acid contents. A decreasing order of sensitivity has been reported for viruses, with single stranded DNA and RNA, double stranded DNA viruses, haploid cells of bacteria and yeast, diploid yeast cells and avian and mammalian cells. In fact, mammals are more sensitive to radiations than birds, reptiles, amphibians and fishes.

Effect or Radiations on Mammals

Among mammals, the tissue of bone marrow, lymphoid organs and the lining of intestine are most radiosensitive. The radiations cause radiation syndromes which may be haemopoietic, gastrointestinal, central nervous system and cardiovascular. They may cause lymphopenia and leukemia diseases in human beings. Due to radiations, the rate of mitosis decreases in the epithelial cells lining the mucosa of small intestine of mammals and results the denude mucosa which ulcerates, haemorrhages and becomes an inflamed gangrenous tissue. Death can ensue.

The ionizing radiations affect the skin by injuring the basal layer, *i.e.*, the stratum germinativum of the stratum malpighii to inhibit mitosis. As a result erythema may occur. Cells forming hair are also radiosensitive. When damaged, a loss of hair occurs. Although spermatozoa are somewhat radioresistant, spermatogonia are not and are very radiosensitive. Oocyte are relatively radiosensitive. Large doses of radiation injure the brain and nervous system as well the heart and blood vessels.

Ionizing Radiation and Human Cancer

In human, heavy medical irradiation, atomic accidents or radioactive substances may produce chromosomal aberrations. An interesting example of this is provided by a study of radium-dial painters, who put luminescent paint on watch dials. They retain small amounts of radioactive radium, which accumulate in their bones. The resulting radiation doubles the frequency of chromosomes breaks observed in their lymphocytes and these individuals have a high risk of developing bone cancer (see **Bodmer** and **Cavallisforza**, 1976).

Further, the survivors of the atom bombs dropped on Hiroshima and Nagasaki at the end of World War II are another group of people who have been studied intensively in order to find out more about the genetic and other effects of radiation.

Biological Effects of Atom Bombs

An atomic bomb explosion lasts about 10^{-6} seconds and develops in its centre a temperature of about 1,000,000°C. Its destructive effects are the result of its extreme heat, the ultraviolet flash, shock waves and ionizing radiation. The temperature of the surface of an object such as human skin at about 400 meters from the bomb may be raised to about 50°C. The ultraviolet flash burn may be severe and surface cells of the organisms are most rapidly damaged. The shock waves rip the capillaries of the sense organs and lungs and cause haemorrhages throughout the entire digestive system. The ionizing radiations consists of about 15 per cent neutrons, about 15 per cent alpha and beta particles and the remainder consists of short gamma rays.

A characteristic set of symptoms in a person injured by ionizing radiations is nausea accompanied by vomiting, prostration, after a latent period, fever, bloody diarrhoea, loss of hair and appearance of purple haemorrhagic spots in the skin. Should the individual survive, various skin lesions appear in areas where the germinative epithelium has been killed by the radiations. Ionizing radiation of atomic bomb also caused chromosomal abnormalities and cancer (*e.g.*, leukemia) in Hiroshima survivors.

Economic Uses of Radiations

Besides, their deleterious biological effects, the radiations have economic value. They are employed in medicine, dentistry and radiotherapy. Various types of diseases and cancer have been treated adequately by the radiations.

In the production of new antibiotics and plants, irradiation is being used to economic advantage.

QUESTIONS

Long Answer Questions

1. What is a radiation? Describe different sources of radiation.
2. How do various radiations affect the cells?
3. Discuss the gross effects of radiation on living orgnisms.

Short Answer Question

1. What are the effects of ultraviolet rays on cells?

Very Short Answer Question

1. What is ionizing radiation?

ANSWER

Very Short Answer Question

1. Ionizing radiation is high-energy radiation that is highly mutagenic, producing free radicals that react with DNA, including X rays and gamma rays.

57 Viruses

Viruses (L., venoum or poisonous fluid) are very small submicroscopic biological entities which though lack cellular organization (*viz.,* plasma membrane and metabolic machinery) possess their own genetic material, genetically determined macromolecular organization and characteristic mode of inheritance. For their multiplication, they essentially require the presence of some host cell, *i.e.,* they are obligate cellular parasites of either bacteria, plants or animals.

57.1. STRUCTURE

Viruses are quite a varied group (Fig. 57.l). They range in between 30 to 300 nm or 300 to 3000 Å in size, so they can be observed only by electron microscopy and X-ray crystallography. They have a

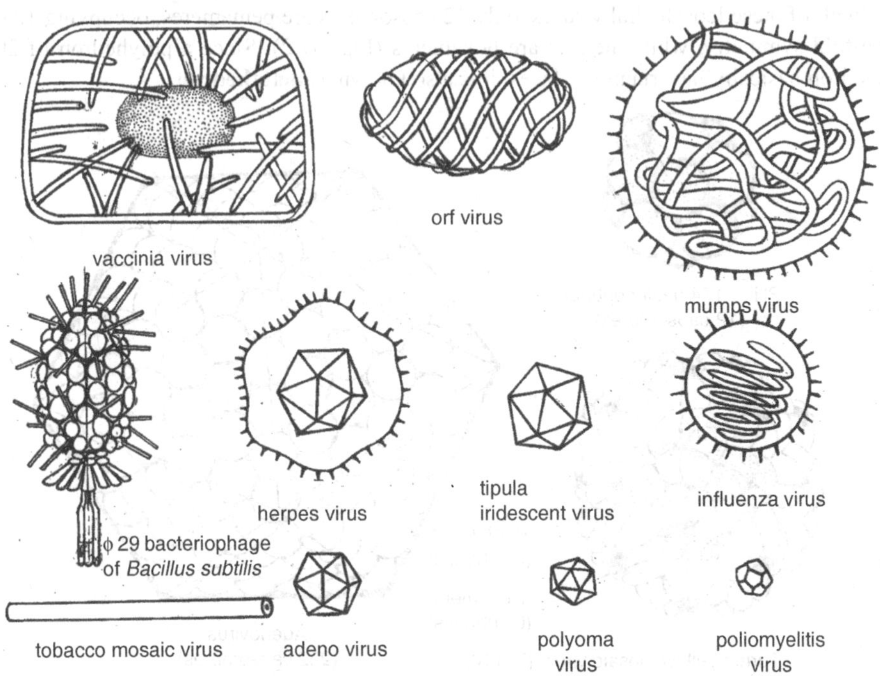

Fig. 57.1. Different kinds of viruses.

regular geometrical and macromolecular organization. Basically an infectious virus particle (called **virion**) is composed of a **core** of only one type of nucleic acid (DNA or RNA) which is wrapped in a protective coat of protein, called **capsid**. The capsid consists of numerous **capsomeres**, each having a few **monomers** or **structural units**. Each structural unit is made up of one or more polypeptide chains. The capsomeres are of different shapes such as hollow prism, hexagonal, pentagonal, lobular or any other shape. The specific arrangement of capsomeres in the capsid determines the shape of a virion. Viruses have the following three different types of symmetry:

1. Icosahedral symmetry. Many viruses have spherical, cubical or polygonal shape which is basically **icosahedral** or 20-sided. Icosahedral symmetry depends on the fact that the assembly of the capsomeres causes the capsid of the virus to be at a state of minimum energy (**Caspar** and **Klug,** 1962). An icosahedral capsid comprises both **pentameres** (*i.e.,* capsomeres containing 5 structural units) and **hexameres** (*i.e.,* capsomeres having 6 structural units). In an icosahedral virus the minimum number of capsomeres is 12 or its multiple such as 32, 42, 72, 92, 162, 252, 362, 492, 642 and 812.

For example, the total number of capsomeres of different icosahedral viruses are:

(*i*) Bacteriophage ϕ (phi) × 174 = 12 pentameres;
(*ii*) Turnip yellow mosaic virus or TYMV = 32 capsomeres;
(*iii*) Poliovirus = 32 capsomeres;
(*iv*) Polyoma virus and papilloma virus = 72 capsomeres;
(*v*) Reovirus = 92 capsomeres;
(*vi*) Herpes virus = 162 capsomeres;
(*vii*) Adenovirus = 252 capsomeres; and
(*viii*) Tipula iridescent virus = 812 capsomeres.

In all of these icosahedral viruses, only 12 capsomeres are pentameres, occupying 12 corners of five-fold symmetry, while the rest are hexameres (Fig. 57.2). Since a polyhedron of 20-sided icosahedron basically has triangular faces, it is also known as **deltahedron.**

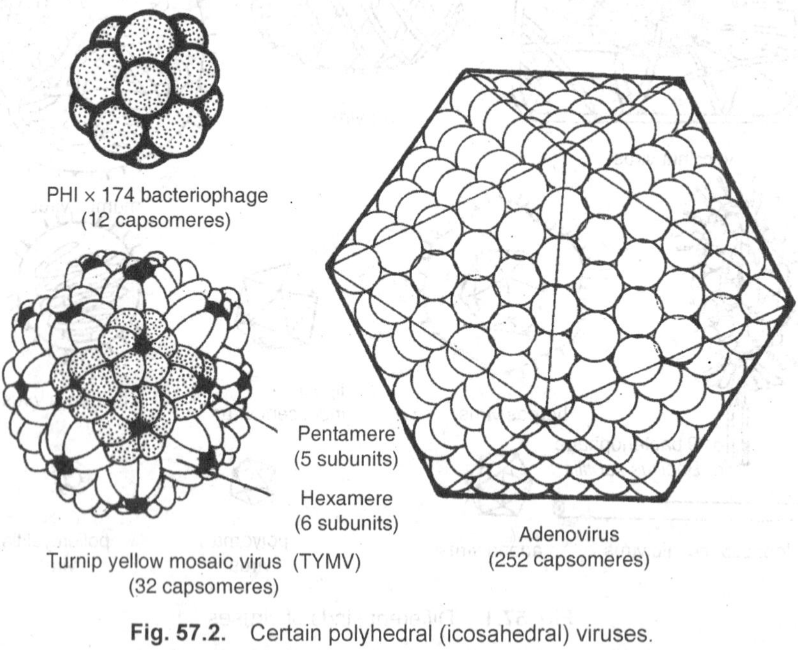

Fig. 57.2. Certain polyhedral (icosahedral) viruses.

2. Helical or cylindrical symmetry. The rod-shaped helical capsid of viruses such as tobacco mosaic virus (TMV), bacteriophage M13 and influenza virus, consists of numerous identical capsomeres arranged into a helix because they are thicker at one end than the other.

3. Complex symmetry. Viruses with complex shaped capsids are of two shapes; those without identifiable capsids (*e.g.*, pox viruses such as vaccinia, cowpox, extromelia and orf viruses) and those with tadpole-shaped structures in which each part has different sort of symmetry (*e.g.*, T-even phages of *E.coli*; T_2 phage has an icosahedral head, helical tail sheath, hexagonal end plate and rod-shaped tail fibres). Some viruses such as rabies virus are bullet-shaped.

Some viruses such as herpes virus, influenza virus, mumps virus and Semliki forest virus are surrounded by a 100–150 Å thick spiked membrane. This membrane contains lipid bilayer of plasma membrane from which projects the virus-specific protein molecules or spikes. It is not made by or specified by the virus itself but is derived from the plasma membrane of the host cell (*i.e.*, animal cell).

57.2. TYPES OF VIRUSES

Viruses are of following three types depending on the host cell.

A. Bacterial Viruses or Bacteriophages

Viruses that parasitize the bacterial cells, are called **bacteriophages** or **phages** (phage means 'to eat'). The phages have specific hosts and they are of variable shapes, sizes and structures. The most widely studied phages are *T*-even bacteriophages such as T_2, T_4, T_6, etc., which infect the colon bacillus, *Escherichia coli* and are also known as **coliphages** (*T* for "type". The plural word phages refers to different species; the word phage is both singular and plural and in the plural sense refers to particles of same type. Thus, T_4 and T_7 are both phages, but a test tube might contain either 1 T_4 phage or 100 T_4 phage).

T_4 bacteriophage is a large-sized tadpole-shaped complex virus (Fig. 57.3). Its capsid comprises of an icosahedral **head** (1250 Å length and 850 Å width; 2000 capsomeres), a short neck with **collar** bearing 'whiskers' and a long helical **tail.** The tail is made up of a thick and hollow **mid-piece,** a hexagonal **base plate** or **end plate** to which are attached six **spikes** and six long **tail fibres.** The mid-piece consists of a central hollow **core** and a spring-like **contractile sheath** which comprises 24 rings of hexameres and remains helically arranged around the core. The T_4 genome or chromosome is a single DNA molecule which is 60 μm long, linear, double-stranded and tightly-packed within the head of the phage. Phage DNA contains more than 1,66,000 nucleotide pairs and encodes more than 200 different proteins (*i.e.*, proteins involved in DNA replication and in the assembly of head and tail). For example, T_4 phage DNA codes for at least 30 different enzymes (*e.g.*, helicases, topoisomerases, DNA polymerases, DNA ligases, etc.) all of which ensure rapid replication of phage chromosome in preference to DNA of *E.coli* (host cell). Further, during DNA replication, an unusual nitrogen base, called **5-hydroxymethylcytosine** is incorporated in place of cytosine in the phage DNA. This unusual base makes phage DNA recognisable from that of host DNA and selectively protects it from the nuclease enzymes. The nucleases are encoded in T_4 phage genome to degrade only the DNA of host cell. Some other phage proteins alter host cell's RNA

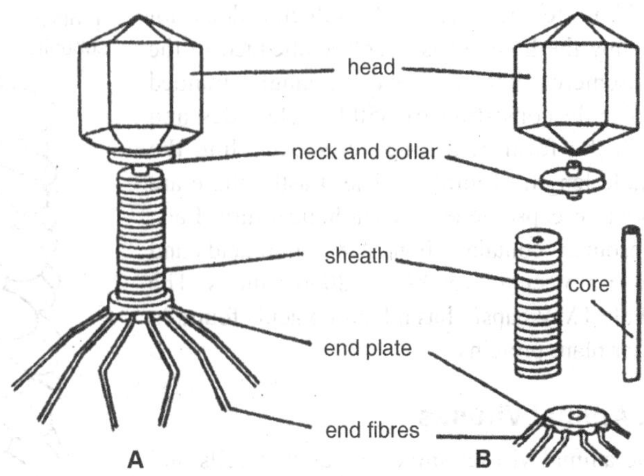

Fig. 57.3. A T_4 bacteriophage. A—External structure; B—Parts.

polymerase enzymes, so that they transcribe different sets of T_4 genes at different stages of viral infection according to the phage's needs.

B. Plant Viruses

The plant viruses parasitize the plant cells and disturb their metabolism and cause severe diseases in them. All plant viruses consist of ribonucleoproteins in their organization. The important plant viruses are tobacco rattle virus (TRV), tobacco mosaic virus (TMV), potato virus, beet yellow virus (BYV), southern bean mosaic virus (SBMV) and turnip yellow virus (TYV). Among plants, few hundred viral diseases are caused, e.g., mosaic diseases of tobacco, cabbage, cauliflower, groundnut and mustard; black-ring spot of cabbage; leaf roll of tomato; leaf curl of papaya, cotton, bean and soyabean; yellow diseases of carrot, peach; little-leaf of brinjal. These diseases are spread mainly by insects such as aphids, leaf hoppers and beetles.

Tabacco mosaic virus (TMV). TMV is the most extensively studied plant virus. It was discovered by **Iwanowski** (1892) and obtained in a pure state (*i.e.,* in paracrystalline form) by **Stanley** (1935). **Bawden** and **Pirie** (1937) extensively purified TMV and showed it to be a nucleoprotein containing RNA. **H. Fraenkel-Conrat** experimentally demonstrated that RNA is the genetic substance of TMV.

TMV is a rod-shaped, helically symmetrical RNA virus (Fig. 57.4). Each virus particle is elongated, cigarette-like in shape having the length of 3000 Å (300 nm) and diameter of 160 Å (16 nm). In each rod of TMV, there are about 2130 identical elliptical protein subunits or capsomeres. The capsomeres are closely packed and arranged in a helical manner around the RNA helix, forming a hollow cylinder. Thus, there is a hollow core (axial hole) of about 40 Å (4nm) diameter which runs the entire length of the rod and contains the RNA molecule. The RNA molecule does not occupy the hole but is deeply embedded in the capsomeres. RNA of TMV is a single-stranded molecule consisting of 6500 nucleotides and is in the form of a long helix extending the whole length of viral particle. Lastly, there are about 16 capsomeres in each helical turn. Each capsomere contains about 158 amino acids and has a molecular weight of 18000 daltons. The whole TMV capsid has all amino acids found in other plant proteins.

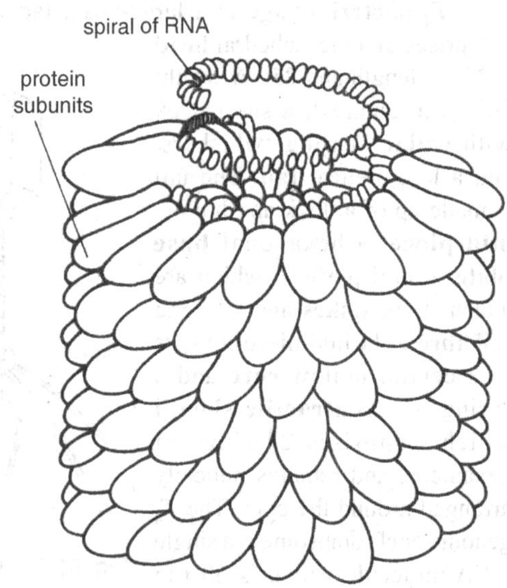

Fig. 57.4. Molecular organization of the tobacco mosaic virus (TMV).

C. Animal Viruses

The animal viruses infect the animal cells and cause different fatal diseases in animals including humans. Generally, they have a polyhedron or spherical shape and genetic material in the form of DNA or RNA. The protein coat or capsid of animal viruses is surrounded by an envelope.

Examples of viruses causing infections in the human beings are: common cold, influenza (Fig. 57.5), mumps, measles, rubella (German measles), chicken pox, small pox, polio, viral hepatitis, herpes simplex, viral encephalitis, fever blisters, warts and some types of cancer. Among livestock and fowl, viruses cause encephalitis, foot and mouth disease, fowl plague, Newcastle disease, pseudorabis, hog cholera and a variety of warts and other tumors. A virus usually displays some specificity for a particular animal group.

Poliomyelitis is a most extensively studied single stranded (ss) RNA-containing animal virus. The polio virus has comparatively very simple organization. It consists of a protein shell built up of 60 structurally equivalent asymmetric protein subunits of approximately 60 Å diameter, packed together in such a way that they form a spherical shell of about 300 Å in diameter. The shell or capsid encloses a single stranded RNA molecule of 5,200 nucleotides.

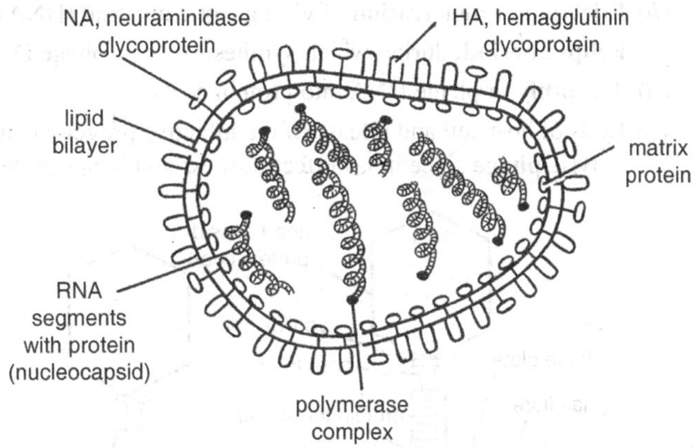

Fig. 57.5. The influenza viron.

Herpes virus is another most extensively studied animal virus. It possesses a DNA containing core embedded in a regular icosahedral **capsid** (162 capsomeres) and an outer **envelope** of lipids, proteins and carbohydrates. The DNA molecule of herpes virus is a single, linear, double-stranded having a molecular weight of 10^8 daltons and codes for about 100 average sized protein molecules.

57.3. VIRUSES ARE LIVING!

Viruses are not considered true cells since they lack two very important characteristics of a cell:

1. They lack a plasma membrane;
2. They lack cytoplasm and a metabolic machinery to sustain life activities. The living and nonliving characters of viruses are summarized below:

1. Nonliving characters of viruses

(*a*) They have the property of crystallization like other inorganic and organic compounds.

(*b*) They lack enzyme system.

(*c*) They lack response to external stimuli.

(*d*) They lack growth.

2. Living characters of viruses

(*a*) Viruses have the capacity to multiply (inside a cellular host).

(*b*) They have the property of undergoing mutations.

(*c*) They have the property of recombination and inheritability.

57.4. LIFE CYCLE OF THE BACTERIOPHAGE

Bacteriophages may have the following two types of life cycles:

1. **Lytic cycles**, in which viral infection is followed by **lysis** (bursting and death) of the host cell and release of new infective phages, *e.g.*, virulent phages such as T_4 and all other *T*-even coliphages.
2. **Lysogenic cycles**, in which infection rarely causes lysis, *e.g.*, temperate phages such as P_1 and lambda (λ) phages.

1. Lytic Cycle of a Virulent Phage

Life cycle of a T_4 bacteriophage (Fig. 57.6) involves the following steps:

(*a*) **Attachment** or **adsorption** of phage to bacterial (host) cell.

(*b*) **Injection** or **penetration** of viral genetic material (DNA) into the host cell.

(*c*) **Eclipse period,** during which synthesis of new phage DNA and protein coats takes place.

(*d*) **Assembly** of phage DNA into protein coats.

(*e*) **Lysis** of host cell and release of the infective progeny phages. Such a phage is called **virulent** or **lytic phage** since it has infectiousness and it causes death of host cell by lysis.

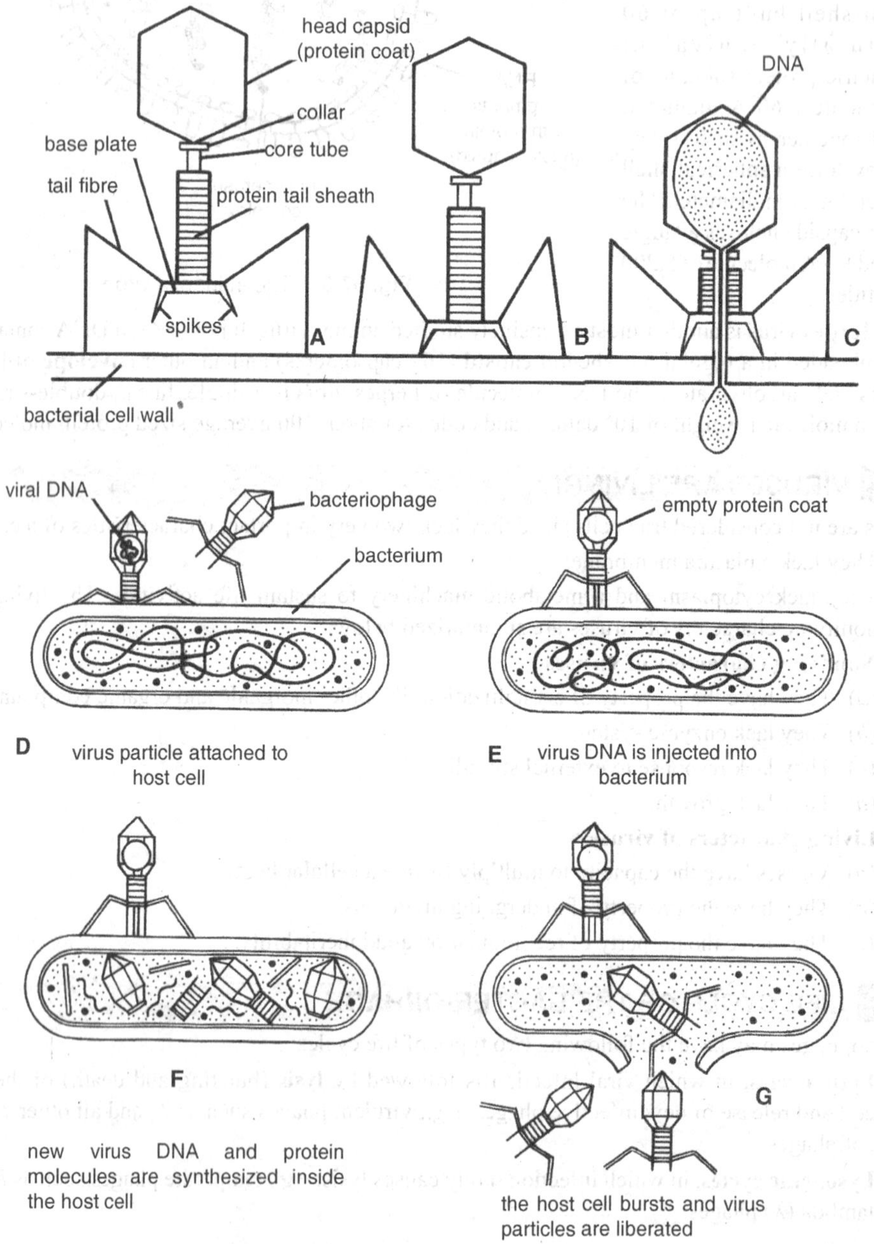

Fig. 57.6. A, B, C—Mode of attachment of a *T*-even phage (bacteriophage) on a bacterial cell wall and injection of DNA into the bacterium (host cell); D, E, F and G—Steps of viral reproduction inside the host cell.

The adsorption of the phage to its host is made possible by a reaction of chemical groups on the two during a random collision. Reactive groups (called **adsorption proteins** or **pilot proteins; Kornberg**, 1974) at the end of the tail of the phage can join with a complementary set of chemical groups (**receptor sites**) in the cell wall of the bacterium. During adsorption, long tail fibres of the phage are first to contact and attach to the cell. They help to position the phage's tail perpendicularly to the cell wall. Once the phage is attached to its prospective host, injection can take place involving a movement of phage DNA from its position inside the head of the phage through the hollow core of the tail into the bacterium. Entry is made possible by a hole punched in the bacterial cell wall, either by contraction of outer sheath of tail or by the action of enzymes carried by phage tail, or both. The protein coat or capsid of the phage remains outside the cell. Once inside the host cell, the phage DNA becomes a **vegetative phage,** *i.e.,* phage genes take over the metabolic machinery of the cell and direct it to produce replicas of the infecting virus. Although the cell continues to procure raw materials and energy from the environment, the phage genes allow only viral components to be built. Further, either the normal ability of the host DNA to control the cell is lost, or the host DNA is completely destroyed by early products of the viral genes. Thus, phage DNA is both replicated and transcribed; first the enzymes needed for synthesis of phage DNA are translated, then the capsid proteins are translated. Phage particles are assembled around condensed cores of the complete phage nucleic acid (by self-assembly method). At last lytic enzymes which have been coded by phage DNA, break open the bacterium and release the new phage particles which diffuse in the surrounding in search of new host.

2. Lysogenic Cycle

Certain bacteriophages such as P_1 and lambda (λ) phages, have entirely different pattern of life cycle than the virulent phages. This pattern is called **lysogeny** and is characterized by delayed lysis after phage infection. A virus with this capacity is called **temperate virus.** The infected host cell is said to be **lysogenic** because dormant virus may at any time become active and begin directing the synthesis of new virus particles.

In lysogeny, the process of adsorption and nucleic acid injection are quite similar to a lytic cycle of virulent phages, although different phages recognise different bacterial cell surface receptors. The next step, however, is unique to lysogeny. The nucleic acid is neither extensively replicated nor extensively transcribed. The virus generally expresses one or a few genes which code for a **repressor protein** that turns off (*i.e.,* represses) the expression of the other genes of the virus. In consequence, virus is not replicated, but phage DNA remains in the bacterium, being replicated in such a way that when the lysogenic bacterium divides, each daughter cell receives at least one phage genome in addition to the bacterial genome. There are two styles to this persistence of phage DNA: the phage chromosome may exist as a fragment of DNA outside the host's chromosome (*i.e.,* in the host's cytoplasm) essentially as a plasmid (*e.g., P_1* bacteriophage) or it may attach itself to the host's chromosome as an episome (*e.g.,* lambda phage) (see Box 57.1). Thus, in case of lambda phage the DNA first becomes circular due to joining of its both cohesive ends and then is integrated into the circular DNA molecule of the bacterium. Such an integrated and dormant viral genome is often termed as **provirus** or **prophage.** The infection of *E.coli* cell with lambda phage and its consequent integration and adoption of lysogeny, renders that cell immune to further attack by phage of the same type.

Box 57.1

Episome is an extra-chromosomal, circular, transposable, closed DNA molecule which can exist either integrated into the bacterial chromosome or separately and autonomously in the cytoplasm. **Plasmid** is that bit of autonomous genetic material of bacteria (i.e., circular DNA) that exists only extra chromosomally and cannot be integrated into the bacterial chromosome (DNA). Now only the term plasmid is used for all kinds of extra-chromosomal autonomous transposable circular DNA fragments.

The calm lysogenic period is ended by some type of shock (*e.g.*, temperature changes, UV irradiation or conjugation of a lysogenic bacterium with a non-lysogenic bacterium) to the lysogenic culture. A shock evidently inactivates the repressor of the phage so that all the phage genes can be expressed. Then the lysogenic bacterium is ruined, for the phage DNA that the host bacterium is harbouring enters the lytic phase. It replicates, transcribes, translates, assembles viron particles and lyses the bacterium (Fig. 57.7).

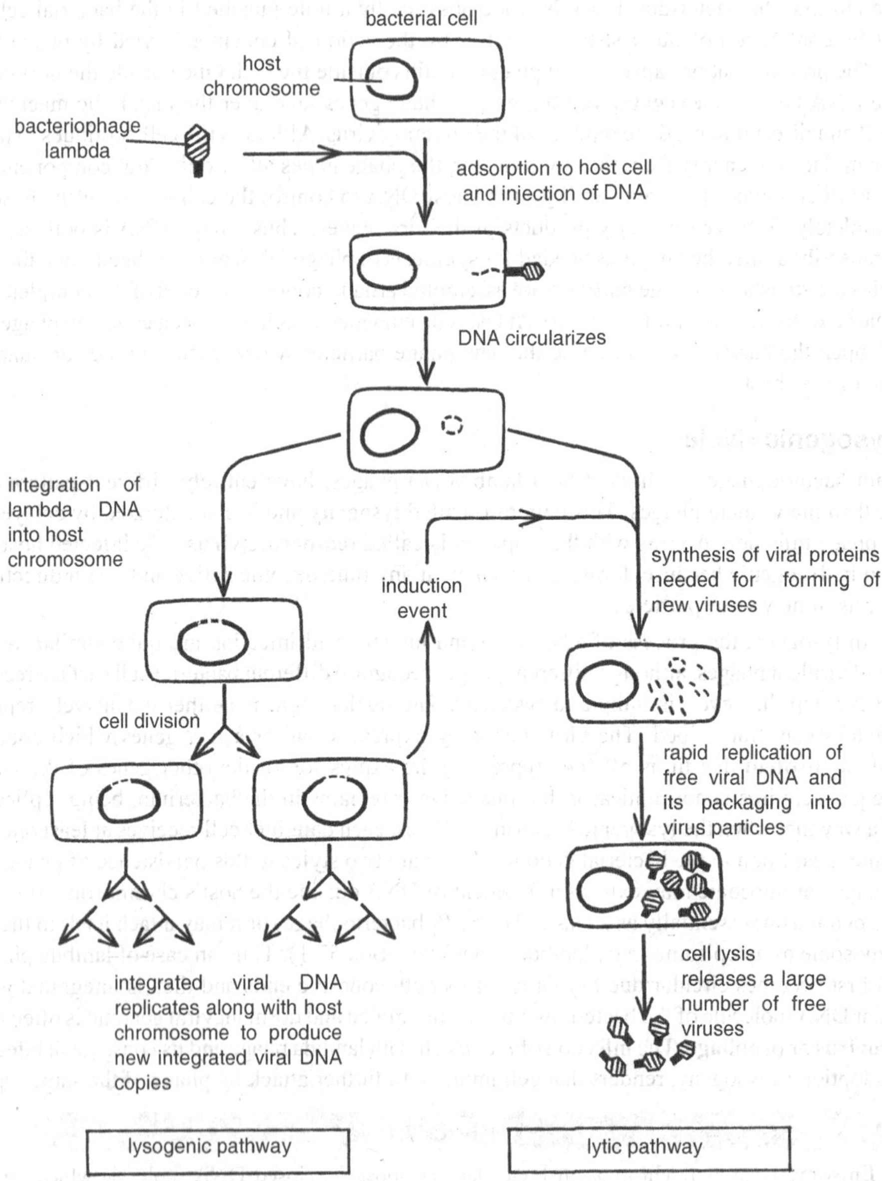

Fig. 57.7. The life cycle of bacteriophage lambda. Its double stranded DNA can exist in both linear and circular forms. The lambda phage can multiply by either a lytic or a lysogenic pathway in the *E.coli*.

3. Recombination in Phages

Recombination is any process generating new combinations of pre-existing genetic material. Recombination in phages was described by **Max** and **Mary Delbruck** (1894). When two different types of phages T_2 and T_4 are placed in a culture of bacteria which is sensitive to both, the new generation of phages consists of the two parental types. But some of them were found to have combination of T_2 and T_4 genes. The recombination in bacteriophages can be equated to the sexual reproduction of plants and animals.

4. Uses of Bacteriophages for Genetic Research

A bacteriophage has many advantages for genetic research. Its simplicity of structure and rapid rate of multiplication and enormous size of population make it very important material for genetic study.

A phage has a genetic system basically like that of a higher organism. However, the genetic material of phage easily mutates, so that mutants of variant phenotype are very frequent.

Phage provides the evidence that DNA is the genetic material and transmits the hereditary characters from one generation to the next. This was proved by **Hershey** and **Chase** by labelling phage DNA with radioactive phosphorus (^{32}P) and its protein with radioactive sulphur (^{53}S). Only ^{32}P was carried to the-next generation. This proved that DNA and not the protein is the carrier of genetic material.

57.5. VIROIDS

As simple as viruses are there are even simpler agents that can infect eukaryotic cells (though apparently not prokaryotic cells; see **Becker** *et al.*, 2007). The **viroids** found in some plant cells represent one class of such agents. Viroids are small, circular RNA molecules, rather like an RNA virus without its capsid. They are, in fact, the smallest known infectious agents. The RNA circles are only about 300–400 nucleotides long and replicated in the host cell even though they do not code for any protein.

It is not yet known how viroids are transmitted from one host to another, since clearly they do not occur in a free form. Most likely, they pass from one plant cell to another only when the surfaces of adjacent cells are damaged so that there is no membrane barrier for the RNA molecule to cross.

Viroids are responsible for diseases of several crop plants such as potatoes and tobacco. A viroid disease of great economic concern is **cadang-cadang disease** of the coconut palm. It is yet not clear how viroids cause disease. They may enter the nucleus and interfere with the transcription of DNA into RNA. Alternatively, they may interfere with the subsequent processing required of most eukaryotic RNA transcripts.

57.6. PRIONS

They represent another class of infectious agents. The term prions was coined to describe *proteinaceous infective particles* that are thought to be responsible for neurological diseases such as **scrapie** in sheep and goats, **Kuru** in humans and **mad cow** disease in cattle. Scrapie is so named because infected animals rub incessantly against trees or other objects, scraping off most of their wool in the process. Kuru is a degenerative disease of the central nervous system originally reported among native peoples in New Guinea patients with this or other prion-based disease suffer initially from mild physical weakness and dementia, but these effects slowly become more severe, and the diseases are eventually fatal. Scrapie like brain-wasting disorder occurs in case of mad cow disease. **Stanley Prusiner**, who received a Nobel Prize in 1997 for his pioneering work in this field, has proposed that these diseases are transmitted by infectious protein-containing particles called **prions**. Because prions do not appear to contain DNA or RNA, **Prusiner** has formulated a unique theory to explain how prions might transmit disease by causing the infections spread of abnormal **protein folding**. According to this theory, a prion protein (designated Pr Psc) is simply a misfolded version of a normal cellular protein (designated Pr Pc). When the misfolded Pr Psc encounters a normal Pr Pc peptide chain in the process of folding, it causes the normal polypeptide to fold improperly. The resulting, abnormally folded protein triggers extensive nerve cell damage in the brain, leading to uncontrolled muscle movements and eventual death. Both the normal and variant forms of prion proteins are found on the surfaces of neurons, suggesting that the protein may somehow affect the receptors that detect nerve signals.

QUESTIONS

Long Answer Questions

1. Describe general organisation of viruses. Why are viruses considered both living as well as non-living?

Short Answer Questions

1. Why viruses are not true cells?
2. Write short notes on the following:
   (*i*) Bacteriophage
   (*ii*) Reproduction in bacteriophage
3. Draw a neat and labelled diagram of the following:
   (*i*) T_4 bacteriophage

Very Short Answer Questions

1. Define the virus.
2. What is viroid?
3. What are prions?

Multiple Choice Questions

1. AIDS causing factors are associated with
   (*a*) protozoa (*b*) DNA virus
   (*c*) bacteria (*d*) RNA virus
2. Viruses are living because they
   (*a*) multiply in the host cells
   (*b*) carry anaerobic respiration
   (*c*) carry metabolic activity
   (*d*) cause infection
3. Virus parasite in bacteria is
   (*a*) bacteriophage (*b*) viroid
   (*c*) cyanophage (*d*) mycoplasma
4. Number of capsomeres in TMV is
   (*a*) 231 (*b*) 1230
   (*c*) 2130 (*d*) 3120
5. Polio virus has
   (*a*) ss DNA (*b*) ds DNA
   (*c*) ss RNA (*d*) ds RNA
6. T_2 has
   (*a*) ds DNA (*b*) ss RNA
   (*c*) ds RNA (*d*) ss DNA
7. When a viral DNA integrates with a host DNA, then the process is referred to as
   (*a*) lysis
   (*b*) prophase
   (*c*) lysogeny
   (*d*) none of these

ANSWERS

Very Short Answer Questions

1. Virus is a subcellular parasite composed of a protein coat and DNA or RNA incapable of independent existence; it invades and infects cells and redirects the host cell's synthetic machinery toward the production of more virus.
2. Viroid is a small, circular RNA molecule that can infect and replicate in host cells even though it does not code for any protein (see **Becker** *et al.*, 2006).
3. Prions are infectious proteinaceous particles.

Multiple Choice Questions

1. (*d*) **2.** (*a*) **3.** (*a*) **4.** (*c*) **5.** (*c*) **6.** (*a*) **7.** (*c*)

58 Archaea (Archaebacteria)

The Archaea or Archaebacteria (Gr., *archaio* = ancient) seem to have diverged very early from the bacteria and the more closely related to eukaryotes than to bacteria. This conclusion comes largely from comparisons of genes that code ribosomal RNAs.

Archaea were initially discovered as inhabitants that we humans avoid, such as bogs, sewage farms, ocean depths, salt brines and hot acid springs, although it is now know that they are also widespread in less extreme and more homely environments, from soils and lakes to the stomach of cattle.

Although they are diverse group, all Archaebacteria share certain key characteristics (Table 58.1). Their cell walls lack peptidoglycan (an important component of the walls of bacteria), the lipids in cell membranes of archaebacteria have a different structure from those in all other organisms; and archaebacteria have distinctive ribosomal RNA (rRNA) sequences. Some of their genes possess introns, unlike those of bacteria.

Table 58.1. Comparison of three domains of life (Source: Raven *et al.*, 2005).

	Feature	Archaea	Bacteria	Eukarya
1.	Amino acid that initiate protein synthesis	Methionine	Formylmethionine	Methionine
2.	Introns	Present in some genes	Absent	Present
3.	Membrane bounded organelles	Absent	Absent	Present
4.	Membrane lipid structure	Branched	Unbranched	Unbranched
5.	Nuclear envelope	Absent	Absent	Present
6.	Number of different RNA polymerases	Several	One	Several
7.	Peptidoglycan in cell wall	Absent	Present	Absent
8.	Response to antibiotic streptomycin and chloramphenicol	Growth not inhibited	Growth inhibited	Growth not inhibited

58.1. CLASSIFICATION

The Archaebacteria are grouped into there general categories—1. methanogens; 2. extremophiles and 3. nonexterme archaebacteria—based primarily on the environments in which they live or their specialized metabolic pathways.

1. Methanogens. They obtain their energy by using hydrogen gas (H_2) to reduce carbon dioxide (CO_2) to methane gas (CH_4). They are strict anaerobes, poisoned by even traces of oxygen. They live in swamps, marshes and the intestines of mammals. Methanogens release about 2 billion tons of methane gas into the atmosphere each year.

2. Extremophiles. They are able to grow under condition that seem extreme to us. There are several types of extermophiles:

***(i)* Thermophiles ("Heat lovers").** They live in very hot places, typically in temperatures ranging from 60°C to 80°C. Many thermophiles are *autotrophs*, and their metabolisms are based on sulphur. Some thermophilic archaebacteria form the basis of food webs around deep-sea thermal vents, where they must withstand extreme temperatures and pressures. Other such as *Sulpholobus* inhabit the hot sulphur spring of Yellowstone National Park (USA) at 70° to 75°C. The recently described *Pyrolobus fumarii* holds the current record for heat stability, with a 106°C temperature optimum and 113°C maximum. It is so heat tolerant that is not killed by a one hour treatment in an autoclave (121°C).

***(ii)* Halophiles ("salt lovers").** They live in very salty places, including the Great Salt Lake in Utah (USA), Mono Lake in California (USA) and the Dead Sea in Israel. Where as salinity of sea water is around 3%, these archaebacteria thrive in, and indeed require water with a salinity of 15 to 20%.

***(iii)* pH-tolerant Archaebacteria.** They grow in highly acidic (pH = 0.7) and very basic (pH = 11) environments.

***(iv)* Pressure-tolerant Archaebacteria.** They are isolated from ocean depths, require at least 300 atmosphere of pressure to survive; they can tolerate upto 800 atmosphere!

3. Nonextreme Archaebacteria. They grow in the same environments as bacteria do. As the genomes of archaebacteria have become better known, microbiologists have been able to identify **signature sequences** of DNA present only in archaebacteria. The newly discovered microorganisms *Nanoarchaeum equitens* was identified as an archaebacterium based on a signature sequence. This odd Icelandic microorganism may have the smallest known genome, only 500 base pairs (**Raven** *et al.*, 2005).

58.2. ORIGIN AND EVOLUTION OF ARCHAEA

As already mentioned, the tree of life has three primary Branches (Domains): Bacteria, Archaea and Eukaryotes.

The classification of living things has traditionally depended on comparisons of their outward appearances: we can see that a fish has eyes, jaws, backbone, brain, and so on, just as we do, and that a worm does not; that a rose bush is cousin to an apple tree, but less similar to a grass. We can readily interpret such close family resemblances in terms of evolution from common ancestors and we can find the remains of many of these ancestors preserved in the fossil record. In this way, it has been possible to begin to draw a family tree of living organisms.

Microbiologists have sought to classify prokaryotes in terms of their biochemistry and nutritional requirements. But this approach also has its pit falls. **Genome analysis** has transformed the problem, giving us a simpler, more direct, and more powerful way to determine evolutionary relationships. The complete DNA sequence of an organism defines the species with almost perfect precision and in exhaustive detail. Moreover, this specification, once we have determined it, is in a digital form—a string of letters—that can be fed directly into a computer and compared with corresponding information for any other living thing. Because DNA is subject to random changes that accumulate over long periods of time, the number of differences between the DNA sequences of two organisms can be used to provide a direct, objective, quantitative indication of the evolutionary distance between them.

This approach has shown that some of the organisms that were traditionally classed together as "bacteria" are as widely divergent in their evolutionary origins as is any prokaryote from any eukaryote. It now appears that prokaryotes comprise two distinct groups that diverged early in the history of life on Earth, either before the ancestors of the eukaryotes diverged as a separate groups or at about the same time. The two groups of prokaryotes are called the **bacteria** (or **eubacteria**) and the **archaea** (or **archaebacteria**). The living world therefore has three divisions or domains: bacteria, archaea and eukaryotes (Fig. 58.1).

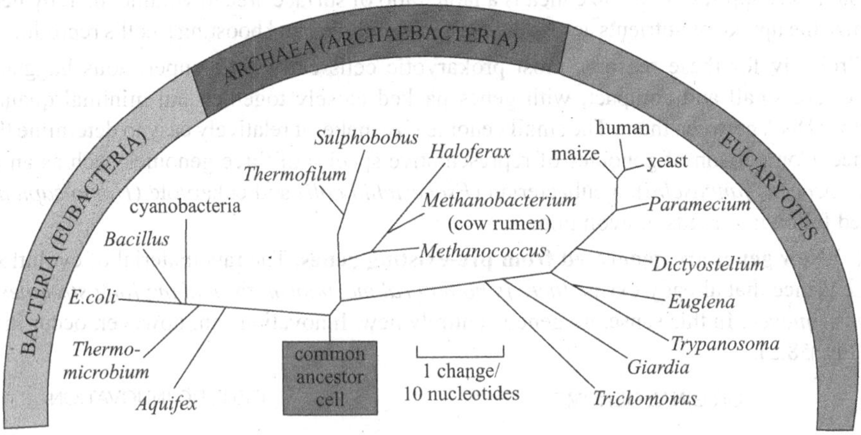

Fig. 58.1. The three major divisions (domains) of the living world. Note that traditionally the word *bacteria* has been used to refer to prokaryotes in general, but more recently has been redefined to refer to *eubacteria* specifically. The life tree is based on comparisons of the nucleotide sequence of a ribosomal RNA subunit (16S rRNA) in the different species. The length of the evolutionary changes that have occurred in this molecule in each lineage (after Alberts *et al.*, 2002).

In outward appearance archaea are not easily distinguished from the more familiar eubacteria. At a molecular level, archaea seem to resemble eukaryotes more closely in their machinery for handling genetic information (replication, transcription and translation), but eubacteria more closely to their apparatus for metabolism and energy conversion. Let us see how this might have occurred!

1. Role of mutation and natural selection. Both in the storage and in the copying of genetic information, random accidents and errors, altering the nucleotide sequence—that is creating **mutations**. Therefore, when a cell divides, its two daughters are often not quite identical to one another or to their parent. On rare occasions, the error may represent a change for the better; more probably, it will cause no significant difference in the cells prospects; and in many cases, the error will cause serious damage—for example, by disrupting the coding sequence for a key protein. Changes due to mistakes of the second type—*selective neutral changes*—may be perpetuated or not: in the competition for limited resources, it is a matter of chance whether the altered cell or its cousins will succeed. But changes that cause serious damage lead nowhere: the cell that suffers, then dies, leaving no progeny. Through endless repetition of this cycle of error and trial—of **mutation** and **natural selection**—organisms evolve: their genetic significations change, giving them new ways to exploit the environment more effectively, to survive in competition with others, and to reproduce successfully.

In the course of evolution, evidently some parts of the genome change more easily than others. A segment of DNA that does not code for protein and has no significant regulatory role is free to change at a rate limited only by the frequency of random errors. In contract, a gene that codes for a highly optimized essential proteins or RNA molecule cannot alter so easily: when mistakes occur, the faulty cells are almost always eliminated. Genes of this latter sort are therefore **highly conserved**. Through 3.5 billion years or more of evolutionary history, many features of the genome have changed beyond all recognition: but the most highly conserved gene remain perfectly recognizable in all living species.

(*i*) Most bacteria and archaea have 1000-4000 genes. Natural selection has generally favoured those prokaryotic cells that can reproduce the fastest by taking up raw materials from their environment; and also replicating themselves most efficiently, at the maximal rate permitted by the available food supplies. Small size means a large ratio of surface area to volume, thereby helping to maximize the uptake of nutrients across the plasma membrane and boosting a cell's reproductive rate.

Probably for these reasons, most prokaryotic cells carry little superfluous baggage; their genomes are small and compact, with genes packed closely together and minimal quantities of regulatory DNA between them. The small genome size makes it relatively easy to determine the DNA sequence. Comparison of genomes of representative species of three genomes such as an archaea (*Methanococcus jannaschii*), a eubacterian (*Escherichia coli*) and eukaryote (*Homo sapiens*) have revealed following trends of evolution:

(*a*) New genes are generated from pre-existing genes. The raw material of evolution is the DNA sequence that already exists: *there is no natural mechanism for making long stretches of new random sequence*. In this sense, no gene is entirely new. Innovation can, however, occur in several ways (Fig. 58.2).

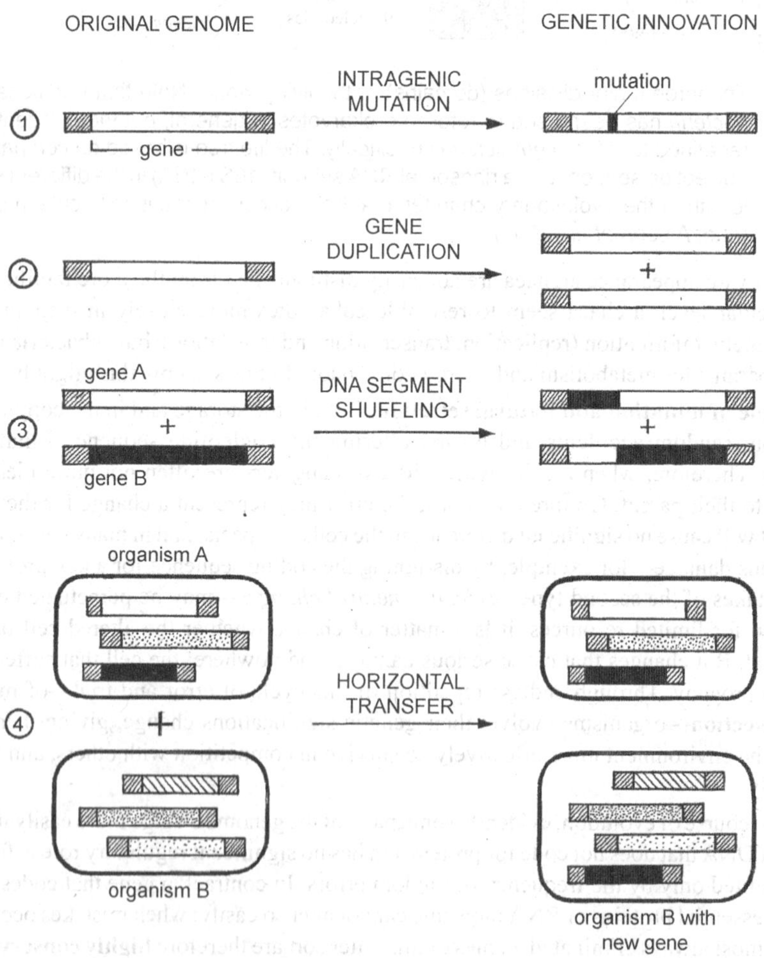

Fig. 58.2. Four Modes of genetic innovation and their effects on the DNA sequence of an organism.

I. Gene duplication. An existing gene can be duplicated so as to create a pair of closely related genes within a single cell.

II. Segment shuffling. Two or more existing genes can be broken and rejoined to make a hybrid gene consisting of DNA segments that originally belonged to separate genes.

III. Horizontal or intercellular transfer. A piece of DNA can be transferred from the genome of one cell to that of another—even to that of other species. This process is in contrast with the usual **vertical transfer** of genetic information from parent to progeny.

Each of these types of changes leaves a characteristic trace in the DNA sequnce of the organism, providing clear evidence that all four processes have occurred.

***(b)* Origin of gene families.** A cell must duplicate its entire genome each time it divides into two daughter cells. However, accidents occasionally result in the duplication of just part of the genome, with retention of original and duplicate segments in a single cell. Once a gene has been duplicated in this way, one of the two gene copies is free to mutate and become specialized to perform a different function within the same cell. Repeated rounds of this process of duplication and divergence, over many millions of years, have enabled one gene to give rise to a whole family of genes within a single genome. Analysis of the DNA sequence of prokaryotic genomes reveals many examples of such gene families: in *Bacillus subtilis*, for example, 47% of the genes have one or more obvious relatives (**Kunst** *et al.*, 1997; Fig. 58.3).

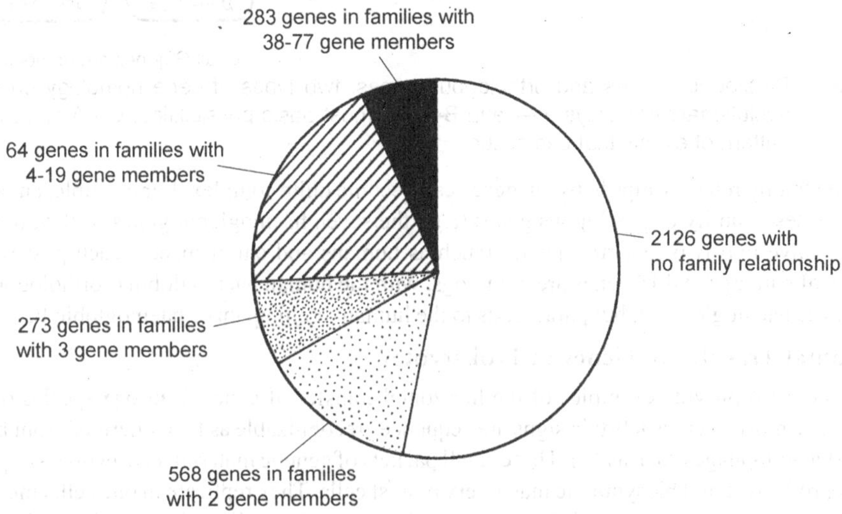

Fig. 58.3. Families of evolutionarily related genes in the genome of *Bacillus subtilis*. The biggest family consists of 77 genes coding for varieties of ABC transporters—a class of membrane transport proteins found in all three domains of the living world.

When genes duplicate and diverge in this way, the individuals of one species become endowed with mutiple variants of a primordial gene. This evolutinonary process has to be distinguished from the genetic divergence that occurs when one species of organism splits into two separate lines of descent at a branch point in the family tree—when the human line of descent became separate from that of chimpanzees, for example. There, the genes gradually become different in the course of evolution, but they are likely to continue to have corresponding functions in the two sister species. Genes that are related in this way—that is, genes in two separate species that derive from the same ancestral gene in the last common ancestor of those two species—are said to be **orthologs**. Related genes that have resulted from a gene duplication event within a single genome—and are likely to have diverged in their function—are said to be **paralogs**. Genes that are related in either way are called **homologs**, a general term used to cover both types of relationships (Fig. 58.4).

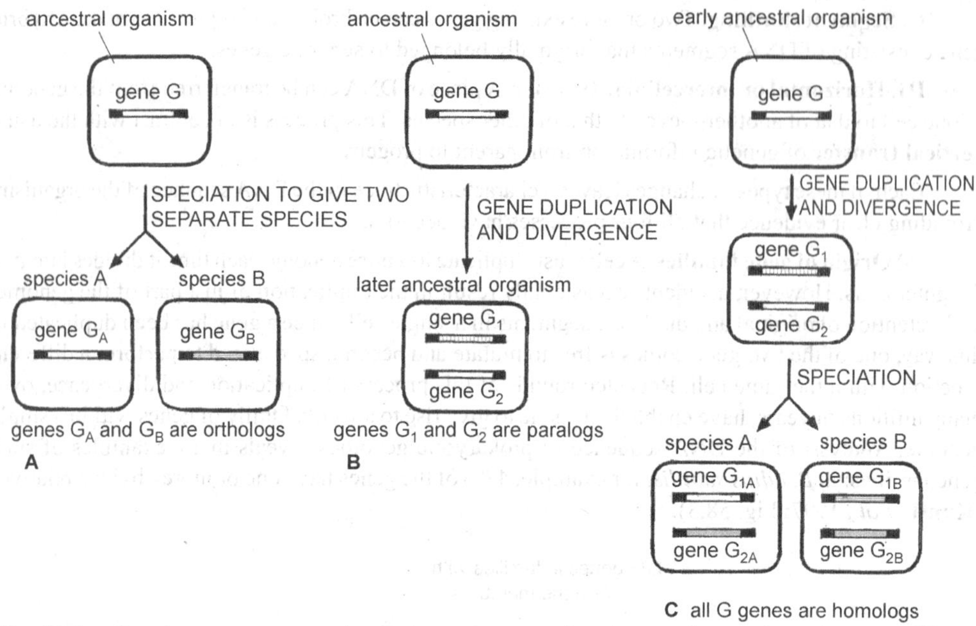

Fig. 58.4. Paralogous genes and orthologous genes: two types of gene homology on different evolutionary pathways. A— and B— the most basic possibilities. C—A more complex pattern of events that can occur.

The family relationships between genes can become quite complex. For example, an organism that possesses a family of paralogous genes (*e.g.*, the seven haemoglobin genes α, β, γ, δ, ε, ξ and θ) may evolve into two separate species (such as humans and chimpanzees) each possessing the entire set of paralogs. All 14 genes are homologs, with the human haemoglobin α orthologous to the chimpanzee haemoglobin α, but paralogous to the human or chimpanzee haemoglobin β, and so on.

Horizontal Transfer of Genes in Prokaryotes

Prokaryotes also provide examples of the horizontal transfer of genes from one species of cell to another. The most obvious tell-tale signs are sequences recognizable as being derived from bacterial viruses (bacteriophages such as T_4). These small packets of genetic material have evolved as parasites on the reproductive and biosynthetic machinery of host cells. They replicate in one cell, emerge from it with a protective wrapping, and then enter and infect another cell, which may be of the same or a different species. Inside a cell, they may either remain as separate fragments of DNA, known as **plasmids**, or insert themselves into the DNA of the host cell and become part of its regular genome. In their travels, viruses can accidentally pick up fragments of DNA from the genome of one host cell and transport them into another cell. Such transfers of genetic material frequently occur in prokaryotes and they are common between eukaryotic cells of the same species.

Horizontal transfers of genes between eukaryotic cells of different species are very rare, and they do not seem to have played a significant part in eukaryote evolution. In contrast, horizontal gene transfers occur much more frequently between different species of prokarotes. Many prokaryotes have a remarkable capacity to take up even nonviral DNA molecules from their surroundings and thereby capture the genetic information these molecules carry. This enables bacteria in the wild to acquire genes from neighboring cells relatively easily. Genes that confer resistance to antibiotics or an ability to produce a toxin, for example, can be transferred from species to species and provide the recipient bacterium with a selective advantage. In this way, new and sometimes dangerous strains of bacteria have been observed to evolve in the bacterial ecosystems that inhabit hospitals or the various niches in the human body. It has been estimated that at least 18% of all the genes in the present day genome of *E.coli* have been acquired by horizontal transfer from another species within the past 100 million years (**Albert** *et al.*, 2002).

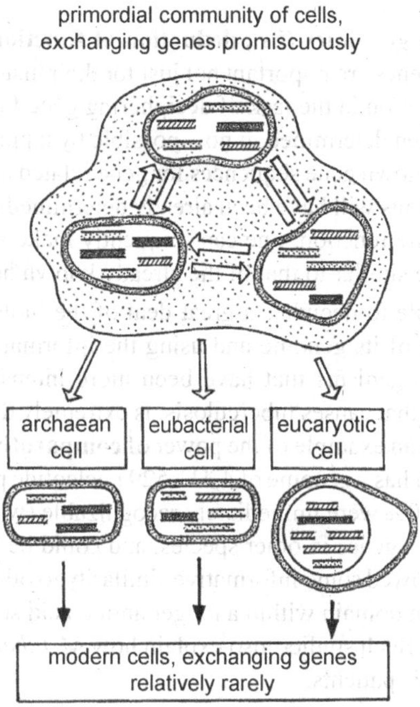

Fig. 58.5. In the early days of life on Earth, cells may have been less capable of maintaining their separate identities and may have exchanged genes much more readily than now. In this way, the archaen, eubacterial and eukaryotic lineages may have inherited different but overlapping subsets of genes from a primordial community of cells that were exchanging genes accidentally.

Horizontal gene transfer in early evolution. Horizontal exchanges of genetic information have an important role in bacterial evolution in todays world, and they may have occurred even more frequently and accidentally in the early days of life on Earth. Indeed, it has been suggested that the genomes of present day eubacteria, archaea and eukaryotes originated not by divergent lines of descent from a single genome in a single ancestral type of cell, but rather as three independent collections of genes that have survived from the pool of genes in a primordial community of diverse cells in which genes were frequently exchanged (Fig. 58.5). This could explain the otherwise puzzling observation that the eukaryotes seem more similar to archaea in their genes for the basic information handling processes of DNA replication, transcription and translation, but more similar to eubacteria in their genes for metabolic processes.

Horizontal gene transfer among bacteria may seem a surprising process, but it has a parallel in a phenomenon familiar to us all: sex. **Sexual reproduction** causes a large-scale horizontal transfer of genetic information between two intially separate cell lineages—those of the father and the mother. A key feature of sex, ofcourse, is that the genetic exchange normally occurs only between individuals of the same species. But no matter whether they occur within a species or between species, horizontal gene transfers leave a characteristic imprint: they result in individuals who are related more closely to one relatives with respect to some genes and more closely to another set of relatives with respect to others.

Sexual reproduction is a widespread (although not universal) phenomenon, especially among eukaryotes. Even bacteria indulge from time to time in controlled sexual exchange of DNA with other members of their own species. Natural selection has clearly favoured organisms that are capable of this behaviour.

Power of comparative genomics (*i.e.*, deduction of function of gene from its sequence). Family relationships among genes are important not just for their historical interest, but because they lead to a spectacular simplification in the task of deciphering gene functions. Once the sequnce of a newly discovered gene has been determined, it now possible by tapping a few key on a computer, to search the entire database of known gene sequences for gene related to it. In many cases, the function of one or more of these homologs will have been already determined experimentally, and thus, since gene sequence determines gene function, one can frequently make a good guess at the function of the new gene: it is likely to be similar to that of the already known homologs.

Thus it becomes possible to decipher a great deal of the biology of an organism simply by analyzing the DNA sequence of its genome and using the information we already have about the functions of genes in other organisms that have been more intensively studied. *Mycobacterium tuberculosis*, the eubacterium that causes tuberculosis, is extremely difficult to study experimentally in the laboratory and provides an example of the power of **comparative genomics**. DNA sequencing has revealed that this organism has a genome of 4,411,529 nucleotide pairs, containing approximately 4000 genes. Of these genes, 40% were immediately recognizable (when the genome was sequenced in 1998) as homolog of known genes in other species, and could be tentatively assigned a function on that basis. Another 44% showed some informative similarity to other known genes—for example, containing a conserved protein domain within a longer amino acid sequence. Only 16% of the 4000 genes were totally unfamiliar. Such studies also explain how *M. tuberculosis* escapes destruction by immune system of tuberculosis patients.

Lastly, given the complete genome sequences of representative organisms from all three domains—archaea, eubacteria and eukaryotes—one can search systematically for homologies that span enormous evolutionary divide. Out of 2264-protein-coding gene families recently defined by comparing the genomes of 18 bacteria, 6 archaeans and eukaryote (yeast), only 76 are truly ubiquitous (*i.e.*, represented in all the genomes analyzed. About 239 gene families are common to all three primary branches of the tree of life).

QUESTIONS

Long Answer Questions

1. Write an essay on Archaea.
2. Describe the classification of Archaebacteria.
3. Give an account of origin and evolution of Archaea.

Short Answer Questions

1. Compare Archaea, Bacteria and Eukarya.
2. Describe horizontal transfer of genes in prokaryotes.
3. Write short notes on:
   (*i*) Methanogens
   (*ii*) Halophile archaea
   (*iii*) Thermophile archaea
   (*iv*) Origin of gene families

Very Short Answer Questions

1. What is archaea?
2. Which microorganism is found to have signature sequence?
3. Which microorganism has smallest known genome?

Multiple Choice Questions

1. The most primitive of monerans are
   (*a*) Rickettsiae
   (*b*) Actinomycetes
   (*c*) Progenote
   (*d*) Archaebacteria
2. Branched chain lipids occur in the plasma membrane of

(*a*) Actinomycetes
(*b*) Mycoplasma
(*c*) Streptomycetes
(*d*) Archaebacteria

3. Which bacteria is utilised in gobar gas plant?
(*a*) methanogens
(*b*) nitrifying bacteria
(*c*) ammonifying bacteria
(*d*) denitrifying bacteria

4. The sulphur dependent archaebacteria are called as
(*a*) thermoacidophiles
(*b*) methanogens
(*c*) ammonifying bacteria
(*d*) nitrifying bacteria

5. Methanogens are strictly
(*a*) aerobic
(*b*) anaerobic
(*c*) parasitic
(*d*) none of these.

6. That group of archaebacteria which live in strong brine are called as
(*a*) methanogens
(*b*) thermoacidophiles
(*c*) halophiles
(*d*) aerobic

7. For production of methane, methanogens
(*a*) oxidise carbon dioxide
(*b*) reduce carbon dioxide
(*c*) reduce alcohol
(*d*) oxidise alcohol

ANSWERS

Very Short Answer Questions

1. Archaea or archaebacteria are one of two main groups of prokaryotes, the other being eubacteria; many archaea thrive under harsh conditions, such as salty, acidic or hot environments, that would be fatal to most other organisms.
2. The archaebacterium *Nanoarchaeum equitens.*
3. The archaebacterium *Nanoarchaeum equitens* (only 500 base pairs).

Multiple Choice Questions

1. (*d*) **2.** (*d*) **3.** (*a*) **4.** (*a*) **5.** (*b*) **6.** (*c*) **7.** (*b*)

59 Bacteria

The bacteria (singular bacterium) are amongst the smallest organisms. They are most primitive, simple, unicellular, prokaryotic and microscopic organisms. All bacteria are structurally relatively homogeneous, but their biochemical activities and the ecological niches for which their metabolic specialization equip them, are extremely diverse.

Bacteria occur almost everywhere: in air, water, soil and inside other organisms. They are found in stagnant ponds and ditches, running streams and rivers, lakes, sea water, foods, petroleum oils from deeper regions, rubbish and manure heaps, sewage, decaying organic matter of all types, on the body surface, in body cavities and in the internal tracts of humans and animals. Bacteria thrive well in warmth, but some can survive at very cold tops of high mountains such as Alps or even in almost boiling hot springs. They occur in vast numbers. A teaspoonful of soil may contain several hundred million bacteria. They lead either an autotrophic (photoautotrophic or chaemoautotrophic), or heterotrophic (saprotrophic or parasitic) mode of existence. The saprophytic or saprotrophic species of bacteria are of great economic significance for humans. Some parasitic species of bacteria are pathogenic (disease producing) to plants, animals and humans.

Bacteria have a high ratio of surface area of volume because of their small size. They show high metabolic rate because they absorb their nutrients directly through cell membranes. They multiply at a rapid rate. In consequence, due to their high metabolic rate and fast rate of multiplication, bacteria produce marked changes in the environment in a short period.

59.1. CLASSIFICATION

The bacteria domain has been classified into following major clades according to *Bergey's Manual of Systematic Bacteriology* (*2002*). (**Note. Clade** (Gr., *Klados* = branch). A group of organisms evolved from a common ancestor).

Bacteria

I. Thermophiles

1. Aquificae, 2. Thermotogae, 3. Chloroflexi and 4. Deinococcus.

Aquificae, *e.g.*, *Aquifex pyrophilus* is a rod shaped hyperthermophile with temperature optimum at 81°C, a chaemoautotroph, it oxidizes hydrogen or sulphur.

II. Gram Negative Bacteria

1. Low G/C: (*i*) Bacilli and (*ii*) Clostridium.

Bacilli. Gram positive bacteria. Largely solitary; many form spores. Responsible for many significant human diseases, *e.g.*, *Bacillus anthracis* (anthrax), *Clostridium botulinum* (botulism), *Staphylococcus*, *Streptococcus*.

2. High G/C

Actinobacteria. Gram-positive bacteria, from branching filaments and produce spores: often mistaken for fungi. Produce many commonly used antibiotics, including streptomycin and tetracycline, *e.g.*, *Streptomyces*, *Actinomyces*.

III. Spirochaetes

Long, coil-shaped cells, common in aquatic environments; a parasitic form is responsible for the disease syphilis. Rotation of internal filaments produce a cork screw movement, *e.g.*, *Treponema pallidum* (syphilis), *Borrelia burgdorferi* (Lyme disease).

IV. Photosynthetic Bacteria

1. Cyanobacteria; 2. Chlamydiae

Cyanobacteria. A form of photosynthetic bacteria. Common in both marine and freshwater environments. Deeply pigmented; often responsible for "blooms" in polluted water. Both colonial and solitary forms are common. Some filamentous cells have cells (called **heterocysts**) specialized in nitrogen fixation, *e.g.*, *Anabaena* and *Chlamydia*.

V. Protobacteria

1. Beta; 2.Gamma; 3. Alpha; 4. Epsilon; 5. Delta

1. Beta protobacteria. A nutritionally diverse group that includes soil bacteria such as *Nitrosomonas* that recycle nitrogen within ecosystem by oxidizing the ammonium (NH_4^+).

2. Gamma protobacteria. They include photosynthetic sulphur bacteria, pathogens, such as *Legionella* (pneumonia-like influenza) and the enteric bacteria that inhabit animal intestines. Enteric bacteria include *Escherichia coli*; *Salmonella* (food poisoning) and *Vibrio cholerae* (Cholera). *Pseudomonas* are a common form of soil bacteria, responsible for many plant diseases.

3. Alpha protobacteria, *e.g.*, *Rickettsia*.

4. Epsilon protobacteria, *e.g.*, *Helicobacter*.

5. Delta protobacteria. The cell of myxobacteria exhibit gliding motility by secreting slimy polysaccharides over which masses of cells glide; when the soil dries out, cells aggregate to form upright multicellular colonies called fruiting bodies, carrying spores. Other delta bacteria are solitary predators that attack other bacteria, *e.g.*, *Chondromyces*, *Bdellovibrio* (see **Raven** *et al.*, 2005).

59.2. MORPHOLOGY

1. Size of bacteria. Typically bacteria range between 1 μm (one micrometre) to 3 μm so they are barely visible under the light microscope. The smallest bacterium is *Dialister pneumosintes* (0.15 to 0.3 μm in length). The largest bacterium is *Spirillum volutans* (13 to 15 μm in length).

2. Forms of bacteria. Bacteria vary in their shapes. Based on their shape, bacteria are classified into the following groups:

(*i*) Cocci (singular coccus). These bacteria are spherical or round in shape. These bacterial cells may occur singly (**micrococci**); in pairs (**diplococci**, *e.g.*, pneumonia causing bacterium, *Diplococcus pneumoniae*); in groups of four (**tetracocci**); in a cubical arrangement of eight or more (**sarcinae**); in irregular clumps resembling bunches of grapes (**staphylococci**, *e.g.*, boil causing bacterium, *Staphylococcus aureus*) or in a bead-like chain (**streptococci**, *e.g.*, sore throat causing bacterium, *Streptococcus pyogenes*; Fig. 59.1).

(*ii*) Bacilli (singular, bacillus). These are rod-like bacteria. They generally occur singly, but may occasionally be found in pairs (**diplobacilli**) or chains (**streptobacilli**). Bacilli cause certain most notorious diseases of humans such as tuberculosis (*Mycobacterium* or *Bacillus tuberculosis*), tetanus (*Clostridium tetani*), typhoid (*Salmonella* or *Bacillus typhosus*), diphtheria (*Corynebacterium diphtheriae*), leprosy (*Mycobacterium leprae*), dysentery and food poisoning (*Clostridium botulinum*).

Certain well known diseases of the animals are also caused by bacilli, *e.g.*, anthrax (*Bacillus anthracis*) and black leg (*Clostridium chauvei*).

(*iii*) Spirilla (singular, spirillum). These are also called **spirochaetes**. These are spiral-shaped and motile bacteria (Fig. 59.1). Spirilla cause human disease such as syphilis (*Treponema pallidum*).

(*iv*) Vibrios (singular vibrio). These are comma-shaped or bent-rod like bacteria (Fig. 59.1). Vibrios cause human disease such as cholera (*Vibrio cholerae*).

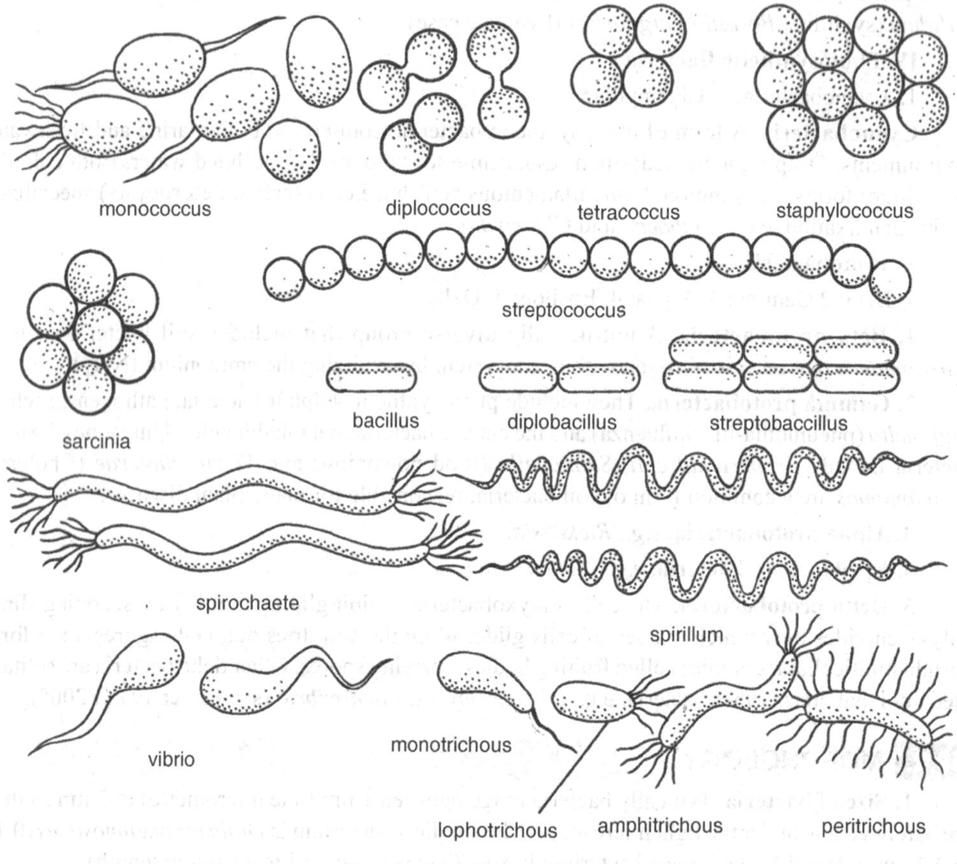

Fig. 59.1. Different forms of bacteria.

3. Gram negative and Gram positive bacteria. On the basis of structure of cell wall and its stainability with the Gram stain, the following two types of bacteria have been recognized: Gram positive and Gram negative bacteria. The Gram staining method is named after **Christian Gram** who developed it in Denmark in 1884. In this technique, when heat-fixed bacteria are treated with the basic dye, crystal violet they become blue or purple. Such blue stained cells are treated with a mordant (*i.e.*, agent that fixes stains to tissues) such as iodine (*i.e.*, potassium iodide or KI solution) and ultimately washed with some organic solvent such as alcohol. Some bacteria retain the blue colour, while others lose it and stay colourless. The former are **Gram positive bacteria** (*e.g.*, *Bacillus subtilis*, *Staphylococcus*, etc.) and the latter are **Gram negative bacteria** (*e.g.*, *Escherichia coli*, *Simonsiella*, cyanobacteria, etc.). Colourless Gram negative bacteria may thereafter be stained pink with safranin stain for their better microscopic visibility.

The long search for the chemical basis of this differentiating staining reaction ended in 1950's when it was detected that cell wall of Gram negative bacteria has high lipid content which tends to be dissolved away by alcohol. The alcohol then can enter the cell and leach out the stain, whereas the cell walls of Gram positive bacteria form a barrier (*i.e.*, peptidoglycan layers) that prevent the penetration of the solvent inside the cell.

4. Structure of bacteria. Structural details of a bacterial cell can only be seen with an electron microscope in very thin sections. A typical bacterial cell has the following components:

A. Outer covering: The outer covering of bacterial cell comprises the following three layers:

I. Plasma membrane. The bacterial protoplast is bound by a living, ultrathin (6 to 8 nm thick) and dynamic plasma membrane. The plasma membrane chemically comprises molecules of lipids and proteins which are arranged in a **fluid mosaic pattern**. That is, it is composed of a bilayer sheet of **phospholipid** molecules with their polar heads on the surfaces and their fatty-acyl chains (tails) forming the interior. The **protein** molecules are embedded within this lipid bilayer, some spanning it, some exist on its inner side and some are located on its external or outer side. These membrane proteins serve many important functions of the cell. For example, the transmembrane proteins act as **carriers** or **permeases** to carry on selective transportation of nutrients (molecules and ions) from the environment to the cell or vice versa. Certain proteins of the membrane are involved in oxidative metabolism, *i.e.*, they act as enzymes and carriers for electron flow in respiration and photosynthesis leading to phosphorylation (*i.e.*, conversion of ADP to ATP). The bacterial plasma membrane also provides a specific site at which the single circular chromosome (DNA) remains attached. It is the point from where DNA replication starts. The first stage in nuclear division involves duplication of this attachment, followed by a progressive bidirectional replication of DNA by two replication forks.

Plasma membrane intrusions. Infoldings of the plasma membrane of all Gram-positive bacteria and some Gram-negative bacteria give rise to the following two main types of structures:

1. Mesosomes (or chondrioids). They are extensions of the plasma membrane within the bacterial cell (*i.e.*, cytoplasm) involving complex whorls of convoluted membranes (Fig. 59.2). Mesosomes tend to increase the plasma membrane's surface and in turn also increase their enzymatic contents. They are seen in chemoautotrophic bacteria with high rates of aerobic respiration such as *Nitrosomonas*, and in photosynthetic bacteria such as *Rhodopseudomonas* where they are the site of photosynthetic pigments. Mesosomes are involved in crosswall (septum) formation during the division of cell.

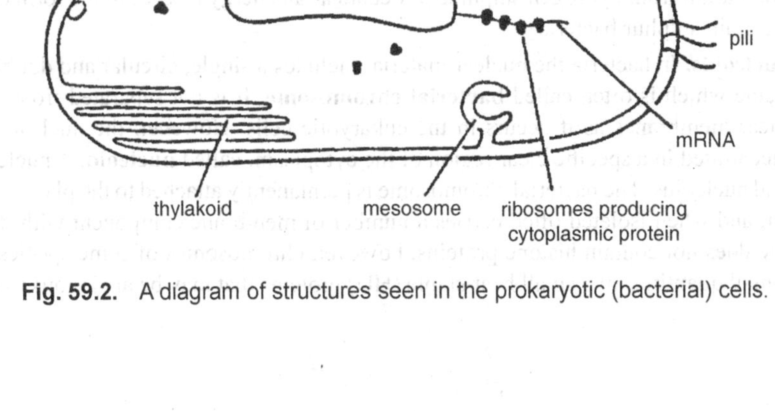

Fig. 59.2. A diagram of structures seen in the prokaryotic (bacterial) cells.

2. Chromatophores. These are photosynthetic pigment-bearing membranous structures of photosynthetic bacteria. Chromatophores vary in form as vesicles, tubes,. bundled tubes, stacks, or thylakoids (as in cyanobacteria).

II. Cell wall. The plasma membrane is covered with a strong and rigid cell wall that renders mechanical protection and provides the bacteria their characteristic shapes (the cell wall is absent in *Mycoplasma*). The cell wall of bacteria differs chemically from the cell wall of plants in that it contains proteins, lipids and polysaccharides. It may also contain chitin but rarely any cellulose.

Electron microscopy has revealed the fact that the cell wall of Gram- negative bacteria comprises the following two layers: 1. Gel, proteoglycan or peptidoglycan (*e.g.*, murein or muramic acid) containing **periplasmatic space** around the plasma membrane and 2. The outer **membrane** which consists of a lipid bilayer traversed by channels of **porin** polypeptide. These channels allow diffusion of solutes. The lipids of lipid bilayer are phospholipids and lipopolysaccharides (LPS). LPS have antigenic property and anchor the proteins and polysaccharides of the surrounding capsule. The cell wall of Gram positive bacteria is thicker, amorphous homogeneous and single layered. Chemically it contains many layers of peptidoglycans and proteins, neutral polysaccharides and polyphosphate polymers such as teichoic acids and teichuronic acids.

III. Capsule. In some bacteria, the cell wall is surrounded by an additional slime or gel layer called capsule. It is thick, gummy mucilaginous and is secreted by the plasma membrane. The capsule serves mainly as a protective layer against attack by phagocytes and by viruses. It also helps in regulating the concentration, and uptake of essential ions and water.

B. Cytoplasm. The plasma membrane encloses a space consisting of **hyaloplam**, **matrix** or **cytosol** which is the ground substance and the seat of all metabolic activities. The cytosol consists of water, proteins (including multifunctional enzymes), lipids, carbohydrates, different types of RNA molecules, and various smaller molecules. The cytosol of bacteria is often differentiated into two distinct areas: a less electron dense nuclear area and a very dense area (or dark region). In the dense cytoplasm occur thousands of particles, about 25 nm in diameter, called **ribosomes**. Ribosomes are composed of ribonucleic acid (RNA) and proteins and they are the sites of protein synthesis. Ribosomes of bacteria are 70S type and consist of two subunits (*i.e.*, a larger 50S ribosomal subunit and a small 30S ribosomal subunit). Non-functional ribosomes exist in the form of separated subunits which are suspended freely in the cytoplasm. During protein synthesis many ribosomes read the codes of single mRNA (messenger RNA) molecules and form **polyribosomes** or **polysomes**.

Reserve materials of bacteria are stored in the cytoplasm either as finely dispersed or distinct granules called **inclusion bodies** or **storage granules**. There are three types of reserve materials. First, there are organic polymers which either serve as reserves of carbon, as does **poly-β-hydroxybutyric acid**, or as stores of energy, as does a polymer of glucose, called **granulose** (*i.e.*, glycogen). Second, many bacteria contain large reserves of inorganic phosphate as highly refractile granules of metaphosphate polymers known as **volutin**. The third type of reserve material is elemental **sulphur**, formed by oxidation from hydrogen sulphide. It occurs as an energy reserve in the form of spherical droplets in certain sulphur bacteria.

C. Nucleoids. In bacteria the nuclear material includes a single, circular and double stranded DNA molecule which is often called **bacterial chromosome**. It is not separated from the cytosol by the nuclear membranes as it occurs in the eukaryotic cells. However, the nuclear material is usually concentrated in a specific clear region of the cytoplasm, called **nucleoid**. A nucleoid has no ribosome and nucleolus. The bacterial chromosome is permanently attached to the plasma membrane at one point, and when isolated often carries a number of membrane component with it. Bacterial chromosome does not contain histone proteins, however, chromosomes of some species are found to contain small quantities of a small heat-stable (HU) proteins that may be analogous to eukaryotic histones.

All three classes of RNA (*i.e.*, mRNA, tRNA, and rRNA) are formed (transcribed) by the activity of the single RNA polymerase (RNAP) species in prokaryotes. The messenger RNA formed at the chromosome is directly available for translation without processing, and so ribosomes may attach to the beginning of the mRNA strand and begin translation, while the other end of the mRNA is still being formed by transcription from DNA. Proteins for use within the cell are synthesized at cytoplasmic ribosomes; but ribosomes responsible for the synthesis of membrane proteins or proteins destined for export from the cell to form either the cell wall or secretory products, are attached to the plasma membrane. The resulting exportable polypeptides are ejected directly into or through the membrane as they are formed.

Plasmids. Many species of bacteria may also carry extra chromosomal genetic elements in the form of small, circular and closed DNA molecules, called **plasmids**. Some plasmids are merely **bacteriophage (viral) DNA** which may alternatively be incorporated within the chromosome. Other plasmids may be separated parts of the normal genome from the same or a foreign cell, and may recombine with the main chromosome. One function of some of these plasmids (called **colcinogenic factors**) is the production of antibiotically active proteins or **colicins** which inhibit the growth of other strains of bacteria in their vicinity. Some plasmids may act as **sex** or **fertility factors (F factors)** which stimulate bacterial conjugation. **R factors** are also plasmids which carry genes for the resistance to one or more drugs (antibiotics) such as chloramphenicol, neomycin, penicillin, streptomycin, sulphonamides and tetracyclines.

D. Flagella and other structures. Many bacteria (*e.g.*, *E.coli*) are motile and contain one or more **flagella** for the cellular locomotion (swimming). Bacterial flagella are smaller than the eukaryotic flagella (*i.e.*, they are 15 to 20 nm in diameter and up to 20 μm long) and are also simpler in organization. A bacterial flagellum consists of a helical tube containing a single type of protein subunit, called **flagellin**. The flagellum is attached at its base, by a short flexible hook that is rotated, like a propeller of ship, by the flagellar rotatory **motor** *i.e.*, basal body; Fig. 59.3). The flagellar motor comprises four distinct parts: rotor (M ring), stator, bearing (S ring) and rod. The '**rotor**' is a protein disc integrated into the plasma membrane. It is driven by energy stored in the transmembrane proton H^+ gradient (not by ATP breakdown; see **Jones**, 1986) and rotates rapidly (~100 revolutions/second) in the-lipid bilayer against another protein disc, called the '**stator**'. A **rod** links the 'rotor' to a hook and flagellum, thereby causing them to rotate. The protein "**bearing**" serves to seal the outer membrane of the cell wall as the rotating rod passes through it. The 'stator' and 'bearing' remains stationary.

According to the number and arrangement of the flagella in a bacterial cell, following four types of flagellation patterns have been recognized: **(1) Monotrichous.** There is a single flagellum at one pole of the cell. **(2) Lophotrichous.** There are several flagella at one pole. **(3) Amphitrichous.** The cell bears at least one flagellum at each pole. **(4) Peritrichous.** There are flagella all over the surface of cell (Fig. 59.1). Flagella-like **axial filaments** are the characteristics of some spirochetes which move like snakes through the environment. The axial filaments do not project away from the cell but are wrapped around the cell surface.

Fimbriae or pili. Some bacteria (mostly Gram negative bacilli) contain non-flagellar, extremely fine, appendages called **fimbriae** or **pili** (singular **pilus**). Pili are non- motile but adhesive structures. They enable the bacteria to stick firmly to other bacteria, to a surface or to some eukaryote such as mold, plant and animal cells including red blood cells and epithelial cells of alimentary, respiratory and urinary tracts. Pili help in conjugation (*e.g.*, long F-pili or **sex pili** of male bacteria); in the attachment of pathogenic bacteria to their host cells (*e.g.*, attachment of gonorrhea-causing coccus, *Neisseria gonorrhoeae*, to the epithelial cells of the human urinary tract) and in acting as specific sites of attachment for the bacteriophages. Pili are known to be coded by the genes of the plasmid.

Spinae. Some Gram positive bacteria have tubular, pericellular and rigid appendages of single protein moiety, called **spinin**. They are called **spinae** and are known to help the bacterial cells to tolerate some environmental conditions such as salinity, pH, temperature, etc.

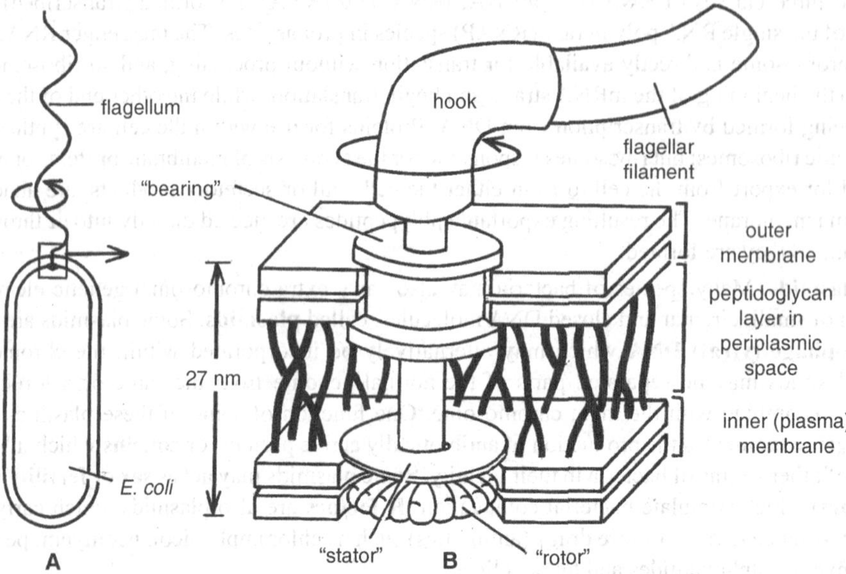

Fig. 59.3. Schematic drawing of the flagellar rotatory 'motor' of *E.coli.*

5. Nutrition in bacteria. Bacteria show wide diversity in their nutrition. Some are chemosynthetic, some are photosynthetic, but most of them are heterotrophic. Heterotrophic bacteria are mostly either saprophytic or parasitic; Parasitic bacteria live on the body of plants and animals and with few exceptions, most bacteria are pathogenic.

Modes of respiration of bacteria are both aerobic and anaerobic. Some of the end products of bacterial anaerobic respiration are useful to humans, so, they are used in the manufacture of various foods such as butter, cheese and vinegar. *Pseudomonas* is a gram negative heterotrophic aerobic form which can decompose (biodegrade) a wide variety of organic compounds such as hydrocarbons. So it is used in reducing water pollution due to petroleum spillage.

59.3. REPRODUCTION

A. Asexual reproduction. Bacteria reproduce asexually by **binary fission** and **endospore** formation and sexually by **conjugation.** In the binary fission, the cell divides into two genetically identical daughter cells. During this process, the single circular chromasome first makes a copy of itself (*i.e.*, it duplicates) and daughter chromosomes become attached to the plasma membrane. They separate as the bacterial cell enlarges and ultimately the formation of a cross wall between the separating daughter chromosames, divides the parent cell into two daughter cells.

Under unfavourable ecalogical conditions, many bacteria (*e.g.*, *Clostridium*, *Bacillus*, etc.) form spores which are not reproductive units but represent an inactive state. In endospore formation, a part of the protoplasmic material is used to form an impermeable coat or cyst wall around the chromosome along with some cytoplasm. The rest af the cell degenerates. The spore being metabolically inert can survive an unsuitable temperature, pH and drought. Under favourable conditions spores imbibe water, become metabolically active again and germinate.

B. Sexual reproduction. Under this heading have been included processes in which **genetic recombination** occurs. Genetic recombination includes following three methods: 1. Conjugation; 2. Transformation and 3. Transduction (Fig. 59.4).

1. Conjugation. In 1946 **Lederberg**, a student of **Tatum**, established in bacteria a process called **sex**. In conjugation there is a transfer of genetic material between two conjugants belonging to sexually different strains.

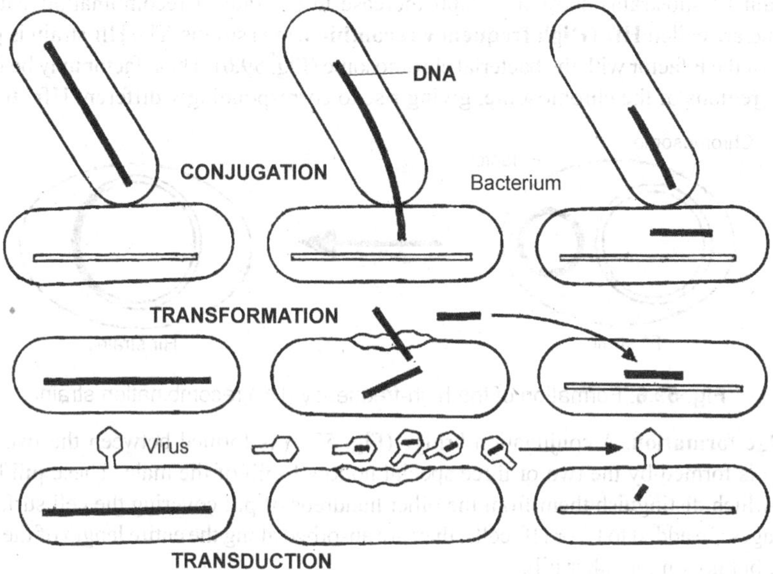

Fig. 59.4. Genetic recombination in bacteria.

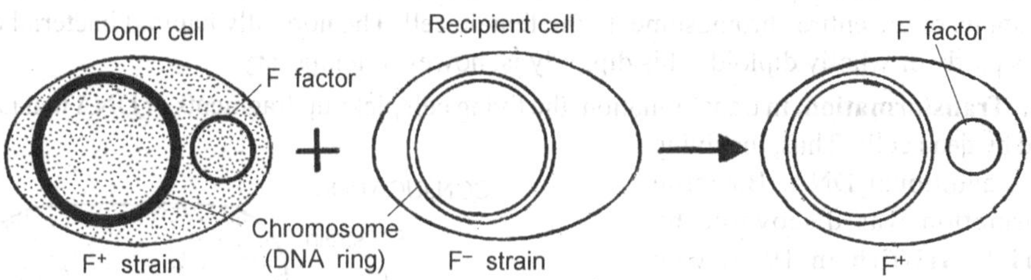

Fig. 59.5. Transfer of F factor from F^+ strain to F^- strain.

The first step of conjugation involves the mutual recognition of "male" and "female" cells. This is achieved by complementary surface molecules, the male component of which is formed by the **F** [= **fertility factor** or **sex factor** (plasmid)]. Depending upon the presence or absence of the F factor two mating types have been found in the K12 strain of the bacterium, *Escherichia coli*. Cells in which the F factor is present are called F^+ (*i.e.*, **male** or **donor cells**). Cells in which the F factor is absent are designated as F^- (*i.e.*, **female** or **recipient cells**). The F factor is cytoplasmic factor (DNA) which is usually separate from the chromosome and it multiplies independently.

The following results are obtained by crossing or growing the two strains together:

(*i*) $F^- \times F^-$ crosses are always sterile

(*ii*) $F^+ \times F^-$ crosses are fertile, but conjugation is a one-in-a-million rarity

(*iii*) $F^+ \times F^+$ crosses are fertile, but at very low level

(*iv*) If F^+ and F^- strains are grown together, upto 70% F^- cells are converted into F^+ by cell contact (Fig. 59.5). The F^+ cells transfer the F factors to the F^- cells, without much chromosomal transfer. The F^+ cells remain F^+.

(*v*) Prolonged incubation with acridine dyes which are mutagens converting F^+ strains into F^-.

High-Frequency Recombination (Hfr recombination)

Certain F^+ substrains show a × 1000 increase in the rate of recombination with F^- strains ($F^+ \times F^-$) and are called **Hfr (High frequency recombination)** strains. The Hfr strain is produced by integration of the F factor with the bacterial chromosome (Fig. 59.6). The F factor may be incorporated in different regions of the chromosome, giving rise to correspondingly different Hfr strains.

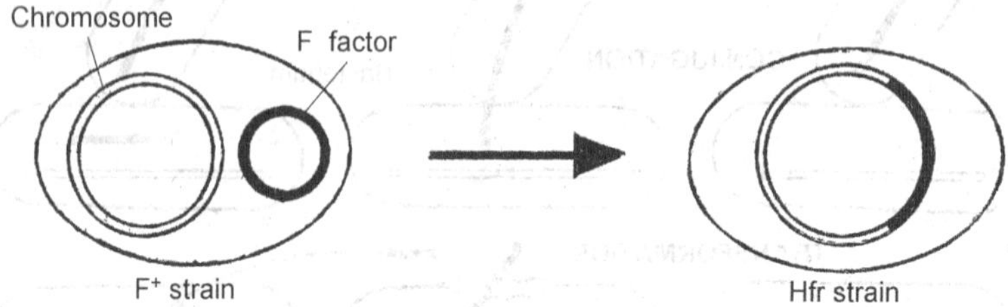

Fig. 59.6. Formation of the high-frequency (Hfr) recombination strain.

Bridge formation. A conjugation bridge (Fig. 59.7) is formed between the two conjugants. This bridge is formed by the two or three special hollow F-pili of the male. These pili have special properties which distinguish them from the other hundreds of pili covering the cell surface. If MS-2 bacteriophages are added to F^+ or Hfr cells, they are absorbed along the entire length of the conjugation bridge pili, but not on the other pili.

DNA transfer takes place slowly from the male cell to the female cell through the groove (passage) in the F-pilus. This transfer is energy-dependent. The male cell transfers a part of its chromosome, or the entire chromosome, to the female cell. The normally haploid bacterial cell now becomes partly or wholly diploid. This diploidy is, however, temporary.

2. Transformation. In transformation, the living cell picks up fragments of DNA that have been released by dead cells. Thus, the living cell gets additional DNA. Bacterial transformation was discovered by **Frederick Griffith** in 1928, who used two strains of the bacterium *Diplococcus pneumoniae* (Fig. 59.8). In the wild type strain (S III; called **smooth**) the cells have a carbohydrate capsule and are **virulent** (pathogenic). If injected into mice, the smooth strain causes severe septicemia and death. A mutant strain (R II; called **rough**) is without capsule and avirulent (non-pathogenic). Heat killed smooth cells are avirulent. Griffith found that when the non-capsulated variety (S III) was injected a mouse along with heat-killed encapsulated (rough) variety (R II) a small fraction of the former acquired the ability to produce a capsule and become virulent (*i.e.*, R II) (Fig. 59.9). Griffith thought that the transforming agent was a protein.

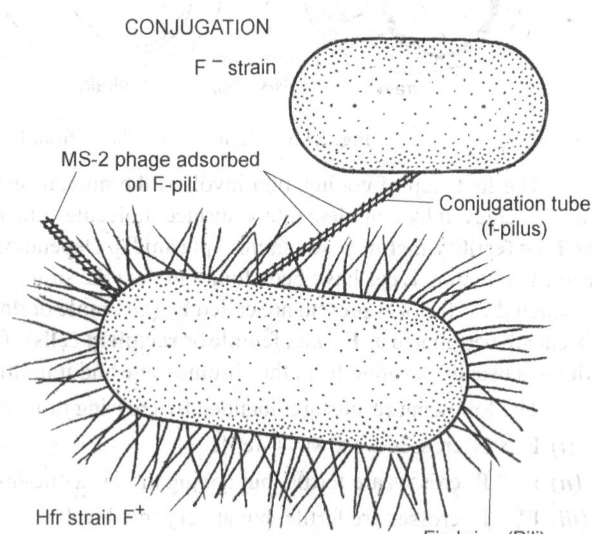

Fig. 59.7. Conjugation tube (or bridge) formation between two conjugants during bacterial conjugation.

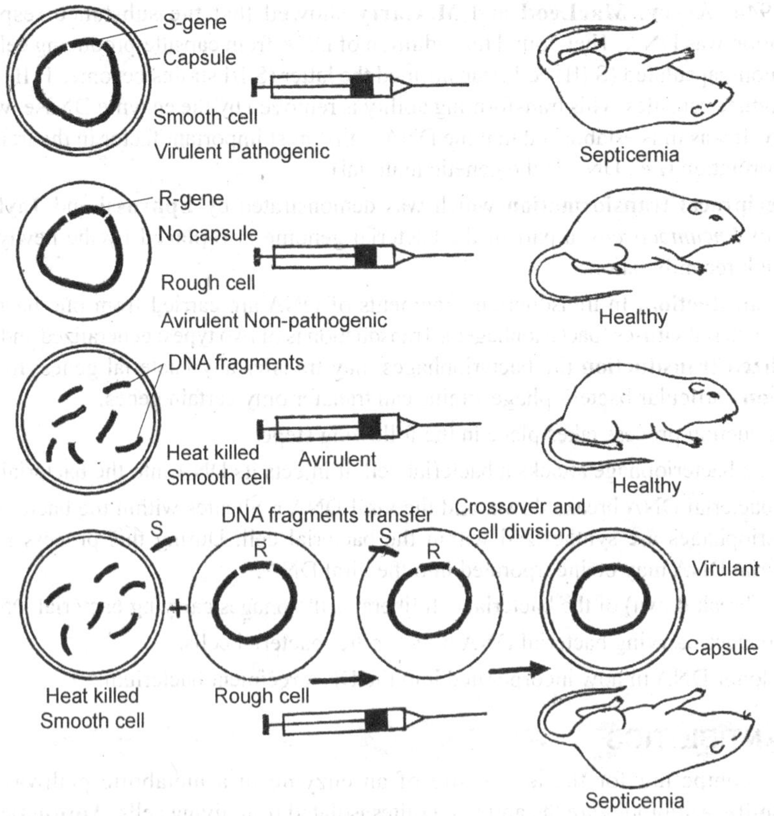

Fig. 59.8. Diagram illustrating Griffith's experiment on bacterial transformation.

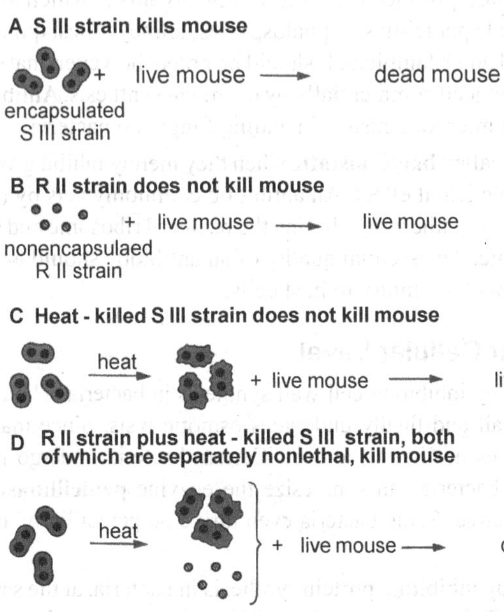

Fig. 59.9. Explanation of Griffith experiments showing conversion of a non-lethal bacterial strain to a lethal form by a cell extract (after Freifelder, 1985).

In 1944, **Avery, MacLeod** and **McCarty** showed that the substance responsible for transformation was DNA. They found that addition of DNA from capsule-producing cells (R II) to a culture of non-capsulated (S III) cells transformed the latter (S III strains becomes R II) and enabled them to produce capsules. This transforming ability is removed by the enzyme DNase, which breaks down DNA. It was thus established that the DNA is the most important factor in the transmission of genetic information (*i.e.*, DNA is the genetic material).

In **reciprocal transformation** which was demonstrated by **Ephrussi** and **Taylor** between 25 strains of *Pneumococcus*, a part of the bacterial genome is **replaced** by the newly introduced DNA through recombination.

3. Transduction. In transduction, fragments of DNA are carried from one bacterial cell to another by bacterial viruses (bacteriophages). Transduction is of two types: generalized and specialized. In **generalized transduction** the bacteriophages may transfer any bacterial genes. In **specialized transduction** particular bacteriophage strains can transfer only certain genes.

Transduction process takes place in the following stepes:

1. When a bacteriophage attacks a bacterial cell, it injects its DNA into the bacterial cell.
2. The bacterial DNA breaks down, and the viral DNA replicates within the bacterium. Several bacteriophages are synthesized within the bacterial cell. During this process a part of the bacterial DNA may be incorporated into the viral DNA.
3. Lysis (break down) of the bacterial cell liberates the phages carrying bacterial DNA.
4. The phages carrying bacterial DNA infects other bacterial cells.
5. The donor DNA in now incorporated into the DNA recipient bacterium.

59.4. ANTIBIOTICS

Substances competing for the active site of an enzyme in a metabolic pathway are called **antimetabolite**. Antibiotics are the antimetabolites isolated from living cells. **Antibiotics** have been defined as substances, produced by microorganisms with antimicrobial activity in minute quantities. Thus, antibiotic is a substance produced by one living organism which is harmful to another. Examples of antibiotics include penicillins, cephalosporins, tetracyclines, quinolones, streptomycin, gentamicin, chloramphenicol and rifampicin. It should be noted, however, that some antibiotics, *e.g.*, chloramphenicol are now produced commercially by chemical synthesis. Antibiotics against bacteria are produced by a number of microorganisms, including fungi and bacteria.

Antibacterial drugs are called **bacteriostatic** when they merely inhibit growth, and **bactericidal** when they have an irreversible lethal effect. An antibiotic commonly acts by (*i*) inhibiting cell wall synthesis, (*ii*) inhibiting protein synthesis by altering the bacterial ribosome and (*iii*) inhibiting mRNA synthesis on the DNA template. An essential quality of an antibiotic should be the ability to destroy the parasitic microorganism without injury to host cells.

Action of Antibiotics at Cellular Level

The antibiotic **penicillin** acts by inhibiting cell wall synthesis in bacteria. The cell continues to grow without the protective cell wall and finally undergoes osmotic lysis. Since mammalian cells do not have cell walls, penicillin does not harm the host. Certain bacteria undergo mutation and become resistant to penicillin. Some bacteria can synthesize the enzyme **penicillinase** which breaks down penicillin and makes it ineffective. Some bacteria even thrive on penicillin by utilizing it as a source of nutritional carbon.

Certain antibiotics act by inhibiting protein synthesis in bacteria, at the same time not affecting protein synthesis in the host. This is because of the differences between the 70S bacterial ribosome and 80S eukaryotic ribosome. The antibiotic **streptomycin** act by inhibiting protein synthesis by altering the structure of the 30S ribosomal subunit. This results in the mis-reading of the mRNA code. Streptomycin is rapidly bactericidal. It is more effective against extracellular bacteria than intracellular bacteria, possibly because Mg^{2+} and $H_3PO_4^{2+}$ ions in the cell reduce streptomycin activity. The antibiotic, **tetracyclin** inhibits protein synthesis by interfering with binding of tRNA with the ribosomes. The antibiotic **chloramphenicol (chloromycetin)** inhibits protein synthesis in bacteria by interacting with the 50S ribosomal subunit. The tetracyclines and chloramphenicol are conventionally called **broad spectrum** antibiotics because of their wide ranging antimicrobial activity.

The antibiotic **puromycin** blocks protein synthesis by acting as an analog to terminal of phenyl tRNAs thus displacing aminoacyl tRNA.

The antibiotic **actinomycin D** acts against bacteria at the level of nucleic acid synthesis. It binds directly to double-stranded (but not single stranded) DNA and inhibits synthesis of mRNA on the DNA template. It has no effect on the host cells because DNA is bound to basic protein (histones) in eukaryotes. In very high concentrations actinomycin D blocks replication of DNA.

59.5. NITROGEN FIXATION

Nitrogen is the fourth most abundant element in living organisms. It constitutes less than 0.1% of the Earth's crust but makes up about 80% of the atmosphere, mostly as dinitrogen (N_2). **Nitrogen fixation** is a biological process that converts N_2 to NH_3 (ammonia). Nitrogen fixation is carried out only by prokaryotic species. The NH_3 produced by N fixation may be assimilated into amino acids and then to protein and other N compounds, or it may be converted by nitrifying bacteria to NO_2^- and NO_3.

Enzymatic nitrogen fixation is limited to prokaryotes. This trait is associated with eubacteria (Cyanobacteria, actinomycetes and α-protobacteria) and some methanogenic archaea. Some nitrogen fixing bacteria live as **endosymbionts** of plants. *Azospirillum* is an example of free living nitrogen fixing bacteria. In the presence of specific rhizobial species, legume roots form unique structure (the **root nodules**) in which nitrogen fixation occur. Rhizobial species are α-protobacteria and include following genera: 1. *Rhizobium*, 2. *Sinorhizobium*, 3. *Azorhizobium*, 4. *Mesorhizobium* and 5. *Bradyrhizobium* (**Buchanan** *et al.*, 2000).

Enzymology of Nitrogen Fixation

The biological reduction of dinitrogen (N_2) to ammonia is carried out by two enzymes: **dinitrogenase** and **dinitrogenase reductase**. Together these enzymes are often referred to as **nitrogenase** and its role has been schematically shown in Figure 59.10.

Nitrogen Fixation is Sensitive to Oxygen

Nitrogen fixation is a unique biochemical reaction that consumes energy in energy-rich compounds while requiring strong biological reduction. Because nitrogenase enzyme and some of proteins supply it with reductants are sensitive to oxygen, many nitrogen-fixing bacteria are **anaerobes**. Neither fermentation nor anaerobic respiration oxidizes reduced carbon compounds as efficiently as aerobic respiration, so anaerobic bacteria must process large quantities of substrate to generate the ATP required for dinitrogen fixation. In contrast, **aerobes** have the advantage of high ATP production from aerobic metabolism but must contend with oxygen-sensitivity of nitrogenase. In some cases, free-living nitrogen-fixing organisms use mechanical or biochemical barriers to keep oxygen away the biological catalysts of nitrogen fixation. In other cases, the nitrogen fixation machinery is segregated spatially in specialized structures. For example, some filamentous cyanobacteria generate **heterocysts**, thick-walled cells that fix nitrogen but cannot complete all the reactions of oxygenic photosynthesis. Heterocysts produce the ATP needed for nitrogen fixation by way of cyclic process that does create oxygen gas. Some nonfilamentous cyanobacteria segregate photosynthesis from nitrogen fixation temporally, performing oxygenic photosynthesis in the light and nitrogen fixation in the dark.

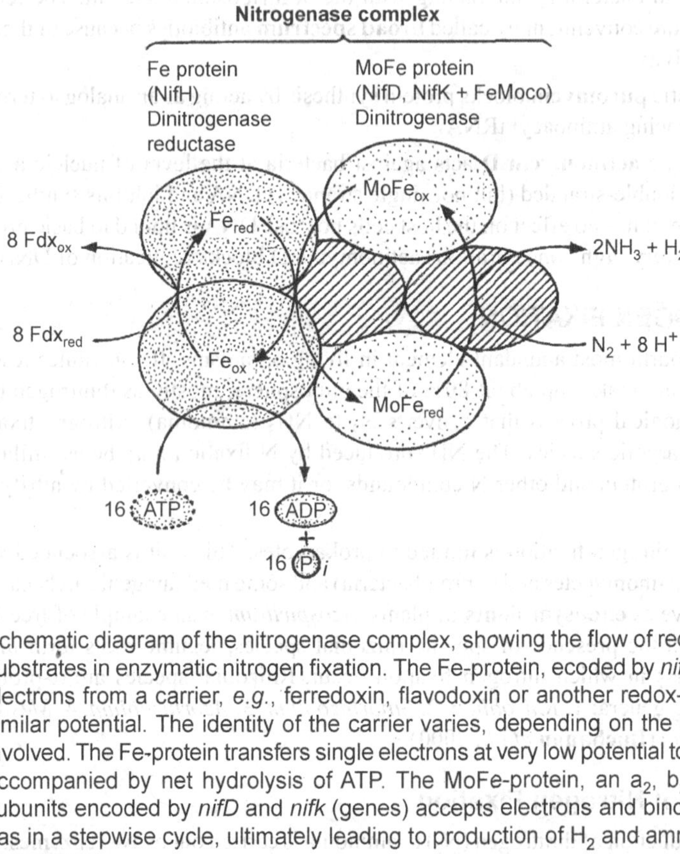

Fig. 59.10. Schematic diagram of the nitrogenase complex, showing the flow of reducing power and substrates in enzymatic nitrogen fixation. The Fe-protein, ecoded by *nifH* (gene) accepts electrons from a carrier, *e.g.*, ferredoxin, flavodoxin or another redox-active species of similar potential. The identity of the carrier varies, depending on the biological system involved. The Fe-protein transfers single electrons at very low potential to the MoF-protein, accompanied by net hydrolysis of ATP. The MoFe-protein, an a_2, b_2, heterodimer of subunits encoded by *nifD* and *nifk* (genes) accepts electrons and binds H^+ ions and N_2 gas in a stepwise cycle, ultimately leading to production of H_2 and ammonia (Buchanan *et al.*,2000).

Oxygen management during nitrogen fixation in aerobes. Nitrogenase enzyme is sensitive to oxygen. Three factors are important in maintaining low oxygen concentrations in nodules. First, the entry of O_2 into the nodule is regulated by a variable permeability barrier in **nodule parenchyma**. Second, **leghaemoglobin (Lb)**, a oxygen-binding plant protein, plays an active role in regulating and delivering oxygen in the infected cells. Third, bacterial respiration constitutes a major sink whereas the free-living rhizobia typically have a **cytochrome oxidase** with *km* for oxygen of around 50 nM, the nodules bacteriod cytochrome oxidase has a very low K_m for oxygen, about 8 nM.

Mechanism of symbiotic nitrogen fixation (in root nodules). A few groups of plants (*e.g.*, legumes) can obtain NH_4 efficiently from nitrogen (N_2) by participating in a **symbiosis** with **nitrogen-fixing bacteria**. The best characterized nitrogen-fixing symbiosis are the interactions between legumes and rhizobia. Bacteria infects the plant roots (Fig. 59.11).

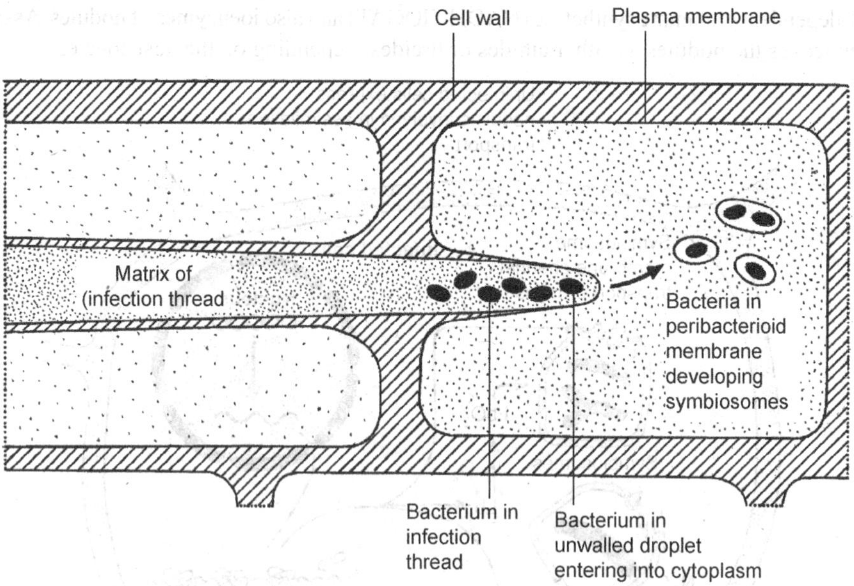

Fig. 59.11. Release of bacterium from a walled infection thread into a target cell. The bacteria are taken up into the host plant cell enveloped in host plasma membrane. The bacteria undergo limited cell division (**indeterminate** or **meristematic nodules**, *e.g.*, alfalfa, pea and clover) or extensive rounds of cell division accompanying host cell division (**determinate** or **spherical nodules**, *e.g.*, soyabean, trefoil and bean-*Phaseolus*).

Infection requires communication between the bacteria and the plant specific plant compounds, such as **flavonoids** are released by developing roots. For example, two compounds that occur naturally in alfalfa, **luteolin** (a flavone) and **4-methoxychalcone** (a chalcone) act as inducers of *nod* gene expression in *Sinorhizobium meliloti*. These are detected by the bacteria and induce the bacteria to make **lipo-oligosaccharide Nod factor** (Nod D, Nod C, Nod A, Nod B, etc.) which alter plant gene expression and cell division in roots of specific host plants. As the nodule grows, bacteria invade some plant cells and develop into nitrogen-fixing forms called **bacteroids** (Fig. 59.12). Bacteroids are surrounded by a plant-derived membrane and form **symbiosome** (Fig. 59.13) which controls the exchange of nutrients between the bacteroid and the plant cytoplasm. A mature nodule is organized to support the energy intensive **nitrogen fixation** reaction by delivering oxygen and carbon source to the bacteroids at low concentrations of oxygen in order to protect **nitrogenase**, a notoriously sensitive enzyme. Ammonia released by bacteroids is first assimilated in the plant by glutamine synthetase (GS) and

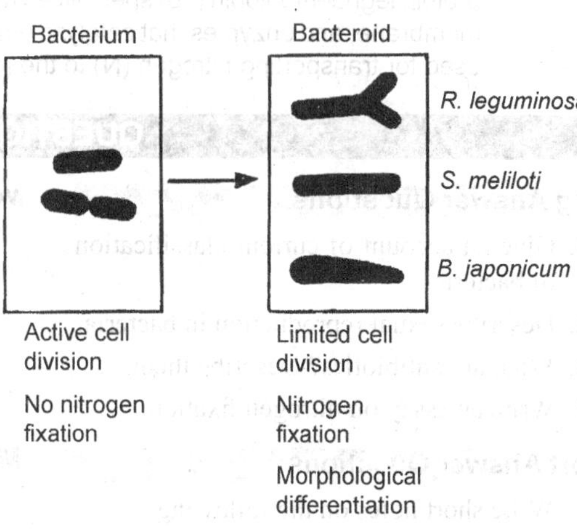

Fig. 59.12. After release bacteria differentiate to form morphologically distinct bacteroids. Different plant bacterial combinations yield distinct bacteroid forms (after Buchanan *et al.*, 2000).

NADH-dependent glutamate synthetase (NADH-GOGAT) and also ioenzymes of nodules. Assimilated nitrogen leaves the nodules as either amides or ureides, depending on the host species.

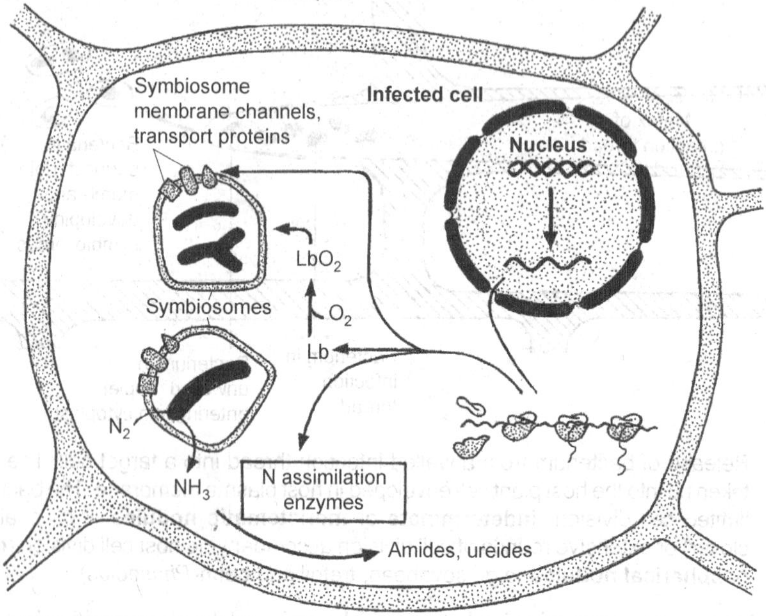

Fig. 59.13. The plant nucleus encodes a series of genes that are expressed only during later stages of nodule development, in response to bacterial differentiation. As a general group, these plant genes are termed late nodulins. Some of these transcripts are specific to infected or uninfected cells. These plant-encoded products include the oxygen-binding protein leghaemoglobin (Lb) specialized membrane proteins targeted to the symbiosome membrane, and enzymes that catalyze ammonia assimilation and synthesis of molecules used for transporting nitrogen (N) to the rest of the plant (after Buchanen *et al.*, 2000).

QUESTIONS

Long Answer Questions

1. Give an account of current classification of bacteria.
2. Describe sexual reproduction in bacteria.
3. What are antibiotics? Describe them.
4. Write an essay on nitrogen fixation.

Short Answer Questions

1. Write short notes on the following:
   (*i*) Gram vegative bacteria
   (*ii*) Symbiotic nitrogen fitation
   (*iii*) Antibiotics
   (*iv*) Symbiosome

Very Short Answer Questions

1. What is bacteria?
2. Define transduction.
3. What is nitrogenase?
4. Give an example of symbiotic N_2 fixation bacteria of root nodules.

Multiple Choice Questions

1. Prokaryotic cells occur in
   (*a*) PPLO (*b*) bacteria
   (*c*) cyanobacteria (*d*) all of these.
2. Muramic acid found in the cell wall of bacterial cell is a derivative of

(*a*) alkaloid (*b*) glucose
(*c*) fat (*d*) protein

3. Infoldings of plasma membrane regions which provide anchors for DNA are called
(*a*) ribosome (*b*) centrioles
(*c*) lomasomes (*d*) mesosomes

4. Apart from DNA in the bacterial chromosome, there is a circular extra chromosomal DNA called
(*a*) chromosome (*b*) mesosome
(*c*) plasmid (*d*) none of these

5. Volutin granules in bacterial cell are present in
(*a*) mitochondria
(*b*) nucleoplasm
(*c*) plasma membrane
(*d*) cytoplasm

6. Flagella with single strand and composed of flagellin is found in
(*a*) prokaryotes (*b*) eukaryotes
(*c*) both (*a*) and (*b*) (*d*) none of these

7. Symbiotic nitrogen fixing bacteria present in root nodules of legumes occur in genus
(*a*) *Xanthomonas* (*b*) *Pseudomonas*
(*c*) *Azotobacter* (*d*) *Rhizobium*

8. Non-legume nitrogen fixing bacteria belong to genus
(*a*) *Rhizobium* (*b*) *Frankia*
(*c*) *Clostridium* (*d*) *Azotobacter*

9. A free-living aerobic soil bacterium capable of fixing nitrogen is
(*a*) *Clostridium* (*b*) *Rhizobium*
(*c*) *Azotobacter* (*d*) *Streptococcus*

10. Gram stain is
(*a*) a stain produced from gram seeds
(*b*) a chemical
(*c*) a staining technique developed by a Danish scientist Christian Gram
(*d*) none of these

11. The term antibiotic was first used by
(*a*) A. Fleming (*b*) S. Waksman
(*c*) J. Listere (*d*) L. Pasteur

12. Conjugation in bacteria was discovered by
(*a*) Watson and Crick
(*b*) Lederberg and Tatum
(*c*) Jacob and Monad
(*d*) Zinder and Lederberg

13. Penicillin kills bacteria by
(*a*) inhibition of RNA synthesis
(*b*) inhibition of DNA synthesis
(*c*) lysis of protoplasm
(*d*) suppression of cell wall synthesis

14. Bacteria reproduce sexually by
(*a*) endospores (*b*) transformation
(*c*) conidia (*d*) exosphores

15. Pili are characteristic appendages of
(*a*) mycoplasma (*b*) virus
(*c*) algae (*d*) bacteria

ANSWER

Very Short Answer Questions

1. Bacteria are most primitive, simple, unicellular, prokaryotic and microscopic organisms of great economic importance.
2. Transduction is a type of sexual reproduction in bacteria in which fragments of DNA are carried from one bacterial cell to another by the bacteriophage.
3. Nitrogenase is the name of two enzymes: dinitrogenase and dinitrogenase reductase which are involved in reduction of N_2 into ammonia during nitrogen fixation.
4. *Rhizobium*.

Multiple Choice Questions

1. (*d*)	2. (*b*)	3. (*d*)	4. (*c*)	5. (*d*)	6. (*a*)	7. (*d*)
8. (*b*)	9. (*c*)	10. (*c*)	11. (*b*)	12. (*b*)	13. (*d*)	14. (*b*)
15. (*d*)						

60 Recombinant DNA Technology

60.1. INTRODUCTION

Technical developments often lead to quantum jumps forward in science (**Malacinski** 2003). In the past three decades, a recently developed technology, called **genetic engineering** or **recombinant DNA technology** has revolutionized genetics in both basic research and its pracital aspects. The discovery of two naturally occurring and quite remarkable classes of biological molecules made possible the development of methods to isolate and manipulate DNA fragments. These two classes of molecules are **plasmid DNA (Vector)** and **restriction enzymes**. Within recombinant DNA technology, a DNA fragment-even and entire gene and its controlling elements—can be isolated, the fragment coupled with a plasmid or phage and the hybrid inserted into a bacterium. In addition, since bacteria can be replicated in vast quantities, this process can be used for large-scale production of foreign DNA inserted itself; if the bacterial host is able to express or synthesize the protein product of the foreign DNA, the hybrid plasmid can direct the production of large quantities of otherwise scarce or expensive proteins, or even of proteins containing specific mutations created in the laboratory.

A. Plasmids

In the early 1950s, it was discovered that bacteria transfer genes to other bacteria by a process called **conjugation** (Fig. 60.1). One bacterium attaches small projections (**pili**) on the surface of those on the surface of an

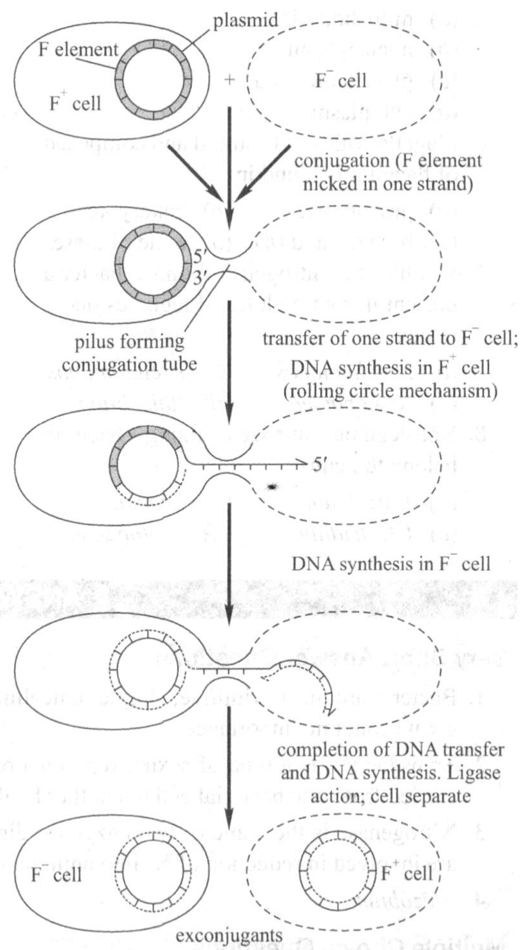

Fig. 60.1. Diagram illustrating transfer of the F element from an F^+ to an F^- cell during conjugation in *E.coli* (after Goodenough and Levine, 1974).

adjacent bacterium. DNA from the donor bacterium is passed to the recipient through the pili. The ability to form pili and to donate genes to neighbors is a genetically controlling trait. The genes controlling this trait are not located on the bacterial chromosome, but rather on separate genetic element called **plasmid** (Fig. 60.5). Plasmids are ideal **vectors** to accept and carry pieces of foreign DNA (including entire genes) into a bacterial host, where the foreign DNA can be replicated along with the plasmid. As vectors, plasmids are, therefore, one of the major tools of recombinant DNA technology.

B. Restriction Enzymes

Restriction enzymes, many of which have been found in bacteria, recognize and degrade DNA from foreign organisms. Bacteria have confronted the invasion of foreign DNAs for million of years, and have evolved these enzymes as protective mechanisms that preserve their own DNA while destroying the invading DNA. Each restriction enzyme recognizes only one short sequence, usually (but not always) 4 to 6 base-pairs long. For example, **EcoRI**, one of the first restriction enzymes to be isolated (from *E.coli*) cuts DNA only at the sequence:

$$\begin{array}{c} \downarrow \\ \text{G–A–A–T–T–C} \\ \text{C–T–T–A–A–G} \\ \uparrow \end{array}$$

In this case, the target sequence is called a **palindrome**, since it reads the same backwards as forwards. Many restriction enzymes recognize palindromic target sequences. If this sequence occurs once in a circular invading plasmid, the enzyme will open the circle by cutting it once. If the sequence occurs at several sites, the DNA will be cut into several pieces. In general, the names of restriction enzymes are derived from the first letter of the genus followed by first two letters of the species name of their bacterial source. For example, EcoR is isolated from *Escherichia coli* (by **Meselson** and **Yuan**), Hind III from *Hemophilus influenzae*, Taq I from *Thermus aquaticus*. Several hundred of these enzymes have been isolated from hundreds of species of microorganisms (Box 60.1).

Box 60.1

The enzymes required in recombinant DNA technology are restriction endonucleases, DNA ligases, DNA polymerases (mainly DNA polymerase III) and reverse transcriptase.

C. Cloning

For the molecular biologist, these restriction enzymes allow plasmids to be opened up *in vitro* so that foreign DNA can be inserted. Restriction enzymes also offer a way of isolating predictable fragments of any DNA molecule. With the discovery of plasmid are restriction enzymes, it became technically simple to obtain large quantities of exact copies of any chosen DNA fragment by **cloning**. This was done by using restriction enzyme to isolate "the fragment" inserting it into a plasmid, and inserting the **hybrid plasmid** into a host bacterium, which then reproduced many copies of plasmids as the bacteria proliferated. The rearrangement of DNA in a living organism is genetic recombination, hence the name **recombinant DNA technology**.

60.2. MECHANISM OF RECOMBINANT DNA TECHNOLOGY

The process of generating recombinant DNA can be divided into following four steps:

1. The production of desired DNA fragments;
2. The insertion of these fragments into a suitable vector such as a plasmid or phage;
3. The introduction of that vector into an appropriate host (usually a strain of *E.coli*);
4. The identification, selection and characterization of recombinant clones.

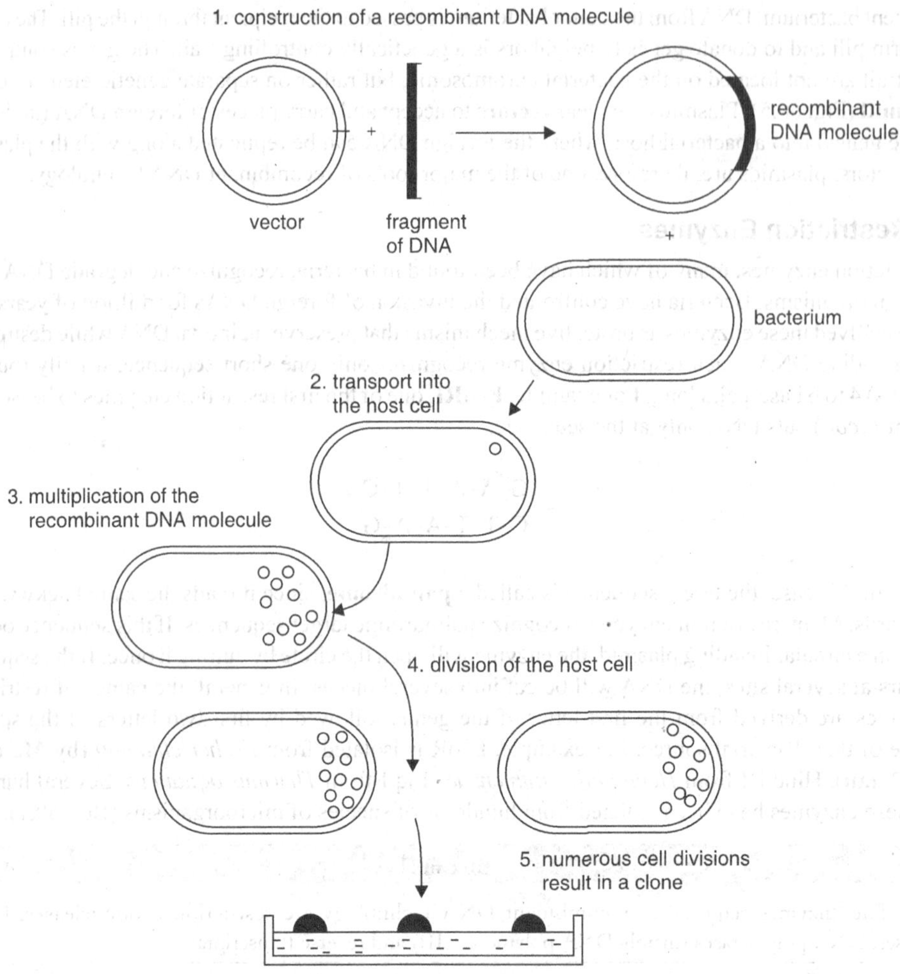

Fig. 60.2. The basic events in recombinant DNA technology.

I. Isolation and characterization of DNA Fragments

Owing to their specificity, **restriction endonucleases** were the first nucleases found to cleave DNA at specific sequences. These enzymes have been proven useful in genetic analyses from the oligonucleotide to the chromosomal levels. At the oligonucleotide level, they are used to provide specific fragments for cloning, sequencing, and custom-made mutations. Segments of DNA thought to contain region that control gene expression can, if desired, be cut away selectively, and the effect of their loss on gene expression can be monitored. At the chromosomal level, they permit analyses of entire chromosomes by "walking" or "jumping".

Most restriction enzymes make two single-strand breaks, one in each strand (close to, but not opposite each other), thus, generating two 3′-OH and two 5′-P group at each restriction enzyme site. A technically useful property of restriction enzymes was detected by electron microscopy: because the complementary 3′-OH to 5′-P base pair over hanging termini (called **cohesive** or **"sticky" ends** because they can overlap one another) produced by many restriction enzymes, fragments **circularized** spontaneously. These circles could be relinearized by heating, but, if after circularization they were also reacted with *E.coli* **DNA ligase** (which covalently joins 3′-OH and 5′-P groups), circularization became permanent. This observation was the first evidence for two important characteristics of restriction enzymes (Fig.60.3):

(*i*) Restriction enzymes make breaks in symmetric sequences.

(*ii*) The breaks are usually not directly opposite one another, therefore sticky ends are produced.

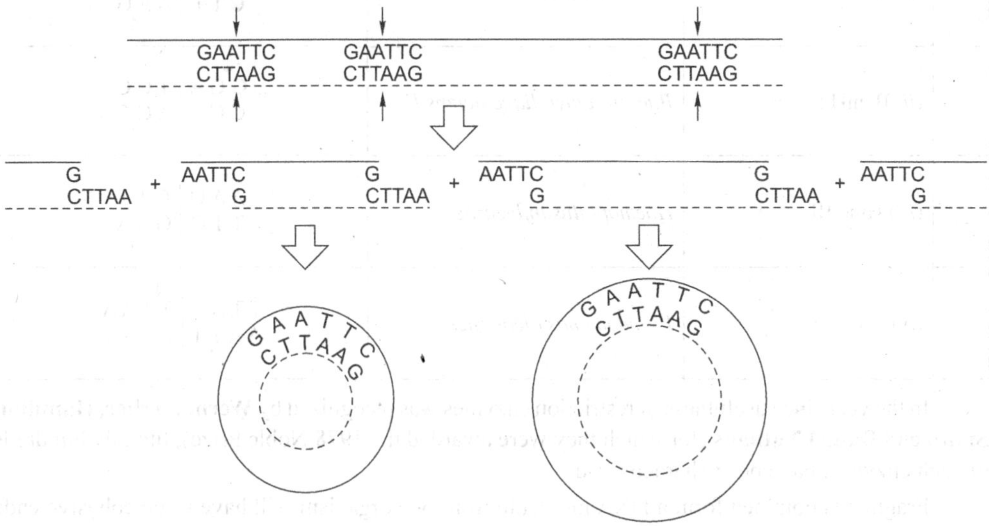

Fig. 60.3. Formation of restriction fragments that have overhanging cohesive ends following digestion with restriction enzymes (here, at the 5′ ends) can circularize. Small arrows indicate cleavage sites of EcoR1; large arrows indicate the next step in the process.

Palindromes. The sequences recognized by many restriction enzymes are **palindromes**, as shown here:

A B C	C′ B′ A′	or	A B	B′ A′	or	A B X	B′ A′
A′ B′ C′	C B A		A′ B′	B A		A′ B′ X′	B A

Capital letter represent bases (A, B and C may be the same or different), the (′) indicates the complementary base, and the vertical line is the axis of symmetry.

Examination of a very large number of restriction enzymes showed that the breaks are usually in one of the two distinct arrangements: **staggered but symmetric** around the line of symmetry, forming two different cohesive ends—a single-stranded extension with a 5′-P terminus and 3′-OH extension; or both cuts at the center of symmetry, forming blunt ("**flush**") **ends**. Fragments with blunt ends cannot circularize spontaneously. Table 60.1 has enlisted the recognition sequences and cleavage sites for several restriction enzymes.

Table 60.1. Some restriction endonucleases and their cleavage sites (source Malacinski 2003).

	Name of Enzyme	Microorganism	Target sequence and Cleavage sites
A. Generate flush/blunt ends			
	Bal I	*Brevibacterium albidum*	↓ TGG | CCA ACC | GGT ↑

	Name of Enzyme	Microorganism	Target sequence and Cleavage sites
B. Generate cohesive ends			
	(*i*) EcoR I	*Escherichia coli*	↓ G A A | T T C C T T | A A G ↑
	(*ii*) BamH I	*Bacillus amyloliquefaciens H*	↓ G G A | T C C C C T | A G G ↑
	(*iii*) Hind III	*Haemophilus influenzae*	↓ A A G | C T T T T C | G A A ↑
	(*iv*) Pac I	*Pseudomonas alcalignes*	↓ T T A A | T T A A A A T T | A A T T ↑

In the year since usefulness of restriction enzymes was recognized by **Werner Arber**, **Hamilton Smith** and **Daniel Nathans** (for which they were awarded the 1978 Noble Prize), literally hundreds of such enzymes have been characterized.

Fragments obtained form a DNA molecule from one organism will have same cohesive ends as the fragments produced by the same enzyme acting on DNA molecules from another organism. This unique fact is one of the corner stones of recombinant DNA technology.

Advantages of DNA of Phages or Plasmids

Since most restriction enzymes recognize a unique sequence, the number of cuts made in the DNA from an organism by a particular enzyme is limited. The DNA of a typical bacterial chromosome, which contains roughly 2×10^6 base pairs, is cut into several hundred to several thousand fragments, and nuclear DNA of mammals ($\sim 3 \times 10^9$ base pairs) is cut into more than a million. These numbers are large but, still relatively small compared to the total number of bases in the DNA of an organism. Of great usefulness in cloning are the less complex DNA molecules, such as **bacteriophage λ (lambda)** or **plasmids** which have only 1 to 10 sites of cutting (or even none) for particular restriction enzymes. Plasmids having a single site for each of a number of restriction enzymes are especially valuable in cloning.

If restriction sites common to two cloning partners (a plasmid and a foreign DNA) are not available, an alternative strategy can be employed. Each of the two DNAs can be cleaved with an enzyme that generates protruding termini identical to those generated by the other restriction enzyme. These **cohesive compatible ends** can then be ligated.

Restriction maps. Maps of enzyme cutting sites, called **restriction maps**, provide a convenient method for comparing DNA fragments. Whether two fragments are identical, or whether they share some of the same nucleotide sequence that can be quickly established by comparing their restriction maps. Thus, DNA fragments exhibit different restriction maps can be said to contain at least some differences in their nucleotide sequences. Without investing a great amount of time and effort to perform a complete sequencing reaction, a quick determination of the extent of relatedness of different DNA fragments can often be made by simply comparing restriction maps (Fig. 60.4).

II. Vectors

If a DNA fragment is to be introduced permanently into a cell, a carrier DNA molecule or a vector is essential. A **vector** is a small DNA molecule capable of self-replication and is used as a carrier of DNA fragment inserted into it for cloning. The vector is also called **cloning vehicle** or **cloning DNA**.

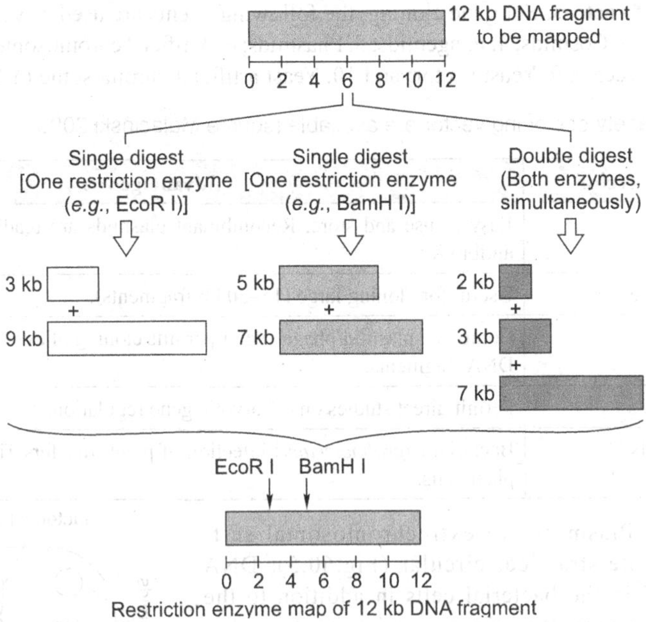

Fig. 60.4. Procedure for preparing a restriction enzyme map for DNA fragment. The cutting sites for the two enzymes (EcoRI and BamHI) used for this simplified example of a 12-kb DNA fragments are established by comparing results of single and double digests. Sizes of the fragments produced by digestion are determined with gel electrophoresis (kb = kilobase = 1000 bases).

Characteristics of a Good Vector

A good vector must have the following properties:

1. It should be able to replicate autonomously, so that it can generate multiple copies of itself along with DNA insert with a single host cell.
2. A vector is supposed to have one **origin of replication** for DNA replication.
3. A vector should be small in size and low molecular weight. Ideally it should be less than 10kb (kilo base) in size, because large DNA molecules are broken during purification and present difficulty during manipulation required for gene cloning.
4. A vector should be easily isolated and purified.
5. It should be easily introduced into host cell.
6. Vector should afford easy transformation of host cell.
7. The vector should contain a suitable marker or markers to permit its detection in the host cell and selection of transformed host cell.
8. The vector should have unique target sites for as many restriction enzymes as possible, so that DNA insert can be integrated without disrupting essential functions of the vector.
9. When objective is gene transfer, the vector should have the ability to integrate either itself of the DNA insert it carries into the genome of the host cell.
10. When objective is expression of the DNA insert, the vector should contain suitable control elements such as promotor, operator and ribosome binding site, etc.
11. The transformed cell containing DNA insert recombinant DNA should be identifiable from those transformed by the vector molecule alone.

Example of vectors. For gene cloning, the following agents are used as vector: 1. Plasmids; 2. Bacteriophages; 3. Cosmids; 4. Phagemids; 5. Phasmids; 6. Artificial chromosome vector; 7. Fosmid vector; 8. Shuttle vector; 9. Yeast vector; and 10. Yeast artificial chromosome (YAC) vector.

Table 60.2. A variety of cloning vector are available (source Malacinski 2003).

	Type of vector	Advantage
1.	Plasmids	Easy to use and store. Recombinant plasmids are readily selected within antibiotics.
2.	Lambda phage	Useful for cloning large (15–20 kb fragments).
3.	Cosmids	Combined plasmid/phage vector permits cloning of even larger (*e.g.*, 45 kb) DNA fragments.
4.	Yeast plasmids	Permit direct studies on eukaryotic gene regulation.
5.	Plant plasmids	Bacterial (*Agrobacterium*) infection of plant transfers Ti plasmid into host plant cells.

Plasmids. Plasmids are extrachromosomal self-replicating, double-stranded, circular (Fig.60.5), DNA molecules found in the bacterial cells in addition to the bacterial chromosomes. Plasmids range in size from 1×10^6 daltons to 200×10^6 daltons. These may exist either independently or may become integrated into bacterial chromosome. Generally plasmids are nonessential for the bacterial cells except under specific environments. Plasmids contain genetic information for their own replication. They also contain gene for antibiotics, heavy metal resistance, nitrogen fixation, pollutant degradation, etc.

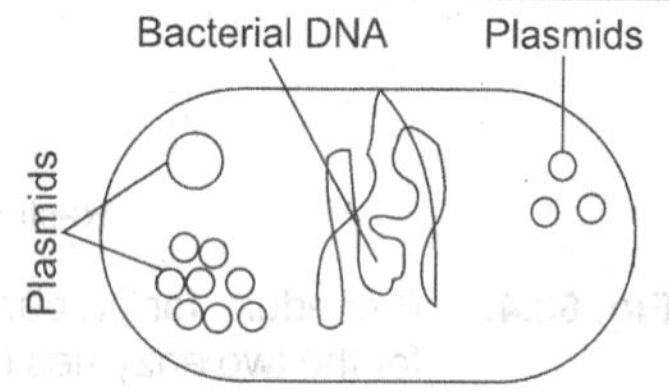

Fig. 60.5. Bacterium cell showing its chromosome and plasmids.

(*i*) Phage lambda (λ) as a vector. A commonly used vector is that of the lambda (λ) phage. Bacteriophage λ, which infects *E.coli* cells, can be used as cloning vector. DNA of λ phage is 48.5 kb in length. At its ends are the *cos* (cohesive) *sites*, which consist of 12 bp cohesive ends. The *cos* ends allow the DNA to be circularized in the host cell (Fig. 60.6).

For the cloning of large DNA fragments, upto about 20 kb, much of the nonessential lambda DNA is removed and replaced by the insert (desired or target gene; Fig. 60.7). The recombinant DNA is then packaged within viral particles *in vitro*, and these are allowed to infect bacterial cells (*E.coli*) which have been plated out on agar.

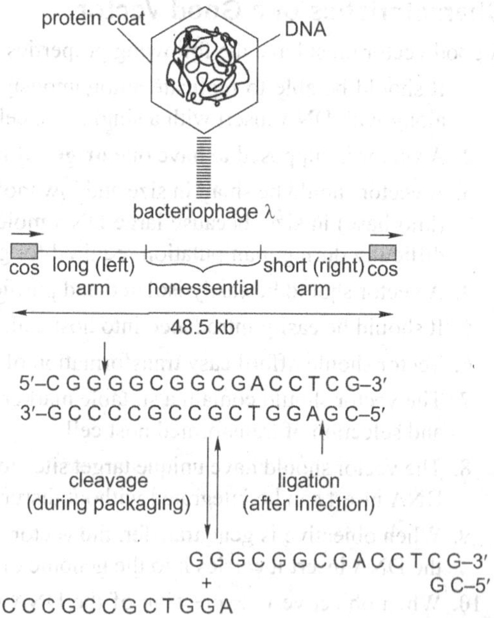

Fig. 60.6. A—Phage λ (= lambda) and its genome; B—The *cos* ends of genome of phage λ.

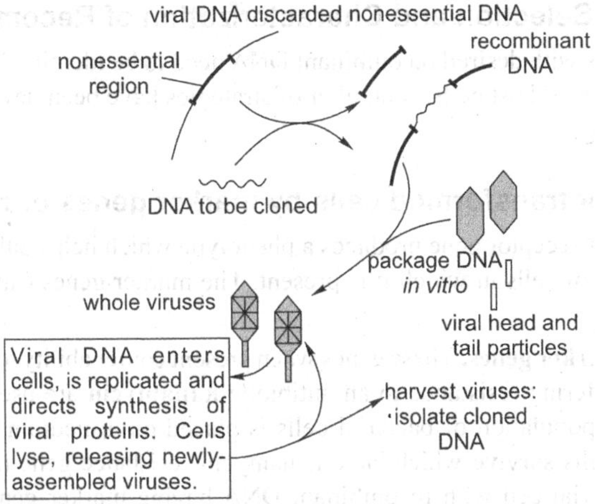

Fig. 60.7. Cloning DNA in a λ bacteriophage. Nonessential DNA is cut out of the λ DNA using restriction enzymes and is separated from essential DNA fragments by electrophoresis. Essential DNA and DNA to be cloned are ligated and packaged in empty head particles of phage and tails are added to get infective bacterial viruses.

Once inside the bacterial cells, the recombinant viral DNA is replicated. All the genes needed for normal lytic growth are still present in the DNA and so multiplication of the virus takes place by cycles of cell lysis and infection of surrounding cells. It gives rise to **plaques** of lysed bacterial cells on a background, or **lawn**, of bacterial cells. Cloned DNA can be recovered from the viruses in these plaques (Fig. 60.8).

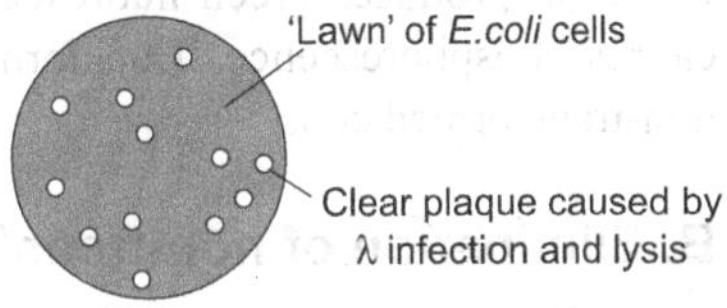

Fig. 60.8. The formation of plaques by λ (lambda phage) infection.

III. Cloning of Genes

Several strains of *E.coli* have been used as host cells or cloning organism (*e.g.*, HB101, h303 and RR1). The cloning organism lacks restriction enzymes and are also deficient in modifying enzymes to protect DNA from degradation. *E.coli* has been largely used for most recombinant studies. In one such experiment a bacterial gene for leucine production was introduced into a yeast strain (eukaryotic cell) that was deficient in leucine. Some *E.coli* strains that bud off mini cells have been found most desirable in transcription and translation of genes to be studied. The *mini cells* contain plasmids and other components required for gene expression, but lack chromosomal DNA. Hence in such systems expression of recombinant DNA can be easily studied, since there is no interference from other DNA fragments.

In order to produce large amounts of a particular DNA fragment (genes) in pure form, DNA cloning has to be carried out. DNA fragments can be introduced in bacterial cells, but they will not be able to replicate and after some time may get lost. Therefore, DNA fragments carried by vector (plasmids or phages) when introduced in bacteria get replicated. A population of recombinant DNA molecules can be made, each recombinant molecule containing one of the foreign DNA fragments in the original mixture. This can then be introduced into a population of bacteria such that each bacterial cell contains a different type of recombinant DNA molecule. If we can identify the bacterial cell that contains recombinant DNA bearing the foreign DNA fragments that we desire, it can be cultured and large amounts of recombinant DNA isolated.

IV. Identification, Selection and Characterization of Recombinant Clones

The clones of host cells with desired recombinant DNA need to be identified and sorted out from the mixture of various types of host cells. A number of strategies have been developed for this purpose. These are the following:

A. Identification of transformed cells by marker genes or reporter genes

A marker gene or receptor gene produces a phenotype which helps either in the easy selection or quick identification of cells in which it is present. The marker genes can be either selectable or scorable.

(*i*) Selectable marker genes. These genes when present confer ability to survive under selective conditions. Genes confering resistance to an antibiotic **kanamycin** are used as selectable marker. For example, when a population of bacterial cells is placed on a medium containing kanamycin, only those bacterial cells survive which have kanamycin resistance gene (*kan'*) and form colony. The transformed bacterial cell with recombinant DNA having marker gene *kan'* survive and are sorted out.

(*ii*) Scorable marker genes. These are those genes that produce distinct phenotypes. These characteristics allow an easy identification of cells with and without them. For example—(*a*) Gene *gus* (β-glucuronidase) when present produces blue color in presence of appropriate substrate, (*b*) Gene *gfp* produces green fluorescence protein, and (*c*) Gene *lux* produces protein luciferase that causes phosphorescence. Transformed bacterial cells with any of these genes are identified from non-transformed cells.

B. Elimination of non-transformed cells

When vectors are introduced into the host cells, following three types of cells are formed: (*i*) Non-transformed host cells; (*ii*) Transformed host cells with unaltered vector, *i.e.*, without recombinant DNA; (*iii*) Transformed host cells with recombinant DNA having DNA inserts.

In the process or sorting, the first step is elimination of non-transformed cells. A good vector has at least two marker genes, one of them is selectable marker. For example, vector pBR322 (an artificial plasmid) has marker gene *tet^r* and *amp^r* (for tetracyclin and ampicillin resistance). The bacterial culture is plated on a medium containing two antibiotics-tetracycline and ampicillin. All the colonies that develop on this medium will consist of antibiotic resistant transformed bacterial cells. However, these cells may be with unaltered vector or with recombinant DNA vector.

C. Identification of clones having recombinant DNA

The identification and isolation of those clones of bacterial cells that are transformed by the presence of recombinant DNA from those cells that have unaltered vector is the next step. Two methods are employed for their identification.

(*i*) Selection of vectors with two selectable markers. In case of vectors that have two selectable markers, the DNA insert is placed within one of these markers. For example, in case of vector pBR322, the two marker genes are *amp^r* and *tet^r*. The DNA insert with one marker *amp^r*. The other marker *tet^r* is used for the elimination of non-transformed cells. The transformed clones are then replica plated on ampicillin containing medium. The clones containing recombinant DNA will be sensitive to ampicillin because the DNA insert inactivates the *amp^r* gene. Such clones are identified and isolated from the master plate.

(*ii*) Selection of vector with a gene complementary to the missing host gene. Some vectors contain a gene or a part of gene, which complements a function missing in their host cells. For example, in vectors pUC (a plasmid) gene *lacZα* complements certain lac Z^- strain of *E.coli*. The same combination is used for some λ vector and M13 phage vectors. In such cases, the DNA insert is placed in such a way that it disrupts function of *lacZα* gene in the vector. Therefore, *E.coli* cells containing the recombinant DNA are deficient in enzyme β-galactosidase. Such colonies form white plaques on a medium containing X-gal. These white colonies are selected. The *E.coli* cells without recombinant DNA (*i.e.*, with unchanged vector) form blue colonies.

D. Selection of clone containing specific DNA insert

The population of recombinant clones obtained this way is heterogenous having different segments of DNA. The clone with DNA insert of interest need to be identified and separated. Following strategies are used for this purpose.

1. Colony hybridization. Recombinant clones with no easily recognized phenotype are most commonly identified by determining if the DNA in the colonies **hybridizes** with a radioactive probe to the foreign DNA. Bacterial colonies are replica-plated onto a solid support, usually nitrocellulose or nylon filters. The bacteria are lysed with alkali that also denatures their DNA, neutralized and allowed to hybridize to the 32P-labelled probe. The location of the 32P-hybridized probe is then determined by exposing X-ray film to the filters. Bacterial colonies releasing DNA hybridizing to the DNA the probe can then be identified by aligning the X-ray film with the original bacterial plates. In this way, it is possible to screen for many hundreds of colonies simultaneously, and recognize colonies that carry recombinant plasmids (Fig. 60.9). This method is called **colony hybridization**.

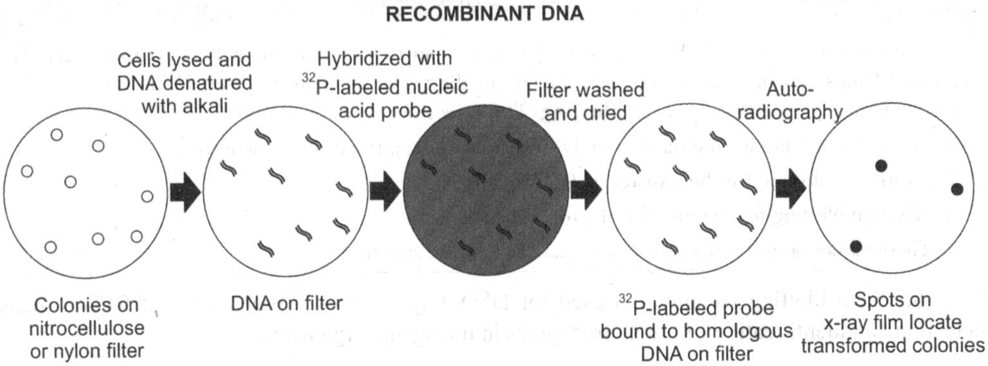

Fig. 60.9. Detection of transformed cells by colony hybridization with a radioactive probe. Only eight of thousand of colonies on a reference plate (not showed) are included.

2. Southern blotting or Southern hybridization. (It was developed by **E.M. Southern, 1975**). In this technique, a DNA molecule is cut into discrete fragments by a restriction enzyme. It is electrophoresed through an agarose gel which separates the various fragments according to size. The DNA is then denatured into single strands by exposing the gel to NaOH. A few pieces of filter paper soaked in buffer are placed under the gel. A large piece of nitrocellulose paper is laid over the agarose gel, followed by several layers of absorbent material such as filter paper. This dry absorbent material pulls the buffer up through the gel from the lower layer (Fig. 60.10). This washes the DNA off the gel and on to the filter, where it covalently binds to the filter.

The positions of the DNA molecules on the filter paper are identical to their position in the gel. The nitrocellulose filter containing the DNA is first dried and then exposed to a solution of ^{32}P labelled mRNA called **molecular probe** from the gene to be isolated. The radioactive mRNA hybridizes (*i.e.*, establishes the hydrogen bonds) only with the single-stranded DNA in restriction fragments that contain complementary sequences. The nitrocellulose filter is then removed and placed in contact with photographic film that when developed will reveal fragments from the original gel containing complementary sequences to the mRNA used in the assay. The procedure allows specific identification of restriction fragments containing DNA sequences to specific RNA molecules.

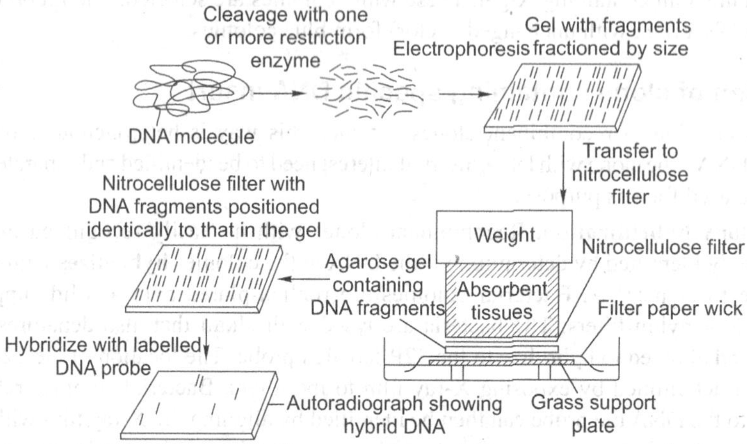

Fig. 60.10. Various steps of Southern blotting technique.

Box 60.2

A mixture of DNA, RNA and protein fragments can be separated by **gel electrophoresis**. The separated bands can be stained and seen directly in the gel or they are transferred in a nitrocellulose membrane through a technique called **blotting**. There are three types of blotting:

1. Southern blotting for the transfer of DNA bands to the nitrocellulose membrane.
2. Northern blotting for the transfer of RNA bands.
3. Western blotting for the transfer of protein bands.

These names do not reflect any functional or historical significance.

Southern blotting technique is used for DNA finger printing, preparation of RFLPs maps, detection and identification of transferred genes in transgenic organisms.

60.3. TECHNIQUES OF INTRODUCTION OF CLONED GENES IN HOSTS

After a recombinant DNA is constructed it is introduced into a suitable host by following methods.

1. Transformation. Mandell and **Higa** (1979) have found that majority of cells including *E.coli*, yeast and mammalian cells when treated with $CaCl_3$ solution, rate of transformation is greatly increased.

2. Transfection. Vectors that have *COS* sequences of lambda phage such as cosmids, plasmids and λ-phage vectors are packed *in vitro* into empty phage heads. The packaging produces completes λ particles with recombinant DNA. These newly synthesized phage particles are caused to infect *E.coli* cells. Thus, introduction of foreign DNA into bacteria by bacteriophages is called **transfection.**

3. Electroporation. Electroporation is the introduction of foreign DNA or recombinant DNA into cell by a very brief exposure to a very high voltage electric pulses ranging from 4000–8000 V/em. This induces transient pores in the plasma membrane. These pores give passage through which foreign DNA can enter the cells, provided DNA is in direct contact with the membrane.

Transfer of desired gene through electroporation is carried out in the protoplasts of rice, wheat, maize, sorghum, petunia and tobacco.

4. Liposome mediated gene transfer. Liposomes are small lipid bags in which large plasmids are enclosed. The liposomes are induced to fuse with the protoplasts and are used for gene transfer. DNA enters the protoplasts (naked plant cells) by endocytosis of liposomes. This technique is used in gene transfer in tobacoo, carrot, petunia, etc., and has the following advantages: (*i*) protection of DNA/RNA from nuclease digestion; (*ii*) stability of nucleic acids; (*iii*) low toxicity; (*iv*) high degree of multiplication; and (*v*) applicability to a wide range of cell types.

5. Microinjection. Microinjection is the technique of delivering foreign DNA directly into the cytoplasm or nucleus of the cells or male pronucleus of the fertilized egg or into one or two celled embryo with the help of micromanipulators. The microinjection assembly consists of a low power **stereoscopic dissecting microscope** and two **micromanipulators**. The microscope is needed to view the cell or ovum during the entire process of DNA introduction. One of the micromanipulator is a glass **micropipette** that is used to hold the ovum by partial suction. This is called **holding pipette.** The other micromanipulator has a glass injection needle and is used to introduce DNA into the ovum or its cytoplasm or in male pronucleus. The microinjection technique can be applied to animal cells and also to the plant protoplast.

The process of microinjection involves following steps:

1. One end of a glass micropipette is heated until the glass become somewhat liquified. It is stretched quickly so that a very fine tip is formed at the heated end. It is about 0.5 mm in diameter and resembles an injection needle.
2. Cells to be microinjected are placed in a container.
3. A holding pipette is placed in the field of view of the microscope and manipulated to hold the target cell at its tip.
4. The tip injecting micropipette is gently pushed through the membrane of the cell.
5. Contents of the injecting micropipette are delivered into the cytoplasm.

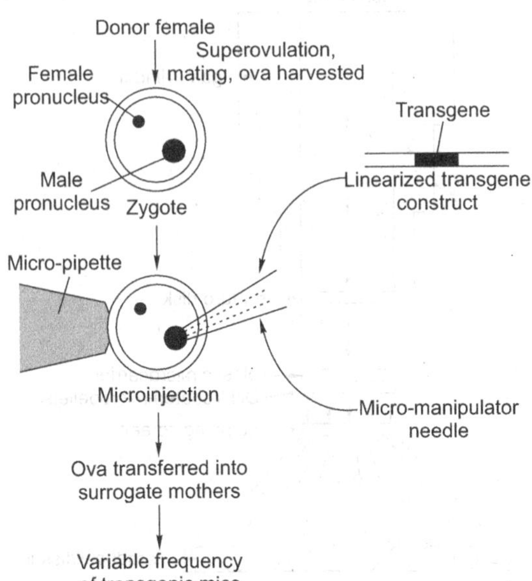

Fig. 60.11. Microinjection of cloned plasmid DNA into the male pronucleus of fertilized ovum.

Amphibian *Xenopus* (toad) oocytes have been widely used for the transcription by microinjection. The injected foreign DNA randomly comes in contact with DNA of the nucleus. Its expression is possible only when it is attached to a suitable promoter. **Rubin** and **Spradling** (1982) successfully introduced *Drosophila* gene for *xenothine hydrogenase* into P element (parental element). Such embryos on further development produced rosy eyes instead of mosaic eyes.

6. Microprojectile or biolistic or particle bombardment gun. It was developed by **Prof. Stanford** and coworkers at Cornell University (USA) in 1987. As the term denotes, it shoots foreign DNA into plant cells or tissues at a very high speed. This technique is also known as **particle gun method, biolistc process, gene gun** and **microprojectile gun**. This technique is most suitable for those plants which hardly regenerate and do not show sufficient response to gene transfer through *Agrobacterium*, for example, rice, wheat, corn, sorghum, chicken pea and pigeon pea.

Recently, transfection has been successful in both mitochondria and chloroplasts (both have been proven to be difficult targets for genetic engineering because they have double-membrane walls) using a biolistic process.

The gene gun consists of a chamber connected to an outlet to create vacuum (Fig. 60.12). At the top, a cylinder is temporarily sealed off from the rest of chamber with a plastic rupture disk. Helium gas flows into the cylinder. A plastic microcarrier is placed close to rupture disk. It contains DNA-coated tungsten particle, the microscopic pellets (*i.e.*, coated microprojectiles). While work the apparatus is placed in Laminar flow just to maintain sterile conditions. The target cells/tissues are placed in the apparatus. A stopping screen is placed between the target cells and microcarrier assembly. Helium gas is flown in the cylinder at high velocity. When pressure of cylinder exceeds the bursting point of plastic disk, it gets ruptured. Helium shock waves propel the plastic microcarrier containing DNA-coated micropellets. The stopping screen allows the micropellets to pass through and deliver DNA into target cells. The transformed cells are regenerated onto nutrient medium. The regenerated plant tissues are selected over culture media containing either antibiotic or herbicides. The selected plants are then analysed for expression of foreign DNA.

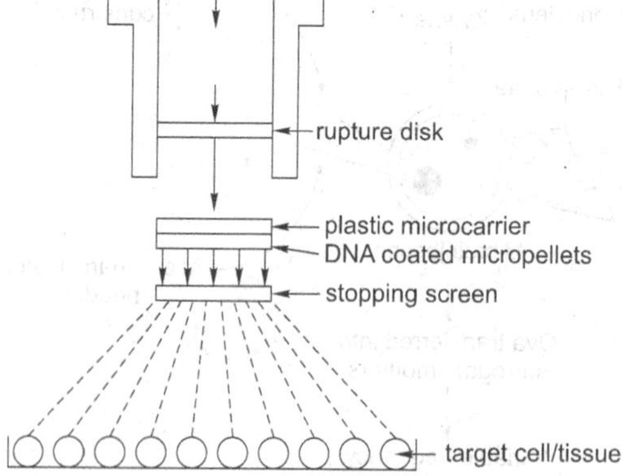

Fig. 60.12. Working system of particle bombardment gun.

The technique of biolistic method is suitable for those plants which hardly regenerate and which do not respond to gene transfer through *Agrobacterium*.

Using gene gun, scientists have got success in delivering foreing DNA into epidermal tissue of *Allium cepa*, scutellar tissue of maize and leaf and cell cultures of many crops (**Peters**, 1993). In addition to bacterial cells, algae, fungi, plants organelles (*e.g.*, cholorplasts and mitochondria), animal or human cells and fruitfly embryos have been successfully transofrmed. **Ramaiah** and **Skinner** (1997) produced transgenic alfalfa through direct delivery of DNA into pollen grains by particle bombardment method. In 1998, scientists of Plant Transformation Group at ICGEB (New Delhi) got success in transforming human interferon gamma gene into chloroplasts of tobacco, maize, etc. using a particle gun mediated gene transformation.

60.4. APPLICATION OF DNA RECOMBINANT TECHNOLOGY

Recombinant DNA technology is being used effectively to genetically engineer transgenic plants and animals, drought resistant plants, crop plants resistant to herbicides, pests and viruses. However, its greatest impact has been on agriculture, biotechnology and medice.

I. Agriculture. Genetic engineering can play a vital role in agriculture and forestery. 1. Seed genotypes can be programmed to yield high protein grains that could have resistance to heat, moisture and disease. 2. Certain crop varieties have been produced that could directly fix up atmospheric nitrogen without dependence on fertilizers. 3. DNA recombinant technology could also lead to the production of such varieties of seeds that may show pest resistance. 4. The transfer of genes by recombinant DNA technology and tissue culture techniques can also produce new combination of cereals, pulses and vegetables that would give a much higher yield.

Recombinant DNA technology has produced **genetically modified organisms** or **GMOs** such as **Bt cotton** (which is a transgenic plant containing *Bt gene* from *Bacillus thuringiensis*).

In the field of forestry, gene transfer techniques can be very effective in afforestation by producing new species of woody plants and tree that may not be browsed by cattle.

II. Biotechnology. Commercial utilization of recombinant DNA technology to obtain beneficial products through manipulation of the genome of living cells is known as **biotechnology**. This term was coined by **Karl Ereky** (1971). Biotechnology has great impact on virtually all domains of human welfare:

1. Fermentation industry. It is possible to engineer biofuels or energy crops that grow rapidly to generate huge biomass and that can be directly used for fuel or processed into alcohols or other products. The crops may be produced to contain high content of oils, carbohydrates, proteins and other products. Wastes from these can be converted into methane by anaerobic digestion. The technology has greatly improved the existing **fermentation process** and **modification** of microorganisms for production of industrial chemicals such as ethylene oxide, ethylene glycol and alcohol (*i.e.*, alcoholic beverages such as beer, wine , brandy, rum, gin, etc.).

2. Chemical industries. A number of chemical industries are dependent upon biological processes that yield biofertilizers, industrial solvents, food flavours, enzymes (*e.g.*, α-amylase, cellulase, lactase, lipase, pectinase, protease, glucoamylase, glucose isomerase, etc.; they are used in detergents, textiles, leather, dairy industries and in medicines), organic acids (*e.g.*, lactic acid, acetic acid, citric acid, gluconic acid, fumaric acid, etc.) and a number of organic compounds. The L-malic acid is produced by microorganisms that contain fumarase immobilised on an organic support. This acid is used in food industry to a great extent.

A number of **amino acids** synthesized industrially include L-asparagine, lysine, L-glutamine, DL-alanine, L-valine and L-glutamic acid, which are produced by genetically engineered microorganisms. L-asparagine is in high commercial demand since it is used as a medicine in certain type of cancers. Lysine is used in animal nutrition and is produced by fermentation process to serve as a supplement to animal feeds. Similarly, glutamic acid is another important amino acid which is produced by certain species of microbes, such as *Arthrobactor* and *Corynebacterium glutamicum*. Glutamic acid is used as a feed supplement and its sodium salt is utilized in food preparations.

3. Biodegradation. Biotechnology has proved to be a useful tool in treating lignocellulosic waste with the help of cellulolytic fungi, such as *Aspergillus niger*, *Trichoderma reesei*, etc. Biodegradation is usually less expensive with a high yield of byproducts. Genetic engineering has been able to enhance microorganisms, enabling them to degrade a huge range of human-made chemicals, wastes and synthetic products.

Bioremediation. Use of living organisms or the genetically engineered microorganisms to degrade environmental pollutants is called **bioremediation**. It is to remove hazardous chemicals accumulated in the cells or to detoxify them into nontoxic forms. The removal of hydrocarbons, dyes, industrial wastes, heavy metals, xenobiotics, etc., by microorganisms is also bioremediation.

4. Pharmaceutical industries. This industry has gained considerably from biotechnology since a number of genetically engineered products are manufactured commercially. Ever increasing demand of antibiotics, hormones (*e.g.*, somatostatin, insulin, etc.), antiviral drugs (inferons such as IFN-β and IFN-γ), blood clotting factors and vaccines (recombinant vaccines and edible vaccines), etc., have pushed recombinant DNA technology in the most enviable position.

III. Medicine. One of the greatest advantage of recombinant DNA technology is its application to medicine and immunotechnology. Plant-based substances and microorganisms are now being manipulated to yield large quantities of useful drugs, intermediates and vaccines at low costs. Besides, the techniques are used in developing diagnostic tools to detect in herited disorders.

Immunodiagnostics. Application of immunoassays in medical diagnosis began with the measurement of insulin in the plasma using anti-insulin antibodies. The use of monoclonal antibodies hold a promise to identify many microorganisms and diseases caused by them. Monoclonal antibodies have been put to innunmerable uses such as serotherapy, tumour and cancer therapy, identification of major histocompatibility antigens and serogenetic classification of infectious microoganisms. More recently, monoclonal antibodies have been used as tools in enzyme genetics in developing fast as a discipline of immunogenetics.

Inherited human disorders. Gene-specific probes for several human diseases have been isolated. For example, human growth hormone genes contain a specific deletion in people suffering from certain forms of pituitary dwarfims; immunoglobin genes are being studied in some who have severe immune deficiency, and the cloned cDNA for collagen is a specific probe for some forms of **osteogenesis imperfecta**, a hereditary disease in which bone becomes fragile.

Inherited disorders of defective haemoglobins pose a serious problems among inherited diseases. These are caused by mutations. Severe anaemia, such as **thalassemia**, **sickle-cell anaemia**, etc., are caused due to mutation in the coding region or defective RNA-splicing. These can be diagnosed with the help of restriction enzymes by making profiles of normal and suspect DNA. A novel method of locating a disease gene has been perfected which is called **chromosome walking**. Ths technique scans the DNA to identify specific regions or genes associated with the disease.

DNA Fingerprinting: The Ultimate Identification Test

Every year in court cases all over the world, the ability to establish a person's identity is essential for a just decision. Genetics has come to rescue of the courts and now the following new questions are routinely asked in the courts: (1) Is the drop of blood found at the crime scene from suspect on trial? Who is the child's father? Until recently, there was no foolproof test. In a criminal case, if there was no identifiable fingerprint left behind at the crime scene, there was no case. Blood tests can determine who is not the parent, not who is. A test has now been developed that provides hundred per cent positive identification. The test is called **DNA fingerprinting**. The test of DNA fingerprinting can show conclusively whether the genetic material in a drop of blood matches that of the suspect, or it can be used to solve paternity cases.

The technique of DNA fingerprinting relies on developments from recombinant DNA technology and allows an examination of each individual's unique genetic blueprint—DNA. The technique was discovered in England by **Alec Jeffreys** (Born 1950). It is based on the fact that DNA of each individual is interrupted by a series of identical DNA sequences called **repetitive DNA** or **tandem repeats** *i.e.*, VNTR or variable-number of tandem-repeats loci. The pattern, length and number of these repeats are unique for each individual. **Jeffreys** developed a series of DNA probes, which are short pieces of DNA that seek out any specific sequences they match, and base pair with that sequence. Such molecular probes are used to detect the unique repetitive DNA patterns characteristic of each individual. The procedure of DNA fingerprinting has the following steps: 1. DNA is purified from a small sample of blood, semen, or other DNA-bearing cells, and digested into smaller fragments with restriction endonucleases. 2. The fragments are separated by agarose gel electrophoresis. 3. The separated fragments are transferred to a nylon membrane by the technique of Southern blotting. 4. The DNA probes labelled with radioactive material are added to a solution containing the nylon membrane. 5. Wherever the probes fit a band containing respective DNA sequences, they attach. 6 The X-ray film is pressed against the nylon filter and exposed at bands carrying the radioactive probes attached to the fragments. 7. The patterns of bands obtained on the film is 100 per cent unique for each person, except for identical twins who would have the same pattern.

The **forensic application** of the DNA fingerprinting technique involves a comparison between the DNA fingerprint obtained from cells at a crime scene (*i.e.*, blood cells, hair, sperm cells, etc.) with a DNA fingerprint from cells provided by the suspect. If the DNA pattern matches exactly, certain identification is made. For **paternity determination**, DNA fingerprints of the mother, child and alleged father are compared. In this cases, one-half of the bands in the child comes from the mother and the other half from the father. All the paternal bands in child's DNA fingerprint must match with the alleged father for positive paternity identification.

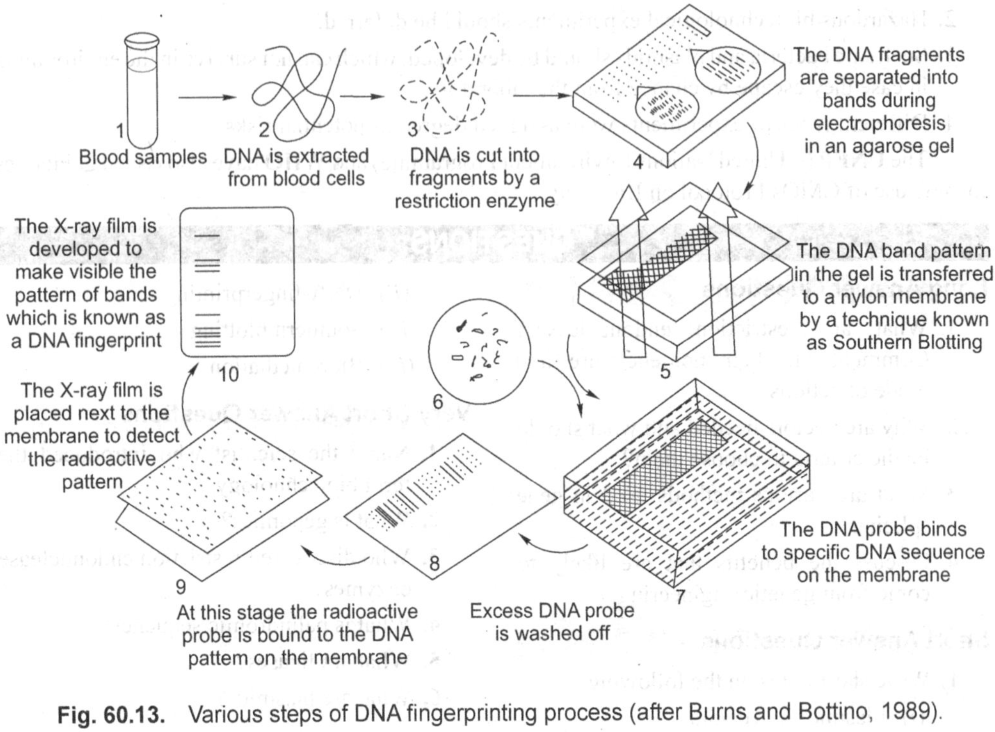

Fig. 60.13. Various steps of DNA fingerprinting process (after Burns and Bottino, 1989).

In India, DNA fingerprinting tests are carried out at the Centre for Cell and Molecular Biology (CCMB), Hyderabad. For this purpose, a test with the **BKM-DNA probe** (=banded krait minor satellite DNA) earlier used for identification of sex chromosomes (by **Dr. Lalji Singh**) has been found to cost one-tenth of the cost of tests used in Europe and U.S.A. Paternity dispute cases are much more common in India and most of them are referred to CCMB for DNA evidence. The first such test on DNA fingerprinting was used in June, 1989 to settle a drawn-out paternity case in Madras.

Biosafety

With the expansion in the field of genetic engineering and commercialization of transgenic organisms, scientists have expressed fear about the use of such genetically engineered microorganisms, transgenic plants and animals, that they might disturb the ecosystem. The **genetically engineered microorganisms (GEMs)** can disturb the ecosystem either (1) by multiplying rapidly and replacing the native microbes or (2) by transferring genes related to virulence or pathogenesis into native microbial population and increasing their virulence or changing harmless microbes into pathogenic microbes. The transgenic plants or genetically modified plants (GM) could pose biological and ecological risks by (*i*) the production of toxic or allergic metabolites, (*ii*) introducing unexpected new susceptibilities to pathogens, (*iii*) by transmission of new traits to sexually compatiable weeds, (*iv*) by disturbing ecosystems through persistence or altered reaction in parasites, symbionts and competitors, (*v*) by the escape of transgenes provide resistance against antibiotics into other neighbouring plants, and (*vi*) by providing resistance in insects against pesticides and in weeds against herbicides.

Realizing the possible hazards of cloning recombinant DNA technology, **National Institute of Heath (NIH)**, U.S.A., established the Recombinant Adivisory Committee (RAC) in 1974 which gave following recommendations:

1. Recombinant DNA work should be carried out with appropriate safety measures.
2. Hazardous biotechnological experiments should be deferred.
3. Such safe bacteria and plasmids should be developed, which can not survive in the environment in case they escape by chance from the laboratory.
4. Different types of experiments were assigned degree of potential risks.

The **UNEP** (= United Nations Environment Programme) and **WHO** have also issued guidelines for safe use of GMOs Protocol on Biosafety.

QUESTIONS

Long Answer Questions

1. What are restriction endonucleases? Comment on their nomenclature and mode of actions.
2. Why are vectors needed and what should be the criteria to choose them?
3. What are various methods used in gene splicing?
4. Discuss the benefits that are likely to come from genetic engineering.

Short Answer Questions

1. Write short notes on the following:
   (*i*) Plasmids as vectors
   (*ii*) DNA-fingerprinting
   (*iii*) Southern blotting
   (*iv*) Bioremediation

Very Short Answer Questions

1. Name the scientist who introduced the term biotechnology.
2. What is genomics?
3. Who discovered restriction endonuclease enzymes?
4. What is palindromic sequence?
5. What are ligases?
6. What are plasmids?

7. Why only type II restriction endonucleases are used in recombinant DNA technology?
8. Why the name restriction endonuclease is used for these endonucleases?
9. Which enzyme act as a polymerizing enzyme during artificial synthesis of DNA?
10. What is proteomics?
11. What are knockout mice?
12. What is superovulation?

Multiple Choice Questions

1. Introduction of foreign gene for improving genotype is called
(a) tissue culture
(b) vernalization
(c) genetic engineering
(d) eugenics
2. What does Bt stand for in popular crop of Bt cotton?
(a) Biotechnology
(b) *Bacillus tomentosa*
(c) *Bacillus thuringiensis*
(d) None of these
3. 'Molecular scissors' used in genetic engineering is
(a) DNA polymerase
(b) DNA ligase
(c) restriction endonuclease
(d) helicase
4. Restriction endonuclease acts on
(a) any fragment of DNA
(b) (+) fragment of DNA
(c) (–) fragment of DNA
(d) double helical DNA
5. One of the following is used to join the segments of DNA during genetic engineering
(a) lipase (b) ligase
(c) gyrase (d) helicase
6. Which of the following is a genetic vector?
(a) plasmid (b) phage
(c) cosmid (d) all of these
7. The transfer of genetic material from one cell to another by a phage is called
(a) transformation (b) conjugation
(c) transduction (d) hybridization
8. Plasmids are autonomously replicating minichromosomes found in
(a) bacteriophage
(b) *Escherichia coli*
(c) *Paramecium*
(d) *Euglena*
9. The Ti plasmid is often used for making transgenic plants. This plasmid is found in
(a) *Azobacter*
(b) *Rhizobium* of the roots of leguminous plant
(c) *Agrobacterium*
(d) Yeast
10. Vectorless gene transfer includes
(a) particle gene
(b) microinjection
(c) electroporation
(d) all the above
11. A collection of organisms, usually viruses, bacterial or yeast, which have been transformed by the addition of extra genes from another species forms
(a) gene library (b) gene pool
(c) gene replication (d) gene cloning
12. Recent technique used for separating fragments of DNA is
(a) northern blotting
(b) southern blotting
(c) eastern blotting
(d) western blotting
13. A technique used to make numerous copies of a specific segment of DNA quickly and accurately
(a) translation
(b) transcription
(c) ligase chain reaction
(d) polymerase chain reaction
14. The main technique involved in agriculture biotechnology is called
(a) tissue culture
(b) transformation
(c) plant breeding
(d) DNA replication

ANSWERS

Very Short Answer Questions

1. **Karl Ereky** (1917) coined the term biotechnology.
2. Genomics is a computer based study and designing of genome of human being or any organism.
3. **W. Arber** (1962) discovered restriction endonuclease enzymes.
4. Palindromic sequences are base sequences in two strands of DNA that read the same in $5' \rightarrow 3'$ direction.
5. Ligases are enzymes found in all living cells that join together either broken ends of DNA or the ends of foreign DNA (insert DNA) and vector DNA.
6. Plasmids are small, circular DNA molecules that occur naturally in bacterial cells and mutliply along with host cell.
7. Only type II restriction endonuclease used for recombinant DNA technology and restriction mapping because they produce cuts within the restriction sites in both the strands of DNA molecule and the cut ends have palindromic sequence of nitrogen bases.
8. The name restriction endonuclease is used for those endonuclease enzymes because they restrict bacteriophages from infecting the bacterial cell.
9. DNA polymerase III.
10. Proteome represents the complete set of proteins expressed during cell's life span and proteomics is the study of proteome.
11. Knockout mice is a strain of mice in which specific genes have been either replaced or disrupted so that they do not produce specific phenotype related to them.
12. When a female mammal is stimulated with specific hormones for the development and maturation of mutiple ovarian follicles, the phenomenon is called **super ovulation** or **stimulated ovarian cycle.**

Multiple Choice Questions

1. (*c*)	**2.** (*c*)	**3.** (*c*)	**4.** (*d*)	**5.** (*b*)	**6.** (*d*)	**7.** (*c*)
8. (*b*)	**9.** (*c*)	**10.** (*d*)	**11.** (*a*)	**12.** (*b*)	**13.** (*d*)	**14.** (*a*)

Glossary

Acentric A part of the whole chromosome which is without the centromere.

Acetyl CoA A metabolic intermediate produced through catabolism of compounds, including fatty acids, and used as the initial substrate for the central respiratory pathway, the TCA (Krebs cycle).

Acrocentric A chromosome with centromere near one end, so that one arm is very short.

Actin A globular cytoskeletal protein that polymerizes to form a flexible, helical filament capable of interacting with myosin. Actin filaments provide mechanical support for eukaryotic cells, determine's the cell's shape, and enable cell movements.

Actin-binding proteins Any of nearly 100 different proteins belonging to numerous families that affect the assembly of actin filament, their physical properties, and their interactions with one another and with other cellular organelles.

Action potential The collective changes in membrane potential, beginning with depolarization to threshold and ending with return to resting potential, that occur with stimulation of an excitable cell and act as the basis for neural communication.

Activation energy The minimal kinetic energy needed for a reactant to undergo a chemical reaction.

Active site The part of an enzyme molecule that is directly involved in binding the substrate.

Active transport It is a energy-requiring process in which a substance binds to a specific transmembrane protein, changing its conformation to allow passage of the substance through the membrane against the electrochemical gradient for that substance.

Adenine A nitrogen having purine base which occur in DNA, and RNA. Pairs normally with thymine in DNA.

Adenosine triphosphate (ATP) It is a nucleotide consisting of adenosine binding to three phosphate groups; it is the principal immediate-energy source for prokaryotic and eukaryotic cells.

Adherens junctions (Zonulae adherens) Adherens junctions are a type of specialized adhesive junctions particularly common in epithelia. The plasma membranes in this region are separated by 20 to 35 nm and are sites where cadherin molecules are concentrated. The cells are held together by linkages between the extra cellular domains of cadherin molecules that bridge the gap between neighbouring cells.

Aerobes Organisms dependent on the presence of oxygen to metabolize energy-rich compounds.

Allele (also allelomorph; adj. allelic or allelomorphic) One member of a pair or a series of genes that can occur at a particular locus on homologous chromosomes.

Allopolyploid A polyploid having whole chromosome sets from different species.

Allosteric modulation Modification of the activity of an enzyme through interaction with a compound that binds to a site (*i.e.*, allosteric site) other than the active site.

Alpha (α) helix One possible secondary structure of polypeptides, in which the backbone of the chain forms a spiral (*i.e.*, helical) conformation.

Alternative splicing Widespread mechanism by which a single gene can encode two or more related proteins.

Amide bond	The chemical bond that forms between carboxylic acids and amines (or acidic and amino functional groups) while producing a molecule of water.
Amino acids	The monomeric units of proteins; each is composed of three functional groups attached to a central, α carbon: an amino group, a defining side chain, and a carboxyl group.
Aminoacyl-tRNA synthetase (AARS)	An enzyme that covalently links amino acids to the 3′ ends of their cognate tRNAS. Each amino acid is recognized by a specific aminoacyl-tRNA synthetase.
Amitosis	Direct nuclear division, without the separation of daughter chromosomes..
Amphidiploid	A tetraploid individual having two sets of chromosomes from each of two known ancestral species.
Amphimixis	To bring together the elements from two gametes in fertilization.
Amphipathic	Molecules having spatially separated hydrophilic and hydrophobic regions.
Amphoteric	Structural property allowing the same molecule to act as an acid or as a base.
Anabolic pathway	A metabolic pathway resulting in the synthesis of relatively complex products.
Anaerobes	Organisms that utilize energy-rich compounds through oxygen-independent metabolic pathways such as glycolysis and fermentation.
Anaphase	The stage of mitosis during which sister chromatids separate from one another.
Anaphase A	The movement of chromosomes toward the poles during mitosis.
Anaphase B	Elongation of the mitotic spindle, which causes the two spindle poles to move farther apart.
Androgenesis	The phenomenon of parthenogenesis by male.
Aneuploidy	A karyotypic abnormality or variation in chromosome number by whole chromosomes, but less than an entire set, *e.g.*, 2n+l (trisomy), 2n–1 (monosomy).
Angiogenesis	The formation of new blood vessels.
Angstrom (A°)	0.1 nanometer (nm) or 0.0000001 millimeter.
Animal models	Laboratory animals that exhibit characteristics of a particular human disease.
Animal pole	That portion of egg which has active cytoplasm and zygote nucleus.
Anion	An ionised atom or molecule with a negative charge.
Anisogamy	The fusion of unequal sized gametes.
Annulate lamellae	The pores and lamellae (*i.e.*, the pore complex of nuclear envelope) of most eukaryotes and endoplasmic reticulum of invertebrate and oocytes and spermatocytes of vertebrates.
Antibody	An immunoglobulin protein produced by plasma cells derived from B lymphocytes that interacts with the surface of a pathogen or foreign substance to facilitate its destruction.
Anticodon	A triplet of nucleotides specific to each tRNA corresponding to a particular amino acid and complimentary to the codon of mRNA.
Antigen	Any substance recognised by an immune system as foreign to the organism.
Antiserum	Fluid containing desired antibodies that remains after the removal of the cells and clotting factors from whole blood that has been exposed to given antigen.
Apoenzyme	The protein component of a holoenzyme which also contains a nonprotein coenzyme or prosthetic group.
Apomixis	The occurrence of the external form of sexual reproduction without the occurrence of meiosis and fertilization.
Apoptosis	A type of orderly or programmed cell death in which the cell responds to certain signals by initiating a normal response that leads to the death of the cell. Death by apoptosis is characterised by overall compaction of the cell and its nucleus, the orderly dissection of the chromatin into pieces at the hands of a special DNA-splitting endonuclease, and the rapid engulfment of the dying cell by phagocytosis.

Artifact A structure seen in a microscopic image that results from the coagulation or precipitaion of materials that had no existence in the living cell.

Asexual reproduction The process of reproduction that does not involve fusion of cells.

Assay Some identifiable feature of a specific protein such as the catalytic activity of an enzyme, used to determine the relative amount of that protein in a sample.

Aster "Sunburst" arrangement of microtubules around each centrosome during mitosis.

ATP synthase The ATP-synthesizing enzyme of the inner mitochondrial membrane, which is composed of two chief components: The F_1 of headpiece and the F_0 basal piece, the latter of which is embedded in the membrane itself.

Auto-antibodies Antibodies against components of the body's own tissues.

Autoimmune diseases Diseases characterized by an attack of the immune system against the body's own tissues. Includes multiple sclerosis, insulin-dependent diabetes.

Autophagy The destruction of organelles and their replacement during which an organelle is surrounded by a double membrane. The membrane surrounding the organelle then fuses with a lysosome.

Autopolyploidy A polyploidy all of whose sets of chromosomes are those of the same species.

Autoradiography A method for localizing radioactive atoms in microscope preparations of biological materials by exposing of a photographic film emulsion to radioactive atoms incorporated in the biological specimen including cell.

Autosome A chromosome not associated with the sex of the individual and therefore possessed in matching pairs by diploid members of both sexes.

Auxotroph An individual unable to carry on some particular synthesis hence requiring supplementing of minimal medium by some growth factor.

Axon A single, prominent extension that emerges from the cell body and conducts outgoing impulses away from the cell body and towards the target cell(s).

Axonal transport Process by which vesicles, cytoskeletal polymers, and macromolecules are moved along microtubules within the axon of a neuron. Anterograde transport moves materials from the cell body toward the synaptic terminals, whereas retrograde transport moves materials in the opposite direction.

Axoneme The central, microtubule-containing core of a cilium or flagellum. Most axonemes consist of nine peripheral doublets, two central microtubules and numerous accessory structures.

Bacillus Rod-shaped bacterium.

Bacteriophage The viruses attacking the bacterial cells.

Barr body (sex chromatin) A mass of chromatin in the nucleus of resting cells, resulting from inactivation of an X chromosome.

Basal body A structure that resides at the base of the cilium or flagellum and which generates their outer microtubules. Basal bodies are identical in structure to centrioles. Both can give rise to one another.

Base excision repair (BER) A cut and patch mechanism for the removal from the DNA of altered nucleotides, *e.g.*, uracil (formed from cytosine) and 8-oxoguanine.

Basement membrane (basal lamina) Thickened layer of approximately 50 to 200 nm of extra-cellular matrix that surrounds muscle and fat cells and underlies the basal surface of epithelial tissues such as the skin, the inner lining of the digestive and respiratory tracts, and the inner lining of blood vessels.

Beta (β) clamps One of the non-catalytic components of the replisome that encircles DNA and keeps the polymerase associated with DNA template.

Beta-oxidation Splitting of fatty acids into two carbon units.

Bioenergetics	The study of the various types of energy transformations that occur in living organisms.
Biogenesis	The production or orgin of livings from non living things.
Bivalent	A pair of synapsed homologous chromosomes, whether or not these have yet replicated to form chromatids.
Budding	A mode of asexual reproduction in which new organisms develops from the parent body in the form of an outgrowth.
C_3 pathway	The metabolic pathway by which carbon dioxide is assimilated into the organic molecules of the cell during photosynthesis. RuBP (= ribulose 1,5 biphosphate) is used by Rubisco as the initial CO_2 acceptor. The product then fragments into two three-carbon PGA molecules.
C_4 pathway	Alternate pathway for carbon fixation utilizing phosphoenolypyruvate as the CO_2 acceptor to produce four carbon compounds (predominantly malate and oxaloacetate).
Cadherins	A family of related glycoproteins that mediate Ca^{2+}-depends cell-cell adhesion.
Calmodulin	A small calcium-binding protein that is widely distributed. Each molecule of calmodulin contains four binding sites for calcium.
Calvin cycle (or Calvin-Benson cycle)	The pathway for conversion of CO_2 into carbohydrates, the cycle occurs in cyanobacteria and all eukaryotic photosynthetic cells.
Carbohydrates (Glycans)	Organic molecules including simple sugars (monosaccharides, *e.g.*, glucose) and multisaccharide polymers, which largely serve as energy-storage (*e.g.*, starch, glycogen) and structural compounds (*e.g.*, cellulose) in cells.
Carcinogen	A chemical substance or physical agent (Such as radiation) that causes cancer.
Carotenoid	Any of a group of yellow, orange or red pigments associated with chlorophylls in plants and in animal fat.
Carrier	A transport protein within the membrane that binds temporarily with a molecule to be transported across the membrane.
Caspases	A family of cysteine proteases that are activated at an early stage of apoptosis and are responsible for the degradative events observed during cell death.
Catabolism	Breakdown of organic molecules within the cell with the release of energy.
Catalyst	A substance that affects the rate of chemical reaction but is not effected itself in the process.
Cation	An ionized atom or molecule with an extra positive charge.
Cell	A structural and physiological unit of living organisms (Bacteria, plants and animals) which have protoplasm.
Cell concept	A generalization of the cell theory, stating that the cell is the ultimate structural unit of the organism.
Cell culture	The technique used to grow cells outside the organisms.
Cell cycle	The sequence of events in dividing cells in which an interphase consisting of G_1, S and G_2 periods separates one mitosis from the mitosis of successive cell cycle.
Cell division	A method by which new cells are originated from preexisting cells.
Cell fractionation	Bulk separation of the various cell organelles by differential centrifugation.
Cell-free system	An experimental system to study cellular activities that do not require whole cell. Such systems typically contain a preparation of purified proteins and/or subcellular fractions and are amenable to experimental manipulation.
Cell fusion	Technique whereby two different types of cells (from one organism or from different species) are joined to produce one cell with one, continuous plasma membrane.
Cell-mediated immunity	Carried out by T lymphocytes, that when activated, can specifically recognize and kill an infected (or foreign) cell.

Cell line	Cells that are commonly used in tissue culture studies that have undergone genetic modifications that allow them to be grown indefinitely.
Cell plate	Structure between the cytoplasm of two newly formed daughter cells that gives rise to a new cell wall in plant cells.
Cell signaling	Communication in which information is relayed across the plasma membrane to the cell interior and often to the cell nucleus by means a series of molecular interactions.
Cell theory	Theory of biological organization, which has three tenets: all organisms are made up of one or more cells; the cell is the structural unit of life, cells only arise from the division of preexisting cells.
Cell wall	A rigid, nonliving (Karp 2010) or dynamic (Buchanan *et al.*, 2000) structure that provides support and protection for the cell it surrounds.
Cellulose	Unbranched glucose polymer with $\beta(1 \rightarrow 4)$ linkages that assembles into cables and serves as a principal structural element of plant cell walls.
Centrioles	Cylindrical structures, about 0.2 μm across and typically about twice as long, that contain nine evenly spaced fibrils, each of which appears in cross section as a band of three microtubules. Centrioles are nearly always found in pairs, with each of the members situated at right angles to one another.
Centromere	Marked indentation on a mitotic chromosome that serves as the site of kinetochore formation.
Centrosome	A complex structure that contains two barrel-shaped centrioles surrounded by amorphous electron-dense pericentriolar material (or PCM) where microtubules are nucleated.
Chaperones	Proteins that bind to other polypeptides, preventing their aggregation and promoting their folding and/or assembly into multimeric proteins.
Chaperonins	Members of the Hsp60 class of chaperones, *e.g.*, GroEl, that form a cylindrical complex of 14 subunits within which the polypeptide folding reaction takes place.
Check point	Mechanism that halt the progress of the cell cycle if (1) any of the chromosomal DNA is damaged, or (2) certain critical processes, such as DNA replication or chromosome alignment during mitosis have not been properly completed.
Chemiosmotic mechanism	The mechanism for ATP synthesis whereby the movement of electrons through the electron-transport chain results in establishment of a proton gradient across the bacterial, thylakoid, or inner mitochondrial membrane, with the gradient acting as a high-energy intermediate, linking oxidation of substrates to the phosphorylation of ADP.
Chiasma (pl. chiasmata)	The point of contact and interchange between the chromatids of two homologous chromosomes during prophases of meiosis.
Chloroplasts	Cytoplasmic organelles containing several pigments, particularly the light absorbing chlorophylls.
Chromatid	One of the two identical, longitudinal halves of a chromosome which share a common centromere with a sister chromatid,; results from the replication of chromosomes during a nuclear division.
Chromatin	Nuclear material (DNA, RNA, histone and non-histone proteins) composing the chromosomes.
Chromatography	A term used for a wide variety of techniques in which a mixture of dissolved components is fractionated as it moved through some type of immobile matrix.
Chromocentre	Irregular mass of heterochromatin.
Chromomere	Small, stainable thickenings arranged linearly along a chromosome.
Chromonema (pl. chromonemata)	One of the delicate, helically coiled, thread-like filaments composing a chromosome.

Chromosome Nucleoprotein structures, generally more or less rod-like during nuclear division, the physical sites of nuclear genes which are arranged in linear order. Each species has a characteristic number of chromosomes although individuals with fewer or more than this characteristic number occur, specially in plants.

Cilia Hair like motile organelles that project from the surface of a variety of eukaryotic cells. Cilia tend to occur in large numbers on a cell's surface.

Ciliary or axonemal dynein A huge protein (MW about 2 million daltons) responsible for conversion of the chemical energy of ATP into the mechanical energy of ciliary locomotion.

***Cis* cisternae** The cisternae of the Golgi complex closer to the endoplasmic reticulum.

Cistron A segment of DNA specifying one polypeptide chain in protein synthesis. Under the concept of triplet code, one cistron must contain three times as many nucleotide pair as amino acids in the chain it specifies.

Clonal selection theory Well supported theory that B and T lymphocytes develop their ability to produce specific antibodies or T cell receptors prior to exposure to antigen.

Clone A line of cells derived from a single cell and presumed to contain same genetic information. The multicellular organisms that are propagated by vegetative (non-sexual) reproduction are also called clones.

Coated vesicles Vesicle that bud from a membrane compartment typically possess a multi-subunit protein coat that promotes the budding process. COPI–, COPII– and clathrin-coated vehicles.

Codon Sequences of three nucleotides (nucleotide triplets) in mRNAs that specify amino acids.

Coenocyte Mass of cytoplasm, containing nuclei, formed as a result of repeated divisions of the cytoplasm.

Coenzyme An organic nonprotein component of an enzyme.

Cofactor The nonprotein component of an enzyme, it can be either inorganic or organic.

Cohesin A multiprotein complex that keeps replicated chromatids associated with one another until they are separated during cell division.

Co-linearity The spatial correlation between codons in DNA and amino acids in the polypeptide translated from the DNA blueprint.

Collagens A family of fibrous glycoproteins known for their high tensile strength that function exclusively as part of the extracellular matrix.

Colloid A chemical substance having particles which range from 1 mm to 100 mm in size.

Complementary base pairing Specific hydrogen bond interactions between a particular purine and a particular pyrimidine component in nucleic acids; for example, guanine and cytosine or adenine and thymine or uracil.

Compound A chemical substance formed by the chemical combination of two or more simple substances.

Conductance The movement of small ions across membranes.

Conformation The three dimensional arrangement of the atoms within a molecule, often important in understanding the biological activity of proteins and other molecules in a living cell.

Congression The movement of duplicated chromosomes to the metaphase plate during prometaphase of mitosis.

Conjugation Side by side temporary association of two bodies, as of synapsed chromosomes in meiosis, or of two organisms during sexual reproduction.

Connexon Multisubunit complex of a gap junction formed from the clustering within the plasma membrane of an integral membrane protein called connexin. Each connexon is composed of six connexin subunits arranged around a central opening (or annulus) about 1.5 nm in diameter.

Consensus sequence	The most common version of a conserved sequence. The TTGACA sequence of a bacterial promoter (known as the—35 element) is an example of a consensus sequence.
Conserved sequences	Refers to the amino acid sequences of particular polypeptides or the nucleotide sequences of particular nucleic acids. If two sequences are similar to one another, *i.e.*, homologous, they are said to be conserved, which indicates that they have not diverged very much from a common an ancestral sequence over long periods of evolutionary time.
Constitutive heterochromatin	Chromatin that remains in the compacted state in all cells at all times and, thus, represents DNA that is permanently silenced. It consists primarily of highly repeated sequences.
Constitutive	Constant or unchanging; for example, a constitutive enzyme which is synthesized at a constant rate and not subject to regulation.
Conventional (or type II) myosins	A family of myosins, first identified in muscle tissue, that are the primary motors for muscle contraction but are also found in a variety of non-muscle cells. Type II myosins are needed for spitting a cell in two during cell division and generating tension at focal adhesions.
Cotransport	A process that couples the movement of two solutes across a membrane, termed symport if the two solutes are moved in the same direction and antiport if the two move in the opposite directions.
Covalent bonds	Strong chemical bonds between atoms created when they share pairs of electrons (*e.g.*, bond between carbon and hydrogen, C—H).
Cristae	The many deep folds that are characteristic of the inner mitochondrial membrane, which contain the molecular machinery of oxidative phosphorylation.
Crossing over (genetic recombination)	1. Reshuffling of the genes on chromosomes (thereby disrupting linkage groups) that occurs during meiosis as a result of breakage and reunion of segments of homologous chromosomes. 2. A process in which genes are exchanged between non-sister chromatids of homologous chromosomes. Chiasmatas are visible evidences of crossing over.
Cyclic AMP (cAMP)	Adenosine monophosphate molecule arranged in cyclic form; its formation within cells is triggered by the action of hormones; cAMP functions to control cellular biochemistry.
Cyclin-dependent kinases (Cdks)	Enzymes that control progression of cells through the cell cycle.
Cytochrome system	Group of iron-bearing enzymes located in mitocbondria through which electrons pass; in the process energy is trapped into ATP molecules.
Cytochromes	A type of electron carrier consisting of a protein bound to a heme group.
Cytogenetics	Study of the cellular structures and mechanisms associated with genetics.
Cytokines	Proteins secreted by cells of the immune system that alter the behaviour of other immune cells.
Cytokinesis	The division of cytoplasm during cell division.
Cytology	The study of structure and function of cell.
Cytoplasm	The portion of cellular protoplasm which occurs between the plasma membrane and nuclear membrane.
Cytosine	A pyrimidine base occurring in DNA and RNA. Pairs with guanine in DNA.
Dalton	A measure of molecular mass, with one dalton equivalent to one unit of atomic mass (*e.g.*, the mass of a H atom).
Dehydrogenase	An enzyme that catalyzes a redox reaction by removing a hydrogen atom from one reactant.
Deletion	At the molecular level, the removal of one or more bases from a DNA sequence. At the cytological level, the absence of segment of a chromosome (also known as deficiency).

Denaturation 1. Separation of DNA double helix into its two component strands. 2. The unfolding or disorganization of a protein from its native or fully folded state.

Dendrites Fine extensions from the cell bodies of most neurons; dendrites receive incoming information from external sources, typically other neurons.

Deoxyribonucleic acid (DNA) A usually double-stranded, helically coiled, nucleic acid molecule, composed of phosphate-deoxyribose "back bones" which are connected by paired nitrogen bases (purines and pyrimidines) attached to the deoxyribose sugar, the genetic material of all living organisms and many viruses.

Depolarization A decrease in the electrical potential difference across a membrane.

Desmosome (maculae adherens) Disc-shaped adhesive junction containing cadherins found in a variety of tissues, but most notably epithelia, where they are located basal to the adherens junction. Dense cytoplasmic plaques on the inner surface of the plasma membranes in this region serve as sites of anchorage for looping intermediate filaments that extend into the cytoplasm.

Diakinesis A stage of prophase-I of meiosis in which paired chromosomes are much shortened and thickened.

Dictyosome The Golgi complex of plant cells which has no secretory function.

Differentiation Specialization of cells to perform specific functions in the organism.

Diffusion Migration or molecules or ions as a result of their own random movements from a region of higher concentration to a region of lower concentration. It requires no energy (also called free diffusion).

Dioecious Individuals producing either sperm or egg, but not both. In dioecious species, the sexes are separate.

Diploid An individual or cell having two complete sets of chromosomes.

Diplotene A stage of prophase–1 of meiosis in which each of the synaptic chromosomes gets doubled by splitting.

Disjunction The separation of homologous chromosomes during anaphase–1 of meiosis.

DNA gyrase A type II topoisomerase that is able to change the state of supercoiling in a DNA molecule by relieving the tension that builds up during replication. It does this by traveling along the DNA and acting like a "swivel", changing the positively supercoiled DNA into negatively supercoiled DNA.

DNA methylation An epigenetic process in which methyl groups are added to cytosine residue in DNA by DNA methyltransferases. In vertebrates, DNA methylation occurs on certain CpG residues in the promoter regions of genes and is associated with inactivation of gene expression.

DNA polymerases The enzymes responsible for constructing new DNA strands during replication or DNA repair.

DNA tumor viruses Viruses capable of infecting vertebrate cells, transforming them into cancer cells.

DNA-dependent RNA polymerases (RNA) The enzymes responsible for transcription in both prokaryotic and eukaryotic cells.

Domain A region within a protein (or RNA) that folds and functions in a semi-independent manner.

Double-strand breaks (DSBs) DNA damage often resulting from ionizing radiation involving a fracture of both strands of the double helix.

Duplication The occurrence of the segment twice in the same chromosome or in the same complement.

Dyad A pair of homologous chromosomes before their duplication in the meiosis.

Dynein An exceptionally large, cargo-carrying multisubunit motor protein that moves along microtubules toward their minus end. This family of proteins occurs as cytoplasmic dyneins and ciliary or axonemal dyneins.

Effector A substance that brings about a cellular response to a signal.

Egg (egg cell) The female gamete.

Electrophoresis A technique for separating molecules, particularly proteins, according to over all electric charge of the molecules.

Embryonic stem (ES) cells A type of cell that has virtually unlimited powers of differentiation, found in the mammalian blastocyst, which is an early stage of embryonic development comparable to the blastula of other animals.

Endergonic reactions Reactions that are thermodynamically unfavorable and cannot occur spontaneously, possessing a $+\Delta G$ value.

Endocytosis Mechanism for the uptake of fluid and solutes into a cell. Can be divided into two types: bulk-phase endocytosis, which is non-specific, and receptor-mediated endocytosis, which requires the binding of solute molecules such as LDL or transferrin to a specific cell-surface receptor.

Endomembrane system Functionally and structurally interrelated group of membranous cytoplasmic organelles including the endoplasmic reticulum, Golgi complex, endosomes, lysosomes and vacuoles.

Endoplasmic reticulum (ER) A tubular double membrane system in the cytoplasm, continuous with the nuclear membrane and bearing many ribosomes. Contain many enzymes and have skeletal, transporting and biosynthetic functions of the cytoplasm.

Endosomes Organelles of the endocytic pathway. Materials taken up by endocytosis are transported to early endosomes where they are sorted, and then on to late endosomes and ultimately to lysosomes. Late endosomes also function as destination sites of lysosomal enzymes transported from the Golgi complex.

Endosymbiont theory Proposal, which is based on considerable evidence, that mitochondria and chloroplasts arose from symbiotic prokaryotes that took up residence within a primitive host cell.

End-product inhibition A sequence of chemical reactions wherein the last product formed inhibits one of the early steps; a form of negative feedback which limits the rate of chemical reactions.

Energy Capacity for doing work. It exists in two forms: potential and kinetic.

Enhancer A regulatory site in the DNA that may be located at considerable distance either upstream or downstream from the promoter that it regulates. Binding of one or more transcriptional factors to the enhancer can dramatically increase the rate of transcription of the gene.

Entropy (S) A measure of the relative disorder of the system or universe associated with random movements of matter; because all movements cease at absolute zero (OK), entropy is zero only at that temperature.

Enzyme inhibitor Any molecule that can bind to an enzyme and decrease its activity. It is of two types: noncompetitive or competitive.

Enzyme Any substance, protein in whole or in part that regulate the rate of a specific biochemical reaction in living organisms.

Epigenetic inheritance Heritable changes, *i.e.*, changes that can be transmitted from one cell to its progeny, that do not involve changes in DNA sequence. Epigenetic changes can result from DNA methylation, covalent modification of histones, and likely other types of chromatin modifications.

Equatorial plate The figure formed at the spindle equator in nuclear division.

Euchromatin Chromatin that returns to its dispersed state during interphase.

Eukaryotes Organisms (*e.g.*, plants, animals, protists and fungi) with a well-defined nucleus enclosed in a nuclear membrane and usually having one or more other membranous subcellular compartments.

Euploidy	Variation in chromosome number by whole sets or exact multiple of the monoploid (haploid) number, *e.g.*, diploid, triploid. Euploids above the diploid level may be referred to collectively as polyploids.
Exocytosis	A mode of transport of substances out of the cell by enclosure in a portion of the plasma membrane and subsequent expulsion to the outside.
Exocytosis	The process of membrane fusion and content discharge during which the membrane of a secretory granule or vesicle comes into contact with the overlying plasma membrane with which it fuses, thereby forming an opening through which the contents of the granule or vesicle can be released.
Exons	Those parts of a split gene that contribute to a mature RNA product.
Exonuclease	A DNA or RNA-digesting enzyme that attaches to either the 5′ or 3′ end of the nucleic acid strand and removes one nucleotide at a time from that shrinking end.
Extracellular matrix (ECM)	An organised network of extracellular matrials that is present beyond the immediate vicinity of the plasma membrane. It may play an integral role in determining the shape and activities of the cell.
Facilitated diffusion	Assisted passage (transport) of molecules across the membrane toward their lower concentration along a gradient.
Feedback inhibition	A mechanism to control metabolic pathways where the end product interacts with an enzyme in the pathway, resulting in inactivation of the enzyme.
Fermentation	Anaerobic decomposition of an organic substance in the body of a living organism; glycolysis.
Fertilization	The fusion of cytoplasm and male and female pronuclei of male gamete (sperm or pollen) and a female gamete (ovum) to form a zygote in sexually reproducing organisms.
Feulgen reaction	Chemical staining process which reacts specifically with genetic material (DNA) in cells.
Fission	A mode of asexual reproduction in which the body of the organism divide transversely or longitudinally into two or more equal parts.
Flagella	Hairlike motile organelles that project from the surface of a variety of eukaryotic cells.
Fluid-mosaic model	Model presenting membranes as dynamic structures in which both lipids and associated proteins are mobile and capable of moving within the membrane to engage in interactions with other membrane molecules.
Fractionated	Disassembly of a preparation into its component ingredients so that the properties of individual species of molecules can be examined.
Free radical	Highly reactive atom or molecule that contains a single unpaired electron.
Gamete	A protoplast (sex cell) which in the process of sexual reproduction, fuses with another protoplast. One of them may be termed as male and another female.
Gametogenesis	The origin, maturation and differentiation of gametes from the generative cells.
Gel	Jelly-like state of protoplasm.
Gene amplification	Differential replication of some genes, producing many copies of these genes.
Gene	The particulate determiner of a hereditary trait; a particular segment of DNA molecule, generally located in the chromosome.
Genetic code	The code that relates nucleotide sequences in nucleic acids to amino acid sequences. Each triplet of nucleotide designates a particular amino acid; thus, the genetic code allows the translation of information stored in DNA and the use of that information in protein synthesis.
Genome	A complete set of chromosomes or of chromosomal genes, inherited as an unit from one parent.
Gluconeogenesis	Synthesis of carbohydrates from noncarbohydrate precursors such as fats or proteins.

Glycolysis	The oxidation of sugar (glucose) or glycogen to pyruvic acid without the use of oxygen; also known as the Embden-Meyerhof pathway.
Golgi complex	A cytoplasmic organelle which have a complex system of tubules and vesicles and which is related with the production of cellular secretions, also called Golgi apparatus.
Guanine	A purine base occurring in DNA and RNA. Pairs normally with cytosine in DNA.
Haemoglobin	An iron protein pigment of blood functioning in oxygen carbon dioxide exchange of living cells.
Haploid	An organism or cell that has one copy of each chromosome.
Hard keratin	A keratin used for production of structures such as hair, nails and horns.
Heat-shock proteins	A highly conserved group of chaperone proteins expressed in cells exposed to elevated temperatures or other forms of environmental stress.
Hedgehog	A secreted signaling molecule that stimulates a pathway regulating cell fate during embryonic development.
Helicase	An enzyme that catalyzes the unwinding of DNA.
Helix-loop-helix	A transcription factor DNA-binding domain formed by the dimerization of two polypeptide chains. The dimerization domains of these proteins consists of two helical regions separated by a loop.
Helix-turn-helix	A transcription factor DNA-binding domain in which three or four helical regions contact DNA.
Hemicellulose	A polysaccharide that cross links cellulose microfibrils in plant cell wall.
Hemidesmosome	A region of contact between cells and the extra cellular matrix at which keratin filaments are attached to integrin.
Hepatitis B viruses	A family of DNA viruses that infect liver cells and can lead to the development of liver cancer.
Hepatitis C viruses	A family of RNA viruses that infect liver cells and can lead to the development of liver cancer.
Heterochromatin	Darkly stained part of chromatin in interphase nucleus the condensed transcriptionally inactive chromatin.
Hetero-chromosomes	Chromosomes with exceptional form or behaviour especially at meiosis, such as a sex chromosomes fragments.
Heterogamy	Union of two unlike gametes.
Heterokaryon	A cell having two different types of nuclei, as a result of fusion of two cell types without nuclear fusion.
Heterozygote	An individual whose chromosomes bear unlike genes of a given allelic pair. Heterozygote produces more than one kind of gametes with respect to a particular locus (gene).
Histone acetylation	The modification of histones by the addition of acetyl groups to specific lysine residues.
Histone	Any of several proteins that can combine with DNA to form chromosomes, having a high content of the basic amino acids arginine and lysine.
Holliday junction	The central intermediate in recombination, consisting of a crossed-strand structure formed by homologous base pairing between strands of two DNA molecule.
Holliday model	A molecular model of genetic recombination involving the formation of heteroduplex regions.
Homeobox	Conserved DNA sequences of 180 base pairs that encode homeodomains.
Homeodomain	A type of DNA binding domain found in transcription factors that regulate gene expression during embryonic development.

Homologous chromosomes	Chromosomes occurring in pairs, one derived from each of two parents, normally (except in case of chromosomes associated with sex such as X and Y chromosomes) morphologically alike and bearing the same gene loci. Each member of such a pair is the homolog of the other.
Homozygote	An individual whose chromosomes bear identical genes of a given allelic pair or series. Homozygotes produce only one kind of gamete with respect to a particular locus and therefore "breed true".
Hormones	Signaling molecules produced by endocrine glands that act on cells at distant body sites.
Hyaloplasm	The ground substance of cytoplasm which is colloidal and contain enzymes for glycolysis and structural material, such as sugars-amino acids, water, vitamins, nucleotides, etc., commonly known as cytoplasmic matrix, nutrient soup or cell sap.
Hybridization molecular	Formation of a double-stranded structure, DNA-DNA, RNA-DNA, or RNA-RNA, by hydrogen bonding of complementary single stranded molecules or parts of molecules.
Hydrolysis	The process by which one molecule is processed into two smaller molecules by the addition of water, usually resulting in the release of energy.
Hydrophilic	Soluble in water.
Hydrophobic	Not soluble in water.
Idiogram	A diagrammatic representation of the complete chromosome set of an individual, arranged according to (decreasing) size and/or accepted numbering system. Idiograms may be drawn from suitable microscope preparations, or these may be photographed and various chromosomes arranged in order (and in homologous pairs in diploid organisms).
Immune system	Physiological system consisting of organs, scattered tissues, and independent cells that protect the body from invading pathogens and foreign materials.
Immunity	State in which the body is not susceptible to infection by a particular pathogen.
Immunoblotting	A method that uses antibodies to detect proteins separated by SDS-polyacrylamide gel electrophoresis.
Importin	A karyopherin that recognizes nuclear localization signals and directs nuclear import.
In utero	(Latin) in the womb.
In vitro	(Latin in glass), pertaining to experiments done on cells grown outside of the organisms.
In vivo	(Latin, in the living) pertaining to experiments on a whole living organism.
Induced fit	A model of enzyme action in which the configurations of both the enzyme and the substrate are altered by substrate binding.
Inducer (Effector)	A substance inactivating a repressor.
Inducer	A compound that binds to a protein repressor and activates transcription of a bacterial operon.
Inducible operon	An operon in which the presence of the key metabolic substance induces transcription of the structural genes.
Initiation codon	The triplet AUG, the site to which the ribosome attaches to the mRNA to assure that the ribosome is in the proper reading frame to correctly read the entire message.
Initiation factors	Soluble proteins (IFs in bacteria and eIFs in eukaryotes) that make initiation of translation possible.
Inositol, 1, 4, 5-triphosphate (IP3)	A second messenger, formed from the hydrolysis of PIP2, that signals the release of calcium ions from the endoplasmic reticulum.
Insulator	A sequence that divides chromatin into independent domains and prevents an enhancer from acting on a promoter in a separate domain.
Integral protein	A membrane-associated protein that penetrates or spans the lipid bilayer.
Integrin	A transmembrane protein that mediates the adhesion of cells to the extracellular matrix.

Interchange	An exchange of non-homologous terminal segments of chromosomes.
Interference	The reduction in likelihood of a second crossover closely adjacent to another. Interference ordinarily increases with decreased distance.
Interkinesis	See interphase, resting stage.
Intermediate filaments (IFs)	Strong, ropelike cytoskeletal fibers about 10 nm in diameter that provide mechanical strength to cells in tissues.
Interphase	1. The stage of cell life during which that cell is not dividing. 2. A period of the cell cycle between mitoses that include G_1, S and G_2 phases.
Introns	Non-coding sequences of split gene that interrupt exons.
Inversion	A chromosomal aberration that results after a chromosome is broken in two places and the resulting center segment is reincorporated into the chromosome in reverse order.
Ion channel	A transmembranous structure (*e.g.*, an integral protein with an aqueous pore) permeable to a specific ion or ions.
Ion pump	A protein that couples ATP hydrolysis to the transport of ions across a membrane.
Ion	A atom or molecule with a net positive or negative charge because it has lost or gained one or more electrons during a chemical reaction.
Ionic bond	A noncovalent bond occurring between oppositely charged ions, also called a salt bridge.
Isochromosome	A chromosome with two morphologically identical chromosomes.
Isoforms	Different versions of a protein. Isoforms may be encoded by separate, closely related genes, or formed as splice variants by alternative splicing from a single gene.
Isogametes	Gametes not differentiated as to sex. Fusing isogametes are physically similar but, in some species are chemically differentiated.
Isotonic	Property of one compartment having the same solute concentration compared with that in a given compartment.
Isozymes	Multiple forms of an enzyme with varying activity.
Junctional complex	A region of cell-cell contact containing a tight junction, an adherens junction and a desmosome.
Karyokinesis	The division of the nucleus during cell division.
Karyology	The biology of the nucleus.
Karyolymph	A clear fluid material within the nuclear membrane; nuclear sap.
Karyopherin	A nuclear transport receptor.
Karyotype	A distribution of photographed metaphase chromosomes from a cell or individual showing chromosomes in pairs and in order of decreasing size, the size shape and appearance of metaphase chromosomes.
Keratin	A type of intermediate filament protein of epithelial cells.
Kilobase (kb)	One thousand nucleotides or nucleotide base pairs.
Kinesin	A motor protein that moves along microtubules toward plus end. It moves membranous vesicles and other organelles along microtubules through the cytoplasm.
Kinetic energy	Energy released from a substance through atomic or molecular movements.
Kinetochore microtubules	Microtubules of the mitotic spindle that attach to condensed chromosones at their centromeres.
Kinetochore	A button-like structure situated at the outer surface of the centromere to which the microtubules of the spindle attach.
Knockout	Inactivation of a chromosomal gene by homologous recombination with a cloned mutant allele.

Krebs cycle Continuation of glycolysis; oxidative metabolism of pyruvic acid which is an end product of glucose fermentation (glycolysis), occurring within mitocbondria, requiring O_2 and yielding large quantities of energy that eventually trapped in the bounds of ATP; also called tricarboxylic acid cycle or citric acids cycle.

Lagging strand The newly synthesized daughter DNA strand that is synthesized discontinuously, so called because the initiation of each fragment must wait for the parent strands to separate and expose additional template.

Lamellipodium The leading actin-base edge of a moving fibroblast, which is extended out from the cell as a broad, flattened, veil-like projection that glides over the substratum.

Laminin The principal adhesion protein of basal laminae.

Lamins Intermediate filament proteins that form the nuclear lamina.

Lampbrush chromosomes Giant bivalents with extensively looped-out regions of chromatin fibres, especially prominent in amphibian oocyte nuclei during the diplotene stage of meiosis I propase.

Lateral gene transfer Transfer of genes from one species to another.

Leading strand The newly synthesized daughter DNA strand that is synthesized continuously in the direction of movement of the replication fork.

Leptotene The first stage of the extended prophase of meiosis I during which homologous chromosomes pair before condensation.

Leucoplast A plastid that stores energy sources in non-photosynthetic plant tissues.

Ligand Any molecule that can bind to a receptor because it has a complementary structure.

Ligand-gated channels Ion channels that open in response to the binding of signaling molecules.

Ligase Enzyme that joins together the parts of single strands of DNA between the 5′ end of one strand and the 3′ end of another.

Locus (pl., loci) The position or place on a chromosome occupied by a particular gene or one of its alleles.

Lysis (n.) v.i. or vt. lyse Disintegration or dissolution; usually the destruction of a bacterial host cell by infecting phage particles.

Lysogenic bacteria Living bacterial cells harbouring temperate phages (viruses).

Lysosomes Cytoplasmic organelles of animal cells which contain digestive enzymes for intracellular digestion of bacteria and other foreign bodies which enter the cells by the process of phagocytosis or pinocytosis. They may cause cell destruction if ruptured.

M phase The part of the cell cycle that includes two events: the process of mitosis, during which duplicated chromosones are separated into two nuclei, and cytokinesis, during which the entire cell is physically divided into two daughter cells.

Macromolecules Large, highly organised molecules crucial to the structure and function of cells; divided into polysaccharides, certain lipids, proteins and nucleic acids.

Major histocompatibility complex (MHC) A region of genome that encode MHC proteins. The genes that encode these proteins tend to be highly polymorphic, being represented by a large number of different alleles. These genetic differences between humans account for the tendency of a person to reject a transplant from another person other than an identical twin.

Mass spectrometry Methodology to identify molecules (including proteins). A proteins or mixture of proteins is fragmented, converted into gaseous ions, and propelled through a tubular component of a mass spectrometer, causing the ions to separate according to their mass/charge (m/z) ratio. Identification of the protein(s) is made by comparison with a computer database of the sequence of proteins encoded by a particular genome.

Matrix One of two aqueous compartments of a mitochondrion; the matrix is located within the interior of the organelle; the second compartment is called the intermembrane space and is located between the outer and inner mitochondrial membrane.

Meiocyte Any cell that undergoes meiosis.

Meiosis Nuclear division in which the diploid or somatic chromosome number is reduced by half. In the first of the two meiotic divisions homologous chromosomes first replicate, then pair (synapse), and finally separate to different daughter nuclei, which thus have half as many chromosomes as the parent nucleus, and one of each kind instead of two. A second division, in which chromosome replicates separate into daughter nuclei, follows so that meiosis produces four monoploid daughter nuclei from one diploid nucleus.

Meiospore In plants, one of the asexual reproductive cells produced by meiosis from a meiocyte.

Membrane fluidity A property of the physical state of the lipid bilayer of a membrane that allows diffusion of membrane lipids and proteins within the plane of the membrane. Inversely related to membrane viscosity. Membrane fluidity increases as the temperature rises and in bilayers with more unsaturated lipids.

Membrane potential The electrical potential difference across a membrane.

Meristem A cellular region in plants characterized by repeated cell division and wherein therefore, growth in size originates.

Mesosome An extensively infolded portion of the prokaryotic plasma membrane which has been implicated in respiratory and cell division functions.

Messenger RNA (mRNA)
1. Ribonucleic acid transcribed by DNA and conferring amino acid specificity on a ribosome.
2. The intermediate molecule between a gene and the polypeptide for which it codes. Messenger RNA is assemblage as a complementary copy of one of the two DNA strands that encodes the gene.

Metabolism The total of the chemical reactions occuring within a cell.

Metaphase The stage of nuclear division in which the chromosomes are located in the equatorial plane of the spindle prior to centromere division.

Metastasis Spread of cancer cells from a primary tumor to distant sites in the body where the formation of secondary tumors may arise.

Micro RNAs (mi RNAs) Small RNAs (20–23 nucleotide long) that are synthesized from many sites in the genome and involved in inhibiting translation or increasing degradation of complementary mRNAs.

Microfibrils Bundles of cellulose molecules that confer rigidity to the cell wall and provide resistance to pulling forces.

Microfilaments Solid 8-nm thick, cytoskeletal structures composed of a double-helical polymer of the protein actin. They play a key role in virtually all types of contractility and motility within cells.

Microscope An instrument that provides a magnified image of a tiny object.

Microsome A heterogenous collection of vesicles formed from the endomembrane system (primarily the endoplastic reticulum and golgi complex) after homogenization.

Microtubule An unbranched proteinous hollow cylindrical assembly of protofilaments which is involved in cell movement phenomena; spindle fibres, ciliary subfibres and centriole-subfibres are microtubules.

Microtubule-associated proteins (MAPs) Proteins other than tubulin contained in microtubules obtained from cells. MAPs may interconnect microtubules to form bundles and can be seen as cross-bridges connecting microtubules to each other. Other MAPs may increase the stability of microtubules, alter their rigidity and influence the rate of their assembly.

Microtubule-organizing centers (MTOCs) A variety of specialized structures that exert a role in initiating microtubule formation.

Microvilli Finger-like projections of plasma membranes of animal ethithelial cells, particularly of the gut.

Mitochondrion (Pl. mitochondria) The small cytoplasmic organelles in which aerobic cellular respiration occur and which are related with the production of energy by Krebs cycle, electron transport chain, beta oxidation of fatty acids, etc.

Mitosis Nuclear division in which replication of chromosomes is followed by separation of the products of replication and their incorporation into two daughter nuclei. Daughter nuclei are normally identical with each other and with the original parent nucleus in both kind and number of chromosomes. Mitosis may or may not involve cytoplasmic division.

Model organisms Organisms that have been widely used for research so that a great deal is known about their biology. These organisms have properties that have made them excellent research subjects. Such organisms include the bacterium, *E.coli*; the budding yeast, *S. cerevisiae*; the nematode, *C. elegans*; the fruit fly, *D. melanogaster*; the mustard plant, *A. thaliana* and the mouse, *M. musculus*.

Molecular chaperons Various families of proteins whose role is to assist the folding the assembly of proteins by preventing undesirable interactions.

Monoecious Individual producing both sperms and eggs.

Monoploid An individual having a single complete set of chromosomes. Synonym, haploid.

Monosomic An individual lacking one chromosome of a set ($2n - 1$).

Motor proteins Proteins that utilise the energy of ATP hydrolysis to generate mechanical forces that propel the protein, as well as attached cargo, along one of the components of the cytoskeleton. Three families of motor proteins are known kinesins and dyneins move along microtubules and myosins move along microfilaments.

Mutation A sudden change in genotype having no relation to the individual's ancestry. Used for changes in single gene itself ("point mutation") and for chromosomal aberrations.

Muton A smallest segment of DNA or subunit of a cistron that can be changed and thereby bring about a mutation; probably as small as one nucleotide pair.

Myeloma A tumour consisting of a clone of antibody-forming cells.

Myosin 1. The muscle protein making up the thick filaments of striated muscle and of a few other cell systems; has ATPase activity.

2. A large family of motor proteins that moves along actin-containing microfilaments. Most myosins are plus-end directed motors. **Conventional myosin (myosin II)** is the protein that mediates muscles contractility as well as certain types of nonmuscle motility, such as cytokinesis. **Unconventional myosins** (myosin I and III–XVIII) have many diverse roles including organelle transport.

Nanometer Measure of length equaling 10^{-9} meters.

Nascent protein A protein in the process of being synthesized, *i.e.*, not yet complete.

Negative staining Procedures in which heavy-metal deposits are collected everywhere on the specimen grid except in the locations of very small particulate materials, including high-molecular-weight aggregates such as viruses, ribosomes, multisubunit enzymes, cytoskeletal elements and protein complexes.

Non-disjunction The failure of homologous chromosomes to separate at anaphase-1 of meiosis.

Nucleosomes	Repeating subunits of chromatin. Each nucleosome of 146 base pairs of supercoiled DNA wrapped almost twice around a disc-shaped complex of eight histone molecules. The nucleosome core particles are connected to one another by a stretch of linker DNA.
Nuclear lamina	A thin meshwork composed of intermediate filaments that lines the inner surface of the nuclear envelope.
Nuclear localization signal (NLS)	Sequence of amino acids in a protein that is recognised by a transport receptor leading to translocation of the protein from the cytoplasm to the nucleus.
Nucleic acid	Polymers composed nucleotides, which in living organisms are based on one of two sugars, ribose or deoxyribose, yielding the terms ribonucleic acid (RNA) and deoxyribonucleic acid (DNA).
Nucleoid	The poorly defined region of a prokaryotic cell that contains its genetic material.
Nucleolar organising region (NOR)	The specific part of the nucleolar organizing chromosome where ribosomal RNA genes are situated and where the nucleolus is produced.
Nucleolus	Deeply staining nuclear body which is involved in rRNA synthesis; it is usually associated with secondary constrictions or nucleolar organizers of certain chromosomes and has important role in the biogenesis of ribosomes.
Nucleoplasm	The unstructured matrix portion of the nucleus in which the chromosomes and nucleoli are bathed.
Nucleoside	Portion of DNA or RNA molecule composed of one deoxyribose molecule (in DNA) or ribose (in RNA) plus a purine or a pyrimidine.
Nucleotide	Portion of DNA or RNA molecule composed of one deoxyribose phosphate unit (in DNA) or one ribose phosphate unit (in RNA) plus a purine or a pyrimidine.
Nucleus	A membrane-bounded cellular organelle having chromatin, nucleolus and nucleoplasm and regulating growth and reproduction of the cell.
Nullisomy	The absence of a particular chromosome.
Okazaki fragments	Small segments of DNA that are rapidly linked to longer pieces that have been synthesized previously, to form the lagging strand.
Oncogenes	Genes that encode proteins that promote the loss of growth control and the conversion of the cell to a malignant state. These genes have the ability to transform cells.
Oncogenic	Having a tendency to "cause" cancer.
Ontogeny	The complete development of the individual from zygote, spore, etc., to adult form.
Oocyte	The egg mother cell.
Oogenesis	The process of development and differentiation of the ovum from the oogonium.
Operator site	A segment of DNA in one operon that affects activity or non-activity of associated cistrons; it may be combined with a repressor and thereby 'turn off' the associated cistrons.
Operon	A system of cistrons operator and promoter sites by which a given genetically controlled metabolic activity is regulated.
Organelle	A structural differentiation of the cell (*e.g.*, mitochondria, chloroplasts, ribosomes, etc.) containing particular enzymes and performing particular functions for the whole cell or individual.
Orientation	The movement of chromosomes so that their centromere lie axially with respect to the spindle.
Ovary	The female gonad in animals or the ovule containing portion of the pistil of a flower.
Ovule	Structure within the ovary of pistil of a flower, which becomes a seed.
Oxidation	Reactions involving loss of electrons or hydrogens or gain of oxygen.

Oxidative phosphorylation In cellular respiration, oxidation in the cytochrome system accompanied by the addition of phosphate molecules to ADP forming ATP.

Pachytene A stage of prophase-I of meiosis in which the chromosomes are visible as long paired threads.

Parthenogenesis Development of a new individual from an unfertilized egg.

Pectins A heterogeneous class of negatively charged polysaccharides that make up the matrix of the plant cell wall. Pectins hold water and form a gel that fills in the spaces between the fibrous elements.

Peptide bond A chemical bond (CONH) linking amino acid residues together in a protein.

Peptide A string of a few amino acids joined by peptide bonds.

Peptidyl transferase That portion of the large ribosomal subunit that is responsible for catalyzing peptide bond formation; peptidyl transferase activity resides in the large ribosomal RNA molecule.

Perosixomes (microbodies) Simple, membrane-bound, multifunctional organelles of the cytoplam that carry out a diverse array of metabolic reactions, including substrate oxidation leading to formation of hydrogen peroxide. For example, peroxisomes are the site of oxidation of uric acid, and the synthesis of plasmalogens. Plant glyoxysomes, which carry out the glyoxylate cycle are a type of peroxisome.

pH Expression of the relative abundance of H^+ in solution (*i.e.*, negative logarithm of H^+ concentration); relative acidity of a solution. Neutral pH is 7; lower levels are acid, higher are basic.

Phagocytosis Process by which particulate materials are taken into cells. Materials are enclosed within a fold of the plasma membrane, which buds into cytoplasm to form a vesicle called a phagosome.

Photon Energy quantum or packet of light energy.

Photorespiration A series of reactions in which O_2 is attached to RuBP and eventually resulting in the release of recently fixed CO_2 from the plant.

Photosynthesis Manufacture of sugar from CO_2 and H_2O in the light and in the presence of chlorophyll in eukaryotic chloroplasts or within prokaryotic photosynthetic membranes.

Phragmoplast Dense material roughly aligned in the equatorial plane of the previous metaphase plate in plant cells, consisting of clusters of interdigitating microtubules oriented perpendicular to the future cell plate, together with vesicles and associated electrondense material.

Phylogeny The evolutionary development of a species or other taxonomic group.

Pinocytosis A process by which the cell engulf droplets of liquids.

Plasma membrane The differentially permeable and limiting membrane of the cell through which extracellular substances may be selectively sampled and cell products may be liberated.

Plasmodesmata Cytoplasmic channels between plant cells.

Plastids Structures for storage of starch, pigment and other cellular products in plant cell.

Ploidy Variation in number of whole chromosomes.

Polar bodies The expelled products of the division of the oocyte nucleus in animals during oogenesis.

Polymer A chemical compound composed of two or more units of the same compound.

Polymerase An enzyme that catalyzes polymerization. In particular, DNA and RNA polymerases catalyze the formation of the nucleic acids from the nucleotide constituents on the basis of a single-stranded DNA template.

Polypeptide A compound containing amino acid residues jointed by peptide bonds. A protein may consists of one or more specific polypeptide chains.

Polyploid An individual having more than two complete sets of chromosomes, *e.g.*, triploid (3n), tetraploid (4n).

Polysome Polyribosome, an aggregation of ribosomes, connected by a strand of messenger RNA, which is active in protein synthesis.

Polytene chromosomes A multistranded endoreplicated chromosomes

Porins Integral proteins found in bacterial, mitochondrial and chloroplast outer membranes that act as large, relatively nonselective channels.

Primary structure The sequence of amino acids in a polypeptide chain.

Primase Type of RNA polymerase that assembles the short RNA primers that begin the synthesis of each Okazaki fragment of the lagging strand.

Prion An infectious agent associated with certain mammalian neurodegenerative disease that is composed solely of protein.

Procentriole An immature centriole.

Processive A term applied to proteins (*e.g.*, kinesin of RNA polymerase) that are capable of moving considerable distances along their track or template (*e.g.*, a microtubule or a DNA molecule) without dissociating from it.

Prokaryotes Organisms of the bacteria and blue-green algae groups lacking a nucleus separated from the cytoplasm by a membrane.

Promoter site Site on DNA at which mRNA synthesis initiated in an operon. May be combined with a repressor to inhibit mRNA synthesis by associated cistrons.

Prophase The first stage of nuclear division, including all events upto arrival of chromosomes at the equator of the spindle.

Proteasome Barrel-shaped, multiprotein complex in which cytoplasmic proteins are degraded. Proteins selected for destruction are linked to ubiquitin molecules and threaded into the central chamber of the proteasome.

Proteome The entire inventory of proteins in a particular organism, cell type, or organelle.

Protoplasm The living material of the cell.

Protoplast A structural unit of protoplasm; all the living (protoplasmic) material of the cell. Its two principal parts are nucleus and cytoplasm.

Puff A region of expanded chromosome undergoing active transcription, usually observed in giant polyene chromosomes.

Pumps, ion Systems that underwrite active transport of molecules across membrane by expelling one substance out of the cell and thereby helping to drive many kinds of molecules into the cell along an energy gradient.

Purine Nitrogen base occurring in DNA and RNA; these are adenine and guanine.

Pyrimidine Nitrogenous base occurring in DNA (thymine and cytosine) or RNA (uracil and cytosine).

Quality control Cells contains various mechanisms that ensure that the proteins and nucleic acids they synthesize have the appropriate structure. For example, misfolded proteins are translocated out of the ER and destroyed by proteasomes in the cytosol; mRNAs that contain premature termination codons are recognised and repaired.

Quantum The energy of a photon, its amount being, inversely proportional to the wavelength of emitted radiation.

Quaternary structure Specific assemblages of different polypeptide chains which, when combined in the protein, do not have the same structural or chemical properties as in the individual chains.

Rabs A family of small G proteins involved in vesicle transfecting. (For example, in humans more than 60 Rabs (rab proteins) have been isolated, each of which become associated with different membrane compartment to give each of them a unique surface identity).

Ran A GTP binding protein that exists in an active GTP-bound form or an inactive GDP-bound form. RAN ragulates nucleocytoplasmic transport of proteins, and works in association with importin proteins.

Ras Ras is a small GTPase (protein) that is anchored at the inner side of the plasma membrane by a lipid group. It acts similar to G proteins but also acts both as a switch and a molecular timer. Mutations in human *RAS* gene that lead to tumor formation prevent the Ras protein from hydrolyzing the bound GTP back to GDP form. As a result, the mutant version of Ras protein remained in the "on" position, sending a continuous message downstream along the signaling pathway, keeping the cell in proliferative mode.

Ras protein is a part of a superfamily of more than 150 small G proteins including Rabs, Sar 1 and Ran. These proteins are involved in the regulation of numerous processes including cell division, differentiation, gene expression, cytoskeletal organization, vesicle trafficking and nucleocytoplasmic transport.

Ras-MAP kinase cascade A cascade that is turned on in response to a wide variety of extracellular signals and plays a keys role in regulating vital activities such as cell proliferation and differentiation.

Receptor Any substance that can bind to a specific molecule (ligand), often leading to uptake or signal transduction.

Recombination The new association in one individual of phenotype traits from both parents (*i.e.*, a non-parental individual). Also random combinations of genes from different linkage groups resulting from meiosis.

Recon The smallest segment of DNA or unit of a cistron that is capable of recombination; probably as one nucleotide pair.

Reduction division See meiosis.

Regulated secretion Discharge of materials synthesized in the cell that have been stored in membrane-bound secretory granules in the peripheral regions of the cytoplasm, occuring in response to an appropriate stimulus.

Regulatory gene A gene responsible for "switching on or off" other genes, usually through production of a repressor that regulates the activity of the other gene.

Replicate To form replicas; applies to synthesis of new DNA from preexisting DNA as part of nuclear division.

Replication fork A site within a replicating duplex DNA molecule at which synthesis of complementary strands is proceeding at that moment.

Repressible enzyme Type of enzyme that is synthesized in the absence of its substrate which represses synthesis of enzyme.

Repressor A protein produced by a regulatory gene that can combine with and repress action of an associated operator gene.

Resolution The ability to see two neighboring points in the visual field as distinct entities.

Restriction endonuclease (restriction enzyme) Nuclease occurring in bacteria that recognize short nucleotide sequences within duplex DNA and cleave the backbone at highly specific sites on both strands of the duplex.

Reverse transcriptase An RNA-dependent DNA polyerase. An enzyme that uses RNA as a template to synthesize a complementary strand of DNA. (An enzyme that is found in RNA-containing viruses and used in the laboratory to synthesize cDNAs).

Ribonucleic acid (RNA) A single-stranded nucleic acid molecule, synthesized principally in the nucleus from deoxyribonucleic acid, composed of a phosphate-ribose backbone with purines (adenine and guanine) and pyrimidines (uracil and cytosine) attached to the sugar ribose. RNA is of several kinds and functions to carry the "genetic message" from nuclear DNA to the ribosomes.

Ribonucleotide Portion of RNA molecule composed of one ribose-phosphate unit plus a purine or a pyrimidine.

Ribosomal RNAs (or rRNAs) The RNAs of a ribosome. rRNAs recognise and bind other molecules, provide structural support and catalyse the chemical reaction in which amino acids are covalently linked to one another

Ribosome A complex structure composed of RNA and protein which is the site of protein synthesis in the cytoplasm, mitochondria and chloroplasts.

Ribozyme An RNA molecule that functions as a catalyst in cellular reactions.

RNA polymerase-I The transcribing enzyme found in eukaryotic cells that synthesizes the large (28S, 18S and 5.8S) ribosomal RNAs.

RNA polymerase-II The transcribing enzymes in the eukaroytic cells that synthesizes messenger RNAs and most small nuclear RNAs.

RNA polymerase-III The transcribing enzyme found in eukaryotic cells that synthesis the various transfer RNAs and the 5S ribosomal RNA.

RNA splicing The process of removing the intervening DNA sequences (*i.e.*, introns) from a primary transcript.

RNA World A proposed stage in the early evolution of life before the appearance of DNA and proteins in which RNA molecules served both as genetic material and catalysts.

Rough endoplasmic reticulum (RER) That part of endoplasmic reticulum that has ribosomes attached. The RER appears as an extensive membranous organelle composed primarily of flattened sacs (cisternae) separated by a cytosolic space. RER functions include synthesis of secretory proteins, lysosomal proteins, integral membrane proteins and membrane lipids.

Satellite region A region of DNA that contains highly repetitive base sequences.

Second messenger A substance that is formed in the cell as the result of the binding of a first messenger—a hormone or other ligand-to a receptor at the surface of the cell.

Secondary structure The local structure of the polypeptide chain, over a distance of only a few amino acid residues; may exist either as an irregular chains or as a regular and repeating shape such as the α-helix.

Secreted Discharge outside the cell.

Secretory granule Large, densely packed, membrane-bound structure containing highly concentrated secretory materials that are discharged into the extracellular space (secreted) following a stimulatory signal.

Sedimentation coefficient (s) A quantitative measure of the rate of sedimentation of a given substance in a centrifugal field, expressed in Svedberg unit.

Self-assembly The property of proteins (or other structures) to assume the correct (native) conformation based on the chemical behaviour dictated by the amino acid sequence.

Self-fertilization Functioning of a single individual as both male and female parent. Plants are "selfed" if sperm and egg supplied by the same individual.

Semiconservative replication The usual mode of duplex DNA synthesis resulting in daughter duplex molecules which contain one parental strand and one newly formed strand.

Sex chromosomes Chromosomes not occurring in individual pairs in both sexes in diploid organism; in human beings and fruit fly these are designated as X and Y chromosomes.

Sexual reproduction Mode of reproduction which is accomplished by the union of male and female gametes.

Sliding filament mechanism A proposal applied to explanation of cellular movement phenomena, particularly to muscle contraction and to ciliary bending.

Signal transduction The overall process in which information carried by extracellular messenger molecules is translated into changes that occur inside a cell.

Smooth endoplasmic reticulum (SER) That part of endoplasmic reticulum that is without attached ribosomes. The membraneous elements of the SER are typically tubular and form an interconnecting system of pipelines curving through the cytoplasm in which they occur. The SER functions vary form cell to cell and include the synthesis of steroid hormones, detoxification of a wide variety of organic compounds, mobilization of glucose from glucose 6-phosphate, and separation of calcium ions.

Sodium-potassium pump (Na^+/K^+-ATPase) Transport protein that uses ATP as the energy source for transporting sodium and potassium ions, with the result that each conformational change transports three sodium ions out of the cell and two potassium ions into the cell.

Sol Colloidal solution.

Soma (adj. Somatic) The body, cells of which in mammals and flowering plants normally have two sets of chromosomes, one derived from each parent.

Spermatocyte The cell that undergoes meiosis to produce four spermatids.

Spermigenesis Differentiation of the spermatozoa.

Spermotagenesis Development of sperms.

Spindle The axially differentiated part of the cytoplasm within which the centromere of the chromosomes are held during metaphase.

Spliceosome A macromolecular complex containing a variety of proteins and a number of distinct ribonucleoprotein particles that functions in removal of introns from a primary transcript.

Split genes Genes with intervening sequences.

Spore An asexual reproductive protoplast capable of developing into a new individual.

Stem cells Cells situated in various tissues of the body that constitute a reserve population capable of giving rise to the various cells of that tissue. Stem cells can be defined as undifferentiated cells that are capable of both 1. self-renewal, that is, production of cells like themselves, and 2. differentiation into two mature cell types.

Stereoisomers Two molecules that structurally are mirror images of each other and may have vastly different biological activity.

Stroma Space outside the thylakoid but within the relatively impermeable inner membrane of the chloroplast envelope.

Structural gene A cistron.

Subcellular fractionation An approach that allows different organelles (*e.g.*, nucleus, mitochondria, plasma membrane, endoplasmic reticulum) having different properties, to be separated from one another.

Substrate The reactant bound by an enzyme.

Synapse The specialized junction of a neuron with its target cell.

Synapsis The pairing of homologous chromosomes occurring in prophase I of meiosis.

Synaptonemal complex (SC) A ladder-like structure composed of three parallel bars with many cross fibers. The SC holds each pair of homologous chromosomes in the proper position to allow the continuation of genetic recombination between strands of DNA.

Syngamy The union of sex cells (gametes) in reproduction.

tDNA The DNA encoding tRNAs.

Telocentric A chromosome with the centromere located at one end.

Telomerase A rare enzyme that can add new repeat units of DNA to the 3′ end of the over hanging strand of a telomere. Telomerase is a reverse transcriptase that synthesizes DNA using an RNA template.

Telophase The concluding stage of nuclear division characterized by the reorganization or interphase nuclei.

Template 1. A model, mold, or pattern; DNA act as a template for RNA synthesis.
2. A single strand of DNA (or RNA) that contains the information (encoded as a nucleotide sequence) for construction of a complementary strand.

Terminalization Expansion of the association of two pairs of chromatids on one side of the chiasma at the expense of that on the other side.

Tertiary structure The three-dimensional folding of the polypeptide chain into a complex structural form brought about by interactions among side chains of amino acids at some distance from one another in the primary structure.

Tetrad The four monoploid (haploid) cells arising from meiosis of a megaspore or microsporocyte in plants; also group of four associated chromatids during synapsis.

Thermodynamics Study of the changes in energy accompanying events in the physical universe.

Thylakoids Flattened, membraneous sacs formed by the chloroplast's internal membrane, which contain the energy-transducing machinery for photosynthesis.

Thymine A pyrimidine base occurring in DNA, pairs normally with adenine.

Tonoplast The membrane that bounds the vacuole of a plane cell. Plant vacuoles store a large variety of molecules including inorganic ions, organic ions, organic acids, sugars, enzymes (some like the lysosomes), storage proteins and many types of secondary metabolites. Proteins in the tonoplast transport all of these molecules except for storage proteins—into the vacuole.

Transcription The formation of a complementary RNA from a DNA template.

Transcriptosome The entire inventory of RNS's transcribed by a particular cell, tissue, or organism.

Transduction The incorporation of a gene into a cellular genome by means of a virus.

Transfer RNA (tRNA) Amino acid-specific RNA which transfers activated amino acids to mRNA where protein synthesis takes place.

Transfer RNAs (tRNAs) A family of small RNAs that translate the information encoded in nucleotide "alphabet" of an mRNA into the amino acid "alphabet" of a polypeptide.

Translation The process by which the sequence of amino acids is assembled into a polypeptide at the ribosome, under the direction of the coded base sequence copied from DNA into messenger RNA.

Translocation The shift of a portion of a chromosome to another part of the same chromosome or to an entirely different chromosome.

Translocation The step in the translation elongation cycle that involves (1) ejection of the uncharged tRNA from the P site and (2) the movement of the ribosome three nucleotides (one codon) along the mRNA in the 3′ direction.

Translocon A protein-lined channel embedded in the ER membrane; the nascent polypeptide is able to move through the translocon in its passage from the cytosol to the ER lumen.

Transport vesicles The shuttles, formed by budding from a membrane compartment, that carry materials between organelles.

Transposons DNA segments capable of moving from one place in the genome to another.

Transverse (T) tubules Membranous folds along which the impulse generated in a skeletal muscle cell is propagated into the interior of the cell.

Tricarboxylic acid cycle (TCA cycle) The circular metabolic pathway that oxidizes acetyl CoA, conserving its energy; the cycle is also known as the Krebs cycle or the citric acid cycle.

Trisomy A chromosome complement that has one extra chromosome, *i.e.*, a third homologous chromosome.

Tritium (^3H) The radioactive isotope of hydrogen.

Tubulin The protein that forms the walls of microtubules. Isoforms include α, β and γ tubulin.

Tumor-suppressor genes Genes that encode proteins that restrain cell growth and prevent cells from becoming malignant.

Turnover The regulated destruction of cellular materials and their replacement.

Ubiquitin A small, highly conserved protein that is linked to proteins targeted for internalization by endocytosis or degradation in proteasomes.

Unfolded protein response (UPR) Comprehensive response that occurs in cells whose ER cisternae contain an excessively high concentration of unfolded or misfolded proteins. Sensors that detect this situation trigger a pathway that leads to the synthesis of proteins (*e.g.*, molecular chaperones such as BiP), that can alleviate the stress.

Unineme Concept of chromosome structure having only one chromatin fibre in the unreplicated chromosome or in each chromatid of a replicated chromosome.

Unit membrane A membrane formed of two layers of lipid molecules sandwiched between the two layers or protein molecules. It forms the outer boundary of cell and almost all cellular organelles.

Untranslated regions (UTRs) Non-coding segments contained at both 5′ and 3′ ends of mRNA.

Unwinding protein Structural protein that binds to single-stranded regions of duplex DNA during replication and recombination.

Uracil A pyrimidine base occurring in RNA.

Vacuole A single membrane-bound, fluid filled cytoplasmic structure that comprises as much as 90% of the volume of many plant cells.

Vesicle A small, spherical membranous element filled with protein in a watery medium.

Viruses Small, obligatory intracellular pathogens that are not considered to be alive because they cannot divide directly, which is required by the cell theory of life.

Vitamin Organic compounds found in small amounts in all living organisms and which are essentially required by body for its normal working.

Wobble hypothesis Crick's proposal that the steric requirement between the anticodon of the tRNA and the codon of the mRNA is flexible at the third position, which allows two codons that differ only at the third position to share the same tRNA during protein synthesis.

Zygote The protoplast resulting from the fusion of two gametes in sexual reproduction; a fertilized egg.

Zygotene A stage of prophase I of meiosis which is characterized by the synapsis or paring of the homologous chromosomes.

Selected Reading

Agarwal, V.K., 2014, *Zoology For Degree Students–1*, S.Chand & Company Pvt. Ltd., New Delhi.

Alberts, Bruce, Johnson, A., Lewis, J., Raff, M., Roberts, K. and Walter P., 2002, *Molecular Biology of the Cell*, Garland Science, New York.

Ambrose E.J. and Dorothy Measty, 1970, *Cell Biology*, The English Language Book Society & Nelson, London.

Avers, C.J., 1976, *Cell Biology*, D. Van Nostrand Company, New York.

Balinsky, B.I., 1970, *An Introduction to Embryology*, W.B. Saunders Company, Philadelphia and London.

Barry, J.M. and E.M. Barry. 1973, *Molecular Biology: An Introduction to Chemical Genetics*, Prentice-Hall of India Private Limited, New Delhi.

Becker, Wayne M., Kleinsmith, L.J., Lewis J. and Hardin, J. 2006, *The World of the Cell*, Pearson Education, Inc., New York.

Berill, N.J., 1971, *Developmental Biology*, McGraw-Hill Book Company, New York.

Bodmer, W.F. and L.L. Cavalli-Sforza 1976, *Genetics, Evolution, and Man*, W H., Freeman and Company, Sanfrancisco, London.

Bostok,. C.J. and A.T. Summer 1978, The *Eukaryotic Chromosomes*, North Holland Publishing Company, Amsterdam.

Bowen. W.R. 1969, *Experimental Cell Biology*, The Macmillan Company. New York.

Brachet. J and A.E. Mirsky 1959-1964. *The Cell* (I-V Volumes), Academic Press, New York.

Bradbury, E.M., N. Maclean and H.R. Matthews 1981, *DNA, Chromatin and Chromosomes*, Blackwell Scientific Publications, Oxford.

Branton, D. and D.Deamer 1972, *Membrane Structure*, Springer-Verlag, New York.

Branton, D. And R. Park 1968, *Papers on Biological Membrane Structure*, Little Brown & Co., Boston.

Buchanan, Bob b., Gruissem, W. and Jones, R.L. 2000, *Biochemistry & Molecular Biology of Plants*, I.K. International Pvt. Ltd., New Delhi.

Burns, G.W. 1980, *The Science of Genetics* (4th ed). The Macmillan Company, New York.

Butler J.A.V., 1959, *Inside the Living Cell*, George Allen and Unwin Ltd., London.

1964, The *Life of the Cell*, George Allen and Unwin Ltd., London.
1968, *Gene Control in the Living Cell*, George Allen and Unwin Ltd., London.

Cockrum, E.L. and W.J. McCauely, 1965, *Zoology*, W.B. Saunders Company, Philadelphia.

Cohn, N.S. 1969, *Elements of Cytology*, Harcourt, Brace and World, Inc., New York.

Cooper, Geoffrey and Hausman, Robert E. 2007, *The Cell: A Molecular Approach*, ASM Press Washington, D.C., USA.

DeRobertis, E.D.P., W.Nowinski and F.A. Saez. 1970, *Cell Biology*, W.B., Saunders Company, Philadephia.

DeRobertis E.D.P., F.A. Saez and E.M.F. DeRobertis 1975, *Cell Biology*, W.B. Saunders Company, Philadelphia.

De Witt, William, 1977, *Biology of the Cell*, W.B. Saunders Company, Philadelphia.

Dowben, R.M. 1971, *Cell Biology*, Harper & Row, Publishers, New York.

Dupraw, E.J. 1969, *Cell and Molecular Biology*, Academic Press, New York.

Dyson R.D., 1978, *Cell Biology*, Allyn and Bacon, Inc. Boston.

Fairley, J.L. 1966, *Essentials of Biological Chemistry*, Reinhold Publishing Corporation, New York.

Fawcett. D.W. 1966, *The Cell: Its Organelles and Inclusions,* W.B. Saunders Philadelphia.

Fraser, Dean 1967, *Viruses and Molecular Biology*, The Macmillan Company, New York.

Freeman, J.A. 1964, *Cellular Fine Structure,* McGraw-Hill Book Company, New York.

Garber, E.D. 1972, *Cytogenetics*, McGraw-Hill Book Company, New York.

Gerking S.D. 1969, *Biological Systems*, W.B. Saunders Company, Philadelphia.

Giese, A.C. 1973, *Cell Physiology*, W.B. Saunders Company, Philadelphia.

Goodenough V. and R.P. Levine 1974, *Genetics*, Holt Reinhard and Winston, Inc., New York.

Gilbert, Scott F., 2010, *Developmental Biology*, Sinauer Associates, Inc. Publishers, Sunderland, USA.

Gunning, B. and M.W. Steer 1975, *Plant Cell Biology*, Crane, Russak, New York.

Harris, H., 1970, *Nucleus and Cytoplasm*, Clarendon Press, Oxford.

Harrison, R. and G. Lunt. 1976, *Biological Membranes: Their Structure and Function*; Halsted Press, New York.

Haggis. C.H., 1964, *Introduction to Molecular Biology,* Longmans, London.

Hall, John, E. 2011, *Guyton and Hall Textbook of Medical Physiology*, Saunders, Elsevier, Philadelphia, USA,

Hanawalt, P.C. and Robert H. Haynes 1970, *The Chemical Basis of Life*, W.H. Freeman and Company, San Francisco.

Herreld II., C.F. 1977, *Biology*, Macmill and Publishing Co; Inc., New York.

Hickman, Jr., Cleveland P., Roberts, L.S. and Larson A. 1996, *Integrated Principles of Zoology*, WCB McGraw-Hill, Boston, USA.

Horton, H.R. Moran, L.A., Scrimgeour, K.G., Perry, M.D. and Rawn, J.D. 2006, *Principles of Biochemistry*, Pearson, Prentice Hall, London.

Karp, Gerald 2010, *Cell Biology*, John Wiley & Sons, Inc., N.J.

Khorana H.G. 1978, *Synthetic Nucleic Acids and the Genetic Code*,. J. Am. Med. Assoc., 205 Lasker Award, 1968.

Kimball, J.W. 1970, *Cell Biology,* Addison-Wesley Publishing Company, California.

Kornberg. A., 1974, *DNA Synthesis*, W.H. Freeman & Co., San Francisco.

Kornberg, R.D. and A. Clug. 1981, *The Nucleosome*, Scientific American, Feb. 1981, 48-60.

Lima-deFaria A., 1969, *Handbook of Molecular Cytology,* North Holland Publishing Company, Amsterdam.

Mays, Laura Livingston 1981, *Genetics*, Macmillan Publishing Co., Inc., New York.

Lehman I. and D. Uyemura 1976, *DNA Polymerase I: Essential Replication Enzymes*, Science, 193: 1963.

Lodish, Harvey, Berk, A. Matsudaira, P., Kaiser, C.A., Krieger M., Scott, M.P., Zipursky, S.L. and Darnell, J. 2004, *Molecular Cell Biology*, W.H. Freeman and Company, New York.

Malacinki, George M. 2005, *Freifelder's Essentials of Molecular Biology*, Narosa Publishing House, New Delhi.

Meglitsch P.A. 1972, *Invertebrate Zoology*, Oxford University Press, Oxford.

Marcer E.H. 1965, *Cell and Cells Structure*, Hutchinson Educational, London.

Murray, Robert K., Granner, D.K. Mayes P.A. and Rodwell, V.W. 2000, *Harper's Biochemistry,* Appleton & Lange, Stamford Connecticut, USA.

Nass, Gisela, 1970, *The Molecules of Life*, Weiden Feld and Nicolson, London.

Nomura, M.A. Tissieres and P. Lengyel. 1974, *Ribosomes*, Cold Spring Harbour Laboratory, New York.

Perry, R. 1976, *Processing of RNA*, Ann. Rev. Biochem; 45: 605.

Purves, William K., Orians, G.H., Heller H.C. and Sadava D. 1997, *Life. The Science of Biology, Vol. I The Cell and Heredity,* W.H. Freeman and Company, Utah, USA.

Quinn, P.J., 1976, *The Molecular Biology of Cell Membranes*, Macmillian Press, London.

Rabinowitch, E.I. and Govindjee. 1965, *The Role of Chlorophyll in Photosynthesis*, Scientific American, July, 1965.

Raven, Peter H., Johnson, G.B., Losos, J.B. and Singer, S.R. 2005, *Biology*, Tata McGraw-Hill Publishing Company Limited, New Delhi.

Reinert, J. and H. Ursprung 1971, *Origin and Continuity of Cell Organelles*, Springer Verlag, Berlin.

Rosenberg, E., 1971, *Cell and Molecular Biology*, Holt, Rinehart and Winston, Inc., New York.

Sager, Ruth, 1972, *Cytoplasmic Genes and Orgenelles*, Academic Press, New York.

Smith, C.U.M., 1968, *Molecular Biology*, Faber and Faber, London.

Spencer, J.H., 1972, *The physics and Chemistry of DNA or RNA*, W.B. Saunders Company, Philadelphia.

Spirin, A.S. and L.P. Gavrilova 1969, *The Ribosomes*, Springer-Verlag Inc., New York.

Stern, H. and D.L. Nanney. 1965, *The Biology of Cells*. Wiley Estern Private Ltd., New Delhi.

Sybenga, J., 1972, *General Cytogenetics*, North-Holland Publishing Company, Amsterdam.

Tamarin, Robert H. 2002, *Principles of Genetics*, Tata McGraw-Hill Publishing Company Limited, New Delhi.

Taylor, J.H., 1974, *Units of DNA Replication in Chromosomes of Eukaryotes*, Int. Rev. Cytol., 37:1.

Tedeschi, H., 1974, *Cell Physiology*, Academic Press, New York.

Toner, P.G. and K.E. Carr 1971, *Cell Structure*, Churchill Livingston, Edinburgh and London.

Verma P.S. and Agarwal V.K. 2014, *Chordate Embryology* (*Developmental Biology*), S.Chand & Co. Ltd., New Delhi.

Verma P.S. and Agarwal V.K. 2014, *Genetics,* S.Chand & Company Ltd. New Delhi.

Verma P.S. and Agarwal V.K. 2014, *Cytology,* S.Chand & Company Pvt. New Delhi.

Watson, J.D. 1976, *Molecular Biology* of the Gene, (3rd ed.), W.A. Benjamin, Inc., New York.

Weisz, P.B. 1971, *The Science of Biology*, Tata McGraw-Hill Publishing Company Ltd., Bombay.

Werner, D.D. 1977, *The Molecular Biology of Plant Cells*, Blackwell Scientific, Oxford.

Literary Sources in Cell Biology

A. Journals

1. Journal of Cell Biology.
2. Experimental Cell Research.
3. Journal of Molecular Biology.
4. Journal of Ultrastructure Research.
5. Journal of Cell Science.
6. Journal of Cellular and Comparative Physiology.
7. Journal of General Physiology.
8. Chromosoma.
9. Cytogenetics.
10. Heredity and Hereditary.
11. Nature.
12. Science.
13. Proceedings of the Royal Society.
14. Proceeding of National Academy of Science (Washington).
15. Biochemica et. Biophysica.
16. Biochemical Journal or Journal of Biological Chemistry.
17. Life Sciences.
18. American Scientist.
19. Bioscience.
20. Hospital Practice.
21. New Scientist.
22. The Science.
23. Science News.
24. Current Biology.
25. Scientific American.
26. Bioessays.
27. Nature Biotechnology.
28. Current Opinions in Cell Biology.
29. Trends Microbiology.
30. Trends Cell Biology.

B. Reviews

1. International Review of Biology.
2. Quarterly Review of Biology.
3. Physiological Review.
4. Biological Review (Cambridge).
5. Advances in Genetics.
6. Plant Physiology.
7. Science Progress (Oxford)
8. Bacteriological Reviews.
9. Biomedicine.
10. Biomembrane.
11. Cell.
12. Life Sciences.
13. New England Journal of Medicine.
14. Reviews of Physiology, Biochemistry and Pharmacology.
15. Sub-Cellular Biochemistry.
16. Nature Reviews Molecular Cell Biology.
17. Annual Reviews Cell Development Biology.
18. Nature Reviews Molecular Cell Biology.
19. Nature Reviews Genetics.
20. Nature Reviews Immunology.
21. Annual Reviews of Immunology.
22. Annual Reviews Biochemistry.

C. Abstracts

1. Biological Abstracts.
2. Index Medicus.
3. Excerpta Medica.
4. Chemical Abstracts.

D. Monographs

1. The Cell (5 volumes) by Brachet and Mirsky.
2. Handbook of Molecular Cytology By A. Lima-de-Faria (1969).
3. Giba Foundation Symposia (Each volume a separate topic).

Index

A

N

O

P

Q

R

S

T

U